UNITS

Quantity	Name of unit	In terms of base units	In other common terms
Capacitance (C)	farad (F)	$kg^{-1} \cdot m^{-2} \cdot s^4 \cdot A^2$	C/V
Electric charge (q)	coulomb (C)	$s \cdot A$	
Electric field (**E**)		$kg \cdot m \cdot s^{-3} \cdot A^{-1}$	N/C or V/m
Electric potential (V) (also EMF [$\mathscr{E}$])	volt (V)	$kg \cdot m^2 \cdot s^{-3} \cdot A^{-1}$	J/C or W/A
Electric resistance (R)	ohm (Ω)	$kg \cdot m^2 \cdot s^{-3} \cdot A^{-2}$	V/A
Energy (E)	joule (J)	$kg \cdot m^2/s^2$	N · m
Force (**F**)	newton (N)	$kg \cdot m/s^2$	
Frequency (ν)	hertz (Hz)	s^{-1}	
Inductance (L)	henry (H)	$kg \cdot m^2 \cdot s^{-2} \cdot A^{-2}$	Wb/A or V · s/A
Magnetic field (**B**)	tesla (T)	$kg \cdot s^{-2} \cdot A^{-1}$	Wb/m^2
Magnetic flux (Φ_B)	weber (Wb)	$kg \cdot m^2 \cdot s^{-2} \cdot A^{-1}$	V · s
Magnetic intensity (**H**)		A/m	
Power (P)	watt (W)	$kg \cdot m^2/s^3$	J/s
Pressure (p)	pascal (Pa)	$kg \cdot m^{-1}s^{-2}$	N/m^2 or J/m^3

PHYSICAL PROPERTIES

AIR (room temperature and average atmospheric pressure at sea level)

Density	$1.20\ kg/m^3$
Specific heat (c_p)	$1.00 \times 10^3\ J\ kg^{-1}\ K^{-1}$
Speed of sound	343 m/s
Index of refraction	1.000293 (visible light)

WATER (room temperature and atmospheric pressure)

Density	$1.00 \times 10^3\ kg/m^3$
Specific heat	$4.18 \times 10^3\ J\ kg^{-1}\ K^{-1}$
Speed of sound	$1.26 \times 10^3\ m/s$
Index of refraction	1.33 (visible light)

EARTH

Density (mean)	$5.49 \times 10^3\ kg/m^3$
Radius (mean)	6.37×10^6 m
Mass	5.97×10^{24} kg
Atmospheric pressure	1.01×10^5 Pa (average sea level)
Mean earth-moon distance	3.84×10^8 m

SOLAR SYSTEM (see Appendix A for more data)

Body	Mean radius of orbit (m)	Mean radius of body (m)	Mass (kg)
Sun		6.96×10^8	1.99×10^{30}
Mercury	5.79×10^{10}	2.42×10^6	3.35×10^{23}
Venus	1.08×10^{11}	6.10×10^6	4.89×10^{24}
Earth	1.50×10^{11}	6.37×10^6	5.97×10^{24}
Mars	2.28×10^{11}	3.38×10^6	6.46×10^{23}
Jupiter	7.78×10^{11}	7.13×10^7	1.90×10^{27}
Saturn	1.43×10^{12}	6.04×10^7	5.69×10^{26}
Moon	3.84×10^8	1.74×10^6	7.35×10^{22}

Physics

Second Edition

Frederick J. Keller
Clemson University

W. Edward Gettys
Clemson University

Malcolm J. Skove
Clemson University

Consulting Editor
Lawrence Coleman

McGraw-Hill, Inc.

NEW YORK ST. LOUIS SAN FRANCISCO AUCKLAND BOGOTÁ CARACAS LISBON LONDON MADRID MEXICO MILAN MONTREAL NEW DELHI PARIS SAN JUAN SINGAPORE SYDNEY TOKYO TORONTO

PHYSICS

2 3 4 5 6 7 8 9 0 VNH VNH 9 0 9 8 7 6 5 4 3

ISBN 0-07-023461-2

This book was set in Garamond by Progressive Typographers Inc.
The editors were Susan J. Tubb, Deena Cloud, Margery Luhrs, and Jack Maisel;
the designer was Armen Kojoyian;
the production supervisor was Janelle S. Travers.
The cover illustration was done by Greg MacNichol.
The photo editor was Safra Nimrod; the photo researcher was Mira Schachne.
New drawings were done by Fine Line Illustrations, Inc.
Von Hoffmann Press, Inc., was printer and binder.

Cover: Artist's representation of a laser fusion reactor with a lithium vortex. The design involves no interior walls. It was developed by workers at the Oak Ridge National Laboratory.

Library of Congress Cataloging-in-Publication Data

Keller, Frederick J., (date).
Physics / Frederick J. Keller, W. Edward Gettys, Malcolm J. Skove. — 2nd ed.
p. cm.
Includes index.
ISBN 0-07-023461-2
1. Physics. I. Gettys, W. Edward. II. Skove, Malcolm J., (date). III. Title.
QC23.K398 1993
530—dc20 92-38761

ABOUT THE AUTHORS

FREDERICK J. KELLER was born in Huntington, West Virginia on May 10, 1934. He served in the United States Air Force for four years as an aircraft navigator. He took his B. S. from Marshall University in 1960, and his M. S. and Ph. D. from The University of Tennessee in 1962 and 1966. In 1966 he joined the faculty of Clemson University where he is now professor of physics. Dr. Keller has been deeply involved with the training of high school physics teachers for many years. He was the director of two summer institutes for high school teachers and the associate director of six others.

W. EDWARD GETTYS was born in Gaffney, South Carolina on March 16, 1939. He received his B. S. and M. S. degrees in physics at Clemson University. He earned his Ph.D. in physics at Ohio University, and joined the faculty at Clemson University in 1963, where he was professor of physics for 28 years. In addition to research in condensed matter physics, he has been active in the advanced placement physics program and has worked toward improving high school physics instruction. He is now Professor Emeritus of Physics at Clemson University.

MALCOLM J. SKOVE was born in Cleveland, Ohio on March 3, 1931. He served as a midshipman in the United States Navy and a corporal in the United States Army. He attended The University of New Mexico, Pennsylvania State University, Jyochi Diagaku (Tokyo), and Clemson University, where he obtained a B. S. in industrial physics in 1956. He was awarded a Ph. D. in physics from the University of Virginia in 1960. He has taught at the University of Virginia, Illinois State University, University of Puerto Rico, Clemson University, Haile Selassie University (Addis Ababa), and the Swiss Federal Institute of Technology. In 1988 he was program director for solid state physics at the National Science Foundation. He is now Alumni Professor of Physics at Clemson University.

CONTENTS IN BRIEF

CONTENTS

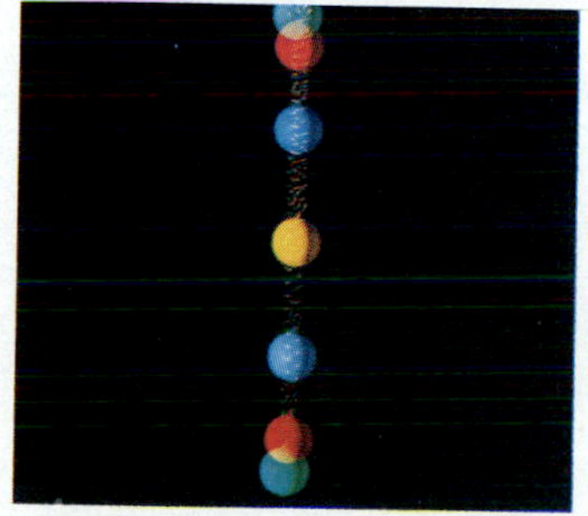

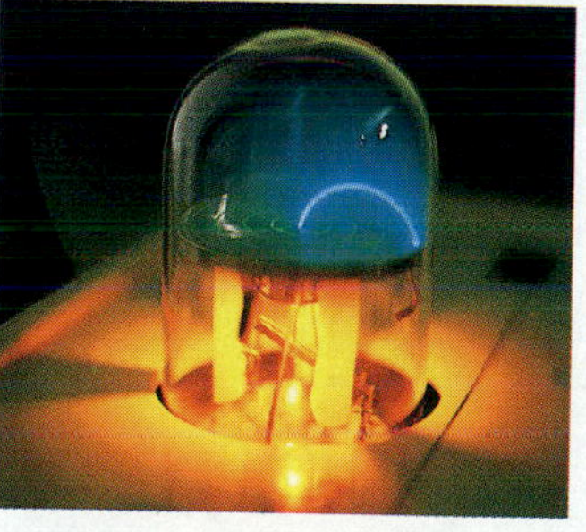

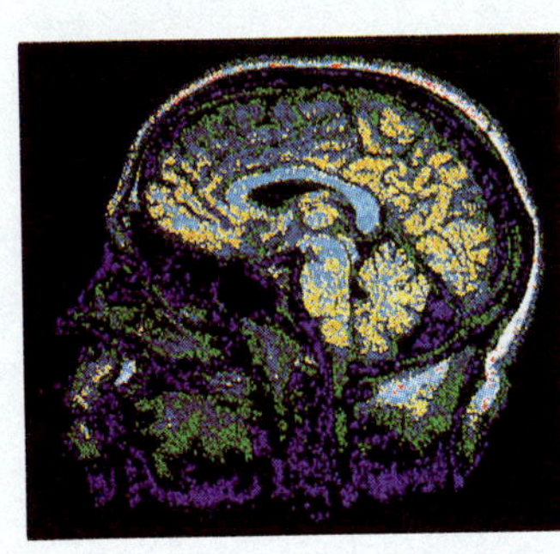

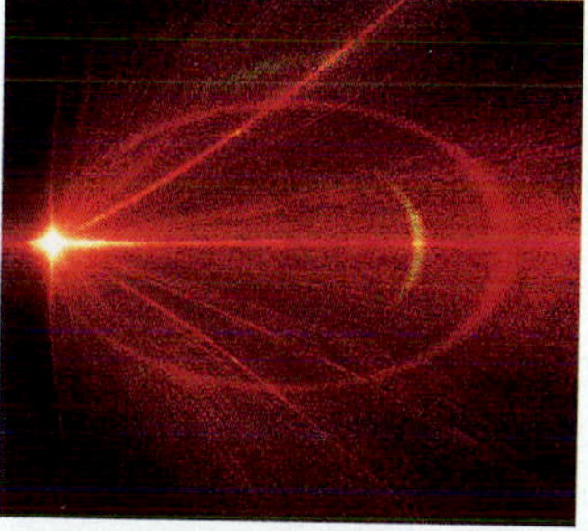

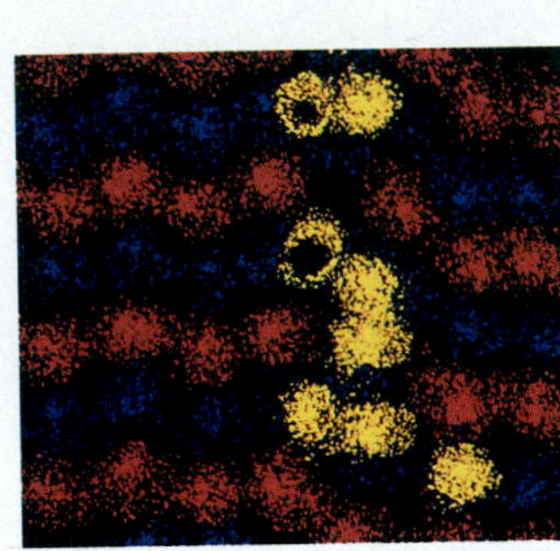

PREFACE

The book is designed for a sequence of courses in physics for science and engineering majors. This sequence usually incorporates two or three three-semester-hour courses (or two or three five-quarter-hour courses). The text is available in two versions: *Physics,* which contains 39 chapters and ends with the Bohr model of hydrogen; and *Physics: Classical and Modern,* which contains an additional six chapters on modern physics. We presume that students beginning in this physics sequence will either have completed or will be concurrently enrolled in a first course in calculus.

Objectives

The main objectives of this textbook are to engage students in the discovery of physics and to provide them with theory and applications in a clear, understandable presentation. To help accomplish this, any new concept or topic is introduced with a familiar example whenever possible. With this specific-to-general approach, we draw on the students' experience and avoid launching into an unfamiliar subject with an abstract discussion. Typically, the discussion flow in a chapter is (familiar example) → (general principle) → (further examples). Similarly, models are frequently used to explain physical phenomena. The technique of constructing and using a model to provide an approximate description of a real process is demonstrated whenever the opportunity arises.

Our experience in teaching undergraduates has proven that only when students appreciate the conceptual nature of a subject can they truly understand how a problem is solved. With this in mind, we have included additional topics that enable students to grasp the broader concepts, thereby making problem solving easier.

FEATURES OF THE SECOND EDITION

Organization

The chapter-by-chapter organization of material is largely traditional: mechanics (Chapters 2 to 15), thermodynamics (Chapters 16 to 19), electricity and magnetism (Chapters 20 to 31), waves (Chapters 32 to 34), optics (Chapters 35 to 37), and modern physics (Chapters 38 and 39).

Nontraditional aspects of the text's organization are the placement of Newton's Law of Universal Gravitation (Chapter 7) and Waves (Chapter 32). The rationale for an early introduction to gravitation is to have the gravitational force law available for the discussion of potential energy (Chapter 9) and to be able to apply a fundamental force law to Newton's Second Law as early as possible.

The rationale for the placement of waves after electricity and magnetism is to provide a cohesive treatment of mechanical, sound, and electromagnetic waves and a stronger demonstration and utilization of the unifying principles in physics. For instructors who wish to treat waves earlier in the course, Chapter 32 is easily placed after Oscillations, Chapter 14.

The breadth of material in each chapter is intended to correspond to that presented in three or four one-hour class meetings. To satisfy the diverse needs of different curricula and different instructors, most chapters contain more information than can be discussed in this time. Chapters 15, 30, 31, 33, 35, and 37 contain material that has little or no bearing on subsequent chapters, and a course sequence can be designed that omits one or more of them.

Major Changes

This second edition contains a number of changes and improvements based on suggestions offered by users of the text and reviewers of the manuscript. The text has been revised thoroughly, updated where appropriate, and rewritten to improve clarity. The major changes are the following.

1. There are over 25 percent more problems than were in the first edition, with a wider range of difficulty.
2. A new student self-test follows each worked example.
3. New modern physics highlights, written by Lawrence Coleman of the American Institute of Physics, are placed at the end of ten chapters.
4. Spreadsheet software is used in the solutions to numerical problems.
5. "Physics at Work," an interactive videodisk, is now available to adopters of the textbook.

Chapter Design

Each chapter begins with a short introduction that is intended to orient and motivate the student toward attaining the chapter's goals and ends with a concise summary. Most chapters include a Commentary that is designed to pique the student's interest. Some of the Commentaries include biographies of famous or especially interesting personalities in physics such as Isaac Newton or Count Rumford. Others have to do with important and rapidly developing technologies such as lasers or high-Tc superconductors or with interesting philosophical points such as determinism or the meaning of a physical law.

Modern Highlights

Many instructors are anxious to provide their students with a glimpse of the excitement of physics at the frontier. However, a number of fundamental concepts are used to discuss such topics, and the course often ends before all these concepts can be defined and mastered. Our modern physics highlights circumvent this problem by dealing with topics in modern physics in a nonrigorous fashion. The treatment of the material and the level of the discussion is similar to that in *Scientific American*. We have selected the topics that we believe are of greatest interest to contemporary physicists, and we have placed these at the earliest juncture in the book that is in keeping with the growing sophistication of the students taking the course.

Problem Solving

Since learning physics goes hand-in-hand with developing students' problem-solving skills, we have given careful attention to the placement and level of examples in each chapter and in end-of-chapter questions, exercises, and problems. Each chapter contains about 10 examples, 10 self-tests, 25 questions, 50 exercises, and 15 problems. Answers to the odd-numbered exercises and problems are given in the back of the book. For the sake of accuracy, these answers have been checked by at least two persons working independently.

Examples. Examples appearing early in a chapter are usually simple and are often used to illustrate the definition and units of a newly introduced physical quantity. Later, examples become more difficult and demonstrate problem-solving techniques. A few examples are used to show how to make order-of-magnitude estimates and to discuss simple models.

Self-Tests. To encourage students to read the text with pencil, paper, and calculator in hand, we follow each worked example with a self-test. Without such a prod, many students might read the text too passively.

Questions. A question can usually be answered without the need of pencil, paper, or calculator. The difficulty level of the questions has a wide range; some can be answered with a simple application of a definition, while others may require an extension of an idea presented in the text. Indeed, in some cases, a question should be regarded as a springboard for a classroom discussion of an advanced topic. For instance, Question 24-13 leads to the conclusion that the carrier drift velocity in some materials is not necessarily parallel to the electric field (that is, the electric conductivity is a tensor quantity), and a marble-and-pegboard model is used to visualize this phenomenon.

Exercises. Most of the exercises are not difficult, many involve a simple one-step calculation. We believe that it is important to provide a mechanism whereby a reasonably industrious student can enjoy quick success. Since exercises are keyed to specific sections of the chapter, a student has a limited amount of material to grasp in order to attack an exercise.

Problems. Problems are usually more difficult than exercises and involve more steps and a wider range of material. Some problems develop material beyond the level of that presented in the text. A few of the problems are very challenging. Lawrence S. Pinsky of the University of Houston contributed the "additional problems" that are at the end of a number of chapters.

Acknowledgments

We gratefully acknowledge the dedicated work of Susan Tubb, Deena Cloud, Safra Nimrod, Margery Luhrs, and Jack Maisel at McGraw-Hill. Loren Winters at the North Carolina School of Science and Mathematics and Danny Overcash at Lenoir-Rhyne College granted us their talents as photographers.

Larry Coleman, education fellow at the American Institute of Physics and Professor of Physics at the University of Arkansas, Little Rock, served as our consulting editor on this edition. In addition to writing the modern physics highlights and the new chapter on astrophysics, "The Physics of Stars," Larry guided this revision with his thoughtful comments. He has helped to make this a better book for professors and students.

We are indebted to the following reviewers for their patience and helpful suggestions: Edward Adelson, Ohio State University; Albert Altman, University of Lowell; Barbara Bates, Lakeland Community College; Carroll Bingham, University of Tennessee; William Coghlan, Grand Canyon College; Jai N. Dahiya, Southeast Missouri State University; J. P. Davidson, University of Kansas-Lawrence; Barry Freidman, Sam Houston State University; Linda Fritz, Franklin and Marshall College; Robert E. Gibbs, Eastern Washington University; Alan Goldman, Iowa State University; Walter Grandy, Jr., University of Wyoming; Michael J. Hones, Villanova University; George Horton, Rutgers University; Alvin Jenkins, North Carolina State University; Walter H. Johnson, University of Minnesota; Edwin Jones, University of South Carolina; Hans Laue, University of Calgary; Roger Ludin, California Polytechnic State University; Robert Marchini, Memphis State University; Paul Marguard, Casper College; Michael J. Mooney, Rose-Hulman Institute of Technology; Eugene

Mosca, U.S. Naval Academy; Captain David Myers, U.S. Air Force Academy; Michael J. Naughton, State University of New York-Buffalo; David Newton, DeAnza Community College; Andrew P. Odell, Northern Arizona University; Norman Pearlman, Purdue University; Harvey Picker, Trinity College; Francis Pinchanick, University of Massachusetts; Lawrence Pinsky, University of Houston; Derek L. Pursey, Iowa State University; Frank A. Rickey, Jr., Purdue University; John Risley, North Carolina State University; Donald F. Ryan, State University of New York-Plattsburgh; Kumar Sharman, University of Manitoba; Robert Siemann, Cornell University; Charles R. Taylor, Pacific Lutheran University; Rev. Clarence M. Wagener, S.J., Creighton University; and David M. Wolfe, The University of New Mexico.

We also gratefully acknowledge the following reviewers of the first edition: Albert Altman, University of Lowell; John P. Barach, Vanderbilt University; Richard G. Barnes, Iowa State University; John H. Broadhurst, University of Minnesota; Richard R. Bukrey, Loyola University of Chicago; Joseph S. Chalmers, University of Louisville; Colston Chandler, University of New Mexico; William R. Cochran, Youngstown State University; Peter R. Fontana, Oregon State University; Anthony P. French, Massachusetts Institute of Technology; J. David Gavenda, University of Texas, Austin; Vince Griffin, Tulsa Junior College; Robert B. Hallock, University of Massachusetts, Amherst; Paul Heckert, Western Carolina University; Virgil L. Highland, Temple University; Robert P. Hurst, SUNY at Buffalo; Mario Iona, University of Denver; Alvin W. Jenkins, Jr., North Carolina State University; Peter B. Kahn, SUNY-Stony Brook; Carl A. Kocher, Oregon State University; Donald Kydon, University of Waterloo; B. A. Logan, University of Ottawa; Oscar Lumpkin, University of California, San Diego; Joseph L. McCauley, University of Houston; Alvin Meckler, University of Maryland, Baltimore County; Ralph C. Minehart, University of Virginia; William J. Mullin, University of Massachusetts, Amherst; Jack H. Noon, University of Central Florida; Benedict Oh, Pennsylvania State University; Lawrence S. Pinsky, University of Houston; Stanley J. Shepherd, Pennsylvania State University; Wilbur C. Thoburn, Iowa State University; James Trefil, George Mason University; Somdev Tyagi, Drexel University; Gordon G. Wiseman, University of Kansas; Lowell Wood, University of Houston; Richard K. Yamamoto, Massachusetts Institute of Technology, and Jens Zorn, University of Michigan.

We especially acknowledge the contributions of Wendy Schaffer and Veneeta Ribeiro, who provided the students' perspective — the most important of all.

Finally, we acknowledge the support of our families. It was they who also endured the many long hours and the occasional short tempers that accompany such a project.

Frederick J. Keller
W. Edward Gettys
Malcolm J. Skove

ANCILLARIES

The following ancillaries are available with this text:

For the Student

Study Guide, Marllin L. Simon and G. Donald Thaxton, Auburn University. Each chapter contains a comprehensive review of the principle concepts presented in that chapter. In addition, we have included over 1600 practice exercises with detailed solutions, 215 example problems with an approach strategy and detailed solution for each, and 925 sample test questions with the answers.

Student Solutions Manual, Jai Dahiya and Giulio Venezian, Southeast Missouri State University. Contains the worked-out solutions to odd-numbered questions, problems, and exercises in the text.

Schaum's 3000 Solved Problems in Physics. A complete and expert source of problems with solutions. Also includes sections on problem solving tips, computer applications and lists of available software.

For the Instructor

Instructor's Solutions Manual, Jai Dahiya and Giulio Venezian, Southeast Missouri State University, and William Stephens. This manual contains the complete worked-out solutions to all questions, problems, and exercises in the text.

Overhead Transparencies. Over 160 4-color acetates of useful figures from the text are included.

Test Bank, John Garlow, Tarrant County Junior College. Over 1000 multiple choice and free response questions and problems are included in this manual. This is approximately four times the amount that can be used in a typical course, and, in a given term, the instructor has a broad variety of test questions from which to choose.

McGraw-Hill Testing System. This computerized version of the examination questions found in the test bank enables the instructor to select questions by section, topic, question type, difficulty level and other criteria. Instructors may add their own criteria and edit their own questions.

Videos. These original videos, created by George Horton of Rutgers University, walk a student through some of each chapter's problems with helpful problem-solving hints and strategies.

Physics at Work. The original videodisc including over 1500 still images plus hundreds of film clips and computer graphics was prepared exclusively for the

McGraw-Hill physics series. The videodisc contains fascinating experiments, demonstrations and multi-screen equations that enable students to understand difficult concepts. The videodisc features custom graphics of hard to visualize phenomena such as molecular orbitals, wave behavior and relative points of reference. Includes bar-coded Image Directory indexed by name, concept, and frame number and Quick Reference Card. English and Spanish narration are included. Optional computer software is available.

This ancillary program was designed to complement both your teaching efforts and your students' learning process. If you would like information and costs on any supplemental materials, please contact your local McGraw-Hill representative.

Physics

INTRODUCTION 1

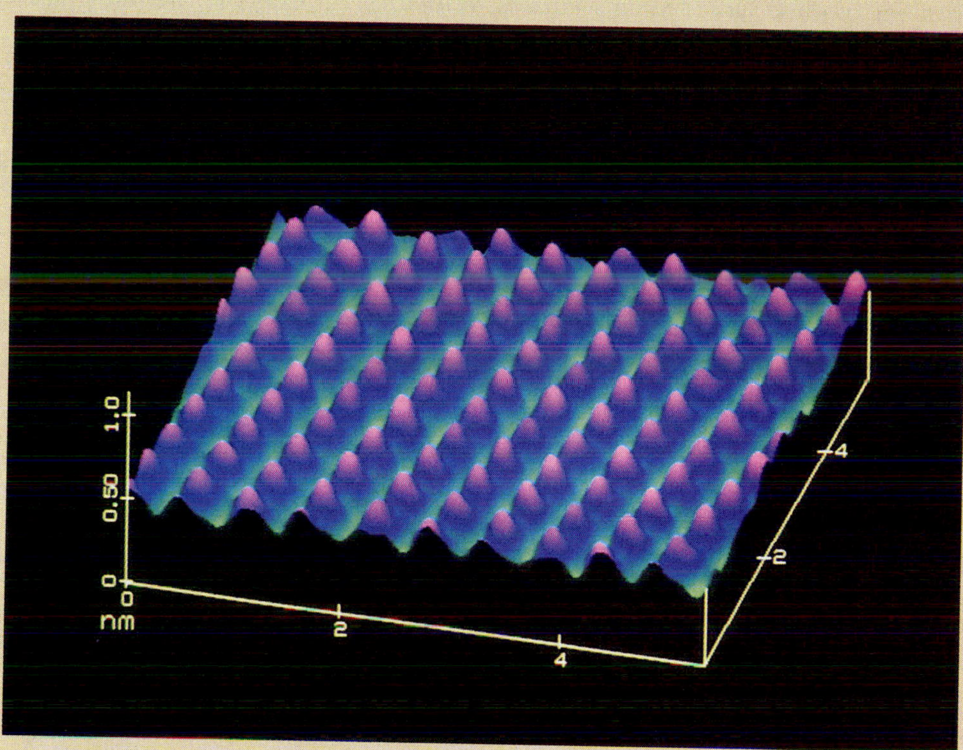

An image of atoms on the surface of table salt. Modern instruments can produce images of individual atoms on the surface of a solid. This "picture" was taken with an atomic force microscope.

Until about 1850, there were texts and courses in what was called natural or experimental philosophy. The name recognized the contrast between subjects that were dependent on experiments and those, such as literature and religion, that were not. As the results and conclusions of experimental philosophy accumulated, it became difficult for a single person to work in the whole field, and subdivisions appeared. Long before 1850, chemistry, astronomy, geology, and other such studies split off into independent disciplines. As this happened, the core that was left came to be known as ***physics.*** Because of its central importance in the sciences, an understanding of physics is required in many other disciplines.

Physics is a quantitative science that includes mechanics, thermal phenomena, electricity and magnetism, optics, and sound. These subjects are part of classical physics. If speeds near that of light or sizes near that of an atom are important in a physical problem, the topics of modern physics, the discoveries of the 1900s, must be discussed. These topics include relativity and quantum mechanics.

Physics is the study of particularly simple systems, such as single atoms. Scientific methods are often expressed more clearly in these simple systems of physics than in many other sciences. Because of this, physics is often regarded as a model for the "scientific method."

1-1 STANDARDS

In physics, the quantities measured are carefully defined. Not only must the number that is measured be precise, but the measurement must be referred to a common standard, such as an inch or a meter. Often the measurements will contain more than one of these agreed-upon standards. Velocities, for example, are often measured in units of meters per second. The meters and the seconds must be traceable to a standard meter and a standard second.

In the first part of this course you will need only three standards, those for time, length, and mass. Additional standards will be needed later for discussing temperature and electricity. All other quantities that are used in this part of the course can be built from the standards for time, length, and mass. Since these three standard quantities are not defined in terms of any others, they are sometimes called ***indefinable.*** The standards for time, length, and mass are laid down in prescriptions that give a method for reproducing them and for comparing them with measured quantities. This is called an ***operational definition*** because it specifies the operations that must be performed to reproduce the standards and to compare them with measured quantities. Thus, although time, length, and mass are indefinable in terms of other quantities, each of them has an operational definition.

Time, length, and mass have operational definitions.

Consider the desirable features of a standard:

1. It should be immutable, so that measurements made today can be compared with those made in the next century.
2. It should be easily available, so that many laboratories can duplicate it.
3. It should be precise, so that the standard is available to whatever precision is technologically possible.
4. It should be universally agreed upon, so that results obtained in different countries are comparable.

As an example, consider the definition of the standard of time. The second was originally defined in terms of the length of the day. Later it was found that the length of the day, as measured by clocks based on other phenomena, varied throughout the year, from year to year, and from century to century. If a second were defined to be 1/86,000 of a day, then such a variation in the length of a day would be impossible

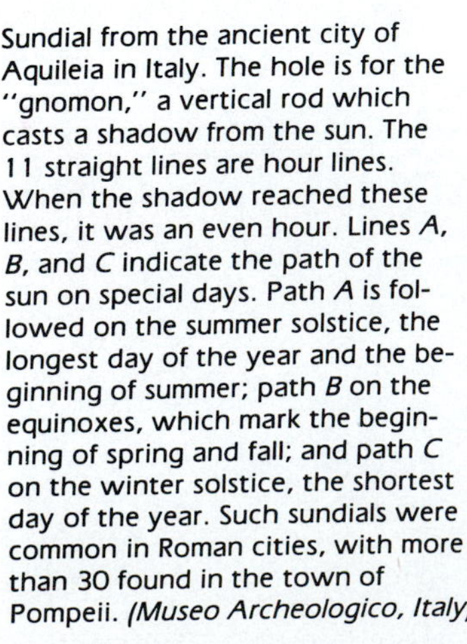

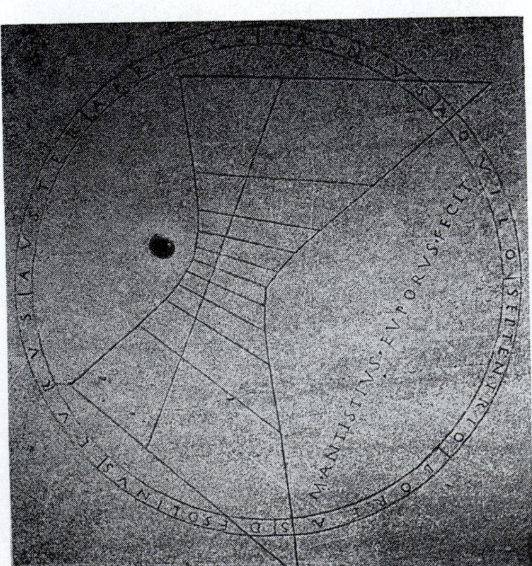

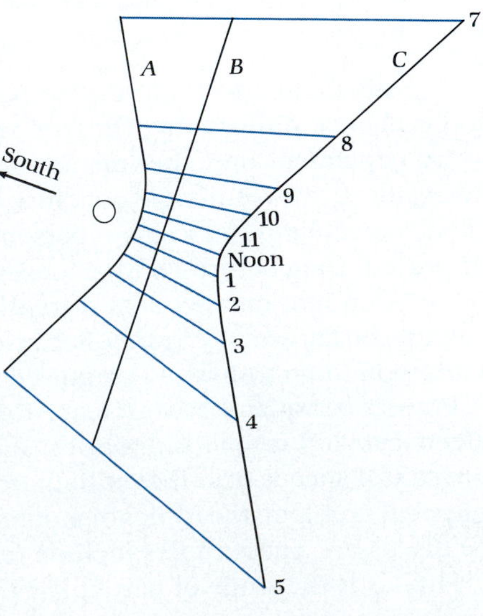

Sundial from the ancient city of Aquileia in Italy. The hole is for the "gnomon," a vertical rod which casts a shadow from the sun. The 11 straight lines are hour lines. When the shadow reached these lines, it was an even hour. Lines *A*, *B*, and *C* indicate the path of the sun on special days. Path *A* is followed on the summer solstice, the longest day of the year and the beginning of summer; path *B* on the equinoxes, which mark the beginning of spring and fall; and path *C* on the winter solstice, the shortest day of the year. Such sundials were common in Roman cities, with more than 30 found in the town of Pompeii. *(Museo Archeologico, Italy)*

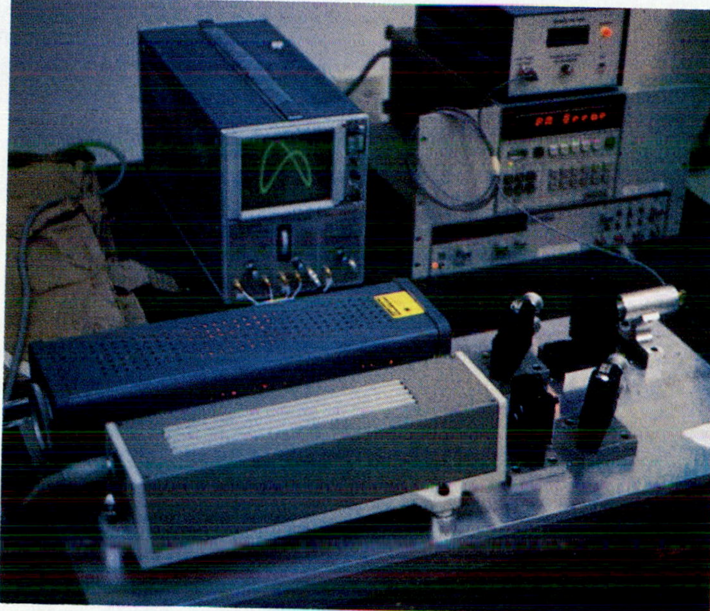

Time and length standards. This cesium clock (*left*) is the primary standard for measuring time at the National Institute of Standards and Technology (NIST, formerly the National Bureau of Standards). This device keeps time with an accuracy of about 0.000 003 s per year. NIST uses this iodine-stabilized helium-neon laser (*right*) as its standard for length. The laser approximates the ideal meter to an accuracy of about 0.000 000 000 1 m.

by definition. But this definition would have consequences that would be disconcerting. We would find that the vibrational period of an atom or the speed of light depended on the date it was measured. Since we believe that the vibrations of atoms and the speed of light do not vary with time, a definition of time based on the properties of atoms or on what are believed to be constants of nature, such as the speed of light, is preferred.

An international organization, the Conférence Générale des Poids et Mesures, or CGPM (in English, the General Conference on Weights and Measures), is internationally recognized as the authority on the definition of units. At present the CGPM definitions of the standards for time, length, and mass are as follows:

Time. ***One second is 9,192,631,770 periods of a certain vibration of the atom*** $\boldsymbol{Cs^{133}}$. You need not be concerned with the details of the atomic behavior or the insides of clocks that use these atoms, except that these clocks are the most reproducible timekeepers that are now known. Two of these clocks will agree with each other to a precision of one part in 10^{13}, or about one second in a million years. These clocks are not cheap nor particularly easy to build, but they are being used in the standards laboratories of several nations, including the National Institute of Standards and Technology (NIST) near Washington, D.C.

As far as being immutable, it is believed that atomic properties are independent of time, but that is only one of the assumptions that go into this definition. You might consider what would happen if the properties of atoms did change over the age of the universe! The observed properties of the universe put severe restrictions on how much the properties of atoms can change in time.

Length. ***One meter is the length of the path traveled by light in vacuum during a time interval of 1/299,792,458 of a second.*** Note that the definition of the meter depends on the definition of the second and on a presumption of the constancy of the speed of light.

Mass. ***One kilogram is the mass of a particular platinum-iridium cylinder kept near Paris, France.*** This was agreed upon at the first meeting of the CGPM in 1889. The reason this standard is not yet based on atomic standards is that the

Mass standard. The NIST standard kilogram arrived in the United States on January 2, 1890. It was fabricated by the International Bureau of Weights and Measures from a platinum-iridium alloy.

measurement of atomic masses and their comparison with large-scale masses is not yet as precise as the measurements that can be made on large-scale objects such as the standard kilogram. The kilogram was defined so that the mass of 10^{-3} cubic meter of water is very nearly one kilogram at a temperature of 20°C.

Secondary and tertiary standards that are compared with the primary standards are kept in standardization laboratories throughout the world. It is possible to trace the length of a common laboratory meter stick through these standards to a measurement of the path covered by light in 1/299,792,458 of a second, the second of a watch to the vibrations of a cesium atom, and the scale in a grocery store to the standard kilogram in France.

1-2 SYSTEMS OF UNITS

Besides the standards on which all measurements are based, we need a system of units. A system of units includes (i) the standards, (ii) a method of forming larger and smaller units, and (iii) definitions of derived quantities such as energy, power, and force. For example, much of the commerce in the United States uses the British system of units in which the unit of distance is the inch (defined to be 0.0254 of the standard meter) and the unit of mass is a pound-mass (defined to be 0.4359237 of the mass of the standard kilogram). Larger and smaller units, such as the foot (12 inches) and the ounce ($\frac{1}{16}$ of a pound), are also part of the British system of units. Derived units, such as that for the horsepower (550 feet squared pound-mass per second cubed) are also defined. Although still in common usage in the United States, these units are not often used elsewhere and are almost never used in scientific work. Engineers in the United States generally must be familiar with both the British system and the SI system, which we now describe.

TABLE 1-1. ***Prefixes for SI Units***

Symbol	Name	Amount
E	exa	10^{18}
P	peta	10^{15}
T	tera	10^{12}
G	giga	10^{9}
M	mega	10^{6}
k	kilo	10^{3}
↓	↓	↓
c	centi	10^{-2}
m	milli	10^{-3}
μ	micro	10^{-6}
n	nano	10^{-9}
p	pico	10^{-12}
f	femto	10^{-15}
a	atto	10^{-18}

The Systême Internationale d'Unités, or SI system of units, was set up in 1960 by the CGPM. This is the system used by most of the world, essentially all science, and this book. It uses the kilogram (kg), the meter (m), and the second (s) as base units and has a general method of forming larger and smaller units. The larger and smaller units are formed with prefixes that modify the base and derived units by factors of various powers of a thousand. These prefixes and their abbreviations are listed in Table 1-1. Derived units, such as the unit for power, the watt [defined as one kilogram meter squared per second cubed (kg m² s⁻³)], are modified in the same manner by the same prefixes. Thus one microwatt (1 μW) is one-millionth of a watt, and one kilowatt (1 kW) is 1000 watts.

The range of measurements that are made in physics is illustrated in Fig. 1-1. The behavior of objects even over these ranges seems to have some order. Physics attempts to describe this order.

1-3 DIMENSIONS, UNITS, AND PRECISION

A measurement has a dimension, a unit, and a precision.

Physics is a quantitative science that deals with the real world. Physicists are concerned with the grubby details of measurement. A measurement of a physical quantity, such as a length of 4.2 m, includes (i) a dimension, (ii) a unit, and (iii) a precision. The "m" tells us that the dimension is length and that the unit of length being used is the meter. The 4.2 (rather than 4.2157) characterizes the precision with which the measurement was made.

Dimensions

The ***dimension*** *of a quantity is the physical property that the quantity describes.* For example, the dimensions of the standard quantities time, length, and mass are

Figure 1-1. An indication of the range over which physical measurements are made. Measurements of time and distance extend over a range of 10^{40}, and measurements of mass extend over a range of 10^{80}.

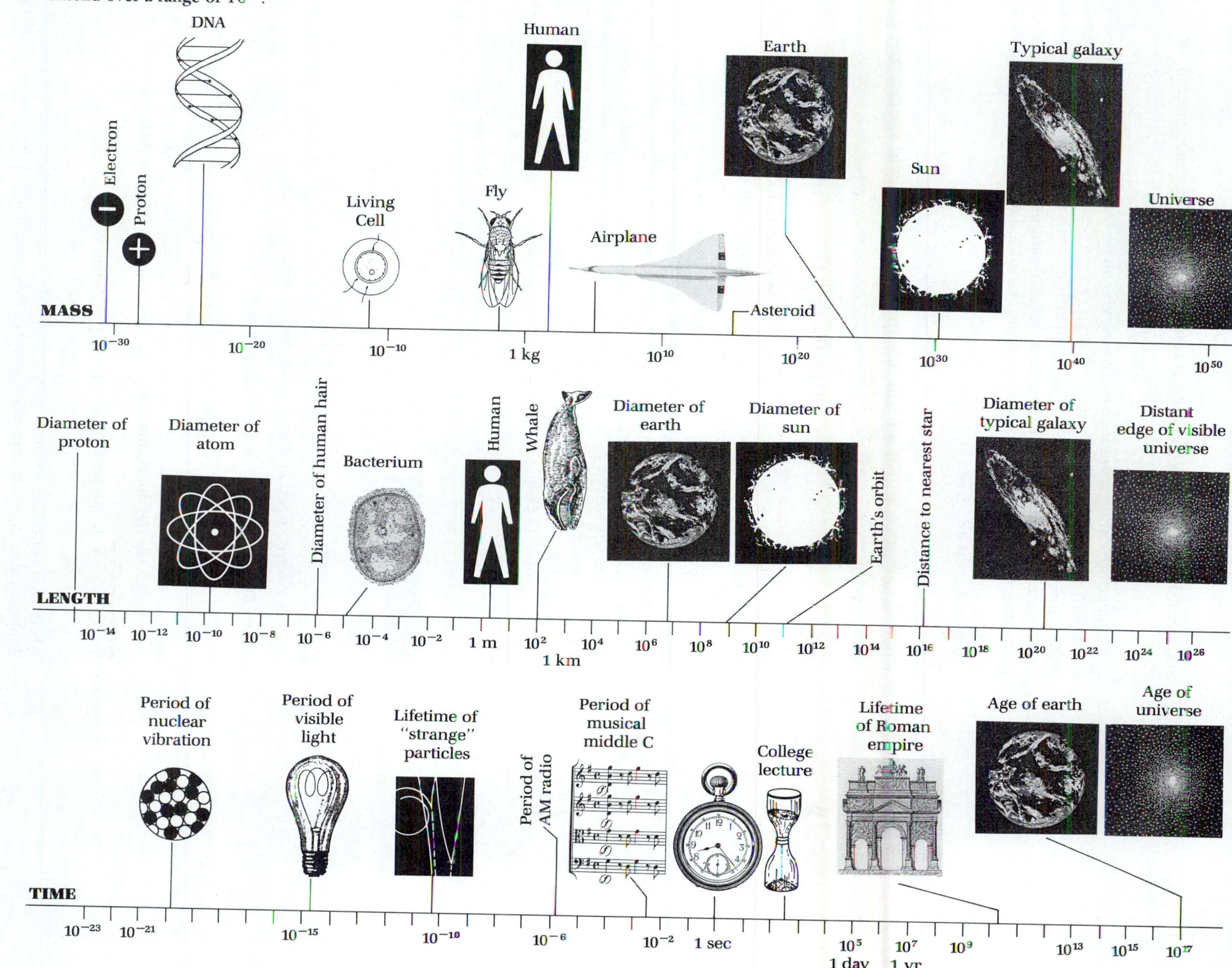

simply time [T], length [L], and mass [M]. The dimensions of other quantities are combinations of these and other standard quantities. For example, the dimension of speed v is length divided by time, or

$$[V] = [L]/[T]$$

and the dimension of acceleration a is length divided by time squared, or

$$[A] = [L]/[T]^2$$

A procedure called ***dimensional analysis*** can be helpful in spotting an error in an equation, or in helping to verify that an error has not been made. Dimensional analysis is based on the fact that only quantities with the same dimension can be equal to one another. For example, a length cannot be equal to a time. Also, quantities with different dimensions cannot be added to one another or subtracted from one another.

Suppose you are given an equation

$$x = x_0 + vt$$

where x and x_0 represent lengths, v represents a speed, and t represents a time. Checking the dimensions, we have

$$[L] = [L] + \frac{[L]}{[T]}[T]$$

On the right-hand side, the dimension of time in the numerator and the denominator of the second term cancel, so that each term has the dimension [L]. We conclude that the dimensionality of the equation is correct.

Rules for dimensional analysis

The rules for dimensional analysis are (i) assign a dimension to each symbol in an equation according to the physical property it describes, (ii) multiply and divide the dimensions using the rules of algebra, and (iii) check the resulting dimension of each term for agreement. An additional consideration is that any quantity contained in the argument of a transcendental function must be dimensionless. For example, sin x is meaningless if x represents a distance. However, sin x/d is meaningful if x and d have the same dimension. Note that dimensional analysis cannot be used to confirm or deny the presence of any pure numbers (such as 2 or π) in an equation.

EXAMPLE 1-1

Dimensions in equations. Suppose you are given the equation

$$x = x_0 + vt + \tfrac{1}{2}at^3$$

where x represents a length, t a time, v a speed, and a an acceleration. Use dimensional analysis to verify whether this equation may be valid.

Solution. Checking the dimensions, we have

$$[L] = [L] + \frac{[L]}{[T]}[T] + \frac{[L]}{[T]^2}[T]^3$$

After cancellation, each term has the dimension [L] except for the last term on the right-hand side, which has the dimension [L][T]. We conclude that the equation cannot be valid.

SELF-TEST 1-1. Suppose you are given the equation

$$x = x_0 + vt + \tfrac{1}{2}at^2$$

where each symbol represents the same physical quantity as it did in the example above. Use dimensional analysis to determine whether this equation may be correct. ***ANSWER:*** The equation may be correct.

Units

A unit is the scale with which a dimension is measured. For example, a unit of length can be a meter, or a foot, or a mile. Since various sets of units are commonly used, we must know how to convert a unit from one set to another set. We must often convert back and forth between SI and British units. Suppose you are given a length of 4.29 feet (ft) and you want to know this length in meters. In the conversion tables on the inside back cover of this book, you find that

Converting from one set of units to another

$$1 \text{ ft} = 0.305 \text{ m}$$

You can construct a conversion factor with this information. A conversion factor is the number 1 written such that it converts a unit from one set to another. In this case,

$$1 = \frac{0.305 \text{ m}}{1 \text{ ft}} = 0.305 \text{ m/ft}$$

Multiplying the length 4.29 ft by this form of the number 1 gives

$$4.29 \text{ ft} = (4.29 \text{ ft})(0.305 \text{ m/ft}) = 1.31 \text{ m}$$

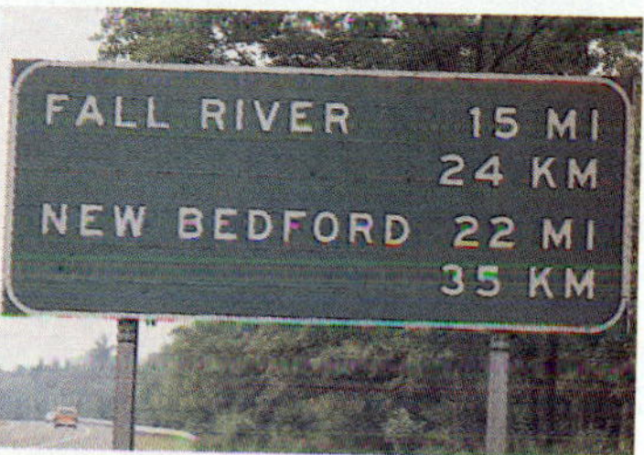

Length units. This road sign is an indication of the gradual shift to SI units in the United States.

Notice that the units in the numerator and denominator cancel one another. That is, the units are treated algebraically as though they were numbers.

Suppose you want to convert a length of 2.85 m to a length in feet. A conversion factor for this process is

$$1 = \frac{1 \text{ ft}}{0.305 \text{ m}} = 3.28 \text{ ft/m}$$

so that

$$2.85 \text{ m} = (2.85 \text{ m})(3.28 \text{ ft/m}) = 9.34 \text{ ft}$$

An alternative procedure for this conversion is to divide by the conversion factor 0.305 m/ft:

$$2.85 \text{ m} = \frac{2.85 \text{ m}}{0.305 \text{ m/ft}} = 9.34 \text{ ft}$$

Therefore the quantity to be converted may be either multiplied or divided by a conversion factor, depending on which procedure gives the quantity the desired unit.

EXAMPLE 1-2

Converting a density. The mass density of a substance is the mass of a given amount of the substance divided by the volume occupied by that amount. For example, 1.0 kg of water (under normal circumstances) occupies a volume of 1.0 liter, or 1.0×10^{-3} m^3. Thus the density of water is

$$\text{Density} = \frac{\text{mass}}{\text{volume}} = \frac{1.0 \text{ kg}}{1.0 \times 10^{-3} \text{ m}^3} = 1.0 \times 10^3 \text{ kg/m}^3$$

Use the relation 1 pound (lb) = 0.453 kg to convert the density of water to the British unit pound per cubic foot (lb/ft^3).

Solution. For converting between pound and kilogram, we have

$$1 = \frac{0.453 \text{ kg}}{1 \text{ lb}} = 0.453 \text{ kg/lb}$$

If 1.0×10^3 kg/m^3 is divided by this conversion factor, then the unit kilogram will be converted to pound. For converting between cubic meters and cubic feet, we have

$$1 = \frac{(0.305\text{ m})^3}{(1\text{ ft})^3} = (0.305)^3\text{ m}^3/\text{ft}^3$$

If we multiply 1.0×10^3 kg/m³ by this conversion factor, then the unit $(1/\text{m}^3) = \text{m}^{-3}$ will be converted to $(1/\text{ft}^3) = \text{ft}^{-3}$. Therefore

$$1.0 \times 10^3\text{ kg/m}^3 = 1.0 \times 10^3\text{ kg/m}^3\,\frac{(0.305)^3\text{ m}^3/\text{ft}^3}{0.453\text{ kg/lb}} = 62.6\text{ lb/ft}^3$$

Mark through each unit that cancels to verify that the units are correct.

SELF-TEST 1-2. In SI units, the acceleration due to gravity is 9.8 m/s². Convert this acceleration to feet per second squared. ***ANSWER:*** 32 ft/s².

Precision and Significant Figures

The precision of a measured value of a physical quantity is reflected in the number of significant figures (or significant digits) used in stating the value. For example, the number 90.2 has three significant figures, while 7.5 has two significant figures. A leading zero in a number is never counted as a significant figure—the number 0.054 has two significant figures. A trailing zero is a significant figure if it is to the right of the decimal point—the number 7.40 has three significant figures. Some difficulty arises from the interpretation of whether a trailing zero to the left of the decimal point is counted as a significant figure. Does the number 6900 have two, three, or four significant figures? If the number of significant figures is not specified in such a case, then it is prudent to treat 6900 as if it has four significant figures.

The definition of a significant figure is the following:

Definition of a significant figure

A ***significant figure*** is a digit in a number, except that leading zeros are not significant figures and trailing zeros are significant figures only if so specified.

The ***least significant figure*** in a number is the significant figure farthest to the right. In the number 8.47, the 7 is the least significant figure.

Suppose you are asked to give your height in SI units and you know that it is 68 inches in British units. You convert the 68 inches to meters by multiplying by the conversion factor 0.0254 m/inch with your calculator:

$$(68\text{ inches})(0.0254\text{ m/inch}) = 1.7272\text{ m}$$

If you give this value as your height, then you are implying that you know your height to a precision of 0.1 mm (the thickness of a sheet of paper). Such a claim is completely unwarranted. To avoid claiming too much precision, the value you state in the SI unit should reflect about the same precision as the original value in the British unit. If you blindly state whatever your calculator reads, you will almost always be quoting nonsense because your calculator treats any number as exact. You must "round off" the value your calculator gives so that the value after rounding off properly reflects the precision of the value. That is, your stated value should have the proper number of significant digits.

Rules for significant figures

The rules used in this book for rounding a value to the proper number of significant digits are the following:

1. **Rounding off** If the digit to the right of the least significant figure in the final answer is 4 or less, the value is rounded *down*. To two significant figures, the number 8.54 is rounded off to 8.5. If the digit to the right of the least significant digit in the final answer is 5 or higher, the value is rounded *up*. To two significant figures, the number 8.55 is rounded off to 8.6.

2. Multiplication and division The result from multiplication or division has the same number of significant figures as the least precisely known number used in the calculation. For example,

$$3.218\ \text{m}/0.53\ \text{s} = 6.1\ \text{m/s}$$

3. Addition and subtraction The least significant figure in the result of addition or subtraction occupies the same position relative to the decimal point as the number in the calculation whose least significant figure is farthest to the left. For example,

$$8.1\ \text{m} + 3.77\ \text{m} = 11.9\ \text{m}$$

or

$$4.84\ \text{kg} - 4.3\ \text{kg} = 0.5\ \text{kg}$$

As you can see, the number of significant figures in the result is not necessarily the same as that in the least precisely known number involved in the calculation.

4. Transcendental functions The result of the evaluation of a transcendental function is given the same number of significant figures as the argument of the function. For example,

$$\sin 34° = 0.56$$

or

$$\ln 9.356 = 2.236$$

The rules introduced here do not have the full sophistication of a thorough statistical analysis, but they do provide a simple way to state an answer that reflects its precision.

EXAMPLE 1-3

Significant figures in a calculation. Consider the area of the sheet of wood shown in Fig. 1-2. *(a)* Determine the area A_r of the rectangular part. *(b)* Determine the area A_s of the semicircular part. *(c)* Determine the area A of the entire sheet.

Solution. *(a)* The area A_r of the rectangular part is

$$A_r = (1.2\ \text{m})(1.37\ \text{m}) = 1.6\ \text{m}^2$$

Although 1.2 multiplied by 1.37 with a calculator gives 1.644, the answer is rounded to two significant digits because that is the precision of the least precise number in the calculation. *(b)* The area of a circle of radius R is πR^2. Thus the area A_s of the semicircular part is

$$A_s = \tfrac{1}{2}\pi(1.37\ \text{m}/2)^2 = 0.737\ \text{m}^2$$

The area is given to three significant digits because the diameter is given to that precision. *(c)* The area A of the entire sheet is

$$A = 1.644\ \text{m}^2 + 0.737\ \text{m}^2 = 2.4\ \text{m}^2$$

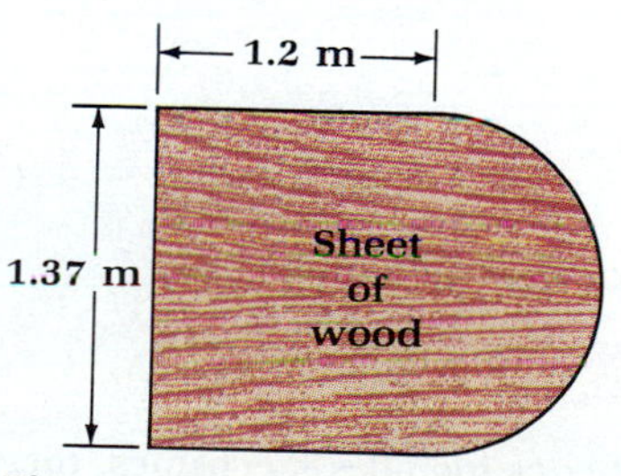

Figure 1-2. Example 1-3.

In finding A, we wrote A_r as 1.644 m² rather than 1.6 m². If we had used 1.6 m², the result would have been $A = 2.3$ m². That is, when performing intermediate calculations, you should keep one or two digits beyond the least significant digit. Indeed, since it requires no extra work on your part, you may as well keep all the digits that your calculator will store. It is only your final answer that must reflect the precision of the numbers you used in the calculation.

SELF-TEST 1-3. The thickness of the sheet of wood in the above example is 16.5 mm. What is the volume occupied by the wood? ***ANSWER:*** 3.9×10^{-2} m³.

1-4 PROBLEM-SOLVING TECHNIQUES

In almost all courses in physics, your progress is judged by how well you can solve problems. A standard approach to solving problems is useful. It can help you organize your thoughts and show you where to begin. The procedures listed below are a distillation of techniques previous students have found to be useful.

- **Draw a diagram.** Use the diagram to make explicit the conditions of the problem. Use it as an extension of your memory; write the given information on the page. Identify the unknown(s) on the diagram if possible.
- **Select a principle.** Investigate the conditions stated in the problem so that you can find a principle or concept that is applicable. Often the application of a principle leads to one or more equations that involve the unknown(s). If there is more than one unknown, you may need more than one equation.
- **Solve the equation.** Perform the algebra that isolates the unknown on the left-hand side of an equation. Avoid inserting known numerical values into the equation until you have solved the equation for the unknown in terms of the symbols used for the known quantities. This usually makes the algebra simpler. After you have solved the equations, insert the values of the known quantities (including their units) and perform the calculation.
- **Check your answer.** Check the dimensions and units of your answer. Use your diagram to check the value of your answer for reasonableness. Consider your answer in terms of intuition and common sense.

Estimating. One way to check the reasonableness of an answer is to perform a rough calculation, or estimate, using powers of 10. Estimating also helps to develop your physical intuition. Suppose you are asked to calculate the volume of a typical adult human. A typical adult has a mass closer to 100 kg than to 10 kg, so we use 100 kg. The mass density of a human is about the same as water, $1 \times 10^3\ \text{kg/m}^3$. If a mass is divided by a mass density, the result will have the dimension of volume. Thus

$$\frac{100\ \text{kg}}{1 \times 10^3\ \text{kg/m}^3} = 0.1\ \text{m}^3$$

As you proceed through your physics course, look for the numerical value of any quantity that can be used in making an estimate. Examples of such quantities encountered so far are the acceleration of gravity ($\approx 10\ \text{m/s}^2$) and the density of water ($\approx 1 \times 10^3\ \text{kg/m}^3$). Remain aware of your personal standards for mass, length, and time: Your mass is about 60 kg; the distance from your foot to your waist is about 1 m; the time to say "one-thousand-and-one" is about 1 s.

SUMMARY

Physics includes mechanics, thermal phenomena, electricity and magnetism, optics, and sound. The standards that are used to define the quantities used in physics were set up by the CGPM. They are the second (given as a certain number of vibrational periods of a cesium atom), the meter (given as the distance that light travels in a certain time), and the kilogram (given as the mass of a cylinder kept near Paris).

The SI system of units consists of three basic units—the second, meter, and kilogram. The SI system also includes a method for naming larger and smaller units that differ by factors of a thousand and a set of derived units, such as those for force and power.

Quantities in physics have dimensions as well as magnitude. When solving problems in physics, you must be sure that the dimensions of the quantities used are consistent, that they are expressed in the same units, and that the precision of the answer is correctly indicated. In this text, the precision is indicated by the number of significant figures.

It is wise to use a standard technique for solving problems in physics. When possible, you should draw a diagram, select relevant principles, apply the principles to obtain an equation or equations, solve the equations, and evaluate your answer.

QUESTIONS

1. What is the dimension of the volume of a cube? Of the volume of a sphere? Of the ratio of the volume of a sphere to the volume of a cube whose side is equal to the diameter of the sphere?

2. What is the dimension of the unit "liter"?

3. What are the dimensions of 60 mi/h, 3 qt, and 2.5 kg/m^3?

4. If you were told that everything in the universe had expanded to twice its former size while you slept last night, how would you check to see if it were true? What if all processes suddenly ran at half speed? What if the mass of everything in the universe doubled? What if all of these things happened simultaneously?

5. Suppose we made contact with a civilization in another galaxy. Could we tell them the size of any of our standards? Of all of our standards?

6. There are really two parts to the standard of time, the size of the unit and the origin of time. How would you tell if your watch had the correct size for the second? How would you tell if it had the correct origin? Which is the answer to the question, "What time is it?"

7. Because civil time is based on the properties of the sun's apparent motion and physical time is based on atomic properties, occasionally "leap seconds" must be inserted as a correction. Do you think the correction is put into physical time or into civil time? Why?

8. Because of relativistic effects, time scales based on atomic properties are different both in size and in synchronization for observers moving with respect to each other. What does this say about the existence of a universal time?

9. Why do we keep a standard kilogram but no standard meter or standard clock?

10. What is the difference between dimensions and units?

11. If an equation is consistent in one set of units, is it dimensionally consistent? Why or why not?

12. If we multiply several numbers with differing precision, which one determines the precision of the answer?

13. If we add a series of numbers with the same precision but differing magnitudes, which one determines the precision of the answer?

14. If you give your mass as 55.6234 kg, in what sense are you being misleading?

15. The speed of light is approximately 3×10^8 m/s. If we define a new unit of time as 1 blink = 30 μs, what is the speed of light in meters per blink? How far does light travel in one blink?

EXERCISES

In Exercises 1 through 7, the symbols represent the following: x, a length; v, a speed; a, an acceleration; t, a time; m, a mass; and k, a dimensionless number.

1. Consider the equation

$$v^2 = v_0^2 + 2b(x - x_0)$$

where the symbol b represents a quantity whose dimension is unknown. Use dimensional analysis to determine the dimension of b. What physical quantity do you know that has this dimension?

2. *(a)* In the equation $v^n = ka^j x$, what numbers must n and j be to make the equation dimensionally correct? *(b)* What, if anything, can you learn about k from dimensional analysis?

3. We will later use the equation $K = \frac{1}{2}mv^2$. What is the dimension of K?

4. Check the following equations for dimensional consistency:
(a) $v^2 + v^3 = 2ax$
(b) $x = v^2/a$
(c) $v = 3at + x/t$
(d) $x = at^2 \sin[(x/t^2)/a]$

5. Find the nonzero integers b, c, and d such that $a^b v^c t^d$ is dimensionless.

6. Suppose an object suspended from a spring is pulled downward and released. In such a case the object oscillates, and the frequency f of the oscillations has the dimension of reciprocal time [1/T]. Experiments show that f depends on the mass m of the object and on the stiffness s of the spring, where s has the dimension [M]/[T]^2. Determine the values of the exponents b and c in the relation

$$f = (\text{dimensionless constant})\ m^b s^c$$

7. What is the conversion factor that can be multiplied by a speed in kilometers per hour to convert that speed to meters per second?

8. Convert a speed of 15 furlongs per fortnight to meters per second. One furlong is the length of a furrow in a 10-acre-square field and is $\frac{1}{8}$ mile. One fortnight is two weeks.

9. Automobile magazines give values of acceleration in units of miles per hour-seconds. *(a)* Determine a conversion factor that, upon multiplication, changes an acceleration in miles per hour·second (mi/h·s) to meter per second squared (m/s^2).

(b) Use your conversion factor to convert 12 mi/h·s to meters per second squared.

10. A carat is a unit of mass equal to 200 mg. How many 1-carat diamonds does it take to make a pound?

11. The speed of light in vacuum is 3.00×10^8 m/s. Convert this speed to miles per second.

12. Find the product of 21.6 m and 5.3 m.

13. Find the sum of 84.626 s and 923.1 s.

14. Do the computation

$$(46.1\ \text{m})(0.231\ \text{s}) + \frac{492\ \text{s}}{13\ \text{m}^{-1}}$$

15. Do the computation

$$\frac{5.47 \times 10^4\ \text{m/s}^2}{(26.67 \times 10^{-8}\ \text{s}^{-1})^2} - (3.63 \times 10^{11}\ \text{m}) \cos 56°$$

16. Evaluate the following:

$$6.02\ \text{m} + (3.7\ \text{m}) \ln (4.870)$$

17. The earth has a radius of 6.4×10^6 m = 6.4 Mm. The density of rocks on the earth's surface is roughly 3 times that of water. Estimate the mass of the earth.

18. Find the sum of the following three time intervals: 3.15 h, 43.5 min, and 21.2 s. First express your answer in units of hours. How many significant figures can you claim? What is the answer expressed in units of seconds? Is the number of significant figures different from your first answer? Explain.

19. Evaluate $(6.00 \times 10^2)(315)$ and give the number of significant figures. Repeat for (600)(315) and explain any difference in the number of significant figures from the first calculation.

20. The length of the hypotenuse of a right triangle is given by $\sqrt{x^2 + y^2}$, where x and y are the lengths of the other two sides of the triangle. Determine the length of the hypotenuse and the number of significant figures in the answer if *(a)* $x = 3.00$ m, $y = 1.00$ m and *(b)* $x = 3.00$ m, $y = 0.100$ m.

21. Estimate *(a)* the mass of Mt. Everest, which is about 10^4 m above sea level and *(b)* the mass of an elephant.

22. Estimate *(a)* the driving time from Chicago to Miami by automobile, *(b)* the volume of water in the oceans, *(c)* the density of a crocodile.

23. The earth's oceans have an average depth of about 1 mi. Estimate the mass of water on earth.

24. Estimate the amount of water that falls on the earth's surface as rain each year.

25. One hectare is the area of a square 100 m on a side. Estimate the mass of wood in a hectare of virgin forest containing oak trees.

VECTORS

Force is a vector quantity. These dogs exert forces on the person's hand in various directions. To determine their net effect, the forces must be added as vectors.

Measurements of wind include both magnitude and direction. We speak of a 5 m/s wind out of the north. The wind's magnitude is 5 m/s and its direction is from north to south. Temperature, on the other hand, is characterized only by a magnitude. We speak of a temperature of 22°C. Temperature is a scalar quantity, and wind is a vector quantity. You are already familiar with how to deal mathematically with scalars. In this chapter we establish the framework for handling vector quantities, and we use this framework to show how vectors are added and subtracted. The multiplication of vectors is investigated in later chapters.

Wind vane and anenometer. The wind vane measures the wind direction and the anenometer measures the wind speed. The combination measures the wind velocity, which is both the speed and the direction.

2-1 SCALARS AND VECTORS

The many different quantities you will encounter in this course can be divided into two types, *scalars* and *vectors*.

Scalars

A ***scalar*** is a quantity that can be represented by a single number, along with the proper unit. A scalar has only a size; it has no directional properties. Some examples are

A distance: The distance around your waistline is 0.85 m.

A mass: Your mass is 58 kg.

A time interval: The time between successive heartbeats is 0.95 s.

Scalars combine according to the rules of algebra.

The rules for combining scalars are the rules of ordinary algebra.

Vectors

A vector has a direction.

A displacement is a vector from one point to another point.

A ***vector*** is a quantity that has both a size (magnitude) and a direction. To specify a vector, more than just a single number with its units is required. An example of a vector is a ***displacement.*** Suppose that you walk from a point P to a point Q. Your displacement can be represented by a directed line segment such as the one shown in Fig. 2-1*a*. The sense, or direction, of the line segment is indicated by the arrowhead. This displacement is a vector that locates point Q relative to point P.

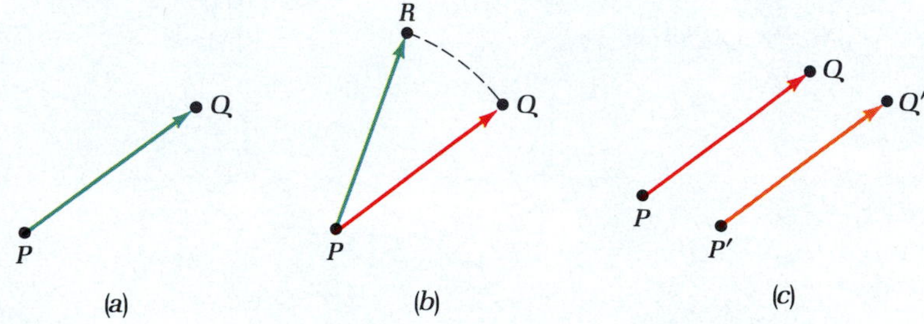

Figure 2-1. *(a)* A displacement locates point Q relative to point P. *(b)* Points Q and R are equidistant from point P, but the two displacements are different because they have different directions. *(c)* Two displacements are equal if they have the same length and the same direction.

Notice that the displacement from P to Q involves more than just the distance between the two points. The orientation, or direction, of the line segment in the plane is also needed. Suppose that you walked from P to a different point R, as shown in Fig. 2-1*b*. The distance between P and R is the same as the distance between P and Q, but the two displacements, the two vectors, are different because they have different directions. That is, ***a displacement is characterized by a distance and a direction.***

Two displacements are equal if they have the same length and the same direction. Two equal displacements are shown in Fig. 2-1*c*. One displacement is from P to Q, and the other displacement is from P' to Q'. You can imagine picking up one of the displacements and moving it, without changing its length or its direction, until it coincides with the other displacement. This procedure, moving a displacement while preserving its length and direction, is often used when dealing with vectors.

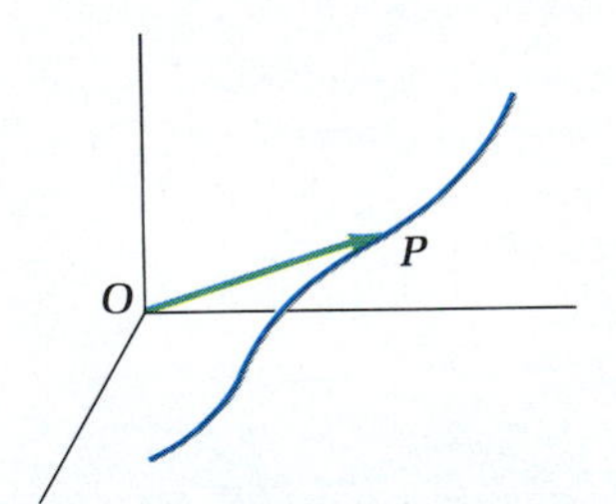

Figure 2-2. A position vector locates a point P on a path relative to an origin O.

Displacements will be used in Chaps. 3 and 4 to discuss the motion of an object. As an object moves in space, a representative point on the object, say the center of a ball, traces out a path. Such a path is shown in Fig. 2-2. A given point P on the path is located relative to the origin O in the figure by a displacement. A displacement that locates a point relative to the origin of a coordinate system is called a ***position vector.***

Position vector

Other vector quantities will be introduced in the following chapters. You will soon be dealing with force, velocity, and acceleration as vectors. Like displacement, other vector quantities are specified by a direction and a ***magnitude.*** The magnitude of a vector is a nonnegative number (with a unit) that indicates the size of the vector without regard to its direction. The magnitude of a displacement is just the distance between the two points. For example, the magnitude of the displace-

The magnitude of a vector is independent of its direction.

ment of the upper-right-hand corner of this page relative to the lower-left-hand corner is about 0.3 m. What is the magnitude of the displacement of the upper-left-hand corner relative to the lower-right-hand corner? ***Remember that the magnitude of a vector is independent of its direction and is never negative.***

To distinguish a vector symbol from that for a scalar, boldface type is used in the text to represent a vector quantity. Thus a displacement can be represented by a symbol such as **D**. Since boldface is difficult to produce when written by hand, a vector symbol is often denoted by an arrow placed over a letter. Thus both **A** and $\vec{A}$ represent a vector. The magnitude of a vector is often called its ***absolute value,*** indicated by $|\mathbf{A}| = A$. That is, the vector is represented by the boldface **A** (or by $\vec{A}$), and its magnitude is represented by the lightface A. A scalar is also indicated by a lightface symbol. Indeed, the *magnitude* of a vector, being independent of the *direction* of the vector, is a scalar.

The vector **A** has magnitude A.

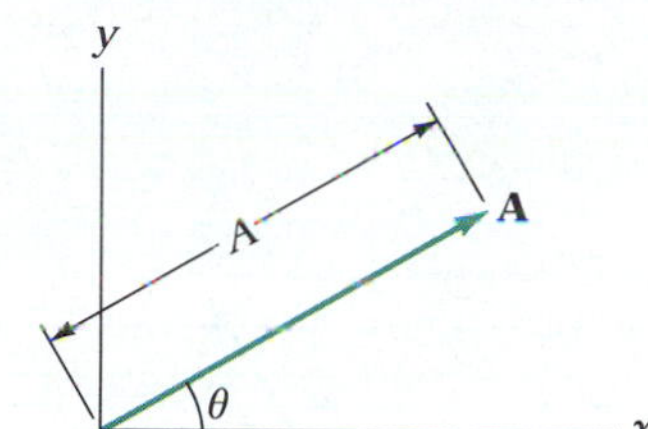

Figure 2-3. A vector **A** in a plane can be specified by its magnitude A and the angle θ.

How many numbers are required to specify a vector? Consider the special case of a vector in a plane, such as the vector **A** in Fig. 2-3. Giving the magnitude A of the vector requires one number. To distinguish the direction of this vector from other directions in the plane, we need an additional number. It is convenient to specify the angle θ between the direction of the vector and a reference direction. Often the reference direction is chosen as the positive x direction, as in the figure. If measured counterclockwise from this direction, the angle θ is taken as positive. Thus the vector **A** in the plane of Fig. 2-3 is specified by the two numbers, A and θ, that give the magnitude and direction of the vector. Two numbers are required to specify a vector in the two spatial dimensions of a plane. For the more general case of a vector in three spatial dimensions, three numbers are required. Such vectors are considered in Sec. 2-3.

EXAMPLE 2-1

Magnitude and direction of a vector. An ant crawls on a tabletop. A position vector **r** locates the ant at a point with coordinates $x = 35$ mm, $y = 45$ mm relative to the origin of the coordinate system shown in Fig. 2-4. Determine the magnitude and the direction of this vector.

Solution. The magnitude r of the position vector is the distance from the origin to the point in question. From the pythagorean theorem, this distance is $r = \sqrt{x^2 + y^2}$, or

$$r = \sqrt{(35 \text{ mm})^2 + (45 \text{ mm})^2} = 57 \text{ mm}$$

The direction of the vector is determined by the angle θ, where $\tan\,\theta = y/x = 45$ mm/35 mm. Then

$$\theta = \tan^{-1}(45/35) = 52°$$

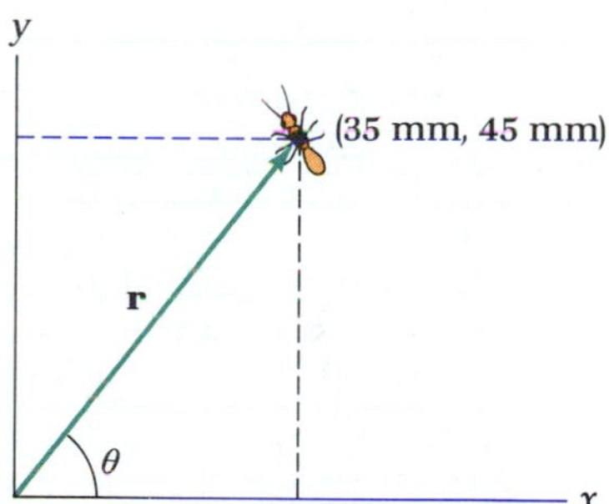

Figure 2-4. Example 2-1: A position vector **r** locates an ant at the point $x = 35$ mm, $y = 45$ mm.

SELF-TEST 2-1. Construct a diagram similar to that in Fig. 2-4 and display a position vector **r** with $r = 57$ mm, $\theta = -52°$. Why is this vector different from the one in Example 2-1? ***ANSWER:*** The vector in the self-test has the same magnitude as the one in the example, but it has a different direction.

2-2 GRAPHICAL ADDITION OF VECTORS

Graphical addition of vectors by the head-to-tail method

Since a vector has both magnitude and direction, vector addition does not obey the rules of ordinary algebra. We must define the procedure for adding vectors. The process is conveniently expressed in graphical terms: Consider two vectors **a** and **b** that lie in the plane of Fig. 2-5a. The lengths of the line segments representing these vectors are proportional to the magnitudes of the vectors. To form the sum **a** + **b**, place vector **b** so that its tail is at the head of vector **a**, as in Fig. 2-5b. Construct the

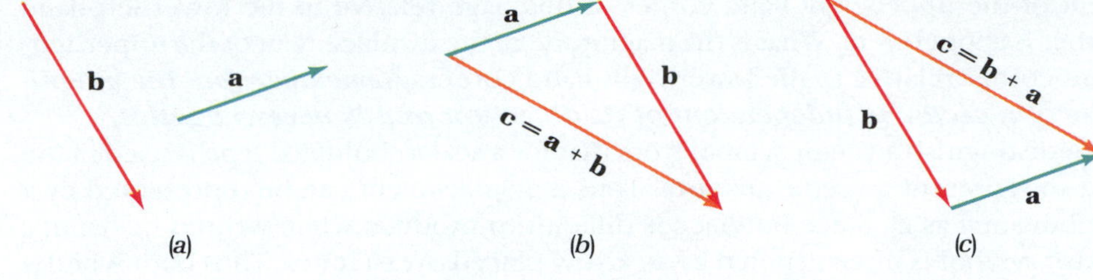

Figure 2-5. Vectors **a** and **b** are added graphically with the head-to-tail method to produce the vector sum **c**. The vector sum **c** is independent of whether the tail of **b** is placed at the head of **a** or the tail of **a** is placed at the head of **b**.

directed line segment from the tail of **a** to the head of **b**. This represents a vector **c** = **a** + **b**, which is the sum, or ***resultant,*** of vectors **a** and **b**. Since the vectors to be added are arranged head to tail, this graphical method is called the ***head-to-tail method.***

If the roles of the two vectors in Fig. 2-5*b* are interchanged so as to form the sum **b** + **a**, as in Fig. 2-5*c*, the same resultant **c** is obtained. Thus vector addition is commutative, and

$$\mathbf{a} + \mathbf{b} = \mathbf{b} + \mathbf{a} \qquad (2\text{-}1)$$

This is in agreement with our experience with displacements. Walking 3 km north and then 2 km east takes us to the same point as walking 2 km east and then 3 km north.

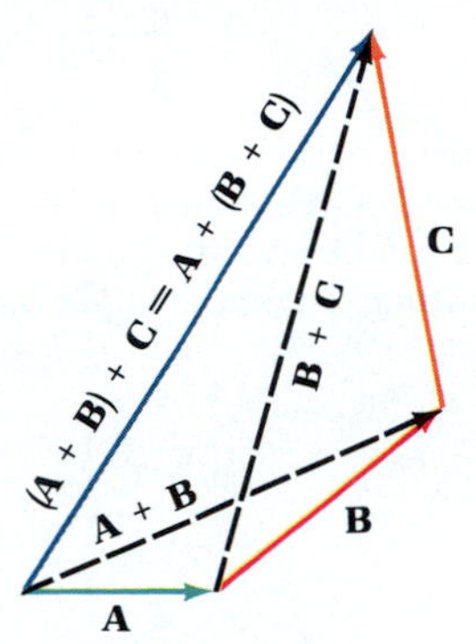

Figure 2-6. A vector sum is independent of the order in which the vectors are added.

The addition of vectors also obeys the associative law illustrated in Fig. 2-6. The result of adding vectors **A** and **B** first and then adding in vector **C** is the same as adding **B** and **C** together first and then adding in vector **A**, or

$$(\mathbf{A} + \mathbf{B}) + \mathbf{C} = \mathbf{A} + (\mathbf{B} + \mathbf{C}) \qquad (2\text{-}2)$$

Because of the associative and commutative laws of vector addition, vectors may be grouped and added in any convenient order.

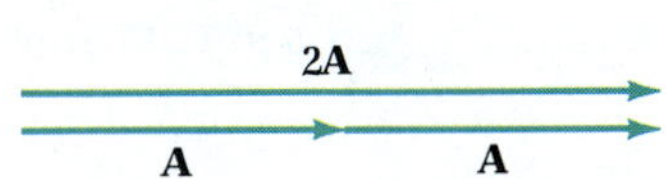

Figure 2-7. Vector **A** added to itself gives the vector 2**A**.

Suppose that a vector is added to itself. The vector sum **A** + **A** is illustrated in Fig. 2-7. The resultant is a vector with the same direction as **A** and with magnitude 2*A*. This vector is denoted 2**A**. More generally, consider the product of a scalar *s* and a vector **A**. We define the combination as a vector, **B** = *s***A**, such that if *s* is positive, then **B** is parallel to **A** and has magnitude $B = sA$. If the scalar *s* is negative, then **B** is opposite in direction to **A** and has magnitude $B = |s\mathbf{A}| = |s|A$. (Remember that the magnitude of a vector cannot be negative.) Two examples are shown in Fig. 2-8.

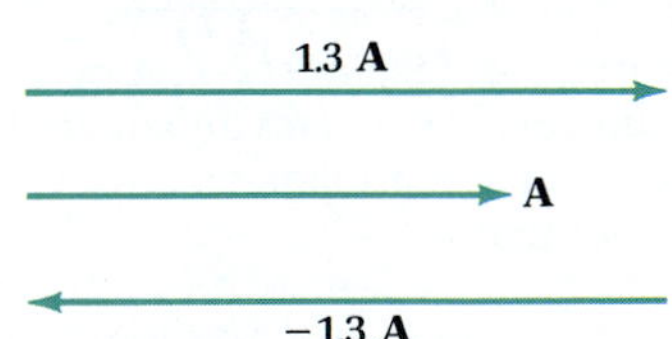

Figure 2-8. The vector 1.3**A** is parallel to **A**; the vector −1.3**A** is opposite to **A**.

Consider now the vector −1**A** or, more simply, −**A**. The vector −**A** has a direction opposite the direction of **A** and a magnitude equal to *A*. We say that the vector −**A** is equal but opposite to the vector **A**. These two vectors are shown in Fig. 2-9. Notice that their sum, **A** + (−**A**), results in a vector of magnitude 0, the ***null vector.*** That is, **A** + (−**A**) = **A** − **A** = 0.

By using the negative of a vector, we have tacitly introduced the idea of subtraction of vectors. In the general case of two vectors **A** and **B**, the difference **A** − **B** is defined as the sum of the vectors **A** and −**B**,

$$\mathbf{A} - \mathbf{B} = \mathbf{A} + (-\mathbf{B}) \qquad (2\text{-}3)$$

An example of subtracting two vectors is shown in Fig. 2-10. The vector −**B** is added to vector **A** to form **D** = **A** − **B**. The vector **C** = **A** + **B** is also shown for comparison.

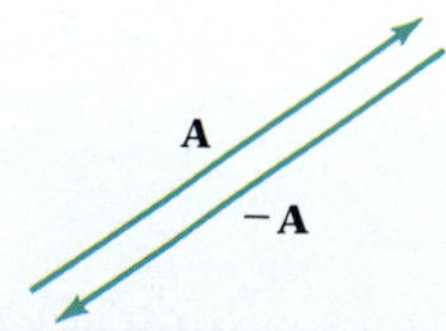

Figure 2-9. Vectors **A** and −**A** have equal magnitudes and opposite directions. Their sum is the null vector 0.

It is often convenient to display two or more vectors with their tails at a common point, as in Fig. 2-11*a*. The graphical addition of two vectors can be performed, without placing them head to tail, by using the ***parallelogram method:*** Given two vectors placed with a common origin, as in Fig. 2-11*a*, complete the parallelogram by constructing sides parallel to the two vectors **A** and **B**, as shown in Fig. 2-11*b*. The resultant vector **C** is directed from the common origin along the diagonal of the parallelogram. Figure 2-11*c* shows the equivalence of the parallelogram method and the head-to-tail method for adding vectors.

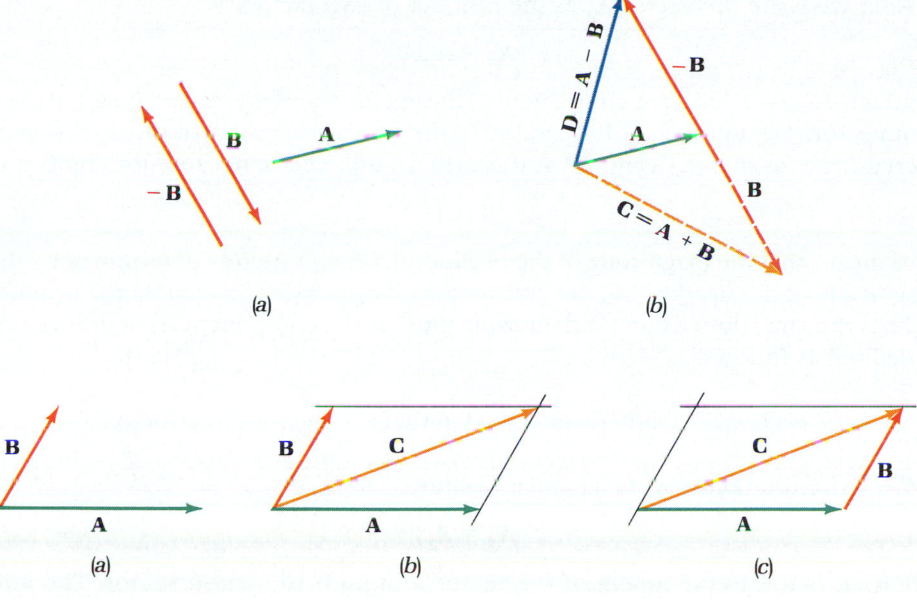

Figure 2-10. The vector $\mathbf{D} = \mathbf{A} - \mathbf{B}$ is obtained by adding $\mathbf{A}$ and $-\mathbf{B}$. For comparison, the vector $\mathbf{C} = \mathbf{A} + \mathbf{B}$ is also shown.

Figure 2-11. *(a)* Two vectors have a common origin. *(b)* The resultant $\mathbf{C} = \mathbf{A} + \mathbf{B}$ is constructed using the parallelogram method. *(c)* The parallelogram method of vector addition is equivalent to the head-to-tail method.

EXAMPLE 2-2

Adding displacements. A helicopter leaves an airport and makes two stops. First, it travels directly east for a distance of 3.0 km, then it travels directly northeast (45° north of east) a distance of 4.5 km. Use the graphical method of adding vectors to determine the magnitude and direction of the helicopter's total displacement. Give the direction in terms of the angle that the total displacement makes with an east-west line.

Solution. The two displacements are labeled $\mathbf{D}_1$ and $\mathbf{D}_2$ in Fig. 2-12, where they are shown added head to tail to give the total displacement $\mathbf{D}$. If you carefully measure the length of $\mathbf{D}$ using the proper scale for this figure, you find that $D \approx 7.0$ km. Measurement of the angle between $\mathbf{D}$ and an east-west line with a protractor gives $\theta = 27°$.

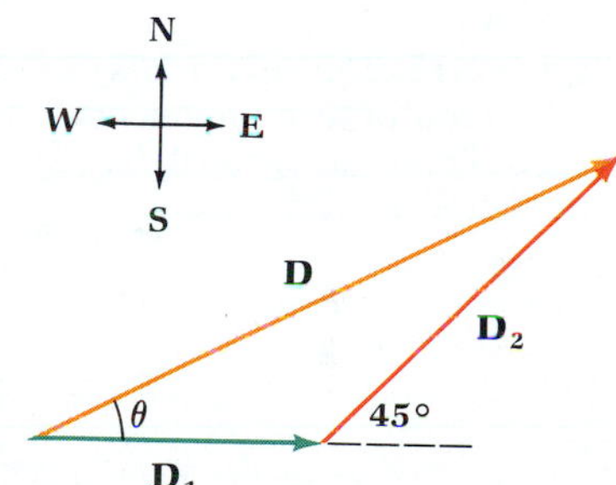

Figure 2-12. Example 2-2.

SELF-TEST 2-2. With a ruler and a protractor, use the graphical method to determine the magnitude of the sum of two displacements, one with magnitude 1.5 km and directed north, the other with magnitude 4.0 km and directed 45° south of east. ***ANSWER:*** 3.1 km.

2-3 UNIT VECTORS AND VECTOR COMPONENTS

From the previous example and self-test, you can see that the graphical method of adding vectors is inconvenient. We need an analytic way to express vectors to deal with them more easily. This is done by using a coordinate frame and unit vectors.

Unit Vectors

Figure 2-13 shows a vector $\mathbf{A}$ against the backdrop of a rectangular coordinate frame. We can write $\mathbf{A}$ as the sum of two vectors $\mathbf{A}_x$ and $\mathbf{A}_y$:

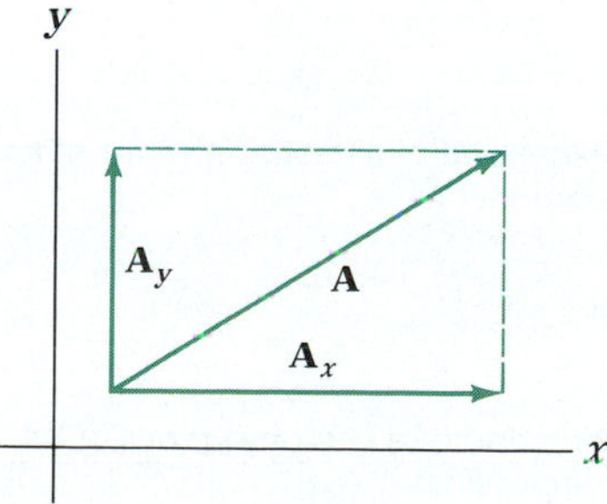

Figure 2-13. Resolving vector $\mathbf{A}$ into two vectors, $\mathbf{A}_x$ parallel to the x axis and $\mathbf{A}_y$ parallel to the y axis.

$$\mathbf{A} = \mathbf{A}_x + \mathbf{A}_y$$

The vector $\mathbf{A}_x$ is obtained from the projection of $\mathbf{A}$ along the x axis, and similarly, the vector $\mathbf{A}_y$ is obtained from the projection of $\mathbf{A}$ along the y axis. Separating a vector into a sum of two or more vectors is called ***resolving the vector.*** The vector $\mathbf{A}$ has been resolved into two vectors, one along the x axis and the other along the y axis.

Resolving a vector

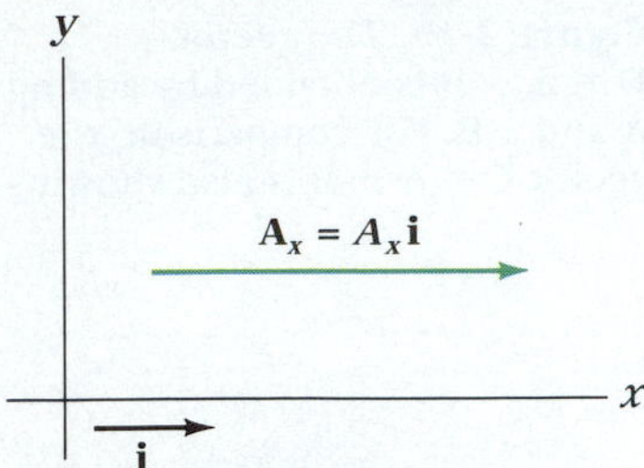

Figure 2-14. The vector $\mathbf{A}_x$ can be written as the product $A_x\mathbf{i}$, where A_x is the x component of vector **A** and **i** is the x unit vector. The unit vector **i** has length 1 and points toward increasing x.

Next we write the vector $\mathbf{A}_x$ as the product of two factors:

$$\mathbf{A}_x = A_x\mathbf{i}$$

The factor A_x is called the x component of the vector **A** and the factor **i** is called the x unit vector. The unit vector **i** is parallel to the x axis and points in the direction of increasing x, as shown in Fig. 2-14. It is called a unit vector because its length is 1:

$$|\mathbf{i}| = 1$$

This means that the magnitude of the product $A_x\mathbf{i}$ is completely determined by the magnitude of A_x. In addition, the unit vector **i** is dimensionless so that the product $A_x\mathbf{i}$ has the same dimension (and the same unit) as the component A_x. A unit vector is defined as follows:

A ***unit vector*** is a mathematical device that specifies a direction only.

We write the vector $\mathbf{A}_y$ in a similar fashion:

$$\mathbf{A}_y = A_y\mathbf{j}$$

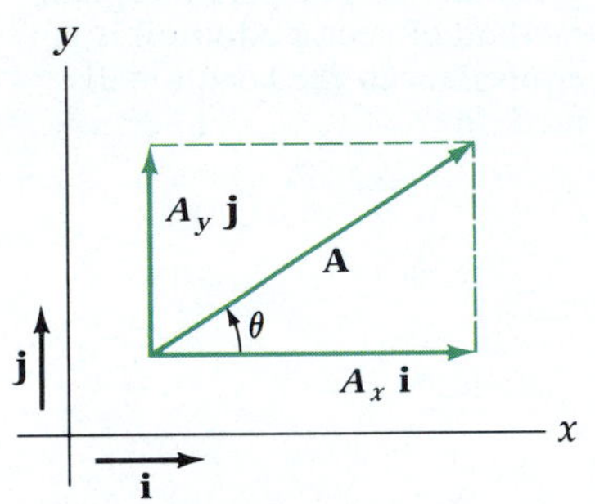

Figure 2-15. The vector **A** written in terms of its components A_x and A_y and in terms of the unit vectors **i** and **j**.

where A_y is the y component of the vector **A** and **j** is the y unit vector. The unit vector **j** is parallel to the y axis and points in the direction of increasing y. As shown in Fig. 2-15, the vector **A** can now be written

$$\mathbf{A} = A_x\mathbf{i} + A_y\mathbf{j} \tag{2-4}$$

This system for writing a vector is very convenient when performing calculations with vectors.

Unit vectors can be used in a number of ways. They can be used to describe the orientation of a flat surface. As we shall see in later chapters, the orientation of a surface is specified by constructing a unit vector directed perpendicular to the surface face.

The symbol for a unit vector is printed in boldface in a book, as are other vectors. As stated earlier, the handwritten symbol for a vector usually has an arrow over it. In the case of unit vectors, it is customary to put a caret over the handwritten symbol rather than an arrow. For example, the handwritten symbol for the **i** unit vector is $\hat{\mathbf{i}}$.

Magnitude and Direction of a Vector

Since $A_x\mathbf{i}$ and $A_y\mathbf{j}$ are perpendicular, they form the sides of a right triangle with **A** as the hypotenuse. Therefore the pythagorean theorem gives the magnitude of **A** as

Magnitude of a vector in terms of its components

$$A = \sqrt{A_x^2 + A_y^2} \tag{2-5}$$

Also, since the tangent of an angle is defined as the ratio of the opposite to the adjacent sides of a right triangle, we have

Direction of a vector in terms of its components

$$\tan\theta = A_y/A_x \qquad \text{or} \qquad \theta = \tan^{-1}(A_y/A_x) \tag{2-6}$$

where θ is the angle between **A** and the x axis, as discussed earlier (Fig. 2-3). The angle θ is often referred to as the *direction of the vector.* For example, suppose $A_x = 4.0$ m and $A_y = 3.0$ m. Then $A = \sqrt{(4.0\text{ m})^2 + (3.0\text{ m})^2} = 5.0$ m, and $\theta = \tan^{-1}(3.0\text{ m}/4.0\text{ m}) = 37°$.

Equations 2-5 and 2-6 give the magnitude and direction of a vector in terms of its components so that if A_x and A_y are given, then we can determine A and θ. Now consider the reverse situation. Suppose that A and θ are given and we wish to find A_x and A_y. These relations come from the definitions of the sine and cosine of an angle.

The sine of an angle in a right triangle is defined as the opposite side divided by the hypotenuse, and the cosine of an angle in a right triangle is defined as the adjacent side divided by the hypotenuse. Applying these definitions to Fig. 2-15, we have

$$A_x = A \cos \theta \tag{2-7}$$

$$A_y = A \sin \theta \tag{2-8}$$

Components of a vector in terms of its magnitude and direction

Suppose $A = 2.8$ m and $\theta = 57°$. Then $A_x = 2.8$ m $\cos 57° = 1.5$ m and $A_y = 2.8$ m $\sin 57° = 2.3$ m.

EXAMPLE 2-3

Resolving a vector into its components. A vector **A** has magnitude 5.1 m and makes an angle of $\theta = 122°$ with the x axis. *(a)* Draw the vector **A** with its tail at the origin. *(b)* Determine the components of **A**, and write **A** in terms of the unit vectors **i** and **j**.

Solution. *(a)* The vector is shown with its tail at the origin in Fig. 2-16. Notice that the angle $\theta = 122°$ is measured counterclockwise from the x axis because it is positive. Since an angle measured clockwise is negative, the direction of this vector **A** is also given by $\theta = 122° - 360° = -238°$. Indeed, we can always add or subtract 360° to any angle θ and not change the direction of the vector. It is customary to give the value of θ in the range from 0° to 360°, or else in the range from −180° to 180°. *(b)* The components of **A** are

$$A_x = 5.1 \text{ m} \cos 122° = -2.7 \text{ m}$$

$$A_y = 5.1 \text{ m} \sin 122° = 4.3 \text{ m}$$

The vector **A** is

$$\mathbf{A} = (-2.7 \text{ m})\mathbf{i} + (4.3 \text{ m})\mathbf{j}$$

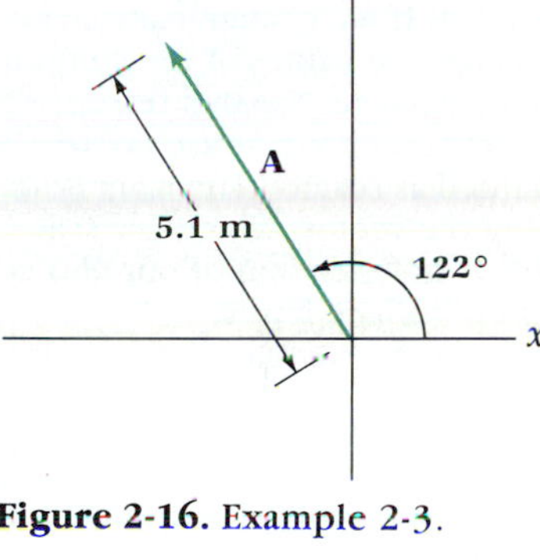

Figure 2-16. Example 2-3.

SELF-TEST 2-3. Consider the vector $\mathbf{A} = (-4.6 \text{ m})\mathbf{i} + (-6.3 \text{ m})\mathbf{j}$. *(a)* Determine the magnitude of **A**. *(b)* Give the direction of **A** with the value of θ in the range from 0° to 360°. *(c)* Give the direction of **A** with the value of θ in the range from −180° to 180°. ***ANSWER:*** (a) 7.8 m; (b) 234°; (c) −126°. Note that 234° − 360° = −126°.

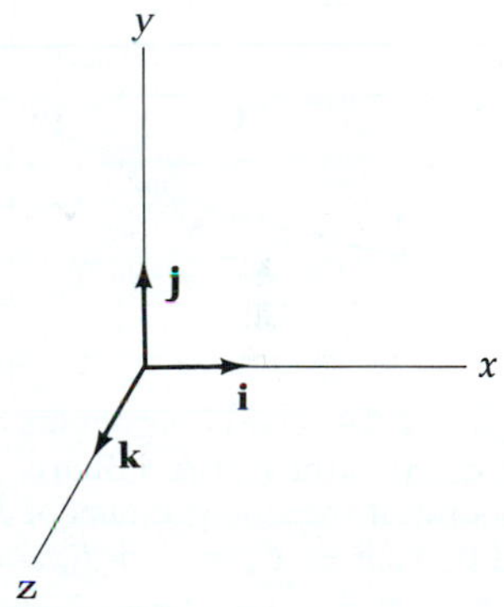

Figure 2-17. Unit vectors **i, j, k** lie along the x, y, z axes.

Extension to Three Dimensions

Vectors in three dimensions are specified in terms of three components and three unit vectors. In cartesian coordinates (or rectangular coordinates), the unit vector that is parallel to the z axis and points in the direction of increasing z is given the symbol **k**. The three mutually perpendicular cartesian unit vectors **i, j,** and **k** are shown in Fig. 2-17. A vector **F** is shown in three dimensions in Fig. 2-18. The vector is then written

$$\mathbf{F} = F_x\mathbf{i} + F_y\mathbf{j} + F_z\mathbf{k}$$

The magnitude of the vector is given by

$$F = \sqrt{F_x^2 + F_y^2 + F_z^2}$$

The direction of the vector can be expressed in terms of angles relative to coordinate axes. However, it is usually more convenient in three dimensions to just give the three components F_x, F_y, F_z. As a special case, let **F** represent a position vector **r** that locates a point with coordinates (x, y, z) relative to the origin of a coordinate system. For a position vector, the coordinates are also the components: $\mathbf{r} = x\mathbf{i} + y\mathbf{j} + z\mathbf{k}$. For example, $\mathbf{r} = (3 \text{ m})\mathbf{i} + (4 \text{ m})\mathbf{j} + (-5 \text{ m})\mathbf{k}$ locates the point (3 m, 4 m, −5 m).

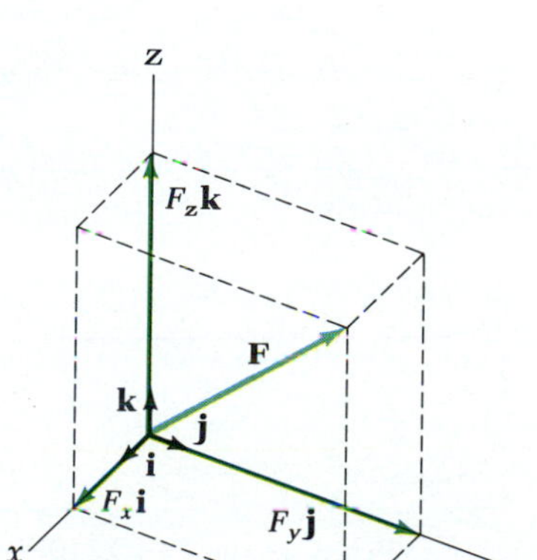

Figure 2-18. A vector **F** in three dimensions has components F_x, F_y, F_z.

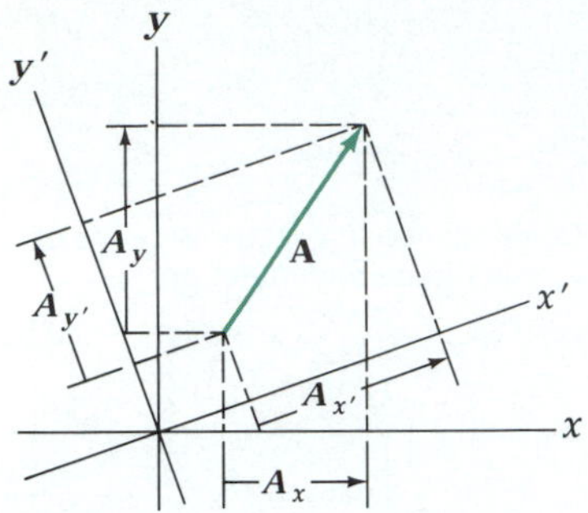

Figure 2-19. The vector **A** is independent of our choice of coordinate frame, but the vector components depend on this choice. If we change coordinate frames, the values of the components change. For this reason, vector components are not regarded as scalars. The magnitude A of the vector, which is a scalar, is independent of our choice of coordinate frame.

More about Vector Components

It is tempting to regard vector components, such as A_x and A_y in the expression $\mathbf{A} = A_x\mathbf{i} + A_y\mathbf{j}$, as scalars. After all, vector components are certainly not vectors because the vector part of **A** is represented by the unit vectors **i** and **j**. But technically, A_x and A_y are not scalars because the value of a vector component depends on the coordinate frame we choose to use, whereas the value of a scalar or a vector does not depend on the coordinate frame.

In Fig. 2-19 a vector **A** is shown against the backdrop of two coordinate frames, with one frame rotated relative to the other. In describing the vector, we may choose either frame, and our choice will not affect the vector **A**. However, the components of **A** are different when we change from one frame to the other: $A_x \neq A_{x'}$ and $A_y \neq A_{y'}$. The magnitude of **A** is a scalar and

$$A = \sqrt{A_x{}^2 + A_y{}^2} = \sqrt{A_{x'}{}^2 + A_{y'}{}^2}$$

Thus the magnitude of a vector is independent of the choice of coordinate frame and can be properly regarded as a scalar. Since a vector component depends on the arbitrary choice of a coordinate frame, it cannot be a scalar; it is simply a vector component.

2-4 VECTOR ADDITION, ANALYTICAL METHOD

The graphical method of vector addition is useful for visualizing a problem involving vectors, but it is not convenient for precise work. This is where unit vectors and vector components are used.

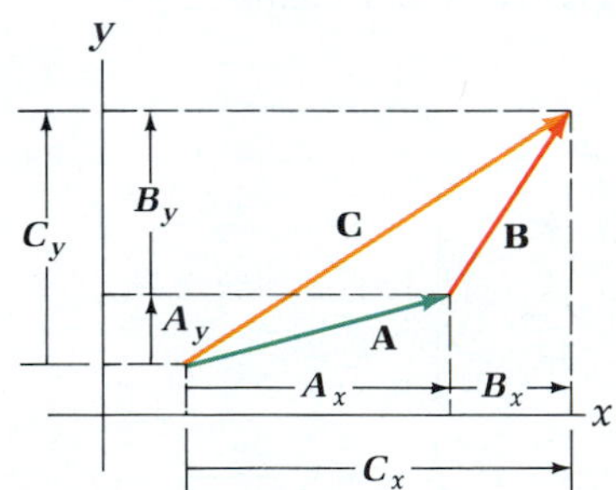

Figure 2-20. The components of the vector sum $\mathbf{C} = \mathbf{A} + \mathbf{B}$ are the sum of the components of **A** and **B**. That is, $C_x = A_x + B_x$ and $C_y = A_y + B_y$.

Figure 2-20 shows vectors **A** and **B** added head to tail to produce the vector **C**. The figure also shows **A**, **B**, and **C** resolved along the x and y axes. From the figure, it is clear that

$$C_x = A_x + B_x \qquad C_y = A_y + B_y$$

We now demonstrate algebraically what Fig. 2-20 shows graphically:

$$\mathbf{C} = \mathbf{A} + \mathbf{B}$$

Writing the vectors in terms of components and unit vectors gives

$$C_x\mathbf{i} + C_y\mathbf{j} = (A_x\mathbf{i} + A_y\mathbf{j}) + (B_x\mathbf{i} + B_y\mathbf{j})$$

Collecting terms on the right-hand side gives

$$C_x\mathbf{i} + C_y\mathbf{j} = (A_x + B_x)\mathbf{i} + (A_y + B_y)\mathbf{j}$$

If the vectors are equal, then each component must be equal, or

$$C_x = A_x + B_x \qquad C_y = A_y + B_y$$

The extension of these results to three dimensions gives

$$C_x = A_x + B_x \qquad C_y = A_y + B_y \qquad C_z = A_z + B_z \tag{2-9}$$

Each component of the resultant vector **C** is the sum of the corresponding components of **A** and **B**.

EXAMPLE 2-4

Adding displacements of a cruise ship. A cruise ship leaves port and sails due east for a distance of 231 km. To avoid a storm, it turns and sails 42.1° south of east for 209 km, and then sails 54.8° north of east for 262 km. Determine the magnitude and direction of the resultant displacement **R**. Neglect the curvature of the earth and assume that all displacements lie in the same plane.

Solution. Calling the successive displacements **E, F,** and **G**, we first determine their components relative to the coordinate axes shown in Fig. 2-21. Thus

$$E_x = 231 \text{ km} \cos 0 = 231 \text{ km}$$

$$F_x = 209 \text{ km} \cos (-42.1°) = 155 \text{ km}$$

$$G_x = 262 \text{ km} \cos 54.8° = 151 \text{ km}$$

$$E_y = 231 \text{ km} \sin 0 = 0$$

$$F_y = 209 \text{ km} \sin (-42.1°) = -140 \text{ km}$$

$$G_y = 262 \text{ km} \sin 54.8° = 214 \text{ km}$$

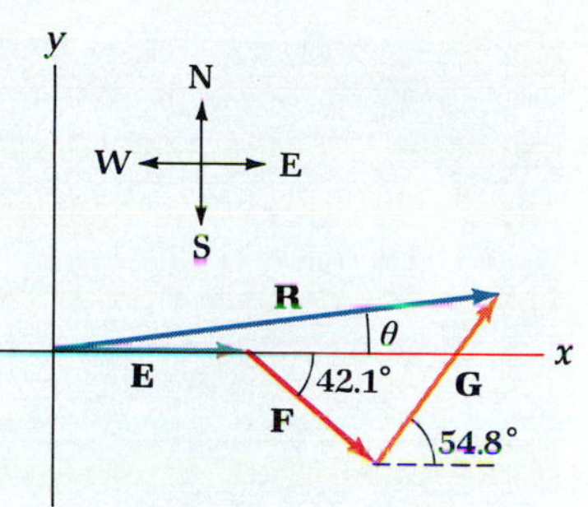

Figure 2-21. Example 2-4.

Notice that $-42.1°$ was used for **F** since an angle is negative if measured clockwise from the positive x axis and positive if measured counterclockwise from that axis. The x and y components of the resultant **R** are given by

$$R_x = E_x + F_x + G_x = 537 \text{ km}$$

$$R_y = E_y + F_y + G_y = 74 \text{ km}$$

Using Eqs. 2-5 and 2-6, we obtain

$$R = \sqrt{(537 \text{ km})^2 + (74 \text{ km})^2} = 542 \text{ km}$$

$$\theta = \tan^{-1} (74/537) = 7.8°$$

SELF-TEST 2-4. Add a fourth displacement **H**, directed due north with $H = 125$ km, to the three displacements given in Example 2-4. What are the magnitude and the direction of the resultant displacement? ***ANSWER:*** 573 km, 20.3° north of east.

EXAMPLE 2-5

The law of cosines. Consider two vectors **A** and **B** with an angle θ between their directions, as shown in Fig. 2-22. *(a)* Show that the magnitude of the resultant vector **C** is given by the *law of cosines,*

$$C = \sqrt{A^2 + B^2 + 2AB \cos \theta}$$

(b) Suppose you walk 350 m south on Broadway, turn 65° toward the east and continue for 280 m on 42nd Street. Use the law of cosines to determine the magnitude of the resultant displacement.

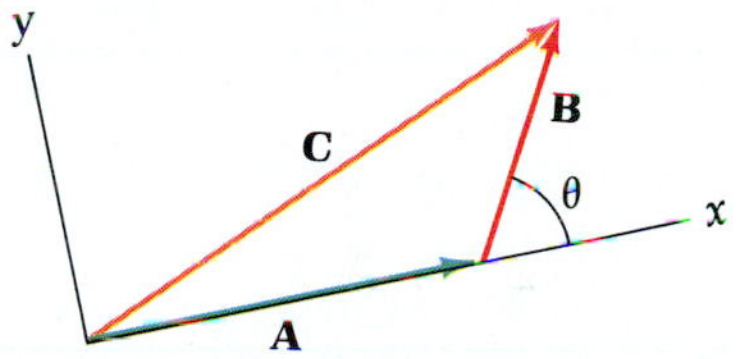

Figure 2-22. Example 2-5.

Solution. *(a)* For simplicity, coordinate axes have been chosen so that **A** is along the x axis, and the components of **A** and **B** are

$$A_x = A \cos 0 = A \qquad A_y = A \sin 0 = 0$$

$$B_x = B \cos \theta \qquad B_y = B \sin \theta$$

Then $C_x = A + B \cos \theta$ and $C_y = B \sin \theta$. From Eq. 2-7, $C = \sqrt{C_x^2 + C_y^2}$ or

$$C = \sqrt{(A + B \cos \theta)^2 + (B \sin \theta)^2}$$

$$= \sqrt{A^2 + 2AB \cos \theta + B^2 \cos^2 \theta + B^2 \sin^2 \theta}$$

$$= \sqrt{A^2 + B^2 + 2AB \cos \theta}.$$

where we have used $\cos^2 \theta + \sin^2 \theta = 1$.

(b) Letting $A = 350$ m, $B = 280$ m, and $\theta = 65°$ in the law of cosines, we obtain

$$C = \sqrt{(350 \text{ m})^2 + (280 \text{ m})^2 + 2(350 \text{ m})(280 \text{ m}) \cos 65°}$$

$$= 530 \text{ m}$$

SELF-TEST 2-5. Consider the vector sum **C** = **A** + **B**, where the magnitude and direction of **A** are 2.9 m and 25°, and the magnitude and direction of **B** are 3.6 m and 155°. *(a)* Make a sketch showing **A** and **B** with their tails at the origin of a coordinate frame. Complete the sides of the parallelogram and include **C** in your sketch. *(b)* Use the law of cosines to determine the magnitude of **C**. *(c)* Use your sketch to verify the result you got from the law of cosines. ***ANSWER:*** (b) 2.8 m.

J. Willard Gibbs.

COMMENTARY: Vectors and J. Willard Gibbs

Much of the vector notation that we now use can be attributed to Josiah Willard Gibbs (1839–1903). Gibbs was a strong advocate of the use of single symbols to represent objects, such as vectors, that consist of several quantities. For example, the vector equation $\mathbf{A} + \mathbf{B} = \mathbf{C}$ is a simple and concise way to summarize three ordinary equations:

$$A_x + B_x = C_x \qquad A_y + B_y = C_y \qquad A_z + B_z = C_z$$

Willard Gibbs was born in New Haven, Connecticut, and spent most of his life at or near Yale College, where his father was a professor. Gibbs was a student at Yale and achieved distinction in mathematics and in Latin. His graduate work there earned him the first doctorate awarded in engineering in the United States. His thesis was entitled "On the Form of the Teeth of Wheels in Spur Gearing." After several years of travel and study in Europe, Gibbs returned to New Haven and in 1871 was appointed professor of mathematical physics at Yale. No salary was provided, and Gibbs worked without pay until 1875 when he was paid at a reduced rate of $2000 a year.

Gibbs had an interest in practical devices and inventions; he held a patent on an improved air brake for railroad cars. However, his major contributions to physics were mathematical rather than experimental. His published papers were abstract and difficult to understand and consequently attracted little attention. But his work was noticed and appreciated by the brilliant Scottish physicist, James Clerk Maxwell (1831–1879) (see the Commentary on Maxwell in Chapter 27). There was a comment on Maxwell's death that only one person (Maxwell) could understand Gibbs's work, and he was now dead.

Gibbs's major work was in thermodynamics and statistical mechanics. (Statistical mechanics provides a theoretical basis for thermodynamics.) He extended these fields to apply to mixtures of chemical substances, and he is often described as the father of physical chemistry. An important concept in statistical mechanics is phase space, a fictitious space with axes labeled by three spatial coordinates and three velocity components for each particle or molecule of a system such as a gas. The phase space for a single particle therefore has six dimensions. The phase space for a system with N particles, say a gas with 10^{22} molecules, has $6N = 6 \times 10^{22}$ dimensions. Just as we deal with a position vector that locates a point in three dimensions, Gibbs considered the motion of a point in the $6N$-dimensional phase space. Many of the techniques and much of the notation that we use for vectors in three dimensions were used by Gibbs in $6N$ dimensions.

Gibbs used mathematics to describe and understand the processes of nature. In an address on the applications of multiple algebra (vectors), Gibbs noted that since position in space is essentially a vector quantity, "Nature herself takes us by the hand, and leads us along by easy steps. . . ." Gibbs maintained that the mathematics used by physicists must always be directed toward the results of experiments. He is reputed to have said, "A mathematician may say anything he pleases, but a physicist must be at least partially sane."

For further reading, see *J. Willard Gibbs* by Raymond J. Seeger (Pergamon Press, New York, 1974) and *Willard Gibbs* by Muriel Rukeyser (Doubleday, Doran & Co., New York, 1942).

SUMMARY

Section 2-1. Scalars and Vectors

A scalar is specified by a single number. A vector has a magnitude and a direction and is specified by three numbers. A vector in a plane is specified by two numbers. A displacement is a vector from one point in space to another.

Section 2-2. Graphical Addition of Vectors

Vectors can be added graphically using the head-to-tail method or the parallelogram method. Vector addition is commutative and associative. A vector multiplied by a scalar is another vector. The vector $-\mathbf{A}$ is equal in magnitude and opposite in direction to the vector **A**. Subtraction of vectors is defined by

$$\mathbf{A} - \mathbf{B} = \mathbf{A} + (-\mathbf{B}) \tag{2-3}$$

Section 2-3. Unit Vectors and Vector Components

A unit vector is a dimensionless vector of magnitude 1 that specifies a direction. Unit vectors **i, j,** and **k** lie along the axes of an *xyz* coordinate system. A vector in a plane can be specified either by its magnitude and direction or by its components. The components are found from the magnitude and direction, using

$$A_x = A \cos\theta \tag{2-7}$$

$$A_y = A \sin\theta \tag{2-8}$$

The magnitude and direction are found from the components, using

$$A = \sqrt{A_x^2 + A_y^2} \tag{2-5}$$

$$\tan\theta = A_y/C_x \tag{2-6}$$

A vector in three dimensions is expressed in terms of its components by $\mathbf{F} = F_x\mathbf{i} + F_y\mathbf{j} + F_z\mathbf{k}$.

Section 2-4. Vector Addition, Analytical Method

Vectors are added analytically using their components. If $\mathbf{C} = \mathbf{A} + \mathbf{B}$, then

$$\begin{aligned} C_x &= A_x + B_x \\ C_y &= A_y + B_y \\ C_x &= A_z + B_z \end{aligned} \tag{2-9}$$

QUESTIONS

1. Can a scalar be negative? Can the magnitude of a vector be negative? Can a vector component be negative? Explain.

2. Consider the displacements that lie along the two diagonals of this page, one from the lower-left to the upper-right corner and the other from the lower-right to the upper-left corner. Are the magnitudes of these vectors equal? Are the vectors equal? Explain.

3. Is there a distinction between a position vector and a displacement? Explain.

4. You and a friend decide to go from the first to the third level of a department store. Leaving from a common point by the elevator, you take the elevator up, while your claustrophobic friend takes the escalator. You and your friend meet at the third-level elevator. How do your path lengths compare? How do your displacements compare? How do the magnitudes of your displacements compare?

5. Two vectors of equal magnitude are added. Depending on the directions of the two, what is the maximum magnitude of the resultant? What is the minimum?

6. Displacement **D** has magnitude 12 m, and displacement **E** has magnitude 9 m. The resultant, $\mathbf{F} = \mathbf{D} + \mathbf{E}$, has magnitude 3 m. What can you say about the directions of **D** and **E**?

7. If $\mathbf{A} \neq -\mathbf{B}$, is it possible for $\mathbf{A} + \mathbf{B}$ to equal zero? Explain.

8. Is it possible for $\mathbf{a} + \mathbf{b} + \mathbf{c}$ to equal zero if the three vectors **a, b,** and **c** have *(a)* unequal magnitudes, *(b)* equal magnitudes? Explain.

9. You overhear a classmate describing a 400-m race, "Since the finish line and the starting line are at the same point on the track, the displacement of a runner for the entire race is zero." Do you agree with this statement? Explain.

10. A square city block is 150 m on a side. If you take the walkway from one corner to the next, what is the magnitude of your displacement? You continue around the block to the next corner diagonally opposite your initial location. What is the magnitude of this second displacement? What is the magnitude of your resultant displacement? What is the total distance traveled?

11. While facing east, you take a giant step that results in a displacement of magnitude 1 m. Is this displacement a unit vector? Explain.

12. Different cartesian coordinate systems can be chosen that have different orientations for the coordinate axes. Does the value of a scalar depend on the orientation of the axes? Do the components of a vector depend on the orientation of the coordinate axes? Does the vector depend on the orientation of the axes? Does the magnitude of the vector depend on the orientation of the axes? Explain.

13. The vector $\mathbf{V} = \mathbf{i} + \mathbf{j} + \mathbf{k}$ has magnitude $V = \sqrt{3}$. What are the magnitudes of the vectors $\mathbf{U} = -\mathbf{i} - \mathbf{j} - \mathbf{k}$ and $\mathbf{W} = \mathbf{i} + \mathbf{j} - \mathbf{k}$?

14. A position vector in the *xy* plane lies in the third quadrant. What is the sign of each component?

15. A vector $\mathbf{F} = F_x\mathbf{i} + F_y\mathbf{j} + F_z\mathbf{k}$ lies in the *yz* plane. What is the value of its *x* component F_x?

16. If you change from one coordinate system to another with a different orientation of axes, do the unit vectors **i, j, k** remain the same? Explain. [See Prob. 4.]

17. What are the components of the unit vector **j**?

18. What range of angles is returned by your calculator by the $\tan^{-1}$ routine? By the $\cos^{-1}$ routine? By the $\sin^{-1}$ routine?

19. Does your calculator have radians or degrees as the default unit for angles? How can you tell?

20. What angles have a tangent equal to $+1, -1, 0, +\infty, -\infty$?

21. At the end of each chapter, the final question will ask you to complete a table that contains some of the symbols used in that chapter to represent physical quantities. Fill in the table entries by stating concisely the meaning of the symbol; stating whether the quantity is a scalar, a vector, or a vector component; and giving the SI unit for the quantity. The table below contains some of the symbols for this chapter. Complete the table:

Symbol	Represents	Type	SI Unit
$\|\mathbf{A}\|$	Magnitude of a vector	Scalar	*
C_x		Component	*
$\mathbf{r}$	Position vector		m
$\mathbf{i}$		Vector	None

* The SI unit for this entry depends on the quantity. If **A** or $C_x\mathbf{i}$ is a displacement, then the SI unit is the meter.

EXERCISES

Section 2-1. Scalars and Vectors

1. Determine the magnitude of the position vector that locates the point with coordinates *(a)* (1.0 m, 2.0 m, 0.0 m); *(b)* (0.0 m, 1.0 m, 2.0 m); *(c)* (1.0 m, 2.0 m, 3.0 m).

2. The origin of a coordinate system is at one corner of a rectangular room, with the coordinate axes along the three edges from that corner. A mosquito starts at the origin and crawls only along edges to reach the corner with coordinates (4.2 m, 3.8 m, 2.6 m). *(a)* What is the minimum path length for the mosquito? *(b)* What is the magnitude of the displacement for the trip? *(c)* What is the answer to part *(b)* if the mosquito flies directly from the origin to the opposite corner? *(d)* What is the answer to part *(a)* if the mosquito crawls along walls? (*Hint:* Imagine the walls form a box which can be unfolded so that the walls lie in the same plane.)

3. Set up an xy coordinate system in the plane of a sheet of paper. Using a ruler and a protractor, display each of the following vectors: *(a)* the position vector locating the point (55 mm, 65 mm); *(b)* the displacement from the point (32 mm, 18 mm) to the point (87 mm, 83 mm); *(c)* the displacement from the point (0.0 mm, 18 mm) that has magnitude 85 mm at a 50° angle from the positive x axis. What do these vectors have in common?

Section 2-2. Graphical Addition of Vectors

4. Use a ruler and protractor to transfer each pair of vectors in Fig. 2-23 to a sheet of paper and determine their sum graphically.

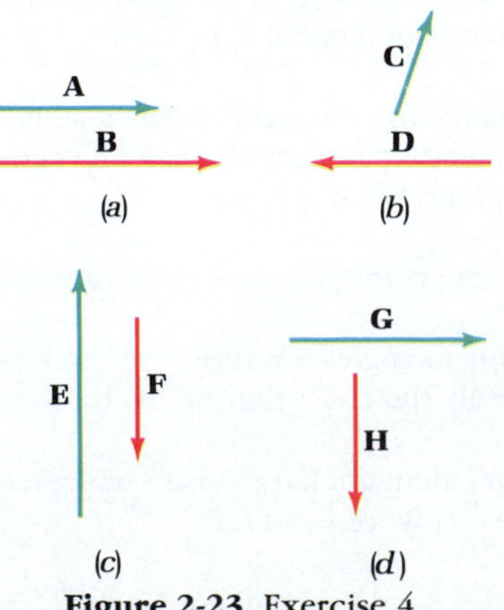

Figure 2-23. Exercise 4.

5. Five vectors, all having the same magnitude and lying in a plane, make angles with the positive x axis of 0, ±72°, and ±144°. Determine graphically the sum of these vectors.

6. Vectors **A**, **B**, and **C** lie in a plane as shown in Fig. 2-24. Using ruler and protractor, determine *(a)* $\mathbf{A}+\mathbf{B}+\mathbf{C}$; *(b)* $\mathbf{A}-\mathbf{B}-\mathbf{C}$; *(c)* $\mathbf{C}+\mathbf{B}+\mathbf{A}$.

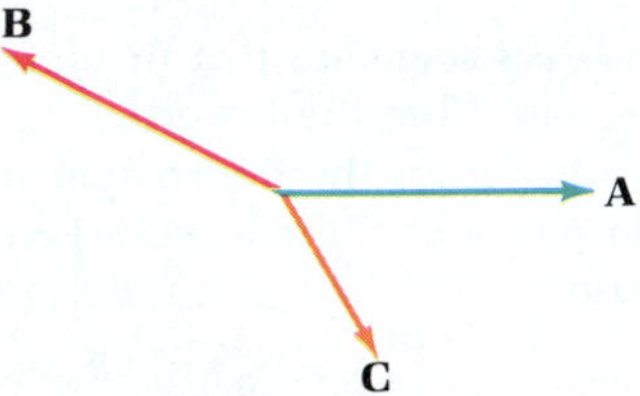

Figure 2-24. Exercise 6.

7. On a diagram, show a pair of vectors **d** and **e** such that
(a) $\mathbf{d}+\mathbf{e}=\mathbf{f}, f=d-e, d>e$
(b) $\mathbf{d}+\mathbf{e}=\mathbf{f}, f=d+e$
(c) $\mathbf{d}-\mathbf{e}=\mathbf{f}, f=d+e$
(d) $\mathbf{d}+\mathbf{e}=\mathbf{f}, f=\sqrt{2}d, d=e$

8. In the parallelogram method of addition illustrated in Fig. 2-11*b*, the sum $\mathbf{A}+\mathbf{B}$ is represented by one diagonal of the parallelogram. Show that the other diagonal, directed from the head of **B** to the head of **A**, represents $\mathbf{A}-\mathbf{B}$.

Section 2-3. Unit Vectors and Vector Components

9. *(a)* On a sheet of paper, construct an xy coordinate system and display the position vector locating the point $x=54$ mm, $y=22$ mm. *(b)* Determine the magnitude and direction of this vector. *(c)* What are the x and y components of this vector?

10. Verify that Eqs. 2-7 and 2-8 correctly give the components A_x and A_y for a vector **A** that can lie in any one of the four quadrants of the xy plane.

11. A coordinate system is oriented so that **i** points east, **j** points north, and **k** points up. The entrance to Physics Hall is 340 m from the entrance to the library along the direction 49° west of north. Both entrances are at the same horizontal level. Determine the components of the displacement *(a)* from the library entrance to the Physics Hall entrance; *(b)* from the Physics Hall entrance to the library entrance; *(c)* from the library entrance to the Physics Hall wind vane, which is 35 m directly above the entrance.

12. *(a)* The unit vector **n** lies in the *xy* plane and makes an angle θ with the positive *x* axis. Express **n** in terms of **i**, **j**, and θ. *(b)* Show that $\mathbf{i}\ 3/\sqrt{14} - \mathbf{j}\ 1/\sqrt{14} + \mathbf{k}\ 2/\sqrt{14}$ is a unit vector.

13. The position vector $\mathbf{r} = x\mathbf{i} + y\mathbf{j}$ locates the point (x, y). *(a)* What is the magnitude *r* of this vector? *(b)* Determine the expression in component form for the unit vector $\hat{\mathbf{r}} = \mathbf{r}/r$.

14. A vector **v** (a velocity) has components $v_x = 34$ m/s, $v_y = -12$ m/s. Determine the magnitude and direction of the vector *(a)* **v**; *(b)* 2**v**; *(c)* -2**v**; *(d)* $(1/v)$**v**. *(e)* If v and θ represent the magnitude and direction of **v**, determine the components of the vector with magnitude $2v$ and direction 2θ.

Section 2-4. Vector Addition, Analytical Method

15. Given the two dimensionless vectors $\mathbf{A} = 3\mathbf{i} + 4\mathbf{j}$ and $\mathbf{B} = -2\mathbf{i} - 6\mathbf{j} + 5\mathbf{k}$, determine *(a)* **A** + **B**; *(b)* **A** − **B**; *(c)* **B** − **A**; *(d)* **B** + **A**.

16. Two displacements are given by **d** = (3 m)**i** + (4 m)**j** + (5 m)**k** and **e** = (2 m)**i** + (−6 m)**j** + (−1 m)**k**. Determine *(a)* the resultant **f** = **e** + **d**; *(b)* A vector **g** such that **d** − **e** + **g** = 0.

17. The position vector $\mathbf{r}_Q = x_Q\mathbf{i} + y_Q\mathbf{j}$ locates a point *Q* with coordinates (x_Q, y_Q). *(a)* What is the expression for the position vector $\mathbf{r}_P$ that locates point *P* with coordinates (x_P, y_P)? *(b)* Determine the components of the displacement from point *P* to point *Q*. *(c)* Explain why these components do not depend on the location of the origin of the coordinate system.

18. Given the displacement **a** = (5 m)**i** + 0**j**, find two other displacements **b** and **c**, also of magnitude 5 m and in the *xy* plane, such that **a** + **b** + **c** = 0. Is the pair **b**, **c** unique? Explain.

19. A delivery truck has successive displacements of 1.37 km southeast, 0.85 km north, and 2.12 km 17° west of north. Determine the magnitude and direction of the resultant displacement.

20. A yacht is tacking into the wind on a zigzag path. On the first leg of the course, the yacht has a displacement of 12 km at 84° east of north. After the second leg has been completed, the yacht's resultant displacement is 15 km at 23° east of north. Determine the magnitude and direction of the second leg of the course.

21. Eight vectors when arranged head to tail form a regular octagon of edge 25 mm. Use the coordinate system shown in Fig. 2-25. *(a)* Determine the components of each of the vectors that form the octagon. *(b)* Determine the magnitude and direction of the vectors labeled **a**, **b**, and **c** in the figure.

22. A goose flies 120 m in a straight line, turns abruptly and flies 160 m in a straight line at 77° from the original course. *(a)* Determine the magnitude of the resultant displacement. (*Hint:* See Example 2-5.) *(b)* What is the total distance traveled by the goose?

23. Use the law of cosines in Example 2-5 to determine the angle between vectors **A** and **B** if $A = 23$ mm, $B = 18$ mm, and $C = 7$ mm ($\mathbf{C} = \mathbf{A} + \mathbf{B}$).

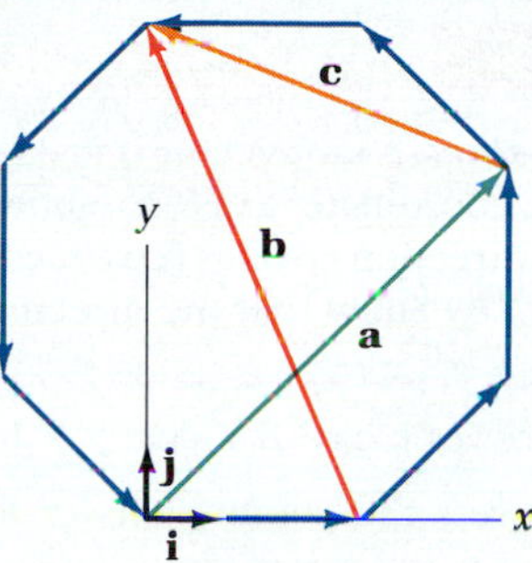

Figure 2-25. Exercise 21.

24. The law of cosines is sometimes expressed using the interior angle ϕ shown in Fig. 2-26 rather than the exterior angle θ. Show that this form is given by

$$C = \sqrt{A^2 + B^2 - 2AB\cos\phi}$$

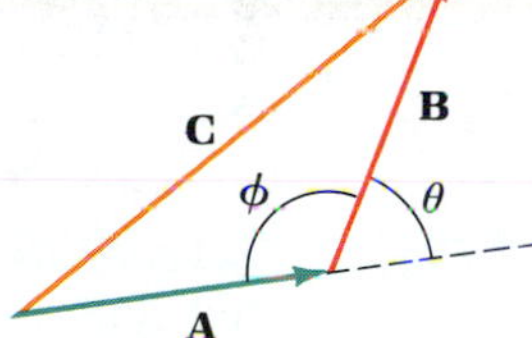

Figure 2-26. Exercise 24.

25. *(a)* If **C** = **A** + **B**, show that $|A - B| \leq C \leq A + B$. *(b)* What is the angle between **A** and **B** if $|A - B| = C$? *(c)* If $A + B = C$?

26. For the vectors **A** and **B** shown in Fig. 2-27, use graphical methods to determine *(a)* **A** + **B**, *(b)* 2**A** + 2**B**, *(c)* 2**A** − 3**B**.

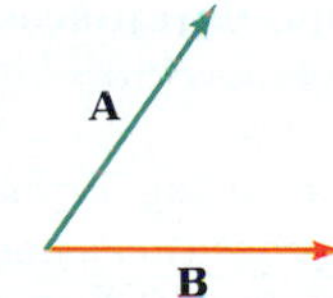

Figure 2-27. Exercise 26.

27. Given the vector **S** = 4**i** + 5**j** in the *xy* plane, determine a unit vector that also lies in the *xy* plane but is perpendicular to **S**. Is there more than one such unit vector? Explain.

28. A hiker has a displacement of (2.7 km)**i** − (7.6 km)**j** on the first day of a two-day hike. If the hiker's resultant displacement is (3.1 km)**i** + (1.2 km)**j**, what is her displacement on the second day of the hike?

29. Point *P* is located by a position vector $\mathbf{r}_P = -(3\text{ m})\mathbf{i} - (4\text{ m})\mathbf{j} - (5\text{ m})\mathbf{k}$. Point *Q* is located by a position vector $\mathbf{r}_Q = (1\text{ m})\mathbf{i} - (1\text{ m})\mathbf{k}$. Determine the displacement of *Q* relative to *P*. Sketch a diagram to show these vectors.

30. Vector **C** in the *xy* plane has magnitude $C = 12.0$ and direction $\theta_C = 66°$. Vector **A**, also in the *xy* plane, has magnitude $A = 6.0$ and direction $\theta_A = 33°$, and **C** = **A** + **B**. Determine the magnitude and direction of vector **B**.

PROBLEMS

1. ***Direction cosines.*** A unit vector **n** makes angles of α, β, γ with the x, y, z coordinate axes, as shown in Fig. 2-28. *(a)* Show that the direction cosines ($\cos\alpha$, $\cos\beta$, $\cos\gamma$) are the components of **n**. *(b)* Show that the direction cosines satisfy the identity

$$\cos^2\alpha + \cos^2\beta + \cos^2\gamma = 1$$

(c) If $\mathbf{F} = F\mathbf{n}$, express the components of **F** in terms of the direction cosines of **n** and the magnitude F.

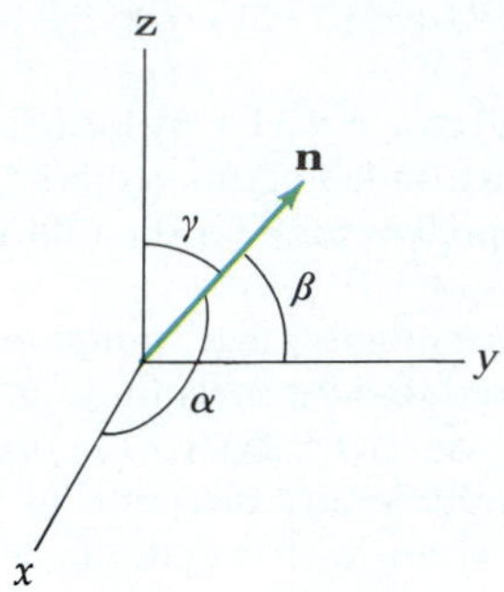

Figure 2-28. Problem 1.

2. ***Adding and subtracting displacements.*** Two displacements are given by $\mathbf{d} = (2.1\ \mathrm{m})\mathbf{i} + (-1.3\ \mathrm{m})\mathbf{j}$ and $\mathbf{e} = (0.8\ \mathrm{m})\mathbf{i} + (1.6\ \mathrm{m})\mathbf{j}$. Determine the angle between *(a)* **d** and **e**; *(b)* **d** and **f** where $\mathbf{f} = \mathbf{d} + \mathbf{e}$; *(c)* **f** and **g** where $\mathbf{g} = \mathbf{d} - \mathbf{e}$.

3. ***Finding the angle between two vectors.*** Let θ represent the angle between two vectors **a** and **b** with $b = \frac{1}{2}a$. If $\mathbf{c} = \mathbf{a} - \mathbf{b}$, determine the value of θ if
(a) $c = \sqrt{a^2 + b^2} = \sqrt{5}a/2$
(b) $c = \sqrt{a^2 - b^2} = \sqrt{3}a/2$
(c) $c = 3a/2$
(d) $c = \frac{1}{2}a$

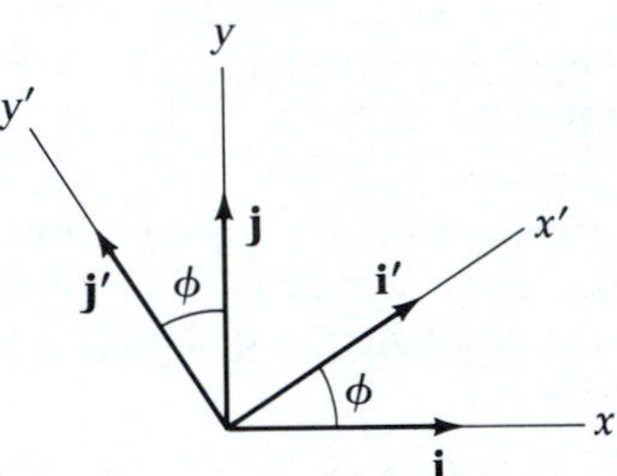

Figure 2-29. Problems 4 and 8.

4. ***Using different coordinate axes.*** Two sets of coordinate axes and associated unit vectors are shown in Fig. 2-29. *(a)* Show that

$$\mathbf{i}' = \mathbf{i}\cos\phi + \mathbf{j}\sin\phi$$

$$\mathbf{j}' = -\mathbf{i}\sin\phi + \mathbf{j}\cos\phi$$

(b) A vector **A** can be expressed as $\mathbf{A} = A_x\mathbf{i} + A_y\mathbf{j}$ or as $\mathbf{A} = A_x'\mathbf{i}' + A_y'\mathbf{i}'$. Use part *(a)* to show that

$$A_x' = A_x\cos\phi + A_y\sin\phi$$

$$A_y' = -A_x\sin\phi + A_y\cos\phi$$

More advanced treatments of vectors use expressions such as these to define vectors.

5. ***Constructing unit vectors.*** Given vector **A** of magnitude A, you can construct a unit vector with the same direction as **A**. We use the symbol $\hat{\mathbf{A}}$ to denote a unit vector formed from the vector **A**. Thus $\hat{\mathbf{A}} = (1/A)\mathbf{A} = \mathbf{A}/A$. *(a)* Prove that $\hat{\mathbf{A}}$ is a unit vector parallel to **A**. *(b)* The position vector **r** locates the point (x, y, z) from the origin of a coordinate system. Determine the x, y, and z components of the unit vector $\hat{\mathbf{r}}$.

6. ***Adding four vectors.*** Four vectors **A, B, C,** and **D** have magnitudes, in arbitrary units, of 1, 2, 3, and 4, respectively. The directions are unspecified and can be considered variable. Determine *(a)* the largest and *(b)* the smallest magnitude of the resultant $\mathbf{E} = \mathbf{A} + \mathbf{B} + \mathbf{C} + \mathbf{D}$. *(c)* Rework with the restriction that three of the vectors lie in a plane and the fourth vector is perpendicular to that plane.

7. ***More on unit vectors.*** A point in the xy plane has coordinates $x = 3$ m, $y = -4$ m. Determine *(a)* the position vector locating the point; *(b)* a unit vector that is parallel to the position vector; *(c)* a unit vector, also in the xy plane, that is perpendicular to the unit vector in part *(b)*; *(d)* a unit vector that is perpendicular to both of the unit vectors found in parts *(b)* and *(c)*.

8. ***Components of vectors in different coordinate systems.*** In one xy coordinate system, two vectors are given by $\mathbf{A} = 2.4\mathbf{i} + 3.0\mathbf{j}$ and $\mathbf{B} = -1.6\mathbf{i} - 1.6\mathbf{j}$. *(a)* Determine the magnitude and direction of $\mathbf{C} = \mathbf{A} + \mathbf{B}$. *(b)* Let $\phi = 30.0°$ in Fig. 2-29 and use the results of Prob. 4 to determine expressions for the vectors **A** and **B** in terms of **i′** and **j′**. *(c)* Use the components from part *(b)* to determine the magnitude and direction, relative to the x' axis, of **C** and compare with the results in part *(a)*.

9. ***Magnitude and direction of a vector.*** *(a)* Determine the magnitude of vector $\mathbf{G} = (1.0\ \mathrm{m})\mathbf{i} + (2.0\ \mathrm{m})\mathbf{j} + (3.0\ \mathrm{m})\mathbf{k}$. *(b)* Determine the angle between **G** and each of the coordinate axes. (*Hint:* See Prob. 1.)

10. ***Mutually perpendicular unit vectors.*** Vector $\mathbf{u} = (3/5)\mathbf{i} - (4/5)\mathbf{j} + (0)\mathbf{k}$. *(a)* Show that **u** is a unit vector. *(b)* Determine two unit vectors, **v** and **w**, that are perpendicular to **u** and to each other.

3 MOTION IN ONE DIMENSION

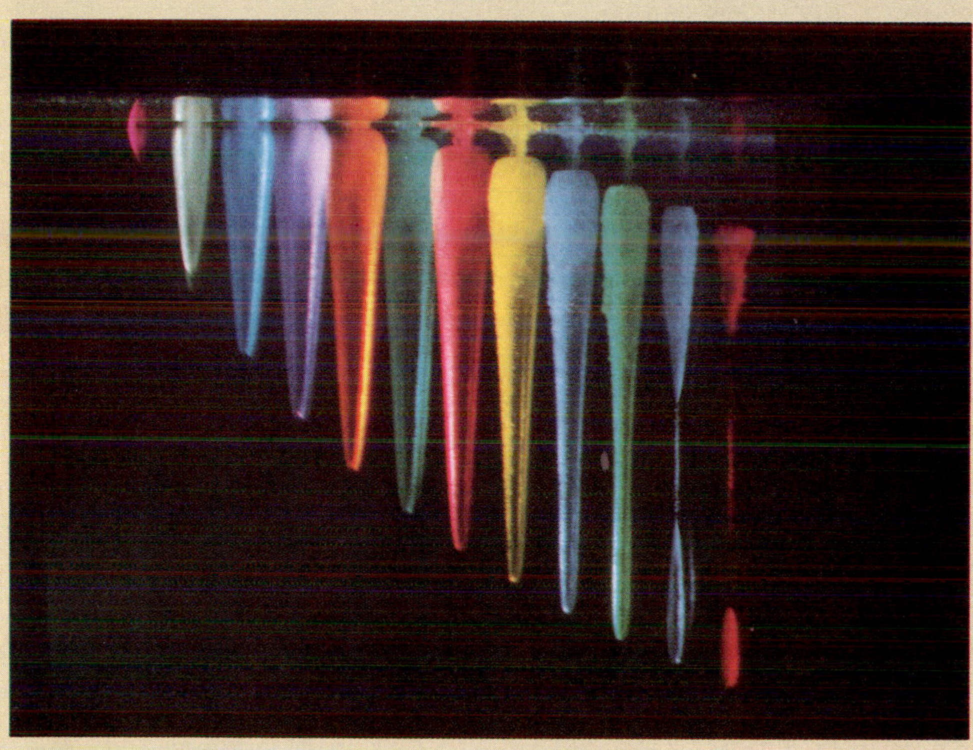

Multicolor stroboscopic photo of a BB as it enters water after being shot downward from above the water surface. The photo was taken with a rotating mirror so that the BB's image is displaced sideways by equal distances for equal time intervals.

Mechanics is the study of motion. Suppose you are interested in the motion of a satellite orbiting the earth. You know the satellite's present position, its speed, and the direction it is traveling. You wish to be able to predict the satellite's path and the time it will reach each point on its path. These are the kinds of questions that are addressed by mechanics.

Mechanics is divided into two parts: kinematics and dynamics. ***Kinematics*** is the study of motion without regard to the cause of the motion. In kinematics, we define some of the quantities that are used in mechanics, quantities such as velocity and acceleration. Then, using the definitions, we establish relationships among these quantities. ***Dynamics,*** which contains the laws of motion, allows us to predict an object's motion from information about the object and its environment. In addition to the kinematical quantities (position, velocity, and acceleration), dynamics deals with concepts such as force and mass.

In preparation for dynamics, this chapter and the next are about kinematics. This chapter is restricted to discussions of objects moving along a straight line — that is, it deals with one-dimensional (1-D) kinematics. Motion in two dimensions (2-D) is discussed in Chap. 4.

A satellite in orbit. The laws of mechanics predict the motion of such objects.

3-1 POSITION VECTOR AND DISPLACEMENT

In these early chapters, discussion of the motion of an object is simplified by treating the object as a particle.

Definition of a particle

A ***particle*** is an idealized entity with no size or internal structure.

Treating an extended object as a particle is a valid approximation if the object's size is irrelevant to the problem at hand. For example, suppose you flip a coin in the air. If you wish to determine the maximum height reached by the coin or the amount of time it stays in the air, you may treat the coin as a particle. However, if you wish to know whether the coin will land heads up or tails up, then you may not treat it as a particle. Then you must take account of the rotation of the coin.

Establishing a frame of reference

To describe the motion of an object, the first step is to establish a coordinate frame, or a ***frame of reference.*** For motion along a straight line, this entails selecting first an origin at some point along the line and then a positive direction.

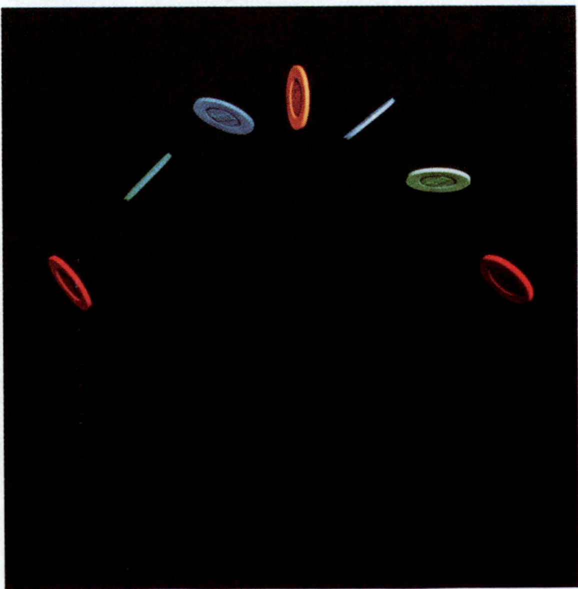
Multicolor strobe photograph of a poker chip flipped into the air. If the rotation of the chip is important, then the chip may not be treated as a particle. In these chapters on mechanics, we often use multicolor strobe photographs to demonstrate motion. By flashing at equal time intervals, a strobe photograph shows a moving object at a number of positions along its path. In a multicolor strobe photograph, the object appears to be of a different color at different positions because the strobe light flashes are different colors. This multicolor feature can be useful in several ways. For example, the photograph can reveal the simultaneous positions of two objects.

Measurements are then made relative to this frame of reference. Consider a car that is traveling along a straight road that runs east-west. Any conveniently located point can serve as the origin, such as a point adjacent to a large tree beside the road. Letting the x axis be along the road, we choose eastward as the positive direction for the unit vector $\mathbf{i}$. The car's *position vector* $\mathbf{r}$ is given by

$$\mathbf{r} = x\mathbf{i} \tag{3-1}$$

Position vector $\mathbf{r}$ in 1-D

The car's coordinate x is the component of its position vector. When the car is 55 m east of the origin, its position vector is $\mathbf{r} = (55 \text{ m})\mathbf{i}$. When it is 25 m west of the origin, its position vector is $\mathbf{r} = -(25 \text{ m})\mathbf{i}$.

A ***position vector*** locates an object relative to a frame of reference.

A ***displacement*** $\Delta\mathbf{r}$ occurs with a change in position. It is the difference between a final position vector $\mathbf{r}_f$ and an initial position vector $\mathbf{r}_i$:

$$\Delta\mathbf{r} = \mathbf{r}_f - \mathbf{r}_i = (x_f - x_i)\mathbf{i} = \Delta x\,\mathbf{i} \tag{3-2}$$

Displacement $\Delta\mathbf{r}$

If the car's initial position was 25 m west of the origin and its final position was 55 m east of the origin, then its displacement was

$$\Delta\mathbf{r} = [(55 \text{ m}) - (-25 \text{ m})]\mathbf{i} = (80 \text{ m})\mathbf{i}$$

The use of the unit vector $\mathbf{i}$ is superfluous in one-dimensional kinematics. The object is always along a straight line, such as the x axis, and the sign of x tells us which side of the origin it is on. This means that we can describe motion in one dimension by using x rather than $\mathbf{r}$, and by using Δx rather than $\Delta\mathbf{r}$. However, this simplicity does not extend to two and three dimensions. Even though it is unnecessary in one-dimensional kinematics, we use this vector notation in this chapter to prepare for the discussions of two- and three-dimensional motion that lie ahead.

EXAMPLE 3-1

A bobsled sliding up, then down a slope. Sometimes the solution to a dynamics problem produces an expression for the coordinate x of an object as a function of the time t, written as $x(t)$. After having found such an expression, we can use it to determine the coordinate at a particular time. Suppose a bobsled is sliding up a straight snowy slope. The sled continuously moves more slowly as it slides up the slope; it comes to a stop momentarily; and then it slides backward down the slope. An analysis of the sled's motion gives its coordinate x as a function of time t as

$$x(t) = 18 \text{ m} + (12 \text{ m/s})t - (1.2 \text{ m/s}^2)t^2$$

where x is measured along the sled's path and the positive x direction is up the slope. *(a)* Construct a graph of the sled's coordinate versus time from $t = 0.0$ s to $t = 8.0$ s by plotting points at each 1.0 s. *(b)* Determine the sled's displacement between $t_i = 1.0$ s and $t_f = 7.0$ s. *(c)* The distance an object travels is the total length of its path. Estimate the distance traveled by the sled between $t_i = 1.0$ s and $t_f = 7.0$ s.

Solution. *(a)* The sled's coordinate at, say 2.0 s, is found by substituting $t = 2.0$ s into the equation for $x(t)$:

$$x = 18 \text{ m} + (12 \text{ m/s})(2.0 \text{ s}) - (1.2 \text{ m/s}^2)(2.0 \text{ s})^2 = 37 \text{ m}$$

Finding the coordinate at the other times yields the data in Table 3-1. Figure 3-1 is a graph of x versus t. Incidentally, notice the way the multiplication of the units in each term of the above equation gives the unit m. For instance, the unit of the third term is $(\text{m/s}^2)(\text{s})^2 = \text{m}$. Any time you perform a calculation like this, make certain the units are consistent. (If you wish to refresh your memory about units, see Sec. 1-3.) *(b)* Using data from Table 3-1, we find that the displacement between $t_i = 1.0$ s and $t_f = 7.0$ s is

$$\Delta\mathbf{r} = (x_f - x_i)\mathbf{i} = (43 \text{ m} - 29 \text{ m})\mathbf{i} = (14 \text{ m})\mathbf{i}$$

(c) From the graph we can estimate the position at which the sled changed its direction of motion. Its coordinate was increasing until $x \approx 48$ m, and then it began decreasing. Begin-

TABLE 3-1

t, s	*x*, m
0.0	18
1.0	29
2.0	37
3.0	43
4.0	47
5.0	48
6.0	47
7.0	43
8.0	37

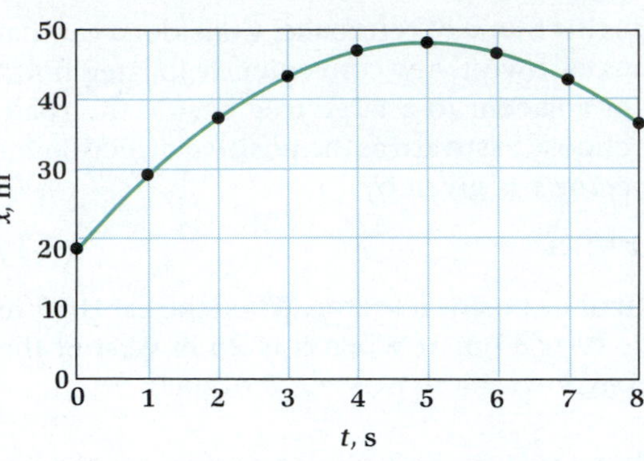

Figure 3-1. Example 3-1: Coordinate x of a bobsled versus time t. Initially sliding up the slope, the sled stopped momentarily at $t \approx 5.0$ s and then slid backward down the slope.

ning at $t = 1.0$ s, it traveled in the $+x$ direction a distance of about 48 m − 29 m = 19 m. Ending at $t = 7.0$ s, it traveled in the $-x$ direction a distance of about 48 m − 43 m = 5 m. The total distance traveled from $t = 1.0$ s to $t = 7.0$ s was about 19 m + 5 m = 24 m. Notice the distinction between the displacement and the distance traveled. Displacement is a vector, but distance is a scalar. The magnitude $|\Delta \mathbf{r}|$ of the displacement is equal to the distance traveled if the object's direction of motion does not change. If the direction does change, then $|\Delta \mathbf{r}|$ does not equal the distance traveled.

SELF-TEST 3-1. Suppose the coordinate of an object is given by the expression

$$x(t) = 9.2 \text{ m} - (0.35 \text{ m/s}^3)t^3$$

(a) Find the value of x at the instant $t = 1.4$ s. That is, determine $x(1.4 \text{ s})$. Make sure the units cancel to give the correct unit to your answer. *(b)* Write down the object's position vector $\mathbf{r}$ at $t = 1.4$ s. Be certain that your answer is expressed as a vector. ***ANSWERS:*** *(a)* $x(1.4 \text{ s}) = 8.2$ m; *(b)* $\mathbf{r}(1.4 \text{ s}) = (8.2 \text{ m})\mathbf{i}$.

3-2 VELOCITY AND SPEED

The velocity $\mathbf{v}$ of an object tells how fast the object is traveling *and* the direction it is heading at some instant of time. The best way to understand the meaning of velocity is to first define and discuss the *average velocity* and then use it to define velocity.

Average Velocity

An object's ***average velocity*** $\overline{\mathbf{v}}$ during a time interval t_i to t_f is

$$\overline{\mathbf{v}} = \frac{\mathbf{r}_f - \mathbf{r}_i}{t_f - t_i} = \frac{\Delta \mathbf{r}}{\Delta t} \tag{3-3}$$

where $\mathbf{r}_f$ and $\mathbf{r}_i$ are the position vectors that locate the object at times t_f and t_i, respectively. A symbol with a bar over it, such as $\overline{\mathbf{v}}$, is the customary way to represent the average of a quantity. In one dimension, the average velocity has only one component:

$$\overline{\mathbf{v}} = \frac{(x_f - x_i)\mathbf{i}}{t_f - t_i} = \frac{\Delta x}{\Delta t}\mathbf{i} = \overline{v}\mathbf{i}$$

The average velocity component is

$$\overline{v} = \frac{x_f - x_i}{t_f - t_i} = \frac{\Delta x}{\Delta t}$$

As examples, we now calculate $\overline{v}$ during two different time intervals for the sled in Example 3-1. Using values from Table 3-1, we find that $\overline{v}$ from $t_i = 1.0$ s to $t_f = 4.0$ s is

$$\bar{v} = \frac{47\ \text{m} - 29\ \text{m}}{4.0\ \text{s} - 1.0\ \text{s}} = \frac{18\ \text{m}}{3.0\ \text{s}} = 6.0\ \text{m/s}$$

For the time interval from $t_i = 1.0$ s to $t_f = 3.0$ s, we have

$$\bar{v} = \frac{43\ \text{m} - 29\ \text{m}}{3.0\ \text{s} - 1.0\ \text{s}} = \frac{14\ \text{m}}{2.0\ \text{s}} = 7.0\ \text{m/s}$$

On a graph of x versus t, the average velocity component is equal to the slope of the straight line that joins two points on the graph. Figure 3-2 shows a straight line that connects the points for the interval from $t_i = 1.0$ s to $t_f = 4.0$ s. The figure shows that the slope of this line is $\bar{v} = 6.0$ m/s. Keep in mind that a graph of the coordinate versus time does not represent the path of the object. In this chapter, the path of the object is always along a straight line.

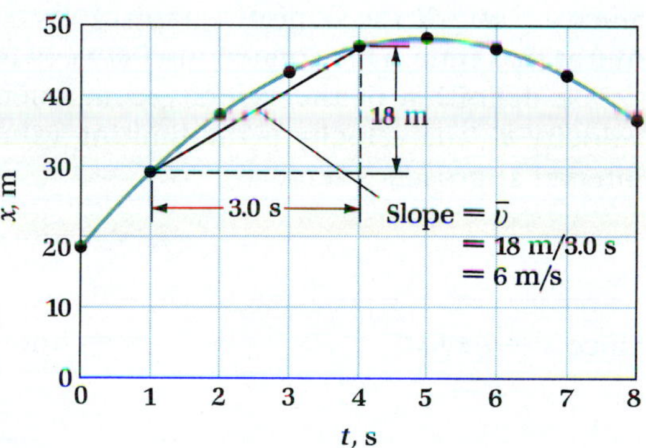

Figure 3-2. The average velocity component between t_i and t_f is equal to the slope of the straight line connecting the points at t_i and t_f on a graph of x versus t. For the bobsled in Example 3-1, $\bar{v} = 6$ m/s between $t_i = 1.0$ s and $t_f = 4.0$ s.

Velocity

We now use average velocity to define the concept of velocity. The *average velocity* characterizes an object's motion *during a time interval,* whereas the *velocity* characterizes its motion *at an instant of time.* To emphasize that the velocity pertains to an instant of time, it is sometimes called the *instantaneous velocity.*

Again consider the bobsled in Example 3-1. In the graph of x versus t, imagine that we plot many points so that the points appear as a smooth continuous curve, as in Fig. 3-3. The slope of the line between any two points gives $\bar{v}$ over the time interval between these points. Now imagine finding $\bar{v}$ for smaller and smaller time intervals. In each succeeding calculation we keep t_i fixed and select t_f closer and closer to t_i. From the figure you can see that as t_f approaches t_i, the slope of each succeeding line approaches that of the line tangent to the curve at t_i. The velocity component v is defined as the limiting value of $\bar{v}$ as the time interval $\Delta t = t_f - t_i$ approaches zero. That is, v is equal to the slope of the line tangent to the x-versus-t curve, or

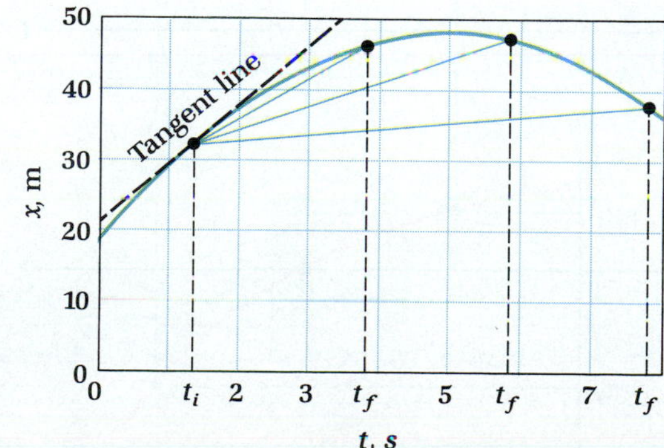

Figure 3-3. As t_f approaches t_i on a graph of x versus t, the slope of each line connecting the points at t_i and t_f approaches the slope of the line tangent to the curve at $t = t_i$.

$$v = \lim_{\Delta t \to 0} \bar{v} = \lim_{\Delta t \to 0} \frac{\Delta x}{\Delta t}$$

As Δt approaches zero, so does Δx. In the limit as both approach zero, their ratio approaches v.

The slope of a line tangent to a curve at a point is usually referred to as the ***slope of the curve*** at that point. This slope is given by the derivative of x with respect to t:

$$\text{Slope of curve} = \lim_{\Delta t \to 0} \frac{\Delta x}{\Delta t} = \frac{dx}{dt}$$

Thus v is defined as the derivative of x with respect to t:

Definition of the velocity component.

$$v = \frac{dx}{dt}$$

Since $v = dx/dt$, the slope of a graph of x versus t at a particular time gives the value of v at that time. This connection between x and v is demonstrated in Fig. 3-4.

The definition of the velocity component leads to the general definition of velocity $\mathbf{v}$. The velocity is the limiting value of the average velocity as the time interval approaches zero:

$$\mathbf{v} = \lim_{\Delta t \to 0} \bar{\mathbf{v}}$$

Since $\bar{\mathbf{v}} = \Delta \mathbf{r}/\Delta t$,

$$\mathbf{v} = \lim_{\Delta t \to 0} \frac{\Delta \mathbf{r}}{\Delta t}$$

or

$$\mathbf{v} = \frac{d\mathbf{r}}{dt} \tag{3-4}$$

The ***velocity*** of an object is the rate of change of its position vector. Velocity tells how fast the object is traveling and the direction it is heading at some instant of time.

Velocity is a vector quantity because it is defined as a displacement (a vector) divided by a time interval (a scalar). As with other vectors, the symbol $\mathbf{v}$ for velocity is shown in boldface in a textbook or with an arrow over it when handwritten. The velocity component v is not a vector; it is a vector component. The symbol v for the

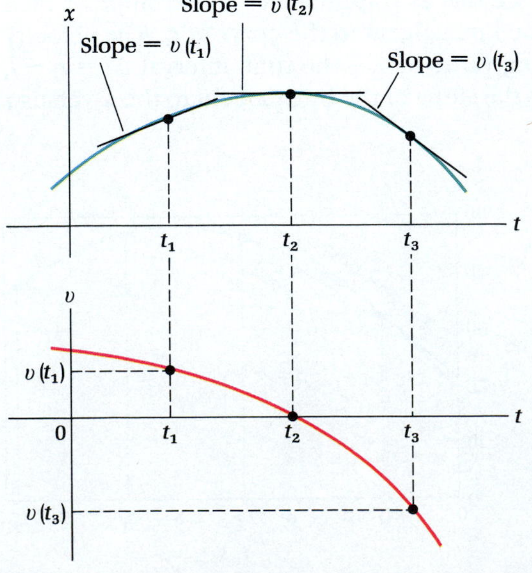

Figure 3-4. For the motion of an object, the graph of x versus t is directly related to the graph of v versus t. The slope of the x-versus-t curve at each point (at each instant of time) is equal to the value of v at that same point.

velocity component is written in italics in a textbook, as are other vector components. For one-dimensional motion along the x axis, the relation between these two quantities is $\mathbf{v} = v\mathbf{i}$. Despite this distinction, it is common usage to refer to the velocity component simply as the velocity. That is, the word "component" is dropped for brevity. This imprecise language is usually quite harmless when discussing motion along one dimension. The SI unit of velocity is meter per second (m/s).

Speed

The speed of an object is the magnitude of its velocity:

$$\text{Speed} = |\mathbf{v}| = \left|\frac{d\mathbf{r}}{dt}\right| \tag{3-5}$$

In one dimension,

$$\text{Speed} = |v\mathbf{i}| = |v| = \left|\frac{dx}{dt}\right|$$

Since speed is the magnitude of a vector, it is a scalar quantity that is never negative. We often select the direction of $+\mathbf{i}$ so that the velocity component v is positive. In such cases, the speed $|v|$ and the velocity component v are the same. Some representative speeds are listed in Table 3-2.

TABLE 3-2. ***A Few Speeds in Meters per Second (Approximate)***

North America (relative to Europe, continental drift)	10^{-9}
Glacier (relative to earth's surface)	10^{-6}
Human walking (relative to sidewalk)	1
Jet taking off (relative to runway)	80
Earth's surface at equator (relative to center of earth)	4.6×10^2
Center of earth (relative to sun)	3.0×10^3
Solar system (relative to center of our galaxy)	2.5×10^5
Fastest known galaxy (relative to earth)	2.4×10^8
Light	3.0×10^8

EXAMPLE 3-2

More about the bobsled. *(a)* For the sled in Example 3-1, determine an expression for the velocity component $v(t)$ as a function of time. *(b)* Construct a graph of the velocity component versus time from $t = 0.0$ s to $t = 8.0$ s by plotting points at each 1.0 s. *(c)* Using data from Table 3-1, show the sled's position along its straight-line path at $t = 0.0$, 2.0, 5.0, and 8.0 s. Use arrows to represent the sled's velocity at each of these times.

Solution. *(a)* From Example 3-1,

$$x(t) = 18\text{ m} + (12\text{ m/s})t - (1.2\text{ m/s}^2)t^2$$

Since $x(t)$ is a polynomial in t, we use the rule for taking the derivative of a power of t:

$$\frac{d}{dt}t^n = nt^{n-1}$$

Therefore

$$\begin{aligned} v(t) &= \frac{d}{dt}[18\text{ m} + (12\text{ m/s})t - (1.2\text{ m/s}^2)t^2] \\ &= 12\text{ m/s} - 2(1.2\text{ m/s}^2)t \\ &= 12\text{ m/s} - (2.4\text{ m/s}^2)t \end{aligned}$$

(b) The value of v at, say $t = 2.0$ s, is

$$v = 12\text{ m/s} - (2.4\text{ m/s}^2)(2.0\text{ s}) = 7\text{ m/s}$$

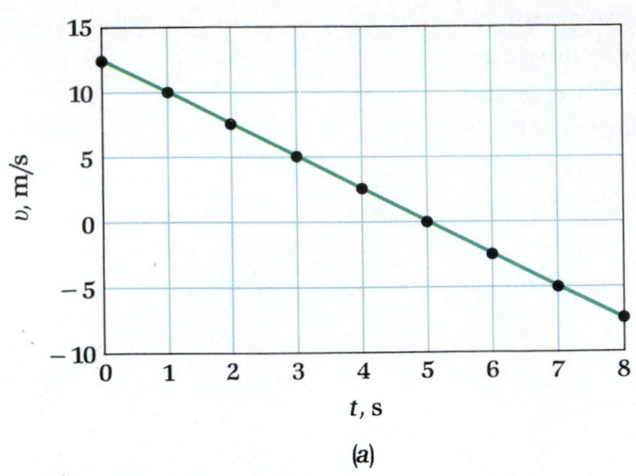

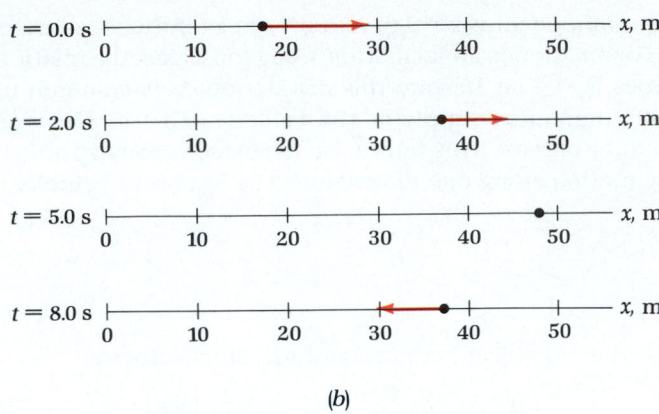

Figure 3-5. Example 3-2: *(a)* Velocity component versus time. *(b)* Position and velocity at four different times. The sled was momentarily at rest at $t = 5.0$ s.

Finding the velocity component at the other times yields the data in Table 3-3, and these data are plotted in Fig. 3-5*a*. Notice that v passes through zero at $t = 5.0$ s. At this instant the sled is momentarily at rest before it begins its backward slide. *(c)* The sled's position and velocity at the four times are shown in Fig. 3-5*b*. The length of each arrow corresponds to the sled's speed at that instant. Compare the motion at $t = 2.0$ s and at $t = 8.0$ s in Fig. 3-5*(b)*. The sled's position and speed are the same at these times, but the velocity is different because the direction of motion is different. This clearly shows the distinction between velocity (a vector) and speed (a scalar).

TABLE 3-3

t, s	v, m/s
0.0	12
1.0	10
2.0	7
3.0	5
4.0	2
5.0	0
6.0	−2
7.0	−5
8.0	−7

SELF-TEST 3-2. Suppose the coordinate of an object is given by the expression

$$x(t) = 2.5\ \text{m} + (8.9\ \text{m/s})t - (1.5\ \text{m/s}^3)t^3$$

(a) Find an expression for $v(t)$. Make certain each factor is written with the proper units. *(b)* Determine the value of v at $t = 2.0$ s. *(c)* What is the object's speed at $t = 2.0$ s? *(d)* Write down the object's velocity $\mathbf{v}$ at the instant $t = 2.0$ s. Be sure your answer is expressed as a vector. ***ANSWERS:*** *(a)* $v(t) = 8.9\ \text{m/s} - (4.5\ \text{m/s}^2)t^2$; *(b)* $v(2.0\ \text{s}) = -9.1$ m/s; *(c)* $|v(2.0\ \text{s})| = 9.1$ m/s; *(d)* $\mathbf{v}(2.0\ \text{s}) = (-9.1\ \text{m/s})\mathbf{i}$.

EXAMPLE 3-3

How to convert the unit of a speed. Speedometers in cars often give the speed in the unit miles per hour (mi/h). *(a)* Determine the conversion factor for changing the unit of a speed from miles per hour to meters per second. *(b)* Convert a speed of 55 mi/h to meters per second.

Solution. *(a)* From the tables on the back cover, 1 mi = 1.61 km. Also, 1 h = 3600 s. Thus 1 mi/1 h = 1.61 km/3600 s = 0.447 m/s, or

$$1\ \text{mi/h} = 0.447\ \text{m/s}$$

We can write the number 1 by dividing both sides by 1 mi/h:

$$1 = \frac{0.447\ \text{m/s}}{1\ \text{mi/h}} = 0.447\ \text{m h mi}^{-1}\ \text{s}^{-1}$$

This way of writing the number 1 is our conversion factor. *(b)* Using the conversion factor from part *(a)*, we have

$$55\ \text{mi/h} = (55\ \text{mi h}^{-1})(0.447\ \text{m s}^{-1}\ \text{mi}^{-1}\ \text{h}) = 25\ \text{m s}^{-1} = 25\ \text{m/s}$$

SELF-TEST 3-3. *(a)* Find the conversion factor that changes a speed expressed in kilometers per hour to a speed in meters per second. *(b)* Convert a speed of 29 km/h to meters per second. ***ANSWERS:*** *(a)* 0.278 m h $\text{km}^{-1}\ \text{s}^{-1}$; *(b)* 8.1 m/s.

Strobe photograph of a toy tractor moving with constant velocity. The length of the tractor is 0.010 m and the time between flashes is 0.5 s. Determine the tractor's speed.

3-3 MOTION WITH CONSTANT VELOCITY

An example of motion with constant velocity is that of a car traveling along a straight road at a steady speed. In this case, the coordinate x as a function of time is given by

$$x(t) = x_0 + vt$$

The symbol x_0 represents the value of x at $t = 0$ and is referred to as the initial coordinate or the initial position.

Suppose the car is traveling along a straight street at a steady speed of 10 m/s and is 20 m past an intersection at the time we select for $t = 0$. If we let our x axis be along the street with $+\mathbf{i}$ in the direction of travel and the origin at the intersection, then the expression for the coordinate is

$$x = 20 \text{ m} + (10 \text{ m/s})t$$

Figure 3-6 shows graphs of x versus t and v versus t during the time interval $t = 0.0$ to 4.0 s. The graph of x versus t is a straight line with slope v, and the graph of v versus t is a straight line with zero slope. What is the expression for $x(t)$ if $+\mathbf{i}$ is chosen to be opposite the direction of travel and the intersection is still the origin? Sketch graphs of x and v versus t for this case.

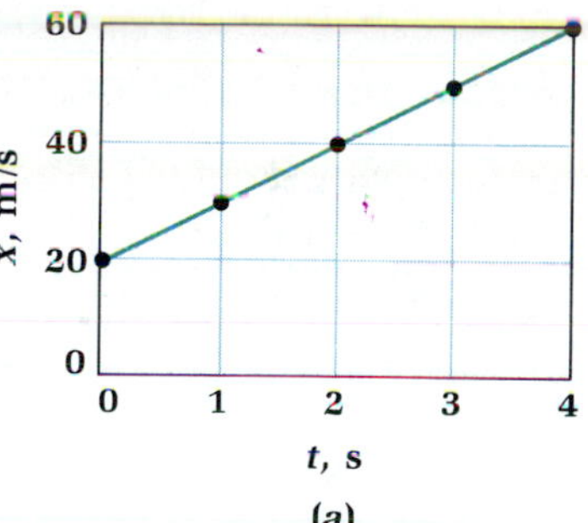

(a)

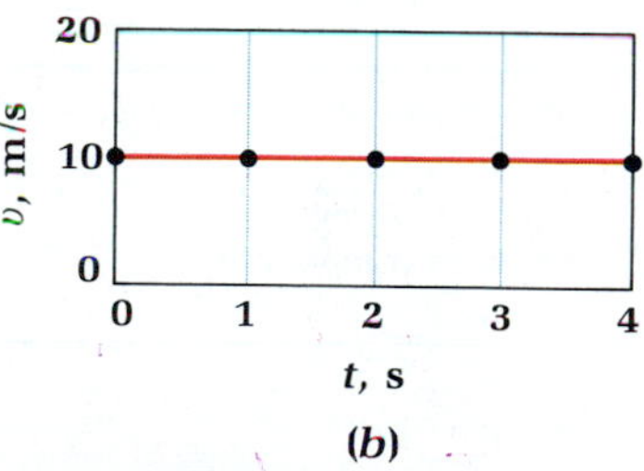

(b)

Figure 3-6. Graphs of *(a)* x versus t and *(b)* v versus t for a car traveling with constant velocity.

3-4 ACCELERATION

The acceleration of an object characterizes how rapidly the object's velocity is changing, both in magnitude and direction. Acceleration is the rate of change of velocity.

Average Acceleration

Similar to the way we used average velocity to define velocity, we now use average acceleration to define acceleration. An object's *average acceleration* $\bar{\mathbf{a}}$ over a time interval from t_i to t_f is

$$\bar{\mathbf{a}} = \frac{\mathbf{v}_f - \mathbf{v}_i}{t_f - t_i} = \frac{\Delta\mathbf{v}}{\Delta t} \qquad (3\text{-}6)$$

Average acceleration

where $\mathbf{v}_f$ and $\mathbf{v}_i$ are the velocities at times t_f and t_i, respectively. In one dimension, the average acceleration has only one component. Since $\mathbf{v}_f - \mathbf{v}_i = v_f\mathbf{i} - v_i\mathbf{i}$,

$$\bar{\mathbf{a}} = \frac{(v_f - v_i)\mathbf{i}}{t_f - t_i} = \frac{\Delta v}{\Delta t}\mathbf{i} = \bar{a}\mathbf{i}$$

where v_f and v_i are the velocity components at times t_f and t_i. The quantity $\bar{a}$ is the average acceleration component:

$$\bar{a} = \frac{v_f - v_i}{t_f - t_i} = \frac{\Delta v}{\Delta t}$$

Acceleration

To determine the acceleration **a**, we find the limiting value of the average acceleration $\bar{\mathbf{a}}$ as the time interval approaches zero:

$$\mathbf{a} = \lim_{\Delta t \to 0} \bar{\mathbf{a}} = \lim_{\Delta t \to 0} \frac{\Delta \mathbf{v}}{\Delta t}$$

Since

$$\lim_{\Delta t \to 0} \frac{\Delta \mathbf{v}}{\Delta t} = \frac{d\mathbf{v}}{dt}$$

acceleration is defined as

Definition of acceleration

$$\mathbf{a} = \frac{d\mathbf{v}}{dt} \tag{3-7}$$

The ***acceleration*** of an object is the rate of change of its velocity.

For one-dimensional motion along the x axis, $\mathbf{a} = a\mathbf{i}$, so that $a = dv/dt$. Also, since $v = dx/dt$, we have

Acceleration component

$$a = \frac{dv}{dt} = \frac{d}{dt}\left(\frac{dx}{dt}\right) = \frac{d^2x}{dt^2}$$

Figure 3-7 shows the connection between a graph of v versus t and a graph of a versus t. Note that the relation between a and v is similar to the relation between v

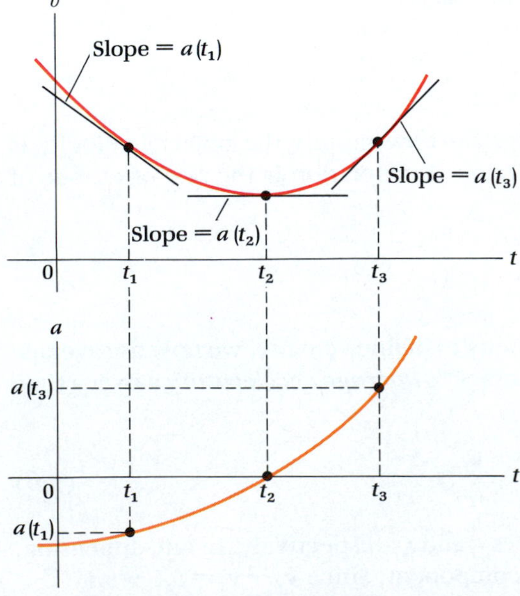

Figure 3-7. The relation between a graph of v versus t and a versus t is similar to the relation between graphs of x versus t and v versus t. The slope of the v curve at each point is equal to the value of a at that same point.

and x. That is, the slope of the curve for v at each instant is equal to the value of a at that same instant.

Further, the acceleration of an object can be found from a graph of x versus t. In a graph of x versus t, $v = dx/dt$ is the slope of the graph. The rate at which the slope v changes is given by $a = dv/dt = d^2x/dt^2$. Thus a is a measure of the rate of change of the slope. In other words, a is related to the bending of the graph. The tighter the bending of the graph, the larger is $|d^2x/dt^2|$ and the larger is the magnitude of the acceleration. If the graph bends upward with increasing t, as shown in Fig. 3-8*a*, then v is increasing with time and a is positive. If the graph bends downward with increasing t, as shown in Fig. 3-7*b*, then v is decreasing with time and a is negative. If the graph is straight, as shown in Fig. 3-8*c*, then v is constant and a is zero.

Acceleration **a** is a vector quantity because it is a velocity change (a vector) divided by a time interval (a scalar). The acceleration component a is a vector component. For one-dimensional motion along the x axis, the relation between them is

$$\mathbf{a} = a\mathbf{i}$$

Just as it is common usage to refer to the velocity component as the velocity, it is also customary to refer to the acceleration component as the acceleration. This abbreviated language rarely causes confusion as long as the motion is restricted to one dimension. The magnitude of the acceleration is a scalar that is never negative. In one dimension,

$$\text{Acceleration magnitude} = |\mathbf{a}| = |a\mathbf{i}| = |a|$$

The SI unit for acceleration is meter per second squared (m/s^2). A few representative accelerations are listed in Table 3-4.

The term ***deceleration*** is often used to describe an object whose speed is decreasing. For example, a car decelerates when the driver applies the brakes. Deceleration is *not* the opposite of acceleration. An object is accelerating when its speed is either increasing or decreasing. Indeed, in the next chapter we shall find that an object following a curved path is accelerating even when its speed is constant. Thus deceleration is a particular kind of acceleration. For an object that is slowing down, the object's deceleration is given by the rate of change of its speed.

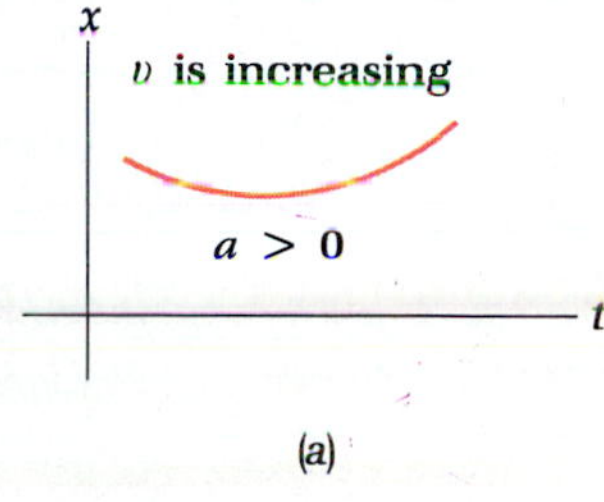

(a)

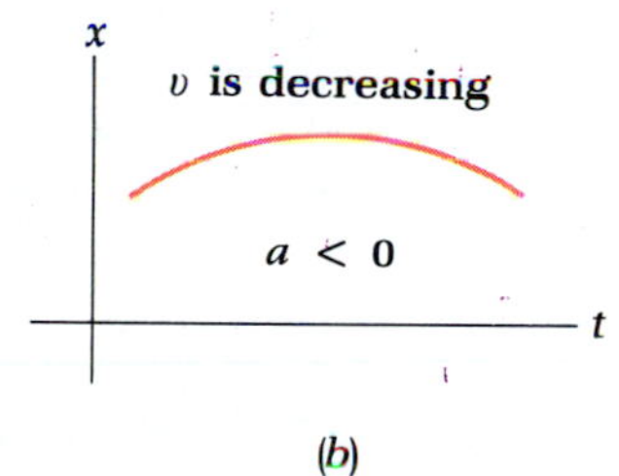

(b)

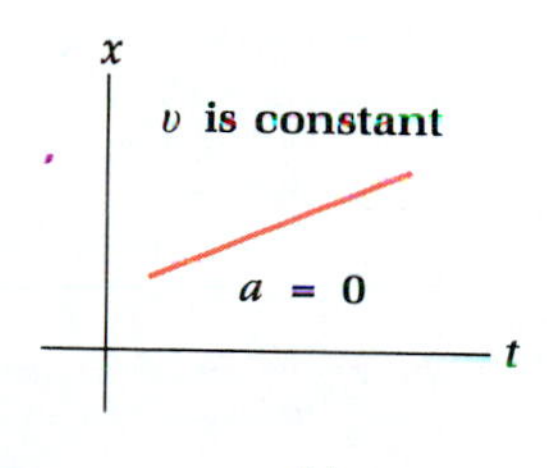

(c)

Figure 3-8. *(a)* A graph of x versus t that is concave upward. In this case v is increasing with time and a is positive. *(b)* A graph of x versus t that is concave downward. In this case v is decreasing with time and a is negative. *(c)* A graph of x versus t that is straight. In this case v is constant and a is zero.

TABLE 3-4. ***A Few Acceleration Magnitudes in Meters per Second Squared (Approximate)***

Solar system (relative to center of our galaxy)	2×10^{-10}
Earth's center (relative to sun)	6×10^{-3}
Earth's surface at equator (relative to earth's center)	3×10^{-2}
Jet airliner on its takeoff roll (relative to runway)	4
Object falling near earth's surface (relative to earth's surface)	10
Proton in a laboratory accelerator (relative to accelerator)	10^{14}

EXAMPLE 3-4

A varying acceleration. Suppose that the coordinate of an object is given by

$$x(t) = (4.0\ \text{m/s})t + (1.1\ \text{m/s}^3)t^3$$

Find expressions for *(a)* v and *(b)* a. *(c)* Construct a graph of v versus t in the time interval from $t = 0.0$ s to $t = 4.0$ s. *(d)* Determine a at $t = 1.0$ s, and show the line tangent to the v-versus-t graph whose slope is equal to this value of the acceleration component.

Solution. *(a)* Since $v = dx/dt$,

$$\begin{aligned} v &= \frac{d}{dt}[(4.0\ \text{m/s})t + (1.1\ \text{m/s}^3)t^3] \\ &= 4.0\ \text{m/s} + 3(1.1\ \text{m/s}^3)t^2 \\ &= 4.0\ \text{m/s} + (3.3\ \text{m/s}^3)t^2 \end{aligned}$$

TABLE 3-5

t, s	v, m/s
0.0	4.0
1.0	7.3
2.0	17
3.0	34
4.0	57

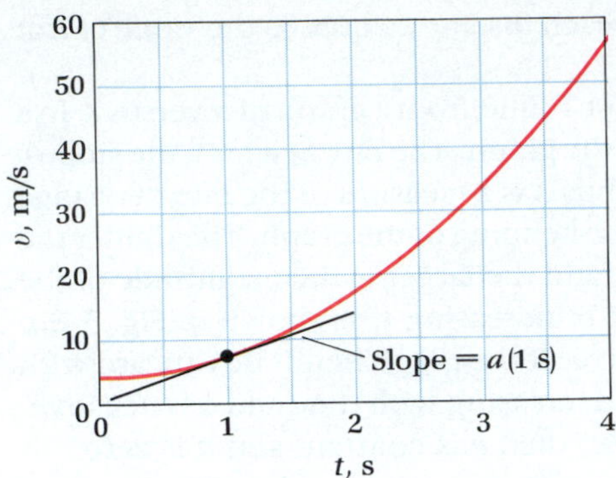

Figure 3-9. Example 3-4.

(b) Since $a = dv/dt$,

$$a = \frac{d}{dt}[4.0 \text{ m/s} + (3.3 \text{ m/s}^3)t^2] = 2(3.3 \text{ m/s}^3)t$$
$$= (6.6 \text{ m/s}^3)t$$

(c) To construct the graph, we use the expression from part *(a)* above to evaluate v at each second from $t = 0.0$ s to $t = 4.0$ s. The value of v at, say 2.0 s is

$$v(2 \text{ s}) = 4.0 \text{ m/s} + (3.3 \text{ m/s}^3)(2.0 \text{ s})^2 = 17 \text{ m/s}$$

The other values of v are listed in Table 3-5, and the graph of v versus t is shown in Fig. 3-9. *(d)* Using the expression for a from part *(b)* above gives

$$a(1 \text{ s}) = (6.6 \text{ m/s}^3)(1.0 \text{ s}) = 6.6 \text{ m/s}^2$$

The line whose slope is equal to this value of the acceleration component is tangent to the graph of v versus t at $t = 1.0$ s, as shown on the graph in Fig. 3-9.

SELF-TEST 3-4. The expression for the coordinate of the sled in Example 3-1 is

$$x(t) = 18 \text{ m} + (12 \text{ m/s})t - (1.2 \text{ m/s}^2)t^2$$

(a) Determine an expression for the sled's acceleration component a. *(b)* Is the acceleration constant or does it vary with time? *(c)* Construct a graph of a versus t from $t = 0.0$ s to $t = 8.0$ s. *(d)* What is the value of the acceleration magnitude $|a|$ at $t = 4.0$ s? ***ANSWERS:*** *(a)* $a = -2.4 \text{ m/s}^2$; *(b)* constant; *(d)* $|a| = 2.4 \text{ m/s}^2$.

EXAMPLE 3-5

Analyzing an entire trip. A barge equipped with an engine travels back and forth along a straight canal that is aligned east to west. Figure 3-10 is a graph of the barge's coordinate versus time where the origin is at the dock and the $+x$ direction is toward the east. Use this graph to describe the barge's motion.

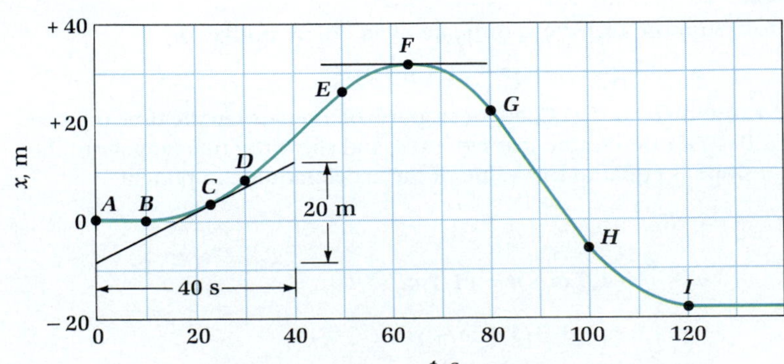

Figure 3-10. Example 3-5: A graph of x versus t for a barge.

Solution. *A* to *B*: The barge remains stationary at the dock from $t = 0$ s to $t = 10$ s.

B to *D*: At $t = 10$ s the barge begins accelerating. During the time interval from $t = 10$ s to $t = 30$ s, a is positive (the graph bends upward); the acceleration is toward the east. This means that v is increasing. Also, v is positive (the slope of the graph is positive) so that the velocity is directed toward the east and the speed is increasing. The slope of the tangent line at $t = 24$ s shows that the velocity at that instant is $\mathbf{v} = (20 \text{ m}/40 \text{ s})\mathbf{i} = (0.5 \text{ m/s})\mathbf{i}$.

D to *E*: From $t = 30$ s to $t = 50$ s, the acceleration is zero (the graph is straight) so that the velocity is constant toward the east with magnitude

$$v = \frac{26 \text{ m} - 8 \text{ m}}{50 \text{ s} - 30 \text{ s}} = 0.9 \text{ m/s}$$

E to *G*: From $t = 50$ s to $t = 80$ s, the acceleration component is negative (the graph bends downward) so that the velocity component is decreasing. At $t = 64$ s the barge is instantaneously at rest as it changes its direction of motion from eastward to westward. From $t = 50$ s to $t = 64$ s the velocity component and the speed are decreasing. Between $t = 64$ s and $t = 80$ s the speed is *increasing* despite the fact that the velocity component is *decreasing*. This is because the velocity component is negative and when a negative quantity is decreasing (becoming more negative), its absolute value is increasing.

G to *H*: During the time interval from $t = 80$ s to $t = 100$ s, the barge travels westward at a constant speed of

$$[(22 \text{ m}) - (-6 \text{ m})]/(100 \text{ s} - 80 \text{ s}) = 1.4 \text{ m/s}$$

H to *I*: Between $t = 100$ s and $t = 120$ s the acceleration component is positive so that the velocity component is increasing (becoming less negative) and the speed is decreasing. The barge comes to rest 18 m west of the dock at $t = 120$ s.

SELF-TEST 3-5. Figure 3-11 shows curves of x versus t for two sprinters racing in a 100-m dash. *(a)* Which sprinter, *A* or *B*, has the larger average acceleration during the time interval from $t = 0$ s to 5 s? *(b)* Which sprinter has the larger acceleration at the instant $t = 5$ s? *(c)* Which sprinter, if either, decelerates during the race? *(d)* Which sprinter seems to be in better condition? ***ANSWERS:*** *(a) A; (b) B; (c) A; (d) B.*

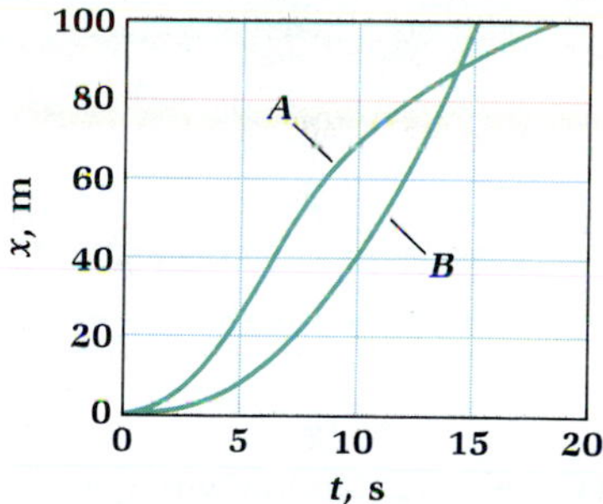

Figure 3-11. Self-test 3-5. Two sprinters in a 100-m race.

3-5 MOTION WITH CONSTANT ACCELERATION

An important type of motion is motion with constant acceleration. The bobsled, whose adventures we followed in Examples 3-1 and 3-2, moved with constant acceleration. When an object moves with constant acceleration, the acceleration is equal to the average acceleration. Thus $a = \bar{a} = \Delta v/\Delta t$, or $a = (v_f - v_i)/(t_f - t_i)$. To find an expression for $v(t)$, we let $t_f = t$ and $t_i = 0$, so that $v_f = v(t)$ and $v_i = v(0) = v_0$. This gives

$$a = \frac{v(t) - v_0}{t - 0}$$

Solving for $v(t)$, we obtain

$$v(t) = v_0 + at \qquad (3\text{-}8)$$

Velocity component v as a function of t, constant a

The object's velocity depends linearly on the time t.

We can find the expression for $x(t)$ from the definition of velocity and Eq. 3-8. Since

$$v(t) = \frac{d}{dt}x(t)$$

and

$$v(t) = v_0 + at$$

we have

$$\frac{d}{dt}x(t) = v_0 + at \qquad (3\text{-}9)$$

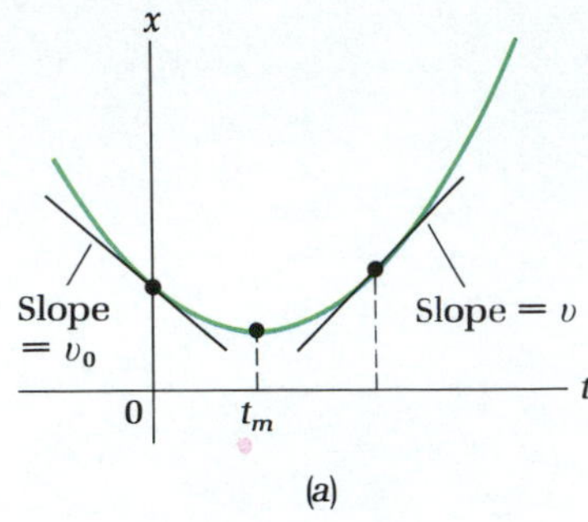

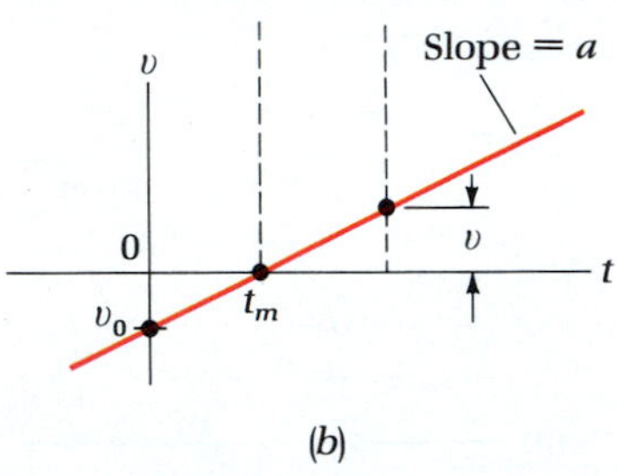

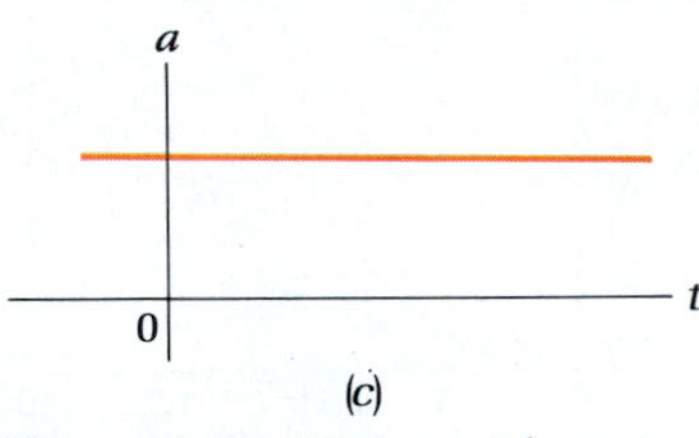

Figure 3-12. Motion with constant acceleration. For the case shown, x_0 is positive, v_0 is negative, and a is positive. *(a)* The graph of x versus t is a parabola whose slope at each point is v. *(b)* The graph of v versus t is a straight line with slope a. *(c)* The graph of a versus t is a horizontal straight line.

That is, $x(t)$ is a function of t such that its time derivative gives $v_0 + at$. From the rule

$$\frac{d}{dt} t^n = nt^{n-1}$$

you can see that if we take the derivative of a polynomial in t, then each exponent is reduced by 1. This means that for dx/dt to give Eq. 3-9, the time dependence of $x(t)$ must be of the form

$$x(t) = C_0 + C_1 t + C_2 t^2 \tag{3-10}$$

where C_0, C_1, and C_2 are constants. To verify that Eq. 3-10 does give the correct time dependence for x, we take its derivative and compare the result to $v = v_0 + at$. Taking the derivative gives

$$\frac{d}{dt} x(t) = \frac{d}{dt}(C_0 + C_1 t + C_2 t^2) = C_1 + 2C_2 t$$

Comparing this expression with $v_0 + at$, we have

$$v_0 + at = C_1 + 2C_2 t$$

If we let $C_1 = v_0$ and $C_2 = \frac{1}{2}a$, then the derivative of Eq. 3-10 gives Eq. 3-9.

We have verified that Eq. 3-10 has the proper time dependence to describe motion with constant acceleration, and we have determined the values of C_1 and C_2. Now we find C_0 by evaluating Eq. 3-10 at $t = 0$. This gives $x(0) = C_0$. Since this initial value of x is customarily designated as x_0, we let $C_0 = x_0$. Substituting the values for the constants into Eq. 3-10, we obtain

$$x(t) = x_0 + v_0 t + \tfrac{1}{2}at^2 \tag{3-11}$$

Thus x depends quadratically on t.

Figure 3-12 shows graphs of x, v, and a versus t for a case where x_0 is positive, v_0 is negative, and a is positive. These graphs reveal the symmetry of the motion about the time t_m. At this time, the object is instantaneously at rest ($v = 0$) and it is changing its direction of travel. At any time *before* t_m, the value of x is the same as it is for the same time *after* t_m. Also, the object's speed is the same for equal times before and after t_m. Since x depends quadratically on t, the graph of x versus t in Fig. 3-12*a* is a parabola.

Equations 3-8 and 3-11 describe the motion in terms of the time t. A third equation can be developed from these two by eliminating t. Solving Eq. 3-8 for t, we find $t = (v - v_0)/a$. Inserting this result into Eq. 3-11 gives

$$x = x_0 + v_0\left(\frac{v - v_0}{a}\right) + \frac{1}{2}a\left(\frac{v - v_0}{a}\right)^2$$

Rearranging and solving for v^2, we obtain

Velocity component v as a function of x, constant a

$$v^2 = v_0^2 + 2a(x - x_0) \tag{3-12}$$

Another equation that is sometimes used in describing an object moving in one dimension with constant acceleration is

$$x = x_0 + \tfrac{1}{2}(v_0 + vt) \tag{3-13}$$

You can obtain this expression by using Eq. 3-8 to eliminate a in Eq. 3-11. (See Prob. 10.) Table 3-6 summarizes the important equations in this section.

TABLE 3-6. *Equations Describing Motion with Constant Acceleration*

$v(t) = v_0 + at$
$x(t) = x_0 + v_0 t + \frac{1}{2}at^2$
$v^2 = v_0^2 + 2a(x - x_0)$
$x = x_0 + \frac{1}{2}(v_0 + v)t$

More About the Graphical Relation between Kinematical Quantities. As we have seen, the slope of a graph of x versus t gives v, and the slope of a graph of v versus t gives a. However, we often wish to proceed in the reverse order. That is, we wish to determine v from a graph of a versus t, and we wish to find x from a graph of v versus t. Figure 3-13 shows a graph of v versus t when a is constant. This graph

(Top) Strobe photograph of a toy car moving to the left and increasing its speed. The car's acceleration is in the same direction as its velocity. *(Bottom)* Strobe photograph of a toy car moving to the right and decreasing its speed. The car's acceleration is opposite in direction to its velocity.

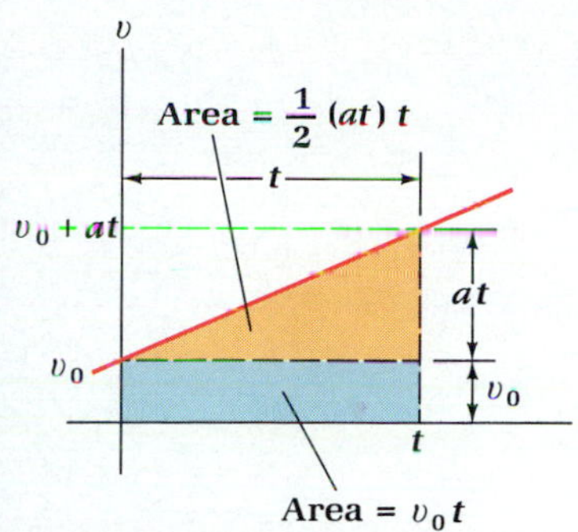

Figure 3-13. Relation between v and x from a graph of v versus t. When acceleration a is constant, the area under the graph of v versus t is equal to the area of the rectangle plus the area of the triangle: Total area = $v_0t + \frac{1}{2}at^2$. From Eq. 3-11, this is equal to $x - x_0$. Thus the area under the graph of v versus t gives $x - x_0$.

describes Eq. 3-8: $v = v_0 + at$. The area between the graph and the t axis, and between the v axis and the vertical line at t, is called the ***area under the graph.*** This area is equal to the area of the rectangle of height v_0 and width t plus the area of the triangle of height at and width t:

$$\text{Area under graph} = v_0t + \tfrac{1}{2}(at)t = v_0t + \tfrac{1}{2}at^2$$

From Eq. 3-11 we see that $v_0t + \frac{1}{2}at^2 = x - x_0$. Therefore we find that the area under the graph v versus t gives the coordinate x if the initial coordinate x_0 is known. Although we have shown this only for the case of constant acceleration, it is valid in general, but the general proof requires integral calculus (see App. I), which we defer until Chap. 8. Figure 3-14 illustrates the general result using the subscripts i and f for initial and final values of the quantities.

The relation between v and a is similar to the relation between x and v. As you would expect, the area under a graph of a versus t gives $v - v_0$, similar to the way the area under the graph of v versus t gives $x - x_0$. This result is shown in Fig. 3-15.

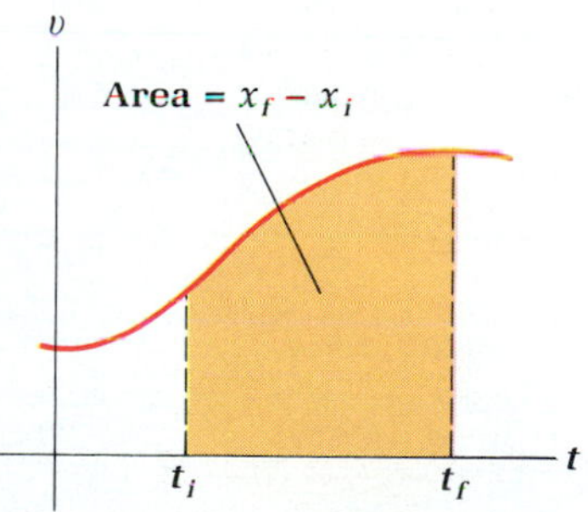

Figure 3-14. In general, the area under a graph of v versus t for a time interval is equal to the change in x during the time interval.

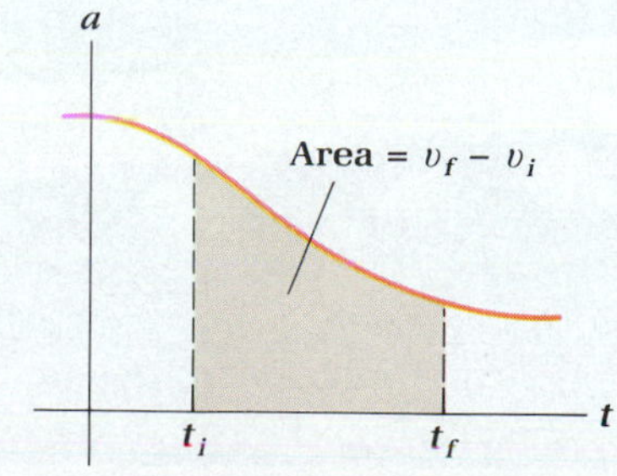

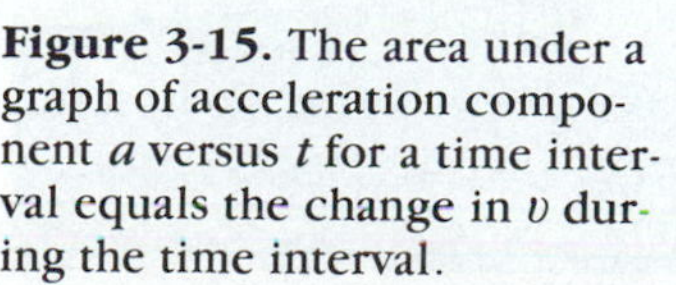

Figure 3-15. The area under a graph of acceleration component a versus t for a time interval equals the change in v during the time interval.

EXAMPLE 3-6

Estimating the acceleration of a tennis ball. During a serve, the speed of a tennis ball may increase from nearly zero to approximately 50 m/s while the racket is in contact with the ball. During the time of contact, the ball may move a distance of about 1 m. Use this information to estimate the acceleration magnitude of the ball during the serve.

Solution. The ball's acceleration during the serve is probably *not* constant. However, the initial speed, the final speed, and the distance moved are given. To proceed with a solution, we must assume that the acceleration is constant. Thus we use Eq. 3-12 and set $v_0 = 0$, $(x - x_0) = d$, and solve for a. This gives

$$a = v^2/2d = (50\ \text{m/s})^2/(2\ \text{m}) = 1 \times 10^3\ \text{m/s}^2$$

This is a rough estimate, but it does give some idea of the ball's acceleration magnitude.

SELF-TEST 3-6. Using information from the above example, estimate the time interval in which a tennis ball is in contact with the racket during a serve. What assumption were you required to make to find an answer? ***ANSWERS:*** 0.02 s. Assume constant acceleration.

EXAMPLE 3-7

TABLE 3-7

t, s	x, m
0.00	−0.030
0.20	0.067
0.40	0.203
0.60	0.377
0.80	0.589
1.00	0.840

Experiment with a rolling ball. A useful device for observing the motion of an object is a stroboscope. A stroboscope is a light that briefly flashes at regular time intervals. If a stroboscope is used in conjunction with a camera whose shutter remains open, the camera records the position of an object at each successive flash. Figure 3-16 shows a stroboscopic picture of a billiard ball rolling down a straight slope. The flashes are 0.20 s apart, and the scale is calibrated in meters. Let the unit vector **i** point in the direction of motion (to the right), and let $t = 0.00$ s be the time when the ball is farthest up the slope. Table 3-7 gives the coordinate of the center of the ball at each time t. *(a)* Assume that the ball moves with constant acceleration, and find values of x_0, v_0, and a. *(b)* Write the expressions for $x(t)$ and $v(t)$. *(c)* Verify that the data in Table 3-7 are consistent with constant acceleration.

Solution. *(a)* If the ball moves with constant acceleration, the data of Table 3-7 can be given by Eq. 3-11. As you can see from the table, $x_0 = -0.030$ m. Therefore we have two unknowns to find; they are a and v_0. Since two unknowns can be found from two equations, we can insert two data pairs from Table 3-7, call them (t_1, x_1) and (t_2, x_2), into Eq. 3-11 and solve for the unknowns. These equations are

$$x_1 = x_0 + v_0 t_1 + \tfrac{1}{2} a t_1^2 \tag{A}$$

$$x_2 = x_0 + v_0 t_2 + \tfrac{1}{2} a t_2^2 \tag{B}$$

Solving Eq. A for v_0, we find

$$v_0 = \frac{x_1 - x_0}{t_1} - \frac{1}{2} a t_1 \tag{C}$$

Inserting this result into Eq. B and solving for a, we obtain

$$a = \frac{2[t_1(x_2 - x_0) - t_2(x_1 - x_0)]}{t_1 t_2^2 - t_1^2 t_2} \tag{D}$$

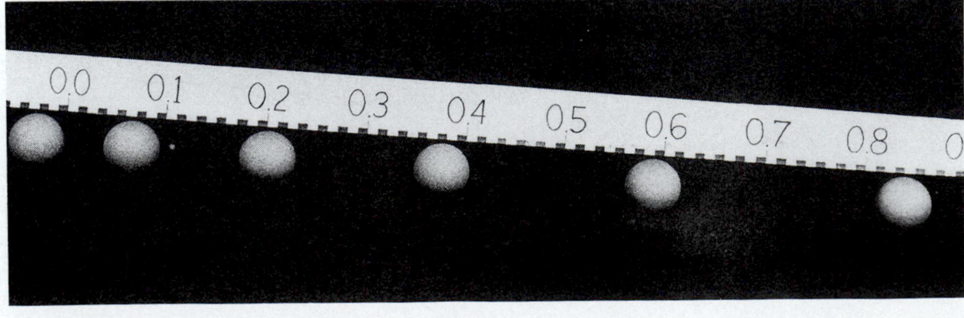

Figure 3-16. Example 3-7: A stroboscopic picture of a billiard ball rolling down a straight slope.

We choose (0.20 s, 0.067 m) and (1.00 s, 0.840 m) as our data pairs from Table 3-7. Substituting these data into Eq. D gives

$$a=\frac{2[(0.20\text{ s})(0.870\text{ m})-(1.00\text{ s})(0.097\text{ m})]}{(0.20\text{ s})(1.00\text{ s})^2-(0.20\text{ s})^2(1.00\text{ s})}=0.96\text{ m/s}^2$$

Using this value of a and a data pair from the table, say (0.20 s, 0.067 m), in Eq. C gives

$$v_0=\frac{0.097\text{ m}}{0.20\text{ s}}-\frac{1}{2}(0.96\text{ m/s}^2)(0.20\text{ s})=0.39\text{ m/s}$$

(b) The ball's coordinate and velocity component as functions of time are

$$x(t)=-0.030\text{ m}+(0.39\text{ m/s})t+(0.48\text{ m/s}^2)t^2$$

$$v(t)=0.39\text{ m/s}+(0.96\text{ m/s}^2)t$$

(c) To verify that the data in Table 3-7 are consistent with the assumption of constant acceleration, each value of t from the table can be inserted into the expression for $x(t)$ to see whether each calculation gives the corresponding value of x. For instance, checking the pair (0.40 s, 0.203 m),

$$x=-0.030\text{ m}+(0.39\text{ m/s})(0.40\text{ s})+(0.48\text{ m/s}^2)(0.40\text{ s})^2=0.20\text{ m}$$

This data pair is consistent with the assumption of constant acceleration. Some of the other data pairs can be checked in the same way.

SELF-TEST 3-7. Without using numerical values, sketch graphs of x versus t and v versus t, similar to Fig. 3-12*a* and *b*, for the following cases of constant acceleration:
(a) $a<0,\ v_0>0,\ x_0>0$
(b) $a>0,\ v_0<0,\ x_0>0$
(c) $a>0,\ v_0>0,\ x_0>0$

3-6 FREE-FALL

We are all familiar with falling objects—for example, a paperweight that is knocked off the edge of a desk. Often in describing the motion of such objects, we can neglect air resistance. If air resistance has a negligible effect on a falling object, then it is valid to assume that the object's acceleration is due entirely to gravity. In this case the motion is called ***free-fall.*** Treating the motion of the paperweight as free-fall is a valid approximation as long as it does not fall too far. Even for short falls, this approximation is poor for an object such as a feather or a badminton birdie.

Galileo Galilei is regarded as the father of modern science. A brief biography of Galileo is contained in Chap. 4.

Galileo Galilei (1564–1642) made quantitative studies of free-fall and determined that the acceleration due to gravity is constant. Indeed, it was Galileo who established the usefulness of the concept of acceleration as it is now defined. (Some of Galileo's results are discussed in Prob. 12.) Modern measurements verify that objects in free-fall have a constant downward acceleration; the acceleration is the same at each instant during the fall. Further, this acceleration is the *same* for *different* objects (Fig. 3-17). This familiar but nevertheless intriguing result will be considered in more detail in Chap. 7.

The magnitude of the acceleration due to gravity is represented by the symbol g. Although g varies slightly from place to place on the earth's surface, a value that is accurate enough for our purposes is

Magnitude of the acceleration due to gravity

$$g=9.8\text{ m/s}^2$$

We shall discuss the slight variation of g on the earth's surface in Chap. 7.

In describing free-fall, we customarily choose the y axis along the direction of motion with the unit vector $\mathbf{j}$ directed upward. Then the acceleration of a freely falling object is

Free-fall acceleration

$$\mathbf{a}=-g\mathbf{j}$$

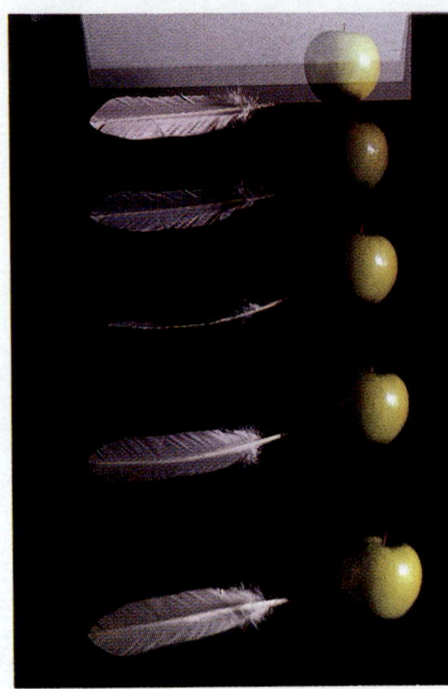

Figure 3-17. An apple and a feather fall together in an evacuated chamber. When the effects of air resistance are negligible, all objects fall with the same acceleration.

A minus sign is explicitly put into the equation because the acceleration is downward, and the symbol g represents a positive number. Since free-fall is motion with constant acceleration, we can use Eqs. 3-8, 3-11, and 3-12 to describe this motion by letting $a = -g$ and by changing the coordinate from x to y:

$$v(t) = v_0 - gt \tag{3-14}$$

$$y(t) = y_0 + v_0 t - \tfrac{1}{2}gt^2 \tag{3-15}$$

$$v^2 = v_0^2 - 2g(y - y_0) \tag{3-16}$$

Figure 3-18 shows graphs of a, v, and y versus time t for an object that undergoes free-fall after it is released from rest at $y = 0$.

If an object, such as a baseball, is thrown vertically upward with initial speed v_0, two quantities that can be readily measured are (i) the time t_m required for the ball to reach its maximum height and (ii) the ball's maximum height h_m. Let the origin of the coordinate frame be at the release point, and let $t = 0$ correspond to the instant the ball is released so that $y_0 = 0$. When the ball reaches its maximum height, its velocity is zero. Assuming the ball is in free-fall, we use Eq. 3-14 and find $v(t_m) = 0 = v_0 - gt_m$, or

$$t_m = \frac{v_0}{g} \tag{3-17}$$

The coordinate evaluated at this time is the maximum height, $h_m = y(t_m)$. Inserting t_m from Eq. 3-17 into Eq. 3-15 gives

$$y(t_m) = h_m = 0 + v_0\left(\frac{v_0}{g}\right) - \frac{1}{2}g\left(\frac{v_0}{g}\right)^2$$

or

$$h_m = \frac{v_0^2}{2g} \tag{3-18}$$

The same result is obtained starting with Eq. 3-16, inserting $v = 0$, $y - y_0 = h_m$, and solving for h_m.

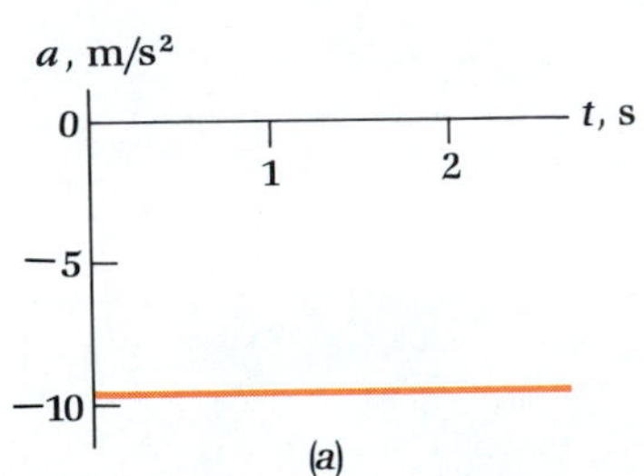

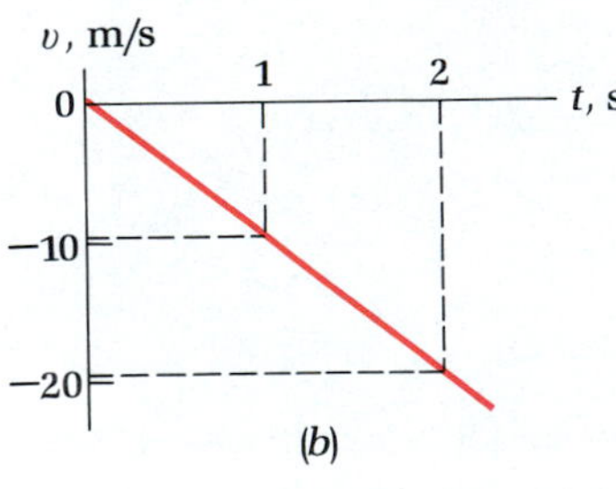

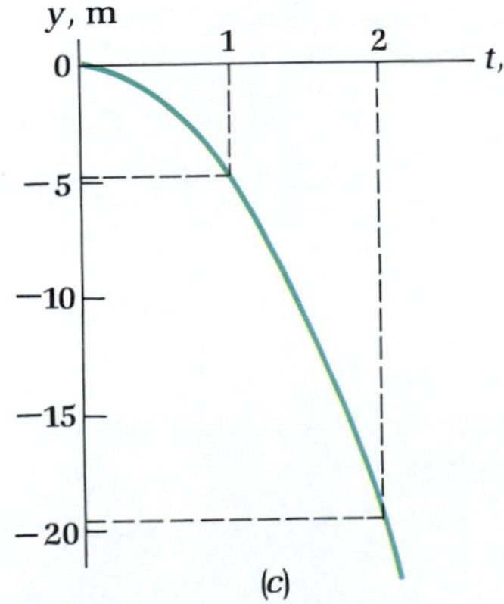

Figure 3-18. Graphs of *(a)* a, *(b)* v, and *(c)* y for free-fall when $v_0 = 0$ and $y_0 = 0$.

EXAMPLE 3-8

Flight of a rock. A rock was thrown vertically upward such that the time required for it to reach its maximum height was 1.2 s. The release point was 1.5 m above the ground. *(a)* Letting $t = 0$ be the instant the rock was released and $y = 0$ correspond to the ground, determine the expressions for the rock's velocity component and coordinate as functions of time. *(b)* Evaluate these expressions at $t = 0.0$, 0.60, 1.2, and 1.8 s. *(c)* Sketch the rock's position, velocity, and acceleration at each of these times. Use arrows to represent the velocity and acceleration. Neglect air resistance.

Solution. *(a)* To find these expressions, we must determine the values of y_0 and v_0 and insert them into Eqs. 3-14 and 3-15. The value of y_0 is given ($y_0 = 1.5$ m), and v_0 can be found from Eq. 3-17.

$$v_0 = t_m g = (1.2 \text{ s})(9.8 \text{ m/s}^2) = 12 \text{ m/s}$$

Thus the expressions for v and y are

$$v = 12 \text{ m/s} - (9.8 \text{ m/s}^2)t$$

$$y = 1.5 \text{ m} + (12 \text{ m/s})t - (4.9 \text{ m/s}^2)t^2$$

(b) At $t = 0.0$ s, we have $v = v_0 = 12$ m/s and $y = y_0 = 1.5$ m. At $t = 0.60$ s, we have

$$v = 12 \text{ m/s} - (9.8 \text{ m/s}^2)(0.60 \text{ s}) = 5.9 \text{ m/s}$$

$$y = 1.5 \text{ m} + (12 \text{ m/s})(0.60 \text{ s}) - (4.9 \text{ m/s}^2)(0.60 \text{ s})^2 = 6.8 \text{ m}$$

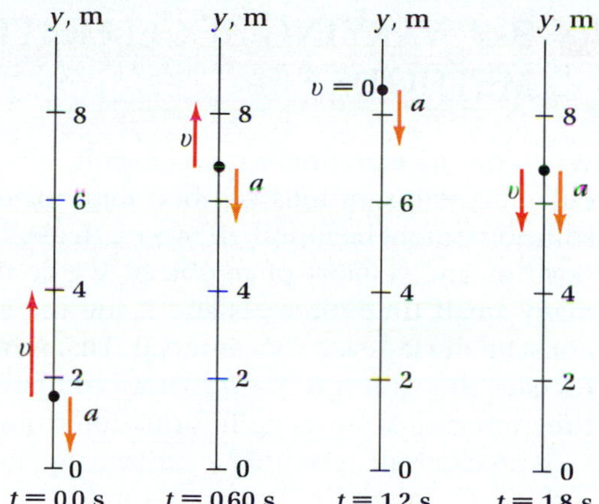

Figure 3-19. Example 3-8: A rock is thrown vertically upward. The rock's position, velocity, and acceleration are shown at four times.

At $t = 1.2$ s, $v = 0.0$ m/s because $t_m = 1.2$ s. The coordinate at this time is

$$y = 1.5\ \text{m} + (12\ \text{m/s})(1.2\ \text{s}) - (4.9\ \text{m/s}^2)(1.2\ \text{s})^2 = 8.6\ \text{m}$$

At $t = 1.8$ s, we have

$$v = 12\ \text{m/s} - (9.8\ \text{m/s}^2)(1.8\ \text{s}) = -5.9\ \text{m/s}$$

$$y = 1.5\ \text{m} + (12\ \text{m/s})(1.8\ \text{s}) - (4.9\ \text{m/s}^2)(1.8\ \text{s})^2 = 6.8\ \text{m}$$

These values at $t = 1.8$ s could have been predicted from those at $t = 0.60$ s and from the symmetry of the motion. *(c)* The sketches of the rock's position, velocity, and acceleration at these times are shown in Fig. 3-19.

SELF-TEST 3-8. *(a)* From the example above, determine the rock's speed when its coordinate is $y = 6.5$ m. *(b)* At what time t did it have this speed while traveling upward? *(c)* At what time t did it have this speed while traveling downward? ***ANSWERS:*** *(a)* 6.3 m/s; *(b)* 0.55 s; *(c)* 1.8 s.

EXAMPLE 3-9

Look out below! A flowerpot falls from a second-floor window. What is its speed just before it hits the ground 3.0 m below? Neglect air resistance.

Solution. We can use Eq. 3-16, $v^2 = v_0^2 - 2g(y - y_0)$, to develop an equation for the speed v of any object at the instant it has fallen a vertical distance h from rest (neglecting air resistance). Since the object falls from rest, we set $v_0 = 0$, and since it falls downward a distance h, we have $(y - y_0) = -h$. Substituting into Eq. 3-16 gives

$$v^2 = (0)^2 - 2g(-h)$$

Or,

$$v = \sqrt{2gh}$$

The flowerpot falls 3.0 m, so its speed just before hitting the ground is

$$v = \sqrt{2(9.8\ \text{m/s}^2)(3.0\ \text{m})} = 7.7\ \text{m/s}$$

SELF-TEST 3-9. *(a)* Consider a ball thrown vertically upward such that its speed is v_0 at the instant of release. Show that the maximum height h_m attained by the ball during its flight is given by

$$h_m = v_0^2/2g$$

where h_m is measured from the point of release. Neglect air resistance. *(b)* A ball is tossed vertically upward with a release speed of 4.4 m/s. What is the maximum height it will reach relative to the point of release? ***ANSWER:*** *(b)* 0.99 m.

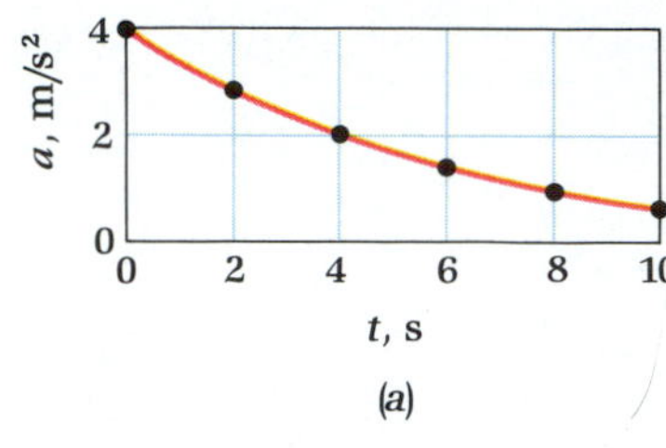

(a)

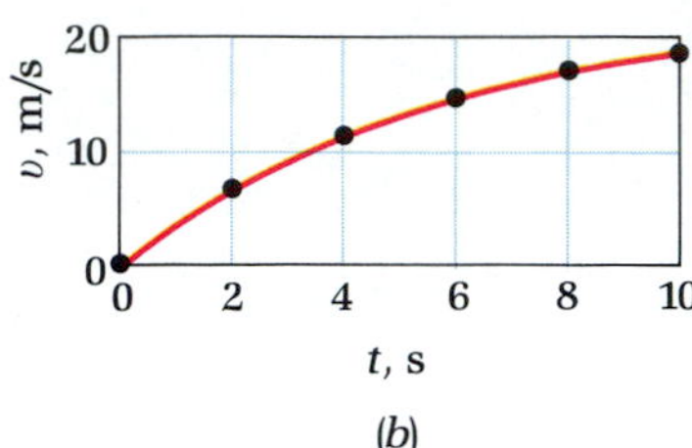

(b)

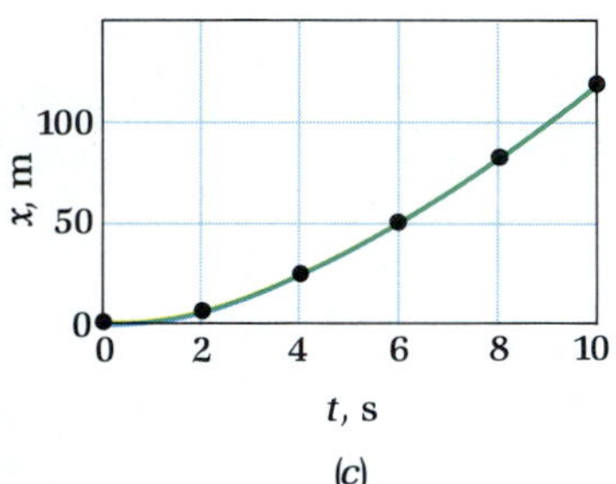

(c)

Figure 3-20. Graphs generated using spreadsheet software. The graphs show a, v, and x for a car starting from rest with an acceleration component given by Eq. 3-19.

3-7 VARYING ACCELERATION, NUMERICAL METHODS

When the acceleration is not constant, solving a kinematics problem is often difficult. Analytic solutions can be found for only a few special cases. If an analytic solution cannot be found, then we can use a numerical procedure to determine the position and velocity of an object. We do this by simply dividing the motion into many small time intervals and using the approximation that the acceleration is constant during each time interval. That is, we use constant acceleration kinematics for each time interval. We instruct a computer to calculate x and v at the end of each time interval Δt by using information from the previous time interval.

As an example, consider a car starting out from a stop sign. Suppose the magnitude of the car's acceleration is 4 m/s² at $t = 0$, then 2 m/s² at $t = 4$ s, then 1 m/s² at $t = 8$ s, and so on. That is, the magnitude of the acceleration becomes smaller by a factor of 2 every 4 s. An equation for a is

$$a(t) = a_0 2^{-bt} \tag{3-19}$$

where $a_0 = 4\ \text{m/s}^2$ and $b = 0.25\ \text{s}^{-1}$. Figure 3-20a shows a graph of a versus t.

Two ways to proceed are to (i) use spreadsheet type software or (ii) use a computer language such as BASIC or Fortran. Each of these procedures has advantages and disadvantages in comparison with the other. For example, a spreadsheet will often graph the data for you with just a few simple commands. On the other hand, the languages are more flexible and more readily applied to a variety of problems, but then you are left to graph the data yourself. In this section, we describe the use of a spreadsheet.

A spreadsheet is designed like a ledger, that is, with columns and rows. As shown in Table 3-8, the columns are labeled with letters and the rows with numbers. The position of an entry within a spreadsheet is called a cell, and a cell is located by its column letter and row number—for example, cell C5. The cells may contain words, numbers, or formulas. The cells in row 2 of Table 3-8 contain words that identify the quantities listed below them. Cell A3 contains the number zero as the first time entry, $t = 0$. Cell B3 contains the formula from Eq. 3-19 for calculating the acceleration, where * stands for multiplication, ^ stands for exponentiation (raising a quantity to a power), and A3 refers to the time given in cell A3. In this example, we have instructed the computer to determine values of t, a, v, and x at each 0.01 s over a time interval of 10 s. Consider cell C5, which has the formula for calculating v at the time given in A5 ($t = 0.02$ s). The formula states that v for the time in A5 is equal to the value of v from cell C4 plus the increase due to the acceleration. This increase is the product of the acceleration times the time interval, or the value of a from cell B5 times 0.01 s. Table 3-9 gives the beginning and ending data for this problem, and Fig. 3-20 shows the curves that were generated.

TABLE 3-8. ***Beginning and ending of a spreadsheet program for the car whose acceleration component is given by Eq. 3-19. Notice that numbers, such as 0.00 in cell A3, are right-justified within a cell, whereas formulas and words are left-justified within a cell.***

	A	B	C	D
1	Varying Acceleration			
2	Time (s)	Acc. (m/s2)	Vel. (m/s)	Pos. (m)
3	0.00	4*2^(−0.25*A3)	0.00	0.00
4	+A3+0.01	4*2^(−0.25*A4)	+C3+B4*0.01	+D3+C4*0.01
5	+A4+0.01	4*2^(−0.25*A5)	+C4+B5*0.01	+D4+C5*0.01
⋮	⋮	⋮	⋮	⋮
1001	+A1000+0.01	4*2^(−0.25*A1001)	+C1000+B1001*0.01	+D1000+C1001*0.01
1002	+A1001+0.01	4*2^(−0.25*A1002)	+C1001+B1002*0.01	+D1001+C1002*0.01
1003	+A1002+0.01	4*2^(−0.25*A1003)	+C1002+B1003*0.01	+D1002+C1003*0.01

TABLE 3-9. *Beginning and ending data from the spreadsheet program listed in Table 3-8.*

	A	B	C	D
1	Varying Acceleration			
2	Time (s)	Acc. (m/s2)	Vel. (m/s)	Pos. (m)
3	0.00	4.00	0.00	0.00
4	0.01	3.99	0.04	0.00
5	0.02	3.99	0.08	0.00
⋮	⋮	⋮	⋮	⋮
1001	9.98	0.71	18.97	120.78
1002	9.99	0.71	18.98	120.97
1003	10.00	0.71	18.99	121.16

SELF-TEST 3-10. Write a spreadsheet program that describes a car starting from rest with an acceleration given by Eq. 3-19, with $a_0 = 3\ \mathrm{m/s^2}$ and $b = 0.25\ \mathrm{s^{-1}}$. Make graphs of your data and compare them with those in Fig. 3-20.

COMMENTARY: Marching Lockstep through Physics

A textbook can be deceptive. In this textbook, you will be introduced to many laws, equations, and rules. These formulations are regarded as successful because they accurately and concisely describe many phenomena. Any theory or experiment that is now viewed as a failure will not be discussed. It is the lack of any discussion of the failures that can be deceiving. It can lead to the mistaken impression that such failures are rare, or that they never existed, or that they are without value.

Studying physics in a textbook such as this is similar to walking along a beaten path. The path has been worn smooth by those before us. There are hardly any bumps or holes that might cause us to stumble. As it exists now, the path at a particular place is nothing like it was when it was first trod. Then it was full of briars and brambles. In times past, one person or another cut a path that later turned out to be off the main trail. Textbooks avoid such diversions.

Where does this beaten path lead? It leads to the frontier of physics. The spirit of physics is at the frontier. That is where the uncertainty and the excitement are. Nearing the frontier, we find that the path becomes more obscure. Now there are lots of holes and bumps and briars and brambles. At the frontier many people are hacking straight ahead. However, real progress often is made by those who back up, step off the beaten path, and start an entirely new approach. For instance, Albert Einstein, in developing the theory of relativity (Chap. 38), returned to the most basic ideas of space and time. He showed that beyond a certain point the beaten path was leading the wrong way.

As you read this textbook, or any textbook, you should remain skeptical. But do not let your skepticism interfere with learning the material. If you are to blaze a new trail, you will probably need to start from somewhere on the beaten path.

SUMMARY

Section 3-1. Position Vector and Displacement

The position vector $\mathbf{r}$ locates an object relative to the origin of a reference frame. In one dimension, $\mathbf{r} = x\mathbf{i}$ where x is the object's coordinate. The displacement $\Delta\mathbf{r}$ is the change in the position vector,

$$\Delta\mathbf{r} = \mathbf{r}_f - \mathbf{r}_i$$

Section 3-2. Velocity and Speed

The average velocity of an object is the object's displacement during a time interval divided by the time interval, $\overline{\mathbf{v}} = \Delta\mathbf{r}/\Delta t$. The velocity is the limiting value of the average velocity as the time interval approaches zero:

$$\mathbf{v} = \lim_{\Delta t\to 0} \overline{\mathbf{v}} = \frac{d\mathbf{r}}{dt} \qquad (3\text{-}4)$$

Speed is the magnitude of the velocity, speed $= |\mathbf{v}|$.

Section 3-3. Motion with Constant Velocity

When the velocity is constant, the coordinate x varies linearly with time t:

$$x = x_0 + vt$$

Section 3-4. Acceleration

The average acceleration of an object during a time interval is the change in the object's velocity divided by the time interval, $\bar{\mathbf{a}} = \Delta\mathbf{v}/\Delta t$. The acceleration is the limiting value of the average acceleration as the time interval approaches zero:

$$\mathbf{a} = \lim_{\Delta t \to 0} \bar{\mathbf{a}} = \frac{d\mathbf{v}}{dt} \qquad (3\text{-}7)$$

Section 3-5. Motion with Constant Acceleration

When an object moves with constant acceleration along one dimension,

$$v(t) = v_0 + at \qquad (3\text{-}8)$$

$$x(t) = x_0 + v_0 t + \tfrac{1}{2}at^2 \qquad (3\text{-}11)$$

$$v^2 = v_0{}^2 + 2a(x - x_0) \qquad (3\text{-}12)$$

Section 3-6. Free-Fall

Free-fall is a particular example of motion with constant acceleration. If a freely falling object moves along a vertical line (the y axis), then we use the espressions above with y as the coordinate and $-g$ as the constant acceleration component.

Section 3-7. Varying Acceleration, Numerical Methods

Numerical methods can be used to describe motion with varying acceleration. Each iteration is taken over a time interval that is small enough so that the acceleration can be treated as constant.

QUESTIONS

1. Suppose we flip a coin vertically upward. If we are interested in the maximum height of the coin, is it valid to treat the coin as a particle? If we are interested in whether the coin lands heads up or tails up, is it valid to treat the coin as a particle?

2. A jogger travels from $x = 0$ to $x = 50$ m between $t = 0$ and $t = 10$ s. Between $t = 10$ s and $t = 15$ s, the jogger travels from $x = 50$ m to $x = 25$ m. Is the distance traveled by the jogger equal to the magnitude of his displacement *(a)* between $t = 0$ and $t = 10$ s, *(b)* between $t = 0$ and $t = 15$ s? Explain.

3. The symbol $v(t)$ usually means "v as a function of time t," but it could be used to represent the product v times t. In the following two equations, which interpretation do you give to $v(t)$?
(a) $v(t) = 7.3$ m
(b) $v(t) = (6.1\ \text{m/s}^2)t$

4. While running in a race of length 24 mi, a hypothetical runner becomes increasingly tired. During the first hour of the race she runs 12 mi, during the second hour she runs 6 mi, during the third hour she runs 3 mi, and so on. Each hour she runs half the remaining distance. How long will it take her to complete the race?

5. In describing the motion of a car traveling west, we let the $+x$ direction be toward the east. Consider the following statements:
(a) The velocity of the car is -32 m/s
(b) The velocity of the car is $(-32\ \text{m})\mathbf{i}$.
(c) The velocity of the car is $(-32\ \text{m/s})\mathbf{i}$.
(d) The speed of the car is -32 m/s.
(e) The speed of the car is 32 m/s.
(f) The velocity component of the car is -32 m/s.
Which, if any, of these statements is meaningless? Explain what is wrong with any statement that is meaningless.

6. Judy says that the average speed of an object is the magnitude of the object's average velocity. Martha says that the average speed of an object is the distance traveled by the object during a time interval divided by the time interval. Will Judy and Martha always agree on the value of the average speed of an object? Describe a case where they will agree. Describe a case where they will disagree. Decide which definition you prefer and give your reasons.

7. Can the speed of an object be negative? If so, give an example. If not, explain why not.

8. We defined speed as the magnitude of the velocity. Does a car's speedometer reading correspond to this definition? Explain.

9. Does a car's odometer measure distance or displacement? Explain.

10. A car travels along a straight east-west street. We let the unit vector $\mathbf{i}$ point toward the east. What is the sign of v if the car is traveling *(a)* toward the east, *(b)* toward the west? What is the sign of a if the car is traveling *(c)* toward the east and slowing down, *(d)* toward the east and speeding up, *(e)* toward the west and slowing down, *(f)* toward the west and speeding up?

11. Consider the definition of deceleration (Sec. 3-4). If $dv/dt < 0$, is it necessarily true that the object is decelerating? Give examples to support your answer.

12. In describing the motion of a rock that is thrown vertically upward, we let the unit vector $\mathbf{j}$ point upward. What is the sign of the rock's velocity component v *(a)* before it reaches its maximum height, *(b)* at the instant it reaches its maximum height, *(c)* after it reaches its maximum height? *(d)* What is the rock's speed at the instant it reaches its maximum height? After the rock has reached its maximum height, *(e)* is its speed increasing, decreasing, or remaining the same? *(f)* Is its velocity component increasing, decreasing, or remaining the same?

13. For the rock in the previous question, what is the sign of the rock's acceleration component *(a)* before it reaches its maximum height, *(b)* at the instant it reaches its maximum height, *(c)* after it reaches its maximum height? *(d)* What is the magnitude of the rock's acceleration at the instant it reaches its maximum height?

14. A girl throws a ball vertically upward with an initial speed of 10 m/s and catches it at the same height when it returns. Neglecting air resistance, what is the ball's speed when it is caught?

15. Reaching out from a balcony, you throw rock A vertically upward. Then you throw rock B vertically downward from the same release point and with the same initial speed as rock A. If you neglect air resistance, which rock has the higher speed just before it hits the ground?

16. Two golf balls are dropped from rest from the top of a tall building. Ball 1 is dropped at $t = 0$, and ball 2 is dropped at $t = 0.5$ s. Ball 1 hits the ground at $t = 3.0$ s. *(a)* Between $t = 0.5$ s and $t = 3.0$ s, does the separation between the two balls increase, decrease, or remain the same? *(b)* When does ball 2 hit the ground?

17. A graph of x versus t for an object is shown in Fig. 3-21. What are the algebraic signs of v and a at times *(a)* t_1; *(b)* t_2; *(c)* t_3?

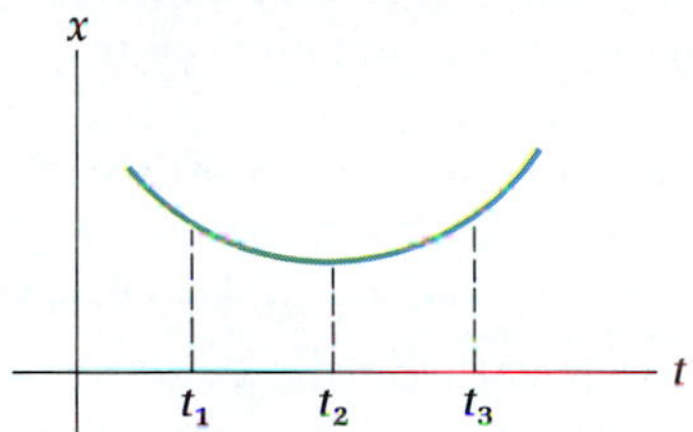

Figure 3-21. Question 17.

18. A graph of v versus t for an object is shown in Fig. 3-22. What is the algebraic sign of a at times *(a)* t_1; *(b)* t_2; *(c)* t_3?

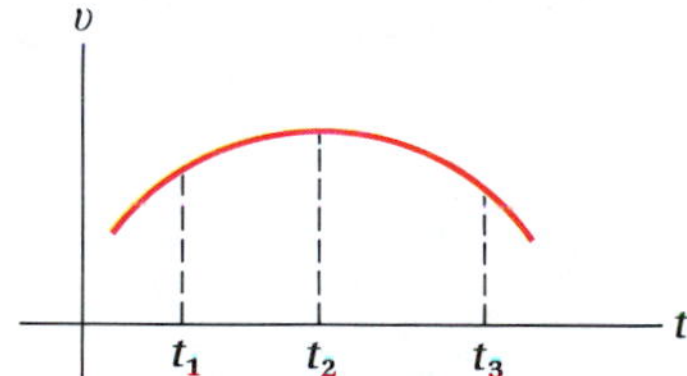

Figure 3-22. Question 18.

19. *(a)* In each part of Fig. 3-23, determine whether the velocity component v is larger at t_1 or at t_2. Keep in mind that -3 is larger than -5. *(b)* In each part of Fig. 3-23, determine whether the speed $|v|$ is larger at t_1 or at t_2.

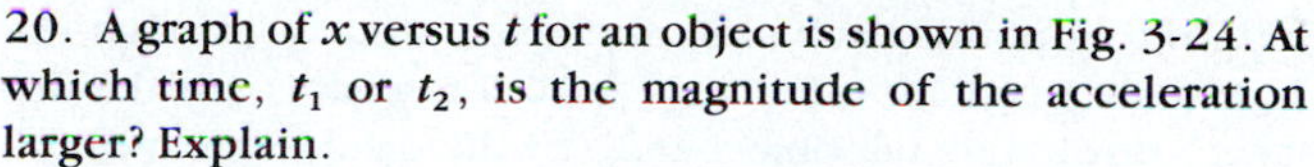

20. A graph of x versus t for an object is shown in Fig. 3-24. At which time, t_1 or t_2, is the magnitude of the acceleration larger? Explain.

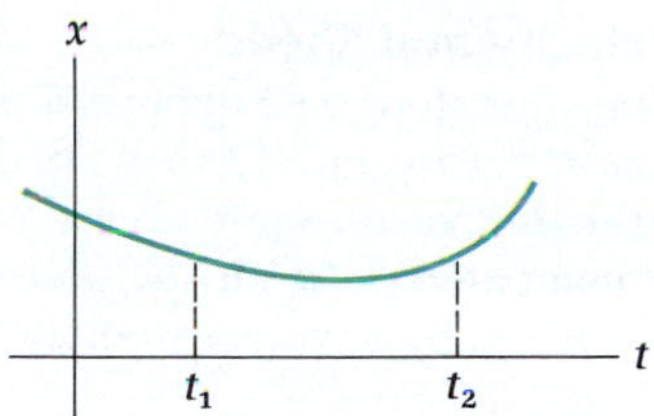

Figure 3-24. Question 20.

21. Suppose the expression that gives a versus t is linear. What is the expression for v versus t?

22. Suppose the expression for the coordinate of an object is of the form $x(t) = C_0 + C_1 t + C_3 t^3$. *(a)* What is the time dependence of a? Is it quadratic? Linear? Constant? Something else? *(b)* What is the time dependence of v?

23. Complete the following table:

Symbol	Represents	Type	SI Unit
$\mathbf{r}$			m
x_0	Initial coordinate		
v		Component	
$\lvert v \rvert$			
a			

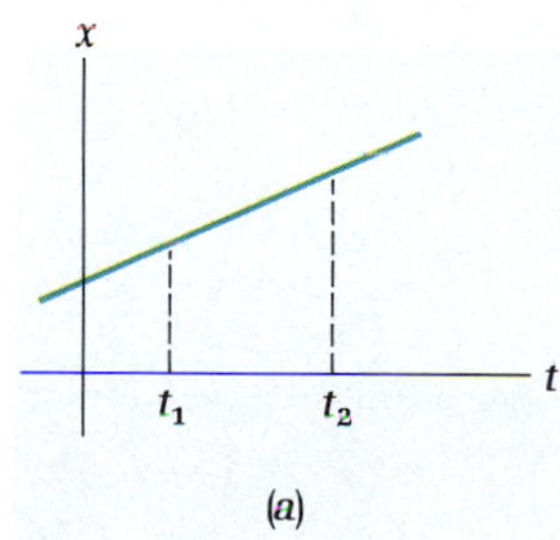

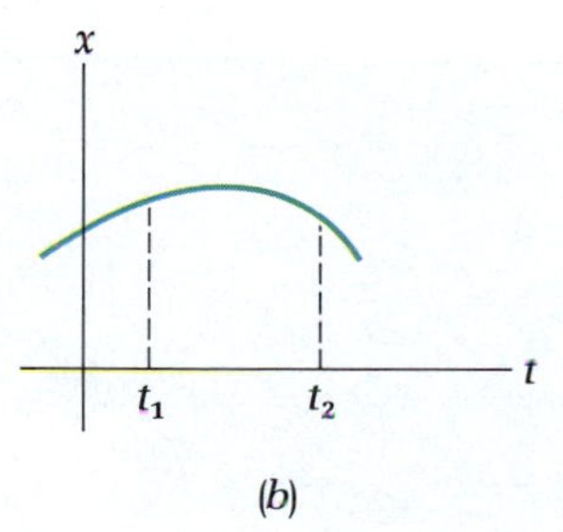

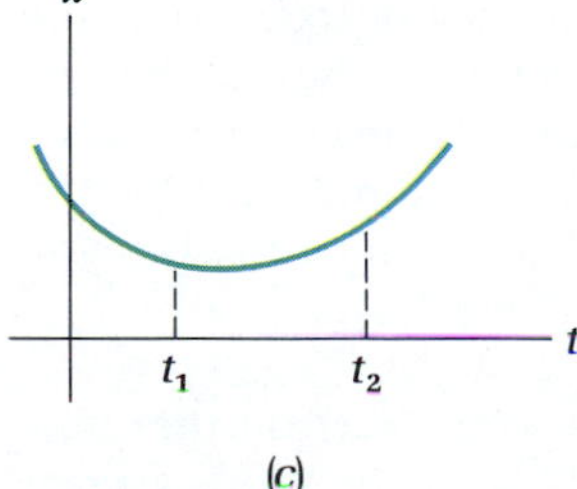

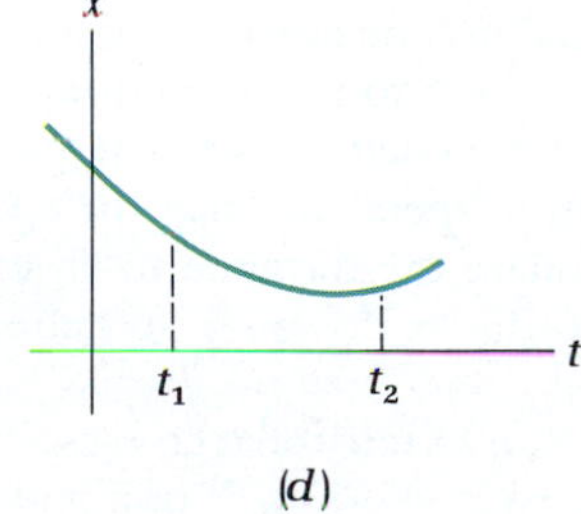

Figure 3-23. Question 19.

EXERCISES

Section 3-1. Position Vector and Displacement

1. A jogger is 16 m west of a stop sign at time t_i and is 37 m east of the stop sign at time t_f. Let the stop sign be the origin and let the unit vector $\mathbf{i}$ point east. Determine *(a)* x_i; *(b)* x_f; *(c)* $\mathbf{r}_i$; *(d)* $\mathbf{r}_f$; *(e)* $\Delta\mathbf{r}$.

2. The coordinate for a bicycle is given by the expression $x(t) = -(14 \text{ m/s})t + 74$ m. Make a graph of x versus t from $t = 0.0$ s to $t = 6.0$ s by plotting points for each second. Sketch a curve through the points.

3. The coordinate of an object is given by the expression $x(t) = 52 \text{ mm} \sin (0.44 \text{ rad/s})t$. (Recall that 2π rad $= 360°$. Be certain you change from degrees to radians on your calculator.) *(a)* Make a graph of x versus t from $t = 0.0$ s to $t = 15.0$ s

by plotting points each second. Sketch a curve through the points. Between $t = 0.0$ s and $t = 10.0$ s, what is *(b)* the distance traveled by the object and *(c)* the displacement of the object?

Section 3-2. Velocity and Speed

4. A car traveling west along a straight road is 81 m east of a manhole cover at $t_i = 15$ s and 13 m west of the cover at $t_f = 22$ s. *(a)* If the unit vector **i** points east, what is the car's average velocity component? *(b)* If the unit vector **i** points west, what is the car's average velocity component?

5. *(a)* Determine a conversion factor between feet per second and meter per second. *(b)* Convert a speed of 25 m/s to feet per second.

6. One light-year (abbreviated ly) is the distance light travels in one year. *(a)* Given that the speed of light is about 3.0×10^8 m/s, determine a conversion factor between meter and light-year. The distance from earth to the star Sirius (the brightest star in the heavens, other than the sun) is about 10 ly. Determine the distance to Sirius *(b)* in meters and *(c)* in miles.

7. The average distance from the earth to the sun is about 100 million mi (1×10^8 mi), and the speed of light is 3.0×10^8 m/s. *(a)* How long does it take the light from the sun to reach the earth? *(b)* One light-minute is the distance light travels in one minute. Determine the distance to the sun in units of light-minutes.

8. The speed of sound in air at ordinary temperatures is about 340 m/s. Suppose you see a lightning flash in an approaching storm and 6.0 s later you hear the thunder. *(a)* Estimate the distance that the storm is from you by assuming that the speed of light is infinite. *(b)* What sort of precision (the number of significant digits) would be required of the time measurement for the assumption in part *(a)* to be invalid?

9. Often, just about a second or two after the lights flicker in your house, you may hear the sound of a distant explosion. The sound may be due to the explosion of a nearby transformer and the flicker to the subsequent destruction of the transformer. Assuming that electric energy propagates along the wires at an infinite speed and that the speed of sound in air is 340 m/s, estimate the distance to an exploded transformer when you hear the explosion 0.50 s after the lights flicker.

10. *(a)* A ferry boat crosses a 550-m-wide river from the east to the west shore in 1 min and 9 s. What is the boat's average velocity? *(b)* The boat makes the return trip in 58 s. What is the boat's average velocity for the return trip? *(c)* What is the boat's average velocity for the entire round-trip? Keep in mind that velocity is a vector quantity.

11. The equation for the coordinate of an object as a function of time is $x(t) = (2.2\ \text{m/s}^3)t^3 - 18$ m. *(a)* What is the object's average velocity component between $t_i = 1.0$ s and $t_f = 3.0$ s? *(b)* What is the object's velocity component at $t = 2.0$ s?

12. A graph of the coordinate versus time for an object is shown in Fig. 3-25. From this graph, determine $\bar{v}$ *(a)* between $t = 1.0$ s and $t = 4.0$ s and *(b)* between $t = 5.0$ s and $t = 9.0$ s. Find *(c)* $v(3.0\ \text{s})$ and *(d)* $v(8.0\ \text{s})$.

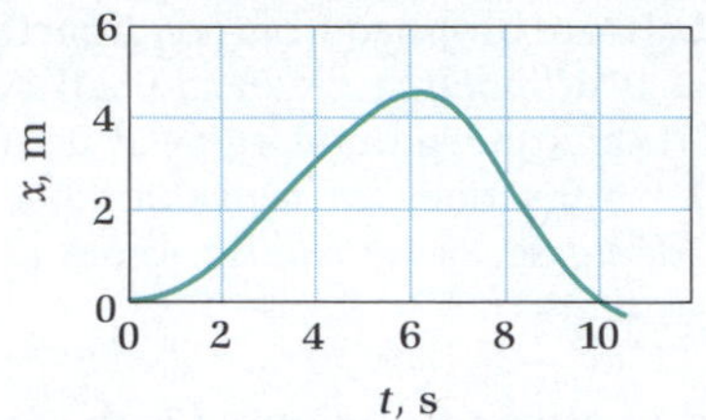

Figure 3-25. Exercise 12.

13. Use the graph of coordinate versus time in Fig. 3-26 to find $\bar{v}$ between $t = 1.0$ s and $t = t_f$ when *(a)* $t_f = 5.0$ s; *(b)* $t_f = 4.0$ s; *(c)* $t_f = 3.0$ s; *(d)* $t_f = 2.0$ s. *(e)* Estimate $v(1.0\ \text{s})$.

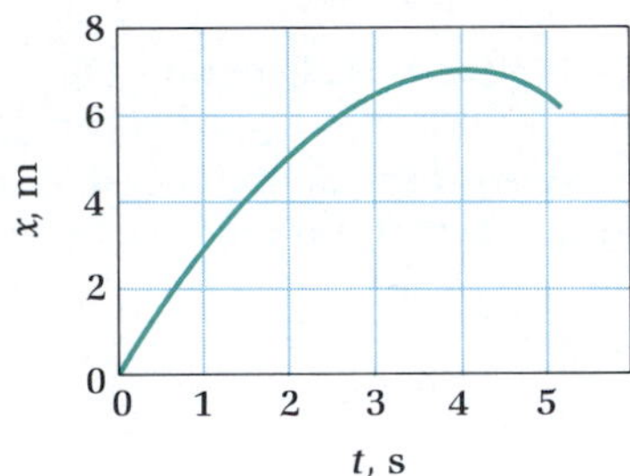

Figure 3-26. Exercise 13.

14. An expression for the coordinate of an object is $x(t) = -(3.5\ \text{m/s}^3)t^3 - (1.8\ \text{m/s})t$. *(a)* Write an expression for $v(t)$. *(b)* What is $v(2.6\ \text{s})$? *(c)* What is v_0?

Section 3-3. Motion with Constant Velocity

15. Figure 3-27 shows a stroboscopic photograph of a golfer driving a ball off a tee. The stroboscopic light flashed 100 times per second. Estimate the speed of the ball as it left the tee.

Figure 3-27. Exercise 15.

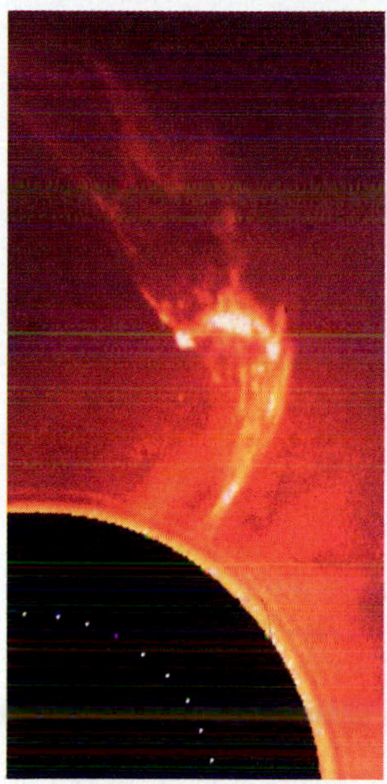

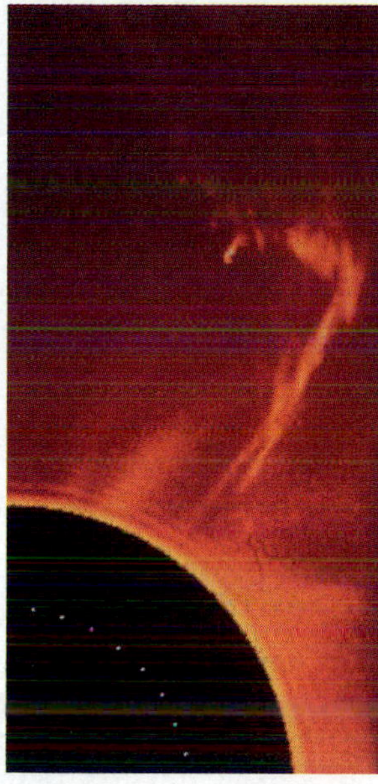

Figure 3-28. Exercise 16.

16. Figure 3-28 shows two photographs of a burst of incandescent gas emitted from the surface of the sun. The two photographs were taken 33 minutes apart. Estimate the speed of the emitted material between the photographs. The radius of the sun is about 7×10^8 m.

17. Figure 3-29 shows a stroboscopic photograph of a billiard ball as it rolls across a horizontal surface. The time between flashes is 0.10 s. Use the meter stick as your reference frame and let the unit vector **i** point to the right. *(a)* Assuming that the ball is moving to the right, write an expression for $x(t)$ with $t = 0.00$ s corresponding to the ball's position at the far left. *(b)* Assuming that the ball is moving to the left, write an expression for $x(t)$ with $t = 0.00$ s corresponding to the ball's position at the far right.

18. A 60-km footrace that was held in New York in 1983 was won with a time of about 4 h and 40 min. Assuming constant speed, what was the speed of the winner?

19. Major-league pitchers typically throw a baseball with a speed of about 90 mi/h. The distance between the pitcher's mound and home plate is about 20 m. Assuming the ball travels at constant speed, estimate the time required for the ball to travel from the mound to the plate.

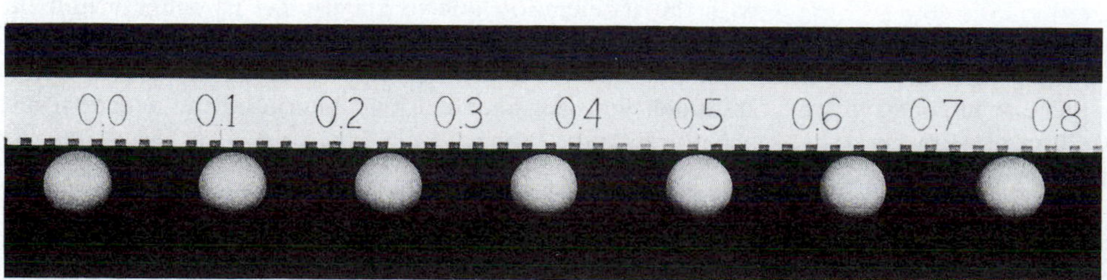

Figure 3-29. Exercise 17.

20. A jogger runs with a constant velocity of 2.2 m/s toward the north. Choose the origin of the coordinate frame at an oak tree, let the unit vector **i** point north, and let $t = 0$ correspond to the instant the jogger is 15 m north of the tree. *(a)* Determine an expression for the jogger's coordinate as a function of time. *(b)* What is the jogger's coordinate at $t = 28$ s? *(c)* When will the jogger's coordinate be 51 m?

Section 3-4. Acceleration

21. Automobile magazines sometimes give the magnitude of the acceleration in units of miles per hour per second ($\text{mi h}^{-1}\,\text{s}^{-1}$). *(a)* Determine a conversion factor for converting miles per hour per second to m s^{-2}. *(b)* Convert an acceleration magnitude of 12 $\text{mi h}^{-1}\,\text{s}^{-1}$ to meters per second squared (m s^{-2}).

22. The speed of a car increases from 18 to 23 m/s in a time interval of 5.8 s. *(a)* Let the $+x$ direction be along the direction of travel and determine the average acceleration component. *(b)* Let the $+x$ direction be opposite the direction of travel and determine the average acceleration component.

23. The speed of a car decreases from 23 to 18 m/s in a time interval of 5.8 s. *(a)* Let the $+x$ direction be along the direction of travel and determine the average acceleration component. *(b)* Let the $+x$ direction be opposite the direction of travel and determine the average acceleration component.

24. A graph of the velocity component versus time for an object is shown in Fig. 3-30. Determine $\bar{a}$ *(a)* between $t = 0.0$ s and $t = 3.0$ s; *(b)* between $t = 3.0$ s and $t = 9.0$ s. *(c)* Find $a(5.0\text{ s})$.

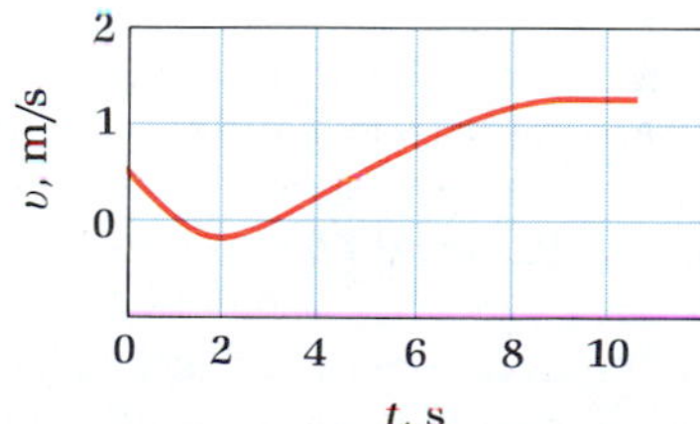

Figure 3-30. Exercise 24.

25. Use the graph of the velocity component versus time in Fig. 3-31 to find $\bar{a}$ between $t = 1.0$ s and $t = t_f$ when *(a)* $t_f = 5.0$ s; *(b)* $t_f = 4.0$ s; *(c)* $t_f = 3.0$ s; *(d)* $t_f = 2.0$ s. *(e)* Estimate $a(1\text{ s})$.

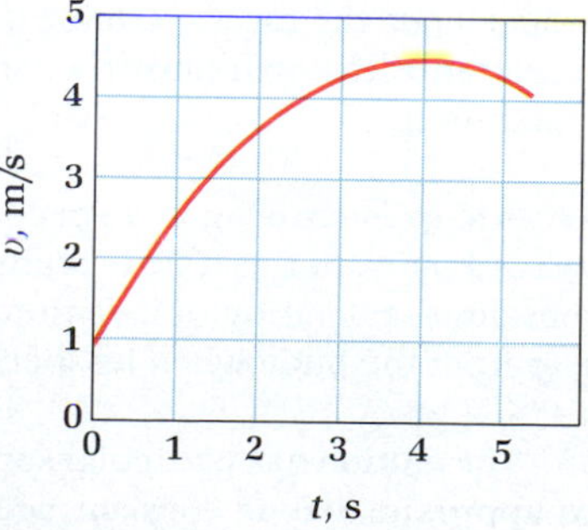

Figure 3-31. Exercise 25.

26. An expression for the coordinate of an object is $x(t) = -(1.6\text{ m/s}^3)t^3 + (2.1\text{ m/s}^2)t^2 - 42\text{ m}$. *(a)* Write an expression for $a(t)$. *(b)* Determine $a(4.1\text{ s})$. *(c)* What is $a(0) = a_0$?

27. An expression for the velocity component of an object is $v(t) = (3.2\text{ m/s}^3)t^2 - 6.1\text{ m/s}$. *(a)* Write an expression for $a(t)$. *(b)* Determine $a(2.7\text{ s})$. *(c)* What is $a(0) = a_0$?

Section 3-5. Motion with Constant Acceleration

28. The acceleration of a sprinter at the beginning of a race can be approximated as constant. Use this approximation and an acceleration magnitude of 3.8 m/s^2 to determine *(a)* the distance traversed by the sprinter in the first 2.0 s of a race and *(b)* the sprinter's speed 2.0 s after the start of the race.

29. Assume that an airliner on its takeoff run moves at constant acceleration with magnitude 3.6 m/s^2. *(a)* Write an expression for the airliner's velocity component as a function of time. *(b)* What is the airliner's speed 24 s after the start of the run? *(c)* Write an expression for the airliner's coordinate as a function of time. *(d)* What is the distance traversed by the airliner during the first 24 s of the run?

30. An object has a constant acceleration of 4.0 m/s^2 toward the south. At a certain instant the object's velocity is 8.4 m/s toward the north, and its position is 47 m north of our origin. *(a)* Establish a convenient reference frame and starting time for describing the motion. Determine an expression for the object's *(b)* coordinate and *(c)* velocity component as functions of time. Determine the object's *(d)* coordinate and *(e)* velocity component 3.0 s after the instant mentioned above.

31. The motion of a particular sprinter can be approximated by constant acceleration with magnitude 3.4 m/s^2 for the first 40 m after she leaves the starting line. What is her speed when she has traveled *(a)* 20 m and *(b)* 40 m?

32. A car is traveling along a straight road at a speed of 22 m/s. At the instant the car passes a "stop ahead" sign, it begins to slow down with a constant acceleration magnitude of 2.9 m/s^2. *(a)* What is the car's speed 30 m beyond the sign? *(b)* What is the car's speed 60 m beyond the sign? *(c)* If the car continues its constant acceleration until it stops just at the stop sign, how far apart are the signs?

33. The driver of a car traveling along a straight road with a speed of 18 m/s observes a sign that gives the speed limit as 25 m/s. The sign is 85 m ahead at the instant the driver begins to accelerate the car. Determine the magnitude of the constant acceleration that will cause the car to pass the sign at the posted speed limit.

34. A ship is cruising at a speed of 6.3 m/s at the instant it passes a buoy. At this time it begins to increase its speed with a constant acceleration of magnitude 0.20 m/s^2. How far is the ship from the buoy when its speed is 8.6 m/s?

35. The motion of a particular sprinter during a 50-m dash can be approximated as constant acceleration of magnitude 3.7 m/s^2. Let $t = 0$ correspond to the beginning of the dash, and determine the time the sprinter has traversed *(a)* 5.0 m and *(b)* 10.0 m.

36. A car is traveling at a speed of 14 m/s at the instant it passes a sign that gives the speed limit as 20 m/s. If the car increases its speed with a constant acceleration of magnitude 1.4 m/s^2, how long after it passes the sign will its speed be at the speed limit?

37. A car is 18 m past the entrance to a restaurant and traveling at a speed of 16 m/s when the driver applies the brakes. The speed of the car decreases with a constant acceleration of magnitude 2.3 m/s^2. How long after the driver applies the brakes will the car be 65 m past the entrance?

38. Major-league pitchers typically throw a baseball with a speed of 90 mi/h. Estimate the acceleration of the ball during the throw.

39. The typical speed of a bullet as it leaves the muzzle of a rifle is about 700 m/s. Estimate the acceleration of the bullet while it is in the rifle's barrel.

40. *(a)* Obtain Eq. C in Example 3-7 by solving Eq. A for v_0. *(b)* Obtain Eq. D in Example 3-7 by inserting Eq. C into Eq. B and solving for a.

41. Table 3-10 gives the coordinate of an object in terms of time. Assume that the acceleration is constant and determine the values of *(a)* x_0; *(b)* a; *(c)* v_0. *(d)* Use your answers to write an expression for $x(t)$. *(e)* Use the expression from part *(d)* to verify that the data are consistent with the assumption of constant acceleration. (*Hint:* See Example 3-7.)

TABLE 3-10

t, s	x, m
0.0	3.0
1.0	7.5
2.0	15.2
3.0	26.1
4.0	40.2
5.0	57.5

42. Two drag racers starting from rest run the same course. Racer *A* gets to the finish line in half the time of racer *B*. Assuming constant acceleration for both racers, find the ratio of the acceleration of racer *A* to that of racer *B*.

Section 3-6. Free-Fall

43. Sometimes it is convenient to compare an acceleration with the acceleration due to gravity. Let us define a unit of acceleration which we call the g: $1\ g = 9.8\text{ m/s}^2$ (exactly). Suppose a car increases its speed from zero to 25 m/s in 4.0 s. Assuming constant acceleration, determine the acceleration magnitude in the unit g.

44. A rock is released from rest at $t = 0$ from the top of an observation tower. Let the y axis be vertical with the unit vector $\mathbf{j}$ pointing upward. Make a table that lists values of a, v, and y at half-second intervals from $t = 0.0$ s to $t = 3.0$ s. Use these data

to make graphs of a, v, and y versus t. Sketch the curves and compare them with Fig. 3-18.

45. An astronaut stands on the steps of her spaceship that is resting on the surface of planet X, and she drops a rock from a height of 3.5 m. The rock hits the surface in 0.83 s. Determine the magnitude of the acceleration due to gravity on the surface of planet X.

46. A flowerpot falls from rest from a window sill that is 6.2 m above the ground. *(a)* What is the pot's speed as it hits the ground? *(b)* How long does it take the pot to hit the ground? *(c)* How far has the pot fallen after 0.50 s? *(d)* What is the pot's speed after 0.50 s? *(e)* What is the pot's acceleration after 0.50 s?

47. A ball is thrown vertically upward with an initial speed of 12 m/s from a release point that is 1.8 m above the ground. Let $t = 0$ correspond to the instant of release and let the origin of the coordinate be at the ground with the unit vector **j** pointing upward. *(a)* Determine h_m and t_m. *(b)* At what time prior to t_m does the ball have a velocity component of +5.0 m/s and at what time after t_m does it have a velocity component of −5.0 m/s? *(c)* What are the coordinates that correspond to the two velocity components in part *(b)*?

48. A rock thrown vertically upward at $t = 0$ reaches a maximum height of 14 m above the release point. *(a)* What is its initial speed? *(b)* At what time does it pass the release point on the way down?

49. A ball thrown vertically upward hits a telephone wire that is 5.1 m above the release point with a speed 0.70 m/s. What was the ball's initial speed?

Section 3-7. Varying Acceleration, Numerical Methods

50. Repeat the calculations for the example of the accelerating car discussed in Sec. 3-7, except let the time interval between calculations be 0.02 s rather than 0.01 s. Keep the overall time interval at 10 s. Compare your results with those in the section. Does this longer time interval give results that are significantly different?

Additional Exercises

51. According to entomologists, a dragonfly is capable of increasing its speed from 0 to 30 km/h over a distance of 3 m. Estimate the acceleration magnitude that a dragonfly is capable of performing. Give the value in the SI unit.

52. Suppose a subway train has a constant acceleration magnitude of 2.5 m/s² both when its speed is increasing and when it is decreasing, and the distance between stations is 500 m. *(a)* What is the minimum time required for the train to make the trip? *(b)* What is the maximum speed the train attains during a trip?

53. On earth, a spring gun shoots a pellet to a maximum height of 8.4 m. When taken to planet X (in another solar system) this same gun shoots a pellet to a maximum height of 12.6 m. What is g on planet X? What assumptions must you make to solve this problem with the information given?

54. A drag racer accelerates from rest with constant acceleration such that it reaches the 10-m mark with a speed of 10 m/s. How far from the 10-m mark will the racer be when its speed is 20 m/s if it continues with this same constant acceleration?

55. From the graph of x versus t shown in Fig. 3-32, estimate v at $t = 3.5$ s. What is the algebraic sign of a at $t = 3.5$ s? Is the object decelerating at $t = 3.5$ s?

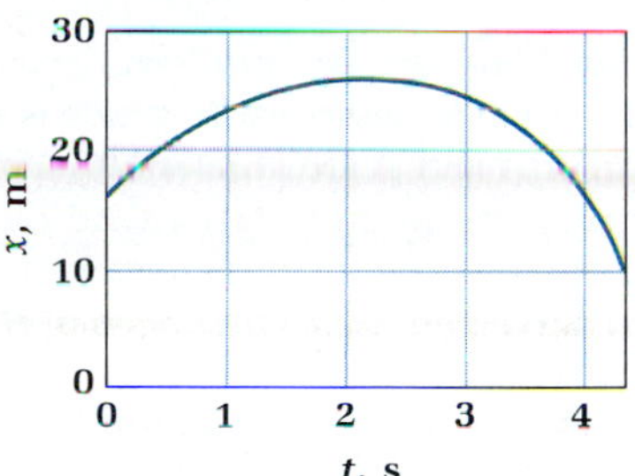

Figure 3-32. Exercise 55.

56. In some magazines about cars, accelerations are given in the unit kilometers per hour per second [km/(h·s)]. *(a)* Determine the conversion factor that changes the unit of acceleration from kilometers per hour per second to meters per second squared. *(b)* What is the value of 7.2 km/(h·s) in meters per second squared?

57. From the graph of x versus t in Fig. 3-33, can you tell whether there is an instant at which the acceleration is zero? If so, estimate t at this instant.

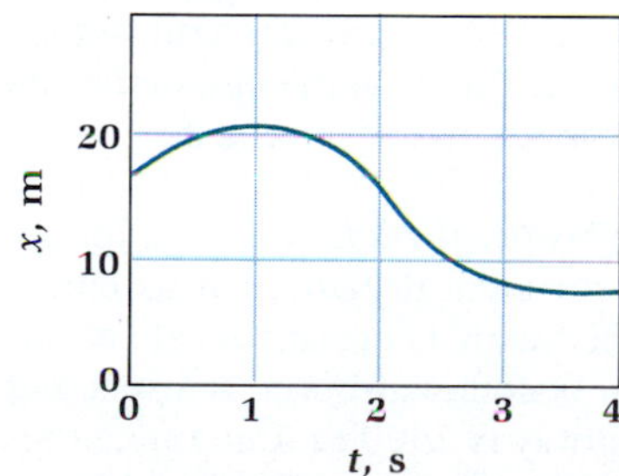

Figure 3-33. Exercise 57.

58. Suppose an archer is capable of shooting an arrow such that its speed at the instant it loses contact with the string is 70 m/s. Estimate the arrow's acceleration magnitude. What assumptions did you make in order to produce this estimate?

PROBLEMS

1. ***A 100-meter dash.*** A sprinter runs the 100-m dash in 10.0 s. Approximate his motion by assuming constant acceleration over the first 15 m and then constant velocity over the remaining 85 m. Determine *(a)* his final speed; *(b)* the time required for the first 15 m; *(c)* the time required for the remaining 85 m; *(d)* the acceleration magnitude for the first 15 m.

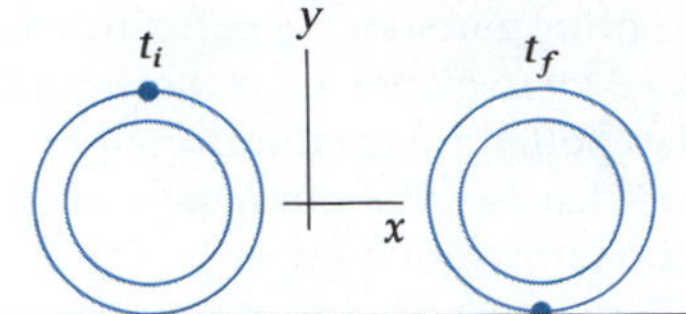

Figure 3-34. Problem 2.

2. ***Finding a displacement vector.*** A wad of chewing gum is stuck on the top of a wheel at time t_i, as shown in Fig. 3-34. The wheel then rolls without slipping in the $+x$ direction, and at time t_f the gum is on the bottom of the wheel. Use the coordinate frame in the figure to give the displacement $\Delta\mathbf{r}$ of the gum in terms of the radius R of the wheel and the unit vectors $\mathbf{i}$ and $\mathbf{j}$.

3. ***One car catching up with another.*** Car A, traveling at a constant speed of 18 m/s, passes car B, which is at rest at a stop sign. At the instant A and B are abreast, B accelerates with a constant magnitude of 4.6 m/s². Determine *(a)* the time required for B to catch A; *(b)* the distance traveled by B during the time required to catch A; *(c)* the speed of B as it passes A.

4. ***Measuring a person's reaction time.*** Consider a method of comparing the reaction times of different people. Have a friend hold her thumb and forefinger about 20 mm apart while you hold a ruler vertically so that the bottom end is between the thumb and finger. Your friend is to catch the ruler the instant she sees it released. By finding the distance the ruler falls before it is caught, we can measure a reaction time. If the ruler falls 200 mm, what is your friend's reaction time? Assume free-fall.

5. ***Designing a runway for jets.*** Suppose you are to design a runway for use by a particular type of jet. On the takeoff run, the speed of this aircraft increases with a constant acceleration of magnitude 4.0 m/s² until it becomes airborne at a speed of 85 m/s. Should the pilot be required to abort the takeoff, the jet's speed decreases with a constant acceleration of magnitude 5.0 m/s². Determine the length of the runway needed to allow the pilot to abort the takeoff at the instant the jet reaches flying speed and still not run out of pavement.

6. ***Designing brakes for a jet.*** Suppose you are to design the braking system for a jet airplane. On its takeoff run, the jet increases its speed with a constant acceleration of magnitude 3.5 m/s² until it becomes airborne at a speed of 95 m/s. The length of the runway is 2500 m. Determine the magnitude of the constant acceleration that will stop the jet at the end of the runway, assuming that the pilot aborted takeoff at the instant the jet attained takeoff speed.

7. ***Average velocity when acceleration is constant.*** *(a)* Show that for motion with constant acceleration $\bar{v} = \frac{1}{2}[v(t_f) + v(t_i)]$. [*Hint:* Use the definition of the average velocity component and note that $t_f^2 - t_i^2 = (t_f + t_i)(t_f - t_i)$.] *(b)* Consider the case where $a = 2.0$ m/s², $v_0 = 1.0$ m/s, $t_i = 1.0$ s, and $t_f = 3.0$ s. Make a graph of v versus t from $t = 0.0$ s to $t = 4.0$ s. Show $v(t_i)$, $v(t_f)$, and $\bar{v}$ on the graph.

8. ***An acceleration that is linear with t.*** Consider an acceleration that varies linearly with time: $a(t) = a_0 + bt$. Find expressions for *(a)* $v(t)$ and *(b)* $x(t)$. *(c)* What is the physical meaning of b?

9. ***Simple harmonic motion with a computer.*** Use spreadsheet software to calculate the motion of an object whose acceleration component is given by the expression $a = -(0.25\ \text{s}^{-2})x$. Let $x_0 = 0$ and $v_0 = 0.50$ m/s, and examine the motion between $t = 0$ and $t = 16$ s. Use a time interval of 0.01 s between successive calculations. Do you recognize the curves on your graphs? They describe a type of motion called *simple harmonic motion,* which will be discussed in Chap. 14.

10. ***Completing Table 3-6.*** Obtain Eq. 3-13 from Eqs. 3-8 and 3-11.

11. ***Flipping a coin.*** A coin was flipped vertically into the air such that it rotated from heads to tails and then back to heads 10 times per second. The coin was released with heads up at a height of 0.49 m above the surface on which it landed and its maximum height above that surface was 1.13 m. Did the coin land heads up or tails up?

12. ***Galileo's law of odd numbers.*** An object undergoes free-fall after having been released from rest. We divide the time of fall into many equal time intervals Δt. Show that the change of coordinate during each successive time interval follows the pattern $\Delta y_2 = 3\Delta y_1$, $\Delta y_3 = 5\Delta y_1$, $\Delta y_4 = 7\Delta y_1$, . . . , $\Delta y_n = (2n - 1)\Delta y_1$. (See Fig. 3-35.)

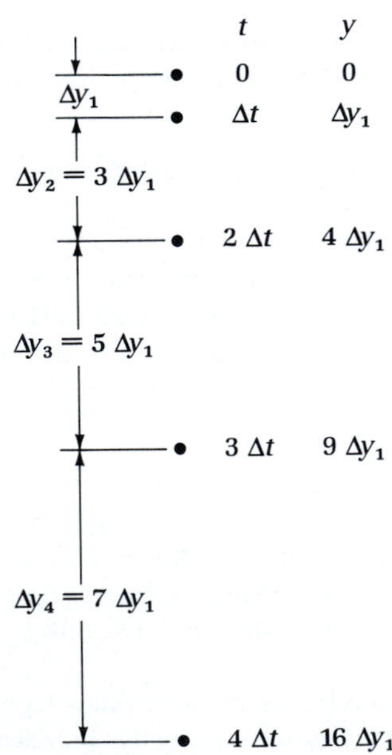

Figure 3-35. Problem 12.

13. ***Simple harmonic motion: A preview.*** Suppose $x(t) = A \sin \omega t$, where A and ω represent constants. Determine expressions for *(a)* $v(t)$ and *(b)* $a(t)$. *(c)* Show that $a = -\omega^2 x$. Let $A = 10.0$ mm and $\omega = 0.628$ rad/s and make graphs of *(d)* x; *(e)* v and a versus time from $t = 0.0$ s to $t = 10.0$ s. Plot points at each 1-s interval and sketch curves through the data. (*Comment:* This type of motion is called *simple harmonic motion* and will be discussed in detail in Chap. 14.)

14. **Air resistance.** An approximate expression that accounts for air resistance during the fall of an object such as a baseball is

$$a = -g - bv|v|$$

where b is a constant. *(a)* Construct a spreadsheet with $b = 0.0055\ \text{m}^{-1}$ for an object falling from rest for a time interval of 10 s. *(b)* Compare $y(10\ \text{s})$ given from your spreadsheet with $y(10\ \text{s})$ given by assuming free-fall.

15. **Finding the parameters in an equation for $x(t)$.** During the time interval from $t = 0.0$ to 4.0 s, the coordinate of an object as a function of time is given by the expression

$$x(t) = bt^2 - ct^3$$

At the instant $t = 4.0$ s, the object comes to rest at $x = 32.0$ m. *(a)* Determine the value of b and of c. Be certain your answers include the proper SI units. *(b)* Determine the time and coordinate at which the acceleration is zero. *(c)* Evaluate x, v, and a at each second in the interval from $t = 0.0$ to 4.0 s and plot the data. Sketch a curve through each set of points.

16. **A cheetah stalking a gazelle.** Suppose a gazelle is capable of accelerating from rest to its top speed of 25 m/s in a distance of 50 m and, after attaining this speed, the animal can maintain it for an extended period of time. In addition, suppose a cheetah can accelerate from rest to its top speed of 30 m/s in a distance of 60 m, but then can maintain this speed for only 4.0 s before it must give up the chase. How close to a gazelle must a cheetah be before it can launch a successful attack?

17. **A way to measure g.** One way to measure g is to project an object, such as a marble, vertically upward in an evacuated chamber, and use photodetectors to measure accurately the time intervals Δt_1 and Δt_2 during the marble's flight, as shown in Fig. 3-36. *(a)* Show that the expression which gives g, in terms of the measured values of Δt_1, Δt_2, and Δy, is

$$g = 8\Delta y/[(\Delta t_2)^2 - (\Delta t_1)^2]$$

(b) Suppose the measurements give $\Delta t_1 = 0.1483$ s, $\Delta t_2 = 0.6554$ s, and $\Delta y = 0.5000$ m. Calculate g with these data.

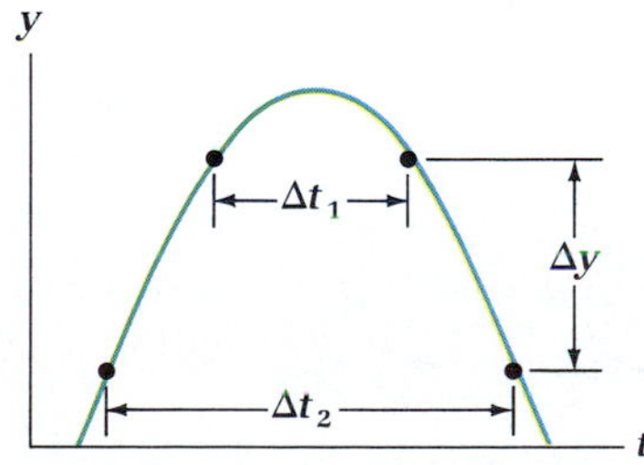

Figure 3-36. Problem 17.

18. **Hanging in the air.** An object is tossed vertically upward such that it travels straight up to its peak height and then falls straight down to the point from which it was launched. Consider the amount of time Δt_L the object spends in the lower half of its flight compared with the amount of time Δt_U it spends in the upper half, as shown in Fig. 3-37. Show that Δt_U is larger than Δt_L by a factor of about 2.4, or $\Delta t_U \approx 2.4(\Delta t_L)$. The object travels the same distance in the lower half as it does in the upper half, but it is moving more slowly in the upper half. This effect partially accounts for the illusion that an athlete, when jumping vertically, seems to hang in the air.

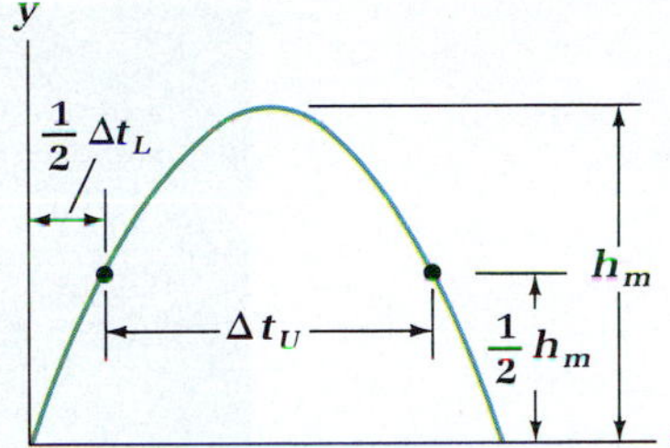

Figure 3-37. Problem 18.

19. **Exponential time dependence.** Suppose the coordinate of an object is given by the expression

$$x = \lambda[(t/\tau) - 1 + e^{-t/\tau}]$$

where λ is a factor that has the dimension of length and τ is a factor that has the dimension of time. Determine expressions for *(a)* v and *(b)* a. *(c)* Let $\lambda = 133$ m and $\tau = 5.77$ s and tabulate x, v, and a at each 2.0 s between 0.0 and 10.0 s. *(d)* Plot these data. Your graphs should be the same as those in Fig. 3-20. *(e)* What is the physical significance of λ/τ? *(f)* What is the physical significance of λ/τ^2?

20. **Flight of a rocket.** A toy rocket is fired vertically upward. The rocket's acceleration magnitude is constant at 14 m/s² until its fuel is exhausted 2.0 s after liftoff. *(a)* What is the rocket's speed at the instant it runs out of fuel? *(b)* What is its maximum height? *(c)* What is its speed just before it hits the ground? Neglect air resistance.

21. **Estimating the acceleration of a ball as it bounces.** If you drop a tennis ball from rest at a height of 2.0 m above the floor, it will bounce to a height of about 1.0 m. *(a)* Determine the speed of the ball just before it hits the floor. *(b)* Determine the speed of the ball just as it loses contact with the floor immediately after the bounce. Assuming that the ball is compressed by a maximum of about 2.0 mm during the bounce, estimate the ball's acceleration magnitude and give its acceleration direction while in contact with the floor, *(c)* as it slows down, and *(d)* as it speeds up. *(e)* What assumptions must you make to provide these estimates with the information given?

22. **Speed of a BB pellet.** Figure 3-38 is a double-flash photograph that shows the blurred image of a BB pellet at two positions as the pellet moves rapidly to the right. Also in the photograph are a meter stick and a disk with a straight line along a radius. The disk makes 50.0 revolutions in 1.00 s. By taking measurements from the photograph and by using the disk as your clock, determine the speed of the BB pellet with a precision of two significant digits.

23. **BB passing through water: Graphical analysis.** The chapter opening photograph (page 27) shows a BB shot from a gun as it passes through water. The BB's speed as it enters the water's surface is $v_0 = 233$ m/s. Let y be the coordinate of the BB with $+\mathbf{j}$ downward and $y = 0$ at the water's surface, and let $t = 0$ correspond to the instant the BB strikes the surface. The following data are taken from the photograph:

Figure 3-38. Problem 22.

t, ms	*y*, mm	*t*, ms	*y*, mm
0.085	31.5	1.885	201.0
0.385	73.5	2.185	216.5
0.685	110.0	2.485	231.0
0.985	137.5	2.785	243.5
1.285	162.5	3.085	254.5
1.585	182.0	3.358	265.0

(a) Plot these data on a graph of y versus t. *(b)* On the same graph, plot the function $y = b\ \ln[(v_0 t/b) + 1]$ where $b = 0.142$ m. Sketch a curve through the plotted points. Your graph should show that this function gives a reasonably accurate expression for y as a function of t.

24. ***BB passing through water: theoretical analysis.*** Problem 23 shows that the coordinate y of the BB in the chapter opening photograph (page 27) as a function of time t is well described by $y = b \ln [(v_0 t/b) + 1]$. *(a)* Show that the velocity component of the BB is given by

$$v = v_0/[(v_0 t/b) + 1]$$

(b) Show that the acceleration component of the BB is

$$a = -\frac{v_0^2}{b[(v_0 t/b) + 1]^2} = -v^2/b$$

Thus the acceleration is opposite the velocity and proportional to the square of the speed.

MOTION IN TWO DIMENSIONS 4

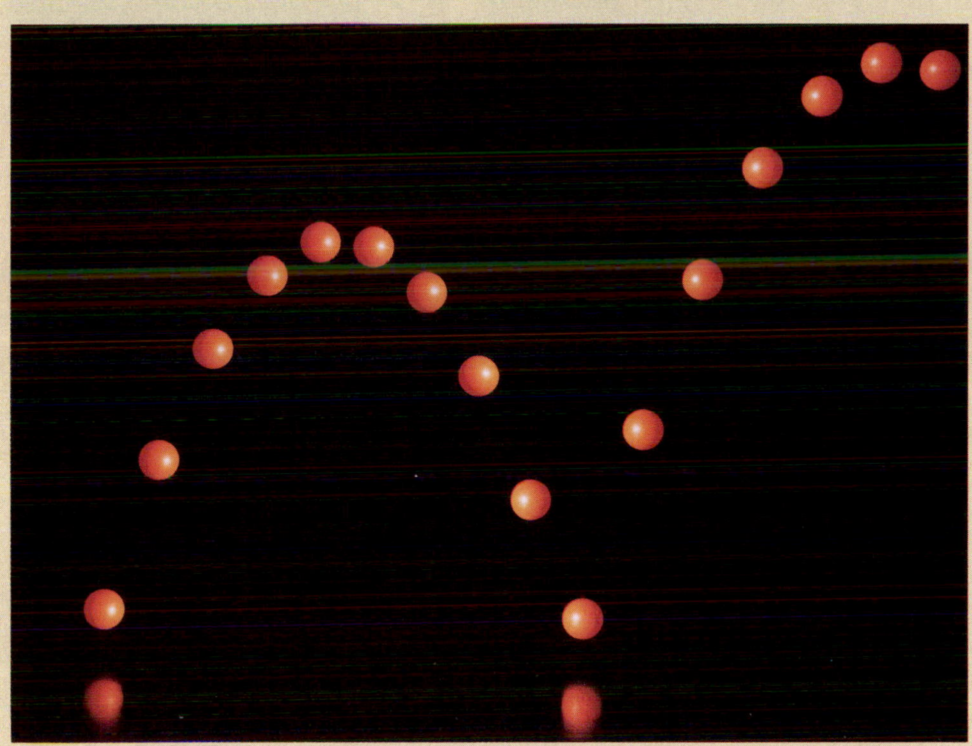
A stroboscopic photograph of a bouncing ball. Can you tell which way the ball is bouncing?

In Chap. 3 we defined the three kinematical quantities — position, velocity, and acceleration — and we used them to describe motion along a straight line. Now we apply these definitions to objects moving in two dimensions, or moving in a plane. The vector nature of the velocity and acceleration is manifested more clearly by motion in two dimensions. Contrary to motion in one dimension, the velocity and acceleration are not necessarily along the same line. Indeed, as you will see, the velocity and acceleration lie along different directions when the path of an object is not straight.

At this stage, it is not necessary to pursue kinematics into three dimensions because the further extension from two to three dimensions is straightforward and can be treated as the need arises. Besides, many of the motions that occur in nature are confined (approximately) to a plane. Two such motions, which we describe in detail, are that of an object launched or thrown into the air (such as a ball tossed from one person to another) and that of an object traveling in a circle (such as a planet orbiting the sun).

Photograph of the planet Saturn taken from *Voyager 2*.

4-1 POSITION, VELOCITY, AND ACCELERATION

The following are definitions of the three kinematical quantities in two dimensions.

Position

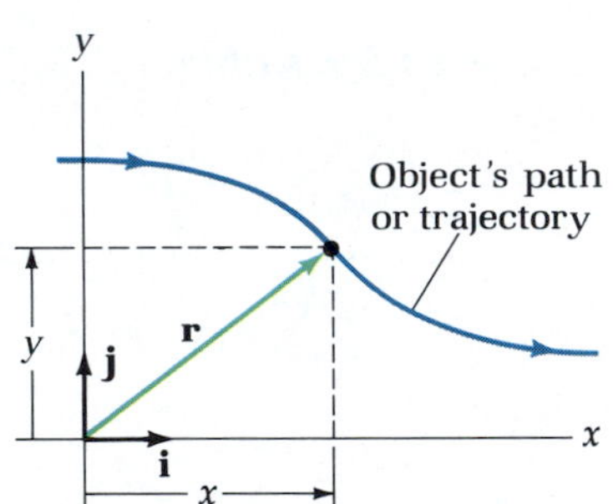

Figure 4-1. Position vector **r** in two dimensions.

The position vector **r** locates an object relative to the origin of a reference frame, as shown in Fig. 4-1. In two dimensions,

$$\mathbf{r} = x\mathbf{i} + y\mathbf{j} \tag{4-1}$$

where x and y are the object's coordinates. Notice the difference between the graph in Fig. 4-1 and the type of graphs used in Chap. 3. Graphs in Chap. 3 were of x versus t or v versus t, as in Fig. 3-12. Figure 4-1, which is typical of the graphs used in this chapter, shows the object's path, or trajectory, in the xy plane.

The term "position of an object" often refers to the object's coordinates written as an ordered pair (x, y). For example, an object located at $x = 3.0$ m and $y = 4.0$ m has a position vector $\mathbf{r} = (3.0 \text{ m})\mathbf{i} + (4.0 \text{ m})\mathbf{j}$ and a position (3.0 m, 4.0 m). The displacement, $\Delta\mathbf{r} = \mathbf{r}_f - \mathbf{r}_i$, is directed from an object's initial position to its final position, as shown in Fig. 4-2. In component form,

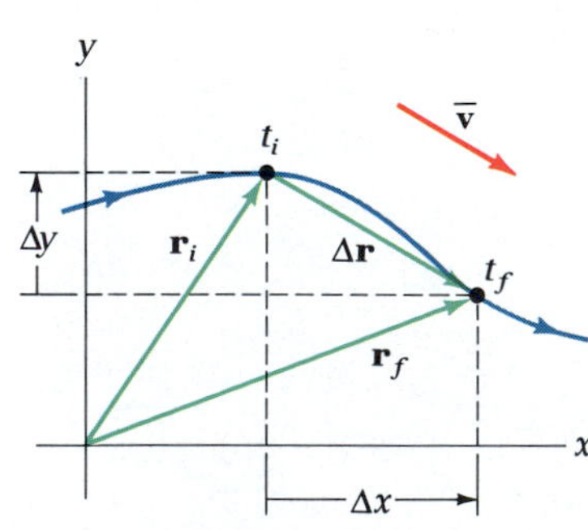

Figure 4-2. Displacement $\Delta\mathbf{r}$ and average velocity $\bar{\mathbf{v}}$; $\bar{\mathbf{v}}$ is parallel to $\Delta\mathbf{r}$. The position-vector triangle shows $\Delta\mathbf{r} = \mathbf{r}_f - \mathbf{r}_i$, or $\mathbf{r}_f = \mathbf{r}_i + \Delta\mathbf{r}$.

$$\Delta\mathbf{r} = (x_f\mathbf{i} + y_f\mathbf{j}) - (x_i\mathbf{i} + y_i\mathbf{j}) = (x_f - x_i)\mathbf{i} + (y_f - y_i)\mathbf{j}$$

Letting $\Delta x = x_f - x_i$ and $\Delta y = y_f - y_i$, we have

$$\Delta\mathbf{r} = \Delta x\mathbf{i} + \Delta y\mathbf{j}$$

Velocity

An object's average velocity $\bar{\mathbf{v}}$ for a time interval Δt is its displacement divided by the time interval, or

$$\bar{\mathbf{v}} = \frac{\Delta\mathbf{r}}{\Delta t} = \frac{\Delta x}{\Delta t}\mathbf{i} + \frac{\Delta y}{\Delta t}\mathbf{j} = \bar{v}_x\mathbf{i} + \bar{v}_y\mathbf{j}$$

Since $\bar{\mathbf{v}} = \Delta\mathbf{r}/\Delta t$, the direction of the average velocity is the same as $\Delta\mathbf{r}$ (Fig. 4-2).

The ***velocity*** is defined as the limiting value of the average velocity as the time interval approaches zero:

$$\mathbf{v} = \lim_{\Delta t \to 0} \overline{\mathbf{v}} = \lim_{\Delta t \to 0} \frac{\Delta \mathbf{r}}{\Delta t} = \frac{d\mathbf{r}}{dt} \tag{4-2}$$

Figure 4-3 shows this limiting process on a graph of the object's path. In passing from Figs. 4-3*a* through *c*, we show $\overline{\mathbf{v}}$ for smaller and smaller time intervals as t_i is held fixed and t_f approaches t_i. As Δt approaches zero (Fig. 4-3*d*), $\Delta \mathbf{r}$ approaches zero and $\overline{\mathbf{v}}$ becomes $\mathbf{v}$, which is parallel to a line tangent to the path and points in the direction of motion.

The ***velocity*** $\mathbf{v}$ at any point on an object's path is directed parallel to the path and points in the direction of motion.

Inserting $\Delta \mathbf{r}$ in terms of its components into the definition of velocity gives

$$\mathbf{v} = \lim_{\Delta t \to 0} \left(\frac{\Delta x}{\Delta t} \mathbf{i} + \frac{\Delta y}{\Delta t} \mathbf{j} \right) = \mathbf{i} \left(\lim_{\Delta t \to 0} \frac{\Delta x}{\Delta t} \right) + \mathbf{j} \left(\lim_{\Delta t \to 0} \frac{\Delta y}{\Delta t} \right) = \frac{dx}{dt} \mathbf{i} + \frac{dy}{dt} \mathbf{j}$$

Since $\mathbf{v} = v_x \mathbf{i} + v_y \mathbf{j}$, we have

$$v_x = \frac{dx}{dt} \quad \text{and} \quad v_y = \frac{dy}{dt} \tag{4-3}$$

If the coordinates x and y are known as functions of time, then the velocity is determined by calculating the derivatives of the expressions $x(t)$ and $y(t)$.

The magnitude of the velocity is the speed v:

$$v = \sqrt{v_x^2 + v_y^2} \tag{4-4}$$

The direction in which the object is headed at any time may be described in terms of the angle θ between the velocity vector and the x axis. Figure 4-4 shows that

$$\tan \theta = \frac{v_y}{v_x} \quad \text{or} \quad \theta = \tan^{-1} \frac{v_y}{v_x} \tag{4-5}$$

where θ is positive when measured counterclockwise from the x axis. From the figure you can see that the velocity components are

$$v_x = v \cos \theta \quad \text{and} \quad v_y = v \sin \theta \tag{4-6}$$

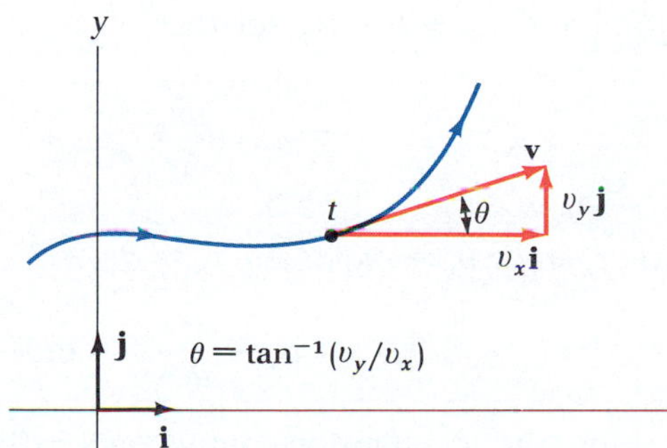

Figure 4-4. Resolving the velocity in terms of its components gives $v_x = v \cos \theta$ and $v_y = v \sin \theta$

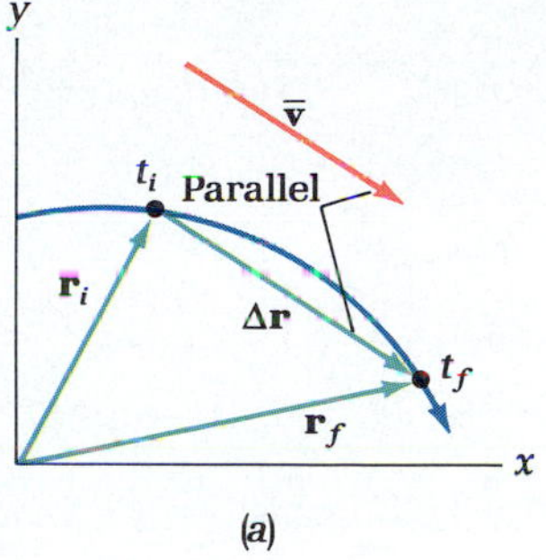

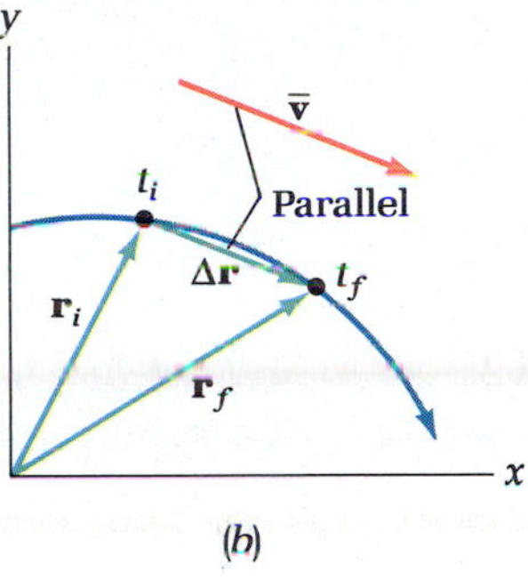

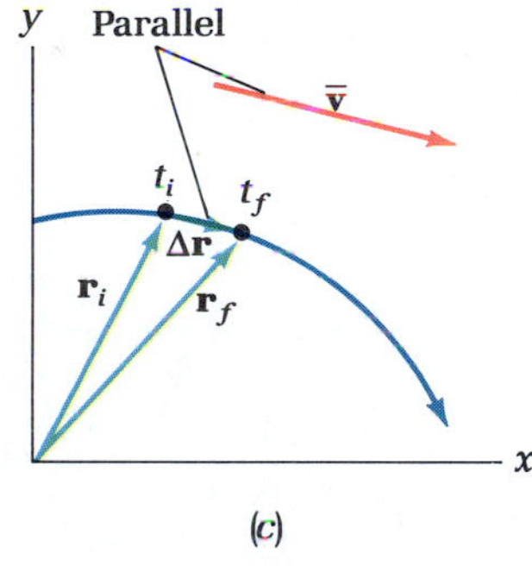

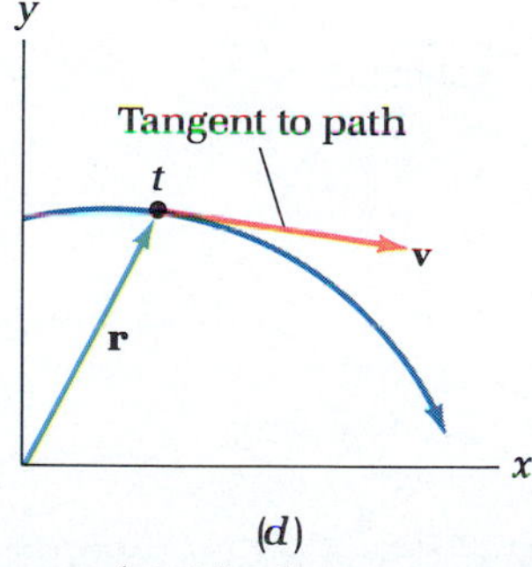

Figure 4-3. The limiting process by which $\Delta \mathbf{r}$ approaches zero as Δt approaches zero. The average velocity $\overline{\mathbf{v}}$ has the same direction as $\Delta \mathbf{r}$, and, as Δt approaches zero, the average velocity $\overline{\mathbf{v}}$ approaches the velocity $\mathbf{v}$. This limiting process shows that $\mathbf{v}$ is parallel to a line tangent to the path.

EXAMPLE 4-1

Resolving a velocity into its components. A motorboat is traveling 52° south of east at a speed of 12 m/s. Establish a coordinate frame and determine the boat's velocity components.

Solution. Let $+x$ be toward the east and $+y$ be toward the north, as shown in Fig. 4-5. Since an angle measured counterclockwise is positive, we have $\theta = -52°$ in this case. From Eqs. 4-6,

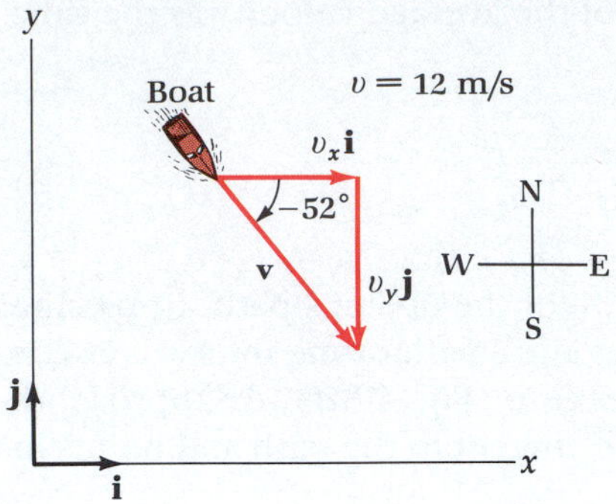

Figure 4-5. Example 4-1: A motorboat traveling 52° south of east. In this case $\theta = -52°$.

$$v_x = 12 \text{ m/s} \cos(-52°) = 7.4 \text{ m/s}$$

$$v_y = 12 \text{ m/s} \sin(-52°) = -9.5 \text{ m/s}$$

SELF-TEST 4-1. Suppose a car is traveling at a speed of 28 m/s in a direction 34° north of east. If we let $+\mathbf{i}$ point east and $+\mathbf{j}$ point north, what are the car's velocity components? Using unit vectors, write down the velocity vector for the car. ***ANSWER:*** $\mathbf{v} = (23 \text{ m/s})\mathbf{i} + (16 \text{ m/s})\mathbf{j}$.

Acceleration

The average acceleration $\bar{\mathbf{a}}$ of an object over a time interval Δt is the object's velocity change divided by the time interval:

$$\bar{\mathbf{a}} = \frac{\Delta \mathbf{v}}{\Delta t}$$

The ***acceleration*** is defined as the limiting value of the average acceleration as the time interval approaches zero:

Acceleration

$$\mathbf{a} = \lim_{\Delta t \to 0} \bar{\mathbf{a}} = \lim_{\Delta t \to 0} \frac{\Delta \mathbf{v}}{\Delta t} = \frac{d\mathbf{v}}{dt} \tag{4-7}$$

Since $\Delta \mathbf{v} = \Delta v_x \mathbf{i} + \Delta v_y \mathbf{j}$, we have

$$\mathbf{a} = \lim_{\Delta t \to 0} \left(\frac{\Delta v_x}{\Delta t} \mathbf{i} + \frac{\Delta v_y}{\Delta t} \mathbf{j} \right) = \mathbf{i} \left(\lim_{\Delta t \to 0} \frac{\Delta v_x}{\Delta t} \right) + \mathbf{j} \left(\lim_{\Delta t \to 0} \frac{\Delta v_y}{\Delta t} \right) = \frac{dv_x}{dt} \mathbf{i} + \frac{dv_y}{dt} \mathbf{j}$$

Writing $\mathbf{a} = a_x \mathbf{i} + a_y \mathbf{j}$, we see that

$$a_x = \frac{dv_x}{dt} \quad \text{and} \quad a_y = \frac{dv_y}{dt} \tag{4-8}$$

Acceleration components

Further, since $v_x = dx/dt$ and $v_y = dy/dt$,

$$a_x = \frac{d^2x}{dt^2} \quad \text{and} \quad a_y = \frac{d^2y}{dt^2}$$

Therefore the acceleration components are determined from derivatives of expressions for $v_x(t)$ and $v_y(t)$ or from second derivatives of $x(t)$ and $y(t)$.

Figure 4-6 shows this limiting process graphically. In Figs. 4-6*a* through *c*, velocity-vector triangles, with sides $\mathbf{v}_i$, $\mathbf{v}_f$, and $\Delta \mathbf{v}$, are used to determine the magnitude and direction of $\Delta \mathbf{v}$ in each succeeding case as t_i is held fixed while t_f approaches t_i. By its definition, $\bar{\mathbf{a}} = \Delta \mathbf{v}/\Delta t$, so that $\bar{\mathbf{a}}$ has the same direction as $\Delta \mathbf{v}$. Thus $\mathbf{a}$ has the same direction as the limiting value of $\Delta \mathbf{v}$ as Δt approaches zero. When the limit is reached in Fig. 4-6*d*, $\bar{\mathbf{a}}$ has become $\mathbf{a}$, and t_i and $\mathbf{v}_i$ are designated simply as t and $\mathbf{v}$. This figure shows that $\mathbf{a}$ is partly directed inward, or toward the

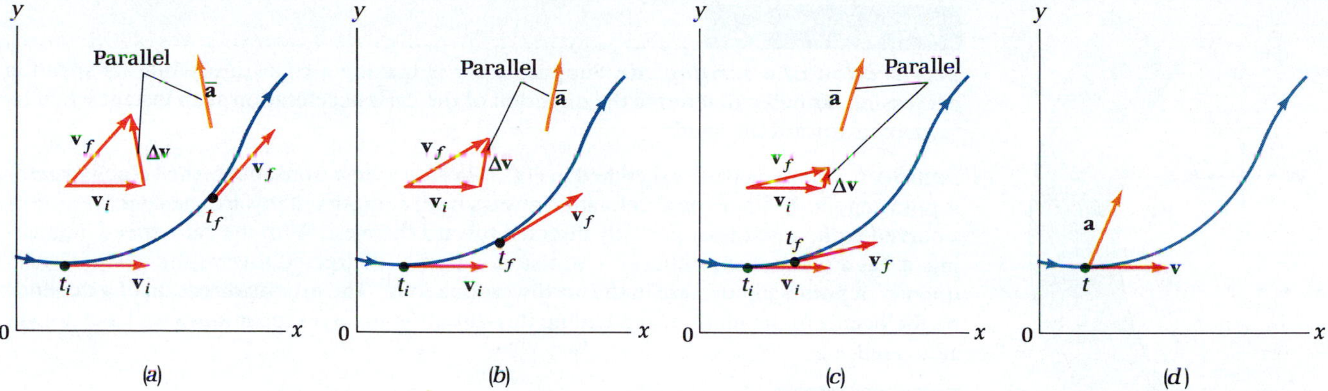

Figure 4-6. The acceleration **a** is equal to the limiting value of $\Delta\mathbf{v}/\Delta t$ as Δt approaches zero. In *(a)* through *(c)*, we let Δt approach zero by holding t_i fixed as t_f approaches t_i. Velocity-vector triangles with sides $\mathbf{v}_i$, $\mathbf{v}_f$, and $\Delta\mathbf{v}$ are used to determine the direction of $\Delta\mathbf{v}$ in each case: $\Delta\mathbf{v} = \mathbf{v}_f - \mathbf{v}_i$, or $\mathbf{v}_f = \mathbf{v}_i + \Delta\mathbf{v}$. Since $\bar{\mathbf{a}} = \Delta\mathbf{v}/\Delta t$, $\bar{\mathbf{a}}$ has the same direction as $\Delta\mathbf{v}$. *(d)* As the limit is reached, $\bar{\mathbf{a}}$ becomes **a**. Note that **a** is not parallel to the path. Indeed, this figure shows that **a** always has a component toward the concave, or inward, side of a curved path.

concave side of the path. This is because the limiting value of $\Delta\mathbf{v}$ is partly directed inward, or toward the concave side of the path.

When an object follows a curved path, the object's acceleration always has a component toward the concave side of the path.

In the case shown in Fig. 4-6, **a** also has a component along the path and in the direction of **v**. This is because we happened to show an object whose speed is increasing ($v_f > v_i$). If we had shown a case where the object's speed is decreasing ($v_f < v_i$), then **a** would have a component along the path but opposite **v**. Or, if we had shown an object with constant speed ($v_f = v_i$), then **a** would have no component along the path and **a** and **v** would be perpendicular.

When an object's speed is increasing, the object's acceleration always has a component in the direction of its velocity. When an object's speed is decreasing, the object's acceleration always has a component opposite its velocity.

Figure 4-7 shows the relation between the direction of **v** and the direction of **a** for an object whose speed is increasing (Fig. 4-7*a*), for an object whose speed is constant (Fig. 4-7*b*), and for an object whose speed is decreasing (Fig. 4-7*c*).

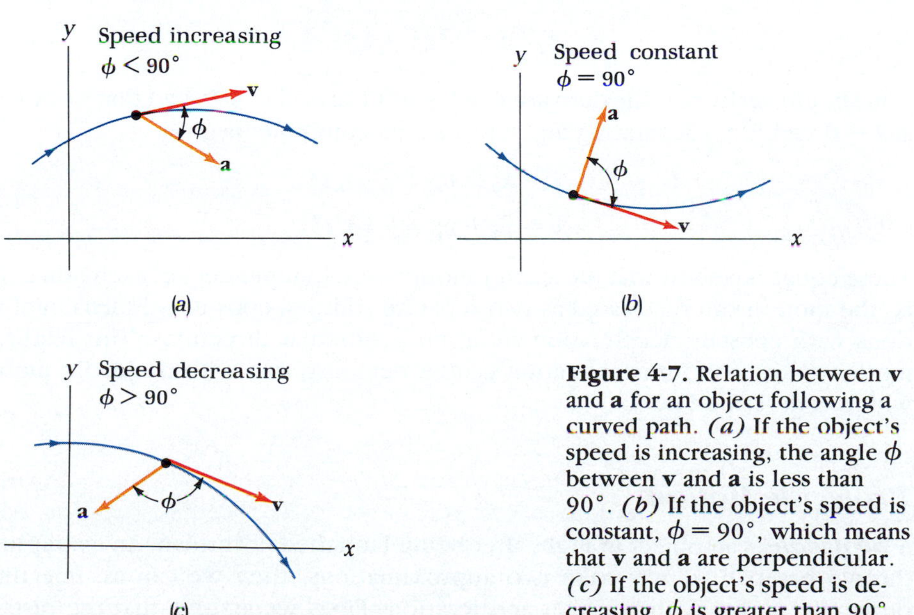

Figure 4-7. Relation between **v** and **a** for an object following a curved path. *(a)* If the object's speed is increasing, the angle ϕ between **v** and **a** is less than 90°. *(b)* If the object's speed is constant, $\phi = 90°$, which means that **v** and **a** are perpendicular. *(c)* If the object's speed is decreasing, ϕ is greater than 90°.

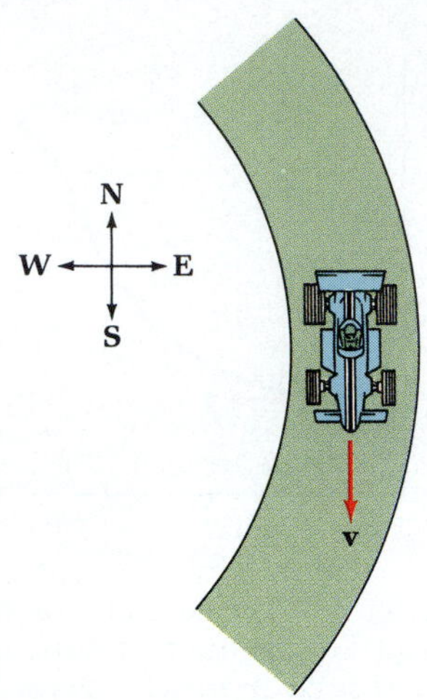

Figure 4-8. Example 4-2.

EXAMPLE 4-2

Acceleration of a turning car. Suppose a car is making a right turn while its speed is decreasing. Roughly determine the direction of the car's acceleration at an instant when its velocity is toward the south.

Solution. The car's path is sketched in Fig. 4-8 using a view from above; the velocity vector is pointing south. Since the acceleration always has a component toward the concave side of a curved path, **a** is at least partially directed toward the west. With the car's speed decreasing, **a** has a component opposite **v** so that **a** is partially directed toward the north. Consequently, **a** points somewhere in the northwest quadrant. The precise direction of **a** depends on the details of the motion, and finding this direction involves techniques we have not yet addressed.

SELF-TEST 4-2. A car is turning left and its speed is increasing. At the instant the car is headed east, into which quadrant is the car's acceleration directed? You may find it helpful to draw a sketch similar to Fig. 4-8. ***ANSWER:*** Northeast.

4-2 CONSTANT ACCELERATION: PROJECTILE MOTION

Now we consider motion with constant acceleration. To find expressions for **v** and **r**, we proceed as we did for constant acceleration in one dimension (Sec. 3-4). When the acceleration is constant, it is equal to its average value: $\mathbf{a} = \bar{\mathbf{a}} = \Delta\mathbf{v}/\Delta t$. If $\mathbf{v}_f = \mathbf{v}$, $\mathbf{v}_i = \mathbf{v}_0$, $t_f = t$, and $t_i = 0$, then $\mathbf{a} = (\mathbf{v} - \mathbf{v}_0)/(t - 0)$, or

$$\mathbf{v} = \mathbf{v}_0 + \mathbf{a}t \tag{4-9}$$

In terms of components,

$$\mathbf{v} = (v_{x0}\mathbf{i} + v_{y0}\mathbf{j}) + (a_x\mathbf{i} + a_y\mathbf{j})t = (v_{x0} + a_x t)\mathbf{i} + (v_{y0} + a_y t)\mathbf{j}$$

so that

$$v_x = v_{x0} + a_x t \quad \text{and} \quad v_y = v_{y0} + a_y t \tag{4-10}$$

The equation for **r** can be determined by finding an expression whose derivative gives Eq. 4-9:

$$\mathbf{r} = \mathbf{r}_0 + \mathbf{v}_0 t + \tfrac{1}{2}\mathbf{a}t^2 \tag{4-11}$$

You should verify that the derivative of Eq. 4-11 gives Eq. 4-9, and that **r** evaluated at $t = 0$ yields $\mathbf{r}_0$. Separating Eq. 4-11 into its components gives

$$\begin{aligned} x &= x_0 + v_{x0}t + \tfrac{1}{2}a_x t^2 \\ y &= y_0 + v_{y0}t + \tfrac{1}{2}a_y t^2 \end{aligned} \tag{4-12}$$

These equations show that the x and y motions are independent of each other. That is, the motion can be treated as two separate simultaneous one-dimensional motions with constant acceleration along perpendicular directions. This feature of motion with constant acceleration can be demonstrated with projectile motion, which is described next.

Projectile Motion

A ***projectile*** is an object in flight after being launched or thrown. An example is a thrown baseball. If we make two approximations, then we can assume that a projectile moves with constant acceleration. First, we assume that the distance traveled is much smaller than the radius of the earth so that the acceleration due to gravity remains essentially fixed. Clearly, this approximation is valid for such cases as the flight of a baseball or golf ball. Second, we assume that air resistance is negligible. Often, this approximation is *not* valid. The effects of air resistance increase with speed, so that neglecting air resistance is invalid for large speeds. This matter is considered further in Sec. 4-5, but for now, we use both of these approximations and assume that the projectile's acceleration is constant. We let the y axis of our coordinate frame be vertical with $+\mathbf{j}$ upward. Then,

$$a_x = 0 \qquad \text{and} \qquad a_y = -g$$

Acceleration components of a projectile

where $g = 9.8\ \text{m/s}^2$. Suppose the projectile is launched such that its initial velocity $\mathbf{v}_0$ is at an angle θ_0 with the x axis, as shown in Fig. 4-9. We call θ_0 the ***angle of projection.*** By resolving the initial velocity, we obtain the initial velocity components: $v_{x0} = v_0 \cos\theta_0$ and $v_{y0} = v_0 \sin\theta_0$, where v_0 is the initial speed. Substituting these values into Eqs. 4-10 gives

$$v_x = v_0 \cos\theta_0 \qquad \text{and} \qquad v_y = v_0 \sin\theta_0 - gt \tag{4-13}$$

Velocity components of a projectile

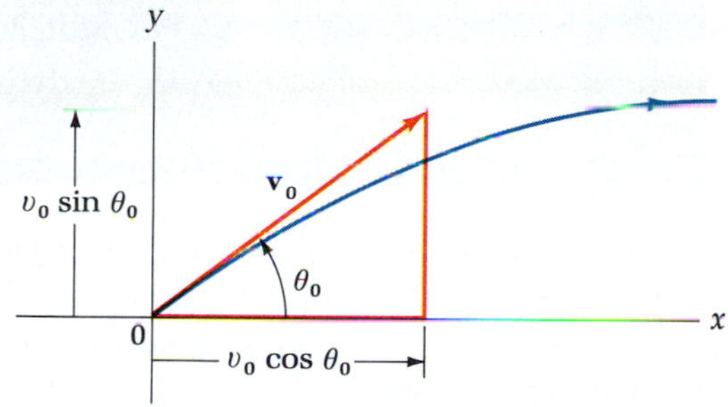

Figure 4-9. Resolving the initial velocity of a projectile into its components: $v_{x0} = v_0 \cos\theta_0$ and $v_{y0} = v_0 \sin\theta_0$, where v_0 is the initial speed and θ_0 is the angle of projection.

If the origin of the reference frame is placed at the initial position, then $x_0 = y_0 = 0$ and Eqs. 4-12 give

$$x = (v_0 \cos\theta_0)t \qquad \text{and} \qquad y = (v_0 \sin\theta_0)t - \tfrac{1}{2}gt^2 \tag{4-14}$$

Coordinates of a projectile

Therefore the x motion can be regarded as one-dimensional motion with constant velocity, and the y motion can be regarded as one-dimensional motion with constant acceleration.

Figure 4-10 demonstrates the independence of the x and y parts of the motion. This figure is a stroboscopic photograph of the simultaneous motion of two golf balls. One ball was released from rest at the same instant the other ball was launched horizontally. The motion of the ball that was released from rest is described by Eqs. 4-14 if $v_0 = 0$. This gives $x = 0$ for all t and $y = -\tfrac{1}{2}gt^2$. The motion of the ball that was launched horizontally is described by Eqs. 4-14 if $\theta_0 = 0$. This gives $x = v_0 t$ and $y = -\tfrac{1}{2}gt^2$. Thus Eqs. 4-14 give the same y coordinate for each ball at each instant of time, and this result is verified by Fig. 4-10.

This observation was first explained by Galileo. Describing a particle projected off the edge of a horizontal surface, he wrote

> . . . then the moving particle, which we imagine to be a heavy one, will on passing over the edge of the plane acquire, in addition to its previous uniform and perpetual motion, a downward propensity due to its own weight; so that the resulting motion, which I call projection *[projectio]*, is compounded of one which is uniform and horizontal and of another which is vertical and naturally accelerated.*

* Galileo Galilei, *Two New Sciences,* Henry Crew and Alfonso de Salvio (trans.), Dover, New York, 1954, p. 244.

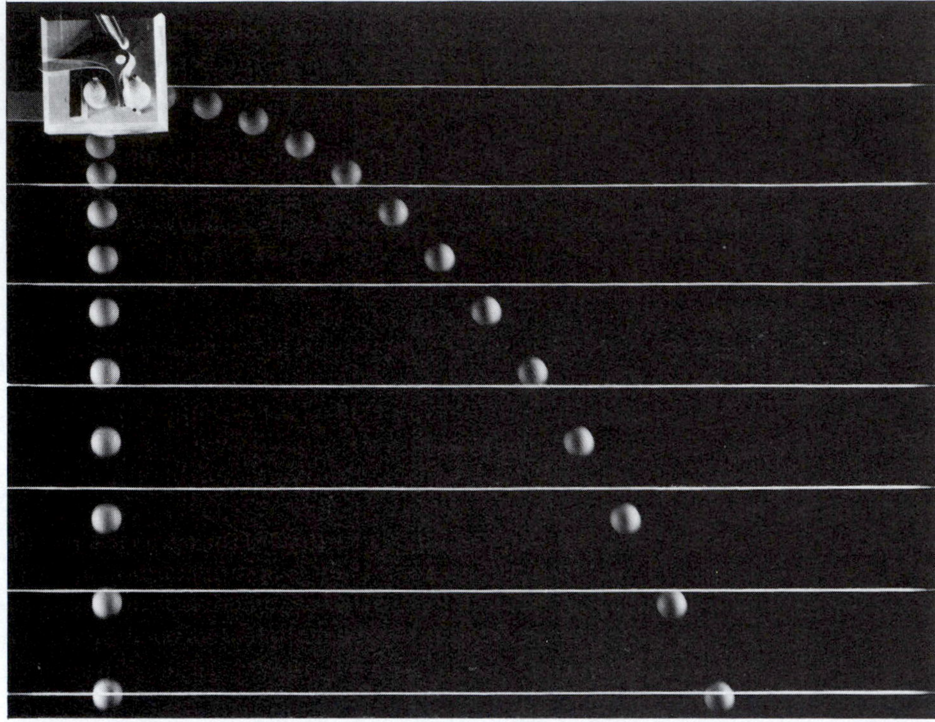

Figure 4-10. A stroboscopic picture of two golf balls. One ball was released from rest at the instant the other was launched with a horizontal velocity. The photograph shows that the simultaneous vertical coordinates of each ball are the same, which is in accord with Galileo's observations.

An equation for the path or trajectory of a projectile can be found by eliminating the time between the expressions for x and y in Eqs. 4-14. Solving $x = (v_0 \cos \theta_0)t$ for the time gives $t = x/(v_0 \cos \theta_0)$. Substituting this result into the expression for y and rearranging gives

Trajectory of a projectile

$$y = (\tan \theta_0)x - \frac{g}{2(v_0 \cos \theta_0)^2} x^2 \tag{4-15}$$

We have found that if the effects of air resistance can be neglected, then the trajectory of a projectile is a parabola. Figure 4-11 shows this parabolic path with the velocity indicated at several points. Notice that v_x remains fixed throughout the motion, whereas v_y decreases in magnitude as the projectile is going up and increases in magnitude as the projectile is going down. That is, v_y continually decreases, corresponding to the fact that a_y is negative. At the instant the projectile reaches its maximum height, $v_y = 0$.

The projectile motion shown in Fig. 4-11 is consistent with the conclusions of the previous section about the direction **a**. There we showed that **a** always has a

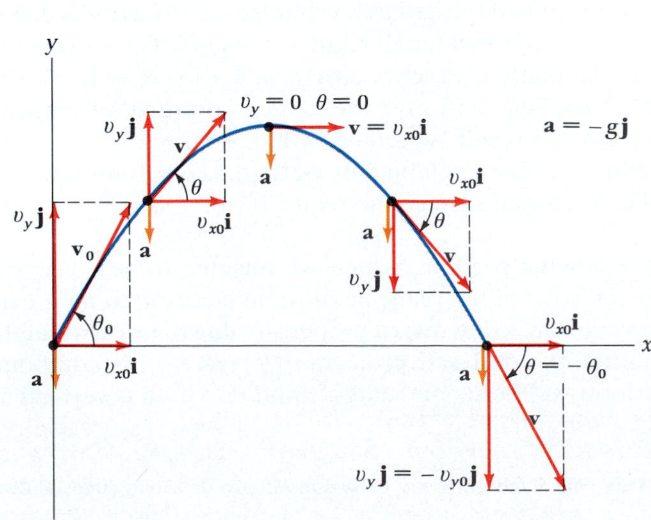

Figure 4-11. If the effects of air resistance are negligible, the trajectory of a projectile is a parabola. Note that the x component of the velocity remains fixed while the y component continually changes.

component directed toward the concave side of the path, and that the angle ϕ between **a** and **v** is greater than 90° when the speed is decreasing and less than 90° when the speed is increasing. In projectile motion, **a** is vertically downward, so that it is directed toward the concave side of the parabolic path. As the projectile rises, its speed decreases and $\phi > 90°$. As the projectile falls, its speed increases and $\phi < 90°$. Each of these conclusions is verified by Fig. 4-11.

EXAMPLE 4-3

Time of maximum height. A rock is thrown with an initial speed $v_0 = 17$ m/s and at a projection angle of 58°. *(a)* Develop an expression for the time t_m required for the rock to reach its maximum height h_m. *(b)* Evaluate t_m for this case.

Solution. *(a)* Since $v_y = 0$ when $t = t_m$, Eq. 4-13 gives

$$0 = v_0 \sin\theta_0 - gt_m$$

Or,

$$t_m = (v_0 \sin\theta_0)/g$$

(b) For this case,

$$t_m = (17 \text{ m/s} \sin 58°)/(9.8 \text{ m/s}^2) = 1.5 \text{ s}$$

SELF-TEST 4-3. *(a)* Show that the expression for the maximum height h_m of the rock in the previous example is

$$h_m = (v_0 \sin\theta_0)^2/2g$$

where h_m is measured relative to the point of release. *(b)* Evaluate h_m for this case. ***ANSWER:*** *(b)* 11 m.

EXAMPLE 4-4

Horizontal range. The horizontal range R of a projectile is the horizontal distance traversed by the projectile from its launch point to where it passes through $y = 0$ on its way down. Determine an expression for the horizontal range.

Solution. Equation 4-15 for the trajectory of a projectile is

$$y = (\tan\theta_0)x - \frac{g}{2(v_0 \cos\theta_0)^2}x^2$$

Note that $y = 0$ at $x = 0$ by our selection of the origin of our coordinate frame. Also, by the definition of the horizontal range R as given above, $y = 0$ at $x = R$. (See Fig. 4-11.) Substituting $y = 0$ and $x = R$ into the trajectory formula and solving for R gives

$$R = 2(v_0^2 \sin\theta_0 \cos\theta_0)/g$$

This can be simplified using the trigonometric identity $\sin 2\alpha = 2\cos\alpha \sin\alpha$:

$$R = (v_0^2 \sin 2\theta_0)/g$$

SELF-TEST 4-4. A golf ball is struck such that its initial speed is 14 m/s and its projection angle is 49°. How far down the fairway does the ball land? Assume that the fairway is level and neglect air resistance. ***ANSWER:*** 20 m.

More about the Horizontal Range

Examination of the expression for the horizontal range given in Example 4-4 allows us to answer an interesting question. For a given initial speed v_0, at what projection angle should we launch a projectile so that its horizontal range is maximum? That is, what angle θ_0 makes R a maximum when v_0 is held fixed? Since $\sin 2\theta_0$ has a maximum value of 1 when $2\theta_0 = 90°$, R is maximum when $\theta_0 = 45°$. Thus the maximum horizontal range is $R_m = v_0^2/g$, and this maximum occurs when the projection angle is $\theta_0 = 45°$. In addition, this maximum is symmetric about $\theta_0 = 45°$, as shown in Fig. 4-12. For example, suppose $v_0 = 10$ m/s and $\theta_0 = 45° + 6° = 51°$. Then the range $R = (10 \text{ m/s})^2 \sin(102°)/(9.8 \text{ m/s}^2) = 10$ m. Now suppose $v_0 = 10$ m/s and $\theta_0 = 45° - 6° = 39°$. Then $R = (10 \text{ m/s})^2 \sin(78°)/(9.8 \text{ m/s}^2) = 10$ m. See Prob. 4 for more about this symmetry.

Table 4-1 summarizes the important equations describing projectile motion. These equations are more general than those discussed above because the initial coordinates are left unspecified as (x_0, y_0) rather than being placed at (0, 0).

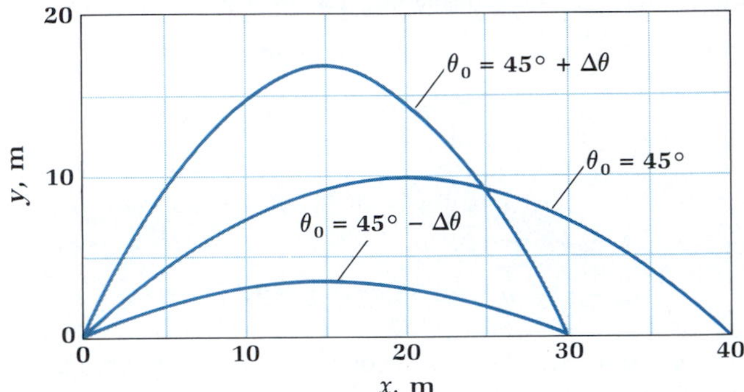

Figure 4-12. The range R is symmetric for changes $\Delta\theta$ in the projection angle about the angle $\theta_0 = 45°$. That is, the range is the same when $\theta_0 = 45° + \Delta\theta$ as it is when $\theta_0 = 45° - \Delta\theta$.

TABLE 4-1. *Projectile Motion*

$a_x = 0$	$a_y = -g$
$v_x = v_0 \cos\theta_0$	$v_y = v_0 \sin\theta_0 - gt$
$x = x_0 + (v_0 \cos\theta_0)t$	$y = y_0 + (v_0 \sin\theta_0)t - \frac{1}{2}gt^2$

$$y = y_0 + (\tan\theta_0)(x - x_0) - \frac{g}{2(v_0 \cos\theta_0)^2}(x - x_0)^2$$

EXAMPLE 4-5

A home run? A baseball bat hits a ball such that the ball's initial speed is 35 m/s and its projection angle is 42° above the horizontal. The outfield fence is 115 m from home plate and is 4 m high. Neglecting air resistance, determine whether the ball clears the fence. Assume that the ball was struck at a height of 1 m above the level playing surface at home plate.

Solution. Letting the origin be at home plate, we have $x_0 = 0$ and $y_0 = 1$ m. Using the equation for the trajectory in Table 4-1, we find the value of y when $x = 115$ m:

$$y = 1 \text{ m} + (\tan 42°)(115 \text{ m}) - \frac{9.8 \text{ m/s}^2}{2(35 \text{ m/s} \cos 42°)^2}(115 \text{ m})^2 = 9 \text{ m}$$

Since the fence is 4 m high, the ball clears the fence by 5 m, and the hit is a home run. We shall return to this example and consider air resistance effects in Sec. 4-5.

SELF-TEST 4-5. Consider a projectile launched over level terrain with an initial velocity $\mathbf{v}_0 = (6.0 \text{ m/s})\mathbf{i} + (8.0 \text{ m/s})\mathbf{j}$. Use the launch position as the origin, neglect air resistance, and let $g = 10 \text{ m/s}^2$ for ease of calculation. Determine *(a)* the angle of projection θ_0; *(b)* the initial speed v_0; *(c)* the time t_m for the projectile to reach its maximum height; *(d)* the projectile's position vector $\mathbf{r}$, velocity $\mathbf{v}$, and acceleration $\mathbf{a}$ at $t = t_m$. Make sure each of your answers has the correct units and that you use unit vectors to write vector quantities. ***ANSWERS:*** *(a)* 53°; *(b)* 10.0 m/s; *(c)* 0.80 s; *(d)* $\mathbf{r} = (4.8 \text{ m})\mathbf{i} + (3.2 \text{ m})\mathbf{j}$; $\mathbf{v} = (6.0 \text{ m/s})\mathbf{i}$; $\mathbf{a} = -(10 \text{ m/s}^2)\mathbf{j}$.

Figure 4-13. Plastic letters undergoing uniform circular motion on a phonograph turntable. The image of the letter farthest from the spindle is more blurred because it is moving faster.

4-3 UNIFORM CIRCULAR MOTION

An object executes ***uniform circular motion*** when it travels in a circle at constant speed. An example is the plastic letters shown riding on a rotating phonograph turntable in Fig. 4-13. In the term "uniform circular motion," the word "uniform" refers to the constant speed. Note that even though the object's speed is constant, its velocity is *not* constant. The velocity is a vector tangent to the path at each point. Thus the velocity continually changes direction as the object travels around the circle. Because the velocity is continually changing, uniform circular motion is accelerated motion. Our main aim in this section is to determine the magnitude and direction of this acceleration.

Velocity

To find the acceleration of an object, we first consider its changing velocity. Since the speed of an object executing uniform circular motion is constant, this speed is simply the distance around the circle divided by the time required for the object to travel this distance, or the time to complete one revolution. If R is the radius of the circle and T is the time for one revolution, then the distance around the circle is $2\pi R$ and

$$v = 2\pi R/T \tag{4-16}$$

The time T is called the ***period*** of the motion.

As an example, suppose one of the letters in Fig. 4-13 is 0.12 m from the turntable's spindle and the turntable is rotating at 33.3 revolutions per minute (rev/min). The period of the rotation is

$$T = (60\ \text{s/min})/(33.3\ \text{rev/min}) = 1.8\ \text{s/rev}$$

The letter's speed is

$$v = 2\pi(0.12\ \text{m})/(1.8\ \text{s}) = 0.42\ \text{m/s}$$

Acceleration

In Sec 4-1, we found that the acceleration of an object following a curved path has a component directed toward the concave side of the path, and if the speed of the object is constant, $\mathbf{a}$ is perpendicular to $\mathbf{v}$. This is illustrated in Fig. 4-14 for the specific case of an object in uniform circular motion. Figure 4-14*a* shows two triangles: a position-vector triangle, made up of $\mathbf{r}_i$, $\mathbf{r}_f$, and $\Delta\mathbf{r}$, and a velocity-vector triangle, composed of $\mathbf{v}_i$, $\mathbf{v}_f$, and $\Delta\mathbf{v}$. Since $\mathbf{v}_f$ is perpendicular to $\mathbf{r}_f$ and $\mathbf{v}_i$ is perpendicular to $\mathbf{r}_i$, $\Delta\mathbf{v}$ is perpendicular to $\Delta\mathbf{r}$. Keep in mind that the direction of $\Delta\mathbf{v}$ is the same as that of $\bar{\mathbf{a}}$. When $\Delta\mathbf{v}$ is placed on the line that bisects the angle between $\mathbf{r}_i$ and $\mathbf{r}_f$, it is directed toward the center of the circle. Figure 4-14*b* shows these

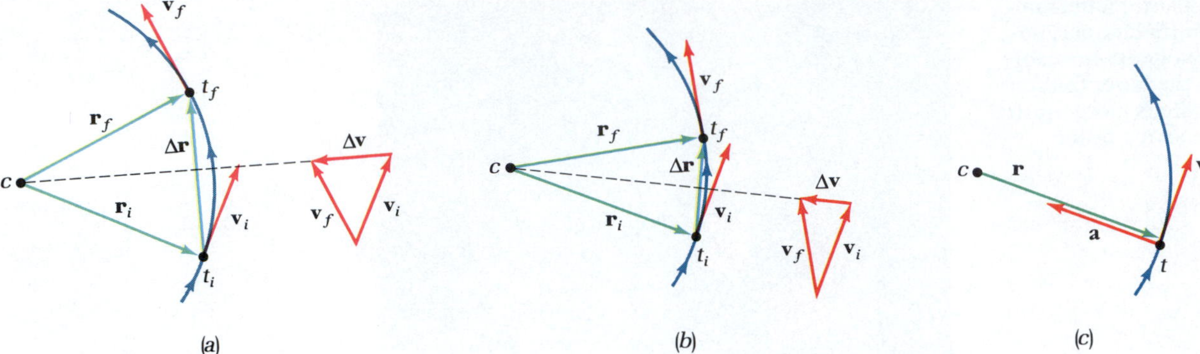

Figure 4-14. By its definition, the direction of $\bar{\mathbf{a}}$ is the same as that of $\Delta\mathbf{v}$. Therefore the direction of $\mathbf{a}$ is the same as that of $\Delta\mathbf{v}$ in the limit as Δt approaches zero. *(a)* When placed midway between the initial and final positions, $\Delta\mathbf{v}$ is directed toward the center of the circle. *(b)* As t_f is made closer to t_i, $\Delta\mathbf{v}$ remains directed toward the center. *(c)* In the limit as Δt approaches zero, $\bar{\mathbf{a}}$ becomes $\mathbf{a}$. Thus $\mathbf{a}$ is directed toward the center of the circle.

triangles for a smaller time interval, and $\Delta\mathbf{v}$ is again directed toward the center of the circle. In Fig. 4-14*c* we have passed to the limit as Δt approaches zero. In this limit, $\bar{\mathbf{a}}$ has become $\mathbf{a}$, and $\mathbf{a}$ is directed toward the center of the circle. Thus the acceleration of an object executing uniform circular motion is toward the center of the circle.

The magnitude of the acceleration is the limiting value of $|\Delta\mathbf{v}|/\Delta t$ as Δt approaches zero:

$$a = \lim_{\Delta t \to 0} \frac{|\Delta\mathbf{v}|}{\Delta t}$$

Again consider Fig. 4-14*a*. Since $r_i = r_f = R$ and $v_i = v_f = v$, both the position-vector and velocity-vector triangles are isosceles (two sides are equal). Also, since $\mathbf{v}_i$ is perpendicular to $\mathbf{r}_i$ and $\mathbf{v}_f$ is perpendicular to $\mathbf{r}_f$, these triangles are similar. Because the triangles are similar, the ratio of base length to side length for one of the triangles is equal to that for the other triangle: $|\Delta\mathbf{v}|/v = |\Delta\mathbf{r}|/R$ or $|\Delta\mathbf{v}| = v|\Delta\mathbf{r}|/R$. Thus

$$a = \lim_{\Delta t \to 0} \frac{|\Delta\mathbf{v}|}{\Delta t} = \lim_{\Delta t \to 0} \frac{v|\Delta\mathbf{r}|/R}{\Delta t} = \frac{v}{R} \lim_{\Delta t \to 0} \frac{|\Delta\mathbf{r}|}{\Delta t}$$

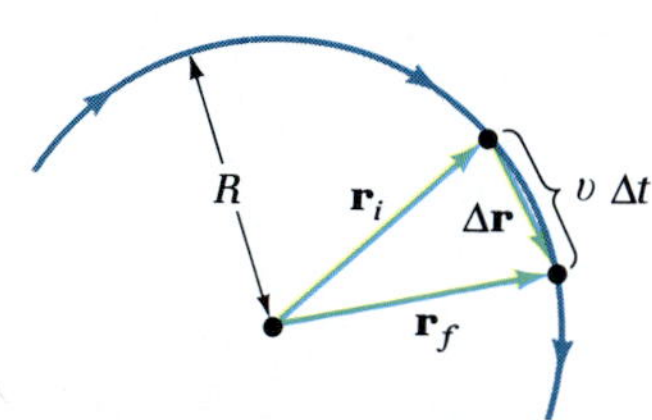

$|\Delta\mathbf{r}| \approx v\,\Delta t$

Figure 4-15. When $|\Delta\mathbf{r}|$ is much smaller than R, the arc length $v\,\Delta t$ is approximately equal to $|\Delta\mathbf{r}|$. In the limit as Δt approaches zero, the approximation becomes exact.

where we have factored v/R out of the limit because neither v nor R depends on Δt. From Fig. 4-15, you can see that if $|\Delta\mathbf{r}|$ is small compared with R, then $|\Delta\mathbf{r}| \approx v\,\Delta t$ or $|\Delta\mathbf{r}|/\Delta t \approx v$. In the limit as Δt approaches zero, this approximation becomes exact:

$$\lim_{\Delta t \to 0} \frac{|\Delta\mathbf{r}|}{\Delta t} = v$$

Thus $a = (v/R)v = v^2/R$. Since the direction of this acceleration is toward the center of the circle, the acceleration is called the ***centripetal acceleration.*** "Centripetal" comes from a Greek term which means center-seeking. Accordingly, we use the symbol a_c to represent the magnitude of this acceleration:

Magnitude of the centripetal acceleration

$$a_c = \frac{v^2}{R} \tag{4-17}$$

Since v and R are constants, the magnitude of the centripetal acceleration is constant. However, as the object follows its circular path, the direction of the acceleration continuously changes because it is always directed from the object toward the center of the circle. Thus the centripetal acceleration is constant in magnitude, but varies in direction.

Since $v = 2\pi R/T$, this acceleration magnitude can also be written as

$$a_c = \frac{(2\pi R/T)^2}{R} = \frac{4\pi^2 R}{T^2} \tag{4-18}$$

For the example with the letter in Fig. 4-13, $R = 0.12$ m and $T = 1.8$ s, so that $a_c = 4\pi^2(0.12 \text{ m})/(1.8 \text{ s})^2 = 1.5 \text{ m/s}^2$.

EXAMPLE 4-6

Acceleration of a car in a circular turn at constant speed. A car, initially traveling west, makes a right turn, following a circular arc of radius $R = 22$ m, and finishes the turn heading north. The car's speed is constant at $v = 8.5$ m/s throughout the turn. Determine the car's acceleration *(a)* at the instant after it begins turning, *(b)* when it is halfway through the turn, and *(c)* at the instant before it finishes the turn.

Solution. *(a)* During the turn the car is in uniform circular motion, so that the magnitude of its acceleration is $a_c = v^2/R = (8.5 \text{ m/s})^2/22 \text{ m} = 3.3 \text{ m/s}^2$. The car's path is shown in Fig. 4-16. As the car enters the turn, the center of the circle is to the north. Thus the acceleration is 3.3 m/s² toward the north. *(b)* At the instant the car is halfway through the turn, the center of the circle is toward the northeast. Consequently, the acceleration is 3.3 m/s² toward the northeast. *(c)* Similarly, just before the turn is completed, the acceleration is 3.3 m/s² toward the east.

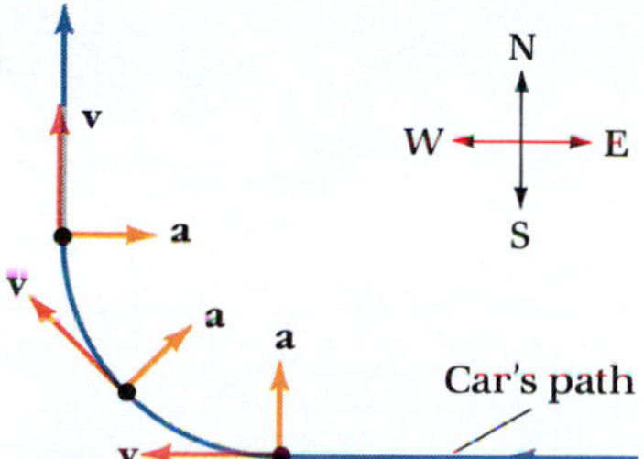

Figure 4-16. Example 4-6: A car initially traveling west makes a right turn of 90° along a circular arc. Just after the beginning of the turn, **v** is toward the west and **a** is toward the north; halfway through the turn, **v** is toward the northwest and **a** is toward the northeast; just before the end of the turn, **v** is toward the north and **a** is toward the east.

SELF-TEST 4-6. A typical satellite in low earth orbit has an altitude of about 300 km above the surface and a period of about $T = 1.5 \text{ h} = 5.4 \times 10^3$ s. Given that the radius of the earth is $6.37 \times 10^6 \text{ m} = 6.37$ Mm, determine the centripetal acceleration magnitude for such a satellite. ***ANSWER:*** 9.0 m/s².

4-4 RELATIVE MOTION

Alvin was standing on the curb on the south side of an east-west street waiting for a bus, as shown in Fig. 4-17. Bill was riding in a car traveling at a speed of 20 m/s toward the east. Just before the car passed close to Alvin, when it was northwest of him, Bill threw a banana peel out of the window with velocity components (according to Bill) of 10 m/s toward the south and 10 m/s toward the west. The banana peel hit Alvin's ear.

Alvin and Bill met later, and Alvin accused Bill of throwing the peel at him. Bill pleaded innocent by truthfully saying that he threw the peel toward the south*west,* and at the time of release Alvin was south*east* of the release point. This episode reveals that two observers moving at constant velocity relative to one another and each measuring the velocity of an object will find that their velocity measurements yield different results.

We use Fig. 4-18 to understand this disagreement between Alvin and Bill. This figure shows two reference frames, the frame for observer A (Alvin) and the frame for observer B (Bill). Each observer is at rest in his frame. The two frames are moving at constant velocity relative to one another along their common x direction. (In the Alvin-Bill dispute, the unit vector **i** is toward the east and the unit vector **j** is toward the north.)

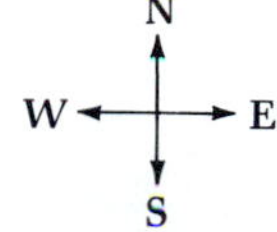

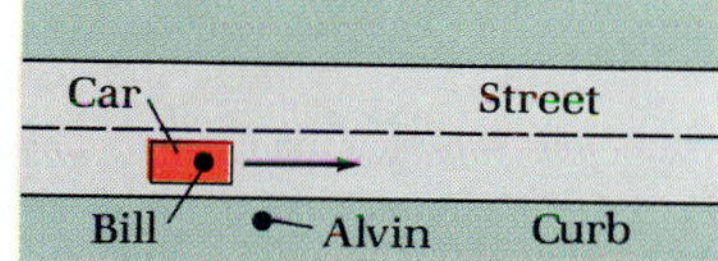

Figure 4-17. Bill was riding in a car traveling east. Alvin was standing on the curb.

To distinguish measurements by different observers, we use a double-subscript notation. We let $\mathbf{r}_{PA}$ be the position vector of a particle P (the peel) according to observer A. That is, the first subscript indicates the object whose position is mea-

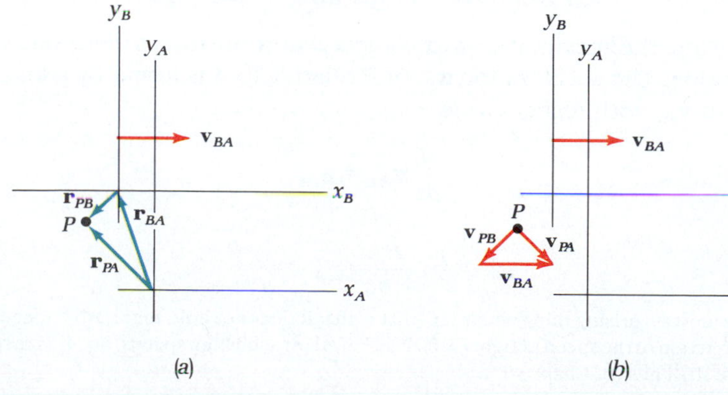

Figure 4-18. Observers in frame A and frame B measure particle P's *(a)* position vector and *(b)* velocity.

sured and the second subscript indicates the observer who made the measurement. In words, $\mathbf{r}_{PA}$ is the position vector of P relative to A. A similar notation is used for velocities. For example, $\mathbf{v}_{BA}$ is the velocity of observer B relative to observer A. If observer B is moving in the $+x$ direction according to A at a speed v, as shown in the figure, then $\mathbf{v}_{BA} = v\mathbf{i}$. This means that A is moving in the $-x$ direction according to B: $\mathbf{v}_{AB} = -v\mathbf{i}$. In this double-subscript notation, reversing the order of the subscripts changes the sign of the quantity:

$$\mathbf{v}_{AB} = -\mathbf{v}_{BA} \tag{4-19}$$

From Fig. 4-18a, the position vector $\mathbf{r}_{PA}$ of P relative to A is given by

Relative position vectors

$$\mathbf{r}_{PA} = \mathbf{r}_{PB} + \mathbf{r}_{BA} \tag{4-20}$$

where $\mathbf{r}_{PB}$ is the position vector of P relative to B and $\mathbf{r}_{BA}$ is the position vector of B relative to A. Notice the arrangement of the subscripts in Eq. 4-20. The first subscripts on each side of the equation are the same and the last subscripts on each side are the same. Also, the "in-between" subscripts on the right-hand side are the same. This subscript rule is easy to remember and is quite useful.

The velocity of P relative to A is found by taking the derivative of $\mathbf{r}_{PA}$ with respect to t:

$$\frac{d}{dt}\mathbf{r}_{PA} = \mathbf{v}_{PA}$$

Similarly,

$$\frac{d}{dt}\mathbf{r}_{PB} = \mathbf{v}_{PB} \qquad \text{and} \qquad \frac{d}{dt}\mathbf{r}_{BA} = \mathbf{v}_{BA}$$

Therefore, if we take the derivative of Eq. 4-20 with respect to t, we find

$$\mathbf{v}_{PA} = \mathbf{v}_{PB} + \mathbf{v}_{BA} \tag{4-21}$$

We find that the relative velocities follow the same subscript rule as the relative positions (Eq. 4-20). This result is shown graphically in Fig. 4-18b.*

Now we apply Eq. 4-21 to the incident involving Alvin and Bill. For simplicity, we consider only the horizontal motion of the banana peel and neglect air resistance. Then, the velocity of the peel relative to Bill is $\mathbf{v}_{PB} = (-10\text{ m/s})\mathbf{i} + (-10\text{ m/s})\mathbf{j}$, and the velocity of Bill relative to Alvin is $\mathbf{v}_{BA} = (20\text{ m/s})\mathbf{i}$. Inserting these values into Eq. 4-21 gives

$$\begin{aligned}\mathbf{v}_{PA} &= (-10\text{ m/s})\mathbf{i} + (-10\text{ m/s})\mathbf{j} + (20\text{ m/s})\mathbf{i} \\ &= (10\text{ m/s})\mathbf{i} + (-10\text{ m/s})\mathbf{j}\end{aligned}$$

That is, the velocity of the peel relative to Alvin was toward the southeast. Since Alvin was southeast of the release point, the peel headed directly toward him. Despite Bill's innocence, the peel hit Alvin's ear with a speed of

$$\sqrt{(10\text{ m/s})^2 + (-10\text{ m/s})^2} = 14\text{ m/s}$$

Now consider the acceleration of an object according to two observers who are in relative motion. The acceleration $\mathbf{a}_{PA}$ of P relative to A is found by taking the time derivative of $\mathbf{v}_{PA}$ with respect to t:

$$\frac{d}{dt}\mathbf{v}_{PA} = \mathbf{a}_{PA}$$

* Perhaps the most surprising thing about Eq. 4-21 is that it does not hold for relative speeds that are an appreciable fraction of the speed of light (3.0×10^8 m/s). At such high speeds, Eq. 4-21 is replaced by a result that we shall give in Chap. 38.

Similarly,

$$\frac{d}{dt}\mathbf{v}_{PB} = \mathbf{a}_{PB} \quad \text{and} \quad \frac{d}{dt}\mathbf{v}_{BA} = \mathbf{a}_{BA}$$

Taking the time derivative of Eq. 4-21 gives

$$\mathbf{a}_{PA} = \mathbf{a}_{PB} + \mathbf{a}_{BA} \tag{4-22}$$

Relative accelerations

The relative accelerations follow the same subscript rule as the relative positions and the relative velocities. In addition, all three quantities follow the same rule for reversing subscripts. In the case of the acceleration,

$$\mathbf{a}_{BA} = -\mathbf{a}_{AB}$$

Reversing the subscripts on any of these quantities changes the sign of the quantity.

A simple application of Eq. 4-22 is finding the connection between the acceleration of particle P relative to observer A and to observer B when the observers move with constant velocity relative to one another. If the two observers move with a constant relative velocity, then $\mathbf{a}_{BA} = 0$ and

$$\mathbf{a}_{PA} = \mathbf{a}_{PB}$$

In this case, each observer measures particle P's acceleration to be the same.

We now summarize the results of this section. When two frames of reference are in motion relative to one another, the velocity of an object measured relative to these frames is different, and the difference is the relative velocity of the frames. Similarly, when two frames accelerate relative to one another, the acceleration of an object measured relative to these frames is different, and the difference is the relative acceleration of the frames.

If the value of a kinematical quantity, such as a velocity or an acceleration, is to have meaning, then the reference frame relative to which the quantity is measured must be specifically stated or clearly implied. Usually, it is clearly implied.

We shall have more to say about reference frames in the next chapter.

Multicolor strobe photograph of a ballistics-cart demonstration. When the ball and the cart are shown at the same instant, they appear with the same color. The cart moves with nearly a constant velocity relative to the table (and to the camera). A spring mechanism in the cart projects the ball vertically upward relative to the cart. The photograph shows that relative to an observer fixed to the table, the ball is projected horizontally as well as vertically, and the horizontal component of the ball's velocity is essentially the same as the cart's.

EXAMPLE 4-7

Parcel dropped from an airplane. Charles is in an airplane flying overhead with a constant horizontal velocity relative to the ground of 75 m/s toward the east. Emily, who is standing on the ground, observes Charles as he drops a parcel out of the airplane's window, as shown in Fig. 4-19. At the instant of release, the parcel is at rest relative to Charles. Describe the motion of the parcel according to *(a)* Charles and *(b)* Emily. Let the origins for Charles and Emily be at the point of release at the instant of release, and let this instant be $t = 0$. Also, let the unit vector $\mathbf{j}$ be directed vertically upward and the unit vector $\mathbf{i}$ be directed toward the east. Neglect air resistance.

Solution. *(a)* According to Charles, the parcel is released from rest so that it falls vertically downward in a straight line and hits the ground directly beneath the airplane. The acceleration of the parcel (P) relative to Charles (C) is $\mathbf{a}_{PC} = -g\mathbf{j}$. The parcel's x component of velocity is zero during the entire motion, and its y component of velocity is $-gt$ so that the velocity of the parcel relative to Charles is $\mathbf{v}_{PC} = -gt\mathbf{j}$. *(b)* Since their relative velocity is constant ($\mathbf{a}_{CE} = 0$), Emily (E) agrees with Charles about the parcel's acceleration ($\mathbf{a}_{PE} = -g\mathbf{j}$), but they measure different velocities. The velocity $\mathbf{v}_{PE}$ of the parcel relative to Emily is $\mathbf{v}_{PE} = \mathbf{v}_{PC} + \mathbf{v}_{CE}$, where $\mathbf{v}_{CE}$ is the velocity of Charles relative to Emily. The velocity of Charles relative to Emily is the same as the velocity of the airplane relative to the ground: $\mathbf{v}_{CE} = (75\text{ m/s})\mathbf{i}$. Thus

$$\mathbf{v}_{PE} = (75\text{ m/s})\mathbf{i} - gt\mathbf{j}$$

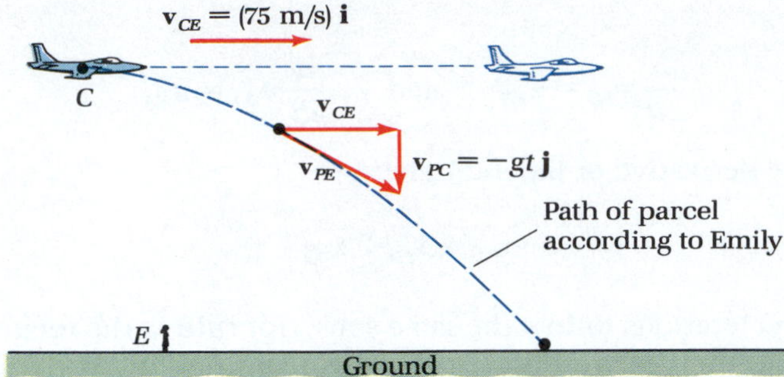

Figure 4-19. Example 4-7: Charles drops a parcel from the plane. From Emily's point of view, the path of the parcel is a parabola.

This velocity expression corresponds to projectile motion. According to Emily, the path of the parcel is a parabola, as shown in Fig. 4-19.

SELF-TEST 4-7. On a radar screen, air controller *C* is measuring the velocity of two jet airliners, *A* and *B*. Controller *C* measures the velocity of *A* as 200 m/s toward the north and the velocity of *B* as 250 m/s toward the southeast. Let **i** point toward the east and **j** toward the north. In terms of these unit vectors, write $\mathbf{v}_{AC}$, $\mathbf{v}_{BC}$, $\mathbf{v}_{AB}$, and $\mathbf{v}_{BA}$. ***ANSWERS:*** $\mathbf{v}_{AC} = (200\text{ m/s})\mathbf{j}$; $\mathbf{v}_{BC} = (177\text{ m/s})\mathbf{i} - (177\text{ m/s})\mathbf{j}$; $\mathbf{v}_{AB} = -(177\text{ m/s})\mathbf{i} + (377\text{ m/s})\mathbf{j}$; $\mathbf{v}_{BA} = (177\text{ m/s})\mathbf{i} - (377\text{ m/s})\mathbf{j}$.

4-5 AIR RESISTANCE, NUMERICAL METHODS

As an example of numerical methods in a two-dimensional problem, we consider the effect of air resistance on the baseball in Example 4-5. The effect of air resistance on a baseball can be approximated by adding a term to the acceleration that depends on the square of the speed of the ball through the air and points in a direction opposite to the velocity. The acceleration then becomes

$$\mathbf{a} = \mathbf{g} - bv\mathbf{v}$$

where **g** is the gravitational acceleration and b is a constant. (Measurements indicate that an approximate value of b for a baseball in air is 0.0055 m^{-1}.) With $+y$ vertically upward and $+x$ horizontal and in the direction of motion, $\mathbf{g} = -g\mathbf{j}$ and

$$\mathbf{a} = (-bvv_x)\mathbf{i} + (-g - bvv_y)\mathbf{j}$$

Writing $v = \sqrt{v_x^2 + v_y^2}$ gives

$$a_x = -bv_x\sqrt{v_x^2 + v_y^2}$$
$$a_y = -g - bv_y\sqrt{v_x^2 + v_y^2}$$

Table 4-2 shows the beginning of a spreadsheet that determines the baseball's motion. The initial values of x, y, v_x, and v_y are taken from Example 4-5. The initial velocity components are $v_{x0} = v_0 \cos\theta_0 = (35\text{ m/s})\cos 42° = 26$ m/s, and $v_{y0} = v_0 \sin\theta_0 = 23$ m/s. We notice that the time of the flight of the ball without air resistance is about 5 s from the fact that $v_x = 26$ m/s and the range of the ball is

TABLE 4-2. ***Beginning spreadsheet for the flight of a baseball***

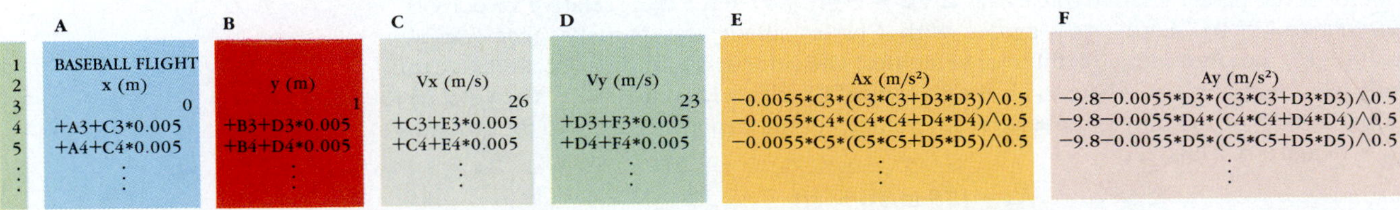

	A	B	C	D	E	F
1	BASEBALL FLIGHT					
2	x (m)	y (m)	Vx (m/s)	Vy (m/s)	Ax (m/s²)	Ay (m/s²)
3	0	1	26	23	−0.0055*C3*(C3*C3+D3*D3)∧0.5	−9.8−0.0055*D3*(C3*C3+D3*D3)∧0.5
4	+A3+C3*0.005	+B3+D3*0.005	+C3+E3*0.005	+D3+F3*0.005	−0.0055*C4*(C4*C4+D4*D4)∧0.5	−9.8−0.0055*D4*(C4*C4+D4*D4)∧0.5
5	+A4+C4*0.005	+B4+D4*0.005	+C4+E4*0.005	+D4+F4*0.005	−0.0055*C5*(C5*C5+D5*D5)∧0.5	−9.8−0.0055*D5*(C5*C5+D5*D5)∧0.5
⋮	⋮	⋮	⋮	⋮	⋮	⋮

about 120 m: (120 m)/(26 m/s) ≈ 5 s. Therefore we let the time intervals be 5 ms, and in that way 1000 iterations will encompass 5 s.

Figure 4-20 shows the baseball's trajectory as it was generated by the spreadsheet. For comparison, another spreadsheet was used to generate the trajectory when air resistance is neglected, and these results are included in the figure. In Exercise 38, you are asked to produce a spreadsheet for the no-air-resistance case. The graphs show that when air resistance is considered the prospective home run in Example 4-5 is nothing more than an ordinary fly ball.

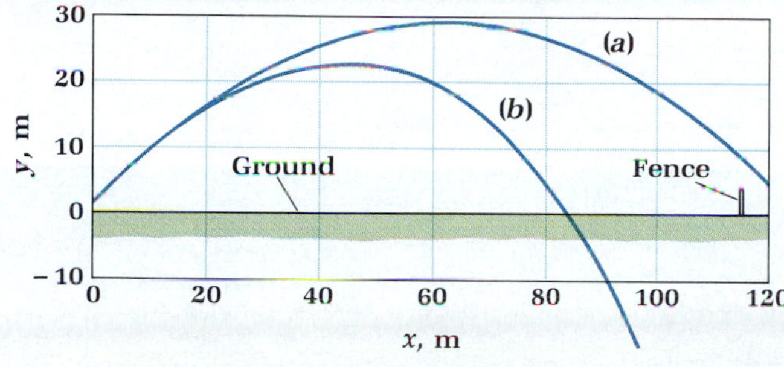

Figure 4-20. Trajectory of a baseball with the initial conditions given in Example 4-5: *(a)* without air resistance, and *(b)* with air resistance.

COMMENTARY: Galileo Galilei

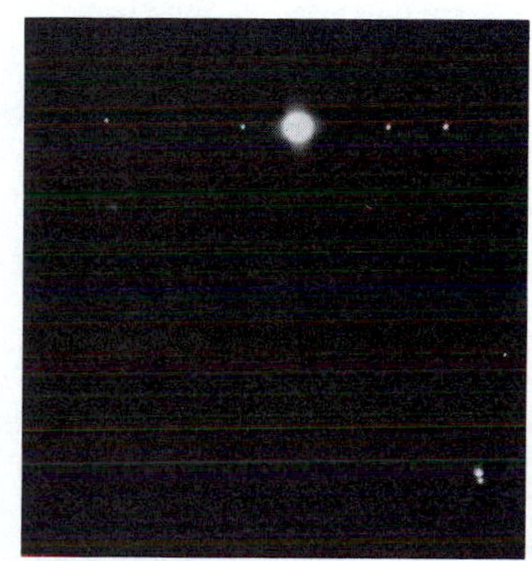

Jupiter with its four brightest moons. As viewed from the earth, the moons appear along a straight line because the orbital planes of the moons are nearly the same. Longitude was very difficult to measure in Galileo's day, and as one of his many practical accomplishments, he devised a means for measuring longitude from observations of Jupiter's moons.

Galileo was born in Pisa, Tuscany (part of present-day Italy), in 1564, the year of William Shakespeare's birth and Michelangelo's death. His father, Vincenzio Galilei, was a musician and merchant with a lively interest in cultural activities. Vincenzio's ancestors were distinguished citizens of Florence, but at the time of Galileo's birth the family was in economic decline. His father sent him, at age 17, to the University of Pisa to study medicine, but Galileo soon became captivated by a passion for science and mathematics. He demonstrated his aptitude for experimentation and measurement when, as a medical student, he devised a pendulum-type clock for measuring pulse rates. His extraordinary abilities were apparent early, and he was appointed professor of mathematics at Pisa at the age of 26.

Since Galileo was the eldest son, his father's death left him with the family's financial burdens, but without means. He incurred huge debts to provide dowries for two of his sisters and partially supported a younger brother for many years. Throughout his life, Galileo never completely escaped the specter of poverty. He never married, but he was the father of three children by Marina Gamba, who was his affectionate companion for 10 years. Their separation was amicable and Marina subsequently married one of Galileo's friends. Galileo consigned his two daughters, Virginia and Livia, to a convent at the ages of 13 and 12, and they later became nuns. He treated his son Vincenzio with somewhat more favor than his daughters, and he eventually took action to make his son legitimate. Late in his life, his daughter Virginia became his close companion. Her death at the age of 36 was an extremely grievous event in Galileo's life.

Because of his exuberance and fiery personality, Galileo's life was filled with turbulence and controversy. He is probably more famous for his difficulties with the Roman Catholic church than for his remarkable scientific achievements. He was tried twice by the Inquisition. He supported the sun-centered, or copernican, theory of the universe, whereas church doctrine held that the earth is stationary at the center of the universe. At his second trial, in 1633, when he was 69 and feeble, Galileo was forced to recant his scientific beliefs to avoid imprisonment.

The phases of Venus. When Venus is on the opposite side of the sun from the earth (superior conjunction), it is at its farthest distance from the earth and its illuminated side faces the earth. When it is on the same side of the sun as the earth (inferior conjunction), it is nearest the earth and its illuminated side faces away from the earth.

Galileo's scientific interests were so broad that they can scarcely be mentioned here. He studied mechanics, astronomy, optics, and the behavior of fluids. He was an early developer of the telescope, and was among the first to use it as an astronomical instrument. He discovered the moons of Jupiter, the rings of Saturn, the phases of Venus, and the existence of surface features on the moon. (Because of the limited resolving power of his telescopes, he mistook the rings of Saturn to be two satellites, one on each side of the planet.) As we have noted in these two chapters on kinematics, he explained free-fall and projectile motion. The description of relative motion that we discussed in Sec. 4-4 is attributed to him and is known as *galilean relativity.* Many historians give Galileo credit for discovering one of the three laws of motion, the one referred to as Newton's first law, or the law of inertia (Sec. 5-2).

Because of his philosophy of science, Galileo is regarded as the father of modern science. The importance he attached to mathematics in describing physical phenomena is evident in the following passage:

> Philosophy is written in this grand book — I mean the universe — which stands continually open to our gaze, but it cannot be understood unless one first learns to comprehend the language and interpret the characters in which it is written. It is written in the language of mathematics, and its characters are triangles, circles, and other geometrical figures, without which it is humanly impossible to understand a word of it; without these, one is wandering about in a dark labyrinth.*

Galileo was primarily concerned with mathematics as the "language" of nature, and he showed little interest in pure mathematics.

After his second trial, Galileo was kept in near-seclusion for the remaining nine years of his life. Despite being sick and blind, he wrote some of his greatest works during this time. His enthusiasm was unflagging to the end. He died in 1642, the year of Isaac Newton's birth.

* Ludovico Geymonat, *Galileo Galilei,* Stillman Drake (trans.), McGraw-Hill, New York, 1965, p. 106.

SUMMARY

Section 4-1. Position, Velocity, and Acceleration

In two dimensions the position vector is

$$\mathbf{r} = x\mathbf{i} + y\mathbf{j} \tag{4-1}$$

and the velocity is

$$\mathbf{v} = \frac{dx}{dt}\mathbf{i} + \frac{dy}{dt}\mathbf{j}$$

The velocity is parallel to a line tangent to the path and points in the direction of motion. The acceleration is

$$\mathbf{a} = \frac{dv_x}{dt}\mathbf{i} + \frac{dv_y}{dt}\mathbf{j}$$

For a curved path, **a** is always toward the concave side of the path. If the speed is increasing, the angle ϕ between **v** and **a** is less than 90°; but if the speed is decreasing, ϕ is greater than 90°.

Section 4-2. Constant Acceleration: Projectile Motion

Projectile motion with negligible air resistance is an example of motion with constant acceleration in two dimensions. The trajectory of a projectile is a parabola. Table 4-1 summarizes the equations that describe a projectile.

Section 4-3. Uniform Circular Motion

An object that moves in a circle at constant speed is in uniform circular motion, and its acceleration, called the centripetal acceleration, is toward the center of the circle. The magnitude of the centripetal acceleration is

$$a_c = \frac{v^2}{R} \tag{4-17}$$

Section 4-4. Relative Motion

Two observers moving at constant velocity with respect to one another measure different values for the velocity of an object. The difference between their measurements is equal to their velocity relative to one another. These two observers measure the same acceleration for an object.

Section 4-5. Air Resistance, Numerical Methods

Numerical methods can be used to account approximately for the effects of air resistance on the motion of a projectile.

QUESTIONS

1. What sort of relationship, if any, always holds between the directions *(a)* of an object's position vector and velocity, *(b)* of its displacement and average velocity, *(c)* of a line tangent to the path and the velocity, *(d)* of the change in velocity and the average acceleration, *(e)* of a line tangent to the path and the acceleration?

2. If an object is traveling in a straight line, what sort of relationship, if any, always holds between the directions of the object's path and its acceleration?

3. If an object has a curved path, what sort of relationship, if any, always holds between the direction in which the path is curving and the direction of the object's acceleration?

4. Is it possible for an object to accelerate and still have *(a)* a constant speed, *(b)* a straight path, *(c)* a constant velocity? Explain.

5. If the velocity of an object is changing and becomes zero at an instant of time, must its acceleration be zero at that instant? Can its acceleration be zero at that instant? Can its acceleration be zero over a time interval that includes that instant? Give an example that corresponds to each of your answers.

6. Is it possible for an object's velocity to be in the direction opposite to its acceleration? If it is possible, give an example. If not, explain.

7. If an object is accelerating during a time interval, can its velocity be zero at an instant during this time interval? Can the velocity be zero during the entire time interval? Give examples to support your answers.

8. We have seen that if the acceleration of an object is constant, then the velocity varies linearly with time and the position vector varies quadratically with time. Suppose the acceleration varies linearly with time. What is the time variation of *(a)* the velocity and *(b)* the position vector?

9. The following list contains various objects and the range of their trajectories when they are launched with a variety of speeds into the air at an angle of 45° over flat terrain. Suppose we wish to predict the positions of the objects during their motion to an accuracy of 10 percent. For which objects is it valid to neglect air resistance?

(a) Golf ball, 5 m
(b) Golf ball, 100 m
(c) Baseball, 5 m
(d) Baseball, 100 m
(e) Badminton birdie, 3 m
(f) Ping-pong ball, 3 m
(g) Frisbie, 10 m
(h) Boomerang, 10 m

10. Judy says that an object tossed into the air is falling if its acceleration is downward. Martha says that the object is falling if its velocity is downward. In what part of the trajectory of a projectile will Judy and Martha agree? In what part will they disagree? Which definition of the term "falling" do you prefer, Judy's or Martha's?

11. A projectile is launched with an initial velocity of $(3\ \text{m/s})\mathbf{i} + (2\ \text{m/s})\mathbf{j}$. Neglect air resistance. *(a)* What is its velocity at the top of its trajectory? **(b)** What is its acceleration at the top of its trajectory? Be certain that each of your answers is in the form of a vector.

12. A projectile is launched with an initial velocity $\mathbf{v}_0 = v_{x0}\mathbf{i} + v_{y0}\mathbf{j}$. Neglect air resistance. Does the range of the projectile depend on *(a)* v_{x0}; *(b)* v_{y0}? Does the maximum height of the projectile depend on *(c)* v_{x0}; *(d)* v_{y0}?

13. A stone is thrown into the air at an angle of, say 40° with respect to the horizontal. Is its velocity ever parallel to its acceleration at any time during its motion? If so, when? If not,

explain. Is its velocity ever perpendicular to its acceleration at any time during the motion? If so, when? If not, explain.

14. Suppose you throw a rock at a bottle that will begin falling from rest at the instant you release the rock. To hit the bottle, should you aim above it, below it, or directly at it? Explain.

15. Is it possible for an object to be moving with constant speed and be accelerating at the same time? If so, give an example of such a case. If not, explain.

16. If an object follows a curved path, can its velocity be constant? Can its speed be constant? Can its acceleration be constant? Can the magnitude of its acceleration be constant? If the answer to any of these questions is yes, give an example.

17. Suppose an object is traveling in a circle at constant speed and the origin is at the center of the circle. What is the relationship, if any, between the directions of the object's *(a)* position vector and velocity, *(b)* position vector and acceleration, *(c)* velocity and acceleration?

18. In uniform circular motion, *(a)* is the speed constant? *(b)* Is the velocity constant? *(c)* Is the magnitude of the acceleration constant? *(d)* Is the acceleration constant? Explain.

19. A penny on a rotating turntable executes uniform circular motion with a speed of 0.8 m/s and an acceleration magnitude of 4 m/s². The penny is repositioned on the turntable such that the radius of its circular path is halved. What is its subsequent speed and acceleration magnitude?

20. A penny on a rotating turntable executes uniform circular motion with a speed of 0.4 m/s and an acceleration magnitude of 2 m/s². The rotational speed of the turntable is doubled so that the period of the penny's motion is halved. What is its subsequent speed and acceleration magnitude?

21. ***Acceleration of a pendulum bob.*** A pendulum is composed of a heavy object, such as a rock, which is allowed to swing back and forth on the end of a string in a vertical plane (Fig. 4-21). The heavy object is called the ***pendulum bob,*** and its path is along the arc of a circle. In the figure, the bob is instantaneously at rest when it is at the far right. Suppose the bob is swinging to the left, which arrow best indicates the direction of its acceleration at each position shown? How about when it is swinging to the right?

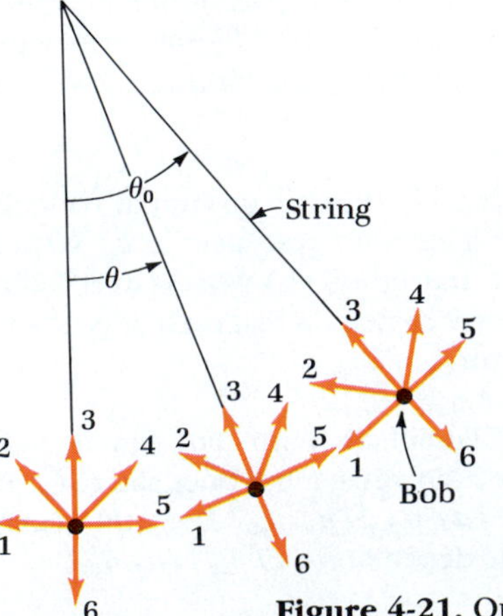

Figure 4-21. Question 21: A pendulum bob is shown at three positions as it swings back and forth in a plane.

22. Consider an object at point P along each trajectory shown in Fig. 4-22. *(a)* In Fig. 4-22*a*, what are the algebraic signs of v_x and v_y? *(b)* In Fig. 4-22*b*, what are the algebraic signs of a_x and a_y if the object's speed is increasing? *(c)* In Fig. 4-22*c*, what are the algebraic signs of a_x and a_y if the object's speed is decreasing?

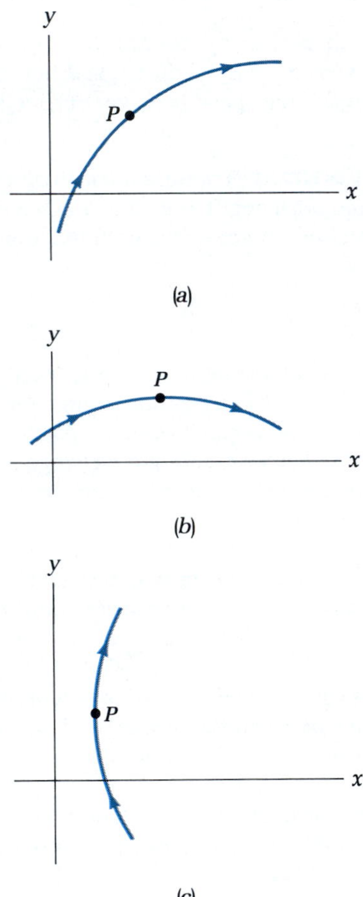

Figure 4-22. Question 22.

23. Suppose an object follows a spiral path while traveling at constant speed (Fig. 4-23). Is the object's velocity constant? Is its acceleration constant? Is its acceleration magnitude constant? If its acceleration magnitude is not constant, is it increasing or decreasing?

24. As you know, the earth circles the sun once each year. From our perspective on earth, the sun returns to its original position relative to the backdrop of distant stars in one year. How would it look to us if the sun circled the earth, instead of the other way around?

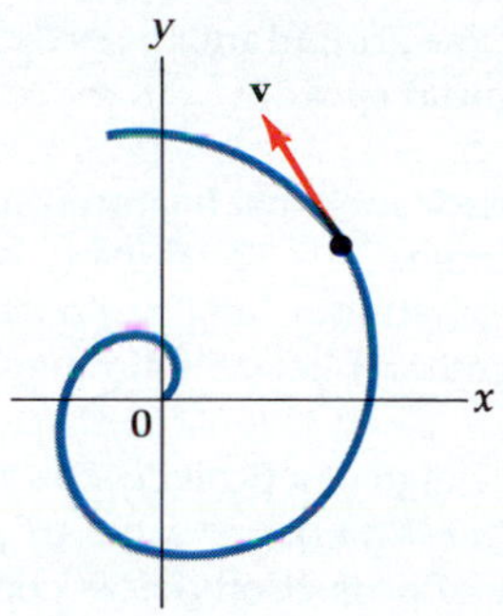

Figure 4-23. Question 23.

25. *(a)* Suppose two observers who are moving at constant velocity relative to one another each measure the velocity of an object. Will they find the same result? If so, explain. If not, how will they differ? *(b)* Suppose the observers measure the object's acceleration. Will they find the same result? If so, explain. If not, how will they differ?

26. When the acceleration is constant, we found that the x and y motions are independent of one another. That is, a_x has no effect on v_y or y, and a_y has no effect on v_x or x. Thus, when air resistance has a negligible effect on the motion of a projectile, the x and y motions are independent. Is this still true when air resistance is not negligible? Explain.

27. Did Bill intend to hit Alvin with the banana peel? (*Hint:* Bill made an A in Physics.)

28. Complete the following table:

Symbol	Represents	Type	SI Unit
$\mathbf{v}$		Vector	
a_x			m/s²
y		Component	
v_0	Initial speed		
a_c		Scalar	
$\mathbf{v}_{BA}$			

EXERCISES

Section 4-1. Position, Velocity, and Acceleration

1. A jogger runs around a circular track of radius 45.0 m (Fig. 4-24). Let the origin of an xy coordinate frame be at the center of the circle, with $+\mathbf{i}$ toward the east and $+\mathbf{j}$ toward the north, and let $t = 0.0$ s correspond to the instant the jogger's coordinates (x, y) are (45.0 m, 0.0 m). *(a)* At $t = 16.8$ s, the jogger is directly northeast of the origin. What is her position vector? *(b)* At $t = 33.6$ s, the jogger is due north of the origin. What is her displacement between $t = 0.0$ s and $t = 33.6$ s? *(c)* What is the distance she traveled between $t = 0.0$ s and $t = 33.6$ s?

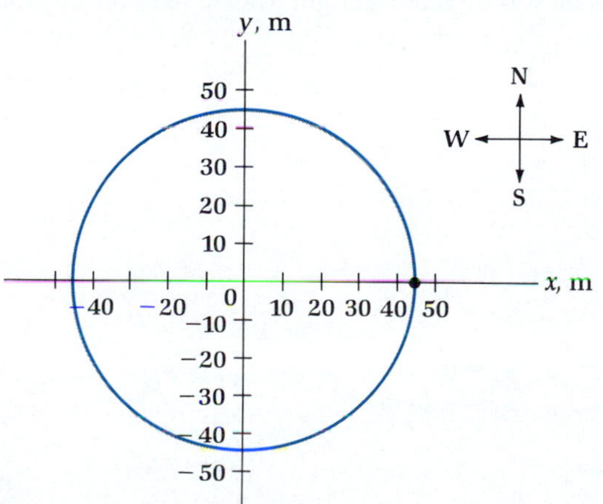

Figure 4-24. Exercise 1.

2. The jogger in the previous exercise is directly northwest of the origin at $t = 50.4$ s. *(a)* What is her position vector? *(b)* What is her displacement between $t = 0.0$ s and $t = 50.4$ s? *(c)* What is the distance she traveled between $t = 0.0$ s and $t = 50.4$ s?

3. *(a)* What is the average velocity of the jogger in Exercise 1 between $t = 0.0$ s and $t = 33.6$ s? *(b)* At $t = 8.4$ s her position is (41.6 m, 17.2 m), and at $t = 25.2$ s her position is (17.2 m, 41.6 m). What is her average velocity during this time interval?

4. The jogger in Exercise 1 completes one revolution in 134.4 s. *(a)* If she travels at constant speed, what is her speed? *(b)* What is her velocity at $t = 16.8$ s? Compare this answer with the answers in the previous exercise.

5. A car travels the horizontal path from A through K, as shown in Fig. 4-25. The paths from C to E and G to I are circular arcs of radii 200 and 300 m, respectively. The car is at rest at A, accelerates uniformly from A to C, travels at a constant speed of 23 m/s from C to I, then slows down with a constant acceleration until it is at rest at K. *(a)* What is the car's velocity at B? *(b)* What is its acceleration at B? *(c)* What is its velocity at D? *(d)* What is its average acceleration between C and E?

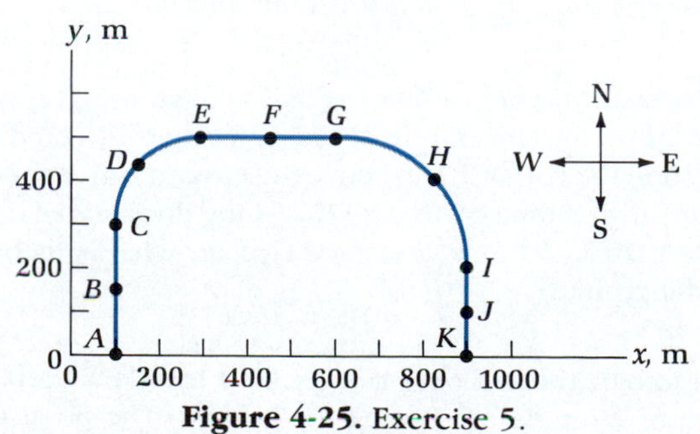

Figure 4-25. Exercise 5.

6. For the car in the previous exercise (Fig. 4-25), what is its velocity at *(a)* F; *(b)* H; *(c)* J? What is its acceleration at *(d)* F and *(e)* J? *(f)* What is its average acceleration between G and I?

7. The position vector of an object is $\mathbf{r} = [(3.5\ \text{m/s})t + 4.2\ \text{m}]\mathbf{i} + [(5.1\ \text{m/s})t]\mathbf{j}$. *(a)* Determine $\mathbf{v}$. *(b)* Make a graph of the object's path from $t = 0.0$ s to $t = 3.0$ s. Plot points each 0.5 s and sketch the path. Show $\mathbf{v}$ at several places on the path to indicate the direction of motion.

8. The coordinates of an object are given by $x = (1.9\ \text{m/s}^2)t^2$ and $y = (0.47\ \text{m/s}^3)t^3 - 5.6$ m. What are the object's *(a)* velocity components and *(b)* acceleration components?

9. At a particular point on a roller-coaster track, the track makes an angle of 29° with respect to the horizontal. If the roller-coaster car passes this point with a speed of 16 m/s, what are the *(a)* horizontal and *(b)* vertical components of the car's velocity?

10. As an airplane climbs after takeoff, the horizontal and vertical components of its velocity are 97 m/s and 22 m/s, respectively. What are *(a)* the airplane's speed and *(b)* the angle between its velocity and the horizontal?

Section 4-2. Constant Acceleration: Projectile Motion

(*Note:* The effects of air resistance are neglected in the exercises in this section.)

11. A billiard ball rolls toward the east down a straight plank that makes an angle of 15° with respect to the horizontal. The magnitude of the ball's acceleration is 1.81 m/s² during the entire motion. Let the x axis be horizontal with $+\mathbf{i}$ toward the east and let $+\mathbf{j}$ point vertically upward. Let $t = 0$ correspond to the instant the ball is released at the origin. *(a)* Write expressions for a_x, a_y, v_x, v_y, x, and y. *(b)* Repeat part *(a)* with $\mathbf{i}$ directed down the ramp and $\mathbf{j}$ perpendicular to it. *(c)* Is this motion in one or two dimensions? Explain.

12. A projectile has an initial speed of 26 m/s and a projection angle of 48° at the launch point. At $t = 2.1$ s after launch, *(a)* what is the projectile's distance from the launch point? *(b)* What is its speed? *(c)* What is the direction it is heading relative to the horizontal?

13. A rock is thrown into the air with an initial speed of 36 m/s and at a projection angle of 62°. Let the origin be at the point of release and let $t = 0$ correspond to the instant of release. Write expressions for v_x, v_y, x, and y as functions of time t.

14. A baseball pitcher throws the ball with an initial speed of 40 m/s. At the instant of release the velocity is directed horizontally and the ball is 2.1 m above the ground and 20 m (horizontally) from home plate. *(a)* How long does it take to pass over the plate? *(b)* As it crosses the plate, what is its height above the ground?

15. A football thrown over a level field travels a horizontal distance of 17 m before hitting the ground. The point of release is 1.5 m above the ground and the projection angle is 16°. What is the ball's initial speed?

16. A golf ball is struck such that its initial speed is 42 m/s and the angle of projection is 34°. The fairway is level. What is the ball's *(a)* time of maximum height, *(b)* time of flight, *(c)* maximum height, and *(d)* horizontal range?

17. The horizontal range of a projectile is 48 m and its initial speed is 33 m/s. *(a)* What is its angle of projection? *(b)* Is there another angle of projection that is consistent with these values of R and v_0? If so, find this angle. If not, explain.

18. A ball is thrown from a balcony with an initial speed of 31 m/s and at a projection angle of 24°. The point of release is 8.2 m above flat terrain. *(a)* What is the horizontal distance from the release point to where the ball hits the ground? **(b)** What is the straight-line distance from the release point to where the ball hits the ground?

19. An air gun fires a pellet with an initial speed of 52 m/s. The gun is fired with its barrel directed 75° above the horizontal, with the end of the barrel 1.9 m above the ground. The terrain is level. *(a)* How long after the gun is fired will the pellet reach its maximum height? *(b)* What is the pellet's maximum height above the ground? **(c)** What is the horizontal distance traveled by the pellet from the gun barrel to the point where it hits the ground?

20. Make a graph of the trajectory of a projectile. Let $x_0 = 0.0$ m, $y_0 = 0.0$ m, $v_0 = 40.0$ m/s, and $\theta_0 = 50.0°$. Plot points at each 0.5 s and label each point with its value of t. Sketch a curve through the points. Determine t_m and show $\mathbf{v}$ on the graph at $t = 0.5t_m$, $1.0t_m$, $1.5t_m$, and $2.0t_m$.

21. Figure 4-26 shows the paths of a rock and a bottle. The bottle was released from rest at the instant the rock was launched. The value of g has been taken to be 10.0 m/s² in order to deal with round numbers. Note that the initial velocity of the rock was directed straight toward the bottle. From data taken from the graph, determine v_0 and θ_0 for the rock.

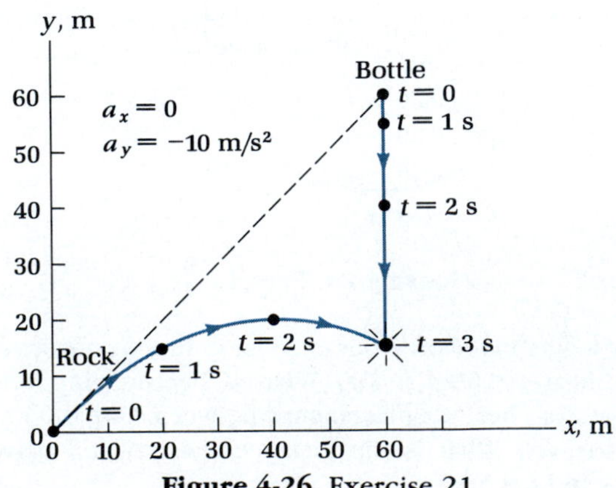

Figure 4-26. Exercise 21.

Section 4-3. Uniform Circular Motion

22. A car makes a circular turn of radius $R = 63$ m while its speed is constant at $v = 12$ m/s. What is the magnitude of its centripetal acceleration?

23. A penny rides on a phonograph turntable at a distance of 130 mm from the spindle. What is the magnitude of the penny's centripetal acceleration when the turntable rotates at *(a)* 33.3 rev/min (revolutions per minute) and *(b)* 45.0 rev/min?

24. The jogger in Exercise 1 (Fig. 4-24) runs at constant speed and completes one revolution in 134.4 s. What is the magnitude and direction of her centripetal acceleration when she is *(a)* north of the origin and *(b)* northwest of the origin? *(c)* Write her acceleration in terms of unit vectors in each case.

25. Determine the magnitude and direction of the velocity of the car in Exercise 5 (Fig. 4-25) at *(a)* *D* and *(b)* *H*. *(c)* Write its acceleration in terms of the unit vectors in each case.

26. An object executes uniform circular motion with the origin of an *xy* coordinate frame at the center of the circle. Make a graph of the circular path and show the velocity and acceleration vectors with their tails at the object's position when the object's coordinates (x, y) are *(a)* $(R, 0)$; *(b)* $(0, R)$; *(c)* $(-R/\sqrt{2}, R/\sqrt{2})$. The object is traveling counterclockwise. *(d)* Repeat with the object traveling clockwise.

27. A boy swings a rock tied to a string around his head in a horizontal circle. The radius of the circle is 0.96 m, and the time for one revolution is 1.1 s. What is the rock's *(a)* speed and *(b)* acceleration magnitude?

28. The orbit of the moon around the earth is nearly circular, with a radius of 3.85×10^8 m and a period of 27.3 days. What is the magnitude of the moon's centripetal acceleration in its motion about the earth?

29. *(a)* The radius of the earth is 6.37×10^6 m. Determine the centripetal acceleration of a point on the earth's surface at the equator relative to the center of the earth in meters per second squared and in *g*'s. *(b)* The radius of the earth's orbit about the sun is 1.5×10^{11} m. Determine the centripetal acceleration of the earth relative to the sun in meters per second squared and in *g*'s. *(c)* Astronomical measurements indicate that our solar system is in nearly a circular orbit about the center of the Milky Way galaxy at a radius of 2.8×10^{20} m and a speed of 2.5×10^5 m/s. Determine the centripetal acceleration of the solar system relative to the center of the galaxy in meters per second squared and in *g*'s. *(d)* Determine the ratios of each pair of these accelerations.

30. In the Fermilab accelerator at Batavia, Illinois, protons travel at nearly the speed of light (3×10^8 m/s) along a circular path of radius 1 km. Find the centripetal acceleration of one of these protons in *(a)* meters per second squared and *(b)* in *g*'s.

31. *(a)* Show that for an object undergoing uniform circular motion, $a_c = 2\pi v/T$. *(b)* A go-cart travels around a circular track once every 5.1 s at a constant speed of 11 m/s. What is the cart's centripetal acceleration?

32. A carnival Ferris wheel has a radius of 7.5 m and makes one revolution every 5.7 s. What are the magnitude and direction of a passenger's acceleration when the passenger is *(a)* at the top and *(b)* at the bottom?

Section 4-4. Relative Motion

33. Becky is driving south in the right lane of a highway at a speed of 22 m/s. Suzie is in the left lane and traveling in the same direction as Becky at a speed of 28 m/s. What is Suzie's *(a)* speed and *(b)* velocity relative to Becky? What is Becky's *(c)* speed and *(d)* velocity relative to Suzie?

34. Observer *A* measures the velocity and acceleration of particle *P* to be $(3 \text{ m/s})\mathbf{j}$ and $(4 \text{ m/s}^2)\mathbf{i}$, respectively. Observer *B* moves with a constant velocity relative to *A*: $\mathbf{v}_{BA} = (2 \text{ m/s})\mathbf{i} + (-1 \text{ m/s})\mathbf{j}$. According to *B*, what is the *(a)* velocity and *(b)* acceleration of *P*?

35. A train is traveling east at a speed of 3.4 m/s. Using a compass for direction, a man on a flatcar of the train walks northeast at a speed of 1.2 m/s relative to the flatcar. What is the man's velocity relative to the ground?

36. An airplane's compass shows that it is headed west, and its airspeed indicator shows that its speed through the air is 175 m/s. Relative to the ground, the velocity of the wind is 42 m/s toward the north. Establish a coordinate frame and determine the velocity of the airplane relative to the ground.

37. The cruising speed of a ferry boat relative to the water is 7.8 m/s. The boat crosses a river to a destination on the other side that is directly north of the departure point. The distance from departure to destination is 1.8 km. The current in the river is 2.3 m/s toward the east. Let $+\mathbf{i}$ point east and $+\mathbf{j}$ point north. *(a)* Determine the velocity of the boat relative to the water such that the boat travels in a straight line to its destination. *(b)* How long does it take the boat to make the crossing? *(c)* Repeat with a current velocity of 4.6 m/s toward the east.

Section 4-5. Air Resistance, Numerical Methods

38. Construct a spreadsheet for the baseball in Example 4-5, where air resistance is neglected. Compare your graph with the no-air-resistance curve shown in Fig. 4-20.

39. Repeat the spreadsheet for the baseball in Sec. 4-5 with air resistance included, except reduce the number of iterations to 100 while keeping the time interval for the entire program at 5 s. Compare your resulting graph with the air-resistance curve of Fig. 4-20. Can you tell any difference?

Additional Exercises

40. Using her radar equipment, a highway patrolwoman *P* measures the speed of a car *C*. The car is traveling east along a straight east-west road. The patrolwoman is traveling west at a speed of 31 m/s and her radar measures the speed of the car relative to her (the patrolwoman) to be 65 m/s. Let the unit vector $\mathbf{i}$ be directed east and use the symbol *R* to denote the road. Determine *(a)* $\mathbf{v}_{PR}$, *(b)* $\mathbf{v}_{CP}$, and *(c)* $\mathbf{v}_{CR}$.

41. The commentary on Galileo includes ten photographs of Venus, all at the same magnification. Use the radius of the earth's orbit around the sun (1.5×10^{11} m) and measurements

from these photographs to make an estimate of the radius of Venus's orbit.

42. An airplane is making a right turn in horizontal flight at a steady speed of 105 m/s. The plane's path is along the arc of a circle of radius $R = 2.7$ km. At a particular instant the plane's velocity is toward the north. Using unit vectors **i** toward the east and **j** toward the north, determine the airplane's *(a)* velocity and *(b)* acceleration at this instant.

43. Suppose an athlete performing a long jump has a speed of 9.3 m/s at the instant she leaves the ground. If her jump distance is 4.7 m, then what is her *(a)* angle of projection and *(b)* time in the air? Treat the athlete as a particle, neglect air resistance, and assume $\theta_0 < 45°$.

44. The expression for the position vector of a particle is $\mathbf{r} = [(-2.3 \text{ m/s})t + (1.9 \text{ m/s}^2)t^2]\mathbf{i} + [(1.2 \text{ m/s}^3)t^3]\mathbf{j}$. Determine expressions for *(a)* the velocity **v** and *(b)* the acceleration **a**. *(c)* At the instant $t = 2.6$ s, evaluate **r**, **v**, and **a**.

45. In some hotels, the elevators are in transparent enclosures so that people inside the elevator can see what is going on outside (and vice versa). While riding upward in such an elevator at a constant speed of 1.3 m/s, Millie (M) tosses a golf ball (G) vertically upward. Bill (B) is standing outside the elevator and observes the action. Let **j** point upward and neglect air resistance. *(a)* At the instant the ball reaches its maximum height in Millie's frame of reference, determine $\mathbf{v}_{GM}$, $\mathbf{v}_{GB}$, $\mathbf{a}_{GM}$, and $\mathbf{a}_{GB}$. *(b)* At the instant the ball reaches its maximum height in Bill's frame of reference, determine $\mathbf{v}_{GM}$, $\mathbf{v}_{GB}$, $\mathbf{a}_{GM}$, and $\mathbf{a}_{GB}$. *(c)* What is the time interval between the instant the ball reaches its maximum height in Millie's frame and when it reaches its maximum height in Bill's frame?

46. *(a)* Determine the magnitude of the centripetal acceleration, relative to the sun, of the planets Mercury, Jupiter, and Pluto. Assume that each orbit is circular. You will find the period T and the orbital radius R for each planet in App. A. *(b)* Multiply each acceleration found in part *(a)* by R^2 for each of these planets and compare the values you obtain for this product a_cR^2. If you performed the calculations correctly, each answer is nearly the same, namely, 1.33×10^{20} m³/s². Indeed, the value of the product a_cR^2 for each of the other planets yields very nearly this same value. We shall investigate this curious result in Chap. 7.

47. In shooting a foul shot, a basketball player releases the ball at a projection angle of 60°. The center of the basket is a horizontal distance of 4.3 m from the release point, and the height of the basket above the release point is 1.5 m. Determine the initial speed required in order to make the shot. Neglect air resistance.

48. A football player wishes to punt the ball such that the ball's "hang time" (the time it is in the air) is 4.5 s. The player's kick is expected to give the ball an initial speed of 27 m/s. Determine the projection angle that will yield this hang time. Neglect air resistance and assume that the ball is caught at the same height as the point at which it was launched. (Neglecting air resistance in this problem leads to a sizable error.)

49. A projectile is launched with an initial speed of 43 m/s and a projection angle of 41°. What is the projectile's height above the release point after it has traveled a horizontal distance of 20 m? Neglect air resistance.

50. A person standing on the earth at a latitude of, say 40° is traveling in a circle centered at a point on the earth's rotational axis. Determine the acceleration magnitude of this person relative to this center of motion.

51. Neutron stars are very dense, have small radii, and rotate rapidly on their axes. Consider a neutron star with a rotational period of 1 s and a radius of 20 km. Determine the magnitude of the centripetal acceleration of a particle on the star's equator.

52. During a football game, a ball carrier is running down the sideline at a constant speed of 9.0 yd/s (Fig. 4-27). At the instant the carrier passes the 50-yd line, a defensive back begins moving toward the sideline by running parallel to the yard markers. At this instant, the defensive back is standing on his 20-yd line at a distance of 20 yd from the sideline. *(a)* Determine the constant acceleration magnitude for the defensive back such that he will collide with the carrier. *(b)* What is the speed of the defensive back at the instant of the collision?

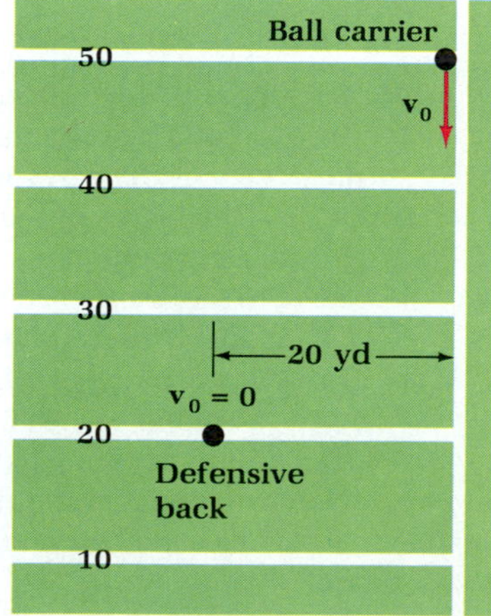

Figure 4-27. Exercise 52.

PROBLEMS

(*Note:* Unless stated otherwise, neglect air resistance in problems dealing with projectile motion.)

1. ***An analytical approach to uniform circular motion.*** The coordinates of an object are $x = R\cos\omega t$ and $y = R\sin\omega t$, where R and ω are constants. *(a)* Show that the velocity components are $v_x = -R\omega\sin\omega t$ and $v_y = R\omega\cos\omega t$. *(b)* Show that the acceleration components are $a_x = -R\omega^2\cos\omega t$ and $a_y = -R\omega^2\sin\omega t$. *(c)* Show that $\mathbf{a} = -\omega^2\mathbf{r}$. *(d)* Make a graph of the path of the object on an xy coordinate frame for the case where $R = 45.0$ m and $\omega = 46.75$ mrad/s. Plot points at each

8.4 s between $t = 0.0$ s and $t = 134.4$ s, and sketch the curve. *(e)* Evaluate v and a at $t = 16.8$ s and draw the vectors on your graph with their tails at the position of the object at this instant.

2. ***Connection between the speed and height of a projectile.*** For an object moving with constant acceleration, show that *(a)* $v_x^2 = v_{x0}^2 + 2a_x(x - x_0)$ and *(b)* $v_y^2 = v_{y0}^2 + 2a_y(y - y_0)$ by eliminating t in Eqs. 4-10 and 4-12. *(c)* Combine the results of parts *(a)* and *(b)* to show that $v^2 = v_0^2 + 2[a_x(x - x_0) + a_y(y - y_0)]$. *(d)* For projectile motion, show that $v^2 = v_0^2 - 2g(y - y_0)$. *(e)* The projectile is launched at coordinates (x, y) of (2.4 m, 1.5 m) with an initial speed of 28 m/s. Use the answer from part *(d)* to find the projectile's speed at the position (37.6 m, 25.4 m).

3. ***Finding θ_0 for a projectile from measurements of h_m and R.*** *(a)* Show that the projection angle θ_0 for a projectile launched from the origin is given by $\theta_0 = \tan^{-1}(4h_m/R)$. *(b)* What is the projection angle for a projectile launched from the origin when the maximum height is 6.2 m and the horizontal range is 32 m? *(c)* What is the projection angle for a projectile whose maximum height is equal to its horizontal range? *(d)* What is the maximum height for a projectile whose range is R_m?

4. ***Symmetry of the horizontal range relative to $\theta_0 = 45°$.*** Show that the horizontal range of a projectile whose projection angle is $\Delta\theta$ less than 45° is the same as that for a projectile whose projection angle is $\Delta\theta$ greater than 45°, v_0 held fixed. That is, show that R is symmetric about $\theta_0 = 45°$. (*Hint:* Consider derivatives of R with respect to θ_0.)

5. ***A ski jump.*** A skier leaves the ski-jump ramp with a velocity of 34 m/s along the horizontal (Fig. 4-28). The ground is a vertical distance of 4.2 m below the launch point and the hill below makes an angle of 25° with respect to the horizontal. Neglecting air resistance, determine the distance from the launch point to the point where the skier touches down. (*Note:* A skilled jumper holds her body such that she obtains "lift" from the air and thus lengthens her flight.)

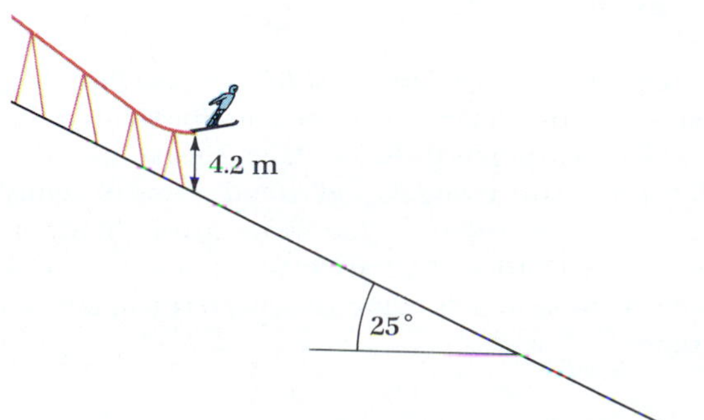

Figure 4-28. Problem 5.

6. ***When to pull out of a dive.*** An airplane is in a vertical dive at a speed of 174 m/s over level terrain. Assume that the plane pulls out of the dive in a circular path and that the maximum acceleration the airplane can tolerate is 8.0 g's. Determine the minimum altitude at which the plane must begin to pull out in order to avoid hitting the ground. Assume constant speed throughout the motion.

7. ***Angle θ for a projectile as a function of time.*** Show that for a projectile the angle between the velocity and the x axis as a function of time is

$$\theta(t) = \tan^{-1}\frac{v_{y0} - gt}{v_{x0}}$$

8. ***A geosynchronous orbit.*** In Sec. 4-3 we showed that the magnitude of the centripetal acceleration can be written $a_c = 4\pi^2R/T^2$, where T is the period (the time to complete one revolution). The centripetal acceleration of an earth satellite is due to the earth's gravitational attraction. An expression for this acceleration is $a_c = gR_e^2/R^2$, where R is the radius of the orbit and R_e is the earth's radius ($R_e = 6.37 \times 10^6$ m). *(a)* Show that the relation between a satellite's orbital radius and its period is

$$R^3 = \frac{gR_e^2}{4\pi^2}T^2 = (1.0 \times 10^{13}\ \text{m}^3/\text{s}^2)T^2$$

(b) A geosynchronous orbit is an orbit over the equator with a 24-h period. In such an orbit a satellite remains directly above a point on the earth's surface. Show that the height of a geosynchronous orbit above a point on the earth's surface at the equator is $h = 5.6\ R_e$.

9. ***Another way to show that $a_c = v^2/R$.*** *(a)* Use the velocity-vector triangle in Fig. 4-14*a* to show that the magnitude of the average acceleration for an object in uniform circular motion can be written

$$|\bar{\mathbf{a}}| = \frac{v^2}{R}\frac{2[\sin(\Delta\theta/2)]}{\Delta\theta}$$

where $\Delta\theta$ is the angle between $\mathbf{v}_i$ and $\mathbf{v}_f$ expressed in radians. Use the expression in part *(a)* to determine $|\bar{\mathbf{a}}|$ when $\Delta\theta$ equals *(b)* $\pi/2$; *(c)* $\pi/4$; *(d)* $\pi/10$; *(e)* $\pi/1000$. *(f)* From the above, deduce the value of

$$\lim_{\Delta\theta \to 0}\frac{\sin(\Delta\theta/2)}{\Delta\theta}$$

(g) Use the results from parts *(a)* and *(f)* to show that $a_c = v^2/R$.

10. ***Fermat's principle.*** A runner is to travel from A to B in Fig. 4-29. Because the footing is different, his speed is different on each side of the x axis. Let v_1 represent his speed above the x axis and v_2 represent his speed below the x axis. Also, let the two straight-line portions of the path be characterized by the angles θ_1 and θ_2, as shown in the figure. Show that his running time will be minimized when the relation between the angles is given by $v_2 \sin\theta_1 = v_1 \sin\theta_2$. (*Hint:* Write an expression for the time required to run each straight-line portion, add these times, and take the derivative of the sum with respect to x and set it equal to zero. Following along this path minimizes the time because the runner travels a longer distance in the region where his speed is greater and a shorter distance in the region where his speed is slower.) *Comment:* This procedure can be used to derive Snell's law of refraction in optics (Chap. 35), in which case it is an application of a rule called *Fermat's principle.*

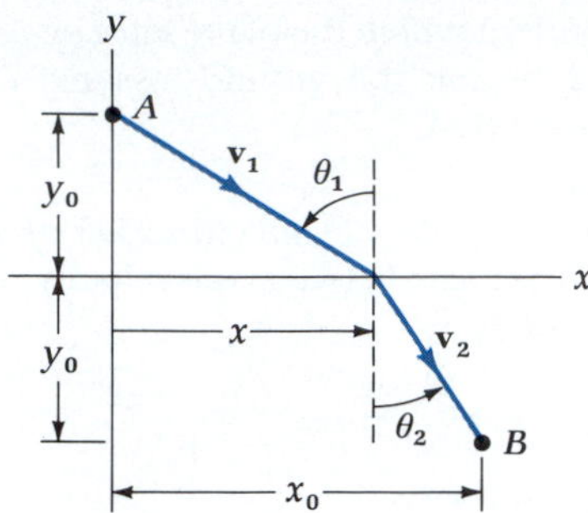

Figure 4-29. Problem 10.

11. ***Catching a baseball on the run.*** A batter hits a baseball such that $v_0 = 33$ m/s and $\theta_0 = 32°$ at a point 1 m above the plate. The horizontal direction of the ball's path is directly toward an outfielder who is 118 m from the plate. At 0.50 s after the ball is struck, the outfielder begins running toward the plate with constant acceleration. What must be the magnitude of his acceleration if he is to catch the ball when it is 1 m above the ground?

12. ***Distinction between*** $|\Delta \mathbf{r}|$ ***and*** Δr. *(a)* Suppose we define $|\Delta \mathbf{r}| = |\mathbf{r}_f - \mathbf{r}_i|$ and $\Delta r = r_f - r_i$. Show that these quantities are different by writing them in terms of coordinates. Make a graph of an object executing uniform circular motion. Consider the object at two points on its path and show $|\Delta \mathbf{r}|$ and Δr. *(b)* Suppose we define $|\Delta \mathbf{v}| = |\mathbf{v}_f - \mathbf{v}_i|$ and $\Delta v = v_f - v_i$. Show that these quantities are different by writing them in terms of velocity components. Make a graph of an object executing uniform circular motion. Consider the object at two points on its path and show $|\Delta \mathbf{v}|$ and Δv.

13. ***Michelson-Morley experiment.*** A girl anchors her boat in the middle of a river (Fig. 4-30). The current in the river is 0.85 m/s toward the east. *(a)* Determine the time required for the girl to swim to a point 50 m east of the boat and return. The girl's swimming speed relative to the water is 1.43 m/s. *(b)* Determine the time required for the girl to swim to a point 50 m north of the boat and return. *(c)* Which trip took the longer time and by how much? *Comment:* This problem is closely associated with a famous experiment called the Michelson-Morley experiment (Chap. 38).

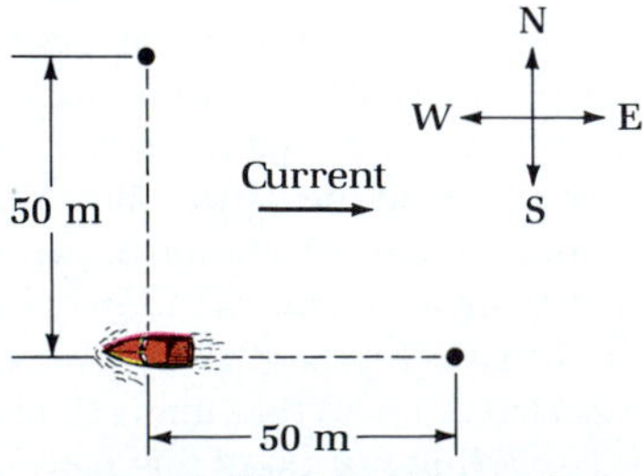

Figure 4-30. Problem 13.

14. ***A projectile as seen by different observers.*** A boy is standing on a flatcar of a train that is moving at a speed of 8.2 m/s relative to the ground. The boy throws a rock straight up (according to him) at a speed of 12.5 m/s. *(a)* What is the initial velocity of the rock according to a woman who is standing on the ground? The boy catches the rock when it returns to the same elevation from which it was released. What is the rock's horizontal range according to *(b)* the boy and *(c)* the woman? What is the time of flight according to *(d)* the boy and *(e)* the woman? Neglect air resistance.

15. ***Terminal speed.*** When an object falls through the air (or other fluid), it asymptotically approaches a terminal speed v_t because of frictional effects. Suppose an object that is falling vertically has an acceleration component given by

$$a_y = -g + bv^2$$

where $b = 0.002\ \text{m}^{-1}$. Determine v_t. (*Hint:* As the object's speed approaches the terminal speed, its acceleration approaches zero.)

16. ***Artillery range-finding.*** *(a)* By firing one shot short of the target and another shot long, an artillery gunner can zero in so that the third shot is on the mark. Suppose that an artillery shot lands a distance ΔR_1 short of the target when the cannon barrel makes an angle of θ_{01} with the horizontal and another round lands ΔR_2 past the target when the angle is θ_{02}. Show that the angle θ_0 which puts the next shot on target is

$$\theta_0 = \frac{1}{2} \sin^{-1} \frac{\Delta R_1 \sin 2\theta_{02} + \Delta R_2 \sin 2\theta_{01}}{\Delta R_2 + \Delta R_1}$$

(b) With the barrel at an angle of 15.20°, an artillery shot lands 130 m short of its target, and with the angle at 15.85°, a second shot lands 160 m long. What is the angle that will cause the third shot to land on the target?

17. ***How to measure the terminal speed of raindrops.*** By the time raindrops are near the surface of the earth, the combined effects of gravity and air resistance have caused them to reach a steady speed called the *terminal speed.* Suppose a steady rain is falling vertically relative to the earth (there is no wind). You get in your car and while moving at a speed of 10 m/s you measure the angle of the rain with the vertical to be 50°. *(a)* What is the terminal speed of the raindrops relative to the earth? *(b)* What is the terminal speed of the raindrops relative to the moving car?

18. ***Acceleration of Venus relative to earth.*** The planets travel around the sun in nearly circular orbits and all the planetary orbits are in nearly the same plane. The radius of the earth's orbit is 1.5×10^{11} m and the radius and period of Venus's orbit are 1.1×10^{11} m and 0.61 year, respectively. What is the acceleration magnitude of Venus relative to the earth when they are *(a)* on the same side of the sun and *(b)* on opposite sides of the sun?

19. ***A cycloid.*** If a wheel rolls at a constant speed without sliding along a level surface, the coordinates of a particle on the edge of the wheel as a function of time t are

$$x = vt - R \sin \frac{vt}{R} \qquad y = R - R \cos \frac{vt}{R}$$

where R is the wheel's radius and v is the speed of the center of the wheel. The coordinate frame is fixed to the surface, with $+x$ along the surface in the direction of motion and $+y$ vertically upward. *(a)* Evaluate x and y at times $t = 0$, $T/4$, $T/2$, . . . , $3T/2$, where $T = 2\pi R/v$, and make a graph of the particle's trajectory. This path is called a *cycloid* and will be discussed again in Sec. 12-8. *(b)* Determine v_x and v_y as functions of time and show $\mathbf{v}$ on your graph at the times $t = 0$, $T/2$, T, and $3T/2$. *(c)* Determine a_x and a_y as functions of time and show $\mathbf{a}$ on your graph at the times $t = 0$, $T/2$, T, and $3T/2$. *(d)* Using a_x and a_y from part *(c)* above, show that $a = v^2/R$.

20. ***More about air resistance.*** Use a spreadsheet to include air resistance in the flight of the baseball in Example 4-5. Use the expression

$$\mathbf{a} = -g\mathbf{j} - cv^{0.7}\mathbf{v}$$

for the baseball's acceleration, where $c = 0.017\ \text{m}^{-0.7}s^{-0.3}$. In other words, assume that the contribution to the acceleration due to air resistance varies with the ball's speed v as $v^{1.7}$. Compare your graph to the air-resistance curve in Fig. 4-20. Can you tell any difference?

21. ***The bouncing plastic ball.*** Examine the strobe photograph of the bouncing plastic ball on the first page of this chapter. *(a)* By using the fact that the diameter of the ball is 0.0254 m and by making measurements from the photograph, determine the height of the bounce shown in the photograph. *(b)* From your answer in part *(a)*, determine the vertical component of the ball's velocity at the instant this bounce began. *(c)* From your answer in part *(b)*, determine the time interval between strobe flashes. *(d)* From your answer in part *(c)* determine the horizontal component of the ball's velocity. *(e)* Determine the range R during the bounce by inserting your answers from parts *(b)* and *(d)* into the expression $R = 2v_{x0}v_{y0}/g$. *(f)* Measure R directly from the photograph and calculate the percent difference between this value of R and the one determined in part *(e)*. Are your measurements precise to one significant digit? To two? To three?

22. ***Golf balls falling simultaneously.*** Examine the strobe photograph of the two golf balls in Fig. 4-10. *(a)* By using the fact that the diameter of a golf ball is 0.0413 m and by making measurements from the photograph, determine the vertical distance that each ball has fallen at the instant it is at its lowest position in the photograph. *(b)* From your answer to part *(a)*, determine the time interval between flashes. *(c)* From your answer to part *(b)*, determine the horizontal velocity component of the ball on the right. *(d)* Using the lowest position in the photograph of the ball on the left as your origin, write expressions for $x(t)$ and $y(t)$ for each ball. *(e)* Write an expression for $y(x)$ for the ball on the right.

23. ***Chipping a golf ball.*** Suppose you wish to chip a golf ball from the fairway to an elevated green (Fig. 4-31). The point where you want the ball to land is a horizontal distance d and a vertical distance h from its launch point. The club you selected launches the ball at a projection angle θ_0. *(a)* Show that, neglecting air resistance, the initial speed v_0 that will place the ball on target is given by

$$v_0{}^2 = \frac{d^2 g}{2\cos^2\theta_0(d\tan\theta_0 - h)}$$

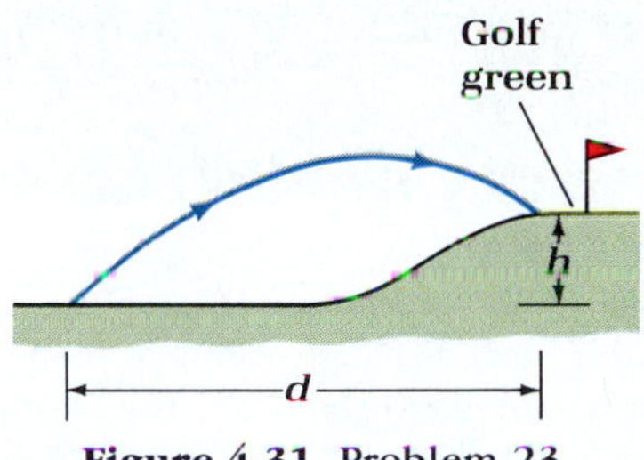

Figure 4-31. Problem 23.

(b) Determine the value of v_0 for the case where $d = 35$ m, $h = 4.0$ m, and $\theta_0 = 41°$.

24. ***Proper heading for an airplane.*** The heading of an airplane is the direction its nose is pointed. Suppose an airplane's destination is northeast of its departure point. That is, the plane's path over the ground is along a line that is 45° north of east. The airplane's speed relative to the air is 135 m/s and the wind is blowing out of the south with a speed of 48 m/s. What is the airplane's heading during the flight?

25. ***Acceleration of the moon relative to the sun.*** To a good approximation, the moon can be considered to be traveling in a circle centered at the earth and the earth can be considered to be traveling in a circle centered at the sun. Determine the magnitude of the moon's acceleration relative to the sun *(a)* when the moon is between the earth and the sun and *(b)* when the earth is between the moon and the sun.

26. ***Mission improbable.*** Your mission, should you choose to accept it, is to guide Secret Agent 007 to her destination in a minimum amount of time. The secret agent is to travel under darkness from point A to B in Fig. 4-32 by swimming the canal and then by creeping along the bank of the canal. You are to place an ultrasonic transmitter at the canal's edge; the secret agent will swim directly toward the transmitter and then creep from the transmitter to B. Determine the placement distance x for the transmitter that will minimize the agent's travel time. Your only information is that the agent's swimming speed is half her creeping speed and that $w = 15$ m and $d = 25$ m. (*Hint:* Develop an expression for the travel time in terms of x and the fixed parameters involved in the problem, then find the value of x that minimizes this time by taking the derivative of the time with respect to x.)

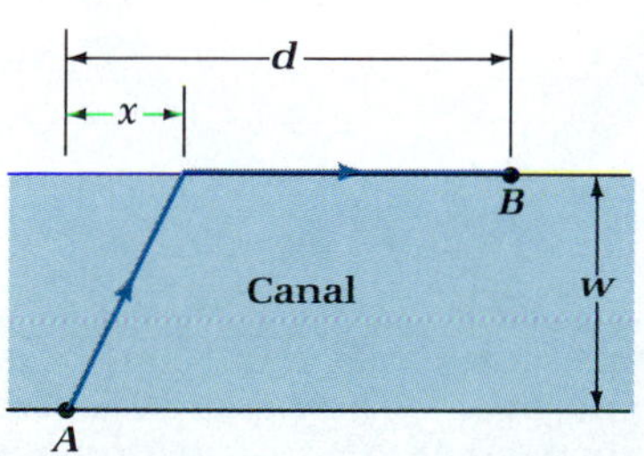

Figure 4-32. Problem 26.

HIGHLIGHTS OF MODERN PHYSICS

The Paradox of the Dark Night Sky

Why is the sky dark at night?

This simple question was first posed by Johannes Kepler in 1610. The copernican universe was a new idea then. Another new idea was that the universe was infinite and was filled with an infinite number of stars. While Kepler embraced the copernican universe, he wrote about the "secret, hidden horror" of an infinite one. He thought that an infinite universe would produce an absurd result, a sky that was everywhere as bright as the sun.

If you stand in the center of a large forest, it would seem that the trees go on forever, since you see tree trunks in every direction. Similarly, if stars go on forever, any line of sight should end on the surface of a star: thus the entire sky should be as bright as the surface of stars. Kepler argued that, since the sky is instead dark, the universe must therefore be finite: stars do *not* go on forever. The forest is replaced by a small grove of trees.

We now sharpen this argument. Consider a large spherical shell of radius r and thickness Δr centered on the earth (Fig. 1).

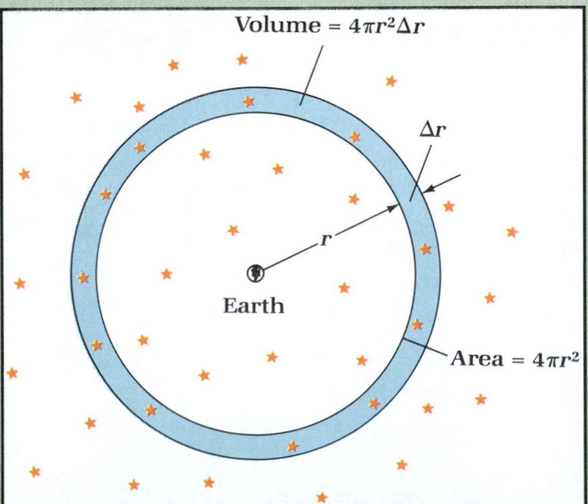

Figure 1. A large spherical shell of radius r and thickness Δr is centered on the Earth. The number of uniformly distributed stars within the shell is proportional to the volume $4\pi r^2\ \Delta r$. In accord with the inverse-square law, all shells of the same thickness will contribute an equal glow to the sky. According to the dark sky paradox, the total light from all shells produces a bright sky, contrary to observation.

The number of stars N inside this shell will be proportional to its volume $4\pi r^2\ \Delta r$. These stars will contribute a weak glow to the sky. A spherical shell of twice the radius but the *same thickness* will have $2^2N = 4N$ stars since the volume of the shell varies as r^2. However, since light intensity decreases as $1/r^2$ (the inverse-square law), each star will contribute exactly $1/2^2 = 1/4$ as much as each star in the smaller shell, so the sky glow is the same as for the smaller shell. More generally, the volume of a shell and the number of stars in it increase as r^2 while the inverse-square law reduces the light from each star by $1/r^2$. The r-dependence cancels, and all spherical shells contribute equally. Summing over an infinity of shells gives infinite brightness.

Of course, the sky would not really be infinitely bright because nearby stars would block out the light from more distant ones (Fig. 2), just as in the forest nearby trees keep us from seeing more distant trees. Instead, the sky would appear to be uniformly filled with the equivalent of 100,000 suns (Exercise 1). It would be as though the earth were in a 6000-K oven. The earth's atmosphere would dissipate in a few seconds, the oceans would evaporate in a few days, and the earth itself would vaporize within a few years.

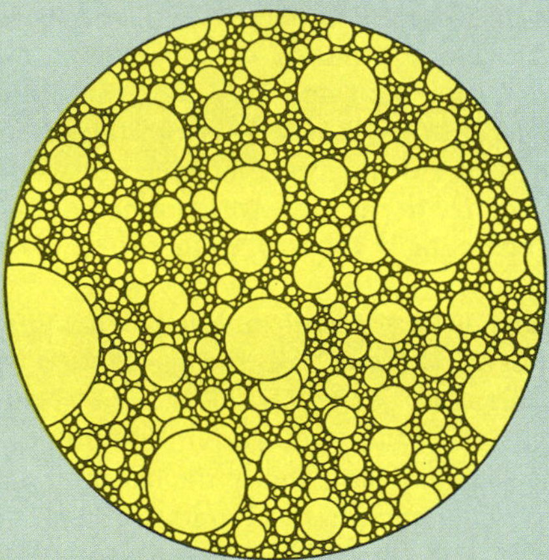

Figure 2. In the bright sky, the entire sky would be covered with stars. The farthest visible stars would be about 10^{23} light-years distant. More distant stars would be blocked from view by the closer ones. In all, about 10^{60} stars would be visible.

Clearly, something is wrong with this picture because the sky is dark at night; it is not filled with 100,000 suns. What is wrong with the argument? Modern astronomy shows that if the universe is finite, it cannot be finite in the sense of having a boundary (a quaint idea that raises the question of what is beyond the boundary). A finite universe must be finite and boundless in the sense that a spherical surface has a finite area but no boundary. If a finite number of trees were planted on the (spherical) earth, then any line of sight would inevitably come to rest on a tree trunk (assuming that light travels only parallel to the ground). The line of sight might have to circle the earth several times before encountering a tree trunk, but eventually it would do so. Similarly, given a finite sphere of curved space, your line of sight might have to travel around the universe several times, but in the end it would encounter the surface of a star. A finite universe with no boundary will not save us from a bright sky. Still, we should admire Kepler's ability to take a simple observation and from it draw a cosmic conclusion.

Since Kepler, many solutions that do not assume a finite universe have been proposed for the paradox of the dark night sky. One was that dark interstellar clouds block the light from distant stars. If we use the forest analogy again, the interstellar clouds would be like a fog that lets you see only nearby trees. This solution fails, too, because the distant stars, which surround the dust clouds, would increase the temperature of the clouds to 6000 K, and they would glow as brightly as the stars themselves.

How far do we have to look until the sky is covered with stars? Assuming that all stars have the same radius R, a line of sight of length d would intercept any star whose center is within R of the line (Fig. 3). The number of stars intercepted equals (volume of cylinder) × (stars/unit volume), or $\pi R^2 d \times n$. Using the sun's radius and a density of about one star per 10^{57} m³ (Exercise 2), the average distance to intercept *one star* is $d/\pi R^2 dn = (\pi R^2 n)^{-1} = 10^{23}$ light-years.

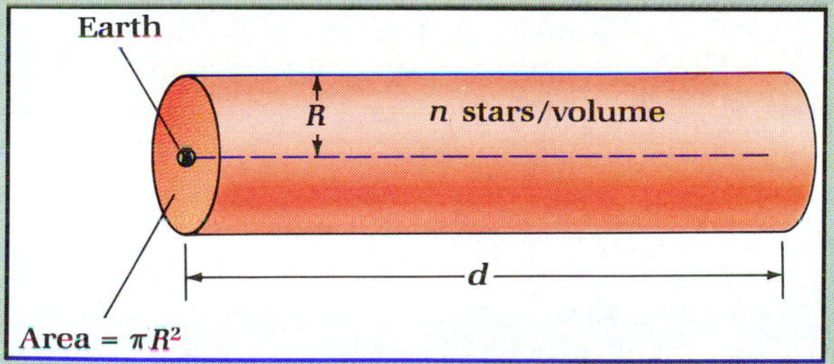

Figure 3. The dashed line represents a line of sight extending out from the earth in an arbitrary direction. Assuming all stars have the same radius R equal to the average stellar value (approximately the radius of the sun), any star whose *center* is within the cylinder intercepts the line of sight. A cylinder of length d will contain $(\pi R^2 d)n$ star centers, where n is the number of stars per unit of volume. Therefore, on average, the line of sight extends out a distance of $1/\pi R^2 n$ before terminating on the surface of a star.

If we look out 10^{23} light-years, the sky should be filled with stars and be everywhere as bright as the surface of the sun. Why isn't it? Of course, the light from these most distant stars must travel for 10^{23} years to reach the earth. If we could see these stars, we would see them as they were 10^{23} years ago, not as they are now. Yet the "big bang" occurred no more than 20 billion years ago. Even if the universe is infinite, we cannot receive light from stars farther than about 10^{10} light-years. The light from stars this close is $10^{10}/10^{23} = 10^{-13}$ times that of a bright sky. Therefore one answer to why the sky is dark at night is very simple: The universe is not old enough to produce a bright sky.

But what happens as the universe ages? Will the sky become progressively brighter as we see more distant stars? No, because a typical stellar lifetime is about 10^{10} years. As the universe ages, the sun and other nearby stars will die out. The outer limit of the most distant visible stars will continue to recede at the speed of light, and an inner, concentric sphere containing burned-out stars will begin to grow outward, also at the speed of light (Fig. 4). Only stars inside this spherical shell, which is

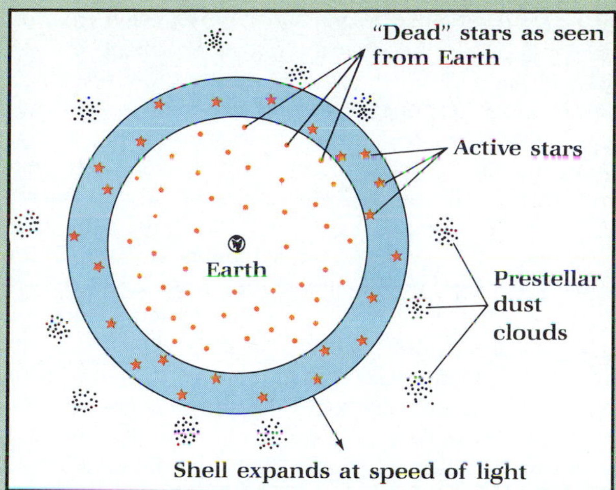

Figure 4. Eventually the sun and other nearby stars will die out. More distant stars will appear to shine because of the time it takes their light to reach earth. Still more distant stars will not have begun to shine because we are looking further into the past. Visible, shining stars will form a spherical shell that moves outward at the speed of light.

expanding at the speed of light, will contribute to the brightness of the sky. From our earlier analysis we know that a shell of a fixed thickness will produce the same amount of light at the earth no matter what its size. The sky will forever remain essentially as bright as it is now. Since it takes 10^5 suns to fill the sky, the collective light from all stars whose light has had time to reach us is equivalent to $10^{-13} \times 10^5 = 10^{-8}$ sun, or about 1 percent of the full moon, a feeble glow.

We get additional insight into the situation by considering this question: How long would it take to fill the universe with enough light to produce the bright sky? Imagine a typical star inside a cell with perfectly reflecting walls (Fig. 5) such that the cell volume equals the average volume of space surrounding a star, 10^{57} m³. The density of radiation inside the cell is the same as that between stars in space. Instead of letting the star's

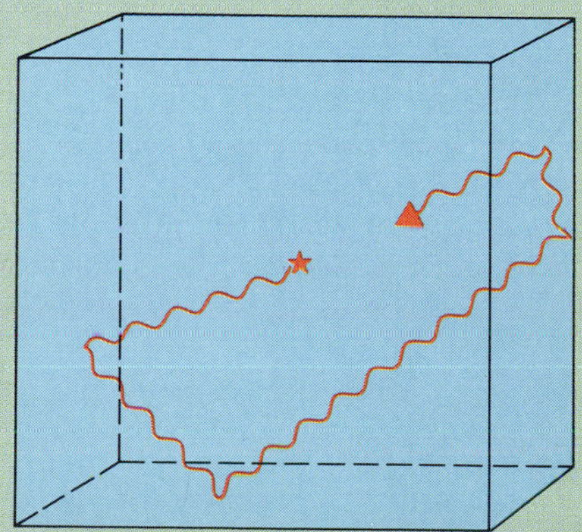

Figure 5. A single star is contained within a cell of volume 10^{57} m³, the average volume per star in the universe. The cell walls are perfectly reflective. It will take an average of 10^{23} years for an emitted photon to be reabsorbed by the star. This is the time required to establish thermodynamic equilibrium within the cell, or the time to fill the cell with radiation to a level corresponding to the 6000-K temperature of the star's surface.

light travel throughout space and mix with that from other stars, we get the same effect by reflecting it back within the walls and keeping other starlight out. So now the question becomes: How long does it take the cell to fill up with light?

At first, the light will stream from the star into the cube, which will begin to be filled with light. After several years, the light will begin to reflect off of the walls of the cell, and much later (because the star is much smaller than the cell) some of the light will begin to strike the star and be reabsorbed. Now, the amount of light in the cube will increase less rapidly than during those first few years when it was not being absorbed. As more light fills the cube, the rate of absorption will increase until finally the rates of absorption and emission become equal. Radiation intensity inside the cube is the same as if the cube walls were at a temperature of 6000 K so the cube will be filled with light as densely as for the bright-sky situation. How long will this take?

This "fill-up time" will approximately equal the average time for an emitted photon to be reabsorbed and is determined exactly as in Fig. 3 where we calculated the average distance for our line of sight to intercept a star. The same procedure gives the average distance a photon will travel before being absorbed by the star. Dividing by the speed of light will give the average time for absorption, which equals the fill-up time. The calculation is the same so the answer is also the same, 10^{23} years. So another answer to why the sky is dark is simply this: Stars do not live long enough to fill the universe with light; to do so they would have to live 10^{23} years, not 10^{10} years. Once again, the light is only 10^{-13} of that needed for a bright sky. For a bright sky, the density of stars would have to be 10^{13} times greater; that is, they would have to be $(10^{13})^{1/3}$ closer together. Instead of being about 1000 light-years apart on average, they would need to be separated by only 0.1 light-years.

Science progresses in part because inspired people ask simple but profound questions. Einstein wondered how light would appear if one were moving along with it at the same speed, a simple question with a nonsimple answer. Why is the sky dark at night? is another such question, and it leads us to profound conclusions about the nature of the universe.

QUESTIONS

1. Refer to Fig. 4. When stars first begin to shine, only those near the earth contribute to the sky brightness. Much later, a spherical shell of stars provides the only contribution, and this shell expands at the speed of light. What is the forest analog of this situation? (*Hint:* How would you plant the forest?)
2. In Fig. 4, what is the thickness of the spherical shell? Is it of constant thickness? Why does it move outward with the speed of light?
3. The text points out that we can see stars only within 10^{10} light-years of earth due to the age of the universe and that we would need to see them to a distance of 10^{23} light-years to have the bright sky. It would seem that the actual sky should be fainter than the bright one by the ratio of *volumes,* $(10^{10}/10^{23})^3 = 10^{-39}$, rather than by the ratio of *radii* $10^{10}/10^{23} = 10^{-13}$, as indicated in the text. Explain.

EXERCISES

1. Calculate the distance in light-years of the farthest stars visible in the bright-sky universe.
2. If stars had been shining for 10^{23} years, the sky would appear to be filled with many suns. How many suns would it take to cover the hemisphere of sky that we can see at any moment? The angular diameter of the sun is 0.5°. (*Hint:* Use the solid angle subtended by the sun and that of the sky.)
3. *(a)* What is the average density of luminous matter in the universe? (There appears to be a lot of dark matter, but it does not contribute to a bright sky so we ignore it.) Near the sun, stars are about 10 light-years apart. Using the fact that an average star like the sun contains about 10^{57} protons plus neutrons (nucleons), calculate the average density of nucleons (nucleons per cubic meter) in the solar neighborhood. *(b)* In the universe as a whole, stars are found in galaxies consisting of about 10^{11} stars. Galaxies are separated from each other by about 10^7 light-years. Estimate the average density of nucleons in the universe. The actual density is about 1 nucleon per cubic meter.
4. *(a)* Show that if the entire mass of all stars were completely converted into radiation, the universe would glow with a temperature of about 20 K. The average density of stellar material spread throughout the universe is equivalent to one hydrogen atom per cubic meter. Therefore consider a cubic meter of space containing one hydrogen atom. The radiation energy density inside a cavity at temperature T is $(4\sigma/c)T^4$, and an energy of mc^2 is produced when a mass m is converted to radiation. *(b)* Stars convert only about 0.1 percent of their mass to energy during their lifetimes. Considering this fact, what would the average density of matter have to be to produce a bright sky at a temperature of 6000 K? *(c)* Based on the result of part *(b)*, estimate the average distance between stars (in light-years) required to produce a bright sky. Use the fact that an average star is equivalent to 10^{57} hydrogen atoms.
5. *(a)* Show that about 10^{60} stars would be visible in a bright sky. *(b)* Estimate the number of stars actually visible in our 20-billion-year-old universe and compare with 10^{60}.

NEWTON'S LAWS OF MOTION

5

Newton's laws explain the motion of this acrobatic person.

Three fundamental principles, called ***Newton's laws of motion,*** form the basis of mechanics. Sir Isaac Newton (1642–1727) presented these principles to the world in his book *Philosophiae Naturalis Principia Mathematica (The Mathematical Principles of Natural Philosophy),* which was published in 1686 and is often referred to as the *Principia.* We shall introduce these principles in this chapter and use them in much of this book. In this chapter the weight of an object and procedures for finding the motion of an object are also discussed.

PHILOSOPHIÆ

NATURALIS

PRINCIPIA

MATHEMATICA.

Autore *J S. NEWTON, Trin. Coll. Cantab. Soc.* Mathefeos Profeffore *Lucafiano,* & Societatis Regalis Sodali.

IMPRIMATUR.

S. PEPYS, *Reg. Soc.* PRÆSES.

Julii 5. 1686.

LONDINI,

Juffu *Societatis Regiæ* ac Typis *Jofephi Streater.* Proftant Venales apud *Sam. Smith* ad infignia Principis *Walliæ* in Cœmiterio *D. Pauli,* aliofq; nonnullos Bibliopolas. *Anno* MDCLXXXVII.

5-1 FORCE AND MASS

Newton's laws are phrased in terms of force and mass. It is helpful to have some understanding of these two concepts before being introduced to the laws. However, because force and mass are defined with Newton's laws, we are confronted with a dilemma. On the one hand, how can we discuss Newton's laws without first knowing the definitions of force and mass? On the other hand, how can we define

Figure 5-1. An equal-arm balance.

force and mass without first stating Newton's laws? We skirt this issue by describing force and mass qualitatively, in terms of everyday experience, and by using these intuitive notions while introducing the laws. Then, after we have stated the laws, we present the formal definitions of force and mass.

Mass

The ***mass*** of an object is a measure of the object's resistance to a change in its velocity. A child's wagon coasting along a horizontal sidewalk is more difficult to stop when it is loaded with bricks than when it is empty. The system, the wagon plus its load, is more massive when the bricks are on board. Mass is a scalar quantity and is additive. That is, if we fasten two objects of mass m_1 and m_2 together, then the mass m_{12} of the composite system is

$$m_{12} = m_1 + m_2$$

In the laboratory, the mass of an object is often measured with an equal-arm balance by comparison with standard "weights" of known mass (Fig. 5-1). As we shall see in Sec. 5-5, an object's weight is proportional to its mass.

Force

While Newton was forming his ideas about mechanics, he struggled mightily with the concept of force. From his notes, one of his early definitions was "Force is ye pressure or crouding of one body upon another." In modern everyday language, a ***force*** is a push or a pull. If you push on an object with your hand, you exert a force on the object. Such a force is the result of direct contact between your hand and the object and is an example of a ***contact force.*** Another familiar force is the weight of an object. The weight of an object is closely related to the ***gravitational force*** exerted by the earth on the object. The gravitational force is examined in detail in Chap. 7. When you take your socks out of the clothes dryer on a dry day and the socks cling to each other, you are observing the effects of ***electric forces.*** A ***magnetic force*** is responsible when you use a small magnet to hold a note on the refrigerator door. Electric and magnetic forces are investigated in Chaps. 20 through 31. ***Nuclear forces*** are outside the realm of direct human experience.* The commentary at the end of Chap. 7 gives an overall view of the forces in nature.

Types of forces in nature

From common experience, we can point out four properties of force:

* In the "extended" edition of this text, nuclear forces are discussed in Chap. 43.

Figure 5-2. A football as it is kicked. The ball is temporarily deformed and it is accelerated.

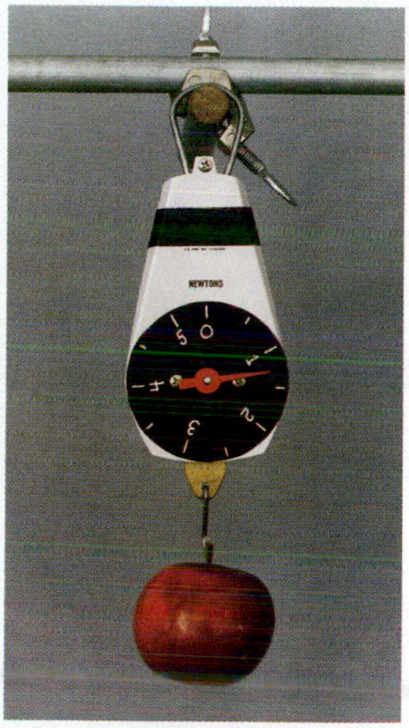

Figure 5-3. Weighing an apple with a spring scale.

Properties of force

1. Since a push or a pull has both magnitude and direction, we expect that force is a vector quantity. In the next section, we shall substantiate this expectation.
2. Forces occur in pairs. If object *A* exerts a force on object *B*, then *B* also exerts a force on *A*. For example, when a foot kicks a football (Fig. 5-2), the foot exerts a force on the ball, but the ball exerts a force on the foot too.
3. A force can cause an object to accelerate. If you kick a football, the ball's velocity changes while your foot is in contact with it.
4. A force can deform an object. As you can see from Fig. 5-2, the ball is deformed by the force exerted on it by the foot.

Property 4, that a force causes an object to be deformed, is often used to measure a force. This is the principle of a spring scale (Fig. 5-3). A spring scale consists of a spring, usually contained in a housing, and a pointer that indicates the amount the spring is stretched or compressed. The magnitude of this force is proportional to the amount the spring is stretched (or compressed), and the direction of the force is along the spring. The scale may be calibrated to read in pounds (lb) or in newtons (N), which is the SI unit of force. A few representative force magnitudes are given in Table 5-1.

The SI unit of force is the newton (N).

TABLE 5-1. ***A Few Representative Forces***

Exerted by	Exerted on	Type	Approximate Magnitude, N
Andromeda galaxy	Milky Way galaxy	Gravitational	7×10^{28}
Sun	Earth	Gravitational	3.5×10^{22}
Saturn V rockets	Apollo spacecraft	Contact	3.3×10^{7}
Earth	You	Gravitational	600
Hydrogen atom nucleus (proton)	Atom's electron	Electric	8×10^{-8}
Hydrogen atom nucleus (proton)	Atom's electron	Gravitational	4×10^{-47}

Environment of an object

Finally, note that the mass of an object is a property of that object alone. In contrast, a force exerted on an object is a result of an interaction between the object and some other object. Further, an object's *environment* consists of other objects that exert forces on that object. For example, if you hold a book in your hand, the important elements of the book's environment are your hand, which exerts an upward force on the book, and the earth, which exerts a downward force on the book (the book's weight).

EXAMPLE 5-1

Converting force units. In the British system of units, force is measured in pounds, and 1.00 lb = 4.45 N. *(a)* What is the weight of a 5.0-lb bag of sugar in newtons? *(b)* If an apple weighs 1.1 N, what is its weight in pounds?

Solution. *(a)* Dividing the equation 1.00 lb = 4.45 N by 1.00 lb, we find that the number 1 can be written as 1 = 4.45 N/1.00 lb = 4.45 N/lb. Multiplying the weight of the bag of sugar by this conversion factor, we obtain

$$5.0 \text{ lb} = (5.0 \text{ lb})(4.45 \text{ N/lb}) = 22 \text{ N}$$

(b) Similarly, the conversion factor can be written as 1 = 1.00 lb/4.45 N = 0.225 lb/N, and the weight of the apple is

$$1.1 \text{ N} = (1.1 \text{ N})(0.225 \text{ lb/N}) = 0.25 \text{ lb}$$

Since an apple plays a prominent role in a famous legend about Newton, it is interesting that a small apple weighs about 1 N.

SELF-TEST 5-1. Is it proper to speak of
(a) The force *exerted by* object *A*?
(b) The mass *exerted by* object *A*?
(c) The force *of* object *A*?
(d) The mass *of* object *A*?
(e) The force *exerted on* object *A*?
(f) The mass *exerted on* object *A*?
ANSWERS: *(a)* yes; *(b)* no; *(c)* no; *(d)* yes; *(e)* yes; *(f)* no.

5-2 NEWTON'S FIRST LAW

As was the custom for a scholarly work in Newton's day, the *Principia* was written in Latin. A translation of Newton's first law is*

Newton's first law

> Law I. Every body continues in its state of rest, or in uniform motion in a right [straight] line unless it is compelled to change that state by forces impressed upon it.

This law is often called the ***law of inertia,*** because "inertia" means resistance to a change, and the law states that an object naturally tends to maintain whatever velocity it happens to have (including a velocity of zero).

We now recast Newton's first law in modern terms. First, if an object is in a state of rest or in uniform motion in a straight line, then its acceleration is zero. Second, we use the term *net force* for Newton's expression "forces impressed upon it":

> The net force $\Sigma\mathbf{F}$ exerted on an object is the vector sum of all the individual forces exerted on it by other objects.

* Isaac Newton, *Philosophiae Naturalis Principia Mathematica* (1686), Andrew Motte (trans., 1729), University of California Press, Berkeley, 1960.

$$\Sigma \mathbf{F} = \mathbf{F}_1 + \mathbf{F}_2 + \mathbf{F}_3 + \cdots$$

Net force

where $\mathbf{F}_1$, $\mathbf{F}_2$, and so on represent the individual forces exerted by other objects. The symbol $\Sigma\mathbf{F}$ is used for the net force because the Greek letter Σ customarily represents the operation of finding a sum. The net force is sometimes called the *resultant force* or the *total force* or the *unbalanced force*. Now, Newton's first law becomes

Newton's first law in modern terms

> If the net force on an object is zero ($\Sigma\mathbf{F} = 0$), then the object's acceleration is zero ($\mathbf{a} = 0$).

Newton's First Law and Common Experience

At first glance, Newton's first law seems to violate common experience. We are inclined to go along with Newton when he tells us that an object at rest tends to remain at rest. But does an object moving at constant velocity tend to maintain that velocity? Suppose you push a grocery cart (exert a force on it) along a supermarket aisle at a constant velocity. If you release the cart (stop exerting the force), it slows down (accelerates) until it comes to rest. This seems to disagree with Newton's first law. That your experience with the cart does not contradict Newton's first law can be seen by recognizing the presence of a *frictional force* and by making a clear distinction between the *net force on the cart* and the *force you exert on it*. While you push the cart at constant velocity, the force you exert is not zero, but the net force is zero because the frictional force on the cart is equal and opposite the force you exert. Frictional forces are discussed in the next chapter, and there you will see in detail how this works. When you release the cart, the net force on it is no longer zero because the frictional force continues to act on it until it comes to rest. Your experience with the cart does not contradict Newton's first law; it is in accord with this law.

A skeptic might say that the frictional force was contrived to make these observations agree with Newton's first law. To see that this is not so, we can effectively eliminate the frictional force by performing experiments on an air track (Fig. 5-4). The glider on an air track rides on a cushion of air so that the frictional force on the glider as it moves along the track is due to air resistance only. Consequently, the frictional force is practically imperceptible, and once the glider is set in motion, it moves along the air track with a velocity that is essentially constant.

Figure 5-4. An air track. Air is blown into the hollow triangular track and escapes through small holes. The escaping air suspends the glider on a cushion of air so that frictional forces on the glider as it moves along the track are negligible.

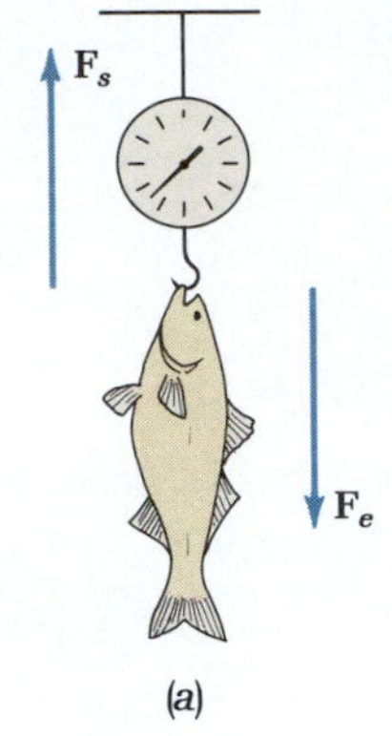

(a)

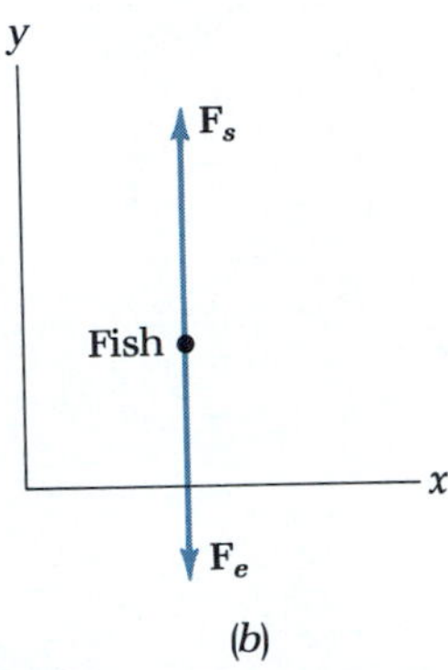

(b)

Figure 5-5. *(a)* A fish is suspended by a spring scale. The reading on the scale gives the magnitude $\mathbf{F}_s$ of the force exerted upward by the scale on the fish. *(b)* Free-body diagram for the fish. The fish is not accelerating so that, by Newton's first law, the net force on it is zero. Thus the scale reads the magnitude of the fish's weight: $\mathbf{F}_s = \mathbf{F}_e$.

Principle of Superposition

A common application of Newton's first law is the weighing of an object. Figure 5-5*a* shows a fish suspended from a spring scale. In the figure, force vectors are used to represent the forces exerted on the fish. To indicate a particular force, a subscript is placed on the symbol **F**, and the subscript refers to the object that exerts the force. The fish's weight is equal to the gravitational force $\mathbf{F}_e$ exerted on it by the earth (directed downward), and $\mathbf{F}_s$ is the force exerted by the scale (directed upward).*

Figure 5-5*b* is a ***free-body diagram*** for the fish. A free-body diagram is an aid used for finding the net force on an object. In the free-body diagram we represent the fish as a dot, and place the tails of the force vectors at the dot. The fish is represented simply as a dot because we assume its extent is of no consequence, which means that it is treated as a particle. (Situations where this assumption is invalid will be discussed in Chaps. 11 through 13.) Using the coordinate frame in the free-body diagram, we find that the net force on the fish is

$$\Sigma\mathbf{F} = \mathbf{F}_s + \mathbf{F}_e = (F_s\mathbf{j}) + (-F_e\mathbf{j}) = (F_s - F_e)\mathbf{j}$$

Since the fish is not accelerating, Newton's first law states that the net force on it is zero. Thus $F_s = F_e$. That is, the scale reads the fish's weight. (Strictly speaking, the weight of an object is a vector and has both magnitude *and* direction, but since weight is always directed downward, the term "weight" often is taken loosely to mean the magnitude of the weight.)

We can use Newton's first law to verify experimentally that force is a vector. Suppose we hang a metal ball from a spring scale and measure the force $\mathbf{F}_a$ exerted by the scale on the ball (Fig. 5-6*a*). By Newton's first law, $\mathbf{F}_a$ is equal and opposite the ball's weight. Next we suspend the ball with two spring scales that make angles of θ and ϕ with the vertical and measure forces $\mathbf{F}_b$ and $\mathbf{F}_c$ (Fig. 5-6*b*). By Newton's first law, $\mathbf{F}_b$ and $\mathbf{F}_c$ combine to give a force that is equal and opposite the ball's weight. Thus the effect of the combined forces $\mathbf{F}_b$ and $\mathbf{F}_c$ is the same as $\mathbf{F}_a$ acting alone. A vector diagram shows that the vector sum $\mathbf{F}_b + \mathbf{F}_c$ is equal to $\mathbf{F}_a$ (Fig. 5-6*c*). That is, the effect of both forces acting simultaneously is the same as their vector sum. ***The property that forces add as vectors is sometimes called the principle of superposition.*** Incidentally, Fig. 5-6*c* is not a free-body diagram. A free-body diagram includes all forces exerted on an object in a particular situation. Figure 5-6*c* is a comparison of forces exerted in two different situations.

(a)

(b)

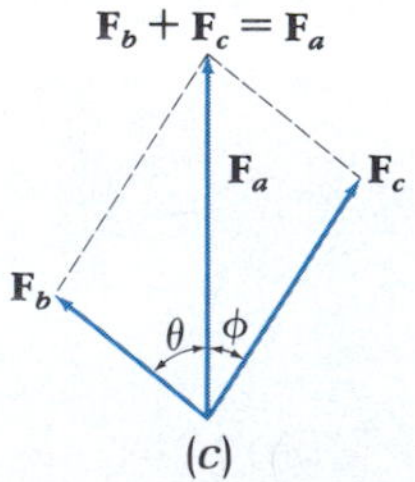

(c)

Figure 5-6. In *(a)*, a spring scale exerts force $\mathbf{F}_a$ on a metal ball. In *(b)*, two spring scales exert forces $\mathbf{F}_b$ and $\mathbf{F}_c$ on the ball. *(c)* Measurement shows that $\mathbf{F}_a$ is the vector sum of $\mathbf{F}_b$ and $\mathbf{F}_c$. To verify that forces add vectorially, read the force magnitudes from the scales and use a protractor to measure the force directions.

* In Sec. 5-5, we shall find that weight is defined such that it is equal to the gravitational force exerted by the earth to an accuracy of better than one-half of 1 percent.

EXAMPLE 5-2

Finding a net force. An object has forces exerted on it by three other objects: $\mathbf{F}_1 = (2.0\ \text{N})\mathbf{j}$, $\mathbf{F}_2 = -(3.0\ \text{N})\mathbf{i}$, and $\mathbf{F}_3 = (5.0\ \text{N})\mathbf{i} - (6.0\ \text{N})\mathbf{j}$. Determine the net force $\Sigma\mathbf{F}$ exerted on the object.

Solution. The net force is

$$\begin{aligned}\Sigma\mathbf{F} &= \mathbf{F}_1 + \mathbf{F}_2 + \mathbf{F}_3 \\ &= (2.0\ \text{N})\mathbf{j} + [-(3.0\ \text{N})\mathbf{i}] + [(5.0\ \text{N})\mathbf{i} - (6.0\ \text{N})\mathbf{j}] \\ &= (2.0\ \text{N})\mathbf{i} - (4.0\ \text{N})\mathbf{j}\end{aligned}$$

SELF-TEST 5-2. Determine the magnitude $|\Sigma\mathbf{F}|$ of the net force in the above example.
ANSWER: 4.5 N.

Inertial Reference Frames

You may be surprised by our next discovery: *Newton's first law is not valid in all reference frames.* Consider a crate sitting on a sidewalk. There are two forces on the crate: the gravitational force $\mathbf{F}_e$ exerted downward by the earth, and an equal and opposite force $\mathbf{F}_s$ exerted upward by the sidewalk. The net force on the crate is zero: $\Sigma\mathbf{F} = 0$. Newton's first law states that the crate's acceleration is zero because the net force is zero. Is the acceleration zero? The answer depends on the reference frame used to measure the acceleration.

In Sec. 4-4, we learned that the acceleration of an object measured by two different observers is different if their reference frames accelerate relative to one another. If you choose a frame fixed to the sidewalk, then the crate is at rest and remains at rest so that its acceleration is zero. But if you choose a frame fixed to a car that accelerates relative to the sidewalk, then the crate's acceleration is not zero. We conclude that Newton's first law is valid in a frame fixed to the sidewalk, but it is not valid in a frame fixed to the car.

Given that Newton's first law is not valid in some reference frames, one might question whether the law is useful. Actually, this very feature leads us to the utilization of the law. We now define a special type of reference frame called an *inertial reference frame.*

Definition of an inertial reference frame

> An ***inertial reference frame*** is a frame in which Newton's first law is valid, or a frame relative to which $\mathbf{a} = 0$ for any object with $\Sigma\mathbf{F} = 0$.

In the discussion above, the frame fixed to the sidewalk was an inertial frame, but the frame fixed to the car was not an inertial frame. A frame that is not inertial is called a *noninertial frame.*

A noninertial frame

> Any frame that accelerates relative to an inertial frame is a ***noninertial frame.***

Suppose the car had been traveling at constant velocity, rather than accelerating relative to the sidewalk. Then the crate's acceleration measured relative to a frame fixed to the car would be zero, and Newton's first law would be valid in that frame.

> Any frame that moves with constant velocity relative to ***an inertial frame*** is itself an inertial frame.

A reference frame we ordinarily use for an object on or near the surface of the earth is a frame whose origin is fixed relative to a nearby point on the earth's surface and one whose axes are fixed relative to the horizontal and vertical (Fig. 5-7). We

Figure 5-7. The origin of an earth-surface frame is fixed relative to a nearby point on the earth's surface, and the frame's axes are fixed relative to the horizontal and vertical at that point.

Figure 5-8. Time-lapse photograph of the night sky with the camera pointed toward Polaris, the North Star. As the earth rotates, the stars appear to travel in circles.

call such a frame an ***earth-surface frame.*** The frame fixed relative to the sidewalk in our discussion above is an example of an earth-surface frame. As you know, the earth rotates on its axis once a day (Fig. 5-8), and it orbits the sun once a year. Because of the earth's rotation, an earth-surface frame at the equator has an acceleration of 0.034 m/s^2 toward the earth's center, and because of the earth's orbital motion, the earth's center has an acceleration of 0.006 m/s^2 toward the sun (see Chap. 4, Exercise 29). Because these accelerations are small, their effects are often insignificant when applying the laws of motion to terrestrial problems. Therefore, unless stated otherwise, we assume that an earth-surface frame can be treated as an inertial frame, as we did in the discussion above. The validity of this assumption will be investigated in Sec. 6-3.

EXAMPLE 5-3

An alternative definition for an inertial reference frame? Suppose you are told that Newton's first law is not needed to define an inertial reference frame. Instead, an alternative definition is put forth: An inertial reference frame is a frame whose acceleration is zero. Can this definition be used? If not, why not?

Solution. Section 4-4 showed that the kinematical quantities, **r**, **v**, and **a**, have meaning only when the frame in which they are measured is explicitly stated or clearly implied. In a stated problem, the frame is usually implied in the description. If we attempt to define a frame by stating that the frame's acceleration is zero, then this immediately prompts us to ask: Zero relative to what? That is, a reference frame is needed to define an acceleration, so a frame cannot be defined by stating that the frame's acceleration is zero unless another frame is defined for measuring this acceleration. This means that the alternative definition given above is useless.

SELF-TEST 5-3. Suppose we measure the forces exerted on an object and the acceleration of the object relative to reference frames 1, 2, and 3. Let the letter *K* denote the object. *(a)* In one instance $\Sigma\mathbf{F} = 0$ and $\mathbf{a}_{K1} \neq 0$. Is frame 1 inertial? *(b)* In another instance $\Sigma\mathbf{F} \neq 0$ and $\mathbf{a}_{K2} = 0$. Is frame 2 inertial? *(c)* In still another case $\Sigma\mathbf{F} = 0$ and $\mathbf{a}_{K3} = 0$. Is frame 3 inertial? *(d)* Frame 4 moves with constant velocity relative to frame 3. Is frame 4 inertial? *(e)* Frame 5 moves with constant acceleration relative to frame 3. Is frame 5 inertial? ***ANSWERS:*** *(a)* no; *(b)* no; *(c)* yes; *(d)* yes; *(e)* no.

5-3 NEWTON'S SECOND LAW

Newton wrote the second law as

> Law II. The change of motion is proportional to the motive force impressed, and is made in the direction of the right line in which that force is impressed.

Newton's second law

By the term "motion," Newton was referring to a quantity now called *momentum*. The ***momentum*** $\mathbf{p}$ of an object of mass m moving with velocity $\mathbf{v}$ is

$$\mathbf{p} = m\mathbf{v}$$

Definition of momentum

By the term "motive force," Newton meant the net force $\Sigma\mathbf{F}$. If the proportionality constant between the "motive force" and the "change of motion" is 1, then Newton's second law is

$$\Sigma\mathbf{F} = \frac{d\mathbf{p}}{dt}$$

Further, if we assume that the object's mass is independent of the time t, then

$$\frac{d\mathbf{p}}{dt} = \frac{d}{dt}(m\mathbf{v}) = m\frac{d\mathbf{v}}{dt} = m\mathbf{a}$$

and Newton's second law can be written

$$\Sigma\mathbf{F} = m\mathbf{a} \qquad (5\text{-}1)$$

Newton's second law in equation form

Usually Eq. 5-1 is called Newton's second law. The equation states:

> An object's acceleration is proportional to the net force exerted on it, and the object's mass is the proportionality factor between the net force and the acceleration.

For a given net force, an object with a larger mass will have a smaller acceleration. ***Mass is that property of an object that causes it to resist any change in its velocity.*** Since *inertia* means resistance to a change, the mass is sometimes called the ***inertial mass.***

Newton's second law provides a definition of the concept of force:

> A ***force*** is that which causes an object to accelerate. A single force acting alone on an object has the same direction as the object's acceleration relative to an inertial reference frame, and the force magnitude is proportional to the acceleration magnitude.

Definition of force

This law connects the unit of force on the one hand and the units of mass and acceleration on the other. The SI unit of force, the newton, is defined with Newton's second law:

Definition of the newton (N)

$$1\ \text{N} = 1\ \text{kg} \cdot \text{m/s}^2 \qquad \text{(exactly)}$$

If an object of mass one kilogram has an acceleration of one meter per second squared relative to an inertial reference frame, then the net force exerted on the object is one newton.

From a comparison of Newton's first and second laws, one is tempted to view the first law as simply a particular case of the second law. Since $\Sigma\mathbf{F} = m\mathbf{a}$, it follows that $\mathbf{a} = 0$ when $\Sigma\mathbf{F} = 0$. However, we used the first law to define the type of reference frame relative to which the acceleration in Newton's second law must be measured, namely, an inertial reference frame. With this interpretation, the first law is a statement about nature that is independent of the second law. The first law states that inertial reference frames exist, and gives a procedure for determining whether a frame is inertial.

EXAMPLE 5-4

Acceleration of the Titanic. At the time it was launched, the ill-fated *H.M.S. Titanic* was the most massive mobile object ever built by humans, having a mass of 6.0×10^7 kg. What would have been the magnitude of the net force required to give the *Titanic* an acceleration of magnitude of $0.1\ \text{m/s}^2$?

Solution. Newton's second law in terms of magnitudes is $|\Sigma\mathbf{F}| = ma$. For the *Titanic* to have an acceleration magnitude of $0.1\ \text{m/s}^2$, the magnitude of the net force should have been

$$|\Sigma\mathbf{F}| = (6.0 \times 10^7\ \text{kg})(0.1\ \text{m/s}^2) = 6 \times 10^6\ \text{N} = 6\ \text{MN}$$

SELF-TEST 5-4. The net force on an object of mass 3.0 kg is $\Sigma\mathbf{F} = -(6.0\ \text{N})\mathbf{i} + (9.0\ \text{N})\mathbf{j}$. Determine the object's acceleration relative to an inertial reference frame. ***ANSWER:*** $-(2.0\ \text{m/s}^2)\mathbf{i} + (3.0\ \text{m/s}^2)\mathbf{j}$.

5-4 NEWTON'S THIRD LAW

A translation of Newton's third law is

> Law III. To every action there is always opposed an equal reaction; or, the mutual actions of two bodies upon each other are always directed to contrary parts.

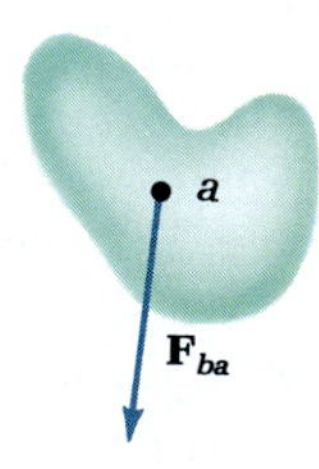

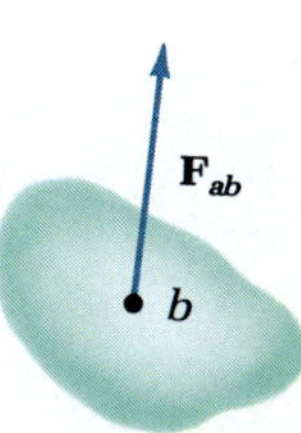

Figure 5-9. Newton's third law. Objects *a* and *b* exert forces on each other. By Newton's third law, these forces are equal and opposite: $\mathbf{F}_{ab} = \mathbf{F}_{ba}$. In this figure the forces are shown as attractive forces: $\mathbf{F}_{ab}$ is toward *a* and $\mathbf{F}_{ba}$ is toward *b*.

The first and second laws are statements about a single object, whereas the third law is a statement about two objects. To discuss the third law, there must be two subscripts on the force symbol $\mathbf{F}$, the first subscript to denote the object that exerts the force and the second subscript to denote the object on which the force acts. Suppose objects *a* and *b* exert forces on each other (Fig. 5-9); $\mathbf{F}_{ab}$ is the force exerted by *a* on *b*, and $\mathbf{F}_{ba}$ is the force exerted by *b* on *a*. Newton's third law states that these two forces are equal and opposite, or

$$\mathbf{F}_{ab} = -\mathbf{F}_{ba} \tag{5-2}$$

> If object *b* exerts a force on object *a*, then object *a* exerts an equal and opposite force on *b*. Forces occur in pairs. A single force cannot exist.

When two objects exert forces on each other, we say that an *interaction* exists between the objects. Newton's third law gives the relation between the two forces that are the result of an interaction. The two forces $\mathbf{F}_{ab}$ and $\mathbf{F}_{ba}$ are often called an ***action-reaction pair.*** One of the forces is called the action force and the other is called the reaction force. Which force is called the action and which is called the reaction is arbitrary. Newton's third law reveals an underlying symmetry in the forces that occur in nature.

EXAMPLE 5-5

Stacking books. Suppose that your physics and history books are lying on your desk, with the history book on top of the physics book (Fig. 5-10). The history and physics books weigh 14 and 18 N, respectively. Identify each force on each book with a double subscript notation and determine the value of each of these forces.

Solution. The free-body diagrams for the books are shown in Fig. 5-10. Since the weight of the history book is equal to the force exerted by the earth on the *h*istory book, it is represented as $\mathbf{F}_{eh}$:

$$\mathbf{F}_{eh} = -(14\ \mathrm{N})\mathbf{j}$$

Other than the earth, the history book interacts only with the physics book. Since the acceleration of the history book is zero, the net force on it is zero by Newton's second law:

$$\mathbf{F}_{ph} + \mathbf{F}_{eh} = 0$$

where $\mathbf{F}_{ph}$ is the force exerted by the *p*hysics book on the *h*istory book. Thus $\mathbf{F}_{ph} = -\mathbf{F}_{eh} = -[-(14\ \mathrm{N})]\mathbf{j}$, or

$$\mathbf{F}_{ph} = (14\ \mathrm{N})\mathbf{j}$$

We find that the physics book exerts an upward force of magnitude 14 N on the history book.

The physics book has three forces exerted on it: $\mathbf{F}_{ep}$ due to the earth, $\mathbf{F}_{hp}$ due to the *h*istory book, and $\mathbf{F}_{dp}$ due to the *d*esktop. Since the physics book weighs 18 N,

$$\mathbf{F}_{ep} = -(18\ \mathrm{N})\mathbf{j}$$

From Newton's third law, $\mathbf{F}_{hp} = -\mathbf{F}_{ph}$, so that

$$\mathbf{F}_{hp} = -(14\ \mathrm{N})\mathbf{j}$$

Newton's second law applied to the physics book gives $\Sigma\mathbf{F} = 0$, or $\mathbf{F}_{dp} + \mathbf{F}_{ep} + \mathbf{F}_{hp} = 0$, or $\mathbf{F}_{dp} = -\mathbf{F}_{ep} - \mathbf{F}_{hp}$, so that

$$\mathbf{F}_{dp} = -[-(18\ \mathrm{N})\mathbf{j}] - [-(14\ \mathrm{N})\mathbf{j}] = (32\ \mathrm{N})\mathbf{j}$$

The desk exerts an upward force of 32 N on the physics book. To arrive at the solution, we have applied Newton's second law twice and Newton's third law once.

SELF-TEST 5-5. Suppose a biology book with a weight of 16 N is placed on top of the history book in Fig. 5-10. Determine *(a)* $\mathbf{F}_{ph}$ and *(b)* $\mathbf{F}_{dp}$. Indicate the direction of each force with unit vectors. ***ANSWERS:*** *(a)* (30 N)**j**; *(b)* (48 N)**j**.

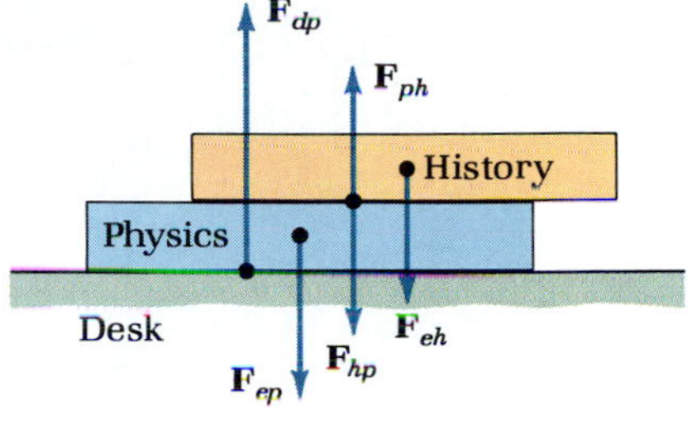

(a)

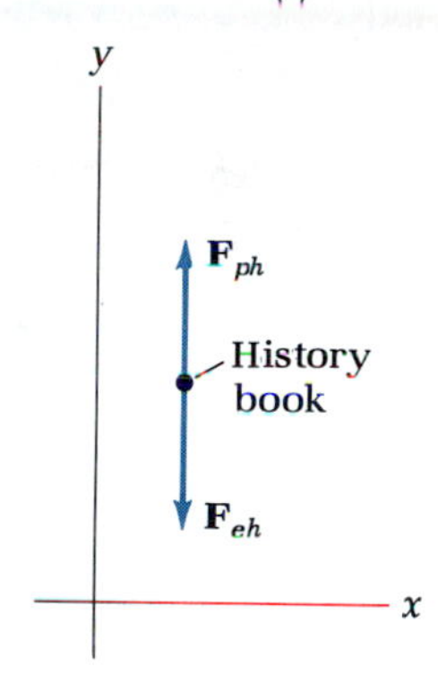

(b)

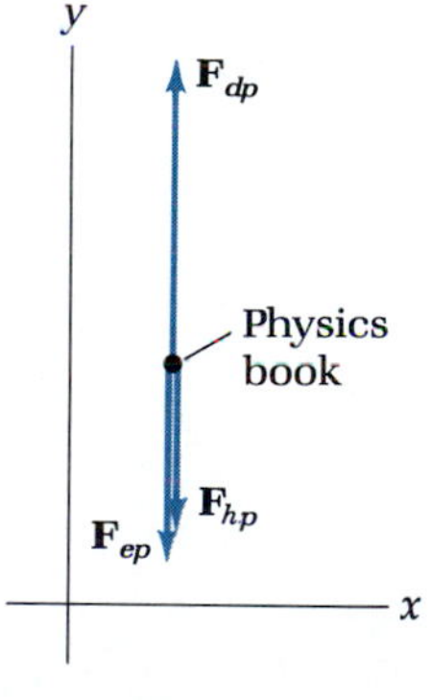

(c)

Figure 5-10. Example 5-5: *(a)* A history book and a physics book rest on a desk. *(b)* Free-body diagram for the history book. *(c)* Free-body diagram for the physics book.

A word of caution is in order. For the particular case in which two forces are exerted on an object with zero acceleration, Newton's second law appears deceptively similar to Newton's third law. For example, Newton's second law applied to the history book above gives $\mathbf{F}_{ph} + \mathbf{F}_{eh} = 0$, or $\mathbf{F}_{ph} = -\mathbf{F}_{eh}$; the forces are equal and opposite. The important thing to note about these two forces is that they are both applied to the *same* object, the history book. This is indicated by the fact that the second subscript on the force symbols is the same—*h* for *h*istory book. On the other hand, Newton's third law applied to the interaction between the physics book and the history book is written $\mathbf{F}_{hp} = -\mathbf{F}_{ph}$; these forces are also equal and opposite. The important thing to note about these two forces is that they are exerted on *different* objects, one on the physics book and one on the history book. This fact is indicated by the reversal of the subscripts on the force symbols. Newton's second

In Newton's third law, each force is applied to a different object.

law applies to a single object, whereas Newton's third law applies to an interaction between two objects. Note that the two forces in Newton's third law never occur in the same free-body diagram. This is because a free-body diagram shows forces acting on a single object, and the action-reaction pair in Newton's third law always act on different objects.

Now we combine Newton's second and third laws to provide a definition for the mass of an object. Suppose objects a and b exert forces on each other, $\mathbf{F}_{ab}$ and $\mathbf{F}_{ba}$. There may be other forces on these two objects besides $\mathbf{F}_{ab}$ and $\mathbf{F}_{ba}$, but we arrange to have the vector sum of these other forces add to zero so that the net force on object a is $\mathbf{F}_{ba}$, and the net force on object b is $\mathbf{F}_{ab}$. Newton's second law applied to each object gives

$$\mathbf{F}_{ba} = m_a\mathbf{a}_a \qquad \text{and} \qquad \mathbf{F}_{ab} = m_b\mathbf{a}_b$$

Inserting this result into Newton's third law, $\mathbf{F}_{ba} = -\mathbf{F}_{ab}$, we have

$$m_a\mathbf{a}_a = -m_b\mathbf{a}_b$$

Or, in terms of acceleration magnitudes,

$$m_a a_a = m_b a_b$$

Now let object b be the standard kilogram (or a replica of it), and let object a be the object whose mass m we wish to determine. Then $m_b = 1$ kg (exactly), $m_a = m$, and

Definition of mass

$$m = 1\text{ kg}\,\frac{a_s}{a}$$

where a_s and a are the acceleration magnitudes of the standard kilogram and the object of mass m, respectively.

EXAMPLE 5-6

Measuring the mass of a cart. Carts A and B, each equipped with a spring bumper (Fig. 5-11a), are pushed together so that their spring bumpers are compressed. When the carts are released, the springs push the carts apart such that $a_A = 0.87\text{ m/s}^2$ and $a_B = 1.42\text{ m/s}^2$. Given that the mass of cart B is 1.0 kg, determine the mass of cart A. The mass of each cart's wheels is much less than that of its body, and the wheel bearings are well lubricated.

Solution. Since the mass of each cart's wheels is much less than that of its body, the effects due to the rotation of the wheels can be neglected and each cart can be treated as a particle (see Chap. 13). Further, since the wheel bearings are well lubricated, frictional effects that tend to slow the carts are negligible, so the force exerted on each cart by the horizontal surface has no horizontal component. This means that each of these forces is vertically upward, as shown in the free-body diagrams in Fig. 5-11b. The vertical component of each cart's acceleration is zero, so that the force upward by the surface on each cart is equal and opposite the cart's weight. Thus the net force on each cart is the force exerted by the other cart, and by Newton's third law, these forces are equal and opposite. This arrangement is a good approximation to the situation we described above in defining the mass of an object.

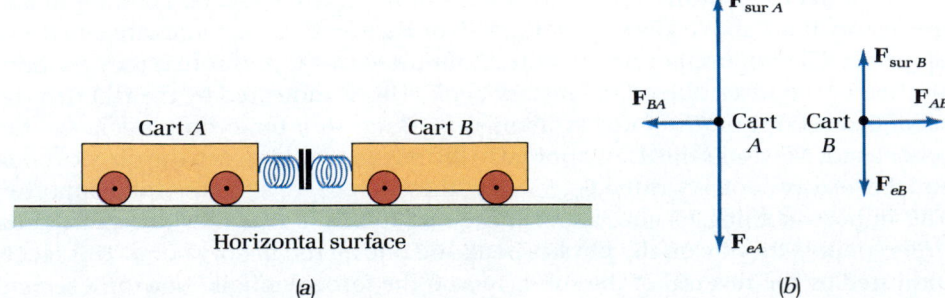

Figure 5-11. Example 5-6: *(a)* Carts A and B, each with a spring bumper, are pushed together and then released. *(b)* Free-body diagrams for the two carts immediately after they are released.

The mass of cart A is

$$m_A = m_B \frac{a_B}{a_A} = 1.0 \text{ kg} \frac{1.42 \text{ m/s}^2}{0.87 \text{ m/s}^2} = 1.6 \text{ kg}$$

SELF-TEST 5-6. Block b is held stationary against a vertical wall (w), as shown in Fig. 5-12. The weight of the block is 5.0 N downward and the hand (h) exerts a horizontal force $\mathbf{F}_{hb} = -(8.0 \text{ N})\mathbf{i}$ on the block. The only other force on the block is the force $\mathbf{F}_{wb}$ exerted by the wall. *(a)* Draw a free-body diagram for the block showing the magnitude and direction of each of the three forces. (*Hint:* Start out by drawing $\mathbf{F}_{hb}$ and $\mathbf{F}_{eb}$, and then estimate $\mathbf{F}_{wb}$.) *(b)* Use your free-body diagram and Newton's second law to find an expression for $\mathbf{F}_{wb}$ in terms of unit vectors. *(c)* Use Newton's third law and your answer to part *(b)* to find an expression for $\mathbf{F}_{bw}$. *ANSWERS: (b)* $(8.0 \text{ N})\mathbf{i} + (5.0 \text{ N})\mathbf{j}$; *(c)* $-(8.0 \text{ N})\mathbf{i} - (5.0 \text{ N})\mathbf{j}$.

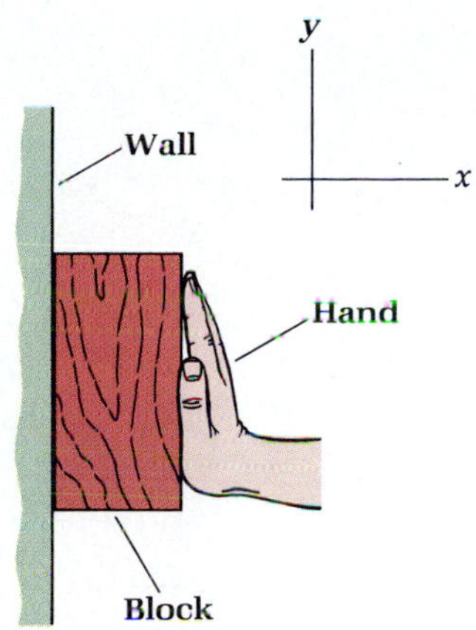

Figure 5-12. Self-test 5-6.

5-5 WEIGHT AND THE GRAVITATIONAL FORCE BY THE EARTH

For an object on or near the earth's surface, two closely connected quantities are (i) the gravitational force exerted on the object by the earth and (ii) the weight of the object.

Gravitational Force by the Earth

When an object is in free-fall, the only significant force on the object is the gravitational force by the earth. For example, forces due to air resistance are negligible. For a feather to execute free-fall, it must fall in vacuum. But a rock is essentially in free-fall if it falls in air, provided it is not allowed to fall too far. If the rock's speed becomes large, air resistance becomes significant and the rock is not in free-fall. Thus in free-fall the net force equals the gravitational force: $\Sigma\mathbf{F} = \mathbf{F}_e$. Applying Newton's second law, $\Sigma\mathbf{F} = m\mathbf{a}$, to an object in free-fall gives

$$\mathbf{F}_e = m\mathbf{g} \tag{5-3}$$

Gravitational force by the earth

where $\mathbf{g}$ is the acceleration of the object measured relative to an inertial frame. Experiment shows that any object in free-fall at some location has the same acceleration as any other object in free-fall at that same location. That is, $\mathbf{g}$ is independent of an object's mass. Equation 5-3 is an example of a ***force law.*** Actually, it is an application of a more general force law, which is the topic of Chap. 7.

Weight

According to the General Conference on Weights and Measures (CGPM), the definition of the weight $\mathbf{F}_w$ of an object of mass m is

$$\mathbf{F}_w = m\mathbf{g}' \tag{5-4}$$

CGPM definition of weight

where $\mathbf{g}'$ is the object's free-fall acceleration measured relative to the frame of reference of the person making the measurement. This means that the weight of an object is proportional to its mass and depends on the frame in which the measurement is made. Also, this definition corresponds to the reading on a spring scale in any reference frame, whether the frame is inertial or noninertial. That is, if an object is at rest on a spring scale fixed in some reference frame, then the scale reads the magnitude F_w of the weight in that frame.

A spring-scale reading gives the weight magnitude.

In particular, when a weight measurement is performed in an inertial frame, then $\mathbf{F}_w = \mathbf{F}_e$ because $\mathbf{g}'$ in an inertial frame is the same as $\mathbf{g}$. As mentioned in Sec. 5-2, we always use the approximation that an earth-surface frame is the same as an inertial frame unless stated otherwise. Thus $\mathbf{F}_w \simeq \mathbf{F}_e$ when the weight measurement is made in an earth-surface frame. In Sec. 6-3, we shall find that this approximation is valid to about one-half of 1 percent in any earth-surface frame. As an example, consider the weight of a 65-kg person measured in an earth-surface frame. Since $\mathbf{g}' \simeq \mathbf{g} = -g\mathbf{j} = -(9.8\ \text{m/s}^2)\mathbf{j}$, we have $\mathbf{F}_w = -(65\ \text{kg})(9.8\ \text{m/s}^2)\mathbf{j} = -(640\ \text{N})\mathbf{j}$, where $\mathbf{j}$ is directed vertically upward.

Comparison of Mass and Weight

The mass of an object is an intrinsic property of that object. That is, an object's mass is a property of that object alone. This means that an object may be characterized by its mass. For example, we may speak of a 9-kg pumpkin. It is also customary to treat the weight of an object as if it is an intrinsic property of the object. For example, we speak of a 20-lb pumpkin. (In SI units, it is a 90-N pumpkin.) However, the weight of an object involves the gravitational force by the earth, so it is a property of the earth as well as of the object. In addition, weight depends on the frame in which it is measured. Therefore, in principle, it is invalid to treat the weight of an object as an intrinsic property of that object.

As a practical matter, we may violate this principle and characterize an object by its weight because (i) the frame used for the measurement is assumed implicitly to be an earth-surface frame and (ii) an object's weight, $\mathbf{F}_w = m\mathbf{g}'$, is proportional to its mass. At any particular point, $\mathbf{g}'$ is the same for any object so that an object's mass determines its weight at that point. However, the acceleration magnitude $\mathbf{g}'$ measured relative to an earth-surface frame varies with position on the earth, even though this variation is small. A 9.00-kg pumpkin weighs $(9.00\ \text{kg})(9.79\ \text{m/s}^2) = 88.1\ \text{N}$ in Florida, and it weighs $(9.00\ \text{kg})(9.82\ \text{m/s}^2) = 88.4\ \text{N}$ in Alaska. Therefore, if we apply our measurements to a limited region on or near the earth's surface, or if we do not require high precision, then the weight of an object may be treated as a property of that object alone.

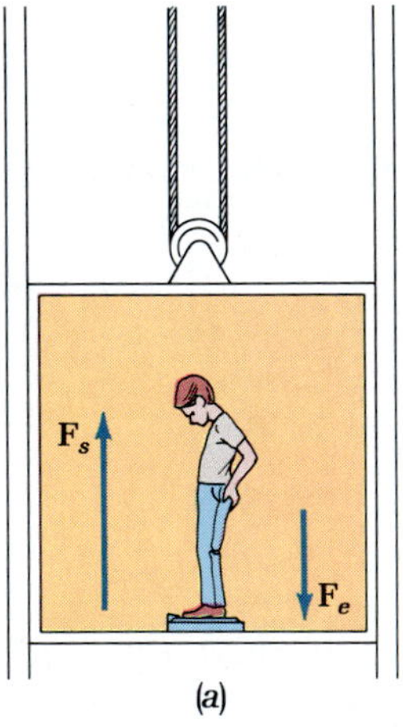

(a)

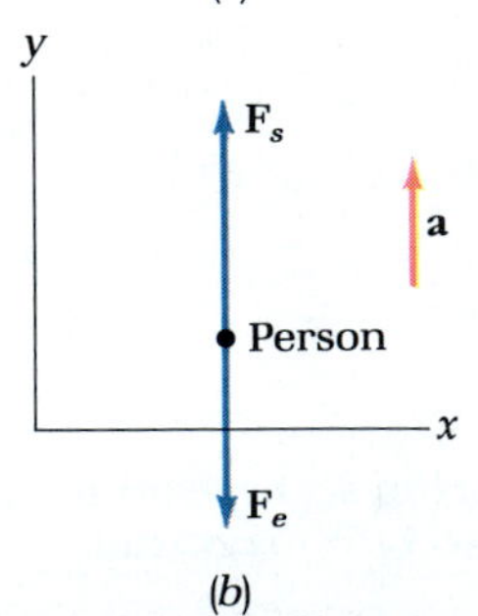

(b)

Figure 5-13. Example 5-7: *(a)* A person standing on a spring scale in an elevator. *(b)* Free-body diagram for the person as the elevator accelerates upward.

EXAMPLE 5-7

Weight measurements in an elevator. A person stands on a spring scale in an elevator (Fig. 5-13). Before the elevator begins to move, the scale reads 651 N (146 lb). As the elevator accelerates upward, the scale reading fluctuates somewhat, but, for simplicity, let us assume that this reading remains steady at 733 N during this time interval. (Incidentally, this larger reading corresponds to the feeling of being pushed down as the elevator accelerates upward.) After the elevator reaches a steady upward speed, the scale again reads 651 N. While the elevator remains stationary, determine the person's *(a)* weight in the elevator frame and *(b)* mass. *(c)* While the elevator accelerates upward, again determine the person's weight in the elevator frame. *(d)* Calculate the elevator's acceleration relative to the earth during the upward acceleration. *(e)* Explain why the spring scale reads the same when the elevator's speed is constant as it does when the elevator is stationary.

Solution. *(a)* The person's weight magnitude is given by the scale reading. Letting $\mathbf{j}$ point upward,

$$\mathbf{F}_w = -(651\ \text{N})\mathbf{j}$$

(b) From Eq. 5-4, the person's mass is $m = F_w/g'$. Before the elevator begins to move, the elevator frame is an earth-surface frame, and we assume that an earth-surface frame is inertial: $\mathbf{g}' \simeq \mathbf{g} = -(9.8\ \text{m/s}^2)\mathbf{j}$. Thus

$$m = (651\ \text{N})/(9.8\ \text{m/s}^2) = 66\ \text{kg}$$

(c) While the elevator accelerates upward, the elevator frame is not an earth-surface frame because of its acceleration relative to the earth. However, a spring scale reads the weight magnitude in any frame. Thus

$$\mathbf{F}_w = -(733\ \text{N})\mathbf{j}$$

in the elevator frame. Note that the person's mass is still 66 kg because the mass of an object is an intrinsic property of the object and is independent of the reference frame. *(d)* The acceleration of the person and the elevator (they accelerate together) can be determined by applying Newton's second law to the person. The free-body diagram in Fig. 5-13*b* shows that the net force on the person is

$$\Sigma\mathbf{F} = \mathbf{F}_s + \mathbf{F}_e = F_s\mathbf{j} - F_e\mathbf{j}$$

where $\mathbf{F}_s$ is the upward force exerted by the scale and $\mathbf{F}_e$ is the downward gravitational force exerted by the earth. The magnitude F_s of the force by the spring scale is equal to the weight magnitude F_w, and it is given by the scale reading: $F_s = 733$ N. Also, since an earth-surface frame is essentially inertial, the scale reading while the elevator is stationary gives the magnitude of the gravitational force by the earth: $F_e = 651$ N. Newton's second law, $\Sigma\mathbf{F} = m\mathbf{a}$, applied to the person is

$$(733\ \text{N})\mathbf{j} - (651\ \text{N})\mathbf{j} = (66\ \text{kg})\mathbf{a}$$

Solving for **a** gives

$$\mathbf{a} = (82\ \text{N}/66\ \text{kg})\mathbf{j} = (1.2\ \text{m/s}^2)\mathbf{j}$$

It is interesting that we have determined an acceleration without measuring distances traveled and time intervals. Instead, we measure the stretch of a spring and use Newton's second law. When used in this way, the scale is an *accelerometer,* a device that measures acceleration. *(e)* The spring scale reads the same in a stationary elevator as it does in an elevator moving at constant velocity because these two frames do not accelerate relative to one another. In each case the person's acceleration relative to an inertial frame is essentially zero ($\mathbf{a} = 0$) so that the force upward by the spring scale is equal and opposite the gravitational force downward by the earth ($\Sigma\mathbf{F} = 0$).

SELF-TEST 5-7. Consider the 66-kg person in the elevator in the above example at an instant when the scale reads 563 N. *(a)* What is the person's weight in the elevator frame? *(b)* What is the gravitational force on the person? *(c)* What is the elevator's acceleration relative to an earth-surface frame? ***ANSWERS:*** *(a)* $-(563\ \text{N})\mathbf{j}$; *(b)* $-(651\ \text{N})\mathbf{j}$; *(c)* $-(1.3\ \text{m/s}^2)\mathbf{j}$.

5-6 PROBLEM-SOLVING TECHNIQUES

In the previous sections you have seen a few examples of mechanics problems. Now we list some procedures for solving such problems and then reinforce these procedures with more examples. Newton's second law, $\Sigma\mathbf{F} = m\mathbf{a}$ provides the fundamental principle for solving a problem. Since the second law is a vector relation, we can separate it into components:

$$\Sigma F_x = ma_x \qquad \Sigma F_y = ma_y \qquad \Sigma F_z = ma_z \qquad (5\text{-}5)$$

Newton's second law, component form

Each component provides an equation that may be used in a problem.

The procedures for solving a problem can be divided into three parts:

1. Draw a sketch of the system and identify the object (or objects) to which you will apply the second law. On your sketch, show force vectors that represent the forces on the object. Introduce a symbol for each quantity by using a notation that helps bring the quantity to mind. For example, if the mass of a block is given, write $m_b = 2.3$ kg, or if an angle is given, write $\theta = 25°$. These are *known quantities*. Also write a symbol for each *unknown quantity* that is to be found. If the problem asks for the acceleration of an object, write $\mathbf{a} = ?$, so that the unknown quantity is clearly stated at the outset.
2. Draw a free-body diagram (or diagrams) with coordinate axes on it. These axes should be oriented so that subsequent calculations will be simplified. Usually this is done by placing the axes along as many of the forces as possible, or by

placing one axis along the acceleration, if its direction is known. This step requires judgment and there is no right way or wrong way of doing it, just one or two easy ways and many difficult ways. Good judgment comes with practice.

3. Using the free-body diagram, write the components of Newton's second law in terms of the known and unknown quantities. Solve these equations for each unknown quantity in terms of the known quantities. Finally, substitute the numerical values of the known quantities (including their units) and calculate each unknown quantity.

EXAMPLE 5-8

Tension in a rope. A bucket with mass $m = 8.4$ kg is suspended by two light ropes, *a* and *b*, as shown in Fig. 5-14*a*. By a "light" rope, we mean one with a mass small enough so that the weight of the rope is much less than the force it exerts. With this approximation, we may assume that the rope is straight. When a rope (or string or cable) is attached to an object, the magnitude of the force exerted by the rope is called the *tension* in the rope. Determine the tension in ropes *a* and *b*.

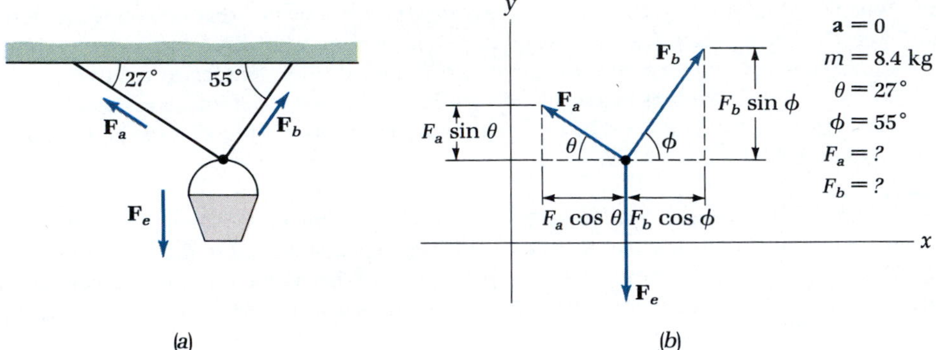

Figure 5-14. Example 5-8: *(a)* A bucket suspended by two ropes. *(b)* Free-body diagram for the bucket.

Solution. The free-body diagram for the bucket is shown in Fig. 5-14*b*. The forces by the ropes on the bucket are represented as $\mathbf{F}_a$ and $\mathbf{F}_b$, and the angles they make with the horizontal are shown as θ and ϕ. Since the bucket remains at rest, its acceleration is zero. Thus Newton's second law gives $\Sigma F_x = 0$, or

$$-F_a \cos\theta + F_b \cos\phi = 0 \qquad \text{(A)}$$

and $\Sigma F_y = 0$, or

$$F_a \sin\theta + F_b \sin\phi - mg = 0 \qquad \text{(B)}$$

where F_a and F_b are the tensions in the ropes, and we have used $F_e = mg$. Equations A and B represent two equations in two unknowns; the unknowns are F_a and F_b. If we solve Eq. A for F_b,

$$F_b = \frac{F_a \cos\theta}{\cos\phi} \qquad \text{(C)}$$

and substitute this result into Eq. B, then we obtain

$$F_a \sin\theta + \frac{F_a \cos\theta \sin\phi}{\cos\phi} - mg = 0$$

Now we have an equation which contains only one unknown. Solving for F_a gives

$$F_a = \frac{mg}{\sin\theta + \cos\theta \tan\phi} \qquad \text{(D)}$$

To obtain a similar expression for F_b, we insert F_a from Eq. D into Eq. C, which gives

$$F_b = \frac{mg}{\sin\phi + \cos\phi \tan\theta} \qquad \text{(E)}$$

(Given Eq. D, could you have predicted Eq. E from symmetry?) Inserting the numerical values of the known quantities, we find

$$F_a = \frac{(8.4\ \text{kg})(9.8\ \text{m/s}^2)}{\sin 27° + \cos 27° \tan 55°} = 48\ \text{N}$$

and

$$F_b = \frac{(8.4\ \text{kg})(9.8\ \text{m/s}^2)}{\sin 55° + \cos 55° \tan 27°} = 74\ \text{N}$$

SELF-TEST 5-8. Suppose the two ropes in the above example are adjusted so that the angles they make with the horizontal are $\theta = 32°$ and $\phi = 61°$. Determine the tension in each rope. ***ANSWER:*** $F_a = 40$ N, $F_b = 70$ N.

EXAMPLE 5-9

A coasting cart. A cart with small wheels and well-lubricated bearings is released from rest at $t = 0$ on a sloping surface, as shown in Fig. 5-15*a*. The cart's mass is $m = 1.3$ kg. *(a)* Determine the magnitude of the force exerted by the surface on the cart. *(b)* Determine the magnitude of the cart's acceleration. At $t = 1.5$ s, determine *(c)* the cart's speed and *(d)* the distance traveled.

Solution. We neglect the effect of the rotation of the small wheels and treat the cart as a particle. Since the bearings are well lubricated, we neglect frictional forces that tend to slow the cart. That is, we assume that the force exerted on the cart by the surface does not have a component parallel to the surface. Therefore this force is normal to the surface and represented as $\mathbf{F}_N$ in the cart's free-body diagram (Fig. 5-15*b*). We let the y axis be perpendicular to the surface because there is no motion in that direction. Thus $a_y = 0$ and the y component of the second law gives $\Sigma F_y = 0$, or

$$F_N - mg\cos\theta = 0 \tag{A}$$

where we have used $F_e = mg$. The x component of the second law, $\Sigma F_x = ma_x$, gives

$$mg\sin\theta = ma \tag{B}$$

where $a = |a_x| = a_x$ is the acceleration magnitude. *(a)* From Eq. A, we have

$$F_N = mg\cos\theta = (1.3\ \text{kg})(9.8\ \text{m/s}^2)\cos 32° = 11\ \text{N}$$

(b) Solving Eq. B for a gives

$$a = g\sin\theta = (9.8\ \text{m/s}^2)\sin 32° = 5.2\ \text{m/s}^2$$

(c) Since the acceleration is constant and the cart started from rest, its speed at $t = 1.5$ s is

$$v = at = (5.2\ \text{m/s}^2)(1.5\ \text{s}) = 7.8\ \text{m/s}$$

(d) The distance traveled after 1.5 s is

$$d = \tfrac{1}{2}at^2 = \tfrac{1}{2}(5.2\ \text{m/s}^2)(1.5\ \text{s})^2 = 5.8\ \text{m}$$

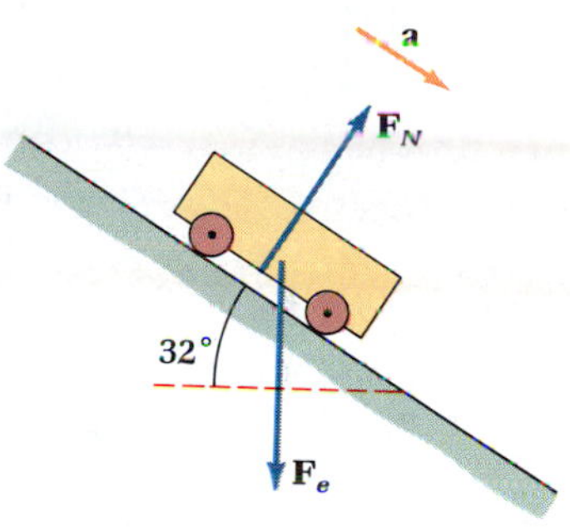

(a)

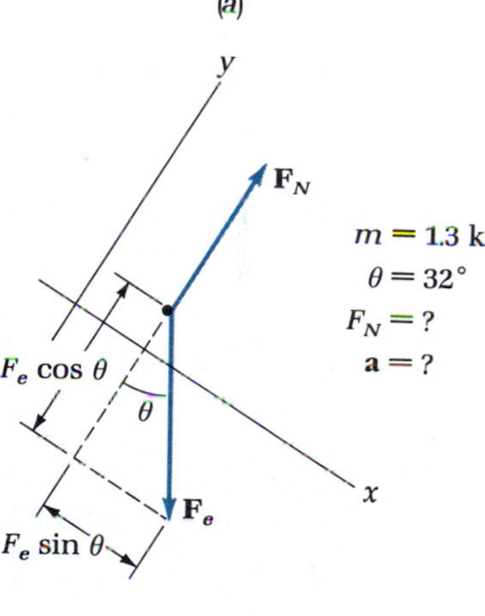

(b)

Figure 5-15. Example 5-9: *(a)* A cart on a sloping surface. *(b)* Free-body diagram for the cart.

SELF-TEST 5-9. Repeat parts *(a)* and *(b)* of the above example using a coordinate frame with the x axis horizontal and the y axis vertical. Now the normal force $\mathbf{F}_N$ and the acceleration $\mathbf{a}$ have both x and y components. This shows how the algebra can become much more burdensome with an inconvenient choice for a coordinate frame.

EXAMPLE 5-10

A cart pulled by a falling block. A cart (mass $m_C = 1.8$ kg) with small wheels and well-lubricated bearings is connected to a block (mass $m_B = 0.50$ kg) by a string which passes over a pulley, as shown in Fig. 5-16*a*. Assume that the pulley rotates freely and that its mass is small enough so that the effect of its rotation is insignificant. That is, the pulley's only effect is to change the direction of the string. Consequently, the tension is the same throughout the string, and the magnitudes of the forces exerted by the string on the cart and on the block are both equal to the tension. Determine *(a)* the acceleration magnitude of the cart (and the block) and *(b)* the tension in the string.

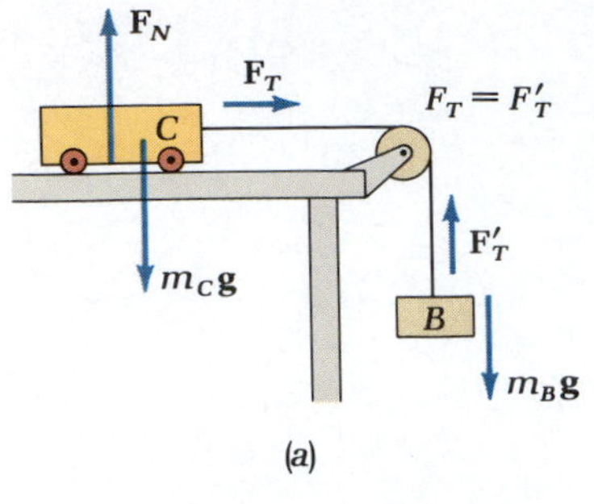

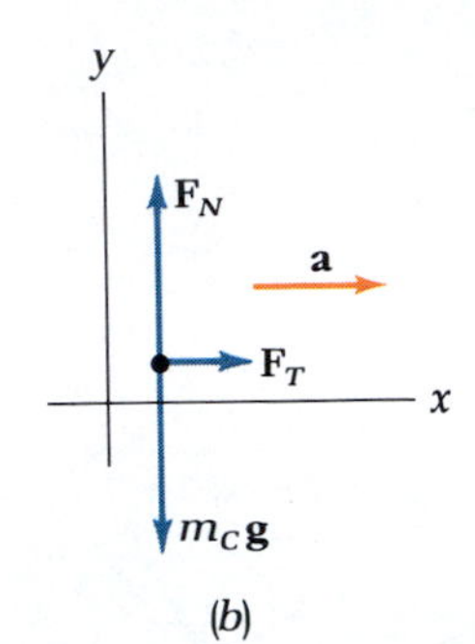

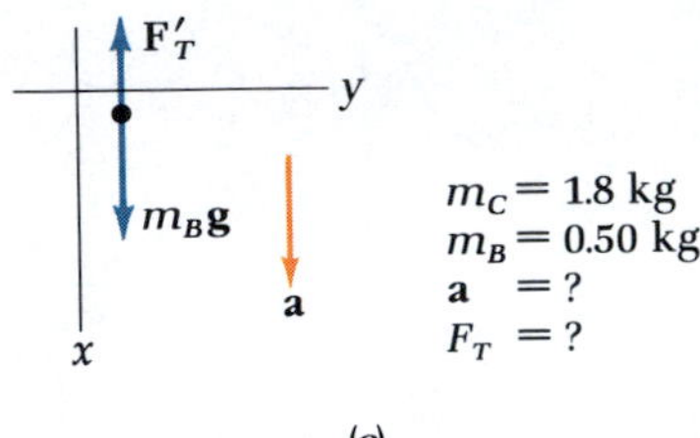

Figure 5-16. Example 5-10: *(a)* A cart being pulled along a horizontal surface by a string tied to a falling block. *(b)* Free-body diagram for the cart. *(c)* Free-body diagram for the block.

Solution. The free-body diagrams for the cart and for the block are shown in Fig. 5-16*b* and *c*, respectively. Because the cart and the block are connected by the string, they have the same acceleration magnitude a. The coordinate frames are oriented so that the acceleration of each object is in the $+x$ direction of its frame. The x component of the second law applied to the cart gives

$$F_T = m_C a \tag{A}$$

where F_T is the tension in the string. For the block, the x component of the second law is

$$m_B g - F_T = m_B a \tag{B}$$

Equations A and B are two equations in two unknowns; the unknowns are a and F_T. *(a)* If we add Eq. A to Eq. B, F_T is eliminated and we obtain

$$m_B g = m_B a + m_C a$$

Solving for a gives

$$a = \frac{m_B}{m_C + m_B} g = \frac{0.50 \text{ kg}}{1.8 \text{ kg} + 0.50 \text{ kg}} 9.8 \text{ m/s}^2 = 2.1 \text{ m/s}^2$$

(b) Inserting the expression above for a into Eq. A gives

$$F_T = \frac{m_C m_B}{m_C + m_B} g = \frac{(1.8 \text{ kg})(0.50 \text{ kg})}{1.8 \text{ kg} + 0.50 \text{ kg}} 9.8 \text{ m/s}^2 = 3.8 \text{ N}$$

SELF-TEST 5-10. Consider the acceleration magnitude a of the system in the above example when the mass of the cart or the mass of the block is changed. *(a)* Evaluate the acceleration magnitude with the mass of the cart doubled from 1.8 to 3.6 kg while the mass of the block remains at 0.50 kg. *(b)* Evaluate the acceleration magnitude with the mass of the block doubled from 0.50 to 1.00 kg while the mass of the cart remains at 1.8 kg. *(c)* Explain the physical reason why a larger mass for the cart decreases the acceleration magnitude while a larger mass for the block increases it. ***ANSWERS:*** *(a)* 1.2 m/s²; *(b)* 3.5 m/s². *(c)* When the mass of the cart increases, the inertia of the system (cart plus block) increases, but the net force exerted on the system does not. When the mass of the block increases, the inertia of the system increases, but the net force exerted on the system increases by a larger factor than does the inertia.

COMMENTARY: Classical Mechanics and Determinism

Now that you have been introduced to Newton's laws of motion, let us stand back and view them in perspective.

1. The first law defines an inertial reference frame, the reference frame for measuring **a** in the second law.
2. The second law, $\Sigma \mathbf{F} = m\mathbf{a}$, connects the forces exerted on an object to the object's acceleration. The second law is often called the equation of motion.
3. The third law, $\mathbf{F}_{ab} = -\mathbf{F}_{ba}$ expresses the relation between the forces that two interacting objects exert on each other.

Force laws

We apply Newton's second law to determine the motion of an object. To find the net force $\Sigma \mathbf{F}$, we need some way to obtain the individual forces exerted on the object. We need force laws. A *force law* is an expression or a rule for determining the force on an object in terms of properties of the object and its environment. Equation 5-3, which gives the gravitational force on (or weight of) an object on or near the surface of the earth, is an example of a force law. Other force laws will follow. In the next chapter we give a force law for frictional forces; in Chap. 7 we develop the gravitational force law; in Chap. 20 we introduce the electric force law; and so on. The combination of the force laws with Newton's laws of motion is called newtonian mechanics, or *classical mechanics.*

Definition of classical mechanics

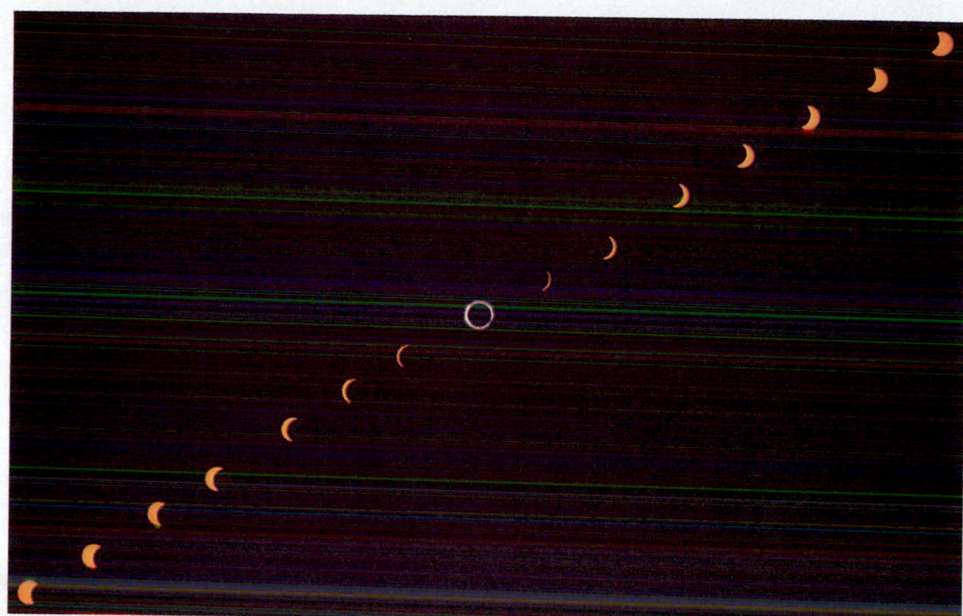

A solar eclipse. Astronomers use classical mechanics to predict such events centuries into the future, or to tell us when they occurred in the distant past.

The procedure for predicting the motion of any object is now clear. From properties of the object and its environment, the force laws are used to obtain the individual forces on the object, and the forces are added to find the net force. Then the acceleration of the object is determined with Newton's second law: $\mathbf{a} = \Sigma\mathbf{F}/m$. If the position and velocity of the object at some instant are known, the methods of kinematics can be used to calculate the velocity and position as functions of time. In principle, if all the forces are known, then the motion can be determined exactly.

Newton's laws of motion imply that an object's future conditions are completely determined by its present conditions, and its present conditions were completely determined by its past conditions. This may be said for every object in the universe. Newton's laws suggest that the evolution of events in the universe is an unfolding of conditions that were determined from some beginning. This idea has had a great influence on philosophy, religion, and the concept of free will. After Newton presented the laws of motion of the world, there arose a branch of philosophical thought called *mechanistic determinism*. This view of the universe was summarized in the words of Pierre Simon de Laplace (1749–1827):

> If an intellect were to know, for a given instant, all the forces that animate nature and the conditions of all the objects that compose her, and were also capable of subjecting these data to analysis, then this intellect would encompass in a single formula the motions of the largest bodies in the universe as well as those of the smallest atom; and the future as well as the past would be present before its eyes.

Indeed, classical mechanics is generally regarded as one of the most successful theories in all of science. Using this theory, engineers have placed astronauts on the moon and sent space probes to the outer reaches of the solar system. With split-second accuracy, astronomers can predict celestial events decades in advance. However, classical mechanics does have limitations. It is not applicable to small objects, those as small as atoms and smaller. Then we must use quantum mechanics (Chap. 39)* With quantum mechanics, we can no longer predict all mechanical quantities to any desired precision, not even in principle. The basis for mechanistic determinism evaporates.

Another limitation of classical mechanics occurs for an object traveling at a speed

* In some cases, quantum mechanics is also applied to macroscopic systems. In the "extended" edition of this text, quantum mechanics is dealt with more fully in Chaps. 40 through 43.

near the speed of light or an event near a massive body, such as a large, dense star. For such cases, we use Einstein's theory of relativity (Chap. 38).

During the 1970s, the 1980s, and to the present time, a new field of study has arisen which bears upon the concept of mechanistic determinism; it is called chaos. Chaos may become a new branch of science, somewhere between mathematics and physics. Its birth and growth have been accompanied by the expanding use of computers in science. Chaos has shown us that a deterministic equation, such as Newton's second law, can produce nondeterministic results. Because it provides a new way to use Newton's laws to attack important problems, chaos is causing a renewed interest in classical mechanics at the cutting edge of physics research.

SUMMARY

Section 5-1. Force and Mass

Mass is a scalar quantity. The mass of an object characterizes the object's resistance to a change in its velocity. Force is a vector quantity. A force can be exerted on an object by some other object. A force can cause an object to become deformed, and it can cause an object to accelerate.

Section 5-2. Newton's First Law

Newton's first law states that if $\Sigma \mathbf{F} = 0$ for an object, then $\mathbf{a} = 0$. Newton's first law is used to define an inertial reference frame: An inertial reference frame is a frame in which Newton's first law is valid.

Section 5-3. Newton's Second Law

The motion of an object is determined with Newton's second law:

$$\Sigma \mathbf{F} = m\mathbf{a} \tag{5-1}$$

Force is defined as that which causes an object to accelerate.

Section 5-4. Newton's Third Law

Newton's third law states that when two objects exert forces on each other, these forces are equal and opposite:

$$\mathbf{F}_{ab} = -\mathbf{F}_{ba} \tag{5-2}$$

If an object interacts with the standard kilogram, and the net force on the object and the standard kilogram is given by their interaction force only, then the mass m of the object is defined as

$$m = 1\ \text{kg}\,\frac{a_s}{a}$$

where a_s and a are the accelerations of the standard kilogram and the object, respectively.

Section 5-5. Weight and the Gravitational Force by the Earth

For an object on or near the earth's surface, the gravitational force $\mathbf{F}_e$ exerted by the earth is

$$\mathbf{F}_e = m\mathbf{g} \tag{5-3}$$

The CGPM definition of weight $\mathbf{F}_w$ is

$$\mathbf{F}_w = m\mathbf{g}' \tag{5-4}$$

In an inertial frame $\mathbf{F}_e = \mathbf{F}_w$, and in an earth-surface frame $\mathbf{F}_e \approx \mathbf{F}_w$.

Section 5-6. Problem-Solving Techniques: Mechanics

The procedures for solving a mechanics problem are (i) draw a sketch, (ii) draw a free-body diagram, and (iii) apply Newton's second law.

QUESTIONS

1. As a corollary to the laws of motion, Newton stated:

> Corollary I. A body, acted on by two forces simultaneously, will describe the diagonal of a parallelogram in the same time it would describe the sides by those forces separately.

In modern terms, what property of force was Newton describing? Explain.

2. By continuously measuring the velocity of an object, can you tell whether the net force on the object is zero? Explain.

3. Is it possible for an object to follow a curved path when the net force on it is zero? Explain.

4. Suppose you are riding in a car at constant velocity when the driver suddenly slams on the brakes so that you are "pushed" forward. Was this "push" exerted on you by some other object? If so, identify the object. If no object exerted the push, then how do you account for your acceleration relative to the car?

5. Suppose you are riding on a steadily rotating merry-go-round and you place a skate on the floor such that it is aligned with its back end toward the center of the merry-go-round and its front end away from the center. When you release the skate from rest (relative to you), it accelerates forward, radially away from the merry-go-round's center. Is there a force on the skate that is directed radially outward? If so, identify the object that exerts this force. If not, explain why the skate accelerates.

6. A block of ice slides down a curved chute, as shown in Fig. 5-17, and then onto a horizontal floor. When the block exits the chute, will it continue its curved path or will it change to a straight path? Explain.

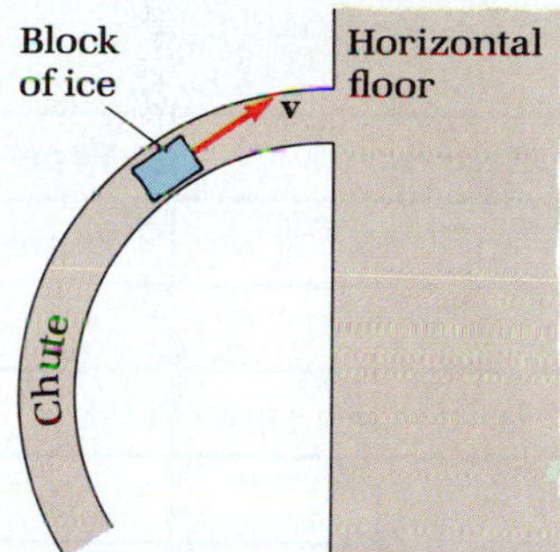

Figure 5-17. Question 6.

7. Suppose you drop objects a and b from a second-floor window. Each object is released from rest at the same instant, and $m_a > m_b$. Assume that at each instant of the time during the fall the force due to air resistance on object a is equal to that on object b. Which object reaches the ground first? Explain.

8. What object exerts the force that holds a compass needle in its north-south orientation?

9. When the driver of a car slams on the car's brakes, what object exerts the force that slows the car? Suppose this happened when the street is covered with ice?

10. Must the acceleration in Newton's second law be measured relative to any particular type of reference frame? If so, what type of reference frame must be used? Are there a limited number of these reference frames? Explain.

11. In the British system of units, mass is measured in slugs (sl), length in feet (ft), and force in pounds (lb), such that 1 lb = (1 sl)(1 ft/s^2). What is your mass in slugs?

12. A friend tells you that Newton's third law cannot be correct because it predicts that an object cannot be moved. The friend says: "Suppose I push on a cart. By Newton's third law the force exerted on me by the cart is equal and opposite the force I exert on the cart. Consequently, the net force is zero and the third law predicts that the cart cannot be moved." Explain what is wrong with this reasoning.

13. Two strings, a and b, can sustain the same maximum tension before breaking. But a force smaller than this maximum tension causes string a to stretch much more than b. If each string has one end tied to a rigid support and the other end is given a quick jerk, which string is more likely to break? Explain.

14. In a free-body diagram, why is it often convenient to orient the axes horizontally and vertically? In Example 5-9, why was it preferable to orient the axes parallel and perpendicular to the surface?

15. Suppose that while standing on a rotating merry-go-round, you hold a rock suspended from the end of a string. Does the string hang vertically downward? Is the net force on the rock zero? Is your frame of reference an inertial frame?

16. Suppose that you drop a marble of mass m into a jar of honey. As the marble sinks, its speed is effectively constant. What is the net force on the marble as it sinks? What are the magnitude and direction of the force exerted by the honey on the marble?

17. If string b in Fig. 5-18 is pulled downward, while the force is gradually increased, string a eventually breaks. However, if string b is given a sharp jerk downward, string b is more likely to break. Explain.

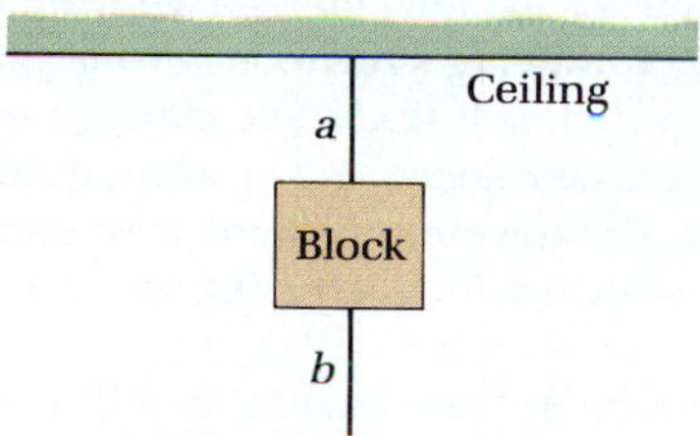

Figure 5-18. Question 17.

18. The force vectors in Fig. 5-19 each have the same magnitude. When exerted on an object, which combinations of these forces, if any, would result in a net force of zero on the object?

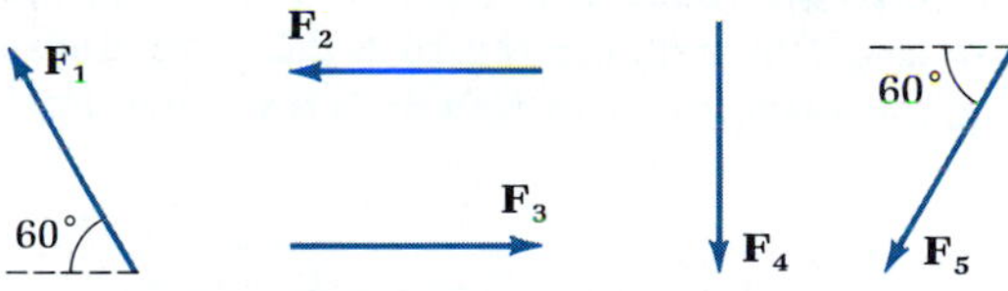

Figure 5-19. Question 18.

19. The strings and the spring scales in Fig. 5-20 have negligible weights, each block weighs 25 N, and the pulleys are essentially frictionless. What is the tension in each string? What is the reading on each spring scale?

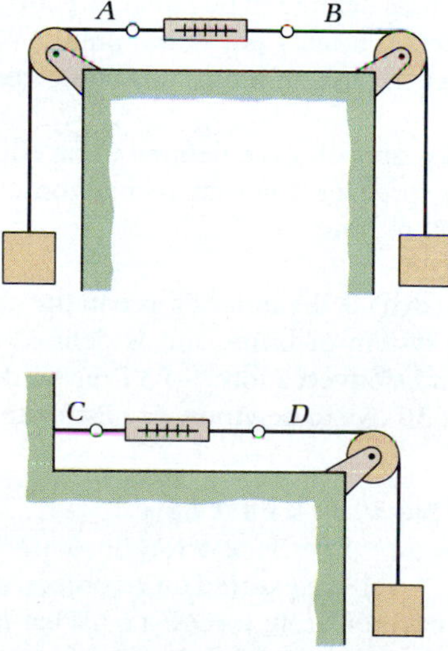

Figure 5-20. Question 19.

20. A rope is stretched tightly between two trees, and a brick is hung by a short string from the middle of the rope. Is it possible to stretch the rope tightly enough so that the rope is straight?

21. A horse is pulling a cart toward town (toward the north) by exerting a horizontal force of magnitude F. What are the magnitude and direction of the force exerted by the cart on the horse?

22. Is there any directional relation between the net force on an object and the object's velocity? If so, what is this relation?

23. What are the magnitude and direction of the gravitational force you exert on the earth?

24. Suppose you are standing on a spring scale in an elevator. In which of the following situations is your apparent weight greatest and in which is it least? The elevator is *(a)* traveling upward with constant speed, *(b)* traveling downward with constant speed, *(c)* traveling upward with increasing speed, *(d)* traveling upward with decreasing speed.

25. If you were on another planet, would you expect your weight to be different from what it is on earth? Would you expect your mass to be different?

26. Write the dimension of mass $[M]$ in terms of force $[F]$, length $[L]$, and time $[T]$.

27. Ecologists are fond of the adage "You can never do just one thing." Discuss the connection between this axiom and Newton's third law. Which do you regard as the more fundamental statement, the adage above or Newton's third law? Explain.

28. Complete the following table:

Symbol	Represents	Type	SI Unit
$\Sigma\mathbf{F}$		Vector	
$\mathbf{F}_{ab}$			N
F_T	Tension in a rope		
F_e			
F_w			
m			

Assumptions to Use in the Following Exercises and Problems

1. An earth-surface frame is the same as an inertial frame.
2. The force exerted by a rope or string is directed along its length, with magnitude equal to the tension.
3. The only effect of a pulley is to change the direction of alignment of the rope or string that is wrapped over the pulley.
4. Any cart can be treated as a particle, and the force exerted on the cart by the surface on which it rolls is directed perpendicular (or normal) to the surface.

EXERCISES

Section 5-1. Force and Mass

1. The unit of mass in the British system is the slug (sl), and 1.000 sl = 14.59 kg. What is the mass of a 72-kg person in slugs?

2. The density of water is about 1000 kg/m^3. *(a)* The tonne (t) is a unit of mass defined to be equal to 1000 kg. What is the density of water in tonnes per cubic meter? *(b)* What is the density of water in units of grams per cubic centimeter?

3. The ton is a unit of force defined to be equal to 2000 lb. *(a)* Convert a force of 1.6 tons to newtons. *(b)* Convert a force of 5.6 MN to tons.

4. The dyne (dyn) is the unit of force in the cgs (centimeter-gram-second) system of units, and is defined to be equal to 1×10^{-5} N. *(a)* Convert a force of 34 mN to dynes. *(b)* Convert a force of 630 dyn to newtons. *(c)* Estimate your weight in dynes.

Section 5-2. Newton's First Law

5. Two forces, $\mathbf{F}_1 = -(2.4\text{ N})\mathbf{i} + (6.1\text{ N})\mathbf{j}$ and $\mathbf{F}_2 = (8.5\text{ N})\mathbf{i} - (9.7\text{ N})\mathbf{j}$, are exerted on an object. *(a)* What is the magnitude of each of these forces? *(b)* What is the angle between each of these forces and the x axis? *(c)* Draw a free-body diagram showing these forces. *(d)* Determine the magnitude and direction of the net force on the object.

6. Two forces, $\mathbf{F}_1$ and $\mathbf{F}_2$, act on an object. If $\mathbf{F}_1 = -(6.1\text{ N})\mathbf{i} + (5.6\text{ N})\mathbf{j} - (4.7\text{ N})\mathbf{k}$ and the net force on the object is $\Sigma\mathbf{F} = -(4.1\text{ N})\mathbf{i} - (2.4\text{ N})\mathbf{j} + (1.1\text{ N})\mathbf{k}$, what is $\mathbf{F}_2$?

7. While sliding a large crate across a floor, Alvin and Bill exert horizontal forces $\mathbf{F}_A$ and $\mathbf{F}_B$ on it. Force $\mathbf{F}_A$ is toward the north with magnitude 130 N and $\mathbf{F}_B$ is 32° east of north with magnitude 180 N. What are the magnitude and direction of the single force that has the same effect as these two forces acting together?

8. In the free-body diagram shown in Fig. 5-21, $F_1 = 22$ N, $F_2 = 18$ N, and $F_3 = 16$ N. *(a)* Determine the components of each of these forces. *(b)* Determine each component, ΣF_x and ΣF_y, of the net force. *(c)* Determine the net force $\Sigma\mathbf{F}$ in terms of unit vectors. *(d)* Determine the magnitude and direction of the net force.

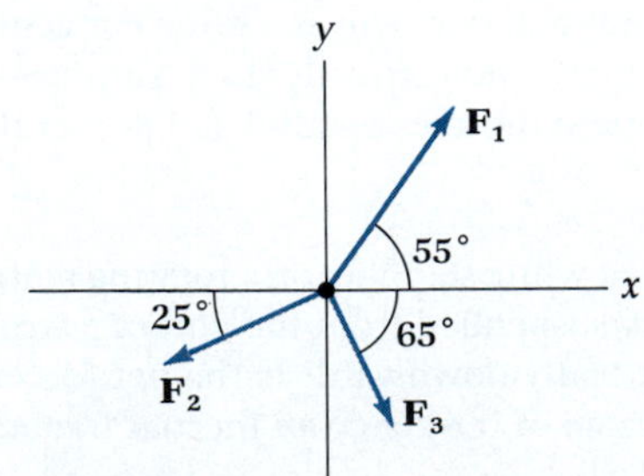

Figure 5-21. Exercise 8.

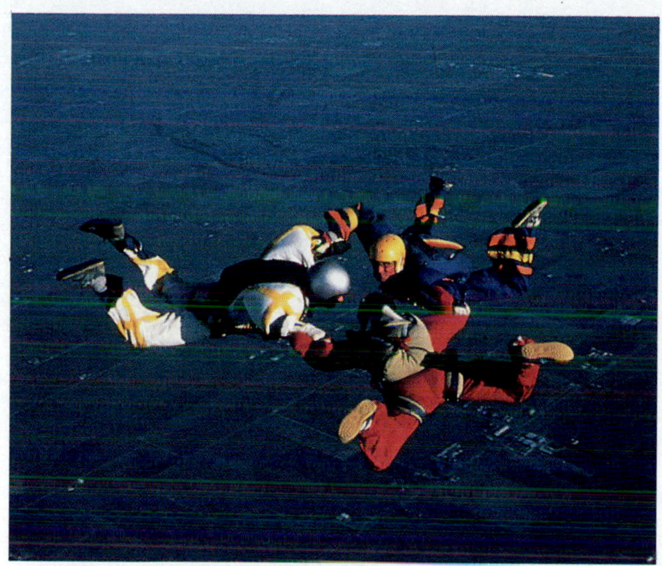

Three skydivers.

9. Shortly after jumping out of an airplane, a skydiver, whose weight is 720 N, reaches a velocity that is essentially constant. In this situation, there are two significant forces exerted on the skydiver. *(a)* What exerts each of these forces? *(b)* What are the magnitude and direction of each of these forces?

10. A box is pulled along a horizontal floor at constant velocity with a spring scale. There are three significant forces on the box: the force by the scale, which is 15 N and horizontal; the weight of the box, which is 25 N; and the force exerted by the floor. Determine the magnitude and direction of the force exerted by the floor.

Section 5-3. Newton's Second Law

11. An 830-kg car starts from rest and reaches a speed of 22 m/s after 10 s. Assuming that this acceleration is uniform, determine the magnitude of the net force on the car.

12. A cathode-ray tube (such as a TV tube) contains an element, called an electron gun, which emits a beam of electrons. Suppose that in an electron gun an electron is accelerated from rest to a speed of 2×10^7 m/s over a distance of 10 mm. Estimate the net force magnitude on an electron in this electron gun. (*Hint:* Use Eq. 3-12 to detrmine an electron's acceleration. The mass of the electron is given on the inside front cover.)

13. The nucleus of an atom is very small, about 10^{-14} m across. Suppose that in a nuclear reaction a neutron with a speed of 1×10^7 m/s impinges upon a nucleus and comes to rest inside it. *(a)* Estimate the net force magintude on the neutron during the reaction. (*Hint:* Use Eq. 3-12 to find the acceleration of the neutron. The mass of the neutron is given in App. F.) *(b)* Estimate the time interval of the reaction.

14. When a tennis ball is served, the ball accelerates from rest (nearly) to a speed of about 50 m/s. The mass of a tennis ball is about 0.06 kg. Estimate the force magnitude exerted by the racket on the ball, assuming that the acceleration is uniform over a distance of 1 m. (*Hint:* Use Eq. 3-12 to find the ball's acceleration.)

15. Estimate the force exerted on a softball ($m = 0.3$ kg) by the pitcher during a typical pitch.

16. Let us make the (unfounded) assumption that the only force on the Milky Way galaxy is the gravitational force by the Andromeda galaxy. Given that the mass of the Milky Way galaxy is 7×10^{41} kg, determine the magnitude of the Milky Way's acceleration. (See Table 5-1.) What is the reference frame in which this acceleration would be measured?

17. A baseball bat strikes a 0.15-kg baseball such that it reverses the ball's velocity from 48 m/s horizontal and eastward to 81 m/s horizontal and westward in a time interval of 0.01 s. Estimate the force by the bat on the ball, assuming the force is uniform and neglecting all other forces on the ball.

18. Estimate the force exerted on a bullet by the expanding gases in a rifle barrel during a shot. The barrel is 0.5 m long, the bullet exits the barrel at a speed of 400 m/s, and the mass of the bullet is 2 g. Assume that the force is constant during the shot and neglect all other forces on the bullet.

Section 5-4. Newton's Third Law

19. Using Table 5-1, determine the magnitude of *(a)* the gravitational force exerted by the Milky Way galaxy on the Andromeda galaxy and *(b)* the gravitational force exerted by the earth on the sun.

20. A 2200-kg truck collides with a 550-kg sports car, and during the collision the net force on each vehicle is essentially the force exerted by the other. If the magnitude of the truck's acceleration is 10 m/s², what is the magnitude of the sports car's acceleration?

21. Carts 1 and 2, each with mass 1.0 kg, are equipped with spring bumpers similar to the carts in Fig. 5-11 (Example 5-6). A block of unknown mass m is fastened to cart 1 and the carts are pushed together, compressing their spring bumpers, and released. The acceleration magnitudes of the carts are $a_1 = 0.51$ m/s² and $a_2 = 1.14$ m/s². The carts have small wheels and well-lubricated bearings. Determine m. State any assumptions that you make.

22. A chemistry book whose weight is 13 N is placed on top of the history book in Fig. 5-10 (Example 5-5) making it a stack of three, not two. Identify each force on each book with an appropriate double-subscript notation and determine the magnitude and direction of each of these forces. How many times do you apply Newton's third law and how many times do you apply Newton's second law in finding these forces?

23. *(a)* Two carts, 1 and 2, are being pushed along by an externally applied force $\mathbf{F}_{a1}$, which is exerted on cart 1, (Fig. 5-22*a*). Treat each cart as a particle and neglect the frictional forces which tend to slow each cart. Given that $F_{a1} = 12$ N, $m_1 = 4.0$ kg, and $m_2 = 2.0$ kg, determine the magnitude and direction of each of the interaction forces $\mathbf{F}_{12}$ and $\mathbf{F}_{21}$. *(b)* Now suppose force $\mathbf{F}_{a1}$ is taken away and an externally applied force $\mathbf{F}_{a2}$ is exerted on cart 2 (Fig. 5-22*b*), where $F_{a2} = 12$ N. Determine the magnitude and direction of each of the interaction forces in this case. *(c)* Explain why the magnitude of the interaction forces is different in the two cases.

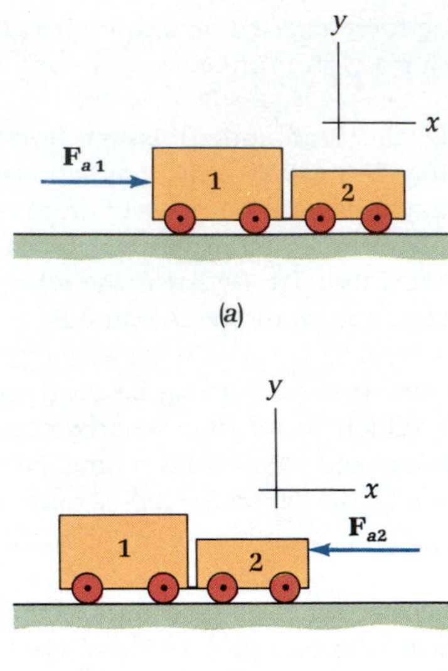

Figure 5-22. Exercise 23.

Section 5-5. Weight and the Gravitational Force by the Earth

24. On Mars, the acceleration magnitude of any object in free-fall relative to a Mars-surface frame is 3.8 m/s². What is the weight of a 68-kg person in a Mars-surface frame?

25. While on the surface of the planet Illocorb, a space traveler stands on a spring scale and the scale reads 950 N. If the space traveler's mass is 71 kg, what is the acceleration of an object in free-fall on Illocorb?

26. A person whose mass is 58 kg stands on a spring scale in an elevator. What are the magnitude and direction of the elevator's acceleration at an instant when the scale reads *(a)* 570 N; *(b)* 420 N; *(c)* 710 N?

27. Determine the weight of an electron in an earth-surface frame. (The mass of the electron is given on the inside front cover.) *(b)* Compare your answer with the net force on an electron in the electron gun of the cathode-ray tube described in Exercise 12.

28. *(a)* Determine the weight of a neutron in an earth-surface frame. (The mass of the neutron is given on the inside front cover.) *(b)* Compare your answer with the net force on a neutron as it is absorbed by the nucleus in the reaction described in Exercise 13.

29. A 77-kg person is standing on a spring scale in an elevator. What is the person's weight while the elevator is *(a)* accelerating upward at 2.8 m/s², *(b)* accelerating downward at 3.1 m/s², *(c)* traveling upward at a constant speed of 4.4 m/s?

Section 5-6. Problem-Solving Techniques: Mechanics

30. A 52-kg skier slides down a straight slope that makes an angle of 24° with the horizontal. *(a)* Neglecting frictional forces, determine the skier's acceleration magnitude. *(b)* What is the skier's speed 1.0 s after starting from rest? *(c)* How far did the skier travel in 1.0 s?

31. A 24-kg box resting on the floor has a rope secured to its top. The maximum tension the rope can withstand without breaking is 310 N. What is the minimum amount of time in which the box can be lifted a vertical distance of 4.6 m by pulling on the rope?

32. Determine the tension in the supporting cable of a 1500-kg elevator while the elevator is accelerating *(a)* upward at 2.1 m/s² and *(b)* downward at 2.1 m/s². Neglect forces other than the cable tension and the weight of the elevator.

33. A 32-kg sled is being pulled along a horizontal icy surface by a rope (Fig. 5-23). The constant tension in the rope is 140 N and frictional forces are negligible. *(a)* Draw a free-body diagram for the sled. *(b)* What are the magnitude and direction of the force exerted by the surface on the sled? *(c)* What is the sled's acceleration magnitude? *(d)* If it starts from rest, how far does the sled travel in 1.3 s?

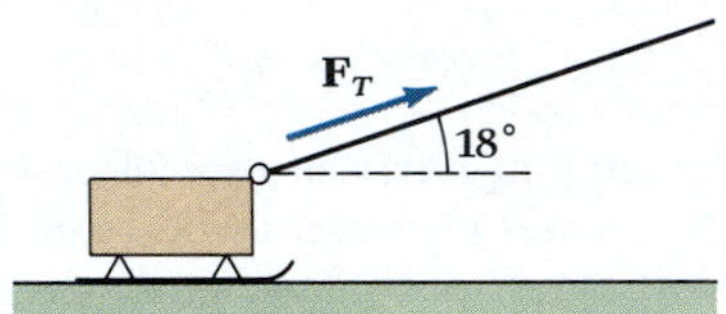

Figure 5-23. Exercise 33.

34. A 1430-kg car accelerates in the same direction as its velocity along a straight horizontal road, and $a = 1.95$ m/s². *(a)* What is the magnitude of the net force on the car? *(b)* Suppose air friction exerts a force of 513 N on the car, directed opposite the velocity. Determine the horizontal and vertical components of the force by the road on the car.

35. The cart in Fig. 5-24 has a mass of 2.4 kg and remains at rest. The cart's axles are well lubricated so that the force exerted on it by the surface has a negligible component parallel to the surface. *(a)* Determine the magnitude of the force by the surface. *(b)* Determine the tension in the string.

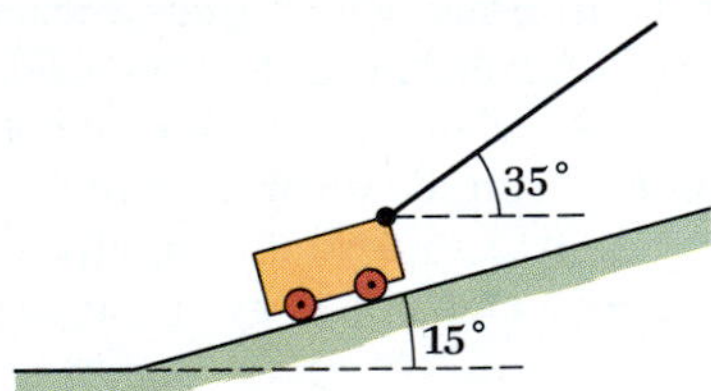

Figure 5-24. Exercise 35.

36. The mass of the suspended block in Fig. 5-25 is 45 kg. Determine the tension in each rope.

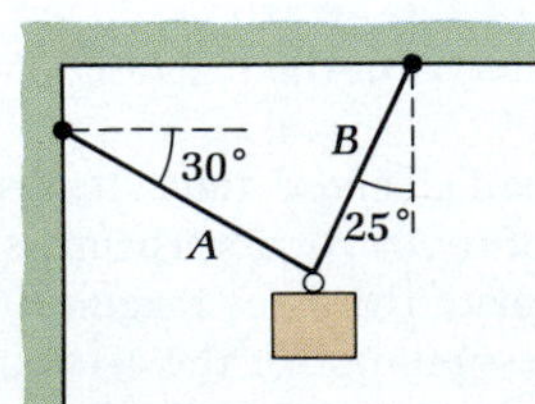

Figure 5-25. Exercise 36.

37. In Fig. 5-26, magnet *A* is directly west of magnet *B*, and *A* exerts a horizontal force of attraction on *B*. Magnet *B*, whose mass is 0.200 kg, is suspended from a string and remains stationary, with $\theta = 27.5°$. *(a)* Determine the tension in the string. *(b)* Determine the magnitude and direction of the force exerted by *A* on *B*.

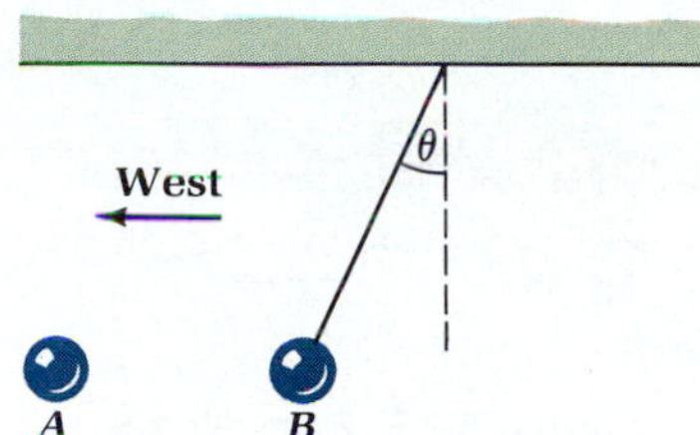

Figure 5-26. Exercise 37.

38. The cart-block system in Fig. 5-16 (Example 5-10) is released from rest with block *B* 1.2 m above the floor. How long will it take for the block to hit the floor?

39. A bird with mass $m = 26$ g perches at the middle of a stretched string (Fig. 5-27) *(a)* Show that the tension in the string is given by $F_T = mg/(2 \sin \theta)$. Determine the tension when *(b)* $\theta = 5°$ and *(c)* $\theta = 0.5°$. Assume that each half of the string is straight.

Figure 5-27. Exercise 39.

40. A girl pushes a 31-kg box sled at constant speed up a straight snowy slope by exerting a horizontal force on the sled, (Fig. 5-28). Neglect the frictional force by the surface on the sled by assuming that the force exerted by the surface on the sled has no component parallel to the surface. Determine the magnitude of the force exerted *(a)* by the girl and *(b)* by the surface.

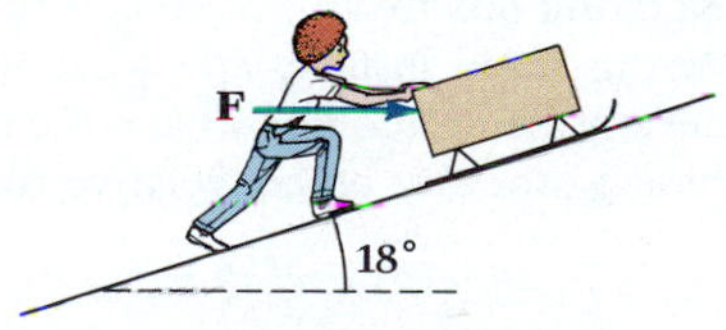

Figure 5-28. Exercise 40.

Additional Exercises

41. A 940-kg car is traveling north along a straight horizontal road, and at a certain instant the car has an acceleration of 1.9 m/s² directed along the line of travel. The car is traveling slowly enough so that air resistance is negligible, and the only significant forces on the car are (i) that exerted by the earth's gravitational attraction and (ii) that exerted by the road due to direct contact with the car. Determine *(a)* the horizontal component and *(b)* the vertical component of the force exerted on the car by the road. *(c)* Find the magnitude and direction of this force.

42. A 4.09-kg block slides along a horizontal surface at constant velocity toward the south (Fig. 5-29). The tension in the string is 25.0 N, and the angle the string makes with the horizontal is 30.0°. *(a)* Determine the magnitude and direction of the force exerted on the block by the surface. *(b)* What are the magnitude and direction of the force exerted by the block on the surface?

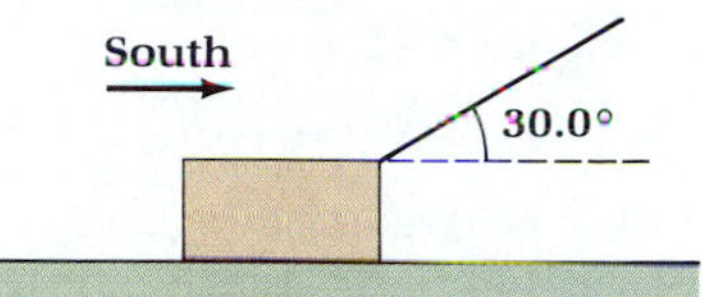

Figure 5-29. Exercise 42.

43. A 3.7-kg cart is being pulled up a 30° slope (Fig. 5-30). The cart is accelerating up the slope with a magnitude of 1.7 m/s². Determine the tension in the string.

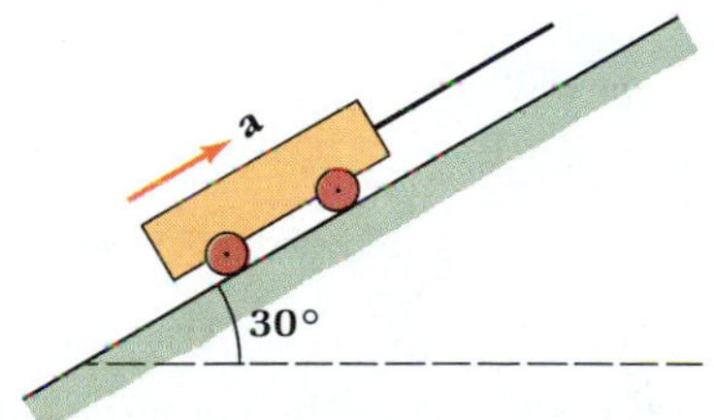

Figure 5-30. Exercise 43.

44. In Fig. 5-31, string *A* is supporting the pulley at its axle; string *B* is secured to the floor, wrapped around the pulley, and connected to the 4.6-kg block. The mass of the pulley is negligible. Determine the tension in each string.

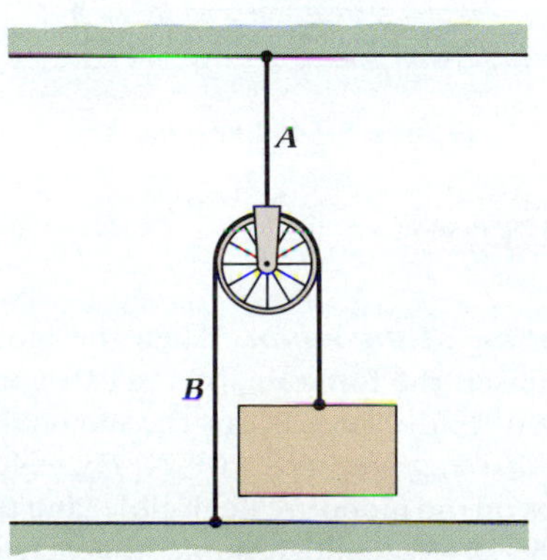

Figure 5-31. Exercise 44.

45. A boat is held fixed in a boat slip by three ropes that are pulled taut (Fig. 5-32). Each rope is horizontal, and the tension in rope 1 is $F_1 = 305$ N. Determine the tension in ropes 2 and 3.

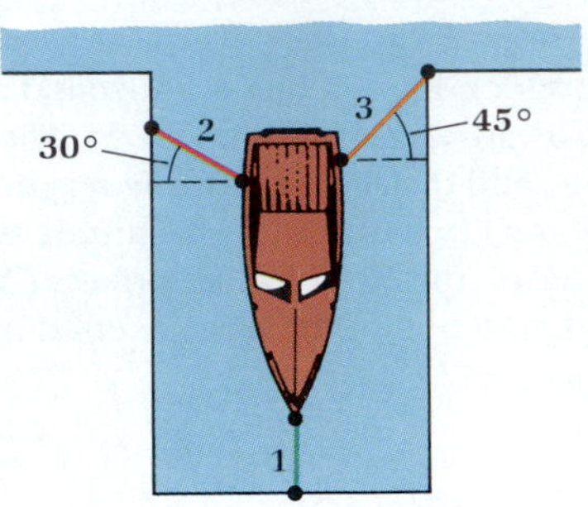

Figure 5-32. Exercise 45.

46. A 1.8-kg block is held stationary against a sloping surface by a force $\mathbf{F}_a$ of magnitude 9.0 N that is directed horizontally and toward the east (Fig. 5-33). *(a)* Determine the magnitude and direction of the force on the block by the surface. *(b)* What are the magnitude and direction of the force on the surface by the block?

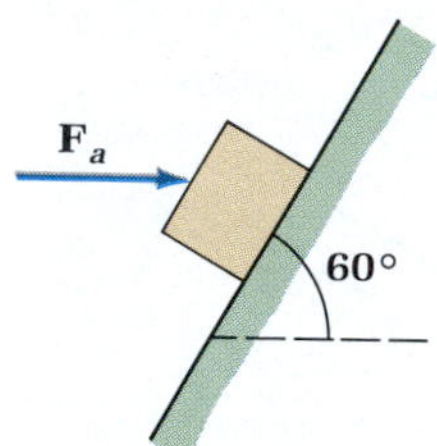

Figure 5-33. Exercise 46.

47. A 55-kg crate is riding on the bed of a truck (Fig. 5-34). As the truck pulls away from a red light, it travels in a straight line on a horizontal road with an acceleration magnitude of 3.4 m/s². Assume that the force on the crate due to the air is negligible. Determine *(a)* the horizontal component and *(b)* the vertical component of the force exerted on the crate by the bed of the truck. *(c)* Find the magnitude of this force.

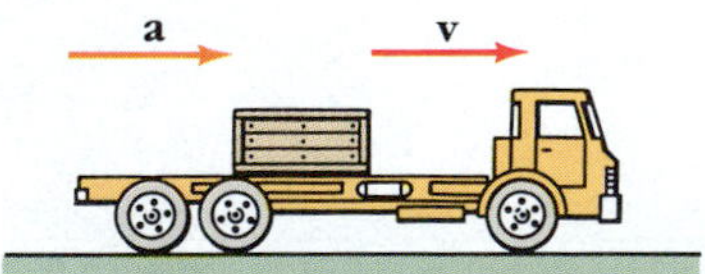

Figure 5-34. Exercise 47.

48. A 2.6-kg cart is held stationary on a 25° slope by a hand that holds the attached string. (Fig. 5-35). Determine *(a)* the tension in the string and *(b)* the magnitude of the force exerted on the cart by the surface. *(c)* If the string is released, what is the acceleration magnitude of the cart?

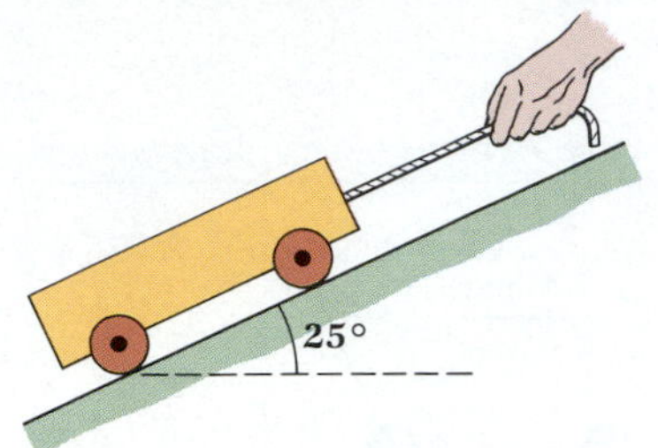

Figure 5-35. Exercise 48.

49. Carts *A* and *B*, each of mass 1.0 kg, are connected by string 1, as shown in Fig. 5-36. A block of mass 1.0 kg is fixed on cart *A*. String 2 is connected to cart *A* and is used to pull the carts such that the acceleration is 4.1 m/s² in the direction of travel. *(a)* Determine the tension in each string. *(b)* Repeat part *(a)* except let the 1.0-kg block be attached to cart *B* rather than cart *A*.

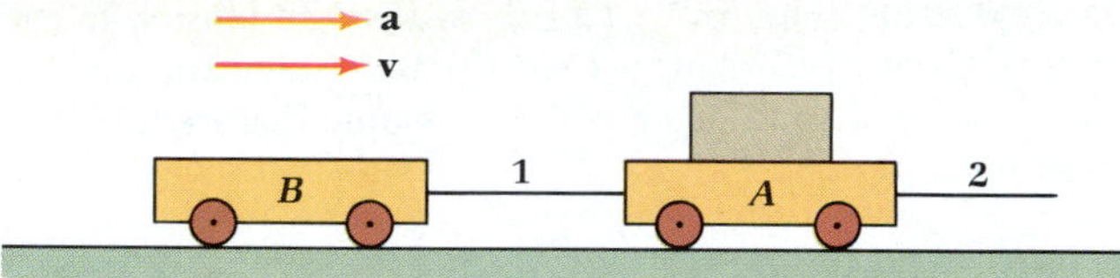

Figure 5-36. Exercise 49.

50. Object *A* exerts a force $\mathbf{F}_{AB} = -(4\text{ N})\mathbf{i} + (7\text{ N})\mathbf{j} - (3\text{ N})\mathbf{k}$ on object *B*. What is the force exerted by *B* on A?

51. Suppose you plan to lower a piano from a balcony to the street below. The piano's weight is 1750 N, and the cable that will support the piano can withstand a maximum tension of 1800 N before breaking. You intend to lower the piano at a constant speed of 0.50 m/s and then, at an instant when the bottom of the piano is at a height *h* above the ground, you intend to reduce the speed at a constant rate such that the speed is zero just as the piano touches the ground. What is the minimum value of *h*?

52. A 58.0-kg person is standing on a spring scale in an elevator that is accelerating downward. At some instant, the scale reads 505 N. At this same instant, *(a)* what is the person's weight in the elevator frame and *(b)* what is the free-fall acceleration magnitude g' for any object relative to this elevator frame?

PROBLEMS

1. ***Acceleration of the moon.*** When the moon is directly overhead at sunset, the force $\mathbf{F}_{em}$ by the earth on the moon is essentially at 90° to the force $\mathbf{F}_{sm}$ by the sun on the moon (Fig. 5-37). Given that $F_{em} = 1.98 \times 10^{20}$ N, $F_{sm} = 4.36 \times 10^{20}$ N, all other forces on the moon are negligible, and the mass of the moon is 7.35×10^{22} kg, determine the magnitude of the moon's acceleration. Is this the acceleration of the moon rela-

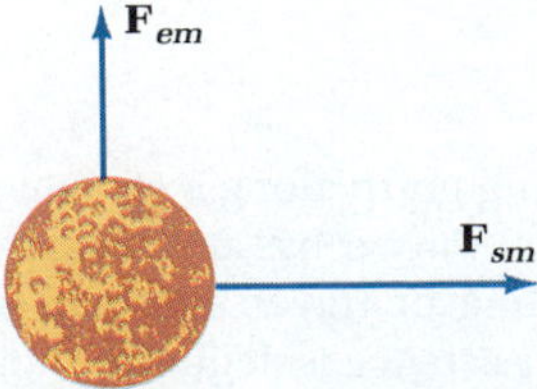

Figure 5-37. Problem 1.

tive to the earth? If not, then relative to what reference frame is the acceleration measured?

2. ***Which way does the block-cart system accelerate?*** In the cart-block system in Fig. 5-38, the cart has mass m_C and the block has mass m_B. Determine an expression for *(a)* the acceleration magnitude a of the cart, *(b)* the tension F_T in the string, *(c)* the force $\mathbf{F}_N$ exerted by the surface on the cart. State the assumptions you must make to work the problem. *(d)* Evaluate a, F_T, and F_N when $\theta = 30°$, $m_C = 4.0$ kg, and $m_B = 2.5$ kg *(e)* Repeat part *(d)* except with $m_B = 2.0$ kg. *(f)* Repeat part *(d)* except with $m_B = 1.5$ kg.

Figure 5-38. Problem 2.

3. ***A hot-air balloon.*** The force that keeps a lighter-than-air aircraft, such as a hot-air balloon or a dirigible, aloft is called a *buoyant force* $\mathbf{F}_B$. This force is related to the displacement of air by the lighter-than-air aircraft. Suppose a hot-air balloon of mass M has a downward acceleration of magnitude a. *(a)* Show that the mass m of ballast that must be dropped overboard to cause the balloon to accelerate upward with magnitude a is $m = 2Ma/(g + a)$. What assumptions must you make to work the problem? *(b)* Evaluate m for the case where $M = 400$ kg and $a = 0.2$ m/s².

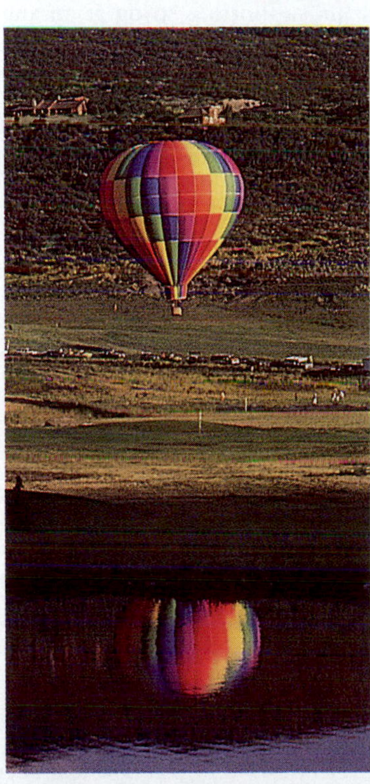

A hot-air balloon.

4. ***Monkey and bananas.*** A 12-kg monkey climbs a light rope (Fig. 5-39). The rope passes over a pulley and is attached to a 16-kg bunch of bananas. Mass and friction in the pulley are negligible so that the pulley's only effect is to reverse the direction of the rope. What is the maximum acceleration the monkey can have without lifting the bananas?

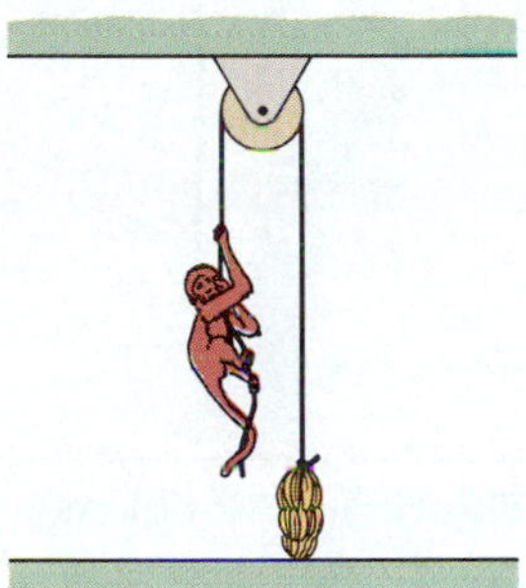

Figure 5-39. Problem 4.

5. ***Atwood's machine.*** Atwood's machine (Fig. 5-40) can be used to measure g. If the two blocks have nearly the same mass, then the acceleration of the system is small and g can be determined without the need to measure short time intervals. Assume that mass and friction in the pulley are negligible (so that the pulley's only effect is to reverse the direction of the light string). *(a)* Show that g can be determined from the expression

$$g = \frac{a(m_2 + m_1)}{m_2 - m_1}$$

In this expression, a is the acceleration magnitude of the blocks, and we have let the mass m_2 of block 2 be greater than the mass m_1 of block 1. *(b)* Suppose you are sent to the planet Norc to measure the magnitude g of the acceleration of free-fall on its surface. Using Atwood's machine with $m_2 = 4.85$ kg and $m_1 = 4.65$ kg, you release the blocks from rest and find that they move a distance of 0.50 m in 2.5 s. What is g on Norc?

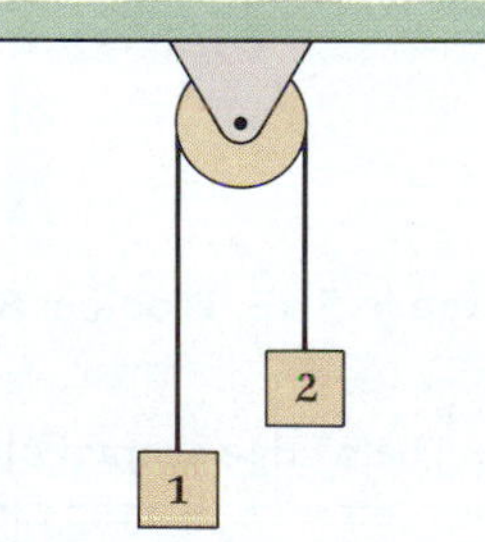

Figure 5-40. Problem 5: Atwood's machine.

6. ***An easier way to shinny up a rope.*** A boy of mass m places one foot in a loop at the end of a light rope that passes over a pulley, and pulls himself upward at constant speed by pulling on the other end of the rope (Fig. 5-41). Neglecting mass and friction in the pulley and the mass of the rope, determine the force exerted on the rope by *(a)* the boy's hands and *(b)* the boy's foot.

Figure 5-41. Problem 6.

7. ***A three-glider train.*** Three air-track gliders, connected by strings, are being pulled along a track by a horizontally applied force **F** such that the acceleration magnitude of the system is 2.0 m/s² (Fig. 5-42). Neglecting friction, determine the tension in each string. The masses of the gliders are $m_a = 2.0$ kg, $m_b = 1.0$ kg, and $m_c = 2.0$ kg.

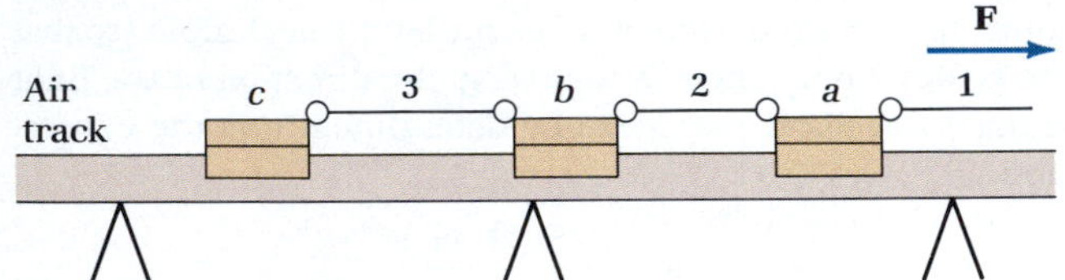

Figure 5-42. Problem 7.

8. ***Cart with a pulley.*** Determine an expression for the acceleration of block *B* in Fig. 5-43 in terms of m_C, m_B, and g. Assume that the cart's wheels are small and have well-lubricated bearings. Neglect mass and friction in the pulleys and neglect the mass of the rope.

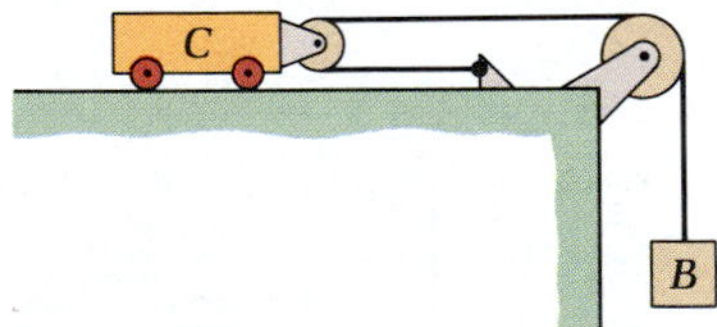

Figure 5-43. Problem 8.

9. ***A flying wedge.*** The wedge-shaped block in Fig. 5-44 has an acceleration to the right such that the cart does not roll up or down its sloping face. The cart's wheels are small and its bearings are well lubricated. *(a)* Show that $a = g \tan \theta$ *(b)* What happens if $a > g \tan \theta$?

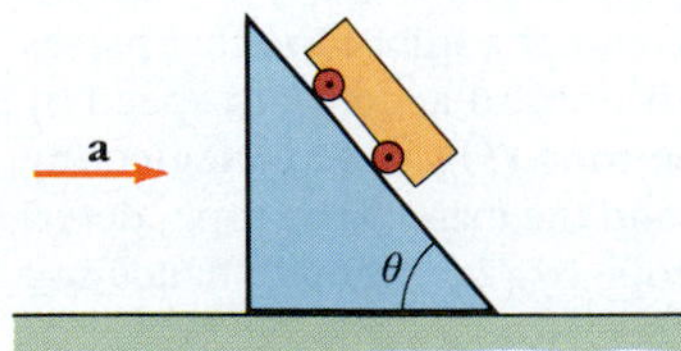

Figure 5-44. Problem 9.

10. ***Forces on football players.*** In football, a 130-kg offensive lineman is blocking a 110-kg defensive linebacker. The acceleration of both players is 0.11 m/s² horizontally toward the north, in the direction the lineman is pushing. The horizontal component of the force exerted by the lineman on the ground is 1335 N. Determine the magnitude and direction of the following forces: *(a)* by the lineman on the linebacker, *(b)* by the linebacker on the lineman, *(c)* by the linebacker on the ground, *(d)* by the ground on the linebacker.

11. ***Two sleds on ice.*** Two sleds (*A* and *B*) face each other on a frozen lake. Sled *A* is equipped with a battery-driven winch that winds in a rope at a constant tension of 5.0 N. The other end of the rope is connected to sled *B*, the winch motor is turned on, and the sleds are released from rest with their front ends a distance of 6.0 m apart. The masses of the sleds are $m_A = 50$ kg and $m_B = 25$ kg. Assume that the sled runners glide across the ice with negligible friction. *(a)* Determine the magnitude of the acceleration of each sled. *(b)* What is the ratio of the speed of *A* to the speed of *B* at each instant during the motion? *(c)* What is the distance traveled by each sled by the time their front ends make contact? *(d)* Which, if any of the answers to parts *(a)*, *(b)*, or *(c)* would be the same if the tension were not held constant by the winch?

12. ***More about Atwood's machine.*** *(a)* Show that the tension in the rope in Atwood's machine (Fig. 5-40) is

$$F_T = \frac{2m_1 m_2 g}{m_1 + m_2}$$

(b) For the case where $m_2 > m_1$, show that $m_1 g < F_T < m_2 g$.

13. ***How to determine an electric force.*** In Fig. 5-45, spheres *A* and *B* are electrically charged in such a way that they attract each other. When *B* is placed at an angle ϕ below *A*, *A* remains stationary, with its supporting string at an angle θ with the vertical. *(a)* Show that magnitude of the force exerted by *B* on *A*, which is an electric force, is

$$F_{BA} = mg/(\cos\phi \cot\theta - \sin\phi)$$

(b) Show the tension in the string is

$$F_T = mg/(\cos\theta - \sin\theta \tan\phi)$$

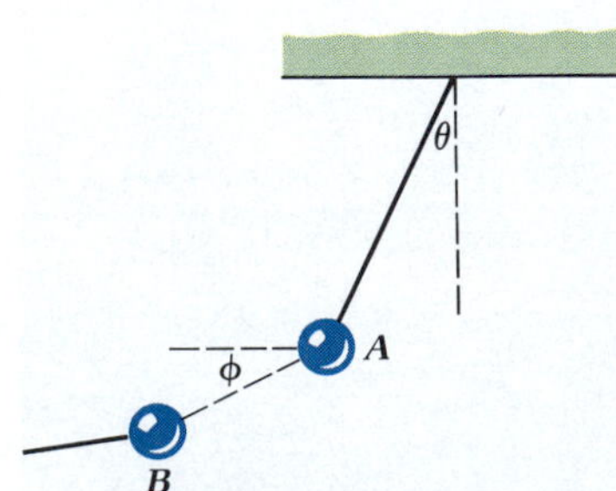

Figure 5-45. Problem 13.

14. ***Variation of tension in a rope due to acceleration.*** Suppose you use a rope to raise a bucket of water from a well. The bucket's trip out of the well goes as follows: *(a)* its speed increases from zero to 0.80 m/s at constant acceleration over the first 0.15 m; *(b)* its speed is constant after the first 0.15 m

and until it is 0.20 m from the top; *(c)* its speed decreases at constant acceleration from 0.80 m/s to zero over the last 0.20 m. The mass of the bucket of water is 4.9 kg. Determine the tension in the rope during each of the three parts of the trip.

15. ***Force on a ball when it strikes a surface.*** A tennis ball of mass $m = 0.06$ kg is released from rest at a height $h = 2$ m above a tennis court's surface. When the ball hits the surface, it is compressed a distance $d = 1$ mm. That is, the center of the ball travels this distance d between the time it makes contact with the surface until the instant it comes to rest before reversing its direction of travel. Assume that the acceleration is constant during this time period. *(a)* Show that the expression for the magnitude of the force exerted on the ball by the surface is

$$F_s = mg[1 + (h/d)]$$

(b) What is the direction of this force? *(c)* Evaluate F_s *(d)* Now suppose a billiard ball is released from the same height. Would the force on the billiard ball by the surface be larger or smaller than that on the tennis ball? Which factors in the equation in part *(a)* account for the difference?

16. ***Estimating the force on an arrow by a bowstring.*** A 0.05-kg arrow is shot vertically upward and reaches a maximum height of 30 m. Estimate the magnitude of the force exerted on the arrow by the bowstring during the shot. List the assumptions you make. In your list, order the assumptions according to their validity with the least valid assumptions listed first.

17. ***A consequence of Newton's third law.*** Two air-track gliders (1 and 2) are equipped with magnets and placed on the air track such that each glider exerts a force of repulsion on the other. The gliders are released from rest with their facing ends in contact and, because of the repulsion, accelerate away from one another. Since the gliders are on an air track, the upward force by the cushioning air is equal and opposite the downward gravitational force by the earth. Let m_1 be the mass of glider 1 and its magnet, and let m_2 be the mass of glider 2 and its magnet. At a particular instant, glider 1 has traveled a distance ℓ_1 and glider 2 has traveled a distance ℓ_2. *(a)* Use Newton's second and third laws to show that $m_1\ell_1 = m_2\ell_2$. *(b)* Let $m_1 = 1.0$ kg, $\ell_1 = 0.40$ m, $\ell_2 = 0.20$ m, and evaluate m_2.

18. ***Carts on opposite slopes.*** Carts *A* and *B* are connected by a string that is placed over a pulley, and then the carts are released from rest on sloping surfaces (Fig. 5-46). *(a)* Assuming that the system accelerates to the right, show that the acceleration magnitude for the system is

$$a = g(m_B \sin\phi - m_A \sin\theta)/(m_A + m_B)$$

(b) Write the expression for the acceleration magnitude assuming the system accelerates to the left. *(c)* Let $\theta = 30°$, $\phi = 60°$, and $m_A = 2.0$ kg. Determine the value of m_B such that $a = 0$.

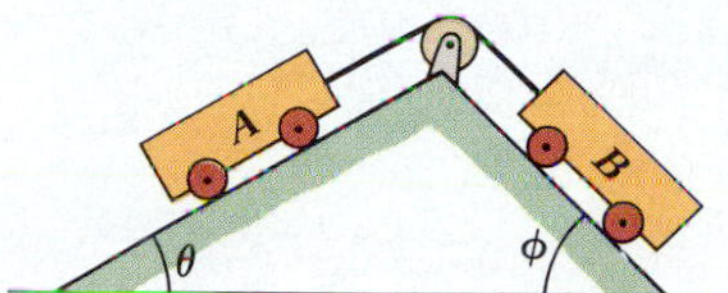

Figure 5-46. Problem 18.

19. ***More about the carts on opposite slopes.*** *(a)* Determine an expression for the tension in the string connecting the carts in the previous problem. Give your answer in terms of the masses, the angles, and g. *(b)* Show that your answer is the same as that for the Atwood machine when $\theta = \phi = 90°$ (see Prob. 12).

20. ***Lifting an object with a cable.*** Suppose you intend to lift an object of mass M a height h with a cable that can sustain a maximum tension F_{max} before breaking. *(a)* Show that the minimum time in which you can lift the object is

$$t_{min} = \sqrt{2hM/(F_{max} - Mg)}$$

(b) Show that the object will arrive at height h with speed

$$v = \sqrt{2h/(F_{max} - Mg)/M}$$

21. ***A block anchors a cart.*** A cart (mass m_C) is connected to a block (mass m_B) by a string, and the system remains stationary after it is placed on a sloping surface (angle θ relative to the horizontal), as shown in Fig. 5-47. *(a)* Determine an expression for the tension F_T in the string in terms of m_B, m_C, g, and θ. Determine the components of the force exerted by the surface on the block *(b)* parallel to the surface and *(c)* perpendicular to the surface.

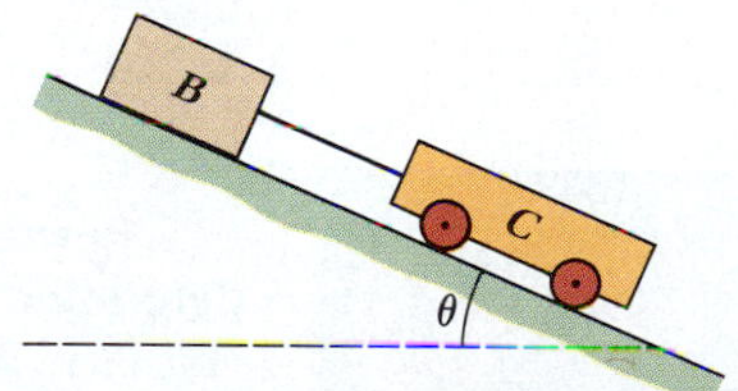

Figure 5-47. Problem 21.

6 APPLICATIONS OF NEWTON'S LAWS OF MOTION

An example of a contact force.

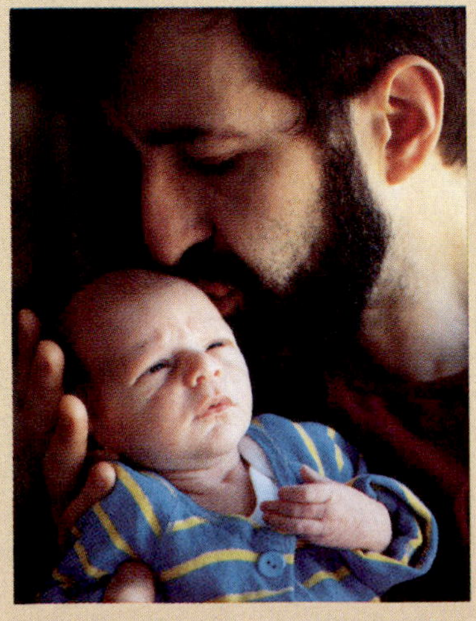

Contact forces are the way we keep in touch with our environment.

Newton's second law is the equation of motion. If we have some way to determine the forces on an object, then, given the object's initial position and velocity, we can apply the second law and determine its subsequent motion. In some cases this procedure is reversed. That is, from measurements of an object's motion, we determine the forces exerted on it. This second method is used in this chapter to investigate some characteristics of contact forces. The dynamics of uniform circular motion and some effects of the earth's rotation are also considered.

6-1 CONTACT FORCES: THE NORMAL FORCE AND THE FRICTIONAL FORCE

Contact forces are so pervasive in our lives that an understanding of their behavior is useful to everyone. One obvious effect of contact forces is to prevent objects from interpenetrating. The fundamental interaction that is responsible for these forces is the electromagnetic force between atoms and molecules, which operates at the level of their constituents, electrons and nuclei. At this microscopic level, contact forces involve many particles and are very complex and incompletely understood. Fortunately, the macroscopic behavior of these forces is much simpler, and we consider contact forces only at this level.

There is a particularly convenient way to describe contact forces between the flat surfaces of two solid objects. The method involves resolving a contact force into two forces, one parallel to the surface of contact and the other perpendicular, and then treating each of these as a separate force. The force parallel to the surface is called the ***frictional force*** and the force perpendicular to the surface is called the ***normal force.***

$\mathbf{F}_N$

mg

(a)

$\mathbf{F}_N$

$2mg$

(b)

Figure 6-1. *(a)* A block of mass m at rest on a surface. The normal force $\mathbf{F}_N$ is exerted by the surface on the block: $F_N = mg$. *(b)* When another block of mass m is placed on top of the first block, the normal force doubles: $F_N = 2\ mg$.

The Normal Force

Suppose a block of mass m is at rest on a horizontal surface, with only its weight and the contact force by the surface exerted on it, as shown in Fig. 6-1*a*. The force by the surface supports the block, holding it at rest. Since the block's acceleration is zero, the net force on it is zero, which means that the contact force is equal and opposite the block's weight. This contact force is the *normal force* $\mathbf{F}_N$ because it is directed perpendicular, or normal, to the surface. For the case shown in Fig. 6-1*a*, $F_N = mg$. Now suppose another block of mass m is placed on top of the original block forming a composite block of mass $2m$ (Fig. 6-1*b*). The weight is now doubled, and to support the composite block, the normal force also doubles: $F_N = 2mg$. That is, the normal force adjusts itself to keep the block from accelerating perpendicular to the surface.

(a)

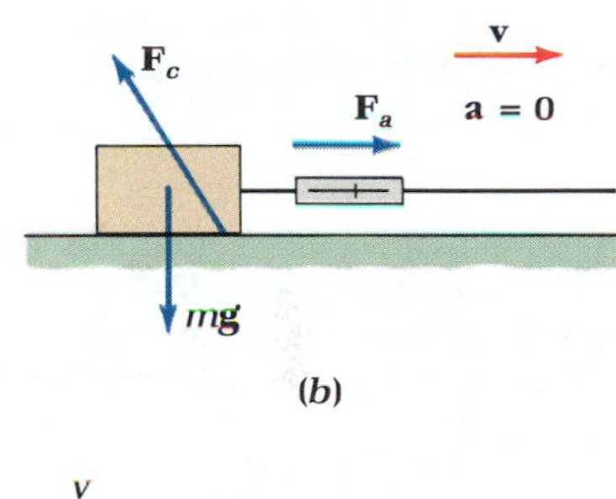

(b)

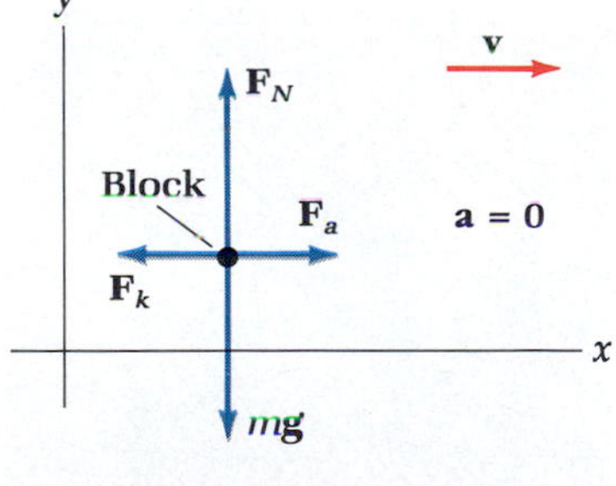

(c)

Figure 6-2. *(a)* A block is pulled at constant velocity by a spring scale. *(b)* Sketch that shows the forces exerted on the block. *(c)* Free-body diagram for the block. The contact force is represented as two forces: the normal force $\mathbf{F}_N$ perpendicular to the surface, and the kinetic frictional force $\mathbf{F}_k$ parallel to the surface and opposite the velocity $\mathbf{v}$.

The Kinetic Frictional Force

In Fig. 6-2*a* we show a block of mass m being pulled at constant velocity by a spring scale along a horizontal surface. There are three forces on the block (Fig. 6-2*b*): $\mathbf{F}_a$, the applied force by the scale; $\mathbf{F}_e = m\mathbf{g}$, the block's weight; and $\mathbf{F}_c$, the contact force by the surface. In the block's free-body diagram (Fig. 6-2*c*), the contact force is represented by two forces: $\mathbf{F}_k$, the frictional force (parallel to the surface and opposite the velocity), and $\mathbf{F}_N$, the normal force (perpendicular to the surface). The subscript on $\mathbf{F}_k$ stands for "kinetic," and $\mathbf{F}_k$ is called the ***kinetic frictional force.*** ("Kinetic" is a Greek term that means motion, and the subscript k refers to the motion between the two surfaces.) Since the block's acceleration is zero, Newton's second law applied to the block gives $\Sigma F_x = 0$ and $\Sigma F_y = 0$. Therefore $F_k = F_a$ and $F_N = mg$, so that the scale reading gives the value of F_k, and F_N equals the block's weight.

To investigate the relation between the normal force and the kinetic frictional

TABLE 6-1. ***A Few Representative Values of Coefficients of Friction***

Surface	μ_k	μ_s
Copper on steel	0.36	0.53
Copper on cast iron	0.29	1.05
Rubber on concrete	0.8	0.9
Wood on leather	0.4	0.5
Steel on ice	0.06	0.10

force, we fasten another block of mass m on top of the sliding block and determine the forces on the composite block of mass $2m$. The normal force exerted by the surface is now double its previous value, $F_N = 2mg$. From the reading on the spring scale, we find that the applied force required to slide the block at constant velocity also doubles, as does F_k, since $F_k = F_a$. Experiments such as this show that, to a good approximation, F_k is proportional to F_N, or

$$F_k = \mu_k F_N \tag{6-1}$$

where the proportionality constant μ_k is a dimensionless number and is called the ***coefficient of kinetic friction.*** Notice that Eq. (6-1) connects only the magnitudes of $\mathbf{F}_k$ and $\mathbf{F}_N$. These forces have perpendicular directions and $\mathbf{F}_k$ is opposite $\mathbf{v}$.

Further experiments show that

1. F_k depends on the nature and condition of the two surfaces, and μ_k usually falls in the range from about 0.1 to about 1.5. (See Table 6-1.)
2. F_k (or μ_k) is nearly independent of speed for low relative speeds of the surfaces, decreasing slightly as the speed increases. We shall use the approximation that F_k is independent of the speed.
3. F_k (or μ_k) is nearly independent of the area of contact for a wide range of areas.

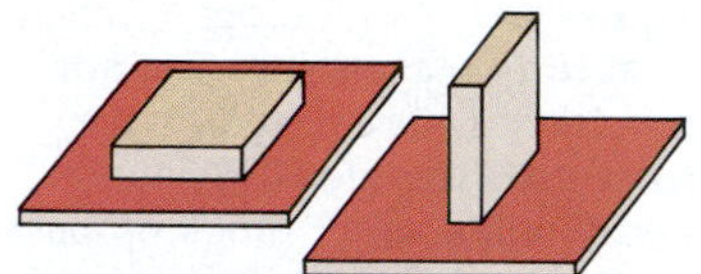

Figure 6-3. A block on a horizontal surface. Upon sliding the block, the frictional force is found to be nearly independent of whether it slides on a face with large area or a face with small area. This demonstrates that, with F_N the same, F_k is nearly the same.

The near independence of μ_k on the area of contact can be demonstrated by sliding a block that has sides with different areas (Fig. 6-3). The surface of each side should consist of the same type of material and should be in the same condition. When the applied force required to slide the block at a given speed on different sides is measured, it is found to be nearly the same. Since F_N is the same in each case, we conclude that μ_k is approximately independent of the area.

A close look at the surface of an object gives an indication of why our description of frictional forces is imprecise. Figure 6-4 shows a photograph, taken with the aid of a scanning electron microscope, of a highly polished surface.

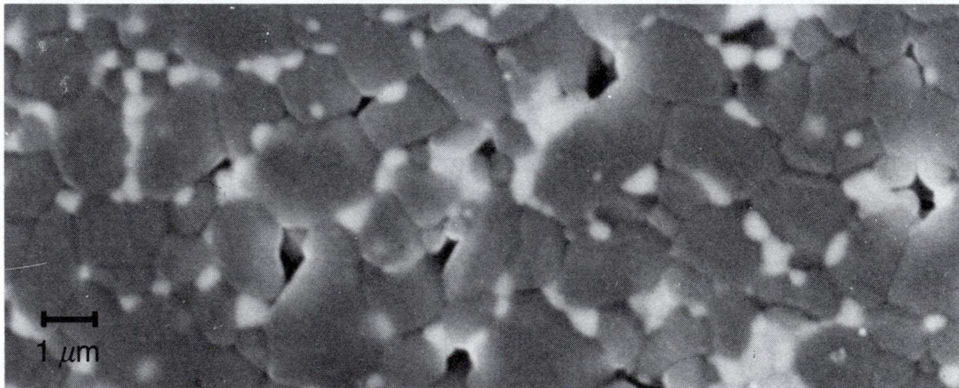

Figure 6-4. Scanning-electron-microscope photograph of a highly polished metal surface. A distance of 1 μm is indicated in the figure.

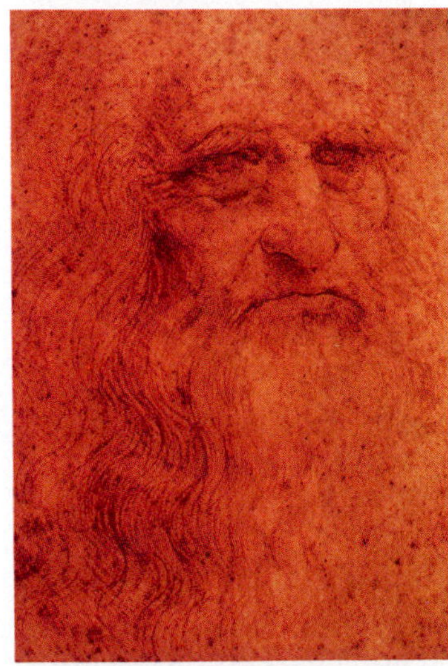

The famous Italian artist, Leonardo da Vinci (1452–1519), was also a scientist, mathematician, engineer, and architect. He studied frictional forces and understood many of their properties.

The Static Frictional Force

A frictional force can also exist between two objects when there is no relative motion. Such a force is called a ***static frictional force*** $\mathbf{F}_s$. In Fig. 6-5a through d, the applied force by the spring scale on the block is gradually increased, but the block remains at rest. Since the acceleration is zero in each case, the applied force $\mathbf{F}_a$ by the scale is equal and opposite the static frictional force $\mathbf{F}_s$ by the surface. The maximum static frictional force $\mathbf{F}_{s,\max}$ occurs just as the block is about to slide. Experiment shows that, to a good approximation, $F_{s,\max}$ is proportional to F_N, or

$$F_{s,\max} = \mu_s F_N \tag{6-2}$$

where the proportionality constant μ_s is called the ***coefficient of static friction.*** Therefore, up to a limit, the static frictional force adjusts to keep one surface from sliding across the other:

$$F_s \leq \mu_s F_N \tag{6-3}$$

Similar to μ_k, the coefficient μ_s depends on the condition and nature of the two surfaces and is nearly independent of the area of contact. Table 6-1 lists μ_k and μ_s for a few representative pairs of surfaces. Normally, for a given pair of surfaces, μ_s is noticeably larger than μ_k.

A way to reduce frictional effects while transporting a load is to use a wheeled vehicle. Ordinarily, it is much easier to move something in a cart than it is in a sledge because the sliding surfaces are at the wheel bearings and can be lubricated. The frictional forces that tend to slow a wheeled vehicle can be treated in much the same way that we treated kinetic friction—namely, by introducing a coefficient of ***rolling friction*** (Exercise 18).

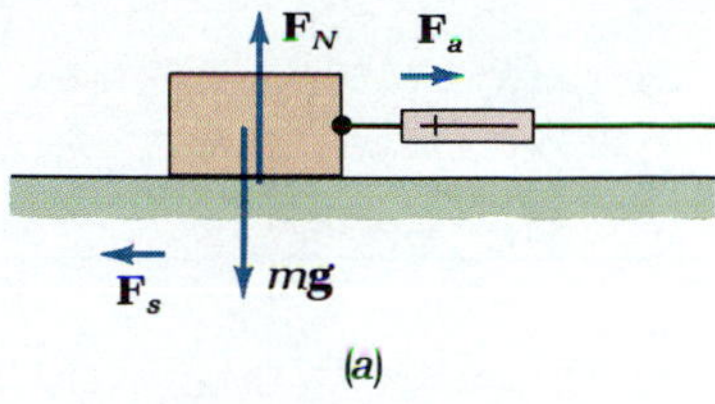

(a)

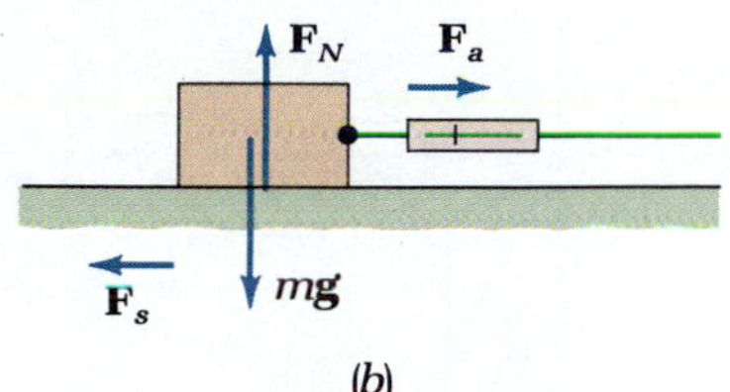

(b)

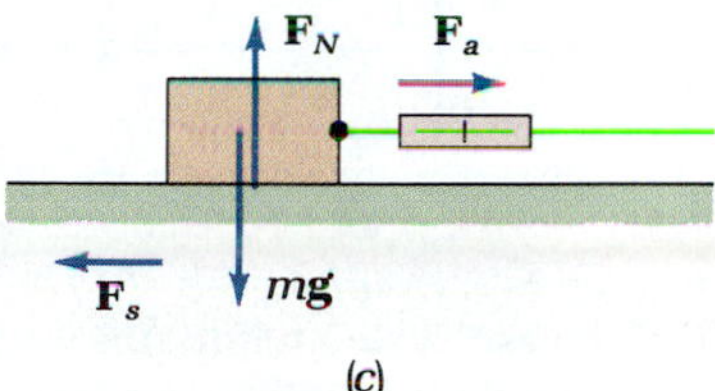

(c)

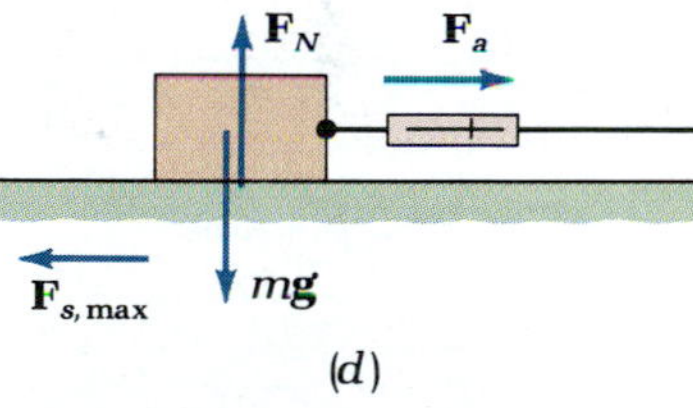

(d)

Figure 6-5. *(a)* through *(c)* The applied force is gradually increased, and in response the static frictional force $\mathbf{F}_s$ gradually increases. *(d)* Just before sliding begins $F_s = F_{s,\max} = \mu_s F_N$.

EXAMPLE 6-1

A gradually increasing applied force. A spring scale is used to exert a horizontal force on a block, as in Figs. 6-2 and 6-5. The block is initially at rest. Let the coefficients of friction be $\mu_s = 0.80$ and $\mu_k = 0.60$ and let the mass of the block be $m = 0.51$ kg. On the same graph, plot F_s and F_k (as appropriate) versus F_a as F_a is increased from 0.0 to 7.0 N in increments of 1.0 N. Sketch a line through the points for F_s versus F_a and F_k versus F_a. Determine the block's approximate acceleration for each value of F_a.

Solution. The surface is horizontal so that $F_N = mg = (0.51\text{ kg})(9.8\text{ m/s}^2) = 5.0$ N. Therefore $F_{s,\max} = \mu_s F_N = (0.80)5.0\text{ N}) = 4.0$ N. At each value of F_a up to this limit, $F_s = F_a$, as shown in the graph in Fig. 6-6, and for each of these values both the velocity and the acceleration of the block are zero.

For the values of F_a greater than 4.0 N, the block slides on the surface and $F_k = \mu_k F_N = (0.60)(5.0\text{ N}) = 3.0$ N. Since $F_a > F_k$, there is a net horizontal force of magnitude $F_a - F_k$. In the case where $F_a = 5.0$ N, the acceleration magnitude is

$$a = \frac{F_a - F_k}{m} = \frac{5.0\text{ N} - 3.0\text{ N}}{0.51\text{ kg}} \approx 4\text{ m/s}^2$$

Similarly, when $F_a = 6.0$ N,

$$a = \frac{6.0\text{ N} - 3.0\text{ N}}{0.51\text{ kg}} \approx 6\text{ m/s}^2$$

and when $F_a = 7.0$ N

$$a = \frac{7.0\text{ N} - 3.0\text{ N}}{0.51\text{ kg}} \approx 8\text{ m/s}^2$$

SELF-TEST 6-1. Using the same arrangement as in the above example, consider decreasing the magnitude of the applied force on the block from 7.0 to 0.0 N in increments of 1.0 N. Make a graph similar to that in Fig. 6-6. Explain any difference between your graph and that of Fig. 6-6 in the region between $F_a = 4.0$ N and $F_a = 3.0$ N.

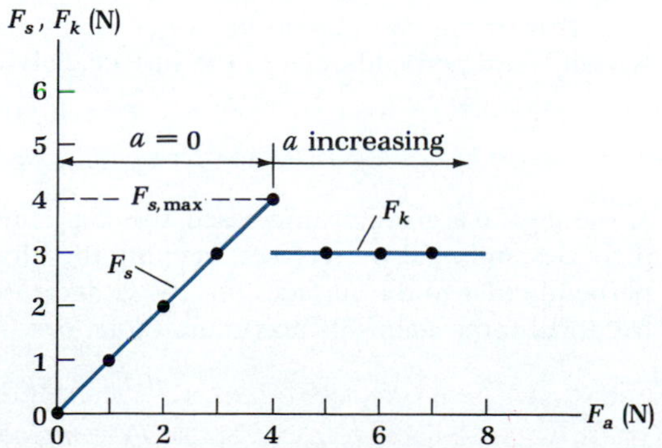

Figure 6-6. Example 6-1: Graph of F_s and F_k versus F_a for a 0.51-kg block on a horizontal surface where $\mu_s = 0.80$ and $\mu_k = 0.60$.

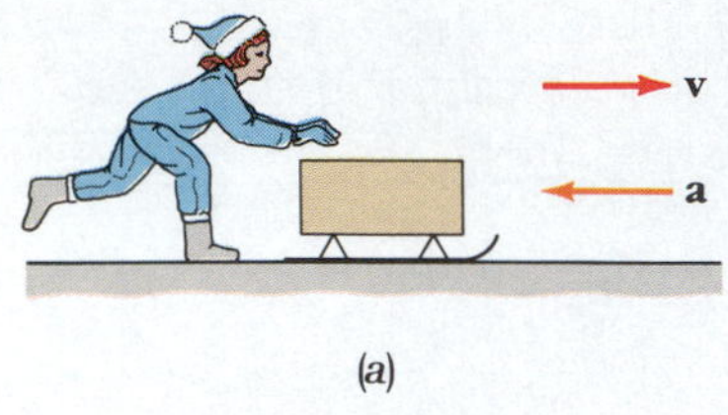

(a)

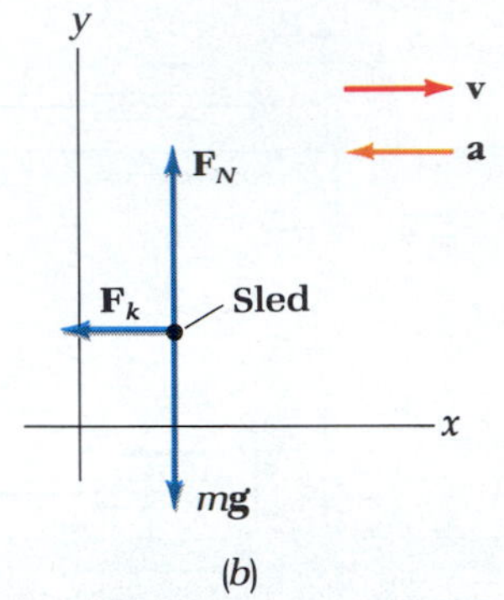

(b)

Figure 6-7. Example 6-2: *(a)* A girl releases a sled while it is traveling with speed v. *(b)* Free-body diagram for the sled after it is released.

EXAMPLE 6-2

A sled sliding to a stop. A girl pushes a sled along a snowy horizontal road. When the sled's speed is $v = 2.5$ m/s (Fig. 6-7*a*), the girl releases it and it slides a distance $d = 6.4$ m before coming to rest. Determine μ_k between the sled runners and the snowy surface.

Solution. The free-body diagram (Fig. 6-7*b*) shows the forces exerted on the sled during the time it is sliding (after it was released by the girl). We let the unit vector **i** point in the direction of the velocity **v**, which is opposite the frictional force: $\mathbf{F}_k = -F_k\mathbf{i}$. The sled's vertical acceleration is zero ($a_y = 0$), so that $\Sigma F_y = 0$, which means that $F_N = mg$. Substituting this value of F_N into Eq. 6-1 gives $F_k = \mu_k F_N = \mu_k mg$. From the free-body diagram, $\mathbf{F}_k$ is the only force along the horizontal so that $\Sigma F_x = -F_k = -\mu_k mg$. The x component of Newton's second law, $\Sigma F_x = ma_x$, is

$$-\mu_k mg = ma_x$$

Solving for a_x gives

$$a_x = -\mu_k g$$

The negative value for a_x shows that the acceleration is directed opposite the velocity. Since a_x is constant, we can use Eq. 3-12:

$$v_x^2 - v_{x0}^2 = 2a_x(x - x_0)$$

Applying this equation to the sled as it slides to a stop from an initial speed v, we have $v_x = 0$, $v_{x0} = v$, $a_x = -\mu_k g$, and $x - x_0 = d$. Thus

$$-v^2 = 2(-\mu_k g)d$$

Solving for μ_k gives

$$\mu_k = \frac{v^2}{2gd} = \frac{(2.5\ \text{m/s}^2)}{2(9.8\ \text{m/s}^2)(6.4\ \text{m})} = 0.050$$

SELF-TEST 6-2. Suppose a sled is released when its speed is 5.0 m/s along a horizontal snowy road and μ_k between the sled's runners and the surface is 0.050. How far will the sled slide before coming to rest? ***ANSWER:*** 26 m.

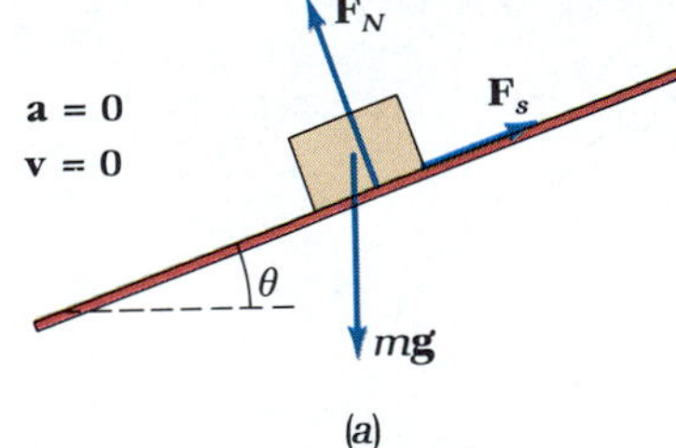

(a)

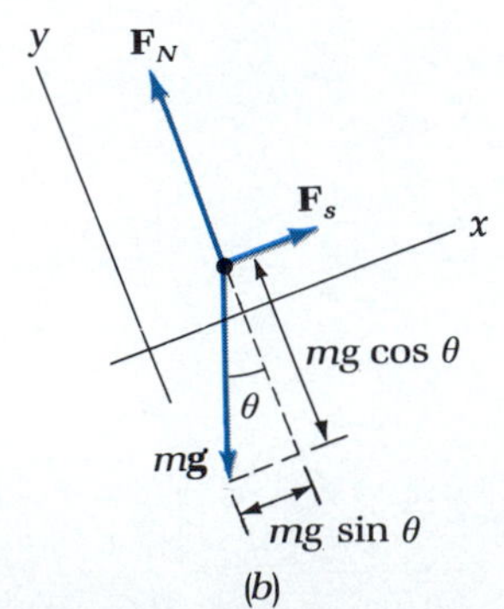

(b)

Figure 6-8. Example 6-3: *(a)* A block on a plank tilted at an angle θ *(b)* Free-body diagram for the block.

EXAMPLE 6-3

Angle of repose. A convenient way to measure the coefficient of static friction between a block and a plank is to place the block on the plank and gradually tilt the plank (Fig. 6-8). The angle θ_s between the plank and the horizontal just before the block begins to slide is called the ***angle of repose,*** or the *critical angle for sliding to begin*. Determine μ_s in terms of θ_s.

Solution. Consider the block at rest on the plank when the plank is tilted at an angle θ, where $\theta < \theta_s$ (Fig. 6-8). Since the block remains at rest, the components of Newton's second law give $\Sigma F_x = 0$ and $\Sigma F_y = 0$, or

$$F_s - mg \sin\theta = 0 \qquad \text{and} \qquad F_N - mg\cos\theta = 0$$

Note that in this case the normal force is not equal and opposite the weight because the weight is not perpendicular to the surface. Solving for F_s and F_N gives

$$F_s = mg \sin\theta \tag{A}$$

$$F_N = mg\cos\theta \tag{B}$$

As the angle θ is gradually increased, the component of the weight down the plane, $mg \sin\theta$, increases and F_s also increases, keeping the block at rest. The component of the weight perpendicular to the surface, $mg\cos\theta$, decreases as θ increases. When $\theta = \theta_s$, the static frictional force attains its maximum value, $F_s = F_{s,\max} = \mu_s F_N$, so that Eqs. A and B become

$$\mu_s F_N = mg \sin\theta_s \tag{C}$$

$$F_N = mg\cos\theta_s \tag{D}$$

When Eq. C is divided by Eq. D, F_N and mg cancel out and we obtain

$$\mu_s = \tan \theta_s$$

Suppose $\theta_s = 38°$ for a block and a plank. Then

$$\mu_s = \tan 38° = 0.78$$

Similarly, you can show that

$$\mu_k = \tan \theta_k$$

where θ_k is the angle between the plank and the horizontal such that the block slides with constant velocity (see Exercise 10).

SELF-TEST 6-3. What is the angle of repose for a copper block on a sloping steel plate? (See Table 6-1.) ***ANSWER:*** 28°.

EXAMPLE 6-4

Sliding a crate. A man slides a 45-kg crate at constant velocity across a horizontal floor by pulling on a rope attached to the crate, as shown in Fig. 6-9*a*. The angle θ between the rope and the horizontal is 33° and the coefficient of kinetic friction between the crate and the floor is 0.63. Determine the tension F_T in the rope.

Solution. The free-body diagram for the crate is shown in Fig. 6-9*b*. Because the crate's acceleration is zero, the components of Newton's second law applied to the crate are $\Sigma F_x = 0$ and $\Sigma F_y = 0$. The free-body diagram shows that $\Sigma F_x = 0$ gives

$$F_T \cos \theta - \mu_k F_N = 0 \qquad (A)$$

and that $\Sigma F_y = 0$ gives

$$F_T \sin \theta + F_N - mg = 0 \qquad (B)$$

Solving Eq. A for F_N gives

$$F_N = (F_T \cos \theta)/\mu_k$$

Substituting this result into Eq. B and solving for F_T gives

$$F_T = \frac{\mu_k mg}{\cos \theta + \mu_k \sin \theta} = \frac{(0.63)(45 \text{ kg})(9.8 \text{ m/s}^2)}{\cos 33° + 0.63 \sin 33°} = 240 \text{ N}$$

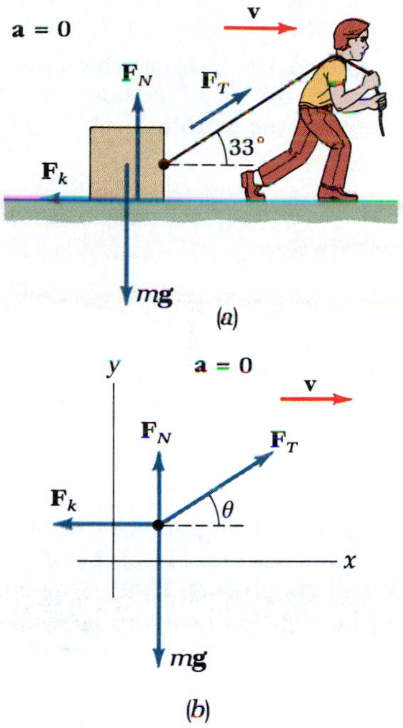

SELF-TEST 6-4. Suppose the man in the above example pulls on the rope such that $F_T = 280$ N. Determine the magnitude of the crate's acceleration. ***ANSWER:*** 1.2 m/s².

Figure 6-9. Example 6-4: *(a)* A man slides a crate across a floor. *(b)* Free-body diagram for the crate.

Frictional Forces due to Fluids

When a solid object, such as a rock, moves in a fluid, such as air or water, the fluid exerts a frictional force on the object. The behavior of this force depends on many things, including the shape of the object, the velocity of the object relative to the fluid, and the nature of the fluid. The property of a fluid responsible for the force is called ***viscosity,*** and the force is called a ***viscous force, drag force,*** or ***retarding force.*** To illustrate numerical methods, we described the effect of such a force on a baseball traveling in air in Sec. 4-5.

Because we wish to show a few of the characteristics of motion in a fluid without becoming bogged down with complicating factors, we shall assume a particularly simple form for the drag force $\mathbf{F}_d$:

$$\mathbf{F}_d = -b\mathbf{v}$$

where b is a proportionality constant. That is, we assume $\mathbf{F}_d$ is directed opposite the object's velocity with a magnitude that increases linearly with the speed. This expression is approximately valid when the fluid is a liquid (such as water) rather than a gas (such as air), and when v is not large.

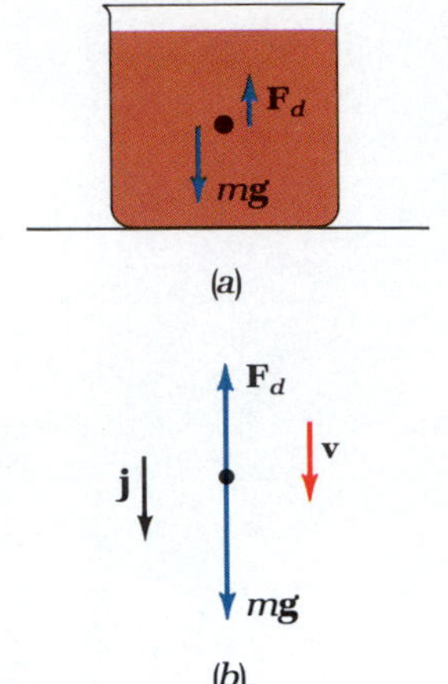

Figure 6-10. *(a)* A marble falls through oil. *(b)* Free-body diagram for the marble.

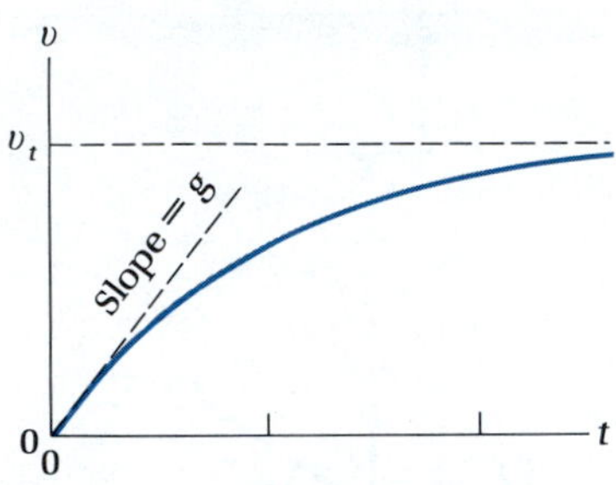

Figure 6-11. A graph of v versus t for the marble. The slope of the graph gives a. At $t = 0$, $v = 0$ and $a = g$. As t becomes large, v approaches v_t and a approaches zero.

Now suppose we release some object, such as a marble of mass m, from rest in some fluid, such as oil (Fig. 6-10). To deal mostly with positive quantities, we let the $+y$ direction be downward (not upward), so that

$$\Sigma F_y = mg - bv$$

(For simplicity, we have neglected a force that is significant in liquids, the ***buoyant force.*** Since the buoyant force is constant, it does not alter the qualitative features of the motion described here. See Sec. 15-4.) As the marble falls, its weight acts as a constant force downward and the fluid exerts a varying force upward. From Newton's second law, $\Sigma F_y = ma_y = ma$, so that

$$a = \frac{mg - bv}{m}$$

Since $a = dv/dt$,

$$\frac{dv}{dt} = g - \frac{b}{m}v$$

This equation, which contains v and its derivative dv/dt, is called a ***differential equation.*** In later chapters, we shall encounter differential equations similar to this, and there we shall proceed to solve them. Here we simply describe the solution.

Figure 6-11 shows a graph of the marble's speed v versus time t as the marble falls. Keep in mind that the slope of this graph at any time gives the acceleration magnitude a. At the instant the marble is released ($v = 0$ at $t = 0$), $a = g - (b/m)(0) = g$, as indicated by the slope of the dashed line. As v increases, $F_d = bv$ increases, causing the net force to decrease. Consequently, a decreases and approaches zero as t becomes large. As a approaches zero, v asymptotically approaches a maximum value called the ***terminal speed*** v_t. Since ΣF_y becomes zero as v approaches v_t, we have $mg - bv_t = 0$, or

$$bv_t = mg$$

This gives

$$v_t = \frac{mg}{b}$$

As an example, suppose the marble's mass is $m = 0.015$ kg and $b = 8$ N·s/m. Then

$$v_t = \frac{(0.015\ \text{kg})(9.8\ \text{m/s}^2)}{8\ \text{N}\cdot\text{s/m}} = 0.02\ \text{m/s}$$

6-2 DYNAMICS OF UNIFORM CIRCULAR MOTION

Circular motion, or nearly circular motion, is common in nature and in mechanical devices. For example, the planets move in nearly circular paths around the sun, and the movements of gears, pulleys, and wheels involve circular motion.

Recall from Sec. 4-3 that an object moving in a circle is accelerating even though the object's speed may be constant. There must be an acceleration during circular motion because the velocity is continuously changing direction. If the speed v of the object is constant, then the motion is called *uniform* circular motion. In this motion the acceleration vector points toward the center of the circle, and its magnitude is $a_c = v^2/R$, where R is the radius of the circle. This acceleration toward the center of the circle is called the ***centripetal acceleration.***

From Newton's second law, we see that an object in uniform circular motion must have a net force exerted on it toward the center of the circle. Since $\Sigma\mathbf{F} = m\mathbf{a}$ and $\mathbf{a}$ points toward the center of the circle with magnitude v^2/R, $\Sigma\mathbf{F}$ must also point toward the center of the circle, and

$$|\Sigma \mathbf{F}| = \frac{mv^2}{R}$$

Centripetal force

This net force toward the center of the circle is called the ***centripetal force.*** Note that the term "centripetal force" does not refer to a type of interaction, like a gravitational force or an electrical force. This term simply indicates that the net force is directed toward the center of the circular motion, without reference to how the force is provided.

EXAMPLE 6-5

A car rounding a curve. A car travels at a constant speed v on a horizontal road while rounding a circular turn of radius R, as shown in Fig. 6-12*a* and *b*. *(a)* The coefficient of static friction between the tires and the road surface is μ_s. Determine an expression for the maximum speed v_m that the car can have without beginning to slide. *(b)* Evaluate v_m for the case where $\mu_s = 0.85$ and $R = 150$ m.

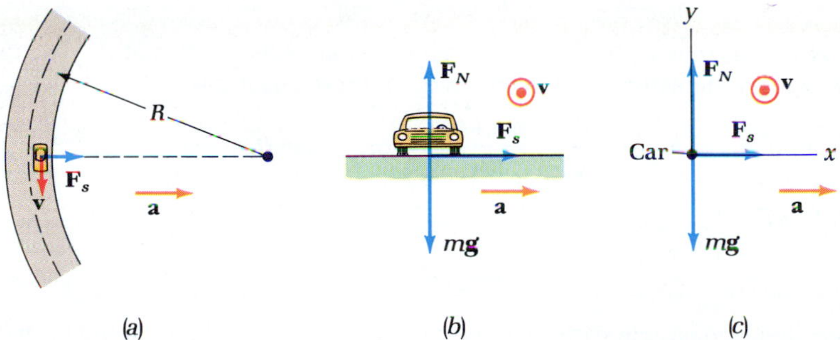

Figure 6-12. Example 6-5: A car rounding a curve. *(a)* Top view. *(b)* Front view. The car's velocity **v** is directed out of the page. A vector directed out of a page is often shown as a dot at the center of a small circle. The dot represents the point of an arrow facing the reader. *(c)* Free-body diagram of the car.

Solution. *(a)* The car's free-body diagram is shown in Fig. 6-12*c*. Since the vertical acceleration is zero, the normal force and the gravitational force are equal and opposite: $F_N = mg$. The net force is horizontal: $\Sigma \mathbf{F} = \mathbf{F}_s$. The static frictional force $\mathbf{F}_s$ (rather than the kinetic frictional force $\mathbf{F}_k$) provides the centripetal force because the car's tire tread does not slide on the road. From Newton's second law, $|\Sigma \mathbf{F}| = F_s = ma = mv^2/R$, and we have

$$F_s = \frac{mv^2}{R}$$

Thus F_s is larger when v is larger, but F_s cannot exceed $F_{s,\max}$. The maximum speed v_m corresponds to $F_s = F_{s,\max}$:

$$\frac{mv_m{}^2}{R} = F_{s,\max} = \mu_s F_N = \mu_s mg$$

Solving for v_m gives

$$v_m = \sqrt{\mu_s gR}$$

Should the car exceed this speed, sliding will begin and the car will no longer travel in a circle. Note that v_m is independent of the mass of the car, but depends on the coefficient of friction and the radius of the curve. *(b)* When $\mu_s = 0.85$ and $R = 150$ m,

$$v_m = \sqrt{(0.85)(9.8\ \text{m/s}^2)(150\ \text{m})} = 35\ \text{m/s}$$

SELF-TEST 6-5. Suppose a horizontal road surface is icy so that μ_s between a car's tires and the road is 0.10. *(a)* What is the maximum speed for the car to round a curve with a radius of $R = 200$ m? *(b)* Determine the car's acceleration magnitude while traveling at this maximum speed. ***ANSWERS:*** *(a)* 14 m/s; *(b)* 0.98 m/s².

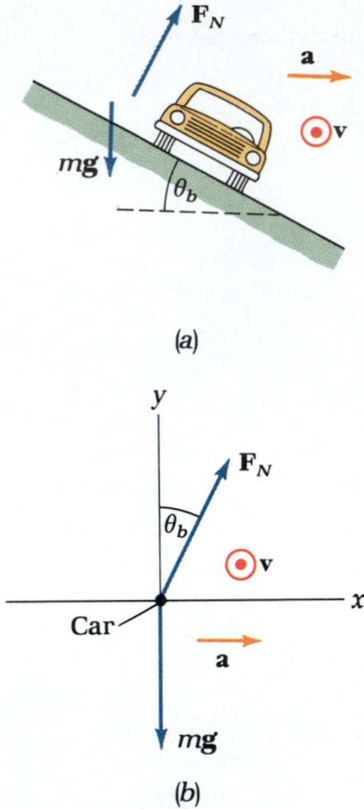

Figure 6-13. Example 6-6: Banking a racetrack curve. *(a)* Car rounding the curve at the design speed v_d. *(b)* Free-body diagram for the car.

EXAMPLE 6-6

Banking a curve on a racetrack. In the design of a racetrack, as well as a highway, the appropriate angle to bank a curve is such that the horizontal component of the normal force exerted by the pavement provides the centripetal force on a car traveling at the design speed v_d. For a car at this speed, a frictional force is not needed to provide the centripetal force, and the car will not tend to slide off the track when the coefficient of friction is reduced by smooth tires or water on the pavement. *(a)* Determine the banking angle for a racetrack curve of radius R such that a car rounding the curve with speed v_d will experience no frictional force perpendicular to its velocity. *(b)* Determine the banking angle for a curve of radius 280 m designed for a speed of 35 m/s.

Solution. *(a)* Figure 6-13*a* shows a sketch of the system where θ_b is the banking angle of the racetrack surface relative to the horizontal. Figure 6-13*b* is a free-body diagram for a car traveling at speed v_d. Since there is no frictional force, the horizontal component of $\mathbf{F}_N$ provides the centripetal force on the car, and the horizontal component of the second law gives

$$F_N \sin \theta_b = \frac{mv_d^2}{R}$$

Since there is no vertical acceleration, the vertical component of $\mathbf{F}_N$ is equal to the car's weight:

$$F_N \cos \theta_b = mg$$

We have two equations with two unknowns; the unknowns are F_N and θ_b. The unwanted unknown, F_N, is eliminated by taking the ratio of these equations:

$$\frac{F_N \sin \theta_b}{F_N \cos \theta_b} = \frac{mv_d^2/R}{mg} \quad \text{or} \quad \tan \theta_b = \frac{v_d^2}{Rg}$$

Solving for θ_b gives

$$\theta_b = \tan^{-1} \frac{v_d^2}{Rg}$$

This expression is consistent with our expectations; it predicts that θ_b should be made larger for a larger value of v_d, and θ_b should be made smaller for a curve with a larger radius of curvature R. *(b)* For a curve with $R = 280$ m designed for $v_d = 35$ m/s,

$$\theta_b = \tan^{-1} \frac{(35\ \text{m/s})^2}{(280\ \text{m})(9.8\ \text{m/s}^2)} = 24°$$

SELF-TEST 6-6. Suppose you are driving a car around a banked curve with $R = 250$ m and $\theta_b = 20°$. At what speed should you drive the car so that a marble placed on the car's dash will tend to roll neither to the left nor to the right? ***ANSWER:*** 30 m/s.

EXAMPLE 6-7

The conical pendulum. A pendulum is composed of a dense object, such as a rock of mass m, suspended from a string of length L. The rock is called the pendulum bob. If the bob is swung in a horizontal circle of radius R, as shown in Fig. 6-14*a*, the system is called a *conical pendulum* because the string sweeps out a cone. The time required for the bob to make one complete revolution is called the *period T*. Newton used a conical pendulum to measure g. Previously, he had accepted a value of g determined by Galileo, but he had reason to believe that Galileo's value was not very accurate. Following Newton, develop an expression for g in terms of T, L, and the angle θ between the string and the vertical.

Solution. The period T is related to the constant speed v of the bob. Since the bob travels a distance $2\pi R$ during one period, $v = 2\pi R/T$. A connection between v and g can be found by applying Newton's second law. The free-body diagram for the bob is shown in Fig. 6-14*b*. The vertical acceleration is zero so that the second law gives $\Sigma F_y = 0$, or

$$F_T \cos\theta = mg \qquad \text{(A)}$$

where F_T is the tension in the string. Since the bob travels in a horizontal circular path, the horizontal acceleration is v^2/R. The centripetal force is the horizontal component of the tension in the string, so that the horizontal component of the second law gives

$$F_T \sin\theta = \frac{mv^2}{R} \qquad \text{(B)}$$

Dividing Eq. B by Eq. A gives

$$\tan\theta = \frac{v^2}{Rg}$$

Substituting $v = 2\pi R/T$ and $R = L \sin\theta$, and then solving for g, we obtain

$$g = \frac{4\pi^2 L \cos\theta}{T^2}$$

Newton swung a pendulum with $L = 81$ inches while keeping $\theta = 45°$ and measured the period. He arrived at a value of g that was accurate to within 4 percent. You should try this experiment and determine the accuracy of your results. It will probably cause you to have great respect for Newton's skill as an experimentalist.

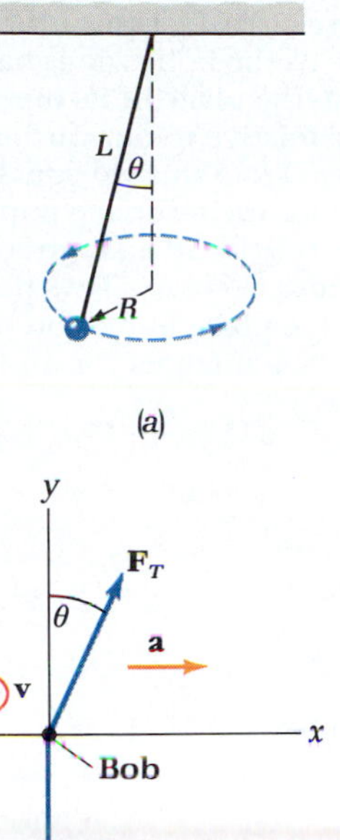

Figure 6-14. Example 6-7: *(a)* A conical pendulum. *(b)* Free-body diagram for the bob.

SELF-TEST 6-7. A conical pendulum is swung such that the angle θ between the string and the vertical is 35°. What is the magnitude of the bob's acceleration? ***ANSWER:*** 6.9 m/s².

6-3 THE ROTATION OF THE EARTH

So far we have used the approximation that an earth-surface frame is an inertial frame, but, as you know, the earth rotates on its axis daily. One consequence of the earth's rotation is the nightly procession of the stars across the sky (Fig. 5-8). However, from looking at the stars, we cannot tell whether the earth is rotating and the stars are fixed, or the earth is fixed and the stars travel in circles around the earth. An experiment that provides convincing evidence that the earth rotates uses a device called a Foucault pendulum.

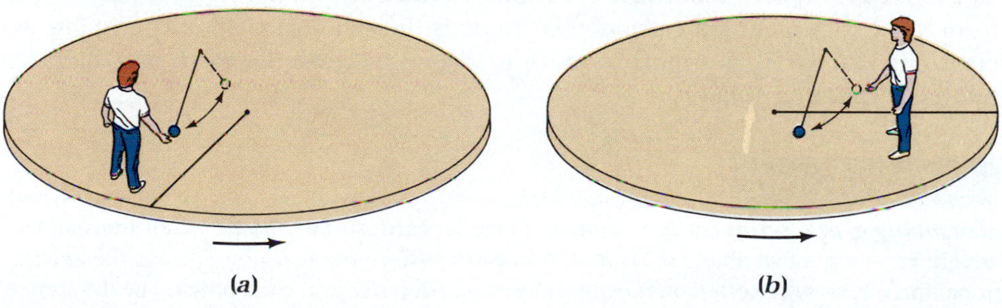

Figure 6-15. A pendulum mounted on a merry-go-round. For simplicity, the pendulum stand is not shown. *(a)* The pendulum bob is set in motion such that the plane of the motion is along a radial line from the center of the merry-go-round. *(b)* After the merry-go-round has rotated 90°, the bob is swinging perpendicular to the radial line, so that the plane of motion has rotated 90° in the merry-go-round frame. From the perspective of an earth-surface frame, the plane of motion maintains a constant alignment. If the bob starts off swinging along north-south, then it remains swinging along north-south as the merry-go-round rotates.

The Foucault Pendulum

To understand the idea behind the Foucault-pendulum experiment, first consider a playground merry-go-round. Suppose you ride on a merry-go-round and you wish to determine whether you are rotating and the trees and bushes in the playground remain stationary, or whether you are stationary and the trees and bushes travel in circles with you at the center. To find out, you set a pendulum swinging (Fig. 6-15), and you observe that, from your merry-go-round perspective, the plane of motion of the swinging pendulum rotates. However, an observer standing on the ground beside the merry-go-round finds the pendulum's plane of motion does not rotate. That is, the plane of swinging does not rotate noticeably in an earth-surface frame. From this we conclude that the merry-go-round rotates while the trees and bushes remain stationary.

Figure 6-16. A Foucault pendulum. As the bob swings back and forth, the plane of its motion rotates relative to our earth-surface frame. From this we conclude that our earth-surface frame rotates relative to an inertial frame because Newton's laws predict that the plane of motion has a constant alignment in an inertial frame.

A Foucault pendulum (Fig. 6-16) uses the same principle as the pendulum on the merry-go-round. However, the mount of a Foucault pendulum is fixed to the earth, and the pendulum is designed so that its plane of swing can be observed for a significant fraction of a day. To explain the result of a Foucault-pendulum experiment, it is convenient to first define another reference frame, an earth-center frame.

Definition of an earth-center frame

An ***earth-center frame*** is one with origin fixed relative to the center of the earth and with axes fixed relative to distant stars.

If the motion of a Foucault-pendulum bob is viewed from our earth-surface frame, the plane of swing gradually rotates. That is, the result is similar to the one we get when riding on a merry-go-round. Calculation reveals that the rotation rate corresponds to a rotating earth-surface frame and a stationary earth-center frame. To put it more precisely, the Foucault-pendulum experiment shows that an earth-surface frame is not inertial because of the earth's rotation and that as far as can be told from this experiment, an earth-center frame is inertial. From this experiment we conclude that the stars appear to move in circles because the earth on which we stand is a spinning sphere.

EXAMPLE 6-8

Measuring* g *at a point on the equator. Since an earth-surface frame is not inertial, the weight $\mathbf{F}_w = m\mathbf{g}'$ of an object measured in an earth-surface frame is not equal to the gravitational force $\mathbf{F}_e = m\mathbf{g}$ exerted on the object by the earth (except at each pole). The difference between $\mathbf{F}_e$ and $\mathbf{F}_w$ is the greatest at points on the equator because these points are farthest from the axis of rotation so that they have the largest centripetal acceleration toward the axis. *(a)* Develop an expression for F_e in terms of F_w for an object of mass m located at a point on the equator. Assume that an earth-center frame is essentially inertial. *(b)* Let the object in part *(a)* be the standard kilogram ($m = 1$ kg exactly) and suppose that its weight is measured to be 9.781 N. To four significant digits, determine g at this point.

Solution. *(a)* Figure 6-17*a* shows a person standing at a point on the equator while holding an object suspended from a spring scale. The reading on the scale gives the weight F_w of the object in the person's earth-surface frame. This reading is also equal to the magnitude F_s of the force exerted upward by the scale: $F_w = F_s$. Figure 6-17*b* shows the object's free-body diagram and its circular path in an earth-center frame. There are two forces exerted on the object: the gravitational force $\mathbf{F}_e$ and the force $\mathbf{F}_s$ by the scale. These two forces are in opposite directions, and $F_e > F_s$ because the object is accelerating toward the center of the earth. The magnitude of the net force is $|\Sigma\mathbf{F}| = F_e - F_s$. If R_e is the earth's radius ($R_e = 6.37 \times 10^6$ m $= 6.37$ Mm) and T_e is the period of the earth's rotation ($T_e =$

8.616×10^4 s $\approx$ 24 h), the object's acceleration magnitude relative to the earth-center frame is $4\pi^2R_e/T_e^2$, and Newton's second law gives $|\Sigma\mathbf{F}| = ma$, or

$$F_e - F_s = m\frac{4\pi^2R_e}{T_e^2}$$

Inserting $F_s = F_w$ and solving for F_e gives

$$F_e = F_w + m\frac{4\pi^2R_e}{T_e^2}$$

Evaluating the centripetal acceleration, we find $4\pi^2R_e/T_e^2 = 0.0339$ m/s², so that

$$F_e = F_w + (0.0339 \text{ m/s}^2)m$$

Since $F_w = mg'$ and $g' \simeq 9.8$ m/s² in any earth-surface frame, this result shows that the difference between F_e and F_w measured in an earth-surface frame at the equator is very small. This difference becomes still smaller with increasing latitude and is zero at the poles.

(b) Now we insert $F_e = mg$ into the result from part *(a)* and then divide by m. This gives

$$g = \frac{F_w}{m} + 0.0339 \text{ m/s}^2$$

Since the measured weight of the standard kilogram at this point is 9.781 N, we find

$$g = \frac{9.781 \text{ N}}{1.0000 \text{ kg}} + 0.0339 \text{ m/s}^2 = 9.815 \text{ m/s}^2$$

Using similar procedures, one can calculate g at other latitudes in terms of measured values of F_w.

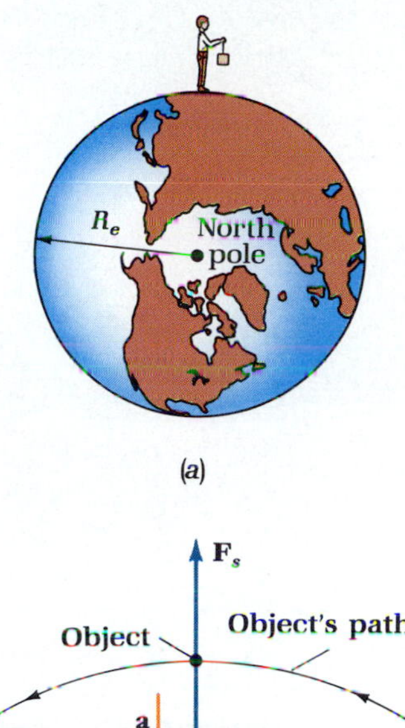

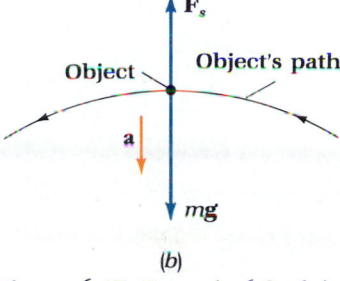

Figure 6-17. Example 6-8: *(a)* A person holding an object suspended from a spring scale at the equator. *(b)* Free-body diagram for the object. The difference between the magnitudes of the two forces is greatly exaggerated for purposes of illustration.

SELF-TEST 6-8. *(a)* To four significant digits, what is the acceleration magnitude relative to an earth-surface frame of an object in free-fall at a point on the equator where the weight of the standard kilogram is $F_w = 9.781$ N? *(b)* To four significant digits, what is the acceleration magnitude of this object relative to an inertial frame? ***ANSWERS:*** *(a)* 9.781 m/s²; *(b)* 9.815 m/s².

COMMENTARY: Isaac Newton

Issac Newton was the most important scientist of all time. His major discoveries included the laws of motion, the law of gravity, and the calculus. Any one of these achievements would be sufficient to grant its discoverer acclaim as a genius of historical proportions. It is not a coincidence that these three discoveries were made by the same person. It was natural for the discoverer of the calculus to also discover the laws of motion, and it was natural for the discoverer of the laws of motion to also discover the law of gravity.

Newton's father, also named Isaac, was illiterate but moderately wealthy. The Newton clan had been rising socially and economically for several generations when Newton's father married his mother, and the marriage represented another step forward. Newton's mother, Hannah Ayscough Newton, could read and write, which was unusual for a woman in those times. However, there is no hint of latent genius to be found in Newton's pedigree.

Newton was born in Woolsthorpe, England, on Christmas day in 1642, within a year of Galileo's death. His father had died the previous October, only six months after he had married Hannah. When Isaac was 3 years old, his mother married Barnabas Smith, an elderly but wealthy rector at a nearby village. Part of the marriage agreement was that young Isaac would be left in the care of his grandmother Ayscough in Woolsthorpe while his mother and stepfather lived some 10 miles away. There is no record of affection between Isaac and his grandmother, and some historians suspect that none existed. When Isaac was 10 years old, his stepfather died, and his mother returned to resume care of Isaac at Woolsthorpe. Many biographers of Newton have speculated on the psychological damage done to the

A portrait of Newton at age 46 by Sir Godfrey Kneller. In the words of biographer Richard Westfall: ". . . an arresting presence, instinct with intelligence, caught when his capacities stood at their height. Without difficulty we recognize the author of the *Principia*."

fatherless boy by the absence of his mother during those formative years. Newton admitted that as a boy he was obstinate and peevish. By most accounts, he became an introspective and humorless adult. He was quick to develop a grudge and was unforgiving.

At 18 years of age, Newton entered Trinity College of Cambridge University. The social structure of Cambridge reflected contemporary English society. An entering student was classified as either a pensioner or a sizar. A pensioner was privileged and given the better accommodations. A sizar performed menial tasks, often as a servant for a pensioner. Newton was a sizar, despite the fact that his mother, who grudgingly allowed him to attend college, easily could have afforded to make him a pensioner. It is disconcerting to imagine the young student, soon to become the intellectual titan of his times, waiting on tables and emptying chamber pots.

Newton's advancement at Cambridge, from sizar to scholar to fellow to Lucasian Professor of Mathematics, played a critical role in his later accomplishments. Residency at Cambridge meant a steady income with few duties, although celibacy was a requirement. These positions gave Newton the freedom to pursue studies of mechanics, optics, mathematics, astronomy, alchemy, and theology. Although his contributions to physics and mathematics account for his enduring fame, he spent a majority of his time and effort on alchemy and theology.

Newton was a solitary scholar. His ability to concentrate on a problem is legendary. He often became so engrossed in his work that he was oblivious to his own human needs, reputedly going for days without food or sleep. When asked on one occasion how he made his discoveries, he said, "By always thinking unto them." On another occasion he said, "I keep the subject constantly before me and wait till the first dawnings open little by little into the full light."

Newton certainly did not suffer fools gladly, and he was an exacting taskmaster toward himself. He was exceedingly cautious about anything he said or wrote, always fearful of a mistake. He explained his reluctance to publish as "fear that disputes and controversies may be raised against me by ignoramuses." It has been said that discoveries by Newton had two phases: he made the discovery and then someone pried it from him so that the rest of the world could share it.

Newton's masterpiece, the *Principia,* was pried from him by Edmund Halley, the astronomer for whom Halley's comet is named. The great unsolved problem at that time was the force on the planets by the sun. Halley, Christopher Wren (the famous architect), Robert Hooke, and possibly others had surmised that the force varies inversely as the square of the separation distance, a so-called inverse-square force (Chap. 7). However, no one was able to prove it. With tongue in cheek, Wren offered a prize worth 40 shillings to the person who could solve the problem. (To these proud and competitive men, the accomplishment itself was worth far more than any prize.) In 1684, Halley visited Newton and asked him what sort of path would be followed by a planet if the force by the sun was an inverse-square force. Newton replied immediately that the path would be an ellipse. Halley was "struck with joy & amazement," and he asked Newton how he knew it. Newton returned a crushingly terse reply: "I have calculated it." With these simple words, Newton announced that he had solved the problem of the centuries. Indeed, Newton was probably the only person of his time who had both the physical insight and the mathematical muscle to solve such a problem. Several months later, Newton sent Halley, who was clerk to the Royal Society (Britain's scientific organization), a treatise that contained the mathematical solution. In those days, only a handful of people would have had the ability to understand this treatise, but fortunately, Halley was one of them. Halley recognized the monumental significance of Newton's work, and he immediately began to press him to publish a book. With Halley as midwife, the *Principia* was born in 1686.

Newton's reluctance to publish caused him to share the credit for some of his discoveries; the most significant example is the calculus. From his notes and letters, it is known that Newton developed the calculus about 10 years before it was published by Gottfried Leibniz in 1684. Possibly, had someone like Halley prodded Newton in the 1670s to publish his findings. Leibniz would not now be considered the codiscoverer of the calculus.

In 1696, Newton moved to London to accept the position of warden of the mint. He played a major role in the revision of the coinage during his term (1696–1699), and in 1699 he was appointed master of the mint. He resigned his professorship at Cambridge in 1701 and was elected a member of Parliament that year. In 1703 he was elected president of the Royal Society and held that post until his death in 1726. In 1705 he was knighted. Upon his death at age 83, he was given a national funeral and buried in Westminster Abbey. Alexander Pope wrote a fitting tribute:

Nature and Nature's Laws lay hid by night;
God said, Let Newton be! And all was light.

For further reading see *Never at Rest* by Richard Westfall (Cambridge University Press, New York, 1980) or *A Portrait of Isaac Newton* by F. E. Manuel (Harvard University Press, Cambridge, Mass., 1968).

SUMMARY

Section 6-1. Contact Forces: The Normal Force and the Frictional Force

A contact force exerted by one solid object on another is resolved into a normal force and a frictional force. The kinetic frictional force on an object as it slides on a surface is

$$F_k = \mu_k F_N \tag{6-1}$$

Up to a maximum value of $F_{s,\max} = \mu_s F_N$, the static frictional force on an object at rest on a surface adjusts itself to keep the object from sliding, so that

$$F_s \leq \mu_s F_N \tag{6-3}$$

When a solid object falls in a fluid, the fluid exerts a retarding or drag force on the object such that the object's acceleration approaches zero and its speed approaches a maximum value called the terminal speed v_t.

Section 6-2. Dynamics of Uniform Circular Motion

For an object to execute uniform circular motion, a net force of constant magnitude, called a centripetal force, must be exerted toward the center of the circle:

$$|\Sigma \mathbf{F}| = \frac{mv^2}{R}$$

Section 6-3. The Rotation of the Earth

A Foucalt pendulum can be used to show that an earth-surface frame is not inertial because of the earth's rotation. The acceleration magnitude of an earth-surface frame relative to an earth-center frame is maximum at the equator, where its value is $0.0339\ \text{m/s}^2 = 0.0035g$.

QUESTIONS

1. Why is it useful to separate a contact force into a normal force and a frictional force? Give at least two reasons.

2. Suppose you wish to prop open a screen door that has a strong closing spring. A brick resting on the floor in front of the door will not keep it from closing; the door simply slides the brick along. However, a wedge-shaped piece of wood can hold the door open if it is sandwiched between the bottom of the door and the floor, even though its weight is much less than the brick's. Explain.

3. You are asked to slide a crate across a horizontal floor. The crate's weight is a few newtons larger than your weight, and the coefficient of friction between the crate and the floor is slightly larger than that between your shoes and the floor. Can you slide the crate by exerting a horizontal force on it? If not, how should you direct the force you exert in order to slide the crate?

4. Suppose you attempted to measure a coefficient of static friction between the surface of water and a block of wood floating on the water. What do you think you would find?

5. Consider a toy boat set in motion across the surface of water in a tub. Suppose we attempt to measure a coefficient of kinetic friction between the boat and the water, using the analysis developed in Example 6-2 for the sled sliding along a snow-covered road. Do you think our value of μ_k would be independent of the initial speed of the boat? If not, do you think that this concept of a coefficient of friction is useful?

6. A crate is placed at the center of a flatbed truck, and as the truck accelerates so does the crate. What type of force exerts the accelerating force on the crate? If the truck's acceleration is larger than a certain maximum value, the crate will slide. Will this maximum acceleration depend on the mass of the crate? Explain.

7. A sidewalk covered with melting ice is much more slippery than one covered with ice that is very cold and not melting. Explain.

8. You may have seen the trick where a tablecloth is jerked from a table, leaving the dishes that were on the cloth-covered table in nearly their original positions. To successfully perform this trick, should the cloth be made of thick burlap or thin silk? Should the dishes have large mass or small mass? Is it better to pull the cloth with a large force or pull it with a gentle and steady force?

9. Cross-country skiers apply a special wax to their skis to allow them to push off with one ski while the other ski slides easily along. What can you say about the ratio μ_s/μ_k between the wax and snow?

10. If you want to stop a car in a minimum distance, why is it better *not* to press so hard on the brakes that the tires slide on the pavement?

11. A carpenter's level placed on the dashboard of a car has the bubble in the center when the car is at rest on a horizontal surface. This indicates that the dashboard is parallel to the surface. Consider observing the bubble as the car makes a right turn on a road banked for cars traveling at speed v_d. Is the bubble centered, right of center, or left of center when the car's speed v is *(a)* $v = v_d$; *(b)* $v > v_d$; *(c)* $v < v_d$?

12. A panel truck is traveling at constant velocity along a horizontal street. A skate, which is aligned with the truck's length, is at rest on the floor and a helium-filled balloon is at rest against the ceiling. Describe the motion of the skate and the balloon when the driver applies the brakes.

13. Given the same initial speed, can a heavier car come to a stop in a shorter distance than a lighter one? Explain.

14. When deciding on the angle at which a highway curve should be banked, does the engineer need to take the mass of the cars into account? Explain.

15. Why is a drag-racing car designed so that most of its mass is over the drive wheels, which are at the rear of the car? (See Fig. 6-18.)

Figure 6-18. Question 15: A drag racer.

16. Explain why a pilot banks an airplane when making a turn.

17. A *turn-and-slip indicator* is an instrument used by a pilot in an airplane. One part of this instrument is a small ball that can roll in a glass tube (Fig. 6-19). The glass tube is circular and oriented in a vertical plane perpendicular to the length of the airplane. The ball rests in the bottom of the tube when the plane is at rest on a horizontal runway, and this position is marked on the tube and called the center position. When the airplane turns, the proper angle of bank is maintained by keeping the ball centered. Suppose the plane is turning right and the ball is left of center. Should the banking angle be increased or decreased? Explain.

Figure 6-19. Question 17: A ball that can roll in a circular glass tube is part of a turn-and-slip indicator on the instrument panel of an airplane.

18. When a jet airliner rapidly accelerates during takeoff, the passengers are "forced" back against their backrests. Does some object exert this force? If not, explain the existence of this force.

19. The gravitational force exerted by the earth on an object is proportional to the object's mass: $\mathbf{F}_e = m\mathbf{g}$. But because an earth-surface frame is not inertial, the gravitational force is not equal to an object's weight in an earth-surface frame. Is an object's weight in an earth-surface frame proportional to its mass?

20. A string supports a plumb bob suspended from a surveyor's tripod. Is the string aligned along the direction of the gravitational force on the bob *(a)* at the equator, *(b)* at the poles, *(c)* at a latitude of 45°? If the string is not along the direction of the force at any of these latitudes, then describe its relation to that direction.

21. The planet Noino orbits its star with the same period and separation distance as the earth orbits the sun. Noino has twice the radius of the earth, and a day on Noino is 48 earth-hours. Is a coordinate frame at rest on Noino's equator more nearly an inertial frame than a frame at rest on the earth's equator? Justify your answer.

22. While exploring the planet Otatop, you are captured and confined to a dungeon. From your cell you cannot see the sky, but you have a spring scale that measures weight very precisely. To pass the time, you place a rock on the scale every day and record the reading. One day you find that the reading is less than before. *(a)* One of your guards tells you that Otatop has changed its rotational speed. If this is true, then is the planet spinning faster or slower? *(b)* Another guard tells you that Otatop has not changed its rotational speed—rather, the gravitational force has somehow become weaker. Can you tell which guard is correct? If you could make your weight measurement at a different latitude, could you tell which guard is correct? Explain.

23. In Example 6-5 we stated that the centripetal force exerted on the car rounding a curve on a horizontal road is provided by the static frictional force $\mathbf{F}_s$. Note that $\mathbf{F}_k$ cannot be a centripetal force because $\mathbf{F}_k$ is directed opposite the relative velocity of the surfaces, and, in uniform circular motion, the velocity is tangent to the circle and perpendicular to $\mathbf{a}$. If the tires should slide on the road, then the car no longer travels in a circle because $\mathbf{F}_k$ cannot have a component perpendicular to the velocity. What is the car's path if its tires should begin to slide?

24. What is the terminal acceleration of an object falling in a fluid?

25. A man is riding inside a boxcar on a train that runs so smoothly that he cannot detect any vibrations. The boxcar has no windows so that he cannot see outside. After awakening from a nap, the man notices that the string supporting a plumb bob from the ceiling of the boxcar is not perpendicular to the floor (Fig. 6-20), even though the boxcar floor was measured to be horizontal when the train was at rest on a horizontal track. Should the man conclude that the train is *(a)* on a hill so that the floor is not horizontal or *(b)* accelerating on level ground so that the floor is horizontal? Are either of these conclusions warranted? Are both warranted? Explain.

Figure 6-20. Question 25: Man and a plumb bob in a boxcar. Is the boxcar on a hill that slopes upward to the left, or is the boxcar accelerating to the left?

26. Complete the following table:

Symbol	Represents	Type	SI Unit
F_N		Scalar	
$\mathbf{F}_k$	Kinetic frictional force		
μ_k			None

EXERCISES

Section 6-1. Contact Forces: The Normal Force and the Frictional Force

1. A 37-kg crate is at rest on a horizontal floor. A rope is attached to the top of the crate, and the rope pulls vertically upward on the crate. What is the normal force exerted on the crate by the floor when the tension in the rope is *(a)* 52 N; *(b)* 170 N; *(c)* 360 N?

2. A 940-kg car is parked on a hill such that the street surface makes an angle of 16° with the horizontal. Determine the normal force and the static frictional force exerted on the car.

3. A 41-kg sofa is to be moved across the room, and $\mu_s = 0.46$ and $\mu_k = 0.39$ between the sofa's legs and the floor. *(a)* What is the minimum horizontal force that will start the sofa sliding? *(b)* What horizontal force is required to keep the sofa sliding at constant velocity?

4. A cord is attached to a 3.9-kg box, and the cord pulls upward on the box at an angle of 32° relative to the horizontal. The tension in the cord is 21 N, but the box remains at rest on a horizontal surface. Determine the magnitudes of *(a)* the nor-

mal force, *(b)* the static frictional force, *(c)* the contact force on the box by the surface.

5. A horizontal force of 28 N is required to start a 2.6-kg block sliding across a horizontal surface. *(a)* What is μ_s between the block and the surface? *(b)* If a horizontal force of 19 N keeps the block sliding at constant velocity, then what is μ_k between the block and the surface?

6. The angle between a plank and the horizontal such that a block just begins to slide down the plank is 26°. What is μ_s between the block and the plank?

7. The speed limit along a certain road is 25 m/s. To stop at a red light, the driver of a car slammed on the brakes and the car slid 57 m before coming to a stop. The coefficient μ_k is 0.80 between the tires and the road. Was the driver exceeding the speed limit?

8. Construct a graph of the minimum stopping distance d versus the speed v of a car at the instant the brakes are applied. Let μ_s between the tires and the pavement be 0.90, and plot v horizontally from zero to 35 m/s in increments of 5 m/s. What is the percent increase in stopping distance for a 20 percent increase in v from 25 to 30 m/s?

9. If the coefficient μ_k is 0.12 between a shuffleboard and a disk, how far will the disk slide when released with an initial speed of 5.2 m/s?

10. *(a)* Show that if a block slides at constant velocity down a plank that makes an angle θ_k with the horizontal, then the coefficient of kinetic friction between the block and the plank is

$$\mu_k = \tan\theta_k$$

(b) What is μ_k between a plank and a block when $\theta_k = 29°$?

11. *(a)* Show that the expression for the magnitude of the normal force exerted by the floor in Example 6-4 (Fig. 6-9) is

$$F_N = \frac{mg\cos\theta}{\cos\theta + \mu_k \sin\theta}$$

(b) Show that the magnitude F_c of the contact force by the floor is

$$F_c = \frac{mg\cos\theta}{\cos\theta + \mu_k\sin\theta}\sqrt{1+\mu_k^2}$$

(c) Using the data from Example 6-4 ($m = 45$ kg, $\theta = 33°$, and $\mu_k = 0.63$), evaluate F_N and F_c.

12. The coefficient μ_s between a sprinter's shoes and the running track is 0.92. What is the sprinter's maximum acceleration?

13. Given that μ_s between the pavement and the tires of a car is 0.85, determine the maximum acceleration of the car on a horizontal road. Assume that half the car's weight is supported by the drive wheels.

14. A 3.4-kg block slides down a sloping surface (Fig. 6-21). The coefficient μ_k between the block and the surface is 0.37. Determine the magnitudes of *(a)* the normal force on the block, *(b)* the frictional force on the block, *(c)* the block's acceleration.

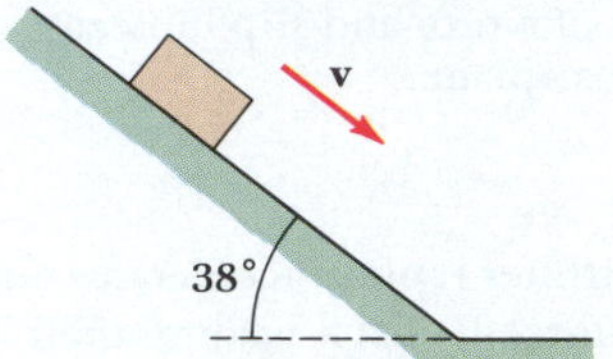

Figure 6-21. Exercise 14.

15. A boy pulls a sled up a snow-covered slope (Fig. 6-22). The mass of the sled is $m = 26$ kg, and μ_s and μ_k between the sled runners and the snow are 0.096 and 0.072, respectively. Determine the magnitude of the force exerted by the boy *(a)* to start the sled sliding and *(b)* to slide the sled at constant velocity.

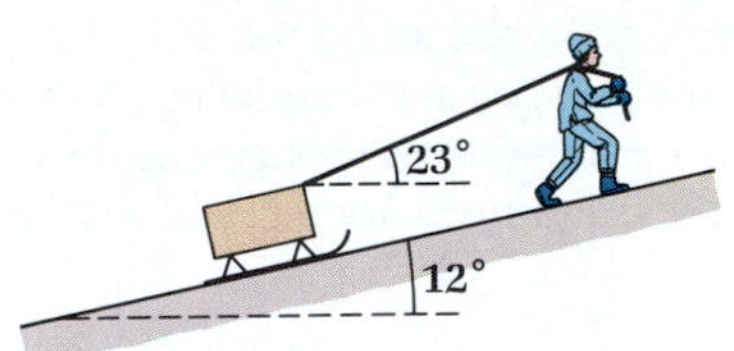

Figure 6-22. Exercise 15.

16. The applied force exerted on the sliding crate shown in Fig. 6-23*a*, *b*, and *c* has the same magnitude in each case, $F_a = 380$ N. The crate's mass is $m = 43$ kg and $\mu_k = 0.47$. Determine the crate's acceleration magnitude in each case. Explain why the acceleration is different in each case even though F_a is the same.

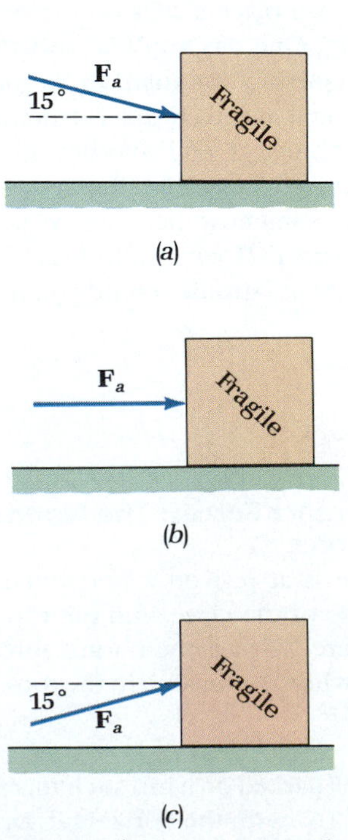

Figure 6-23. Exercise 16.

17. In Fig. 6-24, what is the magnitude of the minimum applied force that will keep the block from sliding down the vertical wall? The mass of the block is $m = 6.4$ kg and μ_s between the block and the wall is 0.76.

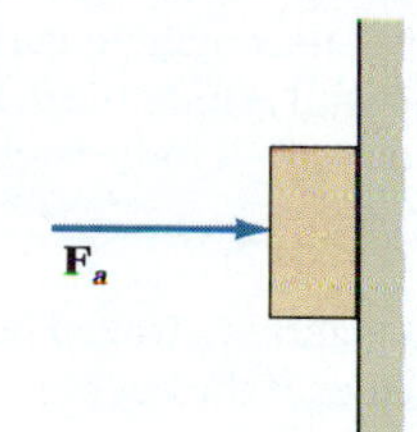

Figure 6-24. Exercise 17.

18. A cart coasts at constant velocity down a straight incline that makes an angle of 3.1° with the horizontal. What is the coefficient of rolling friction μ_r between the cart and the incline's surface?

19. The coefficient of rolling friction μ_r between a cart and a plank is 0.062. What is the cart's acceleration when the cart is placed on the plank and the plank is tilted to an angle of 5.0° relative to the horizontal?

20. A marble with mass $m = 0.012$ kg is found to have a terminal speed of 0.072 m/s when falling in a clear syrup. Assume a drag force of the form $\mathbf{F}_d = -b\mathbf{v}$ and neglect the buoyant force (as we did in the text). *(a)* Determine b. *(b)* Determine the magnitude of the net force on the marble when its speed is 0.050 m/s.

21. The drag force on a 0.081-kg rock falling in oil is given by the expression $\mathbf{F}_d = -(13 \text{ N} \cdot \text{s/m})\mathbf{v}$. What is the rock's terminal speed v_t? Neglect any buoyant forces, as done in the text.

Section 6-2. Dynamics of Uniform Circular Motion

22. What is the maximum speed a car can have while making a turn of radius 130 m on a horizontal road? The coefficient μ_s between the tires and the pavement is 0.91.

23. You are designing a highway in a section where it is to make a turn of radius $R = 310$ m. If cars are expected to travel at a speed of 25 m/s along this section of the highway, what should you make the angle of bank?

24. When Newton used a conical pendulum to determine g, he stated his value by saying that an object released from rest will fall 200 inches in 1 s. Determine Newton's value of g from this statement. Determine the percent error in this value of g.

25. If an aircraft is properly banked during a turn in level flight at constant speed, the force $\mathbf{F}_a$ exerted by the air on the aircraft is directed perpendicular to a plane which contains the aircraft's wings and fuselage (Fig. 6-25). Draw a free-body diagram for such an aircraft. (*Hint:* Note the similarity to the conical pendulum in Example 6-7.) An aircraft traveling at a speed $v = 75$ m/s makes a properly banked turn at a banking angle of 28°. What is the radius of curvature of the turn?

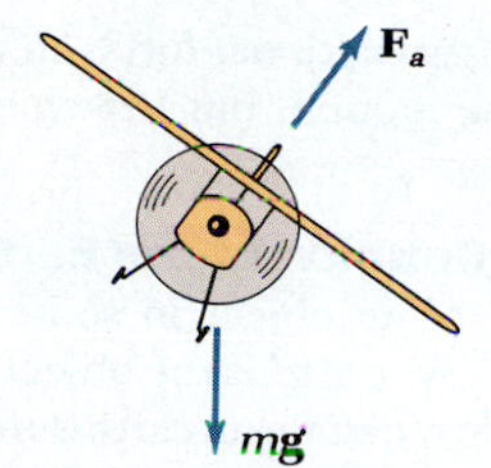

Figure 6-25. Exercise 25.

26. What is the period of a conical pendulum with $L = 1.00$ m, swung at an angle $\theta = 30°$?

27. An 875-kg car turns a corner with a radius of curvature $R = 15$ m. The car's speed is $v = 7.5$ m/s, and the street is horizontal. Determine the magnitudes of *(a)* the car's acceleration, *(b)* the frictional force on the car, *(c)* the normal force on the car, *(d)* the contact force on the car. *(e)* Determine the angle between the contact force and the vertical.

28. Show that two conical pendulums 1 and 2, with different lengths L_1 and L_2, have the same period T if the pendulums are swung such that the vertical distance between the point of suspension and the plane occupied by the circular path of a bob is the same.

29. The period of a phonograph turntable (when set at $33\frac{1}{3}$ rev/min) is $T = 1.8$ s. A penny is placed on a phonograph record, and the turntable is switched on. If the penny is placed at a distance of 0.092 m from the spindle, or less, it stays on the record and travels around in a circle. But if the penny is placed at a distance greater than 0.092 m, it slides off. What is μ_s between the penny and the phonograph record?

30. A bicyclist makes a turn with a radius of curvature of 65 m. The road is banked at an angle of 14°, and the speed of the bicycle is 18 m/s. The combined mass of the bicycle and the bicyclist is 92 kg. Determine the magnitudes of *(a)* the normal force, *(b)* the frictional force, *(c)* the contact force.

31. An interesting trick is to swing a bucket of water in a vertical circle such that the water does not pour out of the bucket while the bucket is inverted at the top of the circle. To successfully perform this trick, the speed of the bucket must be larger than a certain minimum value. *(a)* Determine an expression for the minimum speed v_m of the bucket at the top of the circle in terms of the radius R of the circle. *(b)* Evaluate v_m when $R = 1.0$ m.

32. A car tops a hill as it travels along a road. The road is straight horizontally, but, because of the hill, the road follows the arc of a vertical circle of radius R. *(a)* Determine an expression for the maximum speed v_m the car can have such that its tires remain in contact with the pavement at the crest of the hill. *(b)* Evaluate v_m when $R = 140$ m.

33. A 62-kg girl rides a steadily rotating Ferris wheel. At the top of her circular path, her weight magnitude is 210 N. The distance between the wheel's axis and the seats is 7.1 m. *(a)* What is the girl's weight magnitude at the bottom of her circular path? *(b)* What is her speed? *(c)* What is the period of the

motion? (*Hint:* The gravitational force on the girl remains the same throughout the motion, but her weight varies. See Sec. 5-5.)

Section 6-3. The Rotation of the Earth

34. The weight $\mathbf{F}_w$ of an object in some reference frame is defined as $\mathbf{F}_w = m\mathbf{g}'$, where $\mathbf{g}'$ is the object's acceleration relative to that frame. Show that for an earth-surface frame at a point on the equator, $g' = g - (0.0339 \text{ m/s}^2)$, where g is the object's acceleration magnitude relative to an inertial frame.

35. If the weight magnitude F_w of a fish in an earth-surface frame at a point on the equator is 12.75 N, what is the magnitude F_e of the gravitational force exerted by the earth on this fish?

36. You are on a mission to explore the planet Cilrag in another solar system. You have measured Cilrag's radius to be 7.46 Mm and its rotational period about its axis to be 1.21×10^4 s. Also, you have measured the weight of a replica of the standard kilogram at one of Cilrag's poles to be 8.42 N. Assume that g is the same at each point on Cilrag's surface. *(a)* What is g on Cilrag? *(b)* What is the weight magnitude F_w of the replica measured at Cilrag's equator? *(c)* If you make measurements that are precise to two significant digits, is it valid to use a Cilrag-surface frame at Cilrag's equator as an inertial frame?

37. While exploring the planet Egabbac, you use a spring scale to measure the weight magnitude F_w of a replica of the standard kilogram. At one pole you find $F_w = 12.8$ N, and at the equator you find $F_w = 10.1$ N. The planet's period of rotation about its axis is 8.68×10^3 s. Assume that g is the same at each point on Egabbac's surface. Determine *(a)* g on Egabbac and *(b)* Egabbac's radius.

Additional Exercises

38. Suzie holds a bunch of bananas suspended from a spring scale. While she is standing in the hallway the scale reads 12.0 N. Then she gets on an elevator, and while the elevator accelerates upward, the scale reads 14.8 N. *(a)* What is the mass of the bananas? *(b)* What is the gravitational force by the earth on the bananas? *(c)* What is the weight of the bananas in the accelerating elevator? *(d)* What is the acceleration of the accelerating elevator? Be certain you give the direction of vector quantities.

39. A 75-kg person rides on an elevator. Determine the person's weight magnitude in the elevator frame at an instant when the elevator accelerates downward with a magnitude 1.9 m/s².

40. A 1.6-kg block is pushed against a vertical wall by an applied force that is directed at an angle $\theta = 26.5°$ above the horizontal (Fig. 6-26). The coefficient of static friction between the block and the wall is $\mu_s = 0.50$. Determine the minimum magnitude for the applied force such that the block does not slide down the wall.

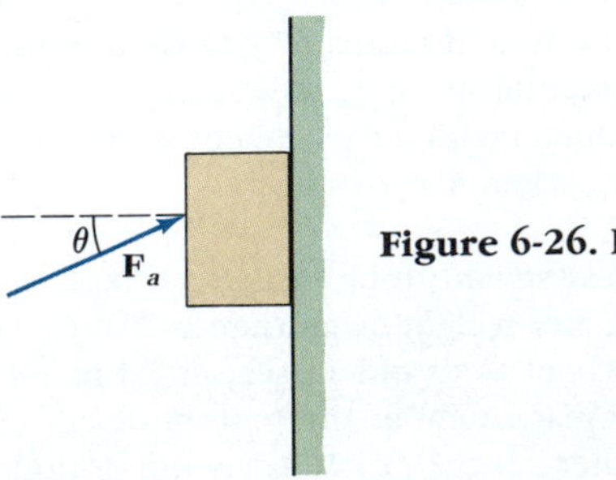

Figure 6-26. Exercise 40.

41. A 1.6-kg block is pushed against a vertical wall by an applied force that is directed at an angle $\theta = 26.5°$ above the horizontal (Fig. 6-26). The coefficient of static friction between the block and the wall is $\mu_s = 0.50$. Determine the maximum magnitude for the applied force such that the block does not slide up the wall.

42. A hockey puck slides across a frozen lake a distance of 11.0 m from the point of release. If the coefficient of kinetic friction between the ice and the puck is 0.16, what is the puck's speed at the instant it is released?

43. While ice skating you find that if you stop propelling yourself when your speed is 3 m/s, you slide to a stop in a distance of 5 m. How far will you slide if you stop propelling yourself when your speed is 6 m/s?

44. If a car is to travel at a speed of 15 m/s on a horizontal road that makes a circular turn of radius 85 m, what is the minimum value for the coefficient of static friction between the road surface and the tires?

45. A conical pendulum with length $L = 0.89$ m has a period of $T = 1.4$ s. Determine the angle θ between the string and the vertical.

46. Your job as part of the space exploration team on the planet Torrac is to measure g on Torrac's surface. You set a conical pendulum of length $L = 1.00$ m in motion at an angle of $\theta = 45.0°$ and measure the period to be 1.55 s. Assuming that your Torrac-surface frame is inertial, determine g on Torrac.

47. You are traveling on an icy road that is very slippery. There is a turn in the road with a radius of 178 m and the road is banked at an angle of 12°. At what speed should you travel such that the road exerts no frictional force on the car?

48. Suppose you tie a 0.41-kg rock to a string and swing the rock in a vertical circle of radius 0.93 m. During one revolution, strobe photography reveals that the rock's speed at the top of its circular path is 3.5 m/s and at the bottom it is 7.0 m/s. Determine the tension in the string when the rock is *(a)* at the top and *(b)* at the bottom.

49. A certain Ferris wheel has a period of 5.8 s, and the circular path of its passengers has a radius of 6.1 m. Suppose a passenger is holding a 1.4-kg purse suspended from a spring scale. What does the scale read at the instant the passenger is *(a)* at the top and *(b)* at the bottom of the circular path?

50. A 13-kg dog is riding in a car that is traveling at a speed of 18 m/s as it moves in a circular path of radius 142 m. Viewed from above, the car is turning clockwise. At the instant the car's velocity is directed toward the west, determine the magnitude and direction of *(a)* the centripetal acceleration of the dog and *(b)* the net force on the dog.

PROBLEMS

1. ***Two blocks sliding as one.*** Two blocks, 1 and 2, are sliding down a plank, (Fig. 6-27). Each block has the same mass m, but the coefficients of kinetic friction between the blocks and the surface are different, with $\mu_2 > \mu_1$. The system accelerates down the slope and the string between blocks remains taut. *(a)* Show that the tension in the string is

$$F_T = \tfrac{1}{2}(\mu_2 - \mu_1)mg \cos \theta$$

(b) Show that the magnitude a of the system's acceleration is

$$a = g[\sin \theta - \tfrac{1}{2}(\mu_2 + \mu_1) \cos \theta]$$

(c) Show that the system slides down the slope with constant velocity when $\theta = \theta_k$, where

$$\theta_k = \tan^{-1}[\tfrac{1}{2}(\mu_2 + \mu_1)]$$

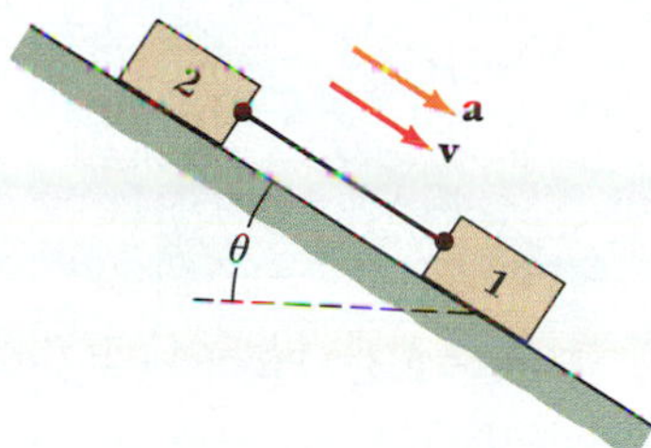

Figure 6-27. Problem 1.

2. ***Maximum acceleration for a truck with an untethered crate.*** A flatbed truck transports a crate (Fig. 6-28). The crate is not fastened to the bed, but the coefficient μ_s between the crate and the bed is 0.70. When the truck starts out from a red light on a horizontal street, what is the maximum acceleration it can have such that the crate will not slide backward relative to the bed?

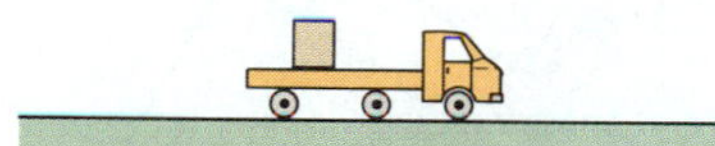

Figure 6-28. Problem 2.

3. ***A block pushed against a cart.*** In Fig. 6-29, block B has mass m, cart C has mass M, and the coefficient of static friction between the block and the cart is μ_s. Neglect frictional effects, which tend to slow the cart, and the rotational effects of the wheels. Determine an expression for the minimum value of F_a such that the block will not slide.

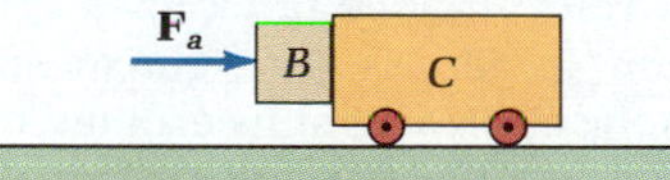

Figure 6-29. Problem 3.

4. ***A rotor.*** A rotor is an amusement park ride that consists of a cylindrical room that rotates about a vertical axis (Fig. 6-30). The passengers enter and place their backs to the wall, and the room begins to rotate. The rotational speed gradually increases, and, when it reaches a certain minimum value, the floor is taken away, exposing the passengers to the (supposed) hazard of falling into a pit below. *(a)* In terms of the coefficient μ_s between the passengers and the wall, and the radius R of the room, determine an expression for the minimum speed v_m of the passengers in their circular path such that they will not slide down the wall. *(b)* Determine an expression for the maximum period of the motion which corresponds to speed v_m.

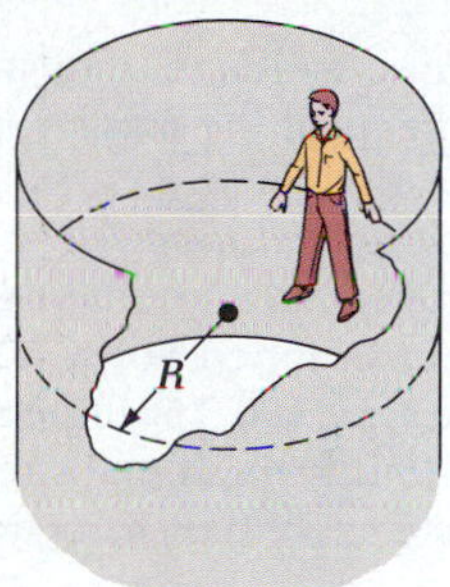

Figure 6-30. Problem 4.

5. ***Minimum force for pulling a crate.*** A crate is pulled along a horizontal surface at constant velocity by an applied force $\mathbf{F}_a$ that makes an angle θ with the horizontal, as shown in Fig. 6-31. The coefficient of kinetic friction between the crate and the surface is μ_k. *(a)* Show that the magnitude F_a is minimum when the angle $\theta = \theta_m$, where

$$\theta_m = \tan^{-1} \mu_k$$

(*Hint:* Find an expression for F_a in terms of θ and take its derivative to find θ_m.) *(b)* Show that the minimum value of F_a is

$$F_{a,\min} = \frac{\mu_k mg}{\sqrt{1 + \mu_k^2}}$$

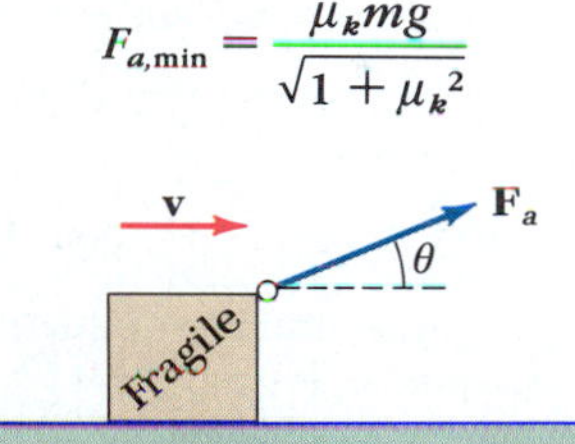

Figure 6-31. Problem 5.

(*Hint:* Determine $\cos \theta_m$ and $\sin \theta_m$ by constructing a right triangle with sides equal to 1 and μ_k so that $\tan \theta_m = \mu_k$.) *(c)* What is the minimum applied force required to slide a 51-kg crate across a horizontal floor at constant velocity when $\mu_k = 0.70$? *(d)* At what angle relative to the horizontal must the force in part *(c)* be directed? *(e)* Compare your answer in part *(c)* with the magnitude of the horizontally applied force that slides the crate at constant velocity.

6. ***Block on a cart.*** In Fig. 6-32, the mass of block B is m and the mass of cart C is M. Show that the maximum applied force $\mathbf{F}_{a,\max}$ such that the block does not slide has a magnitude

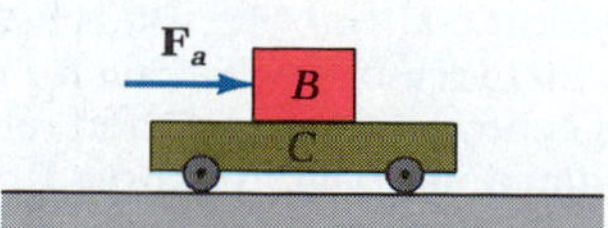

Figure 6-32. Problem 6.

$$F_{a,\max} = \mu_s mg\left(1 + \frac{m}{M}\right)$$

Neglect frictional forces, which tend to slow the cart, and neglect the rotational effects of the cart's wheels.

7. ***Minimum stopping distance for a truck with an untethered crate.*** A crate is placed in the middle of the bed of a flatbed truck and is not strapped down. The coefficient μ_s between the bed and the crate is 0.75. If the truck is traveling at a speed of $v = 22$ m/s along a horizontal street, what is the minimum stopping distance such that the crate will not slide?

8. ***Weight of a fish at latitude λ.*** Consider the weight of a fish in an earth-surface frame at latitude λ (Fig. 6-33*a*). Use the fact that the magnitude F_s of the force on the fish by the scale is equal to the weight magnitude F_w, and assume that an earth-center frame is essentially inertial. *(a)* Show that the centripetal acceleration of the fish relative to an earth-center frame is

$$a_c = \frac{4\pi^2 R_e \cos\lambda}{T_e^2}$$

where T_e is the period of the earth's rotation about its axis ($T_e = 8.616 \times 10^4$ s) and R_e is the earth's radius ($R_e = 6.37$ Mm). *(b)* In the fish's free-body diagram (Fig. 6-33*b*), the x axis is parallel to the equatorial plane and is directed toward the earth's rotational axis. Note that $\mathbf{F}_s$ is *not* directed opposite the gravitational force $\mathbf{F}_e = m\mathbf{g}$ because the centripetal acceleration is directed along $\mathbf{i}$. Use the free-body diagram to show that

$$F_s \cos\theta = mg(1 - \alpha)\cos\lambda$$

and

$$F_s \sin\theta = mg \sin\lambda$$

where $\alpha = 4\pi^2 R_e / gT_e^2$ for brevity. *(c)* Use the results of part *(b)* and $F_s = F_w$ to show that

$$F_w = mg[1 - (2\alpha - \alpha^2)\cos^2\lambda]^{1/2}$$

(*Hint:* Eliminate θ by squaring each equation, adding them, and using the identity $\sin^2\theta + \cos^2\theta = 1$.) *(d)* Show that

$$\tan\theta = \frac{\tan\lambda}{1 - \alpha}$$

(e) Using the numerical data, show that $\alpha = 3.5 \times 10^{-3}$. *(f)* Since $\alpha \ll 1$, show that

$$F_w \approx mg(1 - \alpha\cos^2\lambda)$$

(g) Evaluate F_w and θ at $\lambda = 45.0°$.

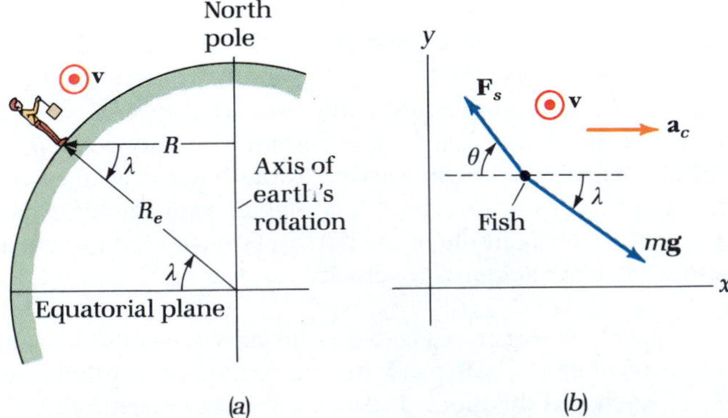

Figure 6-33. Problem 8. The weight of a fish in an earth-surface frame at latitude λ. *(a)* The fish is held suspended by a spring scale. *(b)* Free-body diagram for the fish. Because of the fish's centripetal acceleration relative to an earth-center frame (assumed inertial), the force $\mathbf{F}_s$ exerted by the scale is neither equal in magnitude nor opposite in direction to the gravitational force $\mathbf{F}_e = m\mathbf{g}$ exerted by the earth.

9. ***Sliding block pulled by a falling block.*** In Fig. 6-34, $m_A = m_B = 5.0$ kg, and μ_k between block A and the surface is 0.40. Block A is sliding up the slope. Determine *(a)* the acceleration magnitude of the system and *(b)* the tension in the string. Neglect friction and rotational effects in the pulley.

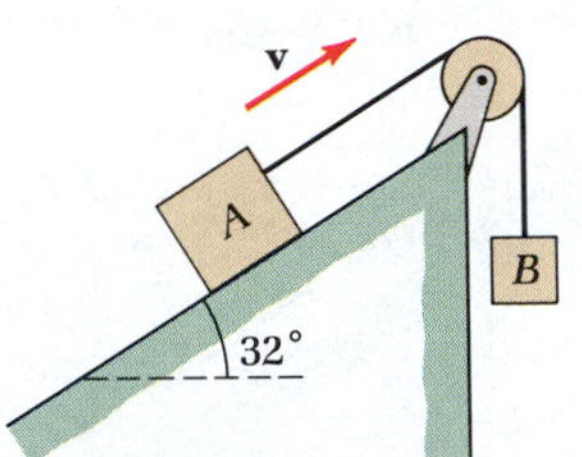

Figure 6-34. Problem 9.

10. ***Blocks on opposite slopes.*** In Fig. 6-35, μ_k is the same between each block and the surface, and $\mu_k = 0.25$. The system is sliding as shown, and $m_A = 7.0$ kg and $m_B = 9.0$ kg. Determine *(a)* the acceleration of the system and *(b)* the tension in the string. Neglect friction and rotational effects in the pulley.

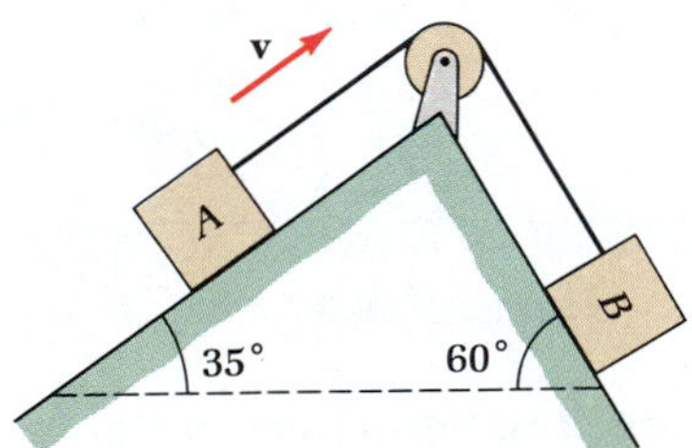

Figure 6-35. Problem 10.

11. ***Weightlessness in an airplane.*** Consider an airplane as it ascends, reaches a maximum altitude, and then descends such that its path is a vertical circle of radius R (Fig. 6-36). The airplane's constant speed v is just right so that the pilot is weightless when the airplane is at its maximum altitude. That is, her weight in the airplane reference frame is zero. Show that, as the airplane moves in this circle, the magnitude F_a of the force exerted by the airplane on the pilot (mostly due to contact with the seat) is given by $F_a = 2mg\sin(\theta/2)$ where m is the pilot's mass and θ is the angle between the airplane's velocity and the horizontal. Note that $F_a = 0$ when $\theta = 0$.

12. ***Marble falling in oil.*** Use a spreadsheet to determine the motion of the marble falling in oil that we discussed in Sec. 6-1. Using about 10 data points, construct a graph of v versus t from $v = 0$ to $v \approx (0.95)v_t$.

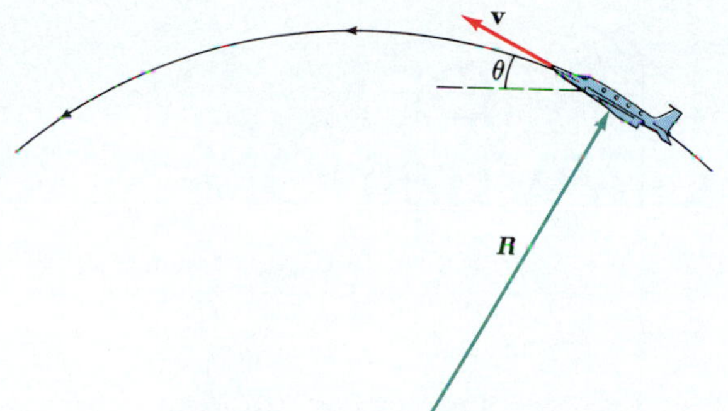

Figure 6-36. Problem 11.

13. ***A skate on a merry-go-round.*** Jim is riding on a steadily rotating merry-go-round that has a period of 8.2 s; the merry-go-round rotates clockwise when viewed from above. He places a 0.69-kg skate at a distance of 3.8 m from the center. The skate has its wheels aligned radially with its front end toward the center of the merry-go-round. The skate has well-lubricated axles so that the force exerted on the skate by the merry-go-round's floor is essentially normal to the floor surface. Jim keeps the skate from rolling backward by holding onto a string tied to the front of the skate. The string is horizontal. At an instant when the skate is directly east of the merry-go-round's center, determine the magnitude and direction of *(a)* the skate's velocity relative to an earth-surface frame; *(b)* the skate's acceleration relative to an earth-surface frame; and *(c)* the force exerted on the skate by the string.

14. ***A suspended rock on a merry-go-round.*** While riding on a steadily rotating merry-go-round, Mary holds a 0.81-kg rock suspended from a string. When the string makes an angle of 15° with the vertical, the rock remains stationary in Mary's frame of reference at a distance of 2.9 m from the merry-go-round's axis. Determine *(a)* the tension in the string; *(b)* the magnitude of the rock's centripetal acceleration relative to an earth-surface frame; and *(c)* the period of the merry-go-round.

15. ***Best angle to stop a downward slide.*** A block of mass m is pushed against a vertical wall by an applied force $\mathbf{F}_a$ directed upward at an angle θ with the horizontal as shown in Fig. 6-29. *(a)* Show that the expression for the minimum value of F_a such that the block will not slide down the wall is

$$F_{a,\min} = mg/(\sin\theta + \mu_s \cos\theta)$$

where μ_s is the coefficient of static friction between the wall and the block. *(b)* Show that $F_{a,\min}$ is further minimized if $\theta = \tan^{-1}(1/\mu_s)$. (*Hint:* Take the derivative of $F_{a,\min}$ with respect to θ and set it equal to zero and solve for this value of θ.) *(c)* Show that for the angle θ in part *(b)*,

$$F_{a,\min} = mg/(1 + \mu_s^2)^{1/2}$$

16. ***Force on the earth by the sun.*** Consider the motion of the earth in a sun-center frame. A *sun-center frame* is a frame whose origin is fixed relative to the center of the sun and whose axes are fixed relative to the distant stars. The radius of the earth's orbit about the sun is $R = 1.50 \times 10^{11}$ m, the period of the orbit is $T = 1$ year $= 3.16 \times 10^7$ s, and the earth's mass is $m = 5.97 \times 10^{24}$ kg. *(a)* Determine the magnitude of the earth's centripetal acceleration relative to a sun-center frame. *(b)* Assuming that the only significant force exerted on the earth is the force by the sun, determine the magnitude of this force.

17. ***Car on a banked curve.*** A 985-kg car travels along a circular turn of radius $R = 162$ m on a road that is banked at an angle of 12° toward the center. Determine the magnitude of the frictional force exerted by the road on the car when *(a)* the car's speed is 24 m/s and *(b)* the car's speed is 12 m/s. Describe the direction of this force in each case.

18. ***Terminal speed of a baseball falling in air.*** The magnitude F_a of the force exerted by the air on a baseball as it falls is nearly proportional to the speed squared: $F_a = cv^2$, where the proportionality constant $c = 0.0013\ \text{N}\cdot\text{s}^2/\text{m}^2$. Determine the terminal speed of a baseball in air.

19. ***Measuring the acceleration of a jet.*** As a passenger on a jetliner, you can measure the plane's acceleration on its takeoff run without measuring a distance traveled or a time interval. Instead, you can use Newton's second law. As the plane begins its run, suspend a dense object, such as a large key, by a thin string, and hold it up between you and the window so you can view it against the vertical walls of buildings in the background. As the jet accelerates, measure the angle between the string and the vertical with a protractor. Suppose you find that the angle is, say, 35° at some instant. What is the magnitude of the acceleration at this instant?

20. ***Measuring the period of a turntable.*** You can measure the period T of a phonograph turntable without a watch. One way is to use a ruler, a protractor, and Newton's second law. First, fix a straight ramp to the turntable surface, as shown in Fig. 6-37. The ramp makes an angle θ with the horizontal, and its bottom edge is at the turntable's spindle. Then, while the turntable is rotating steadily, locate the position where a toy car will roll neither down nor up the ramp. The car's wheels should be capable of moving freely so that friction can be neglected. Show that

$$T = 2\pi[(d \cot\theta \cos\theta)/g]^{1/2}$$

where d is the slant distance up the ramp from the spindle to the car. If $d = 0.31$ m when $\theta = 20.0°$, determine the value of T.

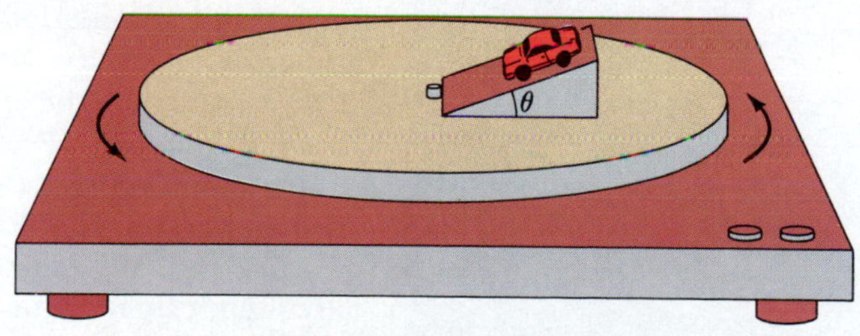

Figure 6-37. Problem 20.

7 NEWTON'S LAW OF UNIVERSAL GRAVITATION

The earth as seen from space.

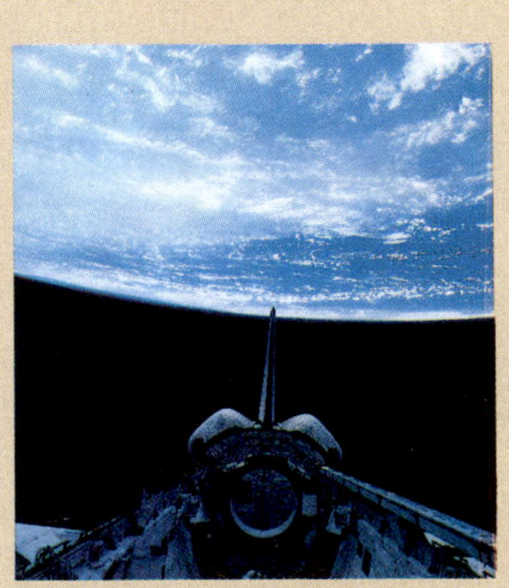

A space shuttle orbiting the earth.

In the *Principia,* Isaac Newton introduced the three laws of motion and the law of universal gravitation. In addition, he demonstrated how to use these laws. Others claimed to have discovered the gravitational law before Newton, but it was his masterful manipulation of this law in the *Principia* that led to uncovering the mysteries of gravity. Newton revealed many of the consequences of the law of gravity, including the elliptical orbits of the planets (Sec. 7-6) and the twice-a-day arrival of the tides (Prob. 14).

Newton recognized the connection between a falling object on earth and a planet orbiting the sun. The motions of these objects, though quite different, are a result of the same type of force, the gravitational force. For example, if you drop a book, the only significant force on the falling book is the gravitational force by the earth. Also, the only significant force on a space shuttle as it orbits the earth is the gravitational force by the earth.

Gravity holds us firmly to the surface of the earth, and it holds the planets in their orbits around the sun. It holds the sun and other stars together in a swirling system of stars we call the Milky Way galaxy (Fig. 7-1). It holds galaxies together in clusters of galaxies (Fig. 7-2). Because of this force, each object in the universe tends to attract all other objects in the universe. This is what we mean when we say that the *gravitational force is universal.*

7-1 THE LAW OF UNIVERSAL GRAVITATION

By using some intuition and some experimental information about the solar system, we now develop the law of universal gravitation. Because the solar system can be described with a simple model, it is almost ideal for the study of the gravitational force. Also, because the nine planets are at greatly different distances from the sun, they provide a broad sampling of the distance dependence of the gravitational force.

A Sun-Center Reference Frame

When describing the motion of an object on earth, we often use an earth-surface frame as an inertial frame. Sometimes, for example, in the case of a Foucault pendulum, we use an earth-center frame. Now if we wish to consider the motion of the planets in the solar system, is either of these frames close enough to being inertial? No. The motion of the planets is very complex from the point of view of either an earth-surface frame or earth-center frame. However, planetary motion is quite simple relative to a sun-center frame.

A ***sun-center reference frame*** is a frame whose origin is fixed relative to the center of the sun and one whose axes are fixed relative to distant stars.

Definition of a sun-center frame

In describing the motion of the planets, we shall assume that a sun-center frame is inertial. We make this assumption because the motion of the planets is simpler

Figure 7-1. Stars cluster together to form galaxies. Any individual star we see in the sky is a member of our own galaxy, the Milky Way, which is named for the appearance of the sky when viewed in the direction toward the center of the galaxy. The stars are so numerous in this direction that their light blends into a continuum. The streak in this photograph is a meteorite passing through the earth's atmosphere.

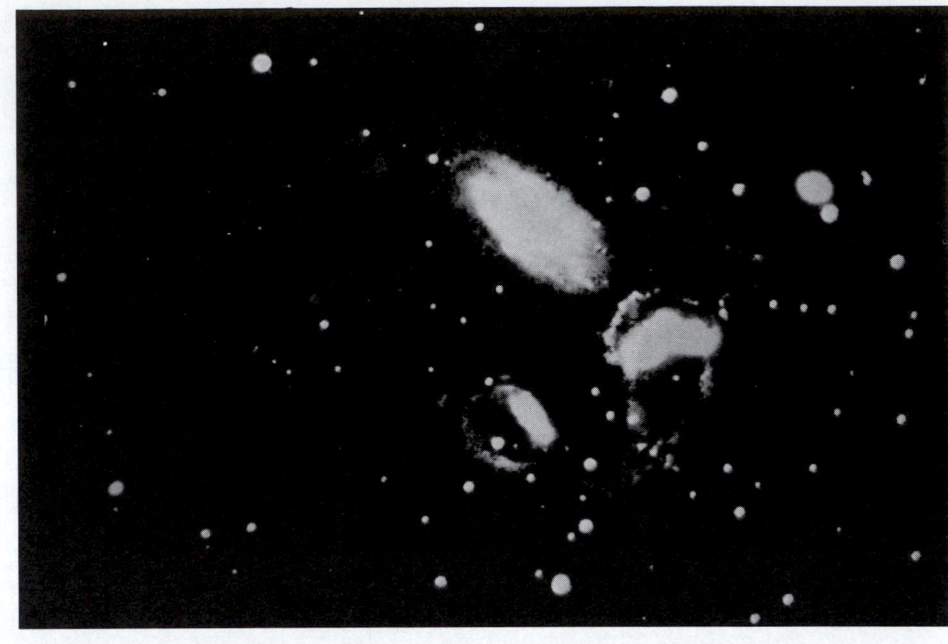

Figure 7-2. A cluster of galaxies. This cluster of four galaxies is called Stephan's Quartet. There is a fifth galaxy in the photograph, but it is much nearer the earth and belongs to another, less compact galaxy cluster. The individual stars seen scattered about in this photograph are within our own galaxy.

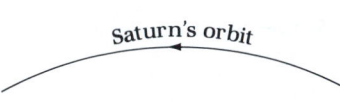

relative to that frame than to any other frame. At the outset of our investigation, this is a sufficiently compelling reason, but at the end of the investigation, we will require more. At the end we require that our findings, based on this assumption, yield a simple, accurate, and consistent theory.

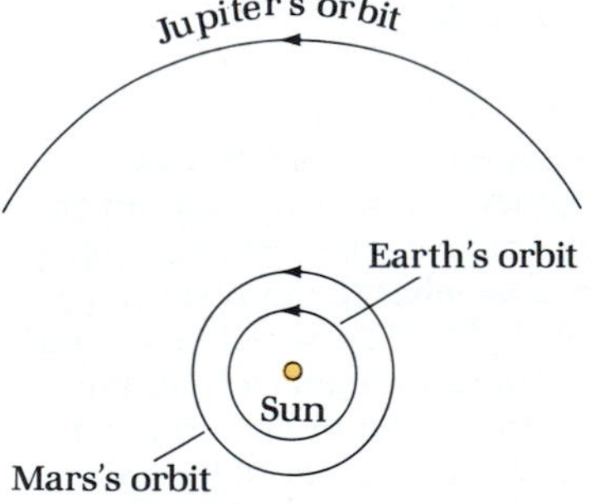

Figure 7-3. A few representative orbits of planets, shown to scale. On this scale the sun's radius is less than one-tenth the thickness of this page.

Model of the Solar System

Our model of the solar system involves three approximations:

1. The sun and the planets are treated as particles. This approximation is justified because the distances between the sun and planets are much greater than their sizes (Fig. 7-3).

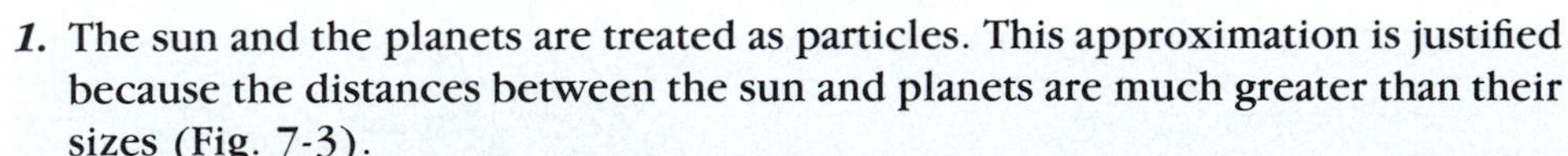

2. Each planet moves around the sun in a circular orbit. Although the orbits are actually elliptical (Sec. 7-6), they are nearly circular. This approximation allows us to use the centripetal acceleration v^2/R relative to the sun as the acceleration relative to an inertial frame in Newton's second law.
3. The only significant force on a planet is the gravitational force exerted by the sun. This approximation is justified by the observation that each planet's orbit is essentially unaffected by the positions of the other planets. Because of this approximation, we use the force exerted by the sun on a planet as the net force on that planet.

Distance Dependence

The model of the solar system described above can be used in finding the distance dependence of the gravitational force. Columns (1) and (2) in Table 7-1 give the periods T and the radii R of the planetary orbits about the sun. Using these data, we can find the acceleration of each planet. Since the speed of an object traveling in a circle of radius R with period T is $v = 2\pi R/T$, the object's centripetal acceleration is

$$a_c = \frac{v^2}{R} = \frac{(2\pi R/T)^2}{R} = \frac{4\pi^2 R}{T^2}$$

TABLE 7-1. ***Solar System Data***

Body	(1) Period T, 10^7 s	(2) Average Orbital Radius R, 10^{11} m	(3) Average Orbital Acceleration $a_c = 4\pi^2 R/T^2$, 10^{-3} m/s²	(4) Acceleration × Orbital Radius Squared $a_c R^2 = 4\pi^2 R^3/T^2$, 10^{20} m³/s²	(5) Mass 10^{24} kg
Sun					1,990,000
Mercury	0.760	0.579	39.6	1.33	0.335
Venus	1.94	1.08	11.3	1.32	4.89
Earth	3.156	1.496	5.929	1.327	5.98
Mars	5.94	2.28	2.55	1.33	0.646
Jupiter	37.4	7.78	0.219	1.33	1900
Saturn	93.5	14.3	0.0646	1.32	569
Uranus	264	28.7	0.0163	1.34	87.3
Neptune	522	45.0	0.00652	1.32	103
Pluto	782	59.1	0.00382	1.33	5.4

Inserting T and R for each planet gives the accelerations listed in column (3). As an example, the earth's acceleration relative to the sun is

$$a_c = \frac{4\pi^2(1.496 \times 10^{11}\ \text{m})}{(3.156 \times 10^7\ \text{s})^2} = 5.929 \times 10^{-3}\ \text{m/s}^2$$

Column (4) of Table 7-1 presents the data that are crucial to our investigation. The product $a_c R^2$ (or $4\pi^2 R^3/T^2$) for each of the planets is nearly the same. Each value is within 1 percent of the average value. Within the realm of our approximations, we can state that the product $a_c R^2$ is the same for each planet. That is,

$$a_c R^2 = k_s$$

where k_s (subscript s for sun) is a constant that is the same for each planet ($k_s = 1.33 \times 10^{20}\ \text{m}^3/\text{s}^2$). Solving for a_c gives

$$a_c = \frac{k_s}{R^2}$$

If we let F_{sp} be the magnitude of the force by the sun on a planet and m_p be the mass of the planet, then Newton's second law gives $F_{sp} = m_p a_c$, or

$$F_{sp} = m_p\left(\frac{k_s}{R^2}\right) = \frac{k_s m_p}{R^2} \tag{7-1}$$

Thus the gravitational force decreases with increasing distance R as $1/R^2$. A force that depends on distance in this fashion is called an ***inverse-square force.***

An inverse-square force

Mass Dependence

In the *Principia,* Newton used the third law of motion to arrive at the mass dependence of the gravitational force law, as indicated by the following passage:

> And since the action of centripetal force upon the attracted body, at equal distances, is proportional to the matter in this body, it is reasonable, too, that it is also proportional to the matter in the attracting body. For the action is mutual, and causes the bodies by mutual endeavor (by law 3) to approach each other, and accordingly it ought to be similar to itself in both bodies. One body can be considered as attracting and the other as attracted, but this distinction is more mathematical than natural. The attraction is really that of either of the two bodies toward the other, and thus is of the same kind in each of the bodies.*

* I. Bernard Cohen, "Newton's Discovery of Gravity," *Scientific American,* March 1981.

Let us follow Newton's reasoning step by step. From Eq. 7-1, the gravitational force by the sun on a planet is proportional to the mass of the planet. Also, from Newton's third law, if the sun exerts a force $\mathbf{F}_{sp}$ on a planet, then the planet exerts a force $\mathbf{F}_{ps}$ on the sun, and the magnitudes are equal: $F_{sp} = F_{ps}$. If a single mathematical expression (or law) is to provide both of these forces, then m_s must enter the expression in the same way as m_p. That is, since $F_{sp} \propto m_p$, we expect that $F_{ps} \propto m_s$. Therefore we let $k_s = Gm_s$, where G is a proportionality constant independent of either mass, and Eq. 7-1 becomes

$$F_{sp} = \frac{Gm_s m_p}{R^2}$$

The expression for F_{ps} is found by exchanging p for s and s for p:

$$F_{ps} = \frac{Gm_p m_s}{R^2}$$

Since these expressions give $F_{ps} = F_{sp}$, they agree with Newton's third law, and the mass of each object enters each expression the same way.

Gravitational Force Law for Particles

To generalize our results, we let $\mathbf{F}_{12}$ represent the gravitational force exerted by particle 1 on particle 2. Since the force is attractive, it is directed from 2 toward 1. The customary way to show this direction is to use a unit vector $\hat{\mathbf{r}}$, which is directed from 1 toward 2, as shown in Fig. 7-4. Thus we have

Newton's law of universal gravitation

$$\mathbf{F}_{12} = -\frac{Gm_1 m_2}{r^2}\,\hat{\mathbf{r}} \qquad (7\text{-}2)$$

where m_1 and m_2 are the masses of the particles and r is the distance between them. The minus sign indicates that the direction of the force is opposite $\hat{\mathbf{r}}$. This is Newton's law of universal gravitation. The symbol G represents a universal constant, and is referred to as "big G" in conversations. Capital "G" is used in order to avoid confusion with the magnitude of the acceleration of gravity near the earth's surface, represented by g and sometimes called "little g."

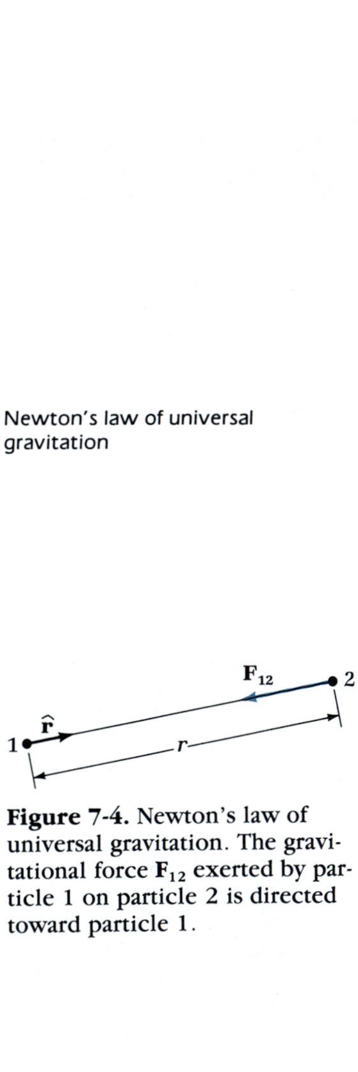

The magnitude of the force exerted by particle 1 on particle 2 is equal to the magnitude of the force exerted by particle 2 on particle 1: $F_{12} = F_{21} = Gm_1m_2/r^2$. Because of this, we often refer to the force *between* particles 1 and 2. Such wording is common, but it misleadingly implies that there is no distinction between the force by 1 on 2 and the force by 2 on 1. This is true only for the *magnitudes* of the forces by Newton's third law. When using such wording you should keep in mind that you are talking about two different forces on two separate particles, and that these two forces are opposite in direction.

Figure 7-4. Newton's law of universal gravitation. The gravitational force $\mathbf{F}_{12}$ exerted by particle 1 on particle 2 is directed toward particle 1.

Force between Extended Objects

Equation 7-2 applies to objects that are separated far enough such that they can be regarded as particles. How do we determine the gravitational force between objects that are not small compared with their separation? For example, how do we find the gravitational force between the earth and a chair near the surface of the earth? To answer this question, Newton applied his newly invented integral calculus. The answer is that we imagine the two objects to consist of a large number of small pieces, each piece being small enough to be regarded as a particle. Then, using integral calculus, we add vectorially the forces due to each piece. The sum (or integral) gives the total gravitational force.

Showing this procedure is beyond our present scope (see Chaps. 20 and 21), but the answer for objects with a spherical distribution of mass, such as a billiard ball and a ping-pong ball, is remarkably simple. Consider two spherically symmetric objects, *A* and *B*, as shown in Fig. 7-5. The gravitational force between *A* and *B* is the same as it would be if *A* and *B* were particles with all the mass of each object concentrated at its center. If you wish to find the gravitational force between two objects with spherical symmetry, such as a billiard ball and the earth, or a billiard ball and a basketball, use Eq. 7-2 with *r* as the distance between centers and $\hat{\mathbf{r}}$ directed along the line between centers. However, the two objects must be outside one another. If the billiard ball is inside the basketball, then Eq. 7-2 is invalid.

Gravitationally, an object with spherical symmetry behaves as a particle with all its mass located at its center.

This result for objects with spherical symmetry is a consequence of two features of the gravitational force: (i) the force is directed along the line joining the particles and (ii) the force is an inverse-square force. The electric force between two charged particles has these same two features. We shall return to this problem of the force between objects with spherical symmetry when discussing electric forces in Chap. 21.

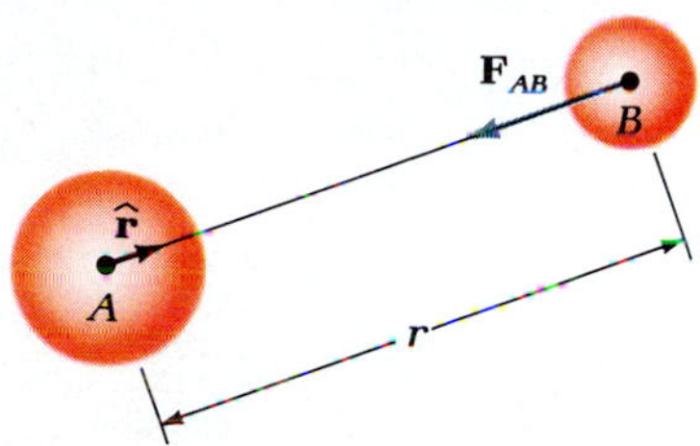

Figure 7-5. Gravitational force $\mathbf{F}_{AB}$ by object *A* with spherical symmetry on object *B* with spherical symmetry. Each object interacts gravitationally as a particle with all the mass of the object located at its center.

Now consider the gravitational force on a chair near the surface of the earth. A chair is shaped nothing like a sphere. What force law do we use in this case? We still use Eq. 7-2. The chair does not have spherical symmetry, but the earth does. In describing the interaction, we may treat the earth as a particle with mass m_e located at the center of the earth. Since the distance from the chair to the center of the earth is $R_e = 6.37 \times 10^6$ m, the separation *r* between the chair and the "earth-treated-as-a-particle" is large compared with the dimensions of the chair (≈ 1 m). This means that the chair may be treated as a particle too. Thus the gravitational force $\mathbf{F}_e$ exerted by the earth on an object of mass *m* at the surface of the earth is

$$\mathbf{F}_e = -\left(\frac{Gmm_e}{R_e^2}\right)\hat{\mathbf{r}}$$

where m_e is the mass of the earth, R_e is the radius of the earth, and $\hat{\mathbf{r}}$ is a unit vector directed away from the center of the earth. Previously, we have expressed this force as

$$\mathbf{F}_e = m\mathbf{g}$$

A comparison of these equations for $\mathbf{F}_e$ provides an expression for the magnitude *g*. Since

$$F_e = \frac{Gmm_e}{R_e^2} \qquad \text{and} \qquad F_e = mg$$

we have

$$g = \frac{Gm_e}{R_e^2} \tag{7-3}$$

Newton's law of universal gravitation shows why all objects falling freely near the surface of the earth have the same acceleration. The acceleration depends on properties of the earth (m_e and R_e) and is independent of any property of the falling object.

7-2 THE GRAVITATIONAL CONSTANT *G*

An accurate measurement of the gravitational constant *G* eluded scientists for many years. The constant cannot be evaluated from the radii and periods of the planetary orbits without first knowing the mass of the sun. We can see why this is so by applying Newton's second law, $\Sigma\mathbf{F} = m\mathbf{a}$, to a planet: $Gm_sm_p/R^2 = m_pa_c$. Solving for *G*, we have

$$G = \frac{a_cR^2}{m_s}$$

As we have seen, the radii and periods of the orbits of the planets give $a_cR^2 \approx 1.33 \times 10^{20}$ m^3/s^2 for each planet (Table 7-1). However, without knowing the mass of the sun, we cannot determine G.

Another way to try to find G is from the measured values of g and R_e. Solving Eq. 7-3 for G gives

$$G = \frac{gR_e^2}{m_e}$$

However, without first knowing the mass of the earth, we cannot determine G.

This problem of finding the value of G can be solved by measuring the magnitude F_{12} of the force between two spherically shaped objects of known masses m_1 and m_2 that are a known distance r apart. Solving Eq. 7-2 for G, we have

$$G = \frac{F_{12}r^2}{m_1 m_2} \tag{7-4}$$

The values of the measurements can then be inserted into Eq. 7-4. However, an accurate measurement of G is very difficult to make. The difficulty is that if m_1 and m_2 are of ordinary size, say a few kilograms, then F_{12} is exceedingly small.

In 1798, 71 years after Newton's death, Henry Cavendish (1731–1810) made the first reasonably accurate measurements from which the value of G can be calculated. The apparatus used by Cavendish, now called a *Cavendish balance,* is shown schematically in Fig. 7-6. He used four lead balls: balls a and b were of equal mass and balls A and B were of equal mass but different from a and b. Balls a and b were placed on each end of a light rod to form a dumbbell, which was suspended by a thin fiber. Balls A and B were placed near a and b on opposite sides, as seen in the figure, so that the gravitational attraction between a and A and between b and B caused the thin fiber to twist by a measurable amount. A mirror was attached to the rod so that a beam of light reflected from the mirror to a scale gave a precise measure of the equilibrium orientation of the rod. Then A and B were repositioned as shown by the dashed circles in the figure. In this second configuration, the gravitational attraction between the balls caused the fiber to twist in the opposite sense. The new equilibrium orientation of the rod was then recorded. From a separate measurement of the forces on a and b that twist the fiber a given amount, the gravitational forces were calculated.

Cavendish described his experiment as "weighing the earth." Indeed, once G is known, the mass of the earth can be determined. For that matter, the mass of the other planets and the mass of the sun can also be found. The presently accepted value of G is

The value of G

$$G = 6.670 \times 10^{-11}\ \text{N} \cdot \text{m}^2/\text{kg}^2$$

Besides finding the value of G, the Cavendish balance has also been used to confirm the mass and distance dependence in Newton's law of universal gravitation.

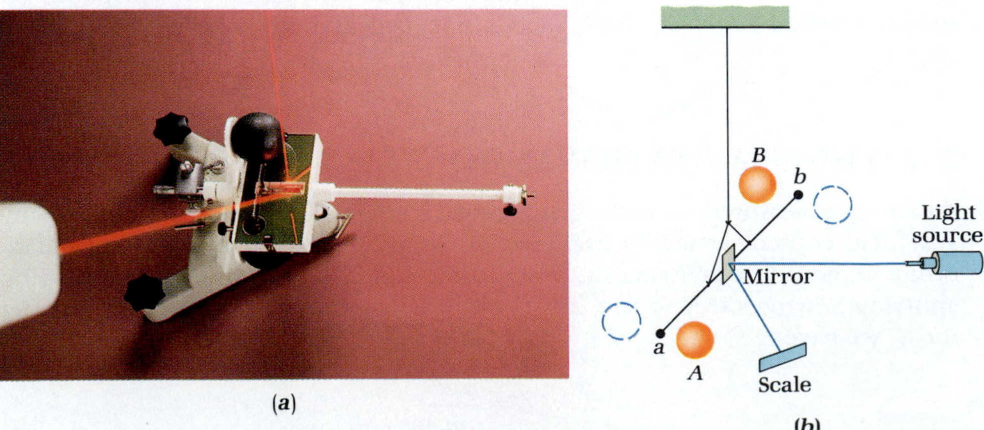

Figure 7-6. *(a)* A Cavendish balance. *(b)* A schematic drawing of a Cavendish balance. The balance can be used to measure G and to provide experimental verification of the distance and mass dependence in Newton's law of universal gravitation.

EXAMPLE 7-1

Size of a force measured by a Cavendish balance. Suppose the lead balls a and A that we use in a Cavendish experiment have radii $r_a = 20.0$ mm and $r_A = 30.0$ mm. Determine the magnitude F_{aA} of the force between these balls when their centers are 60 mm apart. The mass density of lead is $\rho = 11.4 \times 10^3$ kg/m³.

Solution. The mass of an object is given by the product of the mass density and the volume, and the volume of a sphere of radius r is $4\pi r^3/3$. Thus the mass of ball a is

$$m_a = (11.4 \times 10^3 \text{ kg/m}^3)[4\pi(2.00 \times 10^{-2} \text{ m})^3/3] = 0.382 \text{ kg}$$

Similarly, $m_A = 1.29$ kg. From Newton's law of universal gravitation,

$$F_{aA} = \frac{Gm_a m_A}{r^2} = \frac{(6.67 \times 10^{-11} \text{ N}\cdot\text{m}^2/\text{kg}^2)(0.382 \text{ kg})(1.29 \text{ kg})}{(6.0 \times 10^{-2} \text{ m})^2}$$
$$= 9.1 \times 10^{-9} \text{ N}$$

Compared with the weight of the balls, this is a very tiny force indeed. For example, the weight of ball a is $(0.382 \text{ kg})(9.8 \text{ m/s}^2) = 3.7$ N, which is about a billion times larger than F_{aA}.

SELF-TEST 7-1. Determine the magnitude of the gravitational force between a billiard ball of mass 0.2 kg and a basketball of mass 0.6 kg when they are separated by a distance of 0.5 m between their centers. ***ANSWER:*** 3.2×10^{-11} N.

EXAMPLE 7-2

Mass of the sun. Using the value of G and the data of Table 7-1, determine the mass of the sun.

Solution. From Table 7-1 we have $a_c R^2 = 1.33 \times 10^{20}$ m³/s², and we previously noted that $G = a_c R^2/m_s$. Therefore

$$m_s = \frac{a_c R^2}{G} = \frac{1.33 \times 10^{20} \text{ m}^3/\text{s}^2}{6.67 \times 10^{-11} \text{ N}\cdot\text{m}^2/\text{kg}^2} = 1.99 \times 10^{30} \text{ kg}$$

SELF-TEST 7-2. From the numerical values of G, g, and the radius of the earth ($R_e = 6.37 \times 10^6$ m $= 6.37$ Mm), determine the mass m_e of the earth. ***ANSWER:*** 6.0×10^{24} kg.

EXAMPLE 7-3

Variation of F_e with distance from the earth's center. Construct a graph of the distance dependence of the magnitude of the gravitational force by the earth on a 1.00-kg object in the range of distances from R_e to $4R_e$ from the earth's center. Determine the force magnitude at R_e, $2R_e$, $3R_e$, and $4R_e$; plot the points; and then sketch a curve between the points.

Solution. Let e represent the earth and b the object. At R_e,

$$F_{eb} = \frac{Gm_e m_b}{R_e^2}$$
$$= \frac{(6.67 \times 10^{-11} \text{ N}\cdot\text{m}^2/\text{kg}^2)(5.98 \times 10^{24} \text{ kg})(1.00 \text{ kg})}{(6.37 \text{ Mm})^2} = 9.83 \text{ N}$$

At $2R_e$,

$$F_{eb} = \frac{Gm_e m_b}{(2R_e)^2} = \frac{9.83 \text{ N}}{2^2} = \frac{9.83 \text{ N}}{4} = 2.46 \text{ N}$$

At $3R_e$,

$$F_{eb} = \frac{Gm_e m_b}{(3R_e)^2} = \frac{9.83 \text{ N}}{3^2} = \frac{9.83 \text{ N}}{9} = 1.09 \text{ N}$$

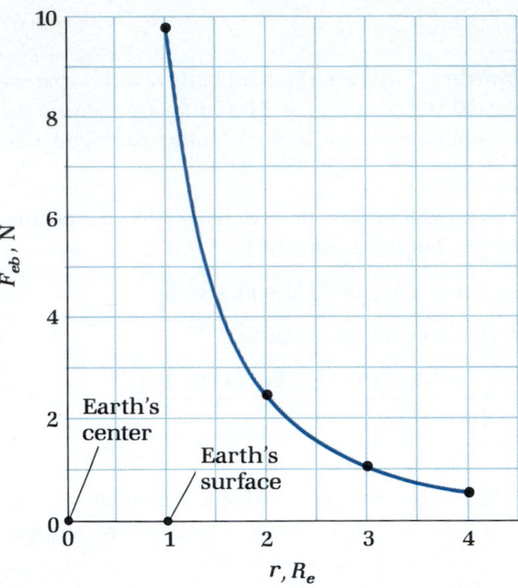

Figure 7-7. Example 7-3. The magnitude F_{eb} of the gravitational force by the earth on object b ($m_b = 1.00$ kg) versus distance r from the center of the earth.

Similarly, at $4R_e$,

$$F_{eb} = \frac{Gm_e m_b}{(4R_e)^2} = \frac{9.83\text{ N}}{4^2} = \frac{9.83\text{ N}}{16} = 0.614\text{ N}$$

The graph is shown in Fig. 7-7.

SELF-TEST 7-3. Determine the magnitude of the gravitational force by the earth on a 1.00-kg object when the object's distance from the earth's center is *(a)* $5R_e$ and *(b)* $10R_e$.
ANSWERS: *(a)* 0.393 N; *(b)* 0.0983 N.

EXAMPLE 7-4

A point where gravitational forces cancel. As with other forces, gravitational forces add vectorially. Consider a spacecraft traveling from the earth to the moon along a straight line between the center of the earth and the center of the moon. At what distance from the center of the earth is the force $\mathbf{F}_{es}$ by the earth on the spacecraft equal and opposite the force $\mathbf{F}_{ms}$ by the moon on the spacecraft? The mass of the moon is $m_m = 7.35 \times 10^{22}$ kg, and the radius of the moon's orbit about the earth is $r_{em} = 3.84 \times 10^8$ m = 384 Mm.

Solution. Figure 7-8 shows a coordinate frame with the center of the earth at the origin and the x axis extending to the moon. Since any gravitational force is attractive, the point P where $\mathbf{F}_{es}$ and $\mathbf{F}_{ms}$ are equal and opposite must be on this axis and between the earth and the moon. In the figure this point is shown at a distance x from the center of the earth and a distance $r_{em} - x$ from the center of the moon. Setting the magnitudes of the two forces equal, we have

$$\frac{Gmm_e}{x^2} = \frac{Gmm_m}{(r_{em} - x)^2}$$

where m is the mass of the spaceship. Rearranging gives

$$\left(1 - \frac{m_m}{m_e}\right)x^2 - (2r_{em})x + r_{em}{}^2 = 0$$

This equation for x is a quadratic equation. Using the solution to a quadratic equation (App. M), we find

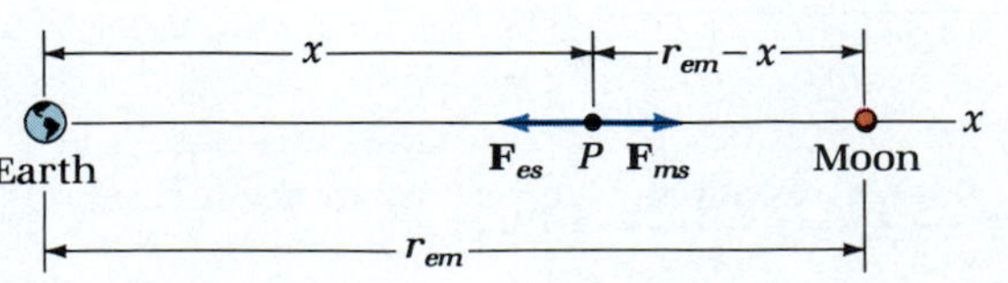

Figure 7-8. Example 7-4. Finding the distance x from the center of the earth to the point P where the gravitational forces on a spaceship due to the earth ($\mathbf{F}_{es}$) and due to the moon ($\mathbf{F}_{ms}$) are equal and opposite.

$$x = \frac{r_{em}}{1 \pm \sqrt{m_m/m_e}}$$

The plus or minus sign indicates that there are two points on the x axis where the force magnitudes are equal. For the minus sign, $x > r_{em}$. This corresponds to a point on the opposite side of the moon from the earth. At this point the two forces are equal in magnitude and in the *same* direction. Since we want the point where the forces are equal in magnitude and *opposite* in direction, we discard the solution with the minus sign. For the plus sign, $x < r_{em}$. This corresponds to a point between the earth and the moon and is the point of interest in this example. The ratio of the mass of the moon to the mass of the earth (see Table 7-1) is

$$\frac{m_m}{m_e} = \frac{0.0735 \times 10^{24}\ \text{kg}}{5.97 \times 10^{24}\ \text{kg}} = 0.0123$$

so that

$$x = \frac{384\ \text{Mm}}{1 + \sqrt{0.0123}} = (0.900)(384\ \text{Mm}) = 346\ \text{Mm}$$

The point between the earth and the moon where the two forces cancel is 90 percent of the way to the moon.

SELF-TEST 7-4. Determine the distance from the earth to the point where the gravitational forces on an object by the earth and by the moon have both the same magnitude and the same direction. ***ANSWER:*** 432 Mm.

7-3 GRAVITATIONAL AND INERTIAL MASS

From Chap. 5, the mass m of an object is the proportionality constant in Newton's second law, $\Sigma\mathbf{F} = m\mathbf{a}$.

Inertial mass

> The ***mass*** of an object is that property of the object that causes it to resist a change in its velocity.

For that reason, the mass that appears in Newton's second law is often called the ***inertial mass.*** Suppose a runaway grocery cart loaded with groceries is rolling out of control down a supermarket aisle and you must stop the cart before it crashes into a stack of cans. The force required to stop the cart depends on its inertial mass.

In this chapter we have seen that the mass of an object also appears in Newton's law of universal gravitation. The magnitude F of the gravitational force on an object of mass m due to another object of mass M is $F = GmM/r^2$. (We assume that the two objects may be treated as particles.) In this expression

Gravitational mass

> The ***mass*** of an object is that property of the object that causes it to be attracted to another object by the gravitational force.

For that reason, the mass that appears in Newton's law of universal gravitation often is called the ***gravitational mass.*** Suppose you are holding a bag of groceries while waiting for your roommate. The force you must exert while holding the bag depends on the gravitational mass of the bag of groceries.

Mass characterizes two different properties of matter.

The difficulty you encounter in stopping the runaway cart has nothing to do with its gravitational mass. The effort you expend in holding the bag of groceries has nothing to do with its inertial mass. The term "mass of an object" characterizes two different properties of the object. On the one hand, it is a measure of an object's resistance to a change of velocity (inertial mass), and on the other hand, it is a measure of an object's gravitational attraction to other objects in its environment (gravitational mass). For the purpose of this discussion, we will distinguish the inertial mass from the gravitational mass by using the symbol m_I for the former and m_G for the latter.

Why are these two different properties of matter, the inertial mass and the gravitational mass, both called "mass"? Because experiment shows that they are proportional to one another. One such experiment is the measurement of the acceleration of different objects during free-fall. During free-fall, all forces on an object are negligible except for the force of gravity; the net force on the object is the gravitational force due to the earth. Consider a golf ball in free-fall near the surface of the earth and let the $+y$ direction be vertically upward. Then $\Sigma F_y = -Gm_Gm_e/R_e^2$, where m_G is the gravitational mass of the ball. The y component of Newton's second law gives

$$-\frac{Gm_Gm_e}{R_e^2} = m_Ia_y$$

where m_I is the inertial mass of the golf ball. Solving for a_y, we have

$$a_y = -\left(\frac{Gm_e}{R_e^2}\right)\left(\frac{m_G}{m_I}\right)$$

Experiment shows that m_I is proportional to m_G, and our units are chosen such that $m_I = m_G$.

The factor (Gm_e/R_e^2) is independent of the object whose motion we are describing (the golf ball), but m_G and m_I depend on this object. As you know, all freely falling objects have the same acceleration; $a_y = -g$. If we drop a rock instead of a golf ball, then $a_y = -g$ for the rock. Thus a_y is independent of the object. This means that the ratio (m_G/m_I) must be independent of the object. In other words, our experiment shows that m_G must be proportional to m_I for each object. Since m_I is proportional to m_G, we may choose our units in such a way that they are made equal. This was tacitly done when G was evaluated from the results of the Cavendish experiment. Thus $m_I = m_G$.

The statement that the inertial mass is the same as the gravitational mass is an experimental statement. The validity of the statement depends on the accuracy of the experiments. Modern experiments have shown that the statement is valid to at least three parts in 10^{11}.

7-4 VARIATION OF g ON THE EARTH'S SURFACE

Newton's law of universal gravitation helps us understand why the earth (as well as the other planets and the sun) is spherical. The earth's own gravitational force tends to shape it as a sphere. Each part of the earth is attracted to every other part. The way to pack all the parts together so that they are as close to each other as possible is to pack them in the shape of a sphere.

The earth is slightly oblate.

However, the earth is not exactly spherical. It is more nearly shaped as an oblate spheroid, slightly flattened at the poles, similar to a beach ball that is compressed by someone sitting on it. This oblateness is a result of the spinning of the earth about its axis. Like a pot being formed on a potter's wheel, the earth bulges slightly around its middle (near the equator). The distance from the center of the earth to one of the poles is about 6.36 Mm, and the distance from the center of the earth to the equator is about 6.38 Mm. A measure of the earth's oblateness is given by

$$\frac{6.38 \text{ Mm} - 6.36 \text{ Mm}}{6.37 \text{ Mm}} = 0.003$$

Therefore the earth is nearly spherical. If a basketball had the same degree of oblateness as the earth, its radius the long way would be $\frac{1}{2}$ mm larger than the short way. There probably are not many basketballs that are as nearly spherical as the earth.

Because of the earth's oblateness and other irregularities, different locations on the earth's surface can be at slightly different distances from the center. This causes g to vary slightly and, in turn, causes the weight magnitude $F_w = mg'$ in an earth-surface frame to vary slightly; this variation depends mainly on latitude. Also, as discussed in Sec. 6-3, the weight of an object depends on the acceleration of the earth-surface frame relative to an inertial frame, and this acceleration depends on latitude (see Prob. 8, Chap. 6). Thus the weight of any object in an earth-surface frame is doubly dependent on latitude, but the combined effect causes a maximum variation of only one-half of 1 percent between a frame at one of the poles and a frame at the equator.

EXAMPLE 7-5

Estimation of the variation of weight from one of the poles to the equator. *(a)* Equation 7-3, $g = Gm_e/R_e^2$, is based on the assumption that the earth is perfectly spherical. Despite this, use Eq. 7-3 with $R_e = 6.36$ Mm to estimate g at one of the poles, and, similarly, use Eq. 7-3 with $R_e = 6.38$ Mm to estimate g at the equator. *(b)* Use your estimates from part *(a)* and the results from Example 6-8 to compare the weight of an object in an earth-surface frame at one of the poles with its weight at the equator.

Solution. *(a)* Using the values of these distances given above, we have

$$g(\text{pole}) = \frac{(6.67 \times 10^{-11}\ \text{N}\cdot\text{m}^2/\text{kg}^2)(5.98 \times 10^{24}\ \text{kg})}{(6.36\ \text{Mm})^2} = 9.86\ \text{m/s}^2$$

Similarly, inserting $R_e = 6.38$ Mm into Eq. 7-3 gives

$$g(\text{equator}) = 9.80\ \text{m/s}^2$$

(b) In Example 6-8 we found that for an object of mass m in an earth-surface frame at the equator

$$mg = F_w + \frac{4\pi^2 R_e}{T_e^2}$$

Inserting the numerical values and solving for F_w gives

$$F_w = (0.9965)mg$$

Since the poles are located on the earth's rotational axis, $F_w = mg$ for a weight measurement in an earth-surface frame at either pole. Using our estimates of g from part *(a)*, we have

$$F_w(\text{pole}) = m(9.86\ \text{m/s}^2) \qquad \text{and} \qquad F_w(\text{equator}) = m(9.77\ \text{m/s}^2)$$

The fractional difference between these values is

$$\frac{9.86 - 9.77}{\frac{1}{2}(9.86 + 9.77)} = 0.009$$

Our estimate of this variation gives a value that is too large by a factor of almost 2. Actual measurements yield $F_w(\text{pole}) = m(9.83\ \text{m/s}^2)$ and $F_w(\text{equator}) = m(9.78\ \text{m/s}^2)$ for a fractional variation of 0.005.

SELF-TEST 7-5. The distance from the center of the earth to sea level at midlatitudes is about 6.37 Mm. Determine g to three significant digits by using this value for R_e in Eq. 7-3. *(b)* Estimate g on top of Mount Everest whose height above sea level is 8.9 km. *(c)* Determine the fractional difference between your answers to parts *(a)* and *(b)*. ***ANSWERS:*** *(a)* 9.83 m/s²; *(b)* 9.80 m/s²; *(c)* 0.003.

7-5 THE GRAVITATIONAL FIELD

Often a convenient way to deal with gravitational forces is to use the concept of a gravitational field. The ***gravitational field*** **g** at a point P is defined as the gravitational force **F** on a particle located at P divided by the mass m of the particle:

$$\mathbf{g} = \frac{\mathbf{F}}{m} \qquad (7\text{-}5)$$

Definition of the gravitational field

Thus the gravitational field is the gravitational force per unit mass.

We previously called **g** the acceleration due to gravity. More specifically, it was used to represent the quantity $-(9.8\ \text{m/s}^2)\mathbf{j}$, where **j** is directed upward (away from the center of the earth). Now we see that this is a particular value of a more general concept. The gravitational field exists everywhere in space, and the value of its magnitude on or near the earth's surface is $9.8\ \text{m/s}^2$.

Consider the gravitational field produced by a particle. From Eq. 7-2, the gravitational force on particle 2 due to particle 1 is $\mathbf{F}_{12} = -(Gm_2m_1/r^2)\hat{\mathbf{r}}$. Let us write this as

$$\mathbf{F}_{12} = m_2\mathbf{g}_1 \tag{7-6}$$

where

Gravitational field produced by a particle

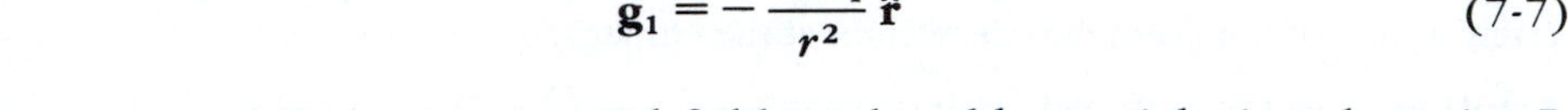

$$\mathbf{g}_1 = -\frac{Gm_1}{r^2}\hat{\mathbf{r}} \tag{7-7}$$

The quantity $\mathbf{g}_1$ is the gravitational field produced by particle 1 at the point P, where P is a distance r from the particle and $\hat{\mathbf{r}}$ is directed from the particle to P (Fig. 7-9).

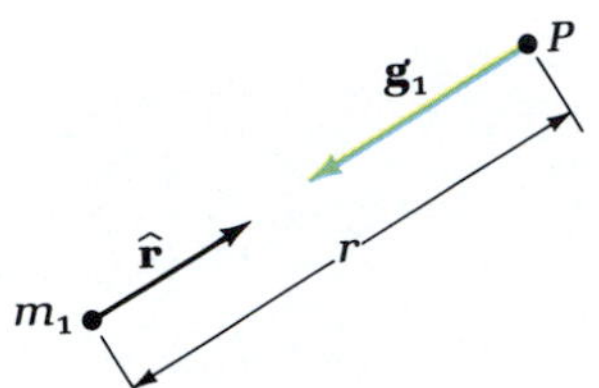

Figure 7-9. Gravitational field $\mathbf{g}_1$ produced by particle 1 at point P. The field has magnitude $g_1 = Gm_1/r^2$ and is directed toward particle 1.

Equations 7-6 and 7-7 taken together give the same information as Eq. 7-2, so that we have not added anything. The convenience comes about because we have divided the calculation of the force on particle 2 into two parts. Equation 7-6 states that the force on 2 is the mass of 2 multiplied by the gravitational field produced by particle 1. Equation 7-7 shows how to determine the gravitational field produced by particle 1. Therefore, in determining the force by 1 on 2, we can start off by calculating the gravitational field produced by 1 while forgetting about 2. Then we find the force on 2 by multiplying the mass of 2 by the value of the field at 2's position. By introducing the gravitational field, the problem has been broken into two parts, which can then be solved one at a time. In dealing with two particles, this procedure is of little use, but in more complex problems it is very helpful.

A spherically symmetric object behaves gravitationally as a particle at points outside the object. Thus the gravitational field produced by a spherically symmetric object of mass m and radius R is

Gravitational field produced by an object with spherical symmetry

$$\mathbf{g} = -\frac{Gm}{r^2}\hat{\mathbf{r}} \qquad (r > R) \tag{7-8}$$

where r is the distance from the center of the object to the point P where the field is evaluated, and $\hat{\mathbf{r}}$ is a unit vector directed away from the center of the object and toward P (Fig. 7-10).

A gravitational field is viewed as a condition established in space by an object with mass.

The concept of a gravitational field introduces more than simply a convenient procedure for calculating gravitational forces; it also provides an alternative view of the gravitational interaction. Now we regard space as being modified by the presence of an object; the object *produces* a gravitational field. A physical quantity, the gravitational field, is associated with each point in space (Fig. 7-11). Later, when we study electricity and magnetism, the concept of a field will be a central feature of our inquiry.

From Eq. 7-5, $\mathbf{g} = \mathbf{F}/m$, the dimension of **g** is force divided by mass. Previously,

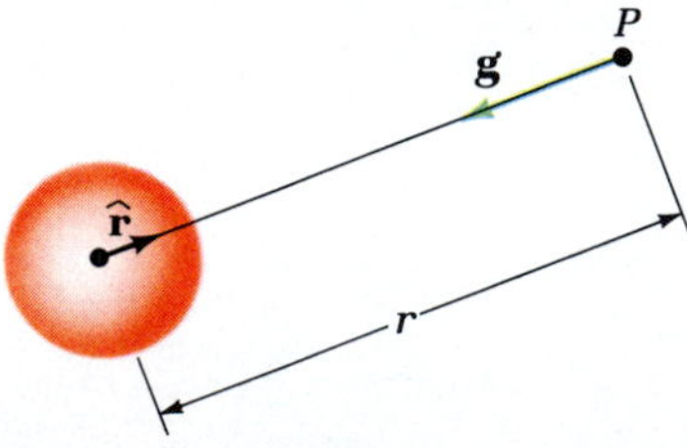

Figure 7-10. Gravitational field **g** produced by an object with spherical symmetry at a point P outside the object.

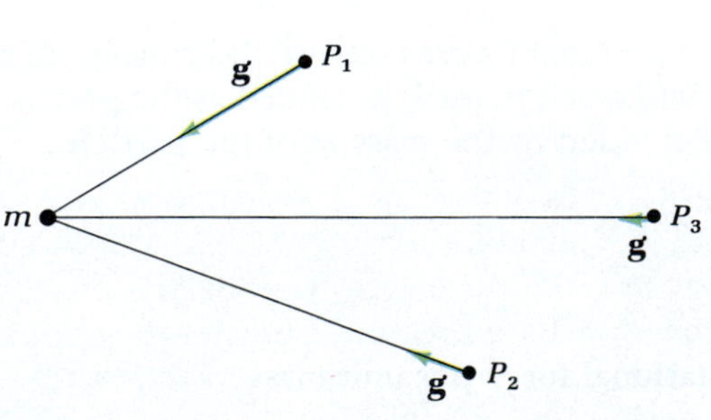

Figure 7-11. The gravitational field produced by a particle of mass m is shown at three representative points. The field is directed toward the particle at each point and decreases with distance r as $1/r^2$. Point P_3 is twice as far from the particle as P_1, so that the field at P_3 has one-fourth the magnitude it has at P_1.

when **g** was viewed as an acceleration, we gave its dimension as length divided by time squared. From Newton's second law, these two dimensions are the same. When we write **g**'s dimension as length/(time)^2, we are treating it as an acceleration, but when we write **g**'s dimension as force/mass, we are treating it as a gravitational field.

EXAMPLE 7-6

Gravitational field at the moon. Since gravitational forces add vectorially, gravitational fields add vectorially too. *(a)* Determine the resultant gravitational field **g** due to the individual fields of the earth ($\mathbf{g}_e$) and of the sun ($\mathbf{g}_s$) at a point P along a straight line between the earth and the sun (Fig. 7-12). The distance from the earth to P is the same as the radius of the moon's orbit around the earth. The radius of the moon's orbit about the earth is 3.84×10^8 m, and the remaining data are given in Table 7-1. *(b)* Determine the force on the moon due to both the earth and the sun when the moon is at P. The moon's mass is 7.35×10^{22} kg.

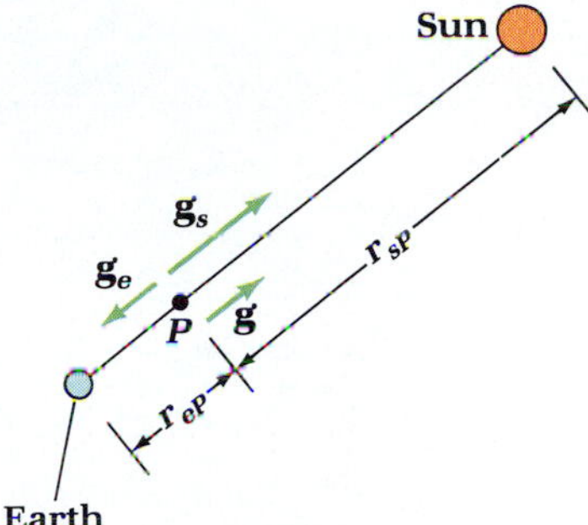

Figure 7-12. Example 7-6. Finding **g** at a point P between the earth and the sun. The distance r_{eP} is exaggerated for purposes of illustration.

Solution. *(a)* Between the earth and the sun the contribution to the gravitational field due to the earth is opposite that due to the sun. The field produced by the earth is toward the earth, and the field produced by the sun is toward the sun. If the unit vector **i** is directed from the earth toward the sun, the resultant field is

$$\mathbf{g} = \mathbf{g}_e + \mathbf{g}_s = -\frac{Gm_e}{r_{eP}^2}\mathbf{i} + \frac{Gm_s}{r_{sP}^2}\mathbf{i}$$

where r_{eP} is the distance from the earth to P and r_{sP} is the distance from the sun to P:

$$r_{eP} = 3.84 \times 10^8 \text{ m}$$

and

$$r_{sp} = 1.496 \times 10^{11} \text{ m} - 3.84 \times 10^8 \text{ m} = 1.492 \times 10^{11} \text{ m}$$

Thus

$$\begin{aligned}\mathbf{g} &= -\frac{(6.67 \times 10^{-11} \text{ N} \cdot \text{m}^2/\text{kg}^2)(5.97 \times 10^{24} \text{ kg})}{(3.84 \times 10^8 \text{ m})^2}\mathbf{i} \\ &\quad + \frac{(6.67 \times 10^{-11} \text{ N} \cdot \text{m}^2/\text{kg}^2)(1.99 \times 10^{30} \text{ kg})}{(1.492 \times 10^{11} \text{ m})^2}\mathbf{i} \\ &= -(2.70 \times 10^{-3} \text{ N/kg})\mathbf{i} + (5.96 \times 10^{-3} \text{ N/kg})\mathbf{i} \\ &= (3.26 \times 10^{-3} \text{ N/kg})\mathbf{i}\end{aligned}$$

The resultant field is directed toward the sun because the contribution to the field by the sun is larger than that by the earth. *(b)* Using the field we determined in part *(a)*, we find that the force **F** on the moon is

$$\begin{aligned}\mathbf{F} = m_m\mathbf{g} &= (7.35 \times 10^{22} \text{ kg})(3.26 \times 10^{-3} \text{ N/kg})\mathbf{i} \\ &= (2.40 \times 10^{20} \text{ N})\mathbf{i}\end{aligned}$$

At point P, the force due to both the earth and the sun is directed toward the sun and away from the earth.

SELF-TEST 7-6. Determine the magnitude g of the gravitational field produced by both the earth and the sun at a point P on the opposite side of the earth from the sun at a distance from the earth equal to the radius of the moon's orbit. ***ANSWER:*** 8.66×10^{-3} N/kg.

7-6 ORBITS, NUMERICAL METHODS

To understand how a satellite continually moves in its orbit, consider launching a projectile horizontally from the top of a high mountain. Because we are interested in satellite motion, we neglect air friction. The distance the projectile travels before hitting the ground depends on the launching speed: the greater the speed, the greater the distance. The distance the projectile travels before hitting the

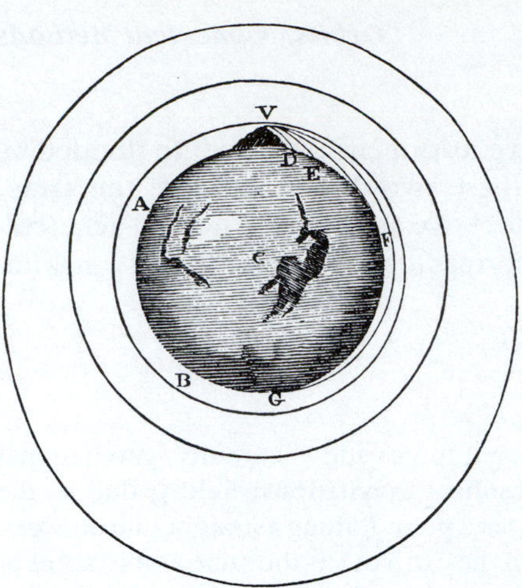

Figure 7-13. A figure used by Newton in the *Principia* to illustrate the trajectory of a projectile launched horizontally from the top of a tall mountain (V). (We neglect friction.) As the projectile's launching speed is increased, its range is increased and its landing point is farther from the mountain (from D to E to F to G). If the launching speed is made large enough, the projectile's path follows the curvature of the earth.

ground is also affected by the curvature of the earth, as illustrated in Fig. 7-13. This figure, used by Newton in the *Principia,* shows different trajectories for different launching speeds. As the launching speed is made greater, a speed is reached whereby the projectile's path follows the curvature of the earth. This is the launching speed which places the projectile in a circular orbit. Thus an object in circular orbit may be regarded as falling, but as it falls its path is concentric with the earth's spherical surface and the object maintains a fixed distance from the earth's center. Since this motion may continue indefinitely, we may say that the orbit is *stable*.

A stable orbit

The dynamics of orbital motion solves a paradox that arises from the ever-attracting nature of the gravitational force. If each object in the universe attracts every other object through the gravitational force, then why doesn't the entire universe collapse on itself? Now we see that orbital motion is an answer to this vexing question. For example, the solar system is stable, with the planets remaining separated from the sun despite the fact that the only significant force on a planet is the gravitational force of attraction by the sun. The reason for the stability is that each planet is in orbital motion and the force of attraction by the sun provides the force that maintains each stable orbit.

Now let us find the speed v of a satellite in circular orbit around the earth. We assume that an earth-center frame is inertial and that the only significant force on the satellite is the gravitational force by the earth. These assumptions are valid if the radius of the orbit is not too large. Newton's second law, $\Sigma\mathbf{F} = m\mathbf{a}$, applied to the satellite gives

$$\frac{Gmm_e}{r^2} = m\left(\frac{v^2}{r}\right) \tag{7-9}$$

where the speed v is relative to the center of the earth. The direction of the force and the acceleration are toward the center of the earth (Fig. 7-14). Solving for v gives

$$v = \sqrt{\frac{Gm_e}{r}}$$

A satellite with this speed continually moves in its circular orbit of radius r.

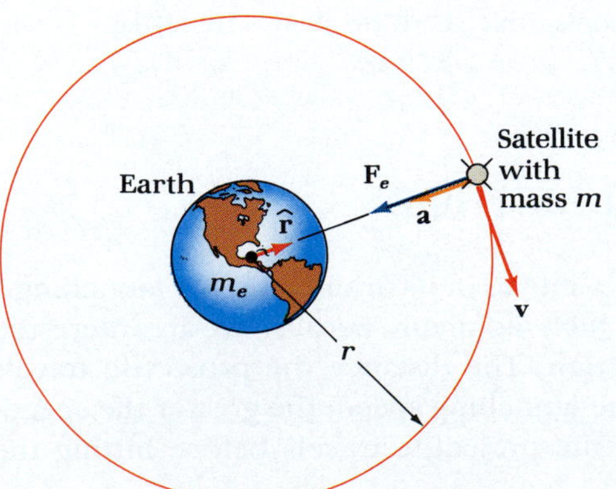

Figure 7-14. A satellite in circular orbit around the earth. The net force on the satellite is the gravitational force due to the earth, $\mathbf{F}_e = -(Gmm_e/r^2)\hat{\mathbf{r}}$. The acceleration $\mathbf{a}$ is parallel to $\mathbf{F}_e$, with magnitude v^2/r. The velocity $\mathbf{v}$ is perpendicular to $\mathbf{F}_e$ and $\mathbf{a}$, and $v = \sqrt{Gm_e/r}$.

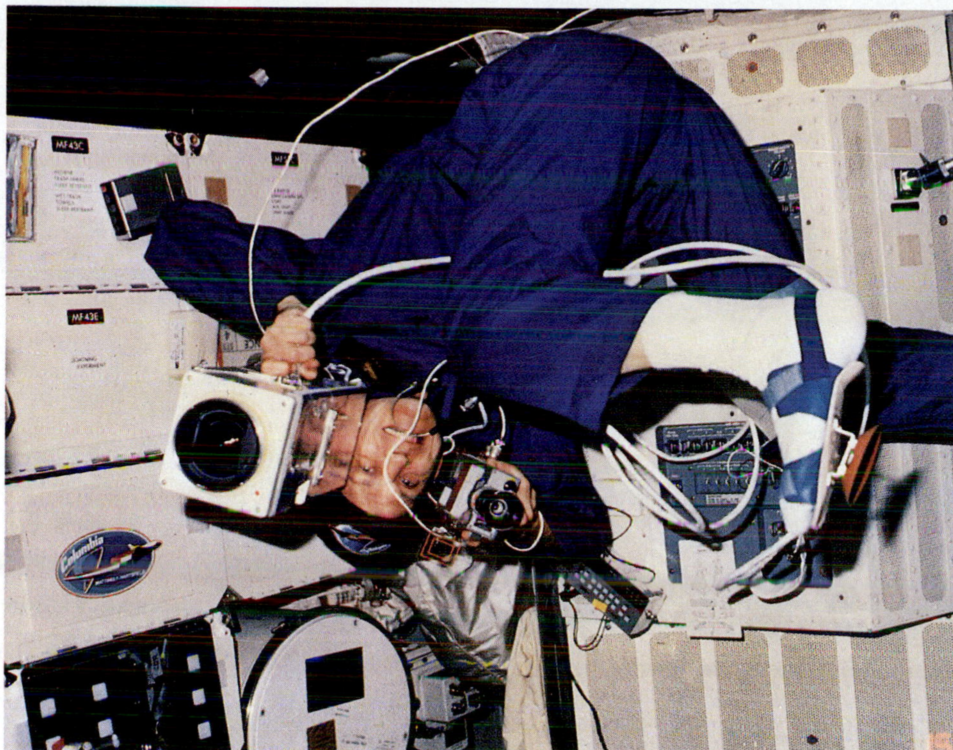

Astronaut in an orbiting spacecraft.

Weightlessness in an Orbiting Spacecraft

We are accustomed to seeing pictures of astronauts and their paraphernalia floating around inside a spacecraft as it orbits the earth. If an astronaut should attempt to weigh something by placing it on a spring scale, the reading would be zero. Indeed, the scale itself would float away if it were not secured. Any object in an orbiting spacecraft is weightless.

Why is this so? Because when Newton's second law is applied to an orbiting object, as in Eq. 7-9, the object's mass appears on both sides of the equation and can be canceled out. Thus quantities such as the acceleration, orbital speed, and orbital period are independent of the mass of the orbiting object. This explains the picture of the spacecraft with everything floating about: All objects in the spacecraft are in orbit along with it. The gravitational force by the earth exerted on each object maintains that object in its orbit and each object follows the same orbital path. Incidentally, this is further evidence of the equivalence of inertial and gravitational mass (Sec. 7-3).

Numerical Methods

In Chap. 4, we constructed a spreadsheet for the trajectory of a baseball (Table 4-2). By modifying this spreadsheet, we can use it to describe the circular orbit of a satellite. To proceed, we (1) select a consistent set of initial conditions, (2) determine the equations for the x and y components of the acceleration, and (3) find a convenient time interval for the iterations.

1. Let the orbital radius be three earth radii: $r = 3R_e = (3)(6.37\text{ Mm}) = 19.11$ Mm. Then the speed of the satellite is

$$v = \sqrt{\frac{Gm_e}{r}} = \sqrt{\frac{(6.67 \times 10^{-11}\text{ N}\cdot\text{m}^2/\text{kg}^2)(5.97 \times 10^{24}\text{ kg})}{19.11\text{ Mm}}} = 4.56\text{ km/s}$$

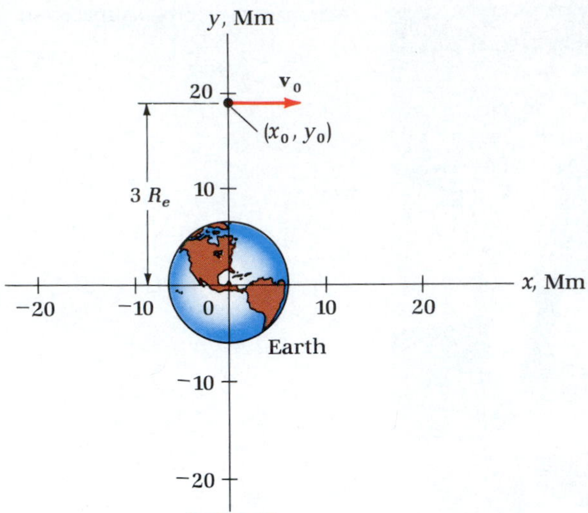

Figure 7-15. Initial conditions used in the spreadsheet in Table 7-2 for an earth satellite.

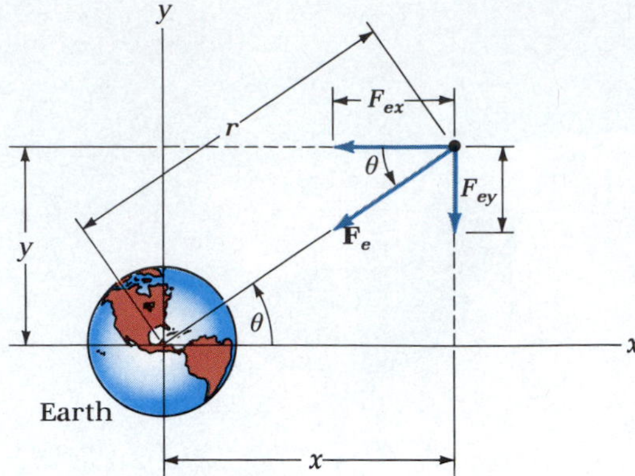

Figure 7-16. Finding the x and y components of the force on a satellite in an earth orbit: $F_{ex} = -F_e(x/r)$ and $F_{ey} = -F_e(y/r)$.

We use the coordinate frame shown in Fig. 7-15. The center of the earth is at the origin, and the initial position of the satellite is at $x_0 = 0$ and $y_0 = 19.11$ Mm. Since the velocity **v** must be perpendicular to a line from the center of the earth to the satellite, we let the initial components of **v** be $v_{x0} = 4.56$ km/s and $v_{y0} = 0$.

2. To determine the acceleration of the satellite, we find the net force on it and use Newton's second law. Figure 7-16 shows the x and y components of the force on the satellite. The x component is

$$F_{ex} = -F_e \cos\theta = -F_e\left(\frac{x}{r}\right) = -\left(\frac{Gm_em}{r^2}\right)\left(\frac{x}{r}\right) = -\frac{Gm_emx}{r^3}$$

From Newton's second law, $a_x = \Sigma F_x/m = F_{ex}/m$. Therefore the x component of the satellite's acceleration is $a_x = -Gm_ex/r^3$. Since $r = \sqrt{x^2 + y^2}$, the expression for a_x in cartesian coordinates is

$$a_x = -\frac{Gm_ex}{(x^2 + y^2)^{3/2}}$$

Inserting the numerical values of G and m_e gives

$$a_x = -\frac{(3.98 \times 10^{14}\ \text{m}^3/\text{s}^2)x}{(x^2 + y^2)^{3/2}}$$

Similarly,

$$a_y = -\frac{(3.98 \times 10^{14}\ \text{m}^3/\text{s}^2)y}{(x^2 + y^2)^{3/2}}$$

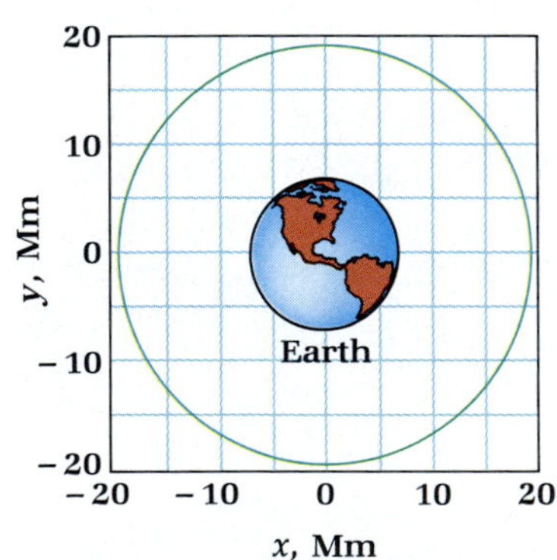

Figure 7-17. Graph of the circular orbit of an earth satellite generated by the spreadsheet in Table 7-2.

3. To determine an appropriate time interval for the iterations, we find the period T of the orbit. For uniform circular motion $v = 2\pi r/T$, or $T = 2\pi r/v$. Inserting $r = 19.11$ Mm and $v = 4.56$ km/s, we find $T = 2.63 \times 10^4$ s. A convenient procedure is to let the computer perform about 1200 iterations with a time interval of 25 s between the iterations [(1200)(25 s) = 3.0×10^4 s]. In that way the spreadsheet will encompass slightly more than one revolution.

Table 7-2 shows the first several rows of our spreadsheet and Fig. 7-17 shows the circular orbit generated by the spreadsheet.

SELF-TEST 7-7. Write the entries in columns C, E, and G that are not displayed in Table 7-2. Let column C correspond to a_y, column E to v_y, and column G to y.

TABLE 7-2. ***First few rows of a spreadsheet for an object in circular orbit around the earth.*** Note that columns C, E, and G for a_y, v_y, and y are not included. **(See Self-Test 7-7.)**

	A	B	D	F
1	A	B	D	F
2	t	ax	vx	x
3	0	0	4560	0
4	25+A3	−3.98E+14*F3/F3*F3+G3*G3)∧1.5	+D3+B4*25	+F3+D4*25
5	25+A4	−3.98E+14*F4/F4*F4+G4*G4)∧1.5	+D4+B5*25	+F4+D5*25
6	25+A5	−3.98E+14*F5/F5*F5+G5*G5)∧1.5	+D5+B6*25	+F5+D6*25
⋮	⋮	⋮	⋮	⋮

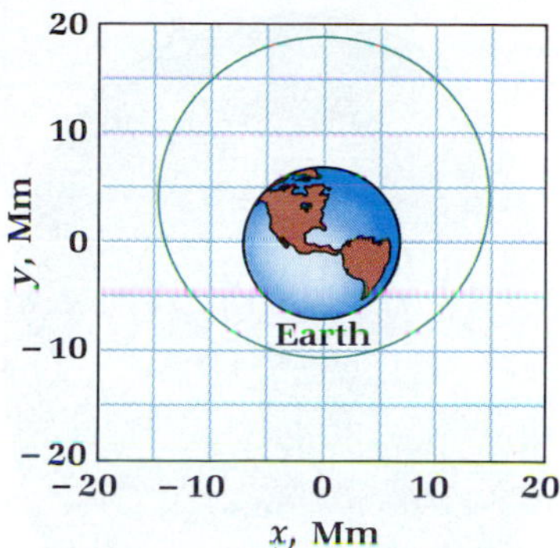

Figure 7-18. Graph of the elliptical orbit of an earth satellite generated by changing the initial velocity in the spreadsheet in Table 7-2.

Elliptical Orbits

The spreadsheet in Table 7-2 can be changed to show what happens if the initial speed is not just the right value for a circular orbit. Suppose the initial velocity is (3.90 km/s)**i** instead of (4.56 km/s)**i**. That is, entry D3 is changed from 4560 to 3900. The graph generated by the adjusted spreadsheet is shown in Fig. 7-18. An analysis of this graph shows that the orbit is an ellipse, with the center of the earth at one focus. The point where the orbit is farthest from the earth is called the *apogee*, and the point where the orbit is nearest the earth is called the *perigee*. From the figure, we estimate that the distance from the center of the earth to the apogee is about 19 Mm, and the distance from the center of the earth to the perigee is about 12 Mm. If we investigate the velocity data generated by the spreadsheet in the neighborhood of the apogee and the perigee, we find that the satellite is traveling slowest at the apogee and fastest at the perigee.

Apogee and perigee

It is instructive to try a few other initial velocities. How small can you make the initial velocity and still have the satellite miss the earth? That is, try to find the value of v_{x0} that gives the distance from the center of the earth to the perigee to be R_e.

We stated above that the orbital path is an ellipse when the velocity does not have just the right value for a circular orbit. This is true as long as the speed is not too great. If the speed is much greater, great enough so that the satellite is just barely able to escape the earth and never return, then the path is a parabola. If the velocity is greater still, then the path is a hyperbola. These types of curves are called *conic sections*. To prove that orbital paths are always one of these conic sections, we would be required to use analytical methods beyond the scope of this book. Such a proof provides an example of some of the advantages of analytical methods as compared with numerical methods. One cannot show with numerical methods that all orbits are conic sections.

Nicolaus Copernicus.

7-7 DISCOVERY OF THE LAW OF GRAVITATION

The way the law of universal gravitation was discovered is often considered the paradigm of modern scientific technique. The major steps involved were (1) the hypothesis about planetary motion given by Nicolaus Copernicus (1473–1543); (2) the careful experimental measurements of the positions of the planets and the sun by Tycho Brahe (1546–1601); (3) the analysis of the data and the formulation of empirical laws by Johannes Kepler (1571–1630); and (4) the development of a general theory by Isaac Newton.

1. Although a sun-centered, or heliocentric, model of the solar system had been proposed by Aristarchus in the third century B.C., for many centuries the western world believed that the earth remains fixed while the sun and planets circle it. This earth-centered, or geocentric, model required elaborate geometrical schemes to account for the observed motion of the planets. In his book *De Revolutionibus Orbium Coelestium (On the Revolution of the Heavenly Spheres),* Copernicus asserted that the geocentric model was not "sufficiently pleasing to the mind." He proposed a system using the following assumptions:

Tycho Brahe.

Johannes Kepler.

(*a*) the earth rotates on its axis once per day, (*b*) the earth revolves around the sun (along with the other planets), and (*c*) the stars are at a much greater distance from the earth than are the sun and the planets.

2. The work of Tycho Brahe exemplifies the fundamental basis of experimental research: *If you want to know how something works, carefully measure its behavior.* He spent the last half of his life, more than 20 years, precisely measuring the positions of the sun and planets. His measurements provided the data for those who followed to solve the mysteries of the motion of celestial objects. During his final years Brahe acquired Johannes Kepler as his assistant.
3. Kepler possessed great mathematical and computational abilities. He used these skills and Brahe's measurements to determine the orbits of the planets, especially those of the earth and Mars. He condensed his findings into three laws:

   All planets move in elliptical orbits with the sun at one focus (Fig. 7-19).

   A line joining any planet to the sun sweeps out equal areas in equal time intervals (Fig. 7-20).

   The square of the period of any planet is proportional to the cube of the planet's average distance from the sun.

   With his three laws, Kepler introduced more than a precise characterization of the solar system. He also initiated a new way of describing natural phenomena. The description takes the form of brief, concentrated, broadly applicable statements, which we now call "laws." Kepler fostered the tenet of modern science that the proper description of natural phenomena is the simplest one that complies with the experimental data. Uncommon for his day, Kepler insisted that a successful theory must conform to the strict details of experimental measurements. Kepler would have greatly appreciated the further generalizations by Newton.
4. By introducing the laws of motion and the law of universal gravitation, Newton provided a general theory that unified the astronomical laws of Kepler and terrestrial experience. One of the tests to which Newton subjected his laws was that they give elliptical orbits for the planets, and thus agree with Kepler's first law. We saw an example of this in the previous section. In Sec. 13-5 we will show that Kepler's second law can be developed from Newton's laws.

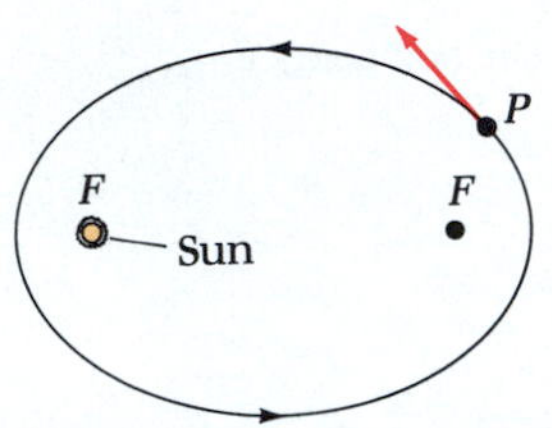

Figure 7-19. Kepler's first law. The orbit of a planet is an ellipse, with the sun at one focus. For the sake of illustration, the eccentricity of the ellipse shown is much greater than that of any of the actual planetary orbits.

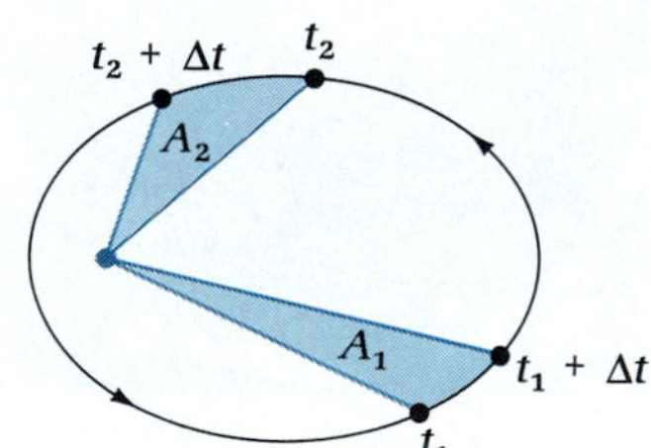

Figure 7-20. Kepler's second law. The time intervals Δt are the same for the two cases shown. Kepler's second law states that $A_1 = A_2$.

We now show that Kepler's third law follows from Newton's second law and the law of universal gravitation applied to a planet in a circular orbit. Newton's second law applied to a planet gives $F_{sp} = m_p a_c$, where F_{sp} is the magnitude of the force exerted by the sun on the planet, m_p is the mass of the planet, and a_c is the magnitude of the centripetal acceleration of the planet around the sun. Since $F_{sp} = Gm_s m_p/r^2$ and $a_c = 4\pi^2 r/T^2$, we have

$$\frac{Gm_s m_p}{r^2} = \frac{m_p 4\pi^2 r}{T^2}$$

Solving for T^2 gives

$$T^2 = \left(\frac{4\pi^2}{Gm_s}\right) r^3$$

The factor in the parentheses is a constant, independent of the planet, so that the square of the period T is proportional to the cube of the distance r to the sun. This is Kepler's third law for the case of a circular orbit.

Kepler's three laws of planetary motion can be derived from Newton's second and third laws in combination with Newton's law of universal gravitation. In retrospect, Kepler's empirical laws represent a first step toward the understanding of natural phenomena that is typical of scientific progress. Newton recognized this when he said: "If I have seen farther than others, it is because I stand on the shoulders of giants."

COMMENTARY: Fundamental Forces and Unification

In recent years exciting discoveries have been made in physics, discoveries that have revolutionized the way we view matter and the forces that determine the behavior of matter. Physicists are attempting to find an ultimate force law which explains all of the many interactions we observe in nature. This quest to unify all forces into one force is based on the concept of a fundamental force. A *fundamental force* is the result of a basic interaction between particles. Such a force explains many phenomena that cannot be attributed to some other force. For example, contact forces between macroscopic objects are not regarded as fundamental forces. These forces are a complex manifestation of the more fundamental electromagnetic force (discussed below). By contrast, the gravitational force is not simply an effect that can be explained as one example of some other force. The gravitational force is a fundamental force. Or is it? How can we be certain? Whether a force is regarded as fundamental at a particular time depends on what we know at that time.

To be specific, let us place ourselves in the year 1967 and list the fundamental forces known then. There are four such forces:

1. The gravitational force
2. The electromagnetic force
3. The strong nuclear force
4. The weak nuclear force

We are already familiar with the gravitational force. After we briefly describe the other three forces and make a comparison of the four fundamental forces, we shall discuss progress toward the unification of the fundamental forces.

The *electromagnetic force* is important in many ways. As mentioned above, it is the underlying cause of the contact forces we experience at the macroscopic scale, such as friction. This interaction gives rise to the electric shock you feel when you touch a metal part of a car after sliding across the seat on a cold and dry winter day. Because it produces electric currents in wires, the electromagnetic force is the basis of a large part of our modern technology. It is regarded as a fundamental force because it is responsible for interactions between some of the elementary particles that compose matter. For example, it provides the attractive force which holds an atom's electrons near the nucleus of the atom. As the name implies, the electromagnetic force includes both electric and magnetic forces. These two forces are intimately connected; both are the result of the same property of matter, the electric charge.

The *strong nuclear force* is the force that binds protons and neutrons together to form an atomic nucleus. Within a nucleus the protons and neutrons are confined to a very small space, about 5×10^{-15} m across. Because of their electric charge, protons repel each other through the electromagnetic force. If it were not for the dominance of the strong nuclear force, the repulsion between the protons would make the nucleus unstable; the protons would fly apart and the nucleus could not exist.

The *weak nuclear force* acts between elementary particles and is responsible for some nuclear reactions. For example, in radioactive decay a nucleus will spontaneously disintegrate into several fragments. The weak nuclear force causes a particular radioactive decay which is called *beta decay*. In addition, the weak nuclear force is important in controlling the rate of some of the nuclear reactions that occur in stars such as the sun. The lifetime of the sun is determined by the characteristics of this force.

A comparison of the fundamental forces is given in Table 7-3. Consider the distance over which each force acts, or the range of the force. The gravitational and the electromagnetic forces are inverse-square forces. The magnitudes of these

TABLE 7-3. *Comparison of the Fundamental Forces*

	Gravitational	Electromagnetic	Strong Nuclear	Weak Nuclear
Example interactions	Binds stars together to form galaxies	Binds electrons to nuclei to form atoms	Binds protons and neutrons together to form nuclei	Responsible for nuclear beta decay
Range	Infinite	Infinite	10^{-15} m	$\ll 10^{-16}$ m
Relative strength	10^{-39}	10^{-2}	1	10^{-5}

forces weaken with distance but never become zero. The range of these two forces is infinite. The strong nuclear force has a very short range; its effects are imperceptible beyond a separation distance of about 10^{-15} m. The range of the weak nuclear force is smaller still, less than 10^{-16} m.

The relative strength of a force is gauged by the magnitude of the force between elementary particles that are within the range of the force. On a strength scale where the strong nuclear force has a value of 1, the electromagnetic force has a value of 10^{-2}, the weak nuclear force a value of 10^{-5}, and the gravitational force a value of 10^{-39}. The gravitational force is by far the weakest of the fundamental forces. Since the gravitational force is often the dominant force on objects you regularly encounter, this may surprise you. In the case of the strong and weak nuclear forces, objects in the macroscopic world of direct human experience are beyond their range. The direct effects of these forces are seen in experiments using high-energy accelerators that probe deeply into matter (Figs. 7-21 and 7-22). In the case of the electromagnetic force, the electric charge on macroscopic objects is often too small to produce noticeable electric forces. Macroscopic objects consist of elementary particles that have both negative and positive charges. Normally the amount of negative charge is almost exactly the same as the amount of positive charge so that the object is nearly electrically neutral and experiences an insignificant electric force.

The historical trend toward unification is shown in Fig. 7-23. As we have seen in this chapter, Newton unified celestial forces with gravitational forces on earth when he discovered the law of universal gravitation. Experimental work during the nineteenth century unified electrical and magnetic phenomena and culminated

Figure 7-21. The Stanford Linear Accelerator Center (SLAC) is one of several large accelerators where experiments are performed on elementary particles. Particles are accelerated along the 2-mi beam tube to very high energies, and then they collide with other particles. Accelerators such as this one help provide the experimental data for the search for an ultimate force law.

with the theory of electromagnetism developed by James Clerk Maxwell. A recent advance toward unification was accomplished independently by Steven Weinberg and Abdus Salam, who showed the underlying connection between the electromagnetic force and the weak nuclear force. This unification gave rise to the so-called *electroweak force*. An approach similar to that of Weinberg-Salam may bring about the unification of the electroweak force with the strong nuclear force in the near future. In Fig. 7-23, we have dared to speculate on the occurrence of an

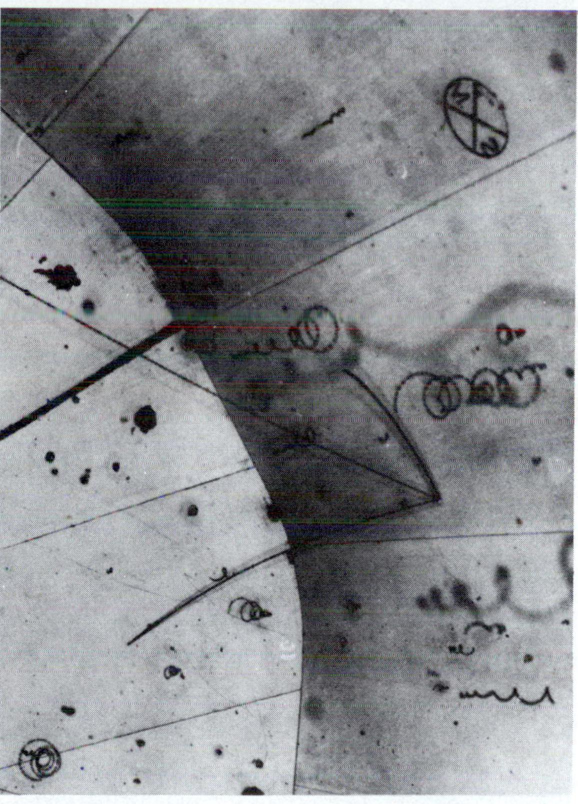

Figure 7-22. The interactions between high-energy elementary particles are revealed in bubble-chamber photographs. As the particles pass through a liquid, such as liquid hydrogen, bubbles are formed along their paths.

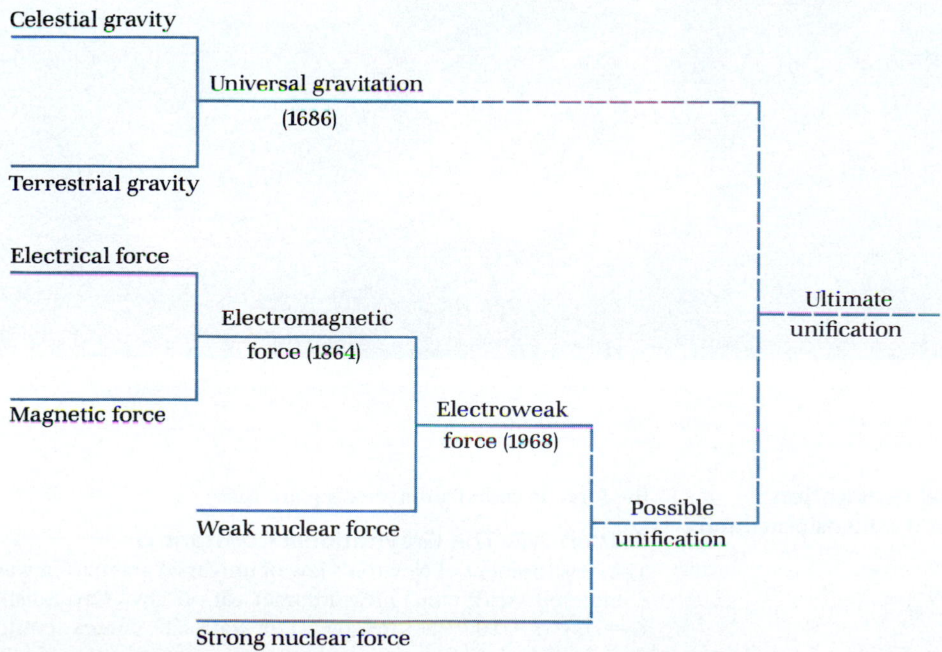

Figure 7-23. Progress toward unification.

ultimate unification. Does this most fundamental force exist? If it does exist, is it simple enough to be knowable by humans? The search for unification is based on our belief that, after removing the cloak of complexity, we shall find that nature laid bare is beautifully simple.

An aerial view shows the main accelerator ring at Fermilab. The diameter of the ring is about 1.9 km.

The collider detector at Fermilab slides into alignment with the Tevatron's particle beams.

SUMMARY

Section 7-1. The Law of Universal Gravitation

Newton's law of universal gravitation is a fundamental force law in nature:

$$\mathbf{F}_{12} = -\frac{Gm_1m_2}{r^2}\,\hat{\mathbf{r}} \qquad (7\text{-}2)$$

The gravitational force is always attractive. Since the force depends on the inverse of the square of the separation distance r, the force is called an inverse-square force.

Section 7-2. The Gravitational Constant *G*

The development of Newton's law of universal gravitation was completed with the measurement of G by Cavendish: $G = 6.670 \times 10^{-11}\ \mathrm{N \cdot m^2/kg^2}$. Gravitational forces could then be calculated.

Section 7-3. Gravitational and Inertial Mass
Mass characterizes two different properties of an object—its resistance to a change in its velocity and its gravitational interaction with other objects. Experiment shows that these two properties are proportional, and our choice of units makes them equal.

Section 7-4. Variation of g on the Earth's Surface
Because the earth does not have exact spherical symmetry, g varies on the surface of the earth.

Section 7-5. The Gravitational Field
The solution to complex problems involving gravitational forces can be facilitated by using the gravitational field, $\mathbf{g} = \mathbf{F}/m$. The gravitational field is viewed as a condition in space produced by an object with mass.

Section 7-6. Orbits, Numerical Methods
A satellite in a circular earth orbit has its velocity directed perpendicular to a line to the center of the earth with magnitude $v = \sqrt{Gm_e/r}$. In general, the path of a satellite is an ellipse.

Section 7-7. Discovery of the Law of Gravitation
The discovery of the law of gravitation involved the work of many great men over a span of hundreds of years. It is a paragon of scientific endeavor.

QUESTIONS

1. Consider the following three prospective force laws for the gravitational force $\mathbf{F}_{sp}$ by the sun on a planet: *(a)* $F_{sp} = Km_sm_p^2/r^2$; *(b)* $F_{sp} = Km_s^2m_p/r^2$; and *(c)* $F_{sp} = Km_sm_p/r^3$, where K is a factor independent of m_s, m_p, or r. Using the periods and orbital radii of the planets as your data base, which of these prospective force laws (if any) can be shown to be inconsistent with the data? Which (if any) are consistent with the data?

2. Two spherically symmetric objects, each of mass m_0, exert a gravitational force of magnitude F_0 on each other when their centers are a distance r_0 apart. What are the magnitudes of these forces when they are separated by *(a)* $2r_0$; *(b)* $3r_0$; *(c)* $4r_0$?

3. Two spherically symmetric objects, each of mass m_0, exert a gravitational force of magnitude F_0 on each other when their centers are a distance r_0 apart. What would the magnitudes of these forces be if each had a mass of *(a)* $2m_0$; *(b)* $3m_0$; *(c)* $4m_0$?

4. If the gravitational force on an object depends linearly on its mass, why is the acceleration of a freely falling object independent of its mass?

5. The inside of a satellite orbiting the earth is often called a *weightless environment.* Is the gravitational force by the earth on objects in this environment zero? How do you account for the zero reading when an astronaut attempts to weigh herself on a spring scale?

6. Describe the way the mass of an astronaut and the gravitational force on the astronaut vary during a trip from the earth to the moon.

7. Suppose you are communicating with an intelligent being from another solar system. Which concept do you think you could define more clearly for this being, mass or weight? Explain.

8. If you are buying gold from a dealer who uses a spring scale to measure the amount of gold, and you wish to get the most gold for your money, do you want the measurement to be made at the equator or at the poles?

9. The planet Egabbac (in another solar system) has a radius twice that of the earth's, but an average mass density which is the same as the earth's. Would the weight of an object on Egabbac's surface be the same as on the earth's, greater than on the earth's, or less than on the earth's? If greater or less than on the earth's, then by how much?

10. Suppose an artificial satellite has a circular orbit around the earth, and we measure its radius and period. Can we use this information and Newton's laws to determine the mass of the satellite? Can we use it to find the mass of the earth?

An astronaut and satellite in orbit together.

11. In the British system of units, the foot (ft) is a unit of length, the second (s) is a unit of time, and the slug (sl) is a unit of mass. What is the unit of G in the British system?

12. Figure 7-24 is a photograph of a planetary nebula. The star at the center is surrounded by a shell of material that was ejected from the star. (The material appears as a ring because its line-of-sight thickness is greater at the edges than in the middle.) Using the approximation that the shell has spherical symmetry, what is the gravitational force on the star due to the shell? Explain.

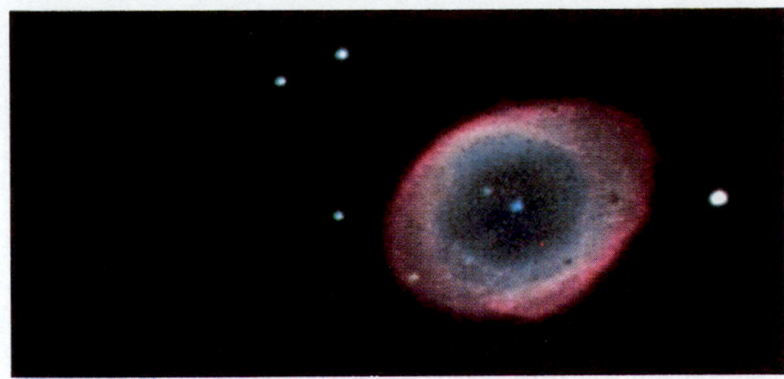

Figure 7-24. Question 12: A planetary nebula.

13. Suppose an artificial satellite is in a circular orbit around the earth at a distance r_0 from the center of the earth. A short burst is fired from its rocket engine in a direction such that its speed quickly increases (but not enough to take it out of earth orbit). *(a)* What is the subsequent path of the satellite? *(b)* Will its perigee distance be greater than, less than, or equal to r_0? *(c)* Will its apogee distance be greater than, less than, or equal to r_0? *(d)* Will its period increase or decrease?

14. Suppose that the rocket engine of the satellite in the previous question is fired in a direction such that the satellite's speed quickly decreases (but not enough to cause the satellite to hit the earth), and rework your answers to that question.

15. Consider two artificial satellites B and C in circular orbits around the earth. The radius of C's orbit is twice that of B's: $r_C/r_B = 2$. What is the ratio of their *(a)* accelerations, *(b)* periods, *(c)* speeds?

16. In its elliptical orbit around the sun, with the sun at one focus, a planet comes closest to the sun at the point called its *perihelion;* at the point called its *aphelion,* a planet is farthest from the sun. At which point is the speed of the planet higher and at which point is it lower?

17. The sun's speed relative to the earth (as measured with respect to background stars) is highest at around January 4 each year and lowest around July 4. When is the earth closest to the sun and when is it farthest from the sun? Does this effect tend to make summers and winters more severe or less severe in *(a)* the northern hemisphere, *(b)* the southern hemisphere?

18. The magnitude of the force on the moon due to the sun is more than twice the magnitude of the force on the moon due to the earth. Would it be more accurate to say that the moon orbits the sun rather than the moon orbits the earth? Explain.

19. If an artificial satellite is orbiting the earth, is it possible for the plane of the orbit not to pass through the center of the earth? On what property of the gravitational force is your answer based?

20. Which planet falls farther toward the sun in 1 s, the earth or Venus?

21. An apple is dropped from rest. An ant on the ground states that the apple accelerates toward the earth and strikes it. A worm in the apple states that the earth accelerates toward the apple and strikes it. Which statement do you consider more appropriate from a dynamical point of view and why?

22. A *geosynchronous orbit* is an orbit in which the satellite remains fixed directly over a point on the earth's surface. *(a)* What must be the period of a geosynchronous orbit? *(b)* Is there a particular plane in which the orbit must be contained? If so, identify the plane.

23. If an object at a distance h above the earth's surface is released from rest, it plummets to the earth's surface. If an object at a distance h above the earth's surface is launched horizontally with a speed $v = \sqrt{Gm_e/(R_e + h)}$, it moves in a stable circular orbit (assuming h is large enough so that air friction is negligible). Explain.

24. Complete the following table:

Symbol	Represents	Type	SI unit
G		Scalar	
$\mathbf{g}$	Gravitational field		
$\mathbf{g}$			m/s^2
m_I			
m_G			

EXERCISES

Section 7-1. The Law of Universal Gravitation

1. Use the data in columns (1) and (2) of Table 7-1 to calculate *(a)* a_c and *(b)* a_cR^2 for Mars and for Neptune.

2. Consider developing Newton's law of universal gravitation by using the data in Table 7-4. This table gives the periods and orbital radii of the four largest moons of Jupiter. *(a)* Determine a_c for each of these moons around Jupiter. *(b)* Determine $k_J = a_cR^2$ for each of the moons. *(c)* If you assume that Jupiter is at rest in an inertial reference frame and that the only significant force on each of the moons is the gravitational force by Jupiter, what can you say about the distance dependence of the force? *(d)* What can you say about the dependence of the force

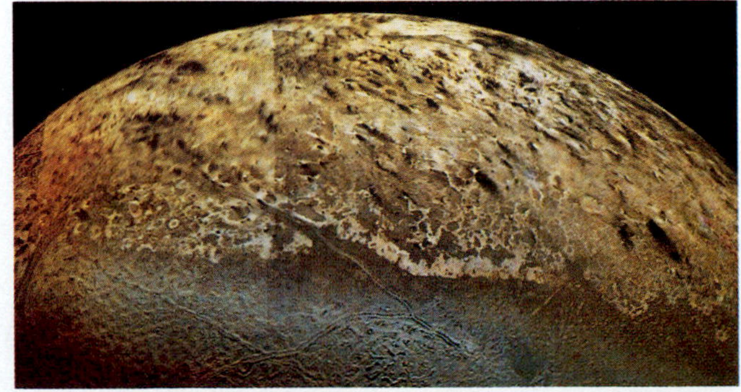

Triton, Neptune's largest moon, viewed from the Voyager spacecraft.

TABLE 7-4. *The Moons of Jupiter*

Name	Orbital Radius, Mm	Period, Days
Io	421.6	1.769
Europa	670.8	3.551
Ganymede	1070	7.155
Callisto	1882	16.689

on the mass of each moon? *(e)* What can you say about the dependence of the force on the mass of Jupiter?

3. You have been sent on a space mission to investigate the solar system of the star Nikpmup, which has three planets in circular orbit. The names of the planets and their orbital radii R about Nikpmup are as follows: Elppa with $R = 1.22 \times 10^{11}$ m, Mulp with $R = 1.89 \times 10^{11}$ m, and Hcaep with $R = 2.45 \times 10^{11}$ m. The orbital period of Elppa is $T = 2.00 \times 10^7$ s. Determine the orbital periods of *(a)* Mulp and *(b)* Hcaep.

Section 7-2. The Gravitational Constant *G*

4. Determine the magnitude of the gravitational force between two billiard balls of mass 0.16 kg when the distance between their centers is 450 mm.

5. Determine the magnitude of the gravitational force between a bowling ball of mass 5.2 kg and a baseball of mass 0.15 kg when the distance between their centers is 640 mm.

6. Determine the magnitude of the gravitational force on the earth *(a)* due to the sun (F_{se}) and *(b)* due to the moon (F_{me}). *(c)* Find the ratio F_{se}/F_{me}. (See Table 7-1 and Example 7-4 for data.)

7. What is the distance between centers of a baseball of mass 0.145 kg and a bowling ball of mass 5.5 kg such that the gravitational force between them is 1.3×10^{-10} N?

8. A billiard ball of mass 0.16 kg exerts a force of 6.2×10^{-10} N on a bowling ball when the distance between their centers is 0.37 m. What is the mass of the bowling ball?

9. At what distance from the center of the earth will a 1-kg object have a weight of 1 N? If released from rest at this distance, what will its initial acceleration be?

10. Use the data of Table 7-4 to determine the mass of Jupiter. What assumptions must you make about the forces on the moons of Jupiter and Jupiter's reference frame?

11. Estimate the magnitude of the gravitational force between the earth and Mars when *(a)* Mars is in the west at sunrise, *(b)* Mars is in the east at sunrise, *(c)* Mars is overhead at sunrise. (See Table 7-1.)

12. *(a)* Using the data of Table 7-1, determine the magnitudes of the forces on the sun due to each of the planets. *(b)* Assuming that all the planets are lined up such that the forces they exert on the sun are all in the same direction, determine the magnitude of the total force on the sun due to the planets. *(c)* From the answer to part *(b)*, determine the maximum acceleration of the sun if the only significant forces on the sun are due to the planets. Compare your answer with the accelerations given in Table 7-1.

13. The radius of Mars is 3.4 Mm and the acceleration of a freely falling object on its surface is 3.7 m/s². Determine the mass of Mars. Neglect any effects that arise from the rotation of Mars about its axis.

14. *(a)* Determine the dimension of G. *(b)* Determine the unit of G in the British system of units.

15. Suppose we invent a unit of mass that we shall call the cavendish (C). One cavendish of mass is defined such that $G = 1.0000\ (\text{AU})^3/(\text{yr}^2 \cdot \text{C})$. Our unit of length is the astronomical unit (AU), the earth-sun distance—1 AU = 1.496×10^{11} m—and our unit of time is the year (yr). *(a)* Determine the conversion factor between the cavendish and a kilogram. *(b)* Find the mass of the sun in cavendish.

16. Estimate the magnitude of the gravitational force on a 1-kg object on the surface of the earth due to *(a)* the sun and *(b)* the moon. *(c)* Compare the forces you find in parts *(a)* and *(b)* with the force on the object due to the earth's gravity.

17. A solar probe is rocketed from earth toward the sun in such a way that it is always between the earth and the sun. At what distance from the center of the earth will the probe be when the force on it due to the sun is equal and opposite the force on it due to the earth? What percentage of the earth-sun distance is this?

18. Reconsider the solar probe of the previous problem. Taking into account the force on the probe due to the moon as well as to the sun and the earth, determine the magnitude of the net force on the probe when it is 264 Mm from the earth for various phases of the moon: *(a)* new moon, *(b)* full moon, *(c)* first quarter. The probe's mass is 100 kg. (*Hint:* The angle between the moon and the sun at the earth is 90° at first quarter.)

19. A 6500-kg lunar probe is rocketed from earth directly toward the moon in such a way that it is always between the earth and the moon. Determine the magnitude of the sum of the gravitational forces on the probe by the earth and by the moon when it is *(a)* one-fourth of the way to the moon, *(b)* one-half of the way, and *(c)* three-fourths of the way.

20. Solve the quadratic equation for x in Example 7-4 and show that $x = r_{em}/(1 \pm \sqrt{m_m/m_e})$. [*Hint:* $1 - u^2 = (1 + u)(1 - u)$.]

21. Determine the ratio of the magnitude of the force by Venus on the earth to that by the sun on the earth at the time when the earth and Venus are nearest each other.

22. Phobos, a satellite of Mars, has a period of 7 h and 39 min and has an orbital radius of 9.4 Mm. From these data determine the mass of Mars. State any assumptions that you make.

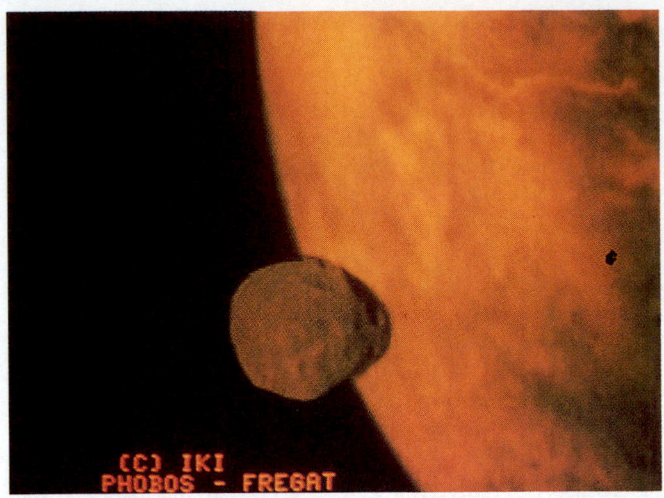

Phobos with Mars in the background.

23. In Fig. 7-25, three particles, each of mass m, occupy the corners of an equilateral triangle of side a. Determine an expression for the magnitude of the force on each particle.

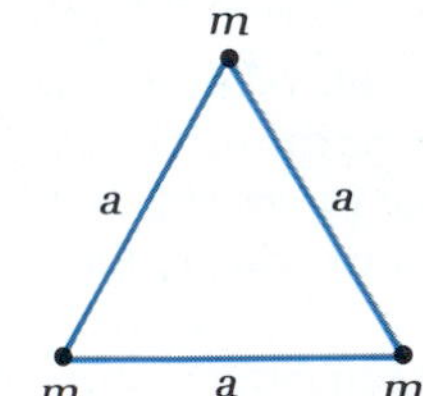

Figure 7-25. Exercise 23.

24. A neighboring galaxy to our Milky Way is the Andromeda galaxy at a distance of 2.1×10^{22} m. The mass of the Milky Way galaxy is 7×10^{41} kg and the mass of Andromeda is 6×10^{41} kg. *(a)* Treating these galaxies as particles, determine the magnitude of the force by Andromeda on the Milky Way. *(b)* Suppose that the magnitude of the net force on our galaxy is equal to the magnitude of the force on it by Andromeda. What is the magnitude of the acceleration of our galaxy relative to an inertial reference frame?

Section 7-4. Variation of *g* on the Earth's Surface

25. Assuming that the earth has exact spherical symmetry, with $R_e = 6.37$ Mm and $m_e = 5.97 \times 10^{24}$ kg, determine g at a height of 0.02 Mm above its surface.

26. We have traveled to another solar system in order to investigate the planet Yrelec. We measured the gravitational field at Yrelec's poles to be 7.69 m/s^2 and at its equator to be 7.52 m/s^2. Assuming this difference is due to the planet's oblateness, estimate the ratio of the distances between the center and the poles and the center and the equator.

27. Determine the fractional reduction of the acceleration of gravity due to an increase in elevation of 10 km near the earth's surface.

Section 7-5. The Gravitational Field

28. *(a)* Determine the magnitude of the gravitational field on the surface of the moon. The moon's radius is 1.74 Mm, and its mass is 7.35×10^{22} kg. *(b)* Determine the weight of a 59-kg astronaut in a moon-surface frame located on the moon's surface.

29. *(a)* Determine the magnitude of the gravitational field on the surface of Mars. The radius of Mars is 3.4 Mm, and its mass is 6.46×10^{23} kg. *(b)* Determine the weight of a 64-kg android in a Mars-surface frame located on the surface of Mars. Neglect any possible effects due to the rotation of Mars about its axis.

30. Three particles, *A, B,* and *C,* each have a mass of 1.9 kg and are placed on the corners of the square shown in Fig. 7-26. *(a)* What is the gravitational field at the empty corner (point *P*)? Give your answer in terms of unit vectors **i** and **j**. *(b)* What is the gravitational force on a particle of mass 2.3 kg placed at the empty corner?

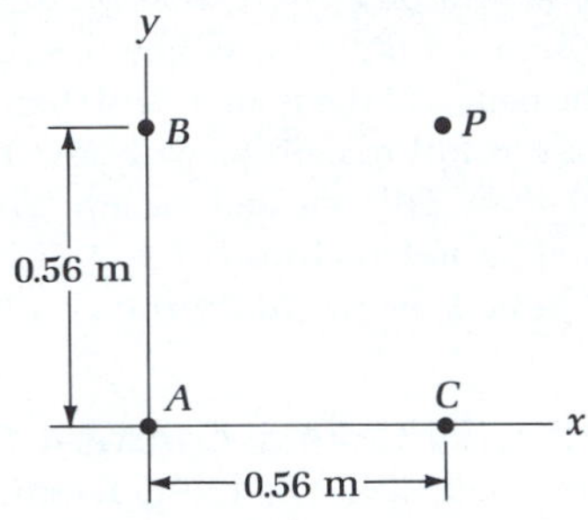

Figure 7-26. Exercise 30.

31. In Fig. 7-27, particle *A* has a mass of 1.4 kg and particle *B* has a mass of 3.1 kg. What is the gravitational field at point *P*?

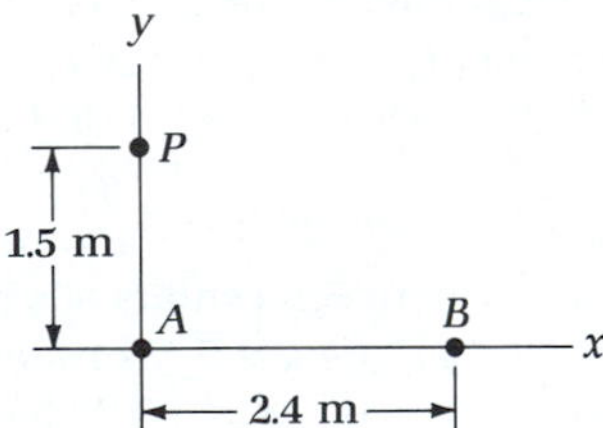

Figure 7-27. Exercise 31.

32. Two planets, Egabbac and Ecuttel, orbit the star Torrac. Curiously, both Egabbac and Ecuttel have a radius twice that of earth, Egabbac has a mass the same as earth, and Ecuttel has an average mass density the same as earth. What is g on the surface *(a)* of Egabbac and *(b)* of Ecuttel?

Section 7-6. Orbits, Numerical Methods

33. An earth satellite is in a circular orbit with $r = 7.19$ Mm. *(a)* What is the satellite's speed? *(b)* What is the period of the orbit?

34. Show that the speed of an earth satellite in circular orbit is given by the expression

$$v = \sqrt{\frac{Gm_e}{R_e + h}}$$

where h is the height of the satellite above the earth's surface.

35. (*a*) Consider an artificial satellite in a circular orbit about the moon with radius $r = 1.5r_m$, where r_m is the radius of the moon (1.74 Mm). Determine the satellites's speed v_c (subscript *c* for circular). (*b*) Use a spreadsheet to determine the motion of the satellite in part (*a*). (*c*) Now use another spreadsheet to determine the satellite's motion when the initial conditions are $r = 1.5r_m$, $v = 2.0v_c$, and the initial velocity is perpendicular to a line to the center of the moon. (*d*) Compare the graphs of the orbits in parts (*b*) and (*c*).

36. (*a*) Use a spreadsheet to determine the motion of an earth satellite. Let the initial conditions be $x = 15.0$ Mm, $y = 0$, $v_x = 0$, and $v_y = 10.0$ km/s. What are the satellite's coordinates at (*b*) the perigee and (*c*) the apogee? What are the satellite's velocity components at (*d*) the perigee and (*e*) the apogee?

37. (*a*) Use a spreadsheet to determine the motion of a sun satellite. Let the initial conditions be $x = 0$, $y = 100$ Gm, $v_x = 18$ km/s, and $v_y = 0$. What are the satellite's coordinates at (*b*) the perihelion (where the satellite is closest to the sun) and (*c*) the aphelion (where the satellite is farthest from the sun)? What are the satellite's velocity components at (*d*) the perihelion and (*e*) the aphelion?

38. While exploring the planet Pilut in another solar system, you place your spaceship in orbit around the planet at a height of 3.54 Mm above the surface. You measure the radius of the planet to be 7.48 Mm and the period of your orbit to be 6.63×10^3 s. Determine Pilut's mass.

Section 7-7. Discovery of the Law of Gravitation

39. The four largest moons of Jupiter have nearly circular orbits, and their periods and orbital radii are given in Table 7-4. Show that the comparison of these orbits conforms to Kepler's third law.

40. Use the data from Table 7-1 to show that a comparison of the orbits of earth and Venus conforms to Kepler's third law.

41. The radius of the orbit of Mars is 1.52 times that of the earth's. Use this information in Kepler's third law to find the period of the orbit of Mars in years.

Additional Exercises

42. While exploring a solar system at the edge of the Andromeda galaxy, you place your spaceship in an orbit about a planet such that the orbital radius is r_0 and the orbital speed is v_0. If you change to an orbit with radius $2r_0$, what is the new orbital speed in terms of v_0?

43. The moon has two significant forces exerted on it, that exerted by the sun and that exerted by the earth. Also, in applying Newton's second law to the moon, it is invalid to use an earth-center frame as inertial. Despite this, Prob. 6 shows that Newton's second law applied to the moon reduces to an equation that appears similar to Newton's second law:

$$\mathbf{F}_{em} \approx m_m \mathbf{a}_{me}$$

to a good approximation. In this expression, $\mathbf{F}_{em}$ is the force exerted by the earth on the moon, m_m is the mass of the moon, and $\mathbf{a}_{me}$ is the acceleration of the moon relative to an earth-center frame. (*a*) Use this expression, the moon's orbital period about the earth ($T = 2.36 \times 10^6$ s), and orbital radius ($r_{me} = 384$ Mm) to estimate the earth's mass. (*b*) Determine the percent error in your answer to part (*a*).

44. A typical neutron star has a mass about the same as the sun (2×10^{30} kg), but a radius of only about 10 km. (*a*) Estimate the magnitude *g* of the gravitational field on the surface of such a star. (*b*) About how long would it take for an object to fall from rest a distance of 1 m on the surface of such a star? Assume free-fall and neglect rotational effects.

45. Consider the gravitational forces exerted on the 2.0-kg particle located at the origin in Fig. 7-28. (*a*) What is the force exerted only by the 4.0-kg particle located at (0.0 m, 4.0 m)? (*b*) What is the force exerted only by the 5.0-kg particle located at (3.0 m, 4.0 m)? (*c*) What is the vector sum of the forces exerted by both the 4.0-kg particle and the 5.0-kg particle? (*d*) What is the gravitational field produced at the origin by both the 4.0-kg and the 5.0-kg particles? Write each of your answers in terms of unit vectors **i** and **j**.

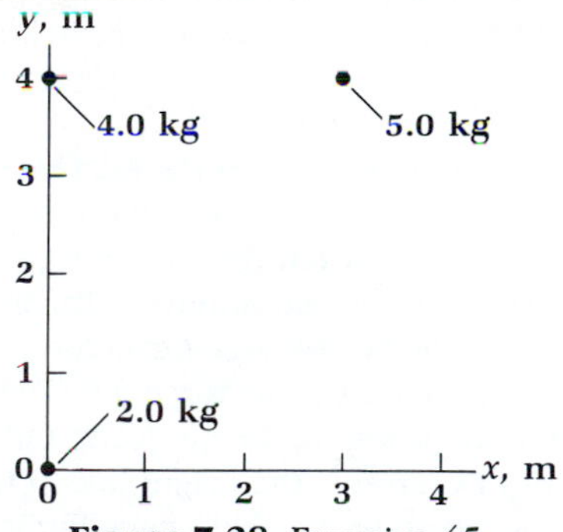

Figure 7-28. Exercise 45.

46. (*a*) Determine the speed of the earth relative to a sun-center frame. (*b*) Determine the speed of a person standing at the equator relative to an earth-center frame. (*c*) Determine the speed of the person in part (*b*) relative to a sun-center frame at noon on June 21. (*d*) Determine the speed of the person in part (*b*) relative to a sun-center frame at midnight on June 21. (*Hint:* Viewed from Polaris, the North Star, the earth moves counterclockwise in its orbit and spins counterclockwise on its axis. Also, the velocity of a point on the earth's equator is parallel to the plane of the earth's orbit at noon and at midnight on June 21.)

47. (*a*) Determine the magnitude of the acceleration of the earth relative to a sun-center frame. (*b*) Determine the magnitude of the acceleration of a person standing at the equator relative to an earth-center frame (*c*) Determine the magnitude of the acceleration of the person in part (*b*) relative to a sun-center frame at noon on March 21. (*d*) Determine the magnitude of the acceleration of the person in part (*b*) relative to a sun-center frame at midnight on March 21. (*e*) Is the weight of a person in an earth-surface frame at the equator greater at noon or at midnight? (*Hint:* Viewed from Polaris, the earth moves counterclockwise in its orbit and spins counterclockwise on its axis. Also, the acceleration of a point on the earth's equator is

parallel to the plane of the earth's orbit at noon and at midnight on March 21.)

48. Two particles, each with mass m, are located on the y axis at positions $(0, a)$ and $(0, -a)$, as shown in Fig. 7-29. Show that the gravitational field $\mathbf{g}$ produced by these particles at points along the x axis is

$$\mathbf{g} = -[2Gmx/(x^2 + a^2)^{3/2}]\mathbf{i}$$

49. A white dwarf is a type of star with a large mass density. A white dwarf may have a mass about the same as the sun (2×10^{30} kg), but a radius of only about 1×10^7 m (about the same as the earth). Estimate the magnitude of the gravitational field at a point on the surface of a white dwarf.

50. Suppose a 4.0-kg particle is located at the origin and a 2.0-kg particle is located on the x axis at $x = 4.0$ m. At what point is the net gravitational field produced by both of these particles equal to zero?

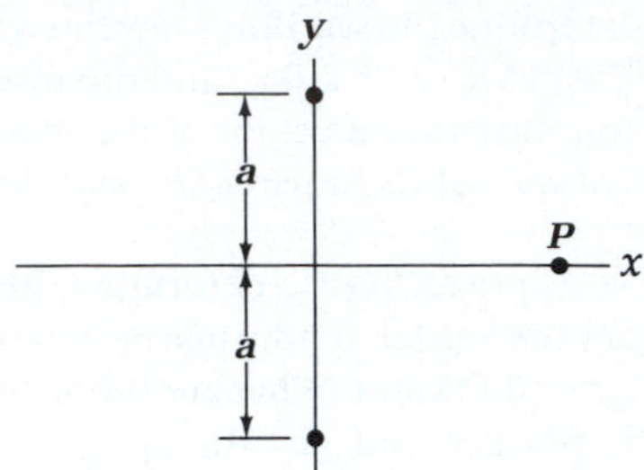

Figure 7-29. Exercise 48.

PROBLEMS

1. ***Estimating G with earth data.*** *(a)* Estimate G by assuming that the average mass density ρ_e of the earth is approximately the same as that of rocks on the surface of the earth (2.7×10^3 kg/m³). Use $g = 9.8$ m/s² and $R_e = 6.37$ Mm. *(b)* Using $m_e = 5.97 \times 10^{24}$ kg, determine the average mass density of the earth.

2. ***Net force on the moon when the earth and sun are at right angles.*** Consider the net force on the moon due to the earth and due to the sun when they are at right angles to the moon (Fig. 7-30). (Notice the different scales on the axes.) *(a)* Assuming these are the only significant forces on the moon, write a vector equation for the net force. *(b)* Draw a free-body diagram of the moon showing the net force and the two individual forces. *(c)* Determine the magnitude of the moon's acceleration relative to an inertial reference frame.

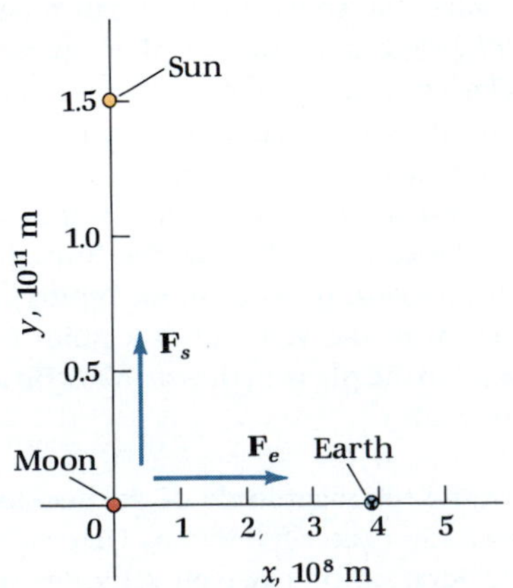

Figure 7-30. Problem 2.

3. ***Exploring the planet Norc.*** While investigating the planet Norc in another solar system, we find that the radius of Norc is 9.54×10^6 m and that the period of a satellite put in circular orbit of radius 1.476×10^7 m is 8.09×10^3 s. Determine *(a)* the mass of Norc, *(b)* the average mass density of Norc, *(c)* the value of the gravitational field on the surface of Norc. *(d)* If the period of Norc's rotation about its axis is 1.04×10^4 s, what will be the reading on a spring scale (calibrated on earth) supporting a 1.0-kg object at Norc's equator?

4. ***Gravitational field along an axis.*** Consider the gravitational field produced by two particles, particle B of mass m_B at the origin of coordinates and particle C of mass m_C at position x_C on the x axis (Fig. 7-31). Develop equations for the gravitational field at points along the x axis for three regions: *(a)* region 1, $x < 0$; *(b)* region 2, $0 < x < x_C$; *(c)* region 3, $x > x_C$. In each case write the field in terms of the unit vector $\mathbf{i}$.

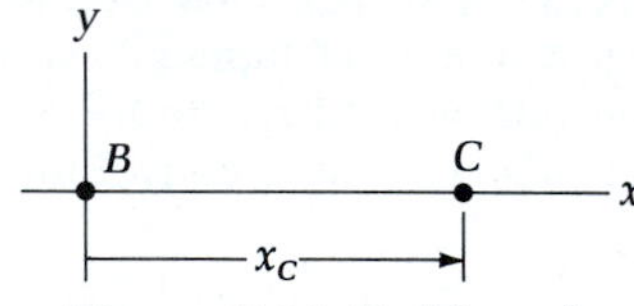

Figure 7-31. Problem 4.

5. ***A null position for*** **g.** *(a)* For the situation described in the previous problem, develop an equation for the position where the field is zero (the null position). *(b)* Find the null position for the case where $m_B = 2$ kg, $m_C = 8$ kg, and $x_C = 6$ m. *(c)* Plot a graph of the gravitational field for this case in the region -6 m $< x < +12$ m.

6. ***Motion of the moon in an earth-center frame.*** Assume that a sun-center frame is inertial and that the only significant forces on the moon are $\mathbf{F}_{sm}$ exerted by the sun and $\mathbf{F}_{em}$ exerted by the earth. (In Example 7-6 we showed that F_{sm} is a little more than twice F_{em}.) Newton's second law, $\Sigma\mathbf{F} = m\mathbf{a}$, applied to the moon is

$$\mathbf{F}_{sm} + \mathbf{F}_{em} = m_m\mathbf{a}_{ms}$$

where m_m is the moon's mass, and $\mathbf{a}_{ms}$ is the acceleration of the moon relative to a sun-center frame (assumed inertial). Show that, because r_{es} (the distance from the earth to the sun) is nearly equal to r_{ms} (the distance from the moon to the sun), this equation reduces to

$$\mathbf{F}_{em} \simeq m_m\mathbf{a}_{me}$$

where $\mathbf{a}_{me}$ is the acceleration of the moon relative to an earth-center frame. (*Hint:* From Sec. 4-4, $\mathbf{a}_{ms} = \mathbf{a}_{me} + \mathbf{a}_{es}$ where $\mathbf{a}_{es}$ is the acceleration of the earth relative to a sun-center frame.) This means that one can describe the motion of the moon relative to the earth using only the force exerted on the moon by the earth. This can be done because the force by the sun on the moon, though more than twice that by the earth, is nearly uniform within the moon's orbit of the earth.

7. ***Using a log-log graph to demonstrate Kepler's third law.*** A log-log graph is useful for showing a relation between two quantities, say x and y, that are related by an expression such as $y^a = kx^b$. The slope of the graph of y versus x on log-log graph paper gives b/a. Use the data in Table 7-1 on the planetary orbits to plot a log-log graph of T versus R. Determine the slope of the line through the data points.

8. ***Orbit of an earth satellite.*** Using an earth-center reference frame, you measure the position of a satellite to be $x = 20.20$ Mm and $y = 18.90$ Mm at a time we designate as $t = 0.0$ s, and then $x = 20.25$ Mm and $y = 18.85$ Mm at time $t = 25.0$ s. Construct a spreadsheet to determine the satellite's orbit and generate a graph of the orbit.

9. ***Orbit of a moon satellite.*** Use a spreadsheet to determine the motion of a satellite in circular orbit around the moon with a radius of 5.000 Mm. Use about 1000 iterations and generate a graph of the orbit.

10. ***Finding the period of a satellite from earth sightings.*** Suppose you are at the equator and are observing an artificial satellite in a circular orbit in the equatorial plane. The satellite passes over you from east to west, and you measure a time interval of 1.055×10^4 s between consecutive sightings of the satellite directly overhead. (*a*) What is the period of the satellite? (*b*) What is the radius of the orbit? (*c*) What is the speed of the satellite? (*d*) What is the acceleration of the satellite? (*Hint:* Be sure to take account of the earth's rotation.)

11. ***An approximate expression for g as a function of height.*** Assuming that the earth has exact spherical symmetry, show that g at a height h above the surface can be written

$$g \approx \left(\frac{Gm_e}{R_e^2}\right)\left(1 - \frac{2h}{R_e}\right)$$

where $h \ll R_e$.

12. ***Orbit of the solar system about the galaxy.*** Measurements indicate that our solar system is in nearly a circular orbit about the center of the Milky Way galaxy at a radius of 2.8×10^{20} m and a speed of 2.5×10^5 m/s. (*a*) Determine the period of the motion. (*b*) Determine the acceleration of our solar system relative to the center of the galaxy. (*c*) Estimate the mass of that part of the galaxy that is inside the orbit of our solar system. State any assumptions that you make. (*d*) Astronomers estimate the mass of the galaxy to be 7×10^{41} kg. Assuming that the sun is a typical star, estimate the number of stars in the galaxy.

13. ***The velocity of Io.*** The orbits of Jupiter's moons listed in Table 7-4 are in nearly the same plane as the orbits of the planets around the sun. In addition, viewed from Polaris, the sense of each orbit is counterclockwise. (*a*) Use the data in Table 7-4 to find the speed of Io in its orbit around Jupiter. (*b*) Now use the data in Table 7-1 to find the speed of Jupiter in its orbit around the sun. (*c*) Consider the velocities of Jupiter and Io in a sun-center frame at a time when Io is between Jupiter and the sun. Are the velocities of Jupiter and Io in nearly the same direction or are they in nearly opposite directions? (*d*) Determine the speed of Io in a sun-center frame at this time.

14. ***The tides.*** In Exercise 6, you are asked to compare the force exerted on the earth by the sun to that by the moon. It turns out that $F_{se}/F_{me} = 178$; the force by the sun is much larger than that by the moon. Despite this, the moon's influence on the earth's tides is more than twice that of the sun's. To investigate this apparent discrepancy, consider celestial objects A and B in Fig. 7-32. Object A has a radius R, object B has a mass M, the distance between centers is r, and $r \gg R$. (*a*) Show that the magnitude of the force exerted by B on particles 1 and 2, each of mass m and located on opposite sides of A, is

$$F_{B1} = GMm/(r - R)^2 \qquad \text{and} \qquad F_{B2} = GMm/(r + R)^2$$

(*b*) Show that the force difference $\Delta F_B = F_{B1} - F_{B2}$ is

$$\Delta F_B \approx 4GMmR/r^3$$

where we have used $r \gg R$. [*Hint:* Use the binomial expansion $(1 + x)^n \approx 1 + nx$ when $x \ll 1$, and let $R/r = x$.] (*c*) Now let object A be the earth and object B be the moon in one case and the sun in another case. Show that

$$\Delta F_{\text{moon}}/\Delta F_{\text{sun}} = 2.2$$

The tides on a planet depend on the difference in the force on one side and the other. Since the moon is much closer than the sun, its effect on the earth's tides is greater than the sun's because the force by the moon, although smaller than that by the sun, varies more from one side of the earth to the other.

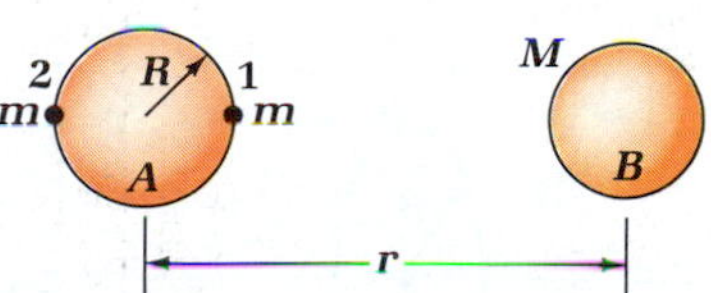

Figure 7-32. Problem 14.

15. ***Gravitational anomalies.*** Local variations in the gravitational field on the surface of the earth can be due to irregularities in density. Consider an effective field $\mathbf{g}_{\text{eff}} = \mathbf{g} + \mathbf{g}_a$ measured at a point P on the earth's surface near a spherical cavity of volume V that is centered a vertical distance d below the surface (Fig. 7-33). Here $\mathbf{g}$ is the field due to a perfectly spherical earth ($g = Gm_e/R_e^2$) and $\mathbf{g}_a$ is the field anomaly due to the cavity. To focus on the effect of the cavity, we disregard any effect due to the earth's rotation and assume that the surrounding rock has a uniform density ρ. (*a*) Explain why $\mathbf{g}_a$ is directed away from the center of the cavity. (*b*) Show that $g_a = G\rho V/(x^2 + d^2)$. (*c*) Show that the horizontal and vertical components of $\mathbf{g}_a$ are $g_{ax} = G\rho Vx/(x^2 + d^2)^{3/2}$ and $g_{ay} = G\rho Vd/(x^2 + d^2)^{3/2}$. (*d*) Show that $g_{\text{eff}} = [(g - g_{ay})^2 + g_{ax}^2]^{1/2}$. (*e*) Use the fact that $g \gg g_{ax}$ and $g \gg g_{ay}$ to show that $g_{\text{eff}} \approx g - g_{ay}$. (*f*) Evaluate the ratio g_{ay}/g for the case where $\rho = 2.7 \times 10^3$ kg/m^3, $V = 1.0 \times 10^6$ m^3, $x = 0$, and $d = 100$ m.

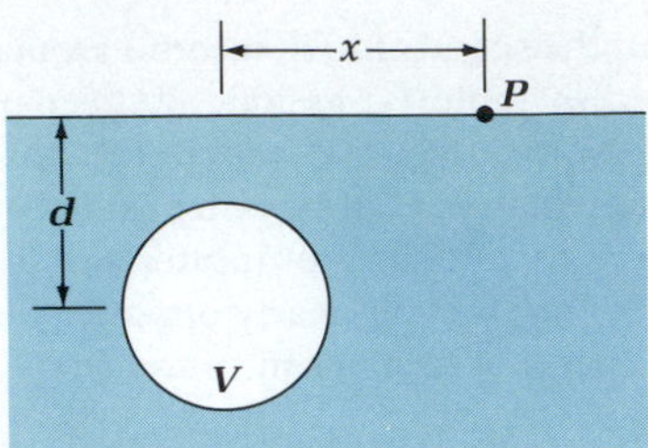

Figure 7-33. Problem 15.

16. ***A geosynchronous orbit.*** The *subpoint* of a celestial object is the point on the earth's surface where a line from the object to the earth's center passes through the earth's surface. If an observer was standing at an object's subpoint, the object would be directly overhead. Communications satellites and weather-monitoring satellites are often placed in a circular orbit such that the satellite's subpoint remains fixed (nearly). In this way the satellite's location in the sky, viewed from earth, remains fixed (nearly). Such an orbit is called *geosynchronous* because the satellite's period is synchronized to the earth's rotational period. Determine the radius of a geosynchronous orbit. What must be the plane of this orbit?

17. ***Gravitational field produced by a ring-shaped mass.*** Consider the gravitational field **g** produced by a thin ring of mass m and radius R. Show that, at points along its axis of symmetry (shown as the x axis in Fig. 7-34),

$$\mathbf{g} = -\frac{Gmx}{(x^2 + R^2)^{3/2}}\,\mathbf{i}$$

(b) Construct a graph of g_x in units of Gm/R^2 versus x in units of R from $x = -2R$ to $x = +2R$ with points plotted at intervals of $R/4$.

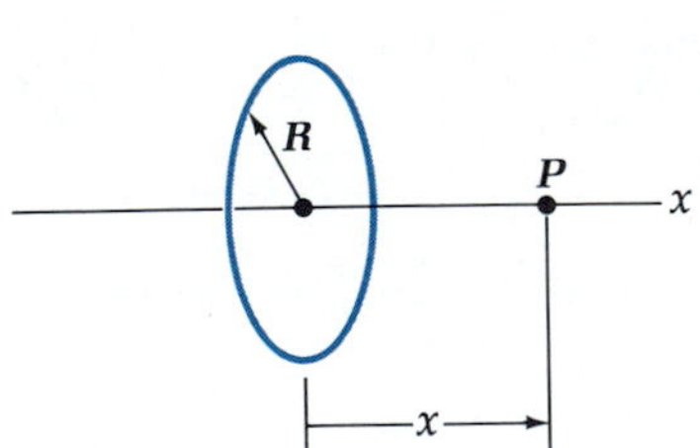

Figure 7-34. Problem 17.

18. ***Time that a satellite remains above the horizon.*** *(a)* An artificial satellite is placed in a circular orbit traveling eastward in the equatorial plane such that its height above the surface is 14.00×10^6 m. What is the time interval the satellite remains above the horizon for an observer stationed at a point on the equator? *(b)* Repeat part *(a)* except let the satellite be traveling westward.

19. ***Minimum rotational period for a celestial object.*** Consider a celestial object, such as a planet, that is composed of particles that are held together only by their mutual gravitational attraction. Such an object has a maximum rotational speed it cannot exceed because, if it did, the gravitational force exerted on the particles near the equator would be insufficient to provide the centripetal force for circular motion. *(a)* Show that for an object with uniform mass density ρ the minimum rotational period is

$$T_{\min} = \sqrt{3\pi/G\rho}$$

(b) Evaluate $T_{\min}$ for a planet with $\rho = 3 \times 10^3$ kg/m³. *(c)* Evaluate $T_{\min}$ for a neutron star with $\rho = 3 \times 10^{17}$ kg/m³. Some neutron stars may have periods of rotation as small as 1 s.

20. ***Orbital speed from Kepler's second law.*** Consider a planet in a elliptical orbit about the sun. The point at which the planet is farthest from the sun is called the ***aphelion*** and the point at which the planet is nearest the sun is called the ***perihelion.*** Let r_a and r_p represent the distance between the sun and the planet at the aphelion and at the perihelion, respectively. *(a)* Use Kepler's second law to show that

$$v_a r_a = v_p r_p$$

where v_a and v_p represent the planet's speed at the aphelion and at the perihelion, respectively. *(b)* For the planet Mercury, $r_a = 6.99 \times 10^{10}$ m, $r_p = 4.60 \times 10^{10}$ m, and $v_a = 3.88 \times 10^4$ m/s. Determine v_p for Mercury.

21. ***Field g produced by a particle.*** Consider the gravitational field **g** produced by a particle with mass m located on the y axis at the point $(0, a)$, as shown in Fig. 7-35. *(a)* Show that **g** at points along the x axis is

$$\mathbf{g} = -\frac{Gmx}{(x^2 + a^2)^{3/2}}\,\mathbf{i} + \frac{Gma}{(x^2 + a^2)^{3/2}}\,\mathbf{j}$$

(b) Show that g_x is minimum at $x = a/\sqrt{2}$ and is maximum at $x = -a/\sqrt{2}$. *(c)* Sketch a graph of g_x between $x = -2a$ and $x = 2a$. *(d)* Show that g_y is maximum at $x = 0$. *(e)* Sketch a graph of g_y between $x = -2a$ and $x = 2a$.

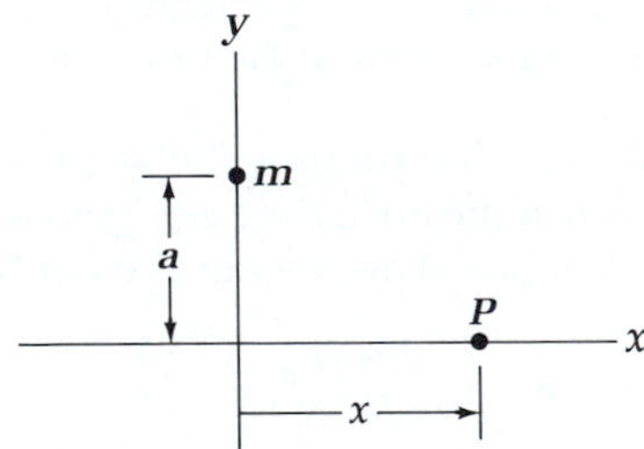

Figure 7-35. Problem 21.

22. ***Field inside a uniform spherical shell.*** Consider the field **g** at a point P inside a thin spherical shell of uniform mass density ρ (Fig. 7-36). The shell has thickness τ and radius r_o, and point P is on the x axis a distance r from the center of the shell. *(a)* Show that the infinitesimal field components produced at P by the thin ring centered on the x axis and with cross-sectional area $\tau(r_o\, d\theta)$ and with radius $r_o \sin\theta$ are

$$dg_y = 0 \quad \text{and} \quad dg_x = \frac{G\rho\tau(r_o\, d\theta)(2\pi r_o \sin\theta)(r_o\cos\theta - r)}{[(r_o\cos\theta - r)^2 + (r_o\sin\theta)^2]^{3/2}}$$

(b) Add the contributions due to the infinite number of infinitesimal rings by integrating dg_x from $\theta = 0$ to $\theta = \pi$. Show that $\mathbf{g} = 0$ at each point inside the spherical shell. [*Hint:* $(r_o\cos\theta - r)^2 + (r_o\sin\theta)^2 = r_o^2 + r^2 - 2rr_o\cos\theta$, and $\sqrt{(r_o - r)^2} = r_o - r$ when $r_o > r$.]

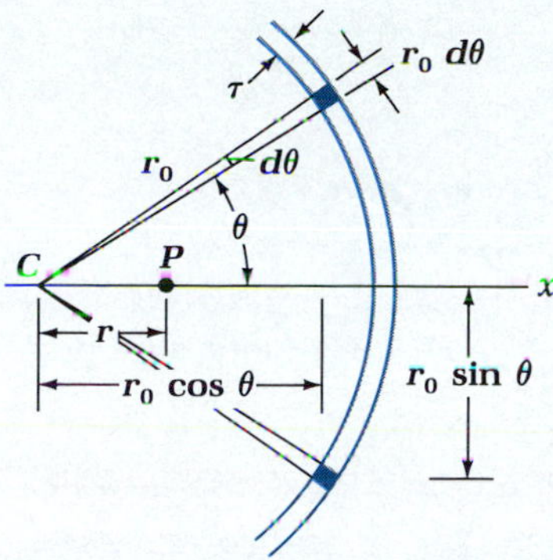

Figure 7-36. Problem 22. Part of the thin spherical shell is shown in cross section. The cross-sectional area $\tau(r_0\,d\theta)$ of the ring is shown darkened.

23. ***Field outside a thin spherical shell.*** Consider the field **g** at a point P outside a thin spherical shell of uniform mass density ρ (Fig. 7-37). The shell has thickness τ and radius r_0, and point P is on the x axis a distance r from the center of the shell. *(a)* Show that the infinitesimal field components produced at P by the thin ring centered on the x axis and with cross-sectional area $\tau(r_0\,d\theta)$ and with radius $r_0 \sin\theta$ are

$$dg_y = 0 \quad \text{and} \quad dg_x = \frac{G\rho\tau(r_0\,d\theta)(2\pi r_0 \sin\theta)(r - r_0\cos\theta)}{[(r_0\cos\theta - r)^2 + (r_0\sin\theta)^2]^{3/2}}$$

(b) Add the contributions due to the infinite number of infinitesimal rings by integrating dg_x from $\theta = 0$ to $\theta = \pi$. Show that

$$g_x = -\frac{Gm}{r^2}$$

where $m = \rho 4\pi r_0^2 \tau$ is the mass of the spherical shell. Thus the field produced outside the shell is the same as that produced by a particle of mass m located at the center. [*Hints:* $(r_0\cos\theta - r)^2 + (r_0\sin\theta)^2 = r_0^2 + r^2 - 2rr_0\cos\theta$, and $\sqrt{(r_0 - r)^2} = r - r_0$ when $r > r_0$.]

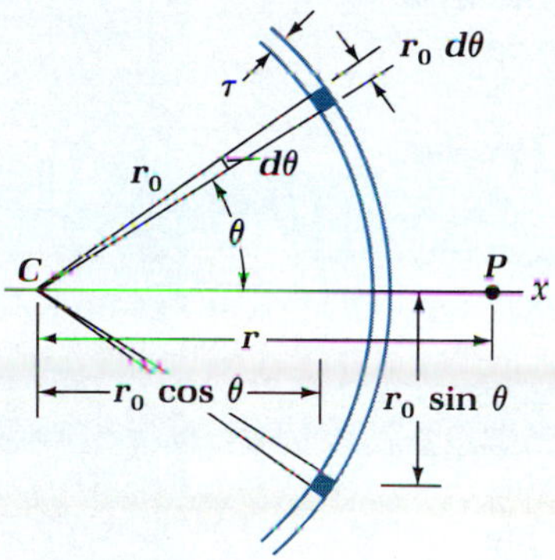

Figure 7-37. Problem 23. Part of the thin spherical shell is shown in cross section. The cross-sectional area $\tau(r_0\,d\theta)$ of the ring is shown darkened.

8 WORK AND ENERGY

Egyptian pyramids. For an analysis of the work required to construct a pyramid, see Problem 9.

This photograph of Venus was taken on March 3, 1979, by the Pioneer Venus orbiter.

In the preceding chapters, we have developed a straightforward method for finding the motion of a particle. For example, we can determine how the position of a planet, such as, Venus, varies with time. The law of gravitation gives the force acting on the planet, and Newton's second law connects the net force to the acceleration. Given the initial position and velocity of Venus, we can then solve for its motion. One method for solving this kind of problem was illustrated in Sec. 7-6. In principle, this procedure can be applied to find the motion of each particle in a complicated system of particles. However, in practice, this may be a cumbersome task. Also, we often do not wish to have such detailed information. Instead, we seek a description that tells us in a broader sense how a system evolves. The concepts of work and energy provide this broader and less detailed perspective on the motion of objects.

In these next two chapters we introduce the concepts of work and energy. The central focus of this chapter is the definition of work and the work-energy theorem. In the following chapter, we discuss one of the fundamental laws of nature, the law of conservation of energy.

8-1 WORK DONE BY A CONSTANT FORCE

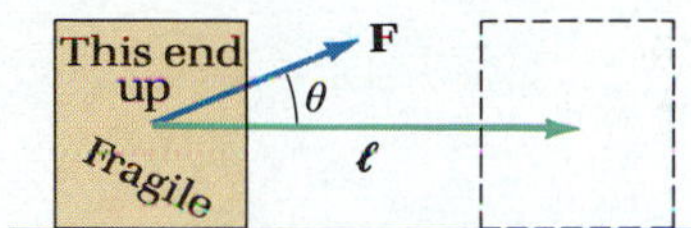

Figure 8-1. A constant force acts on a crate as it undergoes a displacement. The work done by the force is $W = F\ell \cos\theta$.

We introduce the idea of work by considering a constant force acting on an object that moves in a straight line. Suppose you pull a crate in a straight line across a floor. Let **F** represent the force you exert on the crate and let $\Delta\mathbf{r} = \boldsymbol{\ell}$ represent the displacement of the crate (Fig. 8-1). Although there are other forces exerted on the crate besides **F**, we are concerned for the moment with only this one force. The work W done on the crate by you is defined as

$$W = F\ell \cos\theta \tag{8-1}$$

where F and ℓ are the magnitudes of the vectors **F** and $\boldsymbol{\ell}$.

The work done by an agent that exerts a force on an object is equal to the product of the force magnitude times the distance the object is moved along the direction of the force.

The SI unit of work is the joule (J).

The SI unit of work is the newton-meter (N·m), to which we assign the name "joule," abbreviated J and pronounced as jo͞ol. This unit is named after James Prescott Joule (1818–1889), whose experiments helped to clarify the relationship between work and heat.

For work to be done, a force **F** must act on an object and the point at which the force is applied must move. In a sense, this definition of work is consistent with the everyday meaning of the word "work." After all, you provided the muscular effort (the force **F**) to move the crate. On the other hand, Eq. 8-1 states that $W = 0$ if $\ell = 0$. If you push on the crate and it does not move, then, by our definition, you do no work. This aspect of our definition is inconsistent with the everyday meaning of work.

No work is done unless the object moves.

Work depends on angle θ between F and $\boldsymbol{\ell}$.

It is essential to understand that work depends on the relative directions of the force and displacement. This dependence is contained in the factor $\cos\theta$ in Eq. 8-1. For example, the work done by a force parallel to the displacement ($\theta = 0$) is just the product $F\ell$, since $\cos\theta = \cos 0 = 1$. If you push horizontally on a book with a constant force of magnitude 20 N and move it 0.5 m across a tabletop, as seen in Fig. 8-2*a*, then the work done on the book by the force pushing it is $W = (20\text{ N})(0.5\text{ m}) \cos 0 = 10\text{ J}$.

Suppose instead that a force on an object is perpendicular to its displacement. Then the factor $\cos 90° = 0$, and the work done on the object by the force is zero: ***The work done by a force whose direction is perpendicular to the displacement is zero.*** When you carry a book across a room, the supporting force you exert to balance the weight of the book is vertical, while the displacement is horizontal and perpendicular to that force, as shown in Fig. 8-2*b*. The supporting force does no work on the book.

If the angle between the force and displacement vectors is greater than 90°, then $\cos\theta$ is negative and the work done by that force is negative. As you gently lower a 30-N book (Fig. 8-2*c*), the force you exert on the book is upward, opposite to a 0.5-m downward displacement of the book. In this case, the work that you do is $(30\text{ N})(0.5\text{ m}) \cos 180° = -15\text{ J}$. You do negative work on the book because the force that you exert is opposite to the direction of the book's motion.

The work done by a force can be positive, zero, or negative, depending on the angle between force and displacement. Even though the work done by the force that you exert on an object may be zero or negative, the muscular fatigue that you experience may seem much the same as when you do positive work on the object. Muscular fatigue is not a good indicator of the sign or the amount of work that you do on an object.

So far we have looked at the work done by one force acting on an object. If more than one force is present, the work done by each force can be calculated separately as the object undergoes a displacement. The second example below illustrates such a calculation.

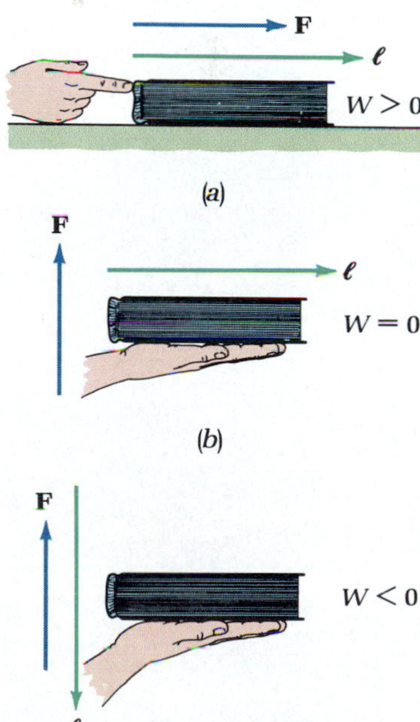

Figure 8-2. *(a)* A force does positive work on a book. *(b)* No work is done on the book because **F** and $\boldsymbol{\ell}$ are perpendicular. *(c)* The force does negative work on the book because **F** and $\boldsymbol{\ell}$ are opposite.

EXAMPLE 8-1

Work done by a rope. A constant force of 17 N is exerted by a rope on a crate as the crate slides 2.0 m across a floor in a straight line. The angle between the force exerted by the rope and the crate's displacement is 25°. What is the work done by the rope?

Solution. Applying Eq. 8-1 gives

$$W = (17\ \text{N})(2.0\ \text{m})\cos 25° = 31\ \text{J}$$

In this case, the force and the displacement were in the same general direction ($\theta < 90°$), so that the work done by the force is positive.

SELF-TEST 8-1. Two ropes are attached to a crate, and the crate slides across a floor in a straight line for a distance of 3.0 m. Rope 1 exerts a force of 25 N at an angle of 20° with the displacement, and rope 2 exerts a force of 11 N at an angle of 115° with the displacement. What is the work done by rope 2? Explain why the algebraic sign of this work is negative. ***ANSWER:*** −14 J.

EXAMPLE 8-2

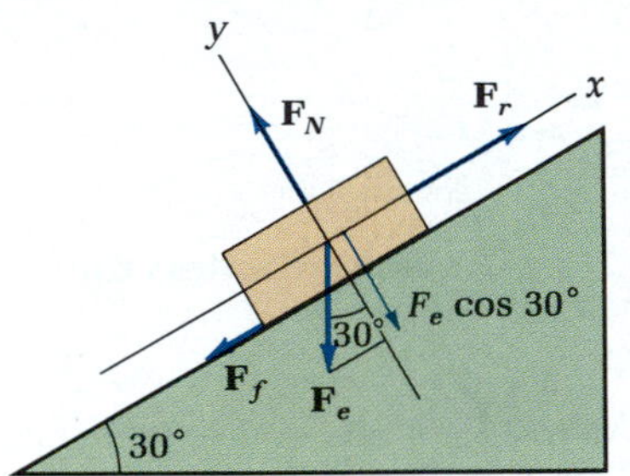

Figure 8-3. Example 8-2.

Work done by several forces. A 48-kg crate is pulled 8.0 m up a 30.0° ramp by a rope with constant tension $F_r = 540$ N (Fig. 8-3). The coefficient of kinetic friction is $\mu_k = 0.40$. Determine the work done by each force acting on the crate.

Solution. Using methods from Chaps. 5 and 6, we resolve forces along coordinate axes, as shown in Fig. 8-3. The force exerted by the rope has magnitude $F_r = 540$ N, while the magnitude of the weight is $F_e = mg = 470$ N. Since there is no motion perpendicular to the plane of the ramp, the normal force must balance the component of weight acting perpendicular to the ramp, $F_N = F_e \cos 30° = 410$ N. The magnitude of the frictional force is then $F_f = \mu_k F_N = 160$ N.

Now we calculate the work done by each force acting on the crate as it moves up the ramp. For the rope force,

$$W_r = (540\ \text{N})(8.0\ \text{m})\cos 0° = 4.3\ \text{kJ}$$

The work done by the gravitational force, or weight, is

$$W_e = (470\ \text{N})(8.0\ \text{m})\cos 120° = -1.9\ \text{kJ}$$

The normal force does no work since its direction is perpendicular to the displacement: $W_N = 0$ J. Finally, the work done by the frictional force is

$$W_f = (160\ \text{N})(8.0\ \text{m})\cos 180° = -1.3\ \text{kJ}$$

SELF-TEST 8-2. When the crate in Example 8-2 reaches the top of the ramp, the rope breaks and the crate slides 8.0 m back down the ramp. What work is done by each of the forces acting on the crate? ***ANSWER:*** $W_f = -1.3$ kJ, $W_e = 1.9$ kJ, $W_N = 0$.

8-2 THE DOT PRODUCT

The definition of work is expressed in Eq 8-1 as a product of the magnitudes F and ℓ of the force and the displacement and the cosine of the angle between their directions: $W = F\ell \cos\theta$. An elegant and useful way to write this expression is in terms of the ***dot product*** of two vectors. Using this product streamlines our notation and simplifies the calculation of work.

In Chap. 2, we concentrated on adding and subtracting vectors. Here we consider one way of multiplying two vectors. The dot product of any two vectors **A** and **B** is defined as

The dot product of two vectors

$$\mathbf{A} \cdot \mathbf{B} = AB\cos\theta \tag{8-2}$$

where θ is the angle between the two vectors, as shown in Fig. 8-4*a*. The right-hand side of Eq. 8-2 is the product of three scalers, the magnitudes A and B and the cosine. Therefore the dot product of two vectors is a scalar quantity. Each of the vectors **A** and **B** has a direction, but the dot product itself does not. Since it is a scalar quantity, the dot product is often called the ***scalar product.***

For a force **F** and a displacement $\boldsymbol{\ell}$, the dot product is $\mathbf{F} \cdot \boldsymbol{\ell} = F\ell \cos \theta$, which is the work done by the force. Thus the work done by a force **F** on an object that moves in a straight line with a displacement $\boldsymbol{\ell}$ is the dot product of these two vectors:

$$W = \mathbf{F} \cdot \boldsymbol{\ell} \tag{8-3}$$

Work as a dot product

The dot product of two vectors can be interpreted geometrically in terms of the projection of one vector onto the other vector. Writing Eq. 8-2 as $\mathbf{A} \cdot \mathbf{B} = A(B \cos \theta)$ and using Fig. 8-4*b*, we see that $\mathbf{A} \cdot \mathbf{B}$ is the product of the magnitude of **A** and the component of **B** along **A**. Alternatively, we can write $\mathbf{A} \cdot \mathbf{B} = B(A \cos \theta)$ and use Fig. 8-4*c*. The equality of these two results expresses the commutativity of the dot product of two vectors; the value is independent of the order of the vectors in the product:

$$\mathbf{A} \cdot \mathbf{B} = \mathbf{B} \cdot \mathbf{A}$$

Commutative and distributive properties of the dot product

The dot product is also distributive:

$$\mathbf{A} \cdot (\mathbf{B} + \mathbf{C}) = \mathbf{A} \cdot \mathbf{B} + \mathbf{A} \cdot \mathbf{C}$$

Another property of the dot product is $\mathbf{A} \cdot (s\mathbf{B}) = s(\mathbf{A} \cdot \mathbf{B})$, where s is a scalar. (Exercise 10 considers a graphical demonstration of these last two results.)

Since vectors are often specified by their components, let us express the dot product of two vectors in terms of their components. Suppose that vectors **A** and **B** are written in component form as $\mathbf{A} = A_x\mathbf{i} + A_y\mathbf{j} + A_z\mathbf{k}$ and $\mathbf{B} = B_x\mathbf{i} + B_y\mathbf{j} + B_z\mathbf{k}$. Then

$$\mathbf{A} \cdot \mathbf{B} = (A_x\mathbf{i} + A_y\mathbf{j} + A_z\mathbf{k}) \cdot (B_x\mathbf{i} + B_y\mathbf{j} + B_z\mathbf{k})$$

In multiplying out the parentheses in this last expression, we see that nine products appear in the sum:

$$\begin{aligned}\mathbf{A} \cdot \mathbf{B} = {} & A_xB_x\mathbf{i} \cdot \mathbf{i} + A_xB_y\mathbf{i} \cdot \mathbf{j} + A_xB_z\mathbf{i} \cdot \mathbf{k} + A_yB_x\mathbf{j} \cdot \mathbf{i} \\ & + A_yB_y\mathbf{j} \cdot \mathbf{j} + A_yB_z\mathbf{j} \cdot \mathbf{k} + A_zB_x\mathbf{k} \cdot \mathbf{i} + A_zB_y\mathbf{k} \cdot \mathbf{j} + A_zB_z\mathbf{k} \cdot \mathbf{k}\end{aligned}$$

Since **i**, **j**, and **k** are mutually perpendicular unit vectors, we can easily evaluate all such products by using Eq. 8-2, the definition of the dot product of any two vectors. For example, $\mathbf{i} \cdot \mathbf{i} = (1)(1) \cos 0 = 1$; $\mathbf{i} \cdot \mathbf{j} = (1)(1) \cos 90° = 0$; and there are similar results for the others. We have

$$\mathbf{i} \cdot \mathbf{i} = \mathbf{j} \cdot \mathbf{j} = \mathbf{k} \cdot \mathbf{k} = 1$$

$$\mathbf{i} \cdot \mathbf{j} = \mathbf{j} \cdot \mathbf{k} = \mathbf{k} \cdot \mathbf{i} = 0$$

$$\mathbf{j} \cdot \mathbf{i} = \mathbf{k} \cdot \mathbf{j} = \mathbf{i} \cdot \mathbf{k} = 0$$

Of the nine products in the equation above for $\mathbf{A} \cdot \mathbf{B}$, only three are nonzero. Thus in terms of the components of **A** and **B**, the dot product is

$$\mathbf{A} \cdot \mathbf{B} = A_xB_x + A_yB_y + A_zB_z \tag{8-4}$$

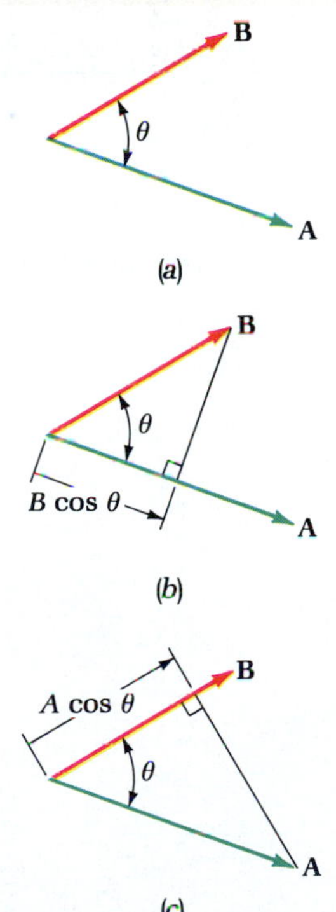

Figure 8-4. *(a)* The dot product of two vectors is a scalar, $\mathbf{A} \cdot \mathbf{B} = AB \cos \theta$. *(b)* $B \cos \theta$ is the projection of vector **B** onto vector **A**. *(c)* $A \cos \theta$ is the projection of vector **A** onto vector **B**.

There are two equivalent ways to evaluate the dot product. If you know or can determine the magnitude of each vector and the angle between them, then you can evaluate the dot product by using Eq. 8-2. Alternatively, if the components of each vector are known, then Eq. 8-4 is simpler to use.

As a special case of Eq. 8-4, suppose that $\mathbf{A} = \mathbf{B}$. Using Eq 8-2 first and then Eq. 8-4, we have

$$\mathbf{A}\cdot\mathbf{A} = (A)(A)\cos 0 = A^2$$

and

$$\mathbf{A}\cdot\mathbf{A} = A_xA_x + A_yA_y + A_zA_z$$

Combining these results yields

$$A^2 = A_x^2 + A_y^2 + A_z^2$$

We have used the dot product to obtain a familiar result: The square of the magnitude of a vector is equal to the sum of the squares of its components. This is the three-dimensional form of the pythagorean theorem.

EXAMPLE 8-3

Work as a dot product. An object moves in a straight line with a displacement given by $\boldsymbol{\ell} = (3.0\text{ m})\mathbf{i} + (4.0\text{ m})\mathbf{j}$. Determine the work done on the object by the constant force $\mathbf{F} = (8.0\text{ N})\mathbf{i} + (-8.0\text{ N})\mathbf{j}$. These vectors are shown in Fig. 8-5.

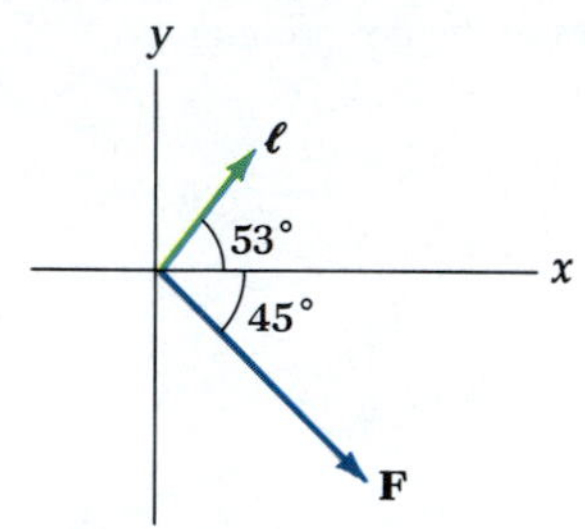

Figure 8-5. Example 8-3.

Solution. Since the components of the vectors are given (note that $F_z = 0$ and $\Delta z = 0$), we use Eq. 8-4 to evaluate Eq. 8-3 for the work:

$$\begin{aligned} W &= \mathbf{F}\cdot\boldsymbol{\ell} = F_x\,\Delta x + F_y\,\Delta y \\ &= (8.0\text{ N})(3.0\text{ m}) + (-8.0\text{ N})(4.0\text{ m}) = -8.0\text{ J} \end{aligned}$$

The dot product can also be evaluated by finding the magnitude of each vector and the angle between the two. Using

$$F = \sqrt{F_x^2 + F_y^2} = 11\text{ N}$$

$$\ell = \sqrt{\Delta x^2 + \Delta y^2} = 5.0\text{ m}$$

and

$$\theta = \tan^{-1}\left(\frac{\Delta y}{\Delta x}\right) - \tan^{-1}\left(\frac{F_y}{F_x}\right) = 53° - (-45°) = 98°$$

we have

$$W = F\ell\cos\theta = (11\text{ N})(5.0\text{ m})\cos 98° = -8.0\text{ J}$$

SELF-TEST 8-3. The object in the above example undergoes the same displacement as before. What work is done by a constant force $\mathbf{F} = (-8.0\text{ N})\mathbf{i} + (8.0\text{ N})\mathbf{j}$ acting on the object? Determine the work by using components, and check your answer by using magnitudes and directions. ***ANSWER:*** $W = 8.0$ J.

EXAMPLE 8-4

Finding the angle between vectors. Suppose you are given two vectors $\mathbf{F} = (4.0\text{ N})\mathbf{i} + (-1.0\text{ N})\mathbf{k}$ and $\boldsymbol{\ell} = (5.0\text{ m})\mathbf{i} + (-2.0\text{ m})\mathbf{j} + (1.0\text{ m})\mathbf{k}$, and you wish to know the angle θ between these vectors. *(a)* Determine an expression that gives θ in terms of the components of **F** and $\boldsymbol{\ell}$. *(b)* Evaluate θ for this case.

Solution. *(a)* Using the vectors **F** and $\boldsymbol{\ell}$ in Eqs. 8-2 and 8-4 gives

$$\mathbf{F}\cdot\boldsymbol{\ell} = F\ell\cos\theta$$

and

$$\mathbf{F}\cdot\boldsymbol{\ell} = F_x\ell_x + F_y\ell_y + F_z\ell_z$$

Setting these two expressions for $\mathbf{F}\cdot\boldsymbol{\ell}$ equal and using $F = \sqrt{F_x^2 + F_y^2 + F_z^2}$ and $\ell = \sqrt{\ell_x^2 + \ell_y^2 + \ell_z^2}$ give

$$\cos\theta = \frac{F_x\ell_x + F_y\ell_y + F_z\ell_z}{\sqrt{F_x^2 + F_y^2 + F_z^2}\,\sqrt{\ell_x^2 + \ell_y^2 + \ell_z^2}}$$

(b) For the values of **F** and $\boldsymbol{\ell}$ given, we find

$$\cos\theta = \frac{(4.0\text{ N})(5.0\text{ m}) + (-1.0\text{ N})(1.0\text{ m})}{\sqrt{(4.0\text{ N})^2 + (-1.0\text{ N})^2}\,\sqrt{(5.0\text{ m})^2 + (-2.0\text{ m})^2 + (1.0\text{ m})^2}}$$

Calculation gives

$$\theta = \cos^{-1}(0.84) = 33°$$

SELF-TEST 8-4. Use the method of the above example to determine the angle between the force **F** and the displacement $\boldsymbol{\ell}$ in Example 8-3. ***ANSWER:*** 98°.

8-3 WORK DONE BY A VARIABLE FORCE

So far we have considered the work done by constant forces. Many of the forces we encounter are not constant. For example, the force exerted on an object by a spring depends on the amount of stretch or compression of the spring.

A Variable Force in One Dimension

For simplicity, we begin by considering the case where the force **F** is along the straight-line motion of the object, along the x axis. Then the component of the displacement is Δx, and the component of the force is F_x. To make clear that the force varies with x, it is written as $F_x(x)$. Figure 8-6*a* shows a typical graph of $F_x(x)$ versus x. How much work is done by such a force as the object moves from x_i to x_f? The work done by the varying force for the full displacement from x_i to x_f can be approximated by adding the work done in each of a large number of small displacements. Each subinterval is taken to be small enough so that the force changes insignificantly as x changes by Δx. We approximate the work ΔW done by the force for the displacement $\Delta\boldsymbol{\ell} = \Delta x\,\mathbf{i}$ by evaluating the force component $F_x(x)$ at the midpoint $\bar{x}$ of the subinterval and writing

$$\Delta W \approx \mathbf{F}\cdot\Delta\boldsymbol{\ell} = [F_x(\bar{x})\mathbf{i}]\cdot(\Delta x\,\mathbf{i}) = F_x(\bar{x})\,\Delta x$$

since $\mathbf{i}\cdot\mathbf{i} = 1$. From Fig. 8-6*a* this value of ΔW is equal to the area of the shaded rectangular region of height $F_x(\bar{x})$ and base Δx.

Adding such contributions for each incremental displacement Δx for the entire displacement from x_i to x_f gives an approximation for the work done:

$$W \approx \Sigma\, F_x(\bar{x})\,\Delta x$$

Suppose N is the number of subintervals into which we have divided the interval $x_f - x_i$; then $N\Delta x = x_f - x_i$. If we make N larger and Δx correspondingly smaller, the accuracy of the approximation for the work will improve because the force $F_x(x)$ varies even less over a smaller subinterval.

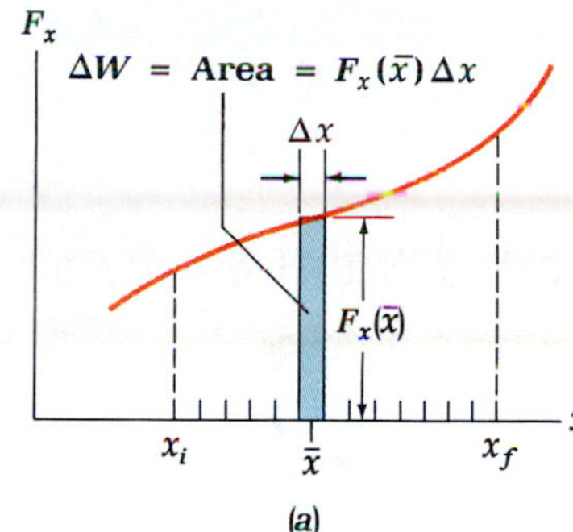

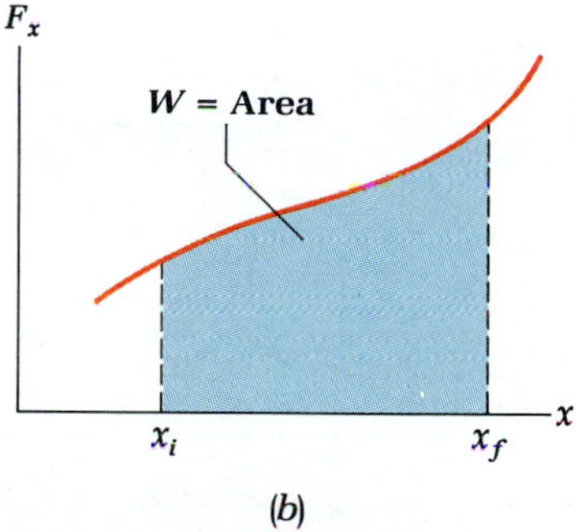

Figure 8-6. *(a)* The area of the rectangle approximately equals the work done for the small displacement, $\Delta W = F_x(\bar{x})\,\Delta x$. *(b)* The work done for the displacement from x_i to x_f equals the area bounded by the curve and the x axis.

Imagine a sequence of such choices of larger N and smaller Δx so that in the limit, as $N \to \infty$ and $\Delta x \to 0$, the result becomes exact. This limiting process makes it possible to express the work as an integral. Since

$$\lim_{\substack{N\to\infty \\ \Delta x\to 0}} \Sigma\, F_x(\bar{x})\,\Delta x = \int_{x_i}^{x_f} F_x(x)\,dx$$

Work done by a varying force in one dimension

the work done by the variable force as the object is displaced from x_i to x_f is

$$W = \int_{x_i}^{x_f} F_x(x)\,dx \tag{8-5}$$

Work equals the area under the graph of $F_x(x)$

The integral for work came from the limiting case of summing the work done for

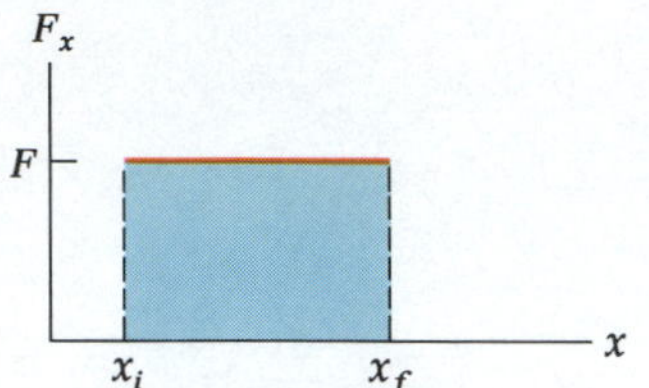

Figure 8-7. A constant force does work on an object. The work done equals the area of the rectangle, $W = F(x_f - x_i)$.

each small step, $\Delta W = F_x(\bar{x})\ \Delta x$. Since $F_x(\bar{x})\ \Delta x$ is the area of the rectangular region in Fig. 8-6*a*, we can give a graphical interpretation to the integral. Between the limits x_i and x_f, the integral for work is equal to the area under the graph of $F_x(x)$. It is the area bounded by the curve and the x axis, as shown in Fig. 8-6*b*. In that figure the sense of the displacement is to the right and the curve lies above the axis. Such an area is positive. If the curve were below the axis (force is opposite displacement), then the work (or area) would be negative. A brief summary of results of integral calculus is given in App. I.

Before investigating a specific force that varies, we first use the area-under-the-curve procedure to find the work done by a constant force. Figure 8-7 shows a graph of F_x versus x for a constant force of magnitude F between $x = x_i$ and $x = x_f$. The area under the curve is then a rectangle whose height is F and whose width is $(x_f - x_i)$, so that the work for a constant force is

$$W = F(x_f - x_i)$$

A Linear Restoring Force

Now we consider the work done by an important type of force, a ***linear restoring force.*** To a good approximation, a spring can exert a linear restoring force. Suppose a spring is fastened at one end to a fixed support and at the other end to a block that can slide on a horizontal surface. In Fig. 8-8*a*, the spring is neither stretched nor compressed; it is in its relaxed state. We let x denote the position of the block from this equilibrium position. Then x is also the amount of stretch of the spring, as shown in Fig. 8-8*b*. For many springs, the magnitude of the force is proportional to the extension x of the spring, as described by ***Hooke's law,***

$$F_x(x) = -kx$$

where k is the ***spring constant*** or the ***force constant*** of the spring.* The stiffer the spring, the larger the spring constant, and for a given x, the larger the force exerted by the spring on the block. The SI unit for the spring constant is newtons per meter (N/m).

The magnitude of the force exerted by the spring increases linearly with the amount the spring is compressed or stretched. If the spring is extended (Fig. 8-8*b*) so that x is positive, F_x is negative and **F** points toward $-\mathbf{i}$. That is, the force tends to restore the block to the equilibrium position ($x = 0$). If the spring is compressed (Fig. 8-8*c*) so that x is negative, then F_x is positive and **F** points toward $+\mathbf{i}$. Again the force tends to restore the block to the equilibrium position. Thus the spring force is a linear restoring force. This behavior for a spring obeying Hooke's law is shown graphically in Fig. 8-8*d*. For most springs the linear dependence of F_x on x is valid for displacements from equilibrium that are small compared with the length of the spring. Large displacements may even cause a spring to become permanently deformed.

The work done by the spring force can be found graphically. Hooke's law, $F_x = -kx$, shows that the force component is negative if x is positive (Fig. 8-8*d*). The work, which is equal to the area bounded by the curve, is negative for an outward displacement since the curve lies below the axis. This area is easily calculated from Fig. 8-8*e* as the sum of the area of the rectangle, $-(kx_i) \cdot (x_f - x_i)$, and the area of the triangle, $-\frac{1}{2}[k(x_f - x_i)] \cdot (x_f - x_i)$:

$$W = -\{(kx_i) \cdot (x_f - x_i) + \tfrac{1}{2}[k(x_f - x_i)] \cdot (x_f - x_i)\} = \tfrac{1}{2}k(x_f^2 - x_i^2)$$

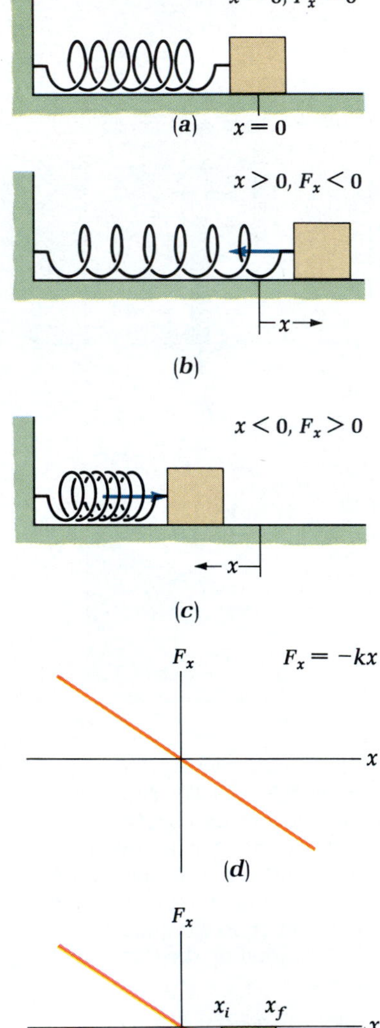

Figure 8-8. *(a), (b), (c)* A spring exerts a restoring force on the block. *(d)* The restoring force is proportional to the displacement but in the opposite direction, $F_x(x) = -kx$. *(e)* The area representing the work done by the spring equals the sum of the area of the rectangle and the area of the triangle.

* Newton's contemporary, Robert Hooke (1635–1703), is credited with first showing that the spring force is proportional to the displacement.

We also determine the work by evaluating the integral in Eq. 8-5. As the block moves from x_i to x_f, the work done by the spring force is

$$W = \int_{x_i}^{x_f} F_x(x)\, dx = \int_{x_i}^{x_f} (-kx)\, dx = -k \int_{x_i}^{x_f} x\, dx$$

where $-k$ has been factored out of the integral because it is constant. Since $\int x\, dx = \frac{1}{2}x^2$, we obtain

$$W = -\tfrac{1}{2}k(x_f^2 - x_i^2) \tag{8-6}$$

Note that the negative sign occurs because the spring force is opposite to the displacement of the block.

EXAMPLE 8-5

Work done by a spring. A block is attached to a spring of spring constant $k = 2200$ N/m and slides on a horizontal surface. Calculate the work done by the spring force on the block as it moves *(a)* from the equilibrium position $x_i = 0$ to $x_f = 0.15$ m and *(b)* from $x_i = 0.15$ m to $x_f = 0.30$ m.

Solution. From Eq. 8-6 the work done by the spring force is given by $-\frac{1}{2}k(x_f^2 - x_i^2)$. Substituting the numerical values, we obtain
(a) $W = -\frac{1}{2}(2200\text{ N/m})[(0.15\text{ m})^2 - (0)^2] = -25\text{ J}$
(b) $W = -\frac{1}{2}(2200\text{ N/m})[(0.30\text{ m})^2 - (0.15\text{ m})^2] = -74\text{ J}$
Notice that the block moves equal distances in parts *(a)* and *(b)*, but the work done by the spring is different because the force depends on x.

SELF-TEST 8-5. What work is done by the spring in the above example if the block moves from an initial position of $x_i = 0.15$ m to a final position of *(a)* 0.20 m; *(b)* 0.00; *(c)* -0.15 m? ***ANSWERS:*** *(a)* -19 J; *(b)* 25 J; *(c)* 0 J.

The General Expression for Work

In addition to dealing with a force that varies as the object moves along a straight line, we must consider a path that is curved rather than straight. A smooth curved path is illustrated in Fig 8-9*a*. While the figure appears to be in the plane of the paper, it could also represent a general curved path that the object traces in three dimensions. To evaluate the work done by a force that can vary in magnitude and direction, we again imagine dividing the path into a number of segments. A given segment can be approximated by a small displacement $\Delta\boldsymbol{\ell}$, as shown in Fig. 8-9*b*. We suppose each displacement is so small that the force vector can be considered unchanging and the path straight as the object undergoes such a minute displacement. For each segment, the element of work done by the force is $\Delta W = \mathbf{F} \cdot \Delta\boldsymbol{\ell}$. We add these contributions for each of the segments that form the path from point i to point f. That sum gives an approximation for the work, $W \approx \Sigma\mathbf{F} \cdot \Delta\boldsymbol{\ell}$.

The same type of limiting process is invoked as in the last section. We consider the limit of such sums as the number of segments tends to infinity while the length of each displacement approaches zero. The limit of that sequence of sums is defined as a *line integral*, $\int_i^f \mathbf{F} \cdot d\boldsymbol{\ell}$. This gives the general expression for the work done by a force on an object as it moves along a path from i to f.

$$W = \int_i^f \mathbf{F} \cdot d\boldsymbol{\ell} \tag{8-7}$$

As an object moves along a path, the work done by a force exerted on the object is equal to the line integral of the force along the path.

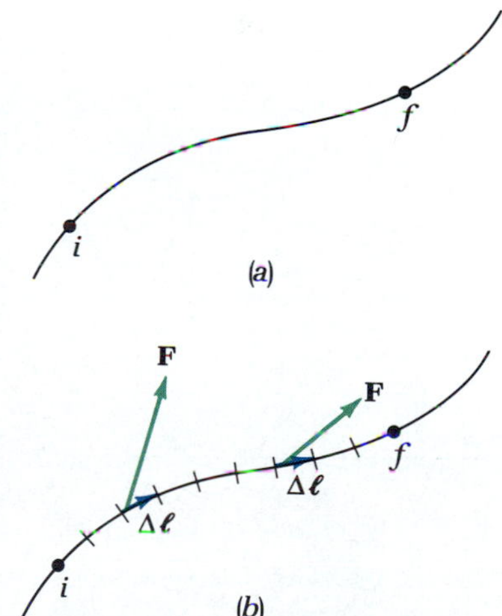

Figure 8-9. *(a)* An object can move along a general path from an initial point i to a final point f. *(b)* The path is divided into small segments. The work done by the force for a small displacement is $\Delta W = \mathbf{F} \cdot \Delta\boldsymbol{\ell}$. The work done by the force for the entire path is the sum of the values for each segment. Notice that the force can vary along the path.

The interpretation of this line integral is that the work done during an infinitesimal displacement $d\ell$ is $dW = \mathbf{F} \cdot d\ell$. The work W is obtained by integrating along the path from an initial point i to a final point f. In Exercise 19, you are asked to show that the previous expressions for work, Eqs. 8-1 and 8-5, are special cases of the general expression above.

Evaluating the line integral in Eq. 8-7 is not always an easy task. If the components of the force are known, then

$$\mathbf{F} \cdot d\ell = (F_x\mathbf{i} + F_y\mathbf{j} + F_z\mathbf{k}) \cdot (dx\,\mathbf{i} + dy\,\mathbf{j} + dz\,\mathbf{k}) = F_x\,dx + F_y\,dy + F_z\,dz$$

and Eq. 8-7 can be expressed as

Using the components to express the line integral for work

$$W = \int_i^f (F_x\,dx + F_y\,dy + F_z\,dz) \tag{8-8}$$

Each component of the force may depend on the coordinates (x, y, z), which vary as the path is traced out.

A simple but important example is to determine the work done by the weight $\mathbf{F}_e$ of an object as it moves along an arbitrary path from the point (x_i, y_i, z_i) to the point (x_f, y_f, z_f). Both points are near the earth's surface and the y axis is chosen vertically upward. A representative path for this calculation is sketched in Fig. 8-10, which shows the weight acting on the object as it undergoes a displacement $d\ell$. Other forces are also acting on the object, but only the work done by the gravitational force is considered. In using Eq. 8-8 we note that the gravitational force has only a y component, $F_{ey} = -mg$ and $F_{ex} = F_{ez} = 0$; then $\mathbf{F}_e \cdot d\ell = (0)\ dx + F_{ey}\ dy + (0)\ dz = (-mg)\ dy$. Since the force has only one component and it is constant, the integral in Eq. 8-8 reduces to a simple one-dimensional one:

$$W = \int_i^f \mathbf{F}_e \cdot d\ell = \int_{y_i}^{y_f} F_{ey}\,dy = \int_{y_i}^{y_f} (-mg)\,dy = -mg \int_{y_i}^{y_f} dy$$
$$= -mg(y_f - y_i)$$

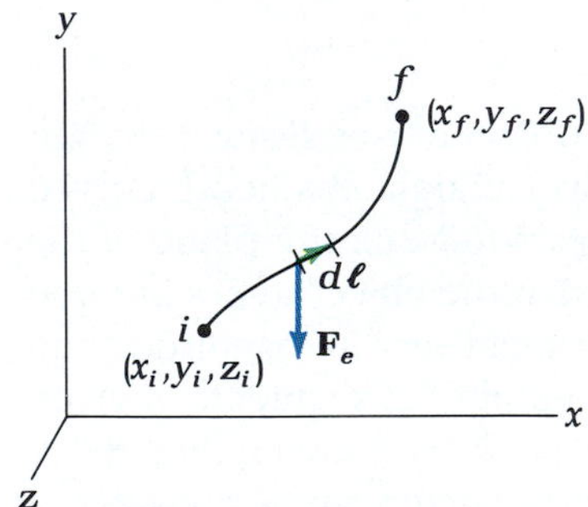

Figure 8-10. An object moves along an arbitrary path from i to f. The work done by the weight for an infinitesimal displacement is $\mathbf{F}_e \cdot d\ell$.

Since the gravitational force acts down, it does negative work on an object that goes to a higher elevation $(y_f > y_i)$ and positive work on an object that goes to a lower elevation $(y_f < y_i)$.

In the above calculation, the path was not specified in detail. The result holds for any path connecting the two endpoints i and f. In other words, ***the work done by the gravitational force is independent of the path connecting the endpoints.***

Work done by the gravitational force is independent of the path: $W = -mg(y_f - y_i)$.

In the next chapter, we shall see how this feature of the work done by the gravitational force is related to gravitational potential energy. For the present just note that the work done by the gravitational force as an object near the earth's surface moves from i to f is independent of the path and is given by

$$W = -mg(y_f - y_i) \tag{8-9}$$

From now on Eq. 8-9 will be used to evaluate the work done by the weight of an object near the surface of the earth.

In determining the work done by the constant gravitational force, it was convenient to use Eq. 8-8 since the force had only a y component and the resulting integral was simple. In other cases, as in the example below, the direct use of Eq. 8-7 for the work can be simpler. There is no clear procedure for determining whether to use Eq. 8-7 or Eq. 8-8 in a particular problem. If the calculation seems exceedingly complex using one of the forms, you should try the other. Of course, there is no guarantee that either equation will be easy to apply.

EXAMPLE 8-6

Work done by friction. A 0.40-kg puck moves in a circular path of radius 0.50 m on a horizontal tabletop, as shown in Fig. 8-11. The coefficient of kinetic friction is $\mu_k = 0.24$. Determine the work done by the frictional force as the puck moves through one-quarter of a revolution.

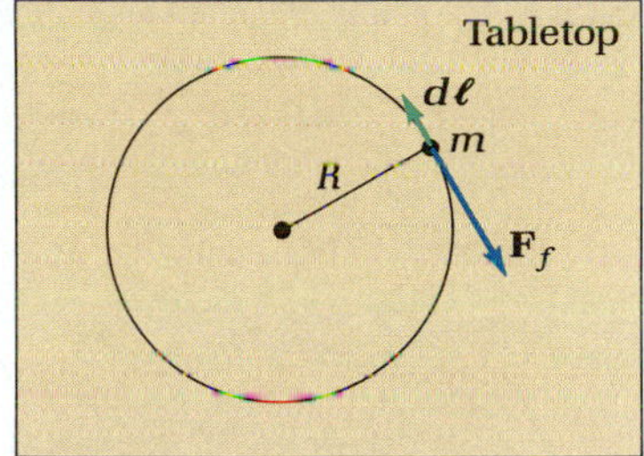

Figure 8-11. Example 8-6: A puck moves counterclockwise in a circle on a horizontal tabletop. The frictional force acts tangent to the circle and opposite to the velocity.

Solution. The normal force exerted by the tabletop on the puck balances the weight of the puck, $F_N = mg$. The magnitude of the frictional force is constant and given by $F_f = \mu_k F_N = \mu_k mg$. The direction of the frictional force changes continuously, always being opposite the velocity of the puck. For an infinitesimal displacement $d\boldsymbol{\ell}$ tangent to the circular path, we have

$$\mathbf{F}_f \cdot d\boldsymbol{\ell} = F_f \, d\ell \cos 180° = -\mu_k mg \, d\ell$$

where $d\ell$ is the magnitude of the displacement. We use Eq. 8-7 to evaluate the work:

$$W = \int_i^f \mathbf{F} \cdot d\boldsymbol{\ell} = \int_i^f (-\mu_k mg) \, d\ell = -\mu_k mg \int_i^f d\ell$$

The integral $\int_i^f d\ell = \frac{1}{4}(2\pi R) = \frac{1}{2}\pi R$ is the distance along the circular arc for one-quarter of a revolution. Thus

$$\begin{aligned} W &= -\mu_k mg \tfrac{1}{2}\pi R = -(0.24)(0.40 \text{ kg})(9.8 \text{ m/s}^2)\tfrac{1}{2}\pi(0.50 \text{ m}) \\ &= -0.74 \text{ J} \end{aligned}$$

The work done by the frictional force is negative because the frictional force is opposite the displacement in each incremental displacement.

SELF-TEST 8-6. Estimate the work done by the gravitational force on you if you *(a)* take an escalator from one level to another level 10 m higher, *(b)* take an elevator from one level to another level 10 m lower, *(c)* walk up a hill 30 m high and descend 20 m on the other side into a valley. ***ANSWERS:*** Suppose your mass is 60 kg. *(a)* −5.9 kJ; *(b)* 5.9 kJ; *(c)* −5.9 kJ.

8-4 THE WORK-ENERGY THEOREM AND KINETIC ENERGY

How do we use work? What is its value in mechanics? In this section we develop a theorem that relates work to kinetic energy, the energy of motion. This work-energy theorem provides a powerful method that connects a particle's speed with its position, no matter how complicated the motion.

Brief Derivation

First, consider the special case of an object moving along a straight line, say, the x axis, with a constant net force acting on it. From Newton's second law, the acceleration component $a_x = \Sigma F_x / m$ is also constant. As the object moves from x_i to x_f, its speed changes from v_i to v_f. From Chap. 3, we have $v_f^2 - v_i^2 = 2a_x(x_f - x_i)$, or

$$a_x(x_f - x_i) = \tfrac{1}{2}v_f^2 - \tfrac{1}{2}v_i^2$$

Now consider the work W_{net} done by the *net* force on the object. Since the motion is along the x axis and the net force is constant, the net work is $W_{net} = (\Sigma F_x)(x_f - x_i)$. Using Newton's second law, $ma_x = \Sigma F_x$, we find that

$$W_{net} = (\Sigma F_x)(x_f - x_i) = (ma_x)(x_f - x_i) = m(\tfrac{1}{2}v_f^2 - \tfrac{1}{2}v_i^2)$$

or

Work-energy theorem

$$W_{net} = \tfrac{1}{2}mv_f^2 - \tfrac{1}{2}mv_i^2 \qquad (8\text{-}10)$$

This result is called the ***work-energy theorem***.

The left-hand side of the work-energy theorem is the work done by the net force, or the net work. We can think of calculating this net work by adding all the forces acting on an object to get the net force and then determining the work done by the net force. Equivalently, we can determine the work done by each force separately and add these individual contributions to get the net work.

The right side of Eq. 8-10 is the difference of the quantity $\tfrac{1}{2}mv^2$ evaluated at initial and final points of the path. *We define $\tfrac{1}{2}mv^2$ as the **kinetic energy** K of an object of mass m with velocity* $\mathbf{v}$:

Definition of kinetic energy

$$K = \tfrac{1}{2}mv^2 \qquad (8\text{-}11)$$

Kinetic energy is the energy of motion. From Eq. 8-11, an object at rest has zero kinetic energy; a moving object has positive kinetic energy. Kinetic energy cannot be negative since its expression contains the square of the speed. According to Eq 8-10, kinetic energy has the same dimension as work, and the SI unit of energy is the joule, the same as the unit of work.

Phrasing the work-energy theorem in words shows the connection between work and kinetic energy:

> The work done by the net force acting on an object is equal to the change in the kinetic energy of that object.

In equation form,

$$W_{net} = \Delta K = K_f - K_i \qquad (8\text{-}12)$$

The kinetic energy increases if the net force on the object does positive work. The kinetic energy decreases if the net force does negative work. If the net work is zero, the kinetic energy does not change.

General Derivation

In developing Eq. 8-10, we established the work-energy theorem for the special case of one-dimensional motion with constant acceleration. We now show that the result is generally valid in three dimensions and with a varying acceleration. Consider the rate of change of the kinetic energy:

$$\frac{dK}{dt} = \frac{d}{dt}(\tfrac{1}{2}mv^2) = \frac{d}{dt}(\tfrac{1}{2}m\mathbf{v}\cdot\mathbf{v}) = \tfrac{1}{2}m\left(\frac{d\mathbf{v}}{dt}\cdot\mathbf{v} + \mathbf{v}\cdot\frac{d\mathbf{v}}{dt}\right)$$

where we used $\mathbf{v}\cdot\mathbf{v} = v^2$. The dot product is commutative, and by definition $\mathbf{a} = d\mathbf{v}/dt$ and $\mathbf{v} = d\boldsymbol{\ell}/dt$. Thus the factor in parentheses on the right can be written

$$\frac{d\mathbf{v}}{dt}\cdot\mathbf{v} + \mathbf{v}\cdot\frac{d\mathbf{v}}{dt} = 2\,\frac{d\mathbf{v}}{dt}\cdot\mathbf{v} = 2\,\mathbf{a}\cdot\mathbf{v} = 2\,\mathbf{a}\cdot\frac{d\boldsymbol{\ell}}{dt}$$

so that

$$\frac{dK}{dt} = \tfrac{1}{2}m\left(2\,\mathbf{a}\cdot\frac{d\boldsymbol{\ell}}{dt}\right) = m\mathbf{a}\cdot\frac{d\boldsymbol{\ell}}{dt}$$

or

$$m\mathbf{a} \cdot d\boldsymbol{\ell} = dK$$

Now we use Newton's second law by inserting $\Sigma\mathbf{F}$ for $m\mathbf{a}$, and we integrate the equation from an initial point i to a final point f:

$$\int_i^f (\Sigma\mathbf{F}) \cdot d\boldsymbol{\ell} = \int_i^f dK$$

The left side of this equation is the work done by the net force, and the right side is the change in the kinetic energy:

$$W_{\text{net}} = K_f - K_i \qquad \text{or} \qquad W_{\text{net}} = \tfrac{1}{2}mv_f^2 - \tfrac{1}{2}mv_i^2$$

Again, we arrive at the work-energy theorem.

The examples at the end of this section illustrate the problem-solving capabilities of the work-energy theorem. With it we are able to treat situations that would otherwise require great computational effort. It should be pointed out, however, that the description of motion obtained from the theorem is a partial one. It essentially connects values of speeds with positions along the path. There is no explicit reference to time. That is, we cannot determine the time dependence of the velocity and position by using the work-energy theorem alone. Even though the description is a partial one, it can increase our understanding of the motion of a system both quantitatively and qualitatively.

EXAMPLE 8-7

A roller coaster. A roller-coaster car starts from rest at the top and moves down the curved track (Fig. 8-12). Determine its speed as it reaches the bottom. Assume that the work done by frictional forces is negligible.

Solution. Since frictional effects are neglected, we consider only two forces acting on the car. One is the gravitational force exerted by the earth; the other force is the normal force exerted by the track that constrains the car to move along the track. The value of the normal force cannot be easily determined since the track is not straight. However, the normal force performs no work in this situation. For any infinitesimal displacement $d\boldsymbol{\ell}$ of the car tangent to the surface, the normal force $\mathbf{F}_N$ is perpendicular to that displacement, $\mathbf{F}_N \cdot d\boldsymbol{\ell} = F_N\, d\ell \cos 90° = 0$. Only the weight of the car does work in this case. Rather than calculating this work explicitly, we recall Eq. 8-9 and the discussion surrounding it. The work done by the gravitational force, $W_g = -mg(y_f - y_i)$, is independent of the path. It depends only on the vertical separation $y_f - y_i$ of the initial and final points. The net work is just the work done by the weight of the car, $W_{\text{net}} = -mg(y_f - y_i)$. Since the car started from rest, $v_i = 0$ and the work-energy theorem gives

$$-mg(y_f - y_i) = \tfrac{1}{2}mv_f^2 - \tfrac{1}{2}m(0)^2 = \tfrac{1}{2}mv_f^2$$

Noting from the figure that the height of the track is $h = y_i - y_f$, we can solve for v_f^2.

$$v_f^2 = -2g(y_f - y_i) = 2gh$$

and

$$v_f = \sqrt{2gh}$$

Note that the final speed at the bottom is independent of the shape of the curved track. In fact, the speed would be the same if there were no track at all and the car were released from rest to fall straight down from a height h. The time required to reach bottom and the direction of the velocity, however, do depend on the shape of the track.

(a)

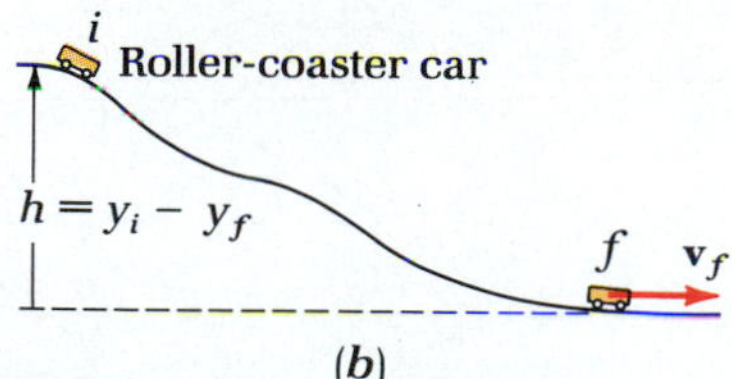

(b)

Figure 8-12. Example 8-7.

SELF-TEST 8-7. Suppose the car in the above example has a speed of 5.0 m/s as it passes the top of the track at a height of 14 m above the ground. Determine its speed as it passes a point that is 10 m above the ground. ***ANSWER:*** 10 m/s.

EXAMPLE 8-8

A braking car. The driver of a 1200-kg automobile cruising at 18 m/s on a level avenue suddenly brakes. The wheels lock and the automobile skids, coming to a stop after traveling 25 m. *(a)* What work is done on the automobile by the frictional force exerted by the road surface? *(b)* Determine the value of the frictional force, assuming it to be constant.

Solution. From the work-energy theorem, we can immediately determine the net work:

$$W_{net} = \tfrac{1}{2}mv_f^2 - \tfrac{1}{2}mv_i^2 = \tfrac{1}{2}m(0)^2 - \tfrac{1}{2}mv_i^2$$
$$= -\tfrac{1}{2}(1200 \text{ kg})(18 \text{ m/s})^2 = -190 \text{ kJ}$$

The three relevant forces are the weight, the normal force, and the frictional force. Of these, only the frictional force does work in this case. (Be sure that you understand why the other two forces do no work!) *(a)* The net work is done by friction, and

$$W_f = W_{net} = -190 \text{ kJ}$$

(b) For a constant frictional force and a displacement along a straight path the force is opposite the displacement, and the work is

$$W_f = \mathbf{F}_f \cdot \Delta\boldsymbol{\ell} = F_f(25 \text{ m}) \cos 180° = F_f(25 \text{ m})(-1)$$

Then,

$$F_f = -\frac{W_f}{25 \text{ m}} = -\frac{-190 \text{ kJ}}{25 \text{ m}} = 7.8 \text{ kN}$$

SELF-TEST 8-8. Determine how much work has been done by the frictional force acting on the car in the above example when the car has *(a)* skidded 12.5 m, which is half the final distance, and *(b)* slowed to 9 m/s, which is half the initial speed. ***ANSWERS:*** *(a)* −97 kJ; *(b)* −150 kJ.

EXAMPLE 8-9

A sled on a hill. A child's sled starts from rest at the top of an icy hill (Fig. 8-13). The portion of the path from f to q is circular, with radius R. Neglect any frictional effects. *(a)* Determine the speed of the sled at f, the lowest point in the path. *(b)* What normal force is exerted by the ice on the sled at this point? *(c)* What are the speed and normal force at point q?

Solution. First we consider the work done by the force exerted by the frictionless icy surface on the sled. That work is zero since the normal force $\mathbf{F}_N$ is perpendicular to a infinitesimal displacement $d\boldsymbol{\ell}$ of the sled tangent to the path, $\mathbf{F}_N \cdot d\boldsymbol{\ell} = 0$. The only other force is the weight of the sled. Its work is given by Eq. 8-9 and is independent of the path. Since the normal force does no work, the net work is the work W_e done by the gravitational force.

(a) For the motion from i to f, $W_{net} = W_e = -mg(y_f - y_i)$ and the difference $y_f - y_i = -2R$. The work-energy theorem gives

$$-mg(-2R) = \tfrac{1}{2}mv_f^2 - \tfrac{1}{2}m(0)^2 = \tfrac{1}{2}mv_f^2$$

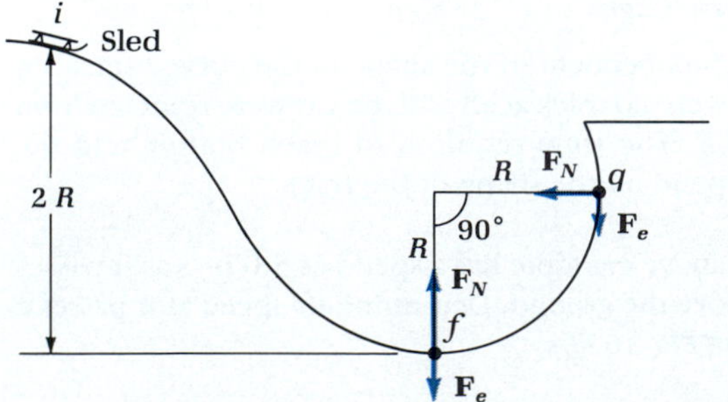

Figure 8-13. Example 8-9: A sled slides on a smooth, icy surface. Frictional forces are negligible.

Solving for v_f gives

$$v_f = \sqrt{4gR}$$

(b) From the figure we can see that the net force on the sled at position f has magnitude $F_N - F_e = F_N - mg$ and is directed toward the center of the circle. This net force is the centripetal force, and from Newton's second law, its magnitude must be mv_f^2/R. (Recall the discussion of circular motion in Chap. 6.) We have then $F_N - mg = mv_f^2/R$, or

$$F_N = mg + \frac{mv_f^2}{R}$$

From part *(a)* above, $v_f^2 = 4gR$ so that $mv_f^2/R = 4mg$. At the bottom of the hill, the normal force exerted on the sled is

$$F_N = mg + 4mg = 5mg$$

or five times the weight of the sled!

(c) As the sled slides from i to q, the work done by the net force is again due only to the gravitational force and is $-mg(y_q - y_i) = mgR$. Applying the work-energy theorem as before gives $mgR = \frac{1}{2}mv_q^2$ so that $v_q = \sqrt{2gR}$. When the sled passes through position q, the normal force alone provides the centripetal force since the weight is tangential at this point, $F_N = mv_q^2/R$. Since $v_q = \sqrt{2gR}$ and $v_q^2 = 2gR$, we have

$$F_N = \frac{mv_q^2}{R} = \frac{m(2gR)}{R} = 2mg$$

SELF-TEST 8-9. In the latter part of the above example, the work-energy theorem was applied to the points i and q in the motion of the sled. What are the answers for the speed of the sled and the normal force acting on it at point q if you apply the work-energy theorem to points f and q? ***ANSWER:*** $v_q = \sqrt{2gR}$, $F_N = 2mg$.

8-5 POWER

"Power" is another of those commonly used words that has a much stricter meaning in physics than in everyday conversation. In physics, power relates work to the time interval in which it is done. ***Power is the time rate at which work is performed.*** In a machine, work is often done at a steady rate, so that the machine is conveniently characterized by its power.

Suppose you do 200 J of work on a box as you slide it across the floor. If the displacement occurs in a time interval $\Delta t = 5$ s, the average rate at which you perform work is 200 J/5 s = 40 J/s. We define the ***average power*** $\bar{P}$ for a time interval Δt during which work ΔW is performed as

$$\bar{P} = \frac{\Delta W}{\Delta t} \tag{8-13}$$

Average power

The SI unit of power is the watt (W), with 1 W = 1 J/s. The watt is named after James Watt (1735–1819), who made significant improvements to the steam engine. Watt introduced the idea of the horsepower as a unit of power to characterize the rate at which these engines performed work. The horsepower (hp) is now defined as 1 hp = 746 W. In the simple calculation above, the average power is 40 W = 0.05 hp.

Rather than dealing with average power, we often use the ***instantaneous power,*** or simply ***power,*** defined as the limit of Eq. 8-13 as Δt approaches zero:

$$P = \frac{dW}{dt} \tag{8-14}$$

Definition of power

Power is the instantaneous rate at which work is performed.

An alternative expression for power can be developed in terms of the force that does the work and the velocity of the object. Suppose in a small time interval Δt, a force $\mathbf{F}$ acts on an object as it is displaced by $\Delta \boldsymbol{\ell}$. Since $\Delta W = \mathbf{F} \cdot \Delta \boldsymbol{\ell}$, the average power is given by

$$\overline{P} = \frac{\mathbf{F} \cdot \Delta \boldsymbol{\ell}}{\Delta t} = \mathbf{F} \cdot \frac{\Delta \boldsymbol{\ell}}{\Delta t}$$

We take the limit as Δt approaches zero and note that $\Delta \boldsymbol{\ell}/\Delta t \rightarrow \mathbf{v}$, the velocity of the object. This gives the power as the dot product of the force and the velocity:

Power in terms of force and velocity

$$P = \mathbf{F} \cdot \mathbf{v} \tag{8-15}$$

Most electrical devices have a power rating. This rating is the rate of conversion of electric energy under normal operating conditions. We can determine from the power rating the consumption of electric energy during a given time interval; the energy used, ΔE, is the product of the power with the time interval: $\Delta E = P\,\Delta t$. A useful energy unit can be introduced here. Suppose the power is expressed in kilowatts (kW) and the time interval in hours (h). The product $P\,\Delta t$ has the dimension of work or energy and the unit kilowatt-hour, abbreviated kW·h. It is the energy unit commonly used by electric utilities.

The energy unit, kW·h

EXAMPLE 8-10

Powering an elevator. An elevator cable pulls a fully loaded elevator upward at a constant speed of 0.75 m/s. The power delivered by the cable is 23 kW. What is the tension in the cable?

Solution. The force $\mathbf{F}_c$ exerted by the cable on the elevator is parallel to the velocity $\mathbf{v}$. Then the dot product in Eq. 8-15 is just the product of the magnitudes:

$$P = \mathbf{F}_c \cdot \mathbf{v} = F_c v \cos 0 = F_c v$$

Solving for F_c, gives

$$F_c = \frac{P}{v} = \frac{23\ \text{kW}}{0.75\ \text{m/s}} = 31\ \text{kN}$$

SELF-TEST 8-10. What work is done on the elevator in the above example by the cable force in a 4.0-s time interval? ***ANSWER:*** 92 kJ.

Exposed elevator.

EXAMPLE 8-11

The cost of electric energy. A 1.0-hp electric motor (1 hp = 746 W) operates a pump continuously. How much work is performed by the motor in one day and at what cost? Let the price of electric energy be \$0.12 per kW·h.

Solution. Since 1 hp = 0.746 kW, the work performed in a 24-h time interval is

$$\Delta W = (0.746 \text{ kW})(24 \text{ h}) = 18 \text{ kW}\cdot\text{h}$$

The cost for one day's operation is

$$(18 \text{ kW}\cdot\text{h})(\$0.12 \text{ kW}^{-1}\text{ h}^{-1}) = \$2.10$$

SELF-TEST 8-11. For the motor in the above example, express the power rating in watts and the work done in a day in joules. What is the cost per joule of the electric energy?
ANSWER: $P = 746$ W, \$$3.3 \times 10^{-8}$ per J.

8-6 INTEGRATION, NUMERICAL METHODS

We have used the concept of the integral in evaluating the work done by a variable force. In simple cases we can perform the integration analytically, as in Sec. 8-3, perhaps with help from App. I. We also encounter integrals that are not so easy to do.

Some integrals can only be evaluated by numerical methods. A simple technique for the numerical evaluation of an integral can be developed by following the discussion of the integral in Sec. 8-3. There we divided the interval into a large number of small subintervals, each of extent Δx. The function $F_x(x)$ to be integrated is evaluated at the midpoint of each subinterval and multiplied by Δx. This gives the area of the rectangle of height $F_x(\bar{x})$ and base Δx. Adding such contributions, we have an approximation to the integral:

$$\int_{x_i}^{x_f} F_x(x)\,dx \approx \Sigma F_x(\bar{x})\,\Delta x \tag{8-16}$$

The approximation should improve as Δx is made smaller. Such a sum of terms can be easily evaluated using a computer. The following problem can be solved in this way:

EXAMPLE 8-12

Work done by a nonlinear force. The rubber band shown in Fig. 8-14 exerts a restoring force given by $F_x(x) = -kx \tanh|x/a|$, with $k = 1100$ N/m and $a = 0.050$ m. The hyperbolic tangent function is

$$\tanh z = \frac{\sinh z}{\cosh z} = \frac{e^z - e^{-z}}{e^z + e^{-z}}$$

This is an example of a restoring force, similar to a spring force, but this rubber-band force is nonlinear. Calculate the work done by this force on your finger as it stretches the band from $x = 0$ m to $x = 0.10$ m.

Solution. We divide the interval from $x = 0.00$ m to $x = 0.10$ m into 50 subintervals of length 0.002 m. The function is to be evaluated at the midpoint of each subinterval and then multiplied by Δx and summed. Table 8-1 shows a portion of the formulas in a spreadsheet that performs this numerical integration. The values in the corresponding cells are shown in Table 8-2. Each entry in column A lists the midpoint $\bar{x}$ of a subinterval, beginning with 0.001 m. The exponential function $e^{x/a}$ is evaluated in column B. Column C evaluates the force component $F_x(x) = -kx \tanh(x/a)$, and column D gives the area of the rectangle,

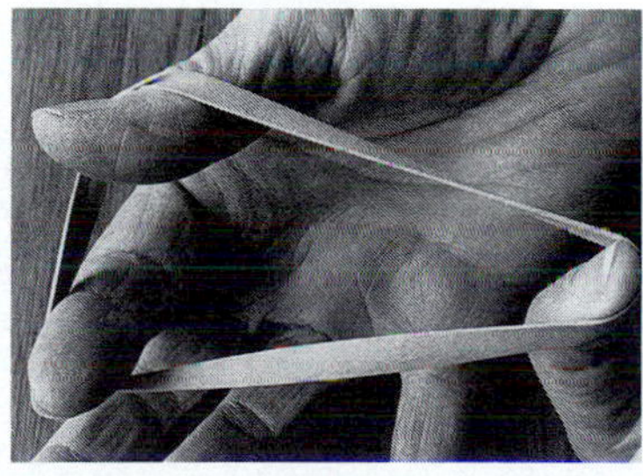
(a)

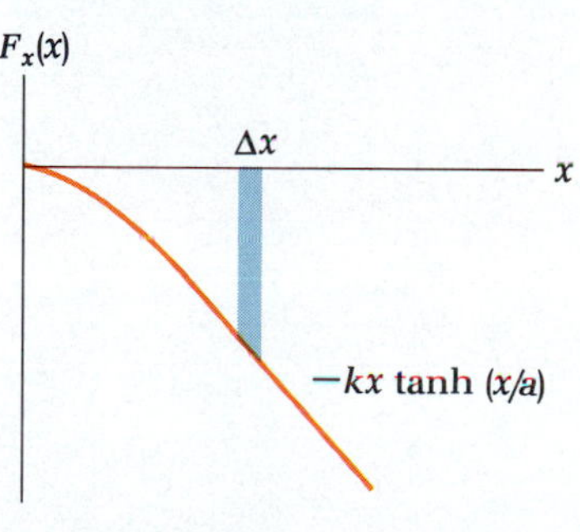

(b)

Figure 8-14. Example 8-12: *(a)* The rubber band exerts a force on the finger. *(b)* A graph of the force exerted by the rubber band.

TABLE 8-1. ***Formulas in a Spreadsheet***

	A	B	C	D
1	X	EXP(X/0.05)	FORCE	WORK
2	0.001	@EXP(A2/0.05)	−1100*A2*(B2−1/B2)/(B2+1/B2)	+C2*0.002
3	0.003	@EXP(A3/0.05)	−1100*A3*(B3−1/B3)/(B3+1/B3)	+C3*0.002
4	0.005	@EXP(A4/0.05)	−1100*A4*(B4−1/B4)/(B4+1/B4)	+C4*0.004
⋮	⋮	⋮	⋮	⋮
51	0.099	@EXP(A51/0.05)	−1100*A51*(B51−1/B51)/(B51+1/B51)	+C51*0.002
52				@SUM(D2..D51)

TABLE 8-2. ***Values in a Spreadsheet***

	A	B	C	D
1	X	EXP(X/0.05)	FORCE	WORK
2	0.001	1.020201	−0.022	−4.4E−05
3	0.003	1.061837	−0.19776	−0.0004
4	0.005	1.105171	−0.54817	−0.0011
⋮	⋮	⋮	⋮	⋮
51	0.099	7.242743	−104.826	−0.20965
52				−4.4938

$\Delta W = F_x(\bar{x})\ \Delta x$. Cell D52 contains the sum of the areas of the rectangles, which is an approximation to the integral. The work done by the rubber band is $W = -4.5$ J.

One way to assess the accuracy of our numerical evaluation of the integral is to modify the spreadsheet to use a larger number of smaller subintervals. If the resulting answer is not significantly different from the previous one, you may be satisfied with the approximation. For example, suppose the interval from $x = 0$ m to $x = 0.010$ m is divided into 100 subintervals of length 0.001 m. If the sum of the rectangular areas is still −4.5 J when rounded to two significant figures, then you can expect that the approximation to the integral is good to this same precision.

SELF-TEST 8-12. Modify the spreadsheet in the above example to use 100 subintervals of length 0.001 m. Compare the approximation for the work done with the result given in the example, which used 50 subintervals of length 0.002 m. What is your estimate of the accuracy of the approximation for the work W?

COMMENTARY: Work and Energy

What is energy? What is work? Can you give a rigorous, one-sentence definition of each of these? Good definitions are not easy to formulate, and seldom is a good definition also a good description. There is the old story of Plato's attempt to define a man as "a two-legged animal without feathers." The inadequacy of this definition became obvious when Diogenes displayed a plucked chicken as an example of Plato's man.

In elementary science books, we often see energy defined as "the ability to do work," and work as "a force acting through a distance." Like Plato's definition of a man, these definitions are descriptive, but they are hardly incisive. We can immediately think of examples, like plucked chickens, that cause us to put aside such definitions. We should not expect the definition of a quantity, such as work, also to serve as a description of it or to give us an intuitive feeling for the concept. Rather, our understanding comes with practice; we apply a rigorous definition to a variety of situations. With each application, the concept becomes clearer—it matures.

Take work, for example. The general expression for work is given by the rather formidable integral in Eq. 8-7. By looking at some simple cases, such as a constant force and a straight path, we see how work depends on the relative directions of force and displacement; the value of the work can be positive, negative, or zero. We find that the work done by the weight of an object is positive if the object moves downward, negative if it moves upward, and zero if it moves horizontally. With these and similar calculations, the concept of work becomes more familiar, a part of our experience.

The real understanding of work comes when we see its connection with kinetic energy, through the work-energy theorem. Here we relate work done by the net force to a change in the object's kinetic energy. For example, if the kinetic energy decreases, then negative work is done by the net force on the object. As the object slows, it can do positive work on some agent in its environment. In a hydroelectric generator, water slows as it does work in turning a turbine blade. In this sense the kinetic energy of an element of water represents its "ability to do work."

We have just begun our look at energy by identifying and defining kinetic energy, an energy of motion. There are other kinds of energy, and we shall identify some of them in the next chapter. It will take some time and effort to develop a good understanding of energy, but the payoff is enormous. The concept of energy extends to all the natural sciences, and, like the common denominator in fractions, allows us to combine and simplify our descriptions of diverse phenomena.

SUMMARY

Sections 8-1 and 8-3. Work Done by a Constant Force and by a Variable Force

The work done on an object by a force acting on it as the object moves along its path is defined in general as the line integral,

$$W = \int_i^f \mathbf{F} \cdot d\boldsymbol{\ell} \qquad (8\text{-}7)$$

In the simple case of a constant force and a displacement $\Delta\boldsymbol{\ell}$ along a straight-line path, the work is given by $\mathbf{F} \cdot \Delta\boldsymbol{\ell}$. For a variable force in one dimension,

$$W = \int_{x_i}^{x_f} F_x(x)\, dx \qquad (8\text{-}5)$$

The work done by the weight of an object near the surface of the earth is

$$W_e = -mg(y_f - y_i) \qquad (8\text{-}9)$$

and is independent of the path connecting the initial and final points. A stretched or compressed spring exerts a restoring force given by Hooke's law, $F_x = -kx$. The work done by the spring is

$$W = -\tfrac{1}{2}k(x_f^2 - x_i^2) \qquad (8\text{-}6)$$

Section 8-2. The Dot Product

The dot product of any two vectors is defined as a scalar quantity:

$$\mathbf{A} \cdot \mathbf{B} = AB\cos\theta \qquad (8\text{-}2)$$

where θ is the angle between the directions of **A** and **B**. The dot product can be expressed in terms of the components of the vectors as

$$\mathbf{A} \cdot \mathbf{B} = A_xB_x + A_yB_y + A_zB_z \qquad (8\text{-}4)$$

Section 8-4. The Work-Energy Theorem and Kinetic Energy

The kinetic energy of an object of mass m with speed v is $K = \frac{1}{2}mv^2$. It is the energy of motion. The work-energy theorem equates the work done by the net force acting on an object to the change in its kinetic energy,

$$W_{\text{net}} = \tfrac{1}{2}mv_f^2 - \tfrac{1}{2}mv_i^2 \qquad (8\text{-}10)$$

Section 8-5. Power

Power is the rate at which work is performed by a force: $P = dW/dt$. The power of a force doing work on an object with velocity **v** is $P = \mathbf{F} \cdot \mathbf{v}$.

Section 8-6. Integration, Numerical Methods

An integral can be approximated by a sum,

$$\int_{x_i}^{x_f} F_x(x)\, dx \approx \Sigma F_x(\bar{x})\, \Delta x \qquad (8\text{-}16)$$

which can be evaluated numerically.

QUESTIONS

1. An object slides on a stationary surface. Can the work done on that object by the kinetic frictional force be positive? Negative? Zero? Does your answer depend on your reference frame? Explain.

2. Can a static frictional force acting on some object perform work? If so, under what circumstances? If not, why not?

3. Can the normal force exerted by a surface on an object do work on that object? If so, under what circumstances?

4. Suppose a crate is pushed across a warehouse floor from one end to the other at constant speed. How does the work done by the frictional force on the crate, if it moves in a straight path, compare with the work for a curved path?

5. If $\mathbf{A} \cdot \mathbf{B} = 0$, is it necessary either that $A = 0$ or that $B = 0$? Explain.

6. Suppose that $-AB < \mathbf{A} \cdot \mathbf{B} < 0$. What can you conclude about the directions of **A** and **B**?

7. Can a force do any work on an object if the force is always perpendicular to the velocity of that object? Explain.

8. Can a force do any work on an object if the force is always perpendicular to the acceleration of that object? (Is it possible for a force to be perpendicular to the acceleration?) Explain.

9. Suppose that the speed of a baseball is doubled when struck by a bat. By what factor does the ball's kinetic energy change?

10. Is it possible for an object to have a negative value of kinetic energy? Explain.

11. A satellite moves in a circular orbit about the earth's center. The gravitational force of the earth provides the centripetal force on the satellite. How much work is done by the gravitational force on the satellite?

12. A satellite moves in an elliptical orbit about the earth. The gravitational force of the earth on the satellite is directed toward the center of the earth. Does the kinetic energy of the satellite change? Explain.

13. Is the work-energy theorem consistent with Newton's first law? Can you expect the work-energy theorem to be valid in a noninertial reference frame? Explain.

14. You may take either the escalator or an elevator from the first to the second floor of a department store. For these two routes, compare the values of the work done on you by the gravitational force.

15. In a British system of units, the foot (ft) is a unit of length and the pound (lb) is a unit of force. What is a unit of work in this system? What is a unit of kinetic energy in this system?

16. If object A exerts a force on object B, then B exerts an equal but opposite force on A (Newton's third law). How can any net work be done on an object if $\mathbf{F}_{BA} = -\mathbf{F}_{AB}$?

17. How much work is done by a spring on the connected block in Fig. 8-8 if the block returns to its starting point? Explain.

18. The horsepower (1 hp = 746 W) is a unit of power that was based on an estimate of the rate at which a horse could do work. While a horse cannot steadily maintain such a power, you can maintain about 5 percent of that for an hour or so. Suppose that we call 37 W a "manpower." Is this how the word is used in commerce? Explain.

19. Does an electric utility sell power or energy, or both? Instead of being called a power bill, should it be called an energy bill? If you don't pay the bill, will the electric utility cut off power or energy, or both?

20. Complete the following table:

Symbol	Represents	Type	SI Unit
W			
K			
$\mathbf{A} \cdot \mathbf{B}$			—
W_{net}	Work done by net force		
P		Scalar	
ℓ			m

EXERCISES

Section 8-1. Work Done by a Constant Force

1. Suppose that you lift a 4-kg book from the floor to a shelf 2 m high. *(a)* What force must you apply to move the book at constant velocity? *(b)* What work is done by this force?

2. *(a)* What force must you apply to a 4-kg book to carry it slowly at constant velocity from one shelf to an adjoining one 3 m away but at the same level? *(b)* How much work is done by this force?

3. Show that the joule is equivalent to $kg \cdot m^2 \cdot s^{-2}$.

4. A 16-kg sled is pulled by a rope over wet snow for a horizontal distance of 3.2 m (Fig. 8-15). The tension in the rope remains constant at 5.8 N, and the rope is at 37° from the horizontal. Determine the work done by the rope on the sled.

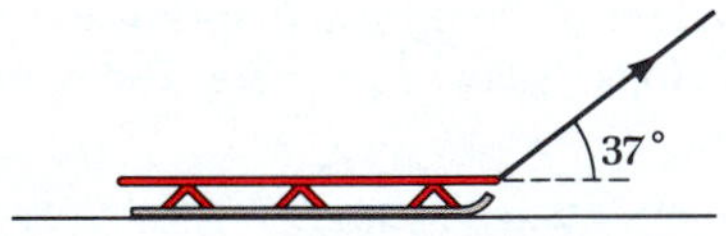

Figure 8-15. Exercise 4.

5. Suppose that the sled in the previous exercise is moving with constant velocity. Determine *(a)* the work done by the frictional force on the sled and *(b)* the coefficient of kinetic friction at the snow-sled interface.

6. A tow plane uses a light cable to pull a glider along a straight path at a constant speed of 280 km/h. If the tension in the cable is 1400 N, how much work is done by the cable force on the glider during 24 min of flight?

Sections 8-2. The Dot Product

7. Prove the inequality, $-AB \leq \mathbf{A} \cdot \mathbf{B} \leq AB$, where **A** and **B** are any two vectors.

8. An object moves 4.2 m along a straight line while a constant force of magnitude 9.8 N acts on it. The work done on the object by this force is -31 J. What is the angle between the force and displacement vectors?

9. An object moving in a straight line has a displacement $(2\text{ m})\mathbf{i} + (3\text{ m})\mathbf{j} - (5\text{ m})\mathbf{k}$ while a constant force $(7\text{ N})\mathbf{i} - (7\text{ N})\mathbf{j} - (2\text{ N})\mathbf{k}$ acts. Evaluate *(a)* the work done by this force and *(b)* the angle between the two vectors.

10. By using the projective interpretation of the dot product, *(a)* construct a graphical or geometric illustration of its distributive property, $\mathbf{A} \cdot (\mathbf{B} + \mathbf{C}) = \mathbf{A} \cdot \mathbf{B} + \mathbf{A} \cdot \mathbf{C}$. *(b)* If s is a scalar, show $\mathbf{A} \cdot (s\mathbf{B}) = s(\mathbf{A} \cdot \mathbf{B})$.

11. Two vectors are arranged head to tail with an angle θ between their directions (Fig. 8-16). Let $\mathbf{C} = \mathbf{A} + \mathbf{B}$, form $\mathbf{C} \cdot \mathbf{C}$, and prove one form of the law of cosines:

$$C = \sqrt{A^2 + B^2 + 2AB \cos \theta}$$

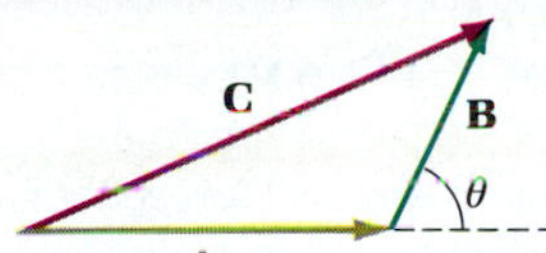

Figure 8-16. Exercise 11.

12. ***Direction cosines.*** A vector in three dimensions is expressed in component form, $\mathbf{F} = F_x\mathbf{i} + F_y\mathbf{j} + F_z\mathbf{k}$. *(a)* Find the projection of $\mathbf{F}$ onto each of the coordinate axes by showing that $\mathbf{i} \cdot \mathbf{F} = F_z$, $\mathbf{j} \cdot \mathbf{F} = F_y$, $\mathbf{k} \cdot \mathbf{F} = F_z$. *(b)* These dot products can also be expressed in terms of the angles that $\mathbf{F}$ makes with the x, y, and z coordinate axes—call them α, β, and γ, respectively. That is, $\mathbf{i} \cdot \mathbf{F} = F \cos \alpha$, etc. Prove the identity: $\cos^2 \alpha + \cos^2 \beta + \cos^2 \gamma = 1$. The quantities $\cos \alpha$, $\cos \beta$, and $\cos \gamma$ are called the ***direction cosines*** of the direction of the vector $\mathbf{F}$ and serve to determine its direction.

Section 8-3. Work Done by a Variable Force

13. A block is attached to a spring of spring constant $k = 2100$ N/m and moves from the equilibrium position out to $x = 0.14$ m. *(a)* How much work is done by the spring force? *(b)* Determine the minimum and maximum magnitudes of the spring force exerted on the block for this motion.

14. An automobile spring is compressed 15 mm by applying a force of magnitude 450 N. *(a)* What is the compression if a force of magnitude 2250 N is applied? *(b)* Determine the spring constant of this spring. *(c)* How much work is done by the spring if it is compressed 15 mm from its relaxed length? *(d)* What additional work is done by the spring if it is compressed an additional 60 mm? *(e)* On what object is work done by the spring?

15. Let the y axis be vertical so that the weight of an object has only a y component, $F_y = -mg$. An object moves along the y axis from y_i to y_f. Show that the work done by the weight is $W_g = -mg(y_f - y_i)$. Compare with Eq. 8-9.

16. A particle moving along the x axis is subjected to a force given by $F_x(x) = F_0\,(e^{x/a} - 1)$, where F_0 and a are constants. *(a)* Determine an expression for the work done by this force as the particle moves from the origin to the point x_1. *(b)* Let $F_0 = 2.5$ N and $a = 0.20$ m. Evaluate the work done if $x_1 = 0.50$ m.

17. Suppose that an object moving along the z axis is acted on by a force given by $F_z(z) = -C/z^2$, where C is a constant. Taking both z_i to z_f as positive, obtain the expression for the work done by this force as the object moves from z_i to z_f.

18. The graph of a coordinate dependent force is shown in Fig. 8-17. Determine from the graph the work done by that force on a particle that moves from 0 to 2.0 m.

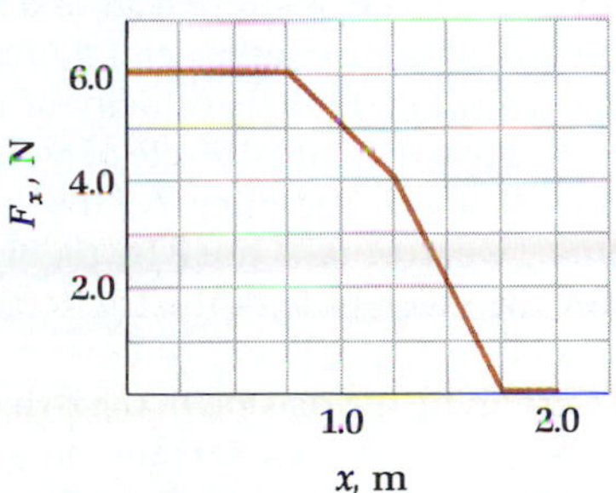

Figure 8-17. Exercise 18.

19. Apply Eq. 8-7 to two special cases: *(a)* If the force $\mathbf{F}$ is constant, show that the line integral gives Eqs. 8-3 and 8-1. *(b)* If the path is along the x axis, show that Eq. 8-5 results.

20. A 1.5-kg ball attached to a light string is whirled in a *horizontal* circle of radius 0.75 m. *(a)* How much work is done by the earth's gravitational force on the ball as it moves halfway around the circle. *(b)* How much work is done by the tension force in the string?

21. A 1.5-kg ball attached to a light string is whirled in a *vertical* circle of radius 0.75 m. *(a)* Evaluate the work done by the earth's gravitational force as the ball moves from the highest point to the lowest point in the circle. *(b)* How much work is done by the tension force in the string?

Section 8-4. The Work-Energy Theorem and Kinetic Energy

22. From $K = \frac{1}{2}mv^2$, show that the SI unit of kinetic energy is the joule.

23. What is the kinetic energy of *(a)* a 1100-kg automobile traveling at 45 km/h, *(b)* a 550-kg subcompact at 90 km/h?

24. A 95-kg crate, given an initial speed of 3.5 m/s, slides across a warehouse floor and comes to rest after traveling 2.3 m. *(a)* Determine the work done by the frictional force. Assume a constant frictional force and determine, *(b)* its magnitude and *(c)* the coefficient of kinetic friction.

25. A 0.015-kg marble is loaded into a spring gun (Fig. 8-18). The spring has a spring constant $k = 120$ N/m and is compressed by 0.12 m. When the spring is released, the marble is shot from the barrel. Neglecting all frictional effects, determine the speed of the marble as it leaves the barrel.

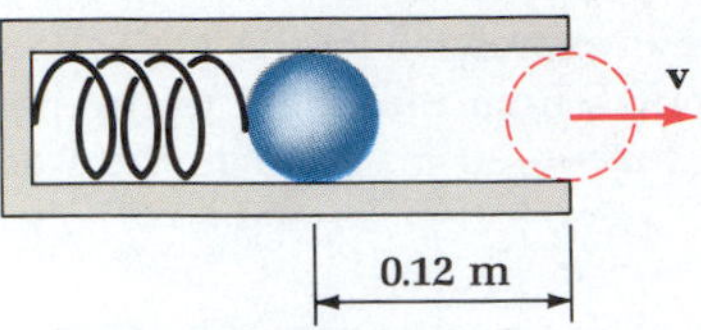

Figure 8-18. Exercise 25.

26. A 70-kg skydiver falls straight down through the air at a constant speed of 140 km/h. During a time interval of 120 s, what work is done by *(a)* the net force, *(b)* the gravitational force, *(c)* the air-resistance force?

27. One end of a light string is slipped around a peg fixed in a horizontal tabletop, while the other end is tied to a 0.50-kg puck. The puck is given an initial velocity of magnitude 3.4 m/s so that it moves in a horizontal circle of radius 0.75 m. The object comes to rest after completing 2.5 revolutions. *(a)* For the entire motion, what work is done by the frictional force? *(b)* Assume that the magnitude of the frictional force is constant and determine the coefficient of kinetic friction at the interface. *(c)* Determine the tension in the string at the instant that the puck completes the first revolution. *(d)* How much work is done by the tension force in the string?

28. A 1.5-g bullet with speed 420 m/s penetrates 0.14 m into a stationary wooden block. *(a)* What work is done by the block in stopping the bullet? *(b)* Estimate the magnitude of the stopping force exerted on the bullet by the wooden block.

29. A 0.37-kg ball is thrown straight up with an initial velocity of 14 m/s. It rises to a maximum height of 8.4 m. *(a)* What work is done by air resistance on the ball? *(b)* Assume that air resistance does about the same work on the downward trip. Estimate the speed of the ball as it returns to its starting point.

30. A 0.35-kg ball, attached to a light string, is initially at position *I* in Fig. 8-19. It is given an initial downward velocity of magnitude 5.0 m/s and swings in a circular arc of radius $R = 0.80$ m in a vertical plane. Neglecting frictional effects, determine the speed of the ball and the tension in the string at position *(a) A; (b) B; (c) C*.

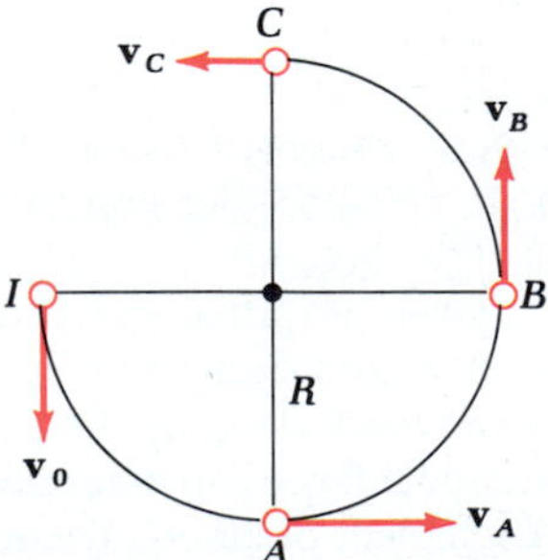

Figure 8-19. Exercise 30.

31. What minimum initial speed must the ball in the previous exercise be given so that it completes the circle in its motion? (*Hint:* The string will just become slack when the ball reaches the highest point in the circle.)

32. A 15-kg toolbox rests on the horizontal bed of a pickup truck. The truck and toolbox begin moving with a constant acceleration of 2.5 m/s² while covering a distance of 18 m. *(a)* Evaluate the kinetic energy of the toolbox at the end of this period. *(b)* What work was done by the static frictional force on the toolbox? *(c)* What is the minimum value of the coefficient of static friction for these surfaces?

33. Suppose that the truck in the previous exercise has a smoother bed. As the truck accelerates from rest at 2.5 m/s², the toolbox slides with a coefficient of kinetic friction of 0.20. It starts from rest at the front and slides until it hits the tailgate of the 2.0-m bed. What work is done by the frictional force on the toolbox for this part of the motion? Use an inertial reference frame at rest relative to the road. (*Caution:* With respect to this roadside reference frame, the displacement of the toolbox is not 2.0 m.)

Section 8-5. Power

34. Show that 1 kW · h = 3.6 MJ.

35. A log is pulled across a level forest floor at a constant speed of 2.3 m/s by a horizontal cable connected to a winch. If the power delivered by the cable is 940 W, what is the tension in the cable?

36. An electric motor performs work on a large compressor at the rate of 1.5 kW. *(a)* How much work is done in 1 month if the motor runs continuously? *(b)* If the electric utility charges $0.12 per kilowatt-hour, estimate the costs of operation for 1 month.

37. Assume that the drag force exerted by the water on a barge is proportional to the speed of the barge relative to the water. A tug delivers 230 hp to the barge when they travel at a constant speed of 0.25 m/s. *(a)* What power is required to move the barge at 0.75 m/s? *(b)* What force does the tug exert on the barge at the lower speed? *(c)* At the higher speed?

38. A horse draws a barge on a canal, as shown in Fig. 8-20. Suppose that the horse does work on the barge at a rate of 0.3 hp (1 hp = 746 W) when the barge has a steady velocity parallel to the canal bank, with a magnitude of 0.7 m/s. *(a)* What is the tension in the (straight) tow rope that makes a 34° angle with the velocity? *(b)* How can the barge move parallel to the bank?

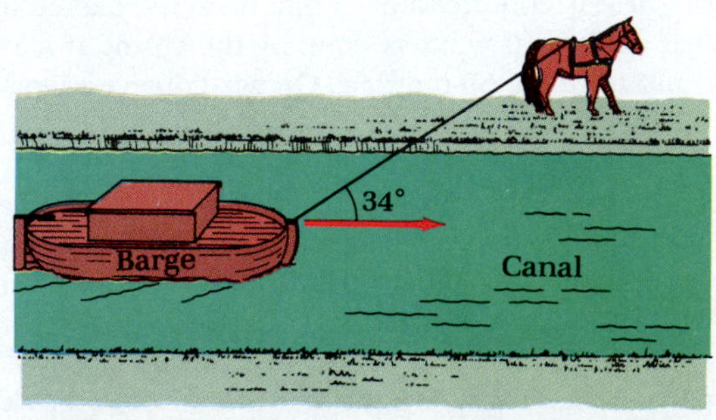

Figure 8-20. Exercise 38: A horse pulls a barge along a canal.

Additional Exercises

39. Estimate the work you perform on *(a)* a loaded grocery cart that you push 10 m on a horizontal floor, *(b)* a physics book that you push 1 m across a tabletop, *(c)* one page of this book as your turn to the next page.

40. A box slides 2 m in a straight line across a horizontal floor. One of the forces acting on the box has magnitude 12 N. At what angle, relative to the direction of the displacement, is this force applied if the work done is *(a)* 24 J; *(b)* 12 J; *(c)* 0; *(d)* −20 J?

41. A force (8.0 N)**i** − (9.0 N)**j** acts on an object whose displacement is $\boldsymbol{\ell}$ = (2.0 m)**i** + (4.0 m)**j** + (4.0 m)**k**. Determine *(a)* the work done by the force and *(b)* the angle between the force and the displacement.

42. Two forces, $\mathbf{F}_1$ = (2.0 N)**i** + (3.0 N)**j** and $\mathbf{F}_2$ = (4.0 N)**i** act on a box that undergoes a displacement $\boldsymbol{\ell}$ = (0.80 m)**i** + (0.40 m)**j**. Determine the work done by *(a)* $\mathbf{F}_1$; *(b)* $\mathbf{F}_2$; *(c)* the net force $\mathbf{F}_1 + \mathbf{F}_2$. *(d)* What is the net work?

43. How much work is done by the gravitational force on a 2-kg physics book that undergoes a displacement of *(a)* (3 m)**i**; *(b)* (3 m)**j**; *(c)* (3 m)**i** − (3 m)**j**? Let **j** be directed straight up.

44. A 35-kg crate is raised by applying a constant tension in a vertical rope that is attached to the crate. The crate starts from rest at floor level and is moving up at 1.5 m/s at the instant its vertical displacement is 2.5 m. Determine *(a)* the work done by the gravitational force on the crate, *(b)* the work done by the tension force, *(c)* the magnitude of the tension force.

45. A simple model for the force exerted on an atom in a linear chain molecule is $F_x(x) = -Ax^3$, where the constant $A = 1 \times 10^{26}$ N/m³. *(a)* Determine an expression for the work done by this force on the atom if the atom moves from x_i to x_f. *(b)* Evaluate the work for $x_i = 0$ and $x_f = 2 \times 10^{-11}$ m.

46. Suppose that the force acting on the atom in the previous exercise is the net force. If the atom's mass is 3×10^{-26} kg and its speed is 2×10^4 m/s at $x_i = 0$, what is its speed at $x_f = 2 \times 10^{-11}$ m?

47. When released in a syrup, a 0.010-kg sphere quickly reaches a terminal (constant) speed downward of 0.12 m/s. At what rate is work done on the sphere when moving at this speed by *(a)* the gravitational force and *(b)* the force due to the syrup?

48. An electron moving along the x direction in a CRT (a cathode-ray tube such as a television picture tube) is subjected to a net force $\mathbf{F} = (1 \times 10^{-12}\ \text{N/m})(0.020\ \text{m} - x)\mathbf{i}$, where x ranges between 0.000 m and 0.020 m. *(a)* Construct a graph showing F_x versus x. *(b)* Measure areas on the graph to determine the work done on the electron as its moves from x_i = 0.000 m to x_f = 0.020 m. *(c)* If the electron starts from rest at x_i, what is its speed at x_f? (The mass of the electron is listed on the inside front cover.)

49. Suppose the electron in the previous exercise is subjected to the force shown graphically in Fig. 8-21. *(a)* Determine the work done on the electron as it moves from x_i = 0.000 m to x_f = 0.020 m. *(b)* The electron starts from rest at x_i; what is its speed at x_f?

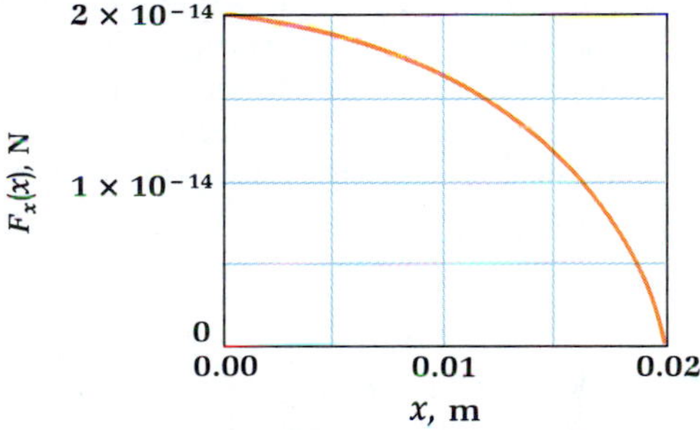

Figure 8-21. Exercise 49.

50. A 5-kg stone is released from rest at a point 5 m above ground level. Determine the rate at which the gravitational force does work on the stone at the instant the stone is *(a)* 5 m; *(b)* 4 m; *(c)* 3 m; *(d)* 2 m; *(e)* 1 m above the ground.

PROBLEMS

1. ***Block projected up a ramp.*** A 15-kg block is projected up a 30.0° ramp, with an initial speed of 4.6 m/s at the bottom of the ramp. The coefficient of kinetic friction for this pair of surfaces is 0.34. Determine the work done on the block as it slides to its highest point on the ramp by *(a)* the net force, *(b)* the weight of the block, *(c)* the normal force, *(d)* the frictional force. *(e)* How far up the ramp does the block slide before coming momentarily to rest?

2. ***The block slides back down.*** Suppose the block in Prob. 1 slides back down the ramp. *(a)* Determine the net work done on the block as it moves to the bottom of the ramp. *(b)* What is its speed as it reaches the bottom?

3. ***Work done along a circular path.*** A ball of mass m is attached to a light string of length L and suspended vertically. A constant horizontal force, whose magnitude F_a equals the weight of the ball, is applied, as shown in Fig. 8-22. Determine the speed of the ball as it reaches the 90° level in terms of L and g. Note that as the ball moves along the circular arc, the dis-

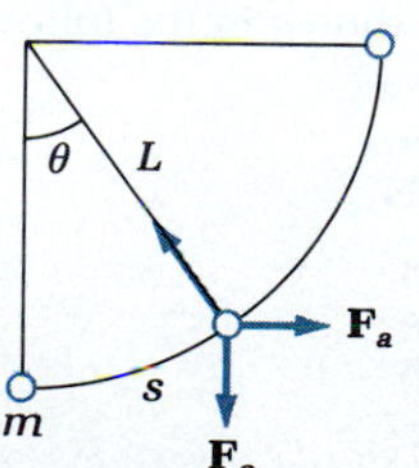

Figure 8-22. Problem 3.

tance moved along the arc is $s = L\theta$, and so $ds = L\,d\theta$. Neglect any frictional effects.

4. ***Work as a line integral.*** As a particle moves in the xy plane, it is acted on by a force whose components are functions of the coordinates of the particle, $F_x = (8.0\ \text{N/m}^2)xy$, $F_y = (6.0\ \text{N/m}^2)y^2$. *(a)* Evaluate the work done by this force as the particle moves from the origin to the point (2.0 m, 2.0 m) along the path $y = x$. *(b)* Repeat the calculation using the path $y = (0.50\ \text{m}^{-1})x^2$. Is the work done by this force independent of path?

5. ***An accelerating elevator.*** Suppose that you are a 60.0-kg passenger in an elevator. The elevator accelerates upward from rest at 1.0 m/s² for 2.0 s, moves at the resulting velocity for 10.0 s, and then decelerates at −1.0 m/s² for 2.0 s. *(a)* For the entire trip, what is the work done by the normal force exerted on you by the elevator floor? *(b)* By your weight? *(c)* What average power is delivered by the normal force for the full 14.0 s? *(d)* What instantaneous power is delivered by the normal force at 7.0 s? *(e)* At 13.0 s?

6. ***Circular motion on a ramp.*** One end of a light string is attached to a 1.2-kg puck which can slide with negligible friction on a 37° ramp. The other end of the string is fixed to a point on the ramp, and the puck moves in a circular path of radius 0.75 m, as seen in Fig. 8-23. At the lowest position, the tension in the string is 110 N. Determine *(a)* the speed of the puck at this lowest point, *(b)* the speed of the puck at the highest point in the circle, *(c)* the tension in the string for this highest position.

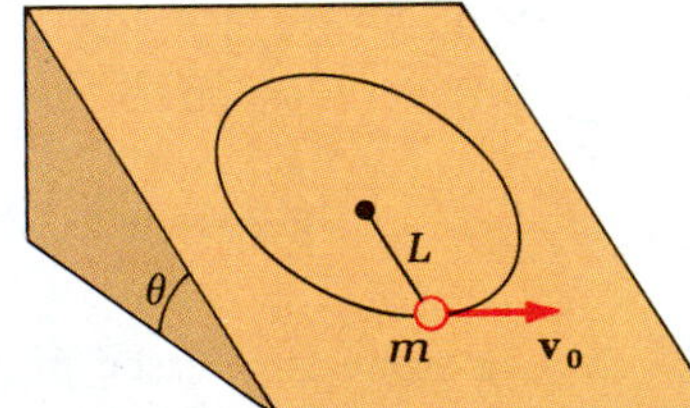

Figure 8-23. Problem 6.

7. ***Minimal conditions for circular motion.*** Suppose that the puck in the preceding problem has speed v_0 at the lowest position. *(a)* Determine the minimum value of v_0 such that the puck can complete the circular path. (*Hint:* The tension in the string will be zero at the highest position.) *(b)* Determine for these conditions the tension in the string at the lowest position. *(c)* Describe the motion qualitatively if the puck has this value v_0 at the lowest postion, but frictional effects, while small, are not negligible.

8. ***Work done by a varying force.*** A 3.00-kg object moving on the x axis experiences a net force that varies with the position of the object as shown in the following table:

x, m	F_x, N
0.00	120
0.20	98
0.40	80
0.60	66
0.80	54

x, m	F_x, N
1.00	44
1.20	36
1.40	30
1.60	24
1.80	20

(a) Assume that F_x varies smoothly with x and construct a graph using the data in the table. *(b)* Estimate the work done by this force if the object moves from $x_i = 0.00$ m to $x_f = 2.00$ m. *(c)* If the object's speed is 4.00 m/s at x_i, what is its speed at x_f?

9. ***Pyramid power.*** The pyramids of ancient Egypt may have been constructed by pulling or pushing large blocks of stone up a gradually rising ramp. Assume that the block was moved along the ramp at constant speed and that friction was minimized by using rollers. Estimate the work required to pull a 10,000-kg block up a ramp to a point 100 m higher than the starting point. *(a)* First assume that there is no frictional force acting on the block. *(b)* Now assume a 5° ramp and a coefficient of sliding friction of 0.2. *(c)* What pulling power is required for part *(b)* if the block moves up the ramp at 0.1 m/s?

10. ***Evaluating work by integration.*** A particle moves along the x axis under the influence of a force with component $F_x(x) = -Ax/(a^2 + x^2)$, where A and a are constants. Determine an expression for the work done as the particle moves from x_i to x_f.

11. ***Ropes and pulleys.*** A constant tension of 600 N is maintained in a rope that passes over a small, freely turning pulley and lifts a 50-kg box as shown in Fig. 8-24*a*, *(a)* If the box starts from rest, determine its speed when it is 2 m higher. *(b)* If a second pulley is used as in Fig 8-22*b*, what tension in the rope will cause the box to move as before? *(c)* What work is done by the tension force in each case?

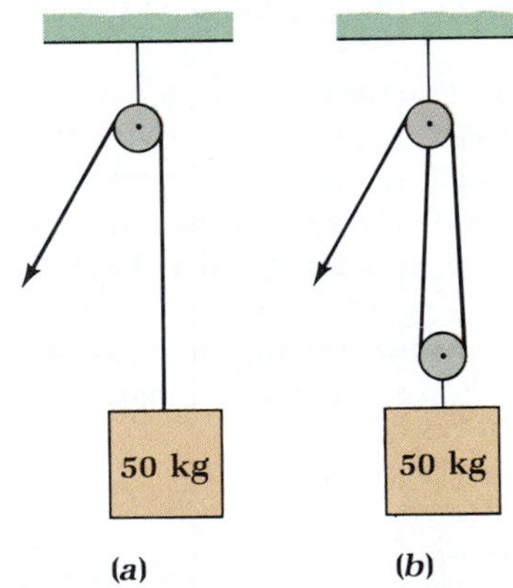

Figure 8-24. Problem 11.

12. ***Elevator estimates.*** A 2000-kg elevator serves a 50-floor hotel. *(a)* Estimate the work done by the elevator lifting mechanism for a trip from the lobby to the top floor. *(b)* Estimate the power requirement for operating the elevator.

13. ***Work done by air resistance.*** A car moving through the air with velocity $\mathbf{v} = v\mathbf{i}$ is subjected to an air-resistance force that is approximately proportional to the square of the speed: $\mathbf{F}_{\text{air}} = -Av^2\mathbf{i}$. For a certain car, the constant $A = 0.6\ \text{N} \cdot \text{s}^2/\text{m}^2$. *(a)* Determine the work done by the air-resistance force if the car travels 1 km at a constant speed of 80 km/h. *(b)* What power must be supplied to maintain this speed against air resistance?

14. ***A conveyer belt.*** A conveyer belt moves luggage up a 20° ramp to the baggage compartment of the aircraft 2 m above. *(a)* If the belt moves at a constant speed of 0.5 m/s, what minimum power is required when the mass of the luggage totals 1000 kg? *(b)* What is the minimum pulling force acting on the belt?

15. ***Work done by a spring force in two dimensions.*** The force exerted by a spring on an object is given by $\mathbf{F} = -kx\mathbf{i} - ky\mathbf{j}$, where k is the spring constant and the object can move in the xy plane. Determine an expression for the work done by the spring force as the object moves from an initial point (x_i, y_i) to a final point (x_f, y_f). Use an arbitrary path and explain why the result is independent of the path connecting the two points.

16. ***Using smaller subintervals.*** Modify the spreadsheet in Table 8-1 to use a larger number of smaller subintervals. For example, try doubling the number of steps while halving the increment Δx. Perform this doubling/halving several times to see if a limiting value of the sum is apparent.

17. ***Speed of a block.*** Suppose that the force exerted by the rubber band in Example 8-12 is the net force acting on a 2.0-kg block. The block starts from rest at $x = 0.10$ m and moves out to $x = 0$. The work-energy theorem can be applied at any point in the motion to calculate the speed of the block at that point. Modify the spreadsheet in Table 8-1 to calculate the speed of the block of each increment of x. (*Hint:* The square root of a variable A can be taken by using the spreadsheet function @ SQRT(A).

18. ***Power from a rubber band.*** Further modify the spreadsheet discussed in Prob. 17 so that it will calculate and display the power supplied by the force at each step of the calculation.

19. ***Work done by a varying force.*** Using Example 8-12 as a guide, determine the work done by a force with $F_x(x) = -Axe^{-x^2/a^2}$ for a particle that moves from x_i to x_f. Let $A = 1 \times 10^3$ N/m, $a = 0.25$ m, $x_i = 0.00$ m and consider three cases: *(a)* $x_f = 0.10$ m; *(b)* $x_f = 1.00$ m; *(c)* $x_f = 1.50$ m. Justify your choice for the value of the increment Δx.

20. ***Speed of a car.*** As in Prob. 17, the work-energy principle can be applied to determine the speed at each incremental step in the motion of an object. Use this technique to determine numerically the work done on the car by the air-resistance force described in Prob. 13. However, let the car start from rest and assume that the engine causes a constant force $\mathbf{F}_e = (1000\ \text{N})\mathbf{i}$ to be exerted on the car in addition to the air-resistance force. That is, the net force is $(1000\ \text{N})\mathbf{i} - Av^2\mathbf{i}$, with $A = 0.6\ \text{N} \cdot \text{s}^2/\text{m}^2$. Take the mass of the car to be 1000 kg and the length of the path to be 1 km. What is the final speed of the car? Graph the speed of the car as a function of position.

9 CONSERVATION OF ENERGY

When the string is released by the archer, potential energy in the bow is converted to kinetic energy of the arrow. See Exercise 42.

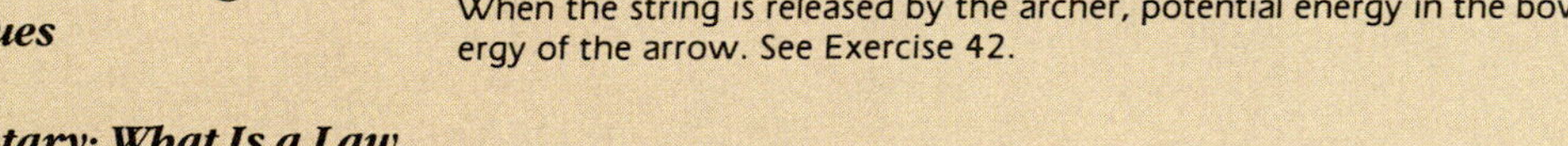

The *Voyager* spacecraft returned video images during its voyage past the planets Jupiter, Saturn, Uranus, and Neptune. The spacecraft is leaving the solar system, never to return.

Energy is eternal delight!
William Blake (1757 – 1827), poet

As we cannot give a general definition of energy, the principle of the conservation of energy simply signifies that there is something which remains constant. Well, whatever new notions of the world future experiments may give us, we know beforehand that there will be something which remains constant and which we shall be able to call energy.
Henry Poincare (1854 – 1912), mathematician-physicist

It is important to realize that in physics today, we have no knowledge of what energy is.
Richard P. Feynman (1918 – 1988), physicist

In this chapter you will encounter the first of the great conservation laws, the ***law of conservation of energy.*** When we say that a quantity is "conserved," we mean that the value of that quantity does not change with time. Its value at some initial instant is the same as its value at some final instant, and at all times in between. If energy is conserved for a system, then the total amount of energy in the system remains the same, although some of the energy may change its form or type. Conservation of energy is similar to the conservation of a fixed amount of financial assets. The form or type may change from money in a checking account to money in a cookie jar, but the total assets remain the same.

A conserved quantity is constant in time.

9-1 ONE-DIMENSIONAL CONSERVATIVE SYSTEMS

The main ideas in energy conservation can be developed for the case of a particle that moves along a straight line. We restrict our attention to this case in this section and generalize the results in Secs. 9-3 and 9-4.

Conservative and Nonconservative Forces

Forces that do work on an object are classified into ***conservative forces*** and ***nonconservative forces.*** Suppose an object undergoes a round-trip. In one dimension, the object moves out along a straight line and then returns along the same line to its starting point.

A force is conservative if it does no net work on an object for any round-trip.

Definition of a conservative force

A nonconservative force is one that is *not* conservative.

The gravitational force is an example of a conservative force. As an object moves up, such as the ball in Fig. 9-1*a*, the gravitational force does negative work (displacement is opposite to force). In Fig. 9-1*b*, as the ball returns downward to its starting point, the gravitational force does positive work. The positive work and negative work add to zero and no work is done for the round-trip. You can also get this result from Eq. 8-9, $W=-mg(y_f-y_i)$. Thus $W=0$ for a round-trip because $y_f=y_i$.

Gravitational force is conservative.

In one dimension, a force is conservative if it depends only on the coordinate that locates the object. An example of such a force is the elastic spring force, $F_x(x)=-kx$, and $W=-\frac{1}{2}k(x_f^2-x_i^2)$ from Eq. 8-6. For a round-trip, $x_f=x_i$, so that $W=0$. The positive and negative contributions to the work cancel for a round-trip. This cancellation occurs because the spring force depends only on where the object is (through x) and not on which way it is moving.

In contrast, the kinetic frictional force usually depends on which way the object moves. If an object slides on a stationary surface, the direction of the frictional force is always opposite the velocity of the object, as illustrated in Fig. 9-2. The frictional force does negative work on the object throughout its motion, and the work cannot be zero for a round-trip. If nonzero work is done by a force on an object as it makes a round-trip, then the force is *nonconservative.* The kinetic frictional force is a nonconservative force.

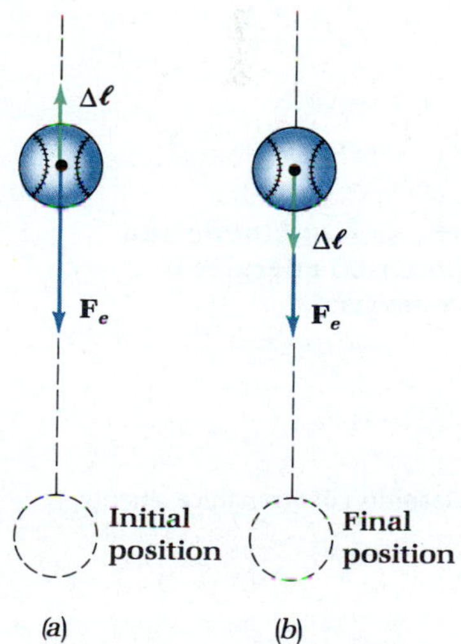

Figure 9-1. A conservative force. The earth's gravitational force does *(a)* negative work on a ball going up and *(b)* positive work on a ball going down. The work for the round-trip is zero.

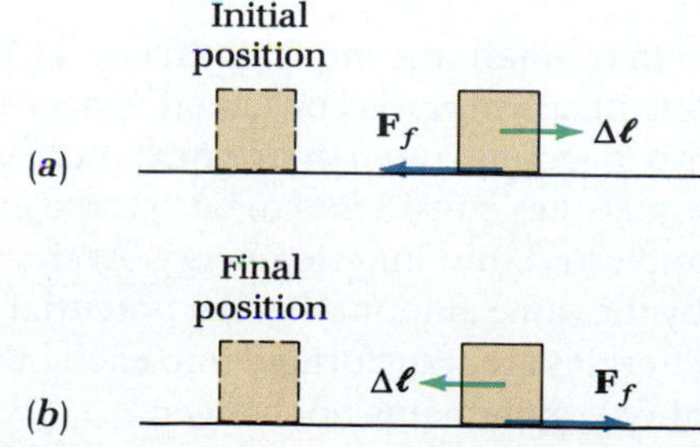

Figure 9-2. A nonconservative force. The frictional force does negative work for both parts of a round-trip. The work for the round-trip is not zero.

Conservative Systems and Mechanical Energy

A ***system*** includes the object under consideration and the parts of the object's environment that interact with it. For a cart traveling down a ramp, the system includes the cart, the ramp, and the earth (interacting with the cart through the gravitational force). If the cart moves rapidly, air resistance may be significant, and the surrounding air is also part of the system.

> A ***conservative system*** is one where only conservative forces do work on the object.

A slowly moving cart with well-lubricated axles is, to a good approximation, a conservative system. In this case, air resistance and friction in the axles are negligible.

Another example of a conservative system is a ball moving vertically in free-fall. We neglect air resistance so that only the conservative gravitational force by the earth does work on the ball. Equation 8-9 gives the work done by gravity as an object moves vertically from y_i to y_f: $W = -mg(y_f - y_i)$. Only the gravitational force does work, so Eq. 8-9 gives the net work:

$$W_{\text{net}} = -mg(y_f - y_i)$$

From the work-energy theorem,

$$W_{\text{net}} = K_f - K_i = \tfrac{1}{2}mv_f^2 - \tfrac{1}{2}mv_i^2$$

Equating these expressions for the net work gives

$$\tfrac{1}{2}mv_f^2 - \tfrac{1}{2}mv_i^2 = -mg(y_f - y_i)$$

The subscripts i for *i*nitial and f for *f*inal refer to any two points in the motion. Placing the terms that pertain to the final point on one side of the equation and the terms that refer to the initial point on the other side gives

$$\tfrac{1}{2}mv_f^2 + mgy_f = \tfrac{1}{2}mv_i^2 + mgy_i \qquad (9\text{-}1)$$

We are already familiar with the kinetic energy $K = \frac{1}{2}mv^2$, which depends on the object's speed. Now we introduce another form of energy—the potential energy U. The formal definition of potential energy is presented later in this section, but for the case of the ball in free-fall, the gravitational potential energy is $U = mgy$. Now Eq. 9-1 becomes

The sum of kinetic and potential energies is conserved.

$$K_f + U_f = K_i + U_i$$

Both $K = \frac{1}{2}mv^2$ and $U = mgy$ change during the motion of the ball, but their sum $K + U$ remains constant.

This leads to the definition of the mechanical energy:

Definition of mechanical energy

> The ***mechanical energy*** E is sum of the kinetic energy K and the potential energy U:

$$E = K + U$$

A decrease of potential energy is offset by an equal gain of kinetic energy and vice versa.

For the ball in free-fall, the mechanical energy does not change during the motion; that is, mechanical energy is conserved. A decrease in the kinetic energy of the ball on its way up is accompanied by an equal increase in the potential energy. Potential energy is sometimes considered to be "stored energy," since it has the "potential" of being converted into kinetic energy. As the ball falls down, the kinetic energy increases by the same amount that the potential energy decreases. Thus kinetic and potential energies are transformed into each other during the motion such that the mechanical energy remains unchanged.

The expression for conservation of mechanical energy, Eq. 9-1, can be written simply as

$$E_f = E_i \tag{9-2}$$

where $E_f = K_f + U_f$ and $E_i = K_i + U_i$. Since the kinetic energy depends on speed and the potential energy depends on position, Eq. 9-2 can be used to connect the speed of an object with its position. It provides us with a partial description of the motion of the object. We can determine how fast the object is moving when it is at a particular position, without explicit reference to the time. Equation 9-2 does not, however, reveal the direction of motion or the time at which the object is at a certain position.

EXAMPLE 9-1

Conservation of mechanical energy. A 2.0-kg stone is thrown straight up with an initial speed $v_i = 8.0$ m/s, as shown in Fig. 9-3. Air-resistance effects may be neglected so that the system consists of the stone and the earth's gravity. *(a)* Evaluate the mechanical energy of the system. *(b)* How high does the stone rise, and what is the potential energy at the maximum height? *(c)* What is the speed of the stone when it reaches one-half the maximum height? *(d)* Describe the changes in kinetic and potential energies during this motion.

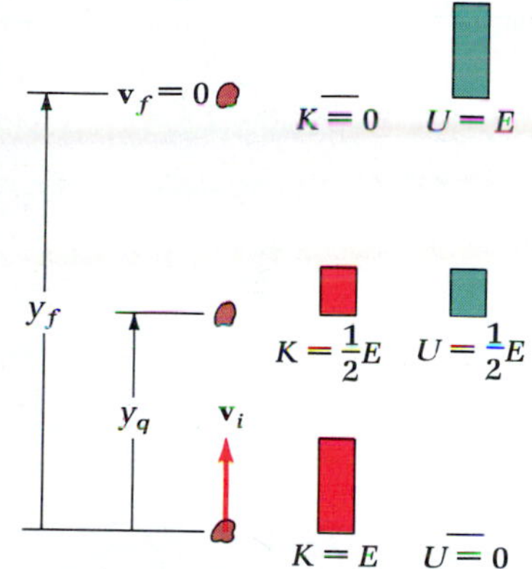

Figure 9-3. Example 9-1: A stone is projected upward. Kinetic energy is converted into potential energy as the stone rises.

Solution. Since only the conservative gravitational force does appreciable work, mechanical energy is conserved. For convenience, vertical positions are measured from the initial location of the stone, so $y_i = 0$.

(a) At the initial point in the motion, $v_i = 8.0$ m/s and $y_i = 0$. The initial kinetic and potential energies are

$$K_i = \tfrac{1}{2}(2.0\text{ kg})(8.0\text{ m/s})^2 = 64\text{ J}$$

$$U_i = mgy_i = mg(0) = 0$$

The mechanical energy of this system at the initial point, and throughout the motion, is

$$E = E_i = 64\text{ J} + 0 = 64\text{ J}$$

(b) Let y_f represent the maximum height of the stone. At this position the speed $v_f = 0$, so that the kinetic energy $K_f = \frac{1}{2}mv_f^2 = 0$. The conserved mechanical energy, $E = K_f + U_f$, must still have the value 64 J, however. So at the maximum height, the potential energy $U_f = 64$ J, and

$$y_f = \frac{U_f}{mg} = \frac{64\text{ J}}{(2.0\text{ kg})(9.8\text{ m/s}^2)} = 3.3\text{ m}$$

(c) Let $y_q = \frac{1}{2}y_f$ represent the y coordinate at half the maximum height. There the mechanical energy is part kinetic energy and part potential energy. The mechanical energy is still 64 J. We have $\frac{1}{2}mv_q^2 + mgy_q = 64$ J. Inserting numerical values gives

$$\tfrac{1}{2}(2.0\text{ kg})v_q^2 = 64\text{ J} - (2\text{ kg})(9.8\text{ m/s}^2)\tfrac{1}{2}(3.3\text{ m})$$

$$v_q^2 = 32\text{ m}^2/\text{s}^2 \qquad \text{and} \qquad v_q = 5.7\text{ m/s}$$

There are two possible velocities for the stone since the stone has this position on the way up ($v_{yq} = +5.7$ m/s) and on the return-trip down ($v_{yq} = -5.7$ m/s). *(d)* During the entire free-fall motion, the mechanical energy is conserved. Initially, when $y_i = 0$, the mechanical energy is entirely kinetic. As the stone moves up, the kinetic energy decreases as the potential energy increases. Halfway up or down, at point q, the kinetic and potential energies become equal. At the highest position, the mechanical energy is entirely potential since $v_f = 0$ there. The division of the constant mechanical energy into kinetic and potential is suggested in the figure at several points in the motion by the bars labeled by the symbols K and U.

SELF-TEST 9-1. For the stone in the above example, determine the potential energy, the kinetic energy, and the speed of the stone when it is at *(a)* $y_1 = 0.8$ m and *(b)* $y_2 = 2.4$ m.
ANSWERS: *(a)* 16 J, 48 J, 7.0 m/s; *(b)* 47 J, 17 J, 4.1 m/s.

Potential Energy and Conservation of Mechanical Energy

In the free-fall example, the change in gravitational potential energy was expressed as the negative of the work done by the conservative gravitational force. By introducing this potential energy, we saw that a change in the kinetic energy was balanced by a change of opposite sign in potential energy so that the mechanical energy remained the same. We can proceed similarly by introducing a potential energy for any conservative force in one dimension. Let $F_x(x)$ represent such a force acting on an object. From Eq. 8-5, the work done by this force is

$$W = \int_{x_i}^{x_f} F_x(x)\, dx$$

The change $U_f - U_i$ in the potential energy is the negative of the work done by a conservative force.

Change in potential energy in one dimension

$$U_f - U_i = -\int_{x_i}^{x_f} F_x(x)\, dx \qquad (9\text{-}3)$$

The negative sign is used so that an increase in potential energy corresponds to a decrease in kinetic energy. For example, the gravitational force does negative work on an object as it goes up, which corresponds to an increase of potential energy and a decrease of kinetic energy.

For any conservative system, conservation of mechanical energy is obtained by combining the definition of potential-energy change with the work-energy theorem. If work is done only by a conservative force, then $W_{\text{net}} = W = -(U_f - U_i)$ and

$$K_f - K_i = W_{\text{net}} = -(U_f - U_i)$$

On rearranging, we have $K_f + U_f = K_i + U_i$, or with $E = K + U$,

Conservation of mechanical energy

$$E_f = E_i \qquad (9\text{-}4)$$

The mechanical energy of a conservative system is conserved.

Gravitational Potential Energy

As an object of mass m moves from a point with vertical coordinate y_i to a point with vertical coordinate y_f, the change in gravitational potential energy is

$$U_f - U_i = mgy_f - mgy_i$$

This is the negative of the work done by the gravitational force on the object.

It is important to notice that only changes or differences in potential energy are relevant. The potential energy can be set equal to zero at a point of our choice, called the *reference point.* For example, we usually select the reference point or level for gravitational potential energy at $y = 0$; that is, $U = 0$ at $y = 0$. Having selected the reference point where $U = 0$, we can then refer to a value of the potential energy at any point as the difference between the potential energy at that point and the reference point. With the reference point at $y = 0$, the gravitational potential energy for an object at a point with vertical coordinate y is

Gravitational potential energy

$$U = mgy \qquad (9\text{-}5)$$

Although the value of the potential energy depends on the choice of the origin (the place where $y = 0$), changes in potential energy do not, as the following example shows.

EXAMPLE 9-2

Changes in gravitational potential energy. Calculate the gravitational potential energy of a 2.1-kg book on the floor and on a shelf 2.0 m above the floor, and evaluate the difference in potential energy of the book between the floor and the shelf. Perform these calculations twice: *(a)* with the origin at the floor and *(b)* with the origin at the shelf.

Solution. Let y_i be the vertical coordinate of the book when on the floor and y_f its coordinate when on the shelf.

(a) If the origin is at the floor, then $y_i = 0$ and $y_f = 2.0$ m. Using Eq. 9-5, we obtain

$$U_i = mgy_i = 0$$

and

$$U_f = mgy_f = (2.1 \text{ kg})(9.8 \text{ m/s}^2)(2.0 \text{ m}) = 41 \text{ J}$$

The difference in potential energy is

$$U_f - U_i = 41 \text{ J} - 0 = 41 \text{ J}$$

(b) With the origin at the level of the shelf, the coordinate of the floor level is $y_i = -2.0$ m, and the shelf is at $y_f = 0$. The values of potential energy at these two locations are different from those in part *(a)* because the origin of coordinates is different. At floor level, the potential energy of the book is

$$U_i = mgy_i = (2.1 \text{ kg})(9.8 \text{ m/s}^2)(-2.0 \text{ m}) = -41 \text{ J}$$

On the shelf the book's potential energy is $U_f = mgy_f = 0$. The difference or change in potential energy is

$$U_f - U_i = 0 - (-41 \text{ J}) = 41 \text{ J}$$

the same as in part *(a)*.

SELF-TEST 9-2. Take the origin of the coordinate system in the above example to be halfway between the floor and the shelf. Determine the gravitational potential energy of the book when it is at the floor and at the shelf. What is the change in potential energy between the floor and the shelf? ***ANSWERS:*** -21 J, 21 J, 41 J.

Elastic Potential Energy

Hooke's law for an ideal spring, $F_x(x) = -kx$, provides another example of a one-dimensional conservative force. The potential energy associated with this force is called the ***elastic potential energy*** of the spring. Figure 9-4*a* shows a typical arrangement of a spring with one end attached to a rigid support and the other end to a movable block. The origin of coordinates is chosen so that $x = 0$ when the spring is relaxed, neither stretched nor compressed. In Fig. 9-4*b*, the x coordinate of the block gives the amount by which the spring is stretched or compressed.

Elastic potential energy of a spring

The work done by the spring was determined in Chap. 8 and is given by Eq. 8-6, $W = -\frac{1}{2}k(x_f^2 - x_i^2)$. From Eq. 9-3, $U_f - U_i = -W$, and we have

$$U_f - U_i = \tfrac{1}{2}kx_f^2 - \tfrac{1}{2}kx_i^2 \qquad (9\text{-}6)$$

By choosing the reference point at $x = 0$, we identify U_f with $\frac{1}{2}kx_f^2$ and U_i with $\frac{1}{2}kx_i^2$. The elastic potential energy of the spring, stretched or compressed by an amount x, is given by

$$U = \tfrac{1}{2}kx^2 \qquad (9\text{-}7)$$

The elastic potential energy of a spring is nonnegative.

This potential energy is never negative because it is proportional to the square of x. If the spring is compressed, making x negative, the potential energy is again positive. The elastic potential energy is zero only if $x = 0$, corresponding to a relaxed spring.

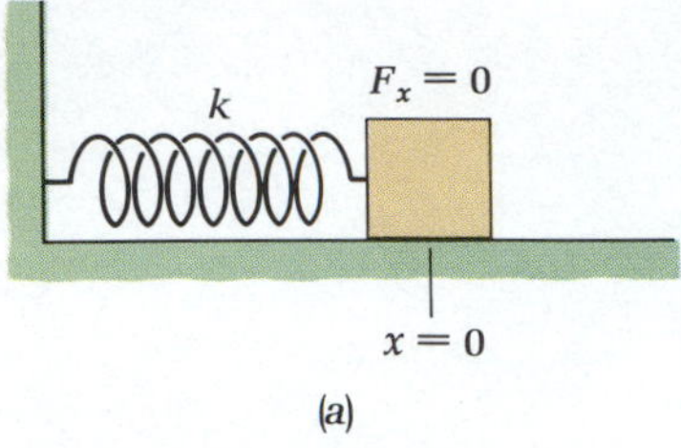

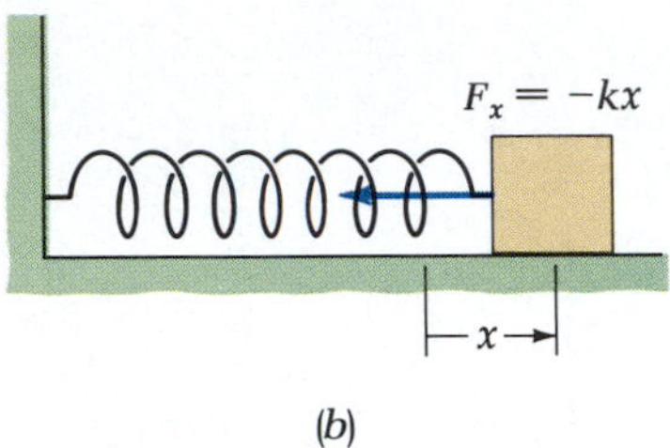

Figure 9-4. Example 9-3: A block is connected to a spring. *(a)* The spring is relaxed. *(b)* The spring exerts a restoring force.

EXAMPLE 9-3

Potential energy of a spring. A 2.5-kg block is attached to the free end of a spring of constant $k = 1100$ N/m, as seen in Fig. 9-4. The block is pulled from $x = 0$ to $x = 0.15$ m and released from rest at an initial instant. The block slides on the surface with negligible friction. *(a)* Evaluate the initial potential energy of this system. *(b)* Determine an expression for the speed of the block as a function of its coordinate. *(c)* What is the speed of the block as it reaches the equilibrium position? *(d)* By how much is the spring compressed as the block comes to rest momentarily to the left of the equilibrium position?

Solution. *(a)* The initial potential energy is

$$U_i = \tfrac{1}{2}kx_i^2 = \tfrac{1}{2}(1100 \text{ N/m})(0.15 \text{ m})^2 = 12 \text{ J}$$

The block is released from rest at this position, so that $K_i = 0$, and the mechanical energy is $E = 12$ J.

(b) Since frictional effects are negligible, only the conservative spring force does work and mechanical energy is conserved. Applying Eq. 9-4, we have

$$\tfrac{1}{2}mv_f^2 + \tfrac{1}{2}kx_f^2 = \tfrac{1}{2}kx_i^2$$

Simplifying, we obtain $v_f^2 = (k/m)(x_i^2 - x_f^2)$, or

$$v_f = \sqrt{\frac{k}{m}(x_i^2 - x_f^2)} \qquad \text{(A)}$$

The values of k, m, and x_i are known.

(c) As the block passes through the equilibrium position, $x_f = 0$, and the speed of the block is

$$v_f = \sqrt{\frac{k}{m}}\, x_i = \sqrt{\frac{1100 \text{ N/m}}{2.5 \text{ kg}}}\, 0.15 \text{ m} = 3.1 \text{ m/s}$$

(d) Now we let f label the point of maximum compression of the spring corresponding to $v_f = 0$. Then from Eq. A we have $x_i^2 - x_f^2 = 0$, or $x_f = \pm x_i = \pm 0.15$ m. The negative sign is selected because the spring is compressed at this instant: $x_f = -0.15$ m.

SELF-TEST 9-3. *(a)* What is the potential energy of the spring in the above example at the instant that the block has speed 2.0 m/s? *(b)* Where is the block at this instant? ***ANSWERS:*** *(a)* 7.4 J; *(b)* 0.12 m.

9-2 GRAPHICAL ANALYSIS OF CONSERVATIVE SYSTEMS

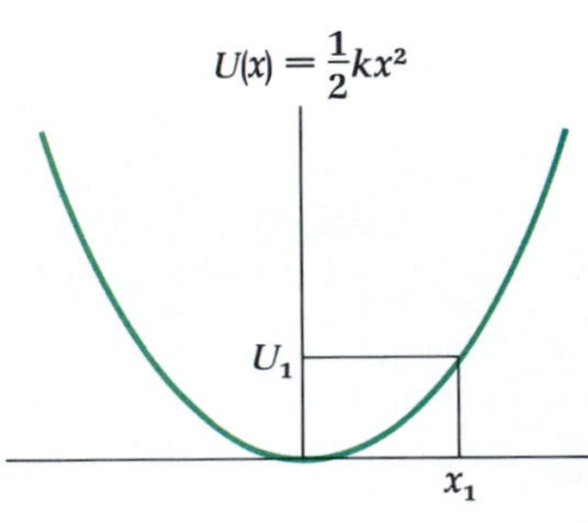

Figure 9-5. *(a)* A block is connected to a spring. The coordinate of the block, x, gives the stretch or the compression of the spring. *(b)* The elastic potential energy of a spring: $U = \tfrac{1}{2}kx^2$.

For a one-dimensional conservative system, the division of the conserved mechanical energy into changing amounts of kinetic and potential energies can be displayed graphically. As a simple example, consider a familiar system, a block connected to a spring as shown in Fig. 9-5*a*. Suppose the block slides with negligible friction on the horizontal surface; its position is determined by the coordinate x with the spring relaxed for $x = 0$. The elastic potential energy of the spring is given by Eq. 9-7, $U = \tfrac{1}{2}kx^2$. The mechanical energy of this system is conserved because only the spring force performs work and it is a conservative force.

Figure 9-5*b* shows the x dependence of the potential energy. The curve is a parabola since the potential energy is proportional to the square of x. From the graph we can determine the value of potential energy U_1 correponding to the coordinate value x_1. Note that this potential-energy function is positive both for extension and for compression of the spring.

Suppose we set the system in motion by pulling the block out a distance x_m and releasing it from rest. In the subsequent motion, the mechanical energy is conserved. The value of the mechanical energy can be expressed in terms of the initial situation. Since the kinetic energy is zero at the initial instant, the mechanical energy is entirely potential energy, $E = U_m = \tfrac{1}{2}kx_m^2$. Later in the motion, that mechanical energy is divided into kinetic and potential energy, $E = K + U$. For example, suppose the block is at position x_1 in Fig. 9-5*b*, moving either to the right or to the left. Part of the mechanical energy is potential energy U_1, and the remainder is kinetic energy K_1, with $E = K_1 + U_1$. This division is shown explicitly in Fig. 9-6. The length of the line up to the curve at point x_1 represents the potential energy U_1. Then the line length from this curve up to E represents the kinetic energy K_1. This graphical construction corresponds to $E = K_1 + U_1$.

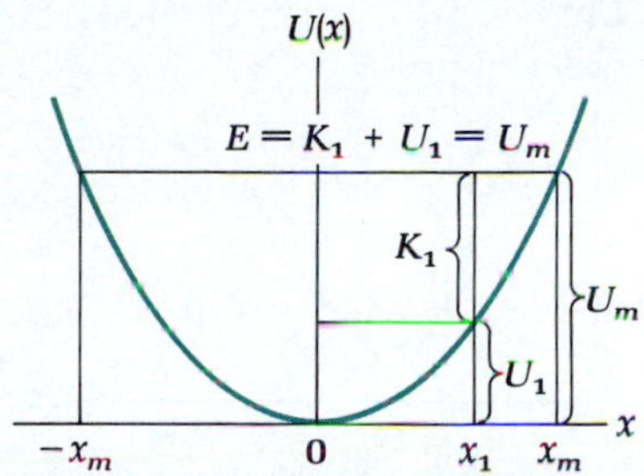

Figure 9-6. The conserved mechanical energy is part kinetic energy and part potential energy: $E = K + U$.

We can follow the motion of this object using the graphical construction outlined in Fig. 9-6. Starting with the initial configuration of this system, we evaluate the mechanical energy. On the vertical axis we mark this value of energy and draw the horizontal line whose height represents the fixed value of the mechanical energy. This line intersects the potential-energy curve at $\pm x_m$. For any value of x between $\pm x_m$, the values of kinetic and potential energies are represented by the line lengths above and below the potential-energy curve, just as at point x_1. Thus, as the block passes through the equilibrium position at $x = 0$, the potential energy is zero and the kinetic energy $K = E$. The mechanical energy is entirely kinetic energy at this point. As a consequence, the speed of the block is greatest at this position. At either of the points x_m and $-x_m$, the mechanical energy is entirely potential energy, and the kinetic energy is (momentarily) zero as the velocity changes direction. These points are called *turning points* for the motion. A turning point is located graphically by the intersection of the potential-energy curve with the horizontal line corresponding to the value of the mechanical energy of the system.

We can also use the potential-energy curve to obtain information about the conservative force acting on the object. Equation 9-3 defines the change in potential energy between two points as the negative of the integral of the conservative force. Because integration is the operation inverse to differentiation, we must have $F_x(x) = -dU/dx$, a purely mathematical fact. More physically we can consider the negative of the work done by the conservative force $F_x(x)$ for a infinitesimal displacement dx. The corresponding infinitesimal change in potential energy is $dU = -F_x(x)\,dx$, or

$$F_x(x) = -\frac{dU}{dx} \tag{9-8}$$

A conservative force is a negative derivative of the potential energy.

The one-dimensional conservative force is the negative of the derivative of the potential-energy function. To see how this works, consider two examples for which we already know the force. For an ideal spring, $U(x) = \frac{1}{2}kx^2$ and $F_x(x) = -d(\frac{1}{2}kx^2)/dx = -kx$. For gravitational potential energy, $U(y) = mgy$ and $F_y(y) = -d(mgy)/dy = -mg$.

Recall from Chap. 3 the graphical interpretation of the derivative of a function, which we now apply to the graph of the potential-energy function. At some point x_1 the derivative of $U(x)$, dU/dx, is equal to the slope of the tangent line to the graph at that point. The force is then the negative of the slope of the tangent line. This interpretation is illustrated in Fig. 9-7 for the mass-spring system we have been discussing. At x_1 the slope of the tangent line is positive and $F_x(x_1)$ is negative; that is, the force is opposite the displacement of the block. At x_2 the slope is negative and $F_x(x_2)$ is positive. Again, the force is opposite the displacement. At the origin, of course, the slope of the tangent line is zero, consistent with the force being zero there.

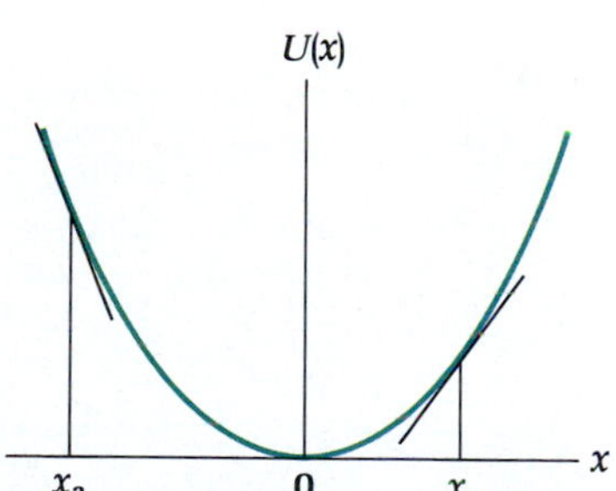

Figure 9-7. The force component is the negative of the slope of the tangent line to the curve: $F_x(x) = -dU/dx$.

Equilibrium and Stability

An ***equilibrium point*** for an object is a point where the net force on the object is zero.

If an object is at an equilibrium point and at rest, then it remains at rest. Consider a roller-coaster car placed at rest at various positions along its track. (i) Suppose the car is placed at a point where the track has a slope. The car would move down the

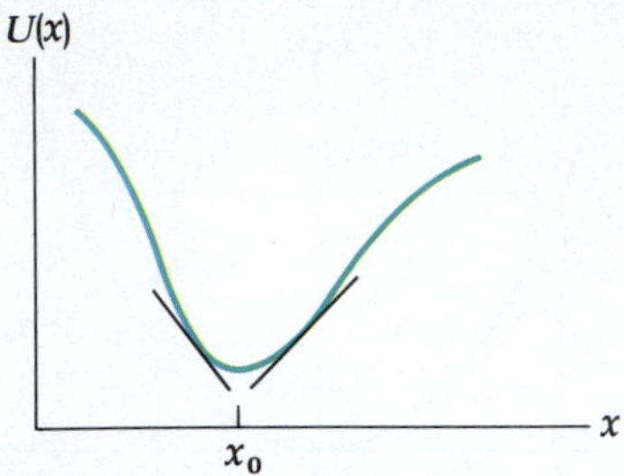

Figure 9-8. The force is zero at x_0, a point of stable equilibrium.

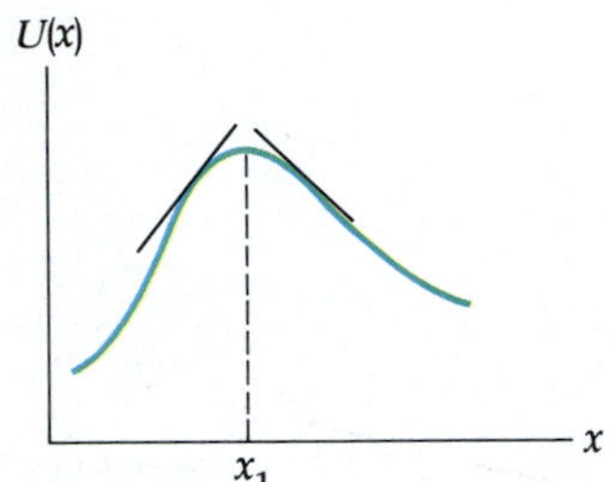

Figure 9-9. The force is zero at x_1, a point of unstable equilibrium.

slope—a point on a slope is not an equilibrium point. (ii) Suppose the car is placed at a trough (or low point) in the tracks. The car would remain at rest—the trough is an equilibrium point. Suppose the car is pushed slightly to one side or the other from the trough. It would tend to return to the trough—the trough is a point of *stable* equilibrium. (iii) Suppose the car is placed at a crest (or high point) in the tracks. The car would remain at rest—a crest is an equilibrium point. Suppose the car is pushed slightly to one side or the other from the crest. It would tend to travel away from the crest—the crest is a point of *unstable* equilibrium. Now we use graphical analysis of potential energy to formalize these ideas.

A graph of a hypothetical potential-energy function is shown in Fig. 9-8. The system is conservative and the potential-energy function represents the net force on the object. The point x_0 is a point of ***stable equilibrium.*** It is an equilibrium point because F_x, equal to the negative slope of the tangent line to the curve, is zero there. Suppose the object, initially at rest at point x_0, is given a small displacement to the right. The force can be estimated from the slope of the tangent line there. The direction of the force is to the left, tending to return the object to the equilibrium position. A similar conclusion results if we displace the object by a small amount to the left of the equilibrium position. The direction of the force is to the right, again to return the object to the equilibrium point.

> A point of equilibrium is stable if, for any small displacement from that point, the net force tends to return the object to the equilibrium point.

It should be clear from Fig. 9-8 that a minimum of the potential-energy function corresponds to a point of stable equilibrium for a one-dimensional conservative system.

Point x_1 in Fig. 9-9 is also an equilibrium point; the force on the object is zero there. However, for this point of ***unstable equilibrium,*** the force on the object, if displaced slightly to either side of x_1, tends to remove the object farther from the equilibrium point.

Unstable equilibrium

> A point of equilibrium is unstable if, for any small displacement of the object from that point, the net force tends to move the object away from the equilibrium point.

For a one-dimensional conservative system, a maximum of the potential-energy function is a point of unstable equilibrium.

9-3 CONSERVATIVE FORCES AND POTENTIAL ENERGY IN THREE DIMENSIONS

Our discussion of conservative forces, potential energy, and the conservation of mechanical energy has been confined to one-dimensional systems. We now extend these concepts to systems in two and three dimensions.

Conservative Force

For a conservative force, the work is zero for any round-trip. In two or three dimensions, a round-trip can be any *closed path,* as seen in Fig. 9-10. We extend the definition of a conservative force to include such paths.

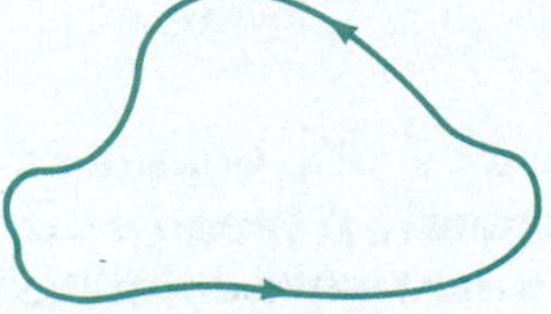

Figure 9-10. A curve that closes on itself forms a closed path.

The work done by a conservative force is zero for any closed path.

Definition of a conservative force

Positive work done by a conservative force for part of a closed path is exactly canceled by negative work done for another part of the path, and this cancellation occurs for any closed path.

The definition of a conservative force can be expressed as an integral. Recall from Chap. 8 that the work done by a force **F** for an infinitesimal displacement $d\boldsymbol{\ell}$ is $\mathbf{F} \cdot d\boldsymbol{\ell}$. The integral $\int \mathbf{F} \cdot d\boldsymbol{\ell}$ for the work done along a path represents the sum of the contributions for each infinitesimal displacement along the path. To indicate a closed path, we put a circle on the integral sign: $\oint$. Thus the work done by a force **F** for a closed path is denoted by $\oint \mathbf{F} \cdot d\boldsymbol{\ell}$. For a conservative force, this work is zero for any closed path, or

For a conservative force, $\oint \mathbf{F} \cdot d\boldsymbol{\ell} = 0$.

$$\oint \mathbf{F} \cdot d\boldsymbol{\ell} = 0 \qquad \text{any closed path} \tag{9-9}$$

Instead of using a closed path, we can consider the work done by a conservative force along alternative paths connecting two endpoints, say, i and f, as shown in Fig. 9-11*a*. Since the work done by a conservative force for a closed path is zero, then the work done along one path from i to f must exactly cancel the work done along the return path from f back to i. In Fig. 9-11*b*, the sense of the return path has been reversed so that both paths go from i to f. Reversing the sense of a path is equivalent to replacing $d\boldsymbol{\ell}$ with $-d\boldsymbol{\ell}$ and so changing the sign of the integral. That is,

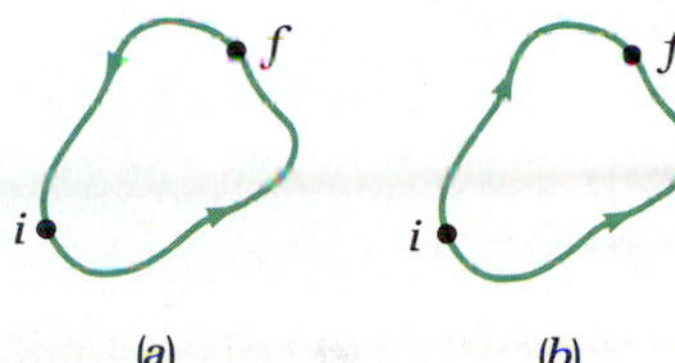

Figure 9-11. The work done by a conservative force is *(a)* zero for a closed path and *(b)* the same for alternative paths going from i to f.

$$\int_f^i \mathbf{F} \cdot d\boldsymbol{\ell} = -\int_i^f \mathbf{F} \cdot d\boldsymbol{\ell}$$

So along any two paths connecting points i and f, a conservative force does the same work.

The work done by a conservature force is independent of the path connecting the endpoints.

Alternative definition of a conservative force

This is an alternative and convenient definition of a conservative force.

We saw in Chap. 8 that the work done by the gravitational force is independent of the path. Thus this force is conservative according to our extended definition. The work done by a frictional force does depend on the path generally and so is nonconservative.

EXAMPLE 9-4

A tale of two paths. In one instance, a 2.0-kg book is lowered along a vertical straight line from point A to point B, where B is 1.5 m directly below A. In another instance, the book is moved from A to B along three straight-line paths, first moving it horizontally 1.0 m to point C, then lowering it vertically 1.5 m to point D, and then horizontally 1.0 m to B. Compare the work done by gravity for these two paths.

Solution. The work done by gravity is

$$W = -mg(y_f - y_i)$$

where y is measured vertically upward. This work depends only on the initial and final values of the vertical coordinate y and is independent of any horizontal coordinate. No work is done by gravity when an object is moved horizontally because a horizontal displacement is perpendicular to the vertical gravitational force, $\mathbf{F} \cdot \Delta\boldsymbol{\ell} = F\Delta\ell \cos 90° = 0$. For both paths,

$$W = -(2.0 \text{ kg})(9.8 \text{ m/s}^2)(-1.5 \text{ m}) = 29 \text{ J}$$

SELF-TEST 9-4. In the above example, determine the work done by gravity when the book is taken around a closed path from point A to C to D to B and finally back to A. ***ANSWER:*** Zero.

Potential Energy

The work done by a conservative force, $W = \int_i^f \mathbf{F} \cdot d\boldsymbol{\ell}$, is independent of the path connecting the endpoints. Since the work does not depend on the path, it can only depend on the two endpoints. This means that the work done by a conservative force can be written as some quantity evaluated at one endpoint minus the quantity evaluated at the other endpoint. This provides the definition of the change in potential energy U.

The change in the potential energy is the negative of the work done by a conservative force

General definition of change in potential energy

$$U_f - U_i = -\int_i^f \mathbf{F} \cdot d\boldsymbol{\ell} \tag{9-10}$$

Again, the negative sign is used so that an increase in potential energy corresponds to a decrease in kinetic energy. With the selection of a convenient reference level, the change in the potential energy gives an expression for the potential energy U.

9-4 CONSERVATION OF MECHANICAL ENERGY

We can obtain conservation of mechanical energy for an object moving in two or three dimensions just as we did for motion in one dimension. From the work-energy theorem, the change in the kinetic energy of the object equals the work done by the net force on the object: $K_f - K_i = W_{\text{net}}$. Suppose that only conservative forces are performing work on the object. The work done by a conservative force is independent of the path and is equal to the negative of the change in potential energy. Combining these results, we have $K_f - K_i = W_{\text{net}} = -(U_f - U_i)$. Rearranging this last expression gives $K_f + U_f = K_i + U_i$, or

$$E_f = E_i \tag{9-11}$$

where $E = K + U$ is the mechanical energy of the system. The mechanical energy is conserved because the only forces performing work are conservative, and the net work done can then be written as the negative change in potential energy.

If only conservative forces do work, the mechanical energy of the system is conserved.

Equation 9-11 expresses the conservation of mechanical energy of a system in one, two, or three dimensions. It connects the speed of the object with its position. Throughout the motion, both the kinetic energy and the potential energy can change, but their sum, $E = K + U$, does not change.

Often a normal force is exerted on an object by a stationary surface on which the object slides. However, this normal force performs no work because the normal force $\mathbf{F}_N$ is perpendicular to any infinitesimal displacement $d\boldsymbol{\ell}$ of the object, and $\mathbf{F}_N \cdot d\boldsymbol{\ell} = 0$. Although the normal force in this case is not zero, the work done by the normal force is zero. It is the *work* performed by the force, and not the force itself, that is important for conservation of mechanical energy.

EXAMPLE 9-5

Changes in gravitational and elastic potential energies. A 2.1-kg block is held against a light spring (of negligible mass) of spring constant $k = 2400$ N/m, which is compressed by 0.15 m. The block is released from rest at point i, and the spring projects the block up the 25° ramp, as shown in Fig. 9-12. The block comes to rest momentarily at point f. Frictional forces are negligible. Assume that the block loses contact with the spring when the spring is relaxed. *(a)* How far up the ramp from point i is point f? *(b)* As the block slides back down the ramp, what is its speed halfway between f and i?

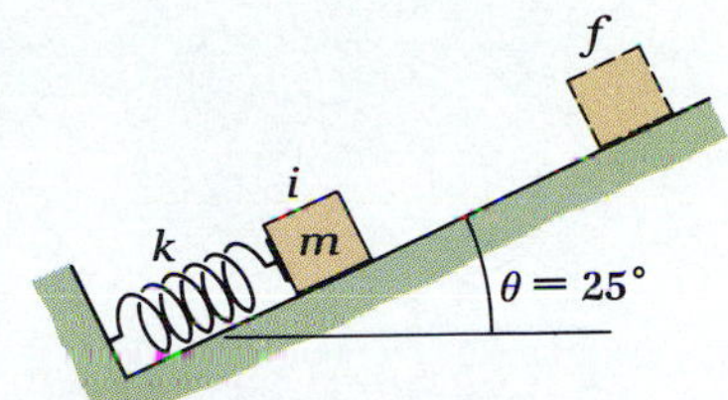

Figure 9-12. Example 9-5: A spring projects a block up a smooth ramp.

Solution. There are three forces acting on the block. Of these three, the spring force and the weight are conservative. The normal force performs no work since the displacement of the block is perpendicular to the normal force. Since only conservative forces perform work, mechanical energy is conserved. There are two contributions to the potential energy, the elastic potential energy of the spring and the gravitational potential energy. We measure the vertical coordinate y from the initial position i, so that $y_i = 0$. The initial value of elastic potential energy is $\frac{1}{2}(2400\text{ N/m})(0.15\text{ m})^2 = 27\text{ J}$.

(a) The block is released from rest at i and comes to rest at f; the kinetic energy is zero at both points. Let y_f be the vertical coordinate of point f; the gravitational potential energy there is mgy_f. The elastic potential energy is zero for point f since the relaxed spring has been left behind. Conservation of mechanical energy gives $mgy_f = 27\text{ J}$. Inserting numerical values, we obtain $y_f = 1.3\text{ m}$ for the vertical coordinate of point f. Measured along the ramp, the distance s is related to y_f by $y_f = s \sin\theta$, or

$$s = \frac{1.3\text{ m}}{\sin 25°} = 3.1\text{ m}$$

(b) As the block comes halfway down on the return trip, the spring remains relaxed, and the 27 J of mechanical energy is divided into kinetic energy and gravitational potential energy. Let h label this point; then

$$\tfrac{1}{2}mv_h^2 + mgy_h = 27\text{ J}$$

with $y_h = \frac{1}{2}y_f = \frac{1}{2}(1.3\text{ m})$. Solving for v_h gives $v_h = 3.6\text{ m/s}$.

SELF-TEST 9-5. Suppose that the spring in the above example is initially compressed by 0.30 m instead of 0.15 m. *(a)* How far up the ramp does the block slide? *(b)* What is the speed of the block when it is halfway up the ramp? ***ANSWERS:*** *(a)* 12 m; *(b)* 7.2 m/s.

EXAMPLE 9-6

A loop-the-loop track. A small ice cube of mass m slides with negligible friction on a "loop-the-loop" track, as shown in Fig. 9-13. The ice starts from rest at a point $y_i = 4R$ above the level of the lowest part of the track. *(a)* What is the speed of the ice cube at point f, the highest point on the circular loop? *(b)* What normal force is exerted on the ice at this point?

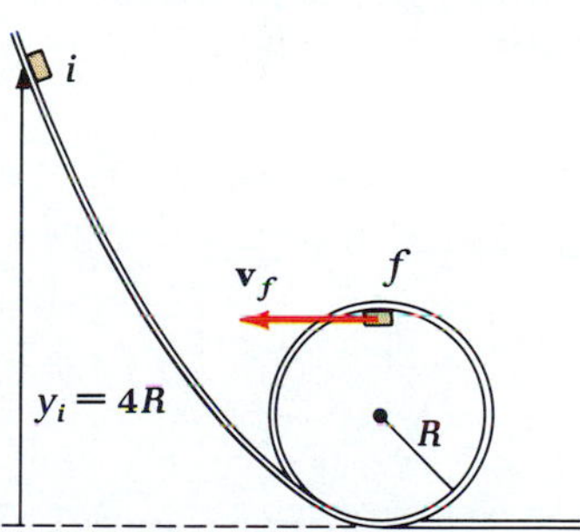

Figure 9-13. Example 9-6: A loop-the-loop track.

Solution. In any infinitesimal displacement of the ice cube along the track, the normal force does no work since it is perpendicular to the displacement. Neglecting frictional effects, we see that only the weight of the ice cube performs work, and it is a conservative force. Mechanical energy is conserved in this motion.

(a) At point i the speed v_i is zero and $y_i = 4R$. At point f the speed v_f is to be determined and $y_f = 2R$. Applying Eq. 9-11, we equate the mechanical energy at point i and f:

$$K_f + U_f = K_i + U_i$$

or,

$$\tfrac{1}{2}mv_f^2 + mg(2R) = 0 + mg(4R)$$

The kinetic energy at point f is then

$$\tfrac{1}{2}mv_f^2 = 2mgR$$

and the speed is $v_f = \sqrt{4gR}$.

(b) At point f both the normal force exerted by the track and the weight of the ice cube are directed downward. Together these provide the centripetal force, of magnitude mv^2/R, necessary for the circular path. From Newton's second law, $\Sigma\mathbf{F} = m\mathbf{a}$:

$$F_N + mg = \frac{mv_f^2}{R}$$

From part *(a)* we have

$$\frac{1}{2}mv_f^2 = 2mgR$$

or $mv_f^2 = 4mgR$ and $mv_f^2/R = 4mg$. The centripetal force here is 4 times the weight! Substituting this value and solving for the normal force gives $F_N = 4mg - mg = 3mg$. In arriving at our answer we can see that the lower the release point i, the slower the speed at point f and the smaller the normal force exerted by the track. An interesting question is: From what minimum height can the ice cube be released and still remain in contact with the track at point f? See Prob. 9-6.

SELF-TEST 9-6. *(a)* What is the speed of the ice cube in the above example at the lowest point of the circular track? *(b)* What is the normal force (magnitude and direction) exerted on the cube by the track at the lowest point of the circular track? ***ANSWERS:*** *(a)* $\sqrt{8gR}$; *(b)* $9mg$ upward.

9-5 NONCONSERVATIVE FORCES AND INTERNAL WORK

In each of the examples in the last section, we were careful to note that only conservative forces were performing work. If a nonconservative force, such as a frictional force, does work on an object, then mechanical energy is not conserved. In this case the mechanical energy can change during the motion of the object.

Nonconservative Forces

To see how the mechanical energy can change, we begin with the work-energy theorem, $K_f - K_i = W_{net}$, which is valid for all forces, nonconservative and conservative. We separate the net work done by all the forces acting on the object into two contributions, $W_{net} = W_{con} + W_{non}$. One contribution is the work done by the conservative forces, which equals the negative of the change in potential energy: $W_{con} = -(U_f - U_i)$. The other contribution is the work done by the nonconservative forces W_{non}. We cannot evaluate this work in general because its value depends on the details of the motion of the object. To evaluate W_{non}, we must know both the path and how the force varies along the path.

Work done by a nonconservative force depends on the path.

By dividing W_{net} into the conservative and nonconservative parts, we can write

$$K_f - K_i = W_{net} = W_{con} + W_{non} = -(U_f - U_i) + W_{non}$$

We rearrange the equation to get $K_f + U_f$ on the left side of the equation and $K_i + U_i$ on the right side:

$$K_f + U_f = K_i + U_i + W_{non}$$

Since $E_f = K_f + U_f$ is the mechanical energy at point f and $E_i = K_i + U_i$ is the mechanical energy at point i, then

Modified work-energy theorem

$$E_f = E_i + W_{non} \tag{9-12}$$

which is a modified form of the work-energy theorem. From this equation, we see that the change in mechanical energy, $E_f - E_i$, is equal to the work done by the nonconservative forces along the path from i to f. Notice as a special case that if $W_{non} = 0$ (no work performed by nonconservative forces), then $E_f = E_i$, and the conservation of mechanical energy is recovered.

The modified work-energy theorem provides a description of the motion of the object if we can evaluate the work done by the nonconservative forces in a specific case. By relating changes in kinetic and potential energies, the theorem connects the values of speed and position of the object. Alternatively, if we know the speed of the object at each of two positions i and f, we can evaluate the work done by the nonconservative forces. The following example illustrates this approach.

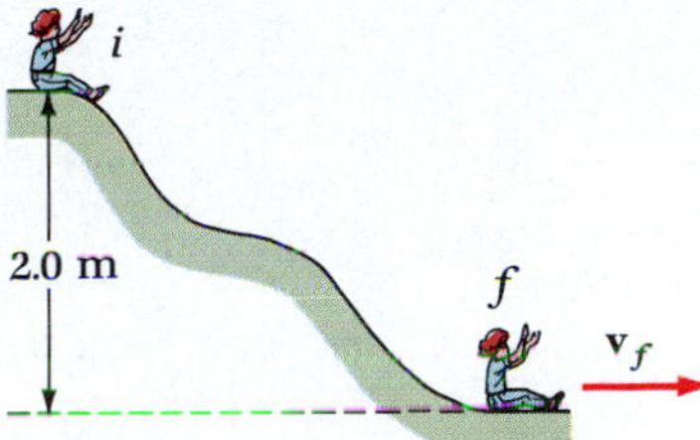

Figure 9-14. Example 9-7: A child on a slide.

EXAMPLE 9-7

Work done by friction. A 17-kg child starts from rest at the top of a 2.0-m slide, as shown in Fig. 9-14. Her speed at the bottom is 4.2 m/s. How much work is done by frictional forces?

Solution. The forces acting on the child are her weight, which is conservative; a normal force exerted by the slide surface, which performs no work; and the nonconservative frictional forces from the slide surface and air-resistance effects. In applying Eq. 9-12, we let $y_f = 0$ and $v_f = 4.2$ m/s so that $y_i = 2.0$ m and $v_i = 0$. This gives, since $U_f = 0$ and $K_i = 0$,

$$\begin{aligned} W_{\text{non}} &= E_f - E_i = K_f - U_i \\ &= \tfrac{1}{2}(17\text{ kg})(4.2\text{ m/s})^2 - (17\text{ kg})(9.8\text{ m/s}^2)(2.0\text{ m}) \\ &= -180\text{ J} \end{aligned}$$

The negative work done by friction corresponds to the decrease in the mechanical energy of the system.

SELF-TEST 9-7. How much work is done by the nonconvervative forces on the child in the above example if she starts from the top of the slide with a speed v_i and has the same speed at the bottom? ***ANSWER:*** 330 J.

Internal Work

In applying methods of work and energy, we have restricted our attention to the case of a single, inert object acted on by external forces. In the situation illustrated by Fig. 9-12, for example, the object we describe is the block. The external forces are exerted by the earth, by the spring, and by the ramp surface. Except for changing its position and speed, the block remains unchanged; it is treated as a particle.

We can also consider more complicated objects, however, which undergo changes in their makeup or composition or shape. Suppose that an athlete climbs hand over hand up a stationary vertical rope at a steady speed. The athlete's gravitational potential energy, $U = mgy$, is increasing. Since the kinetic energy is constant, the mechanical energy, $E = K + U$, is also increasing. Yet the rope does no work on the athlete. Although the rope exerts a force on a hand, the hand *does not move* while grasping the rope.

What is responsible for increasing the mechanical energy of the athlete? What provides the work? In this case, the athlete's contracting muscles perform work. That is, the arms do work in raising the rest of the athlete. These forces are internal to the object or system rather than external. The work done by internal forces exerted by one part of a system on another part is called ***internal work.*** In considering changes in the mechanical energy of a system, we must account for the internal work done by such internal forces.

Internal work can be done by internal forces.

9-6 THE LAW OF CONSERVATION OF ENERGY

Having considered a number of examples for which mechanical energy is conserved, we may have become accustomed to thinking of energy as a conserved quantity generally. If the mechanical energy of a system is not conserved, because nonconservative forces perform the work, then we tend to account for its change. For example, if the mechanical energy increases, we look for the source of this increase. If the mechanical energy decreases, we look for this energy in another form or in another place.

Suppose a crate is given a push so that it slides across a horizontal floor with an initial speed v_i. The initial mechanical energy is $E_i = K_i = \frac{1}{2}mv_i^2$. Because of frictional effects, the crate slows and comes to rest; its mechanical energy has been reduced to zero. We understand that loss of mechanical energy in terms of the work done by the nonconservative frictional force, $E_f - E_i = W_{\text{non}}$ from Eq. 9-12. We can also interpret the decrease of mechanical energy of the crate as an energy-conversion process, mechanical energy being converted into some other kind of energy. With this interpretation, the total amount of energy remains unchanged; only its form changes. What name do we give to this new kind of energy? From our everyday experiences with frictional forces, we expect that the transformed energy is associated with changes of the crate and floor. In particular, measurement shows that the temperature of these surfaces has increased. This energy is called the ***internal energy*** of the system consisting of crate and floor. The increase in internal energy is equal to the decrease in mechanical energy. The total energy of this system is conserved in this way. The internal energy of a system and its connection to temperature and "heat" will be discussed further in Chap. 17. For now we can think of the internal energy of a system as kinetic and potential energy of the molecules of that system.

Internal energy of a system: kinetic and potential energy of molecules

Internal energy, energy at the molecular level, has a different character from the kinetic and potential energy of an object such as a ball. It is a simple matter to convert the potential energy of a ball into kinetic energy; we just let it fall. It is also easy to convert mechanical energy into internal energy of the ball-floor system. After the ball bounces a few times, it comes to rest. Its initial mechanical energy has been converted into internal energy of the ball and floor. The reverse process, converting internal energy into mechanical energy, is not simple, however. We do not expect a ball initially at rest on the floor spontaneously to lower its internal energy and rise from the floor. The conversion from mechanical energy to internal energy as described above is associated with work done by nonconservative forces such as friction. These processes have a one-way character and are often called ***dissipative processes,*** since the mechanical energy is dissipated into the less accessible internal energy of a system.

In dissipative processes, energy becomes less accessible.

We generalize the ideas suggested above. Consider a *closed,* or *isolated system,* one on which no work is done by anything external to the system. No exchange of energy occurs between the system and its environment. We identify the various kinds of energy in the system. There is the kinetic energy of each moving macroscopic component of the system. There may be potential energy present because of elastic springs and gravitational forces. We have already seen the need to include the internal energy of various parts of the system. Other identifiable contributions to the total energy may include acoustic energy, electric energy, chemical energy, nuclear energy, and so on. In short, we should include all energies that could be changing with time. We add together all of these contributions evaluated at a certain time, calling that sum the ***total energy*** of the system.

The law of conservation of energy

> The law of conservation of energy states that the total energy of an isolated system is conserved.

The various contributions to the total energy can change with time, transforming from one type to another, but their sum does not change.

We have not proved the law of conservation of energy. It is a law in the same sense that Newton's second law is a law of nature. We accept it as true or valid so long as no violation is observed. To our knowledge, no violation has ever occurred. Indeed, our acceptance of the law is so strong that when apparent violations are observed, we search for a previously unidentified form of energy to enter into the balance. It was in this way that the existence of the neutrino, a subatomic particle, was proposed.

EXAMPLE 9-8

Sticky clay and energy transformations. A 2.5-kg ball of sticky clay is dropped from a height of 2.0 m above a stationary floor. On hitting the floor, the clay sticks to the floor. Account for energy transformations in the motion.

Solution. We take the clay, the floor, and the earth and its atmosphere as the isolated system. Since the clay is released from rest, $K_i = 0$ and $U_i = mgy_i$. We measure y_i from floor level, and

$$E_i = U_i = (2.5 \text{ kg})(9.8 \text{ m/s}^2)(2.0 \text{ m}) = 49 \text{ J}$$

On coming to rest ($v_i = 0$) on the floor ($y_f = 0$), the clay has zero mechanical energy. We must account for the 49 J of mechanical energy that has been transformed into other kinds of energy of the system. A small amount of the energy is acoustic; we should hear the clay hitting the floor. Because of the dissipative effects of air resistance and the deformation of the clay, most of the 49 J appears as an increase in the internal energy manifested by temperature increases in parts of the system.

SELF-TEST 9-8. A 0.070-kg tennis ball is released from rest at a height of 1.24 m above the floor and after one bounce reaches a maximum height of 0.61 m. Determine the amount of energy dissipated. ***ANSWER:*** 0.43 J.

9-7 SATELLITE MOTION AND ESCAPE SPEED

For an object of mass m close to the earth's surface, where the weight is approximately constant, we have used $U = mgy$ to evaluate the gravitational potential energy. For an object such as a satellite or an interplanetary probe, which is not near the surface, we must use a more general expression for gravitational potential energy.

Newton's law of universal gravitation gives the attractive force exerted by one particle on another particle. The force has the same form for two extended objects if they are spherically symmetric and nonintersecting, as shown in Fig. 9-15 for objects of mass m and M. From Eq. 7-2, the magnitude of the force is

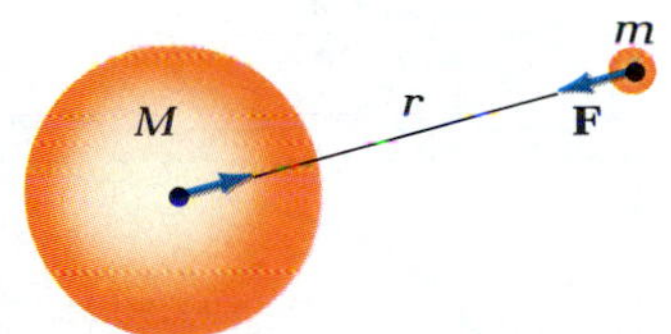

Figure 9-15. Two spherically symmetric objects exert gravitational forces on each other.

$$F = G\frac{Mm}{r^2}$$

where r is the separation of their centers. To determine the gravitational potential energy, we need to calculate the work done by this force on one of the objects as it moves with respect to the other object.

First we note that the gravitational force is conservative, that the work done by this force is independent of the path. Figure 9-16*a* shows a small displacement $\Delta\ell$ (exaggerated for clarity), with the force on the body of mass m directed along the line joining the objects. Figure 9-16*b* shows that the displacement $\Delta\ell$ changes the separation of the two bodies from r to $r + \Delta r$ where

$$\Delta r = |\Delta\ell| \cos(\pi - \phi) = -|\Delta\ell| \cos\phi$$

The work done by $\mathbf{F}$ for this displacement is $\mathbf{F} \cdot \Delta\boldsymbol{\ell} = F|\Delta\ell| \cos\phi = -F\,\Delta r$. The work done for this displacement depends then only on the separation distance r and the change Δr in that separation distance, but not at all on the distance moved perpendicular to the line joining the two bodies. It is this feature that makes the work path-independent. (See Prob. 5.) For a general path the work done is the integral

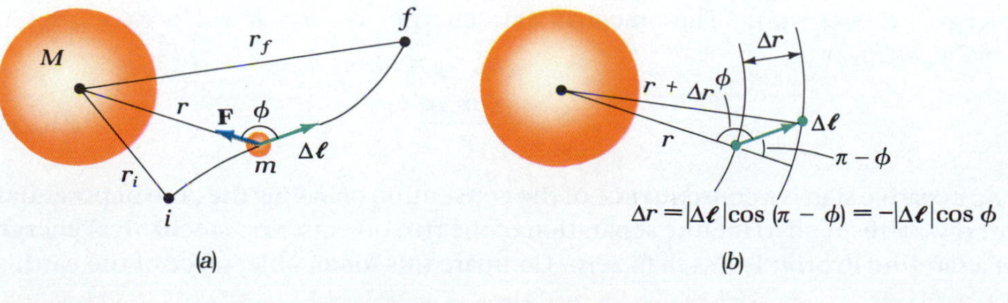

Figure 9-16. *(a)* An object of mass m moves along an arbitrary path connecting i and f. *(b)* The change in the separation distance r is $\Delta r = |\Delta\ell| \cos(\pi - \phi) = -|\Delta\ell| \cos\phi$.

$$W = \int_i^f \mathbf{F} \cdot d\boldsymbol{\ell} = -\int_{r_i}^{r_f} G\frac{Mm}{r^2}\,dr = -GMm\int_{r_i}^{r_f} \frac{1}{r^2}\,dr = GMm\left(\frac{1}{r_f} - \frac{1}{r_i}\right)$$

The change in gravitational potential energy of this system is just the negative of the work done by the gravitational force, $U_f - U_i = -W$, from Eq. 9-10. If the separation of the two objects changes from r_i to r_f, then the potential energy difference is

Gravitational potential energy difference

$$U_f - U_i = -GMm\left(\frac{1}{r_f} - \frac{1}{r_i}\right) \qquad (9\text{-}13)$$

Since the separation r of the two objects can be arbitrarily large, it is conventional to choose the potential energy to be zero for an infinite separation. With this choice, the gravitational potential energy for the two objects separated by a distance r is

Gravitational potential energy for objects separated by a distance r

$$U(r) = -\frac{GMm}{r} \qquad (9\text{-}14)$$

Notice from Eq. 9-14, shown graphically in Fig. 9-17, that as the separation r of the objects tends to infinity, the potential energy $U(r)$ approaches zero. Since the zero of potential energy has been chosen to correspond to infinite separation of the two objects, the potential energy for a finite separation is less than zero, or negative. The potential energy $U(r)$ increases as the separation r increases. See Exercise 40 to relate Eqs. 9-13 and 9-14 to the special case of mgy for an object close to the earth's surface.

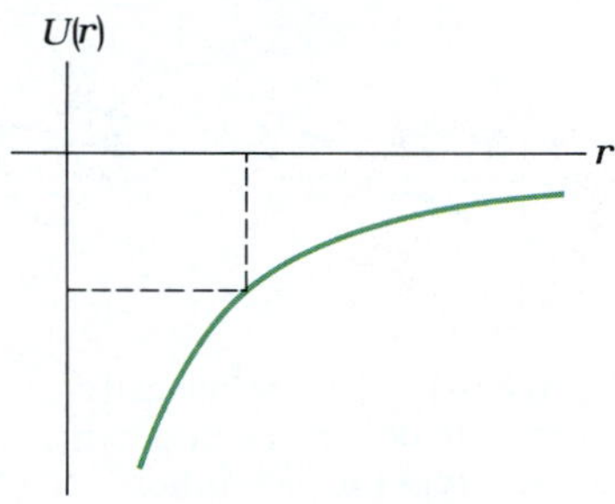

Figure 9-17. Gravitational potential energy $U = -GMm/r$ is negative, approaching zero for infinite separation. As r increases, U increases from more negative to less negative values.

Consider a satellite of mass m in orbit about the earth. Assume that the satellite can be treated as a particle and that the earth is spherically symmetric, with mass m_e. The gravitational potential energy is given by Eq. 9-14 with $M = m_e$, $U(r) = -Gm_em/r$. If the orbit is elliptical, then r and $U(r)$ will be changing. Mechanical energy is conserved because only the gravitational force acts on the satellite, $K + U = E$, or

$$\tfrac{1}{2}mv^2 + \left(-\frac{Gm_em}{r}\right) = E$$

As r increases, the potential energy increases and the kinetic energy decreases. The satellite moves more slowly when farther from the earth, more rapidly when closer to the earth.

For the special case of a circular orbit, the potential energy and the kinetic energy, each remains fixed. We can determine the values of the kinetic energy and the mechanical energy for a circular orbit. The centripetal force is provided by the gravitational force. Newton's second law, $m\mathbf{a} = \Sigma\mathbf{F}$, gives

$$\frac{mv^2}{r} = \frac{Gm_em}{r^2}$$

On multiplying this equation by $\tfrac{1}{2}r$, we obtain

$$\tfrac{1}{2}mv^2 = \frac{\tfrac{1}{2}Gm_em}{r}$$

The kinetic energy of the satellite is one-half the magnitude of the potential energy, $K = \tfrac{1}{2}(-U)$. The mechanical energy is $E = K + U = \tfrac{1}{2}Gm_em/r + (-Gm_em/r)$, or

Energy of a satellite in a circular orbit

$$E = -\frac{\tfrac{1}{2}Gm_em}{r} \qquad (9\text{-}15)$$

The negative sign is a consequence of the convention of having the zero of potential energy correspond to infinite separation of the two objects. The mechanical energy of a satellite in orbit is less than zero. Compare this for an object free of the earth's influence, as $r \rightarrow \infty$. Since $U \rightarrow 0$ and $K \geq 0$ for the free object, the mechanical energy $E = K + U \geq 0$. Thus a satellite in orbit has less energy than if free and so is *bound* to the earth. The absolute value of its mechanical energy is called the ***binding energy*** of the satellite. That much energy would have to be provided to free the satellite from the earth's gravitational influence. For a circular orbit, the binding energy is $|E| = \frac{1}{2}Gm_e m/r$.

A satellite's binding energy

Suppose we give an object sufficient energy to escape from the earth. Depending on its initial location, there is a minimum speed, or ***escape speed,*** that the object must have. We can determine the escape speed using conversation of mechanical energy, $E_f = E_i$. To escape, the object can just come to rest ($K_f = 0$) on achieving infinite separation from the earth ($U_f = 0$). That is, its mechanical energy $E_f = K_f + U_f$ must be at least zero to escape. The object initially at distance r_i ($U_i = -Gm_e m/r_i$) from the earth has escape speed v_i ($K_i = \frac{1}{2}mv_i^2$) at that point if $E_i = \frac{1}{2}mv_i^2 + (-Gm_e m/r_i) = 0$. Thus

$$v_i = \sqrt{\frac{2Gm_e}{r_i}} \tag{9-16}$$

Escape speed from the earth

is the escape speed from that position. Notice that the escape speed depends on the mass of the earth but not on the mass of the escaping object.

EXAMPLE 9-9

Energy and satellites. A 150-kg satellite is in circular orbit of radius 7.3 Mm about the earth. Evaluate *(a)* the potential, kinetic, and mechanical energies and *(b)* the orbital speed. *(c)* What is the escape speed from this altitude?

Solution. *(a)* The potential energy, $U = -Gm_e m/r$, is

$$U = \frac{(6.67 \times 10^{-11}\ \text{N m}^2\ \text{kg}^{-2})(5.98 \times 10^{24}\ \text{kg})(150\ \text{kg})}{7.3\ \text{Mm}}$$
$$= -8.2 \times 10^9\ \text{J} = -8.2\ \text{GJ}$$

The kinetic energy is one-half the magnitude of the potential energy, $K = \frac{1}{2}(-U) = 4.1$ GJ. The mechanical energy is

$$E = K + U = 4.1\ \text{GJ} + (-8.2\ \text{GJ}) = -4.1\ \text{GJ}$$

(b) From the kinetic energy, we can calculate the orbital speed: $v = \sqrt{2K/m} = 7.4$ km/s.
(c) The escape speed from this distance from the earth's center is obtained from Eq. 9-16:

$$v_i = \sqrt{\frac{2Gm_e}{r_i}} = 10\ \text{km/s}$$

SELF-TEST 9-9. A satellite is in an elliptical orbit about the earth. When it is closest to the earth, 9.00×10^6 m from the center, its speed is 6980 m/s. What is the satellite's speed when it is 1.00×10^7 m from the center? ***ANSWER:*** 6660 m/s.

9-8 PROBLEM-SOLVING TECHNIQUES

The concepts of energy and energy conservation provide useful and powerful tools for solving many of the problems that we encounter in this and in succeeding chapters, as well as in science and engineering generally. However, the tools must not be applied blindly. For example, using an equation that expresses conservation of mechanical energy is *wrong* if a nonconservative force such as friction is performing work. Listed below are some steps that will help you to apply energy methods systematically in solving problems.

1. Clearly identify the object or system involved. The system you choose should be an isolated one—that is, one on which no work (or negligible work) is done by any part of its external environment. For example, a system could consist of a

falling ball and the earth, if air-resistance effects are negligible. Often a sketch will help distinguish the system and its environment.

2. Identify the forces acting on various parts of the system and determine which forces are performing work. Again, a sketch can help sort out the forces acting on the system. There are two possibilities:
   a. Only conservative foces are performing work. Then the mechanical energy of the system is conserved. For each conservative force, there is a potential energy. In this chapter we have identified gravitational potential energy and elastic potential energy. For each particle in the system that can move, there is a kinetic energy $\frac{1}{2}mv^2$. The mechanical energy is the sum of the kinetic and potential energies of the system.
   b. Nonconservative forces are performing work. Then the mechanical energy of the system is not conserved and one or more other forms of energy are changing. Identify such other forms of energy and add them to the mechanical energy to obtain the total energy for the system.
3. Identify two appropriate times, say t_i and t_f, for which data such as positions and speeds are known or are to be determined. Use the data to evaluate the expression for the total energy at these two times, E_i and E_f, and set them equal: $E_i = E_f$. This gives one equation that can be solved for one unknown.
4. If there are additional unknowns, then consider other principles that can be applied, such as Newton's second law, to obtain additional equations that contain the unknowns.
5. Solve the equations for the unknown quantities. If possible, solve the equations in terms of the symbols for the given quantities and wait to substitute numerical values until the very end.
6. As usual, check your answer for reasonableness. One useful check is to substitute the now known quantities into the expression $E_i = E_f$ and verify that the equality is satisfied.

COMMENTARY: What Is a Law?

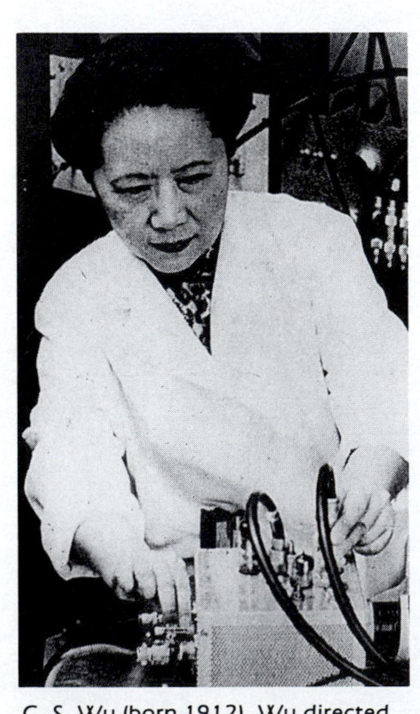

C. S. Wu (born 1912). Wu directed the experiments that showed that parity is not conserved in certain nuclear processes.

If you violate one of society's laws, say a law dealing with fraud, you could face a prison term or a fine or both. The usual role of criminal and civil law in a community is to protect its citizens. What is the role of a law in physics? What happens if you violate one of nature's laws? Is it even possible to break a law of nature?

So far we have encountered five basic laws: Newton's three laws of motion, Newton's law of universal gravitation, and, in this chapter, the law of conservation of energy. We can broaden our understanding of the nature of physical laws by describing some of their general features.

First of all, a law is a fundamental statement that helps form the foundation of a conceptual framework. In this sense, a law is similar to a postulate in plane geometry. In geometry, a postulate is accepted without proof, and various consequences are derived (theorems are proved) from a set of postulates. Similarly in mechanics, Newton's third law is accepted as a fundamental statement about the way forces occur as action-reaction pairs. Newton's third law serves as a postulate.

There is an important difference between a postulate in geometry and a law in physics, however. A mathematical postulate in geometry is not subject to validation. That is, its truth or validity is not open to question: It cannot be tested because mathematics is a pure abstraction and does not have a physical basis. A law in physics does have a physical basis; it should always be regarded as "on trial." Its truth is tested by our insistence that the law, and its consequences, be fully in accord with experiment. If an experiment shows that a new effect is contrary to a law, then the law is abandoned, or at least modified to include the new effect. Thus the laws of physics are not necessarily immutable; they must change in response to

C. N. Yang (born 1922) and T. D. Lee (born 1926). From theoretical work, Yang and Lee predicted a violation of conservation of parity.

newly discovered phenomena. For example, the law of conservation of energy was modified by Einstein to include mass energy, a previously unrecognized form of energy (described in Chap. 38). In doing so, Einstein merged what had been two separate, unrelated conservation laws — conservation of energy and conservation of mass — into a single, more general law.

Despite the word "law" in its name, Hooke's law for a spring, $F_x = -kx$, is not a law in the sense that we are using the term. That is, Hooke's law is not a fundamental statement or principle. Instead, it is a simple and useful rule that characterizes how a material, such as spring steel, responds to an extension or a compression. Hooke's law is an example of an *empirical law,* a rule that summarizes our experience over a limited range of conditions. In contrast, Newton's law of universal gravitation gives the gravitational force between any two particles; and this forms the basis for describing gravitational phenomena, such as the structure of the solar system.

Can you willfully violate one of nature's laws? With what penalty? Suppose that you could devise an ingenious experiment with a result that conflicted with a law. Strangely enough, the penalty could be a Nobel Prize! Violations of nature's laws are richly rewarded. C. N. Yang and T. D. Lee received the 1957 Nobel physics award for suggesting the possibility of observing a violation of what had been accepted as the law of conservation of parity. (Parity has to do with how a description of a system changes if one of the coordinate axes, say the x axis, is inverted through the origin as if it were reflected in a mirror.) The confirming observation was subsequently made by C. S. Wu. She demonstrated that parity is not conserved in certain nuclear decay processes. So go ahead and break a law of physics — if you can.

SUMMARY

Section 9-1. One-Dimensional Conservative Systems

A force is conservative if the work it does is zero for a round-trip. In a one-dimensional conservative system, only conservative forces perform work and mechanical energy, $E = K + U$, is conserved. Potential energy is an energy of position. Two types of potential energy are gravitational potential energy,

$$U = mgy \tag{9-5}$$

and elastic potential energy of a stretched or compressed spring,

$$U = \tfrac{1}{2}kx^2 \tag{9-7}$$

Section 9-2. Graphical Analysis of Conservative Systems

Motion for an object can be displayed on a graph of the potential-energy function. Turning points and equilibrium points can be located on the graph.

Section 9-3. Conservative Forces and Potential Energy in Three Dimensions

In general, a force is conservative if its work is independent of the path connecting the endpoints i and f. The change in potential energy is the negative of the work done by the conservative force.

Section 9-4. Conservation of Mechanical Energy

If only conservative forces do work, then the mechanical energy of the system is conserved. Conservation of mechanical energy can be used to determine features of the motion of the system.

Sections 9-5 and 9-6. Nonconservative Forces and Internal Work and The Law of Conservation of Energy

Work done by nonconservative forces and forces internal to a system can cause the mechanical energy to change. The total energy of a system is the sum of various types of energy, including kinetic, potential, and internal energies. Transformations of energy obey the law of conservation of energy: For an isolated system, the total energy is conserved.

Section 9-7. Satellite Motion and Escape Speed

The gravitational potential energy for two spherically symmetric objects separated by a distance r is given by

$$U(r) = -\frac{GMm}{r} \tag{9-14}$$

where M and m are the masses. A satellite is bound to the earth and its mechanical energy is negative. An object escaping from the earth must have a minimum speed, which is called the escape speed.

Section 9-8. Problem-Solving Techniques

To apply energy-conservation methods to solve problems: identify the isolated system, determine which energies are changing, equate the total energy at two different times, apply additional principles if needed, solve for the unknown quantities.

QUESTIONS

1. Is something conserved by a conservative force? Explain.

2. Is the static frictional force conservative or nonconservative? Explain.

3. What are the differences in the meaning of the phrase "conservation of energy" as used in this text and as used in the news media in connection with energy shortages?

4. Under what circumstances can a normal force, exerted by a surface on some object, perform work? Is this work path-dependent? Explain.

5. Consider an automobile traveling at constant speed on a level road surface. What relevant energy transformations are occurring?

6. As an automobile begins moving, how does it get its kinetic energy?

7. A parachutist leaves a plane at an altitude of 3000 m and immediately opens the parachute. What becomes of her potential energy as she drifts downward at a relatively low speed?

8. If 1.0 kg of water changes its temperature by 1.0°C, its internal energy changes by about 4.2 kJ. Through what vertical height would the 1.0 kg of water have to fall to change its gravitational potential energy by that amount?

9. Can you start a fire by rubbing two sticks together? (Neither stick is a match.) Comment on the energy transformations involved in the attempt.

10. Why is potential energy defined with a negative sign, as in Eq. 9-3? Suppose it were positive instead. What important consequences would result?

11. How can the gravitational potential energy *(a)* be positive (mgy) for an object above but close to the earth's surface and *(b)* be negative ($-Gm_em/r$) for an object above the earth's surface?

12. For an object close to the earth's surface, the y component of the force is always negative, $F_y = -mg$, while the potential energy may be positive or negative, $U = mgy$. What corresponding statements can be made for an object connected to a spring obeying Hooke's law?

13. Mechanical energy is conserved if only conservative forces perform work. Suppose no forces perform work. Is mechanical energy conserved? Explain.

14. Potential energy is often said to be stored energy, as in the gravitational potential energy of the water behind a dam. Can kinetic energy be considered as stored energy? Explain.

15. Gravitational potential energy and kinetic energy can be transformed into each other. Elastic potential energy of a spring and kinetic energy can be transformed into each other. Can elastic potential energy be transformed directly into gravitational potential energy? Explain.

16. Estimate the change in your gravitational potential energy in going from the first to the second floor of a department store by *(a)* riding the escalator, *(b)* riding the elevator, *(c)* running at top speed up the stairs.

17. Identify in each case posed in the previous question the energy transformations that lead to the increase in gravitational potential energy.

18. Develop an analogy between transformations of energy and financial transactions involving coins, currency, checks, loans, and other forms of assets and liabilities.

19. Consider a satellite in each of several possible circular orbits of different radii. To have the satellite speed increase, must its mechanical energy increase or decrease? Explain.

20. How far from earth must an object move before it is *effectively* free of the earth's influence? What do you mean by *effectively* free? Do your answers depend on the location of the object relative to other objects in the solar system? Explain.

21. Consider two satellites of mass m in circular orbits of radii r_1 and r_2, with $r_1 < r_2$. Which satellite has the larger *(a)* kinetic energy, *(b)* potential energy, *(c)* mechanical energy, *(d)* binding energy?

22. The expression for gravitational potential energy in Eq. 9-14 is based on the assumption that both objects are spherically symmetric. Assume that the earth is spherically symmetric. Is a typical satellite spherically symmetric? What about the moon? Explain why the shape of an artificial satellite is not important when considering potential energy.

23. Complete the following table:

Symbol	Represents	Type	SI Unit
U			J
$U_f - U_i$			
E			
$\oint \mathbf{F} \cdot d\boldsymbol{\ell}$			
mgy	Gravitational potential energy		
$-GMm/r$		Scalar	
$\frac{1}{2}kx^2$			

EXERCISES

Section 9-1. One-Dimensional Conservative Systems

1. A 0.55-kg stone is thrown straight up with an initial speed of $v_i = 14$ m/s. Air-resistance effects may be neglected. *(a)* Evaluate the mechanical energy of this system. *(b)* What is the potential energy as the stone reaches the highest point in its motion? *(c)* How high is this point?

2. A 0.22-kg pebble is thrown straight down with an initial speed of 12 m/s from a bridge 15 m above the water surface. Neglect air-resistance effects. *(a)* Determine the mechanical energy of this system. *(b)* With what speed does the pebble reach the water?

3. A 0.75-kg block is connected to a spring of spring constant $k = 2100$ N/m, as shown in Fig. 9-4, and the block is set in motion with a mechanical energy of 47 J. Neglecting frictional effects, determine *(a)* the maximum displacement of the block from its equilibrium position, *(b)* the maximum speed of the block, *(c)* the displacement of the block when its speed is 5.6 m/s.

4. Show that mgy and $\frac{1}{2}kx^2$ have dimensions of energy.

5. A 12,000-kg railroad car rolls at 4.3 m/s with negligible friction on a horizontal track, as shown in Fig. 9-18. Near the end of the track, the car hits and compresses a bumper spring by 0.23 m and comes momentarily to rest. Assuming that only the conservative spring force performs work on the car, determine the spring constant of the spring.

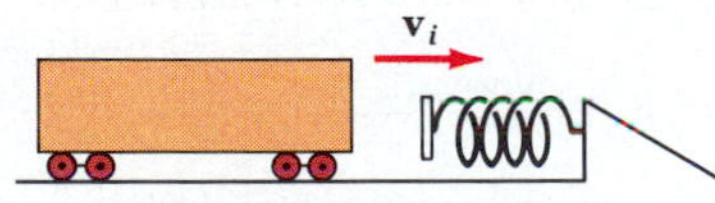

Figure 9-18. Exercise 5.

6. Suppose the force exerted on an object in one dimension is given by $F_x(x) = -\alpha x^3$ where α is a constant with units of newtons per meter cubed (N/m³). Determine the expression for the potential energy corresponding to this conservative force. Let $U = 0$ at $x = 0$.

7. A freight elevator and a 75-kg carton on the elevator floor accelerate upward from rest with a constant acceleration, $a_y = 2.4$ m/s². On reaching the next floor 3.8 m above, the elevator and carton go into free-fall because of an elevator-cable failure. *(a)* What work is done by the normal force exerted on the carton by the elevator floor as they accelerate upward? *(b)* Is this normal force conservative? Explain. *(c)* With what final speed does the carton pass its starting point on the way down?

8. Consider the arrangement shown in Fig. 9-19*a*. Suppose that both springs are relaxed when the block is at $x = 0$. In Fig. 9-19*b* the block is displaced so that x is the amount of stretch for one spring and the amount of compression for the other. Assume that friction is negligible and determine expressions for *(a)* the force component $F_x(x)$ acting on the block and *(b)* the potential energy of this system.

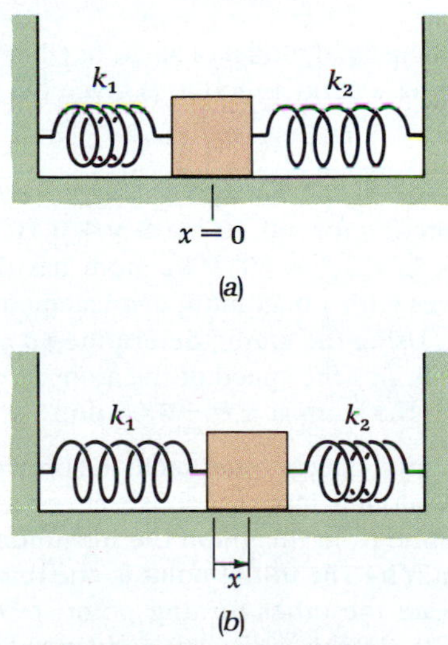

Figure 9-19. Exercise 8.

9. The mass of the block in the previous exercise is 5.0 kg and the springs have constants $k_1 = 1200$ N/m and $k_2 = 1800$ N/m. The block is pulled out and released from rest at $x = 0.20$ m. *(a)* Determine the maximum kinetic energy of the block. Determine the speed of the block at *(b)* $x = 0.0$ and *(c)* $x = -0.10$ m.

10. The block in Example 9-3 is set in motion so that its maximum speed is v_m. *(a)* Show that when the block is at position x, its speed is

$$v = \sqrt{v_m^2 - \frac{k}{m}x^2}$$

Let $k = 1100$ N/m, $m = 2.5$ kg, and $v_m = 3.0$ m/s. Determine *(b)* the maximum distance x_m of the block from equilibrium, *(c)* the speed at $x = -\frac{1}{2}x_m$, *(d)* the position of the block when $v = \frac{1}{2}v_m$.

Section 9-2. Graphical Analysis of Conservative Systems

11. The gravitational-potential-energy function $U(y) = mgy$ is graphed in Fig. 9-20 for a 10.2-kg object close to the earth's surface; $y = 0$ corresponds to ground level. Suppose the mechanical energy of the system is 0.20 kJ. From the graph determine *(a)* the maximum height of the object, *(b)* the maximum kinetic energy and the point where the object has that maximum kinetic energy, *(c)* the location of the object when its kinetic energy equals the potential energy, *(d)* the force on the object at that instant.

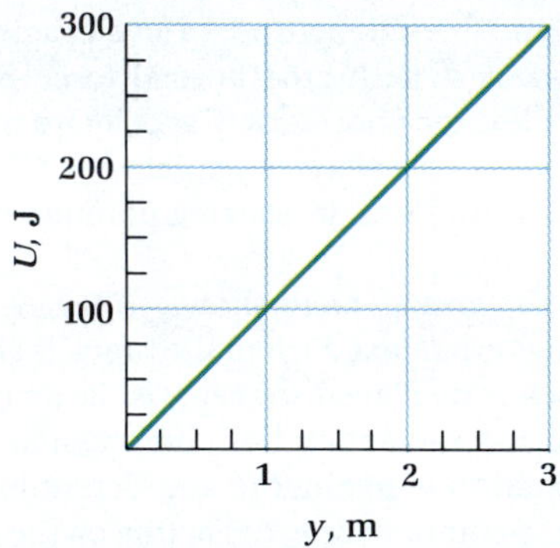

Figure 9-20. Exercise 11.

12. In a one-dimensional model of an atom vibrating in a molecule, the potential-energy function is given by

$$U = \tfrac{1}{4}\alpha x^4 - \tfrac{1}{2}kx^2$$

with $\alpha = 16 \times 10^{20}$ N/m³ and $k = 4.0$ N/m. *(a)* Construct a graph of this function for -0.10 nm $\leq x \leq 0.10$ nm (1 nm $= 10^{-9}$ m). Suppose a 2.3×10^{-26}-kg atom has this potential energy and moves with a maximum displacement of 0.08 nm from the origin. Using the graph, determine *(b)* the speed of the atom at $x = 0$, *(c)* the speed of the atom at $x = 0.05$ nm, *(d)* the force on the atom at $x = -0.05$ nm.

13. For the potential-energy function in the previous exercise, suppose that the atom is instantaneously at rest at $x_i = -0.02$ nm. *(a)* Determine from the graph the maximum kinetic energy of the atom. *(b)* The initial point is one turning point in the motion; locate the other turning point. *(c)* How is the motion for $E < 0$ (in this exercise) fundamentally different from the motion for $E > 0$ (in Exercise 12)?

14. A simple model of a hydrogen molecule uses a one-dimensional potential energy $U(x) = U_0(e^{-2x/a} - 2e^{-x/a})$, where $U_0 = 7.5 \times 10^{-19}$ J and $a = 7.0 \times 10^{-11}$ m. *(a)* Construct a graph of $U(x)$ versus x for $-1.5 \leq x/a \leq 3$. From the graph, determine the turning points of the motion if *(b)* $E = -2.5 \times 10^{-19}$ J and *(c)* $E = +2.5 \times 10^{-19}$ J.

Section 9-3. Conservative Forces and Potential Energy in Three Dimensions

15. Estimate the work done on you by the gravitational force if you *(a)* climb a ladder from the ground to a rooftop 3 m above and *(b)* jump from roof level to the ground and walk back to the bottom of the ladder. *(c)* How much work is done on you by the gravitational force for the round-trip?

16. Consider a constant force $\mathbf{F} = (3\text{ N})\mathbf{i} + (4\text{ N})\mathbf{j}$ which acts on an object moving in the xy plane. Show that the work done by this force as the object moves from the point (x_i, y_i) to the point (x_f, y_f) is independent of the path. Note that $\mathbf{F} \cdot d\boldsymbol{\ell} = F_x\,dx + F_y\,dy$ and evaluate the integral without specifying the path.

17. The force in the previous exercise is conservative. (Why?) *(a)* Determine the expression for the difference in potential energy between points (x_i, y_i) and (x_f, y_f). *(b)* Choose the origin (0, 0) as the reference point at which $U = 0$. What is the potential-energy function $U(x, y)$? Evaluate the potential energy at the point *(c)* (8 m, 6 m) and *(d)* (−8 m, 6 m).

Section 9-4. Conservation of Mechanical Energy

18. Neglecting air resistance, show that the speed of a projectile as it reaches an altitude y depends on the initial speed v_i but is independent of the angle of projection. See Fig. 9-21.

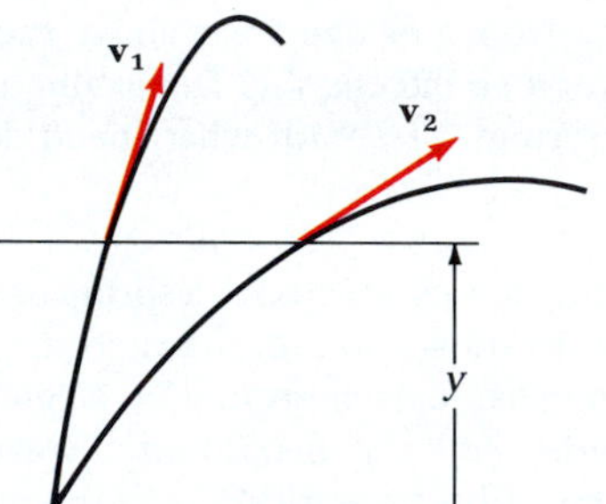

Figure 9-21. Exercise 18: Identical projectiles are fired with the same initial speed but at different angles.

19. A 0.25-kg ball is thrown so that it travels a horizontal distance of 37 m in 2.0 s and reaches a maximum height of 18 m. Air resistance is negligible, so the horizontal velocity component is constant. Determine *(a)* the mechanical energy of the ball, *(b)* the initial speed of the ball, *(c)* the maximum potential energy of the ball.

20. A projectile of mass m is fired with initial speed v_i at an angle of θ_i from the horizontal, as shown in Fig. 9-22. Neglect effects of air resistance and use conservation of mechanical

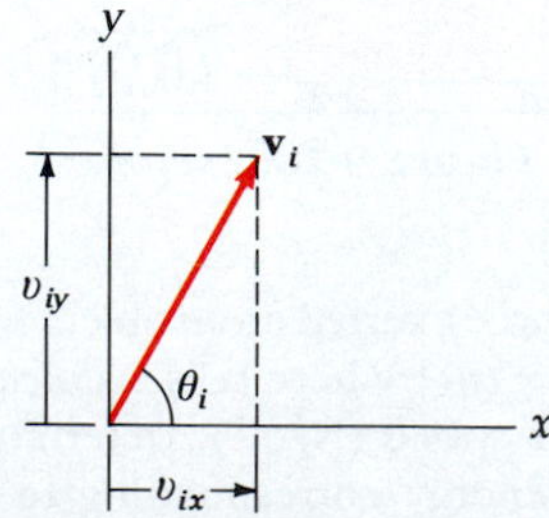

Figure 9-22. Exercise 20.

energy to determine *(a)* the maximum altitude reached and *(b)* the speed of the projectile on returning to ground level. Remember that the x component of the velocity does not change in this motion.

21. A simple pendulum is formed by attaching a ball to one end of a light string; the other end of the string is held fixed, and the ball can swing in a vertical plane. Suppose the ball is released from rest from the position shown in Fig. 9-23 with $L = 450$ mm and $\theta = 30.0°$. Determine *(a)* the speed of the ball and *(b)* the tension in the string when the ball passes through its lowest position.

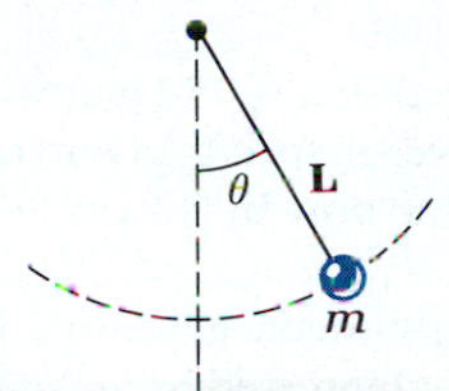

Figure 9-23. Exercise 21.

22. A ball of mass m is attached to a light string and moves in a vertical circle of radius R. It is given an initial velocity of magnitude v_0 at point O, as shown in Fig. 9-24. Determine, in terms of m, g, v_0, and R, the speed of the ball and the string tension at *(a)* point A and *(b)* point B.

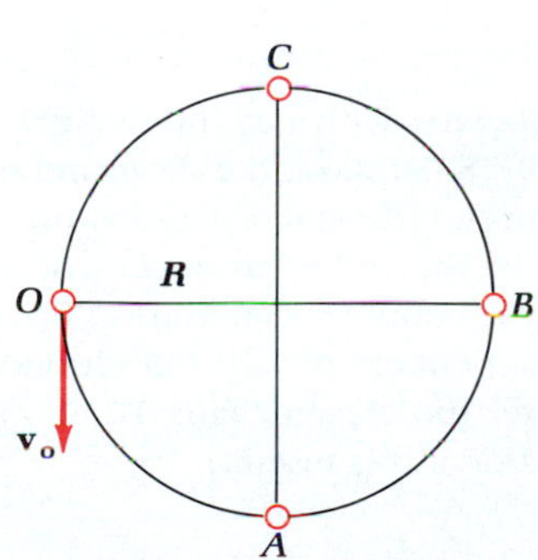

Figure 9-24. Exercise 22.

23. What minimum speed must the ball in the previous exercise be given at point O so that the string does not go slack before the ball reaches point C? See Fig. 9-24.

24. A light spring of constant $k = 1600$ N/m is compressed by 15 mm. A 75-g marble is placed against the spring, as shown in Fig. 9-25. The marble is fired upward when the spring is released. Assume that the marble leaves the spring behind in its relaxed configuration and that frictional effects are negligible. *(a)* What maximum height is attained by the marble? *(b)* With what speed does the marble leave the spring?

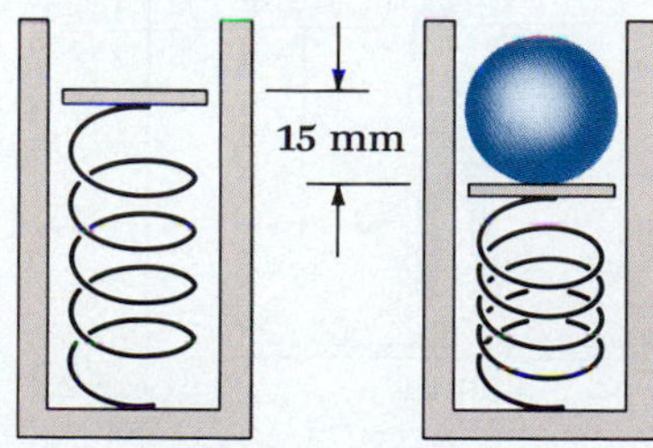

Figure 9-25. Exercise 24.

25. A spring gun similar to that in the previous exercise is used to fire a 75-g marble horizontally from a countertop that is 1.2 m above the floor, as shown in Fig. 9-26. If the spring is compressed by 25 mm, the marble hits the floor 4.2 m from the bottom of the counter, as measured along the floor. Neglect all frictional effects. *(a)* Determine the mechanical energy of the marble in its trajectory. *(b)* Determine the spring constant of the spring. *(c)* How far will the marble travel horizontally if the spring is compressed by 37 mm?

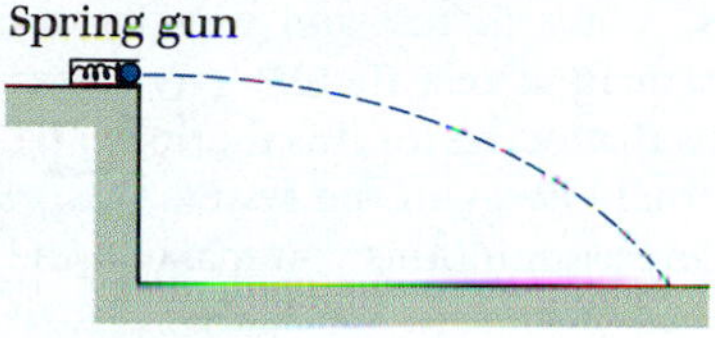

Figure 9-26. Exercise 25.

26. A child sits at the top of a cylindrical tank of radius R, as shown in Fig. 9-27. The surface is very smooth and the child begins to slide with negligible friction. Determine the value of the angle at which the child loses contact with the cylindrical surface.

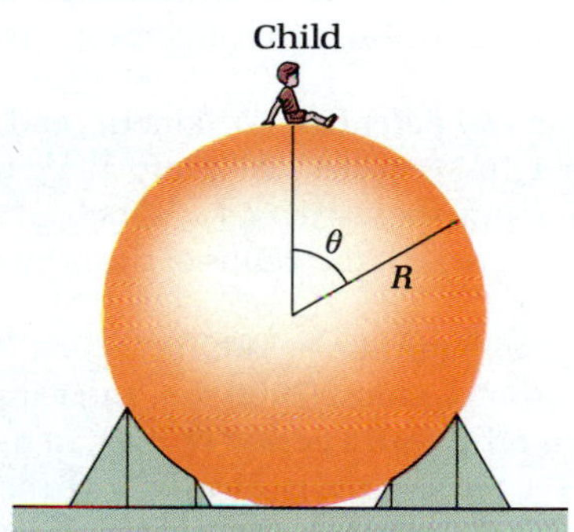

Figure 9-27. Exercise 26.

Section 9-5. Nonconservative Forces and Internal Work

27. Determine the work done by the kinetic frictional force on a 50-kg crate that is pushed horizontally *(a)* around a semicircle of diameter 4 m and *(b)* straight along a diameter. Take $\mu_k = 0.3$.

28. A 0.25-kg rubber ball is released from rest at a height of 1.5 m above the floor. After hitting the floor, the ball rises to a maximum height of 0.8 m. Estimate *(a)* the work done by the floor on the ball and *(b)* the speed of the ball just before and just after it comes into contact with the floor.

29. Suppose that you walk up a flight of stairs at constant speed while changing your vertical coordinate by 10 m. *(a)* Estimate the change in your mechanical energy. *(b)* Explain why the normal force exerted on your shoes by the steps performs no work. *(c)* How do you account for the change in your mechanical energy? What forces perform work?

Section 9-6. The Law of Conservation of Energy

30. *(a)* Estimate the running speed of a pole vaulter who clears the bar at 4 m. *(b)* Describe the energy transformations that occur and explain why the pole is necessary.

31. A 1000-kg automobile coasts down a 5 percent grade at 36 km/h. *(a)* Estimate the magnitude of the net retarding force due to air resistance, friction, and so on. *(b)* Estimate the minimum power provided by the engine if the automobile travels up the grade at a steady 36 km/h.

32. Suppose that frictional effects are not negligible in Example 9-3. The block is released from rest at $x_i = 0.15$ m and after a few seconds comes to rest and remains at $x_f = -0.02$ m. (How can it remain at rest there?) *(a)* Describe the energy transformations that occur for this motion. *(b)* Determine the change in internal energy of the system of spring, block, and surface. State any assumptions you make in the calculation.

33. A 0.33-kg ball is projected upward with an initial speed of $v_i = 23$ m/s. On reaching a level of $y_f = 14$ m, its speed is $v_f = 13$ m/s. Consider the system of projectile, atmosphere, and earth to be isolated. *(a)* Account quantitatively for energy transformations between these two points in the motion by determining the changes in kinetic, potential, and internal energies. *(b)* What can you say about the speed of the projectile as it returns to the 14-m level on its descent?

Section 9-7. Satellite Motion and Escape Speed

34. Show that $U = -GMm/r$ has dimensions of energy.

35. Evaluate the *(a)* potential, *(b)* kinetic, and *(c)* mechanical energies of a 30-Mg space laboratory in a circular orbit of radius 70 Mm about the earth. *(d)* Which, if any, of these energies would increase if the radius of the orbit were smaller?

36. A 125-kg communications satellite is first "parked" in a circular earth orbit of radius 7000 km. Later it is moved to a geosynchronous orbit with a period of 24 h, so that as the earth rotates on its axis, the satellite is always above the same spot on the equator. *(a)* Determine the radius of the geosynchronous orbit. *(b)* What additional energy must be supplied to move the satellite from the parking orbit to the geosynchronous orbit?

37. The planet Mercury has mass 3.3×10^{23} kg and moves in an approximately circular orbit of radius 5.8×10^{10} m about the sun, which has a mass of 2.0×10^{30} kg. *(a)* Determine the mechanical energy of the Mercury-sun system. *(b)* Suppose Mercury were moved to a circular orbit of radius 15×10^{10} m, equal to that of the earth about the sun. How much energy would have to be provided?

38. The mass of the earth is 81 times the mass of the moon. The earth's radius is 3.7 times the moon's radius. Compare the escape speeds of an object from the surfaces of these bodies. Does this comparison help explain why the moon has no atmosphere?

39. A satellite is in an elliptical orbit about the earth. The separation of the satellite from the center of the earth ranges from a minimum at perigee of 7.2 Mm, where its speed is 8.0 km/s, to a maximum of 9.9 Mm at the apogee. Determine the speed of the satellite when *(a)* at the apogee and *(b)* at a separation of 8.4 Mm from the earth's center.

40. For an object close to the earth's surface, the difference in gravitational potential energy is given by $U_f - U_i = mg(y_f - y_i)$. Alternatively, Eq. 9-13 can be used by letting $r_i = R + y_i$ and $r_f = R + y_f$, where R is the earth's radius. Show that Eq. 9-13 approximately reduces to $U_f - U_i = mg(y_f - y_i)$. [*Hint:* Remember that $g = Gm_e/R^2$ and that $(1 + z)^{-1} \approx 1 - z$ for $|z| \ll 1$.]

Additional Exercises

41. A common flea can jump about 20 cm high. Estimate the initial kinetic energy of a 200-mg flea as it leaves the surface.

42. Estimate the speed of a 0.05-kg arrow as it leaves a bow. Treat the bow as if it were a spring, given that a force of 150 N is required to "draw" the bow by 30 cm.

43. When four 50-kg students get into a 1000-kg car, each of the four car springs is depressed by an additional 0.01 m. *(a)* Determine the spring constant for each spring. *(b)* Estimate the potential energy of each spring (i) before and (ii) after the students are in the car.

44. A 1-kg stone is released from rest at a height $y = 4$ m above the ground. Construct a table with values of y ranging in 1-m steps from 4 m to 0. For each value of y, show values of the gravitational potential energy, the kinetic energy, and the speed of the ball.

45. An elevator operates with a counterweight as shown schematically in Fig. 9-28. Suppose the drive and braking mechanisms fail when the elevator is at rest on an upper floor. Neglect frictional effects in the pulley arrangement as the elevator moves down and the counterweight moves up. Determine the change in potential energy of *(a)* the elevator and *(b)* the counterweight after the elevator falls 10 m. *(c)* What is the speed of the elevator at this instant?

Figure 9-28. Exercise 45.

46. The potential energy of an atom vibrating in a so-called soft mode in a crystal is given by $U(x) = \alpha x^4$, where $\alpha = 1 \times 10^{21}$ J/m^4 is a constant. *(a)* Construct a graph of $U(x)$ for $-1.2a \leq x \leq 1.2a$, where $a = 2 \times 10^{-10}$ m. *(b)* Determine the mechanical energy of the atom if one turning point is at $x = \frac{1}{2}a$. *(c)* What is the kinetic energy of the atom at $x = -\frac{1}{4}a$? *(d)* Determine an expression for the force component F_x on the atom as a function of x. *(e)* Evaluate F_x at $x = \frac{1}{2}a$.

47. The "double-well" potential energy function shown in Fig. 9-29 provides a useful model for a molecular vibration. *(a)* Locate any positions of stable equilibrium. *(b)* Suppose the molecule has one turning point at $x = -4 \times 10^{-11}$ m. Where is the other turning point and what is the mechanical energy? Repeat part *(b)* if one turning point is *(c)* at $x = +4 \times 10^{-11}$ m, *(d)* at $x = 4.5 \times 10^{-11}$ m.

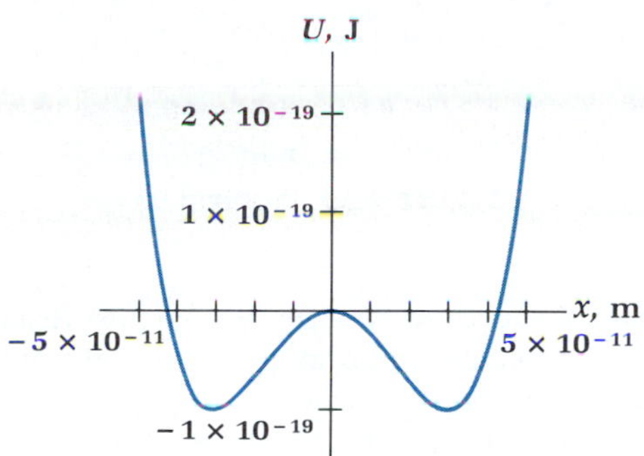

Figure 9-29. Exercise 47.

48. A constant force $\mathbf{F} = F_1\mathbf{i} + F_2\mathbf{j}$ has constant components F_1 and F_2. *(a)* Show that the work done by this force along an arbitrary path connecting points (x_i, y_i) and (x_f, y_f) is given by $W = F_1(x_f - x_i) + F_2(y_f - y_i)$. *(b)* What is an expression for the potential energy associated with this force?

49. A 60-kg skydiver falls with a constant terminal speed. What work is done by the air-resistance force on the skydiver for each 10 m of fall? What can you say about the internal energy of the air and the skydiver?

50. Determine the change in the gravitational potential energy of a 1-kg object that is moved from ground level to a position *(a)* 1 m, *(b)* 1 km, *(c)* 1 Mm, *(d)* 1 Gm above the earth's surface.

51. The Hubble space telescope is in a circular orbit 610 km above the earth's surface. Determine the telescope's orbital *(a)* speed and *(b)* period. *(c)* What would its escape speed be from this altitude?

52. The mass of the Hubble space telescope in the previous exercise is about 9000 kg. Determine its gravitational potential energy *(a)* before launch when the telescope is at the earth's surface and *(b)* when the telescope is in a circular orbit of altitude 610 km. *(c)* What is the telescope's kinetic energy in this circular orbit?

The Hubble space telescope.

PROBLEMS

1. ***Energy transformations.*** A 7.3-kg block begins sliding from the top of a ramp, as shown in Fig. 9-30. At the bottom of the ramp is a light spring of constant $k = 210$ N/m. Neglect frictional effects. *(a)* By how much is the spring compressed as the block comes to rest? *(b)* What is the speed of the block as it reaches the spring?

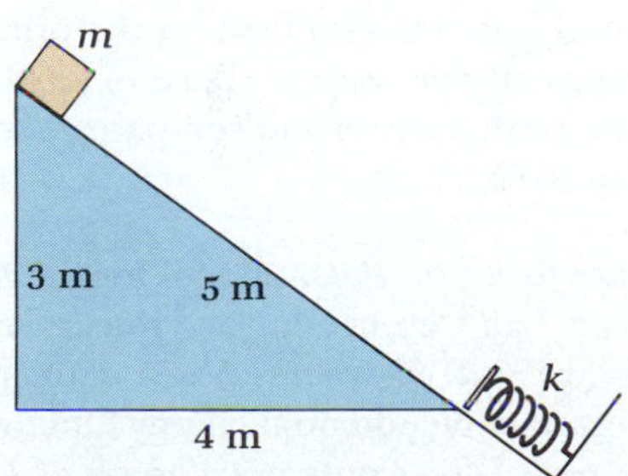

Figure 9-30. Problems 1 and 17.

2. ***Including friction.*** Take kinetic frictional effects into account in the situation posed in the previous problem. Let $\mu_k = 0.10$ and determine *(a)* the speed as the block reaches the spring and *(b)* the maximum compression of the spring.

3. ***Potential energy of a vertical spring.*** A light spring is suspended vertically, as shown in Fig. 9-31*a*. A block of mass m is attached and set in vertical motion. Let y denote the position of the block as seen in Fig. 9-31*b* so that $y = 0$ corresponds to the position where the spring is relaxed. *(a)* Write down the potential-energy function as the sum of elastic and gravitational contributions. *(b)* Determine the net force acting on the block as a function of the coordinate y. *(c)* At what value of y, call it y_1, is the net force zero? *(d)* Show that the potential-energy function found in part *(a)* can be written as $\frac{1}{2}k(y - y_1)^2$ + a constant term. *(e)* By resetting the origin at the equilibrium point, $y - y_1 \rightarrow y$, and resetting the zero of potential energy at this equilibrium point, show that the potential energy of a block connected to a vertical spring can be expressed as $U = \frac{1}{2}ky^2$.

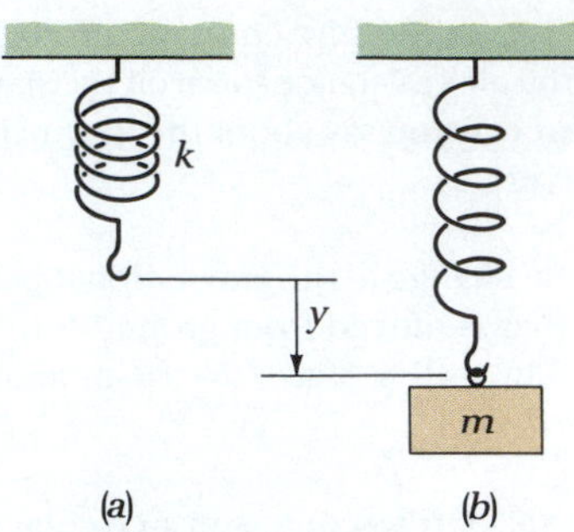

Figure 9-31. Problem 3.

4. ***A 12-6 potential energy.*** The potential energy of interaction between neutral atoms is sometimes approximated by the "12-6" potential-energy function:

$$U(x) = V_0 \left[\left(\frac{a}{x} \right)^{12} - 2 \left(\frac{a}{x} \right)^{6} \right]$$

where x is the separation between centers of the atoms and V_0 and a are constants. *(a)* Construct a graph of this function with $a = 0.30$ nm and $V_0 = 3.2 \times 10^{-21}$ J for 0.24 nm $\leq x \leq$ 0.36 nm. *(b)* Determine from the graph the equilibrium separation where $F_x = 0$. *(c)* Check this value by using the connection $F_x(x) = -dU/dx$. *(d)* Suppose one turning point in the motion is at $x = 0.28$ nm; locate the other turning point. *(e)* Determine the mechanical energy for the motion between these turning points.

5. ***Central forces are conservative.*** If a force on an object is always directed along a line from the object to a given point, which we take as the origin in three dimensions, and the magnitude of the force depends only on the separation of the object from the origin, the force is said to be a *central force.* After reviewing the discussion preceding Eq. 9-13, show that any central force is a conservative force.

6. ***Completing the loop.*** In Example 9-6, a small ice cube slides without friction on a "loop-the-loop" track. Determine the minimum height above the bottom of the circular track from which the cube can be released and still remain in contact with the track at the highest point.

7. ***Using energy methods.*** Two blocks are connected by a light string which passes over a small pulley, as shown in Fig. 9-32. They are released from rest, and frictional effects may be neglected. *(a)* Show that the sum of the work done by the string force on the blocks is zero. *(b)* Using energy methods, determine the common speed of the blocks when m_1 has fallen a distance h. Express your answer in terms of m_1, m_2, g, and h.

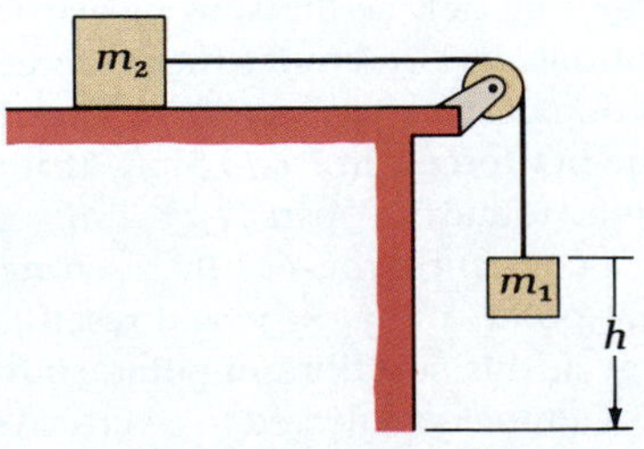

Figure 9-32. Problem 7.

8. ***Energy methods with friction.*** Reconsider the previous problem, but include friction; let μ_k represent the coefficient of kinetic friction. Determine the speed after m_1 has fallen a distance h.

9. ***Analyzing energy graphs.*** Construct a graph of the force component $F_x(x)$ from the potential-energy graph in Fig. 9-33.

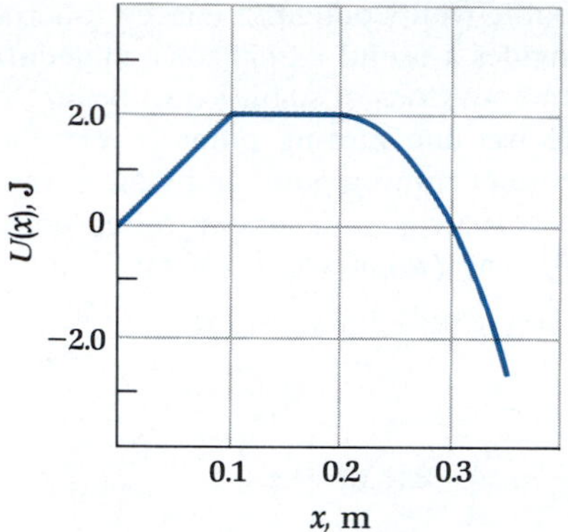

Figure 9-33. Problem 9.

10. ***Potential energy of two springs.*** Suppose that the block in Exercise 8 is in equilibrium at $x = 0$, but that neither spring is relaxed; one spring is stretched by an amount a_1, while the other is stretched by a_2 when $x = 0$. *(a)* By requiring the net force on the block to be zero at $x = 0$, show that $k_1a_1 = k_2a_2$. Determine expressions for *(b)* the elastic potential energy $U(x)$ of this system and *(c)* the force component $F_x(x)$ acting on the block.

11. ***Satellite motion.*** Modify the spreadsheet in Table 7-2 to determine the orbit of the satellite in Exercise 39. *(a)* Print values of the speed along with distance from the earth to check the values calculated in that exercise. Take as initial conditions: X = 0, Y = 7.2E + 6, VX = 8.0E + 3, VY = 0. *(b)* Redo the spreadsheet and display values of the kinetic, potential, and mechanical energies at each iteration. Is mechanical energy conserved?

12. ***Escape speed.*** The spreadsheet in the previous problem can be adapted to investigate escape speeds. For simplicity, take the initial velocity to be directed away from the earth. The initial conditions are X = 0, Y = initial distance from the center of earth, VX = 0, and VY = your estimate of the escape speed. If you underestimate the escape-speed value, the velocity component VY will eventually become negative, indicating that the object is returning to earth. You may have to use a larger incremental time interval than for the orbital problem. Determine numerically the escape speed of an object initially at 7 Mm from the earth's center and compare your answer with that given by Eq. 9-16.

13. ***A molecular force model.*** A model for the force component acting on an atom in a linear molecule is $F_x(x) = -2Ax/(x^2 + a^2)^2$, where A and a are positive constants. *(a)* Determine the potential energy function $U(x)$ such that $U \rightarrow 0$ as $x \rightarrow \infty$. *(b)* Construct a graph of $U(x)$ versus x and identify any positions of stable equilibrium.

14. ***Look out below!*** A 2-kg block is released from rest at the top of an office building. After falling 120 m, the block effectively reaches a constant terminal speed $v_t = 20$ m/s because of air resistance. *(a)* How much work has been done by air resistance? *(b)* How much work is done by air resistance for each subsequent meter of fall?

15. ***Galactic motion.*** The mass of our sun is about 2×10^{30} kg. *(a)* Estimate the sun's gravitational potential energy of interaction with our galaxy. Assume that the galaxy contains 1×10^{11} sunlike stars within the sun's circular orbit of radius 3×10^{20} m about the galactic center. *(b)* Estimate the orbital speed of the sun.

16. ***Motion in a vertical circle.*** A ball of mass m is tied to one end of a string and the other end of the string is held fixed. The ball then swings in a vertical circle of radius L, as shown in Fig. 9-34. The speed of the ball at the lowest point in its path is v_o. Determine expressions for *(a)* the speed of the ball and *(b)* the tension in the string in terms of m, g, v_o, L, and θ. *(c)* What is the minimum value of v_o such that the path is circular?

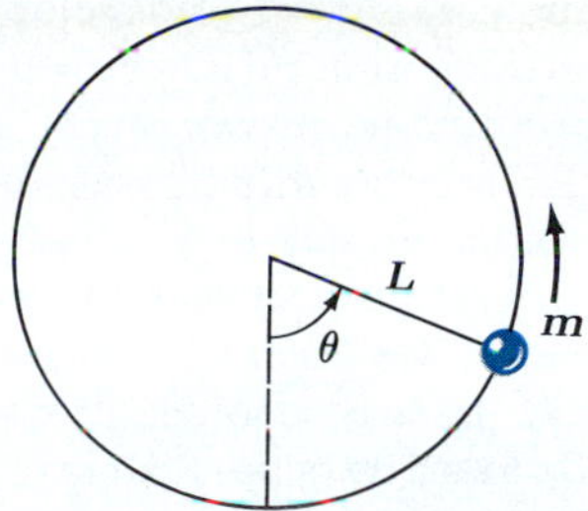

Figure 9-34. Problem 16.

17. ***Moving up the ramp.*** The 7.3-kg block in Prob. 1 is held against the 210-N/m spring compressing it by an amount s. The block is then released and reaches the highest point on the ramp with a speed of 3.4 m/s. Determine the initial compression s of the spring. Neglect friction and the mass of the spring.

18. ***Satellite energies.*** Identical satellites, each of mass m, are put in circular earth orbits of radius R_1 and R_2, respectively, with $R_2 > R_1$. Which satellite has the greater *(a)* gravitational potential energy, *(b)* kinetic energy, *(c)* mechanical energy? *(d)* Satellite 2 is subsequently caused to move in the same orbit as satellite 1. Determine expressions for the change in the potential, kinetic, and mechanical energies of the satellite in terms of G, m_e, m, R_1, R_2.

19. ***Hydroelectric energy conversion.*** A hydroelectric facility produces electric energy at a rate of 100 MW by having water turn a turbine-generator system. *(a)* If the water falls 100 m vertically in reaching the turbine, estimate the amount of water per unit time that exits the facility. *(b)* The water is drawn from a 1-km by 1-km reservoir. At what rate does the water level drop? (Water has a density of 10^3 kg/m^3.)

20. ***Firing a projectile straight up.*** A projectile leaves the earth's surface with an initial speed v_o. Neglect air-resistance effects and show that the projectile reaches a height h above the surface given by

$$h = \frac{v_o^2/(2g)}{1 - v_o^2/(2gR)}$$

where R is the radius of the earth (and $g = Gm_e/R^2$). Why must v_o be less than $\sqrt{2gR}$?

21. ***An electric force in hydrogen.*** In a simple model of a hydrogen atom, an electron moves about a (stationary) proton which exerts an attractive electric force on the electron. The electric force, which is discussed in Chap. 20, is an inverse-square force and can be treated in analogy with the gravitational force. The magnitude of the force on the electron is $F = K/r^2$, where $K = 2.3 \times 10^{-28}$ N·m^2. Determine expressions for the *(a)* electric potential energy, *(b)* kinetic energy, and *(c)* mechanical energy for an electron in a circular orbit of radius R. *(d)* Evaluate each expression for $R = 0.53 \times 10^{-10}$ m.

22. ***Elevator estimates.*** A hotel elevator is designed to carry a load of up to 2000 kg, in addition to its own mass of 2000 kg, at a top speed of 4 m/s. Estimate the power requirements for the operating motor under the following conditions: the elevator is moving *(a)* up at top speed, *(b)* down at top speed, *(c)* accelerating upward, *(d)* accelerating downward. Specify any assumptions that you make about friction, the magnitude of the acceleration, and the effects of a counterweight.

23. ***A change in rest-mass energy.*** An isolated nucleus of an atom of the element Einsteinium (Es) is at rest and spontaneously decays into three products—the nucleus of an atom of Fermium (Fm), an electron, and an antineutrino. One type of energy for this system that changes is *rest-mass energy*—it decreases by 8×10^{-14} J. When the decay products are well separated (their potential energy of interaction is zero), the electron and the Fm atom have final kinetic energies that total 6×10^{-14} J. Determine the energy of the antineutrino.

HIGHLIGHTS OF MODERN PHYSICS

The Anthropic Principle

In 1937 P. A. M. Dirac proposed that the gravitational constant G is not really constant, but decreases with time. He noticed the near-equality of several large numbers obtained by combining various universal constants, such as the speed of light c, the charge of the electron e, the masses of the proton m_p and electron m_e, and G itself. Although the electron is really a point particle, a kind of radius r_e can be associated with it by setting its electrical self-energy e^2/r_e equal to its rest energy mc^2, giving $r_e = e^2/mc^2$. If the present age of the universe t_o (15 billion years) is divided by the time, r_e/c, for light to travel this distance, the result is $N_1 = 6 \times 10^{39}$. On the other hand, the ratio of the electric to gravitational forces between an electron and proton is $N_2 = e^2/Gm_pm_e = 2 \times 10^{39}$. These two large numbers, calculated in such very different ways, are strikingly close to each other. Is this simply a coincidence?

Dirac thought not. He proposed that their approximate equality was due to some underlying law of nature unknown to us, that they might be exactly equal if we knew the correct multiplicative factors. This was a bold assertion because if $N_1 = N_2$, then the expression for the present age of the universe is

$$t_o = e^4/Gm_pm_e^2c^3$$

How can this be right? It equates a number that changes, t_o, to a constant quantity. If the equation is correct today, it will be incorrect in the future. However, Dirac thought it unlikely that we live in a special moment in history when these numbers coincidentally happen to be equal. He postulated that the equality holds for all time, meaning that at least one of the constants on the right side of the equation must change as the universe ages. Letting any of the atomic constants change would have clear, observable consequences, so he decided G must be the variable, in which case, $G \propto 1/t_o$. Dirac concluded that G decreases with time, therefore, it must have been larger in the past.

This idea was widely criticized because a G that decreases with time would have serious, observable consequences. As pointed out in Chap. 44, a higher G in the past implies a brighter sun, and therefore a hotter earth. Going back in time, a point would be reached when the earth's oceans would boil, yet we know life existed at that point. Modern measurements have shown that, if G does vary, it must do so at a *much slower* rate than that postulated by Dirac.

This does not mean, however, that the equality of N_1 and N_2 is a simple coincidence. The Princeton physicist Robert Dicke pointed out that we *do* live in a special time. In fact, this is the only era of time that human beings could possibly live. At times much earlier than the present, there would not have been enough time for stars and planets to form and life to evolve. At much later times, the stars that can support life on their planets will have burned out.

The sun is typical of stars that have significant life-supporting zones surrounding them. Life has only recently developed on one of its planets, and the sun will evolve to a red giant in another five billion years or so. In the entire past and future of the universe, there is a *relatively* brief period of time when physicists can exist to calculate numbers like N_1 and N_2. It is not surprising, then, that a special equality might be satisfied even though it is only temporary (in a cosmic sense). Put another way, if we could do an experiment with a large number of universes identical to this one, at some point, life would evolve in each one, and a physicist would discover Dirac's large number equality.

This is an application of the ***weak anthropic principle: all values we observe or calculate must be consistent with the existence of carbon-based life on earth.*** Most physicists agree with this statement, but there is much disagreement as to whether it has any scientific content. To some, it is tautological, saying only that "we observe what we observe." To others, it has real predictive and scientific value. It has certainly provided a penetrating insight into the near equality of N_1 and N_2.

The most impressive prediction based on the weak anthropic principle was made by the British cosmologist Fred Hoyle. Hoyle predicted the existence of a previously unknown energy state of the ^{12}C nucleus. For life to develop, carbon must be formed inside stars from primordial hydrogen and helium. Carbon is formed from helium in two steps. First, two helium nuclei combine to form beryllium. Beryllium only lasts for 10^{-17} s before breaking up into two helium nuclei again, but this is long enough so that sometimes it collides with another helium to form carbon. However, this will happen only if the reaction probability is large enough. Hoyle knew that this probability would ordinarily be so small at the kinetic energies of the nuclei in the sun that the production of carbon would be negligible. He concluded that ^{12}C must have a state at an energy equal to the total energy of the ^{8}Be and ^{4}He nuclei (7.4 MeV) plus the kinetic energy of the collision, for a total of 7.7 MeV (at the temperature inside the sun). This condition would allow a resonance reaction that would occur with much higher probability. Since carbon-based life *has* formed, he predicted that ^{12}C has a state near 7.7 MeV. Shortly thereafter, a state at 7.66 MeV was found experimentally.

Copernicus eliminated our special status by removing the earth from the center of the universe. In 1973 Brandon Carter of Cambridge suggested that we should not interpret this to mean that we play *no* special role in the universe. The weak anthropic principle is one statement of this special role: the fact that we exist places certain requirements on the way the universe can be. Carter went further by proposing the ***strong anthropic principle: The universe must be such that life can form.*** In the weak form it would be possible for a universe to exist that was not compatible with life; it would probably not look like our universe, but it would be possible. In the strong form, such universes are not allowed. As you might guess, this statement is highly controversial.

Carter noticed that the universe seemed to be fine-tuned to support life. We have already noted that the energy levels of the ^{12}C nucleus are just right to permit the evolution of life. It also turns out that only if ordinary space is *three*-dimensional will planetary orbits be stable; presumably, the earth must have a stable orbit in order that conditions be constant enough for the development of life. Carter believed that the fine structure constant (a dimensionless constant formed from atomic quantities) and electronic mass and charge have the precise values necessary for life to develop. He argued that if G, for example, were slightly smaller, all stars would be small, faint red dwarfs that are unable to sustain life on orbiting planets. If G were slightly larger, only large, bright stars would form, and they would not last long enough for life to develop.

It has often been pointed out that ice's property of being less dense than the liquid state is very unusual and prevents lakes and ponds from completely freezing from top to bottom in the winter. If atomic constants had been different so that ice was denser than liquid water, the ice would sink to the bottom and would accumulate over time since it would not be exposed to sun and air during the summer. Life could probably not evolve or survive in such situations. (This argument is usually made by physicists who live where lakes actually feeze over in the winter.)

Carter suggested that these coincidences (and others not mentioned here) are not really coincidences. The strong anthropic principle says that these situations come about because only a universe in which life could develop is permitted. If so, this could go a long way toward explaining why the various physical constants have their observed values, a long-time goal of physicists.

Dissenters, of whom there are many, point out that there is no demonstration that *any* of the physical constants are constrained to their observed values by the condition that life exists in the universe. Furthermore, it is naive to assume that if the universal constants were altered some rather different life form might not be possible. It is difficult enough to try to deduce what would happen if a single constant were changed; it is far harder to know what would happen if several were changed simultaneously. Finally, the problem is reminiscent of a well-known conundrum: Since the chances that you would end up with your particular set of DNA sequences is vanishingly small, is it not miraculous that you, yourself, were born? Of course, the answer is that *some* specific DNA sequence had to occur; the one that actually occurred, we call "you." Similarly, our universe had to have *some* set of values for its universal constants and, since we exist, they are obviously compatible with life, a consequence of the weak principle. For such reasons, many find that the strong principle is too strong. Its claim that the universe *must* be the way it is is analogous to the claim that you are the inevitable consequence of your parents' DNA, that you could not have been otherwise. Certainly, we are not forced to accept the strong principle even if life is possible only with the actual, observed constants.

QUESTIONS

1. The resonance level for $^4He + {}^{12}C \rightarrow {}^{16}O$ (^{16}O is ordinary oxygen) lies just *below* the total energy of the ^4He and ^{12}C nuclei. Why is this important for the development of life in the universe? (*Hint:* What would be the consequence if ^{12}C were rapidly converted to ^{16}O?)

EXERCISES

1. Estimate the temperature required for the *total* collision energy of the nuclei in Ques. 1 to be 0.3 MeV. Chapter 44 shows that the energy of reacting nuclei is about an order of magnitude above the average thermal energy; use this fact.
2. Calculate the ratio of the gravitational to coulomb force between two protons. Does the ratio depend on the separation between protons?

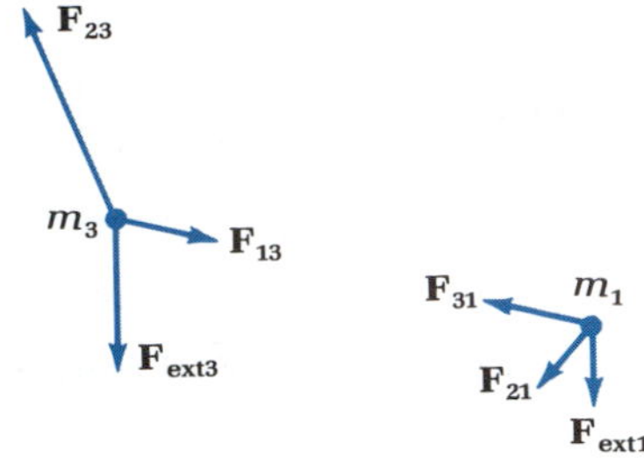

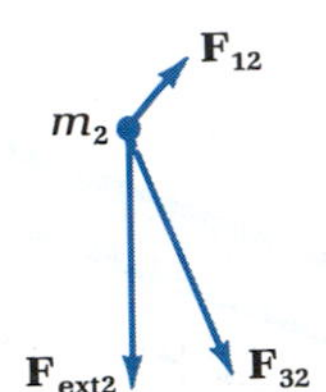

Figure 10-5. Three particles form a system. Each particle exerts a force on the other two particles as shown. $\mathbf{F}_{21}$ is the force that particle 2 exerts on particle 1. There are also forces from sources external to the system, such as that exerted on particle 1, $\mathbf{F}_{ext1}$.

Newton's Laws and Center-of-Mass Motion

Figure 10-5 shows a system of three particles. Each particle has three forces exerted on it. For example, particle 1 has force $\mathbf{F}_{21}$ exerted on it by particle 2, it has force $\mathbf{F}_{31}$ exerted on it by particle 3, and it has force $\mathbf{F}_{ext1}$ exerted on it by an object that is external to our three-particle system. (There may be more than one external agent, but for simplicity only one is considered.) The forces that the particles exert on each other are ***internal forces,*** and the forces exerted by external agents are ***external forces.*** (An external force might be the gravitational force exerted on a particle by the earth.) Newton's second law, $\Sigma\mathbf{F} = m\mathbf{a}$, applied to each particle gives

$$\mathbf{F}_{21} + \mathbf{F}_{31} + \mathbf{F}_{ext1} = m_1\mathbf{a}_1$$

$$\mathbf{F}_{12} + \mathbf{F}_{32} + \mathbf{F}_{ext2} = m_2\mathbf{a}_2$$

$$\mathbf{F}_{13} + \mathbf{F}_{23} + \mathbf{F}_{ext3} = m_3\mathbf{a}_3$$

Now add these three equations, and while doing so, use Newton's third law. Newton's third law states that the action-reaction forces are equal and opposite:

$$\mathbf{F}_{21} = -\mathbf{F}_{12} \qquad \mathbf{F}_{31} = -\mathbf{F}_{13} \qquad \mathbf{F}_{32} = -\mathbf{F}_{23}$$

When the equations are added, the internal forces cancel in pairs so that the sum of all internal forces is zero. This is illustrated in Fig. 10-5, which shows each action-reaction pair as equal in magnitude and opposite in direction. The sum of the three equations is

$$\mathbf{F}_{ext1} + \mathbf{F}_{ext2} + \mathbf{F}_{ext3} = m_1\mathbf{a}_1 + m_2\mathbf{a}_2 + m_3\mathbf{a}_3$$

By Newton's third law, the vector sum of the internal forces is zero for a system with any number of particles. Extending the argument to any number of particles gives

$$\Sigma\mathbf{F}_{ext} = \Sigma m_i\mathbf{a}_i \tag{10-8}$$

where $\Sigma\mathbf{F}_{ext}$ represents the sum of the external forces exerted on each particle in the system. That is, $\Sigma\mathbf{F}_{ext}$ is the ***net external force*** exerted on the system.

Equation 10-7, $\mathbf{a}_{cm} = (\Sigma m_i\mathbf{a}_i)/M$, comes directly from the definition of the center of mass. Rearranging, $M\mathbf{a}_{cm} = \Sigma m_i\mathbf{a}_i$, so that Eq. 10-8 can be written

Newton's second law for a system or an extended object

$$\Sigma\mathbf{F}_{ext} = M\mathbf{a}_{cm} \tag{10-9}$$

> The net external force exerted on a system is proportional to the acceleration of the center of mass of the system, and the system's total mass is the proportionality factor.

This is Newton's second law for the motion of the center of mass of a system (or of an extended object). The motion of the center of mass of a system is the same as that of a single particle with mass M when the net force is $\Sigma\mathbf{F}_{ext}$. As far as the motion of the center of mass is concerned, the system behaves as if all its mass were concentrated at the center of mass; the only forces that affect the motion are external forces. Forces internal to the system have no effect on the motion of the center of mass.

Now it is clear why the red light that marked the center of mass of the baton in Fig. 10-1 moved along a simple parabolic path while other points on the baton moved along more complex paths. The motion of the center of mass was determined by the external gravitational force exerted by the earth and was not influenced by internal forces.

EXAMPLE 10-4

A midair explosion. Two rubber balls (ball A and ball B) with the same mass are squeezed together. While in this compressed condition, the balls are bound with several loops of thread. This two-ball system is tossed into the air. During the first part of its flight, the system follows the parabolic path shown in blue in Fig. 10-6 (air resistance is neglected). Its position is marked with a dot every 0.1 s. Just as it reaches its maximum height, the thread breaks, and the system "explodes." Ball *A* follows a new parabolic trajectory that is contained in the same plane as the original trajectory. Ball *A*'s path is shown in red, and the dots give the ball's position at 0.1-s intervals. *(a)* Mark the position of the center of mass of the system at each 0.1-s interval during the second part of the motion (after the explosion). *(b)* Mark the position of ball *B* at each 0.1 s during the second part of the motion and sketch its trajectory.

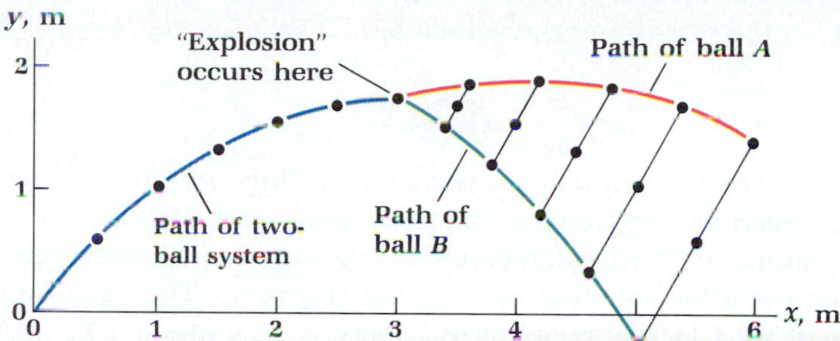

Figure 10-6. Example 10-4: A two-ball system, with the balls initially compressed, explodes as it reaches its maximum height. From the known path of ball *A* after the explosion, we can determine the path of ball *B*.

Solution. *(a)* During the explosion each ball exerts a force on the other, but these forces are internal and do not affect the motion of the center of mass. Therefore the center of mass follows the same path that it would if there were no explosion. Using the known symmetry of the path of a particle in a uniform gravitational field, we mark each position of the center of mass for the second part of the motion by continuing the original parabolic path, as shown by the black dots in Fig. 10-6.

(b) Since the balls have equal mass, the center of mass is midway between the balls at each point. Each position of ball *A* is given, and we found each position of the center of mass in part *(a)*. Now we draw a line from a ball-*A* position to a corresponding center-of-mass position for each 0.1 s and extend this line beyond the center of mass. Then we mark the corresponding ball-*B* position, as shown in the figure. The green curve shows the trajectory of ball *B* during the second part of the motion.

SELF-TEST 10-4. Suppose ball *B* has twice the mass of ball *A*. Keep the motion of the center of mass of the system before the explosion the same as in Fig. 10-6, and keep the motion of ball *A* after the explosion the same as in the figure. *(a)* Mark the position of the center of mass at each 0.1 s during the second part of the motion. *(b)* Mark the position of ball *B* at each 0.1 s during the second part of the motion and sketch the trajectory.

Work and the Motion of a System

Suppose a person standing on roller skates pushes herself away from a wall with her hands (Fig. 10-7). From the point of view of Newton's second law, the external force exerted on the skater by the wall accounts for the motion of her center of mass, and internal forces exerted by her muscles on her bones have no effect on the motion of her center of mass. From the perspective of work and kinetic energy, the force by the wall does no work because the point of application of this force does not move. It is the internal forces exerted by the skater's muscles that do work in this case. This is an example of internal work that was discussed previously at the end of Sec. 9-5. Internal work is work done by internal forces. Internal work can be done when a system changes its shape or composition. Another example of internal work is the expansion of the compressed rubber balls in the "explosion" described in Example 10-4. When applying the work-energy theorem to the motion of a system of particles or to an extended object, one must account for any internal work that may have been performed during the motion.

Figure 10-7. A skater pushes herself away from a wall. The force exerted by the wall on the skater provides the external force that causes the skater to accelerate, but this force does no work because its point of application does not move. Internal forces, exerted by the skater's muscles on her bones, account for the skater's increase in kinetic energy.

10-3 MOMENTUM

When a sportscaster on television describes a team as having momentum, he or she may be alluding to the rate at which the team is scoring points. The definition of momentum in physics is simpler and more specific.

Momentum of a Particle

The ***momentum*** **p** of a particle of mass m and velocity **v** is defined as the product of the mass times the velocity:

Momentum of a particle

$$\mathbf{p} = m\mathbf{v} \tag{10-10}$$

Momentum is a vector quantity because in the product $m\mathbf{v}$, m is a scalar and **v** is a vector. The direction of the momentum is the same as the direction of the velocity, and the magnitude is $p = mv$. Momentum has the dimension [mass]·[length]/[time], and its SI unit is kilogram-meter per second (kg·m/s). The magnitude p of the momentum of a 0.1-kg ball traveling at a speed $v = 20$ m/s is 2 kg·m/s.

Newton originally presented the second law in a form that in our notation becomes

Newton's second law in terms of momentum

$$\Sigma\mathbf{F} = \frac{d\mathbf{p}}{dt} \tag{10-11}$$

This is equivalent to the form used in Chap. 5, $\Sigma\mathbf{F} = m\mathbf{a}$:

$$\Sigma\mathbf{F} = \frac{d\mathbf{p}}{dt} = \frac{d}{dt}(m\mathbf{v}) = m\frac{d\mathbf{v}}{dt} = m\mathbf{a}$$

where we have assumed that the mass is constant. Thus the momentum of a particle is the quantity whose time derivative is the net force on the particle.

Momentum of a System of Particles

The momentum $\mathcal{P}$ of a system of particles is the vector sum of the individual momenta of the particles that compose the system:

Momentum of a system

$$\mathcal{P} = \Sigma\mathbf{p}_i = \Sigma m_i\mathbf{v}_i \tag{10-12}$$

For example, suppose a 50-kg jogger is traveling north at 6.0 m/s ($p_1 = m_1v_1 = 300$ kg·m/s) and a 60-kg jogger is traveling east at 5.0 m/s ($p_2 = m_2v_2 = 300$ kg·m/s). If we treat each jogger as a particle, then the magnitude $\mathcal{P}$ for this two-jogger system is $\sqrt{p_1{}^2 + p_2{}^2} = 420$ kg·m/s, and $\mathcal{P}$ is directed toward the northeast.

The momentum of a system may be written in terms of the velocity $\mathbf{v}_{cm}$ of the center of mass. Equation 10-6, which came from the definition of the center of mass, gives $\mathbf{v}_{cm} = \Sigma m_i\mathbf{v}_i/M$, or $M\mathbf{v}_{cm} = \Sigma m_i\mathbf{v}_i$. Inserting this into Eq. 10-12 gives

Momentum of a system in terms of $\mathbf{v}_{cm}$

$$\mathcal{P} = M\mathbf{v}_{cm} \tag{10-13}$$

The momentum $\mathcal{P}$ of a system or of an extended object is the same as that of a particle of mass M and velocity $\mathbf{v}_{cm}$. As far as the momentum $\mathcal{P}$ of a system is concerned, the system behaves as if all the mass were concentrated at the center of mass and moves with velocity $\mathbf{v}_{cm}$.

Reference Frames

Use of an inertial frame

Problems involving momentum can sometimes be simplified with a judicious choice of a reference frame. Any frame may be used as long as the frame is inertial, or approximately inertial within the precision of the problem. Momentum was developed from Newton's second law, and Newton's second law is valid only if the acceleration is measured relative to an inertial frame. Often the most convenient reference frame is one at rest relative to a system's center of mass because the momentum of the system is zero in such a frame. However, the techniques involved in using a center-of-mass reference frame are beyond the scope of this introductory treatment.

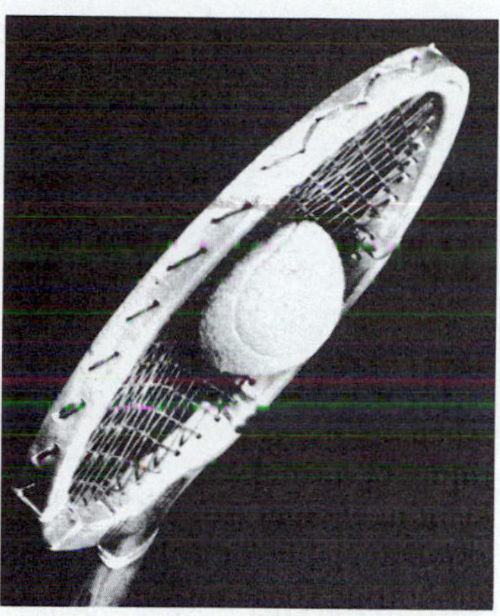

Figure 10-8. A tennis ball being hit by a racket. The distortion of the ball and racket indicates that the force exerted by the racket on the ball is quite large. Such a large force that acts over a small time interval is called an impulsive force.

10-4 IMPULSE

A force of large magnitude that exists for a short time interval is called an ***impulsive force.*** An example of an impulsive force is that exerted by a racket on a tennis ball (Fig. 10-8). While it is being struck, there are other forces exerted on the tennis ball besides that by the racket. There is the gravitational force by the earth, and there is a force exerted by the air. But while it is being struck, these other forces are negligible, and the force by the racket is responsible for drastically altering the motion of the ball. Figure 10-9 shows how the magnitude of the force exerted by the racket on the ball might vary with time. The time interval over which the force acts is very short, about 10 ms, but the force magnitude becomes very large, with a maximum of about 500 N.

The quantity that characterizes the effect of an impulsive force is called the ***impulse,*** and it is represented by the symbol **J**. The magnitude of **J** is given by the area under the curve in a graph of F versus t (for example, the area under the curve in Fig. 10-9), and the direction of **J** is the same as that of the force. Since an integral gives an area under a curve, the impulse of force **F** exerted on an object is

$$\mathbf{J} = \int_{t_i}^{t_f} \mathbf{F}\, dt$$

Impulse of a force

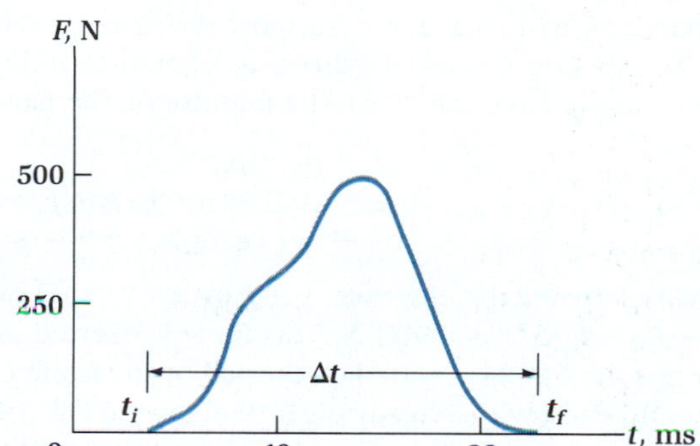

Figure 10-9. A graph of how the impulsive force exerted on the tennis ball in Fig. 10-8 might vary in time.

The dimension of impulse is [force]·[time] so that its SI unit is newton-seconds (N·s).

During the time interval $t_f - t_i = \Delta t$ over which an impulsive force **F** acts, this force is the dominant force, and all others may be neglected. Thus we may use **F** as the net force, and Newton's second law gives

$$\Sigma\mathbf{F} = \mathbf{F} = \frac{d\mathbf{p}}{dt} \qquad \text{or} \qquad d\mathbf{p} = \mathbf{F}\, dt$$

Integrating this equation, we have

$$\Delta\mathbf{p} = \mathbf{p}_f - \mathbf{p}_i = \int_{t_i}^{t_f} \mathbf{F}\, dt$$

Therefore Newton's second law leads to a connection between impulse and momentum:

$$\mathbf{J} = \Delta\mathbf{p} \qquad (10\text{-}14)$$

Impulse is equal to the change in momentum.

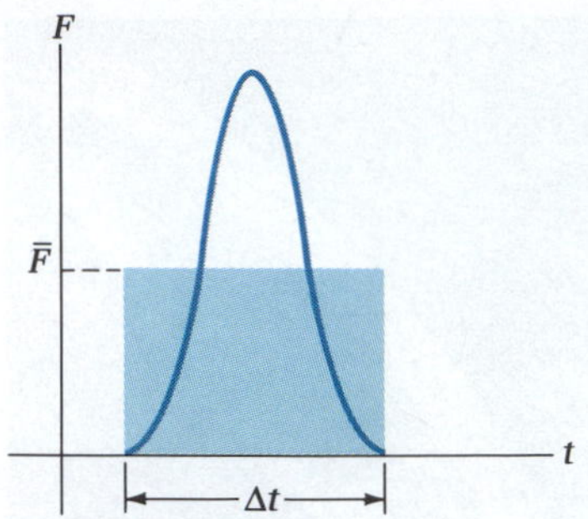

Figure 10-10. The average magnitude $\bar{F}$ of an impulsive force **F** is defined such that the area under the *F*-versus-*t* curve is equal to the area $\bar{F}\ \Delta t$ of the rectangle.

The impulse of a force exerted on an object over a time interval Δt is equal to the object's change of momentum during that time interval. Keep in mind that this expression is valid only if the impulsive force is much larger than the vector sum of all other forces during the time interval Δt.

Sometimes we can measure an object's momentum change due to an impulsive force but are unable to determine the way the force varies with time. We cannot then evaluate the integral $\int \mathbf{F}\, dt$. In such cases we treat the impulsive force as if it were constant at an average magnitude $\bar{F}$ during the time interval Δt, as shown graphically in Fig. 10-10. The value of $\bar{F}$ is such that the area of the rectangle in the figure is the same as the area under the curve. Thus the relation between **J** and $\bar{\mathbf{F}}$ is

$$\mathbf{J} = \int_{t_i}^{t_f} \mathbf{F}\, dt = \bar{\mathbf{F}}\, \Delta t$$

EXAMPLE 10-5

Driving a nail. Estimate the average force magnitude that you exert on a nail when you drive it into a board with a hammer.

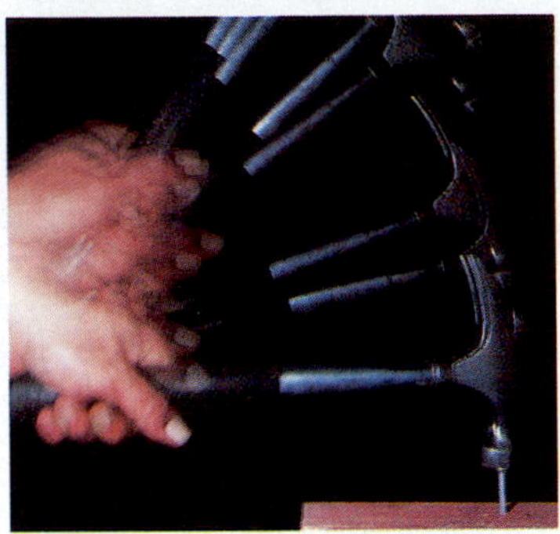

Hammer hitting nail.

Solution. The hammer exerts an impulsive force on the nail because it is in contact with the nail for only a short time but the effect on the nail is large. We will need to estimate the change in momentum of the hammerhead, and the time over which the change occurred. From this we can calculate the average force on the hammer (and through Newton's third law, the average force the hammer exerts on the nail). Let us examine the time the hammer and nail are in contact.

A second is about the time it takes to say "one-one thousand," and the hammerhead isn't in contact with the nail for even a tenth of that time. Consequently, we shall take $\Delta t \approx 0.01$ s. The mass of a hammerhead is about 3.0 kg. The speed of the hammer before it hits the nail can be estimated from the acceleration you give a hammer, perhaps 4 times the acceleration of gravity, or 40 m/s², and the time you take to swing the hammer, about $\frac{1}{2}$ s. This gives $v = at \approx 20$ m/s. The speed of the hammer after the collision depends somewhat on the hammer, nail, board, etc. Suppose the hammer rebounds with half the speed it had before the collision. Choose the direction of motion of the hammer just before it hits the nail as the positive *y* direction. Then the impulse on the hammer is

$$\begin{aligned} J_y &= p_{fy} - p_{iy} \\ &= (0.3\ \text{kg})(-10\ \text{m/s}) - (0.3\ \text{kg})(20\ \text{m/s}) \\ &= -9\ \text{kg}\cdot\text{m/s} = -9\ \text{N}\cdot\text{s} \end{aligned}$$

We estimated that this takes place over 0.01 s. The average force component on the hammer is $\bar{F}_y = J_y/\Delta t$, or -900 N. This force is exerted on the hammer by the nail. An equal and opposite force is exerted on the nail by the hammer. It is quite a large force, more than you could exert with a steady push. As shown in Fig. 10-10, the actual force at its peak might be about twice as large as the average force, possibly several thousand newtons.

SELF-TEST 10-5. A 0.070-kg tennis ball is traveling horizontally toward the north at a speed of 30 m/s when it is struck by a racket. After the encounter the ball is traveling toward the south at a speed of 40 m/s. What are the magnitude and direction of the impulse exerted on the ball by the racket? ***ANSWER:*** 4.9 N·s toward the south.

10-5 CONSERVATION OF MOMENTUM

Newton's second law for a system leads to the principle of conservation of momentum. From Eq. 10-13 we have

$$\mathcal{P} = M\mathbf{v}_{cm}$$

Taking the time derivative of this equation gives

$$\frac{d\mathcal{P}}{dt} = M\frac{d\mathbf{v}_{cm}}{dt} = M\mathbf{a}_{cm}$$

Newton's second law for a system of particles is

$$\Sigma \mathbf{F}_{ext} = M\mathbf{a}_{cm}$$

Combining these results, we obtain

$$\Sigma \mathbf{F}_{ext} = \frac{d\mathcal{P}}{dt} \tag{10-15}$$

Newton's second law for a system in terms of $\mathcal{P}$

which is Newton's second law for a system of particles or for an extended object, written in terms of the momentum $\mathcal{P}$ of the system. The momentum of a system changes at a rate that is equal to the net *external* force exerted on the system.

Now suppose we are dealing with a system for which $\Sigma \mathbf{F}_{ext} = 0$. Then

$$\frac{d\mathcal{P}}{dt} = 0 \quad \text{or} \quad \mathcal{P} = \text{constant} \quad \text{or} \quad \mathcal{P}_i = \mathcal{P}_f$$

That is, if $\Sigma \mathbf{F}_{ext} = 0$ for a system, then the momentum of this system remains the same for as long as $\Sigma \mathbf{F}_{ext}$ remains zero; the momentum $\mathcal{P}_i$ at some time t_i is equal to the momentum $\mathcal{P}_f$ at some time t_f. In other words, the momentum of the system is conserved. We can summarize the ***principle of conservation of momentum:***

If the net external force exerted on a system is zero ($\Sigma \mathbf{F}_{ext} = 0$), then the momentum of the system is constant in time ($\mathcal{P} = constant$).

Principle of conservation of momentum

The key to using the principle of conservation of momentum is in isolating a system for which $\Sigma \mathbf{F}_{ext}$ is zero or is negligibly small.

EXAMPLE 10-6

Two spring-loaded carts. Two carts, each equipped with a spring bumper, are pushed together. While the springs are compressed, the carts are tied together with a cord (Fig. 10-11). When the cord is cut, the springs expand and push the carts apart so that they travel in opposite directions. After the spring bumpers lose contact, cart *A*, whose mass is 1.0 kg, has a speed of 3.0 m/s; cart *B*, whose mass is 3.0 kg, has an unknown speed v_{Bf}. The carts have small wheels and their axles are well lubricated. *(a)* Isolate a system in which you can apply the principle of conservation of momentum, and explain why this principle can be applied to this system. *(b)* Use conservation of momentum to find the value of v_{Bf}.

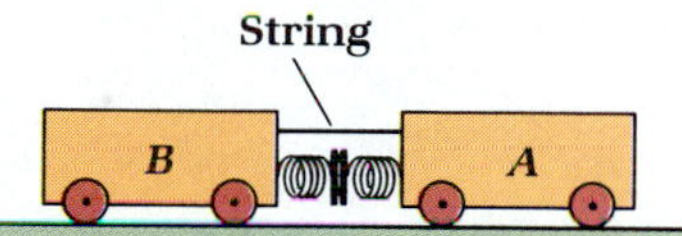

Figure 10-11. Example 10-6: Two carts are tied with a string so that their spring bumpers are compressed.

Solution. *(a)* Since each cart has small wheels, each can be treated as a particle. To apply the principle of conservation of momentum to a system during the time Δt that the carts are being pushed apart, the net external force on the system must be zero: $\Sigma \mathbf{F}_{ext} = 0$. Each cart has a gravitational force exerted on it by the earth and a contact force exerted on it by the table. Since the wheels are small and the axles are well lubricated, we can assume that the contact force is vertically upward, so that it is equal and opposite the downward gravitational force; these external forces add to zero. During Δt, each cart has a horizontal force exerted on it by the other cart so that the net force exerted on either cart is not zero. However, if the system includes both carts, then $\Sigma \mathbf{F}_{ext}$ is zero because the forces exerted by the carts on one another are internal to the system and do not contribute to $\Sigma \mathbf{F}_{ext}$ for the two-cart system. Therefore we choose our system so that it contains both carts.

(b) Now we apply conservation of momentum to the two-cart system. Before the cord was cut, both carts were at rest so that $\mathbf{p}_{Ai} = 0$, $\mathbf{p}_{Bi} = 0$, and $\mathcal{P}_i = \mathbf{p}_{Ai} + \mathbf{p}_{Bi} = 0$. Applying conservation of momentum, we have $\mathcal{P}_f = 0$, or $\mathbf{p}_{Af} + \mathbf{p}_{Bf} = 0$. This means that the momenta of the carts are equal and opposite after Δt. Setting the momenta magnitudes equal gives

$$m_A v_{Af} = m_B v_{Bf}$$

Solving for v_{Bf} and inserting the values gives

$$v_{Bf} = \frac{m_A}{m_B} v_{Af} = \frac{1.0 \text{ kg}}{3.0 \text{ kg}} (3.0 \text{ m/s}) = 1.0 \text{ m/s}$$

Cart *B*, with 3 times the mass of cart *A*, has one-third the speed of cart *A*.

SELF-TEST 10-6. Suppose we perform the spring-loaded carts experiment again, only this time we lock the wheels of cart B so that they will not roll, and the coefficient of sliding friction between the locked wheels and the surface is such that this frictional force is significant. If you measure v_{Af}, can you determine v_{Bf} with conservation of momentum? If you can, identify the system for which $\Sigma\mathbf{F}_{\text{ext}} = 0$. If you cannot, explain why not. ***ANSWER:*** No, because $\Sigma\mathbf{F}_{\text{ext}}$ is significant.

10-6 COLLISIONS IN ONE DIMENSION

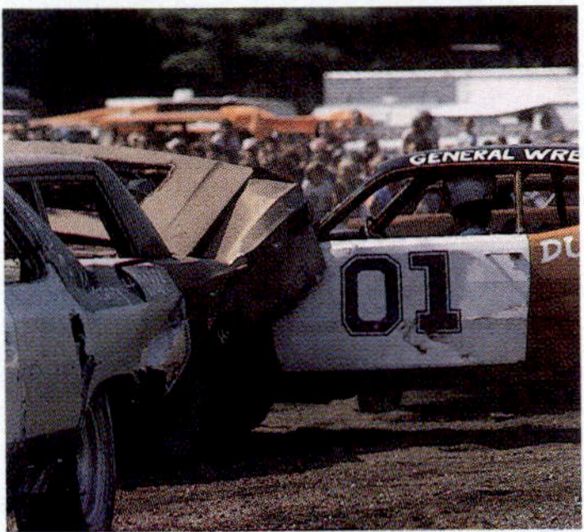

Figure 10-12. A collision in a demolition derby.

In a demolition derby, two cars collide, lock bumpers, then slide off in a heap (Fig. 10-12). If you measure the momentum of each car just prior to the collision, you can use conservation of momentum to estimate the speed and the direction of the two-car heap immediately after the collision.

Collision problems often have characteristics that lend themselves to solution by the application of conservation of momentum. In a collision, two (or more) objects exert impulsive forces on one another. That is, their interaction forces are large but act over a short time interval Δt. If the colliding objects are taken as the system, then during Δt, the external forces exerted on this system are often small enough to be neglected compared with the large forces that the colliding objects exert on each other. This system of colliding objects satisfies the criterion for conservation of momentum—the net external force $\Sigma\mathbf{F}_{\text{ext}}$ is either zero or negligibly small.

Elastic and inelastic collisions

Collisions are often categorized by comparison of the kinetic energy of the colliding objects after the collision (K_f) with that before the collision (K_i). If the kinetic energy after the collision is the same as before the collision ($K_f = K_i$), the collision is called *elastic.* If the kinetic energy after the collision is less than it was before the collision ($K_f < K_i$), the collision is called *inelastic.* A collision between macroscopic objects is inelastic, but some collisions, such as collisions between billiard balls, are nearly elastic. At the opposite extreme from an elastic collision is a *completely inelastic collision.* In a completely inelastic collision, the colliding objects stick together and depart the collision site in unison. It can be shown that a maximum amount of kinetic energy is dissipated during a completely inelastic collision.

A Completely Inelastic Collision in One Dimension

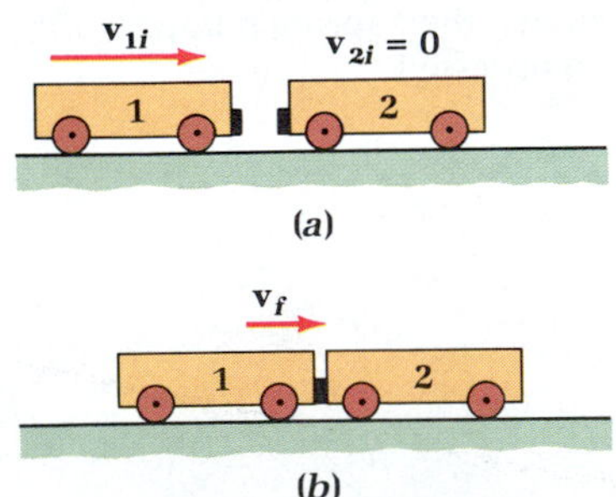

Figure 10-13. Two carts have a head-on, completely inelastic collision. *(a)* Before the collision cart 2 is at rest. *(b)* After the collision the carts move off together.

Cart 1 with mass m_1 has speed v_{1i} just before it collides head-on with cart 2 (Fig. 10-13). Cart 2 with mass m_2 is initially at rest ($v_{2i} = 0$). The carts have bumpers made of a sticky putty so that they adhere and move off together, each with the same speed ($v_{1f} = v_{2f} = v_f$). We assume that the carts have small wheels, and the axles are well lubricated. Therefore we can neglect the horizontal component of the force by the surface compared with the impulsive forces the carts exert on each other during the collision, and the two-cart system satisfies the criterion for conservation of momentum ($\Sigma\mathbf{F}_{\text{ext}} \approx 0$). The x axis is placed along the line of motion with $+\mathbf{i}$ in the direction of cart 1's initial velocity. The momentum of the system before the collision is

$$\mathcal{P}_i = \mathbf{p}_{1i} + \mathbf{p}_{2i} = m_1 v_{1i}\mathbf{i} + m_2(0)\mathbf{i} = m_1 v_{1i}\mathbf{i}$$

The momentum of the system after the collision is

$$\mathcal{P}_f = \mathbf{p}_{1f} + \mathbf{p}_{2f} = m_1 v_f\mathbf{i} + m_2 v_f\mathbf{i} = (m_1 + m_2)v_f\mathbf{i}$$

Applying conservation of momentum, we have $\mathcal{P}_i = \mathcal{P}_f$, or

$$m_1 v_{1i} = (m_1 + m_2)v_f \qquad (10\text{-}16)$$

If we measure three of the quantities in Eq. 10-16, then the fourth can be calculated.

EXAMPLE 10-7

Finding v_f. Often the masses of the colliding objects are known, and the initial speeds are measured. Then conservation of momentum is used to find a final speed. Consider a completely inelastic two-cart collision, similar to the one described above, in which $m_1 = 3.0$ kg, $m_2 = 1.0$ kg, $v_{1i} = 2.0$ m/s, and $v_{2i} = 0.0$ m/s. Determine v_f.

Solution. Since Eq. 10-16 describes a completely inelastic collision with cart 2 initially at rest, we solve it for v_f and insert the numerical values:

$$v_f = \frac{m_1}{m_1 + m_2} v_{1i} = \frac{3.0 \text{ kg}}{3.0 \text{ kg} + 1.0 \text{ kg}} (2.0 \text{ m/s}) = 1.5 \text{ m/s}$$

SELF-TEST 10-7. In collisions between nuclear particles, the speeds are sometimes measured and the mass of one of the colliding particles is then determined with conservation of momentum. A proton of mass $m_p = 1.67 \times 10^{-27}$ kg has a speed of 3.0×10^6 m/s when it is absorbed by a nucleus of unknown mass m that is initially at rest. The new nucleus formed by this reaction has a speed of 1.0×10^6 m/s. Determine the mass m of the nucleus that absorbed the proton. ***ANSWER:*** $3.34 = 10^{-27}$ kg.

An Inelastic Collision in One Dimension

We use two carts with rubber bumpers to perform another collision experiment (Fig. 10-14). In this case the carts do not stick together, so that after the collision they have different speeds. Again, cart 2 is at rest before the collision ($v_{2i} = 0$), and the coordinate frame is the same as above. The momentum of the system before the collision is

$$\mathcal{P}_i = \mathbf{p}_{1i} + \mathbf{p}_{2i} = m_1 v_{1i}\mathbf{i} + m_2(0)\mathbf{i} = m_1 v_{1i}\mathbf{i}$$

The momentum of the system after the collision is

$$\mathcal{P}_f = \mathbf{p}_{1f} + \mathbf{p}_{2f} = m_1 v_{1f}\mathbf{i} + m_2 v_{2f}\mathbf{i}$$

The symbol v_{1f} represents cart 1's velocity component, which may not be equal to its speed $|v_{1f}|$. This is because cart 1's velocity after the collision may be toward either $+\mathbf{i}$ or $-\mathbf{i}$, depending on the nature of the colliding objects. If cart 1 travels toward $+\mathbf{i}$ after the collision, then v_{1f} is positive, but if cart 1 travels toward $-\mathbf{i}$ after the collision, then v_{1f} is negative.

Conservation of momentum gives $\mathcal{P}_i = \mathcal{P}_f$, or

$$m_1 v_{1i} = m_1 v_{1f} + m_2 v_{2f} \qquad (10\text{-}17)$$

If four of the five quantities in Eq. 10-17 are measured, the fifth quantity can be calculated.

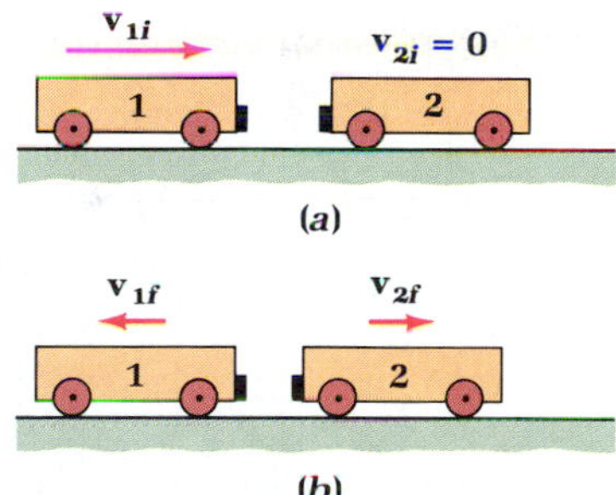

Figure 10-14. Two carts have a head-on inelastic collision: $K_f < K_i$. *(a)* Before the collision cart 2 is at rest. *(b)* After the collision cart 1 may continue to move forward or may move backward, depending on the nature of the collision. Here we show cart 1 moving backward.

An Elastic Collision in One Dimension

Now the carts are equipped with spring bumpers, and the collision experiment is performed again (Fig. 10-15). The spring bumpers are constructed so that they dissipate a negligible amount of energy. Thus mechanical energy is conserved during the collision. Since the potential energy is unchanged, the kinetic energy is also unchanged so that the collision is elastic. Applying conservation of energy, $K_i = K_f$, we find

$$\tfrac{1}{2} m_1 v_{1i}^2 = \tfrac{1}{2} m_1 v_{1f}^2 + \tfrac{1}{2} m_2 v_{2f}^2 \qquad (10\text{-}18)$$

Conservation of momentum applied to this case reproduces Eq. 10-17. Equations 10-17 and 10-18 can be used to calculate two of the quantities they contain if the others are known. However, Eq. 10-18 is difficult to use because the speeds in the

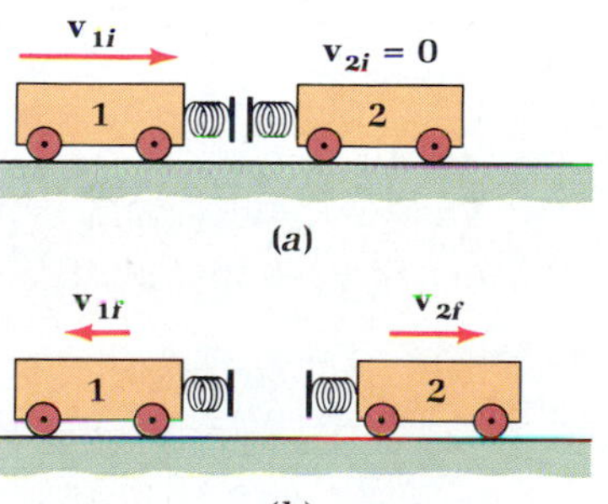

Figure 10-15. Two carts have a head-on elastic collision: $K_f = K_i$. *(a)* Before the collision cart 2 is at rest. *(b)* After the collision cart 1 may continue forward or may move backward. Here we show cart 1 moving backward.

equation are squared. We now develop another equation that can be used in place of Eq. 10-18.

Multiply Eq. 10-18 by 2 and place all quantities that involve cart 1 on one side of the equation and those that involve cart 2 on the other side. This gives

$$m_1 (v_{1i}^2 - v_{1f}^2) = m_2 v_{2f}^2$$

Since $(a^2 - b^2) = (a - b)(a + b)$, this equation may be written as

$$m_1 (v_{1i} - v_{1f})(v_{1i} + v_{1f}) = m_2 v_{2f}^2 \qquad \text{(E)}$$

(The label on Eq. E is a reminder that it came from the energy equation, Eq. 10-18.) Next, take the momentum equation, Eq. 10-17, and write it with quantities involving cart 1 on one side of the equation and those involving cart 2 on the other side. This gives

$$m_1 (v_{1i} - v_{1f}) = m_2 v_{2f} \qquad \text{(M)}$$

(The label on Eq. M is a reminder that it came from the momentum equation, Eq. 10-17.) Now divide Eq. E by Eq. M and obtain

$$v_{1i} + v_{1f} = v_{2f} \qquad \text{(EM)}$$

This is the simplified equation that can be used in place of Eq. 10-18. (The label on Eq. EM is a reminder that it was developed from a combination of conservation of energy and conservation of momentum.)

Rearrangement of Eq. EM reveals an interesting feature of the head-on elastic collision:

$$v_{1i} = v_{2f} - v_{1f}$$

The relative speeds are the same before and after a head-on elastic collision.

The left-hand side of this equation is the relative speed of the carts before the collision, and the right-hand side is the relative speed of the carts after the collision. The equation states that the speed of one cart relative to the other is unchanged by the collision. Although the relative speeds are unchanged, the relative velocities are reversed by the collision. The carts approach one another before the collision and recede from one another after the collision. This result is valid even when both carts are moving before the collision (Prob. 9).

Equations M and EM can be used as two equations to solve for any two of the quantities that are unknown. Equation EM can be used alone even when neither mass is known because it does not contain the masses.

EXAMPLE 10-8

Finding final velocities. Two carts with the following characteristics have a head-on elastic collision: $m_1 = 2.0$ kg, $m_2 = 4.0$ kg, $v_{1i} = 3.0$ m/s, and $v_{2i} = 0.0$ m/s. Determine the velocity of each cart after the collision.

Solution. Equations EM and M both contain the quantities v_{1f} and v_{2f}. Therefore these two equations must be solved to find the two unknowns. First, insert the expression for v_{2f} from Eq. EM into Eq. M and solve for v_{1f}. This gives

$$v_{1f} = \frac{m_1 - m_2}{m_1 + m_2} v_{1i} \qquad \text{(A)}$$

This formula shows that v_{1f} is positive if $m_1 > m_2$, and v_{1f} is negative if $m_2 > m_1$. This means that if cart 1's mass is greater than cart 2's, then after the collision, cart 1 continues to travel in the direction it had before the collision. On the other hand, if cart 2's mass is greater than cart 1's, then cart 1's direction of travel is reversed by the collision. For the numerical values given,

$$v_{1f} = \frac{2.0 \text{ kg} - 4.0 \text{ kg}}{2.0 \text{ kg} + 4.0 \text{ kg}} (3.0 \text{ m/s}) = -1.0 \text{ m/s}$$

Figure 10-16. Two billiard balls suspended by cords from a high ceiling such that they barely touch when at rest.

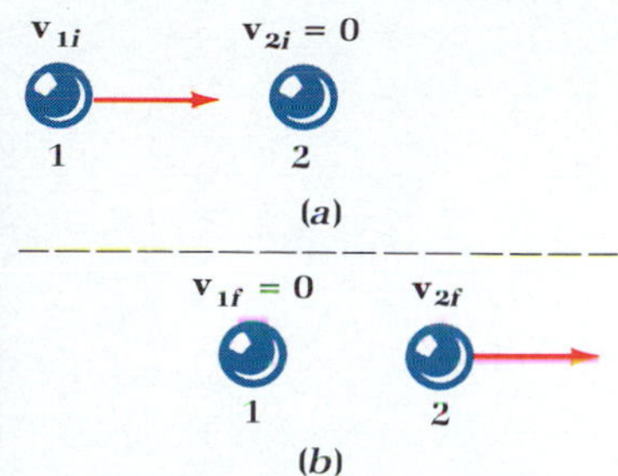

Figure 10-17. A head-on elastic collision between two billiard balls: $m_1 = m_2$. *(a)* Before the collision ball 2 is at rest. *(b)* After the collision ball 1 is at rest and ball 2 has the same speed ball 1 had before the collision.

Now solve Eq. EM for v_{1f} and insert this expression into Eq. M, and then solve for v_{2f}. This gives

$$v_{2f} = \frac{2m_1}{m_1 + m_2} v_{1i} \qquad \text{(B)}$$

From the numerical values,

$$v_{2f} = \frac{2(2.0\text{ kg})}{2.0\text{ kg} + 4.0\text{ kg}} (3.0\text{ m/s}) = 2.0\text{ m/s}$$

The final velocities of the carts are

$$\mathbf{v}_{1f} = (-1.0\text{ m/s})\mathbf{i} \quad \text{and} \quad \mathbf{v}_{2f} = (2.0\text{ m/s})\mathbf{i}$$

SELF-TEST 10-8. Object 1 of unknown mass m_1 has a head-on elastic collision with object 2 of mass 5.0 kg. Object 1's speed before the collision is 55 m/s, and after the collision it is traveling in the opposite direction with a speed of 20 m/s. Object 2 is initially at rest. *(a)* Determine the speed of object 2 after the collision. *(b)* Determine m_1. ***ANSWERS:*** *(a)* 35 m/s; *(b)* 2.3 kg.

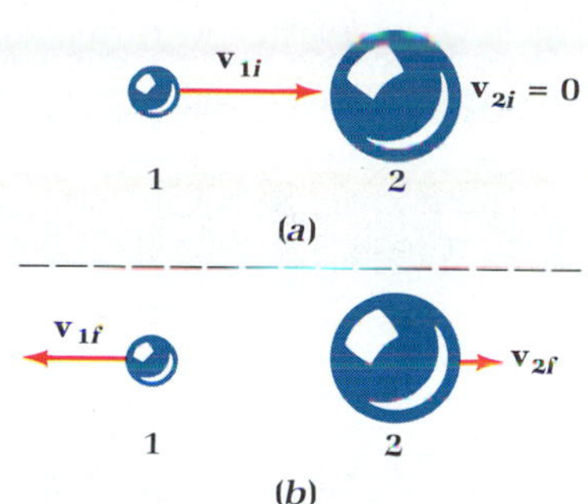

Figure 10-18. A head-on elastic collision between a billiard ball (1) and a bowling ball (2): $m_1 \ll m_2$. *(a)* Before the collision ball 2 is at rest. *(b)* After the collision ball 1 has nearly the same speed as before the collision but its direction of travel is reversed. Ball 2 is hardly affected by the collision.

Further Investigation of Head-on Elastic Collisions

It is instructive to apply Eqs. A and B from the last example to a few special cases. Suppose two billiard balls are hung from a high ceiling, each with two cords that form a "V" (Fig. 10-16). Ball 1 is pulled back and then released so that it collides head-on with ball 2, which is initially stationary. Assume that the collision is essentially elastic so that Eqs. A and B apply. Consider three cases:

1. $m_1 = m_2$ (Fig. 10-17). In the case of equal masses Eq. A shows that $v_{1f} = 0$ and $v_{2f} = v_{1i}$. This means that the balls exchange velocities; ball 2's final velocity is the same as ball 1's initial velocity, and ball 1's final velocity is the same as ball 2's initial velocity.
2. $m_2 \gg m_1$ (Fig. 10-18). Replace billiard ball 2 with a bowling ball whose mass is about 100 times that of a billiard ball. In Eq. A, m_1 is negligible compared with m_2 so that $v_{1f} \approx -v_{1i}$. This means that the billiard ball bounces back with nearly the same speed it had before the collision. Equation B shows that $v_{2f} \approx 0$; the bowling ball is hardly affected by the collision.
3. $m_1 \gg m_2$ (Fig. 10-19). Now let the billiard ball remain stationary and pull the bowling ball back and release it. This means that, in Eqs. A and B, ball 1 is now the bowling ball and ball 2 is now the billiard ball. In Eq. A, m_2 is negligible compared with m_1 and $v_{1f} \approx v_{1i}$; the collision hardly slows the bowling ball at all. From Eq. B, $v_{2f} \approx 2v_{1i}$. Thus the billiard ball gets quite a bump; the collision changes it speed from zero to nearly twice the speed of the bowling ball.

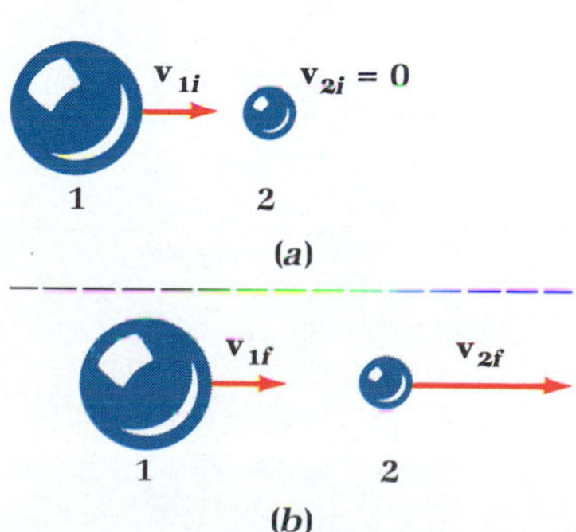

Figure 10-19. A head-on elastic collision between a bowling ball (1) and a billiard ball (2). *(a)* Before the collision ball 2 is at rest. *(b)* After the collision ball 1 has nearly the same velocity it had before the collision and ball 2 has its speed increased from zero to almost twice the speed of ball 1.

EXAMPLE 10-9

Moderating the neutrons in a nuclear reactor. In a nuclear reactor, a neutron released in one fission event must be slowed before it is likely to contribute to an ongoing chain reaction by causing another nucleus to fission. Neutrons can be slowed by collisions with nuclei; this speed-reducing process is called *moderation*. This is similar to the way the cue ball is slowed by collisions with the numbered balls in billiards. In some reactors, hydrogen nuclei (contained in water) are used for this purpose; in other reactors, carbon nuclei (in graphite rods) are used. These collisions are essentially elastic, and compared with the neutron, the moderating nuclei are stationary when struck by a neutron. Determine the fraction by which the speed of a neutron is reduced by a head-on collision with *(a)* a hydrogen nucleus and *(b)* a carbon nucleus. A hydrogen nucleus, which consists of a proton, has about the same mass as a neutron; a carbon nucleus has a mass of about 12 times that of a neutron.

Solution. Since the collision is head-on and elastic, the neutron's velocity component after the collision is given by Eq. A in Example 10-8:

$$v_{1f} = \frac{m_1 - m_2}{m_1 + m_2} v_{1i}$$

where the neutron is object 1 and the struck nucleus is object 2.

(a) Since the mass of the hydrogen nucleus is nearly the same as that of the neutron, $m_1 \approx m_2$ so that $v_{1f} \approx 0$. The neutron is effectively stopped; its speed is reduced to nearly zero by a head-on collision with a proton.

(b) For the carbon-nucleus collision, insert the neutron mass m_n for m_1 and $12m_n$ for m_2:

$$v_{1f} = -(\tfrac{11}{13})v_{1i}$$

Thus the neutron has its speed reduced by a factor of $\frac{2}{13}$, or 15 percent, by this head-on collision. This means that in terms of moderating ability alone, water is much more effective than graphite. However, there are other factors (not discussed here) that weigh in favor of graphite.

SELF-TEST 10-9. Another material that is often used as a moderator is "heavy water." In heavy water, the hydrogen nucleus is a deuteron rather than a proton, and the mass of a deuteron is about twice that of a neutron. Determine the fractional reduction of a neutron's speed when it has a head-on elastic collision with a stationary deuteron. ***ANSWER:*** one-third.

EXAMPLE 10-10

A ballistic pendulum. A ballistic pendulum is a device used to measure the speed of a bullet after it exits the barrel of a rifle (Fig. 10-20). This speed is called the "muzzle velocity." A bullet of mass m is fired horizontally into a wooden block of mass M that forms the bob of a pendulum. The bullet becomes embedded in the block, and the maximum height h that the block rises is measured. Determine an expression for the muzzle velocity v_m in terms of m, M, h, and g. Since the bullet slows very quickly in the block, assume that the block moves an insignificant distance between the time the bullet enters the block and when it becomes embedded.

Solution. There are two time intervals of interest: the time interval of the collision Δt_c and the time interval of the swing Δt_s. Time interval Δt_c begins just as the bullet enters the block

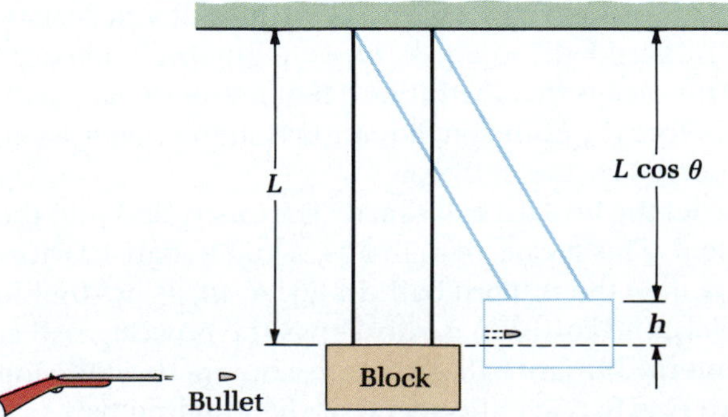

Figure 10-20. Example 10-10: A ballistic pendulum.

with speed v_m, and it ends at the instant the bullet becomes embedded in the block. Time interval Δt_s begins when Δt_c ends, and it ends when the bullet-block system comes instantaneously to rest with the block raised a maximum height h.

The system is the colliding objects—the bullet and the block. During time intervals Δt_c and Δt_s, there are two external forces exerted on this system (neglecting air resistance): There is a downward gravitational force and an upward force by the supporting strings. During Δt_c these forces are essentially equal and opposite so that $\Sigma \mathbf{F}_{ext} \approx 0$. Applying conservation of momentum to the system gives

$$mv_m = (M + m)v_f$$

where v_f is the speed of the block-bullet system at the instant the bullet comes to rest relative to the block. Solving for v_f gives

$$v_f = \frac{mv_m}{M + m}$$

The mechanical energy of the system is not conserved during Δt_c.

During Δt_s, $\Sigma \mathbf{F}_{ext}$ is significant, so the momentum of the system is not conserved. However, during this time interval, only the conservative gravitational force does work on the system. Conservation of mechanical energy can be applied to the system during Δt_s. The system has kinetic energy $\frac{1}{2}(M + m)v_f^2$ at the beginning of Δt_s and zero kinetic energy at the end of Δt_s. From Fig. 10-20, the change in the system's potential energy during Δt_s is $\Delta U = (M + m)gh$. Conservation of mechanical energy states that $\Delta U = -\Delta K$. Thus

$$(M + m)gh = -[0 - \tfrac{1}{2}(M + m)v_f^2]$$

or

$$v_f = \sqrt{2gh}$$

Now the two values of v_f, one from conservation of momentum during Δt_c and the other from conservation of energy during Δt_s, are set equal to one another and solved for v_m. This gives

$$v_m = \frac{M + m}{m}\sqrt{2gh}$$

SELF-TEST 10-10. In a ballistic pendulum experiment, the mass of the bullet is 25.6 g, the mass of the block is 5.42 kg, and the maximum height that the block rises is 0.040 m. Determine the muzzle velocity. ***ANSWER:*** 188 m/s.

10-7 COLLISIONS IN TWO DIMENSIONS

Momentum is a vector quantity. To demonstrate the full problem-solving power of conservation of momentum, we must go beyond the one-dimensional examples considered so far. For simplicity, the discussions are restricted to two-dimensional examples.

Consider a two-particle collision where, before the collision, particle 1 is moving and particle 2 is stationary: $\mathbf{p}_{2i} = 0$ in Fig. 10-21. Particle 1 is the incident particle and particle 2 is the struck particle. If $\Sigma \mathbf{F}_{ext}$ is zero or negligibly small during the collision, then $\mathcal{P}_i = \mathcal{P}_f$ or

$$\mathbf{p}_{1i} = \mathbf{p}_{1f} + \mathbf{p}_{2f} \tag{10-19}$$

Conservation of momentum for this case can be represented with a momentum-vector triangle (Fig. 10-22).

In Fig. 10-23, the coordinate axes are oriented so that $\mathbf{p}_{1i}$ is toward $+\mathbf{i}$. Conservation of momentum in component form is

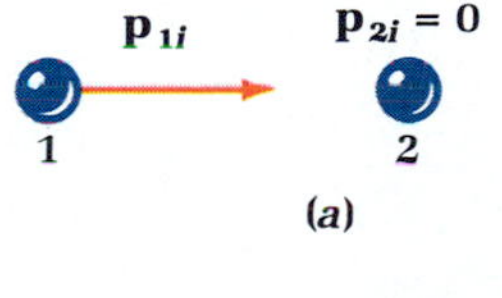

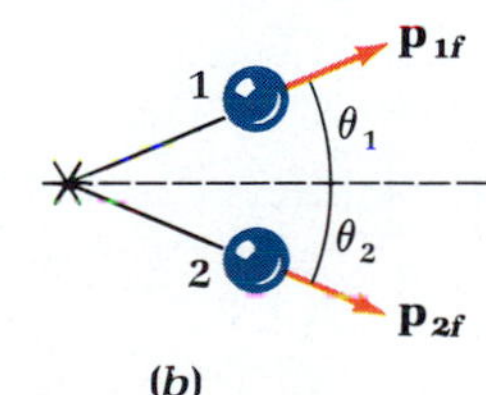

Figure 10-21. *(a)* Incident particle 1 collides with struck particle 2 which is initially stationary. *(b)* After the collision.

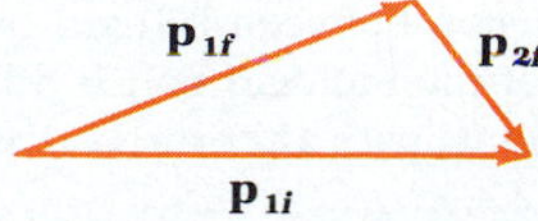

Figure 10-22. When one of the particles is initially stationary, conservation of momentum can be represented by a momentum-vector triangle.

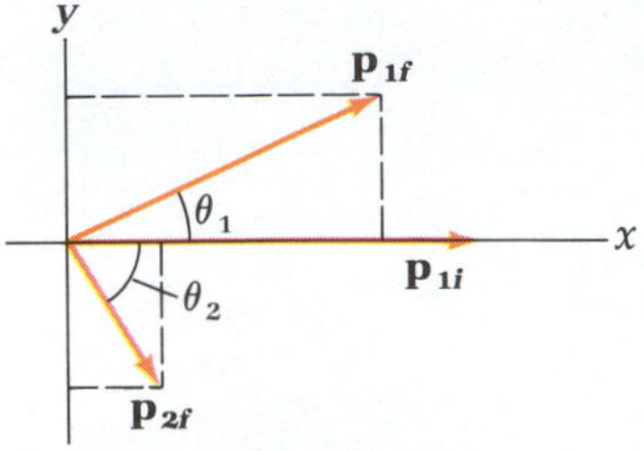

Figure 10-23. With the axes oriented such that $\mathbf{p}_{1i}$ is toward $+\mathbf{i}$, we have $p_{1i} = p_{x1f} + p_{x2f}$ and $p_{y1f} = -p_{y2f}$.

$$\begin{aligned} p_{x1i} &= p_{x1f} + p_{x2f} \\ 0 &= p_{y1f} + p_{y2f} \end{aligned} \tag{10-20}$$

These equations state that the incident particle's x component of momentum before the collision is equal to the sum of the x components for both particles after the collision and that the y components after the collision are equal and opposite (Fig. 10-23).

Sometimes it is convenient to write Eqs. 10-20 in terms of the angles θ_1 and θ_2, and to write $m\mathbf{v}$ for $\mathbf{p}$:

$$\begin{aligned} m_1 v_{1i} &= m_1 v_{1f} \cos\theta_1 + m_2 v_{2f} \cos\theta_2 \\ 0 &= m_1 v_{1f} \sin\theta_1 - m_2 v_{2f} \sin\theta_2 \end{aligned} \tag{10-21}$$

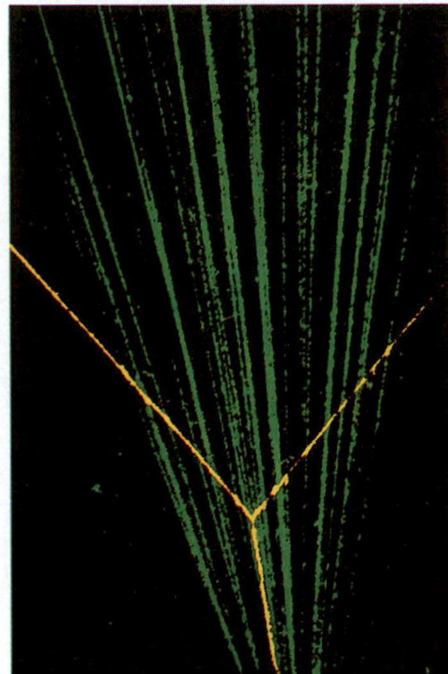

Figure 10-24. Example 10-11: Cloud-chamber photograph shows the tracks of nuclear particles. The two tracks that make an angle of nearly 90° are due to two α particles, one of which was essentially at rest before the collision.

EXAMPLE 10-11

Angle between paths of equal-mass particles in an elastic collision with one particle initially stationary. Figure 10-24 shows a collision between an incident α particle and a struck α particle in a cloud chamber. Figure 10-25 shows a strobe photo of a collision between an incident billiard ball and a struck billiard ball. Note that in each of these cases, the angle between the velocities of the two objects after the collision is nearly 90°. Show that in an elastic collision between particles of the same mass, the angle between their paths after the collision is 90°.

Solution. From Fig. 10-23, the angle between the velocities of the particles after the collision is the angle $\theta_1 + \theta_2$. Since the particles have the same mass, the mass can be divided out of the conservation of momentum equation, giving an expression that contains only velocities:

$$\mathbf{v}_{1i} = \mathbf{v}_{1f} + \mathbf{v}_{2f}$$

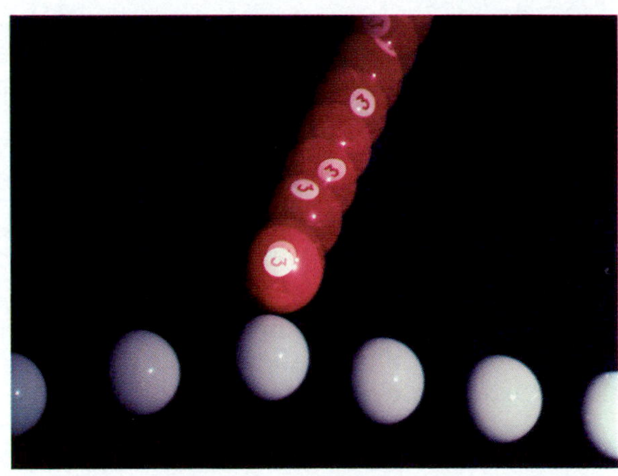

Figure 10-25. Example 10-11: Strobe photo of a billiard-ball collision.

This vector relation is shown with the velocity-vector triangle in Fig. 10-26. The collision is elastic, so kinetic energy is conserved:

$$\tfrac{1}{2}m_1 v_{1i}^2 = \tfrac{1}{2}m_1 v_{1f}^2 + \tfrac{1}{2}m_2 v_{2f}^2$$

Since $m_1 = m_2 = m$, dividing by $\frac{1}{2}m$ gives

$$v_{1i}^2 = v_{1f}^2 + v_{2f}^2$$

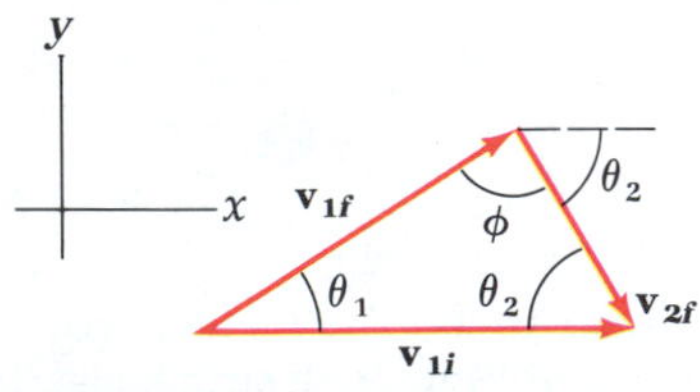

Figure 10-26. Example 10-11: Velocity-vector triangle.

When applied to the velocity-vector triangle in Fig. 10-26, this expression is the pythagorean theorem. Thus the angle between the sides of the triangle labeled $\mathbf{v}_{1f}$ and $\mathbf{v}_{2f}$ must be 90°. The angle $\theta_1 + \theta_2$ between the velocities is 90°.

SELF-TEST 10-11. In a collision between billiard balls, the angle that the struck ball makes with the original path of the incident ball is 55°. Estimate the angle between the incident ball's original path and its path after the collision. ***ANSWER:*** 35°.

EXAMPLE 10-12

A completely inelastic collision between moving objects. Ice skater A of mass m_A is traveling east at speed v_A and ice skater B of mass m_B is traveling north at speed v_B when they meet and grasp one another and move off together. During the encounter the ice skaters quickly reorient their skate blades along the direction of travel so that the horizontal component of the force exerted by the surface on this two-skater system is negligible. Determine the magnitude and direction of the skaters' velocity after the encounter.

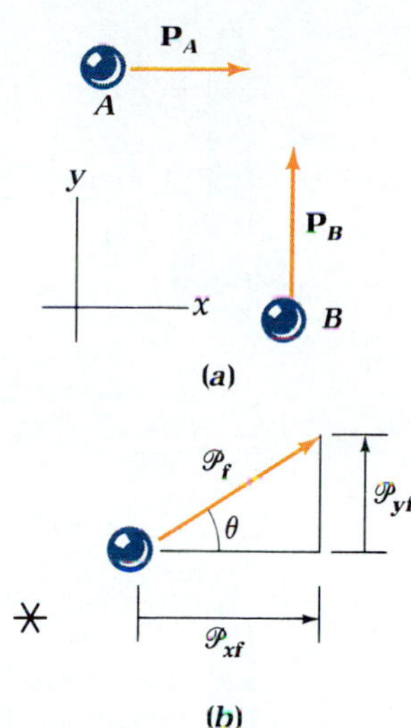

Figure 10-27. Example 10-12: *(a)* Skaters A and B before the encounter. *(b)* The two-skater system after the encounter.

Solution. Since the horizontal component of the force exerted on the skaters by the surface is negligible during the encounter, the criterion for conservation of momentum is satisfied. In Fig. 10-27, the axes are oriented so that $+\mathbf{i}$ is toward the east and $+\mathbf{j}$ is toward the north. Conservation of momentum gives $\boldsymbol{\mathcal{P}}_i = \boldsymbol{\mathcal{P}}_f$, where $\boldsymbol{\mathcal{P}}_i = (m_A v_A)\mathbf{i} + (m_B v_B)\mathbf{j}$ and $\boldsymbol{\mathcal{P}}_f = \mathcal{P}_{xf}\mathbf{i} + \mathcal{P}_{yf}\mathbf{j}$. In component form,

$$m_A v_A = \mathcal{P}_{xf} \qquad \text{and} \qquad m_B v_B = \mathcal{P}_{yf}$$

The speed v_f of the two-skater system after the encounter is $v_f = \mathcal{P}_f / M$, where $\mathcal{P}_f = \sqrt{\mathcal{P}_{xf}^2 + \mathcal{P}_{yf}^2}$ and $M = m_A + m_B$. Using conservation of momentum gives

$$v_f = \frac{\sqrt{(m_A v_A)^2 + (m_B v_B)^2}}{m_A + m_B}$$

The angle θ between the velocity vector and the x axis is $\theta = \tan^{-1}(\mathcal{P}_{yf}/\mathcal{P}_{xf}) = \tan^{-1}(m_B v_B / m_A v_A)$.

SELF-TEST 10-12. In the two-skater encounter above, $m_A = 71$ kg, $v_A = 1.6$ m/s, $m_B = 52$ kg, and $v_B = 3.8$ m/s. Evaluate v_f and θ. ***ANSWER:*** $v_f = 1.9$ m/s; $\theta = 60°$.

10-8 ROCKET MOTION

Rocket motion is unlike that of other vehicles, such as cars or trains. When a car accelerates, the pavement exerts a horizontal frictional force on the tires, and this external force is responsible for the car's acceleration. But a rocket must be capable of accelerating in empty space where there is no external agent to "push against." A rocket moves by ejecting part of itself in the opposite direction from its intended course. When a rocket engine is burning its fuel, the burned fuel material—the exhaust gases—and the rest of the rocket exert forces on each other. The force exerted by the exhaust gases on the rest of the rocket is called the ***thrust*** of the engine, and it is this force that propels the rest of the rocket. A distinctive feature of a rocket is that its mass M changes significantly (it decreases) while its engine is operating.

Thrust of a Rocket Engine

Two important characteristics of a rocket engine are (i) the fuel burn rate $|dM/dt|$ and (ii) the speed v_e of the exhaust gases. The fuel burn rate is represented as $|dM/dt|$ because the rocket's mass M decreases at this same rate. Since the rocket's mass decreases with time, dM/dt is negative; the fuel burn rate is equal to the absolute value of dM/dt. The product $v_e|dM/dt|$ is the magnitude of the thrust $\mathbf{F}_t$ of the engine:

$$F_t = v_e|dM/dt|$$

Thrust of a rocket engine

The direction of the thrust is opposite the velocity of the exhaust gases, or toward the front of the rocket.

Liftoff

Consider a rocket as it lifts off from its pad. At time t the rocket's velocity is $\mathbf{v}$ and its

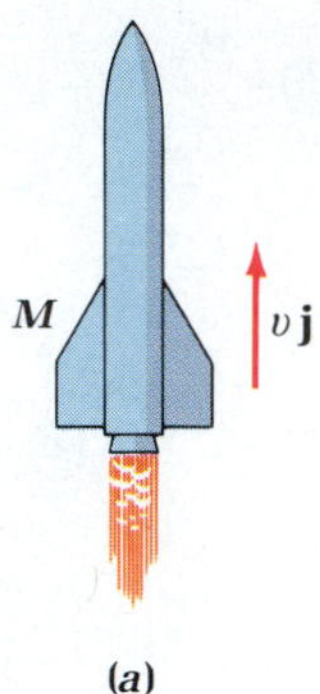

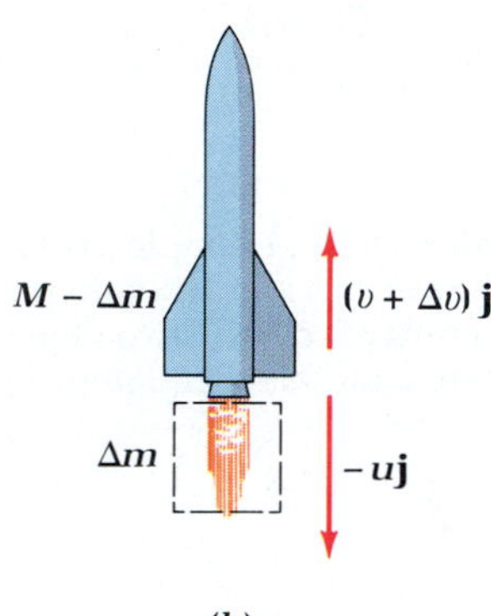

Figure 10-28. A rocket just after liftoff. *(a)* At time t the rocket has mass M and velocity $v\mathbf{j}$. *(b)* At time $t + \Delta t$, the rocket has mass $M - \Delta m$ and velocity $(v + \Delta v)\mathbf{j}$, and the exhaust gases have mass Δm and velocity $(-u)\mathbf{j}$.

mass is M (Fig. 10-28*a*). At a later time $t + \Delta t$, the rocket's velocity is $\mathbf{v} + \Delta\mathbf{v}$, and its mass is $M - \Delta m$, where Δm is the mass of the exhaust gases ejected during Δt (Fig. 10-28*b*). In the figure, the axes are oriented so that $+\mathbf{j}$ is vertically upward. Consider Newton's second law, $\Sigma\mathbf{F}_{ext} = d\mathcal{P}/dt$, applied to our system, which is the rocket plus the exhaust gases ejected during Δt. Thus $\Sigma\mathbf{F}_{ext}\,\Delta t = \Delta\mathcal{P}$, where $\Sigma\mathbf{F}_{ext} = (-Mg)\mathbf{j}$. (Air resistance is neglected.) From the figure,

$$\Delta\mathcal{P} = [(M - \Delta m)(v + \Delta v) + \Delta m(-u)]\mathbf{j} - Mv\mathbf{j}$$

where u is the speed of the exhaust gases relative to the launching pad. The speed u is related to the speed v_e, which is measured relative to the rocket, by

$$u = v_e - v$$

That is, the exhaust gases are moving slower relative to the pad than to the rocket because the rocket is moving upward.

Inserting this value for u and setting $\Sigma\mathbf{F}_{ext}\,\Delta t$ equal to $\Delta\mathcal{P}$ gives

$$(-Mg)\,\Delta t = (M - \Delta m)(v + \Delta v) + \Delta m(v - v_e) - Mv$$

Several terms on the right-hand side cancel one another. Canceling these terms and dividing by Δt yields

$$-Mg = M\frac{\Delta v}{\Delta t} - v_e\frac{\Delta m}{\Delta t} - \frac{\Delta m\,\Delta v}{\Delta t}$$

Now consider the limit as Δt approaches zero. In this limit, $\Delta v/\Delta t$, becomes $dv/dt = a$, $\Delta m/\Delta t$ becomes the burn rate $|dM/dt|$, and $\Delta m\,\Delta v/\Delta t$ becomes negligible because its numerator contains a product of two infinitesimals while its denominator contains only one. These changes give

$$v_e|dM/dt| - Mg = Ma$$

or

$$F_t - Mg = Ma \tag{10-22}$$

The engine's thrust minus the rocket's weight is equal to the rocket's mass times its acceleration. Equation 10-22 shows that the engine's thrust is the force exerted on the rocket by the exhaust gases.

Solving Eq. 10-22 for a gives

Acceleration of a rocket just after liftoff

$$a = \frac{F_t}{M} - g$$

This expression shows that, for the rocket to accelerate in the direction of the thrust, F_t/M must be greater than g. Or, for the rocket to lift off, its mass must be less than F_t/g.

The Rocket Equation

Equation 10-22 can be extended to describe rocket motion in general. In the general case, the external forces may include forces other than gravity, and the equation is a vector relation because the forces may not be along the same line. Therefore $-Mg$ is replaced with $\Sigma\mathbf{F}_{ext}$, F_t with $\mathbf{F}_t$, and a with $\mathbf{a}$:

The rocket equation

$$\mathbf{F}_t + \Sigma\mathbf{F}_{ext} = M\mathbf{a} \tag{10-23}$$

This is the rocket equation, which states that in finding the motion of a rocket, the engine's thrust is added vectorially to the external forces. When using the equation, you must keep in mind that the mass M is not constant; it is decreasing in time.

The rocket equation can be used to find the increase in a rocket's speed when the engine's thrust is in the same direction as the rocket's initial velocity. For simplicity, let the rocket be in interplanetary space where it is far enough from the sun or any planet so that $\Sigma\mathbf{F}_{ext}$ is negligible compared with the engine's thrust. Since we are interested in the rocket's changing speed, the acceleration magnitude is written as dv/dt; to account for the rocket's decreasing mass, F_t is replaced with $v_e|dM/dt| = v_e(-dM/dt)$. This yields

$$v_e(-dM/dt) = M(dv/dt)$$

Separating the variables v and M and integrating gives

$$\int_{v_i}^{v_f} dv = -v_e \int_{M_i}^{M_f} \frac{dM}{M}$$

so that

$$v_f - v_i = -v_e \ln (M_f/M_i) = v_e \ln (M_i/M_f)$$

Speed change of a rocket when $\Sigma\mathbf{F}_{ext}$ is negligible

The change in the rocket's speed is directly proportional to the speed of the exhaust gases and depends logarithmically on the fractional reduction of the mass. Suppose a rocket starts from rest ($v_i = 0$) and performs a burn such that its mass is reduced by a factor of 2 ($M_i/M_f = 2$); also suppose the exhaust speed is $v_e = 2.5 \times 10^3$ m/s. The rocket's speed after the burn is

$$v_f = (2.5 \times 10^3 \text{ m/s}) \ln 2 = 1.7 \times 10^3 \text{ m/s}$$

EXAMPLE 10-13

Blasting off. The engine of a rocket has a burn rate $|dM/dt| = 3.8$ kg/s, and the exhaust-gas speed is $v_e = 2.3 \times 10^3$ m/s. Determine *(a)* the magnitude of the engine's thrust and *(b)* the maximum mass the rocket can have at liftoff from the surface of the earth. *(c)* If the rocket's mass is 900 kg at the instant the engine reaches full power, how long will it take for the rocket to begin to lift off?

Solution. *(a)* The magnitude of the engine's thrust is

$$F_t = v_e|dM/dt| = (3.8 \text{ kg/s})(2.3 \times 10^3 \text{ m/s}) = 8.7 \text{ kN}$$

(b) The maximum mass for liftoff is

$$M_m = F_t/g = 8.7 \text{ kN}/9.8 \text{ m/s}^2 = 890 \text{ kg}$$

(c) Since the rocket's mass is 900 kg at the instant full power is reached, it must eject 10 kg of fuel before it can begin to lift off. With a burn rate of 3.8 kg/s, the time interval between full power and liftoff is

$$\Delta t = 10 \text{ kg}/3.8 \text{ kg/s} = 2.6 \text{ s}$$

SELF-TEST 10-13. You are designing a rocket to leave the surface of Mars ($g = 3.8$ m/s^2). The rocket's mass at liftoff is expected to be 650 kg, and $v_e = 2.5 \times 10^3$ m/s. What is the rocket engine's minimum burn rate? ***ANSWER:*** 0.99 kg/s.

10-9 PROBLEM-SOLVING TECHNIQUES

The central principle in this chapter is Newton's second law for a system of particles:

$$\Sigma\mathbf{F}_{ext} = M\mathbf{a}_{cm} = \frac{d\mathcal{P}}{dt}$$

The key to applying this principle to the solution of a problem is the selection of the objects that compose the system. Some useful problem-solving steps follow:

1. As with any problem, start by drawing a sketch. Problems involving conservation principles often require two sketches, one showing conditions before

some event, such as a collision, and one for conditions after the event. Use a coordinate frame that takes advantage of any symmetry in the problem. Introduce symbols for known and unknown quantities and clearly distinguish between knowns and unknowns.

2. Choose the particles or objects that compose your system, and determine $\Sigma\mathbf{F}_{ext}$ for this system. If, during a time interval Δt, you can select a system for which $\Sigma\mathbf{F}_{ext}$ is zero or negligible compared with the internal forces within the system, then $d\mathcal{P}/dt$ is zero, and the momentum of the system is conserved during Δt (see Example 10-6). You can then use conservation of momentum to determine some of the properties of the objects in your system, such as an unknown mass or an unknown speed. If you are unable to find a system for which $\Sigma\mathbf{F}_{ext}$ is zero, then apply $\Sigma\mathbf{F}_{ext} = M\mathbf{a}_{cm}$ to your system (see Example 10-4).
3. If further information is needed to solve the problem, examine it for the application of other principles. For instance, if only conservative forces do work on the system during Δt, then the mechanical energy of the system is conserved during Δt (see Example 10-10).

COMMENTARY: Symmetry and the Conservation Principles

Newton's laws have led us to two conservation principles: the conservation of energy and the conservation of momentum. Newtonian mechanics leads to a third conservation principle, the conservation of angular momentum, which is discussed in Chap. 13.

The commentary at the end of Chap. 5 discussed the fact that classical mechanics is not always valid. For objects as small as atoms, quantum mechanics replaces classical mechanics. For objects traveling near the speed of light, Einstein's theory of relativity is applicable. However, even when these modern theories replace Newton's classical theory, the conservation principles remain valid. The conservation principles transcend the mechanical theories.

Do the conservation principles have a more fundamental basis than the mechanical theories? Yes. The conservation principles are connected to certain symmetries in the universe. One such symmetry is *spatial symmetry,* or the homogeneity of space. The statement that space is homogeneous means that physical laws are independent of position in the universe. The laws involve separation distances, but they contain no reference to any preferred point in space. When applying the laws, any origin may be chosen for a coordinate frame. Because of spatial symmetry in the universe, momentum is conserved.

Another symmetry is *temporal symmetry,* or the uniformity of time. The uniformity of time means that any instant may be chosen as the zero of time when applying a physical law. Conservation of energy is a consequence of temporal symmetry.

A third symmetry in the universe is *rotational symmetry,* or the isotropy of space. That space is isotropic means that the coordinate axes may be oriented in any way when applying a physical law. Conservation of angular momentum comes from the isotropy of space.

The principle of conservation of electric charge is introduced in Chap. 20. The symmetry associated with conservation of charge is called *gauge symmetry.*

At one time it was thought that the dynamics of a system were independent of the parity of a system. The *parity* of a system refers to whether the system is left-handed or right-handed. For example, an ordinary screw is called right-handed—it is advanced, or "screwed in," by turning the screwdriver clockwise. A left-handed screw is advanced by turning the screwdriver counterclockwise. A left-handed screw appears the same as a mirror image of a right-handed screw. If parity is conserved during an event, then the mirror image of this event should occur with the same probability as the event. An experiment by C. S. Wu and her collaborators in 1957 showed that in some reactions parity is not conserved. Thus the universe does not possess the symmetry that corresponds to conservation of parity.

Theories of the early universe (see the commentary on page 1085), near the time of the "big bang," often use other symmetries, which may have been properties of the universe when it was very young. These additional symmetries are associated with still more conservation principles, which are used to reveal properties of the early universe without knowing the complete laws of physics in the extreme circumstances of that time.

SUMMARY

Section 10-1. Center of Mass

The position of the center of mass of a system of particles is defined by

$$\mathbf{r}_{cm} = \frac{\Sigma m_i \mathbf{r}_i}{M} \tag{10-1}$$

For a continuous system,

$$M\mathbf{r}_{cm} = \int \rho \mathbf{r}\, dV$$

Section 10-2. Motion of the Center of Mass

The velocity and acceleration of the center of mass of a system are

$$\mathbf{v}_{cm} = \frac{\Sigma m_i \mathbf{v}_i}{M}$$

$$\mathbf{a}_{cm} = \frac{\Sigma m_i \mathbf{a}_i}{M}$$

From Newton's second and third laws,

$$\Sigma \mathbf{F}_{ext} = M\mathbf{a}_{cm}$$

Section 10-3. Momentum

The momentum of a particle is defined as

$$\mathbf{p} = m\mathbf{v}$$

Newton's second law can be written

$$\Sigma \mathbf{F} = \frac{d\mathbf{p}}{dt}$$

The momentum of a system of particles is

$$\mathcal{P} = \Sigma m_i v_i = M v_{cm}$$

and Newton's second law for a system is

$$\Sigma \mathbf{F}_{ext} = \frac{d\mathcal{P}}{dt}$$

Section 10-4. Impulse

The impulse of a force is defined as

$$\mathbf{J} = \int_{t_i}^{t_f} \mathbf{F}\, dt$$

When a single impulsive force is applied to an object, the change in momentum of the object is

$$\mathbf{J} = \mathbf{p}_f - \mathbf{p}_i$$

Section 10-5. Conservation of Momentum

If $\Sigma \mathbf{F}_{ext} = 0$ for a system, then $d\mathcal{P}/dt = 0$ and the system's momentum is conserved, $\mathcal{P}_i = \mathcal{P}_f$.

Section 10-6. Collisions in One Dimension

Conservation of momentum is particularly useful in solving collision problems. In one dimension, conservation of momentum provides one equation for determining a property of one of the colliding objects. In an elastic collision, conservation of kinetic energy provides a second relation for finding a second property.

Section 10-7. Collisions in Two Dimensions

The vector nature of momentum is demonstrated with two-dimensional collisions. In two dimensions, conservation of momentum provides two equations.

Section 10-8. Rocket Motion

The magnitude of the thrust of a rocket is

$$F_t = v_e |dM/dt|$$

The rocket equation is

$$\mathbf{F}_t + \Sigma \mathbf{F}_{ext} = M\mathbf{a} \tag{10-23}$$

For a rocket with $\mathbf{F}_t$ in the same direction as $\mathbf{v}_i$ and $\Sigma \mathbf{F}_{ext} = 0$,

$$v_f - v_i = v_e \ln (M_i/M_f)$$

QUESTIONS

1. Estimate the maximum magnitude of momentum you have ever had. In what reference frame?

2. Is there any connection between Newton's first law and the conservation-of-momentum law for a single particle? Explain.

3. Must there be a particle at the center of mass of a system of particles? Explain.

4. Draw a picture of an object that has no mass at its center of mass.

5. Where is the center of mass of a basketball? Of a hula hoop? Of a doughnut? Of this book? Of a horseshoe?

6. Where is the center of mass of a uniform sphere? Of a cube? Of a regular tetrahedron? Of any regular solid polyhedron?

7. What can you say without calculation about the center of mass of a uniform hemisphere? About a D-shaped cylinder? About an isosceles triangle?

8. A truck driver carrying chickens to market is stopped at a weighing station. He bangs on the side of the truck to frighten the chickens so that they will fly up and make the truck lighter. Will his scheme work? Does it make any difference if the truck is open or closed? Explain.

9. Is it possible to conserve kinetic energy but not momentum? Explain. Is it possible to conserve momentum but not kinetic energy? Explain.

10. Is potential energy conserved in an elastic collision? Is it conserved all during the collision, or is the potential energy just the same at the finish as it was at the start?

11. Is it necessary that the forces that do work during an elastic collision be conservative? Is it sufficient that they are conservative for the collision to be elastic?

12. Why are pile drivers used to set piles in the ground rather than just pushing the piles slowly into the ground?

13. How do the air wrenches that mechanics use to remove wheels from your car work? Why do they make a clattering sound?

14. Most of the skid marks left at the scene of an automobile accident are left by the car tires after the collision occurs. How can measuring the direction and length of these skid marks after the collision reveal whether either of the cars involved was speeding before the collision?

15. You are standing still in the middle of a frictionless, flat, iced-over lake. Now how do you get off? Remember that, lacking a net external horizontal force, momentum is conserved.

16. A fat man walks from one end of a light canoe to the other end. If you watch from shore, the man hardly moves at all, but the canoe moves approximately its own length. Why?

17. Make a rough estimate of how long it takes to stop an oceangoing oil tanker. What provides the external force to stop the tanker?

18. Why do energy-absorbing bumpers reduce the death rate in collisions?

19. It has been claimed that a skillful high jumper clears the bar even though her center of mass passes *under* the bar. Show with a diagram that this is possible.

20. An hourglass with a valve that starts the flow of sand is being weighed on a sensitive balance. Compare the momentum of the sand before the valve is turned, when sand is being dropped in a steady stream from the upper to the lower half, and when all the sand is in the bottom. What are the scale readings at these three times? Does the scale read differently when the momentum of the sand is changing?

21. Would a completely inelastic head-on collision between two identical cars with equal but opposite velocities be more or less damaging than a completely inelastic collision between a car and an immovable wall? Explain.

22. Consider a rocket at rest relative to an inertial frame at a point in space where $\Sigma\mathbf{F}_{ext}$ is negligible. Does doubling the engine's thrust by doubling the exhaust speed v_e result in doubling the rocket's speed? Justify your answer.

23. Consider a rocket at rest at a point in space where $\Sigma\mathbf{F}_{ext}$ is negligible. Does doubling the engine's thrust by doubling the burn rate $|dM/dt|$ result in doubling the rocket's speed? Justify your answer.

24. Establish a definition for the geographic center of a country. Introduce symbols and write an equation for the position vector that locates this center.

25. Establish a definition for the population center of a country. Introduce symbols and write an equation for the position vector that locates this center.

26. Complete the following table:

Symbol	Represents	Type	SI Unit
$\mathbf{p}$			
$\mathcal{P}$			kg·m/s
$\mathbf{J}$		Vector	
v_e	Exhaust speed		
F_t			

EXERCISES

Section 10-1. Center of Mass

1. Find the coordinates of the center of mass of the particles of Fig. 10-29.

2. How far is the center of mass of a water molecule from the center of the oxygen atom (Fig. 10-30)? An oxygen atom has a mass about 16 times that of a hydrogen atom.

3. Find the coordinates of the center of mass of the particles of Fig. 10-31. Each particle has a mass of 0.041 kg, and the edge of each square in the grid of the figure represents 0.010 m.

4. Determine the distance from the center of the earth to the center of mass of the earth-moon system. What is the ratio of this distance to the radius of the earth?

5. Two thin rods, one of length ℓ and the other 2ℓ, are joined at their ends to form an "L" (Fig. 10-32). The rods are made of the same material and have the same cross-sectional area. With

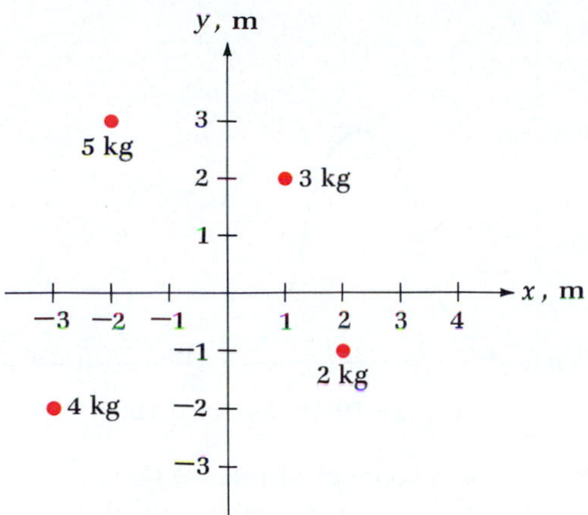

Figure 10-29. Exercise 1.

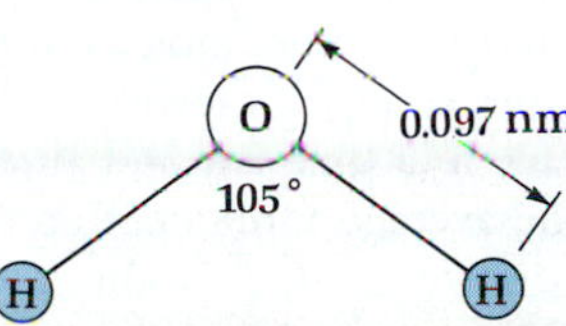

Figure 10-30. Exercise 2: Schematic drawing of a water molecule.

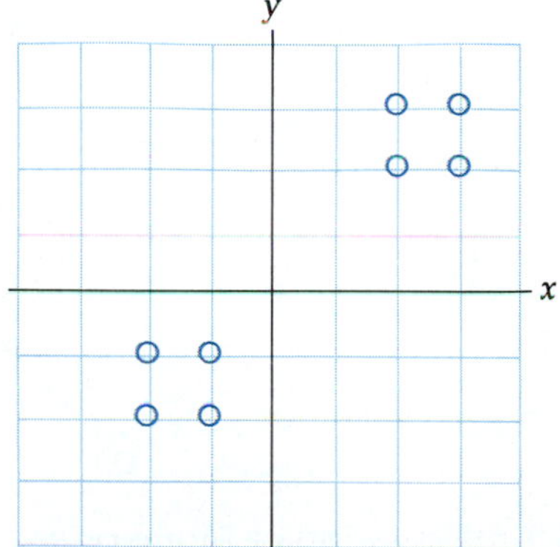

Figure 10-31. Exercise 3.

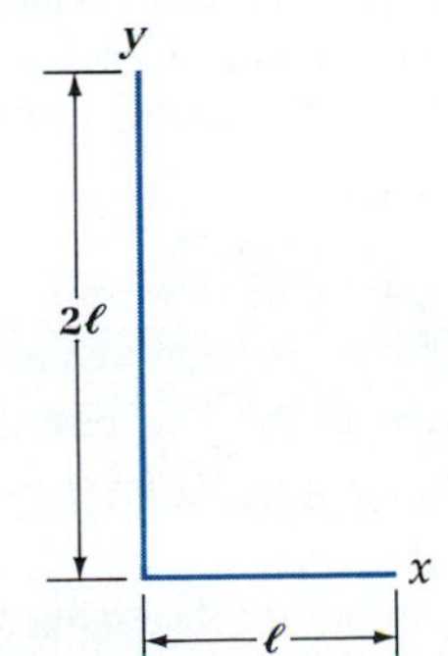

Figure 10-32. Exercise 5.

the corner of the "L" at the origin and the long part along the y axis, as shown in the figure, determine the coordinates of the center of mass in terms of ℓ. Using unit vectors, write an expression for $\mathbf{r}_{cm}$.

6. A thin sheet of plywood is cut into a "U" (Fig. 10-33). What is the distance from the point P at the bottom-center of the "U" to the center of mass?

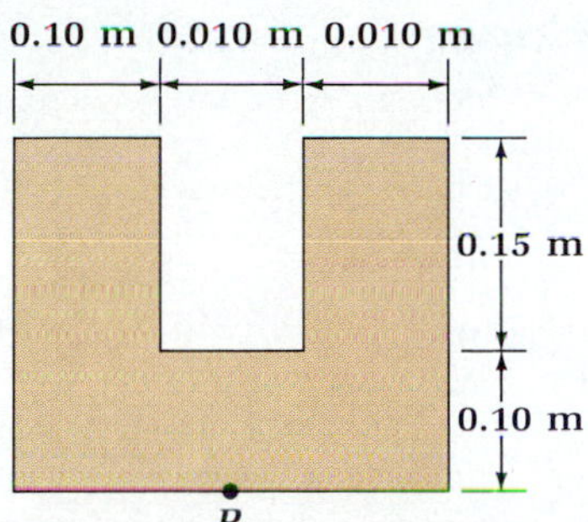

Figure 10-33. Exercise 6.

Section 10-2. Motion of the Center of Mass

7. A 42-kg girl walks along a stationary uniform beam of mass 21 kg. She walks with a speed of 0.75 m/s. What is the speed of the center of mass of the system of girl plus beam?

8. A 125-kg satellite contains two astronauts, Tony and Ben. Tony has a mass of 48 kg and Ben a mass of 52 kg. If the satellite has a velocity of $(32\text{ m/s})\mathbf{j}$, Ben has a velocity of $(30\text{ m/s})\mathbf{i}$, and Tony has a velocity *with respect to the satellite* of $(2\text{ m/s})\mathbf{i}$, what is the velocity of the center of mass of the system of satellite plus astronauts?

9. Two cars are stopped for a red light on opposite sides of an east-west road at an intersection. When the light turns green, car A accelerates toward the south with magnitude 3.4 m/s^2 and car B accelerates toward the north at 1.7 m/s^2. *(a)* If car A and car B have the same mass, what are the magnitude and direction of the acceleration of the center of mass of this two-car system? *(b)* If car A has half the mass of car B, what are the magnitude and direction of the acceleration of the center of mass?

10. A 85-kg man is at the stern of a 12-m-long, 200-kg barge that is free to move in the water (Fig. 10-34). The barge has its center of mass 6 m from either end. *(a)* Where is the center of mass of the system of barge plus man? *(b)* By how much does the center of mass of the system move if the man walks to the bow of the barge? *(c)* By how much has the man moved with respect to the shore? *(d)* By how much has the barge moved with respect to the shore?

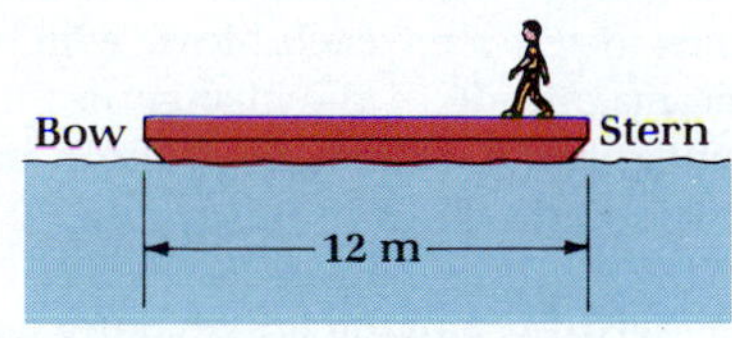

Figure 10-34. Exercise 10.

11. If the mass of the pulley and the string of an Atwood machine (Fig. 10-35) are negligible compared with that of the blocks and if friction is negligible, then the acceleration magnitude of each block is $[g(m_2 - m_1)]/(m_2 + m_1)$, where $m_2 > m_1$. *(a)* Determine an expression for the acceleration magnitude of the center of mass of the two-block system, which is shown inside the dashed lines in the figure. *(b)* Determine an expression for $\Sigma\mathbf{F}_{\text{ext}}$ for this system. *(c)* What external objects are responsible for this net external force?

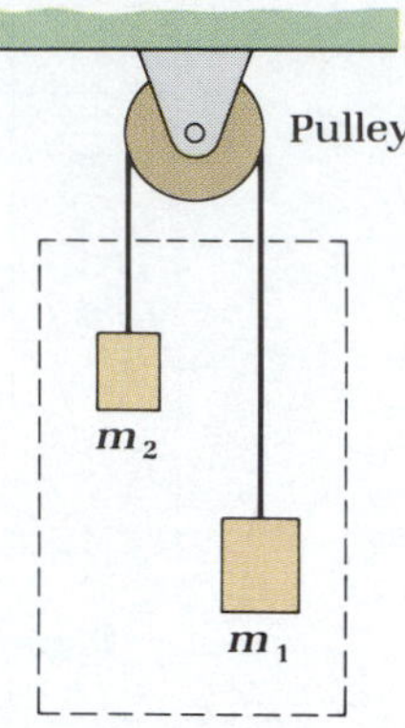

Figure 10-35. Exercise 11.

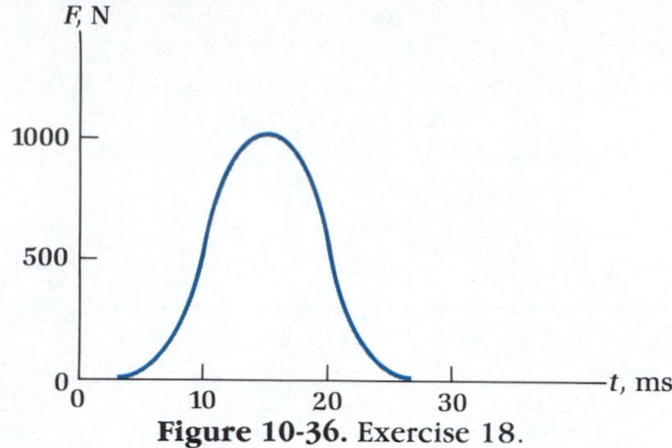

Figure 10-36. Exercise 18.

12. A 1.56-kg fireworks mortar shell is fired straight up, with an initial speed of 31 m/s. It explodes at the peak of its path, breaking into three pieces of different color. The pieces all start out moving horizontally. A 0.78-kg piece lands 212 m north of the mortar, and a 0.26-kg piece lands 68 m east of the mortar. If frictional forces and the wind can be neglected, where does the third piece land?

Section 10-3. Momentum

13. Determine the magnitude of the earth's momentum relative to a sun-center frame.

14. What is the magnitude of the momentum of a 1250-kg car when it is traveling with a speed of 25 m/s?

15. A water molecule is traveling horizontally toward the north at a speed of 350 m/s when it collides with a wall. After the collision the molecule is traveling south with the same speed it had before the collision. The mass of a water molecule is 3.0×10^{-26} kg. *(a)* What are the magnitude and direction of the molecule's momentum before the collision? *(b)* What are the magnitude and direction of the molecule's momentum after the collision? *(c)* What are the magnitude and direction of the change in the molecule's momentum?

16. A 4-Mg truck going straight north at 24 m/s makes a 90° right turn, keeping its speed constant. What is the change in its momentum (magnitude and direction)?

Section 10-4. Impulse

17. A pile driver has a 413-kg block that it drops from a height of 1.4 m onto the pile being driven into the ground. The block essentially comes to rest after each blow, which lasts 43 ms. *(a)* What is the magnitude of the change in the block's momentum? *(b)* What is the magnitude of the average force exerted by the block on the pile?

18. A 0.15-kg baseball is moving horizontally with a speed of 25 m/s when it is hit by a bat such that the magnitude of the impulsive force varies with time, as shown in Fig. 10-36. After the blow, the baseball is again traveling horizontally, but in the opposite direction. Estimate its speed.

19. A 0.071-kg tennis ball is released from rest and falls 1.00 m to the floor. It rebounds to a peak height of 0.48 m. Neglecting air resistance, determine the magnitude and direction of the impulse exerted by the floor on the ball. What are the magnitude and direction of the impulse exerted by the ball on the floor?

20. A ball of mass m bounces off a wall as shown in Fig. 10-37. The angle of incidence θ_i is essentially equal to the angle of reflection θ_r ($\theta_i \approx \theta_r = \theta$), and the speed v_i before the collision is negligibly faster than the speed v_f after the collision ($v_i \approx v_f = v$). *(a)* Write an expression for the magnitude Δp of the ball's momentum change in terms of v and θ. *(b)* If the collision takes place in a time Δt, find an expression for the magnitude $\bar{F}$ of the average force exerted on the ball by the wall. Determine *(c)* Δp and *(d)* $\bar{F}$ for the case where $m = 0.10$ kg, $\theta = 45°$, $v = 22$ m/s, and $\Delta t = 30$ ms.

Figure 10-37. Exercise 20.

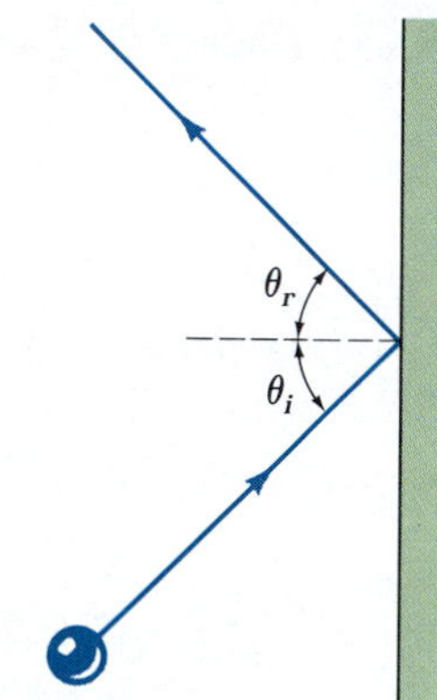

Section 10-5. Conservation of Momentum

21. Two carts, initially at rest, are free to move in the x direction. Cart A has mass 4.52 kg and cart B has mass 2.37 kg. They are tied together, compressing a spring between them, as shown in Fig. 10-38. When the string holding them together is burned and breaks, cart A moves off with a speed of 2.11 m/s. *(a)* With what speed does cart B leave? *(b)* How much energy was stored in the spring before the string was burned?

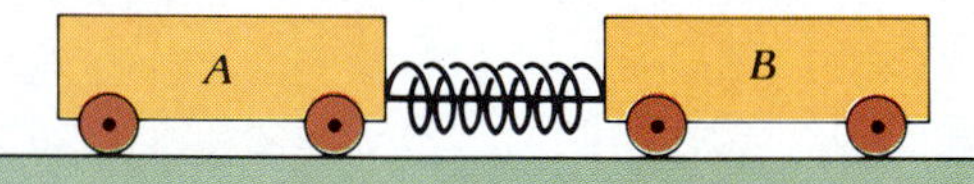

Figure 10-38. Exercise 21.

22. A 62-kg astronaut in free space pushes a 94-kg astronaut. If they were originally stationary, and the 94-kg astronaut leaves with a velocity of (1.7 m/s)**j** with respect to an inertial reference frame, what is the velocity of the 94-kg astronaut with respect to the 62-kg astronaut?

23. A 45-kg girl dives off a 1000-kg boat. She leaves the boat with horizontal speed 5.2 m/s. Assume the boat is originally at rest and free to move in the water. With what speed does the boat start to move off?

24. John and Mary dive off a raft with equal speeds. John has a mass of 75 kg and dives eastward; Mary has a mass of 52 kg and dives southward. In what direction does the raft take off?

25. Two pucks of mass $m_1 = 0.050$ kg and $m_2 = 0.100$ kg are placed on an air table (friction is negligible). The pucks, which are connected by a light rubber band, are pulled apart, stretching the rubber band, and then released simultaneously. At a certain instant before the pucks collide, $v_1 = 0.40$ m/s. *(a)* What is v_2 at this instant? *(b)* What is v_{cm} for this two-puck system at this instant. *(c)* Would the answer to part *(b)* be different if one of the pucks was released a short time after the other?

26. A radioactive nucleus at rest decays into two fragments. One fragment is a small nucleus called an α particle; the larger fragment is called the daughter nucleus. Determine the speed of the recoiling daughter nucleus when the α particle's speed is 5.8×10^6 m/s. The mass of the daughter is 58 times that of the α particle.

Section 10-6. Collisions in One Dimension

27. A 14.2-Mg railroad boxcar with a speed of 1.8 m/s hits and couples to a stationary 23.5-Mg flatcar on a straight and level track. *(a)* Describe the conditions necessary to use conservation of momentum to estimate the speed of the two-car system after the collision. *(b)* Use conservation of momentum to estimate this speed.

28. Astronauts *A* and *B*, each of mass *M*, are floating in interplanetary space at rest relative to each other and to an inertial reference frame. Astronaut *A* tosses a wrench of mass *m* to astronaut *B* such that the speed of the wrench relative to *A* just after the toss is *v*. *(a)* Show that after *B* has caught the wrench, the speed of the astronauts relative to each other is $mv(2M + m)/(M + m)^2$. *(b)* Show that if $m \ll M$, then this speed is approximately $2mv/M$.

29. A girl is running at a speed of 2.5 m/s when she jumps onto a 34-kg sled that is initially stationary on the frozen surface of a lake. If the girl-sled system begins sliding at a speed of 1.5 m/s, what is the mass of the girl?

30. A 100-kg fullback trying to make a touchdown dives directly forward. At the peak of his trajectory he is 1.2 m above the ground and 1.1 m from the goal line; his speed is 4.2 m/s. At that point a 110-kg linebacker at the peak of *his* trajectory, going 2.3 m/s in the opposite direction, hits and holds the fullback tightly. Does the fullback cross the goal and land in the end zone?

31. Astronaut Ann wishes to travel from one side of her spaceship to the other so she pushes herself away from a wall and floats over at a speed relative to the spaceship of 1.00 m/s (Fig. 10-39). Ann's mass is 50 kg and the mass of the spaceship is 500 kg. Astronaut Bob, who is outside the spaceship and was at rest relative to the spaceship and Ann before she pushed off, watches the action. What is Ann's speed relative to Bob as she drifts across the spaceship?

32. A 3.2-kg object with a speed of 15 m/s has a completely inelastic collision with a 4.8-kg object initially at rest. Find the final speed of the combination.

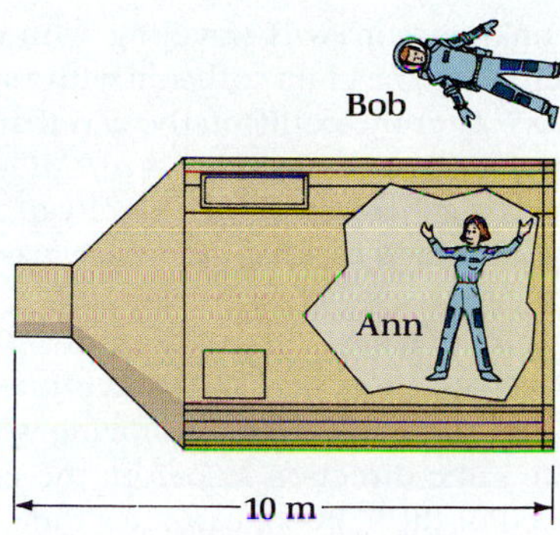

Figure 10-39. Exercise 31.

33. A boy tosses a 3.3-kg beach ball to a 48-kg girl who is initially stationary on roller skates. The ball's velocity is horizontal at the instant she catches it. After the catch, she moves backward with a speed of 0.32 m/s. What is the speed of the ball just before the girl catches it?

34. A 3.2-kg object with a speed of 15 m/s has a head-on elastic collision with a 4.8-kg object initially at rest. Find the speeds of the objects after the collision.

35. An α particle traveling at a speed of 3.5×10^6 m/s has a head-on elastic collision with a uranium nucleus that is at rest. The mass of a uranium nucleus is 59.5 times greater than that of an α particle. What is the speed of the uranium nucleus after the collision?

36. Each of two pendula has a ball as its bob (Fig. 10-40). The balls barely touch when at rest. Ball 1 of mass m_1 is pulled back such that the angle θ between its string and the vertical is 14.5° when the ball is released from rest. Ball 1 then has a head-on collision with ball 2 of mass m_2, which is initially at rest. The collision is elastic and $m_1 = 2m_2$. What is the maximum angle between the string supporting ball 2 and the vertical?

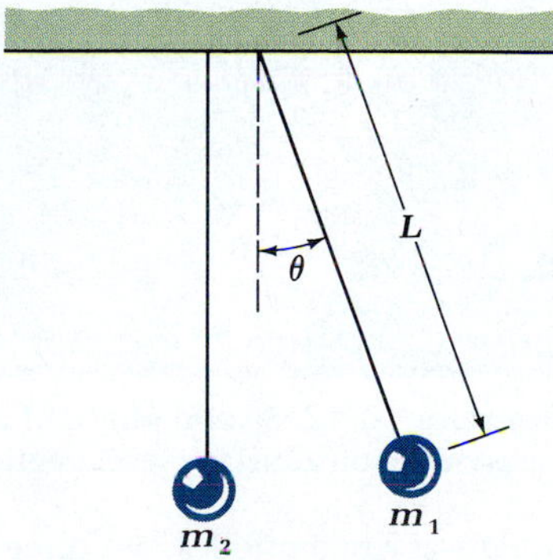

Figure 10-40. Exercise 36.

37. A pistol shoots a 4.5-g bullet into the 1.5-kg block of a ballistic pendulum. The block and bullet then rise 80 mm. What is the muzzle velocity of the bullet fired by this pistol?

38. A 50-g bullet is shot clear through a 1-kg wooden block suspended on a string 2 m long. The center of mass of the block is observed to rise to a maximum height of 50 mm. Find the speed of the bullet as it emerges from the block if its initial speed is 500 m/s. Neglect the loss of mass of the block due to the penetrating bullet.

39. A cart of unknown mass is traveling with a velocity of $(2.8\ \mathrm{m/s})\mathbf{i}$ when it has a head-on collision with a stationary cart of unknown mass. After the collision the cart that was initially stationary moves with a velocity of $(1.4\ \mathrm{m/s})\mathbf{i}$. Assuming that the collision is elastic, determine the velocity of the other cart after the collision. Which cart has the larger mass?

40. A 2.00-kg cart moving with a speed of 3.00 m/s has a head-on collision with a 1.00-kg cart that is initially stationary. After the collision, the 2.00-kg cart is moving with a speed of 1.50 m/s in the same direction as before the collision. *(a)* What is the speed of the 1.00-kg cart after the collision? *(b)* What is the ratio of the kinetic energy of the system before the collision to that after the collision?

Section 10-7. Collisions in Two Dimensions

41. A 1-Mg car going eastward on Main Street at 30 km/h collides with an 8-Mg truck crossing Main Street in a southward direction at 20 km/h. If the vehicles become entangled, how fast and in what direction will they start to move after the collision?

42. Figure 10-41 shows the positions at equal intervals of time of two pucks colliding in two dimensions on a horizontal air table. The diameter of each puck is 60 mm and the time between flashes is 20 ms. Determine the ratio of the mass of the incident puck to that of the struck puck.

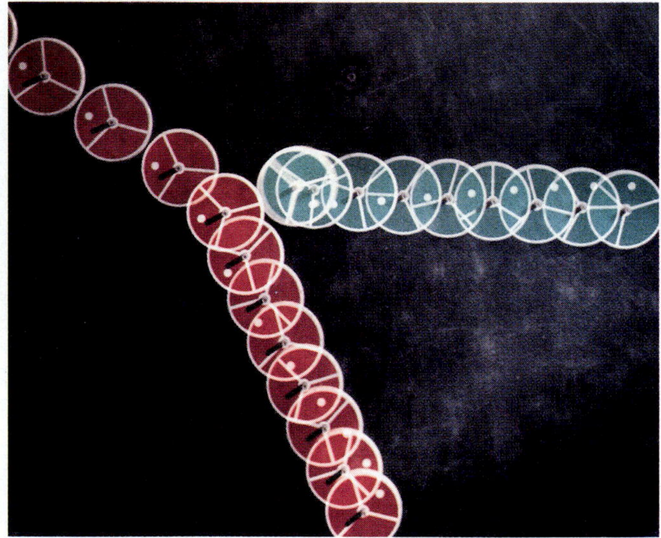

Figure 10-41. Exercise 10-42: Strobe photo of a collision between two pucks riding on a surface with negligible friction.

43. A system initially at rest explodes into three pieces. Piece A has a mass of 2.0 kg, B has a mass of 3.0 kg, and C has a mass of 1.0 kg. After the explosion A's velocity is $(3.0\ \mathrm{m/s})\mathbf{i}$ and B's velocity is $(-1.0\ \mathrm{m/s})\mathbf{j}$. Determine C's velocity after the explosion.

44. Object 1 moving with speed v collides with initially stationary object 2. After the collision, object 1 is moving with speed $\frac{1}{2}v$ at 90° to its original line of motion. The collision is not necessarily elastic and the masses of the objects are unknown and not necessarily the same. Determine the angle between object 2's final velocity and the original line of motion. (*Hint:* Use the identity $\tan\theta = \sin\theta/\cos\theta$.)

45. In a billiard-ball collision between a moving cue ball and an initially stationary numbered ball, the cue ball has an initial speed of 0.88 m/s. After the collision the cue ball's speed is 0.23 m/s. Assume an elastic collision, and determine: *(a)* the numbered ball's speed after the collision, *(b)* the angle between the cue ball's velocity after the collision and its velocity before the collision, and *(c)* the angle between the numbered ball's velocity after the collision and the cue ball's velocity before the collision.

46. A 6.30-kg bowling ball moving at a speed of 2.80 m/s collides with a 1.40-kg pin that is initially stationary. The angle between the ball's initial and final velocities is 16.0°; the angle between the pin's final velocity and the ball's initial velocity is 71.0°. Determine the final speed of *(a)* the ball and *(b)* the pin.

Section 10-8. Rocket Motion

47. Show that the product $v_e|dM/dt|$ has the dimension of a force.

48. What is the acceleration magnitude of a 5860-kg rocket just after liftoff? The rocket's engine has a thrust magnitude of 72.7 kN.

49. A 2000-kg rocket is at rest when its engine is turned on. The rocket is in an interplanetary region of the solar system where $\Sigma\mathbf{F}_{\mathrm{ext}}$ is negligible. What is the rocket's mass at the instant its speed equals v_e?

50. A rocket lifts off from the surface of the earth at time $t = 0$ s. At the instant of liftoff the rocket's mass is 1.00×10^3 kg, the fuel burn rate is constant at 5.00 kg/s, and the speed v_e of the exhaust gases is 2.00×10^3 m/s. *(a)* Evaluate the rocket's mass M at each 25 s from $t = 0$ s to $t = 100$ s, and sketch a graph of M versus t in this time interval. *(b)* Evaluate the rocket's acceleration magnitude a at each 25 s from $t = 0$ s to $t = 100$ s, and sketch a graph of a versus t in this time interval. *(c)* Use your graph from part *(b)* to estimate an average acceleration magnitude during the time interval. Use this value and constant-acceleration kinematics to estimate the height of the rocket at $t = 100$ s.

51. A 10,000-kg spaceship is equipped with a small rocket engine for maneuvering in space. The engine has an exhaust speed of 2.0 km/s and a burn rate of 0.010 kg/s. *(a)* What is the engine's thrust? *(b)* Estimate the length of time the engine must operate to increase the spaceship's speed from zero to 2.0 m/s. *(c)* How much mass is ejected during this time interval?

52. A rocket is in a region of space where $\Sigma\mathbf{F}_{\mathrm{ext}}$ is negligible. The rocket's engine is used to accelerate the rocket along a straight line from speed zero to 5 km/s. The engine's exhaust speed is $v_e = 2.0 \times 10^3$ m/s. What fraction of the rocket's mass is ejected during this time interval?

Additional Exercises

53. A hand-held machine gun fires bullets of mass 0.030 kg with a muzzle velocity of 300 m/s. If the maximum horizontal force the gunner can steadily exert on the gun is 200 N, what is the maximum number of bullets that can be fired horizontally per minute and still have the gun be properly aimed?

54. In a nuclear reactor, a neutron has an elastic collision with an initially stationary proton. The angle between the neutron's velocity after the collision and its velocity before the collision is 45°. The mass of a proton is essentially the same as that of a neutron. Determine the neutron's final kinetic energy K_f in terms of its initial kinetic energy K_i.

55. A 6000-kg railway gondola car is coasting along a level track at a speed of 2.8 m/s when 10,000 kg of sand is dumped vertically downward into it. What is the speed of the car after it is loaded with sand?

56. Show that $y_{cm} = 2h/3$ for the thin, triangular-shaped slab contained in the xy plane in Fig. 10-42. The slab has uniform density ρ, thickness τ, and its shape is an isosceles triangle of height h and base length b. Thus the slab's volume is $bh\tau/2$.

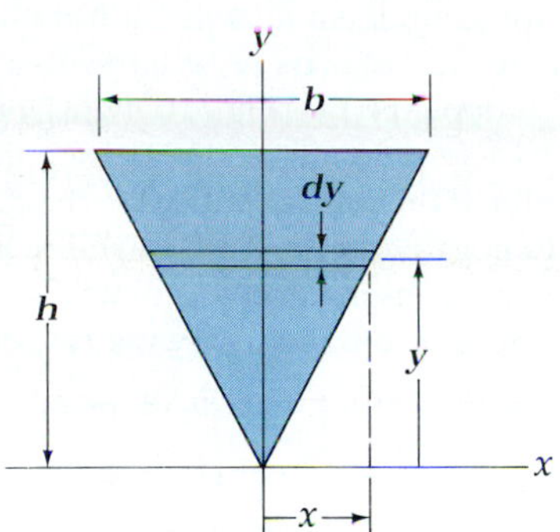

Figure 10-42. Exercise 56.

[*Hint:* From the figure, the relation between the volume element's half-width x and its y coordinate is $x = (b/2h)y$.]

57. A cannon of mass M is mounted on small wheels with well-lubricated axles (Fig. 10-43). The cannon is at rest on the level deck of a ship when it fires a cannonball of mass m with a muzzle speed of v_m relative to the cannon. Write expressions for *(a)* the speed v of the cannonball relative to the ship after the shot and *(b)* the speed V of the cannon relative to the ship after the shot. Each expression should be in terms of v_m, m, and M. Determine *(c)* v, and *(d)* V when $M = 100\ m$, and $v_m = 100$ m/s.

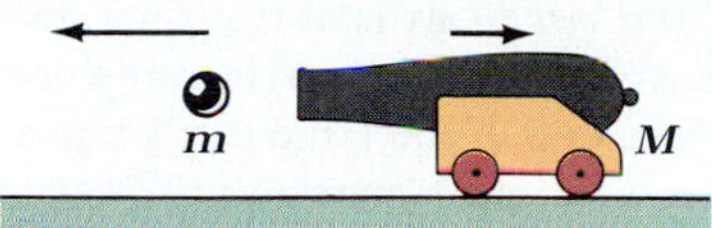

Figure 10-43. Exercise 57.

58. The cannon in the previous problem is secured to the ship and then fired. What is the speed of the cannonball relative to the ship?

59. A stationary fire truck pumps water from its tank and through its hose at a rate of 10 kg/s. The speed of the water as it exits the hose nozzle is 20 m/s. What is the force exerted on the truck by the streaming water?

60. Two thin uniform rods are made of the same material, have the same length ℓ, and have the same mass M. The rods are connected to form a "T." What is the distance between the point where the rods are connected and the center of mass?

PROBLEMS

1. ***Center of mass of a parabolic sheet.*** Consider the center of mass of the thin uniform sheet shown in Fig. 10-44. The sheet edge is parabolic with height h, top width w, and thickness τ. *(a)* Show that the relation between the half-width x of an element and its y coordinate is $y = (4h/w^2)x^2$. *(b)* Show that the volume of the sheet is $2hw\tau/3$. *(c)* Show that $y_{cm} = 3h/5$.

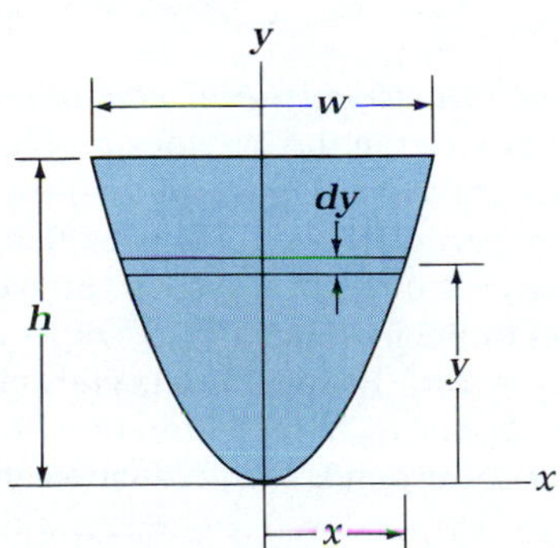

Figure 10-44. Problem 1.

2. ***Center of mass of a system made up of subsystems.*** A system of particles is made up of two subsystems, A and B. Show that the center of mass of the system $\mathbf{r}_{cm}$ is given by

$$M\mathbf{r}_{cm} = M_A\mathbf{r}_A + M_B\mathbf{r}_B$$

where M is the mass of the whole system, M_A and M_B are the masses of the subsystems, and $\mathbf{r}_A$ and $\mathbf{r}_B$ locate the centers of mass of the subsystems.

3. ***Center of mass of a solid cone.*** Consider the center of mass of a solid cone made of a uniform material of mass density ρ (Fig. 10-45). The cone has a height h and base radius R_0. The thin, coin-shaped volume element has radius R and volume $dV = \pi R^2 dy$, and is located at coordinate y. *(a)* Show that the relation between R and y is $R = (R_0/h)y$. *(b)* Show that the volume of the cone is $\pi R_0^2/h/3$. *(c)* Show that $y_{cm} = 3h/4$.

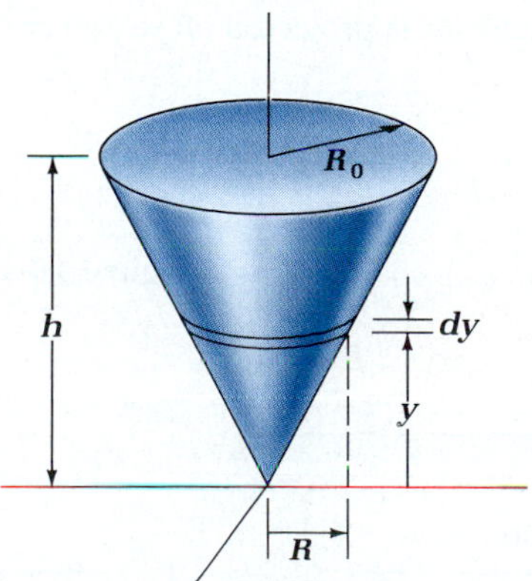

Figure 10-45. Problem 3.

4. ***Center of mass of a solid hemisphere.*** Determine the distance from the center of the flat face of a solid uniform hemisphere to the center of mass. Give your answer in terms of the radius R_0 (*Hint:* Follow the procedure outlined in the previous problem.)

5. ***Center of mass of half a hula hoop.*** Consider the center of mass of a thin, uniform, semicircular ring. The cross-sectional dimensions are negligibly small compared with the radius R_0 of the full ring. The object is shaped like a hula hoop cut into two pieces. Show that the distance from the center of the full circle to the center of mass of the semicircular ring is $2R_0/\pi$. (*Hint:* Divide the ring into infinitesimal elements of arc length $R_0\,d\theta$, cross-sectional area A, and volume $AR_0\,d\theta$. The volume of the entire semicircular ring is πAR_0.)

6. ***A recoiling curved ramp.*** A large cart of mass M is equipped with a curved ramp that is horizontal at its lower end (Fig. 10-46). The large cart is at rest on a level table when a small cart of mass m is released on the ramp at a vertical height h above the lower end of the large cart's ramp. On both carts, the wheels have negligible mass and the axles are well lubricated. *(a)* Show that the speed of the large cart at the instant the small cart is projected off its lower end is

$$V = \sqrt{2m^2gh/[M(m+M)]}$$

(b) Determine an expression for the speed v of the small cart at this instant.

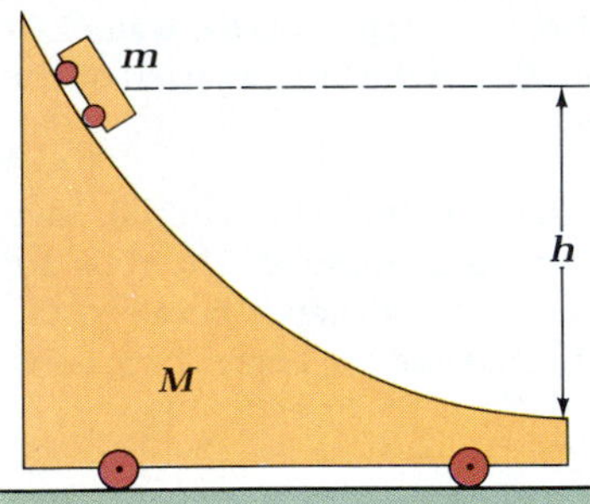

Figure 10-46. Problem 6.

7. ***A recoiling straight ramp.*** A large cart of mass M is equipped with a straight ramp that makes an angle ϕ with the horizontal (Fig. 10-47). The large cart is at rest on a level table when a small cart of mass m is released on the ramp at a vertical height h above the lower end of the large cart's ramp. On both carts, the wheels have negligible mass and the axles are well lubricated. *(a)* Show that the speed of the large cart at the instant the small cart is projected off its lower end is

$$V = \sqrt{2m^2gh\cos^2\phi/[(M+m)(M+m\sin^2\phi)]}$$

(b) Determine an expression for the speed v of the small cart at this instant.

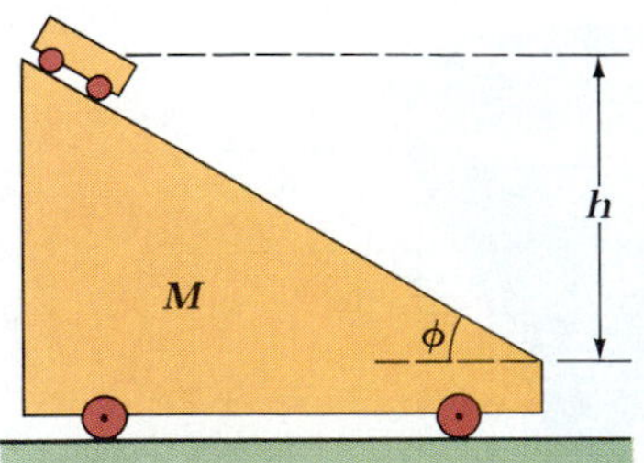

Figure 10-47. Problem 7.

8. ***Multiple collisions.*** Carts A, B, and C, equipped with small wheels and well-lubricated axles, are aligned along a straight level track. Carts A and C have mass $2m$ and cart B has mass m. Cart A is initially moving with speed v toward stationary cart B, and cart C is also stationary and located a distance ℓ from B on the opposite side from A. Cart A collides with B, projecting it forward so that it collides with C. If each collision is elastic, show that carts A and B collide twice and that the time interval between these collisions is $12\ell/7v$.

9. ***A one-dimensional elastic collision with both particles initially moving.*** Particle 1 of mass m_1 is traveling along the x axis with velocity component v_{1i} when it has a head-on elastic collision with particle 2 of mass m_2 as it travels along the x axis with velocity component v_{2i} (Fig. 10-48). Note that each v represents a velocity component rather than a speed, so that it has a positive value if the particle moves in the direction designated as $+\mathbf{i}$ and a negative value if it is traveling in the opposite direction. Show that conservation of momentum and conservation of kinetic energy yield the following two equations:

$$m_1v_{1i} + m_2v_{2i} = m_1v_{1f} + m_2v_{2f}$$

$$v_{1i} - v_{2i} = -(v_{1f} - v_{2f})$$

where v_{1f} and v_{2f} are the velocity components after the collision. (*Hint:* Develop the second equation from a combination of conservation of momentum and conservation of kinetic energy using a procedure similar to that used in Sec. 10-6.)

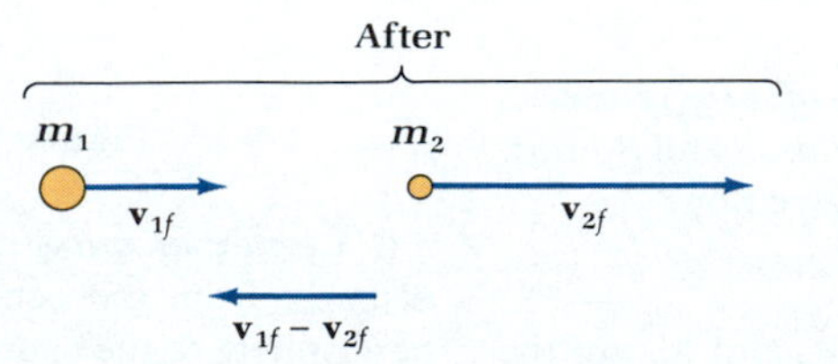

Figure 10-48. Problem 9.

10. ***More about the one-dimensional elastic collision.*** Use the results of the previous problem to show that the velocity components of the particles after the collision are

$$v_{1f} = \frac{m_1 - m_2}{m_1 + m_2}v_{1i} + \frac{2m_2}{m_1 + m_2}v_{2i}$$

$$v_{2f} = \frac{m_2 - m_1}{m_2 + m_1}v_{2i} + \frac{2m_1}{m_2 + m_1}v_{1i}$$

11. ***Examples of one-dimensional elastic collisions.*** Employ the equations given in the previous problem to examine the results of such collisions. Figure 10-49 shows the situation before a collision: particle 1 has velocity (2.0 m/s)$\mathbf{i}$ and particle 2 has velocity (−1.0 m/s)$\mathbf{i}$. Consider three different cases where the masses of the particles are *(a)* $m_1 = m_2$, *(b)* $m_1 = 2m_2$, and *(c)* $m_2 = 2m_1$. For the initial condition given in the figure, determine v_{1f} and v_{2f} in each case. Draw a sketch similar to Fig. 10-48 that corresponds to the situation after each collision.

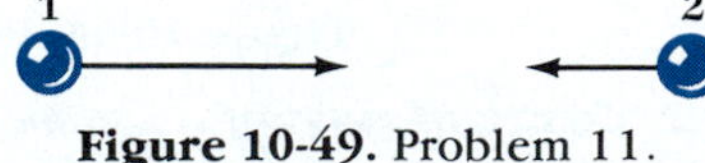

Figure 10-49. Problem 11.

12. ***A railway-car coupling.*** A 20,000-kg railway car is stationary on a slope on a straight track at a location that is 10 m vertically above where the track becomes level. When the car's brakes are released, it rolls down the hill and collides with a 10,000-kg car that is staionary on the level portion of the track. The two cars couple and roll partway up the next hill. Assume that the wheels have negligible mass and that the axles are well lubricated. What is the maximum vertical height of the two-car system above the level portion of the track after the coupling? Is mechanical energy conserved during this entire process? If not, when is it not conserved?

13. ***The ballistic pendulum revisited.*** In the ballistic pendulum discussion in Example 10-10, it was assumed that the block moved a negligible amount during the time Δt_c (the time for the bullet to become embedded in the block). One consequence of this assumption is that the block's vertical rise during this time was neglected. To justify this approximation, estimate Δt_c by assuming that the bullet slows from a speed of 200 m/s to zero with constant acceleration and that it stops after entering the block a distance of 0.010 m. Next, let the speed of the block be 1 m/s at the end of Δt_c, and let the pendulum's length be 1 m. Estimate the block's vertical rise during Δt_c.

14. ***Conservation of momentum in frames moving relative to one another.*** Frame of reference A has a system of n particles with masses m_1, m_2, . . . and velocities $\mathbf{v}_1$, $\mathbf{v}_2$, ... Prove that if total momentum is conserved in frame A, then it is also conserved in frame B, which is moving at constant velocity with respect to A.

15. ***Advantage of a two-stage rocket.*** Consider two rockets with the same total mass ($M = 11{,}000$ kg) and the same payload mass ($M_p = 200$ kg), but one is a one-stage rocket and the other is a two-stage rocket. *(a)* In the one-stage rocket, the mass of the fuel is $M_f = 9700$ kg and the mass of the rocket engine is $M_e = 1100$ kg. (Note that $M_p + M_f + M_e = M$.) Assume $\Sigma\mathbf{F}_{\text{ext}} = 0$, and calculate Δv in terms of v_e when the rocket burns all its fuel. *(b)* In the two-stage rocket, the mass of the fuel and the rocket engine in the first stage are $M_{f1} = 9000$ kg and $M_{e1} = 1000$ kg; in the second stage they are $M_{f2} = 700$ kg and $M_{e2} = 100$ kg. (Note that $M_p + M_{f1} + M_{e1} + M_{f2} + M_{e2} = M$.) In this case the first-stage fuel is completely burned, and then the first-stage engine is discarded and the second stage fuel is completely burned. Assume $\Sigma\mathbf{F}_{\text{ext}} = 0$ and calculate Δv in terms of v_e.

16. ***Energy loss in a completely inelastic collision.*** In a completely inelastic collision between object 1 that is initially moving and object 2 that is initially stationary, a measure of the energy dissipated is the ratio of the kinetic energy of the system after the collision to that before the collision. Show that this ratio is $m_1/(m_1 + m_2)$.

17. ***Energy transfer in a head-on elastic collision.*** In a head-on elastic collision between object 1 that is initially moving and object 2 that is initially stationary, a measure of the energy transferred from 1 to 2 is the ratio of 2's kinetic energy after the collision to 1's kinetic energy before the collision. Show that this ratio is $4m_1m_2/(m_1 + m_2)^2$.

18. ***Speed of a gravitationally attracted object.*** Two objects, one of mass m and the other of mass M, are initially at rest in an inertial reference frame at a great distance r_i from one another. The two exert gravitational forces on one another, but no other significant force is exerted on either of them. Show that as they approach each other to a separation distance r ($r \ll r_i$), the speed v of the object of mass m relative to our inertial reference frame is $\sqrt{2GM^2/[(m + M)r]}$.

19. ***Force exerted by water running through a hose.*** Water passes through a fire hose with a speed of 4.8 m/s and at a rate of 12 kg/s (Fig. 10-50). Consider the section of hose outlined in the figure. What are the magnitude and direction of the force exerted on this section of hose by the streaming water? (Neglect the weight of the water.)

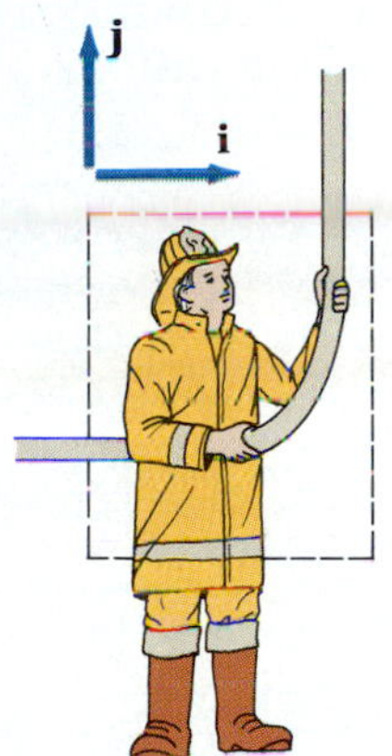

Figure 10-50. Problem 19.

20. ***Inelastic collision between equal-mass objects with the same initial speed.*** Two objects with the same mass m and the same speed v have an inelastic collision (Fig. 10-51). After the collision the two-object system moves with speed $\frac{1}{2}v$. What is the angle θ between the final line of motion and either of the initial velocities?

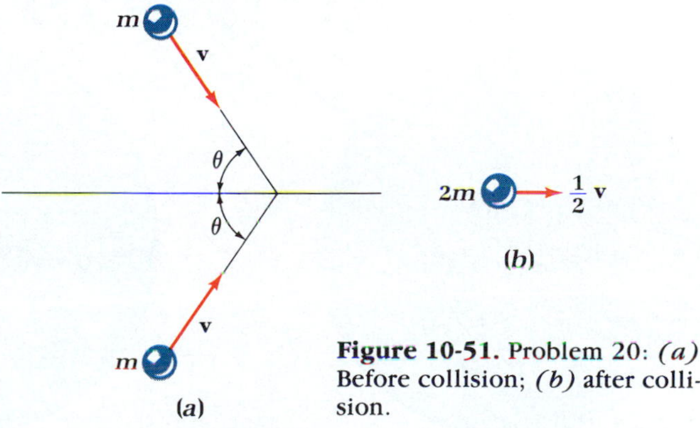

Figure 10-51. Problem 20: *(a)* Before collision; *(b)* after collision.

21. ***A neutron-deuteron elastic collision.*** Show that if a neutron (mass $= m_n$) has an elastic collision with a deuteron (mass $\approx 2m_n$) such that the neutron's direction of motion is changed by 90°, then two-thirds of the neutron's energy is transferred to the deuteron. (*Hint:* Use the identity $\sin^2\theta + \cos^2\theta = 1$.)

HIGHLIGHTS OF MODERN PHYSICS

A Fifth Force?

Physics is the study of the most basic laws of nature. The four fundamental forces—gravitation, electromagnetism, and the strong and weak nuclear forces—form the core of these laws (Commentary, Chap. 7). By the latter part of the twentieth century most physicists were satisfied that these four forces could fully account for the physical world. So it came as a great surprise when Ephraim Fischbach of Purdue University and his colleagues published a paper in 1986 that seemed to show the existence of a fifth force. These physicists had been trying to explain some puzzling data on a strange kind of elementary particle (the neutral K meson) and thought that a yet-unknown force might account for their results.

An independent indication of a fifth force arose from measurements made by other workers indicating that the acceleration of gravity g does not change with depth in mine shafts as predicted by Newton's law of gravitation. The measurements consistently showed that the gravitation constant G was about 1 percent larger than the value obtained from the Cavendish experiment (Sec. 7-2). The data could be explained if gravity were not purely an inverse-square force but also included a medium-range component that dies out within a few hundred meters, as represented by the potential energy

$$U(r) = -\frac{Gm_1m_2}{r}(1 + \alpha e^{-r/\lambda}) \qquad (1)$$

with $\alpha = 7 \times 10^{-3}$ and $\lambda = 200$ m. The first term is the ordinary gravitational potential energy, while the second represents a modification of newtonian gravitation that falls off more rapidly than $1/r^2$. It becomes weaker by $1/e$ for each 200-m increment of distance, becoming negligible for distances of 1 km or more.

Together with the K meson problem, this result hinted at the presence of a new force in the universe. Whether it was a truly new force or merely a modification of Newton's law of gravitation was not obvious, but the latter was unlikely since it would invalidate Einstein's general theory of relativity (Commentary, Chap. 38) which reduces to Newton's inverse-square theory for weak gravitational fields.

Fischbach looked for a third area where the presence of such a force might appear and focused on an experiment by the Hungarian, Roland von Eötvös, that showed that gravitational and inertial masses are equal within one part in 10^9 (Sec. 7-3). In Fig. 1*a*, a pendulum is suspended from a support at an intermediate latitude (neither 0° or 90°) for an idealized nonrotating earth. The tension in the string $\mathbf{F}_T$ and the weight $\mathbf{F}_e$ are equal and opposite so the pendulum bob is in equilibrium. Figure 1*b* is the same, except that the earth's rotation is now included. The pendulum bob must move out from the earth's axis so that the net force on the bob can produce a centripetal acceleration as the bob "orbits" the earth (Fig. 1*c*).

In Fig. 1*b*, will the angle θ by which the pendulum swings out from the radial direction change for a bob of a different mass? Put another way, will the force parallelogram of Fig. 1*b* change its shape? The parallelogram will retain its shape if $\mathbf{F}_e$ and $\Sigma\mathbf{F}$ change proportionally for different masses. For example, if $\mathbf{F}_e$ doubles, then $\Sigma\mathbf{F}$ must also double for the parallelogram to keep its shape.

The weight $\mathbf{F}_e$ depends on the *gravitational mass* of the bob m_G:

$$\mathbf{F}_e = \frac{GMm_G}{R^2}$$

while $\Sigma\mathbf{F}$ depends on the *inertial mass* $\mathbf{m}_I$:

$$|\Sigma\mathbf{F}| = m_I\omega^2R$$

if these two masses are proportional to each other, $m_I = km_G$,

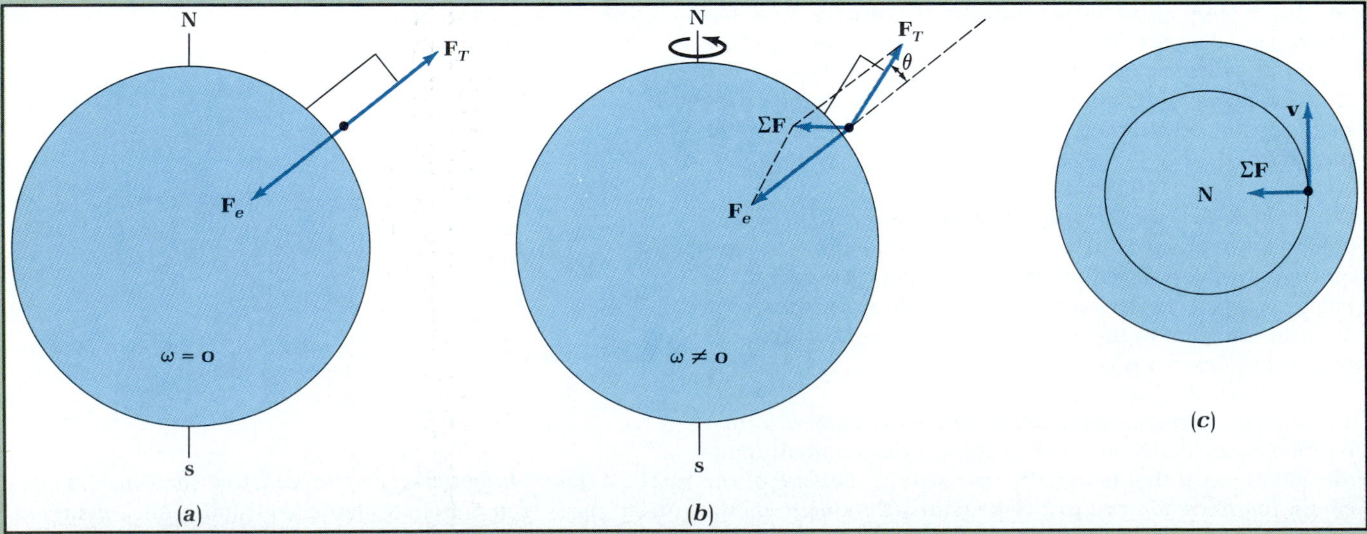

Figure 1. *(a)* On an idealized, nonrotating earth, the plumb bob hangs down directly toward the earth's center. The gravitational force on the bob and the force due to the string exactly cancel. *(b)* On a rotating earth, the bob swings out by the angle θ, so $\mathbf{F}_T$ and $\mathbf{F}_e$ are no longer colinear. The net force, pointing directly toward the axis of rotation, produces the centripetal acceleration for the circular motion of the bob, shown from above the earth's north pole in *(c)*.

then the ratio $F_e/|\Sigma \mathbf{F}|$ will not change when bobs of different masses are used since

$$\frac{F_e}{|\Sigma \mathbf{F}|} = \frac{GMm_G/R^2}{km_G\omega^2 R} = \frac{GM}{k\omega^2 R^3} \qquad (2)$$

is independent of the bob's mass. If the gravitational and inertial masses are *not* proportional, the ratio will change and so will θ.

In principle, the difference could be observed by suspending two bobs of different masses and measuring the two angles, but Eötvös knew the effect could not be large enough to be detected this way. He devised a highly sensitive method based on the Cavendish apparatus described in Sec. 7-2, except that he dispensed with the two large masses and used unequal masses at the ends of the dumbbell (Fig. 2). In effect, two

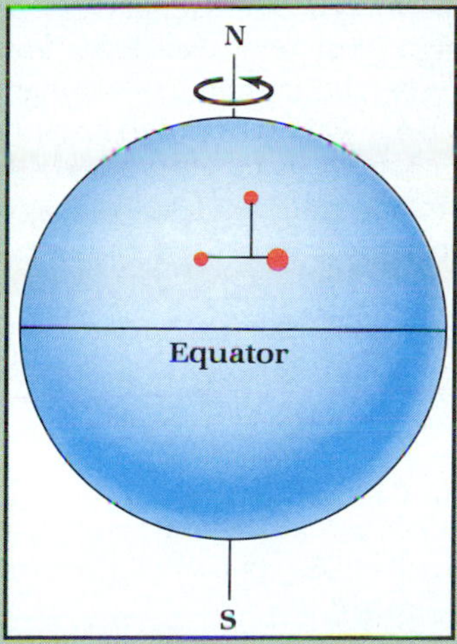

Figure 2. The Eötvös experiment. A rod with unequal masses is suspended in an east-west direction. If m_I and m_G are not proportional, the fiber will twist through an angle α. This is not observable because there is no way to know where the "no-twist" position is. When the apparatus is rotated 180°, the fiber will twist the same amount α in the other direction. The total angular motion 2α will be observable as a movement of the reflected light beam.

plumb bobs (the ends of the dumbbell) were being supported by a single fiber. Extremely small changes in the dumbbell's orientation could be observed with the mirror and light beam arrangement of the Cavendish experiment (Fig. 7-6). Since the dumbbell was placed in an east-west orientation, when the whole apparatus was rotated 180° to reverse the positions of the two masses, any tendency for either mass to swing out farther than the other would cause the supporting fiber to twist in the opposite direction, and the reflected light would move along the scale. Eötvös observed a number of bob pairs made of a variety of materials but saw no deflection of the light beam. He concluded that the gravitational and inertial masses are proportional to each other within experimental accuracy. For simplicity we take them to be equal and are justified in referring to *the* mass of an object.

The new force should act between each object on the ends of the torsion rod and the nearby material in the earth. Gravity acts between *masses*, as shown by the potential energy $U = -Gm_1m_2/r$. Similarly, the electric force acts between *charges*, q_1 and q_2 (Sec. 20-3), with $U = -q_1q_2/4\pi\epsilon_0 r$. Fischbach proposed that the new force acts on *baryon number B*, which for ordinary matter is just the total number of protons and neutrons in the object:

$$U = SB_1B_2\frac{e^{-r/\lambda}}{r}$$

Here, B_1 and B_2 are the baryon numbers of the two objects—for example, one of the ends of the dumbbell and a small volume of material in the earth—and S represents the strength of the force just as G does for gravitation.

The fact that the proposed force couples to something other than mass was important because otherwise the masses would still cancel as in Eq. 2, and the experiment would be insensitive to the fifth force. By coupling to baryon number, the ratio of forces in Eq. 2 can be different for the two torsion objects. It would *not* be different if baryon number were proportional to mass, because the masses would cancel as before. In fact, B is *almost* proportional to mass, but when neutron and protons combine to form nuclei, energy is given off, and by the Einstein formula $E = mc^2$, mass is also carried off (Sec. 43-2), leaving a nuclear mass that is less than the total mass of the protons and neutrons that formed it. This mass loss is different for different nuclei and *is not proportional to the baryon number*. Therefore B for a plumb bob is not proportional to its mass, and $F_e/|\Sigma \mathbf{F}|$ will vary from substance to substance.

Eötvös's data did not show exactly zero deflection for each pair of masses, but he attributed this to normal experimental error and reported a null result. Fischbach proposed that if a force coupling to baryon number was present, what seemed to be random error might actually be due to the different compositions of the torsion objects used by Eötvös. So he was led to plot Eötvös's random torsion balance deflections against the difference between B/m for the two objects, which should give a straight line if such a force exists (Exercise 3). The result (Fig. 3) suggests that his model is correct, for otherwise the points should be scattered randomly on the plot instead of lying so comfortably along a straight line. Furthermore, the values of α and λ obtained from the mine shaft data (Eq. 1)

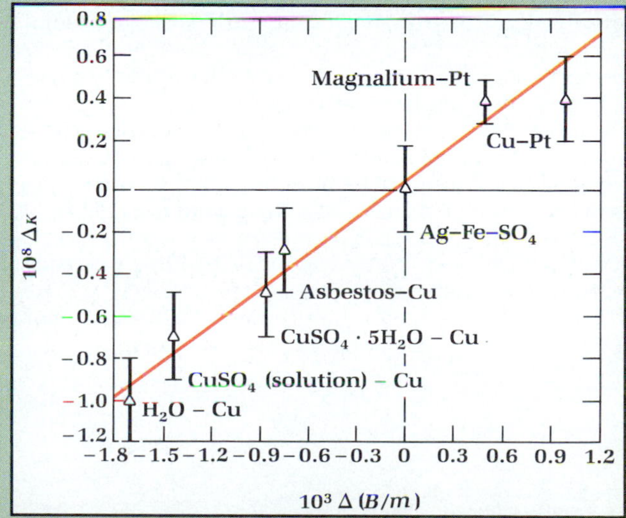

Figure 3. The plot by Fischbach and his coworkers, showing the deflection of the torsion beam as a function of the difference in B/m of the two masses on the ends of the beam. The quantity $\Delta\kappa$ is actually the difference in accelerations of the two masses as a fraction of g, but it is also a measure of the deflection of the torsion beam.

were roughly consistent with the slope of the line in Fig. 3. Very different methods had given similar results.

This paper galvanized work by many others, some critically examining Fischbach's analysis and others designing additional experiments to look for the force. Two early criticisms seemed to kill the idea outright. It was pointed out that the slope of the line in Fig. 3 implied an *attractive* force rather than the *repulsive* force required by the geophysical data. What had at first seemed to be a dramatic confirmation had become a devastating contradiction. Secondly, Fischbach's group and others realized that if one assumed a smooth earth slightly flattened due to rotational effects as Fischbach had done, a plumb bob should hang exactly perpendicular to the surface (Fig. 4). Thus, by symmetry, the net effect of any fifth force must be directed along the supporting fiber, and the force could not produce a deflection anyway. It seemed that the straight line of Fig. 3 was simply a fluke.

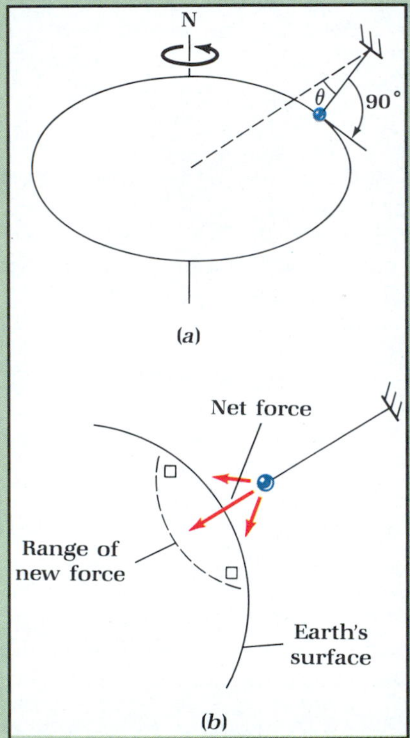

Figure 4. *(a)* The flattening of the earth is exaggerated, showing how the plumb bob hangs down perpendicular to the earth's surface, but at an angle θ with respect to a line to the center of the earth. *(b)* A magnified view of the plumb bob and the local area of the earth. Because of the symmetry, the net force on the plumb bob due to both gravity and the proposed fifth force must be directly in line with the string. The forces due to two symmetrically placed mass elements are shown. The net force is the sum of many such pairs of masses, which clearly add to give a force in line with the fiber. The direction of this force will not change if a fifth force is turned on (or off), so it cannot alter the direction of the plumb bob.

Realizing this, Fischbach's group recalculated the expected deflections based on what was known about the conditions at the time Eötvös made his measurements, such as the location of nearby buildings, cellars, hills—any substantial feature within about a kilometer. They found that not only did the revised data fall on a line once again, but that the force was now repulsive and agreed even more closely with the geophysical data.

The experiments since then are too numerous to describe here, but they give inconsistent results. The first set of experiments split about equally between those that showed no effect and those that did. Most of the results could be brought into agreement by assuming a combination of repulsive and attractive forces, but most physicists thought this was moving too far from the original idea, and that with enough forces—including their respective α's and λ's—it ought to be possible to fit almost any data. Things began to seem contrived. More recent experiments tend to show no effect, and there is evidence that some of the earlier positive results could have come from unknown geologic mass concentrations rather than from a nongravitational force. The final word has not been written on this, but however it turns out, the fifth-force controversy has produced a textbook case of how science works at the frontier of knowledge.

QUESTIONS

1. Why should the line of Fig. 3 pass through the origin? Does the fact that it does not pass *precisely* through the origin invalidate the results? Explain.
2. If a fifth force *did* affect Eötvös's results, explain how this would have increased the apparent error in his data. How could knowledge of the fifth force be used to improve the accuracy of Eötvös's results?
3. If the fifth force exists, will F_e/F_{net} be different for two plumb bobs of different masses but the same material?

EXERCISES

1. Show that $F_e/F_{\text{net}} = $ constant if $B \propto$ mass.
2. Show that $F_e/F_{\text{net}} = $ constant if the mass lost by a nucleus during formation from protons and neutrons is proportional to its mass.
3. Show that if you assume the gravitational and fifth forces act on the bob in the same direction—they don't—then the data of Fig. 3 are expected to lie on a straight line. Do this by using the fact that for small angles, $\theta \approx (F_e + F_5)/|\Sigma \mathbf{F}|$. Write θ in terms of appropriate expressions for F_e, F_5, and $|\Sigma \mathbf{F}|$. Then show that $\theta_1 - \theta_2$, the torsion-rod deflection, depends linearly on $(B_1/m_1 - B_2/m_2)$.

STATIC EQUILIBRIUM OF A RIGID BODY

11

Sculpture in Storm King Park, New York.

Just over a century ago, the Brooklyn Bridge linking Manhattan and Brooklyn was completed. At that time it was the longest suspension bridge on earth. And while it may have been surreptitiously sold many times, it remains in place as a graceful example of an extended object in static equilibrium.

What is the value of studying an extended object in static equilibrium? After all, its motion is trivially simple: It remains at rest. But, analysis reveals information about some of the forces acting on the object. This knowledge is essential when selecting the materials and components of a structure.

Many engineering students take one or more courses in statics. Such courses are based on the principles introduced in this chapter. We focus our attention on simple situations that illustrate how these principles are used to analyze a "statics" problem.

11-1 STATIC EQUILIBRIUM OF A RIGID BODY

In static equilibrium every point on an object remains at rest.

If a particle, a point object, remains at rest in an inertial reference frame, its acceleration is zero, and, from Newton's second law, the net force on the particle is zero, $\Sigma \mathbf{F} = 0$. This is the necessary (and sufficient) condition for the static equilibrium of a particle, as discussed in Chap. 5. But in the real world, we must deal with extended objects instead of particles.

> An extended object is in static equilibrium if every point of the object is at rest and remains at rest.

Of course, if the object is in static equilibrium in one inertial reference frame, every point on the object moves with a common, constant velocity in a different inertial reference frame. We choose the reference frame in which the object is at rest to discuss static equilibrium.

Some extended objects, such as bread dough, a spring, or a pencil eraser, are flexible and can change their shape or size in response to applied forces. On the other hand, a mechanic's wrench remains rigid and nondeformable under ordinary circumstances. This chapter focuses on rigid bodies in static equilibrium.

> A ***rigid body*** (or object) is one for which the distance between any pair of points on the body remains fixed. A rigid body retains its shape or size under the application of forces.

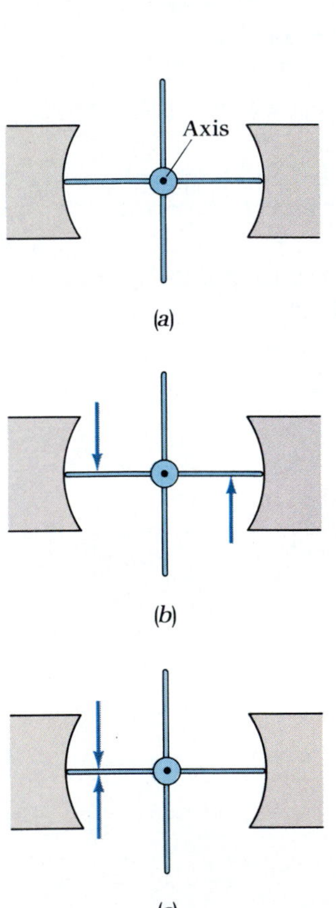

Figure 11-1. A revolving door viewed from above.

The concept of a rigid body is an idealization because every material object undergoes some deformation in response to externally applied forces (Chap. 15). But if these changes are negligible, the object is considered to be a rigid body.

From Chap. 10, the motion of the center of mass of an extended object is determined by the external forces: $\Sigma \mathbf{F}_{ext} = M\mathbf{a}_{cm}$. The object is said to be in ***translational equilibrium*** if the acceleration of the center of mass is zero. If $\mathbf{a}_{cm} = 0$, then

$$\Sigma \mathbf{F}_{ext} = 0 \tag{11-1}$$

This is the condition for translational equilibrium.

Even if the center of mass of a rigid body remains at rest, the object is not necessarily in static equilibrium. It could be changing its spatial orientation by *rotating* about the stationary center of mass. For example, a pulley mounted on a shaft can rotate about an axis passing through its center of mass. Points on the pulley are moving, although the center of mass remains at rest. An object that either does not rotate or that rotates steadily about an axis (Chap. 12) is in ***rotational equilibrium.***

> A rigid body in static equilibrium neither translates nor rotates; it is in both translational and rotational equilibrium.

Consider a revolving door as viewed from above, shown in Fig. 11-1*a*. The door can rotate about a vertical axis which passes through its center of mass. The bearings along the axis will exert a force $\mathbf{F}_b$ (not shown in the figures) on the shaft of the door so that $\Sigma \mathbf{F}_{ext} = 0$ to keep the door in translational equilibrium. Suppose that two pedestrians are in opposite quadrants, exerting equal but opposite forces on the door, as seen in Fig. 11-1*b*. The net external force is still zero, but our experience shows that the door is not in rotational equilibrium. The door actually starts rotating when the pedestrians begin pushing in this way. In Fig. 11-1*c*, however, the door is in static equilibrium. Here the net external force is zero and, by symmetry, the door is in rotational equilibrium.

Experiment shows that the door in Fig. 11-1*d* is also in static equilibrium if the magnitudes of the forces F_1 and F_2 are inversely proportional to the distances from the axis of their points of application: $F_1/F_2 = x_2/x_1$. This result can also be written as

$$x_1F_1 - x_2F_2 = 0 \quad (11\text{-}2)$$

Although $\mathbf{F}_1$ and $\mathbf{F}_2$ do not add to zero, the condition $\Sigma\mathbf{F}_{ext} = 0$ is still satisfied because of the force exerted by the bearings. The force $\mathbf{F}_1$ tends to cause the door to rotate counterclockwise, while $\mathbf{F}_2$ tends to produce a clockwise rotation. These two tendencies balance, or cancel out, for static equilibrium. Equation 11-2 is the application of a condition for rotational equilibrium. The general conditions of static equilibrium of a rigid body will be stated in Sec. 11-3.

A "torque wrench" is equipped with a meter that reads the magnitude of the torque.

EXAMPLE 11-1

A revolving door. One pedestrian exerts a 40-N force F_1 on a revolving door at a point 0.3 m from the axis, as shown in Fig. 11-1*d*. What force F_2 must be exerted at 0.2 m from the axis to keep the door stationary?

Solution. From Eq. 11-2 we have $x_2F_2 = x_1F_1$, so that

$$F_2 = \frac{F_1x_1}{x_2} = \frac{(40\ \text{N})(0.3\ \text{m})}{0.2\ \text{m}} = 60\ \text{N}$$

SELF-TEST 11-1. Suppose that each of the pedestrians described in the above example *pulled* instead of *pushed* on the door with the same force magnitude. Would the door remain stationary? ***ANSWER:*** Yes.

11-2 TORQUE ABOUT AN AXIS

In considering the equilibrium of an object, such as a door, that can rotate about an axis, we have seen the importance of the point of application of a force. The quantity that takes this feature into account is called the torque. ***Torque*** is produced by a force about an axis; it is the torque that tends to cause an object to rotate.

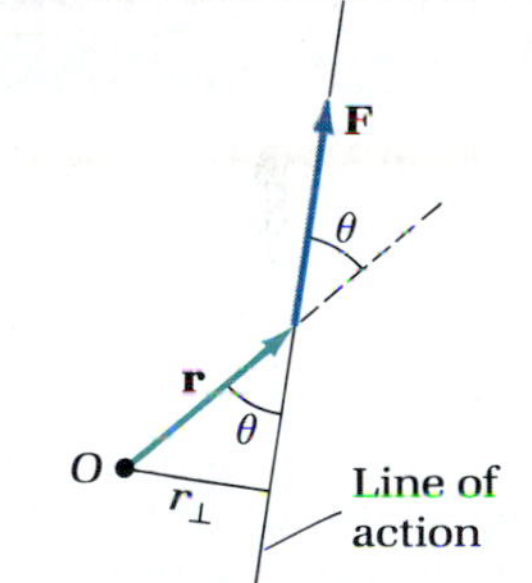

Figure 11-2. A force **F** is applied at a point located by the position vector **r**.

To define torque, consider a force **F** applied at a point on an object, as indicated in Fig. 11-2. The point *O* represents the intersection of a perpendicular axis with the plane containing the vectors **F** and **r**; the vector **r** locates the point of application of the force from this axis. The *line of action* of this force is constructed by extending the line along which **F** lies. The perpendicular distance from the axis to the line of action of the force is $r_\perp = r \sin\theta$, where θ is the angle between the directions of **r** and **F**. The magnitude of the torque about axis *O* produced by this force is defined as

$$\tau = r_\perp F = (r \sin\theta)F \quad (11\text{-}3)$$

The magnitude of the torque about an axis is the product of the force magnitude and the perpendicular distance from the axis to the line of action of the force.

Often the perpendicular distance $r_\perp$ is called the ***moment arm*** of the force, and the torque $r_\perp F$ is called ***moment of the force*** about the axis. Notice from Fig. 11-2 that $r_\perp$ would be the same if the force were applied at any point along the line of action of the force. You can imagine sliding the force vector along that line of action; the magnitude of the torque would be unchanged.

The dimension of the torque is [force]·[length], the same as the dimension of work and energy. The SI unit of torque is newton-meters (N·m) and that of work is joule = newton-meters (J = N·m). Torque is a very different kind of quantity from work, however. While the SI unit of torque is expressed as newton-meters, the unit joule is reserved for work and for energy.

In calculating torques, it is sometimes convenient to use an alternative form of Eq. 11-3, which can be obtained from the construction in Fig. 11-3. From Fig.

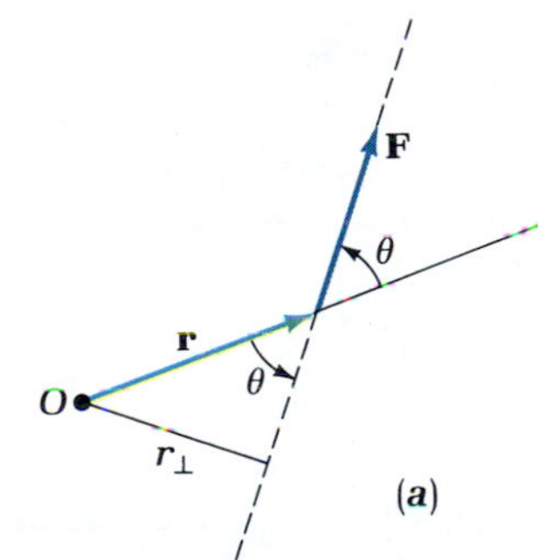

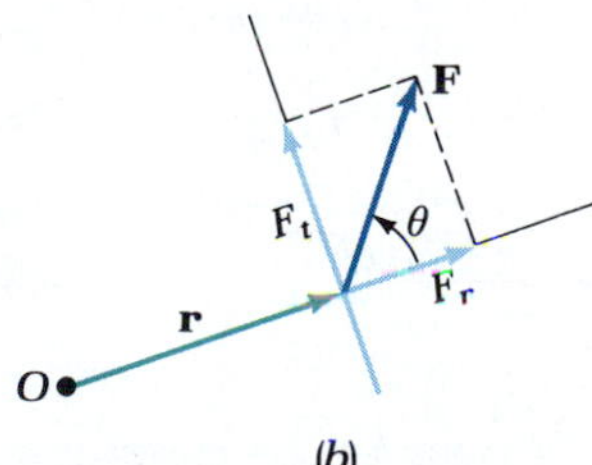

(b)

Figure 11-3. *(a)* $r_\perp$ is the perpendicular distance from the axis to the line of action of the force. *(b)* The force is resolved into components along directions parallel and perpendicular to **r**.

11-3*a* and Eq. 11-3 we have $\tau = r_\perp F = (r \sin \theta)F$. The factor $\sin \theta$ can be associated with the force F rather than with the distance r. In Fig. 11-3*b* the force vector is resolved into a component F_r, parallel to $\mathbf{r}$, and a component F_t, perpendicular to $\mathbf{r}$. If the magnitude of this perpendicular component is denoted by $F_\perp = |F_t| = F \sin \theta$, the magnitude of the torque about axis O is

$$\tau = rF_\perp \tag{11-4}$$

The magnitude of the torque about an axis is the product of the magnitude of the vector $\mathbf{r}$ locating the point of application of the force from the axis and the magnitude of the component of the force perpendicular to $\mathbf{r}$.

Equations 11-3 and 11-4 are equivalent expressions. They are just different ways of expressing the quantity $rF \sin \theta$

Three equivalent expressions for τ

$$rF \sin \theta = r_\perp F = rF_\perp$$

Any one of these three forms may be used to evaluate the magnitude of the torque. Choose the one that seems easiest to apply.

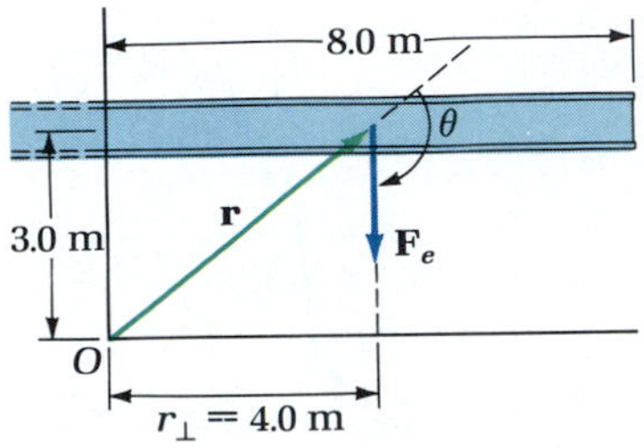

Figure 11-4. Example 11-2: The weight of a section of a cantilevered beam exerts a torque about O.

EXAMPLE 11-2

A cantilevered beam. Part of a steel I beam extends 8.0 m beyond a wall, as shown in Fig. 11-4. Assume that the weight of this uniform section acts at the center of the section, with magnitude $F_e = 2100$ N. Determine the magnitude and direction of the torque about the axis O.

Solution. From the information provided in the figure, the torque can be evaluated most easily from Eq. 11-3:

$$\tau_e = r_\perp F_e = (4.0 \text{ m})(2.1 \text{ kN}) = 8.4 \text{ kN}\cdot\text{m}$$

With somewhat more effort, the expression $rF \sin \theta$ can be applied. The angle θ between $\mathbf{r}$ and $\mathbf{F}$ is given by

$$\theta = 90° + \tan^{-1} \frac{3.0}{4.0} = 127°$$

and

$$r = \sqrt{(3.0 \text{ m})^2 + (4.0 \text{ m})^2} = 5.0 \text{ m}$$

Then

$$\tau_e = rF_e \sin \tau = (5.0 \text{ m})(2100 \text{ N}) \sin 127° = 8.4 \text{ kN}\cdot\text{m}$$

SELF-TEST 11-2. Show that you get the same result for the problem in the above example by using Eq. 11-4.

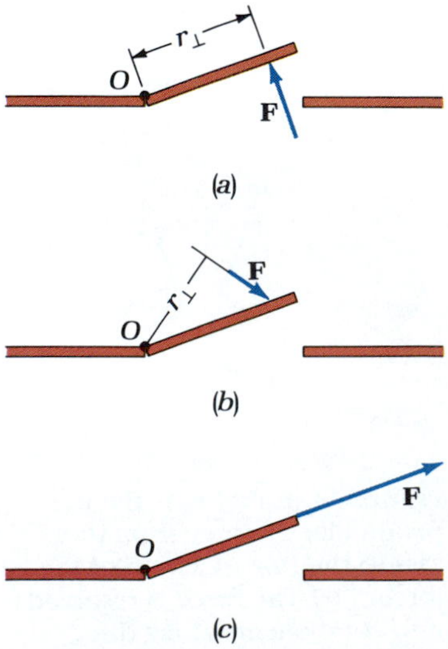

Component of a Torque

A torque about an axis can cause rotation in a clockwise sense or in a counterclockwise sense.* A sign convention is used to distinguish between these two possibilities. Suppose a horizontal force $\mathbf{F}$ is exerted on a door (Fig. 11-5). Let the xy plane be horizontal, and let the z axis be along the hinges with $+\mathbf{k}$ directed out of the page. Imagine that you are viewing the door from above—looking downward along hinges (the z axis). If a force tends to cause the door to rotate in a counterclockwise sense (Fig. 11-5*a*), then the z component τ_z of the torque is positive. If a force tends to cause the door to rotate in a clockwise sense (Fig. 11-5*b*), then the z component τ_z of the torque is negative. This sign convention is consistent with the general definition of torque as a vector quantity (Sec. 11-6). When τ_z is positive, the

Figure 11-5. Forces applied to a door, as viewed from above.

* It is common to speak of rotation "in a clockwise direction" rather than "in a clockwise sense." However, we reserve the term "direction" for vector quantities only, and as it turns out, a rotation cannot be represented as a vector quantity. In keeping with this strict interpretation, we use the term "sense" rather than "direction."

vector τ is directed toward $+\mathbf{k}$, and when τ_z is negative, the vector τ is directed toward $-\mathbf{k}$. Since the sign of τ_z is determined by the sense of the rotation, a positive τ_z is often called a counterclockwise torque, and a negative τ_z is called a clockwise torque. In Fig. 11-5*c*, the line of action of the force passes through the hinges so that $r_\perp = 0$ and $\tau = 0$.

EXAMPLE 11-3

Sign of τ_z. For each force shown in Fig. 11-6, determine the sign of the z component of the torque about the axis O. The z axis comes perpendicularly out of the plane of the page.

Solution. The view in Fig. 11-6 is from a point on the $+z$ axis looking back toward the origin. If a force tends to cause a counterclockwise rotation from this point of view, τ_z is positive. If a force tends to cause a clockwise rotation from this point of view, τ_z is negative. From the figure, $\mathbf{F}_1$ tends to cause a counterclockwise rotation so τ_{z1} is positive; $\mathbf{F}_2$ tends to cause a clockwise rotation so τ_{z2} is negative; $\mathbf{F}_3$ tends to cause a clockwise rotation so τ_{z3} is negative.

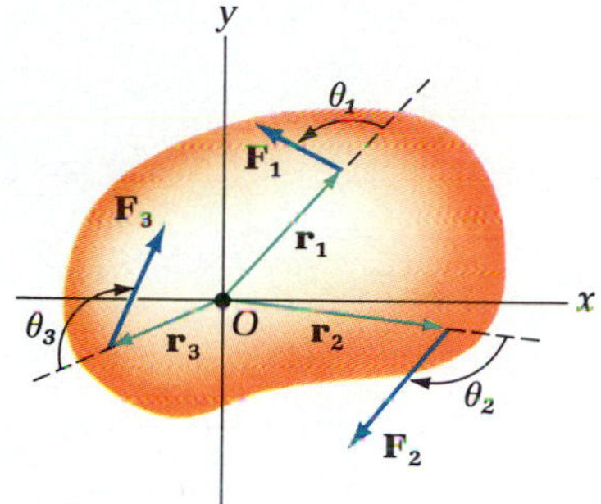

Figure 11-6. Example 11-3: Three forces exert torques about axis O.

SELF-TEST 11-3. Refer back to the revolving door in Fig. 11-1*d*. Take the z axis to be perpendicular to the plane of the figure and out toward you. Determine the sign of the torque components τ_{z1} and τ_{z2} due to forces $\mathbf{F}_1$ and $\mathbf{F}_2$. ***ANSWER:*** $\tau_{z1} > 0$; $\tau_{z2} < 0$.

11-3 CONDITIONS FOR STATIC EQUILIBRIUM

The condition for translational equilibrium of an object is given by Eq. 11-1, $\Sigma\mathbf{F}_{\text{ext}} = 0$. A rigid body in static equilibrium, must be in rotational as well as translational equilibrium. The condition for rotational equilibrium is stated in terms of the torques produced by the external forces acting on the rigid body. Rotational equilibrium requires the balancing of tendencies to rotate clockwise and counterclockwise about any axis because of torques about that axis. This condition can be expressed as a vector equation:

$$\Sigma\tau_{\text{ext}} = 0 \tag{11-5}$$

Condition for rotational equilibrium

The condition for rotational equilibrium of a rigid body is a special case of the more general consideration of rotational dynamics to be developed from Newton's laws in Chap. 13.

Together Eqs. 11-1 and 11-5 constitute the conditions for static equilibrium of a rigid body.

For a rigid body in static equilibrium, the net external force must be zero and the net external torque must be zero:

$$\Sigma\mathbf{F}_{\text{ext}} = 0 \qquad \Sigma\tau_{\text{ext}} = 0 \tag{11-6}$$

Conditions for static equilibrium

These conditions are often referred to as the *first* ($\Sigma\mathbf{F}_{\text{ext}} = 0$) and *second* ($\Sigma\tau_{\text{ext}} = 0$) conditions of equilibrium of a rigid body.

Each of these two vector equations has x, y, and z components, for a total of six equations. However, in many situations, all external forces effectively lie in a given plane, say the xy plane. Such forces are said to be *coplanar,* and we confine our attention to the case of coplanar forces. Then the external forces have only x and y components, and the external torques have only z components. These torque components correspond to clockwise and counterclockwise torques about some axis perpendicular to the xy plane. The conditions for static equilibrium become

Coplanar forces lie in a plane.

$$\Sigma\mathbf{F}_{x,\text{ext}} = 0 \qquad \Sigma\mathbf{F}_{y,\text{ext}} = 0 \qquad \Sigma\tau_{z,\text{ext}} = 0 \tag{11-7}$$

Conditions for static equilibrium with coplanar forces

Equations 11-7 can be solved for at most three unknowns. The equations contain forces, distances, and angles. Up to three of these quantities can be determined by solving the equations simultaneously. In many simple situations, the number of unknown quantities will be three or less. There are also cases in which the number of unknowns is greater than three. In such cases, additional information would have to be provided to determine the unknowns.

11-4 PROBLEM-SOLVING TECHNIQUES

The following steps are useful for solving typical statics problems:

1. Sketch the situation showing the rigid body in static equilibrium. Introduce symbols for the quantities involved. Clearly identify known quantities and unknown quantities. Distinguish between the unknown quantities that you need to know and those that you will not need.
2. Construct the free-body diagram. Accurately represent the shape and orientation of the rigid body, and show the point of application of each external force. Indicate the magnitude and direction of the forces as accurately as you can at the outset. Some of these quantities will be unknown so you will need to estimate them.
3. Select a coordinate frame for the free-body diagram. The choice of the axis about which you evaluate torques is especially crucial because a proper choice allows you to easily evaluate the torques. Often, it is convenient to choose an axis for which the moment arm for a particular force about the axis is zero so that the torque due to this force is zero. This is the way to eliminate an unwanted unknown from the equations. Experience is the most reliable guide for making this choice, and experience comes with practice.
4. Apply the conditions of static equilibrium, Eqs. 11-7. Solve these equations for the wanted unknowns.

The examples below demonstrate these techniques.

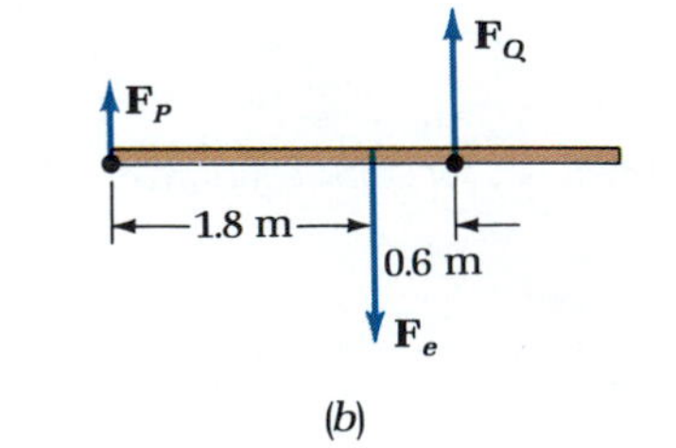

EXAMPLE 11-4

Board and sawhorses. A uniform 48-N board of length 3.6 m rests horizontally on two sawhorses, as shown in Fig. 11-7*a*. What normal forces are exerted on the board by the sawhorses?

Solution. Since the board is uniform, the center of mass is 1.8 m from each end, and the weight is assumed to act at that point. Figure 11-7*b* shows the free-body diagram for the board. The sawhorses exert vertical normal forces of magnitude F_P and F_Q on the board at points P and Q. These two normal forces are the unknowns. Since the board is in static equilibrium, we can apply the conditions of static equilibrium (Eqs. 11-7). Take the x axis as horizontal and the y axis as vertically upward. Then $\Sigma F_{x,\text{ext}} = 0$ is automatically satisfied because all the forces act vertically. Requiring $\Sigma F_{y,\text{ext}} = 0$, we have $F_P + F_Q - F_e = 0$, or

$$F_P + F_Q = F_e$$

The sum of the two normal forces must balance the weight. We must now select an axis about which to calculate torques. Any axis will do, but a convenient choice is the axis through point P because $\mathbf{F}_P$ exerts no torque about this axis. The weight exerts a clockwise torque, and $\mathbf{F}_Q$ exerts a counterclockwise torque about the axis through P. With the z axis perpendicularly out of the page, applying $\Sigma\tau_{z,\text{ext}} = 0$ gives $(2.4\text{ m})F_Q - (1.8\text{ m})F_e = 0$, or

$$(2.4\text{ m})F_Q = (1.8\text{ m})(48\text{ N})$$

$$F_Q = 36\text{ N}$$

The other normal force can be determined immediately since $F_P + F_Q = F_e = 48$ N, or

$$F_P = 12\text{ N}$$

Figure 11-7. Example 11-4: *(a)* A board rests on two sawhorses. *(b)* The free-body diagram of the board.

SELF-TEST 11-4. Work out the above example by choosing an axis through point Q. Repeat for an axis through the center of mass of the board.

EXAMPLE 11-5

A horizontal boom. A uniform boom supports a load weight of magnitude $F_L = 1100$ N, as shown in Fig. 11-8*a*. The boom is attached to the wall by a pin that exerts a force $\mathbf{F}_p$ on the boom. The weight of the boom acts at its midpoint and has magnitude 200 N. Determine the horizontal and vertical components of $\mathbf{F}_p$ and the tension in the guy wire.

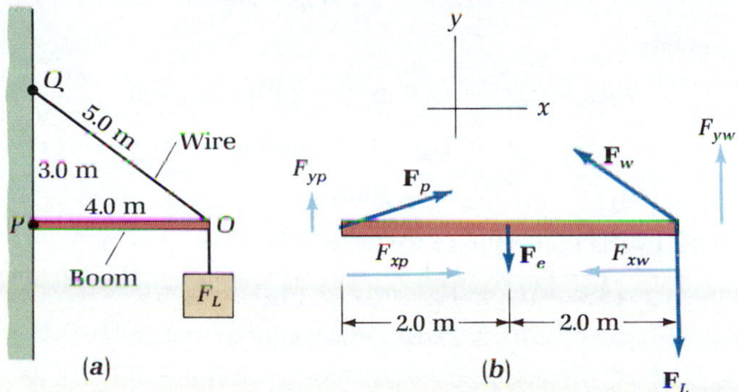

Figure 11-8. Example 11-5: *(a)* A load weight is supported by a boom. *(b)* The free-body diagram of the boom. The components of $\mathbf{F}_p$ and $\mathbf{F}_w$ are shown displaced for clarity.

Solution. We take the boom as a rigid body in static equilibrium and display its free-body diagram in Fig. 11-8*b*. Notice that the load is in equilibrium so that the tension in the vertical wire attached to the load must also be F_L. Neither the magnitude nor the direction of the pin force $\mathbf{F}_p$ is known, but let us assume that it has x and y components as shown. If F_{xp} and F_{yp} come out as positive in our solution, then our assumption about the direction of $\mathbf{F}_p$ is correct. If one or both of the components have negative values, then we can correct the direction of $\mathbf{F}_p$ accordingly. The three unknowns are F_{xp}, F_{yp}, and the guy-wire tension F_w. The tension force can be resolved into x and y components from similar triangles. Thus $F_{yw}/F_w = 3.0\text{ m}/5.0\text{ m}$, or $F_{yw} = \frac{3}{5}F_w$. Similarly, $F_{xw} = -\frac{4}{5}F_w$. Applying the conditions for translational equilibrium, we have

$$F_{xp} - \tfrac{4}{5}F_w = 0 \qquad (A)$$

$$F_{yp} + \tfrac{3}{5}F_w - F_e - F_L = 0 \qquad (B)$$

It is convenient to choose the axis at point O at the right end of the boom. The lines of action of F_w, F_L, and F_{xp} all pass through this axis, and therefore these forces exert no torque about O. The force component F_{yp} causes a clockwise torque, and F_e a counterclockwise torque about this axis. For rotational equilibrium then,

$$(2.0\text{ m})F_e - (4.0\text{ m})F_{yp} = 0$$

so that

$$F_{yp} = \tfrac{1}{2}F_e = \tfrac{1}{2}(200\text{ N}) = 100\text{ N} \qquad (C)$$

Substituting this into Eq. B above gives $\frac{3}{5}F_w = F_e + F_L - 100$ N, or

$$F_w = 2000\text{ N}$$

F_{xp} is obtained from Eq. A above:

$$F_{xp} = \tfrac{4}{5}\,F_w = 1600\text{ N}$$

What is the practical value of this calculation? From the solution we see that the guy-wire tension is 2000 N. We must use a wire substantial enough to support this tension without significant stretching. Similarly, we can see that there are forces exerted on the boom that tend to compress it and to bend it. The boom must be stiff enough to sustain these forces without appreciable deformation.

SELF-TEST 11-5. Use the now-determined force values in the above example to calculate the net torque about an axis passing through point P. Your calculation should give zero, which illustrates our statement that the choice of an axis is arbitrary.

EXAMPLE 11-6

Torques about different axes. In an alternative approach to a statics problem, we can obtain three equations in three unknowns by applying the condition $\Sigma\tau_{z,\text{ext}} = 0$ about three appropriately chosen axes. These equations are not independent of the three conditions in Eqs. 11-7, but are sometimes easier to apply. Reconsider the previous example by requiring $\Sigma\tau_{z,\text{ext}} = 0$ about axes through O, P, and Q.

Solution. About axis O we have, as before,

$$(2.0\text{ m})F_e - (4.0\text{ m})F_{yp} = 0$$

$$F_{yp} = 100\text{ N}$$

About axis P, we obtain

$$(4.0\text{ m})\ \tfrac{3}{5}F_w - (2.0\text{ m})F_e - (4.0\text{ m})F_L = 0$$

$$\tfrac{3}{5}F_w = \tfrac{1}{2}F_e + F_L = 1200\text{ N}$$

$$F_w = 2000\text{ N}$$

And about axis Q, $\mathbf{F}_w$ exerts no torque so we have

$$(3.0\text{ m})F_{xp} - (2.0\text{ m})F_e - (4.0\text{ m})F_L = 0$$

$$F_{xp} = \tfrac{2}{3}F_e + \tfrac{4}{3}F_L = 1600\text{ N}$$

SELF-TEST 11-6. Use the method of the above example, but calculate torques about axes passing through points P, Q, and the center of mass of the boom. Is this calculation more difficult than the one in Example 11-6? ***ANSWER:*** Taking torques about the axis through the center of mass gives $(0.60)(2.0\text{ m})F_w - (2.0\text{ m})F_{yp} - (2.0\text{ m})F_L = 0$. This calculation is more difficult because two unknowns, F_w and F_{yp}, appear in the equation.

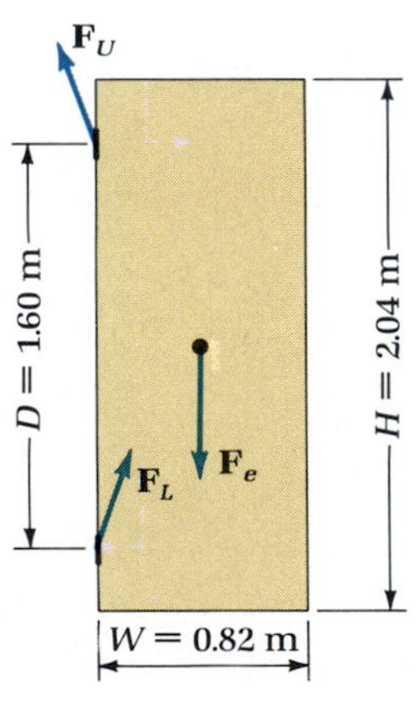

In the examples considered so far, the conditions of static equilibrium, Eqs. 11-7, have been sufficient to determine all of the unknown quantities. There are many situations, however, in which those conditions alone cannot determine all of the unknowns. For example, there may be more than three unknown forces, but only three equations connect them. The additional information that must be provided to solve this type of problem often comes from considering the mechanical properties of materials. We shall not pursue this approach, but the following examples will illustrate this type of problem.

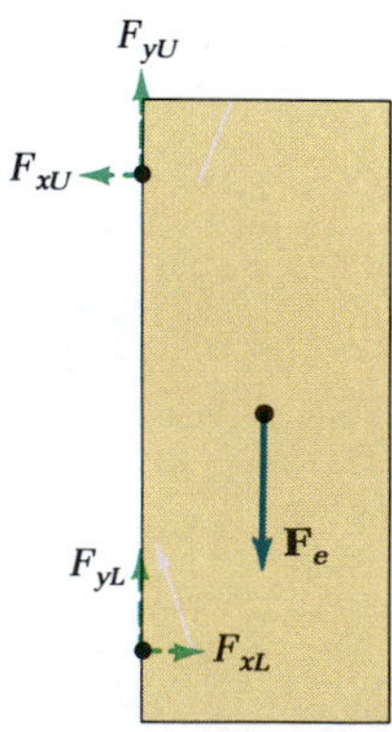

Figure 11-9. Example 11-7: *(a)* A door is supported by two hinges. *(b)* The free-body diagram of the door.

EXAMPLE 11-7

Torques on a door. A uniform door of dimensions 0.82 m by 2.04 m and weight of magnitude 210 N is hung by two hinges placed symmetrically 1.60 m apart, as shown in Fig. 11-9*a*. *(a)* Determine the horizontal component of the force exerted by each hinge on the door. *(b)* Determine the vertical component of the force exerted by each hinge on the door. Comment on the difficulty encountered.

Solution. *(a)* The free-body diagram is shown in Fig. 11-9*b*, with the forces exerted by the upper and lower hinges resolved into horizontal and vertical components. The weight of the uniform door acts at the midpoint of the door. Requiring the x and y components of the external forces to add to zero gives

$$F_{yU} + F_{yL} - F_e = 0$$

$$F_{xU} + F_{xL} = 0$$

So the sum of the vertical components of the hinge forces balances the weight, and the horizontal co nponents of the hinge forces balance each other. Choose an axis through the upper hinge position and perpendicular to the plane of the figure. The weight of the door produces a clockwise torque about this axis. The only other nonzero torque is caused by the component F_{xL}, and it exerts a counterclockwise torque about the axis. Then

$$DF_{xL} - \tfrac{1}{2}WF_e = 0$$

$$(1.60 \text{ m})F_{xL} = (0.41 \text{ m})(210 \text{ N})$$

$$F_{xL} = 54 \text{ N}$$

The upper hinge force component balances this, so

$$F_{xU} = -54 \text{ N}$$

(b) We have determined two unknowns, F_{xU} and F_{xL}, and have only a single equation connecting the other two unknown force components:

$$F_{yU} + F_{yL} = 210 \text{ N}$$

From the information provided we cannot determine how this 210-N load is shared by the two hinges. Notice that F_{yU} and F_{yL} have the same line of action and therefore exert the same net torque about an axis, no matter how the 210-N total is shared. Additional information is required to determine F_{yU} and F_{yL}, and that information depends in detail on how the hinges are mounted and aligned.

SELF-TEST 11-7. Rework the above example by choosing an axis through the lower hinge and perpendicular to the plane of the figure. ***ANSWER:*** Taking torques about the axis through the lower hinge gives $(1.60 \text{ m})F_{xU} + (0.41 \text{ m})F_e = 0$.

EXAMPLE 11-8

A leaning ladder. A uniform ladder leans against a wall, as shown in Fig. 11-10*a*. What minimum value of the coefficient of static friction at the floor-ladder interface will prevent the ladder from slipping?

Solution. There are four unknown forces in this problem. Normal forces are exerted on the ladder by the floor and by the wall. Static frictional forces are also exerted on the ladder by the two surfaces. Assume that the wall surface is very smooth and that the frictional force there is negligible. This assumption reduces the number of unknown forces to three. The free-body diagram is shown in Fig. 11-10*b*. Choose the axis to pass through the point of contact of the ladder and floor. Applying the conditions of static equilibrium gives for the magnitudes of these forces:

$$F_{fF} = F_{NW} \qquad F_{NF} = F_e \qquad HF_{NW} - \tfrac{1}{2}LF_e = 0$$

These equations are easily solved for the unknown forces:

$$F_{NW} = \frac{L}{2H}F_e \qquad F_{fF} = \frac{L}{2H}F_e \qquad F_{NF} = F_e$$

Since F_{fF} is a static frictional force, its value must not exceed the maximum value of the static frictional force, $F_{fF} \leq \mu_s F_{NF}$ from Eq. 6-3. Imposing this condition gives $(L/2H)F_e \leq \mu_s F_e$ or, on simplifying,

$$\mu_s \geq \frac{L}{2H}$$

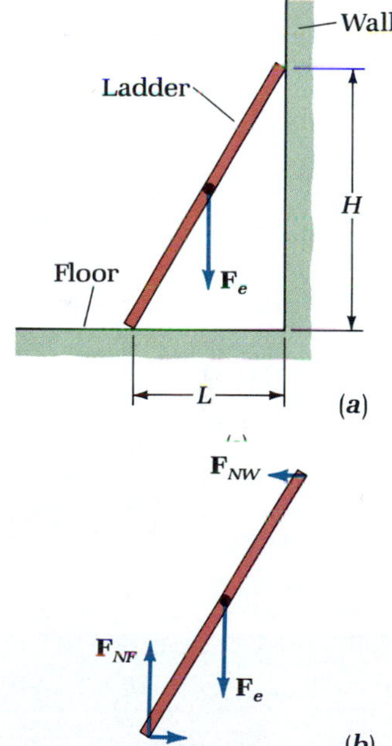

Figure 11-10. Example 11-8: *(a)* A uniform ladder is in static equilibrium. *(b)* The free-body diagram of the ladder. The frictional force at the wall is negligible.

SELF-TEST 11-8. Describe qualitatively the changes in the magnitude of each force on the ladder in the above example if a small, upward frictional force is exerted by the wall on the ladder. ***ANSWER:*** The weight F_e would be unchanged, and all the other forces would be smaller.

11-5 CENTER OF GRAVITY

We have taken the effective point of application of the weight of an object to be at the center of mass of that object. What is the justification for this assumption? In what sense is there a single point at which the full weight of the object acts? What is the full weight of the object?

The weight of an object is the sum of the weight of its parts.

Since an object can be regarded as a large number of small parts, the weight of the object $\mathbf{F}_e$ is just the sum of the weights of these parts. For an object of ordinary size close to the earth's surface, the weight of each part is directed downward, and we can easily add them. Letting m_i represent the mass of one of those elements, we have

$$F_e = \Sigma m_i g = (\Sigma m_i)g = Mg \tag{11-8}$$

where $M = \Sigma m_i$ is the total mass of the object and g is the acceleration due to gravity. Adding together the weights of the individual pieces of the object gives, according to Eq. 11-8, the full weight of the object, independent of the points of application of these forces.

The point of application of each force is important when calculating torques. The weight of each part of the body will exert a torque about some axis, as shown in Fig. 11-11. For the force of magnitude $F_i = m_i g$, the torque component τ_{zi} is

$$\tau_{zi} = -(x_i)(m_i g)$$

The overall torque is the sum of the torques due to the weight of each part.

where x_i is the x coordinate of that part of the object and the negative sign accounts for the clockwise sense of the torque. If the contributions from each piece of the body are added, the torque component due to the full gravitational force acting on the object is

$$\tau_z = \Sigma[-(x_i)(m_i g)] \tag{11-9}$$

This expression shows that the torque due to the weight of an object depends on how the parts of the object are distributed spatially. There is an effective point on the object, the ***center of gravity,*** at which the full weight of the object can be considered to act so as to cause the same torque given by Eq. 11-9. If x_{cg} is the x coordinate of that point, the torque component due to the full weight, of magnitude $F_e = \Sigma m_i g$, is

$$\tau_z = -x_{cg} F_e = -x_{cg} \Sigma m_i g \tag{11-10}$$

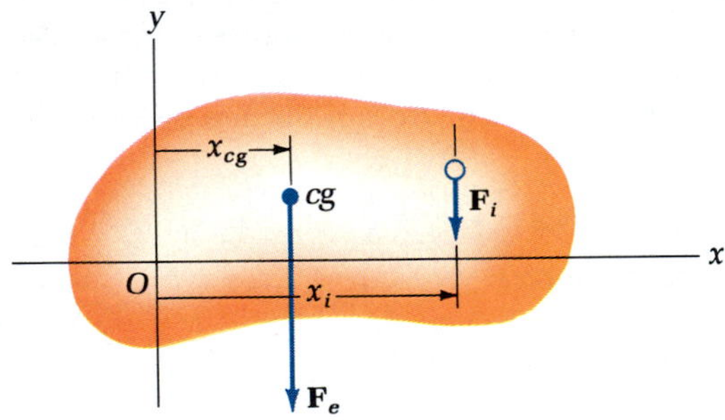

Figure 11-11. The full weight of an object acts at the center of gravity. The weight of one part of the object has magnitude $F_i = m_i g$.

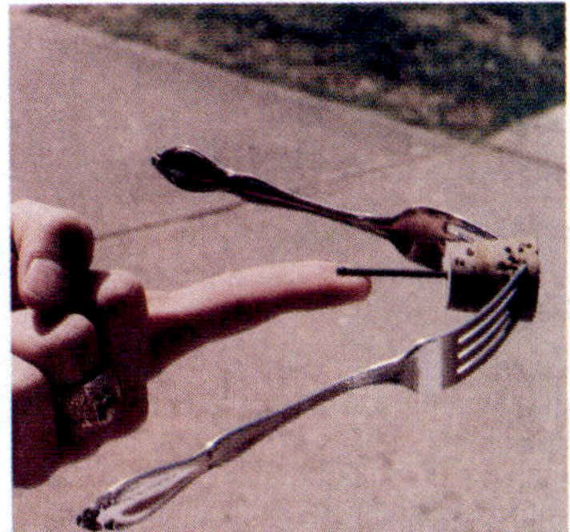

An inexpensive and common demonstration of the center of gravity is with two forks and a cork.

The x coordinate of the center of gravity is determined by equating the right-hand sides of Eqs. 11-9 and 11-10:

$$x_{cg} = \frac{\Sigma(x_i)(m_i g)}{\Sigma m_i g}$$

The center of gravity is given by

$$x_{cg} = \frac{\Sigma x_i m_i g}{\Sigma m_i g} \tag{11-11}$$

In most cases of interest, the value of g is the same for every part of the object and can be divided out of the expressions in Eq. 11-11. Dividing out the common factor gives

$$x_{cg} = \frac{\Sigma x_i m_i}{\Sigma m_i}$$

The right-hand side is the x coordinate of the center of mass of the object. *If the*

acceleration due to gravity is the same for each part of the object, then the center of gravity and the center of mass coincide.

The center of gravity of a rigid body can be determined experimentally by suspending the object from two or more points, as shown in Fig. 11-12. If the object is suspended by a string at point P, the conditions of static equilibrium require that $F_e = F_s$ and that the two forces have the same line of action. The center of gravity must lie directly below point P. Similarly, the center of gravity must lie directly below point Q, another point of suspension. The center of gravity is at the intersection of these two lines.

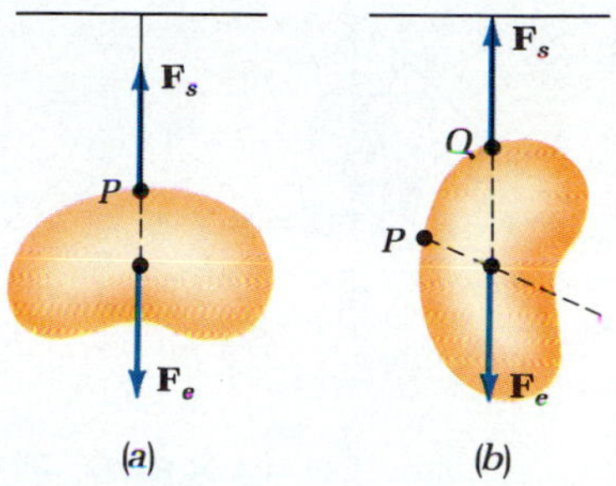

Figure 11-12. The center of gravity must lie directly below the point of suspension.

11-6 TORQUE AND THE CROSS PRODUCT OF VECTORS

The direction of the torque produced by a force has been taken to be along an axis perpendicular to the plane containing the force **F** and the vector **r** that locates the point of application of the force. If that plane is the xy plane, then the torque vector is along the z axis, and the torque component τ_z is referred to as the torque about the z axis. It is desirable to consider the torque as a vector, without reference to a particular coordinate system. The definition of torque can be generalized by introducing a *vector product,* or *cross product of vectors.*

The Cross Product

The dot product or scalar product of two vectors was defined in Chap. 8. The cross product is an entirely different kind of product of two vectors. As the names imply, the vector product, or cross product of two vectors, yields another vector, and the product is denoted by a cross (×) between the factors. Consider two vectors **A** and **B**, as shown in Fig. 11-13*a*, where the angle θ is the smaller of the angles between the two vectors.

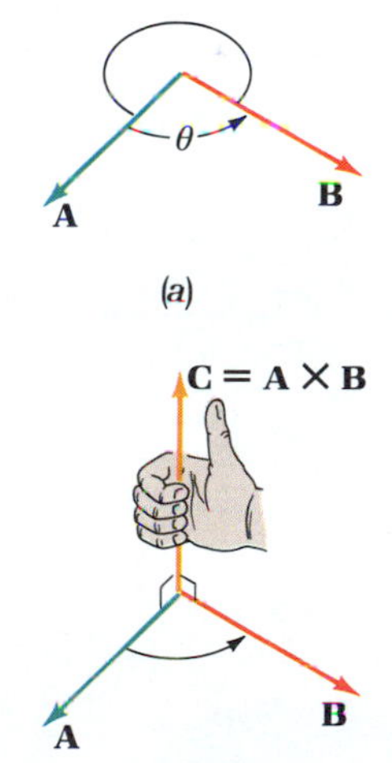

Figure 11-13. *(a)* Two vectors lie in a plane, with θ the smaller of the angles between them, *(b)* Vector **C** = **A** × **B** is perpendicular to both **A** and **B**, with a direction given by the right-hand rule.

The ***cross product*** between vectors **A** and **B** is defined as a vector **C** = **A** × **B**. The direction of **C** is given by the right-hand rule and the magnitude C is given by

$$C = AB \sin \theta \qquad (11\text{-}12)$$

The phrase "right-hand rule" stands for the following convention for obtaining the direction of the vector **C** = **A** × **B**:

> Vector **C** is perpendicular to the plane containing vectors **A** and **B**. Curl the fingers of the right hand in the sense that would rotate the first factor **A** into the second factor **B**. The extended thumb gives the direction of **C**.

The rotation that would carry **A** into **B** is through the smaller of the two angles between them. The right-hand rule is illustrated in Fig. 11-13*b*, where **C** is perpendicular to both **A** and **B**. The definition of the cross product **C** = **A** × **B** consists of two parts: Both the magnitude and the direction of **C** must be specified. Equation 11-12 gives the magnitude, and the right-hand rule gives the direction.

For example, suppose **A** is a displacement of magnitude 2.5 m due north and **B** is a displacement of magnitude 3.0 m directed northeast at the earth's surface. Then **C** = **A** × **B** has magnitude $C = (2.5\text{ m})(3.0\text{ m}) \sin 45° = 5.3\text{ m}^2$, and the direction of **C** from the right-hand rule is downward, perpendicular to the earth's surface.

The cross product is noncommutative.

The cross product of two vectors is *noncommutative*: **A** × **B** ≠ **B** × **A**. Let us

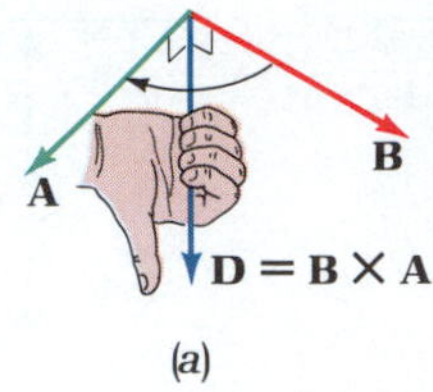

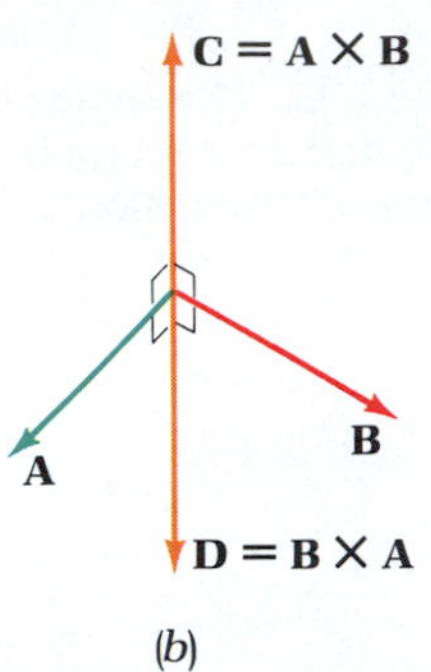

Figure 11-14. *(a)* The direction of $\mathbf{D} = \mathbf{B} \times \mathbf{A}$ is given by the right-hand rule. *(b)* The cross product is noncommutative: $\mathbf{B} \times \mathbf{A} = -\mathbf{A} \times \mathbf{B}$.

change the order of the factors in the cross product; consider the vector $\mathbf{D} = \mathbf{B} \times \mathbf{A}$, with **A** and **B** the same vectors as before. The magnitude is again given by Eq. 11-12, $D = BA \sin\theta$. Applying the right-hand rule, however, requires us to imagine rotating **B** into **A**, as indicated in Fig. 11-14*a*. The direction of $\mathbf{D} = \mathbf{B} \times \mathbf{A}$ is opposite the direction of $\mathbf{C} = \mathbf{A} \times \mathbf{B}$, as shown in Fig. 11-14*b*. The magnitudes of **C** and **D** are the same, $C = D = AB \sin\theta$. Hence

$$\mathbf{B} \times \mathbf{A} = -\mathbf{A} \times \mathbf{B} \tag{11-13}$$

Reversing the order of the factors reverses the direction of the cross product. Although noncommutative, the cross product does obey the distributive rule:

$$\mathbf{A} \times (\mathbf{B} + \mathbf{C}) = \mathbf{A} \times \mathbf{B} + \mathbf{A} \times \mathbf{C}$$

A geometric interpretation can be applied to the magnitude of the cross product of any two vectors. In Eq. 11-12, the factor $\sin\theta$ can be associated with the magnitude of either **A** or **B**. For example, $A \sin\theta$ is the magnitude of the component, $A_\perp = A \sin\theta$, which is perpendicular to **B** (Fig. 11-15*a*). Thus the magnitude $C = |\mathbf{A} \times \mathbf{B}| = A_\perp B$. Similarly, the construction in Fig. 11-15*b* shows that $C = AB_\perp$. *The magnitude of the cross product of vectors* **A** *and* **B** *is the product of the magnitude of one vector and the magnitude of the perpendicular component of the other vector.* The cross product is a measure of how perpendicular two vectors are to each other.

There are some special cases that deserve attention. Suppose two vectors **A** and **B** are perpendicular. Then since $\sin 90° = 1$, $|\mathbf{A} \times \mathbf{B}| = AB$. At the other extreme are two vectors that are either parallel ($\theta = 0$) or opposite ($\theta = 180°$). In either case the factor $\sin\theta = 0$, and the magnitude of the cross product $|\mathbf{A} \times \mathbf{B}| = 0$; thus $\mathbf{A} \times \mathbf{B} = 0$, the null vector. *The cross product of two vectors that are either parallel or opposite is zero.*

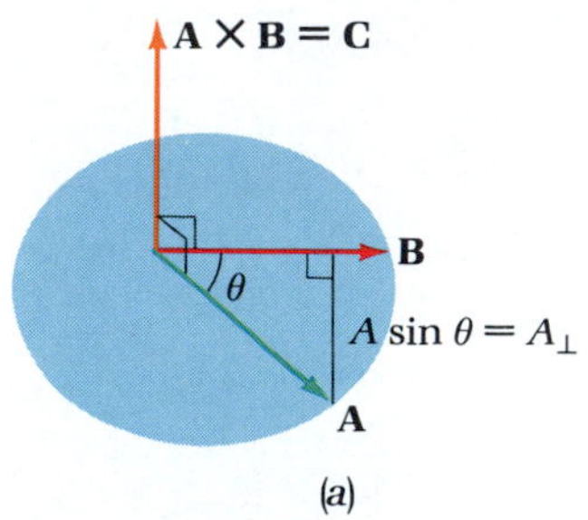

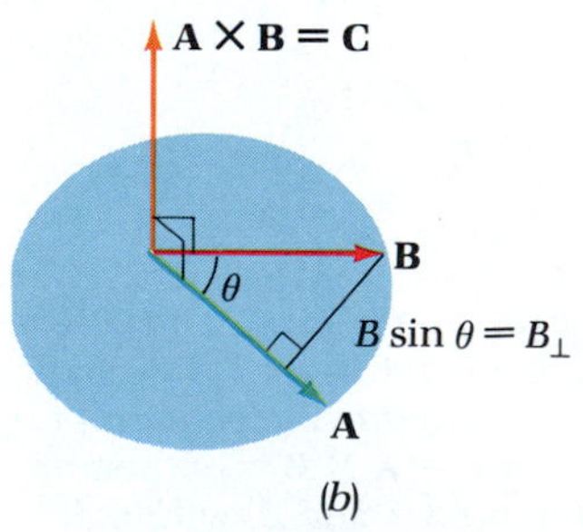

Figure 11-15. The magnitude of the cross product can be expressed as either *(a)* $|\mathbf{A} \times \mathbf{B}| = A_\perp B$ or *(b)* $|\mathbf{A} \times \mathbf{B}| = AB_\perp$.

EXAMPLE 11-9

Area as a cross product. A surveyor marks off two displacements from a pin: **A** of magnitude $A = 204.56$ m due east and **B** of magnitude $B = 188.32$ m at 74.82° north of east, as seen in Fig. 11-16. *(a)* Evaluate the cross product $\mathbf{A} \times \mathbf{B}$ and *(b)* give a geometric interpretation of the result.

Solution. *(a)* Rather than using Eq. 11-12 directly, we find the component of **A** perpendicular to **B**:

$$A_\perp = (204.56 \text{ m}) \sin 74.82° = 197.4 \text{ m}$$

Then the magnitude of $|\mathbf{A} \times \mathbf{B}|$ is

$$A_\perp B = (197.4 \text{ m})(188.32 \text{ m}) = 37{,}180 \text{ m}^2$$

The direction of $\mathbf{A} \times \mathbf{B}$ is obtained from the right-hand rule applied to Fig. 11-16*a*. Rotating the direction of **A** (east) into the direction of **B** (north of east) gives a direction that is perpendicular to the plane of **A** and **B** and straight up, out of the plane of the figure. *(b)* From Fig. 11-16*b*, $A_\perp B$ ($= 37{,}180$ m²) is equal to the area of the parallelogram formed by **A** and **B**, and the geometrical interpretation is that $\mathbf{A} \times \mathbf{B}$ represents an area. With this inter-

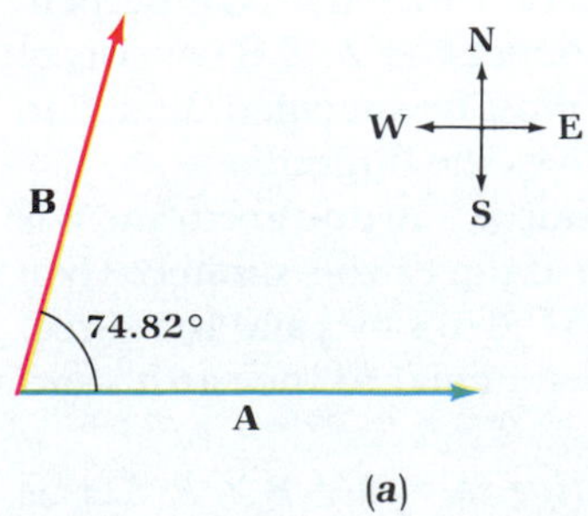

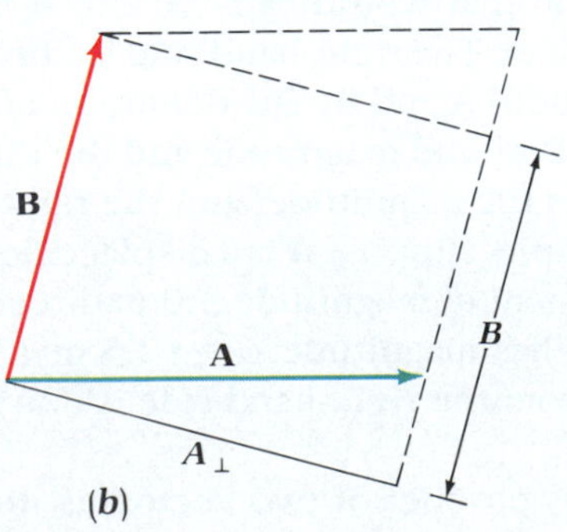

Figure 11-16. Example 11-9: *(a)* Two displacements are laid off from a pin by a surveyor. *(b)* Displacements **A** and **B** form a parallelogram. The area of the parallelogram equals the area $A_\perp B$ of the rectangle with sides $A_\perp$ and B.

pretation, an area is a vector that has a direction that is perpendicular to the plane of the figure.

SELF-TEST 11-9. Evaluate **B** × **A** where **A** is a displacement with $A = 204.56$ m due east and **B** is a displacement with $B = 188.32$ m due west. ***ANSWER:*** zero.

Cross Product in Component Form

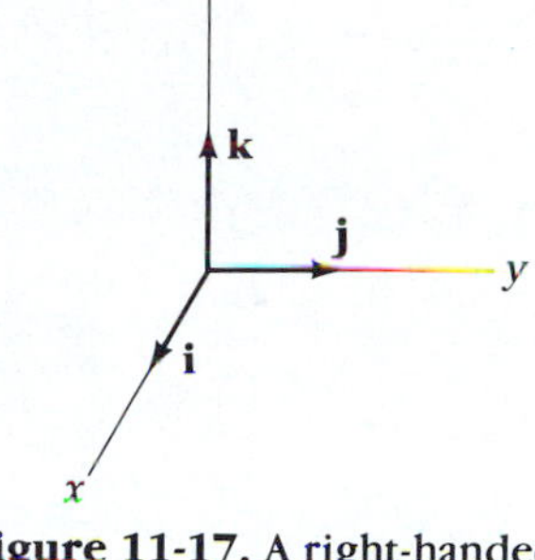

Figure 11-17. A right-handed coordinate system with **i** × **j** = **k**.

The cross product of two vectors can be evaluated in terms of the components of the two vectors. To do so, the cross products of the unit vectors must be considered. Let us evaluate **i** × **i**. The result is zero, or the null vector, because these two vectors are parallel ($\sin\theta = 0$). For the same reason, $\mathbf{j} \times \mathbf{j} = \mathbf{k} \times \mathbf{k} = 0$. To evaluate **i** × **j**, refer to Fig. 11-17, which shows the unit vectors on an *xyz*-coordinate system. The magnitude of each unit vector is 1, or unity, and so the magnitude of **i** × **j** is also 1: $|\mathbf{i} \times \mathbf{j}| = (1)(1)\sin 90° = 1$. The direction of **i** × **j** is given by the right-hand rule and, from Fig. 11-17, is along the positive *z* axis. Thus **i** × **j** is a vector of magnitude 1 directed along the *z* axis; but this is just the unit vector **k**. We have $\mathbf{i} \times \mathbf{j} = \mathbf{k}$. Simply reversing the order of the factors gives $\mathbf{j} \times \mathbf{i} = -\mathbf{k}$.

A right-handed coordinate system

The coordinate system shown in Fig. 11-17 is called a ***right-handed coordinate system***. The unit vectors **i**, **j**, **k** associated with the *x*, *y*, *z* axes, in that order, are connected by the right-hand rule: $\mathbf{i} \times \mathbf{j} = \mathbf{k}$. Only right-handed coordinate systems are used in this text.

Cross products of unit vectors

The cross products of the remaining pairs of unit vectors can be evaluated as above. There are a total of nine such products, and you should convince yourself of the following results:

$$\mathbf{i} \times \mathbf{i} = \mathbf{j} \times \mathbf{j} = \mathbf{k} \times \mathbf{k} = 0$$
$$\mathbf{i} \times \mathbf{j} = \mathbf{k} = -\mathbf{j} \times \mathbf{i} \qquad \mathbf{j} \times \mathbf{k} = \mathbf{i} = -\mathbf{k} \times \mathbf{j} \qquad \mathbf{k} \times \mathbf{i} = \mathbf{j} = -\mathbf{i} \times \mathbf{k} \tag{11-14}$$

Consider now two vectors **A** and **B** given in terms of their components:

$$\mathbf{A} = A_x\mathbf{i} + A_y\mathbf{j} + A_z\mathbf{k} \qquad \mathbf{B} = B_x\mathbf{i} + B_y\mathbf{j} + B_z\mathbf{k}$$

We form the cross product $\mathbf{C} = \mathbf{A} \times \mathbf{B}$, expressing **A** and **B** in terms of their components:

$$\mathbf{A} \times \mathbf{B} = (A_x\mathbf{i} + A_y\mathbf{j} + A_z\mathbf{k}) \times (B_x\mathbf{i} + B_y\mathbf{j} + B_z\mathbf{k})$$

The distributive property allows us to multiply out the parentheses, giving nine terms involving cross products of pairs of unit vectors. For example, the terms $A_xB_x\mathbf{i} \times \mathbf{i}$ and $A_zB_y\mathbf{k} \times \mathbf{j}$ appear. We evaluate the cross products using Eqs. 11-14 and obtain

$$\mathbf{A} \times \mathbf{B} = (A_yB_z - A_zB_y)\mathbf{i} + (A_zB_x - A_xB_z)\mathbf{j} + (A_xB_y - A_yB_x)\mathbf{k} \tag{11-15}$$

The cross product in component form

Another way to express the result is

$$\mathbf{A} \times \mathbf{B} = \mathbf{C} = C_x\mathbf{i} + C_y\mathbf{j} + C_z\mathbf{k}$$

and, by comparison with Eq. 11-15, the components of **C** are given by

$$C_x = A_yB_z - A_zB_y \qquad C_y = A_zB_x - A_xB_z \qquad C_z = A_xB_y - A_yB_x$$

EXAMPLE 11-10

Components of a cross product. Determine the vector representing the area measured by the surveyor in Example 11-9 by evaluating the cross product in component form. Use a coordinate system with **i** directed east and **j** directed north; then **k** is directed vertically upward.

Solution. From Example 11-9 we determine the components of the displacements: $A_x = 204.56$ m, $A_y = 0$, $A_z = 0$, and $B_x = (188.32\text{ m})\cos 74.82° = 49.31$ m, $B_y = (188.32\text{ m})\sin 74.82° = 181.7$ m, $B_z = 0$. Let **C** represent the vector area, $\mathbf{C} = \mathbf{A} \times \mathbf{B}$. From Eq. 11-15, the components of **C** are

$$C_x = A_yB_z - A_zB_y = (0)(0) - (0)(181.7\text{ m}) = 0$$

$$C_y = A_zB_x - A_xB_z = (0)(49.31\text{ m}) - (204.56\text{ m})(0) = 0$$

$$C_z = A_xB_y - A_yB_x$$
$$= (204.56\text{ m})(181.7\text{ m}) - (0)(49.31\text{ m}) = 37{,}180\text{ m}^2$$

The area measured by the surveyor is represented by

$$\mathbf{C} = (37{,}180\text{ m}^2)\mathbf{k}$$

SELF-TEST 11-10. Consider a set of unit vectors directed as follows: $\hat{\mathbf{E}}$ horizontally toward the east, $\hat{\mathbf{N}}$ horizontally toward the north, $\hat{\mathbf{U}}$ vertically upward, $-\hat{\mathbf{E}}$ horizontally toward the west, $-\hat{\mathbf{N}}$ horizontally toward the south, and $-\hat{\mathbf{U}}$ vertically downward. Determine the cross product *(a)* $\hat{\mathbf{N}} \times (-\hat{\mathbf{E}})$; *(b)* $(-\hat{\mathbf{U}}) \times \hat{\mathbf{U}}$; *(c)* $\hat{\mathbf{N}} \times \hat{\mathbf{U}}$; and *(d)* $\hat{\mathbf{E}} \times \hat{\mathbf{U}}$. ***ANSWERS:*** *(a)* $\hat{\mathbf{U}}$; *(b)* 0; *(c)* $\hat{\mathbf{E}}$; *(d)* $-\hat{\mathbf{N}}$.

Torque as a Cross Product

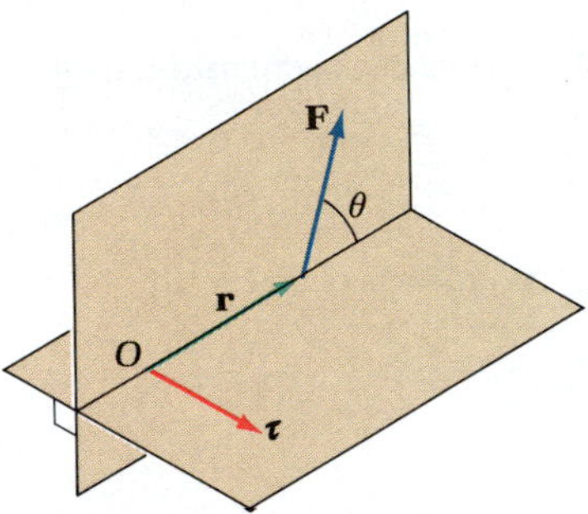

Figure 11-18. A position vector **r** locates the point of application of a force **F** from a reference point *O*. The torque about *O* is defined as $\boldsymbol{\tau} = \mathbf{r} \times \mathbf{F}$.

The cross product can be used to give a general definition of torque as a vector quantity. Suppose a force **F** is applied to an object at a point located by a position vector **r** relative to some origin or reference point *O*, as shown in Fig. 11-18.

The torque $\boldsymbol{\tau}$ exerted by force **F** about reference point *O* is defined as the cross product of **r** and **F**:

$$\boldsymbol{\tau} = \mathbf{r} \times \mathbf{F} \qquad (11\text{-}16)$$

The magnitude of the torque is $\tau = rF\sin\theta$, and the direction is given by the right-hand rule to be perpendicular to the plane containing **r** and **F**. If we choose this plane to be the *xy* plane, then the torque has only a *z* component and is the same as the torque about this *z* axis, as defined in Sec. 11-2. For some other choice of a coordinate system, the torque $\boldsymbol{\tau}$ would have components τ_x, τ_y, τ_z. The component τ_x is the torque about the *x* axis, τ_y is the torque about the *y* axis, and τ_z is the torque about the *z* axis. Equation 11-16 can be used to find the torque in terms of the components of **r** and **F**. Since $\mathbf{r} = x\mathbf{i} + y\mathbf{j} + z\mathbf{k}$ and $\mathbf{F} = F_x\mathbf{i} + F_y\mathbf{j} + F_z\mathbf{k}$,

$$\boldsymbol{\tau} = (yF_z - zF_y)\mathbf{i} + (zF_x - xF_z)\mathbf{j} + (xF_y - yF_x)\mathbf{k}$$

EXAMPLE 11-11

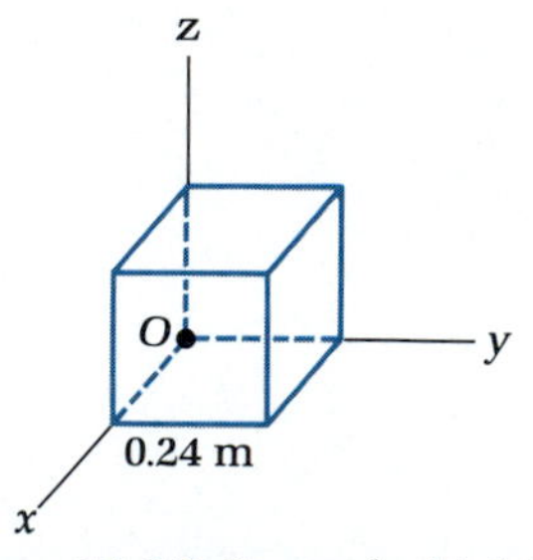

Figure 11-19. Example 11-11: A cubic block of ice has edge 0.24 m. The *z* axis is vertically upward and the weight is directed downward.

Torque as a cross product. A 12-kg block of ice in the shape of a cube of edge 0.24 m lies in the first octant of a coordinate system, as shown in Fig. 11-19. Evaluate the torque due to the weight of the block about the corner at the origin.

Solution. The center of gravity of the cubic block is halfway along the body diagonal and is located by

$$\mathbf{r}_{cg} = (0.12\text{ m})\mathbf{i} + (0.12\text{ m})\mathbf{j} + (0.12\text{ m})\mathbf{k}$$

The weight has only a *z* component (vertical) and is

$$\mathbf{F}_e = -mg\mathbf{k} = (-120\text{ N})\mathbf{k}$$

Using Eq. 11-16, we form the cross product $\boldsymbol{\tau} = \mathbf{r}_{cg} \times \mathbf{F}_e$. Expressing the cross product in component form, we have

$$\tau_x = y_{cg}F_{ez} - z_{cg}F_{ey} = -14\ \text{N}\cdot\text{m}$$
$$\tau_y = z_{cg}F_{ex} - x_{cg}F_{ez} = 14\ \text{N}\cdot\text{m}$$
$$\tau_z = x_{cg}F_{ey} - y_{cg}F_{ex} = 0$$

or

$$\boldsymbol{\tau} = (-14\ \text{N}\cdot\text{m})\mathbf{i} + (14\ \text{N}\cdot\text{m})\mathbf{j}$$

SELF-TEST 11-11. Another force acts on the block of ice in the above example. It acts on the upper face of the block at the point (0.12 m, 0.12 m, 0.24 m); the force **F** is (89 N)**j**. Determine the torque due to this force about point *O*. ***ANSWER:*** $\boldsymbol{\tau} = -(11\ \text{N}\cdot\text{m})\mathbf{i} + (11\ \text{N}\cdot\text{m})\mathbf{k}$

COMMENTARY: Cables and Bridges

Figure 11-20. The Brooklyn Bridge.

In a suspension bridge such as the Brooklyn Bridge or the Golden Gate Bridge, the graceful curve of a main cable is an essential feature that attracts and holds the eye. These bridges are acknowledged as works of art as well as engineering marvels. Cable-supported bridges can span long distances. For example, the Golden Gate Bridge has two main pylons, or towers, with a central span of 1280 m between them. When such lengths must be spanned, a suspension bridge is often the only feasible type.

The deck, or roadway, is typically connected to the curved main cable by smaller, usually vertical, cables. Some bridges have staying cables, or stays, that are essentially straight and are used instead of curved main cables to support the deck. In other cases, stays are used in addition to curved main cables. A portion of the cable structure of the Brooklyn Bridge is shown in Fig. 11-20. The stays radiate fanlike from the pylon and appear to cross the vertical cables that descend from the curved main cable. According to the designer, John Roebling, "The supporting power of the stays alone will be 15,000 tons, ample to hold up the floor, If the [vertical] cables were removed, the bridge would sink in the center but would not fall."

In treating a cable (or a rope or a string) in this text, we have made the simplifying assumption that the weight of the cable is negligible compared with the tension in it. Hence an ideal cable is assumed to be straight, with the same tension everywhere. In a real cable, the tension varies along its length, and the tension and the curvature are determined by the distribution of the load weight. The weight of the cable itself may be a significant part of the total load. The shape is determined mathematically by requiring each element of the cable to be in static equilibrium. For a uniform, flexible cable supporting only its own weight, the curve is called a *common catenary*. This problem was first solved in 1691, largely through the efforts of James Bernoulli. Adding the weight of the deck and including the effects of the rigidity of the spanning truss greatly increase the complexity of the problem. Sophisticated computer programs are now used to perform numerical analyses for building bridges and other structures.

Since a cable must safely support a large load, special requirements are imposed on cable materials and sizes. A typical cable is composed of steel wires that have a composition different from the steel used for structural beams. As a result, cable steel can support, before breaking, about twice the load, for the same cross-sectional area, as high-strength structural steel. A cable with a large cross section is composed of many steel wires. Typically, a main cable is formed from wires of 5 mm diameter. These wires may be strung one or several at a time along the entire span as the cable is assembled in place during the construction of the bridge. Many passes are required for a large main cable. For example, each main cable of the Golden Gate Bridge contains more than 27,000 wires. The total length of wire used was about 80,000 mi, enough to circle the earth more than three times.

For further reading, see *Cable-Supported Bridges* by Niels J. Gimsing (John Wiley & Sons, New York, 1983).

SUMMARY

Section 11-1. Static Equilibrium of a Rigid Body

An extended object is in static equilibrium if every point of that object remains at rest. A rigid body is an object for which the distance between any pair of points on the object remains fixed.

Section 11-2. Torque about an Axis

Let **r** locate the point of application of a force **F** on a body. If the plane containing **r** and **F** is taken as the xy plane, the torque about the z axis is τ_z. The magnitude of the torque is given by $\tau = r_\perp F$. The torque component τ_z is positive if it tends to produce a counterclockwise rotation of the object when viewed from the positive z axis, and τ_z is negative if the tendency of rotation is clockwise.

Section 11-3. Conditions for Static Equilibrium

For a rigid body to be in static equilibrium, both the condition for translational equilibrium, $\Sigma \mathbf{F}_{\text{ext}} = 0$, and the condition for rotational equilibrium about any point, $\Sigma \tau_{\text{ext}} = 0$, must be satisfied. For coplanar forces, these conditions reduce to

$$\Sigma \mathbf{F}_{x,\text{ext}} = 0 \qquad \Sigma \mathbf{F}_{y,\text{ext}} = 0 \qquad \Sigma \tau_{z,\text{ext}} = 0 \tag{11-7}$$

where the coplanar forces lie in the xy plane. These equations can be solved for up to three unknowns.

Section 11-4. Problem-Solving Techniques

There are four steps to follow: Draw a sketch; draw a free-body diagram; choose a coordinate frame; solve the equations of static equilibrium. The third step is crucial because a problem can be simplified by the proper choice of the axis about which the torques are calculated.

Section 11-5. Center of Gravity

The center of gravity of an extended object is that point at which the full gravitational force on the object can be considered to act. The center of gravity and the center of mass coincide for objects of ordinary size close to the earth's surface.

Section 11-6. Torque and the Cross Product of Vectors

The cross product of two vectors **A** and **B** is also a vector, $\mathbf{C} = \mathbf{A} \times \mathbf{B}$. The magnitude of **C** is $C = AB \sin\theta$, with θ the smaller angle between **A** and **B**. The direction of **C** is given by the right-hand rule and is perpendicular to both **A** and **B**. The cross product in component form is

$$\mathbf{A} \times \mathbf{B} = (A_yB_z - A_zB_y)\mathbf{i} + (A_zB_x - A_xB_z)\mathbf{j} + (A_xB_y - A_yB_x)\mathbf{k} \tag{11-15}$$

Torque about a point O is defined as the cross product.

$$\boldsymbol{\tau} = \mathbf{r} \times \mathbf{F} \tag{11-16}$$

QUESTIONS

1. Give examples of some objects that are rigid bodies and some that are not rigid bodies.

2. Consider a block of gelatin at rest in a bowl. Does the distance between any pair of points in the block remain the same? What happens if you press on it with a spoon? Is the gelatin a rigid body?

3. Because of the load it supports, a steel post has its length shortened by 0.1 percent. Is this post a rigid body?

4. Explain how and why an *equal*-arm balance can be used to compare weights. Can it compare masses?

5. Explain how and why an *unequal*-arm balance can be used to compare weights.

6. Archimedes claimed to be capable of moving the earth, given a lever and a place to stand. Explain the principle behind his claim. Is it a practical claim?

7. A playground seesaw has a total length of 3 m and is pivoted at its midpoint. Where and with what force should Uncle press on the board to balance 20-kg Baby at one end of the board?

8. Consider an 18-wheel tractor-trailer at rest on a level surface. Is the normal force on each wheel the same for all 18 wheels? What is the sum of the normal forces on the wheels?

9. Trucks are weighed at a highway weigh station by adding the scale reading when the front half of the truck is on the scale to the reading when the back half is on the scale. Does this procedure give an accurate measure of the truck's weight? Explain.

10. In reference to the preceding question, should a truck be at rest when the scale readings are taken? Should the scale be on level terrain? Explain.

11. Suppose you suspend an object from the ceiling by using two wires. One wire is attached to each end of the object, and the wires may have any orientation when attached to the ceiling. What orientation of the wires corresponds to a minimum in the tension of each?

12. In Example 11-8, the floor exerts a normal force and a frictional force on the ladder. These two forces can be added to obtain the resultant force $\mathbf{F}_F$ exerted on the ladder by the floor. Is this force directed along the ladder?

13. From the equation $\mathbf{A} \times \mathbf{B} = 0$, can you conclude that either $\mathbf{A} = 0$ or $\mathbf{B} = 0$? Explain.

14. Consider the torque you exert on a wrench as you tighten a bolt. Is the direction of the torque about the bolt's axis in the direction that the bolt is advancing, or opposite this direction?

15. Can a rigid body be in translational equilibrium and rotational equilibrium but not in static equilibrium? Explain.

16. Must there be any matter at the center of gravity of an object?

17. Give an example of a situation in which the center of gravity and the center of mass of an object do not coincide.

18. Does the phrase "center of gravity of the earth" have meaning? Explain.

19. The torque exerted by a force about some axis depends on the choice of axis. How can the condition $\Sigma\tau_{z,\text{ext}} = 0$ be satisfied for *any* choice of axis?

20. What are the units of torque and of work in a British system of units in which the pound (lb), the foot (ft), and the second (s) are the basic units?

21. Complete the following table:

Symbol	Represents	Type	SI Unit
τ_z		Component	
τ			
$\Sigma\tau_{\text{ext}}$			
$\mathbf{A}\times\mathbf{B}$	Cross product of vectors		—
$r_\perp$			m

EXERCISES

Section 11-2. Torque about an Axis

1. Before finger holes are drilled, a uniform bowling ball of radius 120 mm has its weight of magnitude 65 N acting at its center. Determine the magnitude and direction of the torque exerted by this force about an axis perpendicular to the plane of Fig. 11-21 and passing through point *(a)* *A*; *(b)* *B*; *(c)* *C*; *(d)* *O*.

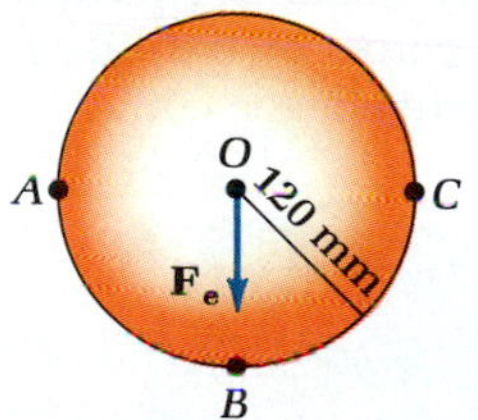

Figure 11-21. Exercise 1: A uniform bowling ball.

2. Show that the SI unit of torque can be expressed as kilogram-meters per second to the minus second power ($\text{kg}\cdot\text{m}^2\cdot\text{s}^{-2}$).

3. Suppose that you hold a 2.0-kg stone in your hand with your arm extended horizontally from your side. *(a)* What force does the stone exert on your hand? *(b)* Estimate the torque due to this force on your hand about a horizontal axis through your nose and perpendicular to your arm.

4. A force **F** acts on an object at a point with coordinates (x, y), as shown in Fig. 11-22. Evaluate the torque produced about the *z* axis by adding the torque components produced separately by F_x and F_y. In this way show that $\tau_z = xF_y - yF_x$.

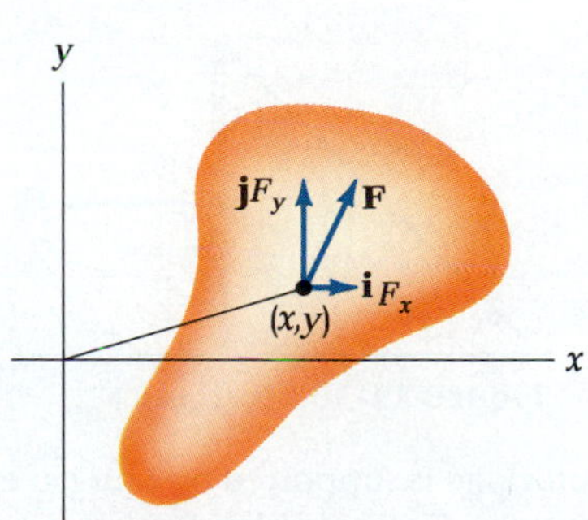

Figure 11-22. Exercise 4.

Section 11-4. Problem-Solving Techniques

5. A uniform 325-g meter stick is in balance in Fig. 11-23, with the knife edge directly below the midpoint. The 478-g mass is suspended by a light string at the 89.1-cm mark. *(a)* From what point is the 642-g mass suspended? *(b)* What force does the knife edge exert on the stick? *(c)* Would the answers be different if this apparatus were on the moon? In interstellar space?

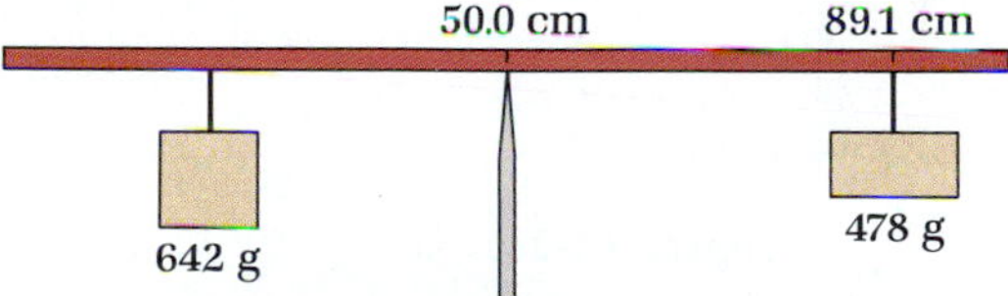

Figure 11-23. Exercise 5.

6. Suppose that you weigh 500 N and that you are standing on the board in Example 11-4. Determine the upward normal forces on the board if you are standing *(a)* at the middle of the board and *(b)* at the left-hand end. *(c)* How close can you stand to the right-hand end without tipping the board?

7. Estimate the force you must apply to the lever in Fig. 11-24 to move the 2000-N boulder.

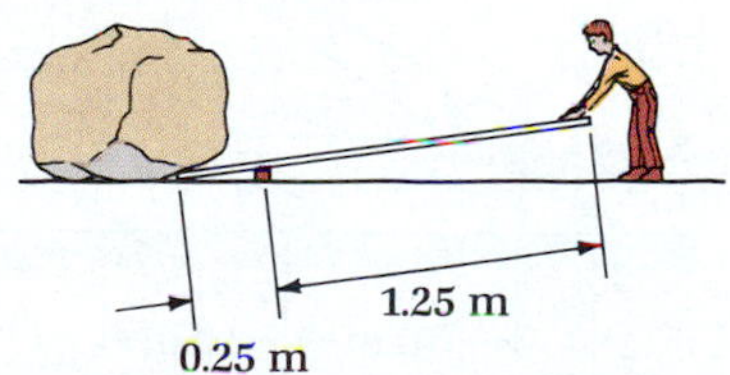

Figure 11-24. Exercise 7.

8. A load weight $F_L = 900$ N is supported by the boom shown in Fig. 11-25. A pin exerts a force with vertical and horizontal

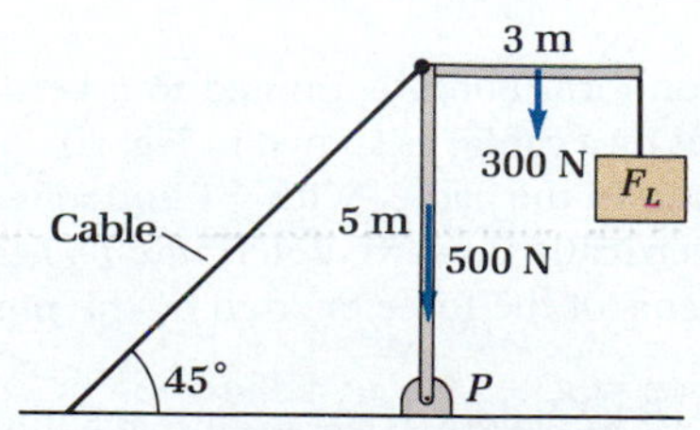

Figure 11-25. Exercise 8.

components at P. The weight of each section of the structure acts at the midpoint of that section. *(a)* Determine the components of the force exerted by the pin and the tension value in the cable. *(b)* What maximum load F_L can be supported if the cable tension is not to exceed 2500 N?

9. An 8.0-m horizontal boom supports a 20.0-kN load, as shown in Fig. 11-26. A pin exerts a force on the boom at the left-hand end. *(a)* Neglecting the weight of the boom itself, find the vertical and horizontal components of the pin force and the tension in the cable. *(b)* What is the direction of the pin force?

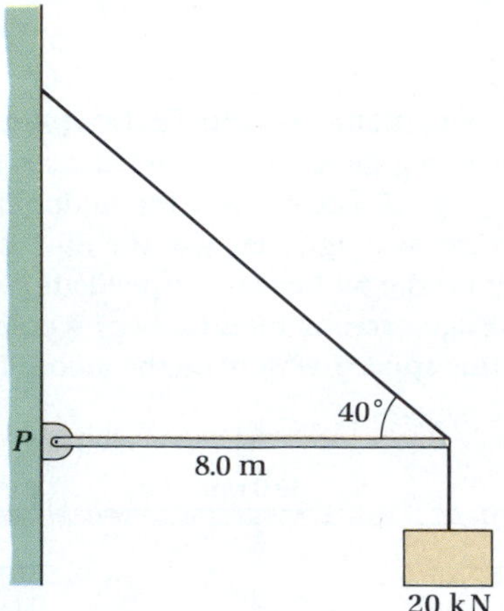

Figure 11-26. Exercise 9.

10. Rework the preceding exercise, taking the boom's weight of magnitude 4.0 kN to be acting at its midpoint. Again determine the direction of the pin force and explain any differences from the earlier result.

11. A 30-kN truck-crane on level ground supports a 20-kN load, as shown in Fig. 11-27. *(a)* Determine the normal forces exerted on the front and rear wheels by the ground. *(b)* What minimum load would cause the crane to tip?

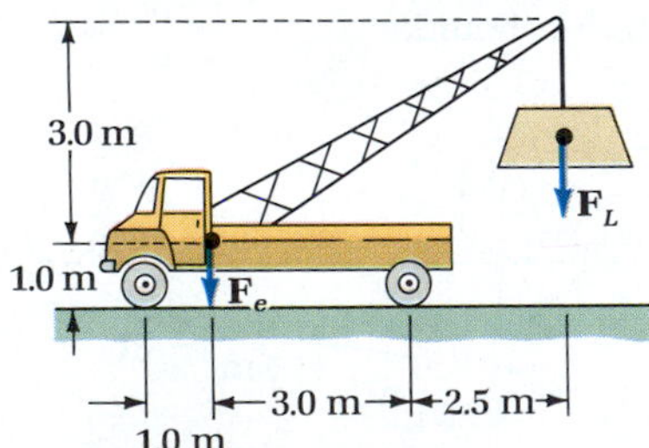

Figure 11-27. Exercise 11.

12. Rework the preceding exercise, but orient the truck-crane to face uphill on a 20° slope. Explain your assumptions concerning the frictional forces exerted on the wheels.

13. A 350-N uniform boom is pinned to a vertical wall and held horizontal by a cable, as shown in Fig. 11-28. *(a)* Determine the tension in the cable. With $+\mathbf{i}$ horizontal toward the right and $+\mathbf{j}$ vertically upward, determine *(b)* the x and *(c)* the y components of the force exerted by the pin.

14. A 4.0-m, 15-kg ladder leans against a smooth wall; the ladder's lower end touches the floor 1.0 m from the wall. The

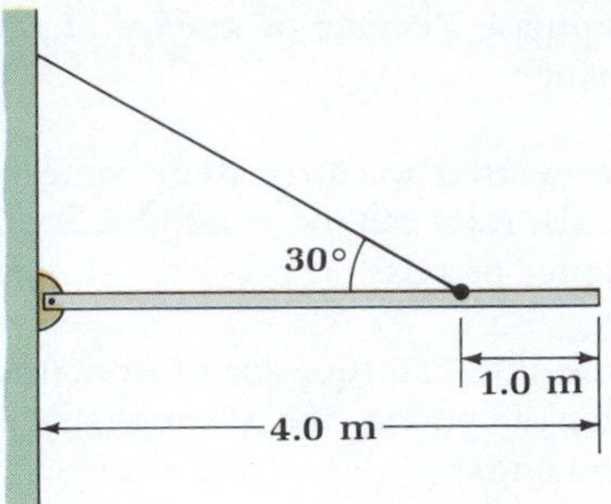

Figure 11-28. Exercise 13.

weight of the ladder acts at its midpoint. A 52-kg painter stands on the ladder at a point 1.5 m from its top. Determine *(a)* the (horizontal) force exerted by the wall, *(b)* the normal and frictional forces exerted by the floor, and *(c)* the minimum coefficient of friction needed at the ladder-floor interface to keep the ladder from sliding.

15. A uniform disk of radius R rests in contact with a curb of height $h = \frac{1}{2}R$, as shown in Fig. 11-29. A horizontal pull force of magnitude $F_p = F_e/3$ is applied at the center of the disk. Determine in terms of F_e *(a)* the normal force exerted by the floor and *(b)* the horizontal and vertical components of the force exerted by the corner on the disk. *(c)* What is the direction of the force exerted by the corner?

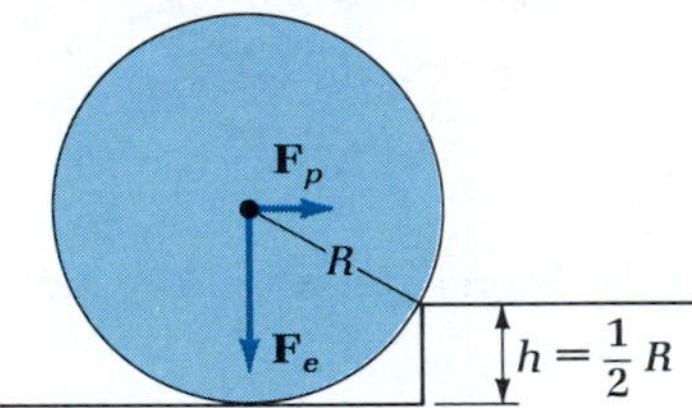

Figure 11-29. Exercise 15.

16. For the conditions given in the previous exercise, determine the minimum magnitude of the horizontal force F_p necessary to raise the disk over the curb. Express this force in terms of F_e. (*Hint:* What happens to the magnitude of the normal force just as the disk leaves the floor surface?)

17. A rigid rod is in static equilibrium as seen in Fig. 11-30, with a horizontal force applied at the midpoint. The weight of the rod may be neglected. The floor exerts normal and frictional forces on the rod. *(a)* Assuming the rod does not slip, determine the tension in the cable, the static frictional force, and the normal force, all in terms of the value of F_a. *(b)* Determine the minimum value of μ_s for which the rod does not slip.

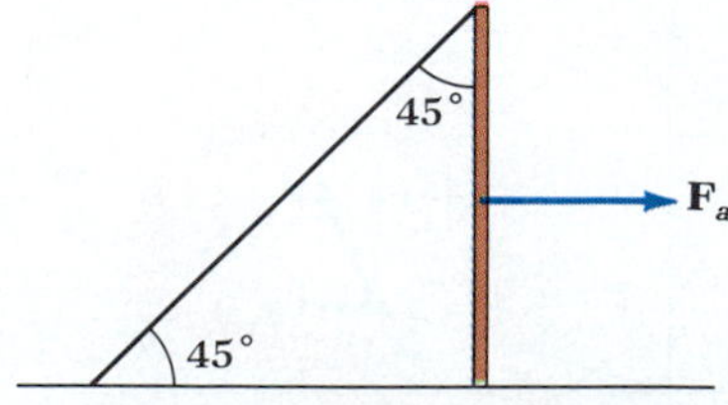

Figure 11-30. Exercise 17.

18. A 2.5-m footbridge is supported by a pier at each end. The center of gravity of the 120-kg bridge is at its midpoint. A 60-kg man moves steadily across the bridge. Determine the magni-

tudes of the upward forces exerted by the piers on the bridge as functions of the position of the man as he moves from one end of the bridge to the other.

Section 11-5. Center of Gravity

19. Locate the center of gravity of the inverted L-shaped structure shown in Fig. 11-25.

20. A uniform, square metal plate of edge 25.0 mm has a square section of edge 5.0 mm cut out, as shown in Fig. 11-31. Locate the center of gravity of this plate.

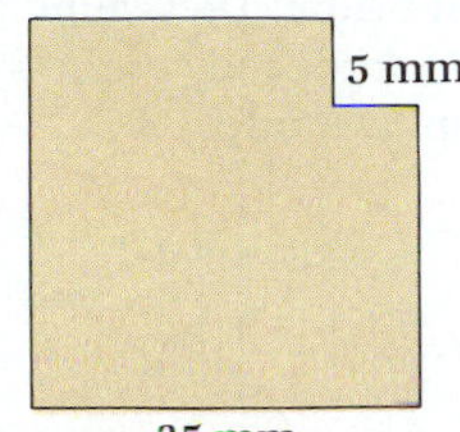

Figure 11-31. Exercise 20.

21. A uniform cubic block is placed on a very rough surface so that when the surface is slowly tilted, the block does not slide. At what angle of the ramp will the block tip over?

Section 11-6. Torque and the Cross Product of Vectors

22. With respect to a coordinate system centered at point O, a force acts at a point located on the y axis. The torque exerted by this force about point O lies in the xz plane. What can you determine about the direction of this force?

23. A force $\mathbf{F} = (174\ \text{N})\mathbf{i} + (203\ \text{N})\mathbf{j} + (-166\ \text{N})\mathbf{k}$ is exerted on an object at a point located by the position vector $\mathbf{r} = (1.35\ \text{m})\mathbf{i} + (-2.22\ \text{m})\mathbf{j}$ from a reference point O. Evaluate the torque exerted by this force about point O.

24. Consider three displacements **a**, **b**, **c**, as shown in Fig. 11-32. Show that $\mathbf{a} \cdot (\mathbf{b} \times \mathbf{c})$ equals the volume of the parallelepiped formed by these vectors.

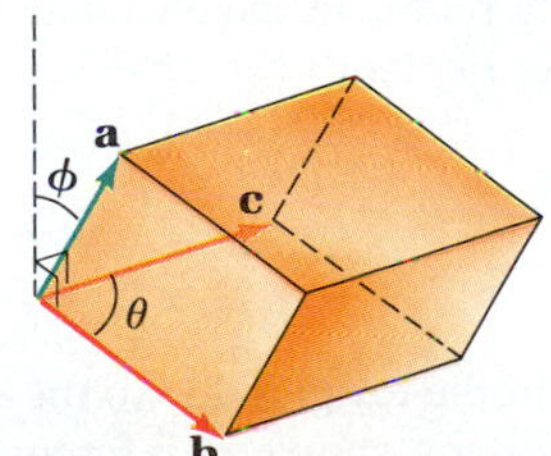

Figure 11-32. Exercise 24.

25. A tetrahedron has one vertex at the origin and the other three vertices at points (1, 0, 0), (1, 1, 0), and (1, 1, 1), where all distances are in meters. Determine the volume of the tetrahedron. (*Hint:* See the previous exercise.)

26. For each face of the tetrahedron in the previous exercise, find a vector perpendicular to that face and directed out of the enclosed volume.

27. *(a)* Using the rules for evaluating an ordinary determinant, show that $\mathbf{A} \times \mathbf{B}$ can be obtained by evaluating

$$\begin{vmatrix} \mathbf{i} & \mathbf{j} & \mathbf{k} \\ A_x & A_y & A_z \\ B_z & B_y & B_z \end{vmatrix}$$

(b) Evaluate the cross product for the case $\mathbf{A} = \mathbf{i} + 2\mathbf{j} + 3\mathbf{k}$ and $\mathbf{B} = 2\mathbf{i} - \mathbf{j} - \mathbf{k}$.

28. *(a)* Establish the so-called BAC-CAB rule:

$$\mathbf{A} \times (\mathbf{B} \times \mathbf{C}) = \mathbf{B}(\mathbf{A} \cdot \mathbf{C}) - \mathbf{C}(\mathbf{A} \cdot \mathbf{B})$$

(*Hint:* Evaluate the products in terms of components.)
(b) If **A** is perpendicular to both **B** and **C**, show that $\mathbf{A} \times (\mathbf{B} \times \mathbf{C}) = 0$.

29. Consider two nonparallel vectors **A** and **B** and their cross product $\mathbf{C} = \mathbf{A} \times \mathbf{B}$. If a and b are any two scalars, show that $a\mathbf{A} + b\mathbf{B}$ is perpendicular to **C**. What is the geometrical interpretation of this result?

30. Verify Eqs. 11-14 by evaluating all nine cross products.

31. Two equal and opposite forces $\mathbf{F}_1$ and $\mathbf{F}_2$ with different lines of action are exerted at points 1 and 2 on the object shown in Fig. 11-33. The lines of action are separated by a perpendicular distance of $d = 2.0$ m, and $|\mathbf{F}_1| = |\mathbf{F}_2| = 5.0$ N. Use a coordinate frame with the xy plane parallel to the plane of the page and the z axis directed out of the page. *(a)* With the origin placed midway between points 1 and 2, determine τ_{z1}, τ_{z2}, and the net torque τ_z. *(b)* With the origin placed at point 1, determine τ_{z1}, τ_{z2}, and the net torque τ_z. *(c)* With the origin placed at point 2, determine τ_{z1}, τ_{z2}, and the net torque τ_z. This force arrangement is an example of a *couple* (see Prob. 1).

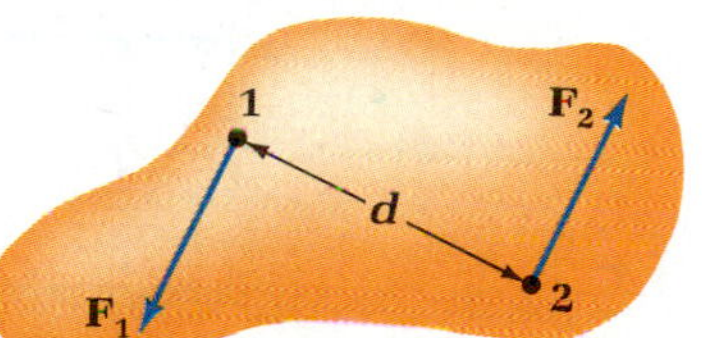

Figure 11-33. Exercise 31.

32. Suppose that the condition for translational equilibrium of a rigid body is satisfied, $\Sigma\mathbf{F}_{\text{ext}} = 0$. Show that the net external torque $\Sigma\tau_{\text{ext}}$ is independent of the reference point about which torques are evaluated. This proves that if $\Sigma\tau_{\text{ext}} = 0$ about one point, then $\Sigma\tau_{\text{ext}} = 0$ about any point, so long as $\Sigma\mathbf{F}_{\text{ext}} = 0$.

Additional Exercises

33. A 50-kg diver stands at one end of a 3-m long diving board (Fig. 11-34). *(a)* Neglect the weight of the board and determine the torque on the board due to the diver about an axis

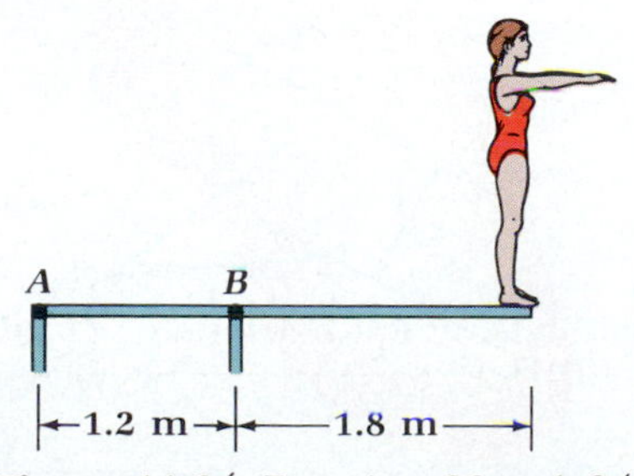

Figure 11-34. Exercises 33 and 34.

through point A perpendicular to the plane of the figure. What force is exerted on the board by the support at *(b)* point B and *(c)* point A?

34. Rework the previous exercise taking the weight of the board into account. Assume that the weight of the 100-kg board acts at the center of the board.

35. Estimate the magnitude of the force exerted on the nail by each jaw of the pliers in Fig. 11-35. Assume that the forces applied at the handles have magnitude 100 N and that the overall length of the pliers is 250 mm.

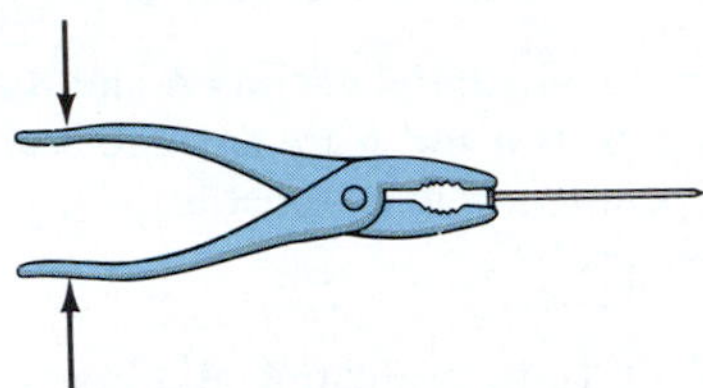

Figure 11-35. Exercise 35.

36. A horizontal, uniform 30,000-kg bridge has supports at each end 30 m apart. A 10,000-kg truck moves across the bridge. Determine the force exerted by each support on the bridge as a function of the position x of the truck. Let $x = 0$ correspond to one end of the bridge.

37. The center of gravity of a nonuniformly packed 40-kg crate is located as shown in Fig. 11-36. The crate is at rest on a horizontal surface. If a force of magnitude $F_a = 300$ N is applied horizontally as shown, the crate just begins to tip over. Determine the height H.

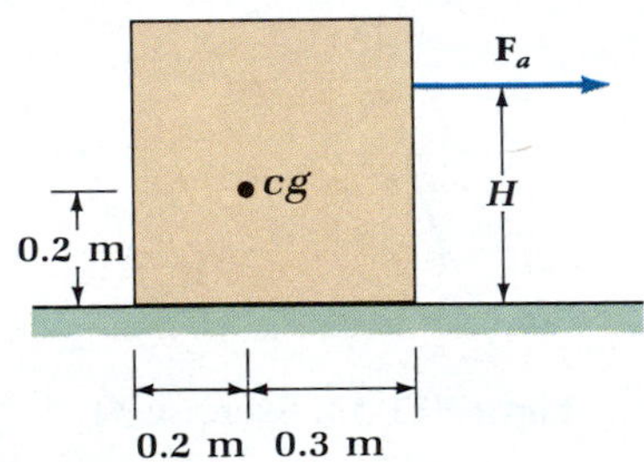

Figure 11-36. Exercise 37.

38. Rework the previous exercise with the force applied to the left instead of to the right. Determine the height H at which the applied force causes the crate just to tip over at the lower-left corner.

39. Three coplanar forces act at corners of a square plate (Fig. 11-37). The forces have magnitudes $F_1 = F_2 = F$ and $F_3 = 2F$. A fourth force is to be applied at point P such that the plate is in static equilibrium. Determine the *(a)* magnitude and *(b)* direction of the force and *(c)* the x coordinate of point P in terms of the given quantities.

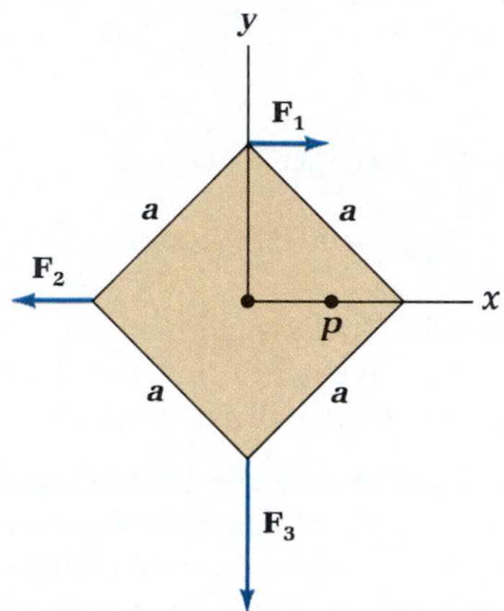

Figure 11-37. Exercise 39.

40. A force $\mathbf{F} = (120 \text{ N})\mathbf{i} + (180 \text{ N})\mathbf{j}$ acts on a body. Determine the torque about the origin O if the point of application of the force is *(a)* $\mathbf{r} = (3.0 \text{ m})\mathbf{i}$; *(b)* $\mathbf{r} = (3.0 \text{ m})\mathbf{j}$; *(c)* $\mathbf{r} = (3.0 \text{ m})\mathbf{k}$; *(d)* $\mathbf{r} = (3.0 \text{ m})\mathbf{i} - (2.0 \text{ m})\mathbf{j}$; *(e)* $\mathbf{r} = -(2.0 \text{ m})\mathbf{i} - (3.0 \text{ m})\mathbf{j}$.

41. Relative to an origin O, a force $\mathbf{F} = (60 \text{ N})\mathbf{i} + (80 \text{ N})\mathbf{j}$ exerts a torque along the positive z direction with magnitude 24 N·m. Determine a possible point of application of this force. Consider only points in the xy plane for simplicity. Is your answer unique? Explain.

PROBLEMS

1. ***A couple.*** A *couple* consists of two forces that are equal in magnitude F and opposite in direction, with a perpendicular distance d between the lines of action, as shown in Fig. 11-38. Show that the sum of the torques due to the couple about any point is the same for any choice of reference point and is of magnitude Fd.

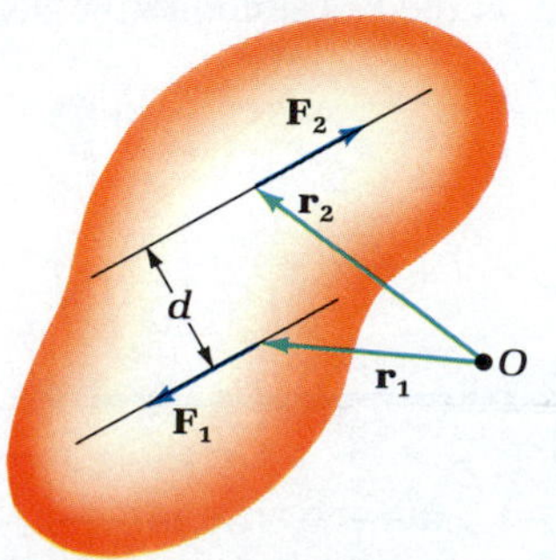

Figure 11-38. Problem 1.

2. ***Where does the normal force act?*** The normal force exerted on an object by a surface is actually a sum of a large number of such forces distributed over the area of contact of the two surfaces. The effective point of application of the full normal force is such that the torque is the same as that produced by the distributed normal forces. A horizontal force of magnitude $F_p = \frac{1}{3}F_e$ is applied at the top of the uniform cubic block shown in Fig. 11-39. *(a)* Locate the effective point of application of the normal force, assuming that the block does not slide. *(b)* What is the minimum value of the coefficient of static friction?

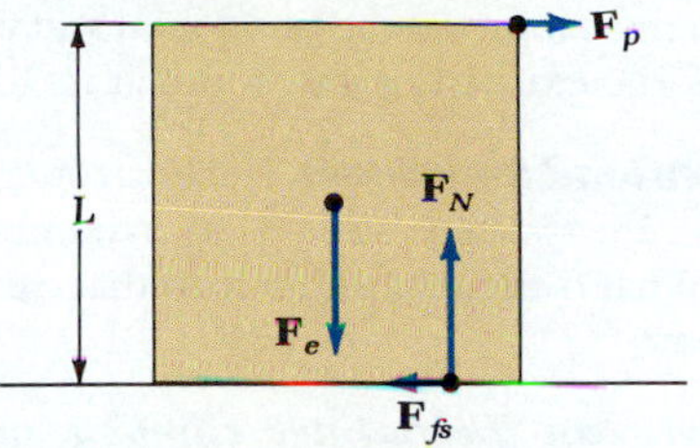

Figure 11-39. Problem 2.

3. ***Tip the block.*** Rework the previous problem, but suppose that the magnitude F_p of the horizontal force is increased. Assuming that the block does not slide, determine the value of F_p that will cause the block to tip over. What is the minimum value of the coefficient of static friction to keep the block from sliding?

4. ***Find the tensions.*** The irregularly shaped object of length L and weight of magnitude F_e shown in Fig. 11-40 is in static equilibrium, with two light strings attached as indicated. Determine the x coordinate of the center of gravity of the object and the tension in each string in terms of F_e, L, θ, and ϕ.

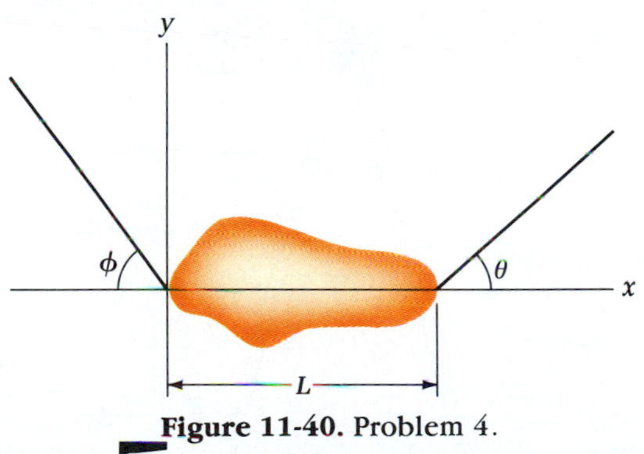

Figure 11-40. Problem 4.

5. ***An asymmetric A-frame.*** Two uniform boards are hinged together at one end to form a right angle as shown in Fig. 11-41. One board has length 3.0 m and weight of magnitude 120 N, and the other board has length 4.0 m and weight of magnitude 160 N. The combination rests in contact with a smooth floor and a light, horizontal cable is strung between the boards at a height of 1.0 m from the floor. Determine *(a)* the normal force exerted by the floor on each board, *(b)* the tension in the cable, *(c)* the force that one board exerts on the other at the hinged apex *A*.

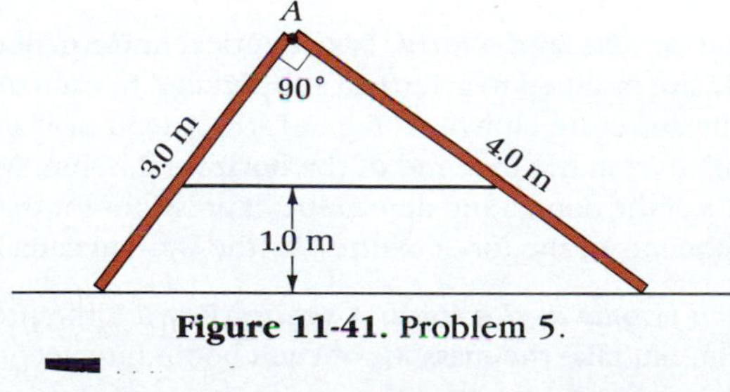

Figure 11-41. Problem 5.

6. ***Estimating a muscular force.*** Figure 11-42 shows a horizontal forearm supporting a 100-N stone. Using your own arm as a model for numerical values, estimate the force exerted on the forearm by *(a)* the biceps muscle and *(b)* the upper arm bone at the elbow. *(c)* How sensitive are your answers to your estimates of where the biceps is attached, where the center of gravity of the hand-forearm is located, and the weight of the hand-forearm?

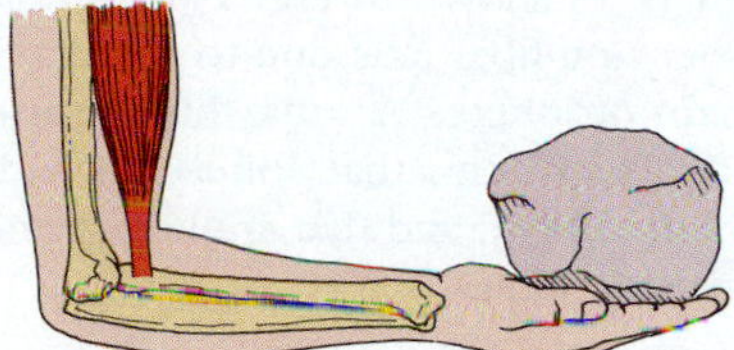
Figure 11-42. Problem 6.

7. ***A crane.*** A simple crane is shown supporting a 15-kN load in Fig. 11-43. The center of gravity of the 7.5-m, 2.5-kN boom is 3.0 m, as measured along the boom from the lower end, which is pinned at *P*. Cable *C* can be drawn up by a winch to change the elevation angle of the boom. *(a)* Determine the tension in cable *C* and the components of the pin force at *P* on the boom for $\theta = 30°$. *(b)* Repeat for $\theta = 60°$ and compare the answers for these two angles.

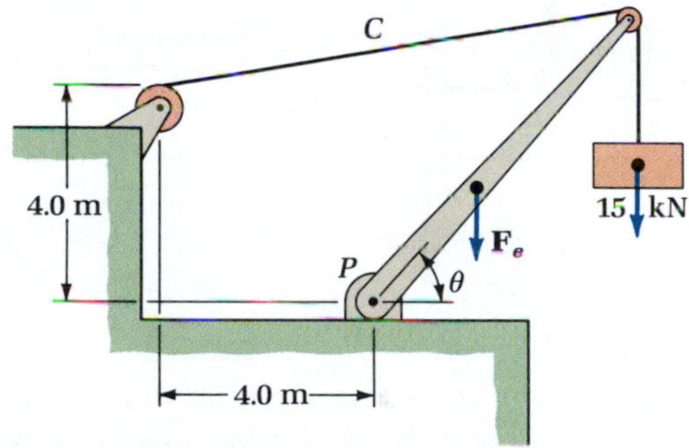

Figure 11-43. Problem 7.

8. ***The law of sines.*** Consider the triangle formed by vectors **A**, **B**, and **C** = **A** + **B**, as seen in Fig. 11-44. By forming the cross products **A** × **C** and **B** × **C**, prove the *law of sines:*

$$\frac{\sin\alpha}{A} = \frac{\sin\beta}{B} = \frac{\sin\gamma}{C}$$

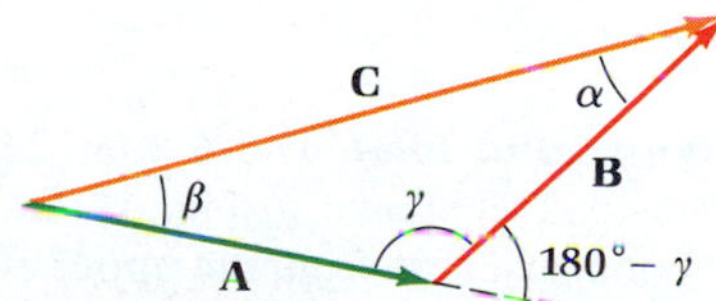

Figure 11-44. Problem 8.

9. ***A vector identity.*** Prove the identity $\mathbf{A} \cdot (\mathbf{B} \times \mathbf{C}) = (\mathbf{A} \times \mathbf{B}) \cdot \mathbf{C}$.

10. ***More unit vectors.*** Consider the three vectors **u**, **v**, **w** with

$$\mathbf{u} = \frac{\mathbf{i}+\mathbf{j}}{\sqrt{2}} \qquad \mathbf{v} = \frac{-\mathbf{i}+\mathbf{j}}{\sqrt{2}} \qquad \mathbf{w} = \mathbf{k}$$

(a) Show that each of these is a unit vector. *(b)* Show that the three vectors are mutually perpendicular. *(c)* Evaluate the cross product of each pair of these vectors. *(d)* If an *xyz*-

coordinate system is set up with x, y, z axes along the directions of **u**, **v**, **w**, in that order, is this coordinate system right-handed?

11. ***Torques and the Cavendish balance.*** A schematic diagram of the Cavendish balance, used for measuring the gravitational constant G, is shown in Fig. 7-6. Estimate the torque about the suspension-fiber axis due to the gravitational force between the pairs of spheres. Assume that the distance between spheres a and b is 15 cm, that spheres a and b have mass 0.25 kg and radius 2 cm, and that spheres A and B have mass 2.0 kg and radius 3.5 cm.

12. ***A boom and a load.*** A 1200-kg load is suspended from a uniform 400-kg boom (Fig. 11-45). The length of the boom is 5.0 m and a horizontal cable is fastened to its midpoint. Determine *(a)* the tension in the cable and *(b)* the magnitude and direction of the force exerted on the boom at its lower end by the pin P.

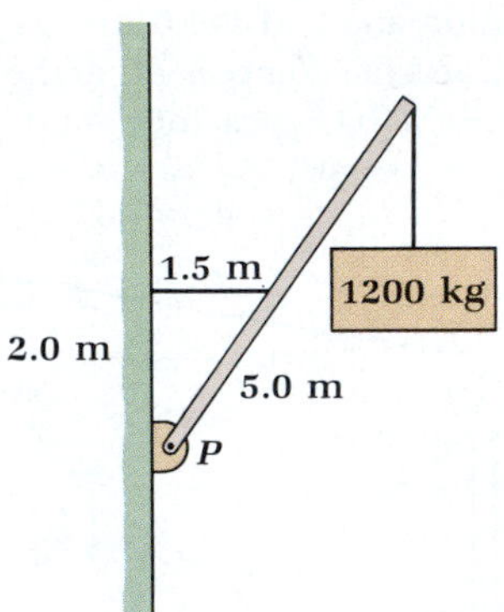

Figure 11-45. Problems 12 and 15.

13. ***Child hanging onto a door.*** A uniform 25-kg door is 2.0 m high and 1.0 m wide. It has three hinges located at the uppermost *(U)*, middle *(M)*, and lowermost *(L)* points on one vertical edge. Suppose that a 15-kg child hangs from the door knob, 0.1 m from the edge, and that the upper hinge has failed and exerts no force on the door. *(a)* Determine the vertical and horizontal components of the force exerted by the other two hinges on the door. Assume that the two hinges equally share the vertical load. *(b)* Repeat for the case of the top and bottom hinges intact and the middle hinge having failed.

14. ***Another boom and load.*** The 6.0-m, 240-kg uniform boom supports a load as shown in Fig. 11-46. The cable is attached to the boom at 1.5 m from the upper end. Determine the tension in the cable and the horizontal and vertical components of the force exerted on the boom at its lower end.

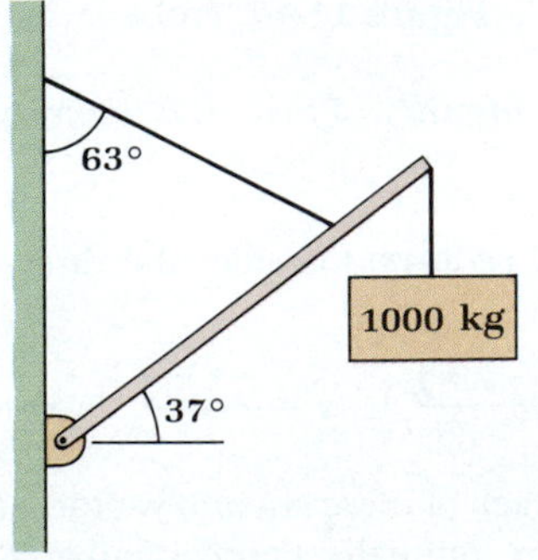

Figure 11-46. Problem 14.

15. ***Maximum load for a boom.*** The horizontal cable in Prob. 12 and Fig. 11-45 can sustain a maximum tension of 16×10^3 N. What is the maximum load that can be suspended from the boom?

16. ***How high can the painter climb?*** A uniform 350-N, 8.0-m ladder leans against a smooth wall, with the ladder's lower end on the floor 2.5 m from the wall. The coefficient of static friction at the floor is $\mu_s = 0.21$. A painter weighing 440 N begins climbing the ladder slowly. How far up the ladder can she climb before the ladder begins to slide?

17. ***Ladders strapped together.*** Two 6.0-m ladders are hinged together at their tops and connected by a strap at their midpoints, as shown in Fig. 11-47, with their bases 3.0 m apart. The floor is very smooth so that friction is negligible. Each ladder weighs 150 N. Determine the x and y components of the force exerted on the ladder *(a)* by the floor at point A, *(b)* by the strap at point B, *(c)* on the left at point C by the ladder on the right, *(d)* by the strap at point D, *(e)* by the floor at point E.

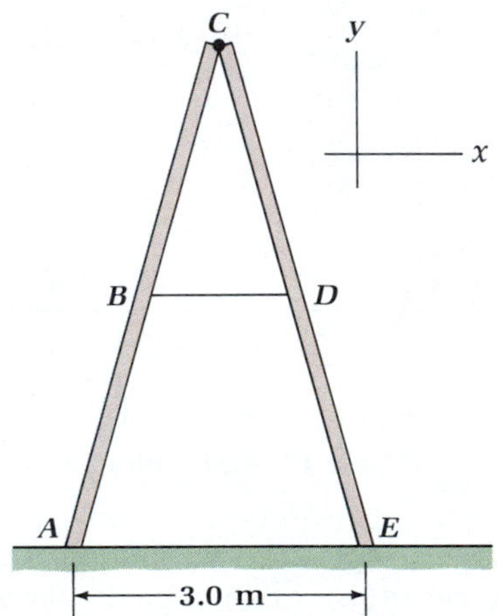

Figure 11-47. Problems 17 and 18.

18. ***Painter on the strapped ladders.*** A 450-N painter climbs the ladder on the left in the previous problem, and she stands at the midpoint where the strap is attached at point B. Determine the x and y components of the force exerted on the ladder *(a)* by the floor at point A, *(b)* by the strap at point B, *(c)* on the left at point C by the ladder on the right, *(d)* by the strap at point D, *(e)* by the floor at point E.

19. ***Two booms and a load.*** Two identical uniform booms of length L are fastened to a vertical support and to each other to form the structure shown in Fig. 11-48. A load of mass M is suspended from the free end of the horizontal boom. Neglect the mass of the booms and determine expressions for the x and y components of the force exerted by the wall on each boom.

20. ***Two booms and a load; a sequel.*** Rework the previous problem, but take the mass M_b of each boom into account.

21. ***A barnyard gate.*** A 480-N gate is fastened by two hinges to a post, as shown in Fig. 11-49. The guy wire is drawn up such that the horizontal component of the force exerted by the

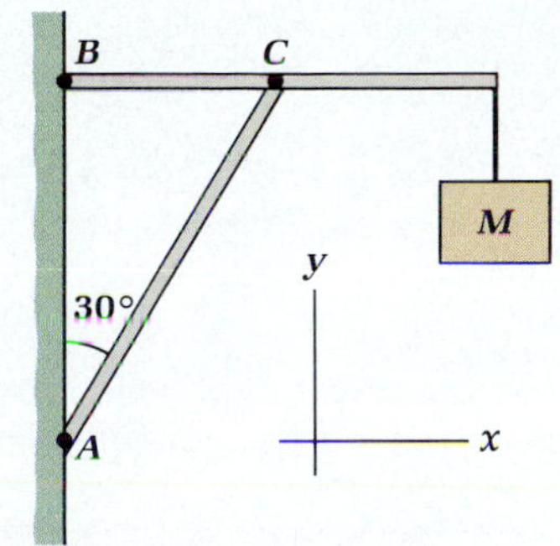

Figure 11-48. Problems 19 and 20.

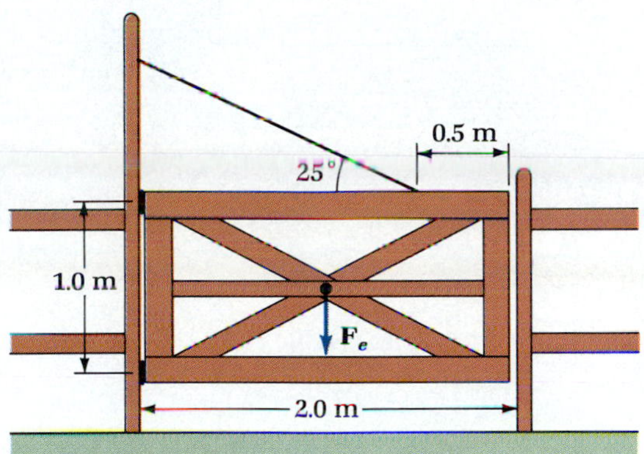

Figure 11-49. Problem 21.

upper hinge is zero. Evaluate the horizontal component of the force exerted by the lower hinge, the tension in the wire, and the sum of the vertical components of the hinge forces.

22. ***Finding two masses from one weight measurement.*** In Fig. 11-50, a spring scale supports a uniform meter stick that has a block attached to it. *(a)* Determine the mass of the block. *(b)* Determine the mass of the meter stick.

Figure 11-50. Problem 22.

12 ROTATION I

A carnival merry-go-round. The camera shutter was left open long enough to demonstrate the merry-go-round's rotational motion.

Objects that rotate are encountered frequently — a door rotates on its hinges, a pulley on its axle, a compact disk in a disk player. The earth has two rotational motions; it spins on its axis once a day, and it orbits the sun once a year. At the level of atoms and molecules, both spin and orbital motions play important roles in the behavior of matter. When these properties are discussed in later chapters, an understanding of rotational motion will be crucial. This chapter is mostly about *rotational kinematics,* and the next chapter discusses *rotational dynamics.*

12-1 TRANSLATION AND ROTATION OF A RIGID OBJECT

It is helpful to state the definition of translational motion before beginning a discussion of rotation. An example of translational motion is the motion of the body of a car as the car travels along a straight road.

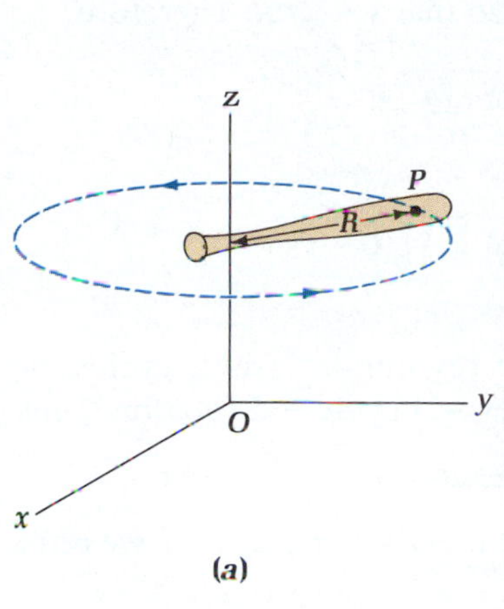

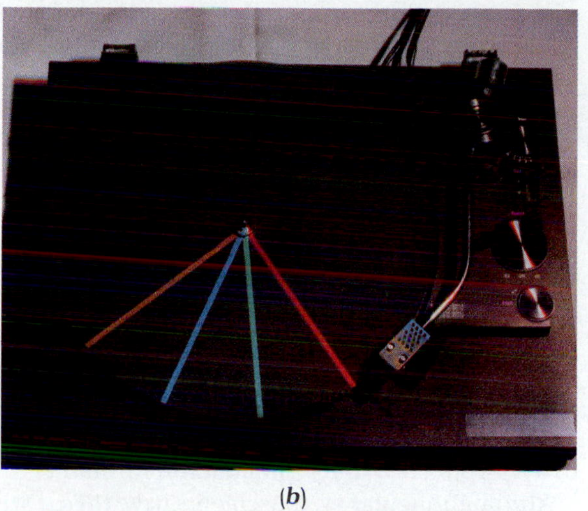

Figure 12-1. Each particle in a rotating rigid object travels in a circle centered at the axis of rotation, except those particles that are on the axis. *(a)* The coordinate frame we usually use to describe a rotating rigid object. *(b)* Multicolor strobe photograph of a rotating turntable.

Definition of translational motion

> A rigid object undergoes ***translational motion*** when each particle of the object has the same displacement in the same time interval.

An example of rotational motion is the motion of a door being opened or closed.

Definitions of rotational motion and axis of rotation

> A rigid object undergoes ***rotational motion*** when each particle of the object travels in a circle, except those particles on the axis of rotation. The ***axis of rotation*** is a straight line that extends through the centers of the circles described by the motion of the particles.

For a rotating rigid object, a perpendicular line drawn from the axis of rotation to any particle of the object sweeps out the same angle in the same time interval as any other such line.

Figure 12-1*a* shows the standard arrangement for describing the rotation of a rigid object about a fixed axis. The z axis is the axis of rotation so that the circular path of each particle is centered on the z axis and parallel to the xy plane. The circle's radius R is the perpendicular distance from the axis to the particle.

An example of translational and rotational motion combined is the motion of a moving car's wheel relative to a coordinate frame fixed to the earth. Relative to this frame the axis of rotation of the wheel (along the wheel's axle) executes translational motion along the road. If the road is straight, the orientation of the wheel's axis remains fixed relative to this coordinate frame.

A completely general motion of a rigid object involves changes in the orientation of the axis of rotation as well as translation of this axis. An example is a wobbly pass of a football. The ball rotates about an instantaneous axis, but since the throw is wobbly, the orientation of the axis, as well as its position, changes. In this chapter, we restrict our discussions to the rotation of a rigid object about an axis that maintains a fixed orientation.

12-2 ANGULAR MEASUREMENT

You are familiar with measuring angles in degrees. However, a more convenient unit to use for rotational motion is the SI unit, the ***radian (rad)***. In Fig. 12-2, the angle θ in radians between the x axis and the line segment OP is defined as the ratio s/R, where s is measured along the arc from the x axis to P and R is the radial distance from O to P:

$$\theta = \frac{s}{R} \qquad (12\text{-}1)$$

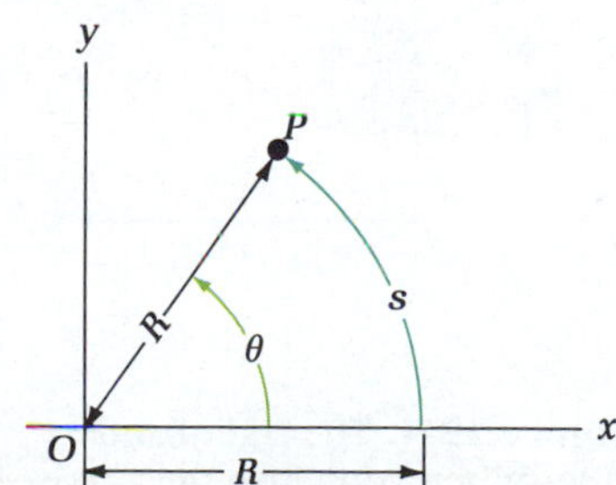

Figure 12-2. Definition of angular measure in radians: $\theta = s/R$. The same length unit is used to measure both s and R.

where s and R are measured in the same unit of length.

For a full circle, s is the circle's circumference, so that $s = 2\pi R$. Therefore

$$\theta(\text{full circle}) = \frac{2\pi R}{R} = 2\pi \text{ rad}$$

Since θ(full circle) = 360°, we have

$$2\pi \text{ rad} = 360° \qquad \text{or} \qquad \pi \text{ rad} = 180°$$

Thus 1 rad = $180°/\pi \approx 57.3°$.

Another way to measure an angle is in terms of revolutions (rev), cycles, or rotations. All three terms refer to one full circle: 1 rev = 1 cycle = 1 rotation. Thus

$$1 \text{ rev} = 360° = 2\pi \text{ rad} \qquad \text{(exactly)}$$

Revolutions are often used when discussing angular speeds. For example, we refer to a 45-rpm (rev/min) phonograph turntable.

Although angular measurements have these three commonly used units (radians, degrees, and revolutions), angle is a dimensionless quantity. It is defined in terms of the ratio of lengths. When performing calculations with angular quantities, you will find it useful to write the angular unit along with the numerical value, as we have been doing for other quantities.

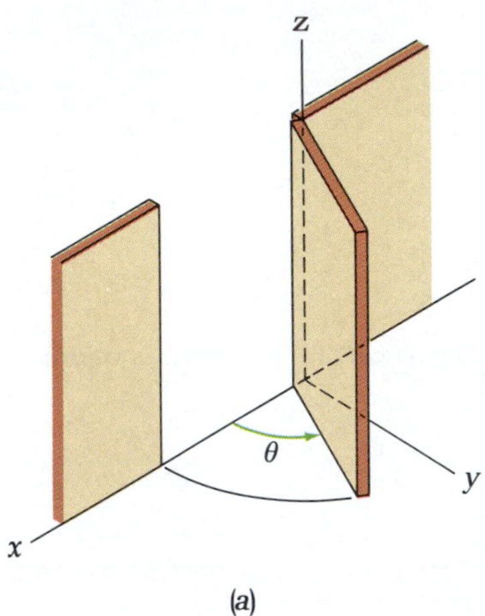

(a)

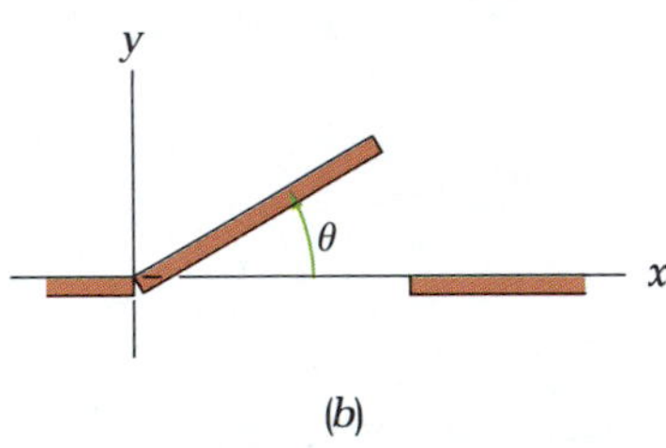

(b)

Figure 12-3. *(a)* Angular coordinate θ of a gate. *(b)* View from above, looking down onto the xy plane. Measured from the x axis, the angular coordinate θ is positive for counterclockwise rotation and negative for clockwise rotation.

EXAMPLE 12-1

Converting angles. Convert an angle of 64° to *(a)* radians and *(b)* revolutions.

Solution. *(a)* Since π rad = 180°, the number 1 can be written as $1 = \pi$ rad/180°. Thus

$$64° = 64°(\pi \text{ rad}/180°) = 1.1 \text{ rad}$$

(b) The number 1 can also be written: 1 = 1 rev/360°. Therefore

$$64° = 64°(1 \text{ rev}/360°) = 0.18 \text{ rev}$$

SELF-TEST 12-1. Show that *(a)* 45° = $(\pi/4)$ rad and *(b)* 60° = $(\pi/3)$ rad.

12-3 ANGULAR COORDINATE, VELOCITY, AND ACCELERATION

The kinematics of a rigid object rotating about a fixed axis is mathematically similar to that of a particle moving along a straight line (Chap. 3). The rotational kinematical quantities are defined in a way to strengthen this analogy.

Angular Coordinate θ

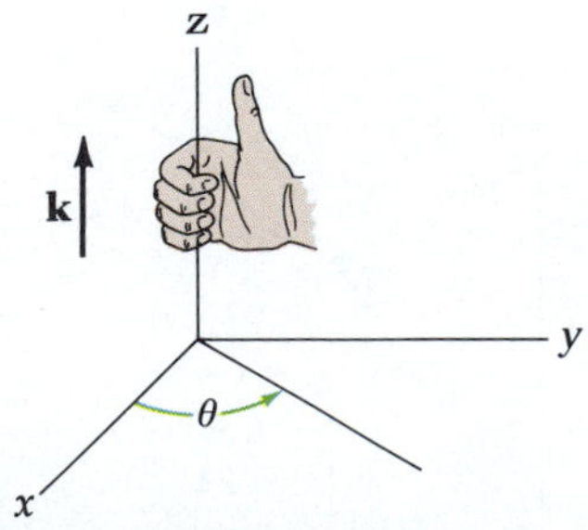

Figure 12-4. The right-hand rule. When you grasp the z axis with your right hand such that your thumb points in the $+\mathbf{k}$ direction, your fingers curl in the positive sense for θ.

The ***angular coordinate*** of a gate can be measured as shown in Fig. 12-3. Let the z axis be along the hinges and the xy plane be the plane of the ground, with the x axis along the fence and the y axis perpendicular to the fence. The angular coordinate θ of the gate is measured from the x axis to the gate. When the gate is closed, $\theta = 0$. When the gate is open, its angular coordinate is measured in a counterclockwise sense from the x axis when viewed from above; that is, an angle measured counterclockwise is positive, and an angle measured clockwise is negative. A right-hand rule gives the positive sense for θ (Fig. 12-4). If you imagine grasping the z axis with your right hand so that your thumb points in the $+\mathbf{k}$ direction, your fingers curl in the positive θ sense.

A major distinction between the angular coordinate θ and the linear coordinate x is that θ is cyclic. That is, the angular coordinates θ and $\theta + 2\pi$ represent the same angular position (Fig. 12-5). More generally, if n is any positive or negative integer, then θ and $\theta + n(2\pi)$ represent the same angular position. Ordinarily values of θ are adjusted so that they fall in the range from 0 to 2π rad or from $-\pi$ rad to π rad.

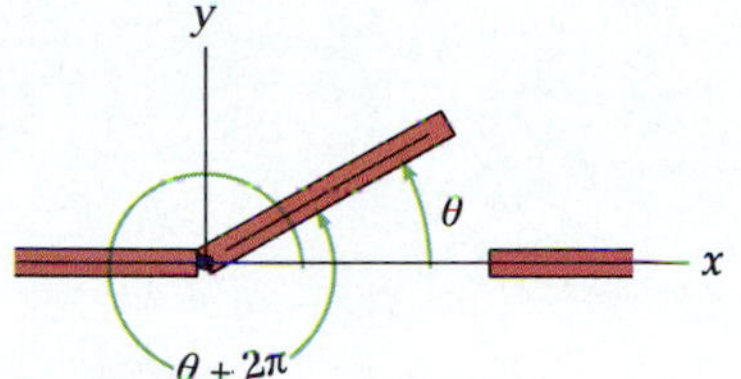

Figure 12-5. The angular coordinate is cyclic; the coordinates θ and $\theta + n(2\pi)$ rad represent the same angular position, where n is any positive or negative integer.

Angular Velocity and Angular Speed

The angular velocity component of an object, such as the swinging gate in Fig. 12-3, is

$$\omega_z = \frac{d\theta}{dt} \qquad (12\text{-}2)$$

The ***angular velocity component*** ω_z is the time rate of change of the angular coordinate θ.

If θ is increasing in time so that ω_z is positive, then the rotation is in the positive θ sense, and the gate is swinging counterclockwise when viewed from above. If θ is decreasing in time so that ω_z is negative, then the rotation is in the negative θ sense, and the gate is swinging clockwise. The magnitude of ω_z is the ***angular speed*** ω:

Angular speed

$$\omega = |\omega_z|$$

Thus the magnitude of ω_z tells how rapidly an object is rotating and its algebraic sign tells the sense of the rotation.

In general, vectors are used to describe angular motion. The angular velocity $\boldsymbol{\omega}$ is a vector whose magnitude is the angular speed and whose direction is parallel to the axis in the direction given by a right-hand rule. The right-hand rule for $\boldsymbol{\omega}$ is the following: Grasp the axis of rotation with the right hand so that the fingers are curled in the sense of the rotation; the extended thumb then points in the direction of $\boldsymbol{\omega}$. For example, the earth rotates from west to east. If you imagine grasping the earth with your right hand and with your fingers curled from west to east, then your extended thumb points northward. The angular velocity of the earth is directed along the axis toward the north. If the z axis is placed along an object's rotational axis, then

Right-hand rule for $\boldsymbol{\omega}$

$$\boldsymbol{\omega} = \omega_z \mathbf{k}$$

because a right-hand rule is used both for the direction of $\boldsymbol{\omega}$ and for the positive θ sense.

It may seem curious to let the vector $\boldsymbol{\omega}$ be directed along the axis of rotation. After all, when an object rotates, none of the particles that compose the object moves along the axis; they all move in circles contained in planes that are perpendicular to the axis. However, no direction in the plane of the motion of a particle can be used for $\boldsymbol{\omega}$ because, during one rotation, the particle's velocity sweeps through all directions in the plane. Thus no single direction in this plane is characteristic of the motion. But the axis of rotation provides a line in space that can be uniquely associated with the rotation. Therefore $\boldsymbol{\omega}$ is taken to be parallel to the axis, and the right-hand rule is used to decide which direction along the axis corresponds to which sense for the rotation.

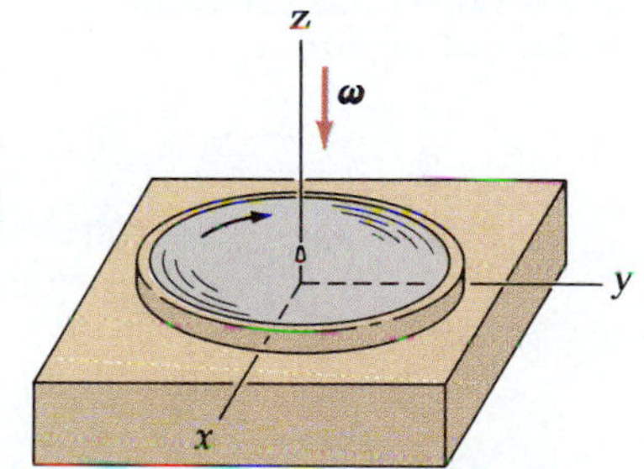

Figure 12-6. Example 12-2: A rotating phonograph turntable. The z axis passes through the spindle with $+z$ upward, and the x and y axes are fixed relative to the desktop. Viewed from above, the turntable rotates in the clockwise sense so that ω_z is negative.

EXAMPLE 12-2

Angular velocity of a turntable. Figure 12-6 shows a phonograph turntable rotating steadily at 45 rev/min in the clockwise sense when viewed from above. The z axis is along the axis of rotation, with $+z$ upward. The x and y axes are horizontal and fixed relative to the desk on which the turntable mechanism rests. Using this coordinate frame, determine ω_z in units of radians per second.

Solution. Converting ω from revolutions per minute to radians per second, we have

$$\omega = 45 \text{ rev/min} \frac{2\pi \text{ rad/rev}}{60 \text{ s/min}} = 4.7 \text{ rad/s}$$

Using the right-hand rule (Fig. 12-4), we see that the turntable is rotating such that the direction of $\boldsymbol{\omega}$ is downward. That is, if you grasp the axis of rotation with your right hand such that your fingers curl in the sense of the rotation, then your thumb points downward. Thus ω_z is negative, so that

$$\omega_z = -4.7 \text{ rad/s}$$

SELF-TEST 12-2. Determine the magnitude and direction of the angular velocity (in radians per second) of a phonograph turntable rotating at 33 rev/min. The turntable is on its side so that the axis is horizontal along an east-west line, and the rotation is clockwise when viewed from the west. ***ANSWER:*** 3.5 rad/s, east.

Angular Acceleration

Immediately after a phonograph turntable is turned on (or off), the angular velocity changes. The turntable's angular acceleration is equal to the rate of change of its angular velocity.

The ***angular acceleration component*** α_z is equal to the time rate of change of the angular velocity component:

Definition of the angular acceleration component

$$\alpha_z = \frac{d\omega_z}{dt} = \frac{d^2\theta}{dt^2} \tag{12-3}$$

Thus α_z is positive when ω_z is increasing, it is zero when ω_z is constant, and it is negative when ω_z is decreasing. The magnitude of the angular acceleration is $\alpha = |\alpha_z|$.

The angular acceleration $\boldsymbol{\alpha}$ is a vector whose direction is defined with a right-hand rule such that

$$\boldsymbol{\alpha} = \alpha_z \mathbf{k}$$

Thus $\boldsymbol{\alpha}$ is directed along the axis in the same direction as $\boldsymbol{\omega}$ when ω_z is increasing but opposite $\boldsymbol{\omega}$ when ω_z is decreasing. Angular acceleration has the dimension of $[\text{time}]^{-2}$ and the SI unit radians per second squared (rad/s^2).

Although the angular kinematical quantities (θ, $\boldsymbol{\omega}$, and $\boldsymbol{\alpha}$) have been defined by considering a rotating rigid object, these quantities can be used to describe the motion of a particle traveling in a circle. In a rotating rigid object, each particle of the object (except those on the axis of rotation) travels in a circle, and each particle has the same $\boldsymbol{\omega}$ and $\boldsymbol{\alpha}$.

Each particle of a rotating rigid object has the same values of ω_z and α_z.

12-4 KINEMATICS OF ROTATION ABOUT A FIXED AXIS

Two important types of rotational motion are (i) constant angular velocity and (ii) constant angular acceleration.

Constant Angular Velocity

Consider an object, such as the turntable seen in Fig. 12-6, rotating with a constant angular velocity. From Eq. 12-2,

$$\frac{d\theta}{dt} = \omega_z$$

Integrating from an initial time zero to a final time t gives θ as a function of t:

$$\theta - \theta_0 = \int_0^t \omega_z \, dt' = \omega_z \int_0^t dt' = \omega_z t$$

where ω_z is factored out of the integral because it is constant. Thus

$$\theta(t) = \theta_0 + \omega_z t \qquad (12\text{-}4)$$

Angle θ as a function of time, constant ω_z

EXAMPLE 12-3

Turntable rotating at a constant angular velocity. ***(a)*** Give the equation that describes the angular position of the turntable in Fig. 12-6 when it is rotating at a constant speed of 45 rev/min. The initial angular coordinate is $\theta_0 = 1.2$ rad. ***(b)*** Determine θ at $t = 2.4$ s.

Solution. *(a)* From Example 12-2, $\omega_z = -4.7$ rad/s. Substituting this value and $\theta_0 = 1.2$ rad into Eq. 12-4 gives

$$\theta(t) = 1.2 \text{ rad} - (4.7 \text{ rad/s})t$$

(b) At $t = 2.4$ s,

$$\theta = 1.2 \text{ rad} - (4.7 \text{ rad/s})(2.4 \text{ s}) = -10 \text{ rad}$$

SELF-TEST 12-3. Write an expression for $\theta(t)$ that describes a turntable rotating at 33 rev/min with $\theta_0 = 1.5$ rad. Use the coordinate frame shown in Fig. 12-6. ***ANSWER:*** $\theta = 1.5 \text{ rad} - (3.5 \text{ rad/s})t$

Constant Angular Acceleration

Suppose the gate in Fig. 12-3 rotates with a constant angular acceleration as it swings open. From Eq. 12-3,

$$\frac{d\omega_z}{dt} = \alpha_z$$

Integrating from an initial time zero to a final time t gives ω_z as a function of t:

$$\omega_z - \omega_{z0} = \int_0^t \alpha_z \, dt' = \alpha_z \int_0^t dt' = \alpha_z t$$

where α_z is factored out of the integral because it is constant. Thus

$$\omega_z(t) = \omega_{z0} + \alpha_z t \qquad (12\text{-}5)$$

Angular velocity component ω_z as a function of time, constant α_z

Substituting this value of ω_z into Eq. 12-2 gives

$$\frac{d\theta}{dt} = \omega_z = \omega_{z0} + \alpha_z t$$

Integrating from an initial time zero to a final time t gives θ as a function of t:

$$\theta - \theta_0 = \int_0^t (\omega_{z0} + \alpha_z t') \, dt' = \omega_{z0} \int_0^t dt' + \alpha_z \int_0^t t' \, dt'$$

where ω_{z0} and α_z are factored out of the integrals. (Why?) The integration gives $\theta - \theta_0 = \omega_{z0} t + \frac{1}{2}\alpha_z t^2$, or

$$\theta(t) = \theta_0 + \omega_{z0} t + \tfrac{1}{2}\alpha_z t^2 \qquad (12\text{-}6)$$

Angle θ as a function of time, constant α_z

By eliminating the time t between Eqs. 12-5 and 12-6 (see Exercise 16), you can show that

$$\omega_z^2 = \omega_{z0}^2 + 2\alpha_z(\theta - \theta_0) \qquad (12\text{-}7)$$

Angular velocity component ω_z as a function of θ, constant α_z

Table 12-1 lists the equations that describe rotation with constant angular velocity and constant angular acceleration. Also given in the table are the analogous expressions for a particle moving along a straight line. This comparison shows the similarity between these two types of motion.

TABLE 12-1. ***Analogy between Translation and Rotation***

Translation (one dimension)	Rotation (fixed axis)
Constant linear velocity	Constant angular velocity
$x = x_0 + v_x t$	$\theta = \theta_0 + \omega_z t$
Constant linear acceleration	Constant angular acceleration
$v_x = v_{x0} + a_x t$	$\omega_z = \omega_{z0} + \alpha_z t$
$x = x_0 + v_{x0} t + \frac{1}{2} a_x t^2$	$\theta = \theta_0 + \omega_{z0} t + \frac{1}{2}\alpha_z t^2$
$v_x^2 = v_{x0}^2 + 2a_x(x - x_0)$	$\omega_z^2 = \omega_{z0}^2 + 2\alpha_z(\theta - \theta_0)$

EXAMPLE 12-4

Turntable rotating with a constant angular acceleration. Suppose that after being switched off, the turntable in Example 12-2 slows to a stop in a time interval of 1.7 s. *(a)* Find an equation for the turntable's angular coordinate as a function of time while it is slowing to a stop, assuming that the angular acceleration is constant. Let $t = 0$ correspond to the instant the turntable is turned off, and let θ_0 be zero. *(b)* Through what angular coordinate does the turntable rotate while coming to a stop?

Solution. *(a)* Using Eq. 12-5 to find α_z gives

$$\alpha_z = \frac{\omega_z - \omega_{z0}}{t}$$

Since the turntable comes to rest in 1.7 s, $\omega_z = 0$ when $t = 1.7$ s, and from Example 12-2, $\omega_{z0} = -4.7$ rad/s. Therefore

$$\alpha_z = \frac{0 - (-4.7 \text{ rad/s})}{1.7 \text{ s}} = 2.8 \text{ rad/s}^2$$

Substituting into Eq. 12-6 gives

$$\theta(t) = -(4.7 \text{ rad/s})t + (1.4 \text{ rad/s}^2)t^2$$

(b) Since $\theta_0 = 0$ and the turntable stops in 1.7 s, the angle through which the turntable rotates in coming to a stop is

$$\theta(1.7 \text{ s}) = -(4.7 \text{ rad/s})(1.7 \text{ s}) + (1.4 \text{ rad/s}^2)(1.7 \text{ s})^2 = -3.9 \text{ rad}$$

SELF-TEST 12-4. A phonograph turntable is turned on and its angular speed increases from zero to 33 rev/min in 2.2 s. Let $t = 0.0$ s be the instant the turntable begins rotating, and use the coordinate frame in Fig. 12-6. With $\theta_0 = 0.0$ rad, write an expression for $\theta(t)$ between $t = 0.0$ s and $t = 2.2$ s, assuming constant angular acceleration. ***ANSWER:*** $\theta(t) = -(0.79 \text{ rad/s}^2)t^2$.

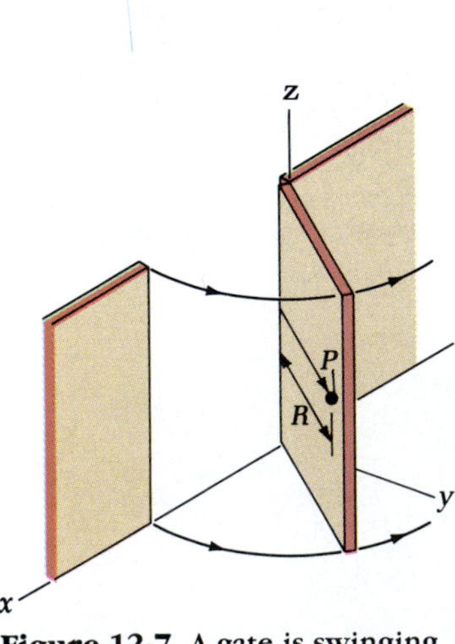

Figure 12-7. A gate is swinging open. Each particle in the gate has the same angular velocity and angular acceleration, but particles with different distances from the axis of rotation have different linear velocities and linear accelerations.

12-5 RELATIONS BETWEEN ROTATIONAL AND TRANSLATIONAL QUANTITIES

Consider a gate swinging open as shown in Fig. 12-7. Imagine that the gate consists of many small pieces, each small enough to be considered a particle. Because the gate is rigid, each particle has the same angular velocity $\boldsymbol{\omega}$ and angular acceleration $\boldsymbol{\alpha}$. Thus $\boldsymbol{\omega}$ and $\boldsymbol{\alpha}$ characterize the motion of the entire gate. However, particles with different distances from the axis of rotation have different linear velocities **v** and linear accelerations **a**. We now establish the connection between the angular velocity and angular acceleration of the gate on the one hand, and the linear velocity and linear acceleration of a particle P in the gate on the other.

As before, let the ground compose the xy plane, and place the origin directly below the hinges so that the z axis is the axis of rotation (Fig. 12-8). Particle P travels in a circle of radius R and can be located relative to the x axis by the ***arc coordinate*** s measured along the arc. The sign convention for s is the same as for θ: Viewed from above, s is positive in the counterclockwise sense and negative in the clockwise sense. The *tangential component of the velocity* of P is defined as

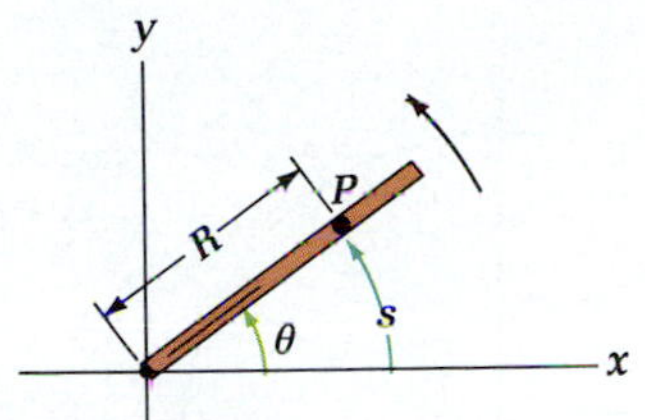

Figure 12-8. Top view of the gate in Fig. 12-7. A particle at radius R can be located with the arc coordinate s. The sign convention for s is similar to that for θ, so $s = R\theta$.

$$v_t = \frac{ds}{dt}$$

Thus v_t is positive when s increases with time (counterclockwise rotation) and is negative when s decreases with time (clockwise rotation). With this definition, v_t has the same algebraic sign as ω_z. Further, since $s = R\theta$,

$$v_t = \frac{ds}{dt} = \frac{d(R\theta)}{dt} = R\frac{d\theta}{dt}$$

a

Or,

$$v_t = R\omega_z \tag{12-8}$$

Since particle P travels in a circle, its velocity $\mathbf{v}$ has a tangential component only and its speed v is $v = |\mathbf{v}| = |v_t| = R|\omega_z|$, or

$$v = R\omega \tag{12-9}$$

y
a_t
a
a_R
x

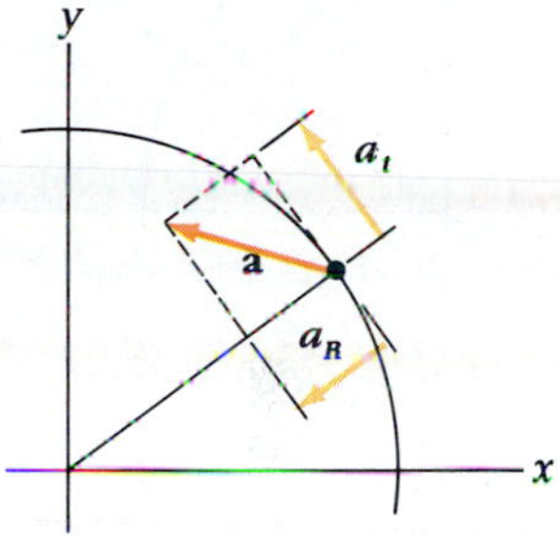

Figure 12-9. Tangential and radial components of the linear acceleration. The radial component $a_R = v^2/R$ corresponds to the projection of a along the radial line. This projection is always inward. The tangential component a_t corresponds to the projection of a tangent to the circle.

For a given angular speed ω, the linear speed of a particle is proportional to its distance R from the axis of rotation. Suppose the gate is rotating at an angular speed of 0.5 rad/s. A particle 0.2 m from the axis of rotation has a linear speed of (0.2 m)(0.5 rad/s) = 0.1 m/s; a particle 0.4 m from the axis of rotation has a linear speed of (0.4 m)(0.5 rad/s) = 0.2 m/s.

Now the linear acceleration $\mathbf{a}$ is resolved into tangential and radial components, as shown in Fig. 12-9. The tangential component a_t of the linear acceleration of a particle traveling in a circle is defined as the time derivative of the tangential component of the linear velocity:

$$a_t = \frac{dv_t}{dt} = \frac{d}{dt}(R\omega_z) = R\frac{d\omega_z}{dt}$$

Since $\alpha_z = d\omega_z/dt$,

$$a_t = R\alpha_z \tag{12-10}$$

Tangential acceleration component

Similar to v_t, a_t for the particle is proportional to its distance R from the axis of rotation.

Section 4-3 showed that a particle traveling in a circle at constant speed v has an acceleration of magnitude v^2/R directed toward the center of the circle (along a radial line). This is the centripetal acceleration. In the present case, where the speed may vary, the quantity v^2/R represents one component of the linear acceleration, the component that corresponds to the projection of $\mathbf{a}$ toward the center of the circle (Fig. 12-9). This component is called the *radial component* a_R of the linear acceleration, $a_R = v^2/R$. We can write a_R in terms of the angular speed ω by using Eq. 12-9 to substitute for v: $a_R = v^2/R = (R\omega)^2/R$, or

$$a_R = R\omega^2 \tag{12-11}$$

Radial acceleration component

As with v_t and a_t, a_R for a particle is proportional to the particle's distance R from the axis of rotation.

Since a_t and a_R are components of $\mathbf{a}$ along perpendicular directions (Fig. 12-9), the pythagorean theorem gives

$$a = \sqrt{a_t^2 + a_R^2}$$

Using Eqs. 12-10 and 12-11, we have $a_t^2 = R^2\alpha_z^2 = R^2\alpha^2$ and $a_R^2 = R^2\omega^4$, so

$$a = \sqrt{R^2\alpha^2 + R^2\omega^4} = R\sqrt{\alpha^2 + \omega^4}$$

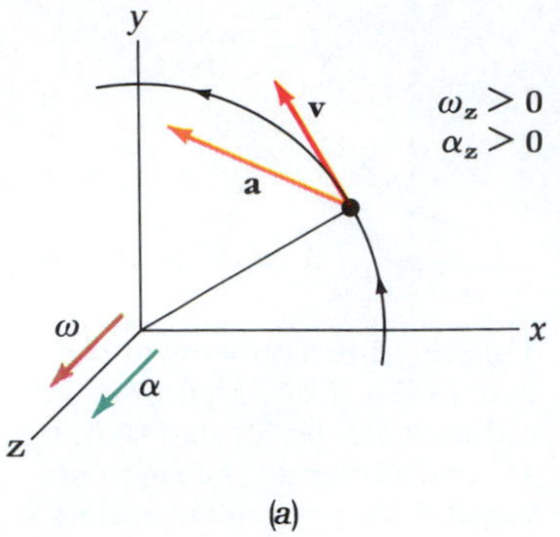

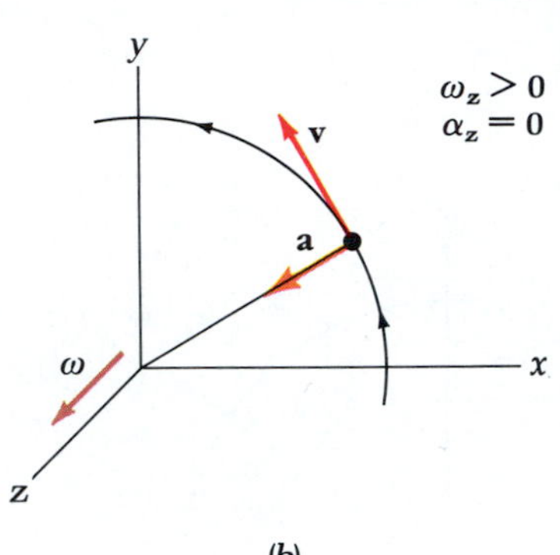

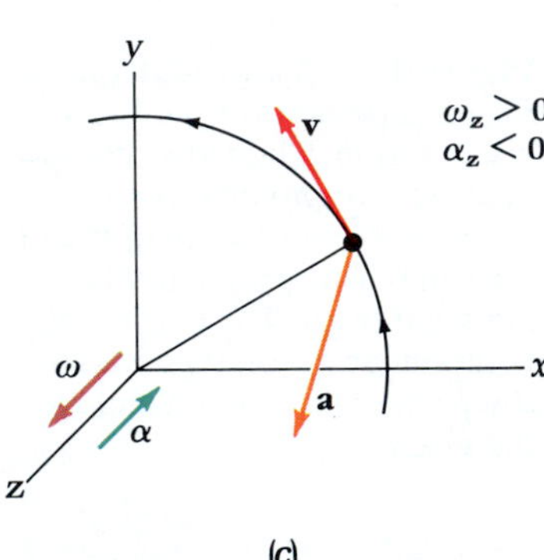

Figure 12-10. The directional relationships between **v**, **a**, $\boldsymbol{\omega}$, and $\boldsymbol{\alpha}$ for a particle in circular motion about the z axis when $\omega_z > 0$. *(a)* $\alpha_z > 0$; *(b)* $\alpha_z = 0$; *(c)* $\alpha_z < 0$.

Thus the magnitude a of the linear acceleration of a particle is proportional to the particle's distance R from the axis of rotation.

Figure 12-10 shows the linear velocity and linear acceleration for the case where ω_z (and v_t) is positive. In Fig. 12-10*a*, α_z (and a_t) is positive; in Fig. 12-10*b*, α_z (and a_t) is zero; and in Fig. 12-10*c*; α_z (and a_t) is negative. You may wish to draw the corresponding figures for the case where ω_z is negative.

When using Eqs. 12-8, 12-9, and 12-11, be certain that you insert the values of ω and ω_z in the SI unit radians per second, and not in degrees per second or revolutions per second. These equations are based on the expression $s = R\theta$. If the same unit is used to measure s and R, then θ is in radians. Similarly, when using Eq. 12-10, you should insert α_z in the SI unit radians per second squared.

EXAMPLE 12-5

Acceleration of a child riding a merry-go-round. A child is riding on a playground merry-go-round at a distance of 2.1 m from the vertical axis of rotation. At a particular instant the merry-go-round is rotating counterclockwise when viewed from above with an angular speed of 0.42 rad/s; this angular speed is decreasing such that $\alpha = 0.14$ rad/s². For the child, determine *(a)* v_t, *(b)* a_t, *(c)* a_R, and *(d)* a. Let the xy plane be horizontal, and let the z axis be along the axis of rotation with $+z$ upward.

Solution. *(a)* Since the merry-go-round is rotating counterclockwise, ω_z is positive: $\omega_z = +0.42$ rad/s. Thus

$$v_t = R\omega_z = (2.1 \text{ m})(0.42 \text{ rad/s}) = 0.88 \text{ m/s}$$

(b) To find a_t, we need α_z. We know that $\alpha = |\alpha_z| = 0.14$ rad/s², but we must find the sign of α_z. The angular speed ω is decreasing and ω_z is positive, so that ω_z is decreasing. Thus α_z is negative: $\alpha_z = -0.14$ rad/s². From Eq. 12-10,

$$a_t = R\alpha_z = (2.1 \text{ m})(-0.14 \text{ rad/s}^2) = -0.29 \text{ m/s}^2$$

(c) From Eq. 12-11,

$$a_R = \omega^2 R = (0.42 \text{ rad/s})^2(2.1 \text{ m}) = 0.37 \text{ m/s}^2$$

(d) The magnitude a of the child's acceleration is

$$a = R\sqrt{\alpha^2 + \omega^4} = (2.1 \text{ m})\sqrt{(0.14 \text{ rad/s}^2)^2 + (0.42 \text{ rad/s})^4}$$
$$= 0.47 \text{ m/s}^2$$

SELF-TEST 12-5. For the child in the previous example, draw sketches, similar to those in Fig. 12-10, showing the child's path, velocity, and acceleration. Is the angle between **a** and **v** less than or greater than 90°? ***ANSWER:*** Greater than 90°.

12-6 ROTATIONAL KINETIC ENERGY: MOMENT OF INERTIA

When a wheel rotates, there is kinetic energy associated with the rotation. The wheel consists of many small particles, and the kinetic energy of a particle i with mass m_i and speed v_i is $\frac{1}{2}m_i v_i^2$. The kinetic energy K of the entire wheel is the sum of the kinetic energies of all the particles that compose the wheel:

$$K = \Sigma \tfrac{1}{2} m_i v_i^2$$

Particles that are different distances from the axis of rotation have different linear speeds v_i, but, because the wheel is rigid, each particle has the same angular speed ω. (This fact is indicated by the lack of a subscript i on the ω.) Using $v_i = R_i\omega$, the kinetic energy of the wheel can be written as

$$K = \Sigma \tfrac{1}{2} m_i R_i^2 \omega^2$$

Two factors contained in this sum are the same for every term; they are the $\frac{1}{2}$ and the ω^2. Therefore these factors can be taken out of the summation:

$$K = \tfrac{1}{2}\omega^2(\Sigma m_i R_i^2)$$

The quantity $\Sigma m_i R_i^2$ is called the ***moment of inertia*** I:

$$I = \Sigma m_i R_i^2 \qquad (12\text{-}12)$$

Definition of the moment of inertia

The moment of inertia has the dimension [mass][length]^2, and its SI unit is kilogram meter squared ($kg \cdot m^2$). We shall discuss the moment of inerta in more detail in the next section.

In terms of the moment of inertia, the kinetic energy of a rotating object is

$$K = \tfrac{1}{2}I\omega^2 \qquad (12\text{-}13)$$

Rotational kinetic energy

If the expression for rotational kinetic energy is compared with $K = \frac{1}{2}mv^2$ for translation, then I is the rotational analog of the mass m, and ω is the rotational analog of the speed v. As a brief example, suppose a door, whose moment of inertia I about its hinges is $I = 8.2\ kg \cdot m^2$, is rotating with an angular speed of $\omega = 0.71$ rad/s. The door's rotational kinetic energy is

$$K = \tfrac{1}{2}(8.2\ kg \cdot m^2)(0.71\ rad/s)^2 = 2.1\ kg \cdot m^2/s^2 = 2.1\ J$$

12-7 MOMENT OF INERTIA

The moment of inertia is the rotational analog of mass. Mass is the property of an object that causes the object to resist a change in its velocity.

> The ***moment of inertia*** of an object about an axis is that property of the object that causes it to resist a change in its angular velocity about that axis.

The moment of inertia is sometimes called the ***rotational inertia.***

Some general features of the moment of inertia can be discovered by considering the array of eight particles shown in Fig. 12-11. Each particle has the same mass m, and each is held fixed relative to the others by rods of negligible mass. First we determine the moment of inertia about an axis through the center of the array and parallel to the side with dimension b, shown as the z axis in the figure:

$$I = \Sigma m_i R_i^2 = m_1 R_1^2 + m_2 R_2^2 + \cdots + m_8 R_8^2$$

Each particle is the same distance from the axis of rotation:

$$R = \sqrt{(\tfrac{1}{2}a)^2 + (\tfrac{1}{2}a)^2} = \frac{a}{\sqrt{2}}$$

Thus

$$I = m\left(\frac{a}{\sqrt{2}}\right)^2 + m\left(\frac{a}{\sqrt{2}}\right)^2 + \cdots + m\left(\frac{a}{\sqrt{2}}\right)^2 = 8m\left(\frac{a}{\sqrt{2}}\right)^2 = 4ma^2$$

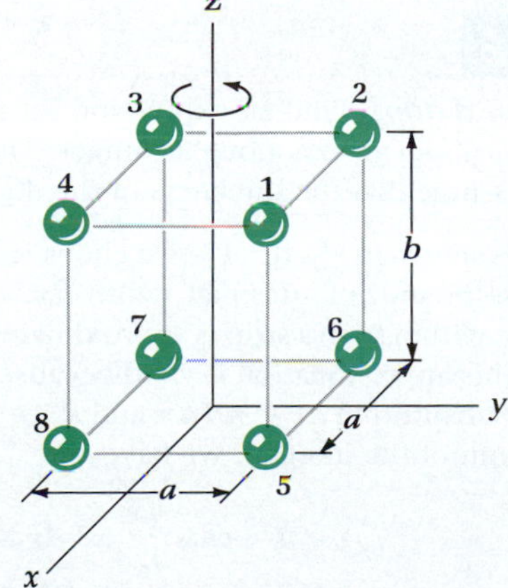

Figure 12-11. An array of eight particles, each with mass m, held together by rigid rods of negligible mass. The z axis passes through the center of the array and is parallel to the edges of length b. About the z axis, $I = 4ma^2$, and about an axis through particles 3 and 7, $I = 8ma^2$. The moment of inertia depends on how the mass is distributed relative to the axis of rotation.

Next consider the moment of inertia of this array about an axis parallel to the z axis that passes through particles 3 and 7. The distance from this axis to particles 1 and 5 is $R_1 = R_5 = \sqrt{a^2 + a^2} = \sqrt{2}a$; to particles 2, 4, 6, and 8 is $R_2 = R_4 = R_6 = R_8 = a$; and to particles 3 and 7 is $R_3 = R_7 = 0$. Thus

$$I = \Sigma m_i R_i^2 = 2m(\sqrt{2}a)^2 + 4m(a)^2 + 2m(0)^2 = 8ma^2$$

For rotation about the axis through the center, $I = 4ma^2$, but for rotation about the axis through particles 3 and 7, $I = 8ma^2$. Thus, unlike the mass, the moment of inertia is not an intrinsic property of a system. Rather, it depends on the mass of the system *and* on the location of the axis of rotation. Notice that I depends strongly on the distribution of mass perpendicular to the axis; if the distance a is doubled, I is quadrupled. However, I is independent of the mass distribution parallel to the axis; it is independent of the distance b in the array in Fig. 12-11.

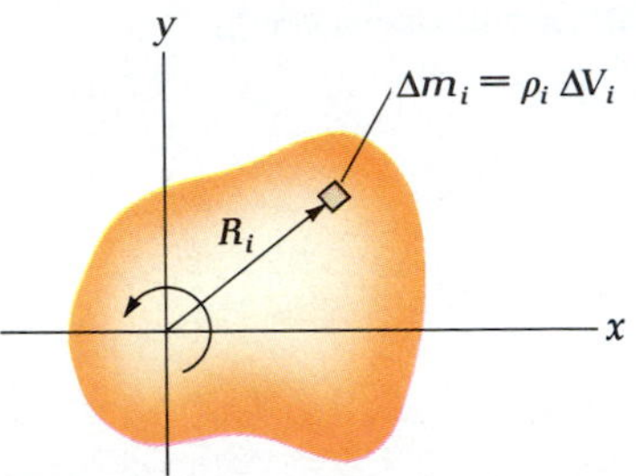

Figure 12-12. Moment of inertia of a continuous object about the z axis. The $+z$ direction is out of the page.

Moment of Inertia of a Continuous Object

In finding the moment of inertia of a continuous object, such as a pulley or a wheel, imagine the object to consist of many small pieces, each small enough to be considered a particle (Fig. 12-12). The mass of a piece is $\Delta m_i = \rho_i \Delta V_i$, where ρ_i is the mass density of the material and ΔV_i is the small volume occupied by the piece. Thus

$$I = \Sigma\, \Delta m_i R_i^2 = \Sigma \rho_i\, \Delta V_i R_i^2$$

As the volume ΔV_i approaches an infinitesimal dV, the sum transforms into an integral:

$$I = \lim_{\Delta V_i \to 0} \Sigma \rho_i\, \Delta V_i R_i^2 = \int_V \rho R^2\, dV$$

Thus the moment of inertia of a continuous object is

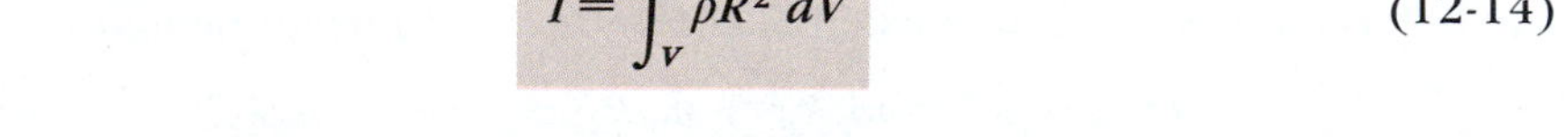

$$I = \int_V \rho R^2\, dV \qquad (12\text{-}14)$$

where the integration is over the volume of the object. If the density is uniform, then ρ can be taken out of the integral:

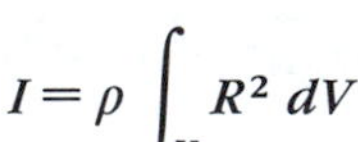

$$I = \rho \int_V R^2\, dV$$

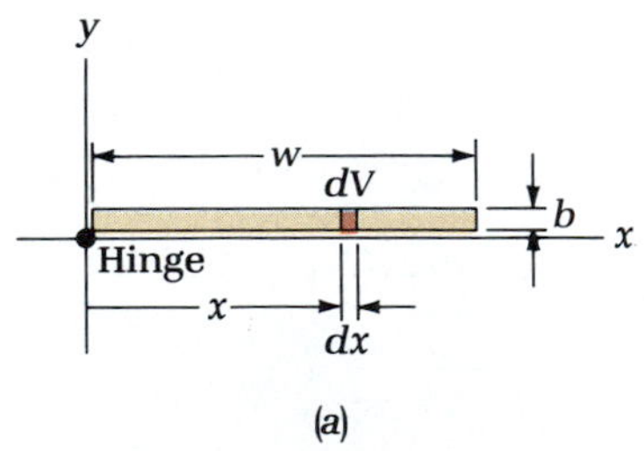

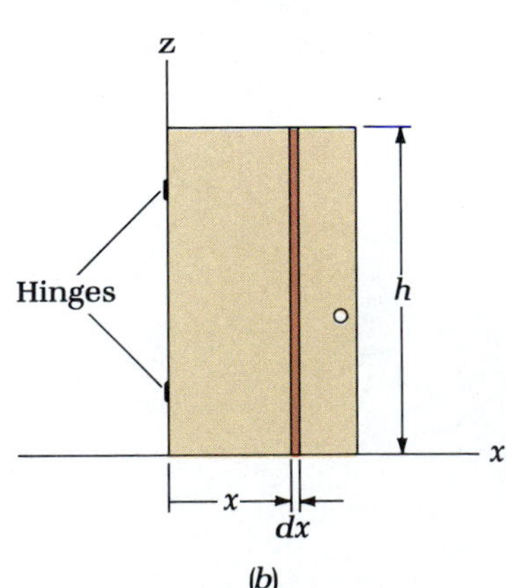

Figure 12-13. Example 12-6: *(a)* Top view of the door. The $+z$ direction is out of the page. *(b)* Front view of the door. Since $w \gg b$, nearly all the mass in dV is the same distance from the axis.

EXAMPLE 12-6

Moment of inertia of a door. Find an expression for the moment of inertia of a door of uniform mass density ρ about an axis along its hinges. The door has mass M, height h, width w, and thickness b. Assume that the thickness of the door is much smaller than its width.

Solution. The door is shown in Fig. 12-13. We choose as our element of volume a thin strip of height h, thickness b, and infinitesimal width dx. This element of volume is chosen because all of the mass within such a strip is approximately the same distance x from the axis through the hinges. This approximation is valid because the door's thickness is small compared with its width. Substituting $dV = hb\, dx$, and $R^2 = x^2$ into Eq. 12-14 and factoring the constants ρ, h, and b out of the integral, we have

$$I = \rho h b \int_0^w x^2\, dx$$

Performing the integration gives

$$I = \frac{\rho hbw^3}{3}$$

Since $\rho = M/V = M/hbw$, we have $\rho hbw = M$. Thus

$$I = \frac{Mw^2}{3}$$

The height of the door does not appear in the final answer. Thus this answer is valid for a door of any height. For that matter, it is also the expression for the moment of inertia of a stick of length w about an axis perpendicular to the stick and passing through one end.

SELF-TEST 12-6. Evaluate the moment of inertia of a uniform door for rotation about an axis through its hinges. The door has a mass of 27.3 kg and a width of 0.95 m. ***ANSWER:*** 8.2 kg · m².

EXAMPLE 12-7

Moment of inertia of a hollow cylinder. Find an expression for the moment of inertia of a hollow right circular cylinder for rotation about its symmetry axis. Let the cylinder have inner radius R_1, outer radius R_2, height h, and uniform mass density ρ, as shown in Fig. 12-14.

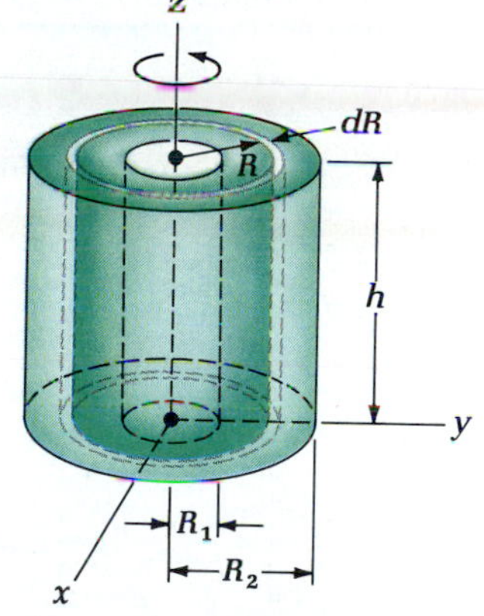

Figure 12-14. Example 12-7: Calculating the moment of inertia of a hollow cylinder. The mass within the volume element $dV = 2\pi Rh\, dR$ is essentially all at a distance R from the z axis. All the mass of the cylinder is contained between the values R_1 and R_2 of the integration variable R.

Solution. Choose as the element of volume a thin cylindrical shell of height h, circumference $2\pi R$, and infinitesimal thickness dR (Fig. 12-14), $dV = h2\pi R\, dR$. This element is chosen because all the mass within such an element is essentially the same distance from the axis of rotation. Substituting this value of dV into Eq. 12-14 and factoring constants out of the integral gives

$$I = 2\pi\rho h \int_{R_1}^{R_2} R^3\, dR$$

The limits on the integral are determined by the values of R within which the mass is contained. Performing the integration gives

$$I = \tfrac{1}{2}\pi\rho h(R_2^4 - R_1^4)$$

To express this in terms of the mass of the cylinder, rewrite I as

$$I = \tfrac{1}{2}\pi\rho h(R_2^2 - R_1^2)(R_2^2 + R_1^2)$$

The volume V of the hollow cylinder is the product of its height times the area of its base:

$$V = h\pi(R_2^2 - R_1^2)$$

Its mass M is

$$M = \rho V = \rho h\pi(R_2^2 - R_1^2)$$

Therefore

$$I = \tfrac{1}{2}M(R_2^2 + R_1^2)$$

SELF-TEST 12-7. A thick-walled pipe has an outer radius of 65 mm, an inner radius of 42 mm, and a mass of 0.28 kg. Determine the moment of inertia of the pipe for rotation about its symmetry axis. ***ANSWER:*** 8.4×10^{-4} kg · m².

The moment of inertia for objects with two other shapes can be found from the answer to the above example.

1. The moment of inertia of a solid cylinder of radius R_0 for rotation about its symmetry axis is found by setting the inner radius R_1 equal to zero and setting the outer radius R_2 equal to R_0. This gives

$$I = \tfrac{1}{2}MR_0^2$$

This equation is also valid for a disk-shaped object such as a phonograph record (neglecting the hole) and for a long thin object such as a log with uniform radius.

2. The moment of inertia of a thin cylindrical shell of radius R_0 for rotation about its symmetry axis is found by using the approximation that $R_1 \approx R_0$ and $R_2 \approx R_0$. This gives

$$I = MR_0{}^2$$

This equation is valid both for a hula hoop and for a thin-walled pipe.

Table 12-2 gives moments of inertia for objects with various shapes about axes that pass through the center of mass.

TABLE 12-2. ***Some Representative Moments of Inertia.*** *In each case the density is uniform and the axis of rotation is through the center of mass. (a) Thin-walled cylinder. (b) Thin-walled cylinder. (c) Thin-walled hollow sphere. (d) Solid cylinder. (e) Solid cylinder. (f) Solid sphere. (g) Thick-walled hollow cylinder. (h) Long, thin rod. (i) Rectangular plate.*

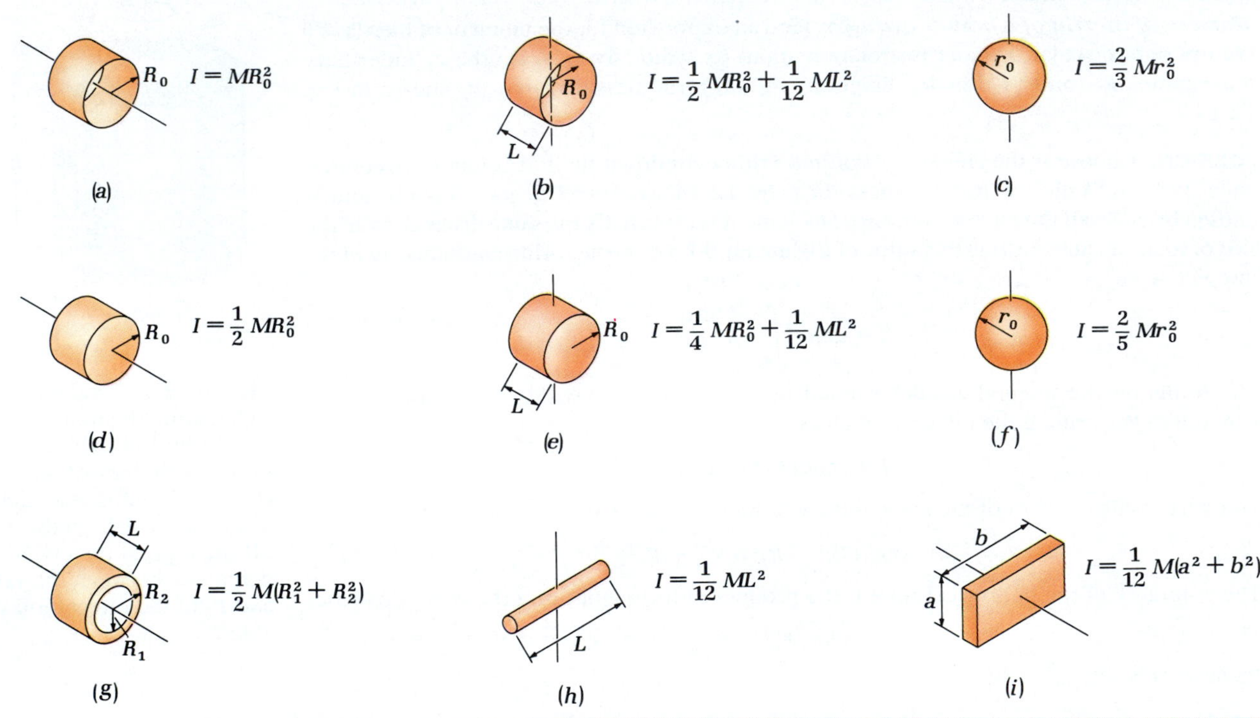

Parallel-Axis Theorem

Calculating the moment of inertia of an object about an axis that is not an axis of symmetry can be very complex. Fortunately, there is an easily applicable theorem that helps do this, the *parallel-axis theorem.* This theorem provides a relation between the moment of inertia I_P about an axis through an arbitrary point P and the moment of inertia I_{cm} about a parallel axis through the object's center of mass.

Consider the moment of inertia of the arbitrarily shaped object shown in Fig. 12-15 about an axis that passes through the point P. In this figure, the z axis passes through the center of mass and is parallel to the rotational axis through P. Since the z axis passes through the center of mass, $x_{cm} = 0$ and $y_{cm} = 0$. The distance between the two parallel axes is $d = \sqrt{a^2 + b^2}$. From Eq. 12-12, the moment of inertia I_P about an axis through P is

$$I_P = \Sigma m_i R_{Pi}{}^2 = \Sigma m_i[(x_i - a)^2 + (y_i - b)^2]$$

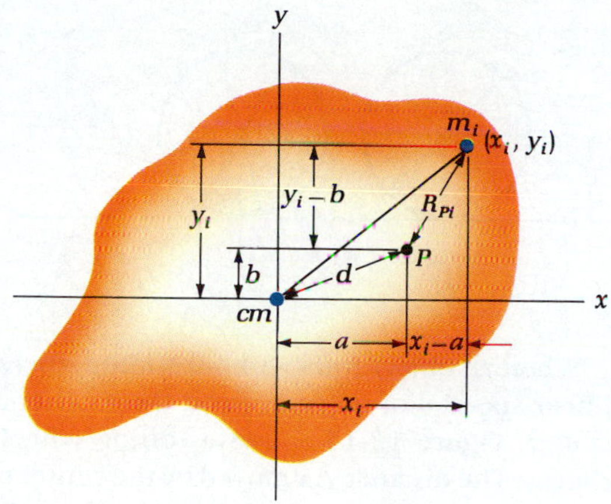

Figure 12-15. The parallel-axis theorem. The z axis passes through the center of mass and is parallel to the rotational axis that passes through point P (the $+z$ direction is out of the page). The moment of inertia I_P about the axis through P is $I_P = I_{cm} + Md^2$, where d is the distance between the axes.

where R_{Pi} is the distance from the axis (through P) to particle i. Multiplying out the squared terms gives

$$I_P = \Sigma m_i(x_i^2 + y_i^2) - 2a\Sigma m_i x_i - 2b\Sigma m_i y_i + (a^2 + b^2)\Sigma m_i$$

where the quantities that are the same in each term of the sums have been factored out of the summations. In the above equation, the term $\Sigma m_i(x_i^2 + y_i^2)$ is the moment of inertia I_{cm} of the object about the axis through the center of mass (the z axis) because x_i and y_i are measured relative to the axis through the center of mass:

$$I_{cm} = \Sigma m_i(x_i^2 + y_i^2)$$

From Sec. 10-1, the center of mass is defined such that $\Sigma m_i x_i = Mx_{cm}$ and $\Sigma m_i y_i = My_{cm}$. This means that the choice of coordinate frame causes the second and third terms to be zero: $\Sigma m_i x_i = Mx_{cm} = 0$ and $\Sigma m_i y_i = My_{cm} = 0$. The fourth term, $(a^2 + b^2)\Sigma m_i$, equals Md^2 because $\Sigma m_i = M$ and $a^2 + b^2 = d^2$. Therefore

$$I_P = I_{cm} + Md^2 \qquad (12\text{-}15)$$

The parallel-axis theorem

which is the ***parallel-axis theorem.***

EXAMPLE 12-8

Moment of inertia of a rod about one end. Table 12-2 gives the moment of inertia of a uniform thin rod about an axis through the center of mass and perpendicular to the rod's long axis. Determine the moment of inertia of the rod about an axis parallel to the axis described above and passing through a point P at one end, as shown in Fig. 12-16.

Solution. Using Eq. 12-15, we have

$$I_P = I_{cm} + Md^2 = \frac{ML^2}{12} + M\left(\frac{L}{2}\right)^2 = \frac{ML^2}{3}$$

Compare this result with the solution in Example 12-6.

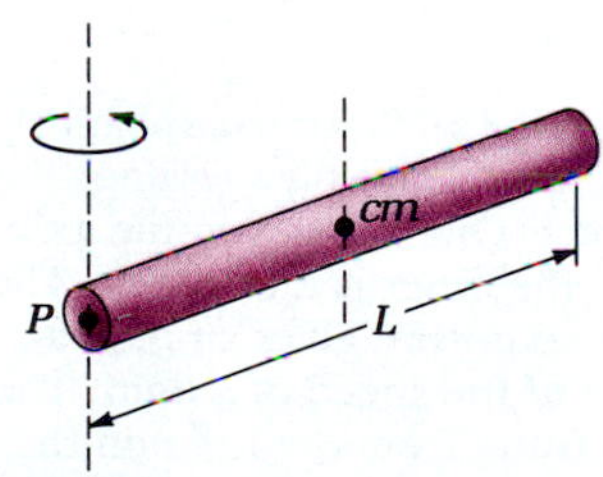

Figure 12-16. Example 12-8: Using the parallel-axis theorem to find the moment of inertia of a uniform thin rod about an axis perpendicular to the rod's length and passing through one end (P).

SELF-TEST 12-8. Table 12-2*d* gives the moment of inertia of a uniform solid cylinder for rotation about its symmetry axis as $I = \frac{1}{2}MR_0^2$. Use the parallel-axis theorem to find an expression for the moment of inertia of a solid cylinder for rotation about an axis parallel to the symmetry axis and at a distance $\frac{1}{2}R_0$ from the symmetry axis. ***ANSWER:*** $\frac{3}{4}MR_0^2$.

12-8 ROLLING OBJECTS

So far we have considered objects that rotate about a fixed axis only. The following discussion deals with an object that rolls, such as a wheel or a ball. For simplicity, the object has a circular cross section and rolls without sliding along a straight line. In this case the axis of rotation undergoes translation but maintains a fixed orientation.

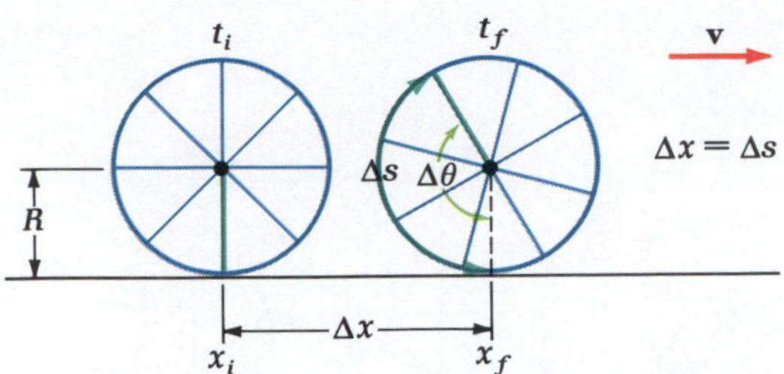

Figure 12-17. Rolling without sliding. In the time interval Δt, the center of the wheel moves a distance Δx, which is equal to the distance Δs that the point of contact moves along the wheel's edge: $\Delta x = \Delta s$. Thus $v = \Delta x/\Delta t = \Delta s/\Delta t = R\,\Delta\theta/\Delta t = R\omega$.

When an object rolls without sliding, there is a simple connection between the linear speed v of its center and the rotational speed ω about an axis through its center. Figure 12-17 shows a rolling wheel of radius R at a time t_i and a later time t_f. The distance Δx moved by the center of the wheel in a given time interval Δt is equal to the distance Δs moved by the point of contact along the wheel's edge: $\Delta x = \Delta s$. If v is the linear speed of the center of the wheel and ω is the angular speed about the axis of rotation, then

$$v = \frac{\Delta x}{\Delta t} = \frac{\Delta s}{\Delta t} = R\frac{\Delta\theta}{\Delta t} = R\omega$$

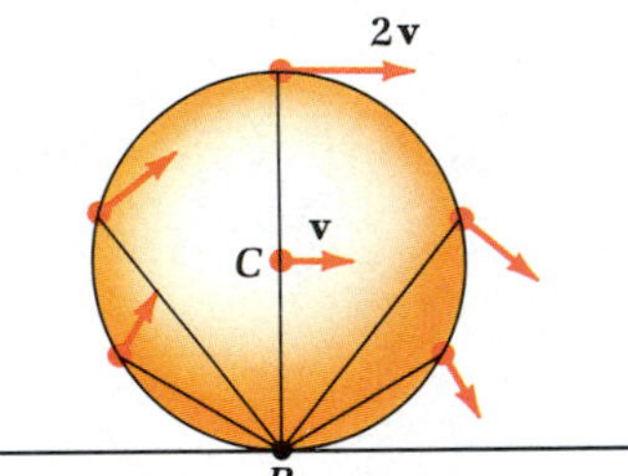

Figure 12-18. A rolling object at any instant is rotating about an axis through the point of contact P. The linear velocity of any particle is perpendicular to a line from P to the particle, and the particle's linear speed is proportional to the length of that line.

For rolling without sliding, we have

$$v = R\omega \tag{12-16}$$

Now we find the kinetic energy of a rolling object. Figure 12-18 shows that such an object rotates about an axis that passes through the point of contact P between the object and the surfaces on which it is rolling. This axis has a fixed orientation parallel to the surface and perpendicular to the direction of motion. Since no sliding occurs, the point of contact P is *instantaneously* at rest, and the axis of rotation of the object instantaneously passes through that point (Fig. 12-19). In Fig. 12-18, the velocities of several points on the object are shown. The velocities are directed perpendicular to a line from P and have magnitudes proportional to the lengths of those lines. For example, the velocity of the point at the top of the object is $2v$, twice the velocity of the center.

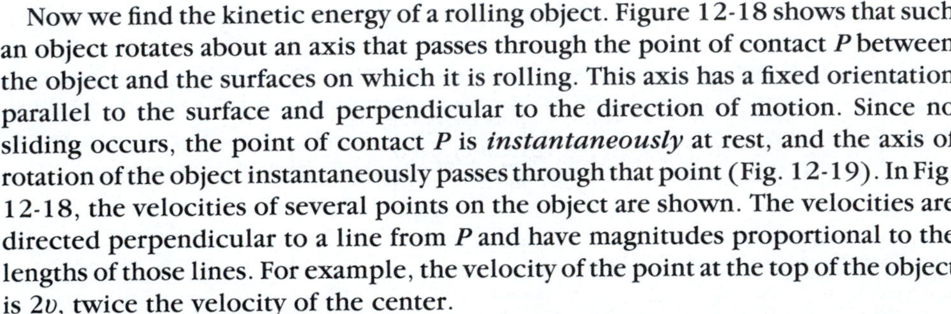

Note that the angular speed ω_P about an axis through P is equal to the angular speed ω_C about an axis through C, because the speed of C relative to P is $\omega_P R$ and

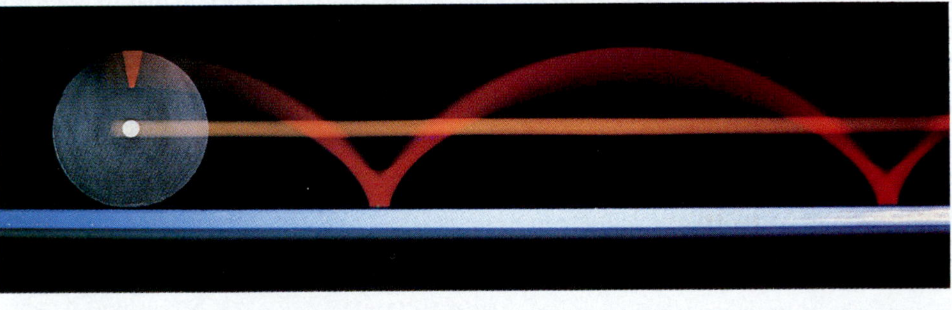

Figure 12-19. Time-exposure photo of lights on a rolling wheel. One light is on the axle and the other is on the rim. The film exposure gives an indication of the speed of a light. The exposure from the light on the axle is uniform, showing that this point moves at constant speed. The light on the rim exposes the film the greatest when the light is near the surface (at the bottom) and the least when it is farthest from the surface (at the top). This is because this point slows as it approaches the surface, stops instantaneously, and then reverses its direction. Also, this point is traveling fastest when it is at the top. The curve traced out by the light on the rim is called a cycloid.

the speed of P relative to C is $\omega_C R$. Thus $\omega_P R = \omega_C R$, and consequently the subscripts are dropped: $\omega = \omega_P = \omega_C$.

The kinetic energy of the object is

$$K = \tfrac{1}{2} I_P \omega^2$$

where I_P is the moment of inertia of the object about an axis perpendicular to the direction of motion, parallel to the surface, and through P. If the object has a symmetric distribution of mass about an axis through C, then C corresponds to the center of mass, and the parallel-axis theorem gives

$$K = \tfrac{1}{2}(I_{cm} + Md^2)\omega^2$$

where I_{cm} is the moment of inertia of the object about an axis perpendicular to the direction of motion, parallel to the surface, and through C. The distance d between these two axes is the radius R. Since the object rolls without sliding, $v = \omega R$, or $\omega = v/R$. This gives $K = \frac{1}{2}I_{cm}\omega^2 + \frac{1}{2}M(R)^2(v/R)^2$, or

$$K = \tfrac{1}{2}I_{cm}\omega^2 + \tfrac{1}{2}Mv^2 \qquad (12\text{-}17)$$

Kinetic energy of a rolling object

Thus the kinetic energy of a rolling object can be expressed as the sum of two terms—one term corresponds to rotation about the center of mass, and the other corresponds to translation of the center of mass.

The velocity of a particle in a rolling object may also be regarded as being the result of a combination of pure translation and pure rotation of the object, as illustrated in Fig. 12-20. In particular, notice how the velocity of the particle at point A on the rolling object is the vector sum of the velocities at A due to pure translation and pure rotation.

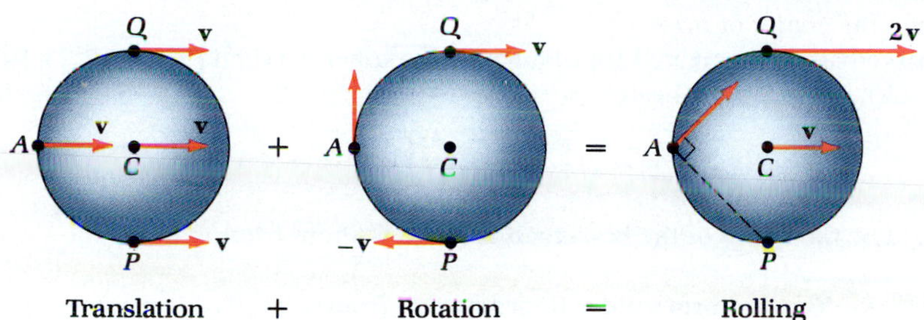

Figure 12-20. The linear velocity of a particle of a rolling object may be regarded as a combination of a linear velocity that is the same as that of the axis (pure translation) and a linear velocity due to the particle's rotation about the axis (pure rotation).

Equation 12-17 is valid when the object rolls with partial sliding, even though it was developed assuming no sliding. However, in this case $v \neq \omega R$. The only restriction on Eq. 12-17 is that the axis of rotation must maintain a fixed orientation.

EXAMPLE 12-9

Speed of an object rolling down a slope. *(a)* Develop an equation for the speed of an object that rolls from rest down a hill without sliding, as shown in Fig. 12-21. Assume that nonconservative forces, such as air resistance, are negligible. *(b)* Using the answer from part *(a),* find the speed of a basketball after it rolls from rest down a hill. The ball's vertical drop is $h = 2.3$ m.

Solution. *(a)* Since nonconservative forces are negligible, we use conservation of mechanical energy (Sec. 9-4):

$$K_f + U_f = K_i + U_i$$

$$\tfrac{1}{2}Mv_f^2 + \tfrac{1}{2}I_{cm}\omega_f^2 + Mgy_f = \tfrac{1}{2}Mv_i^2 + \tfrac{1}{2}I_{cm}\omega_i^2 + Mgy_i$$

Since the object starts from rest, $v_i = 0$ and $\omega_i = 0$. Also, we let $v_f = v$, $\omega_f = \omega$, and $y_i - y_f = h$. Rearranging terms gives

$$Mgh = \tfrac{1}{2}Mv^2 + \tfrac{1}{2}I_{cm}\omega^2$$

This equation states that the decrease in potential energy, as the object rolls down the hill, is

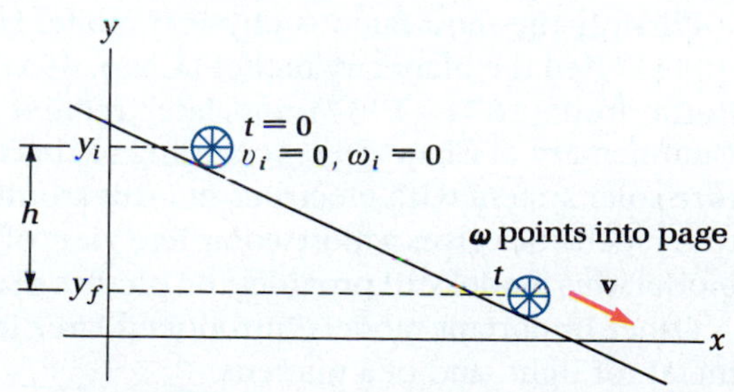

Figure 12-21. Example 12-9: An object rolling from rest down a straight slope.

equal to its increase in kinetic energy. Substituting $\omega = v/R$ and solving for v, we obtain

$$v = \sqrt{\frac{2gh}{1 + (I_{cm}/MR^2)}}$$

From Sec. 3-5, the speed of a particle after falling a vertical distance h from rest, neglecting friction, is $v = \sqrt{2gh}$. Similarly, this is the expression for the speed of a block sliding down a slope without friction. Therefore, since part of the kinetic energy of a rolling object is rotational kinetic energy, its speed after a given vertical drop is less than that of an object that has only a translational part to its kinetic energy. The amount that the speed is decreased due to the rolling depends on the ratio I_{cm}/MR^2: The larger this ratio, the smaller will be the speed for a given vertical drop h. For the objects listed in Table 12-2 that have spherical or cylindrical symmetry, this ratio is independent of both M and R, because I_{cm} about the symmetry axes of these objects is proportional to MR^2. Therefore the speed of such a rolling object is independent of either its mass or its radius, but depends only on how its mass is distributed about an axis perpendicular to the direction of motion, parallel to the surface, and through the center of mass.

(b) A basketball can be treated approximately as a spherical shell (Table 12-2). Substituting $I_{cm} = 2Mr_0^2/3 = 2MR^2/3$ gives

$$v = \sqrt{\frac{6gh}{5}} = \sqrt{\frac{6(9.8\ \text{m/s}^2)(2.3\ \text{m})}{5}} = 5.2\ \text{m/s}$$

What would be the speed of the basketball if it slid without friction down the slope?

SELF-TEST 12-9. A uniform solid cylinder starting from rest rolls down a slope. *(a)* Write an expression for the cylinder's speed v at the instant its vertical height has decreased an amount h. *(b)* Evaluate v when $h = 0.50$ m. ***ANSWERS:*** *(a)* $v = \sqrt{4gh/3}$; *(b)* 2.6 m/s².

COMMENTARY: The Use of Models in Physics

The word "model" evokes a vision of a dollhouse or a scaled-down replica of an airplane. But to a physicist, a model is an idealized mental picture of a physical system or natural phenomenon. Several models have already been used in this book. For example, in the discussion of projectile motion in Sec. 4-2 a ball launched into the air was treated as a particle and air resistance was neglected. In Chap. 7, a model of the solar system was developed, while in this chapter, a rolling ball was treated as a rigid object and dissipative effects were neglected.

A good model has three desirable attributes: simplicity, agreement with experiment, and generality. To ensure simplicity, we comply with a rule of parsimony: ***In developing a model, use the simplest conceivable assumptions that are consistent with observation and logic.*** Therefore a model builder must discard all nonessential details from the description of a phenomenon—details that would cause the calculations to become unwieldy or overly complex. The key to constructing a good model is determining which details are nonessential.

A model is created by using approximations that are valid within some realm. Ordinarily, the validity of a model begins to break down as the model is extended beyond the realm of its origin. Indeed, a model's usefulness depends greatly on how far it can be extended.

Possibly the most famous physical model is the Bohr model of the atom, sometimes called the planetary model (Chap. 39). This model was suggested by Ernest Rutherford (1871–1937) and later refined by Niels Bohr (1885–1962) (see Commentary of Chap. 39). According to this model, an atom is similar to a miniature solar system with electrons in orbit around a central nucleus. Although quantum mechanics gives a more complete view of atoms than is possible with the Bohr model, this model still provides the picture of an atom in the minds of most people.

Other important models introduced later in this book are models of a gas, of a metal, of light, and of a nucleus.

SUMMARY

Section 12-1. Translation and Rotation of a Rigid Object

In translational motion, all the particles in a rigid object have the same displacement in the same time interval. In rotational motion, all the particles in a rigid object execute circular motion about the axis of rotation.

Section 12-2. Angular Measurement

The SI unit of angle, the radian (rad), is defined as $\theta = s/R$, where s and R are measured with the same length unit.

Section 12-3. Angular Coordinate, Velocity, and Acceleration

Angular kinematical quantities are the angular coordinate θ, the angular velocity component $\omega_z = d\theta/dt$, and the angular acceleration component $\alpha_z = d\omega_z/dt$. The right-hand rule is used to define the positive sense for rotation.

Section 12-4. Kinematics of Rotation about a Fixed Axis

The kinematics of a rigid object rotating about a fixed axis is analogous to that of a particle moving in a straight line. The equations that describe constant angular velocity and constant angular acceleration are listed in Table 12-1.

Section 12-5. Relations between Rotational and Translational Quantities

For a particle in a rotating rigid object, the relation between the linear and angular velocity components are

$$v_t = R\omega_z \tag{12-8}$$

The linear acceleration components are

$$a_t = R\alpha_z \tag{12-10}$$

$$a_R = R\omega^2 \tag{12-11}$$

Section 12-6. Rotational Kinetic Energy: Moment of Inertia

The rotational kinetic energy of a rigid object is

$$K = \tfrac{1}{2}I\omega^2 \tag{12-13}$$

where the moment of inertia is

$$I = \Sigma m_i R_i^2 \tag{12-12}$$

Section 12-7. Moment of Inertia

The moment of inertia of a continuous object is

$$I = \int_V \rho R^2\, dV \tag{12-14}$$

The moment of inertia of an object about an axis parallel to an axis through the center of mass is given by the parallel-axis theorem:

$$I_P = I_{cm} + Md^2 \tag{12-15}$$

Section 12-8. Rolling Objects

The axis of rotation of a rolling object undergoes translation. The kinetic energy of a rolling object can be written

$$K = \tfrac{1}{2}I_{cm}\omega^2 + \tfrac{1}{2}Mv^2 \tag{12-17}$$

QUESTIONS

1. When is it particularly useful to measure angles in radians? When is the angular unit of measure unimportant?

2. Are all angular measurements dimensionless, regardless of the units?

3. What is the path of a particle in a rigid object rotating about a fixed axis?

4. How would you define a fixed axis of rotation of a rigid object in terms of the motion of the particles that compose the object?

5. If a rigid object has only translational motion (for example, the body of a car traveling in a straight line on a flat road), are there any points within the object that always have the same velocity as the center of mass? If so, which ones?

6. If a rigid object has only rotational motion about a fixed axis (for example, a swinging door), are there any points within the object that always have the same velocity as the center of mass? If so, which ones?

7. If a rigid object moves with both translation and rotation about an axis with a fixed orientation (for example, a rolling wheel), are there any points within the object that always have the same velocity as the center of mass? If so, which ones?

8. If a rigid object moves with both translation and rotation about an axis that is not fixed (for example, a football during a wobbly pass), are there any points within the object that always have the same velocity as the center of mass? If so, which ones?

9. Suppose you have a meter stick and a dime. How would you measure the angle subtended by the moon at the earth? Does it require less arithmetic to find the angle in degrees or in radians? Which angular unit would you say is easier to use in this case?

10. If we imagine looking down on the solar system from Polaris (the North Star), both the earth's orbital motion around the sun and its rotation on its axes are in a counterclockwise sense. When is your linear speed with respect to the sun the greater, at night or during the day?

11. What is the direction of the angular velocity of a rigid object rotating about a fixed axis? What is the direction of the linear velocity of a particle in a rigid object rotating about a fixed axis?

12. A rigid object rotating about a fixed axis has nonzero angular velocity and angular acceleration. Particle A in the object is twice as far from the axis of rotation as particle B. What is the ratio of the following quantities for A and B: *(a)* the angular speeds, *(b)* the linear speeds, *(c)* the magnitudes of the angu-

lar accelerations, *(d)* the tangential components of the accelerations, *(e)* the radial components of the accelerations, *(f)* the magnitudes of the linear accelerations?

13. Do the angular velocities of the hands of a wall clock point into the wall or out of the wall? At the instant the clock is unplugged, do the angular accelerations of the hands point into the wall or out of the wall?

14. Consider a particle on the end of the second hand of a wall clock at the instant the hand passes the 12 just after the clock is plugged in (such that the hand's angular speed is increasing). Let the origin of coordinates be at the center of the clock, with the z axis pointing out of the wall along the axis of rotation, the x axis pointing horizontally toward the 3, and the y axis pointing vertically upward toward the 12. Which cartesian components of the linear velocity and linear acceleration of the particle are zero? What is the algebraic sign of the nonzero components? Which cartesian components of the angular velocity and angular acceleration are zero? What is the algebraic sign of the nonzero components?

15. A car is moving forward and slowing down. Is the direction of the angular velocity of the wheels toward the driver's left or right? What is the direction of the angular acceleration of the wheels?

16. Can an object have more than one moment of inertia? Other than an object's shape and mass, what information must be given to find its moment of inertia?

17. One side of a door (Fig. 12-22) is made of material with a larger mass density than the other side. To minimize the moment of inertia about an axis of rotation along the hinges, should the hinges be placed at the heavier side or the lighter side? Explain.

Figure 12-22. Question 17: Top view of a door.

18. Consider three rods made of the same material and with the same length and mass, but with different cross-sectional shapes (Fig. 12-23). Which of the three has the largest moment of inertia about an axis through the center of mass and along the rod's long axis? Which rod has the smallest moment of inertia about the axis?

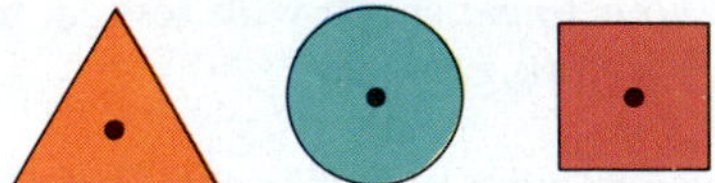

Figure 12-23. Question 18: Cross section of three rods.

19. Is it possible to find an axis of rotation (call the axis A) about which the moment of inertia for an object is smaller than the moment of inertia about an axis through the center of mass and parallel to A?

20. Suppose you allow the following objects, starting from rest, to roll down the same straight slope at the same instant: a basketball, a billiard ball, a hollow can with the ends removed, a soccer ball, and a bowling ball. Which do you expect will reach the bottom first? Which do you expect will reach the bottom last? Do you expect any ties? Assume that mechanical energy is conserved.

21. Suppose you are designing a cart for coasting down a hill. To maximize your coasting speed, should you design the wheels so that their moments of inertia about their rotation axes are large or small, or does it matter? Keeping the moment of inertia of the wheels fixed, will the cart's speed be increased or decreased by increasing the mass of the cart's body? Assume that mechanical energy is conserved.

22. Two similar barrels A and B with the same radius and mass roll down the same slope starting at the same instant. Barrel A is filled with liquid water, and barrel B is filled with ice of equal mass. (Barrel B is slightly longer than barrel A to compensate for the slightly larger density of liquid water.) Which barrel reaches the bottom of the slope first? Explain.

23. In three separate experiments, we roll the same ball down three different slopes that are shaped as shown in Fig. 12-24. The ball starts from rest at point P and rolls without sliding to point Q, which is the same vertical distance h below P in each case. For which slope—*(a)*, *(b)*, or *(c)*—is the ball's speed v the greatest as the ball passes Q? For which is it least? Assume that only conservative forces do work.

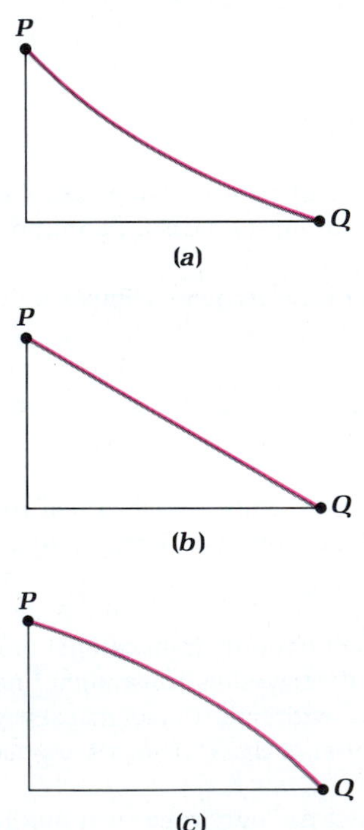

Figure 12-24. Questions 23 and 24.

24. For which slope in the previous question—*(a)*, *(b)*, or *(c)*—is the ball's travel time Δt between P and Q the greatest? For which is it least? Assume that only conservative forces do work. (Newton showed that the curve from P to Q that corresponds to the minimum travel time Δt is a cycloid.)

25. Complete the following table:

Symbol	Represents	Type	SI Unit
θ			
ω		Vector	
ω_z		Component	
α_z			
v_t			m/s
a_t			
a_R			
I	Moment of inertia		

EXERCISES

Section 12-2. Angular Measurement

1. Determine the angle between the 12 and the 4 on the face of a clock in *(a)* radians, *(b)* degrees, and *(c)* revolutions.

2. The diameter of the sun subtends an angle of 0.53° at the earth. *(a)* Convert this angle to radians. *(b)* Given that the radius of the earth's orbit is 1.5×10^{11} m, estimate the diameter of the sun.

3. The knob of a door is 0.84 m from an axis through the hinges. *(a)* Determine the distance the knob moves when the door rotates through an angle of 35°. *(b)* Determine the distance the knob moves when the door rotates through an angle of 0.61 rad.

4. When a dime (diameter = 18 mm) is held at a distance of 2 m from one's eye, it just barely obscures one's view of the moon. What is the angle subtended by the moon's diameter at the earth in *(a)* radians and *(b)* degrees? *(c)* During a lunar eclipse, the shadow of the earth is cast on the moon. From observations of such eclipses, the ancient Greeks knew that the moon's diameter is about $\frac{1}{4}$ that of the earth's. Using this information, estimate the earth-moon distance in units of earth radii.

5. Determine the distance the earth's center of mass travels relative to the sun in 1 day. The radius of the earth's orbit is 1.5×10^{11} m.

Section 12-3. Angular Coordinate, Velocity, and Acceleration

6. In the following list of angular coordinates, which, if any, corresponds to the same angular position? 48.69 rad, −27.38 rad, 36.20 rad, 67.54 rad.

7. Find the angular speed of a 78-rev/min turntable in radians per second.

8. A Ferris wheel rotating steadily makes 1 rev every 7.6 s. The riders face west and are moving forward when at the top. What is the wheel's angular velocity (magnitude and direction)?

9. Suppose $\theta(t) = -(1.4 \text{ rad/s}^3)t^3 + (6.8 \text{ rad})$. Find *(a)* $\omega_z(t)$; *(b)* $\omega_z(2.1 \text{ s})$; *(c)* $\omega(2.1 \text{ s})$; *(d)* $\theta(2.1 \text{ s})$.

10. Suppose $\omega_z(t) = (2.3 \text{ rad/s}^4)t^3 - (7.5 \text{ rad/s}^3)t^2$. Find *(a)* $\alpha_z(t)$; *(b)* $\alpha_z(1.6 \text{ s})$; *(c)* $\alpha(1.6 \text{ s})$; *(d)* $\omega_z(1.6 \text{ s})$; *(e)* $\omega(1.6 \text{ s})$.

Section 12-4. Kinematics of Rotation about a Fixed Axis

11. Write an equation for $\theta(t)$ for a phonograph turntable rotating steadily at 33 rev/min. Use the ordinary coordinate frame with $+\mathbf{k}$ upward and the xy plane horizontal, and let $\theta_0 = 0$.

12. A gear begins rotating from rest with a constant angular acceleration of magnitude 0.21 rad/s². What is its angular speed after completing *(a)* 1 rev, *(b)* 2 rev, *(c)* 4 rev?

13. A wheel rotates about a fixed horizontal axis that is aligned east-west. If the $+\mathbf{k}$ direction is toward the west, the wheel's angular velocity component is given by

$$\omega_z(t) = 5.8 \text{ rad/s} - (2.2 \text{ rad/s}^2)t$$

(a) What is the angular acceleration component? *(b)* Write an equation for $\theta(t)$, with θ_0 set equal to zero. *(c)* Find the time t_q at which the angular velocity is zero. *(d)* What is the direction of a particle's linear velocity at the top of the wheel before t_q and after t_q. *(e)* Write an equation for ω_z^2 as a function of θ.

14. A wheel with a fixed axis of rotation along the vertical rotates with constant angular acceleration. At $t = 0$ the wheel is rotating at 17 rev/s in the counterclockwise direction when viewed from above, and 6.2 s later it is rotating at 11 rev/s in the counterclockwise direction when viewed from above. Develop equations for *(a)* $\omega_z(t)$ and *(b)* $\theta(t)$ using a coordinate frame with the z axis as the axis of rotation and $+\mathbf{k}$ upward. *(c)* Develop an equation for ω_z^2 as a function of θ. Use SI units.

15. The front door of a north-facing house swings inward, and the hinges are on the west side of the door frame. Let $+\mathbf{k}$ point upward along the hinges and let $\theta = 0$ correspond to the closed position. The door is opened starting from rest such that its

angular acceleration is constant. At the instant the opening angle is 0.72 rad, the angular speed is 1.4 rad/s. Develop equations for *(a)* $\omega_z(t)$, *(b)* $\theta(t)$, and *(c)* ω_z^2 as a function of θ.

16. Develop Eq. 12-7 by elminating the time t between Eqs. 12-5 and 12-6.

Section 12-5. Relations between Rotational and Translational Quantities

17. The tip of the second hand on a clock is 93 mm from the axis of rotation. What is the linear speed of the tip?

18. *(a)* Determine the earth's angular speed about the sun in radians per second. *(b)* Determine the linear speed of the center of the earth relative to the sun. The earth-sun distance is 1.5×10^{11} m.

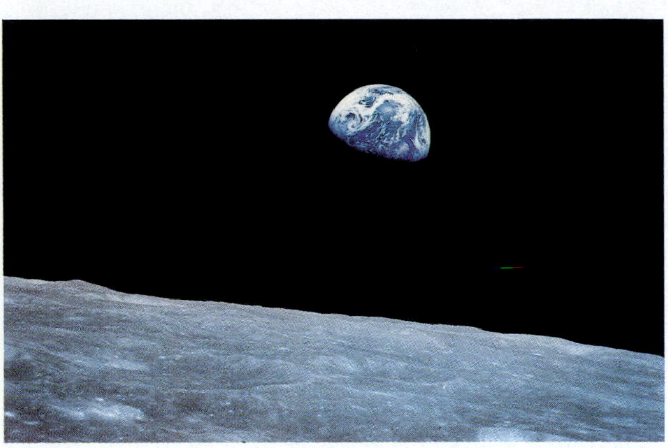

19. A child riding on a merry-go-round is 1.4 m from the axis of rotation. *(a)* Determine the magnitudes of the tangential and radial components of the child's linear acceleration at the instant the merry-go-round's angular speed is 0.34 rad/s and its angular acceleration has a magnitude of 0.18 rad/s². *(b)* What is the child's linear speed and the magnitude of his linear acceleration.

20. Wheels A and C are connected by belt B which does not slip (Fig. 12-25); $R_A = 250$ mm and $R_C = 410$ mm. Find the angular speed of wheel A at the instant the angular speed of wheel C is 1.7 rad/s.

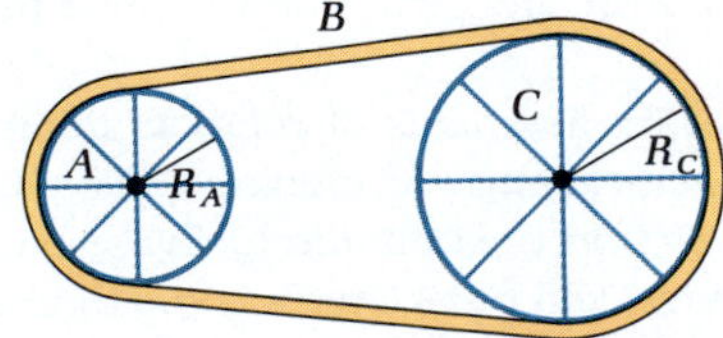

Figure 12-25. Exercise 20: Wheels A and C connected by belt B.

Section 12-6. Rotational Kinetic Energy: Moment of Inertia

21. Show that the quantity $\frac{1}{2}I\omega^2$ has *(a)* the dimension of energy and *(b)* the SI unit of joule (J).

22. The moment of inertia of a grindstone about its axis of rotation is 0.11 kg·m². *(a)* What is the kinetic energy of the grindstone when rotating at an angular speed of 28 rad/s? *(b)* What is the kinetic energy if the speed is doubled to 56 rad/s?

23. A wheel rotating about a fixed axis has a kinetic energy of 29 J when its angular speed is 13 rad/s. What is the wheel's moment of inertia about the axis of rotation?

24. A stone of mass m (=0.68 kg) is tied to a string of negligible mass and swung in a horizontal circle of radius R (=0.84 m) such that the period of the motion is T (=0.39 s). Determine the kinetic energy of the stone two ways: *(a)* Treat the stone as a particle of mass m and linear speed v $(=2\pi R/T)$. *(b)* Treat the stone as a rotating system of moment of inertia I $(=mR^2)$ and angular speed ω $(=2\pi/T)$. *(c)* Compare your answers.

Section 12-7. Moment of Inertia

25. Consider the array of four particles shown in Fig. 12-26. The particles are contained in the xy plane and are connected by rods of negligible mass. Determine the moment of inertia about *(a)* the x axis (I_x), *(b)* the y axis (I_y), and *(c)* the z axis (I_z). (Notice that $I_z = I_x + I_y$. This equation is called the *plane-figure theorem.* As the name implies, the theorem is valid only if the system is of negligible thickness so that it can be contained in a plane, such as the xy plane. The array in Fig. 12-26 satisfies this criterion and obeys the plane-figure theorem. See Prob. 6.)

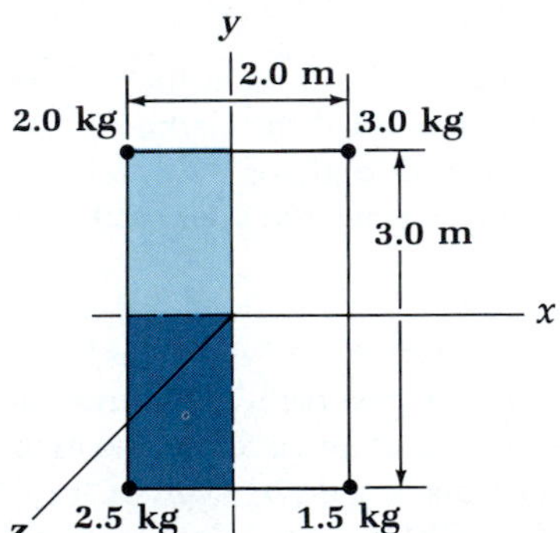

Figure 12-26. Exercise 25: The particles are contained in the xy plane, and the rectangle is centered at the origin.

26. Estimate the moment of inertia of a tennis ball for rotation about a diameter. A tennis ball has a mass of 0.070 kg, an outer radius of 32 mm, and a thickness of 5 mm.

27. A door has a height of 2.1 m, a width of 1.1 m, a thickness of 42 mm, and a uniform density of 0.88×10^3 kg/m³. What is the moment of inertia of the door about an axis along its hinges?

28. *(a)* Estimate the moment of inertia about a vertical axis through the center of mass of a man standing erect with arms at his sides. The man is 1.8 m tall and has a mass of 73 kg. Treat the man as a right circular cylinder with density 1.0×10^3 kg/m³. *(b)* Estimate the man's moment of inertia about the same axis if he holds his arms out horizontally; treat his arms as thin rods.

29. Find the moment of inertia of the thin, triangular-shaped slab of uniform density shown in Fig. 12-27 about the z axis in terms of its mass M and dimensions a and b.

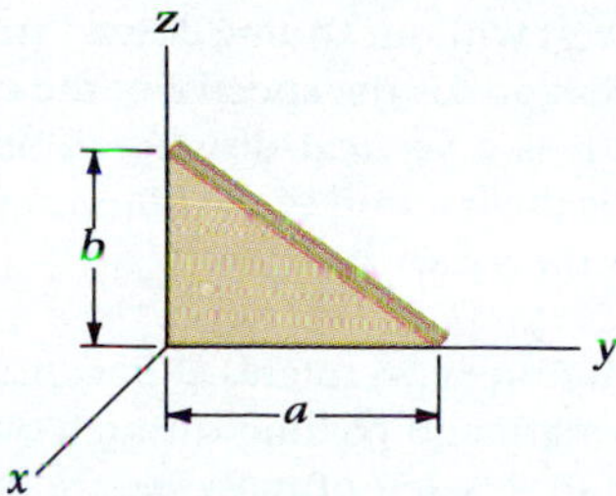

Figure 12-27. Exercise 29.

30. Find the moment of inertia of the solid right circular cylinder of uniform density shown in Fig. 12-28 about the z axis in terms of its mass M and radius R.

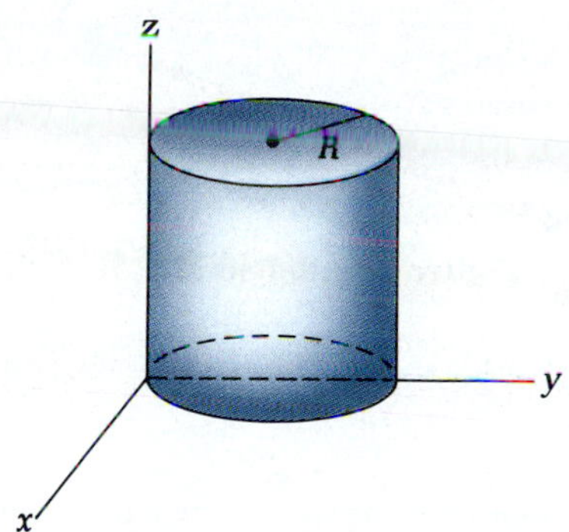

Figure 12-28. Exercise 30.

31. Using Table 12-2, find the moment of inertia of a solid sphere of uniform density, mass M, and radius r_0 about an axis that passes a distance of $\frac{1}{2}r_0$ from the center. Give your answer in terms of M and r_0.

32. *(a)* Determine the mass density of the earth ($m_e = 6.0 \times 10^{24}$ kg, $R_e = 6.4$ Mm), assuming that its density is uniform, and compare your answer with the average density of rocks on the earth's surface ($\rho = 2.7 \times 10^3$ kg/m^3). *(b)* Estimate the earth's moment of inertia about an axis through its center by assuming it has a uniform mass density. *(c)* Based on your answer to part *(a)*, is your estimate in part *(b)* an overestimate or an underestimate? Explain.

Section 12-8. Rolling Objects

33. A wheel of radius 340 mm rolls in a straight line without sliding. At the instant the center of the wheel has a linear speed of 1.4 m/s, determine *(a)* the wheel's angular speed about its center and *(b)* the linear speed of a particle at the top of the wheel.

34. *(a)* What is the kinetic energy of an 8.2-kg bowling ball that is rolling without sliding when the linear speed of its center is 1.7 m/s? *(b)* What fraction of the kinetic energy corresponds to rotation about the center of mass and what fraction corresponds to translation of the center of mass?

35. Find *(a)* the linear speed and *(b)* the angular speed of a hollow cylinder that rolls from rest a distance of 6.7 m down a straight slope that makes an angle of 12° with the horizontal. The cylinder has an outer radius of 96 mm, an inner radius of 75 mm, and a mass of 0.83 kg. Neglect dissipative effects.

36. The following objects are rolling without sliding. In each case, find the ratio of the rotational kinetic energy about the center to the total kinetic energy: *(a)* a hollow cylinder, *(b)* a solid cylinder, *(c)* a hollow sphere, *(d)* a solid sphere. Assume uniform densities.

37. Estimate the kinetic energy of a thrown Frisbee immediately after release. What fraction of your estimate is due to its spinning about its center of mass?

Additional Exercises

38. Our solar system orbits the center of the Milky Way galaxy at a radius of about 2.2×10^{20} m and with a period of about 1.7×10^8 years. What is the solar system's speed in this Milky Way reference frame?

39. A wheel is rotating with an angular speed of 39 rad/s and is slowing with a constant angular acceleration of magnitude 0.63 rad/s^2. *(a)* How long will it take the wheel to come to rest? *(b)* Through how many radians will the wheel rotate in this time?

40. A disk is rotating about the z axis such that

$$\alpha_z = (0.24 \text{ rad/s}^3)t - 0.89 \text{ rad/s}^2$$

(a) Given that $\omega_{z0} = 3.1$ rad/s, determine the expression for $\omega_z(t)$. *(b)* With $\theta_0 = 2.7$ rad, what is the equation for $\theta(t)$?

41. A flywheel is rotating about the z axis according to the relation

$$\theta(t) = 4.2 \text{ rad} - (2.9 \text{ rad/s})t + (0.31 \text{ rad/s}^3)t^3$$

Determine *(a)* $\omega_z(t)$ and *(b)* $\alpha_z(t)$.

42. Consider a reference frame with its origin at the earth's center, with its z axis along the earth's rotational axis, and with its x and y axes fixed relative to distant stars. Relative to this frame, the earth's rotational period T_e is 23 h 56 min. *(a)* Write an expression for the speed v of a point on the earth's surface at latitude λ in terms of T_e, λ, and the earth's radius R_e. *(b)* Write an expression for the acceleration magnitude a of such a point. *(c)* Evaluate v and a at $\lambda = 45°$. *(d)* Sketch a graph of v and a versus λ from $\lambda = 0°$ to $\lambda = 90°$.

43. Consider a reference frame with the center of the sun at the origin, with the z axis perpendicular to the plane of the earth's orbit, and with the x and y axes fixed relative to distant stars. Relative to this frame, what is the earth's *(a)* angular speed, *(b)* linear speed, and *(c)* linear acceleration magnitude?

44. A typical neutron star, or pulsar, has a radius of a few kilometers, a mass about the same as the sun, and a very large angular speed. Estimate the rotational kinetic energy of a neutron star that has a period of 50 ms.

45. An airplane propeller measures 3.2 m from tip to tip and has a mass of 35 kg. What is the propeller's rotational kinetic energy when rotating at 1000 rev/min?

46. Consider the moment of inertia I of a uniform cube of mass M and dimension ℓ along each edge. *(a)* Write an expression for I for rotation about an axis parallel to a cube edge and passing through the center. (*Hint:* See Table 12-2.) *(b)* Write an expression for I for rotation about an axis along a cube edge.

47. A wheel is rolling without sliding in a straight line along a level surface (Fig. 12-29). Let $+\mathbf{i}$ be horizontal in the direction of motion of the center, and let $+\mathbf{j}$ be vertically upward. Consider an instant when the velocity of the center of the wheel is $(2.0\ \text{m/s})\mathbf{i}$. *(a)* What is the velocity of the point T on the wheel's edge at that instant? *(b)* What is the velocity of the point F on the wheel's edge at that instant?

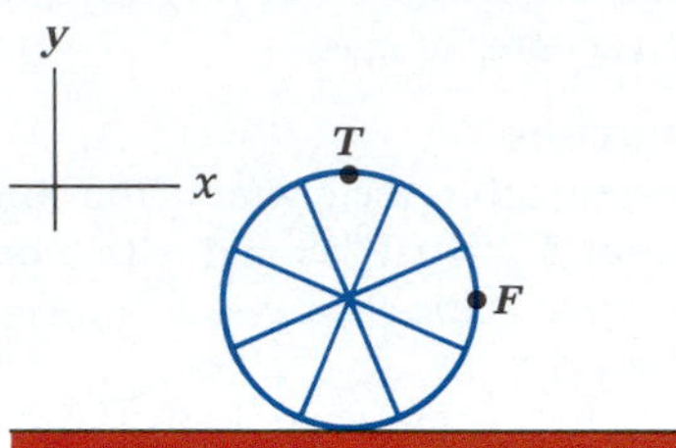

Figure 12-29. Exercise 47.

48. Estimate the moment of inertia of a 5.8-kg automobile tire that has an outer radius 0.31 m.

49. A thick-walled cylinder whose inner radius is half its outer radius rolls from rest without sliding down a straight slope. *(a)* Develop an expression for the speed v of the cylinder's axis at an instant when it is a vertical distance h below its starting point. *(b)* What is the fraction of the kinetic energy that is due to rotation about the axis?

50. A rod of negligible mass rotates about an axis perpendicular to the rod's length and passing through one end (Fig. 12-30). Spheres 1 and 2, each of mass m, are fixed on the rod. Sphere 1 is a distance ℓ from the axis, and sphere 2 is a distance 2ℓ from the axis. *(a)* Determine an expression for the moment of inertia of this arrangement in terms of m and ℓ. *(b)* What fraction of the moment of inertia is due to sphere 2?

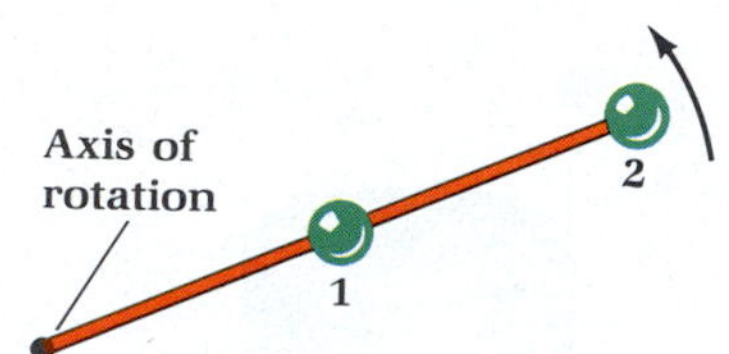

Figure 12-30. Exercise 50.

PROBLEMS

1. ***Constant angular acceleration.*** A wheel rotating about the z axis with constant angular acceleration has the following angular coordinates: $\theta = 0.0$ at $t = 0$ s; $\theta = -3.8$ rad at $t = 1.0$ s; and $\theta = -5.0$ rad at $t = 2.0$ s. *(a)* Find α_z and ω_{z0}. *(b)* Write equations for the $\omega_z(t)$ and $\theta(t)$.

2. ***Moment of inertia of a solid cone.*** Consider the moment of inertia I of a uniform, solid cone for rotation about its symmetry axis (Fig. 12-31). The volume of the cone is $\pi R_0^2 h/3$.

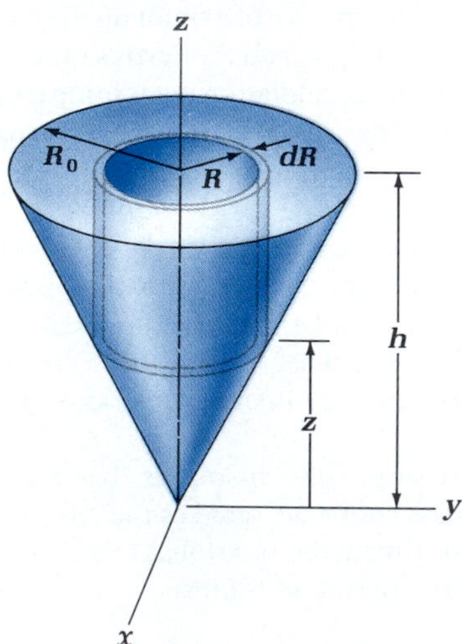

Figure 12-31. Problem 2.

(a) Show that the volume dV of the element in the figure is $2\pi R\ dR(h - z)$. Note that all parts of this element are essentially the same distance from the axis. *(b)* Show that $I = 3MR_0^2/10$. (*Hint:* Both z and R are variables in the moment-of-inertia integral and $z/h = R/R_0$.)

3. ***Kinetic energy of the earth.*** Consider dividing the earth's kinetic energy relative to the sun into two parts: one part for the orbital motion of the center of mass and the other part for the rotation about its axis. Calculate and compare these two energies, assuming the earth's mass density is uniform. The earth's mass and radius are 6.0×10^{24} kg and 6.4×10^6 m, respectively. The radius of the earth's orbit is 1.5×10^{11} m.

4. ***A hinged plank as it falls.*** One end of a uniform plank of length L is held fixed by a frictionless hinge (Fig. 12-32). The plank is given a slight push from the vertical position such that it swings down under the influence of gravity. At the instant the plank's orientation passes through the horizontal, find *(a)* its angular speed, *(b)* the linear speed of its center, *(c)* the radial component of the linear acceleration of its center, *(d)* the linear speed of its free end. Find your answers in terms of L and g.

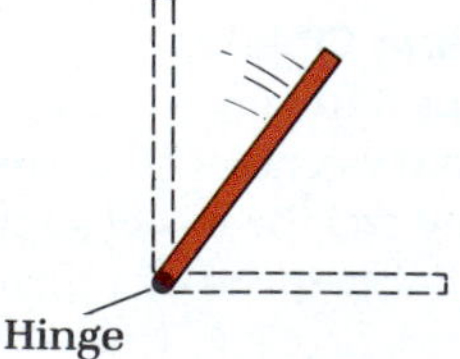

Figure 12-32. Problem 4.

5. ***Radius of gyration.*** The radius of gyration k of an object about an axis is defined as

$$k = \sqrt{\frac{I}{M}}$$

where I is the moment of inertia of the object about the axis and M is its mass. That is, if the entire mass of the object were concentrated at the distance from the axis given by the radius of gyration, the resulting moment of inertia would be the same as that of the object. *(a)* Determine the radii of gyration for the objects with the shapes and axes given in Table 12-2. *(b)* Write the equation for the speed of the rolling object in Example 12-9 in terms of its radius of gyration rather than its mass and moment of inertia.

6. ***Plane-figure theorem.*** Consider an object of negligible thickness contained in the *xy* plane, as shown in Fig. 12-33. Show that

$$I_z = I_x + I_y$$

where I_x, I_y, and I_z are the moments of inertia of the object about the *x*, *y*, and *z* axes, respectively.

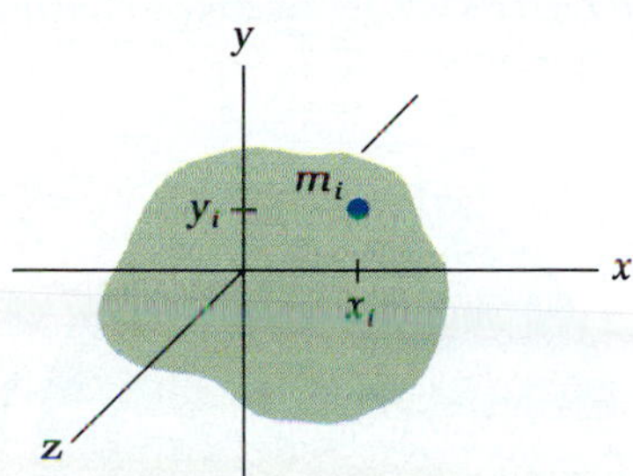

Figure 12-33. Problem 6.

7. ***Moment of inertia of a thin disk about a diameter.*** Find the moment of inertia of a thin circular disk with uniform density about an axis along a diameter in terms of the disk's mass and radius. Do the calculation two ways: *(a)* by direct integration of Eq. 12-14 and *(b)* by using the plane-figure theorem (Prob. 6).

8. ***Using the parallel-axis theorem and the plane-figure theorem.*** Recall that the moment of inertia of a uniform door of height *h*, width *w*, and mass *M* (neglecting its thickness) about an axis along the hinges is $Mw^2/3$. Use the parallel-axis theorem and the plane-figure theorem (Prob. 6) to find the door's moment of inertia about an axis through its center and perpendicular to its face. (See Table 12-2.)

9. ***A rolling sphere.*** A uniform sphere of radius 26 mm and mass 0.175 kg rolls from rest without sliding down a straight slope. After the sphere has undergone a vertical drop of 130 mm, the linear speed of its center is 1.3 m/s. What is the moment of inertia of the sphere about a diameter?

10. ***Designing a soapbox-derby cart.*** Let M_b be the mass of the cart's body, and let *m* be the mass of each wheel so that the cart's total mass is $M = M_b + 4m$. In a soapbox derby, the cart starts from rest. We wish to maximize the coasting speed *v* of the cart after it has undergone a vertical drop *h*. *(a)* Develop an expression for *v* in terms of *h*, *g*, and the ratio m/M. Neglect dissipative effects, assume the wheels roll without sliding, and treat each wheel as a uniform disk. Evaluate *v* when $h = 5.0$ m and m/M is *(b)* 0.05 and *(c)* 0.15. If *M* is fixed, is it better to have more mass in the body and less in the wheels, or is it better to have more mass in the wheels and less in the body?

11. ***A falling block rotates a cylinder.*** A block of mass *m* is tied to a string of negligible mass that is wrapped around a uniform cylinder of mass *M* and radius R_0 (Fig. 12-34). The

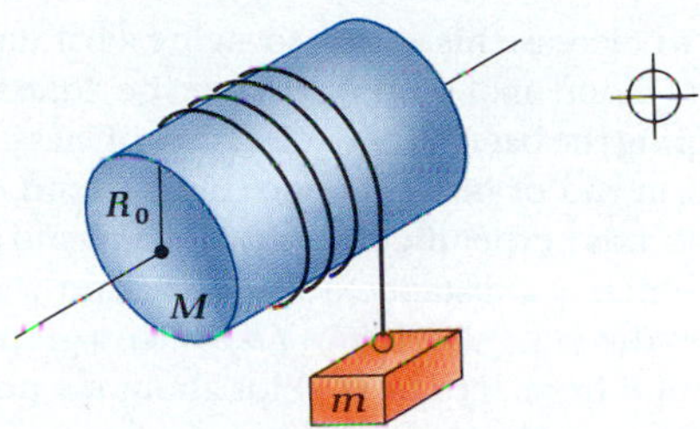

Figure 12-34. Problem 11.

cylinder is free to rotate, with negligible friction, about a fixed horizontal axis through its center. After the block has dropped a vertical distance *h* from rest, find *(a)* the linear speed of the center of the block and *(b)* the angular speed of the cylinder about its axis of rotation.

12. ***A falling yo-yo.*** Estimate the linear speed of the center of a yo-yo of mass *M* and radius R_2 after it has fallen from rest a distance *h* while attached to a string (Fig. 12-35). The string is of negligible mass and thickness and is wound around an axle of radius R_1. Neglect the moment of inertia of the axle, and treat the body of the yo-yo as a uniform disk. Compare your answer for the case where $R_2/R_1 = 5$ with the speed of free-fall from the same height.

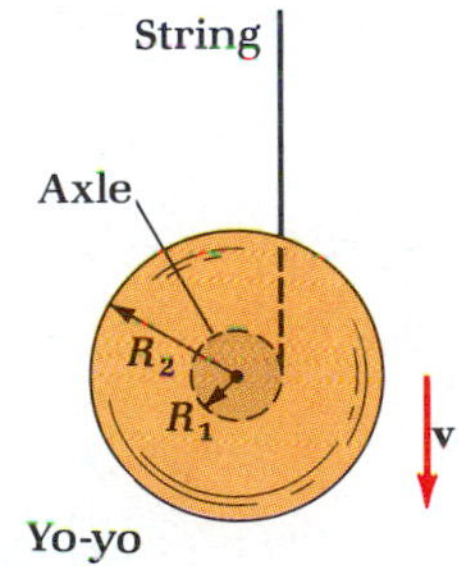

Figure 12-35. Problem 12.

13. ***Roller-coaster car in a loop.*** Consider a roller-coaster car at the instant it is halfway up a loop, as shown in Fig. 12-36. The linear velocity of the car is 13 m/s upward, the radius of the loop is 5.3 m, and frictional forces, which tend to reduce the car's mechanical energy, are negligible. Determine the magnitude and direction of the car's *(a)* angular velocity, *(b)* angular acceleration, *(c)* linear acceleration.

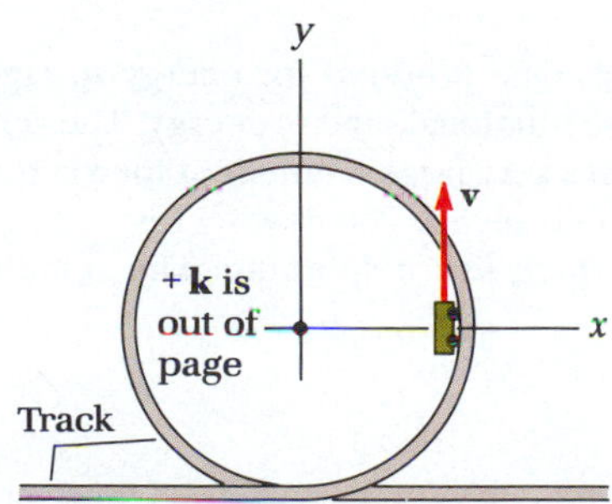

Figure 12-36. Problem 13.

14. ***Choking up on a bat.*** In the next chapter we shall show that the moment of inertia is a measure of the resistance of an object to a change in its angular velocity. A way for a baseball

player to increase his ability to swing a bat rapidly is to reduce the bat's moment of inertia. This can be done by "choking up," or grasping the bat nearer to its center of mass. *(a)* Treat a bat as a uniform rod of mass M and length ℓ, and consider rotation about an axis perpendicular to the length and passing through a point which is a distance h from one end (Fig. 12-37). Show that $I = M(\ell^2/3 - \ell h + h^2)$. *(b)* Make a graph of I versus h for values of h from zero to $\frac{1}{2}\ell$. Plot about six points and sketch a curve through them.

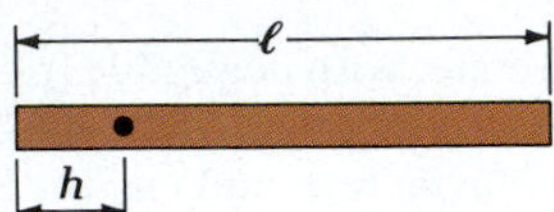

Figure 12-37. Problem 14.

15. ***Marble in a rotating bowl.*** Suppose a hemispherical bowl of radius r_0 is rotating with constant angular velocity ω about its vertical axis of symmetry (Fig. 12-38). If a marble placed on the inside surface of the bowl is at rest relative to the surface at a distance R from the axis, the marble remains at rest relative to the surface. Show that

$$R = \sqrt{r_0^2 - \frac{g^2}{\omega^4}}$$

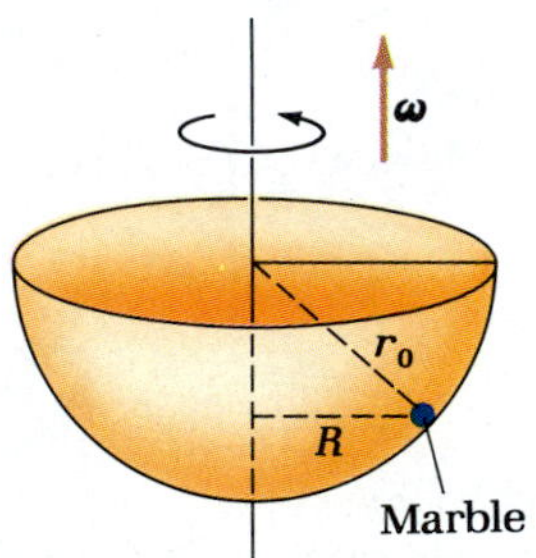

Figure 12-38. Problem 15.

16. ***Moment of inertia of an array of particles.*** In Fig. 12-11, find an expression for the moment of inertia of the array about an axis that passes through particles 3 and 8.

17. ***Moment of inertia of an array of particles II.*** In Fig. 12-11, find an expression for the moment of inertia of the array about an axis that passes through particles 2 and 8.

18. ***Fill 'er up.*** One proposal for energy storage to "fuel" a vehicle is to use rotational kinetic energy. The vehicle contains a flywheel with a very large moment of inertia that is designed to rotate at a high speed. The flywheel is "charged" at an energy station, which sets it spinning. The vehicle then drives about by gradually using energy tapped from the flywheel. *(a)* What is the energy contained in a flywheel that is a uniform cylinder of mass 400 kg and radius 1.0 m when it is rotating at 500 rad/s? *(b)* What is the distance such a vehicle could travel if it requires 250 J to move 1.0 m?

19. ***Artificial gravity.*** One possible problem for space travelers is the deleterious effects on the body from the lack of a gravitational force. A solution is to design a spaceship that supplies an artificial "gravity" by rotating. Suppose the spaceship is a hollow cylinder in which the space travelers walk on the inner walls (Fig. 12-39). If the spaceship's inner radius is 100 m, what must the angular speed ω of the spaceship be such that the space traveler's weight relative to the inner wall is the same as her earth weight?

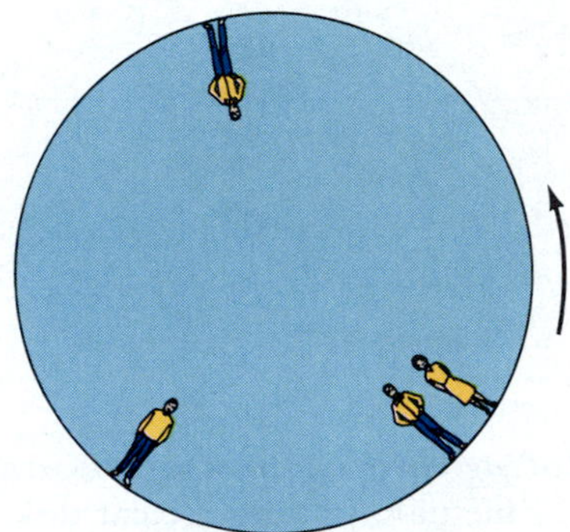

Figure 12-39. Problem 19.

20. ***The Atwood machine revisited.*** An Atwood machine has blocks of mass m_1 and m_2 ($m_2 > m_1$); a pulley of radius R_0, mass M and moment of inertia $\frac{1}{2}MR_0^2$ about its axis; and a string of negligible mass (Fig. 12-40). The system is released from rest, and friction in the bearings is negligible. Show that the expression for the speed v of either block after it has moved a distance h is

$$v = \sqrt{\frac{2gh(m_2 - m_1)}{m_1 + m_2 + \frac{1}{2}M}}$$

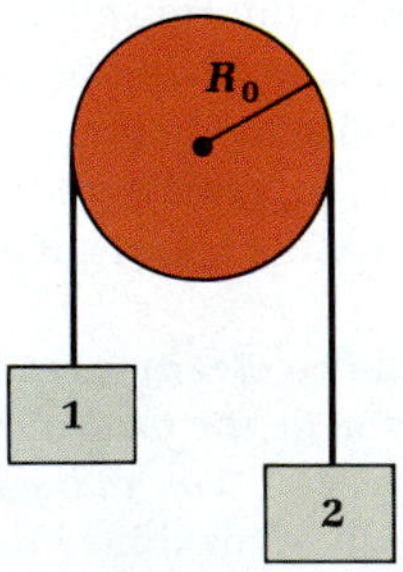

R_0
1
2

Figure 12-40. Problem 20.

ROTATION II 13

A spiral galaxy. The shape of a spiral galaxy strongly suggests that its stars are rotating about its central axis

Electrons, nuclei, and molecules — wheels, gears, and pulleys — planets, stars, and galaxies — all rotate. What causes a wheel to begin to rotate? After the wheel is rotating, what causes it to stop? A torque exerted on the wheel causes its angular velocity to change just as a force causes the linear velocity of an object to change.

Once a wheel is set in rotation, it tends to continue rotating. Eventually torques due to friction in the bearings and air resistance slow the wheel, bringing it to rest. The fact that the wheel tends to continue to rotate is evidence that there is momentum associated with rotational motion — angular momentum. Angular momentum plays a central role in our study of rotational dynamics.

13-1 ANGULAR MOMENTUM OF A PARTICLE

The rotational quantity analogous to the linear momentum **p** is the *angular momentum* $\boldsymbol{\ell}$. In Chap. 10, Newton's laws were used to develop the principle of conservation of momentum. Angular momentum is defined so that Newton's laws lead to another conservation principle—the principle of conservation of angular momentum.

Definition of Angular Momentum

The ***angular momentum*** $\boldsymbol{\ell}$ of a particle relative to a reference point O (Fig. 13-1) is equal to the cross product of the particle's position vector $\mathbf{r}$ measured from O and the particle's linear momentum $\mathbf{p}$:

Definition of angular momentum

$$\boldsymbol{\ell} = \mathbf{r} \times \mathbf{p} \tag{13-1}$$

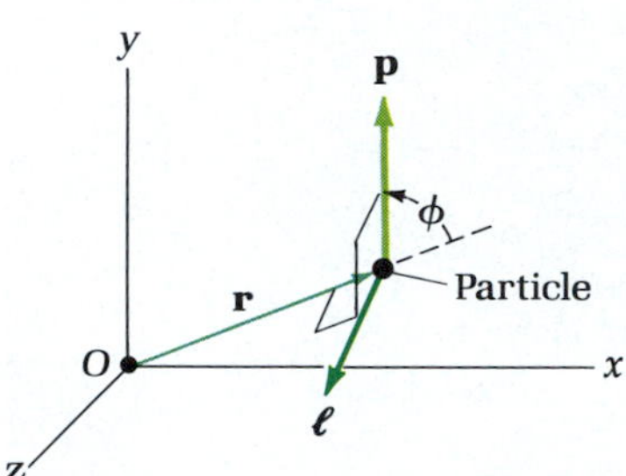

Figure 13-1. The angular momentum of a particle about the origin O is $\boldsymbol{\ell} = \mathbf{r} \times \mathbf{p}$, where $\mathbf{r}$ is the particle's position vector and $\mathbf{p} = m\mathbf{v}$ is its linear momentum. Thus $\boldsymbol{\ell}$ is perpendicular to the plane that contains $\mathbf{r}$ and $\mathbf{p}$.

Since it is defined as a cross product, the angular momentum is a vector quantity. It is perpendicular to the plane containing $\mathbf{r}$ and $\mathbf{p}$, with its direction given by the right-hand rule (Sec. 11-5). The dimension of angular momentum is [mass][length]^2[time]$^{-1}$, and its SI unit is kilogram-meters squared per second (kg·m^2/s). The magnitude ℓ of the angular momentum can be written a number of equivalent ways:

$$\ell = rp \sin \phi = rmv \sin \phi = r_{\perp} p = rp_{\perp} \tag{13-2}$$

where ϕ is the angle between $\mathbf{r}$ and $\mathbf{p}$ (or $\mathbf{v}$), and where $r_{\perp} = r \sin \phi$ and $p_{\perp} = p \sin \phi$.

Since the definition of $\boldsymbol{\ell}$ contains $\mathbf{r}$, the value of $\boldsymbol{\ell}$ depends on the point O about which it is calculated. Selecting this point is an important part of making the angular momentum a useful quantity. Any time an angular momentum is discussed, the point about which it is calculated should be kept clearly in mind.

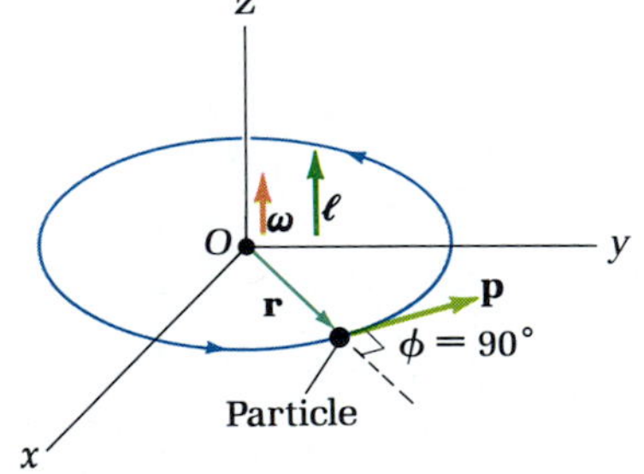

Figure 13-2. For a particle in circular motion about the origin in the xy plane, the angular momentum relative to the origin is $\boldsymbol{\ell} = mR^2\omega_z\mathbf{k}$.

Particle Traveling in a Circle

A simple and important case to consider is a particle traveling in a circle of radius R. The angular momentum is determined about the center of the circle so that $r = R$ (Fig. 13-2). Since in circular motion $\mathbf{v}$ is perpendicular to $\mathbf{r}$ at each point along the path, $\phi = 90°$ at each point. From Eq. 13-2,

$$\ell = Rmv \sin 90° = Rmv$$

The magnitude of angular momentum for a particle in circular motion

For motion in a circle, $v = \omega R$, so $\ell = Rm(\omega R)$, or

$$\ell = mR^2\omega$$

The direction of $\boldsymbol{\ell}$ depends on the sense of the rotation. In Fig. 13-2, vectors $\mathbf{r}$ and $\mathbf{p}$ are contained in the xy plane, which means that $\boldsymbol{\ell}$ is along the z axis. Which way $\boldsymbol{\ell}$ points along the z axis, toward $+\mathbf{k}$ or $-\mathbf{k}$, is determined with the right-hand rule. If you curl the fingers of your right hand so that $\mathbf{r}$ would rotate into $\mathbf{p}$, then your extended thumb gives the direction of $\boldsymbol{\ell}$. When the motion is counterclockwise (viewed from the $+z$ direction), $\boldsymbol{\ell}$ is toward $+z$ (ℓ_z is positive), and when the motion is clockwise, $\boldsymbol{\ell}$ is toward $-z$ (ℓ_z is negative). Since this also corresponds to the sign convention for ω_z, $\boldsymbol{\ell}$ can be written as

$$\boldsymbol{\ell} = \ell_z\mathbf{k} = mR^2\omega_z\mathbf{k} \tag{13-3}$$

Angular momentum of a particle in circular motion

If the motion is uniform circular motion, ω_z is constant, and consequently $\boldsymbol{\ell}$ is constant.

Particle Traveling in a Straight Line

Strange as it may seem at first thought, a particle traveling in a straight line has angular momentum. Figure 13-3 shows a particle in the xy plane moving in a straight line parallel to the x axis. From Eq. 13-2, the magnitude of the particle's angular momentum relative to origin O is

$$\ell = rp \sin\phi = mvr_\perp \tag{13-4}$$

where $r_\perp$ is the perpendicular distance from the particle's path to O. Thus a particle with a straight-line path has a nonzero angular momentum about any point that is not on its path. If the particle's speed is constant, its angular momentum is constant, but if the particle accelerates along a straight line, the angular momentum changes with time. By the right-hand rule, the direction of $\boldsymbol{\ell}$ for the particle in Fig. 13-3 is in the $-\mathbf{k}$ direction.

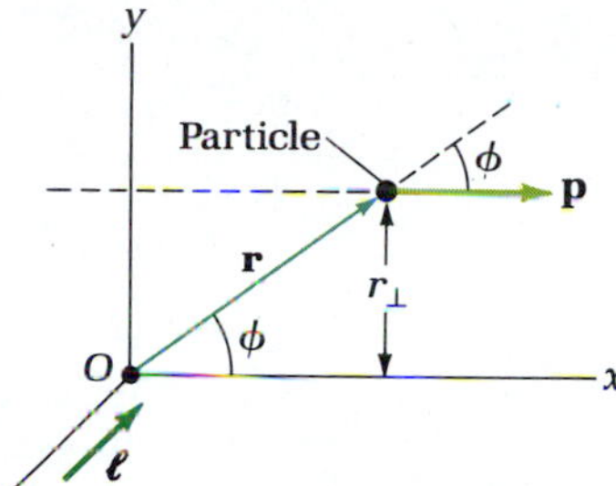

Figure 13-3. For a particle traveling in a straight line, ℓ relative to O is $\ell = r_\perp mv$, where $r_\perp$ is the perpendicular distance from the particle's path to O.

EXAMPLE 13-1

Orbital angular momentum of the earth. Determine the magnitude of the earth's angular momentum due to its orbital motion around the sun. The earth's orbital radius is $R = 1.5 \times 10^{11}$ m, and its mass is $m_e = 6.0 \times 10^{24}$ kg.

Solution. The earth completes 1 rev (2π rad) per year ($T = 365$ days), so that its angular speed $\omega = 2\pi/T$ about the sun is

$$\omega = \frac{2\pi}{(365\ \text{day})(24\ \text{h/day})(3600\ \text{s/h})} = 2.0 \times 10^{-7}\ \text{rad/s}$$

The magnitude ℓ of the earth's orbital angular momentum about the sun is

$$\begin{aligned}\ell &= m_eR^2\omega \\ &= (6.0 \times 10^{24}\ \text{kg})(1.5 \times 10^{11}\ \text{m})^2(2.0 \times 10^{-7}\ \text{rad/s}) \\ &= 2.7 \times 10^{40}\ \text{kg}\cdot\text{m}^2/\text{s}\end{aligned}$$

SELF-TEST 13-1. A 1000-kg car travels south at 25 m/s and passes by at a distance of 10 m from you when it is directly west of your position. What are the magnitude and direction of the car's angular momentum relative to you? ***ANSWER:*** 2.5×10^5 kg·m²/s, upward.

Relation between Angular Momentum and Torque

Newton's second law is the relation between the net force $\Sigma\mathbf{F}$ exerted on a particle and the particle's momentum $\mathbf{p}$:

$$\Sigma\mathbf{F} = \frac{d\mathbf{p}}{dt}$$

Consider the rotational analog of this relation. The time derivative of the angular momentum is

$$\frac{d\boldsymbol{\ell}}{dt} = \frac{d}{dt}(\mathbf{r} \times \mathbf{p})$$

Using the product rule for differentiation gives

$$\frac{d\boldsymbol{\ell}}{dt} = \frac{d\mathbf{r}}{dt} \times \mathbf{p} + \mathbf{r} \times \frac{d\mathbf{p}}{dt}$$

The first of the two cross products on the right-hand side of this equation is zero because $d\mathbf{r}/dt = \mathbf{v}$ and $\mathbf{p} = m\mathbf{v}$. The vectors $\mathbf{v}$ and $m\mathbf{v}$ are parallel, so their cross product must be zero. Therefore

$$\frac{d\boldsymbol{\ell}}{dt} = \mathbf{r} \times \frac{d\mathbf{p}}{dt}$$

Using Newton's second law for a particle, $\Sigma\mathbf{F} = d\mathbf{p}/dt$, we have

$$\frac{d\boldsymbol{\ell}}{dt} = \mathbf{r} \times \Sigma\mathbf{F}$$

The quantity $\mathbf{r} \times \Sigma\mathbf{F}$ is the net torque $\Sigma\boldsymbol{\tau}$ exerted on the particle about the origin (Sec. 11-5): $\Sigma\boldsymbol{\tau} = \mathbf{r} \times \Sigma\mathbf{F}$. Thus

Rotational analog of Newton's second law for a single particle

$$\Sigma\boldsymbol{\tau} = \frac{d\boldsymbol{\ell}}{dt} \tag{13-5}$$

In Newton's second law, $\Sigma\mathbf{F} = d\mathbf{p}/dt$, $\mathbf{p}$ must be measured relative to an inertial reference frame. Thus for Eq. 13-5 to be valid, $\boldsymbol{\ell}$ must be measured relative to an inertial reference frame.

13-2 ANGULAR MOMENTUM OF A SYSTEM OF PARTICLES

From Chap. 10, the combination of Newton's second law for a particle, $\Sigma\mathbf{F} = d\mathbf{p}/dt$, and Newton's third law, $\mathbf{F}_{12} = -\mathbf{F}_{21}$, shows that for a system of particles (or an extended object):

$$\Sigma\mathbf{F}_{\text{ext}} = \frac{d\mathcal{P}}{dt}$$

where $\Sigma\mathbf{F}_{\text{ext}}$ is the net external force on the system and $\mathcal{P}$ is the total momentum of the system. Internal forces—the forces the particles within the system exert on each other—do not contribute to a change in the system's total momentum. This leads to the conservation of momentum: If $\Sigma\mathbf{F}_{\text{ext}} = 0$, then $\mathcal{P}$ is constant. Now consider this same line of reasoning with rotational quantities.

The total angular momentum $\mathbf{L}$ for a system of particles is the vector sum of the individual angular momenta of the particles:

Total angular momentum

$$\mathbf{L} = \boldsymbol{\ell}_1 + \boldsymbol{\ell}_2 + \cdots = \Sigma\boldsymbol{\ell}_i \tag{13-6}$$

where each angular momentum is measured relative to the same reference point. For simplicity, consider a system of two particles: $\mathbf{L} = \boldsymbol{\ell}_1 + \boldsymbol{\ell}_2$. The time derivative of the total angular momentum is

$$\frac{d\mathbf{L}}{dt} = \frac{d\boldsymbol{\ell}_1}{dt} + \frac{d\boldsymbol{\ell}_2}{dt}$$

From Eq. 13-5,

$$\frac{d\mathbf{L}}{dt} = \Sigma\boldsymbol{\tau}_1 + \Sigma\boldsymbol{\tau}_2$$

The torques exerted on the particles can be divided into two categories: the torques they exert on each other (internal torques) and the torques exerted on them by other objects (external torques). Newton's third law requires that the net internal torque be zero. The law states that *the forces of interaction between particles 1 and 2 are equal and opposite,* $\mathbf{F}_{12} = -\mathbf{F}_{21}$, and that *the direction of the two forces is along the line between them* (Fig. 13-4). That is, the two forces are equal and opposite and have the same line of action. Since the line of action is the same, each force has the same moment arm $r_\perp$ about *any* reference point. Consequently, each torque magnitude is the same. The right-hand rule applied to each cross product, $\boldsymbol{\tau} = \mathbf{r} \times \mathbf{F}$, shows that the two torques are in opposite directions: $\boldsymbol{\tau}_{12} = -\boldsymbol{\tau}_{21}$. So it goes for all pairs of torques due to the interactions between all pairs of particles within a system (or an extended object). The vector sum of all internal torques add to zero. Therefore

Figure 13-4. The torques due to the interaction of a pair of particles are equal and opposite because the forces of interaction obey Newton's third law.

$$\Sigma\boldsymbol{\tau}_{\text{ext}} = \frac{d\mathbf{L}}{dt} \qquad (13\text{-}7)$$

Rotational analog of Newton's second law for a system of particles

The net external torque exerted on a system of particles (or an extended object) is equal to the rate of change of the system's total angular momentum.

As an example, the planets and the sun in the solar system exert forces on one another, and the torques from these forces about any point cancel in pairs. The angular momentum of the solar system relative to any point is unaffected by the torques due to forces between planets or between the sun and any planet.

The torques and the angular momentum in Eq. 13-7 may be measured about any point as long as the same point is used for all measurements, and as long as the point is fixed in an inertial reference frame. Although we shall not prove it here, this equation is even more general than that.

Equation 13-7, $\Sigma\boldsymbol{\tau}_{\text{ext}} = d\mathbf{L}/dt$, is valid if the reference point is at the center of mass of the system, even when the center of mass accelerates relative to an inertial reference frame.

Often this makes it convenient to separate the motion of a system into two parts: the translational motion of the center of mass, in which $\Sigma\mathbf{F}_{\text{ext}} = d\mathcal{P}/dt$ is used, and the rotational motion about the center of mass, in which $\Sigma\boldsymbol{\tau}_{\text{ext}} = d\mathbf{L}/dt$ is used.

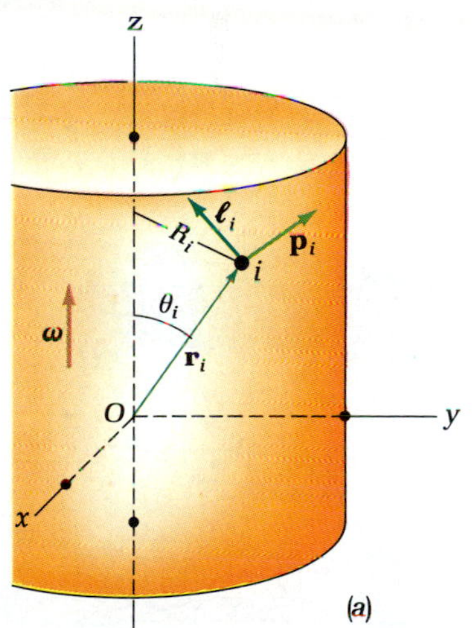

13-3 ROTATIONAL DYNAMICS OF A RIGID OBJECT ABOUT A FIXED AXIS

The rotation of a rigid object about a fixed axis is of great practical interest. We now investigate the angular momentum of such an object and develop its equation of motion.

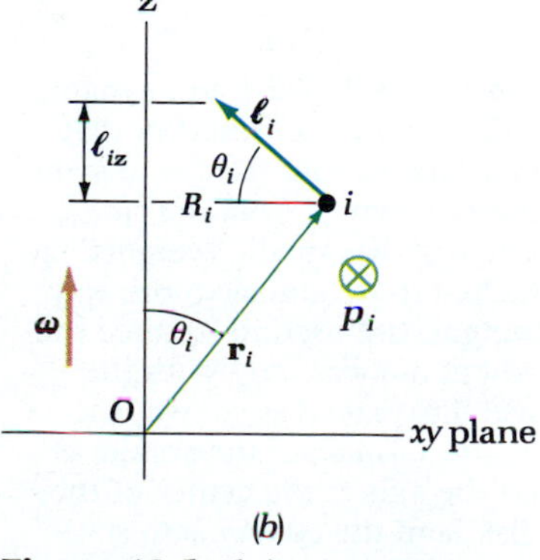

Figure 13-5. *(a)* As a rigid object rotates about the z axis, particle i moves in a circle centered on the axis. The distance from origin O to i is r_i, and the distance from the axis to i is $R_i = r_i \sin\theta_i$. *(b)* Relative to O, the axial component of $\boldsymbol{\ell}_i$ is $\ell_{iz} = \ell_i \sin\theta_i = m_i R_i^2 \omega_z$.

Angular Momentum

Consider the relation between the angular momentum $\mathbf{L}$ of a rigid object relative to a point O on the axis of rotation and the object's angular velocity $\boldsymbol{\omega}$ about the axis. As we proceed, we must take care to distinguish between a distance r from a point (origin O in Fig. 13-5) and a distance R from an axis (z axis in Fig. 13-5). The angular momentum $\mathbf{L}$ is referred to a *point* (such as O), whereas the angular velocity $\boldsymbol{\omega}$ is referred to an *axis* (such as the z axis).

The z (or axial) component of an object's angular momentum is $L_z = (\Sigma\boldsymbol{\ell}_i)_z = \Sigma\ell_{iz}$. To find L_z, we first find ℓ_{iz} for particle i. In Fig. 13-5*a*, θ_i is the angle between $\mathbf{r}_i$ and the z axis, so $R_i = r_i \sin\theta_i$. The angular momentum of particle i, $\boldsymbol{\ell}_i = \mathbf{r}_i \times \mathbf{p}_i$, has both an axial and a radial component. As shown in the figure, $\mathbf{r}_i$ is perpendicular to $\mathbf{p}_i$ so that $\ell_i = r_i p_i \sin 90° = r_i m_i v_i$. Since $v_i = R_i \omega$, $\ell_i = r_i m_i R_i \omega$. From Fig. 13-5*b*, the z component of particle i's angular momentum is

$$\ell_{iz} = r_i m_i R_i \omega_z \sin \theta_i = m_i R_i (r_i \sin \theta_i) \omega_z = m_i R_i^2 \omega_z$$

Note that since R_i and ω_z are referred to the axis, ℓ_{iz} is independent of where the origin is located on the axis. *Therefore this expression is valid for any origin as long as the origin is on the axis.*

Adding ℓ_{iz} for each particle in the object gives

$$L_z = \Sigma m_i R_i^2 \omega_z = \omega_z \Sigma m_i R_i^2$$

where ω_z is factored out of the sum because it does not depend on i. (ω_z is the same for each particle.) The quantity $\Sigma m_i R_i^2$ is the moment of inertia I of the object about the axis of rotation (Sec. 12-7). Thus

$$L_z = I\omega_z \tag{13-8}$$

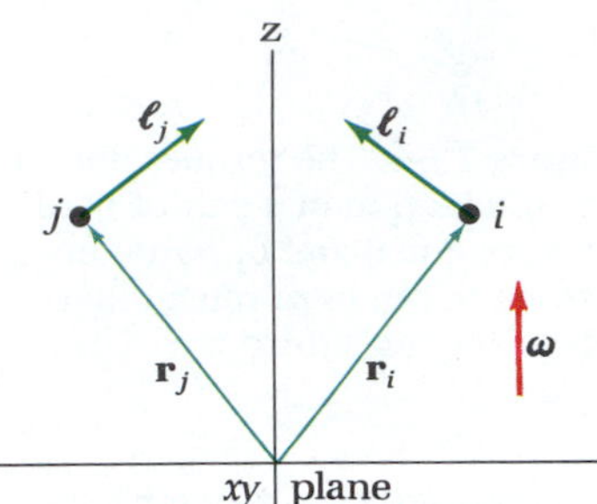

Figure 13-6. When particles i and j, which are on opposite sides of the axis, have the same mass ($m_i = m_j$), the radial components of ℓ_i and ℓ_j are equal and opposite. If the radial components of ℓ for the particles cancel when forming $\mathbf{L} = \Sigma \ell_i$, then $\mathbf{L} = L_z \mathbf{k}$ and $\mathbf{L}$ is parallel to ω.

Equation 13-8 is valid for any rigid object, but it refers only to the axial component of $\mathbf{L}$. It turns out that when an object has sufficient symmetry about the axis, $\mathbf{L}$ is parallel to the axis, or $\mathbf{L} = L_z\mathbf{k}$. For example, consider a homogeneous cylinder rotating about its symmetry axis. Because the cylinder is homogeneous, particles i and j on opposite sides of the axis have the same mass so that the radial components of their angular momenta are equal and opposite (Fig. 13-6). When we form the vector sum $\mathbf{L} = \Sigma \ell_i$, the radial components of the angular momenta of the particles cancel in pairs. Thus $\mathbf{L} = L_z\mathbf{k}$.

Besides this pairwise cancellation, there are other ways an object can have $\mathbf{L}$ parallel to the axis. Since ω is parallel to the axis by definition, $\mathbf{L}$ and ω are parallel to each other in such cases, and

$$\mathbf{L} = I\omega \qquad \text{(sufficient symmetry only)} \tag{13-9}$$

If an object does not have sufficient symmetry about the axis such that $\mathbf{L}$ is parallel to the axis, then the direction of $\mathbf{L}$ will rotate as the object rotates. An example is a car's wheel that is out of dynamic balance. Such a wheel tends to wobble, and the torque required to maintain the rotation about the axle is exerted by the bearings in the axle (Fig. 13-7).

The translational analog of Eq. 13-9 for a system of particles is $\mathcal{P} = M\mathbf{v}_{cm}$, where $\mathcal{P}$ is the total linear momentum, M is the total mass, and $\mathbf{v}_{cm}$ is the velocity of the center of mass (Sec. 10-3).

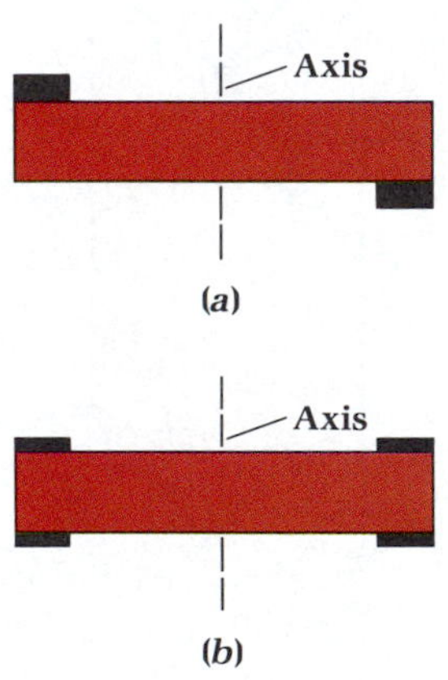

Figure 13-7. Static and dynamic balance of a uniform disk that can rotate about its axis of symmetry (viewed from the side). The disk has small "weights" attached to it, similar to the way weights are used to balance the wheel of a car. *(a)* With the weights placed as shown, the center of mass of the system is on the axis at the center of the disk, and the system is in static balance, but it is not in dynamic balance. When the system rotates, $\mathbf{L}$ is not along the axis. *(b)* The weights are split and the half-weights are attached as shown. Now the system's symmetry about the axis is such that $\mathbf{L}$ is along the axis, and the system is in dynamic balance.

Equation of Motion

Newton's second law for a rigid object undergoing translational motion is $\Sigma\mathbf{F}_{ext} = M\mathbf{a}_{cm}$. The rotational analog of this equation is found by taking the time derivative of Eq. 13-8:

$$\frac{dL_z}{dt} = \frac{d}{dt} I\omega_z = I\frac{d\omega_z}{dt} = I\alpha_z$$

We have factored I out of the derivative (because it is constant for a rigid body rotating about a fixed axis), and we have used $d\omega_z/dt = \alpha_z$. The z component of Eq. 13-7 is $\Sigma\tau_z = dL_z/dt$. From the expression above, dL_z/dt can be replaced with $I\alpha_z$ to give

$$\Sigma\tau_z = I\alpha_z \tag{13-10}$$

This is the equation of motion for a rigid object constrained to rotate about a fixed axis, and it is a consequence of Newton's laws.

Moment of Inertia Revisited

The moment of inertia was introduced in Chap. 12 in connection with the rotational kinetic energy. The development of Eqs. 13-8 and 13-10 provides an alternative introduction for the moment of inertia. In this latter case, the reason for the term "inertia" in its name is more apparent. Equation (13-10) shows that *the moment of inertia is the measure of an object's resistance to a change in its angular velocity*. For a given net external torque, an object with a larger moment of inertia will have a smaller angular acceleration. Consequently, the moment of inertia is sometimes referred to as the ***rotational inertia***.

EXAMPLE 13-2

Spin angular momentum of the earth. Determine the earth's angular momentum magnitude due to its daily rotation. We call this angular momentum the earth's *spin angular momentum* $\mathbf{L}_s$. Assume that $\mathbf{L}_s$ is parallel to the earth's axis and that the earth's mass density is uniform. The mass of the earth is 6.0×10^{24} kg, and its radius is 6.4 Mm.

Solution. The moment of inertia of a uniform sphere about an axis along a diameter is $2Mr_0^2/5$ (Table 12-2), so the earth's moment of inertia (assuming its density is uniform) is $I = [2(6.0 \times 10^{24}\ \text{kg})(6.4\ \text{Mm})^2]/5 = 9.8 \times 10^{37}\ \text{kg}\cdot\text{m}^2$. The angular speed of the earth's daily rotation is

$$\omega = \frac{2\pi}{T} = \frac{2\pi}{(24\ \text{h})(3600\ \text{s/h})} = 7.3 \times 10^{-5}\ \text{rad/s}$$

The magnitude of the earth's spin angular momentum is

$$L_s = I\omega = (9.8 \times 10^{37}\ \text{kg}\cdot\text{m}^2)(7.3 \times 10^{-5}\ \text{rad/s}) = 7.2 \times 10^{33}\ \text{kg}\cdot\text{m}^2/\text{s}$$

SELF-TEST 13-2. What are the magnitude and direction of the spin angular momentum of a phonograph record relative to its center when the record is rotating at 33 rev/min? The record's mass is 0.11 kg, its radius is 0.15 m, and its rotation is clockwise when viewed from above. ***ANSWER:*** $4.3 \times 10^{-3}\ \text{kg}\cdot\text{m}^2/\text{s}$, downward.

EXAMPLE 13-3

Finding the motion of a door. Two forces are exerted on a uniform door, as shown in Fig. 13-8. The door is initially rotating in the sense shown with an angular speed of 0.45 rad/s. Assuming the torques due to these forces remain constant, determine the door's *(a)* angular acceleration component, *(b)* angular velocity component as a function of time, and *(c)* angular coordinate as a function of time. Neglect the torque due to friction in the hinges. The door's mass is $M = 38$ kg, and its width is $w = 0.88$ m. Other data are $\phi = 63°$, $F_b = 15$ N, and $F_c = 12$ N.

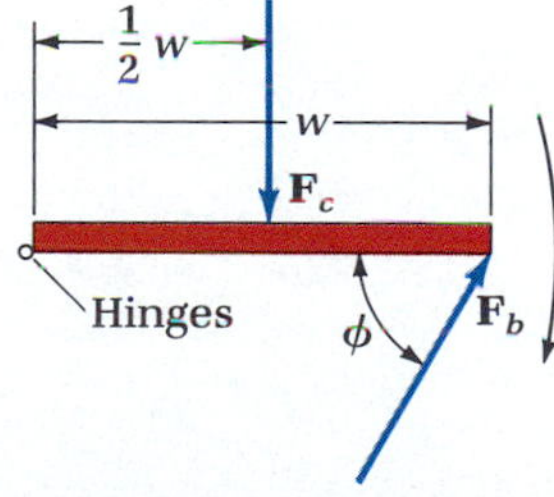

Figure 13-8. Example 13-3: The z axis is perpendicular to the page (along the hinges), and $+\mathbf{k}$ is out of the page.

Solution. *(a)* We let the z axis in Fig. 13-8 be along the hinges, with $+z$ pointing out of the page. The z component of the net torque about any point on the axis is

$$\Sigma\tau_z = wF_b \sin\phi - \tfrac{1}{2}wF_c$$

Solving Eq. 13-10 for α_z gives $\alpha_z = \Sigma\tau_z/I$, where $I = Mw^2/3$ (Example 12-6). Substituting $\Sigma\tau_z$ from above gives

$$\alpha_z = \frac{3(F_b \sin\phi - \frac{1}{2}F_c)}{Mw} = \frac{3[(15\ \text{N}) \sin 63° - \frac{1}{2}(12\ \text{N})]}{(38\ \text{kg})(0.88\ \text{m})} = 0.66\ \text{rad/s}^2$$

(b) The right-hand rule applied to the initial direction of rotation shown in Fig. 13-8 gives $\boldsymbol{\omega}_0$ into the page so that ω_{z0} is negative: $\omega_{z0} = -0.45$ rad/s. Using the result from part *(a)* and the expression $\omega_z = \omega_{z0} + \alpha_z t$, we have

$$\omega_z = -0.45\ \text{rad/s} + (0.66\ \text{rad/s}^2)t$$

(c) Using the expression $\theta = \theta_0 + \omega_{z0}t + \frac{1}{2}\alpha_z t^2$ and letting the door's initial angular position θ_0 be zero, we have

$$\theta = -(0.45 \text{ rad/s})t + (0.33 \text{ rad/s}^2)t^2$$

SELF-TEST 13-3. Determine the time t at which the door in the above example is instantaneously at rest. ***ANSWER:*** 0.68 s.

EXAMPLE 13-4

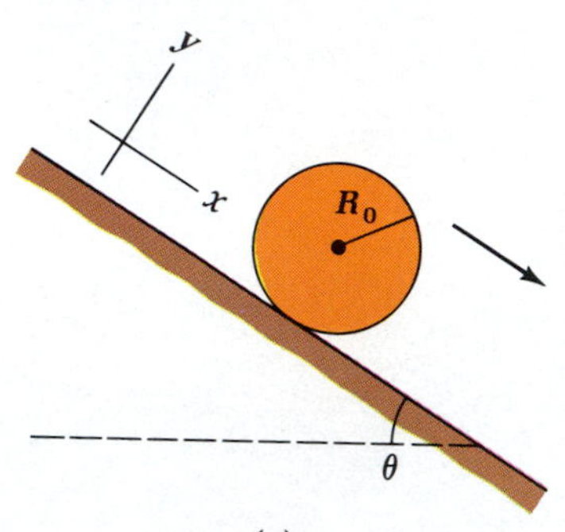

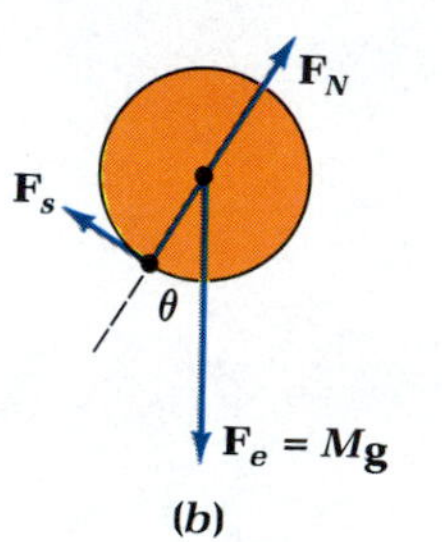

Figure 13-9. Example 13-4: *(a)* Object rolling down a slope. *(b)* Free-body diagram.

Acceleration of a rolling object. A symmetric wheel of mass M and radius R_0 rolls without sliding down a straight slope that makes an angle θ with the horizontal (Fig. 13-9*a*). The wheel's moment of inertia about its symmetry axis is I. Find an expression for the acceleration of the center of mass of the wheel in terms of I, M, R_0, g, and θ.

Solution. Figure 13-9*a* shows a coordinate frame with $+\mathbf{i}$ in the direction of the center-of-mass motion, and with $+\mathbf{k}$ parallel to the rotational axis and directed out of the page. The wheel's free-body diagram is shown in Fig. 13-9*b*. There are three forces exerted on the wheel: its weight $\mathbf{F}_e$, the normal force by the surface $\mathbf{F}_N$, and the static frictional force by the surface $\mathbf{F}_s$. (Static friction is acting because the wheel is not sliding.) Using $F_e = Mg$, we find that the x component of Newton's second law, $\Sigma F_x = Ma_x$, gives

$$Mg \sin\theta - F_s = Ma \qquad \text{(A)}$$

Notice that the lines of action of $\mathbf{F}_e$ and of $\mathbf{F}_N$ each pass through the center of the wheel so that if we take torques about an axis through the center, these forces do not contribute to $\Sigma\tau_z$. Only $\mathbf{F}_s$ contributes to $\Sigma\tau_z$, and its moment arm is R_0. Thus, applying $\Sigma\tau_z = I\alpha_z$ to this axis, the rotational equation of motion gives

$$-F_s R_0 = -I\alpha \qquad \text{(B)}$$

Since $\alpha = a/R_0$, Eq. B can be written as

$$F_s = (I/R_0^2)a$$

Substituting this expression for F_s into Eq. A and solving for a yields

$$a = \frac{g \sin\theta}{1 + (I/MR_0^2)}$$

Again, as in Example 12-9, we find that the ratio I/MR_0^2 is an important characteristic of a rolling object; the larger this ratio, the smaller is the acceleration. For comparison, it is interesting to recall that the acceleration magnitude of a cart rolling down a slope, neglecting the moment of inertia of its wheels (Example 5-9), is $a = g \sin\theta$.

SELF-TEST 13-4. Apply the result of the above example to some specific objects. Write the expression for the acceleration magnitude when the object is *(a)* a thin-walled cylinder, *(b)* a solid cylinder, and *(c)* a solid sphere. ***ANSWERS:*** *(a)* $(g \sin\theta)/2$; *(b)* $2(g \sin\theta)/3$; *(c)* $5(g \sin\theta)/7$.

EXAMPLE 13-5

A falling crate connected to a windlass. A crate tied to a rope that is wrapped around a windlass is released (Fig. 13-10*a*). The mass of the crate is $M_C = 35$ kg, and the mass and radius of the windlass are $M_W = 94$ kg and $R_0 = 83$ mm. Determine *(a)* the magnitude a of the crate's linear acceleration and *(b)* the tension F_T in the rope. The windlass may be treated as a uniform cylinder of radius R_0, and the torque due to friction in the bearings and the mass of the rope both may be neglected.

Solution. The free-body diagrams of the crate and the windlass are shown in Fig. 13-10*b*, where $M_W\mathbf{g}$ is the weight of the windlass, $M_C\mathbf{g}$ is the weight of the crate, $\mathbf{F}_T$ is the force by the rope on the windlass, $\mathbf{F}'_T$ is the force by the rope on the crate, and $\mathbf{F}_B$ is the force by the bearings on the windlass. We choose a coordinate system oriented with $+\mathbf{k}$ along the windlass's axis and out of the page. Since the rope's mass is negligible, $|\mathbf{F}_T| = |\mathbf{F}'_T| = F_T$. *(a)* The vertical component of Newton's second law applied to the crate gives

$$F_T - M_C g = -M_C a \tag{A}$$

The axial component of the torque exerted on the windlass by the rope relative to any point on the axis is $\tau_z = F_T R_0$. The torques due to $\mathbf{F}_B$ and $M_W\mathbf{g}$ are zero because the line of action of each of these forces passes through the axis. Thus $\Sigma\tau_z = F_T R_0$. The rotational analog of Newton's second law, Eq. 13-10, gives

$$F_T R_0 = I\alpha_z = I\alpha \tag{B}$$

The magnitudes of the linear acceleration of the crate and the angular acceleration of the windlass are related by $a = \alpha R_0$. Also, since the windlass is a uniform cylinder, $I = \frac{1}{2}M_W R_0{}^2$. Substituting into Eq. B and dividing both sides by R_0, we obtain

$$F_T = \tfrac{1}{2}M_W a \tag{C}$$

Inserting F_T from Eq. C into Eq. A and solving for a gives

$$a = \frac{M_C g}{M_C + \frac{1}{2}M_W} = \frac{(35\text{ kg})(9.8\text{ m/s}^2)}{35\text{ kg} + \frac{1}{2}(94\text{ kg})} = 4.2\text{ m/s}^2$$

Notice that a is less than g. Why is this so? *(b)* Substituting the above value of a into Eq. C gives

$$F_T = \frac{\frac{1}{2}M_W M_C g}{M_C + \frac{1}{2}M_W} = \frac{\frac{1}{2}(94\text{ kg})(35\text{ kg})(9.8\text{ m/s}^2)}{35\text{ kg} + \frac{1}{2}(94\text{ kg})} = 200\text{ N}$$

Notice that the tension in the rope is less than the weight of the crate. Why is this so?

SELF-TEST 13-5. Suppose the rope in the above example will break if the tension exceeds 230 N. What is the maximum mass that the crate can have such that the rope will not break when the crate is released? ***ANSWER:*** 47 kg.

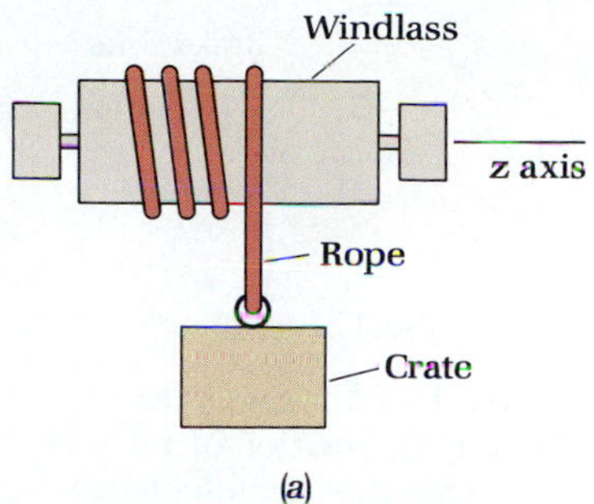

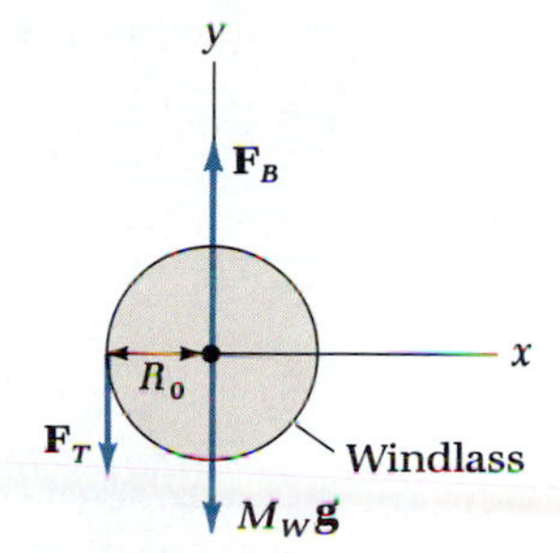

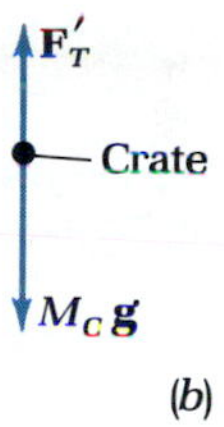

Figure 13-10. Example 13-5: *(a)* Front view. The z axis is in the plane of the page (along the axle). *(b)* Free-body diagrams. The z axis is perpendicular to the page, with $+\mathbf{k}$ out of the page.

13-4 ROTATIONAL WORK AND POWER FOR A RIGID OBJECT

When a force causes a torque on a rotating object, work is done by the agent that exerts the force. In Fig. 13-11, we consider the work done by a force $\mathbf{F}$ during a time interval in which a door's angular position changes by an amount $d\theta$. The work dW done by $\mathbf{F}$ is

$$dW = \mathbf{F} \cdot d\mathbf{s} = F_t(R\,d\theta)$$

where F_t is the tangential component of $\mathbf{F}$. Since $\tau_z = F_t R$,

$$dW = \tau_z\,d\theta \tag{13-11}$$

The work W done by torque τ_z when the door rotates from θ_i to θ_f is

$$W = \int_{\theta_i}^{\theta_f} \tau_z\,d\theta \tag{13-12}$$

For a constant torque, τ_z can be taken out of the integral, and

$$W = \tau_z(\theta_f - \theta_i) = \tau_z\,\Delta\theta \tag{13-13}$$

The translational analog of Eq. 13-12 is $W = \int F_x\,dx$ for one-dimensional motion.

The work done by a torque may be negative. Figure 13-12 shows a rotating wheel being slowed by the torque due to a brake shoe. With the z axis along the axle and $+\mathbf{k}$ out of the page, the angular displacement of the wheel is positive ($\Delta\theta > 0$), and by the right-hand rule τ_z due to the force $\mathbf{F}_f$ is negative. Therefore τ_z and $\Delta\theta$ have opposite signs, which makes W negative. What are the algebraic signs of $\Delta\theta$, τ_z, and W when the wheel rotates in the opposite sense with the brakes on?

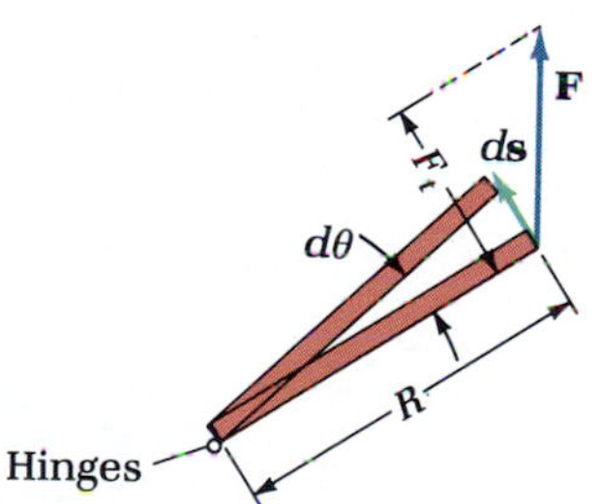

Figure 13-11. Work done in rotating a door. The z axis is perpendicular to the page (along the hinges), and $+\mathbf{k}$ is out of the page.

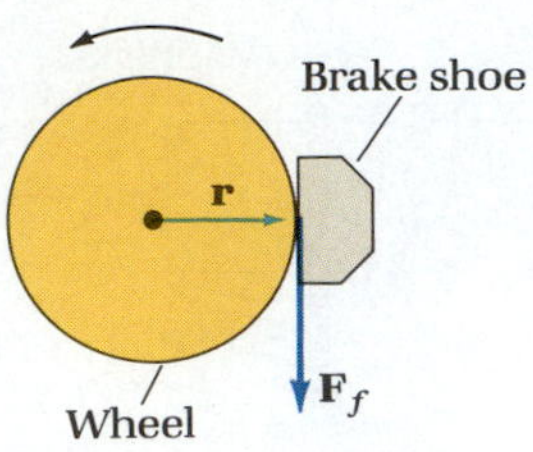

Figure 13-12. A brake shoe slowing the rotation of a wheel. The z axis is perpendicular to the page (along the wheel's axle), with $+\mathbf{k}$ out of the page.

The work-energy theorem gives the relation between the net work on an object and the change in the object's kinetic energy (Sec. 8-4). We now develop the rotational counterpart of the work-energy theorem for a rigid object rotating about a fixed axis. Combining Eqs. 13-10 and 13-11 gives

$$dW_{net} = \Sigma\tau_z\, d\theta = I\alpha_z\, d\theta = I\frac{d\omega_z}{dt}\,\omega_z\, dt$$

Since

$$\frac{d}{dt}\left(\tfrac{1}{2}\omega_z^2\right) = \omega_z\frac{d\omega_z}{dt}$$

we have

$$dW_{net} = I\frac{d}{dt}\left(\tfrac{1}{2}\omega_z^2\right) dt = \tfrac{1}{2}I\frac{d}{dt}\left(\omega_z^2\right) dt = \tfrac{1}{2}I\frac{d}{dt}\left(\omega^2\right) dt$$

This gives

$$W_{net} = \tfrac{1}{2}I\int_{t_i}^{t_f}\frac{d}{dt}\left(\omega^2\right) dt$$

where I has been taken out of the integral because the object is rigid and the axis is fixed. Performing the integration gives

The rotational work-energy theorem for a rigid object rotating about a fixed axis

$$W_{net} = \tfrac{1}{2}I\omega_f^2 - \tfrac{1}{2}I\omega_i^2 \qquad (13\text{-}14)$$

The power P delivered to a rotating rigid object by a torque is the rate at which work is done by the torque. Using Eq. 13-11, we have

Power delivered to a rigid object rotating about a fixed axis

$$P = \frac{dW}{dt} = \tau_z\frac{d\theta}{dt} = \tau_z\omega_z \qquad (13\text{-}15)$$

As with the work, the power due to a torque may be negative. If a torque tends to slow the rotation, then τ_z has the opposite sign from ω_z and P is negative. Such an energy transfer tends to decrease the object's rotational energy.

EXAMPLE 13-6

Work done on the windlass. *(a)* Find the work done on the windlass by the force due to the rope in Example 13-5 (Fig. 13-10) during the time interval in which it rotates through an angle of 45°. *(b)* Assuming the windlass is at rest at the beginning of the 45° rotation, determine its angular speed at the end of the rotation. *(c)* Find the power delivered to the windlass by the rope at the instant the 45° rotation is completed.

Solution. *(a)* From Example 13-5, the torque due to the rope is constant, $\tau_z = F_T R_0$, so that Eq. 13-13 may be used: $W = F_T R_0\,\Delta\theta$. Since 45° corresponds to $\pi/4$ rad,

$$W = (200\text{ N})(0.083\text{ m})(\pi/4\text{ rad}) = 13\text{ J}$$

(b) Since the torque due to the rope is the net torque about the axis of rotation, the solution from part *(a)* above gives $W_{net} = 13$ J. Also, the initial angular velocity was zero, so that $W_{net} = \tfrac{1}{2}I\omega^2$, or

$$\omega = \sqrt{\frac{2W_{net}}{I}} = \sqrt{\frac{2W_{net}}{\tfrac{1}{2}MR_0^2}}$$

$$= \sqrt{\frac{2(13\text{ J})}{\tfrac{1}{2}(94\text{ kg})(0.083\text{ m})^2}} = 9.0\text{ rad/s}$$

(c) Using the coordinate frame of Fig. 13-10*b* ($+\mathbf{k}$ out of the page), τ_z due to the rope is positive:

$$\tau_z = F_T R_0 = (200\text{ N})(0.083\text{ m}) = 17\text{ N}\cdot\text{m}$$

and ω_z at the end of the 45° rotation is positive: $\omega_z = 9.0$ rad/s. Thus

$$P = \tau_z\omega_z = (17\ \text{N}\cdot\text{m})(9.0\ \text{rad/s}) = 150\ \text{W}$$

SELF-TEST 13-6. Repeat the calculation in the above example, but apply it to a 90° rotation rather than a 45° rotation. ***ANSWERS:*** *(a)* 26 J; *(b)* 13 rad/s; *(c)* 210 W.

13-5 CONSERVATION OF ANGULAR MOMENTUM

Chapters 9 and 10 introduced the principles of conservation of energy and conservation of linear momentum, respectively. Now a third conservation principle is presented: the principle of *conservation of angular momentum.*

Suppose the net external torque on a system of particles is zero: $\Sigma\boldsymbol{\tau} = 0$. From Eq. 13-7, $\Sigma\boldsymbol{\tau} = d\mathbf{L}/dt$, so that

$$\frac{d\mathbf{L}}{dt} = 0 \qquad \text{or} \qquad \frac{d}{dt}(\Sigma\boldsymbol{\ell}_i) = 0$$

If the time derivative (or rate of change) of a quantity is zero, then the quantity remains constant. Therefore, when the net external torque on a system is zero, we have

$$\mathbf{L} = \Sigma\boldsymbol{\ell}_i = \text{constant}$$

Although the individual angular momenta of the particles that compose the system may change when $\Sigma\boldsymbol{\tau} = 0$, their sum cannot change. Another way to express this result is

$$\boxed{\mathbf{L}_i = \mathbf{L}_f} \qquad (13\text{-}16)$$

Total angular momentum is conserved when $\Sigma\boldsymbol{\tau} = 0$.

where $\mathbf{L}_i$ and $\mathbf{L}_f$ are the initial and final total angular momenta of the system. Thus the ***principle of conservation of angular momentum*** is

The principle of conservation of angular momentum

> If the net torque on a system of particles (or an extended object) relative to some point is zero, then the total angular momentum of the system relative to that same point is constant.

EXAMPLE 13-7

Boy jumping on a merry-go-round. A boy of mass m running with speed v jumps onto the outer edge of a playground merry-go-round that is initially at rest (Fig. 13-13*a*). What is the angular speed ω of the merry-go-round after the boy is at rest relative to the merry-go-round? Assume the torque due to friction in the axle of the merry-go-round is negligible.

Solution. We treat the boy as a particle, and arrange our coordinate system as shown in Fig. 13-13*b*. We take the system to be the boy and the merry-go-round. Since the torque due to friction in the axle is negligible, there are no external torques on the system, and its total angular momentum is conserved. The boy and the merry-go-round exert forces and torques on each other, but these forces and torques are internal; they are equal and opposite, so their sum is zero.

Since the merry-go-round is initially at rest, the initial angular momentum of the system about the axle (the z axis) is entirely due to the boy:

$$L_{zi} = mvR$$

where R is the radius of the merry-go-round. If I is the moment of inertia of the merry-go-round, then the moment of inertia of the system after the boy is at rest relative to the merry-go-round is $I + mR^2$. Therefore

$$L_{zf} = (I + mR^2)\omega_z$$

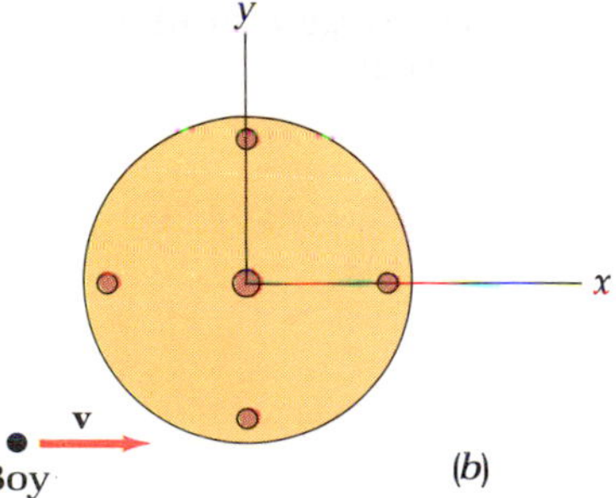

Figure 13-13. Example 13-7: *(a)* A boy jumps onto a playground merry-go-round. *(b)* Top view (+**k** is out of the page).

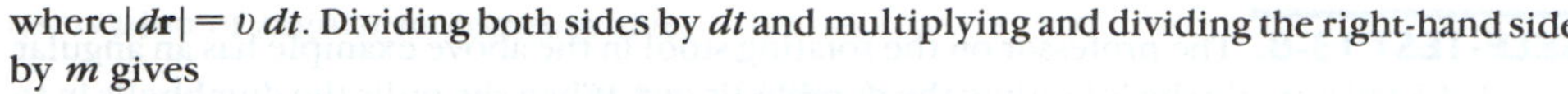

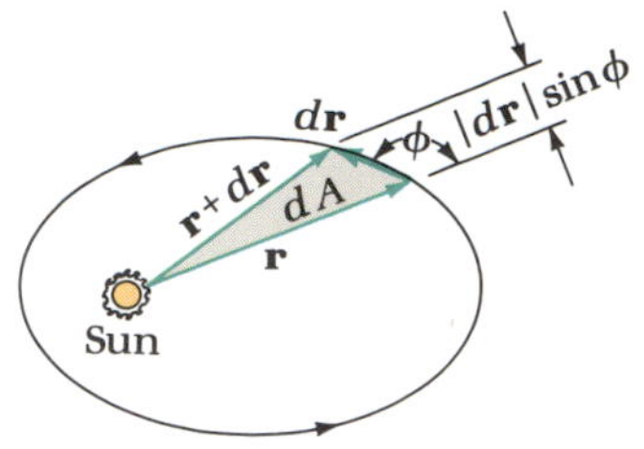

Figure 13-16. Example 13-9: Kepler's second law: The line connecting a planet to the sun sweeps out equal areas in equal time intervals.

where $|d\mathbf{r}| = v\,dt$. Dividing both sides by dt and multiplying and dividing the right-hand side by m gives

$$\frac{dA}{dt} = \frac{mvr\sin\phi}{2m}$$

The numerator on the right-hand side is the magnitude of the planet's orbital angular momentum, which is constant. Thus dA/dt = constant, which is a mathematical statement of Kepler's second law.

SELF-TEST 13-9. The earth's orbit is slightly elliptical so that the earth is nearest the sun (perihelion) in early January and farthest from the sun (aphelion) in early July. *(a)* When is the earth's orbital speed the largest? *(b)* When is it the smallest? ***ANSWERS:*** *(a)* early January; *(b)* early July.

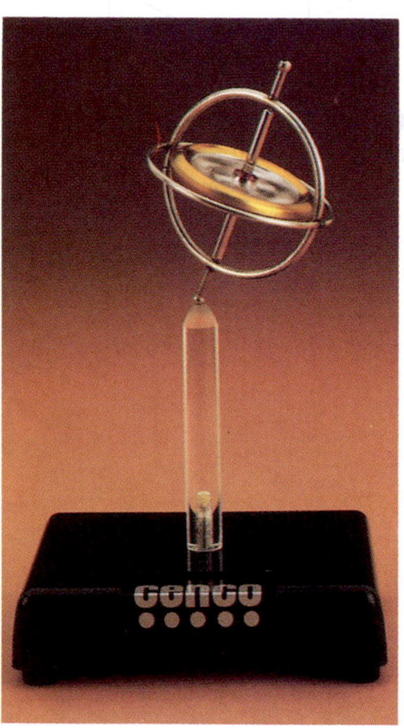

A simple gyroscope.

13-6 MOTION OF A GYROSCOPE

So far Eq. 13-7, $\Sigma\boldsymbol{\tau} = d\mathbf{L}/dt$ has been applied to only a few special cases. In specific examples, we have discussed rigid objects that are symmetric about an axis of rotation that has a fixed orientation. In such cases, $\Sigma\boldsymbol{\tau}$ (and $d\mathbf{L}/dt$) is in the same direction as **L**; both quantities are along the axis, and the problem is essentially one-dimensional. The study of gyroscopic motion affords the opportunity to consider a more general case, where $\Sigma\boldsymbol{\tau}$ and **L** are not along the same line. This provides a clear demonstration of the vector nature of torque and angular momentum.

Gyroscopic motion can be exhibited with a top or with a toy gyroscope. The usual demonstration of this motion in the physics classroom is with a bicycle wheel gyroscope (Fig. 13-17). The wheel spins rapidly with angular speed ω_s about an axle. One end of the axle (end P in Fig. 13-18) forms a ball-and-socket joint with the top of a stand so that the axle pivots freely. When the other end of the axle is released, an astonishing thing happens. The spinning wheel seems to defy gravity. It does not fall to the floor as it would if it were not spinning. Instead, the wheel and axle rotate about a vertical axis, turning out the path shown in Fig. 13-18. This rotation of the axle and wheel is called ***precession;*** ω_p represents the angular speed of precession. For precession to occur, we must have $\omega_s \gg \omega_p$.

Figure 13-17. A bicycle-wheel gyroscope. A precessing wheel is captured in a flash photograph. The wheel had a red light attached to the end of its axle. The camera shutter was open as the wheel executed its motion in a darkened room, and the red light traced out the path of the end of the axle. The camera's flash revealed the wheel's position at an instant during its motion.

As you can see from Fig. 13-17, there is more to this motion than we have mentioned so far. When first released, the axle bobs up and down as the wheel precesses. This bobbing motion is called ***nutation,*** and it becomes damped out due to friction in the pivot. We shall ignore the nutation by considering the motion after the nutation has become negligible.

Figure 13-18 shows a coordinate frame with the origin at the pivot, the y axis vertical, and the xz plane horizontal. We let the axle be horizontal for simplicity, and consider the motion at the instant the axle is along the z axis. There are two external forces on the wheel and its axle (which we regard as the system). They are the weight $\mathbf{F}_e$ acting downward at the center of the wheel and the force $\mathbf{F}_p$, due to the stand, acting at the pivot P. We neglect the torque due to friction in the pivot. The net external torque on the system about P is

$$\Sigma\boldsymbol{\tau} = F_e D\mathbf{i} = MgD\mathbf{i}$$

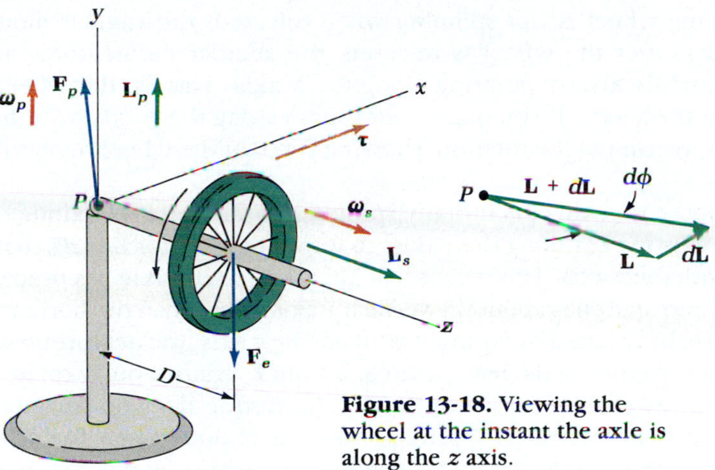

Figure 13-18. Viewing the wheel at the instant the axle is along the z axis.

That is, only the weight contributes to the torque about P because the line of action of $\mathbf{F}_p$ passes through P. Therefore the infinitesimal change $d\mathbf{L}$ in total angular momentum about P during an infinitesimal time interval dt is

$$d\mathbf{L} = \Sigma\boldsymbol{\tau}\, dt = (MgD\, dt)\mathbf{i}$$

The total angular momentum $\mathbf{L}$ of the system about P is the sum of the angular momentum $\mathbf{L}_s$ due to the spinning wheel and the angular momentum $\mathbf{L}_p$ due to the precession of the wheel and axle:

$$\mathbf{L} = \mathbf{L}_s + \mathbf{L}_p$$

The spin angular momentum is

$$\mathbf{L}_s = I_s\omega_s\mathbf{k}$$

where I_s is the moment of inertia of the wheel about its axle. The precessional angular momentum is

$$\mathbf{L}_p = I_p\omega_p\mathbf{j}$$

where I_p is the moment of inertia of the wheel and axle about the y axis. This gives

$$\mathbf{L} = I_s\omega_s\mathbf{k} + I_p\omega_p\mathbf{j}$$

Note that $\mathbf{L}$ is perpendicular to $d\mathbf{L}$, which means that $\mathbf{L}$ will not change in magnitude but will change in direction.

Now we assume ω_s is much larger than ω_p, so that the total angular momentum is very nearly equal to the spin angular momentum. With this simplification, we can see from the triangle in Fig. 13-18 that the infinitesimal precession angle $d\phi$ swept out in time interval dt is

$$d\phi = \frac{dL}{L} = \frac{MgD\, dt}{I_s\omega_s}$$

Since $\omega_p = d\phi/dt$,

$$\omega_p = \frac{MgD}{I_s\omega_s} \tag{13-18}$$

Thus, using reasonable assumptions, we have developed an equation for the angular speed of precession. The equation indicates that as friction in the axle causes ω_s to decrease, ω_p will increase, a feature of the motion that is easily noticeable.

Let us now try to understand why the wheel does not fall if it is spinning, but does fall if it is not spinning. First we describe, in terms of torque and angular momentum, the way a nonspinning wheel does fall. Suppose the wheel in Fig. 13-18 is not spinning, so that $\mathbf{L}_s$ does not exist. The net torque $\Sigma\tau$ and $d\mathbf{L}$ are both along the x axis. Since the wheel is *not* spinning when released, the angular momentum is initially zero; after the wheel is released, the angular momentum increases in magnitude, while always pointing along the x axis. That is, the wheel and axle rotate about the x axis. If you place your thumb along the x axis, your fingers will curl in the direction of the rotation, showing you that the wheel rotates downward or falls.

Now suppose the wheel is initially spinning. In this case $d\mathbf{L}$ (along the x axis) adds to the existing $\mathbf{L}_s$ (along the z axis) to form a new vector $\mathbf{L} + d\mathbf{L}$ that makes an angle $d\phi$ with the z axis. This means that after time dt the axle has precessed by an angle $d\phi$. This rotation is about the vertical y axis rather than the horizontal x axis. Since the axle has rotated by an angle $d\phi$ from the z axis, the net torque now makes an angle $d\phi$ with the x axis. Both vectors, $\Sigma\tau$ and $\mathbf{L}$, continuously rotate about the vertical, with $\Sigma\tau$ staying 90° ahead of $\mathbf{L}$. The torque does not maintain a fixed direction when the wheel is spinning, whereas it does when the wheel is not spinning. The torque causes a spinning wheel to precess, and it causes a nonspinning wheel to fall.

SUMMARY

Section 13-1. Angular Momentum of a Particle

The angular moment $\boldsymbol{\ell}$ of a particle about a point O is

$$\boldsymbol{\ell} = \mathbf{r} \times \mathbf{p} \tag{13-1}$$

where $\mathbf{r}$ is measured relative to O and $\mathbf{p} = m\mathbf{v}$. From Newton's second law, the relation between $\boldsymbol{\ell}$ and the net torque $\Sigma\tau = \mathbf{r} \times \Sigma\mathbf{F}$ is

$$\Sigma\tau = \frac{d\boldsymbol{\ell}}{dt} \tag{13-5}$$

Section 13-2. Angular Momentum of a System of Particles

The total angular momentum of a system of particles is the vector sum of the angular momenta of the particles that compose the system:

$$\mathbf{L} = \Sigma\boldsymbol{\ell}_i \tag{13-6}$$

Using Eq. 13-5 and Newton's third law, we find

$$\Sigma\tau_{\text{ext}} = \frac{d\mathbf{L}}{dt} \tag{13-7}$$

where $\Sigma\tau_{\text{ext}}$ is the net external torque on the system.

Section 13-3. Rotational Dynamics of a Rigid Object about a Fixed Axis

For the case of a rigid object rotating about a fixed axis,

$$L_z = I\omega_z \tag{13-8}$$

and

$$\Sigma\tau_z = I\alpha_z \tag{13-10}$$

Relative to any point on the axis, a torque's axial component is

$$\tau_z = RF_t$$

Section 13-4. Rotational Work and Power for a Rigid Object

The work done by a torque on a rigid object rotating about a fixed axis is

$$W = \int_{\theta_i}^{\theta_f} \tau_z \, d\theta \tag{13-12}$$

Work done by the net torque changes the rotational kinetic energy:

$$W_{\text{net}} = \tfrac{1}{2}I\omega_f^2 - \tfrac{1}{2}I\omega_i^2 \tag{13-14}$$

The power to a rotating rigid object is

$$P = \tau_z\omega_z \tag{13-15}$$

Section 13-5. Conservation of Angular Momentum

If the net external torque on a system is zero, the total angular momentum of the system is conserved:

$$\mathbf{L}_i = \mathbf{L}_f \tag{13-16}$$

Section 13-6. Motion of a Gyroscope

The motion of a gyroscope provides an example of rotational motion where the rotating object is not constrained to rotate about an axis with fixed orientation.

Table 13-1 summarizes the mathematical similarity between translational motion in one dimension and rotational motion about a fixed axis.

TABLE 13-1. *Analogy between Rotation and Translation*

Translation (one dimension)		Rotation (fixed axis)	
Coordinate	x	Coordinate	θ
Velocity component	v_x	Velocity component	ω_z
Acceleration component	a_x	Acceleration component	α_z
Mass	M	Moment of inertia	I
Force component	F_x	Torque component	τ_z
Momentum	$\mathcal{P}_x$	Momentum	L_z
$\Sigma F_{\text{ext},x} = Ma_x$		$\Sigma\tau_{\text{ext},z} = I\alpha_z$	
$\mathcal{P}_x = Mv_x$		$L_z = I\omega_z$	
$W = \int F_x\,dx$		$W = \int \tau_z\,d\theta$	
$W_{\text{net}} = \frac{1}{2}Mv_f^2 - \frac{1}{2}Mv_i^2$		$W_{\text{net}} = \frac{1}{2}I\omega_f^2 - \frac{1}{2}I\omega_i^2$	
$P = F_x v_x$		$P = \tau_z\omega_z$	

QUESTIONS

1. What is the angle between a particle's linear velocity and its angular momentum?

2. If a particle is traveling in a straight line, are there any points about which its angular momentum is zero? Explain.

3. A particle is moving along a straight line with increasing speed, and point P is not on its line of motion. Is the direction of its angular momentum about P constant? Is the magnitude of its angular momentum about P constant?

4. If a particle is in uniform circular motion, is either the direction or the magnitude of the angular momentum about the center of its motion constant? If the particle's speed is changing as it travels in a circle, is either the direction or the magnitude of the angular momentum constant?

5. If the net torque exerted on a particle is in the same direction as the particle's angular momentum, is there a change in the direction of the particle's angular momentum? Is there a change in the magnitude of the particle's angular momentum?

6. If the net torque on a particle is perpendicular to the particle's angular momentum, is there a change in the direction of the particle's angular momentum? Is there a change in the magnitude of the particle's angular momentum?

7. Consider an isolated system of two particles a and b that interact with each other such that $\mathbf{F}_{ab} = -\mathbf{F}_{ba}$, but the direction of the forces is perpendicular to the line joining the particles, as shown in Fig. 13-19. What happens to this system as time goes on? Is total linear momentum conserved? Is total angular momentum conserved? Is such a system possible? Explain.

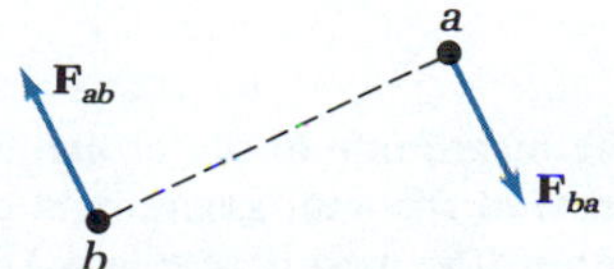

Figure 13-19. Question 7.

8. Explain why heavily muscled calves can be a disadvantage to a sprinter.

9. Explain why it is often advisable for a batter to "choke up" on the bat (Fig. 13-20) when facing a pitcher who can throw a baseball at a very high speed.

Figure 13-20. Question 9: Choking up on a bat.

10. When a billiard ball rolls down a slope without sliding, what force is responsible for the torque that causes the angular acceleration about an axis through the center of mass? What force is responsible for the torque that causes the angular acceleration about an axis through the point of contact with the surface?

11. Legend has it that a cat always lands on its feet. High-speed cameras have shown that when a cat begins a fall with its feet up, its tail rotates rapidly and the cat's body also rotates, so that it does, in fact, land on its feet. Explain the motion in terms of conservation of angular momentum. Include in your explanation a comparison of the sense of the rotation of the cat's body

with that of its tail. How do you think a bobtailed cat might do in a fall that begins with its feet up?

12. A small satellite orbiting the earth has only one window for the astronaut, and the window is facing away from the earth. Explain how the astronaut can rotate the satellite so he can view the earth and not use any rocket fuel in the process.

13. A spinning ice skater rapidly extends his arms. (Neglect friction during the time interval the arms are extended.) Is his kinetic energy conserved? Is his potential energy conserved? Is his mechanical energy conserved? Is his angular momentum conserved? If these quantities are not conserved, do they increase or decrease?

14. A yo-yo with half the string wound on its axle is placed on its edge on the floor (Fig. 13-21). Consider pulling gently on the string in three different directions indicated by $\mathbf{F}_a$, $\mathbf{F}_b$, and $\mathbf{F}_c$ in the figure. The force in each case is gentle enough so that the yo-yo does not slide. In which case, if any, does the string wind onto the yo-yo? In which case, if any, does the string wind off the yo-yo?

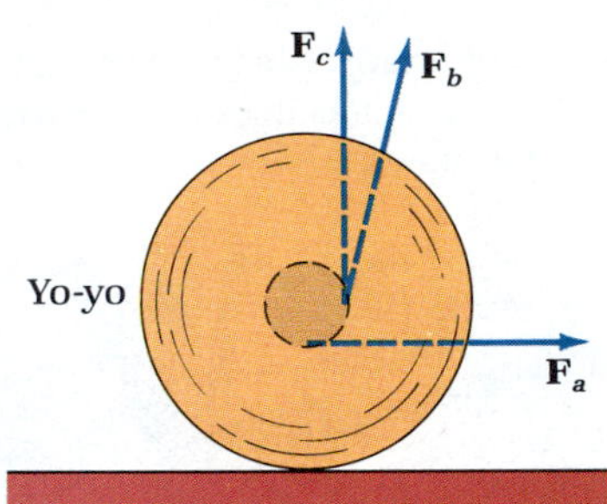

Figure 13-21. Question 14.

15. A particle moves in a straight line at constant speed; origin O is not on the particle's path. Does the line between O and the particle sweep out equal areas in equal times? Explain.

16. Consider the angular momentum $\mathbf{L}$ of a uniform door relative to origin O on the axis through the hinges. Let the z axis be along the hinges and let ω_z be positive. Describe qualitatively the direction of $\mathbf{L}$ when O is *(a)* midway between the top and bottom of the door, *(b)* at the top, *(c)* at the bottom. Reconsider the question, with ω_z negative.

17. For the case considered in Fig. 13-6, ω_z is positive so that $\mathbf{p}_i$ is into the page and $\mathbf{p}_j$ is out of the page. Draw the figure for the case where ω_z is negative. With $m_i = m_j$, are the radial components of $\boldsymbol{\ell}_i$ and $\boldsymbol{\ell}_j$ still equal and opposite?

18. A pencil that was balanced on its eraser end falls over such that the eraser does not slide (Fig. 13-22). Consider the contact force $\mathbf{F}_c$ exerted by the surface on the pencil. *(a)* Is the x component of $\mathbf{F}_c$ positive or negative? Explain. *(b)* Is the y component of $\mathbf{F}_c$ equal to mg, less than mg, or greater than mg? *(c)* Sketch Fig. 13-22 on a scratch sheet and show $\mathbf{F}_c$ and $m\mathbf{g}$ on your sketch.

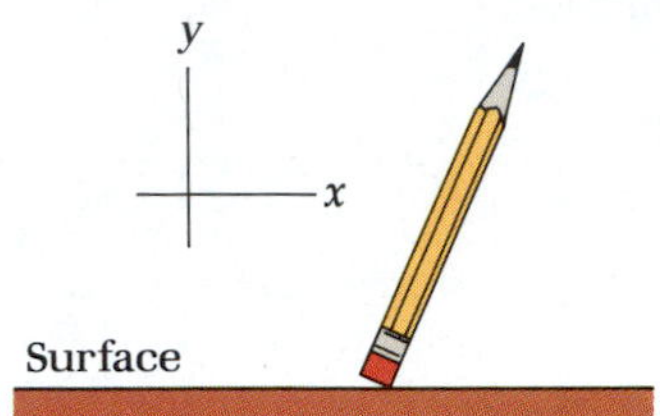

Figure 13-22. Question 18: A falling pencil.

19. Why is it easier to balance a spinning basketball on the end of your finger than a nonspinning baseketball?

20. Suppose the angular momentum of a system about a particular point is zero. Must the net external torque on the system about that point be zero? Must the net external force on the system be zero?

21. Assume that the axle-wheel system in Sec. 13-6 is constructed such that the mass of the axle is much less than the mass of the wheel. If the mass of the wheel is doubled, what will be the approximate effect on the angular speed of precession?

22. Consider the bicycle wheel in Fig. 13-18 spinning in the opposite direction from that shown. What are the directions of ω_p, $\mathbf{L}$, $d\mathbf{L}$, and $\Sigma\tau_{\text{ext}}$ in this case?

23. You hand a porter a suitcase that contains a large spinning rotor. What happens when the porter turns a corner?

24. Complete the following table:

Symbol	Represents	Type	SI Unit
$\boldsymbol{\ell}$		Vector	
$\mathbf{L}$			kg·m²/s
$\Sigma\tau_{\text{ext}}$			
r	Distance from a point		
R			
P			

EXERCISES

Section 13-1. Angular Momentum of a Particle

1. An observer stands 125 m south of an east-west road, and a 1340-kg automobile is traveling east along the road. *(a)* What are the magnitude and direction of the angular momentum of the automobile relative to the observer at the instant it is directly north of the observer and moving at a speed of 36.4 m/s? *(b)* What is the angular momentum after the car has traveled 325 m beyond the position in part *(a)* and is still moving at 36.4 m/s?

2. Determine the magnitude of the orbital angular momentum of Mars relative to the sun, assuming a circular orbit of radius 2.28×10^{11} m. The mass of Mars is 6.46×10^{23} kg, and the period of its orbit is 5.94×10^{7} s.

3. What is the angular momentum relative to the origin for a 4.1-kg particle at the instant its position is $\mathbf{r} = (-3.5\text{ m})\mathbf{i} + (1.4\text{ m})\mathbf{j}$ and its velocity is $\mathbf{v} = (-2.0\text{ m/s})\mathbf{i} + (-6.3\text{ m/s})\mathbf{j}$?

4. A 3.6-kg particle is moving in the xy plane. What is its angular momentum (magnitude and direction) relative to the origin at the instant it crosses the x axis at $x = +4.6$ m and at a speed of 2.4 m/s? The angle between $\mathbf{i}$ and $\mathbf{v}$ at that instant is $+0.76$ rad.

5. A 72-g bead slides without friction on a vertically oriented circular wire of radius 0.93 m, as shown in Fig. 13-23. If the bead is released from rest at $\theta = 0.87$ rad, what is the angular momentum of the bead relative to C at the instant it passes the x axis?

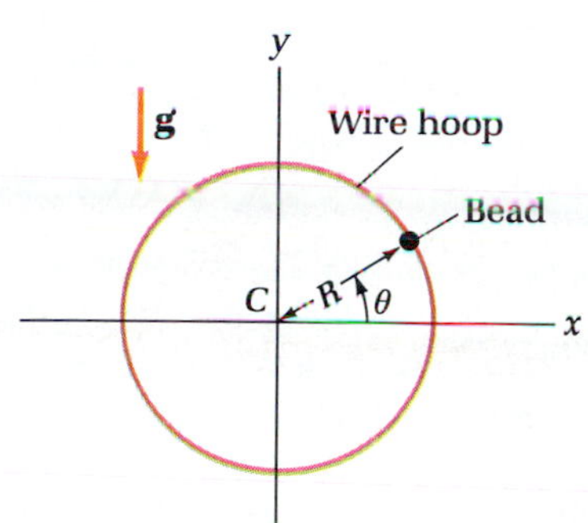

Figure 13-23. Exercise 5.

6. Recall that the kinetic energy of a particle may be written $K = \frac{1}{2}p^2/m$, where $p = mv$. Show that the kinetic energy of a particle moving in a circle of radius R can be written $K = \frac{1}{2}\ell^2/I$, where $I = mR^2$ and ℓ is measured relative to the center of the circle.

Section 13-2. Angular Momentum of a System of Particles

7. A system consists of three particles which have angular momenta, relative to the origin, of $\boldsymbol{\ell}_a = (2.4\text{ kg}\cdot\text{m}^2/\text{s})\mathbf{i}$, $\boldsymbol{\ell}_b = (-6.1\text{ kg}\cdot\text{m}^2/\text{s})\mathbf{k}$, and $\boldsymbol{\ell}_c = (-4.8\text{ kg}\cdot\text{m}^2/\text{s})\mathbf{i} + (1.6\text{ kg}\cdot\text{m}^2/\text{s})\mathbf{j}$. What is the total angular momentum of the system relative to the origin?

8. Two particles A and B exert forces on one another of magnitude 14 N, as shown in Fig. 13-24. *(a)* Determine the magnitude and direction of the torque exerted by B on A about the origin. *(b)* Determine the magnitude and direction of the torque exerted by A on B about the origin. *(c)* Determine the net torque on the two particles due to their interaction.

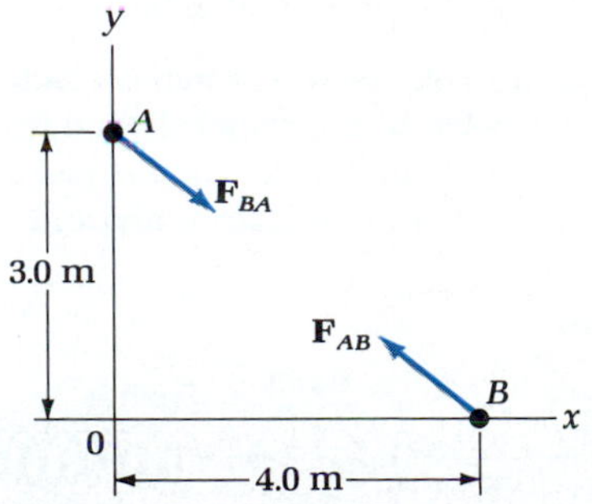

Figure 13-24. Exercise 8: The $+\mathbf{k}$ direction is out of the page.

9. Two particles of equal mass m move in opposite directions along straight-line paths with the same speed v. The paths of the particles are parallel and separated by a distance D. Show that the magnitude of the total angular momentum of the two particles is mvD relative to any point.

10. Two particles of mass M_1 and M_2 are connected to each other and to the axis of rotation by a rigid rod of negligible mass (Fig. 13-25). The two particles have an angular speed ω and are at distances of R_1 and R_2 from the axis. Write an expression for the total angular momentum in terms of M_1, M_2, R_1, R_2, and ω.

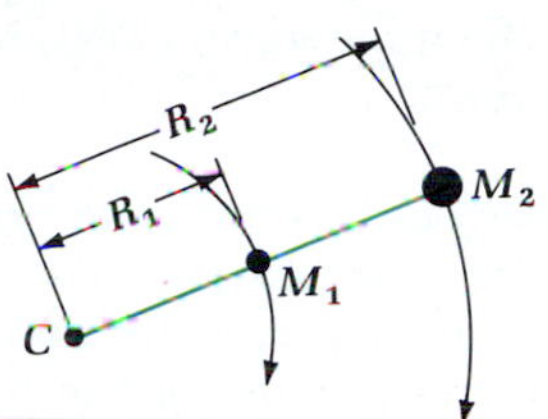

Figure 13-25. Exercise 10.

Section 13-3. Rotational Dynamics of a Rigid Object about a Fixed Axis

11. What is the angular momentum, relative to the center of the record, of an 85-g phonograph record of radius 150 mm when it is rotating on a 78-rev/min turntable?

12. *(a)* What is the z component of the net external torque on a uniform door of mass 22 kg and width 0.95 m at the instant its z component of angular acceleration is 8.2 rad/s²? The z axis is along the hinges and $+z$ is upward. *(b)* If the net torque component in part *(a)* is entirely due to a force perpendicular to the face of the door and applied at the knob, which is 0.89 m from the axis through the hinges, what is the magnitude of this force?

13. A grinding wheel with a moment of inertia about its axle of 0.15 kg·m² has a constant net torque of magnitude 18 N·m exerted on it about its axle. Assuming it starts from rest at $t = 0$, develop equations for the wheel's angular acceleration, angular velocity, and angular position as functions of time.

14. A bicycle wheel with a moment of inertia about its axle of 0.25 kg·m² and an initial angular speed of 12 rad/s slows to a stop because of friction in the bearings in a time interval of 320 s. Determine the magnitude of the torque due to friction, assuming it is constant.

15. *(a)* The pulley in Fig. 13-26*a* has a moment of inertia of 0.085 kg·m² about its axle and a radius of 170 mm. The string

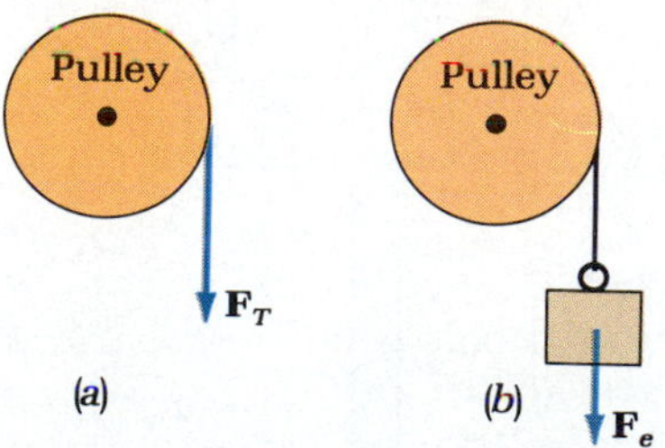

Figure 13-26. Exercise 15.

wrapped around the pulley exerts a constant force of magnitude 32 N. Neglecting the torque due to friction in the bearings, determine the pulley's angular acceleration magnitude. *(b)* The same pulley now has a 32-N block tied to the string and is released (Fig. 13-26*b*). Determine the pulley's angular acceleration magnitude. *(c)* Explain why the answers to parts *(a)* and *(b)* are different.

16. A 3.0-kg pulley with a radius of 120 mm has a 7.3-kg block tied to a string that is wrapped around the pulley, similar to the arrangement shown in Fig. 13-26*b*. The block falls from rest a distance of 450 mm in 0.33 s. Find the moment of inertia of the pulley about its axle, assuming frictional effects are negligible. What can you say about the way the mass of the pulley is distributed about its axis?

17. Rework the previous exercise, taking friction in the pulley bearings into account. Using the procedure of Exercise 14, we find that the torque due to friction between the pulley and its bearings is 0.23 N·m. Compare your answer with that of the previous exercise.

18. Blocks *B* and *C* are tied to a string that passes over a pulley *P*, as shown in Fig. 13-27. Neglecting friction between the pulley and its bearings, neglecting the mass of the string, and assuming that the string does not slip, determine *(a)* the magnitude of the linear acceleration of the blocks, *(b)* the tension in the part of the string connected to *B*, *(c)* the tension in the part of the string connected to *C*, *(d)* the magnitude of the force on the pulley due to its bearings. *(e)* Compare the answer in part *(d)* with the total weight of the blocks and the pulley, and explain why these two results differ. Treat the pulley as a uniform disk of radius R_0 and use these data: $R_0 = 78$ mm, $m_p = 0.74$ kg, $m_B = 0.83$ kg, and $m_C = 0.57$ kg.

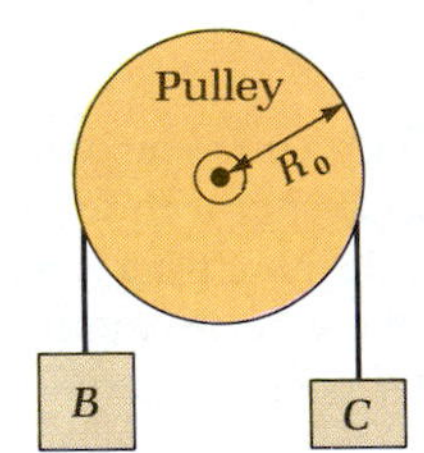

Figure 13-27. Exercise 18.

19. Blocks *B* and *C* are tied together by a light string that passes over a pulley, as shown in Fig. 13-28. Friction between block *B* and the table and friction in the pulley are negligible. Assuming the string does not slip, determine *(a)* the magnitude of the linear acceleration of the blocks, *(b)* the tension in the string between block *B* and the pulley, *(c)* the tension in the string between block *C* and the pulley. Find your answers in terms of m_B, m_C, *I*, R_0, and *g*.

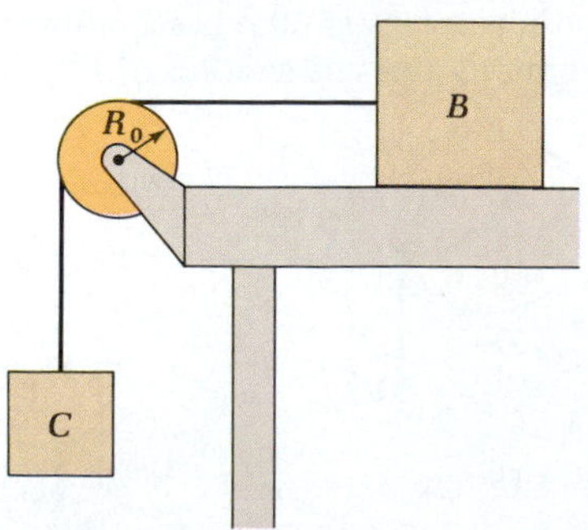

Figure 13-28. Exercise 19.

20. A 4.5-kg block is tied to a string that is wrapped around a pulley, and the block slides down a slope as shown in Fig. 13-29. The coefficient of kinetic friction between the block and the slope is 0.30, and there is a constant torque due to friction, of magnitude 1.3 N·m acting on the pulley. The moment of inertia of the pulley about its axle is 0.016 kg·m². Other data are $\theta = 0.73$ rad and $R_0 = 85$ mm. Determine *(a)* the magnitude of the block's acceleration and *(b)* the tension in the string.

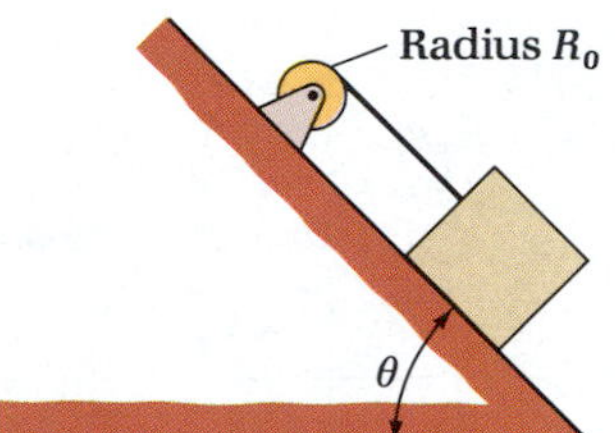

Figure 13-29. Exercise 20.

21. Find the force on the windlass of Example 13-5 that is caused by its bearings.

Section 13-4. Rotational Work and Power for a Rigid Object

22. *(a)* What is the work done by the frictional torque at the bearings of a bicycle wheel of moment of inertia 0.22 kg·m² when the wheel slows to a stop from an angular speed of 14 rad/s? *(b)* The wheel comes to rest in a time interval of 86 s. Assuming the frictional torque is constant, what is the magnitude of the angular acceleration? *(c)* Find the power delivered to the wheel as a function of time where $t = 0$ corresponds to the instant its angular speed was 14 rad/s.

23. Show that the kinetic energy of a rigid body rotating about a fixed axis may be written $K = \frac{1}{2}L^2/I$.

24. What is the power delivered to the pulley by the rope in Fig. 13-26*a* at the instant its angular speed is 2.9 rad/s and $F_T = 32$ N?

25. Determine the linear speed of the two blocks in Fig. 13-27 at the instant block *B* has fallen 12 mm after being released from rest.

26. The pulley *P* in Fig. 13-30 is rotated clockwise until the spring is stretched 52 mm, and then released from rest. Determine *(a)* the speed of the block when it passes through the position at which the spring is unstretched and *(b)* the power

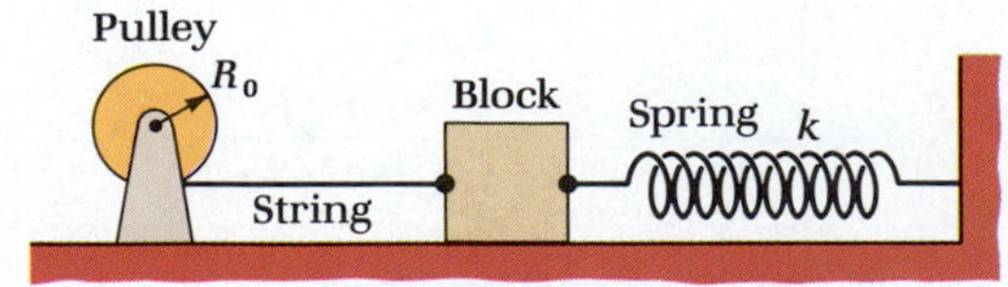

Figure 13-30. Exercise 26.

delivered to the block by the spring when it passes through the position at which the spring is unstretched. Neglect friction between the block and the surface and in the pulley bearings. The data are $m_B = 4.3$ kg, $I_p = 0.016\ \text{kg}\cdot\text{m}^2$, $R_0 = 73$ mm, and $k = 230$ N/m.

Section 13-5. Conservation of Angular Momentum

27. An ice skater spins about a vertical axis at an angular speed of 15 rad/s with outstretched arms, then quickly pulls her arms in to her sides in a time interval so small that the effect of frictional forces due to the ice is negligible. Her initial moment of inertia about the axis of rotation is $1.72\ \text{kg}\cdot\text{m}^2$, and her final moment of inertia is $0.61\ \text{kg}\cdot\text{m}^2$. *(a)* What is her resultant angular speed? *(b)* What is the change in her kinetic energy? *(c)* Explain this change in kinetic energy.

28. A 22-kg child stands halfway between the center and the edge of a merry-go-round that is rotating freely at an angular speed of 1.8 rad/s. *(a)* What is the angular speed of the merry-go-round after the child walks to its edge? *(b)* What is the change in the kinetic energy of the system (child plus merry-go-round)? The radius of the merry-go-round is 3.0 m, and its moment of inertia, when empty, is $610\ \text{kg}\cdot\text{m}^2$. Neglect friction in the axle during the time interval in which the child moves.

29. An open door of mass M is at rest when struck by a thrown ball of mass m ($m \ll M$) at a point that is a distance D from an axis through the hinges (Fig. 13-31). Just before the ball strikes the door, its path is perpendicular to the door face, and, since $m \ll M$, its path is still nearly perpendicular just after the collision. The door has a uniform density and width w. Let v_i and v_f represent the initial and final speeds of the ball. Neglect friction in the hinges during the time interval of the collision. *(a)* Taking the system to be the ball and the door, explain why the total linear momentum of the system is not conserved. *(b)* Is the angular momentum of the system about any axis conserved? If so, identify the axis. *(c)* Use the approximation discussed above and determine an expression for the resulting angular speed ω of the door in terms of the quantities introduced. *(d)* Evaluate ω when $m = 1.1$ kg, $M = 35$ kg, $w = 73$ cm, $D = 62$ cm, $v_i = 27$ m/s, and $v_f = 16$ m/s. (This collision is considered further in Prob. 19.)

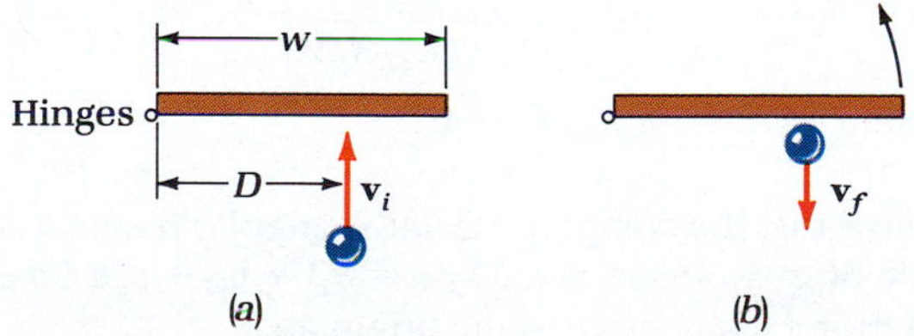

Figure 13-31. Exercise 29: Top view of the door: *(a)* before the collision; *(b)* after the collision.

30. The door described in the previous exercise is at rest when struck by a ball of putty of mass m (Fig. 13-32). The initial velocity of the putty is horizontal and makes an angle θ with a normal to the door, and the putty sticks to the door after the collision. *(a)* Find the resulting angular speed of the door. *(b)* Find the change in kinetic energy of the system (door plus putty). *(c)* Evaluate the answers to parts *(a)* and *(b)* when $m = 1.1$ kg, $M = 35$ kg, $w = 73$ cm, $D = 62$ cm, $\theta = 0.38$ rad, and $v = 27$ m/s.

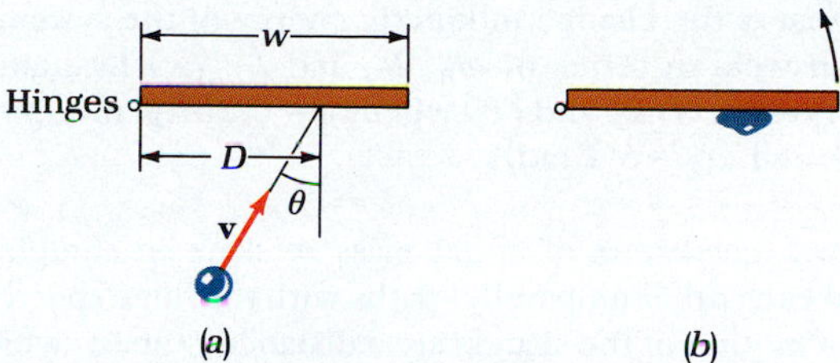

Figure 13-32. Exercise 30: Top view of the door: *(a)* before the collision; *(b)* after the collision.

31. A block of mass m moves in a circle of radius R_i with speed v_i, while sliding without friction on a horizontal tabletop. The block is tied to a string that passes through a hole in the table (Fig. 13-33). There is no frictional force between the table and the string. The end of the string under the table is displaced downward so that after the displacement the block moves in a circle of radius R_f. *(a)* Show that the ratio of the tension in the string after the displacement to that before the displacement equals $(R_i/R_f)^3$. *(b)* Determine the tension in the string after the displacement for the case where the tension before the displacement is 3.4 N and $R_f = \frac{1}{2}R_i$.

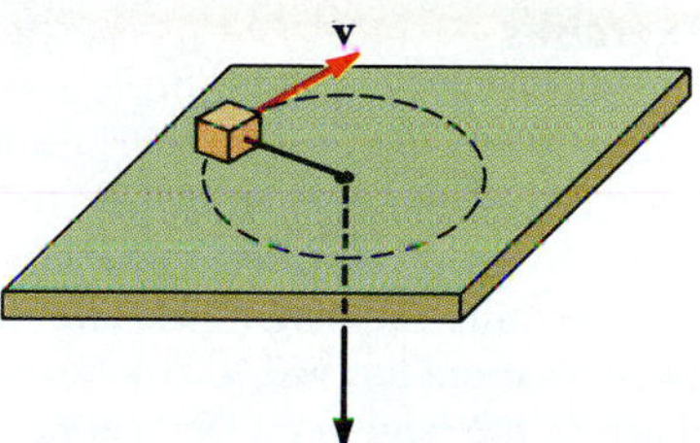

Figure 13-33. Exercise 31.

32. Disk B rotates freely with angular velocity ω_B and has a rod that projects along its axis of rotation. A hole in the center of disk C is fitted over the rod, as shown in Fig. 13-34, and disk C, initially at rest, is dropped onto disk B. *(a)* Determine the angular velocity of the system after the frictional interaction between B and C brings them to a common angular velocity.

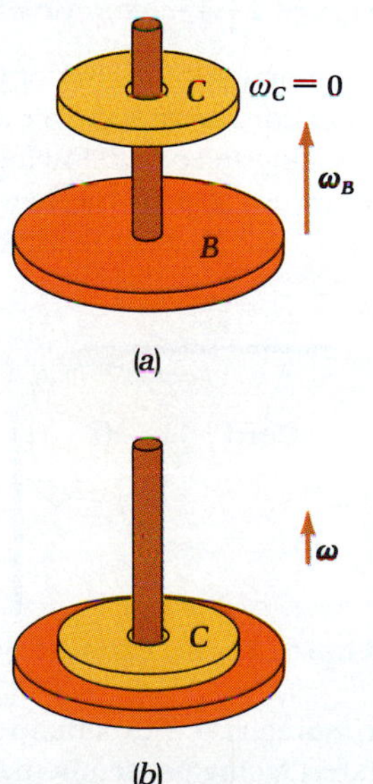

Figure 13-34. Exercise 32: Wheel B is connected to the shaft and wheel C fits loosely over the shaft: *(a)* before C is dropped; *(b)* after C is dropped.

(b) What is the change in kinetic energy of the system? Find your answers in terms of ω_B, I_B, and I_C. *(c)* Evaluate your answers to parts *(a)* and *(b)* when $I_B = 0.20$ kg·m², $I_C = 0.10$ kg·m², and $\omega_B = 6.2$ rad/s.

33. Two ice skaters of equal mass m skate in straight lines toward each other on parallel paths with the same speed v. The lines of motion of the skaters are a distance D apart, which is a little more than an arm's length. When the skaters are abreast, they clasp arms and rotate in a circle of diameter D. *(a)* Determine the resulting angular speed of the skaters in terms of the quantities introduced above. *(b)* Find the change, if any, in the kinetic energy of the system.

Section 13-6. Motion of a Gyroscope

34. A bicycle-wheel gyroscope is precessing without nutation, with its axle horizontal (Fig. 13-17). The wheel is spinning at 58 rad/s about its axle. The moment of inertia of the wheel about its axle is 0.23 kg·m², and the moment of inertia of the wheel and axle about the pivot is 0.14 kg·m². The axle is 280 mm long and the wheel is centered on the axle. *(a)* Find the angular speed of precession, assuming $L_s \gg L_p$. *(b)* Determine L_s and L_p and justify the assumption in part *(a)*.

Additional Exercises

35. The 2.6-kg disk in Fig. 13-35 is uniform and has a radius of 0.38 m. The forces exerted on it at the points shown have magnitudes $F_1 = 12$ N, $F_2 = 24$ N, and $F_3 = 18$ N. In addition to these forces, a fourth force exerted at the axis holds the axis stationary. *(a)* Determine the magnitude and direction of the net torque on the disk about its axis. Let $+\mathbf{k}$ be directed parallel to the axis and out of the page. *(b)* Determine the magnitude and direction of the disk's angular acceleration.

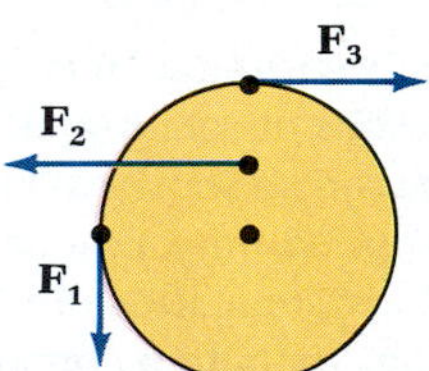

Figure 13-35. Exercise 35.

36. A pulley with a moment of inertia of 0.022 kg·m² and a radius of 0.31 m has a cord wrapped over it (Fig. 13-36). The pulley is rotating clockwise and is slowing at the rate of 4.9 rad/s². The cord does not slip. If the tension in the vertical section of the cord is 1.22 N, what is the tension in the horizontal section?

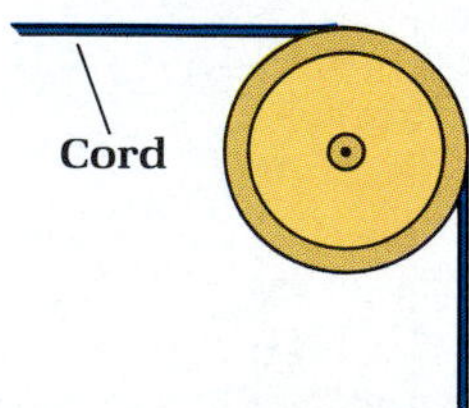

Figure 13-36. Exercise 36.

37. A 20-kg girl running at 4.8 m/s jumps onto an open gate that was initially at rest facing perpendicular to the girl's velocity. The girl holds her feet off the ground by clinging to the latch on the side of the gate opposite the hinges, and, after the encounter, the girl-gate system swings with angular speed ω. The mass of the gate is 30 kg and its width is 1.5 m. Determine ω and state any significant assumptions that you use.

38. A diver dives off a springboard with her body outstretched and rotating about a horizontal axis through her center of mass such that her angular speed is 2.4 rad/s and her moment of inertia about her axis of rotation is 11.2 kg·m². *(a)* If she tucks her arms and legs in close to her body such that her moment of inertia becomes 6.5 kg·m², what is her new angular speed? Determine her rotational kinetic energy *(b)* when her body is outstretched and *(c)* when her body is tucked in.

39. A neutron star is created by the collapse of an ordinary star. A typical neutron star has a mass 1 or 2 times that of the sun, a radius of a few kilometers, and an exceedingly large angular speed of from one thousand to several thousand radians per second. The material that collapsed into a neutron star is believed to originate from the star's core, which had a radius of about 1×10^7 m before the collapse. *(a)* Estimate the initial rotational speed of a star's core that subsequently collapses to a neutron star with an angular speed of 1000 rad/s. *(b)* The sun's rotational period is about 25 days. Determine the sun's angular speed for comparison with your answer to part *(a)*.

40. You are in a spaceship and wish to change its orientation, say, 10° about some axis. You attach a flywheel support to the spacecraft with the flywheel's axle along the chosen axis. The moment of inertia of the flywheel is 1.0×10^{-5} that of the spacecraft. *(a)* Through how many revolutions must you spin the flywheel to achieve the desired spacecraft orientation? *(b)* Should the flywheel be rotated in the same sense or in the opposite sense from the desired rotation of the spacecraft?

41. A string is wound around a yo-yo axle and the end of the string is held fixed. Then the yo-yo is released and the string unwinds as the yo-yo falls. Let the yo-yo's mass be M, its outer radius be R_2, and the radius of its axle be R_1. For simplicity, neglect the thickness of the string and the axle's contribution to the yo-yo's moment of inertia ($I = \frac{1}{2}MR_2^2$). Assume that the string remains vertical as the yo-yo falls. *(a)* Show that the magnitude of the yo-yo's acceleration is

$$a = \frac{2g}{[2 + (R_2/R_1)^2]}$$

(b) Evaluate a when $R_2/R_1 = 4$.

42. Show that the components of the angular momentum of a particle of mass m and velocity $\mathbf{v} = v_x\mathbf{i} + v_y\mathbf{j} + v_z\mathbf{k}$ located at $\mathbf{r} = x\mathbf{i} + y\mathbf{j} + z\mathbf{k}$ relative to the origin are

$$\ell_x = m(yv_z - zv_y)$$
$$\ell_y = m(zv_x - xv_z)$$
$$\ell_z = m(xv_y - yv_x)$$

43. Use the result of the previous exercise to determine the angular momentum relative to the origin of a 2.7-kg particle located at $\mathbf{r} = (-3.5\text{ m})\mathbf{i} + (1.1\text{ m})\mathbf{j} + (2.8\text{ m})\mathbf{k}$ traveling with a velocity $\mathbf{v} = (1.4\text{ m/s})\mathbf{i} + (-3.3\text{ m/s})\mathbf{k}$. Give your answer in terms of unit vectors.

44. A helicopter has three rotor blades. Each blade is 5.5 m

long and has a mass of 250 kg. Determine the angular momentum magnitude of this rotor assembly when its angular speed is 300 rev/min.

45. A thin rod of length 1.0 m and mass 0.12 kg can rotate freely (with negligible friction) about a horizontal axis that is perpendicular to the rod's length and passes through its midpoint. The rod is initially horizontal and at rest. A 0.040-kg wad of putty is dropped from rest at a height of 0.50 m above one end of the rod, and the putty sticks to the rod's end when they collide. Determine the angular speed of the rod-putty system just after the collision.

46. Suppose a star the size of the sun with a radius of 7.0×10^8 m collapses to become a white dwarf with a radius 3.0×10^3 m. The star has an angular speed of 1.2 revolutions per month (the same as the sun's) before the collapse. What is its angular speed after the collapse?

47. A uniform disk of mass M and radius R_0 is constrained to rotate about a vertical axis. Friction in the axle is negligible as the disk rotates freely at an initial angular speed ω_i. A ball of putty of mass m is dropped vertically onto the disk, and it sticks to the disk at a distance $\frac{1}{2}R_0$ from the axis. *(a)* Find an expression for the angular speed ω_f of the disk-putty system after the encounter. *(b)* Evaluate ω_f when $M = 2.0$ kg, $m = 1.0$ kg, and $\omega_i = 5.0$ rad/s.

48. A thin uniform rod of length 1.4 m and mass 0.32 kg is mounted with a horizontal axle perpendicular to its length and passing through its midpoint. A 0.20-kg block is secured to one end of the rod, and the rod is then aligned horizontally and released from rest. What is the magnitude of the rod's angular acceleration just after it is released? Friction in the axle is negligible.

49. A pulley of radius 0.19 m can rotate with negligible friction about a horizontal axle. A cord is wrapped around the pulley and given a steady pull such that the tension in the cord is constant at 7.7 N. What is the power delivered to the pulley at an instant when its angular speed is 2.6 rad/s?

50. While tightening a bolt, you exert a steady force of magnitude 19 N on the wrench handle as the bolt rotates 90°. The force you exert is always directed perpendicular to the handle's long dimension, and it is applied at a distance of 0.21 m from the axis of rotation. What is the work done by you?

PROBLEMS

1. ***Velocity* v *in terms of* ω *and* r.** Consider a particle moving in a circle. Show that the equation $\mathbf{v} = \boldsymbol{\omega} \times \mathbf{r}$ gives the correct magnitude and direction for $\mathbf{v}$, where $\mathbf{r}$ is the position of the particle relative to the center of the circle.

2. ***Torque equals the rate of change of the angular momentum.*** Consider a particle at position $\mathbf{r} = x\mathbf{i} + y\mathbf{j} + z\mathbf{k}$, with linear momentum $\mathbf{p} = p_x\mathbf{i} + p_y\mathbf{j} + p_z\mathbf{k}$ that has force $\mathbf{F} = F_x\mathbf{i} + F_y\mathbf{j} + F_z\mathbf{k}$ exerted on it. *(a)* Determine the components of the angular momentum $\boldsymbol{\ell}$ and torque $\boldsymbol{\tau}$ relative to the origin. *(b)* Show that $d\ell_z/dt = v_x p_y - v_y p_x + xF_y - yF_x$. *(c)* Show that the first two terms in the previous equation add to zero so that $d\ell_z/dt = \tau_z$.

3. ***Starting a ball rolling with no sliding.*** To start a billiard ball of radius r_0 rolling without sliding on a horizontal frictionless surface, a horizontal force is exerted on the ball such that its line of action is a distance h above the center of the ball (Fig. 13-37). Find h in terms of r_0. (*Hint:* Consider both the translational and rotational motion of the ball.)

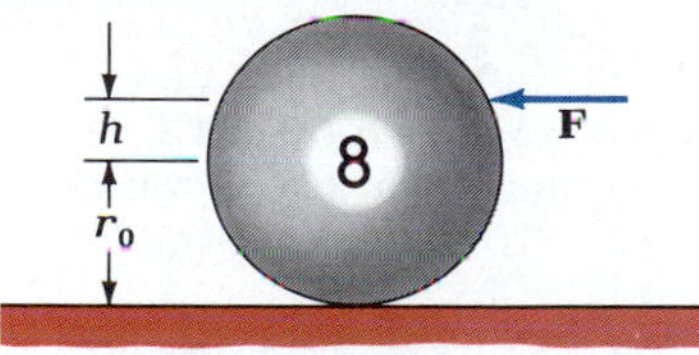

Figure 13-37. Problem 3.

4. ***Minimum coefficient of friction for rolling without sliding down a slope.*** Consider a billiard ball rolling without sliding down a straight slope that makes an angle θ with the horizontal. *(a)* Show that the magnitude of the linear acceleration of the ball is $(5g \sin \theta)/7$. *(b)* Show that the minimum value for the coefficient of friction between a ball and the surface in order for the ball to roll down the slope without sliding is $(2 \tan \theta)/7$. (*Hint:* Consider both the translational and rotational motion of the ball.)

5. ***Sliding with no rolling becomes rolling with no sliding.*** A billiard ball of radius r_0 is initially sliding without rolling at linear speed v_0 on a horizontal surface; the coefficient of friction between the ball and the surface is μ. Show that, at the instant the ball begins to roll without sliding, *(a)* its linear speed is $5v_0/7$; *(b)* the time elapsed is $2v_0/7\mu g$; *(c)* the distance traveled is $12v_0^2/49\mu g$. (*Hint:* During the time the ball is both rolling and sliding, the surface exerts a constant force μmg at the point of contact. Also, the equations $v = \omega r_0$ and $a = \alpha r_0$ are valid at the instant rolling without sliding begins.)

6. ***Rolling a yo-yo with a horizontally directed force.*** Consider the force $\mathbf{F}_a$ exerted on the yo-yo of Fig. 13-21. *(a)* Assuming the yo-yo does not slide, find the magnitude of its linear acceleration in terms of its mass M, its radius R_2, the radius of its axle R_1, and F_a. Assume $I = \frac{1}{2}MR_2^2$. *(b)* Show that the maximum value F_a can have such that the yo-yo does not slide is $3\mu MgR_2/(R_2 + 2R_1)$, where μ is the coefficient of friction between the yo-yo and the surface.

7. ***Minimum release height for completing a loop.*** A billiard ball of radius r_0 rolls without sliding from rest at the top of a loop-the-loop track having a loop of radius R_0, as shown in Fig. 13-38. What is the minimum height h from which the ball can be released and have it not leave the track as it passes the top of the loop? Assume $r_0 \ll R_0$.

8. ***Speed of a rolling basketball.*** A basketball of radius r_0 rolls without sliding down the side of a circular track of radius R_0 (Fig. 13-39). The ball is released from rest at angle θ. What is its speed as it reaches the bottom?

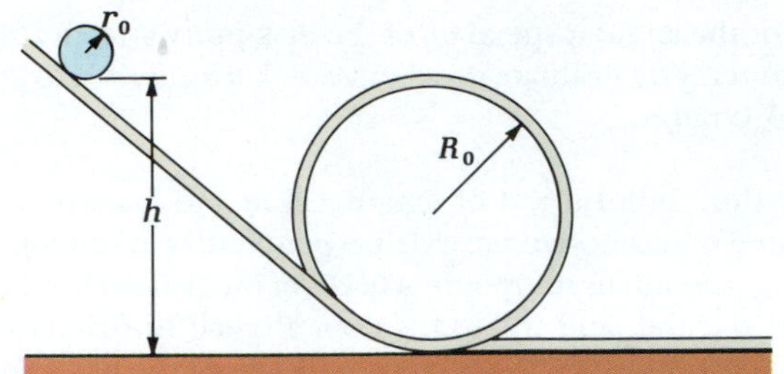

Figure 13-38. Problem 7.

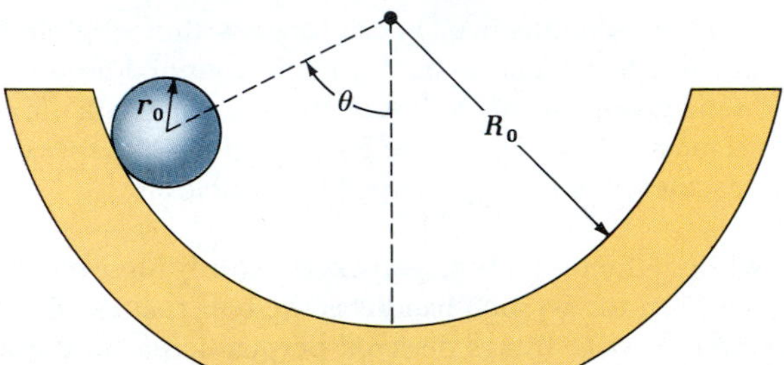

Figure 13-39. Problem 8.

9. ***Angular speed of a falling trapdoor.*** A uniform trapdoor of mass M and length h rotates without friction about a horizontal axis through its hinges, as shown in Fig. 13-40. Show that the angular speed of the door at position θ, after beginning the fall from a vertical orientation and with a negligible initial speed, is $\sqrt{3g(1 - \cos\theta)/h}$.

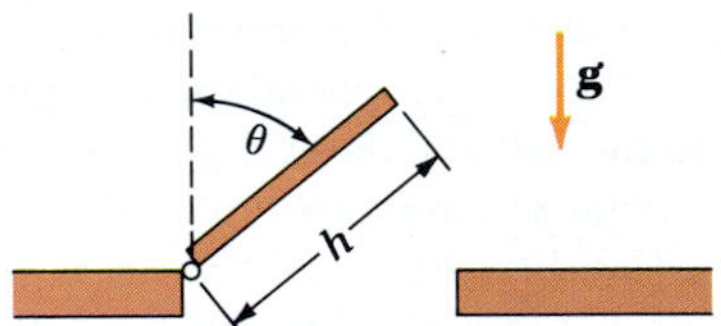

Figure 13-40. Problem 9: Side view of a trapdoor.

10. ***Interacting wheels.*** Two wheels B and C rotate without friction about axles that are parallel but displaced, as shown in Fig. 13-41. The wheels have the same mass M, radius R_0, and moment of inertia I. Wheel C has angular velocity $\omega_0\mathbf{k}$, and wheel B is at rest when their edges are brought into contact.

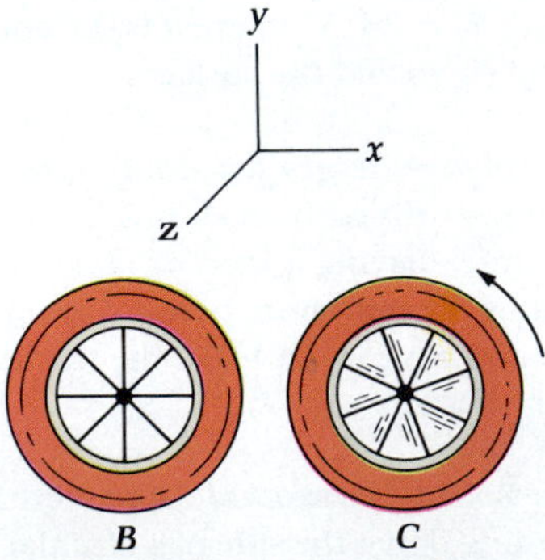

Figure 13-41. Problem 10: The $+\mathbf{k}$ direction is out of the page.

After a while, their angular speeds are the same as a result of their frictional interaction: $\omega_C = \omega\mathbf{k}$ and $\omega_B = -\omega\mathbf{k}$. *(a)* If we take the two wheels (excluding the axles) as the system, is the kinetic energy of the system conserved? *(b)* Is the angular momentum of the system conserved? *(c)* Show that the torque on either wheel is $-FR_0\mathbf{k}$ about any axis parallel to the axles, where F is the magnitude of the vertical component of the frictional force of interaction. *(d)* Show that $\omega = \frac{1}{2}\omega_0$. *(e)* What is the ratio of the final to the initial kinetic energy? *(f)* What is the final angular momentum of the system? (*Hint:* Be sure to take account of the force on the wheels due to their axles.)

11. ***Work done on a circulating block.*** Reconsider the block of Exercise 31 (Fig. 13-33). Suppose the string is pulled down very slowly so that the block can be considered to be nearly moving in a circle of radius R at any instant. *(a)* Show that the tension in the string varies with R as $mv_i^2R_i^2/R^3$, where i refers to the initial values. *(b)* Use your answer from part *(a)* to find the work done on the block by the tension in the string in changing the radius of the block's circular path from R_i to R_f. *(c)* Show that your answer from part *(b)* equals the change in kinetic energy of the block.

12. ***Inverting the axis of a rotating wheel.*** A student is sitting on a stool that can rotate about a vertical axis, and she is holding a bicycle wheel such that the axle is vertical, as shown in Fig. 13-42. The moment of inertia of the wheel about its axle is 0.21 kg·m², and the moment of inertia of the student plus wheel plus stool seat about the stool's axle is 2.8 kg·m². The initial angular velocity of the wheel about its axle is (61 rad/s)**k**, and the initial angular velocity of the student about the stool's axle is zero, where we take **k** as vertically upward. The student rotates the wheel's axle 180° such that it is again vertical and the angular velocity of the wheel is (−61 rad/s)**k**. *(a)* Determine the resulting angular velocity of the student about the stool's axle. *(b)* Determine the work done by the student. Neglect friction in the stool's axle.

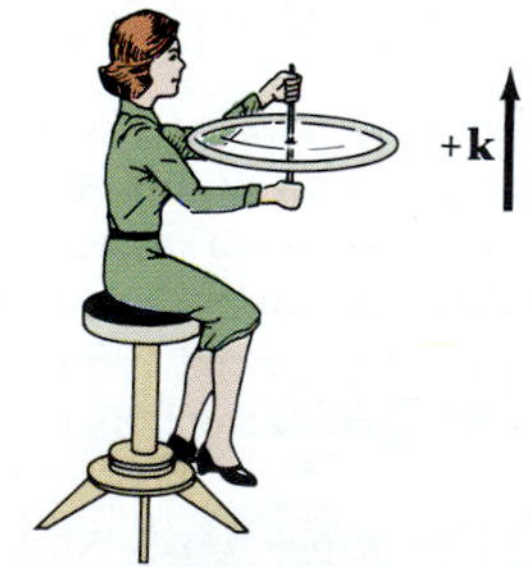

Figure 13-42. Problem 12: The $+\mathbf{k}$ direction is vertically upward.

13. ***Precession of the bicycle-wheel gyroscope.*** Suppose the axle of a steadily precessing bicycle wheel is not horizontal, as it was shown to be in Fig. 13-18; rather, the axle makes an angle θ with the xz plane. Using the assumptions given in Sec. 13-6, show that the angular speed of precession is independent of θ.

14. ***Moment of inertia of the bicycle-wheel gyroscope about a vertical axis.*** Find I_p for the bicycle-wheel gyroscope in Fig. 13-18 with the wheel centered on the axle. Assume

$I_s = M_W R_0^2$, where M_W is the mass of the wheel (tire and rim) and R_0 is its radius. That is, treat the wheel as a hoop. Use the plane-figure theorem (Prob. 6 in Chap. 12) and the parallel-axis theorem to show that $I_p = \frac{1}{2}M_W R_0^2 + M_W D^2 + 4mD^2/3$, where m is the mass of the axle and $2D$ is its length.

15. ***Force exerted by the pivot on the bicycle-wheel gyroscope.*** *(a)* Show that the horizontal component of the force due to the pivot on the bicycle-wheel gyroscope described in Sec. 13-6 is $Mg[M^2gD^3/(I_s\omega_s)^2]$. *(b)* Determine the magnitude of the force due to the pivot. *(c)* Determine the angle between the vertical and the direction of the force due to the pivot.

16. ***Precession of the earth.*** The earth's axis of rotation precesses because the earth is not a perfect sphere. The earth can be considered as a perfect sphere plus a "belt" (Fig. 13-43). *(a)* Show that if we further approximate the belt as a dumbbell, with spheres of mass Δm at the positions shown in the figure, then the magnitude τ_s of the torque exerted by the sun about the center of the earth is

$$\tau_s = \frac{6\ GM_s\ \Delta m R_e^2 \sin 23^\circ \cos 23^\circ}{R_0^3}$$

where M_s is the mass of the sun, R_e is the radius of the earth, and R_0 is the earth-sun distance. *(b)* Averaging over the actual bulge and over the year gives an average torque of 3/8 the above result. Also, comparison of the earth's circumference around the poles and around the equator gives Δm as 8/3000 the mass of the earth. Show that this leads to a precession of the earth's axis with a period of about 80,000 years. (Effects of the moon bring the precessional period down to about 26,000 years.)

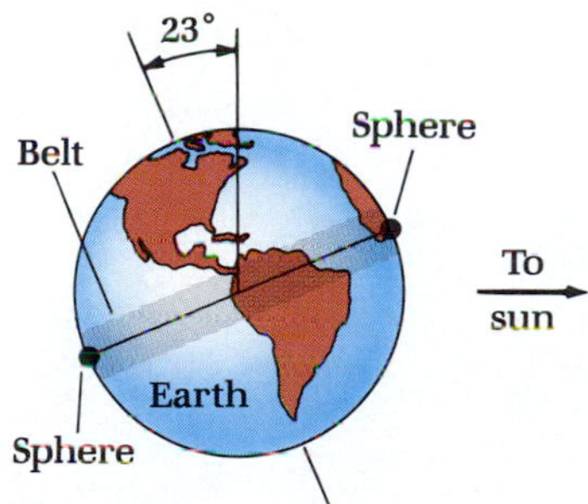

Figure 13-43. Problem 16.

17. ***A case where* L *and* ω *are not parallel.*** Consider a system in which $\boldsymbol{\omega}$ and $\mathbf{L}$ are not parallel. Two spheres of equal mass M are connected by a rod of negligible mass and rotate on an axle, as shown in Fig. 13-44. Show that at the instant the spheres are in the yz plane: *(a)* the angular momentum about the origin is

$$\mathbf{L} = 2Mr^2\omega \sin\theta\,[(\sin\theta)\mathbf{k} + (\cos\theta)\mathbf{j}]$$

(b) the torque on the axle due to the bearings about the origin is

$$\boldsymbol{\tau}_b = -(2Mr^2\omega^2 \sin\theta \cos\theta)\mathbf{i}$$

(*Hint:* Consider the projection of $\mathbf{L}$ and $d\mathbf{L}$ onto the xy plane similar to Fig. 13-18.) *(c)* the horizontal component of the force exerted on the axle by the upper bearings is

$$F_b = \frac{Mr^2\omega^2}{D}\cos\theta \sin\theta$$

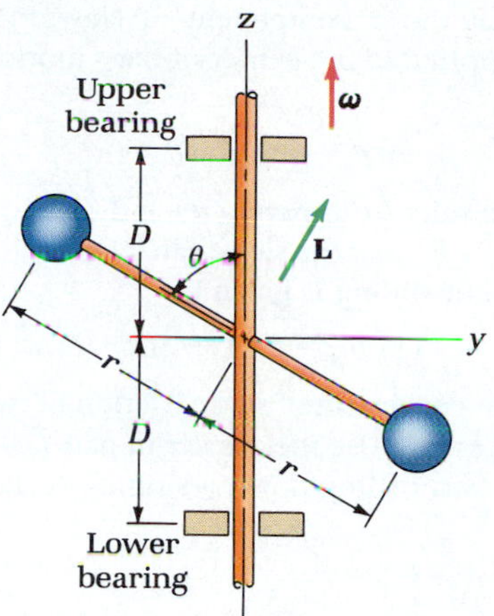

Figure 13-44. Problem 17: The $+\mathbf{i}$ direction is out of the page.

18. ***Angular momentum relative to a point on the circumference.*** The particle of mass m in Fig. 13-45 is traveling in uniform circular motion about the point $(a, 0)$. *(a)* Show that the magnitude of its angular momentum about the origin is

$$\ell = ma^2\omega(1 + \cos\omega t)$$

where ω is its angular speed and $t = 0$ corresponds to the instant the particle is at $(2a, 0)$. *(b)* Make a graph of ℓ versus t from $t = 0$ to $t = 2\pi/\omega$.

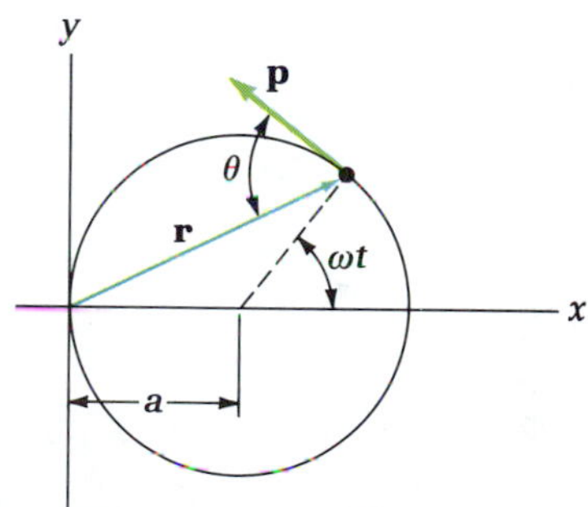

Figure 13-45. Problem 18.

19. ***A ball colliding with a door.*** Consider the collision between the ball and the door discussed in Exercise 29 (Fig. 13-31). *(a)* Assume that the collision is elastic (the kinetic energy of the system is conserved) and show that

$$v_f = v_i \frac{Mw^2 - 3mD^2}{Mw^2 + 3mD^2}$$

$$\omega = v_i \frac{6MD}{Mw^2 + 3mD^2}$$

(b) Use the above expressions and the values from Exercise 29 to evaluate v_f and ω. Was the collision discussed in Exercise 29 elastic?

20. ***Maximum angle for rolling without sliding.*** *(a)* In the case of the wheel rolling without sliding down the slope in Example 13-4, show that the static frictional force exerted by the surface has magnitude

$$F_s = \frac{Mg \sin\theta}{(1 + MR_0^2/I)}$$

(b) Show that the y component of Newton's second law, $\Sigma F_y = Ma_y$, applied to the center-of-mass motion of the wheel gives

$$F_N - Mg \cos \theta = 0$$

(c) Use the results from parts *(a)* and *(b)* to show that the maximum angle θ_m that the slope can have such that the object will roll without sliding is given by

$$\tan \theta_m = \mu_s(1 + MR_0^2/I)$$

where μ_s is the coefficient of static friction between the object and the surface. *(d)* Use the answer in part *(c)* to evaluate θ_m for a billiard ball rolling down a sloping surface where $\mu_s =$ 0.40.

21. ***Yo-yo and its string.*** Consider the yo-yo in Fig. 13-21. Let the yo-yo have a radius of R_0 and let its axle have a radius of R_1. Also, neglect the thickness of the string. Determine an expression for the angle between the string and the horizontal such that the string neither winds onto nor off the yo-yo's axle.

22. ***A critical angle for sliding to begin.*** A uniform ball of radius r_0 is placed on top of a somewhat larger ball of radius R. The larger ball is held fixed and the smaller ball is given a slight nudge so that it starts essentially from rest as it rolls off the larger ball. Let θ be the angle between the vertical and a line between the centers of the balls, and let μ_s be the coefficient of static friction between the balls. Determine an expression for the angle θ_c at which the smaller ball begins to slide.

OSCILLATIONS 14

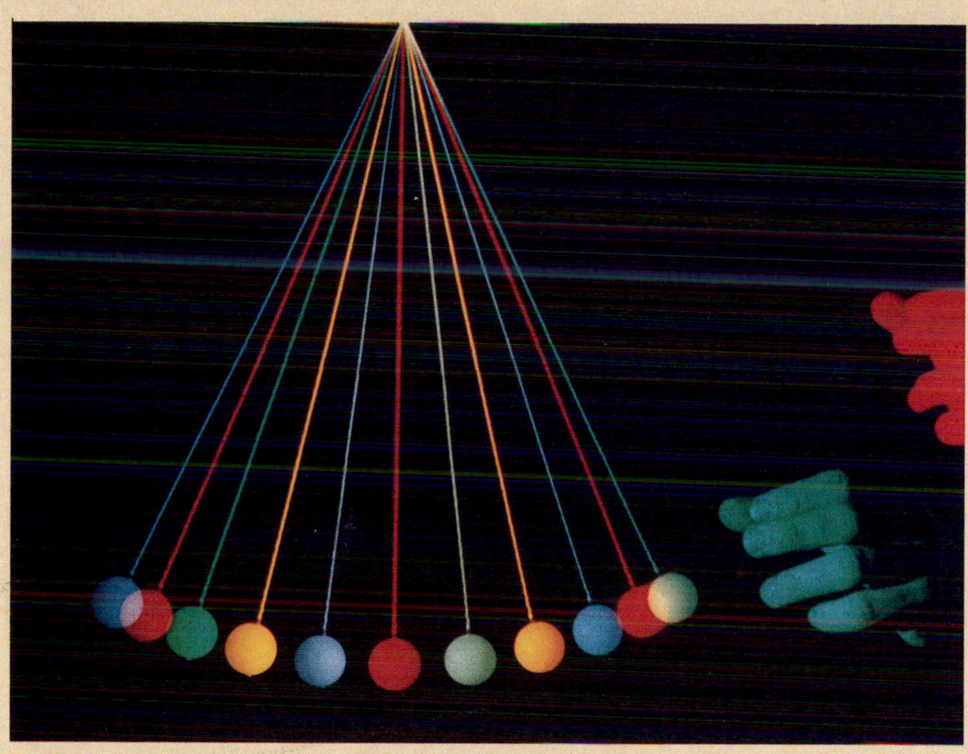

Multicolor strobe photograph of a pendulum during half of a cycle.

In nature, *oscillatory,* or *periodic, motion* is common. Many effects are periodic; examples are the heartbeat of an animal, the seasons of the year, the swinging of a clock pendulum, the vibrations of atoms in solids, the electric current in the wires of the light bulb that illuminates this page, and on and on. On the grandest scale of all, some cosmologists believe that the entire universe may be oscillating with an interval of tens of billions of years between oscillations.

Two types of motion that are closely associated with oscillatory motion are circular (or nearly circular) motion and wave motion. The average temperature in your town changes periodically as the seasons change, and these changes are associated with the earth's circular motion about the sun. As waves of water pass a dock, the water level on the dock pilings oscillates up and down. In the study of waves in Chap. 32, we shall find that their description is based largely on the concepts learned in this chapter.

14-1 KINEMATICS OF SIMPLE HARMONIC MOTION

The simplest type of oscillatory motion is called *simple harmonic motion,* or *SHM*. Simple harmonic motion can be demonstrated with a ball suspended from a spring (Fig. 14-1). When the ball is lifted above its equilibrium position and released, it oscillates vertically with SHM, neglecting small dissipative effects. Figure 14-2 shows photographically that the ball's oscillations are essentially sinusoidal. That is, a graph of the ball's coordinate versus time follows the curve of a sine or a cosine.

Figure 14-1. Multicolor strobe photo of a metal ball oscillating vertically on a spring. Because of the multicolor flash, the ball appears to have a different color at different positions. This photo shows the motion during one-half of a period. The diameter of the ball is 3.50 cm and the time between flashes is 60 ms. To show that this motion is simple harmonic motion (SHM), make a graph of the ball's coordinate (relative to the center) versus time and compare your graph to a cosine curve.

Figure 14-2. Strobe photograph of the oscillating ball in Fig. 14-1. This photograph was taken with a rotating mirror so that the image on the film is horizontally displaced equal distances in equal time intervals. Thus, the photograph shows the vertical displacement as a function of time. Compare the photograph with the curve in Fig. 14-3.

Definition of SHM

An object undergoes ***simple harmonic motion*** if its coordinate varies sinusoidally with time (as a sine or a cosine function).

Let x be the coordinate of an object undergoing SHM; then

Coordinate x of an object executing SHM

$$x = A \cos (\omega t + \phi) \tag{14-1}$$

Definitions of amplitude A, angular frequency ω, phase constant ϕ, and phase $(\omega t + \phi)$

As the object oscillates one way and then the other, x varies sinusoidally with time between $x = A$ and $x = -A$ (Fig. 14-3). Thus A is called the ***amplitude*** because it characterizes the extent of the motion. The symbol ω represents the ***angular frequency,**** and its value determines the rate of the oscillations. The parameter ϕ, called the ***phase constant,*** is selected by our choice of when we begin the measurement ($t = 0$). In Eq. 14-1, the entire quantity in parentheses, the argument of

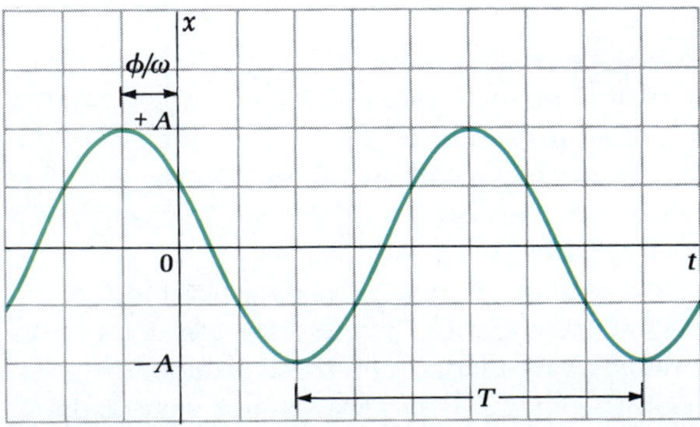

Figure 14-3. Coordinate x versus time t for an object executing SHM.

* It is customary to use ω as the symbol for the angular frequency. Since this is also the symbol for angular speed (Chap. 13), care must be taken to distinguish them.

the cosine, is called the ***phase***, $\omega t + \phi$. In the next section we present a dynamical description of SHM that leads to Eq. 14-1.

A distinguishing feature of any oscillatory motion, including SHM, is that the motion repeats itself after a characteristic time interval, which is called the ***period*** T. That is, *the object undergoes a complete cycle of its motion during the time interval T*, as shown in Fig. 14-3. Thus, for a full cycle, the phase $(\omega t + \phi)$ increases by 2π rad while the time t increases by T, or

Definition of the period T

$$\omega(t + T) + \phi = (\omega t + \phi) + 2\pi$$

Subtracting $\omega t + \phi$ from each side gives $\omega T = 2\pi$, or

$$T = \frac{2\pi}{\omega} \tag{14-2}$$

Relationship between T and ω

The period T is inversely proportional to ω; the larger the angular frequency, the smaller the period and the more quickly the object completes a cycle.

Besides T and ω, there is a third quantity that is used to specify the rate of the oscillations, the ***frequency*** ν:

$$\nu = \frac{1}{T} \tag{14-3}$$

Definition of frequency ν

Since T is the time per cycle, ν is the number of cycles per unit of time. Substituting $T = 2\pi/\omega$ into Eq. 14-3, we find $\nu = \omega/2\pi$, or

$$2\pi\nu = \omega$$

Radians and cycles are dimensionless, so ν and ω have the same dimension, namely, $[\text{time}]^{-1}$. The SI units for these closely related quantities are different: radians per second (rad/s) in the case of ω, and hertz (Hz) in the case of ν. (A frequency of 1 cycle/s = 1 Hz*.) As an example, suppose an object executing SHM has a period $T = 2$ s. Then the frequency is $\nu = 1/(2 \text{ s}) = 0.5 \text{ s}^{-1} = 0.5$ Hz, and the angular frequency is $\omega = 2\pi(0.5 \text{ Hz}) = \pi$ rad/s.

The velocity and the acceleration of an object executing SHM are found by applying the procedures of kinematics from Chap. 3: Given an expression for x, we determine $v_x = dx/dt$ and $a_x = dv_x/dt = d^2x/dt^2$. Differentiating x in Eq. 14-1 with respect to t (see App. D), we find

$$v_x = \frac{dx}{dt} = -\omega A \sin(\omega t + \phi) \tag{14-4}$$

Velocity component during SHM

Differentiating a second times gives

$$a_x = \frac{d^2x}{dt^2} = -\omega^2 A \cos(\omega t + \phi) \tag{14-5}$$

Acceleration component during SHM

The derivative of a sinusoidally varying function is itself a sinusoidally varying function with the same frequency. Thus v_x and a_x oscillate with the same frequency as x, as shown in Fig. 14-4. Note that each differentiation changes the multiplicative factor by the factor ω; x oscillates between A and $-A$, v_x oscillates between ωA and $-\omega A$, and a_x oscillates between $\omega^2 A$ and $-\omega^2 A$. Therefore the maximum speed of the object is $v_{max} = \omega A$, and its maximum acceleration magnitude is $a_{max} = \omega^2 A$.

Another effect of each differentiation is a change in phase by $\frac{1}{2}\pi$ rad, or 90°. From a comparison of the graphs of x and v_x in Fig. 14-4, you can see that v_x passes through each maximum and minimum one-fourth of a period before x. Since one-fourth of a period corresponds to a phase change of $\frac{1}{2}\pi$ rad, or 90°, we sometimes describe this by saying, "v_x leads x by 90°." Similarly, as you can see from the figure, a_x leads v_x by 90°, and a_x leads x by 180°.

* The hertz is named for H. R. Hertz (1857–1894), whose contributions are discussed in Sec. 34-6.

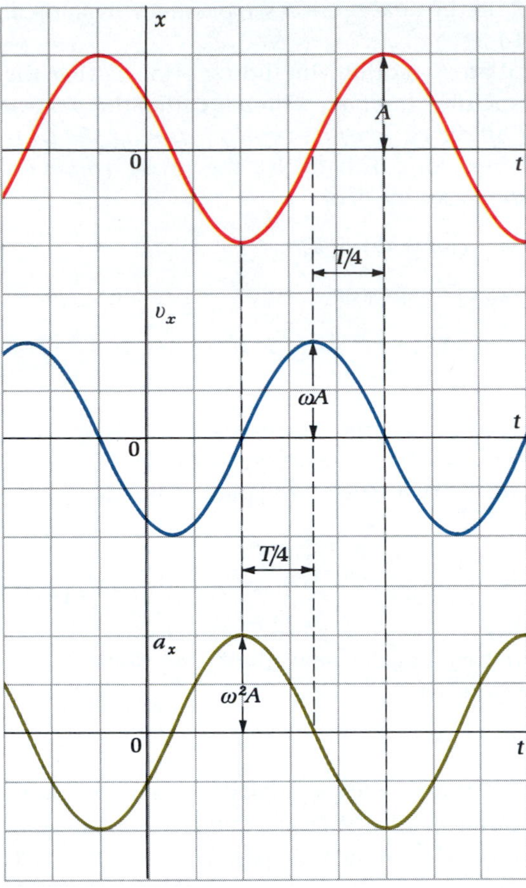

Figure 14–4. The relationships between x, v_x, and a_x for an object executing SHM: x and v_x are out of phase by $\frac{1}{2}\pi$ rad, or 90°; v_x and a_x are out of phase by 90°; and x and a_x are out of phase by 180°.

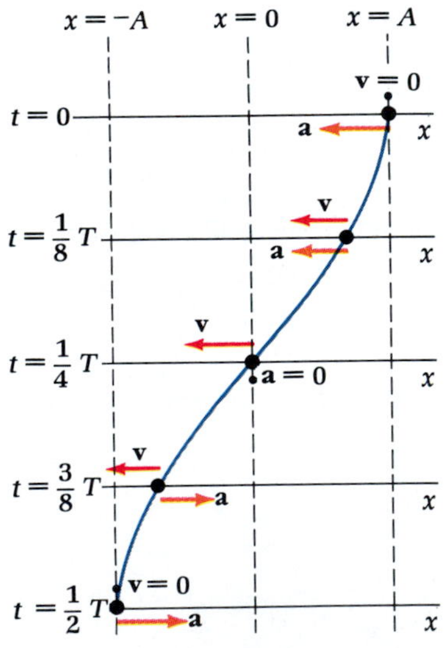

Figure 14–5. An object moving with SHM is shown at five different times during half of a cycle. Note the magnitude and direction of the velocity and acceleration at each time.

Figure 14-5 shows an object at five different times during half a period of its motion from $t = 0$ to $t = T/2$. We have set $\phi = 0$ so that the figure shows the motion occurring between $x = A$ and $x = -A$. As an example of how the figure was constructed, let us determine x, v_x, and a_x at $t = 3T/8$. Substituting $t = 3T/8$ into the phase, we find $\omega t + \phi = (2\pi/T)(3T/8) + 0 = 3\pi/4$, so that $x = A \cos 3\pi/4 \approx -0.7A$; $v_x = -\omega A \sin 3\pi/4 \approx -0.7\omega A$; and $a_x = -\omega^2 A \cos 3\pi/4 \approx 0.7\omega^2 A$. At $t = 3T/8$, the figure shows the object at $x = -0.7A$, with $\mathbf{v} = -0.7v_{max}\mathbf{i}$ and with $\mathbf{a} = 0.7a_{max}\mathbf{i}$. Carefully examine the relationships among x, v_x, and a_x in the figure.

For an object executing SHM, there is a direct relation between the object's displacement ($x\mathbf{i}$) and its acceleration ($a_x\mathbf{i}$). Since

$$a_x = \frac{d^2x}{dt^2} = -\omega^2 A \cos(\omega t + \phi)$$

and

$$x = A \cos(\omega t + \phi)$$

we have

$$a_x = -\omega^2 x \tag{14-6}$$

For any object undergoing SHM, its acceleration and its displacement are always in opposite directions, and their magnitudes are proportional.

A simple harmonic oscillator is a system that undergoes SHM.

This relation is used to identify a system that executes SHM. A system that undergoes SHM is called a ***simple harmonic oscillator.***

EXAMPLE 14-1

Describing SHM when A and T are given. A simple harmonic oscillator has an amplitude of 0.17 m and a period of 0.84 s. Determine *(a)* the frequency and *(b)* the angular frequency of the motion. Write expressions for the time dependence of *(c)* the coordinate, *(d)* the velocity component, and *(e)* the acceleration component.

Solution. *(a)* The frequency is $\nu = 1/T = 1/(0.84\ \text{s}) = 1.2$ Hz.

(b) The angular frequency is $\omega = 2\pi/T = 2\pi/(0.84\ \text{s}) = 7.5$ rad/s.

(c) From Eq. 14-1, $x = A\cos(\omega t + \phi)$. The value of A is given and ω was found in part *(b)*. To use this expression, we must determine a value for ϕ. Since nothing in this problem requires otherwise, let us choose $\phi = 0$ for simplicity. Thus

$$x = (0.17\ \text{m})\cos[(7.5\ \text{rad/s})t]$$

(d) From Eq. 14-4,

$$\begin{aligned} v_x &= -\omega A \sin(\omega t + \phi) \\ &= -(7.5\ \text{rad/s})(0.17\ \text{m})\sin[(7.5\ \text{rad/s})t] \\ &= -(1.3\ \text{m/s})\sin[(7.5\ \text{rad/s})t] \end{aligned}$$

Notice that $v_{max} = 1.3$ m/s.

(e) From Eq. 14-5,

$$\begin{aligned} a_x &= -\omega^2 A \cos(\omega t + \phi) \\ &= -(7.5\ \text{rad})/\text{s})^2(0.17\ \text{m})\cos[(7.5\ \text{rad/s})t] \\ &= -(9.5\ \text{m/s}^2)\cos[(7.5\ \text{rad/s})t] \end{aligned}$$

Notice that $a_{max} = 9.5\ \text{m/s}^2$.

SELF-TEST 14-1. From the above example, determine the values of *(a)* x, *(b)* v_x, and *(c)* a_x at $t = 0.27$ s. (*Hint:* When you evaluate the cosine and the sine of the phase angle with your calculator, be certain the calculator is set to accept the angle in radians.) ***ANSWERS:*** *(a)* −75 mm; *(b)* −1.1 m/s; *(c)* 4.2 m/s².

Finding ϕ and A from the Initial Conditions

Often, when we are dealing with a system that undergoes SHM, the values of ϕ and A are not directly measured, but the values of x_0 and v_{x0} are known. The quantities x_0 and v_{x0} are called the *initial conditions.* Consider finding ϕ and A from the initial conditions. Setting $t = 0$ in Eqs. 14-1 and 14-4, we find

$$x_0 = A\cos\phi \qquad \text{and} \qquad v_{x0} = -\omega A \sin\phi \tag{14-7}$$

These equations give x_0 and v_{x0} in terms of ϕ and A, but we want ϕ and A in terms of x_0 and v_{x0}. First we find ϕ by eliminating A. Dividing the second of these equations by the first, we have $v_{x0}/x_0 = (-\omega A\sin\phi)/(A\cos\phi) = -\omega\tan\phi$. Solving for ϕ, we obtain

$$\phi = \tan^{-1}\frac{-v_{x0}}{\omega x_0} \tag{14-8}$$

Phase constant ϕ in terms of x_0 and v_{x0}

Next we find A by eliminating ϕ. Squaring each of Eqs. 14-7 and forming the sum $\sin^2\phi + \cos^2\phi = 1$, we have $(v_{x0}/\omega A)^2 + (x_0/A)^2 = 1$, or

$$A = \sqrt{x_0{}^2 + \frac{v_{x0}{}^2}{\omega^2}} \tag{14-9}$$

Amplitude A in terms of x_0 and v_{x0}

A simple example is to determine ϕ and A for the case where the object starts from rest. Then $v_{x0} = 0$ so that Eq. 14-8 gives $\phi = \tan^{-1}(0) = 0$, and Eq. 14-9 gives $A = \sqrt{x_0{}^2 + 0} = |x_0|$. This corresponds to the case shown in Fig. 14-5.

14-2 DYNAMICS OF SIMPLE HARMONIC MOTION

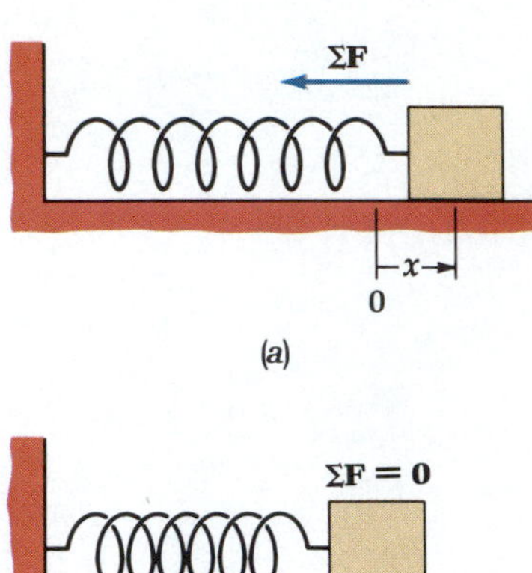

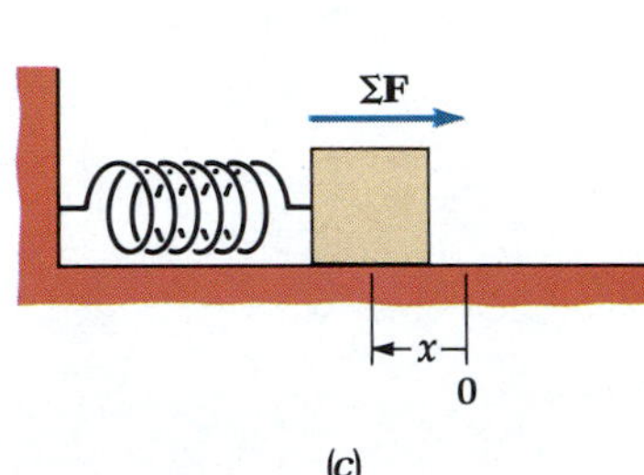

Figure 14-6. A block attached to a spring. In this idealized system, friction is negligible and the mass of the spring is insignificantly small compared with the mass of the block. The net force on the block is due to the spring, $\Sigma\mathbf{F} = \mathbf{F}_s = -kx\mathbf{i}$. *(a)* The block is displaced to the right, and $\Sigma\mathbf{F}$ is directed to the left. *(b)* The block is at the equilibrium position, $\Sigma\mathbf{F} = 0$. *(c)* The block is displaced to the left, and $\Sigma\mathbf{F}$ is directed to the right. If the block is displaced from equilibrium and released, it executes SHM.

The previous section showed how to describe simple harmonic motion; this section discusses the causes of SHM. As a representative simple harmonic oscillator, consider a block of mass m connected to a light spring with spring constant k (Fig. 14-6). In this idealized system, the block slides along the horizontal surface with negligible friction so that the force due to the surface is equal and opposite the block's weight, and the net force on the block is the force $\mathbf{F}_s$ due to the spring: $\Sigma\mathbf{F} = \mathbf{F}_s$. From Sec. 8-3, the force due to the spring is

$$\mathbf{F}_s = -(kx)\mathbf{i} \tag{14-10}$$

where x is the coordinate of the block measured from the position where the spring is relaxed; that is, it is neither stretched nor compressed. This type of force is called a ***linear restoring force.*** It is called "linear" because it is linearly proportional to the displacement $x\mathbf{i}$, and it is called "restoring" because the force is directed opposite the displacement. If x is positive, the force is toward $-x$, and if x is negative, the force is toward $+x$. The force tends to restore the object to the central position ($x = 0$), and the larger the displacement, the larger the force.

Since the spring force provides the net force on the block, Newton's second law, $\Sigma\mathbf{F} = m\mathbf{a}$, gives

$$-kx = ma_x$$

Writing a_x as d^2x/dt^2 and rearranging gives

$$\frac{d^2x}{dt^2} = -\frac{k}{m}x \tag{14-11}$$

Newton's second law becomes a differential equation for the coordinate x. A solution to the equation is an expression for x as a function of time, which satisfies the equation. What function of time has its second derivative proportional to the negative of the function itself? Such a function was presented in the last section. The second derivative of a cosine function is proportional to the negative of the cosine function. (This is also true for a sine function. See Exercise 12.) Thus the solution to Eq. 14-11 can be written

$$x = A\cos(\omega t + \phi) \tag{14-1}$$

As we have seen, the second derivative of x with respect to t is

$$\frac{d^2x}{dt^2} = -\omega^2 A\cos(\omega t + \phi)$$

Substitution into Eq. 14-11 gives

$$-\omega^2 A\cos(\omega t + \phi) = -\frac{k}{m}A\cos(\omega t + \phi)$$

Thus Eq. 14-1 is a solution to Eq. 14-11, provided $\omega^2 = k/m$. This means that the block executes SHM, and the angular frequency is

Angular frequency in terms of the spring constant k and the mass m

$$\omega = \sqrt{\frac{k}{m}} \tag{14-12}$$

A linear restoring force produces SHM.

For a strong spring (large k) or small mass, the oscillations are rapid; for a weak spring (small k) or large mass, the oscillations are slow. These predictions agree with common experience with oscillating systems involving springs. Simple harmonic motion is caused by a net force that is a linear restoring force.

Our conclusion that the block-spring system executes SHM with angular frequency $\omega = \sqrt{k/m}$ can also be seen by comparing Eq. 14-11 with Eq. 14-6, $a_x = -\omega^2 x$. If Eq. 14-11 is written with d^2x/dt^2 replaced by a_x, this comparison is

$$a_x = -\frac{k}{m}x \qquad \text{and} \qquad a_x = -\omega^2 x$$

With $\omega^2 = k/m$, these equations are the same.

EXAMPLE 14-2

Describing SHM when m, k, and the initial conditions are given. Suppose the block in Fig. 14-6 has a mass of 0.31 kg and the spring constant of the spring is 63 N/m. The block is pulled aside such that the spring is stretched 0.074 m and the block is released from rest at $t = 0$. *(a)* Determine ω, T, and ν. *(b)* Write expressions for x, v_x, and a_x.

Solution. *(a)* The angular frequency is

$$\omega = \sqrt{\frac{k}{m}} = \sqrt{\frac{63\ \text{N/m}}{0.31\ \text{kg}}} = 14\ \text{rad/s}$$

The period is

$$T = \frac{2\pi}{\omega} = 2\pi\sqrt{\frac{m}{k}} = 2\pi\sqrt{\frac{0.31\ \text{kg}}{63\ \text{N/m}}} = 0.44\ \text{s}$$

The frequency is

$$\nu = \frac{\omega}{2\pi} = \frac{1}{2\pi}\sqrt{\frac{k}{m}} = \frac{1}{2\pi}\sqrt{\frac{63\ \text{N/m}}{0.31\ \text{kg}}} = 2.3\ \text{Hz}$$

(b) Since the block was released from rest with the spring stretched 0.074 m, $x_0 = 0.074$ m and $v_{x0} = 0$. Therefore $\phi = 0$ and

$$x = (0.074\ \text{m}) \cos\,[(14\ \text{rad/s})t]$$

We have $v_{max} = \omega A = (14\ \text{rad/s})(0.074\ \text{m}) = 1.1\ \text{m/s}$ so that

$$v_x = -(1.1\ \text{m/s}) \sin\,[(14\ \text{rad/s})t]$$

Also, $a_{max} = \omega^2 A = (14\ \text{rad/s})^2(0.074\ \text{m}) = 15\ \text{m/s}^2$ so that

$$a_x = -(15\ \text{m/s}^2) \cos\,[(14\ \text{rad/s})t]$$

SELF-TEST 14-2. Suppose the block in Fig. 14-6 has a mass of 0.50 kg, and the spring constant of the spring is 25 N/m. What is the block's acceleration at an instant when $x = 0.058$ m? (*Hint:* Keep in mind that there is a direct relation between x and a_x.) ***ANSWER:*** $-(2.9\ \text{m/s}^2)\mathbf{i}$.

14-3 THE ENERGY OF A SIMPLE HARMONIC OSCILLATOR

In Chap. 9 we found that the force due to a spring is a conservative force and that the expression for the potential energy of a spring is $U = \frac{1}{2}kx^2$. Using Eq. 14-1, we find that the potential energy of the idealized block-spring harmonic oscillator in Fig. 14-6 is $U = \frac{1}{2}kx^2 = \frac{1}{2}k[A\cos(\omega t + \phi)]^2$, or

$$U = \tfrac{1}{2}kA^2 \cos^2(\omega t + \phi) \qquad (14\text{-}13)$$

Potential energy of a simple harmonic oscillator

Similarly, Eq. 14-4 can be used to find the kinetic energy of the block-spring system: $K = \frac{1}{2}mv^2 = \frac{1}{2}m[-\omega A \sin(\omega t + \phi)]^2$, or

$$K = \tfrac{1}{2}m\omega^2 A^2 \sin^2(\omega t + \phi) \qquad (14\text{-}14)$$

Kinetic energy of a simple harmonic oscillator

The maximum value of the square of a sine or cosine function is 1, so these energies can be expressed as

$$U = U_{max} \cos^2(\omega t + \phi)$$

$$K = K_{max} \sin^2(\omega t + \phi)$$

where $U_{max} = \frac{1}{2}kA^2$ and $K_{max} = \frac{1}{2}m\omega^2 A^2$. Since $\omega^2 = k/m$, $K_{max} = \frac{1}{2}m\omega^2 A^2 = \frac{1}{2}m(k/m)A^2 = \frac{1}{2}kA^2$, or

$$K_{max} = U_{max}$$

In the block-spring oscillator system of Fig. 14-6, only the spring force does work. Consequently, the mechanical energy E of the oscillator is

$$E = K + U = K_{max} \sin^2(\omega t + \phi) + U_{max} \cos^2(\omega t + \phi)$$

Using $K_{max} = U_{max}$ and $\sin^2\theta + \cos^2\theta = 1$, we have $E = K_{max} = U_{max}$, or

Mechanical energy of a simple harmonic oscillator

$$E = \tfrac{1}{2}m\omega^2 A^2 = \tfrac{1}{2}kA^2 \quad (14\text{-}15)$$

The mechanical energy of the oscillator is constant; the simple harmonic oscillator is a conservative system.

Graphs of K and U versus time are shown in Fig. 14-7 ($\phi = 0$ for simplicity). Each function oscillates between zero and E. The energy of the oscillator continuously changes from potential energy to kinetic energy, then back to potential energy, and on and on.

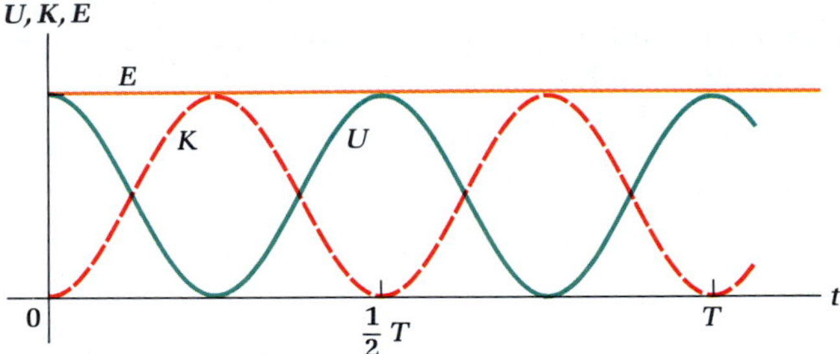

Figure 14–7. Potential energy U and kinetic energy K versus time t for a simple harmonic oscillator ($\phi = 0$). Note that $E = U_{max} = K_{max}$.

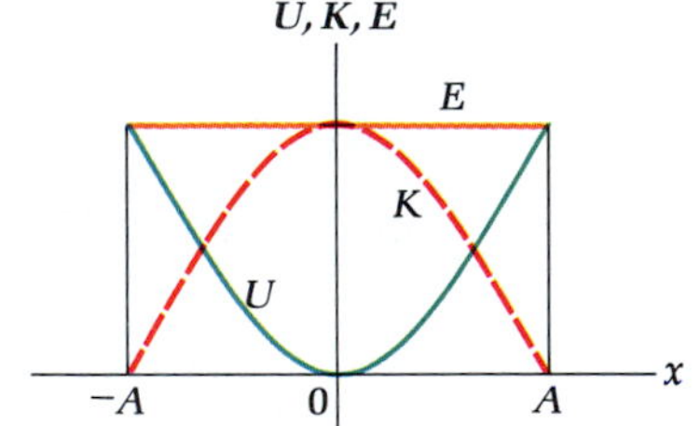

Figure 14–8. Potential energy U and kinetic energy K versus coordinate x for a simple harmonic oscillator.

Equations 14-13 and 14-14 give the potential and kinetic energies as functions of time. Now consider these energies as functions of the coordinate x. The equation for potential energy as a function of x is $U = \frac{1}{2}kx^2$. Conservation of energy is used to find K as a function of x: $E = K + U = K + \frac{1}{2}kx^2$, or

$$K = E - \tfrac{1}{2}kx^2 = \tfrac{1}{2}kA^2 - \tfrac{1}{2}kx^2 = \tfrac{1}{2}k(A^2 - x^2)$$

Fig. 14-8 shows graphs of U and K versus x. Each curve is a parabola centered at $x = 0$. Consider finding the point at which the two curves cross. At this point $U = K$ or $\frac{1}{2}kx^2 = \frac{1}{2}kA^2 - \frac{1}{2}kx^2$. Solving for x, we find $x = \pm A/\sqrt{2} \approx \pm 0.7A$.

EXAMPLE 14-3

Using energy conservation in SHM. Suppose a block-spring system has $k = 18$ N/m and $m = 0.71$ kg. The system is oscillating with an amplitude $A = 54$ mm. *(a)* Determine the angular frequency of the oscillations. *(b)* Develop an expression for the speed v of the block as a function of x, and use the expression to find v at $x = 34$ mm. *(c)* Develop an expression for the block's distance $|x|$ from the central point as a function of the speed v, and use the expression to find $|x|$ when $v = 0.18$ m/s.

Solution. *(a)* The angular frequency is

$$\omega = \sqrt{\frac{k}{m}} = \sqrt{\frac{18\ \text{N/m}}{0.71\ \text{kg}}} = 5.0\ \text{rad/s}$$

(b) Using conservation of energy with $E = \frac{1}{2}kA^2$ gives

$$\tfrac{1}{2}kA^2 = \tfrac{1}{2}mv^2 + \tfrac{1}{2}kx^2$$

Solving for v gives

$$v = \omega\sqrt{A^2 - x^2}$$

where we have used $\omega = \sqrt{k/m}$. The speed at $x = 34$ mm is

$$v = (5.0 \text{ rad/s})\sqrt{(0.054 \text{ m})^2 - (0.034 \text{ m})^2} = 0.21 \text{ m/s}$$

(c) From conservation of energy with $E = \frac{1}{2}m\omega^2A^2$,

$$\tfrac{1}{2}m\omega^2A^2 = \tfrac{1}{2}mv^2 + \tfrac{1}{2}kx^2$$

Solving for $\sqrt{x^2} = |x|$ gives

$$|x| = \sqrt{A^2 - \left(\frac{v}{\omega}\right)^2}$$

Thus, when $v = 0.18$ m/s,

$$|x| = \sqrt{(0.054 \text{ m})^2 - \left(\frac{0.18 \text{ m/s}}{5.0 \text{ rad/s}}\right)^2} = 40 \text{ mm}$$

SELF-TEST 14-3. What is the energy of the oscillator in the above example? ***ANSWER:*** 26 mJ.

14-4 EXAMPLES OF SIMPLE HARMONIC MOTION

As the main example of SHM, we considered an object connected to a light horizontal spring and moving on a horizontal surface. The net force was due to the spring, and dissipative effects of friction were assumed to be negligible. The acceleration of the object was proportional to its displacement but in the opposite direction. There are many other systems in which acceleration is essentially proportional to but opposite the displacement, so that simple harmonic motion occurs. Following are some examples of such systems.

An Object on a Vertical Spring

Suppose that a light spring with spring constant k is suspended vertically from a support (Fig. 14-9*a*). Initially the spring is neither stretched nor compressed. Now a block of mass m is attached to the other end of the spring, and the block is gently lowered until it comes to equilibrium (Fig. 14-9*b*). In this configuration the spring is stretched by an amount ℓ and exerts an upward force of magnitude $k\ell$ on the block. The block is in equilibrium at this position, and the spring force balances the downward weight of the block of magnitude mg:

$$k\ell = mg \tag{14-16}$$

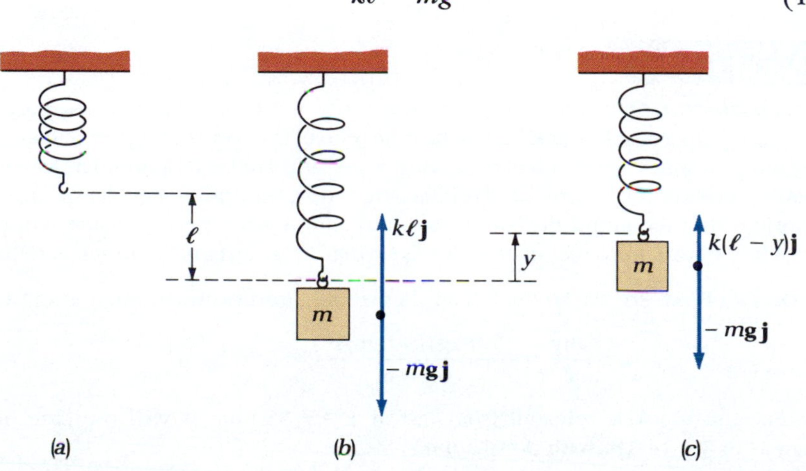

Figure 14-9. *(a)* A light spring is suspended vertically. *(b)* The block is in equilibrium, with $\mathbf{j}(k\ell - mg) = 0$. *(c)* The spring is stretched by $\ell - y$, and the net force is $[k(\ell - y) - mg]\mathbf{j} = -ky\mathbf{j}$, which tends to restore the block to its equilibrium position.

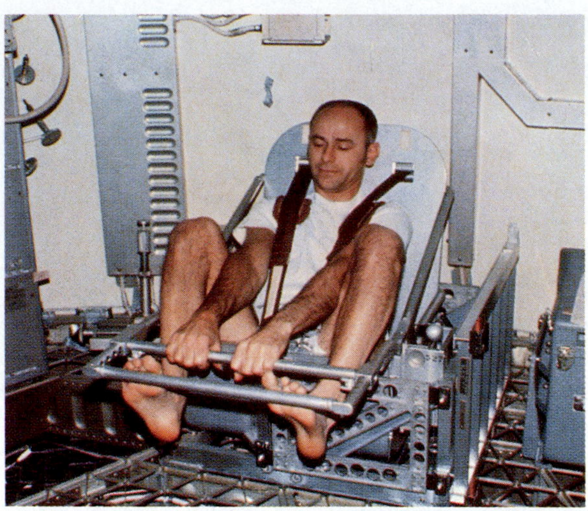

Since an astronaut in an orbiting spacecraft cannot be weighed in the usual way, the astronaut is made part of an oscillating system similar to the block-spring oscillator. The astronaut's mass is determined from a measurement of the period.

If the block is displaced vertically from this equilibrium position, the spring force will not balance the weight, and the block will be accelerated. Let us determine the acceleration. Choose the origin of the y axis at the equilibrium position, as shown in Fig. 14-9c, which also shows the two forces acting on the block when its coordinate is y. Notice that y gives the displacement of the block from the equilibrium position. However, the spring is stretched by an amount $\ell - y$, so that the spring exerts an upward force of magnitude $k(\ell - y)$. The weight of the block is downward and of magnitude mg. Therefore the net force has a y component given by $\Sigma F_y = k(\ell - y) - mg$. Applying Newton's second law gives

Displacement of the block from its equilibrium position is given by y.

$$\Sigma F_y = k(\ell - y) - mg = ma_y$$

Equation 14-16 determines the stretch ℓ of the spring when the block is in equilibrium, so that $k\ell - mg = 0$. Thus the expression above simplifies to $-ky = ma_y$, or

$$a_y = -\frac{k}{m} y \tag{14-17}$$

The acceleration of the block is proportional to but opposite its displacement from equilibrium.

Compare Eq. 14-17 with the standard form for SHM in Eq. 14-6, $a_x = -\omega^2 x$. Except for the use of y instead of x for the coordinate, the equations are the same if $\omega^2 = k/m$. That is, the motion of the block connected to the vertical spring is SHM with angular frequency $\omega = \sqrt{k/m}$. The coordinate of the block is given by

A block on a vertical spring moves with SHM.

$$y = A\cos(\omega t + \phi) \qquad \left(\omega = \sqrt{\frac{k}{m}}\right) \tag{14-18}$$

EXAMPLE 14-4

A vertical oscillator. One end of a light vertical spring is attached to a rigid support, as seen in Fig. 14-9. A 5.0-kg block is attached to the other end of the spring and gently lowered to its equilibrium position. The stretch of the spring is measured to be 180 mm. The block is then pulled down an additional 75 mm and released from rest. Determine *(a)* the spring constant of the spring, *(b)* the amplitude of the motion, and *(c)* the period of the motion. *(d)* Determine the elastic potential energy of the spring at the instant the block is released.

Solution. *(a)* From Eq. 14-16, which applies to the equilibrium configuration, we have

$$k = \frac{mg}{\ell} = \frac{(5.0\text{ kg})(9.8\text{ m/s}^2)}{0.18\text{ m}} = 270\text{ N/m}$$

(b) Since the block is released from rest at $y = -75$ mm, it will oscillate between ± 75 mm, as in Eq. 14-18, with $A = 75$ mm.

(c) Since $\omega = \sqrt{k/m}$, the period $T = 2\pi/\omega$ is

$$T = 2\pi\sqrt{\frac{m}{k}} = 2\pi\sqrt{\frac{5.0 \text{ kg}}{270 \text{ N/m}}} = 0.85 \text{ s}$$

(d) The elastic potential energy of the spring depends on the stretch of the spring. At the instant the block is released, the spring is stretched from its relaxed length by 180 mm plus the additional 75 mm, or by 255 mm. The elastic potential energy U_s is

$$U_s = \tfrac{1}{2}(270 \text{ N/m})(0.255 \text{ m})^2 = 8.8 \text{ J}$$

Mechanical energy is conserved in this motion, and there is a continual interchange of kinetic energy, gravitational potential energy, and elastic potential energy. (See Exercise 25.)

SELF-TEST 14-4. What are the maximum speed and acceleration magnitude of the block in the above example? ***ANSWER:*** $v_{max} = 1.9$ m/s; $a_{max} = 14$ m/s^2.

The Simple Pendulum

The periodic motion of a pendulum has long been used in pendulum clocks to regulate the mechanism that causes the hands to move around the dial. For small displacements from equilibrium, a pendulum undergoes SHM. Here we consider a *simple pendulum,* a pendulum with all the mass concentrated at one end and suspended about the other end, such as the ball-and-cord pendulum in Fig. 14-10. The ball forms the "bob" of the pendulum whose length is L.

Figure 14-10*a* is a strobe photograph that shows positions of the bob at equal time intervals. In this motion there is a back-and-forth exchange of energy between kinetic and potential. The kinetic energy is maximum at the lowest point of the swing, and the gravitational potential energy is maximum at the highest points of the swing.

Although the pendulum swings in the two dimensions of a plane, it swings in a circular arc, and the motion can be analyzed by using a single angular coordinate and applying rotational dynamics.

The pendulum is shown in its equilibrium orientation in Fig. 14-10*b*. There are two external forces: the weight $\mathbf{F}_e$ of the bob and a force $\mathbf{F}_s$ exerted on the cord at the upper end by a fixed support. We choose an axis O at the upper end of the cord and perpendicular to the plane of the figure. For this orientation, the torque about axis O is zero for each external force, and the angular acceleration $\alpha_z = 0$.

If the pendulum is displaced, as seen in Fig. 14-10*c*, there is a net external torque about axis O due to the weight. For a positive value of the angle θ, as shown in the figure, this torque tends to cause a clockwise rotation so as to restore the pendulum to its equilibrium orientation. The perpendicular distance from the axis to the line of action of the weight is $L \sin\theta$. With the z axis out of the plane of the figure, the torque component is $\tau_z = -F_e L \sin\theta = -mgL \sin\theta$. Since the mass of the cord can be neglected, the moment of inertia of the simple pendulum about axis O is due to the mass m of the bob at a distance L from the axis. That is, $I = mL^2$. Now apply Newton's second law for angular motion, Eq. 13-10, $\Sigma\tau_z = I\alpha_z$, or

$$-mgL \sin\theta = mL^2 \alpha_z$$

Solving for the angular acceleration component, we obtain

$$\alpha_z = -\frac{g}{L}\sin\theta \tag{14-19}$$

for the rotational motion.

Now compare Eq. 14-19 with Eq. 14-6 for SHM, $a_x = -\omega^2 x$. The left-hand sides are analogous: $\alpha_z = d^2\theta/dt^2$ is an angular acceleration component, and $a_x = d^2x/dt^2$ is a linear acceleration component. The right-hand sides will likewise correspond, but *only* if the motion is restricted to small displacements from equi-

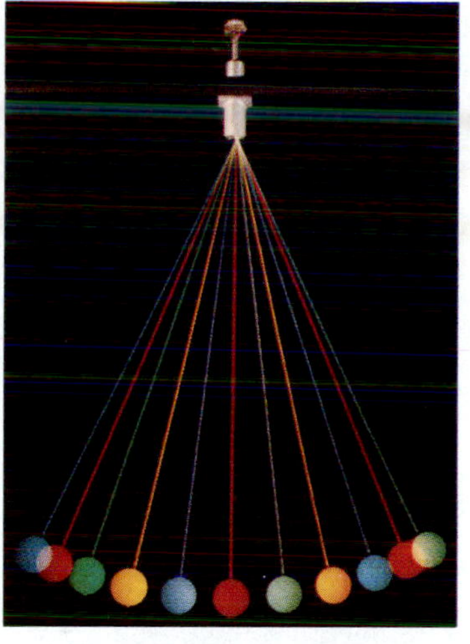

(a)

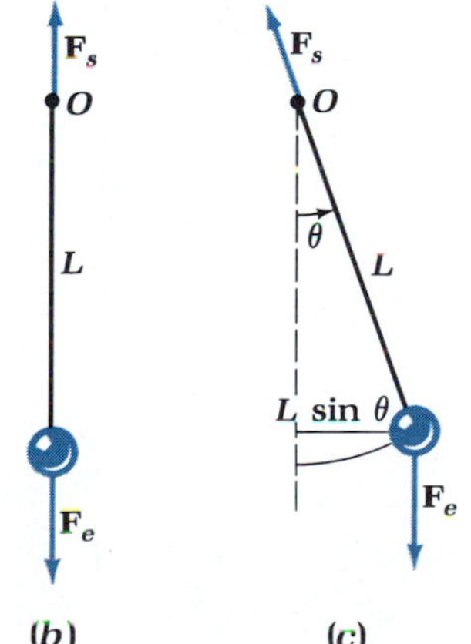

(b) (c)

Figure 14-10. *(a)* A multicolor strobe photo shows how a pendulum swings from one side to the other. *(b)* The net torque about axis O is zero when the pendulum is at the equilibrium orientation. *(c)* The torque about axis O due to the weight tends to restore the pendulum to the equilibrium orientation. The torque about axis O due to the support force $\mathbf{F}_s$ is zero.

librium. Then $\sin\theta \approx \theta$, where θ is expressed in radian measure. [For example, if $\theta = 0.100$ rad (or 5.73°), then $\sin\theta = 0.0998 \approx 0.100$. Set your calculator to radian measure for angles and compare values of $\sin\theta$ and θ for larger and smaller θ.] Replacing $\sin\theta$ with θ in Eq. 14-19 gives

$$\alpha_z = -\frac{g}{L}\theta$$

which is completely analogous to $a_x = -\omega^2 x$, with the replacement $\omega^2 = g/L$. Since α_z is proportional to $-\theta$ for small displacements from equilibrium, the pendulum moves with SHM. Therefore the angular coordinate for the simple pendulum for small displacements is

A pendulum undergoes SHM for small displacements.

$$\theta = A\cos(\omega t + \phi) \qquad (\omega = \sqrt{g/L}) \tag{14-20}$$

Since θ is an angular coordinate, some of the symbols in Eq. 14-20 must be interpreted with care. The amplitude A represents the maximum *angular* coordinate θ_{max}. That is, θ oscillates between $+\theta_{max}$ (to the right in Fig. 14-10) and $-\theta_{max}$ (to the left in Fig. 14-10). The angular *frequency* of the oscillatory motion is $\omega = \sqrt{g/L}$. It must be distinguished from the *angular velocity* component $\omega_z = d\theta/dt$ of the pendulum. Taking the derivative of θ in Eq. 14-20 gives

$$\omega_z = -\omega A\sin(\omega t + \phi)$$

This expression for the angular velocity component of the pendulum is analogous to $v_x = dx/dt = -\omega A \sin(\omega t + \phi)$ for the linear velocity component for the linear motion.

Since the angular frequency is $\omega = \sqrt{g/L}$, the period $T = 2\pi/\omega$ of a simple pendulum is given by

$$T = 2\pi\sqrt{\frac{L}{g}} \tag{14-21}$$

The period of a simple pendulum is determined by its length.

Notice that the period is independent of the mass of the bob; it depends only on the length of the pendulum and g. If the period of a pendulum is determined by precise time measurements, the pendulum can be used to measure g. Precision measurements made with a physical pendulum, described below, are capable of determining local variations in g due to density variations in the earth's upper surface and are useful in locating deposits of natural resources.

EXAMPLE 14-5

Measuring g with a pendulum. A moon explorer sets up a simple pendulum of length 860 mm and measures its period for small displacements to be 4.6 s. Determine the acceleration due to gravity at this location on the surface of the moon.

Solution. Solving Eq. 14-21 for g, we obtain

$$g = \frac{4\pi^2 L}{T^2} = \frac{4\pi^2(0.86\ \text{m})}{(4.6\ \text{s})^2} = 1.6\ \text{m/s}^2$$

SELF-TEST 14-5. What is the length of a simple pendulum that has a period of 4.6 s on earth? ***ANSWER:*** 5.3 m.

The Physical Pendulum

A simple pendulum, with all of its mass concentrated at one end, is a special case of the more general physical pendulum. ***A physical pendulum*** is a rigid body pivoted to rotate about a fixed horizontal axis O, as shown in Fig. 14-11. The mass is distributed along the length of the pendulum, and the center of gravity is at point C, which is a distance L from the axis O. The equilibrium orientation is shown in Fig. 14-11*a*, with the center of gravity directly below the point of suspension. If the pendulum is displaced from equilibrium (Fig. 14-11*b*), the torque component τ_z due to the weight tends to cause a clockwise rotation so as to restore the pendulum to the equilibrium orientation.

The motion of a physical pendulum can be analyzed in the same way as that of a simple pendulum, by applying Newton's second law for angular motion. For the situation shown in Fig. 14-11*b*, $L \sin \theta$ is the perpendicular distance from the axis to the line of action of the weight $\mathbf{F}_e$. The torque component is $\tau_z = -F_e L \sin \theta = -mgL \sin \theta$, where the positive z axis is taken to be out of the plane of the figure. We let α_z represent the angular acceleration component of the pendulum and I its moment of inertia about the axis O. Then neglecting any frictional torques so that the only torque is due to the weight, we have

$$\Sigma\tau_z = -mgL \sin \theta = I\alpha_z$$

Now, just as for the simple pendulum, small angular displacements are considered, so that $\sin \theta \approx \theta$. Then

$$\alpha_z = -\frac{mgL}{I}\theta \qquad (14\text{-}22)$$

which again is analogous to $a_x = -\omega^2 x$, with $\omega^2 = mgL/I$. Therefore, for small displacements, the physical pendulum undergoes SHM, and the angular coordinate is given by

$$\theta = A \cos (\omega t + \phi) \qquad \left(\omega = \sqrt{\frac{mgL}{I}}\right) \qquad (14\text{-}23)$$

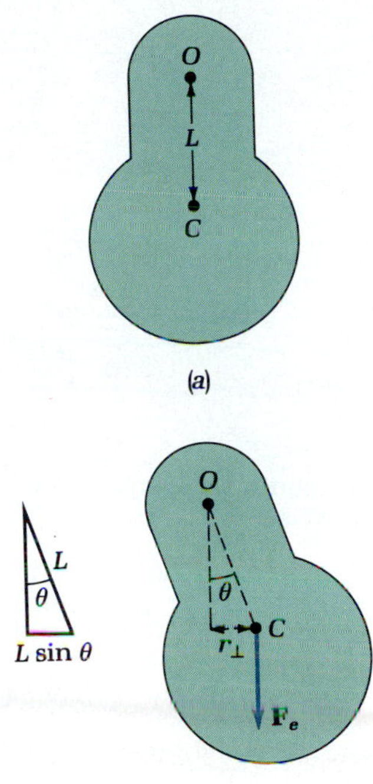

Figure 14–11. *(a)* A rigid body is in rotational equilibrium, with the center of gravity C directly below the axis of rotation O. *(b)* The torque component due to the weight $\mathbf{F}_e$ is $\tau_z = -r_\perp F_e = -(L \sin \theta)(mg)$. The positive z axis passes through O and comes out of the plane of the figure.

EXAMPLE 14-6

A thin-rod pendulum. A thin, uniform rod of mass m and length D is pivoted to rotate freely about a horizontal axis at one end. Determine the period of this pendulum for small displacements from equilibrium.

Solution. For small displacements from equilibrium, the rod oscillates in SHM with angular frequency $\omega = \sqrt{mgL/I}$. Since L represents the distance from the axis to the center of gravity, located at the center of the uniform rod, $L = \frac{1}{2}D$. The moment of inertia of the rod of length D about an axis through one end is $mD^2/3$. (See Example 12-8, but note that L is used there for the length of the rod.) Making these substitutions, we have

$$\omega = \sqrt{\frac{mgL}{I}} = \sqrt{\frac{mg\frac{1}{2}D}{mD^2/3}} = \sqrt{\frac{3g}{2D}}$$

The period is

$$T = \frac{2\pi}{\omega} = 2\pi\sqrt{\frac{2D}{3g}}$$

SELF-TEST 14-6. What must be the length of a simple pendulum to have the same period as the physical pendulum of length D in the above example? ***ANSWER:*** $2D/3$.

The Torsional Oscillator

Two examples of a ***torsional oscillator***, or ***torsional pendulum***, are shown schematically in Fig. 14-12. A vertical fiber is attached to a rigid object such as a plate or a rod. The plate in Fig. 14-12*a* can rotate in a horizontal plane about an axis along the vertical fiber. The angle θ gives the orientation of the plate from its equilibrium orientation. At equilibrium, $\theta = 0$ and the fiber is not twisted. When twisted through angle θ, the fiber exerts a restoring torque on the plate that tends to

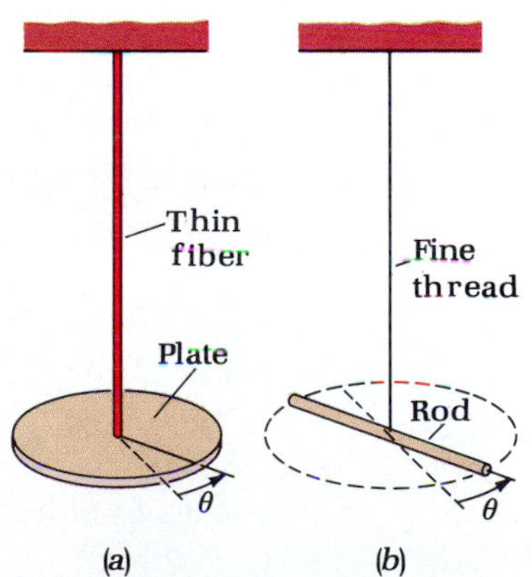

Figure 14–12. A fiber twisted through angle θ exerts a restoring torque on *(a)* a horizontal plate and *(b)* a horizontal rod.

return the plate to its equilibrium orientation. For many fibers this restoring torque is proportional to the angle of twist θ. Let the z axis be vertically up along the fiber. Then

$$\tau_z = -\kappa\theta \tag{14-24}$$

gives the torque component, where κ is called the *torsion constant* of the fiber.

Suppose that only the fiber exerts a torque on the plate, $\Sigma\tau_z = -\kappa\theta$. If I is the moment of inertia of the plate about the fiber axis, then Eq. 13-10 gives $-\kappa\theta = I\alpha_z$, or

$$\alpha_z = -\frac{\kappa}{I}\theta$$

A torsion pendulum undergoes SHM.

Since this has the form $\alpha_z = -\omega^2\theta$, the plate undergoes SHM in the angular coordinate θ and $\omega^2 = \kappa/I$:

$$\theta = A\cos(\omega t + \phi) \qquad \left(\omega = \sqrt{\frac{\kappa}{I}}\right) \tag{14-25}$$

EXAMPLE 14-7

A torsional oscillator. A fine thread is attached to the midpoint of a 10-g pencil of length $L = 200$ mm (Fig. 14-12*b*). The system is set in motion as a torsional oscillator, and the period is observed to be 4 s. *(a)* Estimate the torsion constant of the thread. *(b)* If the amplitude of the motion is 3 rad, determine the maximum magnitude of the restoring torque on the pencil.

Solution. *(a)* Since $\omega = \sqrt{\kappa/I}$ for a torsional oscillator, we have $\kappa = \omega^2 I = 4\pi^2 I/T^2$, where $T = 2\pi/\omega$ is the period. From Table 12-2, the pencil has a moment of inertia $I = mL^2/12$:

$$\kappa = \frac{4\pi^2 mL^2}{12T^2} = \frac{m\pi^2L^2}{3T^2}$$

$$= \frac{(0.01\text{ kg})\pi^2(0.2\text{ m})^2}{(3)(4\text{ s})^2} = 8\times10^{-5}\text{ N}\cdot\text{m}\cdot\text{rad}^{-1}$$

(b) The amplitude is the maximum value of θ: $\theta_{max} = A = 3$ rad. (Such a torsional oscillator can undergo SHM even if the angle θ is not small.) From Eq. 14-24,

$$\tau_{max} = |\tau_z|_{max} = \kappa\theta_{max} = (8\times10^{-5}\text{ N}\cdot\text{m}\cdot\text{rad}^{-1})(3\text{ rad})$$

$$= 2\times10^{-4}\text{ N}\cdot\text{m}$$

SELF-TEST 14-7. Determine the maximum angular speed ω_{max} of the pencil in the above example if the maximum amplitude is $\theta_{max} = 3$ rad. ***ANSWER:*** 5 rad/s.

14-5 SIMPLE HARMONIC MOTION AND UNIFORM CIRCULAR MOTION

There is an intimate connection between the simple harmonic motion of an object moving along a line, such as the x axis, and the motion of a particle at constant speed in a circle. Exploring this connection may help you to understand each type of motion better and to see how some other types of motion are related to SHM.

Consider a particle or reference point Q that moves at constant speed v around a circle of radius A, as shown in Fig. 14-13*a*. The radial line OQ from the origin to the point Q makes an angle θ with the positive x axis. Since point Q moves with constant speed, the angle θ changes uniformly, and $d\theta/dt = \omega_z$. (See Sec. 12-5.) For the case shown in the figure, ω_z is positive, and we shall drop the z subscript and use the angular speed $\omega = v/A$. Since ω is constant, $\theta = \omega t + \phi$, where the phase constant ϕ is the initial value of θ.

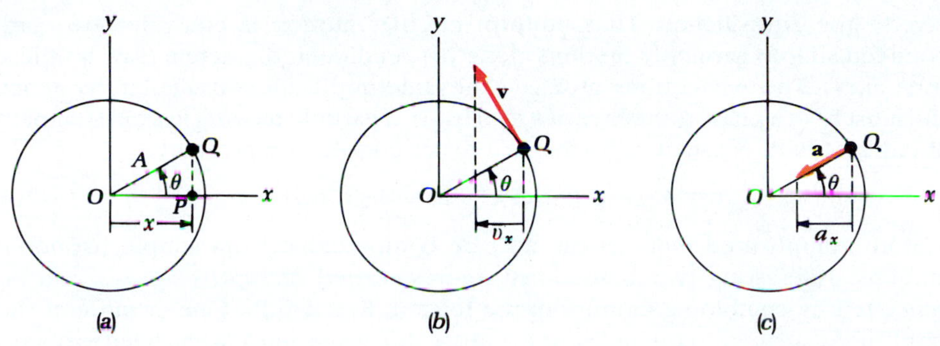

Figure 14–13. Point Q moves counterclockwise, around a circle of radius A, with constant angular speed ω so that $\theta = \omega t + \phi$. *(a)* The x coordinate of Q and of P is $x = A\cos\theta = A\cos(\omega t + \phi)$. *(b)* The x component of the velocity of Q and of P is $\upsilon_x = \upsilon\cos(\theta + \pi/2) = -\omega A\sin(\omega t + \phi)$. *(c)* The x component of the acceleration of Q and of P is $a_x = a\cos(\theta + \pi) = -\omega^2 A\cos(\omega t + \phi)$.

Now the x coordinate of the reference point Q is determined. This is also the x coordinate of point P on the x axis. From the figure we have $x = A\cos\theta$, or with $\theta = \omega t + \phi$,

$$x = A\cos(\omega t + \phi)$$

which is Eq. 14-1 for SHM. Thus as point Q moves around the circle at constant speed, point P moves along the x axis in SHM.

Point P can be thought of as the *projection* of point Q onto the x axis. The projection can be realized by using light from a distant slide projector to observe on a screen the shadow of a particle moving in a circle. Such an arrangement is shown schematically in Fig. 14-14.

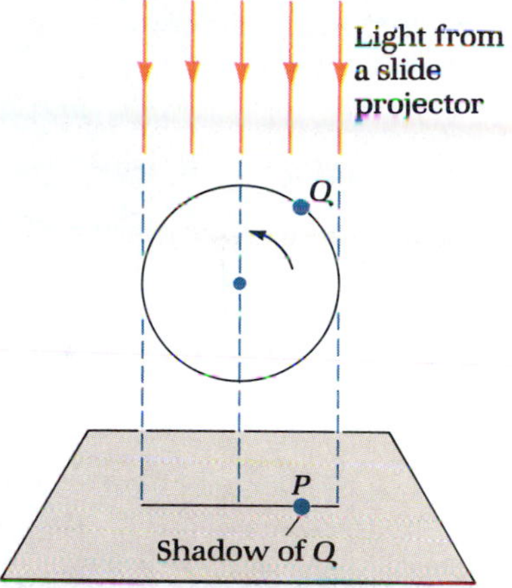

Figure 14–14. Light from a slide projector illuminates an object Q rotating on a turntable at constant angular speed. The shadow of Q on a screen moves along a line in SHM.

The x components of the velocity and acceleration of point Q in its uniform circular motion can also be examined. The velocity is tangent to the circular path (Fig. 14-13*b*). The angle between the velocity and the positive x axis is $\theta + \pi/2$. The velocity component is

$$\upsilon_x = \upsilon\cos\left(\theta + \frac{\pi}{2}\right) = -\upsilon\sin\theta$$

But $\upsilon = \omega A$ and $\theta = \omega t + \phi$, so that

$$\upsilon_x = -\omega A\sin(\omega t + \phi)$$

which is Eq. 14-4 for the velocity component of a point P in SHM.

The acceleration for uniform circular motion is the centripetal acceleration. It is directed toward the center, and its magnitude is $a = \upsilon^2/A = \omega^2 A$, from Sec. 12-5. Notice from Fig. 14-13*c* that the direction of the acceleration is opposite the direction of the position vector from O to Q. Therefore **a** makes an angle $\theta + \pi$ with the positive x axis. The acceleration component is $a_x = a\cos(\theta + \pi) = -a\cos\theta$. Substituting $a = \omega^2 A$ and $\theta = \omega t + \phi$ gives

$$a_x = -\omega^2 A\cos(\omega t + \phi)$$

which is Eq. 14-5 for the acceleration component of a point P in SHM.

In terms of the coordinate, velocity, and acceleration, the x part of the motion of a particle in a circle of radius A with constant angular speed ω is equivalent to the simple harmonic motion of a particle with amplitude A and angular frequency ω. Similar conclusions hold for the y part of the motion. For example, the y coordinate of point Q in Fig. 14-15 is

$$y = A\sin(\omega t + \phi)$$

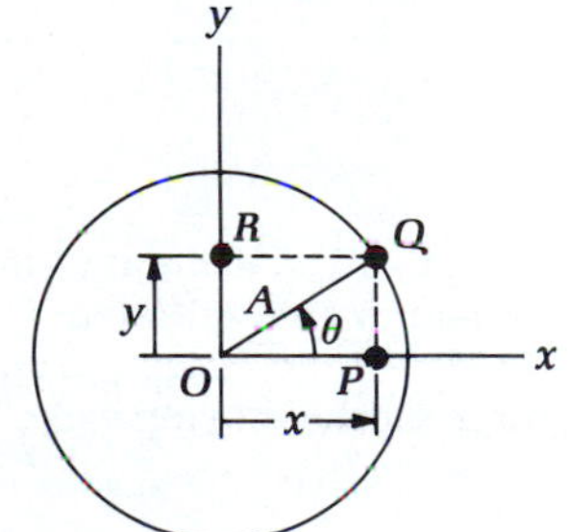

Figure 14–15. Point P is the projection of Q onto the x axis, and point R is the projection of Q onto the y axis.

which is also the y coordinate of point R, the projection of point Q onto the y axis. This motion along the y axis is also SHM, with amplitude A and angular frequency ω. The use of a sine for y and a cosine for x means that the x and y parts of the motion have a phase difference of $\pi/2$.

We have seen that SHM is equivalent to the projection of uniform circular motion onto the x axis or onto the y axis, or onto any diameter of the circle. We can also

Uniform circular motion is compounded from two simple harmonic motions.

reverse the equivalence. Thus uniform circular motion is equivalent to compounded simple harmonic motions along perpendicular diameters (say, on the x and y axes). The two motions must have the same amplitude and angular frequency and must have a phase difference of $\pi/2$. That is, a particle moving in a circular path of radius A with constant angular speed ω has coordinates given by

$$x = A\cos(\omega t + \phi) \qquad y = A\sin(\omega t + \phi) \tag{14-26}$$

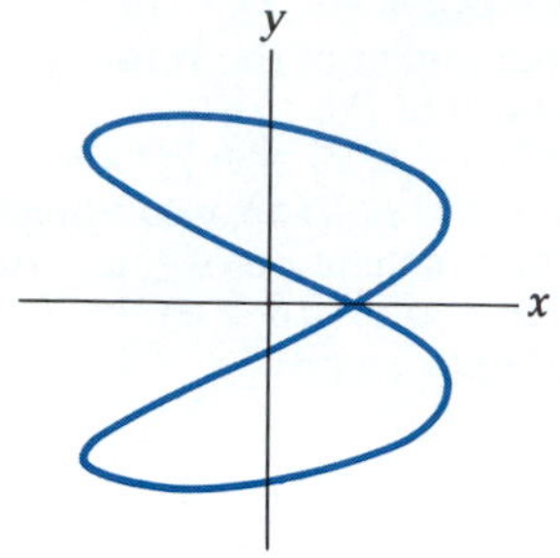

Figure 14-16. A Lissajous figure with ω for the x part of the motion twice that for the y part of the motion.

More complicated motions can also be compounded from simple harmonic motions. Interesting two-dimensional figures, called ***Lissajous figures,*** can be generated by combining motions of the form in Eqs. 14-26. One or more of the amplitudes, angular frequencies, or the phase difference must be changed to obtain noncircular figures. One such example is shown in Fig. 14-16. In this example, the frequency of the x part of the motion is twice the frequency of the y part.

The complicated vibrations of atoms in a crystal can also be represented by compounding simple harmonic motions. In the simplest model, an atom moves with three simple harmonic motions, one each along three mutually perpendicular directions. We shall use this idea in Chap. 18 to understand the response of a substance to a change in its temperature.

14-6 DAMPED HARMONIC MOTION

In our discussion of oscillations, dissipative effects such as those due to frictional forces have been neglected. These effects are almost always present, and often they cannot be neglected. Suppose that a pendulum is set in motion. Although it may swing through many cycles before there is a noticeable decrease in its amplitude, the motion will eventually cease or be "damped out" unless the mechanical energy dissipated by friction is replenished.

Consider a block connected to a spring and oscillating vertically about the equilibrium position at $y = 0$. As we saw in Sec. 14-4, the net force component is $\Sigma F_y = -ky$ and the motion is SHM if dissipative forces are neglected. We can investigate the effects of dissipation by adding a damping mechanism, as shown schematically in Fig. 14-17. A vane, a part of the oscillating object, is in a fluid which exerts a resistive or damping force on the vane. This force will act in a direction opposite the velocity of the object. By varying the shape of the vane and by using different fluids, this damping force can be made large or small.

Figure 14-17. A damping vane is attached to an oscillating block. The liquid exerts a damping force on the moving vane.

A simple model for the damping force is one that is proportional to the velocity of the block but with the opposite direction: $\mathbf{F}_D = -b\mathbf{v}$, where b is a constant that depends on the fluid and the shape of the vane. Since the force is opposite the velocity, it does negative work for every displacement of the block. That is, this force causes the mechanical energy of the oscillator to decrease.

Including this force in Newton's second law for the block, we have

$$\Sigma F_y = -ky - bv_y = ma_y$$

or

$$a_y = -\frac{k}{m}y - \frac{b}{m}v_y \tag{14-27}$$

This is the equation of motion for the ***damped harmonic oscillator.*** Obtaining the solution of this equation involves mathematical techniques beyond the level of this course. We quote the solution, which you can verify as outlined in Exercise 43, for the case of relatively small damping:

Solution for the underdamped oscillator

$$y = e^{-\gamma t} A\cos(\omega_D t + \phi) \tag{14-28}$$

where $\gamma = b/2m$ and $\omega_D = \sqrt{k/\text{m} - (b/2m)^2}$. This solution is valid for $(b/2m)^2 < k/m$. Except for the exponential factor $e^{-\gamma t}$, the solution would be SHM with an angular frequency ω_D that is less than the "normal" angular frequency

$\omega = \sqrt{k/m}$, the frequency with no damping (if $b = 0$). The exponential factor decreases continuously and approaches zero as t increases. In effect, the oscillation has a continuously decreasing amplitude. The motion is said to be *underdamped*. An example is shown graphically in Fig. 14-18 for a case with $(b/2m)^2 = 0.0050k/m$, so that $\gamma = 0.071\omega_D$.

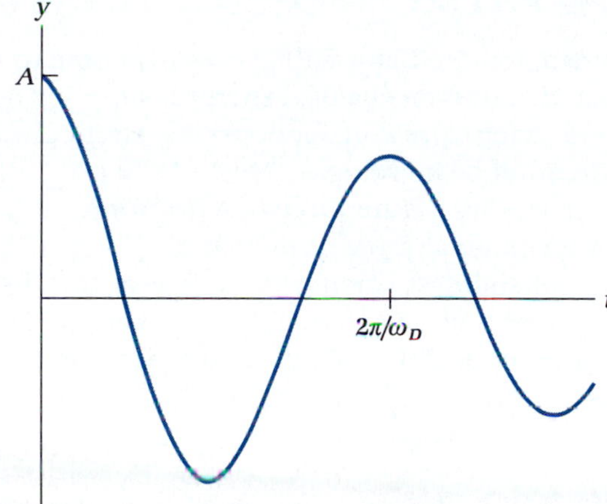

Figure 14-18. The solution for an underdamped harmonic oscillator with $(b/2m)^2 = 0.005k/m$ so that $\gamma = 0.07\omega_D$. The oscillator is released from rest at $t = 0$, with $y_0 = A$.

The solution of Eq. 14-27 is qualitatively different for large damping. If $(b/2m)^2 > k/m$, then the damping force effectively prevents oscillations, and the motion is called *overdamped*. As shown in Fig. 14-19, an overdamped oscillator initially displaced from equilibrium slowly approaches the equilibrium position without passing through it. The form of the overdamped solution is considered in Exercise 44.

Solution for the overdamped oscillator

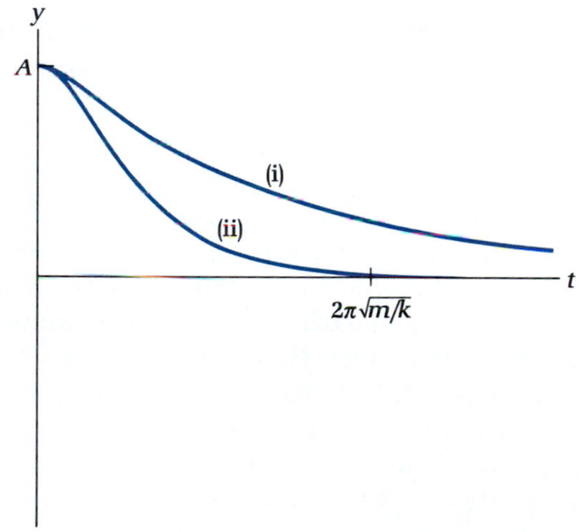

Figure 14-19. An oscillator is released from rest with $y = A$ at $t = 0$. Two cases are shown: (i) overdamped motion with $(b/2m)^2 = 6k/m$ and (ii) critically damped motion with $(b/2m)^2 = k/m$.

If $(b/2m)^2 = k/m$, then the motion is called *critically damped*. As suggested in Fig. 14-19, a critically damped oscillator does not oscillate but approaches the equilibrium position more rapidly than the overdamped oscillator. The form of the critically damped solution is considered in Exercise 45.

Solution for the critically damped oscillator

EXAMPLE 14-8

An underdamped oscillator. For the underdamped motion shown in Fig. 14-18, determine the ratio of the coordinate at $t = T_D = 2\pi/\omega_D$ to the coordinate at $t = 0$.

Solution. The underdamped solution is given by Eq. 14-28, and the motion in Fig. 14-18 corresponds to $\gamma = 0.071\omega_D$. The sinusoidal part of the equation, $A \cos(\omega_D t + \phi)$, has the same value at $t = 0$ and at $t = T_D$ because $\cos(\omega_D T_D + \phi) = \cos(2\pi + \phi) = \cos\phi$. Therefore the ratio of the coordinates at these times is given by the ratio of the exponential parts:

$$\frac{e^{-\gamma T_D}}{e^{-\gamma \cdot 0}} = \frac{e^{-\gamma 2\pi/\omega_D}}{1} = e^{-2\pi(0.071)} = 0.64$$

SELF-TEST 14-8. For the oscillator in the above example, determine the ratio of the coordinate at time $t = 2T_D$ to the coordinate at time $t = 0$. ***ANSWER:*** 0.41.

14-7 FORCED OSCILLATIONS AND RESONANCE

Forced oscillations are driven by an external agent.

A damped oscillator will eventually come to rest as its mechanical energy is dissipated, unless mechanical energy is supplied by a driving force. For example, a child on a swing can swing for hours if a parent gives the swing an occasional push in the direction of its velocity. Most of the oscillations that occur in machinery and in electric circuits are *forced oscillations,* oscillations that are created and sustained by an external force or influence.

The simplest driving force is one that itself oscillates as a sine or a cosine. Suppose such an external force $\mathbf{F}_E$ is applied to an oscillator that moves along the x axis, such as a block connected to a spring. We write the external force component as

An external driving force with angular frequency ω_E

$$F_{Ex} = F_0 \cos \omega_E t \tag{14-29}$$

where F_0 is the maximum magnitude of the force and the force component oscillates sinusoidally with angular frequency ω_E. The frequency of the external force is generally different from the natural angular frequency $\omega = \sqrt{k/m}$ of the oscillator, which is its angular frequency only if there is no damping and no driving force.

If we include the force component in Eq. 14-29 in Newton's second law for a damped harmonic oscillator, then

$$\Sigma F_x = F_0 \cos \omega_E t - kx - bv_x = ma_x$$

That is, three forces are acting: an external force, a restoring force, and a damping force. Dividing by the mass gives the equation of motion,

$$a_x = -\omega^2 x - 2\gamma v_x + \left(\frac{F_0}{m}\right) \cos \omega_E t \tag{14-30}$$

where $\omega^2 = k/m$ and $\gamma = b/2m$ as before.

The techniques for solving Eq. 14-30 are beyond the scope of this course, but some interesting features of the solution are described here. The general solution consists of the sum of two terms. One of these is called the *transient solution,* and it is the solution for a damped harmonic oscillator discussed in the last section. This solution depends in detail on the initial conditions but will be damped out eventually. That leaves the other term, which is called the *steady-state solution.* It is the solution due to the external driving force and persists after the transient solution has died away. We suppose the motion began in the distant past so that for $t \geq 0$, only the steady-state solution remains.

The transient solution damps out and leaves only the steady-state solution.

The steady-state solution oscillates sinusoidally with the same frequency as the external force. It has a steady or fixed amplitude A_0 and has a definite phase difference ϕ_E with the external force. This solution, which you can verify in Prob. 10, is

$$x = A_0 \cos (\omega_E t - \phi_E) \tag{14-31}$$

where

$$A_0 = \frac{F_0/m}{\sqrt{(\omega_E{}^2 - \omega^2)^2 + 4\gamma^2\omega_E{}^2}}$$

$$\tan \phi_E = \frac{2\gamma\omega_E}{\omega^2 - \omega_E{}^2}$$

The amplitude of the motion A_0 is proportional to the amplitude of the driving force F_0.

The amplitude also depends on the driving frequency ω_E. That is, the oscillator responds differently to a driving force of the same magnitude but a different driving frequency. To understand this response, think of the natural frequency ω as fixed and the driving frequency ω_E as variable. For each value of ω_E, we can determine the amplitude A_0 of the motion. This dependence is shown in Fig. 14-20 for an oscillator with small damping. Notice that the amplitude of the motion is small if ω_E is either much larger or much smaller than the natural angular frequency ω. The amplitude is largest if $\omega_E \approx \omega$. In this case the driving force is approximately in phase with the velocity, so that positive work is done by this force on the block during most of the cycle. Thus the oscillator can obtain more energy from the driving force and the amplitude is large.

Maximum amplitude occurs if $\omega_E \approx \omega$.

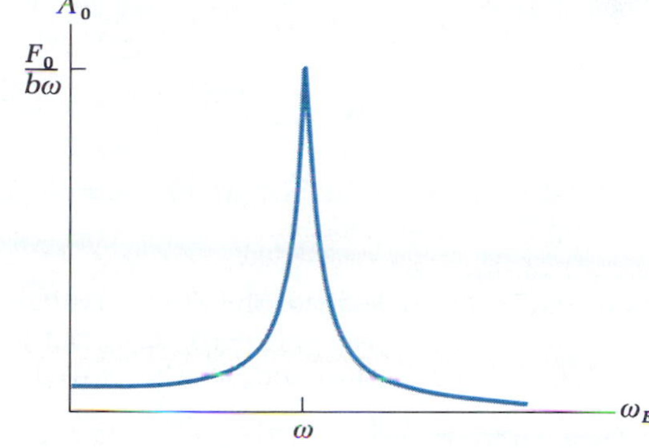

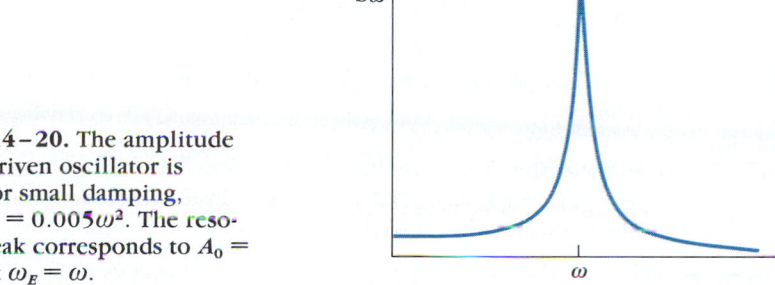

Figure 14–20. The amplitude A_0 of a driven oscillator is shown for small damping, $(b/2m)^2 = 0.005\omega^2$. The resonance peak corresponds to $A_0 = F_0/b\omega$ at $\omega_E = \omega$.

The dramatic increase in the amplitude of the motion for $\omega_E \approx \omega$ is called ***resonance.*** Resonance also can occur when any oscillating system is driven by or coupled to another oscillating system, if their frequencies are about the same. In effect, the coupling between the systems is enhanced if the frequencies are equal. Depending on the circumstances, resonance can be desirable or undesirable. For example, the characteristic shape of a guitar allows a resonant coupling between the vibrating string and the vibrating air in the sound box of the instrument. A radio or television receiver is tuned to be in resonance with the frequency of the signals to be received. On the undesirable side, unwanted vibrations in a mechanical system may occur with a large amplitude if the system is driven at resonance. A car has a characteristic frequency that depends on its mass and the spring constants of its springs. Suppose the car has a wheel that is "out of balance." At a particular speed, the wheel can vibrate the car at its resonant frequency, causing an annoyance for the passengers.

Resonance occurs between coupled systems with nearly equal frequencies.

Low tide and high tide at the Bay of Fundy. Unusually large tides such as this occur in a number of bays and sounds around the world. The size of these tides can be explained in terms of resonance. The natural period of oscillation of water back and forth in the bay is very nearly the same as the driving frequency of the tide.

EXAMPLE 14-9

Resonance amplitude. Show, as indicated in Fig. 14-20, that the amplitude of oscillation for resonance at $\omega_E = \omega$ is given by $A_0 = F_0/b\omega$.

Solution. The frequency dependence of the amplitude A_0 is given by the expression just after Eq. 14-31,

$$A_0 = \frac{F_0/m}{\sqrt{(\omega_E{}^2 - \omega^2)^2 + 4\gamma^2\omega_E{}^2}}$$

At $\omega_E = \omega$, the denominator becomes

$$\sqrt{(0)^2 + 4\gamma^2\omega^2} = 2\gamma\omega$$

Thus

$$A_0 = \frac{F_0/m}{2\gamma\omega} = \frac{F_0/m}{(b/m)\omega} = \frac{F_0}{b\omega}$$

Notice that a smaller value of b, which corresponds to less damping, leads to a larger value of the amplitude A_0 at resonance.

SELF-TEST 14-9. Suppose the driven oscillator in the above example is described by $m = 0.050$ kg, $F_0 = 1.0$ N, $\omega = 25$ rad/s, and $\gamma = 1.8\ \text{s}^{-1}$. Evaluate the amplitude of the oscillator *(a)* at resonance and *(b)* when driven at $\omega_E = 50$ rad/s. ***ANSWERS:*** *(a)* 0.22 m; *(b)* 0.011 m.

COMMENTARY: Chaos

Chaos has been called, by some, the third revolution of the twentieth century in physics, after relativity and quantum physics. Examples of chaos have been observed in countless systems and situations. What is chaos, and what are its implications?

An Example of Chaos

Figure 14-21 shows a simple system that can illustrate chaos. A thin rod has a washer-shaped magnet at its lower end, and the rod and magnet can swing as a pendulum. Two other magnets are fixed in a horizontal plane just below the swinging magnet. They are placed so that each attracts the magnet on the rod and there are just two configurations of stable equilibrium for this system. The magnet on the rod can be in equilibrium at a position just above one of the magnets or just above the other magnet.

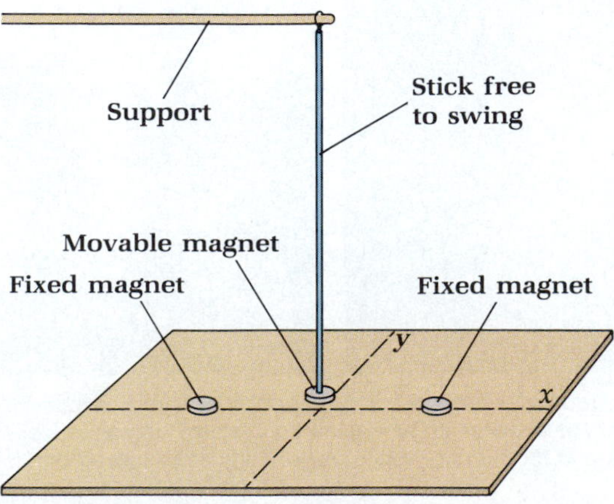

Figure 14-21. A magnet at the end of a swinging stick is attracted by two fixed magnets.

Suppose the pendulum is set in motion by pulling it aside and releasing it. The motion can be rather complex because the magnetic force depends strongly on the relative positions of the magnets. Since there is some damping, the system will lose mechanical energy and eventually end up at one of the equilibrium configurations. A possible path of the pendulum tip is sketched in Fig. 14-22*a*. In this case it ends up near magnet M_1. If the pendulum starts out from a different point, it may end up near the other magnet M_2, as shown in Fig. 14-22*b*.

So far, there is nothing surprising about this system. Where is the chaos? After all, the system always ends up in one of the two equilibrium configurations. But which one? Clearly, that depends on how the pendulum starts out—that is, on the initial conditions. Suppose the pendulum is always released from rest. The other initial conditions are determined by the initial position, which can be specified by the x and y coordinates of the pendulum tip as projected into the plane containing the fixed magnets.

Figure 14-23 shows schematically how a portion of this plane is divided into parts. The white parts show the set of initial release points for which the pendulum tip ends up over magnet M_1. The blue parts show initial release points for which the pendulum tip ends up over magnet M_2. That is, each color represents the "basin of attraction" for each magnet. Although the boundary between the blue and white regions appears smooth and sharp, a magnified view of the boundary would show a structure, called a ***fractal,*** that is exceedingly complex. Figure 14-24 shows an example of a fractal structure. It is for a different vibrating system and uses three colors to distinguish three basins of attraction.

The fractal structure vividly shows regions that are mainly white but that also contain points with the other colors. Magnification of this region shows a structure that is similar to the original mix of swirling colors. This "self-similarity" is a characteristic of a fractal. The complexity extends to all levels of magnification.

Now we can see the nature of chaos for the magnetic pendulum system. The slightest change in the initial point of release can lead to a very different final state, with the tip of the pendulum over magnet M_2 instead of over magnet M_1. The mechanical system is still deterministic in that its motion can be determined, in principle, from a knowledge of the initial position. However, in practice it is not possible to determine the initial position with perfect precision—or to predict which magnet will ultimately capture the pendulum tip.

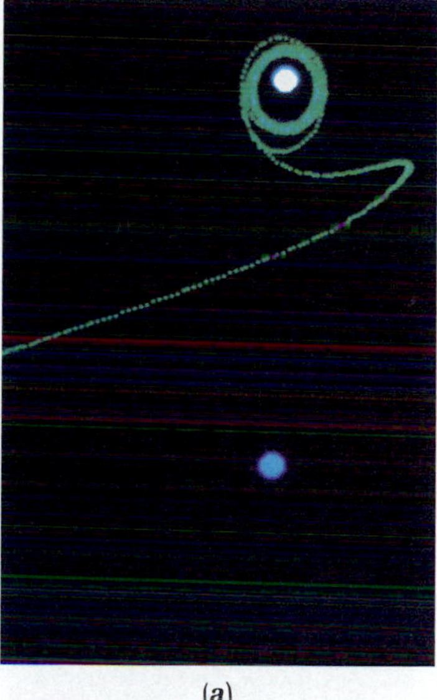

(*a*)

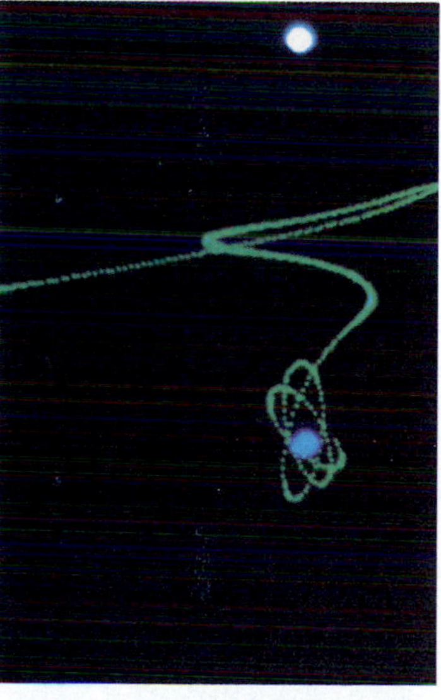

(*b*)

Figure 14–22. *(a)* When released from rest at one position, the pendulum tip ultimately settles over magnet M_1. *(b)* When released from rest at a neighboring point, the pendulum tip settles over magnet M_2.

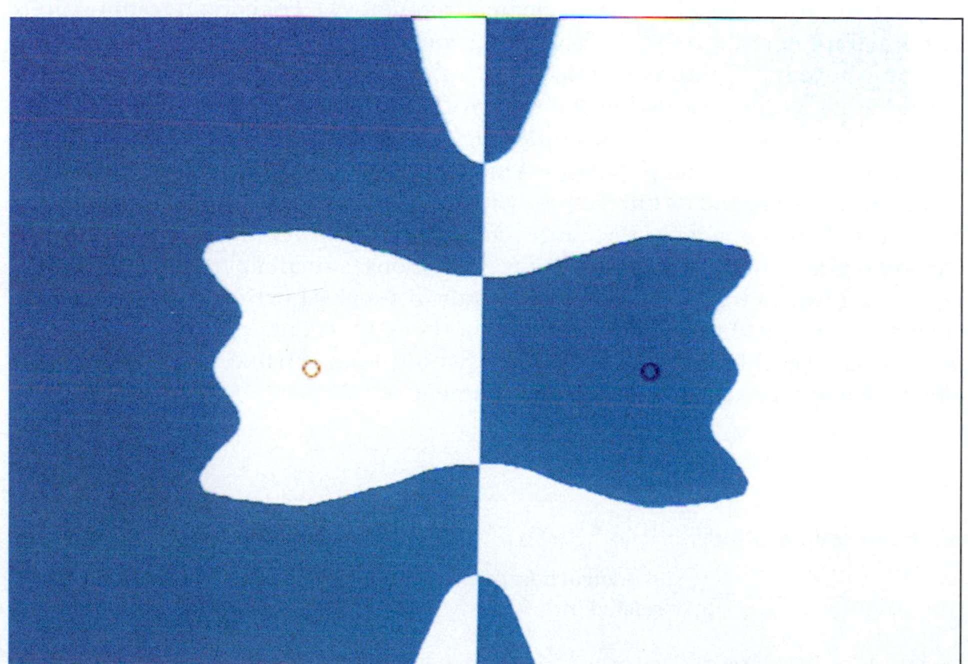

Figure 14–23. The white regions correspond to release points for which the pendulum tip settles over magnet M_1. Blue regions correspond to release points for which the pendulum tip settles over magnet M_2.

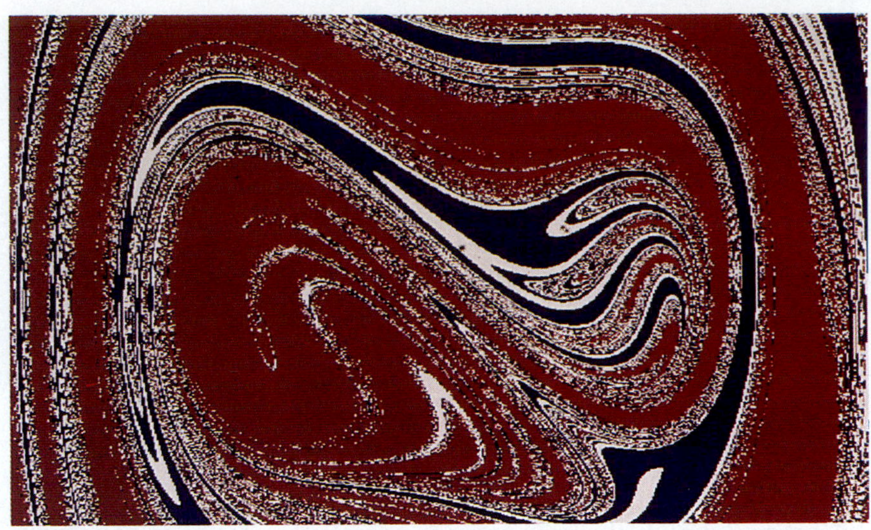

Figure 14–24. A different oscillating system has three basins of attraction; a portion of the fractal structure is shown.

Chaos and Nonlinearity

The oscillators that we considered in the earlier sections of this chapter—simple, damped, and forced oscillators—have all been *linear oscillators.* That is, the equation of motion for a linear oscillator contains the coordinate x, the velocity v_x and the acceleration a_x to the first power only; see Eq. 14-30, for example. Linear oscillators and linear systems do not exhibit chaos. A system that shows chaos must be *nonlinear.* One example of a nonlinear oscillator is a damped, driven pendulum. The equation of motion is

$$I\alpha_z = -b\omega_z - mgL \sin \theta + \tau_o \cos \Omega t$$

where the term $\sin \theta$ is the nonlinear term. For *small* oscillations, $\sin \theta \approx \theta$, and the equation is approximately linear. The next approximation is $\sin \theta \approx \theta - \theta^3/6$, which is clearly nonlinear. In the above equation of motion, I is the moment of inertia of the pendulum of mass m with the center of mass a distance L from the pivot, b is a damping constant, and τ_o is the maximum magnitude of an external torque that varies sinusoidally with angular frequency Ω. For certain combinations of the damping and driving constants, the pendulum has chaotic motion—the motion is essentially unpredictable.

Most systems encountered in the real world are nonlinear and, under suitable circumstances, exhibit chaos. Examples include mechanical and electrical oscillators, the flow of fluids, and population kinetics. Other complex systems, such as the earth's atmosphere and its interactions with the earth's surface and the sun, are also nonlinear. If this system—the earth's atmosphere and its interactions—is in a chaotic regime, then the implications are enormous: a small fluctuation, perhaps a leaf falling from a tree, will ultimately result in a major hurricane that would not have developed had the leaf held on for another day. Accurate long-term weather prediction would be an impossibility, as would be a reliable prediction of the effects of our interaction with the environment.

SUMMARY

Section 14-1. Kinematics of Simple Harmonic Motion

For an object undergoing SHM,

$$x = A \cos (\omega t + \phi) \quad (14\text{-}1)$$

$$v_x = -\omega A \sin (\omega t + \phi) \quad (14\text{-}4)$$

$$a_x = -\omega^2 A \cos (\omega t + \phi) \quad (14\text{-}5)$$

The angular frequency ω, the frequency ν, and the period T are related by

$$\omega = 2\pi\nu = \frac{2\pi}{T}$$

SHM can be identified by the relation

$$a_x = -\omega^2 x \tag{14-6}$$

Section 14-2. Dynamics of Simple Harmonic Motion
SHM is caused by a net force that is a linear restoring force. Newton's second law applied to such an object can be cast as a differential equation:

$$\frac{d^2x}{dt^2} = -\frac{k}{m}x \tag{14-11}$$

A solution to this equation is $x = A\cos(\omega t + \phi)$, with $\omega = \sqrt{k/m}$.

Section 14-3. The Energy of a Simple Harmonic Oscillator
The potential and kinetic energies of a simple harmonic oscillator are

$$U = \tfrac{1}{2}kA^2\cos^2(\omega t + \phi) \tag{14-13}$$

$$K = \tfrac{1}{2}m\omega^2A^2\sin^2(\omega t + \phi) \tag{14-14}$$

and the mechanical energy is constant:

$$E = \tfrac{1}{2}m\omega^2A^2 = \tfrac{1}{2}kA^2 \tag{14-15}$$

Section 14-4. Examples of Simple Harmonic Motion
Systems which were shown to undergo SHM are (i) a block suspended from a vertical spring, $\omega = \sqrt{k/m}$; (ii) a simple pendulum, $\omega = \sqrt{g/L}$; (iii) a physical pendulum, $\omega = \sqrt{mgL/I}$; and (iv) a torsional oscillator, $\omega = \sqrt{k/I}$.

Section 14-5. Simple Harmonic Motion and Uniform Circular Motion
For a particle executing uniform circular motion in the xy plane, x and y components of the motion are each simple harmonic motion.

Section 14-6. Damped Harmonic Motion
The equation of motion for a damped harmonic oscillator is

$$a_y = -\frac{k}{m}y - \frac{b}{m}v_y \tag{14-27}$$

For the underdamped case

$$y = e^{-\gamma t}A\cos(\omega_D t + \phi) \tag{14-28}$$

The amplitude of the oscillations decreases exponentially with time. If the oscillator is critically damped or overdamped, no oscillations occur.

Section 14-7. Forced Oscillations and Resonance
The equation of motion for a forced oscillator is

$$a_x = -\omega^2 x - 2\gamma v_x + \left(\frac{F_0}{m}\right)\cos\omega_E t \tag{14-30}$$

The amplitude of the oscillations is

$$A_0 = \frac{F_0/m}{\sqrt{(\omega_E^2 - \omega^2)^2 + \gamma^2\omega_E^2}}$$

the amplitude is maximum when resonance occurs: $\omega_E \approx \omega$.

QUESTIONS

1. Give three examples of oscillating systems. Are any of these systems simple harmonic oscillators? If not, why not?

2. What is the distance traveled during one period by an object executing SHM with amplitude A?

3. Suppose the angular frequency ω of a simple harmonic oscillator is doubled. By what factor does this change *(a)* the frequency ν, *(b)* the period T, *(c)* the amplitude A, *(d)* the phase constant ϕ?

4. Suppose the amplitude A of a simple harmonic oscillator is doubled. By what factor does this change *(a)* the angular frequency ω, *(b)* the frequency ν, *(c)* the period T, *(d)* the maximum speed v_{max}, *(e)* the maximum acceleration magnitude a_{max}, *(f)* the mechanical energy E?

5. During SHM, are the displacement and velocity ever in the same direction? The velocity and acceleration? The displacement and the acceleration?

6. An astronaut is to stay in earth orbit for several months. Devise a procedure for keeping track of the astronaut's mass.

7. In finding $\omega = \sqrt{k/m}$ for the block-spring system, we neglected the mass of the spring. Suppose we used a spring with significant mass. Describe its effect on the motion of the system.

8. Can you find the amplitude or the phase constant for a simple harmonic oscillator if you know the initial coordinate but not the initial velocity? If you know the initial velocity but not the initial coordinate? Explain.

9. A block of mass m is suspended from two light springs in two different ways (Fig. 14-25). Each spring has the same spring constant k and same unstretched length. What is the angular frequency of the oscillations, in terms of k and m, when the springs are side by side, as in Fig. 14-25*a*? What is the angular frequency when the springs are end to end, as in Fig. 14-25*b*?

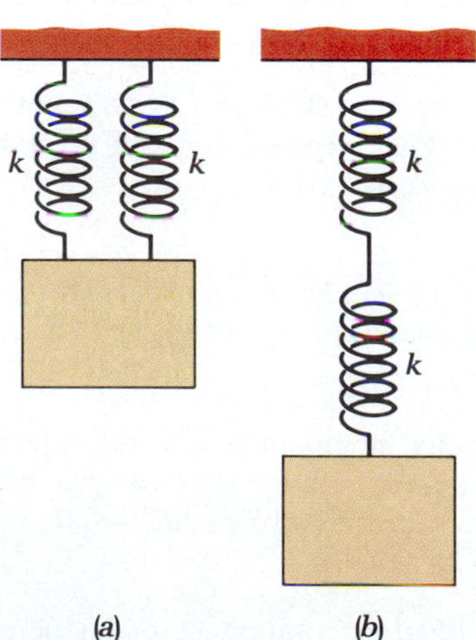

Figure 14–25. Question 9.

10. A block-spring simple harmonic oscillator has $E = 4$ J when $x = A$. At $x = A$, what are *(a)* U and *(b)* K? At $x = 0$, what are *(c)* E; *(d)* U; *(e)* K? At $x = -A$, what are *(f)* E; *(g)* U; *(h)* K?

11. Estimate the spring constant of the springs of an automobile by estimating its mass and its period of oscillation if the springs are pushed down and released.

12. Why do automobiles have shock absorbers? Estimate a value of γ in Eq. 14-28 from the way a typical automobile bounces.

13. Two common lengths for a pendulum in pendulum clocks are about 1 m and about $\frac{1}{4}$ m. What is the advantage of each of these lengths?

14. Would a pendulum clock have the same period at the equator as at one of the poles? What about a windup watch that uses a torsional pendulum? Explain.

15. Does the period of a pendulum depend on its temperature? Did you think of this in answering the previous question?

16. A child on a swing is swinging very high, almost to the point where the ropes are horizontal. If the child stops pumping and lets the amplitude of the swing die down, how will the period of the swing change?

17. Why do soldiers break the cadence of their march when crossing a bridge? It is sometimes said that a cat can cause a bridge to collapse. Is that possible? Explain.

18. The earth moves about the sun in an almost circular orbit. Is this motion compounded of two simple harmonic motions? If so, what are the angular frequencies?

19. Figure 14-26 shows a conical pendulum with the bob swinging in a horizontal circle, say in the xy plane. How are the period and the amplitude of the x motion related to the period and the amplitude of the y motion? What are some other figures that can be traced out by such a pendulum? Try it.

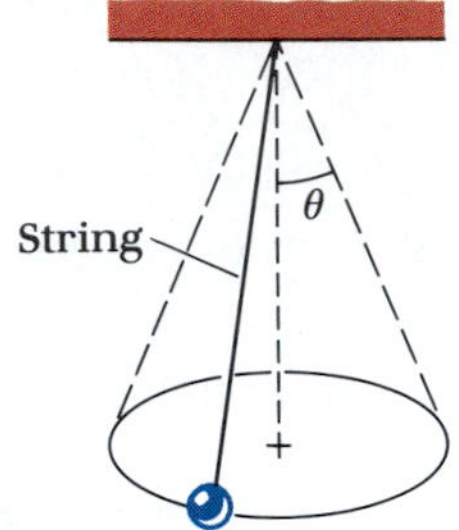

Figure 14-26. Question 19.

20. Complete the following table:

Symbol	Represents	Type	SI Unit
ω			
T	Scalar		
ν			
A			
ϕ			rad
κ	Torsion constant		
ω_E			

EXERCISES

Section 14-1. Kinematics of Simple Harmonic Motion

1. The angular frequency of an object undergoing SHM is 5.8 rad/s. Determine the period and the frequency of the motion.

2. *(a)* Show that the dimension of ωA is the same as velocity, [length][time]$^{-1}$. *(b)* Show that the dimension of $\omega^2 A$ is the same as acceleration, [length][time]$^{-2}$.

3. An object executes SHM with $A = 63$ mm, $\omega = 4.1$ rad/s, and $\phi = 0$. *(a)* Write expressions for x, v_x, and a_x. *(b)* Determine x, v_x, and a_x at $t = 1.7$ s.

4. The coordinate of an object is given by $x = (0.057 \text{ m}) \cos [(3.9 \text{ rad/s})t]$. *(a)* What are A, ω, ν, T, and ϕ? *(b)* Write expressions for v_x and a_x. *(c)* Determine x, v_x, and a_x at $t = 2.3$ s.

5. The velocity component of an object is given by $v_x = (1.8 \text{ m/s}) \sin [(7.1 \text{ rad/s})t]$. *(a)* What are ω, A, ν, T, and ϕ? *(b)* Write expressions for x and a_x. *(c)* Determine x, v_x, and a_x at $t = 0.25$ s.

6. The acceleration component of an object is given by $a_x = -(16.8 \text{ rad}^2/\text{s}^2)x$. The object's initial coordinate is $x_0 = 24$ mm and its initial velocity component is $v_{x0} = 0.71$ m/s. Determine expressions for the object's *(a)* coordinate, *(b)* velocity component, *(c)* acceleration component as functions of time.

7. An object executes SHM with $a_{max} = 13 \text{ m/s}^2$, $T = 0.94$ s, and $\phi = \frac{1}{2}\pi$. *(a)* Write expressions for x, v_x, and a_x. *(b)* Determine x, v_x, and a_x at $t = 0.54$ s.

8. An object is executing SHM with a period $T = 1.8$ s, an initial coordinate $x_0 = 0$, and an initial velocity component $v_{x0} = -0.33$ m/s. *(a)* Write expressions for x, v_x, and a_x as functions of time. *(b)* Make graphs of x, v_x, and a_x versus time from $t = 0$ s to $t = 3.0$ s.

9. An object undergoes SHM with a frequency $\nu = 0.42$ Hz. The initial coordinate is $x_0 = 0.021$ m, and the initial velocity component is $v_{x0} = 1.3$ m/s. Determine A, v_{max}, and a_{max}.

Section 14-2. Dynamics of Simple Harmonic Motion

10. Show that the dimension of $\sqrt{k/m}$ is [time]$^{-1}$.

11. A block-spring simple harmonic oscillator has $k = 45$ N/m and $m = 0.88$ kg. Determine *(a)* ω; *(b)* ν; *(c)* T.

12. *(a)* Show that $x = A \sin(\omega t + \delta)$ is a solution to Eq. 14-11,

$$\frac{d^2x}{dt^2} = -\frac{k}{m}x$$

(b) Determine the relation between δ in the expression above and ϕ in Eq. 14-1, $x = A \cos(\omega t + \phi)$.

13. The period of a block-spring simple harmonic oscillator is $T = 0.87$ s. If the mass of the block is 0.62 kg, what is the spring constant of the spring?

14. A block-spring simple harmonic oscillator has $k = 27$ N/m and $m = 0.46$ kg. Assuming the block was released from rest at $x = 29$ mm, write expressions for x, v_x, and a_x as functions of time.

15. The frequency of a block-spring harmonic oscillator is 1.4 Hz. If the spring constant of the spring is 26 N/m, what is the mass of the block?

16. A block of mass 1.4 kg oscillates with SHM due to a net force exerted by a spring. The amplitude of the motion is $A = 0.14$ m, and the maximum speed of the block is $v_{max} = 2.3$ m/s. What is the spring constant of the spring?

Section 14-3. The Energy of a Simple Harmonic Oscillator

17. A block-spring harmonic oscillator has a force constant $k = 22$ N/m and oscillates with an amplitude $A = 87$ mm. What is the mechanical energy of the oscillator?

18. The mass of the block in a block-spring harmonic oscillator is $m = 0.49$ kg. If the oscillator has a period $T = 0.91$ s and an amplitude $A = 62$ mm, what is the mechanical energy of the oscillator?

19. For a block-spring harmonic oscillator, the spring constant of the spring is $k = 31$ N/m and the mass of the block is $m = 0.74$ kg. The block is released from rest at $t = 0$ and $x_0 = 39$ mm. Write expressions for *(a)* U and *(b)* K as functions of time. *(c)* On the same sheet of graph paper, plot U and K versus t for one period of the motion.

20. A block-spring harmonic oscillator oscillates such that the block's amplitude is A and its maximum speed is v_{max}. *(a)* At what distance from the central point, in terms of A, is the block when its speed is $\frac{1}{2}v_{max}$? *(b)* What is the block's speed, in terms of v_{max}, when its distance from the central point is $\frac{1}{2}A$?

21. A block-spring harmonic oscillator with $k = 23$ N/m and $m = 0.47$ kg has a mechanical energy of 25 mJ. *(a)* What is the amplitude of the motion? *(b)* What is the maximum speed of the block? *(c)* What is the block's speed when $x = 11$ mm? *(d)* What is the block's distance from the central point when its speed is 0.25 m/s?

Section 14-4. Examples of Simple Harmonic Motion

22. A 2.0-kg block is attached to a light vertical spring and gently lowered, so that the block is in equilibrium after the spring extends by 450 mm. The same block is then attached to the unstretched spring and released from rest. Determine *(a)* the period and *(b)* the amplitude of the resulting vertical oscillation.

23. When a standard 1.000-kg mass is connected to a vertical spring of negligible mass, the period of oscillations is 1.43 s. When an object of unknown mass replaces the standard, the period is 1.85 s. Determine *(a)* the unknown mass and *(b)* the spring constant of the spring.

24. A 3.0-kg block is attached to a light vertical spring of spring constant 240 N/m. The system oscillates vertically, and the maximum speed of the block is 1.0 m/s. Determine for this motion *(a)* the amplitude, *(b)* the period, and *(c)* the maximum magnitude of the acceleration of the block. *(d)* How much is the spring stretched if the block is at rest at the equilibrium position?

25. A 5.0-kg block is attached to a 1200-N/m spring and gently lowered to its equilibrium position at $y = 0$. Take the potential energy of the unstretched spring to be zero and the gravitational potential energy of the block to be zero at $y = 0$. *(a)* What is the mechanical energy of the system in this configuration? *(b)* The block is pulled down an additional 25 mm and released from rest. Determine the initial mechanical energy of the system. *(c)* Where is the block when its speed is maximum and what is this maximum speed? *(d)* What is the maximum value of the gravitational potential energy for this motion?

26. The periods for simple, physical, and torsional pendula are given by $2\pi\sqrt{L/g}$, $2\pi\sqrt{I/mgL}$, and $2\pi\sqrt{I/\kappa}$. Show that each of these expressions has dimensions of time.

27. A 20-kg child swings on a 3-m swing with a 0.07-rad amplitude. Determine *(a)* the period and the frequency ν and *(b)* the maximum linear speed of the child for this motion.

28. Tarzan swings from one tree to another on a vine that is joined to a branch 15 m above his head and midway between the two trees. *(a)* Estimate the time interval for his swing if he just lets go of one tree and swings on the vine to the next. *(b)* Does the answer depend on the distance between the trees? *(c)* Would the time interval be different if Tarzan pushed off from one tree with an initial speed? Explain.

29. A uniform meter stick is pivoted about a horizontal axis at one end. *(a)* Determine the period of small oscillations of this stick. *(b)* Estimate the period of small oscillations if the stick is suspended at a small hole drilled through the 250-mm mark.

30. Two small identical 1.0-kg balls are attached at each end of a 1.0-m rigid rod of negligible mass. Determine the period of small oscillations of this pendulum if it is suspended at *(a)* one end, *(b)* the other end, *(c)* one-third of the way from one end to the other, *(d)* the midpoint.

31. A mechanical windup clock is timed by the period of a torsional oscillator called a balance wheel. The balance wheel has moment of inertia 4.20×10^{-8} kg·m^2 and is designed to oscillate with a period of 0.250 s. *(a)* Determine the torsion constant of the spiral spring that acts as a torsional fiber. *(b)* What is the maximum magnitude of the angular acceleration of the wheel if the amplitude of the oscillation is 0.45 rad?

32. A uniform circular disk of radius 180 mm and mass 0.75 kg is attached to a thin fiber of torsion constant

$\kappa = 24$ N·m/rad, as seen in Fig. 14-12. The disk is turned through 1.5 rad from its equilibrium orientation and released from rest. Determine *(a)* the period, *(b)* the angular frequency, *(c)* the maximum angular speed, *(d)* the maximum rotational kinetic energy of the disk for the motion.

33. A 150-mm-long, 25-g pencil is suspended as seen in Fig. 14-12*b*. If equal but opposite horizontal forces of magnitude 2.2 mN are applied at each end perpendicular to the pencil, the pencil is in a new equilibrium position that is rotated by $\pi/4$ rad from its original equilibrium position. Determine *(a)* the period of oscillation and *(b)* the maximum rotational kinetic energy of the pencil after the forces are removed.

Section 14-5. Simple Harmonic Motion and Uniform Circular Motion

34. A particle moves with constant speed $v = 12$ m/s in a circular path of radius $A = 0.50$ m. For the motion of its projection onto the x axis, determine *(a)* the period, *(b)* the amplitude, *(c)* the maximum speed, *(d)* the maximum magnitude of acceleration.

35. The object Q in Fig. 14-14 is 150 mm from the center of a turntable that rotates at $33\frac{1}{3}$ rev/min. Consider the motion of its shadow on the screen. Write expressions as functions of time t for x, v_x, and a_x. Assume that x is a maximum at $t = 0$.

36. As the bob of the conical pendulum in Fig. 14-26 traces out a circular path, the light string of length L sweeps out a cone of half-angle θ. Show that the period of the circular motion of the bob is given by $2\pi\sqrt{(L/g)\cos\theta}$.

37. Suppose that the circle traced out by the bob in the previous exercise is in the xy plane, with the center at the origin. *(a)* Show that the x and y coordinates undergo SHM. *(b)* Determine the amplitude and the angular frequency for the x and y motions.

Section 14-6. Damped Harmonic Motion

38. Determine the SI units of *(a)* the damping constant b that appears in Eq. 14-27 and *(b)* the constant $\gamma = b/2m$ in Eq. 14-28.

39. A block attached to a spring is set in oscillation, with an initial amplitude of 120 mm. After 2.4 min the amplitude has decreased to 60 mm. *(a)* When will the amplitude be 30 mm? *(b)* Determine the value of γ for this motion.

40. A 2.5-kg block is connected to a 1250-N/m spring. The block is released from rest at $t = 0$ at 28 mm from the equilibrium position, and the motion is damped with $b = 50$ kg/s. *(a)* Determine the angular frequency ω_D of the damped harmonic motion. *(b)* Determine the initial amplitude A and the phase constant ϕ in Eq. 14-28 for this motion. (*Caution:* ϕ is not zero.) *(c)* Determine x, v_x, and a_x at $t = \pi/5$ s.

41. The pendulum of a clock normally swings through an arc of length 135 mm. The mechanism, which is driven by a falling weight, stops if the arc length is less than 50 mm. After the weight reaches the bottom of the clock at the time 9:17, the arc length of the pendulum decreases to 95 mm at 9:22. *(a)* What is the constant γ for this damped harmonic motion? *(b)* What time does the clock stop?

42. A steady force of 120 N is required to push a 700-kg boat through the water at a constant speed of 1.0 m/s. Assume that the damping force exerted by the water is given by $\mathbf{F}_D = -b\mathbf{v}$. *(a)* Determine the value of b. *(b)* The boat is fastened by springs on two posts, as shown in Fig. 14-27, and held at 2.0 m from its equilibrium position by a 450-N horizontal force. Write the expression for the motion of the boat after it is released at $t = 0$.

Figure 14–27. Exercise 42: The boat is displaced 2.0 m to the right from its equilibrium position and released from rest.

43. By direct substitution, show that Eq. 14-28 is a solution of Eq. 14-27. [*Hint:* The functions $\cos(\omega_D t + \phi)$ and $\sin(\omega_D t + \phi)$ are independent. That is, $C\cos(\omega_D t + \phi) + D\sin(\omega_D t + \phi) = 0$ if and only if $C = D = 0$.]

44. Verify by direct substitution that the solution of Eq. 14-27 for the case of the overdamped oscillator is given by

$$y = e^{-\gamma t}(Ae^{pt} + Be^{-pt})$$

where $\gamma = b/2m$ and $p = \sqrt{(b/2m)^2 - k/m}$ and A and B are constants.

45. For critical damping, $b/2m = \sqrt{k/m} = \omega$. Verify that the critically damped solution of Eq. 14-27 is given by

$$y = e^{-\omega t}(A + Bt)$$

where A and B are constants.

Section 14-7. Forced Oscillations and Resonance

46. Show that the meter is the SI unit of A_0 in the expression following Eq. 14-31.

47. The block in Exercise 40 is subjected to a sinusoidal driving force of angular frequency $\omega_E = 25$ rad/s and maximum magnitude $F_0 = 12$ N. Determine *(a)* the amplitude A_0 and *(b)* the phase constant ϕ_E for the steady-state motion. *(c)* Determine the amplitude if the system is driven at resonance.

48. Construct graphs similar to that in Fig. 14-20 for $b/2m =$ *(a)* 0.07ω; *(b)* 0.10ω; and *(c)* 0.70ω. Show all three cases on the same graph and comment on the effect of increased damping.

Additional Exercises

49. Suppose SHM is defined with a sine function instead of a cosine function: $x = A\sin(\omega t + \delta)$. Write expressions for *(a)* v_x and *(b)* a_x. Develop expressions for *(c)* δ and *(d)* A in terms of x_0, v_{x0}, and ω.

50. While exploring planet X you set in motion a simple pendulum of length $L = 1.000$ m and find that for small amplitude oscillations $T = 1.96$ s. Determine g on planet X.

51. The coordinate of an object as a function of time is

$$x = (0.27 \text{ m}) \cos [(9.3 \text{ rad/s})t - 2.4 \text{ rad}]$$

What are *(a)* the amplitude and *(b)* the angular frequency of this object's SHM? Determine *(c)* the frequency and *(d)* the period of the motion. At $t = 0.33$ s, determine *(e)* the phase and *(f)* the magnitude and direction of the displacement. (Use a unit vector to show the direction.)

52. For the object in the previous exercise, determine the magnitude and direction of *(a)* the velocity and *(b)* the acceleration at $t = 0.33$ s. (Use a unit vector to show the direction.)

53. Figure 14-28 shows a graph of the potential energy U versus x for a simple harmonic oscillator of mass $m = 0.87$ kg. The oscillator is released from rest at $x = 15$ mm. At the instant the oscillator is at $x = 0$ mm, determine *(a)* its kinetic energy and *(b)* its speed. At the instant the oscillator is at $x = 10$ mm, determine *(c)* its kinetic energy and *(d)* its speed.

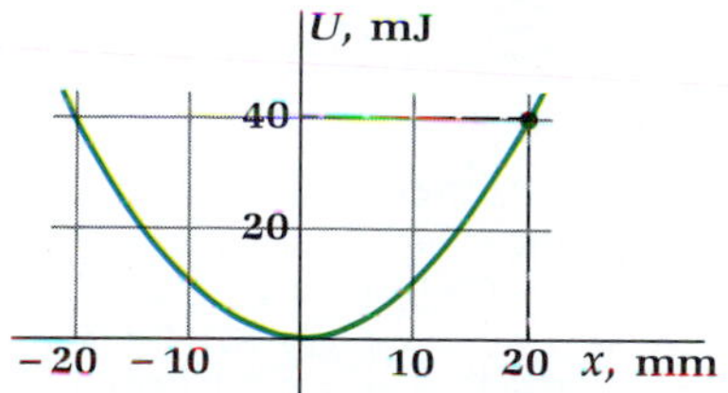

Figure 14–28. Exercise 53.

54. For the oscillator in the previous exercise, determine *(a)* the spring constant k and *(b)* the angular frequency ω. *(c)* Write an expression for x as a function of time where $t = 0$ corresponds to the instant the oscillator is released.

55. A flat circular disk, similar to a phonograph record without the hole, is suspended from a point on its edge so that it forms a physical pendulum that oscillates in the plane of the face of the disk. Show that the angular frequency for small oscillations is $\omega = \sqrt{2g/3R}$, where R is the disk's radius.

56. A flat circular disk, similar to a phonograph record without the hole, is suspended from a point midway between its center and its edge so that it forms a physical pendulum that oscillates in the plane of the face of the disk. Show that the angular frequency for small oscillations is $\omega = \sqrt{2g/3R}$, where R is the radius of the disk.

57. A uniform, 34-kg kitchen door can swing either into or out of the kitchen. The spring that restores the door to the closed position is linear. For simplicity, we neglect friction so that the net restoring torque can be written as $\Sigma\tau_z = -\kappa\theta$. The z axis is along the rotational axis, and θ is measured from the closed position. The door has a width of 0.92 m and oscillates with a period of 1.4 s. Determine κ.

58. Construct a spreadsheet that will produce the graph of A_0 versus ω_E shown in Fig. 14-20. Let $\omega = 1.00$ rad/s, $m = 1.00$ kg, $F_0 = 10.0$ mN, $b = 0.141$ N·s/m, and let ω_E range from 0.00 to 2.00 rad/s.

59. In an experiment, one end of a vertical spring is fixed to a rigid support and various blocks of different mass are attached in turn to the free end of the spring. For each block the period of oscillation is measured. The data are graphed, as shown in Fig. 14-29, with the *square* of the period *versus* the mass of the block. Determine the spring constant of the spring.

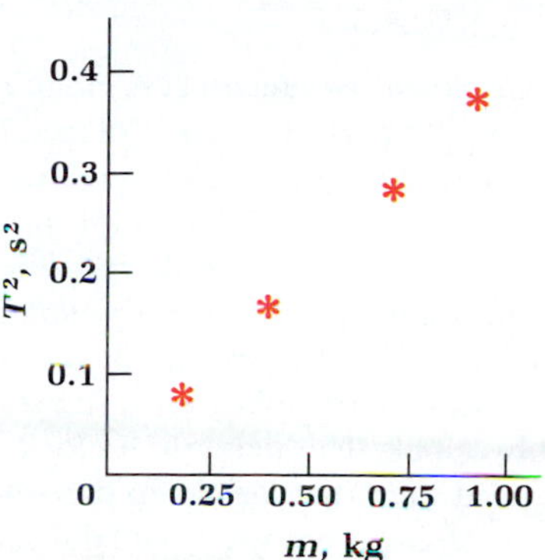

Figure 14–29. Exercise 59.

60. The bob of a simple pendulum of length 1.75 m swings through the lowest point of its motion with a *linear* speed of 0.50 m/s. Determine *(a)* the maximum angular speed $|\omega_z|_{max}$ and *(b)* the amplitude θ_{max} for this motion.

61. Suppose that you grasp a fixed horizontal bar and swing as a pendulum. Estimate the period of small oscillations.

62. A rigid rod of length 2.0 m is suspended at one end and swings as a physical pendulum with small amplitude. Determine the length of a simple pendulum that would have the same period.

63. A rigid rod of length D and negligible mass is pivoted to rotate about one end. A small ball of mass M is threaded onto the rod and fixed to the midpoint, and a second identical ball is fixed to the free end of the rod. Determine an expression for the period of small oscillations.

64. A torque of magnitude 0.278 N·m is required to twist a torsional fiber through 2.25 rad. When this fiber is used with a plate to form a torsional oscillator, the period of oscillation is 0.77 s. What is the moment of inertia of the plate about the axis of rotation?

65. The earth moves about the sun in an approximately circular path with a radius of 1.5×10^{11} m and a period of 1 year. Relative to an xy coordinate system centered on the sun and in the plane of the earth's orbit, write expressions for *(a)* coordinates, *(b)* velocity components, and *(c)* acceleration components for the earth's motion. Choose time t such that $x = 1.5 \times 10^{11}$ m and $y = 0$ at $t = 0$.

66. Solve the equation of motion, Eq. 14-30, for the case of negligible damping ($\gamma = 0$). (*Hint:* Try a solution of the form $x = A_0 \cos \omega_E t$ and determine A_0. Why must $\omega_E \neq \omega$?)

PROBLEMS

1. ***Travel time for an oscillator.*** In terms of the period T, determine the time required for a simple harmonic oscillator to travel from $x = A$ to $x = \frac{1}{2}A$.

2. ***Position of an oscillator.*** Consider a block-spring simple harmonic oscillator with $k = 200$ N/m and $m = 2.4$ kg. The initial conditions of the oscillator are $x_0 = 0.15$ m and $v_{x0} = 0.45$ m/s. Determine the position of the block at $t = 3.0$ s.

3. ***Pendulum clock on an elevator.*** A clock regulated by a simple pendulum with period 0.11 s is placed in an elevator. *(a)* What is the period when the elevator is accelerating upward with magnitude $\frac{1}{2}g$? *(b)* What is the period when the elevator is traveling upward at a steady speed of 5.1 m/s? *(c)* What is the period when the elevator is accelerating downward with magnitude $\frac{1}{2}g$? *(d)* At the start of the day, the pendulum clock is on the elevator at the lobby level and is set to the same time as the lobby clock. The pendulum clock and the lobby clock are calibrated such that they keep the same time if they remain stationary. After many trips up and down in the elevator during the day, the pendulum clock is back in the lobby. Do the clocks agree? If not, which one reads the earlier time?

4. ***Temperature and the timing of a clock.*** A clock regulated by a simple pendulum is accurate when its temperature remains fixed at 20°C. The pendulum increases its length by 0.0010 percent for each 1.0°C increase in temperature. What is the temperature if the clock loses 2.0 s in 1 day?

5. ***Including the mass of a spring.*** When we found that the angular frequency of the oscillations of the idealized block-spring system was $\omega = \sqrt{k/m}$, we neglected the mass of the spring. Let us now take account of the mass m_s of the spring. Assume that the extension of the spring is uniform over its length ℓ and that its linear mass density μ is uniform. With these assumptions, $\mu = m_s/\ell$, and the relation between the speed v of the block and the speed v_ξ of an element of the spring located at ξ (see Fig. 14-30) is $v_\xi = (\xi/\ell)v$. *(a)* Show that the kinetic energy K_s of the spring is

$$K_s = \frac{m_s v^2}{6}$$

(b) Write the expression for the mechanical energy of the block-spring system, including the kinetic energy of the spring. By comparison with the expression for the mechanical energy of a block-spring system having a spring of negligible mass, $E = \frac{1}{2}mv^2 + \frac{1}{2}kx^2$, show that the angular frequency taking the spring's mass into account is

$$\omega = \sqrt{\frac{k}{m + m_s/3}}$$

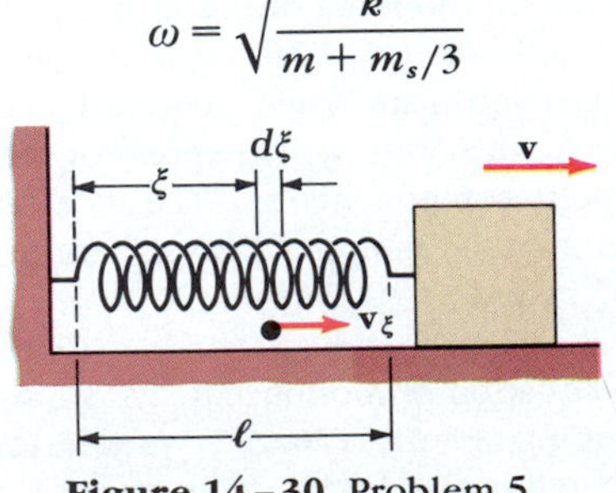

Figure 14–30. Problem 5.

6. ***A swinging stick.*** *(a)* A uniform meter stick swings freely from one end. What is the period for small oscillations? *(b)* A lump of wax whose mass is 12 percent of the mass of the meter stick is placed at the 500-mm mark. What is the period for small oscillations now: *(c)* Where could the lump be placed on the stick such that the period is not affected?

7. ***Ice in a bowl.*** A small, slippery ice cube is placed near the bottom of a bowl with radius of curvature $R = 140$ mm. What is the period of small oscillations of the ice cube?

8. ***Oscillating weights.*** A 100.0-g "weight" holder is attached to a light spring and oscillates vertically with a period of 0.33 s (Fig. 14-31). Suppose a 10.0-g weight is put on the holder. What is the maximum amplitude of oscillation of this system such that the 10-g weight remains in contact with the holder throughout the motion?

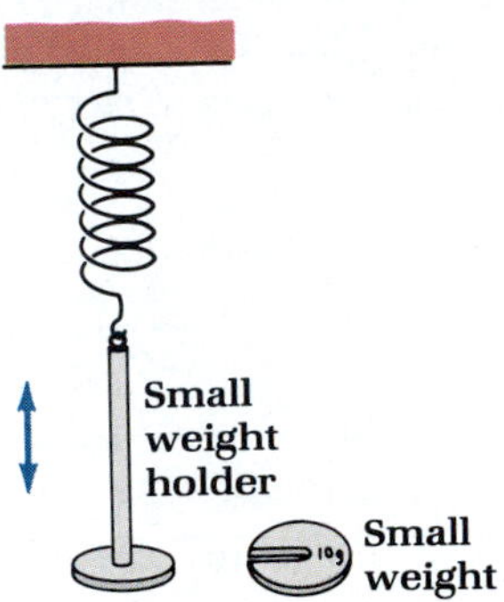

Figure 14–31. Problem 8.

9. ***Driving an oscillator at maximum power.*** If an external force $\mathbf{F}_E$ is applied to an object, the rate at which work is done by the force is $P = \mathbf{F}_E \cdot \mathbf{v}$, where $\mathbf{v}$ is the velocity of the object. (See Sec. 8-5.) *(a)* Determine an expression for the power delivered by the driving force to an oscillator described by Eq. 14-31. *(b)* Your answer to part *(a)* should contain the term $\cos(\omega_E t)\sin(\omega_E t - \phi_E)$. The average of this term over one cycle is

$$-\tfrac{1}{2}\sin\phi_E = \frac{-\gamma\omega_E}{\sqrt{(\omega^2 - \omega_E^2)^2 + 4\gamma^2\omega_E^2}}$$

Show that the average power $\overline{P}$ is a maximum if $\omega_E = \omega$.

10. ***Solution for the forced oscillator.*** Verity that Eq. 14-31 is a solution of Eq. 14-30. Note that

$$\cos(\omega_E t - \phi_E) = \cos\omega_E t\cos\phi_E + \sin\omega_E t\sin\phi_E$$

and

$$\sin(\omega_E t - \phi_E) = \sin\omega_E t\cos\phi_E - \cos\omega_E t\sin\phi_E$$

11. ***Nonlinear damping.*** A damping force can be represented as a force with a direction opposite to the velocity and proportional to the square of the speed: $F_x = -c|v_x|v_x$ where c is a positive quantity. Using such a force in Newton's second law for an oscillator leads to the equation of motion,

$$a_x = -\frac{c}{m}|v_x|v_x - \frac{k}{m}x$$

A solution can be found numerically by adapting the spreadsheet in Table 4-2. Let $c/m = 1.0\ \text{m}^{-1}$; $k/m = 39.48\ \text{s}^{-2}$; $x_0 = 0.50$ m; and $v_{x0} = 0$. Determine x at intervals $\Delta t = 0.05$ s from $t = 0$ s to $t = 2$ s and construct a graph of x versus t. Repeat for $c/m = 5.0\ \text{m}^{-1}$.

12. ***Nonlinear restoring force.*** All springs have some degree of nonlinear behavior, so that the restoring force is not strictly proportional to x. Consider a spring force component of the form

$$F_x = -kx - \alpha x^3$$

Adapt the spreadsheet in Table 4-2 (or from the previous problem) to determine the motion of an undamped oscillator subject to this force. Let $m = 1.0$ kg; $k = 39.48$ N/m; $\alpha = 1000\ \text{N/m}^3$ $v_{x0} = 0$; $\Delta t = 0.05$ s. Consider the motion from $t = 0$ s to $t = 2.0$ s for three cases: *(a)* $x_0 = 0.02$ m; *(b)* $x_0 = 0.05$ m; *(c)* $x_0 = 0.1$ m. Is the period independent of the amplitude? Explain.

13. ***Three circular pendula.*** A physical pendulum has a circular shape of radius R and is pivoted about a point on its circumference as shown in Fig. 14-32. Determine the period of small oscillations if the shape is *(a)* a uniform disk, *(b)* a thin hoop, *(c)* a ring with inner radius $\frac{1}{2}R$. (*Hint:* Use Table 12-2 and the parallel-axis theorem.)

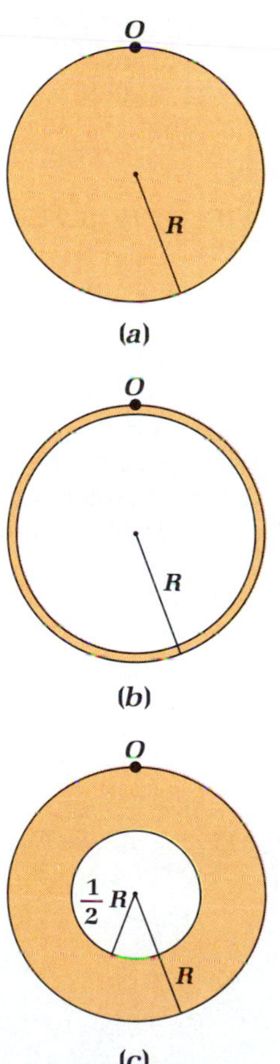

Figure 14-32. Problem 13.

14. ***Adding springs to a pendulum.*** A uniform rod of length L and mass M is pivoted about a fixed axis at one end. It is connected to identical horizontal springs of spring constant k and negligible mass, as shown in Fig. 14-33. Determine an expression for the period of small oscillations of the rod.

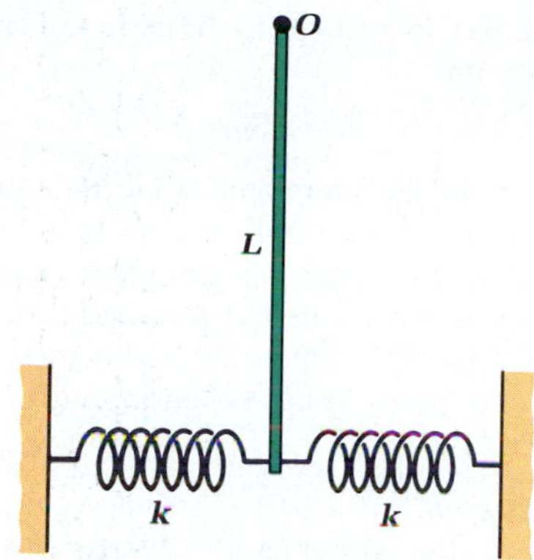

Figure 14-33. Problem 14.

15. ***The Henon map.*** A fractal set can be generated by repeatedly calculating pairs of coordinates (x_n, y_n) from the relations

$$x_{n+1} = 1 - ax_n^2 + y_n$$

$$y_{n+1} = bx_n$$

with $a = 1.4$, $b = 0.3$. This is called a *Henon mapping*. Use a spreadsheet or other application to calculate and plot successive xy points. Begin the iteration with $x_1 = 1$, $y_1 = 0$ but begin the plotting after n has reached 20 or so to mask the choice of the starting point.

16. ***An oscillating disk.*** Figure 14-34 shows the top view of a uniform disk that can rotate with negligible friction about a fixed vertical axis. The disk has mass M and radius R_0 and is symmetric about its axis. Two springs, each with spring constant k, are attached to the disk at a point that is a distance $\frac{1}{2}R_0$ from the center. Determine the angular frequency ω for small oscillations about the equilibrium position.

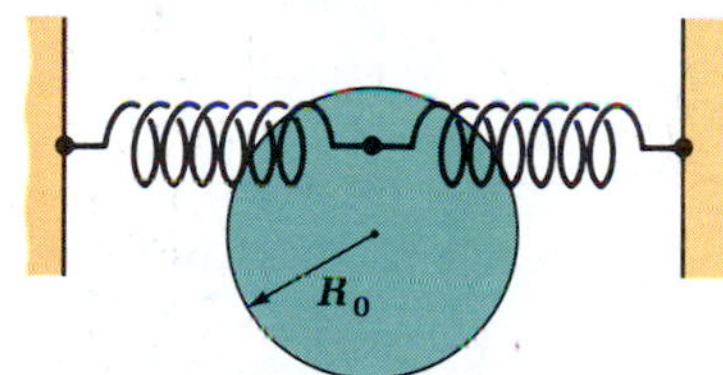

Figure 14-34. Problem 16.

17. ***Center of oscillation.*** The center of oscillation of a physical pendulum is the point P at which the entire mass of the pendulum would be concentrated to form a simple pendulum of the same frequency (Fig. 14-35). *(a)* Show that the distance

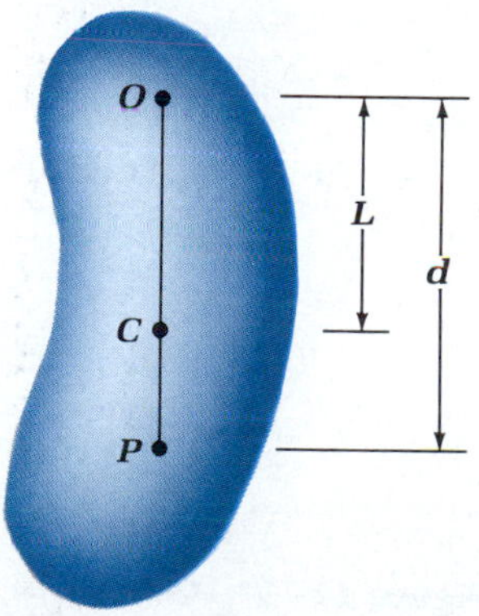

Figure 14-35. Problem 17.

d from the point of suspension O to the center of oscillation of a physical pendulum is

$$d = I/mL$$

where I is the pendulum's moment of inertia about its axis of rotation, m is its mass, and L is the distance from the axis to the center of gravity. *(b)* Use the answer in part *(a)* to show that for a physical pendulum formed by suspending a uniform rod from one end, $d = 2D/3$, where D is the rod's length. *(c)* Verify the answer to part *(b)* by comparing the result of Example 14-6 with Eq. 14-21.

18. ***The "sweet spot."*** If a ball is struck by a bat such that contact is made at the "sweet spot," the batter experiences a minimum reaction from the bat due to the collision. The sweet spot for a physical pendulum is the center of oscillation described in the Prob. 17. To see how this works, consider a simple case where the bat is represented by a uniform rod suspended from one end so that it forms a physical pendulum (Fig. 14-36). Show that if **F** is applied at the center of oscillation ($d = 2D/3$), no additional force is needed to keep point O at rest. (*Hint:* First determine the angular acceleration component α_z of the rod about point O. Then determine the acceleration **a** of the center of mass of the rod. Show that $\mathbf{a} = \mathbf{F}/M$, where M is the mass of the rod and **F** is the force exerted by the ball. That is, **F** is the net force on the rod.)

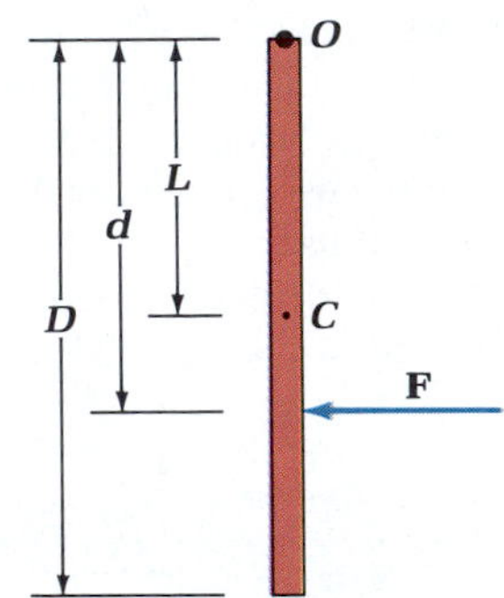

Figure 14-36. Problem 18.

19. ***Finding the minimum period.*** A uniform rod of length D forms a physical pendulum that rotates about the point O, which is a distance x from the center of gravity C (Fig. 14-37). *(a)* Develop an expression for the period T in terms of x and D. *(b)* Show that T has its minimum value when $x = D/\sqrt{12}$. *(c)* Find the expression for the minimum value of T.

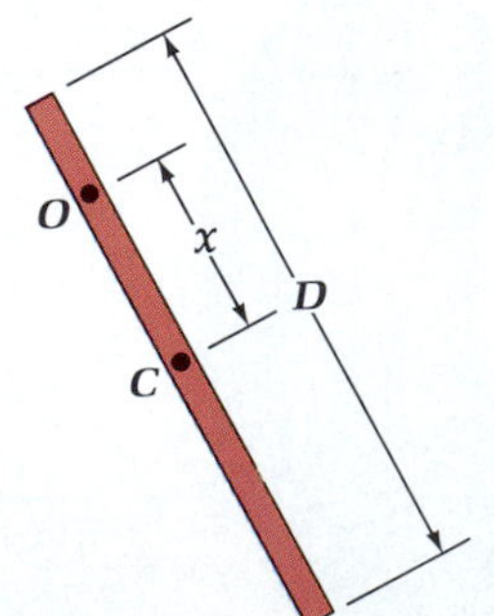

Figure 14-37. Problem 19.

20. ***Phase difference between two oscillators.*** Two simple harmonic oscillators a and b move with the same frequency and amplitude along lines parallel to the x axis; each oscillator has its central position at $x = 0$. The phase difference $\Delta\phi$ between such oscillators is the difference between their phase angles. *(a)* If oscillator b is at its central position, with v_x positive at the instant $x_a = A$, what is $\Delta\phi$? *(b)* If oscillator b is at $x_b = \frac{1}{2}A$, with v_x negative at the instant $x_a = A$, what is $\Delta\phi$?

21. ***Length of an equivalent pendulum.*** A uniform solid sphere of radius r_1 rolls without sliding inside a spherical bowl of radius r_2 (Fig. 14-38). The motion is confined to a vertical plane. Show that for small oscillations, the period and frequency of the motion is the same as for a simple pendulum of length $L = 7(r_2 - r_1)/5$. (*Hint:* Use a sketch to determine the relation between the angular speed of rotation of the rolling sphere to the angular speed of the center of the sphere about the center of the spherical bowl.)

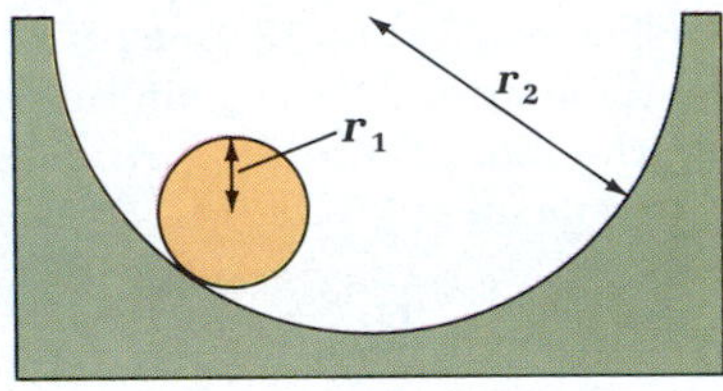

Figure 14-38. Problem 21. A uniform sphere rolls back and forth (in a vertical plane) inside a spherical bowl.

22. ***A tunnel through the earth.*** If we assume that the earth has a uniform mass density, it can be shown that a particle of mass m within the earth's surface has a gravitational force of magnitude $F = mgr/R$ exerted on it by the rest of the earth. In this expression, $g = 9.8\ \text{m/s}^2$, r is the distance from the earth's center to the particle, and R is the radius of the earth. Since the particle is inside the earth, $r < R$. The force is directed toward the center of the earth. Suppose a tunnel is dug through the earth so that a train can travel in a straight line from city A to city B (Fig. 14-39). Assume that frictional forces on the train are negligible; the only forces on it are the normal force by the track and the gravitational force by the earth given above. Show that if the train starts from rest, the time required for the trip is 42 min between any two cities, independent of the distance between them.

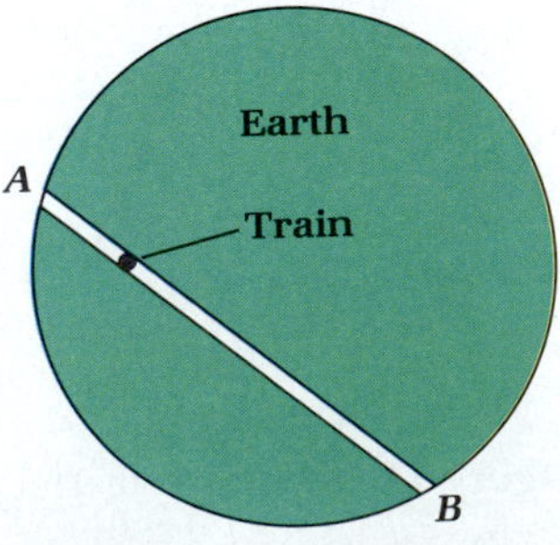

Figure 14-39. Problem 22. Starting from rest, a train travels without friction from city A to city B through a straight tunnel in the earth (assumed uniform). Independent of the distance between the cities, the travel time is 42 min.

SOLIDS AND FLUIDS 15

Although a glacier consists mostly of ice, it flows—albeit quite slowly.

Most substances can be classified into one of three *phases:* solid, liquid, or gas. When acted on by outside forces, solids tend to keep their volume and shape, liquids tend to keep their volume but not their shape, and gases tend to keep neither their volume nor their shape.

The dividing lines between solids, liquids, and gases are not sharp, and many objects cannot be definitively categorized. Glass gradually changes from what is clearly a liquid into what is clearly a solid as it is cooled through the narrow range of temperatures in which it hardens. The "solid" rocks that make up the mantle of the earth slowly change shape in response to the forces that have separated the Americas from Africa and Europe. Bread dough and Silly Putty flow like liquids if acted upon by small forces but fracture in response to large forces. By suitable changes of pressure and temperature, it is possible to gradually change a liquid into a gas, with no sharp dividing line to mark the change of phase. Further, some people speak of a fourth phase of matter, the plasma phase, which has properties distinct from liquids, gases, and solids.

Despite these ambiguities, the classification of solids, liquids, and gases is a useful system. A crystal of salt and a diamond are typical solids, water is a typical liquid, and the air we breathe is a typical gas.

15-1 STRESS AND STRAIN

If you squeeze a rubber ball, the ball becomes deformed. The quantity that characterizes the force exerted by you while squeezing the ball is called the *stress,* and the quantity that characterizes the amount the ball is deformed is called the *strain.* That is, when a stress is applied to an object by an external agent, the object responds by undergoing strain. Stress causes strain.

Stress

It is easier to deform a solid if the force is applied over a small area. For example, a force applied to a stake over the small area at the sharp end causes more deformation than the same force applied over the larger area of the blunt end. Therefore stress involves the ratio of the force to the area of the surface over which the force is applied:

$$\text{Stress} = \text{force/area}$$

Stress is a force per unit area. The SI unit for stress is newtons per meter squared (N/m^2), which is given the name ***pascal (Pa):***

$$1\ \text{Pa} = 1\ \text{N/m}^2$$

In the United States, the British system, for which the unit of stress is pounds per square inch ($lb/inch^2$ or psi), is still sometimes used.

Since a force may be applied to a surface several different ways, there are several kinds of stress, as illustrated in Fig. 15-1. Each force shown in Fig. 15-1 is distributed uniformly over the surface on which it is applied.

The type of stress shown in Fig. 15-1*a* is called a ***tensile stress.*** With a tensile stress two forces, each of magnitude F_n, are applied to opposite faces of a solid, such as the cube-shaped object shown. Each force is normal (perpendicular) to its face, and each is directed away from the face on which it acts. A tensile stress tends to stretch the object. The tensil stress σ_t is defined as

$$\sigma_t = \frac{F_n}{A} \tag{15-1}$$

where A is the area of a single face. Suppose that $F_n = 50$ N and $A = (10\ \text{mm})^2 =$

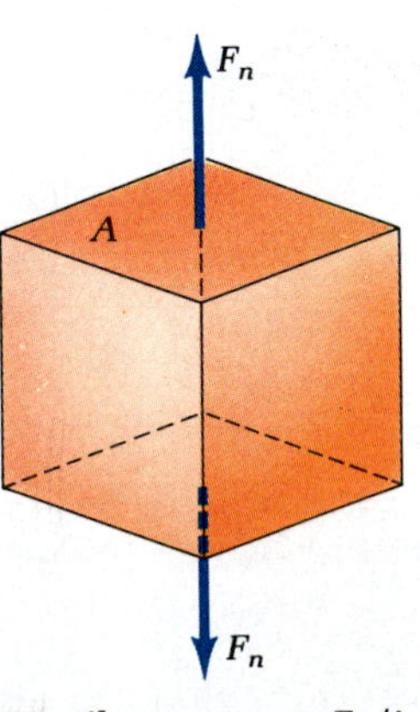

Tensile stress $\sigma_t = F_n/A$

(a)

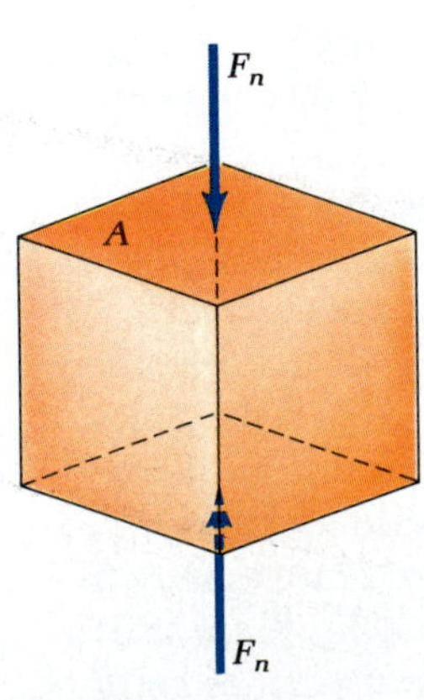

Compressive stress $\sigma_c = F_n/A$

(b)

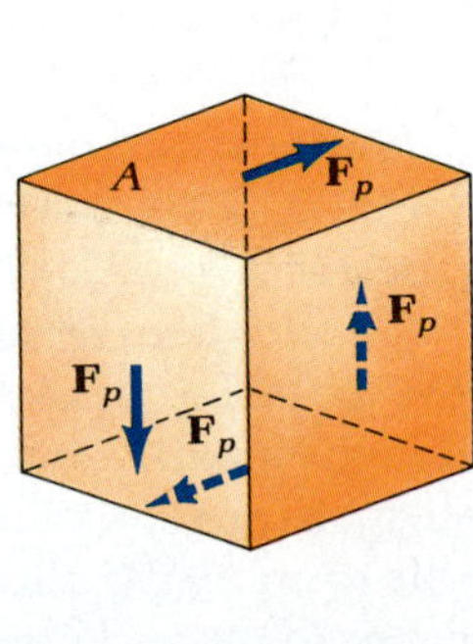

Shear stress $\sigma_s = F_p/A$

(c)

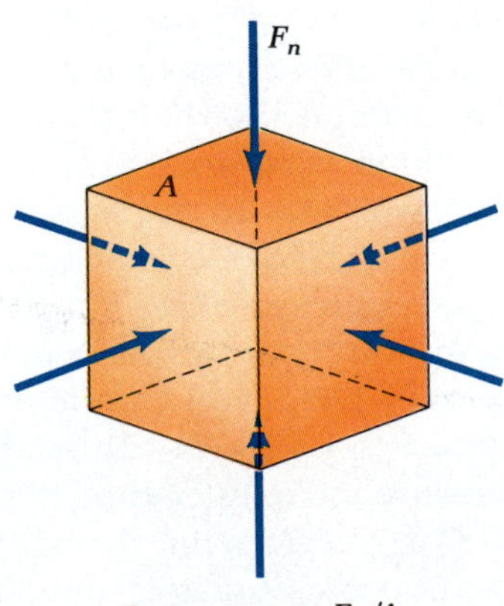

Pressure $p = F_n/A$

(d)

Figure 15-1. Stress is defined as the force per unit area on a surface. *(a)* and *(b)* If the force is normal to the surface, the stress is a tensile or compressive stress. *(c)* If the force is parallel to the surface, the stress is a shear stress. *(d)* If the same force is applied normal to all surfaces, it is called a pressure.

1.0×10^{-4} m². Then $\sigma_t = 50$ N$/1.0 \times 10^{-4}$ m² $= 5.0 \times 10^5$ N/m² $= 5.0 \times 10^5$ Pa $= 500$ kPa.

If each normal force is directed toward the face over which it is applied (Fig. 15-1*b*), then the stress is a ***compressive stress.*** A compressive stress tends to squeeze the object. The compressive stress σ_c is

$$\sigma_c = \frac{F_n}{A}$$

A ***shear stress*** is shown in Fig. 15-1*c*. In this case four forces, each of magnitude F_p, are applied parallel to the cube faces. At least four forces are required to keep the object in equilibrium ($\Sigma \mathbf{F} = 0$ and $\Sigma \tau = 0$). The shear stress σ_s is

$$\sigma_s = \frac{F_p}{A} \tag{15-2}$$

The type of stress called ***pressure*** is shown in Fig. 15-1*d*. On each face of the solid the force is directed perpendicularly inward, and the pressure p is

$$p = \frac{F_n}{A} \tag{15-3}$$

A common pressure is the pressure of the atmosphere on any exposed surface. At sea level, the value of atmospheric pressure is 1.01×10^5 Pa $= 101$ kPa. The force magnitude due to the atmosphere on a square patch of your skin with dimension 10 mm on a side is $F_n = pA = (101 \text{ kPa})(10 \text{ mm})^2 = 10$ N.

Strain

Strain is a measure of the deformation of a solid when a stress is applied to it.

The solid deforms in a way that depends on the stress and on the type of material. An isotropic (same properties in all directions) material, such as glass, responds to stress in the same way in all directions. In contrast, wood responds differently if a tensile stress is applied along the grain rather than across the grain.

Figure 15-2*a* shows the deformation of a solid when a tensile stress is applied. The solid elongates or stretches, and the original length ℓ is increased by $\Delta\ell$. The ***tensile strain*** ϵ_t is defined as the *fractional* change in length,

$$\epsilon_t = \frac{\Delta\ell}{\ell} \tag{15-4}$$

Since the strain is the ratio of lengths, it is dimensionless and has no units. If a cable of length 12 m is stretched by 6 mm, then the tensile strain is $\epsilon_t = 0.006 \text{ m}/12 \text{ m} = 5 \times 10^{-4}$.

The response of a uniform isotropic solid to a shear stress (Fig. 15-2*b*) is called a ***shear strain*** and is defined as

$$\epsilon_s = \phi \approx \frac{\Delta x}{\ell}$$

where Δx is the displacement of the cube corner from its right-angle position, as shown in Fig. 15-2*c*. The shear strain is also dimensionless.

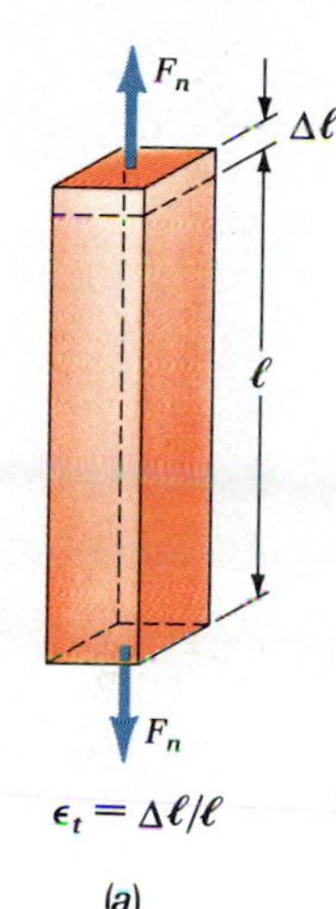

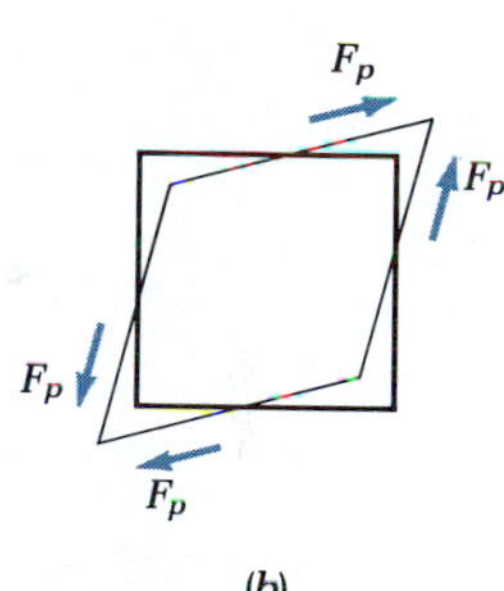

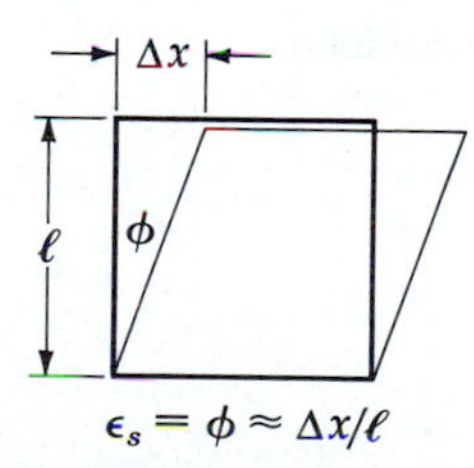

Figure 15-2. *(a)* The tensile strain is defined as the increase, due to the applied forces, in length $\Delta\ell$, divided by the unstressed length ℓ. *(b)* When shear stresses are applied, the body deforms. *(c)* Rotating the stressed body so that one side is coincident with the position of the unstressed body allows the shear stress to be defined as the angle ϕ. For usual shear strains, $\phi \approx \Delta x/\ell$.

How is the strain in a solid related to the applied stress? This is a question that only experiment can answer. The magnitude of the strain as a function of the applied stress for a common material, such as copper, is shown in Fig. 15-3. For strains less than about $\frac{1}{2}$ percent, the strain is proportional to the stress within experimental accuracy. This behavior is known as ***Hooke's law.*** (Note that Hooke's law does not have the same generality as Newton's laws; it merely expresses a property of many, but not all, materials.) At higher strains, experiments show that the strain is not exactly proportional to the stress. The value of the stress at which this becomes noticeable is called the ***proportional limit*** and is shown at A in Fig. 15-3.

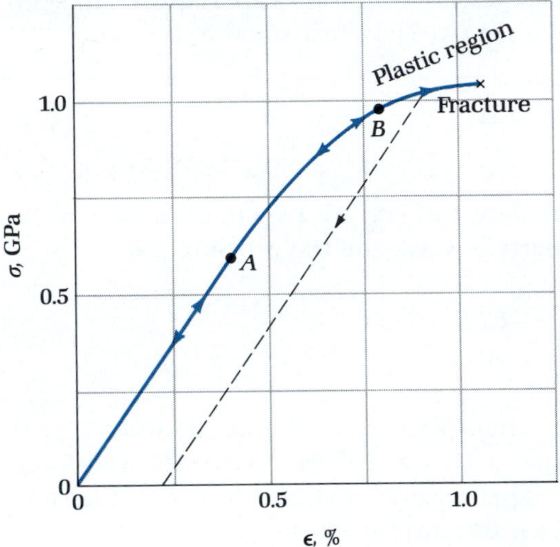

Figure 15-3. A stress-strain curve that might be measured for a typical solid. For stresses larger than that at point A, the stress-strain curve can no longer be considered linear. This point is called the proportional limit. If the stress is released after having been higher than that at point B, called the elastic limit, the solid is permanently elongated; as the stress is released, a line such as the dashed one is followed.

If the stress is large enough, the solid will be permanently deformed. The value of the stress at which this occurs is called the ***elastic limit,*** or the ***yield point,*** and is shown at B in Fig. 15-3. The region between the yield point and fracture is called the *plastic region.*

The elastic properties of an isotropic solid are described by two quantities, one for normal stress and one for shear stress. In the proportional region of the stress-strain curve, the ratio of the stress to the strain is a constant whose value depends on the material of the solid. ***Young's modulus*** Y expresses this linear relation between normal stress and strain for a solid:

Young's modulus

$$Y = \frac{\sigma_t}{\epsilon_t} = \frac{F_n/A}{\Delta\ell/\ell}$$

or $\sigma_t = Y\epsilon_t$. Young's modulus has the dimensions of force per unit area and SI units of pascal (Pa). A steel wire has a large value of Y, while a soft rubber band stretches easily and has a small value of Y. See Table 15-1.

The ***shear modulus*** S expresses the relation between the shear stress and shear strain:

Shear modulus

$$S = \frac{\sigma_s}{\epsilon_s} = \frac{F_p/A}{\Delta x/\ell}$$

or $\sigma_s = S\epsilon_s$. The shear modulus also has dimensions of force per unit area and SI units of pascal. A solid that strongly resists twisting has a large value of S. As you can see in Table 15-1, large values of Y and S usually occur in the same solid.

The ***bulk modulus*** B relates the fractional change in volume to the change in the applied pressure:

Bulk modulus

$$B = -V\frac{dp}{dV} \approx -\frac{\Delta p}{\Delta V/V}$$

It measures how much pressure is required to compress a substance by a given fraction. The volume of liquids and gases can also be changed if subjected to pressure changes. Table 15-1 contains entries for the bulk modulus for some liquids and gases, as well as for solids. Notice that a typical gas (which is easy to compress) has a much smaller bulk modulus than a liquid (which is more difficult to compress).

Unfortunately, the terms "stress" and "strain" are used almost interchangeably in everyday conservation, particularly in a psychological context such as "job-related stress" or "emotional strain." To help keep the concepts straight, it is useful to think of stress (a force per unit area) as the cause of a strain (a deformation).

TABLE 15-1. ***Elastic Constants for a Few Materials****

Material	*Y*, GPa	*S*, GPa	*B*, GPa
Aluminum	70	30	70
Brass	91	36	61
Copper	110	44	140
Lead	15	5.6	7.7
Steel	200	84	160
Tungsten	390	150	200
Bone	15	80	
Concrete	25		
Diamond	1120	450	540
Glass	55	23	31
Ice	14	3	8
Wood			
(‖ grain)	10		
(⊥ grain)	1		
Mercury	0	0	27
Water	0	0	2.2
Most gases at constant room temperature and atmospheric pressure	0	0	10^{-4}

* All values are given in Gpa = 10^9 Pa.

EXAMPLE 15-1

Stress and strain in a steel wire. A steel logging wire is fastened to a log and then pulled by a tractor. The wire is 12.5 mm in diameter, and its length is 10.5 m between the tractor and the log. A force of magnitude 9500 N is required to pull the log. Determine *(a)* the stress in the wire and *(b)* the strain in the wire. *(c)* How much does the wire stretch when the log is pulled?

Solution. *(a)* From the definition of tensile stress:

$$\sigma_t = \frac{F_n}{A} = \frac{9500\ \text{N}}{\pi(6.25\ \text{mm})^2} = 77\ \text{MPa}$$

(b) The strain is found by using the result of part *(a)* and the value Young's modulus for steel in Table 15-1. Solving $Y = \sigma_t/\epsilon_t$ for ϵ_t gives

$$\epsilon_t = \frac{\sigma_t}{Y} = \frac{77\ \text{MPa}}{200\ \text{GPa}} = 3.9 \times 10^{-4}$$

(c) The change in the length of the wire is found from the definition of strain, $\epsilon_t = \Delta\ell/\ell$. Thus

$$\Delta\ell = \epsilon_t\ell = (3.9 \times 10^{-4})(10.5\ \text{m}) = 4.1\ \text{mm}$$

Although the change in length is measurable, it is not enough to be noticeable to a logger.

SELF-TEST 15-1. Rework the above example, except let the diameter and length of the steel wire be 18.0 mm and 21 m, respectively. ***ANSWERS:*** *(a)* 37 MPa; *(b)* 1.9×10^{-4}; *(c)* 3.9 mm.

15-2 DENSITY

When we say that iron is "heavier" than aluminum, what do we mean? We do not mean that every piece of iron weighs more than every piece of aluminum; we mean that given equal volumes of iron and of aluminum, the iron weighs more and thus has more mass. This is quantified in the mass per unit volume, or ***density*** ρ, of a substance:

Density

$$\rho = \frac{m}{V}$$

where m is the mass and V the volume. The dimension of density is mass divided by volume, and the SI unit is kilograms per meter cubed (kg/m^3). Approximate densities of a few materials are given in Table 15-2, and the densities of the elements are given in the periodic table (App. P). As we will see in the next section, the pressure in a fluid at a given depth is determined by the density of the fluid.

Because water plays such a large role in ordinary life, it is common to compare the density of a material with the density of water by dividing the former by the latter and calling the ratio the ***specific gravity*** of the material.* Because the specific gravity is the ratio of two quantities with the same dimension, it is dimensionless. For example, the specific gravity of wood can be determined from the entries for wood and water in Table 15-2: $(0.7 \times 10^3\ kg/m^3)/(1.00 \times 10^3\ m^3) = 0.7$. Specific gravity is used more in commerce than in science. The conditions of battery fluid and of wine are commonly characterized by specific gravity.

TABLE 15-2. ***Approximate Density of Some Materials****

Material	Density ρ, kg/m^3	Material	Density ρ, kg/m^3
Aluminum	2.7×10^3	Blood	1.05×10^3
Copper	8.9×10^3	Ethyl alcohol	0.81×10^3
Gold	19.3×10^3	Mercury	13.6×10^3
Iridium	22.6×10^3	Water	1.00×10^3
Iron or steel	7.8×10^3	Seawater	1.03×10^3
Lead	11.3×10^3		
Platinum	21.4×10^3	Air	1.29
Tungsten	19.3×10^3	Helium	0.179
		Hydrogen	0.090
Bone	1.8×10^3	Steam (100°C)	0.6
Concrete	2.4×10^3	Uranium hexafluoride	15
Diamond	3.5×10^3		
Glass	2.6×10^3	Interstellar space	3×10^{-22}
Ice	0.92×10^3	Sun (average)	1.4×10^3
Wood	0.7×10^3	Earth (average)	5.5×10^3
		Neutron Star	10^{17}

* Unless noted otherwise, values are for room temperature and atmospheric pressure, and for extraterrestrial locations (the last grouping).

EXAMPLE 15-2

Density of silicon. A rectangular block of silicon is 120 mm by 165 mm by 255 mm and has a mass of 11.8 kg. *(a)* What is the density of silicon? *(b)* What is the specific gravity of silicon?

Solution. *(a)* The volume of the silicon is

$$V = (0.120\ m)(0.165\ m)(0.255\ m) = 5.05 \times 10^{-3}\ m^3$$

Thus the density of silicon is

$$\rho = \frac{m}{V} = \frac{11.8\ kg}{5.05 \times 10^{-3}\ m^3} = 2.34 \times 10^3\ kg/m^3$$

* Specific gravity is actually a misnomer, since it has nothing to do with gravity.

(b) The specific gravity of silicon is the ratio of the density of silicon to the density of water, or

$$\text{sp. gr.} = \frac{2.34 \times 10^3\ \text{kg/m}^3}{1.00 \times 10^3\ \text{kg/m}^3} = 2.34$$

SELF-TEST 15-2. A computer memory chip measures 7 mm × 20 mm × 1 mm. Estimate the mass of the chip, which is composed essentially of silicon. ***ANSWER:*** 0.3 g.

15-3 PRESSURE IN A STATIC FLUID

If a shear stress is applied to a liquid or a gas, the material will not come to equilibrium but will continually deform or flow. Such substances are called ***fluids.*** Under static conditions, there cannot be a shear stress in a fluid. It follows that on any surface bounding a fluid at rest, the force must be normal to the surface. That is, the stress in a fluid is a pressure p.

The pressure in a fluid at rest is independent of the orientation of the surface on which it acts. By way of illustration, suppose a small piece of curved tubing is suspended horizontally in a fluid (Fig. 15-4). If there were any difference in the pressure at the two ends, the fluid would flow, contrary to our supposition that the fluid was static. Since the ends of the tube may have any orientation, we conclude that the fluid pressure at a given level is the same in all directions.

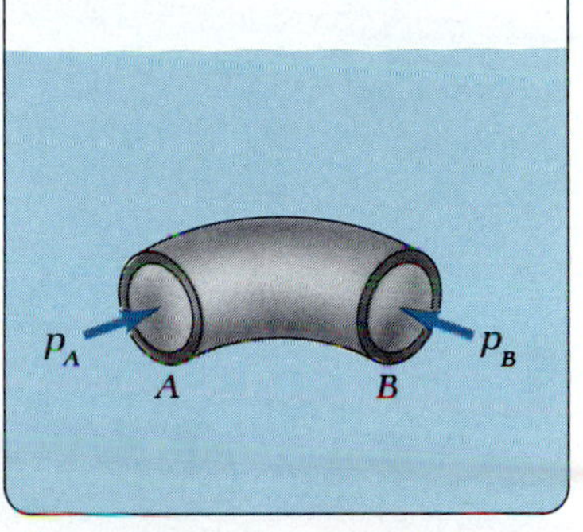

Figure 15-4. If a short, hollow, bent tube is immersed in a static fluid, no flow occurs. This implies that the pressure is equal at the ends of the horizontal tube and that the pressure is the same in all directions in a fluid.

Pressure as a Function of Depth

Although the pressure in a static fluid is the same for a given horizontal level, the pressure does vary with vertical position because of the weight of the fluid. Consider a small element of fluid as shown in Fig. 15-5. The element is located vertically by a coordinate y and is of infinitesimal thickness dy. The pressure of the

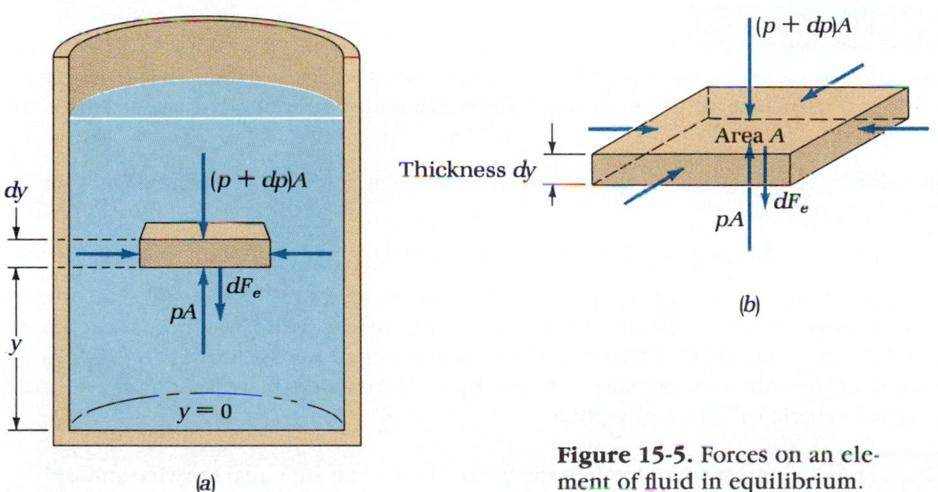

Figure 15-5. Forces on an element of fluid in equilibrium.

surrounding fluid exerts forces on the element. Since the pressure is the same horizontally, the horizontal forces add to zero. Vertically, there are three forces acting on the element: the gravitational force exerted by the earth $d\mathbf{F}_e$, the force of magnitude pA exerted upward at the lower face, and the force of magnitude $(p + dp)A$ exerted downward at the upper face of the element. Since the fluid is static, these forces must add to zero, or

$$pA - (p + dp)A - dF_e = 0$$

so that

$$A\,dp = -dF_e$$

That is, this pressure difference is such as to balance the weight of the element of fluid.

Now substituting $dF_e = (dm)g = (\rho\, dV)g = \rho A\, dy\, g = \rho g A\, dy$, where dm is the mass of the element of fluid of volume $dV = A\, dy$, we have

$$dp = -\rho g\, dy \tag{15-5}$$

The cause of this pressure difference is the necessity of holding the element of fluid up against the force of gravity; the pressure difference supports the weight of the element of fluid. Since y is measured vertically upward, the negative sign indicates that the pressure decreases as the height above the bottom increases. Conversely, the pressure increases as the depth increases.

To determine the difference in pressure between two levels in the fluid, we need to integrate Eq. 15-5. If p_1 and p_2 refer to the pressures at height y_1 and y_2, respectively, then

$$\int_{p_1}^{p_2} dp = -\int_{y_1}^{y_2} \rho g\, dy'$$

A typical liquid has a large bulk modulus B and so is nearly incompressible. This means that the density of the liquid is essentially independent of pressure and therefore is independent of position in the fluid. Then ρ, along with g, can be taken outside the integral, which yields

$$p_2 - p_1 = -\rho g(y_2 - y_1)$$

If the top surface of the liquid is at y_2, where the pressure has a value $p_2 = p_0$, then the pressure $p_1 = p$ at a depth $h = y_2 - y_1$ satisfies $p_0 - p = -\rho g h$, or

Pressure increases linearly with depth in an incompressible fluid.

$$p = p_0 + \rho g h \tag{15-6}$$

Thus the pressure increases linearly with depth in an incompressible fluid.

EXAMPLE 15-3

Crushing pressure. What is the pressure at the bottom of the ocean in a place where it is 3.00 km deep? The pressure at the top of the ocean is atmospheric pressure, or 1.01×10^5 Pa.

Solution. From Table 15-2, the density of seawater is $1.03 \times 10^3\ \mathrm{kg/m^3}$. Thus from Eq. 15-6,

$$p = 1.01 \times 10^5\ \mathrm{Pa} + (1.03 \times 10^3\ \mathrm{kg/m^3})(9.80\ \mathrm{N/kg})(3.00 \times 10^3\ \mathrm{m}) = 3.0 \times 10^7\ \mathrm{Pa}$$

or about 300 times atmospheric pressure. The density of water 3 km deep in the ocean is about $1\frac{1}{2}$ percent higher than at the top, and thus our results, which assume an incompressible fluid, are slightly off. (That is why we have rounded off the answer to two places.) Because of the enormous pressure, exploration of deep undersea areas requires a specially designed vehicle called a bathysphere.

SELF-TEST 15-3. At what depth in the ocean is the pressure equal to twice atmospheric pressure? ***ANSWER:*** 10 m.

Pascal's Principle

If the pressure at the surface of a liquid p_0 is increased by Δp, then Eq. 15-6 shows that the pressure at an arbitrary point a distance h below the surface also increases by Δp. This result is called ***Pascal's principle:***

Pascal's principle

> Pressure applied to an enclosed incompressible static fluid is transmitted undiminished to all parts of the fluid.

Multicolor strobe photo of the splash of a milk drop into a small pool formed by the previous drop. Different flash colors make the milk appear with different colors at different times. The first two images captured the drop just before it hit the pool and the second two show the formation and expansion of a coronet of milk.

The hydraulic press illustates Pascal's principle nicely. Figure 15-6a, from a nineteenth-century physics text, shows a press for cotton bales. A large pressure can be created at r with a small force, because the area at r is small. The same pressure is applied to every surface of the fluid, including the piston P. Since the area at P is large, the force exerted by the fluid there is large. The ratio of the magnitude of the force on the cotton bale to the force applied at r is equal to the ratio of the areas of the pistons at P and r. That is, with reference to Fig. 15-6b, $F/f = A/a$, or $F = f(A/a)$.

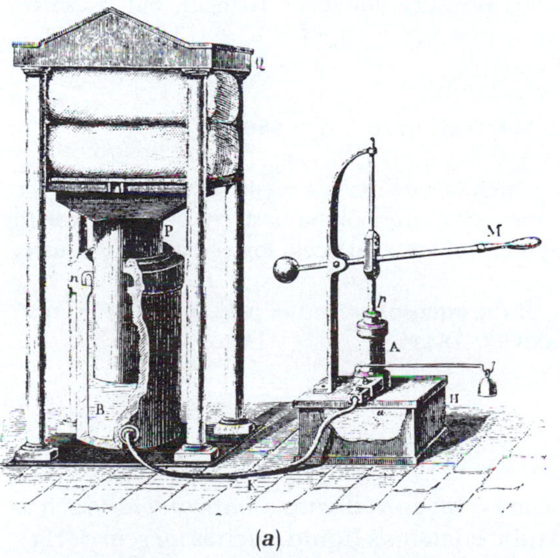

(a)

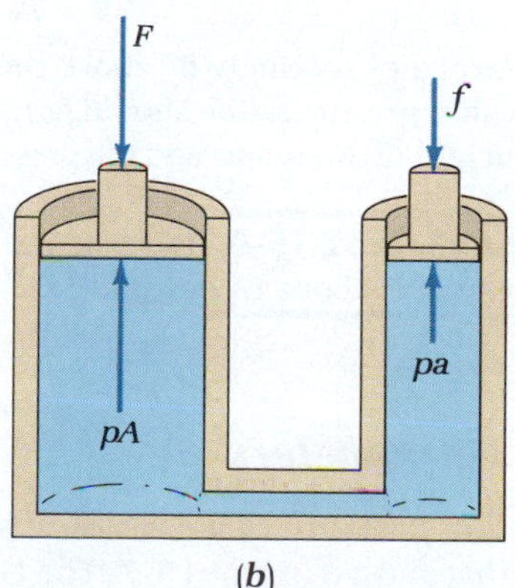

(b)

Figure 15-6. The hydraulic press. *(a)* An illustration of a cotton press. *(b)* A schematic drawing of the press.

Compressible Fluids

If the fluid is noticeably compressible, the relation between the density and the height in the fluid must be known before Eq. 15-5 can be integrated. For many gases, such as air, there is a relation between the pressure p and the density ρ when the temperature is constant: $p/p_0 = \rho/\rho_0$, where p_0 and ρ_0 are the pressure and density at a reference point, say at $y = 0$. Equation 15-5 can then be written $dp = -\rho g\, dy = (p\rho_0/p_0)g\, dy$, or

$$\frac{dp}{p} = -\frac{\rho_0 g}{p_0}\, dy$$

This expression can be integrated from the reference level $y = 0$ to a level h:

$$\ln p(h) - \ln p_0 = -\frac{\rho_0 g}{p_0}(h - 0)$$

or

$$\ln \frac{p(h)}{p_0} = -\frac{\rho_0 g}{p_0}\, h$$

or

Static pressure of the atmosphere, compressible fluid

$$p(h) = p_0 e^{-(\rho_0 g/p_0)h} \qquad (15\text{-}7)$$

The pressure decreases exponentially with height.

EXAMPLE 15-4

Pressure changes with altitude. *(a)* What is the pressure difference between the floor and the ceiling of a 4.0-m-high room? *(b)* How high above the earth's surface is the pressure half that at the surface? Assume that the temperature of the atmosphere is constant.

Solution. *(a)* From Table 15-2

$$\frac{\rho_0 g}{p_0} = \frac{(1.29\ \text{kg/m}^3)(9.80\ \text{N/kg})}{1.01 \times 10^5\ \text{Pa}} = 1.25 \times 10^{-4}\ \text{m}^{-1}$$

Thus

$$p(\text{ceiling}) = p(\text{floor})e^{-(1.25\times10^{-4}\,\text{m}^{-1})(4\,\text{m})} = 0.99950p(\text{floor})$$

or $p(\text{floor}) - p(\text{ceiling}) = (5.0 \times 10^{-4})p(\text{floor}) = 50$ Pa. [Another method is to note that $e^{-x} \approx 1 - x$ when $x \ll 1$, so that for small height differences, $p - p_0 = p_0(1 - \rho_0 g/h) = 50$ Pa.] Since the density of air is low, this pressure difference is small, but it can be measured.

(b) Solving Eq. 15-7 for h,

$$h = -\frac{p_0}{\rho_0 g}\ln\frac{p}{p_0} = (-8.0 \times 10^3\ \text{m})\ln(0.50) = 5500\ \text{m}$$

Airplanes routinely fly above this height, which is why oxygen masks are necessary if the cabin pressurization should fail. Actually, the temperature of the air decreases slightly with height in this range, and the pressure at 5500 m is about 15 percent lower than this estimate.

SELF-TEST 15-4. Estimate the pressure of the atmosphere at the peak of Mount Everest, which is about 14 km above sea level. ***ANSWER:*** 18 kPa.

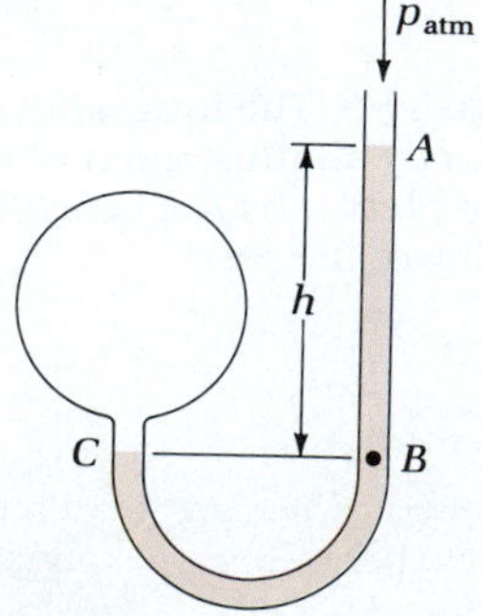

Figure 15-7. A manometer.

Manometers

The pressure of a gas inside a vessel can be measured with a ***manometer*** such as that shown in Fig. 15-7. The U-shaped tube contains a liquid, such as mercury (Hg), which rises to different levels on either side. The pressure at point A is atmospheric pressure p_{atm}, since the tube is open above A. From Eq. 15-6, the pressure in the liquid of the manometer at B is $p_{\text{atm}} + \rho gh$, where ρ is the density of the fluid in the manometer. The pressure at point C is the same as the pressure at point B, since they are at the same level. Thus the pressure p in the bulb of Fig. 15-7 is

$$p = p_{\text{atm}} + \rho gh$$

The total pressure p is called the ***absolute pressure,*** while $p_g = p - p_{atm} = \rho gh$ is called the ***gauge pressure.***

EXAMPLE 15-5

Gauge pressure. A mercury manometer is connected to an air tank, as shown in Fig. 15-8. What is the gauge pressure at points a and b and at any point inside the air tank?

Solution. At point a, the pressure is that of the atmosphere, and thus the gauge pressure is zero. At point b, the pressure is $p_b = p_{atm} + \rho gh$, and the gauge pressure $p_{bg} = p_b - p_{atm} = \rho gh$ is

$$p_{bg} = (13.6 \times 10^3 \text{ kg/m}^3)(9.8 \text{ N/kg})(0.044 \text{ m}) = 5.9 \text{ kPa}$$

Inside the tank, the pressure is essentially the same as it is at b, because the density of air is so small, and there is a negligible change of pressure from point b to the top of the tank.

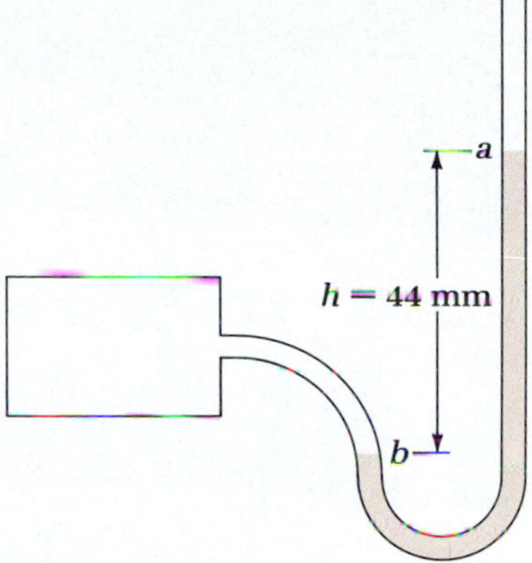

Figure 15-8. Example 15-5.

SELF-TEST 15-5. Suppose that some air is added to the tank in the above example so that the height $h = 56$ mm. Determine *(a)* the gauge pressure and *(b)* the absolute pressure of the air in the tank. ***ANSWERS:*** *(a)* $p_g = 7.5$ kPa; *(b)* $p = 108$ kPa.

Barometers

A barometer measures the pressure exerted by the atmosphere.

A ***barometer*** is a device for measuring the pressure of the atmosphere. A simple barometer consists of a filled tube of mercury that is inverted, with the open end immersed in a pool of mercury (Fig. 15-9). The mercury level drops in coming to equilibrium, so that the top of the mercury column is at a height h above the level of the pool. The space in the tube above the mercury is nearly a vacuum. The small amount of mercury vapor there has a pressure of less than a millionth of an atmosphere. Therefore $p_0 \approx 0$ in Eq. 15-6, and $p_{atm} = \rho_{Hg} gh$. That is, the pressure of the atmosphere is capable of supporting a column of mercury of height h. Put another way, the weight of the column of mercury is equal to the weight of a column of the air of the same cross section that extends to the top of the atmosphere.

Since $p_{atm} = \rho_{Hg} gh$, the pressure is proportional to the height h of the mercury column. At sea level $p_{atm} = 1.01 \times 10^5$ Pa and $h = p_{atm}/(\rho_{Hg} g) = 760$ mm on average, depending on the weather. In years past, pressures were often measured this way, and pressure measurements are still often expressed in units of millimeters of mercury (mmHg), or torr, in terms of the height of a mercury column that the pressure can support. Blood pressures, for example, are routinely expressed in millimeters of mercury.

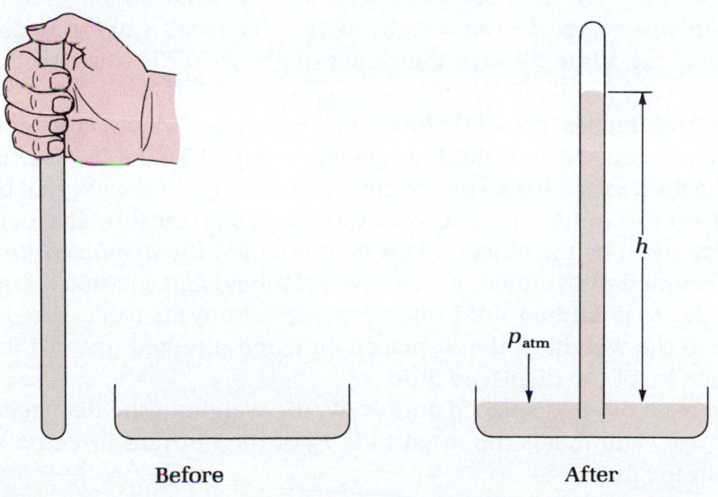

Figure 15-9. A barometer. A tube is filled with mercury, the bottom of the tube is immersed in mercury, and the thumb released. The mercury level in the tube falls to height h, which depends on the pressure of the atmosphere. Since standard atmospheric pressure corresponds to 760 mmHg, the tube must be longer than 760 mm.

EXAMPLE 15-6

Pressure in torr. *(a)* What is the absolute pressure at the bottom of a 6.2-m-deep fresh-water lake? *(b)* What is the gauge pressure at the bottom? *(c)* What is the gauge pressure at the bottom measured in the non-SI unit torr?

Solution. *(a)* The absolute pressure p is

$$p = p_{atm} + \rho gh = 1.01 \times 10^5 \text{ Pa} + (1.00 \times 10^3 \text{ kg/m}^3)(9.8 \text{ m/s}^2)(6.2 \text{ m}) = 1.62 \times 10^5 \text{ Pa}$$

It is worth noting that ρg for water is about 10^4 N/m³, so that for every extra 10 m of depth of water, the pressure increases by 10^5 Pa, or about 1 atm. *(b)* The gauge pressure p_g is the absolute pressure minus the atmospheric pressure: $p_g = p - p_{atm} = 6.1 \times 10^4$ Pa. *(c)* The gauge pressure in torr is the height h (in millimeters) of the column of mercury the gauge pressure will support:

$$h = \frac{p_g}{\rho_{Hg} g} = \frac{6.1 \times 10^4 \text{ Pa}}{(13.6 \times 10^3 \text{ kg/m}^3)(9.8 \text{ N/kg})} = 460 \text{ mm}$$

Thus the gauge pressure can also be expressed as 460 torr.

SELF-TEST 15-6. The second number in a blood pressure reading, the 70 in a 120 over 70 reading, is called the diastolic pressure and is measured in millimeters of mercury. Determine the value of this gauge pressure in pascals. ***ANSWER:*** 9.3 kPa.

15-4 ARCHIMEDES' PRINCIPLE

According to a story, Hiero, the king of Syracuse, ordered a new crown. On receiving the crown, he wasn't satisfied. He suspected the goldsmith of adulterating the gold in the crown with silver. The king asked his mathematician friend, Archimedes, if it were possible to determine if the crown was pure gold without cutting into it. Archimedes was contemplating the problem as he went to his bath. As he stepped into the full tube, he recognized that the water he displaced was equal to the volume of his body under the water. In the story, Archimedes shouted, "Eureka!" ("I found it!"), jumped out of the bath, and ran down the street. What he had discovered was that the volume of the irregularly shaped crown could be found by immersing it in water. By comparing the weight of the crown with the weight of an equal volume of pure gold, he could determine if the crown was pure gold. According to the legend, the goldsmith had cheated and was executed! Although the story may not be true, Archimedes did go on to write *Περι Οχουμενων (On Floating Bodies),* which established the general principles of hydrostatics.

The principle, which is named after Archimedes, is this:

> A body that is partly or entirely submerged in a fluid is buoyed up by a force equal in magnitude to the weight of the displaced fluid and directed upward along a line through the center of gravity of the displaced fluid.

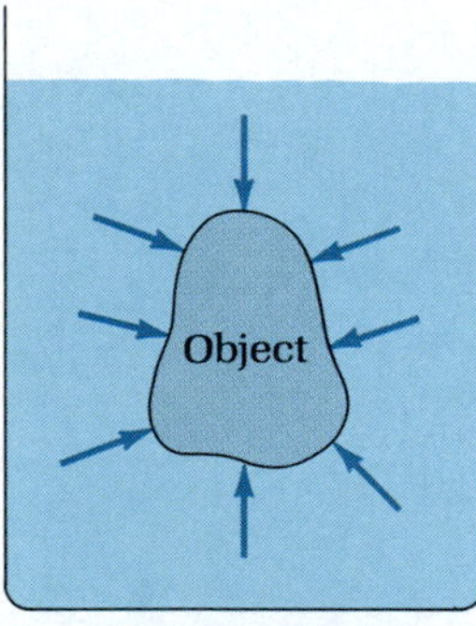

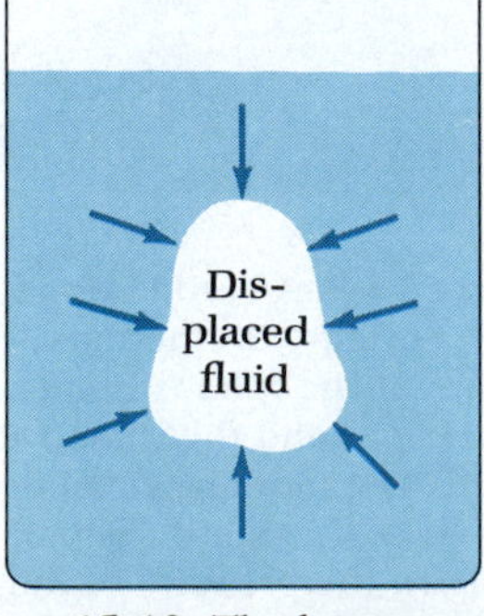

Figure 15-10. The buoyant force on an object is equal to the force on the fluid that the object displaces.

To see how Archimedes' principle follows from Newton's laws, consider an object at rest while immersed in a fluid, as shown in Fig. 15-10. The pressure of the surrounding fluid exerts forces on the object, and the net force exerted by the fluid is the *buoyant force.* But this same force was exerted on the fluid that occupied the volume now taken by the object. (This fluid is called the *displaced fluid.*) When the fluid occupied the volume, it was in translational and rotational equilibrium. Therefore the surrounding fluid must provide a buoyant force that is equal in magnitude to the weight of the displaced fluid and directed upward through the center of gravity of the displaced fluid.

Since the mass of the displaced fluid is ρV, the weight of the displaced fluid has magnitude $\rho g V$, and this is the magnitude F_B of the buoyant force on any object immersed in the fluid:

$$F_B = \rho g V$$

Notice that ρ is the density of the displaced fluid and V is the volume of the displaced fluid.

EXAMPLE 15-7

Buoyant force on a treasure chest. A box of treasure with a mass of 92 kg and a volume of 0.031 m³ lies at the bottom of the ocean. How much force is needed to lift it?

Solution. The forces acting on the chest are shown in Fig. 15-11. The applied force **F** and the buoyant force $\mathbf{F}_B$ acting upward on the chest must balance the downward-acting weight so that

$$F = mg - \rho g V = (92\,\text{kg})(9.8\,\text{N/kg}) - (1.03 \times 10^3\,\text{kg/m}^3)(9.8\,\text{N/kg})(0.031\,\text{m}^3) = 590\,\text{N}$$

Since 590 N is the weight of 60 kg (in air), it would be like supporting a 60-kg box that was above water.

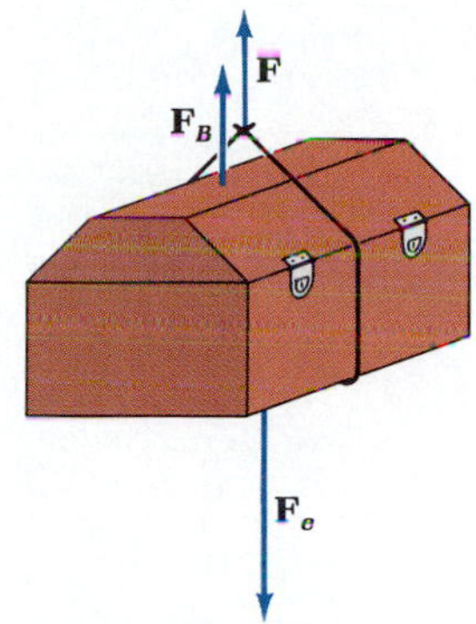

Figure 15-11. Example 15-7.

SELF-TEST 15-7. What applied force **F** is necessary to hold the chest in the above example part way out of the water so that half of its volume is immersed? ***ANSWER:*** 750 N.

EXAMPLE 15-8

A helium balloon. What volume of helium is needed to float a balloon if the empty balloon and its equipment have a mass of 390 kg?

Solution. The buoyant force on the balloon due to the displaced air must be equal and opposite to the weight of the balloon and equipment plus the helium. Thus $F_B = (390\text{ kg} + m_{\text{He}})g = \rho_{\text{air}} g V$. But the mass of the helium depends on the volume of the balloon: $m_{\text{He}} = \rho_{\text{He}} V$. Notice that we have assumed that the volume of the empty balloon and equipment is negligible. Thus $\rho_{\text{air}} g V = (390\text{ kg} + \rho_{\text{He}} V)g$, or

$$V = \frac{390\text{ kg}}{\rho_{\text{air}} - \rho_{\text{He}}} = 350\text{ m}^3$$

where the air and helium are assumed to be at atmospheric pressure.

SELF-TEST 15-8. Rework the above example but consider a balloon filled with hydrogen instead of helium. Why is hydrogen not used for such balloons? ***ANSWER:*** 330 m³; because hydrogen is flammable.

15-5 BERNOULLI'S EQUATION

Fluids in motion are much more complex than fluids at rest. A description of a moving fluid involves knowing the velocity of the fluid, as well as the pressure and density, at all points. Figure 15-12 illustrates some of the complexity. The fine smoke particles provide a visualization of the moving air. Near the tip of the burning cigarette, the pattern is smooth and nearly constant over time. Higher up, the pattern is more complicated and changes with time. This time-varying kind of flow is called *turbulent*. Such flow is not completely understood. Therefore our discussion is restricted to nonturbulent flow and to steady-state conditions. By "steady state," we mean that the pressure, density, and velocity of the fluid do not change with time at a given point, although they may vary with position in the fluid.

An example of nonturbulent, steady-state flow is shown in Fig. 15-13*b*. The flow is generally from left to right in the figure. The dark lines that illustrate the flow pattern are formed by dye injected into the fluid. Such lines of flow are called *streamlines*. Since a streamline shows how the fluid particles move, a streamline can be drawn so as to be everywhere parallel to the velocity of the fluid, as shown in Fig. 15-14.

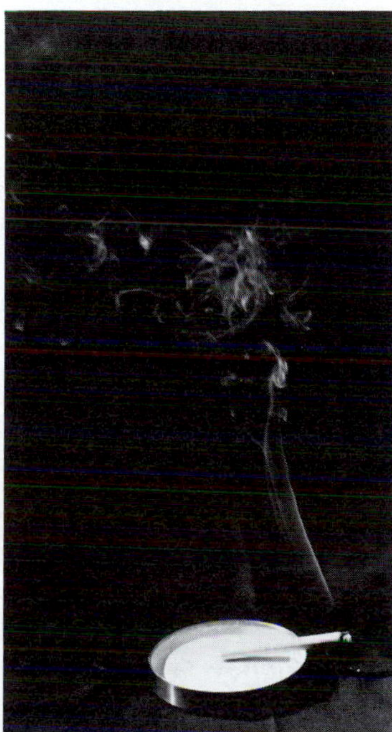

Figure 15-12. The flow of smoke from a cigarette. At first the flow is smooth and steady, but as the smoke rises, the flow becomes unstable and forms vortices that vary in time.

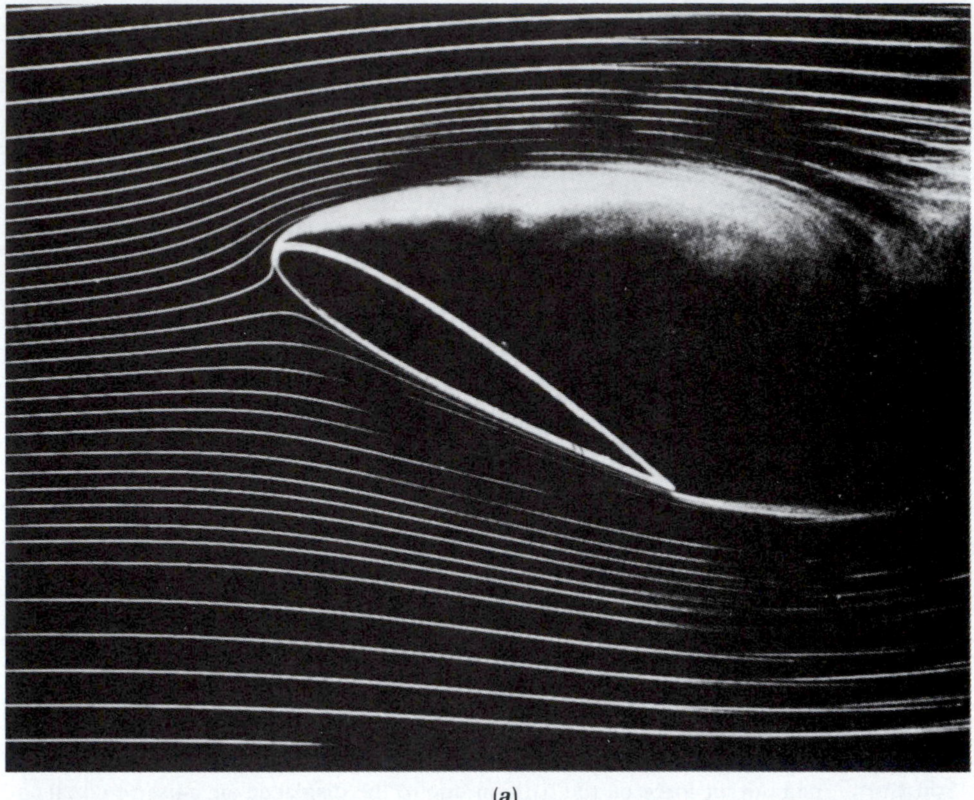

(a)

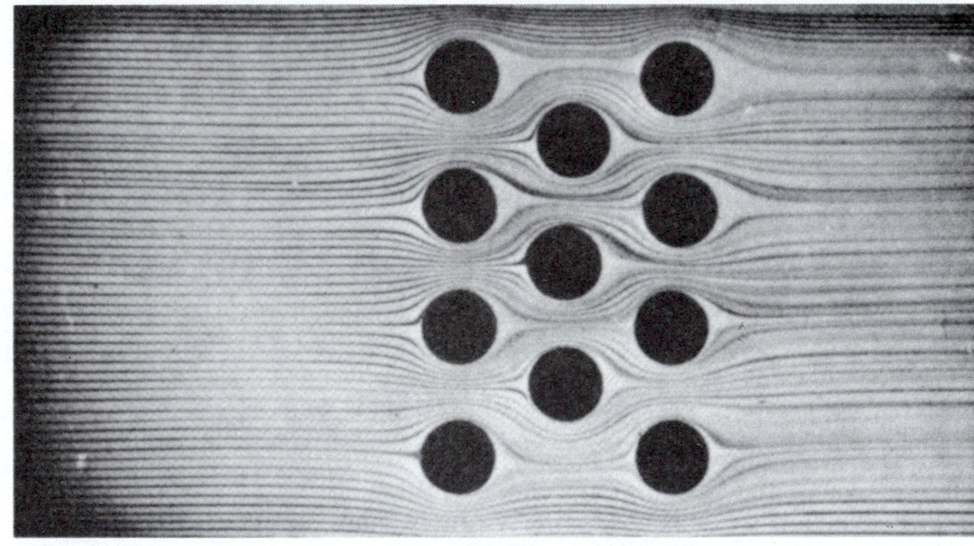

(b)

Figure 15-13. (a) Lines of flow around an airfoil. (b) A fluid flows through an array of cylinders. The streamlines show the paths followed by dye particles that have been injected into the flow.

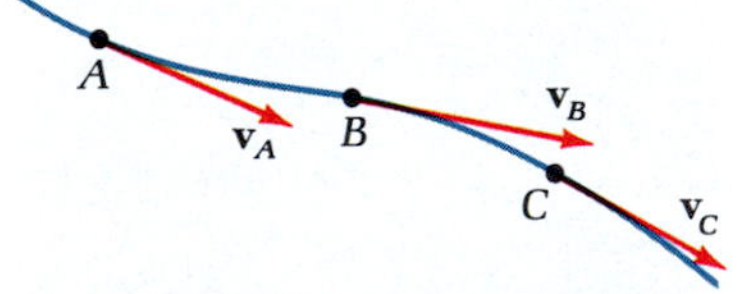

Figure 15-14. A streamline in a moving fluid. At each point the streamline points in the direction of the velocity of the fluid.

Mass Conservation

A tube of flow is bounded by streamlines.

Lines of flow can be drawn so that they enclose a *tube of flow* (Fig. 15-15). Since a streamline is parallel to the fluid velocity, fluid flows along the tube, and no fluid can pass through a wall of the tube. We can use this feature to develop an expression for conservation of mass for the fluid. Consider the element of fluid entering the lower portion of the tube in Fig. 15-15. The volume of this element is the area A_1 times the length $\Delta\ell = v_1\,\Delta t$ of that section of the tube, $\Delta V = A_1 v_1\,\Delta t$. We have chosen the length of the element to be the distance $v_1\,\Delta t$ traveled in a time Δt by the fluid at that end of the tube. Consequently, the mass of fluid that enters the lower

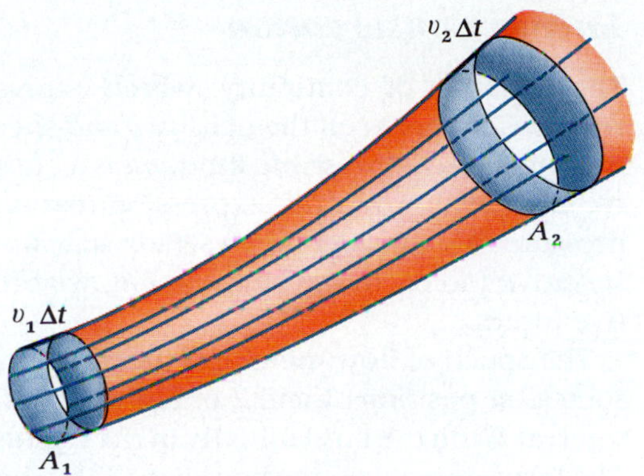

Figure 15-15. A tube of flow is surrounded by streamlines.

end of the tube during a time interval Δt is $\rho_1 A_1 v_1 \Delta t$, where ρ_1 is the density of the fluid there. Similar reasoning gives $\rho_2 A_2 v_2 \Delta t$ as the mass of fluid that leaves the other end of the tube during the same time interval. Since no fluid accumulates in the tube for steady-state flow, the mass entering must equal the mass leaving, $\rho_1 A_1 v_1 \Delta t = \rho_2 A_2 v_2 \Delta t$, or

$$\rho_1 v_1 A_1 = \rho_2 v_2 A_2 \tag{15-8}$$

Equation of continuity

This is called the ***equation of continuity.*** It expresses conservation of mass in a steady flow.

Further, if we assume that the fluid is incompressible or, equivalently, that the density is constant, then $\rho_1 = \rho_2$ and

$$v_1 A_1 = v_2 A_2 \tag{15-9}$$

Equation of continuity for an incompressible fluid

This is a good assumption for water, and even works fairly well for airflow around wings or in heating and cooling ducts, where the pressure changes are small. The product vA gives the volume flow rate, which is represented by the symbol $Q = vA$. Multiplying the volume flow rate vA by the density ρ gives the mass flow rate, $\rho v A$, which is the mass per unit time passing through area A.

EXAMPLE 15-9

Flow rates in a pipe. *(a)* A water line necks down from a pipe with a 12.5-mm radius to a pipe with a 9-mm radius. If the speed of the water in the 12.5-mm pipe is 1.8 m/s, what is the speed in the smaller pipe? See Fig. 15-16. *(b)* What is the volume flow rate? *(c)* What is the mass flow rate? Assume that the water is incompressible.

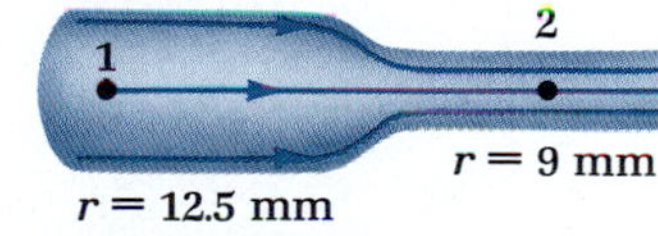

Figure 15-16. Example 15-9.

Solution. *(a)* Using the equation of continuity for an incompressible fluid, Eq. 15-9, we have

$$v_2 = \frac{v_1 A_1}{A_2} = v_1 \frac{r_1^2}{r_2^2} = 3.5 \text{ m/s}$$

(b) The volume flow rate Q is

$$\begin{aligned} v_1 A_1 = v_2 A_2 &= [\pi(12.5 \times 10^{-3} \text{ m})^2](1.8 \text{ m/s}) \\ &= 8.8 \times 10^{-4} \text{ m}^3/\text{s} \end{aligned}$$

(c) The mass flow rate is

$$\begin{aligned} \rho v_1 A_1 = \rho v_2 A_2 &= (1.0 \times 10^3 \text{ kg/m}^3)(8.8 \times 10^{-4} \text{ m}^3/\text{s}) \\ &= 0.88 \text{ kg/s} \end{aligned}$$

SELF-TEST 15-9. Suppose that the radius of the pipe in the above example is 6.0 mm instead of 9.0 mm at position 2. Rework the example for this case. Which answers are different? ***ANSWERS:*** *(a)* 7.8 m/s; *(b)* 8.8×10^{-4} m³/s; *(c)* 0.88 kg/s.

Energy Conservation

The equation of continuity, which expresses conservation of mass, provides a relationship between the density ρ and speed v of the fluid along the flow. For the case of an incompressible fluid, vA is a constant. Using work and energy methods, we can develop another expression that further connects these variables and the pressure. We consider here a steady-state flow in which no work is done by nonconservative forces. In the next section, we briefly describe an effect of nonconservative forces.

The speed of flow, height of the fluid, and pressure may vary along a flow line, such as at positions 1 and 2 in Fig. 15-17. Consider the work done in a short time interval Δt on the fluid initially in the region bounded by A_1, A_2, and the flow tube. The force exerted on the boundary A_1 by the fluid behind it is p_1A_1. The work done by this force in time Δt is the product of this force and the distance through which it moves, $v_1\,\Delta t$. Thus $W_1 = p_1A_1v_1\,\Delta t$. Similarly at A_2, the pressure does work $W_2 = -p_2A_2v_2\,\Delta t$. Note that at A_1 there is work done *on* the fluid, while at A_2 there is work done *by* the fluid; hence the difference in sign. The net work $W = W_1 + W_2$ done on the fluid is

$$W = p_1A_1v_1\,\Delta t - p_2A_2v_2\,\Delta t$$

From the equation of continuity for an incompressible fluid, $v_1A_1 = v_2A_2$. Thus $\Delta V = v_1A_1\,\Delta t = v_2A_2\,\Delta t$ is the volume between A_1 and A_1' or between A_2 and A_2'. It is the volume of fluid entering one end and leaving the other end of the tube in the time interval Δt. The net input work on the fluid can now be written: $W = (p_1 - p_2)\,\Delta V$.

The work-energy theorem shows that this work is equal to the change in the mechanical energy of the fluid initially bounded by A_1 and A_2. Since the flow is steady, the properties of the fluid in the region between A_1' and A_2 are constant. There is a change in mechanical energy in the newly occupied region between A_2 and A_2' and in the region between A_1 and A_1', which is left behind. Since the volume of each of these regions is ΔV and the density is constant, the mass contained in either region is $\Delta m = \rho\,\Delta V$. The potential energy is given by $(\Delta m)gy$ and the kinetic energy by $\frac{1}{2}(\Delta m)v^2$. The change in mechanical energy is thus

$$\Delta E = [(\Delta m)gy_2 + \tfrac{1}{2}(\Delta m)v_2^2] - [(\Delta m)gy_1 + \tfrac{1}{2}(\Delta m)v_1^2]$$

The work-energy theorem then yields

$$(p_1 - p_2)\,\Delta V = [(\Delta m)gy_2 + \tfrac{1}{2}(\Delta m)v_2^2] - [(\Delta m)gy_1 + \tfrac{1}{2}(\Delta m)v_1^2]$$

or, dividing by ΔV with $\rho = \Delta m/\Delta V$, we find

$$p_1 - p_2 = \rho gy_2 + \tfrac{1}{2}\rho v_2^2 - \rho gy_1 - \tfrac{1}{2}\rho v_1^2$$

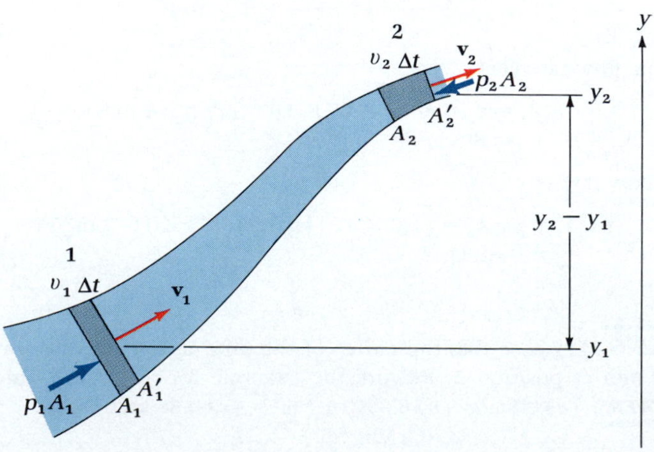

Figure 15-17. Flow along a tube of flow. Volumes ΔV in the shaded regions are equal: $\Delta V = A_1v_1\,\Delta t = A_2v_2\,\Delta t$. The mass of each shaded region is $\Delta m = \rho\,\Delta V$.

Collecting terms so that the same subscript appears on each side of the equation, we obtain

$$p_1 + \rho g y_1 + \tfrac{1}{2}\rho v_1^2 = p_2 + \rho g y_2 + \tfrac{1}{2}\rho v_2^2 \qquad (15\text{-}10)$$

Bernoulli's equation

This is known as ***Bernoulli's equation*** and was originally stated in his *Hydrodynamica* in 1738. Note that we have assumed steady, nonturbulent, energy-conserving flow of an incompressible liquid in the derivation of Bernoulli's equation.

Although our derivation of Eq. 15-10 relates the pressure, speed, and height along a flow line, it is shown in more advanced texts that if the flow is "irrotational," Bernoulli's equation holds even across flow lines. You can visualize what "irrotational" means by imagining a small paddle wheel inserted in the fluid. If the paddle wheel does not have a tendency to rotate, then the flow is irrotational.

Bernoulli's equation holds throughout the fluid.

Notice that if the fluid is at rest, then $v_1 = v_2 = 0$, and Bernoulli's equation is equivalent to Eq. 15-6 for the variation of pressure with depth in a static fluid.

Bernoulli's equation, coupled with the equation of continuity, Eq. 15-9, can be used to solve several types of fluid-flow problems. The following examples illustrate this method.

EXAMPLE 15-10

Pressure variations along a pipe. Water (treated as an incompressible fluid whose viscosity is negligible) flows through the horizontal pipe of Fig. 15-16. At point 1 the gauge pressure is 51 kPa and the speed is 1.8 m/s. What are the speed and gauge pressure at point 2?

Solution. The speed at point 2 can be found from the equation of continuity for an incompressible fluid:

$$v_2 A_2 = v_1 A_1$$

$$v_2 = v_1 \frac{\pi r_1^2}{\pi r_2^2} = 3.5 \text{ m/s}$$

The pressure at point 2 can then be found from Bernoulli's equation. Since $y_1 = y_2$,

$$\begin{aligned} p_2 &= p_1 + \tfrac{1}{2}\rho v_1^2 - \tfrac{1}{2}\rho v_2^2 \\ &= 5.1 \times 10^4 \text{ Pa} + \tfrac{1}{2}(1.00 \times 10^3 \text{ kg/m}^3)[(1.8 \text{ m/s})^2 - (3.5 \text{ m/s})^2] \\ &= 4.7 \times 10^4 \text{ Pa} \end{aligned}$$

At first it may seem paradoxical that the pressure at 2 is lower than that at 1. But if you consider that the fluid must be accelerated between 1 and 2, it is clear that the pressure at 1 must be higher than the pressure at 2. Even if the fluid has a small viscosity, the forces due to viscosity may be small compared with the forces necessary to accelerate the fluid, and p_2 may be less than p_1. We see from this example that when the height is constant, higher speed in the fluid implies lower pressure, and lower speed implies higher pressure.

SELF-TEST 15-10. Explain why the two pressures that appear in Bernoulli's equation can be either gauge pressures or absolute pressures, as long as both pressures are the same. ***ANSWER:*** Bernoulli's equation involves the difference in two pressures; thus any pressure added to both pressures will not affect the result.

EXAMPLE 15-11

A leak in a standpipe. A vertical standpipe with an inner diameter of 1.5 m has a hole of diameter 15 mm in its side that is 2.5 m below the level of water in the standpipe (Fig. 15-18). What is the speed of the water in the jet streaming out of the hole?

Solution. Apply Bernoulli's equation to points 1 and 2:

$$p_1 + \rho g y_1 + \tfrac{1}{2}\rho v_1^2 = p_2 + \rho g y_2 + \tfrac{1}{2}\rho v_2^2$$

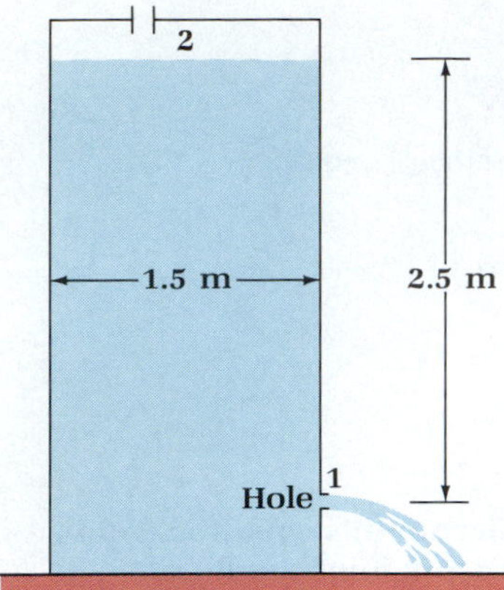

Figure 15-18. Example 15-11: Water leaks from a standpipe.

Point 1 is just outside the hole with the water streaming at speed v_1, and point 2 is any point at the water surface in the standpipe. Both points are at atmospheric pressure so that $p_1 = p_2$ or $p_1 - p_2 = 0$. Substituting this result into Bernoulli's equation and dividing out the density ρ gives

$$\tfrac{1}{2}v_1^2 + gy_1 = \tfrac{1}{2}v_2^2 + gy_2$$

Since $v_2A_2 = v_1A_1$, the speed v_2 of the water at point 2 is negligible compared with the speed v_1 of the streaming water because A_2 is so much larger than A_1:

$$v_2 = (A_1/A_2)v_1 = [(0.015\ \text{m})^2/(1.5\ \text{m})^2]v_1 = (0.00010)v_1 \approx 0$$

Thus the speed of the streaming water is given by

$$v_1^2 = 2g(y_2 - y_1) = 2(9.8\ \text{m/s}^2)(2.5\ \text{m}) = 49\ \text{m}^2/\text{s}^2$$

Taking the square root, we have $v_1 = 7.0$ m/s.

SELF-TEST 15-11. Suppose the diameter of the vertical pipe in the above example is only 45 mm instead of 1.5 m. Determine *(a)* the speed of the water streaming from the hole and *(b)* the speed at which the water surface drops in the standpipe. ***ANSWERS:*** *(a)* 7.4 m/s; *(b)* 0.82 m/s.

EXAMPLE 15-12

Pressure in a water line. A water-supply system uses a water tank for storage so that water will be available when needed. If the water level in the tank at point *A* in Fig. 15-19 is 12 m above the water main, and the speed in the main at point *B* is 16 m/s, what is the gauge pressure at points *A* and *B*?

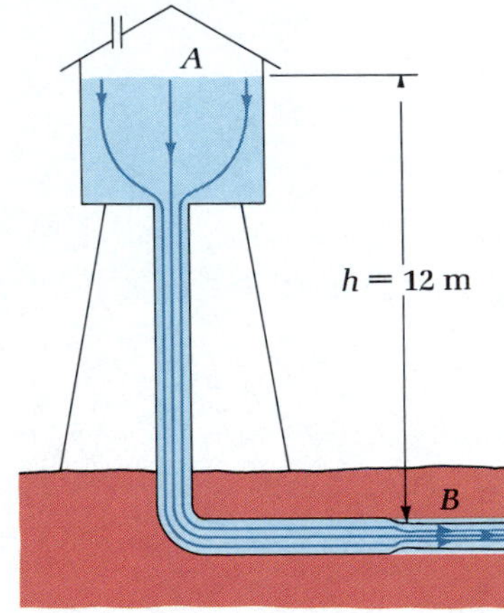

Figure 15-19. Example 15-12.

Solution. At point *A*, the gauge pressure is zero, because the tank is open to the atmosphere. Applying Bernoulli's equation at *A* and *B*,

$$p_B + \rho g y_B + \tfrac{1}{2}\rho v_B^2 = p_A + \rho g y_A + \tfrac{1}{2}\rho v_A^2$$

The speed of water in the large tank v_A is essentially zero (can you show why?), and $y_A - y_B = 12$ m. Thus

$$\begin{aligned} p_B &= \rho g h - \tfrac{1}{2}\rho v_B^2 \\ &= (1.00 \times 10^3\ \text{kg/m}^3)[(9.8\ \text{N/kg})(12\ \text{m}) - \tfrac{1}{2}(16\ \text{m/s})^2] \\ &= -1.0 \times 10^4\ \text{Pa} \end{aligned}$$

The gauge pressure in the line is negative! This means the absolute pressure in the line is less than atmospheric pressure. (The absolute pressure is not less than zero, however.) In the design of water systems, such situations are to be avoided, since if there were a hole in the pipe, contaminated groundwater might be sucked in. In real water systems the flow is often turbulent, and water seldom has speeds over 3 m/s.

SELF-TEST 15-12. If the gauge pressure in the water main in the above example is to be at least 10 kPa, what is the maximum speed of water in the main? ***ANSWER:*** 15 m/s.

The operation of many practical devices can be explained with Bernoulli's equation. Figure 15-20*a* shows a perfume atomizer. Similar devices are used to spray paint or insecticide. When the bulb in the atomizer is squeezed, air rushes through the narrow neck of the atomizer. If the perfume has the density of water, an air speed of about 17 m/s in the narrow channel will lower the pressure at point *A* so that atmospheric pressure will push the perfume 2 cm up the tube leading to the narrow channel. The suction tube that dentists use to clear the mouth of saliva uses a narrow channel with a high-speed stream of water to create a pressure lower than atmospheric.

This lowering of presssure due to the speed of the fluid is the basis for the venturi meter shown in Fig. 15-20*b*. The pressure difference between *A* and *B* is a measure of the speed of the fluid. Many automobile carburetors have a narrow channel called the ***venturi tube,*** shown in Fig. 15-20*c*. Its purpose is to lower the pressure

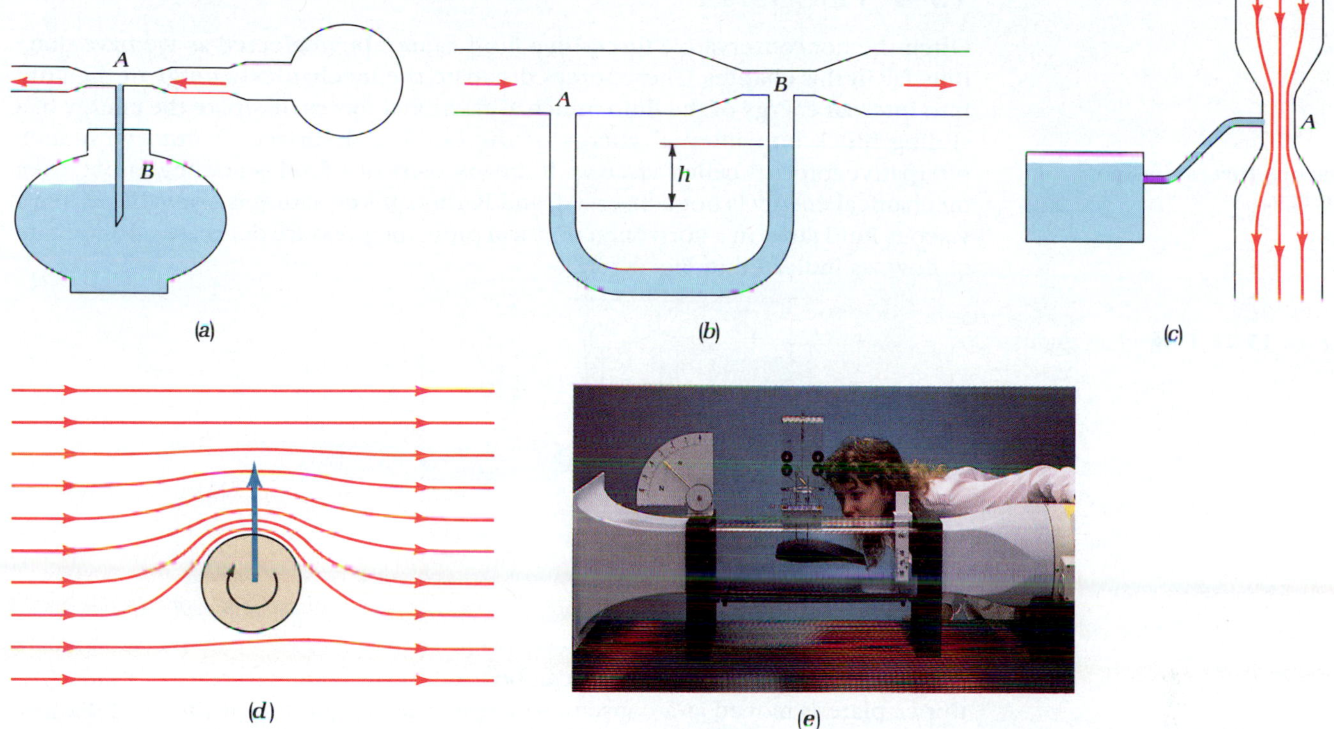

Figure 15-20. Illustrations of Bernoulli's principle.

at point A so that gasoline will enter the airflow leading to the combustion chambers of the engine.

In storms and high winds, the air rushing by a building leads to a lower pressure on the outside of the building than in the still air inside the building. This can cause windows to pop out of the building. Note that windows are almost always blown *out,* not *in*! Similarly, chimneys usually have a small draft even without a fire because wind blows across the top of the chimney, while the air inside the house is stationary. If you watch a fireplace on a windy night, you can easily see the Bernoulli effect.

Figure 15-20*d* shows a rotating cylinder in a moving fluid. If the cylinder drags some fluid around with it, the net flow of fluid is as shown. This makes the speed of the air higher above the cylinder than below it, and pressure lower above the cylinder than below it. Thus there is a net force on the cylinder perpendicular to the flow direction of the fluid (up the page in this case). This is called the *Magnus effect.* Ships have been built with rotating cylinders rather than sails, and although the effect works, the ships have not been of lasting practical use (Fig. 15-21). Figure 15-20*e* is an airfoil demonstration.

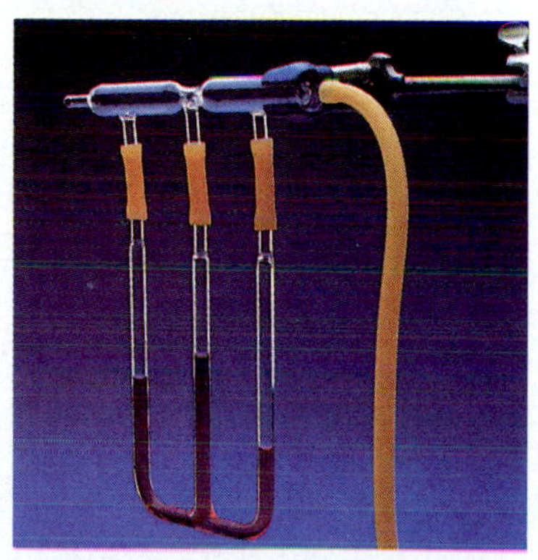

Venturi-tube demonstration. A colored-water manometer reveals the relative pressure at three positions along the tube. As air passes through the horizontal tube, the speed of the air is greater where the tube cross-section is smaller. From Bernoulli's equation, the pressure is lower where the speed is greater. Consequently, the colored water in the manometer arm attached to the smaller section is higher than in the other arms.

Figure 15-21. A Flettner ship. The two tall cylinders were rotated and in a wind drove the ship forward.

15-6 VISCOSITY

Viscous forces dissipate energy.

Often the nonconservative forces in a fluid cannot be neglected as we have done thus far in this chapter. These forces dissipate the mechanical energy of the fluid into internal energy of the fluid, much as frictional forces dissipate the energy of a sliding block into internal energy of the block and surface. A fluid with such dissipative forces is called ***viscous***. If the viscosity of a fluid is not negligible, then mechanical energy is not conserved, and Bernoulli's equation is not valid. When a viscous fluid flows in a horizontal uniform pipe, the pressure decreases along a line of flow, as indicated in Fig. 15-22.

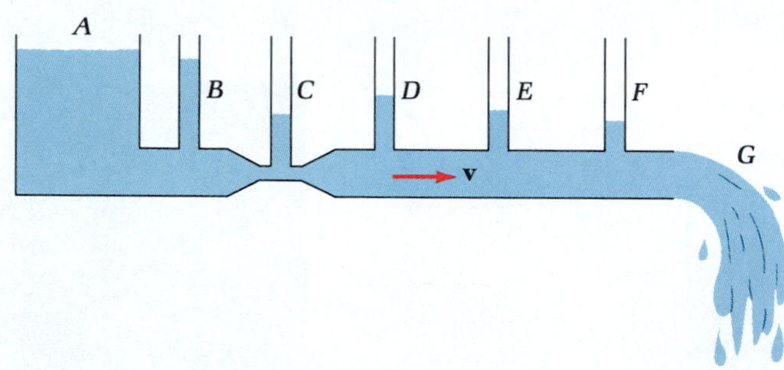

Figure 15-22. Flow of a viscous fluid.

The arrangement in Fig. 15-23 can be used to study the viscosity of fluids. The upper plate is moved at a constant low speed across the top of the fluid. Experiments shows that, for most fluids, the speed of the fluid at points between the two plates of Fig. 15-23 varies linearly with the distance away from the moving plate. Fluids for which the horizontal force component required to move the plate is proportional to the speed of the plate are called *newtonian fluids.* Water and air are examples of nearly newtonian fluids. Certain plastics and suspensions, such as blood and water-clay mixtures, are examples of quite nonnewtonian fluids in which the magnitude of the force required to move the plate might be proportional to the square of the speed. At high speeds, the flow becomes turbulent and very complex in all fluids.

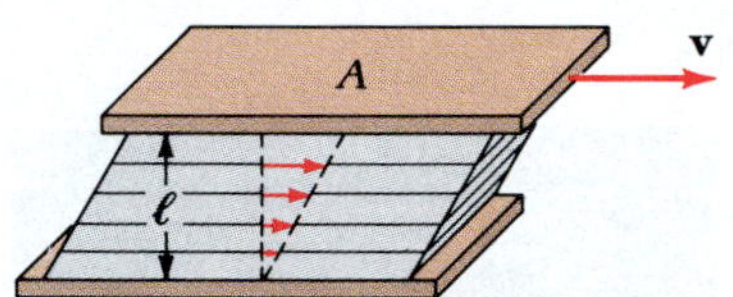

Figure 15-23. When the upper plate is pulled slowly, the viscous fluid between the plates flows in laminae whose speed is proportional to their distance from the stationary plate at the bottom, as indicated by the length of the arrows in the figure.

The magnitude of the force F on the moving plate is found experimentally to depend not only on the speed v of the moving plate, but also to be proportional to the area of the plate A and inversely proportional to the distance ℓ between the moving plate and the stationary plate:

Viscous force

$$F = \frac{\eta A v}{\ell} \tag{15-11}$$

where η is a constant of proportionality called the ***viscosity.*** The SI unit of viscosity is newton-seconds per meter squared ($N \cdot s \cdot m^{-2}$). A common non-SI unit for viscosity is the poise (P), equal to $0.1\ N \cdot s \cdot m^{-2}$. The viscosities of most fluids depend on temperature. For example, the viscosity of water decreases from about $2 \times 10^{-3}\ N \cdot s/m^2$ at the freezing point (0°C) to about $3 \times 10^{-4}\ N \cdot s/m^2$ at the boiling point (100°C). Common gases, such as air, have viscosities of around $2 \times 10^{-5}\ N \cdot s/m^2$ over that same temperature range.

In laminar flow, the layers of fluid do not mix.

The pattern of motion of the fluid shown in Fig. 15-23 is called ***laminar flow.*** Each layer (lamina) of fluid exerts a force on the layer beside it, but since the flow is not turbulent, the layers do not mix.

The volume flow rate Q, due to a pressure difference Δp, of a viscous fluid through a circular pipe of radius R and length L can be shown to be

$$Q = \frac{\pi(\Delta p)R^4}{8\eta L} \tag{15-12}$$

The derivation of this result is indicated in Prob. 14.

EXAMPLE 15-13

Effects of arteriosclerosis. Suppose that arteriosclerosis decreases the radius of the channel in an artery in the heart by a factor of 2. By what factor must the heart increase the pressure difference in the artery to keep the flow rate constant? Assume that blood is a newtonian fluid and that the flow is laminar.

Solution. Since the flow rate is proportional to the fourth power of the radius, decreasing the radius by a factor of 2 decreases the flow rate by a factor of 2^4, or 16. The flow rate is proportional to the first power of the pressure difference, and thus the heart must increase the pressure difference by a factor of 16.

Since blood is not, in fact, a newtonian fluid and the flow of blood is not without turbulence, this result is approximate. Nevertheless, it illustrates the difficulties imposed by arteriosclerosis.

SELF-TEST 15-13. Suppose arteriosclerosis decreases the radius of an artery to 75 percent of its original value. By what factor must the heart increase the pressure difference to maintain the original flow rate? ***ANSWER:*** 3.2.

COMMENTARY: Archimedes

The story of Archimedes jumping out of his bath with the solution to Hiero's problem (see Sec. 15-4) is part of the story of the origins of modern science. Archimedes was born about 287 B.C. and died in the sack of his native Syracuse by the Romans under Marcellus in 212 B.C.

When Rome threatened the Greek-speaking settlements in southern Italy and Sicily, Archimedes helped defend Syracuse by using his scientific knowledge for military applications. According to some accounts, he designed mirrors that set fire to part of a Roman fleet as it approached the city. For their part, the Romans admired the abilities of Archimedes, and Marcellus directed that he be unharmed when the city fell. But legend says that a Roman soldier, coming upon Archimedes in deep study over some mathematical figures and being brushed off by the preoccupied man, peremptorily drew his sword and ran the mathematician through.

In this chapter we have seen that buoyant forces are associated with Archimedes. But he is also credited with a device for pumping water, Archimedes screw, and he constructed an organ that used water to force air across pipes. His father may have been an astronomer; in any case, Archimedes wrote a text on the construction of spheres, and his descriptions were used in making astronomical models. One of these models used water power to simulate the earth and other planets going around the sun, a model proposed by his friend Eratosthenes. When Syracuse fell, Marcellus took one of the planetary spheres in his share of the booty.

Archimedes shunned fame for his engineering prowess. Plutarch wrote that Archimedes thought that "every kind of act connected with daily needs was ignoble and vulgar," and much preferred pure mathematics. He derived expressions for the

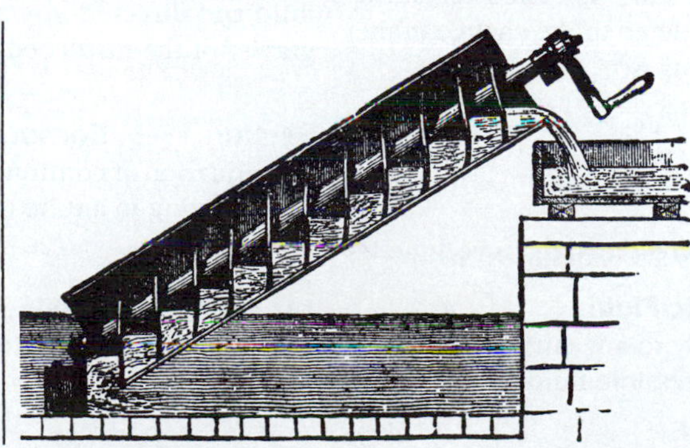

Archimedes' screw. This ancient device is still used for irrigation in some regions.

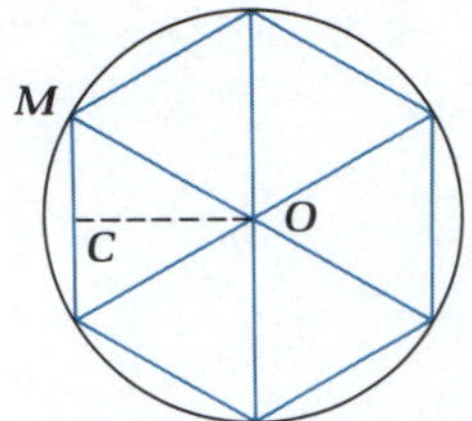

Figure 15-24. The area of a circle is approximated by the area of an inscribed hexagon. Line OC is called the apothem.

volume of figures such as spheres and cones and for the center of gravity of circular cones and figures of revolution generated by hyperbolas. In establishing such proofs, he used a process not far removed from the calculus that Newton invented.

Among his mathematical proofs is one that showed that the area of a circle is equal to πr^2. We can give the flavor of Archimedes' work with a quick and dirty version of this proof. Consider the hexagon inscribed inside the circle of Fig. 15-24. The area of the hexagon is the area of 12 right triangles congruent to triangle COM, and thus the area of the hexagon is equal to that of a right triangle in which the two sides about the right angle are equal to (i) the apothem OC (the perpendicular distance from the center of the circle to the side of the hexagon) and (ii) the perimeter of the hexagon. Similarly, a polygon of n sides inscribed in the circle has an area equal to that of a right triangle in which the sides about the right angle are the apothem and the perimeter of the polygon. Now as n becomes large, the apothem approaches the radius of the circle and the perimeter of the polygon approaches the circumference of the circle. In the limit of large n, the area of the polygon is equal to that of the circle (Archimedes was more careful here), and the area of the circle is equal to the area of a right triangle in which the sides are equal to the circle's radius r and to its circumference $2\pi r$. Thus the area of the circle is $\frac{1}{2}(r)(2\pi r) = \pi r^2$. (The limiting process of calculus was foreshadowed by the limit in which the polygon approaches a circle.) Archimedes' proof involved polygons which circumscribed as well as inscribed the circle, and by working with polygons of up to 96 sides, he was able to show that $(3 + \frac{10}{71}) < \pi < (3 + \frac{1}{7})$, or $3.1408 < \pi < 3.1429$.

The importance of Archimedes' work in the history of science can scarcely be exaggerated. When Alexandria was conquered by the Arabs and the library burned, much of his work was lost. But the Arabs preserved and expanded some of it, and other parts were preserved in the Byzantine empire. When Europe emerged from the dark ages, Archimedes' work was translated into Latin and greatly influenced the beginnings of the scientific revolution. Galileo mentions Archimedes over 100 times, using such expressions as *superhumanus* Archimedes, *inimitabilis* Archimedes, and *divinissimus* Archimedes. The connection between mathematics and the description of experiment that is the core of modern science was nascent in the work of this great mathematician and inventor.

SUMMARY

Section 15-1. Stress and Strain

The force per unit area on a solid is called the stress σ, and the resulting deformation of the solid is called the strain ϵ. The stress and strain in a solid are related to each other by Hooke's law:

$$\sigma_t = Y\epsilon_t$$

$$\sigma_s = S\epsilon_s$$

where Y is called Young's modulus and S is called the shear modulus. The change in volume is related to the change in the applied pressure by the bulk modulus B:

$$\Delta p = -B\left(\frac{\Delta V}{V}\right)$$

Section 15-2. Density

The density ρ of an object is its mass m divided by its volume V.

Section 15-3. Pressure in a Static Fluid

The pressure in a fluid acts normally to any surface. It varies with vertical position in an incompressible fluid by

$$p = p_0 + \rho gh \qquad (15\text{-}6)$$

In most gases, when the temperature is constant, the pressure is given as a function of height of

$$p(h) = p(0)e^{-(\rho_0 g/p_0)h} \qquad (15\text{-}7)$$

Section 15-4. Archimedes' Principle

Archimedes discovered the principle behind buoyant forces: A body which is partly or entirely submerged in a fluid is buoyed up by a force equal in magnitude to the weight of the displaced fluid and directed upward along a line through the center of gravity of the displaced fluid.

Section 15-5. Bernoulli's Equation

The equation of continuity expresses conservation of mass for a fluid flowing in a tube of variable area A:

$$\rho_1 v_1 A_1 = \rho_2 v_2 A_2 \qquad (15\text{-}8)$$

For an incompressible nonviscous fluid, conservation of mechanical energy leads to Bernoulli's equation along a streamline:

$$p_1 + \rho g y_1 + \tfrac{1}{2}\rho v_1{}^2 = p_2 + \rho g y_2 + \tfrac{1}{2}\rho v_2{}^2 \qquad (15\text{-}10)$$

Section 15-6. Viscosity
For newtonian fluids the viscous force F on a surface of area A is

$$F = \frac{\eta A v}{\ell} \tag{15-11}$$

Applying this to the flow in a pipe shows that the total flow rate in the pipe is

$$Q = \frac{\pi(\Delta p)R^4}{8\eta L} \tag{15-12}$$

QUESTIONS

1. Classify the following as solid, liquid, gas, or other: ice, a glacier, gelatin, water vapor, fog, a cloud, taffy, bread dough, sugar, honey.

2. Give an example of a substance that could be considered a solid or a liquid, depending on the time scale of the experiment.

3. Two blocks are glued together to form a cube, as shown in Fig. 15-25, where the blue diagonal line represents the glue joint. If forces are applied as shown, is the stress on the glue joint tensile or shear, or a combination of both? If a combination, what is the ratio of the tensile stress to the shear stress?

Figure 15-25. Question 3.

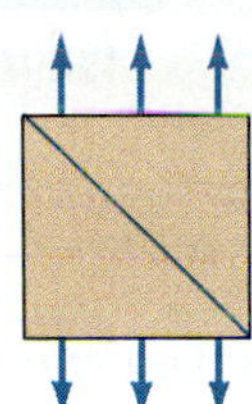

4. When we say a substance is "strong," do we usually mean that it has a high Young's modulus or a high yield point or a high shear modulus? Is a wooden rod stronger than a glass rod?

5. For very high "strength" applications, composite solids such as fiberglass-epoxy are often used. Can you explain why in terms of Young's modulus and the yield point?

6. Is it true that the higher an element is in the periodic chart, the higher its density?

7. What is the least dense gas (at atmospheric pressure and room temperature)? What is the most dense substance at atmospheric pressure and room temperature?

8. In which vertical tube of the apparatus shown in Fig. 15-26 will a fluid rise highest when the fluid is at rest? Ignore any effects of surface tension.

Figure 15-26. Question 8.

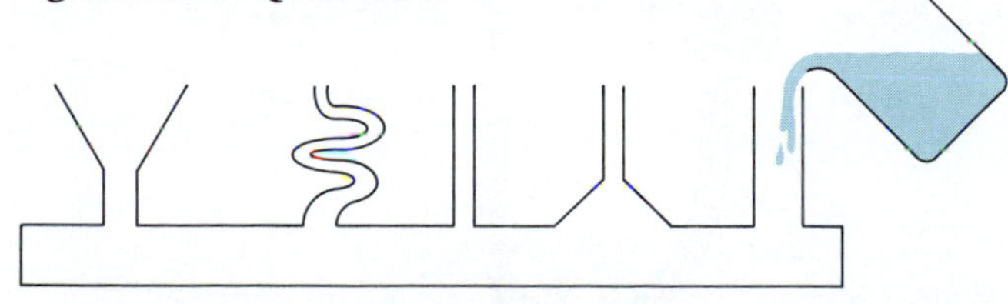

9. How does the "cartesian diver" of Fig. 15-27 work? When the membrane at the top is pushed down, the diver descends, when the membrane is released the diver ascends. If the diver is stationary in the middle of the apparatus, what is its overall density? What properties of a fluid are important in operating the diver?

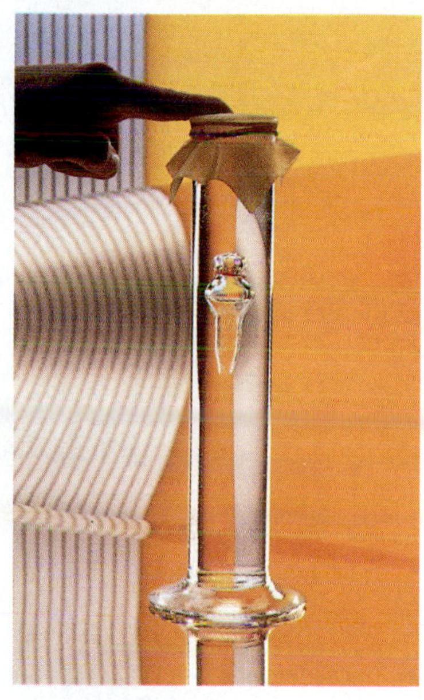

Figure 15-27. Question 9: cartesian diver.

10. Why is it easier to swim in seawater than in fresh water?

11. After oil tankers discharge their cargo of oil, they put water in their tanks for the return voyage, even though water isn't needed at their destination. Why?

12. What is the magnitude of the buoyant force on the block immersed in water, as shown in Fig. 15-28*b*?

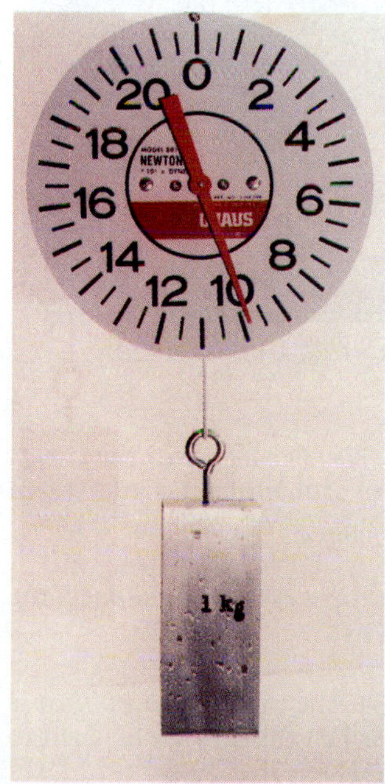

(*a*)

(*b*)

Figure 15-28. Question 12.

13. If you look at the stream of water issuing from a faucet, you will see that it narrows as it descends. Relate this effect to the equation of continuity.

14. How high will the jet of water in Fig. 15-29 go? Explain.

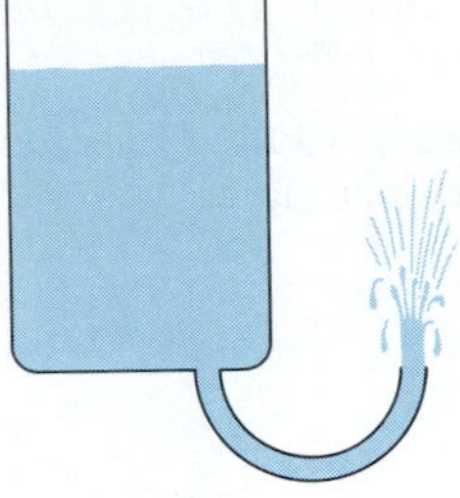

Figure 15-29. Question 14.

15. What is wrong with the Escher print in Fig. 15-30? (Physically, not aesthetically!)

Figure 15-30. Question 15: *Waterfall*, a lithograph by Escher.

16. If you blow between two sheets of paper, will the sheets tend to separate or to come toward each other? Try it! Explain your results.

17. When a car with a convertible top travels at high speed, the top always seems to bulge out. Why?

18. In the pitot tube of Fig. 15-31, the pressure at point A is higher than at point B. Why? How can this device be used to measure the airspeed of an airplane?

19. Is viscous flow in a pipe irrotational?

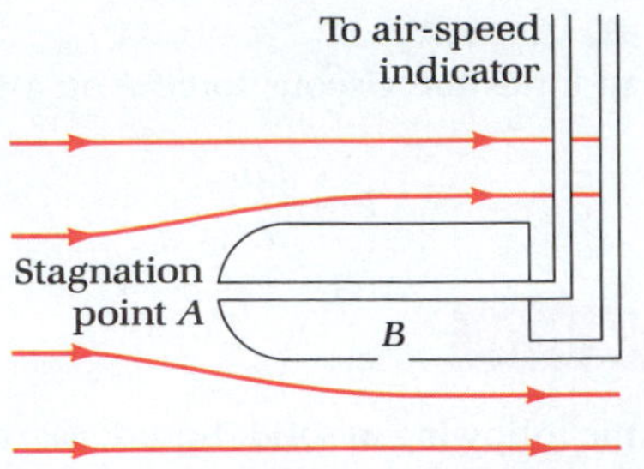

Figure 15-31. Question 18: A pitot tube.

20. The apparatus of Fig. 15-32 is used in an experiment on water flow, and the water levels are as shown. Show that the value of $p + \rho gh + \frac{1}{2}\rho v^2$ is not the same at points A and B. What is the reason for this disagreement with Bernoulli's equation?

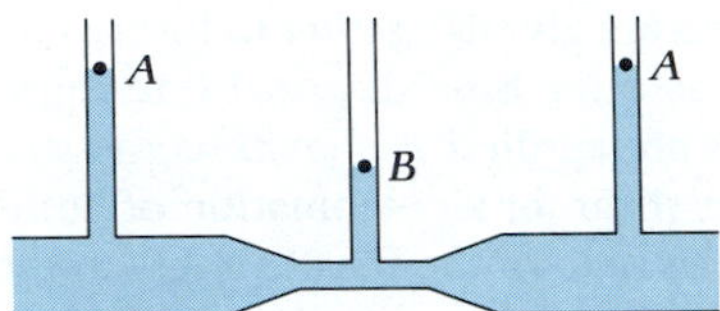

Figure 15-32. Question 20.

21. If a large truck passes you on the highway, does the airflow blow your car toward or away from the truck? Explain.

22. A glass of water is full to the brim, and an ice cube floats in the water. When the ice cube melts, does the glass run over, is the water level lowered, or does the water level stay the same? Explain.

23. The device of Fig. 15-33 has been proposed as a perpetual-motion machine. The buoyant force on the rope in the water forces the rope to go up on the left side. If all frictional forces are negligible, why won't it work? Explain (beyond citing conservation of energy).

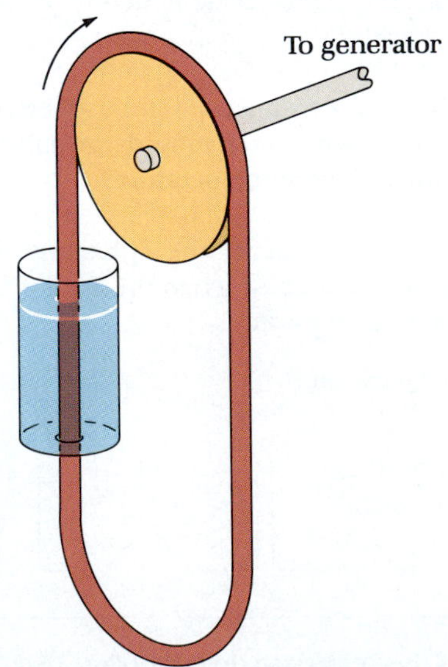

Figure 15-33. Question 23.

24. A "suction" pump, which operates by lowering the pressure in a pipe that extends to the bottom of a well, can only pump water from a well 10 m deep or less. Why?

25. Explain how the siphon of Fig. 15-34 is able to transfer water from A to B. Why doesn't the water in the tube merely run back into the two reservoirs? What is the maximum value that h can have if the siphon is to work?

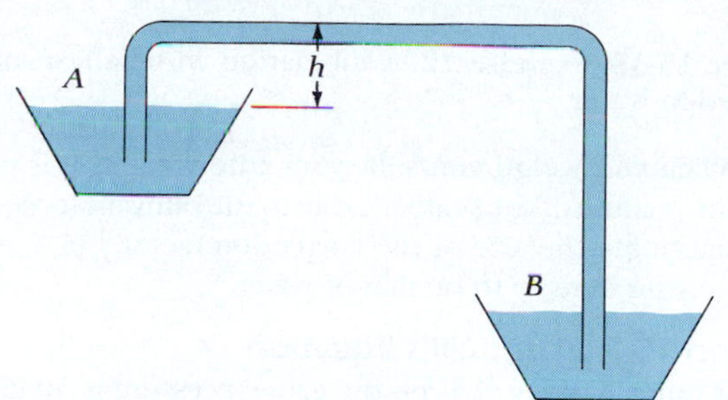

Figure 15-34. Question 25.

26. Complete the following table:

Symbol	Represents	Type	SI Unit
σ			Pa
ϵ	Strain		
p		Scalar	
Y			
B			
S			
ρ			
η			

EXERCISES

Section 15-1. Stress and Strain

1. A 12.7-mm cube of copper is put into a vise and squeezed by a force of 215 N exerted by either side of the vise. What is the stress in the copper cube?

2. An object of mass 13.4 kg is held up by a 1.33-m-long steel wire of cross section 1.63 mm². What is the stress in the wire?

3. A 22.4-kg block is hung from a brass cube by a 91-mm-diameter aluminum wire that is 750 mm long; the cube has 85.0-mm-long sides (Fig. 15-35). What is the stress *(a)* in the aluminum and *(b)* in the brass? Assume that the brass is under a pure shear stress.

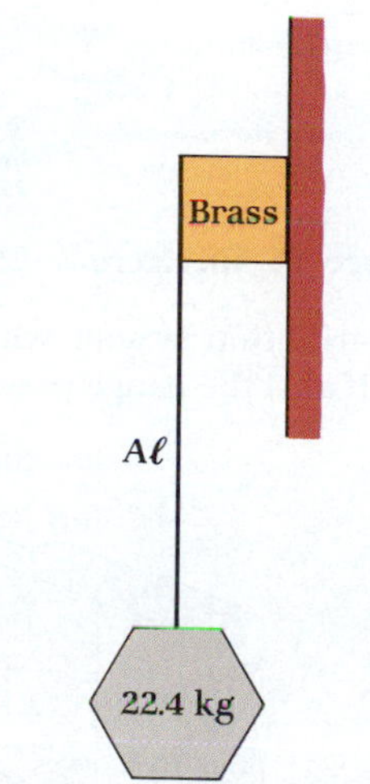

Figure 15-35. Exercise 3.

4. A 130-Mg building is on a foundation of 160 concrete blocks, each of effective cross section 6000 mm². Estimate the stress in the concrete blocks.

5. *(a)* What is the strain in the copper cube of Exercise 1? By how much does the side of the copper cube decrease?

6. *(a)* What is the strain in the wire of Exercise 2? *(b)* What is the increase in length of the wire as the weight is attached?

7. The block of Exercise 3 is attached to the wire but held up entirely by your hand. *(a)* When you remove your hand, what is the strain in the brass cube? *(b)* By how much is the angle between the sides of the brass cube changed? *(c)* How far downward does the block move?

8. The highest steady pressure available in the laboratory is about 20 GPa in a diamond anvil. If liquid mercury stayed in the same phase when this much pressure was applied, by what fraction would its volume be decreased as this pressure was applied?

Section 15-2. Density

9. A 73.0-mm-radius sphere of plutonium has a mass of 32.3 kg. What is the density of plutonium?

10. In the past, the meter was defined by the length of a platinum bar of rectangular cross section 25.3 mm × 4.0 mm. What is the mass of this old standard?

11. What is the density of water at the bottom of the sea where the pressure is 30 MPa? What is the density of lead at the bottom of the sea?

12. When a tensile stress is applied to an object, not only is the object's length increased along the direction of the applied force, its width is decreased perpendicular to the direction of the applied force. For many substances the ratio of the fractional contraction perpendicular to the applied force to the fractional elongation parallel to the applied force is nearly $\frac{1}{3}$. This ratio is called the Poisson ratio ν. Assuming ν is $\frac{1}{3}$ for steel, by what fraction does the density of a 15-mm steel cube change when a tensile stresss of 1 GPa is applied?

Section 15-3. Pressure in a Static Fluid

13. The deepest parts of the ocean are about 10 km below the surface. What is the pressure at this depth?

14. Pressures in air ducts are frequently given in "inches of water." What is the conversion factor between the pressure measured in inches with a water manometer and the pressure measured in pascals?

15. The highest mountains are about 8 km high. What is the pressure at the top? Assume the atmosphere has a constant temperature.

16. What is the force on a circular eardrum of diameter 7 mm if the inside of the ear is at the pressure at sea level and the outside of the ear at the pressure inside an airplane, that of the atmosphere 1500 m above sea level? (This force causes the pain that occurs when you cannot equalize the pressure on the two sides of the eardrum through the eustachian tube that connects the inside of the ear to the atmosphere.)

17. In Fig. 15-36, how much force must be exerted at A to support the 0.85-Mg automobile at B? The piston at A has a diameter of 17 mm and the piston at B a diameter of 300 mm.

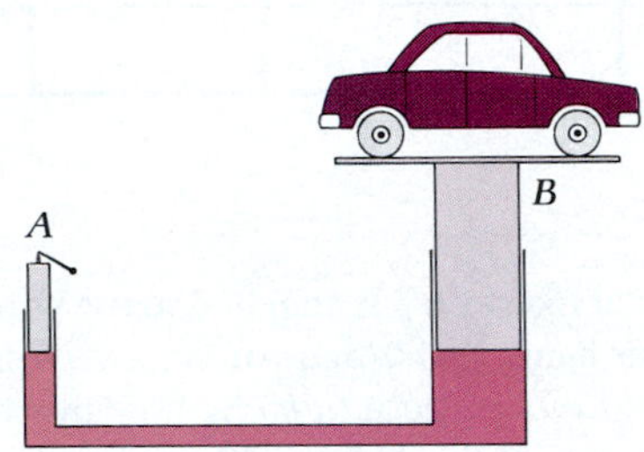

Figure 15-36. Exercises 17 and 42.

Section 15-4. Archimedes' Principle

18. An iceberg has a mass of 13 Gg. What volume of water does it displace?

19. An iceberg has 100 m^3 of its volume above water level. How much water does it displace?

20. A crown is weighed as in Fig. 15-37. In air it weighs 28.24 N, and in water it weighs 26.36 N. What is the density of the crown?

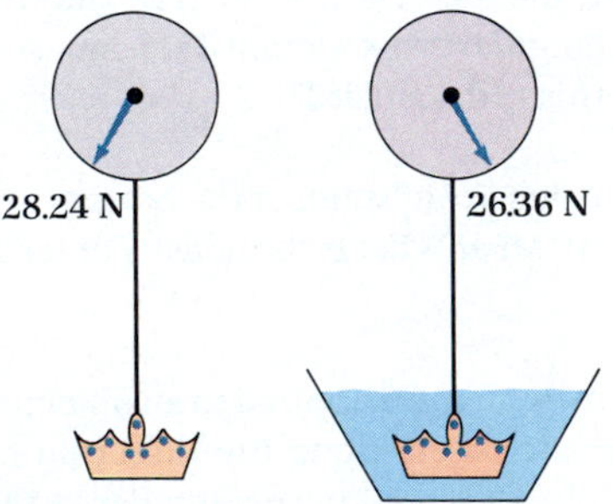

Figure 15-37. Exercise 20.

21. Estimate the volume of your body without making any length measurements or assumptions.

22. Suppose a submarine has a cross section which is a section of a cylinder, as seen in Fig. 15-38. *(a)* Where is the center of gravity of the displaced water? *(b)* Suppose the submarine carries ballast fixed in the bottom of its hold so that the center of gravity of the submarine is at point A, $3r/4$ below the center of the cylinder. If the submarine rolls over by 10°, what is the magnitude of the restoring torque about the center of mass that tends to right the submarine? Give the torque in terms of the mass of the submarine m, the density of water ρ, the radius of the cylindrical section r, and the acceleration of gravity g.

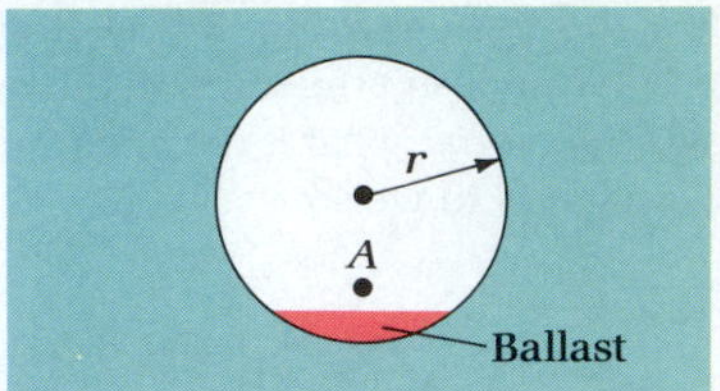

Figure 15-38. Exercise 22: A submarine with ballast submerged in water.

23. When you weigh yourself, your true weight F_e is greater than the reading of the scale F_e', due to the buoyant force of the air. Determine the size of the correction factor f in $F_e = fF_e'$. Assume your density to be that of water.

Section 15-5. Bernoulli's Equation

24. At point A of Fig. 15-39, the gauge pressure is 50 kPa and the speed of the water flowing in the circular pipe is 2.4 m/s. The pipe is horizontal. *(a)* What is the flow rate at points A and B? *(b)* What is the speed at point B? *(c)* What is the gauge pressure at point B? Assume streamline flow.

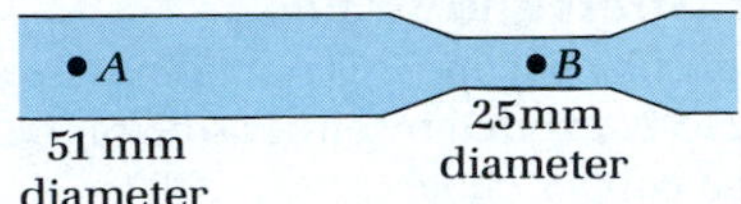

Figure 15-39. Exercise 24.

25. At point A of Fig. 15-40 the gauge pressure is 75 kPa and the speed of the water flowing in this 50-mm-diameter pipe is 1.7 m/s. The pipe splits into two smaller pipes, each of 25-mm diameter. *(a)* What are the flow rates at A and at B? *(b)* What is the speed at point B? *(c)* What is the gauge pressure at point B? Assume streamline flow and constant height.

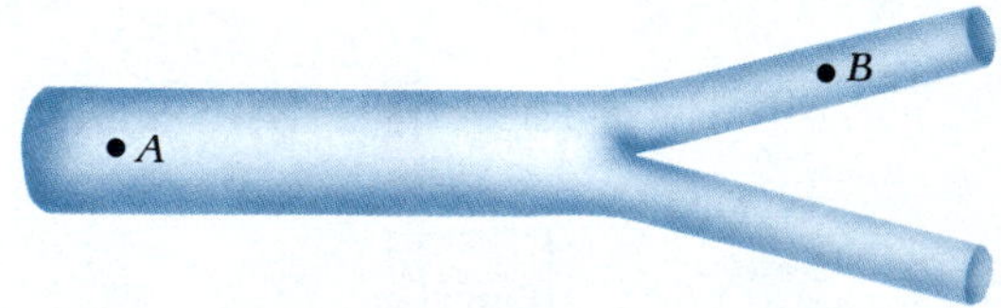

Figure 15-40. Exercise 25.

26. Part of an air-distribution system with square ducts is shown in Fig. 15-41. If at A the gauge pressure is 320 Pa and

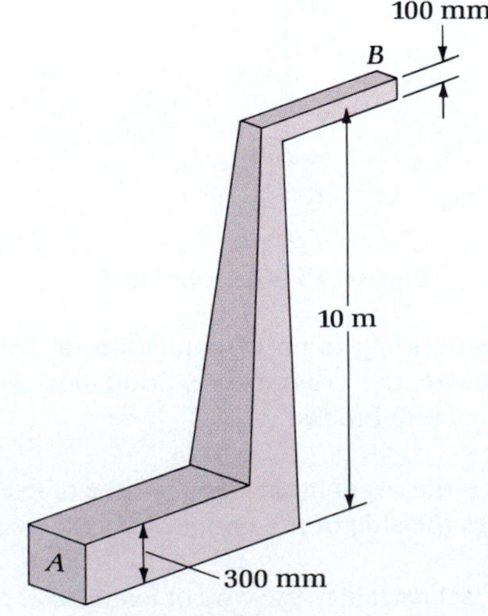

Figure 15-41. Exercise 26.

the flow rate is 2.2 m³/s, what is the air pressure at *B*? Assume incompressible streamline flow.

27. Show that Bernoulli's equation reduces to Eq. 15-6 when the flow speed is the same at both points.

28. A fire truck has a pump that must supply enough pressure to enable the water to reach the top floor of a 30-m-high building in a uniform hose at a speed of 10 m/s. Ignoring viscosity, what is the minimum gauge pressure the pumper must supply? What is the minimum power?

29. An automobile engine has a displacement of 1.6 L. Each time the engine makes two complete revolutions, an amount of air approximately equal to the displacement is sent through the engine. At highway speeds the engine turns at 3500 rev/min. The venturi tube of the engine's carburetor has a radius of 9.1 mm. Estimate the gauge pressure in the venturi under these conditions.

30. The insecticide sprayer of Fig. 15-42 has a pump with a diameter of 60 mm. The insecticide level is 90 mm below the inlet tube at *A*. The tube at *A* has a diameter of 2 mm. Estimate the minimum speed with which the plunger should be pushed if the air jet at the end is to contain insecticide. Assume that the insecticide has the density of water and that the airflow is incompressible and streamline.

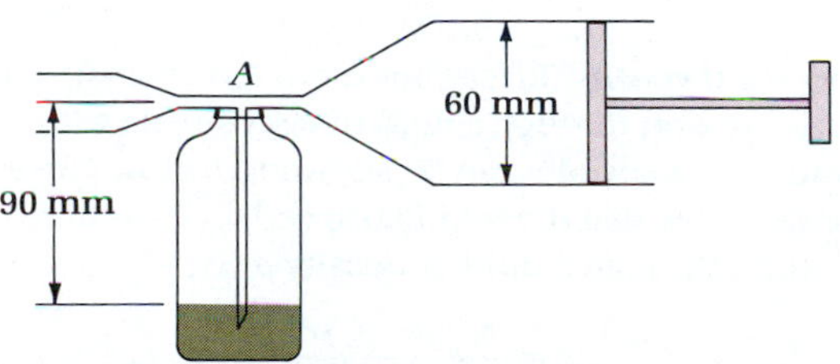

Figure 15-42. Exercise 30.

31. Water flows steadily from a holding tank, as seen in Fig. 15-43. The cross section of the pipe at point *A* is 0.055 m², at point *B* it is 0.040 m², and the cross section of the discharge stream at *C* is 0.025 m². Neglect viscosity and turbulence. *(a)* What is the speed at *C*? *(b)* What is the flow rate? *(c)* What are the gauge pressures at points *A* and *B*?

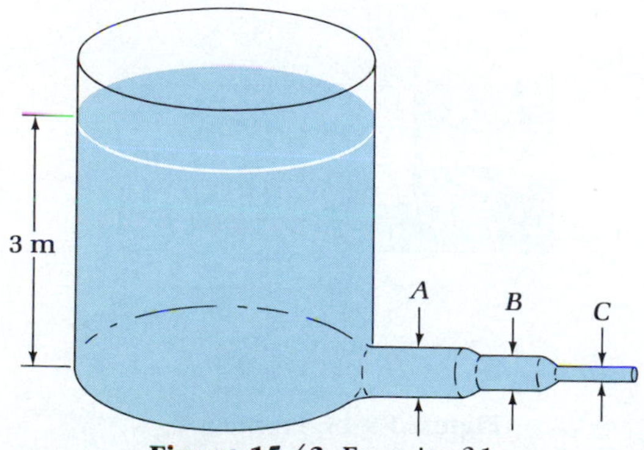

Figure 15-43. Exercise 31.

32. A window of area 5.0 m² has still air on its inside and air of speed 15 m/s on its outside. If you suppose that the flow is irrotational so that the two sides can be connected by Bernoulli's equation, how much force would be exerted on the window? In which direction?

Section 15-7. Viscosity

33. A flat-bottomed canal boat with a bottom area of 30 m² is dragged along a canal at 1.5 m/s. The bottom of the boat is 140 mm above the bottom of the canal. The viscosity of canal water is 1×10^{-3} N·s/m². What is the viscous force on the boat? (Normally the forces that waves exert on a canal boat are much greater than the viscous force.)

34. A cylindrical viscometer is shown in Fig. 15-44. When the inner cylinder is rotated at constant speed, the stationary outer cylinder experiences a torque. The magnitude of the torque on the outer cylinder is a measure of the viscosity of the fluid between the cylinders. When the distance between the cylinders ΔR is small compared with the radius R_1 and R_2 of either cylinder, then the force applied to the outer cylinder by the fluid is

$$F = \frac{\eta A v}{\ell} \approx \eta(2\pi RL)\frac{v}{\Delta R}$$

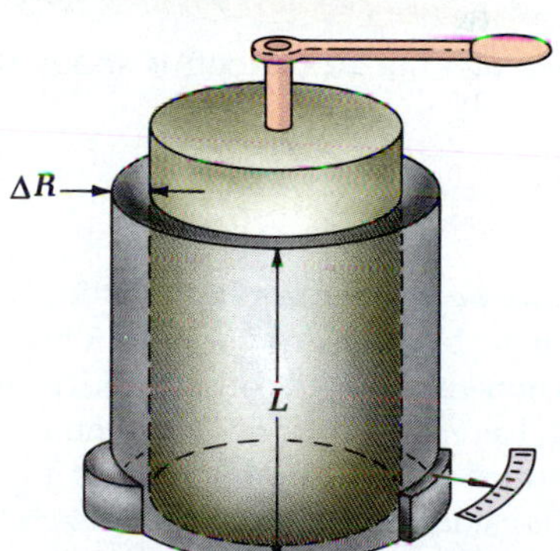

Figure 15-44. Exercise 34.

where $R = \frac{1}{2}(R_1 + R_2)$ and v is the tangential speed of the inner cylinder. *(a)* What is the torque applied to the outer cylinder? *(b)* A cylindrical viscometer has radii of 92 and 93 mm and a length of 170 mm. When the inner cylinder is rotated at a constant angular speed of 20 rev/min, the stationary outer cylinder experiences a torque of 0.54 N·m. What is the viscosity of the fluid in the viscometer?

35. Oil of specific gravity 0.765 and viscosity 2.5×10^{-3} Pa·s is sent through a small oil hole of length 10 mm to lubricate a bearing. If the pressure drop along the tube is 0.305 mPa and the hole diameter is 0.843 mm, what volume of oil will pass through the bearing in an hour? Assume laminar flow.

Additional Exercises

36. A 35-kg load is suspended from a metal wire of diameter 2.0 mm, and the wire elongates by 0.06 percent of its original length. Determine *(a)* the stress in the wire and *(b)* Young's modulus for this metal.

37. A composite block is to be formed from wood of density 710 kg/m³ and copper. What fraction of the total volume of the block must be wood in order for the block to remain suspended when completely immersed in water?

38. A 240-mm layer of oil of density 840 kg/m³ floats on a 360-mm layer of water in a tank. Determine the absolute pressure and the gauge pressure at *(a)* the oil-water interface and *(b)* the bottom of the water layer.

39. A solid wooden ball floats in the layered fluid of the previous exercise. It is in equilibrium with its top half in the oil and its bottom half in the water. What is the density of the wood?

40. A rectangular box of dimensions 0.25 m × 0.50 m × 0.75 m is partially evacuated so that the air pressure inside is 90 percent of the air pressure outside, which is 1.01×10^5 Pa. Determine the net force on each face of the box due to this pressure difference.

41. Estimate the difference in your blood pressure between your feet and your head when you are standing. Assume that your blood is a static fluid (not a good idea!).

42. If the automobile in Exercise 17 is raised by a height H, through what distance does the piston at A move? Assume that the fluid is incompressible.

43. The volume of a hot-air balloon is about 3000 m³ when inflated. Suppose that the balloon fabric, the cargo, and suspension system have a total mass of 200 kg. Estimate the average density of the hot air in the balloon at liftoff. Assume that the density of the surrounding air is 1.29 kg/m³.

44. A water hose has an inner diameter of 12 mm, and water emerges from a nozzle at the end in a stream of diameter 6 mm. When the stream is directed straight up, the water rises to a maximum height of 2 m. Determine *(a)* the volume flow rate and *(b)* the water pressure in the hose.

45. A 5-m/s wind blows parallel to the outside surface of a 2-m by 3-m office window. Estimate the net force on the window due to the air pressure difference between the inside and outside surfaces. Assume that the pressure difference is due to the streamline flow of a nonviscous, incompressible fluid.

46. When (viscous) crude oil flows through a 0.4-m diameter pipe, a pressure difference per unit length of 0.3 Pa/m is needed to maintain a flow rate of 0.25 m³/s. What pressure difference per unit length would be needed to maintain the same flow rate in a 0.6-m diameter pipe?

47. Determine the viscosity of the crude oil in Exercise 46.

PROBLEMS

1. ***A composite bar.*** Three bars are welded together end to end to form a single bar. The first bar is 0.55 m long, has a cross section of 420 mm², and is made of copper. The second section is 0.75 m long, has a cross section of 390 mm², and is made of cast iron. The third bar is 0.45 m long, has a cross section of 405 mm², and is made of aluminum. *(a)* What is the mass of the complete bar? *(b)* What is its average density? *(c)* If tensile forces of 10 kN are applied at either end of the complete bar, by how much does it elongate?

2. ***Oscillations of a cork.*** A cork of density ρ_c has a cylindrical cross section of radius r and length ℓ. It is floating in water of density ρ. Show that, if it is given a small push down into the water from its equilibrium floating position and then released, it will vibrate up and down with a period $2\pi\sqrt{\rho_c\ell/(\rho g)}$.

3. ***Cork, oil, and water.*** Oil of density $\rho_0 < \rho_{H_2O}$ floats on water. The cork described in the previous problem has density $\rho_c > \rho_0$ and floats between the two layers so that its top is in oil and its bottom in water. What fraction of the cork is below the water? Give your answer in terms of ρ_0, ρ_c, and ρ_{H_2O}.

4. ***Buoyant force on the kilogram.*** The standard kilogram is made of an alloy that is 90 percent platinum and 10 percent iridium. What is the magnitude of the buoyant-force correction (because of the atmosphere) that must be made when the standard kilogram is weighed in air?

5. ***Heights of buildings.*** The maximum compressive strain that concrete can safely withstand is about 0.1 percent. Estimate the maximum height of a concrete building of constant cross section. Using the same criterion, estimate the maximum height of a constant-cross-section steel building.

6. ***Terminal speed.*** The viscous force F_v on a small sphere of radius r in motion through a fluid of viscosity η (with laminar flow) and with a speed v can be shown to follow Stokes' law: $F_v = 6\pi r\eta v$. Show that the terminal speed v_t of a sphere of density ρ_s falling through a fluid of density ρ_0 is

$$v_t = \frac{2(\rho_s - \rho_0)r^2 g}{9\eta}$$

7. ***A venturi meter.*** A large tank has a pipe that leads out of its bottom. (See Fig. 15-45.) Along the pipe is a constriction that has a diameter one-third the diameter of the rest of the pipe. In this constriction is a tube which leads to a second tank containing the same fluid as the first tank. When liquid flows out of the first tank, to what height h_2 does the fluid rise in the tube? Express your answer in terms of h_1. Assume laminar flow of a nonviscous fluid.

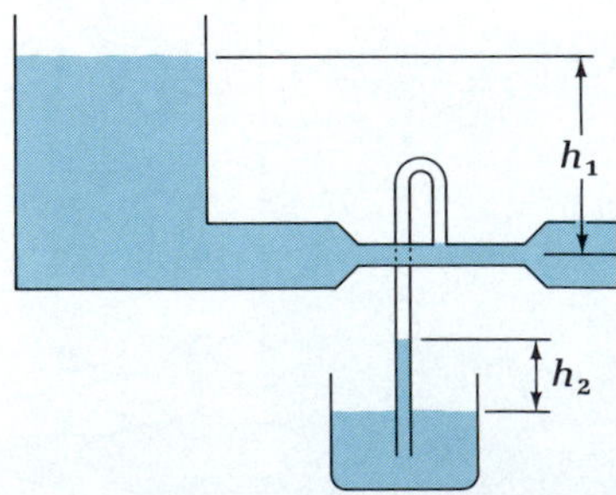

Figure 15-45. Problem 7.

8. ***Leaks in a tank.*** Several holes are drilled in a cylindrical tube containing water. Water issues from the holes horizontally, as seen in Fig. 15-46. At what height should a hole be

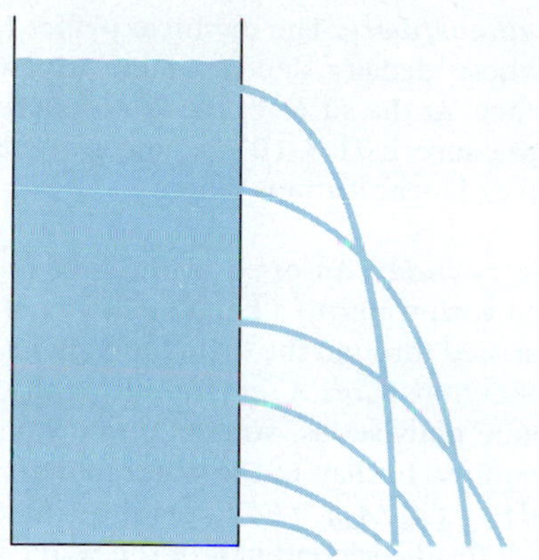

Figure 15-46. Problem 8.

drilled so that water spewing from it hits the ground farthest from the cylinder? Assume the fluid is nonviscous.

9. ***Draining a keg.*** A keg of beer may be approximated by a cylinder 750 mm high and 250 mm in radius. If a hole is drilled near the bottom, and a 1.0-m-long tube with inner diameter 3.8 mm is attached, how long will it take for half the beer to drain out? There is a hole in the top of the keg to keep the top of the beer at atmospheric pressure. Assume that beer has the density and viscosity of water and is kept at 5°C.

10. ***Designing a hydrometer.*** A hydrometer is made of glass with a hollow spherical bottom of radius r and an upright tube of radius R, as shown in Fig. 15-47. It is to measure specific gravity in the range from 1.100 to 1.300, the specific gravity being given by the depth to which the hydrometer sinks in the fluid. The hydrometer's mass is adjusted by adding lead shot. Find the total mass m of the hydrometer and the length ℓ of the tube to measure this range of specific gravities. Give your answer in terms of r, R, and the density of water ρ.

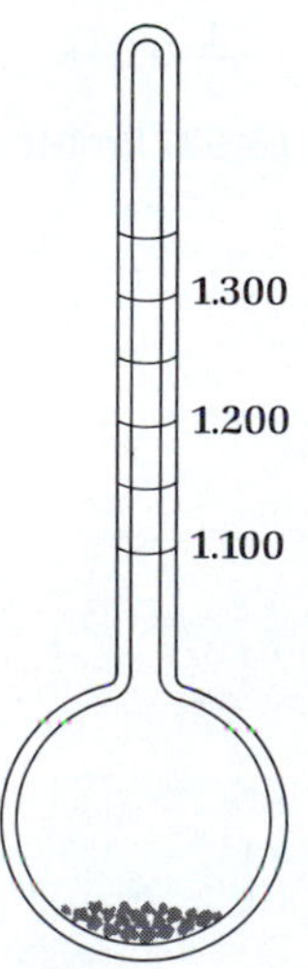

Figure 15-47. Problem 10.

11. ***Stretching a wire.*** A steel wire of cross section A and Young's modulus Y is stretched horizontally between two posts. Initially the tension in the wire is negligible. An object of mass m is hung from the middle of the wire, causing it to sag, as seen in Fig. 15-48. *(a)* Show that the wire stretches until θ is given by the transcendental equation

$$2 \sin \theta(\sec \theta - 1) = \frac{mg}{YA}$$

(b) What is the value of θ when $mg/YA = 1$ percent?

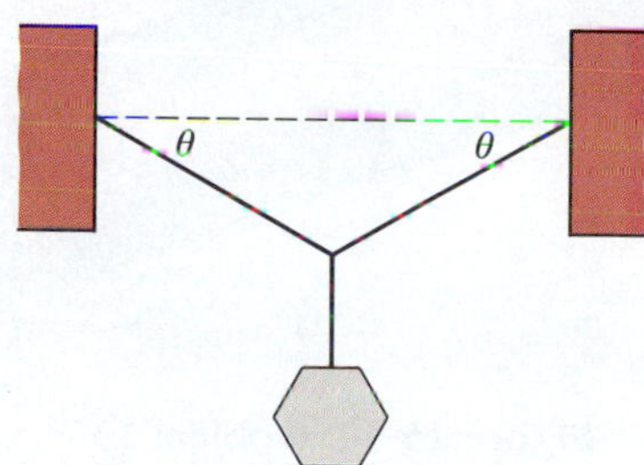

Figure 15-48. Problem 11.

12. ***Water over a spillway.*** The flow of water over a weir or spillway is illustrated in Fig. 15-49. Assuming negligible viscosity, shown that the flow rate Q of water over the weir is

$$Q = \tfrac{2}{3} w \sqrt{2gy^3}$$

where y is the height of the water level above the weir. Actual flows are about one-half of this because of viscosity and the drop in level of the water between the reservoir and the weir.

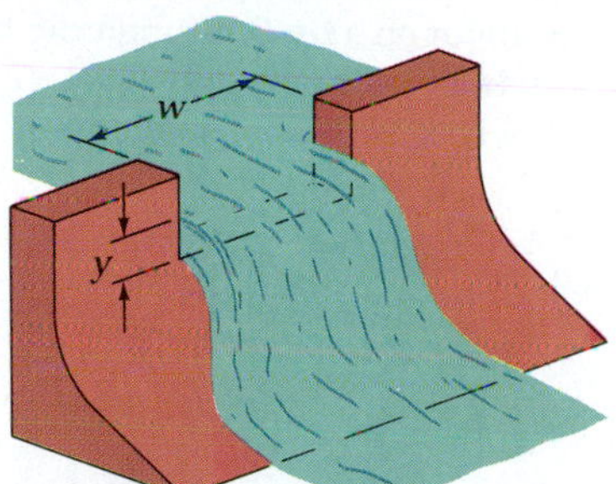

Figure 15-49. Problem 12.

13. ***Viscous flow in a pipe.*** When a viscous fluid flows in a pipe, the speed of the fluid is greatest along the axis and decreases to zero at the wall of the pipe. To determine how the speed varies with distance R from the axis, consider a cylinder of fluid of length L and radius R, as shown in Fig. 15-50. The speed of the fluid at radius R is v. *(a)* Show that the magnitude of the force on the cylinder of fluid of radius R due to the fluid outside is $F = \eta(2\pi RL)(dv/dR)$. *(b)* For constant speed, show that $dv/dR = -\Delta pR/(2\eta L)$, where $\Delta p = |p_1 - p_2|$ is the pressure difference along a length L of the pipe. *(c)* Integrate this expression from a radius R' to the inner radius R_{pipe} of the pipe (where $v = 0$) to show that the speed of the fluid at a radius R' is

$$v(R') = \frac{|p_1 - p_2|}{4\eta L} (R_{\text{pipe}}^2 - R'^2)$$

14. ***Viscous flow rate.*** Show that Eq. 15-12 follows from the results of Prob. 13 by integrating the flow rate,

$$dQ = v(R')dA = v(R')(2\pi R'\, dR')$$

from $R' = 0$ to $R' = R_{\text{pipe}}$.

15. ***Pressure in a water main.*** A water distribution system has a main line of diameter 0.25 m that serves a user with a

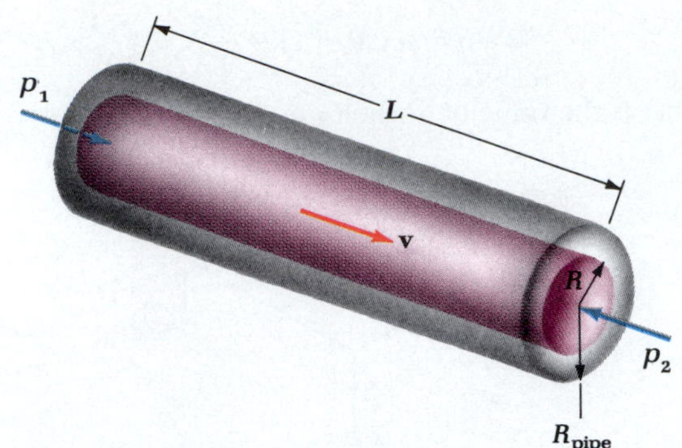

Figure 15-50. Problem 13.

maximum flow rate of 0.20 m³/s. If the user's line has diameter 0.10 m and the elevation is 5.0 m above the level of the main line, what must be the minimum main-line gauge pressure?

16. ***Force and torque on a dam.*** A dam of height H and width W is exposed to the pressure of the water as shown in Fig. 15-51. *(a)* Show that the magnitude of the horizontal force exerted on the dam by the water is $F_W = \frac{1}{2}\rho g W H^2$. (*Hint*: Since the pressure varies with depth, you must set up and perform an integration.) *(b)* Show that the magnitude τ of the torque exerted by the water about an axis O through the bottom of the dam along its width is given by $\tau = \rho g W H^3/6$. *(c)* Evaluate F_W and τ for a dam of height 15 m and width 40 m.

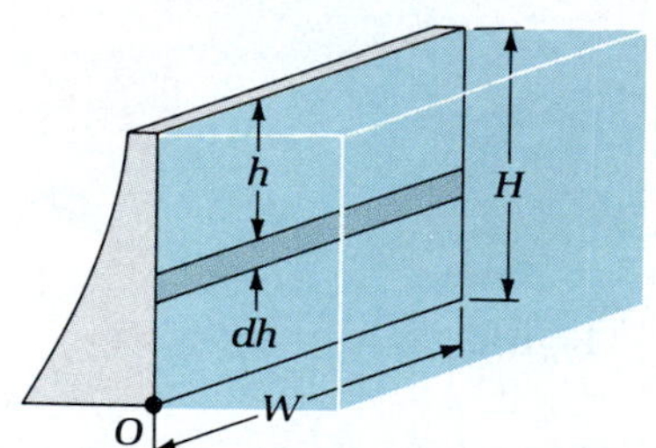

Figure 15-51. Problem 16.

17. ***Another atmosphere.*** The mythical planet Lineac has an atmosphere whose density decreases linearly with altitude above the surface. At the surface, the atmosphere has density 1.29 kg/m³, pressure 1.01×10^5 Pa, and $g = 9.8$ m/s². How high is the top of Lineac's atmosphere?

18. ***A leaking cylinder.*** An open cylinder of cross-sectional area A_1 is filled with water to a height H above the bottom. A small hole is drilled through the cylinder wall, and a stream of water of cross-sectional area A_2 emerges from the hole ($A_2 \ll A_1$). *(a)* Assume nonviscous, streamline flow and show that the time $T_{1/2}$ required for half of the water to drain out is given by $(\sqrt{2} - 1)\sqrt{H/g}\,(A_1/A_2)$. *(b)* Explain qualitatively why it will take longer for the second half of the water to drain out.

19. ***Technique for measuring density.*** Equal volumes of two liquids are poured into a U-shaped tube that has a uniform cross section. The liquids have different densities, ρ_1 and ρ_2, and they do not mix. The liquids are in equilibrium, as shown in Fig. 15-52. If $\rho_2 = 1200$ kg/m³, determine the value of ρ_1.

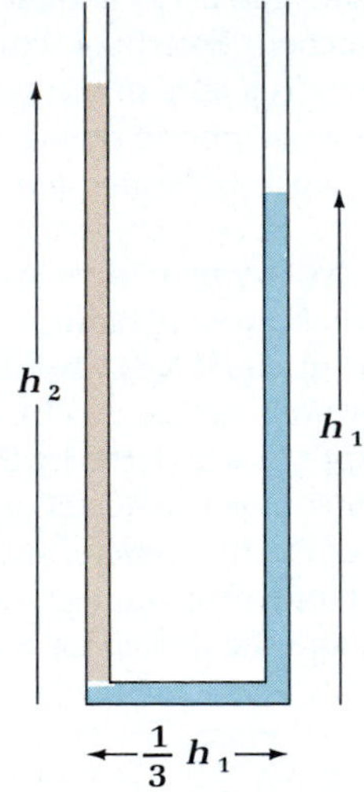

Figure 15-52. Problem 19.

TEMPERATURE AND HEAT TRANSFER

16

A thermogram in color. Special film was used to show heat loss from these structures.

Suppose an ice cube is taken from a freezer and placed on a level table. For a short time, the response of this object to its surroundings can be described by the methods of mechanics. It is in static equilibrium and the net force acting on the ice cube is zero: The earth exerts a downward gravitational force on the ice cube which is balanced by the upward normal force exerted by the table (along with a small buoyant force due to the atmosphere). Before long, however, the ice begins to melt, and eventually we see a puddle of water on the table.

We are not able to describe or understand this melting process on the basis of mechanics alone. Some new concepts, independent of mechanics, must be developed to discuss situations of this sort. In these next four chapters, we shall consider some of the ideas that are central to thermodynamics and to statistical mechanics. The concepts of temperature and of heat transfer are introduced in this chapter.

16-1 MICROSCOPIC AND MACROSCOPIC DESCRIPTIONS

A microscopic description is at the molecular level.

Suppose that a fixed amount of a gas, say 5 g of oxygen, occupies a container. A microscopic description of this system would begin with the recognition that the gas consists of molecules. Is it reasonable to attempt to determine the motion of each molecule by applying Newton's laws? With a large, high-speed computer, a very small system consisting of several thousand molecules can be simulated to provide useful information. In these simulations the motion of each molecule is followed in detail. Any property of the system that depends on the positions and velocities of the molecules can be calculated. This approach is known as ***molecular dynamics.***

However, the number of molecules in 5 g of oxygen is of the order of 10^{23}, far too large to be treated by molecular dynamics. Just listing the instantaneous values of the position and the velocity of each individual molecule would be an overwhelming task. That amount of information is too vast to assimilate and therefore would be of little value in describing this system and its interaction with its environment. Instead, it is more useful to consider averages. The molecular dynamics approach also evaluates quantities that involve averages over molecular motions. The methods of statistical mechanics connect averages of molecular properties to quantities such as temperature and pressure that are part of our everyday world. We shall see connections of this sort in Chap. 18.

Thermodynamics is the larger-scale, or macroscopic, description.

A macroscopic description deals with properties on a scale much larger than the molecular one. On this larger scale there is no direct reference to the molecular properties of the system. The macroscopic description of the interaction of a system with its surroundings is called ***thermodynamics.*** Although thermodynamics does not depend on our knowledge of the molecular structure of matter, interpreting the thermodynamic description in terms of molecular averages often helps us to visualize what is going on.

State Variables

Some of the quantities used in thermodynamics have already been discussed. One of these is the pressure of a gas (Sec. 15-1). The ***pressure*** p is the force per unit area exerted by a gas on a surface.

$$p = F/A$$

Another quantity is the ***volume*** V occupied by a gas. The pressure p and the volume V are examples of *state variables.*

State variables characterize the state, or condition, of a system.

Another state variable for a system is the amount of substance in the system, or the *number of moles* that compose a system. This quantity is represented by the symbol n and abbreviated mol. One ***mole*** of a substance is the amount of the substance that contains Avogadro's number of molecules. ***Avogadro's number*** is

Avogadro's number

$$N_A = 6.022 \times 10^{23} \text{ molecules/mol}$$

Thus, the number of moles in a sample of some substance is the number of molecules N contained in the sample divided by Avogadro's number.

$$n = N/N_A$$

Variables of state include *p*, *V*, *n*, *T*, *U*, and *S*.

Additional state variables that will be defined later include *temperature T, internal energy U,* and *entropy S.*

Molar Mass

The ***molar mass*** M_0 of a pure substance is the mass of Avogadro's number of molecules. Thus, the mass m of a sample of a substance is equal to the product of the number of moles n in the sample and the molar mass of the substance.

$$m = nM_0$$

It is customary to give molar masses in grams rather than in kilograms, in which case the units of M_0 are g/mol.

For a monatomic substance, such as helium (He), the molar mass is the same as the atomic mass (also called atomic weight). (See App. P for atomic masses.) Since the atomic mass of He is 4.0 g/mol, the molar mass of He is $M_0 = 4.0$ g/mol. For a substance whose molecules contain more than one atom, such as carbon dioxide (CO_2), the molar mass is the sum of the atomic masses of the atoms in each molecule. The atomic mass of carbon (C) is 12.0 g/mol, and the atomic mass of oxygen (O) is 16.0 g/mol, so the molar mass of CO_2 is

$$M_0 = [12.0 \text{ g/mol} + 2(16.0 \text{ g/mol})] = 44.0 \text{ g/mol}$$

EXAMPLE 16-1

Number of moles and number of molecules in a sample. Water (H_2O) has two hydrogen atoms and one oxygen atom in each molecule. Determine *(a)* the number of moles and *(b)* the number of molecules in a 0.14-kg sample of water.

Solution. From App. P, the atomic mass of hydrogen is 1.0 g/mol, so the molar mass of H_2O is

$$M_0 = 2(1.0 \text{ g/mol}) + 16.0 \text{ g/mol} = 18.0 \text{ g/mol}$$

(a) The number of moles in the sample is

$$n = m/M_0 = (0.14 \text{ kg})(1000 \text{ g/kg})/18.0 \text{ g/mol} = 7.8 \text{ mol}$$

(b) The number of molecules in the sample is

$$N = nN_A = (7.8 \text{ mol})(6.022 \times 10^{23} \text{ molecules/mol}) = 4.7 \times 10^{24} \text{ molecules}$$

SELF-TEST 16-1. Under normal conditions, oxygen is diatomic (two atoms per molecule). Determine the mass of a sample that contains 2.82×10^{23} oxygen molecules. ***ANSWER:*** 15.0 g.

EXAMPLE 16-2

Gas in a cylinder. A gas occupies a cylindrical container with circular cross section of radius $R = 0.22$ m and length $L = 0.35$ m. The pressure of the gas is 2.00 atm, or 2.02×10^5 Pa. Determine *(a)* the volume in liters (L) occupied by the gas and *(b)* the force exerted by the gas on one of the circular faces of the cylinder.

Solution. *(a)* The volume of the cylinder is given by the product of its length L and its cross-sectional area πR^2: Since 1.0 L = 0.001 m³ (exactly),

$$V = \pi R^2 L = 0.053 \text{ m}^3$$

$$= 0.053 \text{ m}^3 \frac{1.0 \text{ L}}{0.001 \text{ m}^3} = 53 \text{ L}$$

(b) The force exerted on an area $A = \pi R^2$ is

$$F = pA = (202 \text{ kPa})(0.15 \text{ m}^2) = 31 \text{ kN}$$

SELF-TEST 16-2. A gas consists of a mixture of two components: 2.2 mol of He and 1.8 mol of neon (Ne). What is the mass of the gas? ***ANSWER:*** 26.8 g.

16-2 THERMAL EQUILIBRIUM AND THE ZEROTH LAW OF THERMODYNAMICS

If a single value of a state variable is to characterize a system, then the state variable must have this single value throughout the entire system, and it must remain at this value. When this is the case, the system is in ***equilibrium.***

Definition of equilibrium

> A system is in equilibrium when its state variables are constant in time and uniform throughout the system.

For example, when a system is in equilibrium, its pressure is the same at each point in the system, and it does not change during the time involved.

Thermal Equilibrium

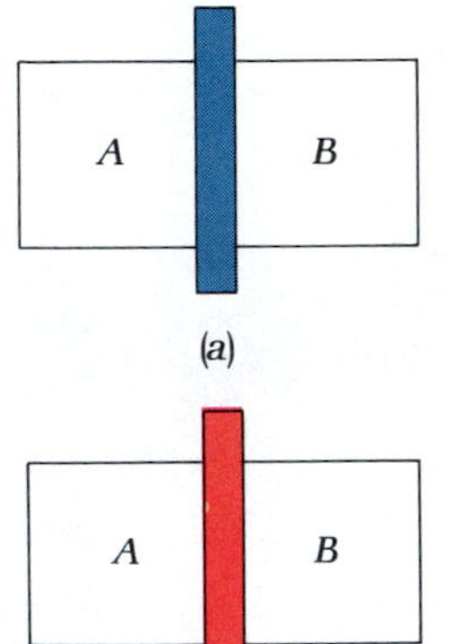

Figure 16-1. Two systems are separated *(a)* by a stationary adiabatic wall which inhibits their interaction and *(b)* by a stationary diathermic wall which allows them to interact.

The concept of thermal equilibrium can be illustrated by considering two systems that are separated first with one kind of barrier and then with another kind. If the systems are separated with a so-called *adiabatic wall,* then the systems are insulated from each other—they are *thermally isolated* (Fig. 16-1*a*). When we change the variables of state of one of the systems, the variables of state of the other remain unaffected. On the other hand, if the systems are separated with a *diathermic wall,* then the systems interact. That is, when the state variables of one system are changed, the state variables of the other also change in response (Fig. 16-1*b*). The systems are in *thermal contact.* In practice, an adiabatic wall can be approximated by a thick layer of insulating material, such as Styrofoam, and a diathermic wall can be constructed with a thin layer of metal, such as copper.

Suppose two systems *A* and *B* are allowed to interact through a diathermic wall, as seen in Fig. 16-2. The enclosing adiabatic wall prevents any thermal contact with systems outside the enclosure. As a result of the interaction between *A* and *B*, some of the variables of each system will change. Eventually, however, these variables will settle down to constant values, and each system will achieve an equilibrium state. The two systems *A* and *B* are then said to be in thermal equilibrium.

> Two systems are in ***thermal equilibrium*** if, when put in contact through a diathermic wall, their variables of state do not change.

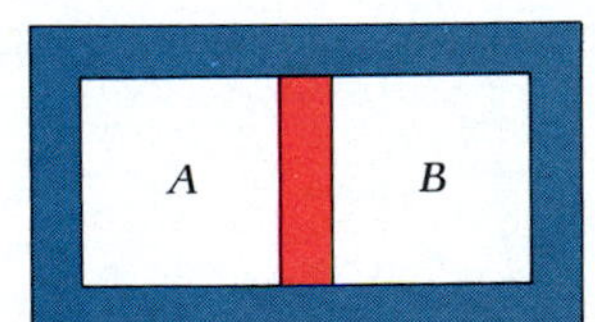

Figure 16-2. Two systems interact through a diathermic wall. The adiabatic cover insulates *A* and *B* from the outside.

The Zeroth Law

Two systems can be in thermal equilibrium even if they are not in direct contact. Figure 16-3 shows an adiabatic wall separating systems *A* and *B*, though each is in contact with a third system *C* through a diathermic wall. After a sufficiently long time, the variables of each system become constant. Therefore systems *A* and *C* are in thermal equilibrium, and systems *B* and *C* are in thermal equilibrium. Experiment indicates that systems *A* and *B* are also in thermal equilibrium. This result is contained in the statement of the ***zeroth law of thermodynamics:***

> Two systems in thermal equilibrium with a third system are in thermal equilibrium with each other.

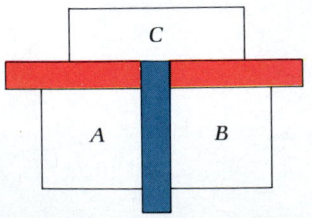

Figure 16-3. An adiabatic wall separates *A* and *B*. Each interacts with *C* through a diathermic wall.

This statement may seem obvious. That it is not logically necessary is suggested by the lovers' triangle: Heathcliff loves Shirley and Garfield loves Shirley, but Heathcliff and Garfield do not love each other. The need for the zeroth law was recognized only after the first law of thermodynamics had been given its name. The unusual term "zeroth" was therefore used to indicate that this law precedes the first law.

In the next section, we shall define temperature by specifying a procedure for its measurement. However, it is often helpful to think of the temperature of a system as a quantity that is related to the random motion of the molecules of the system. This relation will be developed in Chap. 18. For now, we just associate an increase in the temperature of a system with an increase in average molecular speed and in average molecular kinetic energy.

Temperature and thermal equilibrium

The concept of temperature is intimately related to the state of thermal equilibrium of two systems. *Two systems in thermal equilibrium have the same temperature.* If two systems are placed in contact and their variables change, then the systems are not at the same temperature, but they will come to a common temperature as thermal equilibrium is achieved. Suppose that one of these systems is a thermometer used to measure the temperature. After the thermometer comes to thermal equilibrium with the other system, the thermometer has the same temperature as the other system. We actually measure the temperature of the thermometer! The other system has the same temperature because it is in thermal equilibrium with the thermometer.

16-3 THERMOMETERS AND THE IDEAL GAS TEMPERATURE SCALE

Temperature is described subjectively by terms such as "hot," "warm," and "cold." That is, "hot" corresponds to a higher temperature than "cold." These terms do not have consistent objective meaning. For example, suppose that you are served a bowl of hot soup and a glass of cold water. After they have been sitting for a while, you complain that your soup is cold and that your water is warm. Even so, a thermometer will show that the soup still has a higher temperature than the water.

Temperatures are expressed quantitatively using a scale that is established arbitrarily by specifying the procedure by which the temperature measurement is performed—in short, by specifying what thermometer is used.

Thermometers

Some desirable properties of a thermometer

A thermometer is any device or system that is used to connect a value of one of its variables to the temperature. There are some desirable features that a thermometer should possess. It should have a property, such as a length, a pressure, or an electrical resistance, that varies with temperature in an easily measured way. Its readings should be reproducible. Its construction should be easily duplicated so that like thermometers can be used throughout the world. It should be capable of reading temperatures over a wide range. Real thermometers have these features in varying degrees.

Mercury thermometer

A familiar example of a thermometer is the *mercury-in-glass thermometer* in which the length of the column of mercury in the glass capillary indicates the temperature. The temperature is taken to have a linear dependence on the length of the mercury column. Graduations can be marked at equal-length intervals on the glass to indicate values of temperature between two fixed points. The normal melting point and normal boiling point of water, for example, could be chosen as the fixed points. (The adjective "normal" implies that the system is at normal atmospheric pressure of 101 kPa.)

Thermocouple thermometer

Resistance thermometer

A *thermocouple* is another type of thermometer. It consists of two junctions of two different metallic wires. If one junction is maintained at a reference temperature (such as in a liquid-water-and-ice bath) and the other junction is at another temperature, then an electric potential difference, or voltage (to be discussed in Chap. 22), exists between the junctions. This potential difference is used to measure the temperature. Another type of electrical thermometer is a *resistance thermometer.* The electrical resistance of the thermometer varies with temperature and is used to indicate the temperature.

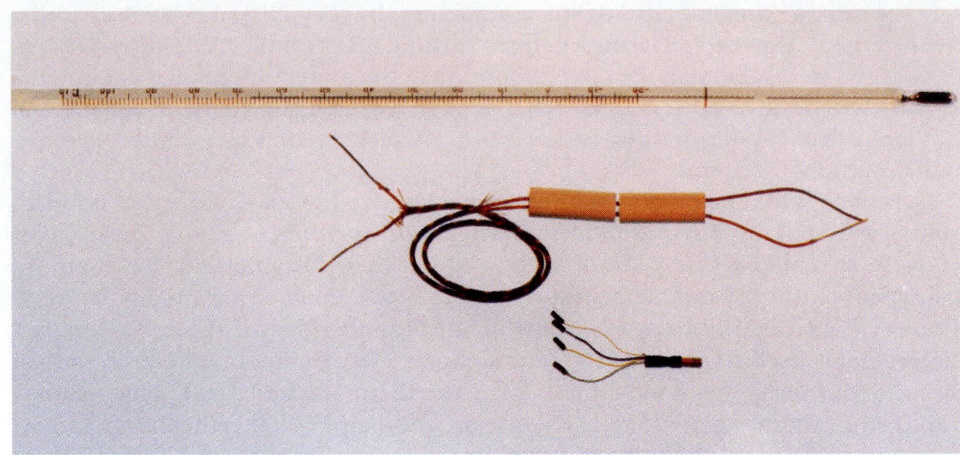

Three common types of thermometers. A mercury-in-glass thermometer uses the difference in expansity of mercury relative to glass. A thermocouple uses the potential difference between junctions of wires of different types of metals. A resistance thermometer uses the temperature-dependence of electrical resistance.

Temperature is proportional to pressure in a constant-volume gas thermometer.

A *constant-volume gas thermometer* is shown in Fig. 16-4. It consists of a bulb containing a gas with a mechanism for ensuring that the volume occupied by the gas remains constant — hence the name "constant-volume gas thermometer." The pressure of the gas is measured and is used to determine the temperature by choosing the temperature to be proportional to the pressure of the gas.

Suppose that the thermometers listed above are calibrated to give identical readings at some fixed point, and then the thermometers are used to measure the temperature of some other system. It is likely that each thermometer would give a slightly different reading. In such a case, the value of the measured temperature depends on the thermometer used to measure it, and this is an unsatisfactory result. In Chap. 19 we shall define a temperature, called the *thermodynamic temperature,* which is independent of the properties of any material. Until then, we work with temperatures as measured with a constant-volume gas thermometer. As we shall see, the use of this thermometer will lead to a temperature measurement that is virtually independent of which gas is used in these thermometers.

Constant-Volume Gas Thermometer. Suppose we compare the temperatures of two systems A and B with a constant-volume gas thermometer. First we put the

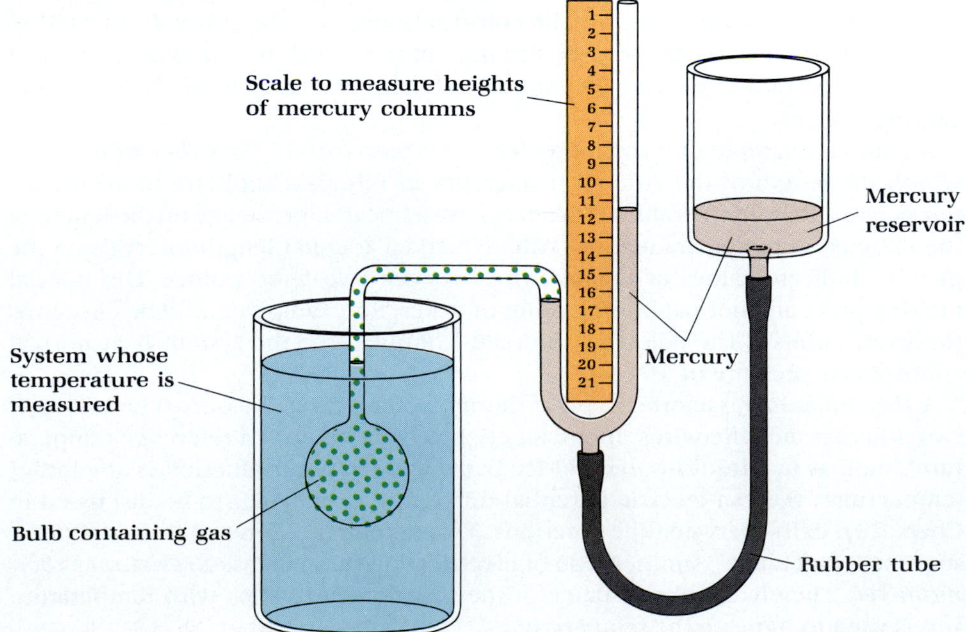

Figure 16-4. A constant-volume gas thermometer. One way to construct such a thermometer is in conjunction with a mercury manometer. A bulb containing the gas is placed in thermal contact with the system whose temperature is measured and they are allowed to come into thermal equilibrium. The mercury level is adjusted to assure that the gas occupies the same volume during each measurement, and the pressure is determined from the difference in the mercury levels on each side of the manometer.

thermometer in contact with system A and allow it to come to thermal equilibrium with A. System A and the gas in the thermometer then have the same temperature, call it T_A. The pressure of the gas in the thermometer bulb is measured; let p_A denote its value. Next the thermometer is allowed to come to thermal equilibrium with system B. The temperature T_B corresponds to the value p_B of the gas pressure of the thermometer. Since the temperature has been chosen to be proportional to the pressure of the gas in the thermometer, we have

$$\frac{T_B}{T_A} = \frac{p_B}{p_A}$$

The kelvin temperature scale is named in honor of Lord Kelvin (1824 – 1907). Born as William Thomson in Belfast, Ireland, Kelvin made significant contributions both to thermodynamics and to electromagnetism. A famous remark by Kelvin is the following: "I often say that when you can measure what you are speaking about, and express it in numbers, you should know something about it, but when you cannot express it in numbers, your knowledge is of a meager and unsatisfactory kind."

For example, if the pressure ratio is $p_B/p_A = 2.1$, then the temperature of system B is 2.1 times the temperature of system A. Note, however, that we do not have a value for either temperature. The scale has not yet been established.

To establish a temperature scale, a numerical value of the temperature is assigned to some fixed point, a definite state of some system. By convention, the ***triple point of water*** is chosen as that fixed point. We are all familiar with a system consisting of ice and liquid water. Two phases, solid and liquid, can coexist in equilibrium at a temperature called the ***melting point.*** Similarly, the liquid and vapor (steam) phases are in equilibrium at the ***boiling point.*** The ***sublimation point*** corresponds to the solid and vapor phases in equilibrium. The temperatures of these points depend on the pressure of the liquid or vapor phases, as indicated in Fig. 16-5, which shows temperatures on the Celsius scale. Lowering the pressure lowers the temperature of the boiling point, for example. There is a unique temperature of the system of water at which all three phases are in equilibrium. For this state the melting point, boiling point, and sublimation point all coincide. This state is the ***triple point*** of water. It occurs at a pressure of 610 Pa and at a temperature of 0.01°C.

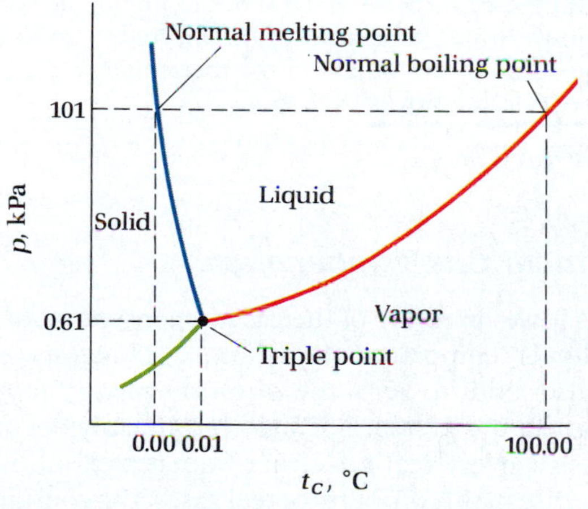

Figure 16-5. A portion of the phase diagram for water is shown. The scales are *not* linear. At the triple point the solid, liquid, and vapor phases are all in equilibrium.

Solid, liquid, and vapor phases are in equilibrium at the triple point.

The temperature scale obtained from the constant-volume gas thermometer corresponds to an arbitrary but convenient choice of the triple-point temperature. The triple-point temperature T_3 is chosen to be

$$T_3 = 273.16 \text{ K} \tag{16-1}$$

Triple-point temperature

where the unit abbreviation "K" stands for kelvin, the SI unit of temperature. The convenience of this particular numerical value, $T_3 = 273.16$ K, is that a temperature interval of 1 K corresponds to one division, or degree, on the Celsius scale. Thus the Celsius degree and the Kelvin "degree" are the same size. The normal melting point of water on the Kelvin scale is 273.15 K (0.00°C), and the normal boiling point is 100 K higher at 373.15 K (100.00°C).

The kelvin (K) and the Celsius degree have the same size.

To calibrate the constant-volume gas thermometer at the triple point of water, we allow the thermometer to come to thermal equilibrium with water at its triple point. The pressure p_3 of the gas in the thermometer is measured at this temperature T_3. To measure the temperature of some system, we allow the thermometer to come to thermal equilibrium with that system. Letting T denote the temperature of that system and p denote the pressure of the gas in the thermometer at that temperature, we have from Eq. 16-1

$$\frac{T}{T_3} = \frac{p}{p_3}$$

or

$$T = (273.16 \text{ K}) \frac{p}{p_3} \tag{16-2}$$

Once the pressure of the gas has been determined when it is in thermal equilibrium with a system of water at its triple point, the temperature of some other system is determined from Eq. 16-2 by measuring the pressure of the gas when it is in thermal equilibrium with that system.

EXAMPLE 16-3

Measuring the boiling point of nitrogen. When in thermal equilibrium at the triple point of water, the pressure of He in a constant-volume gas thermometer is 1020 Pa. The pressure of the He is 288 Pa when the thermometer is in thermal equilibrium with liquid nitrogen at its normal boiling point. What is the normal boiling point of nitrogen as measured using this thermometer?

Solution. Substituting the pressure values in Eq. 16-2 gives

$$T = (273.16 \text{ K}) \frac{288}{1020} = 77.1 \text{ K}$$

SELF-TEST 16-3. In the above example, the constant-volume gas thermometer is used to measure the temperature of an "antifreeze" solution for an automobile radiator. What is the pressure of the He gas in the thermometer if the temperature of the antifreeze is 248 K ($-25°$C)? ***ANSWER:*** 926 Pa.

Ideal Gas Temperature

The absolute zero is the lowest temperature.

Real gases change phase above the absolute zero.

A lower pressure of the gas in a constant-volume gas thermometer corresponds to a lower temperature. Equation 16-2 suggests that, as p tends to zero, the temperature also tends to zero, the *absolute zero of temperature.* At this lowest of temperatures, the pressure of a gas presumably becomes zero. As a practical matter, however, every real gas either liquefies or solidifies before this point is reached. Because of this property of real gases, the constant-volume gas thermometer cannot be used to measure temperatures below about 1 K.

There is another problem with the constant-volume gas thermometer: Identically constructed thermometers give different temperature readings if different gases or different amounts of the same gas are used. These differences in temperature readings are usually very small. To assign a single value of the temperature, an extrapolation technique is employed. Suppose that measurements are made with a set of constant-volume gas thermometers containing smaller and smaller amounts of gas. A smaller amount of gas corresponds to a smaller value of the pressure p_3 of the gas at the triple point. The results of measuring a temperature with this set of thermometers are shown graphically in Fig. 16-6. The thermometers are distinguished by the type of gas and by the amount of gas as indicated by the value of p_3. If the measurements for a given gas are extrapolated down to a zero value of p_3, these extrapolations yield the same value of temperature, independent of the type of gas. *In the limit of infinitely dilute gas ($p_3 \to 0$), the temperature determined by the constant-volume gas thermometer is the same for all gases.*

Figure 16-6. The temperature of the normal boiling point of water is measured with a set of constant-volume gas themometers. The thermometers contain different gases or different amounts of a given gas. Each thermometer is characterized by the pressure of its gas p_3 at the triple point. By extrapolation to $p_3 = 0$, the temperature is found to be 373.15 K.

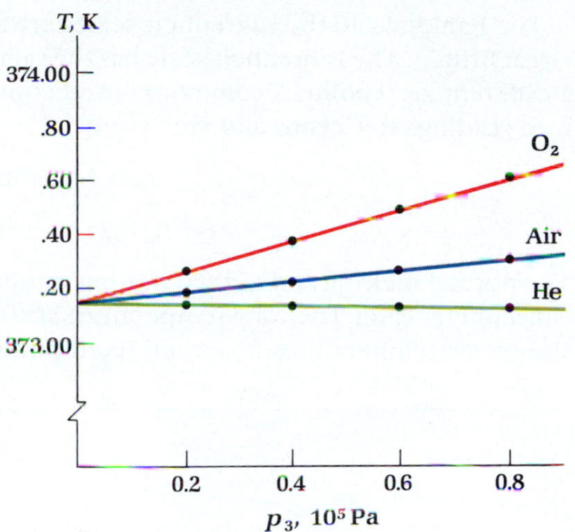

Thus Eq. 16-2 is modified to include this result: The ***ideal gas temperature*** is defined by

$$T = \lim_{p_3 \to 0} \left(\frac{p}{p_3}\right)(273.16 \text{ K}) \qquad (16\text{-}3)$$

Ideal gas temperature

where the limit is taken by performing the extrapolation process described above. Over the temperature range in which the gas thermometer can be used, the ideal gas temperature is independent of any particular gas and is identical with the thermodynamic temperature that we discuss later.

The Ideal Gas. In this dilute limit, real gases occupying the same volume have the same dependence of temperature on pressure; T and p are proportional, with the same constant of proportionality for all gases at a constant volume. It is useful to consider a fictitious gas whose proportional dependence of temperature on pressure holds not only in the dilute case but at any pressure. The ***ideal gas*** is the name given to this imaginary gas. We shall learn more about the ideal gas in the next three chapters. For now we note that its properties are approximated by real dilute gases and that a temperature scale, based on the common properties of dilute gases, bears its name.

For the ideal gas at constant volume, T is proportional to p.

16-4 OTHER TEMPERATURE SCALES

Historically the Celsius temperature scale was based on the properties of water. The temperature of the normal melting point corresponded to 0°C and that of the normal boiling point to 100°C. Now this scale is defined by

$$t_C = T - 273.15 \text{ K} \qquad (16\text{-}4)$$

Celsius temperature

where the Celsius temperature is denoted by t_C. Notice from Eq. 16-4 that the Celsius and Kelvin scales differ only in their zero point. The absolute zero ($T = 0$ K) is at $t_C = -273.15$°C. The normal melting point of water ($t_C = 0.00$°C) is at $T = 273.15$ K. Since the size of the "degree" is the same on the two scales, temperature differences have the same numerical value.

The Rankine temperature scale has the same zero point as the Kelvin scale, but the size of the degree is smaller. With T_R representing a temperature on the Rankine scale, its connection with the Kelvin temperature is

$$T_R = \tfrac{9}{5}T \qquad (16\text{-}5)$$

Rankine temperature

For example, the normal boiling point of water on the Rankine scale is $T_R = (9/5)(373 \text{ K}) = 671$°R.

The Rankine and the Fahrenheit scales are still in use in the United States and in Great Britain. The Fahrenheit scale has the same degree size as the Rankine scale but a different zero point. A commonly used conversion changes Fahrenheit temperature readings to Celsius and vice versa:

Fahrenheit temperature

$$t_F = \tfrac{9}{5}t_C + 32°\text{F}$$
$$t_C = \tfrac{5}{9}(t_F - 32°\text{F}) \qquad (16\text{-}6)$$

The normal melting and normal boiling points of water are at 32 and 212°F on the Fahrenheit scale. The four temperature scales are displayed in Fig. 16-7, which shows the temperatures of several fixed points.

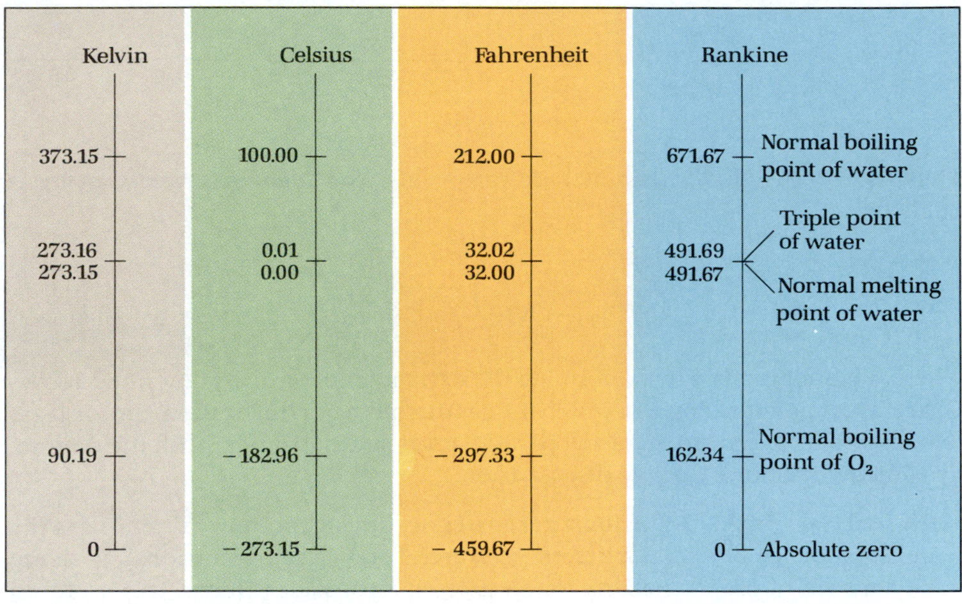

Figure 16-7. A comparison of some temperature values on four temperature scales.

16-5 THERMAL EXPANSION

Most substances expand with increasing temperature and contract with decreasing temperature. This *thermal expansion* is usually quite small, but it can be an important effect. Suppose the length of a solid rod is L_0 at some reference temperature T_0. If the temperature is changed by an amount $\Delta T = T - T_0$, then the length changes by an amount $\Delta L = L - L_0$. Experiment shows that under usual circumstances the change in length is proportional to the temperature change, at least for a small temperature change. We expect that the change in length should be proportional to the reference length L_0. That is, if the change in length of a rod 2 m long is 0.4 mm, then the change in length of a 1-m rod should be 0.2 mm. The change in length also depends on the type of material. For example, copper and iron rods of equal length at one temperature have different lengths at other temperatures.

These features can be put into equation form by introducing a coefficient that is characteristic of the material. The average ***coefficient of linear expansion*** is denoted by α. The change in length ΔL for a temperature change ΔT is given by

$$\Delta L = \alpha L_0 \, \Delta T \qquad (16\text{-}7)$$

Although α depends on the temperature interval ΔT and the reference temperature T_0, that dependence is usually negligible for moderate temperature changes. The coefficient α does not depend on the length L_0. The dimension of α is reciprocal temperature, and the commonly used unit is reciprocal degrees Celsius ($°\text{C}^{-1}$). Note that this unit is the same as the SI unit, reciprocal kelvin (K^{-1}), because we are using temperature changes. Table 16-1 lists the values of α for several common substances.

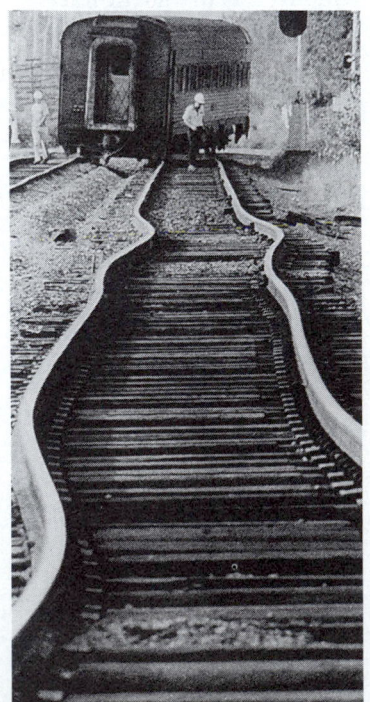

Railroad tracks expand to the point of buckling on a hot summer day.

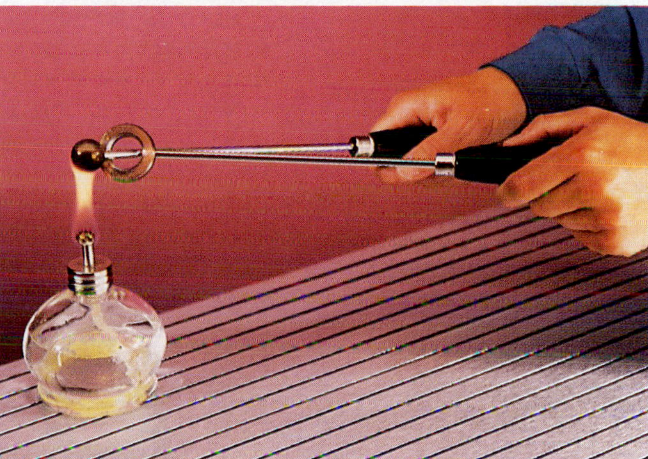

(Left) A bimetallic strip bends as its temperature is increased. The strip is a composite of two strips of different metals bonded together. Why does it bend? *(Right)* Ball-and-ring thermal expansion demonstration. The ball barely fits through the ring when both are at room temperature. If the temperature of the ball alone is increased, it will not fit through the ring. If the temperatures of both the ball and the ring are increased, the ball again fits through the ring. This shows that when the ring expands, the size of the hole increases.

Our discussion of thermal expansion has been based on the change in length of a rod, but Eq. 16-7 applies to any linear dimension, such as the diameter of a cylinder or even the radius of a circular hole in a plate. You can think of thermal expansion as analogous to a photographic enlargement in which every linear feature of an isotropic substance changes proportionally. (An isotropic substance has the same properties in all directions.)

We can understand why, on a microscopic level, a typical solid expands with an increase in temperature. In a solid, neighboring atoms exert springlike forces on each other and undergo vibrational motions. At a given temperature, a typical molecule oscillates about its average position as indicated schematically in Fig. 16-8. An effective-potential-energy function of two adjacent atoms separated by a distance r is shown. This function is asymmetric about its minimum, and the average separation depends on the energy of the molecule. We associate an increase in temperature with an increase in the average molecular energy. With an increase in energy, the average separation of molecules increases. This effect, when applied to the atoms in a solid, gives rise to thermal expansion.

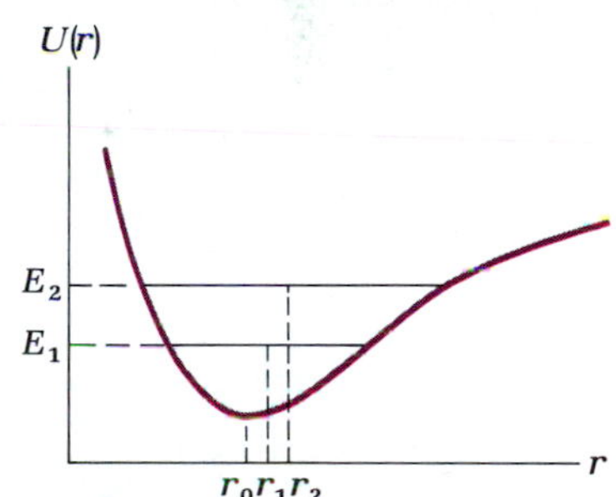

Figure 16-8. The potential energy of interaction of two neighboring atoms is asymmetric about the minimum at r_0. The average separation is r_1 when the vibrational energy is E_1. The average separation increases to r_2 if the vibrational energy increases to E_2 because of a temperature increase.

For liquids, as well as for solids, it is convenient to consider volume changes that correspond to temperature changes. If V_0 is the volume of a substance at a reference temperature T_0, then the change in volume ΔV that accompanies a temperature change ΔT is given by

$$\Delta V = \beta V_0 \, \Delta T \tag{16-8}$$

Coefficient of volume expansion β

where β is the average ***coefficient of volume expansion.*** Its value is characteristic of the particular substance. Values of β for some liquids are listed in Table 16-1.

TABLE 16-1. ***Some Expansion Coefficient Values***

Linear expansion		Volume expansion	
Substance (solid)	**α, 10^{-5} °C^{-1}**	**Substance (liquid)**	**β, 10^{-5} °C^{-1}**
Aluminum	2.4	Methanol	113
Copper	1.8	Glycerin	49
Steel	1.1	Mercury	18
Glass	0.1 – 1.3	Turpentine	90
Concrete	0.7 – 1.4	Acetone	132

Since the product of three linear dimensions gives a volume, it is not surprising that linear expansion and volume expansion are related. The result of Prob. 2 shows that $\beta = 3\alpha$ for an isotropic substance.

Notable by its absence from Table 16-1 is liquid water. The positive values of α and β for the substances in that table indicate that they expand with increasing temperature. Water also expands (but not linearly) with a temperature increase in the temperature range from about 4 to 100°C. However, between 0°C and about 4°C, water contracts with a temperature increase. This behavior is shown in Fig. 16-9, in which the volume of 1 kg of water is plotted versus temperature. The inset shows the region of smallest volume (largest density) round 4°C. This variation of volume or of density with temperature is responsible for the stratification that sometimes occurs in large bodies of fresh water. The anomalous thermal expansion of water is ultimately due to the interaction of the unusually shaped water molecules.

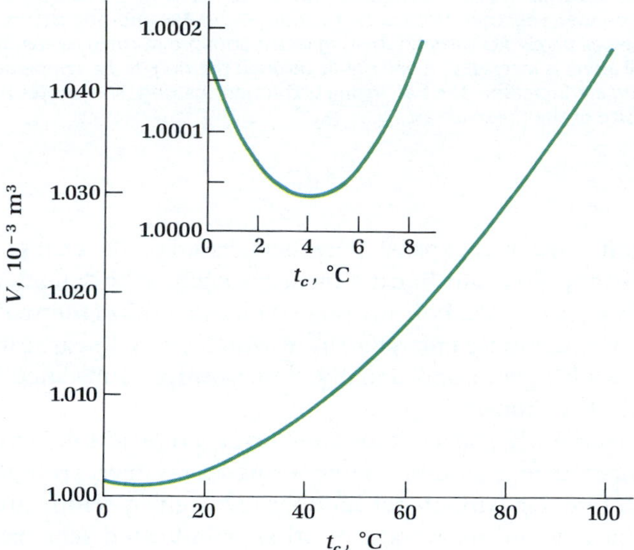

Figure 16-9. The temperature dependence of the volume of 1.000 kg of water is shown. The inset shows the region around 4°C where the density of water is a maximum.

EXAMPLE 16-4

Expanding concrete. A concrete slab has length of 12 m at −5°C on a winter day. What change in length occurs from winter to summer, when the temperature is 35°C?

Solution. From Table 16-1, the coefficient of linear expansion for concrete is around $1 \times 10^{-5}\,°C^{-1}$. Using Eq. 16-7, we have

$$\begin{aligned}\Delta L &= \alpha L_0\, \Delta T \\ &= (1 \times 10^{-5}\,°C^{-1})(12\ \text{m})(40°C) \\ &= 5\ \text{mm}\end{aligned}$$

Adjacent slabs in highways and in sidewalks are often separated by pliable spacers to allow for this kind of expansion.

SELF-TEST 16-4. A copper rod lengthens by 5 mm when its temperature increases by 40°C. What is the original length of the rod? ***ANSWER:*** 7 m.

EXAMPLE 16-5

Volume expansion of a sphere. An aluminum sphere has a radium R of 3.000 mm at 100.0°C. What is its volume at 0.0°C?

Solution. The volume of the sphere ($4\pi R^3/3$) at 100°C is $V = 113.1\ \text{mm}^3$. From Table 16-1, $\alpha = 2.4 \times 10^{-5}\,°C^{-1}$ and $\beta = 3\alpha = 7.2 \times 10^{-5}\,°C^{-1}$. Applying Eq. 16-8, we obtain

$$\Delta V = (7.2 \times 10^{-5}\,°C^{-1})(113.1\text{ mm}^3)(-100°C)$$
$$= -0.81\text{ mm}^3$$

The volume at 0°C is $113.1\text{ mm}^3 - 0.8\text{ mm}^3 = 112.3\text{ mm}^3$.

An alternative approach is to evaluate the radius of the sphere (a linear dimension) at 0°C and calculate the volume from $V = 4\pi R^3/3$. (In Exercise 21, you will be asked to show that you get the same answer this way.)

SELF-TEST 16-5. What temperature change would cause the volume of mercury to change by 0.1 percent? ***ANSWER:*** 6 K.

16-6 HEAT TRANSFER

Suppose we wish to increase the temperature of a flask of water. The procedure would be, in everyday language, to "heat it up." We could, for example, apply a flame to the flask, or we could drop a hot object into the water. In either case there is a transfer of energy to the water, and the energy transfer occurs because there is a temperature difference between the water and some part of its surroundings. This observation is the basis of our definition of the term "heat."

> ***Heat*** is the energy transferred between a system and its surroundings due solely to a temperature difference between that system and some part of its surroundings.

Definition of heat

The reference to a temperature difference is an essential part of the definition of heat because energy can also be transferred in other ways (to be described in the next chapter). To qualify as heat, the energy added to or removed from a system must have been transferred directly and solely because of a difference of temperature between the system and its environment.

How is this energy transferred at the molecular level? In the process of ***heat conduction,*** described more fully below, energy is transferred in the collisions of the randomly moving molecules of a substance. Consider a rod with a temperature difference between its ends. Molecules at the higher temperature end will be moving faster on average than molecules at the lower temperature end. In a typical collision, the slower molecule will gain energy, and the faster molecule will lose energy. Averaged over many collisions involving the molecules all along the rod, there is a net transfer of energy—heat—because of this temperature difference.

The proper use of the term "heat" is as an amount of energy transferred to or from a system. It is not an energy that resides in a system or belongs to a system as potential energy does. It is therefore *incorrect* to speak of the "heat in a system" or the "heat of a system." Rather we speak of the "heat added to a system" or the "heat extracted from a system." The symbol Q is used to represent heat. As an energy transfer, heat has dimensions of energy. The SI unit of heat is the joule (J). Other heat units are also in common usage, and they will be introduced as the need arises.

How is heat transferred between a system and its environment or, for simplicity, between two systems? The processes by which heat is transferred are classified into three categories: *conduction, convection,* and *radiation.* In some situations, only one of these mechanisms may be significantly operative. But often two or all three processes may be contributing significantly. We shall discuss them separately, beginning with heat conduction.

Heat Conduction

In the conduction process, heat is transferred between two systems through a connecting medium. We assume that no part of the medium is moving. Thus the medium must be a rigid solid, or if fluid, it must have no circulating currents. Consider the situation shown schematically in Fig. 16-10. A uniform rod of cross-

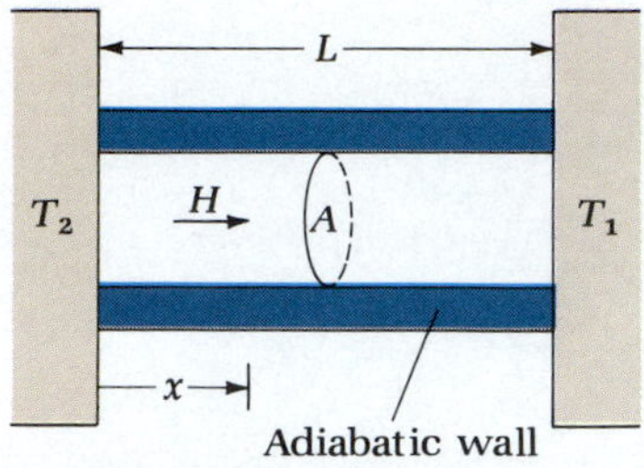

Figure 16-10. A uniform rod conducts heat from a higher temperature T_2 to a lower temperature T_1. The lateral surface of the rod is insulated.

sectional area A and length L is the medium separating two systems maintained at temperatures T_1 and T_2. Heat is transferred through the medium from the higher temperature, say T_2, to the lower temperature. An adiabatic wall covering the lateral surface of the rod prevents any flow of heat from the surface.

We can expect the temperature to vary along the length of the rod. The temperature should be T_2 at the left end ($x = 0$) and T_1 at the other end ($x = L$). At an intermediate point in the rod, located by x, the temperature should be between T_1 and T_2 and generally is changing with time. Some temperature profiles are sketched in Fig. 16-11. Experiment shows that after *steady-state* conditions are achieved (the temperature at any given point no longer is changing), the temperature varies linearly along the length of the rod if T_1 and T_2 are not greatly different.

For steady-state heat flow in the rod, the heat Q flowing through a cross section of the rod in a time interval Δt is the same all along the rod. Thus, under steady-state conditions, the energy is transferred through the medium, and no part of the medium is gaining or losing energy. We let the ***heat current*** H be the heat per unit time flowing through a cross section:

Heat current

$$H = \frac{Q}{\Delta t}$$

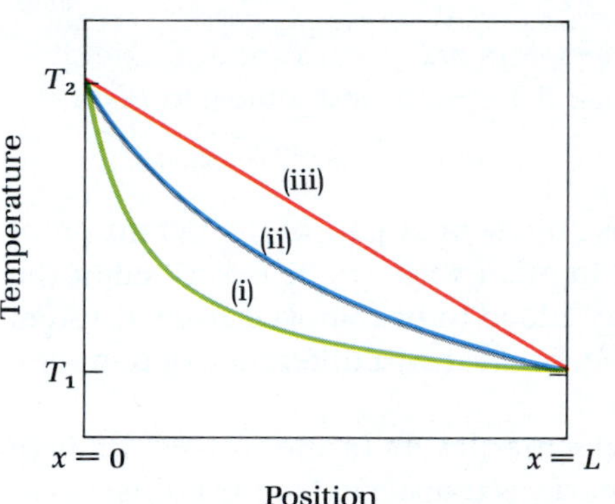

Figure 16-11. Temperature profiles are shown at different times for the rod of Fig. 16-10, where the rod's temperature was initially uniform at T_1. Profile (i) is for shortly after the heat conduction begins and profile (ii) is somewhat later. After a sufficiently long time, the steady-state profile (iii) is achieved.

The steady-state heat current H then has the same value everywhere along the uniform rod. From experiment we find that the steady-state heat current H in the rod is (i) proportional to the temperature difference $T_2 - T_1$, (ii) proportional to the cross-sectional area A, (iii) inversely proportional to the length L, and (iv) dependent on the material. These features are summarized by the expression

$$H = kA\frac{T_2 - T_1}{L} \tag{16-9}$$

Thermal conductivity

where k is the ***thermal conductivity*** characteristic of the material (but not the size or shape) of the rod. The temperature dependence of the thermal conductivity of most substances is usually slight. The SI unit of heat current is the watt (W), and the SI unit of thermal conductivity is $W \cdot m^{-1} \cdot K^{-1}$. Table 16-2 lists values of the thermal conductivity for some common materials.

TABLE 16-2. ***Some Values of Thermal Conductivity and Some R-Values***

Substance	k, $W \cdot m^{-1} \cdot K^{-1}$	Material of given thickness	R-value, $°F \cdot ft^2 \cdot h \cdot Btu^{-1}$
Aluminum	237	Asphalt shingles	0.44
Copper	401	Plasterboard, $\frac{1}{2}$ inch	0.45
Concrete	0.9–1.3	Plywood, $\frac{3}{4}$ inch	0.95
Wood	0.05–0.36	Brick veneer	0.6–1.3
Fiberglass batt	0.04	Styrofoam panel, $\frac{3}{4}$ inch	6.3
Air	0.02	Fiberglass roll, 3.5 inch	11
Styrofoam	0.01		

From Eq. 16-9, we see that the heat current can be large for a material with a large value of thermal conductivity. Such a substance is called a "good" heat conductor and can be used to approximate a diathermic wall. Metals typically are good heat conductors; copper is one of the best. In contrast, a substance with a small value of thermal conductivity would be a poor heat conductor and a good insulator. A thick insulating layer approximates an adiabatic wall. Styrofoam is an excellent insulator (below its melting point of course).

A good heat conductor has a large thermal conductivity; a good insulator has a small thermal conductivity.

Equation 16-9 can be generalized to describe heat conduction in non-steady-state conditions and in a variety of geometries. At a point in a medium characterized by thermal conductivity k, let H represent the instantaneous heat current through a small area element A. If x is a coordinate perpendicular to the plane of the area and the temperature varies with that coordinate, then dT/dx is the ***temperature gradient*** at that position. These quantities are related by

Temperature gradient

$$H = -kA\frac{dT}{dx} \qquad (16\text{-}10)$$

Heat-conduction equation

The negative sign in Eq. 16-10 corresponds to the fact that heat flows from higher to lower temperature. Thus, if the temperature is decreasing in the direction of increasing x, then $-dT/dx$ is positive and heat flows in the positive x direction.

EXAMPLE 16-6

Heat current in a refrigerator wall. Estimate the heat current in the Styrofoam insulation in the walls of a kitchen refrigerator. How much heat flows through these walls in an hour?

Solution. The total wall area of a refrigerator is around 4 m², and the temperature difference between inside and outside is about $25°C - 5°C = 20°C$. We estimate the Styrofoam thickness at 30 mm. Using the thermal conductivity of Styrofoam from Table 16-2, we have from Eq. 16-9,

$$H = \frac{(0.01\ \mathrm{W\cdot m^{-1}\cdot K^{-1}})(4\ \mathrm{m^2})(20°\mathrm{C})}{0.03\ \mathrm{m}}$$

$$= 30\ \mathrm{W}$$

In an hour the heat would be

$$Q = H\,\Delta t = (30\ \mathrm{W})(3600\ \mathrm{s}) = 100\ \mathrm{kJ}$$

SELF-TEST 16-6. Suppose that the refrigerator in the above example uses a 30-mm thickness of fiberglass insulation instead of Styrofoam. What is the heat current in the fiberglass?
ANSWER: 100 W.

EXAMPLE 16-7

Temperature gradient. Evaluate the temperature gradient in the Styrofoam insulation of Example 16-6.

Solution. Select the positive x axis to be perpendicular to one of the walls and directed from outside the refrigerator to the inside. The temperature profile for this steady-state heat flow is illustrated by line (iii) in Fig. 16-11. The temperature gradient is the slope of the straight line in the figure:

$$\frac{dT}{dx} = \frac{-20°\mathrm{C}}{0.03\ \mathrm{m}} = -700°\mathrm{C\cdot m^{-1}}$$

Since $dT/dx < 0$, the heat curent is in the positive x direction, or into the refrigerator.

SELF-TEST 16-7. The temperature of the air in an enclosed room varies linearly from

24°C at the floor to 29°C at the ceiling, which is 2.5 m above the floor. *(a)* Let x be directed upward and determine the temperature gradient. *(b)* Which way would heat flow by conduction? ***ANSWERS:*** *(a)* 2°C/m; *(b)* down.

R-Value. In the building-construction industry, the insulating value of materials under conditions of steady-state heat conduction is expressed as the R-value, or R. We apply Eq. 16-9 to a slab of thickness $\Delta x = L$ with a (positive) temperature difference of $\Delta T = T_2 - T_1$ across it:

$$H = \frac{kA\,\Delta T}{\Delta x}$$

We solve for the temperature difference ΔT in terms of the heat current per unit area H/A:

$$\Delta T = \frac{H}{A}\frac{\Delta x}{k}$$

R-value

The *R-value* is defined as $\Delta x/k$, and the equation above becomes

$$\Delta T = \frac{H}{A} R \qquad (16\text{-}11)$$

British thermal unit

In a system of units used in the building industry in the United States, the unit of heat is the ***British thermal unit (Btu),*** and 1 Btu = 1055 J; the unit of area is the square foot (ft^2), the unit of time is the hour (h); the unit of length (for the thickness) is the inch; and temperature difference is expressed in degrees Fahrenheit (°F). The unit of R-value, whether explicitly expressed or just understood, is $°\text{F}\cdot\text{ft}^2\cdot\text{h}\cdot\text{Btu}^{-1}$. For example, if a roll of fiberglass insulation has R 11 stamped on it, the R-value is $11\,°\text{F}\cdot\text{ft}^2\cdot\text{h}\cdot\text{Btu}^{-1}$. We shall often follow the convention of omitting this unwieldy unit when quoting an R-value. R-values for some building materials are listed in Table 16-2.

The R-value for a composite equals the sum of R-values for the layers.

Equation 16-11 and the R-value are useful in determining heat conduction through composite slab-type structures such as walls or floorings. Under steady-state conditions, the heat current per unit area is the same all along the thickness (why?), the temperature difference across the composite is the sum of temperature differences across the individual layers, and the effective R-value of the composite is simply the sum of R-values of the layers.

EXAMPLE 16-8

Temperature distribution in a composite wall. A wall consists of ½-inch plasterboard, 3½ inches of fiberglass insulation, and ¾-inch plywood. *(a)* Determine the effective R-value of the composite wall. If the inside temperature is 70°F and the outside temperature is 10°F, evaluate *(b)* the heat current per unit area in the wall and *(c)* the temperature distribution in the wall.

Solution. The composite wall is shown in cross section in Fig. 16-12*a*, with temperatures at the interfaces between layers indicated. Applying Eq. 16-11 to each layer in turn, we have

$$T_2 - T_1 = \frac{H}{A} R_c$$

$$T_3 - T_2 = \frac{H}{A} R_b$$

$$T_4 - T_3 = \frac{H}{A} R_a$$

The heat current per unit area (H/A) is the same in all layers under steady-state conditions, and there is no net gain or loss of energy in any layer. The sum of these three equations is

$$T_4 - T_1 = \frac{H}{A}(R_a + R_b + R_c)$$

$$= \frac{H}{A} R_{\text{eff}}$$

So the effective R-value R_{eff} is the sum of the individual R-values. From Table 16-2, we obtain

$$R_{\text{eff}} = 0.45 + 11 + 0.95 = 12.4\,°\text{F}\cdot\text{h}\cdot\text{ft}^2\cdot\text{Btu}^{-1}$$

(b) Using this value, we can evaluate the heat current per unit area:

$$\frac{H}{A} = \frac{T_4 - T_1}{R_{\text{eff}}} = 4.8\ \text{Btu}\cdot\text{h}^{-1}\cdot\text{ft}^{-2}$$

(c) The temperature values T_2 and T_3 can now be determined from the first two equations above:

$$T_2 = T_1 + \frac{H}{A} R_c$$

$$= 10°\text{F} + (4.8)(0.95)°\text{F} = 15°\text{F}$$

In a similar way, T_3 can be determined to be 68°F. The temperature profile is shown in Fig. 16-12*b*. Note that the largest temperature drop is across the fiberglass insulation, the component with the largest R-value.

SELF-TEST 16-8. The plywood in the above example is replaced with a $\frac{3}{4}$-inch Styrofoam panel. For this composite wall under the same conditions, determine *(a)* the effective R-value and *(b)* the heat current per unit area. ***ANSWERS:*** *(a)* 17.7°F·h·ft²·Btu⁻¹; *(b)* 3.4 Btu·h⁻¹·ft⁻².

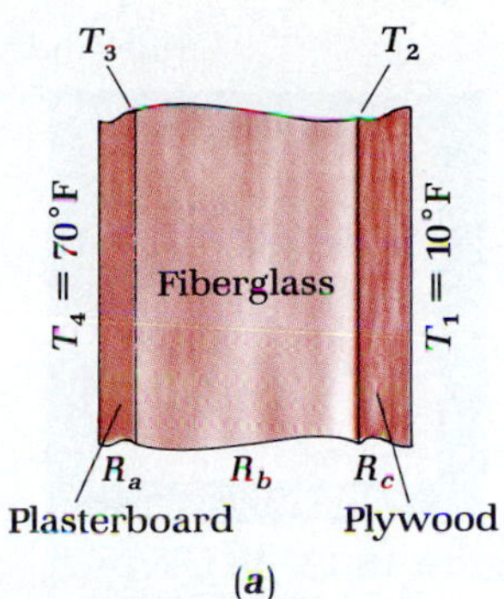

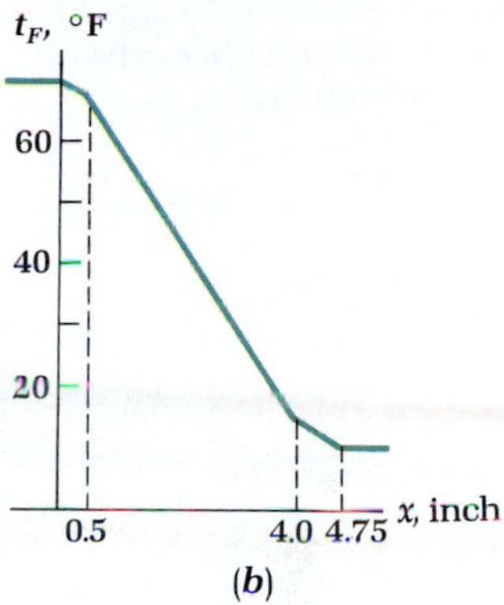

Figure 16-12. Example 16-8: *(a)* A composite wall in cross section consists of three layers: $\frac{1}{2}$-inch plasterboard, $3\frac{1}{2}$-inch fiberglass roll, and $\frac{3}{4}$-inch plywood. *(b)* The temperature profile for the wall is shown. The largest temperature drop is across the fiberglass insulation.

Heat Convection and Radiation

Convection currents

In ***convection,*** energy is transferred by macroscopic movement of matter in the form of convection currents. Such currents can occur spontaneously in fluids whose density varies with temperature. In air, for example, because of the earth's gravitational field, convection currents are established as higher-temperature (lower-density) air rises and lower-temperature (higher-density) air sinks. Forced convection is produced using blowers. Calculations of heat flow by convention are complex and will not be treated here.

An object emits energy from its surface.

Radiation, the third mechanism of heat transfer, can be the dominant mechanism of heat transfer in some situations. All objects emit energy from their surfaces. A portion of this radiant energy may easily be seen if the surface is at a high-enough temperature (such as a glowing ember). Even at much lower temperatures, a surface still emits energy, although an insignificant amount is visible. You can feel the radiation coming from a warm stove, for example.

A surface at temperature T (on the Kelvin scale) will emit radiant energy at a rate proportional to the surface area A and to the fourth power of the temperature. The expression for the radiated power P is called the *Stefan-Boltzmann law:*

The Stefan-Boltzmann law

$$P = e\sigma A T^4 \tag{16-12}$$

The Stefan-Boltzmann constant

where e is the *emissivity,* which characterizes the emitting properties of the particular surface ($0 \leq e \leq 1$), and σ is the *Stefan-Boltzmann constant,* $\sigma = 5.67 \times 10^{-8}\ \text{W}\cdot\text{m}^{-2}\cdot\text{K}^{-4}$, which is the same for all objects.

Surfaces absorb as well as emit radiation. Consider an object at temperature T_2 surrounded by walls at temperature T_1, as seen in Fig. 16-13. Experiment shows that the temperatures will become equal. Then the surface of the enclosed object and the surrounding wall surface must each emit and absorb energy at the same rate at that temperature to maintain thermal equilibrium. That is, at temperature T the surface of the enclosed object must emit and absorb energy at the same rate. Thus a

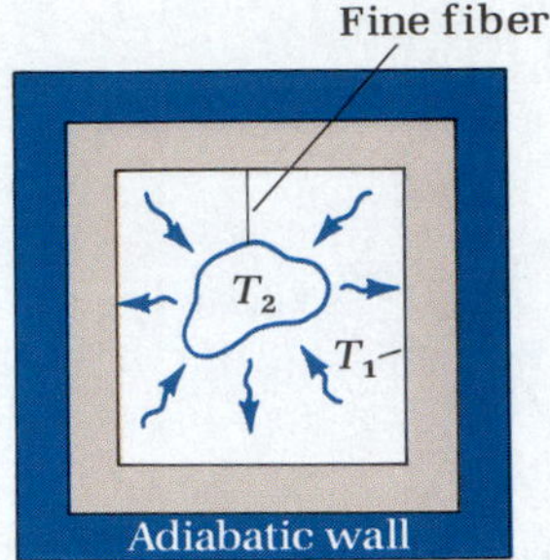

Figure 16-13. An insulated object at temperature T_2 is surrounded by walls at temperature T_1. Both surfaces absorb and emit energy.

good absorbing surface is also a good emitting surface ($e \approx 1$), and a poor absorbing (but a good reflecting) surface is a poor emitting surface ($e \approx 0$).

Suppose the object in the enclosure is maintained at a somewhat higher temperature than the surrounding walls; $T_2 > T_1$. Then its rate of energy emission is greater than its absorption rate, and the net rate of heat transfer (or heat current) H is given by

$$H = e\sigma A(T_2^4 - T_1^4) \tag{16-13}$$

Since the contributions to the heat-transfer rate depend on the fourth power of temperature, the effect can be large at high temperatures and for large temperature differences.

EXAMPLE 16-9

Radiative heat transfer. Estimate the rate of heat transfer between the bare head of a bald man (at 37°C) and the surroundings at *(a)* 20°C and *(b)* −40°C.

Solution. Approximate the head by a sphere of radius $a = 120$ mm and of emissivity $e = 1$. Since $0 \leq e \leq 1$, we are assuming that the skin is an excellent emitter and absorber. The surface area is then $A = 4\pi a^2 = 0.2\ \text{m}^2$. The temperatures must be expressed on the Kelvin scale. Equation 16-13 gives

$$H = (1)(5.67 \times 10^{-8}\ \text{W}\cdot\text{m}^{-2}\cdot\text{K}^{-4})(0.2\ \text{m}^2)[(310\ \text{K})^4 - T_1^4)]$$

(a) With $T_1 = 293$ K, we obtain

$$H = 20\ \text{W}$$

(b) For $T_1 = 233$ K, a similar calculation gives

$$H = 70\ \text{W}$$

SELF-TEST 16-9. On a clear night, the earth's surface, at about 300 K, radiates heat into dark sky, which is at 3 K. Estimate the heat current per unit area in watts per square meter leaving the earth's surface. ***ANSWER:*** 500 W/m².

SUMMARY

Section 16-1. Microscopic and Macroscopic Descriptions

The macroscopic description of a system uses thermodynamic variables of state such as pressure, volume, and temperature.

Section 16-2. Thermal Equilibrium and the Zeroth Law of Thermodynamics

Two systems in contact are in thermal equilibrium if their properties are no longer changing. The zeroth law of thermodynamics states that two systems in thermal equilibrium with a third system are in thermal equilibrium with each other. If two systems are in thermal equilibrium, they have the same temperature.

Section 16-3. Thermometers and the Ideal Gas Temperature Scale

The ideal gas temperature is defined by a limiting process using real dilute gases in constant-volume gas thermometers. The scale is set by assigning to the triple point of water the temperature 273.16 K.

Section 16-4. Other Temperature Scales

Temperatures can be expressed on different scales. The relation between the Celsius and Kelvin scales is

$$t_C = T - 273.15\ \text{K} \tag{16-4}$$

Section 16-5. Thermal Expansion

The length L_0 of an object changes with a change in temperature ΔT by an amount

$$\Delta L = \alpha L_0\, \Delta T \tag{16-7}$$

Volume changes are described by

$$\Delta V = \beta V_0\, \Delta T \tag{16-8}$$

Section 16-6. Heat Transfer

Heat Q is defined as the energy transferred between a system and its surroundings solely because of a temperature difference between the system and its surroundings. Heat conduction is one of the mechanisms of heat transfer and is described by the heat-conduction equation:

$$H = -kA\frac{dT}{dx} \tag{16-10}$$

Convection is a mechanism of heat transfer in which macroscopic convection currents are present. For heat transferred by radiation, the radiated power from a surface is given by the Stefan-Boltzmann law

$$P = e\sigma AT^4 \tag{16-12}$$

QUESTIONS

1. Describe some situations in a typical kitchen for which it would be desirable to use *(a)* an adiabatic wall and *(b)* a diathermic wall.

2. A liquid is a fluid and a gas is a fluid. How are liquids and gases different? How are they similar?

3. Is pressure a variable of state for a liquid? For a solid? Explain.

4. What is the mass of 1 mol of diatomic hydrogen H_2? How many molecules are in 1 mol? How many atoms are in 1 mol? Answer these same questions for He and for CO_2.

5. In a constant-volume gas thermometer, the pressure increases with increasing temperature. How does volume vary with temperature in a constant-pressure gas thermometer?

6. If both glass and mercury expand with increasing temperature, how does a mercury-in-glass thermometer work?

7. What difficulties are encountered in measuring very low temperatures with a constant-volume gas thermometer? What about very high temperatures?

8. Suppose a nice round number like 300 were assigned as the triple-point temperature. What changes would occur in the ideal gas temperature scale? Explain the advantage of choosing the triple-point temperature to be 273.16 K.

9. If two systems are in thermal equilibrium, they have the same temperature. Is the converse true? That is, if two systems have the same temperature, are they in thermal equilibrium? What can you say about two systems that have different temperatures?

10. Is it possible for a container of water to be freezing and boiling at the same time? Explain.

11. The outside diameter of a hollow aluminum sphere increases with increasing temperature. What happens to the inside diameter? Explain.

12. Describe the temperature variation with depth of the water in a lake *(a)* in summer and *(b)* in winter.

13. Pressure, volume, and temperature are variables of state for a thermodynamic system. Is heat a variable of state? Explain.

14. As a practical matter, there is always a temperature difference between a system and some part of its environment, however remote. Must there always be some heat transferred because of that temperature difference? Explain.

15. Suppose a wooden rod and a metal rod are both at room temperature. Which one feels cooler to the touch and why?

16. Explain why a copper teakettle is commonly fitted with a wooden handle.

17. What is the SI unit of R-value?

18. What are the significant mechanisms of heat transfer for a single-pane window? For a double-pane window?

19. Glass has a relatively high thermal conductivity. Account for the good insulating properties of fiberglass batts or rolls.

20. A thermos is double-walled, with the space between walls evacuated. What is the advantage of this construction?

21. Why are the walls of a thermos silvered, that is, highly reflecting?

22. Why are nights with clear skies usually colder than nights with cloudy skies during the same season?

23. Complete the following table:

Symbol	Represents	Type	SI Unit
p	Pressure		
V			
n			
T			K
t_C			
α			
β			
Q		Scalar	
H			
k			
R			
σ			

EXERCISES

Section 16-1. Microscopic and Macroscopic Descriptions

1. Suppose that the instantaneous values of the position and velocity of a molecule in a system can be determined from an application of Newton's laws to the system, and that the calculation time per molecule is 1 ns on a high-speed computer. Estimate the time in years that would be required to calculate the positions and velocities for all of the molecules (10^{23}) of that system.

2. If 2.4 mol of He gas occupies a volume of 82 L (1 L $= 10^{-3}$ m^3), *(a)* what is the mass of the gas? *(b)* How many molecules

are in this system? *(c)* Estimate the average separation between molecules and compare that with the size of a helium atom (about 50 pm).

3. One mole of a gas at standard temperature and pressure (STP corresponds to $T = 273$ K and $p = 101$ kPa) occupies a volume of 22.4 L. Suppose the container is in the shape of a cube. *(a)* Determine the length of the cube edge. *(b)* What force is exerted by the gas on each face of the container?

Section 16-3. Thermometers and the Ideal Gas Temperature Scale

4. Suppose that the gas in Exercise 3 comes to thermal equilibrium with water at its normal boiling point. If the volume is fixed at 22.5 L, *(a)* what is the pressure of the gas and *(b)* what force does the gas exert on each face of the container?

5. Helium gas in a constant-volume thermometer is at pressure 1439 Pa when in thermal equilibrium with water at its triple point. *(a)* What is the pressure of this gas when in thermal equilibrium with zinc at its normal melting point (693 K)? *(b)* The pressure of the gas is 406 Pa when in thermal equilibrium with a liquid at its normal boiling point. What is the temperature of this boiling point?

6. Four constant-volume gas thermometers are used to measure the temperature of the normal melting point of zinc. Each thermometer contains a different amount of the same gas, and these gases have different pressures p_3 at the triple point of water, as shown in the table below. The pressure readings p when in thermal equilibrium with zinc at its melting point are also shown in the table. From these data, construct a graph similar to Fig. 16-5 and perform the extrapolation to determine the ideal gas temperature of the normal melting point of zinc.

Thermometer	1	2	3	4
p_3, kPa	217.12	123.01	84.09	49.83
p, kPa	551.01	312.08	213.31	126.39

7. The pressure of the gas in a constant-volume gas thermometer is 24.5 mmHg when in thermal equilibrium with water at its normal boiling point. What is the pressure (in mmHg) when the gas is in thermal equilibrium with water at *(a)* its triple point, *(b)* its normal melting point, *(c)* 37°C?

8. Temperatures can be determined using a resistance thermometer in which the electrical resistance is measured. Suppose the temperature of the thermometer is proportional to its resistance. [The unit of resistance is the ohm (Ω), defined in Chap. 24.] The thermometer is calibrated by measuring the resistance to be 100.000 Ω at the triple point of water and to be 104.783 Ω at the normal boiling point of water. What is the resistance-thermometer temperature if the resistance is *(a)* 102.445 Ω? *(b)* 98.729 Ω? *(c)* What is the resistance at the normal melting point of water?

9. The length of a column of mercury in a glass capillary is 43 mm when in thermal equilibrium with the underside of a healthy person's tongue (37°C). For a person with "two degrees of fever" (39°C), the length is 67 mm. What is the temperature of a tepid bath for which the length of the column is 16 mm? State any assumptions that you make.

Section 16-4. Other Temperature Scales

10. Express the normal body temperatures of 37°C on *(a)* the Fahrenheit scale, *(b)* the Kelvin scale, *(c)* the Rankine scale.

11. At what temperature (if any) are the readings the same on *(a)* the Celsius and Fahrenheit scales, *(b)* the Kelvin and Fahrenheit scales, *(c)* the Kelvin and Rankine scales, *(d)* the Kelvin and Celsius scales?

12. The normal boiling point of helium is 4.2 K; a comfortable room temperature is 295 K; the surface of the sun is at about 6000 K; the interior of a star is about 10 MK. Express these temperatures on *(a)* the Celsius scale, *(b)* the Fahrenheit scale, *(c)* the Rankine scale.

Section 16-5. Thermal Expansion

13. A steel rule is calibrated at 22°C against a standard so that the distance between numbered divisions is 10.00 mm. *(a)* What is the distance between these divisions when the rule is at −5°C? *(b)* If a nominal length of 1 m is measured with the rule at this lower temperature, what percent error is made? *(c)* What absolute error is made for a 100-m length?

14. A copper plate at 0°C has thickness of 5.00 mm and a circular hole of radius 75.0 mm. Its temperature is raised to 220°C. Determine the values at this temperature of *(a)* the thickness of the plate, *(b)* the radius of the hole, *(c)* the circumference of the boundary of the circular hole, *(d)* the area of the hole in the plate.

15. A steel shaft has diameter 42.51 mm at 28°C. It is to be fitted to a steel pulley with a circular hole of diameter 42.50 mm at that temperature. *(a)* By how much must the temperature of the shaft be reduced so that it can fit in the hole? *(b)* Suppose the temperature of the entire structure is reduced to −5°C after the shaft has been fitted. Will the shaft come loose? Explain.

16. Rework Exercise 15, with an aluminum pulley replacing the steel pulley.

17. A simple pendulum consists of a bob attached to a fine steel wire so that the length of the pendulum is 0.2482 m at 27°C. *(a)* What is the change in the period of the pendulum (Sec. 14-4) if its temperature is changed by −5°C? *(b)* If this pendulum is used as a clock, accurate at 27°C, how many seconds does the clock gain or lose in one day because of this temperature change?

18. The density of aluminum is 2692 kg/m^3 at 20°C. *(a)* What is the mass of an aluminum sphere ($V = 4\pi R^3/3$) whose radius R at this temperature is 25.00 mm? *(b)* What is the mass of the aluminum at 100.0°C? *(c)* What is the density of aluminum at 100.0°C? *(d)* What are the answers to parts *(b)* and *(c)* if the aluminum is in the shape of a cube?

19. A glass ($\beta = 2.2 \times 10^{-5}$°C^{-1}) bulb is completely filled with 176.2 mL of mercury ($\beta = 18 \times 10^{-5}$°C^{-1}) at 0.0°C. The bulb is fitted, as illustrated in Fig. 16-14, with a glass tube of

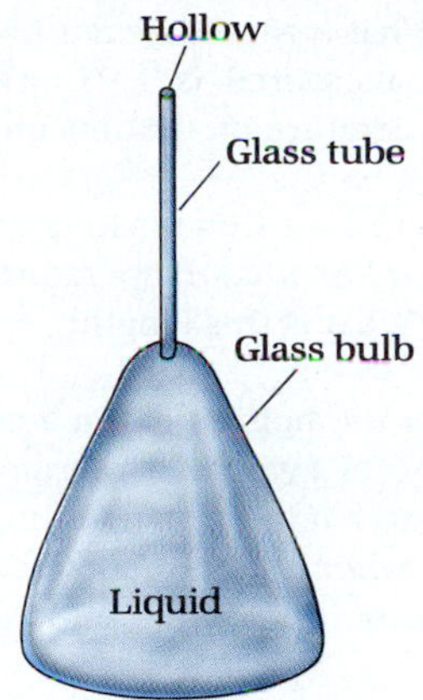

Figure 16-14. Exercise 19.

inside diameter 2.5 mm at 0.0°C. How high does the mercury rise in the tube if the temperature of the system is raised to 50.0°C? The change in diameter of the glass tube may be neglected. Why?

20. Suppose the glass bulb of Exercise 19 is filled with an oil, occupying the 176.2-mL volume at 0.0°C. At a temperature of 8.0°C, the oil has risen in the glass tube to a height of 190 mm. Evaluate the volume coefficient of expansion for this liquid.

21. Work Example 16-5 as suggested there by determining the radius of the sphere at 0.0°C and calculating the volume at that temperature.

22. A certain type of plastic with $\beta = 2 \times 10^{-7}$ °C^{-1} remains at rest if released when completely immersed in water at 6°C. What happens to the plastic if it is released when completely immersed in water at *(a)* 8°C, *(b)* 4°C, *(c)* 1°C? See Fig. 16-9.

Section 16-6. Heat Transfer

23. An aluminum pot contains water that is kept steadily boiling (100°C). The bottom surface of the pot, which is 12 mm thick and 1.5×10^4 mm² in area, is maintained at a temperature of 102°C by an electric heating unit. The remaining part of the surface is well insulated from the surroundings. Evaluate the heat current entering the water through the bottom surface.

24. Determine the SI units of R-value and use those units to express the R-value of a roll of fiberglass insulation of thickness 3.5 inch (90 mm).

25. A piece of wood, in the shape of a 350-mm by 350-mm slab of thickness 15 mm, conducts heat through this thickness under steady-state conditions. The heat current in the slab is measured to be 14.3 W when a temperature difference of 25°C is maintained across the slab. *(a)* Evaluate the temperature gradient in the slab. *(b)* Determine the thermal conductivity of this wood. *(c)* Would this material be classified as a good heat conductor or as a good heat insulator?

26. Approximate the living space of a residence by a box having a 40-ft by 40-ft floor and ceiling and 8-ft-high walls. Suppose the interior is maintained at 70°F, while the exterior surfaces of walls and ceilings are exposed to a steady 10°F and the exterior surface of the floor remains at 40°F. The wall structure has an effective R-value $R_w = 10$, the ceiling has $R_c = 15$, and the floor has $R_f = 8$, all in building-industry units. Evaluate the heat current in *(a)* ceiling, *(b)* walls, *(c)* floor. *(d)* Suppose these conditions are maintained for a 24-h period. What is the heat loss to the outside in this period? *(e)* The inside temperature is maintained by consuming fuel costing \$0.05 per 1000 Btu. What is the heating cost for the 24-h period?

27. The R-value of a building material is determined experimentally by constructing a box from the material and measuring the electric power input to a heater inside the box which maintains the inside at a given temperature. Suppose a box of total area 96 ft² is constructed of $\frac{3}{4}$-inch-thick particleboard and that a power input of 1100 W will maintain a temperature difference of 30°F between inside and outside. *(a)* What is the total heat current in the walls in units of Btu·h^{-1}? *(b)* Calculate the R-value of this particleboard. *(c)* What is the thermal conductivity of particleboard in SI units?

28. Two rods with R-values R_1 and R_2 have the same area and are joined in series end to end, as shown in Fig. 16-15*a*. The lateral surfaces are insulated and opposite ends are maintained at different temperatures. For steady-state heat conduction, *(a)* show that the series combination has an effective R-value R_{eff} given by

$$R_{eff} = R_1 + R_2$$

(b) Suppose $R_1 < R_2$; which rod has the larger temperature drop across its length?

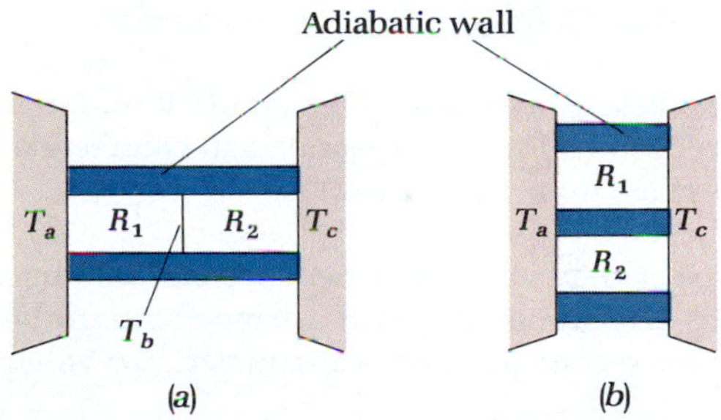

Figure 16-15. Exercise 28 and 29.

29. Suppose the two rods from the previous exercise are arranged in parallel, as shown in Fig. 16-15*b*. For steady-state heat conduction, show that the parallel combination has an effective R-value R_{eff} given by

$$\frac{2}{R_{eff}} = \frac{1}{R_1} + \frac{1}{R_2} \quad \text{or} \quad R_{eff} = \frac{2R_1R_2}{R_1 + R_2}$$

Note that the combined heat current $H = H_1 + H_2$.

30. A wall consists of two layers: plasterboard with R-value $R_p = 0.45$ and brick veneer with R-value $R_b = 1.2$. If a temperature difference of 50°F exists across the wall, determine *(a)* the heat current per unit area in the wall and *(b)* the temperature at the plasterboard-brick interface if the inside temperature is 70°F and the outside temperature is 20°F.

31. A 250-mm-long wooden ($k = 0.19$ W·m^{-1}·K^{-1}) tube has circular cross section with inner radius $a = 10$ mm and outer radius $b = 20$ mm. Fitting snugly within the tube is a circular aluminum rod of the same length. A temperature difference of

150°C is maintained across the ends of this compound bar, and heat loss at the lateral surface is negligible. For steady-state conditions, *(a)* what is the total heat current in the compound bar and *(b)* how much energy is transferred through the bar in an hour?

32. A metal sphere of radius 150 mm has a surface of emissivity 0.40. *(a)* At what rate does it emit energy if its temperature is maintained at 900°C? *(b)* Suppose the sphere at 900°C is in an evacuated enclosure whose walls are maintained at 500°C. At what rate must energy be supplied to the sphere under these steady conditions?

33. Compare the radiated power per unit area emitted by the surface of *(a)* the sun at 6000 K, *(b)* the earth at 300 K, *(c)* the dark side of the moon at 200 K, *(d)* a neutron star at 3 K. For simplicity, take the emissivity of each surface to be 1.

34. Consider a body of emissivity e at temperature T_2 which is surrounded by walls at temperature T_1. Suppose that the temperature difference $\Delta T = T_2 - T_1$ is small compared with T_1. Show that $T_2^4 - T_1^4 \approx 4T_1^3\,\Delta T$ and that the net heat current H is

$$H = 4e\sigma A T_1^3\,\Delta T$$

Additional Exercises

35. A sample of gas consists of 1.2 mol of He and 0.8 mol of Ne. Determine *(a)* the mass of the gas and *(b)* the total number of molecules in the sample.

36. *(a)* Which of the following temperatures is lowest: −234°F, −165°C, 85 K? *(b)* Which is highest?

37. A steel bridge spans a total length of 40 m. Estimate the length needed to allow for expansion and contraction if temperatures range from −20 to +40°C.

38. A copper-constantan thermocouple has one junction at 0.0°C; the second junction is at a temperature to be determined. If the second junction is at 100.0°C, the voltage measured is 4.28 mV. What is the temperature of the second junction if the voltage measured is 1.91 mV? Assume a linear dependence of temperature on thermocouple voltage.

39. At 0.0°C a steel rod is 1.0041 m long and an aluminum rod is 1.0038 m long. *(a)* At what temperature will the rods have equal lengths? *(b)* What is this length?

40. The pressure of the helium gas in a constant-volume gas thermometer is 12560 Pa when the thermometer is in thermal equilibrium with water at its normal boiling point. What is the pressure of the gas when the thermometer is in contact with oxygen at its normal boiling point (90.2 K)?

41. A rectangular aluminum plate has dimensions 364 mm × 448 mm at 0.0°C. *(a)* Determine the area of the plate at 0.0°C. Determine *(b)* the dimensions and *(c)* the area of the plate at 150°C.

42. At 0.0°C a square copper plate of edge 240 mm has at its center a circular hole of radius 80 mm. Determine the edge dimension and the radius of the hole if the plate is at 100°C.

43. Estimate the total energy per second radiated by the sun. The sun's radius is about 7×10^8 m and its surface temperature is about 6000 K.

44. A double-glazed window consists of two sheets of 3-mm-thick glass separated by a 12-mm layer of air. Determine the heat current per unit area for such a window if the inside and outside temperatures differ by 20°C. The thermal conductivities are $k_{glass} = 1\ \text{W}\cdot\text{m}^{-1}\cdot\text{K}^{-1}$, $k_{air} = 0.02\ \text{W}\cdot\text{m}^{-1}\cdot\text{K}^{-1}$. Assume a steady-state heat flow due to conduction and neglect convection.

45. Determine the magnitude of the temperature gradient in the glass sheets and in the air for the window in the previous exercise.

PROBLEMS

1. ***Area expansion.*** Consider a rectangular plate of length L_0 and width W_0 at some reference temperature T_0. The area of the plate at this temperature is $A_0 = L_0 W_0$. With a change in temperature ΔT, each linear dimension changes by an amount determined by the coefficient of linear expansion α. Show that the change in area ΔA is given by

$$\Delta A = 2\alpha A_0\,\Delta T$$

where the small area element $\Delta L\,\Delta W$ is neglected.

2. ***Relation between linear and volume expansion.*** A rectangular block at temperature T_0 has dimensions of L_0, W_0, and H_0 and a volume of V_0. Show that the coefficient of volume expansion $\beta = 3\alpha$. What approximations have been made? What would be the relation between β and α for a sphere?

3. ***Heat conduction with cylindrical symmetry.*** Apply the heat-conduction equation, Eq. 16-10, to steady-state radial heat flow corresponding to cylindrical symmetry, as shown in Fig. 16-16*a*. Suppose a long inner cylinder of radius a is main-

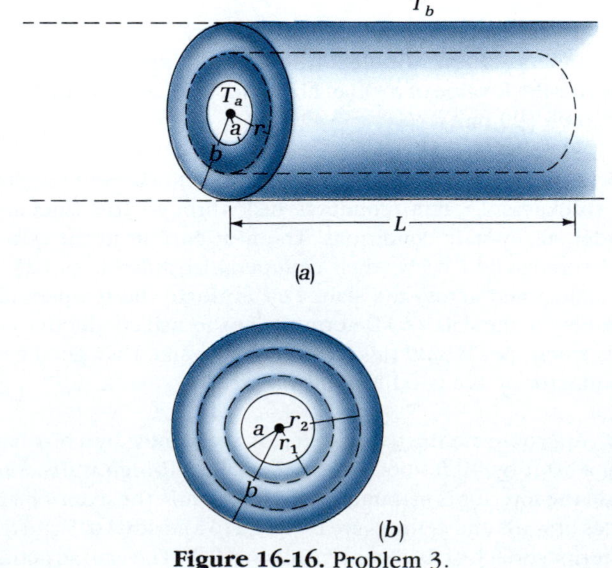

Figure 16-16. Problem 3.

tained at temperature T_a. Surrounding the inner cylinder is a cylindrical medium of thermal conductivity k and outer radius b. The outer surface is maintained at a lower temperature T_b. The radial coordinate r is the perpendicular distance from the axis to a cylindrical surface, and we consider a length L of the cylinder. The area A through which the heat flows is $2\pi rL$. *(a)* By requiring that energy be conserved, show that the heat current H has the same value through concentric cylinders of radii r_1 and r_2 (see Fig. 16-16*b*). *(b)* Show that the temperature gradient at distance r from the axis is given by

$$\frac{dT}{dr} = \frac{-H}{2\pi rLk}$$

(c) Integrate this equation to obtain the temperature distribution

$$T(r) = -\left(\frac{H}{2\pi Lk}\right) \ln r + \text{constant}$$

(d) The constant of integration and the heat-current value are determined from the temperature values at the boundaries. Show that

$$H = \frac{2\pi kL(T_a - T_b)}{\ln (b/a)}$$

$$T(r) = T_a + \frac{(T_b - T_a) \ln (r/a)}{\ln (b/a)}$$

4. ***Heat conduction with spherical symmetry.*** Consider a medium between two concentric *spheres* of radius a at temperature T_a and radius b at temperature T_b. Using the previous problem as a guide (but now r is the radial distance from the center of the sphere), show that

$$\frac{dT}{dr} = \frac{-H}{4\pi r^2 k}$$

and that

$$H = \frac{4\pi abk(T_a - T_b)}{b - a}$$

$$T = \frac{bT_b - aT_a}{b - a} + \frac{ab(T_a - T_b)}{(b - a)r}$$

5. ***An insulated pipe.*** A cylindrical metal pipe of outside radius 12 mm carries high-pressure steam at temperature 140°C. It is in contact with and surrounded by a cylindrical insulating sleeve of outside radius 28 mm and of thermal conductivity $k = 0.11\ \text{W}\cdot\text{m}^{-1}\cdot\text{K}^{-1}$. The outside surface is exposed to a fixed temperature of 35°C. For each meter of length, determine *(a)* the heat current and *(b)* the temperature distribution in the insulating medium. *(c)* Construct a graph of T versus r to show the temperature distribution. *(d)* Evaluate the temperature gradient at a point 20 mm from the axis. (*Hint:* See Prob. 3.)

6. ***Insulating a sphere.*** A radioactive copper sphere of radius 25 mm is insulated from its 25°C surroundings by a spherical Styrofoam blanket of inner radius 25 mm and of thickness 15 mm. If the heat current into the surroundings is 600 mW (equal to the power provided by the radioactivity), determine *(a)* the temperature of the copper sphere and *(b)* the temperature distribution in the Styrofoam insulation. *(c)* Show the temperature distribution graphically. *(d)* Explain why the copper sphere can be considered to have the same temperature throughout. (*Hint:* See Prob. 4.)

7. ***A bimetallic strip.*** A bimetallic strip consists of two metal strips with different coefficients of linear expansion α_1 and α_2; the two strips are both of thickness d and length L_0 at T_0. They are bonded together and, with a change in temperature ΔT, will curve in a circular arc, as shown in Fig. 16-17. Show that the radius of curvature R is given approximately by

$$R = \frac{d}{(\alpha_2 - \alpha_1)\ \Delta T}$$

(*Hint:* Let $R_1 = R - \frac{1}{2}d$ and $R_2 = R + \frac{1}{2}d$ represent the mean radii of the two strips and equate the angles subtended by each strip.)

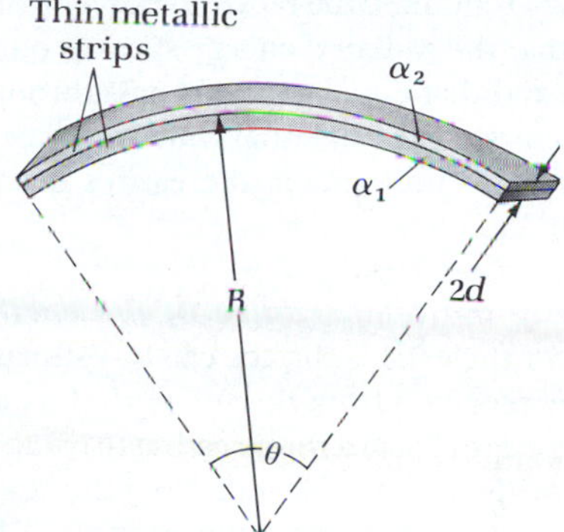

Figure 16-17. Problem 7.

8. ***Conduction and radiation.*** A 90-mm-thick roll of fiberglass insulation separates two wall surfaces whose temperatures are 290 and 270 K. *(a)* Determine the heat current per unit area (in SI units). *(b)* Compare with the net heat current per unit area that would be transferred between these surfaces by radiation if the space between these walls were evacuated. Take the emissivity of each wall to be 0.5. *(c)* Repeat the above calculations, taking the surfaces temperatures to be 490 and 470 K. *(d)* Repeat for surface temperatures of 90 and 70 K. Assume that the thermal conductivity and the emissivities are independent of temperature.

9. ***Laplace's equation.*** For steady-state heat conduction in one dimension, the temperature distribution $T(x)$ is the solution of *Laplace's equation* for a medium with no sources of heat:

$$\frac{d^2T}{dx^2} = 0$$

This equation is to be solved [integrated to determine $T(x)$] with the temperature specified at the boundaries. At $x = 0$, the temperature is T_2; at $x = L$, the temperature is T_1. Solve Laplace's equation for the situation illustrated in Fig. 16-10.

10. ***Laplace's equation for cylindrical symmetry.*** For problems with cylindrical symmetry, Laplace's equation (see Prob. 9) is

$$\frac{d}{dr}\left(r\frac{dT}{dr}\right) = 0$$

Solve this equation subject to the boundary conditions:

$$T = T_a \quad \text{at} \quad r = a$$

$$T = T_b \quad \text{at} \quad r = b$$

Compare the solution with the one displayed in Prob. 3.

11. ***Laplace's equation for spherical symmetry.*** Laplace's equation (see the previous two problems) for problems with spherical symmetry is

$$\frac{d}{dr}\left(r^2\frac{dT}{dr}\right)=0$$

Show that the temperature distribution displayed in Prob. 4 is a solution of Laplace's equation.

12. ***Power radiated from the sun.*** The sun's radius is 7×10^8 m, its surface temperature is 6000 K, and its emissivity is almost 1. *(a)* Calculate the radiated power from the surface. *(b)* Assume that the radiated energy spreads out uniformly in all directions and that equal amounts pass through concentric spheres of different radii in equal time intervals. Evaluate the radiated power per unit area at the earth's distance from the sun, 1.5×10^{11} m.

13. ***Estimating the temperature of the earth.*** An average temperature for the earth's surface can be estimated by balancing the energy received from the sun with the energy radiated away from the earth. Form a rough estimate using the following information: At the earth's distance from the sun, the incident energy per unit area per unit time is about 1400 W/m^2, of which a fraction, $A \approx 0.36$, is reflected. The earth radiates energy from its surface at a rate per unit area that corresponds to an average temperature T_e and emissivity $e = 1 - A$. Estimate T_e and compare it with the mean surface temperature of 285 K.

14. ***Temperature distribution in two dimensions.*** For steady-state heat flow, the temperature variation with position can be determined by solving Laplace's equation. In Probs. 9, 10, and 11, only one spatial variable was considered. In two dimensions, Laplace's equation can be solved approximately by using an iterative technique that is ideally suited for a spreadsheet calculation. The technique is based on the idea that the temperature at a given point is equal to the average of the temperatures at surrounding points. Consider the rectangular region shown in Fig. 16-18. The upper boundary is maintained at 100°C and the other three boundaries are kept at 0.0°C. If the region is divided into cells, then each cell can correspond to a cell in a spreadsheet, and the number assigned to the cell represents the temperature at the center of that cell. In the spreadsheet, the cells in the top row contain the value 100 while those in the bottom row and the leftmost and rightmost columns contain 0, as seen in Table 16-3. An interior cell contains a formula that averages the four surrounding cells. For example, cell B2 contains the formula

$$(\mathrm{B1} + \mathrm{A2} + \mathrm{B3} + \mathrm{C2})/4.$$

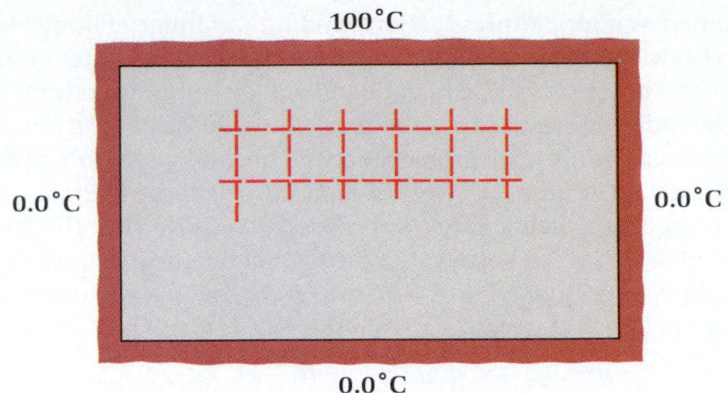

Figure 16-18. Problem 14.

The iteration is performed by repeatedly recalculating the spreadsheet. (A common spreadsheet uses function key F9 to recalculate.) After a number of iterations, the values in the interior settle down to give an approximate temperature distribution. Try this for several choices of the number of rows and columns. You can display the temperature distribution by constructing contours of equal temperatures.

15. ***Heat flow lines in two dimensions.*** Adapt the spreadsheet described in the previous problem to a rectangular region with the uppermost and leftmost boundaries at 100°C and the lowermost and rightmost boundaries at 0.0°C. You can visualize the heat flow by drawing lines from high to lower temperatures that are perpendicular to contours of equal temperature. These heat flow lines, sketched in Fig. 16-19, are analogous to the streamlines for fluid flow in Chap. 15.

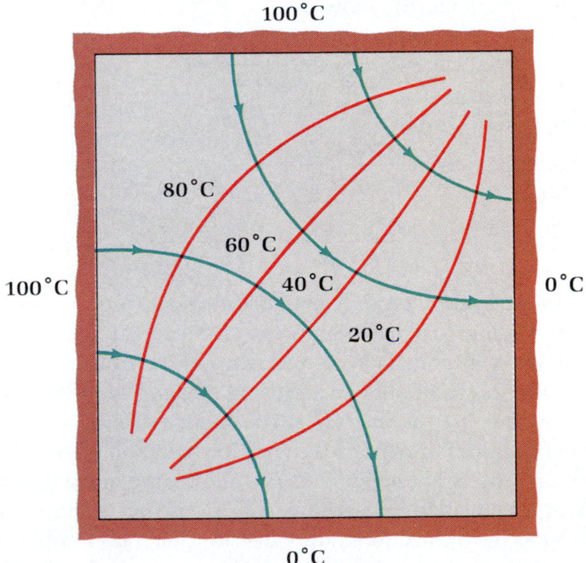

Figure 16-19. Problem 15.

TABLE 16-3. ***Part of a Spreadsheet for Laplace's Equation in Two Dimensions***

	A	B	C	D	E	F	G	H	I
1	100	100	100	100	100	100	100	100	100
2	0								0
3	0				Formulas				0
⋮	⋮								⋮
9	0								0
10	0	0	0	0	0	0	0	0	0

16. ***A composite slab.*** Two slabs of the same size but of different materials are joined as shown in Fig. 16-20*a*. Opposite faces of area 0.25 m^2 are maintained at temperatures of 120°C and 20°C. The temperature profile is shown in Fig. 16-20*b*. If the thermal conductivity $k_1 = 4.8\ W \cdot m^{-1} \cdot K^{-1}$, determine *(a)* the value of k_2 and *(b)* the heat current in the composite slab.

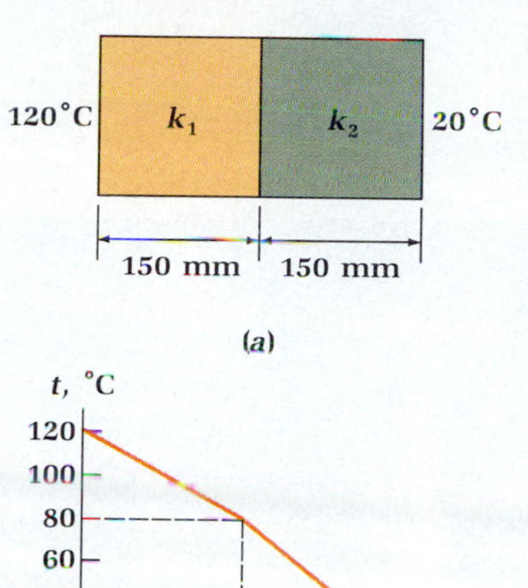

Figure 16-20. Problem 16.

17. ***Temperature of a light bulb.*** The effective emitting area of the tungsten filament in a light bulb is about $1 \times 10^{-5}\ m^2$. Estimate the operating temperature of the filament in a 100-W light bulb. Compare that with the melting point of tungsten, which is 3650 K.

18. ***Adding insulation to a ceiling.*** A house with no insulation has a 400-W/m^2 heat current per unit area in the ceiling when the temperature difference is 25°C between inside and outside. *(a)* What thickness of fiberglass batt insulation is necessary to reduce the heat current per unit area to 40 W/m^2? *(b)* If the cost of installed insulation is $10 per square meter and the cost of energy for heating is $30 per gigajoule (GJ), how many operating hours are needed under these conditions to retrieve the cost of the insulation?

19. ***Plate tectonics.*** The thermal conductivity of the rock that makes up the earth's crust and upper mantle is about 2 $W \cdot m^{-1} \cdot K^{-1}$. There is a heat flow from the interior to the surface of the earth, and the heat current per unit area near the surface is about 20 mW/m^2. *(a)* Estimate the temperature at depths of 1, 10, and 100 km. Assume a surface temperature of 300 K. *(b)* Estimate the depth at which the temperature is 1600°C. At this temperature, the mantle becomes ductile and allows for the (slow) motion of the upper-lying plate.

20. ***Keeping warm.*** Metabolic processes in the human body maintain the surface temperature at about 300 K if the heat current at the body's surface is about 100 W. Estimate the thickness of a down covering that provides comfort against a surrounding temperature of 270 K. The thermal conductivity of loose down is about 3 $mW \cdot m^{-1} \cdot K^{-1}$.

17 THE FIRST LAW OF THERMODYNAMICS

Replica of the steam locomotive Jupiter. The original Jupiter represented the West at the completion of the transcontinental railroad.

Hero's engine. Heat from the flame boils water to produce steam, which escapes from the "jets" and rotates the flask. The engine, which converts heat to work, was discovered about 150 B.C.

The origins of thermodynamics and its laws are in the very practical inventions of the industrial revolution, particularly the steam engine. These inventions eventually replaced the labor performed by human and beast with the mechanical work performed by heat engines. Obtaining mechanical work from an engine, such as a steam engine, required the burning of fuel and the accompanying heat transfer between the flame and the working substance in the engine (usually water).

Not until the second half of the last century did it become widely recognized that heat and mechanical work are energy transfers and that energy is a conserved quantity. Among those who contributed to the gradual development of these ideas were Benjamin Thompson (1753–1814) and James Joule (1818–1889). It was Thompson (see the Commentary) who recognized the inadequacy of treating heat as a fluid ("caloric") that could flow from one body to another. From his observations of the high temperatures produced in the boring of cannon, he proposed a connection between heat and the work done by friction. In a series of experiments, Joule, after whom the SI unit of energy is named, determined the amount of mechanical work that is equivalent to heat in raising the temperature of water. These were crucial steps in arriving at the statement of the first law of thermodynamics.

17-1 EQUATIONS OF STATE

What determines the value of a variable of state, say the pressure p of a gas? Suppose we put a certain amount of gas (n moles) in a container of fixed size (volume V) and maintain it at a constant temperature T. Experimentally, we find that the pressure of the gas cannot now be adjusted; fixing the values of n, V, and T determines the pressure. However, if one or more of n, V, and T are changed, then the pressure p may also change to a value that is determined by the new values of the other variables. That is, these variables of state are related by a mathematical equation. This relation is called the *equation of state* of that substance. Using the equation of state, we can evaluate any one of the variables of state, the pressure for example, in terms of the values of the remaining variables of state. The equation of state of a substance can be an extremely complicated function of the variables of state. But measurement can give their interdependence over all experimentally accessible conditions. We can, for example, hold n and T fixed and determine how the pressure p of a gas varies with changes in volume V.

The equation of state relates the variables of state.

The Ideal Gas Equation of State

Experiments of this sort for gases at low densities lead to the following conclusions:

1. For fixed n and T, p and V are inversely proportional. Thus, if the volume is doubled, the pressure is halved. This relationship, long known as ***Boyle's law,*** can also be written as

Boyle's law

$$pV = \text{constant} \qquad (n,\ T \text{ fixed})$$

2. For fixed n and V, p and T (on the Kelvin scale) are proportional. This result follows from our discussion of the constant-volume gas thermometer in the last chapter. The mathematical relation is

$$\frac{p}{T} = \text{constant} \qquad (n,\ V \text{ fixed})$$

3. For fixed V and T, p and n are proportional. Injecting an additional amount of gas into the container increases the pressure proportionately,

$$\frac{p}{n} = \text{constant} \qquad (V,\ T \text{ fixed})$$

All of these results, along with the results of varying other pairs of variables, are summarized in the relation

$$\frac{pV}{nT} = \text{constant}$$

where the constant in this expression is essentially independent of the variables p, V, n, and T so long as the density of the gas is low. From experiment the constant is found to have about the same value for all gases. In the limit of dilute gases, or for the ideal gas, the value of the constant is the same for all gases and is called the ***universal gas constant*** R. Its value in SI units is

$$R = 8.31\ \text{J}\cdot\text{mol}^{-1}\cdot\text{K}^{-1}$$

Universal gas constant

Thus the variables of state of a real dilute gas are related by an equation of state which is the same for all such gases. It is called the ***ideal gas equation of state:***

$$pV = nRT \tag{17-1}$$

Ideal gas equation of state

The ideal gas equation of state is obeyed approximately by a real gas whose pressure is not too large and whose temperature is not too low—that is, a dilute gas.

Equation 17-1 can be used only for a real gas of this sort. (See Prob. 2 for an example of another equation of state.) Unless stated otherwise, we shall assume that any gas may be treated as an ideal gas.

EXAMPLE 17-1

Using the ideal gas equation of state. *(a)* What is the pressure of $n = 0.85$ mol of He occupying a volume V of 0.012 m³ at a temperature T of 273 K? *(b)* What is the volume of this gas at that same pressure but at temperature $T = 580$ K?

Solution. *(a)* From the ideal gas equation of state, Eq. 17-1, we have

$$p = \frac{nRT}{V}$$

$$= \frac{(0.85 \text{ mol})(8.31 \text{ J}\cdot\text{mol}^{-1}\cdot\text{K}^{-1})(273 \text{ K})}{0.012 \text{ m}^3} = 160 \text{ kPa}$$

(b) It is often convenient to form ratios of state variables for each of two states. Let the state of the gas in part *(a)* be denoted by a and the state for part *(b)* be denoted by b. The variables for each state satisfy (n remains the same)

$$p_aV_a = nRT_a \qquad \text{and} \qquad p_bV_b = nRT_b$$

Dividing one equation by the other and canceling the common factors n and R, we obtain

$$\frac{p_bV_b}{p_aV_a} = \frac{T_b}{T_a}$$

In this example, $p_a = p_b$, so

$$V_b = V_a\frac{T_b}{T_a}$$

Because we are dealing with ratios here, we may express the volume in any unit. Let us use the liter (1 L = 0.001 m³) as a convenient unit of volume. Then

$$V_b = 12 \text{ L}\frac{580 \text{ K}}{273 \text{ K}} = 25 \text{ L}$$

SELF-TEST 17-1. What is the volume (in liters) occupied by 1.00 mol of a gas whose temperature is $T = 295$ K and whose pressure is $p = 101$ kPa? Assume that the gas may be treated as an ideal gas. ***ANSWER:*** 22.4 L.

The p-V Diagram

The ideal gas equation of state allows us to determine the value of one of the variables of state of a gas in terms of the others. If we deal with a fixed amount of gas (n remains the same), then any two of the remaining variables serve to determine the third. For convenience in displaying certain kinds of information, we often select p and V to be the independent variables; then the value of T is determined from Eq. 17-1. In this way the values of p and V determine the state of the gas.

A point on a p-V diagram represents a state of a system.

A state of a system such as a gas can be represented on a ***p-V diagram***. The axes in Fig. 17-1 are scaled to indicate values of pressure and volume, and a point in the plane corresponds to definite values of p and V. Each point on the p-V diagram represents a state of the system with a specified number of moles n. From the values of p and V, the temperature T for that state can be evaluated. Thus a value of the temperature is associated with each point on the p-V diagram. The state labeled a in Fig. 17-1, for example, has temperature $T_a = 300$ K.

States with the same temperature lie on an isotherm.

There is a set of states on the p-V diagram (and of the system) that have the same temperature. The state labeled b in Fig. 17-1 has twice the pressure and half the volume of state a, but it has the same temperature. The set of all states with that value of the temperature forms a curve on a p-V diagram called an ***isotherm.*** In Fig. 17-1 the curve containing points a and b is the 300-K isotherm.

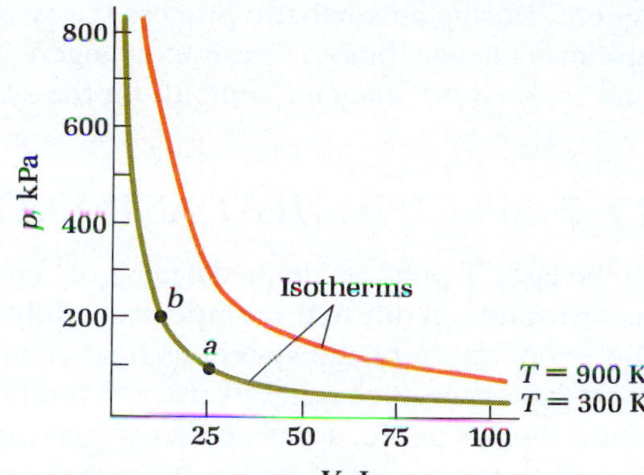

Figure 17-1. Isotherms on a p-V diagram. State a and state b, represented by points on a p-V diagram, have the same temperature. The 300- and 900-K isotherms are shown.

A Quasi-Static Process

In thermodynamics, the term "process" refers to changing the state variables of a system. Strictly speaking, many of the formulas are valid only for so-called *quasi-static processes.*

A ***quasi-static process*** is one that occurs slowly enough such that the system may be regarded as being in equilibrium as it arrives at each successive state during the change.

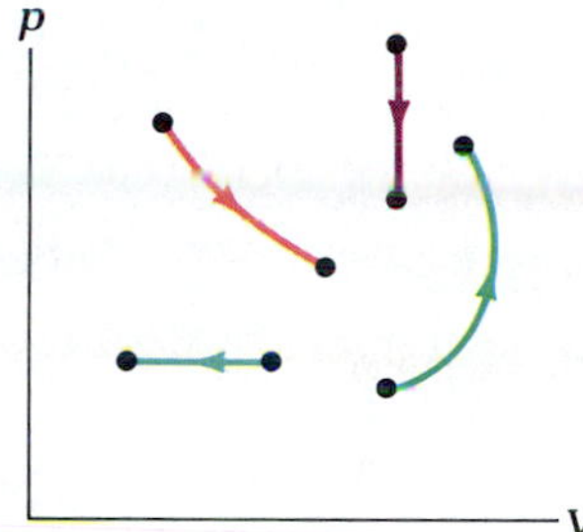

Figure 17-2. Quasi-static processes. A quasi-static process occurs slowly enough that the system proceeds through a series of equilibrium states, and the process can be represented by a curve on a p-V diagram. Several quasi-static processes are shown here.

A quasi-static process can be represented by a curve on a p-V diagram, as shown in Fig. 17-2. Each point on a curve representing a process corresponds to a different state of the system. The direction of the process is indicated by the arrowhead.

A quasi-static process is an idealization that may be approximated by a real process. But a system may undergo a process that is far from being quasi-static. For example, a system may rapidly or violently change its state. Figure 17-3 shows a process on a p-V diagram where the volume of a gas is rapidly increased from V_i to V_f. Since the change is rapid, the pressure of the gas is different in different parts of the system. Consequently, the "pressure of the system" is not well defined, and the process cannot be represented by a curved line on a p-V diagram. Unless stated otherwise, we shall consider only processes that can be regarded as being quasi-static.

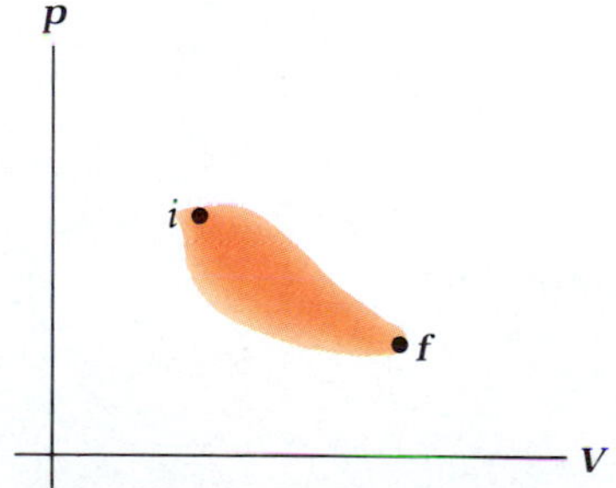

Figure 17-3. A process that is not quasi-static. As the volume of the system rapidly changes from V_i to V_f, the pressure is different in different parts of the system so that one value of p does not characterize the system. The process cannot be described by a curve on a p-V diagram.

Names of Processes

Some processes are simple enough to be given names. Those that are important to us are listed below:

1. An ***isochoric process*** is one in which the system's volume is fixed ($\Delta V = 0$). Since the volume does not change during an isochoric process, the process is represented by a vertical line on a p-V diagram. Which of the processes in Fig. 17-2 is isochoric?

2. An ***isobaric process*** is one in which the pressure of the system is held constant ($\Delta p = 0$). Because the pressure is constant, an isobaric process is represented by a horizontal line on a p-V diagram. Which of the processes in Fig. 17-2 is isobaric?

3. An ***isothermal process*** is one in which the system's temperature remains fixed ($\Delta T = 0$). On a p-V diagram, an isothermal process follows along an isotherm, such as one of those shown for the ideal gas in Fig. 17-1. Thus the curve that represents an isothermal process on a p-V diagram depends on the equation of state of the system. Which of the processes shown in Fig. 17-2 could possibly represent an isothermal process for the ideal gas?

4. An ***adiabatic process*** is one in which no heat is transferred to or from the system. During an adiabatic process the system's pressure, volume, and temperature may change, but no heat is exchanged. The curve that represents an adiabatic process on a p-V diagram depends on the system.

17-2 SPECIFIC HEAT AND LATENT HEAT

In the last chapter, we defined heat as the energy transferred between a system and its surroundings due to the temperature difference between them. What happens to the temperature of the system as heat is added to it from its surroundings? Our everyday experience suggests that the temperature of the system can increase as a result. For example, suppose that we monitor the temperature of water in a pot placed on the burner of a stove. A graph showing how the temperature of the water changes with time is shown schematically in Fig. 17-4. As heat is transferred from the burner to the water, the temperature of the water increases steadily. However, when the water is at its boiling point, its temperature no longer changes, even though heat is still being added. In the first case, the heat added causes a temperature change, which corresponds to the molecules moving faster on average. In the second case, the added heat produces a change in phase from liquid to vapor without changing the temperature. At the molecular level, such a phase change corresponds to increasing the average separation of the molecules and decreasing their interaction.

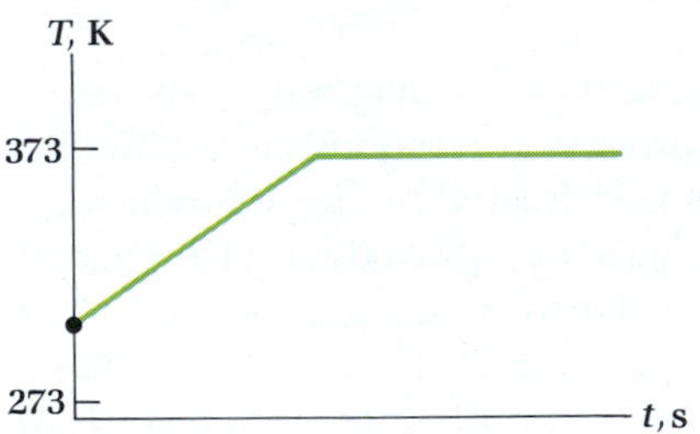

Figure 17-4. Temperature versus time for a sample of water as heat is added. The temperature increases until it reaches the boiling point, but then it no longer changes.

Specific Heat

Suppose we measure the temperature change ΔT of a system of mass m as we add heat Q to the system. For the moment we consider a system that does not change phase—if it is a liquid, it remains a liquid. For small ΔT, experiment shows that Q is directly proportional to ΔT. Further, the heat required to raise the temperature depends linearly on the mass of the system. Thus the specific heat is defined as follows:

Specific heat

> The ***specific heat*** of a substance is the heat required to increase the temperature of 1 kg of the substance by 1 K.

The specific heat depends on the nature of the substance, but it is independent of the mass of the system.

The heat required to change the temperature of a system depends on the process. That is, in general an isochoric process requires a different amount of heat from an isobaric process for the same temperature change. For an isobaric process, these features can be put into equation form with

Specific heat c_P at constant pressure

$$dQ_p = mc_p\, dT \tag{17-2}$$

where c_p is the *specific heat capacity at constant pressure.* It is customary to drop the word "capacity" and to refer to c_p simply as the specific heat at constant pressure. The subscript p is a reminder that the process is isobaric.

If the heat is added isochorically, then

Specific heat c_V at constant volume

$$dQ_V = mc_V\, dT \tag{17-3}$$

where c_V is the specific heat at constant volume and the subscript V indicates that the process is isochoric. The SI unit for c_p and c_V is $\text{J}\cdot\text{kg}^{-1}\cdot\text{K}^{-1}$. Since a temperature change in kelvin is the same as in degrees Celsius, the value of a specific heat in the unit $\text{J}\cdot\text{kg}^{-1}\cdot\text{K}^{-1}$ is the same as in the unit $\text{J}\cdot\text{kg}^{-1}\cdot{}^\circ\text{C}^{-1}$. Table 17-1 lists values of c_p for various substances.

For many substances, and particularly for gases, c_p and c_V are significantly different. Indeed, an important parameter for a gas is the ratio

$$\gamma = c_p/c_V$$

Ratio of specific heats γ

For example, the value of γ for air is 1.4. That is, 40 percent more heat is required to raise the temperature of a given amount of air by, say 1 K, at constant pressure than is required at constant volume. On the other hand, for most solids and liquids under ordinary conditions, c_p and c_V are nearly equal, or $\gamma \approx 1$.

Strictly speaking, the specific heats c_p and c_V are defined in the limiting case of infinitesimal temperature changes dT, and each is a function of T in general. However, for many substances c_p and c_V are nearly independent of T for moderate temperature changes ΔT so that we may use

$$Q_p = mc_p\,\Delta T \qquad \text{and} \qquad Q_V = mc_V\,\Delta T \tag{17-4}$$

Specific heat connects heat added and temperature change.

We avoid using the symbol Δ with heat Q because heat is an energy transfer and not a change in a state variable. In contrast, the temperature change is denoted by ΔT. We will have more to say about this distinction later.

EXAMPLE 17-2

Heating a skillet. How much heat must be added to a 2-kg cast-iron skillet to raise its temperature by 120°C?

Solution. The specific heat of iron, from Table 17-1, is $c_P = 447\ \text{J}\cdot\text{kg}^{-1}\cdot\text{K}^{-1}$. Equation 17-4 gives

$$Q_p = (2\ \text{kg})(447\ \text{J}\cdot\text{kg}^{-1}\cdot{}^\circ\text{C}^{-1})(120^\circ\text{C})$$
$$= 100\ \text{kJ}$$

This represents the minimum heat required, assuming no losses to other parts of the environment.

SELF-TEST 17-2. A 0.50-kg block of material is increased in temperature by 5.0 K when 2.08 kJ of heat is added to it at constant pressure. What is c_P for this material? ***ANSWER:*** 830 $\text{J}\cdot\text{kg}^{-1}\cdot\text{K}^{-1}$.

Molar Heat Capacity. It is sometimes preferable to specify the amount of material in a system by the number n of moles rather than by the mass m; $n = m/M_O$ where M_O is the molar mass of the material. When this is the case, we use the ***molar heat capacities*** $\mathscr{C}_p$ and $\mathscr{C}_V$ rather than the specific heats c_p and c_V. The molar heat capacities are defined by

$$dQ_p = n\mathscr{C}_p\,dT \tag{17-5}$$

Molar heat capacity $\mathscr{C}_p$

and

$$dQ_V = n\mathscr{C}_V\,dT \tag{17-6}$$

Molar heat capacity $\mathscr{C}_V$

The SI unit for the molar heat capacity is $\text{J}\cdot\text{mol}^{-1}\cdot\text{K}^{-1}$. Some molar-heat-capacity values are listed in Table 17-1.

A comparison between Eq. 17-2 and Eq. 17-5 shows that $mc_p = n\mathscr{C}_p$. Since $M_o = m/n$, the relation between c_p and $\mathscr{C}_p$ is

$$M_o c_p = \mathscr{C}_p$$

TABLE 17-1. ***Some Specific-Heat and Molar-Heat-Capacity Values at 25°C and Atmospheric Pressure***

	c_p		
Substance	$\mathrm{J \cdot kg^{-1} \cdot K^{-1}}$	$\mathrm{cal \cdot g^{-1} \cdot {}^\circ C^{-1}}$	$\mathscr{C}_p$, $\mathrm{J\ mol^{-1}\ K^{-1}}$
Aluminum	910	0.215	24.4
Copper	386	0.092	24.5
Iron	447	0.107	25.0
Lead	128	0.031	26.8
Mercury	140	0.033	28.0
Tungsten	136	0.032	25.0
Helium	5200	1.24	20.8
Nitrogen	1040	0.25	29.1
Oxygen	920	0.22	29.4
Carbon (diamond)	509	0.121	6.1
Water	4180	0.998	75.3
Ice (−10°C)	2100	0.50	38
Alcohol (ethyl)	2500	0.60	91.5
Glass (crown)	67	0.016	

Similarly, $M_o c_V = \mathscr{C}_V$. However, some caution must be exercised when using these relations because the units we are using involve both grams and kilograms. If we use the units gram per mole (g/mol) for M_o, $\mathrm{J \cdot kg^{-1} \cdot K^{-1}}$ for c_p, and $\mathrm{J \cdot mol^{-1} \cdot K^{-1}}$ for $\mathscr{C}_p$, then the relation becomes

$$M_o c_p = (1000\ \mathrm{g \cdot kg^{-1}})\ \mathscr{C}_p$$

A similar adjustment must be made in the relation between c_V and $\mathscr{C}_V$.

The Specific Heat of Water and the Calorie. The ***calorie*** (cal) is a unit of energy that was originally based on the properties of water. It was defined as the amount of heat required to raise the temperature of 1 g of water by 1°C. Thus the specific heat of water would be $c_p = 1\ \mathrm{cal \cdot g^{-1} \cdot {}^\circ C^{-1}}$ exactly. The specific heat of liquid water actually depends very slightly on temperature between 0 and 100°C. The calorie is now defined in terms of the joule. The conversion is

Definition of the calorie

$$1\ \mathrm{cal} = 4.186\ \mathrm{J}$$

The calorie is a convenient energy unit to use for a system of liquid water, since $c_p = 1.00\ \mathrm{cal \cdot g^{-1} \cdot {}^\circ C^{-1}}$ to three significant figures over the 100°C range. Another similar unit of heat is the Btu (British thermal unit). It was originally defined such that 1 Btu raised the temperature of 1 lb of water by 1°F. The conversion is 1 Btu = 1055 J.

British thermal unit, Btu

Latent Heat

At atmospheric pressure, the temperature of the melting point of water (ice) is 0.00°C. As water melts, changing its phase from solid to liquid, heat must be added, even though the temperature remains fixed. Similarly, as the liquid freezes to the solid phase, heat must be removed. The amount of heat per unit mass that is added to or removed from a substance undergoing a phase change is called the ***latent heat*** *L*:

Latent heat

$$Q = mL \tag{17-7}$$

The SI unit of latent heat is joules per kilogram (J/kg). The latent heat for a substance undergoing a liquid-solid phase change is denoted by L_f, called the ***latent heat of fusion.*** For a liquid-vapor phase change, the ***latent heat of vapor-***

ization is denoted by L_v. Some latent-heat values for fusion and vaporization are listed in Table 17-2. We note that in addition to the type of phase changes described here, there are different types of phase changes that involve no latent heat.

TABLE 17-2. ***Some Latent-Heat Values at Atmospheric Pressure***

Substance	L_f (fusion), $MJ \cdot kg^{-1}$	L_v (vaporization), $MJ \cdot kg^{-1}$
Aluminum	0.400	12.3
Copper	0.205	4.80
Iron	0.275	6.29
Lead	0.023	0.87
Mercury	0.011	0.29
Tungsten	0.192	4.35
Nitrogen	—	0.20
Oxygen	—	0.21
Water	0.335	2.260
Alcohol (ethyl)	—	1.1

EXAMPLE 17-3

Converting ice to steam. At atmospheric pressure how much heat must be added to 0.50 kg of water in the form of ice at 0°C to convert it to steam (vapor) at 100°C?

Solution. There are three contributions—heat added to melt the ice, heat added to raise the temperature of the liquid from 0 to 100°C, and heat added to change the phase from liquid to vapor. The specific-heat and latent-heat values are taken from Tables 17-1 and 17-2:

$$\begin{aligned} Q &= (0.50\text{ kg})(0.335\text{ MJ/kg}) + (0.50\text{ kg})(4180\text{ J}\cdot\text{kg}^{-1}\cdot{}^\circ\text{C}^{-1})(100^\circ\text{C}) \\ &\quad + (0.50\text{ kg})(2.26\text{ MJ/kg}) \\ &= 0.17\text{ MJ} + 0.21\text{ MJ} + 1.1\text{ MJ} = 1.5\text{ MJ} \end{aligned}$$

Note the large contributions to the total heat added to the water that are made during the phase changes.

SELF-TEST 17-3. Use the value of c_p for copper listed in Table 17-1 to calculate its value of $\mathscr{C}_p$. Carefully check the units of each quantity in the relation. ***ANSWER:*** $24.5\text{ J}\cdot\text{mol}^{-1}\cdot\text{K}^{-1}$.

17-3 WORK

Heat is energy transferred between a system and its environment because of a temperature difference between them. Another way to transfer energy is work.

> ***Work*** is energy transferred between a system and its environment by means independent of the temperature difference between them.

There are a number of ways energy can be transferred as work. For example, work can be done with electric forces or with magnetic forces. However, we shall be concerned with the mechanical work done by a system due to the contact force the system exerts on its surroundings.

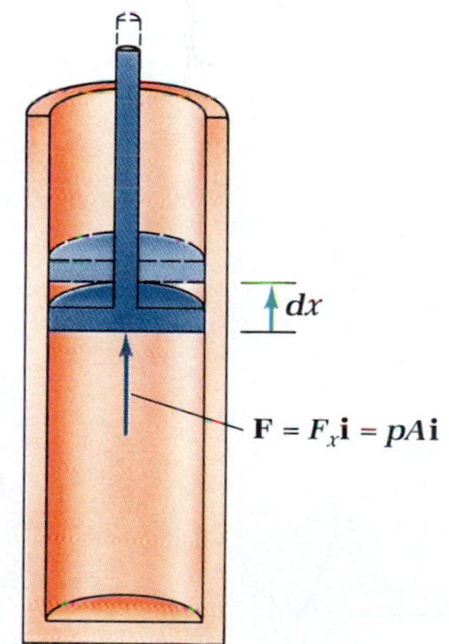

Figure 17-5. A system does work on its surroundings. The enclosed gas at pressure p exerts a force $\mathbf{F} = F_x\mathbf{i} = (pA)\mathbf{i}$ on the piston. The piston is displaced an amount $(dx)\mathbf{i}$ so that the work done by the gas is $dW = \mathbf{F}\cdot(dx)\mathbf{i} = pA\,dx = p\,dV$.

An example of work done by a system is the work done by a gas (our system) in a cylinder as the gas expands and causes a displacement of a movable piston (part of the surroundings), as shown in Fig. 17-5. The piston has face area A and the pressure of the gas is p so that the force exerted by the gas on the piston is $\mathbf{F} = F_x\mathbf{i} = (pA)\mathbf{i}$. If the displacement of the piston is $d\boldsymbol{\ell} = dx\mathbf{i}$, then the work dW done by the gas is $dW = \mathbf{F}\cdot d\boldsymbol{\ell} = pA\,dx$. Since $A\,dx$ is the infinitesimal change dV in the volume of the system, the work done by the system is

$$dW = p\,dV \tag{17-8}$$

Infinitesimal work

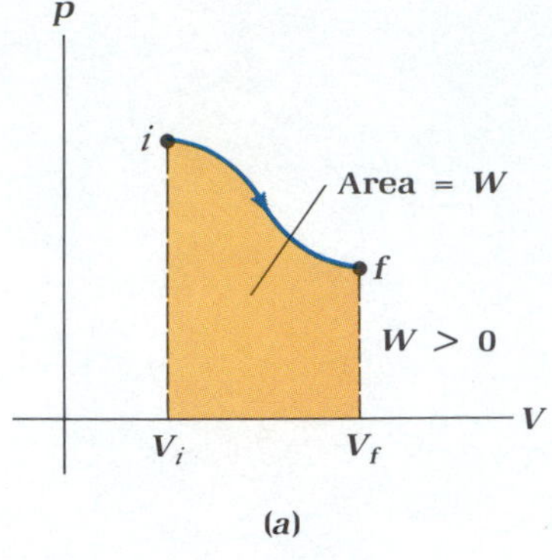

(a)

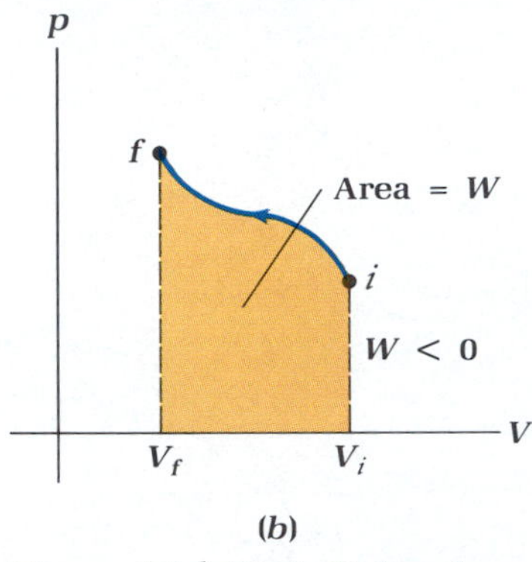

(b)

Figure 17-6. Work W done by a system during a process is equal to the area under the curve representing the process on a p-V diagram. *(a)* The work done by the system is positive ($W > 0$) when the system expands ($V_f > V_i$). *(b)* The work done by the system is negative ($W < 0$) when the system is compressed ($V_f < V_i$).

For a finite process in which the volume changes quasi-statically from V_i to V_f, Eq. 17-8 must be integrated to determine the work done for the process:

$$W = \int_{V_i}^{V_f} p\, dV \tag{17-9}$$

The work is the integral of the function p with respect to the volume V. This means that the work is given by the area under the curve representing the process on a p-V diagram, shown in light orange in Fig. 17-6*a*. To evaluate the integral, we must know how the pressure varies during the process.

From Eq. 17-9 we can see that, since p is positive, W will always be positive when $V_f > V_i$, as shown in Fig. 17-6*a*. On the other hand, W will always be negative when $V_f < V_i$, as shown in Fig. 17-6*b*. That is, if the system expands ($\Delta V > 0$), then $W > 0$, and we say that work is done *by* the system. If the system contracts or is compressed ($\Delta V < 0$), then $W < 0$, and we say that work is done *on* the system.

Work Done during an Isochoric Process. During an isochoric process ($\Delta V = 0$), the work done by the system is zero because there is no displacement. For work to be done, the volume of the system must change. If we use a subscript to denote the quantity that is held fixed during a process, then W_V represents the work done during an isochoric process, and we have

$$W_V = 0$$

Work Done during an Isobaric Process. Let p_0 represent the constant pressure during an isobaric process. As shown in Fig. 17-7, the work W_P is given simply by the area of a rectangle of height p_0 and width $V_f - V_i = \Delta V$, so that

$$W_p = p_0\, \Delta V \tag{17-10}$$

Another way to arrive at this result is to perform the integral in Eq. 17-9. Since the pressure is constant, it can be factored out of the integral, which immediately leads to Eq. 17-10.

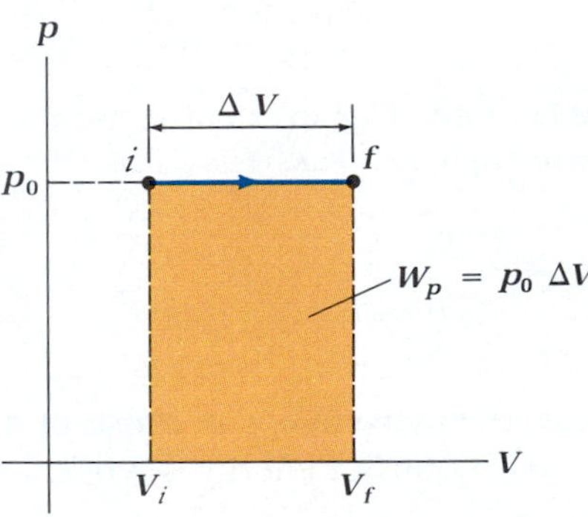

Figure 17-7. Work done during an isobaric process. The work done is equal to the area of the rectangle of height p_0 and width ΔV: $W_p = p_0\, \Delta V$.

Work Done during an Isothermal Process for an Ideal Gas. The curve representing an isothermal process on a p-V diagram depends on the system, and, consequently, the work done depends on the system. With the equation of state for an ideal gas, we can determine an expression for the isothermal work W_T for this system. An isothermal expansion of an ideal gas is shown in Fig. 17-8. To evaluate the work done by the gas as it expands, solve the equation of state for the pressure: $p = nRT/V$. Then insert this expression into Eq. 17-9, which gives

$$W_T = \int_{V_i}^{V_f} \frac{nRT}{V}\, dV = nRT \int_{V_i}^{V_f} \frac{dV}{V} = nRT(\ln V_f - \ln V_i)$$

In evaluating the integral, we have factored out the constants nRT. (Remember, T is constant for an isothermal process.) Since $\ln V_f - \ln V_i = \ln (V_f/V_i)$, we have

$$W_T = (nRT) \ln\left(\frac{V_f}{V_i}\right) \tag{17-11}$$

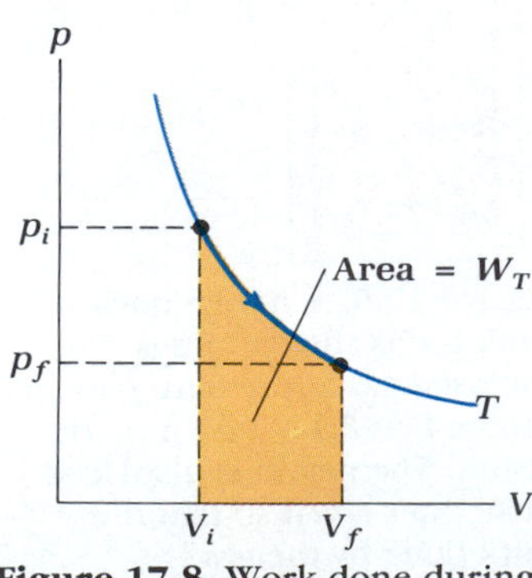

Figure 17-8. Work done during an isothermal process. Integration of the function $p = nRT/V$ with T constant gives $W_T = nRT \ln(V_f/V_i)$.

Note that if $V_f > V_i$, then $\ln(V_f/V_i) > 0$ and the work done is positive; but if $V_f < V_i$, then $\ln(V_f/V_i) < 0$ and the work done is negative.

Work Done during an Adiabatic Process for an Ideal Gas. For an adiabatic process, the relation between p and V depends on the ratio of specific heats γ. (Recall that $\gamma = c_p/c_V = \mathscr{C}_p/\mathscr{C}_V$.) The calculation of the work W_Q done by an ideal gas during an adiabatic process is outlined in Prob. 15. We quote the result here for completeness:

Adiabatic work done by an ideal gas

$$W_Q = \frac{p_i V_i}{\gamma - 1}\left[1 - \left(\frac{V_i}{V_f}\right)^{\gamma - 1}\right]$$

where p_i and V_i are the pressure and volume of the initial state and V_f is the volume of the final state.

EXAMPLE 17-4

Work done along two paths. In two separate experiments, a sample of helium gas is taken from the same initial state (state i) to the same final state (state f), but along two separates paths on a p-V diagram (Fig. 17-9). Path a has two parts: first an isochor and second an isobar. Path b is an isotherm. Determine the work done by the gas *(a)* along path a and *(b)* along path b. The pressure and volume at the initial and final states are $p_i = 202$ kPa, $V_i = 48$ L; $p_f = 91$ kPa, $V_f = 106$ L.

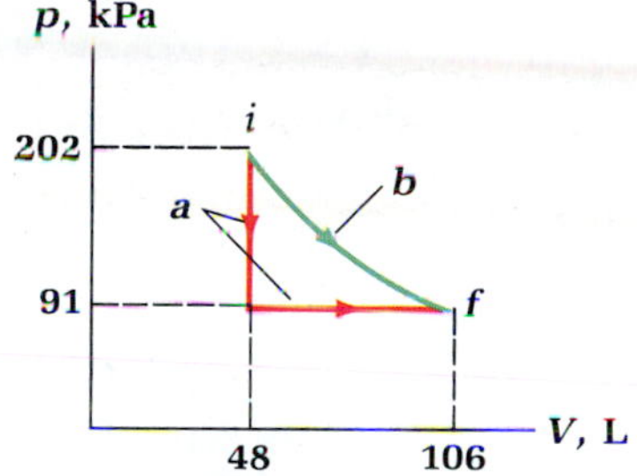

Figure 17-9. Example 17-4.

Solution. *(a)* The work done during the isochoric part of path a is zero because $W_V = 0$ for any system. The work done during the isobaric part of path a is

$$W_p = p_0\,\Delta V = (91\text{ kPa})(0.106\text{ m}^3 - 0.048\text{ m}^3) = 5.3\text{ kJ}$$

Thus the work done by the system along path a is

$$W_a = W_V + W_p = 0 + 5.3\text{ kJ} = 5.3\text{ kJ}$$

[Incidentally, notice that the product pV has the dimension of energy and that (1 kPa) (1 L) = 1 J.] *(b)* Equation 17-11 gives the work done during an isothermal process: $W_T = (nRT)\ln(V_f/V_i)$. For an isothermal process, $nRT = p_iV_i = p_fV_f$, so that two alternative forms of W_T are $W_T = (p_iV_i)\ln(V_f/V_i)$ and $W_T = (p_fV_f)\ln(V_f/V_i)$. Using the last of these expressions, we find

$$W_b = W_T = (91\text{ kPa})(106\text{ L})\ln\left(\frac{106\text{ L}}{48\text{ L}}\right) = 7.7\text{ kJ}$$

Despite the fact that the initial and final states are the same for the two paths, the work along path a (5.3 kJ) is different from the work along path b (7.7 kJ). That the work along b is greater than along a is apparent from the p-V diagram in Fig. 17-9, and from the realization that the work is given by the area under the curve. Indeed, it is evident that the work along two different paths between the same initial and final states on a p-V diagram is, in general, different. The work done depends on the details of the process; it is "path-dependent." Our use of process-labeling subscripts on the work symbol (such as W_T) emphasizes this path dependence.

SELF-TEST 17-4. For the system in the above example consider the work done along path c between states i and f, where path c is first an isobar at pressure $p_i = 202$ kPa and second an isochor at volume $V_f = 106$ L. *(a)* Draw a p-V diagram that shows path a, b, and c. *(b)* Evaluate W_c and compare it with W_a and W_b. ***ANSWER:*** *(b)* 11.7 kJ.

17-4 THE FIRST LAW OF THERMODYNAMICS

We have identified the two types of energy transfer between a system and its surroundings: heat Q and work W. Before introducing the first law of thermodynamics, we must emphasize the sign convention for Q and W. This sign convention is illustrated schematically in Fig. 17-10, where the direction of the energy flow is represented by arrows. Heat Q is positive when heat energy *enters* the system, but

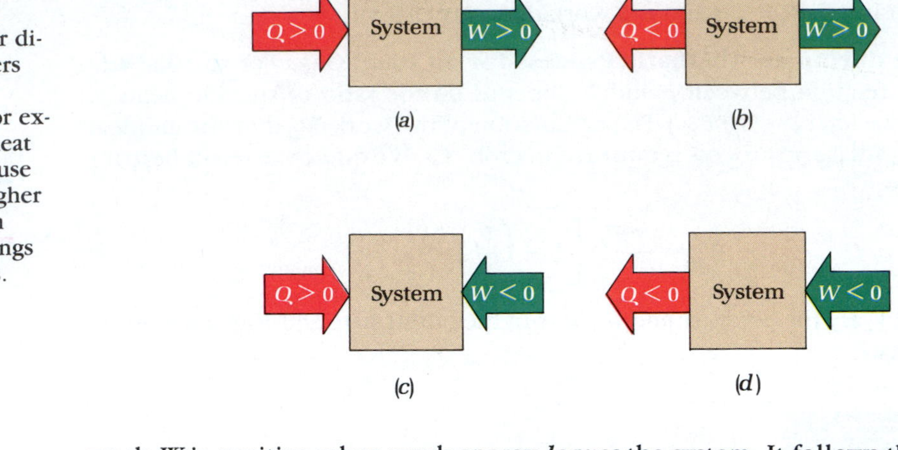

Figure 17-10. Sign convention for heat Q and work W. The arrows illustrate the sense or direction of the energy transfers between a system and its surroundings. In part *(a)* for example, the system absorbs heat from the surroundings because the surroundings are at a higher temperature, and the system does work on the surroundings because the system expands.

work W is positive when work energy *leaves* the system. It follows that heat Q is negative when heat energy *leaves* the system and that work W is negative when work energy *enters* the system. Therefore, during a process that involves both heat and work, the combination $Q - W$ represents the *net* energy *entering* the system.

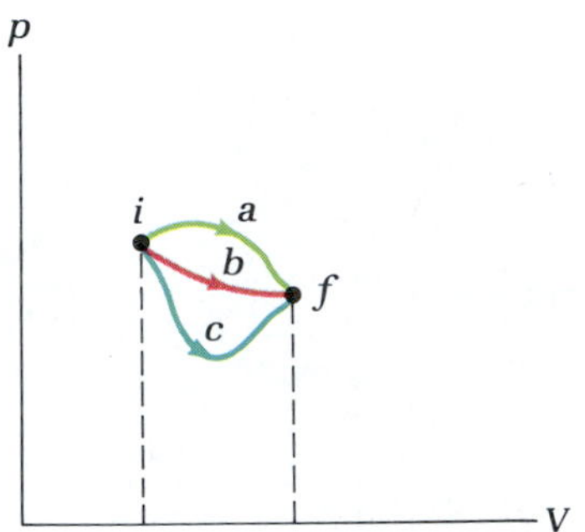

Figure 17-11. Work depends on the process. Three different processes are shown connecting an initial and a final state. Since W equals the area under the curve on a p-V diagram, W is different for each of these processes.

Work Depends on the Process and Heat Depends on the Process. As we saw in Example 17-4, when a system undergoes a process between two states, the work done depends on the path of the process on a p-V diagram. This result is evident from Fig. 17-11, which shows three different processes between the same initial and final states. For each process, the work done by the system—equal to the area under the curve—has a different value. That is,

$$W_a \neq W_b \neq W_c$$

The work done by a system that undergoes a process from an initial state to a final state depends on the details of that process.

Measurements show that the heat added to the system also depends on how the process is performed. For the three processes displayed in Fig. 17-11 the heat added has different values (although this result is not apparent from the figure). Using the same labels as before, we have

$$Q_a \neq Q_b \neq Q_c$$

Heat added to a system that undergoes a process from an initial state to a final state depends on the details of that process.

James Prescott Joule (1818–1889) was the first to establish the first law of thermodynamics quantitatively with a series of precise experiments.

The First Law of Thermodynamics

Now consider the net energy transferred during each process, namely, $Q - W$. Experiment shows that the net energy transferred is the same for *any* process. In particular, for the processes shown in Fig. 17-11,

$$Q_a - W_a = Q_b - W_b = Q_c - W_c$$

That is, the combination $Q - W$ is independent of the path of the process on a p-V diagram.

The fact that the net energy transferred is independent of the process leads to the ***first law of thermodynamics:***

For any process where heat Q is added to a system and work W is done by the system, the net energy transferred, $Q - W$, equals the change ΔU in the internal energy of the system.

In equation form,

$$\Delta U = Q - W \tag{17-12}$$

The first law of thermodynamics

The ***internal energy*** U is a state variable whose change ΔU is given by the net energy, $Q - W$, that is transferred.

To demonstrate the significance of this, consider a specific example (Fig. 17-12). Suppose we assign a value of U to a sample of gas when its volume and pressure are $V_1 = 1.00$ L and $p_1 = 100$ kPa, say, $U = U_1 = 150$ J. Now the gas undergoes a process to a second state where $V_2 = 2.00$ L and $p_2 = 200$ kPa, and measurements give $Q - W = 450$ J for the process. Thus $U_2 = U_1 + \Delta U = U_1 + (Q - W) = 150\text{ J} + 450\text{ J} = 600$ J at this second volume and pressure. If we change the state back to $V_1 = 1.00$ L and $p_1 = 100$ kPa, then measurements give $Q - W = -450$ J and $U = U_1 = 150$ J again—no matter what process we used on the return. Also, if we go back to $V_2 = 2.00$ L and $p_2 = 200$ kPa along any path whatsoever, then $U = U_2 = 600$ J again. That is, U depends only on the system's state and not on the process that placed the system in that state. That is what we mean by the term "state variable." A state variable depends only on the state and not at all on how the system arrived at that state.

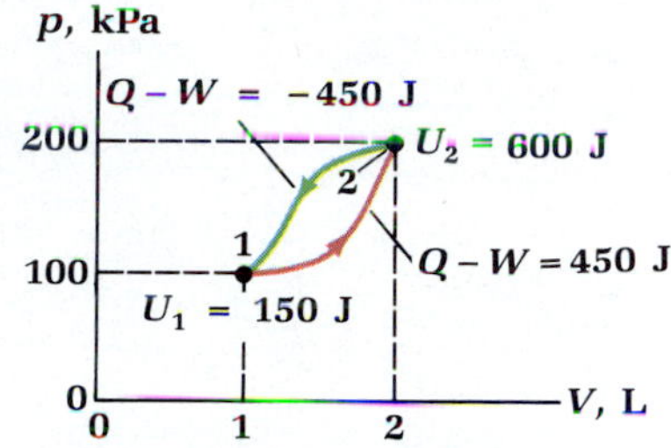

Figure 17-12. Internal energy U is a state variable. As a system undergoes a process from i to f, the net energy transferred is $\Delta U = Q - W$. When the system returns to i, experiment shows that $\Delta U = Q - W$ is equal and opposite what it was for i to f, no matter what process is used. Thus U depends only on the state of the system, and not on how the system was placed in that state.

The first law embodies the principle of conservation of energy. The very fact that the internal energy U is a state variable implies that energy is conserved. Suppose a system is isolated so that it cannot interact with its surroundings—that is, suppose $Q = 0$ and $W = 0$. This means that $\Delta U = 0$, or that the internal energy of an isolated system remains fixed.

EXAMPLE 17-5

Using the first law. During a process (process *a*) that takes a system from state *i* to state *f*, 16 kJ of heat is absorbed by the system and 12 kJ of work is done by the system. During a return process (process *b*), which takes the system from *f* back to *i*, 18 kJ of heat is rejected by the system. What is the work done by the system during the return process?

Solution. We use process *a* to find ΔU between states *i* and *f*. For process *a*, $Q_a = 16$ kJ and $W_a = 12$ kJ so that

$$\Delta U_a = U_f - U_i = Q_a - W_a = 16\text{ kJ} - 12\text{ kJ} = 4\text{ kJ}$$

For process *b*, $Q_b = -18$ kJ (negative because the heat is rejected) and, since this process is from state *f* to state *i*,

$$\Delta U_b = U_i - U_f = -\Delta U_a = -4\text{ kJ}$$

From the first law, $\Delta U_b = Q_b - W_b$, we have

$$W_b = Q_b - \Delta U_b = -18\text{ kJ} - (-4\text{ kJ}) = -14\text{ kJ}$$

Since the work is negative, work is done on the system by its surroundings—the system is compressed during process *b*.

SELF-TEST 17-5. Suppose the system in the above example is taken from *i* to *f* with a process *c* where $W_c = 15$ kJ. What is Q_c? ***ANSWER:*** 19 kJ.

EXAMPLE 17-6

A tale of two processes. ***(a)*** The temperature of 0.25 kg of water is gradually raised from 1.1 to 7.7°C at atmospheric pressure with a resistance heater. During this process, the volume of the water changes insignificantly. Determine the change in internal energy of the water. ***(b)*** An amount 0.25 kg of water initially at 1.1°C in a thermos is vigorously stirred until the temperature rises to 7.7°C. The initial and final pressures are atmospheric. Determine the change in internal energy of the water and the work done by the water.

Solution. *(a)* For this process, the heat added to the water at constant pressure is given by

$$Q = mc_p\,\Delta t_c = (0.25\text{ kg})(4180\text{ J}\cdot\text{kg}^{-1}\cdot{}^\circ\text{C}^{-1})(6.6^\circ\text{C}) = 6.9\text{ kJ}$$

where we have assumed that the specific heat of water from Table 17-1 is essentially constant over this temperature range. Since the volume change is negligible, the work done by the water in this quasi-static process is also negligible, $W = 0$. The first law of thermodynamics, Eq. 17-12, gives

$$U_f - U_i = Q - W = 6.9\text{ kJ} - 0 = 6.9\text{ kJ}$$

(b) In this second process, which is not quasi-static, the initial state and the final state are the same as for the process in part *(a)*. The change in internal energy must also be the same, because it depends only on the initial and final states and not on the process:

$$U_f - U_i = 6.9\text{ kJ}$$

Since this process occurs with the system insulated by the thermos, the heat added is negligible, $Q = 0$. The first law applied to this process gives

$$U_f - U_i = Q - W$$

$$6.9\text{ kJ} = 0 - W$$

or $W = -6.9$ kJ. The water does negative work on the stirring mechanism.

SELF-TEST 17-6. Give a physical interpretation that explains why the work done by the system is negative during the second process discussed in the above example.

17-5 SOME APPLICATIONS OF THE FIRST LAW

The first law of thermodynamics describes the energy exchanges for any process that takes a system from an initial equilibrium state to a final equilibrium state. The first law is applied here to several types of processes. In each case we evaluate two of the three quantities—Q, W, ΔU—and use the first law to determine the third quantity.

Isochoric Process

Since no work is done by the system during an isochoric process, the first law of thermodynamics gives

$$\Delta U = Q_V$$

That is, if the volume remains fixed, then the only way that the system can interact with its surroundings is by heat transfer. If heat is added to the system, the internal energy increases, and if heat is rejected by the system, the internal energy decreases. The subscript V is a reminder that the heat is transferred isochorically.

EXAMPLE 17-7

Change of ΔU during an isochoric process. The temperature of 2.50 mol of He is raised at constant volume from 275.0 to 325.0 K. The initial state is at atmospheric pressure, and the molar heat capacity of He is $\mathscr{C}_V = 12.5\text{ J}\cdot\text{mol}^{-1}\cdot\text{K}^{-1}$ over this range of temperatures and pressures. Evaluate the change in internal energy of this gas.

Solution. Since the volume remains fixed, the heat added to the system is, from Sec. 17-2,

$$Q_V = n\mathscr{C}_V\,\Delta T = (2.50\text{ mol})(12.5\text{ J}\cdot\text{mol}^{-1}\cdot\text{K}^{-1})(50.0\text{ K}) = 1.56\text{ kJ}$$

Since the process occurs at constant volume, $W = 0$. The first law gives for the change in internal energy,

$$U_f - U_i = Q_V - W = 1.56\text{ kJ} - 0 = 1.56\text{ kJ}$$

SELF-TEST 17-7. For the process in the above example, sketch the path of the process on a p-V diagram. Label the points that represent the initial and final states with i and f and show the direction of the process with an arrowhead.

Adiabatic Process

No heat is exchanged between a system and its surroundings during an adiabatic process, so the first law becomes

$$\Delta U = -W_Q$$

This means that the system's internal energy increases when the system is adiabatically compressed. That is, if $\Delta V < 0$, then $W < 0$ (as we saw earlier) and $\Delta U > 0$. Also, the system's internal energy decreases when the system adiabatically expands. (If $\Delta V > 0$, then $W > 0$, and $\Delta U < 0$.)

In practice, an adiabatic process can be performed by surrounding the system with a good insulator or by perfoming the process rapidly enough so that a negligible amount of heat is transferred.

EXAMPLE 17-8

Compression stroke in an automobile engine. During the compression stroke for a cylinder in an experimental engine, the volume decreases by a factor of 8 (the compression ratio). The work done by the air-fuel mixture for this compression is measured to be $W = -200$ J. Evaluate the change in internal energy of the air-fuel mixture.

Solution. Because the compression occurs quickly, this process can be considered to be adiabatic: $Q = 0$. The work done by the mixture is $W_Q = -200$ J and is negative since the volume decreased. Applying the first law, we obtain for the change in internal energy

$$U_f - U_i = Q - W = 0 - (-200 \text{ J}) = 200 \text{ J}$$

SELF-TEST 17-8. A nail is quickly withdrawn from a board and its internal energy increases by 60 J. What work is done by the nail on its surroundings? ***ANSWER:*** -60 J.

Isobaric Process

For an isobaric process, both heat and work may be transferred between the system and its surroundings. The work done by the system for a quasi-static isobaric process is given by Eq. 17-10. The heat added to the system can be evaluated using specific-heat or latent-heat data.

EXAMPLE 17-9

Change in U during an isobaric process. The volume occupied by 1.00 kg of water at 100°C and atmospheric pressure changes from 1.0 L in the liquid phase to 1700 L in the vapor (steam) phase. Evaluate the difference in internal energy of 1.00 kg of water vapor and 1.00 kg of liquid water at the normal boiling point.

Solution. Consider a process in which heat is added at constant atmospheric pressure to convert 1.00 kg of water from liquid at 100°C to vapor at 100°C. The water is contained in a cylinder fitted with a piston that moves outward to keep the pressure fixed. The volume change is 1700 L − 1L = 1.7 m³. From Eq. 17-10 the work done by the system is

$$W_p = p\,\Delta V = (101 \text{ kPa})(1.7 \text{ m}^3) = 170 \text{ kJ}$$

The heat added during the process is evaluated using the latent heat of vaporization from Table 17-2:

$$Q_p = mL_V = (1.00 \text{ kg})(2260 \text{ kJ/kg}) = 2260 \text{ kJ}$$

The difference in internal energy for these two states of water is

$$U_f - U_i = Q_p - W_p = 2260 \text{ kJ} - 170 \text{ kJ} = 2090 \text{ kJ} \approx 2100 \text{ kJ}$$

The calculation shows that of the 2260 kJ of heat added to the system, about 170 kJ of work is performed by the system, and the remainder, about 2100 kJ, appears as an increase in internal energy. Although we used a specific process to calculate the difference in internal energy of these two states of water, note that the 2100-J difference in internal energy is independent of the process.

SELF-TEST 17-9. Suppose 0.35 mol of neon gas undergoes an isobaric process from an initial state $p_i = 101$ kPa, $V_i = 8.64$ L to a final state $p_f = 101$ kPa, $V_f = 11.52$ L. Determine the change in the internal energy. For neon, $\mathscr{C}_p$ is constant at 18.8 $\text{J}\cdot\text{mol}^{-1}\cdot\text{K}^{-1}$ in this temperature range. ***ANSWER:*** 370 J.

Free Expansion

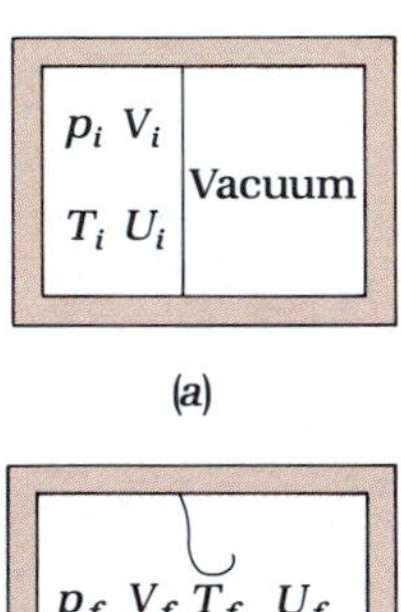

Figure 17-13. A free expansion. *(a)* A membrane confines a gas to one side of an insulated container. *(b)* The membrane ruptures spontaneously, and the gas expands freely to occupy the entire volume.

Consider a gas that initially occupies one compartment of a two-chambered container as illustrated in Fig. 17-13. A membrane separates the two chambers, and the one on the right is evacuated. The entire assembly is insulated from the exterior. The initial state of the gas is characterized by the values of some of its variables of state: p_i, V_i, T_i, and U_i. Suppose now that the membrane separating the two chambers spontaneously breaks and the gas expands freely to fill the entire container. This process is called a *free expansion.* During the free expansion, the gas is not in equilibrium, so the process is not quasi-static and cannot be represented on a *p-V* diagram. But the gas will eventually come to a final equilibrium state in which the variables of state have values p_f, V_f, T_f, and U_f. How have these final values changed from the initial values? Certainly the volume has increased, and measurement shows that the pressure has decreased. The temperature change can also be measured (although not easily for a very small change).

The change in the internal energy of the gas can be calculated by applying the first law of thermodynamics to the free-expansion process. The process is adiabatic because of the insulation, so $Q = 0$. No part of the surroundings moves (we consider the rupturing membrane to be an inert part of the system), so the system does no work on its surroundings: $W = 0$. Therefore the internal energy does not change:

$$U_f - U_i = Q - W = 0 - 0 = 0 \qquad (17\text{-}13)$$

The initial and final states of this gas have the same internal energy.

It is customary to think of the internal energy of a gas as a function of the independent variables *V* and *T*: $U(V, T)$. Experiments in which a gas undergoes a free expansion give information about this functional dependence. If, in a free expansion, the volume changes from V_i to V_f and the temperature changes from T_i to T_f, then Eq. 17-13 gives

$$U(V_f, T_f) = U(V_i, T_i) \qquad (17\text{-}14)$$

For real gases, measurements show that the temperature changes slightly in a free expansion. Thus Eq. 17-14 indicates a dependence of internal energy on the volume. However, in the limit of dilute gases or for an ideal gas, the temperature drop tends to zero for a free expansion. Let $T_i = T_f = T$; then Eq. 17-14 gives

$$U(V_f, T) = U(V_i, T)$$

U depends only on T for an ideal gas.

which implies that the internal energy of an ideal gas does not depend on the volume at all. The free-expansion process has led us to the following conclusion: *The internal energy U(T) of an ideal gas depends only on the temperature.*

Isothermal Process

From our above discussion of the free expansion, the internal energy of an ideal gas depends only on the temperature. Therefore, if the temperature of an ideal gas remains fixed, so does its internal energy, and the first law becomes $0 = Q - W$, or

$$Q_T = W_T$$

for an ideal gas undergoing an isothermal process. If the gas isothermally expands ($\Delta V > 0$ so that $W_T > 0$), then heat is absorbed ($Q_T > 0$). If the gas isothermally contracts ($\Delta V < 0$ so that $W_T < 0$), then heat is rejected. This expression is also valid for a real gas, provided it is sufficiently dilute.

EXAMPLE 17-10

Isothermal expansion of a gas. A metal cylinder fitted with a movable piston contains 0.24 mol of N_2 gas at an initial pressure of 140 kPa. The piston is slowly withdrawn until the volume of the gas has doubled. The gas remains in good thermal contact with its surroundings at 310 K during the process. How much heat is added to the gas for this process?

Solution. In experiments at these temperature and pressure values, nitrogen behaves as an ideal gas. The process is isothermal at 310 K, and the internal energy of the gas, depending only on T, does not change: $\Delta U = 0$. The work done by an ideal gas for an isothermal expansion is given by Eq. 17-11: $W_T = nRT \ln(V_f/V_i)$. Applying the first law to the process, we have

$$\begin{aligned} Q &= \Delta U + W_T = 0 + (nRT)\ln\left(\frac{V_f}{V_i}\right) \\ &= (0.24 \text{ mol})(8.31 \text{ J}\cdot\text{mol}^{-1}\cdot\text{K}^{-1})(310 \text{ K})(\ln 2) \\ &= 430 \text{ J} \end{aligned}$$

SELF-TEST 17-10. Suppose the gas in the above example undergoes a free expansion; that is, it expands freely into an insulated, evacuated region as seen in Fig. 17-13. The final volume is twice the initial volume, and the final and initial temperatures are the same, 310 K. What are the values of Q, W, and ΔU? Treat the gas as an ideal gas. Explain any differences from the answers in the above example. ***ANSWER:*** $Q = 0$; $W = 0$; $\Delta U = 0$.

Throttling Process

The *throttling process* is of practical importance because it is used in most refrigerating machines. The process occurs when a fluid at higher pressure streams through a porous wall or through a small valve into a region of lower pressure. Usually the process is a continuous one, as illustrated in Fig. 17-14, with the pressure difference across the valve maintained by a pump or compressor and the valve region insulated. Problem 5 shows that as a given amount of gas undergoes the throttling process, the quantity $U + pV$, called the ***enthalpy,*** remains fixed.

If the fluid on the high-pressure side is a liquid close to evaporating and is partially vaporized on the low-pressure side, then the temperature is substantially lowered on the low-pressure side. It is this property that is used in many refrigeration systems.

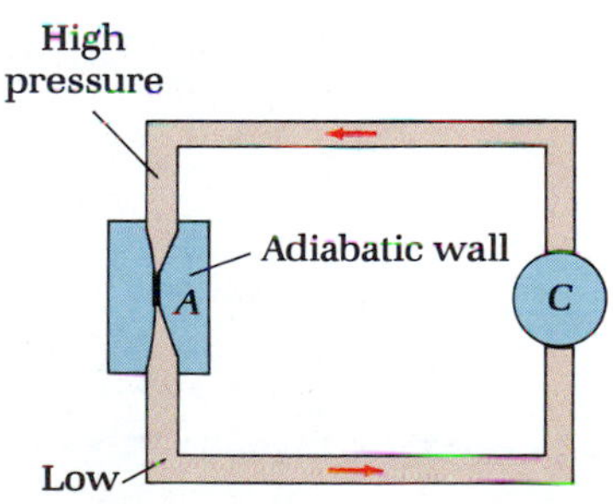

Figure 17-14. A throttling process. Gas at a higher pressure streams through a valve A to a lower-pressure side. The pressure difference is maintained by the compressor C.

Cyclic Process

A *cycle* is a process in which the system is returned to the same state from which it started, as shown in Fig. 17-15. That is, the initial and final states are the same. The cycle has great practical importance because engines and refrigerators use cyclic processes. Also, the cycle has theoretical importance because it can reveal some fundamental properties of nature. These aspects of a cycle will be considered in Chap. 19.

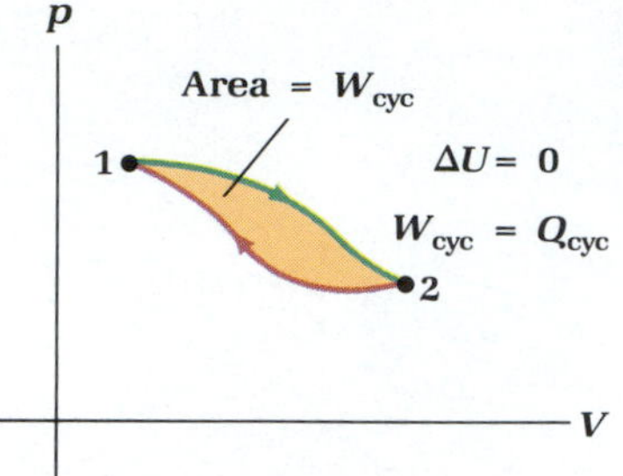

Figure 17-15. A cycle. The area enclosed by the path of a cycle on a p-V diagram equals the net work W_{cyc} done by the system. If the sense of the cycle is clockwise (as in this figure), then $W_{cyc} > 0$. If the sense of the cycle is counterclockwise, then $W_{cyc} < 0$. Since $\Delta U = 0$, $Q_{cyc} = W_{cyc}$.

A cycle can be separated into two parts: an expansion part (from state 1 to state 2 in Fig. 17-15) and a compression part (from state 2 to state 1). The work done by the system during the expansion part is positive, whereas the work done during the compression part is negative. As you can see from the figure, the net work done by the system, W_{cyc}, equals the area enclosed by the path of the cycle in the p-V diagram.

The net work is positive when the pressure is higher during the expansion and lower during the compression, which is the case shown in the figure. In this case, the path proceeds in a clockwise sense. As we shall see in Chap. 19, this clockwise cycle is characteristic of a heat engine. On the other hand, if the pressure is lower during the expansion and higher during the compression, then the path proceeds in a counterclockwise sense and W_{cyc} is negative. A counterclockwise cycle is a feature of a refrigerator.

When a system completes a cycle, each state variable returns to its original value. State variables we know of so far are V, p, T, and U. That is, the change in any state variable after a complete cycle is zero — in particular, $\Delta U = 0$. If we let Q_{cyc} denote the net heat absorbed by the system during a cycle, then the first law gives

$$\Delta U = 0 = Q_{cyc} - W_{cyc}$$

or

$$Q_{cyc} = W_{cyc}$$

For a cycle, the net work done by the system equals the net heat added to the system.

Some of the conclusions from this section are summarized in Table 17-3.

TABLE 17-3. ***Some Results from the First Law***

Process	Features of the process
Isothermal ($\Delta T = 0$)	$\Delta U = 0$ for an ideal gas, $Q_T = W_T$
Isochoric ($\Delta V = 0$)	$W_V = 0$, $\Delta U = Q_V$
Isobaric ($\Delta p = 0$)	$W_p = p_0\,\Delta V$, $\Delta U = Q_p - p_0\,\Delta V$
Adiabatic ($Q = 0$)	$\Delta U = -W_Q$
Cycle	$\Delta U = 0$, $Q_{cyc} = W_{cyc}$

EXAMPLE 17-11

A heat-engine cycle. Suppose 0.0401 mol of an ideal gas undergoes the cycle shown in Fig. 17-16, where $p_0 = 100$ kPa and $V_0 = 1.00$ L. Path a is isochoric, along path b the pressure decreases linearly with the volume, and path c is isobaric. The molar heat capacities of the gas are $\mathscr{C}_V = 12.46\,\text{J}\cdot\text{mol}^{-1}\cdot\text{K}^{-1}$ and $\mathscr{C}_p = 20.77\,\text{J}\cdot\text{mol}^{-1}\cdot\text{K}^{-1}$. Calculate each of the values in Table 17-4.

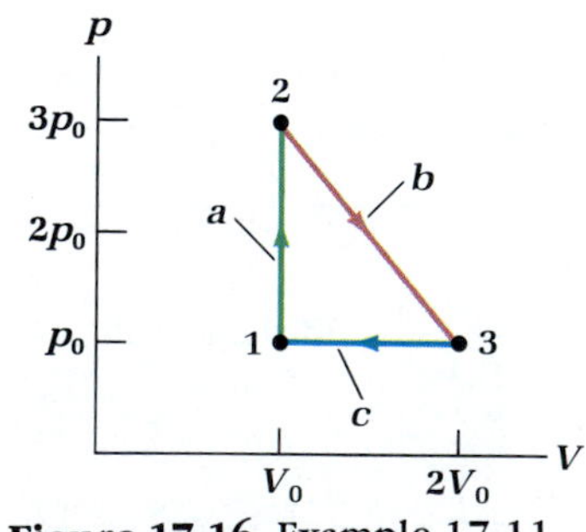

Figure 17-16. Example 17-11.

Solution. First we calculate the work done by the system during each process. Since process a is isochoric,

$$W_a = 0$$

The work done during process b is the area of a trapezoid of average height $2p_0$ and width V_0:

$$W_b = 2p_0V_0 = 2(100\text{ kPa})(1.00\text{ L}) = 200\text{ J}$$

Since process c is an isobaric compression,

$$W_c = -p_0V_0 = -(100\text{ kPa})(1.00\text{ L}) = -100\text{ J}$$

Second, we attempt to calculate the heat absorbed during each process. To do this, we use the ideal gas equation of state, $pV = nRT$, to determine the temperatures of states 1, 2, and 3. For T_1,

$$T_1 = \frac{p_0V_0}{nR} = \frac{(100\text{ kPa})(1.00\text{ L})}{(0.0401\text{ mol})(8.31\text{ J}\cdot\text{mol}^{-1}\cdot\text{K}^{-1})} = 300\text{ K}$$

Similarly, $T_2 = 900$ K and $T_3 = 600$ K. The heat absorbed during process a is

$$Q_a = n\mathscr{C}_V(T_2 - T_1) = (0.0401\text{ mol})(12.46\text{ J}\cdot\text{mol}^{-1}\cdot\text{K}^{-1})(600\text{ K}) = 300\text{ J}$$

and

$$Q_c = n\mathscr{C}_p(T_1 - T_3) = -250\text{ J}$$

We are unable to calculate Q_b in this way because we have no molar heat capacity for that process, but the first law will allow us to find Q_b after we determine ΔU for process *b*.

Third, we use the first law, $\Delta U = Q - W$, to find the internal energy changes. For process *a*,

$$\Delta U_a = Q_a - W_a = 300\text{ J} - 0 = 300\text{ J}$$

and for process *c*,

$$\Delta U_c = Q_c - W_c = -250\text{ J} - (-100\text{ J}) = -150\text{ J}$$

We cannot find ΔU_b in this way because we do not know Q_b, but since U is a state variable, ΔU for the cycle is zero. That is, the column of numbers under ΔU in Table 17-4 adds to zero,

$$0 = \Delta U_a + \Delta U_b + \Delta U_c$$

or

$$\Delta U_b = -\Delta U_a - \Delta U_c = -300\text{ J} - (-150\text{ J}) = -150\text{ J}$$

Now that we have ΔU_b, we can use the first law to find Q_b:

$$Q_b = \Delta U_b + W_b = -150\text{ J} + 200\text{ J} = 50\text{ J}$$

This completes the table. We were able to solve the problem by using the *p-V* diagram, the equation of state, two of the molar heat capacities, and the first law.

SELF-TEST 17-11. Suppose we reverse the path of the cycle in the above example and make it run counterclockwise instead of clockwise. Draw the *p-V* diagram for this cycle and fill in the table of energy values similar to Table 17-4. ***ANSWER:*** All signs are the opposite of those in Table 17-4.

TABLE 17-4. ***Example 17-11***

Process	Energies in J		
	W	*Q*	ΔU
a	0	300	300
b	200	50	−150
c	−100	−250	−150
Cycle	100	100	0

COMMENTARY: Benjamin Thompson, Count Rumford

Benjamin Thompson, Count Rumford.

The flow of heat from a body at a higher temperature to one at a lower temperature is somewhat analogous to the flow of a fluid, say water, from a higher to a lower elevation. It is not surprising that the early theory of heat flow treated heat as a fluidlike substance called caloric. As a body lost caloric, its temperature would drop, and the temperature of a body gaining caloric would increase. While many features of heat flow can be explained with the idea of such a fluid, the caloric theory turned out to be inconsistent with experiment.

Credit for seriously challenging the concept of caloric is usually given to Benjamin Thompson, also known as Count Rumford of Bavaria. Fearing the spread of the French Revolution, the ruling elector of Bavaria assigned Count Rumford to overseeing the construction of cannons for the defense of the borders. As a cannon was bored, Rumford noticed that the cannon stock, the metal chips, and the boring tool all had an increase in temperature. That is, heat seemed to be continuously generated instead of being conserved as the caloric theory supposed.

Rumford conducted quantitative experiments by measuring the temperature change that occurred when a blunt tool was used in the boring process. In one experiment, water was used to cool the boring tool and the cannon stock. Rumford measured the increasing temperature of the water and noted "the surprise and astonishment expressed in the countenances of the by-standers, on seeing so large a quantity of cold water heated, and actually made to boil without any fire." He concluded that heat was not a material substance since there seemed to be no limit to it. Rather, it was the result of friction, or of the work done by friction.

Count Rumford was born Benjamin Thompson in Woburn, Massachusetts, in 1753. His early life showed little promise of eventual nobility. He served two unfulfilled apprenticeships with shopkeepers. One shopkeeper complained to Thompson's mother that Benjamin spent more time under the counter making little machines and reading science books than he did behind the counter serving customers. Thompson's fortunes improved when, at age 19, he married a wealthy 33-year-old widow in Concord, New Hampshire, a region that was also known as Rumford.

Thompson was a loyalist in the disputes between Britain and her American colonies. He served as a major in a company of militia. When his loyalist sentiments became known, a group of colonists, dressed as Indians, arrived at his doorstep, threatening a tar-and-feathers treatment. Thompson escaped to Boston with a horse, $20, and his life.

By his own account, Thompson served the British as a courageous and inventive officer during the American Revolution. After one of his horses drowned crossing a river, he invented a cork life preserver for a horse carrying a cannon on its back. He also designed a gun carriage that could be carried by three horses and could be assembled and fired in 75 seconds.

After being knighted by England's King George III, Thompson was introduced to the court of Theodor, Elector of Bavaria. There he conducted experiments on the properties of silk, an important product of Bavaria at the time. He entertained the court with calculations such as, "If a silk gown worn by a lady weighs 28 ounces, it is very certain that she carries upon her back upwards of 2000 miles in length of silk, as spun by the worm. . . ."

Appointed a major general by the elector, Thompson improved the lot of the Bavarian soldiers. While investigating materials that could provide the greatest warmth for his soldiers, he discovered the great insulating value of a layer of entrapped air. Thompson also provided opportunities for the soldiers that would allow them to earn money for necessities. In experiments to determine the best lighting conditions for workhouses for the poor, Thompson established the candle as a standard for measuring illumination.

Thompson's benefactor, the Elector Theodor, enjoyed a brief reign as vicar of the Holy Roman Empire between the death of Emperor Leopold II and the coronation of Emperor Francis II. As vicar, Theodor had limited powers, but he did have the privilege of elevating a person to the nobility. Thus, on May 9, 1792, Theodor exercised this privilege, and Benjamin Thompson became Count Rumford.

The Count established two large awards for scientific discoveries in the subjects of heat and light. The awards were to be medals of silver or gold equal in value to the interest that had accrued on the original principal. One of the awards was to be handled by the Royal Society in London. When no medal was awarded after six years, Count Rumford had himself appointed to the selecting committee, and in 1802 he became the first recipient of the Rumford Medal. His accomplishments were not recognized by his contemporaries, however, and when he died of "nervous fever" in 1814, few people witnessed his burial.

For further reading, see *Benjamin Thompson, Count Rumford* by Sanborn C. Brown (M.I.T. Press, Cambridge, Mass., 1979).

SUMMARY

Section 17-1. Equations of State
The variables of state p, V, T, and n for a substance are connected by a mathematical equation called the equation of state. For an ideal gas, or a real dilute gas, the equation of state is $pV = nRT$. An equilibrium state of a system such as a gas can be represented by a point on the p-V diagram. A quasi-static pro-

cess, in which the system passes through a succession of equilibrium states, is represented by a curve on a p-V diagram.

Section 17-2. Specific Heat and Latent Heat

The heat added to a system at constant pressure is related to the temperature change by the specific heat.

$$dQ_p = mc_p\, dT \tag{17-2}$$

if the mass m is specified, or by the molar heat capacity

$$dQ_p = n\mathscr{C}_p\, dT \tag{17-5}$$

if the number of moles n is specified. The specific heat at constant volume c_V and the molar heat capacity at constant volume $\mathscr{C}_V$ are similarly defined. The heat added to a system of mass m undergoing a phase change is

$$Q = mL \tag{17-7}$$

where L is the latent heat for the phase change.

Section 17-3. Work

Work is an energy transfer between a system and its surroundings because of the motion of some part of the surroundings. The work done by a fluid on its surroundings is

$$W = \int_{V_i}^{V_f} p\, dV \tag{17-9}$$

The work done by the system is positive if energy is transferred from the system to the surroundings.

Section 17-4. The First Law of Thermodynamics

The first law of thermodynamics gives the relation between the energy transferred between a system and its surroundings and the change in internal energy of the system:

$$Q - W = U_f - U_i \tag{17-12}$$

The heat added to the system and the work done by a system depend on the details of the process. The internal energy is a variable of state; the change in internal energy $U_f - U_i$ depends only on the two states and not on the process connecting them.

Section 17-5. Some Applications of the First Law

The first law of thermodynamics can be easily applied to some special processes such as the isochoric, adiabatic, isobaric, isothermal, throttling, and cyclic processes. From free-expansion experiments on dilute gases, the first law implies that the internal energy of an ideal gas depends only on temperature.

QUESTIONS

1. Are there any circumstances in which the equation of state $pV = nRT$ correctly describes He in *(a)* its gaseous phase? *(b)* A liquid phase? Explain.

2. Are there any circumstances in which the equation of state $pV = nRT$ correctly describes H_2O in *(a)* its liquid phase? *(b)* One of its solid phases? Explain.

3. Air at ordinary temperatures and pressures is a mixture of several different gases, principally nitrogen and oxygen. Is its equation of state given by $pV = nRT$? What is the meaning of n in this case?

4. What is the value of the universal gas constant R in units of $\mathrm{cal \cdot mol^{-1} \cdot K^{-1}}$?

5. Can a process which is not quasi-static be shown on a p-V diagram? Explain.

6. Heat can be added to a system while holding the pressure fixed. Heat can also be added at constant volume. Can heat be added while holding the temperature fixed? Explain.

7. Work can be done by a system while holding the pressure fixed or while holding the temperature fixed. Can work be done by the system while holding the volume fixed? Explain.

8. Must the internal energy of a system increase if heat is added to the system? Explain.

9. Must the internal energy of a system increase if its temperature increases? Explain.

10. Suppose that a system does work W on its surroundings in a process. How much work is done by the surroundings on the system? Can you think of any exceptions to your answer?

11. The outside surface of a metallic container of gas is buffed vigorously by a polishing wheel. Is the energy transferred to the gas inside the container called heat or work? Explain.

12. A gas in a cylinder expands as the piston is withdrawn. A frictional force acts between the piston and the walls of the cylinder. Is the energy transferred to the gas due to the friction heat or work? Explain.

13. Can you "heat up" a bowl of soup without adding heat? Must the temperature of a system change if heat is added?

14. Suppose a system undergoes a process in which the final state has the same volume as the initial state. Can you determine how much work the system has done? Use a p-V diagram to explain your answer.

15. An ice cube is placed in a well-insulated beaker of lukewarm water. Take the system to consist of the ice cube and the water. The ice melts, and the final state of this system is liquid water at some final lower temperature. Is this process adiabatic? What provided the energy to melt the ice?

16. Explain why your skin would be more severely burned if put in contact with 1 g of steam at 100°C than with 1 g of liquid water at 100°C.

17. Work can be represented conveniently on a p-V diagram. What useful information can be shown on a P-T diagram? (See Chap. 16.)

18. Show an isobaric process on a p-T diagram. Show an isothermal process on a p-T diagram. Show, for an ideal gas, an isochoric process on a p-T diagram.

19. Given the initial and final states of a system on a p-V dia-

gram and the value of the change in internal energy, can you determine how much heat was added and how much work was done? Explain.

20. In a cycle the final state of the system is the same as the initial state. Given the heat added in a cycle, can you determine how much work is done by the system? Explain.

21. From the properties or condition of a system undergoing a cycle, can you determine how many cycles have been performed previously? Explain.

22. A small magnet immersed in a liquid stirs the liquid by being driven by an external rotating magnet. Explain why this energy transfer to the liquid is work even though there is no change in volume.

23. Which, if any, of the following types of processes must be quasi-static? Explain your answer.
(a) Isobaric
(b) Isothermal
(c) Isochoric

24. An electric water heater has resistive elements immersed in the water in the tank. *(a)* If the system is considered to be the whole unit, is the energy transferred to the system heat or work? *(b)* If the system is considered to be just the water, is the energy transferred to the system heat or work? *(c)* In each case above, what happens to the internal energy of the system?

25. Draw an analogy between the first law of thermodynamics and your personal finances, using the three concepts: income q, expenditures w, and cash on hand u. Does it make sense to speak of the amount of expenditures on hand? What equation connects the three quantities q, w, u? Why is the analogy imperfect?

26. Nutritionists (and dieters) work with the "large Calorie," 1 Calorie = 1 kcal. Suppose that you are on a strict 1600-Calorie-a-day diet. At what average rate in joules per day does your body metabolize your food? Express this rate in watts also.

27. Complete the following table:

Symbol	Represents	Type	SI Unit
R		Scalar	
c_p			
c_V			
$\mathscr{C}_V$			
L	Latent heat		
W			
ΔU			J
U			

EXERCISES

Section 17-1. Equations of State

1. An ideal gas undergoes a process in which the temperature is doubled and the pressure is tripled. *(a)* By what factor is the volume changed? *(b)* Show the initial and final states on a p-V diagram.

2. Determine the volume in liters (1 L = 10^{-3} m^3) occupied by 1 mol of an ideal gas at atmospheric pressure (101 kPa) and 0.0°C.

3. Helium gas initially is in a state characterized by $p = 0.73$ kPa, $V = 12$ L, and $T = 320$ K. *(a)* Determine the amount of gas present. *(b)* The gas expands isothermally until the volume is 18 L. Determine the pressure of the He in this state. *(c)* Show the process on a p-V diagram.

4. An ideal gas is initially in the state labeled by p_i, V_i, T_i. It undergoes an isothermal expansion to an intermediate state m in which the pressure is $p_m = \frac{1}{2}p_i$. The gas is then compressed to a final state f at that constant value of pressure p_m until the volume is returned to its initial value. *(a)* Show these processes on a p-V diagram. *(b)* Determine the values of the variables p, V, T for states m and f. Express these in terms of the values for the initial state.

5. Gas pressure is often expressed in atmospheres (1 atm = 101 kPa) and volume in liters (L). Determine the value of the universal gas constant in units of L·atm·mol^{-1}·K^{-1}. [*Note:* (1 Pa)(1 m^3) = 1 J.]

6. A cylinder contains 2.54 mol of O_2 at 113 kPa and 325 K. The gas is compressed isothermally to half its original volume. The moving piston does not fit tightly, and 0.26 mol escapes in the process. *(a)* Determine the final pressure of the gas. *(b)* Can this process be represented meaningfully on a p-V diagram? Explain.

7. Determine the number of molecules per cubic meter for an ideal gas at standard conditions of 1.0 atm and 0.0°C.

8. The coefficient of volume expansion (see Chap. 16) for a gas at constant pressure is defined by $\beta = (1/V)(dV/dT)$, where the derivative is taken with the pressure regarded as a constant. *(a)* Show that $\beta = 1/T$ for an ideal gas. *(b)* Evaluate β at 0.0°C for an ideal gas.

Section 17-2. Specific Heat and Latent Heat

9. How much heat must be added at constant pressure to a 3-g iron nail to raise its temperature by 20°C?

10. How much heat must be added at constant pressure to 1.5 mol of iron to raise its temperature from 280 to 320 K?

11. An insulated vessel contains 0.75 kg of water at 20°C, and 1.24 kg of lead, initially at 95°C, is added. *(a)* Assuming no energy exchanges with the surroundings, determine the final temperature of the water-lead system. *(b)* Considering only the water as the system, how much heat was added to the water in the process?

12. Suppose that the water-lead system in the previous exercise is initially at 15°C and that 1800 J of heat is added using an immersion heater. *(a)* What is the final temperature of the system? *(b)* How much heat is added to the lead?

13. How much water, initially at 25°C, must be added to 0.35 kg of ice at 0.0°C to completely melt the ice? The final state consists of liquid at 0.0°C.

14. A 0.35-kg lump of ice at 0.0°C is placed in an insulated container of water initially at 25°C. *(a)* If the original amount of water is 2.0 kg, determine the final temperature and composition of the system. *(b)* Repeat the calculation for 1.0 kg of water present initially.

15. The temperature of 1.2 kg of H_2O is measured with a thermometer of mass 0.033 kg and of specific heat 1070 J·kg^{-1}·°C^{-1}. The thermometer reads 23.5°C before it is inserted in the water. After coming to thermal equilibrium with the water, the thermometer reads 57.9°C. *(a)* Neglecting other energy exchanges with the surroundings, determine the temperature of the water before the thermometer is inserted. *(b)* Suppose this thermometer is used to measure the temperature of 0.012 kg of water. Comment on the effect of the measurement process on the measured value.

16. The specific heat of Si is measured by dropping a 1.50-kg Si slug, initially at 40.0°C, into 3.00 kg of water, initially at 25.0°C. The system comes to a final temperature of 26.2°C. Neglecting energy transfers to the outside, determine the specific heat of Si from these data.

Section 17-3. Work

17. A fluid expands at constant atmospheric pressure (101 kPa) from an initial volume of 0.344 m³ to a final volume of 0.424 m³. *(a)* Determine the work done by the fluid. *(b)* Determine the work done by the fluid if the first process is reversed—that is, the fluid is compressed at atmospheric pressure back to the original volume.

18. A cylinder with a movable piston contains 96 g of O_2 initially at 150 kPa pressure and at 290 K. *(a)* Determine the volume occupied by the gas. *(b)* The gas expands at constant pressure, performing 7.2 kJ of work in the process. What is the volume of the final state? *(c)* The pressure is then increased isochorically to 300 kPa. How much work is done by the gas for the entire process?

19. An ideal gas undergoes a process, shown in Fig. 17-17, which consists of an isobaric expansion followed by an isothermal compression. Determine the work done by the gas for *(a)* the isobaric expansion, *(b)* the isothermal compression, *(c)* the entire process. *(d)* Check your answers by estimating areas under the process curves in the figure.

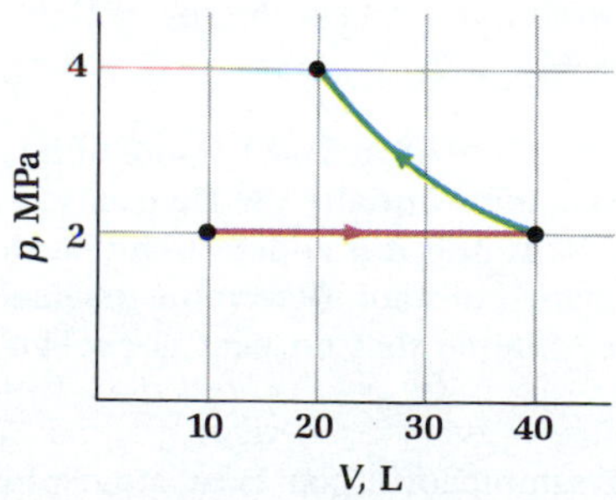

Figure 17-17. Exercise 19.

20. It is possible to have a process similar to that shown in Fig. 17-17 for which the net work is zero. Suppose the isobaric process is as shown and the isothermal compression takes the system to some final volume V_f. Determine V_f such that $W = 0$ for the entire process.

21. An ideal gas is initially in the state specified by p_i, V_i. It undergoes a process in which the pressure changes linearly with volume to the final state specified by p_f, V_f, as shown in Fig. 17-18. *(a)* Determine the work done by the gas for this process in terms of the initial and final pressure and volume values. *(b)* Evaluate the work done by the gas for this process if $p_i = 140$ kPa, $V_i = 0.064$ m³, $p_f = 108$ kPa, $V_f = 0.096$ m³. *(c)* Determine the temperatures of the initial and final states if the system consists of 3.0 mol of gas.

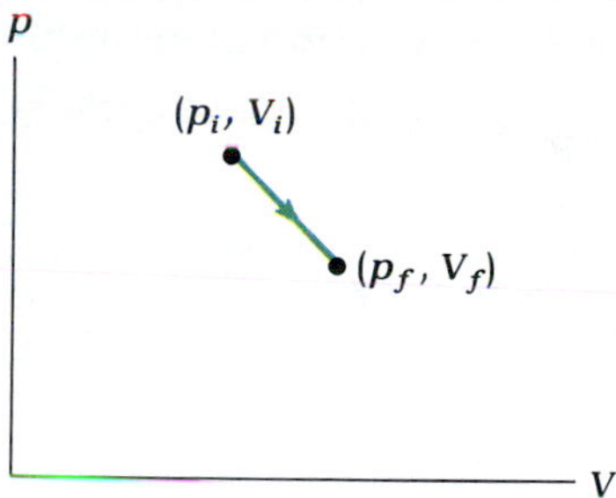

Figure 17-18. Exercise 21.

22. A gas undergoes a quasi-static process from state *i* to state *f* where the pressure depends on the volume according to the relation

$$p = (p_i V_i^{3/2}) V^{-3/2}$$

Show that the work done by the gas for this process is given by the expression

$$W = 2p_i V_i (1 - \sqrt{V_i/V_f}$$

(Note: For an ideal gas with $\gamma = \frac{3}{2}$, these relations describe an adiabatic process.)

23. The value of γ for air is 1.40. Suppose 1.0 mol of air is initially in a state such that $p_i = 202$ kPa and $V_i = 45$ L. The air expands adiabatically to a volume $V_f = 65$ L. Determine *(a)* the work done by the air and *(b)* the final pressure.

Section 17-4. The First Law of Thermodynamics

24. A system undergoes a process in which 27 J of heat is added to the system while it performs 8 J of work. *(a)* What is the change in internal energy of the system? *(b)* If the internal energy of the initial state is 304 J, what is the internal energy of the final state? *(c)* Suppose the system is taken from the same initial state to the same final state by a different process in which the system performs 12 J of work. What is the internal energy of the final state and how much heat is added?

25. In Fig. 17-19, a fluid undergoes first an isobaric process from states 1 to 2 where $Q_p = 10.0$ kJ and then an isochoric process where $Q_V = 11.0$ kJ. Given the $U_1 = 5.0$ kJ, determine

(a) U_2 and *(b)* U_3. *(c)* If the fluid undergoes process *c* with the curved path where $W_c = -6.6$ kJ, determine Q_c.

26. A nail is placed in a water bath and both are initially at 20°C. The water and the nail are gradually heated to 90°C. The internal energy of the nail increases by 45 J. An identical nail initially at 20°C is quickly withdrawn from a wooden block by a hammer. As a result, the temperature of the nail is 90°C. For each of the two processes, state whether heat was added or work was done. By how much did the internal energy of the nail change in the second process?

Section 17-5. Some Applications of the First Law

27. One mole of He undergoes the process 1-2-3 shown in Fig. 17-19. The molar heat capacity values are $\mathscr{C}_V = 12.5\ \text{J}\cdot\text{mol}^{-1}\cdot\text{K}^{-1}$ and $\mathscr{C}_p = 20.8\ \text{J}\cdot\text{mol}^{-1}\cdot\text{K}^{-1}$. Evaluate the difference in internal energy between *(a)* states 1 and 2, *(b)* states 2 and 3, *(c)* states 1 and 3.

28. At atmospheric pressure and 0.0°C, the volume of 1.00 kg of ice is $V_S = 0.917$ L, while the volume of the same mass of liquid water is $V_L = 1.000$ L. Evaluate for 1.00 kg of water the difference in internal energy of the liquid and solid states, $U_L - U_S$.

29. Determine the difference in internal energy for 1 mol of H_2O between liquid at the normal melting point and liquid at the normal boiling point. The small changes in the volume of liquid water as its temperature is changed at atmospheric pressure may be neglected.

30. Consider the cycle shown in Fig. 17-16 with $p_0 = 150$ kPa and $V_0 = 0.50$ L. For the entire cycle determine *(a)* the work done by the system and *(b)* the heat added to the system.

31. For each process listed in the table below, supply the sign (+, −, or 0) for each missing entry:

Process description	Q	W	ΔU
Isochoric pressure drop			−
Isobaric compression of an ideal gas			
Adiabatic expansion			
Isothermal expansion of an ideal gas			
Isothermal compression of an ideal gas			
Free expansion of a gas			
Cyclic process		+	

32. Each kilogram of water drops about 50 m at Niagara Falls. By how much does the temperature of the water rise as a result? Assume that the increase in internal energy of the water in the process is due to the change in its gravitational potential energy.

Additional Exercises

33. A lead brick is released from rest and falls 4 m to the floor below. Estimate the maximum increase in temperature of the brick as a result. Why is it not necessary to know the mass of the brick?

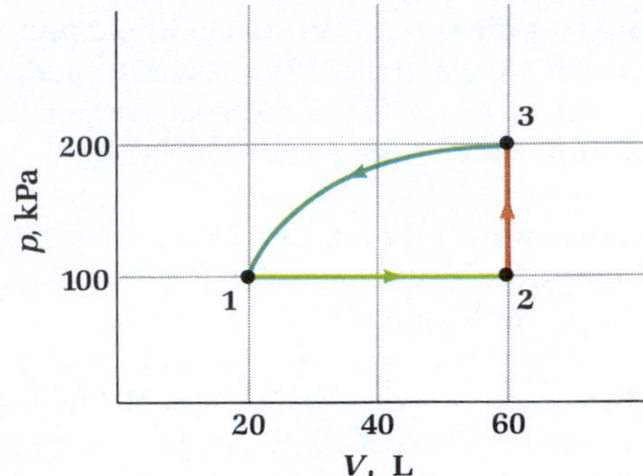

Figure 17-19. Exercise 25.

34. A 2000-kg automobile is traveling at 20 m/s when the driver applies brakes and brings the car to rest. Estimate the change in the internal energy of the surrounding air.

35. Determine the minimum amount of water, initially at 50°C, that is needed to melt a 1 kg of ice initially at 0°C. Assume that the water and ice are placed together in an insulated container and neglect energy exchanges with the surroundings.

36. A cylinder of volume 0.044 m³ contains He gas at 345 K and 160 kPa. *(a)* Determine the number of moles present. *(b)* The gas undergoes an isothermal compression to a final volume of 0.018 m³. What is the final pressure? Show this process on a *p*-*V* diagram.

37. *(a)* Show that the work done by an ideal gas undergoing an isothermal process can be written $W_T = (nRT)\ln(p_i/p_f)$. *(b)* Evaluate W_T for 0.080 mol of a gas whose pressure increases isothermally from 150 to 300 kPa at $T = 350$ K.

38. Determine the difference in internal energy between the solid and the liquid phase of 1 kg of iron at its melting point at atmospheric pressure. Neglect any change in volume on melting.

39. Water enters a hot-water heater at 15°C, and the temperature is raised to 60°C by an electric heating element. Estimate the cost per day for heating water if the average usage is 1 m³ per day and electric energy sells for $0.15 per kilowatt-hour.

40. A substance goes from an initial state *i* to a final state *f* by a process *a*, for which $Q_a = 1250$ J, $W_a = 380$ J. Suppose the substance goes from the same initial state *i* to the same final state *f* by a *different* process *b*, for which $W_b = 440$ J. Determine $U_f - U_i$ and Q_b.

41. Two gases, 2.0 mol of He and 1.0 mol of N_2, are mixed in an insulated container. Initially the He gas is at 320 K in the container. The N_2 at 380 K is added such that the pressure of the mixture remains constant. Determine the final temperature of the mixture. Assume that no heat is exchanged with the surroundings.

42. Suppose a sample of argon is at atmospheric pressure ($p_i = 101$ kPa) and occupies a volume $V_i = 1.00$ L. What is the

change of the internal energy of this gas when it is compressed adiabatically to a volume $V_f = 0.75$ L? The ratio of specific heat for argon is $\gamma = 1.65$.

43. Suppose a 2-kg iron skillet contains 1 kg of water. What is the temperature change of the skillet and the water if 100 kJ of heat is added? Assume that the skillet and the water have the same temperature change.

44. How much heat must be extracted at atmospheric pressure from 0.50 kg of water that is initially liquid at 37°C to change it to ice at −10°C?

45. Helium gas at 310 K is contained in a cylinder with a movable piston. The gas undergoes the following steps: *(a)* the pressure increases from 91 to 202 kPa at a constant volume of 48 L and *(b)* the volume increases from 48 to 106 L at a constant pressure of 202 kPa. Determine the work done by the gas for each step and for the entire process. Sketch the process on a p-V diagram.

46. Suppose that 0.25 kg of water undergoes a process where 7.5 kJ of heat is gradually added and the work done by the water in vigorous shaking is −7.5 kJ. *(a)* What is the change in the internal energy? *(b)* If the initial temperature is 1.1°C, what is the final temperature? Assume that the initial and final pressures are atmospheric.

47. Determine the change in internal energy of 2.50 mol of helium gas if the temperature changes from 275 to 375 K while the volume is held constant. Over this temperature range $\mathscr{C}_V = 12.5\ \text{J}\cdot\text{mol}^{-1}\cdot\text{K}^{-1}$ for helium.

48. Determine the change in the internal energy of 1.00 kg of water at atmospheric pressure if it is initially ice at 0.00°C with volume 1.1×10^{-3} m^3 and changes to liquid at 0.00°C with volume 1.0×10^{-3} m^3. Explain why the work can be neglected in comparison with the heat added.

49. A system undergoes the cycle shown in Fig. 17-20 where $p_0 = 100$ kPa and $V_0 = 1.00$ L. Heat is absorbed by the system during process a such that $Q_a = 450$ J, and heat is absorbed by the system during process b such that $Q_b = 200$ J. Also $U_1 = 200$ J. Determine *(a)* U_2, *(b)* U_3, *(c)* W_{cyc}, and *(d)* Q_c. *(e)* Did the system absorb heat or reject heat during process c?

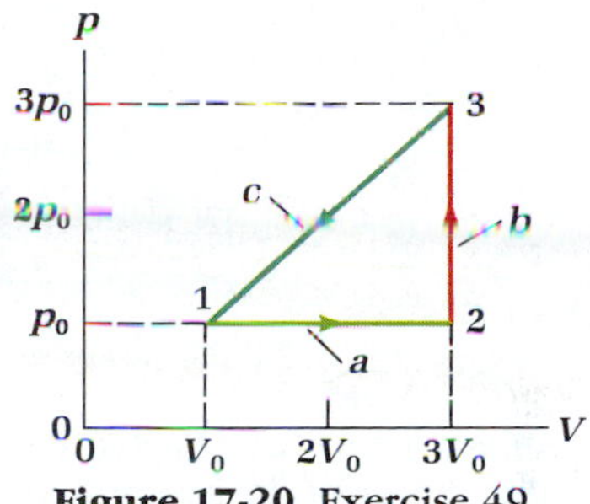

Figure 17-20. Exercise 49.

50. Repeat the previous exercise except with the processes occurring in the opposite sense. That is, the path of the cycle is counterclockwise rather than clockwise.

PROBLEMS

1. ***A temperature-dependent molar heat capacity.*** The molar heat capacity $\mathscr{C}_p$ for Al changes nearly linearly with temperature from $24.4\ \text{J}\cdot\text{mol}^{-1}\cdot\text{K}^{-1}$ at 300 K to $28.1\ \text{J}\cdot\text{mol}^{-1}\cdot\text{K}^{-1}$ at 600 K. *(a)* Construct a mathematical expression for $\mathscr{C}_p$ of the form $\mathscr{C}_p = A + BT$ by evaluating the constants A and B from the given data. *(b)* Construct a graph showing this temperature dependence of the molar heat capacity. *(c)* Determine the amount of heat added at constant pressure to 2.50 mol of Al as its temperature is raised from 300 to 500 K.

2. ***The van der Waals equation of state.*** The *van der Waals equation of state* for 1 mol of a gas is

$$\left(p + \frac{a}{V^2}\right)(V - b) = RT$$

where a and b are constants that are experimentally determined for a particular gas. The equation describes gases at high densities and pressures more accurately than does the ideal gas equation of state. Use this equation of state to evaluate the work done by 1 mol of a gas for a quasi-static isothermal process at temperature T from an initial state p_i, V_i to a final state p_f, V_f. Express the answer in terms of V_i, V_f, T, and the constants a and b.

3. ***Isothermal compressibility.*** The isothermal compressibility of a substance is defined as (see Chap. 15) $-(1/V)(dV/dp)$, with the understanding that T is held fixed. Evaluate the isothermal compressibility for *(a)* an ideal gas and *(b)* a van der Waals gas (see the previous problem). *(c)* In the latter case, show that as the parameters a and b approach zero, the answer to part *(a)* is obtained.

4. ***A volume-dependent internal energy.*** The internal energy of a system can be considered as a function of V and T: $U(V, T)$. (a) Use the first law of thermodynamics to show that

$$c_V = \frac{1}{m}\frac{dU}{dT}$$

where V is held fixed in taking the derivative with respect to T. *(b)* Explain why a corresponding expression does not hold for c_p; that is,

$$c_p \neq \frac{1}{m}\frac{dU}{dT}$$

where the derivative is taken with p fixed.

5. ***Enthalpy and a throttling process.*** A continuous throttling process is sketched in Fig. 17-21*a* in which a fluid streams adiabatically through a valve from a high-pressure side (p_2) to a low-pressure side (p_1). For a given amount of fluid, the process can be simulated by the process illustrated in Fig. 17-21*b* and *c*. The pistons on either side of the valve move in such a way that an amount of fluid at pressure p_2 is forced through the valve to the region at pressure p_1. Evaluate the work done by the fluid, apply the first law, and show that the enthalpy $U + pV$ is constant for the process: $U_2 + p_2V_2 = U_1 + p_1V_1$.

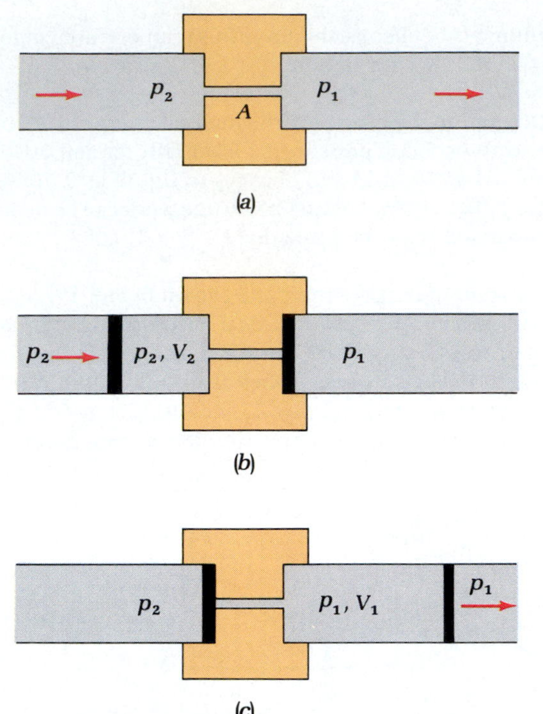

Figure 17-21. Problem 17-5: *(a)* A throttling process is simulated by *(b, c)* the motion of the two pistons.

6. ***Low temperature molar heat capacity.*** At low temperatures the molar heat capacity of a solid such as NaCl has a temperature dependence given approximately by

$$\mathscr{C}_V = 234\ R\left(\frac{T}{\Theta}\right)^3$$

where R is the universal gas constant and Θ is the Debye temperature, a parameter characteristic of the material. For NaCl the Debye $\Theta = 300$ K, and the above expression is valid for T small compared with Θ. Determine the amount of heat added at constant volume to 1.0 mol of NaCl in raising its temperature from 1.0 to 9.0 K.

7. ***Adiabatic process for an ideal gas.*** An ideal gas undergoes a quasi-static adiabatic process. For such a process, show that *(a)* $TV^{\gamma-1} =$ constant and *(b)* $p^{\gamma-1}/T^{\gamma} =$ constant. (See Exercise 22.)

8. ***Isothermal and adiabatic processes for an ideal gas.*** One mole of an ideal gas is initially at 300 K and occupies a volume of 24 L. It undergoes an isothermal expansion in which the volume doubles, followed by an adiabatic compression in which the temperature rises to 600 K. Take $\mathscr{C}_p = 2.5R$ and $\mathscr{C}_V = 1.5R$ for this gas. *(a)* Evaluate the final volume and pressure of the gas. *(b)* Show the process on a *p-V* diagram. *(c)* Evaluate the heat added for the entire process. *(d)* Evaluate the work done by the gas for the entire process (see Exercise 22). *(e)* What is the change in internal energy of the gas? (See the previous problem also.)

9. ***A cycle with two isothermal and two adiabatic processes.*** An ideal gas undergoes the quasi-static process shown in Fig. 17-22. Processes a and c are isothermal and processes b and d are adiabatic. Use the first law and the fact that the internal energy of an ideal gas depends only on the temperature to show that the work done by the gas during the two adiabatic processes is related by $W_b = -W_d$.

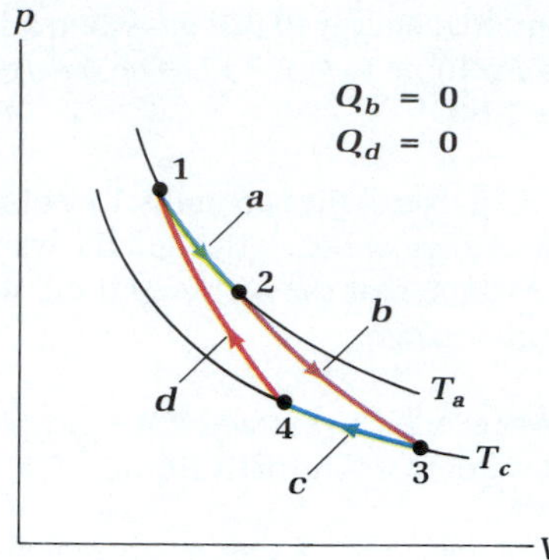

Figure 17-22. Problem 9. An ideal gas undergoes a cycle consisting of two isothermal and two adiabatic processes.

10. ***Internal energy change for a polyatomic gas.*** One model of a polyatomic gas has the following temperature dependence of the molar heat capacity:

$$\mathscr{C}_V = R\left[\frac{5}{2} + \frac{(\Theta/T)^2 e^{\Theta/T}}{(e^{\Theta/T} - 1)^2}\right]$$

where R is the universal gas constant and Θ is a parameter characterizing this gas. Adapt the integrating program from Table 8-1 to determine for a particular gas ($\Theta = 85$ K) the difference in internal energy between two states of 1.0 mol of this gas having the same volume but at temperatures $T_i = 10$ K and $T_f = 200$ K. Since $\mathscr{C}_V = (1/n)(dU/dT)$, note that

$$U_f - U_i = \int_{T_i}^{T_f} n\mathscr{C}_V(T)\ dT$$

11. ***Ideal gas undergoing a cycle I.*** Suppose 0.0963 mol of an ideal gas undergoes the cycle shown in Fig. 17-23, where process b is isothermal. *(a)* Determine the temperatures of states 1 and 2. *(b)* Given that $\mathscr{C}_V = 15.0$ J/mol·K, calculate the heat Q_a absorbed during the isochoric process a. *(c)* Use the first law to find Q_c. *(d)* Evaluate W_{cyc}.

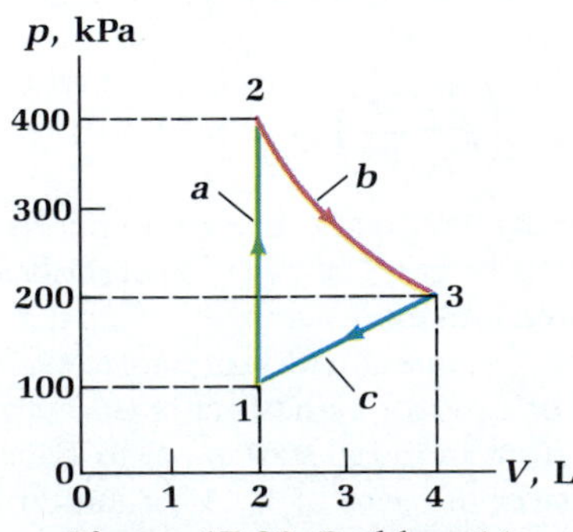

Figure 17-23. Problem 11.

12. ***Ideal gas undergong a cycle II.*** Repeat the previous problem except with the processes reversed so that the cycle proceeds in a counterclockwise sense.

13. ***Ideal gas undergoing a cycle III.*** Suppose 0.0782 mol of an ideal gas undergoes the cycle shown in Fig. 17-24. Process *a* is isochoric at $V = 2.08$ L, process *b* has the pressure varying linearly from 200 to 300 kPa as the volume decreases to 1.00 L, and process *c* is adiabatic. The molar heat capacities are $\mathscr{C}_V = 16.62\ \mathrm{J \cdot mol^{-1} \cdot K^{-1}}$ and $\mathscr{C}_p = 24.93\ \mathrm{J \cdot mol^{-1} \cdot K^{-1}}$. Construct a table similar to Table 17-4 in Example 17-11, with each energy transfer given in joules.

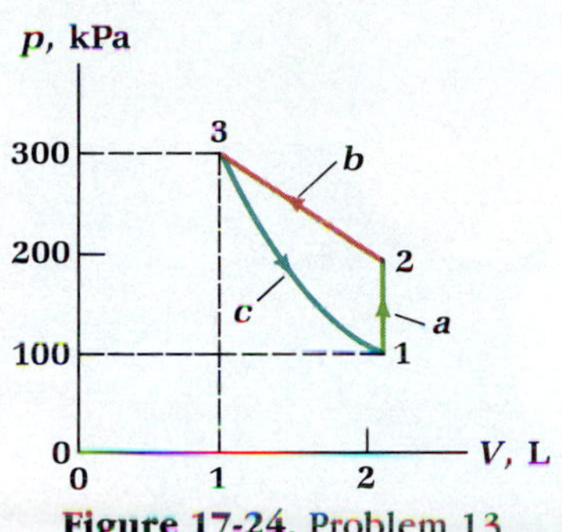

Figure 17-24. Problem 13.

14. ***Ideal gas undergoing a cycle IV.*** Repeat the previous problem except with the processes reversed so that the cycle proceeds in a clockwise sense.

15. ***Adiabatic work done by an ideal gas.*** In Chap. 18 we shall find that an ideal gas undergoing a quasi-static adiabatic process follows the relation

$$p = KV^{-\gamma}$$

where $\gamma = \mathscr{C}_p/\mathscr{C}_V$ is the ratio of molar heat capacities and K is a constant that may be written as $K = p_i V_i^\gamma$. *(a)* Show that the work done by the gas as its volume changes from V_i to V_f is given by

$$W_Q = \frac{p_i V_i}{\gamma - 1}\left[1 - \left(\frac{V_i}{V_f}\right)^{\gamma - 1}\right]$$

(b) For air, $\gamma = 1.40$. Determine W_Q for a sample of air, initially at $p_i = 202$ kPa, when it adiabatically expands from $V_i = 45$ L to $V_f = 65$ L.

18 KINETIC THEORY OF GASES

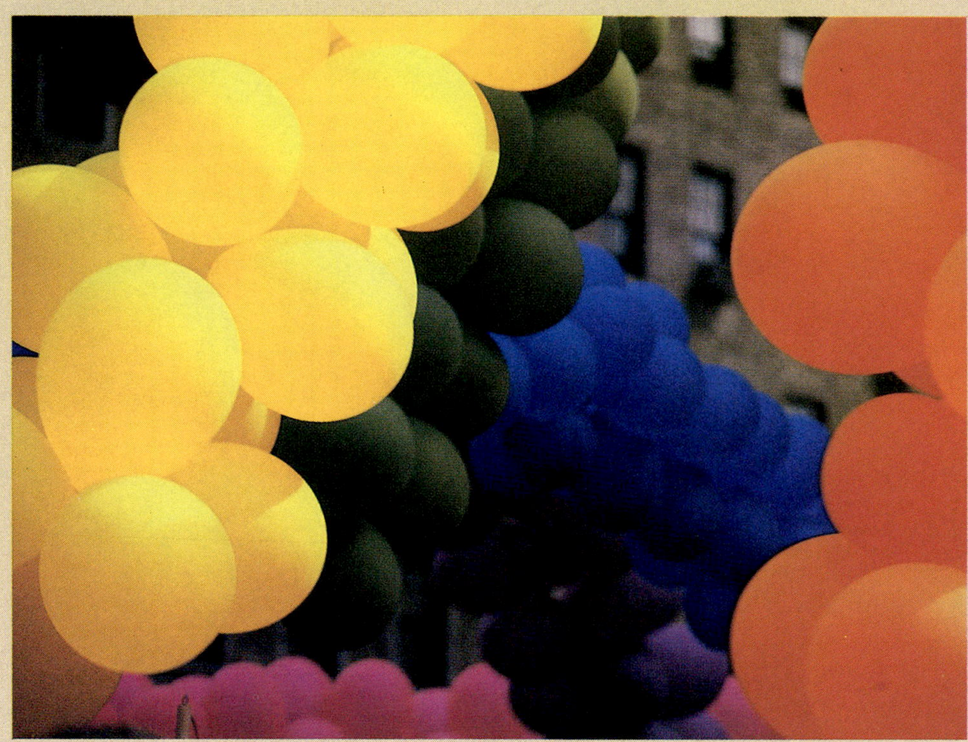

The stretched rubber of an inflated balloon is evidence of the collisions of gas molecules with the walls of their enclosure. The pressure of the gas is caused by these molecular collisions.

"Your chances of winning the jackpot are 1 in 2,300,000,000." "There is safety in numbers." "If at first you don't succeed, try, try again." These clichés express the conventional wisdom that comes from averaging over the large numbers of entities in a complex system. In a similar way, we seek to relate the macroscopic behavior of a system such as a gas to its constituents, a large number of molecules.

The number of molecules in a typical system is indeed large. A coin such as a penny contains about 10^{23} atoms. Even the "vacuum" of interstellar space may have 10^8 molecules in 1 m^3. Because of these large numbers, we do not attempt to follow the motion of individual molecules, but consider averages over molecular motions instead. *Kinetic theory,* a special branch of statistical mechanics, allows us to express some of the macroscopic quantities in terms of averages over molecular motions. A much deeper understanding of the thermodynamic properties of a system comes in this way.

The motion of small spheres is used to demonstrate the random motion of gas molecules.

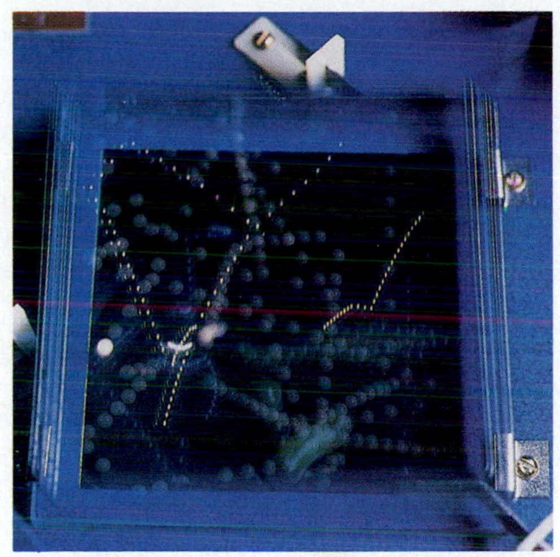

18-1 MOLECULAR MODEL OF AN IDEAL GAS

To formulate a microscopic or molecular model of an ideal gas, we make several simplifying assumptions about the behavior of molecules in the gas and the system that they form:

Assumptions about the molecular model of an ideal gas

1. *Large numbers.* The gas consists of N molecules, where N is a very large number. Each molecule has mass m and a size that is negligible compared with the average distance between molecules.

2. *Mechanics.* The motions of the individual molecules are adequately described by newtonian mechanics.

3. *Collisions.* A molecule moves freely, with negligible forces acting on it, except when it contacts and collides with another molecule or with a wall of the container. All collisions are elastic, and in the case of a collision of a molecule with a wall, only the velocity component perpendicular to the wall changes. That component changes its sign on collision but not its magnitude.

4. *Randomness.* The molecules of the gas are in random motion, and the gas is in equilibrium. The methods of elementary probability theory may be applied to the system.

Some of these assumptions are more restrictive than is really necessary. However, the connections between the microscopic and the macroscopic descriptions that we seek are more easily obtained using the assumptions. Also, for simplicity, we consider the gas to be enclosed in a stationary, rectangular box with edges L_1, L_2, L_3, and of volume $V = L_1L_2L_3$. Our final results are independent of the shape of the container.

Averages and Probability

In what sense do we speak of the "average motion of a molecule"? To illustrate, we consider some simple but useful averages over molecular motions, beginning with the average velocity. Since the molecules are moving randomly, a given molecule is equally likely to be moving in any direction and can have almost any speed. We use a coordinate system with axes parallel to the edges of the box and select the x component of the velocity of a molecule of the system, as shown in Fig. 18-1. We label this molecule with an index j and denote its velocity component by v_{jx}. Now we average over all N molecules in the system:

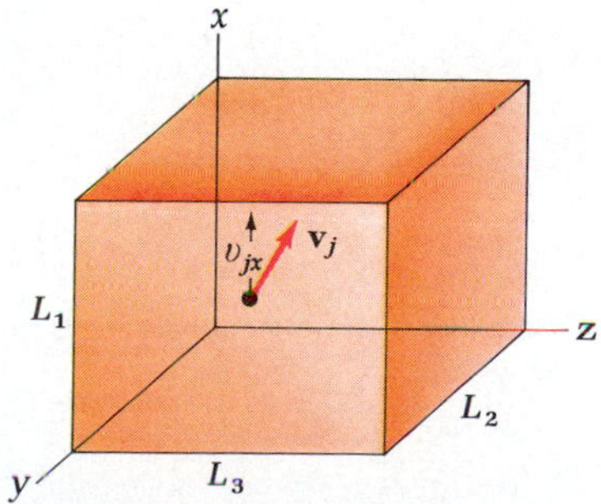

Figure 18-1. One of the molecules of a gas is shown in a container of volume $V = L_1L_2L_3$. This molecule has velocity $\mathbf{v}_j$, with x component v_{jx}.

$$\langle v_x \rangle = \frac{\Sigma v_{jx}}{N} \tag{18-1}$$

where the symbol $\langle\,\rangle$ denotes the average or mean value and the sum is over all the molecules of the system. Since a component of the velocity can have negative as well as positive values, there is a molecule with a negative value of v_x for each molecule with a positive value of v_x. Positive and negative contributions to the sum in Eq. 18-1 occur with the same frequency, or with the same *probability*. Thus the sum Σv_{jx} is zero and $\langle v_x \rangle = 0$. The same reasoning applies to the y and z components of velocity, and we have

The average molecular velocity is zero, but the average speed is *not* zero.

$$\langle v_x \rangle = \langle v_y \rangle = \langle v_z \rangle = 0$$

That is, the average molecular velocity is zero:

$$\langle \mathbf{v} \rangle = \langle v_x \rangle \mathbf{i} + \langle v_y \rangle \mathbf{j} + \langle v_z \rangle \mathbf{k} = 0$$

Mean Square Speed. Although the average velocity (a vector) is zero, this result does *not* mean that the molecules are at rest! Indeed the average molecular speed (a scalar) is not zero. Consider the square of the speed. For the molecule labeled by j,

$$v_j^2 = v_{jx}^2 + v_{jy}^2 + v_{jz}^2$$

The average of this quantity, called the ***mean square speed*** $\langle v^2 \rangle$, will be of primary importance to us. It is the average of the square of the molecular speed and is given by

Mean square speed

$$\langle v^2 \rangle = \frac{\Sigma v^2}{N}$$

Later we shall see how $\langle v^2 \rangle$ is related to the temperature of a gas.

The mean square speed can be expressed in terms of the average of the square of one of the components. Since the x, y, and z directions are equivalent, we have

$$\langle v_x^2 \rangle = \langle v_y^2 \rangle = \langle v_z^2 \rangle$$

and $\langle v^2 \rangle = \langle v_x^2 \rangle + \langle v_y^2 \rangle + \langle v_z^2 \rangle$, or

$$\langle v^2 \rangle = 3\langle v_x^2 \rangle \tag{18-2}$$

The factor of 3 corresponds to the three equivalent spatial directions, and for later convenience we have chosen to use the square of the x component.

Position Probability. We have discussed averages involving the velocity and speed. Now consider averages of position. What is the average value $\langle x \rangle$ of the x coordinate for the molecules of the system? From Fig. 18-1 we can see that $\langle x \rangle = (\Sigma x_j)/N$ must have a value between 0 and L_1. Consistent with the random molecular motion, a gas in equilibrium is spatially homogeneous. That is, all values of x between 0 and L_1 are equally likely, and the average value of x is $\frac{1}{2}L_1$:

$$\langle x \rangle = \frac{\Sigma x_j}{N} = \tfrac{1}{2}L_1$$

Put another way, the probability, or chance, is 1/2 that a particular molecule is in one side, say $0 \leq x \leq \frac{1}{2}L_1$, of the box. It is equally likely to be in either half. More generally, the probability $\mathcal{P}$ that a given molecule has an x coordinate within a range ℓ_1 is given by the fraction that distance is of the total range L_1 of x.

Probability that an x coordinate is in a range ℓ_1

$$\mathcal{P} = \frac{\ell_1}{L_1} \tag{18-3}$$

Suppose $L_1 = 0.50$ m so that x ranges from $x = 0.00$ m to $x = 0.50$ m. The probability that a particular molecule has its position in the range of, say, $x = 0.30$ m to $x = 0.40$ m is $\mathcal{P} = 0.10\text{ m}/0.50\text{ m} = 0.20$.

Calculation of the Pressure

A gas at pressure p exerts a force of magnitude $F = pA$ on a wall of area A. From the microscopic point of view, this force is caused by the continual rain of molecules colliding with the wall. Consider the force on the upper wall of the container in Fig. 18-1:

$$\mathbf{F} = \begin{pmatrix} \text{force exerted by the rain of} \\ \text{molecules on the upper wall} \end{pmatrix}$$

By Newton's third law, the force exerted by the molecules on the wall is equal and opposite the force exerted by the wall on the molecules. Thus

$$\mathbf{F} = -\begin{pmatrix} \text{force exerted by the upper wall} \\ \text{on the raining molecules} \end{pmatrix}$$

Newton's second law states that the force on a particle is equal to the rate of change of the particle's momentum. Consequently,

$$\mathbf{F} = -\frac{\begin{pmatrix} \text{change in momentum of molecules} \\ \text{that strike the wall during } \Delta t \end{pmatrix}}{\Delta t} \tag{18-4}$$

To find $\mathbf{F}$, the momentum change of those molecules that collide with the wall during Δt must be determined. Figure 18-2 shows the collision of a representative molecule (molecule j). According to our model, only the x component of the molecule's velocity undergoes a change; it is v_{jx} before the collision and $-v_{jx}$ after the collision. The molecule's x component of momentum change is

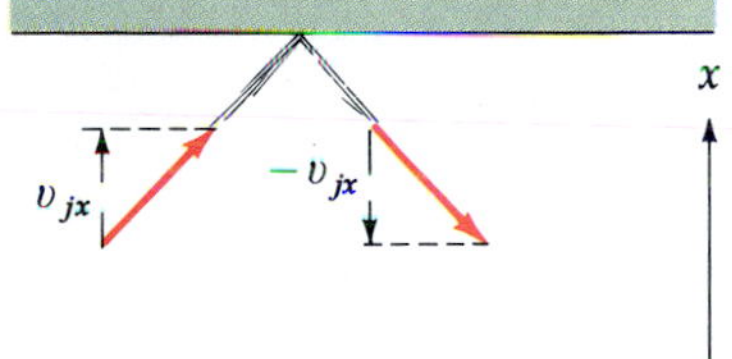

Figure 18-2. A molecule collides elastically with a wall. The velocity component perpendicular to the wall changes from v_{jx} before the collision to $-v_{jx}$ after the collision.

$$m(-v_{jx}) - m(v_{jx}) = -2mv_{jx} \tag{18-5}$$

What is the probability molecule j will collide with the wall during Δt and undergo this momentum change? To collide with the wall during Δt, molecule j must be within the distance $|v_{jx}|\Delta t$ of the wall—otherwise it is too far away to hit the wall during Δt. Equation 18-3 states that the probability that molecule j is within a distance $\ell_1 = |v_{jx}|\Delta t$ of the wall is $\mathcal{P} = |v_{jx}|\Delta t/L_1$. But molecule j may be going the wrong way; it may be going downward (v_{jx} negative) rather than upward (v_{jx} positive). Since the probability that molecule j has positive v_{jx} (rather than negative) is 1/2, the probability that molecule j will hit the wall during Δt is given by

Probability that the molecule will hit the wall

$$\mathcal{P} = \frac{\frac{1}{2}v_{jx}\Delta t}{L_1} \tag{18-6}$$

where the absolute value sign is taken away because v_{jx} is now positive. In addition, the time interval Δt must be small enough so that the probability that molecule j collides with another molecule during Δt is negligible.

The change in momentum of the molecules that strike the wall during Δt is equal to the sum over all molecules in the container of the product of the probability a molecule will hit the wall during Δt (from Eq. 18-6) multiplied by the molecule's momentum change (from Eq. 18-5). Inserting this into Eq. 18-4, we have

$$F = -\frac{\Sigma\frac{1}{2}[(v_{jx}\,\Delta t)/L_1](-2mv_{jx})}{\Delta t}$$

Simplifying gives

$$F = \frac{m\,\Sigma v_{jx}^2}{L_1}$$

From Fig. 18-1, the area of the upper wall is $A = L_2L_3$, so that the pressure on the wall is

$$p = \frac{F}{A} = \frac{m \, \Sigma v_{jx}^2}{L_1 L_2 L_3} = \frac{m \, \Sigma v_{jx}^2}{V}$$

where we have used the fact that $L_1 L_2 L_3$ is equal to the volume V of the container. Therefore

$$pV = m \Sigma v_{jx}^2 \qquad (18\text{-}7)$$

The sum that appears in Eq. 18-7 is closely related to the mean square speed through Eq. 18-2, $\langle v^2 \rangle = 3\langle v_x^2 \rangle$. By multiplying and dividing by the number of molecules in the system, we have

$$\Sigma v_{jx}^2 = N \frac{\Sigma v_{jx}^2}{N} = N\langle v_x^2 \rangle = \frac{N\langle v^2 \rangle}{3}$$

Equation 18-7 then becomes

Pressure is proportional to mean square speed.

$$pV = \frac{Nm\langle v^2 \rangle}{3} \qquad (18\text{-}8)$$

which shows that for a given volume, ***the pressure of a gas is proportional to the mean square speed of the molecules.*** The faster the molecules move, the larger their mean square speed and the greater the pressure. By averaging over the molecular motions, we have obtained the connection between a macroscopic quantity, the pressure, and a microscopic quantity, the mean square speed of the molecules. This is the central point of this section.

EXAMPLE 18-1

Root-mean-square speed. The square root of the mean square speed is called the ***root-mean-square speed*** v_{rms}:

$$v_{\text{rms}} = \sqrt{\langle v^2 \rangle}$$

This quantity will be discussed in more detail in the next section and in Sec. 18-6. Determine the root-mean-square speed of the molecules of a 1.0-mol sample of neon gas, which occupies 22.4 L at a pressure of 101 kPa. Neon is monatomic under normal conditions.

Solution. Solving Eq. 18-8 for $\langle v^2 \rangle$ and taking the square root gives

$$v_{\text{rms}} = \sqrt{\langle v^2 \rangle} = \sqrt{3pV/Nm}$$

Note that the product Nm is the mass of the 1-mol sample. From App. P, the atomic mass of neon is $M_0 = 20.2$ g/mol. Since neon is monatomic, the mass of the sample is $Nm = 0.0202$ kg. Inserting the numerical values into the expression, we have

$$v_{\text{rms}} = \sqrt{3(101 \text{ kPa})(22.4 \text{ L})/(0.0202 \text{ kg})} = 580 \text{ m/s}$$

This speed is about a factor of 20 times the speed limit on an interstate highway.

SELF-TEST 18-1. Determine v_{rms} for a 1.0-mol sample of nitrogen gas, which occupies 22.4 L at a pressure of 101 kPa. Nitrogen is diatomic under these conditions. Explain why your answer is different from the one above for neon. ***ANSWER:*** 490 m/s.

18-2 THE MICROSCOPIC INTERPRETATION OF TEMPERATURE

From macroscopic considerations in Sec. 17-1, we saw that the equation of state of an ideal gas is

$$pV = nRT$$

From microscopic considerations with the molecular model of a gas from the previous section, we found

$$pV = \frac{Nm\langle v^2\rangle}{3}$$

A comparison of these relations leads to

$$\frac{Nm\langle v^2\rangle}{3} = nRT$$

The average ***translational kinetic energy*** $\langle K\rangle$ of the gas molecules is $\langle K\rangle = \frac{1}{2}m\langle v^2\rangle$. (We emphasize that this is the translational kinetic energy because, as we see in the next section, molecules can have other forms of kinetic energy.) Also, since Avogadro's number N_A is the number of molecules per mole and n is the number of moles, the number N of molecules is $N = nN_A$. Inserting this into the above expression and solving for T, we find

$$T = \frac{2N_A}{3R}\,\frac{1}{2}m\langle v^2\rangle = \frac{2N_A}{3R}\langle K\rangle \qquad (18\text{-}9)$$

Ludwig Boltzmann (1844–1906), along with J. C. Maxwell and J. W. Gibbs, was one of the pioneers of statistical mechanics. His correlation of probability and entropy was crucial to the development of the molecular theory of thermodynamics.

This gives the molecular interpretation of temperature:

> Temperature is proportional to the average translational molecular kinetic energy.

When dealing with the energies of molecules, we denote the ratio R/N_A by the symbol k, which is called the ***Boltzmann constant:***

Boltzmann constant k

$$k = \frac{R}{N_A} = 1.38 \times 10^{-23}\ \text{J/K}$$

The Boltzmann constant is one of the more useful physical constants (see inside front cover). In terms of the Boltzmann constant, the relation between temperature and the translational kinetic energy of the molecules is

Molecular interpretation of temperature

$$\langle K\rangle = \frac{3}{2}kT \qquad (18\text{-}10)$$

This equation embodies the molecular interpretation of temperature. It connects the average translational molecular kinetic energy (a microscopic quantity) with the temperature (a macroscopic quantity).

Internal Energy of a Monatomic Ideal Gas

The molecules of a monatomic gas consist of single atoms. The "noble gases"—He, Ne, Ar, Kr, Xe, and Rn—are monatomic under ordinary conditions. With our model, we assume that such molecules behave as particles and that the potential energy of their interaction may be neglected. The only form of energy these molecules may possess is translational kinetic energy. They may not possess, for example, rotational or vibrational energies. Therefore the internal energy U of an ideal monatomic gas is the sum of the translational kinetic energies of its molecules:

$$U = \Sigma \tfrac{1}{2}mv_i^2 = N\tfrac{1}{2}m\langle v^2\rangle$$

Or

$$U = N\langle K\rangle \qquad (18\text{-}11)$$

From Eq. 18-10, we have

$$U = \tfrac{3}{2}NkT = \tfrac{3}{2}nRT \tag{18-12}$$

where we have used $Nk = nN_Ak = nR$.

The internal energy of a monatomic ideal gas is proportional to T.

The internal energy of a monatomic ideal gas is proportional to the Kelvin temperature.

In the previous chapter we saw that the application of the first law to a free expansion led to the conclusion that the internal energy of an ideal gas (not necessarily monatomic) depends only on the temperature. Now we find that our model, applied to a monatomic ideal gas, corroborates this conclusion and predicts that the temperature dependence is linear.

Root-Mean-Square Speed

If the temperature of a system increases, then the molecules move faster on the average. We can phrase this in terms of speed by using the root-mean-square speed v_{rms},

$$v_{\text{rms}} = \sqrt{\langle v^2 \rangle} \tag{18-13}$$

The root-mean-square speed is one of the quantities that characterize the distribution of molecular speeds, which we discuss later in Sec. 18-6. It is not the average speed but is the square root of the mean square speed. To see how v_{rms} depends on temperature, we use $\langle K \rangle = \frac{1}{2}m\langle v^2 \rangle$ so that Eq. 18-10 becomes

$$\tfrac{1}{2}m\langle v^2 \rangle = \tfrac{3}{2}kT$$

Solving for $\langle v^2 \rangle$ and taking a square root gives

$$v_{\text{rms}} = \sqrt{\frac{3kT}{m}} \tag{18-14}$$

Thus the root-mean-square speed is proportional to the square root of the Kelvin temperature and inversely proportional to the square root of the molecular mass.

Quantum Effects. Equations 18-10 and 18-14 imply incorrectly that molecular motion ceases as $T \rightarrow 0$. One of the assumptions on which these results are based becomes invalid for a system of molecules at sufficiently low temperatures: Newton's laws do not adquately describe the motion of a molecule under these conditions, and the methods of quantum statistical mechanics must be used. These quantum effects are important in the behavior of systems such as electrons in condensed matter and the superfluid phases of liquid He.

EXAMPLE 18-2

Internal energy of a monatomic gas. A system consists of 2.21 mol of Ar at 273 K. Determine for this gas *(a)* the average molecular kinetic energy, *(b)* the internal energy, and *(c)* the root-mean-square speed.

Solution. *(a)* The average molecular kinetic energy is given by Eq. 18-10:

$$\langle K \rangle = \tfrac{3}{2}(1.38 \times 10^{-23}\ \text{J/K})(273\ \text{K})$$
$$= 5.65 \times 10^{-21}\ \text{J}$$

(b) The internal energy of this monatomic ideal gas can be calculated using Eq. 18-12. Alternatively, we can determine the number of molecules in the system:

$$N = nN_A = (2.21\ \text{mol})(6.02 \times 10^{23}\ \text{mol}^{-1}) = 1.33 \times 10^{24}$$

The internal energy is then

$$U = N\langle K \rangle = (1.33 \times 10^{24})(5.65 \times 10^{-21}\ \text{J}) = 7.52\ \text{kJ}$$

(c) The mass m of a molecule is $m = M_0/N_A$, where M_0 is the molecular mass and N_A is

Avogadro's number. From App. P, the molar mass of monatomic argon is $M_0 = 39.9$ g/mol, so that

$$m = \frac{0.0399 \text{ kg/mol}}{6.02 \times 10^{23} \text{ mol}^{-1}} = 6.63 \times 10^{-26} \text{ kg}$$

The root-mean-square speed for Ar at this temperature is from Eq. 18-14:

$$v_{\text{rms}} = \sqrt{\frac{(3)(1.38 \times 10^{-23} \text{ J/K})(273 \text{ K})}{6.63 \times 10^{-26} \text{ kg}}} = 413 \text{ m/s}$$

SELF-TEST 18-2. *(a)* Determine an expression for the internal energy of a monatomic ideal gas in terms of its pressure and volume rather than its temperature. *(b)* Use the expression from part *(a)* to evaluate the internal energy of an ideal monatomic gas that occupies 1.0 L at a pressure of 100 kPa. ***ANSWER:*** *(a)* $U = 3pV/2$; *(b)* 150 J.

18-3 EQUIPARTITION OF ENERGY

According to Eq. 18-10, the average translational molecular kinetic energy for a system at temperature T is given by

$$\langle K \rangle = \tfrac{1}{2}m\langle v^2 \rangle = \tfrac{3}{2}kT$$

It is useful to trace back through the last two sections to determine the origin of the factor of 3 in this expression. The first appearance of this factor is in Eq. 18-2 for the mean square speed $\langle v^2 \rangle$. It came about because of the equivalence of the three spatial directions, or the equivalence of the average of square velocity components:

$$\langle v^2 \rangle = \langle v_x^2 \rangle + \langle v_y^2 \rangle + \langle v_z^2 \rangle = 3\langle v_x^2 \rangle$$

Because of this equivalence, which has its roots in the fact that the properties of the system do not depend on our choice of orientation for the xyz-coordinate system, we can write

$$\tfrac{1}{2}m\langle v_x^2 \rangle = \tfrac{1}{2}m\langle v_y^2 \rangle = \tfrac{1}{2}m\langle v_z^2 \rangle = \tfrac{1}{2}kT$$

The sum of these three equal contributions gives Eq. 18-10.

The factor of 3 is connected to the three translational ***degrees of freedom*** of a monatomic molecule. For our purposes, each degree of freedom corresponds to the ability of a molecule to participate in a one-dimensional motion that contributes to the mechanical energy of that molecule. This is best illustrated by a translational degree of freedom: A molecule can have a velocity component in the x direction with a contribution to the mechanical energy of $\tfrac{1}{2}mv_x^2$). [Note that although we can speak of a one-dimensional contribution to the kinetic energy, the kinetic energy of a molecule is $\tfrac{1}{2}m(v_x^2 + v_y^2 + v_z^2)$.] Since there are three spatial directions in which the molecule can move, the monatomic ideal gas has three degrees of freedom, and the average molecular mechanical energy $\langle E \rangle = \langle K \rangle$ is

Degrees of freedom

$$\langle E \rangle = 3(\tfrac{1}{2}kT)$$

A general statement of this result is referred to as the ***equipartition of energy theorem.*** *For a system of molecules at temperature T, with each molecule having ν degrees of freedom the average molecular mechanical energy* $\langle E \rangle$ is

Equipartition of energy theorem

$$\langle E \rangle = \nu(\tfrac{1}{2}kT) \tag{18-15}$$

Equation 18-15 implies that, on the average, an amount $\tfrac{1}{2}kT$ of mechanical energy is associated with each degree of freedom. For the monatomic ideal gas once again, there are only three translational degrees of freedom, $\nu = 3$, and Eq. 18-10 results.

Equation 18-16 is used to express $dU = n\mathscr{C}_V\,dT$ in terms of the temperature change dT, so that $0 = n\mathscr{C}_V\,dT + p\,dV$. Solving this expression for dT gives

$$dT = -\frac{p\,dV}{n\mathscr{C}_V} \qquad \text{(adiabatic process)}$$

Another expression for dT can be obtained from the ideal gas equation of state $pV = nRT$ by taking its differential, $p\,dV + V\,dp = nR\,dT$, or

$$dT = \frac{p\,dV + V\,dp}{nR} \qquad \text{(ideal gas)}$$

On eliminating dT (and n) from these two expressions, we get

$$p\,dV + V\,dp = -\frac{R}{\mathscr{C}_V}p\,dV = -\frac{\mathscr{C}_p - \mathscr{C}_V}{\mathscr{C}_V}p\,dV$$
$$= -(\gamma - 1)p\,dV$$

where we have used both $R = \mathscr{C}_p - \mathscr{C}_V$ and $\gamma = \mathscr{C}_p/\mathscr{C}_V$. Rearranging the equation gives

$$V\,dp = -\gamma p\,dV$$

or

$$\frac{dp}{p} = -\gamma\frac{dV}{V}$$

for the infinitesimal adiabatic process. For larger changes in p and V, we integrate (using the indefinite integral with a constant of integration):

$$\int\frac{dp}{p} = -\int\gamma\frac{dV}{V} + \text{constant}$$

or

$$\ln p + \gamma\ln V = \text{constant}$$

where we have assumed that γ does not change over the range of integration. The properties of the logarithm allow us to write this as

$$\ln pV^{\gamma} = \text{constant}$$

or

$$pV^{\gamma} = K$$

where K is a constant. This is Eq. 18-18.

During an adiabatic process, the pressure and volume of the ideal gas change, but in such a way that Eq. 18-18 is satisfied. We may select any two states of the system connected by an adiabatic process—let one of them be the initial state i—and apply Eq. 18-18, $pV^{\gamma} = K = p_iV_i^{\gamma}$, or

$$pV^{\gamma} = p_iV_i^{\gamma}$$

This equation may be solved for one of the variables, p for example:

$$p = \frac{p_iV_i^{\gamma}}{V^{\gamma}} \qquad (18\text{-}19)$$

This expression shows how the pressure of the ideal gas varies with its volume for a quasi-static adiabatic process. This expression is represented by a curve on a p-V diagram, as shown in Fig. 18-8 for a gas with $\gamma = 1.40$. Also shown for reference are two isotherms ($pV = nRT =$ constant for an isothermal process). The adiabatic curve at a point on the diagram is steeper than an isotherm at that point (see Prob. 1).

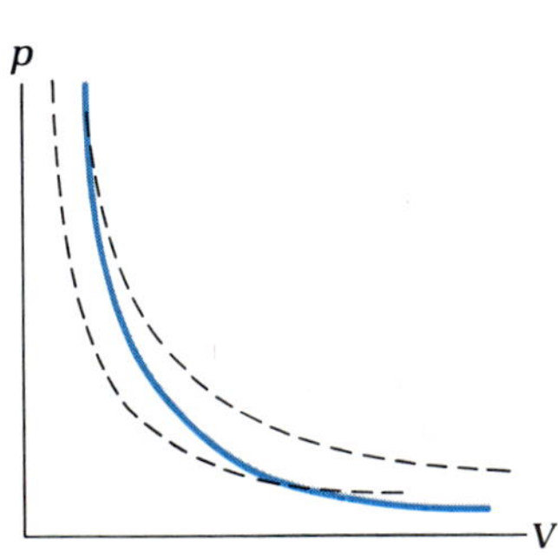

Figure 18-8. The solid curve represents a quasi-static, adiabatic process for an ideal gas. Also shown as dashed lines are two isotherms.

EXAMPLE 18-5

Diesel engine. In one cylinder of a diesel engine, air initially at atmospheric pressure and 310 K occupies a volume of 0.420 L. It is compressed, quasi-statically and adiabatically, to a volume of 0.028 L (a compression ratio of 15). Determine the final *(a)* pressure and *(b)* temperature.

Solution. *(a)* Using Eq. 18-19 and the value of $\gamma = 1.40$ for air (see the entries for N_2 and O_2 in Table 18-1), we obtain

$$p_f = p_i\left(\frac{V_i}{V_f}\right)^{\gamma}$$

$$= (101\ \text{kPa})\left(\frac{0.420}{0.028}\right)^{1.40}$$

$$= 4500\ \text{kPa}$$

(b) The temperature can be determined using the ideal gas equation of state:

$$T_f = T_i\frac{p_f V_f}{p_i V_i}$$

$$= 310\ \text{K}\,\frac{(4500\ \text{kPa})(0.028\ \text{L})}{(101\ \text{kPa})(0.420\ \text{L})} = 920\ \text{K}$$

During this adiabatic process, the volume decreased by a factor of 15, but the pressure increased by a factor of about 45. Consequently, the temperature increased by a factor of about 3 ($45/15 = 3$).

SELF-TEST 18-5. In the above example, we saw that the temperature increased during the adiabatic compression because the factor by which the pressure increased is larger than the factor by which the volume decreased. Thus the product pV increased, and T is proportional to pV. Suppose the gas undergoes an adiabatic expansion. Does its temperature change? If so, does the temperature increase or decrease? ***ANSWER:*** The temperature changes; it decreases.

EXAMPLE 18-6

Work done during an adiabatic compression. *(a)* Develop an expression for the work done by a gas during a quasi-static adiabatic process in terms of p_i, V_i, V_f, and γ. This expression was introduced in the previous chapter (Sec. 17-3). *(b)* Use the expression found in part *(a)* to evaluate the work done by the air in the previous example.

Solution. *(a)* The work done during an adiabatic process is found by inserting the expression for the pressure as a function of volume (Eq. 18-19) into the work integral and performing the integration:

$$W = \int p\,dV = \int_{V_i}^{V_f}\frac{p_i V_i^{\gamma}}{V^{\gamma}}\,dV = p_i V_i^{\gamma}\int_{V_i}^{V_f}\frac{dV}{V^{\gamma}}$$

$$= p_i V_i^{\gamma}\frac{1}{1-\gamma}\left(\frac{1}{V_f^{\gamma-1}} - \frac{1}{V_i^{\gamma-1}}\right)$$

$$= p_i V_i\frac{1}{1-\gamma}\left(\frac{V_i^{\gamma-1}}{V_f^{\gamma-1}} - \frac{V_i^{\gamma-1}}{V_i^{\gamma-1}}\right)$$

$$= \frac{p_i V_i}{\gamma-1}\left[1 - \left(\frac{V_i}{V_f}\right)^{\gamma-1}\right]$$

(b) Inserting the numerical values from the previous example into the expression in part *(a)* gives

$$W = \frac{(101\ \text{kPa})(0.42\ \text{L})}{0.40}\left[1 - \left(\frac{0.42}{0.028}\right)^{0.40}\right]$$

$$= -210\ \text{J}$$

The value of the work is negative because the gas is compressed so that work is done on the gas by its surroundings.

SELF-TEST 18-6. Use the first law to determine the change ΔU in the internal energy of the air during the adiabatic compression described in the above example. ***ANSWER:*** $\Delta U = Q - W = 0 - (-210\text{ J}) = 210\text{ J}$.

18-6 THE DISTRIBUTION OF MOLECULAR SPEEDS

The mean square speed $\langle v^2 \rangle$ and its square root, the root-mean-square speed v_{rms}, have been related to the macroscopic properties of gases. We have considered average or mean values because experiment shows that there is a distribution of speeds for the molecules of a gas. Some molecules have speeds less than v_{rms}, while others have speeds greater than v_{rms}. There exists a range of speeds from zero up to rather large values.

The distribution function $f(v)$ gives the distribution of molecular speeds.

The distribution of speeds is described by a ***distribution function*** $f(v)$. It is defined so that, of the N molecules in the system, the number ΔN that have a value of the speed between v and $v + \Delta v$ is given by

$$\Delta N = f(v)\ \Delta v$$

That is, $f(v)$ is the number of molecules per unit speed range. Since we usually consider the total number of molecules to be very large, we work in the limit as Δv tends to zero, so that

$$dN = f(v)\ dv$$

is the number of molecules with speed between v and $v + dv$. Integrating over all speeds yields the total number of molecules:

$$N = \int_0^\infty f(v)\ dv$$

James Clerk Maxwell (1831 – 1879). A brief biography of Maxwell is given in the commentary at the end of Chap. 27.

The distribution function for molecular speeds was first obtained by James Clerk Maxwell (1831 – 1879), who applied statistical concepts to the random motion of gas molecules. (See the Commentary in Chap. 19 and in Chap. 27.) That distribution of molecular speeds turns out to be

$$f(v) = Av^2 \exp \frac{-\frac{1}{2}mv^2}{kT} \tag{18-20}$$

where k is the Boltzmann constant and m is the molecular mass. The factor A is independent of the molecular speed v:

$$A = 4\pi N \left(\frac{m}{2\pi kT} \right)^{3/2}$$

(Problem 6 considers one experimental technique for measuring the distribution of molecular speeds.)

The form of the distribution function is shown graphically in Fig. 18-9*a* for Ar at 200 K and at 600 K in Fig. 18-9*b*. At either temperature the number of molecules with very low speeds is small. Nor are there many molecules with very high speeds. With increasing temperature, the distribution shifts to higher speeds and broadens. The area under the entire curve, equal to the integral of $f(v)$ over all speeds, is the total number of molecules N. Thus the area is the same for both graphs in Fig. 18-9; it corresponds to a choice of $N = 10^{23}$ molecules, about 1/6 of a mol.

The broadening of the distribution with increasing temperature comes from the factor T in the denominator in the argument of the exponential function, $\exp(-\frac{1}{2}mv^2/kT)$. A smaller value of the molecular mass in the numerator of the argument would have the same result on the appearance of the distribution function as an increase in temperature. That is, the distribution of molecular speeds is broader for a gas with smaller molecular mass than for a gas with larger molecular

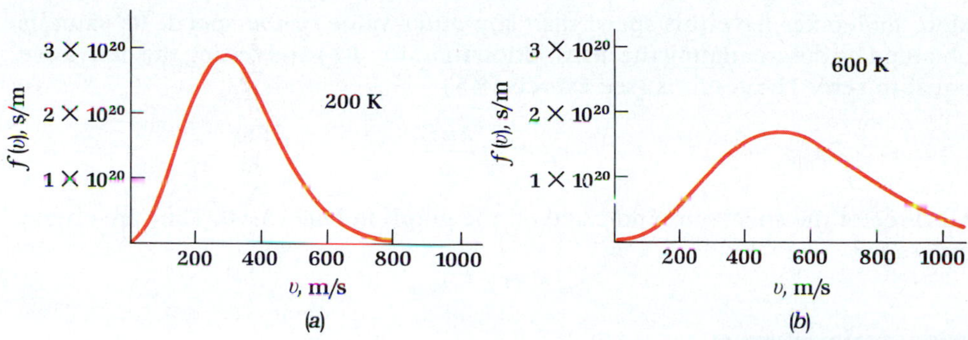

Figure 18-9. The distribution of molecular speeds is shown for Ar at *(a)* 200 K and *(b)* 600 K. The total number of molecules is 1×10^{23}.

mass at the same temperature. For example, Fig. 18-9*b* could also describe the distribution of speeds for He, with 1/10 the molecular mass of Ar, at a temperature of 60 K.

The value of the root-mean-square speed v_{rms} is indicated on the graph in Fig. 18-10. We have already determined its value in terms of the molecular mass m and the temperature, Eq. 18-14:

$$v_{rms} = \sqrt{\langle v^2 \rangle} = \sqrt{\frac{3kT}{m}}$$

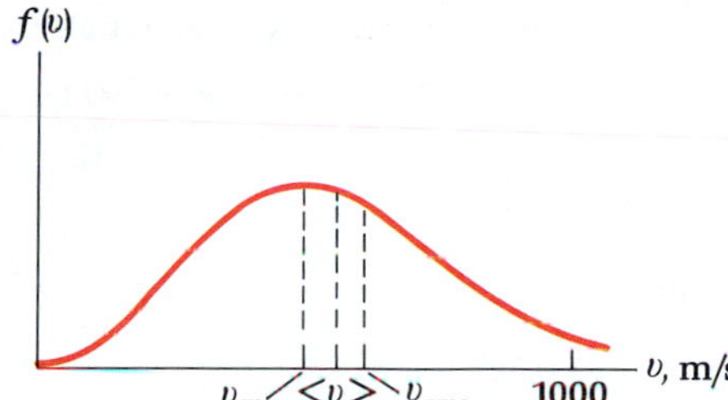

Figure 18-10. Three characteristic speeds are shown for the case of Fig. 18-9*b*. The values are $v_m = 500$ m/s, $\langle v \rangle = 560$ m/s, $v_{rms} = 610$ m/s.

Its value can also be obtained by using the distribution function to average the square of the speed. Since $f(v)\,dv$ is the number of molecules with speed v in the range dv, the mean square speed (the average of v^2) is given by

$$\langle v^2 \rangle = \frac{\int_0^\infty v^2 f(v)\,dv}{N}$$

Integrals of this sort can be found in most tables of integrals after making the change of variables $x = \frac{1}{2}mv^2/kT$. The result is $\langle v^2 \rangle = 3kT/m$, so that $v_{rms} = \sqrt{3kT/m}$ as expected.

Root-mean-square speed: $v_{rms} = \sqrt{3kT/m}$

We identify two other characteristic speeds. One is the mean speed $\langle v \rangle$. Notice that the (scalar) mean *speed* $\langle v \rangle$ is different from the (vector) mean *velocity* $\langle \mathbf{v} \rangle$, which is zero. We obtain the mean speed by averaging the speeds of the molecules. Each value of the speed v is multiplied by the number of molecules with that speed, $f(v)\,dv$. The result is then summed or integrated and then divided by the total number of molecules:

$$\langle v \rangle = \frac{\int_0^\infty v f(v)\,dv}{N}$$

The integral can be evaluated from tables, with the result,

$$\langle v \rangle = \sqrt{\frac{8kT}{\pi m}}$$

Mean speed: $\langle v \rangle = \sqrt{8kT/\pi m}$

Comparing the mean speed $\langle v \rangle$ with $v_{rms} = \sqrt{3kT/m}$, we find that $\langle v \rangle < v_{rms}$, since $8/\pi < 3$. The third characteristic speed is called the most probable speed v_m, which corresponds to the peak, or maximum value, of the distribution function.

More molecules have this speed than any other value of the speed. Its value is obtained by differentiating the distribution function $f(v)$ and setting the derivative equal to zero. The result is (see Exercise 35)

Most probable speed: $v_m = \sqrt{2kT/m}$

$$v_m = \sqrt{\frac{2kT}{m}}$$

All three of the speeds are indicated on the graph in Fig. 18-10. They are always ordered as

$$v_m < \langle v \rangle < v_{\rm rms}$$

EXAMPLE 18-7

Molecular speeds. Air is a mixture of gases, chiefly N_2 and O_2 with smaller amounts of other gases. Each component gas at temperature T has a distribution function given by Eq. 18-20. *(a)* Determine the most probable speed, the mean speed, and the root-mean-square speed of O_2 molecules in air at 300 K. *(b)* Repeat for H_2.

Solution. *(a)* The mass of an O_2 molecule is $m = M/N_A$ where $M = 32$ g/mol is the molar mass:

$$m = \frac{0.032 \text{ kg/mol}}{6.02 \times 10^{23} \text{ mol}^{-1}}$$
$$= 5.3 \times 10^{-26} \text{ kg}$$

The combination kT/m is needed for all three speeds:

$$\frac{kT}{m} = \frac{(1.38 \times 10^{-23} \text{ J/K})(300 \text{ K})}{5.3 \times 10^{-26} \text{ kg}} = 7.8 \times 10^4 \text{ m}^2/\text{s}^2$$

Then

$$v_m = \sqrt{2\left(\frac{kT}{m}\right)} = 390 \text{ m/s}$$

$$\langle v \rangle = \sqrt{\frac{8}{\pi}\left(\frac{kT}{m}\right)} = 450 \text{ m/s}$$

$$v_{\rm rms} = \sqrt{3\left(\frac{kT}{m}\right)} = 480 \text{ m/s}$$

(b) For H_2 we only need to use the correct mass, about 1/16 the mass of the O_2 molecule. Since the speeds are proportional to the inverse square root of the molecular mass, the speeds for H_2 are 4 times those for O_2:

$$v_m = 1.6 \text{ km/s}$$
$$\langle v \rangle = 1.8 \text{ km/s}$$
$$v_{\rm rms} = 1.9 \text{ km/s}$$

SELF-TEST 18-7. Determine each of the three speeds, v_m, $\langle v \rangle$, and $v_{\rm rms}$, for N_2 at 300 K. Do this by using the answers to part *(a)* of the above example and the mass-ratio technique demonstrated in part *(b)*. ***ANSWER:*** 420 m/s; 480 m/s; 510 m/s.

SUMMARY

Section 18-1. Molecular Model of an Ideal Gas

A molecular model of an ideal gas includes the following assumptions: there are a large number of molecules, the molecules behave as particles, Newton's laws apply to molecular motions, collisions are elastic, and the molecules move randomly.

The mean square speed is defined by $\langle v^2 \rangle = \Sigma v_i^2/N$ and is related to the pressure of the gas by

$$pV = \tfrac{1}{3}Nm\langle v^2 \rangle \qquad (18\text{-}8)$$

Section 18-2. The Microscopic Interpretation of Temperature

The internal energy of a monatomic ideal gas is given by

$$U = \tfrac{3}{2}nRT \qquad (18\text{-}12)$$

Temperature is interpreted at the molecular level in terms of the average translational kinetic energy of molecules:

$$\langle K \rangle = \tfrac{3}{2}kT \qquad (18\text{-}10)$$

The root-mean-square speed, $v_{\rm rms} = \sqrt{\langle v^2 \rangle}$, depends on the temperature and the molecular mass:

$$v_{rms} = \sqrt{\frac{3kT}{m}} \quad (18\text{-}14)$$

Section 18-3. Equipartition of Energy

The average energy of a molecule with ν active degrees of freedom is

$$\langle E \rangle = \nu(\tfrac{1}{2}kT) \quad (18\text{-}15)$$

On the average, $\frac{1}{2}kT$ of energy is associated with each degree of freedom.

Section 18-4. Heat Capacities of Ideal Gases and Elemental Solids

The molar heat capacities $\mathscr{C}_V$ and $\mathscr{C}_p$ for an ideal gas with ν active degrees of freedom are given by

$$\mathscr{C}_V = \tfrac{1}{2}\nu R \qquad \mathscr{C}_p = \mathscr{C}_V + R$$

At higher temperatures, the molar heat capacity of an elemental solid is given by a simple model to be $\mathscr{C}_V = 3R$, and $\mathscr{C}_p \approx \mathscr{C}_V$. This model assumes three translational and three vibrational degrees of freedom for each atom.

Section 18-5. Adiabatic Process for an Ideal Gas

An ideal gas undergoing a quasi-static adiabatic process takes on pressure and volume values satisfying

$$pV^{\gamma} = K \quad (18\text{-}18)$$

where K is a constant and $\gamma = \mathscr{C}_p/\mathscr{C}_V$.

Section 18-6. The Distribution of Molecular Speeds

The distribution of molecular speeds for an ideal gas is

$$f(v) = Av^2 \exp(-\tfrac{1}{2}mv^2/kT) \quad (18\text{-}20)$$

where $A = 4\pi N(m/2\pi kT)^{3/2}$. Three speeds which help characterize the distribution are the most probable speed, $v_m = \sqrt{2kT/m}$; the mean speed, $\langle v \rangle = \sqrt{8kT/\pi m}$; and the root-mean-square speed, $v_{rms} = \sqrt{3kT/m}$.

QUESTIONS

1. Explain how $\langle v \rangle$ can be different from zero while $\langle v_x \rangle = \langle v_y \rangle = \langle v_z \rangle = 0$, and so $\langle \mathbf{v} \rangle = 0$.

2. For an ideal gas, $\langle v \rangle < v_{rms}$, or $\langle v \rangle^2 < \langle v^2 \rangle$. That is, the *square of the mean speed* is less than the *mean square speed.* Is it possible for $\langle v \rangle^2$ to equal or exceed $\langle v^2 \rangle$ under any circumstances? Explain.

3. In a pure gas, all molecules are identical and have the same mass. Is the average translational kinetic energy still given by $\frac{3}{2}kT$ for a mixture of gases such as air? Explain.

4. Show that the ideal gas equation of state can be written as $pV = NkT$ and identify N in this equation.

5. The walls of a container of gas also consist of molecules. Thus a gas molecule colliding with a wall having the same temperature as the gas actually collides with one or more molecules. How can we justify, in an average sense, our assumption that the perpendicular component of velocity of a molecule merely reverses on a collision with the wall?

6. Can the gravitational potential energy of interaction of the molecules of a gas with the earth be neglected in comparison with the kinetic energy of the molecules? What about the gravitational potential energy of interaction of the molecules with each other? Explain.

7. Consider the air in a basketball during a game. Is (the vector) $\langle \mathbf{v} \rangle = 0$ for the air? Explain.

8. A molecule in a liquid must have a minimum, characteristic amount of kinetic energy in order to escape. How can a liquid such as water in an open container cool on evaporation?

9. Why must $\mathscr{C}_p$ be greater than $\mathscr{C}_V$ for a gas? Is this result true for liquids and solids as well? Explain.

10. If a molecule can change its speed on colliding with another molecule, can the distribution function $f(v)$ be independent of time? Explain.

11. If the distribution function is time-dependent, $f(v, t)$, can the gas be in equilibrium? Explain.

12. What happens to the temperature of an ideal gas if the gas is compressed adiabatically?

13. How can the temperature of a gas change during an adiabatic process, since no heat is exchanged with the surroundings? (*Hint:* Consider the change in velocity of a molecule colliding with a moving wall.)

14. Suppose that an opposite pair of walls of a container of gas are maintained at different temperatures. By what mechanism involving molecular collisions is heat conducted through this gas? Note that the gas is not at a uniform temperature.

15. Why is it improper to speak of the temperature of a molecule? Of a system of 100 molecules? How many molecules must a system have before we can speak meaningfully of the temperature of that system?

16. Imagine a huge interstellar spaceship with billions of humans drifting randomly through the interior of the ship. The humans occasionally collide with each other and with the walls of the ship. Is it meaningful to speak of a gas of humans (anthropogas)? If so, estimate the human v_{rms}.

17. To what values do the differences $v_{rms} - \langle v \rangle$ and $\langle v \rangle - v_m$ tend as the temperature of a gas drops? Describe the change in appearance of the graph of the distribution function as the temperature drops.

18. The speed of sound in He is greater than the speed of sound

in air at the same temperature and pressure. How can this be understood at the molecular level?

19. Molecular speeds range in value from zero to arbitrarily high values (but less than the speed of light). Explain on this basis why the distribution of molecular speeds can be asymmetric about the most probable speed.

20. Use features of the distribution of molecular speeds to give a possible explanation of the fact that the moon has virtually no atmosphere. Why is there almost no He in the earth's atmosphere?

21. Radon is a (radioactive) monatomic gas. Predict the values of $\mathscr{C}_V$ and $\mathscr{C}_p$ for this gas.

22. Complete the following table:

Symbol	Represents	Type	SI Unit
$\langle v^2 \rangle$	Mean square speed		
v_{rms}			
$\langle v \rangle$		Scalar	
v_m			
$\langle K \rangle$			
k			J/K
ν			
$\mathscr{C}_p$			
$\mathscr{C}_V$			
γ			
$f(v)$			

EXERCISES

Section 18-1. Molecular Model of an Ideal Gas

1. The following numbers, in units of meters per second, represent a small sample of 10 molecular speeds: 290, 47, 182, 439, 330, 268, 302, 372, 344, 410. Determine for this sample *(a)* the mean speed $\langle v \rangle$ and *(b)* the mean square speed $\langle v^2 \rangle$. *(c)* Compare the values of $\langle v^2 \rangle$ and $\langle v \rangle^2$.

2. Estimate the average spacing between the molecules of 1 mol of a gas at atmospheric pressure and 300 K. Compare this with the size of an O_2 molecule (about 0.2 nm).

3. Consider a pure gas consisting of identical molecules of mass m. Show that the velocity components of the center of mass of this gas are related to the average velocity components of the molecules by

$$(v_{cm})_x = \langle v_x \rangle \qquad (v_{cm})_y = \langle v_y \rangle \qquad (v_{cm})_z = \langle v_z \rangle$$

In what reference frame is our description of the ideal gas based?

4. On a coin toss the probability of a heads is 1/2 and the probability of a tails is 1/2. What is the probability of getting *(a)* three heads in a row? *(b)* Heads, tails, tails, in that order, in three tosses? *(c)* One head (and one tail) in two tosses? *(d)* Two heads in four tosses? (*Hint:* List all 16 possible outcomes and find the fraction which result in two heads.)

5. Suppose a coin is tossed 1000 times. What is the expected number of heads? Do you *really* expect to get precisely that number of heads? See the previous exercise.

6. Estimate the impulse exerted on your skin by a collision with a typical molecule of the air at room temperature. Assume that $\langle v^2 \rangle = 2 \times 10^5\ m^2/s^2$.

7. Show that Eq. 18-8 can be written as

$$pV = \frac{nM\langle v^2 \rangle}{3}$$

where M is the molar mass.

8. One mole of a gas at $p = 101$ kPa occupies a volume $V = 28.8$ L. Determine the mean square speed if the gas is *(a)* He, *(b)* H_2, *(c)* CO_2, *(d)* UF_6.

Section 18-2. The Microscopic Interpretation of Temperature

9. Determine the root-mean-square speed for each of the following molecular species in air at 300 K: *(a)* N_2, *(b)* O_2, *(c)* CO_2. *(d)* What is the average molecular translational kinetic energy for each of these species?

10. Show that Eq. 18-14 can be written in terms of the molar mass M as

$$v_{rms} = \sqrt{\frac{3RT}{M}}$$

11. By what factor must the temperature of a gas change so that its root-mean-square speed changes by *(a)* 10 percent, *(b)* −10 percent, *(c)* 50 percent?

12. A gas mixture of 0.80 mol of He and 0.15 mol of Ne is at 400 K. Evaluate *(a)* the root-mean-square speed for each type of molecule, *(b)* the average translational kinetic energy for each type, *(c)* the internal energy of this mixture.

13. The electron volt (eV) is a convenient energy unit to use for atoms and molecules. The conversion is

$$1\ \text{eV} = 1.60 \times 10^{-19}\ \text{J}$$

Express the average translational kinetic energy in electron volts for molecules of a gas at *(a)* 90 K (normal boiling point of O_2), *(b)* 300 K (temperature of the earth's surface), *(c)* 6000 K (temperature of the sun's surface).

14. Justify the following statement: The value of kT for a system at room temperature is $\frac{1}{40}$ eV. (See the previous exercise.)

15. A sealed tank car moves by you on a railroad track at 80 km/h in the positive x direction. It contains N_2 gas at 300 K. *(a)* What are the values of $\langle v_x \rangle$, $\langle v_y \rangle$, and $\langle v_z \rangle$ for the gas? (See Exercise 3.) *(b)* What are the values according to the railroad engineer? *(c)* Is the expression $\langle K \rangle = \frac{3}{2}kT$ for the average translational kinetic energy valid in every reference frame? Explain.

Section 18-3. Equipartition of Energy

16. A nonlinear polyatomic gas at 650 K has molecules with 12 active degrees of freedom. *(a)* Evaluate the average molecular mechanical energy. *(b)* What is the value of v_{rms}? Take $m = 1.3 \times 10^{-25}$ kg. *(c)* How many vibrational degrees of freedom are active?

17. The temperature of 2.2 mol of O_2 is raised from 10 to 140°C. *(a)* Assuming that vibrational degrees of freedom are not active over this temperature range, calculate the change in internal energy of the gas. *(b)* Can you determine how much heat was added and how much work was done? Explain.

Section 18-4. Heat Capacities of Ideal Gases and Elemental Solids

18. Show that both the universal gas constant R and the Boltzmann constant k have dimensions of molar heat capacity. Note that the mole is dimensionless.

19. *(a)* Suppose that the process in Exercise 17 occurred at constant volume. How much heat was added to the gas and how much work was done by the gas? *(b)* What is the value of $\mathscr{C}_p$ for this range of temperatures? *(c)* What are the answers to part *(a)* if the temperature change occurred at constant pressure?

20. Given that F_2 is diatomic and gaseous at room temperature, estimate the values of $\mathscr{C}_V$, $\mathscr{C}_p$, and γ for F_2 at 25°C. Compare with the values in the next exercise.

21. The measured values of $\mathscr{C}_p$ and γ for F_2 at 25°C are $\mathscr{C}_p = 31.4\ \mathrm{J \cdot mol^{-1} \cdot K^{-1}}$ and $\gamma = 1.36$. *(a)* Determine the value of $\mathscr{C}_V$. *(b)* Are the vibrational degrees of freedom active at this temperature? Explain.

22. From the graph in Fig. 18-5, estimate the number of active degrees of freedom for H_2 at *(a)* 100 K, *(b)* 600 K, *(c)* 2000 K. *(d)* Estimate the value of $\mathscr{C}_p$ at 600 K.

23. Estimate the value of γ for H_2 at each of the three temperatures listed in the previous exercise.

24. A cylinder with a movable piston contains 18 g of Ar at $p_i = 180$ kPa, $V_i = 12.0$ L. The gas is compressed to a final state given by $p_f = 480$ kPa, $V_f = 6.0$ L. *(a)* Determine the initial and final temperatures. *(b)* What is the change in internal energy?

25. The Ar gas in the previous exercise undergoes the process shown in Fig. 18-11. *(a)* Determine the net heat added to the gas, using values of $\mathscr{C}_p$ and $\mathscr{C}_V$. *(b)* Calculate the work done by the gas by finding the area under the curve. *(c)* Check this answer by using the first law and the result of the previous exercise.

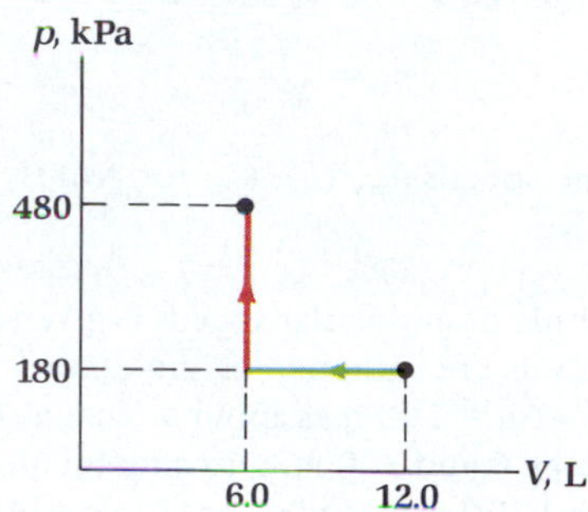

Figure 18-11. Exercise 25.

26. *(a)* Estimate the molar heat capacity of solid KCl at room temperature. Note that 1 mol consists of $2N_A$ atoms. *(b)* Estimate the value of c_V, the specific heat at constant volume, for KCl at room temperature.

Section 18-5. Adiabatic Process for an Ideal Gas

27. A monatomic ideal gas with $\gamma = 1.67$ undergoes an adiabatic expansion from the initial state $p_i = 320$ kPa, $V_i = 12$ L to a final volume $V_f = 18$ L. *(a)* Determine the final pressure of the gas. *(b)* Determine the initial and final temperatures, given that $n = 1.4$ mol. *(c)* Show the process on a p-V diagram.

28. *(a)* Evaluate the work done by the gas for the adiabatic expansions described in the previous exercise by performing the integral $\int p\, dV$. *(b)* Check the result by determining the change in internal energy of the gas and using the first law.

29. Prove for an ideal gas that $T_i V_i^{\gamma-1} = T_f V_f^{\gamma-1}$ for a quasi-static adiabatic process. What is the corresponding expression for an isobaric process?

30. Helium is initially at $p_i = 101$ kPa in a cylinder of volume $V_i = 2.25$ L and in equilibrium with surroundings which remain at 300 K. The gas is quasi-statically, but quickly (adiabatically), compressed to $V_f = 1.64$ L and held at that volume as it comes to equilibrium with the surroundings. *(a)* Evaluate the highest temperature and highest pressure attained by the He. *(b)* Show the process on the p-V diagram for the He. For the entire process, determine *(c)* the work done by the He and *(d)* the heat added to the He.

31. The temperature of 64 g of O_2 is raised from 15 to 45°C. What is the change in the internal energy of the O_2 if the process is *(a)* isochoric, *(b)* isobaric, *(c)* adiabatic? Assume that the gas is ideal.

32. The O_2 in the previous exercise is initially at atmospheric pressure. For each of the three quasi-static processes set forth above, determine the heat added to the gas and the work done by the gas.

33. One mole of an ideal gas is initially at $p_i = 200$ kPa, $V_i = 20$ L. On the same p-V diagram, show the quasi-static adiabatic process leading to a final volume $V_f = 30$ L if the ideal gas is *(a)* He, *(b)* Ne, *(c)* O_2, *(d)* CO_2.

34. Show that the number of active degrees of freedom ν is related to γ by $\nu = 2/(\gamma - 1)$.

Section 18-6. The Distribution of Molecular Speeds

35. Using the distribution function for molecular speeds, show that the most probable speed is given by

$$v_m = \sqrt{\frac{2kT}{m}}$$

36. Evaluate the speeds v_m, $\langle v \rangle$, v_{rms} for *(a)* H_2 at 300 K and *(b)* O_2 at 300 K.

37. A small sample of molecular speeds is given in the table at right, where ΔN is the number of molecules with a speed within a range of $\Delta v = 100$ m/s about v. That is, one molecule has speed between 0 and 100 m/s, two molecules have speeds between 100 and 200 m/s, and so on. Determine for this sample *(a)* the mean speed, *(b)* the root-mean-square speed, *(c)* the most probable speed.

ΔN	v, m/s
1	50
2	150
4	250
3	350
1	450

38. The speed of sound v_s in an ideal gas is given by (see Sec. 33-1 and also Prob. 4)

$$v_s = \sqrt{\frac{\gamma RT}{M}}$$

where $\gamma = \mathscr{C}_p/\mathscr{C}_V$ and M is the molar mass. Calculate the speed of sound at 300 K in *(a)* He, *(b)* N_2, *(c)* air.

39. Prove that the speed of sound in a gas (see previous exercise) is related to the mean speed by

$$\frac{v_s}{\langle v \rangle} = \sqrt{\frac{\gamma\pi}{8}}.$$

Evaluate this ratio for He and for air.

PROBLEMS

1. ***Slope of various processes on a p-V diagram.*** Consider a process represented by a curve on a p-V diagram, as shown in Fig. 18-12. At a given point (V, p) the slope of a line tangent to the curve is dp/dV. Evaluate dp/dV at a point for an ideal gas undergoing *(a)* an isobaric process, *(b)* an isothermal process, *(c)* an adiabatic process. *(d)* Which process has the steepest slope?

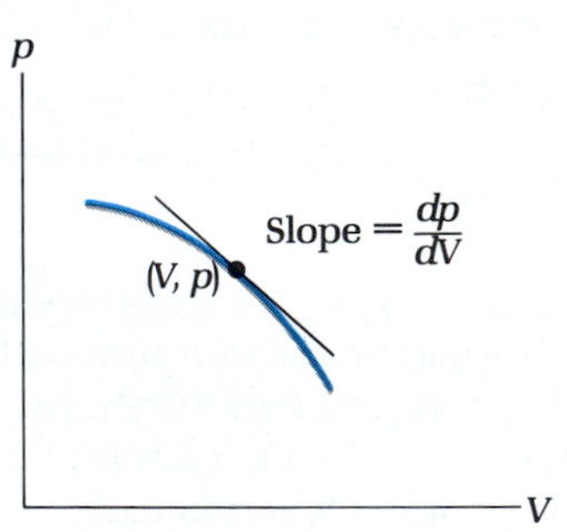

Figure 18-12. Problem 1.

2. ***Expression for*** $\langle v^n \rangle$. Using a table of integrals or otherwise, show that the mean value of v^n, $\langle v^n \rangle$, is given by (n is a positive integer)

$$\langle v^n \rangle = B_n \left(\frac{2kT}{m}\right)^{n/2}$$

where B_n depends on whether n is even or odd and is given by

$$B_1 = \frac{2}{\sqrt{\pi}}$$

$$B_2 = \frac{3}{2}$$

$$B_n = \frac{n+1}{2}\,\frac{n-1}{2}\cdots(2)(1)\frac{2}{\sqrt{\pi}} \qquad n \text{ odd, } n > 1$$

$$= \frac{n+1}{2}\,\frac{n-1}{2}\cdots\frac{3}{2} \qquad n \text{ even, } n > 2$$

3. ***Escape speed.*** The escape speed v_i of a molecule from above the earth's surface was calculated in Chap. 9. Its square is given by

$$v_i^2 = \frac{2GM_e}{r_i}$$

where M_e is the mass of the earth and r_i is the initial separation of the molecule from the center of the earth. The temperature of the earth's upper atmosphere can be 1000 K at an altitutde of about 150 km during a period of maximum sunspot activity. *(a)* Determine the mean square speed of an N_2 molecule under these conditions. *(b)* Determine the ratio of the number of molecules at this location that have the escape speed to the number that have the root-mean-square speed. *(c)* Would this ratio be different for H_2? Explain.

4. ***Speed of sound.*** Ordinary sound waves in a gas have a speed v_s given in terms of the density ρ and the bulk modulus B (see Chap. 15) by

$$v_s = \sqrt{\frac{B}{\rho}}$$

The pressure and density variations that constitute the sound wave occur so rapidly that the adiabatic bulk modulus, $B = -V(dp/dV)$, is to be used, and the derivative is taken appropriate to an adiabatic process. Show that the speed of sound in an ideal gas is given by

$$v_s = \sqrt{\frac{\gamma RT}{M}}$$

5. ***Molecular rotation.*** Oxygen nuclei in an O_2 molecule are about 0.2 nm apart, and virtually all of the mass of the molecule is concentrated in the two nuclei. *(a)* Estimate the moment of inertia of the molecule about an axis which is a perpendicular bisector of the line joining the two nuclei. *(b)* Estimate the root-mean-square *angular speed* ω_{rms} about this axis for O_2 at 400 K.

6. ***Measuring a molecular speed distribution.*** One method of measuring molecular speeds is to use a rotating-drum apparatus, shown schematically in Fig. 18-13. Molecules escaping from an oven at temperature T enter the drum only when the slit S passes. The molecules stick on a glass plate G on the opposite side, with a spatial distribution along the plate depending on the molecular speed distribution. Suppose the drum is 120 mm in diameter and rotates at 1200 rev/min and that Ag atoms emerge from the oven maintained at 1000 K. Determine the distance along the plate G between points struck by atoms with speeds $\frac{1}{2}v_{rms}$ and $\frac{3}{2}v_{rms}$.

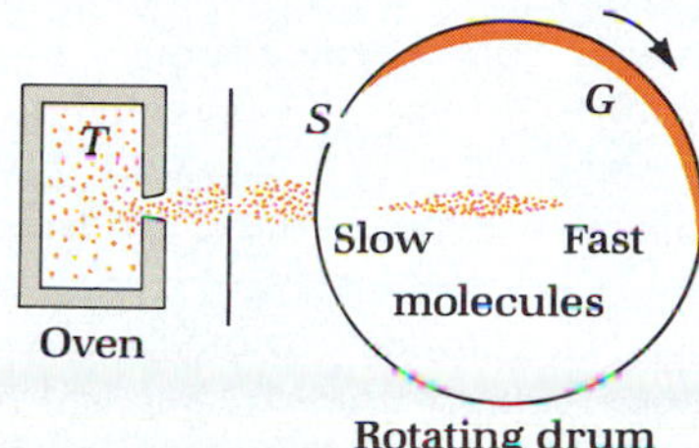

Figure 18-13. Problem 18-6: Silver atoms from an oven pass through a slit in a rotating drum.

7. ***Einstein model of vibrations in a solid.*** In the Einstein model of an elemental solid, all atoms are assumed to vibrate with the same frequency ν_0. Each atom can oscillate along each of the three spatial directions, and the (quantum) average mechanical energy of each oscillator is given by

$$\langle E \rangle = \frac{h\nu_0}{e^{h\nu_0/kT} - 1}$$

where $h = 6.63 \times 10^{-34}$ J·s is Planck's constant (see Chap. 39). *(a)* Show that the molar heat capacity from this model is

$$\mathscr{C}_V = 3R \frac{(h\nu_0/kT)^2 \, e^{h\nu_0/kT}}{(e^{h\nu_0/kT} - 1)^2}$$

(b) Construct a graph of this function for T ranging from 1 to 300 K, taking $\nu_0 = 10^{12}$ Hz as a typical frequency. *(c)* What value does $\mathscr{C}_V$ approach as T becomes large?

8. ***Specific heat of water.*** Estimate the specific heat c_p of water vapor at low pressure. Assume three translational and three rotational degrees of freedom. Compare with the measured value $c_p = 2000$ J·kg^{-1}·K^{-1} at room temperature. Are vibrational degrees of freedom active? Explain.

9. ***Dalton's law of partial pressures.*** Consider a mixture of several gases with N_1 molecules of mass m_1. N_2 molecules of mass m_2,. . . . By extending the development in Sec. 18-1, prove ***Dalton's law of partial pressures:*** *The total pressure of a mixture of gases is equal to the sum of the partial pressures of the component gases.* The ***partial pressure*** is the pressure of a gas if it alone were present.

10. ***Fraction of molecules in a range of speeds.*** The fraction of molecules with speed v within dv is the distribution function in Eq. 18-20 divided by the total number of molecules in the system: $[f(v)\,dv]/N$. The fraction with speeds between v_1 and v_2 is therefore given by the integral

$$\frac{1}{N}\int_{v_1}^{v_2} f(v)\,dv$$

This integral must be evaluated numerically. *(a)* Write or adapt a program to do this, choosing an interval size for Δv which is appropriate for O_2 at 300 K. *(b)* Determine the fraction of molecules with speeds between zero and v_m, between v_m and $\langle v \rangle$, and between $\langle v \rangle$ and v_{rms}.

11. ***The median speed.*** Use the program developed in the preceding problem to determine the *median speed,* the speed such that half the molecules have speeds below this value and the other half have speeds above this value. Note that you must integrate the distribution function from $v = 0$ out to a speed such that half the molecules are accounted for. Compare the median speed with the most probable, the mean, and the root-mean-square speeds.

12. ***Random numbers.*** The BASIC function RND(1) returns a (pseudo) random number whose value lies between 0 and 1. Develop a program to generate N random numbers within this range, calculate the mean of these numbers, and calculate the root-mean-square value. Do this for $N = 10$, 100, and 1000. What do you expect for the mean of these randomly distributed numbers?

13. ***The gaussian distribution.*** A different kind of distribution of numbers can be obtained by using the random-number function from the previous problem. Consider a number obtained from the following expression:

$$v = 300 + \sum_{i=1}^{50} 10*(\text{RND}(1) - .5)$$

That is, added to 300 is the sum of 50 random numbers, each lying between -5 and $+5$. Write a program to generate $N = 100$ values for v of this type. Choose $\Delta v = 5$ and count the number of values that lie between v and $v + \Delta v$. Inspect these values and construct a graph of values similar to that in Fig. 18-9. This approximates a *normal,* or *gaussian, distribution.*

14. ***Investigating our model of a gas.*** The first assumption listed for the model of a gas in Sec. 18-1 is that the number of molecules is very large and that the size of a molecule is negligible compared with the average distance between molecules. A typical atom has a diameter of about 3×10^{-10} m. *(a)* Determine the volume available to an atom in a gas when $T = 300$ K and $p = 101$ kPa. *(b)* From your answer in part *(a)*, estimate the average distance between the gas molecules. *(c)* Find the ratio of your answer to part *(b)* and the diameter of a typical atom.

HIGHLIGHTS OF MODERN PHYSICS

Statistical Physics

Air molecules are in continuous chaotic motion. What is the chance that all the air molecules in a room could, for an instant, randomly fly into the wastepaper basket in the corner of the room? Is such a thing possible? Is it probable?

Perhaps surprisingly, the probability of this happening is far greater than the probability that all the air molecules in the room where you are sitting have the actual positions that they have right NOW. Although this may seem contrary to experience, we will prove it.

Let the volume of the room be V_R and that of the wastepaper basket be V_{WB}. What is the chance that all n molecules in the room will randomly end up in the wastebasket? The probability of any molecule being found in the wastebasket is obviously V_{WB}/V_R. The probability of any *two* molecules being found in the wastebasket is the *product of the separate probabilities,* $(V_{WB}/V_R)^2$. For all n molecules the probability is $(V_{WB}/V_R)^n$. This is an exceedingly small number since $V_{WB}/V_R \ll 1$ and n is large. For a 0.3-m³ wastebasket and a 3000-m³ room, $V_{WB}/V = 10^{-4}$, so for $n = 10^{24}$ molecules, the probability is $(10^{-4})^n = 10^{-4,000,000,000,000,000,000,000,000}$, a very small number indeed.

Can this possibly be larger than the probability of a specific distribution of molecules spread evenly throughout the room? Imagine the room divided into many small cells (Fig. 1), each equal to the smallest volume a molecule can be forced to occupy. This is an ill-defined number but for our purposes we will use the volume of a typical air molecule, roughly a cube 2×10^{-10} m on a side, or a volume of about 10^{-29} m³. In our room of volume 3000 m³, there are $N = 3000/10^{-29} = 3 \times 10^{32}$ cells. The probability that a single molecule is found in a specific cell is $1/N$. The probability that all n molecules are in their specific cells at any instant is $(1/N)^n = (\frac{1}{3} \times 10^{32})^n = 10^{-3.2\times10^{25}}$, or $10^{-2.8\times10^{25}}$ times the probability of all 10^{24} molecules being found in the wastebasket. It is incomparably smaller.

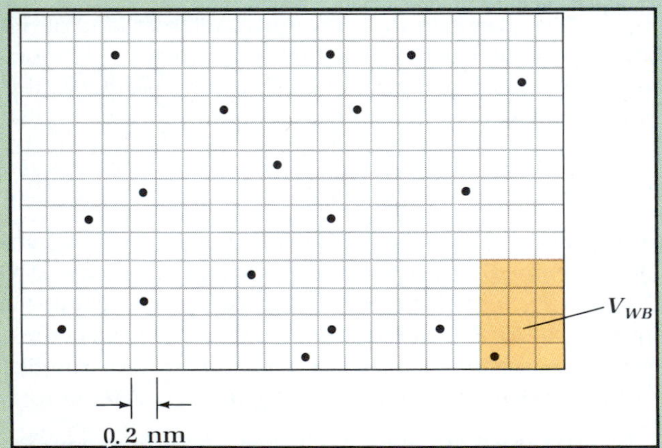

Figure 1. A room of volume V_R is divided into N small cubic cells of edge 0.2 nm. The shaded area is the wastebasket of volume V_{WB}. The figure shows n molecules occupying various cells throughout the room. The probability of the molecules being in those locations equals the probability of selecting those particular cells in a random drawing. This equals the probability of selecting any one cell, $1/N$, raised to the nth power. N equals the total number of cells.

The difference is that in the second calculation the molecular positions were exactly specified whereas in the first one they could be *anywhere* in the wastebasket. Had we specified the positions in the first case also, the probabilities would have been the same. This is analogous to the fact that a royal flush in poker is just as likely as *any* specific hand, no matter how nondescript the cards. For the same reason all lottery numbers are equally good: 01 02 03 04 05 06 is as likely as 23 18 07 12 37 15.

The distinction here is between the microstate of the system and its macrostate. A *microstate* is specified by giving a complete atomic description of each particle. It corresponds to our second calculation. A *macrostate* is specified by giving the values of macroscopic quantities such as pressure, temperature, and volume, without regard to what is happening at the atomic level. In the first calculation we specified only the macrostate: the air was entirely in the wastebasket. We did not know, or care about, the precise locations of the individual molecules.

These two concepts are connected by the *principle of equal probabilities: All possible microstates consistent with a given macrostate are equally likely.* All poker hands, lottery numbers, or molecular distributions are equally likely. A microstate in which all air molecules are spread uniformly, but in precise positions, is no more probable than a microstate in which all molecules are in the wastebasket, again with precise positions. *A macrostate is the average over all possible microstates.* To see what is meant by "averaging over all possible microstates" we examine a simple case that will lead to a powerful and important result.

The Boltzmann Distribution

Consider a four-particle system in which the particles are restricted to discrete, equally spaced energies (Fig. 2). This assumption is not necessary, but it simplifies the discussion. The particles are distinguishable from each other, a seemingly inconsequential factor that will turn out to be important. You can think of them as having different colors. We imagine the system to be immersed in a water bath so that it has the temperature T of the water. The system is removed from the water bath and immediately insulated so that its temperature and energy remain as they were when it was submerged. The question we pose is this: What are the states of the particles? Are they all in the lowest $E = 0$ state, or are they distributed among a number of states? If the latter, *how* are they distributed? The answer is found by tabulating all of the possible distributions (the microstates), then calculating the average number of particles in each state by averaging over the microstates.

Suppose the total energy of the system is 3ε. Figure 2 shows the three ways of arranging the four particles so the total energy is 3ε. Actually, there are more configurations than shown because the particles are distinguishable. There are four different

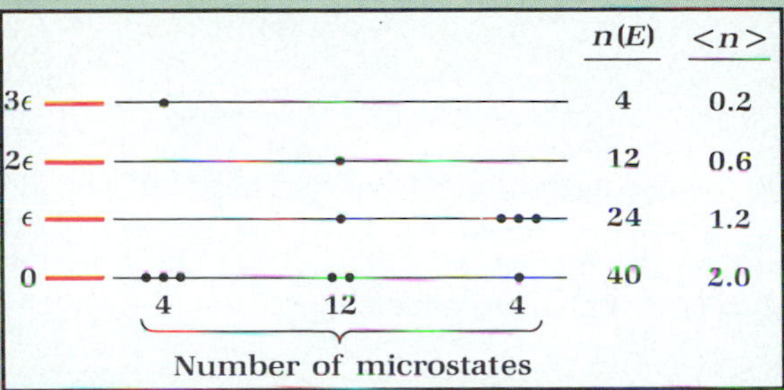

Figure 2. A system of four particles that occupy evenly spaced states. The states are separated by an energy ε, and the total energy of the system is 3ε. The three arrangements of the particles shown are the only ones for which the total energy is 3ε. However, each arrangement can be produced several different ways since the particles are distinguishable; the number of distinct configurations corresponding to each arrangement is shown directly below it. The column $n(E)$ lists the total number of particles in each particle state summed over all 20 microstates. The column $\langle n \rangle$ lists the average number of particles in each state averaged over the 20 microstates; it is obtained by dividing the first column by the number of microstates.

configurations corresponding to the leftmost arrangement because any one of the four (distinguishable) particles can occupy the $E = 3\varepsilon$ state, and each such situation would be a distinguishable microstate. Similarly, there are four microstates corresponding to the rightmost arrangement because, again, a different microstate is produced for each of the four possible particles in the $E = 0$ state. Finally, there are 12 microstates corresponding to the middle arrangement because there are four ways to put a particle in the $E = 2\varepsilon$ and, for each choice in that state, three choices remain for the $E = \varepsilon$ state, for a total of $4 \times 3 = 12$ states. After these first two states are filled, there are no additional choices for the zero-energy state; the two remaining particles have to go there, no matter what. The total number of microstates is $4 + 12 + 4 = 20$.

The first column of numbers along the right-hand side of Fig. 2 shows the total number of particles in each state summed over all 20 microstates. Thus the average number of particles per microstate in the $E = 0$ level is $40/20 = 2.0$, and similarly for the other states. These average values are shown in the far right column and are plotted in Fig. 3. Note how the curve falls off for larger energies, reminiscent of an exponential. If a similar analysis is done for a large number of particles, the curve is indeed found to be exponential, namely,

$$\langle n(E) \rangle = Ae^{-E/kT}$$

This is called the ***Boltzmann distribution*** and it gives the average number of particles in a state of energy E for a system at a temperature T. The proportionality constant A is determined by requiring that the sum over all states equal the total number of particles in the system, but it is often not necessary to know its value, as the following example shows.

Example 1. A two-state system. Consider a system consisting of only two states separated by the energy ε (Fig. 4). Although this is a simple situation, it is a common one: an electron in a magnetic field is an important example.

What is the proportion of particles in the two states for a temperature T? You might expect that at $T = 0$ K all the particles would be in the lower state. What do you expect for very large T?

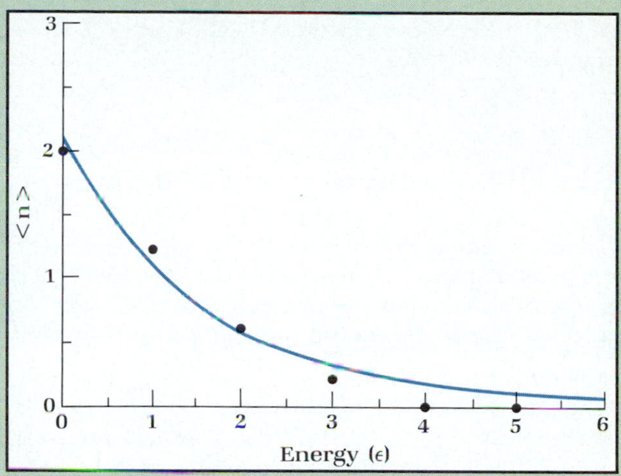

Figure 3. The data from Fig. 2 are represented by the individual data points. The Boltzmann distribution for this case ($kT = 1.49\varepsilon$ and $A = 2.1$) is shown by the solid curve. The parameters kT and A are obtained by requiring that the total number of particles be 4 and the total energy be 3ε. The Boltzmann distribution is normally valid only when the number of particles is large, but it works reasonably well even in this case.

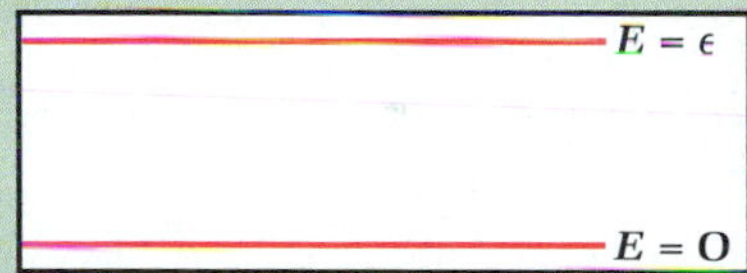

Figure 4. A two-state system. The states are separated by the energy ε.

Solution. The fraction of the particles in the upper state is given by the Boltzmann distribution,

$$\frac{\langle n(\varepsilon) \rangle}{N} = \frac{\langle n(\varepsilon) \rangle}{\langle n(0) \rangle + \langle n(\varepsilon) \rangle} = \frac{Ae^{-\varepsilon/kT}}{Ae^{0} + Ae^{-\varepsilon/kT}}$$

where $\langle n(0) \rangle$ and $\langle n(\varepsilon) \rangle$ are the numbers of particles in the lower and upper states, respectively, and N is the total number of particles. Notice that the normalization factor A cancels out, giving

$$\frac{\langle n(\varepsilon) \rangle}{N} = \frac{e^{-\varepsilon/kT}}{1 + e^{-\varepsilon/kT}} \quad \text{and} \quad \frac{\langle n(0) \rangle}{N} = \frac{1}{1 + e^{-\varepsilon/kT}} \qquad (1)$$

for the lower state. The sum of these fractions is unity, as it must be. It is so common for the factor A to cancel out that the remaining quantity, $e^{-\varepsilon/kT}$, is given its own name: the ***Boltzmann factor.***

In deriving the Boltzmann distribution we assumed that the particles were distinguishable, but this is not generally valid. Consider the fact that all hydrogen atoms, all oxygen atoms, and all electrons are identical. However, in many cases identical particles *are* effectively distinguishable. For example, carbon atoms in diamond are distinguishable by virtue of their different positions. Teachers sometimes place identical twins in specific seats so they can be distinguished by where they are. Therefore the Boltzmann distribution is far more widely applicable than its assumption of distinguishability implies.

Self Test: Derive the expression in Eq. 1 for the fraction of particles in the lower state.

Maxwell-Boltzmann Speed Distribution

The Boltzmann distribution provides an easy derivation of the Maxwell-Boltzmann speed distribution (Eq. 18-20). Recall that $f(v)\ dv$ is the number of particles with speeds between v and $v + dv$. Clearly, $f(v)\ dv$ is proportional to three quantities:

1. The number of particles N
2. The volume in velocity space between v and $v + dv$ (Fig. 5)
3. The probability that a particle will have the speed v. This probability is proportional to the Boltzmann factor with $E = \frac{1}{2}mv^2$, $e^{-E/kT} = e^{-mv^2/2kT}$.

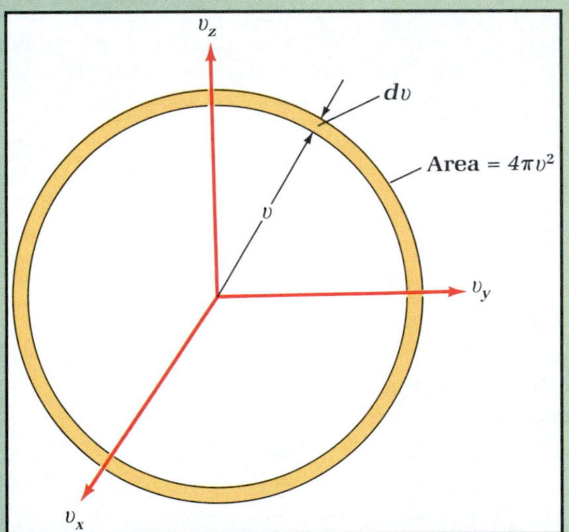

Figure 5. The Maxwell-Boltzmann distribution $f(v)\ dv$ gives the number of molecules within the spherical shell of radius v and thickness dv. The volume of the shell equals its surface area $4\pi v^2$ times its thickness dv. The axes are the x, y, and z components of the particle's velocity. This coordinate system is called *velocity space* for obvious reasons.

Therefore

$$f(v)\ dv = CN(4\pi v^2\ dv)e^{-mv^2/2kT}$$

where C is the overall proportionality constant. The integral over all speeds equals the total number of particles: $\int_0^\infty f(v)\ dv = N$. Integrating and solving for C gives $C = (m/2\pi kT)^{3/2}$, which reproduces Eq. 18-20.

QUESTIONS

1. Why is it valid to use the Boltzmann distribution in calculating the Maxwell-Boltzmann speed distribution? The gas molecules are not fixed in position so why can they be considered distinguishable? Under what conditions might they no longer be considered distinguishable?
2. Justify the following expression for $\langle E\rangle$:

$$\langle E\rangle = \frac{1}{N}\sum_i \langle n(E_i)\rangle E_i$$

EXERCISES

1. Redo the calculation illustrated in Fig. 2, but this time let the total energy be 5ε. Keeping $N = 4$, show that the average numbers of particles in the individual states are 1.5, 1.1, 0.71, 0.43, 0.21, and 0.07, starting from the lowest state.
2. *(a)* The average energy of a particle is given by

$$\langle E\rangle = \frac{1}{N}\sum_i \langle n(E_i)\rangle E_i$$

where the sum is over all states of the system. Use this to calculate the average energy of the particles in the example of Fig. 2. *(b)* There is a much easier way to calculate the average energy in this case. What is it? (This method is not available in large-particle situations where the Boltzmann distribution must be used.)
3. Integrate $f(v)\ dv$ to show that the proportionality constant C is $(m/2\pi kT)^{3/2}$.
4. Redo the calculation of Fig. 2, but let the four particles be indistinguishable. Calculate $\langle n\rangle$ for each state. This distribution is called the *Bose-Einstein distribution* and applies to such things as light and low-temperature superconductivity.
5. Show that both expressions in Eq. 1 go to $\frac{1}{2}$ as $T \to \infty$. Draw graphs showing how the fraction of particles in each state varies with temperature.

THE SECOND LAW OF THERMODYNAMICS

19

Lava flow

Occasionally we see a film clip or videotape that is being run backward. It is amusing to see the customary order of a sequence of events reversed: Views of people walking backward, water flowing uphill, or a previously demolished building arising from the rubble all accentuate our bias toward a one-way perception of time, flowing from past to future. Of the laws of nature that we have encountered so far — Newton's laws, the law of gravitation, the conservation laws, the first law of thermodynamics — none depends on the sense or direction of time. That is, these laws remain the same if time t is replaced by $-t$. The motion of a ball up and down in free-fall (no friction), for example, would look the same if time were reversed. If all of these laws are obeyed, why then does a time-reversed sequence of some events seem unnatural, improbable, even impossible to us? Indeed some processes *are* impossible, and it is the *second law of thermodynamics* which states that such processes do not occur.

There are many different but equivalent ways to phrase the second law. Much of the language reflects the law's origins in attempts to improve the efficiency of steam engines. We shall consider several statements of the second law of thermodynamics in this chapter and see that it applies to much more than just steam engines.

19-1 HEAT ENGINES AND THE SECOND LAW

The first law of thermodynamics, $\Delta U = Q - W$, deals with energy transfers between a system and its environment. In a typical process, heat Q is added to the system and work W is done by the system. The process can be regarded as one that transforms energy: Energy enters the system as heat and leaves the system as work. A gas expanding at constant pressure in a cylinder is a simple example. Or in a different process, the transformation or conversion of energy can be in the opposite sense, with energy entering the system as work and leaving the system as heat.

The conversion of work into heat occurs spontaneously when work is done by dissipative forces such as friction. In an automobile-braking mechanism, for example, energy enters the disk as work done by frictional forces between brake pads and rotating disk. The temperature of the disk increases, and because of the temperature difference, energy is transferred to the surroundings as heat. Thus the energy is dissipated.

A Heat Engine

The conversion of heat into work is highly desirable from an economic point of view. Energy transferred as work can be put to such practical uses as lifting a weight or turning a shaft to operate machinery or an electric generator. Energy transferred as heat cannot be directly used to lift a weight or turn a shaft. Simply burning gasoline to transfer energy as heat to an automobile will not propel the automobile. It is necessary first to convert heat into work. A ***heat engine*** is a device that converts heat to work.

Consider the process illustrated in Fig. 19-1: A metallic cylinder fitted with a movable piston contains a compressed ideal gas in an initial state characterized by p_i, V_i, with a temperature T the same as the surroundings and $p_i > p_o$. The gas is allowed to expand isothermally, with heat entering the gas from the surroundings. The gas does work on the moving piston as the volume of the gas increases. Since the process is isothermal and the system is an ideal gas, the change in internal energy $\Delta U = 0$. From the first law, $\Delta U = 0 = Q - W$. In this process, an amount of heat Q has been converted to work W, and $W = Q$.

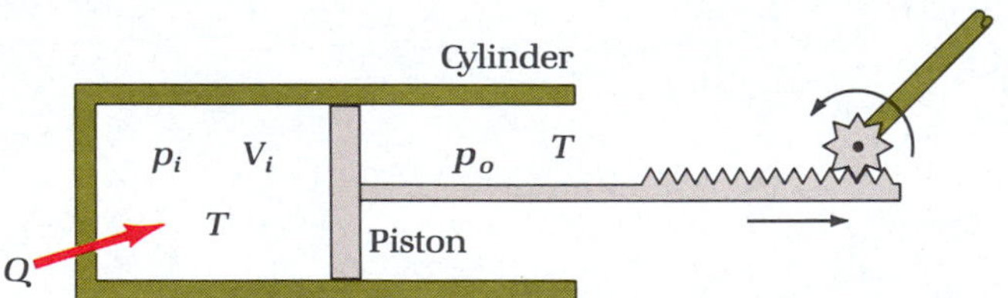

Figure 19-1. A compressed gas expands isothermally. Initially the pressure of the gas inside the cylinder is greater than the pressure of the gas outside ($p_i > p_o$).

As a practical method for converting heat to work, the single expansion process is not very satisfactory: It is a one-time affair. Once the pressure of the gas drops to atmospheric pressure p_o, no further expansion will occur. We were only able to convert heat into work with this process because the gas was originally compressed. To return the gas to its initial state would require that some work be done *on the gas* to compress it. Part of the work obtained during the expansion must be reinvested to prepare the gas for a subsequent expansion. Suppose that we return the gas to its initial state by a different path, one with lower pressure values so that less work has to be done on the gas. Then the net result is that (i) heat has been added to the gas; (ii) the gas has been returned to its initial state, ready for another expansion; and (iii) more work was done by the gas during the expansion than was reinvested to complete the cycle. Thus we are led to the idea of a cycle operating as a heat engine.

A cyclic heat engine continually converts heat to work.

In a cycle the system returns to its initial state, and the cycle may be repeated any number of times. This repetitive feature makes the cycle attractive as a heat engine. We shall usually analyze a single cycle so that all quantities refer to one cycle of operation. A general, quasi-static cycle is shown in Fig. 19-2 on a p-V diagram. For convenience the initial state has been chosen to correspond to the beginning of an expansion. The system does positive work on its surroundings during the expansion. The system does negative work on the surroundings during the compression as the system is returned to the initial state. The net work W_{cyc} done by the system for the entire cycle is equal to the area enclosed by the cycle of the p-V diagram, as we saw in Chap. 17. It is positive if the cycle proceeds in the clockwise sense because the work done during the expansion is greater than the magnitude of the negative work done during the compression. Let Q_{cyc} represent the net heat added to the system for the entire cycle. Since $\Delta U = 0$ for a cycle, the first law gives

$$Q_{cyc} = W_{cyc} \tag{19-1}$$

That is, the net work done for the cycle equals the net heat added for the cycle because the system is returned to its initial state ($\Delta U = 0$).

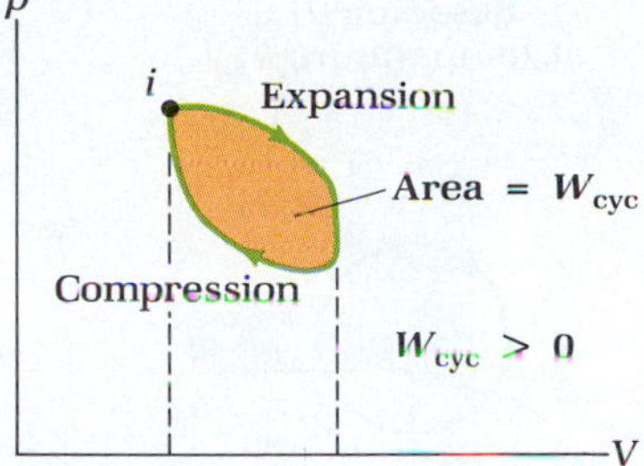

Figure 19-2. A p-V diagram for a heat engine. The net work W_{cyc} done by a system that undergoes a cycle is the area enclosed on a p-V diagram. For a heat engine, the pressure is higher during the expansion than it is during the compression. Consequently, the work done by the system during the expansion is greater than the work done on the system during the compression, and the net work is positive, $W_{cyc} > 0$.

In a modern steam engine, such as one used in the generation of electrical energy, work is done in turning a steam turbine rather than in moving a piston. As shown in Fig. 19-3, heat from fossil or nuclear fuel is added to water to form high-pressure steam, which performs work by expanding against the turbine blades. To be compressed at low pressure, the steam is condensed, which requires the extraction or the exhaust of heat from the water. This heat exchange is accomplished in cooling towers or by using cooling water from a large reservoir such as a river. Of the amount of heat obtained from the burning of fuel in a modern power plant, 60 percent or more is exhausted in completing the cycle. Thus 40 percent, at most, of the heat provided by the fuel is converted to mechanical work (and subsequently to electric energy, with small losses).

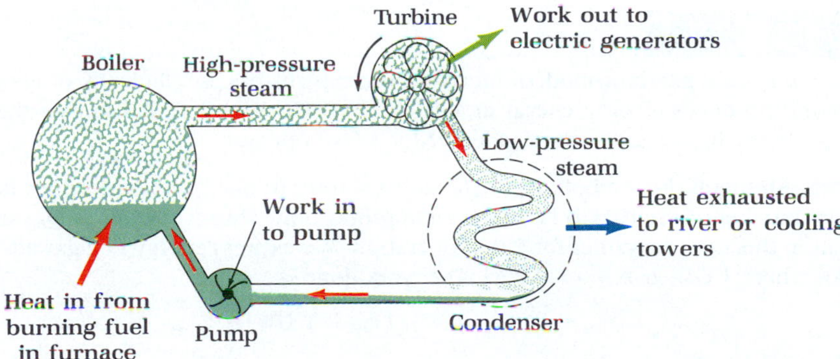

Figure 19-3. Schematic diagram of an electric generating steam plant. Water, in the form of liquid water and steam, is the working substance (or system). High-pressure steam is produced in the boiler by heat absorbed from the furnace. The steam does work by turning the turbine blades. (This work is subsequently used to produce electrical energy.) Lower pressure is maintained on the other side of the turbine by extracting heat and condensing the steam. This heat is exhausted to a river or to cooling towers. A pump forces water back into the high-pressure boiler to complete the cycle.

Generalized Representation of a Heat Engine. All cyclic heat engines have in commom some of the features described above for the steam engine. Some substance, called the ***working substance,*** undergoes a cyclic process. Heat is exchanged with the surroundings by the working substance at two or more different temperatures. Heat is added to the system at the higher temperature and must be exhausted from the system at the lower temperature to complete the cycle. It is the ***net*** heat added that equals the work for the cycle.

For simplicity, we restrict our attention to cycles operating between only two heat reservoirs, called H and L in Fig. 19-4. Reservoir H is at the higher temperature T_H, and reservoir L is at the lower temperature T_L. We let Q_H represent the heat transferred to the system (the working substance) from reservoir H and let Q_L represent the heat transferred to the system from reservoir L. For the working substance of a heat engine, heat Q_H is positive because the working substance absorbs heat from reservoir H, but heat Q_L is negative because the working substance rejects heat to reservoir L. That is, $Q_L < 0$ so that $Q_L = -|Q_L|$. To deal with positive values, we usually replace Q_L with $-|Q_L|$ in our formulas. The net heat added to the working substance during one cycle is

$$Q_{cyc} = Q_H + Q_L = Q_H - |Q_L|$$

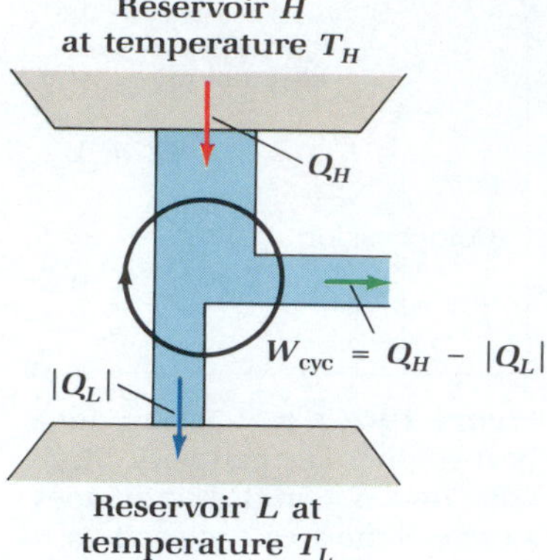

Figure 19-4. A schematic diagram of a heat engine. Heat Q_H is absorbed from reservoir H by the engine's working substance, and heat $|Q_L|$ is rejected to reservoir L. Work W_{cyc} is done by the working substance on the engine's surroundings, where $W_{cyc} = Q_H - |Q_L|$ by the first law. (Q_L is negative.) The values of W_{cyc}, Q_H, and $|Q_L|$ are proportional to the widths of the pathways so that the diagram illustrates the first law. The engine is represented as a circle, and the arrow in the clockwise sense reminds us that this is the sense of the path of the cycle on a p-V diagram (Fig. 19-2).

Equation 19-1, which is the first law applied to a cycle, becomes

$$W_{cyc} = Q_H - |Q_L| \tag{19-2}$$

In an idealized version of the electric generating steam plant, reservoir H replaces the furnace (or nuclear reactor), and heat Q_H is added to the water in the boiler (or steam generator). Also, reservoir L replaces the river or cooling towers, and heat $|Q_L|$ is extracted from the water in the condensing coils.

Efficiency of a Heat Engine. We regard heat Q_H as the required input of energy and work W_{cyc} as the useful output of energy for each cycle. The exhausted heat $|Q_L|$ is called "waste heat" because this energy is no longer useful to this engine. As we shall later see, this heat is not wasted in the sense that it is wantonly or carelessly discarded. Rather, it is necessarily rejected because of constraints placed upon us by nature. The ***efficiency*** η of a heat engine is defined as the ratio [useful output]/[required input], or

$$\eta = \frac{W_{cyc}}{Q_H} \tag{19-3}$$

The efficiency can also be expressed in terms of the two heat-exchange values. Substituting Eq. 19-2 into Eq. 19-3 gives $\eta = (Q_H - |Q_L|)/Q_H$, or

$$\eta = 1 - \frac{|Q_L|}{Q_H} \tag{19-4}$$

From this form, we see that the efficiency of a heat engine increases as $|Q_L|$ is reduced relative to Q_H.

EXAMPLE 19-1

An electric power plant. A modern electric power plant has an efficiency of about 35 percent and produces electric energy at a rate of $P = 10^9$ W = 1 GW. Estimate the heat exchanges in the boiler and in the condenser for 1 h of operation.

Solution. Although the symbols W_{cyc}, Q_H, and $|Q_L|$ refer to one cycle, the values for any particular number of cycles will be in the same proportion. Thus we let W_{cyc}, Q_H, and $|Q_L|$ represent in this case the values for 1 h of operation. We express energy in gigawatt-hours (GW·h), where 1 GW·h = 3.6×10^{12} J. The work done is

$$W_{cyc} = Pt = (1\ \text{GW})(1\ \text{h}) = 1\ \text{GW}\cdot\text{h}$$

From Eq. 19-3, the heat Q_H is

$$Q_H = \frac{W_{cyc}}{\eta} = \frac{1\ \text{GW}\cdot\text{h}}{0.35} = 3\ \text{GW}\cdot\text{h}$$

and from the first law

$$Q_H - |Q_L| = W_{cyc}$$

$$|Q_L| = Q_H - W_{cyc} = 3\ \text{GW}\cdot\text{h} - 1\ \text{GW}\cdot\text{h} = 2\ \text{GW}\cdot\text{h}$$

The Four Corners Power Plant at Farmington, NM.

We have found that one-third of the heat absorbed by the working substance (water) is converted to work and the other two-thirds is exhausted to the environment at the site of the power plant.

SELF-TEST 19-1. If a power plant has an efficiency of 36.0 percent and produces electric energy at the rate of 385 MW, what is the rate at which the power plant exhausts heat into the river used to cool its condenser? ***ANSWER:*** 684 MW.

The Kelvin-Planck Statement of the Second Law

Equation (19-4) combines the definition of efficiency and the first law of thermodynamics applied to a cycle. The maximum efficiency that would be allowed *mathematically* from that equation is 1, or 100 percent, corresponding to $|Q_L| = 0$. Imagine a cycle with no heat exhausted or wasted: $|Q_L| = 0$; then the first law would give $W_{cyc} = Q_H$. The working substance, having completed a cycle, would be unchanged, and an amount of heat extracted from a reservoir at a single temperature would have been completely converted to work.

There have been many attempts to construct a heat engine with 100 percent efficiency. All atempts have failed. That a 100 percent efficient heat engine is impossible is one way of stating the second law of thermodynamics:

> There exists no cycle that extracts heat from a reservoir at a single temperature and completely converts it into work.

{K-P} statement of the second law

This form of the second law is called the *Kelvin-Planck statement* of the second law and, as an abbreviation, we sometimes refer to this statement as {K-P}.

It is important to recognize that the 100 percent efficient heat engine would obey the first law of thermodynamics. It is the second law that denies the possibility of a cycle with no heat exhausted at a lower temperature. Imagine, for example, a cycle that extracts heat Q_H from the ocean at a single temperature T_H and converts it completely into work W (to propel a ship). The first law would require $W_{cyc} = Q_H$, but the second law denies the existence of such a cycle.

The second law is not a strictly quantitative law. The Kelvin-Planck statement is qualitative. It states the impossibility of certain types of processes. Nevertheless, the second law is just as rigorous as the first law and has equal standing with the first law. As we encounter other ways of phrasing the second law, we shall see that it applies to other processes in addition to those suggested by the Kelvin-Planck statement.

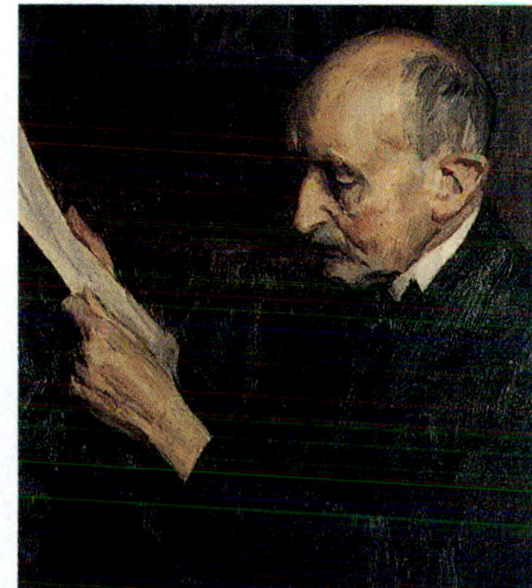

Max K. E. L. Planck (1858–1947). Celebrated for introducing his "quantum of action" known as Planck's constant *h* (Chap 39). Of Planck's many contributions, his study of heat radiation is the most notable. Planck suffered the loss of his son, who was executed by the Nazis for his participation in a plot to assassinate Adolf Hitler.

A thermoelectric converter. When the two copper legs of the converter are maintained at different temperatures, an electric current is produced that operates an electric motor that turns a fan. The converter changes heat to work and is an example of a heat engine.

19-2 REFRIGERATORS AND THE SECOND LAW

The Kelvin-Planck statement of the second law of thermodynamics is expressed in terms appropriate to a heat engine. The second law can also be stated in a way that relates to the operation of a refrigerator.

The Refrigerator

A refrigerator is a device operating in a cycle designed to extract heat from its interior so as to achieve or to maintain a lower temperature inside. During the refrigeration cycle, heat is exhausted to the outside, which is usually at a higher temperature than the inside. A net amount of work, provided typically by an electric motor, is done on the system for a cycle. A common refrigeration cycle uses a throttling process as described in Sec. 17-5. Work is done on the working substance, a fluid such as freon, by a compressor C, as shown in Fig. 19-5. The compressor maintains a high pressure difference across the throttling valve A. As the liquid evaporates on the low-pressure, low-temperature side, heat Q_L is added to the fluid from the inside of the refrigerator, which causes the inside temperature to drop. On the high-pressure, high-temperature side, heat $|Q_H|$ is exhausted from the fluid to the outside as the fluid condenses.

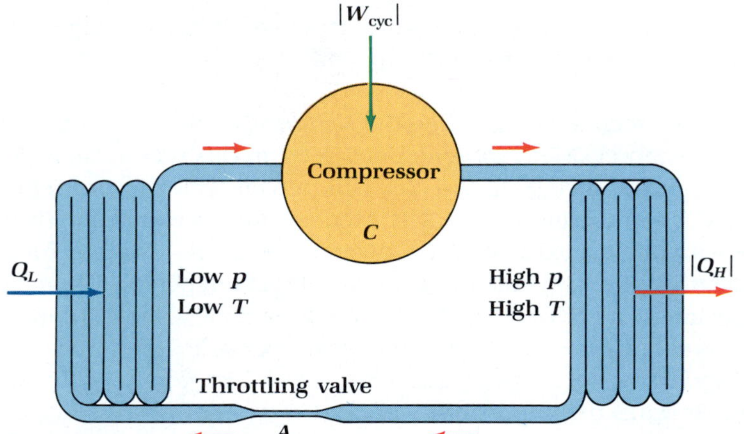

Figure 19-5. A refrigeration cycle. A typical refrigerator uses a compressor to pump freon (which is the system or refrigerant) through a throttling process. The throttling valve A is placed at the point where the freon enters the inside of the freezer compartment. Heat Q_L is absorbed on the low-pressure, low-temperature side, which is the evaporator inside the freezer. Heat $|Q_H|$ is rejected on the high-pressure, high-temperature side, which is the condenser outside of the refrigerator. (Q_H is negative.)

The p-V diagram for a system undergoing a refrigeration cycle is shown in Fig. 19-6, where the point i represents the state of the system just prior to expansion. The system's pressure is lower during the expansion than it is during the compression. Since W for the system is positive during the expansion and negative during the compression, W_{cyc} is negative. That is, when a cycle on a p-V diagram is counterclockwise, $W_{cyc} < 0$, as we saw in Chap. 17.

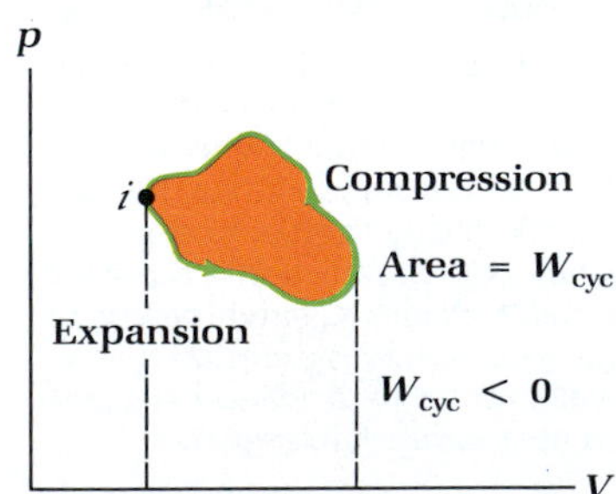

Figure 19-6. A p-V diagram for a refrigeration cycle. Since the pressure is lower during the expansion than during the compression, W_{cyc} is negative. That is, more work is done on the system by its surroundings than is done by the system on its surroundings. Compare this cycle with the heat-engine cycle shown in Fig. 19-2.

The energy transfers taking place in a refrigeration cycle are shown schematically in Fig. 19-7. Heat Q_L is added to the system from the low-temperature reservoir, representing the inside of the refrigerator, at T_L. A positive amount of heat $|Q_H|$ is exhausted to the high-temperature reservoir, representing the outside surroundings, at T_H. Negative work W_{cyc} is done by the system, or positive work $|W_{cyc}|$ is done on the system by the compressor motor. The first law of thermodynamics applied to the cycle gives $W_{cyc} = Q_H + Q_L$ or (on multiplying by -1) $-W_{cyc} = -Q_H - Q_L$. The result can be written in terms of positive quantities, $|W_{cyc}| = -W_{cyc}$ and $|Q_H| = -Q_H$:

$$|W_{cyc}| = |Q_H| - Q_L \qquad (19\text{-}5)$$

A comparison of Eqs. 19-5 and 19-2 and of Figs. 19-7 and 19-4 suggests that a refrigerator is like a heat engine running backward. The energy transfers are reversed for the heat engine and refrigerator cycles. But we should not expect the magnitudes of Q_L, Q_H, and W_{cyc} to be the same for the two cycles. There are usually mechanical constraints that actually prevent a particular heat engine from being run backward.

One characteristic of a refrigeration cycle (similar to the efficiency of a heat engine) is its *coefficient of performance K*. It is defined by

$$K = \frac{Q_L}{|W_{cyc}|} \tag{19-6}$$

It is the ratio of the useful quantity for a refrigerator, the heat extracted from the inside, to the cost, the work performed by the motor.

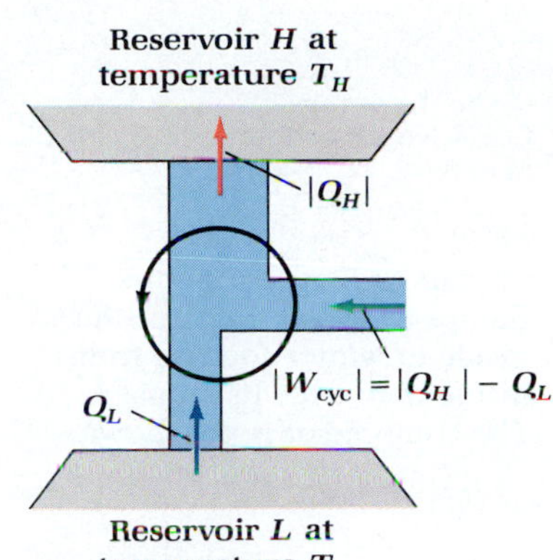

Figure 19-7. Schematic diagram of a refrigerator. Work $|W_{cyc}|$ is done on the system (the refrigerant) by its surroundings (a compressor). Work W_{cyc} is negative. Heat Q_L is absorbed by the system from reservoir L and heat $|Q_H|$ is rejected to reservoir H. Heat Q_H is negative. From the first law, $|W_{cyc}| = |Q_H| - Q_L$. The widths of the pathways are proportional to the energy transfers. The circle which represents the refrigerator has an arrow that reminds us of the counterclockwise path of the cycle on a p-V diagram (Fig. 19-6).

EXAMPLE 19-2

A kitchen refrigerator. A refrigerator with a 480-W compressor motor is rated by a coefficient of performance $K = 2.8$. Estimate the rate of heat exchange *(a)* at the condenser and *(b)* at the evaporator.

Solution. *(a)* The rate at which heat is absorbed by the system (typically freon) at the condenser (inside the freezer compartment) is dQ_L/dt. Also, for a steadily running refrigerator, the rate of energy transfer is proportional to the energy transferred per cycle, so that Eq. 19-6 may be written

$$K = \frac{Q_L}{|W_{cyc}|} = \frac{dQ_L/dt}{d|W_{cyc}|/dt}$$

Solving for dQ_L/dt, we find

$$\frac{dQ_L}{dt} = K\frac{d|W_{cyc}|}{dt} = (2.8)(480\text{ W}) = 1.3\text{ kW}$$

(b) The rate at which heat is rejected by the system at the evaporator (typically at the back or underneath the refrigerator) is $|dQ_H|/dt$. From the first law, $|Q_H| = Q_L + |W_{cyc}|$, so that

$$\frac{d|Q_H|}{dt} = \frac{dQ_L}{dt} + \frac{d|W_{cyc}|}{dt} = 1.3\text{ kW} + 0.48\text{ kW}$$
$$= 1.8\text{ kW}$$

SELF-TEST 19-2. A refrigerator absorbs heat at its condenser (inside the freezer compartment) at a rate of 1.65 kW, and its electric motor uses 540 W of electric power. What is the refrigerator's coefficient of performance? ***ANSWER:*** 3.1.

The Heat Pump

The refrigeration cycle extracts heat Q_L from a low-temperature reservoir (inside a refrigerator) which tends to reduce the inside temperature. The cycle exhausts heat $|Q_H|$ to the outside, and the outside temperature tends to increase. In effect, heat is "pumped" by the refrigeration cycle from lower- to higher-temperature regions.

A heat pump "pumps" heat from lower to higher temperatures.

The ***heat pump,*** which is essentially a refrigerator, is in common use in moderate climates for space heating and cooling. For space heating, the evaporator heat exchanger is outside the structure, as shown in Fig. 19-8*a*, and extracts heat Q_L from surrounding air. The condenser heat exchanger is inside the structure, and heat $|Q_H|$ is exhausted to the inside air. The cycle pumps heat from outside to the inside. For cooling, the inside heat exchanger becomes the evaporator and the outside heat exchanger becomes the condenser, as seen in Fig. 19-8*b*. Heat is pumped from inside to the outside. In either case, the cycle can pump heat from a lower-temperature reservoir to a higher-temperature one.

A heat pump used for space heating can be characterized by a coefficient of performance different from that of a refrigerator. The useful quantity for heating purposes is $|Q_H|$, the heat exhausted to the inside. The quantity we pay for is still the work $|W_{cyc}|$. The heat-pump coefficient of performance K_{hp} is defined as

Coefficient of performance K_{hp} for a heat pump

$$K_{hp} = \frac{|Q_H|}{|W_{cyc}|} \tag{19-7}$$

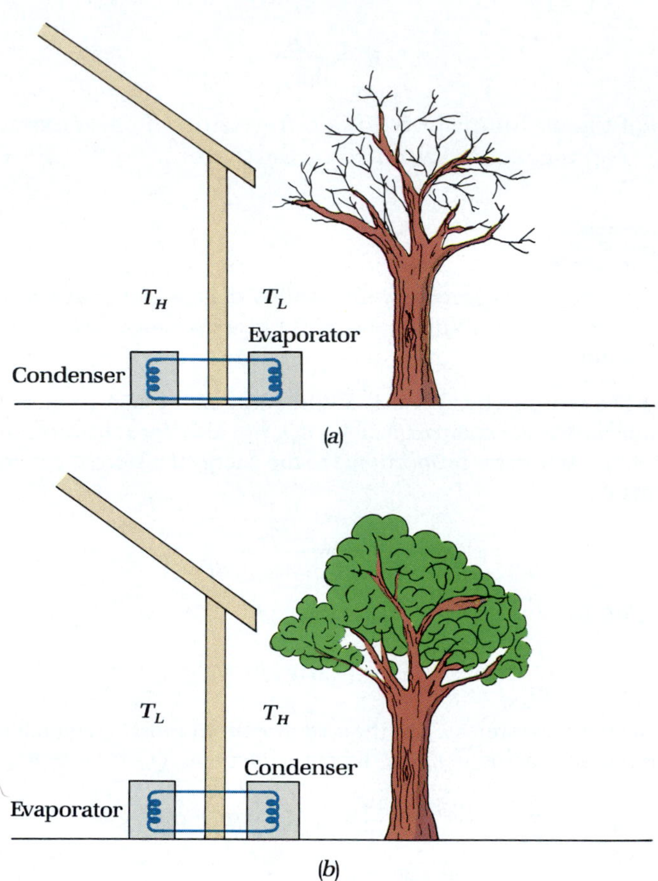

Figure 19-8. A heat pump pumps heat *(a)* from outside to inside in winter and *(b)* from inside to outside in summer. The compressor is not shown.

EXAMPLE 19-3

A heat pump. During an hour of operation, a heat pump uses 1.4 kW·h of electric energy while supplying 1.1×10^4 Btu to the inside of a home. Determine *(a)* the coefficient of performance and *(b)* the heat extracted from the outside in an hour.

Solution. First express the energy transfers in the same unit, converting Btu to kW·h:

$$|Q_H| = \frac{(1.1 \times 10^4 \text{ Btu})(1055 \text{ J}\cdot\text{Btu}^{-1})}{3.6 \times 10^6 \text{ J}\cdot\text{kW}^{-1}\cdot\text{h}^{-1}}$$
$$= 3.2 \text{ kW}\cdot\text{h}$$

(a) The coefficient of performance is

$$K_{hp} = \frac{|Q_H|}{|W_{cyc}|} = \frac{3.2 \text{ kW}\cdot\text{h}}{1.4 \text{ kW}\cdot\text{h}} = 2.3$$

(b) The heat extracted from the outside in 1 h is obtained using the first law, applying it to the operation of the heat pump for this time period:

$$Q_L = |Q_H| - |W_{cyc}|$$
$$= 3.2 \text{ kW}\cdot\text{h} - 1.4 \text{ kW}\cdot\text{h} = 1.8 \text{ kW}\cdot\text{h}$$

SELF-TEST 19-3. A heat pump has a coefficient of performance $K_{hp} = 2.5$, and its compressor is rated at 2.2 kW. What is the rate at which the heat pump transfers heat to the house? ***ANSWER:*** 5.5 kW.

The Clausius Statement of the Second Law

Rudolf J. E. Clausius (1822–1888) was a theoretician with the ability to make significant progress without requiring complicated mathematical analysis. Clausius introduced many concepts that were later expanded by J. C. Maxwell.

From the last two examples, and generally from Eqs. 19-6 and 19-7, we see that a coefficient of performance can have a value greater than 1. Indeed, for refrigerating or heating purposes, larger values of a coefficient of performance mean greater economy. Mathematically, a larger value of a coefficient of performance is achieved by decreasing $|W_{cyc}|$ relative to Q_L or $|Q_H|$. The value could be increased without bound by having $|W_{cyc}|$ approach zero. If $|W_{cyc}| = 0$, then the first law applied to the cycle would give $|Q_H| = Q_L$, and the net result would be the transfer of heat from a reservoir at lower temperature to one at higher temperature, with no change in any other system. No such process has ever been observed. Its impossibility is another way to phrase the second law of thermodynamics.

> No process is possible whose sole net result is the transfer of heat from a lower to higher temperature.

This form of the second law is called the *Clausius statement* and, as an abbreviation, we sometimes refer to this statement as {C}.

The second law denies the possibility of a refrigeration cycle where *no* work is provided. This means that the second law prohibits the spontaneous flow of heat from lower to higher temperature. Stated in this way, our everyday experience is in full accord with the second law. Whenever heat flows, it is always from higher temperature to lower temperature.

The Equivalence of {K-P} *and* {C}

The Clausius statement of the second law deals with heat transfer from lower to higher temperature. The Kelvin-Planck statement deals with the conversion of heat into work. How can these very differently phrased statements correspond to one and the same law of nature? They are equivalent statements because any process prohibited by one statement can also be shown to be prohibited by the other statement. Figure 19-9 illustrates this equivalence. Suppose, as seen in Fig. 19-9*a*, that there were a 100 percent efficient heat engine, a violation of {K-P}. Then the

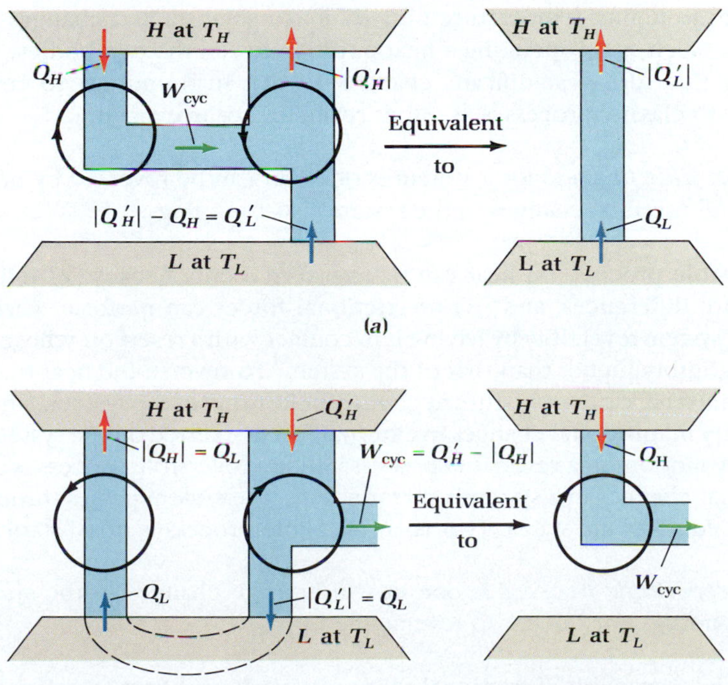

Figure 19-9. *(a)* The work from a hypothetical 100 percent efficient heat engine is used to run a refrigerator. The engine violates the Kelvin-Planck statement. Although the refrigerator by itself does not violate the Clausius statement, the combination [hypothetical engine + refrigerator] does violate it. Thus a violation of {K-P} implies a violation of {C}. *(b)* A hypothetical refrigerator that requires no work is used in combination with an engine. The refrigerator violates {C}. Although the engine by itself does not violate {K-P}, the combination [engine + hypothetical refrigerator] does violate it. That is, a violation of {C} implies a violation of {K-P}.

work from that engine could be used to run a refrigeration cycle. Taking the two cycles together, the net and only result would be a transfer of heat from lower to higher temperature, a violation of {C}. Thus a process forbidden by the Kelvin-Planck statement is also forbidden by the Clausius statement.

In a similar way, the converse is illustrated in Fig. 19-9*b*. A refrigeration cycle violating the Clausius statement could be run in conjunction with a heat engine. The combination constitutes a cycle that could extract heat at a single temperature and completely convert it into work, thus violating the Kelvin-Planck statement.

Alternative, equivalent statements of the second law have equal stature in thermodynamics. It is mainly for convenience (and a consequence of the history of the development of the second law) that we have more than one statement. We may find it much easier to apply one statement of the second law, rather than the other, to some proposed process.

19-3 REVERSIBILITY AND THE CARNOT CYCLE

The second law of thermodynamics tells us that no heat engine can be 100 percent efficient ($|Q_L|$ cannot be zero in Eq. 19-4) and that no refrigerator can have an infinite coefficient of performance ($|W_{cyc}|$ cannot be zero in Eq. 19-6). But the second law does not tell us how great the efficiency of a heat engine or the coefficient of performance of a refrigerator can be. Still, our experience suggests that frictional effects and heat transfer through large temperature differences, for example, tend to reduce the efficiency of engines.

Reversible and Irreversible Processes

The flow of heat from a body at higher temperature to one at lower temperature is an example of an irreversible process. It is irreversible in the sense that the reversed process, one with the order of events reversed in time, is impossible according to the second law. Here is an instance of the connection between the second law and the sense or direction of time. The time-reversed process, the spontaneous flow of heat from lower to higher temperature, does not occur. Heat can be transferred from lower to higher temperature only by making substantial changes in the arrangement, such as by operating a heat pump between the two bodies.

We use this idea—significant changes in the surroundings to reverse the process—to classify processes as either reversible or irreversible.

Definition of a reversible process

A ***reversible process*** for a system is one that can be reversed by making only infinitesimal changes in the system's surroundings.

In a reversible process, (i) heat can be transferred only because of infinitesimal temperature differences, and (ii) no frictional forces can perform work. Heat is added to a system reversibly by having it in contact with a reservoir whose temperature is negligibly higher than that of the system. To reverse the heat transfer, the reservoir must have a temperature negligibly below that of the system. Similarly, by making only infinitesimal changes in external forces exerted on the system, we can reverse any motion for a reversible process. Since a reversible process is a result of infinitesimal changes in a system's surroundings, the system passes through a succession of equilibrium states. That is, a reversible process is quasi-static.

A reversible process is quasi-static.

An irreversible process

An ***irreversible process*** is one in which finite changes in the system's surroundings must occur to reverse the process.

A process is irreversible if frictional forces do work or if heat transfers through a finite temperature difference. Suppose a block slides on a surface and the coefficient of friction is nonnegligible. Then the force exerted on the block by its surroundings must be changed by a finite amount to reverse the process. This process is irreversible because frictional forces do work. Suppose heat flows through a wall as a result of a nonnegligible temperature difference from one side of the wall to the other. Then a finite amount of work must be provided to a heat pump to reverse the process. This process is irreversible because heat transfer occurs as a result of a finite temperature difference.

As stated above, any reversible process is quasi-static. The converse is not true: *A process can be quasi-static but not reversible.* A block can slide across a rough surface at an infinitesimal speed (quasi-statically), but the nonnegligible frictional force does work and the process is irreversible. Or heat may be transferred at an infinitesimal rate (quasi-statically) through an insulating wall, but if there is a finite temperature difference across the wall, the process is irreversible.

A little reflection shows that every *real* process is irreversible. A reversible process is a useful idealization, similar in spirit to a frictionless interface or a spherically symmetric earth or a massless string in mechanics. A particular process can be considered nearly reversible to the extent that only very small changes in the surroundings of the system will reverse the process. The smaller the required changes, the more nearly reversible the process. We classify a process as reversible in this limiting sense. We assume that a process can be found or developed which is as close to being reversible as desired to connect any two states of a system. The concept of a reversible process plays an essential role in our further analysis of heat engines.

A reversible process is an idealization.

The Carnot Cycle

Consider all conceivable heat engines operating between two reservoirs at temperatures T_H and T_L. Each engine has an efficiency less than 100 percent, according to the second law. Which of these engines has the greatest efficiency? The answer to this theoretical question that has practical importance was first obtained by the French engineer Sadi Carnot (1796–1832). Carnot considered an idealized heat engine that would achieve the maximum efficiency for engines working between temperatures T_H and T_L. The cycle is called the *Carnot cycle* and constitutes the *Carnot engine.*

The Carnot cycle is idealized in that it is a reversible cycle. The cycle can be reversed (from operating as an engine to operating as a refrigerator) by making only infinitesimal changes in external conditions. The cycle, operating as an engine, consists of the following four steps (note the emphasis on *reversible* for the Carnot cycle):

1. A reversible isothermal expansion at T_H: Heat Q_H is added to the system.
2. A reversible adiabatic process: The temperature of the system drops from T_H to T_L.
3. A reversible isothermal compression at T_L: Heat $|Q_L|$ is exhausted from the system.
4. A reversible adiabatic process to complete the cycle: The temperature of the system increases from T_L back to T_H.

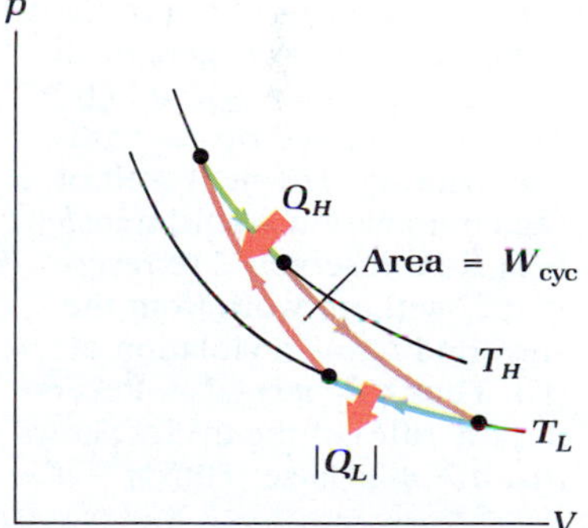

Figure 19-10. The Carnot engine cycle for an ideal gas has the following reversible steps: (i) Heat is added during the isothermal expansion at T_H. (ii) The temperature drops from T_H to T_L as the gas expands adiabatically. (iii) Heat $|Q_L|$ is extracted from the gas during the isothermal compression at T_L. (iv) The temperature increases from T_L to T_H as the gas is compressed adiabatically back to the initial state. The net work, $W_{cyc} = Q_H - |Q_L|$, for the cycle equals the enclosed area.

The Carnot engine cycle is shown on a *p-V* diagram in Fig. 19-10 for the case of an ideal gas as the working substance. The energy transfers are shown schematically, with the net work W_{cyc} equal to the area enclosed by the cycle.

Carnot's Theorem. The Carnot engine is a special case of a more general heat engine called a ***reversible engine,*** a cycle consisting entirely of reversible steps. It is the reversible engine, *any* reversible engine, which is the most efficient heat engine operating between two reservoirs. This result is contained in ***Carnot's theorem:***

Sadi Carnot (1796–1832). Carnot's masterpiece was *Reflexions sur la puissance motrice du feu (Reflections on the motive power of fire)*, published in 1824. In *Reflexions*, Carnot established the second law of thermodynamics without knowledge of the first law.

All reversible engines operating between temperatures T_H and T_L have the same efficiency, and no engine operating between these temperatures can have an efficiency greater than this.

Carnot's theorem contains two conclusions:

1. The efficiency of a reversible engine operating between the two reservoirs is independent of the nature of the working substance or of the details of the mechanism. All reversible engines have the same efficiency when operating between these two temperatures.

2. The efficiency of a reversible engine is the maximum efficiency for engines operating between these two temperatures. The efficiency of any engine operating between these reservoirs must be less than or equal to that of a reversible engine.

For our purposes, the distinction between the general reversible engine and the Carnot engine is unimportant. We shall use the Carnot engine, because of its simplicity, as a particular type of reversible engine.

The proof of Carnot's theorem consists in showing that if Carnot's theorem were not true, then the second law of thermodynamics could be violated. (Carnot's theorem can be taken as yet another way to phrase the second law of thermodynamics.) We briefly sketch the proof of the first part of Carnot's theorem by assuming that two Carnot engines C and C' have different efficiencies, η and η', when operating between the same two reservoirs. Specifically, assume that $\eta' > \eta$. Then, as seen in Fig. 19-11, we can use the more efficient one, C', to drive the other, C, as a refrigerator (it is reversible), having as a net result the transfer of heat from lower to higher temperature. This result is impossible according to the second law; thus our assumption that $\eta' > \eta$ is false. By interchanging the roles of the two cycles, we can show that the assumption $\eta > \eta'$ is also false. Therefore, $\eta = \eta'$. The second conclusion of Carnot's theorem is addressed in Prob. 3.

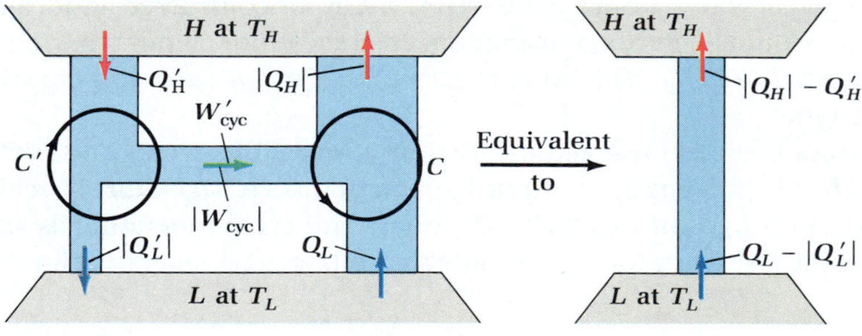

Figure 19-11. First conclusion of Carnot's theorem. Carnot cycle C' is operated as an engine and its work is used to run Carnot cycle C as a refrigerator. If $\eta' > \eta$, then $|Q_H| > Q'_H$ and $Q_L > |Q'_L|$ (as illustrated by the pathway widths). The net result of the combination would transfer heat from reservoir L to reservoir H without work from the surroundings—a violation of {C}. Thus $\eta' > \eta$ is false. Reversing the roles of the cycles shows that $\eta > \eta'$ is false. Thus $\eta = \eta'$.

We shall use the two conclusions of Carnot's theorem in the next section to obtain some useful and remarkable results.

19-4 THE KELVIN, OR THERMODYNAMIC, TEMPERATURE

The reversible engine is the most efficient heat engine operating between two reservoirs at temperatures T_H and T_L. We now determine the value of this highest efficiency. Since all reversible engines operating between these reservoirs have the same efficiency, we choose a Carnot engine whose working substance is an ideal gas. This choice makes the calculation simple, but the result is the same for any working substance in any reversible engine.

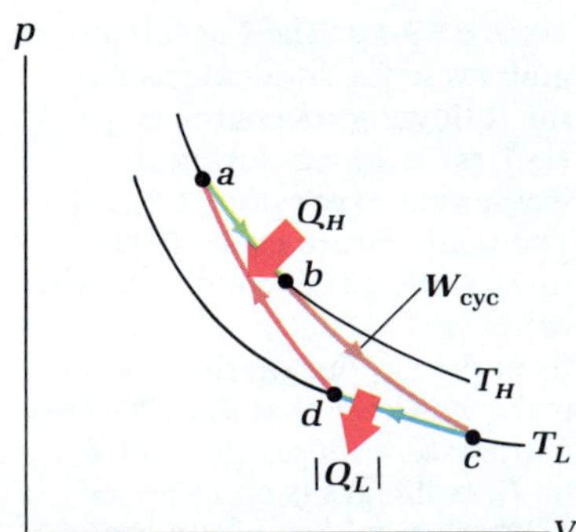

Figure 19-12. An ideal gas undergoes a Carnot cycle between reservoirs at T_H and T_L.

Consider n moles of an ideal gas undergoing a Carnot cycle between reservoirs at T_H and T_L, as shown in Fig. 19-12. We need to evaluate the heat exchanges Q_H and $|Q_L|$ to determine the efficiency from Eq. 19-4. For the isothermal process connecting states a and b, the change in internal energy is zero ($\Delta U = 0$ for the ideal gas because $\Delta T = 0$). From the first law and the ideal gas equation of state,

$$Q_H = W_{ab} = \int_a^b p\,dV$$
$$= \int_{V_a}^{V_b} \frac{nRT_H}{V}\,dV = nRT_H \ln \frac{V_b}{V_a}$$

In a similar way, $|Q_L|$ is evaluated for the isothermal compression from c to d:

$$|Q_L| = nRT_L \ln \frac{V_c}{V_d}$$

The efficiency of the cycle is given by Eq. 19-4, $\eta = 1 - |Q_L|/Q_H$, or

$$\eta = 1 - \frac{nRT_L \ln (V_c/V_d)}{nRT_H \ln (V_b/V_a)} \tag{19-8}$$

This result can be simplified by considering the volume changes for the adiabatic processes in the cycle. For an ideal gas, pV^γ is a constant for a quasi-static adiabatic process. Using $pV = nRT$, we write this as $pV^\gamma = pVV^{\gamma-1} = nRTV^{\gamma-1}$, so that $TV^{\gamma-1}$ is also a constant. Thus $T_H V_b^{\gamma-1} = T_L V_c^{\gamma-1}$ for the adiabatic expansion, and $T_H V_a^{\gamma-1} = T_L V_d^{\gamma-1}$ for the adiabatic compression. On dividing the first of these equations by the second, we obtain $(V_b/V_a)^{\gamma-1} = (V_c/V_d)^{\gamma-1}$, or

$$\frac{V_b}{V_a} = \frac{V_c}{V_d}$$

The logarithms in Eq. 19-8 are therefore equal. On canceling common factors in Eq. 19-8, the efficiency of the Carnot cycle operating between T_H and T_L is

$$\eta = 1 - \frac{T_L}{T_H} \tag{19-9}$$

Efficiency of a Carnot cycle

The efficiency of this heat engine (an ideal gas operating in a Carnot cycle) depends only on the ideal gas temperatures of the reservoirs. According to Carnot's theorem, the efficiency is the same for all reversible engines operating between these reservoirs.

The Thermodynamic Temperature

Suppose some other reversible engine operates between these same two reservoirs. Perhaps the working substance is a very complex material; it can even change phase during part of the cycle. Carnot's theorem assures us that the efficiency of this cycle has exactly the same value as that for the ideal gas Carnot cycle. All reversible engines operating between these reservoirs have the same efficiency. The only common features of the set of all reversible engines operating between the two reservoirs are the reservoirs themselves and the ratio of heat exchanges of the cycle with these reservoirs. That is, since $\eta = 1 - |Q_L|/Q_H$, the ratio $|Q_L|/Q_H$ has the same value for all reversible engines operating between this pair of reservoirs. This observation forms the basis of our final definition of temperature.

The *thermodynamic,* or *Kelvin, temperature* is defined by the operation of a reversible engine between two reservoirs or systems. The ratio of the temperatures of these two systems is defined to be equal to the ratio of heat exchanges of a reversible engine operated between these systems:

$$\frac{T_H}{T_L} = \left|\frac{Q_H}{Q_L}\right| \tag{19-10}$$

Just as described in Chap. 16, the Kelvin scale is established by choosing 273.16 K as the temperature of the triple point of water. The temperature T of some system is

$$(0.30\text{ kg})c_p(90°\text{C} - T_f) = (0.70\text{ kg})c_p(T_f - 10°\text{C})$$

The final temperature is $T_f = 34°\text{C} = 307$ K. We calculate separately the entropy change of each subsystem by reversibly changing its temperature to T_f. For the hotter water, call it subsystem 1:

$$\Delta S_1 = \int_i^f \frac{dQ}{T} = m_1 c_p \int_{T_1}^{T_f} \frac{dT}{T} = m_1 c_p \ln \frac{T_f}{T_1}$$
$$= (0.30\text{ kg})(4.2\text{ kJ}\cdot\text{kg}^{-1}\cdot\text{K}^{-1})\left(\ln \frac{307\text{ K}}{363\text{ K}}\right)$$
$$= -210\text{ J/K}$$

The entropy of the hotter water decreases on cooling, and the minus sign comes from the logarithm. The entropy of the cooler water increases, according to a similar calculation:

$$\Delta S_2 = (0.70\text{ kg})(4.2\text{ kJ}\cdot\text{kg}^{-1}\cdot\text{K}^{-1})\left(\ln \frac{307\text{ K}}{283\text{ K}}\right)$$
$$= 240\text{ J/K}$$

The entropy change for the system is the sum of these two contributions:

$$\Delta S = \Delta S_1 + \Delta S_2 = -210\text{ J/K} + 240\text{ J/K}$$
$$= 30\text{ J/K}$$

Notice that the entropy of the system increases. This result is generally valid for a mixing process. While the entropy of part of a system can decrease, the entropy of the other part increases by a greater amount.

SELF-TEST 19-10. Two 0.50-kg samples of water are mixed together in a well-insulated chamber. One of the samples was initially at 290 K and the other at 310 K so that the mixture comes to equilibrium at 300 K. *(a)* What is the entropy change of the system due to the mixing? *(b)* Explain why this entropy change is not zero. ***ANSWERS:*** *(a)* 2.3 mJ/K; *(b)* the entropy change of the surroundings is zero because the container is well insulated. But the process is irreversible so that the entropy change of the system must increase.

19-6 ENTROPY AND THE SECOND LAW

What is entropy? What is its conceptual value in increasing our understanding of thermodynamics? These questions can be addressed in part by evaluating entropy changes, such as we did in several examples in the last section. We can see from Example 19-10 that, unlike energy, the entropy of an isolated system is not necessarily conserved. The entropy of one part of the system described in that example decreased while the entropy of the other part increased by a greater amount; the entropy of the isolated system increased. Let us generalize this observation.

A system and its surroundings are called the universe.

Any process can be described in terms of the changes in the system of interest to us and the changes in the surroundings of that system. Together our system and the relevant part of the surroundings form a larger, isolated system that we call the *universe*. Consider the entropy changes that occur for a process. We denote the change in entropy of our system by ΔS_{sys} and the change in entropy of its surroundings by ΔS_{sur}. The sum of these changes is the change in entropy of the universe ΔS_{univ}:

$$\Delta S_{\text{univ}} = \Delta S_{\text{sys}} + \Delta S_{\text{sur}}$$

In every calculation, we find that the entropy of the universe either increases or remains the same. The entropy of the universe never decreases. This result is in accord with yet another statement of the second law of thermodynamics:

Entropy statement of the second law

For any process, the entropy of the universe either increases (if the process is irreversible) or remains the same (if the process is reversible).

In equation form,

$$\Delta S_{\text{univ}} \geq 0 \tag{19-17}$$

Since it is expressed in terms of entropy, we call this the *entropy statement* of the second law. Recall that a reversible process is an idealization; all real processes are irreversible. For any real process, the entropy of the universe increases.

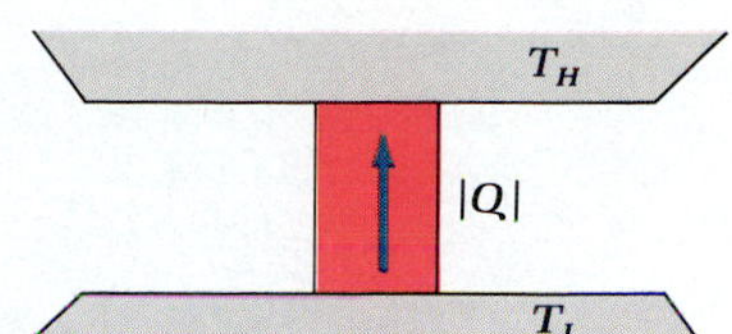

Figure 19-15. The spontaneous flow of heat from lower to higher temperatures does not occur. The entropy of the universe would decrease in violation of the second law.

A process for which the entropy of the universe decreases is an impossible process according to this statement. For example, consider the spontaneous flow of heat $|Q|$ from lower temperature T_L to higher temperature T_H, as shown in Fig. 19-15. This is an impossible process because it violates the Clausius statement of the second law. If it could occur, the entropy changes, ΔS_L and ΔS_H, of the low- and high-temperature reservoirs would be

$$\Delta S_L = \frac{-|Q|}{T_L} \qquad \Delta S_H = \frac{|Q|}{T_H}$$

The entropy change of the universe would be the sum of these:

$$\Delta S_{\text{univ}} = \frac{-|Q|}{T_L} + \frac{|Q|}{T_H} = -|Q|\left(\frac{1}{T_L} - \frac{1}{T_H}\right)$$

The quantity in parentheses is positive because $T_L < T_H$, and the entropy change of the universe would be negative. Thus the spontaneous flow of heat from lower to higher temperature also violates the entropy statement of the second law.

The spontaneous transfer of heat from higher to lower temperature does not violate the second law. By a calculation similar to that above, you can show that the entropy of the universe increases for this commonly occurring process. Thus the second law allows the spontaneous flow of heat from higher to lower temperature but forbids the reverse process. Indeed, the flow of heat is irreversible.

Heat flows irreversibly from higher to lower temperature.

Other irreversible processes can be analyzed with the same result. Suppose the second law allows a process to occur (increasing the entropy of the universe). But the reverse process does not occur because it would violate the second law (decreasing the entropy of the universe). The second law, or the increase in entropy of the universe, explains in this way the one-way nature of macroscopic processes. The "arrow of time" is in a sense, or direction, corresponding to increasing entropy. That is, the procession of events in time is always toward states of the universe with equal or greater entropy.

The "arrow of time" is toward increasing entropy.

Another connection between irreversibility, increasing entropy, and the second law concerns the idea of "loss of opportunity to do work" or the "unavailability of work." Consider the irreversible, spontaneous flow of heat from higher to lower temperature. As a result of this flow, the opportunity is lost to use that heat to perform work by operating a heat engine between the two temperatures. If the flow of heat continues until the two bodies come to the same temperature, then no work can be done (according to the Kelvin-Planck statement of the second law). Of the work that could have been performed by operating a heat engine between the two bodies at different temperatures, none can be obtained after they have come to the same temperature. The opportunity has been lost because of the irreversible heat flow. None of that work is available any longer.

Work becomes unavailable as entropy increases.

Entropy and the second law can also be interpreted at the microscopic level. There the description is in statistical terms. Instead of forbidding a particular process, the second law describes the process as very highly improbable. For example, the probability is virtually zero that all of the gas molecules are in one part of a container. After the (irreversible) free expansion in Example 19-9 has occurred, it is highly unlikely that the gas will spontaneously return to the initial state, with all of the molecules occupying one side of the container.

Increasing entropy can be interpreted at the microscopic level as corresponding to a change from a more orderly situation or configuration to a less orderly one. That

Increasing entropy corresponds to increasing disorder.

is, natural processes lead to a more disorderly or chaotic state of the universe. It is often easy to decide which of two configurations is the more disordered. For example, a liquid phase of a substance, with molecules moving about, is usually more disordered than the solid phase, with the molecules arranged on a lattice (Example 19-7). A gas confined to a smaller volume becomes more disordered if it expands freely into a larger volume (Example 19-9). An isolated system with parts at different temperatures is more ordered than that system when all parts have come to the same temperature (Example 19-10). In this last example, a heat engine can be run between parts at different temperatures, but not between parts at the same temperature.

Heat death of the universe

Considerations of this sort can lead to speculation on the ultimate fate of the universe. In every natural (irreversible) process, the entropy of the universe increases. The universe evolves from more highly ordered states to more disordered ones. The so-called heat death of the universe would correspond to that maximum entropy state of uniform composition and temperature, with no opportunity to perform work. On a less cosmological level, we can comprehend (perhaps imperfectly) the aging process, the spontaneous flow of heat, the inevitable deterioration of any machine, and the inability of any machine to spontaneously repair itself.

COMMENTARY: Maxwell's Demon

Our understanding of a subtle concept can often be sharpened by constructing and considering a paradox. The second law of thermodynamics provides sufficient subtlety, so there is ample opportunity for sharpening our understanding. A paradox is often stated in an amusing way, perhaps with the use of caricatures, talking animals, or the like. In one well-known paradox, described below, the principal character is Maxwell's demon.

James Clerk Maxwell (see the biographical sketch in the Commentary in Chap. 27) introduced most of the ideas of probability that we used in the last chapter to describe a system at the molecular level. Seeking to make the use of statistical methods respectable, Maxwell noted its practicality by saying, "This branch of Mathematics, which is generally thought to favor gambling, dicing, and wagering, and therefore highly immoral, is the only 'Mathematics for Practical Men,' as we ought to be."

Maxwell helped in developing a bridge of understanding between macroscopic and microscopic aspects of the second law of thermodynamics. At the macroscopic level, the second law deals with irreversibility by using terms such as "heat," "temperature differences," and "friction." In contrast, the second law is expressed in statistical terms at the microscopic level. There, irreversibility is seen as the likely outcome of virtually countless random processes that involve vast numbers of molecules.

Consider this simple situation: An insulated container has two identical chambers that are separated by a fixed wall. Identical samples of gas occupy the two chambers, and they are each in equilibrium at temperature T. No work is done on any part of the system and no heat is added to it. According to the second law of thermodynamics, no temperature difference between the two gases can spontaneously develop. Put another way, the entropy of the isolated system of two gases cannot decrease.

Maxwell, in his text, *Theory of Heat,* introduced a paradox along with a tiny creature who has become known as Maxwell's demon. The demon is stationed at the wall separating the two gases described above. A small trapdoor is fitted into the wall, and the demon can open or close it with negligible work being done. The demon can determine a molecule's speed as it approaches the trapdoor, and his quick reflexes enable him to open and close the trapdoor selectively, allowing faster molecules to pass from the left-hand chamber to the right-hand chamber and slower molecules to pass the other way. As a result, without any work being done, a temperature difference develops between the two gases, and the entropy of the system of two gases decreases. Thus Maxwell's demon has paradoxically brought about a violation of the second law.

Often a paradox is resolved by a new discovery or a broadened interpretation. Maxwell's paradox stimulated much critical thinking, and its resolution involved phenomena that are described in later chapters. For example, Brillouin (1854–1948) argued that the enclosures are bathed in a background of thermal (or blackbody) radiation, as described in Chap. 39. Because of this background, the demon would not be able to distinguish individual molecules or sort them according to their speeds. Brillouin also developed a connection between entropy and information. If information about a system is assembled by some process, then the degree of disorder of the system (and its entropy) is correspondingly reduced. If the demon used some device to sort out the speeds of the molecules, say a light not in equilibrium with the remainder of the system, then the decrease in entropy of the gases would be more than offset by the increase in entropy of the demon and his apparatus. In other words, taking all entropy changes into account, the total entropy would not decrease.

For further reading, see *James Clerk Maxwell—A Biography* by Ivan Tolstoy (University of Chicago Press, Chicago, 1981); "Maxwell's Demon" in *Mr. Tompkins in Paperback* by G. Gamow (Cambridge University Press, New York, 1965); and "Demons, Engines and the Second Law" by Charles H. Bennett in *Scientific American,* November 1987.

SUMMARY

Section 19-1. Heat Engines and the Second Law

The efficiency η of a heat engine is

$$\eta = \frac{W_{cyc}}{Q_H} \tag{19-3}$$

According to the Kelvin-Planck statement of the second law of thermodynamics, there exists no cycle which extracts heat from a reservoir at a single temperature and completely converts it into work.

Section 19-2. Refrigerators and the Second Law

The coefficient of performance of a refrigerator is

$$K = \frac{Q_L}{|W_{cyc}|} \tag{19-6}$$

and of a heat pump is

$$K_{hp} = \left|\frac{Q_H}{W_{cyc}}\right| \tag{19-7}$$

According to the Clausius statement of the second law, no process is possible whose sole, net result is the transfer of heat from lower to higher temperature.

Section 19-3. Reversibility and the Carnot Cycle

A reversible process is one which can be reversed by making only infinitesimal changes in the surroundings of a system. If a process is not reversible, it is an irreversible process. A Carnot engine is a reversible engine and obeys Carnot's theorem: All reversible engines operating between temperatures T_H and T_L have the same efficiency and no engine can have a greater efficiency than this.

Section 19-4. The Kelvin, or Thermodynamic, Temperature

The thermodynamic temperature T of a system is defined by

$$T = 273.16\text{ K}\left|\frac{Q}{Q_3}\right| \tag{19-11}$$

where Q and Q_3 are the heat exchanges of a Carnot cycle operating between the system and water at its triple point. The efficiency of a Carnot engine operating between T_L and T_H is given by

$$\eta = 1 - \frac{T_L}{T_H}$$

and is the upper limit on the efficiency of real engines operating between these temperatures.

Section 19-5. Entropy

Entropy is a state variable whose change dS is given by the reversibly transferred heat dQ divided by the temperature T. The difference in the entropy of two states is determined by

$$\Delta S = S_f - S_i = \int_i^f \frac{dQ}{T} \tag{19-15}$$

where the integral is for any reversible process connecting the two states.

Section 19-6. Entropy and the Second Law

The change in entropy of the universe is the sum of the entropy changes of a system and of its surroundings. According to the entropy statement of the second law, for any process the entropy of the universe either increases (if the process is irreversible) or remains the same (if the process is reversible). The increase in entropy of an isolated system for an irreversible process is connected to the loss of an opportunity of the system to perform work. On the microscopic level, the irreversible process is from a more ordered state to a more disordered state.

QUESTIONS

1. What are some advantages of a cycle as a heat engine? Are there any disadvantages? Explain.

2. Is every cycle a heat engine? Does every heat engine operate in a cycle? Explain.

3. Is it possible to cool a room, such as a kitchen, by leaving the refrigerator door open? Explain.

4. In what sense does a heat pump pump heat? Is there a useful analogy with a water pump? Describe similarities and differences.

5. What are some factors that cause real heat engines to have efficiencies lower than the Carnot efficiency?

6. To increase the efficiency of a Carnot engine, is it more effective to increase T_H by an amount ΔT or decrease T_L by ΔT? Explain.

7. Describe some processes that are nearly or approximately reversible. Describe some that are highly irreversible.

8. How can the gasoline internal-combustion engine be considered as a cycle when fresh air is taken in at each intake stroke?

9. What plays the role of the high-temperature reservoir in the gasoline engine? What plays the role of the low-temperature reservoir?

10. In a hydroelectric plant, electric energy is generated when falling water turns a turbine. Is a heat engine involved in this energy conversion process? Explain. Is there an upper limit on the efficiency of this process? If so, what is that limit? If not, why not?

11. Consider the following two processes performed on identical ice cubes initially at 0°C:

1. You hold the ice tightly in your hand. The ice melts and the liquid comes to thermal equilibrium at body temperature.
2. You first smash the ice with a hammer and then you hold the pieces tightly in your hand. The pieces melt and the liquid comes to thermal equilibrium at body temperature.

Is either of these processes reversible? How do the entropy changes of the system (H_2O) compare for the two processes?

12. Does the entropy of a gas increase, decrease, or remain the same if its expands *(a)* reversibly and isothermally, *(b)* reversibly and adiabatically?

13. Does the entropy of a gas increase, decrease, or remain the same if it is compressed *(a)* reversibly and isothermally, *(b)* reversibly and adiabatically?

14. Explain the distinction between a quasi-static process and a reversible process.

15. Which is a more orderly state of matter—liquid or vapor? Which has the greater entropy at the boiling point? Must heat be added to change from liquid to vapor or from vapor to liquid?

16. Explain why the loss of opportunity for work as a result of an irreversible process does not violate the first law of thermodynamics.

17. Many living organisms are characterized by growth and development to highly differentiated (ordered) structures. Explain why this feature of life is not in conflict with the second law.

18. Suppose that you spend the weekend cleaning your room, making it neat and orderly. Has the entropy of the universe increased, decreased, or remained the same? Explain.

19. On freezing, a substance changes from a liquid to a solid, a more orderly structure, and the entropy of the substance decreases. Explain why this process does not violate the second law.

20. Consider the following two statements: "You can't get something for nothing," and "You can't even break even." Discuss the sense of these statements in relation to the first and second laws of thermodynamics.

21. How can you increase the entropy of 1 kg of water? How can you decrease the entropy of 1 kg of water?

22. The term "quality" is sometimes used to describe the heat extracted from a reservoir and is related to the potential for economically converting it to work. Of the heat extracted from a variety of reservoirs at different temperatures, which do you think has the highest quality? Where would mechanical energy be placed on this quality scale?

23. Complete the following table:

Symbol	Represents	Type	SI Unit
Q_H			
Q_L			
T_H			K
η	Efficiency of a heat engine		
K			
K_{hp}			
S		Scalar	

EXERCISES

Section 19-1. Heat Engines and the Second Law

1. In each cycle of operation, a heat engine takes in 440 J of heat and performs work at 28 percent efficiency. Determine for a cycle *(a)* the work done, *(b)* the heat exhausted from the engine, *(c)* the change in internal energy of the working substance.

2. A heat engine operating steadily between two reservoirs has a heat input of 20 MJ and a heat exhaust of 14 MJ each hour. *(a)* What is the efficiency of this engine? *(b)* How much work is done in 1 h? *(c)* What is the output power?

3. In one cycle of operation, a heat engine takes in 2200 J of heat and performs 620 J of work. *(a)* What is the efficiency of this engine? *(b)* How much heat is exhausted in each cycle? *(c)* If the engine completes a cycle each 0.033 s, then at what rate is heat added, heat exhausted, work done?

4. Electric energy is produced at a steam plant at a rate of 500 MW, with an overall efficiency of 34 percent. At what rate is *(a)* energy released from the burning coal and *(b)* heat added to the river water used for cooling? *(c)* The heat of combustion of coal is about 3×10^{10} J/ton. Estimate the amount of coal burned each day at this plant.

5. One mole of He, considered to be an ideal gas, is the working substance in a heat engine that operates in the cycle shown in Fig. 19-16. State *a* has pressure and volume values $p_a = 101$ kPa, $V_a = 22.4$ L. *(a)* Determine the temperatures of states *a*, *b*, *c*, and *d*. *(b)* How much heat is added in one cycle? *(c)* How much work is done in one cycle? *(d)* How much heat is exhausted in one cycle? *(e)* What is the efficiency of this engine?

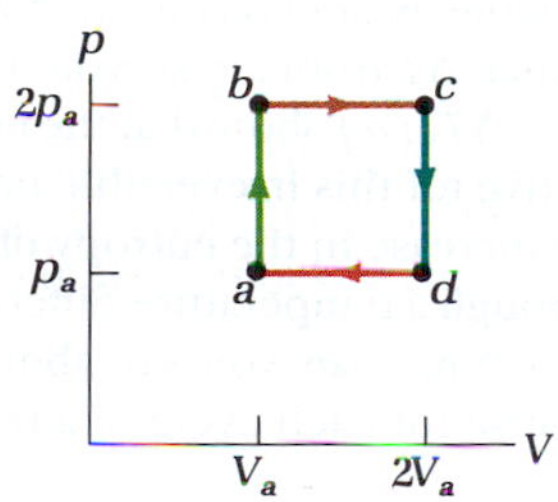

Figure 19-16. Exercise 5.

6. A heat engine carries 2.2 mol of air ($\gamma = 1.4$) through the cycle shown in Fig. 19-17, where process *bc* is adiabatic. State *a* has pressure $p_a = 150$ kPa and volume $V_a = 38$ L. Determine *(a)* the pressure of state *b* and the temperatures of *b* and of *c*, *(b)* the heat added to the air in one cycle, *(c)* the heat extracted from the air in one cycle, *(d)* the efficiency of the engine. Assume the air is an ideal gas.

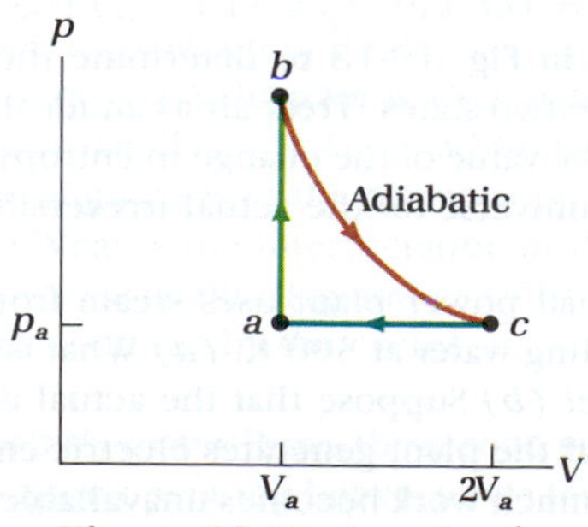

Figure 19-17. Exercise 6.

Section 19-2. Refrigerators and the Second Law

7. A refrigeration cycle exhausts 250 J of heat to a room while the motor provides 80 J of work. *(a)* How much heat is extracted from the interior of the refrigerator? *(b)* What is the coefficient of performance of the refrigerator?

8. A refrigerating unit at a food-processing plant cools food by extracting 2×10^8 J of heat each hour. It operates with a coefficient of performance of 4.3. Determine for each hour of operation *(a)* the work done on the refrigerator and *(b)* the heat exhausted to the outside. *(c)* What power is required by the refrigerator?

9. A heat pump operates with a coefficient of performance $K_{hp} = 2.2$. Work is provided by an electric motor rated at 3.5 kW. At what rate is heat *(a)* exhausted at the higher temperature and *(b)* extracted at the lower temperature? *(c)* What is the cost for 1 h of operation if electric energy costs \$0.10 per kW·h?

10. On a typical winter night, the rate of heat loss from the inside of a house averages 6 kW. Suppose that these losses are balanced by a heat pump with a 3-kW motor which runs about 30 min each hour. *(a)* What is the coefficient of performance of the heat pump under these conditions? *(b)* At what rate is heat extracted from the outside? *(c)* Estimate the cost for a full day of operation under these conditions. Assume electric energy costs \$0.10 per kW·h.

Section 19-3. Reversibility and the Carnot Cycle

11. On a *p-V* diagram, carefully sketch a Carnot cycle for air (an ideal gas with $\gamma = 1.4$) operating between 250 and 350 K. Take $n = 1.0$ mol and the isothermal expansion at 350 K to be from 22 to 32 L. Estimate the work done in a cycle from the area enclosed on the diagram.

12. Consider the Carnot cycle described in the previous exercise. Determine by direct calculation for one cycle *(a)* the heat added at 350 K, *(b)* the heat exhausted at 250 K, *(c)* the work done from the first law (compare with the estimate from the graph). *(d)* What is the efficiency of this engine?

Section 19-4. The Kelvin, or Thermodynamic, Temperature

13. Suppose a reversible cycle, with He as the working substance, operates between water at its normal melting point and water at its normal boiling point. *(a)* What is the ratio of heat exchanges at these two reservoirs? *(b)* If 0.0125 J of heat is extracted from the higher-temperature water in one cycle, how much is exhausted at the lower temperature? *(c)* Would the above answers be different if a Carnot cycle with freon as a working substance was operated between the two reservoirs? Explain.

14. A nearly reversible cycle is operated between Hg at its triple point and water at its triple point. Suppose the heat exchanges with these two reservoirs are $|Q_{Hg}| = 22.1\ \mu$J and

4. ***Heat engines in series.*** Suppose that the heat exhaust from one engine is used as the heat input for another engine, as shown schematically in Fig. 19-20. One engine operates between T_H and T_I with efficiency η, while the other engine operates between T_I and T_L with efficiency η'. The overall efficiency is $\eta_{\text{net}} = (W_{\text{cyc}} + W'_{\text{cyc}})/Q_H$. *(a)* Show that $\eta_{\text{net}} = \eta + (1 - \eta)\eta'$. *(b)* Suppose that each engine is reversible; show that $\eta_{\text{net}} = 1 - T_L/T_H$.

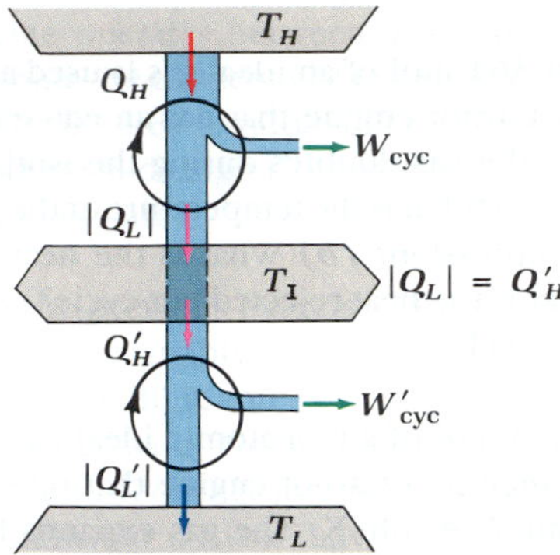

Figure 19-20. Problem 4.

5. ***The gasoline engine and the Otto cycle.*** A useful approximation to the cycle for the gasoline internal-combustion engine is provided by an idealized cycle called the *Otto cycle,* shown in Fig. 19-21. The working substance in a real gasoline engine is mostly air with a small admixture of gasoline vapor. The Otto cycle has air as the working substance and utilizes the same air over and over. Starting a point a on the p-V diagram, the air is compressed adiabatically, corresponding to the compression stroke, from volume V_1 to volume V_2 at point b. The compression ratio $r = V_1/V_2$ characterizes the engine and is determined by the motion of the piston in the cylinder. The pressure increases at constant volume from b to c, corresponding to the rapid burning of fuel which is ignited by a spark. An adiabatic expansion, corresponding to the power stroke, follows from c to d, and the pressure is reduced at constant volume from d back to a, completing the cycle. The intake and exhaust strokes for the gasoline engine are not shown for the Otto cycle. *(a)* Show that the efficiency of the Otto cycle is given in terms of the compression ratio by

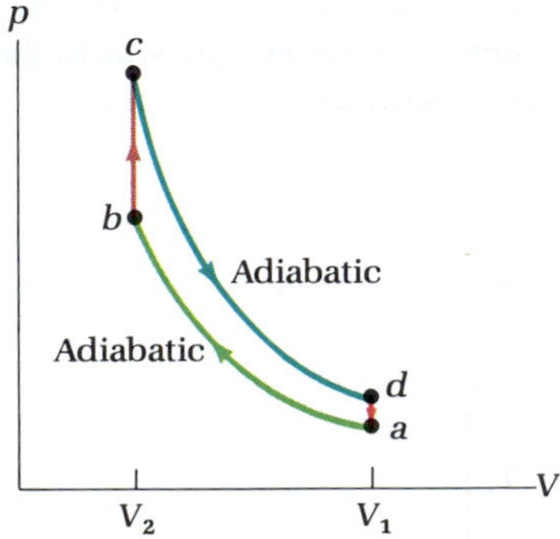

Figure 19-21. Problem 5: The Otto cycle approximates an internal-combustion engine. The compression ratio $r = 3$ for this figure.

(b) The compression ratio for a typical automobile engine is around $r = 8$ and $\gamma = 1.4$ for air. Evaluate the Otto-cycle efficiency for this case. The efficiency of a real engine is more like 20 percent.

6. ***The diesel engine and the air-standard cycle.*** The operation of a diesel engine can be approximated by the *air-standard diesel cycle,* shown schematically in Fig. 19-22. The adiabatic compression ab corresponds to the compression stroke of the air. This stroke is characterized by the compression ratio $r_C = V_a/V_b$. The isobaric expansion bc corresponds to the burning of injected fuel and is followed by the adiabatic expansion cd, with an expansion ratio $r_E = V_a/V_c$. The isochoric process da completes the cycle. *(a)* Show that the efficiency of the air-standard diesel cycle is given by

$$\eta = 1 - \frac{r_E^{-\gamma} - r_C^{-\gamma}}{\gamma(r_E^{-1} - r_C^{-1})}$$

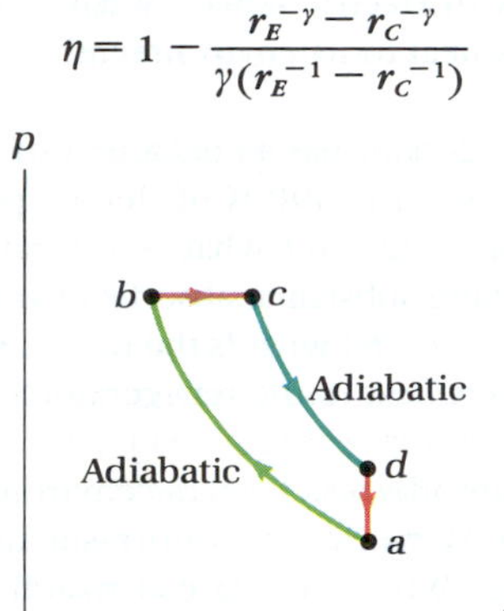

Figure 19-22. Problem 6: The air-standard diesel cycle.

(b) Estimate the efficiency of a diesel engine with $r_C = 15$ and $r_E = 5$. Assume $\gamma = 1.4$. Do you expect a real diesel engine to have this efficiency? Explain.

7. ***An irreversible engine.*** An irreversible engine operates between reservoirs at $T_H = 550$ K and $T_L = 350$ K, with an efficiency of 25 percent. In each cycle, heat $Q_H = 1200$ J is extracted from the reservoir at T_H and added to the working substance in the engine. Heat $|Q_L|$ is exhausted from the engine into the reservoir at T_L. *(a)* Determine the change in entropy of the universe for one cycle of operation. *(b)* How much more work could a reversible engine operating between these reservoirs perform with the same heat input for each cycle? *(c)* Show that for each cycle the amount of work that is unavailable because of the irreversible process is equal to $T_L\,\Delta S_{\text{univ}}$.

8. ***The first law in terms of entropy.*** A system undergoes an infinitesimal, reversible process with $dW = p\,dV$. *(a)* Show that the *first* law of thermodynamics applied to the process can be written

$$T\,dS = dU + p\,dV$$

(b) What is the corresponding integrated form for a reversible process? *(c)* Is either of these expressions valid for an irreversible process? Explain.

9. ***An entropy change for helium.*** Determine for 1.0 mol of He the entropy difference between the state with $V = 22$ L, $T = 280$ K and the state with $V = 44$ L, $T = 1120$ K. Treat He as an ideal gas.

10. ***The Stirling engine.*** An ideal gas is the working substance

for an engine operating in the cycle shown in Fig. 19-23. Determine the efficiency of this cycle in terms of the temperatures T_H and T_L and volumes V_1 and V_2. Notice that heat exchanges occur for all four processes that make up the cycle. This engine is an idealized version of the (irreversible) Stirling engine.

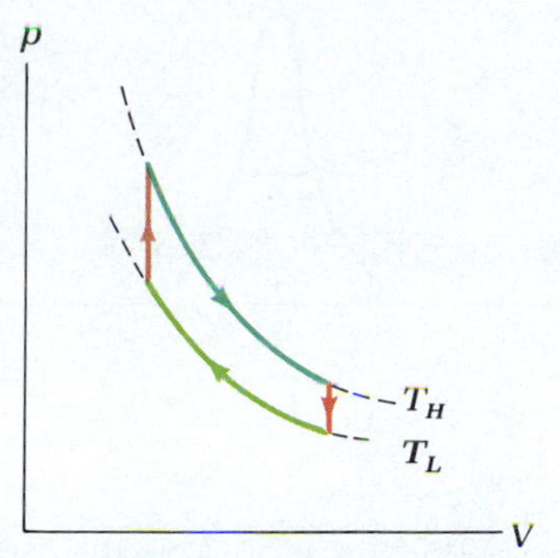

Figure 19-23. Problem 10.

11. **_Entropy change for a polyatomic gas._** Consider the model of a polyatomic gas described in Prob. 10 in Chap. 17. Adapt the integrating program used in that problem to determine the difference in entropy for that gas ($\Theta = 85$ K) between two states of 1.0 mol having the same volume but at temperatures $T_i = 10$ K and $T_f = 200$ K. Note that

$$S_f - S_i = \int_{T_i}^{T_f} \frac{n\mathscr{C}_V(T)\, dT}{T}$$

where $dQ = n\mathscr{C}_V\, dT$ is the heat added reversibly at constant volume for an infinitesimal step in the isochoric process.

12. **_Entropy of mixing two equal amounts._** A sample of mass m of a liquid substance is separated into two equal parts of mass $\frac{1}{2}m$ each. One part is placed at a temperature T_1 and the other part is placed at a higher temperature T_2. Next the two parts are mixed in an insulated container at constant pressure. *(a)* Show that the change of entropy of the system as a result of the mixing is

$$\Delta S = mc_p \ln\left(\frac{T_2 + T_1}{2\sqrt{T_2 T_1}}\right)$$

(b) Show that $\Delta S > 0$.

HIGHLIGHTS OF MODERN PHYSICS

Cold Atoms

Most discussions of atoms begin with a *single* atom, which until recently was something that had never been observed. Other discussions treat the light emitted by a *motionless* atom, something that, strictly speaking, does not exist. There are pedagogic reasons for proceeding this way, but physically, the situation is a highly artificial one. In real life, atoms are present in unimaginably large numbers and fly around at high speed. However, if physicists could study single, slow-moving atoms, it would allow experiments of unprecedented precision. In recent years a remarkable series of advances have made this possible.

Let us see how "cold atoms" can permit a better standard of time. From Chap. 1, the second is defined as the time taken for 9,192,631,770 oscillations of radiation from cesium atoms undergoing a transition between two energy states. Any periodic phenomenon can be used to measure time: a swinging pendulum, a rotating earth, or the vibrations of light. In using light as a standard, the trick is to produce a very narrow range of frequencies, the narrower the better.

A red-hot piece of iron at 1000 K would make a terrible time standard because the frequency distribution is so broad (Fig. 1*a*). The second might be defined in terms of the peak frequency, but it would be difficult to measure the peak precisely. A gas of hot, glowing atoms would make a better timekeeper because the emitted light produces a series of much narrower peaks whose frequencies can be determined more precisely than the broad blackbody distribution (Fig. 1*b*).

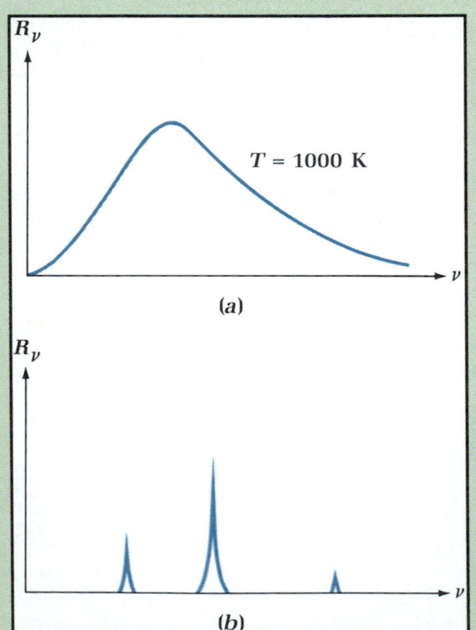

Figure 1. *(a)* The broad frequency distribution of light from a heated solid is very different from *(b)* the series of peaks produced by a hot gas.

The narrow peak in Fig. 2*a* represents the frequency distribution of a typical atomic emission line. A line is produced when an atom that has been excited to a higher state falls back down to the ground state, emitting light that carries off the energy given up by the atom (Fig. 2*b*). The frequency spread of the light $\Delta\nu_0$ depends on the time τ the atom stays in the excited state before returning to the ground state,

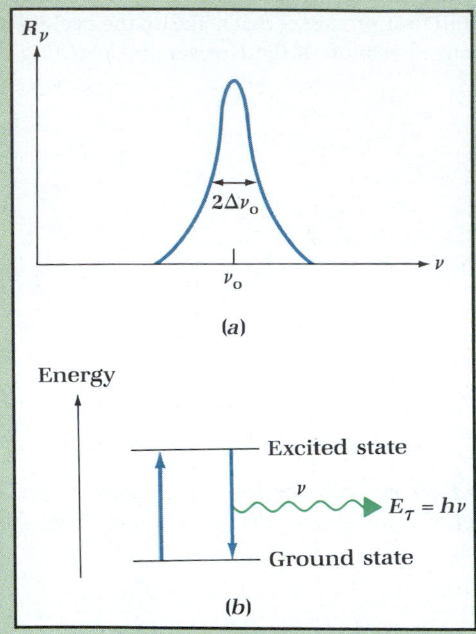

Figure 2. *(a)* A discrete spectral line is characterized by its peak frequency ν_0 and its half-width $\Delta\nu_0$, measured as half its full width at half the peak's maximum value. *(b)* The spectral line is produced when an atom falls from a higher (more energetic) state to a lower state, in this case the ground state.

$$\Delta\nu_0 \approx 1/\tau$$

For a typical atom, $\tau \approx 10^{-8}$ s, so $\Delta\nu_0 \approx 10^8$ Hz, or about 1 part in 10^7 for 500-nm light.

The frequency spread $\Delta\nu_0$ is called the ***natural linewidth*** because it is characteristic of the isolated, stationary atom. However, thermal motion of the atoms produces a much broader line because of the Doppler effect, the same effect that causes a train whistle to have a higher (or lower) pitch when the train is approaching (or receding) (Sec. 33-6). Similarly, light emitted when the atom is moving toward or away from the observer will be shifted toward higher or lower frequencies, respectively (Fig. 3). The fractional frequency shift $\Delta\nu/\nu$

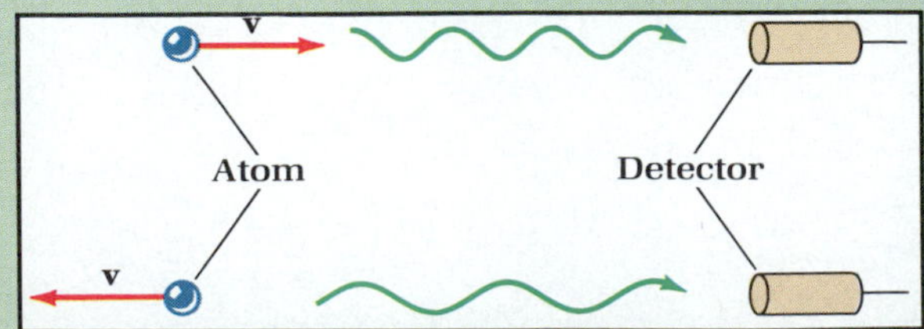

Figure 3. An atom moving toward a detector emits a photon. For the detector, the atom is an approaching light source, so the wavelength of the photon will be Doppler-*shortened* relative to its $v = 0$ value. The wavelength of light from a receding atom is lengthened by the Doppler effect.

equals the ratio of the atom's speed to the speed of light v/c, when $v \ll c$. Since the observer receives light from atoms moving in many different directions and with all speeds represented by the Maxwell-Boltzmann speed distribution (Sec. 18-6), the narrow line becomes much broader. For a gas at 800 K, $v \approx 600$ m/s and

$$\frac{\Delta\nu}{\nu} \approx \frac{v}{c} = \frac{600 \text{ m/s}}{3 \times 10^8 \text{ m/s}} \approx 20 \times 10^{-7}$$

or 20 times the natural width.

New techniques use lasers to slow atoms enough to recover the natural width. A slowing force is exerted on the atoms using the *resonance absorption of light*. Resonance absorption can only occur if the laser frequency is close to the resonance frequency ν_0, but the probability of being absorbed decreases on both sides of ν_0 (Fig. 4). The laser frequency ν used is slightly less than ν_0 so that if the atom is moving *toward* the laser (Fig. 4*a*), the laser light will be Doppler-shifted toward the resonance frequency ν_0, and it will have a high probability of being absorbed.

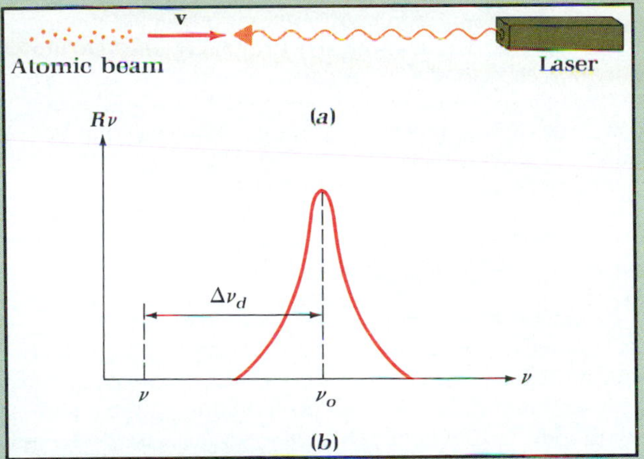

Figure 4. *(a)* A beam of atoms of speed v is illuminated by a laser beam of frequency ν, equal to the resonance frequency of the atoms ν_0 minus the Doppler shift of the light approaching the atoms $\Delta\nu_d$. *(b)* Because of the motion of the atoms, the laser frequency ν is increased by just the amount necessary for it to be in resonance, and the probability of absorption is maximum.

In a typical experiment, a beam of vaporized sodium atoms is struck head on by a laser beam (Fig. 5). The sodium atoms are slowed by the force exerted by the laser beam. In the modern viewpoint, the force comes from treating the light beam as a stream of particles, or photons, each with a mass of $m_\gamma = h\nu/c^2$ ($h = 6.67 \times 10^{-34}$ J·s is *Planck's constant*). When a photon is absorbed by an atom of mass M moving with speed v_i, momentum must be conserved:

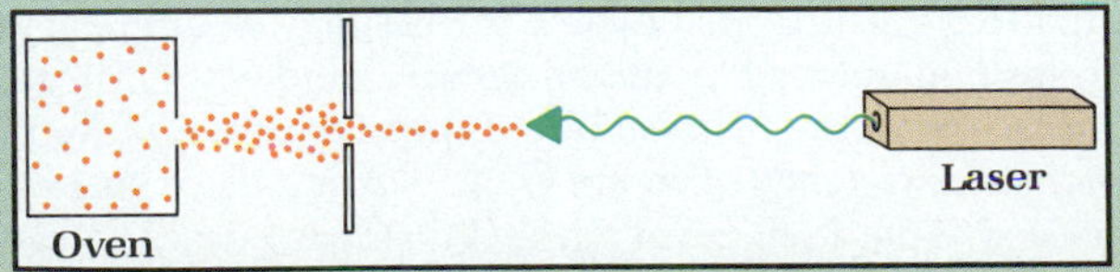

Figure 5. Sodium atoms are vaporized in an oven and allowed to stream out a hole, forming a beam. They are further constricted to a narrow beam by one or more collimating apertures before encountering a counterpropagating laser beam.

$$Mv_i - m_\gamma c = Mv_f$$

$$\Delta v = v_i - v_f = \frac{m_\gamma c}{M}$$

For the yellow D line in the sodium spectrum ($\lambda = 590$ nm), $\Delta v = 0.03$ m/s, which is quite small compared with the initial speed of 600 m/s. It will take $600/0.03 = 20{,}000$ photons to stop an atom.

When a photon is absorbed, the sodium atom jumps from its ground state to the next higher state. It stays in the excited state for only about 10^{-7} s, then emits a 590-nm photon and returns to the ground state. Now the atom can absorb another photon, as many as 10^7 photons per second. (Why?)

As the atom slows, the laser is no longer properly turned to excite it because for the slower atom, the Doppler shift $\Delta\nu$ is smaller, and the laser frequency no longer overlaps the resonance absorption peak (Fig. 4*b*). For sodium, a speed decrease of only 10 m/s will change the Doppler shift by 17 MHz, which exceeds the 10-MHz width of the resonance. Thus, after only 1000 photons are absorbed, the laser will no longer be tuned for absorption. Various techniques have been applied to maintain resonance. In one, a magnetic field is used to shift the atom's energy levels. The field strength decreases in the direction of the beam velocity so the resonance condition is continually maintained as the atoms slow down.

The atoms do not come to a complete standstill under these conditions because each time an atom emits a photon, it recoils. The recoils do not quite average out to zero so that after absorbing 20,000 photons, the atoms move at about 3 m/s in random directions, corresponding to a temperature of 75 mK. Additional absorptions increase the temperature and begin to push the atom back in the opposite direction, two undesirable results. Since "atom traps" are unable to retain any atom above about 20 mK, the temperature needs to be reduced even further. To do this the atoms are put into "optical molasses." Six laser beams are focused on the slowly moving atoms (Fig. 6*a*).

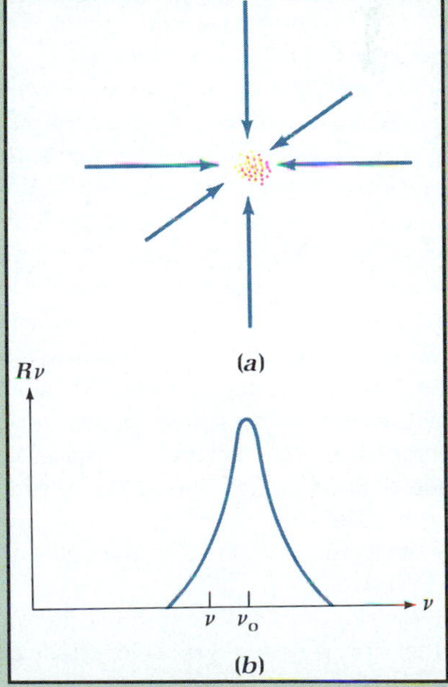

Figure 6. *(a)* "Optical molasses" produced by three pairs of oppositely directed laser beams. The frequency of each beam is slightly below the resonance frequency of the absorption line, as shown in *(b)*. The optimum effect is obtained if ν is slightly more than $\Delta\nu_0$ below the resonance frequency.

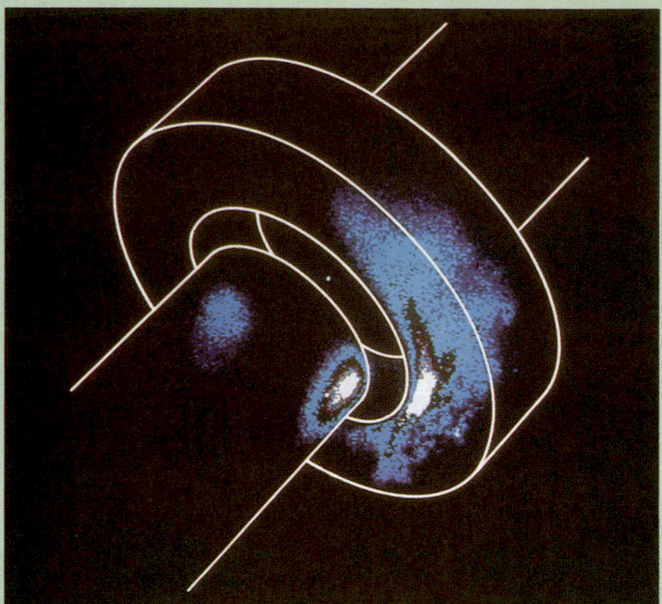

Trapped atoms.

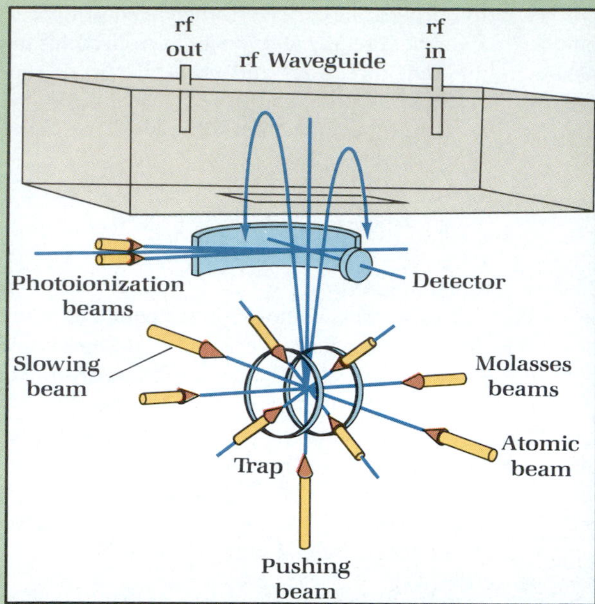

Figure 7. The atomic fountain. Atoms are injected toward the trap from the right, are nearly stopped by the "slowing beam," are cooled by the "molasses beams," and then are given an upward velocity from the "pushing beam." When the microwave frequency is just right, transitions will be produced between atomic states, and these transitions will be detected by the photoionization beam/detector components. Because of the relatively long time the atoms spend in the waveguide (0.25 s), the resonant frequency can be measured with an accuracy of 2 Hz.

The laser frequency is tuned slightly below the center of the resonance (Fig. 6*b*) so that for any atom moving toward one of the lasers, the laser frequency is Doppler-shifted closer to the peak, and absorption is enhanced. The opposing laser beam will be shifted *further* from resonance and absorption will be decreased. Thus an atom will always feel a force opposing its motion. The net effect is that the random motion of each atom is reduced to a very small level, to as little as 3 μK, or about 1 cm/s for cesium.

When atoms are moving so slowly, measurements of high precision are possible because of the inverse relationship between time and frequency, $\Delta \nu = 1/\Delta t$. The NIST atomic clock uses this principle to obtain its high accuracy by producing a transition between two energy states of a beam of cesium atoms at opposite ends of a 4-m tube. Here, Δt is the time for atoms in the beam to drift the length of the tube, which is made as long as practical in order to maximize Δt and minimize $\Delta \nu$. By slowing the atoms from hundreds of meters per second to a few centimeters per second, the same precision can be obtained over a much shorter distance, making a more compact clock. However, a horizontal beam is no longer possible because an atom with $v = 1$ cm/s will take 100 s to travel 1 m and hence will fall a great distance.

Therefore Stephen Chu and his colleagues at Stanford University have devised an "atomic fountain" in which cold atoms are given a small upward velocity by a pulse from a laser below (Fig. 7). The cold atoms rise several centimeters into a microwave cavity before falling back. Here, Δt is the time spent in the microwave cavity, and this can be much longer than the time for cesium atoms to traverse the 4-m tube, giving correspondingly greater precision.

The Global Positioning Satellite system relies on highly precise atomic clocks in satellites. A receiver on the ground can be accurately located by detecting the time difference to receive signals from several satellites; each satellite is kept in step with the others by an internal cesium clock. The clock accuracy of 10^{-8} s allows the receiver to be located within a few meters. Clocks based on atomic fountains would be much smaller than conventional cesium clocks and promise improved accuracy by two to three orders of magnitude, reducing the positional error to a few millimeters. While that is unnecessarily accurate for navigation, it could greatly facilitate the determination of continental drift and other geodetic measurements.

QUESTIONS

1. Explain why a sodium atom can absorb only 10^7 photons per second from the ground state.
2. Why will the apparatus shown in Fig. 6 not slow atoms down if the speed of the atoms is large enough that the Doppler shift of the laser light is much more than $\Delta \nu_0$?

EXERCISES

1. Calculate the deceleration experienced by sodium atoms due to laser slowing. Assume 10^7 photons with $\lambda = 590$ nm are absorbed per second. Express your answer in terms of the acceleration due to gravity, g.
2. Calculate the speed of a 10-mK sodium atom (corresponding to the maximum energy allowing capture in an atom trap).
3. What initial upward speed should an atom in an atomic fountain have in order that $\Delta \nu = 2$ Hz? Assume the frequency of the transition induced by the microwave field is 9×10^9 Hz.
4. An atom traveling with an initial speed of 600 m/s absorbs photons of wavelength 590 nm at the maximum possible rate. How far will the atom travel while it is slowing down? What is its average deceleration?

COULOMB'S LAW AND THE ELECTRIC FIELD

20

Lightning is a common but spectacular example of an electrical phenomenon.

Electromagnetic interactions hold electrons and nuclei together to form atoms, they hold atoms together to form molecules, and they hold molecules together to form macroscopic objects. The constituents of your body, its atoms and molecules, are held together by electromagnetic forces. Many of the effects we see going on around us are, at root, the result of electromagnetic forces. For example, green plants absorb sunlight, an electromagnetic wave, and convert the energy to electromagnetic potential energy in the form of carbohydrate molecules, the basis of nearly all of the life on earth.

You may wonder why we use the word "electromagnetic" here, always combining "electric" with "magnetic." The reason is that both electric and magnetic effects involve the same property of matter, a property we call *electric charge.* Although electric and magnetic effects are intimately connected, they are not inseparably connected. If we confine our study to charges that are at rest, and that remain at rest (electrostatics), then we can separate electricity from magnetism. This we shall do in these beginning chapters.

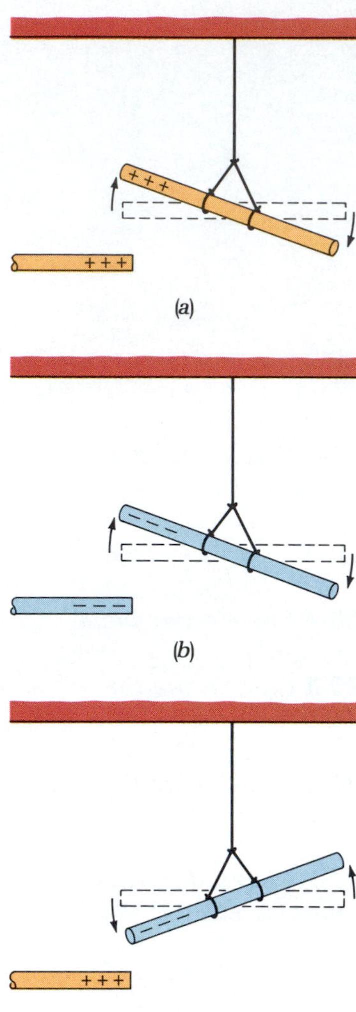

Figure 20-1. Suspended charged rods. *(a)* Two glass rods charged by rubbing with silk repel each other. *(b)* Two plastic rods charged by rubbing with fur repel each other. *(c)* A charged plastic rod is attracted to a charged glass rod.

20-1 ELECTRIC CHARGE AND MATTER

When the weather is cool and dry, it is easy to "charge" an object. You can pass a plastic comb through your hair and then pick up tiny bits of paper with the comb. Or you can rub a balloon on your sweater and the balloon will adhere to the wall, or possibly to the ceiling. If you slide across a car seat and then touch a metallic part of the car, you can give yourself quite a shock. These effects are manifestations of one of the fundamental forces in nature, the electric force.

Electric Charge

Suppose we rub the end of a glass rod with a piece of silk cloth and suspend the rod with string. The rubbed end of the rod is shown in Fig. 20-1*a*, with plus signs on it. Next we rub the end of another glass rod with a piece of silk and bring the rubbed end of the second rod near that of the first. As shown in the figure, the first rod swings away from the second, indicating a force of repulsion between them. Rods treated in this fashion are said to be "electrically charged" or to "possess electric charge," and the force they exert on one another is called the *electric force.*

A similar experiment using plastic rods rather than glass and a piece of fur rather than silk gives similar results (Fig. 20-1*b*). In this case the rubbed ends of the rods are shown with minus signs. When the rubbed ends of the plastic rods are brought near one another, the suspended rod swings away; the rods repel.

If the rubbed end of the glass rod is brought near the rubbed end of the suspended plastic rod, the plastic rod swings toward the glass rod (Fig. 20-1*c*), indicating a force of attraction. Furthermore, a glass rod that has been charged by rubbing with a silk cloth is attracted by the cloth. That is, the silk cloth is charged as well as the glass rod. Similarly, a plastic rod that has been charged by rubbing with fur is attracted by the fur.

Suppose we perform such experiments with rods made of many different materials and rubbed with many different types of cloth. We learn two important results from these experiments. First, if rods A and B attract, then all other rods fall into one of two categories: (i) those that repel A and attract B, or (ii) those that repel B and attract A. Any rod in the first category repels any other rod in the same category and attracts any rod in the second category. Similarly, any rod in the second category repels any other rod in the same category and attracts any rod in the first category. We conclude that there are two types of charge: Rods in the first category have one type of charge and rods in the second have the other, or opposite type.

The second important result of these experiments is that when a rod is charged by rubbing with a cloth, the type of charge the rod obtains is opposite to the type of charge the cloth obtains. Quantitative measurements show that the amount of these opposite types of charge is the same. Thus, if one object becomes charged, another object also becomes charged, but with the opposite type of charge.

The Model

We now present a model that "explains" these electrical effects:

1. Matter contains two types of electric charge, called *positive* and *negative.* Uncharged objects have equal amounts of each type of charge, and when objects are charged by rubbing, electric charge is transferred from one object to the other. After the charging process is completed, one of the objects has excess positive charge and the other has excess negative charge.

2. Objects with like charge repel each other.

3. Objects with unlike charge attract each other.

The law of conservation of charge

This model is nearly the same as that proposed by Benjamin Franklin (1706–1790). Inherent in the model is the ***law of conservation of charge:*** *Electric*

A suspended charged rod swings away from a similarly charged rod brought nearby. The two rods have excess charge of the same sign.

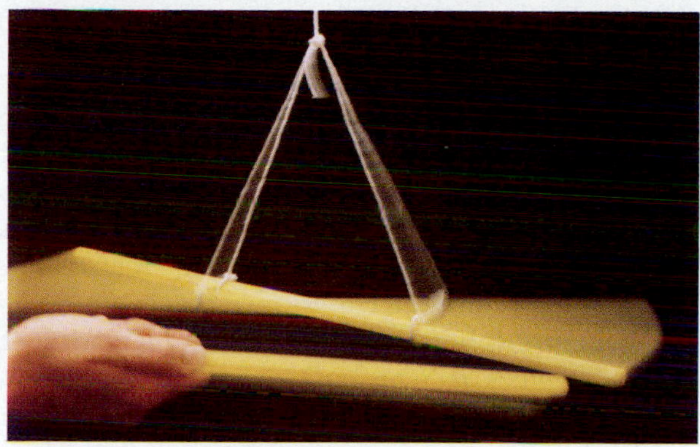

charge can neither be created nor destroyed; it can only be transferred. When an object is said to "possess charge," we mean that it possesses *excess* charge.

Atomic Structure

Our present model of atoms contains Franklin's model of electricity. It also has features that are the result of other experiments, which show that atoms consist of three types of particles: electrons, protons, and neutrons. Electrons and protons possess charge, but neutrons are electrically neutral. The magnitude of an electron's charge is the same as that of a proton, but of the opposite sign. Neutral atoms have equal numbers of protons and electrons. Franklin's choice of sign for charged objects was adopted worldwide and survives to this day. This choice assigns negative charge to electrons and positive charge to protons.

The protons and neutrons of an atom are held together in a small nucleus, which is surrounded by an electron cloud (Fig. 20-2). A nucleus can contain from 1 to about 100 protons, depending on the chemical element, and usually contains about the same number of neutrons. For example, the nucleus of an aluminum atom contains 13 protons and 14 neutrons. A proton and a neutron have about the same mass, and each has about 2000 times the mass of an electron. Therefore, an atom's nucleus usually has a mass that is about 4000 times the mass of its electrons.

The size of the nucleus of an atom is much less than the extent of the electron cloud, as shown in Fig. 20-2. A nucleus is typically about 5×10^{-15} m across, whereas the electron cloud has a diameter of about 2×10^{-10} m. Thus the linear dimension of an atom is about 40,000 times that of its nucleus.

Electron cloud

Nucleus

$\sim 5 \times 10^{-15}$ m

$\sim 2 \times 10^{-10}$ m

Figure 20-2. Model of the atom. The extent of the electron cloud and an enlarged view of the nucleus are shown.

Charge is quantized.

An aspect of electric charge contained in our atomic model is the *quantization of charge.* When something is said to be "quantized," it means that the something comes in lumps, or units of discrete size, and cannot be divided into smaller and smaller pieces. For example, cash is quantized in units of one penny in the monetary system in the United States. If charge is possessed by particles—electrons and protons—and if these particles cannot be divided into smaller pieces, then charge cannot be divided into smaller pieces either.

20-2 INSULATORS AND CONDUCTORS

Most materials can be classified as one of two types—as a conductor or as an insulator. A ***conductor*** is a material that readily allows charge to flow through it, and an ***insulator*** is a material that does not readily allow charge to flow. Metals, such as copper and aluminum, are good conductors (poor insulators), whereas glass and rubber are good insulators (poor conductors).

We can demonstrate the contrasting electrical behavior of conductors and insulators by performing an experiment that relies on this behavior. Figure 20-3*a* shows a conducting sphere supported by an insulating stand. In Fig. 20-3*b* a negatively charged rod has been brought close to the sphere. While the charged rod is near the sphere, the sphere is "grounded," as shown schematically in Fig. 20-3*c*. The term "grounding" means that a conducting path has been provided between the sphere and the earth. This is often done by connecting a wire between the sphere and a water faucet, but in this case a sufficient grounding procedure is simply to touch the sphere with your finger because your body is an adequate conductor for this purpose. With the rod still nearby, the ground connection is taken away (Fig. 20-3*d*). Finally, when the rod is taken away, we find that the sphere is left with excess charge (Fig. 20-3*e*). We can verify that the sphere is charged and determine the sign of its charge by bringing it near a charged suspended rod of known sign, as shown in Fig. 20-1. This procedure is called ***charging by induction*** because the sphere is charged without actually coming in contact with the charged rod.

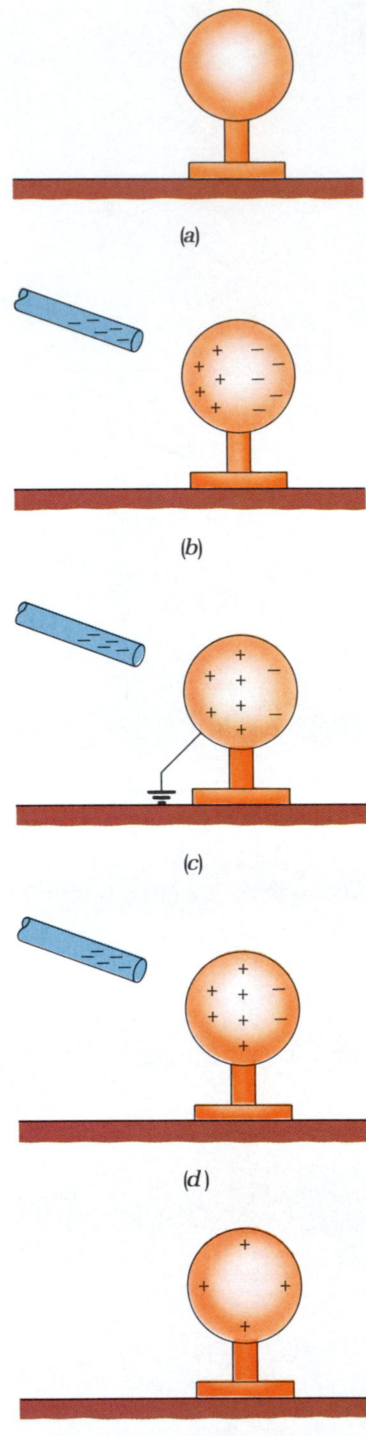

Figure 20-3. Charging by induction. *(a)* A neutral conducting sphere on an insulating stand. *(b)* A negatively charged rod is brought near the sphere. *(c)* The sphere is grounded. *(d)* The ground connection is taken away while the charged rod is still nearby. *(e)* After the charged rod is taken away, excess positive charge remains trapped on the sphere.

Charging by induction is illustrated in Fig. 20-3 with the plus and minus signs. When the charged rod is near the sphere (Fig. 20-3*b*), the side of the sphere nearest the rod has excess positive charge and the side farthest from the rod has excess negative charge because (i) unlike charges attract and like charges repel and (ii) because charge is able to move readily through the conducting sphere. When the sphere is grounded (Fig. 20-3*c*), the negative charges can move still farther from the negatively charged rod by passing through the grounding wire. After the wire is removed, the excess positive charge remains on the sphere because charge will not readily flow through the insulating stand; the excess charge is trapped. Of course, the surrounding air must also be an insulator. Air is a reasonably good insulator, depending on weather conditions; dry air is a better insulator than humid air. For charging by induction to work, the sphere must be a good conductor and the stand must be a good insulator, and for it to work well, the humidity must be low.

Ordinarily, when charge moves through a material, it is electrons that do the moving. We say that these electrons are the ***charge carriers.*** A conducting material, such as a metal, can be thought of as an array of fixed positive ions* interspersed with "free" electrons capable of moving through the material. The number of free electrons in a conductor depends on the material, but is on the order of one per atom. An insulator is a material in which there are almost no free electrons. An electron in an insulator is held near the position of a particular atom or molecule and is not allowed to pass from one molecular site to the next.

In a conducting fluid, such as saltwater, there are practically no free electrons, but the fluid can still be an excellent conductor. In this case the salt dissolves in the water as positive and negative ions, and the ions, which are capable of moving through the fluid, become the charge carriers.

20-3 COULOMB'S LAW

The force law for stationary charged particles was determined in 1784 by Charles Augustin Coulomb (1736–1806). Using a torsion balance (Fig. 20-4), he established the distance and charge dependence of the electric force. In recognition of his work, the SI unit of charge is called the ***coulomb*** (C).

* A positive ion is an atom or molecule with a deficiency of one or more electrons. A negative ion has an excess of one or more electrons.

To describe the electrical interaction between two particles a and b, which possess charge q_a and q_b, we use the coordinate frame shown in Fig. 20-5. Particle a is at the origin a distance r from b, and a unit vector $\hat{\mathbf{r}}$ points away from a along the line joining a and b. Experiment shows that the expression for the force $\mathbf{F}_{ab}$ exerted by a on b is

$$\mathbf{F}_{ab} = \frac{1}{4\pi\epsilon_0}\frac{q_a q_b}{r^2}\hat{\mathbf{r}} \tag{20-1}$$

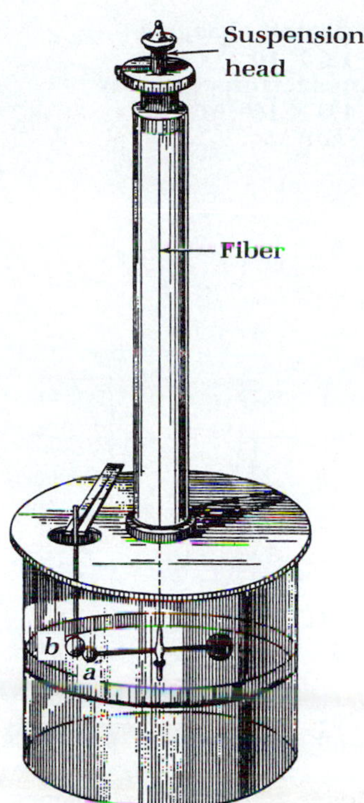

Figure 20-4. Coulomb's torsion balance. a and b are charged spheres.

where $1/4\pi\epsilon_0$ is a proportionality constant that is independent of the separation distance and of the amount of charge on either particle. Equation (20-1) is called ***Coulomb's law.*** Note that the electric force between charged particles is an inverse-square force: $F_{ab} \propto 1/r^2$. If the distance between the particles is doubled, then the magnitude of the force is reduced by a factor of 4 (Fig. 20-6). You are familiar with an inverse-square force from our discussion of Newton's law of universal gravitation (Chap. 7).

Coulomb's law contains the result that particles of like sign repel and particles of unlike sign attract. If q_a and q_b have the same sign, then the product $q_a q_b$ is positive and the direction of $\mathbf{F}_{ab}$ in Eq. (20-1) is the same as $\hat{\mathbf{r}}$; the equation says that b is repelled by a. If q_a and q_b have opposite signs, then the product $q_a q_b$ is negative and $\mathbf{F}_{ab}$ is directed opposite $\hat{\mathbf{r}}$, which describes attraction between b and a. The magnitude of the force depends on the magnitude of the product of the charges: $F_{ab} \propto |q_a q_b|$. That is, the force magnitude depends linearly on the magnitude of each charge.

As with other forces, the electric force is found from experiment to be a vector quantity. The effect of two or more electric forces acting simultaneously on a particle is determined by adding the forces vectorially. In addition, the electrical interaction obeys Newton's third law. If $\mathbf{F}_{ab}$ is the force by a on b, then the force $\mathbf{F}_{ba}$ by b on a is $\mathbf{F}_{ba} = -\mathbf{F}_{ab}$.

The proportionality constant in Coulomb's law, $1/4\pi\epsilon_0$, appears complex, but it is nevertheless only a proportionality constant. The factor 4π is contained in this constant to simplify other equations encountered in the next chapter. The factor ϵ_0 represents a constant called the ***permittivity*** of free space, or vacuum. The value of ϵ_0 is

$$\epsilon_0 = 8.854 \times 10^{-12}\ \mathrm{C^2/N\cdot m^2}$$

Permittivity of free space

This gives

$$\frac{1}{4\pi\epsilon_0} = 8.987 \times 10^9\ \mathrm{N\cdot m^2/C^2} \approx 9 \times 10^9\ \mathrm{N\cdot m^2/C^2}$$

The proportionality factor $1/4\pi\epsilon_0$ is a large number because the coulomb is a large unit of charge. Suppose two particles, each with charge 1 C, are separated a distance of 1 m. From Coulomb's law, the magnitude F of the force between them is

$$F = \frac{1}{4\pi\epsilon_0}\frac{|q_a q_b|}{r^2} = (9 \times 10^9\ \mathrm{N\cdot m^2/C^2})\frac{(1\ \mathrm{C})(1\ \mathrm{C})}{(1\ \mathrm{m})^2} = 9 \times 10^9\ \mathrm{N}$$

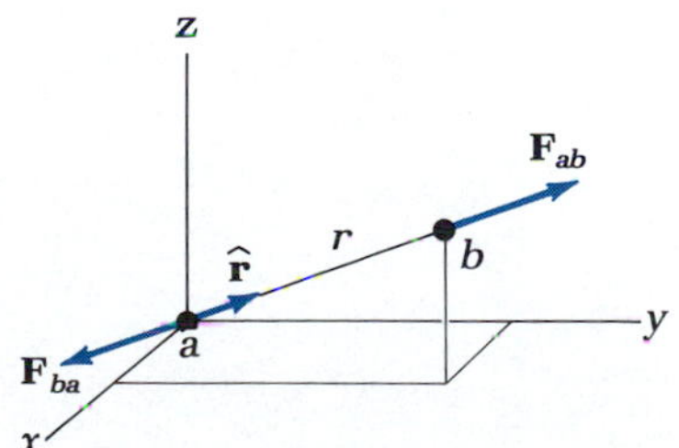

Figure 20-5. If particles a and b have charge of like sign (which is the case shown), then the force $\mathbf{F}_{ab}$ by a on b is directly away from a. The force $\mathbf{F}_{ba}$ by b on a is directly away from b, in agreement with Newton's third law. On a scratch sheet, draw a similar figure for the case where the particles have charge of opposite sign.

The magnitude of this force is huge, equivalent to the weight of about 15 million adult humans. This calculation shows that 1 C is a very large amount of charge. A rod charged by rubbing, as we described in Sec. 20-1, will typically possess a charge of about 10 nC (10^{-8} C). Measuring the charge on a glass rod rubbed with silk in units of coulombs (C) is similar to measuring the thickness of this page in units of kilometers. The coulomb is defined by an experimental procedure involving the measurement of a magnetic force (Chap. 27), and its size is a consequence of this definition.

A proton and an electron have equal and opposite charge. The magnitude of this fundamental charge is represented by the symbol e, and its value is

A proton charge is $+1.6 \times 10^{-19}$ C. An electron charge is -1.6×10^{-19} C.

$$e = 1.60207 \times 10^{-19}\ \text{C}$$

The charge of a proton is $+e$, and the charge of an electron is $-e$. If N_e is the number of electrons in an object and N_p is the number of protons, then the charge q on the object is

$$q = (N_p - N_e)e$$

Thus $N_p > N_e$ for a positively charged object, $N_p = N_e$ for a neutral object, and $N_p < N_e$ for a negatively charged object.

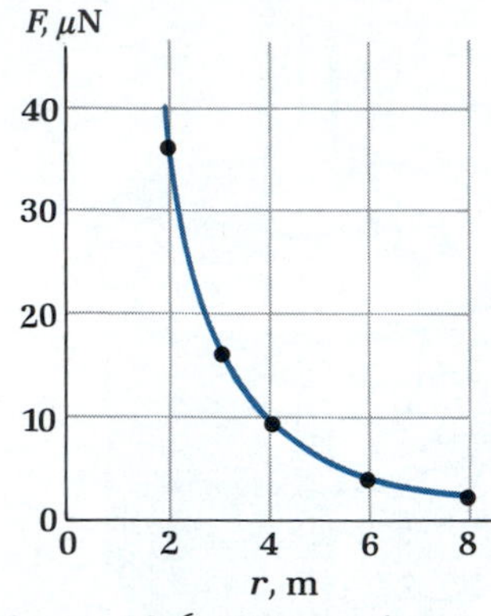

Figure 20-6. Distance dependence of the electric force between two charged particles. The force is an inverse-square force. Data from this graph are used in Example 20-1.

EXAMPLE 20-1

Population imbalance of electrons. Suppose the graph in Fig. 20-6 corresponds to an interaction between two baseballs with equal positive charge. For each baseball, *(a)* determine the charge, *(b)* determine the number of missing electrons, and *(c)* estimate the fraction of missing electrons.

Solution. *(a)* Since each baseball has the same charge, we let q represent both q_a and q_b in Eq. 20-1 and solve for q:

$$q = \sqrt{4\pi\epsilon_0 r^2 F}$$

We may select values of F and r from any of the points on the curve in Fig. 20-6; suppose we choose $F = 9.0\ \mu\text{N}$ and $r = 4.0$ m:

$$q = \sqrt{\frac{(4.0\ \text{m})^2(9.0\ \mu\text{N})}{9.0 \times 10^9\ \text{N}\cdot\text{m}^2/\text{C}^2}} = 130\ \text{nC}$$

(b) Letting n represent the number of electrons missing from each baseball ($n = N_p - N_e$), we have $q = ne$, or

$$n = \frac{q}{e} = \frac{130\ \text{nC}}{1.6 \times 10^{-19}\ \text{C}} = 7.9 \times 10^{11}\ \text{electrons}$$

For a macroscopic object to have a significant charge, the number of missing or extra electrons must be very large because the charge of a single electron is very small. *(c)* The fraction of electrons missing is n/N_p because N_p (the number of protons) is equal to the number of electrons in a neutral object. A baseball has a mass of about 0.15 kg, and about half the mass is attributed to protons and about half to neutrons. Dividing the mass of a baseball by the mass of a proton-neutron pair gives an estimate of N_p:

$$N_p \approx \frac{M}{m_p + m_n} = \frac{0.15\ \text{kg}}{2(1.67 \times 10^{-27}\ \text{kg})} \approx 5 \times 10^{25}\ \text{protons}$$

Thus

$$\frac{n}{N_p} \approx \frac{7.9 \times 10^{11}\ \text{missing electrons}}{5 \times 10^{25}\ \text{protons}} \approx 2 \times 10^{-14}$$

This means that about one out of every 5×10^{13} [or $1/(2 \times 10^{-14})$] electrons is missing from each baseball. For comparison, the earth's human population is about 5×10^9, so that 5×10^{13} is about 10,000 times the earth's population of humans. You can see that a very tiny imbalance in the electron population causes a macroscopic object to have a significant charge.

SELF-TEST 20-1. *(a)* Suppose a penny has a charge of -3.2 nC. How many excess electrons does this penny have? *(b)* The mass of a penny is about 3.3 g. Estimate the fraction of excess electrons in this penny. ***ANSWERS:*** *(a)* 2.0×10^{10} electrons; *(b)* 2.0×10^{-14}.

EXAMPLE 20-2

Comparison with gravitation. From Chap. 7 we know that the gravitational force between two particles is an attractive inverse-square force that depends on the product of the masses of the particles. Compare the magnitudes of the electric and gravitational forces between an electron and a proton by calculating the ratio of these forces.

Solution. The magnitudes of the electric force and the gravitational force between a proton (charge = e, mass = m_p) and an electron (charge = $-e$, mass = m_e) are

$$F_E = \frac{1}{4\pi\epsilon_0}\frac{e^2}{r^2} \quad \text{and} \quad F_G = G\frac{m_p m_e}{r^2}$$

respectively. The ratio is

$$\frac{F_E}{F_G} = \frac{1}{4\pi\epsilon_0 G}\frac{e^2}{m_p m_e}$$

Since both forces are inverse-square forces, r^2 cancels out of the ratio. This means that the comparison is valid at any separation distance:

$$\frac{F_E}{F_G} = \frac{(9.0 \times 10^9\ \text{N}\cdot\text{m}^2/\text{C}^2)(1.6 \times 10^{-19}\ \text{C})^2}{(6.7 \times 10^{-11}\ \text{N}\cdot\text{m}^2/\text{kg}^2)(1.7 \times 10^{-27}\ \text{kg})(9.1 \times 10^{-31}\ \text{kg})} \approx 2 \times 10^{39}$$

Between an electron and a proton, the gravitational force is negligible in comparison with the electric force.

SELF-TEST 20-2. Determine the ratio of the magnitudes of the electric and gravitational forces that two electrons exert on one another. ***ANSWER:*** 4×10^{42}.

EXAMPLE 20-3

Vector sum of electric forces. Determine the force on particle c in Fig. 20-7 due to particles a and b. The charges on the particles are $q_a = 3.0\ \mu\text{C}$, $q_b = -6.0\ \mu\text{C}$, and $q_c = -2.0\ \mu\text{C}$.

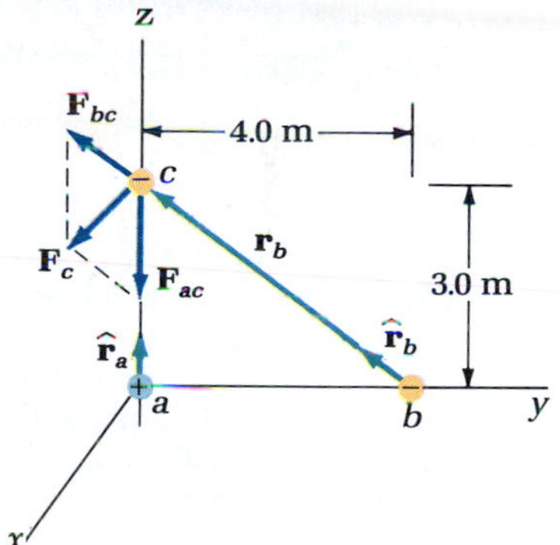

Figure 20-7. Example 20-3: Finding the force on charged particle c due to charged particles a and b.

Solution. Using Coulomb's law, we find that the force on particle c due to particle a is

$$\mathbf{F}_{ac} = \frac{1}{4\pi\epsilon_0}\frac{q_c q_a}{r_a^2}\hat{\mathbf{r}}_a$$

From the figure, $r_a = 3.0$ m. Also, the coordinate system has been arranged such that the unit vector $\hat{\mathbf{r}}_a$ that points from a to c is $\mathbf{k}$. Thus

$$\mathbf{F}_{ac} = (9.0 \times 10^9\ \text{N}\cdot\text{m}^2/\text{C}^2)\frac{(-2.0\ \mu\text{C})(3.0\ \mu\text{C})}{(3.0\ \text{m})^2}\mathbf{k}$$
$$= (-6.0 \times 10^{-3}\ \text{N})\mathbf{k} = -(6.0\ \text{mN})\mathbf{k}$$

The direction of the force is attractive because q_a and q_c have opposite signs. Similarly, the force on particle c due to particle b is

$$\mathbf{F}_{bc} = \frac{1}{4\pi\epsilon_0}\frac{q_c q_b}{r_b^2}\hat{\mathbf{r}}_b$$

From the figure, $r_b = \sqrt{(3.0\ \text{m})^2 + (4.0\ \text{m})^2} = 5.0$ m. The unit vector $\hat{\mathbf{r}}_b$ is found by dividing the vector $\mathbf{r}_b$, which extends from b to c, by its length r_b. Since $\mathbf{r}_b = (-4.0\ \text{m})\mathbf{j} + (3.0\ \text{m})\mathbf{k}$, we have

$$\hat{\mathbf{r}}_b = \frac{\mathbf{r}_b}{r_b} = \frac{(-4.0\ \text{m})\mathbf{j} + (3.0\ \text{m})\mathbf{k}}{5.0\ \text{m}} = -(0.80)\mathbf{j} + (0.60)\mathbf{k}$$

Substituting these values into the equation for $\mathbf{F}_{bc}$ gives

$$\mathbf{F}_{bc} = \left(9.0 \times 10^9\ \frac{\text{N}\cdot\text{m}^2}{\text{C}^2}\right)\frac{(-2.0\ \mu\text{C})(-6.0\ \mu\text{C})}{(5.0\ \text{m})^2}(-0.80\mathbf{j} + 0.60\mathbf{k})$$
$$= (-3.5\ \text{mN})\mathbf{j} + (2.6\ \text{mN})\mathbf{k}$$

The net force $\mathbf{F}_c$ on the particle c is the sum of the two individual forces:

$$\mathbf{F}_c = \mathbf{F}_{ac} + \mathbf{F}_{bc} = (-6.0\ \text{mN})\mathbf{k} + [(-3.5\ \text{mN})\mathbf{j} + (2.6\ \text{mN})\mathbf{k}]$$
$$= (-3.5\ \text{mN})\mathbf{j} + (-3.4\ \text{mN})\mathbf{k}$$

The magnitude of the force is

$$F_c = \sqrt{(-3.5\ \text{mN})^2 + (-3.4\ \text{mN})^2} = 4.9\ \text{mN}$$

The vectors $\mathbf{F}_{ac}$, $\mathbf{F}_{bc}$, and $\mathbf{F}_c$ are shown in Fig. 20-7. Examine the figure to verify that the components calculated above are consistent with the graphical results. It is often useful to check your numerical results with a reasonably accurate diagram.

SELF-TEST 20-3. *(a)* For the arrangement of particles in the above example, determine the force exerted on particle *a* by particles *b* and *c*. *(b)* What is the magnitude of this force?
ANSWERS: *(a)* $\mathbf{F}_a = (10\text{ mN})\mathbf{j} + (6.0\text{ mN})\mathbf{k}$; *(b)* $F_a = 12$ mN.

20-4 THE ELECTRIC FIELD

A ***field*** is a quantity that can be associated with position. For example, the temperature of the air in a room has a specific value at each point in the room. If T represents the temperature, then there exists a function $T(x, y, z)$ that gives the temperature at each point (x, y, z). Further, the temperature may change with time t, in which case it is a function of t as well, $T(x, y, z, t)$. Since temperature is a scalar quantity, $T(x, y, z, t)$ is an example of a *scalar field.*

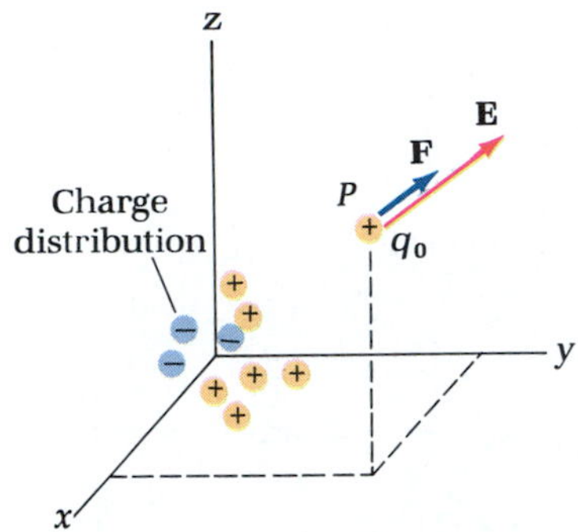

Figure 20-8. At point P the electric field **E** due to a distribution of charge is defined as the electric force **F** exerted by the distribution on a test particle placed at P divided by the charge q_0 of the test particle: $\mathbf{E} = \mathbf{F}/q_0$.

In addition to scalar fields, there are *vector fields*—that is, vector quantities that exist at points in space. Wind in the earth's atmosphere is an example of a vector field. At each point in the earth's atmosphere there is a velocity **v** of the air. The three components of this vector field are functions of position and time. Using cartesian coordinates, we can write these components as $v_x(x, y, z, t)$, $v_y(x, y, z, t)$, and $v_z(x, y, z, t)$.

In Chap. 7 we introduced the gravitational field $\mathbf{g} = \mathbf{F}/m$, where **F** is the gravitational force on an object of mass m. At each point near its surface, the earth produces a gravitational field of magnitude 9.8 N/kg that points toward the center of the earth. If a 3.6-kg object is near the earth's surface, then there is a force of magnitude $F = mg = (3.6\text{ kg})(9.8\text{ N/kg}) = 35$ N exerted on it because of the earth's gravitational field. The gravitational field is an example of a vector field.

The definition of the electric field is similar to that of the gravitational field. Consider the electric field **E** produced by the array of charged particles clustered around the origin in Fig. 20-8. A group of charged particles like this is called a ***charge distribution.*** A test particle with charge q_0 is placed at point P.

The field **E** at point P is defined as the electric force **F** exerted by the charge distribution on the test particle divided by the charge q_0 on the test particle.

Definition of the electric field **E**

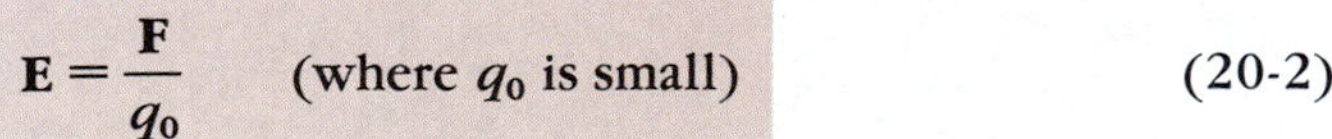

$$\mathbf{E} = \frac{\mathbf{F}}{q_0} \qquad \text{(where } q_0 \text{ is small)} \tag{20-2}$$

Charge distribution

The field **E** depends on the value and location of the charges that compose the charge distribution, and it depends on the point at which the field is measured. We view the electric field as a condition set up in space by the charge distribution. We assume that this electric field exists whether or not a test particle is there to measure it, just as the temperature exists at each point in a room whether or not a thermometer is there to measure it.

The value of **E** is independent of the test particle used to measure it. The force **F** on the test particle is the vector sum of the individual forces exerted on it by the particles in the charge distribution. Since each of these forces is proportional to q_0, the ratio $\mathbf{E} = \mathbf{F}/q_0$ is independent of q_0. To ensure that **E** is independent of the test particle, the charge q_0 is made small enough such that its effect on the particles in the charge distribution is negligible. That is why Eq. 20-2 is qualified by the condition that q_0 be small. (See Ques. 14.) The purpose of the test particle is to test the field, just as the purpose of a thermometer is to test the temperature.

The direction of E is that of the force on a positively charged test particle.

Since force **F** is a vector quantity and $\mathbf{E} = \mathbf{F}/q_0$, **E** is a vector field and obeys the principle of superposition. The direction of **E** is the same as the direction of the force on a positively charged test particle, or opposite the force on a negatively charged test particle. The dimension of the electric field is force divided by electric charge, and its SI unit is newtons per coulomb (N/C).

Suppose a test particle with charge $q_0 = 81$ nC has a force $\mathbf{F} = (2.7\ \mu\text{N})\mathbf{i} + (1.1\ \mu\text{N})\mathbf{j} + (1.3\ \mu\text{N})\mathbf{k}$ exerted on it by a charged object (Fig. 20-9). Then the electric field **E** produced by the object at the position of the test particle is found to be

$$\mathbf{E} = \frac{\mathbf{F}}{q_0} = \frac{(2.7\ \mu\text{N})\mathbf{i} + (1.1\ \mu\text{N})\mathbf{j} + (1.3\ \mu\text{N})\mathbf{k}}{81\ \text{nC}}$$
$$= (33\ \text{N/C})\mathbf{i} + (14\ \text{N/C})\mathbf{j} + (16\ \text{N/C})\mathbf{k}$$

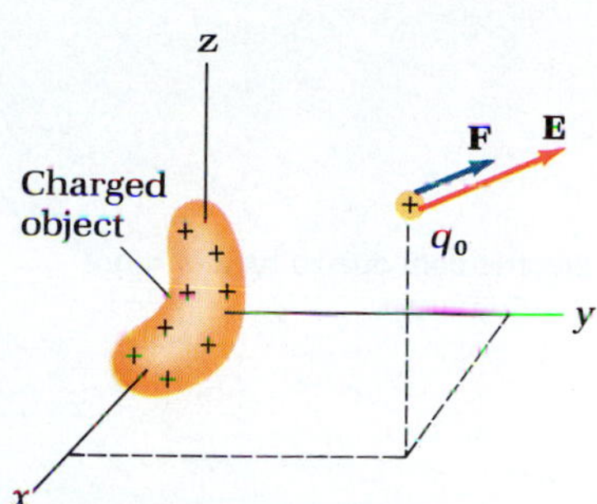

Figure 20-9. Force on a charged particle due to the electric field produced by a charged object.

If **E** is known at a point, then the force on a particle with charge q_0 placed at that point is found by solving Eq. 20-2 for **F**:

$$\mathbf{F} = q_0\mathbf{E} \tag{20-3}$$

Suppose a particle with charge $q_0 = -1.0$ nC is placed at a point where $\mathbf{E} = (56\ \text{N/C})\mathbf{i} + (-37\ \text{N/C})\mathbf{j} + (14\ \text{N/C})\mathbf{k}$. The electric force on the particle is

$$\mathbf{F} = q_0\mathbf{E} = (-1.0\ \text{nC})[(56\ \text{N/C})\mathbf{i} + (-37\ \text{N/C})\mathbf{j} + (14\ \text{N/C})\mathbf{k}]$$
$$= (-56\ \text{nN})\mathbf{i} + (37\ \text{nN})\mathbf{j} + (-14\ \text{nN})\mathbf{k}$$

Note that the force on the negatively charged particle is directed opposite **E**.

20-5 CALCULATING THE ELECTRIC FIELD

The electric field produced by a charge distribution can be found from Coulomb's law and the principle of superposition.

The Electric Field Due to Charged Particles

Consider the field produced by a particle with charge q located at the origin (Fig. 20-10). From Coulomb's law, the force exerted by this particle on the test particle of charge q_0 is

$$\mathbf{F} = \frac{1}{4\pi\epsilon_0}\frac{q\,q_0}{r^2}\,\hat{\mathbf{r}}$$

Dividing by q_0 gives the electric field at the position of q_0:

$$\mathbf{E} = \frac{1}{4\pi\epsilon_0}\frac{q}{r^2}\,\hat{\mathbf{r}} \tag{20-4}$$

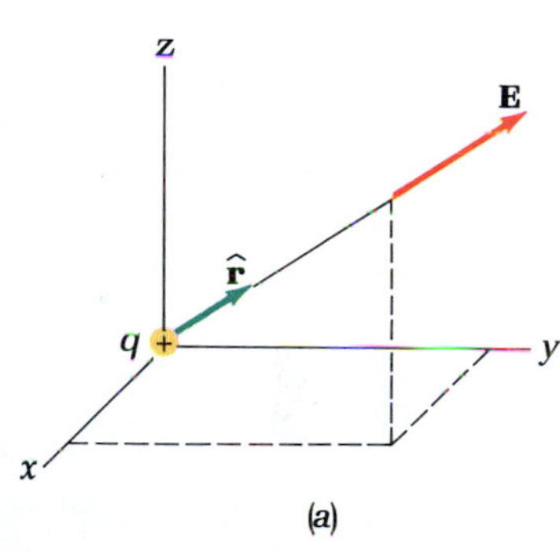

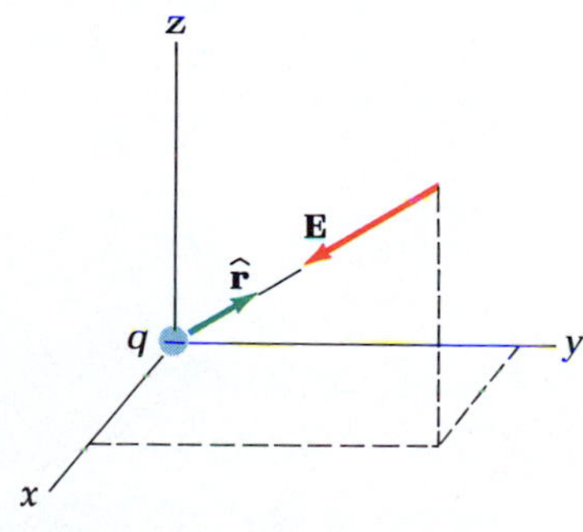

Figure 20-10. Electric field due to a point charge at the origin. *(a)* Positive q. *(b)* Negative q.

A charged particle is often called a ***point charge,*** and Eq. 20-4 is the field produced by a point charge q. The important features of this field are:

1. The field magnitude E is proportional to $|q|$.

2. E is proportional to $1/r^2$. If $E = 80$ N/C at $r = 10$ cm, then at $r = 20$ cm, $E = (80\ \text{N/C})/2^2 = 20$ N/C.

3. The vector **E** points directly away from a positive charge (Fig. 20-10*a*) or directly toward a negative charge (Fig. 20-10*b*).

Now consider the field produced by two or more charged particles. From the principle of superposition, the force **F** on the test particle is the vector sum of the individual forces:

$$\mathbf{F} = \frac{1}{4\pi\epsilon_0}\frac{q_0q_1}{r_1^2}\,\hat{\mathbf{r}}_1 + \frac{1}{4\pi\epsilon_0}\frac{q_0q_2}{r_2^2}\,\hat{\mathbf{r}}_2 + \cdots$$
$$= \frac{q_0}{4\pi\epsilon_0}\left(\frac{q_1}{r_1^2}\,\hat{\mathbf{r}}_1 + \frac{q_2}{r_2^2}\,\hat{\mathbf{r}}_2 + \cdots\right)$$
$$= \frac{q_0}{4\pi\epsilon_0}\sum\frac{q_i}{r_i^2}\,\hat{\mathbf{r}}_i$$

where q_i is the charge on particle i, r_i is the distance from particle i to point P, and $\hat{\mathbf{r}}_i$ is a unit vector that points from particle i to point P. Dividing by q_0 gives the field at P:

Electric field due to two or more point charges

$$\mathbf{E} = \frac{1}{4\pi\epsilon_0} \sum \frac{q_i}{r_i^2} \hat{\mathbf{r}}_i \tag{20-5}$$

The electric field due to two or more point charges is the vector sum of the individual contributions to the field produced by each charge separately. Finding the electric field due to a distribution of point charges is a problem that requires vector addition. We show two equivalent procedures in our examples, one using unit vectors and one using components. Example 20-3 has already provided an illustration of the first procedure, and the following example shows the second procedure.

EXAMPLE 20-4

Principle of superposition. Two particles 1 and 2, with charges $q_1 = +16$ nC and $q_2 = +28$ nC, are at positions (x, y, z) of (0, 0, 0) and (0, −2.0 m, 0), respectively (Fig. 20-11). Find **E** *(a)* at point P_a (0, 1.0 m, 0) and *(b)* at P_b (0, 0, 1.5 m).

Solution. *(a)* The magnitudes of the two contributions to **E** at P_a are

$$E_1 = (9.0 \times 10^9\ \mathrm{N \cdot m^2/C^2}) \frac{16\ \mathrm{nC}}{(1.0\ \mathrm{m})^2} = 140\ \mathrm{N/C}$$

$$E_2 = (9.0 \times 10^9\ \mathrm{N \cdot m^2/C^2} \frac{28\ \mathrm{nC}}{(3.0\ \mathrm{m})^2} = 28\ \mathrm{N/C}$$

Since $\mathbf{E}_1$ and $\mathbf{E}_2$ are both in the y direction at P_a,

$$\mathbf{E} = \mathbf{E}_1 + \mathbf{E}_2 = (140\ \mathrm{N/C})\mathbf{j} + (28\ \mathrm{N/C})\mathbf{j} = (170\ \mathrm{N/C})\mathbf{j}$$

(b) Before finding the field at P_b, it is convenient to note that the distance from q_2 to P_b is $\sqrt{(1.5\ \mathrm{m})^2 + (2.0\ \mathrm{m})^2} = 2.5$ m. The magnitudes of the two contributions to **E** at P_b are

$$E_1 = (9.0 \times 10^9\ \mathrm{N \cdot m^2/C^2}) \frac{16\ \mathrm{nC}}{(1.5\ \mathrm{m})^2} = 64\ \mathrm{N/C}$$

$$E_2 = (9.0 \times 10^9\ \mathrm{N \cdot m^2/C^2}) \frac{28\ \mathrm{nC}}{(2.5\ \mathrm{m})^2} = 40\ \mathrm{N/C}$$

In this case the two contributions to the field are not parallel: $\mathbf{E}_1 = E_1\mathbf{k}$ and $\mathbf{E}_2 = (E_2 \cos\theta)\mathbf{j} + (E_2 \sin\theta)\mathbf{k}$. From the figure, $\cos\theta = 2.0\ \mathrm{m}/2.5\ \mathrm{m} = 0.80$ and $\sin\theta = 1.5\ \mathrm{m}/2.5\ \mathrm{m} = 0.60$. (Notice that $\cos\theta$ and $\sin\theta$ are the components of the unit vector $\hat{\mathbf{r}}_2$ that points from q_2 to P_b, $\hat{\mathbf{r}}_2 = 0.80\mathbf{j} + 0.60\mathbf{k}$.) The field at P_b is

$$\begin{aligned}\mathbf{E} = \mathbf{E}_1 + \mathbf{E}_2 &= (64\ \mathrm{N/C})\mathbf{k} + [(40\ \mathrm{N/C})(0.80)\mathbf{j} + (40\ \mathrm{N/C})(0.60)\mathbf{k}] \\ &= (32\ \mathrm{N/C})\mathbf{j} + (88\ \mathrm{N/C})\mathbf{k}\end{aligned}$$

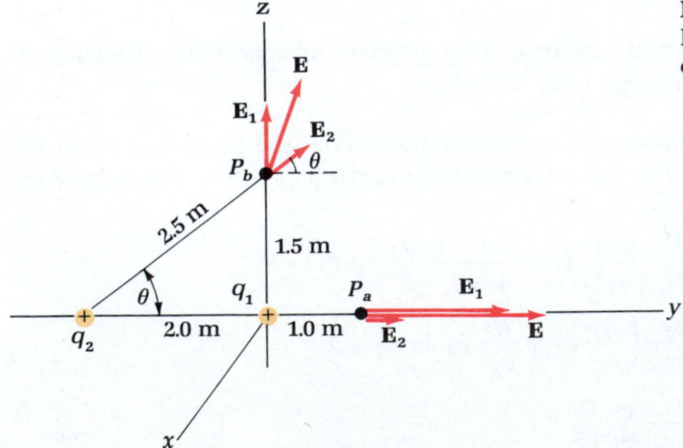

Figure 20-11. Example 20-4: Finding **E** at P_a and P_b due to charged particles 1 and 2.

SELF-TEST 20-4. Determine $\mathbf{E}$ midway between the particles in the above exercise, that is, at the point $(0, -1.0\text{ m}, 0)$. Be certain your answer is in the form of a vector. ***ANSWER:*** $\mathbf{E} = (110\text{ N/C})\mathbf{j}$.

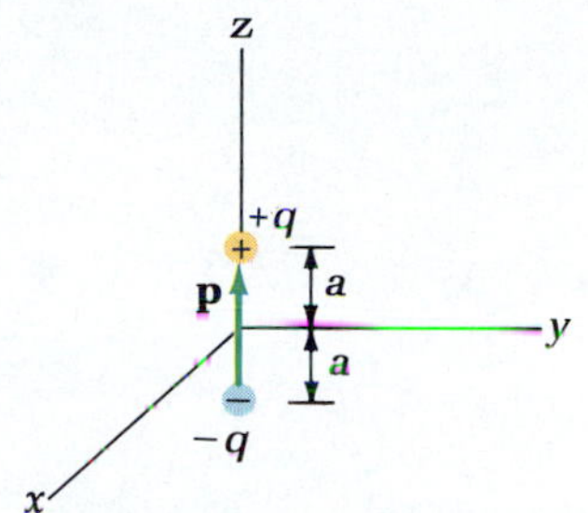

Figure 20-12. An electric dipole at the origin and pointing in the $+z$ direction. The dipole moment of this dipole is $\mathbf{p} = (2aq)\mathbf{k}$.

The Electric Dipole

An important charge distribution is an electric dipole. A ***dipole*** consists of two point charges of equal magnitude and opposite sign. Ordinarily, the symbol q representing the charge on an object may be positive or negative. But when referring to a dipole, we customarily let q represent the magnitude of either charge so that one particle has charge $+q$ and the other has charge $-q$. In Fig. 20-12, the particle with positive charge is at position $(0, 0, a)$ and the particle with negative charge is at $(0, 0, -a)$. The *electric dipole moment* $\mathbf{p}$ is a vector whose magnitude is the product of the charge magnitude q and the separation distance $2a$: $p = 2aq$. The direction of the dipole moment is from the particle with negative charge and toward the particle with positive charge. In the case shown in the figure, $\mathbf{p} = (2aq)\mathbf{k}$.

EXAMPLE 20-5

Field of a dipole at points in its bisector plane. *(a)* Find an expression for the electric field in the perpendicular bisector plane of a dipole. *(b)* Apply the result found in part *(a)* to the case where $\mathbf{E}$ is evaluated at distances from the dipole that are much greater than the separation between the dipole's two point charges.

Solution. *(a)* Figure 20-13 shows the dipole centered at the origin with the xy plane as the perpendicular bisector plane. Consider $\mathbf{E}$ at the point P on the y axis. The two contributions to $\mathbf{E}$ are $\mathbf{E}_+$ produced by the positive charge and $\mathbf{E}_-$ produced by the negative charge: $\mathbf{E} = \mathbf{E}_+ + \mathbf{E}_-$, where

$$\mathbf{E}_+ = \frac{1}{4\pi\epsilon_0}\frac{q}{r_+^2}\hat{\mathbf{r}}_+ \quad \text{and} \quad \mathbf{E}_- = \frac{1}{4\pi\epsilon_0}\frac{-q}{r_-^2}\hat{\mathbf{r}}_-$$

In the equation for $\mathbf{E}_-$, the negative sign for the charge has been explicitly included because q is a positive number. The distance r_+ from $+q$ to P is the same as r_- from $-q$ to P; $r_+ = r_- = r = \sqrt{y^2 + a^2}$. The vector $\mathbf{r}_+$ from $+q$ to P is $(y\mathbf{j} - a\mathbf{k})$ so that the unit vector $\hat{\mathbf{r}}_+ = (y\mathbf{j} - a\mathbf{k})/r$. This gives

$$\mathbf{E}_+ = \frac{1}{4\pi\epsilon_0}\frac{q}{r^2}\frac{y\mathbf{j} - a\mathbf{k}}{r} = \frac{q}{4\pi\epsilon_0 r^3}(y\mathbf{j} - a\mathbf{k})$$

Similarly,

$$\mathbf{E}_- = \frac{1}{4\pi\epsilon_0}\frac{-q}{r^2}\frac{y\mathbf{j} + a\mathbf{k}}{r} = \frac{q}{4\pi\epsilon_0 r^3}(-y\mathbf{j} - a\mathbf{k})$$

When we add $\mathbf{E}_+$ and $\mathbf{E}_-$ to form $\mathbf{E}$, the y components cancel, and the z components give

$$\mathbf{E} = \mathbf{E}_+ + \mathbf{E}_- = \frac{1}{4\pi\epsilon_0}\frac{-2aq}{r^3}\mathbf{k}$$

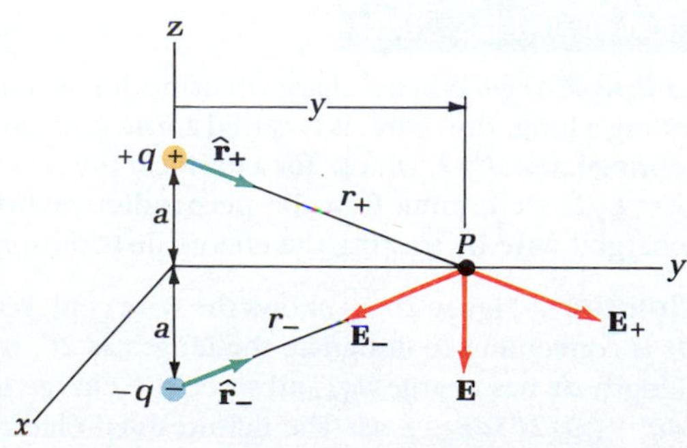

Figure 20-13. Example 20-5: Finding $\mathbf{E}$ in the perpendicular bisector plane of a dipole. The two contributions to the field are $\mathbf{E}_+$ and $\mathbf{E}_-$.

Now we note that the field must have azimuthal symmetry about the z axis. That is, E is the same at all points in the xy plane that are the same distance R from the origin, where $R = \sqrt{x^2 + y^2}$. Therefore we can generalize the result by replacing y^2 with R^2 in the expression for r, which means that $r = \sqrt{R^2 + a^2}$. Thus the field at any point in the dipole's bisector plane is

$$\mathbf{E} = -\frac{1}{4\pi\epsilon_0} \frac{\mathbf{p}}{(R^2 + a^2)^{3/2}}$$

where $\mathbf{p} = (2aq)\,\mathbf{k}$ is the dipole moment. The minus sign indicates that the field is directed opposite the dipole moment $\mathbf{p}$, as shown in the figure. *(b)* At points far from the dipole compared with the separation of the point charges, $R \gg a$ and $\sqrt{R^2 + a^2} \approx R$, or

$$\mathbf{E} \approx -\frac{1}{4\pi\epsilon_0} \frac{\mathbf{p}}{R^3}$$

In the perpendicular bisector plane of a dipole, we find that (i) the direction of $\mathbf{E}$ is opposite $\mathbf{p}$, (ii) E is proportional to p, and (iii) E falls off as $1/R^3$ (when $R \gg a$). As a numerical example of this third result, suppose $a \ll 10$ cm and $E = 80$ N/C at $R = 10$ cm. Then at $R = 20$ cm, $E = (80 \text{ N/C})/2^3 = 10$ N/C.

SELF-TEST 20-5. Two particles with equal and opposite charges of magnitude 2.0 nC are separated by 0.10 m, with the negative particle directly above the positive particle. *(a)* Treating the particles as a dipole, find the field magnitude E in the horizontal bisector plane at a distance of 1.5 m from the dipole. *(b)* Is the direction of the field at this point upward or downward? ***ANSWERS:*** *(a)* $E = 0.53$ N/C; *(b)* upward.

The Electric Field Due to Continuous Charge Distributions

The charge on macroscopic objects, such as the charged rods discussed earlier, is due to an imbalance of electron and proton populations. Since the charge on the electron or on the proton is small compared with ordinary-sized macroscopic charges, such a charge must comprise a large number of extra or missing electrons. Therefore we may treat the charge as a continuous distribution of infinitesimal charge elements dq. Applying Eq. 20-4 to such a situation, we have that the infinitesimal electric field $d\mathbf{E}$ due to dq is

Infinitesimal electric field $d\mathbf{E}$ due to charge element dq

$$d\mathbf{E} = \frac{1}{4\pi\epsilon_0} \frac{dq}{r^2} \hat{\mathbf{r}} \qquad (20\text{-}6)$$

where r is the distance from the element of charge dq to the point P at which the electric field is evaluated, and $\hat{\mathbf{r}}$ is a unit vector that points from dq to P. The electric field $\mathbf{E}$ due to all charge elements is found by integration: $\mathbf{E} = \int d\mathbf{E}$, or

Electric field due to a continuous charge distribution

$$\mathbf{E} = \frac{1}{4\pi\epsilon_0} \int \frac{dq}{r^2} \hat{\mathbf{r}} \qquad (20\text{-}7)$$

where the limits on the integral are determined by the extent of the charge distribution.

EXAMPLE 20-6

A line charge. When a charge distribution is long and narrow, as it is when charge is spread along a long, thin wire, it is called a ***line charge.*** A line charge is characterized by its ***linear charge density*** λ, which, for a uniform line charge, is the charge Q divided by the length L: $\lambda = Q/L$. Determine $\mathbf{E}$ in the perpendicular bisector plane of a long, straight, uniformly charged wire by treating the charge distribution as a uniform line charge.

Solution. Figure 20-14 shows the wire centered at the origin and oriented along the z axis. It is convenient to designate the length as 2ℓ, which means that $\lambda = Q/2\ell$. The element of length dz has charge dq, and since the charge is uniformly distributed, $dq/Q = dz/2\ell$, or $dq = (Q/2\ell)dz = \lambda\, dz$. The infinitesimal electric field $d\mathbf{E}$ due to dq is

Figure 20-14. Example 20-6: Electric field **E** in the perpendicular bisector plane of a straight uniform line charge. By symmetry, the axial component is zero, or $E_z = 0$, so that in the *xy* plane the field is radial. We show later that if the charge distribution is very long, $\ell \gg R$, then $E \approx \lambda/2\pi\epsilon_0 R$, where $R = \sqrt{x^2 + y^2}$.

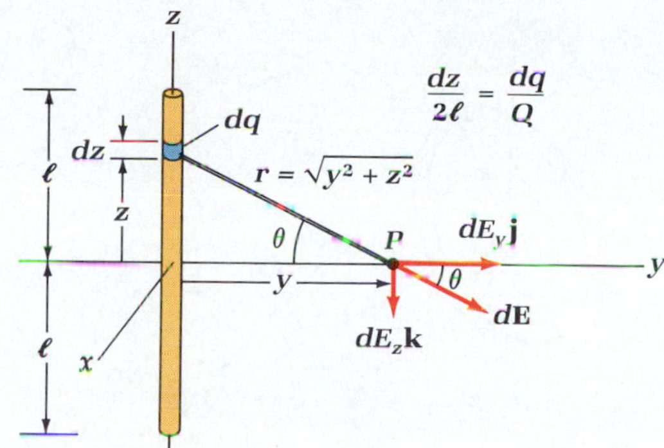

$$d\mathbf{E} = dE_y\mathbf{j} + dE_z\mathbf{k} = (dE\cos\theta)\mathbf{j} - (dE\sin\theta)\mathbf{k}$$

where the magnitude of $d\mathbf{E}$ is

$$dE = \frac{1}{4\pi\epsilon_0}\frac{\lambda\,dz}{(y^2 + z^2)}$$

From the figure we see that $\cos\theta = y/\sqrt{y^2 + z^2}$ and $\sin\theta = z/\sqrt{y^2 + z^2}$. [Notice that a unit vector $\hat{\mathbf{r}}$ that points from dz to P is $\hat{\mathbf{r}} = (\cos\theta)\mathbf{j} - (\sin\theta)\mathbf{k}$.]

First consider the y component of the field:

$$E_y = \int dE_y = \frac{\lambda y}{4\pi\epsilon_0}\int_{-\ell}^{+\ell}\frac{dz}{(y^2 + z^2)^{3/2}}$$

where the constants λ and y have been taken out of the integral. You may be surprised that y is regarded as a constant here, but keep in mind that the integral corresponds to adding up the infinitesimal electric fields $d\mathbf{E}$ due to all the infinitesimal charge elements dq along the z axis. During this procedure y, the coordinate of P, is held fixed.

The limits on the integral from $-\ell$ to $+\ell$ correspond to the region occupied by the charge. From integral tables,

$$\int_{-\ell}^{+\ell}\frac{dz}{(y^2 + z^2)^{3/2}} = \left.\frac{z}{y^2\sqrt{y^2 + z^2}}\right|_{-\ell}^{+\ell} = \frac{2\ell}{y^2\sqrt{y^2 + \ell^2}}$$

Substituting this into our expression for E_y gives

$$E_y = \frac{1}{2\pi\epsilon_0}\frac{\lambda}{y}\frac{\ell}{\sqrt{\ell^2 + y^2}}$$

Performing the integration of the z component of **E** gives $E_z = 0$ (Exercise 27). This result is expected from the symmetry of the charge distribution with respect to the xy plane. For every element of charge with a positive z coordinate there is a corresponding element with a negative z coordinate, and their contributions to E_z cancel.

We may generalize our result to include any point in the xy plane by noting that E must have azimuthal symmetry about the z axis, similar to that of the dipole in Example 20-5. Thus

$$E = \frac{1}{2\pi\epsilon_0}\frac{|\lambda|}{R}\frac{\ell}{\sqrt{\ell^2 + R^2}}$$

where $R = \sqrt{x^2 + y^2}$. The direction of **E** at any point in the xy plane is directly away from the origin (assuming λ is positive).

SELF-TEST 20-6. A thin rod of length 1.6 m has a positive charge of 32 nC uniformly distributed along its length. *(a)* What is the magnitude E of the field at a point in the rod's perpendicular bisector plane and at a distance of 2.0 m from the center of the rod? *(b)* Is the direction of this field toward or away from the rod? ***ANSWERS:*** *(a)* $E = 67$ N/C; *(b)* away.

EXAMPLE 20-7

A charged ring. Determine **E** at points along the axis of a charged circular ring of radius a and charge Q. The charge on the ring is distributed uniformly and is narrow enough to be considered a line charge, similar to the way the mass is distributed on a hula hoop.

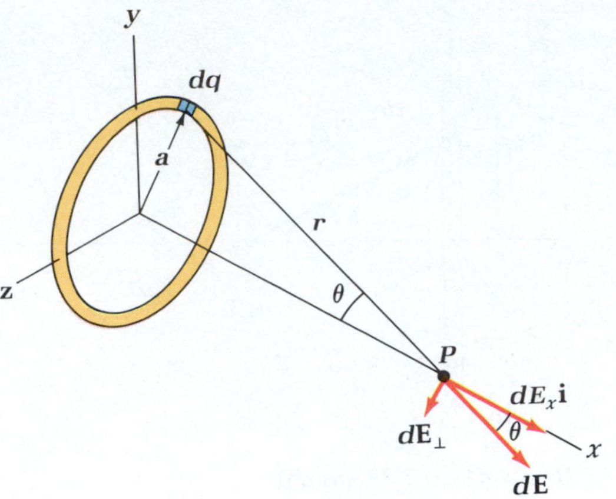

Figure 20-15. Example 20-7: Electric field **E** on the axis of a uniformly charged ring. The ring is centered at the origin and contained in the *yz* plane. By symmetry, the component of **E** perpendicular to the *x* axis is zero.

Solution. Figure 20-15 shows the ring centered at the origin and contained in the *yz* plane. The infinitesimal electric field $d\mathbf{E}$ due to charge dq can be divided into two components, dE_x parallel to the *x* axis and $dE_\perp$ perpendicular to the *x* axis. The symmetry of the charge distribution requires that $\int d\mathbf{E}_\perp = 0$ because elements of charge on opposite sides of the circle produce infinitesimal components of electric field $dE_\perp$ that cancel one another. That is, the electric field must point along the *x* axis. The field's axial component is

$$E = \int dE_x = \int dE \cos\theta = \frac{1}{4\pi\epsilon_0}\int \frac{dq}{r^2}\cos\theta$$

The factor $\cos\theta/r^2$ can be factored out of the integral because θ and r are the same for each charge element dq. This gives

$$E_x = \frac{\cos\theta}{4\pi\epsilon_0 r^2}\int dq = \frac{Q\cos\theta}{4\pi\epsilon_0 r^2}$$

where we have used $\int dq = Q$. We can write E_x in terms of x and a by noting that from the figure, $\cos\theta = x/\sqrt{x^2 + a^2}$ and $r^2 = x^2 + a^2$. Thus

$$E_x = \frac{Qx}{4\pi\epsilon_0 (x^2 + a^2)^{3/2}}$$

Figure 20-16 shows a graph of E_x versus x for the case where $a = 0.50$ m and $Q = +11.1$ nC. Are you surprised to see E_x negative when x is negative? The reason E_x is positive for positive x and negative for negative x is that the field is directed away from the positively charged ring. Note that the field is zero at the center of the ring. For comparison, we also show E_x for a point charge of $+11.1$ nC at the origin. The field of the ring approaches that of the point charge when $|x|$ becomes much greater than a.

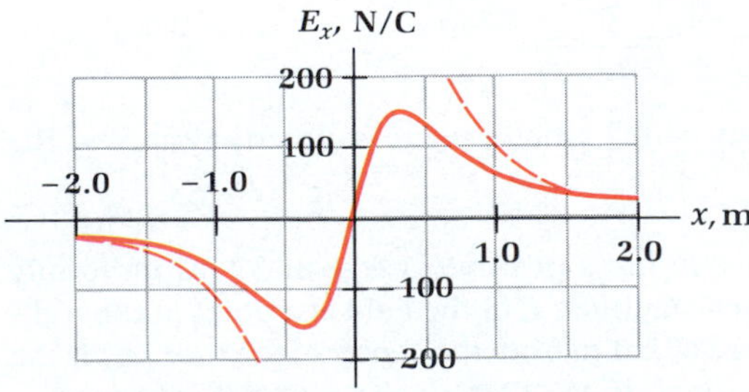

Figure 20-16. Example 20-7: Field component E_x versus x on the axis of the charged ring shown in Fig. 20-15, with $Q = 11.1$ nC and $a = 0.50$ m. The dashed curve is E_x for a charged particle with the same charge as the ring and located at the origin. These two field components approach the same value as $|x|$ becomes large compared with a.

SELF-TEST 20-7. A ring of radius 0.40 m has a positive charge of 20 nC uniformly distributed around it. What is the magnitude E of the field at a point on the ring's axis at a distance of 0.30 m from the center? ***ANSWER:*** $E = 430$ N/C.

EXAMPLE 20-8

A disk-shaped surface charge. Sometimes the charge on an object is spread over the object's surface as a thin layer, like a coat of paint. Such a charge distribution is called a ***surface charge.*** The ***surface charge density*** σ of a uniform surface charge is the charge Q divided by the area A of the surface on which the charge resides: $\sigma = Q/A$. Use the result of the previous example to find **E** on the axis of a thin, disk-shaped charge distribution of radius R_0. The charge distribution is uniform, similar to the mass distribution for a phonograph record (without the hole).

Solution. Since the charge distribution is uniform and shaped as a thin disk, we treat it as a surface charge with $\sigma = Q/\pi R_0^2$, where πR_0^2 is the area of the disk. We divide the disk into rings with infinitesimal width da (Fig. 20-17), so that the area of a ring of radius a and width da is its circumference $2\pi a$ times da, and its charge is $dq = \sigma 2\pi a\, da$. From the result of the previous example, dE_x on the axis of a ring with radius a and charge $dq = \sigma 2\pi a\, da$ is

$$dE_x = \frac{(\sigma 2\pi a\, da)x}{4\pi\epsilon_0(x^2 + a^2)^{3/2}}$$

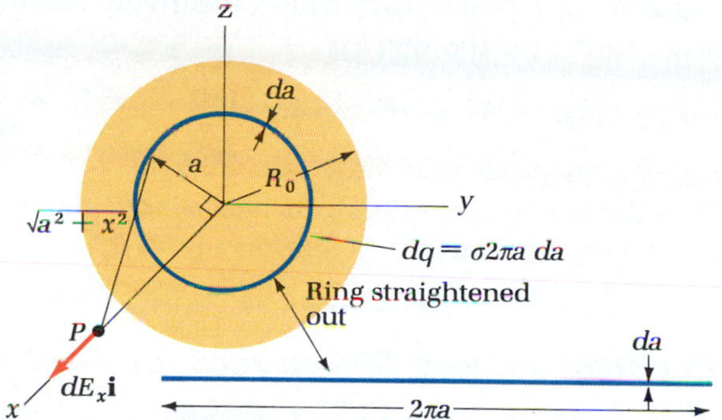

Figure 20-17. Example 20-8: Electric field **E** on the axis of a uniformly charged disk. The disk is centered at the origin and contained in the yz plane. The field is the sum of the infinite number of infinitesimal fields due to charged rings, with $dq = \sigma 2\pi a\, da$. We show later that if the radius of the disk is large compared with the distance from its center to P, $R_0 \gg |x|$, then $E \approx \sigma/2\epsilon_0$.

Integrating this expression from $a = 0$ to $a = R_0$ adds all the contributions to E_x due to each ring from radius $a = 0$ to $a = R_0$:

$$E_x = \frac{2\pi\sigma x}{4\pi\epsilon_0}\int_0^{R_0} \frac{a\, da}{(x^2 + a^2)^{3/2}}$$

Evaluating the integral gives

$$E_x = \frac{\sigma x}{2\epsilon_0}\left(\frac{1}{\sqrt{x^2}} - \frac{1}{\sqrt{x^2 + R_0^2}}\right)$$

Since $1/\sqrt{x^2}$ is always larger than $1/\sqrt{x^2 + R_0^2}$, the algebraic sign of E_x is the same as x (assuming σ is positive). Thus E_x is positive for positive x and negative for negative x. That is, **E** points away from a positively charged disk.

SELF-TEST 20-8. A positive charge of 20 nC is distributed uniformly in the shape of a thin disk of radius 0.40 m. What is the magnitude E of the field at a point on the disk's axis at a distance of 0.30 m from the center? Compare your answer with that of the previous self-test. Explain why E produced by the disk is larger. ***ANSWER:*** 900 N/C.

Useful Approximations for the Line Charge and the Surface Charge

Some useful approximations can be developed from the solutions to Examples 20-6 and 20-8. First, from Example 20-6, E in the bisector plane of a uniform line charge is

$$E = \frac{1}{2\pi\epsilon_0}\frac{|\lambda|}{R}\frac{\ell}{\sqrt{\ell^2 + R^2}}$$

Consider the field due to a very long line charge at a point near the charge but far from its ends. Then $\ell \gg R$, so that $\ell/\sqrt{\ell^2 + R^2} \approx \ell/\sqrt{\ell^2} = 1$, and

Electric field near a long, straight line charge, far from its ends

$$E \approx \frac{1}{2\pi\epsilon_0}\frac{|\lambda|}{R} \qquad (R \ll \ell) \qquad (20\text{-}8)$$

The field is directed away from the line charge (assuming λ is positive). An important feature of this field is that it falls off as $1/R$. If $E = 80$ N/C at $R = 10$ cm, then at $R = 20$ cm, $E \approx 40$ N/C.

Similarly, the solution to Example 20-8 can be used to find the approximate field near a large plane sheet with uniform surface charge density by finding E_x at points where $|x| \ll R_0$. In Exercise 33, you are asked to show that this gives

$$E_x \approx \frac{\sigma}{2\epsilon_0}\frac{x}{\sqrt{x^2}} \qquad (|x| \ll R_0)$$

In this expression, the factor $x/\sqrt{x^2}$ is $+1$ for $x > 0$ and -1 for $x < 0$. This means that for positive σ, E_x is positive when x is positive and E_x is negative when x is negative, or **E** points away from a positively charged sheet. The approximate field magnitude near the sheet is

Electric field near a large plane sheet of charge, far from its edges

$$E \approx \frac{|\sigma|}{2\epsilon_0} \qquad (|x| \ll R_0) \qquad (20\text{-}9)$$

Thus the field is almost uniform near a large plane sheet of charge. If $E = 80$ N/C at $x = 10$ cm, then at $x = 20$ cm, $E \approx 80$ N/C.

Problem-Solving Techniques for Finding the Field E

Using Coulomb's law to find the field **E** produced by a charge distribution is essentially a problem in vector addition.

1. Draw a diagram showing the charge distribution on the backdrop of a coordinate frame. If you are to find **E** at a particular point, include that point in your diagram (see Fig. 20-11). If you are to find **E** as a function of coordinates, select a representative point and include it in your diagram (see Fig. 20-13). If the charge distribution is continuous, select a representative charge element (see Fig. 20-14). Also, show the values of separation distances on your sketch.

2. Use the Coulomb's-law form of the electric field to write each contribution to the field: $\mathbf{E} = (q/4\pi\epsilon_0 r^2)\hat{\mathbf{r}}$ for a particle and $d\mathbf{E} = (dq/4\pi\epsilon_0 r^2)\hat{\mathbf{r}}$ for a continuous distribution of charge.

3. Perform the vector sum. If the charge distribution is an array of particles, this is simply a matter of collecting terms for each component. If the charge distribution is continuous, you will need to perform an integral. Before performing the integral, be certain you clearly distinguish between quantities that vary during the integration and those that do not.

20-6 LINES OF THE ELECTRIC FIELD

Lines of the electric field provide an aid for visualizing the field; they are essentially a map of the field. Although we draw field lines on a two-dimensional sheet of paper or a blackboard, we visualize their existence in three-dimensional space. Later we shall use field lines to describe magnetic fields as well. The concept was originated by the great British experimental physicist, Michael Faraday (1791–1867).

Figure 20-18 shows electric field lines for a few cases. A line is drawn such that **E** is tangent to the line at each point on the line, and arrowheads indicate the direction of the field. For example, near a point charge the lines are radial (Fig. 20-18*a*

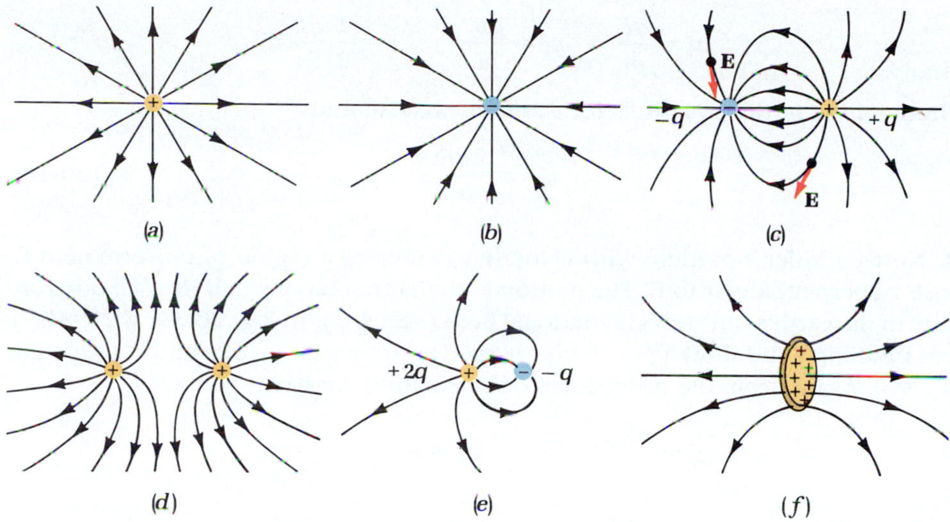

Figure 20-18. Electric field lines. *(a)* Particle with positive charge. *(b)* Particle with negative charge. *(c)* Dipole. *(d)* Two particles with equal positive charge. *(e)* Two particles with charges $+2q$ and $-q$. *(f)* Uniformly charged disk.

and *b*); they are directed away from a positive charge and toward a negative charge. To illustrate how **E** is tangent to field lines, **E** is shown at two points on the field lines of a dipole in Fig. 20-18*c*.

In a given drawing, the spacing of the lines indicates the magnitude of the field. In regions where the lines are close together, or dense, E is large, and where they are far apart, E is small. As it turns out, the density of lines is proportional to E, a fact which can be proved with Gauss's law, the subject of the next chapter.

Since the density of lines is proportional to E, the number of lines directed away from a positive charge or directed toward a negative charge is proportional to $|q|$. As an exercise, count the lines directed away from and directed toward each particle in Fig. 20-18*e*. Are the numbers of these lines correct for the stated values of the charges?

A uniform field is represented by field lines that are equally spaced, straight, and parallel. In the previous section we found that the field near a uniformly charged disk, but not near its edges, is almost uniform. The field lines of a uniformly charged disk are shown in Fig. 20-18*f*, and you can see that near the disk and away from its edges the lines are drawn so that they appear approximately equally spaced, straight, and parallel.

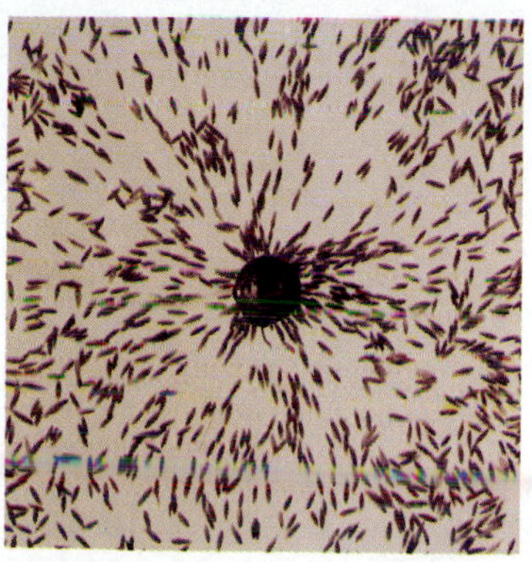

Grass seeds floating on the surface of oil tend to align with an electric field. Thus, the seeds provide a two-dimensional representation of field lines in three dimensions. Here the field is produced by a small charged sphere, and the seeds demonstrate that the field around an isolated point charge is spherically symmetric.

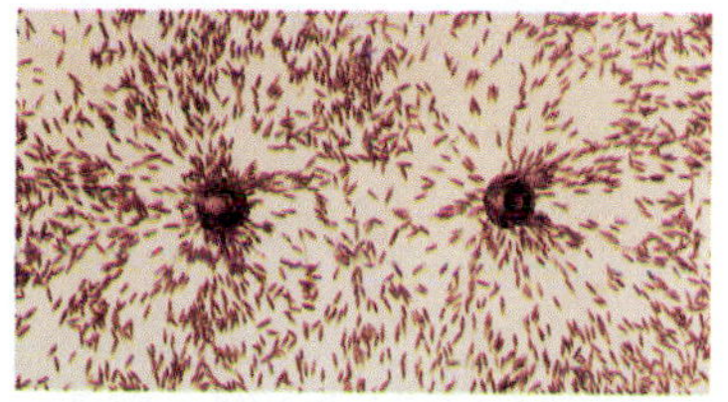

Grass-seed demonstration of the appearance of electric field lines around two equal point charges. The point charges are represented by two small rings with charge of the same sign and magnitude. As you can see, the field in the region midway between the charges is small. Why?

20-7 CHARGED PARTICLE IN A UNIFORM ELECTRIC FIELD

In Sec. 20-5 we determined the electric field produced by several charge distributions. Once the field is known, we can use Eq. 20-3, $\mathbf{F} = q\mathbf{E}$, to examine the effect of the field on a particle. If the electric force is the only significant force on the particle, then $q\mathbf{E}$ is the net force and Newton's second law gives

$$q\mathbf{E} = m\mathbf{a} \qquad \text{or} \qquad \mathbf{a} = \frac{q\mathbf{E}}{m}$$

We consider two specific cases: (1) a particle that is initially at rest in a uniform field and (2) a particle that is projected with velocity $\mathbf{v}_0$ into a uniform field, with $\mathbf{v}_0$ perpendicular to **E**.

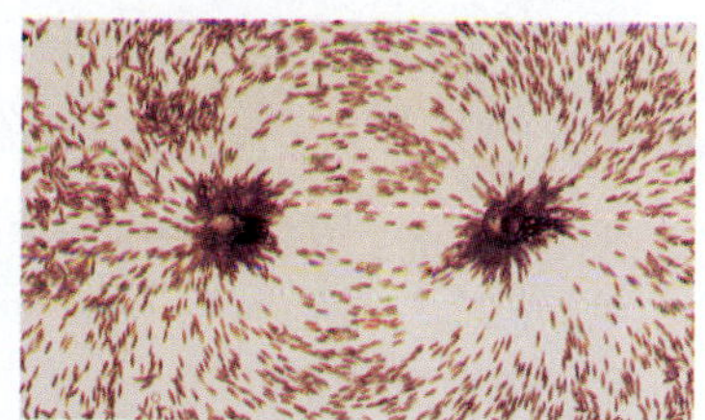

Grass-seed demonstration of the appearance of electric field lines around a dipole. Two small rings with charge of the same magnitude but opposite sign represent the dipole. The field in the region midway between the rings is large. Why?

1. A charged particle released from rest in a uniform electric field will move with constant acceleration along a line parallel to **E** in the same way a rock released from rest in a uniform gravitational field falls vertically downward along a line parallel to **g**. If we place the origin at the release point with the x axis in the direction of **E** and set $t = 0$ when $x = 0$, then the procedures of kinematics (Chap. 3) give

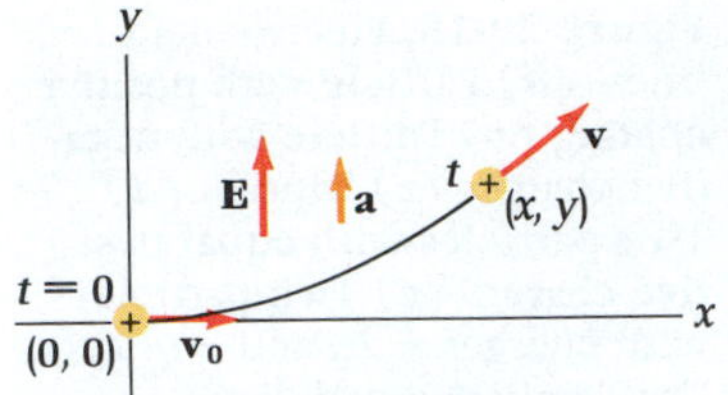

Figure 20-19. A charged particle in a uniform electric field travels in a parabolic path. The particle shown has a positive charge.

$$a_x = \frac{qE}{m} \qquad v_x = \frac{qE}{m}t \qquad x = \frac{1}{2}\frac{qE}{m}t^2$$

Eliminating t in the equations for v_x and x, we also find

$$v_x^2 = \frac{2qE}{m}x \tag{20-10}$$

2. Now consider a particle with velocity $\mathbf{v}_0$ entering a region of uniform field $\mathbf{E}$, with $\mathbf{v}_0$ perpendicular to $\mathbf{E}$. The motion is similar to that of a ball thrown horizontally in the earth's uniform gravitational field (Sec. 4-2). In Fig. 20-19, we let the y axis be along $\mathbf{E}$ and show a positively charged particle with velocity $v_0\mathbf{i}$ at the origin at $t = 0$. Again, from the procedures of kinematics, we have

$$a_y = \frac{qE}{m} \qquad a_x = 0 \qquad a_z = 0$$

$$v_y = \left(\frac{qE}{m}\right)t \qquad v_x = v_0 \qquad v_z = 0$$

$$y = \frac{1}{2}\left(\frac{qE}{m}\right)t^2 \qquad x = v_0 t \qquad z = 0$$

Thus the motion is contained in the xy plane. Eliminating t in the equations for y and x gives the parabolic path of the particle:

$$y = \frac{1}{2}\frac{qE}{mv_0^2}x^2 \tag{20-11}$$

The path shown in Fig. 20-19 is that of a particle with positive charge. If the charge is negative, then a_y is negative and the path curves downward rather than upward.

EXAMPLE 20-9

An electron linearly accelerated from rest. Determine the speed of an electron that travels a distance of 8.3 mm after starting from rest in a uniform field of 4.0×10^3 N/C.

Solution. We let the x axis point along $\mathbf{E}$ so that $\mathbf{E} = E\mathbf{i}$. From Eq. 20-10, the speed v is

$$v = \sqrt{v_x^2} = \sqrt{\frac{2qE}{m}x}$$

In this equation, x is negative because $q\ (= -e)$ is negative. The direction of the force on the negatively charged electron is opposite the field direction. Since the electron started at rest from the origin, its x coordinate is negative throughout the motion. Substituting the numerical values gives

$$v = \sqrt{\frac{2(-1.6 \times 10^{-19}\text{ C})(4.0 \times 10^3\text{ N/C})}{9.1 \times 10^{-31}\text{ kg}}(-8.3\text{ mm})}$$
$$= 3.4 \times 10^6\text{ m/s}$$

SELF-TEST 20-9. Determine the speed of a proton after it travels the same distance (starting from rest) as the electron in the above example and in a uniform field of the same magnitude. Compare your answer with that for the electron. What is the ratio of these speeds? ***ANSWER:*** $v = 8.0 \times 10^4$ m/s; the ratio is $43 \approx \sqrt{1800}$.

EXAMPLE 20-10

A deflected electron. An electron traveling horizontally with a speed of 3.4×10^6 m/s enters a region of uniform electric field that is directed upward, with $E = 520$ N/C. The field extends horizontally for a distance of about 45 mm (Fig. 20-20). Determine *(a)* the vertical displacement and *(b)* the velocity of the electron as it emerges from the region of the field.

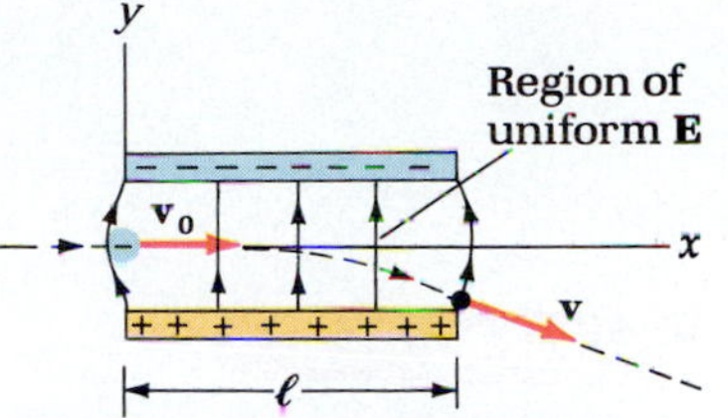

Figure 20-20. Example 20-10: An electron is deflected downward while passing through a region of uniform **E** that points upward.

Solution. *(a)* Figure 20-20 shows the origin of coordinates at the point where the electron enters the field and it leaves the field at $x = \ell$, where $\ell = 45$ mm. Using Eq. 20-11, we find that the electron's y coordinate as it emerges from the field is

$$y = \frac{\frac{1}{2}qE\ell^2}{mv_0^2} = \frac{\frac{1}{2}(-1.6 \times 10^{-19}\ \text{C})(520\ \text{N/C})(45 \times 10^{-3}\ \text{m})^2}{(9.1 \times 10^{-31}\ \text{kg})(3.4 \times 10^6\ \text{m/s})^2}$$
$$= -8.0 \times 10^{-3}\ \text{m} = -8.0\ \text{mm}$$

The electron is deflected downward 8.0 mm by the electric field. *(b)* The x component of the velocity remains constant at 3.4×10^6 m/s, but the y component changes. The time t_1 required for the electron to pass through the field is given by

$$\ell = v_0 t_1 \qquad \text{or} \qquad t_1 = \frac{\ell}{v_0}$$

Substituting this into the equation for v_y gives

$$v_y = \frac{qEt_1}{m} = \frac{qE\ell}{mv_0} = \frac{(-1.6 \times 10^{-19}\ \text{C})(520\ \text{N/C})(4.5 \times 10^{-2}\ \text{m})}{(9.1 \times 10^{-31}\ \text{kg})(3.4 \times 10^6\ \text{m/s})}$$
$$= -1.2 \times 10^6\ \text{m/s}$$

The electron emerges from the electric field with a downward component to its velocity, so that its downward displacement increases as it travels beyond the region of the field in a straight line.

SELF-TEST 20-10. Consider a proton traveling at the same speed as the electron in the above example as it enters the field. Would the proton be deflected more, less, or the same as the electron? Would it be deflected in the same direction or in the opposite direction?
ANSWER: A proton would be deflected less than an electron and in the opposite direction.

Cathode-Ray Tube

Figure 20-21 shows schematically a device called a cathode-ray tube (CRT). In a CRT, electrons are first accelerated (as in Example 20-9) and then deflected (as in Example 20-10). A CRT is used for the video display in television sets, computer monitors, oscilloscopes, and so forth.

The electrons are emitted from a heated filament and accelerated by a horizontal electric field set up by charged plates in the "electron gun." The electrons emerging from a hole in one of the plates form a beam, like bullets projected from the muzzle of a machine gun. The beam then passes through a region of uniform electric field perpendicular to the beam direction. This deflecting field is produced by charged metal plates called *deflection plates.* The deflecting field directs the beam to the intended point on the face of the fluorescent screen, producing a luminous spot. The deflection plates shown in the figure are the vertical deflection plates; they control the vertical position of the luminous spot. Similarly, there are

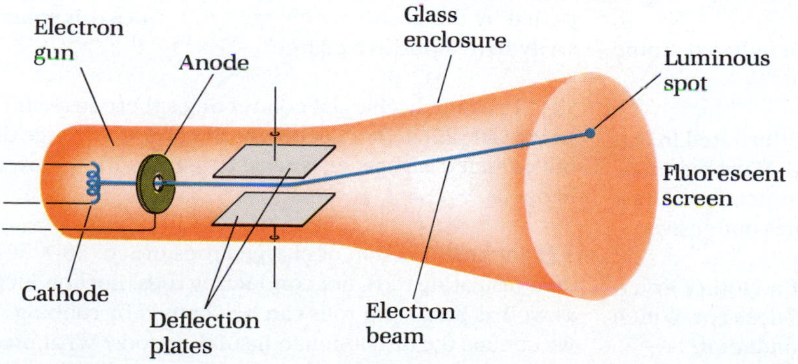

Figure 20-21. Schematic diagram of a CRT, showing the electron gun and a set of deflection plates.

horizontal deflection plates, which are not shown. Some CRTs, such as those used in television sets, deflect the beam, with a magnetic field rather than an electric field. We discuss magnetic deflection in Chap. 26.

SUMMARY

Section 20-1. Electric Charge and Matter

Electric charge is a fundamental property of matter. The charge of a proton is $+e$ and the charge of the electron is $-e$, where $e = 1.602 \times 10^{-19}$ C. A macroscopic object possesses charge q if it has an imbalance in its proton and electron populations: $q = (N_p - N_e)e$.

Section 20-2. Insulators and Conductors

Most materials can be categorized as either conductors or insulators. A conductor readily allows charge carriers to flow, while an insulator does not readily allow charge carriers to flow.

Section 20-3. Coulomb's Law

Coulomb's law gives the electric force $\mathbf{F}_{ab}$ exerted by charged particle a on charged particle b:

$$\mathbf{F}_{ab} = \frac{1}{4\pi\epsilon_0} \frac{q_a q_b}{r^2} \hat{\mathbf{r}} \qquad (20\text{-}1)$$

This force is an inverse-square force that is attractive if the charges have opposite sign and repulsive if the charges have the same sign.

Section 20-4. The Electric Field

The electric field $\mathbf{E}$ due to a charge distribution is the electric force $\mathbf{F}$ exerted by the distribution on a test particle divided by the charge of the test particle:

$$\mathbf{E} = \frac{\mathbf{F}}{q_0} \qquad (\text{small } q_0) \qquad (20\text{-}2)$$

Section 20-5. Calculating the Electric Field

For a point charge,

$$\mathbf{E} = \frac{1}{4\pi\epsilon_0} \frac{q}{r^2} \hat{\mathbf{r}} \qquad (20\text{-}4)$$

where $\hat{\mathbf{r}}$ is a unit vector directed from q to the point P at which $\mathbf{E}$ is evaluated. The field due to a continuous distribution of charge is

$$\mathbf{E} = \frac{1}{4\pi\epsilon_0} \int \frac{dq}{r^2} \hat{\mathbf{r}} \qquad (20\text{-}7)$$

Section 20-6. Lines of the Electric Field

The spatial characteristics of an electric field can be illustrated with electric field lines. By their direction and spacing, field lines indicate both the direction and magnitude of the field.

Section 20-7. Charged Particle in a Uniform Electric Field

If the electric force is the only significant force on a particle, then Newton's second law gives $\mathbf{a} = q\mathbf{E}/m$. When a charged particle moves in a uniform field, its motion is described by constant-acceleration kinematics.

QUESTIONS

1. Suppose the magnitudes of the charge on the electron and proton were not the same, but differed by, say, 0.1 percent. Would the world be much different? Explain.

2. Suppose the signs of the charge on the proton and electron were reversed, positive for the electron and negative for the proton. Would the world be much different? Explain.

3. If you charge a balloon by rubbing it with wool, it will adhere to a wall. Why? Is there a net charge on the wall? (*Hint:* Examine Fig. 20-3.) The balloon will eventually fall. Why?

4. The net charge on the sphere in Fig. 20-3*b* is zero. Is there a net electric force on the sphere? If so, explain why.

5. The sphere in Fig. 20-3*b* is said to have an induced dipole moment. Explain why this name is appropriate.

6. In the charging-by-induction procedure illustrated in Fig. 20-3, suppose the rod were positively charged. What would be the sign of the charge on the sphere after the procedure? Draw figures similar to those in Fig. 20-3 to support your answer.

7. After two pairs of socks are taken out of a clothes dryer, pair *A* sticks together for a long time but pair *B* does not. Which pair is made of a material that is the better conductor?

8. Compare the mass in Newton's law of universal gravitation with the charge in Coulomb's law. How are they similar and how are they different?

9. Suppose you have two metal spheres on insulating stands similar to the one shown in Fig. 20-3, and you wish to give them equal and opposite charge using a rod with positive charge. Using sketches, describe how to proceed. How do you give them equal positive charge? How do you give them equal negative charge?

10. A conducting sphere suspended from a string is attracted to a positively charged rod. Does the sphere necessarily have a negative charge? Another suspended conducting sphere is repelled by the positively charged rod. Does this sphere necessarily have a positive charge?

11. A positively charged conducting sphere suspended from a string is repelled by a positively charged rod at large distances, but attracted at short distances. Using sketches, show how this happens.

12. In the discussion of charged rods in Sec. 20-1, we considered insulating rods, not conducting rods. But conducting rods as well as insulating rods can be charged by rubbing. Why did we confine the discussion to insulating rods? What precautions

would you need to take in order to perform the charging experiments with a conducting rod?

13. Can the following quantities be described as fields? If so, are they scalar fields or vector fields?
(a) Money in the bank
(b) The water velocity in a stream
(c) The mass density of concrete

14. To determine **E** due to a positively charged conducting sphere at a point *P* near the sphere, we measure the force $\mathbf{F}_1$ on a test particle with positive charge q_1 placed at *P*. However, q_1 is not sufficiently small, and consequently the charge distribution on the sphere is significantly affected by the presence of particle 1. How will F_1/q_1 compare with *E*? If we measure the force $\mathbf{F}_2$ on test particle 2 with positive charge $q_2 = \frac{1}{2}q_1$, how does $\frac{1}{2}F_1$ compare with F_2? Which is closer to *E*, F_1/q_1 or F_2/q_2? Reconsider these questions when the test particles have negative charge rather than positive.

15. Suppose $E = 100$ N/C at $r = 40$ cm from a point charge. What is *E* at $r = 20$ cm?

16. Consider *E* in the perpendicular bisector plane of a dipole at points such that $R \gg a$, where *R* is the distance from the dipole and $2a$ is the charge separation. If $E = 100$ N/C at $R = 40$ cm, what is *E* at $R = 20$ cm?

17. At a distance $R = 40$ cm from a long, straight, uniform line charge, far from its ends, $E = 100$ N/C. What is *E* at $R = 20$ cm?

18. At a distance $|x| = 40$ cm from a large plane sheet of uniform surface charge, far from its edges, $E = 100$ N/C. What is *E* at $|x| = 20$ cm?

19. Figure 20-16 shows a graph of E_x versus *x* on the axis of a positively charged ring centered at the origin. Suppose a proton, constrained to move along the *x* axis, is released from rest at $x = 1.0$ m. *(a)* What sort of motion will the proton undergo? Similarly consider *(b)* a proton released at $x = -1.0$ m, *(c)* an electron released at $x = 1.0$ m, *(d)* an electron released at $x = 0.1$ m.

20. Since protons are crowded together in atomic nuclei, there must be another type of force in addition to gravitational and electric. This is evidence for the existence of a nuclear force that acts on protons (and, as we shall see later, neutrons). Do you expect that this nuclear force falls off more strongly with distance than $1/r^2$ or less strongly with distance than $1/r^2$? Explain. (*Hint:* The nuclear force between protons is much weaker than the electric force when the protons are separated by distances much greater than a nuclear diameter.)

21. When released from rest in an electric field, a positively charged particle will begin to move along a field line (assuming the electric force is the net force). *(a)* Will the path of the particle follow the field line if the field line is straight? *(b)* Will the path of the particle follow the field line if the field line is not straight? If it does not follow the field line, then will its path bend more or less than the field line?

22. If we draw 10 field lines directed away from a point charge of $+2.5\ \mu$C, how many lines should we draw directed toward a $-1.5\ \mu$C point charge in the same diagram?

23. Explain why lines of the electric field cannot cross. Suppose the lines do cross at some point; what would this say about the electric force on a charged particle at that point?

24. Sometimes the term "quantum" is used to denote the smallest value that a quantized quantity can have. For example, the quantum of cash in the monetary system of the United States is the penny. What is the quantum of electric charge? Give some examples of quantities that are quantized, and some examples of quantities that are not quantized. Is there a quantum of mass? Can you ever be sure that a seemingly unquantized quantity is really unquantized?

25. Can the electric field contributions due to two point charges of the same sign separated by a distance ℓ cancel each other to give $\mathbf{E} = 0$ at any point? If so, describe the location of this point. If so, and if the charges are of different magnitude, is the point of zero field nearer the large charge or the small one?

26. Can the electric field contributions due to two point charges of opposite sign and different magnitude separated by a distance ℓ cancel each other to give $\mathbf{E} = 0$ at any point? If so, describe the location of this point. If so, is the point of zero field nearer the charge of large magnitude or the one of small magnitude?

27. Consider a dipole in a uniform field **E**. Is there a net electric force on the dipole? If so, what is the direction of the force relative to **E**? If the dipole moment **p** is oriented perpendicular to **E**, is there a torque on the dipole? If so, does this torque tend to align the moment parallel to **E** or opposite **E**?

28. Suppose a dipole is in an electric field and its moment **p** points in the direction of **E**. Further, suppose **E** is nonuniform such that its magnitude increases in the direction of **E**. Is there a net electric force on the dipole? If so, is the net electric force in the direction of **E** or opposite **E**? Reconsider this question for the case where the magnitude of **E** decreases in the direction of **E**.

29. Complete the following table:

Symbol	Represents	Type	SI Unit
q	Electric charge		
ϵ_0			
e			C
$\mathbf{E}$		Vector	
$\mathbf{p}$			
λ			
σ			

EXERCISES

Section 20-2. Insulators and Conductors

1. Draw a figure similar to Fig. 20-3 for the case where the rod has a positive charge.

2. Two uncharged conducting spheres on insulating stands are brought into contact (so that they touch one another) and then a charged rod is brought nearby (Fig. 20-22). *(a)* Draw a figure similar to Fig. 20-22 and use positive and negative signs to show where excess charge resides on the spheres. *(b)* Suppose the spheres are separated while the rod is nearby, and the rod is removed far away. One of the spheres is thereafter found to have a charge of −4 nC. Which sphere has this negative charge? What is the charge on the other sphere? What would the charge on the spheres be if the rod was far removed before they were separated?

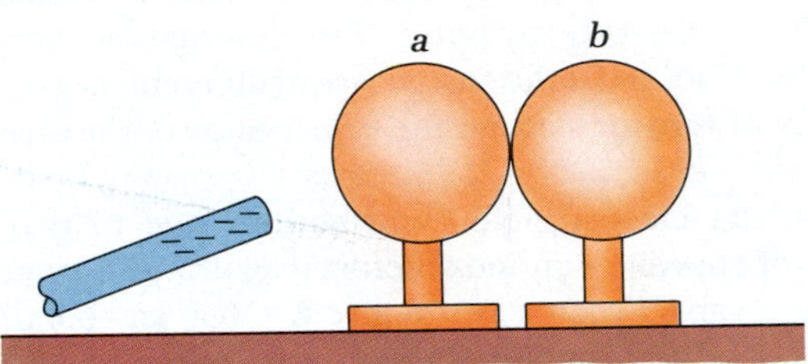

Figure 20-22. Exercise 2.

Section 20-3. Coulomb's Law

3. Two charged particles exert an electric force of magnitude 4.0 μN on each other at a distance of 0.10 m. Construct a table listing r and F for values of r between 0.10 and 0.50 m at intervals of 0.10 m. Plot your data on a graph similar to Fig. 20-6.

4. Two charged particles exert a 3.6-μN electric force of attraction on one another at a separation of 120 mm. If one of the particles has a charge of +1.2 nC, what is the charge on the other?

5. *(a)* If an object has a charge of +1.0 nC, how many of its electrons are missing? *(b)* If an object has a charge of −3.0 nC, how many extra electrons does it contain?

6. Estimate the total number of electrons in a 0.5-kg solid block by assuming the material has about the same number of protons and neutrons. If the block has an excess charge of +10 nC, what is the fraction of missing electrons?

7. A unit of charge called a *faraday* is the charge on Avogadro's number ($N_0 = 6.02 \times 10^{23}$) of protons. *(a)* Determine the conversion factor (C/faraday) for changing a charge from faradays to C. *(b)* Convert a charge of 0.04 faraday to C.

8. *(a)* The two protons in the H_2 molecule are about 10^{-10} m apart. What is the magnitude of the electric force one of them exerts on the other? *(b)* The two protons in the helium nucleus are about 10^{-15} m apart. What is the magnitude of the electric force one of them exerts on the other? *(c)* What is the ratio of the force magnitude found in part *(b)* to that found in part *(a)*?

9. Three particles with charge $q_a = +14$ nC, $q_b = -26$ nC, and $q_c = +21$ nC are arranged along a straight line (Fig. 20-23). Particle b is between a and c, 120 mm from a and 160 mm from c. *(a)* Determine F_{ab}, F_{cb}, and the magnitude of the net force on b. Is the net force on b toward a or c? *(b)* Determine F_{ac}, F_{bc}, and the magnitude of the net force on c. Is the net force on c toward or away from b?

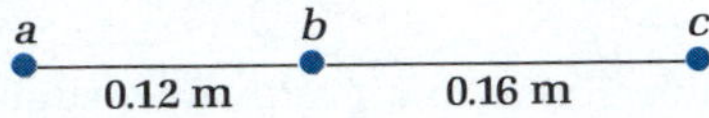

10. Three particles with charge $q_a = +14$ nC, $q_b = -26$ nC, and $q_c = +21$ nC are arranged on the corners of a right triangle (Fig. 20-24). Particle a is at the 90° corner a distance of 18 cm from b and 24 cm from c. Arrange a coordinate system with a at the origin, b along the x axis, and c along the y axis, and then determine the cartesian components and the magnitude of the net force on c.

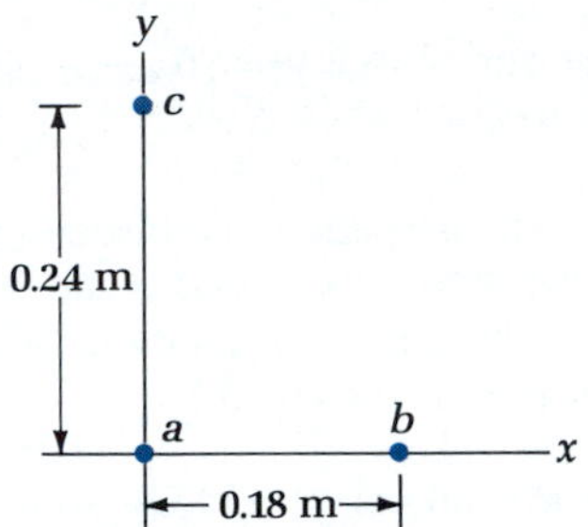

Figure 20-24. Exercise 10.

11. Three particles with equal charge q are at the corners of an equilateral triangle of side d (Fig. 20-25). What is the force on each of the particles?

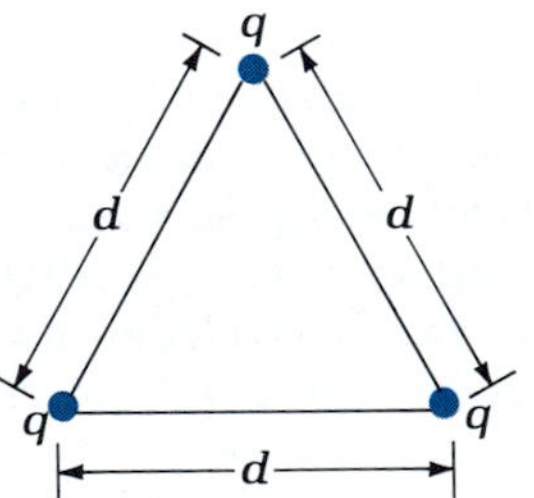

Figure 20-25. Exercise 11.

12. Three particles with equal positive charge q are at the corners of a square of side d (Fig. 20-26). *(a)* What is the magnitude of the force on a fourth particle of positive charge q_0 when it is placed at the center of the square? *(b)* What is the magnitude of the force on the fourth particle when it is placed at the vacant corner of the square?

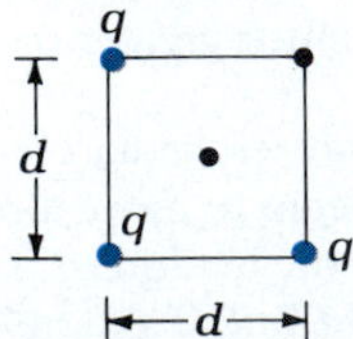

Figure 20-26. Exercise 12.

13. Two particles a and b have equal mass of 2.6 g and charge of equal magnitude q but opposite sign. Particle a is suspended from the ceiling on a thread of length 0.35 m and negligible

mass (Fig. 20-27). When a and b are separated by a horizontal distance of 0.25 m, a is in static equilibrium, with the string at an angle of 45° with the vertical. Determine q.

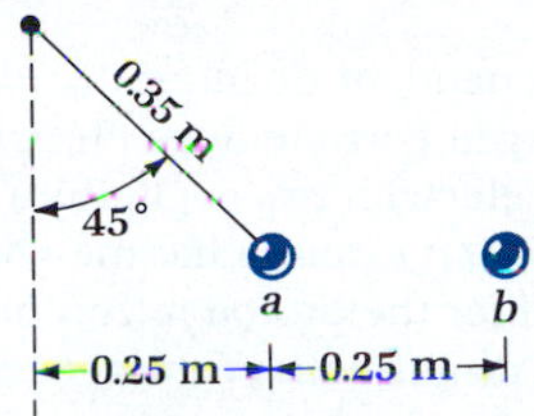

Figure 20-27. Exercise 13.

14. Two particles a and b are 0.20 m apart and have charge $q_a = +2.0$ nC and $q_b = +1.0$ nC. At what position relative to a and b can we place a third particle with charge q such that the force on it due to a and b is zero? *(b)* Rework part *(a)* for the case where $q_b = -1.0$ nC.

Section 20-4. The Electric Field

15. The electric force exerted by a charge distribution on a particle of charge +2.6 nC when placed at position P is vertically upward, with $F = 0.58\ \mu$N. *(a)* At P what is **E** due to the distribution? *(b)* What is the electric force exerted by this field on a particle placed at P with charge −13 nC? (*Note:* To work this problem with the information given, you must assume the charge on the particles is not large enough to disturb the distribution of charges.)

16. The electric force exerted by a distribution of charges on a particle of charge +1.4 nC at a particular position is $\mathbf{F} = (-0.24\ \mu\text{N})\mathbf{i} + (0.55\ \mu\text{N})\mathbf{j}$. *(a)* What is the electric field due to the distribution at that position? *(b)* What is the electric force exerted by this field on a particle with charge −25 nC placed at that position? (See the note at the end of Exercise 15.)

17. Within the earth's atmosphere there exists an electric field, of average magnitude 150 N/C, that points downward. *(a)* Determine the charge-to-mass ratio (in C/kg) an object must have in order to be suspended in midair by electric and gravitational forces. *(b)* Assume that the number of protons and neutrons are equal and determine the fraction of excess electrons within the object. Comment on the feasibility of such an experiment.

18. In attempting to evaluate at point P the electric field due to a system of charges, we measure the electric force on particle a ($q_a = +134$ nC) placed at P and find it to be $\mathbf{F}_a = (2.86\ \text{mN})\mathbf{i} + (3.41\ \text{mN})\mathbf{j}$. Particle a is then far removed, particle b ($q_b = +66.9$ nC) is placed at P, and the force on b is measured to be $\mathbf{F}_b = (1.46\ \text{mN})\mathbf{i} + (1.73\ \text{mN})\mathbf{j}$. *(a)* Did particle a's presence at point P have a measurable effect on the charge distribution of the system? *(b)* Estimate the electric field at P when both a and b are far removed.

Section 20-5. Calculating the Electric Field

19. A particle with charge +5.8 nC is placed at the origin. *(a)* Determine the cartesian components of the electric field due to the particle at positions (x, y, z) of (15 cm, 0, 0); (15 cm, 15 cm, 0); (15 cm, 15 cm, 15 cm); (10 cm, 20 cm, 0). *(b)* Determine E at the positions given in part *(a)*.

20. Rework the previous exercise for a particle with charge −5.8 nC at the origin.

21. *(a)* Determine at the position of particle b the electric field that is due to particles a and c ($q_a = +14$ nC and $q_c = +21$ nC) in Fig. 20-23. *(b)* Determine at the position of particle c the electric field that is due to particles a and b ($q_b = -26$ nC) in Fig. 20-23.

22. *(a)* Determine at the position of particle a the electric field that is due to particles b and c ($q_b = -26$ nC and $q_c = +21$ nC) in Fig. 20-24. *(b)* Determine at the position of particle c the electric field that is due to particles a and b ($q_a = +14$ nC) in Fig. 20-24.

23. Determine the electric field *(a)* at the center of the square and *(b)* at the vacant corner in Fig. 20-26.

24. A dipole centered at the origin consists of two particles, one with charge $+1.6 \times 10^{-19}$ C at $z = +0.41 \times 10^{-10}$ m and the other with charge -1.6×10^{-19} C at $z = -0.41 \times 10^{-10}$ m. *(a)* Determine **p**. *(b)* Determine the electric field in the xy plane that is due to the dipole at a distance of 1.0 μm from the origin. *(c)* Rework part *(b)* when the distance is 2.0 μm.

25. A dipole of moment $\mathbf{p} = 2aq\mathbf{k}$ is centered at the origin. Determine **E** along the z axis at points far from the dipole, $|z| \gg a$. (*Hint:* Use the binomial expansion.)

26. A uniform line charge with $Q = 24$ nC and length 120 mm ($\ell = 60$ mm) is centered at the origin and oriented along the z axis (Fig. 20-14). *(a)* Determine the linear charge density λ. *(b)* Determine E at (20 mm, 0, 0), and at (200 mm, 0, 0). *(c)* E is due to a very long uniform line charge ($\ell \gg 200$ mm) along the z axis, with the same linear charge density as found in part *(a)*. Find E at (20 mm, 0, 0) and at (200 mm, 0, 0) and compare your answers with those found in part *(b)*.

27. *(a)* Show that the integral that gives E_z for the line charge in Example 20-6 is

$$E_z = \frac{-\lambda}{4\pi\epsilon_0} \int_{-\ell}^{+\ell} \frac{z\,dz}{(y^2 + z^2)^{3/2}}$$

(b) Show that $E_z = 0$.

28. Determine E due to a very long line charge of linear charge density $\lambda = 300$ nC/m in the range of distances from $R = 0.10$ m to $R = 0.80$ m, using intervals of 0.10 m. Make a graph of E versus R with your data and sketch a continuous curve through the points.

29. A uniform ring of charge as shown in Fig. 20-15, with charge $Q = 11.1$ nC and radius $a = 0.50$ m, is centered at the origin and contained in the yz plane. Determine E_x at (1.0 m, 0, 0) and (−0.5 m, 0, 0). Check your answer with Fig. 20-16.

30. *(a)* E is a field due to a point charge, with charge Q located at the origin. Write an expression for E_x at points along the x axis. *(b)* Determine the ratio of your expression for E_x in part *(a)* and that due to the charged ring of Example 20-7. *(c)* Evaluate the ratio found in part *(b)* at $x = a$, $5a$, and $10a$. *(d)* Use the binomial expansion to show that E_x due to the charged

ring of Example 20-7 approaches that due to a point charge in the limit as $x \gg a$.

31. The dipole moment of a water molecule has a magnitude of 6.2×10^{-30} C·m. If we let the water molecule be replaced by a dipole with $p = 2ae$, where e is the magnitude of the electronic charge, what is a?

32. A thin, disk-shaped uniform surface charge with $Q = 28$ nC and radius $R_0 = 200$ mm is centered at the origin and contained in the yz plane (Fig. 20-17). *(a)* Determine the surface charge density σ. *(b)* Determine E_x along the x axis in both the positive and negative directions between $x = 40$ mm (-40 mm) and $x = 200$ mm (-200 mm) at intervals of 40 mm. *(c)* Make a graph of E_x versus x with these data and sketch a continuous curve through the points. *(d)* On the same graph, plot E_x versus x for a disk with the same charge density as found in part *(a)* but with a very large radius ($R_0 \gg |x|$).

33. Use the solution to Example 20-8 to show that for points near a large sheet with uniform surface charge density,

$$E_x = \frac{\sigma}{2\epsilon_0} \frac{x}{\sqrt{x^2}}$$

34. Use the binomial expansion to show that the electric field due to a disk-shaped surface charge (Example 20-8) approximates the field due to a point charge at distances large compared with the radius of the disk, $|x| \gg R_0$.

35. A rod in the shape of a semicircle with a uniform linear charge density is shown in Fig. 20-28. Develop an equation for the magnitude of the electric field at the center P in terms of the rod's charge Q and radius a.

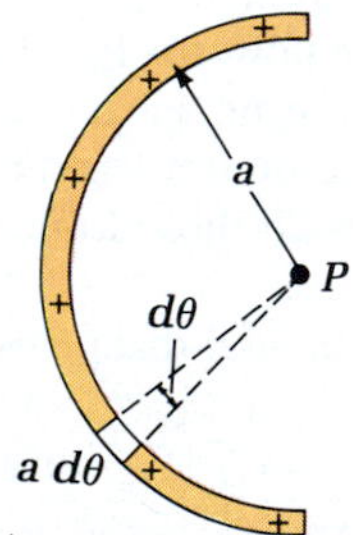

Figure 20-28. Exercise 35.

Section 20-6. Lines of the Electric Field

36. The negatively charged particle in Fig. 20-29 has a charge of 62 nC. What is the charge on the positively charged particle?

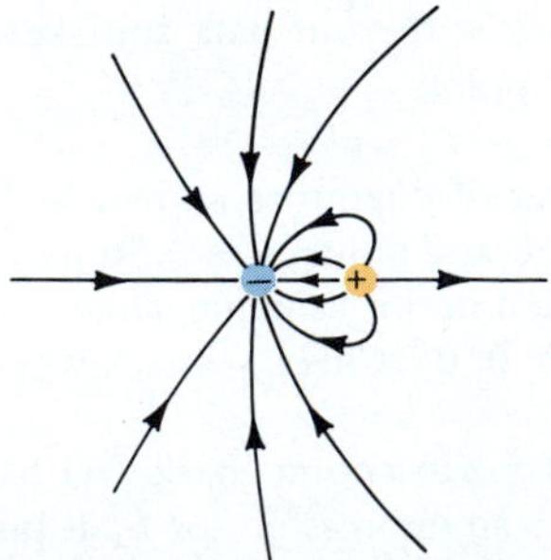

Figure 20-29. Exercise 36.

37. Draw a figure similar to Fig. 20-25 and sketch the field lines. Each of the three particles has the same positive charge.

38. Draw a figure similar to Fig. 20-14 and sketch the field lines in the yz plane. Assume that λ is positive.

Section 20-8. Charged Particle in a Uniform Electric Field

39. The particle-dependent quantity that determines the acceleration of a charged particle in an electric field (assuming forces other than electric are negligible) is the particle's charge-to-mass ratio. *(a)* Determine the charge-to-mass ratio for the electron and for the proton in coulombs per kilogram. *(b)* Determine the magnitudes of the acceleration of an electron and of a proton in a field of magnitude 1 N/C. *(c)* According to your findings in part *(b)*, what is the ratio of the proton's acceleration to the electron's?

40. An electron accelerates from rest in a uniform electric field to a speed of 2.9×10^6 m/s in a distance of 14 mm. What is the magnitude of the electric field?

41. A charged particle accelerates from rest in a uniform electric field of magnitude $E = 5.6 \times 10^3$ N/C to a speed of 5.7×10^5 m/s after traveling a distance of 0.30 m. *(a)* What is the particle's charge-to-mass ratio? *(b)* Is this particle a proton or an electron?

42. What is the time interval required for the electron of Example 20-9 to travel the 8.3 mm?

43. Show that the angle between the x axis and the electron's subsequent straight-line path as it emerges from the region of the field in Fig. 20-20 is given by $\theta = \tan^{-1}(qE\ell/mv_0^2)$. Evaluate this angle using the numerical values given in Example 20-10.

44. An electron enters a region of uniform electric field $\mathbf{E} = -(360 \text{ N/C})\mathbf{j}$ with a velocity $\mathbf{v} = (1.6 \times 10^6 \text{ m/s})\mathbf{i}$. Determine the electron's velocity as it emerges from the region of the field if the field has a horizontal extent of $\ell = 29$ mm (Fig. 20-20).

Additional Exercises

45. Particle a with charge $+65$ nC is located at (0.0 m, 3.0 m), and particle b with charge -88 nC is located at the origin. *(a)* Determine **E** at the point P located at (4.0 m, 3.0 m). Write your answer in terms of unit vectors. *(b)* If a test particle with charge $+2.0$ nC is placed at P, what is the force **F** on it?

46. A dipole has its moment directed toward $+\mathbf{i}$ with magnitude 20 nC·m. *(a)* What is the magnitude of the field in the perpendicular bisector plane of the dipole at a distance of 0.44 m from it? *(b)* What is the direction of this field?

47. A straight, thin, uniformly charged rod has a length of 1.6 m and a charge of -640 nC. *(a)* What is the magnitude of the field at a point in the rod's perpendicular bisector plane at a distance of 1.4 m from the center of the rod? *(b)* What is the direction of the field?

48. A thin, circular, uniformly charged ring has a radius of 0.31 m and a charge of $+420$ nC. *(a)* Determine the field magnitude E at a point on the ring's axis that is 0.97 m from its center. *(b)* What is the direction of the field?

49. A thin, circular, uniformly charged disk has a radius of 0.28 m and a charge of -580 nC. *(a)* Find the magnitude E of the field at a point on the disk's axis that is a distance of 0.41 m from its center. *(b)* What is the direction of the field?

50. An electron gun in a CRT accelerates the electrons over a distance of 4.1 mm in a field of magnitude 5.9 kN/C. What is the speed of the emitted electrons?

51. In a CRT, the deflecting plates have a dimension of 57 mm along the direction of the entering electron beam, the field magnitude between the plates is 280 N/C, and electrons enter the deflecting field with a speed of 2.9×10^6 m/s. What is the angle through which the beam is deflected?

PROBLEMS

1. ***Zero force on particles with charge Q.*** Four charged particles are arranged at the corners of a square of side a (Fig. 20-30). The particles at opposite corners have equal charge. *(a)* Find the relationship between Q and q such that the force on each particle with charge Q is zero. *(b)* With the relationship between Q and q given by the result of part *(a)*, determine the magnitude of the force on each of the two particles with charge q.

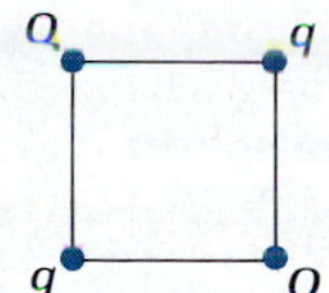

Figure 20-30. Problem 1.

2. ***Field of a dipole in the plane of the dipole moment.*** *(a)* Show that at a point P in the yz plane the components of **E** due to a dipole of moment $\mathbf{p} = 2aq\mathbf{k}$ at the origin (Fig. 20-31) are

$$E_y = \frac{1}{4\pi\epsilon_0} \frac{3pyz}{(y^2 + z^2)^{5/2}}$$

$$E_z = \frac{1}{4\pi\epsilon_0} \frac{p(2z^2 - y^2)}{(y^2 + z^2)^{5/2}}$$

where the distance from the origin to P is much greater than a. *(b)* Show that this general result agrees with the results of Example 20-5 and Exercise 25, which applied to points on the axis only.

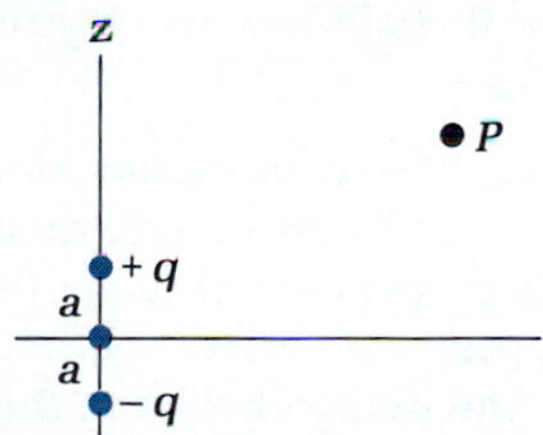

Figure 20-31. Problem 2.

3. ***A quadrupole.*** A quadrupole is an arrangement of charges that, among other characteristics, has zero net charge (monopole moment) and a zero net dipole moment. The charge distribution shown in Fig. 20-32 is an example of a quadrupole; it can be viewed as two dipoles pointing in opposite directions. For this quadrupole show that E varies with distance as r^{-4} at points along the y axis and z axis when $r \gg a$. (Note that E varies with distance as r^{-2} for a monopole, r^{-3} for a dipole, and r^{-4} for a quadrupole at distances far from the charge distribution.)

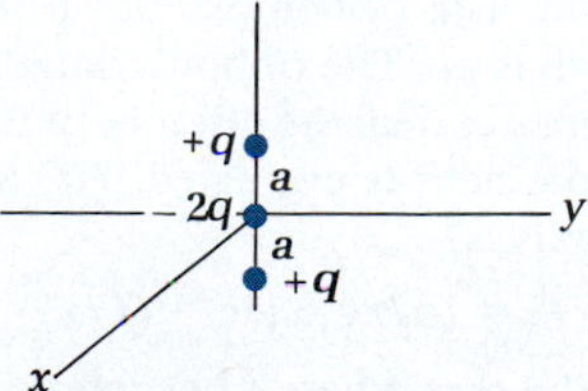

Figure 20-32. Problem 3.

4. ***Points of maximum and minimum E_x on the axis of a charged ring.*** Show that E_x on the axis of a positively charged ring of radius a has its maximum value at $x = +a/\sqrt{2}$ and its minimum value at $x = -a/\sqrt{2}$. Use Fig. 20-16 to check your answer.

5. ***Field along the axis of a straight line charge.*** Show that E_z along the z axis for the line charge in Fig. 20-14 is given by $\lambda\ell/[2\pi\epsilon_0(z^2 - \ell^2)]$, where $z > \ell$. (*Hint:* Let z' be the coordinate of the infinitesimal charge dq and z to be the coordinate of the point P at which the field is determined. The integration variable is z', and z is held fixed during the integration.)

6. ***Field in the plane of a straight line charge.*** Determine the cartesian components of **E** for the line charge in Fig. 20-14 at a point P in the yz plane. (See the hint in the preceeding problem.)

7. ***Field due to both a line charge and a surface charge.*** Determine the approximate electric field at a point near two charge distributions: a long, straight line charge and a large, plane surface charge (Fig. 20-33). The surface charge is due to a thin, uniformly charged disk ($\sigma = +42$ nC/m²) with a very large radius. The disk is contained in the yz plane and centered at the origin. The long line charge is uniform ($\lambda = +15$ nC/m) and is parallel to the y axis, with its center on the x axis at the point (+39 mm, 0, 0). Estimate the cartesian components of **E** at (55 mm, 0, 62 mm).

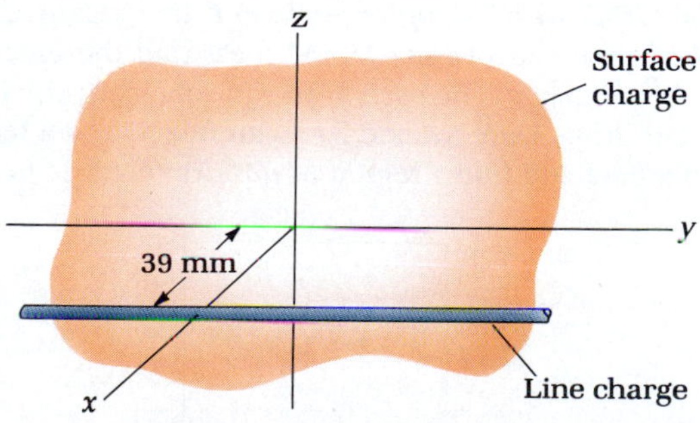

Figure 20-33. Problem 7.

8. ***Field produced by a charged parallel-plate capacitor.*** A parallel-plate capacitor is an important electrical device that is discussed in later chapters. During normal operation, a parallel-plate capacitor maintains a charge distribution that can be ap-

proximated as two parallel sheets of uniform surface charge density of equal magnitude σ but of opposite sign. That is, one sheet has charge density $+\sigma$ and the other has charge density $-\sigma$. (Note that in the context of a parallel-plate capacitor, σ represents a positive number.) Each lateral dimension of these sheets of charge is much larger than the separation between the sheets, so that the magnitude of the contribution to the field by each sheet is approximately $\sigma/2\epsilon_0$. Use the principle of superposition to show that *(a)* the magnitude of the field between the sheets is approximately σ/ϵ_0 and *(b)* the magnitude of the field outside the sheets is essentially zero.

9. ***Field on the axis of a square line charge.*** Determine **E** at points on the axis of a square line charge with side 2ℓ and uniform linear charge density λ (Fig. 20-34). Compare your answer with the results of the charged-ring example (Example 20-7).

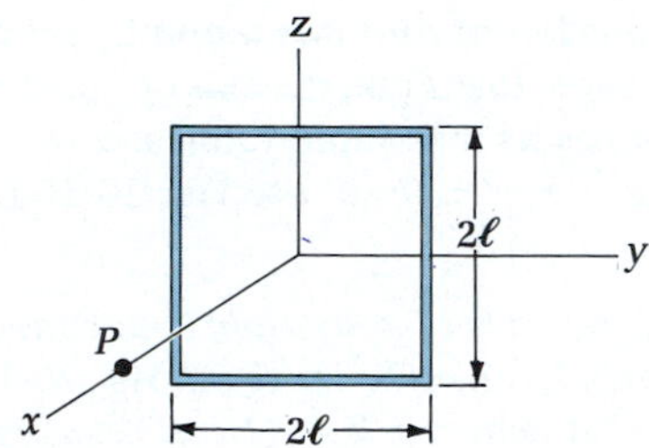

Figure 20-34. Problem 9.

10. ***Field due to a semi-infinite line charge.*** *(a)* At points along the y axis find **E** due to a uniform line charge that extends along the z axis from the origin to $z = \ell$. *(b)* Generalize your answer to part *(a)* to include points in the xy plane. *(c)* Use your answer to part *(b)* to find **E** at points in the xy plane when **E** is due to a semi-infinite line charge that occupies the positive half of the z axis ($z = 0$ to $z = +\infty$).

11. ***Motion of a particle along the axis of a charged ring.*** Consider the motion of a negatively charged particle (charge q and mass m) that is constrained to move only along the x axis and at the same time experiences an electric force due to the (positively) charged ring discussed in Example 20-7. *(a)* What sort of motion will the particle execute when released at some point other than the origin? *(b)* Find an expression for the period T of the particle's motion if it is released at $x = x_0$, where $x_0 \ll a$.

12. ***The Millikan oil-drop experiment.*** In a classic experiment R. A. Millikan (1868–1953) measured the electronic charge. The apparatus he used is shown schematically in Fig. 20-35. Oil drops were formed by an atomizer, and a few fell through a hole and into a region of uniform electric field between charged plates. He could observe a particular drop with the microscope and determine its mass by measuring its terminal speed. He then charged the drop by irradiating it with x-rays and adjusted the electric field so that the drop would be in static equilibrium because of equal and opposite gravitational and electric forces. *(a)* What is the charge on a drop of mass 2.32×10^{-14} kg that remains suspended in an electric field of 2.03×10^5 N/C? Assume $g = 9.80$ N/kg. *(b)* How many electronic charges does the answer to part *(a)* represent?

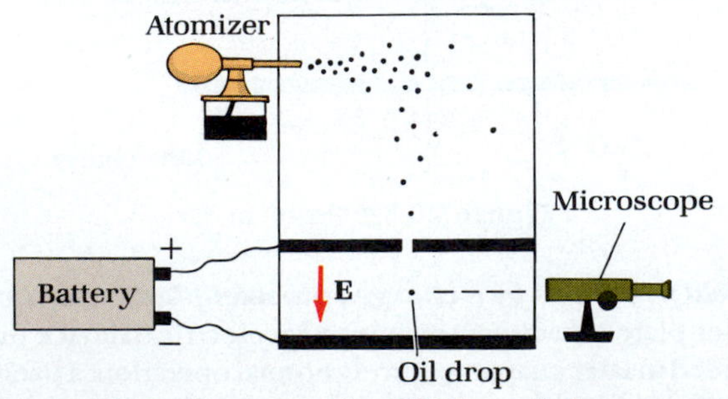

Figure 20-35. Problem 12: The Millikan oil-drop experiment.

13. ***Field on the axis of a square sheet of charge.*** Determine the behavior of the electric field on the axis of a square sheet of charge with a uniform surface charge density (Fig. 20-36). Let the square distribution of charge be represented by 10 parallel line charges, with proper spacing to approximate a square. Construct a spreadsheet that determines the field at about 10 equally spaced points from about 0.2ℓ to 2.0ℓ, where 2ℓ is the length of the sides of the square. Graph your results and compare them with those given for a circular disk with the same charge and with area $(2\ell)^2$. Use numerical values that are both convenient and realistic. You may wish to rework the problem using 20 line charges rather than 10 in order to estimate the accuracy of your answer.

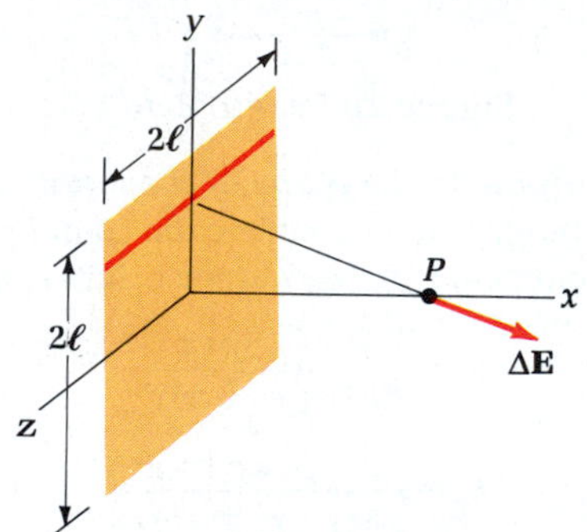

Figure 29-36. Problem 13.

14. ***Zero force on two particles.*** Two particles, one with positive charge Q_0 and the other with positive charge $3Q_0$, are separated a distance d. Determine the location and charge on a third particle that will result in a net electric force of zero on the first two particles.

15. ***Field on the axis of a semicircular ring.*** For the uniform line charge shown in Fig. 20-28, determine an expression for **E** along an axis that is perpendicular to the plane that contains the semicircle and passes through P. Note that **E** has components both parallel and perpendicular to this axis.

16. ***Field produced by a long, straight, ribbon-shaped surface charge.*** Consider a long, straight, ribbon-shaped charge distribution with a uniform surface charge density σ. Let the yz plane contain the flat plane of the ribbon with the y axis along the ribbon's length, the z axis along the ribbon's width, and the origin at the center so that the x axis is perpendicular to the plane of the ribbon. The ribbon extends from $z = -\ell$ to $z = +\ell$ so that its width is 2ℓ. The ribbon's length is much greater than ℓ and much greater than the distance $|x|$ from the ribbon to the point where the field is evaluated. *(a)* Show that E_x at a point P on the x axis is

$$E_x = (\sigma/\pi\epsilon_0)\tan^{-1}(\ell/x)$$

(b) Show that in the case where ℓ becomes much greater than $|x|$, $E \approx |\sigma|/2\epsilon_0$.

GAUSS'S LAW 21

The spherical symmetry of these marbles in their colorful setting is pleasing to the eye. The mathematical symmetry of Gauss's Law is pleasing to the intellect.

The electric field produced by stationary charged objects can be obtained in two equivalent ways: with Coulomb's law or with Gauss's law. The previous chapter described the first way, and this chapter presents the second. Coulomb's law gives a simple and direct way of expressing the electric force. Gauss's law is more subtle, more elegant, and sometimes more useful. Gauss's law requires more mathematical sophistication than Coulomb's law, and the reward for its mastery is a deeper insight into the electrical interaction.

21-1 FLUX

Gauss's law is expressed in terms of the *flux* of the electric field, or the *electric flux*. Before learning about Gauss's law, you must understand the concept of flux. The word "flux" comes from the Latin *fluere,* which means to flow. The concept originated in the theory of fluids, where flux represents the rate at which a fluid passes through an imaginary surface. As you will see later, flux is also useful when dealing with magnetic fields.

The genius of Karl Friedrich Gauss (1777–1855) was apparent at an early age, and his prowess for involved mental calculations is legend. His interests included astronomy and physics, but his most important works were in mathematics. He laid the mathematical foundations for much of the theoretical physics that came in the latter part of the nineteenth and early twentieth centuries, including Einstein's theory of gravitation.

Flux of a Uniform Field for a Plane Surface

Surface vector $\Delta\mathbf{S}$ for a plane surface

The flux Φ of a vector field involves (i) the field and (ii) a surface for which the flux is evaluated. To find the flux for a surface, we represent the surface with a *surface vector.* The surface vector $\Delta\mathbf{S}$ for a plane surface has a magnitude ΔS equal to the area of the surface and is directed perpendicular to the surface. Suppose you take a sheet of notebook paper with length a and width b and hold the sheet horizontally, as shown in Fig. 21-1. Then $\Delta\mathbf{S} = \Delta S\mathbf{j} = (ab)\mathbf{j}$. The flux for such a surface is the dot product between the field and the surface vector. For example, the flux Φ_g of the gravitational field $\mathbf{g}$ for this surface is

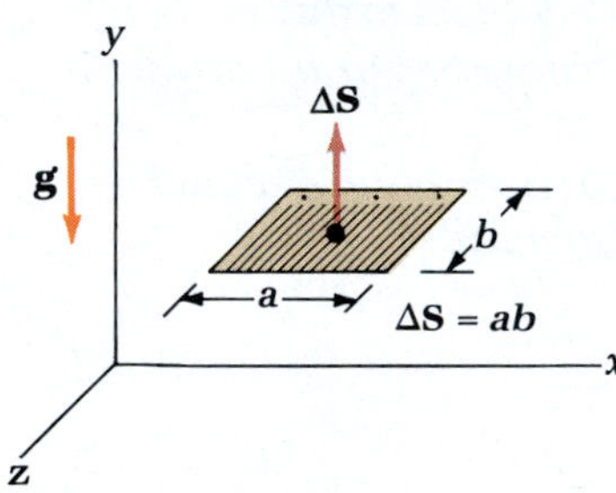

Figure 21-1. A sheet of notebook paper held horizontally. The flux of the gravitational field for the sheet is $\Phi_g = \mathbf{g}\cdot\Delta\mathbf{S} = -g\,\Delta S = -gab$.

$$\Phi_g = \mathbf{g}\cdot\Delta\mathbf{S} = (-g\mathbf{j})\cdot(ab\mathbf{j}) = -gab$$

where we have used $\mathbf{j}\cdot\mathbf{j} = 1$. A typical sheet of notebook paper has dimensions $a = 0.27$ m and $b = 0.22$ m, so that $\Phi_g = -(9.8\text{ N/kg})(0.27\text{ m})(0.22\text{ m}) = -0.58\text{ N}\cdot\text{m}^2/\text{kg}$.

Now suppose the sheet of notebook paper is held vertically so that it faces toward the x direction. Then $\Delta\mathbf{S} = (ab)\mathbf{i}$, and

$$\Phi_g = \mathbf{g}\cdot\Delta\mathbf{S} = (-g\mathbf{j})\cdot(ab\mathbf{i}) = 0$$

because $\mathbf{j}\cdot\mathbf{i} = 0$. Suppose the sheet is held at a 45° angle to the vertical so that $\Delta\mathbf{S}$ is parallel to the unit vector $(1/\sqrt{2})\mathbf{i} + (1/\sqrt{2})\mathbf{j}$. What is the flux in this case?

The surface vector's direction has a twofold ambiguity because a plane surface has *two* directions perpendicular to the surface, one opposite the other. The direction of $\Delta\mathbf{S}$ for the sheet in Fig. 21-1 could be given as $-\mathbf{j}$ rather than $+\mathbf{j}$. This twofold ambiguity can be resolved when the surface is a *closed* surface. By a closed surface, we mean a surface that encloses a volume, like the closed box in Fig. 21-2. In this case, we can define the direction of $\Delta\mathbf{S}$ for each side of the box as either into or out of the enclosed volume. Following custom, we choose the direction of $\Delta\mathbf{S}$ such that it points *out* of the enclosed volume. This means that $\Delta\mathbf{S}$ for the top surface is toward $+\mathbf{j}$, $\Delta\mathbf{S}$ for the right-hand face is toward $+\mathbf{i}$, and so on. What is the direction of $\Delta\mathbf{S}$ for the left-hand face? How about the bottom face?

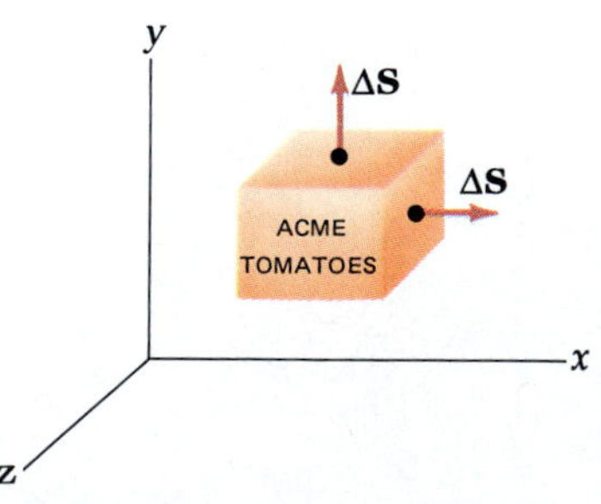

Figure 21-2. When a surface is closed (that is, it encloses a volume), the twofold ambiguity of $\Delta\mathbf{S}$ is resolved by letting $\Delta\mathbf{S}$ point out of the enclosed volume.

Now consider the flux of the electric field.

The flux Φ_E of a uniform electric field $\mathbf{E}$ for a plane surface $\Delta\mathbf{S}$ is the dot product of $\mathbf{E}$ and $\Delta\mathbf{S}$.

$$\Phi_E = \mathbf{E}\cdot\Delta\mathbf{S} = E\,\Delta S\cos\theta \qquad \text{(plane surface and uniform } \mathbf{E} \text{ only)}$$

The dot product takes into account the orientation of the surface with respect to the field direction. As shown in Fig. 21-3, the factor $\Delta S\cos\theta$ is the area of the projection of the surface onto a plane perpendicular to $\mathbf{E}$ (Fig. 21-3c), and may be viewed as the effective area for the flux. Since it is defined as a dot product, flux is a scalar

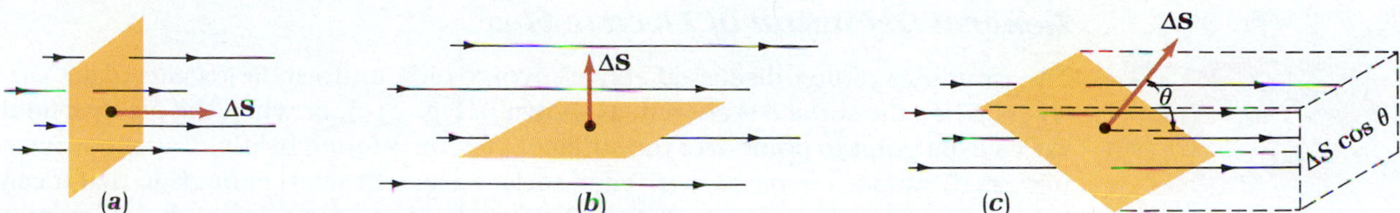

Figure 21-3. Flux Φ_E for plane surfaces with various orientations in a uniform field **E**, shown with field lines. *(a)* **E** parallel to $\Delta\mathbf{S}$, $\Phi_E = E\,\Delta S$. *(b)* **E** perpendicular to $\Delta\mathbf{S}$, $\Phi_E = 0$. *(c)* The general case where **E** and $\Delta\mathbf{S}$ are at an angle θ. Note that in the expression $\Phi_E = E\,\Delta S\cos\theta$, the factor $\Delta S\cos\theta$ can be regarded as the surface's effective area.

quantity. The SI unit of electric flux is newtons per coulomb times square meter $([\mathrm{N/C}](\mathrm{m}^2))$ or $\mathrm{N\cdot m^2/C}$.

It is often helpful to use field lines to form a mental picture of flux (Fig. 21-3). As we shall discuss in the next section, the number of lines that cross a surface is proportional to the flux for the surface. This field-line characterization provides an aid for visualizing flux, but it is not useful for calculations because of the discrete nature of the lines.

EXAMPLE 21-1

Flux for a wedge-shaped surface. The wedge-shaped surface in Fig. 21-4 is in a region of uniform field $\mathbf{E} = (600\ \mathrm{N/C})\mathbf{i}$. *(a)* Determine the electric flux for each of the five surfaces. *(b)* Find the net flux for the entire closed surface.

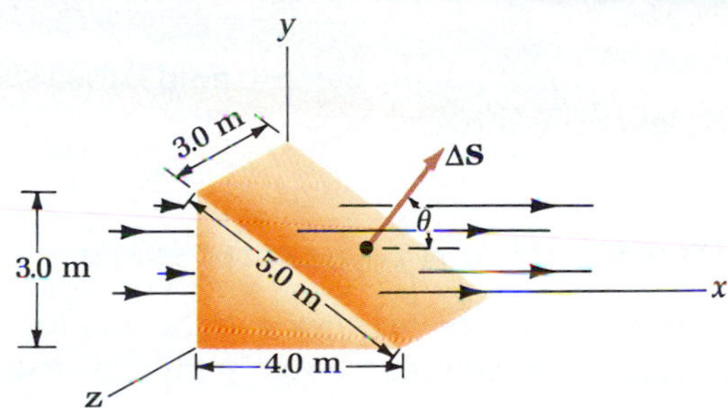

Figure 21-4. Example 21-1: The uniform field **E** is represented by field lines, and the surface vector for the slant surface is shown.

Solution. *(a)* The flux is zero for the two triangular side surfaces and for the bottom surface because the direction of the surface vector for each of these surfaces is perpendicular to **E** so that $\cos\theta = \cos 90° = 0$. For the square surface on the left-hand side, $\Delta\mathbf{S} = (3.0\ \mathrm{m})(3.0\ \mathrm{m})(-\mathbf{i}) = -(9.0\ \mathrm{m}^2)\mathbf{i}$, and

$$\Phi_E = \mathbf{E}\cdot\Delta\mathbf{S} = [(600\ \mathrm{N/C})\mathbf{i}]\cdot[-(9.0\ \mathrm{m}^2)\mathbf{i}]$$
$$= -5400\ \mathrm{N\cdot m^2/C}$$

For the slanted surface,

$$\Phi_E = \mathbf{E}\cdot\Delta\mathbf{S} = E\,\Delta S\cos\theta$$

From the figure, $\Delta S = (3.0\ \mathrm{m})(5.0\ \mathrm{m}) = 15\ \mathrm{m}^2$, and $\cos\theta = 3.0\ \mathrm{m}/5.0\ \mathrm{m} = 0.60$. Therefore the flux for the slanted surface is

$$\Phi_E = (600\ \mathrm{N/C})(15\ \mathrm{m}^2)(0.60) = 5400\ \mathrm{N\cdot m^2/C}$$

(b) For three of the five surfaces the flux is 0; for the other two the values are $5400\ \mathrm{N\cdot m^2/C}$ and $-5400\ \mathrm{N\cdot m^2/C}$. Thus the net flux for the closed surface is zero. Notice that the contribution to the flux for a closed surface is positive for the portion of the surface where the field is directed out of the enclosed volume and negative for the portion of the surface where the field is directed into the enclosed volume.

The net flux for this closed surface can also be seen to be zero from examination of the field lines. No lines cross the triangular sides or the bottom. Figure 21-4 shows four lines directed into the volume where they cross the square surface and the same four lines directed out of the volume where they cross the slant surface. The net number of lines directed out of the volume where they cross the closed surface is the number out minus the number in. For this case the net number of lines is $4 - 4 = 0$.

SELF-TEST 21-1. A uniform electric field of magnitude 400 N/C is directed vertically upward. What is the magnitude of the electric flux for a flat square surface measuring 0.50 m on each side and oriented such that a normal to the surface makes an angle of 38° with the vertical? ***ANSWER:*** $79\ \mathrm{N\cdot m^2/C}$.

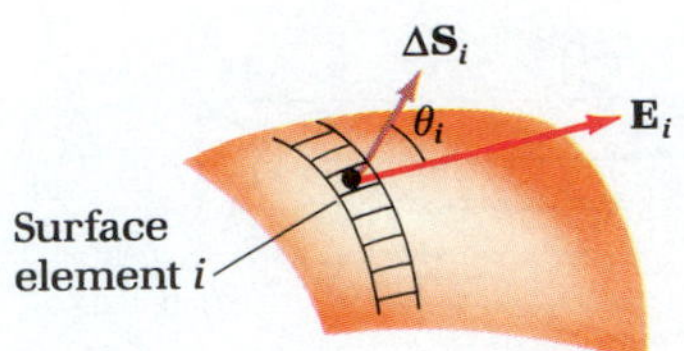

Figure 21-5. To find the flux for a curved surface and/or a nonuniform field, the surface is divided into a large number of small surface elements and the flux for each element is added. In the limit as the size of the elements approaches zero and their number approaches infinity, the sum approaches an integral. Such an integral is called a *surface integral.*

General Definition of Electric Flux

The examples of flux discussed above involve only uniform fields and plane surfaces. When the surface is curved, as shown in Fig. 21-5, or when the electric field varies from point to point over the surface, the flux is found by dividing the surface into small surface elements, with each surface element small enough so that it can be considered a plane and so that the electric field variation across the element is negligible. The flux for the entire surface is then the sum of the individual contributions to the flux from each of the small surface elements. In the limit as the size of each element approaches zero and their number approaches infinity, the sum becomes an integral:

$$\Phi_E = \lim_{\Delta S_i \to 0} \sum_i \mathbf{E}_i \cdot \Delta \mathbf{S}_i = \int \mathbf{E} \cdot d\mathbf{S}$$

Or

$$\Phi_E = \int \mathbf{E} \cdot d\mathbf{S} = \int E \cos\theta \, dS \qquad (21\text{-}1)$$

The integral in Eq. 21-1 extends over the surface for which the flux is calculated and is called a *surface integral.* Thus the general definition of electric flux is

Definition of the flux of an electric field

The **flux** Φ_E of the electric field **E** for a surface **S** is the surface integral of **E** over the surface.

Definition of a Gaussian Surface

Mostly we shall be interested in the flux for a closed surface. When the surface of integration is closed, the sign $\oint$ for a closed integral is used:

Electric flux for a closed surface

$$\Phi_E = \oint \mathbf{E} \cdot d\mathbf{S} \qquad (21\text{-}2)$$

The closed surface for which the flux is calculated is ordinarily an imaginary or hypothetical surface, called a *gaussian surface.*

A gaussian surface

A **Gaussian surface** is a hypothetical closed surface for which the flux is evaluated.

A gaussian surface does not necessarily correspond to the surface of an object. Whenever you use Gauss's law, you may devise a surface of any size and shape to use as your gaussian surface. Selecting the proper size and shape for a gaussian surface is one of the key elements in using Gauss's law.

21-2 GAUSS'S LAW

Gauss's law can be stated this way:

The electric flux for any closed surface is equal to the net charge enclosed by the surface divided by ϵ_0.

In equation form,

Gauss's law

$$\Phi_E = \frac{\Sigma q}{\epsilon_0} \qquad \text{or} \qquad \oint \mathbf{E} \cdot d\mathbf{S} = \frac{\Sigma q}{\epsilon_0} \qquad (21\text{-}3)$$

where the closed surface (the gaussian surface) for which the flux is calculated can be of *any shape or size* and the symbol Σq represents the *net* charge contained within the volume enclosed by the surface.

Using Gauss's Law to Find the Field Produced by a Point Charge

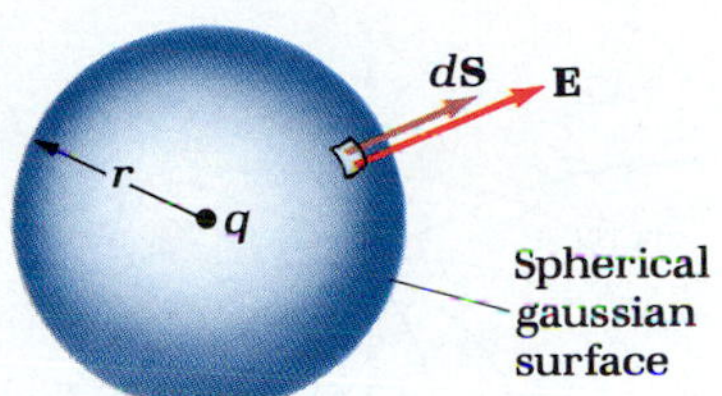

Figure 21-6. Finding the flux due to the field produced by a point charge at the center of a spherical gaussian surface.

As a first example of Gauss's law, we use this law to find the expression for the field produced by a point charge. Figure 21-6 shows a spherical gaussian surface of radius r with a point charge q at its center. In performing the flux integral $\oint \mathbf{E} \cdot d\mathbf{S}$, we note from symmetry that $\mathbf{E}$ is directed radially away from q, which means that $\mathbf{E}$ is parallel to $d\mathbf{S}$ at each point on the sphere's surface, or $\mathbf{E} \cdot d\mathbf{S} = E\,dS$. Also from symmetry we can see that E depends only on distance r from q, so that E is the same at each point on the sphere, or E is constant with respect to the integration. Thus the calculation of the flux for the spherical surface proceeds as follows:

$$\Phi_E = \oint \mathbf{E} \cdot d\mathbf{S} = \oint E\,dS = E \oint dS = E(4\pi r^2)$$

where the integral $\oint dS$ is simply the surface area $4\pi r^2$ of the sphere. Since the total charge contained inside our gaussian sphere is $\Sigma q = q$, Gauss's law gives

$$E(4\pi r^2) = \frac{q}{\epsilon_0} \qquad \text{or} \qquad E = \frac{q}{4\pi\epsilon_0 r^2}$$

This is the same result we obtained with Coulomb's law. Using symmetry arguments, we have shown that Gauss's law gives the same expression for the field produced by a point charge as Coulomb's law.

More about Field Lines

In the previous section, we mentioned that the flux for a surface is proportional to the number of field lines that cross the surface. For the case of a spherical surface centered at a point charge, we can now demonstrate this with Gauss's law. Figure 21-7 shows two such surfaces and the field lines around the point charge q. Since the lines emanate from a positive charge, terminate on a negative charge, and are continuous in between, the number of lines that cross each sphere is the same. By Gauss's law, the flux Φ_E for each sphere is the same because the enclosed charge $\Sigma q = q$ is the same. Thus the flux for each surface is proportional to the number of lines that cross the surface.

When field lines were introduced in Sec. 20-6, we stated that the density of the lines (or their spacing) indicates the field magnitude. Again, this can now be demonstrated for the case of a point charge. In Fig. 21-7, the radius of the larger sphere is twice that of the smaller sphere: $r_2 = 2r_1$. Thus the area of the larger sphere is 4 times that of the smaller sphere: $A_2 = 4\pi r_2{}^2 = 4\pi(2r_1)^2 = 4(4\pi r_1{}^2) = 4A_1$. If N is the number of lines crossing a sphere of area A, then the density of lines is N/A. Since the number of lines crossing each sphere is the same, the density of lines at the larger sphere is one-fourth that at the smaller sphere. Also, since the field is an inverse-square field, E at the larger sphere is one-fourth that at the smaller sphere. Thus E is proportional to the density of lines.

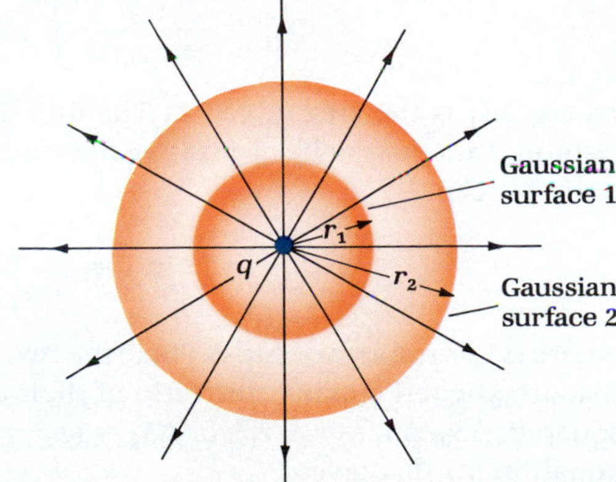

Figure 21-7. Two gaussian spheres with different radii are centered at the same charged particle. Field lines extend continuously from positive to negative charges so that the number of lines crossing each sphere is the same. Also, by Gauss's law the flux for each surface is the same. Therefore the flux for such a surface is proportional to the number of lines crossing the surface.

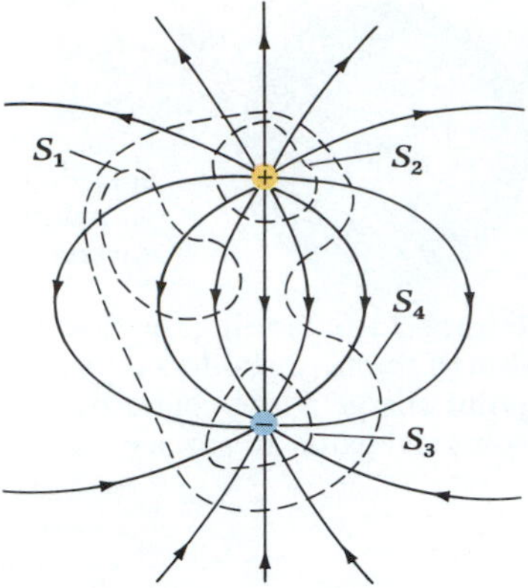

Figure 21-8. Four gaussian surfaces in the field of a dipole. The gaussian surfaces are shown in cross section, and the dashed lines represent the intersection of the surfaces with the plane of the figure. From the field lines, you can verify that $\Phi_{E1} = 0$, $\Phi_{E2} > 0$, $\Phi_{E3} < 0$, and $\Phi_{E4} = 0$.

Interpretation of Σq *in Gauss's Law*

When using Gauss's law, you should keep in mind that Σq represents the *net enclosed* charge, that is, the positive charge minus the negative charge inside the surface. We can explore this feature of Gauss's law by considering the flux for the four different gaussian surfaces in the field of a dipole shown in Fig. 21-8. Surface S_1 contains no charge, so that from Gauss's law the flux for surface S_1 is $\Phi_{E1} = 0$. The field lines in Fig. 21-8 substantiate the fact that the flux for S_1 is zero. Three field lines are directed into the volume enclosed by S_1 and three field lines are directed out, so that the net number of lines directed out is zero. Surface S_2 contains charge q so that Gauss's law gives $\Phi_{E2} = q/\epsilon_0$. From the field lines in Fig. 21-8, Φ_{E2} is positive because the lines are directed out of the enclosed volume at each point on S_2. Similarly, the flux for S_3 is $\Phi_{E3} = -q/\epsilon_0$ from Gauss's law, and the field lines show that the flux is negative because they are directed into the enclosed volume at each point on S_3. Surface S_4 contains both particles, so the net charge enclosed is $\Sigma q = q - q = 0$. Consequently, Gauss's law requires $\Phi_{E4} = 0$. As an exercise, count the number of lines directed into and out of S_4. Is your count in agreement with Gauss's law?

21-3 DEVELOPING GAUSS'S LAW FROM COULOMB'S LAW

In Chap. 20, we presented Coulomb's law as the result of experiment, and used it to write the field produced by a point charge as

$$\mathbf{E} = \frac{q}{4\pi\epsilon_0 r^2}\,\hat{\mathbf{r}}$$

We now give Gauss's law this same experimental foundation by developing Gauss's law from Coulomb's law and the principle of superposition. The above expression for the field produced by a point charge is the form of Coulomb's law we shall use.

Flux for an Arbitrary Surface, Charged Particle Outside

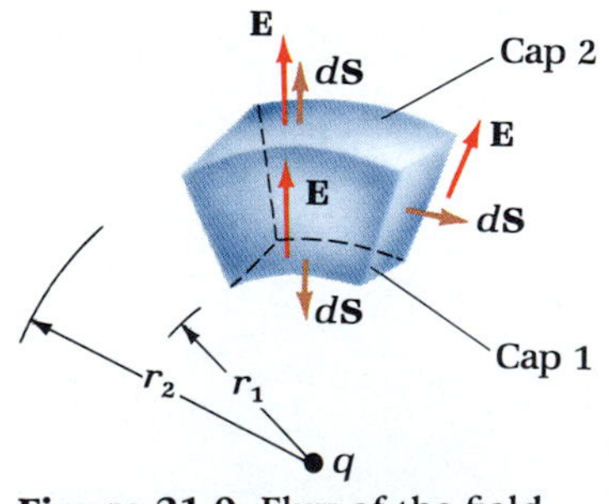

Figure 21-9. Flux of the field due to a charged particle for a rounded-block gaussian surface. The surface is bounded by two spherical caps and four flat sides. The flux for each flat side is zero, and the flux for the two spherical caps is equal and opposite, so the flux for the entire surface is zero.

Consider the flux for the rounded-block gaussian surface shown in Fig. 21-9. The field is due to a charged particle, and the surface is bounded by four flat sides and two spherical caps. Each flat side is aligned radially with the particle, and each cap is a patch of a sphere centered at the particle. Therefore the flux for the sides is zero because $\mathbf{E}$ is perpendicular to $d\mathbf{S}$ at each point on the sides. The flux Φ_{E1} for cap 1 is negative (assuming q is positive) because the direction of $\mathbf{E}$ is opposite $d\mathbf{S}$ at each point on cap 1: $\int \mathbf{E}\cdot d\mathbf{S} = -\int E\,dS$. Further, E is the same at each point on the surface and can be factored out of the flux integral: $E = q/4\pi\epsilon_0 r_1{}^2$. Thus

$$\Phi_{E1} = -\frac{q}{4\pi\epsilon_0 r_1{}^2}\int dS = -\frac{q}{4\pi\epsilon_0 r_1{}^2}\,\Delta S_1$$

where ΔS_1 is the area of cap 1. The flux Φ_{E2} for cap 2 is calculated in a similar fashion, except that Φ_{E2} is positive because $\mathbf{E}$ is in the same direction as $d\mathbf{S}$ at each point on cap 2:

$$\Phi_{E2} = \frac{q}{4\pi\epsilon_0 r_2{}^2}\,\Delta S_2$$

where ΔS_2 is the area of cap 2. Since the two spherical caps are bounded by flat sides that are aligned radially, the ratio of their areas is equal to the ratio of their radii squared: $\Delta S_2/\Delta S_1 = r_2{}^2/r_1{}^2$, or $\Delta S_2 = (r_2{}^2/r_1{}^2)\Delta S_1$. Substituting this result into the equation for Φ_{E2} gives

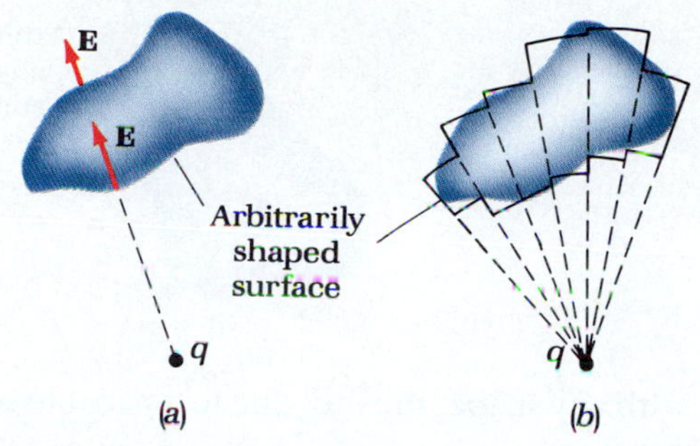

Figure 21-10. *(a)* A charged particle outside the enclosed volume of an arbitrarily shaped gaussian surface shown in cross section. *(b)* Superimposed approximation to the surface in *(a)* consisting of a number of spherical caps and flat sides. The flux for each flat side is zero, and the flux for each pair of spherical caps is equal and opposite: $\Phi_E = 0$.

$$\Phi_{E2} = \frac{q}{4\pi\epsilon_0 r_2^2}\frac{r_2^2}{r_1^2}\Delta S_1 = \frac{q}{4\pi\epsilon_0 r_1^2}\Delta S_1 = -\Phi_{E1}$$

The net flux for the closed surface is

$$\Phi_E = \Phi_{E2} + \Phi_{E1} = -\Phi_{E1} + \Phi_{E1} = 0$$

The net flux is zero because the flux for cap 1 is the negative of the flux for cap 2.

Now we introduce another point: *A surface of any shape can be constructed from an infinite number of infinitesimal spherical caps and flat sides.* Figure 21-10*a* shows a cross section of an arbitrarily shaped, closed surface with a particle of charge q outside the enclosed volume. Figure 21-10*b* shows this same surface with a superimposed approximation to the surface that consists of a number of spherical caps centered at the particle and flat sides aligned with the particle. You can see that the surface can be viewed as the limit of an infinite number of infinitesimal spherical caps and flat sides. Since the flux for the caps cancels in pairs and the flux for the sides is zero, for the arbitrarily shaped surface in Fig. 21-10*a*, the flux due to the charged particle outside the enclosed volume is zero:

$$\Phi_E = 0 \qquad \text{(arbitrarily shaped closed surface, } q \text{ outside)}$$

Flux for an Arbitrary Surface, Charged Particle Inside

Now consider the flux due to a point charge q at the center of a spherical gaussian surface of radius r (Fig. 21-6). At each point on the surface, $\mathbf{E}$ is parallel to $d\mathbf{S}$ ($\mathbf{E}\cdot d\mathbf{S} = E\,dS$) and E has the same value and can be factored from the integral ($E = q/4\pi\epsilon_0 r^2$). Thus

$$\Phi_E = \oint \mathbf{E}\cdot d\mathbf{S} = \oint E\,dS = E\oint dS = \frac{q}{4\pi\epsilon_0 r^2}4\pi r^2 = \frac{q}{\epsilon_0}$$

Since Φ_E does not contain r, the flux is the same for a sphere of any radius.

We can use this result to find the flux for the gaussian surface shown in Fig. 21-11. This surface is mostly a sphere centered at the particle, except that a cap of area ΔS_1 is cut out and replaced with a raised cap of area ΔS_2. The volume directly beneath the raised cap is enclosed by flat sides (aligned with the particle) so that the surface is a closed surface. The flux for this surface is the same as for a sphere because the flux for the flat sides is zero and the flux for cap 2 is equal to the flux missing because of the absence of cap 1; the argument is similar to the previous discussion of the rounded block. Thus $\Phi_E = q/\epsilon_0$.

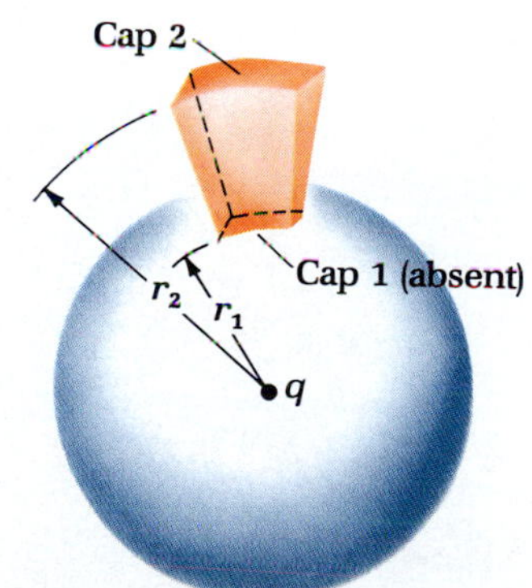

Figure 21-11. Flux of the field produced by a point charge at the center of a gaussian surface that is mostly a sphere except for the raised cap. The flux for cap 2 is equal to the flux missing because of the absence of cap 1, so the flux for the entire surface is the same as that for a sphere: $\Phi_E = q/\epsilon_0$.

Further, any arbitrarily shaped surface may be regarded as the limit of an infinite number of infinitesimal spherical caps and flat sides. Figure 21-12*a* shows the cross section of an arbitrarily shaped closed surface with a charged particle inside, and Fig. 21-12*b* shows a number of spherical caps and flat sides centered at the particle. As before, the arbitrarily shaped surface can be viewed as the limit of an infinite number of infinitesimal spherical caps and flat sides. Therefore, for a closed surface

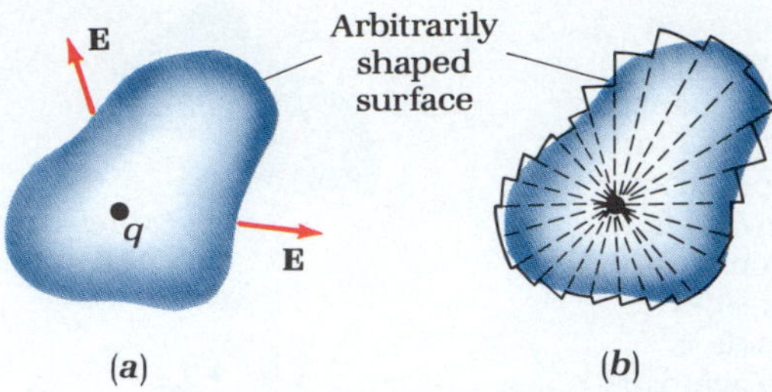

Figure 21-12. *(a)* A charged particle inside the enclosed volume of an arbitrarily shaped gaussian surface shown in cross section. *(b)* Superimposed approximation to the surface in *(a)*, consisting of a number of spherical caps and flat sides. The flux for each flat side is zero, and the sum of the fluxes for the spherical caps is the same as for a sphere: $\Phi_E = q/\epsilon_0$.

with any shape, the flux due to a particle of charge q inside the enclosed volume is q/ϵ_0:

$$\Phi_E = \frac{q}{\epsilon_0} \qquad \text{(arbitrarily shaped closed surface, } q \text{ inside)}$$

Flux for an Arbitrary Surface, Charged Particles Inside and Outside

So far we have considered the electric flux due to a single charged particle. Suppose there is more than one particle to consider. To be specific, consider the surface in Fig. 21-13, where the flux is due to the three charged particles. From the principle of superposition, the field is the vector sum of the individual contributions to the field: $\mathbf{E} = \mathbf{E}_1 + \mathbf{E}_2 + \mathbf{E}_3$. The flux is

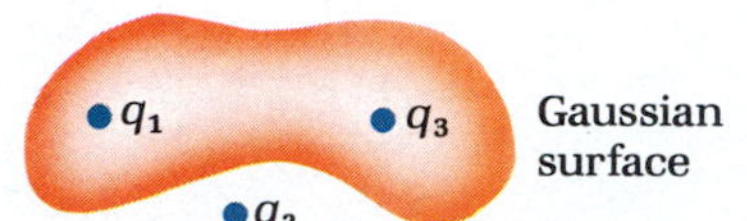

Figure 21-13. Of the three charged particles, particles 1 and 3 are enclosed by the gaussian surface and particle 2 is not. Using the principle of superposition, $\mathbf{E} = \mathbf{E}_1 + \mathbf{E}_2 + \mathbf{E}_3$, we find that the flux for the surface is $\Phi_E = (q_1 + q_3)/\epsilon_0$. That is, particle 2 does not contribute to the flux.

$$\Phi_E = \oint \mathbf{E} \cdot d\mathbf{S} = \oint (\mathbf{E}_1 + \mathbf{E}_2 + \mathbf{E}_3) \cdot d\mathbf{S}$$

Since the integral of a sum is the sum of the integrals, we may write this as

$$\Phi_E = \oint \mathbf{E}_1 \cdot d\mathbf{S} + \oint \mathbf{E}_2 \cdot d\mathbf{S} + \oint \mathbf{E}_3 \cdot d\mathbf{S}$$

Particles 1 and 3 are inside the surface so that their contributions to the flux are q_1/ϵ_0 and q_3/ϵ_0, respectively. Particle 2 is outside the surface so that its contribution to the flux is zero:

$$\Phi_E = \frac{q_1}{\epsilon_0} + 0 + \frac{q_3}{\epsilon_0}$$

Generally, for any number of charged particles, the flux for an arbitrarily shaped closed surface is

$$\Phi_E = \frac{\Sigma q_i}{\epsilon_0} \qquad \text{or} \qquad \oint \mathbf{E} \cdot d\mathbf{S} = \frac{\Sigma q_i}{\epsilon_0}$$

which is Gauss's law. Our development shows clearly that *the field* **E** *in the flux integral is the field due to all charged particles, both inside and outside the enclosed volume, but the charges included in the sum* Σq_i *are only the charges of the particles that are inside the enclosed volume.*

Comparison of Gauss's Law and Coulomb's Law

Since Gauss's law can be derived from Coulomb's law [in the form $\mathbf{E} = (q/4\pi\epsilon_0 r^2)\hat{\mathbf{r}}$] and the principle of superposition, Gauss's law is the same statement about nature as Coulomb's law (in electrostatics). Each is a consequence of the electric force being an inverse-square force that depends linearly on the charge and is directed along a line between the particles. However, Gauss's law is more general than Coulomb's law, and its validity is broader. The electric field **E** that comes from Coulomb's law is due to stationary charges only. In Chap. 28, we shall find that a time-changing magnetic field produces an electric field. The electric field **E** in Gauss's law represents this field as well as the field of stationary charges.

EXAMPLE 21-2

Flux for gaussian surfaces. Find the flux for *(a)* gaussian surface *A* and *(b)* gaussian surface *B* that enclose the charged particles shown in Fig. 21-14. The values of the charges are $q_1 = -41$ nC, $q_2 = +73$ nC, and $q_3 = -65$ nC.

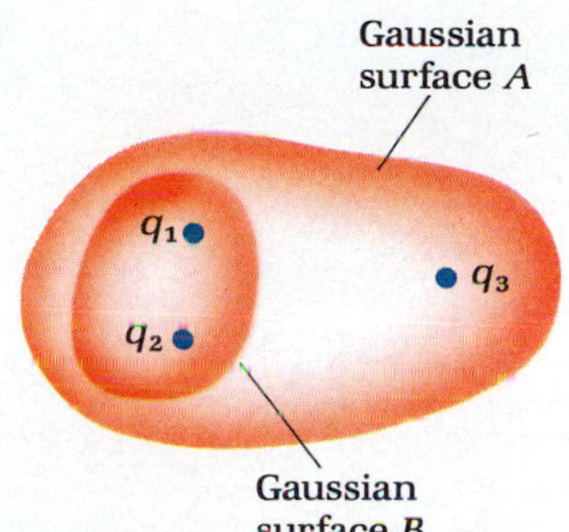

Figure 21-14. Example 21-2: Gaussian surface *A* encloses all three charged particles. Gaussian surface *B* encloses only particles 1 and 2.

Solution. *(a)* The flux for surface *A* is

$$\Phi_E = \frac{\Sigma q}{\epsilon_0} = \frac{q_1 + q_2 + q_3}{\epsilon_0}$$

$$= \frac{-41\ \text{nC} + 73\ \text{nC} - 65\ \text{nC}}{8.85 \times 10^{-12}\ \text{C}^2/(\text{N}\cdot\text{m}^2)}$$

$$= -3.7 \times 10^3\ \text{N}\cdot\text{m}^2/\text{C}$$

because all three particles are enclosed by *A*. *(b)* Because only q_1 and q_2 are enclosed by *B*, the flux for surface *B* is

$$\Phi_E = \frac{\Sigma q}{\epsilon_0} = \frac{q_1 + q_2}{\epsilon_0}$$

$$= \frac{-41\ \text{nC} + 73\ \text{nC}}{8.85 \times 10^{-12}\ \text{C}^2/(\text{N}\cdot\text{m}^2)}$$

$$= +3.6 \times 10^3\ \text{N}\cdot\text{m}^2/\text{C}$$

SELF-TEST 21-2. *(a)* In Fig. 21-14, suppose a gaussian surface encloses particles 1 and 3, but not 2. What is the flux for this surface? *(b)* Now suppose a gaussian surface encloses particles 2 and 3, but not 1. What is the flux for this surface? ***ANSWERS:*** *(a)* -12×10^3 N·m²/C; *(b)* 900 N·m²/C.

21-4 USING GAUSS'S LAW TO FIND *E*

Gauss's law can be used to find the electric field due to a charge distribution that has a high degree of symmetry. If the charge distribution is highly symmetric, then some features of the field, such as its direction, can be deduced by inspection of the symmetry of the distribution, without need of calculation. You can then (i) select a gaussian surface that capitalizes on the symmetry, (ii) determine the flux for this gaussian surface in terms of *E*, and (iii) solve Gauss's law for *E*. The first step is the most crucial. The surface chosen to be the gaussian surface should be one for which the flux can be readily determined. The following examples illustrate the technique.

In finding **E**, the choice of a gaussian surface is crucial.

EXAMPLE 21-3

Field near a long line charge. Find an approximate expression for *E* near a long, straight, uniformly charged wire (linear charge density λ) at a point *P* that is far from either end of the wire.

Solution. The first step in finding *E* with Gauss's law is the selection of the gaussian surface. To do this the symmetry of the field must be determined by inspection of the charge distribution. In Fig. 21-15, the wire is along the *z* axis, and *P* is in the *xy* plane at a distance *R* from the *z* axis ($R = \sqrt{x^2 + y^2}$). Since *P* is far from either end, we expect from symmetry that **E** will point directly away from the *z* axis (assuming λ is positive) and be parallel to the *xy* plane. Also, we expect that *E* will depend only on distance *R* from the wire. That is, we expect that the field has cylindrical symmetry about the *z* axis. The gaussian surface that takes advantage of these symmetrical features of the field is a right circular cylinder with axis along the *z* axis (Fig. 21-15). For this gaussian surface (i) the flux for both the top and bottom is zero because **E** is perpendicular to $d\mathbf{S}$ at each point on the top and bottom, and (ii) the flux for the cylindrical surface is simply *E* times the area of the cylindrical surface because **E** is parallel to $d\mathbf{S}$ at each point on the cylindrical surface and its magnitude is the same at each point on the cylindrical surface. Therefore

$$\Phi_E = \oint \mathbf{E}\cdot d\mathbf{S} = E(2\pi Rh)$$

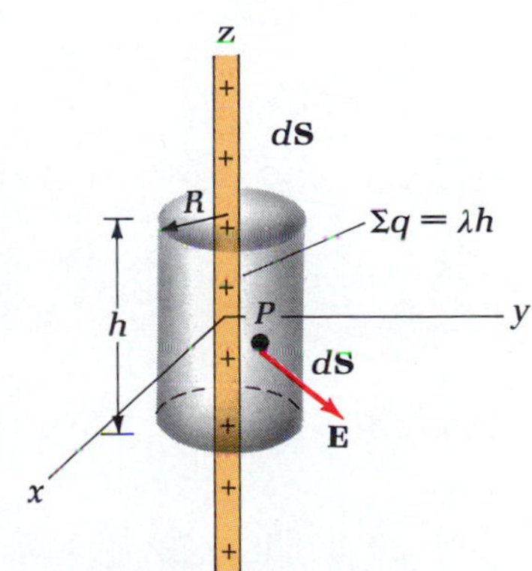

Figure 21-15. Example 21-3: We use a cylinder as a gaussian surface to find the field near a uniform line charge far from its ends. The charge enclosed by the cylinder is $\Sigma q = \lambda h$.

where $2\pi Rh$ is the surface area of the cylindrical surface of radius R and height h.

From the figure, the charge inside the cylinder is the product of the linear charge density λ and the height h of the cylinder: $\Sigma q = \lambda h$. Thus Gauss's law gives

$$\Phi_E = \frac{\Sigma q}{\epsilon_0} \quad \text{or} \quad E2\pi Rh = \frac{\lambda h}{\epsilon_0}$$

Solving for E gives

$$E = \frac{\lambda}{2\pi\epsilon_0 R}$$

This is an approximate result for points far from the ends of a very long line charge; it is strictly valid only for an infinitely long line charge. This is the same result obtained with Coulomb's law (Eq. 20-8).

SELF-TEST 21-3. The magnitude of the electric field at a distance of 0.15 m from a long uniform line charge (far from its ends) is 200 N/C. What is the magnitude of the field at a distance of 0.30 m? ***ANSWER:*** 100 N/C.

EXAMPLE 21-4

Field near a large plane sheet of charge. Find an approximate expression for E produced by a large plane sheet of charge with a uniform surface charge density σ at a point near the sheet but far from its edges.

Solution. First determine the symmetry of the field and select a gaussian surface. In Fig. 21-16, the charged sheet occupies the yz plane, and point P is near the x axis. Since P is far from the edges of the sheet, we expect from symmetry that the field must point directly away from the sheet (assuming σ is positive) and along the x axis. Further, if E depends on position at all, then it can only depend on x. We take advantage of this symmetry by using a gaussian surface that is a right circular cylinder, centered at the origin and with axis along the x axis. For this surface (i) the flux for the cylindrical surface is zero because **E** is perpendicular to $d\mathbf{S}$ at each point on the cylindrical surface, and (ii) the flux for each end is simply E times the area ΔS of that end because **E** is uniform and parallel to $d\mathbf{S}$ at each point on either end. Therefore

$$\Phi_E = \oint \mathbf{E} \cdot d\mathbf{S} = E\,\Delta S + E\,\Delta S = 2E\,\Delta S$$

From the figure, the charge enclosed by the gaussian cylinder is the product of the surface charge density σ and the cross-sectional area of the cylinder, which is the same as the area ΔS of an end. Thus $\Sigma q = \sigma\,\Delta S$. Gauss's law gives

$$\Phi_E = \frac{\Sigma q}{\epsilon_0} \quad \text{or} \quad 2E\,\Delta S = \frac{\sigma\,\Delta S}{\epsilon_0}$$

Solving for E, we have

$$E = \frac{\sigma}{2\epsilon_0}$$

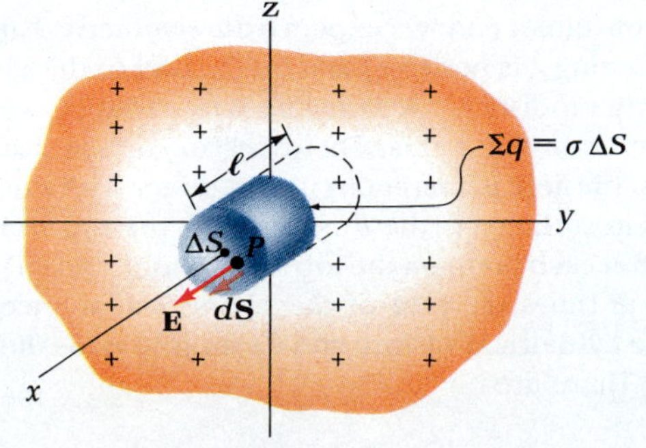

Figure 21-16. Example 21-4: We use a cylinder as a gaussian surface to find the field near a uniform planar sheet of charge far from its edges. The gaussian cylinder extends through the sheet so that it encloses charge $\Sigma q = \sigma\,\Delta S$.

Note that this is an approximation for E at points near the charged sheet and far from its edges. This is the same result obtained with Coulomb's law (Eq. 20-9).

SELF-TEST 21-4. If the electric field near a large plane sheet of uniform surface charge is 500 N/C, and the field is directed toward the sheet, what is the surface density of the sheet?
ANSWER: $-8.85\ \mathrm{nC/m^2}$.

EXAMPLE 21-5

A charged spherical shell. Determine **E** at points both inside and outside a thin, uniformly charged spherical shell of radius r_0 and charge Q. The charge distribution is similar to the mass distribution of a Ping-Pong ball.

Solution. To select a gaussian surface, we determine the symmetry of the field. Since the charge distribution is spherical, **E** has only a radial component and its magnitude depends only on distance r from the center of the charge distribution. The gaussian surface that capitalizes on this symmetry is a spherical surface with the same center as the spherical shell of charge. First consider the field at points inside the shell by finding the flux for a spherical gaussian surface whose radius r is less than the radius r_0 of the shell (Fig. 21-17*a*). Since **E** must be radial and can only depend on r, the flux for the gaussian sphere is

$$\Phi_E = E(4\pi r^2)$$

From Fig. 21-17*a* the charge inside the gaussian sphere is zero because the gaussian sphere is completely inside the shell of charge. Gauss's law gives

$$\Phi_E = \frac{\Sigma q}{\epsilon_0} \qquad \text{or} \qquad E(4\pi r^2) = 0$$

The field at any point on the gaussian sphere must be zero. This is true for any gaussian sphere as long as its radius is less than the radius of the charged shell. Thus the field is zero at all points inside the spherical shell of charge:

$$E = 0 \qquad (r < r_0)$$

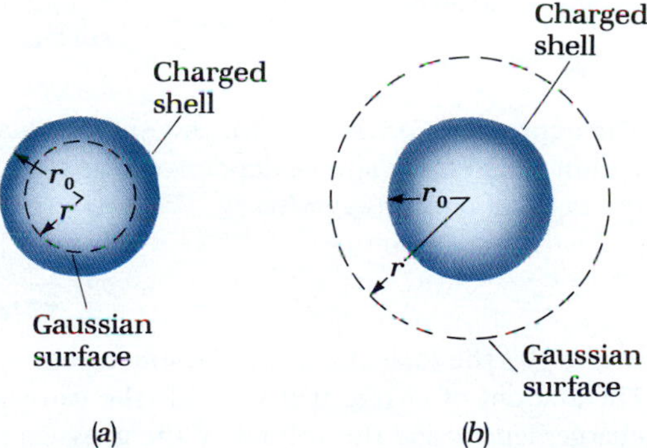

Figure 21-17. Example 21-5: *(a)* Spherical gaussian surface inside the charged spherical shell: $\Sigma q = 0$. *(b)* Spherical gaussian surface outside the charged spherical shell: $\Sigma q = Q$.

Now consider the field at points outside the charged shell by finding the flux for a spherical gaussian surface whose radius r is larger than the radius r_0 of the shell (Fig. 21-17*b*). Again, because of the spherical symmetry, the expression for the flux for the gaussian surface is $\Phi_E = E(4\pi r^2)$. This surface encloses the entire charge Q of the shell, $\Sigma q = Q$, so that Gauss's law gives

$$\Phi_E = \frac{\Sigma q}{\epsilon_0} \qquad \text{or} \qquad E(4\pi r^2) = \frac{Q}{\epsilon_0}$$

Solving for E, we have

$$E = \frac{Q}{4\pi\epsilon_0 r^2} \qquad (r > r_0)$$

This expression for E is familiar; it is the same as that due to a point charge Q located at the center of the shell.

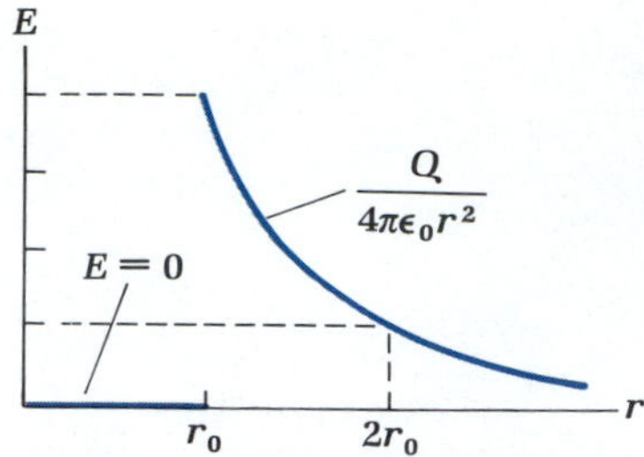

Figure 21-18. Example 21-5: A graph of E versus r for a uniformly charged spherical shell.

Figure 21-18 shows a graph of E for a uniformly charged spherical shell as a function of distance r from its center. Notice the remarkable simplicity of this result. The field at all points inside the charged shell is zero, and the field outside the shell is the same as the field due to a particle with charge Q located at the center of the shell. From field measurements outside a uniform spherical shell of a charge, one cannot determine whether the field is due to a charged shell or a charged particle with the same charge.

SELF-TEST 21-5. Suppose you are investigating a planet in another solar system. You bore a hole into the planet's surface and find that it is hollow inside; its mass distribution is similar to that of a basketball. Next you throw a rock into the hole directly toward the center of the planet. Describe the rock's motion. ***ANSWER:*** The rock will travel in a straight line at a constant speed.

EXAMPLE 21-6

A uniformly charged sphere. Determine **E** at points both inside and outside a uniform spherical distribution of charge of radius r_0 and charge Q. The charge distribution is similar to the mass distribution of a billiard ball; it is continuous and uniform throughout the volume of the sphere.

Solution. As in the previous example, **E** has only a radial component and depends only on distance from the center. Therefore we use spherical gaussian surfaces with the same center as the charge distribution. To find the field inside the charge distribution, we use a gaussian surface whose radius is less than the radius of the charge distribution, $r < r_0$ (Fig. 21-19*a*).

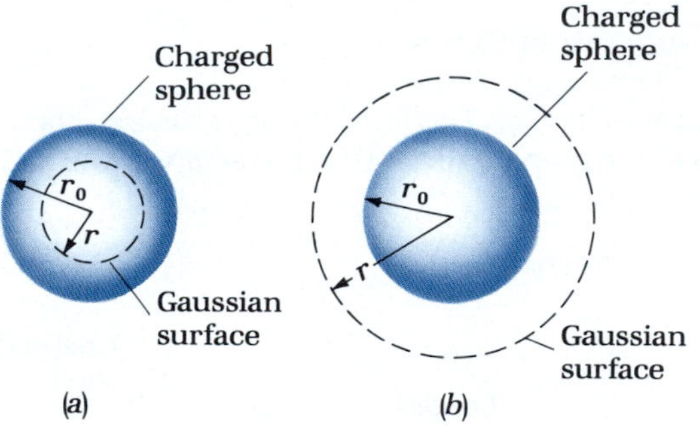

Figure 21-19. Example 21-6: *(a)* A spherical gaussian surface inside a spherical distribution of uniform volume charge density. $\Sigma q = (r^3/r_0^3)Q$. *(b)* A spherical gaussian surface outside the spherical charge distribution $\Sigma q = Q$.

The expression for the flux for this surface is again $\Phi_E = E(4\pi r^2)$. The charge contained within this gaussian sphere depends on the radius r of the gaussian sphere. We let ρ represent the ***volume charge density:***

$$\rho = \frac{Q}{4\pi r_0^3/3}$$

where Q is the total charge on the sphere and $4\pi r_0^3/3$ is the volume of the charged sphere. The amount of charge that is inside the gaussian sphere of radius r is the product of the charge density and the volume of the gaussian sphere:

$$\Sigma q = \rho \frac{4\pi r^3}{3} = \frac{Q}{4\pi r_0^3/3}\,\frac{4\pi r^3}{3} = Q\frac{r^3}{r_0^3}$$

Gauss's law, $\Phi_E = \Sigma q/\epsilon_0$, for this case is

$$E(4\pi r^2) = \frac{Q(r^3/r_0^3)}{\epsilon_0}$$

Solving for E gives

$$E = \frac{Qr}{4\pi\epsilon_0 r_0^3} \qquad (r < r_0)$$

The electric field increases linearly with r at points inside the charged sphere.

Finding the electric field outside the charged sphere is quite similar to the previous example. We use a gaussian sphere with $r > r_0$, as shown in Fig. 21-19*b*. The expression for the flux is again $E(4\pi r^2)$, and the charge enclosed by the gaussian sphere is the total charge Q on the charged sphere. Gauss's law gives

$$E(4\pi r^2) = \frac{Q}{\epsilon_0}$$

so that

$$E = \frac{Q}{4\pi\epsilon_0 r^2} \qquad (r > r_0)$$

The expression for the field at a point outside the sphere is the same as for the spherical shell of charge and for a point charge. A graph of E produced by a uniform spherical charge distribution is shown in Fig. 21-20.

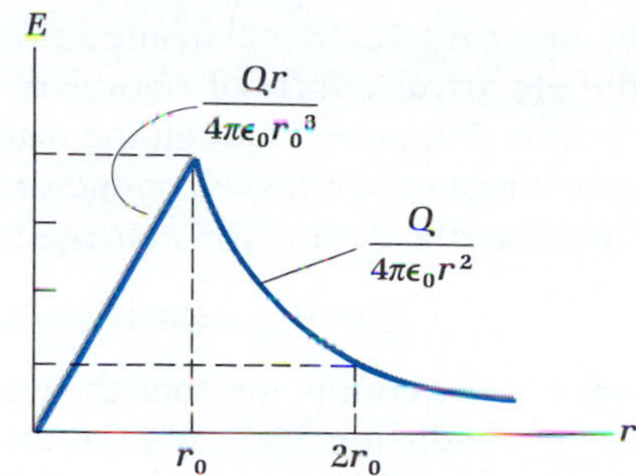

Figure 21-20. Example 21-6: A graph of E versus r for a spherical distribution of uniform volume charge density.

SELF-TEST 21-6. What is the volume charge density of a uniformly charged spherical distribution of charge of radius 0.18 m and charge +65 nC? ***ANSWER:*** 2.7 μC/m³.

Spherical Charge Distributions

The uniformly charged spherical shell in Example 21-5 and the uniformly charged spherical volume in Example 21-6 exhibit a feature that is common to spherically symmetric charge distributions:

> The field outside a spherically symmetric distribution of charge is directed radially, and its magnitude is
>
> $$E = \frac{Q}{4\pi\epsilon_0 r^2}$$
>
> where r is the distance from the center of the distribution.

This result is independent of how the charge is distributed radially. The field inside a spherically symmetric distribution depends on how the charge is distributed radially. The two examples cited above represent two possibilities.

Problem-Solving Techniques

Gauss's law can be used to find an expression for the electric field produced by a charge distribution when the distribution is highly symmetric. Also, if the distribution is nearly symmetric, then Gauss's law may be used to provide an approximate view of the spatial behavior of the field. Here are some guidelines:

1. Examine the symmetry of the charge distribution with an eye toward selecting a gaussian surface. If the charge distribution has cylindrical symmetry, like that of a straight-line charge, then choose a cylindrical gaussian surface. If the charge distribution has spherical symmetry, like that of a spherical shell of charge, then choose a spherical gaussian surface. The flux for such surfaces, in terms of E, can be evaluated by inspection.

2. Determine an expression for the charge enclosed by your gaussian surface in terms of the appropriate charge density—λ, σ, or ρ.

3. Insert your results from steps 1 and 2 into Gauss's law and solve for E.

21-5 ELECTROSTATIC PROPERTIES OF A CONDUCTOR

Some general properties of a conductor in an electric field can be determined with Gauss's law.

Field and Charge Inside a Conductor

In electrostatics, $\mathbf{E} = 0$ inside a conductor because (i) electrostatics is the study of the electrical effects of stationary charges and (ii) a conductor contains charge carriers that move through the material when an electric field exists in that material. When we discuss a conductor in the context of electrostatics, the situation is such that the conductor's charge carriers are not moving, which requires that

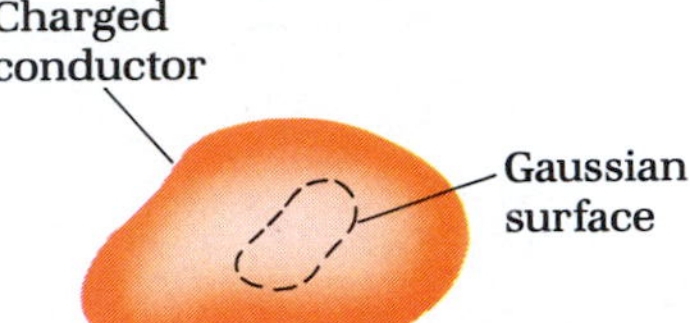

Figure 21-21. Conductor with a gaussian surface completely inside.

$$\mathbf{E} = 0 \qquad \text{(inside a conductor under static conditions)}$$

If $\mathbf{E} \neq 0$ in a conductor, then the charge carriers move and an electric current exists in the conductor. Electric currents are discussed in Chap. 24.

Given that $\mathbf{E} = 0$ inside a conductor, we can use Gauss's law to determine where a conductor's excess charge resides. Figure 21-21 shows a conductor with a gaussian surface completely inside it. Since $\mathbf{E}$ is zero everywhere inside the conductor, Φ_E is zero for this surface. Therefore Gauss's law requires that the net enclosed charge be zero. This is true for any closed surface as long as the surface is completely contained within the conductor. Thus no excess charge can exist anywhere inside a conductor—that is, the volume charge density ρ must be zero for a conductor. A net volume charge density can only exist in an insulator.

No excess charge can exist inside a conductor under static conditions.

If none of a conductor's excess charge can reside inside the conductor, then where does the charge reside? It must reside on the conductor's surface. Figure 21-22 shows a charged conductor with a gaussian surface just inside the actual surface of the conductor. The field is zero at each point on the gaussian surface, so that $\Phi_E = 0$ and the charge enclosed by the gaussian surface is zero. Therefore the conductor's excess charge resides outside the gaussian surface. This means that it must reside on the conductor's actual surface. If a conductor possesses excess charge, then this charge is distributed as a surface charge density σ.

In electrostatics, excess charge resides on the surface of a conductor.

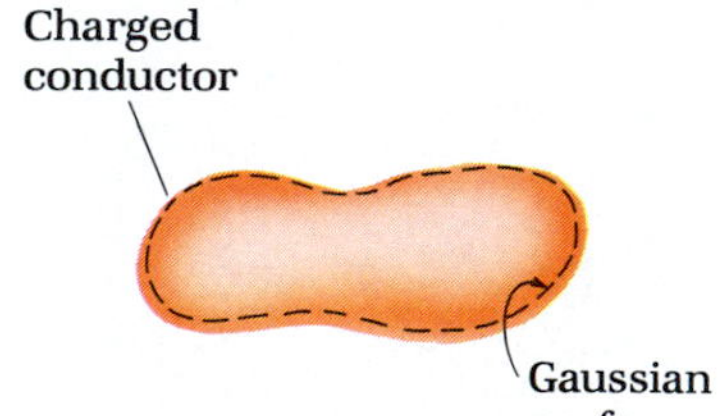

Figure 21-22. Conductor with a gaussian surface just inside its surface.

In general, the surface charge density σ on a conductor varies with position on the surface of the conductor. For example, a neutral conductor can have a positive charge density over part of its surface and a negative charge density over another part such that the total charge is zero.

Field Just Outside a Conductor

Gauss's law can be used to investigate the field just outside a conductor. Consider a point that is so close to the conductor's surface that the surface can be considered planar. The field can then be divided into two components: the component E_t tangent to the surface and the component E_n normal to the surface. With the techniques introduced in the next chapter, you can show $E_t = 0$ (see Prob. 19 in Chap. 22). That is, just outside a conductor the field is directed perpendicular to the conductor's surface, either directly toward it or directly away from it.

Now we use Gauss's law to find E_n just outside the surface of a conductor. Since $\mathbf{E}$ is perpendicular to the conductor's surface, we choose a gaussian surface that is a small cylinder that penetrates the conductor's surface (Fig. 21-23). The cylinder's ends are parallel to the conductor's surface and the cylinders sides are perpendicular to the conductor's surface. The flux for the cylinder's sides is zero because $E_t = 0$. The flux for the cylinder's end that is inside the conductor is zero because $\mathbf{E} = 0$ inside a conductor. The only contribution to the flux for this gaussian cylinder comes from the field just outside the surface. Letting ΔS be the area of the end of

Figure 21-23. Charged conductor with a small cylindrical gaussian surface intersecting its surface.

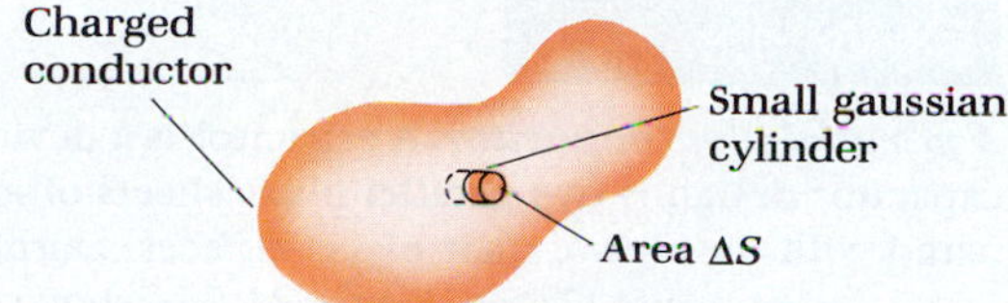

the cylinder that is outside the conductor, we have

$$\Phi_E = E_n \, \Delta S$$

The charge enclosed by the gaussian cylinder is $\sigma \, \Delta S$, where σ is the charge density on the conductor's surface. Gauss's law gives

$$E_n \, \Delta S = (\sigma \, \Delta S)/\epsilon_0 \qquad \text{or} \qquad E_n = \sigma/\epsilon_0$$

Thus the characteristics of the field just outside a conductor are

$$E_n = \frac{\sigma}{\epsilon_0} \qquad \text{and} \qquad E_t = 0 \tag{21-4}$$

Electric field just outside the surface of a conductor

At points on the surface where σ is positive, the field just outside the surface points away from the surface (E_n is positive), and at points on the surface where σ is negative, the field just outside the surface points toward the surface (E_n is negative).

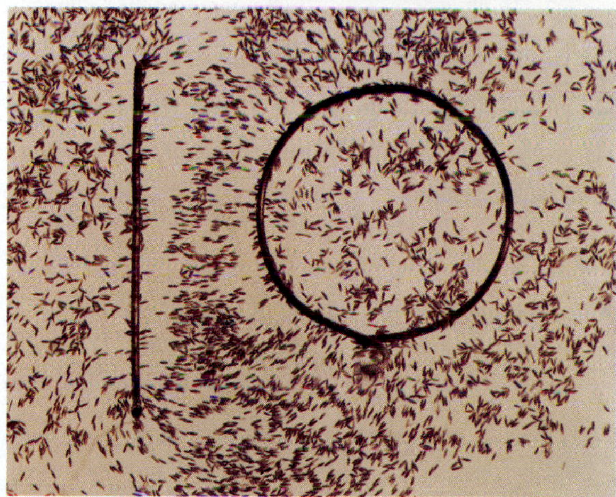

Grass-seed demonstration of electric field lines around a charged plate and a charged sphere of the opposite sign. A ring and a bar with charge of equal magnitude and opposite sign provide a two-dimensional representation of the sphere and the plate. Notice that there is no alignment of the seeds inside the ring.

EXAMPLE 21-7

A charged spherical conductor. Find the magnitude of the field just outside an isolated spherical conductor of radius r_0 and charge Q.

Solution. Since the surface charge density of the sphere is uniform by symmetry, it is simply the charge Q divided by the surface area $4\pi r_0^2$ of the sphere: $\sigma = Q/4\pi r_0^2$. Using Eq. 21-4, we find that the magnitude of the field is

$$E = \frac{|\sigma|}{\epsilon_0} = \frac{|Q|}{4\pi\epsilon_0 r_0^2}$$

Since a conductor carries its charge as a surface charge density, a charged spherical conductor must correspond to a uniform spherical shell of charge. In Example 21-5 we found that a uniformly charged spherical shell has $\mathbf{E} = 0$ inside. Therefore Fig. 21-18 gives the magnitude of the field due to a charged spherical conductor as a function of distance r from the center, and our answer above corresponds to E evaluated at $r = r_0$.

SELF-TEST 21-7. The electric field has a magnitude of 430 N/C at a point very near an aluminum sphere of radius 0.055 m. The field is directed away from the sphere. What is the charge on the sphere? ***ANSWER:*** 0.14 nC.

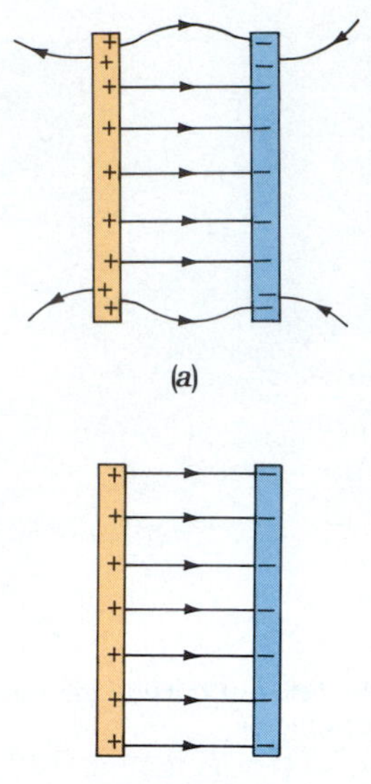

Figure 21-24. Example 21-8: *(a)* A parallel-plate capacitor showing a fringing field. *(b)* An ideal parallel-plate capacitor with its uniform charge densities and uniform field.

EXAMPLE 21-8

A parallel-plate capacitor. A capacitor is a device used in electric circuitry, and a simple capacitor design is two parallel plane sheets of metal whose separation is very small compared with the area of their plane surfaces. During normal operation, the two plates carry equal and opposite charge. Figure 21-24*a* shows the plates of a capacitor in cross section. The separation of the plates is greatly exaggerated for purposes of illustration, and the charge distribution and field lines are shown schematically. The charge on each plate is nearly uniformly distributed because of the attraction by the charge of opposite sign on the other plate. Also, the field between the plates is nearly uniform except for the "fringing field" near the edges. If the plates are brought closer and closer together, then the two charge distributions and the field become more nearly uniform. For simplicity, assume each plate has a uniform charge distribution and that the field between the plates is uniform (Fig. 21-24*b*). Find **E** between the plates of a capacitor in which the area A of each plate is 0.19 m² and the magnitude of the charge Q on each plate is 1.3 μC.

Solution. Since the field between the plates is uniform, its magnitude in the entire region is the same as just outside each surface: $|\sigma|/\epsilon_0$ from Eq. 21-4. The magnitude of the uniform surface charge density on each plate is $|\sigma| = Q/A$ so that

$$E = \frac{|\sigma|}{\epsilon_0} = \frac{Q}{A\epsilon_0}$$

$$= \frac{1.3\ \mu\text{C}}{(0.19\ \text{m}^2)[8.85 \times 10^{-12}\ \text{C}^2/(\text{N}\cdot\text{m}^2)]}$$

$$= 7.7 \times 10^5\ \text{N/C}$$

The direction of the field is away from the positive plate and toward the negative plate.

SELF-TEST 21-8. The field between the plates of a parallel-plate capacitor is 1.4×10^4 N/C. What is the magnitude of the surface charge density on each plate? ***ANSWER:*** 0.12 μC/m².

The Faraday Ice-Pail Experiment

Neither an electric field nor a volume charge density exists inside a conductor under static conditions. Suppose there is a hollow region inside a charged conductor. Does an electric field exist in this hollow region? Does excess charge reside on

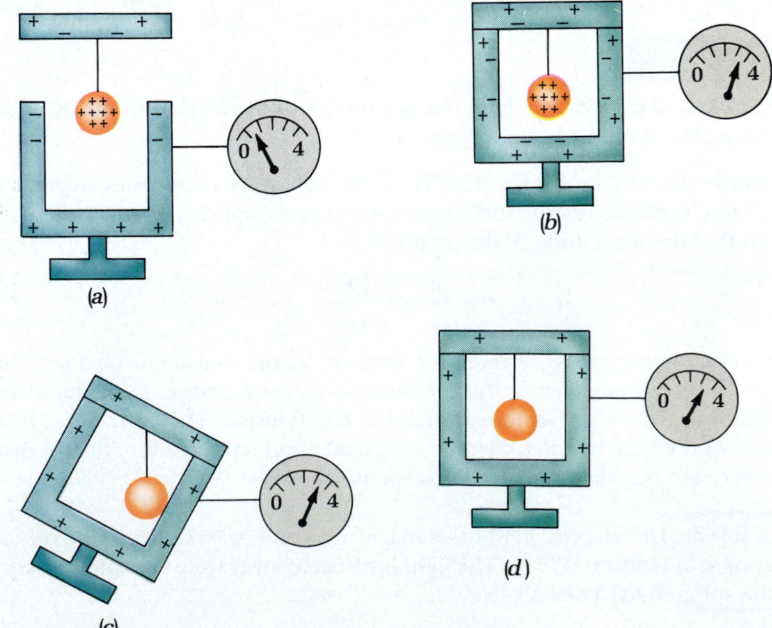

Figure 21-25. Faraday ice-pail experiment. *(a)* Charged metal sphere is lowered into a neutral metal bucket on an insulating stand. An electrometer shows that the outside of the can becomes charged as the sphere is lowered. *(b)* Bucket completely encloses the sphere. *(c)* Bucket is tilted so that the sphere touches the inside of the can. *(d)* Sphere has zero charge after contact, and the outside of the can has a charge equal to the original charge on the sphere.

the inside surface of the conductor? To answer these questions we consider an operation called the Faraday ice-pail experiment.

Suppose we hang a charged metal sphere by an insulating thread inside a closed metal can. The can possesses no net charge and is mounted on an insulating stand (Fig. 21-25). An electrometer, a device that measures charge, indicates that the outside of the can becomes charged as the sphere is lowered into the can (Fig. 21-25*a*) and is a maximum when the sphere is completely enclosed (Fig. 21-25*b*). When the can is tilted (Fig. 21-25*c*) so that the sphere and can make contact, the can and sphere compose a single conductor, but the electrometer reading is unaffected. If the sphere is removed from the can and tested for the presence of charge, it is found to be uncharged.

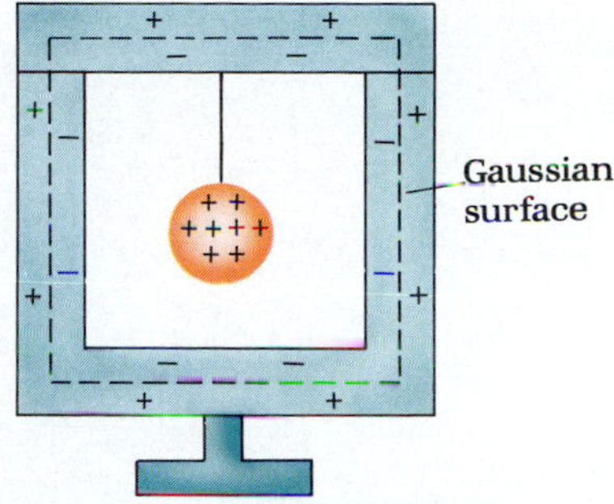

Figure 21-26. For the situation in Fig. 21-25*b*, a gaussian surface is shown within the metal bucket. Since **E** = 0 at each point on the gaussian surface, the charge enclosed is zero. This means that the charge on the inner surface of the bucket is equal and opposite the charge on the sphere. Since the bucket is neutral, the charge on its outer surface is equal to the charge on the sphere.

Gauss's law predicts that the charge induced on the inner surface of the can in Fig. 21-25*b* is the negative of the charge on the sphere. To be specific, suppose the initial charge on the sphere is +26 nC. Figure 21-25*b* is repeated in Fig. 21-26, with the addition of a gaussian surface that is contained inside the conducting material of the metal can. Since **E** = 0 at each point on this gaussian surface, the flux for the gaussian surface is zero. Thus the gaussian surface encloses no net charge. The gaussian surface surrounds the sphere with its charge of +26 nC and the inner surface of the metal can with its induced charge. Therefore the induced charge on the inner surface of the can must be −26 nC. Since the can was originally neutral and is insulated from ground, its net charge is zero. Therefore the charge on its outer surface must be +26 nC.

An electric field does exist in the hollow region of the can in Fig. 21-25*b*. You should use Gauss's law to verify this for yourself.

Since the electrometer reading was unaffected by the contact between the sphere and the inner surface of the can in Fig. 21-25*c*, the charge on the outer surface of the can in Fig. 21-25*d* is still +26 nC. Also, since no charge exists on the sphere after contact, we conclude that no net charge exists on the inner surface of the can after contact (Fig. 21-25*d*). As it turns out, not only is the net charge on the inner surface zero, but the surface charge density at every point on the inner surface is also zero, and the field in the hollow region of the can is zero. To prove this requires another property of the electric field—it is a conservative field. This is the subject of the next chapter.

COMMENTARY: Michael Faraday

Michael Faraday

Throughout our study of electromagnetism, the name Michael Faraday appears often. The laws of electricity and magnetism owe more to the experimental discoveries of Faraday than to those of any other person. He originated the concept of a field; he discovered electromagnetic induction, which led him to invent the dynamo, the forerunner of the electric generator; and he explained electrolysis in terms of electric forces at the molecular level.

Faraday was born at Newington, Surrey, England, in 1791, the son of a blacksmith. His only formal education was in reading, writing, and arithmetic as a young child. He became an apprentice bookbinder at the age of 14 and developed an insatiable appetite for reading. His lifelong fascination with science began when he happened to read an article on electricity in a copy of the *Encyclopaedia Britannica* which had been left for rebinding.

Normally, the world of science in the nineteenth century would be closed to a person with Faraday's education, but fortune intervened. Sir Humphry Davy, the famous chemist, was temporarily blinded in a laboratory accident at the Royal Institution in 1812, and Davy hired Faraday as his laboratory assistant. So Faraday began his scientific career as a chemist's assistant, but the assistant soon surpassed the master. During this time Faraday discovered and described benzene and was the first to discover compounds of chlorine and carbon.

Besides Davy, another great scientist played a major role in Faraday's life, the Scottish theoretician James Clerk Maxwell (see the Commentary in Chap. 27). Despite the fact that most of their contemporaries rejected Faraday's idea of electric and magnetic fields, Maxwell seized on it and made it mathematically legitimate. As you will see, the laws of electricity and magnetism, called *Maxwell's equations,* are normally stated in terms of the electric and magnetic fields. Maxwell showed mathematically that these fields contain energy and momentum, as Faraday had anticipated. Maxwell published these works during the 1860s, but by then Faraday was senile; he was to die a few years later.

Faraday had a great talent for explaining scientific results to the public and for instilling an interest in science. He instigated the Friday Evening Discourses at the Royal Institution, which are still used as a channel of communication between scientist and layperson. He was reknowned for his lectures to the young. His book for children entitled *The Chemical History of a Candle* is a classic and is still in print.

For further reading see *Michael Faraday, a Biography,* by L. Pearce Williams (1965).

SUMMARY

Section 21-1. Flux

The flux of the electric field for a surface is defined as the surface integral of the electric field over the surface. The electric flux for a closed surface is

$$\Phi_E = \oint \mathbf{E} \cdot d\mathbf{S} \qquad (21\text{-}2)$$

The closed surface for which the flux is evaluated is an imaginary or mathematical surface called a gaussian surface. Field lines can be used to visualize the flux for a surface.

Section 21-2. Gauss's Law

Gauss's law states that the flux for any closed surface is equal to the net charge enclosed by that surface divided by ϵ_0:

$$\Phi_E = \frac{\Sigma q}{\epsilon_0} \quad \text{or} \quad \oint \mathbf{E} \cdot d\mathbf{S} = \frac{\Sigma q}{\epsilon_0} \qquad (21\text{-}3)$$

Section 21-3. Developing Gauss's Law from Coulomb's Law

In electrostatics Gauss's law is equivalent to Coulomb's law.

Section 21-4. Using Gauss's Law to Find *E*

Gauss's law can be used to find the electric field produced by some highly symmetric charge distributions. The crucial step in such a calculation is the selection of the gaussian surface. Table 21-1 lists the results we obtained.

Section 21-5. Electrostatic Properties of a Conductor

Under electrostatic conditions, (i) $\mathbf{E} = 0$ everywhere inside a conductor and (ii) $E_t = 0$ just outside a conductor. With these conditions, Gauss's law requires that (i) no excess charge can exist anywhere inside a conductor and (ii) $E_n = \sigma/\epsilon_0$ just outside a conductor.

TABLE 21-1. *The Electric Field for Various Charge Distributions*

Charge distribution	Field magnitude	
Near a long, straight, uniform line charge, far from the ends	$E \approx \dfrac{\lambda}{2\pi\epsilon_0 R}$	
Near a large planar sheet of uniform surface charge, far from the edges	$E \approx \dfrac{\sigma}{2\epsilon_0}$	
Inside and outside a spherical shell of uniform surface charge density	$E = 0$	$(r < r_0)$
	$E = \dfrac{Q}{4\pi\epsilon_0 r^2}$	$(r > r_0)$
Inside and outside a sphere of uniform volume charge density	$E = \dfrac{Qr}{4\pi\epsilon_0 r_0^3}$	$(r < r_0)$
	$E = \dfrac{Q}{4\pi\epsilon_0 r^2}$	$(r > r_0)$
Inside a conductor	$E = 0$	
Just outside a conductor	$E_n = \dfrac{\sigma}{\epsilon_0} \quad E_t = 0$	

QUESTIONS

1. If $\mathbf{E} = 0$ at every point on a surface, is the flux for the surface necessarily zero? Suppose the surface is closed; what can you say about the charge enclosed by the surface?

2. If the flux for a surface is zero, is it necessarily true that $\mathbf{E} = 0$ at every point on the surface? Explain.

3. If no excess charge exists at any point inside a closed surface, is the field at each point on this surface necessarily zero? Is the flux for the surface necessarily zero?

4. If the flux for a closed surface is zero, can excess charge exist at points inside the surface? Explain.

5. If the net charge inside a closed surface is zero, can field lines cross the surface? If field lines do cross the surface, what can you say about the number of lines directed into the enclosed volume compared with the number directed out?

6. What sort of analogy can you draw between the flow of water and electric flux? What are the similarities and what are the differences?

7. In analogy to water flow, positively charged objects are often referred to as "sources" of electric field, and negatively charged objects are referred to as "sinks" of electric field. Explain the usefulness of this terminology in view of electric flux and field lines.

8. A Möbius strip (Fig. 21-27) is a one-sided surface that can be constructed by giving a strip of paper a twist and pasting the two ends together. Can you find the flux for a Möbius strip in a uniform field? Explain.

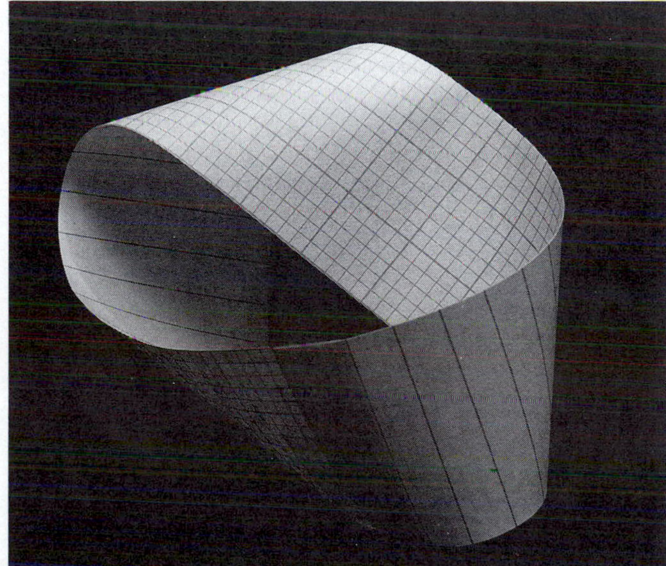

Figure 21-27. Question 8: A Möbius strip.

9. A Klein bottle (Fig. 21-28) can be considered a closed surface in the sense it has no edges. Can you apply Gauss's law to this "closed surface"? If not, why not?

Figure 21-28. Question 9: A Klein bottle.

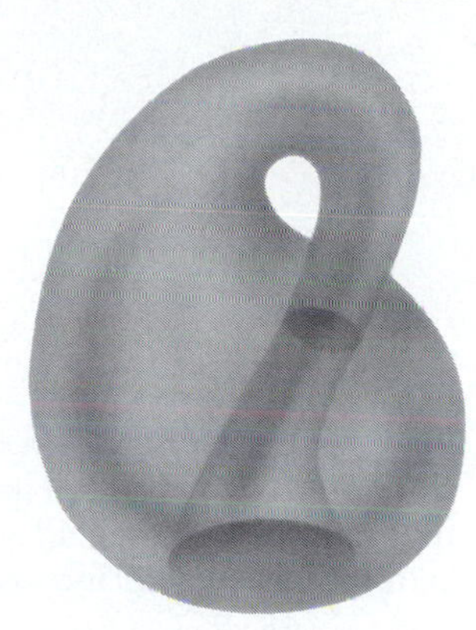

10. Is the field $\mathbf{E}$ in the flux integral of Gauss's law due (i) only to charges inside the gaussian surface, (ii) only the charges outside the gaussian surface, or (iii) to all charges everywhere?

11. Are the charges Σq in Gauss's law (i) only the charges inside the gaussian surface, (ii) only the charges outside the gaussian surface, or (iii) all charges everywhere?

12. To find the electric flux for a gaussian surface, do you need any information other than the net charge enclosed by the surface? To find the electric field at points on a gaussian surface, do you need any information other than the net charge enclosed by the surface? Explain.

13. What is the flux for a surface that encloses an electric dipole?

14. Would Gauss's law be useful in finding the electric field due to a shell shaped as a cube and having uniform surface charge? If so, what sort of gaussian surface would you use? Suppose you use a cube-shaped gaussian surface with the same center and orientation as the charged cubical shell. From symmetry, what can you say about the field direction at points on the gaussian cube? From symmetry, what can you say about the variation of E at points on the gaussian cube? Do you think you can find an expression for the flux in terms of E?

15. Could a cube-shaped gaussian surface be used to find the approximate electric field at points near the middle of a long, uniformly charged wire (Fig. 21-29)? Explain.

Figure 21-29. Question 15.

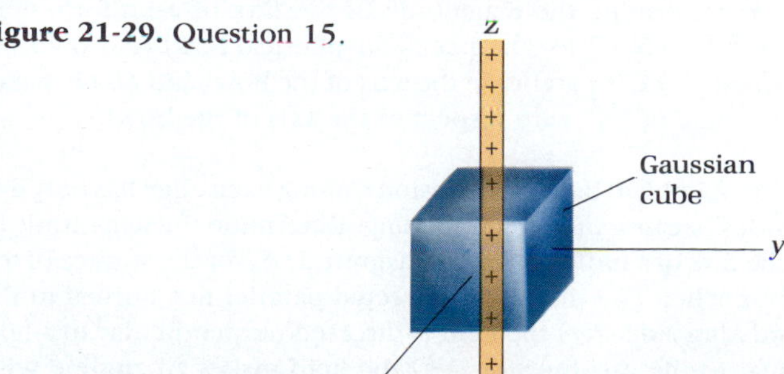

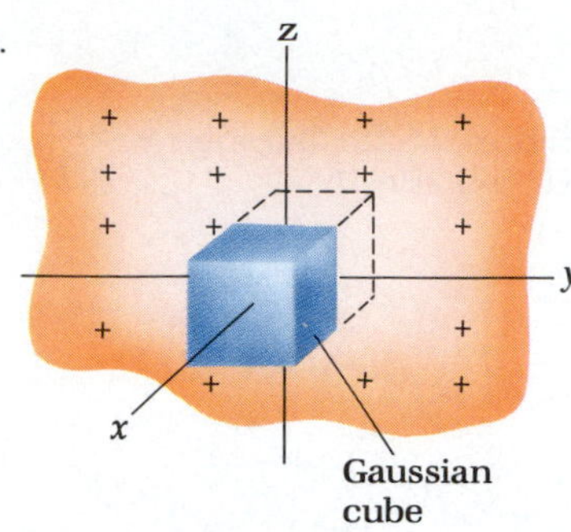

Figure 21-30. Question 16.

16. Could a cube-shaped gaussian surface be used to find the approximate electric field at points near the center of a large, uniformly charged plane disk (Fig. 21-30)? Explain.

17. Suppose two particles with different charge are near one another and are enclosed by a gaussian surface. If the particles exchange positions, does the flux for the surface change? Does the field at points on the surface change?

18. A metal sphere with no excess charge is suspended from an insulating string in a region where an electric field exists. Before the sphere is placed in position the field is uniform and pointed vertically upward: $\mathbf{E}_0 = E_0\mathbf{j}$. Will excess charge accumulate anywhere on the sphere? If so, where will it accumulate and what is its sign? If charge does accumulate, what is the contribution, due to this induced charge alone, to the electric field inside the sphere? Would a thin spherical shell of metal behave differently from a solid metal sphere?

19. An irregularly shaped neutral conductor encloses an irregularly shaped hollow region that contains a particle with charge $q = +10$ nC. What is the charge on the inner surface of the conductor? What is the charge on the outer surface of the conductor? Suppose the particle is moved to a different position inside the hollow region. Will the value of the charge on the inner or outer surfaces be affected? Will the surface charge densities at points on the inner or outer surfaces be affected? Will the field inside or outside the conductor be affected?

20. Suppose **E** is constant in direction but changes in magnitude in a given region. What can you conclude about charge in the region?

21. Coulomb's law and Newton's law of universal gravitation are both inverse-square-force laws, with the force directed along the line between the interacting particles. Can we apply Gauss's law to gravitational fields as well as to electric fields? If an object is contained inside a gaussian surface, what property of this object is proportional to the gravitational flux for the surface? Explain.

22. Consider using the concept of flux to describe rainfall. What would determine the amount of water collected in a bucket resting on a horizontal surface during a rainfall accompanied by a steady wind? What should be the SI units of rain flux?

23. Complete the following table:

Symbol	Represents	Type	SI Unit
Φ_E		Scalar	
$d\mathbf{S}$			m^2
Σq			
ρ	Volume charge density		
$\oint$		N/A	N/A

EXERCISES

Section 21-1. Flux

1. Determine the magnitude of the flux of a uniform field, $E = 660$ N/C directed vertically upward, for a plane rectangular surface (dimensions 1.5 m by 2.1 m) when *(a)* the surface is horizontal, *(b)* the surface is vertical, *(c)* a normal to the surface makes an angle of 32° with respect to the vertical.

2. The magnitude of the flux for a plane circular surface of radius 0.54 m in a uniform field is 340 $N \cdot m^2/C$ when the surface is oriented with its normal parallel to the field direction. What is E at each point on the surface?

3. Determine the magnitude of the flux of a uniform field ($E = 840$ N/C) for an open hemispherical bowl ($r = 0.41$ m) when *(a)* **E** is parallel to the axis of the bowl and *(b)* **E** makes an angle of 63° with respect to the axis of the bowl.

4. A cubical box of dimension ℓ along each edge has only five sides because one side is missing. Determine the magnitude of the flux of a uniform field of magnitude E_0 for the surface of the box when *(a)* the field is directed parallel to a normal to the missing side, *(b)* the field is directed perpendicular to a normal to the missing side, *(c)* the field makes an angle θ with respect to a normal to the missing side.

5. What are the cartesian components of $\Delta\mathbf{S}$ for the slanted surface in Fig. 21-4?

6. What is the magnitude of the flux of a vertical field $\mathbf{E}_0$ for the surface of a cuspidor (Fig. 21-31) with opening radius a and neck radius b?

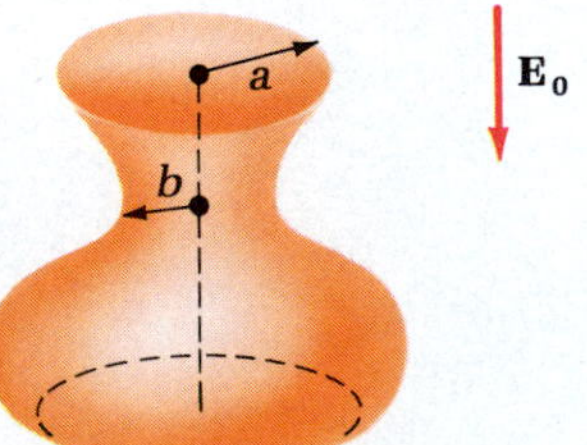

Figure 21-31. Exercise 6.

7. A plane surface of area 2.8 m^2 is oriented such that its surface vector is parallel to a uniform field and 98 field lines cross the surface. What is the angle between the field direction and the surface vector when the surface is oriented such that 38 field lines cross the surface?

8. *(a)* What is the flux of a uniform field $\mathbf{E} = (-240\text{ N/C})\mathbf{i} + (-160\text{ N/C})\mathbf{j} + (390\text{ N/C})\mathbf{k}$ for a plane sur-

face $\Delta\mathbf{S} = (-1.1\ \text{m}^2)\mathbf{i} + (4.2\ \text{m}^2)\mathbf{j} + (2.4\ \text{m}^2)\mathbf{k}$? *(b)* What is the projection of $\Delta\mathbf{S}$ on a plane perpendicular to **E**? *(c)* What is the angle between $\Delta\mathbf{S}$ and **E**?

Section 21-2. Gauss's Law

9. A spherical gaussian surface of radius 1.0 m is centered at a particle with charge of 1.0 nC. *(a)* What is the area of the gaussian sphere? *(b)* What is E at each point on the gaussian sphere? *(c)* Determine the flux for the gaussian sphere from your answers to parts *(a)* and *(b)*. *(d)* Repeat parts *(a)*, *(b)*, and *(c)* for a gaussian sphere of radius 2.0 m.

10. For an open hemispherical surface, what is the magnitude of the flux due to a particle with charge $q = -26$ nC located at the center of the corresponding sphere?

11. A cube-shaped gaussian surface has one corner at the origin of coordinates and the diagonally opposite corner at (ℓ, ℓ, ℓ) such that the edges of the cube are aligned with the coordinate axes. Charged particles and their locations (x, y, z) are $q_1 = 33$ nC at $(\ell/2, 0, 2\ell)$, $q_2 = -54$ nC at $(\ell/3, \ell/4, \ell/3)$, and $q_3 = 28$ nC at $(\ell/4, \ell/2, \ell/3)$. What is the flux for the gaussian surface?

12. A particle with charge $q = -72$ nC is at the center of a cubical gaussian surface of dimension ℓ along each edge. *(a)* What is the flux for the closed gaussian cube? *(b)* What is the magnitude of the flux for one of the six sides? *(c)* If the charge were not at the center, would the answer to either part *(a)* or part *(b)* be different?

Section 21-3. Developing Gauss's Law from Coulomb's Law

13. A solid angle $\Delta\Omega$ subtended at a point is defined as

$$\Delta\Omega = \frac{\Delta S}{r^2}$$

where ΔS is the area of a spherical cap of radius r centered at the point. A solid angle is dimensionless, and its SI unit is the steradian (sr). Given that $\Delta S_2/\Delta S_1 = r_2^2/r_1^2$, show that spherical caps 1 and 2 in Fig. 21-9 subtend the same solid angle at the position of the charged particle.

14. Near the surface of the earth there exists an average electric field of about 150 N/C directed downward. Estimate the net charge on the earth. Assuming that this charge is distributed uniformly over the surface of the earth, estimate the surface charge density.

15. The downward-directed electric field in the earth's atmosphere is found to decrease in magnitude with increasing altitude above the surface. Suppose E is 100 N/C at 200 m above the surface of the earth and 50 N/C at 300 m above the surface of the earth. Estimate the average volume charge density in the earth's atmosphere in the altitude range from 200 to 300 m.

16. Consider a very large slab of plastic that has a thickness of 5.4 mm and a uniform charge density ρ. Just outside the slab and near its center, **E** is directed away from the slab on each side and has a magnitude of 940 N/C. Determine ρ.

17. A uniformly charged spherical shell with total charge $Q = -34$ nC is centered at the origin (Fig. 21-32). Find the flux for a gaussian cube with edges aligned with the axes and one corner at the origin. The length of the cube edges is larger than the sphere's radius. What is the flux for each face of the cube?

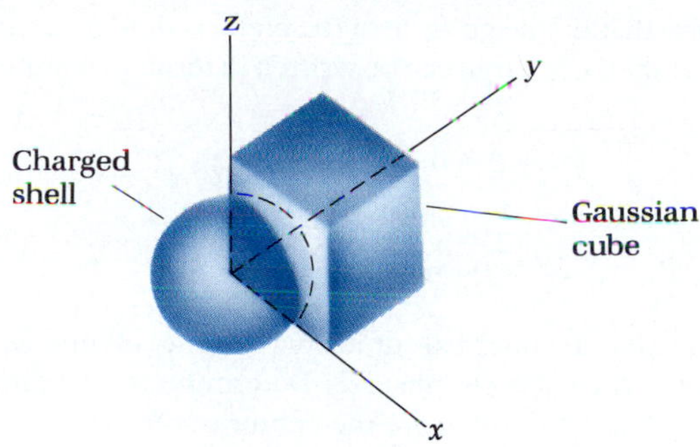

Figure 21-32. Exercise 17.

18. Make a cross-sectional sketch of the rounded-block gaussian surface in Fig. 21-9 with the point charge and its field lines in the plane of the drawing. Use your sketch to show that the field-line representation of the field is consistent with $\Phi_E = 0$ for the surface.

19. Make a cross-sectional sketch of the gaussian surface in Fig. 21-11 such that the raised cap is shown. Also show the point charge and its field lines in the plane of the drawing. Use your sketch to show that the field-line representation of the field is consistent with $\Phi_E = q/\epsilon_0$ for this surface.

20. Show that Gauss's law for gravitation

$$\Phi_G = \oint \mathbf{g}\cdot d\mathbf{S} = -4\pi G\,\Sigma m_i$$

is consistent with Newton's law of universal gravitation. What is the meaning of Σm_i? Why is the minus sign necessary? Why is the 4π present?

Section 21-4. Using Gauss's Law to Find *E*

21. A thin straight rod has a charge of -230 nC uniformly distributed along its 6.3-m length. *(a)* Determine the linear charge density. *(b)* Estimate **E** near the middle of the rod at a perpendicular distance of 25 mm.

22. The electric field at a point that is a perpendicular distance of 18 mm from the middle of a long, thin, straight, uniformly charged rod is directed toward the rod and has a magnitude of 3.5×10^4 N/C. *(a)* What is the linear charge density of the rod? *(b)* What is the magnitude of the field at a point that is a perpendicular distance of 9 mm from the middle of the rod?

23. Consider a thin, square planar sheet of uniform charge density, with $Q = +79$ nC and area 1.2 m². *(a)* Determine the surface charge density. *(b)* Estimate E near the middle of the sheet at perpendicular distances of 10 and 20 mm.

24. The electric field at a point that is a perpendicular distance of 12 mm from the middle of a thin, uniformly charged planar disk is directed toward the disk with magnitude 7.4×10^3 N/C. The radius of the disk is 0.91 m. *(a)* What is the surface charge density of the disk? *(b)* What is the charge on the disk? *(c)* Estimate the field at a perpendicular distance of 24 mm from the center of the disk.

25. A thin, uniformly charged spherical shell has $Q = -87$ nC and $r_0 = 55$ mm. *(a)* What is the surface charge density σ of the shell? *(b)* Find E at $r = 25$, 50, 75, and 100 mm from the center of the shell.

26. Show that the magnitude of the electric field for a uniform spherical shell of charge can be written in terms of σ rather than Q:

$$E = 0 \qquad (r < r_0)$$

$$E = \frac{\sigma r_0^2}{\epsilon_0 r^2} \qquad (r > r_0)$$

27. Consider a spherical uniform volume charge ρ, with $Q = 61$ nC and $r_0 = 48$ mm. *(a)* Determine ρ. *(b)* Find E at $r = 24$, 48, and 96 mm from the center of the sphere.

28. Show that the magnitude of the electric field for a sphere of uniform volume charge can be written in terms of ρ rather than Q:

$$E = \frac{\rho r}{3\epsilon_0} \qquad (r < r_0)$$

$$E = \frac{\rho r_0^3}{3\epsilon_0 r^2} \qquad (r > r_0)$$

29. Suppose you measure the electric field to be 284 kN/C radially outward at a distance $r = 15$ mm from the center of a uniform spherical distribution of volume charge. *(a)* With this information alone, determine any of the following quantities that you can: the charge Q, the radius r_0, the charge density ρ. *(b)* With the additional information that $E = 370$ kN/C at $r = 30$ mm, determine any of the remaining quantities from part *(a)* that you can.

Section 21-5. Electrostatic Properties of a Conductor

30. A solid metal ball of radius 62 mm has a charge of +46 nC. *(a)* How is this charge distributed? *(b)* Find E at $r = 30$, 60, and 90 mm, where r is the distance from the center of the sphere to the point where E is evaluated. *(c)* Does the answer to part *(a)* or part *(b)* change if the ball is hollow?

31. When a thin planar conducting sheet is placed in a uniform electric field that is directed perpendicular to the sheet, the field outside the conductor near its middle is nearly unchanged. Suppose such a sheet is placed in a field of magnitude 940 N/C. What is the magnitude of the surface charge density that is induced on each side of the sheet near its middle?

32. An irregularly shaped conductor has a surface charge density of -52 nC/m² at point P on its surface. *(a)* What are E_t and E_n at a point adjacent to P just inside the conductor surface? *(b)* What are E_t and E_n at a point adjacent to P just outside the conductor surface?

33. The electric field just outside point P on the surface of an irregularly shaped conductor is 620 N/C directed away from the surface. What is the surface charge density at point P?

34. The square plates of a parallel-plate capacitor are 260 mm on a side and are separated by 1.2 mm. The magnitude of the electric field midway between the plates is 1.4 kN/C. What are the magnitudes of *(a)* the charge and *(b)* the charge density on the plates?

35. Use Gauss's law to show that **E** is nearly zero in the region near the middle just outside the plates of a parallel-plate capacitor (see Fig. 21-24).

36. The *dielectric strength* of an insulating material is the maximum electric field that can exist in the material without dielectric breakdown. When an insulator breaks down, some of its constituent molecules ionize and the material becomes a conductor. Lightning is an example of the dielectric breakdown of air. The dielectric strength of dry air at room temperature is about 3×10^6 N/C. Estimate the maximum charge that can be placed on a conducting sphere of radius 20 mm without breakdown of the surrounding air.

Additional Exercises

37. A surface is shaped like a soup bowl and its opening has a radius of 97 mm. The surface is oriented as a soup bowl would be if it were placed on a horizontal table. What is the magnitude of the electric flux for this surface in a uniform electric field of 120 N/C directed generally upward at an angle of 28° to the vertical?

38. A closed surface is near three point charges; $q_1 = -7.9$ nC, $q_2 = -3.6$ nC, and $q_3 = 5.3$ nC. The surface encloses charges 1 and 3, but charge 2 is outside the surface. What is the electric flux for this surface?

39. The surface vector for an open surface has a two-fold ambiguity. Consider a flat, rectangular surface of dimension 0.42 m by 0.61 m that has the edge of its short dimension horizontal and the edge of its long dimension at an angle of 58° to the horizontal. Let **i** be directed horizontal and perpendicular to the edge with the short dimension and let **j** be directed vertically upward. In terms of **i** and **j**, what are the two surface vectors that could represent this surface?

40. The radius of a uniform spherical volume charge is 100 mm and the magnitude of the electric field at its surface is 1000 N/C. What is the magnitude of the field *(a)* at the center, *(b)* 50 mm from the center, *(c)* 150 mm from the center, and *(d)* 200 mm from the center?

41. The radius of a uniform spherical shell of charge is 100 mm and the magnitude of the electric field just outside its surface is 1000 N/C. What is the magnitude of the field *(a)* at the center, *(b)* 50 mm from the center, *(c)* 150 mm from the center, and *(d)* 200 mm from the center?

42. A uniform spherical shell of charge with radius 999 mm and charge -111 nC is centered at the origin and a long thin uniform line charge with linear charge density 27.8 nC/m is along the z axis. Determine **E** at points along the x axis with an x coordinate of *(a)* 0.50 m; *(b)* 1.0 m, *(c)* 2.0 m, and *(d)* 3.0 m.

PROBLEMS

1. ***Field produced by a charged cylindrical shell.*** Consider a very long, straight cylindrical shell of uniform surface charge density σ and with radius R_0. The charge is distributed similar to the way mass is distributed for a long, thin-walled pipe. *(a)* Show that the approximate magnitude of the electric field at points far from the ends of the shell is given by

$$E = 0 \qquad (R < R_0)$$

$$E = \frac{\sigma R_0}{\epsilon_0 R} \qquad (R > R_0)$$

where R is the perpendicular distance from the axis of the cylindrical shell to the point where the field is evaluated. *(b)* Show that this result is consistent with Example 21-3 when $R > R_0$. *(c)* Make a graph of E versus R.

2. ***Field produced by a charged conducting rod.*** When a long, straight conducting rod is given excess charge, the charge is distributed approximately as a uniform surface charge density. Suppose a solid, straight metal rod of radius 11 mm and length 5.4 m has a charge of -47 nC. *(a)* Estimate the surface charge density on the rod. *(b)* Use the answer to the previous problem to find E in the perpendicular bisector plane at $R = 5$, 15, and 30 mm, where R is the perpendicular distance from the axis of the rod to the point where E is evaluated. *(c)* Is the answer to part *(b)* changed if the rod is hollow?

3. ***A line charge coaxial with a conducting pipe.*** A straight, thin wire filament 12 m long with $Q = -74$ nC and uniform linear charge density is coaxial with a neutral conducting pipe of the same length; the inner radius is 6.0 mm and outer radius is 9.0 mm (Fig. 21-33). *(a)* Estimate the induced surface charge densities on the inner and outer surfaces of the pipe. *(b)* For points in the perpendicular bisector plane, make a graph of E versus R in the range from $R = 1.0$ to 15 mm, where R is the perpendicular distance from the filament.

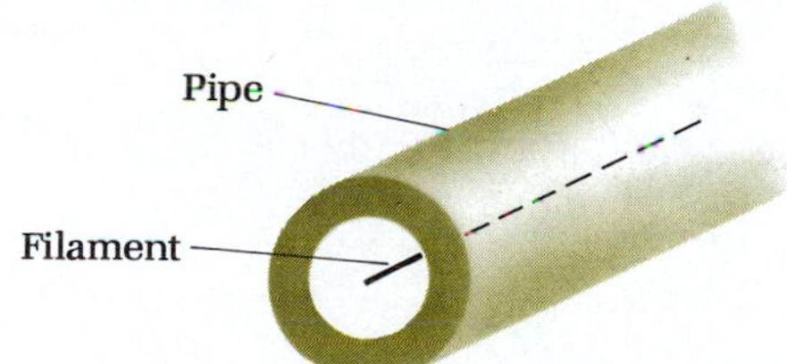

Figure 21-33. Problem 3.

4. ***A cylinder with a uniform volume charge.*** Consider a very long cylinder of uniform volume charge density ρ with radius R_0. The charge is distributed similarly to the mass of a long solid rod with circular cross section, such as a broom handle. *(a)* Show that the magnitude of the electric field at points far from the ends of the charge distribution is given by

$$E = \frac{R\rho}{2\epsilon_0} \qquad (R < R_0)$$

$$E = \frac{R_0^2\rho}{2\epsilon_0 R} \qquad (R > R_0)$$

where R is the perpendicular distance from the axis of the cylindrical charge to the point where E is evaluated. *(b)* Show that this result is consistent with Example 21-3 when $R > R_0$. *(c)* Make a graph of E versus R.

5. ***Oppositely charged spherical shells.*** Consider two concentric spherical shells of uniform surface charge densities σ_a and σ_b and radii a and b, $b > a$, as shown in cross section in Fig. 21-34. The two spherical shells have equal and opposite charge: $q_a = -Q$ and $q_b = +Q$, where $Q > 0$. *(a)* Develop expressions for E in all three regions of space: $r < a$, $a < r < b$, and $r > b$, where r is the distance from the center of the spheres to the point where E is evaluated. *(b)* Make a graph of E versus r from $r = 0$ to $3a$ for the case in which $b = 2a$.

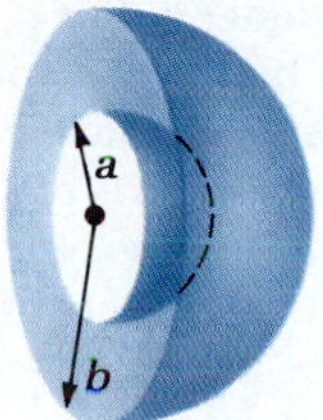

Figure 21-34. Problem 5.

6. ***Charged spherical shells.*** Repeat the previous problem, except let $q_b = +2Q$.

7. ***Thick conducting shell concentric with a point charge.*** A thick, spherical conducting shell with inner radius $a = 30$ mm, outer radius $b = 50$ mm, and zero net charge has a particle with charge $q = 28$ nC at its center (shown in cross section in Fig. 21-35). *(a)* Determine the charge densities on the inner and outer surfaces of the conductor. *(b)* Find E as a function of distance r from the particle and make a graph of E from $r = 20$ to 70 mm.

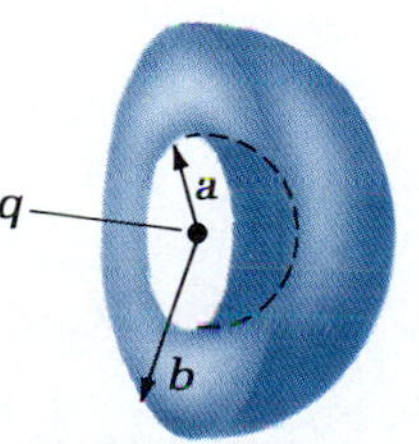

Figure 21-35. Problem 7.

8. ***Thick spherical shell of uniform volume charge.*** Consider a thick spherical shell of uniform volume charge density ρ, charge Q, inner radius a, and outer radius b. *(a)* Show that

$$E = 0 \qquad (r < a)$$

$$E = \frac{Q}{4\pi\epsilon_0 r^2}\frac{r^3 - a^3}{b^3 - a^3} \qquad (a < r < b)$$

$$E = \frac{Q}{4\pi\epsilon_0 r^2} \qquad (r > b)$$

where r is the distance from the center of the sphere to the point where E is evaluated. *(b)* Make a graph of E versus r.

9. ***A spherically symmetric nonuniform volume charge.*** A spherically symmetric distribution of charge has radius r_0 and charge Q. The volume charge density increases linearly from the center, $\rho = Ar$, where A is constant. Show that

$$A = \frac{Q}{\pi r_0^4}$$

$$E = \frac{Qr^2}{4\pi\epsilon_0 r_0^4} \qquad (r < r_0)$$

$$E = \frac{Q}{4\pi\epsilon_0 r^2} \qquad (r > r_0)$$

(*Hint:* An appropriate volume element for a spherically symmetric charge distribution is a thin spherical shell of thickness dr: $dV = 4\pi r^2\, dr$.)

10. ***A shell of nonuniform volume charge.*** A thick spherical shell has a charge Q, an inner radius a, and an outer radius b. The charge distribution between a and b is spherically symmetric but varies with distance from the center: $\rho = A/r$, where A is a constant. A point charge q is placed at the center of the sphere. *(a)* Determine q in terms of Q, a, and b such that the field between a and b is independent of r. *(b)* What is the field for $r < a$? *(c)* What is the field for $r > b$?

11. ***Uniform spherical volume charge with a nonconcentric spherical cavity.*** An otherwise uniform spherical distribution of charge has a spherical cavity completely inside it (Fig. 21-36). The position vector from the center of the sphere to the center of the cavity is $\mathbf{r}_0$. Show that the field inside the cavity is uniform and parallel to $\mathbf{r}_0$: $\mathbf{E} = \rho\mathbf{r}_0/3\epsilon_0$. Notice that this result is independent of the radius of either the sphere or the cavity. (*Hint:* Treat the cavity as a negative uniform charge density superimposed onto the positive uniform charge density of the sphere.)

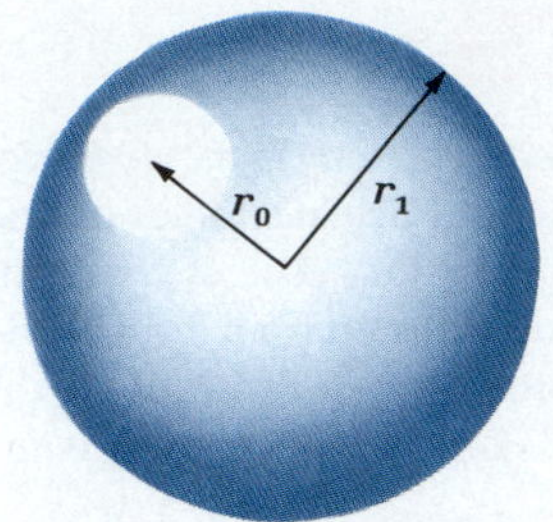

Figure 21-36. Problem 11.

12. ***A simple atomic model.*** Ernest Rutherford, who guided the experiments that established the modern view of atoms, used a simple model of the atom to explain the experimental results. He introduced the concept of a nucleus by representing an atom with atomic number Z as a particle with charge $+Ze$ (the nucleus) at the center of a uniform spherical distribution of charge with charge $-Ze$ and radius r_a (the electrons). Use Gauss's law to show that E at a distance r from the center of this charge distribution is

$$E = 0 \qquad (r > r_a)$$

$$E = \frac{Ze}{4\pi\epsilon_0}\left(\frac{1}{r^2} - \frac{r}{r_a^3}\right) \qquad (r < r_a)$$

ELECTRIC POTENTIAL

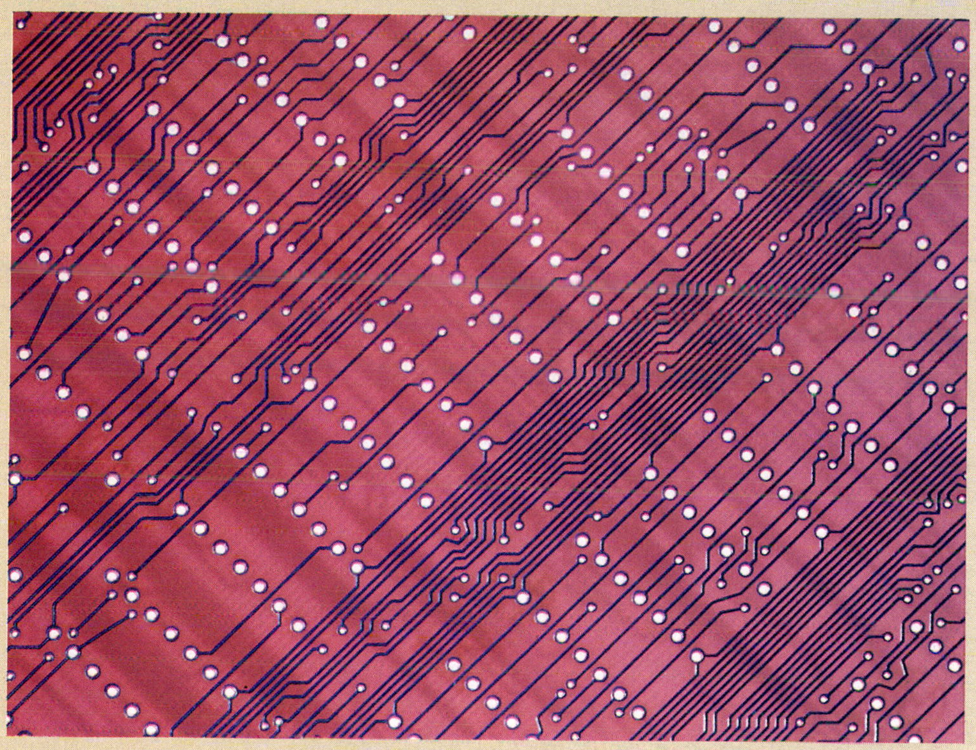

Conducting paths on a circuit board.

As discussed in the two previous chapters, the electrical effect of a charge distribution can be described in terms of the electric field produced by the distribution. In this chapter we introduce another kind of field, called the *electric potential,* or simply the *potential.* The electric field **E** is defined as a *force per unit charge,* and since force is a vector, **E** is a vector field. The potential $\mathcal{V}$ is defined as *potential energy per unit charge,* and since potential energy is scalar, $\mathcal{V}$ is a scalar field. Because $\mathcal{V}$ is a scalar, it is often more convenient to use than **E**, but each is derivable from the other. Indeed, the relation between **E** and $\mathcal{V}$ is analogous to that between a conservative force and its associated potential energy.

You may be familiar with electric potential by another name; it is often called the *voltage.* This term for potential comes from the common unit that is used in measuring potential, the *volt.* Therefore, referring to a potential as voltage is similar to referring to distance as mileage; the name for a concept arises from its unit of measure. The volt is named for Count Alessandro Volta (1745–1827), professor of physics at the University of Pavia, Italy. Volta invented the electric cell, or battery, which was the first device to provide steady electric currents.

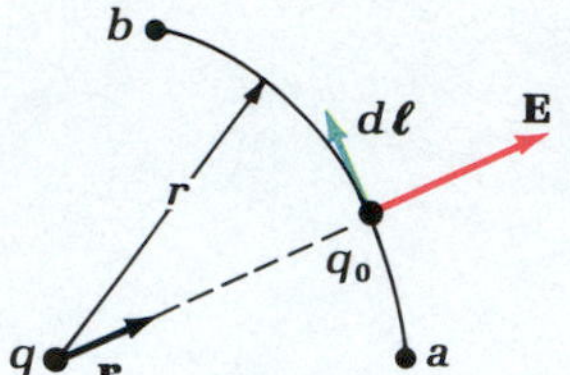

Figure 22-1. A test particle with charge q_0 is moved along a circular arc centered at a fixed point charge q. Since $d\boldsymbol{\ell}$ is perpendicular to $\mathbf{E}$ at each point on the path, the work done by the electric force is zero.

22-1 ELECTRIC POTENTIAL ENERGY

Electric potential is electric potential energy per unit charge. Before presenting a formal definition of potential, we first develop an expression for electric potential energy. The principal ingredients in this discussion are (i) the field due to a point charge and (ii) the principle of superposition.

Potential Energy of a Test Particle in the Field of a Point Charge

Consider the work done by the electric force when a test particle of charge q_0 is moved in the field of a fixed point charge q. In Fig. 22-1, the test particle is moved along a circular arc centered at the point charge q. The electric force on the test particle is $\mathbf{F} = q_0\mathbf{E}$, where $\mathbf{E}$ is the field produced by q. Since $\mathbf{E}$ is perpendicular to $d\boldsymbol{\ell}$ at each point along the circular arc path from a to b, the work W done by the electric force is zero. That is, $\mathbf{E}\cdot d\boldsymbol{\ell} = E\,d\ell\cos 90° = 0$, so that

$$W = \int_a^b \mathbf{F}\cdot d\boldsymbol{\ell} = q_0\int_a^b \mathbf{E}\cdot d\boldsymbol{\ell} = 0$$

This is true for any circular arc path centered at the point charge, and for any path on the surface of a sphere centered at the point charge.

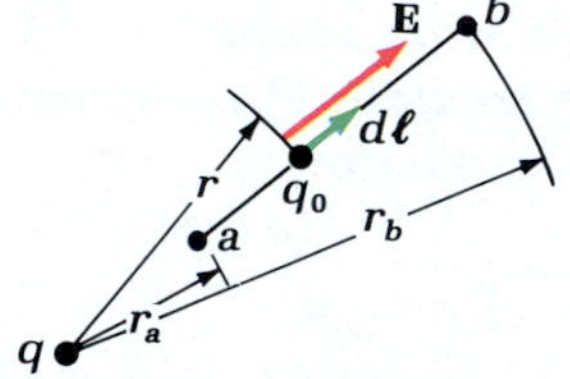

Figure 22-2. The test particle is moved along a radial path, $d\boldsymbol{\ell} = dr\,\hat{\mathbf{r}}$. The work done by the electric force is $(qq_0/4\pi\epsilon_0)(1/r_a - 1/r_b)$.

In Fig. 22-2, the test particle is moved along a radial path from a to b. Because the path is radial, the infinitesimal displacement $d\boldsymbol{\ell}$ can be written as $dr\,\hat{\mathbf{r}}$, where dr is an infinitesimal change in distance between the particles and $\hat{\mathbf{r}}$ is a unit vector that points away from the point charge. The field produced by q is $\mathbf{E} = (q/4\pi\epsilon_0 r^2)\hat{\mathbf{r}}$ so that

$$\begin{aligned} W &= \int_a^b \mathbf{F}\cdot d\boldsymbol{\ell} = q_0\int_a^b \mathbf{E}\cdot d\boldsymbol{\ell} \\ &= q_0\int_{r_a}^{r_b}\left(\frac{q}{4\pi\epsilon_0 r^2}\hat{\mathbf{r}}\right)\cdot(dr\,\hat{\mathbf{r}}) \\ &= \frac{q_0 q}{4\pi\epsilon_0}\int_{r_a}^{r_b}\frac{dr}{r^2} = \frac{q_0 q}{4\pi\epsilon_0}\left[-\frac{1}{r}\right]_{r_a}^{r_b} \\ &= \frac{q_0 q}{4\pi\epsilon_0}\left(\frac{1}{r_a} - \frac{1}{r_b}\right) \end{aligned}$$

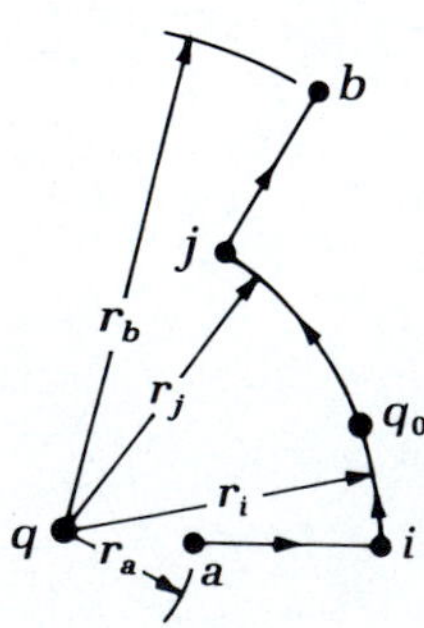

Figure 22-3. The test particle is moved along a radial-arc-radial path. Since the path from i to j is a circular arc, the work done by the electric force is independent of r_i and r_j.

Now suppose the test particle is moved along the path *aijb* in Fig. 22-3. Path a to i is radial, path i to j is a circular arc, and path j to b is radial. The work done by the electric force is

$$\begin{aligned} W = \int_a^b \mathbf{F}\cdot d\boldsymbol{\ell} &= \int_a^i \mathbf{F}\cdot d\boldsymbol{\ell} + \int_i^j \mathbf{F}\cdot d\boldsymbol{\ell} + \int_j^b \mathbf{F}\cdot d\boldsymbol{\ell} \\ &= \frac{q_0 q}{4\pi\epsilon_0}\left(\frac{1}{r_a} - \frac{1}{r_i}\right) + 0 + \frac{q_0 q}{4\pi\epsilon_0}\left(\frac{1}{r_j} - \frac{1}{r_b}\right) \end{aligned}$$

Since $r_i = r_j$, the two terms containing r_i and r_j, namely, $-q_0 q/4\pi\epsilon_0 r_i$ and $q_0 q/4\pi\epsilon_0 r_j$, cancel one another, and

$$W = \int_a^b \mathbf{F}\cdot d\boldsymbol{\ell} = \frac{q_0 q}{4\pi\epsilon_0}\left(\frac{1}{r_a} - \frac{1}{r_b}\right)$$

For this combination radial-arc-radial path, the work done by the electric force depends only on the separation distance between the particles before the displacement (r_a) and after the displacement (r_b). This is true no matter how many radial and arc segments we wish to use because terms that involve distances other than r_a and r_b always cancel out.

Figure 22-4 shows an arbitrarily shaped path between a and b with a number of small arc paths and radial paths superimposed on it. An arbitrarily shaped path can be viewed as an infinite number of infinitesimal arc paths and radial paths. As a test particle is moved from a to b along the arbitrarily shaped path, the work done for each infinitesimal arc path is zero, so that the work is the sum of the individual contributions from each of the infinitesimal radial paths. When these contributions are added, the terms involving distances other than r_a and r_b cancel out, as shown in the previous paragraph. Thus

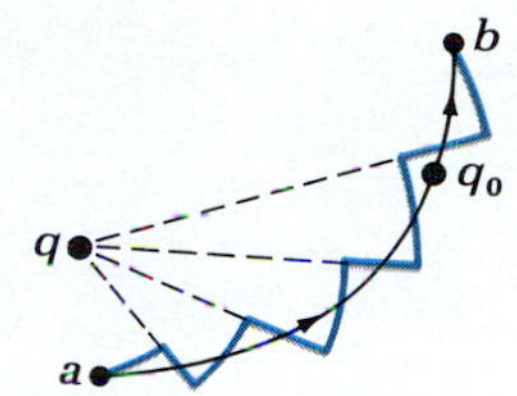

Figure 22-4. An arbitrarily shaped path with a number of small arc paths and radial paths superimposed on it. An arbitrary path may be viewed as an infinite number of infinitesimal arc paths and radial paths. The work done along each arc path is zero, so that the work done along the arbitrarily shaped path is the sum of the individual contributions along each radial path.

$$W = \int_a^b \mathbf{F} \cdot d\boldsymbol{\ell} = \frac{q_0 q}{4\pi\epsilon_0}\left(\frac{1}{r_a} - \frac{1}{r_b}\right) \tag{22-1}$$

for the arbitrarily shaped path between a and b.

The arc paths in Fig. 22-4 are contained in the plane of the page because we are constrained to show figures on a two-dimensional sheet, but they could just as well be out of the plane. Therefore, Eq. 22-1 is valid for any path whatsoever between a and b. That is, the work done by the electric force is independent of the path. If the work done by a force is independent of the path, the force is called a *conservative force* (Chap. 9), and the change in potential energy is the negative of the work done by the conservative force:

$$U_b - U_a = -W = -\int_a^b \mathbf{F} \cdot d\boldsymbol{\ell} \tag{22-2}$$

Substituting from Eq. 22-1, we find that the change in the test particle's potential energy is

$$U_b - U_a = \frac{q_0 q}{4\pi\epsilon_0}\left(\frac{1}{r_b} - \frac{1}{r_a}\right) \tag{22-3}$$

Although only changes in potential energy have physical significance, it is convenient to select a reference position at which the potential energy is defined as zero. In this way we can speak of the potential energy of a charged particle at a particular point, always remembering that its value depends on the reference position selected. From examination of Eq. 22-3, you can see that a simple choice is to identify the term containing $1/r_b$ with U_b and the term containing $1/r_a$ with U_a. This gives the potential energy $U(r)$ for a separation distance r as

Potential energy of a test particle at a distance r from a point charge

$$U(r) = \frac{q_0 q}{4\pi\epsilon_0 r} \tag{22-4}$$

By substitution into Eq. 22-4, you can see that this corresponds to setting $U = 0$ at $r = \infty$. That is, the potential energy is chosen to be zero when the two particles are far removed from one another such that the electrical effect they have on one another is negligible. We often indicate this convention by writing $U_\infty = 0$.

To reinforce our understanding of the test particle's potential energy U, we can apply the reference-level selection ($U_\infty = 0$) to Eq. 22-2, and then give the result a physical interpretation. Let the point b in Eq. 22-2 correspond to a separation distance r [$r_b = r$, $U_b = U(r)$], and let the point a correspond to a very large separation ($r_a = \infty$, $U_\infty = 0$). Equation 22-2 then becomes

$$U(r) = -\int_\infty^r \mathbf{F} \cdot d\boldsymbol{\ell}$$

From this expression you can see that $U(r)$ is the negative of the work done by the electric force when the test particle is moved from a distant point to where it is a distance r from the point charge. The negative of the work done by a force is the same as the work done *against* that force. Thus we can view $U(r)$ *as the work done against the electric force (by an external agent) when the test particle is moved from a distant point to a separation distance r from the point charge.*

Potential Energy of a Test Particle in the Field of a Number of Point Charges

Suppose the test particle is in the field of two point charges q_1 and q_2. By the principle of superposition, the electric force **F** on the test particle is

$$\mathbf{F} = q_0\mathbf{E} = q_0(\mathbf{E}_1 + \mathbf{E}_2)$$

where $\mathbf{E}_1$ and $\mathbf{E}_2$ are the contributions to the field due to q_1 and q_2. The work done by **F** when the test particle is moved from a to b is

$$\int_a^b \mathbf{F}\cdot d\boldsymbol{\ell} = \int_a^b q_0(\mathbf{E}_1 + \mathbf{E}_2)\cdot d\boldsymbol{\ell}$$

$$= q_0\left[\int_a^b \mathbf{E}_1\cdot d\boldsymbol{\ell} + \int_a^b \mathbf{E}_2\cdot d\boldsymbol{\ell}\right]$$

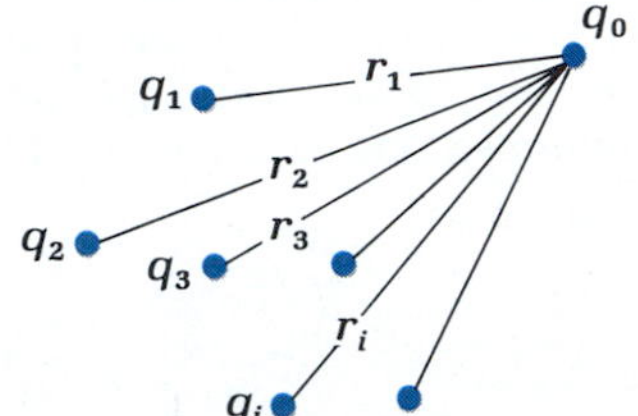

Figure 22-5. The potential energy of the test particle depends on its distance from each of the other particles.

The work can be divided into two contributions, each of which is independent of the path from a to b. Thus the sum of the two contributions to the work is also independent of the path, and the force **F** is conservative. Following the procedures used above when the test particle was in the field of a single point charge (with $U_\infty = 0$), we have

$$U = \frac{q_0}{4\pi\epsilon_0}\left(\frac{q_1}{r_1} + \frac{q_2}{r_2}\right)$$

where r_1 and r_2 are the distances between the test particle and point charges 1 and 2, respectively. Extending to a case where the test particle is in the field of a number of point charges, we find

Potential energy of a test particle in the presence of a number of point charges

$$U = \frac{q_0}{4\pi\epsilon_0}\sum\frac{q_i}{r_i} \qquad (22\text{-}5)$$

where r_i is the distance between the test particle and point charge i (Fig. 22-5).

22-2 ELECTRIC POTENTIAL

In Chap. 20, we found that the electric force **F** on a test particle of charge q_0 in the proximity of a charge distribution is proportional to q_0. The ratio $\mathbf{F}/q_0$ depends on the charge distribution, but is independent of q_0. This ratio was defined as the electric field **E**. Equation 22-5 suggests a similar procedure for the electric potential energy U. If U is divided by q_0, then the resulting quantity is independent of q_0. Thus the *electric potential* $\mathcal{V}$ is defined as

Definition of electric potential

$$\mathcal{V} = \frac{U}{q_0} \qquad \text{(small } q_0\text{)} \qquad (22\text{-}6)$$

The **electric potential** $\mathcal{V}$ at a point P is equal to the electric potential energy U of a test particle placed at point P divided by the charge q_0 of the test particle.

Similar to **E**, $\mathcal{V}$ is a field quantity; it has a value at each point in space. Since U is a scalar, $\mathcal{V}$ is a scalar field. From Eq. 22-6, the SI unit of potential is the joule per coulomb (J/C), which is called the volt (V).

The charge q_0 on the test particle that is used to measure the potential must be small, as indicated in Eq. 22-6. If the charge on the test particle is not small, then its presence may alter the charge distribution that produces the potential, which would change the potential to be measured.

Potential Due to Charged Particles

Dividing Eq. 22-5 by q_0 gives, at a point P, the potential produced by a distribution of charged particles:

$$\mathcal{V} = \frac{1}{4\pi\epsilon_0} \sum \frac{q_i}{r_i} \qquad (22\text{-}7)$$

Potential due to a distribution of charged particles

where r_i is the distance between particle i and point P. The simplest application of Eq. 22-7 is in determining the potential $\mathcal{V}$ a distance r from a single particle with charge q:

$$\mathcal{V} = \frac{q}{4\pi\epsilon_0 r}$$

Potential due to a point charge

Once the potential due to a charge distribution is known, the potential energy of a charged particle (relative to $U_\infty = 0$) can be determined. Suppose a particle, say particle 1 with charge q_1, is placed at a point where the potential is $\mathcal{V}$. Then the electric potential energy U_1 of the particle at that point is $U_1 = q_1\mathcal{V}$.

In atomic and nuclear physics, the charged particles of interest are protons and electrons, in which case the magnitude of the charge is $e = 1.60 \times 10^{-19}$ C. This makes it convenient to define an energy unit, called the *electron volt* (eV), which is equal to the product of the magnitude of the electronic charge times one volt:

$$1 \text{ eV} = (1.6 \times 10^{-19} \text{ C})(1 \text{ V}) = 1.6 \times 10^{-19} \text{ J}$$

Definition of the electron volt

Thus the conversion factor between J and eV is numerically equal to the magnitude of the electronic charge.

EXAMPLE 22-1

The Bohr model of hydrogen. An important early model of an atom is the so-called *Bohr model,* named for its discoverer Niels Bohr (Chap. 39). In the ground state, or lowest energy state, of the hydrogen atom, the electron is in a circular orbit of radius 0.529×10^{-10} m about the nucleus (a proton). *(a)* Determine the electric potential due to the nucleus at a point on the electron's path. Determine the electron's potential energy *(b)* in electron volts and *(c)* in joules.

Solution. *(a)* Each point on the electron's circular path is at a distance $r = 0.529 \times 10^{-10}$ m from the proton. Thus the electric potential due to the proton at any point on the electron's path is

$$\mathcal{V} = \frac{1}{4\pi\epsilon_0}\frac{e}{r}$$

$$= (8.99 \times 10^9 \text{ N}\cdot\text{m}^2/\text{C}^2)\frac{1.60 \times 10^{-19} \text{ C}}{0.529 \times 10^{-10} \text{ m}} = 27.2 \text{ V}$$

(b) The potential energy U of a particle of charge q located at a point where the potential is $\mathcal{V}$ is $U = q\mathcal{V}$. Thus the electron's potential energy in electron volts is

$$U = (-e)(27.2 \text{ V}) = -27.2 \text{ eV}$$

(c) The electron's potential energy in joules is

$$U = (-1.60 \times 10^{-19} \text{ C})(27.2 \text{ V}) = -4.35 \times 10^{-18} \text{ J}$$

When discussing an electron or a proton, it is easier to calculate U in electron volts than in joules, and the size of the numerical value is often more convenient because it is nearer unity.

SELF-TEST 22-1. According to the Bohr model, when the electron is in the first excited state of the hydrogen atom, the radius of its orbit is 2.116×10^{-10} m. *(a)* Determine the electric potential due to the proton at a point on the electron's circular path. *(b)* Determine the electron's potential energy in electron volts. ***ANSWERS:*** *(a)* $\mathcal{V} = 6.80$ V; *(b)* $U = -6.80$ eV.

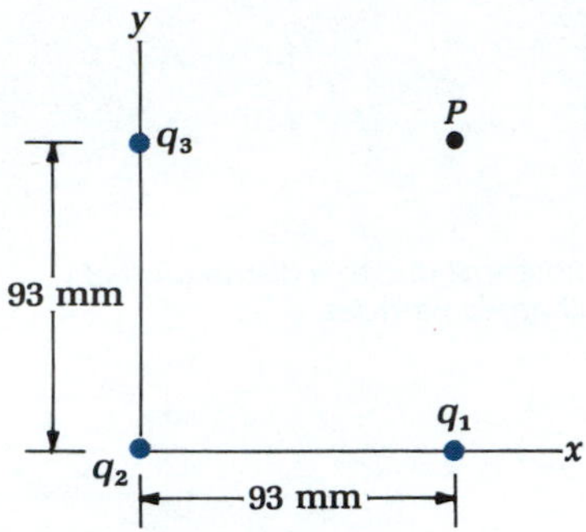

Figure 22-6. Example 22-2.

EXAMPLE 22-2

Potential produced by point charges. Find the potential at point P in Fig. 22-6, where $q_1 = 33$ nC, $q_2 = -51$ nC, and $q_3 = 47$ nC.

Solution. From Eq. 22-7,

$$\mathcal{V} = \frac{1}{4\pi\epsilon_0}\left(\frac{q_1}{r_1} + \frac{q_2}{r_2} + \frac{q_3}{r_3}\right)$$

$$= (9.0 \times 10^9\ \text{N}\cdot\text{m}^2/\text{C}^2)\left(\frac{33\ \text{nC}}{93\ \text{mm}} + \frac{-51\ \text{nC}}{130\ \text{mm}} + \frac{47\ \text{nC}}{93\ \text{mm}}\right)$$

$$= 4.2 \times 10^3\ \text{V}$$

where we have used $\sqrt{(93\ \text{mm})^2 + (93\ \text{mm})^2} = 130$ mm. Notice that it is easier to calculate $\mathcal{V}$ than **E** because the calculation involves a scalar sum rather than a vector sum.

SELF-TEST 22-2. In Fig. 22-6, find the potential produced by the three point charges at a point on the x axis midway between q_1 and q_2. ***ANSWER:*** 580 V.

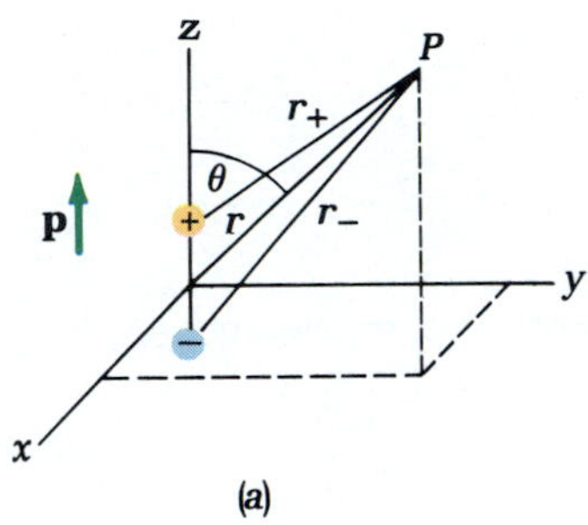

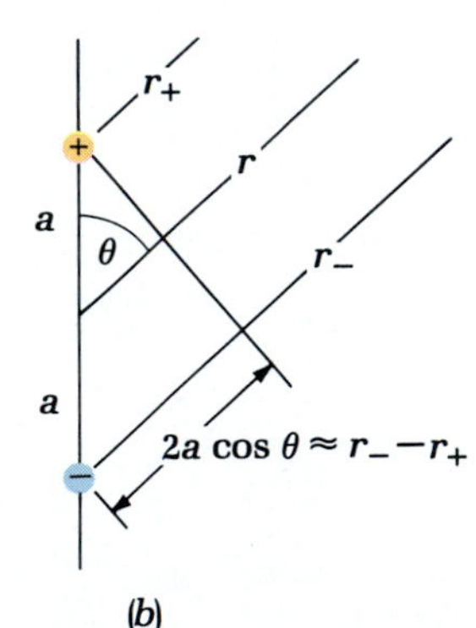

Figure 22-7. Example 22-3: *(a)* Potential at point P due to a dipole with moment $\mathbf{p} = (2aq)\mathbf{k}$ located at the origin. *(b)* When $r \gg a$, $r_- - r_+ \approx 2a \cos\theta$.

EXAMPLE 22-3

Potential due to a dipole. Figure 22-7*a* shows a dipole with moment $\mathbf{p} = (2aq)\mathbf{k}$ located at the origin. Determine the potential produced by the dipole at distances much greater than the separation $2a$ of the charged particles that compose the dipole.

Solution. Letting $\mathcal{V}_+$ and $\mathcal{V}_-$ represent the contributions to the potential due to the particles with positive and negative charge respectively, we have

$$\mathcal{V} = \mathcal{V}_+ + \mathcal{V}_- = \frac{q}{4\pi\epsilon_0 r_+} + \frac{-q}{4\pi\epsilon_0 r_-} = \frac{q}{4\pi\epsilon_0}\left(\frac{1}{r_+} - \frac{1}{r_-}\right)$$

$$= \frac{q}{4\pi\epsilon_0}\frac{r_- - r_+}{r_+ r_-}$$

If $r \gg a$, then from Fig. 22-7*a*, $r_+ \approx r_- \approx r$, and from Fig. 22-7*b*, $r_- - r_+ \approx 2a\cos\theta$. Therefore

$$\mathcal{V} \approx \frac{2aq\cos\theta}{4\pi\epsilon_0 r^2} = \frac{p\cos\theta}{4\pi\epsilon_0 r^2}$$

Using a unit vector $\hat{\mathbf{r}}$ that points away from the origin toward point P, we can write this as

$$\mathcal{V} \approx \frac{\mathbf{p}\cdot\hat{\mathbf{r}}}{4\pi\epsilon_0 r^2}$$

The potential due to the dipole is zero at all points in the xy plane ($\theta = 90°$); it is positive at all points above the xy plane ($0 \le \theta < 90°$); and it is negative at all points below the xy plane ($90° < \theta \le 180°$). Figure 22-8 shows a grid-surface graph of the potential at points in

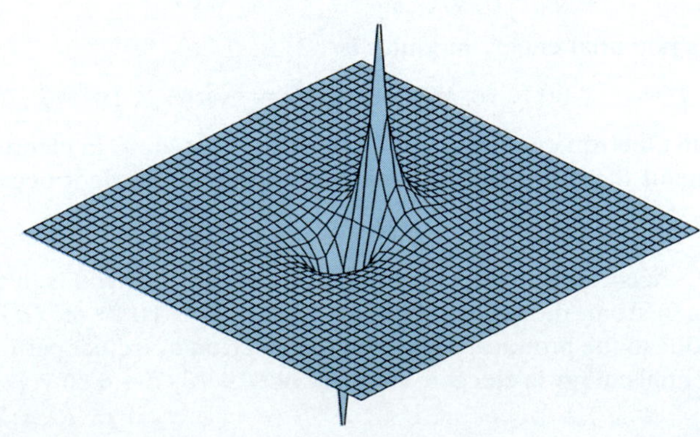

Figure 22-8. Example 22-3: A grid-surface graph of the potential due to a dipole. This is a graph of the potential at points in the xz (or yz) plane. The grid lines show a surface whose height above (or depth below) the plane is proportional to the potential. The potential approaches $+\infty$ at the position of the positive point charge ($z = +a$), and it approaches $-\infty$ at the position of the negative point charge ($z = -a$).

the xz (or yz) plane. The potential falls off as $1/r^2$ along any radial line ($r \gg a$). How does the potential due to a point charge depend on r?

SELF-TEST 22-3. A straight line from the center of a dipole passes first through point P_1 and then through P_2. Point P_1 is 0.20 m from the dipole, point P_2 is 0.40 m from the dipole, and the separation of the dipole's charges is much less than 0.20 m. If the potential produced by the dipole at P_1 is 80 V, what is the potential at P_2? ***ANSWER:*** 20 V.

Potential Due to Continuous Charge Distributions

Equation 22-7 can be transformed into an equation for the potential due to a continuous charge distribution. The continuous charge distribution is divided into an infinite number of infinitesimal charges dq, and in this limit the sum in Eq. 22-7 becomes an integral:

$$\mathcal{V} = \frac{1}{4\pi\epsilon_0} \lim_{\substack{N\to\infty \\ q_i\to 0}} \sum_{i=1}^{N} \frac{q_i}{r_i} = \frac{1}{4\pi\epsilon_0} \int \frac{dq}{r}$$

Or

$$\mathcal{V} = \frac{1}{4\pi\epsilon_0} \int \frac{dq}{r} \qquad (22\text{-}8)$$

where the integration is over the extent of the charge distribution and r is the distance from dq to the point P at which the potential is evaluated.

EXAMPLE 22-4

Potential produced by a charged ring. Determine the potential at points along the axis of a uniformly charged circular ring of radius a and total charge Q. The ring is thin enough to be considered a line charge.

Solution. Figure 22-9 shows the ring contained in the yz plane and centered at the origin so that its axis is the x axis. The potential at point P is

$$\mathcal{V} = \frac{1}{4\pi\epsilon_0} \int \frac{dq}{r} = \frac{1}{4\pi\epsilon_0} \int \frac{dq}{\sqrt{x^2 + a^2}}$$

where the integration involves adding the contributions to the potential due to each charge element dq around the ring. Since x and a are constant with respect to this integration,

$$\mathcal{V} = \frac{1}{4\pi\epsilon_0\sqrt{x^2 + a^2}} \int dq = \frac{Q}{4\pi\epsilon_0\sqrt{x^2 + a^2}}$$

Figure 22-10 shows a graph of $\mathcal{V}$ versus x for the case where $a = 0.50$ m and $Q = +11.1$ nC. It is instructive to compare this example with the corresponding calculation of the electric field, Example 20-7.

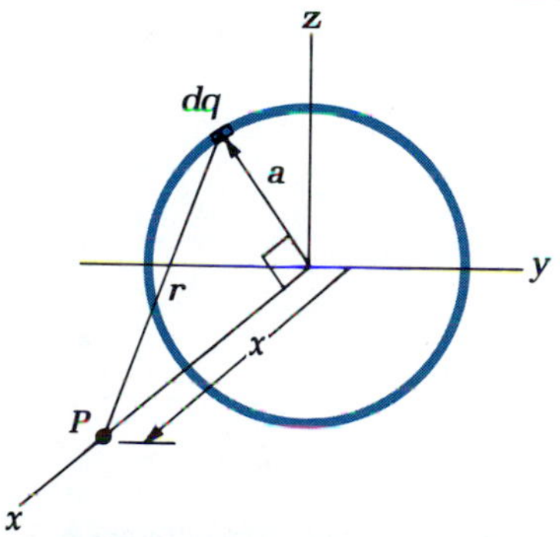

Figure 22-9. Example 22-4: A charged ring contained in the yz plane and centered at the origin.

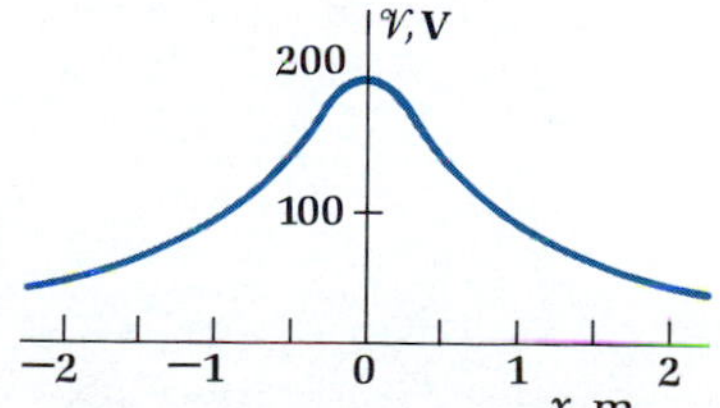

Figure 22-10. Example 22-4: $\mathcal{V}(x)$ along the axis of the charged ring in Fig. 22-9; $Q = 11.1$ nC, $a = 0.50$ m.

SELF-TEST 22-4. Determine the potential produced by a charged ring at a point on the ring's axis at a distance of 0.40 m from the center. The ring has a radius of 0.30 m and a charge of 45 nC. ***ANSWER:*** 810 V.

EXAMPLE 22-5

Potential produced by a charged disk. Determine the potential at points along the axis of a thin, uniformly charged disk of radius R_0 and surface charge density σ.

Solution. As seen in Fig. 22-11, the disk is contained in the yz plane and centered at the

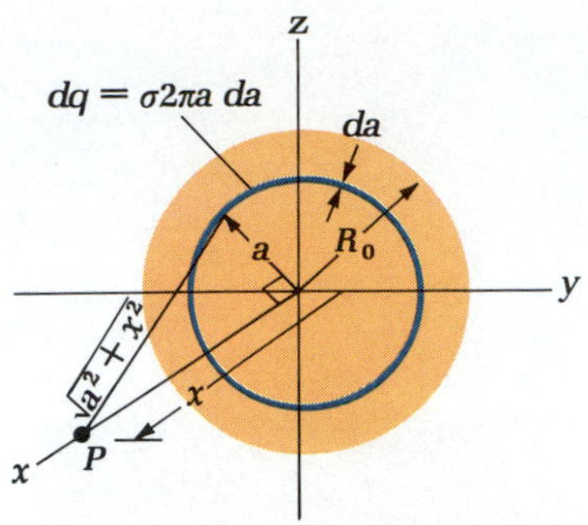

Figure 22-11. Example 22-5: A thin disk with uniform surface charge.

origin so that its axis is the x axis. The disk is divided into infinitesimal rings of area $2\pi a\, da$, and the infinitesimal charge of each ring is $dq = \sigma 2\pi a\, da$. Using the result of the previous example, we have

$$d\mathcal{V} = \frac{dq}{4\pi\epsilon_0 r} = \frac{\sigma 2\pi a\, da}{4\pi\epsilon_0 \sqrt{x^2 + a^2}}$$

$$\mathcal{V} = \int d\mathcal{V} = \frac{2\pi\sigma}{4\pi\epsilon_0} \int_0^{R_0} \frac{a\, da}{\sqrt{x^2 + a^2}} = \frac{\sigma}{2\epsilon_0} \left[\sqrt{x^2 + a^2} \right]_0^{R_0}$$

$$\mathcal{V} = \frac{\sigma}{2\epsilon_0} (\sqrt{x^2 + R_0^2} - \sqrt{x^2})$$

Since $\sigma = Q/\pi R_0^2$, where Q is the charge on the disk, we can also write the potential as

$$\mathcal{V} = \frac{Q}{2\pi\epsilon_0 R_0^2} (\sqrt{x^2 + R_0^2} - \sqrt{x^2})$$

SELF-TEST 22-5. Determine the potential produced by a thin, uniformly charged disk at a point on the disk's axis at a distance of 0.40 m from the center. The disk has a radius of 0.30 m and a charge of 45 nC. Compare your answer with that of the previous self-test.
ANSWER: 900 V.

22-3 POTENTIAL DIFFERENCE

Our definition of potential is based on the reference position we chose for potential energy: $U = 0$ at $r = \infty$. Therefore our reference position for potential $\mathcal{V}$ is also at $r = \infty$. In other words, $\mathcal{V}$ is taken to be zero at points far removed from the charge distribution ($\mathcal{V}_\infty = 0$). However, the selection of a reference position is simply a matter of convenience. When dealing with electric circuits (Chap. 25), a convenient reference position is the earth, or "ground." Only a change in potential energy has physical significance, and, correspondingly, only a change in potential, or *potential difference,* has physical significance.

Let $U_b - U_a$ represent the difference in electric potential energy of a test particle with charge q_0 when it is moved from point a to point b. Then, the potential difference $\mathcal{V}_b - \mathcal{V}_a$ between points a and b is defined as

Definition of potential difference

$$\mathcal{V}_b - \mathcal{V}_a = \frac{U_b - U_a}{q_0} \quad \text{(small } q_0\text{)} \tag{22-9}$$

Sometimes we shall abbreviate the notation by representing $\mathcal{V}_b - \mathcal{V}_a$ with $\Delta\mathcal{V}$.

The potential difference between two points in a region can be determined from the electric field in the region. Since the difference in the electric potential energy of a test particle with charge q_0 is the negative of the work done by the electric force, we have

$$U_b - U_a = -q_0 \int_a^b \mathbf{E} \cdot d\boldsymbol{\ell}$$

Dividing by q_0 gives the potential difference in terms of the electric field:

$$\mathcal{V}_b - \mathcal{V}_a = -\int_a^b \mathbf{E} \cdot d\boldsymbol{\ell} \tag{22-10}$$

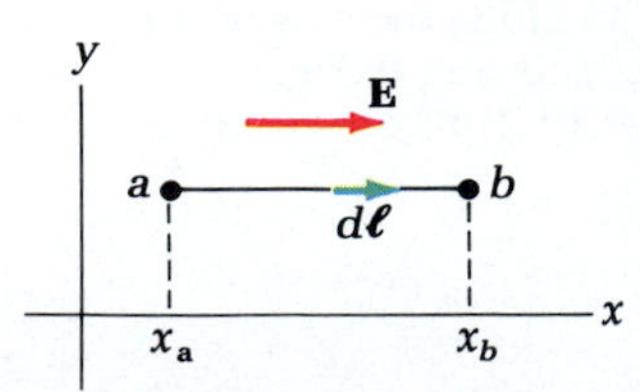

Figure 22-12. Finding $\mathcal{V}_b - \mathcal{V}_a$ in a uniform field pointing in the $+x$ direction.

Since the electric force is conservative, any path connecting points a and b may be used to evaluate the line integral in Eq. 22-10.

A simple application of Eq. 22-10 is to determine the potential difference between points in a uniform electric field. In Fig. 22-12, we let the x axis be along the direction of the field so that $\mathbf{E} = E\mathbf{i}$. Since $\mathbf{E}$ is perpendicular to the yz plane, the potential is constant in planes parallel to the yz plane. To find the potential difference between different planes parallel to the yz plane, we perform the line integral in Eq. 22-10 along a straight line parallel to the x axis, that is, along a field line. In this case $d\boldsymbol{\ell} = dx\,\mathbf{i}$, so that

$$\mathcal{V}_b - \mathcal{V}_a = -\int_{x_a}^{x_b} (E\mathbf{i}) \cdot (dx\,\mathbf{i}) = -\int_{x_a}^{x_b} E\,dx$$

Since E is constant, it may be factored out of the integral so that

$$\mathcal{V}_b - \mathcal{V}_a = -E(x_b - x_a) \qquad \text{or} \qquad \Delta\mathcal{V} = -E\,\Delta x$$

Let $\mathcal{V}_0$ represent the potential of points in the yz plane ($x = 0$), and let $\mathcal{V}(x)$ be the potential of points in planes parallel to the yz plane with coordinate x. This gives

$$\mathcal{V}(x) - \mathcal{V}_0 = -Ex \qquad (22\text{-}11)$$

Potential difference in a uniform field

In a uniform field, the potential difference varies linearly with x and decreases along the field direction.

A gravitational analogy to the uniform electric field is the uniform gravitational field near the surface of the earth. As an object is lifted (moved opposite the field direction), its potential energy relative to some horizontal reference plane (such as the floor of a room) increases linearly with increasing elevation.

EXAMPLE 22-6

Potential difference between the plates of a parallel-plate capacitor. Make a graph of the potential difference $\mathcal{V}(x) - \mathcal{V}_0$ in the space between the plates of the parallel-plate capacitor shown in cross section in Fig. 22-13, and determine $\Delta\mathcal{V}$ between the plates. The lateral dimensions of each plate are much larger than the plate separation, which is $d = 0.50$ mm, and the magnitude of the surface charge density on each plate is $|\sigma| = 1.8\ \mu\text{C/m}^2$.

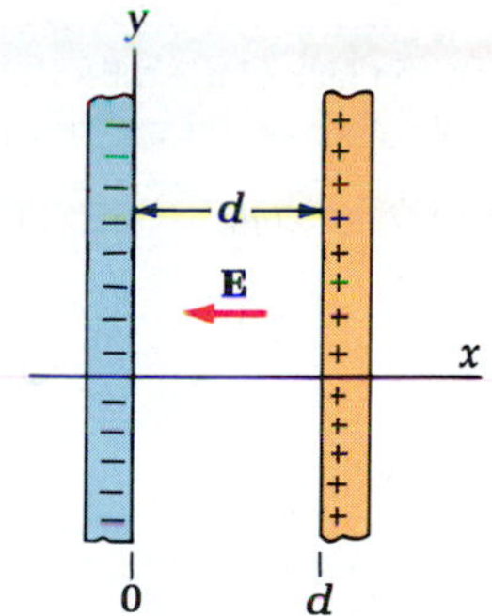

Figure 22-13. Example 22-6: Cross section of a parallel-plate capacitor.

Solution. Since the lateral dimensions of each plate are much larger than the separation between the plates, we may assume that **E** is uniform. In Fig. 22-13 the surface charge of the negative plate occupies the yz plane ($x = 0$), and the surface charge of the positive plate occupies a parallel plane at $x = d$; the x axis passes through the center of the capacitor. With this arrangement, **E** points in the $-x$ direction. Also, from Example 21-8, $E = |\sigma|/\epsilon_0$. Thus $\mathbf{E} = -E\mathbf{i} = -(|\sigma|/\epsilon_0)\mathbf{i}$, and from Eq. 22-11,

$$\mathcal{V}(x) - \mathcal{V}_0 = \frac{|\sigma|}{\epsilon_0}x = \frac{1.8\ \mu\text{C/m}^2}{8.85 \times 10^{-12}\ \text{C}^2/(\text{N}\cdot\text{m}^2)}x$$
$$= (2.0 \times 10^5\ \text{N/C})x$$

A graph of the potential difference is shown in Fig. 22-14. The potential difference $\Delta\mathcal{V}$ between the plates is

$$\Delta\mathcal{V} = \mathcal{V}_d - \mathcal{V}_0 = Ed = (2.0 \times 10^5\ \text{N/C})(0.50\ \text{mm}) = 100\ \text{V}$$

What is the potential difference between the negative plate and the plane at $x = 0.25$ mm?

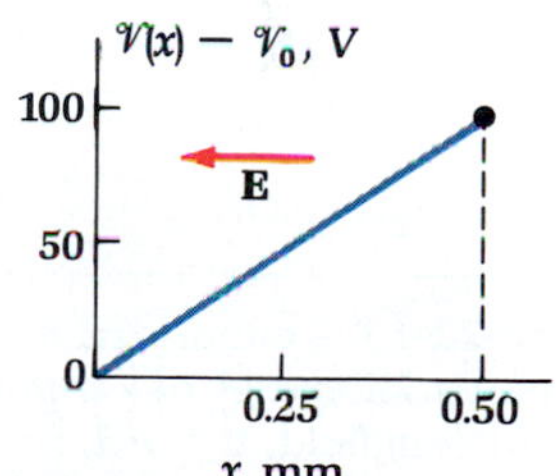

Figure 22-14. Example 22-6: In the space between capacitor plates, $\mathcal{V}(x) - \mathcal{V}_0$ varies linearly from one plate to the other and increases in the direction opposite **E**.

SELF-TEST 22-6. For the capacitor in the above example, suppose we choose to let the potential of the negative plate be zero: $\mathcal{V}_0 = 0$. *(a)* What is the potential at a point 0.15 mm from the negative plate? *(b)* What is the potential at a point 0.35 mm from the negative plate? ***ANSWERS:*** *(a)* 30 V; *(b)* 70 V.

EXAMPLE 22-7

Electron gun in a CRT. The charged plates of an electron gun in a cathode-ray tube, or CRT (Fig. 20-21), are called the anode and the cathode. The anode is maintained at a high potential relative to the cathode and their potential difference, $\Delta\mathcal{V} = \mathcal{V}_A - \mathcal{V}_C$, is called the *accelerating voltage.* Electrons are emitted from the heated cathode with negligible speed and then accelerate toward the anode. By gravitational analogy, the electrons are said to "fall through" the potential difference between the anode and cathode. Some of the electrons emerge from a hole in the anode, and these electrons form the electron beam in the tube. Using an accelerating voltage of $\Delta\mathcal{V} = 2.5$ kV, determine *(a)* the kinetic energy and *(b)* the speed of an electron in the beam.

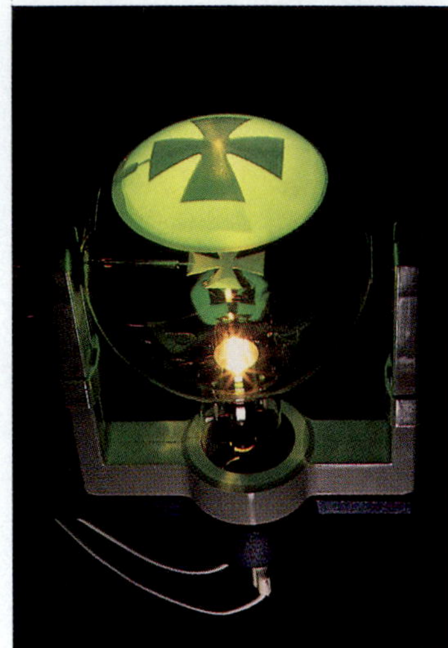

An electron beam produced by an electron gun in a demonstration CRT. The straight-line paths of the electrons casts a "shadow" of a cross on the face of the tube.

Solution. *(a)* The mechanical energy $(K+U)$ of the electrons is conserved as they accelerate from cathode to anode. Letting subscripts C and A refer to quantities evaluated at the cathode and anode, we have $K_A + U_A = K_C + U_C$. Since the electrons start from rest at the cathode, $K_C = 0$, and $K_A + U_A = U_C$, or

$$K_A = U_C - U_A$$

The potential energy of an electron (charge $-e$) at the cathode is $U_C = (-e)\mathcal{V}_C$, and at the anode is $U_A = (-e)\mathcal{V}_A$. The kinetic energy of each electron as it emerges from the hole in the anode is

$$K_A = (-e)\mathcal{V}_C - (-e)\mathcal{V}_A = e(\mathcal{V}_A - \mathcal{V}_C) = e\,\Delta\mathcal{V}$$

Thus, if an electron starts from rest and accelerates through a potential difference $\Delta\mathcal{V}$, then its kinetic energy is $e\,\Delta\mathcal{V}$. In this example, $\Delta\mathcal{V} = 2.5$ kV, so

$$K_A = e(2.5\text{ kV}) = 2.5 \times 10^3\text{ eV} = 2.5\text{ keV}$$

Often an electron with this kinetic energy is described as a "2.5-keV electron." *(b)* Since $K_A = \frac{1}{2}mv^2 = e\,\Delta\mathcal{V}$, the speed v of an electron as it emerges from the hole in the anode is

$$v = \sqrt{\frac{2e\,\Delta\mathcal{V}}{m}}$$

$$= \sqrt{\frac{(2)(1.6 \times 10^{-19}\text{ C})(2.5\text{ kV})}{9.1 \times 10^{-31}\text{ kg}}} = 3.0 \times 10^7\text{ m/s}$$

SELF-TEST 22-7. *(a)* What is the kinetic energy in electron volts of an electron as it emerges from an electron gun with an accelerating voltage of 3.0 kV? *(b)* What is the speed of this electron? ***ANSWERS:*** *(a)* $K = 3.0$ keV; *(b)* $v = 3.2 \times 10^7$ m/s.

EXAMPLE 22-8

Potential energy of a dipole in a uniform field. Determine the potential energy of a dipole in a uniform electric field.

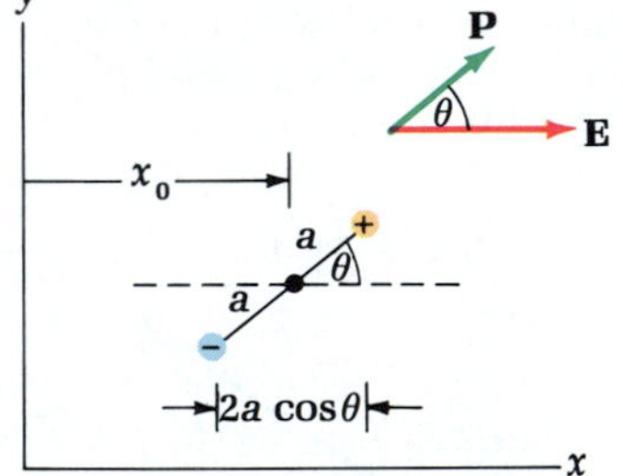

Figure 22-15. Example 22-8: The potential energy of a dipole in a uniform field, $\mathbf{E} = E_x\mathbf{i}$.

Solution. Figure 22-15 shows a dipole oriented at an angle θ with respect to a uniform electric field directed along the x axis. The x coordinates of the positive and negative point charges are $x_0 + a\cos\theta$ and $x_0 - a\cos\theta$, respectively. For a uniform field in the x direction (Eq. 22-11), $\mathcal{V}(x) = -Ex + \mathcal{V}_0$, where $\mathcal{V}_0$ is the potential at points on the yz plane ($x = 0$). Therefore the potential energy U_+ of the positive point charge is

$$U_+ = q[-E(x_0 + a\cos\theta) + \mathcal{V}_0]$$

and the potential energy U_- of the negative point charge is

$$U_- = -q[-E(x_0 - a\cos\theta) + \mathcal{V}_0]$$

We do not include the potential energy of interaction of the two particles because it depends on their separation, which is assumed to be fixed. Thus the potential energy U of the dipole is

$$U = U_+ + U_-$$
$$= q[-E(x_0 + a\cos\theta) + \mathcal{V}_0] - q[-E(x_0 - a\cos\theta) + \mathcal{V}_0]$$
$$= -2aqE\cos\theta = -pE\cos\theta$$

$$U = -\mathbf{p}\cdot\mathbf{E}$$

where $\mathbf{p}$ is the dipole moment. The potential energy of a dipole in a uniform field is independent of its position x_0. Instead it depends on the orientation of the dipole moment with respect to the field direction. The potential energy is $-pE$ when the dipole is aligned parallel to the field ($\theta = 0$, $\cos\theta = +1$); it is zero when the dipole points perpendicular to the field ($\theta = 90°$, $\cos\theta = 0$); and it is $+pE$ when the dipole is aligned opposite the field ($\theta = 180°$, $\cos\theta = -1$). (See Ques. 17 and Exercise 19. Problem 2 discusses additional features of a dipole in a uniform field.)

SELF-TEST 22-8. What is the potential energy of a dipole whose dipole moment makes an angle of 131° to a uniform electric field of magnitude 1200 N/C? The magnitude of the dipole moment is 5.9×10^{-12} C·m. ***ANSWER:*** 4.6×10^{-9} J.

22-4 RELATION BETWEEN E AND $\mathscr{V}$

The potential, with its reference position at a distant point, can be developed from Eq. 22-10 by letting point a correspond to $r = \infty$ ($\mathscr{V}_a = \mathscr{V}_\infty = 0$), and by letting point b correspond to the position P at which the potential $\mathscr{V}$ is evaluated ($\mathscr{V}_b = \mathscr{V}$). This gives

$$\mathscr{V} = -\int_\infty^P \mathbf{E} \cdot d\boldsymbol{\ell} \qquad (22\text{-}12)$$

The potential $\mathscr{V}$ in terms of the field **E**

Thus the potential is the negative of the line integral of the electric field from the distant reference position to the point P where the potential is evaluated. If an expression for **E** due to a charge distribution is known, then $\mathscr{V}$ can be determined, at least in principle, by using Eq. 22-12.

Now we investigate the reverse operation. We wish to find an expression for **E** from a known expression for $\mathscr{V}$. Since $\mathscr{V}$ is the negative of the line integral of **E**, you might expect that **E** is related to the negative of some type of derivative of $\mathscr{V}$. This expectation is correct.

Suppose we use Eq. 22-10 to find the potential difference between two nearby points, $a = (x, y, z)$ and $b = (x + \Delta x, y, z)$. Since the two points have the same y and z coordinates, we integrate along a line parallel to the x axis: $d\boldsymbol{\ell} = dx'\,\mathbf{i}$. (We use x' for the variable of integration because x is used in the limits.) Thus

$$\mathbf{E} \cdot d\boldsymbol{\ell} = (E_x\mathbf{i} + E_y\mathbf{j} + E_z\mathbf{k}) \cdot (dx'\,\mathbf{i}) = E_x\,dx'$$

From Eq. 22-10.

$$\mathscr{V}(x + \Delta x, y, z) - \mathscr{V}(x, y, z) = -\int_x^{x+\Delta x} E_x\,dx'$$

We will presently consider the limit as Δx approaches zero. In anticipation of this, we assume E_x is nearly constant from x to $x + \Delta x$. This means that E_x can be factored out of the integral. With this approximation, the right-hand side of the above expression becomes

$$-E_x \int_x^{x+\Delta x} dx' = -E_x[(x + \Delta x) - (x)] = -E_x\,\Delta x$$

Thus

$$\mathscr{V}(x + \Delta x, y, z) - \mathscr{V}(x, y, z) \approx -E_x\,\Delta x$$

Dividing by Δx and taking the limit as $\Delta x \rightarrow 0$ gives

$$\lim_{\Delta x \rightarrow 0} \left[\frac{\mathscr{V}(x + \Delta x, y, z) - \mathscr{V}(x, y, z)}{\Delta x}\right] = -E_x$$

In the limit as $\Delta x \rightarrow 0$, the approximation becomes exact. The quantity on the left-hand side of the above equation is the derivative of $\mathscr{V}$ with respect to x, with y and z held constant. That is, only the coordinate x varies during the limiting process; the y and z coordinates are fixed. This quantity is represented by the symbol $\partial\mathscr{V}/\partial x$ and is called the *partial derivative of* $\mathscr{V}$ with respect to x.* Thus

* In finding the partial derivative of a function f with respect to one of the coordinates, we treat the other coordinates as constants during the operation. As an example, suppose $f = azx^2/y$, where a is a constant. Then the partial derivative of f with respect to x is

$$\frac{\partial f}{\partial x} = \frac{\partial}{\partial x}\frac{azx^2}{y} = \frac{az}{y}\frac{d}{dx}x^2 = \frac{2azx}{y}$$

Similarly, the partial derivative of f with respect to y is

$$\frac{\partial f}{\partial y} = \frac{\partial}{\partial y}\frac{azx^2}{y} = azx^2\frac{d}{dy}\frac{1}{y} = -\frac{azx^2}{y^2}$$

What is the partial derivative of f with respect to z?

$$E_x = -\frac{\partial \mathcal{V}}{\partial x}$$

Infinitesimal changes of potential in the y and z directions give similar results. Therefore

The electric field in terms of the potential

$$E_x = -\frac{\partial \mathcal{V}}{\partial x} \qquad E_y = -\frac{\partial \mathcal{V}}{\partial y} \qquad E_z = -\frac{\partial \mathcal{V}}{\partial z} \tag{22-13}$$

We find that the components of **E** are given by the negative of the partial derivatives of $\mathcal{V}$. If an expression for $\mathcal{V}$ due to a charge distribution is known, then Eqs. 22-13 can be used to find **E**.

Results similar to Eqs. 22-13 can be developed for other types of coordinates besides cartesian. In particular, if a charge distribution has spherical symmetry, then $\mathcal{V}$ depends only on the radial coordinate r and **E** has only a radial component E_r. In this case,

$$E_r = -\frac{d\mathcal{V}}{dr} \tag{22-14}$$

The potential and the electric field are directly related to one another. Either may be determined from the charge distribution, and either may be determined from the other.

Notice that Eqs. 22-13 and 22-14 show that the dimension of electric field is potential divided by distance. Therefore the SI unit for the electric field may be written as volts per meter (V/m) as well as newtons per coulomb (N/C). Indeed, V/m is probably used more often than N/C.

EXAMPLE 22-9

Field and potential due to a uniform spherical volume charge. Example 21-6 showed that the electric field at a distance r from the center of a uniform spherical volume charge of charge Q and radius r_0 is

$$E = \frac{Qr}{4\pi\epsilon_0 r_0^3} \quad (r < r_0) \qquad \text{and} \qquad E = \frac{Q}{4\pi\epsilon_0 r^2} \quad (r > r_0)$$

In Prob. 21, you are asked to show that the electric potential produced by this charge distribution is

$$\mathcal{V} = \frac{Q}{8\pi\epsilon_0 r_0^3}(3r_0^2 - r^2) \qquad (r < r_0)$$

and

$$\mathcal{V} = \frac{Q}{4\pi\epsilon_0 r} \qquad (r > r_0)$$

A graph of this potential is shown in Fig. 22-16. Since $E_r = -d\mathcal{V}/dr$ for a spherical charge distribution, the negative of the slope of the graph at each point gives the field component at that point. Use this potential to determine the electric field due to a uniform spherical volume charge, and compare your result with the expression for the field that was determined in Example 21-6.

Solution. For the field outside the charge distribution $(r > r_0)$,

$$E_r = -\frac{d\mathcal{V}}{dr} = -\frac{d}{dr}\frac{Q}{4\pi\epsilon_0 r} = -\frac{q}{4\pi\epsilon_0}\frac{d}{dr}\frac{1}{r} = -\frac{Q}{4\pi\epsilon_0}\left(-\frac{1}{r^2}\right) = \frac{Q}{4\pi\epsilon_0 r^2}$$

For the field inside the charge distribution $(r < r_0)$,

$$E_r = -\frac{d}{dr}\frac{Q}{8\pi\epsilon_0 r_0^3}(3r_0^2 - r^2) = \frac{Qr}{4\pi\epsilon_0 r_0^3}$$

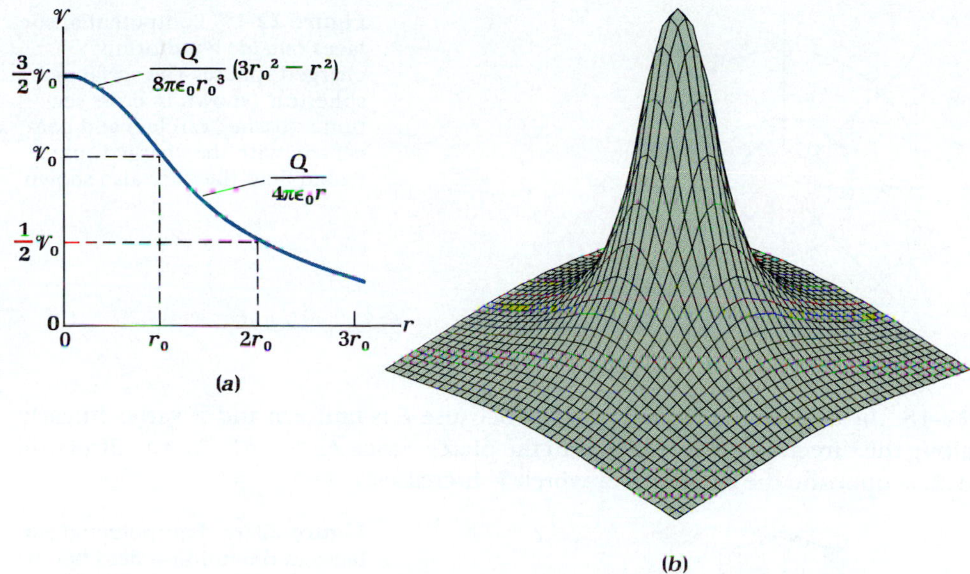

Figure 22-16. Example 22-9: Potential produced by a uniform spherical volume charge. *(a)* Graph of $\mathcal{V}(r)$ versus r. Inside the charge distribution ($r < r_0$), the potential decreases parabolically from the center to the edge, and outside the charge distribution ($r > r_0$), the potential falls off as $1/r$. *(b)* Grid-surface graph of the potential at points in a plane that passes through the center of the distribution.

Since the derivative of this potential gives the same expression for the electric field as that found in Example 21-6, we have verified that the potential is correct.

SELF-TEST 22-9. An early model of the atom, proposed by Ernest Rutherford, was a uniform spherical volume charge due to the electrons with charge $-Ze$ and a point charge $+Ze$ at the center due to the nucleus. Consider such a model for hydrogen ($Z = 1$), where the radius of the negative volume charge due to the electron is $r_0 = 0.53 \times 10^{-10}$ m. *(a)* Determine the value of the potential at the center of a uniform spherical volume charge with charge $Q = -e = -1.6 \times 10^{-19}$ C and radius $r_0 = 0.53 \times 10^{-10}$ m. *(b)* What is the potential energy of the atom's nucleus in electron volts? ***ANSWERS:*** *(a)* -41 V; *(b)* -41 eV.

22-5 EQUIPOTENTIAL SURFACES

An ***equipotential surface*** is a surface on which the potential is constant or equal. No work is done by electric forces when a charged particle moves along an equipotential surface. As with field lines for the electric field, equipotential surfaces are useful for visualizing the spatial behavior of the potential.

Figure 22-17 shows equipotential surfaces and field lines outside a uniformly charged sphere. From Example 22-9, $\mathcal{V} = Q/4\pi\epsilon_0 r$, so that $\mathcal{V}$ is constant if r is constant. Since a surface with constant r is a spherical surface, the equipotential surfaces are spherical, and since r is measured from the charged sphere's center, the equipotential surfaces are concentric with the sphere.

From our discussion in Sec. 22-3, we know that equipotential surfaces in a uniform field are parallel planes that are perpendicular to **E**. Figure 22-18 shows in cross section a parallel-plate capacitor where **E** is uniform, with equipotential surfaces and field lines between the plates.

The field lines in Figs. 22-17 and 22-18 are perpendicular to the equipotential surfaces where they cross. This must be true in all cases because if **E** has a component tangent to a surface, then work is done by the electric force when a charged particle moves along the surface. Therefore **E** cannot have a component tangent to an equipotential surface; **E** is perpendicular to an equipotential surface at each point on the surface.

In a given drawing in which the potential difference between successive pairs of equipotential surfaces is the same, their spacing indicates the magnitude of **E**. The surfaces are spaced closer together in a region where E is larger, similar to the way closely spaced contour lines on a map indicate a steep hill. In Fig. 22-17, the spacing increases with increasing r because E decreases with increasing r. In Fig.

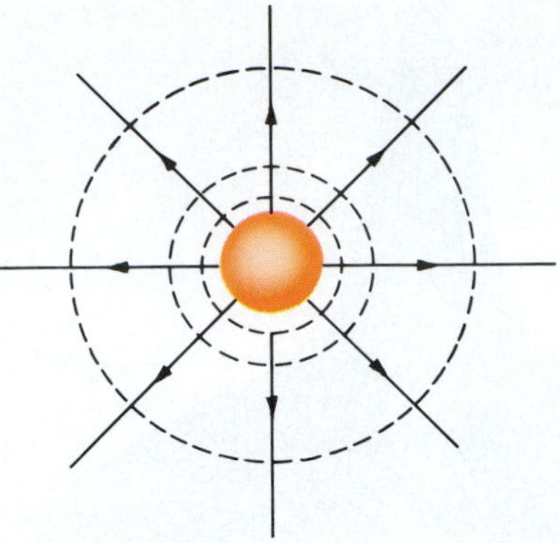

Figure 22-17. Equipotential surfaces outside a uniformly charged sphere. The surfaces are spherical (shown in cross section as dashed circles) and concentric with the charged sphere. Radial field lines are also shown.

22-18, the surfaces are equally spaced because E is uniform and $\mathscr{V}$ varies linearly along the direction perpendicular to the plates. Since $E_x = -\partial\mathscr{V}/\partial x$, the direction of $\mathbf{E}$ is opposite the direction in which $\mathscr{V}$ increases.

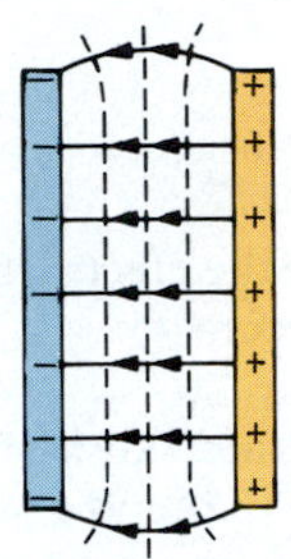

Figure 22-18. Equipotential surfaces in the uniform field between capacitor plates are planes (shown in cross section as dashed lines). Field lines are also shown.

22-6 MORE ABOUT ELECTROSTATIC PROPERTIES OF A CONDUCTOR

In Sec. 21-5, we found that $\mathbf{E} = 0$ inside a conductor under static conditions. Also, we showed that because $\mathbf{E} = 0$, no excess charge can exist inside a conductor; a conductor's charge resides on its surface. Further, the field just outside a conductor is perpendicular to the surface and its component is $E_n = \sigma/\epsilon_0$. Using the concept of potential, we now extend our investigation of the electrostatic properties of a conductor.

Since $\mathbf{E} = 0$ inside a conductor, the volume occupied by conducting material must be a region of uniform potential. To prove this, we apply Eq. 22-10 to two points a and b inside a conductor:

$$\mathscr{V}_b - \mathscr{V}_a = -\int_a^b \mathbf{E} \cdot d\boldsymbol{\ell} \tag{22-10}$$

In electrostatics, all points in a conductor are at the same potential.

In performing the line integral, we choose the path of integration to lie entirely inside the conductor. Then $\mathbf{E} = 0$ at each point along the path, the integral is zero, and $\mathscr{V}_b = \mathscr{V}_a$. This is true for any two points inside the conductor, so that all points in the conductor are at the same potential. In particular, it is often helpful to remember that the surface of a conductor is an equipotential surface.

Previously we spoke of the potential at a point in space. Since all points in a conductor are at the same potential, we may assign a value of the potential to an entire conductor, as long as electrostatic conditions prevail. It is meaningful to speak of a "150-V metal plate." But an insulator cannot be assigned a particular value of potential because the potential may be different at different points inside and on an insulator.

Consider an isolated solid metal sphere of radius r_0 and charge Q. Such a sphere constitutes a spherical shell of uniform surface charge density, and E versus r is shown in Fig. 21-18; E is zero inside the sphere and falls off as $1/r^2$ outside the sphere. Figure 22-19 shows $\mathscr{V}$ versus r; $\mathscr{V}$ is uniform inside and falls off as $1/r$

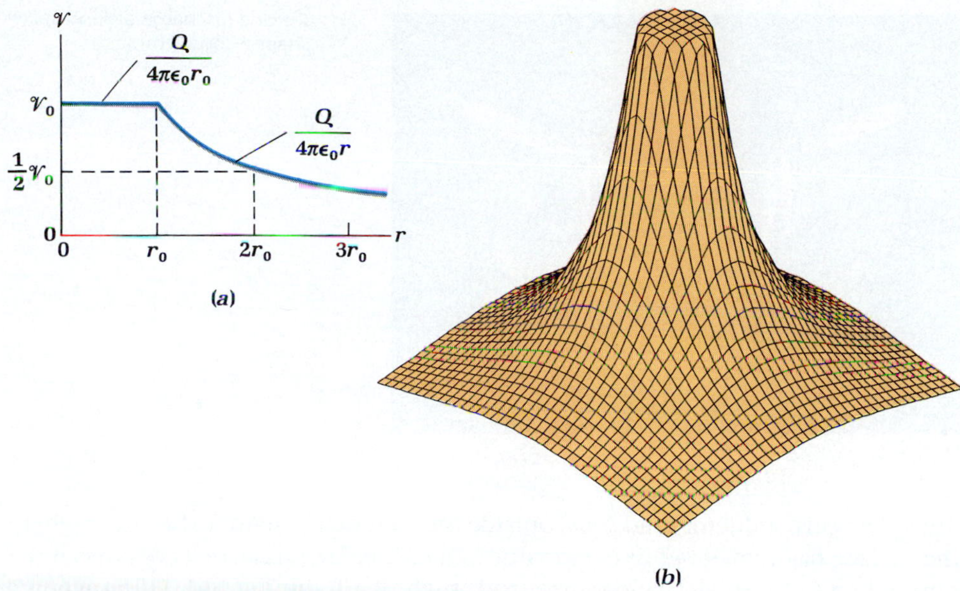

Figure 22-19. Potential produced by a uniform spherical shell of charge. *(a)* Graph of $\mathcal{V}(r)$ versus r. Inside the charge distribution ($r < r_0$), the potential is uniform. Outside the charge distribution ($r > r_0$), the potential falls off as $1/r$. *(b)* Grid-surface graph of the potential at points in a plane that passes through the center of the distribution.

outside. At the sphere's surface and within its volume, $\mathcal{V} = \mathcal{V}(r_0) = Q/4\pi\epsilon_0 r_0$. Suppose $Q = 100$ nC and $r_0 = 0.1$ m, then

$$\mathcal{V} = (9 \times 10^9\ \mathrm{N \cdot m^2/C^2})\frac{100\ \mathrm{nC}}{0.10\ \mathrm{m}} = 9000\ \mathrm{V}$$

The potential "of the sphere" is 9 kV.

Not only is the field zero and the potential uniform within a conductor, but this is true as well for a cavity in a conductor (assuming there are no charged objects inside the cavity). In Sec. 21-5 we showed that if there are no charged objects inside a cavity in a conductor, then there is no net charge on the cavity surface. But this does not exclude the possibility that $\sigma > 0$ over part of the cavity surface and $\sigma < 0$ over another part such that the total charge is zero.

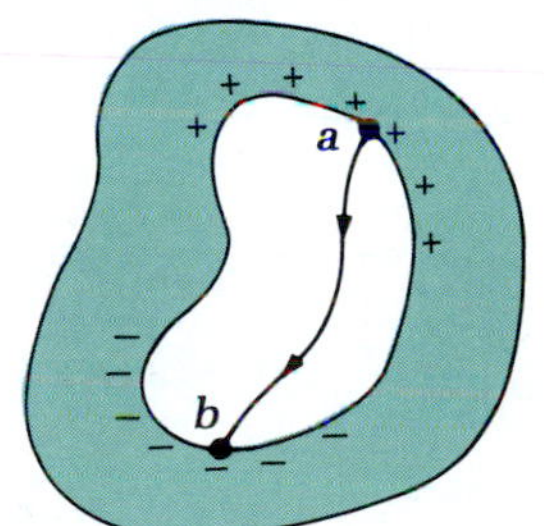

Figure 22-20. Showing that $\mathbf{E} = 0$ inside a cavity in a conductor and that $\sigma = 0$ on the inside surface. The alleged field line drawn between a and b cannot, in fact, exist.

To show that $\sigma = 0$ at each point on the cavity surface, we assume that $\sigma \neq 0$ and see where it leads us. In Fig. 22-20 we show an alleged field line that originates at a positive charge on the cavity surface and terminates at a negative charge on the cavity surface. Suppose we let point a be at the positive charge and point b be at the negative charge and apply Eq. 22-10. If the path for the line integral is taken along the field line, then the line integral cannot be zero because $E > 0$ and $\mathbf{E}$ is parallel to $d\boldsymbol{\ell}$ at each point on the path. This predicts that $\mathcal{V}_b \neq \mathcal{V}_a$. However, we know that $\mathcal{V}_b = \mathcal{V}_a$ because both points are on the conductor's surface. The prediction that $\mathcal{V}_b \neq \mathcal{V}_a$, which is based on the supposition that $\mathbf{E} \neq 0$ inside the cavity and that $\sigma \neq 0$ on the cavity surface, is false. Therefore we conclude that $\mathbf{E} = 0$ inside the cavity and that $\sigma = 0$ at each point on the cavity surface. A field-free region of space can be maintained by surrounding the region with a conductor. Such a procedure is called *electrostatic screening*.

Electrostatic screening

Suppose we wish to give a spherical conductor a very large charge and potential. What determines the maximum charge and potential the conductor can acquire? To answer this question, we must consider the insulating medium, such as air, that surrounds the conductor. The relevant property of the surrounding medium is its **dielectric strength.** The dielectric strength of an insulating material is the maximum magnitude E_{max} of the electric field that can exist in the material without electrical breakdown. When an insulator breaks down, its constituent molecules ionize and the material begins to conduct. In a gas such as air, visible light is emitted (the gas glows) as the electrons recombine with the ionized molecules, a phenomenon called a ***corona discharge.*** A corona discharge can sometimes be observed at night around high-voltage transmission lines.

Dielectric strength E_{max} of an insulator

Corona discharge

By symmetry, the charge on an isolated spherical conductor is distributed uniformly on its surface, but what if the conductor is not spherical? Usually $|\sigma|$ on a

Corona discharge around spike-shaped conductors.

nonspherical conductor (and E just outside the conductor) tends to be larger where the surface has a small radius of curvature. In particular, $|\sigma|$ can be very large at the pointed end of a spike-shaped metal rod, such as a lightning rod. Often when a highly charged thundercloud is overhead, corona discharge can be observed at the end of a lightning rod. Indeed, this corona discharge serves to extend the effective length of a lightning rod and contributes to the protection the rod affords.

EXAMPLE 22-10

Van de Graaff generator. In the physics classroom, a Van de Graaff generator (Fig. 22-21) is used to demonstrate electrical effects. A belt continuously delivers charge to the inside of a metal dome where it is conducted to the outer surface. The ultimate charge and potential of the nearly spherical dome depend on its radius and on the properties of the surrounding insulators (air, the belt, and the supporting tube). Large Van de Graaff generators are used in research laboratories to accelerate subatomic particles to high kinetic energies to study the effects of their collisions. Assume that the maximum magnitudes for the charge (Q_{max}) and potential ($\mathcal{V}_{max}$) of the dome of a Van de Graaff generator are determined by the dielectric strength of the surrounding air ($E_{max} = 3 \times 10^6$ V/m). The radius of the dome is $r_0 = 0.13$ m. Find Q_{max} and $\mathcal{V}_{max}$.

Solution. Treating the dome as an isolated charged spherical conductor, we evaluate the expressions for E and $\mathcal{V}$ at the surface of the sphere of radius r_0, and find

Figure 22-21. Example 22-10: These boys are getting a hair-raising experience from a small Van de Graaff generator.

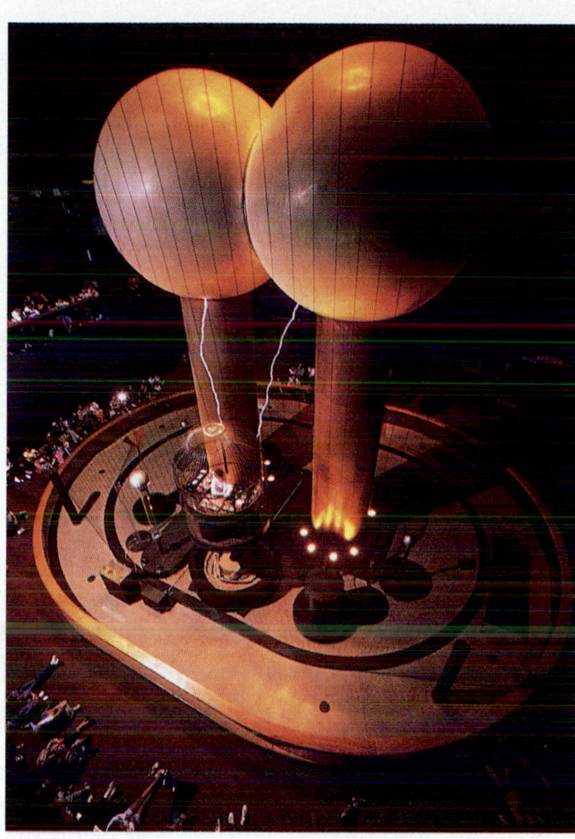

These huge demonstration Van de Graaff generators can produce a large electric potential.

$$E = \frac{|Q|}{4\pi\epsilon_0 r_0^2} = \frac{|\mathcal{V}|}{r_0}$$

Therefore the maximum charge Q_{max} and the maximum potential $\mathcal{V}_{max}$ for a conducting sphere surrounded by an insulator with dielectric strength E_{max} are

$$Q_{max} = E_{max}\, 4\pi\epsilon_0 r_0^2 \qquad \text{and} \qquad \mathcal{V}_{max} = E_{max} r_0$$

In air a sphere of radius 0.13 m has

$$Q_{max} = \frac{(3 \times 10^6\ \text{V/m})(0.13\ \text{m})^2}{9 \times 10^9\ \text{N}\cdot\text{m}^2/\text{C}^2} = 6\ \mu\text{C}$$

and

$$\mathcal{V}_{max} = (3 \times 10^6\ \text{V/m})(0.13\ \text{m}) = 400\ \text{kV}$$

Notice that a dome with twice this radius could contain four times as much charge and could be at twice the potential.

SELF-TEST 22-10. Estimate the maximum charge and potential for a conducting sphere of radius 0.25 m. The sphere is in air and is supported by an insulator whose dielectric strength is greater than that of air. ***ANSWER:*** $Q_{max} = 21\ \mu\text{C}$ and $\mathcal{V}_{max} = 750$ kV.

SUMMARY

Section 22-1. Electric Potential Energy

The electric force is a conservative force. The potential energy of a test particle in the field of a number of fixed particles is

$$U = \frac{q_0}{4\pi\epsilon_0} \sum \frac{q_i}{r_i} \qquad (22\text{-}5)$$

The reference position for U is where the test particle is far removed from the fixed particles ($U_\infty = 0$).

Section 22-2. Electric Potential

The electric potential is defined as

$$\mathcal{V} = \frac{U}{q_0} \quad \text{(small } q_0\text{)} \qquad (22\text{-}6)$$

For a system of charged particles

$$\mathcal{V} = \frac{1}{4\pi\epsilon_0} \sum \frac{q_i}{r_i} \qquad (22\text{-}7)$$

For a continuous distribution of charge

$$\mathcal{V} = \frac{1}{4\pi\epsilon_0} \int \frac{dq}{r} \qquad (22\text{-}8)$$

Section 22-3. Potential Difference

The definition of the potential difference between points a and b is

$$\mathcal{V}_b - \mathcal{V}_a = \frac{U_b - U_a}{q_0} \quad \text{(small } q_0\text{)} \qquad (22\text{-}9)$$

The potential difference in terms of the field **E** is

$$\mathcal{V}_b - \mathcal{V}_a = -\int_a^b \mathbf{E} \cdot d\boldsymbol{\ell} \qquad (22\text{-}10)$$

Section 22-4. Relation between E and $\mathcal{V}$

If an expression for **E** is known, then $\mathcal{V}$ at point P can be found from the line integral of **E**:

$$\mathcal{V} = -\int_\infty^P \mathbf{E} \cdot d\boldsymbol{\ell} \qquad (22\text{-}12)$$

If an expression for $\mathcal{V}$ is known, then **E** can be found from the partial derivatives of $\mathcal{V}$. In particular, when $\mathcal{V}$ is written in terms of cartesian coordinates,

$$E_x = -\frac{\partial \mathcal{V}}{\partial x} \qquad E_y = -\frac{\partial \mathcal{V}}{\partial y} \qquad E_z = -\frac{\partial \mathcal{V}}{\partial z} \qquad (22\text{-}13)$$

Section 22-5. Equipotential Surfaces

An equipotential surface is a surface on which the electric potential is constant. As with field lines for the electric field, drawings of equipotential surfaces help one visualize the spatial dependence of $\mathcal{V}$. Field lines are perpendicular to equipotential surfaces at points where they cross.

Section 22-6. More about Electrostatic Properties of a Conductor

Under static conditions the electric potential is uniform inside a conductor, and consequently the surface of a conductor is an equipotential surface.

QUESTIONS

1. What property (or properties) causes the electric force to be conservative? That it is an inverse-square force? That it is directed along the line between the particles? That it is proportional to the charge magnitude of each particle?

2. If a positively charged particle is moved in the direction of an electric field, does its electric potential energy increase, decrease, or remain the same? How about a negatively charged particle?

3. Consider a point at which $\mathbf{E} = E\mathbf{i}$. From that point give a direction (in terms of a unit vector) in which the potential *(a)* increases, *(b)* decreases, and *(c)* remains the same.

4. A charged particle is moved in the direction of an electric field and its potential energy increases. What is the sign of the charge of the particle?

5. Why is electric potential sometimes called voltage?

6. Consider the proper terminology to use when discussing electric potential energy on the one hand and the electric potential on the other. *(a)* Is it proper to speak of the potential energy of an electron or the potential of an electron? *(b)* Is it proper to speak of the potential energy of a point in space or the potential of a point in space? *(c)* Is it proper to speak of the potential energy produced by an electron or the potential produced by an electron?

7. Suppose someone tells you that no life can exist on Mars because the surface of the planet is at a voltage of 20,000 V. Could this person be correct? Explain.

8. If the potential difference between two points is zero, is there necessarily a path between those two points on which the field is zero at each point? Give an example that supports your answer.

9. If there exists a path on which the field is zero at each point, is the potential difference between two points on this path necessarily zero? Give an example that supports your answer.

10. If you know the numerical value of the electric field at a single point in space, can you use that information to find the potential at that point? If so, how?

11. If you know an expression for the electric field in terms of coordinates in a region of space, can you use that information to find the potential difference between two points within the region? If so, how?

12. If you know the numerical value of the potential at a single point in space, can you use that information to find the electric field at that point? If so, how?

13. Suppose you know the numerical value of the potential difference between two nearby points in space. Can you use this to estimate a component of the electric field between the points? Explain. What assumptions must you make?

14. If you know an expression for the potential throughout a region of space, can you use that information to find the electric field in that region? If so, how?

15. Suppose we interchange the positions of particles 1 and 3 in Fig. 22-6 (Example 22-2). Would the potential $\mathcal{V}$ at P be changed? Would the field **E** at P be changed?

16. Suppose the charge distribution of the ring in Fig. 22-9 is not uniform. Would this change the expression found for the potential at points along the axis in Example 22-4? Would this change the expression found for the electric field at points along the axis (Example 20-6)? Do your answers conform to

Figure 22-22. Question 18.

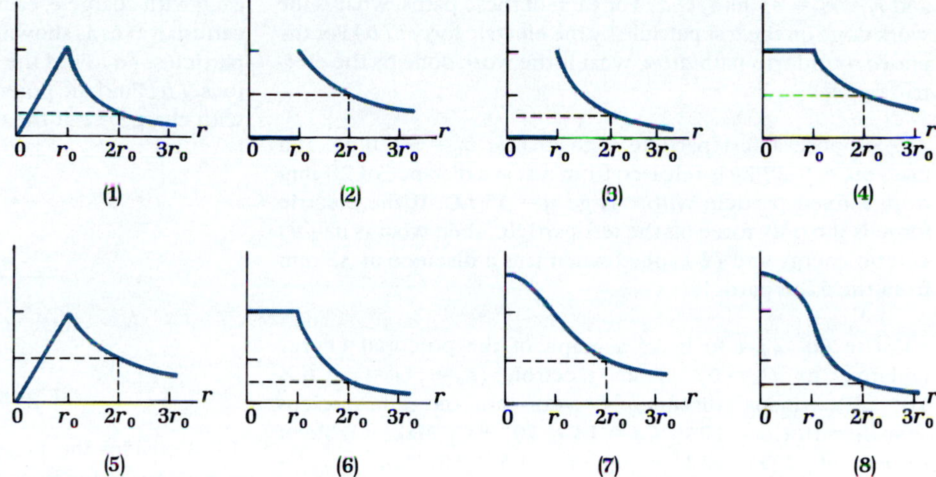

the relation between **E** and $\mathcal{V}$ (Eqs. 22-13)? Explain the apparent discrepancy here.

17. Explain why the potential energy of a dipole in a uniform field depends on its orientation with respect to the field, but not on its position. Why is the potential energy a minimum when the dipole is aligned parallel to the field direction and a maximum when aligned opposite the field direction? If a dipole were in a nonuniform field, would you expect its potential energy to depend on its position?

18. Figure 22-22 shows graphs of a number of functions versus distance r from the center of a spherically symmetric charge distribution of radius r_0. Which graph best represents *(a)* E due to a uniform volume charge, *(b)* $\mathcal{V}$ due to a uniform volume charge, *(c)* E due to a uniform shell of charge, *(d)* $\mathcal{V}$ due to a uniform shell of charge?

19. Can equipotential surfaces intersect one another? Explain.

20. In electrostatics, why is it meaningful to speak of the potential of a conducting object but not of the potential of an insulating object?

21. Suppose the potential difference between the plates in Fig. 22-18 is 100 V. What is the potential difference between the successive pairs of equipotential surfaces shown in the figure?

22. Is it possible for a conductor that possesses a net positive charge to be at a negative potential? If so, describe such a situation. If not, why not?

23. Your employer instructs you to store a delicate instrument such that it will not be exposed to electric fields. Explain how you can do this.

24. Integrated circuit devices are often wrapped in a conducting material when stored or shipped. Why?

25. Complete the following table:

Symbol	Represents	Type	SI Unit
U			J
$\mathcal{V}$		Scalar	
$\Delta\mathcal{V}$	Potential difference		
$d\boldsymbol{\ell}$			
$\partial/\partial x$		N/A	N/A

EXERCISES

Section 22-1. Electric Potential Energy

1. A test particle with charge $q_0 = 4.0$ nC is moved from position a, where it is 35 mm from a fixed particle with charge $q = 85$ nC, to position b, where it is 76 mm from the fixed particle. *(a)* What is the potential energy (relative to $U_\infty = 0$) of the test particle at position a and at position b? *(b)* What is the change in the potential energy of the test particle as a result of the displacement? *(c)* What is the work done by the electric force on the test particle during the displacement?

2. In Fig. 22-23 a test particle with charge $q_0 = 4.0$ nC is shown taken around a closed path in the field of a fixed particle with charge $q = 85$ nC. Path ai is radially aligned with the fixed particle, path ib is a circular arc centered at the fixed particle, and path ba is a circular arc centered at i ($r_a = 45$ mm

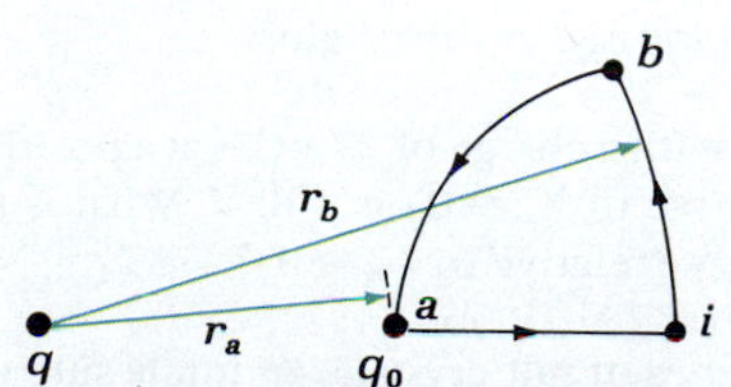

Figure 22-23. Exercise 2.

and $r_i = r_b = 83$ mm). *(a)* For each of these paths, what is the work done on the test particle by the electric force? *(b)* For the entire round-trip path *aiba*, what is the work done by the electric force?

3. Suppose a test particle with charge $q_0 = -6.0$ nC and mass $m_0 = 0.22$ kg is released from rest at a distance of 78 mm from a fixed particle with charge $q = 55$ nC. If the electric force is the only force on the test particle, then what is its *(a)* kinetic energy and *(b)* speed when it is a distance of 32 mm from the fixed particle?

4. Use Eq. 22-4 to make a graph of the potential energy (relative to $U_\infty = 0$) of an electron ($q_0 = -e = -1.6 \times 10^{-19}$ C) versus distance r from an oxygen nucleus [$q = 8e = 8(1.6 \times 10^{-19}\text{ C}) = 13 \times 10^{-10}$ C]. Make a table of values of $U(r)$ and r for $r = 0.5 \times 10^{-10}$ m to $r = 3.0 \times 10^{-10}$ m at intervals of 0.5×10^{-10} m. Plot the tabulated values and sketch the curve.

5. *(a)* Find the potential energy U (relative to $U_\infty = 0$) of particle 1 in Fig. 22-24 *(a)* when $q_1 = 2.0$ nC and *(b)* when $q_1 = 4.0$ nC. The values of the other charges are $q_2 = 25$ nC, $q_3 = 38$ nC, and $q_4 = 32$ nC.

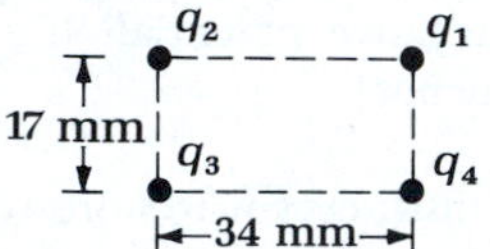

Figure 22-24. Exercise 5.

6. In a classic experiment that helped establish the structure of atoms, high-energy α particles (helium nuclei) were directed at a thin sheet of gold. *(a)* What is the closest distance that an α particle (charge $2e$) can come to a gold nucleus (charge $79e$) if the α particle's kinetic energy is 4 MeV at a great distance from the gold nucleus and its direction of travel is straight toward the gold nucleus? Assume that the position of the gold nucleus remains fixed. *(b)* What must the α particle's initial kinetic energy be in order for it to make contact with the edge of the gold nucleus? Assume that contact is made when the centers of the α particle and the gold nucleus are about 10×10^{-15} m from each other.

Section 22-2. Electric Potential

7. *(a)* Use Eq. 22-6 and the result of Exercise 5, part *(a)*, to find the potential due to particles 2, 3, and 4 at the position of particle 1 in Fig. 22-24. *(b)* Use Eq. 22-6 and the result of Exercise 5, part *(b)*, to find the potential due to particles 2, 3, and 4 at the position of particle 1 in Fig. 22-24. *(c)* Use Eq. 22-7 to find the potential due to particles 2, 3, and 4 at the position of particle 1 in Fig. 22-24.

8. What is the potential (relative to $\mathcal{V}_\infty = 0$) at the point (0.4 m, 0.2 m, −0.5 m) due to a particle with charge $q = 75$ nC and located at the origin?

9. A particle with a charge of 27 nC is at a position where the potential (relative to $\mathcal{V}_\infty = 0$) is 450 V. What is the particle's potential energy (relative to $U_\infty = 0$)?

10. In some types of salt crystals, an ion is surrounded by six ions of opposite sign as nearest neighbors. Consider six ions, each with charge e, each a distance of 1.5×10^{-10} m along the cartesian axes as shown in Fig. 22-25. Treat the ions as charged particles. *(a)* Find the potential at the origin due to these six ions. *(b)* Find the potential energy (in electron volts) of an ion with charge $-e$ at the origin.

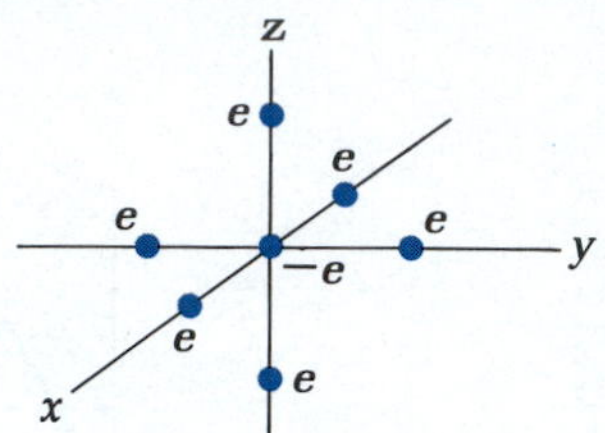

Figure 22-25. Exercise 10.

11. Consider the potential $\mathcal{V}$ due to a dipole with moment $\mathbf{p} = (2aq)\mathbf{k}$ located at the origin, with $q = e$ and $a = 0.4 \times 10^{-10}$ m. Evaluate $\mathcal{V}$ along the z axis from $z = -1.2 \times 10^{-10}$ to 1.2×10^{-10} m at intervals of 0.2×10^{-10} m. (The result of Example 22-3 cannot be used here. Why?) Plot your results on graph paper and sketch in $\mathcal{V}(z)$. How would a graph of $\mathcal{V}$ versus x along the x axis look?

12. Consider a particle with charge Q located at $(0, 0, a)$. Find an expression for the potential due to this particle at points along the x axis and compare your answer with the result of Example 22-4.

13. Graph the results of Example 22-5. Plot $\mathcal{V}/(\sigma R_0/2\epsilon_0)$ versus x/R_0 for $x/R_0 = -2.5$ to 2.5 in intervals of 0.5. Sketch in $\mathcal{V}(x)$.

14. From Prob. 7, the electric potential in the perpendicular bisector plane (xy plane in Fig. 22-26) of a uniform line charge ($\lambda = Q/2\ell$) is

$$\mathcal{V} = \frac{\lambda}{2\pi\epsilon_0} \ln \frac{\ell + \sqrt{\ell^2 + R^2}}{R}$$

where $R = \sqrt{x^2 + y^2}$. Make a graph of $\mathcal{V}$ (in units of $\lambda/2\pi\epsilon_0$) versus R (in units of ℓ). Plot points for $R = \frac{1}{2}\ell, \ell, \ldots, 5\ell$, and sketch the curve.

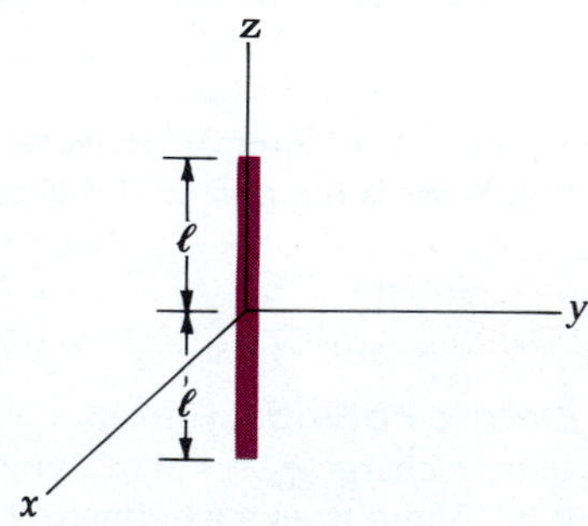

Figure 22-26. Exercise 14.

Section 22-3. Potential Difference

15. Consider a uniform electric field: $\mathbf{E} = -(220\text{ V/m})\mathbf{i}$. *(a)* What is the potential difference between the origin and (1.5 m, 0, 0)? *(b)* What is the potential difference between the origin and (1.5 m, 1.0 m, 0)? *(c)* What is the potential difference between (1.5 m, 0, 0) and the origin? *(d)* What is the potential difference between the origin and (3.0 m, 0, 0)? *(e)*

What is the potential difference between (1.5 m, 0, 0) and (3.0 m, 0, 0)?

16. The potential difference between point a (1.2 m, −2.6 m, 1.8 m) and point b (2.3 m, 1.4 m, −0.8 m) in a uniform field, $\mathbf{E} = E_x\mathbf{i}$, is 730 V. What is E_x?

17. The magnitude of the charge possessed by each plate of a parallel-plate capacitor is 260 nC, the separation of the plates is 0.32 mm, and the lateral area of each plate is 2.2×10^{-2} m². Assuming the plates are close enough together to approximate the field as uniform, *(a)* find the magnitude E of the field between the plates and *(b)* find the potential difference $\Delta\mathcal{V}$ between the plates.

18. The plates of a parallel-plate capacitor have a separation of 0.25 mm, and each plate has a lateral area of 4.6×10^{-2} m². The potential difference between the plates is $\Delta\mathcal{V} = 540$ V. Assuming that the field between the plates is uniform, find *(a)* E, *(b)* $|\sigma|$, and *(c)* $|Q|$.

19. Make a graph of the result of Example 22-8. Evaluate the potential energy U of a dipole in a uniform field when the angle between the dipole moment and the field is 0, 30°, . . . 180°. Make a table of your results and plot U in terms of pE versus θ in degrees. Sketch the curve.

20. *(a)* What is the kinetic energy in electron volts at which an electron emerges from an electron gun with a potential difference between cathode and anode of 970 V? *(b)* What is the speed of an electron as it emerges from the electron gun?

21. A spherically symmetric uniform volume charge distribution with total charge 78 nC and radius 53 mm is centered at the origin. *(a)* What is the potential (relative to $\mathcal{V}_\infty = 0$) at (−12 mm, 17 mm, 22 mm)? *(b)* What is the potential at (28 mm, −45 mm, −42 mm)?

22. Consider a spherically symmetric uniform volume distribution of charge with $Q = 45$ nC and $r_0 = 74$ mm. Determine the potential (relative to $\mathcal{V}_\infty = 0$) at the following distances from the center: *(a)* $r = 0$; *(b)* $r = 37$ mm; *(c)* $r = 74$ mm; *(d)* $r = 148$ mm.

23. Make a rough estimate of the potential at points on the surface of the earth (relative to $\mathcal{V}_\infty = 0$). Assume that the magnitude of the electric field is 150 V/m at the surface of the earth, that it decreases linearly with height above the surface to zero at a height of 50 km, and that it is zero beyond 50 km. The field is directed toward the center of the earth. (The magnitude of your answer will be about a factor of 10 too large because of the crudeness of our assumptions.)

24. Recall that the magnitude of the field far from the ends of a long, uniformly distributed line charge of linear charge density λ is approximately $E \approx \lambda/2\pi\epsilon_0 R$, where R is the perpendicular distance from the line charge. The field is directed away from the line charge (assuming λ is positive). Show that the approximate potential difference between two points a and b far from the ends of the line charge is

$$\mathcal{V}_b - \mathcal{V}_a \approx \frac{\lambda}{2\pi\epsilon_0} \ln \frac{R_a}{R_b}$$

where R_a and R_b are both much smaller than the length of the line charge.

Section 22-4. Relation between E and $\mathcal{V}$

25. From Fig. 22-19, the potential (relative to $\mathcal{V}_\infty = 0$) at all points inside a conducting sphere of radius r_0 and charge Q is $\mathcal{V} = Q/4\pi\epsilon_0 r_0$. Use this result to find the electric field at points inside the sphere.

26. The potential difference between the origin and points with coordinates (x, y, z) is given by $\mathcal{V}(x, y, z) - \mathcal{V}_0 = (740 \text{ V/m})x + (-230 \text{ V/m})y + (-690 \text{ V/m})z$. What is $\mathbf{E}$?

27. The potential in a region of space is given by the expression

$$\mathcal{V} = \frac{\mathrm{A}}{(x^2 + a^2)^2}$$

(a) Given that $\mathrm{A} = 200 \text{ V}\cdot\text{m}^4$ and $a = 0.5$ m, determine $\mathcal{V}$ at $x = 0.5$ m. *(b)* Find an expression for $\mathbf{E}$ in the region. *(c)* Evaluate $\mathbf{E}$ at $x = 0.5$ m.

28. In Example 22-4 we showed that the potential (relative to $\mathcal{V}_\infty = 0$) along the axis of a charged ring is $\mathcal{V}(x) = Q/4\pi\epsilon_0\sqrt{x^2 + a^2}$. Determine E_x along the axis.

29. In Example 22-5 we showed that the potential along the axis of a uniformly charged circular disk is $\mathcal{V}(x) = \sigma(\sqrt{x^2 + R^2} - \sqrt{x^2})/2\epsilon_0$. Determine E_x along the axis.

30. The potential as a function of x is plotted in Fig. 22-27. Estimate E_x at *(a)* $x = -2$ m; *(b)* $x = 1$ m; *(c)* $x = 4$ m.

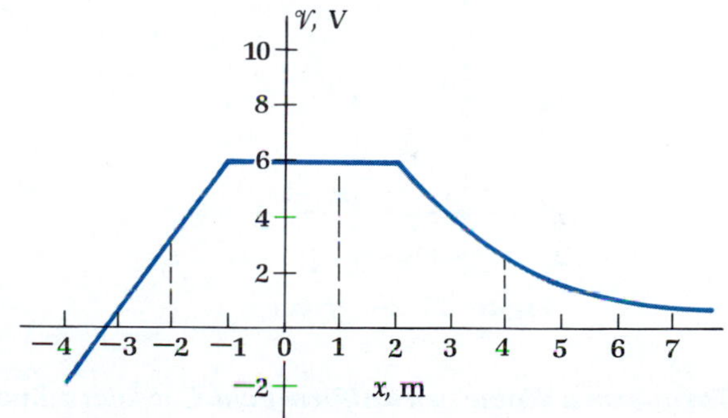

Figure 22-27. Exercise 30.

31. The electric field in a region of space between $x = 1.0$ m and $x = 3.0$ m is given by the expression

$$\mathbf{E} = [(380 \text{ V/m})\, e^{-x/a}]\mathbf{i}$$

where $a = 2.0$ m. Determine the potential difference between $x = 3.0$ m and $x = 1.0$ m; $\Delta\mathcal{V} = \mathcal{V}(3.0 \text{ m}) - \mathcal{V}(1.0 \text{ m})$.

32. In cartesian coordinates the potential due to a point charge at the origin is $\mathcal{V} = q/4\pi\epsilon_0\sqrt{x^2 + y^2 + z^2}$. Use Eqs. 22-13 to show that this gives $\mathbf{E} = (q/4\pi\epsilon_0 r^2)\hat{\mathbf{r}}$, where the unit vector $\hat{\mathbf{r}}$ in cartesian coordinates is $\hat{\mathbf{r}} = \mathbf{r}/r = (x\mathbf{i} + y\mathbf{j} + z\mathbf{k})/\sqrt{x^2 + y^2 + z^2}$.

Section 22-5. Equipotential Surfaces

33. The plates of a parallel-plate capacitor are separated by 0.64 mm, have a lateral area of 0.33 m², and carry equal and opposite charges of magnitude 390 nC. If we wish to show six equipotential surfaces between the plates, including the ones at the plate surfaces, what is the potential difference between successive pairs of equipotential surfaces?

34. A spherically symmetric charge distribution has a radius of 41 mm and a potential (relative to $\mathcal{V}_\infty = 0$) at its surface of 600 V. *(a)* What is the radius of the 300-V equipotential surface? *(b)* What is the radius of the 150-V equipotential surface?

35. Sketch field lines and equipotential surfaces for a dipole. Show four equipotential surfaces: one that is midway between the charges, one surrounding each of the charges, and two others that extend beyond your graph.

36. Consider a thin, uniformly charged disk centered at the origin and contained in the yz plane such that its axis is the x axis (Fig. 22-10). Sketch field lines in the xy plane and the intersection of equipotential surfaces with the xy plane. Show four or five equipotential surfaces.

37. Consider a long, uniformly charged rod centered at the origin and oriented along the z axis (Fig. 22-26). Sketch field lines in the xy plane and the intersection of equipotential surfaces with the xy plane. Assume that the rod is long enough such that it may be treated as infinitely long. Show four equipotential surfaces and make their radii correspond to the same potential difference between each successive pair of surfaces (see Exercise 24).

38. Suppose the drawing in Fig. 22-17 corresponds to a charged sphere with $Q = 88.9\ \mu\text{C}$ and $r_0 = 0.10$ m. *(a)* Determine the potential $\mathcal{V}_0$ at the sphere's surface. *(b)* By making measurements directly from the figure, determine the actual radius of each of the equipotential surfaces shown. *(c)* Determine the potential of each of the surfaces. What is the potential difference between successive pairs of surfaces?

Section 22-6. More about Electrostatic Properties of a Conductor

39. *(a)* What is the charge on an isolated conducting sphere of radius 76 mm when its potential (relative to $\mathcal{V}_\infty = 0$) is 530 V? *(b)* What is the surface charge density of the sphere? *(c)* What is the electric field just outside the surface?

40. What are the maximum charge and maximum potential in dry air on a metal sphere of radius 45 mm?

PROBLEMS

1. ***A nonconservative field.*** Show that the field $\mathbf{E}' = Ay\mathbf{i} + Bx\mathbf{j}$ is *not* conservative if $A \neq B$. (*Hint:* Consider the line integral of $\mathbf{E}'$ from a to b along the two paths shown in Fig. 22-28.)

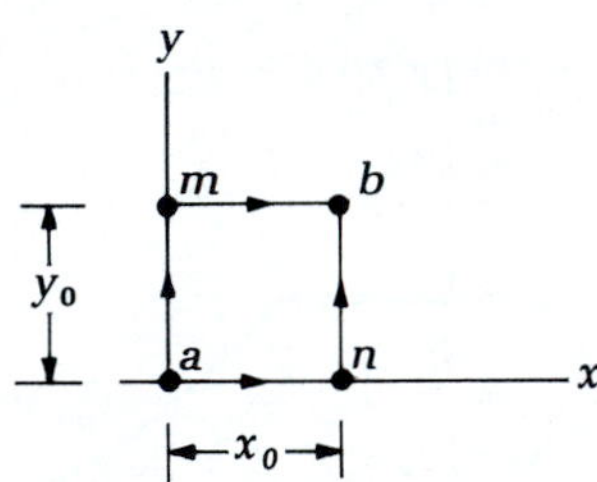

Figure 22-28. Problem 1.

2. ***Torque on a dipole in a uniform field.*** Consider a dipole in a uniform electric field (Fig. 22-15). *(a)* Show that the net electric force on the dipole is zero. *(b)* Show that the torque on the dipole (about an axis perpendicular to the dipole moment and through the center of the dipole) which tends to align it with $\mathbf{E}$ is

$$\tau = \mathbf{p} \times \mathbf{E}$$

(c) Use the result of Example 22-8 to show that the magnitude of the torque may be written

$$\tau = \frac{dU}{d\theta}$$

3. ***Bohr model of hydrogen.*** The Bohr model of the hydrogen atom has the electron orbiting the proton similar to the way a planet orbits the sun. Assume the electron has a circular orbit and that the much more massive proton remains fixed. *(a)* Use Newton's second law applied to uniform circular motion ($\Sigma F = mv^2/r$) and Coulomb's law to show that the relation between the kinetic energy K and the potential energy U is $2K = -U$. (Note that $U < 0$.) *(b)* Show that the mechanical energy ($K + U$) for this circular orbit in terms of the electron-proton separation distance r is $-e^2/8\pi\epsilon_0 r$. *(c)* The ionization energy is the amount of energy required to separate the electron a great distance from the proton. That is, it is the difference in mechanical energy between the bound state of this system [from part *(b)* above] and the state where the potential energy and the kinetic energy are both zero. Given that the ionization energy of the hydrogen atom is 13.6 eV, determine the radius of the electron's orbit. *(d)* Find the speed of the electron in its orbit.

4. ***Energy of α decay.*** Some of the heaviest atomic nuclei are radioactive and decay by emitting an α particle. These α particles usually have a kinetic energy of about 5 MeV (depending on the type of nucleus) after they are far from the nucleus. Use this information to estimate the size of a nucleus by assuming that the α particle leaves the edge of the nucleus with zero kinetic energy and that the electrostatic potential energy is the only potential energy it has at that position. Because of the error introduced by these assumptions, your answer will be about a factor of 4 too large.

5. ***Energy of nuclear fission.*** When a uranium nucleus fissions (splits into two smaller nuclei and other fragments), the combined charge of the two smaller nuclei is the same as the charge of the original uranium nucleus ($92e$) and they have a combined kinetic energy of about 200 MeV when separated a great distance. Assuming these two nuclei have about the same radius r_0 and charge $46e$, make an estimate of r_0.

6. ***Potential at points near a charged disk.*** In Example 22-5 we found that the potential (relative to $\mathcal{V}_\infty = 0$) at points along the axis of a uniformly charged circular disk of radius R_0 is

$$\mathcal{V} = \frac{\sigma}{2\epsilon_0}\left(\sqrt{x^2 + R_0^{\,2}} - \sqrt{x^2}\right)$$

(a) Show that at points on the axis near the center of the disk ($|x| \ll R_0$) the potential is approximately

$$\mathcal{V} \approx \frac{\sigma R_0}{2\epsilon_0}\left(1 - \frac{|x|}{R_0}\right)$$

(b) Use the answer to part *(a)* to find E_x at points on the axis ($|x| \ll R_0$). (*Hint:* When $x > 0$, $|x| = x$, and when $x < 0$, $|x| = -x$.)

7. ***Potential in the bisector plane of a uniformly charged rod.*** Consider a uniformly charged rod of length 2ℓ and linear charge density λ (which is $Q/2\ell$) centered at the origin and oriented along the z axis (Fig. 22-26). Show that the potential at points in the perpendicular bisector plane (the xy plane) is

$$\mathcal{V} = \frac{\lambda}{2\pi\epsilon_0}\ln\frac{\ell + \sqrt{\ell^2 + R^2}}{R}$$

where $R = \sqrt{x^2 + y^2}$.

8. ***Field in the bisector plane of a uniformly charged rod.*** Use the answer to the previous problem to find E in the perpendicular bisector plane of a uniform line charge. (*Hint:* When $\mathcal{V}$ has cylindrical symmetry about the z axis, $E_R = -\partial\mathcal{V}/\partial R$.)

9. ***Potential along the axis of a uniformly charged rod.*** Figure 22-26 shows a uniform line charge ($\lambda = Q/2\ell$) of length 2ℓ extending along the z axis and centered at the origin. Show that the potential at points along the z axis ($|z| > \ell$) is

$$\mathcal{V}(z) = \frac{\lambda}{4\pi\epsilon_0}\ln\frac{|z| + \ell}{|z| - \ell}$$

10. ***Potential in a perpendicular plane through the end of a charged rod.*** Figure 22-29 shows a line charge that extends along the z axis from the origin to $(0, 0, \ell)$. The linear charge density increases with z: $\lambda = cz$, where c is a constant. *(a)* Find an expression for $\mathcal{V}$ along the x axis. *(b)* Modify your answer to part *(a)* so that it is valid for all points in the xy plane.

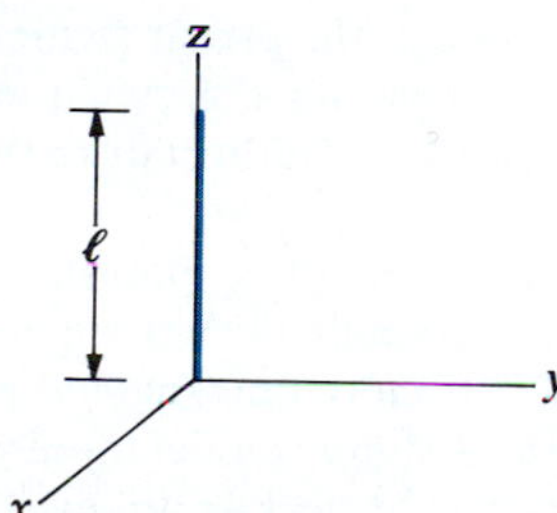

Figure 22-29. Problem 10.

11. ***Potential produced by a linear quadrupole.*** Show that at point $P(r \gg a)$ the potential due to the linear quadrupole shown in Fig. 22-30 is

$$\mathcal{V} \approx \frac{qa^2}{4\pi\epsilon_0 r^3}(3\cos^2\theta - 1)$$

12. ***Field produced by a linear quadrupole.*** *(a)* Express the answer to the previous problem in cartesian coordinates. *(b)* Use the result of part *(a)* and Eqs. 22-13 to find E_x, E_y, and E_z due to the linear quadrupole.

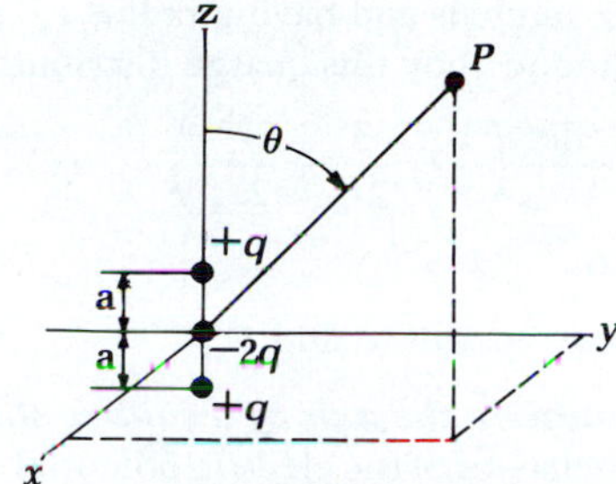

Figure 22-30. Problem 11.

13. ***Potential of concentric spheres.*** A thin-walled conducting spherical shell of outer radius r_{bo} and inner radius r_{bi} is concentric with a solid conducting sphere of radius r_a, as shown in Fig. 22-31. Sphere b has a net charge Q_b, sphere a has a net charge Q_a, and both charges are of the same sign. *(a)* What is the potential of sphere b? *(b)* What is the potential difference between sphere b and sphere a? *(c)* What is the potential of sphere a?

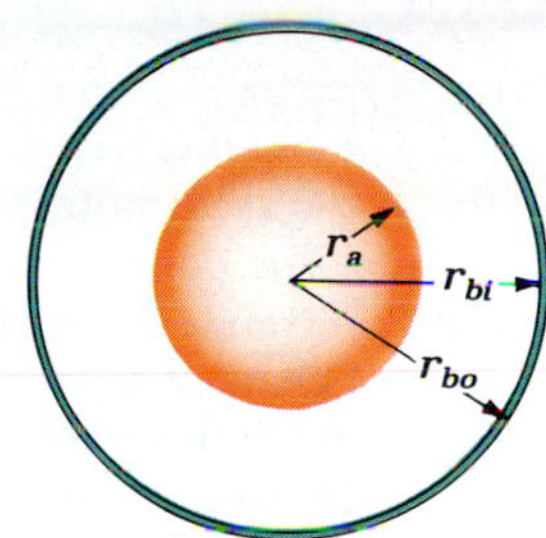

Figure 22-31. Problem 13.

14. ***More about the potential of concentric spheres.*** Reconsider the previous problem for the case where Q_a is positive, Q_b is negative, and $|Q_a| = |Q_b|$.

15. ***Potential produced by a thick shell of volume charge.*** Consider a uniform spherical distribution of net charge Q between an inner radius r_i and outer radius r_o, as shown in cross section in Fig. 22-32. Find the potential in all three regions of space: $r > r_o$, $r_o > r > r_i$, and $r < r_i$.

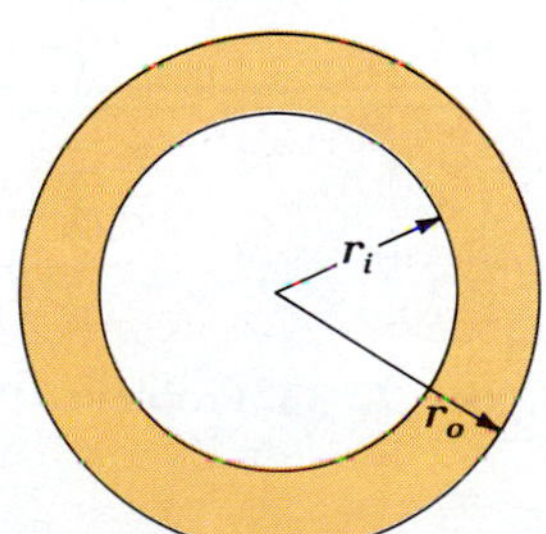

Figure 22-32. Problem 15.

16. ***Field on the axis of a charged rod.*** Use the answer to Prob. 9 to determine E_z on the z axis in Fig. 22-26 for *(a)* $z > \ell$ and *(b)* $z < -\ell$. (*Hint:* When $z > 0$, $|z| = z$, and when $z < 0$, $|z| = -z$.)

17. ***More about Rutherford's atomic model.*** In Prob. 12 of Chap. 21, we discussed a simple atomic model where the nucleus is treated as a point charge $+Ze$ and the electrons are treated as a uniform spherical distribution of charge $-Ze$ con-

centric with the nucleus and having radius r_a. *(a)* Show that the potential produced by this charge distribution is

$$\mathscr{V} = \frac{Ze}{4\pi\epsilon_0}\left(\frac{1}{r} - \frac{3}{2r_a} + \frac{r^2}{2r_a^3}\right) \qquad (r < r_a)$$

$$\mathscr{V} = 0 \qquad (r > r_a)$$

(b) Use the above answer to find E.

18. **Potential on the axis of a square sheet of charge.** Determine the behavior of the electric potential on the axis of a square sheet of charge with uniform charge density. Let the square distribution of charge be represented by 10 parallel line charges, each of length 2ℓ (see Prob. 7), with proper spacing to approximate a square with uniform charge density. Write a program that determines the potential at about 10 equally spaced points from about 0.2ℓ to 2.0ℓ, where 2ℓ is the length of the sides of the square. Graph your results and compare them with those for a circular disk with the same charge and with radius ℓ. Use numerical values that are both convenient and realistic. You may wish to rework the problem using 20 line charges rather than 10 in order to estimate the accuracy of your answer.

19. **Proving that $E_t = 0$ just outside a conductor.** In Section 21-5, we pointed out that $\mathbf{E} = 0$ at any point inside a conductor in electrostatic equilibrium. For points just outside the conductor, we (i) stated that $E_t = 0$ and (ii) proved that $E_n = \sigma/\epsilon_0$. Using the techniques of this chapter, you can prove item (i). Consider performing the line integral of the electric field around the closed rectangular path shown in Fig. 22-33. We choose the rectangle to be small enough such that the conductor's surface can be treated as a plane surface and the electric field outside the conductor is essentially uniform from *a* to *b*. Since we are interested in the field just outside the conductor, the distance δ between the surface and the path from *a* to *b* is vanishingly small. Use the fact that, in electrostatics, the line integral of $\mathbf{E}$ around any closed path is zero to show that the tangential component of the field just outside the conductor is zero: $E_t = 0$.

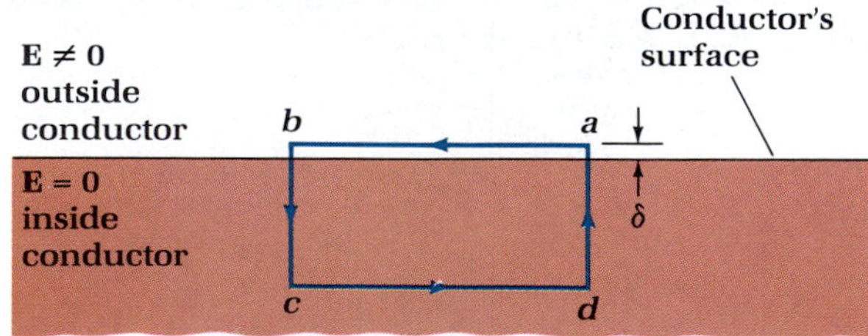

Figure 22-33. Problem 19.

20. **Inflection points in $\mathscr{V}$ versus x for the charged ring.** In Example 22-4, we found that the expression for $\mathscr{V}$ along the axis of a charged ring centered at the origin is

$$\mathscr{V} = \frac{Q}{4\pi\epsilon_0\sqrt{x^2 + a^2}}$$

Determine the inflection points for this function and check your answer by examining Fig. 22-10. These inflection points correspond to the points at which the electric field components reach a maximum on one side and a minimum on the other.

21. **Potential due to a uniform spherical volume charge.** In Example 22-9, the expression for the potential $\mathscr{V}$ produced by a uniform spherical volume charge was differentiated to determine the expression for the electric field due to this charge: $E_r = -d\mathscr{V}/dr$. Perform the reverse procedure. That is, integrate the expression for the field to determine the expression for the potential.

22. **Field produced by a dipole.** From Example 22-3, the potential due to a dipole of moment $\mathbf{p} = (2aq)\mathbf{k}$ located at the origin is $\mathscr{V} = (p\cos\theta)/4\pi\epsilon_0 r^2$, where r is the distance from the origin to the point where $\mathscr{V}$ is evaluated and θ is the angle between the z axis and a line from the origin to this point (Fig. 22-7). *(a)* Write this expression for $\mathscr{V}$ in terms of cartesian coordinates x, y, and z. That is, replace $\cos\theta$ and r with cartesian coordinates. *(b)* Use your result from part *(a)* and Eqs. 22-13 to show that the cartesian components of the electric field produced by this dipole are

$$E_x = \frac{p}{4\pi\epsilon_0}\frac{3xz}{r^5} \qquad E_y = \frac{p}{4\pi\epsilon_0}\frac{3yz}{r^5}$$

and

$$E_z = \frac{p}{4\pi\epsilon_0}\left(\frac{3z^2}{r^5} - \frac{1}{r^3}\right)$$

Additional Problems

23. A uranium nucleus may be thought of as a collection of 92 protons of charge $+e$ and a greater number of neutrons, which are neutral. The coulomb repulsion of the protons for one another has a significant destabilizing effect. In fact, if only a small amount of energy is added, the nucleus tends to break apart (fission) into several pieces. Neglecting any consideration of nuclear binding forces, what is the maximum amount of energy that can be released if a uranium nucleus breaks into two pieces? Note that an integral number of protons must be in each piece, and that the total between the two daughters must be 92. The initial charge radius of the uranium nucleus is 8×10^{-13} cm, and you may assume that the charge density of the daughter is the same as the parent (which is not generally true). Further, you may assume that when inside the nucleus the proton charge contributes uniformly to the charge density.

24. The charge radius r_q of a proton is approximately 0.9×10^{-13} cm. The present evidence suggests that each proton is composed of still more fundamental pointlike charged objects called *quarks*. A simple model represents the proton as being composed of two charge $+\frac{2}{3}e$ up quarks and one charge $-\frac{1}{3}e$ down quark. Let us consider a somewhat naive model where each of these quarks occupies the vertex of an equilateral triangle a distance r_p from its center. How much energy is required to assemble such a proton, assuming we could put it together one quark at a time? Does it matter in which order the quarks are brought in? If $E = mc^2$, where m is the rest mass and c is the speed of light, what mass would the proton have from the electrostatic forces in this model? Express your answer in mega-electron volts (MeV). The proton's measured rest mass is about 938 MeV.

25. There has been a suggestion that a large space station could be built in orbit around the earth using material from the

moon. Assume that this material is to be delivered to the construction site by shooting it up from the moon's surface with an electrostatic gun and assume the maximum mass per shot is limited to 1 gm and the accelerating potential will be 10^6 V. Further assume that the material must just reach escape velocity from the moon. How much charge should each individual projectile be given?

26. Calculate the energy required to assemble a sphere of radius R and uniform charge density ρ. (*Hint:* Consider an assembly process where the sphere is put together in a sequence of infinitesimal concentric shells.)

27. You can use the result of Prob. 4 to estimate the energy yield from nuclear fission processes in which a nucleus such as uranium splits into several pieces. The daughter pieces are more tightly bound together than the original parent nucleus. The nuclear binding forces counter the electrostatic repulsion of the like-charged protons in the nucleus, so you can estimate the binding energy by calculating the energy necessary to contain the protons. Assume a uranium nucleus is a sphere of radius 8×10^{-13} cm and charge $92e$. How much energy can potentially be liberated when this nucleus fissions into two nuclei of equal size, charge $46e$, and radius 6.1×10^{-13} cm?

28. Dust particles in the interplanetary media tend to pick up stray electrons. What is the potential at the surface of a 1-μm dust particle that has picked up 10 electrons?

29. Use the information in Exercise 23 to estimate the net charge on (and within) the earth's surface.

30. A sphere of radius R with center at the origin has similar charges located in the xy plane at 60° intervals, beginning with one on the x axis. If we define $V(r=\infty)=0$, where else is $V=0$?

31. Machines are now being conceived that can accelerate charged particles (electrons and protons usually) with 1 GeV of energy per meter. *(a)* What is the potential difference required to obtain such performance per meter? *(b)* Is it reasonable to try to use static charge distributions to produce these fields? (For example, if charged parallel plates were used to produce the required **E** field, even for 1 m, what would be the force between the plates?)

32. A fine 25-μm wire is suspended 2 mm above a grounded conducting plane. If a potential difference of 2000 V is to be maintained between the wire and the plane, what charge per unit length is required on the wire? (*Hint:* Consider the electric field about two parallel, oppositely charged wires 4 mm apart.)

33. How many kilograms of electrons would have to be removed from the earth (assuming the remaining positive charge is uniformly distributed over the surface) to change the potential at the earth's surface by 1 V?

34. Consider four charges of equal magnitude positioned on the x and y axes, each a distance a from the origin. The sign of the charge corresponds to the sign of the coordinate (the charge on the $+x$ axis is positive and on the $-x$ axis is negative, similarly for the y axis. *(a)* Find an approximate expression for the potential as a function of r and θ for $r \gg a$. *(b)* Find the **E** field from the above potential. *(c)* For which values of θ is **E** radially inward? Does this make sense, given the charge distribution?

35. A certain 12-V automobile battery can provide 6000 W·h of energy. Assuming that it maintains the full 12 V during this effort (which is not true), how much charge must be transferred from one terminal to the other?

36. Consider an alternative universe where the coulomb force varies universally as the *cube* of the distance

$$E = \frac{1}{4\pi\varepsilon_o}\frac{q}{r^3}$$

Energy and work have the same meaning as in our universe. Is it possible to define a meaningful potential analogous to the one we use, and if so, what is its form?

37. Using a potential, which is a *scalar* function, as a shortcut to represent the full directional information in a *vector* field somehow seems like cheating. *(a)* Is there anything missing in the information contained in the potential that is present in the full vector **E** field? *(b)* Is it possible in an abstract sense to always find a scalar function that can represent an arbitrary vector field using a prescription similar to the one we have employed to relate the **E** field to the potential? This question is limited to functions and fields that depend only on the same spatial coordinates and are explicitly time-independent.

38. A long hollow cylinder has a length ℓ and radius r. It has been charged uniformly with a total charge Q. Find the potential at a point on the axis a distance x $(<\frac{\ell}{2})$ from the center of the cylinder.

23 CAPACITANCE, ELECTRIC ENERGY, AND PROPERTIES OF INSULATORS

Large insulators provide electrical isolation for high-voltage transmission lines.

A *capacitor* is one of several kinds of devices used in the electric circuits of radios, computers, and other such equipment. Capacitors provide temporary storage of energy in circuits; they can be made to store and release electric energy in concert with the functions of the circuit. The properties of a capacitor that characterizes its ability to store energy is its *capacitance.*

When energy is stored in a capacitor, an electric field exists within the capacitor. This stored energy can be associated with the electric field. Indeed, energy can be associated with the existence of any electric field. The study of capacitors and capacitance leads us to an important aspect of electric fields, ***the energy of an electric field.***

The study of capacitors and capacitance also provides the background for learning about some of the properties of insulators. Because of their behavior in electric fields, insulators are often referred to as ***dielectrics.***

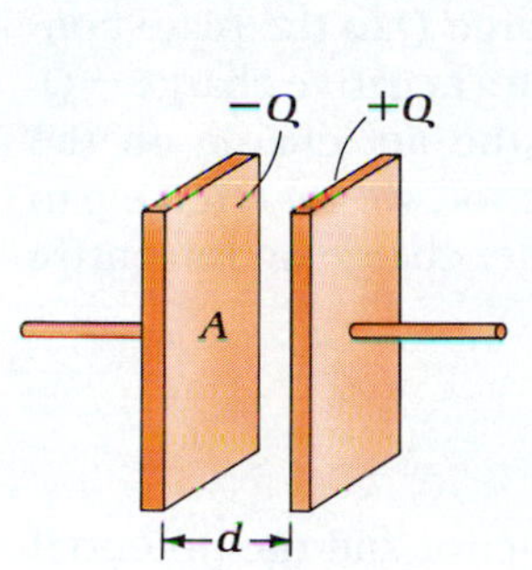

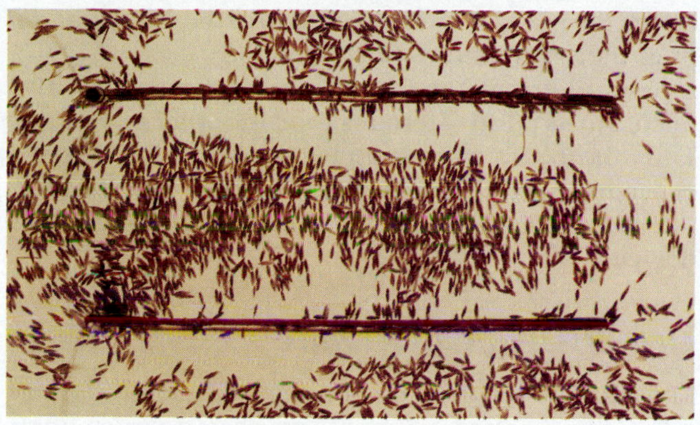

Figure 23-1. *(Left)* A parallel-plate capacitor. *(Right)* Grass-seed demonstration of the electric field between the plates of a parallel-plate capacitor. Rod-shaped electrodes with equal and opposite charge represent the plates. The seeds are well aligned in the region between the rods, but their orientation is nearly random outside the rods and far from the ends. In the region at the ends of the rods, the orientation of the seeds suggests the direction of the "fringing field" at the edges of capacitor plates.

23-1 CAPACITORS AND CAPACITANCE

A ***capacitor*** is a device that consists of two conductors that are close together but are insulated from one another.

Definition of a capacitor

Regardless of their shape, these conductors are referred to as "plates." Figure 23-1 shows a parallel-plate capacitor, which we discussed earlier (Examples 21-8 and 22-6 and Sec. 22-5). The figure also shows conducting wires that are used to connect the plates to other circuit elements. During normal operation the two plates possess charge of equal magnitude but opposite sign. The charge is distributed as a surface charge mostly over the two facing surfaces.

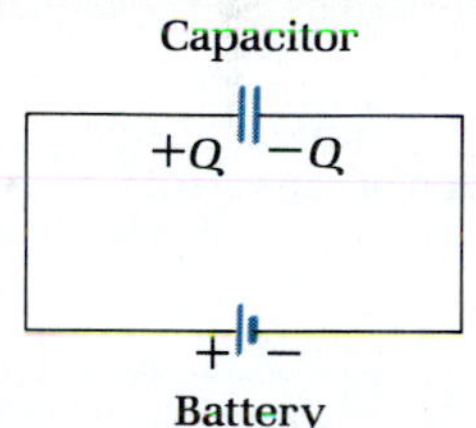

Figure 23-2. A schematic diagram of a capacitor connected to a battery. The longer line at the battery represents its positive terminal and the shorter line its negative terminal. The straight lines represent wires that connect the capacitor to the battery.

Charging a Capacitor

A capacitor can be charged by connecting the wires from the plates to the terminals of a battery, as shown schematically in Fig. 23-2. Batteries are discussed in more detail in Chap. 25. For now we simply note that when a battery is connected to a capacitor, it moves charge carriers from one plate to the other. If the battery remains connected until equilibrium is established (that is, the charge carriers cease to flow), then the potential difference $\mathcal{V}$ between the negative plate and the positive plate is the same as that between the terminals of the battery. (We used the symbol $\Delta\mathcal{V}$ for potential difference in the last chapter. Here we deal only with potential difference, so for brevity we drop the Δ.) The potential difference be-

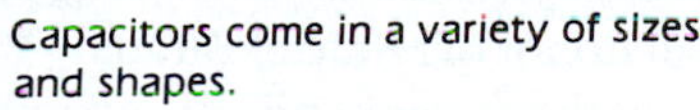
Capacitors come in a variety of sizes and shapes.

tween the plates is often referred to as the *potential difference* across the capacitor or as the *voltage* across the capacitor.

At equilibrium, the battery has transferred a positive charge Q to the plate connected to its positive terminal, and the other plate is left with a negative charge $-Q$. Thus the plates possess equal but opposite charge and the net charge on the capacitor is zero. When we speak of the charge Q on a capacitor, we are referring to the magnitude of the charge on each plate and not to the net charge on the entire device.

Definition of Capacitance

Consider the relationship between the charge Q on a capacitor and the potential difference $\mathcal{V}$ across it. Suppose we charge a capacitor with a 1.5-V battery and find that $Q = 4.5$ nC. If we charge the same capacitor with a 3.0-V battery, then we find that $Q = 9.0$ nC. That is, the ratio $Q/\mathcal{V}$ is the same in both cases: $Q/\mathcal{V} = 4.5\text{ nC}/1.5\text{ V} = 9.0\text{ nC}/3.0\text{ V} = 3.0\text{ nC/V}$. Further investigation shows that the ratio $Q/\mathcal{V}$ is characteristic of a given capacitor; when we increase $\mathcal{V}$ by some factor, Q increases by the same factor. We call this ratio the *capacitance* C of the capacitor:

The **capacitance** of a capacitor is the ratio of the magnitude Q of the charge on each plate to the magnitude $\mathcal{V}$ of the potential difference between the plates.

Definition of capacitance

$$C = \frac{Q}{\mathcal{V}} \tag{23-1}$$

By convention, all quantities in Eq. 23-1 are positive; Q is defined as the magnitude of the charge on each plate, and $\mathcal{V}$ is the magnitude of the potential difference between the plates. Consequently, the capacitance C is always positive.

The term "capacitance" for the quantity C implies that it is a measure of the amount of something that the capacitor can hold or contain. What is it that a capacitor holds? A capacitor holds electric charges, equal but opposite charges on each plate. The capacitance of a capacitor is a measure of its ability to hold these charges. From Eq. 23-1 we see that a larger value of capacitance corresponds to a larger amount of charge for a given potential difference. In Sec. 23-3 we will find that a charged capacitor also holds energy.

From Eq. 23-1, the dimension of capacitance is charge divided by potential and the SI unit of capacitance is coulombs divided by volts (C/V). This SI unit is called a farad (F) in honor of Michael Faraday: 1 F = 1 C/V. One farad is a rather large unit of capacitance; the capacitance of capacitors typically found in electric circuits is in the range from 10^{-12} F, or 1 pF (picofarad), to 10^{-6} F, or 1 μF (microfarad).

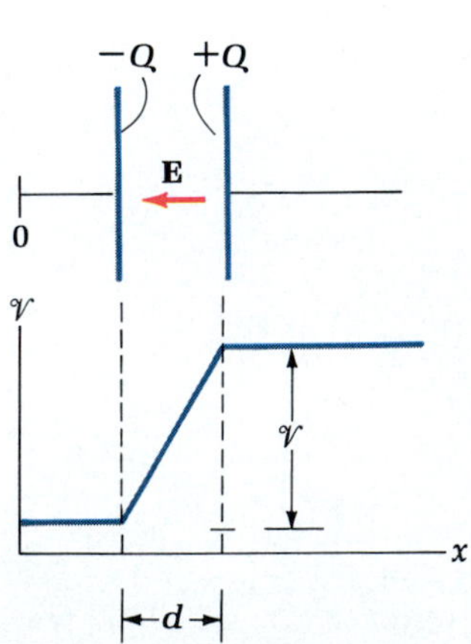

Figure 23-3. Potential difference between the plates of a capacitor.

Capacitance of a Parallel-Plate Capacitor

We now determine an expression for the capacitance of a parallel-plate capacitor. To find the capacitance, we first calculate the potential difference $\mathcal{V}$ across the capacitor for a given charge Q and then divide Q by the expression for $\mathcal{V}$. As we have seen (Examples 21-8 and 22-6), if the lateral dimensions of the plates of a parallel-plate capacitor are much larger than the plate separation, then (i) the surface charge density on the facing surfaces is uniform ($|\sigma| = Q/A$), (ii) the field in the space between the plates is uniform ($E = |\sigma|/\epsilon_0 = Q/\epsilon_0 A$), and (iii) the potential varies linearly with distance from one plate to the other (Fig. 23-3). Thus the potential difference across the capacitor is

$$\mathcal{V} = Ed = \frac{Qd}{\epsilon_0 A}$$

As expected, the potential difference is proportional to Q so that Q cancels out of the ratio $C = Q/\mathcal{V}$. Therefore

$$C = \frac{Q}{\mathcal{V}} = \frac{Q}{Qd/\epsilon_0 A}$$

or

$$C = \frac{\epsilon_0 A}{d} \qquad (23\text{-}2)$$

Capacitance of a parallel-plate capacitor

The capacitance of a parallel-plate capacitor depends on the plate area and the plate separation. To design a parallel-plate capacitor so that its capacitance is large, we make the area large and the separation small. The expression for C also contains ϵ_0, the permittivity of free space (or vacuum). This implies (correctly) that C depends on the medium between the plates, which we have assumed to be a vacuum. We discuss the effect of insulating material between the plates in Sec. 23-4. Notice that Eq. 23-2 shows that the SI unit for ϵ_0 can be written as farads per meter (F/m) as well as coulombs squared per newton-meter squared $[\mathrm{C^2/(N \cdot m^2)}]$.

EXAMPLE 23-1

Parallel-plate capacitor. *(a)* What is the capacitance of a parallel-plate capacitor that has square plates with lateral dimensions of 122 mm on a side, a plate separation of 0.24 mm, and a vacuum between the plates? *(b)* What is the charge on this capacitor if the potential difference across it is 45 V?

Solution. *(a)* Using Eq. 23-2, we have

$$C = \frac{(8.85 \times 10^{-12}\ \mathrm{F/m})(0.122\ \mathrm{m})^2}{2.4 \times 10^{-4}\ \mathrm{m}}$$

$$= 5.5 \times 10^{-10}\ \mathrm{F} = 0.55\ \mathrm{nF} = 550\ \mathrm{pF}$$

(b) Solving Eq. 23-1 for Q, we obtain

$$Q = C\mathcal{V} = (0.55\ \mathrm{nF})(45\ \mathrm{V}) = 25\ \mathrm{nC}$$

SELF-TEST 23-1. *(a)* If the charge on a capacitor is 50 μC when the voltage across it is 25 V, what is the capacitor's capacitance? *(b)* If the voltage across this capacitor is increased to 100 V, what is the charge on the capacitor? ***ANSWERS:*** *(a)* 2.0 μF, *(b)* 200 μC.

EXAMPLE 23-2

The coaxial cable (cylindrical capacitor). Coaxial cables are often used to transmit electric signals, and an important property of a coaxial cable is its capacitance. The cable consists of a conducting wire that is encircled by a coaxial conducting cylinder with an insulator in between (Fig. 23-4*a*). A capacitor of this design is called a cylindrical capacitor. The length of the cable is ordinarily much longer than its radius. Determine the capacitance of a length L of coaxial cable. At this stage, we must assume vacuum is between the wire and the cylinder.

Solution. We treat the wire as the positive plate and the cylinder as the negative plate of a cylindrical capacitor (Fig. 23-4*b*). To use Eq. 23-1 for the capacitance, we first find the expression for $\mathcal{V}$ across the capacitor in terms of Q and then divide Q by $\mathcal{V}$. Since the cable is very long compared with its radius, we can neglect end effects and use the solution to Example 21-3 for E in the space between the plates: $E = \lambda/2\pi\epsilon_0 R$. The linear charge density λ on the positive plate (the wire) is Q/L. The charge on the negative plate (the cylinder) has no effect on the field inside it. (You may wish to use Gauss's law to verify this.) Using Eq. 22-10 and integrating along a field line from the negative plate to the positive plate gives

R_b
R_a

(a)

Outer insulator
$+Q$
Coaxial conducting cylinder
R_a
R_b
Inner insulator
$-Q$

(b)

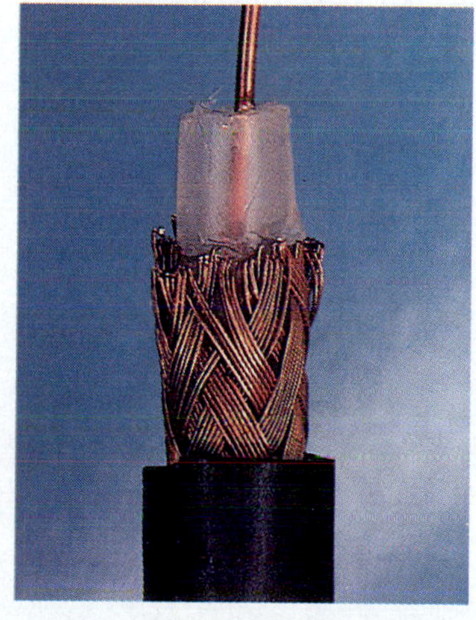

(c)

Figure 23-4. Example 23-2: A coaxial cable is a cylindrical capacitor. An inner solid wire runs through the middle of a plastic insulator that is surrounded by a sheath of braided wire. The braided wire is covered by an outer rubber coating.

$$\mathcal{V} = -\int_b^a \mathbf{E}\cdot d\boldsymbol{\ell} = \int_{R_a}^{R_b} E\,dR$$

$$= \int_{R_a}^{R_b} \frac{Q/L}{2\pi\epsilon_0 R}\,dR = \frac{Q}{2\pi\epsilon_0 L}\int_{R_a}^{R_b} \frac{dR}{R} = \frac{Q}{2\pi\epsilon_0 L}\Big[\ln R\Big]_{R_a}^{R_b}$$

$$= \frac{Q}{2\pi\epsilon_0 L}\ln\frac{R_b}{R_a}$$

Since $C = Q/\mathcal{V}$,

$$C = \frac{Q}{(Q/2\pi\epsilon_0 L)\ln(R_b/R_a)} = \frac{2\pi\epsilon_0 L}{\ln(R_b/R_a)}$$

Again we see that the charge on the capacitor cancels out. The capacitance of a cylindrical capacitor depends on the length of the cylinders and the ratio of the radii (the ratio of the inner radius of the outer cylinder to the outer radius of the inner cylinder). The capacitance also depends on the properties of the insulator between the two cylinders, but we will learn about that in Sec. 23-4.

SELF-TEST 23-2. The radius of the wire in a coaxial cable is 0.86 mm, and the inner radius of the coaxial conducting cylinder is 1.75 mm. Determine the capacitance of a 1.0-m length of the cable. Assume vacuum is between the wire and the cylinder. ***ANSWER:*** 78 pF.

23-2 CAPACITORS IN SERIES AND PARALLEL

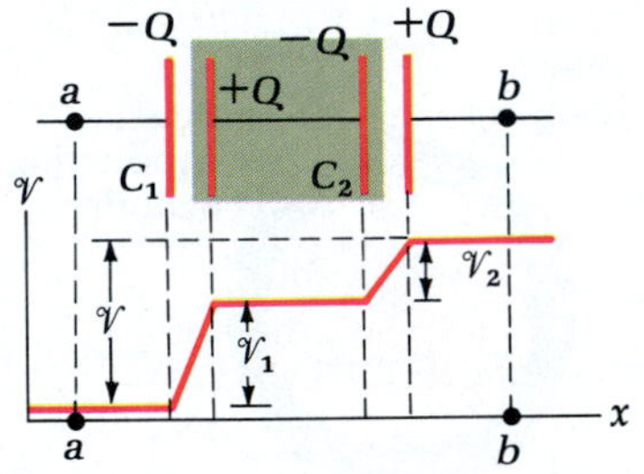

Figure 23-5. Two capacitors connected in series. Since the region inside the shaded area is electrically isolated and initially neutral, the charge on each capacitor is the same.

Circuit elements can be connected in many different ways. Two simple arrangements correspond to the elements being connected in series and in parallel.

Capacitors in Series

Figure 23-5 shows two capacitors with capacitance C_1 and C_2 connected in series, one after the other. Also shown is the variation of the potential along the connecting wires and through the capacitors. Under electrostatic conditions, the potential is uniform along the conducting wires. Notice that the potential difference across both capacitors, $\mathcal{V} = \mathcal{V}_b - \mathcal{V}_a$, is equal to the sum of the potential differences across each capacitor: $\mathcal{V} = \mathcal{V}_1 + \mathcal{V}_2$. This is an example of a general rule:

> The potential difference across a number of electrical devices connected in series is the sum of the potential differences across the individual devices.

We assume that the capacitors were initially uncharged before they were connected together and charged by a battery. Thus the section of insulated conductor enclosed by the shaded region in Fig. 23-5 has no net charge, so that $Q_1 = Q_2 = Q$. Indeed, for any number of capacitors in series (initially uncharged), each has the same charge.

We now find the *equivalent capacitance* C_{12} of the series combination of capacitors 1 and 2.

Definition of equivalent capacitance

> The ***equivalent capacitance*** of a combination of capacitors is the capacitance of a single capacitor which, when used in place of the combination, provides the same external effect.

To provide the same external effect as capacitors 1 and 2 in series, this single capacitor must possess charge of magnitude Q on each of its plates when the potential difference across it is $\mathcal{V}$. That is,

$$C_{12} = \frac{Q}{\mathcal{V}} \quad \text{or} \quad \mathcal{V} = \frac{Q}{C_{12}}$$

Also, $\mathcal{V}_1 = Q/C_1$ and $\mathcal{V}_2 = Q/C_2$. Substitution into $\mathcal{V} = \mathcal{V}_1 + \mathcal{V}_2$ gives

$$\frac{Q}{C_{12}} = \frac{Q}{C_1} + \frac{Q}{C_2}$$

Thus

$$\frac{1}{C_{12}} = \frac{1}{C_1} + \frac{1}{C_2}$$

In general, the equivalent capacitance C_{eq} of any number of capacitors connected in series is

$$\frac{1}{C_{eq}} = \sum \frac{1}{C_i} \qquad (23\text{-}3)$$

Equivalent capacitance of capacitors in series

EXAMPLE 23-3

Capacitors in series. Two capacitors with capacitances of $C_1 = 2.3\ \mu F$ and $C_2 = 4.6\ \mu F$ are connected in series, and the potential difference across the combination is 35 V. Determine *(a)* the equivalent capacitance of the series combination, *(b)* the charge on each capacitor, and *(c)* the potential difference across each capacitor.

Solution. *(a)* For two capacitors in series

$$\frac{1}{C_{12}} = \frac{1}{C_1} + \frac{1}{C_2} = \frac{C_2}{C_1C_2} + \frac{C_1}{C_1C_2} = \frac{C_2 + C_1}{C_1C_2}$$

or

$$C_{12} = \frac{C_1C_2}{C_2 + C_1} = \frac{(2.3\ \mu F)(4.6\ \mu F)}{2.3\ \mu F + 4.6\ \mu F} = 1.5\ \mu F$$

(b) Each capacitor has the same charge, and this charge is equal to the charge on a capacitor with the equivalent capacitance when the potential difference is $\mathcal{V}$. Thus

$$Q_1 = Q_2 = Q = C_{12}\mathcal{V} = (1.5\ \mu F)(35\ V) = 53\ \mu C$$

(c) The potential difference across capacitor 1 is

$$\mathcal{V}_1 = \frac{Q}{C_1} = \frac{53\ \mu C}{2.3\ \mu F} = 23\ V$$

The potential difference across capacitor 2 is

$$\mathcal{V}_2 = \frac{Q}{C_2} = \frac{53\ \mu C}{4.6\ \mu F} = 12\ V$$

Notice that $\mathcal{V}_1 + \mathcal{V}_2 = 23\ V + 12\ V = 35\ V$.

SELF-TEST 23-3. Suppose a third capacitor of capacitance $3.8\ \mu F$ is placed in series with the two capacitors in the above example. Determine the equivalent capacitance of the three in series. ***ANSWER:*** $1.1\ \mu F$.

Capacitors in Parallel

Figure 23-6 is a schematic illustration of two capacitors with capacitances C_1 and C_2 connected in parallel, one beside the other. From the figure, you can see that the potential difference across each capacitor is the same: $\mathcal{V}_b - \mathcal{V}_a = \mathcal{V}_1 = \mathcal{V}_2 = \mathcal{V}$. This is an example of a general rule:

> The potential difference across circuit elements arranged in parallel is the same.

To provide the same external effect as capacitors 1 and 2 in parallel, a single capacitor with capacitance C_{12} must possess charge $Q = Q_1 + Q_2$ when the potential difference across it is $\mathcal{V}$. That is,

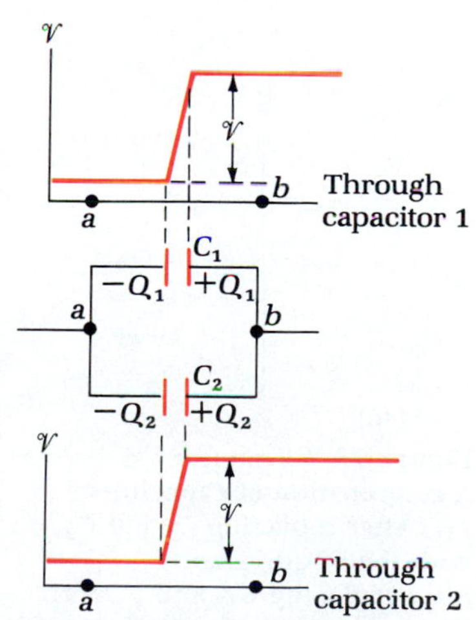

Figure 23-6. Two capacitors connected in parallel. Since the potential difference between *a* and *b* is independent of the path between *a* and *b*, the potential difference across each capacitor is the same.

$$C_{12} = \frac{Q}{\mathcal{V}} = \frac{Q_1 + Q_2}{\mathcal{V}} = \frac{Q_1}{\mathcal{V}} + \frac{Q_2}{\mathcal{V}}$$

Since $C_1 = Q_1/\mathcal{V}$ and $C_2 = Q_2/\mathcal{V}$, we have

$$C_{12} = C_1 + C_2$$

In general, the equivalent capacitance C_{eq} of any number of capacitors connected in parallel is

Equivalent capacitance of capacitors in parallel

$$C_{eq} = \Sigma\, C_i \qquad (23\text{-}4)$$

EXAMPLE 23-4

A combination of capacitors. Consider the combination of capacitors in Fig. 23-7*a*, where $C_1 = 2.9\ \mu F$, $C_2 = 1.8\ \mu F$, and $C_3 = 2.4\ \mu F$. The potential difference across the combination is $\mathcal{V}_b - \mathcal{V}_a = \mathcal{V} = 53$ V. Determine *(a)* the equivalent capacitance C_{123} of the entire combination, *(b)* the potential difference across each capacitor, and *(c)* the charge on each capacitor.

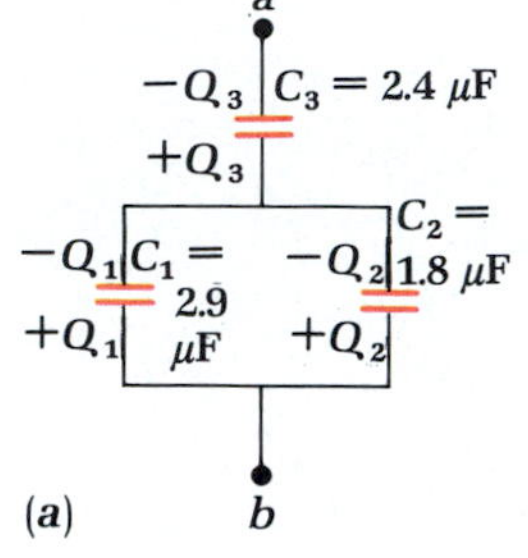

Solution. *(a)* Since capacitors 1 and 2 are in parallel, their equivalent capacitance C_{12} is

$$C_{12} = 2.9\ \mu F + 1.8\ \mu F = 4.7\ \mu F$$

Thus the combination can be represented as shown in Fig. 23-7*b*. The capacitance C_{12} is in series with C_3 so that

$$C_{123} = \frac{C_{12}C_3}{C_3 + C_{12}} = \frac{(4.7\ \mu F)(2.4\ \mu F)}{2.4\ \mu F + 4.7\ \mu F} = 1.6\ \mu F$$

The entire combination is equivalent to the single capacitor shown in Fig. 23-7*c*. *(b)* Since capacitors 1 and 2 are in parallel, the potential difference across each of them is the same: $\mathcal{V}_1 = \mathcal{V}_2$. Also, capacitor 3 is in series with the parallel arrangement of 1 and 2 so that the potential difference $\mathcal{V}_3$ across capacitor 3 plus $\mathcal{V}_1$ (or $\mathcal{V}_2$) gives $\mathcal{V}$:

$$\mathcal{V} = \mathcal{V}_1 + \mathcal{V}_3$$

To find $\mathcal{V}_1$ or $\mathcal{V}_3$ we must first find some of the charges. From the figure you can see that the positive plate of capacitor 3 must possess a charge equal and opposite to the charges on the negative plates of capacitors 1 and 2: $Q_3 = Q_1 + Q_2$. Thus the equivalent capacitance of the entire combination possesses this same charge when the potential difference across it is $\mathcal{V}$:

$$C_{123} = \frac{Q_3}{\mathcal{V}} = \frac{Q_1 + Q_2}{\mathcal{V}}$$

$$Q_3 = Q_1 + Q_2 = C_{123}\mathcal{V} = (1.6\ \mu F)(53\ V) = 84\ \mu C$$

This gives

$$\mathcal{V}_3 = \frac{Q_3}{C_3} = \frac{84\ \mu C}{2.4\ \mu F} = 35\ V$$

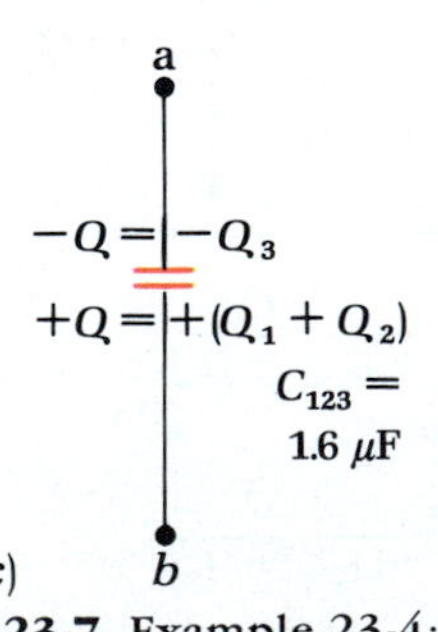

Figure 23-7. Example 23-4: *(a)* A combination of capacitors. *(b)* After replacing C_1 and C_2 with their equivalent C_{12}. *(c)* After replacing C_{12} and C_3 with their equivalent C_{123}.

and

$$\mathcal{V}_1 = \mathcal{V}_2 = \frac{Q_1 + Q_2}{C_{12}} = \frac{84\ \mu C}{4.7\ \mu F} = 18\ V$$

Notice that $\mathcal{V}_1 + \mathcal{V}_3 = 18\ V + 35\ V = 53$ V. *(c)* We already found Q_3 in part *(b)*. The charges on capacitors 1 and 2 are

$$Q_1 = C_1\mathcal{V}_1 = (2.9\ \mu F)(18\ V) = 52\ \mu C$$

and

$$Q_2 = C_2\mathcal{V}_2 = (1.8\ \mu F)(18\ V) = 32\ \mu C$$

Notice that $Q_1 + Q_2 = 52\ \mu C + 32\ \mu C = 84\ \mu C = Q_3$.

SELF-TEST 23-4. *(a)* Capacitors 2 and 3 in the above example are connected in parallel. What is their equivalent capacitance? *(b)* Capacitor 1 from the above example is now connected in series with the two-capacitor arrangement in part *(a)*. What is the equivalent capacitance of this three-capacitor arrangement? ***ANSWERS:*** *(a)* 5.3 μF; *(b)* 1.3 μF.

23-3 ELECTRIC ENERGY AND ENERGY DENSITY

We now develop the concept of electric energy by considering the potential energy of the charges on the plates of a charged capacitor. Once developed, we discuss the electric energy in terms of electric fields, and we introduce the concept of electric energy density.

Electric Energy Stored in a Capacitor

When a battery charges a capacitor, the battery does work as it transfers charge carriers from one plate to the other, raising the potential energy of the carriers. This increased potential energy of the charge carriers constitutes the electric energy stored in a capacitor.

Let U represent the energy of a capacitor after it has been charged to a final charge Q and final potential difference $\mathcal{V}$, and let U', Q', and $\mathcal{V}'$ represent these quantities as they vary during the charging process. At any instant during the charging, the change dU' in the potential energy of the system of charges when charge dQ' is transferred by the battery is

$$dU' = \mathcal{V}'\, dQ'$$

because $\mathcal{V}'$ is the potential energy per unit charge. To find U due to the capacitor being charged from zero to Q, we integrate dU':

$$U = \int_0^Q \mathcal{V}'\, dQ'$$

The potential difference $\mathcal{V}'$ across the capacitor cannot be factored out of the integral because it varies as Q' increases. Indeed, $\mathcal{V}'$ (which equals Q'/C) increases linearly with the charge on the capacitor, as shown in Fig. 23-8. Therefore

$$U = \int_0^Q \frac{Q'}{C}\, dQ' = \frac{1}{C}\int_0^Q Q'\, dQ' = \frac{1}{2}\frac{Q^2}{C}$$

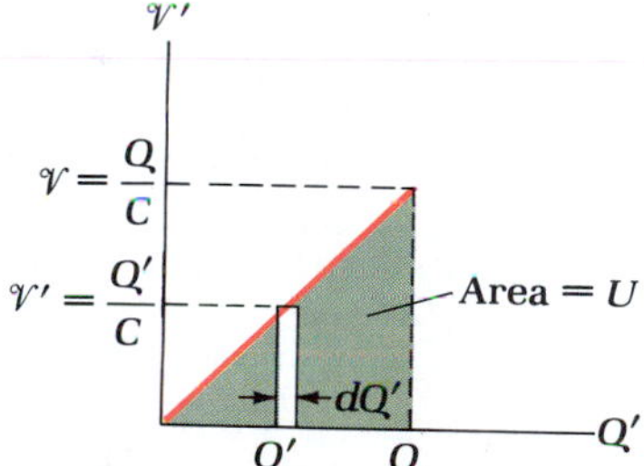

Figure 23-8. The energy U of a capacitor with final charge Q and final potential difference $\mathcal{V}$ is given by the area under the graph of the varying potential difference $\mathcal{V}'$ versus the varying charge Q'. The area is that of a triangle of height Q/C and base Q; $U = \frac{1}{2}(Q/C)Q = \frac{1}{2}Q^2/C$.

The energy of a capacitor depends on the square of its charge. How this integration leads to this result is shown graphically in Fig. 23-8. The energy, equal to $\int \mathcal{V}'\, dQ'$, is given by the area under the graph of $\mathcal{V}'$ versus Q'. Since $\mathcal{V}'$ increases linearly with Q', the area is that of a triangle with height Q/C and with base Q, so that the area (the energy of the capacitor) is $\frac{1}{2}(Q/C)(Q) = \frac{1}{2}(Q^2/C)$.

Charging a capacitor is similar to digging a well (or any deep hole in the ground). It is easier to dig the first part of the well than the last part because the soil taken from the well during the first part does not have to be lifted as far as during the last part. Similarly, the increase in potential energy of the first carriers transferred by the battery is less than that of the last carriers because the potential difference across the capacitor is larger when the last carriers are transferred. This accounts for the variation of $\mathcal{V}'$ as the charge transfer takes place and is the physical reason $\mathcal{V}'$ may not be factored out of the integral.

Using the definition of capacitance, $C = Q/\mathcal{V}$, we can express the energy of a charged capacitor in terms of any two of the three quantities Q, C, and $\mathcal{V}$:

$$U = \frac{Q^2}{2C} \qquad U = \frac{C\mathcal{V}^2}{2} \qquad U = \frac{Q\mathcal{V}}{2} \tag{23-5}$$

Energy of a charged capacitor

As a brief example, suppose a 5.5-μF capacitor has a potential difference of 42 V across it. The capacitor's energy is $U = \frac{1}{2}C\mathcal{V}^2 = \frac{1}{2}(5.5\ \mu\text{F})(42\ \text{V})^2 = 4.9$ mJ.

Energy Density of an Electric Field

In the discussion above, we associated the energy of the capacitor with the potential energy of the charges. An alternative view is to attribute this energy to the electric field that exists between the plates. For a parallel-plate capacitor (with small plate separation and large plate area), $C = \epsilon_0 A/d$ and $\mathcal{V} = Ed$, so that

$$U = \frac{1}{2} C\mathcal{V}^2 = \frac{1}{2}\left(\frac{\epsilon_0 A}{d}\right)(Ed)^2 = \frac{1}{2}\epsilon_0 E^2 (Ad)$$

The factor Ad is the volume between the plates, which corresponds to the volume occupied by the electric field (neglecting edge effects). Since the energy is proportional to the volume occupied by the field, we introduce the energy density (or energy per unit volume) u in the space that contains the field:

$$u = \frac{U}{Ad} = \frac{\frac{1}{2}\epsilon_0 E^2 (Ad)}{Ad} = \tfrac{1}{2}\epsilon_0 E^2$$

Thus

Energy density of an electric field

$$u = \tfrac{1}{2}\epsilon_0 E^2 \qquad (23\text{-}6)$$

Equation 23-6 is more than simply another way to express the energy of a charged capacitor. It suggests that we may view the electric energy of a charge distribution as being attributed to the electric field produced by the charge distribution. Although we do not prove it here, Eq. 23-6 is generally valid; it gives at points in space the energy density due to an electric field produced by any charge distribution. The electric energy density is an example of a scalar field.

EXAMPLE 23-5

Energy density produced by a charged sphere. What is the energy density at a point 0.15 m from the center of a spherically symmetric charge distribution of radius 55 mm and charge 18 nC?

Solution. Since the point in question is outside the charge distribution, the field is the same as the field due to a point charge located at the center of the charge distribution: $E = Q/4\pi\epsilon_0 r^2$. Equation 23-6 gives

$$u = \frac{1}{2}\epsilon_0 E^2 = \frac{1}{2}\epsilon_0\left(\frac{Q}{4\pi\epsilon_0 r^2}\right)^2 = \frac{Q^2}{32\pi^2\epsilon_0 r^4}$$

$$= \frac{(18\text{ nC})^2}{32\pi^2(8.85\times10^{-12}\text{ F/m})(0.15\text{ m})^4} = 230\ \mu\text{J/m}^3$$

SELF-TEST 23-5. Consider the way the energy density produced by a spherical distribution of charge falls off with distance r from the center at points outside the charge distribution. Suppose the energy density outside such a charge distribution has a value of 800 μJ/m^3 at a distance of 1.0 m from the center. What is the energy density at a point 2.0 m from the center? ***ANSWER:*** 50 μJ/m^3.

EXAMPLE 23-6

Energy density and the cylindrical capacitor. Equation 23-6 was developed using the (nearly) uniform field of a parallel-plate capacitor. Show that this equation also gives the energy of a charged cylindrical capacitor, where the field between the plates is not uniform.

Solution. From Example 23-3, E between the plates of a cylindrical capacitor is $E = \lambda/2\pi\epsilon_0 R = Q/2\pi\epsilon_0 RL$, where L is the length of the capacitor. Thus the field is the same at each point in an infinitesimal volume element that is a cylindrical shell of radius R, length L, thickness dR, and volume $2\pi RL\ dR$ (Fig. 23-9). Since this field exists only between the plates, the energy U due to the field is

$$U = \int_{R_a}^{R_b} u(2\pi RL\ dR)$$

The energy density u is

$$u = \frac{1}{2}\epsilon_0 E^2 = \frac{1}{2}\epsilon_0\left(\frac{Q}{2\pi\epsilon_0 RL}\right)^2 = \frac{Q^2}{8\pi^2\epsilon_0 L^2 R^2}$$

Substituting this into the equation for U and factoring constants out of the integral gives

$$U = \frac{Q^2}{4\pi\epsilon_0 L}\int_{R_a}^{R_b}\frac{dR}{R} = \frac{Q^2}{4\pi\epsilon_0 L}\ln\frac{R_b}{R_a}$$

For a cylindrical capacitor (Example 23-2),

$$C = \frac{2\pi\epsilon_0 L}{\ln(R_b/R_a)}$$

Therefore the energy of the field between the plates is $U = \frac{1}{2}Q^2/C$, which corresponds to the potential energy of the charges on the plates given by the first of Eqs. 23-5.

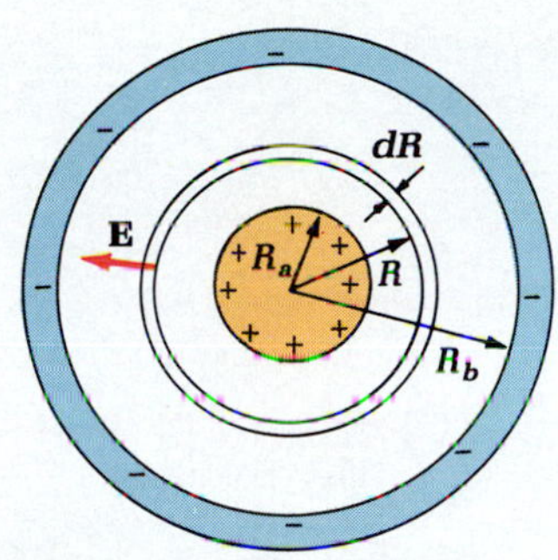

Figure 23-9. Example 23-6: Cross section of a cylindrical capacitor. The field magnitude E is the same at each point in a cylindrical shell of infinitesimal thickness dR and volume $2\pi RL\, dR$.

SELF-TEST 23-6. Consider the way the energy density falls off with perpendicular distance R from the axis of a very long uniform line charge. Suppose the value of the energy density is 400 μJ/m^3 at a distance of 1.0 m from the axis. What is the energy density at a point 2.0 m from the axis? ***ANSWER:*** 100 μJ/m^3.

EXAMPLE 23-7

Energy of a charged conducting sphere. *(a)* Determine an expression for the electric energy of an isolated metal sphere of radius r_0 and charge Q. *(b)* In Example 22-10, we showed that the dome of a demonstration Van de Graaff generator with $r_0 = 0.13$ m can have a charge $Q = 6\ \mu$C. Evaluate U for such a generator.

Solution. *(a)* The electric energy of the charged sphere can be attributed to the field produced by the charge. To find the energy, we integrate (sum) the energy density over the volume occupied by the field. The electric energy inside the metal sphere is zero because $E = 0$ inside a conductor. Outside the sphere, $E = Q/4\pi\epsilon_0 r^2$, so that the field occupies all space outside the sphere. Since E is spherically symmetric, it is the same at each point in an element of volume that is a thin spherical shell of radius r, infinitesimal thickness dr, and volume $4\pi r^2\, dr$ (Fig. 23-10). Therefore

$$\begin{aligned} U &= \int_{r_0}^{\infty} u\, 4\pi r^2\, dr = \int_{r_0}^{\infty}\tfrac{1}{2}\epsilon_0 E^2\, 4\pi r^2\, dr \\ &= \tfrac{1}{2}\epsilon_0\int_{r_0}^{\infty} E^2\, 4\pi r^2\, dr \\ &= \tfrac{1}{2}\epsilon_0\int_{r_0}^{\infty}\left(\frac{Q}{4\pi\epsilon_0 r^2}\right)^2 4\pi r^2\, dr \\ &= \frac{Q^2}{8\pi\epsilon_0}\int_{r_0}^{\infty}\frac{dr}{r^2} = \frac{Q^2}{8\pi\epsilon_0}\left[-\frac{1}{r}\right]_{r_0}^{\infty} \\ &= \frac{Q^2}{8\pi\epsilon_0 r_0} \end{aligned}$$

(b) For the Van de Graaff generator dome,

$$U = (9\times10^9\ \text{N}\cdot\text{m}^2/\text{C}^2)\frac{(6\ \mu\text{C})^2}{2(0.13\ \text{m})} = 1\ \text{J}$$

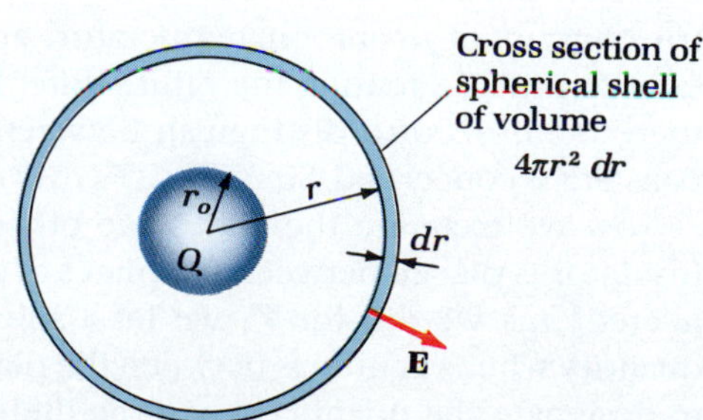

Figure 23-10. Example 23-7. Finding the electric energy of the field produced by a charged metal sphere. The field magnitude E is the same at each point in a spherical shell with infinitesimal thickness dr and volume $4\pi r^2\, dr$.

SELF-TEST 23-7. Two conducting spheres have radii of 0.10 m and 0.20 m, and each sphere is given the same charge. If the energy of the larger sphere is 0.4 J, what is the energy of the smaller sphere? ***ANSWER:*** 0.8 J.

23-4 ELECTROSTATIC PROPERTIES OF INSULATORS

So far we have only studied cases where no material exists in the space between capacitor plates. Now we consider the effect of filling this space with an insulator. These considerations provide a means for investigating the properties of insulators.

To examine the behavior of an insulating material, such as glass or plastic, we first arrange a parallel-plate capacitor with vacuum between its plates. Next we charge the capacitor by connecting it to a battery, and then we disconnect the charging battery. We measure the potential difference across the plates, as shown in Fig. 23-11*a*, and refer to this value as $\mathcal{V}_0$. Now we insert the insulating material we wish to study into the space between the plates and again measure the potential difference (Fig. 23-11*b*). Such experiments reveal that the potential difference changes to a value we call $\mathcal{V}$ and that $\mathcal{V} < \mathcal{V}_0$ in every case.

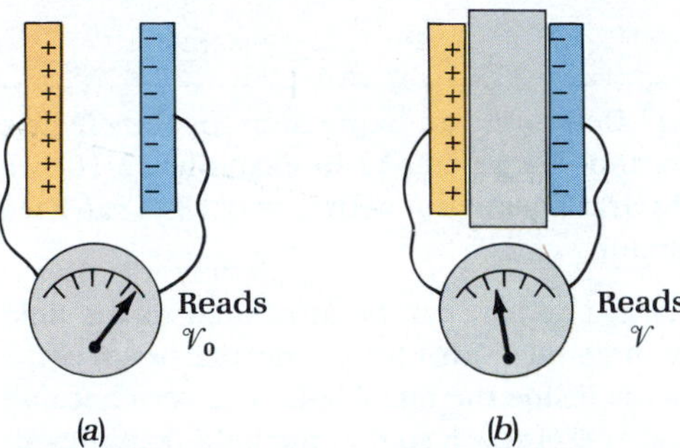

Figure 23-11. *(a)* With vacuum between the plates of the charged capacitor, the voltmeter reads $\mathcal{V}_0$. *(b)* With a dielectric between the plates of the charged capacitor, the voltmeter reads $\mathcal{V}$, $\mathcal{V}_0 > \mathcal{V}$ always.

The reduction of potential difference from $\mathcal{V}_0$ to $\mathcal{V}$ due to the insertion of the insulator cannot be attributed to a reduction of charge on the plates, because if the insulator is removed, the value of the potential difference increases from $\mathcal{V}$ back to $\mathcal{V}_0$. The potential difference would not return to its original value if the charge on the plates had been altered by the insertion of the insulator.

When this experiment is performed with different types of insulating materials, we find that the ratio $\mathcal{V}_0/\mathcal{V}$ depends on the type of material. An insulator is often called a *dielectric,* and this ratio is the *dielectric constant* κ:

> The ***dielectric constant*** of a material is the ratio of the potential difference between the plates of a capacitor when vacuum is between the plates to that when the material is between the plates.

The dielectric constant κ

$$\kappa = \frac{\mathcal{V}_0}{\mathcal{V}} \tag{23-7}$$

Table 23-1 lists values of κ for some representative materials. (Also listed are values of the dielectric strength E_{max}, which was discussed in Sec. 22-6.) For vacuum, κ is exactly 1 because the dielectric constant is defined relative to vacuum. The dielectric constant of air at room temperature and atmospheric pressure is very nearly the same as that of vacuum; the values differ by only about 0.0006. For most purposes there is no need to distinguish between air and vacuum as far as the dielectric constant is concerned. Since $\mathcal{V} < \mathcal{V}_0$ in every case, $\kappa > 1$ for all insulating materials.

Now we examine the way some other quantities (E, C, and U) change as an insulator is placed between the plates of a charged capacitor (the battery is disconnected). As we did for $\mathcal{V}$, we let a subscript zero on the symbol designate the quantity when vacuum is between the plates, and use the symbol with no subscript to designate the quantity when the dielectric is inserted. First consider the field

TABLE 23-1. *Properties of Some Dielectric Materials (20° C)*

Material	Dielectric Constant κ	Dielectric Strength E_{max}, 10^6 V/m
Vacuum	1	
Gases		
Dry air (1 atm)	1.00059	3
Carbon dioxide (1 atm)	1.00098	
Helium (1 atm)	1.00007	
Ethanol (100°C, 1 atm)	1.0061	
Liquids		
Benzene	3.1	
Glycerol	43	
Water	80	
Solids		
Teflon	2.1	60
Polystyrene	2.6	25
Nylon	3.4	14
Paper	3.6	15
Fused quartz	3.8	8
Bakelite	4.9	24
Pyrex glass	5	14
Neoprene	6.8	12
Aluminum oxide	10.3	
Strontium titanate	≈250	8
Barium strontium titanate	≈10^4	

between the plates. Since $\mathcal{V} = Ed$ and $\mathcal{V}_0 = E_0 d$,

$$\kappa = \frac{\mathcal{V}_0}{\mathcal{V}} = \frac{E_0 d}{Ed} = \frac{E_0}{E}$$

or

$$E = \frac{E_0}{\kappa} \tag{23-8}$$

As with the potential difference, the field magnitude is reduced by a factor of $1/\kappa$ when a dielectric is inserted.

Next we look at the effect of a dielectric on the capacitance. Since $\mathcal{V} = Q/C$ and $\mathcal{V}_0 = Q/C_0$,

$$\kappa = \frac{\mathcal{V}_0}{\mathcal{V}} = \frac{Q/C_0}{Q/C} = \frac{C}{C_0}$$

Therefore

$$C = \kappa C_0$$

The insertion of the dielectric causes the capacitance to be increased by a factor κ.

In the case of a capacitor's energy, $U = \frac{1}{2}Q\mathcal{V}$, or $\mathcal{V} = 2U/Q$, so that

$$\kappa = \frac{\mathcal{V}_0}{\mathcal{V}} = \frac{2U_0/Q}{2U/Q} = \frac{U_0}{U}$$

Thus

$$U = \frac{U_0}{\kappa}$$

The energy of the capacitor is reduced by a factor of $1/\kappa$ because of the insertion of the dielectric. Since U is reduced as the dielectric is inserted, there is an electric force that tends to pull the dielectric into the space between the plates (see Prob. 11).

Several factors are involved in choosing a dielectric for a practical capacitor (Fig.

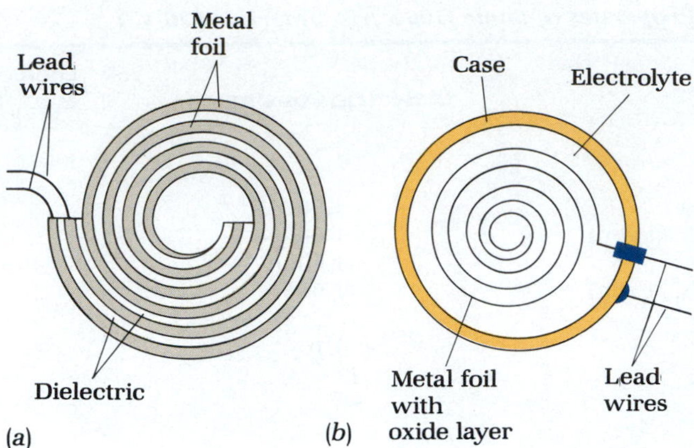

Figure 23-12. Two designs of practical capacitors. *(a)* Two sheets of dielectric and two sheets of metal foil are sandwiched together and rolled into the shape of a cylinder. *(b)* An electrolytic capacitor is one that uses an electrolyte (a conducting solution) as one "plate" and a metal foil as the other. The dielectric is a thin oxide layer on the metal foil.

23-12). First, since C is proportional to κ, a large dielectric constant is preferred so that C can be made large without making the plate area inordinately large. Second, a large dielectric strength E_{max} allows the capacitor to be subjected to a large field without dielectric breakdown. For a parallel-plate capacitor, $\mathcal{V} = Ed$ so that $\mathcal{V}_{max} = E_{max}d$. Therefore a large E_{max} permits d to be small without restricting the maximum operating potential difference, and a smaller value of d means a larger capacitance. Therefore a dielectric with a large κ is required if $\mathcal{V}$ is expected to be large, or if d must be small. Third, a solid insulator provides rigid support between the plates so that the plates cannot make a conducting contact with each other.

EXAMPLE 23-8

Capacitor with a dielectric between the plates. A parallel-plate capacitor is constructed by tightly sandwiching a sheet of paper with thickness 0.14 mm between sheets of aluminum foil. The lateral dimensions of the sheets are 15 mm by 480 mm. Determine *(a)* the capacitance of the capacitor and *(b)* the maximum potential difference that may be placed across it without dielectric breakdown. Neglect edge effects.

Solution. *(a)* Since $C = \kappa C_0$ and $C_0 = \epsilon_0 A/d$,

$$C = \frac{\kappa \epsilon_0 A}{d}$$

From Table 23-1, $\kappa = 3.6$ for paper so that

$$C = \frac{(3.6)(8.85 \times 10^{-12}\ \mathrm{F/m})(1.5 \times 10^{-2}\ \mathrm{m})(0.48\ \mathrm{m})}{1.4 \times 10^{-4}\ \mathrm{m}}$$

$$= 1.6\ \mathrm{nF}$$

(b) For a parallel-plate capacitor, $\mathcal{V}_{max} = E_{max}d$, and from Table 23-1, $E_{max} = 15 \times 10^6\ \mathrm{V/m}$ for paper. Thus

$$\mathcal{V}_{max} = (15 \times 10^6\ \mathrm{V/m})(1.4 \times 10^{-4}\ \mathrm{m}) = 2.1\ \mathrm{kV}$$

SELF-TEST 23-8. Repeat the calculation given in the above example, except use polystyrene as the dielectric. ***ANSWERS:*** *(a)* 1.2 nF; *(b)* 3.5 kV.

23-5 ATOMIC DESCRIPTION OF THE PROPERTIES OF INSULATORS

When a dielectric is placed between the plates of a charged capacitor that is disconnected from the charging battery, the field between the plates is reduced even though the charge on the plates remains fixed. What charges are responsible for this reduction in the field and where do they reside? The charges that cause the reduction of the field are called *bound charges,* or *polarization charges,* and they

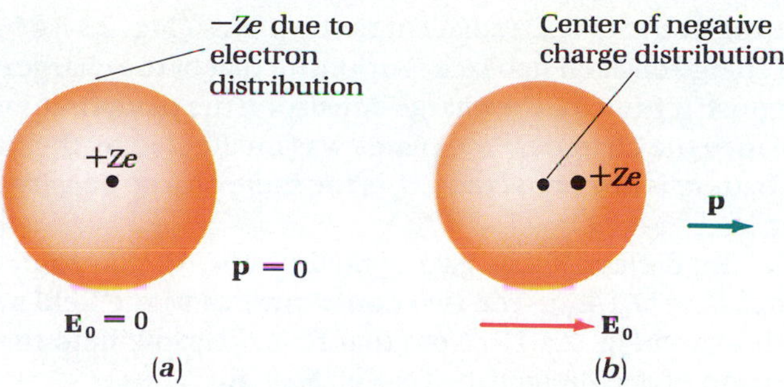

Figure 23-13. A simple model of an atom. The nucleus with charge $+Ze$ is at the center of the spherical distribution of charge $-Ze$ due to the electrons. *(a)* In zero field the center of negative charge coincides with the positive point charge (the nucleus). *(b)* In an external electric field $\mathbf{E}_0$, the center of negative charge and the positive point charge are displaced; the atom has an induced dipole moment.

reside on the surface of the dielectric. We now describe the origin of these charges in terms of the molecules that compose the dielectric.

Figure 23-13*a* shows a simple model of an atom: a nucleus that is effectively a point charge ($+Ze$) surrounded by a spherically symmetric distribution of negative charge ($-Ze$) due to the electrons. Because of the electric attraction between the nucleus and its electrons, the center of negative charge coincides with the position of the nucleus. When the atom is placed in an external electric field $\mathbf{E}_0$ (Fig. 23-13*b*), the force exerted by the field on the nucleus is opposite in direction to the force exerted by the field on the electrons. At equilibrium, the position of the nucleus and the center of negative charge are displaced because of two sets of forces: (i) the forces on the nucleus and the electrons due to $\mathbf{E}_0$ (which tend to separate the nucleus from its electrons) and (ii) the forces between the nucleus and its electrons (which tend to superimpose the nucleus and the center of negative charge). As a result of the external field, the atom has an induced dipole moment, and such an atom is said to be *polarized.*

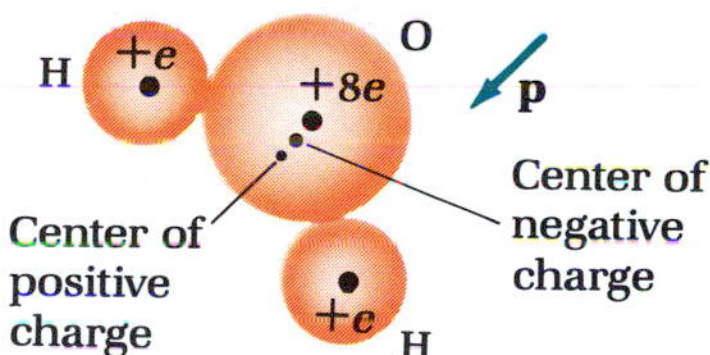

Figure 23-14. The water molecule has a permanent dipole moment.

Some molecules possess permanent dipole moments and, as a consequence, are called *polar molecules.* For a polar molecule, the centers of positive and negative charge do not coincide, even when the molecule is in zero electric field. Water molecules are polar molecules (Fig. 23-14). When no external electric field exists, these molecular dipoles have a random orientation (Fig. 23-15*a*). If a dipole of moment $\mathbf{p}$ is placed in a field $\mathbf{E}$, the field tends to align $\mathbf{p}$ along the direction of $\mathbf{E}$ (see Prob. 2 in Chap. 22). In the case of molecular dipoles, the alignment is not

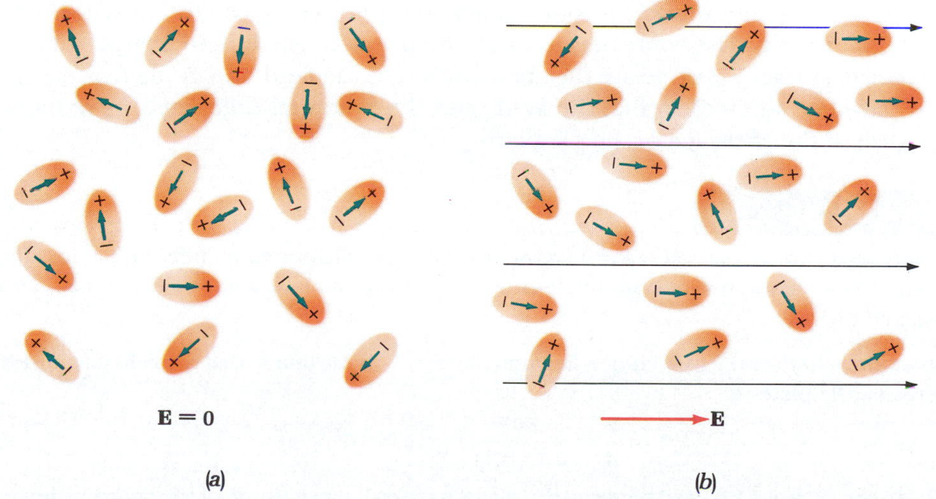

Figure 23-15. *(a)* Dipole moments of polar molecules are randomly oriented in zero electric field. *(b)* Molecules tend to align in an external electric field, but the alignment is not complete because of thermal agitation.

complete because of thermal agitation of the molecules (Fig. 23-15*b*). Thus a characteristic of this type of polarization is its relatively strong temperature dependence; the effect of the polarization decreases with increasing temperature.

Suppose a dielectric slab is placed in the uniform field $\mathbf{E}_0$ between the plates of a parallel-plate capacitor (Fig. 23-16*a*). The dielectric becomes polarized as dipoles are induced by the field and permanent dipoles, if present, are aligned by the field. Because of this polarization, a surface charge density σ_b (*b* for bound) forms on the

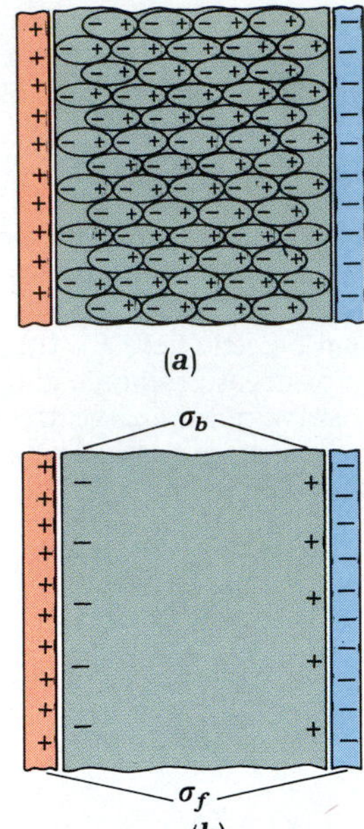

Figure 23-16. *(a)* A dielectric slab is polarized in the uniform field of a parallel-plate capacitor. *(b)* A surface charge density σ_b is formed on the faces of the dielectric adjacent to the plates.

two faces of the dielectric that are adjacent to the plates (Fig. 23-16*b*), and the sign of this induced charge on each slab face is opposite that of the charge on its adjacent plate. To distinguish it from σ_b, the charge density on the plates is now referred to as σ_f (f for free). Since the charge on the plates was unaffected by the insertion of the dielectric (the battery is disconnected), σ_f is the same charge density we previously designated as σ.

The field **E** in the dielectric has two contributions, $\mathbf{E}_f$ due to σ_f ($E_f = |\sigma_f|/\epsilon_0$) and $\mathbf{E}_b$ due to σ_b ($E_b = |\sigma_b|/\epsilon_0$). The two contributions to the field are in opposite directions, as shown in Fig. 23-17. Note that $\mathbf{E}_f$ is the same field that was present before the insertion of the dielectric. That is, $\mathbf{E}_f = \mathbf{E}_0$. Thus

$$\mathbf{E} = \mathbf{E}_0 + \mathbf{E}_b$$

If we place our x axis along **E**, then

$$E\mathbf{i} = E_0\mathbf{i} + (-E_b)\mathbf{i} = (E_0 - E_b)\mathbf{i}$$

So the relation between the field magnitudes is

$$E = E_0 - E_b$$

In terms of the charge densities, we have

$$E = \frac{|\sigma_f| - |\sigma_b|}{\epsilon_0} \tag{23-9}$$

Next we express $|\sigma_b|$ in terms of $|\sigma_f|$ and κ. From Eq. 23-8, $E = E_0/\kappa = |\sigma_f|/\kappa\epsilon_0$, so that

$$\frac{|\sigma_f|}{\kappa\epsilon_0} = \frac{|\sigma_f| - |\sigma_b|}{\epsilon_0}$$

Solving for $|\sigma_b|$ gives

$$|\sigma_b| = \frac{\kappa - 1}{\kappa}|\sigma_f| \tag{23-10}$$

Since the factor $(\kappa - 1)/\kappa$ is less than 1, $|\sigma_b|$ is always less than $|\sigma_f|$.

Now we can understand why the potential difference decreases when a dielectric is inserted into a capacitor with fixed charge on its plates. The field due to the free charge on the plates induces a bound charge of opposite sign on the adjacent faces of the dielectric. The contribution to the field due to the bound charge (on the dielectric surface) is opposite the contribution to the field due to the free charge (on the plate surface) so that the field (and the potential difference) is reduced because of the presence of the dielectric.

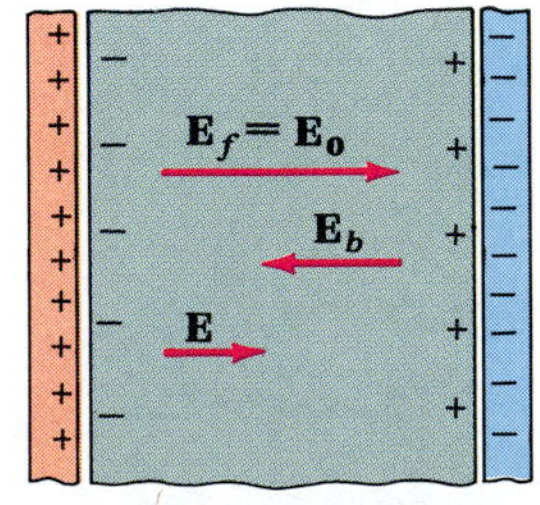

Figure 23-17. The field inside the dielectric has two contributions, $\mathbf{E}_b$ and $\mathbf{E}_f$ (which equals $\mathbf{E}_0$), and they are opposite in direction. In terms of field magnitudes, $E = E_0 - E_b$.

EXAMPLE 23-9

Fields and charge densities for a capacitor. Suppose the potential difference across the parallel-plate capacitor in Example 23-8 is 180 V. Determine *(a)* E, *(b)* E_0, *(c)* E_b, *(d)* $|\sigma_f|$, and *(e)* $|\sigma_b|$.

Solution. *(a)* For the capacitor in Example 23-8, $d = 0.14$ mm so that the field magnitude between the plates is

$$E = \frac{\mathcal{V}}{d} = \frac{180\text{ V}}{0.14\text{ mm}} = 1.3 \times 10^6\text{ V/m}$$

(b) Paper ($\kappa = 3.6$) is used between the plates so that the magnitude of the contribution to the field due to the free charge is

$$E_0 = \kappa E = (3.6)(1.3 \times 10^6\text{ V/m}) = 4.6 \times 10^6\text{ V/m}$$

(c) Since $E = E_0 - E_b$, the magnitude of the contribution to the field due to the bound charge is

$$E_b = E_0 - E = (4.6 \times 10^6\text{ V/m}) - (1.3 \times 10^6\text{ V/m})$$
$$= 3.3 \times 10^6\text{ V/m}$$

(d) The magnitude of the free charge density is

$$|\sigma_f| = \epsilon_0 E_0 = (8.85 \times 10^{-12}\ \text{F/m})(4.6 \times 10^6\ \text{V/m})$$
$$= 4.1 \times 10^{-5}\ \text{C/m}^2$$

(e) The magnitude of the bound charge density is

$$|\sigma_b| = \epsilon_0 E_b = (8.85 \times 10^{-12}\ \text{F/m})(3.3 \times 10^6\ \text{V/m})$$
$$= 3.0 \times 10^{-5}\ \text{C/m}^2$$

We may also find $|\sigma_b|$ from Eq. 23-10:

$$|\sigma_b| = \frac{\kappa - 1}{\kappa}|\sigma_f| = \frac{3.6 - 1}{3.6}(4.1 \times 10^{-5}\ \text{C/m}^2)$$
$$= 3.0 \times 10^{-5}\ \text{C/m}^2$$

SELF-TEST 23-9. Repeat the calculation given in the above example, except use polystyrene as the dielectric. ***ANSWERS:*** *(a)* 1.3×10^6 V/m; *(b)* 3.3×10^6 V/m; *(c)* 2.1×10^6 V/m; *(d)* 3.0×10^{-5} C/m²; *(e)* 1.8×10^{-5} C/m².

EXAMPLE 23-10

Model of a dielectric. Consider a relatively crude model for the polarization of a dielectric (Fig. 23-18). Suppose the dielectric consists of two interpenetrating slabs of uniform charge—one positive (the atomic nuclei) and the other negative (the electrons). Superimposed on one another, the two slabs compose a neutral medium. When placed in a uniform external field $\mathbf{E}_0$, the two slabs are displaced a distance δ at equilibrium. *(a)* Develop an expression that approximates δ in terms of E and κ. *(b)* Apply the result of part *(a)* to estimate δ for the case where paper is the dielectric and $E = E_{max}$ (see Table 23-1).

Solution. *(a)* Consider the side of the slab that has a positive bound charge density on its surface (on the right in Fig. 23-18*b*). On this surface σ_b is given by the product of the volume charge density ρ_p due to the protons in the nuclei of the dielectric and the distance δ: $\sigma_b = \rho_p \delta$. Using Eq. 23-10, we obtain

$$\rho_p \delta = (\kappa - 1)\frac{|\sigma_f|}{\kappa}$$

Since $E = E_0/\kappa = |\sigma_f|/\kappa\epsilon_0$, we have

$$\delta = \frac{(\kappa - 1)\epsilon_0 E}{\rho_p}$$

To use this relation, we should estimate the charge density ρ_p in terms of quantities that are more readily known. If the slab has cross-sectional area A and thickness d (volume Ad), then

$$\rho_p = \frac{N_p e}{Ad}$$

where N_p is total number of protons in the slab. Note that (i) the mass M of the slab is almost entirely due to the protons and neutrons, (ii) the mass m_p of a proton is nearly the same as that of a neutron, and (iii) the number of protons and neutrons in a material is nearly the same. Therefore $M \approx 2N_p m_p$, or $N_p \approx M/2m_p$, and

$$\rho_p \approx \frac{Me}{2Adm_p}$$

Since M/Ad is the mass density ρ_M of the dielectric, we have

$$\rho_p \approx \frac{\rho_M e}{2m_p}$$

Substitution into the equation for δ gives

$$\delta \approx \frac{2(\kappa - 1)\epsilon_0 E m_p}{\rho_M e}$$

The displacement depends on the factor $(\kappa - 1)$. Thus $\delta = 0$ if $\kappa = 1$, which corresponds to

(a)

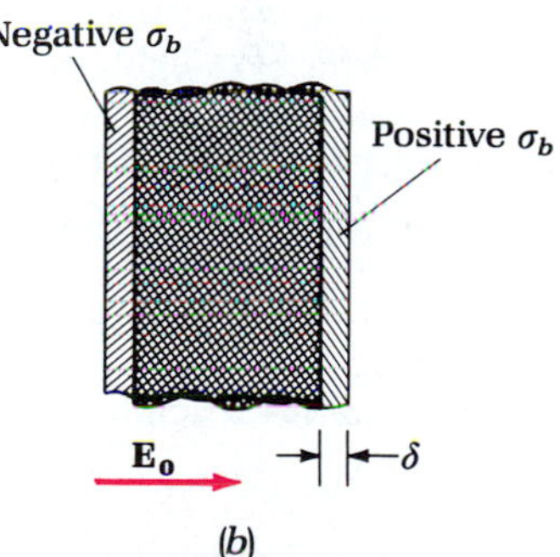

(b)

Figure 23-18. Example 23-10: A crude model of a dielectric. *(a)* A slab of dielectric in zero field. The positive charges (shown as crosshatching sloping downward to the right) are superimposed on the negative charges (shown as crosshatching sloping upward to the right). *(b)* In an electric field the positive and negative charges are displaced a distance δ, which creates equal and opposite surface charge densities of thickness δ on each face: $\sigma_b = \rho_p \delta$.

vacuum. Our model also predicts that δ is directly proportional to E. *(b)* The displacement is δ_{max} when $E = E_{max}$, so that

$$\delta_{max} \approx \frac{2(\kappa - 1)\epsilon_0 E_{max} m_p}{\rho_M e}$$

For paper $\rho_M \approx 800$ kg/m³ and $E_{max} = 15 \times 10^6$ V/m, which gives

$$\delta_{max} \approx \frac{2(3.6 - 1)(8.8 \times 10^{-12}\ \text{F/m})(15 \times 10^6\ \text{V/m})(1.7 \times 10^{-27}\ \text{kg})}{(800\ \text{kg/m}^3)(1.6 \times 10^{-19}\ \text{C})}$$

$$\approx 9 \times 10^{-15}\ \text{m}$$

The diameter of a typical atom is about 3×10^{-10} m, so that this displacement is only about 3×10^{-5} atomic diameters. As we said at the outset, this model is very crude. But it does serve to show that even at the largest field that the dielectric can bear, the displacement of the charges is very small. However, the effect of the displacement is not small, because a very large number of charges are displaced.

SELF-TEST 23-10. The mass density of polystyrene is $\rho_M = 904$ kg/m³. Estimate δ_{max} for polystyrene. ***ANSWER:*** 8×10^{-15} m.

SUMMARY

Section 23-1. Capacitors and Capacitance

A capacitor is an electrical device used in circuits to store charge and electric energy; it consists of two conducting plates separated by an insulator. The capacitance C of a capacitor is

$$C = \frac{Q}{\mathcal{V}} \tag{23-1}$$

The capacitance depends on the geometrical design of the capacitor and the nature of the dielectric between the plates. For a parallel-plate capacitor with vacuum between the plates,

$$C = \frac{\epsilon_0 A}{d} \tag{23-2}$$

Section 23-2. Capacitors in Series and Parallel

The equivalent capacitance of a combination of capacitors is the capacitance of a single capacitor which, when used in place of the combination, provides the same external effect. The equivalent capacitance of a number of capacitors connected in series is

$$C_{eq} = \frac{1}{\Sigma(1/C_i)} \tag{23-3}$$

For a number of capacitors connected in parallel,

$$C_{eq} = \Sigma C_i \tag{23-4}$$

Section 23-3. Electric Energy and Energy Density

The energy of a capacitor is the potential energy of the charges on the capacitor plates:

$$U = \frac{1}{2}\frac{Q^2}{C} = \tfrac{1}{2}C\mathcal{V}^2 = \tfrac{1}{2}Q\mathcal{V} \tag{23-5}$$

When this energy is associated with the electric field, the energy density u in the space occupied by the field (in vacuum) is

$$u = \tfrac{1}{2}\epsilon_0 E^2 \tag{23-6}$$

Sections 23-4 and 23-5. Electrostatic Properties of Insulators; Atomic Description of the Properties of Insulators

When an insulator (or dielectric) is placed in an electric field, the atoms and molecules that compose the insulator become polarized and a bound surface charge density is induced. This bound charge produces an electric field directed opposite the external field, so that both $\mathcal{V}$ and E are reduced. This polarization of a dielectric is characterized by the dielectric constant κ:

$$\kappa = \frac{\mathcal{V}_0}{\mathcal{V}} \tag{23-7}$$

QUESTIONS

1. Explain the meaning of the phrase "charge Q on a capacitor." Does the capacitor as a whole possess this charge?

2. Explain the meaning of the phrase "potential difference $\mathcal{V}$ across a capacitor." In terms of the line integral

$$\mathcal{V} = \mathcal{V}_b - \mathcal{V}_a = -\int_a^b \mathbf{E} \cdot d\boldsymbol{\ell}$$

Where is point a and where is point b? Do we have more than one choice for the position a (or b)? Explain.

3. Suppose a fellow student tells you that since $C = Q/\mathcal{V}$, the capacitance of a capacitor is proportional to the charge on it. How do you respond?

4. Suppose the potential difference across a capacitor is doubled. By what factor does the ratio $Q/\mathcal{V}$ change?

5. The capacity of a bucket to hold a liquid is expressed in terms of its volume. Explain any analogies you can draw between a bucket and a capacitor. Where do your analogies break down?

6. Explain any analogies you can draw between the heat capacity of an object and the capacitance of a capacitor. Where do your analogies break down? Do you consider these better analogies than the ones you devised for the previous question? Explain.

7. When a battery charges a capacitor, the charges on the plates are of equal magnitude but opposite sign. Why? If the plates have different sizes, will they still have charge of the same magnitude?

8. Suppose each plate of a capacitor initially has positive charge q on it and then the capacitor is connected to a battery. Will the subsequent charge on the plates be equal and opposite? What is the charge on each plate in terms of q, C, and $\mathcal{V}$?

9. Suppose a parallel-plate capacitor is charged and then disconnected from the charging battery. Next the plate separation is doubled. Describe any change in each of the following quantities as a result of the change in plate separation (neglect edge effects): *(a)* the charge of the plates, *(b)* the capacitance, *(c)* the field between the plates, *(d)* the potential differences across the capacitor, *(e)* the energy of the capacitor, *(f)* the energy density between the plates, *(g)* the energy in the electric field.

10. Suppose a parallel-plate capacitor is charged and kept connected to the charging battery. Next the plate separation is doubled. Describe any change in each of the following quantities as a result of the change in plate separation (neglect edge effects): *(a)* the potential difference across the capacitor, *(b)* the capacitance, *(c)* the charge on the plates, *(d)* the field between the plates, *(e)* the energy of the capacitor, *(f)* the energy density between the plates, *(g)* the energy in the electric field.

11. Suppose a thin sheet of metal is placed midway between the plates of a parallel-plate capacitor (Fig. 23-19). The metal sheet is insulated from all other objects, and its thickness is negligible compared with the plate separation. By what factor, if any, is the capacitance changed by the insertion of the metal sheet? Does your answer depend on whether the metal sheet is in the middle?

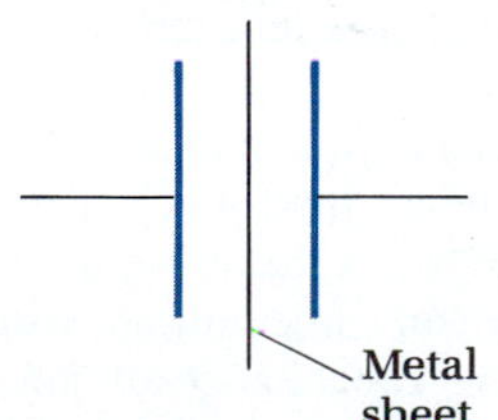

Figure 23-19. Question 11.

12. Suppose a parallel-plate capacitor is charged and then disconnected from the battery. A metal sheet is next inserted between the plates, as shown in Fig. 23-20; the thickness of the sheet is half the plate spacing. Determine the factor by which each of the following is changed: *(a)* the capacitance, *(b)* the charge on the plates, *(c)* the potential difference between the plates, *(d)* the field between the plates (excluding the volume of the metal sheet), *(e)* the energy density between the plates (excluding the volume of the metal sheet), *(f)* the electric energy of the system. *(g)* Based on energy considerations, do you expect an electric force on the metal sheet as it is being inserted? If so, what is the direction of this force? Explain.

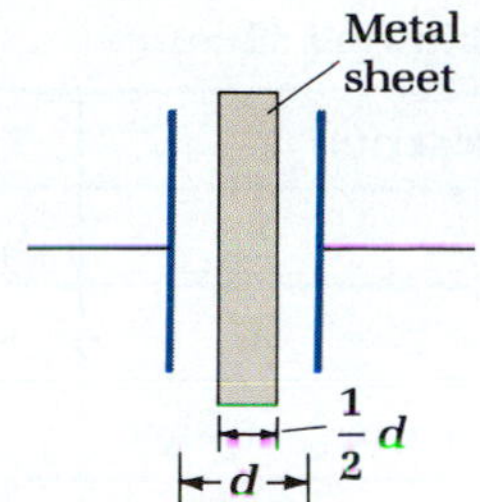

Figure 23-20. Question 12.

13. From examination of Fig. 23-5, which capacitance is larger, C_1 or C_2?

14. The energy of a capacitor is 3.0 μJ after having been charged by a 1.5-V battery. What is the energy of the capacitor after it is charged by a 3.0-V battery?

15. The potential $\mathcal{V}$ may be described as the potential energy per unit charge. Why then is the energy of a charged capacitor $\frac{1}{2}Q\mathcal{V}$ rather than simply $Q\mathcal{V}$?

16. Since an external electric field exerts forces on nuclei and electrons in opposite directions, how do atoms hold together in an electric field?

17. The dielectric constant of water decreases continuously with increasing temperature: $\kappa = 88$ at 0°C and $\kappa = 55$ at 100°C. Explain this behavior in terms of the molecules that compose the water.

18. Suppose you are asked to construct a capacitor that occupies little space, has a large capacitance, and will operate at a high potential. What properties of the construction materials are important? What compromises will you need to make?

19. You can pick up bits of paper with a charged comb even though the paper is neutral. Assuming paper behaves as an ideal insulator (that is, it does not allow charge carriers to move through it), explain how this can happen in terms of the polarization of the atoms in the paper. Will the effect depend on the sign of the charge on the comb? (*Hint:* The electric field due to the charges on the comb falls off with distance from the charged end of the comb.)

20. If the bound charge on a dielectric that fills the space between the plates of a parallel-plate capacitor is to be half that of the free charge, what must be the value of the dielectric constant?

21. When a dielectric of constant κ fills the space between the plates of a charged capacitor, the magnitude of the field is reduced by a factor of $1/\kappa$ and the energy density is reduced by $1/\kappa$. Given that u is proportional to E^2, how can this be so?

22. Complete the following table:

Symbol	Represents	Type	SI Unit
C		Scalar	
C_{eq}			F
u	Electric energy density		
σ_b			
$\mathbf{E}_b$			
σ_f			

EXERCISES

Section 23-1. Capacitors and Capacitance

1. Show that the SI unit for capacitance, the farad (F), is equivalent to $C^2 \cdot s^2/(kg \cdot m^2)$.

2. How much charge is transferred from one plate to the other by a 6.0-V battery when it charges a 3.0-nF capacitor?

3. If the charge on a capacitor is 14.5 μC when the potential difference across it is 25 V, what is its capacitance?

4. A parallel-plate capacitor has circular plates of radius 136 mm that are separated by 1.5 mm in vacuum. What is its capacitance?

5. Suppose you wish to build a 1-F parallel-plate capacitor in which the plate separation is 10 mm. If the plates are square and have vacuum between them, what must the length of their sides be?

6. What is the capacitance of a cylindrical capacitor of length 220 mm with vacuum between its plates? The outer radius of the inner cylinder is 33 mm, and the inner radius of the outer cylinder is 45 mm.

7. Consider a cylindrical capacitor in which the spacing between the plates, $d = R_b - R_a$, is very small compared with the two radii, $R_b \approx R_a \gg d$. Using the result of Example 23-2, show that the expression for the capacitance is nearly the same as for a parallel-plate capacitor with $A = 2\pi R_a L \approx 2\pi R_b L$. (*Hint:* $\ln x \approx x - 1$ when $x \approx 1$.)

8. The capacitance of a single isolated conductor is defined as $C = Q/\mathcal{V}$, where Q is the charge on the conductor when $\mathcal{V}$ is its potential relative to $\mathcal{V}_\infty = 0$. Show that the capacitance of an isolated spherical conductor of radius r_0 in vacuum is $C = 4\pi\epsilon_0 r_0$.

9. A spherical capacitor consists of a conducting sphere of radius r_a surrounded by a concentric spherical conducting shell of inner radius r_b (Fig. 23-21). The capacitance of a spherical capacitor is $C = 4\pi\epsilon_0 r_b r_a/(r_b - r_a)$. (See Prob. 2.) *(a)* Show that if the spacing between the plates of a spherical capacitor, $d = r_b - r_a$, is much smaller than the two radii, then the capacitance is nearly that of a parallel-plate capacitor of area $A = 4\pi r_a^2 \approx 4\pi r_b^2$. *(b)* The surface of the earth (radius = 6400 km) and the ionosphere at an altitude of 100 km may be considered to be the plates of a spherical capacitor. Using the result of part *(a)*, determine the capacitance of this capacitor, assuming vacuum between the plates.

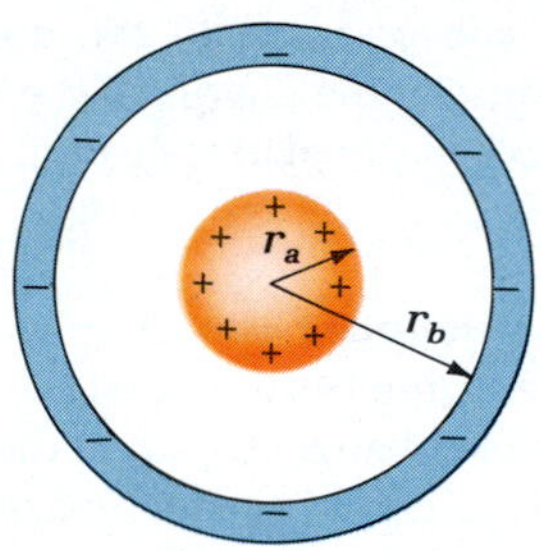

Figure 23-21. Exercise 9: Cross section of a spherical capacitor.

Section 23-2. Capacitors in Series and Parallel

10. A 2.4-μF capacitor is connected in series with a 3.1-μF capacitor and then the combination is charged with a 6.1-V battery. *(a)* What is the equivalent capacitance of the combination? *(b)* What is the charge on each capacitor? *(c)* What is the potential difference across each capacitor?

11. A 2.4-μF capacitor is connected in parallel with a 3.1-μF capacitor and then the combination is charged with a 6.1-V battery. *(a)* What is the equivalent capacitance of the combination? *(b)* What is the potential difference across each capacitor? *(c)* What is the charge on each capacitor?

12. Suppose you need a capacitance of 3.6 μF in a circuit but all you have available is a boxful of capacitors of capacitance 2.4 μF and lower. What is the value of the capacitance of a single capacitor that you can combine with a 2.4-μF capacitor to give an equivalent capacitance of 3.6 μF? How will you connect the two, in series or parallel?

13. Suppose you need a capacitance of 1.7 μF in a circuit but all you have available is a boxful of capacitors of capacitance 2.4 μF and higher. What is the value of the capacitance of a single capacitor that you can combine with a 2.4-μF capacitor to give an equivalent capacitance of 1.7 μF? How will you connect the two, in series or parallel?

14. A 62-nF capacitor is charged with a 12-V battery and then disconnected from the battery. The capacitor's lead wires are

then connected to those of an initially uncharged 38-nF capacitor. *(a)* What is the final charge on each capacitor after they are connected? *(b)* What is the final potential difference across each capacitor after they are connected?

15. A 62-nF capacitor and a 38-nF capacitor are separately charged with a 12-V battery and then disconnected from the battery. Suppose the two lead wires from the positive plates of the capacitors are connected together, and the two lead wires from the negative plates of the capacitors are connected together. *(a)* What is the final charge on each capacitor? *(b)* What is the final potential difference across each capacitor? Now suppose the capacitors' lead wires are reversed such that plates with charge of opposite sign are connected. *(c)* What is the final charge on each capacitor? *(d)* What is the final potential difference across each capacitor?

16. The equivalent capacitance of a number of capacitors connected in parallel is $C_{eq} = \Sigma C_i$. Thus the equivalent capacitance must be larger than the capacitance of any of the individual capacitors simply because a sum of positive numbers must be larger than any single term in the sum. The equivalent capacitance of a number of capacitors connected in series is $C_{eq} = 1/[\Sigma(1/C_i)]$. Show that the equivalent capacitance must be smaller than the capacitance of any of the individual capacitors.

17. In Example 23-3 we showed that the equivalent capacitance of two capacitors connected in series may be written

$$C_{eq} = \frac{C_1 C_2}{C_2 + C_1}$$

(a) Show that the equivalent capacitance of three capacitors connected in series may be written

$$C_{eq} = \frac{C_1 C_2 C_3}{C_2 C_3 + C_1 C_3 + C_1 C_2}$$

(b) Write a similar expression for the equivalent capacitance of four capacitors connected in series.

18. For the arrangement in Fig. 23-7, $C_1 = 2.0\ \mu F$, $C_2 = 3.0\ \mu F$, and $C_3 = 5.0\ \mu F$. *(a)* What is the equivalent capacitance of the combination? Suppose $\mathcal{V}_b - \mathcal{V}_a = 25$ V. *(b)* What is the potential difference across each capacitor? *(c)* What is the charge on each capacitor?

19. For the arrangement in Fig. 23-22, $C_1 = 4.0\ \mu F$, $C_2 = 6.0\ \mu F$, and $C_3 = 5.0\ \mu F$. *(a)* What is the equivalent capacitance of the combination? Suppose $\mathcal{V}_b - \mathcal{V}_a = 65$ V. *(b)* What is the potential difference across each capacitor? *(c)* What is the charge on each capacitor?

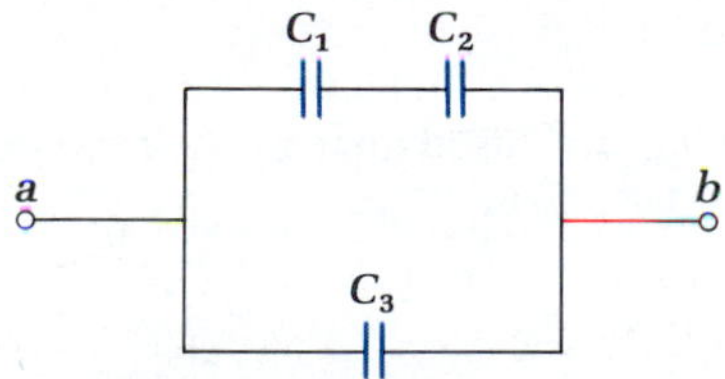

Figure 23-22. Exercise 19.

20. In Fig. 23-23, capacitor 1 is initially charged to a potential difference of $\mathcal{V}_0$ by throwing switch S to the left. *(a)* What is the charge on capacitor 1? Suppose the switch is now thrown to the right. *(b)* What is the final charge on each capacitor and what is the final potential difference across each capacitor? Express your answers in terms of $\mathcal{V}_0$, C_1, C_2, and C_3. *(c)* Evaluate your answers for the case where $\mathcal{V}_0 = 35$ V, $C_1 = 4.0\ \mu F$, $C_2 = 6.0\ \mu F$, and $C_3 = 2.0\ \mu F$.

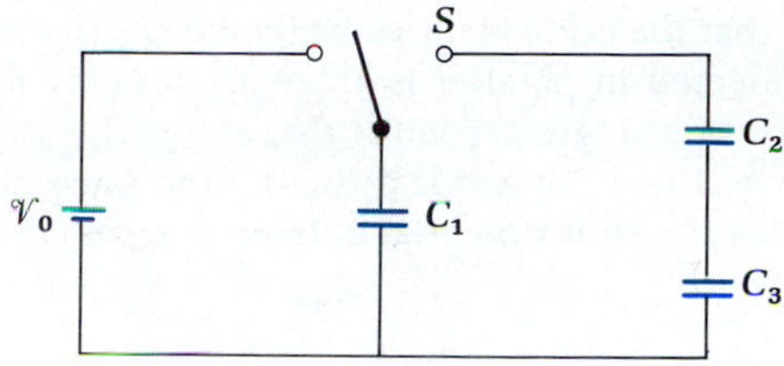

Figure 23-23. Exercise 20.

Section 23-3. Electric Energy and Energy Density

21. A 1.0-μF capacitor is charged with a 10-V battery. *(a)* What is the energy of the capacitor? *(b)* What is the capacitor's energy if it is charged with a 20-V battery?

22. A 0.25-μF parallel-plate capacitor is charged with a 96-V battery. *(a)* What is the energy of the capacitor? *(b)* If the plates are separated by 0.12 mm, what is the field between the plates? *(c)* What is the energy density between the plates? (Neglect edge effects and assume vacuum is between the plates.)

23. As a rough approximation of the electric field in the earth's atmosphere, assume that it is uniform and of magnitude 100 V/m in the region between the surface and the ionosphere, and then zero above the ionosphere. *(a)* What is the electric energy density in the atmosphere? *(b)* Given that the radius of the earth is 6400 km and that the altitude of the ionosphere is about 100 km, estimate the electric energy contained in the earth's atmosphere.

24. A typical lead-acid storage battery used in an automobile stores about 1 kW·h (3.6×10^6 J) of electric energy, and the potential difference between its terminals is 12 V. *(a)* Determine the capacitance of a capacitor that can store this amount of energy when the potential difference across it is 12 V. *(b)* If this capacitor is a parallel-plate capacitor with square plates separated by 1.0 mm, determine the length of the sides of the plates. Neglect edge effects and assume vacuum between the plates. Do you think capacitors are likely to replace batteries for use in automobiles?

25. Determine the electric energy of the system of capacitors discussed in Exercise 19 and illustrated in Fig. 23-22.

26. *(a)* Determine the electric energy of the capacitors of Exercise 14 before and after their lead wires are connected together. *(b)* Determine the electric energy of the capacitors of Exercise 15 before and after their lead wires are connected together. (As we shall see in the next chapter, the lost electric energy is transferred as heat to the environment.)

27. Show that when two capacitors are connected in series the sum of their individual energies $U = U_1 + U_2$ is the same as that of a single capacitor whose capacitance is the equivalent capacitance of the two and whose charge is the same as that for each individual capacitor.

28. Show that the equivalent capacitance C_{12} of capacitors 1 and 2 connected in parallel is $C_{12} = C_1 + C_2$ by finding the capacitance of a single capacitor that stores the same energy as capacitors 1 and 2 when it possesses the same charge as 1 and 2 and has the same potential difference across it as is across 1 and 2.

29. A spherical capacitor consists of a conducting sphere of radius r_a surrounded by a concentric spherical conducting shell of inner radius r_b (Fig. 23-21). The capacitance of a spherical capacitor is $C = 4\pi\epsilon_0 r_a r_b/(r_b - r_a)$. (See Prob. 2.) Show that the energy of the capacitor is $U = \frac{1}{2}Q^2/C$ by integrating the energy density, $u = \frac{1}{2}\epsilon_0 E^2$. (*Hint:* See Example 23-7. The approximate volume element is a spherical shell of volume $4\pi r^2\, dr$.)

30. The capacitance of a single isolated spherical conductor with radius r_0 is given in Exercise 8 as $C = 4\pi\epsilon_0 r_0$, and in Example 23-7 we found that the energy U of an isolated spherical conductor with charge Q is $U = Q^2/8\pi\epsilon_0 r_0$. Show that these two results give the first of Eqs. (23-5): $U = \frac{1}{2}Q^2/C$.

31. Assume that the proton can be represented as a charged conducting sphere of radius 1×10^{-15} m. Using this model, estimate the electric energy of a proton in MeV. (*Hint:* See Example 23-7.)

32. For the situation described in Exercise 20 (Fig. 23-23), show that the final energy stored in the system of capacitors is reduced by the factor $[C_1(C_2 + C_3)]/(C_1C_2 + C_1C_3 + C_2C_3)$ after the switch is thrown to the right.

33. Consider a parallel-plate capacitor in which the plate separation x can be varied while the charge on the plates remain fixed. *(a)* Show that the force on each plate by the other has magnitude $Q^2/2\epsilon_0 A$. (*Hint:* Recall Eq. 9-9: $F_x = -dU/dx$.) *(b)* Use the answer to part *(a)* to find the work done by electric forces when the spacing is changed from d to $3d$. *(c)* Compare your answer to part *(b)* with the change in the energy of the capacitor due to the spacing change.

Section 23-4. Electrostatic Properties of Insulators

34. A capacitor is charged with a 9.6-V battery and then the battery is disconnected. A dielectric is fitted snugly between the plates and the potential then measures 3.2 V. What is the dielectric constant of the dielectric material?

35. What is the capacitance of a parallel-plate capacitor, with plate area 0.024 m² and plate separation 0.26 mm, that has neoprene (see Table 23-1) in the space between the plates? Neglect edge effects.

36. What is the capacitance of a 1-m length of coaxial cable in which the wire radius is 0.91 mm and the inner radius of the coaxial cylinder is 1.22 mm? The insulator between the two is nylon. Neglect end effects.

37. A parallel-plate capacitor with plate area 0.087 m² and plate separation 1.8 mm has a capacitance of 2.4 nF when a certain dielectric fills the space between the plates. What is the dielectric constant of this dielectric?

38. A parallel-plate capacitor has a plate separation of 0.97 mm and a capacitance of 1.4 nF when vacuum is between the plates. The capacitor is charged with a 9.6-V battery and then disconnected from the battery. *(a)* What is the field between the plates? *(b)* What is the charge on the plates? Now, with the battery still disconnected, the space between the plates is filled with a dielectric that has a dielectric constant of 8.2. *(c)* What is the field between the plates? *(d)* What is the charge on the plates?

39. What is the maximum potential difference that can be placed across an air-filled parallel-plate capacitor with plate separation 1.0 mm?

40. What is the maximum electric energy density that can exist in air at room temperature and atmospheric pressure?

41. Suppose you must design a parallel-plate capacitor that has a capacitance of 3.6 nF and a maximum operating potential difference of 4×10^4 V. Further, the dielectric material between the plates must be polystyrene. What is the minimum plate area that you can use?

42. Show that the energy density of the electric field between the plates of a parallel-plate capacitor with a dielectric between the plates is $u = \frac{1}{2}\kappa\epsilon_0 E^2$. (*Hint:* Repeat the development presented in Sec. 23-3, except with a dielectric between the plates rather than vacuum.)

Section 23-5. Atomic Description of the Properties of Insulators

43. A parallel-plate capacitor with Bakelite between the plates has a plate area of 0.070 m² and a capacitance of 4.0 nF. The capacitor is charged with a 10.0-V battery. *(a)* What is the free charge density on the plates? *(b)* What is the bound charge density on the dielectric? *(c)* What is the magnitude of the electric field? *(d)* What is the magnitude of the contribution to the electric field due to the free charge? *(e)* What is the magnitude of the contribution to the electric field due to the bound charge?

44. For a charged parallel-plate capacitor with Teflon as the dielectric in the space between the plates, what is the ratio of the bound charge on the Teflon to the free charge on the plates?

45. Using the model of a dielectric presented in Example 23-10, determine δ_{max} for fused quartz. The mass density of fused quartz is 2.6×10^3 kg/m³.

PROBLEMS

1. ***Capacitor composed of a number of parallel sheets.*** Consider a capacitor that consists of a number n of equally spaced, parallel conducting sheets, as shown in Fig. 23-24. Alternate sheets connected together compose the positive plate, and the other alternate sheets compose the negative plate. Show that the capacitance of this arrangement is

$$C = \frac{(n-1)\epsilon_0 A}{d}$$

Figure 23-24. Problem 1.

where A is the area of each sheet and d is the sheet spacing. Assume vacuum is between the sheets and neglect edge effects.

2. ***A spherical capacitor.*** A spherical capacitor consists of a conducting sphere (radius r_a) as one plate surrounded by a concentric conducting shell (inner radius r_b) as the other plate (Fig. 23-21). Show that the capacitance of this capacitor with vacuum between the plates is

$$C = 4\pi\epsilon_0 \frac{r_b r_a}{r_b - r_a}$$

3. ***A symmetric arrangement of capacitors.*** What is the equivalent capacitance of the arrangement shown in Fig. 23-25?

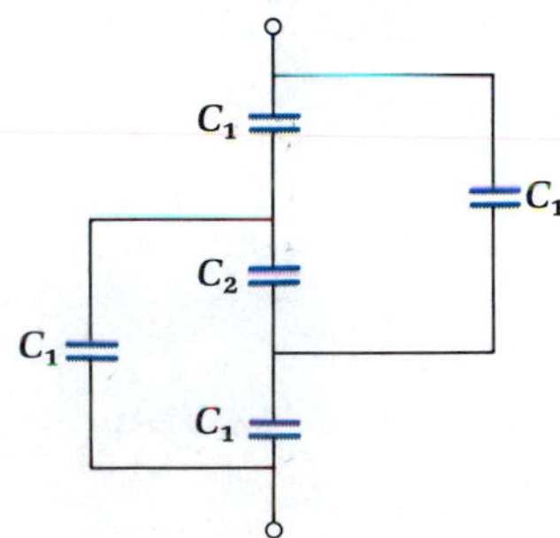

Figure 23-25. Problem 3.

4. ***A type of variable capacitor.*** The section of the circuit that includes the right plate of capacitor a (plate separation d_a) and the left plate of capacitor b (plate separation d_b) in Fig. 23-26 can be rigidly translated back and forth to vary simultaneously the capacitance of both capacitors. During the translation, $d = d_a + d_b$ remains fixed. *(a)* Show that the equivalent capacitance of the arrangement is

$$C_{eq} = \frac{\epsilon_0 A d}{d_a(d - d_a)}$$

where A is the area of the plates of each capacitor. *(b)* Make a graph of C_{eq} versus d_a. At what value(s) of d_a is C_{eq} a minimum? Is there a maximum value of C_{eq}? Explain.

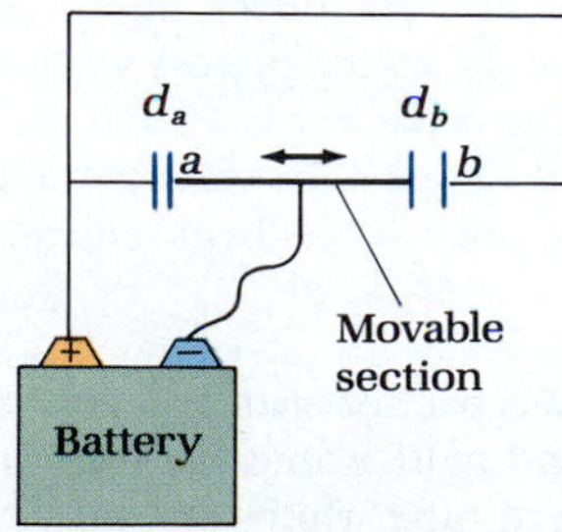

Figure 23-26. Problem 4.

5. ***Capacitance of two distant conducting spheres.*** Consider two conducting spheres a and b, each with a radius (r_a and r_b) that is much smaller than the distance d between their centers. *(a)* Treating the spheres as capacitor plates, show that the capacitance of the spheres is approximately

$$C \approx \frac{4\pi\epsilon_0}{(1/r_a) + (1/r_b) - (2/d)}$$

(*Hint:* Since $d \gg r_a$ and $d \gg r_b$, you may assume that the surface charge density on each sphere is uniform.) *(b)* Estimate the capacitance of two conducting spheres with $r_a = r_b = 20$ mm that are separated by 1 m.

6. ***Stray capacitance.*** Capacitance is exhibited by systems of conductors in electric circuits, whether the capacitance is purposely included in the circuit or not. The wires used to connect circuit devices have some capacitance, although quite small in most cases. For example, the two wires in an ordinary lamp cord behave as a capacitor. Capacitance that is inadvertently contained in a circuit is called *stray capacitance.* Consider estimating the capacitance of two long, straight, parallel wires, each of length L and radius R, separated by a perpendicular distance D between their centers. *(a)* Assuming $L \gg D \gg R$, show that the capacitance of the wires is approximately

$$C \approx \frac{\pi\epsilon_0 L}{\ln (D/R)}$$

(*Hint:* Since $L \gg D$, you may neglect end effects, and since $D \gg R$, you may assume that the surface charge density on each wire is nearly uniform.) *(b)* Estimate the capacitance of two such wires with $R = 1$ mm, $D = 10$ mm, and $L = 1$ m.

7. ***Geiger-Müller tube.*** A Geiger-Müller tube is a device used to measure ionizing radiation. The tube is essentially a cylindrical capacitor with a wire as one plate, a coaxial conducting cylinder as the other plate, and with a gas ($\kappa \approx 1$) as the dielectric. The tube is operated so that the electric field near the wire is very large. Thus an ionizing particle moving through the gas causes the onset of breakdown, and the particle is detected by a burst of electric current from one plate to the other. *(a)* Show that the potential difference $\mathcal{V}$ between the plates can be written

$$\mathcal{V} = E R_a \ln \frac{R_b}{R_a}$$

where E is the magnitude of the electric field just outside the wire, R_a is the radius of the wire, and R_b is the inner radius of the cylinder. Neglect end effects. *(b)* Determine $\mathcal{V}$ for the case where $R_a = 0.30$ mm, $R_b = 20.0$ mm, and $E = 2.0 \times 10^6$ V/m. *(c)* What is the magnitude of the electric field inside the tube and very close to the cylinder?

8. ***Energy of a sphere of uniform charge density.*** *(a)* Show that the electric energy of a spherically symmetric uniform volume charge of radius r_0 and charge Q is

$$U = \frac{3Q^2}{20\pi\epsilon_0 r_0}$$

Assume $\kappa = 1$ everywhere. *(b)* What fraction of the total energy is due to the field inside the charge distribution? *(c)* Estimate the electric energy of a proton (in MeV), assuming that it can be represented as a spherically symmetric uniform volume charge of radius 1×10^{-15} m. Assume $\kappa = 1$ and compare your answer with that found for Exercise 31.

9. ***Finding the radius of a wire with a capacitance measurement.*** A single straight wire of length 1.6 m and unknown radius is insulated with neoprene such that the outer radius of the neoprene is 6 mm. A cylindrical capacitor is formed by painting the outside of the neoprene with a conducting layer of silver paint, and the capacitance is measured to be 0.6 nF. What is the radius of the wire? Neglect end effects.

10. ***Capacitor partially filled with a dielectric.*** A parallel-plate capacitor with plate area A and plate separation d is partially filled with a dielectric of thickness x and dielectric constant κ, as shown in Fig. 23-27. Develop an expression for the capacitance. Neglect edge effects.

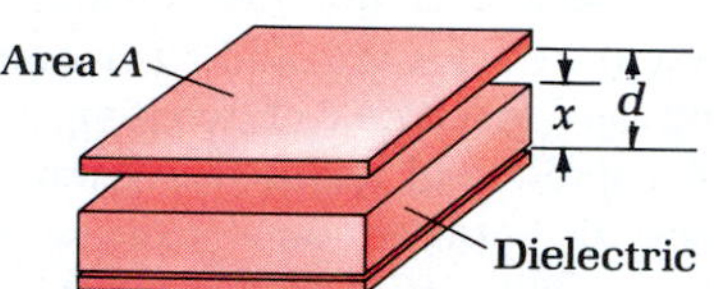

Figure 23-27. Problem 10.

11. ***Capacitor with a partially inserted dielectric.*** A dielectric slab is partially inserted a distance x between the plates of a parallel-plate capacitor of plate separation d and lateral dimensions w_1 and w_2, as shown in Fig. 23-28. *(a)* Show that the capacitance is $C = \epsilon_0 w_2(\kappa x + w_1 - x)/d$, where κ is the dielectric constant of the dielectric. *(b)* Suppose the capacitor is charged and then disconnected from the charging battery before the dielectric is inserted. Develop an expression for the electric force on the slab. What is the direction of this force? Neglect edge effects.

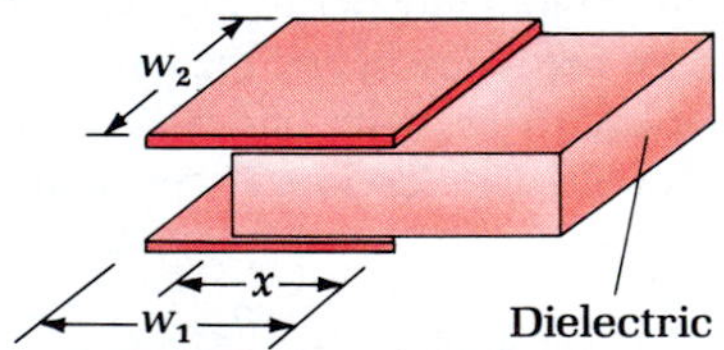

Figure 23-28. Problem 11.

12. ***Displacement using Rutherford's atomic model.*** In Example 23-10 we developed a rough estimate of the displacement of the relative positions of positive and negative charges in a dielectric due to the action of an external electric field. Consider another approach. Assume that the nucleus of an atom can be treated as a point charge with charge Ze and that the electrons can be treated as a uniform volume charge of radius r_a and charge $-Ze$. *(a)* Show that the force on the nucleus is zero when it is displaced a distance δ from the center of negative charge with

$$\delta = \frac{4\pi\epsilon_0 r_a{}^3 E}{Ze}$$

where E is the contribution to the field, at the position of the nucleus, due to charges outside the atom. (*Hint:* See Example 21-6.) *(b)* Determine δ when $E = 15 \times 10^6$ V/m, $Z = 10$, and $r_a = 2 \times 10^{-10}$ m. Compare your answer with that found in Example 23-10.

Additional Problems

13. If the volume of a nucleus of charge Z is equal to KZX (where X is a number between 0.5 and 0.8), what is the capacitance of such nuclei in terms of K, Z, and X? You may assume they are uniform spherical charge distributions for the purpose of this problem.

14. What is the capacitance per unit length of a 25-μm-diameter wire suspended 2 mm above a uniform conducting plane? (See Prob. 10 in Chap. 22.)

15. A conventional 1.5-V D-cell battery may be thought of as a cylindrical capacitor in parallel with a generator (a source of energy that will carry charge across the capacitor to maintain the 1.5 V). What must Q be when the battery is sitting in an isolated condition if the outer radius of the inner conductor is 0.50 cm, the inner radius of the outer conductor is 1.75 cm, and the dielectric constant of the intevening material is 6.1?

16. Consider a capacitor with a mixed dielectric (Fig. 23-29). What is the capacitance of the capacitor?

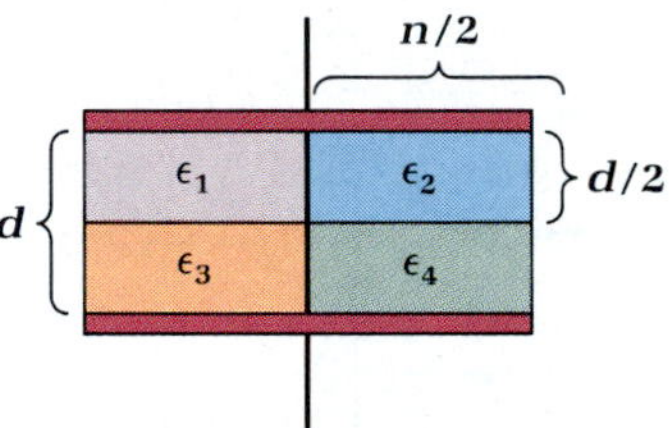

Figure 23-29. Problem 16.

17. What is the capacitance of the earth? Assume the surface is conducting.

18. We can define a gravitational field vector **g** (force per unit mass) in the same sense as we define the **E** field (force per unit charge). Find the energy density in the gravitational field by analogy with the energy density in the electric field. What is the total energy in the earth's gravitational field? (*Hint:* The behavior of the gravitational potential is similar to but not identical with that of the electric potential. Consider building up the mass in a sphere uniformly from 0 to some final value.)

19. A parallel-plate capacitor has a gap d between the plates. There is a block of material of dielectric constant κ and density ρ filling the volume between plates, but the material is free to move laterally (parallel to the plates) without friction. If the initial voltage on the capacitor is V_0 and the block is given a slight displacement to one side, what is the final velocity the block will acquire once it has been entirely expelled from within the plates?

20. *(a)* Consider a capacitor made from two concentric cylinders of length ℓ and radii r_1 and r_2. The outer cylinder is a self-supporting metal tube while the inner cylinder is a thin-walled conducting balloon. Find the effective radial pressure,

as a function of applied voltage V, felt by the inner balloon if the entire device is in a vacuum.

21. A particular capacitor design uses a dielectric that will work up to a maximum applied voltage V_{max}. If 10 of these capacitors are available to make an energy storage bank, is the maximum energy stored obtained with the capacitors in parallel or in series? What is that value? What other practical considerations are there when the bank is charged to maximum energy capacity?

22. Two identical drops of mercury of radius r are also identically charged with a charge Q each. Assume they are in a weightless vacuum environment and initially very far from each other. With what velocity must they mutually accelerate toward one another so that they meet and merge to form a single charged spherical drop? (*Hint:* Consider the initial and final states.)

23. Consider the two networks shown in Fig. 23-30. If C_1 and C_2 are equal and fixed, is it ever possible to select C_3 so that both networks have the same effective total capacitance? If so, find the value of C_3 in terms of C.

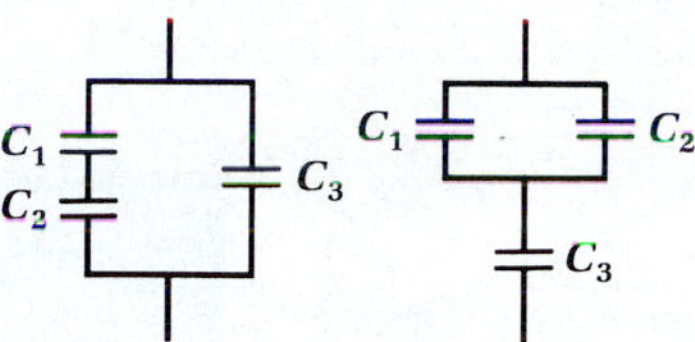

Figure 23-30. Problem 23.

24 CURRENT AND RESISTANCE

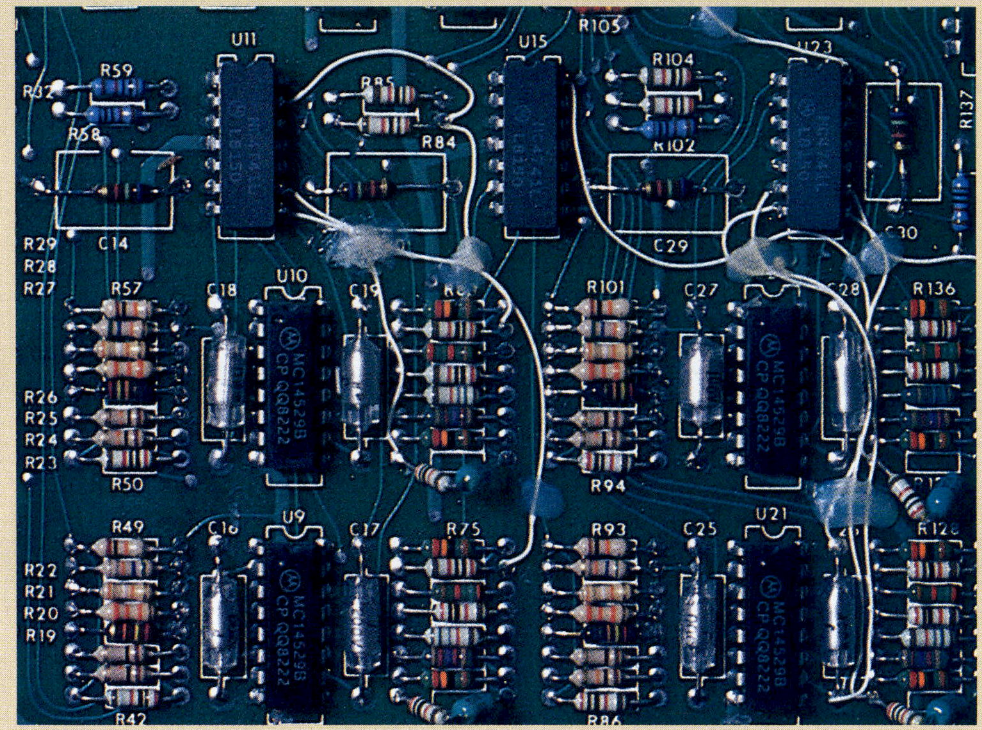

Resistors and integrated circuits on a circuit board. The value of the resistance of a resistor is indicated by the colored bands (Sec. 24-2). The electrical behavior of semiconductors is responsible for the remarkable properties of integrated circuits (Sec. 24-6).

In the last few chapters, we dealt mostly with electrostatics, the effects of stationary charges. Now we begin to consider the motion of charge carriers, or ***electric conduction.*** In electrostatics, $\mathbf{E} = 0$ inside a conductor. However, if we maintain a nonzero field in a conductor, say by connecting it to a battery, then the conductor's charge carriers will flow, and an *electric current* will exist. In this chapter, we describe the effects of steady currents and investigate models that aid in the understanding of electric conduction in matter.

24-1 THE FLOW OF CHARGE

Electrons are the charge carriers in a metal.

A conductor is a material in which some of the charged particles are free to move; these particles are the conductor's charge carriers. For example, a metal can be regarded as an array of positive ions located at fixed lattice sites and interspersed with free electrons. The charge on the free electrons is equal and opposite the charge on the ions to give a neutral medium. The free electrons can move through the lattice; they are the charge carriers in a metal.

Electric Current

The electric current characterizes the flow of charge through a material. Figure 24-1*a* shows a section of conducting wire with positive charge carriers moving to the right. Let dQ be the magnitude of the charge that passes through the plane cross-sectional surface labeled S in time dt.

The ***electric current*** I in the wire is the rate at which charge passes through this surface:

$$I = \frac{dQ}{dt} \qquad (24\text{-}1)$$

Definition of electric current

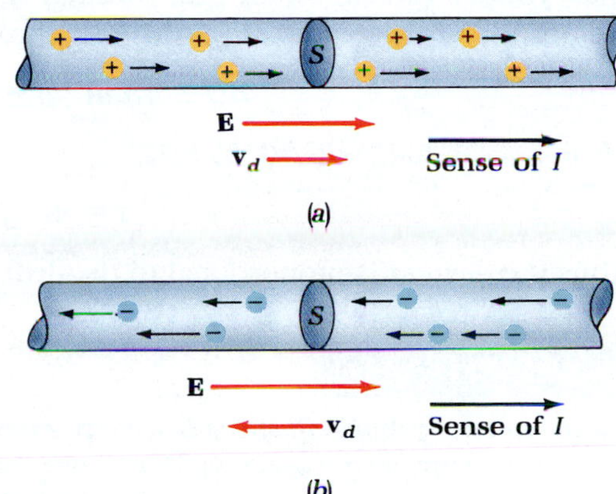

Figure 24-1. *(a)* Current in a wire with positive charge carriers. *(b)* Current in a wire with negative charge carriers. The sense of the current is to the right in each case.

The SI unit of electric current is the ***ampere*** (A), equal to one coulomb per second: 1 A = 1 C/s. The ampere is named for André-Marie Ampere (1775–1836).

Electric current I is a scalar quantity. Even though the electric current is not a vector quantity, it is common practice to speak of the "direction" of the current. This direction corresponds to the direction of flow of positive charge carriers. To emphasize that current is a scalar, we shall refer to the *sense* of the current. The sense of the current in a conductor is given by the direction of motion of positive charge carriers. For example, the current in Fig. 24-1*a* is to the right.

The sense of a current is given by the direction of motion of positive carriers.

Consider the effect of the sign of the charge carriers on the sense of the current. For comparison with the positive charge carriers moving to the right in Fig. 24-1*a*, we show negative charge carriers moving to the left in Fig. 24-1*b*. In Fig. 24-1*a*, positive carriers moving to the right tend to cause the region to the right to become more positive and the region to the left to become more negative. In Fig. 24-1*b*, negative carriers moving to the left also tend to cause the region to the right to become more positive and the region to the left to become more negative. That is, the carrier motion shown in both Fig. 24-1*a* and 24-1*b* gives the same result. Thus the sense of the current is the same in Fig. 24-1*a* as it is in Fig. 24-1*b*; in each case it is to the right. *This means that we need not be concerned with the sign of the carriers when dealing with the external effects of a current; these effects are the same for carriers having either sign.**

The sense of a current is given by the direction opposite the motion of negative carriers.

Drift Velocity

When an externally applied electric field exists in a conductor, it exerts a force on each of the conductor's charge carriers and causes them to move through the material. (Particles other than carriers are displaced slightly, but are confined to

* An exception to this rule is a magnetic phenomenon, called the *Hall effect,* which we discuss in Chap. 26.

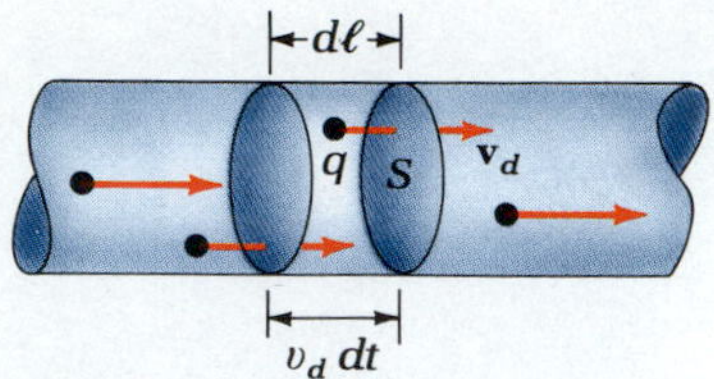

Figure 24-2. Finding the relation between I and v_d. Assuming each carrier has speed v_d, we find that all the carriers in the cylinder of volume $A\,d\ell = Av_d\,dt$ pass through the surface labeled S in time dt. Thus $I = dQ/dt = (nAv_d\,dt|q|)/dt = nAv_d|q|$.

their respective lattice sites.) If the charge carriers were free of other forces, then a constant electric field would cause them to have a constant acceleration. However, the charge carriers interact with the other particles of the material. The combined effect of this interaction and the applied electric field causes the charge carriers to move with a constant average velocity called the ***drift velocity*** $\mathbf{v}_d$.

Now we find the relationship between the current I and the drift speed v_d in a wire with cross-sectional area A. Let n be the number density of charge carriers in the wire (number of carriers per unit volume), and let q be the charge on each carrier. In Fig. 24-2, we assume that each carrier is traveling with speed v_d so that all carriers in the cylinder of length $d\ell = v_d\,dt$ pass through the surface labeled S in time dt. (For a steady flow, these carriers are replaced by those in a neighboring cylinder on the left so that the net charge in this section of wire is unchanged, as it is for all other sections of the wire.) Since $(nA\,d\ell)$ is the number of charge carriers in the cylinder (all of which pass through the surface labeled S in time dt), the magnitude of the charge dQ that passes through the surface in time dt is

$$dQ = nA\,d\ell\,|q| = nAv_d\,dt\,|q|$$

From Eq. 24-1, $I = dQ/dt$, so that

Current in terms of drift speed

$$I = nAv_d|q| \qquad (24\text{-}2)$$

Thus the current is proportional to the drift speed.

EXAMPLE 24-1

Number density and drift speed in a copper wire. *(a)* Determine the number density n of carriers in a copper wire assuming there is one carrier (electron) per copper atom. *(b)* The maximum recommended current in a 14-gauge copper wire (radius = 0.81 mm, $A = 2.1 \times 10^{-6}\ \text{m}^2$) used in household circuits is 15 A. Use your answer to part *(a)* to determine the drift speed of the carriers in such a case.

Solution. *(a)* With one free electron per atom, the number density of carriers is the same as the number density of atoms. Thus $n = N_A\rho_m/M$ where N_A is Avogadro's number, ρ_m is the mass density of copper ($=8.95 \times 10^3\ \text{kg/m}^3$), and M is the atomic mass of copper ($=63.5$ g/mol):

$$n = \frac{(6.02 \times 10^{23}\ \text{mol}^{-1})(8.95 \times 10^6\ \text{g/m}^3)}{63.5\ \text{g/mol}}$$
$$= 8.48 \times 10^{28}\ \text{carriers/m}^3$$

(b) Solving Eq. 24-2 for v_d and using $|q| = e$, we have

$$v_d = \frac{I}{nAe}$$
$$= \frac{15\ \text{A}}{(8.48 \times 10^{28}\ \text{carriers/m}^3)(2.1 \times 10^{-6}\ \text{m}^2)(1.6 \times 10^{-19}\ \text{C})}$$
$$= 5.3 \times 10^{-4}\ \text{m/s} \approx 2\ \text{m/h}$$

The drift speed is literally slower than a snail's pace.

SELF-TEST 24-1. Determine the drift speed of the carriers in a 14-gauge copper wire carrying a current of 5.0 A. ***ANSWER:*** 1.8×10^{-4} m/s.

Electric Current Density

Electric current I characterizes the flow of charge through the entire cross section of a conductor. To describe the flow of charge at points within a conductor, we use the ***current density*** $\mathbf{J}$, which is a vector quantity. If the current density is uniform, the magnitude J of the current density is the current I divided by the cross-sectional area A of the wire:

$$J = \frac{I}{A} \qquad \text{(uniform } \mathbf{J}) \qquad (24\text{-}3)$$

Substitution of Eq. 24-2 into Eq. 24-3 gives J in terms of the drift speed v_d:

$$J = \frac{nAv_d|q|}{A} = nv_d|q|$$

This result can be expressed as a vector equation using the drift velocity $\mathbf{v}_d$:

Current density

$$\mathbf{J} = nq\mathbf{v}_d \qquad (24\text{-}4)$$

Notice that the absolute value sign has been taken from $|q|$ in Eq. 24-4. Thus the current density points in the direction of $\mathbf{v}_d$ for positive carriers, and it points opposite $\mathbf{v}_d$ for negative carriers. Consequently, the direction of $\mathbf{J}$ coincides with the sense of the current in a wire.

If the conductor contains more than one type of charge carrier, then there is a contribution to $\mathbf{J}$ from each type of carrier. Suppose there are two types of charge carriers, a and b. Then

$$\mathbf{J} = n_a q_a \mathbf{v}_{da} + n_b q_b \mathbf{v}_{db} \qquad (24\text{-}5)$$

where the subscripts a and b designate the quantities for each type of charge carrier.

Equations 24-4 and 24-5 are valid for any sort of current distribution, but Eq. 24-3 applies only when the current density is uniform. If the drift velocity of the carriers varies from point to point within a material, as shown in Fig. 24-3, then the current density varies correspondingly. In this case the current I through a surface can be found from the surface integral of the current density $\mathbf{J}$:

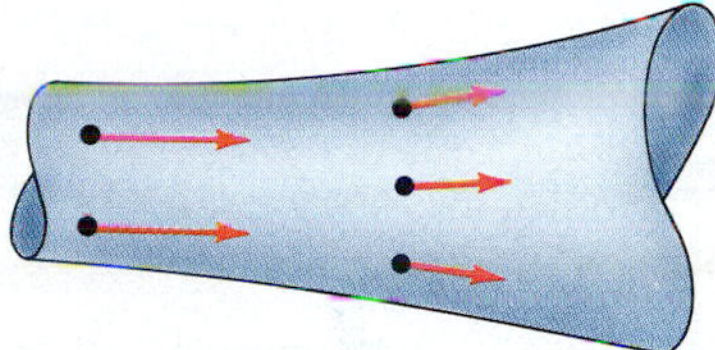

Figure 24-3. The drift velocity is shown as it varies in a conductor with varying cross section. The arrows indicate values of $\mathbf{v}_d$ at a few representative points. If the carriers are positive, these arrows can be used to represent $\mathbf{J}$ also. What if the carriers are negative?

$$I = \int \mathbf{J} \cdot d\mathbf{S}$$

The current through a surface is the flux of the current density for that surface. (See Prob. 2.)

EXAMPLE 24-2

Current density in a 14-gauge copper wire. Assuming $\mathbf{J}$ is uniform, determine the current density in the wire in the previous example.

Solution. For the wire in the example, $I = 15$ A and $A = 2.1 \times 10^{-6}$ m². From Eq. 24-3, the magnitude of the current density is

$$J = \frac{I}{A} = \frac{15\text{ A}}{2.1 \times 10^{-6}\text{ m}^2} = 7.1 \times 10^6 \text{ A/m}^2$$

Since the charge carriers are negative electrons, the direction of $\mathbf{J}$ is opposite the drift velocity.

SELF-TEST 24-2. A wire with cross-sectional area $A = 3.5 \times 10^{-6}$ m² carries a current $I = 10.0$ A. Determine the current density in the wire, assuming it is uniform. ***ANSWER:*** 2.9×10^6 A/m².

24-2 RESISTANCE AND OHM'S LAW

If a potential difference $\mathcal{V}$ is applied across a section of conductor, such as a metal wire, then a current I will be produced in the conductor. The amount of potential difference required to produce a given current depends on a property of the particular section of conductor, a property called its *resistance*. The resistance R is defined as

Definition of resistance

$$R = \frac{\mathscr{V}}{I} \tag{24-6}$$

The ***resistance*** of a section of a conductor is the ratio of the potential difference across the section to the current in the section.

The resistance is appropriately named; for a given section of conductor, it is a measure of that section's opposition to the flow of charge. Since $R = \mathscr{V}/I$ (or $I = \mathscr{V}/R$), a larger resistance for a section of conductor means that a given potential difference will produce a smaller current. Resistance is often added to a circuit to limit or control the current. The circuit element or component used for this purpose is called a *resistor,* and is shown schematically as -/\/\/\/- in circuit diagrams.

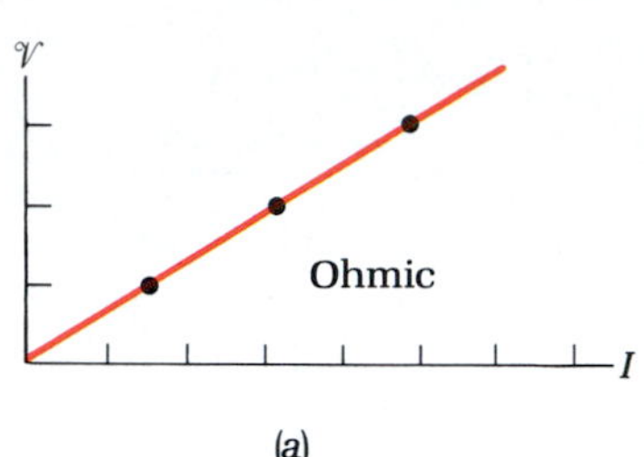

(a)

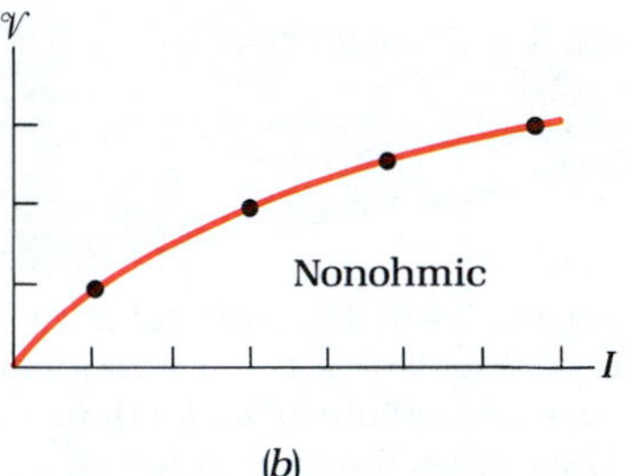

(b)

Figure 24-4. *(a)* The graph of $\mathscr{V}$ versus I for a resistor composed of an ohmic material. The slope of the line gives R. *(b)* The graph of $\mathscr{V}$ versus I for a resistor composed of a nonohmic material. This shows only one of any number of possible relationships.

For many conductors, the current in a section of the conductor is directly proportional to the potential difference across the section, so that the resistance is independent of $\mathscr{V}$ (or I). For example, if the potential difference across the section of conductor is doubled, then the current is doubled. For this case we can write

$$\mathscr{V} = IR \qquad (R \text{ independent of } \mathscr{V} \text{ or } I) \tag{24-7}$$

Equation 24-7 is called ***Ohm's law*** for George Simon Ohm (1787 – 1854), and the SI unit of resistance is the ohm (Ω): $1\ \Omega = 1$ V/A.

The name "Ohm's law" for Eq. 24-7 is possibly misleading because the realm of validity of this equation may be too limited to warrant using the term "law." It is not a fundamental statement about nature as, for example, Coulomb's law is. Rather, the equation is an empirical expression that accurately describes the behavior of many materials over the range of values of $\mathscr{V}$ typically encountered in electric circuits. In these important circumstances, Ohm's law is quite useful.

Materials that "obey" Ohm's law are called *ohmic,* and materials that do not obey Ohm's law are called *nonohmic.* An ohmic conductor is characterized by a single value cf resistance, as shown in Fig. 24-4*a*. That is, its graph of $\mathscr{V}$ versus I is a straight line, so that the slope at each point on the graph is the same. A nonohmic conductor is not characterized by a single value of resistance, and the graph of $\mathscr{V}$ versus I for a nonohmic conductor is not a straight line (Fig. 24-4*b*). A resistor used as a circuit element will usually have its resistance marked on it (often coded in terms of colored bands of rings). Therefore such a resistor is assumed to be ohmic.

Resistivity

The resistance of a section of conductor depends on its size, shape, and composition. Consider a section of conductor with length ℓ and uniform cross-sectional area A, as shown in Fig. 24-5. If a potential difference $\mathscr{V}$ is applied across the section, then a current I exists in it, and the resistance R of the section is $R = \mathscr{V}/I$. Suppose we now apply the same potential difference $\mathscr{V}$ across a section of conductor that has

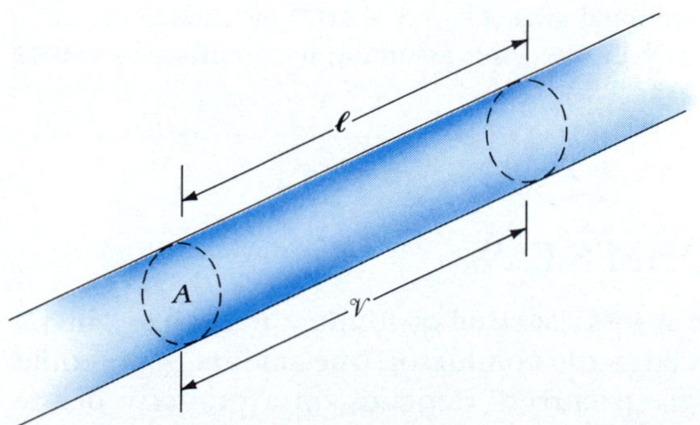

Figure 24-5. The resistance of a section of conductor is directly proportional to ℓ ($R \propto \ell$) and inversely proportional to A ($R \propto 1/A$); $R = \rho\ell/A$.

twice the length of the original one, but is the same in all other respects. We find the current is now half its previous value so that the resistance has doubled. Such measurements indicate that ***the resistance is directly proportional to the length of a section:*** $R \propto \ell$. Suppose we now apply the same potential difference across a section of conductor that has twice the cross-sectional area of the original one, but is the same in all other respects. We find that the current is now twice its previous value so that the resistance is halved. Such measurements indicate that ***the resistance is inversely proportional to the cross-sectional area of a section:*** $R \propto 1/A$.

In addition to this size dependence, the resistance of a section of conductor depends on the material that composes the section. The material-dependence of the resistance is represented by a proportionality factor called the ***resistivity.*** Thus the resistance of a section of conductor of length ℓ, cross-sectional area A, and resistivity ρ is

$$R = \frac{\rho \ell}{A} \tag{24-8}$$

The resistivities of some representative materials are given in Table 24-1.

TABLE 24-1. ***Resistivities at 20° C***

Substance	ρ, Ω·m
Metals	
Silver	1.59×10^{-8}
Copper	1.673×10^{-8}
Gold	2.35×10^{-8}
Aluminum	2.655×10^{-8}
Tungsten	5.65×10^{-8}
Nickel	6.84×10^{-8}
Iron	9.71×10^{-8}
Platinum	10.6×10^{-8}
Lead	20.65×10^{-8}
Semiconductors	
Silicon	4.3×10^{3}
Germanium	0.46
Insulators	
Glass	$10^{10}-10^{14}$
Quartz	7.5×10^{17}
Sulfur	10^{15}
Teflon	10^{13}
Rubber	$10^{13}-10^{16}$
Wood	$10^{8}-10^{11}$
Carbon (diamond)	10^{11}

EXAMPLE 24-3

Resistance of a copper wire. The maximum recommended current in a 12-gauge copper wire (radius = 1.03 mm, $A = 3.31 \times 10^{-6}$ m²) used in household circuits is 20 A. *(a)* What is the resistance of a section of 12-gauge copper wire with length $\ell = 1.00$ m? *(b)* What is the potential difference across this section when the current is 20 A?

Solution. *(a)* Using Eq. 24-8 and the resistivity of copper from Table 24-1, we find

$$R = \frac{\rho \ell}{A} = \frac{(1.673 \times 10^{-8}\ \Omega\cdot\text{m})(1.00\ \text{m})}{3.31 \times 10^{-6}\ \text{m}^2}$$
$$= 5.05 \times 10^{-3}\ \Omega = 5.05\ \text{m}\Omega$$

(b) For a current of 20 A,

$$\mathcal{V} = IR = (20\ \text{A})(5.05\ \text{m}\Omega) = 100\ \text{mV}$$

The resistance of this 1-m length of wire is small, and the potential difference across the section is correspondingly small, even for this relatively large current. The small resistivity of some metals, such as copper (see Table 24-1), accounts for their extensive use in electric circuits.

SELF-TEST 24-3. A wire of length $\ell = 1.70$ m and cross-sectional area $A = 1.80 \times 10^{-6}$ m² has a resistance of $R = 6.46 \times 10^{-2}\ \Omega$. *(a)* Determine the resistivity of the material that composes the wire. *(b)* This material is one of those listed in Table 24-1. Which one?
ANSWERS: *(a)* $6.84 \times 10^{-8}\ \Omega\cdot$m; *(b)* nickel.

Temperature Dependence of the Resistivity of Metals

The resistivity of many pure metals varies almost linearly with temperature over a wide temperature range, as shown in Fig. 24-6 for copper. Since there is usually only a slight amount of curvature in the graph of ρ versus T for metals, we can write

$$\rho \approx \rho_0[1 + \alpha(T - T_0)] \tag{24-9}$$

where ρ is the resistivity at temperature T, ρ_0 is the resistivity at a reference temperature T_0, and α is called the ***temperature coefficient of resistivity.*** That is, for a limited temperature range we can approximate the slightly curved graph of ρ versus T with a straight line. Notice that in Fig. 24-6, the straight line given by Eq. 24-9 is hardly discernible from the curved line at temperatures near T_0. The temperature coefficients of resistivity for some representative materials are given in Table 24-2.

TABLE 24-2. ***Temperature Coefficients of Resistivity at 20° C***

Substance	α, K⁻¹
Metals	
Silver	3.8×10^{-3}
Copper	3.9×10^{-3}
Gold	3.4×10^{-3}
Aluminum	3.9×10^{-3}
Tungsten	4.5×10^{-3}
Nickel	6×10^{-3}
Iron	5×10^{-3}
Platinum	3.93×10^{-3}
Lead	4.3×10^{-3}
Semiconductors	
Silicon	-7.5×10^{-2}
Germanium	-4.8×10^{-2}

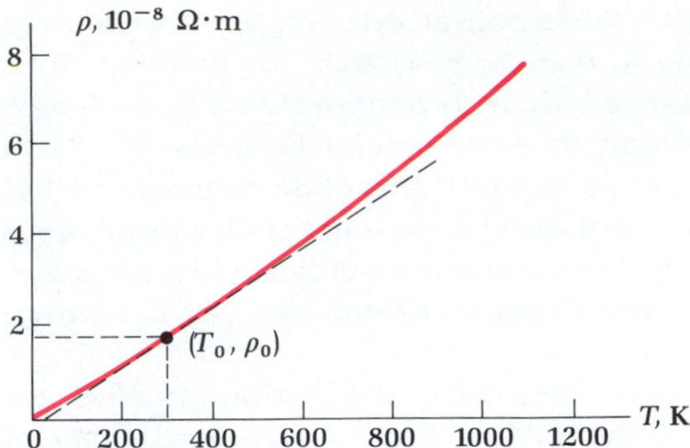

Figure 24-6. Graph of ρ versus T for copper. In the region of the graph near (T_0, ρ_0) the curve is approximately a straight line.

The values of ρ given in Table 24-1 are for 20°C (293 K), so these values may be used as ρ_0 in Eq. 24-9 if 20°C (or 293 K) is used for T_0. The values of α given in Table 24-2 also correspond to 20°C.

The temperature dependence of the resistivity of metals diverges markedly from linearity at low temperatures, below about 20 K. The typical behavior is shown in Fig. 24-7. At these low temperatures the resistance of a metal depends greatly on trace amounts of impurities. Indeed, measurements of resistivity at low temperature are often used to estimate the amount of impurity in a metal.

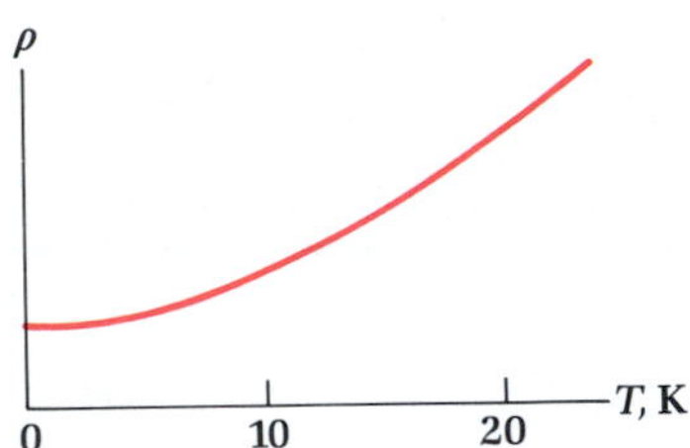

Figure 24-7. Typical low-temperature behavior of the resistivity of a metal.

For some metals a remarkable thing happens as they are cooled to very low temperatures. The resistance totally disappears! This behavior is shown in Fig. 24-8. The phenomenon, called *superconductivity,* was discovered in 1911 by H. Kamerlingh Onnes. Today superconductivity is an active area of research in physics and is increasingly important in engineering.

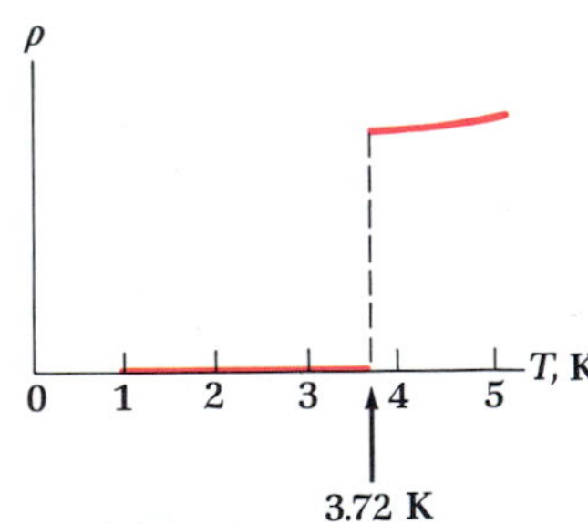

Figure 24-8. The resistivity of tin at low temperatures. The material becomes superconducting below 3.72 K.

EXAMPLE 24-4

Resistance thermometer. A metal wire is often used as a thermometer, and the thermometric property that is measured is the wire's resistance. Platinum is commonly used for this purpose. Suppose we measure the resistance of a platinum resistance thermometer to be 107.9 Ω at a temperature of 20°C. Then, when the thermometer is submersed in a boiling liquid, its measured resistance is 139.3 Ω. Estimate the temperature of the boiling liquid.

Solution. Writing Eq. 24-9 in terms of resistance rather than resistivity (see Ques. 10), we have

$$R \approx R_0[1 + \alpha(T - T_0)]$$

where R (= 139.3 Ω) is the resistance at the unknown temperature T, R_0 (= 107.9 Ω) is the resistance at T_0 (= 293 K), and α (= $3.93 \times 10^{-3}\ \text{K}^{-1}$ from Table 24-2) is the temperature coefficient of resistivity for platinum. Solving for T gives

$$T \approx \frac{R - R_0}{\alpha R_0} + T_0$$

$$\approx \frac{139.3\ \Omega - 107.9\ \Omega}{(3.93 \times 10^{-3}\ \text{K}^{-1})(107.9\ \Omega)} + 293\ \text{K} = 367\ \text{K}$$

Actual resistance thermometers are calibrated at several temperatures, and a regression analysis is used to interpolate between the calibration temperatures.

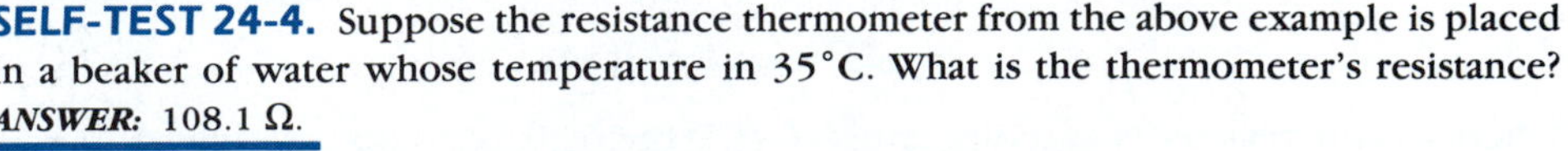

SELF-TEST 24-4. Suppose the resistance thermometer from the above example is placed in a beaker of water whose temperature in 35°C. What is the thermometer's resistance? ***ANSWER:*** 108.1 Ω.

Ohm's Law in Terms of J and E

If an electric field **E** is applied to a conducting material, a current density **J** is produced in the material. The current density at a point in the material depends on the electric field at that point. That dependence is expressed in terms of a property of the material called the ***conductivity*** σ:

$$\mathbf{J} = \sigma \mathbf{E} \qquad (24\text{-}10)$$

(Previously we used the symbol σ to represent a surface charge density. Take care that you do not confuse the two.) From Eq. 24-10, a material that has a larger conductivity than another will have a larger current density for the same electric field. Therefore the conductivity of a material is a measure of the material's ability to allow charge carriers to flow through it.

If the conductivity of the material is independent of **E**, then the material is ohmic and Eq. 24-10 is a vectorial expression of Ohm's law. This can be shown by consideration of the section of conductor in Fig. 24-5. Assume **J** and **E** are uniform within the section so that $I = JA$ and $\mathcal{V} = E\ell$. Substituting for $\mathcal{V}$ and I in Eq. 24-7, $\mathcal{V} = IR$, gives

$$E\ell = JAR$$

Solving for J gives

$$J = \frac{E\ell}{AR} = \frac{\ell}{AR} E$$

Since $J = \sigma E$,

$$\sigma = \frac{\ell}{AR} \qquad (24\text{-}11)$$

For a material in which R is independent of $\mathcal{V}$, σ is also independent of **E** because $\mathcal{V}$ and **E** are directly related ($\mathcal{V} = E\ell$). Thus when σ is independent of **E**, Eq. 24-10 is a form of Ohm's law. Equation 24-10 expresses Ohm's law at a point within the material, whereas Eq. 24-7, $\mathcal{V} = IR$, expresses Ohm's law for a section of material.

From Eq. 24-8, $\rho = RA/\ell$, and from Eq. 24-11, $\sigma = \ell/RA$. Thus

$$\rho = \frac{1}{\sigma}$$

The relation between **E** and **J** can be written in terms of the resistivity:

$$\mathbf{E} = \rho \mathbf{J} \qquad (24\text{-}12)$$

EXAMPLE 24-5

E inside a current-carrying wire. Assuming **J** is uniform, find E inside the copper wire of Examples 24-1 and 24-2.

Solution. Assuming a uniform **J** in Example 24-2, we found that $J = 7.1 \times 10^6 \text{ A/m}^2$. From Table 24-1, $\rho = 1.673 \times 10^{-8}\ \Omega \cdot \text{m}$ for copper:

$$E = \rho J = (1.673 \times 10^{-8}\ \Omega \cdot \text{m})(7.1 \times 10^6 \text{ A/m}^2) = 0.12 \text{ V/m}$$

Notice that this electric field is much smaller than the fields typically found between capacitor plates

SELF-TEST 24-5. What is the conductivity of copper at 20°C? ***ANSWER:*** $5.98 \times 10^7\ \Omega^{-1} \cdot \text{m}^{-1}$.

24-3 RESISTORS IN SERIES AND PARALLEL

Electric circuits usually contain combinations of resistors. The concept of the equivalent resistance of a combination of resistors is useful in finding the current in the various branches of a circuit.

Definition of equivalent resistance

The ***equivalent resistance*** of a combination of resistors is the resistance of a single resistor which, if used in place of the combination, would carry the same current as the combination when it has the same potential difference across it.

In equation form,

$$R_{eq} = \mathcal{V}/I$$

where R_{eq} is the equivalent resistance of the combination, $\mathcal{V}$ is the potential difference across the combination, and I is the current that enters (and leaves) the combination.

Resistors in Series

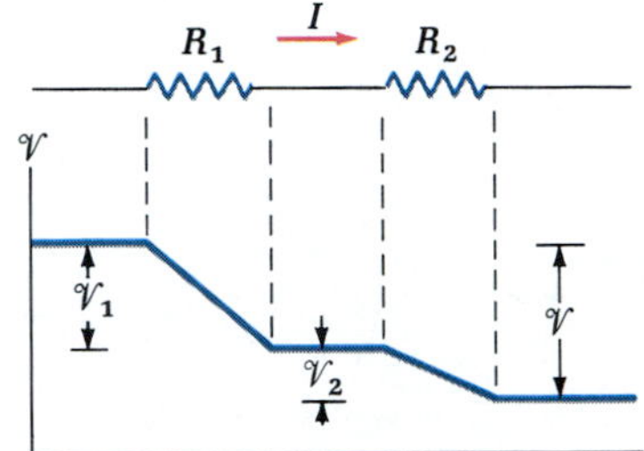

Figure 24-9. Two resistors in series. The potential difference $\mathcal{V}$ across both resistors is equal to the sum of the potential differences across each resistor; $\mathcal{V} = \mathcal{V}_1 + \mathcal{V}_2$. The current I in each resistor is the same.

Figure 24-9 shows two resistors with resistances R_1 and R_2 connected in series. The straight connecting lines indicate wires of negligible resistance. Also shown is the variation of potential along the direction, which corresponds to the sense of the current. Notice that the potential difference $\mathcal{V}$ across the combination of resistors is equal to the sum of the potential differences across each resistor: $\mathcal{V} = \mathcal{V}_1 + \mathcal{V}_2$. Because they are in series, the same current I exists in each resistor so that $\mathcal{V}_1 = IR_1$ and $\mathcal{V}_2 = IR_2$. Therefore

$$\mathcal{V} = IR_1 + IR_2 = I(R_1 + R_2)$$

Thus the equivalent resistance R_{12} is

$$R_{12} = \frac{\mathcal{V}}{I} = R_1 + R_2$$

A single resistor with resistance R_{12} can replace these two and maintain the same external effect. For example, if two series resistors of resistances 4.0 and 2.0 Ω are replaced in a circuit by a single resistor of resistance 6.0 Ω, then the rest of the circuit will be unaffected.

Similarly, for a number of resistors connected in series,

$$R_{eq} = \Sigma R_i \tag{24-13}$$

Resistors in Parallel

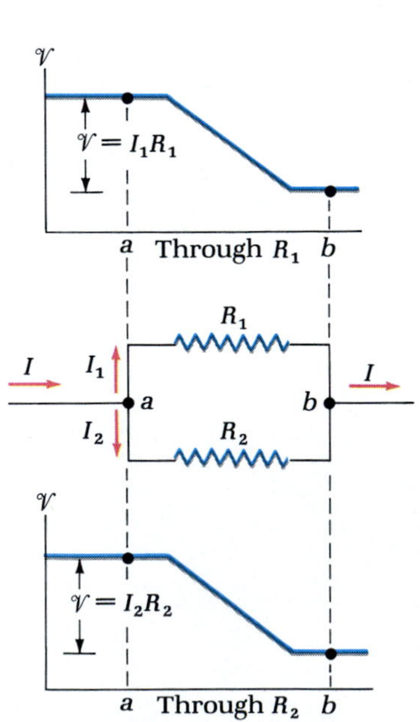

Figure 24-10. Two resistors in parallel. The potential difference $\mathcal{V}$ across each resistor is the same. The current I entering the combination is equal to the sum of the currents in each resistor: $I = I_1 + I_2$.

Figure 24-10 shows two resistors with resistances R_1 and R_2 connected in parallel. Notice that the potential difference $\mathcal{V}$ must be the same for each path, $\mathcal{V}_1 = \mathcal{V}_2 = \mathcal{V}$. Since no charge accumulates at the points a or b, called *branch points*, the current I in the main branch is equal to the sum of the currents I_1 and I_2 in resistors 1 and 2. Or

$$I = I_1 + I_2$$

Since $\mathcal{V} = I_1R_1$ and $\mathcal{V} = I_2R_2$, we have $I_1 = \mathcal{V}/R_1$ and $I_2 = \mathcal{V}/R_2$. Therefore

$$I = \frac{\mathcal{V}}{R_1} + \frac{\mathcal{V}}{R_2} = \mathcal{V}\left(\frac{1}{R_1} + \frac{1}{R_2}\right)$$

The equivalent resistance R_{12} of resistors 1 and 2 in parallel is given by

$$\frac{1}{R_{12}} = \frac{I}{\mathcal{V}} = \frac{1}{R_1} + \frac{1}{R_2}$$

Similarly, for a number of resistors connected in parallel,

$$\frac{1}{R_{eq}} = \sum \frac{1}{R_i} \tag{24-14}$$

Returning to the case of two resistors in parallel and solving for R_{12} gives

$$R_{12} = \frac{R_1 R_2}{R_1 + R_2} \tag{24-15}$$

For example, the equivalent resistance of two parallel resistors of resistances 3.0 and 6.0 Ω is

$$\frac{(3.0\ \Omega)(6.0\ \Omega)}{3.0\ \Omega + 6.0\ \Omega} = \frac{18.0\ \Omega^2}{9.0\ \Omega} = 2.0\ \Omega$$

If these two resistors are replaced by a single resistor of resistance 2.0 Ω, then the rest of the circuit will be unaffected.

Consider the way the current in the main branch is divided into I_1 and I_2 at the branch point a in Fig. 24-10. Since $I_1 = \mathcal{V}/R_1$ and $I_2 = \mathcal{V}/R_2$, the current in each resistor is inversely proportional to its resistance. Suppose a current of 3.0 A is in the main branch in Fig. 24-10 and $R_1 = 3.0\ \Omega$ and $R_2 = 6.0\ \Omega$. Then $R_{12} = 2.0\ \Omega$ and $\mathcal{V} = IR_{12} = (3.0\ \text{A})(2.0\ \Omega) = 6.0\ \text{V}$. Thus $I_1 = 6.0\ \text{V}/3.0\ \Omega = 2.0\ \text{A}$, and $I_2 = 6.0\ \text{V}/6.0\ \Omega = 1.0\ \text{A}$. The larger current is in the resistor with the smaller resistance. That is, most of the current takes the path of least resistance.

EXAMPLE 24-6

A combination of resistors. *(a)* Determine the equivalent resistance of the combination of resistors shown in Fig. 24-11*a*. *(b)* Given that the potential difference across the combination is 36 V, determine the potential difference across each resistor, the total current in the combination, and the current in each resistor.

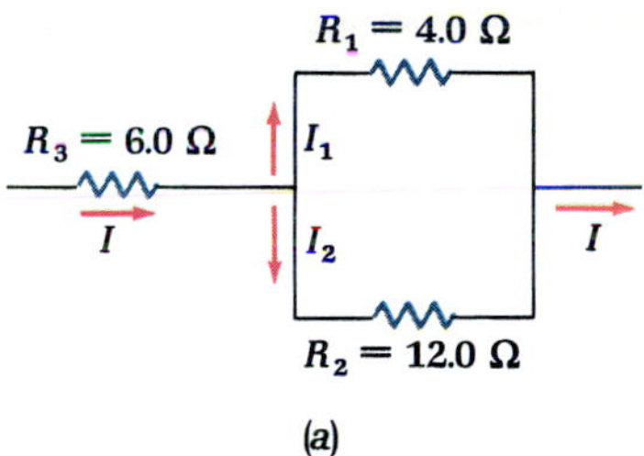

(a)

$R_3 = 6.0\ \Omega$ $R_{12} = 3.0\ \Omega$

I

(b)

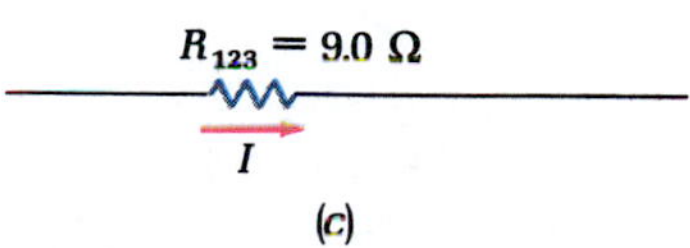

(c)

Figure 24-11. Example 24-6: *(a)* A combination of resistors. *(b)* After replacing R_1 and R_2 with their equivalent R_{12}. *(c)* After replacing R_{12} and R_3 with their equivalent R_{123}.

Solution. *(a)* Using Eq. 24-15 gives the equivalent resistance R_{12} of the parallel combination of resistors 1 and 2 (Fig. 24-11*b*) as

$$R_{12} = \frac{(4.0\ \Omega)(12.0\ \Omega)}{4.0\ \Omega + 12.0\ \Omega} = 3.0\ \Omega$$

This resistance is in series with R_3 so that the equivalent resistance R_{123} of the entire combination is

$$R_{123} = R_{12} + R_3 = 6.0\ \Omega + 3.0\ \Omega = 9.0\ \Omega$$

(b) The current in the equivalent resistance R_{123} is

$$I = \frac{\mathcal{V}}{R_{123}} = \frac{36\ \text{V}}{9.0\ \Omega} = 4.0\ \text{A}$$

This is the current in the main branch of the combination, and it is also the current in resistor 3. Therefore the potential difference $\mathcal{V}_3$ across resistor 3 is

$$\mathcal{V}_3 = IR_3 = (4.0\ \text{A})(6.0\ \Omega) = 24\ \text{V}$$

Since $\mathcal{V} = \mathcal{V}_{12} + \mathcal{V}_3$, where $\mathcal{V}_{12}$ is the potential difference across the parallel resistors 1 and 2, we have

$$\mathcal{V}_{12} = \mathcal{V} - \mathcal{V}_3 = 36\ \text{V} - 24\ \text{V} = 12\ \text{V}$$

[Alternatively, $\mathcal{V}_{12} = IR_{12} = (4.0\ \text{A})(3.0\ \Omega) = 12\ \text{V}$.] The currents I_1 and I_2 in resistors 1 and 2 are

$$I_1 = \frac{\mathcal{V}_{12}}{R_1} = \frac{12\ \text{V}}{4.0\ \Omega} = 3.0\ \text{A}$$

and

$$I_2 = \frac{\mathcal{V}_{12}}{R_2} = \frac{12\ \text{V}}{12.0\ \Omega} = 1.0\ \text{A}$$

SELF-TEST 24-6. Resistors 1 and 2 have resistances of $R_1 = 8.0\ \Omega$ and $R_2 = 12.0\ \Omega$. What is the equivalent resistance of 1 and 2 when they are connected in *(a)* series and *(b)* parallel? ***ANSWERS:*** *(a)* 20.0 Ω; *(b)* 4.8 Ω.

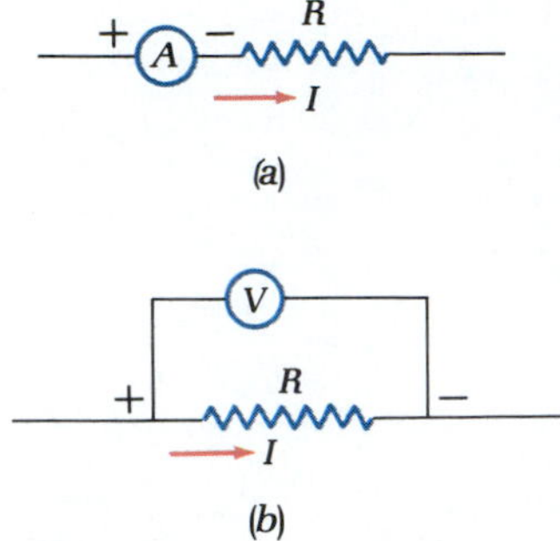

Figure 24-12. *(a)* To measure the current in a circuit element (a resistor in this case), an ammeter is placed in series with the element. *(b)* To measure the potential difference across an element, a voltmeter is placed in parallel with (across) the element.

24-4 AMMETERS AND VOLTMETERS

For an element in an electric circuit, such as a resistor, the two quantities of continuing interest are the current I in the element and the potential difference $\mathcal{V}$ across it. It is instructive to consider the way these two quantities are measured. As the names imply, current is measured with an ammeter and potential difference is measured with a voltmeter.

An *ammeter* measures the current in itself. To measure the current in a circuit element, an ammeter must be placed in series with the element so that the current in the element is the same as the current in the ammeter (Fig. 24-12*a*).

The principal component of an ammeter is some type of current-detecting device, such as a ***galvanometer*** (described in Example 26-6). In addition to a current-detecting device, which includes a dial or digital display, an ammeter usually contains several resistors that are used to change the scale of the meter. The scale of the ammeter is selected with a switch that places one of these resistors in parallel with the current-detecting device. Problem 10 gives further details about the operation of an ammeter.

A *voltmeter* measures the potential difference across itself. To measure the potential difference across an element, a voltmeter is placed in parallel with the element (that is, *across* the element) so that the potential difference across the element is the same as the potential difference across the voltmeter (Fig. 24-12*b*).

As with an ammeter, the principal component of a voltmeter is some type of current-detecting device. Again, as with an ammeter, a voltmeter usually contains several resistors that are used to set the scale of the meter. The scale of the voltmeter is selected with a switch that places one of these resistors in series with the current-detecting device. Thus the major distinction between an ammeter and a voltmeter is that an ammeter has a resistor (usually with a small resistance) in parallel with its current-detecting device, and a voltmeter has a resistor (usually with a large resistance) in series with its current-detecting device. Indeed, an ammeter and a voltmeter are often incorporated in the same instrument so that the same current-detecting device is used for both. Problem 11 gives further details about the operation of a voltmeter.

An important feature of any measuring device is that its interjection into a system should not significantly alter the quantity to be measured. For an ammeter to have a negligible effect on the current it measures, the resistance of the ammeter must be insignificant compared with the rest of the resistance of the branch into which it is placed. In this way its presence will not significantly alter the current in the branch (see Exercise 32). An ideal ammeter is one whose resistance is zero.

For a voltmeter to have a negligible effect on the potential difference it is used to measure, its resistance must be much greater than the resistance of the element it is placed across. If the resistance of the voltmeter in Fig. 24-12*b* is much greater than R, then a very small current will exist in the voltmeter and the voltmeter will have a negligible effect on the potential difference across the resistor (see Exercise 33). An ideal voltmeter is one whose resistance is infinite.

24-5 DRUDE MODEL OF A METAL

A model proposed by P. K. Drude (1863–1906) in 1900 provides some insight into the nature of electric conduction in metals. The principal characteristic of conduction in metals that a successful microscopic model must yield is Ohm's law. We now show that when Newton's second law is used to describe the average motion of the charge carriers in Drude's model of a metal, the metal obeys Ohm's law.

First, consider an important implication of our discussion so far. From combining Eq. 24-4, $\mathbf{J} = nq\mathbf{v}_d$, and Eq. 24-10, $\mathbf{J} = \sigma\mathbf{E}$, we see that $\mathbf{v}_d$ and $\mathbf{E}$ are proportional. That is, an applied field causes the carriers to move with a constant average velocity $\mathbf{v}_d$. But if the field were to furnish the only force on a carrier, then the carrier's

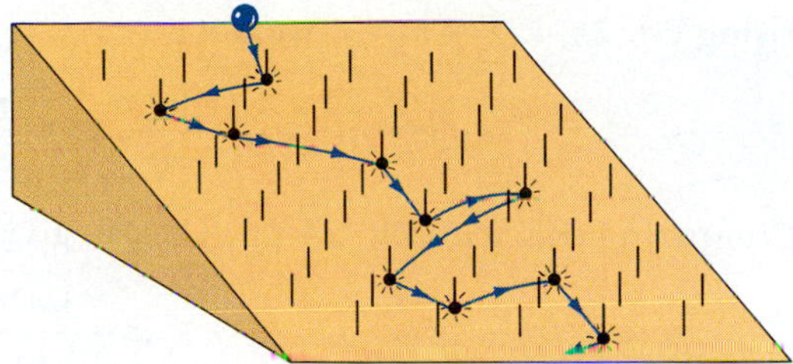

Figure 24-13. A marble rolling down a pegboard. Over a long time interval the motion is characterized by a constant drift velocity down the board.

acceleration would be constant, not its velocity. This means that when a carrier moves through the metal, there must be other forces on it. Indeed, since the average velocity is constant, the sum of all forces on a carrier must be zero on the average.

The situation is analogous to a marble rolling down a pegboard (Fig. 24-13). When first released, a marble will accelerate down the board because of the unbalanced component of the gravitational force down the board. As it collides with the pegs, we notice that its motion, averaged over many collisions, is characterized by a constant average velocity down the board. Averaged over many collisions, the force on a marble due to the pegs is equal and opposite the component of the gravitational force down the board. In Drude's model, a free electron is similar to a marble, the lattice ions are similar to the pegs, and the applied field is similar to the component of the gravitational field down the board.

Drude assumed that the free electrons in metals are the valence electrons that are weakly bound to the atoms when the atoms are isolated (not part of a metal). When the atoms are side by side in a solid, these electrons are free to move through the material. Thus the number density n of carriers is the product of a small integer and the number density of atoms in the material.

The average velocity $\langle \mathbf{v} \rangle$ of the free electrons is the carrier drift velocity $\mathbf{v}_d$: $\langle \mathbf{v} \rangle = \mathbf{v}_d$. Since $\mathbf{E} \propto \mathbf{v}_d$, the average velocity is zero when there is no applied field. That is, if $\mathbf{E} = 0$, then $\langle v_x \rangle = \langle v_y \rangle = \langle v_z \rangle = 0$. The behavior of the free electrons is similar to the behavior of the molecules of a gas in that their velocities are randomly directed.

The force by the applied field $\mathbf{E}$ on an electron is $\mathbf{F} = -e\mathbf{E}$. Since this is the only force on a free electron between collisions, Newton's second law, $\Sigma \mathbf{F} = m\mathbf{a}$, gives the acceleration between collisions as $\mathbf{a} = -e\mathbf{E}/m$. If we align our x axis along $\mathbf{E}$, then the x component of a free electron's velocity at a time t after a collision is

$$v_x = v_{x0} + a_x t = v_{x0} - \left(\frac{eE}{m}\right) t$$

where v_{x0} is the x component of the electron's velocity immediately after the collision. On the average, we have

$$\langle v_x \rangle = \langle v_{x0} \rangle - \left(\frac{eE}{m}\right) \tau$$

where τ characterizes the time interval between collisions. This time interval is often called the *mean free time,* or the *relaxation time.*

For a rather sizable current, the drift speed v_d is only about 10^{-4} m/s (Example 24-1), whereas the average speed $\langle v \rangle$ of the free electrons is about 10^6 m/s. Since $\langle v \rangle$ is a factor of about 10^{10} larger than v_d, the contribution to the motion of the free electrons due to the applied field is negligible at the microscopic level. Therefore it is valid to assume that the velocity of an electron immediately after each collision is randomly directed relative to $\mathbf{E}$, or $\langle v_{x0} \rangle = 0$. Thus

$$\langle v_x \rangle = \frac{-eE\tau}{m}$$

The drift velocity is $\mathbf{v}_d = \langle \mathbf{v} \rangle = \langle v_x \rangle \mathbf{i}$, so that

$$\mathbf{v}_d = \frac{-eE\tau}{m}\mathbf{i}$$

Using Eq. 24-4, $\mathbf{J} = nq\mathbf{v}_d$, we obtain

$$\mathbf{J} = n(-e)\left(\frac{-eE\tau}{m}\mathbf{i}\right) = \frac{ne^2\tau}{m}E\mathbf{i} = \frac{ne^2\tau}{m}\mathbf{E}$$

Comparing this with Eq. 24-10, $\mathbf{J} = \sigma\mathbf{E}$, gives

$$\sigma = \frac{ne^2\tau}{m} \qquad (24\text{-}16)$$

Since $\rho = 1/\sigma$, we also have

$$\rho = \frac{m}{ne^2\tau} \qquad (24\text{-}17)$$

Eureka! The model gives Ohm's law.

If $\sigma = ne^2\tau/m$ is independent of $\mathbf{E}$, then the model yields Ohm's law. The factors n, e, and m are plainly independent of $\mathbf{E}$, but what about τ? We expect τ to depend on $\langle v \rangle$, and $\mathbf{E}$ may change $\langle v \rangle$ by no more than v_d. But we noted earlier that $\langle v \rangle \approx 10^6$ m/s and $v_d \approx 10^{-4}$ m/s. Because of this vast difference, we expect τ to be essentially independent of $\mathbf{E}$, so Drude's model does give Ohm's law.

We can express σ and ρ in terms of the average speed $\langle v \rangle$ by introducing the average distance an electron travels between collisions, the *mean free path* λ:

$$\lambda = \langle v \rangle \tau$$

Substitution for τ into Eqs. 24-16 and 24-17 gives

$$\sigma = \frac{ne^2\lambda}{m\langle v \rangle} \qquad \text{and} \qquad \rho = \frac{m\langle v \rangle}{ne^2\lambda}$$

The Drude model is obviously rather crude. For example, the interaction between an electron and an ion involves the long-range Coulomb force, and in the model this interaction is treated as an abrupt collision. Then, during the time interval between the collisions, the force on an electron due to the ions is neglected. It may seem tempting to try to improve the model by making it more realistic, but there is little to be gained from such an improvement for a very fundamental reason. *It is inappropriate to apply Newton's second law to the motion of an electron in a metal.* An electron in a metal must be described according to *quantum mechanics.* We must learn more about the fundamental laws of physics before we can significantly improve our understanding of conduction in metals.

EXAMPLE 24-7

Mean free time in copper. *(a)* Estimate the mean free time in copper at 20°C (=293 K), assuming one free electron per copper atom. *(b)* Assuming that the average speed of the free electrons is about 10^6 m/s, estimate the mean free path in copper.

Solution. *(a)* Solving Eq. 24-17 for τ gives

$$\tau = \frac{m}{ne^2\rho}$$

From Example 24-1, $n = 8.48 \times 10^{28}$ carriers/m³, so that

$$\tau = \frac{9.11 \times 10^{-31}\ \text{kg}}{(8.48 \times 10^{28}\ \text{carriers/m}^3)(1.60 \times 10^{-19}\ \text{C})^2(1.673 \times 10^{-8}\ \Omega\cdot\text{m})}$$
$$= 2.51 \times 10^{-14}\ \text{s}$$

(b) The mean free path is

$$\lambda = \langle v \rangle \tau = (10^6\ \text{m/s})(2.51 \times 10^{-14}\ \text{s}) = 10^{-8}\ \text{m}$$

This is roughly 100 times the distance between nearest-neighbor atoms in copper.

SELF-TEST 24-7. Estimate the mean free time in copper at $T = 600$ K. ***ANSWER:*** 1.23×10^{-14} s.

24-6 CONDUCTION IN SEMICONDUCTORS

We have divided materials into two classes according to their electric conductivity: conductors and insulators. There is a third class, called ***semiconductors,*** whose conductivity is intermediate between conductors and insulators. Semiconductors play a central role in modern technology. These are the materials that are used in such devices as diodes, transistors, and integrated circuits. The Drude model showed that σ is directly proportional to the number density n of carriers: $\sigma = (e^2\tau/m)n$. The carrier density n is the key factor in controlling the conductivity of a semiconductor.

Conduction in Pure Semiconductors

Semiconductors consist of some of the elements in the middle columns in the periodic table, and an element often used is silicon. To be specific, we discuss silicon as our representative semiconductor.

Silicon ($Z = 14$) has a valence of 4, and when the atoms are together in the form of a solid, each has four nearest-neighbors. In Fig. 24-14, we show a two-dimensional portrayal of the three-dimensional silicon lattice; each atom is shown as an ion core of charge $+4e$ accompanied by four valence electrons. In pure silicon at room temperature, almost all of these valence electrons are bound to their respective ion cores, but thermal fluctuations in energy cause some of them to be free. That is, a small fraction of the silicon atoms are *thermally ionized.* (At room temperature, about one silicon atom out of every 5×10^{12} is ionized, a very small fraction indeed.) The electrons that are released by this thermal ionization become negative charge carriers.

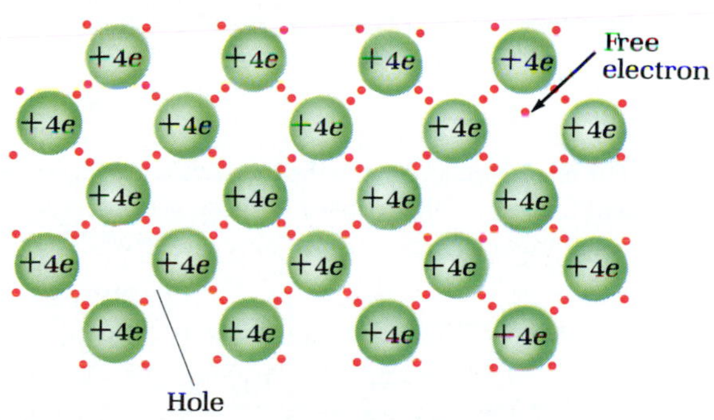

Figure 24-14. A two-dimensional representation of the three-dimensional silicon lattice showing a free electron and a hole.

A free electron is a negative charge carrier in a semiconductor.

In addition to free electrons, semiconductors possess positive charge carriers. Figure 24-14 shows that if an electron is released from its site in a solid, it leaves behind a position at which an electron is missing. This "lack-of-an-electron" is

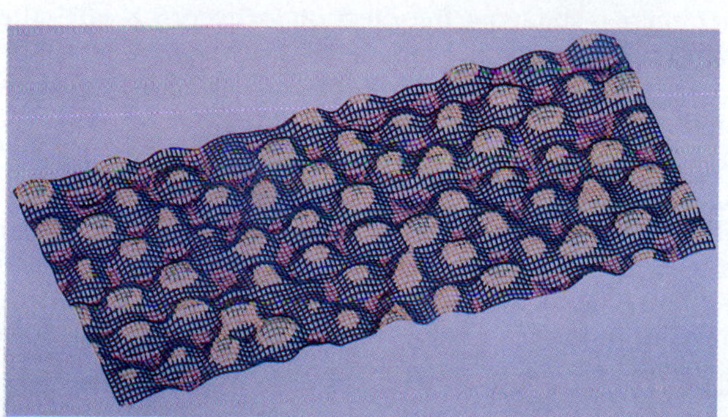

Atoms on the surface of silicon are revealed with an atomic force microscope.

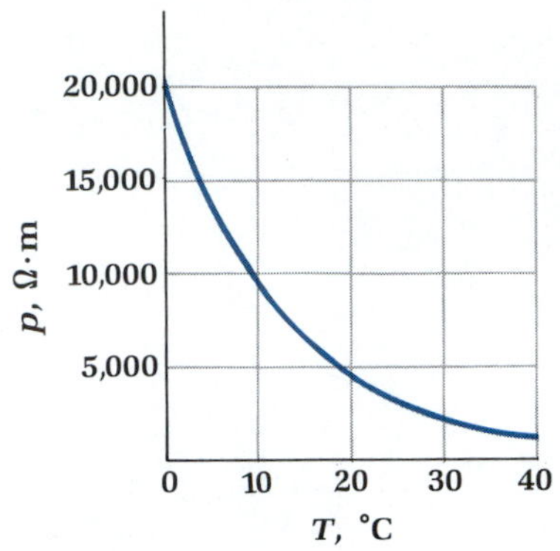

Figure 24-15. Temperature dependence of the resistivity of pure silicon near room temperature.

called a ***hole,*** and an applied electric field can cause a hole to move through the solid in the direction of the field. Thus a hole is a positive charge carrier.

A hole moving through a semiconductor because of an applied electric field is similar to a bubble moving upward from the bottom of a swimming pool because of the gravitational field of the earth; the bubble rises through the water, but water is actually falling as the bubble rises. Instead of describing the water as falling, we find it more convenient to describe the bubble as rising. Our attention is fixed on the lack of water (the bubble) rather than on the water.

The Drude model can be used to describe qualitatively the resistivity of semiconductors. Consider the temperature dependence of the resistivity of silicon shown in Fig. 24-15. From Eq. 24-17, the Drude model predicts that the resistivity ρ is inversely proportional to the number density n of carriers: $\rho \propto 1/n$. Since the existence of the carriers is a consequence of thermal ionization, n increases sharply with temperature. Thus ρ decreases with temperature, in qualitative agreement with Fig. 24-15.

EXAMPLE 24-8

Number density of free electrons in silicon. *(a)* Using the Drude model and the rough assumption that the mean free time τ for the carriers in silicon is the same as we found for those in copper in Example 24-7 ($\tau = 2.44 \times 10^{-14}$ s), estimate the number density of free electrons in silicon at room temperature from the resistivity given in Table 24-1. *(b)* From the answer to part *(a)*, estimate the fraction of ionized silicon atoms.

Solution. *(a)* Solving Eq. 24-17 for n gives

$$n = \frac{m}{\tau e^2 \rho}$$

These carriers consist of both holes and electrons. Thus $n = n_e + n_h$, where the subscripts e and h refer to electrons and holes, respectively. Let us assume that $n_e = n_h = \frac{1}{2}n$. Using the resistivity of silicon from Table 24-1, we have

$$n_e = \frac{\frac{1}{2}(9.11 \times 10^{-31}\ \text{kg})}{(2.44 \times 10^{-14}\ \text{s})(1.60 \times 10^{-19}\ \text{C})^2(4.3 \times 10^3\ \Omega\cdot\text{m})}$$
$$= 1.7 \times 10^{17}\ \text{m}^{-3}$$

(b) To find the fraction of ionized silicon atoms, we must first find the number density n_{Si} of silicon atoms in solid silicon:

$$n_{Si} = \frac{N_A \rho_m}{M}$$

where N_A is Avogadro's number, ρ_m is the mass density of silicon ($= 2.33 \times 10^3$ kg/m^3), and M is the atomic mass of silicon ($= 28.1$ g/mol). This gives $n_{Si} = 4.99 \times 10^{28}\ \text{m}^{-3}$. The fraction of ionized silicon atoms is

$$\frac{n_e}{n_{Si}} = \frac{1.7 \times 10^{17}\ \text{m}^{-3}}{4.99 \times 10^{28}\ \text{m}^{-3}} = 3.4 \times 10^{-12}$$

As we stated earlier, there is about one free electron for every 5×10^{12} silicon atoms, so that the fraction of ionized atoms is actually about 2×10^{-13}. Our calculation gives an answer that is too large by a factor of about 20. In view of our crude approximations, this is about the accuracy we should expect.

SELF-TEST 24-8. Use the same technique as in the above example to estimate the fraction of ionized atoms in germanium at room temperature. The mass density and the atomic mass of germanium are 5.32×10^3 kg/m^3 and 72.6 g/mol, respectively. (Because of the crudeness of the approximations, your answer will be too large by a factor of about 40. See Exercise 35.) ***ANSWER:*** 3.6×10^{-8}.

n-Type and p-Type Semiconductors

From the previous example, the number density of carriers is very low in silicon compared with that of metals. There is about one carrier per atom in a metal, but only about one carrier for every 10^{12} atoms in silicon at room temperature. Consequently, the room temperature resistivity of silicon is about a factor of 10^{11} higher than most metals. However, the number density of carriers in a semiconductor can be increased greatly by the introduction of certain impurities.

Consider the effect of incorporating phosphorus into silicon. Phosphorus ($Z = 15$) has five valence electrons, one more than silicon. If a small amount of phosphorus is introduced into solid silicon, then here and there a site normally occupied by a silicon ion core (charge $+4e$) will be occupied by a phosphorus ion core (charge $+5e$), as shown in Fig. 24-16*a*. Four of the five valence electrons of the phosphorus atom are bound to the phosphorus ion core (in the arrangement normally occupied by electrons around a silicon ion core), and, at room temperature, the one remaining electron is nearly always free. Thus essentially every phosphorus impurity atom contributes (or donates) a negative charge carrier to the material in the form of a free electron. When an electron is called "free," we mean that it is capable of moving through the solid and that it does not remain associated with any particular ion core. An impurity atom that donates a free electron to the host material, as phosphorus does in silicon, is called a *donor*.

A donor atom

Now consider the effect of aluminum in silicon. Aluminum ($Z = 13$) has three valence electrons, one fewer than silicon. If a small amount of aluminum is introduced into solid silicon, then here and there a site normally occupied by a silicon ion core will be occupied by an aluminum ion core (charge $+3e$), as shown in Fig. 24-16*b*. At room temperature, the hole formed by the one fewer valence electron from aluminum is nearly always free, so that nearly all the aluminum ions have four electrons around them, as do almost all the silicon ions. The aluminum atom "accepts" an electron from the host material and the material now contains a charge carrier in the form of a hole. This hole is capable of moving through the solid, and it does not remain associated with any particular ion core. Essentially every aluminum impurity atom contributes a hole to the material. An impurity atom that accepts an electron from the host material, as aluminum does in silicon, is called an *acceptor*.

An acceptor atom

When an impurity is purposely incorporated into an otherwise pure material, we say that the material is *doped*. For example, when phosphorus is introduced into silicon, the resulting material is phosphorus-doped silicon. As we have seen, phosphorus-doped silicon contains extra negative charge carriers. Such a material is called an *n*-type semiconductor; the "*n*" refers to the *n*egative charge of the carriers (electrons). Aluminum-doped silicon is a *p*-type semiconductor; the "*p*" refers to the *p*ositive charge of the carriers (holes).

An *n*-type semiconductor contains negative electrons as carriers, and a *p*-type semiconductor contains positive holes as carriers.

To qualify as an *n*-type semiconductor, the donor concentration should be high enough so that the number density of free electrons introduced by the donor is

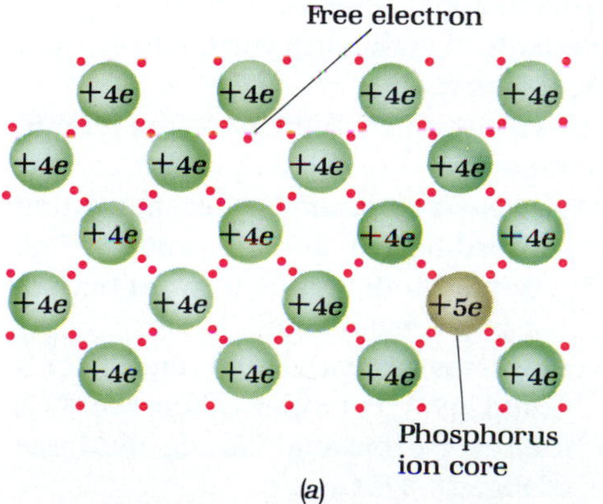

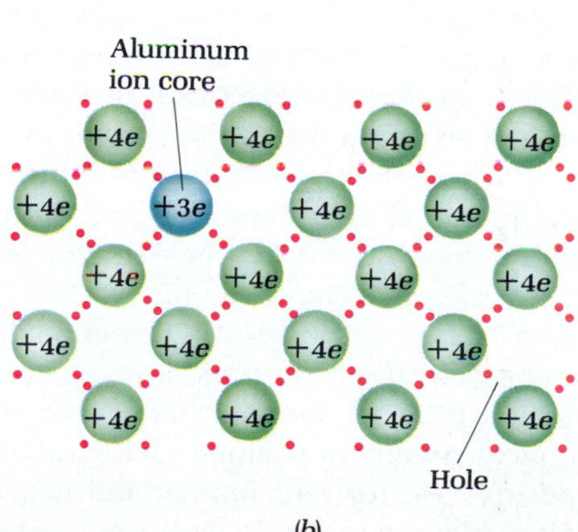

Figure 24-16. *(a)* Phosphorus-doped silicon. At room temperature, four of the five valence electrons of the phosphorus atom remain bound to the ion core while the fifth electron is nearly always free. The material contains free electrons as negative carriers so that it is an *n*-type semiconductor. *(b)* Aluminum-doped silicon. Aluminum has three valence electrons, but at room temperature, four electrons are nearly always bound to the aluminum ion core, and the hole that is formed is free. These holes are positive charge carriers so that the material is a *p*-type semiconductor.

much higher than the number density of carriers in the pure material. In this way the free electrons greatly outnumber the holes in *n*-type material. We say that electrons are the majority carriers in *n*-type material.

Similarly, for a material to be *p*-type, its acceptor concentration should be high enough such that the number density of holes is much higher than that of the carriers in the pure material. Then holes greatly outnumber free electrons and the material is *p*-type. Holes are the majority carriers in *p*-type material.

We previously saw that the carrier concentration in pure silicon at room temperature corresponds to a fraction of about 2×10^{-13} ionized silicon atoms. Therefore, to be an *n*-type (or *p*-type) semiconductor, the fractional concentration of donor (or acceptor) atoms in silicon has only to be much greater than 2×10^{-13}. Typical doping levels in silicon run around 10^{-10} to 10^{-6}. Consequently, the number density of carriers in a doped semiconductor depends on the impurity concentration, and the sign of the charge carriers depends on the type of the impurity. Both are controlled by the manufacturer during the production of the material.

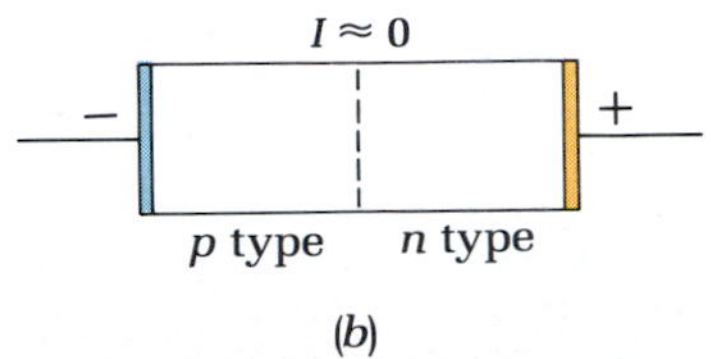

(a)

$I \approx 0$

−
+
p type
n type

(b)

Figure 24-17. *(a)* A forward-biased *pn*-junction diode. *(b)* A reverse-biased *pn*-junction diode. Practically no carriers are available to maintain a current from right to left.

The pn-Junction Diode

Many of the electronic devices that are so useful in circuits are junction devices. These have one or more junctions, which abruptly separate *n*-type material from *p*-type material. The simplest of these devices is the *pn*-junction diode. A ***diode*** is a circuit element that readily allows charge carriers to flow in one direction, but not in the other.

The operation of a *pn*-junction diode can be understood from the following simplified discussion. Suppose a potential difference is applied to a diode in the sense shown in Fig. 24-17*a*. The contact on the *p* side of the diode is at a higher potential than the contact on the *n* side. This tends to produce a current from left to right. For such a current to exist in the *p* region, holes must flow from left to right, and for such a current to exist in the *n* region, electrons must flow from right to left. This readily occurs because the electrons can combine with the holes at the junction, mutually eliminating one another, which allows the carriers to flow continuously.

Now suppose a potential difference is applied to a diode in the sense shown in Fig. 24-17*b*. The contact on the *p* side of the diode is at a lower potential than the contact on the *n* side, which tends to produce a current from right to left. For such a current to exist in the *p* region, holes must flow from right to left, and for such a current to exist in the *n* region, electrons must flow from left to right. For this to occur continuously, some mechanism for producing free electrons and holes at the junction is required. No such mechanism exists in an ordinary diode. (In photodiodes such a mechanism does exist, but that is a different case altogether.)

Another way a current from right to left could exist in the diode is for holes in the *n* region to flow from right to left and electrons in the *p* region to flow from left to right. Then the electrons and holes could combine at the junction. However, there are practically no holes in the *n* region and practically no electrons in the *p* region. Therefore there are practically no carriers available to maintain a current from right to left, so only a negligible current exists in this case.

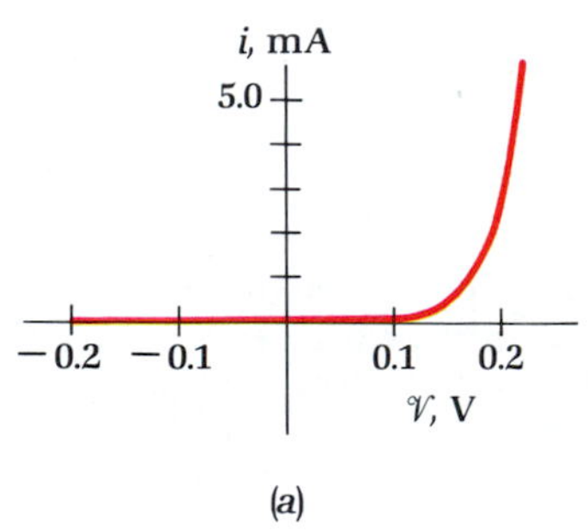

(a)

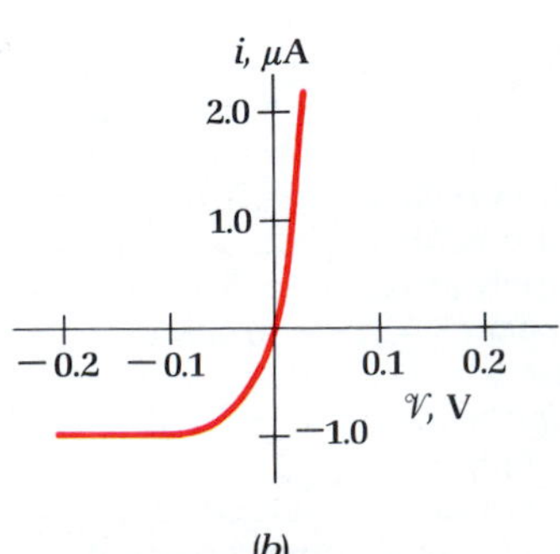

(b)

Figure 24-18. *I-𝒱* curve for a typical *pn*-junction diode. *(a)* Current axis marked in milliamps (10^{-3} A). *(b)* Same data as in *(a)*, with current axis marked in microamps (10^{-6} A).

When a potential difference is applied across a circuit element, we often refer to the potential difference as a *bias*. If the element has directional properties, as a diode does, then it may be *forward-biased* or *reverse-biased* (sometimes called *back-biased*). The diode in Fig. 24-17*a* is forward-biased, and the diode in Fig. 24-17*b* is reverse-biased. Current readily flows in a diode that is forward-biased, and practically no current flows in a diode that is reverse-biased.

Figure 24-18 shows an *I𝒱* curve (current versus potential difference) for a typical *pn*-junction diode (see Exercises 36 and 37). Potential difference $\mathcal{V}$ is positive for forward biasing and negative for reverse biasing. Notice the large differences in the scale on the current axes of Fig. 24-18*a* and *b*.

SUMMARY

Section 24-1. The Flow of Charge

Charge flow through a circuit element is characterized by the electric current:

$$I = \frac{dQ}{dt} \tag{24-1}$$

The sense of the current corresponds to the direction of the drift velocity $\mathbf{v}_d$ of positive charge carriers. The current density $\mathbf{J}$ describes the flow of charge at a point within a conducting medium:

$$\mathbf{J} = nq\mathbf{v}_d \tag{24-4}$$

If $\mathbf{J}$ is uniform, then its magnitude is

$$J = \frac{I}{A} \qquad \text{(uniform } \mathbf{J}\text{)} \tag{24-3}$$

Section 24-2. Resistance and Ohm's Law

Ohm's law is

$$\mathcal{V} = IR \tag{24-7}$$

where the resistance R is independent of $\mathcal{V}$ (or I). When this is the case, the conductor is called ohmic. For a conductor of length ℓ and cross-sectional area A,

$$R = \frac{\rho\ell}{A} \tag{24-8}$$

where ρ is the resistivity of the material. The relation between $\mathbf{E}$ and $\mathbf{J}$ is

$$\mathbf{J} = \sigma\mathbf{E} \tag{24-10}$$

where $\sigma = 1/\rho$.

Section 24-3. Resistors in Series and Parallel

For resistors connected in series,

$$R_{eq} = \Sigma R_i \tag{24-13}$$

and for resistors connected in parallel,

$$\frac{1}{R_{eq}} = \sum \frac{1}{R_i} \tag{24-14}$$

Section 24-4. Ammeters and Voltmeters

To measure the current in a circuit element, an ammeter is placed in series with the element. To measure the potential difference across a circuit element, a voltmeter is placed in parallel with the element.

Section 24-5. Drude Model of a Metal

The Drude model of a conductor yields Ohm's law with

$$\sigma = \frac{ne^2\tau}{m} \tag{24-16}$$

where τ is the mean time between collisions.

Section 24-6. Conduction in Semiconductors

For a pure semiconductor, σ is much smaller than for metals because n is much lower than in metals. Doping a semiconductor with certain impurities can greatly increase n and σ. A semiconductor doped with a donor impurity is *n*-type (carriers are free electrons), and a semiconductor doped with an acceptor impurity is *p*-type (carriers are holes).

QUESTIONS

1. Wires *a* and *b* are made of the same materials and carry the same current, but the radius of wire *a* is half that of *b*. What is the ratio of the carrier drift speeds in the wires? What is the ratio of the current densities in the wires?

2. When you turn on a light switch, the light comes on almost instantaneously, even though the drift speed of the carriers in the wire is only about 10^{-4} m/s. Draw an analogy between this effect and the prompt flow of water from the open end of a long garden hose (initially full of water) after the water flow is turned on at the valve.

3. In developing Eq. 24-2, $I = nAv_d|q|$, we considered only one type of charge carrier. Write a similar equation for the case where there are two types of carriers, *a* and *b*. Is your equation valid if the carriers have opposite sign?

4. Is $\mathbf{J}$ parallel to $\mathbf{v}_d$ for positive carriers? Is $\mathbf{J}$ parallel to $\mathbf{v}_d$ for negative carriers?

5. Is the expression $\mathcal{V} = IR$ for a conductor inconsistent with our requirement in electrostatics that $\mathbf{E} = 0$ inside a conductor?

6. Suppose you measure the current I in a resistor when the potential difference across it is $\mathcal{V}$, and you determine that $R = \mathcal{V}/I = 100\ \Omega$. If the resistor is known to be nonohmic, is it proper to speak of it as a 100-Ω resistor? Explain.

7. Can you apply the expression $\mathcal{V} = IR$ to circuit elements that are not ohmic? If so, what can be said of R?

8. Discuss the distinction between resistance and resistivity. Which is proper, to speak of the resistance of a penny or the resistivity of a penny? Which is proper, to speak of the resistance of copper or the resistivity of copper?

9. If the resistance of a copper wire is 3 Ω at a temperature of 300 K, then what is a reasonable estimate of its resistance at 100 K?

10. Justify the equation used for the resistance as a function of temperature in Example 24-4. What assumption(s) must you make?

11. Explain the difficulty in determining whether a conductor that is well insulated thermally (such as the filament of a light bulb) is ohmic.

12. In the analogy between marbles on a pegboard and charge carriers in a conductor, increasing the tilt of the board corresponds to changing what physical quantity for the conductor? To what does adding more marbles correspond?

13. In some crystalline solids, $\mathbf{J}$ is not parallel to $\mathbf{E}$ except when $\mathbf{E}$ is along particular crystalline directions, called the *principal axes*. In this case, the conductivity σ in the expres-

sion $\mathbf{J} = \sigma\mathbf{E}$ is no longer a scalar; σ is then a *tensor* quantity. Discuss a marble-and-pegboard analogy to conduction in such a solid. Let the pegs have an elliptical cross section, as shown in Fig. 24-19. Suppose the down-slope direction makes an angle of, say 30°, with respect to the major axes of the elliptical cross sections. Will the direction of the average velocity of the marbles be parallel to the unbalanced component of the gravitational force on them? Find two directions along the pegboard which correspond to principal axes. Which of these principal-axis directions offers the larger resistance to the flow of marbles?

Figure 24-19. Question 13: Top view of a section of pegboard that has pegs with elliptical cross sections.

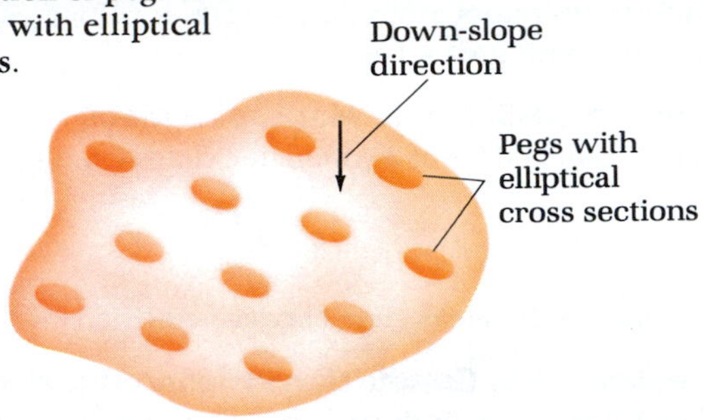

14. In the Drude model, how do we know that $\langle v_{x0} \rangle = 0$ when the applied electric field is zero? How do we justify assuming $\langle v_{x0} \rangle = 0$ when there is an applied electric field $\mathbf{E} = E\mathbf{i}$? Is $\langle v_0 \rangle = 0$ in either case? (Recall that the subscript 0 refers to the instant immediately following a collision.)

15. In applying the Drude-model expression for conductivity $(\sigma = ne^2\tau/m)$, which factor mainly accounts for the fact that conductivity in a pure semiconductor is much less than it is in typical metals?

16. A potential difference exists across three series resistors of resistances R_1, R_2, and R_3. If $R_1 < R_2 < R_3$, then which resistor has the largest potential difference across it? Which resistor has the smallest potential difference across it?

17. A current exists in three parallel resistors of resistances R_1, R_2, and R_3. If $R_1 < R_2 < R_3$, then which resistor carries the largest current? Which carries the smallest current?

18. Compare the expressions for the equivalent resistance of resistors in series and parallel with the equivalent capacitance of capacitors in series and parallel. Explain why the expressions do not compare directly, series to series and parallel to parallel.

19. Why is it preferable to have an ammeter whose resistance is low? Low compared with what? Why is it preferable to have a voltmeter whose resistance is high? High compared with what?

20. It is often difficult to measure a very large resistance by directly measuring the current in it with an ammeter and the potential difference across it with a voltmeter. Why?

21. In an integrated circuit, all the elements—diodes, resistors, capacitors, and others—are made of silicon and the insulator is silicon dioxide (SiO_2). Explain how each element might be constructed from these materials.

22. Complete the following table:

Symbol	Represents	Type	SI Unit
I			
$\mathbf{J}$			
$\mathbf{v}_d$		Vector	
R			Ω
ρ			
σ			
α			
n	Carrier number density		
τ			
$\langle v \rangle$			m/s
λ			

EXERCISES

Section 24-1. The Flow of Charge

1. A steady current of 2.5 A exists in a metal wire. *(a)* What amount of charge passes through a cross section of the wire in 5.0 min? *(b)* How many electrons pass through this surface in 5.0 min?

2. Suppose an electron beam in a television picture tube has a flow of 8.1×10^{15} electrons per second. What is the beam current in A?

3. The amount of charge that passes through a cross section of a wire is given by $Q(t) = (6.5\ \text{C/s}^2)t^2 + 3.5\ \text{C}$ for t varying between 0.0 and 8.0 s. *(a)* What expression gives the current $I(t)$ in this time interval? *(b)* What is the current at the instant $t = 3.4$ s?

4. A 10-gauge aluminum wire (radius 1.30 mm) carries a current of 20 A. Assuming three free electrons per aluminum atom ($\rho_m = 2.7 \times 10^3\ \text{kg/m}^3$, $M = 27.0$ g/mol), determine the drift speed of the electrons.

5. Consider a salt solution that carries, in a long insulating tube of inner radius 12 mm, a current of 0.86 A along the tube axis. The carriers in the solution are singly charged positive ions and negative ions with equal number densities: $n_+ = n_- = 5.7 \times 10^{25}$ ions per cubic meter. Assume that the drift speed of the positive ions is 3 times that of the negative ions. *(a)* Determine the contribution to the current from each type of ion. *(b)* Assuming $\mathbf{J}$ is uniform, determine the contribution to J from each type of ion. *(c)* Determine the drift speed of each type of ion.

6. A 16-gauge copper wire (radius 0.65 mm) is connected in series with an 18-gauge copper wire (radius 0.51 mm) so that they both carry the same current of 2.4 A. Assuming a uniform current density, *(a)* what is the current density in each wire and *(b)* what is the carrier drift speed in each wire?

Section 24-2. Resistance and Ohm's Law

7. Show the ohm can be written $kg \cdot m^2/(s^3 \cdot A^2)$.

8. A potential difference of 12 V produces a current of 16 mA in a resistor. *(a)* What is the resistance of the resistor? *(b)* Assuming the resistor is ohmic, what is the current when the potential difference is 24 V?

9. A 1.0-m length of wire has a resistance of 14 Ω, and the potential difference between its ends is 4.1 V. *(a)* What is the current in the wire? *(b)* Assuming **E** is uniform, determine E inside the wire.

10. Consider the following data taken on two wires:

Wire *a*		Wire *b*	
I, A	$\mathscr{V}$, V	I, A	$\mathscr{V}$, V
0.0	0.0	0.0	0.0
1.0	1.8	1.0	0.6
2.0	2.9	2.0	1.2
3.0	3.7	3.0	1.8
4.0	4.2	4.0	2.4

(a) Make graphs of the data and determine if either wire is ohmic. *(b)* If either wire is ohmic, what is its resistance? *(c)* If either wire is nonohmic, make for it a graph of R versus I. (Notice that we cannot speak of *the* resistance of a nonohmic conductor.)

11. *(a)* What is the resistance of a 1.0-m length of 10-gauge (radius 1.3 mm) copper wire at 20°C? (See Table 24-1.) *(b)* What is the resistance of a 1.0-m length of 10-gauge aluminum wire at 20°C?

12. A 1.3-m length of 8-gauge (radius 1.64 mm) metal wire has a resistance of 8.6 mΩ at 20°C. The metal is one of those in Table 24-1. Which one is it?

13. A very thin layer of metal can be formed by vaporizing the metal that deposits by condensation on a cool substrate of glass. The thickness of the layer can be determined by a resistance measurement. Suppose the resistance of a rectangular deposit of aluminum (31 mm by 5.6 mm) at 20°C is 19 Ω when the potential difference is applied along the layer's larger dimension. *(a)* What is the thickness of this layer? *(b)* What is the resistance of a layer of the same material with the same thickness but with its two rectangular dimensions each twice the value of the previous layer? Can you make a general statement about the resistance of a layer of material of given thickness with rectangular dimensions of a given ratio?

14. A tungsten wire with a circular cross section has a length of 58 mm and a resistance of 6.2 mΩ at 20°C. What is the radius of the wire?

15. The current-carrying rail of an electric train has a cross section of 5.3×10^{-3} m² and is made of steel with a resistivity of about $3 \times 10^{-7}\ \Omega \cdot m$. What is the resistance of 1 km of track?

16. A 12-gauge copper wire is inside the wall of a house so that its length cannot be directly measured. If the wire has a resistance of 23 mΩ at 20°C, what is its length?

17. A metal wire has a room-temperature resistance of 40 mΩ. The wire is melted and all the metal is used to reform it into a wire 3 times its original length. What is the room-temperature resistance of the new wire?

18. Suppose we wish to compare aluminum and copper as materials for a cable that will carry 100 A and have a resistance per unit length of 80×10^{-6} Ω/m. For the two materials, compare *(a)* the cross-sectional areas, *(b)* the current densities, *(c)* the mass per unit length of cable. The mass densities of aluminum and copper are 2.7×10^3 kg/m³ and 8.9×10^3 kg/m³, respectively. *(d)* What is the potential difference across a 1-km length of such a cable?

19. What is the resistivity of iron at 40°C? (See Tables 24-1 and 24-2.)

20. The resistance of a silver wire is 46 mΩ at 20°C. What is its resistance at 45°C?

21. The resistance of a copper wire at 20°C is 130 mΩ. At what temperature is its resistance 110 mΩ?

22. At 20°C the resistance of a copper wire is 8.2 mΩ and the resistance of a gold wire is 7.8 mΩ. Determine the temperaure at which the wires have the same resistance.

23. A 1.8-m length of metal wire with cross-sectional area 2.3×10^{-6} m² carries a current of 65 mA when the potential difference between its ends is 11 mV. *(a)* Assuming **E** is uniform, what is E in the wire? *(b)* What is J in the wire? *(c)* What is σ for the metal? *(d)* What is ρ for the metal?

Section 24-3. Resistors in Series and Parallel

24. Three resistors of resistances 23, 45, and 31 Ω are connected in series. *(a)* What is their equivalent resistance? *(b)* If the potential difference across the combination is 36 V, what is the current in each resistor and what is the potential difference across each resistor?

25. Three resistors of resistances 16, 25, and 31 Ω are connected in parallel. *(a)* What is their equivalent resistance? *(b)* If the potential difference across the combination is 14 V, what is the current in each resistor and what is the potential difference across each resistor?

26. *(a)* If $R_1 = 12\ \Omega$, $R_2 = 21\ \Omega$, and $R_3 = 28\ \Omega$ in Fig. 24-20, what is the equivalent resistance of the combination? *(b)* If a 32-V potential difference is applied across the combination, what is the potential difference across each resistor and what is the current in each resistor?

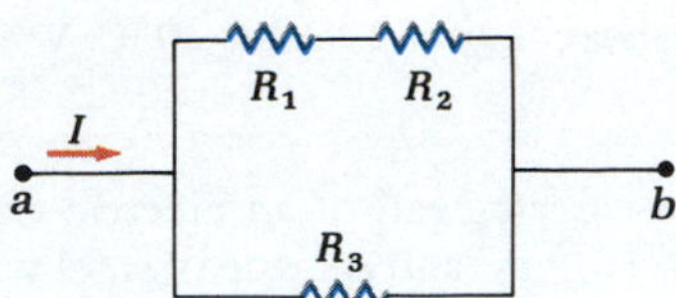

Figure 24-20. Exercise 26.

27. Show that when a number of resistors are connected in parallel, the equivalent resistance of the combination is always less than the resistance of the resistor with the smallest resistance.

28. In Fig. 24-21, $I = 10.0$ A, $R_1 = R_2 = 4.0\ \Omega$, and $R_3 = 8.0\ \Omega$. What is the current in each resistor?

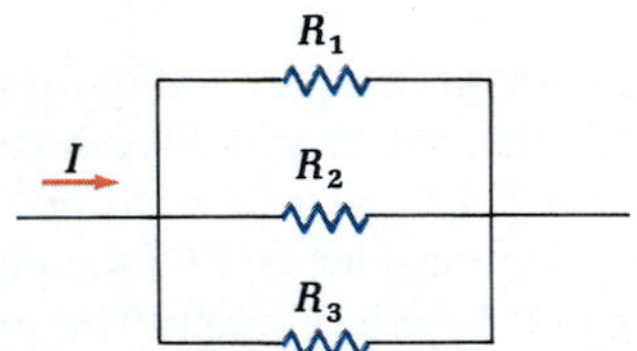

Figure 24-21. Exercise 28.

29. In the arrangement shown in Fig. 24-22, $R_1 = 5.0\ \Omega$, $R_2 = 3.0\ \Omega$, and $R_3 = 4.0\ \Omega$. *(a)* What is the potential difference between points *c* and *d*? *(b)* What is the equivalent resistance from *a* to *b*? *(c)* If the potential difference between *a* and *b* is 8.0 V, what is the potential difference across each resistor and what is the current in each resistor?

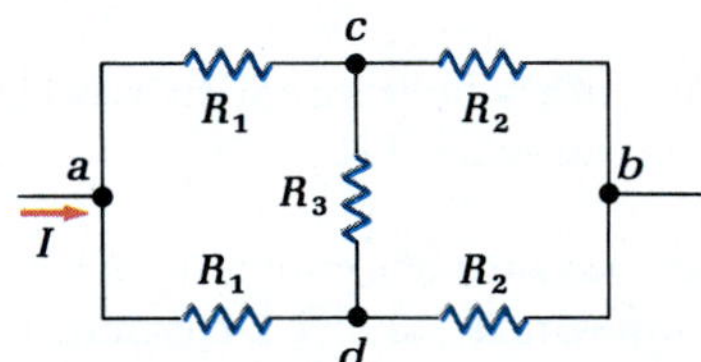

Figure 24-22. Exercise 29.

30. The equivalent resistance of two resistors in parallel may be written

$$R_{eq} = \frac{R_1 R_2}{R_2 + R_1}$$

(a) Show that the equivalent resistance of three resistors connected in parallel may be written

$$R_{eq} = \frac{R_1 R_2 R_3}{R_2 R_3 + R_1 R_3 + R_1 R_2}$$

(b) Write a similar expression for the equivalent resistance of four resistors connected in parallel.

31. For some purposes it is useful to define the *conductance S* of a resistor, $S = 1/R$. *(a)* Show that the equivalent conductance S_{eq} of a series combination of resistors is

$$\frac{1}{S_{eq}} = \sum \frac{1}{S_i}$$

(b) Show that the equivalent conductance of a parallel combination of resistors is

$$S_{eq} = \sum S_i$$

[*Note:* The SI unit of conductance is called the siemens (S); $1\ \text{S} = 1\ \Omega^{-1}$. The unit is named for E. W. von Siemens (1816–1892).]

Section 24-4. Ammeters and Voltmeters

32. To measure the resistance R of a resistor, an ammeter of resistance R_A is placed in series with the resistor and voltmeter is placed across the series combination, as shown in Fig. 24-23. *(a)* Show that the resistance R in terms of the measured readings on the ammeter I_{meas} and voltmeter $\mathcal{V}_{meas}$ is given by

$$R = \frac{\mathcal{V}_{meas}}{I_{meas}} - R_A$$

Notice that if $\mathcal{V}_{meas}/I_{meas} \gg R_A$, then $R \approx \mathcal{V}_{meas}/I_{meas}$. *(b)* Given that $\mathcal{V}_{meas} = 23$ V, $I_{meas} = 62$ mA, and $R_A = 14\ \Omega$, determine R.

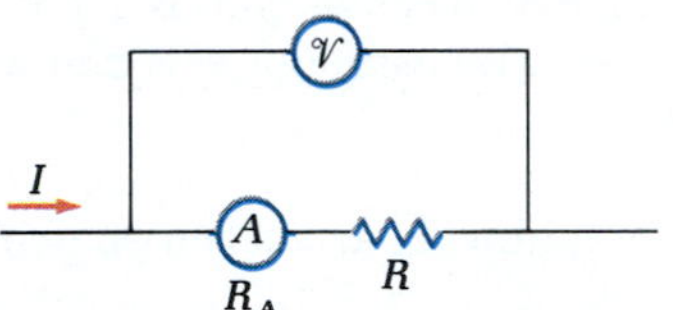

Figure 24-23. Exercise 32.

33. To measure the resistance R of a resistor, a voltmeter of resistance $R_{\mathcal{V}}$ is placed across the resistor and an ammeter is placed in series with the combination, as shown in Fig. 24-24. *(a)* Show that the resistance R in terms of the measured readings on the ammeter I_{meas} and voltmeter $\mathcal{V}_{meas}$ is given by

$$R = \frac{\mathcal{V}_{meas}}{I_{meas} - (\mathcal{V}_{meas}/R_{\mathcal{V}})}$$

(b) Show that if $R_{\mathcal{V}} \gg \mathcal{V}_{meas}/I_{meas}$, then $R \approx \mathcal{V}_{meas}/I_{meas}$. *(c)* Given that $I_{meas} = 16$ mA, $\mathcal{V}_{meas} = 43$ V, and $R_{\mathcal{V}} = 62$ MΩ, determine R.

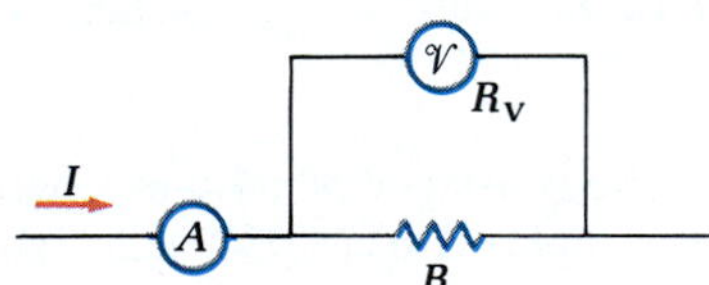

Figure 24-24. Exercise 33.

Section 24-5. Drude Model of a Metal

34. *(a)* Estimate the mean free time in aluminum at 20°C, assuming three free electrons per aluminum atom. *(b)* Using $\langle v \rangle \approx 10^6$ m/s, estimate the mean free path λ in aluminum.

Section 24-6. Conduction in Semiconductors

35. The number density of free electrons n_e and the number density of holes n_h in pure germanium (a semiconductor) at 20°C is about 2×10^{19} carriers per cubic meter for each type of carrier. Assuming free electrons and holes have the same mass and mean free time, estimate the mean free time for these carriers in germanium.

36. Using the data of Fig. 24-18, find the resistance of that *pn*-junction diode for *(a)* a forward-bias potential difference of 0.2 V and *(b)* a reverse-bias potential difference of 0.2 V. *(c)* What is the ratio of the reverse-bias resistance to the forward-bias resistance at 0.2 V?

37. The theoretical expression that quite accurately gives the current i in a *pn*-junction diode in terms of the potential difference $\mathcal{V}$ across it is

$$i = I_0 (e^{e\mathcal{V}/kT} - 1)$$

where I_0 is a parameter that depends on the particular diode, k is the Boltzmann constant, T is the temperature (in kelvins), and e is the magnitude of the electronic charge. [We use i rather than I for this current because it may be negative. Indeed, the current is negative for negative $\mathcal{V}$ (reverse bias).] Using $I_0 = 1.0 \times 10^{-6}$ A, evaluate i at 20°C (293 K) for *(a)* $\mathcal{V} = 0.20$ V and *(b)* $\mathcal{V} = -0.20$ V. *(c)* Determine the forward-bias and reverse-bias resistances at 0.20 V, and find the ratio of the reverse-biased resistance to the forward-bias resistance.

Additional Exercises

38. Suppose a metal has 7.5×10^{28} carriers/m³ (electrons) and the mean free time for the metal is 1.7×10^{-14} s. Use the Drude model to determine the metal's resistivity.

39. A section of a circuit has two resistors of resistance 120 Ω and 370 Ω in series with each other and a 630-Ω resistor in parallel with these two. The potential difference across the section is 29 V. *(a)* What is the section's equivalent resistance? *(b)* What is the current in the section?

40. A section of a circuit has two resistors of resistance 720 Ω and 510 Ω in parallel with each other and a 460-Ω resistor in series with these two. The potential difference across the section is 41 V. *(a)* What is the section's equivalent resistance? *(b)* What is the current in the section?

41. A 16-gauge zinc wire ($A = 1.33 \times 10^{-6}$ m²) carries a current of 5.0 A. Determine *(a)* the number density of carriers and *(b)* the carrier drift speed. (*Hint:* Use the periodic table in Appendix P for information about zinc.)

42. Determine the drift speed of the carriers in a chromium wire that has a current density of 2.5×10^6 A/m². (*Hint:* Use the periodic table in Appendix P for information about chromium.)

PROBLEMS

1. ***Dependence of resistance on length and area.*** *(a)* Show that the length dependence in the expression $R = \rho\ell/A$ is consistent with the expression for the equivalent resistance of two resistors in series, $R_{eq} = R_1 + R_2$, by considering a wire of length ℓ and cross-sectional area A as two wires, each with cross-sectional area A and of lengths ℓ_1 and ℓ_2, connected in series: $\ell = \ell_1 + \ell_2$. *(b)* Show that the area dependence in the expression $R = \rho\ell/A$ is consistent with the expression for the equivalent resistance of two resistors in parallel, $1/R_{eq} = 1/R_1 + 1/R_2$, by considering a wire of length ℓ and cross-sectional area A as two wires of length ℓ and cross-sectional areas A_1 and A_2 connected in parallel: $A = A_1 + A_2$.

2. ***A radial current density.*** Figure 24-25 shows a thin-walled metal tube (inner radius b and outer radius c), with a thin metal disk (thickness d) covering its end. A thin metal wire (radius a) coaxial with the tube is connected to the center

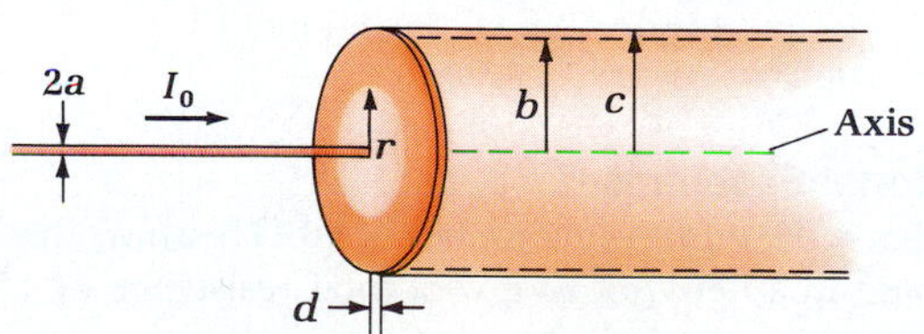

Figure 24-25. Problem 2.

of the disk and a steady current I_0 exists in the wire, disk, and tube as shown. *(a)* In the region of the disk well away from the connection to the wire and the tube, find an expression for the magnitude J of the current density as a function of the perpendicular distance r from the axis. *(b)* In this same region of the disk, find an expression for the current density **J** in terms of **i** and **j** unit vectors. Use a coordinate system with origin on the axis and xy plane parallel to the plane sides of the disk.

3. ***Temperature dependence of resistance of a semiconductor.*** An expression, based on theory, that accurately gives the temperature dependence of the resistivity ρ of a pure semiconductor is

$$\rho = \rho_1 e^{E_g/(2kT)}$$

where ρ_1 is a parameter that depends weakly on the temperature T (compared with the exponential), E_g is the so-called energy gap of the material, and k is the Boltzmann constant. For pure silicon near room temperature, $\rho_1 = 1.4 \times 10^{-6}\ \Omega\cdot$m and $E_g = 1.1$ eV. *(a)* Using the above expression, evaluate ρ at 5° intervals between 0 and 40°C for silicon and make a graph of resistivity versus temperature. Sketch a curve through the data points and compare your graph with Fig. 24-15. *(b)* Develop an expression for α to be used in Eq. 24-9. Evaluate your expression for α for silicon at 20°C and compare your result with the value in Table 24-2. *(c)* Using the linear approximation of Eq. 24-9 with your values of α and ρ_0 for silicon at 20°C, evaluate ρ at 5° intervals from 0 to 40°C and plot the data on the graph you constructed in part *(a)*. Sketch the straight line through the points.

4. ***Constructing a temperature-independent resistance.*** Suppose we wish to construct a composite resistor whose resistance is nearly temperature-independent by connecting two resistors a and b in series. The two resistors have the same cross-sectional area and are assumed to have the same temperature. Show that the effective temperature coefficient of resistivity of the composite resistor is zero at temperature T_0 if the ratio of the lengths of the resistors is $\ell_a/\ell_b = -\rho_{0b}\alpha_b/\rho_{0a}\alpha_a$, where the subscripts a and b refer to the quantities for each resistor. (Why must α_a and α_b be of opposite sign?)

5. ***Resistance of a tapering conductor.*** The radius of the section of conductor shown in Fig. 24-26 varies linearly from a to b along its axis. Show that the resistance of the section for current along the axis is $R = \rho\ell/\pi ab$, where ρ is the resistivity

Figure 24-26. Problem 5.

of the material. (The taper is small, $b - a \ll \ell$, so that you may assume **J** is uniform across any cross section.)

6. ***Dynamic resistance.*** A concept that is sometimes useful when dealing with nonohmic circuit elements is the *dynamic resistance*, $R_{dyn} = d\mathcal{V}/di$. Determine the expression for the dynamic resistance of a *pn*-junction diode from the relation for $i(\mathcal{V})$ given in Exercise 37.

7. ***Resistance of an infinite network.*** Show that the equivalent resistance of the infinite network of resistors shown in Fig. 24-27 is $R_{eq} = (1 + \sqrt{3})R$.

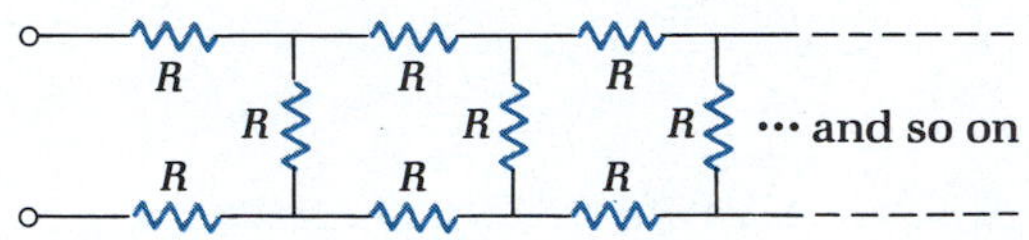

Figure 24-27. Problem 7.

8. ***Resistors along the edges of a cube.*** Twelve resistors of equal resistance R are arranged along the edges of a cube, as shown in Fig. 24-28. *(a)* Show that the equivalent resistance R_{ab} between corner a and corner b is $5R/6$. *(b)* Show that the equivalent resistance R_{ac} between corner a and corner c is $7R/12$. (*Hint:* Use symmetry considerations to reduce the array to resistors in series and parallel.)

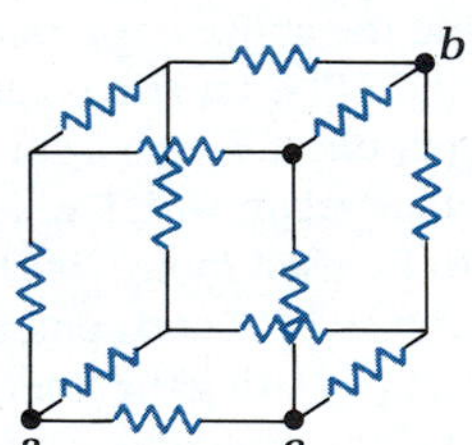

Figure 24-28. Problem 8.

9. ***Resistance of a capacitor.*** *(a)* Show that the resistance R_c of a parallel-plate capacitor of capacitance C is $R_c = \rho\epsilon_0\kappa/C$, where ρ and κ are the resistivity and dielectric constant of the insulator between the plates. *(b)* Determine R_c for the case where $\rho = 2 \times 10^{13}\ \Omega\cdot$m, $\kappa = 5$, and $C = 1\ \mu$F.

10. ***Constructing an ammeter.*** Suppose we wish to construct an ammeter that reads a current I_m when the pointer on the current-detecting device is at its full-scale position (Fig. 24-29). We do this by placing a resistor (called a ***shunt resistor***) of resistance R_{para} in parallel with the current-detecting device so that part of the current in the ammeter is "shunted" through the parallel resistor. The current-detecting device has a resistance R_d, and a current I_{fs} in the device causes a full-scale deflection of its pointer. *(a)* Show that the resistance of the shunt resistor must be

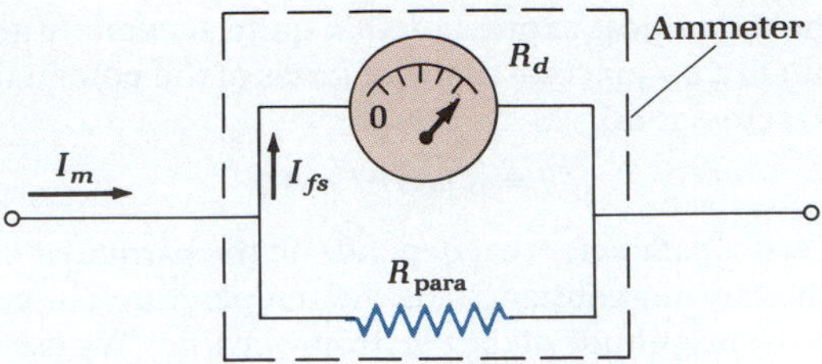

Figure 24-29. Problem 10.

$$R_{para} = \frac{R_d I_{fs}}{I_m - I_{fs}}$$

Notice that we must have $I_m > I_{fs}$. *(b)* Evaluate R_{para} for the case where $I_m = 10$ mA, $R_d = 23\ \Omega$, and $I_{fs} = 63\ \mu$A. What is the resistance of the ammeter for this case?

11. ***Constructing a voltmeter.*** Suppose we wish to construct a voltmeter that reads a potential difference $\mathcal{V}_m$ when the pointer on the current-detecting device is at its full-scale position (Fig. 24-30). We do this by placing a resistor of resistance R_{series} in series with the current-detecting device so that part of the potential difference across the voltmeter is across the series resistor. The current-detecting device has a resistance R_d, and a current I_{fs} in the device causes a full-scale deflection of its pointer. *(a)* Show that the resistance of the series resistor must be

$$R_{series} = \frac{\mathcal{V}_m}{I_{fs}} - R_d$$

Notice that we must have $(\mathcal{V}_m/I_{fs}) > R_d$. *(b)* Evaluate R_{series} for the case where $\mathcal{V}_m = 10$ V, $R_d = 23\ \Omega$, and $I_{fs} = 63\ \mu$A. What is the resistance of the voltmeter for this case?

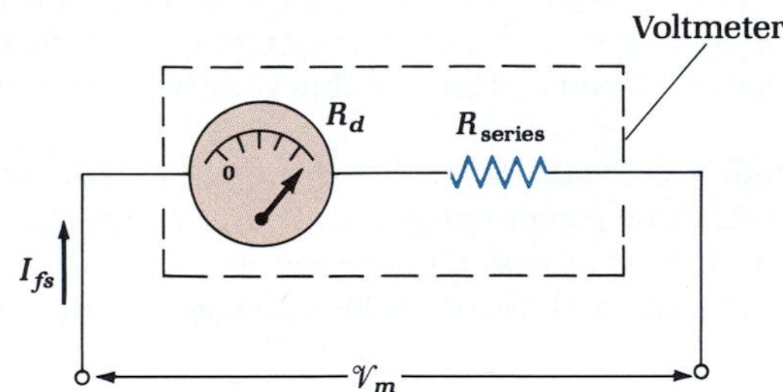

Figure 24-30. Problem 11.

Additional Problems

12. What is the minimum number of 10-Ω resistors that can be assembled in a network to give a total resistance of 15 Ω? Is there any maximum limit?

13. Consider an infinite square grid of resistors, all of resistance R. What is the effective resistance between *any* two adjacent vertices?

14. Consider a fine wire of length ℓ, radius r, and resistivity ρ connected to ground at either end. If an amount of charge Q is placed at a point $\frac{1}{4}\ell$ from one end and allowed to flow, how much net charge will flow in each direction?

15. Many electric cables are made up of multiple strands of fine wire. If a particular cable has 150 strands and the total current drawn by the cable is 0.75 A when the potential difference is 1 V, what is the resistance of each strand individually?

16. If 100 mA can be fatal, at what voltage might electrocution occur if a person were perspiring and his resistance was $\approx 10^3\ \Omega$? How low would resistance have to drop to make electrocution from a 12-V car battery a possibility?

17. Find the resistance (per meter) of a 25-μm insulating wire coated uniformly with a 1-μm layer of gold.

18. The proton beam at the Los Alamos Meson Physics Facility (LAMPF) accelerator has produced 1-mA beams. Assume these protons are moving at $0.7c$, where c is the speed of light. How many protons per second emerge from the accelerator? If the effective potential accelerating a 1-mA beam is 800 MV, what is the effective resistance when the current is being produced?

19. Calculate the equivalent resistance from point A to point B in the network in Fig. 24-31.

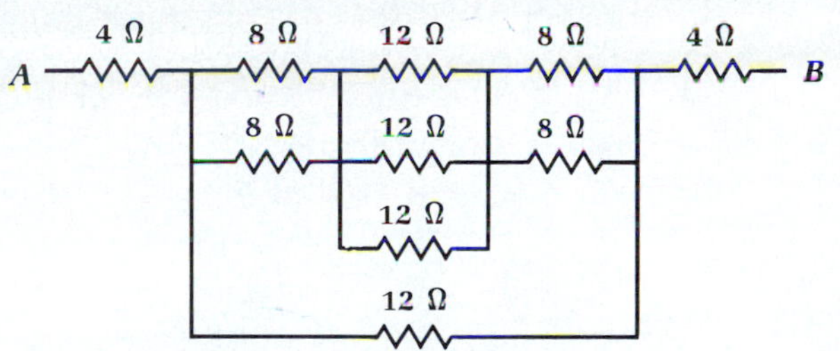

Figure 24-31. Problem 19.

20. A student wishes to measure the resistance of a particular unknown resistor, but she only has a voltmeter. However, she happens to have a 1.5-V battery and a number of resistors of known value. She proceeds to make the following measurements (Fig. 24-32). Ignore the internal battery resistance. *(a)* What is the resistance of the unknown resistor? *(b)* What is the internal resistance of the meter? *(c)* What would the meter have read if the student had made the following measurement?

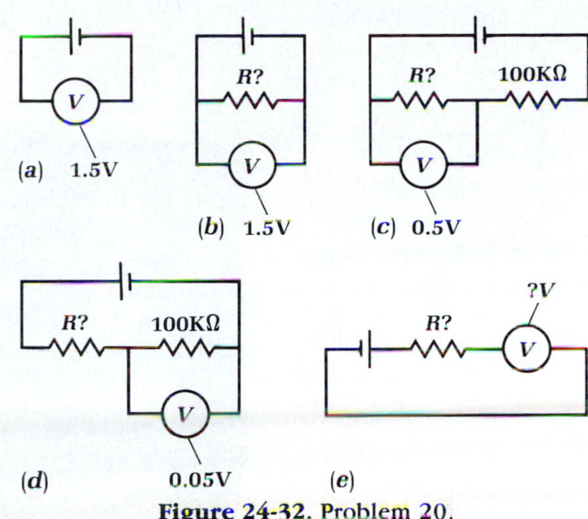

Figure 24-32. Problem 20.

21. What is the resistance per unit length of a cylindrical resistor with an outer metal shell of radius R and an axial conducting wire of radius r, if the volume within the shell is filed with a material of resistivity ρ?

22. Consider an infinite three-dimensional cubic lattice of resistors of equal value R. What is the equivalent resistance across any single resistor in this lattice? What is the resistance between two vertex points in the lattice along a particular line of resistors 100 vertices apart?

25 ENERGY AND CURRENT IN DC CIRCUITS

The design of an electric circuit may be quite complex. The principles you will learn in this chapter provide the framework for analyzing any circuit.

Electric circuits are the bloodstreams in the equipment of the scientist and engineer. In this chapter we meet the simplest of circuits and learn the procedures for analyzing circuits. We limit our discussion to cases where the sense of the current is continuous along one direction, *direct-current (dc) circuits.* Circuits in which the sense of the current oscillates back and forth, called *alternating-current (ac) circuits,* are considered in Chap. 31.

25-1 EMF AND INTERNAL RESISTANCE OF A BATTERY

For a steady current in a circuit, the circuit must contain a source of emf.

For an electric circuit to have a continuous current, the circuit must contain an element that is a source of electric energy. Such an element is called a *source of emf.* (The term "emf," pronounced ee-em-ef, is a contraction of an older expression, the *e*lectro*m*otive *f*orce.) A source of emf provides electric energy to the charge carriers in their trip around a circuit.

A battery is a familiar source of emf. In Chap. 23, we saw that when a battery is connected to the plates of a capacitor, the battery transfers charge carriers from one plate to the other. The motion of these carriers constitutes an electric current. When the terminals of a battery are connected to a resistor, as shown in Fig. 25-1, the battery establishes a current in the circuit. As you know from experience with a flashlight, this current tends to be quite steady for some time. The battery produces this steady current by maintaining a nearly constant potential difference across its terminals. The terminal that is at the higher potential is called the *positive terminal* and the terminal that is at the lower potential is called the *negative terminal.* Thus the sense of the current outside the battery (through the resistor) is from the positive terminal toward the negative terminal, and the sense of the current inside the battery is from the negative terminal toward the positive terminal.

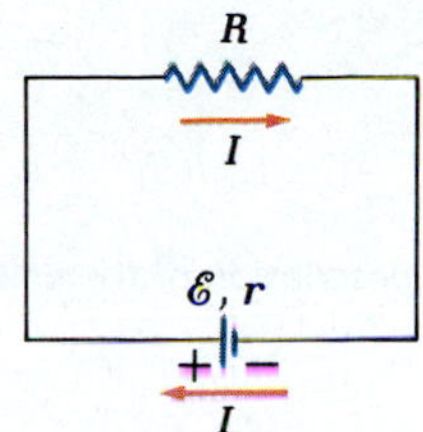

Figure 25-1. A resistor is connected to a battery. The sense of the current outside the battery, through the resistor, is from the positive terminal toward the negative terminal. The sense of the current inside the battery is from the negative terminal toward the positive terminal.

Measuring Emf and Internal Resistance

Two important characteristics of a battery are its emf $\mathscr{E}$ and its internal resistance r. The emf characterizes the energy that the battery provides the charge carriers, and the internal resistance is the battery's own resistance. Figure 25-2 shows how the emf and internal resistance can be determined. A voltmeter placed across the battery measures the battery's *terminal potential difference* $\mathscr{V}$ and an ammeter measures the current I. The current can be changed by changing the resistance of the *variable resistor* (shown as -⩘⩗-). Figure 25-3 shows a graph of $\mathscr{V}$ versus I that is typical for such measurements. The equation that gives this graph is

$$\mathscr{V} = \mathscr{E} - Ir \tag{25-1}$$

The graph's intercept equals the value of the emf $\mathscr{E}$, and its slope gives the internal resistance r. Thus the battery's emf $\mathscr{E}$ is its terminal potential difference when the current in the battery is zero:

$$\mathscr{E} = \mathscr{V} \qquad \text{(when } I = 0) \tag{25-2}$$

The emf of a battery can be measured by placing a high-resistance voltmeter across its terminals while the terminals are not connected to anything else. In this way the current is so small that the Ir term in Eq. 25-1 is negligible compared with $\mathscr{E}$.

The emf is a scalar quantity. However, a battery does have *polarity;* its terminals are distinguishable. We account for a battery's polarity by assigning a sense to its emf. The sense of a battery's emf is from its negative terminal toward its positive terminal (Fig. 25-4).

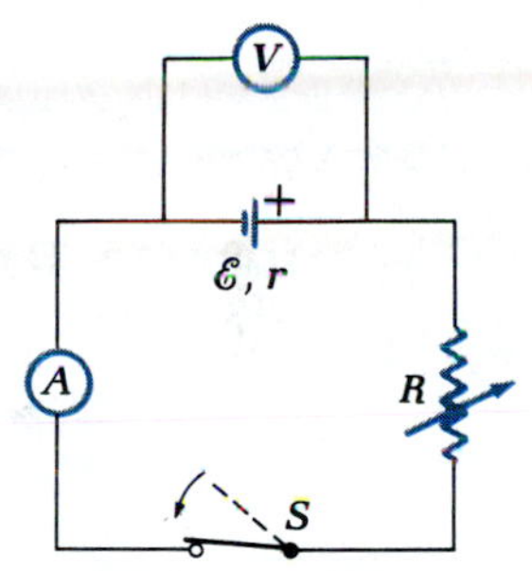

Figure 25-2. Arrangement for finding $\mathscr{E}$ and r. The voltmeter reads the battery's terminal potential difference $\mathscr{V}$ and the ammeter reads the current I. The current is varied by changing the value of the resistance R of the variable resistor.

A toy that demonstrates mechanical analogies to emf, current, and resistance in an electric circuit. Toy penguins equipped with small wheels are lifted to the top of a curved track by an escalator. The penguins then roll down the track with nearly constant speed to the base of the escalator, where they are again lifted. The penguins play the role of charge carriers, and their motion corresponds to a current; the escalator is analogous to a battery as it increases the potential energy of carriers passing through it; friction in the penguins' wheels is similar to resistance.

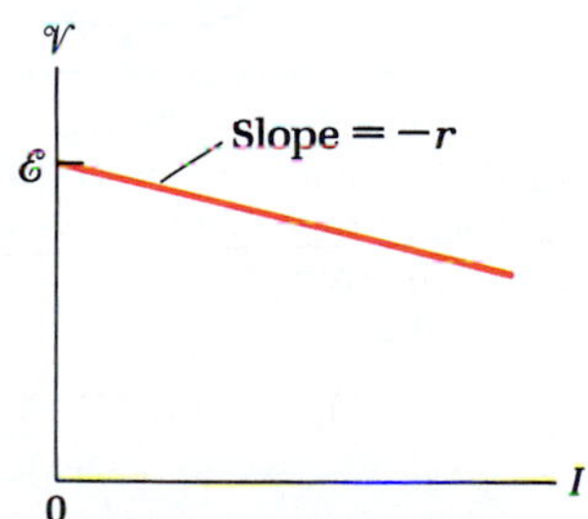

Figure 25-3. A graph of $\mathscr{V}$ versus I from measurements made for the circuit in Fig. 25-2.

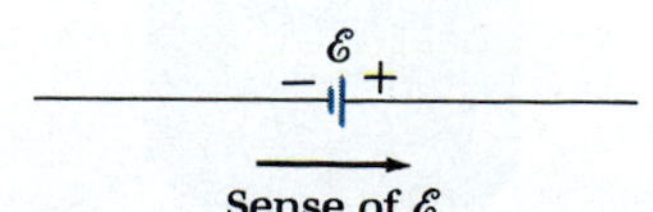

Figure 25-4. The sense of the emf of a battery is from its negative terminal toward its positive terminal.

Physical Description of Emf

Physical description of the emf of a battery

From Eqs. 25-1 and 25-2, you can see that emf has the same dimension as electric potential, namely, energy per unit charge. The dimension energy per unit charge indicates the physical nature of emf. If we consider a small current so that we can neglect the Ir term in Eq. 25-1, then we can describe the emf of a battery as the electric potential energy per unit charge given to the charge carriers by nonelectrostatic forces in the battery as the carriers pass from one terminal to the other. These forces are a result of the chemical action of the battery.

In Fig. 25-5, we show the internal resistance separate from the emf, even though they cannot be physically separated. Traversing the battery along the sense of the current, we find that the potential increases by the amount $\mathscr{E}$ because of the chemical action of the battery and decreases by Ir because of the resistance of the battery, which illustrates the relation $\mathscr{V} = \mathscr{E} - Ir$.

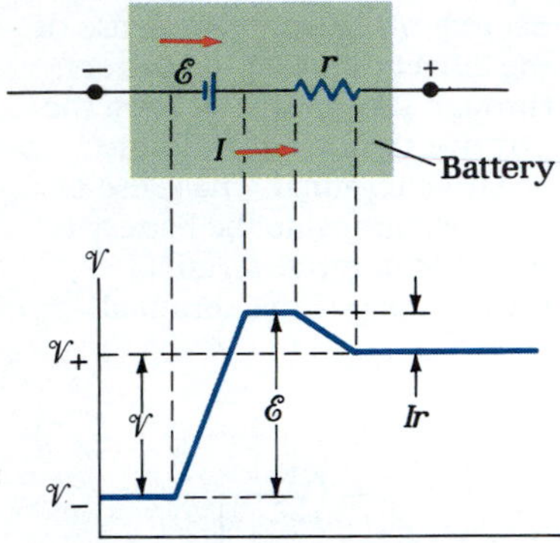

Figure 25-5. A battery with current I in it. The sense of I is the same as the sense of $\mathscr{E}$. The terminal potential difference is $\mathscr{V} = \mathscr{V}_+ - \mathscr{V}_-$.

Properties of Common Batteries

Often a battery is designated by the approximate value of its emf. For example, the emf of an automobile battery is about 12 V and we speak of a 12-V automobile battery; the emf of a flashlight battery is about 1.5 V and we speak of a 1.5-V flashlight battery. Under ordinary operating conditions, the emf of a battery is nearly independent of its condition or its "state of charge." However, the internal resistance of a battery tends to increase with battery usage (unless the battery is recharged). A worn-out flashlight battery may have essentially the same emf as a fresh flashlight battery, but the worn-out battery will have a significantly larger internal resistance. This means that the terminal potential difference of a battery depends on the battery's condition through the Ir term.

The internal resistance of an automobile battery in good condition is around 0.005 Ω and the internal resistance of a fresh flashlight battery is about 0.1 Ω. In many cases, the internal resistance is quite low compared with the rest of the resistance in a circuit and can be neglected.

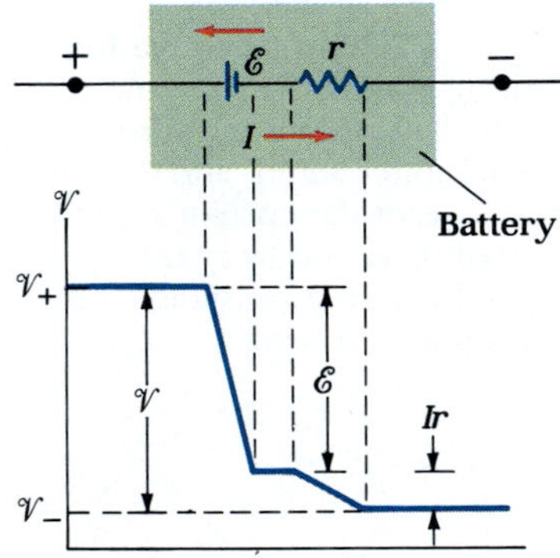

Figure 25-6. A battery being charged. The sense of I is opposite the sense of $\mathscr{E}$. The terminal potential difference is $\mathscr{V} = \mathscr{V}_+ - \mathscr{V}_-$.

If the sense of the current in a battery is the same as the sense of its emf, as seen in Figs. 25-1, 25-2, and 25-5, then the battery is said to be "discharging," and Eq. 25-1 is valid for this case. On the other hand, if the sense of the current is opposite the emf, as shown in Fig. 25-6, then the battery is said to be "charging" or is "being charged." From the figure we see that if we traverse a charging battery along the sense of the current, then the potential *decreases* by the amount $\mathscr{E}$ because of chemical reactions in the battery. As with a discharging battery, the potential decreases by Ir because of the battery's resistance. Thus the terminal potential difference $\mathscr{V}$ across a battery that is being charged is

$$\mathscr{V} = \mathscr{E} + Ir$$

Construction of a Battery

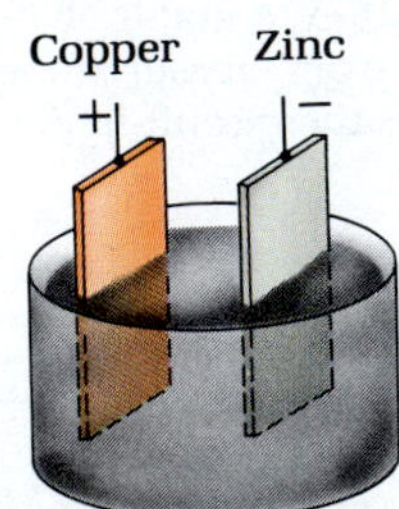

Figure 25-7. A voltaic cell. For these metals a water solution of $CuSO_4$ and $ZnSO_4$ is used as the electrolyte.

A battery is a number of ***voltaic cells*** connected in series, even though we often refer to a single voltaic cell as a battery. A voltaic cell can be made from two plates of different metals that are immersed in an electrolyte (Fig. 25-7). The plate with the higher potential is called the cathode, and a connection to it is the positive terminal. The plate with the lower potential is called the anode, and a connection to it is the negative terminal. In a battery having several cells, the cells are connected in series, with the sense of each cell along the same direction. The battery's emf is the sum of the emf's of its cells, and its internal resistance is the sum of the internal resistances of its cells.

A battery is but one example of a source of emf. Other examples are generators (described in Sec. 28-3), thermocouples, and solar cells. These devices transform energy from some other form to electric energy. For example, a battery transforms chemical energy to electric energy, a generator transforms mechanical energy to electric energy, and a solar cell transforms the energy of light (electromagnetic radiation) to electric energy.

EXAMPLE 25-1

Find the emf and internal resistance of a battery. In Fig. 25-2 the high-resistance voltmeter reads 1.53 V when switch S is open. When switch S is closed, the voltmeter reads 1.41 V and the ammeter reads 0.59 A. What are the battery's *(a)* emf and *(b)* internal resistance? *(c)* What does the voltmeter read when the resistance of the variable resistor is changed so that the ammeter reads 0.86 A?

Solution. *(a)* With switch S open, there is no current in the variable resistor. Since the voltmeter has a high resistance, the current in the battery is negligibly small. Thus the voltmeter reads the battery's emf: $\mathscr{E} = 1.53$ V. *(b)* Solving Eq. 25-1 for r gives

$$r = \frac{\mathscr{E} - \mathscr{V}}{I} = \frac{1.53\text{ V} - 1.41\text{ V}}{0.59\text{ A}} = 0.20\ \Omega$$

(c) When $I = 0.86$ A, we have

$$\mathscr{V} = \mathscr{E} - Ir = 1.53\text{ V} - (0.86\text{ A})(0.20\ \Omega) = 1.36\text{ V}$$

Note that the terminal potential difference is smaller for larger currents because the Ir term is larger for larger currents.

SELF-TEST 25-1. Two 9-V calculator batteries 1 and 2 have the same emf ($\mathscr{E}_1 = \mathscr{E}_2 =$ 9.2 V), but different internal resistances ($r_1 = 2.8\ \Omega$, $r_2 = 1.2\ \Omega$). *(a)* What is the terminal potential difference across each battery when it is required to produce a current of 1.0 A? *(b)* Which battery is in better condition? ***ANSWERS:*** *(a)* $\mathscr{V}_1 = 6.4$ V and $\mathscr{V}_2 = 8.0$ V; *(b)* battery 2 is in better condition.

25-2 ELECTRIC ENERGY AND POWER

When a current exists in a circuit element, energy is transformed. We now investigate energy transformations due to currents in circuit elements.

Energy Dissipated in a Resistor

Consider the energy transformed when a resistor of resistance R carries a current I, as shown in Fig. 25-8. The potential difference across the resistor is $\mathscr{V} = \mathscr{V}_a - \mathscr{V}_b$ and the sense of the current is from a to b. In a time interval Δt, a number of carriers with total charge ΔQ enter the resistor at point a where the potential is $\mathscr{V}_a$, and a number of carriers with an equal total charge ΔQ leave the resistor at point b where the potential is $\mathscr{V}_b$ ($\mathscr{V}_a > \mathscr{V}_b$). The change in the electric point energy ΔU of the carriers is

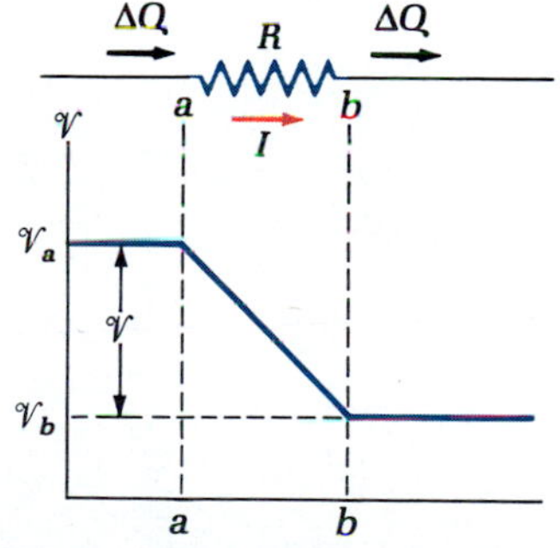

Figure 25-8. Charge carriers lose electric potential energy when they pass through a resistor.

$$\Delta U = \mathscr{V}_b\,\Delta Q - \mathscr{V}_a\,\Delta Q = -(\mathscr{V}_a - \mathscr{V}_b)\,\Delta Q = -\mathscr{V}\,\Delta Q$$

The quantity ΔU is negative because the potential decreases along the sense of the current. The rate at which the carriers lose electric potential energy is $-\Delta U/\Delta t$, and we call this rate the *power P_R dissipated* in the resistor. Thus

$$P_R = -\frac{\Delta U}{\Delta t} = -\frac{-\mathscr{V}\,\Delta Q}{\Delta t} = \mathscr{V}\frac{\Delta Q}{\Delta t}$$

Since $I = \Delta Q/\Delta t$,

$$P_R = \mathscr{V}I \tag{25-3}$$

Equation 25-3 can be quickly brought to mind from the dimensions of $\mathcal{V}$ and I; the dimension of $\mathcal{V}$ is energy per unit charge and the dimension of I is charge per unit time. Their product has the dimension of energy per unit time, or power:

$$\frac{\text{Energy}}{\text{Charge}}\,\frac{\text{charge}}{\text{time}} = \frac{\text{energy}}{\text{time}} = \text{power}$$

We can write P_R in terms of the resistance R of the resistor by using $\mathcal{V} = IR$. We have $P_R = I\mathcal{V} = I(IR)$, or

Joule's law

$$P_R = I^2R \tag{25-4}$$

Alternatively, $P_R = \mathcal{V}I = \mathcal{V}(\mathcal{V}/R)$, or

$$P_R = \frac{\mathcal{V}^2}{R} \tag{25-5}$$

Equations 25-3, 25-4, and 25-5 are equivalent, and Eq. 25-4 is called *Joule's law*.

What happens to the lost electric potential energy of the carriers in a resistor? The carriers lose energy in the collisions that are responsible for the resistance of the resistor. When a resistor carries a current, the temperature of the resistor tends to increase as a result of these collisions. If the temperature of the resistor rises above that of its surroundings, then the resistor transfers heat to its surroundings. (Recall that heat is the transfer of energy due to a temperature difference.) Under steady conditions the energy is continuously transferred as heat to the surroundings. We tersely describe this chain of events by saying that the electric energy is *dissipated* as heat. These effects are sometimes called I^2R (read as "*I*-squared-*R*") heating, sometimes called *Joule heating*, or sometimes called *ohmic heating*.

Electric energy is dissipated as heat in a resistor.

EXAMPLE 25-2

Properties of a light bulb. Household electric light bulbs are rated according to the power P_R dissipated in the filament of the bulb when the potential difference across the bulb is 120 V. (Part of this dissipated energy is in the form of visible light.) *(a)* What is the current in a 75-W light bulb when it is operated at a potential difference of 120 V? *(b)* What is the resistance of the filament of the bulb when it is operated at 120 V? *(c)* If the bulb is operated continuously for 24 h, how much energy is dissipated in the bulb in kilowatt-hour (kW·h)?

Solution. *(a)* Solving Eq. 25-3 for I, we have

$$I = \frac{P_R}{\mathcal{V}} = \frac{75\ \text{W}}{120\ \text{V}} = 0.62\ \text{A}$$

(b) Solving Eq. 25-5 for R, we have

$$R = \frac{\mathcal{V}^2}{P_R} = \frac{(120\ \text{V})^2}{75\ \text{W}} = 190\ \Omega$$

(c) Since the power P_R is constant in time, the energy dissipated in a time interval Δt is $P_R\,\Delta t$:

$$P_R\,\Delta t = (75\ \text{W})(24\ \text{h}) = (0.075\ \text{kW})(24\ \text{h}) = 1.8\ \text{kW}\cdot\text{h}$$

SELF-TEST 25-2. Suppose a light bulb that is designed to dissipate 75 W when operated at 120 V (as in the above example) is operated at 60 V. Assuming the bulb's resistance is the same at 60 V as it is at 120 V, determine the power dissipated. ***ANSWER:*** $(75\ \text{W})/4 = 19\ \text{W}$.

Energy to or from a Battery

Now consider the transformation of energy when a current exists in a battery. If charge carriers pass through a battery in the direction such that the sense of the current is the same as the sense of the emf (the battery is discharging as in Figs. 25-2 and 25-5), then their electric potential energy increases. Let ΔQ be the amount of charge that passes through the battery in time Δt. The change in the electric potential energy of the carriers is $\Delta U = \mathscr{V}\,\Delta Q$, where $\mathscr{V}$ is the terminal potential difference across the battery. In this case, ΔU is positive because the electric potential increases through the battery along the sense of the current (see Fig. 25-5). The rate at which the carriers gain electric potential energy is $\Delta U/\Delta t$, and we call this rate the *power output* P_o from the battery:

$$P_o = \frac{\Delta U}{\Delta t} = \mathscr{V}\frac{\Delta Q}{\Delta t} = \mathscr{V} I$$

Using the expression for the terminal potential difference across a discharging battery, $\mathscr{V} = \mathscr{E} - Ir$, we have $P_o = I\mathscr{V} = I(\mathscr{E} - Ir)$, or

$$P_o = \mathscr{E}I - I^2 r \qquad (25\text{-}6)$$

Power output from a discharging battery

The term $\mathscr{E}I$ in Eq. 25-6 represents the rate at which the electric potential energy of the carriers is increased by chemical reactions in the battery. We shall call this the *power $P_{\mathscr{E}}$ expended by the emf* of the battery: $P_{\mathscr{E}} = \mathscr{E}I$. We recognize the I^2r terms as the power P_r dissipated in the battery due to its resistance r. (The temperature of the battery tends to increase.) This term represents a rate of loss of electric potential energy for the carriers and properly enters the expression with a minus sign. Thus Eq. 25-6 states that the power output P_o of a battery is equal to the power $P_{\mathscr{E}}$ expended by the emf minus the power P_r dissipated as heat: $P_o = P_{\mathscr{E}} - P_r$.

Equation 25-6 is valid for a discharging battery. How is energy transformed in a charging battery? If a battery is being charged, then the sense of the current is opposite the sense of the battery's emf. In this case the potential *decreases* along the sense of the current (see Fig. 25-6), and the electric potential energy of the carriers decreases as they pass through the battery. The power input P_i to the battery is equal to the rate at which the carriers lose electric potential energy in passing through the battery: $P_i = I\mathscr{V}$. Since the terminal potential difference across a charging battery is $\mathscr{V} = \mathscr{E} + Ir$, we have $P_i = I\mathscr{V} = I(\mathscr{E} + Ir)$, or

$$P_i = \mathscr{E}I + I^2 r$$

Power input to a charging battery

In this case the product $\mathscr{E}I$ represents the power delivered to the emf of the battery by the charge carriers.

EXAMPLE 25-3

Output of a flashlight battery. Suppose a flashlight battery of emf 1.5 V and internal resistance 0.61 Ω carries a current of 1.4 A while delivering power to a flashlight bulb. *(a)* Determine the power expended by the battery's emf. *(b)* Determine the power dissipated in the battery. *(c)* Determine the power output of the battery.

Solution. *(a)* The power expended by the battery's emf is

$$P_{\mathscr{E}} = \mathscr{E}I = (1.5\ \text{V})(1.4\ \text{A}) = 2.1\ \text{W}$$

(b) The power dissipated in the battery is

$$P_r = I^2 r = (1.4\ \text{A})^2(0.61\ \Omega) = 1.2\ \text{W}$$

(c) The power output of the battery is

$$P_o = P_{\mathscr{E}} - P_r = 2.1\ \text{W} - 1.2\ \text{W} = 0.9\ \text{W}$$

SELF-TEST 25-3. Suppose the flashlight battery in the above example is being charged, and the current in it is 0.58 A. *(a)* What is the power delivered to the battery's emf? *(b)* What is the power dissipated in the battery? *(c)* What is the power input to the battery? ***ANSWERS:*** *(a)* 0.87 W; *(b)* 0.21 W; *(c)* 1.08 W.

Energy to or from any Circuit Element

Next we consider the rate of energy transformation P in any type of circuit element. If $\mathcal{V}$ is the potential difference across the element and I is the current in the element, then

Rate at which energy is transformed in a circuit element

$$P = I\mathcal{V} \tag{25-7}$$

because $\mathcal{V}$ is the change in the potential energy per unit charge for carriers that pass through the element and I is rate at which charge passes through the element. The product $I\mathcal{V}$ gives the rate at which the electric potential energy of the carriers changes as they pass through the element. If the sense of I is along the direction the potential decreases (as in a resistor or a charging battery), then P gives the rate at which the carriers lose electric potential energy. If the sense of I is along the direction the potential increases (as in a discharging battery), then P gives the rate at which the carriers gain electric potential energy. Equations 25-3 and 25-6 were specific applications of the general relation given by Eq. 25-7.

25-3 KIRCHHOFF'S RULES

In designing a circuit to perform some task, one ordinarily has batteries (or other sources) of known emf and resistors of known resistance. Often the problem is to determine how a given current can be produced in a particular circuit element. Two rules, called *Kirchhoff 's rules* and named for G. R. Kirchhoff (1824–1887), guide us in finding the currents. These rules are referred to as the loop rule and the point rule.

The Loop Rule

The ***loop rule*** states:

The loop rule

> The sum of the potential differences encountered in a round-trip around any closed loop in a circuit is zero.

Since the potential is directly related to the potential energy of the carriers, the loop rule is a statement of conservation of energy. We can write the loop rule as

$$\Sigma\mathcal{V} = 0 \tag{25-8}$$

The loop rule comes from conservation of energy.

As we consider the potential in going around a loop in a circuit, the potential increases through some elements and decreases through others; the sum of the potential differences for a complete round-trip is zero.

Before using the loop rule to find the current in a circuit, let us introduce a new symbol for the current, small i in place of capital I. The symbol I was defined such that it is always positive. But in dealing with circuits, it is convenient to let the current be negative in some cases. Therefore we let the symbol i represent a current that may be negative and I is simply the magnitude of i.

Sign convention for current i

> By convention, i has a positive value when the sense of the current corresponds to the direction of motion of positive carriers.

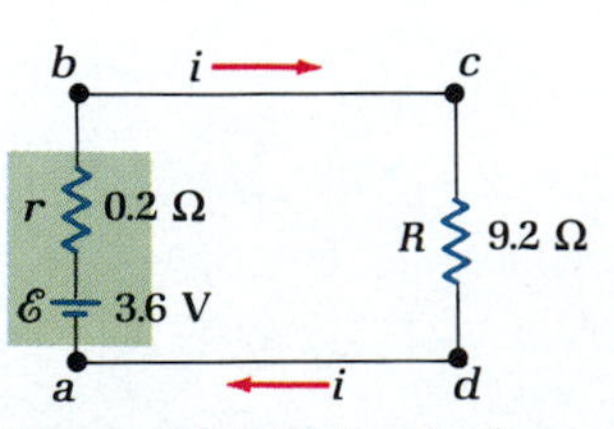

Figure 25-9. Using the loop rule.

Consider using the loop rule to find the current in the circuit of Fig. 25-9. The sense of the current i is shown in the figure. We begin at point a and traverse the loop in the clockwise sense. The loop rule gives

$$(\mathcal{V}_b - \mathcal{V}_a) + (\mathcal{V}_c - \mathcal{V}_b) + (\mathcal{V}_d - \mathcal{V}_c) + (\mathcal{V}_a - \mathcal{V}_d) = 0$$

The potential difference across the section from a to b is the terminal potential difference across the battery: $\mathcal{V}_b - \mathcal{V}_a = \mathcal{E} - ir$. The connecting wires have negligible resistance, so the potential differences $\mathcal{V}_c - \mathcal{V}_b$ and $\mathcal{V}_a - \mathcal{V}_d$ are each zero.

The current through R is from c and d so that $\mathcal{V}_c > \mathcal{V}_d$. Therefore $\mathcal{V}_d - \mathcal{V}_c = -iR$. Substitution into the loop rule gives

$$(\mathscr{E} - ir) + (0) + (-iR) + (0) = 0$$

Solving for i, we have

$$i = \frac{\mathscr{E}}{r + R}$$

With the numerical values given in Fig. 25-9, the current is $i = 3.6\ \text{V}/(0.2\ \Omega + 9.2\ \Omega) = 0.38\ \text{A}$.

In the above analysis we chose to traverse the loop clockwise, but that choice is arbitrary. Suppose we traverse the loop counterclockwise, beginning at point a. The loop rule gives

$$(\mathcal{V}_d - \mathcal{V}_a) + (\mathcal{V}_c - \mathcal{V}_d) + (\mathcal{V}_b - \mathcal{V}_c) + (\mathcal{V}_a - \mathcal{V}_b) = 0$$

The potential difference $(\mathcal{V}_c - \mathcal{V}_d)$ is $+iR$ because $\mathcal{V}_c > \mathcal{V}_d$. Also, $\mathcal{V}_a - \mathcal{V}_b = -(\mathcal{V}_b - \mathcal{V}_a) = -(\mathscr{E} - ir) = -\mathscr{E} + ir$. This gives

$$iR - \mathscr{E} + ir = 0$$

Solving for i we have

$$i = \frac{\mathscr{E}}{r + R}$$

This is the same result as before. The answer is independent of which way we go around the loop.

Our analysis of the circuit in Fig. 25-9 shows that there are two rules we can use to give the algebraic sign of terms we enter into the loop-rule equation:

Rules for the algebraic sign of terms in the loop-rule equation

1. In traversing a resistance R along the sense of the current i, the potential difference across the resistance is entered as $-iR$. In traversing a resistance R opposite the sense of the current i, the potential difference is entered as $+iR$.

2. In traversing a source of emf along the sense of the emf, the potential difference across the source is entered as $+\mathscr{E}$. In traversing a source of emf opposite the sense of the emf, the potential difference across the source is entered as $-\mathscr{E}$.

In using these rules, we treat the internal resistance of the source as a separate resistance.

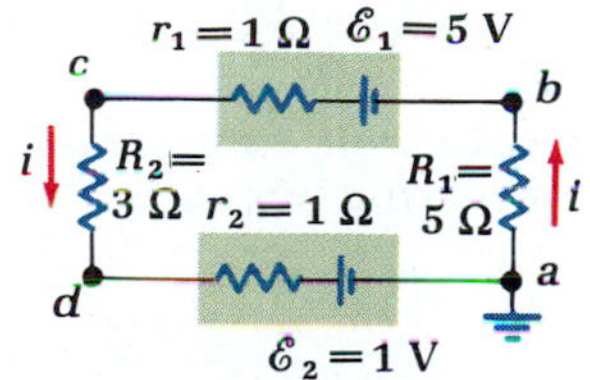

Figure 25-10. In this circuit, $\mathscr{E}_1 > \mathscr{E}_2$ so that the sense of the current is counterclockwise.

Now consider applying the loop rule to the single-loop circuit shown in Fig. 25-10. Since the senses of the emf's of the two batteries are opposite one another, we are not certain about the sense of i. Let us assume that the sense of i is counterclockwise, as shown in the figure. Starting at point a and going counterclockwise around the loop, we write the sum of the potential differences as

$$(-iR_1) + (+\mathscr{E}_1) + (-ir_1) + (-iR_2) + (-ir_2) + (-\mathscr{E}_2) = 0$$

Solving for the current i, we have

$$i = \frac{\mathscr{E}_1 - \mathscr{E}_2}{R_1 + R_2 + r_1 + r_2} \tag{25-9}$$

From the figure, $R_1 + R_2 + r_1 + r_2 = 10\ \Omega$, $\mathscr{E}_1 - \mathscr{E}_2 = 5\ \text{V} - 1\ \text{V} = 4\ \text{V}$, so that $i = 4\ \text{V}/10\ \Omega = 0.4\ \text{A}$.

Notice that Eq. 25-9 yields a positive value for i if $\mathscr{E}_1 > \mathscr{E}_2$ and a negative value for i if $\mathscr{E}_2 > \mathscr{E}_1$. If $\mathscr{E}_2 > \mathscr{E}_1$, then the sense of the current is clockwise, opposite our assumed sense for i in Fig. 25-10. Therefore, if we assume a particular sense for the current at the outset of a problem and the value of the current turns out to be negative, then the actual sense of the current is opposite the assumed sense. That is, the equation automatically gives the sense of the current. This feature of the equations will be useful for dealing with more complex circuits. In more complex circuits we often cannot predict the sense of the current with certainty at the outset

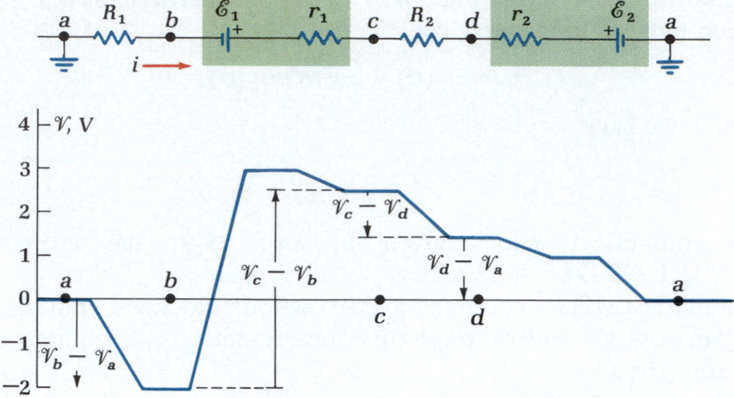

Figure 25-11. Showing the loop rule, $\Sigma \mathscr{V} = 0$, for the circuit in Fig. 25-10.

of the analysis. This uncertainty is of no consequence because we can assume the current has a particular sense, and if this assumption turns out to be wrong, then i has a negative value. In the case of Fig. 25-10, the actual sense of the current is counterclockwise, the same as our assumed sense.

Figure 25-11 shows a graph of the loop rule for the circuit in Fig. 25-10. In our mind's eye, we break the circuit at point *a* and string it along a straight line. Then we show the variation of the potential along the sense of the current. The potential of point *a* has been arbitrarily set to zero (that is, point *a* is "grounded").

The Point Rule

In analyzing circuits with two or more loops, we use the point rule in combination with the loop rule. The point rule states:

The point rule

> The sum of the currents toward a branch point is equal to the sum of the currents away from the same branch point.

The point rule comes from conservation of charge.

Points *a* and *b* in Fig. 25-12 are examples of branch points. Since charge does not accumulate at any point along the connecting wires, the point rule is simply a statement of the conservation of charge. The point rule can be written as

$$\Sigma i_{\text{toward}} = \Sigma i_{\text{away}} \tag{25-10}$$

For example, the point rule applied to point *a* in the circuit in Fig. 25-12 gives

$$i_1 = i_2 + i_3 \tag{25-11}$$

because current i_1 is toward *a* and currents i_2 and i_3 are away from *a*.

We now use the loop rule and the point rule to find the currents i_1, i_2, and i_3 in the circuit of Fig. 25-12. Let loop *abcda* be loop 1 and let loop *aefba* be loop 2. From the figure, our assumed sense for i_1 is chosen to be from *b* to *c* to *d* to *a* (counterclockwise); our assumed sense for i_2 is chosen to be from *a* to *b*; and our assumed sense for i_3 is chosen to be from *a* to *e* to *f* to *b*. Also, for simplicity we include the internal resistance of each battery in the resistance that is in series with

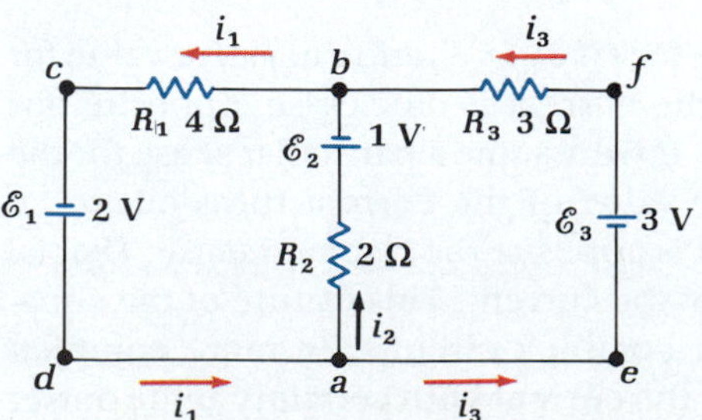

Figure 25-12. A two-loop circuit.

that battery. For example, the internal resistance of battery 1 is included in R_1.

The loop rule applied to loop 1 beginning at point a and going around counterclockwise gives

$$(-i_2R_2) + (-\mathscr{E}_2) + (-i_1R_1) + (\mathscr{E}_1) = 0$$

Rearranging, we have

$$\mathscr{E}_1 - \mathscr{E}_2 = i_1R_1 + i_2R_2 \tag{25-12}$$

The loop rule applied to loop 2 beginning at point a and going around counterclockwise gives

$$(\mathscr{E}_3) + (-i_3R_3) + (\mathscr{E}_2) + (i_2R_2) = 0$$

Rearranging, we have

$$\mathscr{E}_3 + \mathscr{E}_2 = i_3R_3 - i_2R_2 \tag{25-13}$$

We have three equations in three unknowns. The three equations are the point-rule equation (Eq. 25-11) and the loop-rule equations (Eqs. 25-12 and 25-13). The three unknowns are the currents i_1, i_2, and i_3. The point-rule equation is simpler than the other two because each of the coefficients is 1. Using the point-rule equation to eliminate i_1 in Eq. 25-12 gives

$$\mathscr{E}_1 - \mathscr{E}_2 = i_2(R_1 + R_2) + i_3R_1 \tag{25-14}$$

Equations 25-13 and 25-14 represent two equations in two unknowns. Substitution of the numerical values from Fig. 25-12 into these equations gives

$$4\text{ V} = (3\ \Omega)i_3 - (2\ \Omega)i_2$$

and

$$1\text{ V} = (6\ \Omega)i_2 + (4\ \Omega)i_3$$

If we multiply the first of these equations by 3 and add the resulting equation to the second equation, then i_2 is eliminated and we have an equation that contains only i_3. This gives $i_3 = 1$ A. If we now substitute this value of i_3 into either of the above equations, then we find $i_2 = -0.5$ A. Substitution of these values of i_2 and i_3 into Eq. 25-11 gives $i_1 = 0.5$ A. From the signs of our answers we see that the actual senses for i_1 and i_3 are the same as their assumed senses, but the actual sense of i_2 is opposite its assumed sense.

EXAMPLE 25-4

A two-loop circuit. In the circuit of Fig. 25-13*a*, determine the value of $\mathscr{E}$ such that a current of 0.5 A exists in the 8-Ω resistor with sense from a to b.

Solution. First, we notice that the two 2-Ω resistors are in series, so they can be replaced by a single 4-Ω resistor. Also, the 3-Ω resistor is in parallel with the 6-Ω resistor, so they can be replaced by a single 2-Ω resistor. That is, $[(3\ \Omega)(6\ \Omega)]/(3\ \Omega + 6\ \Omega) = 2\ \Omega$. This reduction gives the circuit shown in Fig. 25-13*b*, where we have shown our assumed senses for currents i_1 and i_2. Let loop *abcda* be loop 1 and let loop *abfea* be loop 2. Starting at point a and adding potential differences going counterclockwise around loop 1, we have

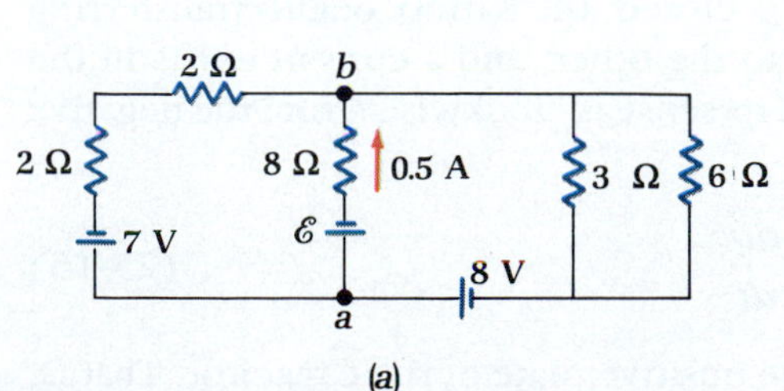

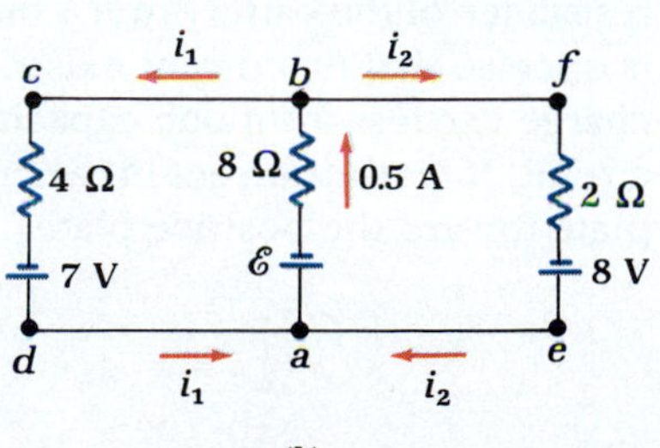

Figure 25-13. Example 25-4: *(a)* Circuit. *(b)* Circuit reduced.

$$(-\mathscr{E}) + [-(0.5\text{ A})(8\ \Omega)] + [-i_1(4\ \Omega)] + (7\text{ V}) = 0$$

or

$$\mathscr{E} = 3\text{ V} - i_1(4\ \Omega) \qquad \text{(loop 1 equation)}$$

Starting at point a and adding potential differences going clockwise around loop 2, we have

$$(-\mathscr{E}) + [-(0.5\text{ A})(8\ \Omega)] + [-i_2(2\ \Omega)] + (8\text{ V}) = 0$$

or

$$\mathscr{E} = 4\text{ V} - i_2(2\ \Omega) \qquad \text{(loop 2 equation)}$$

The point rule applied at point a gives

$$i_1 + i_2 = 0.5\text{ A} \qquad \text{(point-rule equation)}$$

We have three equations in three unknowns; the unknowns are $\mathscr{E}$, i_1, and i_2. We can eliminate i_2 by solving the point-rule equation for i_2 and substituting the result into the loop 2 equation. This gives

$$\mathscr{E} = 3\text{ V} + i_1(2\ \Omega)$$

If we multiply this equation by 2 and add the resulting equation to the loop 1 equation, then i_1 is eliminated and we find $\mathscr{E} = 3$ V.

SELF-TEST 25-4. What is the potential difference, $\mathscr{V}_a - \mathscr{V}_b$, across the section from a to b in the circuit in the above example? ***ANSWER:*** 7 V.

We have shown the procedures for solving one-loop and two-loop circuit problems. Useful circuits often contain many more loops than this. The solution to these problems can be quite complex, but the fundamental principles governing the solutions are the same as for two-loop circuits: conservation of energy (Kirchhoff's loop rule) and conservation of charge (Kirchhoff's point rule).

25-4 *RC* CIRCUITS

Our circuits so far have contained only two types of elements: resistors and batteries. Now we add a third type, a capacitor. From Chap. 23, a capacitor is a device that can hold or contain charge on its plates, charge $+Q$ on one plate and $-Q$ on the other. The potential difference across a charged capacitor is $\mathscr{V} = Q/C$, where C is the capacitance of the capacitor.

We examine two specific cases: (i) An uncharged capacitor is charged by connecting its terminals to a battery in series with a resistor, and (ii) a charged capacitor is discharged by connecting its terminals to a resistor. Our previous discussions have been about steady currents, but now the current varies with time. Indeed, it is the time dependence of the current (and charge) that is our major interest.

Charging a Capacitor

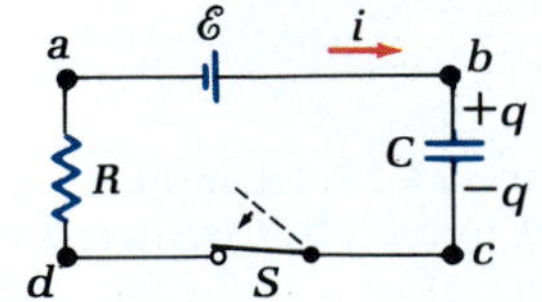

Figure 25-14. A battery charging a capacitor.

Consider a capacitor of capacitance C placed in series with a switch S, resistor of resistance R, and battery of emf $\mathscr{E}$, as shown in Fig. 25-14. (We include the internal resistance of the battery in R.) Initially the capacitor is uncharged and the switch S is open so that no current exists. When S is closed, the battery begins transferring charge carriers from one capacitor plate to the other, and a current exists in the circuit. If i is the current in the circuit and its sense is clockwise (from the negative plate toward the positive plate), then

$$i = \frac{dq}{dt} \tag{25-15}$$

where q is the instantaneous charge on the positive plate of the capacitor. That is, the current in the circuit corresponds to the rate at which charge is transferred from one plate to the other. Consequently, the current is equal to the rate at which the capacitor is charged.

The sum of the potential differences in going clockwise around the loop, beginning at point a, is

$$(\mathcal{V}_b - \mathcal{V}_a) + (\mathcal{V}_c - \mathcal{V}_b) + (\mathcal{V}_d - \mathcal{V}_c) + (\mathcal{V}_a - \mathcal{V}_d) = 0$$

$$(\mathscr{E}) + \left(\frac{-q}{C}\right) + (0) + (-iR) = 0 \qquad (25\text{-}16)$$

Notice that the potential difference $\mathcal{V}_c - \mathcal{V}_b$ across the capacitor is $-q/C$ because the positive plate is on the side with point b and the negative plate is on the side with point c. Thus $\mathcal{V}_b > \mathcal{V}_c$. Substitution of Eq. 25-15 into Eq. 25-16 gives $\mathscr{E} - q/C - R(dq/dt) = 0$. Rearranging, we find

$$\frac{-dq}{\mathscr{E}C - q} = -\frac{1}{RC}\,dt$$

where $\mathscr{E}$, C, and R are constant and q depends on t. To solve this differential equation, it is convenient to let $u = \mathscr{E}C - q$, which gives $du = -dq$. The equation then becomes $du/u = -(1/RC)\,dt$. The indefinite integral is $\ln u = -(t/RC) +$ constant. Replacing u with $\mathscr{E}C - q$, we have

$$\ln(\mathscr{E}C - q) = -\frac{t}{RC} + \text{constant}$$

We evaluate the integration constant by using the initial conditions; when $t = 0$, $q = 0$, so that $\ln \mathscr{E}C =$ constant. Substitution of the integration constant gives $\ln(\mathscr{E}C - q) = -(t/RC) + \ln \mathscr{E}C$, and $\ln(\mathscr{E}C - q) - \ln \mathscr{E}C = \ln[(\mathscr{E}C - q)/\mathscr{E}C]$. Thus

$$\ln \frac{\mathscr{E}C - q}{\mathscr{E}C} = -\frac{t}{RC}$$

or

$$\frac{\mathscr{E}C - q}{\mathscr{E}C} = e^{-t/RC}$$

Solving for q gives

$$q(t) = \mathscr{E}C(1 - e^{-t/RC}) \qquad (25\text{-}17)$$

Charge on a charging capacitor

Figure 25-15 shows a graph of q versus t. We note a few representative points of interest: When $t = 0$, $q = \mathscr{E}C(1 - e^{-0}) = 0$; when $t = RC$, $q = \mathscr{E}C(1 - e^{-1}) = (0.63)\mathscr{E}C$; and as $t \to \infty$, $q \to \mathscr{E}C(1 - e^{-\infty}) = \mathscr{E}C$. The charge asymptotically approaches $\mathscr{E}C$. We let Q_∞ represent the final charge on the capacitor: $Q_\infty = \mathscr{E}C$.

Notice that the product RC characterizes the rate at which the capacitor is charged. From Eq. 25-17, the quantity RC must have the dimension of time because the argument of an exponential must be dimensionless (also see Exercise 30). The SI unit for the product RC is the second ($1\ \Omega \cdot \text{F} = 1$ s). This product is called the *RC time constant* of the circuit and is given the symbol τ: $\tau = RC$. If τ is large, the

The *RC* time constant

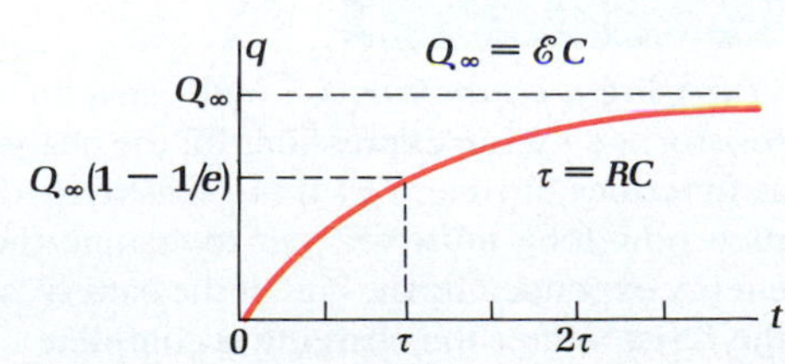

Figure 25-15. Charge on a charging capacitor versus time.

capacitor charges slowly, whereas if τ is small, the capacitor charges rapidly. The time required for the capacitor to reach a given fraction of its final charge is determined solely by τ (or RC).

The current is found by taking the time derivative of Eq. 25-17:

$$i = \frac{dq}{dt} = \frac{d}{dt}[\mathscr{E}C(1 - e^{-t/RC})] = \mathscr{E}C\left(-\frac{1}{RC}\right)(-e^{-t/RC})$$

or

Current in the circuit of a charging capacitor

$$i = \frac{\mathscr{E}}{R}e^{-t/RC} = i_0 e^{-t/\tau} \tag{25-18}$$

where $i_0 = \mathscr{E}/R$ is the initial current. Notice that i_0 is the same as the steady current that would exist if the capacitor were replaced by a connecting wire. Figure 25-16 shows a graph of i versus t. We note a few representative points of interest: When $t = 0$, $i = i_0 e^{-0} = i_0$; when $t = \tau$, $i = i_0 e^{-1} = (0.37)i_0$; when $t = 2\tau$, $i = i_0 e^{-2} = (0.37)^2 i_0 = (0.14)i_0$; and as $t \to \infty$, $i \to i_0 e^{-\infty} = 0$. The current decreases exponentially with time and asymptotically approaches zero as time goes on. Notice that in each time interval equal to one RC time constant, the current is reduced by a factor $e^{-1} = 1/e \approx 0.37$.

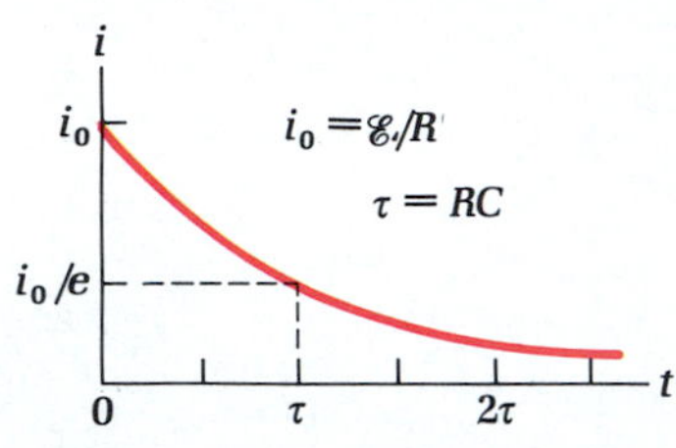

Figure 25-16. Current versus time in the circuit of a charging capacitor.

Now consider the way energy is exchanged during the charging of a capacitor. The energy transferred to the charge carriers of the circuit by the emf of the battery during the entire charging process (as $t \to \infty$) is $\mathscr{E}Q_\infty = \mathscr{E}(\mathscr{E}C) = \mathscr{E}^2C$ because $\mathscr{E}$ is the energy per unit charge transferred by the emf of the battery and a charge $Q_\infty = \mathscr{E}C$ passes through the battery during the entire charging process. Recall from Sec. 23-3 that $\frac{1}{2}Q^2/C$ is the energy stored in a capacitor with charge Q. The energy stored in the capacitor after the charging is completed is $\frac{1}{2}Q_\infty{}^2/C = \frac{1}{2}(\mathscr{E}C)^2/C = \frac{1}{2}\mathscr{E}^2C$. Thus half of the energy expended by the emf of the battery is stored in the capacitor. What happened to the other half of this energy? It is dissipated as heat in the resistor. The rate at which energy is dissipated in the resistor is $P = -(dU/dt) = i^2R$, where U is the electric potential energy of the carriers. The energy $-\Delta U$ dissipated in the resistor during the charging is

$$-\Delta U = -\int dU = \int_0^\infty (i^2R)\,dt$$

$$= \int_0^\infty \left(\frac{\mathscr{E}}{R}e^{-t/RC}\right)^2 R\,dt = \left(\frac{\mathscr{E}}{R}\right)^2\left(\frac{RC}{2}\right)R\int_0^\infty e^{-x}\,dx$$

where $x = (2/RC)t$ and $dx = (2/RC)\,dt$. Since

$$\int_0^\infty e^{-x}\,dx = 1$$

we have

$$-\Delta U = \tfrac{1}{2}\mathscr{E}^2C$$

Half of the energy expended by the battery's emf is stored in the capacitor and half is dissipated as heat in the resistor.

EXAMPLE 25-5

Charging a capacitor. A 5.6-μF capacitor is charged by a 4.2-V battery through a 380-Ω resistor. *(a)* Write expressions for the charge on the capacitor and the current in the circuit as functions of time. *(b)* If the sensitivity of our current measurements is 1 percent of i_0, then how long must we wait to assume the current is effectively zero? *(c)* Evaluate the energy expended by the emf of the battery, stored in the capacitor, and dissipated as heat in the resistor after the charging is complete.

Solution. *(a)* The RC time constant of the circuit is $\tau = (380\ \Omega)(5.6\ \mu\text{F}) = 2.1$ ms. (Recall that $1\ \Omega \cdot \text{F} = 1$ s.) The final charge is $Q_\infty = \mathscr{E}C = (4.2\ \text{V})(5.6\ \mu\text{F}) = 24\ \mu\text{C}$, and the initial current is $i_0 = \mathscr{E}/R = 4.2\ \text{V}/380\ \Omega = 11$ mA. The expressions for the charge and current are

$$q(t) = (24\ \mu\text{C})(1 - e^{-t/(2.1\ \text{ms})})$$

and

$$i(t) = (11\ \text{mA})e^{-t/(2.1\ \text{ms})}$$

(b) To find the time required for the current to fall from i_0 to some value $i(t)$, we solve Eq. 25-18 for t:

$$t = \tau \ln \frac{i_0}{i(t)}$$

Since the sensitivity of our current measurements is 1 percent of i_0, the current is effectively zero when $i(t) < (0.01)i_0$. The time required for the current to become 1 percent of i_0 is $t = \tau \ln [i_0/(0.01)i_0] = \tau \ln 100 = 4.6\tau$. A safe rule of thumb is that for 1 percent sensitivity the charging is complete after about 5τ, or about 10 ms (5×2.1 ms) in this example. Suppose we increase our sensitivity by a factor of 100, to 0.01 percent of i_0. Then how long would we need to wait for the current to be effectively zero? (The answer is twice as long, or about 20 ms, not 100 times as long.) *(c)* The energy expended by the emf of the battery in charging the capacitor is $\mathscr{E}^2C = (4.2\ \text{V})^2(5.6\ \mu\text{F}) = 100\ \mu\text{J}$. The capacitor stores $50\ \mu\text{J}$, and $50\ \mu\text{J}$ is dissipated as heat in the resistor.

SELF-TEST 25-5. In the circuit in the above example, determine *(a)* the charge on the capacitor and *(b)* the current in the circuit at the time $t = 5.0$ ms. ***ANSWERS:*** *(a)* 21 μC; *(b)* 1.1 mA.

Discharging a Capacitor

Consider a capacitor of capacitance C placed in series with a switch S and resistor of resistance R, as shown in Fig. 25-17. Initially the capacitor has charge Q_0 and the switch S is open so that no current exists. At the instant S is closed, charge carriers begin to flow through the circuit to neutralize the charge on the plates. The flow of these carriers constitutes a current in the circuit. Let $q(t)$ be the charge on the capacitor at time t, where $t = 0$ corresponds to the instant S is closed. If i is the current with counterclockwise as its sense (from the positive plate toward the negative plate), then

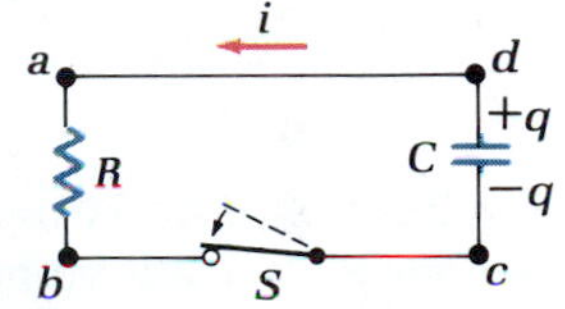

Figure 25-17. A capacitor is discharged through a resistor.

$$i = -\frac{dq}{dt} \tag{25-19}$$

The minus sign must be included because i is positive and dq/dt is negative. Our assumed sense for i has made i positive, and dq/dt is negative because the charge on the plates is decreasing.

Starting at point a, we add potential differences in going counterclockwise around the loop. The loop rule gives

$$-iR + \frac{q}{C} = 0$$

or

$$iR = \frac{q}{C} \tag{25-20}$$

Substituting Eq. 25-19 into Eq. 25-20, we have

$$-\frac{dq}{dt}R = \frac{q}{C} \qquad \text{or} \qquad \frac{dq}{q} = -\frac{1}{RC}\,dt$$

Since R and C are constant, the indefinite integral gives

$$\ln q = -\frac{t}{RC} + \text{constant}$$

The integration constant can be evaluated by noting that when $t = 0$, $q = Q_0$. Therefore $\ln Q_0 =$ constant. Since $\ln q - \ln Q_0 = \ln (q/Q_0)$, we have $\ln (q/Q_0) = -(t/RC)$, or

Charge on a discharging capacitor

$$q(t) = Q_0 e^{-t/RC} = Q_0 e^{-t/\tau} \qquad (25\text{-}21)$$

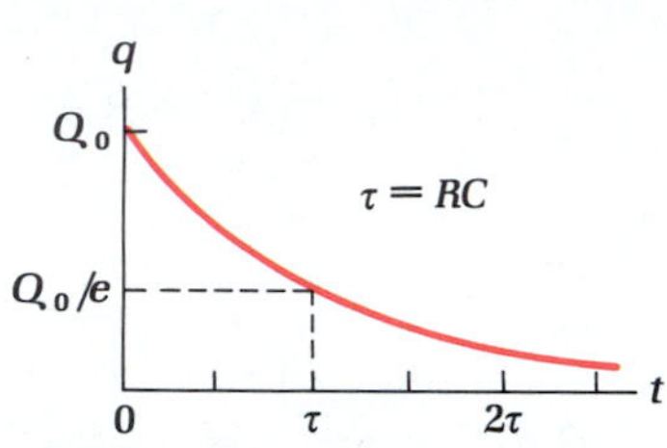

Figure 25-18. Charge versus time on a discharging capacitor.

Figure 25-18 shows a graph of q versus t. Notice the similarity between this graph and the one in Fig. 25-16. The charge on the capacitor decreases exponentially with time and asymptotically approaches zero as time goes on. The time required for the charge to decrease by any given fraction of Q_0 is determined by the product $\tau = RC$.

Now consider the current in the circuit. Since $i = -dq/dt$

$$i = -\frac{d}{dt}(Q_0 e^{-t/RC}) = \frac{Q_0}{RC} e^{-t/RC}$$

The initial potential difference $\mathcal{V}_0$ across the capacitor is $\mathcal{V}_0 = Q_0/C$, so

Current in the circuit of a discharging capacitor

$$i(t) = \frac{\mathcal{V}_0}{R} e^{-t/RC} = i_0 e^{-t/\tau} \qquad (25\text{-}22)$$

where $i_0 = \mathcal{V}_0/R$ is the initial current. Similar to the charge on the capacitor plates, the current decreases exponentially with time, and the RC time constant $\tau = RC$ characterizes the decay of the current.

SUMMARY

Section 25-1. Emf and Internal Resistance of a Battery

The important electrical properties of a battery are its emf $\mathcal{E}$ and its internal resistance r. The terminal potential difference across a discharging battery is

$$\mathcal{V} = \mathcal{E} - Ir \qquad (25\text{-}1)$$

where the sense of the current I in the battery is the same as the battery's emf. For a charging battery,

$$\mathcal{V} = \mathcal{E} + Ir$$

where the sense of the current in the battery is opposite the battery's emf.

Section 25-2. Electric Energy and Power

The power dissipated as heat in a resistor is

$$P_R = I\mathcal{V} = I^2 R = \frac{\mathcal{V}^2}{R}$$

The power output from a discharging battery is

$$P_o = \mathcal{E}I - I^2 r \qquad (25\text{-}6)$$

The power input to a charging battery is

$$P_i = \mathcal{E}I + I^2 r$$

The rate at which energy is transformed in any circuit element is

$$P = I\mathcal{V} \qquad (25\text{-}7)$$

Section 25-3. Kirchhoff's Rules

Kirchhoff's rules are the loop rule:

$$\Sigma\mathcal{V} = 0 \qquad (25\text{-}8)$$

and the point rule:

$$\Sigma i_{\text{toward}} = \Sigma i_{\text{away}} \qquad (25\text{-}10)$$

These rules are expressions of conservation of energy and conservation of charge, respectively. The rules facilitate the analysis of electric circuits.

Section 25-4. *RC* Circuits

The charge on a capacitor that is being charged by a battery with emf $\mathcal{E}$ is

$$q(t) = \mathcal{E}C(1 - e^{-t/\tau}) \qquad (25\text{-}17)$$

where $\tau = RC$ is the RC time constant. The current in the circuit is

$$i(t) = \left(\frac{\mathcal{E}}{R}\right) e^{-t/\tau} \qquad (25\text{-}18)$$

The charge on a discharging capacitor is

$$q(t) = Q_0 e^{-t/\tau} \qquad (25\text{-}21)$$

and the current in the circuit is

$$i(t) = i_0 e^{-t/\tau} \qquad (25\text{-}22)$$

QUESTIONS

1. Suppose the resistance R in Fig. 25-2 is increased. Will the terminal potential difference across the battery increase or decrease?

2. Explain how you would measure the emf and internal resistance of a battery. What apparatus would you need? If you require the use of a resistor, what must the approximate value of its resistance be? What about the internal resistance of any meters you use?

3. Explain the distinction between emf and potential difference. What can you say about the nature of the forces involved with these two quantities?

4. When we refer to a 6-V battery, are we characterizing the battery by its terminal potential difference or by its emf?

5. A battery has two important characteristics, its emf and its internal resistance. Why do we not refer to a battery as a 6-V, 0.5-Ω battery rather than just as a 6-V battery?

6. Can the sense of the current in a battery ever be opposite the sense of its emf? If so, explain how this could happen. If this is the case, which terminal is at the higher potential, the positive terminal or the negative terminal?

7. Can the terminal potential difference across a battery be larger than the emf of the battery? Explain.

8. Explain how a "worn-out" battery can have about the same emf as a fresh battery, but still not produce enough current to be useful. Can a battery that is too worn-out to be used in one circuit still be useful in another circuit?

9. A battery can be rated according to ampere-hours (A·h). For example, an automobile battery might be rated as a 12-V, 90-A·h battery. What are the dimensions of the product (12 V)(90 A·h)? From these dimensions, can you guess the meaning of this product?

10. Two batteries with the same emf $\mathscr{E}$ and internal resistance r are connected in series such that the sense of their emf's is the same. That is, the positive terminal of one is connected to the negative terminal of the other, and the two remaining terminals are the terminals of their combination. What are the emf and internal resistance of this combination?

11. Two batteries with the same emf $\mathscr{E}$ and internal resistance r are connected in parallel. That is, the two positive terminals are connected together and the two negative terminals are connected together, and lead wires from these connections form the terminals of their combination. What are the emf and internal resistance of this combination? Why should such connections only be made with sources which have nearly the same emf? (That is what is done when a car with a worn-out battery is "jump-started" with the battery of another car.)

12. Consider the contrasting properties of the metal used as a heating element wire on the one hand and the metal used in a fuse wire on the other. What physical property (or properties) should be greatly different for the two?

13. Which household light bulb has the larger resistance, a 75-W, 120-V bulb or a 40-W, 120-V bulb?

14. A group of small batteries is connected in series so that the combination has an emf of 120 V. When this source is connected to a 25-W, 120-V light bulb, the bulb glows at its normal brightness. However, when this source is connected to a 300-W, 120-V bulb, the bulb glows much more dimly than it ordinarily does. Explain.

15. Two resistors with resistances R_1 and R_2 are connected in parallel and a current is in the combination. If $R_1 < R_2$, which resistor dissipates the larger amount of energy? Which expression, $P = I^2R$ or $P = \mathscr{V}^2/R$, is more useful in answering this question?

16. Two resistors with resistances R_1 and R_2 are connected in series and a current is in the combination. If $R_1 < R_2$, which resistor dissipates the larger amount of energy? Which expression, $P = I^2R$ or $P = \mathscr{V}^2/R$, is more useful in answering this question?

17. Which battery in Fig. 25-10 is discharging and which is being charged?

18. If the current in a discharging battery is small (so that $\mathscr{E}I \gg I^2r$), then the product $\mathscr{E}I$ is interpreted as the rate at which chemical energy in the battery is being transformed into electric potential energy of the charge carriers. If the current in a charging battery is small, then what is the interpretation of the product $\mathscr{E}I$?

19. In Fig. 25-19, is $\mathscr{V}_b - \mathscr{V}_a$ equal to $+iR$ or $-iR$? What is $\mathscr{V}_a - \mathscr{V}_b$?

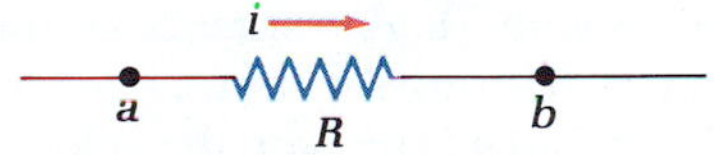

Figure 25-19. Question 19.

20. In Fig. 25-20, is $\mathscr{V}_b - \mathscr{V}_a$ equal to $+Q/C$ or $-Q/C$? What is $\mathscr{V}_a - \mathscr{V}_b$?

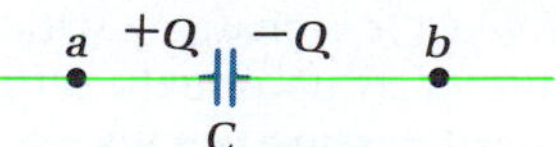

Figure 25-20. Question 20.

21. In Fig. 25-21, is $\mathscr{V}_b - \mathscr{V}_a$ equal to $+(\mathscr{E} - ir)$, $-(\mathscr{E} - ir)$, $+(\mathscr{E} + ir)$, or $-(\mathscr{E} + ir)$? What is $\mathscr{V}_a - \mathscr{V}_b$?

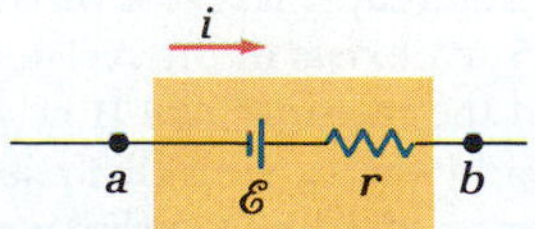

Figure 25-21. Question 21.

22. In Fig. 25-22, is $\mathcal{V}_b - \mathcal{V}_a$ equal to $+(\mathcal{E} - ir)$, $-(\mathcal{E} - ir)$, $+(\mathcal{E} + ir)$, or $-(\mathcal{E} + ir)$? What is $\mathcal{V}_a - \mathcal{V}_b$?

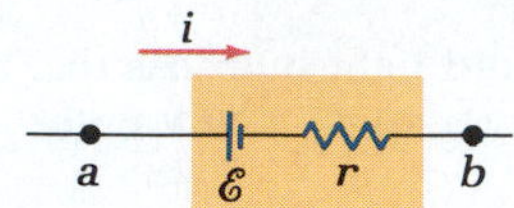

Figure 25-22. Question 22.

23. Consider the discharge of the capacitor in Fig. 25-17. Note that, because the charges on the plates are equal in magnitude but opposite in sign, the net charge on the plates of the capacitor is zero before it is discharged, during the time it is being discharged, and after the discharge is complete. In view of this, does the capacitor lose net charge? If it does not lose net charge, are there any quantities associated with the capacitor that do have a net decrease? Explain.

24. Explain how you would use an *RC* circuit with an ammeter, a battery, a capacitor of known capacitance, and a stopwatch to measure a large resistance. If the unknown resistance is in the range of $10^6\ \Omega$, what would be a convenient value for the capacitance of the capacitor?

25. Complete the following table:

Symbol	Represents	Type	SI Unit
$\mathcal{E}$			V
r	Internal resistance		
i		Scalar	
P			
τ			

EXERCISES

Section 25-1. Emf and Internal Resistance of a Battery

1. A battery with an emf of 1.5 V and an internal resistance of 0.4 Ω has a current of 230 mA in it. The sense of the current is the same as the sense of the battery's emf. What is the battery's terminal potential difference?

2. The terminal potential difference across a 12.5-V battery is 11.9 V when the current in it is 7.8 A. What is meant by the term "12.5-V battery"? Is the battery charging or discharging? What is the internal resistance of this battery?

3. When a high-resistance voltmeter is placed across the terminals of a battery, the voltmeter reads 6.3 V. With the voltmeter leads still in place across the terminals of the battery, the terminals are connected to a resistor in series with an ammeter. The ammeter reads 150 mA and the voltmeter reads 5.9 V. Determine the emf and internal resistance of the battery.

4. When a battery with a high-resistance voltmeter across its terminals is connected in series with an ammeter (the ammeter resistance is 3 Ω) and a 19-Ω resistor, the voltmeter reads 33 V and the ammeter reads 1.50 A. Next the 19-Ω resistor is replaced by a 41-Ω resistor, and the meter readings are 37 V and 0.84 A. What are the emf and the internal resistance of the battery?

5. Suppose the terminal potential difference across a particular battery is 6.5 V when it is charging with a current of 1.9 A (the sense of the current is opposite the sense of the emf), and the terminal potential difference is 5.8 V when it is discharging with a current of 1.2 A (the sense of the current is the same as the sense of the emf). What are the emf and internal resistance of the battery?

Section 25-2. Electric Energy and Power

6. The potential difference across a resistor is 22 V, and a steady current of 65 mA exists in the resistor. *(a)* What is the power dissipated in the resistor? *(b)* How much electric potential energy is lost by the carriers that pass through the element in a time interval of 12 h? Express your answer both in kilowatt-hours (kW·h) and joules (J).

7. The current in a 450-Ω resistor is 32 mA. *(a)* What is the power dissipated in the resistor? *(b)* If this current persists for 60 min, how much energy is dissipated as heat in the resistor? Express your answer both in kilowatt-hours (kW·h) and joules (J).

8. The potential difference across an 880-Ω resistor is 31 V. *(a)* What is the power dissipated in the resistor? *(b)* If this potential difference remains steady for 30 min, how much energy is dissipated as heat in the resistor during that period? *(c)* If the carriers are electrons, how much electric potential energy is lost by each carrier as it passes through the resistor? Express your answer both in joules (J) and electron volts (eV).

9. *(a)* What is the resistance of a 60-W, 120-V household light bulb when the potential difference across it is 120 V? *(b)* What is the current in a 60-W, 120-V bulb when the potential difference across it is 120 V? *(c)* What is the power dissipated in a 60-W bulb when the potential difference across it is 110 V? (Assume the resistance of the filament when operated at 110 V is only negligibly different from when it is operated at 120 V. Do you expect the resistance to be higher when the bulb is operated at 110 V or at 120 V?)

10. Suppose you intend to heat a room with several 800-W electric heaters designed for 120 V. The heaters will be connected to a single circuit (in parallel) which has a circuit breaker that is designed to trip if the current exceeds 15 A. How many heaters can you operate simultaneously without tripping the breaker?

11. Resistors used in electronic circuits have maximum recommended power ratings. *(a)* What is the maximum current you should allow in a 1000-Ω, 0.25-W resistor? *(b)* What is the maximum potential difference you should allow across a 500-Ω, 0.50-W resistor?

12. Resistors used in electronic circuits have maximum recommended power ratings. Suppose a 200-Ω, 0.50-W resistor is placed in series with a 400-Ω, 0.50-W resistor. What is the

maximum allowable current in and potential difference across this combination? Under these conditions, what is the power dissipated in each resistor?

13. Resistors used in electronic circuits have maximum recommended power ratings. Suppose a 200-Ω, 0.50-W resistor is placed in parallel with a 400-Ω, 0.50-W resistor. What is the maximum allowable current in and potential difference across this combination? Under these conditions, what is the power dissipated in each resistor? Compare your answers with those found in the previous exercise.

14. A 6.8-mA current in a battery has the same sense as the battery's emf. The terminal potential difference across the battery is 3.1 V. *(a)* What is the power output of the battery? *(b)* If the charge carriers are electrons, what is the increase in the electric potential energy of each carrier as it passes through the battery? Express your answer both in joules (J) and in electron volts (eV).

15. In the description of a 12-V automobile battery in a sales catalog, the battery is rated at 90 A·h. *(a)* Find the charge in coulombs that corresponds to 90 A·h. *(b)* Estimate the total electric energy (in joules) you might expect this battery to provide (without recharging) before it is discharged.

16. A battery with an emf of 9.0 V and an internal resistance 1.2 Ω carries a current of 260 mA. The sense of the current is the same as the sense of the battery's emf. *(a)* What is the power output of the battery? *(b)* What is the power expended by the emf of the battery? *(c)* What is the power dissipated as heat in the battery?

17. A 12-V battery with an internal resistance of 0.011 Ω is being charged with a current of 7.3 A. *(a)* What is the power input to the battery? *(b)* What is the power delivered to the emf of the battery? *(c)* What is the power dissipated as heat in the battery?

18. A circuit element with a potential difference of 4.8 V across it carries a steady current of 78 mA. The sense of the current is along the direction that the potential decreases. What is the rate at which the electric potential energy of the carriers changes as they pass through the element? Does the electric potential energy of the carriers increase or decrease?

Section 25-3. Kirchhoff's Rules

19. Calculate the potential difference across each element in Fig. 25-10 and check your answers with Fig. 25-11.

20. For the circuit in Fig. 25-23, determine the value of R such that the current in the circuit is 0.5 A.

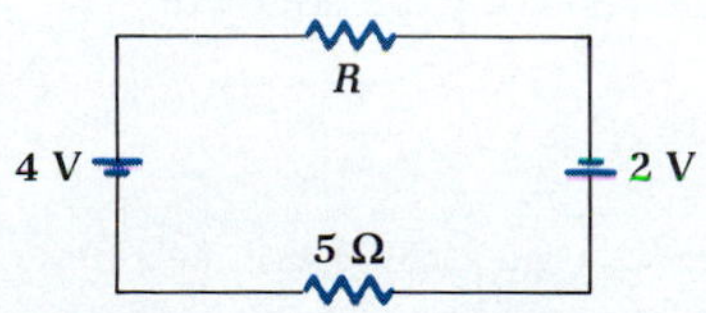

Figure 25-23. Exercise 20.

21. For the circuit in Fig. 25-24, *(a)* determine the value of $\mathscr{E}$ such that the current in the circuit is 2 A, with a counterclockwise sense. *(b)* Determine the value of $\mathscr{E}$ such that the current in the circuit is 2 A, with a clockwise sense.

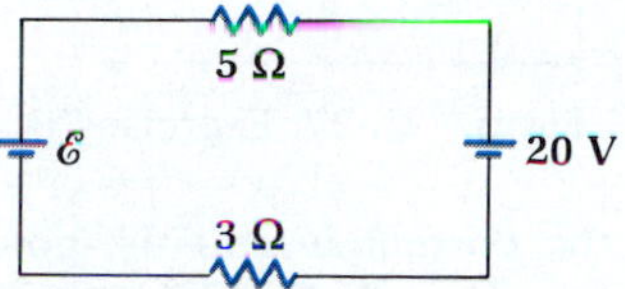

Figure 25-24. Exercise 21.

22. A voltmeter with an internal resistance of 43,000 Ω is connected in series (rather than in parallel) with a resistor of resistance R and a battery of emf 92 V and negligible internal resistance. The voltmeter reads 4.1 V. What is R? (This is useful procedure for measuring large resistances.)

23. For the circuit shown in Fig. 25-12, we found that $4\ \text{V} = (3\ \Omega)i_3 - (2\ \Omega)i_2$ and $1\ \text{V} = (6\ \Omega)i_2 + (4\ \Omega)i_3$. Solve these equations for i_2 and i_3.

24. For each of the two loops in Fig. 25-12 (as defined in Sec. 25-3), construct graphs similar to Fig. 25-11 to illustrate the loop rule. Let point a be grounded.

25. Determine *(a)* current i_1 and *(b)* current i_2 in the circuit of Fig. 25-13 (Example 25-4). *(c)* What is the current in the 3-Ω resistor in Fig. 25-13*a*? *(d)* What is the current in the 6-Ω resistor in Fig. 25-13*a*?

26. *(a)* Determine the current in each of the resistors in Fig. 25-25. *(b)* Determine the potential difference across each of the resistors. *(c)* Determine the power dissipated in each of the resistors.

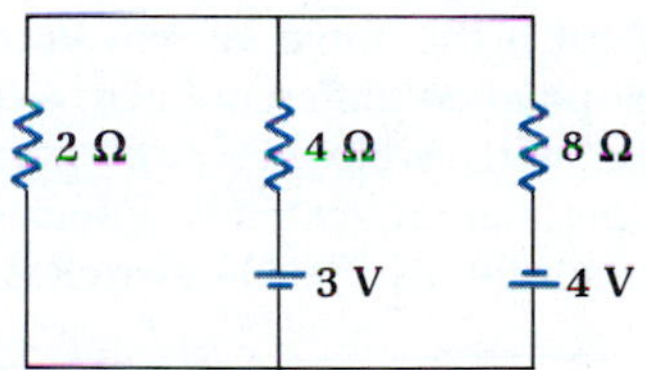

Figure 25-25. Exercise 26.

27. In Fig. 25-26, determine the value of the resistance R such that the current in R is 0.50 A, with sense from a to b.

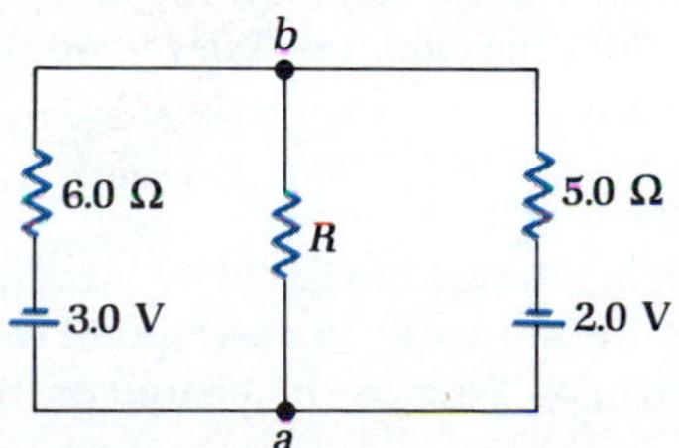

Figure 25-26. Exercise 27.

28. In Fig. 25-27, determine the emf and the sense of the emf of a battery that can be placed at the empty box such that the current in the 6-Ω resistor is 1 A, with sense from a to b.

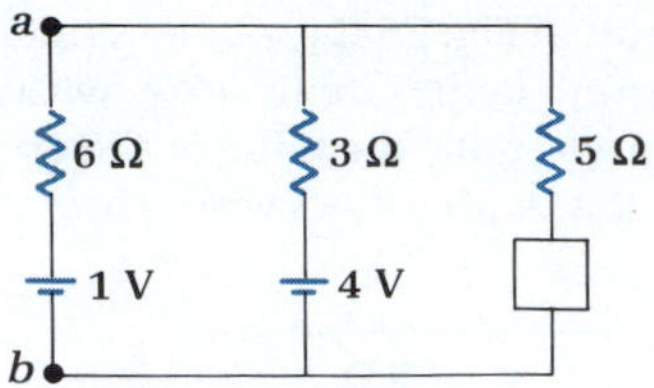

Figure 25-27. Exercise 28.

29. Determine the current in and the potential difference across each of the resistors in Fig. 25-28).

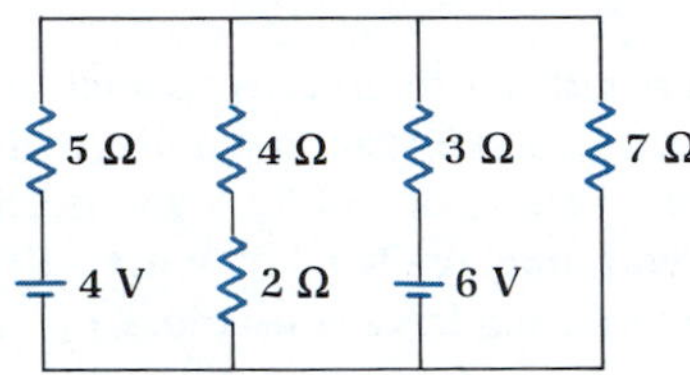

Figure 25-28. Exercise 29.

Section 25-4. *RC* Circuits

30. From Exercise 1 in Chap. 23 the SI unit for capacitance C can be written $s^2 \cdot C^2/(kg \cdot m^2)$, and from Exercise 7 in Chap. 24 the SI unit of resistance R can be written $kg \cdot m^2/(s^3 \cdot A^2)$. Show that the SI unit of the product RC is the second(s).

31. In the circuit shown in Fig. 25-14, let $\mathscr{E} = 14$ V, $R = 75$ kΩ, and $C = 0.84$ μF. *(a)* What is the RC time constant of the circuit? *(b)* What is the charge on the capacitor 50 ms after the switch is closed? *(c)* What is the initial current in the circuit? *(d)* What is the current in the circuit 50 ms after the switch is closed? *(e)* What is the final charge on the capacitor?

32. In the circuit shown in Fig. 25-17, let $Q_0 = 61$ μC, $R = 58$ kΩ, and $C = 1.9$ μF. *(a)* What is the RC time constant of the circuit? *(b)* What is the initial potential difference across the capacitor? *(c)* What is the initial current in the circuit? *(d)* What is the initial potential difference across the resistor (immediately after the switch is closed)? *(e)* What is the charge on the capacitor 50 ms after the switch is closed? *(f)* What is the current in the circuit 50 ms after the switch is closed?

33. In the circuit shown in Fig. 25-14, $\mathscr{E} = 21$ V, $R = 33$ kΩ, and $C = 2.7$ μF. Let $t = 0$ correspond to the instant the switch is closed and assume the internal resistance of the battery is negligible. *(a)* What is the charge on the capacitor at $t = 60$ ms? *(b)* What is the energy stored in the capacitor at $t = 60$ ms? *(c)* What is the energy transferred from the battery to the charge carriers during the time from $t = 0$ to $t = 60$ ms? *(d)* What is the energy dissipated in the resistor during the time from $t = 0$ to $t = 60$ ms?

34. For the circuit shown in Fig. 25-17, let $Q_0 = 45$ μC, $R = 58$ kΩ, $C = 1.6$ μF and let $t = 0$ correspond to the instant the switch is closed. *(a)* What is the charge on the capacitor at $t = 60$ ms? *(b)* What is the energy stored in the capacitor at $t = 0$. *(c)* What is the energy stored in the capacitor at $t = 60$ ms? *(d)* What is the energy dissipated in the resistor between $t = 0$ and $t = 60$ ms?

35. For the circuit shown in Fig. 25-14, let $\mathscr{E} = 35$ V, $R = 64$ kΩ, $C = 1.7$ μF and let $t = 0$ correspond to the instant the switch is closed. The internal resistance of the battery is negligible. *(a)* What is the current in the circuit at $t = 60$ ms? *(b)* What is the rate at which the battery is transferring energy to the carriers at $t = 60$ ms? *(c)* What is the rate at which the resistor is dissipating energy at $t = 60$ ms? *(d)* What is the rate at which energy is being stored in the capacitor at $t = 60$ ms?

36. For the circuit shown in Fig. 25-14, let $\mathscr{E} = 100$ V, $R = 2.0$ kΩ, and $C = 1.0$ μF. On a graph plot the potential difference across the resistor $\mathscr{V}_R$ and on the same graph plot the potential difference across the capacitor $\mathscr{V}_C$ versus the time t from $t = 0$ to $t = 5.0$ ms. Evaluate $\mathscr{V}_R$ and $\mathscr{V}_C$ at each 1.0 ms between $t = 0$ and $t = 5.0$ ms; plot the points and sketch the curves. Show the asymptotic value of $\mathscr{V}_C$ as a horizontal dashed line. At what instant is $\mathscr{V}_R = \mathscr{V}_C$?

37. For the circuit shown in Fig. 25-17, let $R = 2.0$ kΩ, $C = 1.0$ μF, and the initial potential difference across the capacitor $\mathscr{V}_0 = 20$ V. Make a graph of the current in the circuit versus time t from $t = 0$ to $t = 5.0$ ms. Evaluate i at each 1.0 ms between $t = 0$ and $t = 5.0$ ms; plot the points and sketch the curve.

38. Show that the energy initially stored in a capacitor is dissipated as heat in a resistor when the capacitor is discharged through the resistor. Do this by evaluating the time integral of the power dissipated in the resistor, $\int_0^\infty (i^2R)\,dt$, where i is given by Eq. 25-22.

39. A capacitor is charged by a 26-V battery through a 6.2-kΩ resistor. At 3.1 ms after the switch is closed, the potential difference across the capacitor is 13 V. What is the capacitance of the capacitor?

40. A charged capacitor is discharged through a 1.0-MΩ resistor while the current is measured. At a particular instant the current is 150 mA and 0.10 s later the current is 85 mA. What is the capacitance of the capacitor?

41. A 1.5-μF capacitor is discharged through a 2.5-MΩ resistor. If the current at a certain instant is 100 mA, how much longer will it take for the current to reach 10 mA?

42. A 10.0-V battery charges a 2.8-μF capacitor through a 20.0-kΩ resistor. After the charging is completed, what is *(a)* the energy stored in the capacitor, *(b)* the energy dissipated by the resistor, and *(c)* the energy expended by the battery? *(d)* Would the answer to any of these questions be different if the resistance of the resistor were doubled?

PROBLEMS

1. ***Energy transfers in a circuit.*** Consider the energy transfers in the circuit shown in Fig. 25-2. *(a)* Show that the power $P_{\mathscr{E}}$ expended by the emf of the battery is given by $P_{\mathscr{E}} = \mathscr{E}^2/(r + R)$. *(b)* Show that the power P_R dissipated in the resistor is given by $P_R = \mathscr{E}^2 R/(r + R)^2$. *(c)* Show that the power P_r dissipated in the battery is given by $P_r = \mathscr{E}^2 r/(r + R)^2$. *(d)* Use your answers to parts *(a)*, *(b)*, and *(c)* to show that $P_{\mathscr{E}} = P_R + P_r$. *(e)* Consider maximizing P_R by varying R while $\mathscr{E}$ and r are held constant. Show that P_R has a maximum value of $\mathscr{E}^2/4r$ when $R = r$. *(f)* On the same graph, plot $P_{\mathscr{E}}$, P_R, and P_r versus R. Evaluate each power expression between $R = 0$ and $R = 3r$ at intervals of $\frac{1}{2}r$. Plot the points and sketch the curves through them. Note that when $R \gg r$, $P_{\mathscr{E}} \approx P_R \gg P_r$.

2. ***Batteries in parallel.*** Two batteries 1 and 2 with emf's $\mathscr{E}_1$ and $\mathscr{E}_2$ and internal resistances r_1 and r_2 are connected in parallel. That is, the two positive terminals are connected together and the two negative terminals are connected together, and the lead wires from these connections form the terminals of their combination. Show that the effective emf $\mathscr{E}_{12}$ of this combination is

$$\mathscr{E}_{12} = \frac{r_1\mathscr{E}_2 + r_2\mathscr{E}_1}{r_1 + r_2}$$

3. ***A hydroelectric power plant.*** The turbogenerators of a hydroelectric power plant produce a steady current of 9.8 kA for 4.3 h, with a potential difference of 14 kV across the terminals. During this time the level of the lake above the turbogenerators falls from 41.32 to 40.91 m. A negligible amount of water enters the lake from streams, and a negligible amount of water is lost by evaporation and transpiration. The area of the lake is 14×10^6 m^2. Estimate the efficiency of the turbogenerators.

4. ***Heating element in a coffeepot.*** The heating element in an electric coffeepot is designed to carry a current of 5 A when operated at 120 V. *(a)* What is the power dissipated by the pot's heater when it is operated at 120 V? *(b)* The pot increases the temperature of 0.63 L of water from 20°C to the boiling point in 450 s. What is the net heat transferred to the water in this time interval? *(c)* What fraction of the energy dissipated by the heater contributes to the increased internal energy of the water during the 450-s time interval? *(d)* Account for the remainder of the dissipated energy.

5. ***A two-loop circuit.*** Consider a solution to the two-loop circuit problem discussed in Sec. 25-3 with the particular current assignments shown in Fig. 25-29. Show that the currents are given by

$$i_1 = \frac{\mathscr{E}_1(R_2 + R_3) - \mathscr{E}_3 R_2 - \mathscr{E}_2 R_3}{R_1R_2 + R_1R_3 + R_2R_3}$$

$$i_2 = \frac{\mathscr{E}_2(R_1 + R_3) - \mathscr{E}_3 R_1 - \mathscr{E}_1 R_3}{R_1R_2 + R_1R_3 + R_2R_3}$$

$$i_3 = i_1 + i_2$$

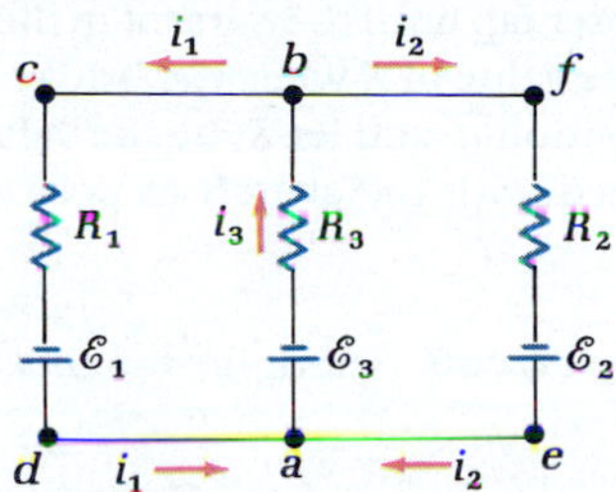

Figure 25-29. Problem 5.

6. ***Kinetic energy of charge carriers.*** The contribution, due to an electric current, to the total kinetic energy of the carriers in a section of conductor can be written as $\Sigma K = \frac{1}{2}\Sigma m v_d^2$, where m is the mass of each carrier, v_d is the carrier drift speed associated with the current, and the sum is over all the carriers in the section of the conductor. *(a)* Using the Drude model of conduction (Sec. 24-3), show that for a section of conductor, $\Sigma K = \frac{1}{2}\tau I \mathscr{V}$, where I is the current in the section, $\mathscr{V}$ is the potential difference across the section, and τ is the mean time between collisions. *(b)* Assuming $\tau \approx 10^{-14}$ s (Example 24-7), estimate ΣK for a 1000-Ω resistor carrying a current of 1 A. What is the power dissipated in this resistor? What is the time interval required for the energy dissipated in the resistor to equal ΣK?

7. ***The Wheatstone bridge.*** A Wheatstone bridge, shown in Fig. 25-30, is a circuit used to measure resistance. In the figure, R_x is the unknown resistance we wish to measure, R_1 is an accurately known variable resistance, R_2 and R_4 are accurately known fixed resistances, and G is a sensitive current-detecting device such as a galvanometer. The bridge is balanced by varying R_1 until the current in the galvanometer is zero. Show that when the bridge is balanced, $R_x = R_1R_4/R_2$.

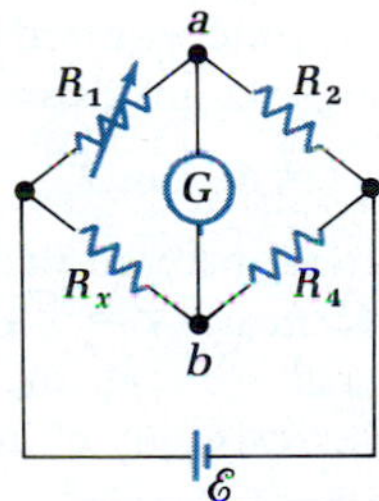

Figure 25-30. Problem 8. A Wheatstone bridge.

8. ***The potentiometer.*** A potentiometer is a circuit used to measure emf (Fig. 25-31). In the figure, $\mathscr{E}_x$ is the unknown emf we wish to measure, $\mathscr{E}_s$ is an accurately known standard emf, and G is a current-detecting device such as a galvanometer. The resistance R_0 has a variable center tap (shown with an arrow) which divides it into two sections, one with resistance R and

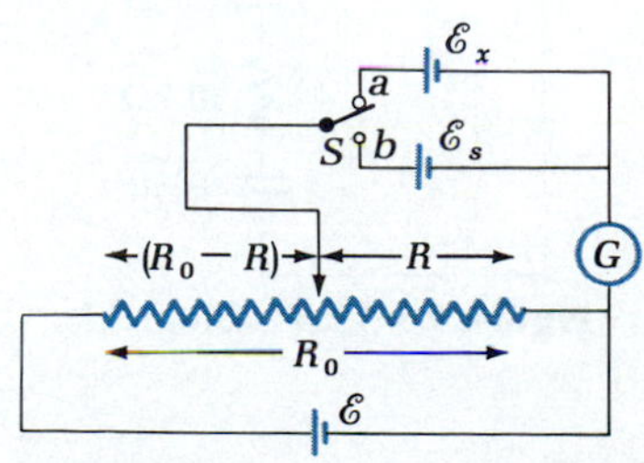

Figure 25-31. Problem 8. A potentiometer.

the other with resistance $R_0 - R$. The switch S is used to connect either $\mathscr{E}_x$ or $\mathscr{E}_s$ into the circuit. The bridge is balanced by adjusting the center tap until the current in the galvanometer is zero. Let R_x be the value of R when the bridge is balanced with the switch in position a, and let R_s be the value of R when the bridge is balanced with the switch in position b. Show that $\mathscr{E}_x = \mathscr{E}_s R_x / R_s$.

9. ***A three-loop circuit.*** Determine the six currents shown in Fig. 25-32.

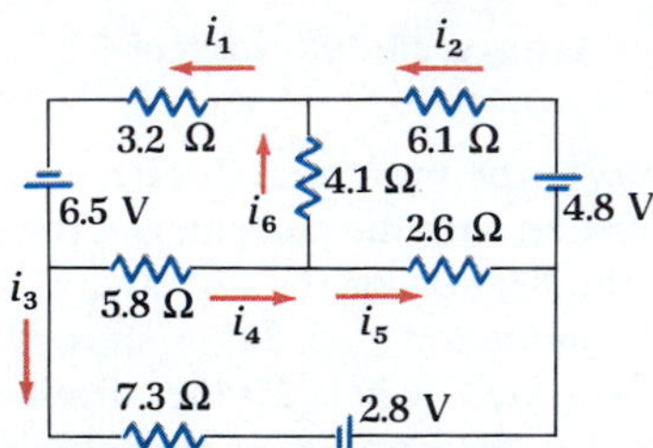

Figure 25-32. Problem 9.

10. ***Charge leaking through a capacitor.*** The resistance of the insulating material between the plates of a parallel-plate capacitor is $R_C = \rho\epsilon_0\kappa/C$ where ρ is the resistivity of the material, κ is the dielectric constant of the material, and C is the capacitance of the capacitor (see Prob. 9 in Chap. 24). Since this resistance is not infinite, a charged capacitor on open circuit discharges slowly because of charge leaking through the insulator. Suppose a capacitor is charged by a battery and then its leads are disconnected from the battery and left open. *(a)* Show that the time dependence of the charge on the capacitor is given by

$$Q(t) = Q_0 e^{-t/\rho\epsilon_0\kappa}$$

where Q_0 is the charge on the capacitor at time $t = 0$. *(b)* Show that the expression for the time required for half the charge on a capacitor to leak through the insulator is

$$t = \rho\epsilon_0\kappa \ln 2$$

(c) Suppose the insulating material between the plates of a capacitor has $\rho \approx 10^{13}\ \Omega\cdot\text{m}$ and $\kappa = 5$. How long does it take for half the charge to leak through the insulator? Discuss a procedure for finding the resistivity of insulators.

11. ***Charging two capacitors in series.*** Consider the circuit in Fig. 25-33. *(a)* What is the *RC* time constant of this circuit? *(b)* Plot the potential difference across each capacitor, and on the same graph, plot the potential difference across the resistor versus time *t*. Evaluate these potential differences at each 1 ms between $t = 0$ ms and $t = 5$ ms. Plot the points and sketch the curves.

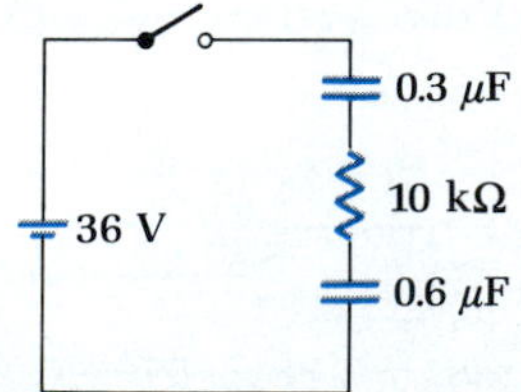

Figure 25-33. Problem 11.

12. ***Energy transfers in an RC circuit.*** Consider the energy transfers in the circuit in Fig. 25-14. *(a)* Show that the power $P_{\mathscr{E}}$ expended by the emf of the battery is given by $P_{\mathscr{E}} = (\mathscr{E}^2/R)e^{-t/\tau}$. *(b)* Show that the power P_R dissipated in the resistor is $P_R = (\mathscr{E}^2/R)e^{-2t/\tau}$. *(c)* Show that the rate at which electric energy is stored in the capacitor is $P_C = (\mathscr{E}^2/R)e^{-t/\tau}(1 - e^{-t/\tau})$. *(d)* Use your answers to parts *(a)*, *(b)*, and *(c)* to show that $P_{\mathscr{E}} = P_R + P_C$. *(e)* Show that P_C maximizes at $\mathscr{E}^2/4R$ when $t = \tau \ln 2$. *(f)* Plot $P_{\mathscr{E}}$, P_R, and P_C versus t on the same graph. Evaluate each power expression between $t = 0$ and $t = 3\tau$ at intervals of $\frac{1}{2}\tau$. Plot the points and sketch the curves through them.

Additional Problems

13. A typical flashlight using two 1.5-V D-cell batteries in series produces 5 W of light for about 2 h. Assume the voltage and output are constant and the efficiency for light production is 50 percent (an equal amount of heat is produced at the same time). *(a)* What is the current flow? *(b)* What is the total net charge that flows through the bulb's filament over the 2-h battery life?

14. If a tea kettle boils in 10 min when the dc voltage to a burner is at a normal level, how long will the same kettle filled with the same amount of water take to boil if the voltage is 10 percent lower?

15. Consider a simple *RC* circuit as shown in Fig. 25-34, where the switch is closed at time $t = 0$. The final (eventual) potential across the capacitor will approach *E*. When will the energy stored in the capacitor reach $\frac{1}{2}$ of its final value? How much energy will have been dissipated in the resistor by that time?

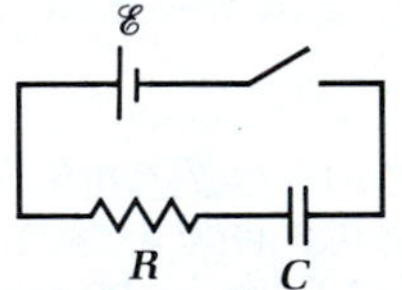

Figure 25-34. Problem 15.

16. Consider the circuit in Fig. 25-35. What is the current measured by the ammeter as a function of time if at $t = 0$ the switch is moved from position 1 to position 2?

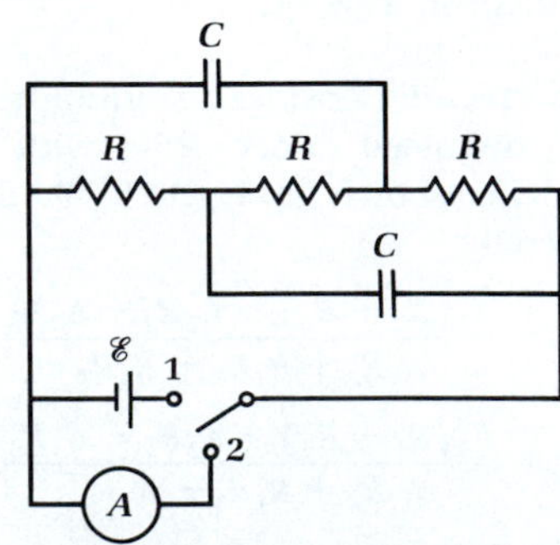

Figure 25-35. Problem 16.

17. Older aircraft use 12-V batteries because of the available automobile electrical components. Newer planes use a 24-V system. With the 24-V system instead of the 12-V system, by what fraction can the weight of the conductors in the electrical wiring be reduced if the power requirements for the electrical equipment remains the same?

18. A solar cell provides a current that is roughly proportional to the intensity of the light illuminating it. This is true over a wide range of voltage. The cell itself has an internal resistance. If the efficiency of a particular cell is 50 percent and the illumination is 10^3 W/m² (sunlight), what potential appears across a 1-cm² cell of 0.1-Ω internal resistance and 2.9-Ω external load?

19. If your electricity use is billed at 10 cents/kWh, how much does it cost to recharge a camcorder battery that is rated at 1.2 A·h at 16 V? Ignore the efficiency of the charger, but consider the internal resistance of the battery to be 0.2 Ω. The recharging time is 1.5 h.

20. Some accelerator electromagnets have coils made of copper pipes through which cooling water flows. The water enters at 20°C. It cannot be allowed to boil, so for safety, its temperature cannot be higher than 90°C at the outflow. If the coil has a resistance of 2.00 Ω and the required current for the needed magnetic field is 400 A, what is the minimum water flow rate that the circulation pumps of the cooling system must supply?

21. The power switch on a portable radio causes an audible click in the headphones when the power is turned on. If the headphones have a lower frequency response cutoff at 100 Hz, what size capacitor should you consider putting from the switch to ground (Fig. 25-36) to eliminate the click? Assume that the radio draws a steady 100 mA in normal operation.

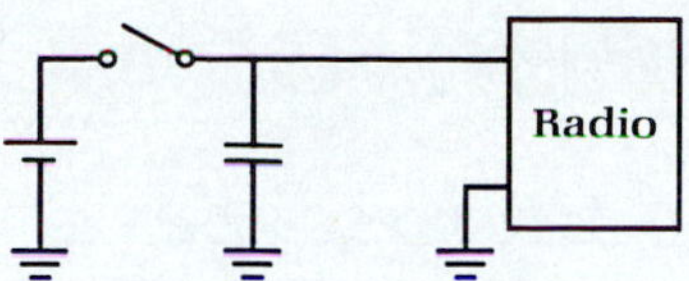

Figure 25-36. Problem 21.

22. A particular notebook computer has a power-saving feature that automatically shuts off power to the LCD (liquid-crystal display) after 3 min of inactivity. If the display normally draws 105 mA from the 12-V (2.0 A·h) battery, and represents $\frac{1}{3}$ of the normal operating power load, how long will the computer run if, after a program had been started, the "shift" key is touched once every 10 min? Assume the program was started during 10 min of continuous use, beginning with the initial turn-on.

23. Design a circuit that will supply a constant current I_0 to a constant external resistive load R_L of 10 Ω, given two fully charged batteries each with E_0 (12 V) and initial internal resistance R_0 of 0.2 Ω. A two-lead device acts as a resistor whose resistance is a function of the applied voltage R (V). You can think of this device as a variable resistor with an infinite resistance voltmeter across it; assume that an operator varies the resistance according to the function as he watches the voltmeter. You must specify the nature of the function to be employed. How long will this circuit be able to supply the constant current if the batteries can supply 10 A·h each and the internal resistance increases linearly (as a function of A·h) to a value of 10 Ω over the 10 A·h draw?

HIGHLIGHTS OF MODERN PHYSICS

The New Microscopes

The concept of atoms goes back millennia, at least to Democritus, but for most of that time atoms have been hypothetical, rather than observable, objects. Einstein wrote a famous paper on Brownian motion—the phenomenon in which dust particles are seen to be knocked around by colliding air molecules—in an attempt to convince skeptics that atoms are real. Recent remarkable advances now allow us to "see" and manipulate individual atoms so that it is hard to imagine anyone not believing in their reality.

The ***scanning tunneling microscope*** was developed in 1982 by the Swiss physicists Gerd Binnig and Heinrich Rohrer. They were awarded the Nobel Prize for this discovery a scant four years later. The microscope uses the principle of quantum-mechanical tunneling described in Sec. 40-9, but we will use a close analog, frustrated total internal reflection (Sec. 35-1), to understand the process.

You have noticed that light reflects off surfaces such as glass and water. You probably have also noticed that the reflection is stronger at glancing angles. The same is true when the light strikes the surface from *inside* the glass, except in this case there is a critical angle beyond which the reflection is 100 percent (Fig. 35-9). This process is called *total internal reflection.* Actually, the light wave does not stop abruptly at the surface, but instead fades out gradually outside the material, decreasing exponentially with distance from the surface. This can be demonstrated by bringing a second piece of glass close to the first one (Fig. 35-10) so that the exponentially decaying light wave is picked up and transmitted by the second glass plate. The intensity of the transmitted light equals that of the decaying wave as it encounters the second plate so the intensity of the transmitted beam increases exponentially as the two plates are brought closer together (Fig. 1).

One of the most important discoveries of the twentieth century is that *particles* act like waves, much as light does. Just as light can tunnel through the "forbidden region" between the glass plates, particles can tunnel through regions that are classically forbidden. A simple example of tunneling occurs when two metals are brought close together, but not so close as to touch, and a voltage is applied between them (Fig. 2). The space between them is classically forbidden because the electrons do not have enough energy to escape from the left-hand metal to flow to the right-hand one. However, just as for light, the electron wave does not stop abruptly at the metal's surface, but falls off exponentially outside the metal. If the second metal is close enough to intercept the wave before it dies out, an electron can tunnel through the forbidden region, and a

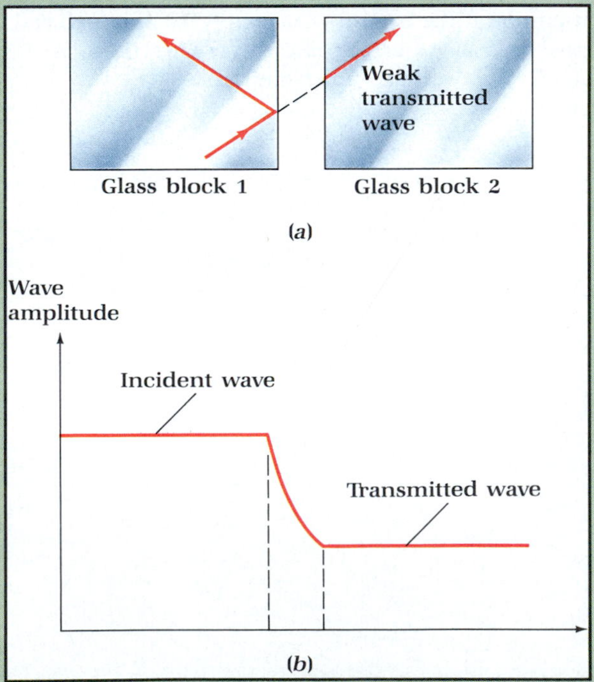

Figure 1. *(a)* A wave is internally incident on the surface of the glass block on the left, at an angle appropriate for total internal reflection. Its amplitude decreases exponentially outside the block until it reaches the second glass block, where it becomes a conventional but weaker transmitted wave. This phenomenon has been used to produce beam splitters to separate a single incident beam into two, one reflected and one transmitted. The relative intensities of the two beams can be adjusted by varying the separation between plates. *(b)* A graph of the amplitude in *(a)*.

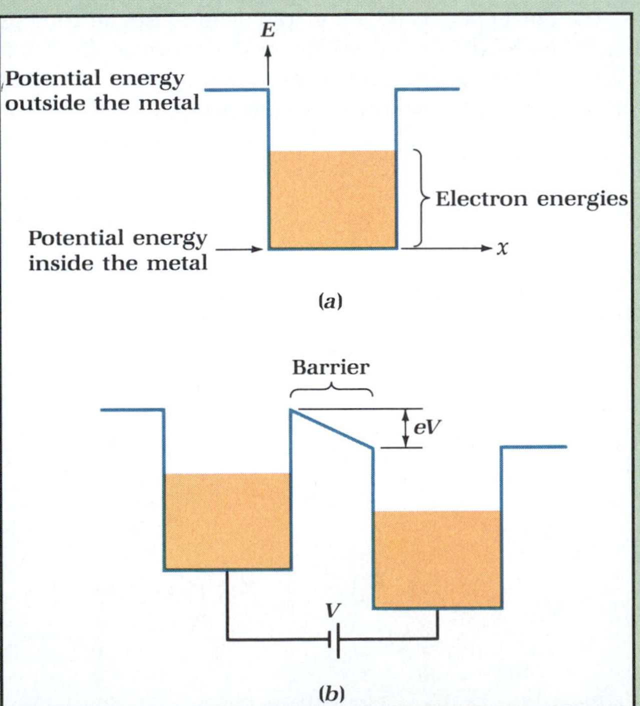

Figure 2. *(a)* The electrons in a metal are bound to it because they are in a potential well produced by the attraction of the positive nuclei. The energies of the electrons are shown by the shaded region. It is apparent that they do not have enough energy to escape from the metal. *(b)* Applying a voltage between two nearby metal blocks shifts the electron potential energy wells relative to each other by eV. Classically, a potential barrier still prevents electrons from passing from one block to the other. Quantum-mechanically, the electrons can tunnel through the barrier.

current will flow between the two metal electrodes. The current increases exponentially as the two metals are brought closer together.

Binnig and Rohrer discovered how to use this effect to produce a microscope. The method presented many difficulties, and it is a tribute to their ingenuity and perseverance that they were able to carry it off. The idea is similar to how one might determine the texture of a rough tabletop in a dark room by running one's fingers back and forth over its surface. Suppose, as shown in Fig. 3, in place of a finger, we use a sharply pointed probe brought near a conducting sample. By applying a small voltage (a few millivolts to a few volts) between the metal probe and the metal sample, we can produce a tunneling current of a few nanoamps. If the probe is moved parallel to the sample surface, the current increases and decreases in response to the hills and valleys of the surface. What is usually done instead is to move the probe toward and away from the sample in order to keep the current—and therefore the distance—constant (Fig. 4*a*). By knowing how much the probe was moved, we know the surface profile along the line of motion across the sample. The surface profile can be determined for a small area of the sample by making multiple scans. Figure 4*b* is an image of individual carbon atoms in graphite.

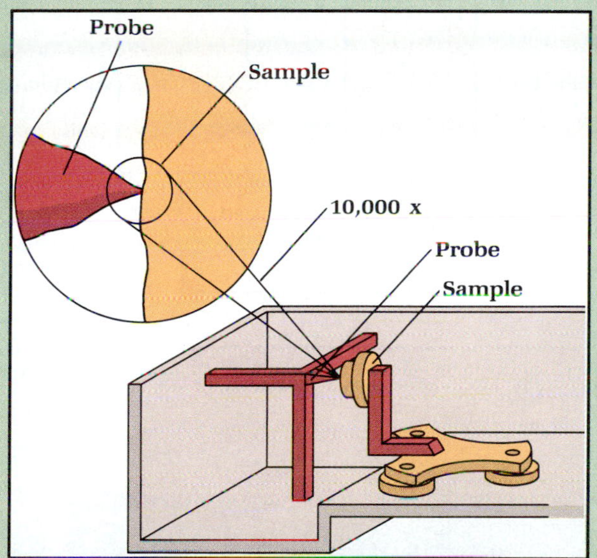

Figure 3. A fine-tipped probe mounted on *xyz* piezoelectric crystals is brought near a sample *S* so a tunneling current can flow.

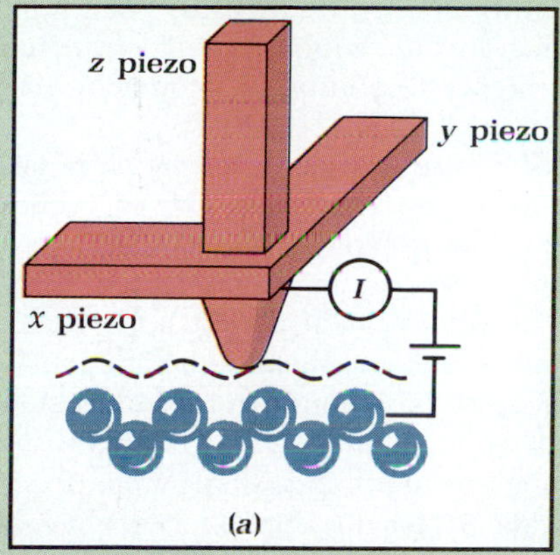

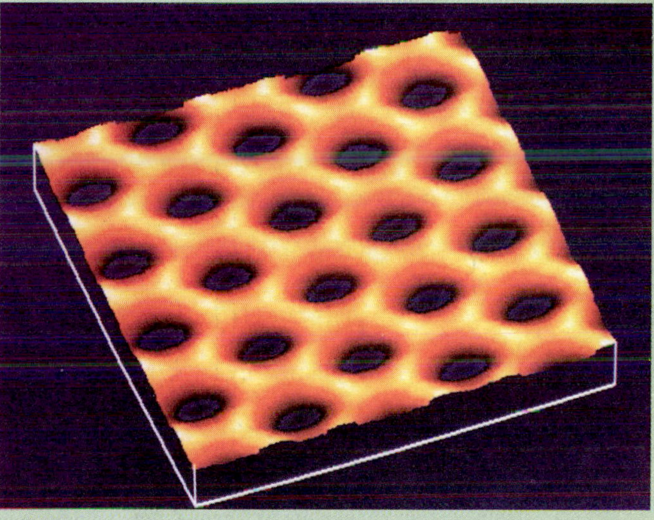

(*b*)

Figure 4. *(a)* The scanning tunneling microscope (STM) probe is moved parallel to the sample surface with individual sample atoms shown. A probe-sample voltage is applied, producing a tunneling current that is very sensitive to the probe-sample distance. A feedback circuit keeps the current constant by activating the *z* piezo to move the probe up and down, following the contour of the atoms on the surface. *(b)* An STM image of individual carbon atoms in graphite.

There is the practical question of how to move the probe back and forth smoothly. Clearly, a mechanical method involving screws and gears would not work because the irregularity of motion would far exceed the atomic dimensions being imaged. In addition, it would be slow to respond and scans would take a long time. Binnig and Rohrer used piezoelectric crystals to provide both the scanning motion (*xy*) and the *z* motion toward and away from the surface, which kept the current constant. These crystals produce an electric voltage when they are compressed, and they expand or contract if an external voltage is applied. It is this feature that allows a vibrating quartz crystal to drive the electronic circuit in a quartz watch. If we apply appropriate voltages to the *x* and *y* piezos in Fig. 4*a*, the probe can be scanned across the desired surface area. Scanning speeds are as large as 10 nm/s, so an entire scan similar to Fig. 4*b* can be done in a few minutes.

As the scan proceeds, a feedback circuit senses any change in the tunneling current and produces a corresponding change in the voltage applied to the *z* piezo until the current returns to its original value. The height information at each point of the scan is obtained by recording the *z*-piezo voltage required to keep the tunneling current constant. Since the crystal movement is calibrated as a function of voltage, this translates into height of the sample surface as a function of the *xy* position.

When Binnig and Rohrer began their work, they did not expect to attain subångstrom resolutions that would show individual atoms. The end of their probe was roughly a sphere with a radius of 100 nm. They calculated that the exponential decay of tunneling current with separation would limit the current to a region about 4.5 nm across. They were quite surprised when the scan showed individual atoms since that requires a resolution of about 0.2 nm. Apparently, there is a local group of atoms forming a peak closest to the sample, and the tunneling takes place at this asperity. It turns out to be simpler to make a sharp point than anyone thought, but efforts are being directed toward even higher resolution by producing the perfect tip consisting of a single atom.

In a kind of bootstrap process it may be possible to use a scanning tunneling microscope (STM) to produce a single-atom tip. In a recent spin-off, STM technology has been used to manipulate individual atoms. The electric field between a probe and a flat surface can be used to apply forces to individual atoms or molecules on a metallic surface and move them to desired locations. It is obvious that the atoms in Fig. 5 are not randomly located. This method may be useful in rearranging atoms on probe tips to produce single-atom tips.

Obviously, the STM is not able to image nonconducting objects. This has given impetus to the development of the *scanning force microscope* (SFM), which relies on the repulsion between any two atoms if they are forced closely together. It is precisely this force that gives hardness to solids: they feel hard because their atoms resist the encroachment of the surface atoms of other objects such as your hand. A stylus on a cantilever spring traces the same kind of xy pattern as in the STM, but now the question is how to measure its up-and-down motion since there is no tunneling current. In early models, an STM probe just behind the SFM probe measured the latter's motion by setting up a tunneling current between the two probes. Later models use laser beams reflected off the back of the SFM probe to follow its up-and-down motion, a method that is less susceptible to contamination of the spring. At first, a typical stylus would exert a force of about 10^{-9} N on the sample atoms, and this was large enough to distort or damage relatively fragile biological samples. Further work has reduced the force to about 10^{-10} N so biological imaging is becoming possible.

QUESTIONS

1. Sketch the shape of the voltage-vs.-time curves for x and y motion that will give an STM scan pattern similar to that shown in Fig. 4. Imagine that the scan is done in the same way you read a book—across the area, then back to the starting point but one line down, and so on.

2. When both the probe and the sample are grounded in an STM (Fig. 6), their potential energy wells are level with each other. The most energetic electrons in the probe are most likely to tunnel through to the sample, but assume the separation is large enough that the electron wave dies out exponentially to essentially zero before reaching the sample and no tunneling occurs, as shown. Sketch the potential-energy curve analogous

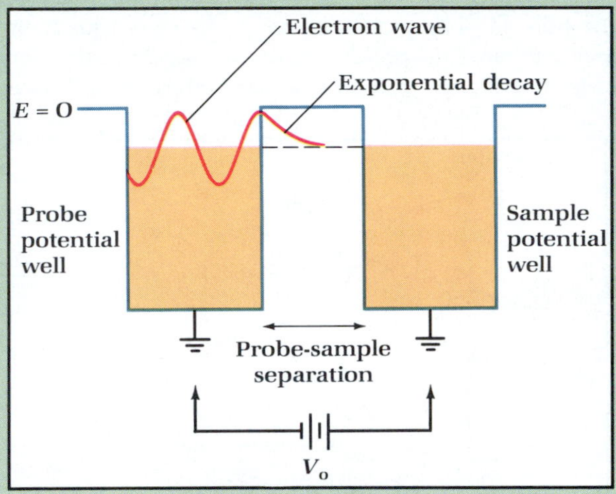

Figure 6. Question 2.

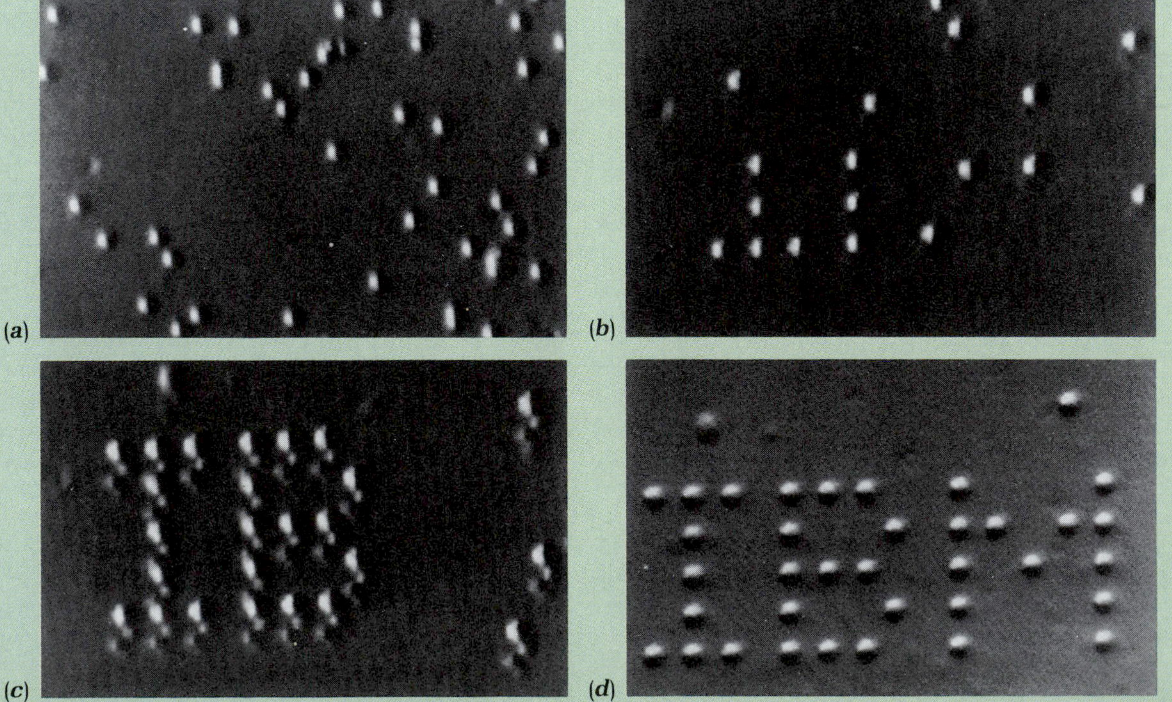

Figure 5. Xenon atoms on a nickel surface are shown in various stages as they are moved from their initial random position a to final position d spelling out the initials of the corporation owning the laboratory.

to the figure if a voltage source between the probe and sample replaces the grounding wires, as shown in the lower part of the figure. What will this do to the height and width of the potential-energy barrier encountered by electrons tunneling through to the sample, compared with the zero-voltage case shown in Fig. 6?

EXERCISES

1. *(a)* Write an expression for the tunneling current as a function of separation if the current increases from 2 nA to 8 nA when the probe is moved .05 nm closer to the sample. *(b)* If the current can be held constant to within 2 percent, estimate the error of the height measurement of the sample's surface.

2. An electron feels an attractive force when it is just outside an STM probe. If we imagine the probe's surface to be a plane, the electric field will have to be perpendicular to the surface, as shown in Fig. 7; otherwise surface currents would be set up in the probe. Therefore the actual field is the same as if there were a positive "image charge" an equal distance on the opposite side of the surface. By symmetry, this would produce a field everywhere normal to the surface as required, so it must be the same as the actual field. *(a)* Calculate the potential energy of the electron as a function of its distance x from the surface. *(b)* Draw a sketch similar to that you produced for Ques. 2, but include this force, showing that the barrier is further reduced.

3. Show that the force between molecules is approximately 10^{-10} N by considering the water molecule. Use the values of the heats of fusion and vaporization and assume that water molecules must be moved about 1 nm apart to break the intermolecular bonds. Use the relation between force and potential energy, Eq. 9-8. This shows why reducing the stylus force in the SFM is important for molecular imaging.

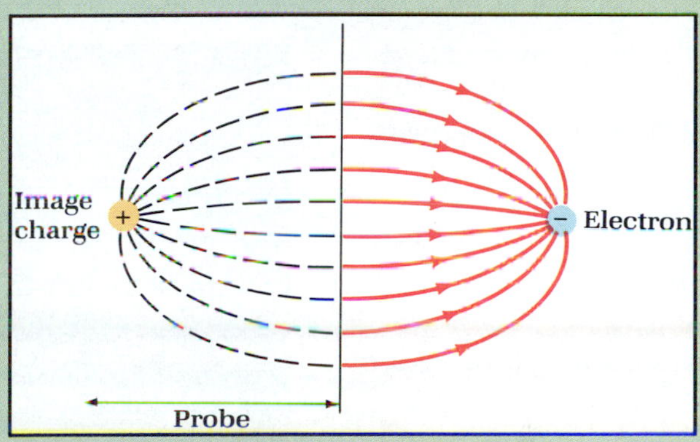

Figure 7. Exercise 2.

26 THE MAGNETIC FIELD

The aurora borealis (northern lights) is a beautiful display of undulating light emanating from the sky. It occurs in far northern and far southern latitudes. The light is emitted in the upper atmosphere by charged particles from the sun. The particles are channeled by the earth's magnetic field so that they usually enter the earth's atmosphere in the region of the magnetic poles.

Observations of the magnetic properties of the mineral magnetite (Fe_3O_4), or lodestone, probably began several thousand years ago when the early Greeks found that these naturally occurring magnets attracted small pieces of iron.* The use of a magnet (as a compass needle) in navigation began about 1000 A.D., although the Chinese may have known of the north-south alignment effect of a magnet much earlier.

Despite these ancient origins, magnetism has become well understood only in the last two centuries. In 1819 Hans Christian Oersted (1777–1851) discovered that an electric current is a source of magnetism. The experiments of Faraday in England and of Joseph Henry (1797–1878) in the United States led to the synthesis of electricity and magnetism by James Clerk Maxwell in the 1860s. A microscopic theory of magnetic materials came with the development of quantum theory in this century, and magnetism in matter is still an area of intensive research.

* Magnetite was found near the city of Magnesia, hence its name.

The effect of a *magnetic field* on electric charges and currents is discussed in this chapter. The sources or causes of magnetic fields are discussed in the next chapter.

26-1 THE MAGNETIC FIELD

Our treatment of electrostatics in Chaps. 20 and 21 made extensive use of the electric field **E**. The field exists everywhere in a region of space and exerts a force **F** on a charge q placed at some point in the region:

$$\mathbf{F} = q\mathbf{E}$$

Magnetic phenomena can be treated similarly by introducing a *magnetic field* **B**. This vector field exerts a force on a *moving* charge. For example, the magnetic field due to a small magnet can deflect the electron beam in a cathode-ray tube, as shown in Fig. 26-1. The deflecting force is observed to be always perpendicular to the velocity of the moving charge. Consider a particle of charge q with velocity **v** at point P:

> The ***magnetic field*** **B** at point P is the vector field that exerts a force **F** on the charged particle given by

$$\mathbf{F} = q\mathbf{v} \times \mathbf{B} \tag{26-1}$$

Definition of the magnetic field **B**

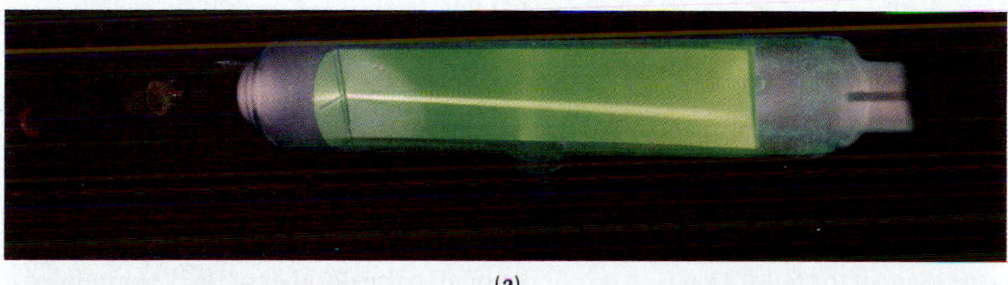

(*a*)

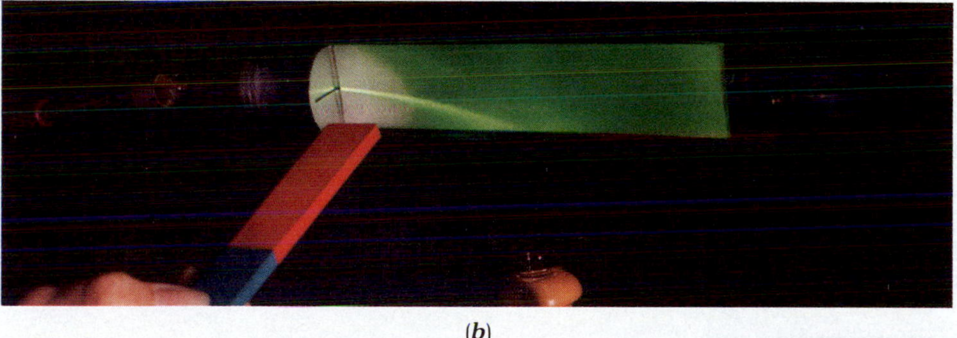

(*b*)

Figure 26-1. *(a)* The path of electrons, moving from left to right in a tube, appears on a fluorescent screen. *(b)* The magnetic field of a magnet deflects the moving electrons in the tube.

For a magnetic force to be exerted on a particle, the particle (i) must be charged and (ii) must be moving. The magnetic field **B** is the vector field that produces the magnetic force. From Chap. 11, the direction of the cross product $\mathbf{v} \times \mathbf{B}$ is perpendicular to the plane containing **v** and **B**, with a right-hand sense. Thus the direction of the force on a charge q is perpendicular both to the velocity of the charge and to the magnetic field at that point. If q is positive, then the force is in the direction of $\mathbf{v} \times \mathbf{B}$, as shown in Fig. 26-2*a*. The magnetic force on a negative charge is opposite the direction of $\mathbf{v} \times \mathbf{B}$, as shown in Fig. 26-2*b*. For either sign of the charge, the magnitude of the force is given by

The magnetic force F is perpendicular to v and B.

$$F = |qvB \sin \theta| \tag{26-2}$$

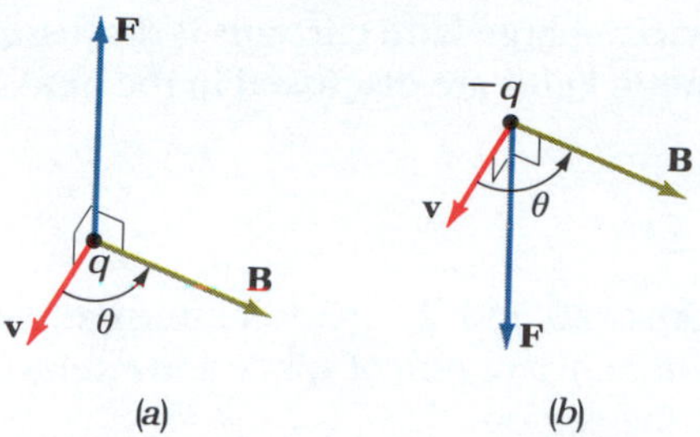

Figure 26-2. The magnetic force on a charged particle is perpendicular to **v** and **B** for *(a)* a positive charge and *(b)* a negative charge.

where θ is the angle between **v** and **B**. Note that the magnetic force is zero if the charge is stationary ($v = 0$) or if **v** and **B** are either parallel or opposite ($\theta = 0$ or $\theta = \pi$).

The SI unit of magnetic field is, from Eq. 26-2, (N/C)/(m/s) or (N)(A^{-1})(m^{-1}), which is called the ***tesla*** (T). Another unit of magnetic field is the ***gauss*** (G); the conversion is $1\ \text{G} = 10^{-4}\ \text{T}$. The tesla is a fairly large unit of magnetic field. For example, the magnitude of the magnetic field of the earth at points near its surface varies, but is around 3×10^{-5} T, or 0.3 G. The largest steady magnetic fields that have been produced in the laboratory are in the range of 30 T.

Since the force exerted by a magnetic field on a moving charged particle is always perpendicular to the velocity, the work done by this force is zero. Consider an infinitesimal displacement $d\boldsymbol{\ell}$ of the charge as it moves with velocity **v**. The work done is $dW = \mathbf{F} \cdot d\boldsymbol{\ell} = \mathbf{F} \cdot \mathbf{v}\, dt$, where $\mathbf{v}\, dt$ is the displacement $d\boldsymbol{\ell}$ for a time interval dt. But the dot product of perpendicular vectors is zero ($\mathbf{F} \cdot \mathbf{v} = 0$) and $dW = 0$. A static magnetic field does no work on a charge. An electric field, of course, *can* do work on a charge.

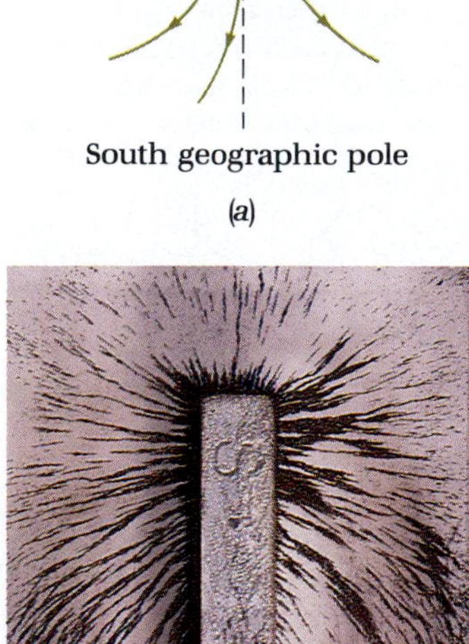

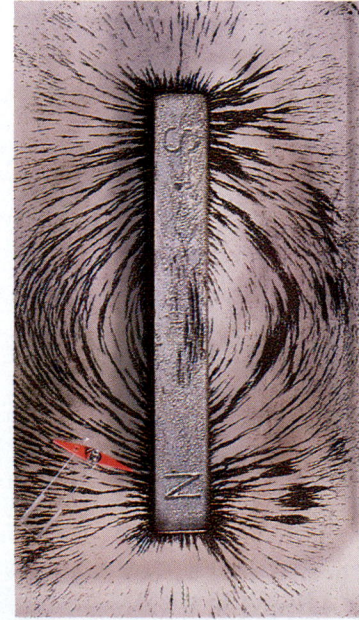

(*b*)

Figure 26-3. *(a)* A portion of the lines of magnetic field are shown outside the earth's surface. *(b)* Iron filings align with the field of a bar magnet.

The spatial distribution of the magnetic field in a region can be represented by lines of magnetic field in the same way that lines of electric field were used. As with the electric field, the lines for a magnetic field are drawn such that the direction of **B** is tangent to the line at a point, and the spacing of the lines indicates the magnitude of the field. That is, B is proportional to the number of lines per unit area crossing a surface perpendicular to the lines. As an example of the use of magnetic field lines, Fig. 26-3*a* shows schematically a portion of the lines for the earth's magnetic field outside its surface. The lines are shown in a plane containing the earth's axis of rotation. Although some of the lines are cut off at the edge of the figure, each line from near the south pole curves around continuously to intersect the surface near the north pole.

Field patterns for a magnetic field can be displayed, as shown in Fig. 26-3*b*, with the use of iron filings, which align with the field. Notice the similarity of the patterns in Fig. 26-3*a* and *b*. The earth's magnetic field pattern is similar to that for a bar magnet.

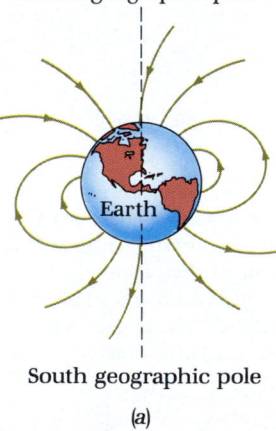

EXAMPLE 26-1

Moving proton in a magnetic field. A proton has a velocity of magnitude 4.4×10^6 m/s at an angle of 62° with a magnetic field of magnitude 18 mT. Determine *(a)* the magnitude and *(b)* the direction of the magnetic force on the proton. *(c)* If this is the only force, what is the proton's acceleration? *(d)* At what rate does the kinetic energy of the proton change?

Solution. *(a)* The magnitude of the force is given by Eq. 26-2:

$$F = |(1.6 \times 10^{-19}\ \text{C})(4.4 \times 10^6\ \text{m/s})(0.018\ \text{T}) \sin 62°|$$
$$= 1.1 \times 10^{-14}\ \text{N}$$

(b) The direction of the force on the (positively charged) proton is perpendicular to the plane of **v** and **B**, with a right-hand sense. The relative directions would be as shown in Fig. 26-2*a*.

(c) From Newton's second law, the acceleration of the proton has the same direction as the net force. Its magnitude is

$$a = \frac{F}{m} = \frac{1.1 \times 10^{-14}\ \text{N}}{1.7 \times 10^{-27}\ \text{kg}}$$
$$= 6.5 \times 10^{12}\ \text{m/s}^2$$

(d) The work done by the magnetic force (which is the only force in this case) is zero because the magnetic force is perpendicular to the velocity. The kinetic energy of the proton does not change.

SELF-TEST 26-1. Rework the above example for the case where the charged particle is an electron rather than a proton. ***ANSWERS:*** *(a)* 1.1×10^{-14} N; *(b)* direction is shown in Fig. 26-2*b*; *(c)* 1.2×10^{16} m/s²; *(d)* zero.

26-2 FORCE ON A CURRENT-CARRYING CONDUCTOR

The magnetic field is defined in Eq. 26-1 in terms of the force on a moving charged particle. If the magnetic field is known in a region, that equation can also be used to determine the magnetic force on a moving charged particle. Since a current in a conductor consists of a collection of moving charge carriers, Eq. 26-1 can be used to determine the magnetic force on a current-carrying conductor.

Consider for simplicity a section of length ℓ of a thin, straight wire of cross-sectional area A, with current I in a uniform magnetic field $\mathbf{B}$, as shown in Fig. 26-4. We evaluate the sum of the magnetic forces on the charge carriers by using their average, or drift, velocity $\mathbf{v}_d$ (see Sec. 24-1). The number of charge carriers in the length ℓ is $N = nA\ell$, where n is the number of charge carriers per unit volume and $A\ell$ is the volume of the length of conductor. If q represents the charge of each carrier, then the total magnetic force on the total charge of Nq is

$$\mathbf{F} = Nq\mathbf{v}_d \times \mathbf{B} = nA\ell q\mathbf{v}_d \times \mathbf{B}$$

It is more convenient to express this result in terms of the current I and a displacement $\boldsymbol{\ell}$, a vector whose direction is the same as the drift velocity of positive charge carriers. (We are assuming that the charge carriers are positively charged. You should be able to show that the force $\mathbf{F}$ is the same if the charge carriers are negatively charged.) The current density has magnitude $j = nqv_d$ and the current $I = jA = nqv_dA$ (from Eq. 24-2). And since $\boldsymbol{\ell}$ has the same direction as $\mathbf{v}_d$, we can write $\ell\mathbf{v}_d = \boldsymbol{\ell}v_d$. Making these substitutions in the expression for the total magnetic force, we have

$$\mathbf{F} = nqA\ell\mathbf{v}_d \times \mathbf{B} = nqv_dA\boldsymbol{\ell} \times \mathbf{B}$$

or

$$\mathbf{F} = I\boldsymbol{\ell} \times \mathbf{B} \qquad (26\text{-}3)$$

Magnetic force on a straight conductor in a uniform field

The magnetic force on this section of the conductor has the direction of $\boldsymbol{\ell} \times \mathbf{B}$ and thus is perpendicular to the plane of $\boldsymbol{\ell}$ and $\mathbf{B}$. In Fig. 26-4, $\boldsymbol{\ell}$ and $\mathbf{B}$ lie in the plane of the paper, and the direction of the force is perpendicularly out of the plane of the paper. Notice that the direction of the displacement $\boldsymbol{\ell}$ is in accord with the conventional sense of the current: in the direction of motion of positive charge carriers. The magnitude of the force is

$$F = I\ell B \sin\theta$$

where θ is the angle between the directions of $\boldsymbol{\ell}$ and $\mathbf{B}$, as shown in Fig. 26-4. The magnitude of the force on the length ℓ of the conductor is proportional to the

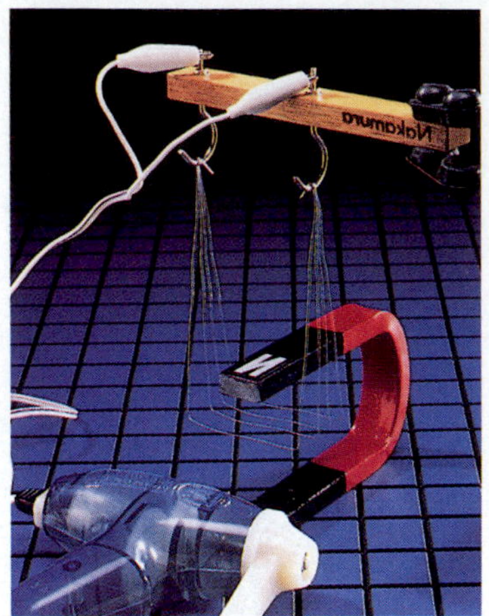

Demonstration of the magnetic force on a current-carrying wire. A wire is arranged with a section that can swing freely in a magnetic field. When a current flows in the wire, the wire swings away from the vertical.

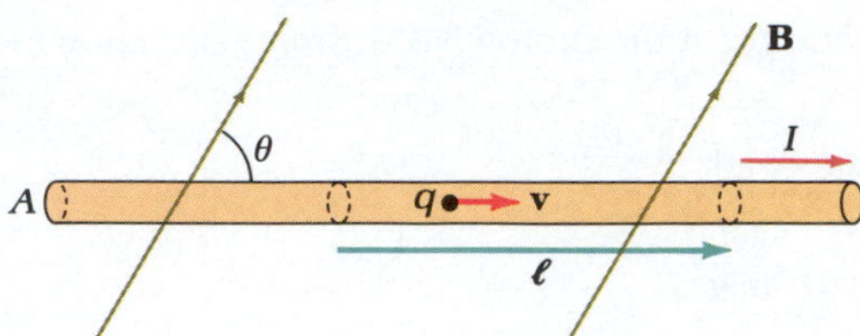

Figure 26-4. A section of a current-carrying wire is in a uniform magnetic field.

length as well as to the current in the conductor and to the magnitude of the magnetic field.

Equation 26-3 applies to a straight section of a thin wire in a uniform magnetic field. We must also deal with conducting wires that are not straight and magnetic fields that are not uniform. We first apply Eq. 26-3 to an infinitesimal length $d\boldsymbol{\ell}$ of a current-carrying conductor. It is convenient to define an infinitesimal *current element* $I\,d\boldsymbol{\ell}$ as the product of the current I and a displacement $d\boldsymbol{\ell}$ whose direction has the sense of the current for that element (Fig. 26-5). The section of wire containing the current element can be considered as straight, and **B** does not vary significantly along the small length $d\boldsymbol{\ell}$. From Eq. 26-3 the magnetic force $d\mathbf{F}$ on the current element is

$$d\mathbf{F} = I\,d\boldsymbol{\ell} \times \mathbf{B} \tag{26-4}$$

The direction of the force is shown in Fig. 26-5, and the magnitude depends on the angle θ between $I\,d\boldsymbol{\ell}$ and **B**:

$$dF = I\,d\ell\,B \sin\theta$$

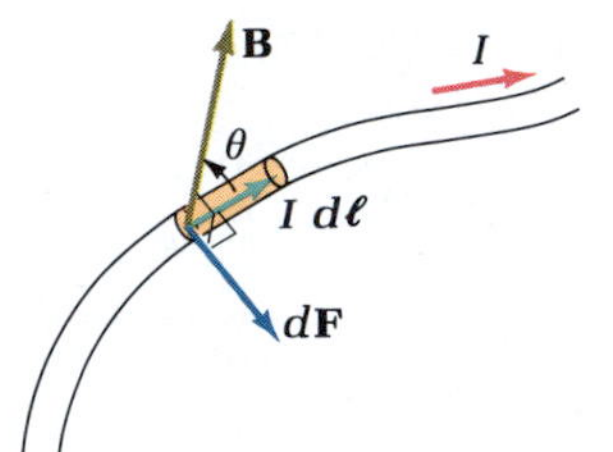

Figure 26-5. A magnetic field exerts an infinitesimal force on a current element. The force is perpendicular to the plane containing $I\,d\boldsymbol{\ell}$ and **B**.

The magnetic force on a longer section of a current-carrying conductor is determined by adding the force on each element of the conductor; that is, Eq. 26-4 is integrated along the conductor:

$$\mathbf{F} = \int I\,d\boldsymbol{\ell} \times \mathbf{B} \tag{26-5}$$

In applying Eq. 26-5, remember that a vector is being integrated and that a cross product is involved.

EXAMPLE 26-2

Magnetic levitation of a wire. A horizontal, straight wire carrying a 16-A current from west to east is in the earth's magnetic field at a place where **B** is parallel to the surface and points north with a magnitude of 0.04 mT. *(a)* Determine the magnetic force on a 1-m length of the wire. *(b)* If the mass of the wire is 50 g, then what current will allow the wire to be magnetically supported (that is, magnetic force balances the weight)?

Solution. *(a)* The magnitude of the force is

$$F = I\ell B \sin\theta = (16\text{ A})(1\text{ m})(4 \times 10^{-5}\text{ T}) \sin 90^\circ$$
$$= 0.6\text{ mN}$$

The direction of the force is given by $\boldsymbol{\ell} \times \mathbf{B}$. Since $\boldsymbol{\ell}$ is directed east and **B** is directed north, the force is up, perpendicular to the earth's surface. What is the force if the wire is oriented so that the sense of the current is from north to south?

(b) With the sense of the current I from west to east, the magnetic force is up, opposite the weight of the wire. To balance the weight with the magnetic force, we equate their magnitudes, $I\ell B = mg$, or

$$I = \frac{mg}{\ell B} = \frac{(0.05\text{ kg})(9.8\text{ m/s}^2)}{(1\text{ m})(4 \times 10^{-5}\text{ T})}$$
$$= 10\text{ kA}$$

While it is not feasible for an ordinary thin wire to carry such a large current, this value indicates that the magnitude of a typical magnetic force on a current-carrying wire is small.

SELF-TEST 26-2. *(a)* If a straight section of wire carries a current horizontally toward the west in a magnetic field that is directed vertically downward, what is the direction of the magnetic force on this section of wire? *(b)* If a straight section of wire carries a current vertically upward in a magnetic field that is directed horizontally toward the north, what is the direction of the magnetic force on this section of wire? *(c)* If a straight section of wire carries a current horizontally toward the south in a magnetic field that is directed horizontally toward the east, what is the direction of the magnetic force on this section of wire? ***ANSWERS:*** *(a)* south; *(b)* west; *(c)* up.

EXAMPLE 26-3

Magnetic force on a straight section of wire. Apply Eq. 26-5 to a straight section of a conductor of length ℓ with current I in a uniform magnetic field and show that the result is Eq. 26-3.

Solution. The current I is constant along the conductor and can be factored from the integral. Since the conductor is straight and the field is uniform, **B** can be factored from the integral—but only to the right because the cross product is not commutative. Equation 26-5 becomes

$$\mathbf{F} = I\left(\int d\boldsymbol{\ell}\right) \times \mathbf{B}$$

and the integral $\int d\boldsymbol{\ell}$ is just $\boldsymbol{\ell}$, the displacement from one end to the other of that section of the conductor. Thus

$$\mathbf{F} = I\boldsymbol{\ell} \times \mathbf{B}$$

which is Eq. 26-3. See Exercise 13 also.

SELF-TEST 26-3. A 0.50-m section of wire is oriented perpendicular to a uniform magnetic field. If the magnitude of the magnetic force on this section of the wire is 14 mN when the current in the wire is 9.6 A, what is the magnitude of the magnetic field? ***ANSWER:*** 2.9 mT.

EXAMPLE 26-4

Force on a semicircular section of wire. A U-shaped conductor carrying current I has its plane perpendicular to a uniform magnetic field, as shown in Fig. 26-6. The curved portion is a semicircle of radius R. Determine the magnetic force on this semicircular portion.

Solution. In Fig. 26-6, the magnetic field is out of the plane of the figure, perpendicular to each current element, and Eq. 26-4 gives the force on a typical current element as shown. The direction of $d\mathbf{F}$ is radially outward, and its magnitude is

$$dF = I\, d\ell\, B \sin 90° = IB\, d\ell$$

We resolve $d\mathbf{F}$ into its x and y components and separately integrate these components of Eq. 26-5:

$$dF_x = dF \cos\phi = IB\, d\ell \cos\phi$$
$$dF_y = dF \sin\phi = IB\, d\ell \sin\phi$$

and

$$F_x = \int IB \cos\phi\, d\ell \qquad F_y = \int IB \sin\phi\, d\ell$$

where the integrals are around the semicircle, with ϕ ranging from 0 to π. The length $d\ell$ subtends an angle $d\phi$ at the center and $d\ell = R\, d\phi$. Making this substitution in the integrals and factoring out the constants I, R, and B gives

$$F_x = IRB \int_0^{\pi} \cos\phi\, d\phi = IRB\,(\sin\pi - \sin 0) = 0$$

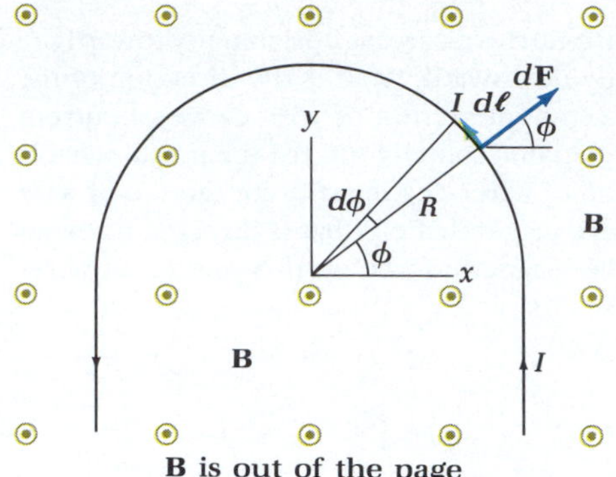

Figure 26-6. Example 26-4: The plane of a U-shaped, current-carrying conductor is perpendicular to a uniform magnetic field.

$$F_y = IRB \int_0^{\pi} \sin\phi \, d\phi = -IRB(\cos\pi - \cos 0) = 2IRB$$

Since $F_x = 0$, the force on the semicircular portion of the conductor is in the positive y direction, and its magnitude is $F = 2IRB$. See Prob. 1 for another approach to this calculation.

SELF-TEST 26-4. Suppose the current in the wire in the above example is 5.0 A, the magnitude of the magnetic field is 60 mT, and the radius of the semicircle is 0.25 m. Determine the magnetic force on the section. ***ANSWER:*** (0.15 N)**j**.

26-3 TORQUE ON A CURRENT LOOP

Since a magnetic field exerts a force on a current-carrying wire, it can also produce a torque. Of particular interest is the torque on a loop of wire pivoted on an axis and carrying a current. The rotational motion caused by such a torque is the basis for an electric motor.

The right-hand rule for the area of a current loop

Consider the rectangular current loop shown in two views in Fig. 26-7. The loop carries current I and is in a uniform magnetic field **B**. The rectangular dimensions of the loop are ℓ and w, so that the area of the plane of the loop is $S = \ell w$. It is convenient to use the vector area **S** to specify the loop's orientation, as illustrated in Fig. 26-7*b*. The direction of **S** is perpendicular to the plane of the loop, with a right-hand sense. To determine the sense, curl the fingers of your right hand to follow the sense of the current around the circuit. Then the extended thumb gives the direction of the area.

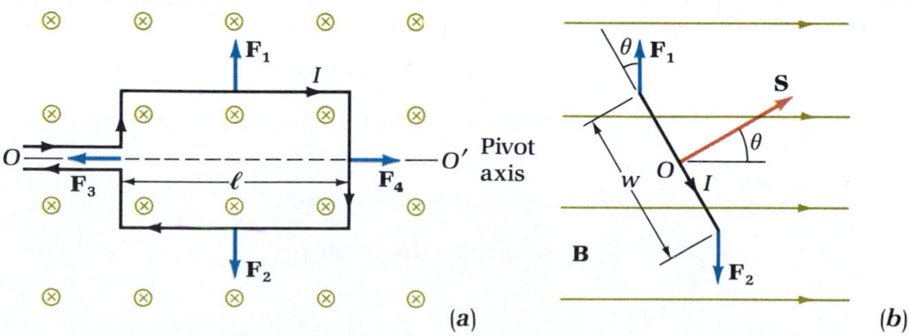

Figure 26-7. A rectangular current loop is in a uniform magnetic field. The area vector **S** is perpendicular to the plane of the loop and has a right-hand sense with respect to the current in the loop.

The magnetic force on each straight segment of the loop can be determined from Eq. 26-3. The force $\mathbf{F}_1$ on the upper element in Fig. 26-7*a* is directed upward and has magnitude

$$F_1 = I\ell B$$

The force $\mathbf{F}_2$ on the lower element has the same magnitude but the opposite direction. These two forces add to zero. Similarly, the forces $\mathbf{F}_3$ and $\mathbf{F}_4$ on the other two segments of length w are equal in magnitude and opposite in direction. Thus the net magnetic force on the current loop is zero. (What is the net magnetic force on the pair of lead wires that feed current to the loop?)

Although the net magnetic force on this loop is zero, the forces do have some effects. The outward-directed forces shown in Fig. 26-7*a* tend to change the shape of the current loop. We assume either that the wires are stiff enough or that mechanical constraints exist so that no appreciable distortion of the loop occurs. Another effect is that these forces produce a torque on the current loop about an axis. (See Chap. 11 for a discussion of torque.) A convenient choice for an axis is one in the plane of the loop and perpendicular to **B**, such as the axis labeled OO' in Fig. 26-7. Notice that if the loop is pivoted about the axis, the torque tends to cause the loop to rotate about the axis.

From Fig. 26-7*a*, we can see that forces $\mathbf{F}_3$ and $\mathbf{F}_4$, acting parallel to axis OO', produce no torque about that axis. The torque produced by $\mathbf{F}_1$ about axis OO' can be determined using Fig. 26-7*b*. The perpendicular distance from the axis to the line of action of this force is $r_\perp = \frac{1}{2}w \sin\theta$; from Eq. 11-3, the magnitude of the torque is

$$\tau_1 = \tfrac{1}{2}wF_1 \sin\theta = \tfrac{1}{2}wI\ell B \sin\theta$$

This torque has a clockwise sense in Fig. 26-7*b*, which corresponds to the direction into the page. Force $\mathbf{F}_2$ also produces a torque with the same sense about this axis, and its magnitude is the same as that produced by $\mathbf{F}_1$. Specifically, $\tau_2 = \tau_1$. Therefore the net magnetic torque has magnitude

$$\tau = \tau_1 + \tau_2 = wI\ell B \sin\theta = ISB \sin\theta$$

where the area of the loop $S = w\ell$ has been used.

The sense of the net torque is clockwise for the arrangement shown in Fig. 26-7*b*. That is the same as the sense of rotation that would carry the direction of the area vector **S** through angle θ into the direction of **B**. This clockwise sense, from a right-hand rule, corresponds to the direction of the torque vector $\boldsymbol{\tau}$, which is perpendicularly into the plane of the paper in Fig. 26-7*b*. (Curl the fingers of your right hand in the sense of the rotation that carries **S** into **B**; the extended thumb then gives the direction into the plane of the figure.) This is the direction of the cross product $\mathbf{S} \times \mathbf{B}$ whose magnitude is $SB \sin\theta$. Thus the torque vector produced on the current loop by a uniform magnetic field is

The right-hand rule gives the direction of $\mathbf{S} \times \mathbf{B}$.

$$\boldsymbol{\tau} = I\mathbf{S} \times \mathbf{B} \qquad (26\text{-}6)$$

Torque on a current loop

You should apply this equation to configurations similar to that seen in Fig. 26-7*b* but with the loop rotated about the axis OO' so that the angle θ is in each of the other three quadrants. In each case the cross product gives the correct direction for the torque on the loop.

Equation 26-6 gives the torque on a single current loop with a rectangular shape. The result also is valid for a plane current loop with any shape in a uniform magnetic field. (See Prob. 2.) Instead of a single loop or turn of wire, we can consider a coil with N turns, as seen in Fig. 26-8. If the coil is closely wound, then each turn lies essentially in a plane. These planes containing the turns are parallel and have the same area **S**, so that the torque on each turn is $\boldsymbol{\tau} = I\mathbf{S} \times \mathbf{B}$. The torque on a coil with N turns carrying current I is just N times the torque on a single current loop. Thus the magnetic torque on the coil in a uniform magnetic field is

$$\boldsymbol{\tau} = NI\mathbf{S} \times \mathbf{B} \qquad (26\text{-}7)$$

Magnetic torque on a coil

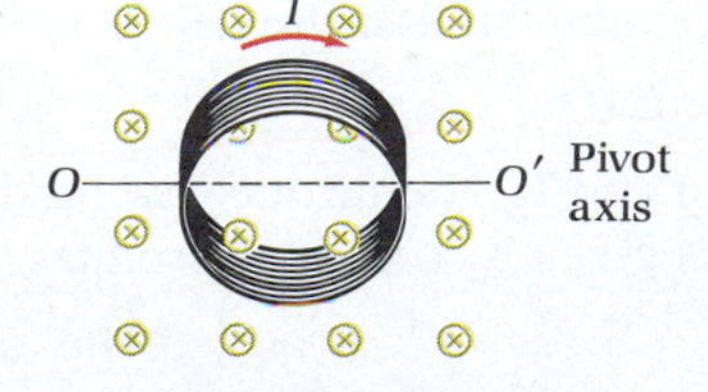

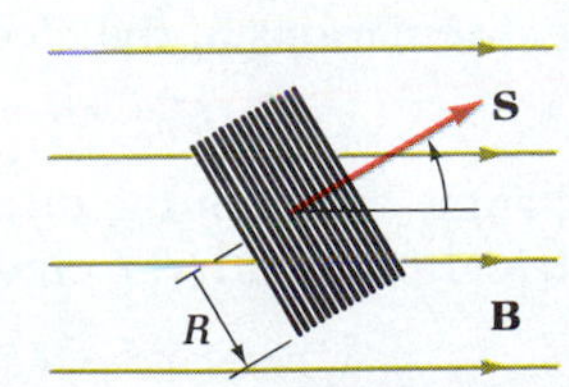

Figure 26-8. A circular coil of radius R is in a uniform magnetic field.

EXAMPLE 26-5

Torque on a current-carrying coil. A simple electric motor has a 100-turn circular coil of radius 15 mm that carries a 65-mA current in a uniform magnetic field of magnitude $B = 23$ mT. At one instant the coil is oriented so that the direction of the area is at $\theta = 25°$ to the field, as shown in Fig. 26-8. The coil is pivoted about an axis through its center perpendicular to **S** and to **B**. *(a)* Determine the magnitude and direction of the magnetic torque on the coil. *(b)* What are your findings if the current sense is reversed? *(c)* For what orientations of the coil is the magnitude of the torque largest and what is this largest value?

Solution. *(a)* The torque is given by Eq. 26-7. The magnitude is

$$\tau = |NI\mathbf{S} \times \mathbf{B}| = NISB \sin \theta$$

The circular coil has face area $S = \pi r^2 = 7.1 \times 10^{-4}\ \text{m}^2$, so

$$\tau = (100)(0.65\ \text{A})(7.1 \times 10^{-4}\ \text{m}^2)(23 \times 10^{-3}\ \text{T}) \sin 25°$$
$$= 4.5 \times 10^{-5}\ \text{N}\cdot\text{m}$$

The direction of the torque is given by the direction of $\mathbf{S} \times \mathbf{B}$, which is into the plane of the paper in Fig. 26-8*b*. The torque tends to produce a clockwise rotation of the coil in that figure.

(b) If the sense of the current is reversed, then **S** is reversed and so is the direction of $\mathbf{S} \times \mathbf{B}$. The angle between the directions of **S** and **B** becomes $180° - 25° = 155°$. The magnitude of the torque is the same as calculated above [$\sin(180° - \theta) = \sin\theta$], but the direction of the torque is reversed.

(c) The maximum value of the magnitude of the torque corresponds to $\sin\theta = \pm 1$, or $\theta = \pm 90°$ in Fig. 26-8*b*. This maximum value is

$$\tau = NISB = 1.1 \times 10^{-4}\ \text{N}\cdot\text{m}$$

What is the direction of the torque for each of these two orientations?

SELF-TEST 26-5. A circular coil of wire is oriented with its windings horizontal; when viewed from above, its current is counterclockwise. *(a)* What is the direction of the area vector **S** for the loop? *(b)* If the loop is in a magnetic field that is directed horizontally toward the west, what is the direction of the magnetic torque on the loop? *(c)* If you are located north of the loop and view it by looking horizontally toward the south, does the magnetic torque tend to cause clockwise or counterclockwise rotation? ***ANSWERS:*** *(a)* up; *(b)* south; *(c)* clockwise.

(a) Fiber S m B I

(b) Fiber m B

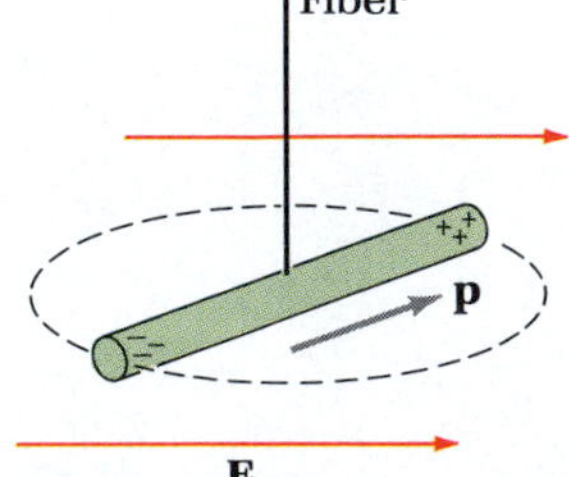

(c)

Figure 26-9. Three objects are suspended by vertical fibers in a horizontal field. *(a)* A current-carrying coil in a magnetic field. *(b)* A bar magnet in a magnetic field. *(c)* An insulating rod charged as a dipole in an electric field.

Magnetic Dipole Moment

If a current-carrying coil is oriented in a uniform magnetic field such that **S** and **B** are parallel ($\theta = 0$), then the magnetic torque is zero. In the absence of torques due to other forces, the coil is in rotational equilibrium with this orientation. However, for any other orientation (except $\theta = \pi$), there is a magnetic torque that tends to align the coil so that **S** and **B** are parallel again. In Fig. 26-9*a*, the coil is suspended in a horizontal magnetic field by a vertical fiber. The coil tends to rotate toward alignment with the field (**S** parallel with **B**) because of the magnetic torque. This same kind of behavior is shown by a bar magnet in a uniform magnetic field (Fig. 26-9*b*) and by an electric dipole in a uniform electric field (Fig. 26-9*c*).

The orientation of an electric dipole in a uniform electric field was considered in Chap. 22. The equilibrium orientation—the electric dipole moment **p** aligned with electric field **E**—corresponds to the minimum value of the electric potential energy U of the electric dipole in an external field. This potential energy depends on the relative orientations of the electric dipole moment and the field:

$$U = -\mathbf{p}\cdot\mathbf{E} = -pE\cos\theta$$

where θ is the angle between the directions of **p** and **E**. The torque that tends to align the electric dipole with the electric field is given by the cross product

$$\boldsymbol{\tau} = \mathbf{p} \times \mathbf{E}$$

This expression for the torque on an electric dipole has the same mathematical form as Eq. 26-7, the torque on a current-carrying coil in a uniform *magnetic* field. In analogy with the electric dipole moment **p**, we can define a *magnetic dipole moment* **m** of a coil carrying a current. Equation 26-7 can be written as

$$\boldsymbol{\tau} = \mathbf{m} \times \mathbf{B} \tag{26-8}$$

Torque on a magnetic dipole and the magnetic dipole moment of a coil

where

$$\mathbf{m} = NI\mathbf{S} \tag{26-9}$$

is the magnetic dipole moment of a coil of face area **S** and with N turns and current I. Equation 26-9 shows that magnetic dipole moment has dimensions of current times area, and the SI unit of magnetic dipole moment is ampere-meters squared ($\text{A}\cdot\text{m}^2$).

Also in analogy with the electric dipole, there is a potential energy for a magnetic dipole in a magnetic field. That potential energy is

$$U = -\mathbf{m}\cdot\mathbf{B} = -mB\cos\theta \tag{26-10}$$

where θ is the angle between **m** and **B** (see Prob. 9). The potential energy is a minimum when **m** and **B** are aligned ($\theta = 0$).

Another look at Fig. 26-9*a* and *b* suggests a connection between a current-carrying coil and a bar magnet. The magnet is also characterized by a magnetic dipole moment. We shall explore the connection in Chap. 30 (but see Prob. 3 at the end of this chapter).

The magnetic torque on a coil, given by Eq. 26-7, can be interpreted by combining Eqs. 26-8 and 26-9. A coil with a current has a magnetic dipole moment $\mathbf{m} = NI\mathbf{S}$, and a uniform magnetic field produces a torque $\boldsymbol{\tau} = \mathbf{m} \times \mathbf{B}$ on a magnetic dipole. Since rotational motion results from a net torque on a coil, many practical applications follow from the effect. It is the basis of operation of electric motors, for example. And since the torque depends on the current, a galvanometer, a meter that measures current, uses the effect.

EXAMPLE 26-6

The moving-coil galvanometer. The essential features of a galvanometer are sketched in Fig. 26-10. A permanent magnet and a soft iron core cause the magnetic field to be approximately radial and uniform in magnitude in the space between each pole and the core. In this space, the wires of a rectangular coil are always perpendicular to the field, so that the effective angle between the directions of **S** and **B** is always 90°. The coil, pivoted about an axis through its center, has an attached pointer, and the scale is calibrated to give current values. A spring exerts a restoring torque on the coil that is proportional to the angular displacement ϕ from the equilibrium orientation at zero current. The magnitude of the restoring torque τ_r depends on the torsional constant κ of the spring: $\tau_r = \kappa\phi$. Show that if a current I exists in the galvanometer, then the coil is in rotational equilibrium with an angular displacement ϕ, which is proportional to the current.

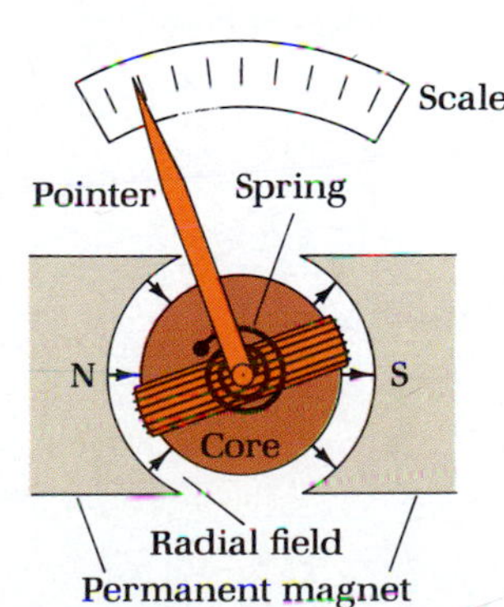

Figure 26-10. Example 26-6: A galvanometer is shown schematically. In the space between the poles of a magnet and the soft iron core, the magnetic field is approximately radial and uniform in magnitude. As a result, the magnetic torque on the current-carrying coil is independent of the orientation of the coil.

Solution. With a current I in the coil, the magnitude of the magnetic torque from Eq. 26-7 is $\tau = NISB$. The coil will rotate from the equilibrium position with no current ($\phi = 0$) to a new equilibrium position where the restoring torque balances the magnetic torque:

$$\kappa\phi = NISB$$

or

$$I = \frac{\kappa}{NSB}\phi$$

Thus the current in the galvanometer is directly proportional to the angular deflection of the pointer. The mechanical galvanometer with a moving coil has been largely replaced by the electronic meter, which has no macroscopic moving parts.

SELF-TEST 26-6. Suppose the current in the coil in Fig. 26-10 is counterclockwise when viewed from above (looking downward at the pointer). That is, the current in the straight

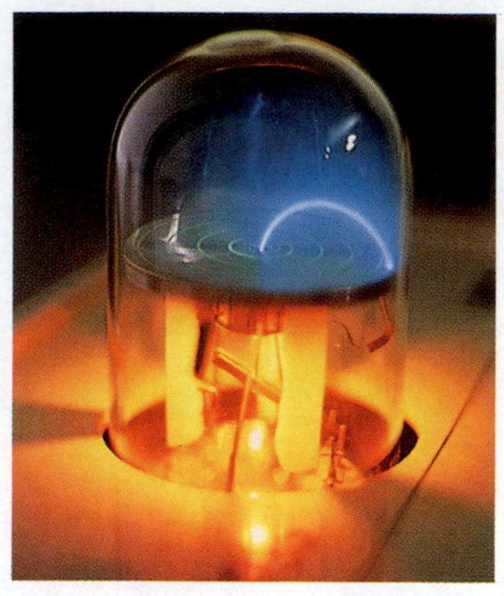
A beam of electrons follows a circular path in a uniform magnetic field.

sections nearest the reader is from left to right. *(a)* What is the direction of the area vector **S** for the coil? *(b)* What is the direction of the magnetic torque on the coil? *(c)* From your view of Fig. 26-10, will the magnetic torque tend to cause clockwise or counterclockwise rotation? ***ANSWERS:*** *(a)* in the same direction as the pointer; *(b)* into the page; *(c)* clockwise.

26-4 MOTION OF CHARGES IN ELECTROMAGNETIC FIELDS

To understand the operation of many modern devices and instruments, we must consider the motion of electrons, protons, and other ions in electric and magnetic (electromagnetic) fields. Electromagnetic forces dominate the motion of charged particles at the atomic level. If an electric field **E** and a magnetic field **B** exist in a region, then the combined force **F** on a particle with charge q and velocity **v** is given by

The Lorentz force

$$\mathbf{F} = q\mathbf{E} + q\mathbf{v} \times \mathbf{B} \tag{26-11}$$

This force is often called the *Lorentz force* after H. A. Lorentz (1853 – 1928), who made many contributions to the understanding of electromagnetic phenomena.

We first consider the motion of a charged particle in a uniform magnetic field with no electric field present. Suppose that the magnitude of the field is B and that the direction is out of the plane of the page, as shown in Fig. 26-11. For simplicity, take the initial velocity to be perpendicular to **B** (but see Prob. 4). From the cross product in Eq. 26-11, the magnetic force is perpendicular to **B** and to the velocity, as shown in Fig. 26-11 for a positive charge. That is, the force and the velocity are perpendicular to each other and lie in a plane perpendicular to **B**. If the magnetic force is the only force acting on the particle, then, from Newton's second law, the acceleration of the particle is perpendicular to the velocity and also lies in the plane perpendicular to **B**. Since the acceleration is perpendicular to the velocity, only the direction of the velocity changes, and the path of the particle is in the plane perpendicular to **B**. The charged particle moves in a circular path of radius r with constant speed v, and the acceleration is the centripetal acceleration of magnitude $a = v^2/r$. The centripetal force is provided by the magnetic force of magnitude $|q\mathbf{v} \times \mathbf{B}|$ so that Newton's second law gives

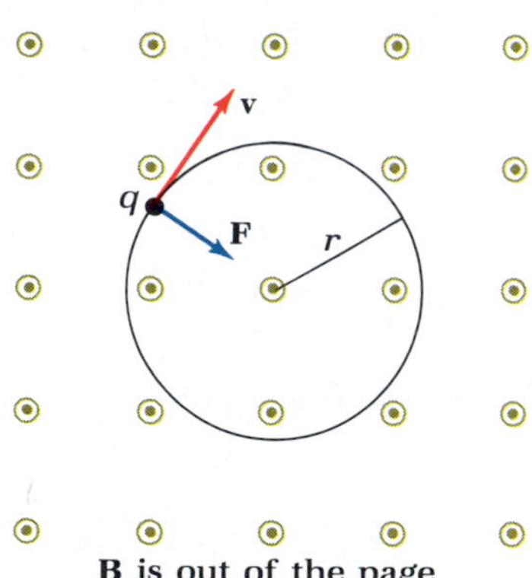

B is out of the page

Figure 26-11. A positively charged particle moves in a circular path perpendicular to a uniform magnetic field.

$$\frac{mv^2}{r} = |q\mathbf{v} \times \mathbf{B}| = |q|vB \sin 90° = qvB$$

where m is the mass of the particle and we have assumed that q has a positive value ($|q| = q$). We can simplify this expression and solve for one of the quantities, for example, the radius of the circular path, in terms of the others:

$$r = \frac{mv}{qB} \tag{26-12}$$

Equation 26-12 shows that the radius of the circular path and the speed are proportional for a charged particle in a uniform magnetic field. Thus for a given type of particle, those with higher speeds have larger radii of curvature. Notice from Eq. 26-12, however, that the angular speed ω is the same for all such particles:

$$\omega = \frac{v}{r} = \frac{q}{m} B \tag{26-13}$$

In a uniform magnetic field, the angular speed, or the *angular frequency,* of the circular motion depends only on the field B and on the charge-to-mass ratio q/m for that type of particle. Since the operation of one of the early particle accelerators, the *cyclotron,* is based on this property, the angular frequency is often called the *cyclotron frequency,* $\omega_c = qB/m$.

The cyclotron frequency depends on the charge-to-mass ratio of the particle.

EXAMPLE 26-7

The bubble chamber. As a charged particle moves through a liquid-hydrogen bubble chamber, it ionizes some of the molecules along its path. Small bubbles that form at the ionized sites make the track of the particle visible. A magnetic field in the region causes the path to be curved, and the momentum of the particle can be determined from the radius of curvature of the path. (See Fig. 26-12.) Determine the magnitude of the momentum and the speed of a proton ($q = e = 1.60 \times 10^{-19}$ C, $m = 1.67 \times 10^{-27}$ kg) at a point where the radius of curvature of the path is 2.67 m and the magnitude of the magnetic field is 0.140 T.

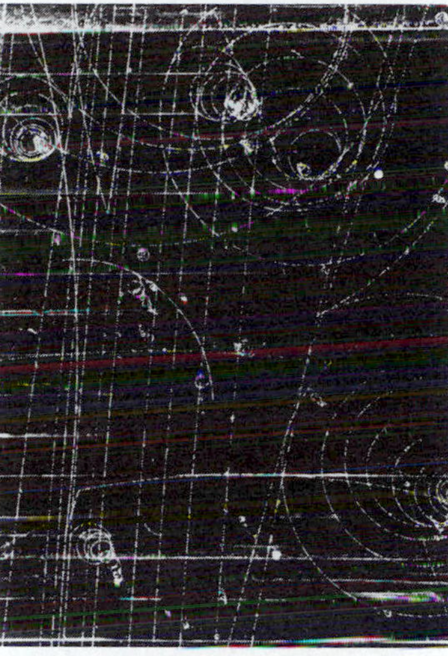

Figure 26-12. Photograph of the paths of charged particles in a bubble chamber. As a particle slows due to interactions with the liquid hydrogen, its speed decreases. As its speed decreases, so does the radius of each infinitesimal section of arc path, and the path is a spiral.

Solution. From Eq. 26-12, the magnitude of the momentum is

$$p = mv = qBr = (1.60 \times 10^{-19}\ \text{C})(0.140\ \text{T})(2.67\ \text{m})$$
$$= 5.98 \times 10^{-20}\ \text{kg}\cdot\text{m/s}$$

The speed of the proton is

$$v = \frac{p}{m} = \frac{5.98 \times 10^{-20}\ \text{kg m/s}}{1.67 \times 10^{-27}\ \text{kg}}$$
$$= 3.58 \times 10^{7}\ \text{m/s}$$

This speed is about $0.1c$, where c is the speed of light. Relativistic effects (Chap. 38) become increasingly important for speeds approaching the speed of light.

SELF-TEST 26-7. At a particular point along its track in a bubble chamber, an α particle ($m = 6.64 \times 10^{-27}$ kg, $q = 2e = 3.20 \times 10^{-19}$ C) has a radius of curvature of 3.15 m. The magnitude of the bubble chamber's magnetic field is 0.19 T. Determine the speed of the α particle at this point. ***ANSWER:*** 2.88×10^{7} m/s.

EXAMPLE 26-8

The cyclotron. Charged particles are accelerated repeatedly in a cyclotron by an alternating potential difference applied across the gap between two hollowed, D-shaped conductors, or "dees," as shown in Fig. 26-13. A uniform magnetic field is perpendicular to the plane of the figure and out of the page. A positive charge with speed v in a dee moves in a circular arc of radius $r = mv/qB$. The particle is only accelerated as it crosses the gap between the dees because the electric field is zero within each dee. The speed and radius increase each time the particle is accelerated in the gap, but the angular frequency ω_c remains the same. The key to having the speed increased each time is to have the accelerating potential difference applied in phase with the circulating charge. Thus the potential difference is applied with an oscillator tuned to the cyclotron frequency. As the beam (a group of charges) reaches the outer edge of the cyclotron, a deflector plate directs the beam out to a target area. Suppose the magnetic field has magnitude 1.4 T in a cyclotron of radius 0.50 m. *(a)* What frequency oscillator must be used to accelerate deuterons? (A *deuteron,* with $q = e$ and $m = 3.3 \times 10^{-27}$ kg, is the nucleus of heavy hydrogen, or deuterium.) *(b)* Determine the speed and kinetic energy of deuterons emerging from the cyclotron.

The first cyclotron was constructed by E. O. Lawrence in 1932.

Solution. *(a)* The angular frequency of the oscillator must match the cyclotron frequency of a deuteron in this magnetic field. From Eq. 26-13,

$$\omega_c = \frac{qB}{m} = \frac{(1.6 \times 10^{-19}\ \text{C})(1.4\ \text{T})}{3.3 \times 10^{-27}\ \text{kg}}$$
$$= 6.8 \times 10^{7}\ \text{rad/s}$$

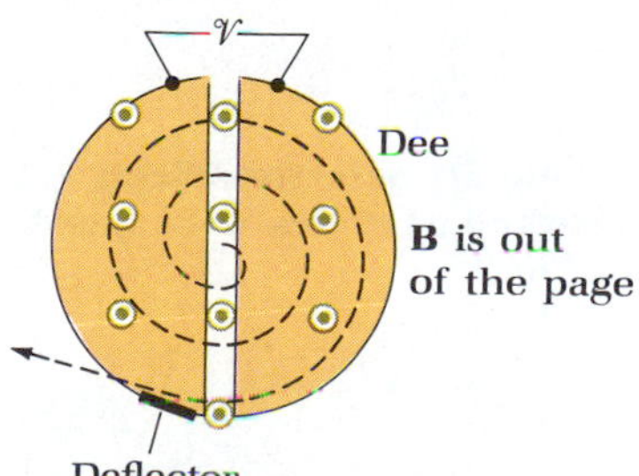

Figure 26-13. Example 26-8: An alternating potential difference accelerates positively charged particles in a beam across the gap between the dees in a cyclotron. The magnetic field is out of the plane of the figure and is uniform in the region of the dees.

Ordinarily, the frequency ν in hertz (Hz; cycles per second) is specified, and

$$\nu = \frac{\omega_c}{2\pi} = 11\ \text{MHz}$$

Electronic oscillators with this frequency are readily available.

(b) The speed of a deuteron at the outer edge of the dee is

$$v = \omega r = (68 \times 10^{6}\ \text{rad/s})(0.50\ \text{m}) = 3.4 \times 10^{7}\ \text{m/s}$$

and the kinetic energy is

$$K = \tfrac{1}{2}mv^2 = \tfrac{1}{2}(3.3 \times 10^{-27}\ \text{kg})(3.7 \times 10^7\ \text{m/s})^2$$
$$= 1.9 \times 10^{-12}\ \text{J} = 12\ \text{MeV}$$

You should be able to show that the kinetic energy of a charged particle emerging from a cyclotron of radius R is given by

$$K = \frac{(qBR)^2}{2m}$$

The maximum kinetic energy is limited for a cyclotron because of relativistic effects, which make the frequency of the circular motion depend on speed for speeds approaching that of light. In the *synchrocylotron,* the frequency is changed to synchronize with the speed-dependent frequency of the charges. See the Commentary for a description of a *synchrotron,* which is a different kind of accelerator.

SELF-TEST 26-8. What is the cyclotron frequency of a proton in a field of magnitude 1.0 T? ***ANSWER:*** 9.6×10^7 rad/s = 15 MHz.

A Velocity Selector

We now consider a configuration of electric and magnetic fields that serves as a *velocity selector* for charged particles. Suppose that uniform electric and magnetic fields exist in a region of space and that these fields are perpendicular, as shown in Fig. 26-14. The force on a charged particle moving in that region is given by Eq. 26-11. For the case illustrated in the figure for a positively charged particle, there is a particular velocity for which the net force is zero. The upward electric force balances the downward magnetic force, so that the net force is zero. For negatively charged particles with this velocity, the directions of the forces would be reversed, and the net force is still zero. Charges with this velocity will pass through the region undeviated. Since the magnetic force depends on the velocity of the particle but the electric force does not, the net force will not be zero for a particle with a different velocity. For a charge with a larger speed, the magnetic force will have a magnitude larger than that of the electric force. Such positively charged particles with a larger speed will be deflected downward. Similarly, slower positively charged particles will be deflected upward.

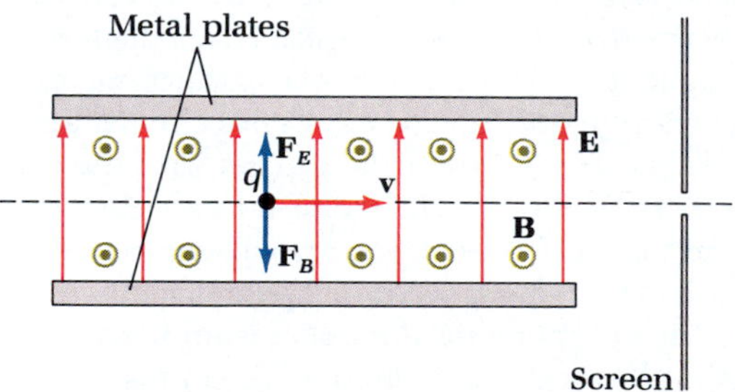

Figure 26-14. A positively charged particle moves in crossed electric and magnetic fields.

A velocity selector selects charged particles with speed $v = E/B$.

The value of the "selected" velocity is obtained by requiring that the combined force in Eq. 26-11 be zero:

$$\mathbf{F} = 0 = q(\mathbf{E} + \mathbf{v} \times \mathbf{B})$$

Taking the magnitude of this expression, we determine the speed of the charged particles passing undeviated through the selector:

$$v = \frac{E}{B}$$

A velocity (or speed) selector is used in devices such as the *mass spectrometer* (Prob. 5) and in experiments such as the Thomson experiment (Prob. 6). A type of "natural" velocity selector is at work in an important effect in conductors called the *Hall effect.*

EXAMPLE 26-9

The Hall effect. Consider a section of a current-carrying conductor in a uniform magnetic field, as shown in Fig. 26-15*a*. With the sense of the current in the positive x direction, positive charge carriers would move in that direction and negative charge carriers in the opposite direction. Each type of charge would be deflected by the magnetic field to the lower surface. We suppose now that only positive charge carriers are present; then the lower surface would become positively charged, leaving the top surface with a negative charge, a deficiency of positive charge. This charge separation produces an electric field in the conductor, as seen in Fig. 26-15*b*. In the steady-state case, the electric field component E_y, called the *Hall field,* exerts an electric force on the moving charge carriers that tends to balance the magnetic force due to the magnetic field component B_z. These crossed electric and magnetic fields act, in an average sense, as a velocity selector for the drift velocity $\mathbf{v}_d$.

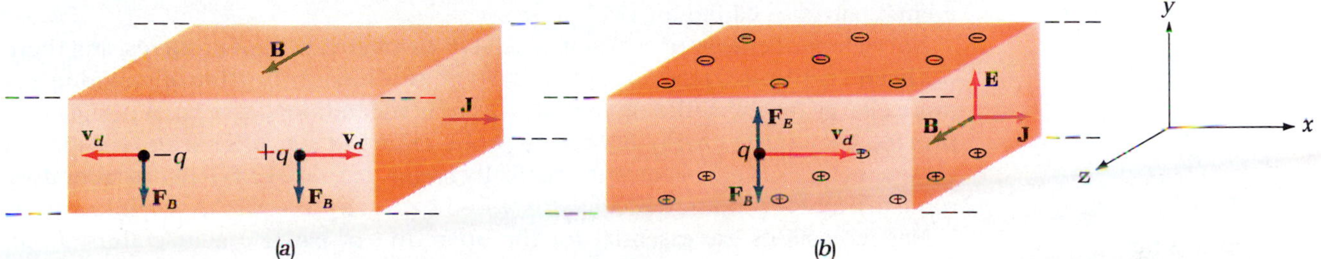

Figure 26-15. Example 26-9: *(a)* A magnetic field is perpendicular to the current density **J** in a conductor. The magnetic force is downward on both positive and negative charge carriers. *(b)* The electric and magnetic forces balance for a positive charge carrier moving with the drift velocity.

Show that the Hall field E_y is proportional to J_xB_z and that the *Hall coefficient* $R_H = E_y/J_xB_z$ is given by

$$R_H = \frac{1}{nq}$$

where n is the density of carriers, each with charge q.

Solution. Consider a particle of charge q moving with the drift velocity $\mathbf{v}_d = v_{dx}\mathbf{i}$, where $\mathbf{i}$ is a unit vector along the x axis. The current density $\mathbf{J}$ in the conductor is

$$\mathbf{J} = J_x\mathbf{i} = nqv_{dx}\mathbf{i}$$

Using Eq. 26-11, we require that the y component of the force be zero: $F_y = q(E_y - v_{dx}B_z) = 0$, or

$$E_y = v_{dx}B_z$$

Multiplying by the density of charge carriers n and using $J_x = nqv_{dx}$ gives

$$nqE_y = nqv_{dx}B_z = J_xB_z$$

or

$$E_y = \frac{J_xB_z}{nq}$$

Thus E_y is proportional to J_xB_z, and the Hall coefficient is the constant of proportionality,

$$R_H = \frac{E_y}{J_xB_z} = \frac{1}{nq}$$

Notice that the Hall coefficient has a positive value, corresponding to our assumption that all the charge carriers were positive. You should be able to show that the Hall coefficient reverses sign if you assume that all the charge carriers are negative. The Hall coefficient gives, in this way, information about the signs of charge carriers in conductors. Since E_y depends on B_z, notice also that the Hall effect can be used to determine the magnetic field in a region by measuring the Hall field E_y in a previously calibrated conductor (How?)

SELF-TEST 26-9. A charged particle moves along a straight path that is perpendicular both to a uniform magnetic field **B** and to a uniform electric field **E**, and **B** and **E** are perpendicular to each other. If $B = 0.28$ T and $E = 160$ V/m, what is the particle's speed? ***ANSWER:*** 5.7×10^3 m/s.

COMMENTARY: Magnetic Fields and Particle Accelerators

There seems to be a general rule in physics that the smaller the object of study, the larger the instruments must be. Nowhere is this rule more evident than in particle physics, the study of the fundamental structure of matter. This structure is revealed in collisions of particles, say of a proton with an antiproton (a particle with the same mass but opposite electric charge of a proton). If the energies of the colliding particles are large enough, a multitude of particles is produced in these collisions. The properties of these particles give clues about the structure of matter on the smallest scale, which is currently expressed in terms of a model involving fundamental particles called quarks.*

Now a higher energy of collision produces a richer variety of particles, and there has been a continued push toward developing accelerators capable of providing yet more energy for the colliding particles. As new accelerators have been designed to provide the higher energy needed to probe matter at a smaller scale, the size of these accelerators has become dramatically larger, from those that fit in a laboratory room to a newly proposed accelerator whose size is expressed in tens of kilometers.

Magnetic fields are essential for the operation of particle accelerators. In an accelerator, the charged particles in a beam are accelerated (given energy) in an evacuated tube or region by an electric field, as described for the cyclotron in Sec. 26-4. A magnetic field is often used to deflect the beam or to cause it to move in a curved path. In the cyclotron, which is a relatively small accelerator, the magnetic field is essentially uniform over the cross section. As the particles in the beam gain energy, the radius of the path increases, $r = mv/qB$, from a very small initial radius to a final radius at the edge of the device. Thus the size of the region of the magnetic field limits the energy gained by a particle in this type of accelerator.

The much larger *synchrotron* type of accelerator is designed to have the particle beam travel along a fixed path in a ringlike tube. Since magnetic fields are needed only at the tube to deflect and guide the beam, it is feasible to have accelerator rings of large diameter. As particles move around the ring, the position of a group of particles in the beam is synchronized with the accelerating stages that increase the particle energy. The currents that produce the magnetic fields are also synchronized with the group, so that the magnetic force on the charged particles bends the beam and causes the group to remain aligned near the center of the tube.

Among the large synchrotrons is one at Fermilab at Batavia, Illinois, which has a 1.9-km ring diameter (see page 160). In the *Tevatron* ring at Fermilab, protons are accelerated to a kinetic energy of 1 TeV (10^{12} eV) and a speed that is close to the speed of light. Antiprotons are accelerated to the same energy, but, since they have negative charge, they circulate in the opposite direction. When the proton and antiproton paths are caused to cross, they collide with 2 TeV of energy. This is the great advantage of a collider: Since none of the energy is associated with the motion of the center of mass, all of the energy is available for processes that occur in the collision.

A much larger synchrotron is presently being planned. The Superconducting SuperCollider (SSC) is designed as a proton-proton collider. It will use superconducting coils on magnets placed around a ring that measures about 80 km around. Two groups of protons will be accelerated to about 20 TeV and caused to collide, so that 40 TeV will be available for collision processes. With the much higher energy provided by this accelerator, it is hoped that some fundamental questions about the structure of matter will be answered. No one will be surprised if the experiments also raise some new questions that call for still higher energies.

* In the "extended" edition of this text, quarks are discussed in Chap. 43.

SUMMARY

Section 26-1. The Magnetic Field

The magnetic field **B** at a point in space is defined in terms of the force exerted by the field on a particle with charge q and velocity **v** at that point:

$$\mathbf{F} = q\mathbf{v} \times \mathbf{B} \tag{26-1}$$

Section 26-2. Force on a Current-Carrying Conductor

The magnetic force on a straight section of a current-carrying conductor in a uniform field is given by

$$\mathbf{F} = I\boldsymbol{\ell} \times \mathbf{B} \tag{26-3}$$

Generally the magnetic force on a section of a current-carrying conductor is

$$\mathbf{F} = \int I\, d\boldsymbol{\ell} \times \mathbf{B} \tag{26-5}$$

Section 26-3. Torque on a Current Loop

If a coil with N turns and current I is in a uniform magnetic field, then the torque exerted by the field on the coil is

$$\boldsymbol{\tau} = NI\mathbf{S} \times \mathbf{B} \tag{26-7}$$

where **S** is the area vector of the plane of a loop of the coil. The direction of **S** is perpendicular to the plane of the area, with a sense given by a right-hand rule.

The magnetic dipole moment **m** of a coil with current I is

$$\mathbf{m} = NI\mathbf{S} \tag{26-9}$$

and the torque on a magnetic dipole in a uniform magnetic field is

$$\boldsymbol{\tau} = \mathbf{m} \times \mathbf{B} \tag{26-8}$$

The potential energy of the dipole is

$$U = -\mathbf{m} \cdot \mathbf{B} \tag{26-10}$$

Section 26-4. Motion of Charges in Electromagnetic Fields

Electric and magnetic fields each exert a force on a particle of charge q. The combined force is given by

$$\mathbf{F} = q(\mathbf{E} + \mathbf{v} \times \mathbf{B}) \tag{26-11}$$

If the velocity is perpendicular to a uniform magnetic field and no other forces are acting, then the charged particle moves in a circular path. The speed of a particle of mass m and the radius of its circular path are connected by

$$r = \frac{mv}{qB} \tag{26-12}$$

The cyclotron frequency ω_c is independent of the values of v and r.

$$\omega_c = \frac{v}{r} = \frac{qB}{m} \tag{26-13}$$

Crossed electric and magnetic fields act as a velocity selector for charged particles entering the region with velocities perpendicular to the crossed fields. Only those with speed

$$v = \frac{E}{B}$$

pass through the region undeviated.

QUESTIONS

1. Is it possible for the magnetic force on a moving charge in a magnetic field to be zero? Explain.

2. Is it possible for the electric force on a moving charge in an electric field to be zero? Explain.

3. Is it possible for the electromagnetic force on a moving charge in an electromagnetic field to be zero? Explain.

4. Is the magnetic force on a moving charge, $\mathbf{F} = q\mathbf{v} \times \mathbf{B}$, always perpendicular to **v**? To **B**? Is **v** always perpendicular to **F**? To **B**? Explain.

5. Suppose the magnetic field were defined so that the field is parallel to the force on a moving positive charge. Explain why such a definition would be unsatisfactory.

6. You are given the following information: (i) The force on a charge $q = 2.4 \times 10^{-14}$ C at rest at a point is 3.7×10^{-12} N in the positive z direction; (ii) the force on the charge at that point when moving in the positive x direction with speed 2.2×10^3 m/s is 3.1×10^{-12} N in the positive z direction; (iii) only electric and magnetic forces are significant. Can you completely determine the electric field at that point? Can you completely determine the magnetic field at that point? If you cannot completely determine either field, what additional information is necessary?

7. The electric field is defined in terms of the electric force **F** on a small test charge q: $\mathbf{E} = \mathbf{F}/q$. Can Eq. 26-1 be similarly solved for **B**? Explain.

8. The charge carriers (electrons) in a metal wire at temperature T are in random thermal motion although the current is zero. The wire is in a magnetic field. Is there a magnetic force on each charge carrier? Is there a magnetic force on the wire? Explain.

9. In a current-carrying conductor, a magnetic field exerts a force on the moving electrons in the conductor. What exerts the force on the conductor?

10. A simple electric motor consists of a current-carrying coil turning in a static magnetic field which exerts a torque on the coil. Explain how the motor can deliver mechanical energy if a static magnetic field can do no work on a moving charge.

11. Explain how you can use a battery, a coil of wire, and a thread as a compass.

12. Equation 26-10 gives the potential energy of a magnetic dipole in a uniform magnetic field. What orientation of the dipole corresponds to the lowest energy? The highest energy? For what orientation is the potential energy zero?

13. For what orientation(s) of a current-carrying coil in a uniform magnetic field is the magnitude of the magnetic torque a maximum? A minimum?

14. The torque on a magnetic dipole in a uniform field, $\tau = \mathbf{m} \times \mathbf{B}$, is zero if **m** and **B** are parallel or if they have opposite directions. Does either of these orientations correspond to stable equilibrium? Explain.

15. An electron in a cyclotron has a cyclotron frequency of $\omega_c = eB/m$. Can an electron have a cyclotron frequency if it is in a magnetic field but not in a cyclotron? Explain.

16. Show that the motion of a charged particle in electric and magnetic fields depends on the charge and mass of the particle only through their ratio q/m. Thus neither q nor m, only their ratio, can be independently determined by observations of the particle's motion in only electromagnetic fields.

17. Does a velocity selector pass both protons and alpha (α) particles ($q_\alpha = 2e$, $m_\alpha = 4m_p$) with the same velocity? What about protons and electrons? Explain.

18. Can a magnetic field be used to separate protons and electrons with the same velocity? What about protons and alpha particles? Explain.

19. Suppose that two charged particles move in circular paths of the same radius in a uniform magnetic field. Must these particles have the same speed? The same charge? The same mass? What can you say about these two particles if they have the same cyclotron frequency? Explain.

20. A proton is released from rest in a region with electric and magnetic fields which are parallel and uniform. How does the proton move? Would an electron behave differently? Explain.

21. If a conductor had both positive and negative charge carriers, could the Hall coefficient be positive? Negative? Zero? Explain.

22. The electron beam in Fig. 26-16 is from left to right. What is the direction of the magnetic field inside the tube?

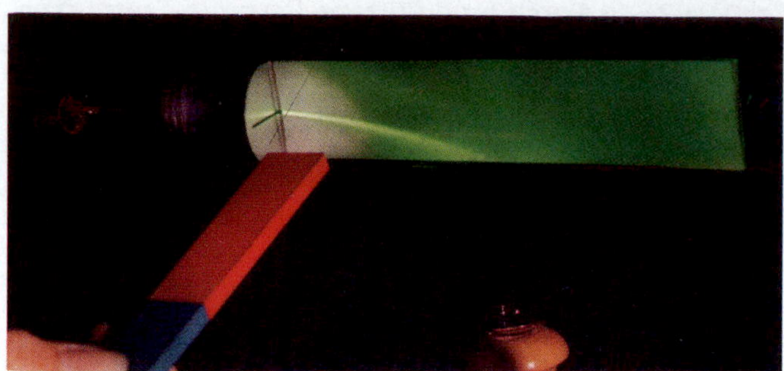

Figure 26-16. Question 22.

23. Complete the following table:

Symbol	Represents	Type	SI unit
B			
$I\,d\boldsymbol{\ell}$		Vector	
τ	Torque		
m			
ω_c			rad/s

EXERCISES

Section 26-1. The Magnetic Field

1. A proton has a velocity of magnitude 2.45×10^6 m/s at a point where $B = 0.117$ T. The direction of the velocity is at 134° from the direction of the magnetic field. *(a)* Show in a diagram the directions of **v**, **B**, and the magnetic force **F** on the proton. *(b)* Determine the magnitude of the force. *(c)* Repeat for an electron with the same velocity.

2. Show that the tesla, the SI unit of magnetic field, can be expressed as $\text{kg}\cdot\text{A}^{-1}\cdot\text{s}^{-2}$.

3. *(a)* Determine the magnetic force on an electron with velocity components $v_x = 4.4 \times 10^6$ m/s, $v_y = -3.2 \times 10^6$ m/s, $v_z = 0$ at a point where the magnetic field has components $B_x = 0$, $B_y = -12$ mT, $B_z = 12$ mT. *(b)* Show the directions of these vectors in a diagram.

4. The magnetic field at a point is determined by projecting protons of speed $v = 3.0 \times 10^7$ m/s through the point from several directions and measuring the force, assumed to be magnetic. It is noticed that if the velocity is parallel to the z direction, the force is zero. But if the velocity is parallel to the y direction, the force is in the negative x direction and has magnitude $F = 1.8 \times 10^{-13}$ N. *(a)* Determine the magnitude and direction of the magnetic field at that point. *(b)* What force (magnitude and direction) is exerted on a proton with that speed if the direction of the velocity is parallel to the x direction?

5. Compare the magnitudes near the earth's surface of the weight of an electron and a typical magnetic force exerted by the earth's magnetic field ($B = 10^{-5}$ T) on an electron with speed 10^6 m/s.

6. Compare the magnitudes near the earth's surface of a typical magnetic force exerted by the earth's magnetic field ($B = 10^{-5}$ T) on an electron with speed 10^6 m/s and the electric force exerted on an electron by an atmospheric electric field ($E = 100$ N/C).

7. A 2-g hailstone with charge -7×10^{-12} C falls vertically downward with speed 80 m/s. In this region there are gravita-

tional, electric, and magnetic fields whose magnitudes are $g = 9.8\ m/s^2$, $E = 120$ N/C, and $B = 40\ \mu T$, respectively. The fields **g** and **E** are directed vertically downward, while the direction of **B** is horizontal and to the north. *(a)* Determine the magnitude and direction of the force exerted on the hailstone by each of these fields. *(b)* Are any of these forces negligible? *(c)* List any other significant forces acting on the hailstone.

Section 26-2. Force on a Current-Carrying Conductor

8. A section of straight wire of length 0.40 m and with current $I = 7.0$ A is oriented at angle $\theta = 27°$ to a uniform magnetic field of magnitude $B = 1.2$ T, as shown in Fig. 26-17. *(a)* Determine the magnitude and direction of the magnetic force on the section of wire. *(b)* How would the answers to part *(a)* change if I is doubled? If B is doubled? If θ is doubled?

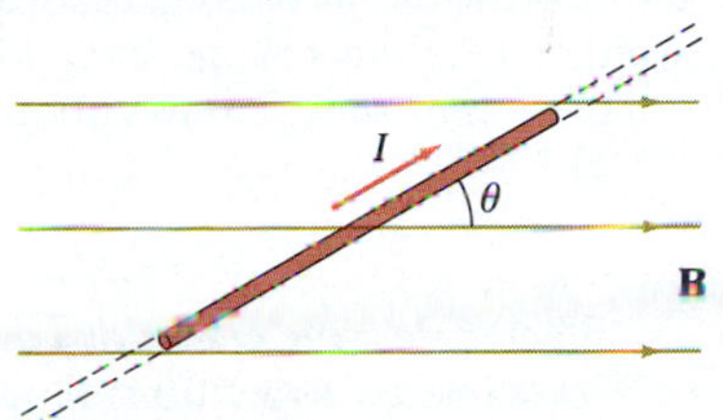

Figure 26-17. Exercise 8.

9. A force of 2.2 mN is found to act on a 250-mm length of current-carrying wire that is perpendicular to a magnetic field of magnitude 340 mT. *(a)* What current exists in the wire? *(b)* Show the sense of the current and the directions of the force and the field in a diagram.

10. Consider a horizontal conducting rod of density ρ and cross-sectional area A that is perpendicular to a horizontal magnetic field **B**, as shown in Fig. 26-18. Flexible lead wires are connected to the ends of the rod, which carries a current I such that the magnetic force balances the weight of the rod. *(a)* Determine the current I in terms of ρ, A, g, and B. *(b)* What is the sense of the current? *(c)* Evaluate the current for $\rho = 2.7 \times 10^3\ kg/m^3$, $A = 100\ mm^2$, $B = 200$ mT, $g = 9.8\ m/s^2$.

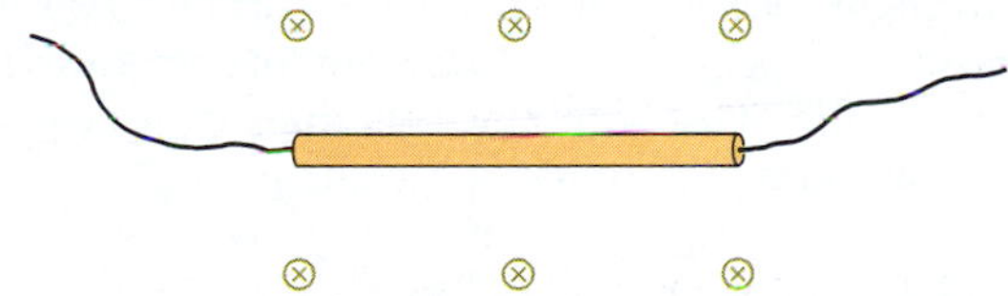

Figure 26-18. Exercise 10.

11. A wire of length ℓ forms a section of a circuit with current I. The section lies along the x axis, as shown in Fig. 26-19. At points along this axis, the magnetic field has only a y component, which varies as $B_y = A/x$, where A is a positive constant. Determine *(a)* the direction and *(b)* the magnitude of the magnetic force on the wire in terms of I, A, and ℓ.

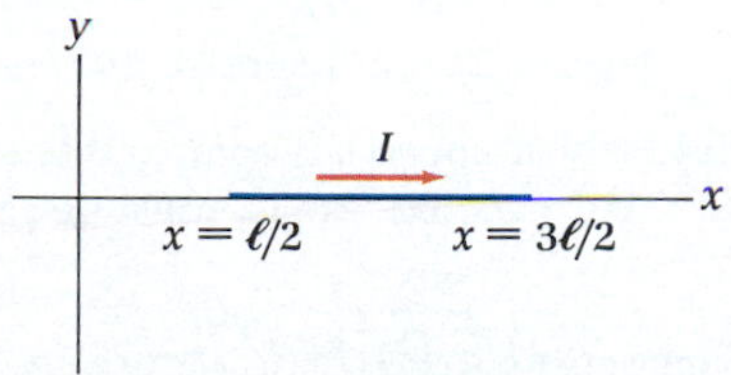

Figure 26-19. Exercise 11.

12. Consider a current-carrying wire of arbitrary shape lying in a plane perpendicular to a uniform magnetic field. Choose a coordinate system so that the magnetic field is in the z direction and the endpoints of the wire lie on the y axis, as shown in Fig. 26-20. A current element can be expressed as

$$I\,d\boldsymbol{\ell} = I\,dx\,\mathbf{i} + I\,dy\,\mathbf{j}$$

Prove that the net magnetic force on this wire is the same as would be exerted on a straight section of wire connecting the two endpoints. See the next exercise for a generalization.

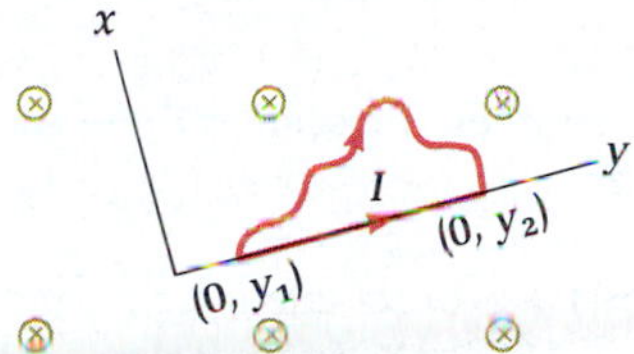

Figure 26-20. Exercise 12.

13. *(a)* Review the discussion in Example 26-3 and prove that the net magnetic force on a current-carrying wire of arbitrary shape in a uniform magnetic field is given by

$$\mathbf{F} = I\boldsymbol{\ell} \times \mathbf{B}$$

where $\boldsymbol{\ell}$ is the displacement from one endpoint of the wire to the other endpoint, with the same sense as the current in the wire. Why must the field be uniform? *(b)* What is the net magnetic force on a current *loop* of arbitrary shape in a uniform magnetic field?

Section 26-3. Torque on a Current Loop

14. A rectangular current loop of dimensions 20 mm by 30 mm is oriented so that the plane of the loop is at 40° to the direction of a uniform magnetic field of magnitude 380 mT. *(a)* Show the arrangement on a diagram and assign a sense to the current. *(b)* What angle does the (vector) area of the loop make with the field? *(c)* Determine the magnetic torque on the loop about an axis perpendicular to **B**, as seen in Fig. 26-7, if the current is 1.5 A.

15. A circular coil has 100 turns, each of radius 15 mm. The coil carries 250-mA current and is pivoted, as shown in Fig. 26-8, to rotate about an axis perpendicular to a uniform magnetic field of magnitude 0.40 T. Determine the magnetic torque on the coil if the direction of the area makes an angle with the field of *(a)* 60°, *(b)* 90°, *(c)* 120°.

16. The rectangular loop shown in Fig. 26-21 is pivoted to rotate about the z axis. A uniform magnetic field of magnitude 300 mT is in the y direction. *(a)* Determine the magnitude and direction of the force on each side of the loop. *(b)* Determine the torque about the z axis. *(c)* Determine the torque about an axis parallel to the z axis but through the center of the loop. *(d)* For what orientation(s) of the loop is the magnitude of the magnetic torque a maximum?

17. The rotating coil in the galvanometer in Example 26-6 has 12 turns and face area $4.0 \times 10^{-4}\ m^2$. The magnitude of the radial magnetic field is 140 mT, and the spring has a torsional

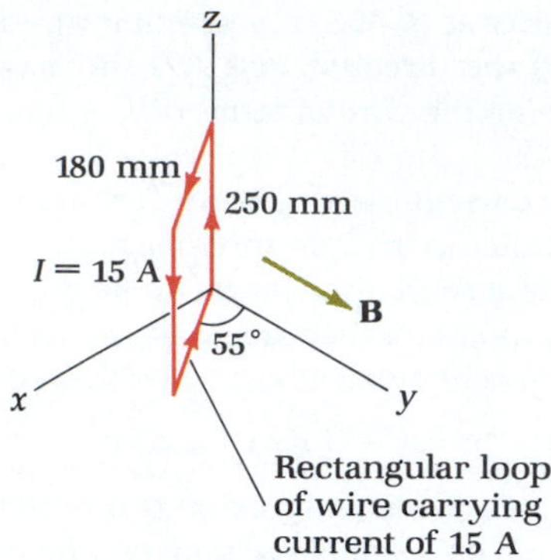

Figure 26-21. Exercise 16.

constant of 6.7×10^{-7} N·m·rad^{-1}. What current causes a deflection of one division on the scale if these divisions are separated by 0.10 rad?

18. One of the coils in a direct-current (dc) motor has 78 turns and face area 1.1×10^{-2} m^2. If the design of the motor specifies the maximum value of the magnitude of the torque to be 4.2 N·m in a field of magnitude 0.34 T, what current must the motor coil draw?

19. A 1200-turn coil has a square cross section of edge 12 mm. It carries a 150-mA current in a uniform magnetic field of magnitude 1.2 T. *(a)* Determine the magnetic dipole moment of the coil. *(b)* Determine the maximum magnitude of the magnetic torque on the dipole. *(c)* For what orientation of the dipole does the torque have one-half of the maximum magnitude?

20. Show that the SI unit of magnetic dipole moment can be expressed as joules per tesla (J/T).

21. A compass needle has a magnetic dipole moment of magnitude 0.1 A·m^2. It points north at a place where the earth's magnetic field points north with a magnitude 5×10^{-5} T. *(a)* How much work must be done to change the orientation of the needle from north to east in this field? *(b)* What is the magnetic torque on the needle when it points east? *(c)* Determine the torque and potential energy if the needle points south.

22. Consider the compass needle in the previous exercise. Given 40 m of wire, design a circular coil that can have the same magnetic dipole moment as the needle, subject to the restriction that the current does not exceed 0.5 A. Specify the radius of the circular coil, the number of turns, and the current.

23. The torque on a small, 400-turn coil carrying 150 mA has a maximum magnitude of 1.6×10^{-3} N·m in a uniform field of magnitude 400 mT. *(a)* Determine the magnetic dipole moment of the coil. *(b)* Determine the face area of the coil. *(c)* If the dipole moment is at 40° to the field direction, what is the potential energy of the dipole? *(d)* How much work must an external agent perform to rotate the dipole moment through 180° from alignment with the field to opposite the field?

24. The electron has an intrinsic magnetic dipole moment of magnitude 9.3×10^{-24} A·m^2. *(a)* Express the value in units of electron volts per tesla (eV/T). *(b)* An electron is in a uniform magnetic field of magnitude 1.5 T. What change in potential energy (in eV) occurs if the orientation of the magnetic dipole changes from alignment with the field to opposite the field? *(c)* What are the answers to parts *(a)* and *(b)* for a proton ($m = 1.4 \times 10^{-26}$ A·m^2)?

Section 26-4. Motion of Charges in Electromagnetic Fields

25. An α particle ($q = 2e$, $m = 6.7 \times 10^{-27}$ kg) moves in a plane perpendicular to a magnetic field of magnitude 0.55 T. *(a)* Determine the magnitude of the momentum of the α particle if the radius of its path is 0.27 m. *(b)* Determine the speed and *(c)* the kinetic energy in electron volts of the α particle.

26. Protons with kinetic energy of 7 MeV are injected into a region perpendicular to a uniform magnetic field of magnitude 0.60 T. *(a)* Determine the radius of the path of a proton. *(b)* By what factor would the radius change if the kinetic energy were doubled? *(c)* Determine the cyclotron frequency for both cases.

27. Deuterons ($q = e$, $m = 3.3 \times 10^{-27}$ kg) with kinetic energy of 12 keV are injected near the center of a cyclotron with $B = 1.7$ T and a 1.1-m diameter. *(a)* What should be the angular frequency of the oscillator used to accelerate the deuterons? *(b)* Determine the initial radius of curvature of the path. *(c)* Determine the final kinetic energy of a deuteron in electron volts. *(d)* How many revolutions are required if the accelerating potential difference across the dees is 500 V?

28. The Hall coefficient for Cu is $R_H = -6 \times 10^{-11}$ V·m·A^{-1}·T^{-1}. *(a)* Estimate the density of charge carriers (electrons) in Cu. *(b)* The mass density of Cu is 8.9×10^3 kg/m^3, and the mass of a Cu atom is 1.06×10^{-25} kg. How many free or mobile electrons does each atom contribute on the average? *(c)* What fraction is this of the 29 electrons in the neutral atom?

29. Figure 26-22 shows the trace of the path of a charged particle in a bubble chamber. Assume that the magnetic field is into the plane of the paper, with magnitude 0.4 T. The smooth spiral path occurs because the particle loses energy in ionizing molecules along the path. *(a)* Which part of the path corresponds to higher kinetic energy for the particle? *(b)* Is the charge positive or negative? *(c)* The radius of curvature ranges from 70 to 10 mm. What is the range of values of the magnitude of momentum if the magnitude of the charge is *e*?

Figure 26-22. Exercise 29.

30. For a particle whose speed v is comparable to the speed of light, $c = 3.00 \times 10^8$ m/s, Eq. 26-12 must be rewritten as

$$\frac{mv}{\sqrt{1 - v^2/c^2}} = qBr$$

where m is the rest mass of the particle (to be discussed in

Chap. 38). *(a)* Determine the speed of a proton (rest mass $m = 1.67 \times 10^{-27}$ kg) in a path with radius of curvature 6.43 m in a magnetic field of magnitude 800 mT. *(b)* Determine the angular frequency of this circular motion and compare with the cyclotron frequency $\omega_c = qB/m$. This relativistic effect limits the cyclotron to accelerating particles to speeds small compared with the speed of light.

31. An electron beam is deflected by an electric field of magnitude $E = 4$ kN/C in the region between charged conducting plates, as shown in Fig. 26-23. The kinetic energy of an electron in the beam is 7 keV. *(a)* What is the direction of the electric field? *(b)* What uniform magnetic field (magnitude and direction) applied in this region would allow the beam to pass undeviated through the region? *(c)* Determine the speed and kinetic energy of a proton that this arrangement of fields would pass.

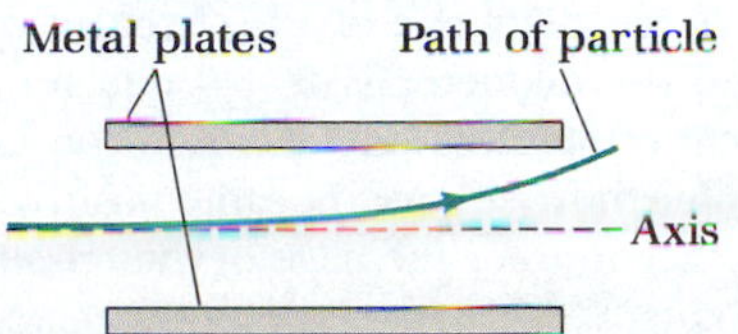

Figure 26-23. Exercise 31.

32. A velocity selector with fields **E** and **B** passes charged particles with velocity **v**. *(a)* Prove that the vectors satisfy the relation

$$\mathbf{E} = \mathbf{B} \times \mathbf{v}$$

(b) Which pairs of vectors *must* be perpendicular?

33. Very large magnetic fields exist in the vicinity of neutron stars. *(a)* Determine the cyclotron frequency of a proton in a region where $B = 10^5$ T. *(b)* Determine the speed of a proton moving in a circular path of radius 1 μm in this field. *(c)* Determine the speed if $r = 1$ m. (See Exercise 30.)

34. A proton travels at 6.5×10^6 m/s horizontally toward the north in a magnetic field of 1.8 T directed horizontally toward the west. What is the magnitude and direction of the magnetic force on the proton?

35. A particle travels in a circle of radius 93 mm at a speed of 5.8×10^6 m/s. The plane of the circle is perpendicular to a magnetic field of magnitude 1.3 T. What is the particle's charge to mass ratio?

36. A charged particle is traveling horizontally toward the north at a speed of 35 km/s in a uniform electric field of 420 V/m directed vertically upward. What magnetic field gives a net force on the particle that is zero? Assume that the electric and magnetic forces are the only significant forces exerted on the particle.

37. A full-scale deflection of a galvonometer rotates the pointer 35°. The magnitude of the galvonometer's magnetic field is 83 mT, the torsion constant of the spring is 3.0 mN/rad, and the area of each loop in the coil is 51×10^{-4} m^2. Determine the number of loops in the coil such that a current of 10 mA will produce a full-scale deflection.

38. A circular coil, similar to the one in Fig. 26-8, has 380 turns, and its axis is in a vertical north-south plane. The angle between the axis and the horizontal is 28°. The coil carries a current of 71 mA, and when viewed by an observer located north of the coil, the sense of the current is clockwise. The area of each loop is 61×10^{-4} m^2. What are the magnitude and direction of the magnetic torque on the coil due to a 0.22-mT field directed horizontally toward the north?

39. A magnetic dipole of magnitude 7.6×10^{-4} A·m^2 is initially directed perpendicular to a magnetic field of magnitude 97 mT. If the dipole's final orientation is in the same direction as the field, what is the work done on the dipole by the field during the rotation?

40. A straight section of wire that is 3.0 m long carries a 5.5-A current horizontally 39° north of east. The magnetic field is directed horizontally toward the north with magnitude 75 mT. What is the magnitude and direction of the magnetic force on the section?

41. The SI unit for magnetic dipole moment is A·m^2, which has the dimension (charge) (length)^2/(time). *(a)* Show that this is the same dimension as (energy)/(magnetic field). *(b)* Determine the conversion factor that changes the unit eV/T to the unit A·m^2.

42. In the Bohr model of the hydrogen atom, an electron orbits its nucleus (a proton) the way a planet orbits the sun. In the atom's *ground state,* the orbit is circular with a radius of 0.053 nm and the electron's speed is 2.2×10^6 m/s. *(a)* Determine the magnitude of the atom's magnetic dipole moment due to this orbital motion in A·m^2. *(b)* Use the conversion factor 1.60×10^{-19} A·m^2·T/eV to convert the result of part *(a)* to eV/T. This value of magnetic dipole moment is called a *Bohr magneton.*

PROBLEMS

1. ***Semicircular wire revisited.*** Determine the magnetic force on the semicircular wire in Example 26-4 by evaluating the infinitesimal force $d\mathbf{F} = I\, d\boldsymbol{\ell} \times \mathbf{B}$ in component form. Note that $\mathbf{B} = B\mathbf{k}$ and that $d\boldsymbol{\ell} = dx\,\mathbf{i} + dy\,\mathbf{j}$. The integrals extend from the point $(R, 0)$ to the point $(-R, 0)$.

2. ***Magnetic moment of an arbitrarily shaped planar loop.*** Consider a plane loop of arbitrary shape carrying current I in a uniform magnetic field. The actual loop can be approximated by a number of small rectangular loops, each with current I, as shown in Fig. 26-24. Along the edges where two

loops join with currents in opposite senses, the net current is zero, as is the force on these imagined conductors. Use this construction to show that the magnetic dipole moment of the plane current loop is given by $\mathbf{m} = I\mathbf{S}$, where $\mathbf{S} = \int d\mathbf{S}$ is the area of the plane of the loop.

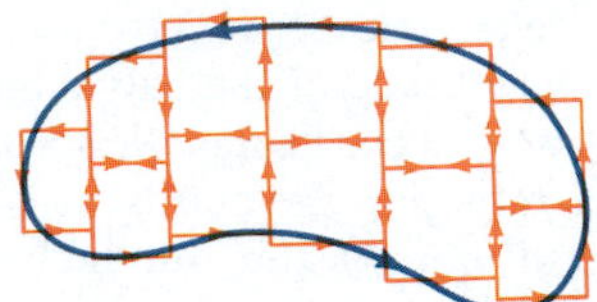

Figure 26-24. Problem 2.

3. ***Orientation of a bar magnet oscillating in a magnetic field.*** Suppose that a bar magnet (or a coil) of magnetic dipole moment $\mathbf{m}$ is suspended, as shown in Fig. 26-9, by a vertical fiber whose torsional constant is negligible. The magnet is in rotational equilibrium when aligned with a uniform magnetic field. If the magnet is rotated in a horizontal plane through a small angle θ, the field produces a torque on the magnet. *(a)* Show that the period of small oscillations of the magnet is given by

$$T = 2\pi \sqrt{\frac{I_0}{mB}}$$

where I_0 is the moment of inertia of the magnet about the axis of rotation. This method can be used in a known magnetic field to measure m for a magnet or a coil—or to measure B if m is known ($m = NIS$ for a coil). *(b)* A long, thin coil (or solenoid) of mass 24 g, with 450 turns of wire in a 85-mm length, has cross-sectional area 0.75 mm^2 and carries current 140 mA. Determine the magnitude of the magnetic field if the solenoid oscillates as described above with a period of 5.4 s.

4. ***Helical path for a charged particle in a magnetic field.*** A particle of mass m and charge q moves in a uniform magnetic field $\mathbf{B} = B\mathbf{k}$. Suppose that its initial velocity is $\mathbf{v}_0 = v_{0x}\mathbf{i} + v_{0z}\mathbf{k}$. The path of the particle is a helix (like the windings of a bedspring) whose axis is along the magnetic field direction. Show that the helix has a radius given by

$$r = \frac{mv_{0x}}{qB}$$

and that the pitch of the helix corresponds to a distance

$$d = \frac{2\pi mv_{0z}}{qB}$$

traveled parallel to the axis for each revolution.

5. ***The mass spectrometer.*** One type of mass spectrometer, used for the precise determination of atomic masses, is diagrammed in Fig. 26-25. Ions with velocity $\mathbf{v}$ from a source pass through a velocity selector, with fields $\mathbf{E}_0$ and $\mathbf{B}_0$, into a region of uniform magnetic field $\mathbf{B}$ and zero electric field. The radius of curvature of the path depends on the charge-to-mass ratio q/m of the ion. *(a)* Show that

$$m = \frac{qBB_0D}{2E_0}$$

where D is the diameter of the path for a particular ion species. Masses are customarily expressed on a scale such that the ^{12}C atom (carbon 12 is the isotope of carbon with six protons and

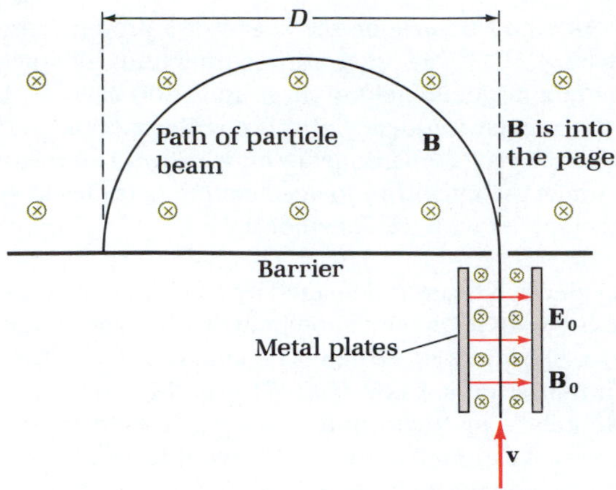

Figure 26-25. Problem 5.

six neutrons) has a mass of exactly 12 atomic mass units (12 u). *(b)* If the diameter of the circular path for ^{12}C is 732.4 mm and the diameter is 671.9 mm for an isotope of boron, determine the mass of the boron ion in atomic mass units (u). Assume the ions have the same positive charge. *(c)* Is it necessary to account for the mass of the missing electron(s) of the ions? Explain.

6. ***The Thomson experiment.*** By measuring the charge-to-mass ratio e/m for electrons, J. J. Thomson in 1897 established quantitatively the nature of cathode rays—rays coming from the cathode of a tube similar to a television picture tube. (Refer to Fig. 20-21.) Figure 26-26 shows that, with an electric field $\mathbf{E}$ but no magnetic field in the region between deflecting plates, the electron beam is deflected by an amount y on the detecting screen. *(a)* Show that the deflection is given by

$$y = \frac{eE}{mv^2}\left(DL + \frac{1}{2}L^2\right)$$

where v is the speed of electrons entering the plate region. Thomson then adjusted the applied magnetic field $\mathbf{B}$ until the beam was undeviated in this velocity selector. *(b)* Show that the charge-to-mass ratio is given in terms of measurable quantities by

$$\frac{e}{m} = \frac{yE}{B^2(DL + \frac{1}{2}L^2)}$$

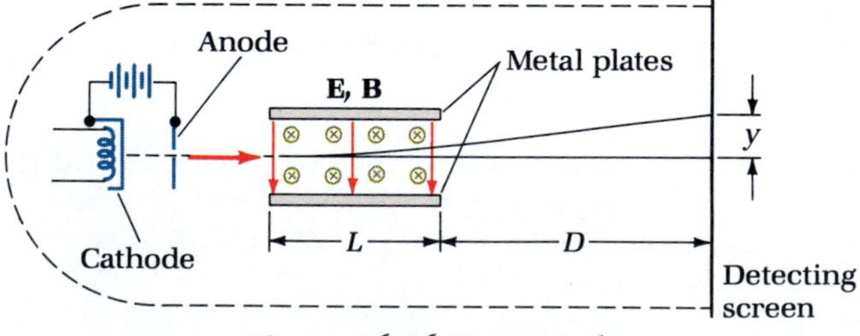

Figure 26-26. Problem 6.

7. ***Coil on a tube.*** A coil with N turns and current I is wound on a thin tube of diameter D and length L, as shown in Fig. 26-27. Light strings are wound around the circumference and tied to a rigid support which is not shown. Assume that the tube-and-coil system is in static equilibrium. *(a)* Show that the strings must be vertical. *(b)* Determine the *minimum* current I_0 for which the tube-and-coil system (of combined mass m)

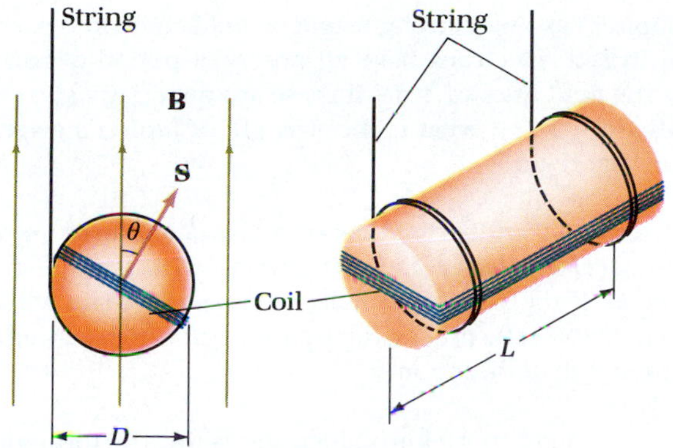

Figure 26-27. Problem 7.

can be in static equilibrium in a uniform magnetic field **B** directed vertically upward. *(c)* What is the angle θ in this case? *(d)* For what angle θ is the system in equilibrium if $I = 2I_0$? *(e)* Is this second configuration one of stable or unstable equilibrium? Explain.

8. ***Loop in a nonuniform field.*** A rectangular current loop with dimensions a and b carries current I in a *nonuniform* magnetic field. The loop is in the xy plane, as shown in Fig. 26-28, perpendicular to a magnetic field which is out of the plane of the figure. At points in the plane, the field is given by $B_x = B_y = 0$ and $B_z = B_0 a/x$, where B_0 is a constant. Thus the magnetic field is in the z direction and it depends on the coordinate x. Determine the magnetic force *(a)* on each section of the current loop and *(b)* on the entire loop. *(c)* Determine the magnetic torque on the current loop about an axis through the center of the loop parallel to the y axis. Note that Eq. 26-7 applies only for a uniform field.

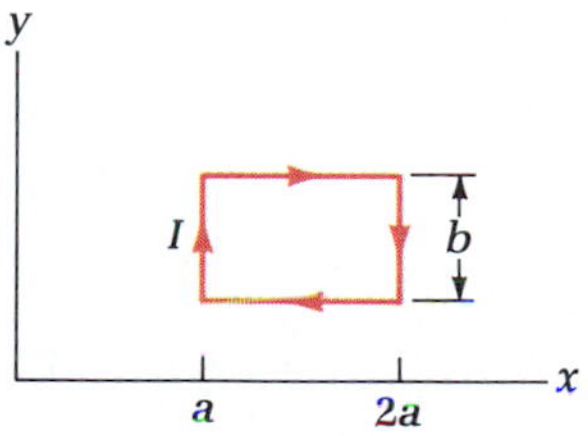

Figure 26-28. Problem 8.

9. ***Potential energy of a magnetic dipole.*** Show that Eq. 26-10 correctly gives the potential energy of a magnetic dipole in a uniform magnetic field. Begin by considering the work done by the magnetic force on a moving segment of a rectangular current loop as it rotates about an axis, as shown in Fig. 26-7. Then show that for a rotation through an infinitesimal angle $d\theta$, the work done by the magnetic force is $dW = -\tau\, d\theta$, where $\tau = |\mathbf{m} \times \mathbf{B}|$ is the magnitude of the magnetic torque on the current loop. Express the change in potential energy as the negative of the work done. $U_f - U_i = -\int dW$, and let $U = 0$ if **m** is perpendicular to **B**.

10. ***Acceleration of a particle in crossed* E *and* B *fields.*** A uniform electric field **E** in the y direction and a uniform magnetic field **B** in the z direction exist in a region of space. A charged particle starts from rest at the origin of a coordinate system. Show that the motion of the particle is determined by the following equations of motion:

$$a_x = \omega_c v_y$$

$$a_y = \frac{qE}{M} - \omega_c v_x$$

$$a_z = 0$$

where $\omega_c = qB/m$ is the cyclotron frequency.

11. ***Motion of a particle in crossed* E *and* B *fields.*** Write or adapt a spreadsheet to integrate the equations of motion in the previous problem. Let the charged particle be a proton and take $E = 2.0 \times 10^5$ N/C, $B = 0.50$ T. The motion takes place in the xy plane. (Why?) Note that the cyclotron frequency sets a time scale for the numerical integration. That is, Δt should be small compared with the period $T = 2\pi/\omega_c$. Graph the path of the particle for a time which is at least twice the period. (This problem can also be solved analytically.)

Additional Problems

12. How fast must an electron be moving if it is to orbit the earth above the equator at an altitude of 250 km if the earth's magnetic field **B** is parallel to the surface of the earth and $\mathbf{B} \approx 10^{-5}$ T? Can you neglect gravity?

13. In an experiment at an accelerator, a large solenoidal magnet has a 2-m "free bore" (central cylindrical opening diameter) and a 2.2-T uniform magnetic field (parallel to the axis). There are detectors in the central region close to the axis that require dc electric power. The current in each power bus cable is 10 A. If the two lines from each circuit are connected to a power supply with one side at ground potential and the other at a $+V$ sufficient to maintain the 10 A, and the cables are routed radially to within 30 cm of the central axis, what is the force (magnitude and direction) on each cable? What would happen if the cables were tightly twisted together for the portion of their radial route through the magnetic field?

14. Consider designing a magnetic-field meter that consists of a beam of electrons moving inside a small CRT (cathode-ray tube) at a known velocity v. The spot where the beam strikes the screen is the pointer. When no field is present, the beam is undeflected and strikes the center of the screen. When a magnetic field is present, the beam spot moves. Assume no **E** fields are present. Assume too that the relative orientations of these devices are fixed with respect to each other. What is the minimum number of devices required that will allow the simultaneous determination of the magnitude and direction of the **B** field? (Assume the device itself does not affect the ambient **B** field.) Could you tell from this device if an **E** field was also present?

15. It is possible to make a magnetic pump if the fluid is a conductor. If the pipe containing the conducting fluid has a square cross section with sides of 10 cm and a magnetic field of 2 T is maintained along the axis of the pipe by an external magnet, what pumping pressure results from a uniformly distributed 100-A current flowing from one side of the square to the opposite side? Does the effective pumping pressure depend on the cross-sectional shape of the pipe?

16. Assume that in your lab the earth's **B** field is horizontal and points due north. *(a)* If a square copper bar is placed horizontally heading east/west with the positive terminal of a battery connected to the eastern end and the negative terminal to the western end, which sides of the bar (top and bottom, or north and south) should be used to attach a meter to measure the Hall voltage? *(b)* What sign will that voltage have with respect to the specific probes?

17. An alternative form of a mass spectrometer involves circulating an ion in a fixed **B** field and measuring the circulation frequency to calculate the mass if the charge is known. If the field in such a device is 90.0 mT and a singly charged ion circulates in 92.1 μs, what element is this most likely to be?

18. A bar of mass m slides on frictionless horizontal rails separated by a distance d. The bar is constrained to remain perpendicular to the rails during its movement. A constant current generator is attached across the rails and maintains a current I in the bar. The bar begins from rest at $t = 0$. Find the velocity of the bar as a function of time if there is a magnetic field **B** (vertically upward) permeating the entire region.

19. Jupiter has a very strong magnetic field compared with the earth. In fact, electrons have an observed period of rotation about the field lines of 1 ns! If these are nonrelativistic (their speeds are $<0.1c$), what is the strength of Jupiter's magnetic field in that region?

20. A thin nonconducting disk of radius R and mass m has a uniform surface charge ρ and is rotating with an angular frequency ω. If the ambient **B** field makes a 45° angle with the initial ω, what is the precession frequency for this rotating disk about the **B** field direction?

21. Aircraft tend to pick up electrons as they fly through the air. If the earth's **B** field is essentially parallel to the surface and points north, which wing of a westbound plane should have a higher potential?

22. An electron in a TV tube is accelerated by a 10-kV potential. If a horizontal external field of 10^{-3} T is present, what is the extent of the deflection if the total path length is 30 cm and the field is perpendicular to the beam initially?

SOURCES OF THE MAGNETIC FIELD

27

The Tokamak Fusion Test Reactor at the Princeton Plasma Physics Laboratory. Fusion nuclear reactors may one day produce electrical energy. Such reactors require large magnetic fields.

The last chapter dealt mainly with the force exerted by a magnetic field on a moving charge or on a current-carrying conductor. We now consider the source or cause of a magnetic field. Most of us have experience with the forces that magnets exert on each other. A magnet can also exert a magnetic force on a current-carrying conductor and on a moving charge. For example, the electron beam in a TV picture tube can be deflected by bringing a magnet close to the screen. (But don't try this with a strong magnet!) The magnet is the source of the magnetic field that deflects the electron beam.

In this chapter the emphasis is on an electric current as the source of a magnetic field. Oersted discovered in 1819 that a compass needle near a conducting wire was deflected when the wire carried a current. The current was the source of the magnetic field that exerted a torque on the compass needle.

Oersted's observation was the first that indicated a connection between electricity and magnetism. Prior to that time, electricity and magnetism were considered to be unrelated. Further connections between these fields will be revealed in the chapters that follow. We begin with the magnetic field produced by a steady distribution of electric current.

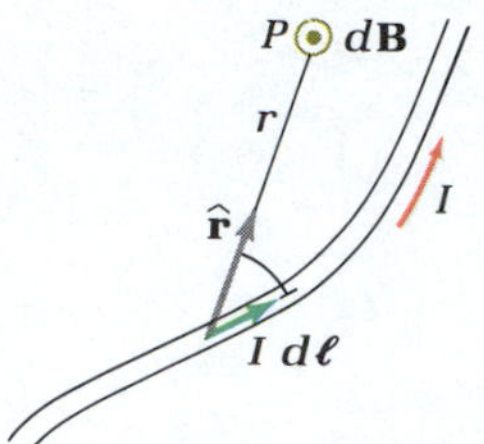

Figure 27-1. A current element $I\,d\boldsymbol{\ell}$ produces a contribution $d\mathbf{B}$ to the magnetic field at point P. If $I\,d\boldsymbol{\ell}$ and $\hat{\mathbf{r}}$ lie in the plane of the figure, then $d\mathbf{B}$ is directed perpendicularly out of the plane for this case.

27-1 THE BIOT-SAVART LAW

Immediately following Oersted's discovery that an electric current is a source of a magnetic field, experiments by A. M. Ampère (1775–1836) and by J. B. Biot (1774–1862) and F. Savart (1791–1841) led to what we now call the ***Biot-Savart law.*** It determines, at a point in space, the magnetic field due to a distribution of electric currents.

The Biot-Savart law is analogous to Coulomb's law for electrostatics. One way of expressing Coulomb's law is to give the electric field produced by a distribution of charge. If dq is a differential or infinitesimal element of charge, then the electric field $d\mathbf{E}$ produced at a point P by that charge is given by Eq. 20-6:

$$d\mathbf{E} = \frac{1}{4\pi\epsilon_0}\frac{dq}{r^2}\hat{\mathbf{r}}$$

where r is the distance from the element of charge to the point P and $\hat{\mathbf{r}}$ is the unit vector directed from the charge to the point P. Integrating over the charge distribution gives the electric field at the point P, $\mathbf{E} = \int d\mathbf{E}$.

Now consider a current distribution such as that shown in Fig. 27-1. A current element $I\,d\boldsymbol{\ell}$ produces a contribution $d\mathbf{B}$ to the magnetic field at a point P. Let r represent the distance from the current element to the point P, with the unit vector $\hat{\mathbf{r}}$ pointing from the current element to point P. Then the Biot-Savart law for an infinitesimal current element is

The Biot-Savart law for a current element

$$d\mathbf{B} = \frac{\mu_0}{4\pi}\frac{I\,d\boldsymbol{\ell}\times\hat{\mathbf{r}}}{r^2} \tag{27-1}$$

The direction of $d\mathbf{B}$ is given by the direction of the cross product $I\,d\boldsymbol{\ell}\times\hat{\mathbf{r}}$ and is perpendicular to the current element $I\,d\boldsymbol{\ell}$ and to the unit vector $\hat{\mathbf{r}}$, with a sense given by a right-hand rule. That is, the extended thumb of the right hand gives the direction of $d\mathbf{B}$ if the fingers curl in the sense of the rotation that carries the direction of $I\,d\boldsymbol{\ell}$ into the direction of $\hat{\mathbf{r}}$. The magnitude of $d\mathbf{B}$ is

$$dB = \frac{\mu_0}{4\pi}\frac{I\,d\ell\sin\theta}{r^2} \tag{27-2}$$

The permeability constant for vacuum, μ_0

where θ is the angle between the directions of $I\,d\boldsymbol{\ell}$ and $\hat{\mathbf{r}}$. The constant μ_0 is called the *permeability constant* for vacuum and is analogous to ϵ_0, the permittivity constant for vacuum in electrostatics. Because of the connections between electricity and magnetism, the values of ϵ_0 and μ_0 are not independent. The value of μ_0 in SI units is determined from the definition of the ampere (A) to be exactly $\mu_0 = 4\pi\times10^{-7}\ \mathrm{T\cdot m\cdot A^{-1}}$ (see Sec. 27-4), or since the combination $\mu_0/4\pi$ often occurs,

$$\frac{\mu_0}{4\pi} = 1\times10^{-7}\ \mathrm{T\cdot m\cdot A^{-1}} \qquad \text{(exactly)}$$

The magnetic properties of air are so nearly the same as for vacuum that the permeability constant μ_0 for vacuum may be used when air is present.

There are some similarities between the Biot-Savart law for the magnetic field and the corresponding Coulomb's law for the electric field:

Similarities of Coulomb's law and the Biot-Savart law

1. Each contains the inverse-square dependence $1/r^2$ on the distance from the point source, with $I\,d\boldsymbol{\ell}$ as the source for $d\mathbf{B}$ and dq as the source for $d\mathbf{E}$.

2. The constant $1/4\pi\epsilon_0$ gives the strength of the electrical interaction, and the constant $\mu_0/4\pi$ gives the strength of the magnetic interaction.

But there are also some significant differences between the two laws:

1. The direction of $d\mathbf{E}$ is radial with respect to the charge dq, while the direction of $d\mathbf{B}$ is perpendicular to the plane containing $I\,d\boldsymbol{\ell}$ and $\hat{\mathbf{r}}$.

Differences between Coulomb's law and the Biot-Savart law

2. Although the simplest distribution of charge is a single isolated point charge, the single isolated current element does not exist for a steady current. Instead, charge must flow into the element from one end and out of the element at the other end, and Eq. 27-1 must always be integrated along the line(s) of the current distribution. Thus at a point P, the magnetic field due to a current distribution is given by the integral form of the Biot-Savart law:

$$\mathbf{B} = \int \frac{\mu_0}{4\pi} \frac{I\,d\boldsymbol{\ell} \times \hat{\mathbf{r}}}{r^2} \qquad (27\text{-}3)$$

Integral form of the Biot-Savart law

where the line integral extends along the entire current distribution. That is, the magnetic field at a point is the linear superposition of the vector contributions due to each of the infinitesimal current elements.

Magnetic Field Due to a Current in a Long, Straight Wire

To illustrate the use of Eqs. 27-1 and 27-3, we determine the magnetic field produced by a steady current I in a long, straight wire. Unless otherwise stated, we shall assume that current-carrying wires have negligible thickness so that they can be represented by lines. A typical current element $I\,d\boldsymbol{\ell}$ is shown in Fig. 27-2, where the point P is at a perpendicular distance R from the wire. The contribution $d\mathbf{B}$ to

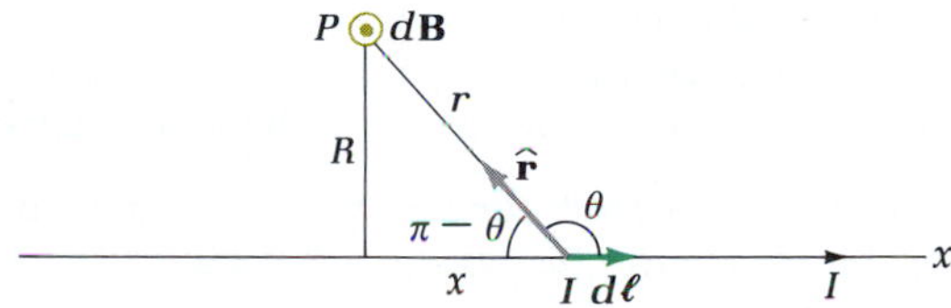

Figure 27-2. A long, straight wire carries a current I. The current element $I\,d\boldsymbol{\ell}$ makes a contribution $d\mathbf{B}$ to the magnetic field at point P. The direction of $d\mathbf{B}$ is out of the plane of the figure.

the magnetic field at that point is given by Eq. 27-1. Applying the right-hand rule to the cross product shows that the direction of $d\mathbf{B}$ is perpendicularly out of the plane of the figure. The magnitude dB is given by Eq. 27-2, with $r^2 = x^2 + R^2$ and $I\,d\ell = I\,dx$. Thus

$$dB = \frac{\mu_0}{4\pi} \frac{I\,dx \sin\theta}{x^2 + R^2}$$

$$= \frac{\mu_0}{4\pi} \frac{IR\,dx}{(x^2 + R^2)^{3/2}}$$

where we have used (see Fig. 27-2) $\sin\theta = \sin(\pi - \theta) = R/(x^2 + R^2)^{1/2}$.

The integral Biot-Savart law in Eq. 27-3 contains the integral of a vector. Particular care must be exercised in handling the direction when integrating a vector. For this case the integral involves the sum of contributions $d\mathbf{B}$ for each current element along the x axis. For any position, such as the one shown in Fig. 27-2, the direction of $d\mathbf{B}$ is perpendicularly out of the plane of the page. The integral represents the sum of these infinitesimal contributions, all with the same direction. Therefore the sum or integral also has that direction, and its magnitude is the sum or integral of the magnitude dB of the contributions. Thus the direction of $\mathbf{B}$ at point P is perpendicularly out of the plane of the figure, and the magnitude is given by

$$B = \int dB = \frac{\mu_0}{4\pi} \int_{-\infty}^{\infty} \frac{IR}{(x^2 + R^2)^{3/2}}\,dx$$

Notice from the limits on the integral that we are considering the ideal case of an

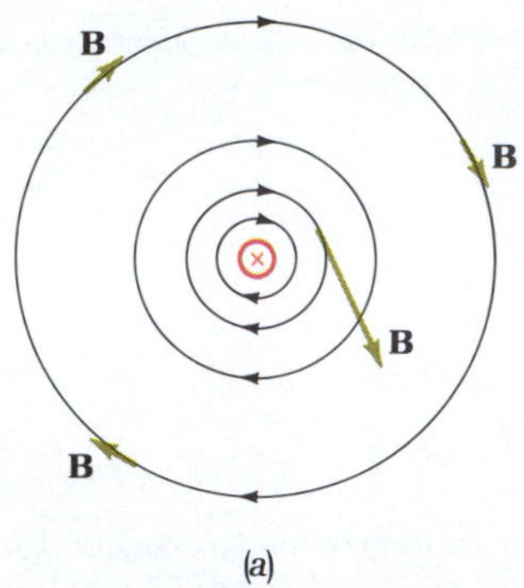

Figure 27-3. *(a)* Continuous lines represent the magnetic field due to a long, straight wire carrying a current into the page. The magnitude of the field decreases with increasing distance from the wire. *(b)* Iron filings form a pattern due to the magnetic field of a current in a wire.

infinitely long wire. This integral can be found in tables or evaluated by using the substitution $x = R \cot(\pi - \theta)$, as suggested by Fig. 27-2. The result is

$$B = \frac{\mu_0 IR}{4\pi}\left[\frac{x}{R^2(x^2+R^2)^{1/2}}\right]_{-\infty}^{+\infty} = \frac{\mu_0 IR}{4\pi}\frac{2}{R^2}$$

Simplifying, we obtain the magnitude of the magnetic field at a perpendicular distance R from a long, straight wire carrying a current I:

$$B = \frac{\mu_0 I}{2\pi R} \tag{27-4}$$

The direction of the field at a point is perpendicular to the plane containing the wire and the point in accord with a right-hand rule:

Grasp the wire with the right hand such that the extended thumb points in the sense of the current; the curled fingers then give the sense of the field.

The lines representing the magnetic field are shown in Fig. 27-3*a* in a plane perpendicular to the wire. These lines are circular and close on themselves. Compare with the pattern of iron filings near a current-carrying wire in Fig. 27-3*b*.

EXAMPLE 27-1

Graph of the field produced by a long, straight wire. A long, straight wire carries a 20-A current. Determine the magnitude of the magnetic field at points 10, 20, and 50 mm from the wire and construct a graph of the dependence of B on distance R from the wire.

Solution. The magnitude of the field is given by Eq. 27-4. For $R = 10$ mm, the field has magnitude (notice that $\mu_0/2\pi = 2 \times 10^{-7}\ \mathrm{T\cdot m\cdot A^{-1}}$)

$$B = \frac{(2 \times 10^{-7}\ \mathrm{T\cdot m\cdot A^{-1}})(20\ \mathrm{A})}{0.01\ \mathrm{m}} = 4 \times 10^{-4}\ \mathrm{T}$$

Since B varies inversely with R, the magnitude of the field at 20 mm from the wire is 1/2 its value at 10 mm from the wire, or $B = 2 \times 10^{-4}$ T. Similarly, $B = 0.8 \times 10^{-4}$ T at 50 mm from the wire. These values are shown on the graph in Fig. 27-4, which illustrates the dependence of B on R.

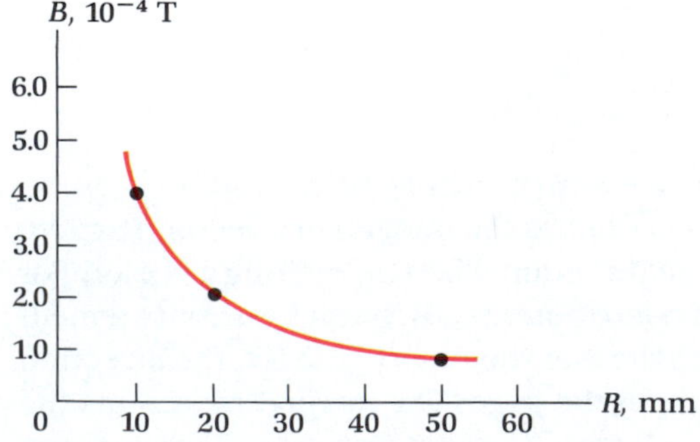

Figure 27-4. Example 27-1.

SELF-TEST 27-1. A long, straight wire is horizontal and carries a current toward the west. What is the direction of the magnetic field produced by the current *(a)* at a point directly above the wire, *(b)* at a point directly south of the wire and at the same elevation as the wire, and *(c)* at a point directly north of the wire and at the same elevation? ***ANSWERS:*** *(a)* north; *(b)* up; *(c)* down.

EXAMPLE 27-2

Field produced by two parallel currents. Two long, straight, parallel wires 240 mm apart carry currents of $I_1 = 20.0$ A and $I_2 = 30.0$ A. The currents have the same sense as shown in Fig. 27-5. The Biot-Savart law implies and experiment shows that the magnetic field due to several currents is the sum of the fields that would be produced by each current separately. Determine the magnetic field in the plane of the two wires at a point P halfway between them.

Solution. The magnetic field at a point is the vector sum of the fields $\mathbf{B}_1$ due to the current I_1 and $\mathbf{B}_2$ due to the current I_2. The magnitude of each field is given by Eq. 27-4:

$$B_1 = \frac{\mu_0 I_1}{2\pi R_1} \qquad B_2 = \frac{\mu_0 I_2}{2\pi R_2}$$

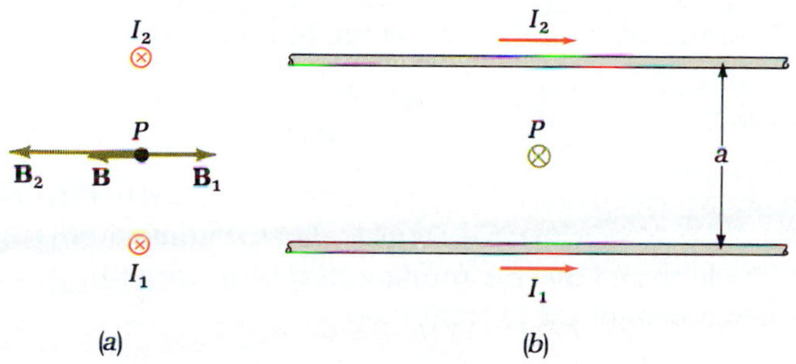

Figure 27-5. Example 27-2: The magnetic field at point P is the sum of contributions due to each current. *(a)* End view. *(b)* Side view.

The direction of each, shown in Fig. 27-5*a*, is determined from the right-hand rule. At point P the two field contributions have opposite directions. The vector sum $\mathbf{B} = \mathbf{B}_1 + \mathbf{B}_2$ has the direction of $\mathbf{B}_2$, the larger of the two contributions, and the magnitude of the sum is given by

$$\begin{aligned} B = B_2 - B_1 &= \frac{\mu_0}{2\pi}\left(\frac{I_2}{R_2} - \frac{I_1}{R_1}\right) \\ &= (2 \times 10^{-7}\ \mathrm{T \cdot m \cdot A^{-1}})\left(\frac{30.0\ \mathrm{A}}{0.12\ \mathrm{m}} - \frac{20.0\ \mathrm{A}}{0.12\ \mathrm{m}}\right) \\ &= 1.7 \times 10^{-5}\ \mathrm{T} \end{aligned}$$

Suppose the sense of I_2 were opposite the sense of I_1. Show that $B = 8.3 \times 10^{-4}$ T at point P.

SELF-TEST 27-2. For the arrangement in the above example, at what perpendicular distance from wire 1 is the magnetic field produced by one of the wires equal and opposite that produced by the other, so that $\mathbf{B} = 0$. ***ANSWER:*** 0.096 m.

Magnetic Field of a Current Loop

As another application of the Biot-Savart law, consider the magnetic field produced by a circular loop of radius a carrying a current I. We confine our calculation to points along the axis of the loop. Such points are equidistant from points on the loop. A typical current element $I\,d\boldsymbol{\ell}$ is shown in Fig. 27-6, where the x axis has been chosen along the axis of the loop. The contribution $d\mathbf{B}$ to the field at a point on the axis is given by Eq. 27-1 and is shown in the figure. For any point on the loop, the current element $I\,d\boldsymbol{\ell}$ and the unit vector $\hat{\mathbf{r}}$ are perpendicular. The magnitude of the field contribution is ($\theta = 90°$ and $\sin\theta = 1$ in Eq. 27-2)

$$dB = \frac{\mu_0 I\,d\ell}{4\pi r^2}$$

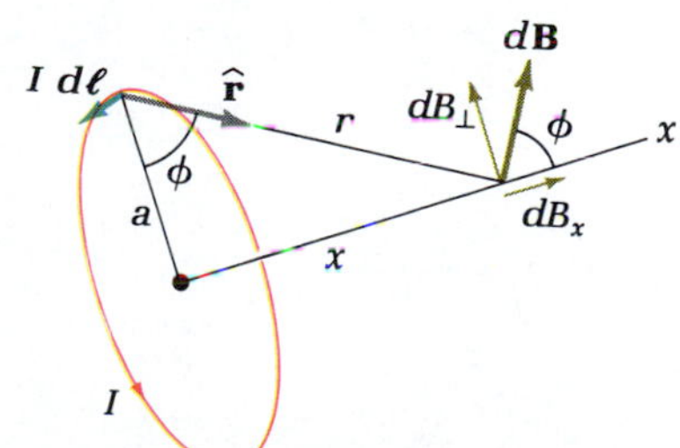

Figure 27-6. A current element of a circular loop produces a magnetic field contribution $d\mathbf{B}$ at a point on the axis of the loop. The vectors $I\,d\boldsymbol{\ell}$ and $\hat{\mathbf{r}}$ are perpendicular.

For each current element around the loop, the magnitude of $d\mathbf{B}$ remains the same but the direction changes. We resolve $d\mathbf{B}$ into components $dB_x = dB\cos\phi$ along the axis of the loop, and $dB_\perp = dB\sin\phi$ perpendicular to the axis. From the symmetry of the problem, the perpendicular component $dB_\perp$ adds or integrates to zero for the loop. We only need to integrate the component along the axis, dB_x.

Thus the magnetic field at a point on the axis of the loop points along the axis, and its component is

$$B_x = \int dB_x = \int \frac{\mu_0 I\, d\ell \cos\phi}{4\pi(x^2 + a^2)}$$

where we have used $r^2 = x^2 + a^2$.

As we integrate $d\ell$ around the loop, each factor in the integral remains the same and can be taken out of the integral, leaving $\int d\ell = \ell = 2\pi a$, which is just the circumference of the loop. From the geometry in Fig. 27-6, $\cos\phi$ can be expressed in terms of a and x:

$$\cos\phi = \frac{a}{r} = \frac{a}{(x^2 + a^2)^{1/2}}$$

Making these substitutions in the equation for B_x gives

$$B_x = \frac{\mu_0 I\, 2\pi a^2}{4\pi(x^2 + a^2)^{3/2}}$$

This result can be expressed more simply in terms of the magnetic dipole moment, $\mathbf{m} = I\mathbf{S}$ from Eq. 26-9. Notice that the factor πa^2 in the numerator is the magnitude S of the area of the loop and that the product $I\pi a^2$ is the magnitude, $m = IS$, of the magnetic dipole moment of the current loop. Notice also that the directions of $\mathbf{B}$ and of $\mathbf{m}$ are the same, along the positive x direction. We have

Magnetic field on the axis of a current loop

$$\mathbf{B} = \frac{\mu_0}{2\pi} \frac{\mathbf{m}}{(x^2 + a^2)^{3/2}} \tag{27-5}$$

For points along the axis of the loop, the magnitude of the magnetic field is a maximum at $x = 0$, the center of the loop. The magnitude of the field decreases with distance along the axis. At points far from the loop, where $|x| \gg a$ so that $(x^2 + a^2)^{3/2} \approx (x^2)^{3/2} = |x|^3$, the magnitude of the field is given approximately by

$$B \approx \frac{\mu_0 m}{2\pi |x|^3}$$

The dependence of B on the inverse cube of the distance from the current loop (far from the loop) is characteristic of a dipole field. (See Exercise 25 in Chap. 20 for the corresponding form of the electric dipole field.)

A current loop behaves as a magnetic dipole in two ways.

We saw in Chap. 26 that a loop with current I and (vector) area $\mathbf{S}$ behaves as a magnetic dipole with dipole moment $\mathbf{m} = I\mathbf{S}$. That is, the torque exerted on such a current loop by an *external* magnetic field $\mathbf{B}_{ext}$ is given by Eq. 26-8, $\boldsymbol{\tau} = \mathbf{m} \times \mathbf{B}_{ext}$. Now we can see that a current loop behaves as a magnetic dipole in another respect: The magnetic field produced by a current loop is a magnetic dipole field at points far from the loop.

EXAMPLE 27-3

Field on the axis of a magnetic dipole. A circular current loop has radius 25 mm and current 750 mA, with a sense as shown in Fig. 27-6. Determine the magnetic field produced by the current loop at $x = 0$, $x = 25$ mm, and $x = -2.0$ m, where the origin of the coordinate is at the center of the loop.

Solution. The magnetic field at any point on the axis of the current loop in Fig. 27-6 is in the positive x direction. This can be seen from the figure for the case of a positive value of x. Sketch similar diagrams to verify this result for a negative value and a zero value of x. The magnitude of the field at a point on the axis is given by Eq. 27-5. The magnetic dipole moment of the current loop has magnitude

$$m = IS = I\pi a^2 = (0.750\ \text{A})\pi(0.025\ \text{m})^2$$
$$= 1.5 \times 10^{-3}\ \text{A}\cdot\text{m}^2$$

At $x = 0$, the magnitude of the field is

$$B = (2 \times 10^{-7}\ \text{T}\cdot\text{m}\cdot\text{A}^{-1})\ \frac{1.5 \times 10^{-3}\ \text{A}\cdot\text{m}^2}{[(0)^2 + (0.025\ \text{m})^2]^{3/2}}$$
$$= 1.9 \times 10^{-5}\ \text{T}$$

Similar calculations give $B = 6.7 \times 10^{-6}$ T at $x = 0.020$ m and $B = 3.7 \times 10^{-11}$ T at $x = -2.0$ m.

SELF-TEST 27-3. From the above example, perform the calculations to evaluate B at $x = 25$ mm and $x = -2.0$ m.

27-2 AMPERE'S LAW

The magnitude of the magnetic field produced at a point by a current I in a long, straight wire is given by Eq. 27-4:

$$B = \frac{\mu_0 I}{2\pi R}$$

where R is the perpendicular distance from the wire to the point. The direction of the magnetic field is tangent to a line of magnetic field, as shown in Fig. 27-7, where the sense of the current is out of the page. Although the magnetic field produced by a steady current is static and the lines representing the magnetic field distribution do not move, each line closes on itself and encircles the current-carrying conductor. This encircling feature of the magnetic filed due to a current can be expressed in geometric terms. Consider a circular path of radius R seen in Fig. 27-7. The circle is in a plane perpendicular to the long, straight, current-carrying wire and centered at the axis of the wire. As a simple closed path, this circle forms the boundary of a surface which the current crosses or pierces. The current piercing such a surface is said to *thread* or to *link* the closed path, just as a filament threads the eye of a needle or as one loop in a chain links another.

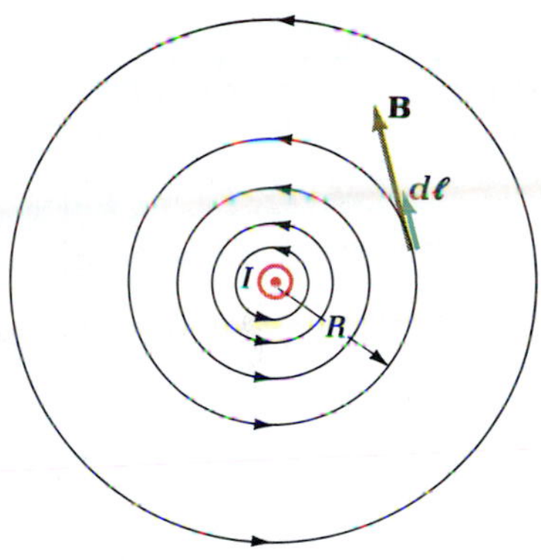

Figure 27-7. The lines of magnetic field for a long, straight, current-carrying wire encircle the wire. At each point on the circular path, **B** and $d\boldsymbol{\ell}$ are parallel.

The encircling relation between a magnetic field and the current linking a closed path can also be expressed quantitatively in a general result known as *Ampere's law.* To obtain the result for the simple case of the circular path illustrated in Fig. 27-7, let $d\boldsymbol{\ell}$ represent an infinitesimal displacement along the closed path. Form the dot product of the displacement $d\boldsymbol{\ell}$ and the magnetic field **B** at a point on the path. Add or integrate these contributions $\mathbf{B}\cdot d\boldsymbol{\ell}$ around the circle to form the line integral of **B** around a closed path:

$$\oint \mathbf{B}\cdot d\boldsymbol{\ell}$$

where the circle on the integral sign denotes a closed path.

For the case illustrated in Fig. 27-7, we see that **B** and $d\boldsymbol{\ell}$ are parallel at each point on the circular path. Thus the dot product is just the product of the magnitudes, $\mathbf{B}\cdot d\boldsymbol{\ell} = B\,d\ell \cos 0 = B\,d\ell$. Further, the magnitude of the magnetic field has the same value at every point on the circle, and B can be factored out of the integral. We have

$$\oint \mathbf{B}\cdot d\boldsymbol{\ell} = \oint B\,d\ell = B \oint d\ell = B(2\pi R)$$

where the integral of $d\ell$ around the circle is just the circumference $2\pi R$. From Eq. 27-4 we see that $B(2\pi R) = \mu_0 I$. Thus, for the circular path shown in Fig. 27-7, the line integral of **B** around the closed path depends only on the current I that links the path and the permeability constant μ_0:

$$\oint \mathbf{B}\cdot d\boldsymbol{\ell} = \mu_0 I \qquad (27\text{-}6)$$

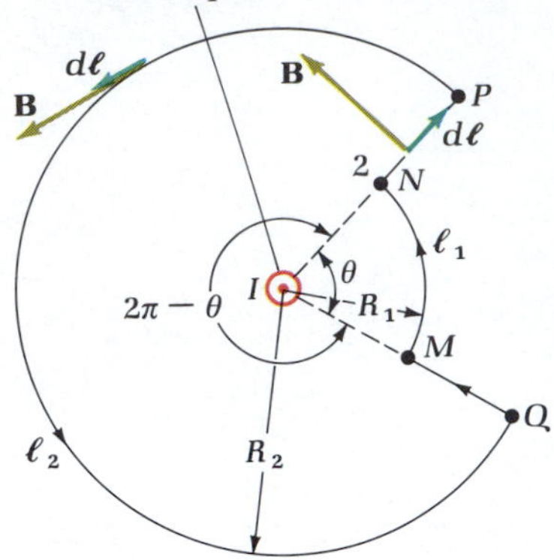

Figure 27-8. A closed path consists of two circular arcs *PQ* and *MN* and two radial lines *QM* and *NP*.

Notice that the result in this equation is independent of the radius R of the circular path.

The result expressed in Eq. 27-6 is even more generally applicable. Consider some other closed paths that are linked by the current in a long, straight wire and that lie in a plane perpendicular to the axis of the wire. The closed path shown in Fig. 27-8 consists of two circular arcs centered on the axis of the wire and two radial lines that connect the arcs. The line integral of **B** around this closed path is the sum of four line integrals—a line integral along each of the two circular arcs and a line integral along each of the two radial lines. Along a radial line, such as *NP*, the magnetic field **B** is perpendicular to the displacement $d\boldsymbol{\ell}$, as shown in Fig. 27-8, and the dot product is zero, $\mathbf{B}\cdot d\boldsymbol{\ell}=0$. Thus there is no contribution to $\oint \mathbf{B}\cdot d\boldsymbol{\ell}$ from the radial lines *NP* and *QM*. Along the arc *MN* of radius R_1, the line integral is easily evaluated following the procedure used above for the full circle:

$$\int_M^N \mathbf{B}_1\cdot d\boldsymbol{\ell}=\int_M^N B_1\,d\ell=B_1\int_M^N d\ell=B_1\ell_1$$

where ℓ_1 is the arc length subtending the angle $\theta=\ell_1/R_1$. Inserting the value of $B_1=\mu_0 I/2\pi R_1$ at a distance R_1 from the wire gives for this contribution to the closed path integral,

$$B_1\ell_1=\frac{\mu_0 I}{2\pi}\frac{\ell_1}{R_1}=\frac{\mu_0 I}{2\pi}\theta$$

Similarly, a calculation of the contribution from the circular arc *PQ* of radius R_2 gives

$$B_2\ell_2=\frac{\mu_0 I}{2\pi}(2\pi-\theta)$$

Adding the four contributions, we have

$$\oint \mathbf{B}\cdot d\boldsymbol{\ell}=\int_M^N \mathbf{B}\cdot d\boldsymbol{\ell}+\int_N^P \mathbf{B}\cdot d\boldsymbol{\ell}+\int_P^Q \mathbf{B}\cdot d\boldsymbol{\ell}+\int_Q^M \mathbf{B}\cdot d\boldsymbol{\ell}$$

$$=\frac{\mu_0 I}{2\pi}\theta+0+\frac{\mu_0 I}{2\pi}(2\pi-\theta)+0=\mu_0 I$$

Therefore, the line integral of **B** around this closed path depends only on the current linking the path: $\oint \mathbf{B}\cdot d\boldsymbol{\ell}=\mu_0 I$.

Another, more general closed path in a plane perpendicular to the wire and linked by the current I is shown in Fig. 27-9*a*. This closed path is approximated in Fig. 27-9*b* by a set of circular arcs and radial lines, and the approximation becomes exact in the limit of an infinite number of infinitesimal circular arcs and radial lines. The procedure used in evaluating the line integral of **B** around the closed path in Fig. 27-8 can be used for the path in Fig. 27-9*b*. The contribution along each radial line is zero, and the sum of the contributions for the circular arcs gives Eq. 27-6 again. This equation is also valid even if the closed path linked by the current I does not lie in a plane. (See Prob. 1.)

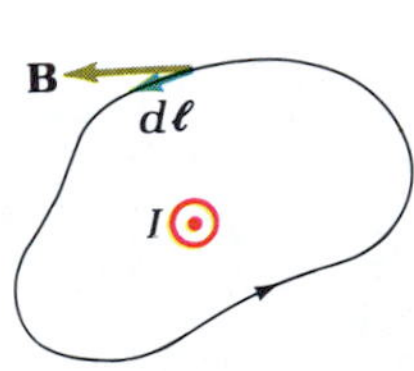

(a)

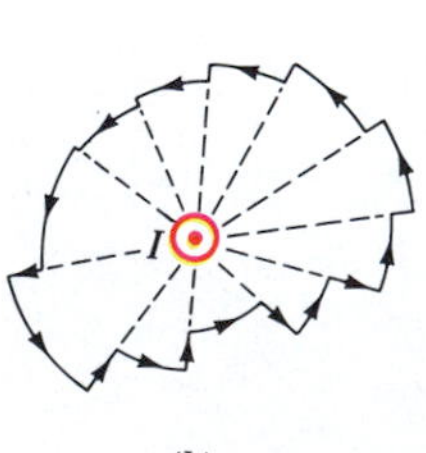

(b)

Figure 27-9. *(a)* A closed path. *(b)* The path is approximated by a set of circular arcs and radial lines.

Now we consider a closed path that is not linked by the current in the long, straight wire. The closed path in Fig. 27-10*a* has two circular arcs, which subtend the same angle θ at their common center, and two radial lines. Again, the contribution to $\int \mathbf{B}\cdot d\boldsymbol{\ell}$ is zero along a radial line. Along the arc of radius R_2, the contribution is $B_2\ell_2$, where $\ell_2=R_2\theta$ and $B_2=\mu_0 I/2\pi R_2$. But along the arc of radius R_1, $\mathbf{B}_1$ and $d\boldsymbol{\ell}$ have opposite directions, so that $\mathbf{B}_1\cdot d\boldsymbol{\ell}=B_1\,d\ell\cos 180^\circ=-B_1\,d\ell$. The contribution of this part of the path is then $-B_1\ell_1$, where $\ell_1=R_1\theta$ and $B_1=\mu_0/2\pi R_1$. The line integral of **B** around this closed path is the sum of the two contributions $B_2\ell_2$ and $-B_1\ell_1$ from the two circular arcs:

$$B_2\ell_2-B_1\ell_1=\frac{\mu_0 I}{2\pi R_2}R_2\theta-\frac{\mu_0 I}{2\pi R_1}R_1\theta=0$$

Thus, for the path that is not linked by the current,

$$\oint \mathbf{B} \cdot d\boldsymbol{\ell} = 0 \tag{27-7}$$

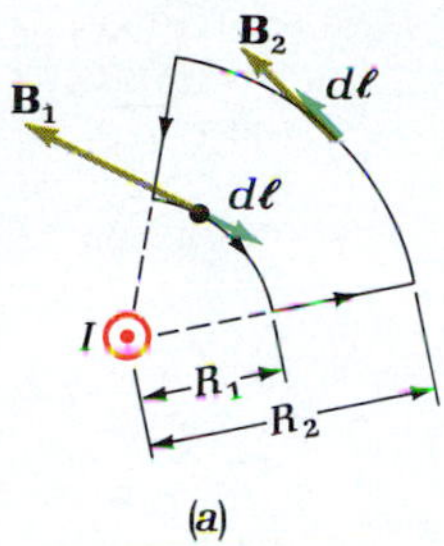

(a)

Equation 27-7 is also valid for a more general path such as that shown in Fig. 27-10*b*. That is, the line integral of **B** around a closed path is zero for a path that is not linked by the current *I*. (See Exercise 20.)

Having calculated the line integral of **B** around a closed path for several cases, we now state the general result. Consider an arbitrary closed path such as that shown in Fig. 27-11. Some of the currents link the closed path and others do not. In addition, these currents may have a general form, not necessarily that of long, straight wires. Let Σi represent the sum of the currents that link the closed path. Then

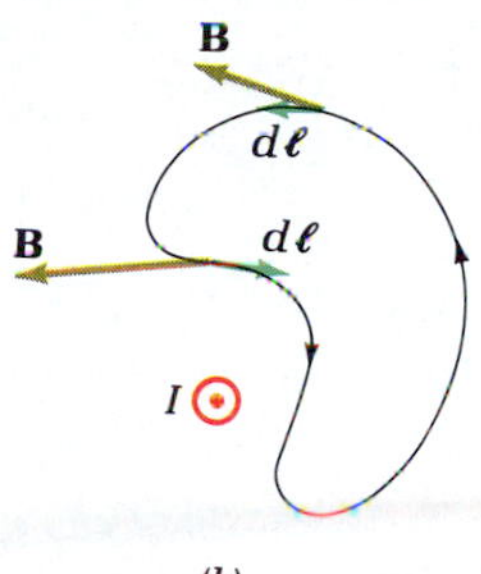

(b)

Figure 27-10. Two closed paths that are not linked by the current in a wire.

$$\oint \mathbf{B} \cdot d\boldsymbol{\ell} = \mu_0 \Sigma i \tag{27-8}$$

which is Ampere's law.

Notice that only the currents that link the closed path are to be included in the sum Σi. For the case illustrated in Fig. 27-11, the currents labeled I_4 and I_5 do not link the closed path and are not included in the sum. The currents labeled I_1, I_2, and I_3 link the path and are added algebraically to give, for this case,

$$\Sigma i = I_1 + I_2 - I_3 \tag{27-9}$$

The sign for each current is determined by a right-hand rule:

> Curl the fingers of the right hand to follow the sense of integration around the closed path; the extended thumb gives the sense of a positive current contribution.

Thus in Fig. 27-11, the currents I_1 and I_2 have the sense of the extended right thumb and are entered with a positive sign in Eq. 27-9. The current I_3 has the opposite sense and is entered with a negative sign in the sum. Currents I_4 and I_5, while contributing to the magnetic field at every point in space, do not contribute to the value of the line integral in Ampere's law. That is, the magnetic field **B** at a point depends on all the currents, but the line integral of **B** around a closed path depends only on the currents that link that path.

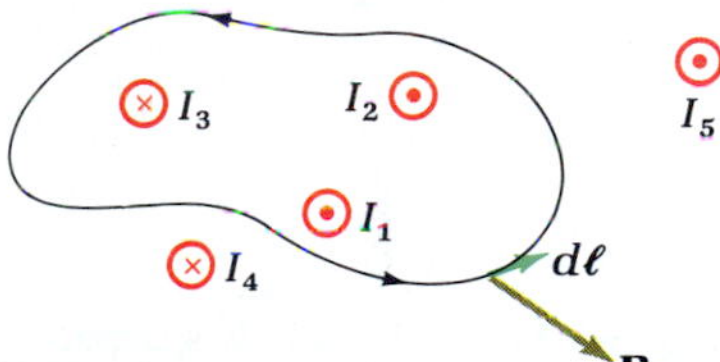

Figure 27-11. Currents I_1, I_2, and I_3 link the closed path; currents I_4 and I_5 do not link the closed path.

A given closed path can form the boundary for an infinite number of surfaces. If the closed path in Fig. 27-11 lies in the plane of the figure, then one choice of a surface is the portion of the plane within the path. But you can also imagine a surface bounded by the closed path that bulges out from the plane of the figure. Any choice of a surface bounded by a simple closed path is valid. (See Ques. 22 for an example of a closed path that is not simple.) The sum of the currents Σi piercing each surface is the same if we consider only steady current distributions. We shall consider the case of nonsteady currents in Sec. 27-6.

Ampere's law is equivalent to the Biot-Savart law.

Ampere's law for magnetic fields can be viewed in analogy with Gauss's law for electric fields. Ampere's law is a general statement about the fields that are produced by steady currents. Any such magnetic field must obey Ampere's law. A more advanced mathematical analysis shows that any magnetic field obtained from the Biot-Savart law must also satisfy Ampere's law. The Biot-Savart law and Ampere's law are equivalent in the same sense that Coulomb's law and Gauss's law are equivalent.

Gauss's law was used to determine the electric field due to certain types of charge distributions having a high degree of symmetry. Ampere's law can be similarly used to determine the magnetic field due to a current distribution having appropriate symmetry. We shall consider some examples of this procedure in the next section.

Static magnetic fields are different from static electric fields.

The analogy between Gauss's law and Ampere's law is not complete. It is essential to note that Ampere's law involves a line integral around a closed path, while Gauss's law involves a very different kind of integral, a surface integral over a closed surface. Static magnetic fields are therefore quite unlike static electric fields. For example, there is no distribution of electric charge that can produce an electrostatic field that is similar to the magnetic field illustrated by the lines in Fig. 27-7.

EXAMPLE 27-4

***Evaluating the line integral of* B.** Determine the value of the line integral of **B** around the closed path in Fig. 27-11, given the current values $I_1 = 1.6$ A, $I_2 = 1.4$ A, $I_3 = 1.7$ A, $I_4 = 4.0$ A, $I_5 = 0.8$ A.

Solution. The value of the line integral is determined from Ampere's law, Eq. 27-8. The currents linking the path are I_1 and I_2 in a positive sense and I_3 in a negative sense. Therefore

$$\oint \mathbf{B}\cdot d\boldsymbol{\ell} = \mu_0\Sigma i = \mu_0(1.6\text{ A} + 1.4\text{ A} - 1.7\text{ A})$$
$$= \mu_0(1.3\text{ A}) = 1.6 \times 10^{-6}\text{ T}\cdot\text{m}$$

SELF-TEST 27-4. Suppose the closed path in Fig. 27-11 is replaced with one that is linked by currents I_1, I_2, and I_5, but not I_3 or I_4. Also, the sense of the new path is opposite the one shown in the figure; it is clockwise rather than counterclockwise. Using the current values from the above example, determine the line integral of **B** around this new path. ***ANSWER:*** -2.8×10^{-6} T·m.

27-3 APPLICATIONS OF AMPERE'S LAW

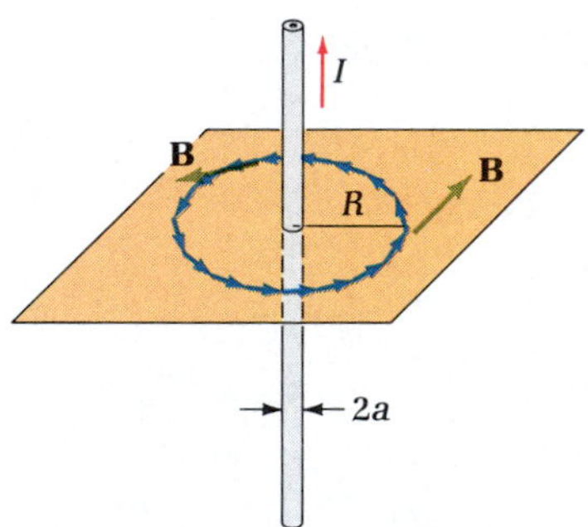

Figure 27-12. **B** and $d\boldsymbol{\ell}$ are parallel, and B is constant at each point on the circular path in a plane perpendicular to the axis of the wire.

Ampere's law, $\oint \mathbf{B}\cdot d\boldsymbol{\ell} = \mu_0\Sigma i$, can be used to determine the magnetic field produced by certain current distributions with a high degree of symmetry. As a simple first example of the procedure, we determine the magnitude of the field produced by a long, straight wire carrying a current I. The answer is already known from the Biot-Savart law for a point outside the wire and is given by Eq. 27-4. In addition to reproducing this equation by using Ampere's law, we shall obtain a new result, the magnetic field at a point inside the wire.

A long, straight wire of radius a carries a current I (Fig. 27-12). Since Ampere's law applies to any closed path, we choose a path that takes advantage of the symmetry. Thus we select a circular path of radius R lying in a plane perpendicular to the axis of the wire and centered on the axis. We apply Ampere's law to this path. From the symmetry of the problem, we observe that (i) the magnetic field **B** is parallel to the displacement $d\boldsymbol{\ell}$ so that $\mathbf{B}\cdot d\boldsymbol{\ell} = B\,d\ell$ and (ii) the magnitude B of the field has the same value at each point on the circular path. Then Eq. 27-8 becomes

$$\oint \mathbf{B}\cdot d\boldsymbol{\ell} = \oint B\,d\ell = B\oint d\ell = B\,2\pi R = \mu_0\Sigma i$$

where the integral of $d\ell$ around the closed circular path equals the circumference $2\pi R$, and, since $R > a$, the single current I links the path, or $\Sigma i = I$. Solving for B, we obtain

$$B = \frac{\mu_0 I}{2\pi R} \qquad (R > a) \qquad (27\text{-}4)$$

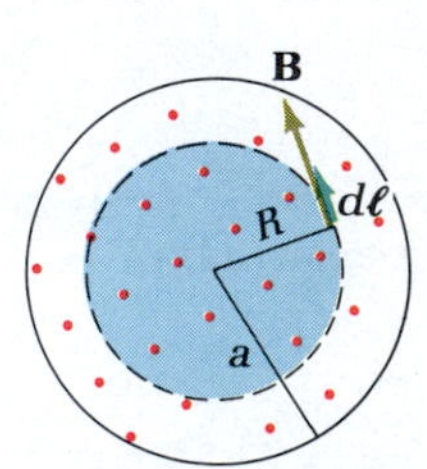

Figure 27-13. Only part of the current links the circular path of radius R, where $R < a$. The current is distributed uniformly over the cross section of the wire, and its sense is out of the plane of the figure.

which is Eq. 27-4, valid for any point outside the long, straight wire. Notice that in the procedure above, we are able to factor the symbol B, which is constant along the circular path, out of the integral in Ampere's law and then solve for B.

Now we consider a point inside the wire a perpendicular distance R from the axis of the wire, with $R < a$ (Fig. 27-13). Again we choose a circular path of radius R in a plane perpendicular to the wire and centered on the axis. Making the same symmetry arguments as above, we note that $\mathbf{B}\cdot d\boldsymbol{\ell} = B\,d\ell$ and that B is constant along the path and can be factored out of the integral. Thus

$$\oint \mathbf{B}\cdot d\boldsymbol{\ell} = \oint B\,d\ell = B\oint d\ell = B\,2\pi R = \mu_0\Sigma i$$

and

$$B = \frac{\mu_0 \Sigma i}{2\pi R}$$

To determine Σi, note that only part of the current I links this circular path. If the current is distributed uniformly over the cross section of the wire, then the current density has magnitude $J = I/\pi a^2$. Since the cross section of the wire within the circle has area πR^2, the current linking the closed circular path is

$$\Sigma i = J\pi R^2 = \frac{I}{\pi a^2}\pi R^2 = I\frac{R^2}{a^2}$$

Substituting this result into the above equation for B gives

$$B = \frac{\mu_0 IR}{2\pi a^2} \qquad (R < a) \qquad (27\text{-}10)$$

Magnetic field inside a long, straight wire

According to Eq. 27-10, the magnetic field is zero at a point on the axis of the wire ($R = 0$). The magnitude of the field increases linearly with distance R from the axis, and at the surface of the wire, the value is $B = \mu_0 I/2\pi a$. This value is obtained from each of Eqs. 27-4 and 27-10 at $R = a$. Outside the wire the magnitude of the magnetic field is given by Eq. 27-4 and decreases inversely with the distance R. These features are summarized graphically in Fig. 27-14.

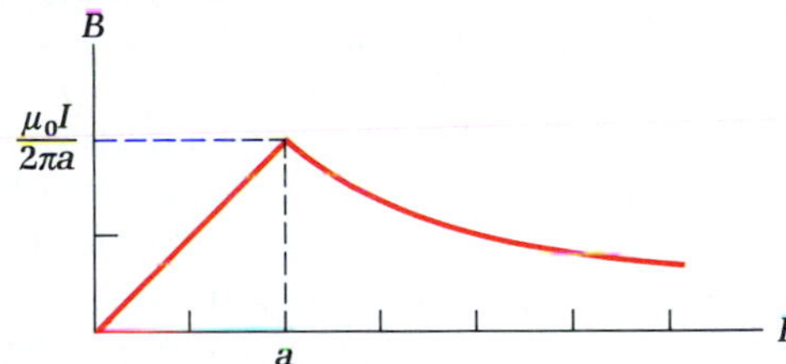

Figure 27-14. The magnitude of the magnetic field due to a current in a long, straight wire of radius a is shown for points inside the wire ($R < a$) and points outside the wire ($R > a$).

EXAMPLE 27-5

Field inside and outside a long, straight wire. A long, straight wire of radius 3.0 mm carries a 25-A current. *(a)* Determine the magnitude of the magnetic field at the surface of the wire. *(b)* At what distances from the axis of the wire is the magnitude of the field equal to one-half its value at the surface?

Solution. *(a)* At the surface of the wire, $R = a$, both Eq. 27-4 and Eq. 27-10 give the same value for B:

$$B = \frac{\mu_0 I}{2\pi a} = \frac{(4\pi \times 10^{-7}\ \text{T}\cdot\text{m}\cdot\text{A}^{-1})(25\ \text{A})}{2\pi(3.0 \times 10^{-3}\ \text{m})} = 1.7\ \text{mT}$$

(b) Outside the wire the magnitude of the magnetic field falls off inversely with distance according to Eq. 27-4. The field is reduced from its value at the surface by one-half at $R = 2a = 6.0$ mm. Inside the wire, Eq. 27-10 holds, and the field has one-half the surface value at $R = \frac{1}{2}a = 1.5$ mm. These results are apparent in Fig. 27-14.

SELF-TEST 27-5. A current in a long, straight wire (wire *A*) produces a field of magnitude 4.0 mT at the surface of the wire. Now wire *A* is replaced with wire *B* whose radius is half that of wire *A*. If wire *B* carries the same current, what is the magnitude of the field at the surface of wire *B*? ***ANSWER:*** 8.0 mT.

The Magnetic Field in a Solenoid

A ***solenoid*** is formed by winding a very long wire onto a cylinder, usually a circular cylinder. The windings, or turns of wire, form a helical coil whose length, measured along the axis of the solenoid, is typically somewhat larger than the diameter. An important characteristic of a solenoid is the number of turns per unit length n.

Number of turns per unit length characterizes a solenoid.

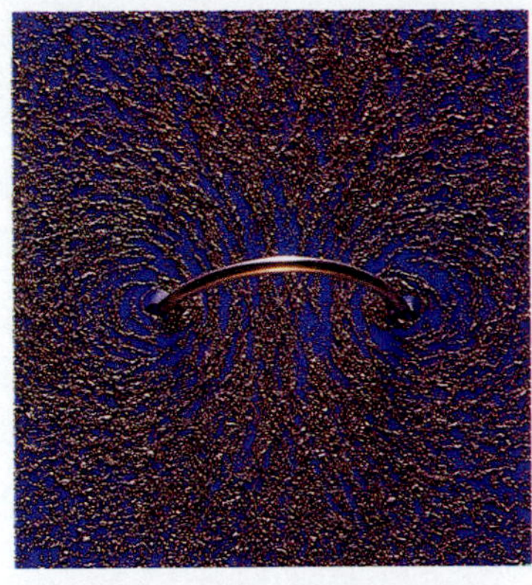

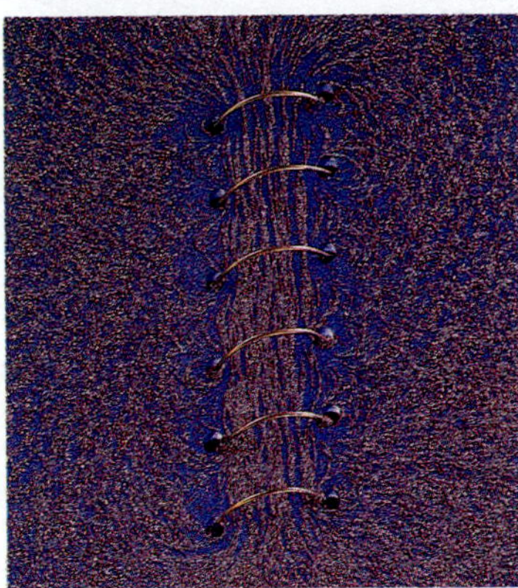

Iron filings indicate the spatial characteristics of the magnetic field around a single current loop (above) and around a loosely wound solenoid (below).

For a solenoid of length L_0 with N turns, the number of turns per unit length is $n = N/L_0$.

To understand the magnetic field produced by the current in a solenoid, we first look qualitatively at the magnetic field produced by a single circular current loop. Magnetic field lines are sketched in a plane perpendicular to the plane of the loop in Fig. 27-15*a*. The magnetic field lines for a loosely wound solenoid are sketched in Fig. 27-15*b*. For a more tightly wound solenoid, the pitch of the helical windings is less and each turn is approximately a current loop. Then each turn will produce a contribution to the magnetic field similar to that of a current loop. Inside the solenoid the contributions to the field from each turn tend to reinforce each other. The resultant field is approximately uniform and parallel to the axis of the solenoid. Outside the solenoid, the contributions to the field from each turn tend to cancel out, and the resultant field is relatively small.

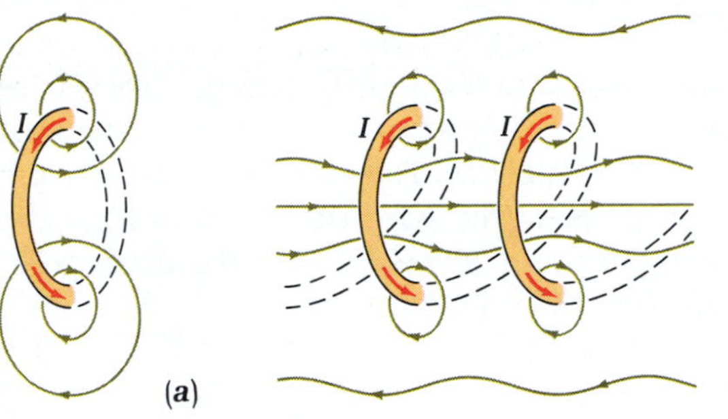

Figure 27-15. Some field lines are shown in a plane for *(a)* a circular current loop and *(b)* a loosely wound solenoid.

These tendencies—toward a uniform field inside the solenoid and a zero field outside the solenoid at points far from either end—become more pronounced for a very long and tightly wound solenoid. In the ideal case of Fig. 27-16, the current distribution in the windings is equivalent to a cylindrical current sheath, and the solenoid has an effectively infinite length. The magnetic field inside the ideal solenoid is parallel to the axis, and the field outside the solenoid is zero.

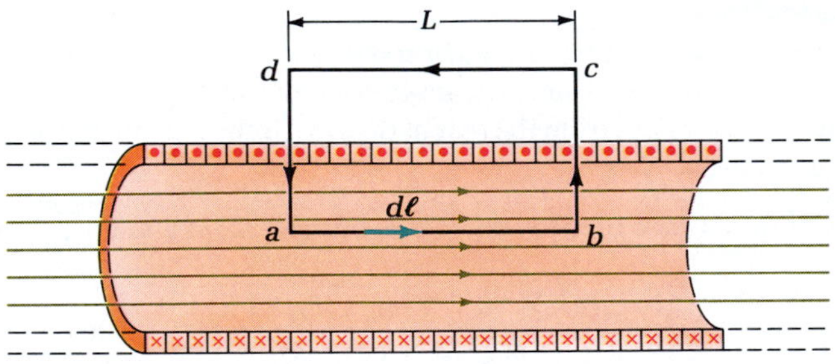

Figure 27-16. The current distribution for an ideal solenoid is equivalent to a cylindrical current sheath.

The magnetic field inside the ideal solenoid can be determined by applying Ampere's law to the closed path shown in Fig. 27-16. The integral around the closed path is the sum of integrals along each of the four straight-line segments:

$$\oint \mathbf{B}\cdot d\boldsymbol{\ell} = \int_a^b \mathbf{B}\cdot d\boldsymbol{\ell} + \int_b^c \mathbf{B}\cdot d\boldsymbol{\ell} + \int_c^d \mathbf{B}\cdot d\boldsymbol{\ell} + \int_d^a \mathbf{B}\cdot d\boldsymbol{\ell}$$

The integral along the segment *bc* is zero because $\mathbf{B}$ and $d\boldsymbol{\ell}$ are perpendicular at each point on that segment, making $\mathbf{B}\cdot d\boldsymbol{\ell} = 0$. For the same reason, the integral along segment *da* is also zero. Along segment *cd*, which is outside the ideal solenoid, $\mathbf{B} = 0$ so that the integral is zero for that segment. The only nonzero contribution to the closed path integral is from segment *ab*. On this segment, $\mathbf{B}$ and $d\boldsymbol{\ell}$ are parallel so that $\mathbf{B}\cdot d\boldsymbol{\ell} = B\,d\ell$, and B has the same value at each point. Thus

$$\oint \mathbf{B}\cdot d\boldsymbol{\ell} = \int_a^b B\,d\ell = B\int_a^b d\ell = BL$$

where L is the length of segment *ab*. For a solenoid with n turns per unit length, the number of turns within the closed path is nL. Since each of these turns carries current I, the net current linking this closed path is

$$\Sigma i = nLI$$

Using the above results in Ampere's law, we have

$$\oint \mathbf{B} \cdot d\boldsymbol{\ell} = BL = \mu_0 nLI$$

or

Magnetic field inside a solenoid

$$B = \mu_0 nI \tag{27-11}$$

While Eq. 27-11 was obtained for the ideal solenoid, it gives a good approximation to the field inside a tightly wound solenoid at points near the axis and far from the ends. The magnetic field is uniform in this region and is determined by the number of turns per unit length n and the current I in the solenoid. A solenoid is often used in the laboratory to provide a region of nearly uniform magnetic field. A ***toroid***, which can be viewed as a long solenoid bent to form a doughnut-shaped coil, is also used in the laboratory. (See Prob. 2.)

EXAMPLE 27-6

Field inside a solenoid. A solenoid has 250 turns on a cylinder of diameter 15.0 mm and length 125 mm. If the current in the solenoid is 0.320 A, determine the magnitude of the magnetic field inside the solenoid.

Solution. Since the length of the solenoid is fairly large compared with its diameter, the field inside the solenoid at points near the axis and far from the ends is approximately given by Eq. 27-11. The number of turns per unit length is (the unit "turn" is dimensionless)

$$n = \frac{250 \text{ turns}}{0.125 \text{ m}} = 2.00 \times 10^3 \text{ turns} \cdot \text{m}^{-1}$$

and the magnitude of the field is

$$B = \mu_0 nI = (4\pi \times 10^{-7} \text{ T} \cdot \text{m} \cdot \text{A}^{-1})(2.00 \times 10^3 \text{ m}^{-1})(0.320 \text{ A})$$
$$= 8.04 \times 10^{-4} \text{ T}$$

SELF-TEST 27-6. The axis of the solenoid is aligned horizontally along east-west. From the perspective of an observer located directly west of the solenoid, the current in each of the solenoid's loops is clockwise. What is the direction of the field inside the solenoid? ***ANSWER:*** east.

27-4 FORCE BETWEEN CURRENTS

In the last chapter, we determined the force exerted on a straight section of a current-carrying wire by a magnetic field (Eq. 26-3). In this chapter, we determined the magnetic field due to the current in a long, straight wire (Eq. 27-4). By combining these two results, we can determine the magnetic force that one current-carrying wire exerts on another.

Consider the arrangement of two long, parallel wires carrying currents I_1 and I_2 and separated by a distance R, as shown in Fig. 27-17. We assume that the separation distance R is small compared with the length of the wires so that the length is effectively infinite. The magnetic force on a segment of one of the wires, the wire with current I_1 and segment length ℓ, can be viewed as due to the magnetic field produced by the current I_2 in the other wire. Let $\mathbf{B}_2$ represent the magnetic field produced by the current I_2 at the position of the wire with current I_1. The direction of this field is as indicated in Fig. 27-17, and the magnitude is given by Eq. 27-4:

$$B_2 = \frac{\mu_0 I_2}{2\pi R}$$

The force **F** exerted by this field on the current element $I_1\ell$ is obtained from Eq. 26-3, $\mathbf{F} = I_1\boldsymbol{\ell} \times \mathbf{B}_2$. The direction of the force is shown in the figure, and the magnitude is

$$F = I_1 \ell B_2$$

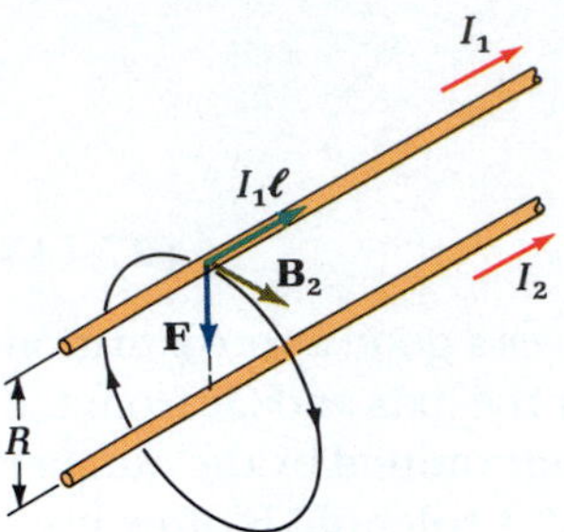

Figure 27-17. The magnetic field of a current I_2 exerts a force on a length of wire with current I_1.

Substituting for the value of B_2 in terms of the current I_2, we obtain the magnitude of the force exerted by the current I_2 in a long, straight wire on a length ℓ of a parallel wire with current I_1:

Force between parallel currents

$$F = \frac{\mu_0 I_1 I_2}{2\pi R}\,\ell \tag{27-12}$$

In arriving at Eq. 27-12, we considered the force on a segment of the wire with current I_1 exerted by the magnetic field of the current I_2. Suppose that the roles of the two currents are interchanged. That is, consider the force on a section of the wire with current I_2 exerted by the magnetic field of the current I_1. A similar analysis shows that the magnitude of the force on a length ℓ of wire 2 with current I_2 is also given by Eq. 27-12 and that the direction of the force on wire 2 is opposite the direction of the force on wire 1. Thus Eq. 27-12 gives the magnitude of the force on a length ℓ of either current-carrying wire due to the current in the other wire. This result for the magnitude of the force is also valid if the two currents have opposite senses instead of the same sense. (What is the direction of the force on each wire if the currents have opposite senses?)

Suppose that both current values were 1 A in Eq. 27-12 and that the wires were 1 m apart. Then the magnitude of the force on a 1-m length of each wire would be

$$F = \frac{\mu_0 (1\text{ A})(1\text{ A})}{2\pi(1\text{ m})}\,1\text{ m}$$
$$= 2 \times 10^{-7}\text{ N}$$

This type of calculation is used to define the ampere (A), the unit of electric current, in terms of the mechanical quantities force and length. (We have been waiting for Eq. 27-12 to give this definition.)

Definition of the ampere

The ***ampere*** is that current which, existing in each of two long (infinite), straight parallel wires separated by exactly 1 m, corresponds to a force per unit length between the wires of exactly 2×10^{-7} N/m.

We note a consequence of this definition of the unit of current.

Definition of the coulomb

The unit of charge, the ***coulomb*** (C), is defined as the amount of charge that flows past a point in one second in a circuit with a steady current of one ampere: 1 C = 1 (A)(s).

This procedure of defining the unit of current and then defining the unit of charge comes about because the magnetic force between steady currents can be more conveniently measured in practice than can the electric force between known charges.

EXAMPLE 27-7

A current balance. **In a current balance, the force per unit length exerted by one wire on another is measured by balancing, on a section of one of the wires, this magnetic force with a mechanical force. Suppose that the two long, straight wires are separated by 15.0 mm and that the balancing force per unit length on a segment of the wire is 7.11×10^{-6} N/m when the same current I exists in both wires. Determine the value of the current.**

Solution. Since $I_1 = I_2 = I$, we have $I_1 I_2 = I^2$ in Eq. 27-12. Solving for I^2,

$$I^2 = \frac{2\pi R}{\mu_0}\frac{F}{\ell} = \frac{2\pi(0.0150\text{ m})}{4\pi \times 10^{-7}\text{ T}\cdot\text{m}\cdot\text{A}^{-1}} 7.11 \times 10^{-6}\text{ N/m}$$

$$= 0.533\text{ A}^2$$

Taking the square root gives the current: $I = 0.730$ A.

Although the ampere is defined in terms of a configuration of two long, straight, parallel wires, a modern current balance, such as that used at the National Institute of Science and Technology, achieves greater precision by measuring the magnetic force between current-carrying coils.

SELF-TEST 27-7. Wires A and B are parallel. *(a)* If wire A carries a current to the north and wire B carries a current to the south, do the wires attract or repel one another? *(b)* If both wires carry a current to the north, do the wires attract or repel one another? ***ANSWERS:*** *(a)* repel; *(b)* attract.

27-5 MAGNETIC FLUX AND GAUSS'S LAW FOR MAGNETIC FIELDS

In analogy with the electric flux Φ_E, introduced in Chap. 21, we define a *magnetic flux* Φ_B of the magnetic field for a surface. Imagine dividing a mathematical surface into infinitesimal area elements. The direction of an area element $d\mathbf{S}$ at a point on the surface is perpendicular to the surface at that point, and a typical element for a surface is shown in Fig. 27-18*a*, along with the magnetic field **B** at a point. The magnetic flux $d\Phi_B$ for the area element $d\mathbf{S}$ is

The area element $d\mathbf{S}$ is perpendicular to the surface.

$$d\Phi_B = \mathbf{B}\cdot d\mathbf{S}$$

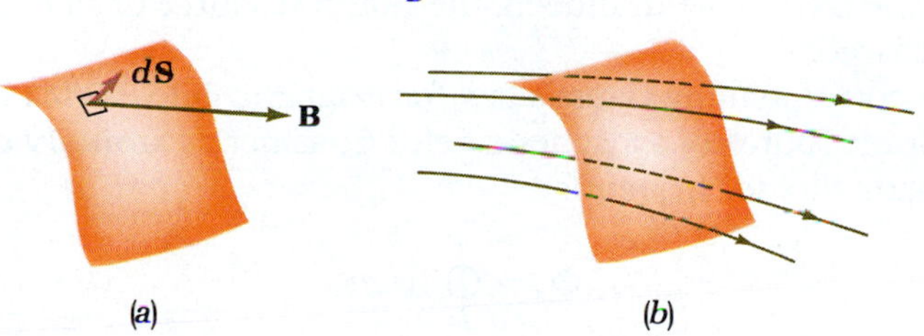

Figure 27-18. *(a)* The magnetic flux for an infinitesimal area $d\mathbf{S}$ is given by $d\Phi_B = \mathbf{B}\cdot d\mathbf{S}$. *(b)* The magnetic flux for a surface is proportional to the number of lines intersecting the surface.

The magnetic flux for a general surface is obtained by integrating (summing) the contributions $d\Phi_B$ as the area element $d\mathbf{S}$ ranges over the surface. Thus

$$\Phi_B = \int \mathbf{B}\cdot d\mathbf{S} \qquad (27\text{-}13)$$

Definition of a magnetic flux for a surface

The ***magnetic flux*** for a surface is the surface integral of the magnetic field over the surface.

The SI unit of magnetic flux is the ***weber*** (Wb), with 1 Wb = 1 T·m².

The magnetic flux for a surface can be interpreted in terms of the magnetic lines that represent the spatial distribution of the magnetic field. In analogy with electric lines and as suggested in Fig. 27-18*b*, the number of magnetic lines intersecting a surface is proportional to the magnetic flux for the surface.

EXAMPLE 27-8

Flux for a cross-sectional surface inside a solenoid. Determine the magnetic flux for a circular cross section of the ideal solenoid shown in Fig. 27-16. Let the inside radius of the solenoid be 7.5 mm, the number of turns per unit length be $2.00 \times 10^3\ \text{m}^{-1}$, and the current in the solenoid be 320 mA.

Solution. The magnetic field in the solenoid is parallel to the axis, and its magnitude (see Example 27-6) is given by Eq. 27-11:

$$B = \mu_0 nI = 8.0 \times 10^{-4}\ \text{T}$$

Each area element $d\mathbf{S}$ on a circular cross section of radius R of the solenoid is parallel to the axis and to $\mathbf{B}$ so that $\mathbf{B} \cdot d\mathbf{S} = B\,dS$, and B is uniform inside the solenoid. Applying these results in Eq. 27-13, we have

$$\Phi_B = \int \mathbf{B} \cdot d\mathbf{S} = \int B\,dS = B \int dS = B(\pi R^2)$$
$$= (8.0 \times 10^{-4}\ \text{T})\pi(7.5 \times 10^{-3}\ \text{m})^2 = 1.4 \times 10^{-7}\ \text{Wb}$$

SELF-TEST 27-8. For the solenoid in the above example, consider a flat circular surface inside the solenoid, with its diameter aligned with the solenoid's axis. What is the magnetic flux for this surface? ***ANSWER:*** zero.

Gauss's Law for B

Gauss's law for the electric field involves the electric flux for a *closed* surface. The direction of the area element for a closed surface is conventionally chosen to be directed outward from the enclosed volume. Gauss's law for **E** states that the electric flux for a closed surface depends only on the charge inside that surface. Thus

$$\Phi_E = \oint \mathbf{E} \cdot d\mathbf{S} = \frac{\Sigma q}{\epsilon_0} \tag{21-3}$$

where Σq is the algebraic sum of the charge within the volume enclosed by the surface. The form of Gauss's law reminds us that, in electrostatics, electric charge is the source of the electric field. Indeed, the simplest source of an electric field is a single point charge.

What is the corresponding Gauss's law for magnetic fields? What interpretation will it give for the source of a magnetic field? Consider an arbitrary *closed* surface and the magnetic flux for that surface:

$$\Phi_B = \oint \mathbf{B} \cdot d\mathbf{S}$$

Imagine this closed surface drawn anywhere in Fig. 27-3, 27-15, or 27-16. The magnetic flux for any closed surface is zero for any of these magnetic fields. Each magnetic line that pierces into a closed surface at one point also pierces out of this closed surface at some other point. The net number of lines crossing the closed surface is zero. Gauss's law for magnetic fields is the statement that this is true for any closed surface. Gauss's law for **B** is

The magnetic flux for any closed surface is zero.

In equation form,

Gauss's law for magnetic fields

$$\oint \mathbf{B} \cdot d\mathbf{S} = 0 \tag{27-14}$$

for any closed surface whatsoever. Note that this result does not apply to an open surface, one that does not enclose a volume. The magnetic flux for an open surface may have any value.

According to Gauss's law for **E**, the electric flux for a closed surface depends on the electric charge inside, and $\oint \mathbf{E} \cdot d\mathbf{S} = \Sigma q/\epsilon_0$ is not zero if the surface encloses a net charge. In contrast, from Gauss's law for **B**, the magnetic flux is zero for any closed surface, $\oint \mathbf{B} \cdot d\mathbf{S} = 0$. That is, there seems to be no magnetic counterpart to the electric charge. If the so-called magnetic charge did exist, it would correspond to a *magnetic monopole,* an isolated magnetic pole (an isolated north pole, for example). There has not been a confirmed observation of a magnetic monopole. In the absence of a magnetic monopole, the simplest source of a magnetic field is a magnetic dipole.

A magnetic monopole has not been observed.

The nonexistence of the magnetic monopole (or magnetic charge) can also be illustrated by the magnetic field lines. A line representing the magnetic field **B** always closes on itself, having no beginning and no end. A bar magnet, which possesses a magnetic dipole moment, has a north pole at one end and a south pole at the other end. The lines representing the field both inside and outside the magnet are sketched in Fig. 27-19*a*. Each line closes on itself so that the magnetic flux is zero for any closed surface. Notice the essential difference between the magnetic dipole field in Eq. 27-19*a* and the electric dipole field in Fig. 27-19*b*. The electric flux for the closed surface in Fig. 27-19*b* is not zero, but the magnetic flux for the corresponding surface in Fig. 27-19*a* is zero. This comparison illustrates a fundamental difference between static electric and magnetic fields, a difference in Gauss's law for **E** and Gauss's law for **B**. For any closed surface,

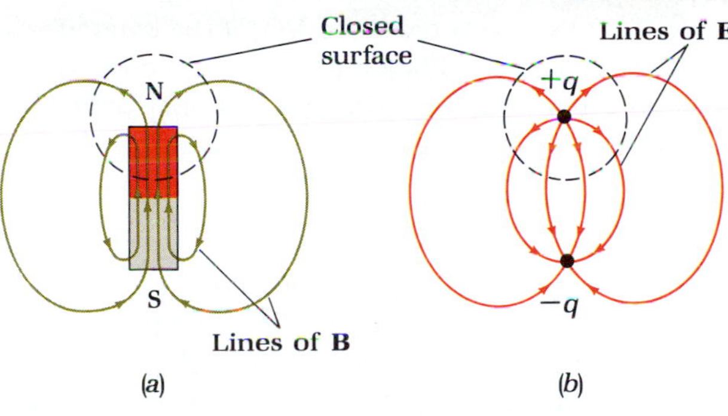

Figure 27-19. *(a)* The lines representing **B** for a bar magnet have no beginning and no end. The magnetic flux for the closed surface is $\oint \mathbf{B} \cdot d\mathbf{S} = 0$. *(b)* The lines representing **E** for an electric dipole begin on the positive charge and end on the negative charge. The electric flux for the closed surface is $\oint \mathbf{E} \cdot d\mathbf{S} = q/\epsilon_0$.

$$\oint \mathbf{E} \cdot d\mathbf{S} = \frac{\Sigma q}{\epsilon_0}$$

$$\oint \mathbf{B} \cdot d\mathbf{S} = 0$$

27-6 THE DISPLACEMENT CURRENT AND AMPERE'S LAW

Our use of Ampere's law, Eq. 27-8, has been limited so far to magnetic fields produced by currents of the sort that exist in continuous conducting wires. More general distributions of current exist, however, and Ampere's law must be modified. This generalization, discovered by Maxwell, represented a major advance in developing a deeper understanding of electromagnetism, including an understanding of the nature of light. It is a tribute to Maxwell's genius that at that time no experiment pointed to the necessity of modifying Ampere's law. To see why a modification is necessary, we return to the idea of a current linking a closed path.

Consider the closed path that encircles the wire carrying a steady current I in Fig. 27-20. Recall that the current linking a closed path is the current that pierces a surface that has the closed path as its boundary. Two such surfaces S_1 and S_2 are shown in the figure. Each surface is bounded by the given closed path. If the current

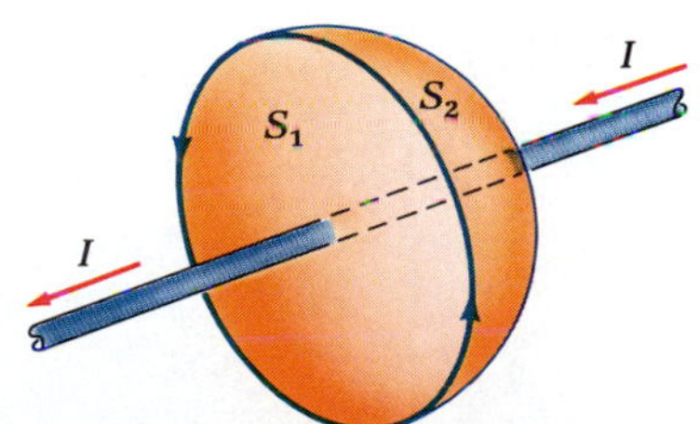

Figure 27-20. The closed path forms the boundary of two surfaces S_1 and S_2. Each surface is pierced by the current I.

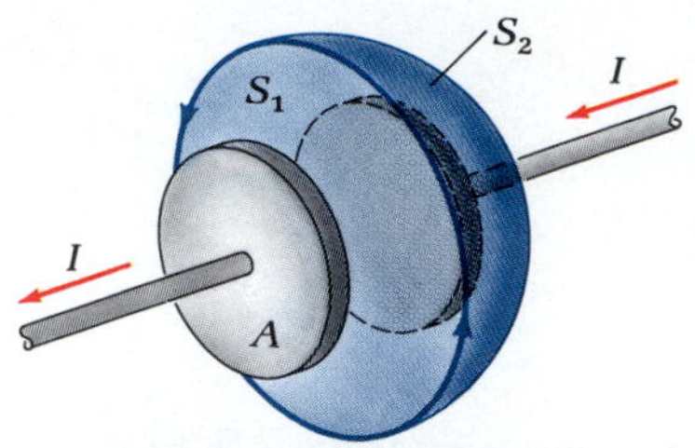

Figure 27-21. Surface S_1 l the space between the pla where there is no current. current I pierces only surf

piercing S_1 were different from the current piercing S_2, then there would be an accumulation of charge in the wire between the two surfaces. Since there is no accumulation of charge in this situation, each surface is pierced by the same current I. Either surface may be used in determining the current linking the closed path. That is, Ampere's law is valid for any surface bounded by the closed path.

A fundamentally different situation is shown in Fig. 27-21. There a parallel-plate capacitor is being charged, and I is the instantaneous value of the current in the connecting wires. The surface S_2 is pierced by the current I as before. The surface S_1, however, is not pierced by this current because this surface is in the space between the capacitor plates. There is an accumulation of charge on the plate between S_1 and S_2 as the capacitor is being charged. The rate dQ/dt at which charge accumulates on the plate is just equal (by conservation of charge) to the current piercing the surface S_2. Thus

$$I = \frac{dQ}{dt}$$

where Q is the instantaneous value of the charge on the capacitor. Since the current linking the closed path seems to depend on which surface we select, there is an inconsistency in Ampere's law for this case.

Maxwell's modification of Ampere's law, applied to the situation in Fig. 27-21, consists of considering a mathematically equivalent current to pierce surface S_1. Then the current linking the closed path will be the same for any surface bounded by that path. Figure 27-22 shows a cross section of the capacitor and the electric field in the region between plates. For simplicity we neglect fringing effects near the edges of the plates and assume vacuum between the plates. From Eq. 21-4, the magnitude of the electric field is

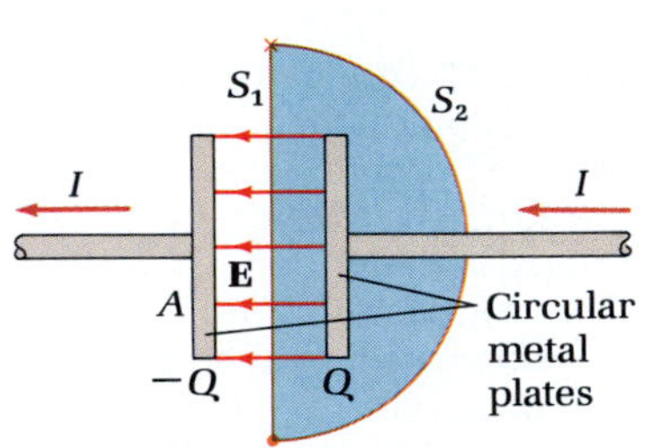

Figure 27-22. The electric flux for the plane surface S_1 is $\Phi_E = EA$, where A is the area of a plate of the capacitor. Edge effects have been neglected.

$$E = \frac{|\sigma|}{\epsilon_0} = \frac{Q}{\epsilon_0 A}$$

where $|\sigma| = Q/A$ is the magnitude of the surface charge density on the capacitor plate whose area is A. The magnitude Q of the charge on the plate can be expressed in terms of the electric flux, $\Phi_E = \int \mathbf{E} \cdot d\mathbf{S}$, for the surface S_1. The flux is $\Phi_E = EA$ since the field exists only in the region between plates. Solving the above equation for Q gives

$$Q = \epsilon_0 EA = \epsilon_0 \Phi_E$$

Taking the time derivative of Q, we find that the current I is related to the time derivative of the electric flux for surface S_1:

$$I = \frac{dQ}{dt} = \epsilon_0 \frac{d\Phi_E}{dt}$$

The right-hand side of the above equation contains the derivative of the electric flux piercing surface S_1, while I is the current piercing surface S_2. That is, $\epsilon_0 d\Phi_E/dt$ is mathematically equivalent for surface S_1 to the current I, which pierces surface S_2. We define this effective current, called the ***displacement current*** I_d, to be

Displacement current

$$I_d = \epsilon_0 \frac{d\Phi_E}{dt} \tag{27-15}$$

We emphasize that the current I pierces surface S_2 in Fig. 27-22, that the displacement current I_d pierces surface S_1, and that $I_d = I$.

The general form of Ampere's law, as modified by Maxwell, can now be stated. To the term for true currents Σi linking a closed path, we add the displacement current I_d linking the path; Ampere's law is

$$\oint \mathbf{B} \cdot d\boldsymbol{\ell} = \mu_0(\Sigma i + I_d)$$

or

$$\oint \mathbf{B} \cdot d\boldsymbol{\ell} = \mu_0 \left(\Sigma i + \epsilon_0 \frac{d\Phi_E}{dt} \right) \qquad (27\text{-}16)$$

Modified form of Ampere's law

By including the displacement current—treating the displacement current as a true current—the total current linking the closed path is the same for any surface bounded by the closed path.

The displacement current and the modified form of Ampere's law will be an essential part of our study of electromagnetic waves in Chap. 34. Except for situations similar to that described in the following example, the effect of the displacement current is negligible in circuits with slowly varying currents and fields.

EXAMPLE 27-9

Magnetic field between the plates of a charging capacitor. A parallel-plate capacitor with circular plates of radius a is being charged. Determine the magnetic field at a point in the space between plates a distance R from the axis of the capacitor with $R \leq a$. (See Fig. 27-23.) Express the result in terms of the instantaneous value of the current I in the charging circuit. Assume that the capacitor's lead wires are long and straight and that they are connected at the center of the plates (Fig. 27-21) so that the arrangement has azimuthal symmetry.

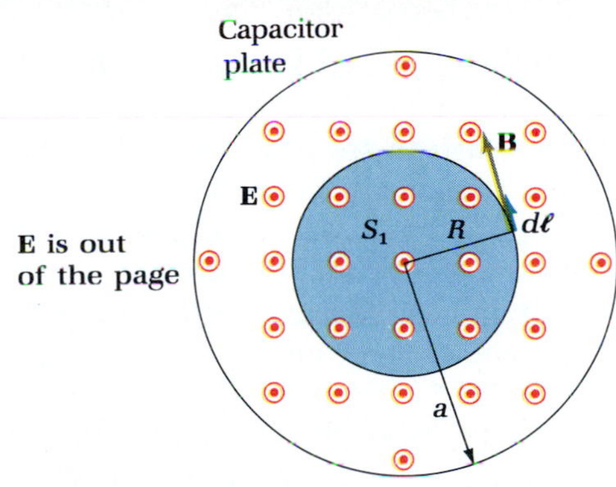

Figure 27-23. Example 27-9: One plate of a capacitor is shown, and the electric field is out of the plane of the figure. The electric flux for the surface S_1 is $\Phi_E = E\pi R^2$.

Solution. Select a closed circular path with radius R centered on the axis of the capacitor and with the plane of the circle parallel to the plates. The electric flux for the plane surface S_1 of area πR^2 is

$$\Phi_E = E\pi R^2$$

The magnitude of the electric field is given by

$$E = \frac{|\sigma|}{\epsilon_0} = \frac{Q}{\epsilon_0 \pi a^2}$$

where πa^2 is the area of a plate of the capacitor and Q is the instantaneous charge on the capacitor. The electric flux is then

$$\Phi_E = \frac{Q}{\epsilon_0 \pi a^2} \pi R^2 = \frac{QR^2}{\epsilon_0 a^2}$$

The displacement current linking the circular path of radius R is given by

$$I_d = \epsilon_0 \frac{d\Phi_E}{dt} = \frac{d}{dt}\left(Q \frac{R^2}{a^2} \right) = \frac{R^2}{a^2} \frac{dQ}{dt} = I \frac{R^2}{a^2}$$

because $I = dQ/dt$.

Now we apply Ampere's law to the closed path. From symmetry considerations, note that $\mathbf{B}$ and $d\boldsymbol{\ell}$ are parallel at each point on the circular path so that $\mathbf{B} \cdot d\boldsymbol{\ell} = B\,d\ell$. Further, the magnitude of the magnetic field is the same at each point on the path and

$$\oint \mathbf{B} \cdot d\boldsymbol{\ell} = \oint B\,d\ell = B \oint d\ell = B2\pi R$$

Since the plane surface lies in the space between capacitor plates, only the displacement current pierces this surface. The right-hand side of Ampere's law becomes

$$\mu_0(\Sigma i + I_d) = \mu_0(0 + I_d) = \frac{\mu_0 I R^2}{a^2}$$

Substituting these results into the modified form of Ampere's law (Eq. 27-16) gives

$$B 2\pi R = \frac{\mu_0 I R^2}{a^2}$$

Solving for B yields

$$B = \frac{\mu_0 I R}{2\pi a^2} \qquad (R \leq a)$$

The magnitude of the magnetic field is zero on the axis of the capacitor ($R = 0$) and increases linearly with distance from the axis (for $R \leq a$).

SELF-TEST 27-9. From the above example, consider a plane parallel to the capacitor plates and midway between them. What is the expression for the magnitude of the magnetic field at points in this plane with $R > a$? ***ANSWER:*** $B = \mu_0 I / 2\pi R$.

COMMENTARY: James Clerk Maxwell

James Clerk Maxwell

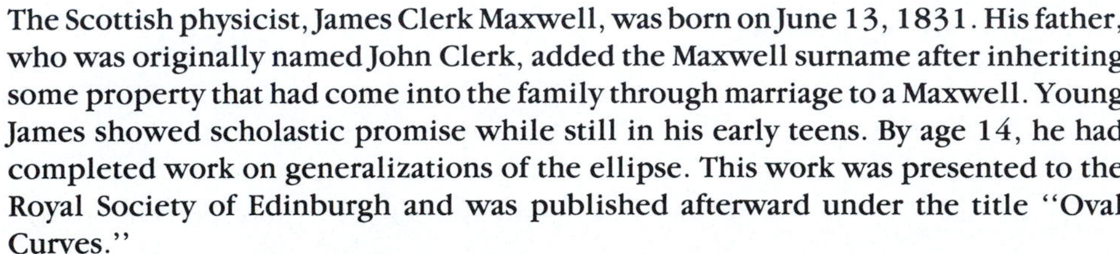

The Scottish physicist, James Clerk Maxwell, was born on June 13, 1831. His father, who was originally named John Clerk, added the Maxwell surname after inheriting some property that had come into the family through marriage to a Maxwell. Young James showed scholastic promise while still in his early teens. By age 14, he had completed work on generalizations of the ellipse. This work was presented to the Royal Society of Edinburgh and was published afterward under the title "Oval Curves."

In 1847, Maxwell attended the University of Edinburgh where, under the tutelage of a physics professor, he received permission to use some of the apparatus after hours. He spent many evenings experimenting and learning. Even his vacations were used for experimenting. He set up a makeshift laboratory "above the wash-house," and for a lab table, he used "an old door set on two barrels, and two chairs, of which one is safe. . . ." By 1849 he had finished two more papers, one titled "Rolling Curves" and the other, "The Equilibrium of Elastic Solids."

Maxwell was largely responsible for developing the mathematical description of electric and magnetic fields. The use of field lines, as conceived by Faraday, to understand electric and magnetic phenomena was mainly qualitative until 1855 when Maxwell provided a useful mathematical model. In his paper "On Faraday's Lines of Force," Maxwell developed the mathematical analogy between the lines representing a field and the flow of an incompressible fluid. The field strength, say of an electric field **E**, corresponded in the analogy to the velocity **v** of a fluid. This analogy, while imperfect, is still reflected in the language that we use to describe fields. Thus we speak of the flux Φ_E of an electric field as if it were related to the flow of a fluid.

Maxwell's use of a fluid analogy continued in his famous four-part paper, "On Physical Lines of Force," published in 1861. Here the fluid model contained vortices; the magnetic field lines were represented by the axes of the vortices, and the magnetic force was related to the pressure of the vortices. The model was rather complex since neighboring vortices had to revolve with the same sense. Thus Maxwell had to introduce rolling contact particles between these vortices. He obtained the mathematical relationships between currents and magnetic fields by considering the motion of the contact particles in the model fluid. This model also led Maxwell to regard light as an electromagnetic phenomenon: "Light consists in the transverse undulations of the same medium which is the cause of electric and magnetic phenomena.''

The mathematical relations between the electric and magnetic fields that Maxwell developed (known as Maxwell's equations and discussed in Chap. 34) provide a complete theoretical basis for all electromagnetic phenomena. These equations were considered by Boltzmann to be so beautiful in their simplicity and elegance that he asked, quoting Goethe, "Was it a god who wrote these lines. . . ?''

In 1871 Maxwell was named the first professor of experimental physics at Cambridge University and was appointed as director of the Cavendish Laboratory, which was then under construction. There he provided, until his death in 1879, the early guidance that allowed the Cavendish Laboratory to become one of the premier centers of basic research in physics.

Maxwell's work in electricity and magnetism merged those two seemingly separate disciplines into a unified theory that encompassed all of electromagnetism. In addition to his synthesis of electricity and magnetism, Maxwell also made essential contributions to thermodynamics and statistical mechanics (see the Commentary, Chap. 19). For all of these contributions, he is generally put in the company of Galileo, Newton, and Einstein—as one on whose shoulders we stand to see nature's more distant horizons.

For further reading, see *James Clerk Maxwell—A Biography* by Ivan Tolstoy (University of Chicago Press, Chicago, 1981).

SUMMARY

Section 27-1. The Biot-Savart Law

At a point in space the contribution $d\mathbf{B}$ to the magnetic field due to an infinitesimal current element $I\,d\boldsymbol{\ell}$ is given by the Biot-Savart law:

$$d\mathbf{B} = \frac{\mu_0}{4\pi}\frac{I\,d\boldsymbol{\ell} \times \hat{\mathbf{r}}}{r^2} \qquad (27\text{-}1)$$

The resultant magnetic field at a point is obtained from the integrated form of the Biot-Savart law:

$$\mathbf{B} = \int \frac{\mu_0}{4\pi}\frac{I\,d\boldsymbol{\ell} \times \hat{\mathbf{r}}}{r^2} \qquad (27\text{-}3)$$

A current in a long, straight wire produces a magnetic field. At a distance R from the axis of the wire, the magnetic field has magnitude

$$B = \frac{\mu_0 I}{2\pi R} \qquad (27\text{-}4)$$

The direction of the magnetic field is given by a right-hand rule. The Biot-Savart law can be used to determine the magnetic field due to other simple current distributions, such as a circular current loop.

Section 27-2. Ampere's Law

According to Ampere's law, the line integral of the magnetic field for any closed path depends only on the sum of the steady currents Σi piercing a surface bounded by the closed path:

$$\oint \mathbf{B} \cdot d\boldsymbol{\ell} = \mu_0 \Sigma i \qquad (27\text{-}8)$$

Section 27-3. Applications of Ampere's Law

Ampere's law can be used to determine the magnetic field due to certain highly symmetric current distributions. The field inside a long, tightly wound solenoid with current I and n turns per unit length is given approximately by

$$B = \mu_0 n I \qquad (27\text{-}11)$$

Section 27-4. Force between Currents

The magnitude of the force on a length ℓ between two long, straight, parallel wires a distance R apart and carrying currents I_1 and I_2 is given by

$$F = \frac{\mu_0 I_1 I_2}{2\pi R}\ell \qquad (27\text{-}12)$$

The ampere, the unit of electric current, is defined in terms of the force per unit length between the wires.

Section 27-5. Magnetic Flux and Gauss's Law for Magnetic Fields

Gauss's law for magnetic fields is

$$\oint \mathbf{B} \cdot d\mathbf{S} = 0 \qquad (27\text{-}14)$$

The magnetic flux for any closed surface is zero, corresponding to the absence of magnetic monopoles. A line representing $\mathbf{B}$ closes on itself.

Section 27-6. The Displacement Current and Ampere's Law

The modified form of Ampere's law is given by

$$\oint \mathbf{B} \cdot d\boldsymbol{\ell} = \mu_0 \left(\Sigma i + \epsilon_0 \frac{d\Phi_E}{dt} \right) \qquad (27\text{-}16)$$

where $I_d = \epsilon_0 d\Phi_E/dt$ is the displacement current.

QUESTIONS

1. Discuss similarities and differences between the electric field $d\mathbf{E}$ due to an element of charge dq and the magnetic field $d\mathbf{B}$ due to a current element $I\,d\boldsymbol{\ell}$.

2. Discuss the meaning of the following statement: "The Biot-Savart law, Eq. 27-1, shows that the magnetic field is an inverse-square field."

3. Discuss similarities and differences for the electric field due to a long line of charge (see Eq. 20-8) and the magnetic field due to the current in a long, straight wire (see Eq. 27-4).

4. Reconcile the statement in Ques. 2 with the expression for the magnitude of the magnetic field due to the current in a long, straight wire, Eq. 27-4.

5. To have a steady current in a wire, the wire must be part of a closed circuit. How can we speak of the magnetic field due to the current in a long, straight wire without considering the rest of the circuit?

6. Draw an analogy between a steady current linking a closed path and two links of a chain.

7. Suppose that the sense of each of the currents is reversed in Fig. 27-11. Would the magnetic field **B** change at a given point? How would the value of the line integral change in Example 27-4?

8. What would the answers in Ques. 7 be if the senses of I_4 and I_5 only were reversed?

9. Often the two lead wires to an electrical device are twisted together (forming a double helix). Use Ampere's law to describe qualitatively the features of the magnetic field due to a twisted pair of wires carrying equal currents with opposite senses.

10. Use the Biot-Savart law qualitatively and the right-hand rule to convince yourself that the lines sketched in Fig. 27-15*a* and *b* are representative of the magnetic field distribution.

11. Using only words and planar sketches, could you explain the distinction between right- and left-handedness to an intelligent extraterrestrial being? Could you explain how to find the cross product of two vectors? How? (It's not as easy as you may think. Can you explain this to a college student?)

12. Figure 27-14 shows the dependence on distance R of the magnetic field inside and outside a long, straight wire carrying a current. Does the magnetic field inside the wire exert a net force on that wire? Explain.

13. Equation 27-4 was obtained from the Biot-Savart law by assuming that the long, straight wire had negligible thickness. The same equation was also obtained by using Ampere's law. What assumptions were made about the current distribution in the latter calculation?

14. Consider the magnetic field at points in between two current-carrying, long, straight wires. Is the magnitude of the field greater for currents with the same sense or with opposite senses in the wires? Explain. Where is the magnitude of the field largest in the two cases? Where is it smallest?

15. A circular loop of fine wire carries a current. At which point in the plane of the loop do you expect the magnetic field to be stronger, at the center of the loop or at a point close to the wire? Does your answer depend on the thickness of the wire? Explain.

16. What is the direction of the magnetic force that a long, straight wire exerts on a parallel wire if the currents in the wires have the same sense? What if they have opposite senses?

17. How can a solenoid whose length is 100 mm have 2000 turns per meter?

18. A solenoid is mistakenly wound in the following way: One layer of windings forms a right-handed helix, which is then covered with a returning second layer. If the solenoid carries current I, the magnetic field at points inside and near the axis of the solenoid is virtually zero. Explain this result.

19. Compare the two sketches in Fig. 27-19. How is a north magnetic pole similar to a positive electric charge? How is it different?

20. How can the magnetic flux for an open surface be nonzero if the magnetic flux for any closed surface is zero?

21. For the situation shown in Fig. 27-22, explain why the displacement current I_d piercing the surface S_1 must equal the current I in the lead wires.

22. A simple closed path forms the boundary of a smooth surface which is *orientable*. That is, if you imagine traversing the path so that the surface always is on your left, then that procedure defines a particular side of the surface. There are nonsimple paths for which the above procedure fails. Try applying it to *(a)* an ellipse, *(b)* a polygon, *(c)* a figure eight, *(d)* the Möbius strip in Fig. 21-27.

23. Complete the following table:

Symbol	Represents	Type	SI Unit
$\hat{\mathbf{r}}$			—
μ_0		Scalar	
$\oint \mathbf{B}\cdot d\boldsymbol{\ell}$			
Σi			
Φ_B			
I_d			A
$\oint \mathbf{B}\cdot d\mathbf{S}$	Magnetic flux for a closed surface		

EXERCISES

Section 27-1. The Biot-Savart Law

1. Using Eq. 27-2, show that the permeability constant μ_0 has SI units of $T \cdot m \cdot A^{-1}$.

2. The magnitude of the magnetic field of the earth averages around 2×10^{-5} T. What current must exist in a long, straight wire in order that its magnetic field at a point 10 mm from the axis of the wire have a magnitude comparable to the earth's magnetic field?

3. A long, straight wire carries a 15-A current. *(a)* Determine the magnitude of the magnetic field at a point 35 mm from the wire. *(b)* On a diagram show the direction of **B** at that point and the sense of the current in the wire.

4. Two long, straight, parallel wires carry currents out of the plane of Fig. 27-24. Determine the magnitude and direction of the magnetic field at *(a)* point *P* and *(b)* point *Q*. Express the magnitude in terms of μ_0, I, and a. *(c)* Evaluate B at points *P* and *Q* for $I = 12$ A and $a = 250$ mm.

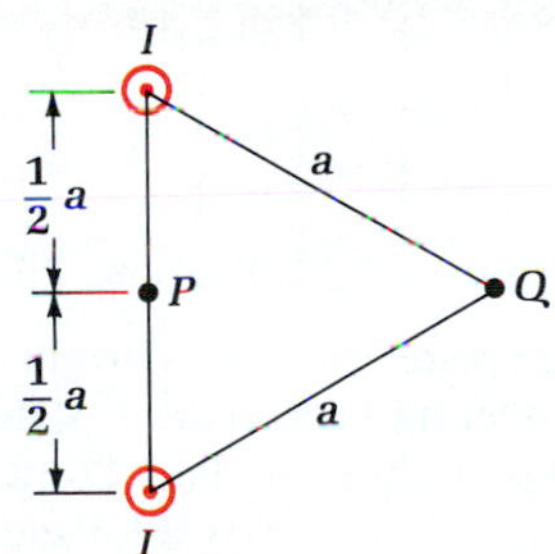

Figure 27-24. Exercise 4.

5. Rework the previous exercise with the sense of the current in the lower wire reversed.

6. Use the Biot-Savart law to determine the magnetic field at the center of a circular current loop. Indicate on a diagram the sense of the current I in the loop and the direction of the magnetic field at the center.

7. Two long, straight, parallel wires separated by a distance D carry currents I_1 and $I_2 = 2I_1$ with the same sense. *(a)* At what distance from the wire with current I_1 is the magnetic field zero between the two wires? *(b)* Locate other points, if any, at which $B = 0$.

8. Suppose the two currents in the previous exercise have opposite senses. Locate those points, if any, at which $B = 0$.

9. Consider the contribution to the magnetic field due to a current in a straight length of wire at a point *P* lying on the axis of the wire. See Fig. 27-25. *(a)* Show that this contribution is zero. *(b)* Is your argument valid even if the point *P* is inside the wire rather than outside? Explain.

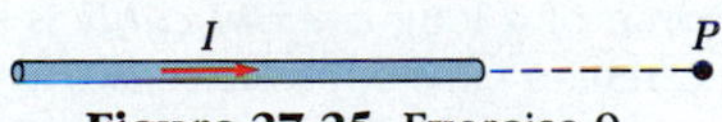

Figure 27-25. Exercise 9.

10. A 5.0-A current exists in two long, straight, parallel wires joined by a semicircular wire of radius 75 mm, as shown in Fig. 27-26. Determine the magnetic field at the center of the semicircle.

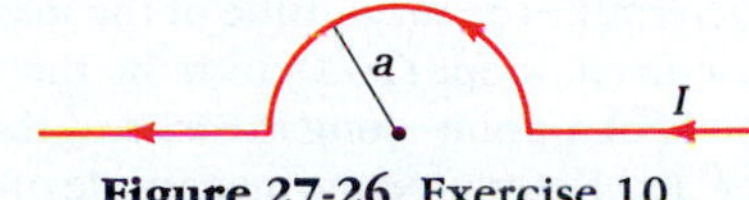

Figure 27-26. Exercise 10.

11. A current loop consists of two concentric circular arcs and two perpendicular radial lines, as shown in Fig. 27-27. *(a)* Determine the magnetic field at the center. *(b)* Evaluate the magnitude of the magnetic field for $I = 20$ A, $a = 30$ mm, $b = 50$ mm.

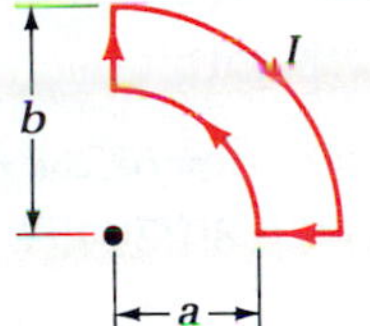

Figure 27-27. Exercise 11.

12. A straight section of conductor of length L carries a current I. *(a)* Show that at point *P* in Fig. 27-28 the contribution to the magnetic field due to this current has magnitude

$$B = \frac{\mu_0 I}{4\pi R} \frac{L}{\sqrt{R^2 + L^2/4}}$$

(*Hint:* See the discussion leading to Eq. 27-4.) *(b)* Show that this expression gives $B \approx \mu_0 I / 2\pi R$ if $L \gg R$.

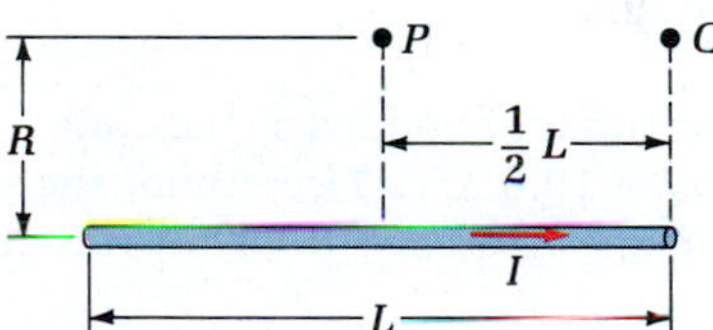

Figure 27-28. Exercise 12.

13. *(a)* Determine the contribution to the magnetic field at point *Q* in Fig. 27-28 due to the current in the straight conductor. *(b)* By taking an appropriate limit, determine the magnetic field at a point a perpendicular distance R from one end of a long (semi-infinite), straight, current-carrying wire. (*Hint:* See the previous exercise.) *(c)* Two long, straight wires carrying current I intersect as shown in Fig. 27-29. Show that the magnetic field at point *P* is directed perpendicularly into the plane of the page and has magnitude $B = \mu_0 I / 4\pi R$.

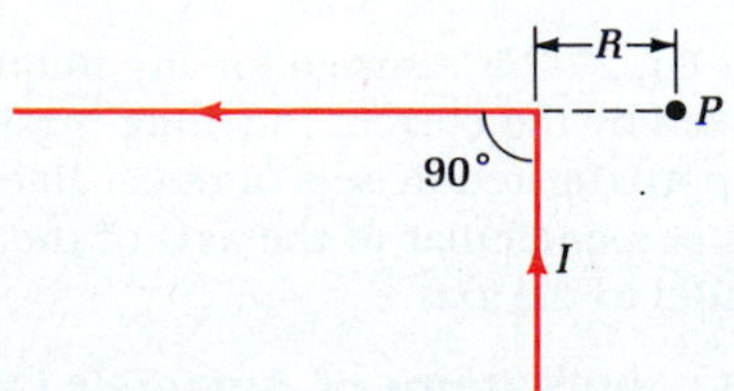

Figure 27-29. Exercise 13.

14. *(a)* Determine the expression for the magnitude of the magnetic field at the center of a square loop of edge a with current I. (See Exercise 12.) *(b)* Evaluate numerically for $a = 150$ mm, $I = 6.0$ A. *(c)* Show the sense of the current and the direction of **B** at the center on a diagram.

15. A circular current loop of radius 2.5 mm carries a 7.4-mA current. *(a)* Determine the magnitude of the magnetic dipole moment of the current loop. *(b)* Determine the magnitude of the magnetic field at a point along the axis of the loop 1.0 m from the center. *(c)* Determine the magnitude of the magnetic field at the center of the loop.

16. The magnitude of the magnetic field due to a current loop is 13 μT at a point on the axis of the loop 250 mm from the center. The circular loop has a 15-mm radius. Determine *(a)* the magnitude of the magnetic dipole moment of the loop and *(b)* the current in the loop. *(c)* Show on a diagram the sense of the current, the direction of the dipole moment, and the direction of **B** at a point on the axis.

17. The electric field at a point on the perpendicular bisector plane of an electric dipole is given approximately (far from the dipole) by

$$\mathbf{E} = \frac{-\mathbf{p}}{4\pi\epsilon_0 r^3}$$

where **p** is the electric dipole moment and r is the distance from the dipole. By analogy, write down the expression for the magnetic field **B** at a point in the plane of and far from a current loop with magnetic dipole moment **m**.

Section 27-2. Ampere's Law

18. *(a)* Determine Σi in Ampere's law for the case illustrated in Fig. 27-11. Take $I_1 = 1.0$ A, $I_2 = 2.0$ A, $I_3 = 3.0$ A, $I_4 = 4.0$ A, $I_5 = 5.0$ A. *(b)* Explain how you determine the sign of each term in the sum.

19. Consider the surface bounded by the closed path shown in Fig. 27-30, with $I = 10.0$ A. *(a)* Determine the sum ΣI for this case. *(b)* What is the value of $\oint \mathbf{B} \cdot d\boldsymbol{\ell}$ for this closed path?

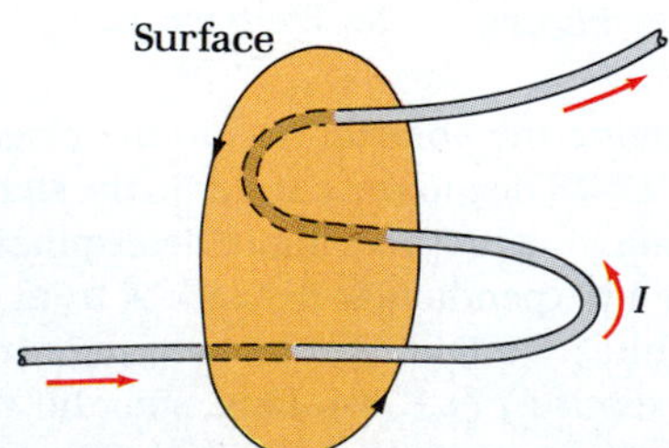

Figure 27-30. Exercise 19: Kinky wire punches through the surface three times.

20. Show that Eq. 27-7 is satisfied for any simple closed path that is not linked by the current in a long, straight wire. The path can be approximated by sets of radial lines and circular arcs in planes perpendicular to the axis of the wire and line segments parallel to the axis.

Section 27-3. Applications of Ampere's Law

21. A long, straight wire of diameter 2.5 mm carries a uniformly distributed 12-A current. *(a)* At what distance from the axis of the wire is the magnitude of the magnetic field a maximum? *(b)* Construct a graph of the magnitude of the magnetic field B versus the radial distance R from the axis of the wire for the region $0 \leq R \leq 3.0$ mm.

22. Determine the magnetic field due to a 320-mA current at a point 2.50 mm from the axis of a long, straight wire of radius *(a)* 4.00 mm, *(b)* 2.50 mm, *(c)* 2.00 mm.

23. Idealized sketches of magnetic lines often show a region of uniform magnetic field ending abruptly; that is, a uniform field region is adjacent to a field-free region. Use Ampere's law and paths such as that shown in Fig. 27-31 to show that the magnetic field must change gradually in regions where there are no currents.

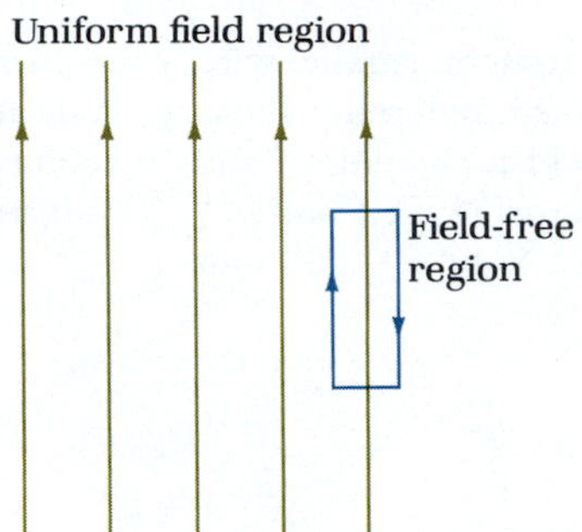

Figure 27-31. Exercise 23.

24. A long, straight wire carries a current I_0, which is uniformly distributed over the cross section of the wire of radius a. *(a)* For the closed path shown in Fig. 27-32, explicitly evaluate $\oint \mathbf{B} \cdot d\boldsymbol{\ell}$ using one or both of Eqs. 27-4 and 27-10. *(b)* Show that the current linking this closed path is given by $\Sigma i = I_0\theta/2\pi$.

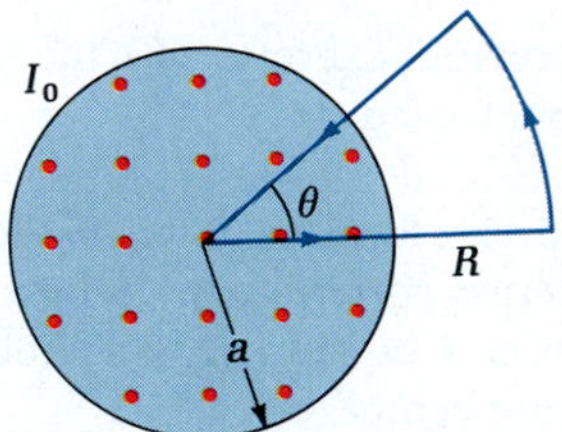

Figure 27-32. Exercise 24.

25. A long, hollow conducting cylinder carries a current I_0, which is uniformly distributed over the cross section, as shown in Fig. 27-33. Determine the magnitude of the magnetic field at a point a distance R from the axis of the cylinder for *(a)* $R \leq b$; *(b)* $b \leq R \leq c$; *(c)* $c \leq R$.

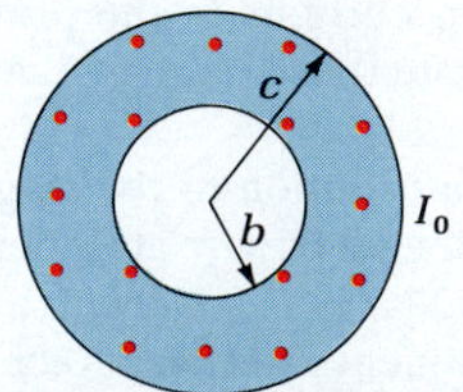

Figure 27-33. Exercise 25.

26. A cross section of a long *coaxial cable* is shown in Fig. 27-34. Equal currents I have opposite senses in the inner and outer conductors. Assume that the current density **J** is uniform

in each conductor. Determine the magnitude of the magnetic field at a point a distance R from the axis of the cable for *(a)* $R \leq a$; *(b)* $a \leq R \leq b$; *(c)* $b \leq R \leq c$; *(d)* $c \leq R$.

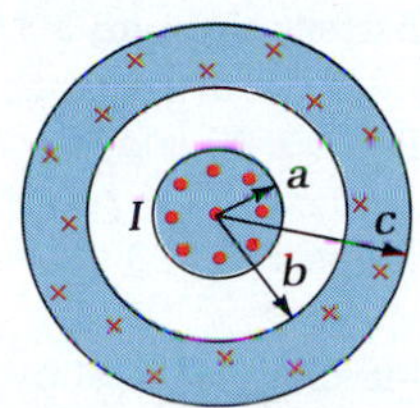

Figure 27-34. Exercise 26.

27. A long solenoid with 850 turns per meter has a 2.4-A current. *(a)* Determine the magnitude of the magnetic field near the center of the solenoid. *(b)* How many turns of wire are on the solenoid if its length is 200 mm? *(c)* Estimate the diameter of the wire.

28. By applying Ampere's law and symmetry arguments to the closed path shown in Fig. 27-35 for an ideal solenoid, show that the magnetic field is uniform inside the solenoid.

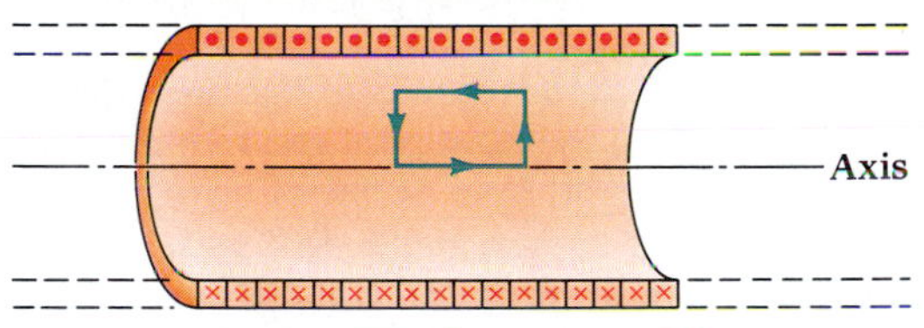

Figure 27-35. Exercise 28.

29. A circular solenoid is designed with a 260-mm² cross-sectional area and a 150-mm length. *(a)* How many turns are required if the magnetic field near the center of the solenoid is to have a maximum field magnitude of 1.8 mT and a maximum current of 0.75 A? *(b)* What length of wire is required? *(c)* If the solenoid is tightly wound with a single layer of copper wire of resistivity $1.7 \times 10^{-8}\ \Omega\cdot$m, what is the resistance of the solenoid? Neglect the thickness of the insulation. *(d)* What potential difference must be applied across the leads to produce a steady magnetic field of magnitude 1.8 mT?

30. Two long, straight, parallel wires are shown in cross section in Fig. 27-36. Each wire carries the same current I with a sense out of the plane of the page. Let x locate a point in the plane containing the axes of the wires. Determine the magnetic field as a function of x for *(a)* $0 \leq x \leq a$; *(b)* $a \leq x \leq D - a$; *(c)* $D - a \leq x \leq D$.

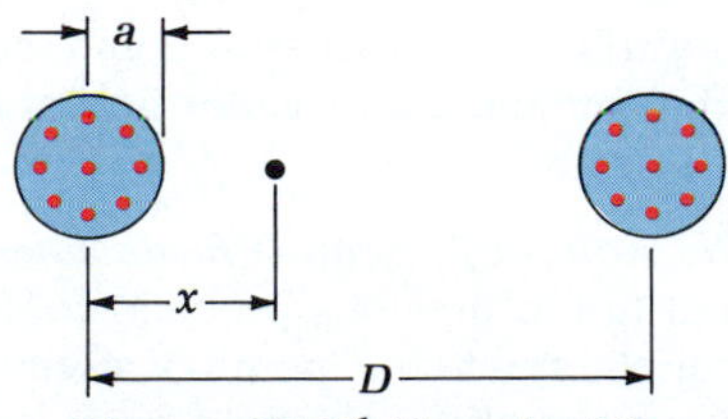

Figure 27-36. Exercise 30.

31. Rework the previous exercise with the current in the wire on the right reversed.

Section 27-4. Force between Currents

32. Two long, straight, parallel wires 30 mm apart carry currents of 12 and 15 A with opposite senses. *(a)* Show on a diagram the direction of the force each wire exerts on the other. *(b)* Determine the magnitude of the force per unit length between the wires. *(c)* How do the answers to parts *(a)* and *(b)* change if the currents have the same sense?

33. Equal currents exist in two long, straight, parallel sections of wire separated by 15 mm. *(a)* If the magnetic force on a 250-mm length of one of the wires is measured to be 0.93 mN, what current exists in the wires? *(b)* By what factor does the force between the wires change if the currents are halved?

34. With the definition of the ampere as the SI unit of current, the four basic units are conventionally taken to be kilogram, meter, second, and ampere. Units for all other quantities in mechanics and electromagnetism are defined in terms of these four. *(a)* Express the unit of magnetic field, the tesla (T), in terms of the four basic units. *(b)* Express μ_0 in these units.

Section 27-5. Magnetic Flux and Gauss's Law for Magnetic Fields

35. The plane surface within a circle of radius 250 mm is in a uniform magnetic field of magnitude 320 mT such that the axis of the circle is at 28° to the field direction. Determine the magnetic flux for this surface.

36. Two surfaces form a closed surface, as shown in Fig. 27-37. Surface S_1 is the plane within a circle of radius a, and surface S_2 is a hemisphere of radius a. Suppose that a uniform magnetic field is at angle θ relative to the axis of the hemisphere. *(a)* Determine the magnetic flux for the plane surface. *(b)* Determine the magnetic flux for the hemispherical surface.

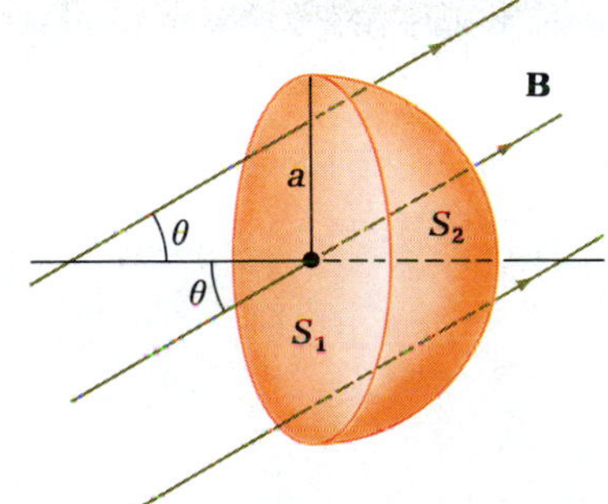

Figure 27-37. Exercise 36.

37. A long, straight wire carries current I. *(a)* Evaluate the magnetic flux for the plane surface bounded by the rectangle shown in Fig. 27-38. *(b)* Evaluate the magnetic flux for a bubblelike surface bulging out of the plane of the figure but having the rectangle as its boundary.

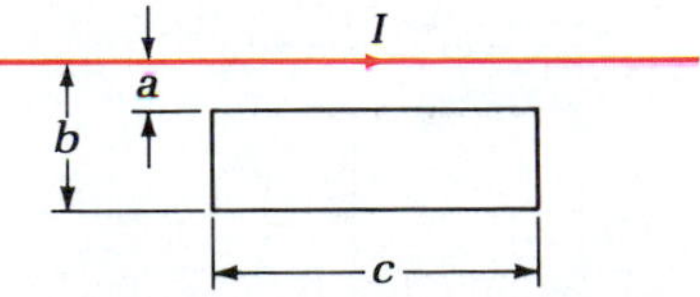

Figure 27-38. Exercise 37.

Section 27-6. The Displacement Current and Ampere's Law

38. A capacitor is being charged as shown in Fig. 27-22. At a certain instant the current in the lead wires is 1.45 A. *(a)* What

is the displacement current for the surface S_1 at the same instant? *(b)* After the capacitor is fully charged, there is no current in the wires. What is the displacement current for surface S_1 in this case? *(c)* Suppose the capacitor is being discharged and the current in the wires is 2.33 A, with a sense opposite that shown in the figure. Now what is the displacement current for surface S_1?

39. A current I is uniformly distributed over the circular cross section S_1 of a long, straight wire. The current density has magnitude $J = I/A$, where A is the area of the surface S_1. An electric field exists at points in the wire whose resistivity is ρ. (Review the discussion connecting **J**, **E**, and ρ in Chap. 24.) *(a)* Determine the electric flux for the surface S_1 in terms of the current I piercing that surface. *(b)* Under what conditions is there a displacement current piercing this surface? *(c)* Evaluate the displacement current for a copper wire ($\rho = 1.7 \times 10^{-8}$ $\Omega \cdot$m) in which the current changes from 320 to 340 mA in 5.0 μs. *(d)* What fraction of the total current is this displacement current in the wire?

40. A 30-nF capacitor in an RC circuit (see Chap. 25) is charged with a 12-V battery through a 10-kΩ resistor, beginning at $t = 0$ ms. Consider the displacement current piercing a surface such as S_1 in Fig. 27-22. *(a)* At what value of t is the displacement current a maximum? *(b)* Evaluate the displacement current at $t = 1.0$ ms and *(c)* at $t = 15$ ms.

41. A capacitor is being charged as shown in Fig. 27-22. At a certain instant the current in the lead wires is 500 μA, and the current leaking from one plate to the other is 40 μA. (This is a "leaky" capacitor. The current would be zero in an ideal capacitor.) *(a)* Determine the displacement current piercing the surface S_1. *(b)* If each plate has area 240 mm^2, at what rate is the electric field changing at that instant? (Neglect edge effects.)

PROBLEMS

1. ***Ampere's law for a nonplanar path.*** A general closed path, not necessarily in a plane, forms the boundary of a surface that is pierced by a current in a long, straight wire. The closed path can be approximated by a large number of lines of three types: circular arcs centered on the axis of the wire, radial lines, and lines parallel to the axis. Show that Eq. 27-6 is satisfied for this type of path.

2. ***The toroid.*** A toroid is a doughnut-shaped coil with N turns of wire wound about the doughnut, or *torus*. A typical toroid is illustrated in Fig. 27-39. For an ideal toroid the magnetic field exists only inside the torus, but the field is not uniform over the cross section. *(a)* Apply Ampere's law to the circular path in the figure to show that the magnitude of the magnetic field is given by

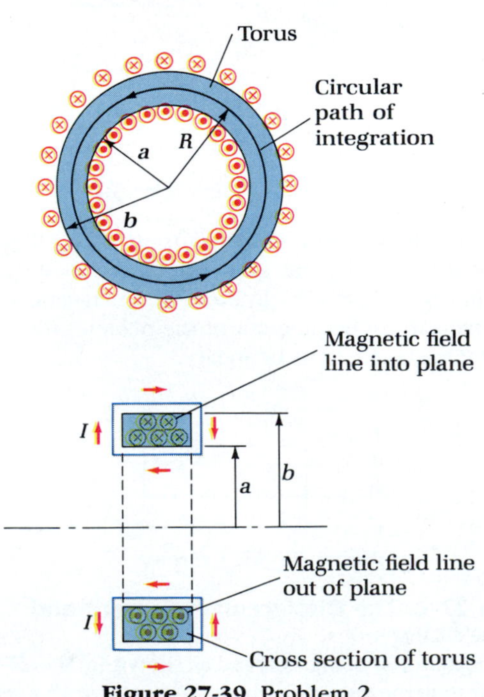

Figure 27-39. Problem 2.

$$B = \frac{\mu_0 NI}{2\pi R}$$

where I is the current in the toroid and R is the radial distance from the axis. *(b)* Find the maximum and minimum values of the magnitude of the field for a 500-turn toroid with current 300 mA and with $a = 75$ mm and $b = 90$ mm. *(c)* Compare the values above with the magnitude of the magnetic field in a solenoid with N turns in a length $2\pi R_{av}$, where $R_{av} = \frac{1}{2}(R_1 + R_2)$.

3. ***Field on the axis of an infinitely long solenoid.*** An ideal solenoid with n turns per unit length can be considered as a continuous set of current loops, as illustrated in Fig. 27-40. For

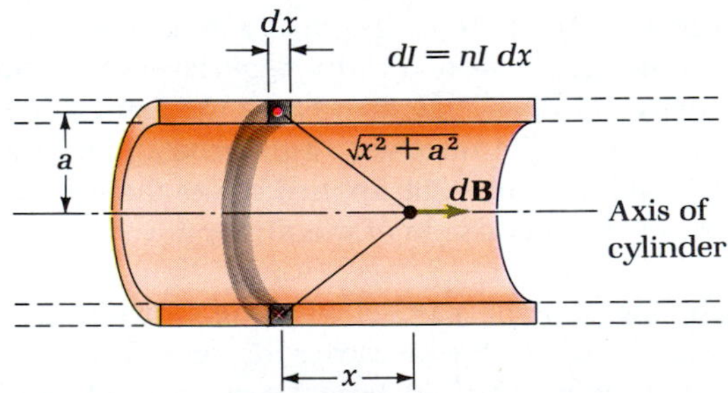

Figure 27-40. Problem 3.

a current loop of width dx, the current is $dI = nI\,dx$. The magnitude dB of the magnetic field of a circular loop for a point on the axis of the loop is given by Eq. 27-5, with $m = \pi a^2\,dI = \pi a^2 nI\,dx$. Use these results and the principle of superposition to determine the magnetic field at a point on the axis of an ideal solenoid.

4. ***Field at the mouth of a semi-infinite solenoid.*** Use the method outlined in the previous problem to determine the magnetic field on the axis at the mouth of a semi-infinite solenoid. Let the open end of the solenoid be at $x = 0$, with the solenoid extending along the negative x axis.

5. ***Field at the center of a solenoid of finite length.*** Use the method of the previous two problems to determine the mag-

netic field at the center of a solenoid of length L and radius a. Let $x = 0$ be at the center of the solenoid, with the ends at $x = \pm\frac{1}{2}L$. Show that the result agrees with the field at the center of a current loop if $a \gg L$ and with the field inside an ideal solenoid if $a \ll L$.

6. ***Field produced by a current sheet.*** Consider the idealized *sheet current* shown in Fig. 27-41. The current exists everywhere in the plane perpendicular to the plane of the page and is described by a *surface current density* K. The current I within a length L is $I = KL$, so that K is the current per unit length on the sheet. Use symmetry arguments and Ampere's law to show that the magnetic field is directed as shown, with a magnitude $B = \frac{1}{2}\mu_0 K$ on each side of the sheet.

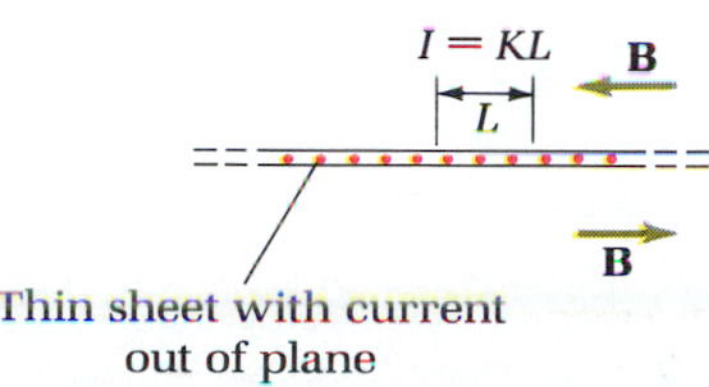

Figure 27-41. Problem 6.

7. ***Field produced by two parallel current sheets.*** Using the result of the previous problem, determine the magnetic field above, below, and in between the two sheet currents shown in Fig. 27-42.

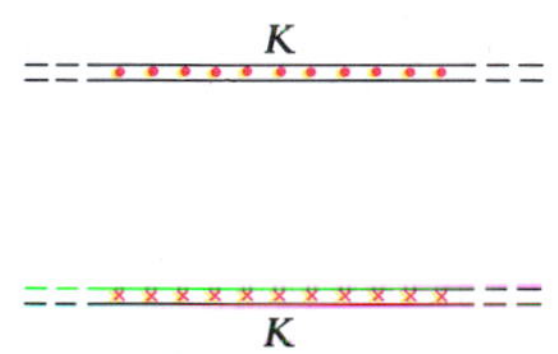

Figure 27-42. Problem 7.

8. ***Sheet of current revisited.*** Use the Biot-Savart law to verify the result in Prob. 6 by considering the sheet current as a collection of long, straight wires of width dx and carrying current $dI = K\,dx$.

9. ***Field produced by a moving charged particle.*** A particle of charge q moves with constant velocity $\mathbf{v}$. *(a)* Justify the interpretation of the product $q\mathbf{v}$ as a current element $I\,d\boldsymbol{\ell}$ located at the instantaneous position of the charged particle. *(b)* If the unit vector $\hat{\mathbf{r}}$ gives the direction from the particle to a point a distance r from the charge, then show that the magnetic field at that point due to the moving charge is given by

$$\mathbf{B} = \frac{\mu_0 q\mathbf{v} \times \hat{\mathbf{r}}}{4\pi r^2}$$

10. ***Numerical calculation of the field by a solenoid—I.*** Consider a solenoid approximated by 101 parallel current loops of radius $a = 10.0$ mm and spaced $d = 0.50$ mm apart. Each loop carries a 1.00-A current and the loop centers lie on the axis of the solenoid. The field at a point on the axis of a current loop has a magnitude given by Eq. 27-5. Choose the x axis to lie along the axis with the origin in the plane of the fifty-first loop. Then the end loops are at $\pm 50d$. Write a spreadsheet to evaluate the magnetic field at *(a)* $x = 0$; *(b)* $x = 50d$; *(c)* $x = 100d$. *(d)* Compare the answers for parts *(a)* and *(b)* with the values for an ideal solenoid and a semi-infinite solenoid.

11. ***Numerical calculation of the field by a solenoid—II.*** Modify the spreadsheet in the previous problem to consider 201 loops, but with the same number of turns per unit length. Evaluate the field at *(a)* $x = 0$; *(b)* $x = 50d$; *(c)* $x = 100d$; *(d)* $x = 200d$. *(e)* Compare with the values for the previous problem.

Additional Problems

12. Consider two electrons separated by a distance d and moving with the same velocity v. What is the total force between these two particles? Does this differ from the case where you consider the same two particles in their initial common rest frame? By how much would the effective mass have to increase (as a function of v) in the frames where these particles appear to be moving to make the resulting accelerations the same as in the rest frame?

13. Find the net force on the rectangle in Fig. 27-38 if it also carries a current I with the side nearest the straight wire having the current in the same direction as that in the wire.

14. Find the initial torque on a rectangular loop of wire suspended between two parallel straight wires as shown in Fig. 27-43.

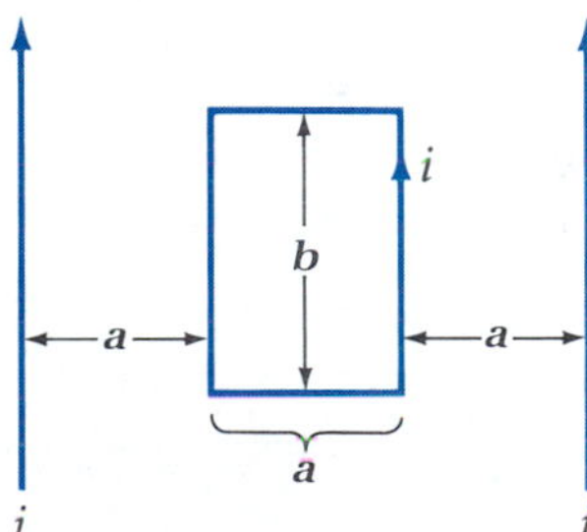

Figure 27-43. Problem 14.

15. Find the magnetic field at the center of an insulating sphere of radius R rotating with angular frequence ω and charged with a uniform surface charge density σ.

16. The earth's magnetic field is $\approx 10^{-5}$ T and we wish to make it null in a small region by creating an opposing field. To do this we arrange two identical circular wire loops of 1-m diameter whose centers are spaced 1 m apart along the direction of the local field, with the planes of the loops perpendicular to the field. What are the magnitude and the sense of the current flow required to null the field exactly at the central point between these two loops? What is the resulting net field at the center of each loop?

17. Suppose a capacitor is filled with a dielectric of constant κ. Find the modification required to Eqs. 27-15 and 27-16 in this case.

18. Magnetic monopoles have fascinated physicists since the time of Maxwell because there is no good reason why they do not exist. Their existence would not conflict with any other known law of nature except Eq. 27-14, which is simply a statement that they have never been seen. Assume that they do exist as a magnetic charge g, which by analogy with normal charge q, gives rise to a radial **B** field diminishing as $1/r^2$. What would be the new form of Eq. 27-14? Find the constants if a magnetic dipole of two opposite-signed monopoles has a magnetic dipole moment with units of g times length (the separation between the monopoles) by analogy with an electric dipole.

19. Consider a circular loop of wire and a square loop with the same wire length. Compare the strength of the **B** field at the center of these loops when the same current flows. Which is stronger? Next consider deforming the square loop into a rectangle with the longer side twice the shorter side. How does this compare? Express each as a function of the area enclosed. Since the circle encloses the maximum area for a fixed circumference, what do you conclude?

20. Consider two concentric, long, thin solenoids, each with the same number of turns per unit length but carrying currents in the opposite sense from one and other. If the outer solenoid has radius a and the inner one has radius b, find the **B** fields in all 3 regions.

FARADAY'S LAW 28

Electric generators at a hydroelectric power plant.

If a current I exists in a circuit with resistance R, then energy is dissipated through Joule heating at a rate given by $P = I^2R$. What is the source of this energy? In the simple circuits considered so far, a battery has usually supplied the energy. The emf $\mathscr{E}$ of the battery, interpreted as the energy per unit charge transferred to a charge carrier through chemical processes, provides energy to the charge carriers in the circuit, at the rate $P = I\mathscr{E}$.

There are also other types of emf, other ways to transfer energy to and from the charge carriers in a circuit. The energy transfer need not occur through chemical reactions. The conversion of mechanical energy (from a rotating steam turbine) to electric energy at a generating station involves a different type of emf, one in which a magnetic field plays an essential role. The development of this important process is based on principles discovered more than 150 years ago. The independent and almost simultaneous observations of magnetically induced currents by Michael Faraday in England and Joseph Henry in the United States led to what is now called *Faraday's law of induction*.

Demonstration of a current generated in a circuit by a moving magnet.

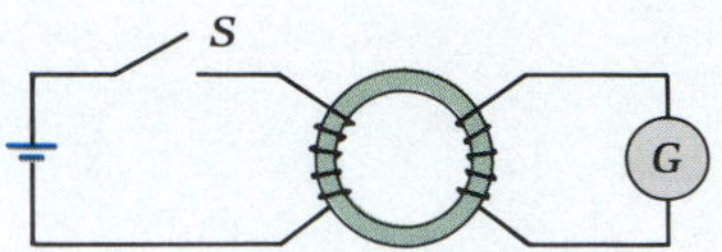

Figure 28-1. Two coils are wrapped around an iron ring. The galvanometer G deflects momentarily when the switch is opened or closed.

28-1 FARADAY'S LAW

Since a steady current in a wire produces a steady magnetic field, Faraday initially (and mistakenly) thought that a steady magnetic field could produce a current. Some of Faraday's investigations of magnetically induced currents utilized an arrangement similar to that shown in Fig. 28-1. A current in the coil on the left produces a magnetic field concentrated in the iron ring. The coil on the right is connected to a galvanometer G, which indicates the presence of any induced current in that circuit. There is no induced current for a steady magnetic field. But an induced current does appear momentarily in the circuit on the right when switch S is closed in the circuit on the left. When switch S is opened, an induced current with the opposite sense appears momentarily. Thus the induced current exists only when the magnetic field, due to the current in the circuit on the left, is *changing*.

The importance of a change is also demonstrated by the arrangement shown in Fig. 28-2. If the magnet is at rest relative to the coil, then no induced current exists. But if the magnet is moved toward the coil, then a current is induced with a sense as indicated in Fig. 28-2*a*. If the magnet is moved away from the coil, then a current is induced with the opposite sense, as shown in Fig. 28-2*b*. Notice that in either case the magnetic field is changing in the vicinity of the coil. An induced current also exists in the coil if it is moved relative to the magnet.

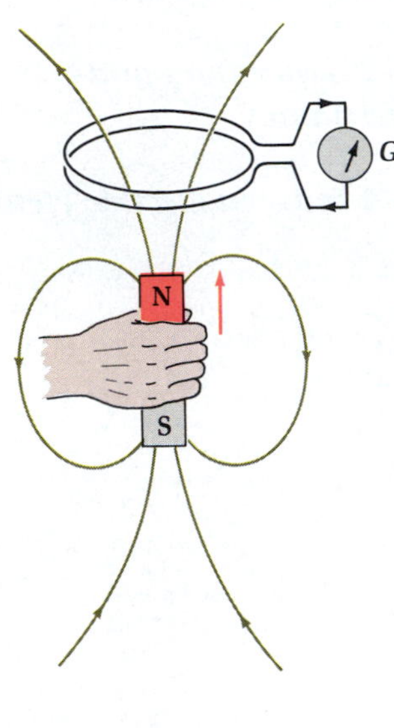

(*a*)

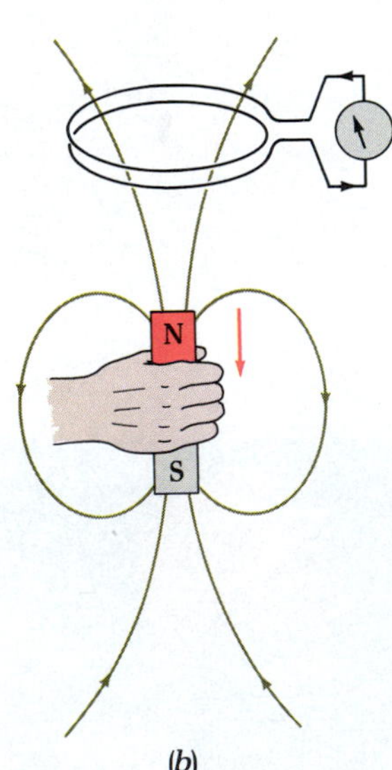

(*b*)

Figure 28-2. *(a)* A current is induced in the coil if the magnet moves toward the coil. *(b)* The induced current has the opposite sense if the magnet moves away from the coil.

The presence of such currents in a circuit implies the existence of an ***induced emf*** $\mathscr{E}$. That is, energy must be supplied to the charge carriers that constitute the current, and emf is the energy per unit charge given to a charge carrier that traverses the circuit. This induced emf is present when the magnetic field is changing, as described above.

The quantitative connection between the changing magnetic field and the induced emf is expressed in terms of the magnetic flux Φ_B for a surface. (Magnetic flux was introduced in Sec. 27-5.) For simplicity, consider a fine loop of conducting wire and an open, mathematical surface bounded by the loop such as the one shown in Fig. 28-3. The magnetic flux for a surface bounded by the loop is given by the surface integral

$$\Phi_B = \int \mathbf{B} \cdot d\mathbf{S}$$

where $d\Phi_B = \mathbf{B} \cdot d\mathbf{S} = B\, dS \cos\theta$ is the flux for the surface element $d\mathbf{S}$. The magnetic flux Φ_B is said to *link* the loop.

A changing magnetic flux linking a loop and the induced emf in the loop are related by **Faraday's law:**

> An emf is induced in a loop of wire when the magnetic flux for a surface bounded by the loop changes in time. The induced emf is given by

$$\mathscr{E} = -\frac{d\Phi_B}{dt} \qquad (28\text{-}1)$$

The emf $\mathscr{E}$ depends on the rate of change of the magnetic flux. The negative sign in this form of Faraday's law relates to the sense of the induced emf in the circuit. (See Lenz's law below.) From Faraday's law, we obtain the relation between the weber (Wb), the unit of magnetic flux, and the volt (V), the unit of emf: 1 V = 1 Wb/s.

Consider the induced emf in a closely wound coil. Each turn in such a coil behaves approximately as a single loop, and we can apply Faraday's law to determine the emf induced in each turn. Since the turns are in series, the total induced emf $\mathscr{E}_T$ in a coil is the sum of the emf's induced in each turn. We suppose that the coil is so closely wound that the magnetic flux linking a turn of the coil at a given instant has the same value for each turn. Then the same emf $\mathscr{E}$ is induced in each turn, and the total induced emf for a coil with N turns is given by

$$\mathscr{E}_T = N\mathscr{E} = N\left(-\frac{d\Phi_B}{dt}\right) = -N\frac{d\Phi_B}{dt} \tag{28-2}$$

where Φ_B is the magnetic flux linking a single turn (or loop) in the coil.

The magnetic flux linking a loop or a turn of a coil in Eq. 28-1 is the flux of the *total* magnetic field for a surface bounded by the loop. There is a contribution to the magnetic flux for a loop due to the loop's own current in addition to the contribution due to an external source, such as a magnet, or the current in another circuit. In this chapter, we shall always assume that loops or coils are part of a circuit with a large resistance so that the induced current is small. Then the flux contribution due to the small induced current is assumed to be negligible compared with the flux due to the other sources. Thus we shall neglect the effect of the induced current in determining the magnitude of the induced emf (but see Lenz's law below). The next chapter will concentrate on the effect that an induced current has on itself.

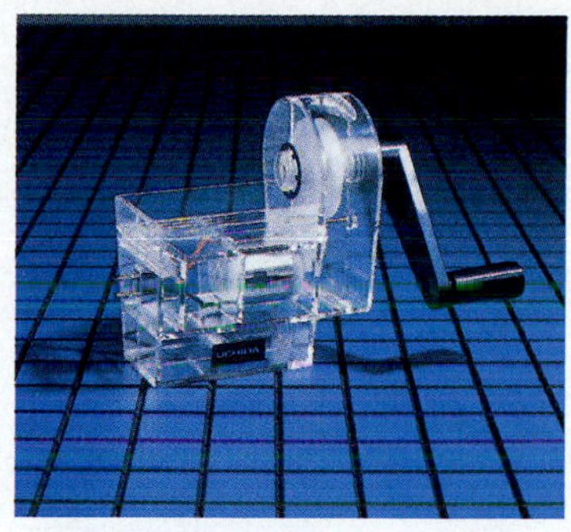

A hand-driven electric generator.

Lenz's Law

The negative sign that appears in Faraday's law (Eq. 28-1) refers to the sense of the induced current. In determining the magnitude of an induced emf from Faraday's law, we ignore the negative sign in the calculation. The sense of the induced emf is determined by applying ***Lenz's law,*** a principle attributed to Heinrich Friedrich Lenz (1804–1865). One way to state Lenz's law is in terms of the contribution of the induced current to the magnetic field:

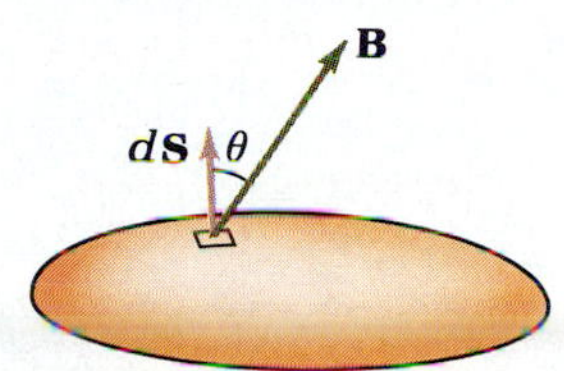

Figure 28-3. A conducting loop forms the boundary of a surface. The magnetic flux for the surface is $\Phi_B = \int \mathbf{B}\cdot d\mathbf{S}$.

> The sense of an induced emf is such that its contribution to the magnetic field opposes the change in the magnetic flux that produces the induced current.

Lenz's law gives the sense of the induced current.

An understanding of Lenz's law and its application can best be obtained by considering examples. Suppose that the magnetic field, and therefore the flux for the surface in Fig. 28-4*a*, is increasing. The sense of the induced current opposes the increasing flux. That is, the contribution to the magnetic field due to the induced current tends to reduce the increasing value of B at points on the surface.

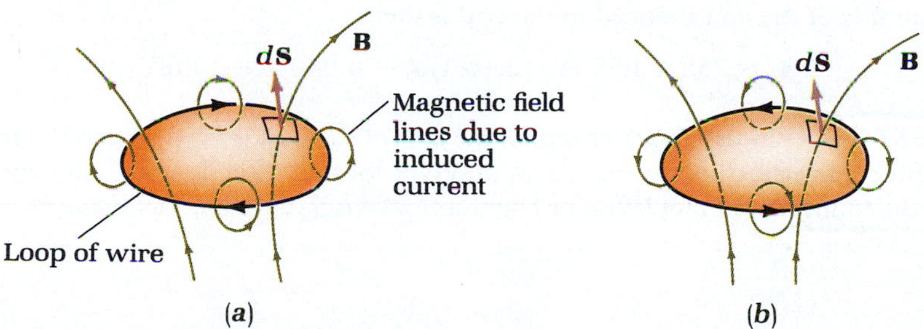

Figure 28-4. The magnetic field contribution due to the induced current opposes the change in flux. The magnetic flux for the surface is *(a)* increasing and *(b)* decreasing.

In contrast, if the induced current had the opposite sense, its field contribution would further increase the flux for the surface, in violation of Lenz's law.

In Fig. 28-4*b*, the magnetic field is assumed to be decreasing at points on the surface. Thus the magnetic flux linking the loop is decreasing, and the field contribution of the induced current tends to increase the decreasing value of B. The induced current has the sense shown in Fig. 28-4*b*. If the current had the opposite sense, the induced current contribution would further decrease the decreasing flux for the surface, again a violation of Lenz's law.

Figure 28-5 shows another interpretation of Lenz's law applied to the situation of Fig. 28-4*a*. The flux linking the loop due to the field of the magnet is increasing as the magnet is moved toward the loop. The induced current in the loop produces a magnetic field similar to the field of a bar magnet (as outlined in Fig. 28-5). The north pole of the equivalent bar magnet repels the real magnet. That is, the mag-

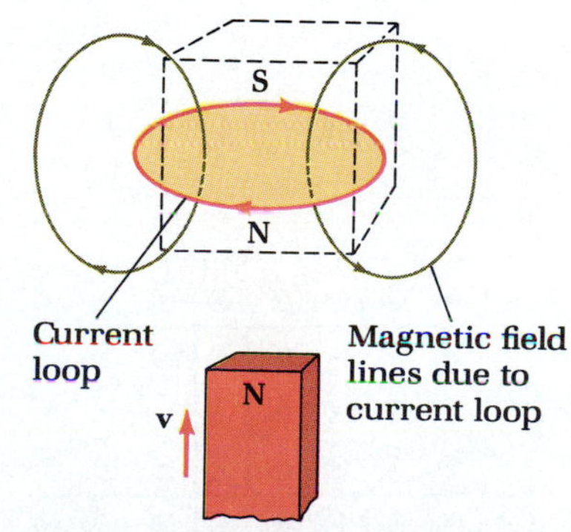

Figure 28-5. The induced current in the loop produces a magnetic field contribution similar to the field of a bar magnet. The two north poles repel.

netic field of the induced current exerts a force on the moving magnet that is opposite the motion of the magnet. In this way the induced current opposes the change in flux that produces it.

Since the induced current always opposes the change in flux that produces it, Lenz's law prohibits runaway situations. Suppose that the magnetic flux linking a loop is increasing. If the induced current had a sense that tended to increase the increasing flux, then the flux could increase at a greater rate, which in turn could induce a larger current to increase the flux at a still greater rate. . . . There is no source of energy for such an unstable growth of current, and the runaway effect does not occur.

In applying Lenz's law to a loop or a coil, you assume the sense of the current to be either one way or the other. One of those senses will be such as to oppose the change in flux that produces the induced current. That is the correct sense. The other sense of the current would aid the change in flux and represents the incorrect choice. Thus you make an initial guess for the sense of the current and then apply Lenz's law to determine if the choice is correct.

EXAMPLE 28-1

Emf induced by a time-changing magnetic field. A 75-turn circular coil of radius 35 mm has its axis parallel to a spatially uniform magnetic field. The magnitude of the field changes at a constant rate from 18 to 43 mT in 240 ms. Determine the magnitude of the induced emf in the coil during this time interval.

Solution. Since the magnetic field is spatially uniform and parallel to the axis of the coil, the flux linking each turn is given by

$$\Phi_B = B\pi R^2$$

where R is the radius of a turn. From Eq. 28-2 the induced emf in the coil is

$$\mathscr{E}_T = -N\frac{d\Phi_B}{dt} = -N\frac{d(B\pi R^2)}{dt} = -N\pi R^2\frac{dB}{dt}$$

The magnitude of the magnetic field changes at a constant rate given by

$$\frac{dB}{dt} = \frac{0.043\ \text{T} - 0.018\ \text{T}}{0.24\ \text{s}} = 0.10\ \text{T/s}$$

The magnitude of the emf induced in the coil is then

$$\mathscr{E}_T = 75\pi(0.035\ \text{m})^2(0.10\ \text{T/s}) = 0.030\ \text{V} = 30\ \text{mV}$$

SELF-TEST 28-1. In the above example, the axis of the coil is vertical and the magnetic field is directed vertically upward. To an observer located directly above the coil, is the sense of the induced emf clockwise or counterclockwise? ***ANSWER:*** clockwise.

28-2 MOTIONAL EMF'S

An emf is induced in a stationary loop or coil if the magnetic flux linking it changes. The magnetic flux, $\Phi_B = \int \mathbf{B}\cdot d\mathbf{S}$, linking a loop involves three things: the field, the area, and the orientation. A change in any one of these can change the flux and lead to an induced emf. One way to change the flux is to change the magnetic field in the region. An emf can also be induced in a conducting circuit if part or all of the circuit moves in a region of magnetic field. These motionally induced emf's have great practical importance. This type of emf occurs in an electric generator, a device that converts mechanical energy to electric energy.

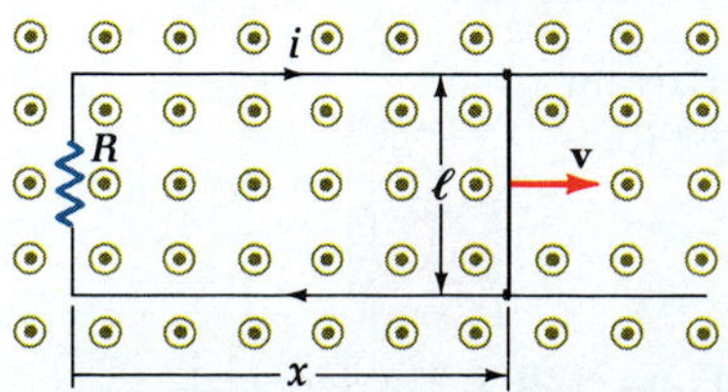

Figure 28-6. A uniform magnetic field is out of the plane of the figure, and an emf is induced in the sliding-wire circuit.

As an example, consider the sliding-wire circuit shown in Fig. 28-6. A U-shaped conductor is at rest in a region of uniform magnetic field perpendicular to the plane of the U. The circuit is completed by a conducting wire of length ℓ that slides on the rails of the U with a velocity **v** that is perpendicular to **B**. For simplicity we assume that the wire is caused to slide with constant velocity. Because the sliding wire moves, the charge carriers in the wire also move. Consequently, the magnetic field exerts a force on each carrier, $\mathbf{F} = q\mathbf{v} \times \mathbf{B}$. The direction of the magnetic force on a positive charge carrier is from the top to the bottom of the sliding wire in the figure.

The magnetic field acting on the charge carriers in the sliding wire tends to make them move around the circuit, thus giving an induced current with the sense indicated in the figure. There are, of course, other forces acting on the charge carriers. (For example, there is a Hall field in the sliding wire; see Example 26-9.) Experiment shows that the net effect of these forces can be expressed in terms of an induced emf determined by applying Faraday's law to the circuit. Since the magnetic field is perpendicular to the plane of the circuit, the magnetic flux for that plane is given by

$$\Phi_B = \int \mathbf{B} \cdot d\mathbf{S} = B\ell x$$

where $S = \ell x$ is the instantaneous area of the circuit loop. The sliding wire is moving with speed v, $v = |dx/dt|$, so that the area and therefore the magnetic flux linking the circuit are changing. The magnitude of the change in flux linking the circuit is

$$\left|\frac{d\Phi_B}{dt}\right| = \left|\frac{d(B\ell x)}{dt}\right| = B\ell\left|\frac{dx}{dt}\right| = B\ell v$$

From Faraday's law, the induced emf in the sliding-wire circuit is

Induced emf for a sliding-wire circuit

$$\mathscr{E} = B\ell v \tag{28-3}$$

This result applies for **B** perpendicular to a plane circuit, with the velocity of the sliding wire perpendicular to the wire's length. See Exercises 11 and 12 for extensions to other cases.

For the case shown in the figure, the sense of the induced current is in accord with Lenz's law. The flux $\Phi_B = B\ell x$ is increasing, and the contribution of the induced current to the magnetic field over the plane surface is opposite the applied magnetic field. Thus the sense of the induced current is such as to oppose the increasing flux. What is the sense of the induced current in the circuit of Fig. 28-6 if the sliding wire moves in the opposite direction?

The expression for the induced emf given in Eq. 28-3 for the sliding-wire circuit can also be obtained from energy considerations. Let i represent the instantaneous current in the circuit and $\mathscr{E}$ the induced emf. The rate at which electric energy is supplied to the circuit is given by $P = i\mathscr{E}$ (Sec. 25-2). The source of this energy can be identified as follows: There is a magnetic force acting on the current-carrying sliding wire in the uniform magnetic field. From Eq. 26-3 this force has magnitude $F = i\ell B$ and its direction, from Fig. 28-6, is to the left, opposite the velocity of the sliding wire. Since the wire moves with constant velocity, the net force acting on it must be zero. To balance the magnetic force (we neglect friction), an applied force of magnitude $F_a = i\ell B$ must act to the right in the figure. This applied force, parallel to the velocity, does work on the wire at the rate $P = \mathbf{F}_a \cdot \mathbf{v} = F_a v = i\ell Bv$ (see Eq. 8-15). Equating the rate at which this mechanical work is done to the rate at which it is transformed to electric energy, we obtain

$$i\mathscr{E} = i\ell Bv$$

or, on dividing by the current i,

$$\mathscr{E} = B\ell v$$

in agreement with Eq. 28-3. Notice that the source of the electric energy associated with this induced emf is the work done by the applied force. Problem 4 considers a case in which no applied force is present.

Induced emf's occur in circuits in a variety of situations involving the motion of a

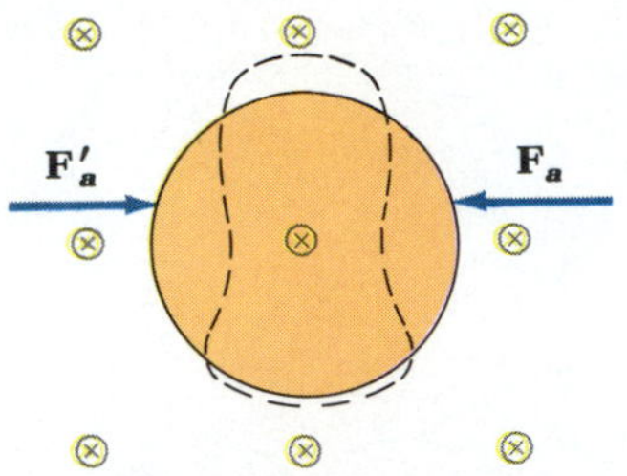

Figure 28-7. A loop in a magnetic field is distorted by applied forces that reduce the area so that the flux changes.

conductor in a static, uniform magnetic field. One example is illustrated in Fig. 28-7. A circular loop whose plane is perpendicular to a uniform magnetic field is distorted by applied forces pushing in the sides. As a result, the area of the plane of the loop decreases, and an induced emf exists. The magnetic flux linking the loop is $\Phi_B = \int \mathbf{B} \cdot d\mathbf{S} = BS$, where S is the instantaneous area of the plane of the loop. Applying Faraday's law gives

$$\mathscr{E} = \left|\frac{d\Phi_B}{dt}\right| = \left|\frac{d(BS)}{dt}\right| = B\left|\frac{dS}{dt}\right|$$

The emf depends on the rate dS/dt at which the area of the loop changes. What is the sense of the induced current in Fig. 28-7? What supplies the electric energy?

The magnetic flux linking a loop or a coil also changes if the loop or coil rotates in a static uniform field. We shall consider this important case in the next section.

In a more general case, an induced emf exists in a circuit if the flux linking it is changing. That change in flux can be due to a combination of a changing magnetic field and the motion of the circuit in a magnetic field. Further, the magnetic field need not be spatially uniform. Faraday's law gives the emf whether the flux changes because of each process alone or because of a combination of processes.

EXAMPLE 28-2

Emf induced by a moving wire. Determine *(a)* the induced emf and *(b)* the induced current in the sliding-wire circuit of Fig. 28-6 if $\ell = 450$ mm, $B = 0.50$ T, and $v = 1.6$ m/s. The 250-Ω resistance R of the circuit is assumed to be concentrated in the base of the U; the rails and sliding wire have negligible resistance. Thus there is a negligible change in the resistance of the circuit as the wire slides along the rails.

Solution. *(a)* The magnitude of the emf is given by Eq. 28-3:

$$\mathscr{E} = B\ell v = (0.50\ \mathrm{T})(0.450\ \mathrm{m})(1.6\ \mathrm{m/s}) = 0.36\ \mathrm{V}$$

(b) Since R represents the total resistance in the circuit, the current is

$$i = \frac{\mathscr{E}}{R} = \frac{0.36\ \mathrm{V}}{250\ \Omega} = 1.4\ \mathrm{mA}$$

Would either of these answers change if the velocity of the wire were reversed? Would the sense of the induced current change?

SELF-TEST 28-2. A sliding-wire circuit similar to the one shown in Fig. 28-6 is contained in a horizontal plane, and the external field is vertically downward. If the sliding wire is moved so that the area enclosed within the circuit is decreasing, what is the sense of the induced current to an observer located directly above the circuit? ***ANSWER:*** clockwise.

28-3 GENERATORS AND ALTERNATORS

Faraday's law provides the basis, or principle, for the conversion of mechanical energy into electric energy. The practical importance of this energy conversion in a technological society is evident. For example, it occurs in an electric generator or dynamo at a commercial generating plant and in the alternator of an automobile. These devices are the result of extensive engineering development; however, the basic principles of their operation can be understood by considering a conducting loop rotating in a magnetic field.

Generators

A conducting loop is in a region of magnetic field, as shown in Fig. 28-8. For simplicity we assume that the magnetic field is uniform and that the loop is caused to rotate by some external agent about an axis O. The axis is in the plane of the loop

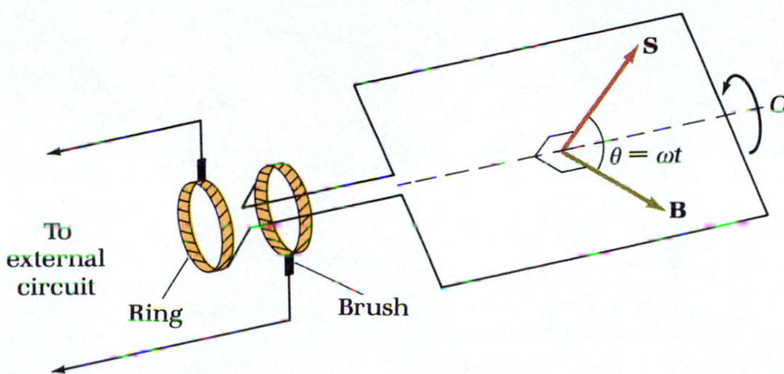

Figure 28-8. A rectangular loop rotates in a uniform magnetic field. Electrical contact is made with brushes sliding on rotating rings.

and perpendicular to the field. The plane loop may have any shape (not necessarily rectangular as in the figure). Let θ represent the angle between the magnetic field and the area vector **S**, which is perpendicular to the plane of the loop. The flux linking this loop is

$$\Phi_B = \int \mathbf{B}\cdot d\mathbf{S} = \mathbf{B}\cdot\mathbf{S} = BS\cos\theta$$

Since the loop is rotating, the angle θ is changing, which leads to a changing flux and, therefore, to an induced emf. Suppose that the loop rotates with a constant angular speed ω about the axis. Then $\theta = \omega t$ and the flux linking the loop is

$$\Phi_B = BS\cos\omega t$$

The induced emf from Faraday's law, $\mathscr{E} = -d\Phi_B/dt$, is given in this case by

$$\mathscr{E} = BS\omega\sin\omega t \tag{28-4}$$

Induced emf in a rotating loop

For a coil with N loops or turns, an induced emf exists in each turn (in series), and the induced emf in the rotating coil is just N times the emf in one loop:

$$\mathscr{E} = NBS\omega\sin\omega t \tag{28-5}$$

Induced emf in a rotating coil

An emf given by Eq. 28-4 or 28-5 oscillates sinusoidally with angular frequency ω or with frequency $\nu = \omega/2\pi$. The maximum, or peak, value of the emf is $\mathscr{E}_{max} = NBS\omega$, which occurs when $\sin\omega t = 1$. Thus the emf oscillates between $+\mathscr{E}_{max}$ and $-\mathscr{E}_{max}$ as the sine function ranges between +1 and −1. The associated current also oscillates, or *alternates,* at this frequency and is called an ***alternating current*** (abbreviated ac). A generator giving an emf of the form of Eq. 28-5 is called an *ac generator.* Notice that the angular frequency ω appears twice in Eq. 28-5. It is contained in the oscillatory term $\sin\omega t$, and the maximum emf is proportional to ω: $\mathscr{E}_{max} = NBS\omega$.

To act as a generator for an external circuit, a rotating coil must be connected to the circuit through lead wires. One connecting arrangement is shown schematically in Fig. 28-8. The wires from the loop are joined to rings which rotate with the loop on a shaft (the shaft is not shown). Electrical contact with the external circuit is through conducting *brushes* which slide on the rotating rings. The emf produced across the brushes is the output voltage V, which is essentially the induced emf. For the ac generator described above, the output voltage is oscillatory, with a time dependence shown in Fig. 28-9.

A different type of connection with a rotating loop is shown schematically in Fig. 28-10. The two brushes contact the halves of a *split-ring commutator.* During one part of the rotation, the output voltage from the coil corresponds to the positive part of the cycle in Fig. 28-9. But when the negative part of the cycle begins, the brushes contact the opposite halves of the commutator. By switching in this way, the commutator causes the sense of the output voltage to remain the same, as shown in Fig. 28-11, rather than alternating. The use of such a commutator leads to a ***direct current*** (dc), in which the current maintains the same sense in the circuit. (Often the symbol "dc" implies a constant value, as well as a constant sense, of the

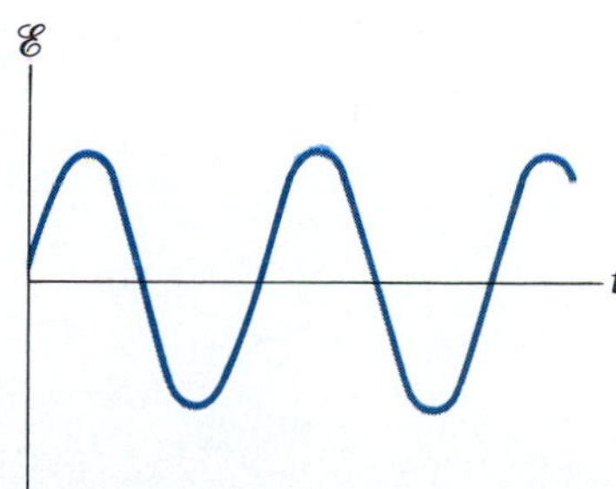

Figure 28-9. The output voltage (or emf) from an ac generator oscillates.

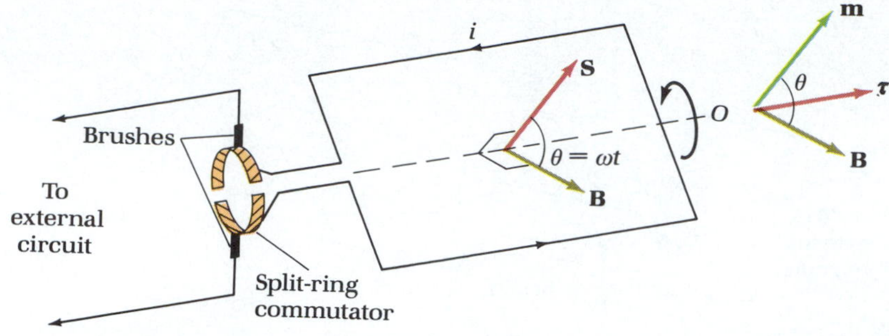

Figure 28-10. Brushes make electrical contact with a rotating split-ring commutator.

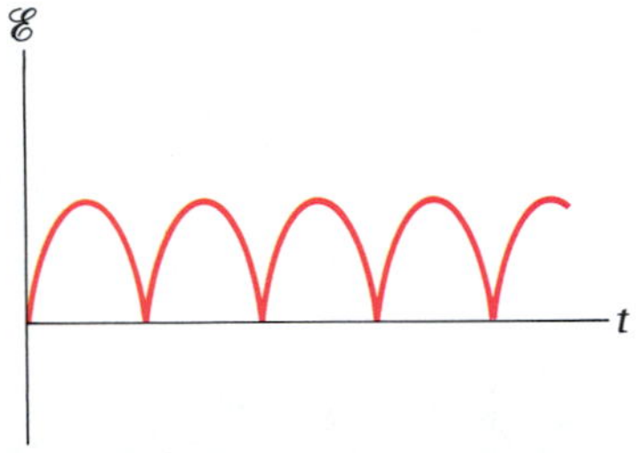

Figure 28-11. The output voltage from a dc generator maintains the same sense.

current.) A generator with a commutator to maintain the sense of the output current is called a *dc generator.*

Lenz's law gives the sense of the induced current in a generating loop. For example, the sense of the induced current i in the loop in Fig. 28-10 is shown at an instant when θ is increasing between 0 and 90°. The flux $\Phi_B = BS\cos\theta$ is decreasing, and the induced current opposes that decrease. This result can also be expressed in terms of the torque produced by the magnetic field on the loop which has the induced current i. In a uniform field the torque on a current loop of area $\mathbf{S}$ with magnetic dipole moment $\mathbf{m} = i\mathbf{S}$ is given by Eq. 26-8: $\boldsymbol{\tau} = \mathbf{m} \times \mathbf{B}$. Notice from Fig. 28-10 that this torque tends to cause a rotation that is opposite the sense of rotation of the loop. This torque would tend to bring the rotating loop to rest. That is, the magnetic torque opposes the externally applied torque that causes the loop to rotate. The mechanical work done by this external agent in maintaining the constant angular speed of the loop is the source of the electric energy "generated" in this generator.

EXAMPLE 28-3

A generator. A 25-turn circular coil of radius $a = 140$ mm rotates at frequency $\nu = 60$ Hz about an axis that is perpendicular to a uniform magnetic field of magnitude 420 mT. The coil is connected to an external circuit with brushes and rings, as shown in Fig. 28-8. *(a)* Write an expression for the output voltage of this generator as a function of time. *(b)* Determine the maximum value of the induced emf in the coil. *(c)* Determine the maximum current in the circuit whose total resistance is 35 kΩ. *(d)* What is the orientation of the coil with respect to the field when the current is maximum? *(e)* Estimate the magnitude of the external torque that must be provided to keep the coil rotating.

Solution. *(a)* The brushes-and-rings connection of Fig. 28-8 corresponds to an ac generator, and the output voltage is essentially given by the induced emf in Eq. 28-5:

$$\mathscr{E} = NSB\omega \sin \omega t$$

(b) The maximum value of the induced emf in the coil occurs when $\sin \omega t = 1$ and is $\mathscr{E}_{max} = NBS\omega = NBS(2\pi\nu)$, where $\omega = 2\pi\nu$. The area of each turn is $S = \pi a^2$ and

$$\mathscr{E}_{max} = (25)(420 \text{ mT})(\pi)(0.14 \text{ m})^2(2\,\pi)(60 \text{ Hz}) = 240 \text{ V}$$

(c) With no other emf's in the circuit, the maximum current is

$$i_{max} = \frac{\mathscr{E}_{max}}{R} = \frac{240 \text{ V}}{35 \text{ k}\Omega} = 7.0 \text{ mA}$$

(d) The maximum current and maximum emf occur when the flux is changing most rapidly. From part *(a)* above, the emf is maximum when $|\sin \omega t| = 1$, corresponding to $\omega t = \theta = \pm 90°$. Since θ is the angle between the field and the area vector, this orientation corresponds to the plane of the coil being parallel to the field. *(e)* From Eq. 26-8, the torque, $\boldsymbol{\tau} = Ni\mathbf{S} \times \mathbf{B}$, produced by the magnetic field on the coil has a maximum magnitude when the current is a maximum: $\tau_{max} = Ni_{max}SB$, or

$$\begin{aligned}\tau_{max} &= (25)(7.0 \text{ mA})(\pi)(0.14 \text{ m})^2(420 \text{ mT}) \\ &= 4.5 \times 10^{-3} \text{ N}\cdot\text{m}\end{aligned}$$

To keep the coil rotating uniformly, the external torque must have a maximum magnitude of at least this value. There will also be frictional torques to counter. Can you estimate the minimum mechanical energy that must be provided to operate this generator for an hour?

SELF-TEST 28-3. What is the maximum induced emf from a 120-Hz generator with 75 turns of area 0.043 m^2 and a magnetic field of magnitude 270 mT? ***ANSWER:*** 660 V.

Alternators

The value of the current in the example above was small (7 mA maximum) because of the large resistance of the circuit. In the lighting circuits of an automobile, currents are typically 10 A or greater. If a generator with brushes is used to provide large currents and emf's, arcing often occurs at the commutator, which leads to deterioration of the electrical contacts. An alternator avoids this problem because the emf is induced in a nonrotating coil (or coils). Thus there are no sliding contacts or brushes in the coil carrying a large current. Instead the brushes are used to supply a smaller current to a rotating electromagnet, as shown schematically in Fig. 28-12. (Some alternators use a rotating permanent magnet.) An emf is induced in the stationary coil as the flux linking the coil changes because of the rotation of the electromagnet. The energy required to keep the electromagnet rotating is supplied by the automobile engine. Although the output current from the coil is ac (can you explain why?), a direct current (dc) is then obtained with the use of diodes of the type described in Sec. 24-4.

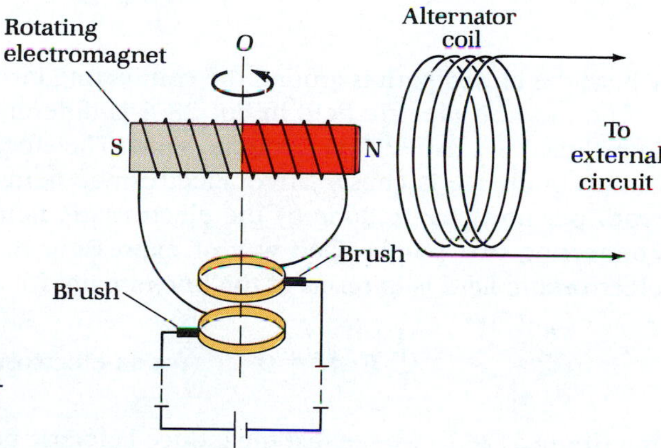

Figure 28-12. A rotating electromagnet induces an emf in a stationary coil of an alternator.

28-4 INDUCED ELECTRIC FIELDS

When a conductor moves in a uniform magnetic field, the charge carriers—moving along with the conductor with average velocity $\mathbf{v}$—experience a magnetic force ($q\mathbf{v} \times \mathbf{B}$). It is this force on the charge carriers in the moving conductor that leads to induced currents in circuits such as those described in Sec. 28-2. Consider, however, a coil or loop that is stationary and has a fixed shape and orientation in a region of magnetic field. That is, no part of the conducting circuit is moving. The flux linking the circuit can change if the magnetic field changes. The induced emf is given by Faraday's law, and Sec. 28-1 dealt with this induced emf. What force leads to the induced current in this case? It cannot be the magnetic force $q\mathbf{v} \times \mathbf{B}$ because the average velocity $\mathbf{v}$ of the charge carriers is zero before the current is induced. Equation 26-11 gives the electromagnetic force on a charged particle: $\mathbf{F} = q(\mathbf{E} + \mathbf{v} \times \mathbf{B})$. If an electric field is present, then an electric force acts on a charge carrier, even if it is initially at rest. We conclude that an electric force acts on the charge carrier and that an electric field is present when the magnetic field changes.

An electric field causes an induced current.

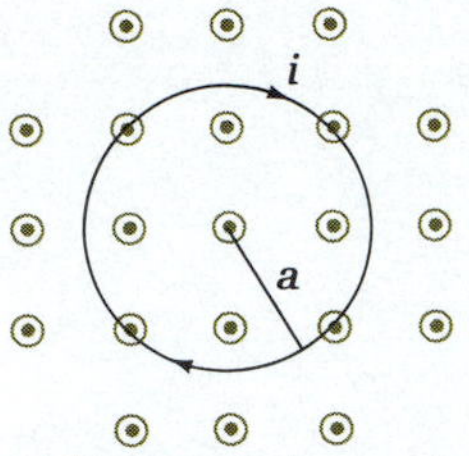

B is out of the page

Figure 28-13. A magnetic field with increasing magnitude induces a current in a circular loop. A tangential electric field drives charge carriers around the loop.

To investigate the nature of the electric field, we consider a configuration with cylindrical symmetry. A fine wire loop of radius a lies inside an ideal solenoid. The axes of the loop and the solenoid coincide so that the plane of the loop is perpendicular to a spatially uniform magnetic field, as shown in Fig. 28-13. Suppose that the magnitude $B(t)$ of the magnetic field is time-dependent. The flux linking the loop is $\Phi_B = B(t)S$, where $S = \pi a^2$ is the area of the loop. The induced emf is then $\mathscr{E} = -S\,dB/dt$. The sense of the induced current is obtained from Lenz's law and is as shown in the figure for the case of $B(t)$ increasing. From the symmetry of the system, we can see that there must be a force **F** acting on each charge carrier, with a component of the force tangent to the loop to cause the induced current. A force with a tangential component must be an electric force, not a magnetic force. (What is the direction of the magnetic force on a charge carrier moving around the loop?) Symmetry also shows that the force has the same magnitude at each point around the conducting wire. Further experiments confirm that an electric field **E** exists in this region and exerts a force $\mathbf{F} = q\mathbf{E}$ on any charge q in the region.

The induced emf is interpreted as the work per unit charge done by this electric force on a charge carrier as it completes a circuit around the loop; thus

$$\mathscr{E} = \frac{W}{q} = \oint \frac{\mathbf{F}\cdot d\boldsymbol{\ell}}{q} = \oint \mathbf{E}\cdot d\boldsymbol{\ell}$$

Induced emf and induced electric field

where the electric field $\mathbf{E} = \mathbf{F}/q$ is the force per unit charge. We think of the induced emf in the circuit as a direct consequence of this electric field, which is an *induced electric field.* The relation between them is

$$\mathscr{E} = \oint \mathbf{E}\cdot d\boldsymbol{\ell} \qquad (28\text{-}6)$$

where the closed path is around the conducting circuit.

The induced electric field in Eq. 28-6 is different from the electrostatic field introduced in Chap. 20 via Coulomb's law. The electric field due to a static distribution of charge is conservative. Electrostatic fields are conservative in that the work per unit charge done by the electrostatic field is independent of the path connecting two points. One way of expressing the conservative nature of the electrostatic field is in terms of the line integral for any closed path:

$$\oint \mathbf{E}\cdot d\boldsymbol{\ell} = 0 \qquad \text{(for an electrostatic field)}$$

An induced electric field is nonconservative.

But from Eq. 28-6, we see that the induced electric field has a nonzero closed-path integral. An induced electric field is a *nonconservative* electric field, a field that cannot be produced by a static distribution of charge. *The nonconservative electric field is produced by a changing magnetic field.*

Using Eq. 28-6, we can express Faraday's law in terms of the nonconservative induced electric field. Substituting the closed-path integral (around the conducting circuit) in Eq. 28-6 for the emf in Faraday's law gives

$$\oint \mathbf{E}\cdot d\boldsymbol{\ell} = -\frac{d\Phi_B}{dt}$$

and $\Phi_B = \int \mathbf{B}\cdot d\mathbf{S}$ is the flux for any surface bounded by the closed path.

This form of Faraday's law was obtained for the closed path that coincided with the conducting loop. Experiment shows that the relation is more general. An induced electric field also exists outside the wire in Fig. 28-13. In fact, the above form of Faraday's law is valid even if no conductor is present in the region.

A form of Faraday's law expressed in terms of the fields is customarily taken to be the fundamental form. It is valid whether conductors are present or absent. (Emf is not a useful concept in the absence of conductors.) The general form, or *integral form,* of Faraday's law is stated as

$$\oint \mathbf{E}\cdot d\boldsymbol{\ell} = -\frac{d}{dt}\int \mathbf{B}\cdot d\mathbf{S} \qquad (28\text{-}7)$$

Integral form of Faraday's law

where the line integral is for any closed path and the magnetic flux $\Phi_B = \int \mathbf{B}\cdot d\mathbf{S}$ is for any surface bounded by the closed path. The sense of the closed path (given by the direction of $d\boldsymbol{\ell}$) and the orientation of the surface (given by the direction of the area element $d\mathbf{S}$) are connected by a right-hand rule: Orient the right hand so that the fingers curl in the sense of $d\boldsymbol{\ell}$. The extended thumb gives the orientation of the area element, as shown in Fig. 28-14. This connection between the sense of the closed path for **E** and the orientation of the surface for **B** is consistent with Lenz's law.

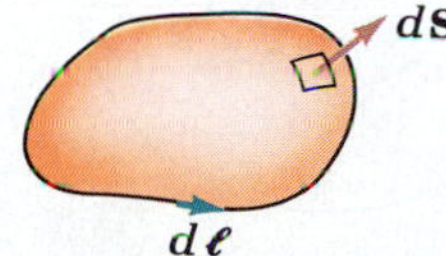

Figure 28-14. The closed path forms the boundary of the oriented surface.

Eddy Currents

Suppose that a changing magnetic field is perpendicular to a face of an extended conductor such as a conducting plate. The induced electric field causes circulating currents called ***eddy currents*** in the plate. Such eddy currents are also produced if a conductor moves through a region of magnetic field. These currents dissipate energy through Joule heating (at a rate $P = i^2R$). A conducting material can be "heated" by the eddy currents induced by changing the magnetic field in the substance, a process called ***induction heating.***

Eddy currents are induced in conductors.

In other cases, the dissipation of energy that accompanies eddy currents may be undesirable. To reduce the eddy currents in the iron core of a transformer (see Sec. 29-5), the core is laminated. That is, thin layers of conducting iron are separated by insulating layers. The insulating layers effectively increase the resistance of the path for circulating charges so that the current is reduced.

Laminated fabrication reduces eddy currents.

Faraday's Law and Ampere's Law

There are some similarities between Faraday's law and Ampere's law with the displacement current included (Eq. 27-16). These two laws are

$$\oint \mathbf{B}\cdot d\boldsymbol{\ell} = \mu_0\left(\Sigma i + \epsilon_0 \frac{d\Phi_E}{dt}\right) \qquad \text{(Ampere's law)}$$

$$\oint \mathbf{E}\cdot d\boldsymbol{\ell} = -\frac{d\Phi_B}{dt} \qquad \text{(Faraday's law)}$$

The equations are most similar if $\Sigma i = 0$. We can interpret Ampere's law by regarding a changing electric field as a cause or source of a magnetic field. And we can interpret Faraday's law by regarding a changing magnetic field as a cause or source of an electric field. These laws, along with Gauss's law for electric fields and for magnetic fields, are obeyed everywhere, even in regions far from charges and currents. The possibility that these laws allow electric and magnetic fields that can sustain each other is explored in Chap. 34. There we shall see that light is an electromagnetic wave.

Changing electric and magnetic fields are sources of each other.

EXAMPLE 28-4

An induced electric field inside a solenoid. The magnitude of the spatially uniform magnetic field in a long solenoid increases at a constant rate dB/dt. Determine the distribution of the induced electric field in this region.

Solution. The electric field must be symmetric with respect to the axis of the solenoid. (See Exercise 29.) To apply Faraday's law, Eq. 28-7, we use a circle of radius R centered on the axis, with its plane perpendicular to the axis. This path forms the boundary for the plane surface within the circle. We choose the sense of the path integral to be as shown by the

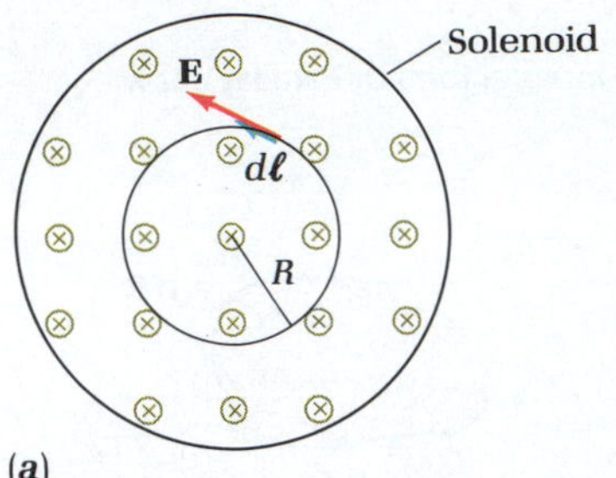

(a)

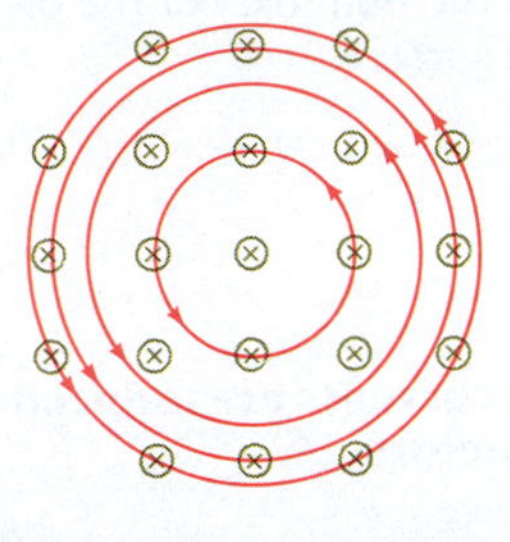

(b)

Figure 28-15. Example 28-4: *(a)* The closed path integral $\oint \mathbf{E}\cdot d\boldsymbol{\ell}$ is evaluated for a circle of radius *R*. *(b)* The lines representing **E** are circular. The field **E** also exists outside the solenoid (see Exercise 26).

direction of $d\boldsymbol{\ell}$ in Fig. 28-15*a*. With this choice, the area element $d\mathbf{S}$ is out of the plane of the figure, opposite to **B**. Since $\mathbf{B}\cdot d\mathbf{S} = B\,dS\cos 180° = -B\,dS$, the flux $\Phi_B = -BS = -B\pi R^2$. From symmetry the electric field is tangent to the path ($\mathbf{E}\cdot d\boldsymbol{\ell} = E\,d\ell$) and has the same magnitude *E* at each point on the path. Thus

$$\oint \mathbf{E}\cdot d\boldsymbol{\ell} = \oint E\,d\ell = E\oint d\ell = E(2\pi R)$$

where $2\pi R$ is the distance around the circular path. Substituting these results in Eq. 28-7 gives

$$E(2\pi R) = -\frac{d}{dt}(-B\pi R^2) = \pi R^2\frac{dB}{dt}$$

or

$$E = \frac{1}{2}R\frac{dB}{dt}$$

The electric field is zero on the axis ($R = 0$), and its magnitude increases linearly with distance *R* from the axis. (Note that dB/dt is independent of *R* inside the solenoid.) The electric field distribution is shown schematically in Fig. 28-15*b*. See Exercise 26 for the electric field outside the solenoid.

SELF-TEST 28-4. The axis of a solenoid is vertical, and an observer is located directly above the solenoid. As viewed by this observer, the sense of the current in the loops of the solenoid is counterclockwise. What is the sense of the induced electric field when the current in the solenoid is *(a)* increasing and *(b)* decreasing? ***ANSWERS:*** *(a)* clockwise; *(b)* counterclockwise.

SUMMARY

Section 28-1. Faraday's Law

A current is induced in a conducting loop if the magnetic flux linking the loop changes. The induced emf in a single loop is given by Faraday's law:

$$\mathscr{E} = -\frac{d\Phi_B}{dt} \tag{28-1}$$

According to Lenz's law, the sense of the induced current is such as to oppose the change in flux which produces it. A small coil with *N* turns may be treated as *N* single loops, and the total emf induced is given by

$$\mathscr{E}_T = -N\frac{d\Phi_B}{dt} \tag{28-2}$$

Section 28-2. Motional emf's

A motion emf is induced if part or all of a circuit moves through a region of magnetic field. In a sliding-wire circuit, the emf is

$$\mathscr{E} = B\ell v \tag{28-3}$$

Section 28-3. Generators and Alternators

From Faraday's law, the emf induced in a rotating coil generator is

$$\mathscr{E} = NBS\omega\sin\omega t \tag{28-5}$$

In an alternator, an emf is induced in stationary coils by a rotating magnet.

Section 28-4. Induced Electric Fields

If a magnetic field is changing, then an induced electric field exists. The electric field is nonconservative. The induced emf due the induced electric field is

$$\mathscr{E} = \oint \mathbf{E}\cdot d\boldsymbol{\ell} \tag{28-6}$$

Faraday's law can be expressed in terms of the fields as

$$\oint \mathbf{E}\cdot d\boldsymbol{\ell} = -\frac{d}{dt}\int \mathbf{B}\cdot d\mathbf{S} \tag{28-7}$$

QUESTIONS

1. What are some similarities and some differences between an emf induced by a changing magnetic field and the emf of a battery?

2. Suppose that an emf is induced in a conducting loop by a changing magnetic field. Is there an internal resistance similar to that in a battery? Explain.

3. Can you think of any way to distinguish between a current induced magnetically in a conducting loop and a current in a loop produced by a battery? Explain.

4. Under what circumstances is the induced emf in a coil with *N* turns equal to *N* times the emf induced in one turn of the coil?

5. Which would cause the larger induced emf in a loop perpendicular to a spatially uniform magnetic field: the magnitude of the field changes linearly from 200 mT to 0 in 1.0 ms or from 1.20 to 1.30 T in 1.0 ms? Explain.

6. Suppose that in Fig. 28-2*a* the coil moves toward the magnet, which is held stationary. In what direction does the pointer deflect on the galvanometer? What if the coil moves away from the stationary magnet? Explain, using Lenz's law.

7. The bar magnet in Fig. 28-16 moves to the right. What is the sense of the induced current in the stationary loop *A*? In loop *B*?

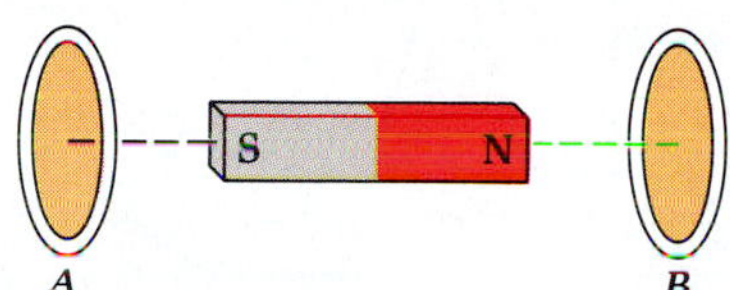

Figure 28-16. Question 7.

8. Suppose that (negatively charged) electrons are the charge carriers in the circuit in Fig. 28-6. What is the direction of the magnetic force on a typical electron in the sliding wire? What is the sense of the current in this case?

9. Determine the sense of the induced current in Fig. 28-6 if the wire slides to the left at constant speed v. What is the magnitude of the induced emf in this case?

10. Suppose that the wire of length ℓ in Fig. 28-6 is at rest, but that the U-shaped wire moves to the left with speed v. Is there an induced current in the circuit? If so, give its sense. If not, explain why not.

11. If the wire of length ℓ and the U-shaped conductor in Fig. 28-6 move together with speed v in the uniform magnetic field, is there an induced current? If so, give its sense. If not, explain why not.

12. What is the sense of the induced current in Fig. 28-8 at an instant when $0 < \theta < 90°$? Explain.

13. Compare the simple generator in Fig. 28-8 with the simple motor in Fig. 26-7. Is there any distinction in principle between a generator and an electric motor? Explain.

14. Imagine measuring the force on a small stationary test charge placed at a point in the changing magnetic field shown in Fig. 28-15. Explain why the force must be exerted by an electric field and not by the magnetic field.

15. What are some similarities and some differences between Faraday's law (Eq. 28-7) and Ampere's law (Eq. 27-16)?

16. A cross section of an ideal solenoid is shown in Fig. 28-17. The magnitude of the spatially uniform field is increasing inside the solenoid and $B = 0$ outside the solenoid. In which of the conducting loops is there an induced current? What is the sense of each current?

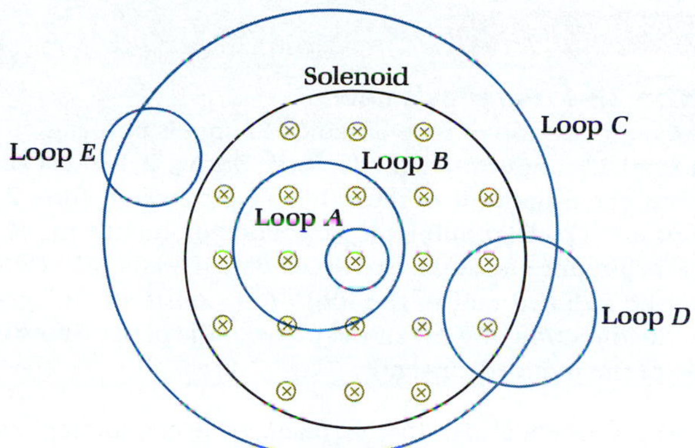

Figure 28-17. Question 16.

17. Suppose that the magnetic field in Fig. 28-17 is not changing. Loop *D* fits through a slot in the wall of the solenoid. If this loop is pulled out of the solenoid, what is the sense of the induced current? In what direction must you exert a force to pull the loop steadily out of the solenoid?

18. Suppose that loop *C* in Fig. 28-17 is moved in the plane of the figure, but each point on the loop remains outside the solenoid. If the magnetic field in the solenoid is not changing, is there an induced current in the loop? Explain. Must you exert a force to move this loop about very slowly? Explain. What would the answers to the above questions be if B is increasing in the solenoid as the loop is moved?

19. A bar magnet, aligned with its axis along the axis of a copper ring, is moved along its length toward the ring. Is there an induced current in the ring? Is there an induced electric field in the ring? Is there a magnetic force on the bar magnet? Explain.

20. The bar magnet in the previous question is moved toward a glass (insulating) ring. Is there an induced current in the ring? Is there an induced electric field in the ring? Is there a magnetic force on the bar magnet? Explain.

21. Is it possible to have an induced emf in a conducting loop (loop *C* in Fig. 28-17) even though the magnetic field is zero at each point on the loop? Explain.

22. Complete the following table:

Symbol	Represents	Type	SI Unit
$d\Phi_B/dt$		Scalar	
$\mathcal{E}$	emf		
$d\mathbf{S}$			
$d\boldsymbol{\ell}$			
$\oint \mathbf{E} \cdot d\boldsymbol{\ell}$			
ω			rad/s

EXERCISES

Section 28-1. Faraday's Law

1. A circular loop of wire of radius 45 mm is perpendicular to a spatially uniform magnetic field. During a 120-ms time interval, the magnitude of the field changes steadily from 240 to 360 mT. *(a)* Determine the magnetic flux linking the loop at the beginning and at the end of the time interval. *(b)* Determine the induced emf in the loop. *(c)* Construct a diagram with the direction of **B** out of the plane of the paper. Show the sense of the induced current.

2. The perpendicular to the plane of a conducting loop makes a fixed angle θ with a spatially uniform magnetic field. If the loop has area S and the magnitude of the field changes at a rate dB/dt, show that the magnitude of the induced emf in the loop is given by $\mathscr{E} = |(dB/dt)S\cos\theta|$. For what orientation(s) of the loop is $\mathscr{E}$ a maximum? A minimum?

3. A 25-turn coil with face area 78 mm² is placed inside a long solenoid, near its center. The axes of the coil and of the solenoid coincide. The current in the solenoid is changed so that the magnetic field in the solenoid changes at a constant rate from 150 mT in one direction to 150 mT in the opposite direction in 75 ms. *(a)* Evaluate the change in flux $\Delta\Phi_B$ linking each turn of the coil for this time interval. *(b)* Determine the induced emf in the coil. *(c)* Rework parts *(a)* and *(b)*, with the axis of the coil making an angle of 70° with the axis of the solenoid.

4. The magnetic flux linking each loop of a 250-turn coil is given by the expression $\Phi_B(t) = A + Dt^2$, where $A =$ 3.0 mWb and $D =$ 15 mWb/s² are constants. *(a)* Show that the magnitude of the induced emf in the coil is given by $\mathscr{E} = (2ND)t$. *(b)* Evaluate the flux linking each turn at $t =$ 0.0, 1.0, 2.0, and 3.0 s. *(c)* Evaluate the induced emf in the coil at each of these instants.

5. Each loop in a 250-turn coil has face area $S = 9.0 \times 10^{-2}$ m². *(a)* What is the rate of change of the flux linking each turn of the coil if the induced emf in the coil is 7.5 V? *(b)* If the flux is due to a uniform magnetic field at 45° from the axis of the coil, what must be the rate of change of the field to induce that emf?

6. The plane of a conducting ring is placed perpendicular to a spatially uniform magnetic field (**B** and **S** parallel, with $S = 10^{-3}$ m²), as shown in Fig. 28-18*a*. The magnitude of the field has a time dependence shown graphically in Fig. 28-18*b*. *(a)* For each 10-ms time interval on the graph, determine the magnitude of the induced emf in the ring and give the sense of the induced current. *(b)* For which 10-ms time interval is the induced current largest? *(c)* Is there any interval for which the emf is obviously not constant? Ignore the behavior near the endpoints of each interval. *(d)* Construct a graph of $\mathscr{E}$ versus t.

7. If the magnetic flux linking a loop changes by $\Delta\Phi_B$ in a *finite* time interval Δt, then the average emf $\overline{\mathscr{E}}$ induced in the loop is $\overline{\mathscr{E}} = \Delta\Phi_B/\Delta t$. *(a)* Determine the average emf induced in the coil of Exercise 4 for the time interval between 0.0 s and 3.0 s. *(b)* At what time is the instantaneous emf equal to this average value? *(c)* What is the current in the coil at this time if the coil's resistance is 15 kΩ?

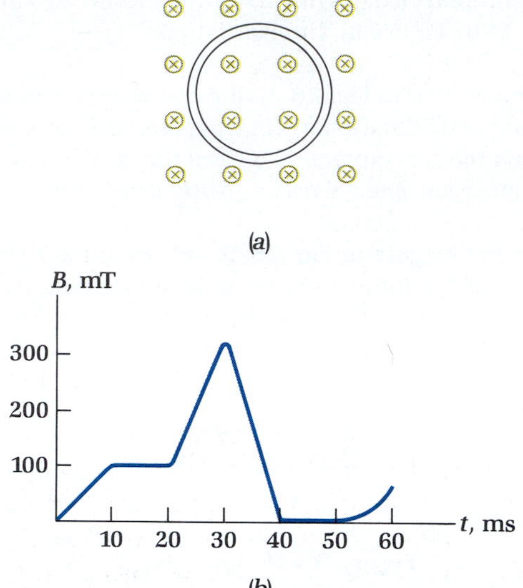

Figure 28-18. Exercise 6.

8. Suppose that the magnetic flux linking a loop changes by $\Delta\Phi_B = \Phi_B(t_2) - \Phi_B(t_1)$ during the time interval between t_1 and t_2. *(a)* Show that the amount of charge that flows past a point in the loop during this time interval is given by $\Delta Q = \Delta\Phi_B/R$, where R is the resistance of the loop. *(b)* The coil in Exercise 4 has resistance 15 kΩ. What charge flows past a point in the coil during the 3.0-s time interval beginning at $t_1 = 0.0$ s?

Section 28-2. Motional emf's

9. *(a)* What emf is induced in the sliding-wire circuit of Fig. 28-6, given that $B = 430$ mT, $\ell = 150$ mm, and $v = 2.6$ m/s? *(b)* Assume that the sliding wire and the rails of the U have negligible resistance and that the 750-Ω resistance of the circuit is concentrated at the left in the figure. What current exists in the circuit? *(c)* Determine the magnitude and direction of the magnetic force acting on the sliding wire.

10. The rolling wheels and axles of railroad cars maintain electrical contact with the rails and form a circuit similar to that shown in Fig. 28-6. (Imagine that a resistor connects the rails at a distant point along the track.) Estimate the emf induced in such a circuit by a typical moving freight train in a region where the vertical component of the earth's magnetic field has magnitude 0.1 mT.

11. Suppose that the direction of **B** in the sliding-wire circuit of Fig. 28-6 is not perpendicular to the plane of the loop. Determine the emf and construct a diagram for each of the following cases: *(a)* **B** is parallel to **v**. *(b)* **B** is perpendicular to **v** but parallel to the plane of the loop. *(c)* **B** is at angle θ with the area **S** of the loop.

12. In the circuit of Fig. 28-19, the velocity of the sliding wire is parallel to the rails so that the wire maintains contact with the rails as it slides. Assume that **B** is uniform and perpendicular to the plane of the circuit and determine the expression for the induced emf in terms of B, ℓ, v, and θ.

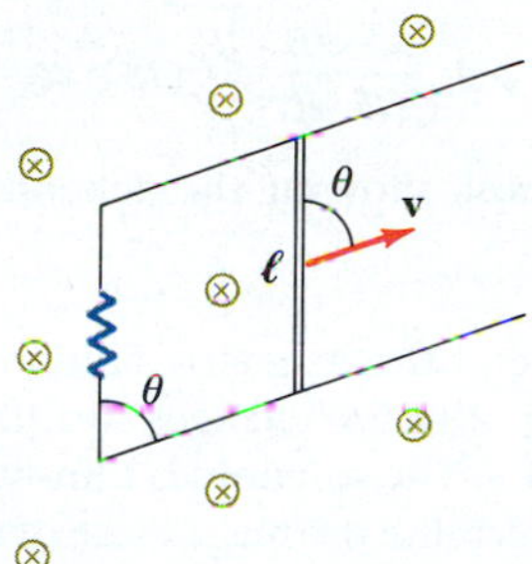

Figure 28-19. Exercise 12.

13. One conducting U-tube slides inside another, as shown in Fig. 28-20. The arrangement is similar to the slide on a trombone. Assume that electrical contact is maintained between the tubes and that a uniform magnetic field is perpendicularly into the plane of the figure. *(a)* If each tube moves toward the other at constant speed, determine the emf induced in the circuit in terms of B, ℓ, and v. *(b)* What is the sense of the induced current?

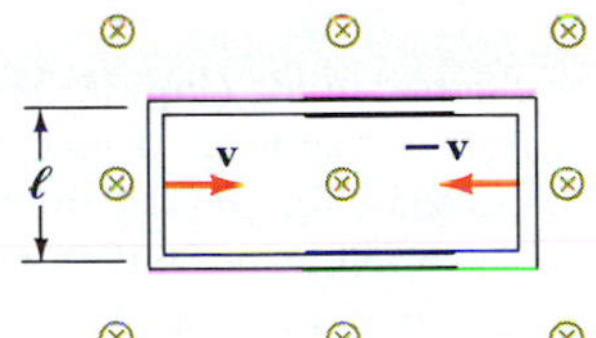

Figure 28-20. Exercise 13.

14. Two thin, flexible tubes, similar to those in the previous exercise, slide together in a plane perpendicular to a magnetic field **B**. In the case illustrated in Fig. 28-21, the tubes constitute an expanding conducting circle. *(a)* If the radius R of the circle increases at a constant rate $v_R = dR/dt$, determine the emf induced in the circuit in terms of B, R, and v_R. *(b)* What is the sense of the induced current?

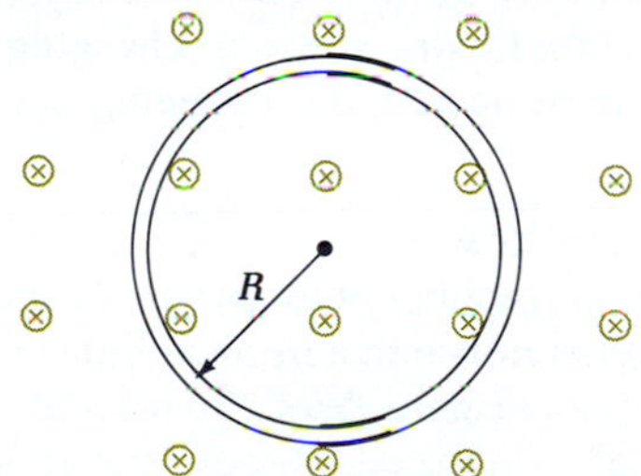

Figure 28-21. Exercise 14.

15. The rectangular conducting loop shown in Fig. 28-22, of dimensions $w = 0.40$ m and $\ell = 0.20$ m, moves perpendicularly into a region of uniform field with constant speed $v = 5.6$ m/s. The leading edge of the loop enters the field region at $t = 0$ and $B = 0.15$ T. *(a)* At what time t_1 does the trailing edge enter the field region? *(b)* Evaluate the induced emf in the loop for $0 < t < t_1$. *(c)* What is the sense of the induced current? *(d)* Determine the net magnetic force acting on the loop during this time interval if the loop's resistance is 1200 Ω. *(e)* Can the magnetic field have an abrupt spatial variation, as assumed in this exercise? See Exercise 23 in Chap. 27.

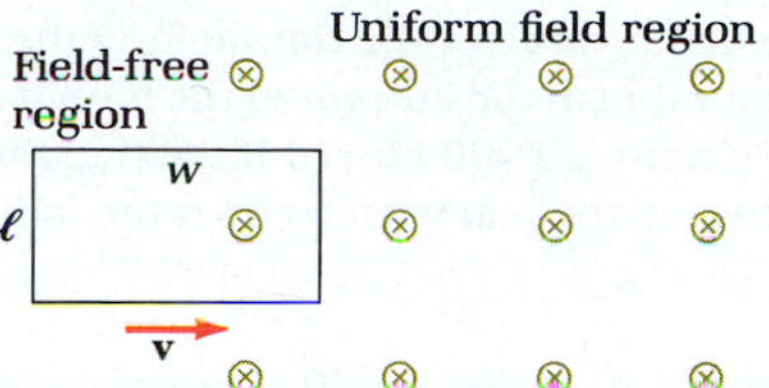

Figure 28-22. Exercise 15.

16. A rectangular current loop moves with constant velocity from a field-free region into a region of uniform magnetic field, as shown in Fig. 28-22. Determine an expression for the induced emf in the moving loop when *(a)* the entire loop is in the field-free region, *(b)* part of the loop is in the region of uniform field, and *(c)* the entire loop is the region of uniform field. *(d)* In each case give the sense of the induced current if one exists in the loop.

17. A conducting rod of length $\ell = 120$ mm is pivoted at one end as the other end slides on a circular conductor perpendicular to a uniform magnetic field $B = 400$ mT, as shown in Fig. 28-23. The rod rotates counterclockwise with constant angular speed $\omega = 370$ rad/s. Assume that all of the $R = 1200$-Ω resistance of the circuit is contained in the resistance symbol in the figure. *(a)* Determine an expression for the induced current in the circuit in terms of ℓ, B, ω, and R. *(b)* Evaluate the current using the values above. *(c)* What is the sense of the induced current? *(d)* Evaluate the magnitude of the magnetic torque on the rotating rod about an axis parallel to **B** and through the pivot point. How can the rod rotate with constant angular speed?

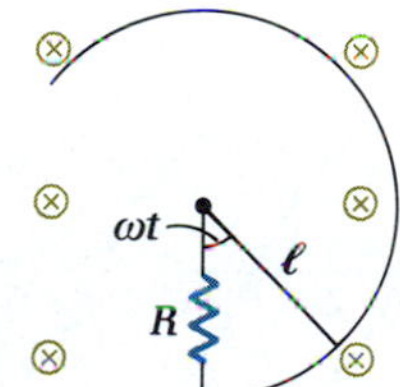

Figure 28-23. Exercise 17.

18. Suppose that the sliding wire in Fig. 28-6 starts from rest at $t = 0$ and is caused to move with a constant acceleration to the right with magnitude a. *(a)* Show that the induced emf is given by $\mathscr{E} = B\ell at$. *(b)* What is the sense of the induced current?

Section 28-3. Generators and Alternators

19. Show that the SI units for each side of Eq. 28-5 are consistent. That is, show that the SI unit of the product $NBS\omega \sin \omega t$ equals the volt.

20. A plane conducting loop has face area of 5×10^{-2} m² and rotates about an axis perpendicular to a uniform magnetic field of magnitude $B = 0.4$ T, as shown in Fig. 28-8. The loop rotates with a constant frequency $\nu = 60$ Hz. *(a)* Determine the angular frequency ω of rotation of the loop. *(b)* What maximum emf is induced in the loop?

21. For the rotating loop in the previous exercise, determine

(a) the maximum value of the flux linking the loop and *(b)* the maximum value of the current in the loop if the total resistance of the circuit is 1500 Ω. *(c)* If the flux linking the loop has a maximum at $t = 0$, at what times is the induced current a maximum?

22. A generator coil rotates at 480 Hz about an axis perpendicular to a uniform magnetic field such as shown in Fig. 28-8. *(a)* If the coil has face area 2.5×10^{-3} m² and $B = 37$ mT, then what maximum emf is induced in each turn of the coil? *(b)* How many turns must this coil have if the maximum emf in the coil is to be 170 V?

23. The 25-turn coil of a generator rotates with angular frequency $\omega = 377$ rad/s in a *nonuniform* but constant magnetic field. The magnetic flux linking each turn of the coil is given by $\Phi_B(t) = C_1 \cos \omega t + C_3 \cos 3\omega t$, where $C_1 = 2.4 \times 10^{-4}$ Wb and $C_3 = 7.1 \times 10^{-6}$ Wb are constants. *(a)* Determine an expression for the induced emf in each turn of the coil. *(b)* What is the maximum value of the output voltage of this generator? *(c)* Evaluate the output voltage of the generator at $t = 2.1$ ms.

24. A generator has six coils spaced symmetrically at 60° around a rotating armature. A multiple commutator connects each coil in turn to an external circuit so that a given coil is connected for only $\frac{1}{6}$ of a revolution of the armature. The flux linking the connected coil has a time dependence that is shown graphically in Fig. 28-24. Disregard the abrupt changes in flux that correspond to one coil being disconnected from the circuit at the commutator while the next coil is being connected. *(a)* Construct a graph showing qualitatively the time dependence of the emf of this generator. *(b)* Estimate the maximum emf. *(c)* Is this generator ac or dc?

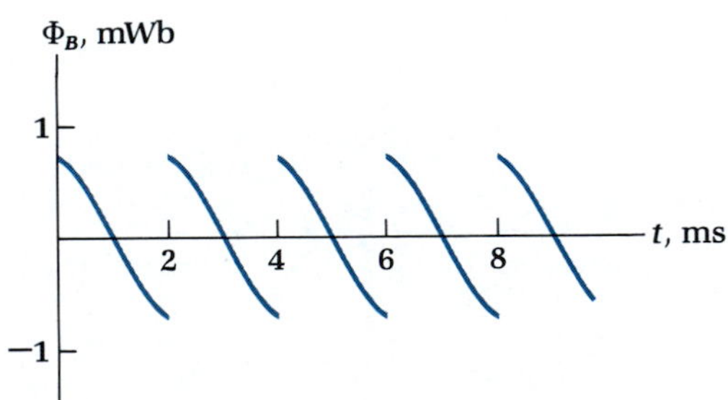

Figure 28-24. Exercise 24.

25. The flux linking each turn of a 300-turn alternator coil is approximately given by $\Phi_B \approx C \cos \omega_1 t$, where $C = 3 \times 10^{-4}$ Wb and $\omega_1 = 2\pi\nu_1$ is the angular frequency of rotation of the alternator shaft. The alternator shaft is connected by a drive belt to an automobile engine and makes three revolutions for each revolution of the engine. *(a)* Determine an expression for the induced emf in the alternator. Determine the maximum emf if the engine speed is *(b)* $\nu_0 = 600$ rev/min and *(c)* $\nu_0 = 4000$ rev/min.

Section 28-4. Induced Electric Fields

26. Assume that the wall of the ideal solenoid in Fig. 28-15*a* is very thin and has radius *a*, so that **B** is spatially uniform for $R < a$ and $B = 0$ for $R > a$. *(a)* Extend Example 28-4 to show that the induced electric field at a point outside the solenoid has magnitude

$$E = \frac{a^2}{2R}\frac{dB}{dt} \qquad (R > a)$$

(b) Construct a graph showing the dependence of E on R for $0 \leq R \leq 2a$.

27. The magnitude of the magnetic field inside an ideal solenoid, as seen in Fig. 28-15*a*, changes steadily from 0 at $t = 0$ s to 0.40 T at $t = 1.4$ s. The solenoid is thin-walled, with a diameter of 15 mm. Determine the magnitude of the electric field at $t = 0.8$ s at a point *(a)* on the axis of the solenoid, *(b)* 5.0 mm from the axis, and *(c)* outside the solenoid 10.0 mm from the axis. *(d)* What are the answers to parts *(a)*, *(b)*, and *(c)* at $t = 1.2$ s?

28. A circular loop of wire is placed with its plane perpendicular to the axis of the solenoid described in the previous exercise. The center of the loop is on the solenoid axis. Evaluate $\oint \mathbf{E} \cdot d\boldsymbol{\ell}$ at $t = 0.8$ s around a circular path coinciding with the loop if the loop has radius *(a)* 5.0 mm and *(b)* 10 mm. *(c)* In each case determine the induced emf in the loop. *(d)* What is the answer to part *(a)* if the center of the loop is 1.0 mm off the solenoid axis?

29. Gauss's law for electric fields (Eq. 21-3) holds generally, even if the magnetic field is changing. Give an argument, based on symmetry and Gauss's law for electric fields, to support the assertion that the induced electric field shown in Fig. 28-15 has only a tangential component. That is, the electric field has no component directed toward or away from the axis of the solenoid. Assume that there is no excess charge inside the solenoid and that the electric field is perpendicular to the axis of the solenoid.

30. The axis of a bar magnet, which produces an axially symmetric magnetic field, coincides with the axis of a circular conducting loop of radius 0.25 m. The magnet moves along this axis toward the loop. At a certain instant the induced emf in the loop is 0.17 V. *(a)* Determine the magnitude of the induced electric field in the loop. Assume that **E** is tangent to the loop at each point (see the previous exercise). *(b)* At what rate is the magnetic flux linking the loop changing at this instant? *(c)* Can you determine how **B** is changing at any given point? Explain.

Additional Exercises

31. Determine the number of loops you should use in the coil of a 60-Hz ac generator so that its maximum emf is 170 V. The magnetic field has a magnitude of 250 mT and each loop has an area of 0.033 m².

32. The magnitude of the magnetic field inside a long solenoid is changing at a rate of 3.9 mT/s. What is the magnitude of the induced electric field at a point inside the solenoid at a perpendicular distance of 15 mm from its axis?

33. The plane of a loop of wire is vertical, and you view the loop from a position on its axis about one meter away. You hold a bar magnet between you and the loop with the magnet aligned along the loop's axis such that the magnet's north pole is toward the loop and its south pole is away. What is the sense of the induced current in the loop when you move the magnet *(a)* toward the loop, and *(b)* away from the loop?

34. A long solenoid of radius 0.15 m has a current decreasing in time. Consider a coordinate frame with its origin at the center of the solenoid, with its z axis along the solenoid's axis, and with $+\mathbf{k}$ in the direction of the magnetic field. (*a*) In terms of unit vectors, what is the direction of the induced electric field at the point (0.10 m, 0.00 m, 0.00 m)? (*b*) What is the direction of this field at the point (0.00 m, 0.10 m, 0.00 m)? (*Hint:* Be certain you use a right-handed coordinate frame; $\mathbf{i} \times \mathbf{j} = \mathbf{k}$.)

35. The plane of a square loop of wire (100 mm × 100 mm) is horizontal and the loop is in a uniform magnetic field directed vertically. (*a*) What is the magnitude of the induced emf in the loop at an instant when the field magnitude is 100 mT and increasing at the rate of 10 mT/s? (*b*) What is the magnitude of the induced emf at a later instant when the field magnitude is 200 mT and still increasing at 10 mT/s?

36. During a certain time interval, the flux of the magnetic field linking a loop of wire is given by

$$\Phi(t) = (4.0\ \text{mWb/s}^2)t^2 + 6.5\ \text{mWb}$$

Determine the magnitude of the induced emf in the loop at the time $t = 2.0$ s.

37. During a certain time interval, the flux of the magnetic field linking a loop of wire is given by

$$\Phi(t) = (9.5\ \text{mWb}) \sin[(377\ \text{rad/s})t]$$

Determine the magnitude of the induced emf in the loop at the time $t = 3.0$ ms.

38. You are to design an ac generator to produce a maximum emf of 54 V at a frequency of 240 Hz. Your coil has 260 turns and each loop has an area of 4.3×10^{-4} m². What should you choose for the magnitude of the magnetic field?

39. Suppose an external agent causes the sliding wire in the circuit of Fig. 28-6 to move at a constant speed of 0.55 m/s. The sliding wire has a length of 0.26 m, the magnitude of the magnetic field is 74 mT, and the induced current is 130 mA. What is the rate at which this external agent is doing work?

40. A long solenoid with a radius of 40 mm is centered at the origin with its axis along the z axis. A current in the wires of the solenoid produces a nearly uniform magnetic field inside the solenoid directed toward $+\mathbf{k}$ and the magnitude of this field is increasing at 20 mT/s. (*a*) Use the expression in Exercise 26 to determine the magnitude of the induced electric field at a point on the x axis 80 mm from the origin. (*b*) What is the direction of the induced electric field?

PROBLEMS

1. ***Another sliding-wire circuit.*** Consider the sliding-wire circuit shown in Fig. 28-25. The wire slides at constant speed and the plane of the circuit is perpendicular to a uniform magnetic field. Show that the induced emf is given by $\mathscr{E} = B\ell v^2 t/D$ for $0 < t < D/v$. What is the expression for the emf for $t > D/v$?

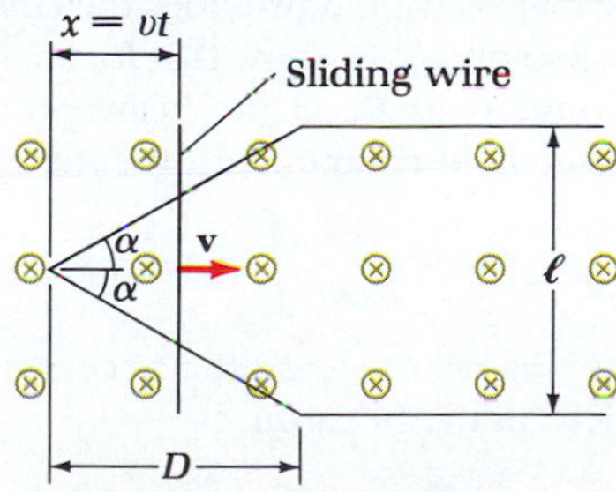

Figure 28-25. Problem 1.

2. ***Wire loop near a long, straight wire.*** A rectangular conducting loop lies in a plane which contains the axis of a long, straight wire carrying a current $i(t)$, as shown in Fig. 28-26. Determine expressions for (*a*) the flux linking the loop and (*b*) the induced emf in the loop in terms of the current in the wire.

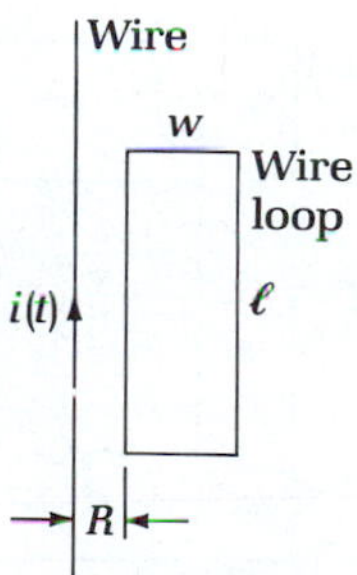

Figure 28-26. Problem 2.

3. ***Loop entering a time-varying magnetic field.*** A rectangular conducting loop moves at constant velocity from a field-free region into a region with a *time-dependent* but spatially uniform magnetic field, as shown in Fig. 28-22. The leading edge enters the uniform field region at $t = 0$, and the trailing edge enters that region at $t_1 = w/v$. During this time interval, the magnetic field changes at a constant rate from B_0 to $2B_0$. Determine expressions, valid for $0 < t < t_1$, for (*a*) the magnetic flux linking the loop and (*b*) the induced emf in the loop. (*c*) Evaluate the emf at $t = 0.10$ s for the case $B_0 = 0.30$ T, $w = 0.50$ m, $\ell = 0.20$ m, $v = 4.0$ m/s.

4. ***Motion of a sliding wire.*** Suppose that the sliding-wire circuit in Fig. 28-6 lies in a horizontal plane, and the wire of mass m slides with negligible friction on the rails. The rails and the sliding wire have negligible resistance, and the resistance R

of the circuit is concentrated in the symbol to the left in the figure. The magnetic field is uniform and constant, but the speed $v(t)$ of the wire is not constant. *(a)* Show that the emf in the circuit is given by $\mathscr{E}(t) = B\ell v(t)$. *(b)* If $i(t)$ represents the current in the circuit, determine the magnetic force acting on the wire. *(c)* Suppose that no other horizontal force acts on the wire, which has speed v_0 at $t = 0$. By applying Newton's second law, show that the speed of the sliding wire is given by $v(t) = v_0 e^{-t/\tau}$, where $\tau = mR/B^2\ell^2$ is a damping time for the motion. *(d)* How far does the wire slide in coming to rest?

5. ***Emf in a sliding-wire circuit with a time-varying magnetic field.*** Suppose that the sliding wire in Fig. 28-6 and in the previous problem is initially at rest at $x = x_0 > 0$. Further, the magnetic field is not constant, but its magnitude increases from B_0 at $t = 0$ at a constant rate $dB/dt = C$ (with $C > 0$). An induced current will exist in the wire, and the wire will move in response to the magnetic force acting on it. *(a)* Show that the induced emf is given by $\mathscr{E} = \ell[B(t)v_x(t) + Cx(t)]$, where $v_x(t)$ is the x component of the velocity of the wire. *(b)* What is the sense of the induced current at $t = 0$? *(c)* In what direction does the wire begin to move at $t = 0$? *(d)* What are the answers to parts *(b)* and *(c)* if B decreases ($C < 0$)?

6. ***Terminal speed of a sliding wire as it falls.*** A sliding-wire circuit is mounted on a board whose plane is vertical, as shown in Fig. 28-27. Assume that all of the resistance R in the circuit is concentrated at the bottom and that the wire slides with negligible friction on the rails. A uniform magnetic field **B** is perpendicular to the plane of the circuit. *(a)* Show that the speed of the sliding wire will approach a terminal speed given by $v_T = mgR/B^2\ell^2$, where m is the mass of the wire. *(b)* What is the expression for the terminal speed if the top of the board is tilted back so that the plane of the circuit makes an angle θ with the horizontal magnetic field?

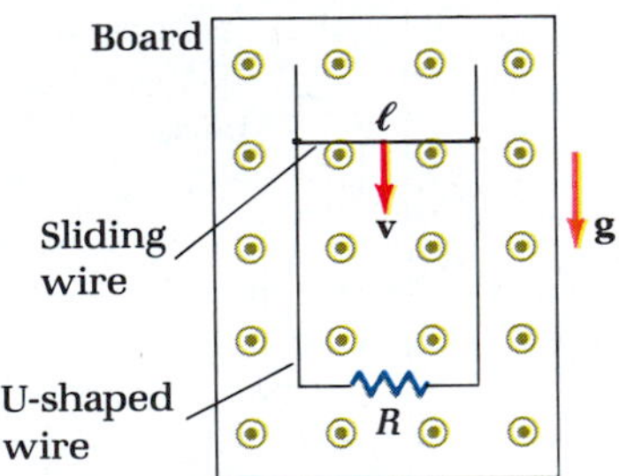

Figure 28-27. Problem 6.

7. ***Moving loop in a spatially varying field.*** Suppose that the rectangular loop in Fig. 28-26 is moving to the right at constant speed v so that $R = R_0 + vt$. If a steady current I exists in the long, straight wire, determine expressions for *(a)* the magnetic flux linking the loop and *(b)* the induced emf. *(c)* What is the sense of the induced current in the loop?

8. ***An induced electric field.*** A magnetic field directed along the z axis has axial symmetry about the axis. If R represents the perpendicular distance from the axis, then at points in the xy plane, the z component of the magnetic field at time t is given by

$$B_z(t) = \frac{Ct/\tau}{a^2 + R^2}$$

where $C = 15$ mWb, $a = 64$ mm, and $\tau = 2.0$ s are constants. *(a)* Determine an expression for the magnitude of the electric field E at a point in the xy plane. *(b)* At what distance from the axis is E a maximum? *(c)* Construct a graph showing the dependence of E on R for $0 \leq \ \leq R \leq 2a$. Assume that $E = 0$ for $R = 0$.

9. ***A search coil.*** A *search coil* (also called a *flip coil*) is sometimes used to measure the magnitude of the magnetic field B in a region. Such a coil has N turns, each with face area S. The coil is small so that the magnetic field is essentially uniform over the dimensions of the coil. The search coil is initially placed with its axis parallel to the field to be measured. It is then rotated, or "flipped," through 180° so that the axis is again parallel to the field but with the opposite sense. During this time, the coil is connected to a *ballistic galvanometer,* a device that measures the total charge $\Delta Q = \int i(t)\,dt$ passing through it. Show that B is given by

$$B = \frac{R\,\Delta Q}{2NS}$$

where R is the resistance of the circuit.

10. ***The betatron.*** The *betatron* uses a changing magnetic field in accelerating electrons (beta particles) to energies around 100 MeV. The electrons circulate in a tube of mean radius R, as shown schematically in Fig. 28-28. They are accelerated by an induced electric field tangent to the circular path. *(a)* Show that the magnitude of the tangential force on an electron is given by

$$F = eE = \frac{e}{2\pi R}\frac{d}{dt}\int B_z(t)\,dS$$

where $B_z(t)$ is the axial component of the magnetic field and the flux is evaluated over the plane bounded by the circular path. *(b)* The magnetic field at the position of an electron in the beam exerts a force directed toward the center. Show that the magnitude of the momentum of an electron in the tube is given by $p = eRB_{0z}(t)$, where $B_{0z}(t)$ is the magnetic field component at the position of the electron beam. *(c)* Parts *(a)* and *(b)* show that the induced electric field provides the tangential force and the magnetic field provides the centripetal force on an electron in a stable orbit. Show that for such an orbit, the magnetic field component B_{0z} at the beam position must be one-half the average field component over the circle; that is,

$$B_{0z}(t) = \frac{1}{2\pi R^2}\int B_z(t)\,dS$$

(d) Explain how this result shows the necessity of a nonuniform magnetic field in the betatron.

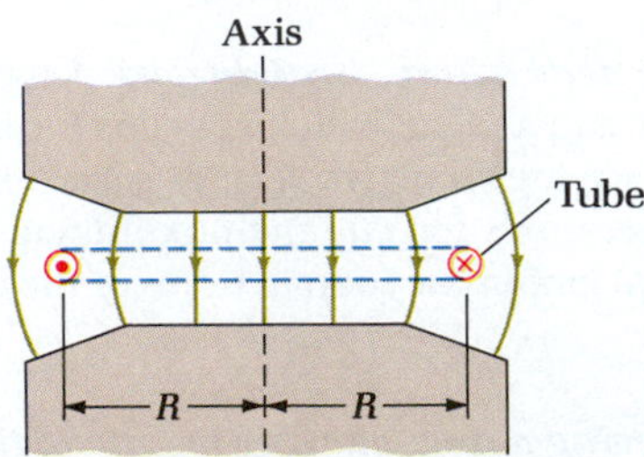

Figure 28-28. Problem 10: Cross section of a betatron.

11. ***More about the betatron.*** The nonuniform magnetic field in the betatron in the previous problem is obtained by using the beveled pole pieces shown schematically in Fig. 28-28. The time dependence of the field is due to an alternating current in the coils (not shown) of the magnet. Suppose that the magnetic field varies sinusoidally, with a frequency $\nu = 60$ Hz, $B_z(t) = B_z(0) \sin 2\pi\nu t$, and the magnetic flux for the circle of radius $R = 0.80$ m has a maximum value of 1.6 Wb. *(a)* What is the sense of the accelerating electron beam if the magnitude of the magnetic field in the figure is increasing? *(b)* Explain why the entire acceleration process must take place over one-fourth of a cycle ($\frac{1}{240}$ s). *(c)* Estimate the kinetic energy gained by an electron in one revolution. *(d)* An electron accelerated in this betatron may have a final kinetic energy of 100 MeV. Estimate the number of revolutions of an electron during the $\frac{1}{240}$-s time interval.

Additional Problems

12. A ***homopolar generator*** uses a solid conducting disk rotating on its' axis. A fixed magnetic field from an external magnet (either a permanent magnet or an electromagnet) is applied parallel to the axis passing through the disk. As the disk is rotated, a potential difference appears between the axle and outside edge of the disk. If the field is of constant strength B between r_1 and r_2 and effectively zero elsewhere, what emf results when the disk is rotating with an angular frequency ω?

13. Consider a homopolar generator as described in Prob. 12. The external **B** field is constant over the entire area of a large thin disk of radius R and mass m, and the disk is initially rotating at an angular frequency ω_0 in frictionless bearings (with no other mechanical attachments). Find the emf in an external load R_L attached between the edge and axle of the disk as a function of time. How long does it take for half of the initial kinetic energy to be removed from the disk? What is the emf at that time?

14. Another use of the homopolar generator as described in the problems above is as a brake. Consider designing an aircraft braking system where a 50,000-kg aircraft has two homopolar disk brakes, one in each main landing gear. Each disk has a radius of 20 cm; the plane's tires have twice that radius. If the touchdown speed on landing is 200 km/h, what constant B field (as a function of the load resistors) should be applied across the brake disks to slow the aircraft to 50 km/h in 5 s? Assume the traction is sufficient to keep the tires from slipping. If the maximum reasonable field available is 5 T, what is the required load resistance?

15. The space shuttle has a wing planform that is roughly an isosceles triangle of height 37.2 m (nose to tail) and base 28.8 m (wingspan). Assume it is a continuous conductor. If the shuttle is in a polar orbit at an altitude of 250 km, heading nose first, right side up, and passing through the earth's 10^{-5} T magnetic field as the field is normal to the wing's surface, what are the potentials of the nose and right and left wing tips with respect to one and other? Could one use this as a source of (emergency) electric power?

16. Consider a flexible wire constrained to lie in a plane permeated by a uniform and constant magnetic field of strength B. The wire is fashioned into a loop with a rectangular shape. The rectangle is deformable with the constraints that each side remain parallel to the original direction and that the total length of the wire be a constant L. Find the emf in this circuit as a function of time if one side is initially ≈ 0 length and increases uniformly in a period T to length $\approx L/2$.

17. Consider the wire and loop shown in Fig. 28-26. If one end of the loop is turned into a sliding bar and the two sides parallel to the wire become extended rails, find the velocity that the bar must have to maintain a current of 1 A if the wire carries a current of 100 A, R is 1 mm, and w is 10 cm.

18. Consider a static **E** field lying in the xy plane having a value E_{0j} if $x \leq 0$ and 0 if $x > 0$. Is this possible, given Faraday's law? Prove that the **E** field at the edge of a parallel-plate capacitor cannot be perpendicular to the plates everywhere, including near the edges of the capacitor.

19. Refer to Ques. 8 concerning the possible existence of magnetic monopoles. What other changes in the laws of electromagnetism would be required? (*Hint:* Consider the fact that a net change in magnetic flux through a loop occurs when a monopole passes through the loop—in other words, monopole currents.)

HIGHLIGHTS OF MODERN PHYSICS

The Handedness of Nature

Is nature ambidextrous? About 85 percent of the people reading this page are right-handed, 13 percent are left-handed, and only 2 percent are ambidextrous. Most kinds of climbing plants wind in right-hand helices as they climb trees and posts (Fig. 1). Similarly, most snail and mollusk shells are in the form of right-handed helices, although both forms occur, depending on the species. If you view these asymmetric organisms in a mirror, their handedness, or ***chirality,*** is changed from left to right or from right to left. In many cases the object seen in the mirror does not exist in the real world. In nature, one chirality predominates.

Figure 1. Morning glories always twine in a right-handed helix. The helix is said to be right-handed because the fingers of the right hand curl in the direction the plant winds as you move along the vine in the direction of the thumb. The mirror image of the plant is never seen.

At a molecular level, the DNA molecule is a right-handed double helix, while many proteins are right-handed single helices. Nineteen of the twenty amino acids that make up the proteins in living things come in two forms, which are mirror images of each other. With a few rare exceptions, all of the amino acids appearing in living organisms are of the L (left-handed) form. The asymmetry of biological molecules is crucially important to living organisms. The birth defects caused by the drug thalidomide, used to treat morning sickness in the early 1960s, were caused by one form of the molecule, while morning sickness was cured by its mirror image. One form of the chemical limonene smells like oranges and the other like lemons.

How did these asymmetries arise? Are they embedded in the fundamental laws of physics? Does nature itself, through these laws, favor one orientation over another? If it does, what does that mean? There are certainly many striking examples of symmetry in nature, such as the sixfold symmetry of snowflakes and the bilateral symmetry of the human body. Do these reflect an underlying symmetry of nature? Until 1957 physicists thought that the basic laws of physics were completely symmetric and that at the most fundamental level, the mirror image of any physical process was also a possible process.

By 1950 four kinds of fundamental forces were known (Commentary, Chap. 7). Gravitation and electromagnetism were known to be completely symmetric, meaning that the equations of these forces have the same form when the coordinate system is inverted. Figure 2 shows that a right-handed cartesian coordinate system is changed into a left-handed one either in a mirror image or when the coordinate system is inverted. Let us see how to verify that the equations of gravitation and electromagnetism are unchanged under inversion.

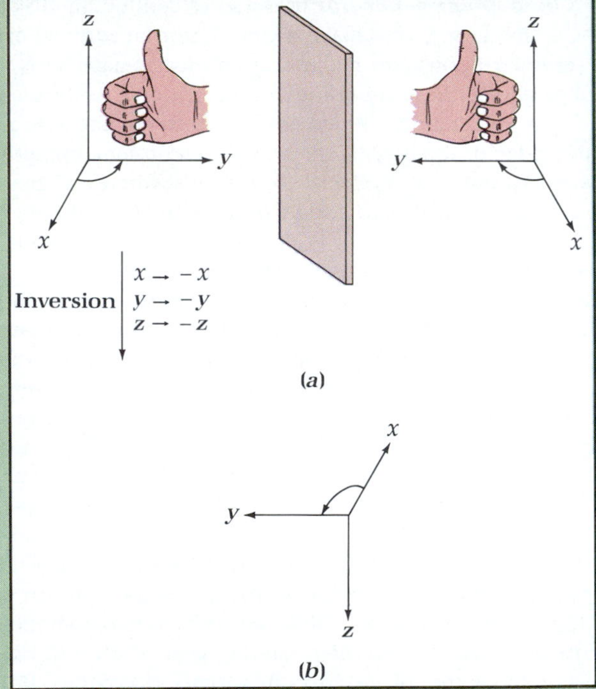

Figure 2. Mirror reflection *(a)* and coordinate inversion *(b)* are equivalent because each turns a right-handed coordinate system into a left-handed one. Mirror reflection reverses only one axis, while inversion reverses all three but they are equivalent. The handedness of a coordinate system is that of the hand for which the thumb points along the positive z axis while the fingers curl in the direction of rotation when the x axis is rotated toward the y axis.

First we determine the effect of inversion on various vectors, starting with the position vector **r**. Since inversion involves changing the sign of each coordinate,

$$\mathbf{r} = x\mathbf{i} + y\mathbf{j} + z\mathbf{k} \rightarrow -x\mathbf{i} - y\mathbf{j} - z\mathbf{k} = -\mathbf{r}$$

By taking successive derivatives of **r**, we see that $\mathbf{v} \rightarrow -\mathbf{v}$ and $\mathbf{a} \rightarrow -\mathbf{a}$. Furthermore, since $\mathbf{F} = m\mathbf{a}$ and mass does not change when the coordinates are inverted, $\mathbf{F} \rightarrow -\mathbf{F}$. In fact, all *true* vectors change sign under coordinate inversion.

What happens to Newton's law of gravitation under inversion? From the above it is easy to see that

The Taj Mahal in Agra, India, has a beautiful symmetrical design.

$$\mathbf{F} = -\frac{Gm_1m_2}{r^2}\hat{\mathbf{r}} \rightarrow (-\mathbf{F}) = -\frac{Gm_1m_2}{r^2}(-\hat{\mathbf{r}})$$

and the negative signs cancel, giving the original equation unchanged (Fig. 3). The coordinate inversion is called the *parity operation,* and we say that gravitation conserves parity. Coulomb's law has the same form as the law of gravitation so it will also conserve parity. We can also show that Ampere's and Faraday's laws do not change, so that like gravitation, electromagnetism conserves parity. Thus, if a process involves only electromagnetic and gravitational forces, the mirror image process is also possible. Experiments have shown that the strong nuclear force also conserves parity.

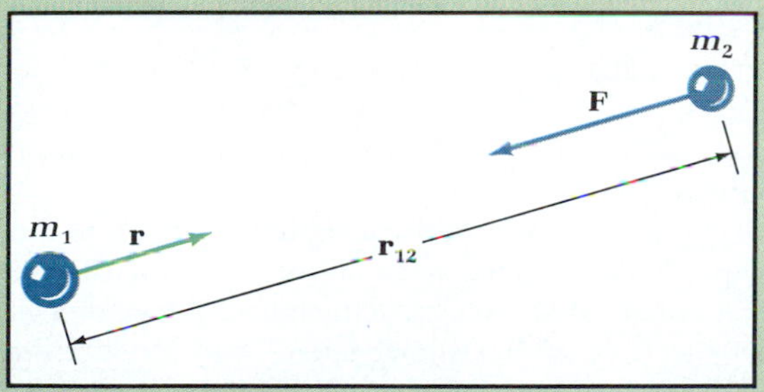

Figure 3. **F** is the gravitational force exerted on m_2 by m_1, and $\hat{\mathbf{r}}$ is the unit vector pointing from m_1 to m_2.

How can the asymmetries of plants, DNA, and proteins arise if all of the fundamental forces are symmetric? In 1956 T. D. Lee and C. N. Yang pointed out that there was in fact no evidence that the *weak interaction* conserves parity. It was known that nuclei emit electrons via the weak interaction, so they suggested an experiment using the electron emitter ^{60}Co. A sample was placed in a magnetic field produced by a current loop, thereby forcing the magnetic moments of the nuclei to align themselves along **B** (Fig. 4). In the mirror-reversed process the current loop reverses, and by the right-hand rule, so does the **B** field. The experiment showed that more electrons were emitted opposite the **B** field than along it. Therefore, in the mirror-reversed experiment more electrons would be emitted downward (the direction of the **B** field in that experiment): the mirror image of the original experiment does not occur.

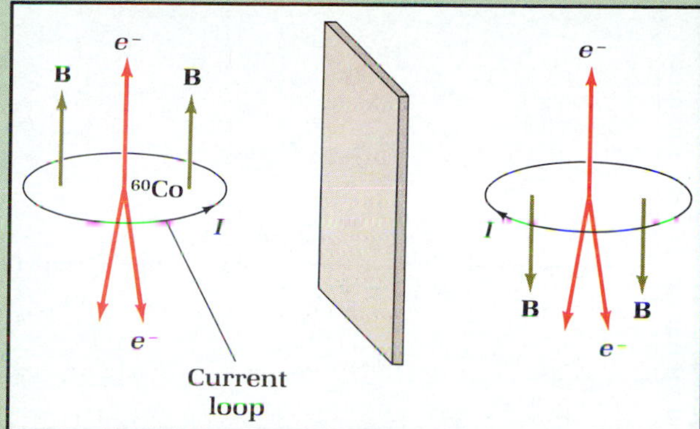

Figure 4. More electrons are emitted by ^{60}Co in the direction opposite the applied **B** field than along the field. In the mirror image the current reverses so **B** also reverses. However, the majority of electrons are still emitted downward in the mirror image. Since experiment shows that most electrons are emitted opposite the **B** field, the mirror-image result is not possible.

Surprisingly, nature distinguishes between right-handedness and left-handedness: one process is possible but the mirror image is not. At a fundamental level the weak interaction changes form when the coordinates are inverted.

Since the weak force acts between electrons and nuclei, although at a very low level, is it possible that this is the origin of nature's asymmetries? Did the action of this force during the formation of amino acids 3 billion years ago produce the observed preponderance of left-handed amino acids? Calculations have shown that the weak force leads to a very slightly lower energy for the left-handed version of several amino acids so they would be slightly favored during formation, but by only one molecule in 10^{17}. This is far too small to measure and is consistent with the observation that the two forms are produced with equal abundance in laboratory reactions. If the weak force is the origin of the asymmetry, clearly some kind of amplification process must be operating.

Communications engineers use signal averaging to detect a weak signal buried in a noisy background. Noise fluctuates randomly between positive and negative values so that in the long term it averages out to zero, whereas the signal, which may be smaller, is always in the same direction and eventually adds to larger values than the noise. Mathematically, if n samples of noise S_n are added, the sum increases as $S_n\sqrt{n}$, while if n samples of signal S_s are added, the sum increases as nS_s. So even if S_s is very small compared with S_n, eventually the averaged signal dominates the averaged noise because n increases more rapidly than $\sqrt{n}$.

In the same way, in a solution of chemicals reacting to produce left- and right-handed amino acids, random fluctuations in the reactions between molecules will tend to average out to zero net production of either chirality, while any asymmetric force will always push the system in the same direction, toward a majority of one chirality over the other. Over time this tendency, always acting in the same direction, can determine the handedness of the system.

More specifically, the amplification might work like this. Even if there is no asymmetrical force at work, if the concentration of reacting chemicals is large enough, the system is unstable and will spontaneously jump to one chirality or the other (time t_{SSB} in Fig. 5) with equal probability, a process called *spontaneous symmetry breaking.* A plot of the numerical dif-

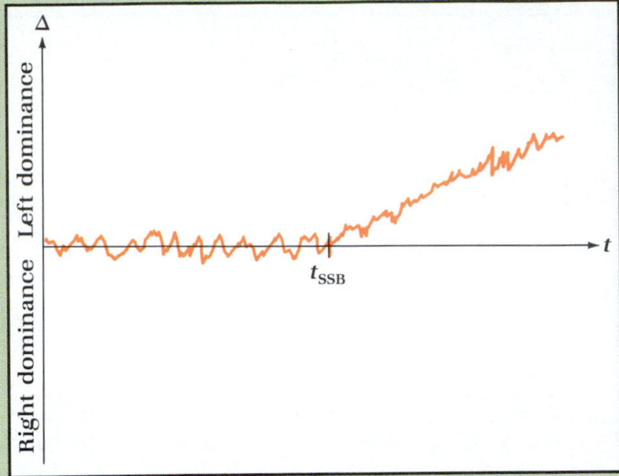

Figure 5. A plot of the difference in concentrations of a left- and right-handed molecule as a function of time. The model assumes the two molecules are formed with equal probability by the reaction of two "nutrient" molecules. If the product of the concentrations of the nutrient molecules exceeds a certain value, the system becomes unstable and will jump toward dominance by one chiral molecule or the other, again with equal probability. If a small bias toward the formation of one type of molecule is introduced, the jump is almost always toward that molecule.

ference between the two chiralities, Δ, will look like Fig. 5. However, if a small weak-force asymmetry is introduced, simulations have shown that it will almost always push the system in the favored direction, for example, toward left-handed amino acids. It is quite possible that the observed amino acid asymmetry arose from such a spontaneous process, nudged toward left-handedness by the weak force. Calculations show that the weak force will be 98 percent certain to produce a dominance of left-handed amino acid molecules in about 100,000 years. No one is sure that this is what actually happened, only that it might have happened this way.

This leaves unanswered the question of the origin of the chirality of snail shells and climbing vines, as well as the question of the handedness of humans. The connections between the chirality of biological molecules and, for example, the chirality of snail shells, remains far too tenuous and mysterious to allow any kind of conclusion . . . yet.

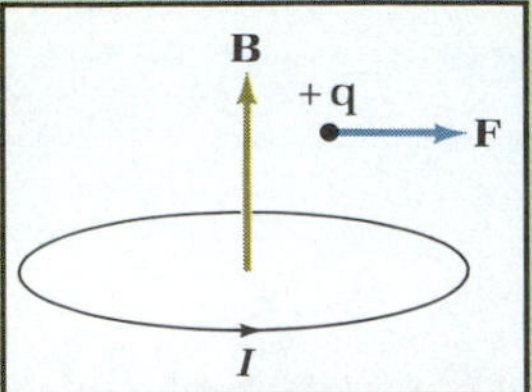

Figure 6. Question 1.

QUESTIONS

1. In Fig. 6, what would the mirror images look like if the current loop were rotated parallel to the mirror with **r** and **B** perpendicular to it?
2. In the ^{60}Co experiment, why was it necessary to align the nuclei using a magnetic field? Why not observe the emitted electrons from an ordinary, unaligned sample of ^{60}Co?
3. What kind of result should the ^{60}Co experiment have given if the weak interaction were to conserve parity?
4. Suppose that in the ^{60}Co experiment all charges change sign when the mirror reflection occurs. Is the reflected experiment a possible process in this case?
5. Explain how you could view a detailed film of the ^{60}Co experiment and determine whether the film was being run forward or backward. Suppose that the film shows the count rates at detectors above and below the ^{60}Co and, rather amazingly, also shows the electrons flowing in the electromagnets' coils.

EXERCISES

1. Suppose a signal of average strength S is buried in noise of average strength $1000S$. How many samples of the total signal have to be added together so that the net signal becomes three times the net noise?
2. *(a)* Figure 6 shows a positive charge moving into the page in the presence of a magnetic field. Show that this process conserves parity by showing that the mirror-image process also follows the laws of electromagnetism. *(b)* Draw a diagram similar to Fig. 6 and its mirror image, assuming the **B** force on a charged particle is similar to the **E** force, namely $\mathbf{F} = q\mathbf{B}$. Does this force conserve parity?

INDUCTANCE 29

Coils of wire at a production facility.

The last chapter introduced Faraday's law, a fundamental law of electromagnetism. The emphasis was on induced emf's and induced electric fields. We discussed generators and alternators as examples of practical devices based on Faraday's law. In this chapter the emphasis is on the application of Faraday's law to determine the behavior of simple circuit elements in circuits with currents that change. An induced emf can appear in a circuit element called an *inductor,* and the circuit is then said to contain *inductance.* We shall also encounter an important principle of electromagnetism — that energy is associated with (or stored in) a magnetic field.

29-1 SELF-INDUCED EMF'S AND SELF-INDUCTANCE

If the magnetic flux linking a circuit loop changes, then an induced emf exists in the circuit and is given by Faraday's law, $\mathscr{E} = -d\Phi_B/dt$ (Eq. 28-1). In the last chapter we assumed that the flux linking a loop was due to an external magnetic

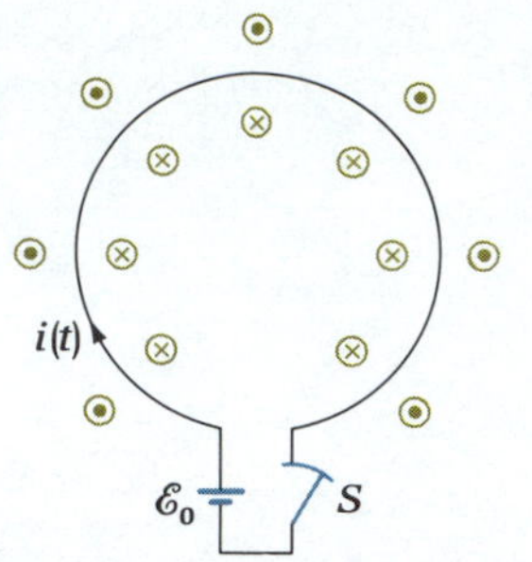

Figure 29-1. A changing current in a loop produces a changing flux linking the loop and a self-induced emf.

field—one produced by a magnet, for example. We now consider a case for which the magnetic field is due to the current in the loop itself.

A circuit with a single plane loop is shown in Fig. 29-1. When the switch S is closed (say, at $t = 0$), the battery causes charge carriers to begin moving. That is, a nonzero current $i(t)$ exists at any time after the switch is closed. The current in the loop produces a magnetic field in the vicinity of the loop, as shown schematically in the plane of the figure. Thus there is a magnetic flux linking the loop, and the flux is due to the loop's own current. Since this current is changing, the magnetic flux is also changing, and from Faraday's law an induced emf, a ***self-induced emf,*** exists in the loop. Faraday's law and Lenz's law apply generally to a magnetically induced emf. A self-induced emf is a special case: The self-induced emf in a circuit is due to the changes that occur in that circuit; in a sense, the changing current acts back on itself.

To see how the self-induced emf in the circuit of Fig. 29-1 depends on the changing current, note that at a point the magnetic field due to the current is proportional to the current. This result can be seen from the linear relationship between **B** and i in the Biot-Savart law, Eq. 27-3, or in Ampere's law, Eq. 27-8. Since the magnetic flux linking the loop is $\Phi_B = \int \mathbf{B} \cdot d\mathbf{S}$, we see that the flux is also proportional to the current in the loop, or

$$\Phi_B = Li \tag{29-1}$$

Self-inductance L and self-induced emf $\mathscr{E}_L$

where the constant of proportionality L is defined as the ***self-inductance*** of the loop. Applying Faraday's law, $\mathscr{E} = -d\Phi_B/dt$, we obtain the self-induced emf $\mathscr{E}_L$ in the loop,

$$\mathscr{E}_L = -L\frac{di}{dt} \tag{29-2}$$

Notice that the self-induced emf depends on the rate of change of the current. The negative sign in Eq. 29-2 determines the sense of the induced emf in accord with Lenz's law. If the magnitude of the current is increasing, then the sense of ξ_L is opposite the sense of i. On the other hand, if the magnitude of the current is decreasing, then the sense of ξ_L is the same as the sense of i. For example, suppose that the current in the loop in Fig. 29-1 is increasing. Then the magnetic flux is also increasing, and the sense of the self-induced emf is opposite the sense of the increasing current.

Although Eq. 29-2 applied to a simple loop circuit, experiments show that it is valid for more complicated circuit arrangements. Self-inductance is purposely introduced into a circuit with a circuit element called an ***inductor.*** An inductor is constructed by forming a coil of wire with many turns or loops; a common inductor design is a solenoid. In the absence of magnetic materials, such as iron, the induc-

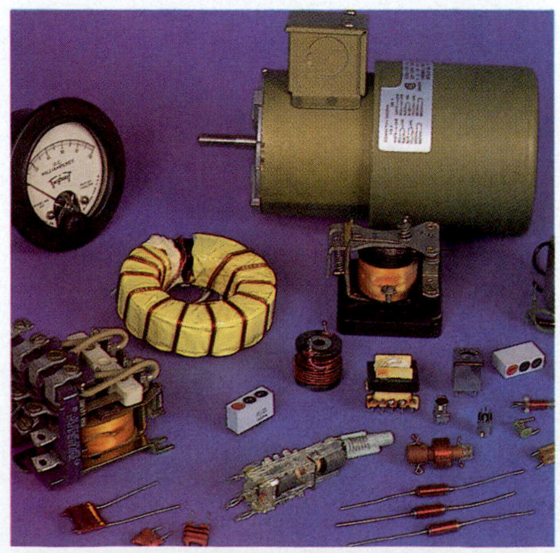

Several types of inductors.

tance L of an inductor depends only on its geometrical design. The SI unit of inductance is the ***henry*** (H), named after Joseph Henry (see the Commentary in this chapter). From Eq. 29-2, we have $1\ \text{H} = 1\ \text{V}\cdot\text{s}\cdot\text{A}^{-1}$. The henry is a fairly large unit of inductance. In electronic circuits, values of self-inductance of typical inductors are usually in the range from 1 μH to 1 mH.

The turns or loops in an inductor are tightly wound so that each loop has essentially the same flux Φ_B linking it. Thus, for an N-turn coil, Eq. 29-1 is recast as

$$N\Phi_B = Li \tag{29-3}$$

where Φ_B is the flux that links one of the N turns in the coil. As an example, we now determine the expression for the inductance of an ideal solenoid. An ideal solenoid is one that is long enough that we may assume that the field is uniform within the cylindrical volume of the solenoid. From Eq. 27-11, the magnitude of this field is $B = \mu_0 ni$, where $n = N/\ell$ is the number of turns per unit length. Since this field is directed along the axis, the flux linking a single turn of area S is

$$\Phi_B = BS = \mu_0 niS$$

The flux linking all N turns of the solenoid is

$$N\Phi_B = (n\ell)(\mu_0 niS) = (\mu_0 n^2 S\ell)i$$

From Eq. 29-3, $L = N\Phi_B/i$, so that the self-inductance of a long, tightly wound solenoid of length ℓ and cross-sectional area S is

$$L = \mu_0 n^2 S\ell \tag{29-4}$$

Inductance of a long solenoid

Notice that the inductance depends, apart from the constant μ_0, only on the length, area, and number of turns per unit length—that is, only on geometrical quantities. We emphasize that Eq. 29-4 is an approximate result, based on the field inside an ideal (infinitely long) solenoid; end effects have been neglected.

Calculating the inductance of a circuit element is feasible for only a few configurations (see Prob. 1). Generally, the inductance can be measured by observing its effect on a circuit containing the element. In the next section we consider a simple circuit with inductance.

EXAMPLE 29-1

A self-induced emf. A coil with self-inductance $L = 23\ \mu$H is in a circuit. Determine the self-induced emf in the coil if the current changes steadily from 1.7 to 2.9 mA in a 50-μs time interval.

Solution. Since the current changes steadily in the time interval,

$$\frac{di}{dt} = \frac{2.9\ \text{mA} - 1.7\ \text{mA}}{0.050\ \text{ms}} = 24\ \text{A/s}$$

From Eq. 29-2, the value of the self-induced emf is

$$\mathscr{E}_L = (23\ \mu\text{H})(24\ \text{A/s}) = 550\ \mu\text{V}$$

The sense of the induced emf, from Lenz's law, is opposite the sense of the increasing current in the coil.

SELF-TEST 29-1. What rate of change of current in a 9.7-mH solenoid produces a self-induced emf of 35 mV? ***ANSWER:*** 3.6 A/s.

EXAMPLE 29-2

Constructing a solenoid. A solenoid is constructed by winding fine wire on a cylindrical frame of radius 25 mm and length 120 mm. How many turns must be wound on the frame if the inductance is to be 0.67 mH? Neglect effects at the ends of the solenoid.

Solution. The number of turns per unit length can be obtained from Eq. 29-4. Solving for n^2,

$$n^2 = \frac{L}{\mu_0 S\ell}$$
$$= \frac{0.67 \text{ mH}}{(4\pi \times 10^{-7} \text{ T}\cdot\text{m}\cdot\text{A}^{-1})[\pi(0.025 \text{ m})^2](0.12 \text{ m})}$$
$$= 2.3 \times 10^6 \text{ m}^{-2}$$

Taking a square root gives $n = 1500 \text{ m}^{-1}$. The number of turns in the 120-mm length is then

$$N = n\ell = (1500 \text{ m}^{-1})(0.12 \text{ m}) = 180 \text{ turns}$$

SELF-TEST 29-2. Suppose the number of turns in the solenoid in the above example is doubled by placing turns on top of the turns already there. The length of the solenoid remains the same, and the wire is thin enough to make the cross-sectional area of these additional loops essentially the same as that of the loops already there. What is the inductance of this modified solenoid? ***ANSWER:*** $2^2(0.67 \text{ mH}) = 4(0.67 \text{ mH}) = 2.68 \text{ mH}$.

29-2 *LR* CIRCUITS

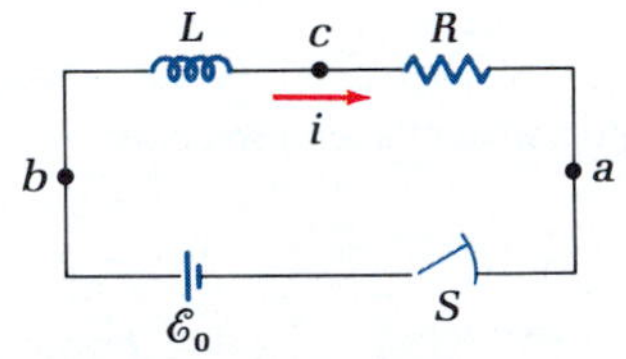

Figure 29-2. An inductance L and resistance R are in series with a battery of emf $\mathscr{E}_0$. Switch S is closed at $t = 0$.

Suppose that a solenoid is connected to a battery through a switch. Beginning at $t = 0$, when the switch is closed, the battery causes charge to move in the circuit. A solenoid, such as the one in Example 29-1, has inductance L and resistance R, and each of these influences the current in the circuit. The inductive and resistive effects of a solenoid are shown schematically in Fig. 29-2. The symbol for inductance (—ᴍᴍᴍ—) is shown in series with the resistance symbol. For simplicity we assume that all of the resistance in the circuit, including the internal resistance of the battery, is represented by R. Similarly, L includes the self-inductance of the connecting wires. A circuit, such as that seen in Fig. 29-2, containing resistance and inductance in series is called an ***LR circuit.***

The role of the inductance in determining the current in the circuit can be understood qualitatively: As the current $i(t)$ in the circuit increases (from $i = 0$ at $t = 0$), there is a self-induced emf $\mathscr{E}_L = -L\, di/dt$ in the inductance whose sense is opposite the sense of the increasing current. That is, since the current is increasing, the sense of the induced emf opposes that increase. This opposition to the increase in current prevents the current from rising abruptly (prevents di/dt from becoming very large). On the other hand, if the current changed negligibly (that is, di/dt is very small), there would be little opposition to a change in the current. The net effect of the inductance is to moderate between these extremes so that the current changes, but not too abruptly. The instantaneous value of the current depends on the values of L, R, and the emf of the battery, $\mathscr{E}_0$.

LR Circuit with a Battery

The time dependence of the current in the LR circuit of Fig. 29-2 can be determined by applying the loop rule (Sec. 25-3). Beginning at point a in the figure and proceeding clockwise, we sum the voltage changes encountered and equate the sum to zero:

$$(\mathscr{V}_b - \mathscr{V}_a) + (\mathscr{V}_c - \mathscr{V}_b) + (\mathscr{V}_a - \mathscr{V}_c) = 0$$

$$\mathscr{E}_0 - L\frac{di}{dt} - Ri = 0$$

Notice that the emf across the inductance, $\mathscr{V}_c - \mathscr{V}_b = -L\, di/dt$, is negative in accord with Lenz's law if i is increasing ($di/dt > 0$). Rearranging the above equation shows that the time dependence of the current in the *LR* circuit is governed by the differential equation

$$L\frac{di}{dt}+Ri=\mathscr{E}_0 \qquad (29\text{-}5)$$

LR circuit equation

This *LR* circuit equation is mathematically equivalent to the equation for the *RC* circuit in Chap. 25, which was solved by separating variables. That is, Eq. 29-5 can be rearranged to give

$$\frac{di}{i-\mathscr{E}_0/R}=-\frac{R}{L}\,dt$$

Integrating each side of the equation gives

$$\ln\left(i-\frac{\mathscr{E}_0}{R}\right)=-\left(\frac{R}{L}\right)t+\ln K$$

where ln K is a constant of integration. The exponential form of this result is $i-\mathscr{E}_0/R=Ke^{-(R/L)t}$. The constant of integration K is evaluated by applying the initial condition, $i(0)=0$. Thus $K=-\mathscr{E}_0/R$. We introduce a characteristic time parameter, $\tau_L=L/R$, called the *inductive time constant,* so that the exponential function is written e^{-t/τ_L}. Then the current in the circuit at time t is given by

Inductive time constant

$$i(t)=\frac{\mathscr{E}_0}{R}\,(1-e^{-t/\tau_L}) \qquad (29\text{-}6)$$

Notice that as $t\rightarrow\infty$, the current approaches the asymptotic value $I=\mathscr{E}_0/R$, which is the value we would obtain by neglecting inductance altogether.

The inductive time constant τ_L in the exponential e^{-t/τ_L} sets the time scale for the *LR* circuit. That is, the current cannot change significantly over a time interval much smaller than τ_L. For example, the current changes from zero at $t=0$ to $(\mathscr{E}_0/R)(1-e^{-1})\approx 0.63\mathscr{E}_0/R$ at $t=\tau_L$. The exponential function e^{-t/τ_L} decreases with time and approaches zero. At $t=5\tau_L$, for instance, $e^{-5}\approx 6.7\times10^{-3}$ and $i\approx 0.993\mathscr{E}_0/R$. In this way the current asymptotically approaches the steady value $I=\mathscr{E}_0/R$. These features of the time dependence of the current in an *LR* circuit are shown graphically in Fig. 29-3.

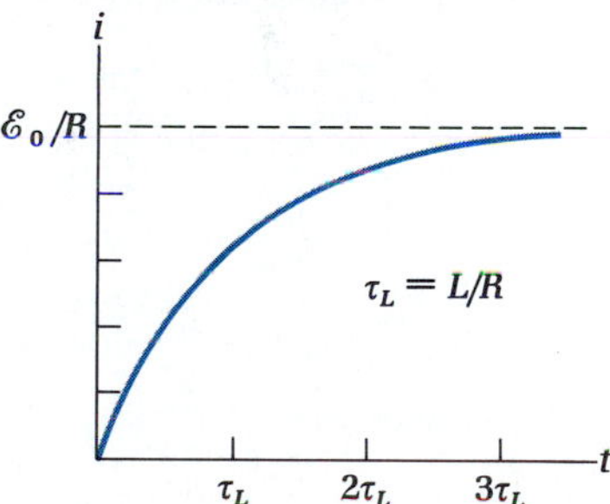

Figure 29-3. The current in the *LR* circuit increases from zero to the asymptotic value $\mathscr{E}_0/R$. The inductive time constant τ_L sets the time scale for changes in current.

LR Circuit without a Battery

The inductance in the *LR* circuit of Fig. 29-2 opposed the increasing current in the circuit. The sense of the self-induced emf was opposite the sense of the increasing current. There is also an *LR* circuit in which the current decreases, and the self-induced emf in the inductor then has the same sense as the decreasing current. The induced emf again opposes the change that produces it.

Figure 29-4*a* shows a circuit with a switching arrangement that removes the battery from the original circuit and forms a new circuit. We suppose that the current in the new circuit, shown as an equivalent circuit in Fig. 29-4*b*, with $R=R_1+R_2$, has an initial value i_0 when the switching occurs at $t=0$. The current will decrease from this initial value, and the sense of the self-induced emf, $\mathscr{E}_L=-L\,di/dt$, is the same as the sense of the decreasing current $(di/dt<0)$. Application of the loop rule (Exercise 15) gives $-L\,di/dt-iR=0$, or

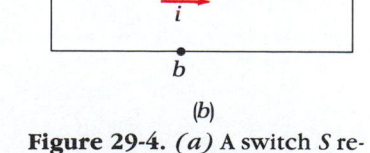

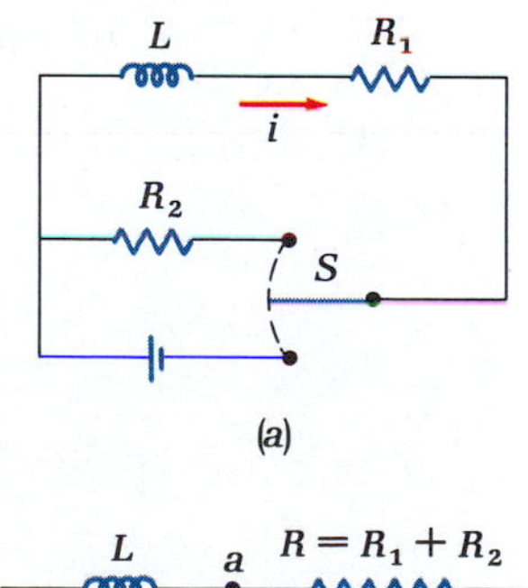

Figure 29-4. *(a)* A switch S removes the battery from the circuit and forms a new circuit containing L, R_1, and R_2. *(b)* The equivalent circuit has $R=R_1+R_2$.

$$L\frac{di}{dt}+Ri=0 \qquad (29\text{-}7)$$

which is the same as the *LR* circuit equation of Eq. 29-5 with no battery in the circuit $(\mathscr{E}_0=0)$.

Equation 29-7 can be solved with the approach used for the *LR* circuit with a battery (see Exercise 16). The solution is

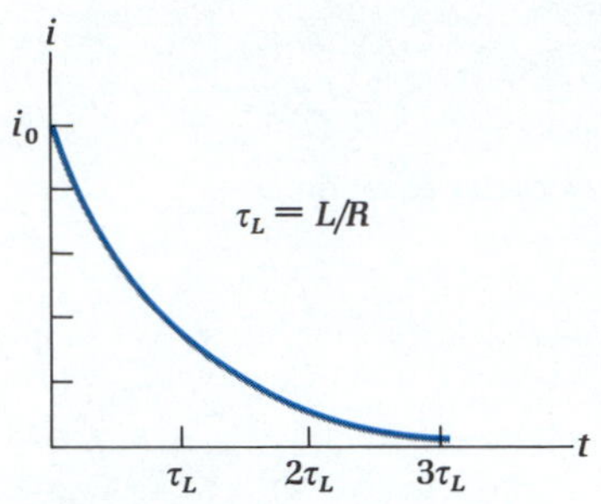

Figure 29-5. The current decreases exponentially from i_0 initially to zero asymptotically.

$$i(t) = Ke^{-t/\tau_L}$$

where K is a constant of integration. We apply the initial condition, $i = i_0$ at $t = 0$, to evaluate the constant K. Since $e^0 = 1$, we have $K = i_0$, and the current in the circuit at time t is given by

$$i(t) = i_0 e^{-t/\tau_L} \tag{29-8}$$

The current decreases exponentially from the initial value i_0 to zero. At $t = \tau_L$ the current has dropped to $i_0 e^{-1} \approx 0.37\, i_0$, and at $5\tau_L$ the current has dropped to about $6.7 \times 10^{-3} i_0$. Once again the inductive time constant, $\tau_L = L/R$, sets the time scale for substantial current changes in the LR circuit. The time dependence of the current is shown graphically in Fig. 29-5.

In some circuits an external emf, such as a battery, is periodically switched in and out of an LR circuit. The time dependence of the current alternates between increasing, as in Fig. 29-3, and decreasing, as in Fig. 29-5. The behavior of the current is shown graphically in Fig. 29-6 for a switching period $T = 6\tau_L$.

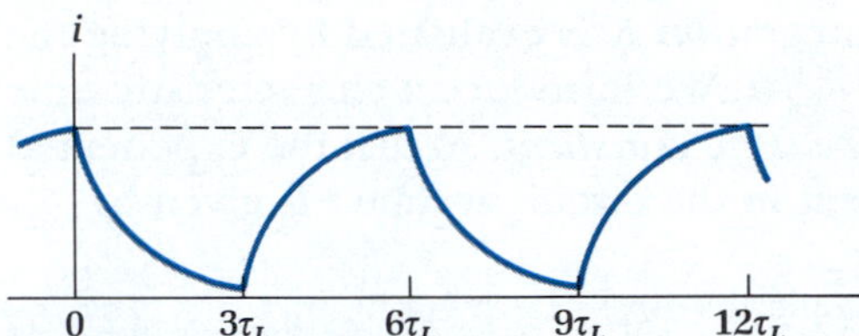

Figure 29-6. The current in an LR circuit is alternately increasing and decreasing as a battery is switched in and out of the circuit. This switching is periodic, with period $T = 6\tau_L$.

EXAMPLE 29-3

Measuring inductance. Inductance can be measured by examining the time dependence of the current in an LR circuit. Suppose we measure the time t_h for the current in the circuit of Fig. 29-2 to reach half the steady value of $I = \xi_0/R$. Determine an expression for L in terms of measured values of t_h and R.

Solution. Equation 29-6 describes the time dependence of the current in the circuit of Fig. 29-2, and the inductance L is contained in the inductive time constant $\tau_L = L/R$. From Eq. 29-6, the current i evaluated at $t = t_h$ is

$$i(t_h) = \frac{\xi_0}{R}(1 - e^{-t_h/\tau_L}) = \frac{\xi_0}{R}\left(\frac{1}{2}\right)$$

because t_h is the time for the current to reach half its steady value of ξ_0/R. Therefore

$$(1 - e^{-t_h/\tau_L}) = \tfrac{1}{2} \qquad \text{or} \qquad e^{-t_h/\tau_L} = \tfrac{1}{2}$$

Taking the logarithm and solving for τ_L, we find $\tau_L = t_h/\ln 2$. Since $\tau_L = L/R$, and $\ln 2 = 0.693$, we have

$$L = \frac{t_h R}{0.693}$$

Thus measured values of t_h and R can be used to determine L. Usually the time t_h is of the order of milliseconds or less, so that an oscilloscope is often used to measure this time.

SELF-TEST 29-3. In an LR circuit like the one in Fig. 29-2, the current reaches 36 mA in 2.2 ms after the switch is closed, and after some time the current becomes steady at 72 mA. If the resistance in the circuit is 68 Ω, what is the inductance in the circuit? ***ANSWER:*** 220 mH.

29-3 ENERGY TRANSFERS IN *LR* CIRCUITS

The electric potential energy of charge carriers changes as carriers move through a potential difference in a circuit element. The power, or rate of energy transformation, in a circuit element with current i and potential difference $\mathscr{V}$ is given by Eq.

25-7, $P = i\mathcal{V}$. What energy transformations occur in an inductor? Joule heating occurs at the rate $P = i\mathcal{V} = i(iR) = i^2R$ because of the ohmic resistance of an inductor. An additional energy transformation takes place while the current is changing in an inductor. That is, across the inductor there is a self-induced emf $\mathcal{V} = L\, di/dt$ and the power for this transformation is $P = i\mathcal{V} = i(L\, di/dt)$, or

$$P = Li\frac{di}{dt} \tag{29-9}$$

Power in an inductance

Whether the charge carriers gain or lose electric potential energy depends on how the current changes. If the current is increasing in the inductor, the sense of the self-induced emf is opposite the current, and the carriers lose potential energy. If the current is decreasing, the sense of the self-induced emf is the same as the sense of the current, and the carriers gain potential energy. Since the charge carriers can either gain or lose potential energy because of the inductance, the potential energy is transformed to and from a *stored energy* in the inductance. That is, this energy is stored as the carriers lose potential energy when the current is increasing in the inductor (as in the circuit in Fig. 29-2). The energy is *recovered* as the carriers gain potential energy when the current is decreasing in the inductor (as in the circuit in Fig. 29-4*b*). We shall call this energy the energy stored in the inductor, and later we shall associate this stored energy with the magnetic field of the inductor. Notice that energy is not stored (is not recoverable) in the resistance of a circuit. Charge carriers can only lose potential energy through Joule heating at the rate $P = i^2R$ in the resistance.

Energy stored in an inductor is recoverable.

The amount of energy stored in an inductor with current i depends on the current. To determine the amount, consider the circuit in Fig. 29-2. At some time t, the current in the circuit is $i(t)$ and it is changing (increasing in this case) at the rate di/dt. From Eq. 29-9, the stored energy U in the inductor is increasing at the rate $dU/dt = P = Li\, di/dt$. Equivalently, the infinitesimal change dU in the stored energy corresponding to a change di in current is $dU = P\, dt = Li(di/dt)\, dt$, or

$$dU = Li\, di$$

Integrating this expression, we obtain the energy stored in the inductor:

$$U = \tfrac{1}{2}Li^2 \tag{29-10}$$

Energy stored in an inductor

The constant of integration has been chosen so that the stored energy in an inductor with no current is zero. For an inductor with current i, the stored energy is positive and independent of the sense of the current.

EXAMPLE 29-4

Energy stored in an inductor. *(a)* Determine the energy stored in a 23-mH coil carrying a 2.5-A current. *(b)* By what factor must the current be increased to have twice that stored energy?

Solution. *(a)* Applying Eq. 29-10 gives

$$U = \tfrac{1}{2}Li^2 = \tfrac{1}{2}(23\ \text{mH})(2.5\ \text{A})^2 = 72\ \text{mJ}$$

(b) Since the stored energy is proportional to the square of the current, the energy is doubled if i^2 is doubled: $i'^2 = 2i^2$. Thus the current i' must be larger by a factor of $\sqrt{2}$, or $i' = \sqrt{2}(2.5\ \text{A}) = 3.5$ A.

SELF-TEST 29-4. *(a)* What is the power in a 25-mH inductor at an instant when the current in it is 380 mA and is increasing at the rate of 150 mA/s? *(b)* At this instant, is the potential energy of the carriers in the inductor increasing or decreasing? ***ANSWERS:*** *(a)* 1.4 mW; *(b)* decreasing.

Magnetic Energy and Energy Density

The stored energy in an inductor carrying a current, $U = \frac{1}{2}Li^2$, is analogous to the energy stored in a charged capacitor, $U = \frac{1}{2}Q^2/C$ (Eq. 23-5). In Chap. 23 we saw that the stored energy in a capacitor could be considered as energy stored in the electric field. The general expression for the electric energy density u_E (energy per unit volume stored in the electric field E) is given by $u_E = \frac{1}{2}\epsilon_0 E^2$ (Eq. 23-6). There is a similar interpretation of the energy stored in an inductor as a special case of energy stored in the magnetic field.

To see how the energy stored in an inductor is related to a magnetic field, consider a long, tightly wound solenoid carrying a current i. The solenoid has length ℓ, cross-sectional area S, and n turns per unit length. We treat it as an ideal solenoid so that the self-inductance is given by Eq. 29-4, $L = \mu_0 n^2 S\ell$. In obtaining this result we took **B** to be parallel to the axis and uniform inside the solenoid, with $B = \mu_0 ni$. Outside the solenoid, B is assumed to be negligibly small. Substituting the above expression for L into Eq. 29-10 gives for the stored energy

$$U = \tfrac{1}{2}Li^2 = \tfrac{1}{2}\mu_0 n^2 S\ell i^2$$

Since $B = \mu_0 ni$, the current $i = B/\mu_0 n$ can be eliminated from the expression above; this gives, after simplifying,

$$U = \frac{B^2}{2\mu_0} S\ell \qquad (29\text{-}11)$$

The energy stored in this solenoid depends on the square of the magnetic field in the solenoid and a geometrical factor $S\ell$, which is the volume of the space inside the solenoid where the magnetic field exists. Dividing the energy U stored in the inductor by the volume $S\ell$ gives the energy per unit volume u_B:

Energy density of the magnetic field

$$u_B = \frac{B^2}{2\mu_0} \qquad (29\text{-}12)$$

This energy density is associated with energy stored in the magnetic field. Experiment shows that the expression applies to any magnetic field. That is, the energy per unit volume stored in the magnetic field **B** at a point in space is given by Eq. 29-12.

Returning to the simple case of the solenoid, we can interpret the energy stored in the inductor, $U = \frac{1}{2}Li^2$, as energy stored in the magnetic field, $U = u_B(S\ell)$. In this instance u_B is uniform; it has the same value at each point inside the solenoid, a region of volume $S\ell$. Notice that if there is no current in the solenoid initially ($i = 0$), then $B = 0$, $u_B = 0$, and $U = 0$. As the current increases to a value i, an increasing magnetic field is produced in the solenoid, and energy is stored in the magnetic field. If the current decreases, then the magnetic field decreases, and the energy is recovered from the magnetic field by the charge carriers in the circuit. This energy may be dissipated in Joule heating in the resistance. Exercise 23 explores this process quantitatively.

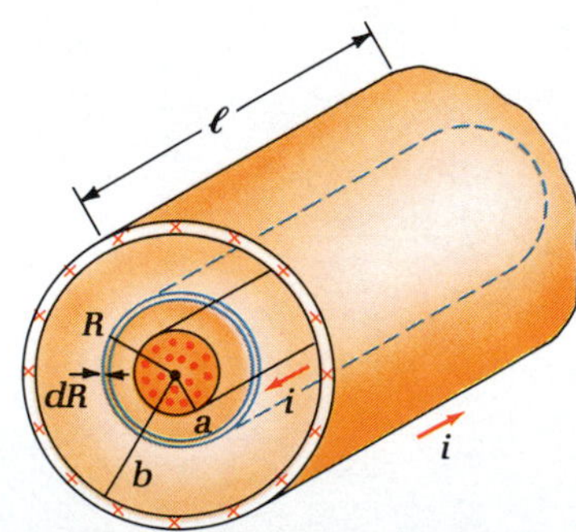

Figure 29-7. Example 29-5: A magnetic field exists in the space between conductors of a long, coaxial cable.

EXAMPLE 29-5

Magnetic energy in a coaxial cable. A long, coaxial cable with a circular cross section is shown in Fig. 29-7. The inner and outer conductors carry current i with opposite senses. *(a)* Determine the energy stored in the magnetic field in the space between conductors for a length ℓ of the cable. *(b)* Estimate the self-inductance per unit length of this coaxial cable.

Solution. *(a)* Ampere's law can be used as in Sec. 27-3 (also see Exercise 26 in Chap. 27) to determine the magnetic field at a point between the conductors at a perpendicular distance R from the axis. The magnitude is $B = \mu_0 i/(2\pi R)$. From Eq. 29-12, the magnetic energy density at this point is

$$u_B = \frac{B^2}{2\mu_0} = \frac{\mu_0 i^2}{8\pi^2 R^2}$$

Since the energy density depends on the distance R, we must set up an integral to determine the energy in this space. An appropriate volume element is a cylindrical shell of radius R, thickness dR, and length ℓ, as shown in the figure. Its volume is $dV = (2\pi R)(\ell)(dR)$, and the energy stored in this volume element is $dU = u_B\, dV = u_B(2\pi R\ell\, dR)$. We integrate this expression from the inner radius a to the outer radius b to obtain

$$U = \int_a^b \frac{\mu_0 i^2}{8\pi^2 R^2}\, 2\pi R\ell\, dR = \frac{\mu_0 i^2 \ell}{4\pi} \int_a^b \frac{1}{R}\, dR = \frac{\mu_0 i^2 \ell}{4\pi} \ln \frac{b}{a}$$

The energy stored in this space is proportional to the square of the current. A magnetic field also exists inside the inner conductor and in the outer conductor. The magnetic energy stored in these regions is small for the case shown in the figure. There the volume of the conductors is small compared with the volume of the space between them. Notice that the magnetic field is zero outside the cable.

(b) If we consider a section of length ℓ of the cable as an inductor of self-inductance L, then the energy stored in the inductor is given by Eq. 29-10, $U = \frac{1}{2}Li^2$. Neglecting the energy stored in the magnetic field inside the conductors (see Prob. 2), we equate $\frac{1}{2}Li^2$ to the expression above for the stored energy and solve for the inductance per unit length L/ℓ. We obtain

$$\frac{L}{\ell} = \frac{\mu_0}{2\pi} \ln \frac{b}{a}$$

The inductance per unit length is an important property of a cable used to transmit electromagnetic signals, such as a TV cable.

SELF-TEST 29-5. A 500-turn solenoid with a length of 40 mm and a cross-sectional area of 3.6×10^{-5} m^2 carries a current of 300 mA. Treating the solenoid as an ideal solenoid, determine its magnetic energy. ***ANSWER:*** 13 μJ.

29-4 MUTUAL INDUCTANCE

When the current changes in a coil in a circuit, there is a self-induced emf in the coil, $\mathscr{E}_L = -L\, di/dt$. As we have seen, the magnetic flux linking each turn of the coil is due to the magnetic field of the current in that circuit—hence the use of the term "self-induced." The magnetic field also extends outside the coil and may influence another nearby circuit.

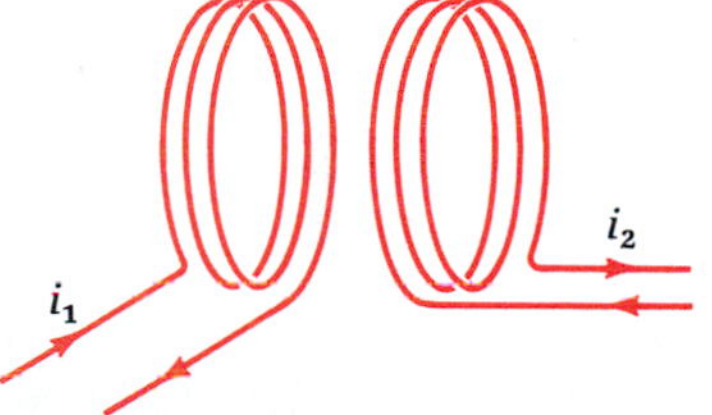

Figure 29-8. A changing current in each coil induces an emf in the other coil.

Figure 29-8 shows stationary coils belonging to separate circuits. Each coil carries a current, and these currents and the magnetic field they produce can be changing. Thus the flux linking each turn of coil 2 changes because of the changing current in coil 1. Similarly, the flux linking each turn of coil 1 changes because of the changing current in coil 2. These contributions to the flux changes and the corresponding induced emf's are in addition to the self-induced contributions. Since induced emf's appear in each coil because of a change in the other coil, the interaction is mutual between the coils, and the effect is called *mutual induction.*

Mutual induction

Consider the induced emf in one of the coils, say coil 2, due to a change in the current i_1 in coil 1. The flux linking each turn of coil 2 has a contribution due to the magnetic-field contribution B_1 of coil 1. We let the symbol Φ_{21} represent the contribution of the magnetic field of coil 1 to the magnetic flux linking a turn in coil 2. It is this flux that changes when the current i_1 changes. We assume that the flux linking each of the N_2 turns is the same. The product $N_2\Phi_{21}$ is called the *number of flux linkages* for the coil. If no magnetic material such as iron is around, the field contribution B_1 is proportional to the current i_1 which produces it (from the Biot-Savart law). Then both Φ_{21} and $N_2\Phi_{21}$ are proportional to i_1. We write the linear relation between $N_2\Phi_{21}$ and i_1 by introducing a constant M_{21}, a *coefficient of mutual inductance:*

Number of flux linkages

$$N_2\Phi_{21} = M_{21} i_1 \qquad (29\text{-}13)$$

We now apply Faraday's law to determine the induced emf due to coil 1 in a turn of coil 2. The induced emf in each turn is $\mathscr{E} = -d(\Phi_{21})/dt$, and the total emf $\mathscr{E}_{21}$ induced in coil 2 is $N_2\mathscr{E}$:

$$\mathscr{E}_{21} = -N_2 \frac{d}{dt}\Phi_{21} = -\frac{d}{dt}(N_2\Phi_{21}) = -\frac{d}{dt}(M_{21}i_1) = -M_{21}\frac{di_1}{dt}$$

Thus the induced emf $\mathscr{E}_{21}$ in coil 2 is proportional to the rate of change of the current di_1/dt in coil 1. The coefficient M_{21} depends on geometrical factors such as the shapes of the coils, the way they are wound, and their relative separation and orientation. The minus sign is used to determine the sense of the induced emf from Lenz's law.

By reversing the roles of the two coils, we can consider the emf $\mathscr{E}_{12}$ induced in coil 1 by the changing current i_2 in coil 2. The number of flux linkages for coil 1 is proportional to the current i_2, $N_1\Phi_{12} = M_{12}i_2$, where M_{12} is a coefficient determined by the geometry. The induced emf in coil 1 is given by

$$\mathscr{E}_{12} = -M_{12}\frac{di_2}{dt}$$

The coefficients M_{12} and M_{21} can be calculated easily for only a few simple arrangements. The values can be generally determined from measurements on the circuits. It turns out that the dependence on geometry of M_{12} and M_{21} is also mutual—that is, $M_{12} = M_{21}$. Thus we can drop the subscripts and let M represent the mutual inductance of the pair. Then the emf's induced in each coil by the changing current in the other are given by

Mutual inductance M

$$\mathscr{E}_{12} = -M\frac{di_2}{dt} \qquad \mathscr{E}_{21} = -M\frac{di_1}{dt} \tag{29-14}$$

This result applies generally to any pair of circuit elements. A changing current in each element induces an emf in the other element. The mutual inductance depends only on the geometry if no magnetic materials are nearby. The SI unit of mutual inductance is the henry (H), the same as the unit of self-inductance.

EXAMPLE 29-6

Mutually induced emf's. A coil in one circuit is close to another coil in a separate circuit. The mutual inductance of the combination is 340 mH. During a 15-ms time interval, the current in coil 1 changes steadily from 23 to 57 mA, and the current in coil 2 changes steadily from 36 to 16 mA. Determine the emf induced in each coil by the changing current in the other coil.

Solution. During the 15-ms time interval, the currents in the coils change at the constant rates of

$$\frac{di_1}{dt} = \frac{57\text{ mA} - 23\text{ mA}}{15\text{ ms}} = 2.3\text{ A/s}$$

$$\frac{di_2}{dt} = \frac{16\text{ mA} - 36\text{ mA}}{15\text{ ms}} = -1.3\text{ A/s}$$

From Eqs. 29-14, the magnitudes of the induced emfs are

$$\mathscr{E}_{21} = (340\text{ mH})(2.3\text{ A/s}) = 0.77\text{ V}$$

$$\mathscr{E}_{12} = (340\text{ mH})(1.3\text{ A/s}) = 0.45\text{ V}$$

Remember that the minus signs in Eqs. 29-14 refer to the sense of each induced emf.

SELF-TEST 29-6. Consider the sense of the mutually induced emf's in Fig. 29-8 according to an observer located to the right of the coils. *(a)* At an instant when the current i_1 is increasing, is the sense of emf ξ_{21} clockwise or counterclockwise? *(b)* At an instant when i_2 is decreasing, is the sense of ξ_{12} clockwise or counterclockwise? ***ANSWERS:*** *(a)* counterclockwise; *(b)* clockwise.

EXAMPLE 29-7

Calculating a mutual inductance. A circular loop of wire of radius a is inside and near the center of a long, tightly wound solenoid with n turns per unit length. The axis of the loop is parallel to the axis of the solenoid. *(a)* Determine an expression for the mutual inductance of the loop and solenoid in terms of geometrical quantities. *(b)* Evaluate the mutual inductance for $n = 2200$ turns per meter and $a = 12$ mm. *(c)* What emf is induced in the loop if the solenoid current changes at a rate of 1.4 A/s? *(d)* What is the sense of the induced current in the loop for the arrangement shown in Fig. 29-9?

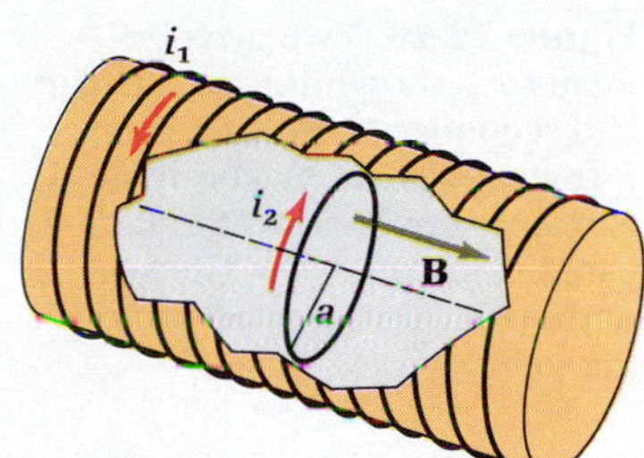

Figure 29-9. Example 29-7: A current is induced in the loop by a changing current in the solenoid.

Solution. *(a)* The magnetic field inside the solenoid due to its current i_1 is given approximately by Eq. 27-11, $B = \mu_0 n i_1$. Since this field is perpendicular to the plane of the loop with area πa^2, the flux linking the loop is $\Phi_{21} = B(\pi a^2) = \pi \mu_0 n a^2 i_1$. There is only one loop or turn so that $N_2 = 1$ in Eq. 29-13. Thus $M i_1 = N_2 \Phi_{21} = \pi \mu_0 n a^2 i_1$, or $M = \pi \mu_0 n a^2$, is the mutual inductance for the loop and solenoid.

(b) The value of the mutual inductance is

$$M = \pi(4\pi \times 10^{-7}\ \mathrm{T \cdot m \cdot A^{-1}})(2200\ \mathrm{m^{-1}})(0.012\ \mathrm{m})^2$$
$$= 1.3\ \mu\mathrm{H}$$

(c) The induced emf in the loop is

$$\mathscr{E}_{21} = M\frac{di_1}{dt} = (1.3\ \mu\mathrm{H})(1.4\ \mathrm{A/s}) = 1.8\ \mu\mathrm{V}$$

(d) Since the current i_1 is increasing $(di_1/dt > 0)$, the flux linking the loop is increasing. From Lenz's law, the induced current must oppose the increase of the flux for a surface bounded by the loop. Thus the sense of the induced current is as shown in the figure.

SELF-TEST 29-7. At a particular instant, the current in the circular loop of wire inside the solenoid in the above example is changing at 2.0 A/s. What is the emf induced in the solenoid as a result of this changing current? ***ANSWER:*** 2.6 μV.

29-5 TRANSFORMERS

Mutual induction, an emf induced in one circuit by a changing current in another, is a process that can sometimes be a problem. For example, unwanted emf's may be induced in a sensitive electronic circuit by other nearby circuits. In other cases mutual induction can be put to useful purposes. Some pacemakers operate with a emf induced by a changing current in a circuit external to the patient's body. This technique avoids the use of an internal power supply which would require surgery for replacement or maintenance. (But can you think of a disadvantage of using mutual induction to power a pacemaker?)

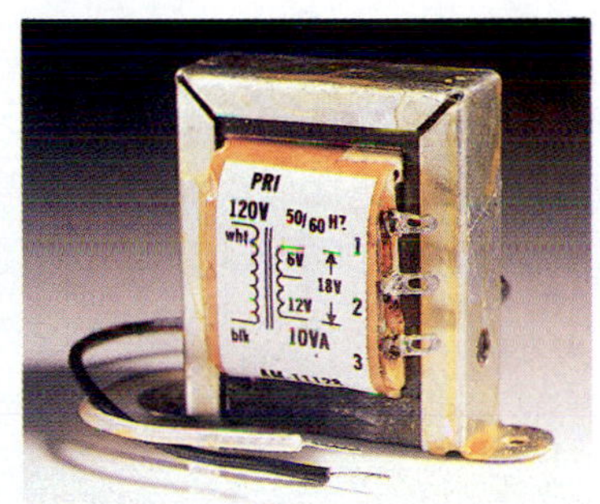

A demonstration transformer.

A device of great practical importance is the ***transformer,*** which uses mutual induction to change (or transform) the voltage from one circuit to another. A simple type of transformer consists of two coils wrapped around an iron core or ring, as shown in Fig. 29-10*a*. The magnetic field produced by the currents in the coils is largely concentrated in the iron. Another arrangement has the turns of one coil wound directly on top of the other coil, as shown in Fig. 29-10*b*. In each case the magnetic flux linking each turn in each coil is virtually the same. One of the coils of the transformer, is designated as the *primary coil;* it has N_p turns. The other coil is the *secondary coil* with N_s turns. Often we consider the primary as the input coil to the transformer and the secondary as the output coil. However, the reference to input and output is not always useful.

Primary and secondary coils of a transformer

The iron core causes the flux Φ_B to be essentially the same for each turn in both the primary and the secondary, and the same emf $\mathscr{E} = -d\Phi_B/dt$ will be induced in each turn if the flux changes. Thus the induced emf, or voltage $\mathscr{V}_s$, in the secondary with N_s turns is

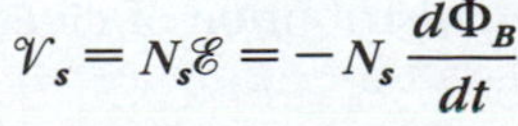

$$\mathscr{V}_s = N_s\mathscr{E} = -N_s\frac{d\Phi_B}{dt}$$

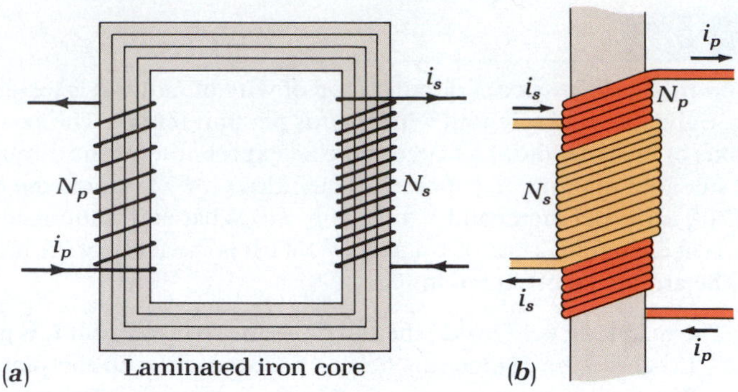

Figure 29-10. Two arrangements are shown for the primary and secondary coils of a transformer. *(a)* The magnetic field is mainly confined to the laminated iron core. *(b)* The secondary coil is wrapped around the primary.

The induced emf, or voltage $\mathcal{V}_p$, in the primary with N_p turns is

$$\mathcal{V}_p = N_p\mathcal{E} = -N_p \frac{d\Phi_B}{dt}$$

Taking the ratio of the voltages gives

$$\frac{\mathcal{V}_s}{\mathcal{V}_p} = \frac{N_s}{N_p} \qquad (29\text{-}15)$$

Step-up and step-down transformers

For example, if the secondary has $N_s = 5N_p$, then $\mathcal{V}_s = 5\mathcal{V}_p$. Such a transformer is called a ***step-up transformer***. That is, the secondary voltage is greater than, or "stepped up" from, the primary voltage. The secondary of a ***step-down transformer*** has fewer turns than the primary, and the secondary voltage is less than the primary voltage.

The detailed operation of a transformer depends on the properties of the other elements in the two circuits. We shall discuss a simple case in which a sinusoidally changing emf is applied to the primary coil and the primary circuit has negligible resistance. We also assume that the secondary circuit has a large resistance. Further, we neglect eddy currents and other losses in the (laminated) iron core. Under these conditions the transformer acts to transfer energy from the primary circuit to the secondary circuit. The power for the primary is $P_p = i_p\mathcal{V}_p$, where i_p is the current in the primary. Similarly, $P_s = i_s\mathcal{V}_s$ is the power in the secondary, where the current is i_s. For the ideal case of no losses (typical transformers have high efficiencies, up to 99 percent), we equate the power in the secondary to that in the primary, by conservation of energy, to get $i_s\mathcal{V}_s = i_p\mathcal{V}_p$ or $i_s/i_p = \mathcal{V}_p/\mathcal{V}_s$. Using Eq. 29-15, we express the ratio of currents in terms of the number of turns on the coils:

By energy conservation, $i_s\mathcal{V}_s = i_p\mathcal{V}_p$.

$$\frac{i_s}{i_p} = \frac{N_p}{N_s} \qquad (29\text{-}16)$$

Comparing Eqs. 29-15 and 29-16, we see that the voltage ratio $\mathcal{V}_s/\mathcal{V}_p$ and the current ratio i_s/i_p behave inversely in a transformer. For example, in a step-up transformer with $N_s > N_p$, the secondary voltage is larger than the primary voltage, but the secondary current is smaller than the primary current.

Transformers are used by electric utilities to reduce Joule heating losses in the distribution of electric energy over long distances. At a generating station, a generator is on the primary side of a step-up transformer. The transmission lines leading to distant points are on the secondary side of the transformer. A relatively large current i_p can exist in the primary circuit, with a moderate voltage $\mathcal{V}_p$ determined by the generator. The voltage is stepped up in the secondary ($\mathcal{V}_s$ may be in the range of 500,000 V), and the secondary current is correspondingly smaller. Reducing the current i_s in the transmission lines reduces the i_s^2R Joule heating loss in the lines. Step-down transformers are used at the other end to reduce the voltage (and increase the current) to safe and convenient levels. Without the use of transformers, the large-scale transmission and distribution of electric energy would not be feasible.

EXAMPLE 29-8

Transformer in a radio. A transformer is used to convert 120-V voltage from a wall receptable to the 9.0-V voltage required for a radio. *(a)* Is a step-up or a step-down transformer used? *(b)* If the primary has 480 turns, how many turns are on the secondary? *(c)* Determine the current in the primary coil if the radio operates with 400 mA.

Solution. *(a)* We regard the primary side of the transformer as the side that supplies energy. In this case it is the 120-V side. A step-down transformer is used to reduce the voltage to 9.0 V.

(b) From Eq. 29-15,

$$N_s = N_p \frac{\mathcal{V}_s}{\mathcal{V}_p} = 480 \frac{9.0 \text{ V}}{120 \text{ V}} = 36 \text{ (turns)}$$

(c) The primary current is determined from Eq. 29-16:

$$i_p = i_s \frac{N_s}{N_p} = 400 \text{ mA} \frac{36}{480} = 30 \text{ mA}$$

SELF-TEST 29-8. What is the power in the transformer in the above example? ***ANSWER:*** 3.6 W.

Transformers at a substation.

EXAMPLE 29-9

Transformer at a power plant. A generator at a hydroelectric station operates at 12 kV and 12 A. (These are root-mean-square values; see Chap. 31.) A step-up transformer is used to step up the transmission-line voltage to 140 kV. *(a)* Determine the current in the transmission line. *(b)* If the resistance of the transmission line is 170 Ω, determine the rate of Joule heating in the line. *(c)* What would be the Joule heating rate if the line voltage were 14 kV?

Solution. *(a)* Since $i_s\mathcal{V}_s \approx i_p\mathcal{V}_p$, we have

$$i_s = \frac{(12 \text{ A})(14 \text{ kV})}{140 \text{ kV}} = 1.2 \text{ A}$$

(b) The rate of Joule heating in the line is

$$P = i_s^2 R = (1.2 \text{ A})^2(170 \ \Omega) = 240 \text{ W}$$

(c) If the line voltage were 14 kV, the current would be 12 A and the heating loss rate would be

$$P = (12 \text{ A})^2(170 \ \Omega) = 24 \text{ kW}$$

SELF-TEST 29-9. For the transformer in the above example, what is the ratio of the number of turns in the secondary to the number of turns in the primary? ***ANSWER:*** 10.

Joseph Henry.

COMMENTARY: Joseph Henry

Joseph Henry, born in Albany, New York, in 1797, is often considered the independent codiscoverer, along with Michael Faraday, of electromagnetic induction. Henry began his study of magnetism as a teacher at the Albany Academy, where he had earlier been a student. Trying to keep abreast of research being done in Europe, he refined an electromagnet described by William Sturgeon of London. Sturgeon's magnet was a soft iron horseshoe wrapped with only a few turns of bare copper wire. These turns were spaced apart so as to prevent a short circuit. Because of the small number of turns, the electromagnet was relatively weak. Henry used silk threads from his wife's petticoat to insulate copper wire. He was able to wrap a horseshoe with 400 overlapping turns, thereby obtaining a much stronger electromagnet.

In 1831, Henry described a primitive electric motor, powered "by magnetic attraction and repulsion." A bar electromagnet was pivoted about a horizontal axis through its center of gravity, with the north pole of a permanent magnet placed under each end of the electromagnet. To reverse the current in the electromagnet and cause it to oscillate, Henry devised a mechanism for dipping the ends of the connecting wires from the electromagnet in and out of the acid of the voltaic cells with each cycle. Of this motor, Henry remarked that "not much importance, however, is attached to the invention, since the article in its present state can only be considered a philosophical toy; although . . . some modification of it on a more extended scale may hereafter be applied to some useful purpose." Henry refused to patent any of his inventions because he "did not consider it compatible with the dignity of science to confine the benefits which might be derived from it to the exclusive use of any individual."

As to the discovery of induction, Faraday in England had "established the general fact, that when a piece of metal is moved in any direction . . . between the poles of a horseshoe magnet, electrical currents are developed in the metal. . . ." This note by a Peter M. Roget was dated December 12, 1831, and appeared in an issue of the *Library of Useful Knowledge.* Henry's account of his own work on induction appeared shortly thereafter (July 1832) in *Gilliman's Journal.* It is possible that Henry's discovery may have predated Faraday's, but it is not clear who had discovered what nor exactly when it had been discovered in the period prior to publication.

Henry later wrote, "Before having any knowledge of the method given in the above account [Faraday's], I had succeeded in producing electrical effects in the following manner, which differs from that employed by Mr. Faraday. . . ." He then described his experiments on self-induction and concluded, "we have . . . electricity converted into magnetism and this magnetism again into electricity. . . ." In any event, the law of induction is known as Faraday's law, and the unit of inductance is the henry.

Henry had a long and productive career in science. He taught at Princeton and was an advisor to the United States government. He served for 32 years as the first secretary (director) of the Smithsonian Institution. During this period of service, he refused to accept an increase in his salary. He is often described as the father of meteorology for his work in that field. Joseph Henry remained active until his death at age 82 in 1878.

For further reading, see "Joseph Henry" in *Famous American Men of Science* by J. G. Crowther (Books for Libraries Press, Freeport, N.Y., 1937) and *Joseph Henry — His Life and Work* by Thomas Coulson (Princeton University Press, Princeton, N.J., 1950).

SUMMARY

Section 29-1. Self-Induced emf's and Self-Inductance

If the current changes in a circuit element such as a coil, a self-induced emf exists and is given by

$$\mathscr{E} = -L\frac{di}{dt} \qquad (29\text{-}2)$$

The self-inductance L depends on the geometry of the element. For a coil,

$$Li = N\Phi_B \qquad (29\text{-}3)$$

where Φ_B is the flux linking each of the N turns in the coil. The self-inductance of a long, tightly wound solenoid is

$$L = \mu_0 n^2 S\ell \qquad (29\text{-}4)$$

Section 29-2. *LR* Circuits

In an LR circuit the time scale for appreciable changes in the current is set by the inductive time constant $\tau_L = L/R$. For an LR circuit with a battery, the current is

$$i(t) = \frac{\mathscr{E}_0}{R}(1 - e^{-t/\tau_L}) \qquad (29\text{-}6)$$

For a circuit without a battery, the current decreases according to

$$i(t) = i_0 e^{-t/\tau_L} \qquad (29\text{-}8)$$

Section 29-3. Energy Transfers in *LR* Circuits

The energy stored in an inductor with current i is

$$U = \tfrac{1}{2}Li^2 \qquad (29\text{-}10)$$

This energy is stored in the magnetic field due to the current. The energy density of magnetic field energy is

$$u_B = \frac{B^2}{2\mu_0} \qquad (29\text{-}12)$$

Section 29-4. Mutual Inductance

The changing currents in two nearby coils mutually induce emf's in each other:

$$\mathscr{E}_{12} = -M\frac{di_2}{dt}$$
$$\mathscr{E}_{21} = -M\frac{di_1}{dt} \qquad (29\text{-}14)$$

where M is the mutual inductance of the pair.

Section 29-5. Transformers

In a transformer the voltages and currents in the primary and secondary coils depend on the number of turns in each:

$$\frac{\mathscr{V}_s}{\mathscr{V}_p} = \frac{N_s}{N_p} \qquad (29\text{-}15)$$

$$\frac{i_s}{i_p} = \frac{N_p}{N_s} \qquad (29\text{-}16)$$

QUESTIONS

1. Why is the term "self" used for a self-induced emf?

2. Suppose the current in a circuit element is increasing. Can the self-induced emf cause the current to increase further? Explain.

3. Which would you expect to have the larger self-inductance—a length of wire forming a single loop or the same length wound to form a small coil? Explain.

4. Does the earth's magnetic field affect the value of the self-inductance of a solenoid? Explain.

5. Suppose that two solenoids carry the same current and are identical except that one solenoid has a laminated iron core. The magnetic field in the iron core is proportional to the current in the solenoid but has a much larger magnitude than in the solenoid without the iron core. Which solenoid has the larger inductance? Explain.

6. A resistor in the form of a solenoid is wound with two layers of wire. The first layer is wound from one end to the other, and the second layer returns so that the currents have opposite senses in the two layers. The manufacturer of the resistor claims it has a negligible self-inductance. Why?

7. If end effects were included, would the self-inductance of a solenoid be larger or smaller than the approximate value given by Eq. 29-4? Explain.

8. How would the value of τ_L for an LR circuit change if $\mathscr{E}_0$ were doubled? If L were doubled? If R were doubled? If L and R were doubled?

9. How long after the switch in Fig. 29-2 is closed does the current equal $\mathscr{E}_0/R$? Explain.

10. Suppose the current in Fig. 29-2 is increasing. Which point is at the higher potential, b or c? What about points a and c? Explain.

11. The current in Fig. 29-4b is decreasing. Which point is at the higher potential, a or b? Explain.

12. Compare and contrast the expressions $\frac{1}{2}q^2/C$ and $\frac{1}{2}Li^2$ for the energy stored in a capacitor and in an inductor.

13. In what sense can we speak of the energy stored in an inductor? Where is the energy? Could part of the energy be outside the inductor (outside a solenoid, for example)? Explain.

14. What provides the magnetic energy stored in an inductor in a circuit such as that shown in Fig. 29-2?

15. Why is the term "mutual" used for mutual induction?

16. Two dissimilar coils are close together, with changing currents in each. Must the currents change at the same rate? Must the mutually induced emf's be the same? Can there be a self-induced emf in either coil? Explain.

17. Suppose that one of the coils in Fig. 29-8 is rotated so that the axes of the coils are perpendicular rather than parallel. Would the mutual inductance increase, decrease, or remain the same? Explain.

18. Some residential areas are served by electric distribution lines at 22 kV. A step-down transformer is used to provide 220-V service to each residence. Why not have the distribution lines at 220 V and eliminate the costly transformers? Why not have househol appliances operate at 22 kV and eliminate the transformers?

19. The primary coil for an automobile ignition system is connected to the 12-V automobile battery. The secondary coil provides several kilovolts to give a spark. If you disassembled an ignition-system unit, how could you determine which coil was the primary and which was the secondary?

20. A hand calculator operates with a 9-V battery. Alternatively it can be used with an adapter that plugs into a 120-V wall outlet. What is inside the adapter? Explain.

21. Step-up transformers are used to reduce the i^2R (Joule heating) losses in electric distribution lines serving distant points. (Recall that a step-up transformer has a smaller current in the secondary.) Why not reduce the resistance R of the lines instead?

22. Transformers are used to change voltages in ac circuits. Can you think of a way to step up a voltage in dc (constant-current) circuits? Explain.

23. Complete the following table:

Symbol	Represents	Type	SI Unit
L			
τ_L		Scalar	
U			J
u_B	Magnetic energy density		
M			
N_p			—
N_s			—

EXERCISES

Section 29-1. Self-Induced emf's and Self-Inductance

1. Show that the unit of inductance, the henry (H), can be expressed as $1\ \text{H} = 1\ \text{kg}\cdot\text{m}^2\cdot\text{s}^{-2}\cdot\text{A}^{-2}$.

2. The current in a 17-μH coil changes at the constant rate of 82 mA/s. Determine the self-induced emf in the coil.

3. A solenoid has inductance $L = 23$ mH. Determine the self-induced emf in the solenoid when *(a)* the current is 125 mA and increasing at the rate 37 mA/s, *(b)* the current is zero and increasing at the rate 37 mA/s, *(c)* the current is 125 mA and decreasing at the rate 37 mA/s, *(d)* the current is 125 mA and not changing.

4. A graph of the time dependence of the current in a 100-mH coil is shown in Fig. 29-11. Construct the corresponding graph of the self-induced emf in the coil. Make estimates of the maximum and minimum values of the emf and explain any change in the sense of the emf.

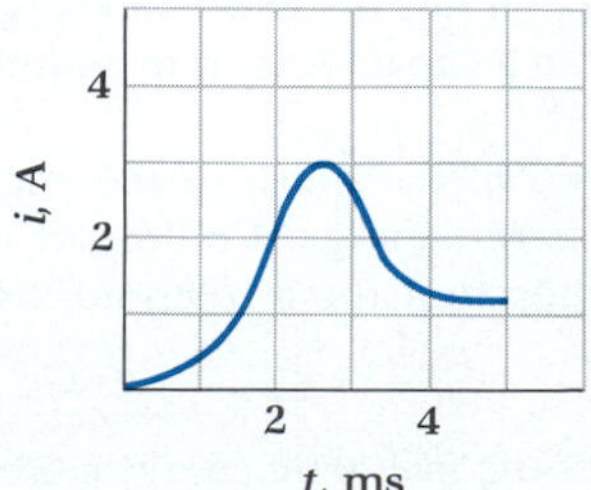

Figure 29-11. Exercise 4.

5. *(a)* At what rate must the current change in a 65-mH coil to have a 1.0-V self-induced emf? *(b)* Should the current be increasing, decreasing, or either? Explain.

6. A solenoid with 1200 turns per meter has length 150 mm and radius 16 mm. *(a)* Determine the self-inductance of the solenoid. *(b)* If the solenoid current increases at a constant rate from 0 to 20 μA in 50 ms, what is the self-induced emf?

7. Show that the SI unit for the permeability constant μ_0 can be expressed as henrys per meter (H/m).

8. A 250-turn coil has self-inductance $L = 65\ \mu$H. *(a)* At an instant when $i = 25$ mA, what is the magnetic flux linking each turn on the coil? *(b)* If the current changes at a rate of 96 mA/s, what emf is induced in the coil?

9. Two long, straight, parallel wires carry a current i, as shown in Fig. 29-12. Each wire has radius a and the axes are separated by distance D. *(a)* Use Ampere's law to show that, at a point between the wires in the plane containing the axes, the magnitude of the magnetic field is (see Chap. 27)

$$B = \frac{\mu_0 i}{2\pi}\left(\frac{1}{R} + \frac{1}{D - R}\right)$$

(b) In this plane consider a surface of length ℓ and width $D - 2a$ between the wires and determine the magnetic flux for the area $\ell(D - 2a)$. Note that the flux for a strip of area $\ell\ dR$ is $B\ell\ dR$. *(c)* Show that the inductance per unit length for this two-conductor line is

$$\frac{L}{\ell} = \frac{\mu_0}{\pi} \ln \frac{D-a}{a}$$

This calculation neglects the effect of the magnetic field inside the wires and is valid for $D \gg a$.

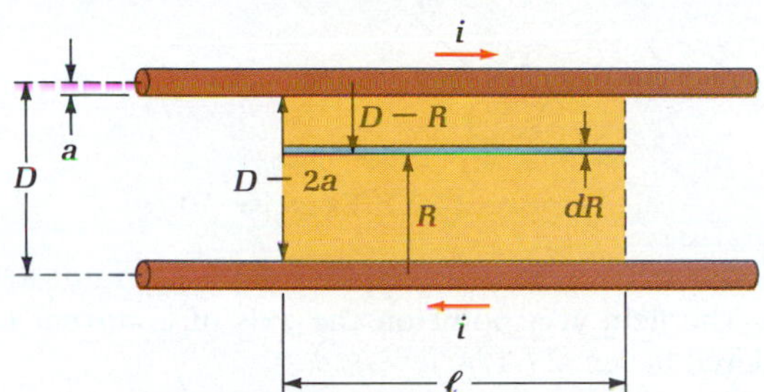

Figure 29-12. Exercise 9.

Section 29-2. *LR* Circuits

10. Show that the SI unit of L/R is the second (s).

11. The circuit in Fig. 29-13 has $\mathscr{E}_0 = 12$ V, $R = 25\ \Omega$, $L = 0.48$ H. The switch is closed at $t = 0$. Determine *(a)* the inductive time constant, *(b)* the current at $t = 25$ ms, *(c)* the current at 1.0 s. *(d)* What is the asymptotic value of the current?

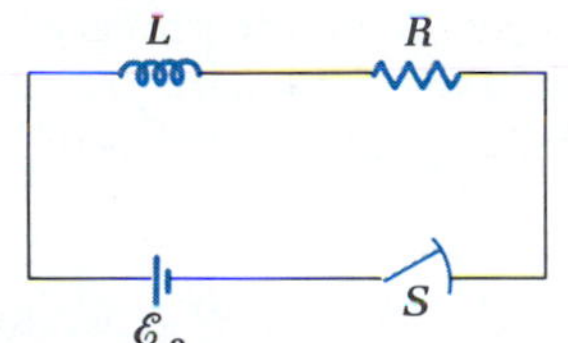

Figure 29-13. Exercise 11.

12. Switch S in the circuit of Fig. 29-14 is closed at $t = 0$. Determine the current i_1 and i_2 *(a)* immediately after the switch is closed, *(b)* after the switch has been left closed for several minutes, *(c)* immediately after the switch is opened. (*Hint:* The current in the inductor cannot change discontinuously.)

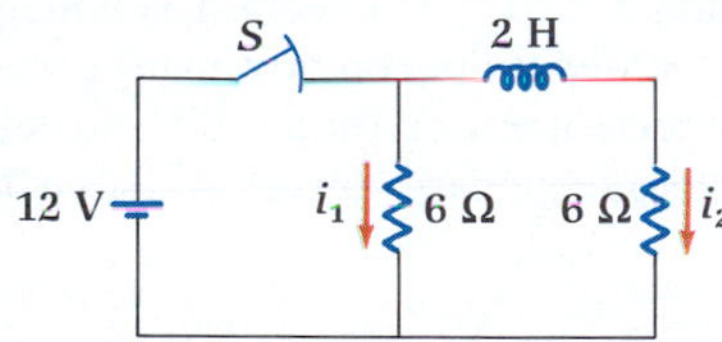

Figure 29-14. Exercise 12.

13. You are given a 23-mH coil with resistance 0.15 Ω and a wide choice of batteries and resistors. Design an *LR* circuit (specify values of R and $\mathscr{E}_0$) with an asymptotic current of 0.80 A that can be reached within about 10 ms.

14. The *LR* circuit in Fig. 29-13 has $R = 2.3\ \Omega$ and $\mathscr{E}_0 = 12$ V; the switch is closed at $t = 0$. At $t = 33$ ms the current is 3.6 A. *(a)* What is the asymptotic current? *(b)* What is the time constant for this circuit? *(c)* At what time is the current 5.0 A? *(d)* What is the value of L?

15. Apply the loop rule to the *LR* circuit without a battery in Fig. 29-4*b* to obtain Eq. 29-7.

16. Solve Eq. 29-7 by direct integration. Use the initial condition $i = i_0$ at $t = 0$ to obtain Eq. 29-8.

17. An *LR* circuit has $\mathscr{E}_0 = 9.2$ V, $R = 72\ \Omega$, $L = 250\ \mu$H. If the switch is closed at $t = 0$, determine *(a)* the current in the circuit, *(b)* the potential difference across the resistor, and *(c)* the potential difference across the inductance at $t = 0$, $t = 3.0\ \mu$s, $t = 7.5\ \mu$s, and $t = 35\ \mu$s.

18. *(a)* Show that if two inductors L_1 and L_2 are in series in a circuit, as shown in Fig. 29-15*a*, the combination is equivalent to an inductance $L = L_1 + L_2$. Assume that no flux from either inductor links the other inductor. *(b)* What is the equivalent inductance if the two are in parallel, as shown in Fig. 29-15*b*?

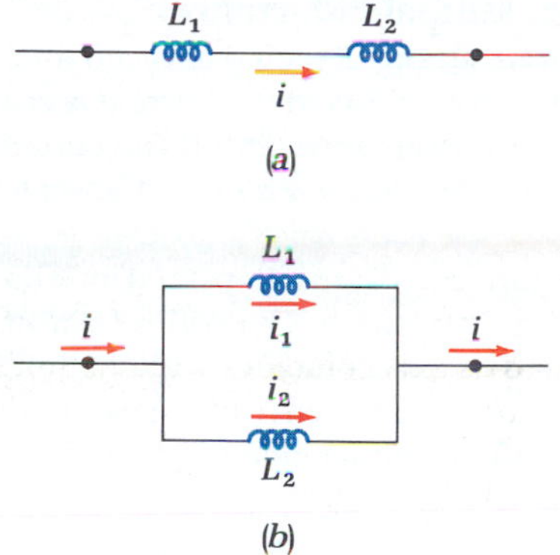

Figure 29-15. Exercise 18: *(a)* Inductors in series. *(b)* Inductors in parallel.

19. Suppose that the initial value of the current is 3.0 A in the circuit shown in Fig. 29-4*b*. The current drops to 1.5 A in 65 ms. *(a)* Determine the inductive time constant for the circuit. *(b)* If $R = 0.50\ \Omega$, what is the inductance L? *(c)* Construct a graph of i versus t for $0 \leq t \leq 200$ ms.

Section 29-3. Energy Transfers in *LR* Circuits

20. A 45-mH solenoid can carry a maximum current of 6.0 A without overheating. *(a)* What maximum energy can be stored in the solenoid? *(b)* For what current is the stored energy half the maximum value?

21. *(a)* For the *LR* circuit in Fig. 29-4*b*, obtain an expression for the stored energy $U(t)$ in the inductor as a function of time. The expression should be in terms of L, i_0, τ_L, and t. *(b)* Let $L = 1.0$ H, $i_0 = 1.0$ A, $\tau_L = 1.0$ s and prepare graphs of $i(t)$ versus t and $U(t)$ versus t. Use the same time scale for both graphs and discuss similarities and differences between the two graphs.

22. Consider the circuit described in Exercise 11. *(a)* How much energy is ultimately stored in the inductor? *(b)* How long after the switch is closed is the stored energy half the ultimate value?

23. At $t = 0$ an initial energy $U_0 = \frac{1}{2}Li_0^2$ is stored in the inductor in the circuit in Fig. 29-4*b*. *(a)* Determine the explicit time dependence of the instantaneous rate $P(t)$ of Joule heating in the resistor. *(b)* Compare the total energy dissipated in the resistor, $\int_0^\infty P(t)\,dt$, with the initial energy stored in the inductor.

24. At a point inside a long solenoid, far from either end, the magnetic field has magnitude $B = \mu_0 ni$. *(a)* Determine the energy per unit volume stored in the magnetic field at such a point if $n = 2000$ turns per meter, $i = 0.50$ A. *(b)* Assuming that the energy density is the same throughout the solenoid, which has length 0.25 m and area 800 mm², determine the energy stored in the solenoid. *(c)* Evaluate the self-inductance of the solenoid using Eq. 29-4.

25. A typical solenoid has 1000 turns per meter and a 1-A current. A typical parallel-plate capacitor has a plate separation of 0.1 mm and a potential difference of 10 V. Compare the magnetic energy density in a typical solenoid with the electric energy density in a typical capacitor.

Section 29-4. Mutual Inductance

26. Two coils are arranged such that the mutual inductance is 75 mH. Coil 1 is in a circuit whose current is changing at a rate of 12 A/s. Coil 2 is in an open circuit (no current). Determine the mutually induced emf in *(a)* coil 1 and *(b)* coil 2.

27. A small N-turn coil of area S is placed inside a long solenoid with n turns per unit length. The coil is far from either end of the solenoid and its axis coincides with the solenoid axis. *(a)* Show that the mutual inductance of the pair is $M = \mu_0 nNS$. *(b)* Evaluate the mutual inductance for $N = 75$, $n = 2000\ \text{m}^{-1}$, and $S = 300\ \text{mm}^2$.

28. Solenoid 1 with n_1 turns per unit length, cross-sectional area S_1, and length ℓ_1 lies inside solenoid 2, which is characterized by n_2, S_2, and ℓ_2. The two solenoids have the same axis. *(a)* Determine an expression for the mutual inductance of the pair in terms of the above quantities. *(b)* Determine an expression for the mutual inductance if the axis of the inner solenoid makes an angle θ with the axis of the outer solenoid.

29. Two coils are wrapped around insulating cylinders with senses as shown in Fig. 29-16. *(a)* Just after the switch is closed in circuit 1, what is the sense of the current in resistor R_2? *(b)* If coil 2 were wound with the opposite sense, what would be the answer to part *(a)*? *(c)* What if both coils were wrapped with senses opposite those in the figure?

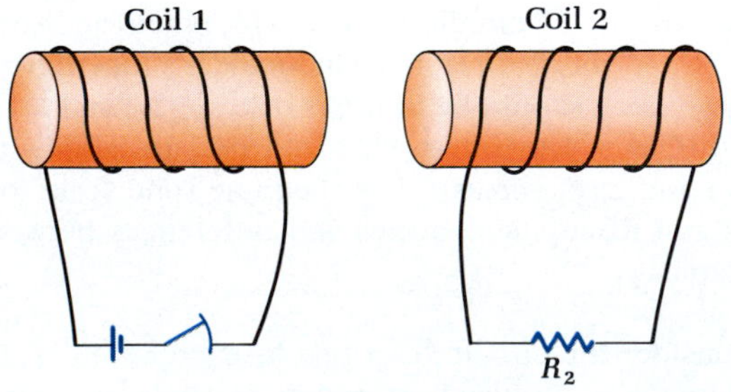

Figure 29-16. Exercise 29.

30. Two circular loops are separated by a distance D that is large compared with their radii R_1 and R_2. The center of one loop lies on the axis of the other and θ represents the angle between their axes, as shown in Fig. 29-17. Show that the mutual inductance of the loops is given approximately by

$$M = \frac{\mu_0 \pi R_1^2 R_2^2}{2D^3} \cos\theta$$

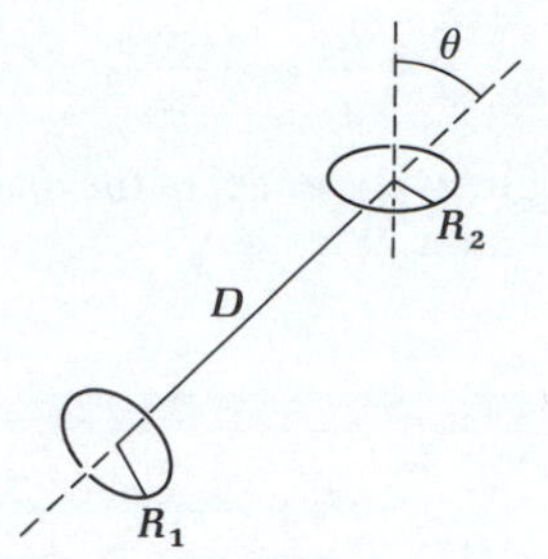

Figure 29-17. Exercise 30.

(*Hint:* The field at a point on the axis of a current loop is considered in Sec. 27-1.)

Section 29-5. Transformers

31. A transformer is used to supply up to 100 A at 240 V to a residence in a neighborhood. The distribution-line voltage is 22 kV. *(a)* What is the ratio of turns in the secondary to turns in the primary? *(b)* If the residence is "drawing" 100 A, what is the minimum primary current? *(c)* At what rate is the transformer supplying energy to the residence?

32. An old vacuum-tube radio plugs into a 115-V wall outlet. For proper operation, the filament in each tube requires 6.3 V, which is supplied by a transformer. *(a)* If the secondary has 28 turns, how many turns are on the primary? *(b)* If the primary current is 60 mA and the secondary provides current for five tube filaments wired in parallel, what is the average current in each filament?

33. The 13-kV output of a 2.0-MW generator is stepped up to 140 kV for transmission to an industrial user 40 km away. *(a)* Determine the transmission-line current. *(b)* Determine the Joule heating loss rate if the transmission line has resistance 20 Ω. *(c)* What percentage is this loss rate of the transmitted power? *(d)* Estimate the diameter of the aluminum transmission line.

34. A variable transformer has a sliding contact that changes the number of turns in the secondary coil. The primary has 600 turns and operates at 120 V. The secondary voltage ranges from near zero to 9.0 V, depending on how many turns are engaged. *(a)* How many turns are used for a 9.0-V output? *(b)* What is the lowest nonzero secondary voltage available from this transformer?

35. An oil furnace uses a transformer to provide an arc for igniting the fuel. The secondary voltage must be around 6 kV, while the primary operates at 120 V and 4.6 A. *(a)* What are reasonable values for N_s and N_p for such a transformer? *(b)* What is the secondary current? *(c)* Estimate the resistance of the secondary coil. (Assume that the coil has all of the resistance in the secondary circuit.)

Additional Exercises

36. How much energy is required to establish a uniform magnetic field of magnitude 1.0 T in a volume of 1.0 m³?

37. A 1.2-mH coil is removed from the ignition circuit of a car for testing. The coil is connected in series with a 13.6-V storage battery and a switch is closed at $t = 0$. After several seconds the current in the circuit is steady at 1.60 A. Determine *(a)* the

resistance in the circuit, *(b)* the inductive time constant, and *(c)* the time at which the current is 0.80 A.

38. As in Fig. 29-4, a battery is switched out of a circuit that contains a 1.2-mH coil in series with a resistor. The total resistance of the circuit is 24 Ω. *(a)* Determine the inductive time constant. *(b)* At 75 μs after the battery was switched out of the circuit, the current was 38 mA. What was the value of the current at the instant the battery was switched out?

39. Large superconducting solenoids have been proposed as energy storage devices. Typically, superconductors cannot tolerate fields in excess of 25 T. Estimate the maximum energy that can be stored by a superconducting solenoid of radius 5 m and length 100 m.

40. A long straight wire carries a 150-mA current. What is the magnetic energy density at a point 200 mm from the wire?

41. Determine the magnetic energy stored inside a 1.0-m section of a coaxial cable. The cable's inner conductor has a radius of 1.2 mm and carries a 10-mA current. The inner radius of the cable's outer conductor is 3.1 mm.

PROBLEMS

1. ***Self-inductance of a toroid.*** In an *N*-turn toroid with current *i* and rectangular cross section, the magnetic field varies with distance *R* from the axis (Prob. 2 in Chap. 27):

$$B = \frac{\mu_0 N i}{2\pi R}$$

(a) For the toroid shown in Fig. 29-18, evaluate the magnetic flux for the rectangular cross section of area *ab*. *(b)* Determine the self-inductance of the toroid.

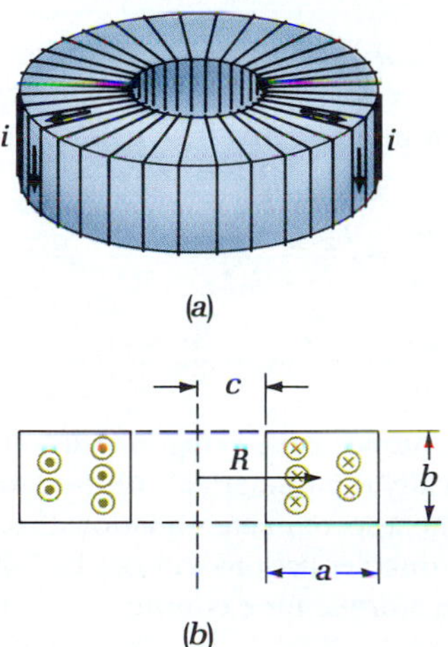

Figure 29-18. Problem 1: *(a)* Toroid in perspective. *(b)* In cross section.

2. ***Magnetic energy inside a wire.*** A long, straight wire of radius *a* carries a current *i* uniformly distributed over its cross section. *(a)* Determine the energy in a length ℓ stored in the magnetic field inside the wire (see Eq. 27-10). *(b)* Apply this result to the coaxial cable of Example 29-5 and determine the inductance per unit length of the cable. Assume that the outer conductor has negligible thickness.

3. ***Self-inductance of a coaxial cable.*** Consider the plane of length ℓ between the conductors of the coaxial cable in Fig. 29-19. The outer conductor has negligible thickness. *(a)* Evaluate the magnetic flux for this plane. *(b)* Apply Eq. 29-1 to the loop that forms the boundary of this plane to estimate the self-inductance per unit length. *(c)* Compare with the result given by Example 29-5.

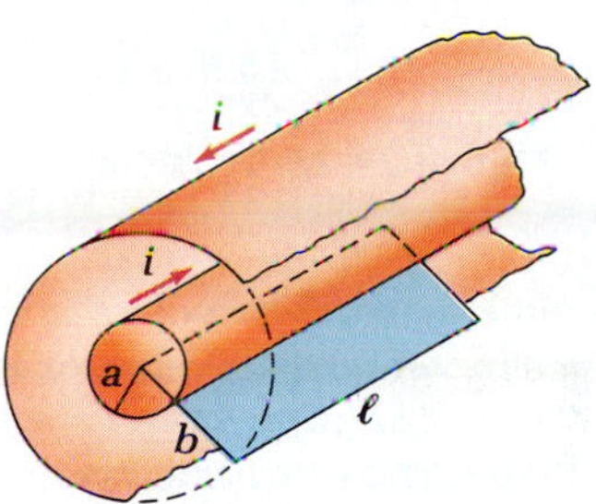

Figure 29-19. Problem 3.

4. ***Mutual inductance between a long, straight wire and a rectangular loop.*** A rectangular loop has an edge parallel to a long, straight wire, as shown in Fig. 29-20. *(a)* Determine the mutual inductance of the system. *(b)* What is the sense of the mutually induced emf in the loop if i_1 is decreasing and i_2 is increasing?

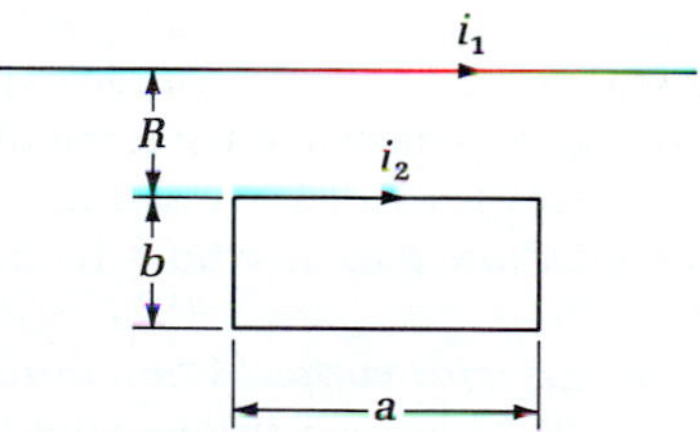

Figure 29-20. Problem 4.

5. ***Mutual inductance between a long, straight wire and a small loop.*** The linear extent of the loop shown in Fig. 29-21 is small compared with its mean distance *R* from the long, straight wire. *(a)* Show that the mutual inductance is approximately given by $M = \mu_0 S/(2\pi R)$, where *S* is the area of the loop whose plane lies in the plane of the figure. *(b)* Use the approximation $\ln(1 + x) \approx x$, valid for $|x| \ll 1$, to show that the answer to part *(a)* is consistent with the answer to the previous problem for the case $b \ll R$. *(c)* How would the expression in part *(a)* change if the plane of the loop made angle θ with the plane of the figure?

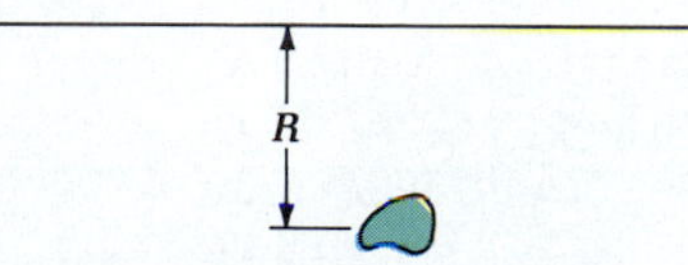

Figure 29-21. Problem 5.

6. ***An LR circuit.*** The current in the *LR* circuit of Fig. 29-22 is zero at $t = 0$ when the switch first closes at position *A*. The switch remains at position *A* for 5.0 s and then is quickly changed to position *B* for the next 5.0 s. *(a)* Determine the current in the circuit at $t = 5.0$ s just before the switch is changed to *B*. *(b)* Determine the current at $t = 10$ s. *(c)* If the switch is changed back to *A* at this instant, determine the current at $t = 15$ s and explain why this answer is different from the answer to part *(a)*.

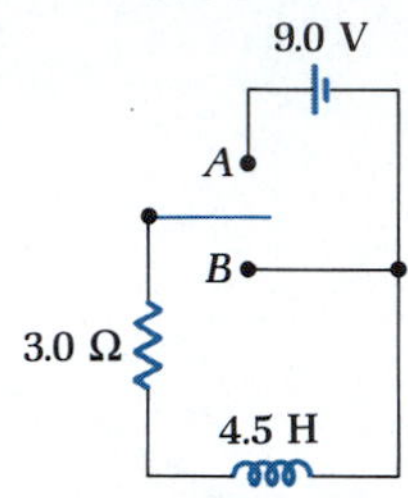

Figure 29-22. Problem 6.

7. ***Resistance and self-inductance of a solenoid.*** A solenoid is constructed by winding 200 turns of 1.0-mm-diameter copper ($\rho = 1.7 \times 10^{-8}\ \Omega \cdot \text{m}$) wire in a single layer on a 25-mm-diameter cylinder. The wire has a very thin insulating coat so that the consecutive turns, which touch, are insulated from each other. *(a)* What is the length of the solenoid? *(b)* What length of wire is required? Determine *(c)* the resistance and *(d)* the self-inductance of the solenoid.

8. ***Magnetic energy in a cylindrical volume with a long, straight wire along its axis.*** A long, straight wire of radius *a* carries a current *i* uniformly distributed over its cross section. Consider a region of space, a cylinder of length ℓ and radius *R* whose axis coincides with the axis of the wire and which is far from the ends of the wire. *(a)* Determine an expression for the magnetic energy stored in this region in terms of μ_0, *i*, *a*, ℓ, and *R*. Separately consider the cases $R < a$ and $R > a$. *(b)* Explain why you should not allow *R* to be arbitrarily large.

9. ***Circuits coupled with mutual inductance.*** The two circuits shown in Fig. 29-23 interact through their mutual inductance *M*. *(a)* Show that the loop rule applied to each circuit gives

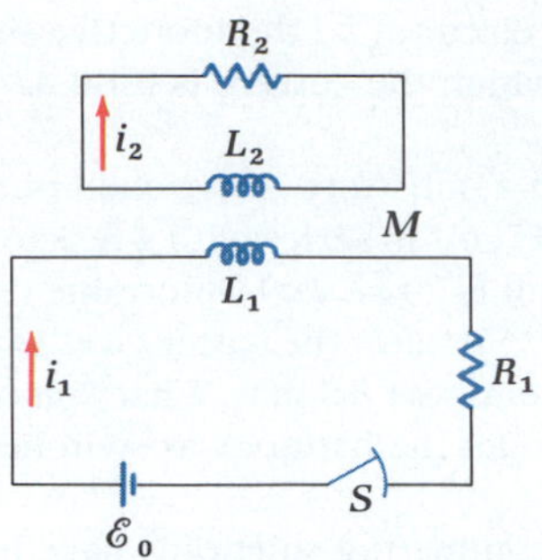

Figure 29-23. Problem 9.

$$L_1 \frac{di_1}{dt} + M \frac{di_2}{dt} + R_1 i_1 = \mathscr{E}_0$$

$$L_2 \frac{di_2}{dt} + M \frac{di_1}{dt} + R_2 i_2 = 0$$

(b) If L_1, L_2, and *M* are all comparable with $R_2 \gg R_1$, then di_2/dt may be neglected in comparison with di_1/dt. Solve the equations in this approximation. Use the initial conditions $i_1(0) = i_2(0) = 0$. *(c)* Construct graphs of $i_1(t)$ and $i_2(t)$ versus *t* for $0 \leq t \leq 1.0$ s. Use the values $L_1 = 1.2$ H, $R_1 = 6.0\ \Omega$, $M = 0.80$ H, $L_2 = 1.4$ H, $R_2 = 600\ \Omega$, $\mathscr{E}_0 = 48$ V. *(d)* Compare the values of the maximum potential difference across R_1 and across R_2.

10. ***Numerically solving coupled equations.*** Write or adapt a spreadsheet to solve the coupled equations in the previous problem numerically. The equations can be solved for the derivatives:

$$\frac{di_1}{dt} = \frac{L_2 \mathscr{E}_0 - L_2 R_1 i_1 + M R_2 i_2}{L_1 L_2 - M^2}$$

$$\frac{di_2}{dt} = -\frac{R_2}{L_2} i_2 - \frac{M}{L_2} \frac{di_1}{dt}$$

and these can be numerically integrated in turn for i_1 and i_2. Compare graphically the numerical solution for i_1 and i_2 with the solution of the approximate equations from the previous problem. The coupled equations can also be solved exactly, by using Laplace transforms, for example.

MAGNETIC FIELDS IN MATTER 30

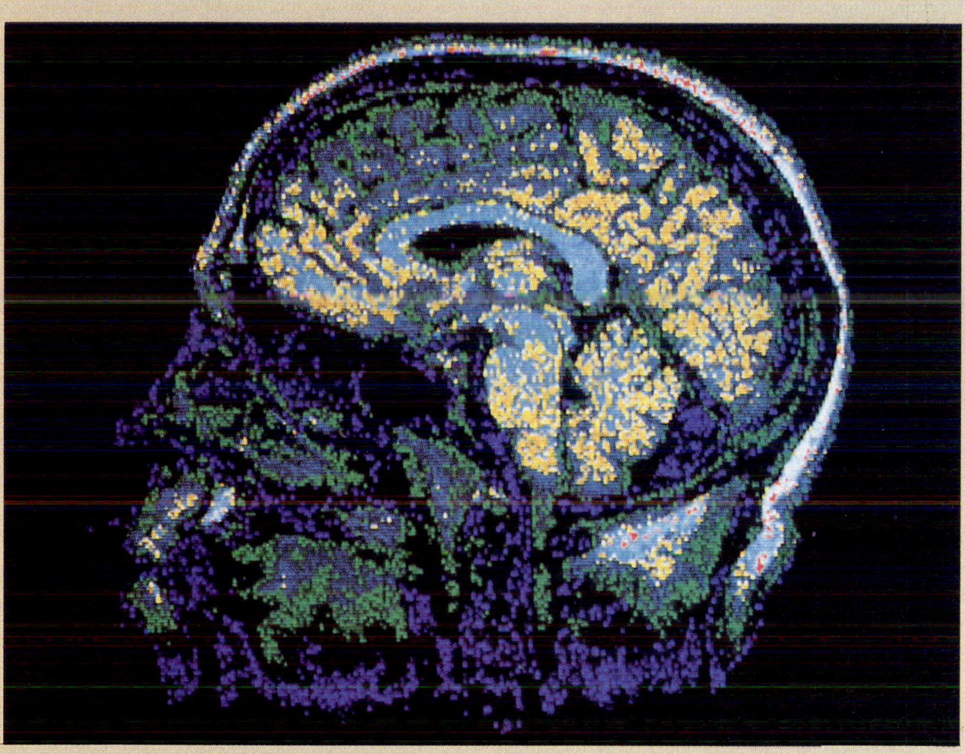

A nuclear magnetic resonance (NMR) image of a cross section of a person's head. With NMR, different chemical elements can be distinguished from one another by the magnetic dipole moments of their nuclei. This information is used to produce a false color image.

Small bits (filings) of iron scattered in the vicinity of an ordinary bar magnet are strongly attracted to the magnet. Sawdust particles, on the other hand, have a much weaker interaction with the magnetic field of the magnet. If the magnet is passed over a mixture of sawdust and iron filings, the filings are pulled from the mixture, leaving the sawdust behind. Why do wood and iron have such different magnetic properties? And what causes the magnetic field of the permanent magnet?

We have approached our study of magnetism chiefly in terms of electric currents — currents as sources of magnetic fields, current-carrying conductors on which the magnetic field exerts forces and torques, currents induced by a changing magnetic flux. The magnetic properties of matter are also described in terms of currents. In many materials these currents are due to the motion of electrons at the atomic level. The effect of these microscopic currents in various materials can range from the almost insignificant interaction of wood with the magnet to the strong attraction of iron filings to the magnet. Such currents are also responsible for the magnetic field of the magnet. In this way the description of magnetic phenomena is ultimately expressed in terms of currents.

The aurora borealis as seen from space.

30-1 ATOMIC CURRENTS, MAGNETIC DIPOLES, AND MAGNETIZATION

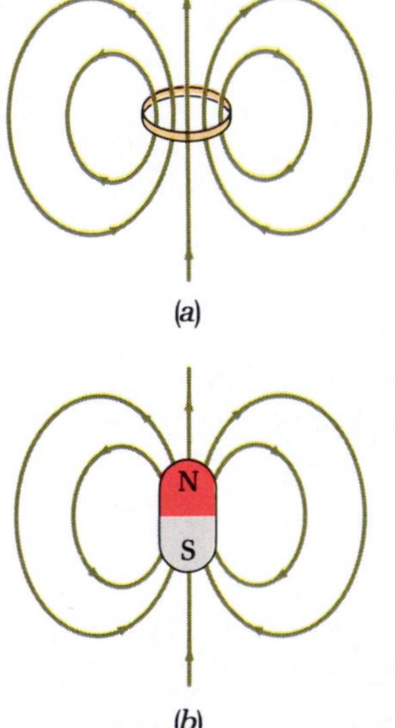

Figure 30-1. A magnetic dipole field exists at points distant from *(a)* a current loop and *(b)* a bar magnet.

In earlier chapters we pointed out the similarity between a permanent bar magnet and a localized current distribution such as in a loop or a coil. Each has features that depend on its magnetic dipole moment **m**. The similar magnetic fields produced by a current loop (Chap. 27) and by a bar magnet are sketched in Fig. 30-1. Further, the torque $\boldsymbol{\tau}$ on a magnetic dipole placed in a uniform field **B** (see Fig. 26-9) is given by Eq. 26-8, $\boldsymbol{\tau} = \mathbf{m} \times \mathbf{B}$. The similarity in magnetic behavior for localized currents and magnets is not coincidental. Indeed, our description of magnetism in matter is based on a picture of currents (and hence magnetic dipole moments) at the molecular level.

Currents and Moments

In a classical model of an atom, negatively charged electrons circulate in orbits about the nucleus. Such circulating charges constitute a localized current distribution that contributes to the atom's magnetic dipole moment. Of course, we do not claim validity for any classical model of an atom. The electronic structure of an atom must be described using quantum theory.* But some of the results from a simple classical theory are the same as those obtained from the quantum theory. One such result is the connection between the orbital contribution to the magnetic moment and the orbital angular momentum **L** of an electron in an atom.

Consider an electron of charge $-e$ and mass m_e in a circular orbit with speed v and radius r about a fixed nucleus, as seen in Fig. 30-2. The orbital period T is the time interval for the electron to travel a distance $2\pi r = vT$ for each revolution. The average electric current for the orbital motion corresponds to the electronic charge passing a section (such as PP' in the figure) in a time interval T; that is, $i = e/T = e/(2\pi r/v) = ev/(2\pi r)$. Notice that the sense of this current is opposite the sense of the orbital motion because the electron bears a negative charge. The magnetic moment of a loop or a coil is given by Eq. 26-9, $\mathbf{m} = i\mathbf{S}$. For a circular

* In the "extended" edition of this text, electronic structure of atoms is discussed in Chap. 41.

current loop with area $S = \pi r^2$ and current i, the magnitude of the magnetic moment is $m = i\pi r^2$. Substituting $i = ev/(2\pi r)$ from above gives

$$m = \tfrac{1}{2}evr$$

The direction of **m** is perpendicular to the plane of the loop. Applying the right-hand rule relative to the sense of the current in Fig. 30-2, we see that **m** is directed into the plane of the figure.

The orbital angular momentum of the electron, $\mathbf{L} = \mathbf{r} \times \mathbf{p} = \mathbf{r} \times (m_e\mathbf{v})$, has magnitude

$$L = m_e vr$$

Its direction in Fig. 30-2 is perpendicularly out of the plane of the figure, opposite the direction of the magnetic moment **m**. Since the expressions for both m and L contain the product vr, we can eliminate this factor. Thus $m = \frac{1}{2}evr = \frac{1}{2}e(L/m_e)$. Noting that **m** and **L** have opposite directions, we write the relation between these two vectors as

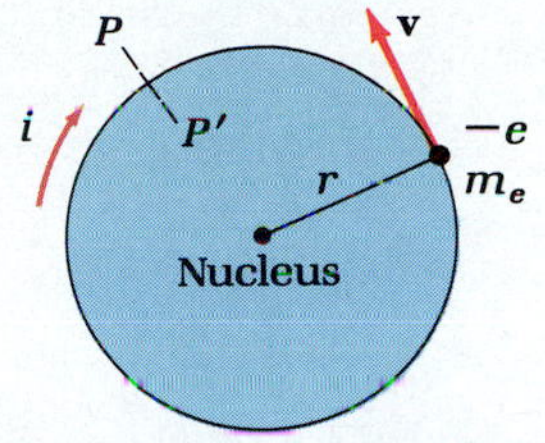

Figure 30-2. An electron orbiting counterclockwise gives a clockwise current. The orbital angular momentum **L** is out of the plane of the figure. The magnetic dipole moment **m** is into the plane of the figure, opposite **L**.

Magnetic moment and orbital angular momentum

$$\mathbf{m} = -\frac{e}{2m_e}\mathbf{L} \qquad (30\text{-}1)$$

This result is valid in general, not just for circular orbits. That is, the orbital contribution of an electron to the magnetic moment is proportional to the orbital angular momentum of the electron. The constant of proportionality $e/2m_e$ depends only on the electronic charge and mass. The two vectors have opposite directions because of the negative charge on the electron.

In addition to an orbital contribution to the magnetic moment, there is a contribution to the magnetic moment of each electron due to its *intrinsic,* or *spin,* angular momentum **S**.* The spin contribution to **m** is proportional to **S**, with a proportionality constant that is approximately twice that for the orbital case. It is given by an expression similar to Eq. 30-1, with $e/2m_e$ and **L** replaced by $2(e/2m_e)$ and **S**:

Magnetic moment and spin angular momentum

$$\mathbf{m} = -\frac{e}{m_e}\mathbf{S} \qquad (30\text{-}2)$$

Together, Eqs. 30-1 and 30-2 give the contribution of an electron to the magnetic moment of an atom. The magnetic moment for an atom (or a molecule) is obtained by adding the contributions of all the electrons. (Some nuclei have magnetic moments, but the values are negligible compared with the electronic contribution.) Many types of molecules have zero magnetic moment if no external field is imposed. We can think of the electronic contributions as canceling because of the different directions of the angular momentum vectors of individual electrons. In the case of the spin angular momentum, most electrons in a molecule pair off with opposite spins, so that the pair gives no net spin contribution to the magnetic moment. For molecules in which the pairing is incomplete, the magnetic moment is due to the few (usually only one) unpaired electrons. These molecules have permanent magnetic moments.

Permanent magnetic moment due to unpaired electrons

Magnetization

So far we have restricted our discussion to isolated atoms or molecules and their magnetic dipole moments. Now consider a large collection of molecules that composes a macroscopic object. At the macroscopic level we deal with quantities that involve averages over many molecules. A useful quantity that is related to an average magnetic dipole moment for many molecules is the *magnetization* **M**.

Consider a volume element ΔV in a material. We assume that ΔV is small on a

* In the "extended" edition of this text, electron spin is discussed in Chap. 41.

macroscopic scale but is large enough to contain a large number of molecules. If $\mathbf{m}_i$ represents the magnetic moment of a molecule labeled by i in the volume element, then the net average magnetic moment for this volume is $\langle \Sigma \, \mathbf{m}_i \rangle$, where the vector sum is over all molecules in the element.

The ***magnetization*** is defined as the magnetic dipole moment per unit volume in the medium:

Magnetization defined

$$\mathbf{M} = \frac{\langle \Sigma \, \mathbf{m}_i \rangle}{\Delta V} \tag{30-3}$$

Thus if the magnetization is known at points in a medium, then the magnetic moment $\mathbf{m}$ of a region in the medium with volume ΔV is $\mathbf{m} = \boldsymbol{M} \, \Delta V$. Notice that the magnetization is a vector quantity. The SI unit of magnetization is amperes per meter (A/m).

The magnetization describes the magnetic state of a medium or a material. For example, if $\mathbf{M} = 0$ everywhere in a medium, then no part of the medium has a magnetic dipole moment. In a magnetized piece of steel on the other hand, the magnitude of the magnetization is large throughout the sample. It is observed to change, for example, if an external magnetic field is imposed or if the temperature is changed. Various materials respond in different ways to changes in their surroundings. Most materials fall into one of three categories of magnetic behavior. We shall discuss *diamagnetism, paramagnetism,* and *ferromagnetism* in the following sections.

EXAMPLE 30-1

Magnetization in a gas. A typical molecule with one unpaired electron has a permanent magnetic moment with a magnitude of around $m_i = 1 \times 10^{-23} \, \text{A} \cdot \text{m}^2$. Suppose that an external magnetic field is applied to a gas of these molecules at temperature $T = 273$ K and pressure $p = 1.01 \times 10^5$ Pa. Assume that the component of $\mathbf{m}_i$ along the direction of the magnetic field averages to 1 percent of m for a molecule. *(a)* Determine the magnetization at a point in this gas, assumed to be ideal. *(b)* Determine the magnetic dipole moment for a 1-mm³ region in the gas.

Solution. *(a)* Since all molecules are the same, the sum over molecules in Eq. 30-3 is $\Delta N \langle \mathbf{m}_i \rangle$, where ΔN is the number of molecules in a volume ΔV. The ideal gas equation of state, $p \, \Delta V = \Delta N \, kT$, where k is the Boltzmann constant, connects ΔN and ΔV. Taking the z axis along the external field direction, we have $\langle m_{iz} \rangle = 0.01 m_i = 1 \times 10^{-25} \, \text{A} \cdot \text{m}^2$, and for the z component of Eq. 30-3,

$$M_z = \frac{\Delta N \langle m_{iz} \rangle}{\Delta V} = \frac{p \langle m_{iz} \rangle}{kT}$$

$$= \frac{(1.01 \times 10^5 \text{ Pa})(1 \times 10^{-25} \text{ A} \cdot \text{m}^2)}{(1.38 \times 10^{-23} \text{ J} \cdot \text{K}^{-1})(273 \text{ K})} = 3 \text{ A/m}$$

That is, $\mathbf{M}$ is along the z axis, with magnitude $M = 3$ A/m.

(b) Since the magnetization is the magnetic moment per unit volume, the magnetic moment of the 1-mm³ region is $\mathbf{m} = \mathbf{M} \, \Delta V$. The magnetic moment has only a z component, which is

$$m_z = M_z \, \Delta V = (3 \text{ A/m})(10^{-9} \text{ m}^3) = 3 \times 10^{-9} \text{ A} \cdot \text{m}^2$$

SELF-TEST 30-1. Determine the magnitude of the magnetization in a gas where each molecule has a magnetic moment of $1 \times 10^{-23} \, \text{A} \cdot \text{m}^2$ and the component of $\mathbf{m}_i$ along the direction of an external magnetic field averages to 2 percent of m for a molecule. The gas has a pressure of 202 kPa and its temperature is 295 K. ***ANSWER:*** 10 A/m.

30-2 DIAMAGNETISM

There are many materials whose individual molecules have no magnetic moment because of their electronic structure. Even if the molecules form a relatively dense liquid or solid, the magnetization **M** of most materials is zero if no external magnetic field is present. That is, even if the molecules are close together, the electrons remain paired so that the substance has no net magnetic moment. However, in the presence of an external magnetic field, molecules do have a small magnetic moment. These molecular magnetic moments are *induced* by the external field. The direction of the induced magnetic moment is opposite the magnetic field direction so that the magnetization of the material is also opposite the magnetic field direction. Such materials are called ***diamagnetic***.

M and B are opposite in diamagnetic materials.

A diamagnetic material can be distinguished from a paramagnetic material (discussed in the next section) by its behavior in a *nonuniform* magnetic field. To understand the effect, consider a magnetic dipole in a nonuniform magnetic field. Recall that **m** for a bar magnet is directed from its south pole to its north pole. The nonuniform field exerts forces of different magnitudes on the two poles. (See Prob. 3.) Thus the dipole experiences a net force in the nonuniform field. Figure 30-3 shows a magnetic dipole in a nonuniform field near the end of a current-carrying solenoid. The direction of the dipole moment in Fig. 30-3*a* is parallel to the magnetic field at points on the axis of the solenoid. In this case the nonuniform field attracts the dipole toward the strong field region, that is, toward the solenoid. With the direction of the dipole moment opposite the direction of the field, as shown in Fig. 30-3*b*, the dipole is repelled from the strong field region, away from the solenoid. Thus the direction of the force depends on the direction of the dipole moment.

A nonuniform magnetic field exerts a net force on a magnetic dipole.

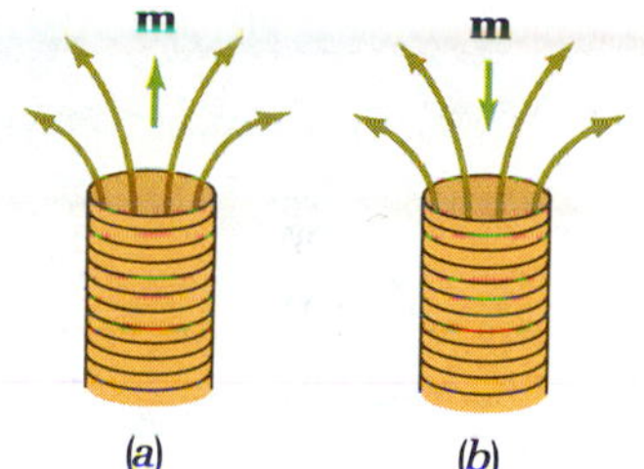

Figure 30-3. *(a)* The magnetic dipole moment **m** of a small magnet is parallel to **B** near the mouth of a solenoid. The nonuniform field attracts the dipole to the stronger field region because the attractive force on the south pole of the magnet has a larger magnitude than the repulsive force on the north pole. *(b)* The magnetic dipole moment **m** is opposite **B**. The nonuniform field repels the dipole from the stronger field region.

Suppose a small, needle-shaped sample of diamagnetic material is placed in a nonuniform magnetic field. The direction of the magnetization in a diamagnetic sample is opposite the magnetic field, and the arrangement is that of Fig. 30-3*b*. The diamagnetic sample is repelled (but very weakly) from the strong field region. (This effect was discovered by Faraday in 1845, and he termed the behavior "diamagnetic.") The magnitude of the repelling force is so small that the interaction of a diamagnetic material with a magnetic field is difficult to observe. You have probably never seen the effect.

The relation between the magnetization **M** in a diamagnetic material and the field **B** can be determined by quantitative measurement. Such measurements usually show that the magnitude of the magnetization is proportional to the magnitude of the magnetic field. That is, the magnetization induced by the magnetic field depends linearly on that field and, for this reason, the material is said to be *linear*.

In isotropic diamagnetic materials, **M** and **B** have opposite directions because the induced magnetic dipole moment is directed opposite **B**, even at the atomic level. While quantum theory must be used to treat the electronic structure of atoms and molecules, the diamagnetic behavior of an atom in an external magnetic field is suggested by applying Faraday's law and Lenz's law to a simple classical model.

Consider an electron orbiting a nucleus, as shown in Fig. 30-4. The orbital contribution to the magnetic moment is opposite the orbital angular momentum and is given by Eq. 30-1, $\mathbf{m}_1 = -e\mathbf{L}_1/2m_e$. An orbit with the opposite orientation is shown in Fig. 30-4*b*, $\mathbf{m}_2 = -e\mathbf{L}_2/2m_e$. These two parts of the figure are shown separately for clarity. You should picture these orbits as having the same center. Before a magnetic field is applied, these orbits are paired if $\mathbf{L}_2 = -\mathbf{L}_1$ so that $\mathbf{m}_1 + \mathbf{m}_2 = 0$.

As a uniform magnetic field is applied in the direction indicated in the figure, an induced electric field (see Faraday's law in Eq. 28-7) changes the orbital speed or frequency differently for the two orbits. To apply Lenz's law, remember that the sense of the current is opposite the sense of the motion of an electron because the charge is negative. In Fig. 30-4*a*, the current increases to oppose the changing magnetic flux. Therefore the angular frequency increases for this orbit. As a conse-

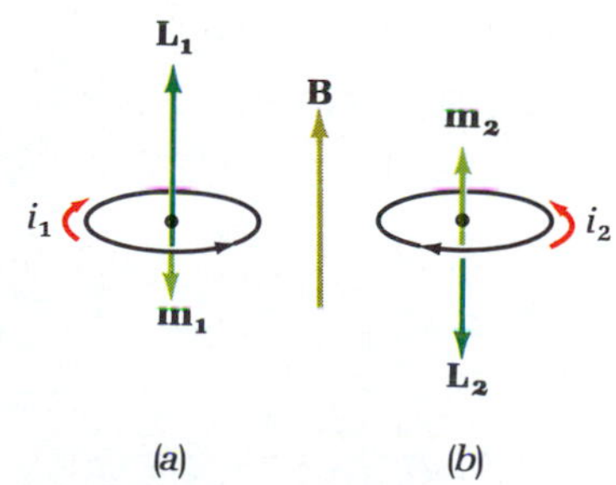

Figure 30-4. If a magnetic field is applied, the magnetic moments of paired electrons do not cancel. Since $m_1 > m_2$ for the case shown, a diamagnetic effect results.

quence, the magnitudes of both $\mathbf{L}_1$ and $\mathbf{m}_1$ increase. Note that $\mathbf{m}_1$ and $\mathbf{B}$ have opposite directions. The same reasoning applied to the orbit in Fig. 30-4*b* shows that the magnitudes of $\mathbf{L}_2$ and $\mathbf{m}_2$ decrease. Now we see that $\mathbf{m}_1$ and $\mathbf{m}_2$ do not completely cancel if a magnetic field is applied. Since $\mathbf{m}_1$ has the larger magnitude, the vector sum $\mathbf{m}_1 + \mathbf{m}_2$ is directed opposite the magnetic field. Thus the direction of the induced magnetic moment is opposite **B**, and this is the diamagnetic effect.

The model used above to illustrate diamagnetic behavior is exceedingly simplistic. A more realistic treatment shows that (i) the induced magnetic moment of such an atom is directed opposite the applied field, (ii) the magnetic moment is usually proportional to the field (leading to a linear material), (iii) the effect is very small, and (iv) diamagnetic properties of materials are essentially independent of temperature.

We have associated diamagnetism with the response of atoms or molecules with paired electrons to an applied magnetic field. (In a substance such as a metal, the free electrons also contribute to the magnetic properties.) It turns out that all molecules possess diamagnetic behavior. However, diamagnetism is completely masked in materials that exhibit paramagnetism or ferromagnetism.

30-3 PARAMAGNETISM

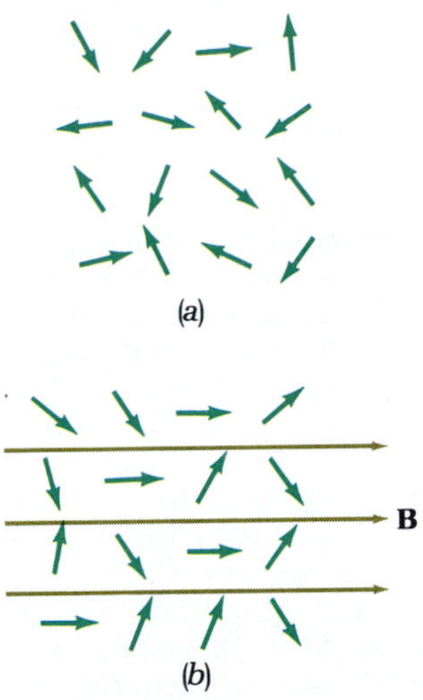

Figure 30-5. *(a)* Magnetic moments of molecules are randomly oriented, and the magnetization $\mathbf{M} = 0$. *(b)* Magnetic moments tend to align with the applied field so that **M** is parallel to **B**.

Molecules with one or more unpaired electrons—Al, O_2, and Fe are examples—possess a permanent magnetic moment. In many materials containing such molecules, the molecular magnetic moments are oriented randomly if no magnetic field is applied (Fig. 30-5*a*). The magnetization, the magnetic moment per unit volume, is zero in this situation because it involves a sum over many molecules.

When a magnetic field is applied to the material, the potential energy of a magnetic dipole is lowered if it can change its orientation and align with the field. To see this, recall that the potential energy of a magnetic dipole in a uniform magnetic field is given by Eq. 26-10, $U = -\mathbf{m} \cdot \mathbf{B}$. This energy is lowest ($U = -mB$) if **m** and **B** are aligned and highest ($U = +mB$) if **m** and **B** are oppositely directed. Thus the potential energy is lowered if the dipole tends toward alignment with the field. (Notice that we are discussing the orientation of the *permanent* magnetic moment of a molecule and not the induced magnetic moment considered in the previous section.)

Opposing the tendency to alignment with the magnetic field are the randomizing thermal motions of the molecules. The average, or overall, alignment is a balance between these influences, as shown schematically in Fig. 30-5*b*. There is a partial alignment of the molecular magnetic moments with the field, and as a consequence, for isotropic materials the magnetization of the material is parallel to the field. If the applied field is removed, the randomness in orientation returns and the magnetization is again zero. Materials with these properties are called ***paramagnetic***.

Diamagnetic and paramagnetic materials behave differently in a nonuniform magnetic field, as previously mentioned. If a needle-shaped sample of a paramagnetic substance is oriented along the axis of a solenoid near one end, the magnetization is parallel to the magnetic field. This configuration is shown in Fig. 30-3*a*, and the paramagnetic needle is weakly attracted to the stronger field region. This behavior is just opposite that of a diamagnetic needle, which is repelled from the strong field region.

Paramagnetism is due to the partial alignment of the permanent magnetic moments with the applied magnetic field. The aligning tendency should increase with an increase in the magnitude of the magnetic field. On the other hand, the randomizing effect of thermal motions should increase with an increase in temperature. These dependencies were first observed by Pierre Curie (1859–1906) and are summarized in *Curie's law,* which relates the magnetization **M** of an isotropic paramagnetic substance with the applied magnetic field **B** and the Kelvin temperature T:

$$\mathbf{M} = \frac{C\mathbf{B}}{\mu_0 T} \tag{30-4}$$

Curie's law

The constant C, called *Curie's constant,* is characteristic of the material and depends on the molecular magnetic moment. The permeability constant μ_0 appears in Eq. 30-4, so Curie's constant has dimensions of temperature. Curie's law is valid except for large magnetic fields and/or low temperatures. It shows that, at a given temperature, **M** and **B** are proportional and the paramagnetic material is linear. Further, the magnetization decreases with increasing temperature. For very large magnetic fields or very low temperatures, the magnetization *saturates* (approaches an ultimate limit) as all molecular magnetic moments approach alignment with the field. Paramagnetic materials are not linear under these extreme circumstances.

Paramagnetic materials are linear.

Some materials, notably the metals, have free electrons that are not bound to any particular atom or molecule. These electrons also contribute to the magnetization of the substance through the magnetic moment associated with their angular momentum. This contribution is typically small and has essentially no dependence on temperature. No classical model can describe the effect adequately, and we shall not discuss it further.

EXAMPLE 30-2

Curie's constant for a crystal. A small amount of doubly ionized manganese (Mn^{++}) is distributed uniformly throughout a crystal of NaCl so that the sample is isotropic and paramagnetic. The magnitude of the magnetization is 6.1 A/m at 310 K in a magnetic field of magnitude 0.87 T. Determine Curie's constant for this sample.

Solution. Since **M** and **B** are parallel in the isotropic paramagnetic material, we take the magnitude of Eq. 30-4 and solve for Curie's constant C:

$$C = \frac{\mu_0 MT}{B}$$

$$= \frac{(4\pi \times 10^{-7}\ \mathrm{T\cdot m\cdot A^{-1}})(6.1\ \mathrm{A/m})(310\ \mathrm{K})}{0.87\ \mathrm{T}}$$

$$= 2.7 \times 10^{-3}\ \mathrm{K}$$

SELF-TEST 30-2. What is the magnitude of the magnetization of the sample in the above example when it is placed in a magnetic field of magnitude 0.45 T and the temperature of the crystal is 280 K? ***ANSWER:*** 3.5 A/m.

30-4 FERROMAGNETISM

For both diamagnetic and paramagnetic materials, the magnetization is nonzero only if an applied magnetic field is present. If the applied field is reduced to zero, the magnetization also becomes zero. There are some substances for which the magnetization persists after the applied field is removed. In ***ferromagnetic materials*** all of the molecular magnetic moments tend to align spontaneously in the same direction. Iron (Fe) is a prime example of a ferromagnetic substance, as the prefix "ferro" suggests. Other elemental solids such as Co, Ni, Gd, and Dy, as well as alloys and compounds containing some of these elements, exhibit ferromagnetic behavior.

Magnetic moments align spontaneously in a ferromagnetic material.

A permanent magnet is made of ferromagnetic material. Even in the absence of an applied magnetic field, the magnetization is nonzero inside the magnet. There is, of course, a magnetic field produced by the magnet itself. The magnetization and the magnetic field of the permanent magnet are due to the alignment of magnetic dipoles.

In paramagnetic materials, an applied magnetic field provides the dominant influence toward partially aligning the magnetic dipoles. A different mechanism is responsible for ferromagnetic behavior. This mechanism involves a quantum phenomenon called *exchange coupling* between neighboring atoms or molecules that cannot be described in classical terms. The effect, however, is quite simple: Large numbers of magnetic dipoles cooperate by having their magnetic moments aligned together. That is, the interaction energy of a particular magnetic moment with its nearby neighbors is lower if they all have the same orientation. The alignment is typically much more complete than is achievable in a paramagnetic substance. Consequently, the magnetization in a ferromagnetic material can be large. In an isolated, needle-shaped carbon-steel magnet, for example, the magnitude of the magnetization can be around $M = 8 \times 10^5$ A/m.

Countering the cooperative alignment of the dipole moments in a ferromagnetic substance is the tendency toward random orientations that increases with increasing temperature. At temperatures above a critical temperature characteristic of the material, the ferromagnetic state is unstable. The critical temperature for iron is 1043 K. At higher temperatures, the magnetic dipole moments are not spontaneously aligned, and iron is not ferromagnetic but is paramagnetic.

Magnetic Domains

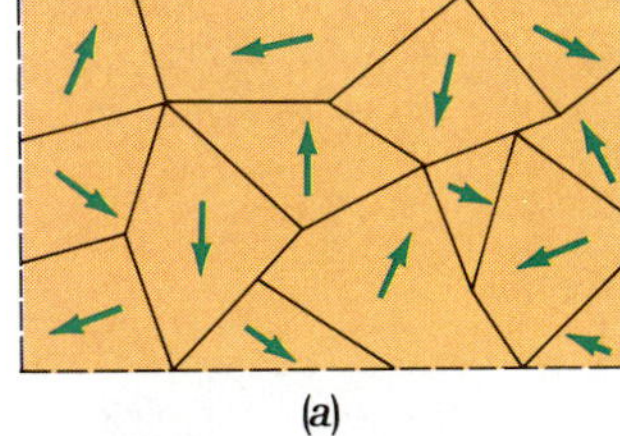

(a)

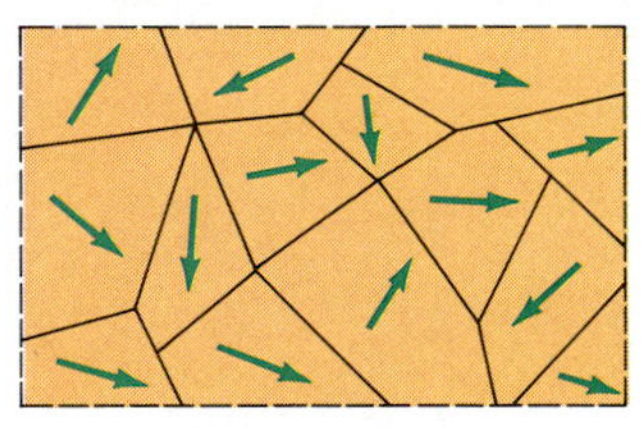

(b)

Figure 30-6. Magnetic domains are shown schematically. *(a)* The domains are randomly oriented in unmagnetized material. *(b)* After a magnetic field is applied, the domains are preferentially oriented.

Although we think if a magnet as having a permanent magnetization, the magnetization can be changed. For example, a common sewing needle will typically be "unmagnetized." But after being exposed to a strong magnetic field, it will attract small bits of iron. The needle has been "magnetized." There are two essential points to consider in this process: (i) How can the needle be unmagnetized initially if the magnetic dipoles are aligned in ferromagnetic materials? (ii) How did applying a magnetic field magnetize the needle?

To understand these features, we must recognize that all of the magnetic dipoles in a ferromagnetic solid may not be aligned in a single direction. Rather, the sample typically consists of a large number of regions, with the magnetic dipoles aligned differently in each region. These regions are called ***magnetic domains***. In a given domain the magnetic dipoles are aligned in a particular direction, which is the direction of the magnetization in this domain. In an adjacent domain the magnetization has a different direction, and the boundary between these domains is called a ***domain wall***. The domain structure is shown schematically in Fig. 30-6*a* for a portion of an unmagnetized sample. In this case the magnetic domains have randomly distributed directions of magnetization. For the whole sample, the average magnetization is nearly zero.

If a magnetic field is applied, then a net magnetization results, as sketched in Fig. 30-6*b*. In some domains the direction of the magnetization switches toward closer alignment with the applied field. In some materials the size of a domain may change because of the motion of domain walls. That is, the direction of the magnetization in some of the domains is close to the applied field direction, and some of these domains can grow at the expense of neighboring domains that are not so closely aligned. These changes are usually irreversible in that some of the domains retain a preferential orientation after the applied field is removed. In this way a net permanent magnetization results; the sample becomes a permanent magnet.

Soft and hard ferromagnetic materials

The "permanent" magnetization of typical ferromagnetic substances is not really permanent. The domains tend to relax back toward the unmagnetized state. Materials are classified as magnetically "soft" or "hard" according to the time required for a significant relaxation. For example, a common iron nail is magnetically soft. It can be magnetized in an applied field, but the magnetization becomes small almost immediately after the applied field is removed. On the other hand, many different types of steel and other alloys and compounds are magnetically hard. The magnetization can persist with little change for years. This "permanence" is important for reliability in a magnetic data-storage medium such as a cassette tape or a floppy disk.

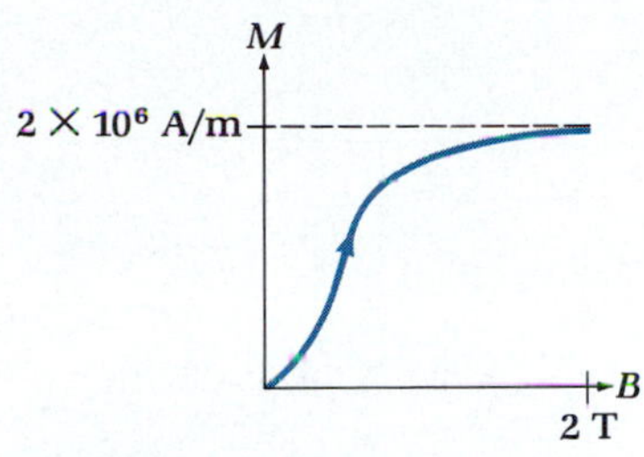

Figure 30-7. In a ferromagnetic material such as Fe, **M** and **B** are not proportional. The magnetization **M** saturates as the magnitude of **B** increases.

The magnetization can be large in a ferromagnetic material if many of the magnetic domains have nearly the same orientation. The magnetic field produced by the oriented domains is correspondingly large. However, the magnetization is not proportional to the magnetic field in ferromagnetic materials. There is an upper limit on the magnetization that corresponds to all magnetic domains having the same alignment. The magnetization approaches this saturation value as the applied field increases. The magnitude of the saturation magnetization for Fe is around $M = 2 \times 10^6$ A/m, corresponding to $B = 2.2$ T in the material. Figure 30-7 shows the nonlinear saturation effect. Another nonlinearity occurs because of the irreversibility of magnetic domain changes. Thus the magnetization at some instant depends not only on the present value of B but also on the previous treatment of the sample. No linear equation connecting **M** and **B** can describe this relationship.

30-5 MAGNETIC INTENSITY H

In previous chapters we assumed that **B** was due solely to a known macroscopic current distribution. We neglected the effect of nearby materials in obtaining expressions for the magnetic field due to these currents. In this chapter we have seen that the magnetic field in a material medium can have two types of contributions. One of these is the contribution due to known macroscopic currents, such as in the windings of a solenoid or a toroid. In some cases we think of this field contribution as an applied field. The other contribution to **B** is from the medium. We describe the effect in terms of the magnetization **M** in the substance. The current in a coil can usually be adjusted, but the magnetization in a sample both depends on and contributes to **B**. Therefore it is not always easy to determine or control **B**, particularly for ferromagnetic materials. In a ferromagnetic substance, **M** and **B** depend on the previous treatment of the sample.

To determine **B** and **M**, it is often convenient to introduce another field. This vector field is represented by the symbol **H** and is called the ***magnetic intensity.*** It is defined by the expression $\mathbf{H} = \mathbf{B}/\mu_0 - \mathbf{M}$, or equivalently,

$$\mathbf{B} = \mu_0(\mathbf{H} + \mathbf{M}) \qquad (30\text{-}5)$$

Definition of magnetic intensity **H**

Notice that **H** and **M** have the same dimensions; the SI unit for **H** is amperes per meter (A/m).

According to Eq. 30-5, **H** and **M** (when multiplied by μ_0) are the two contributions to **B**. (We take **B** as the fundamental field that exerts a force on a moving charged particle: $\mathbf{F} = q\mathbf{v} \times \mathbf{B}$.) Consider these contributions for the simple case of the medium inside a long, closely wound solenoid carrying current i. We assume that end effects can be neglected as well as the magnetization of the windings of the solenoid. Suppose first that the core of the solenoid is vacuum. Since $\mathbf{M} = 0$ for vacuum (why?), Eq. 30-5 shows that $\mathbf{B} = \mu_0\mathbf{H}$, or $\mathbf{H} = \mathbf{B}/\mu_0$ in this case. Both the magnetic field **B** and the magnetic intensity **H** are uniform inside the solenoid and directed along the axis. From Eq. 27-11, $B = \mu_0 ni$, where n is the number of turns per unit length. Thus the magnitude of the magnetic intensity is $H = B/\mu_0 = ni$. We conclude that **H** inside the solenoid is due to the current distribution in the solenoid windings. Notice that H can be adjusted in an experiment by changing the solenoid current.

Now suppose that the space inside the solenoid is filled with some material, and the solenoid current i is adjusted to have the same value as before. For this geometry, it turns out that **H** in the material is the same as with vacuum in the space. That is, the magnetic intensity **H** inside the ideal solenoid is determined solely by the solenoid current. The magnetic field **B** in the material, however, is different from the vacuum case because of the contribution in Eq. 30-5 from the magnetization **M**.

Since the magnetic intensity **H** can be determined from the solenoid current i, the magnetic field, $\mathbf{B} = \mu_0(\mathbf{H} + \mathbf{M})$ can be calculated if **M** is known. In a typical linear diamagnetic or paramagnetic material, **M** and **B** are proportional. Then from Eq. 30-5, **H** and **B** are also proportional in such materials. The linear relation between **H** and **B** is expressed as

$$\mathbf{B} = \mu\mathbf{H} \tag{30-6}$$

Permeability constant μ

where μ is called the ***permeability constant*** of the linear material. For vacuum, $\mathbf{M} = 0$, so that $\mathbf{B} = \mu_0(\mathbf{H} + 0)$ and $\mu = \mu_0$.

Suppose the material in the solenoid is diamagnetic. Then the directions of **M** and **B** are opposite, so that, from Eq. 30-5, **H** and **B** are parallel. Eliminating **B** from Eqs. 30-5 and 30-6, we obtain $\mu\mathbf{H} = \mu_0(\mathbf{H} + \mathbf{M})$, or $(\mu - \mu_0)\mathbf{H} = \mu_0\mathbf{M}$. Since **H** and **M** have opposite directions, we see that $\mu - \mu_0$ is negative; $\mu < \mu_0$ for a diamagnetic material. The permeability constant μ for normal diamagnetic materials is only ***slightly*** less than μ_0. Bismuth, one of the "most diamagnetic" substances, has a permeability constant $\mu = 0.99983\mu_0$. For most practical purposes, diamagnetic effects can be neglected.

In diamagnetic materials, μ is slightly less than μ_0.

For paramagnetic materials, $\mu > \mu_0$.

For a typical paramagnetic substance, **M**, **H**, and **B** are all parallel so that $\mu > \mu_0$. Since the magnetization **M** of a paramagnetic substance is temperature-dependent, the permeability constant μ likewise changes with temperature. For many paramagnetic substances over a wide range of temperatures, μ is slightly greater than μ_0. For example, $\mu = 1.00026\mu_0$ for Pt at 293 K. Since $\mu \approx \mu_0$ under such conditions, paramagnetic effects can often be neglected in determining **B**; that is, $\mathbf{B} = \mu\mathbf{H} \approx \mu_0\mathbf{H}$. Under other circumstances, particularly at low temperatures, paramagnetic effects are important.

In a ferromagnetic material, there is no linearity among **M**, **H**, and **B**. Although Eq. 30-6 can be used to connect **H** and **B**, the value of μ is not characteristic of the material but depends on the previous treatment of the sample. The detailed relationship between the magnetic field **B** and the magnetic intensity **H** can be measured by using the ferromagnetic material as the core of a toroid called a *Rowland ring*. The arrangement is shown in Fig. 30-8. The toroid is like a long solenoid that has been bent, with its ends joined to form a doughnut shape. If the toroid has N_T turns and a mean ring radius R, the number of turns per unit length is $n = N_T/2\pi R$. With this geometry, both **H** and **B** are essentially confined to the ferromagnetic core of the toroid. The magnetic intensity has magnitude $H = ni$ and can be controlled by adjusting the current i. A secondary or search coil is used to monitor a change in B, which causes an induced emf, $\mathscr{E} = NS\, dB/dt$, in the N-turn search coil.

Figure 30-8. A Rowland ring is used to measure B in a ferromagnetic core.

Suppose that initially the core is unmagnetized and that there is no current in the toroid; thus $\mathbf{M} = 0$, $\mathbf{H} = 0$, and $\mathbf{B} = 0$. The current is increased to some value i so that H increases to ni. During this process, B and the flux $\Phi_B = BS$ linking each turn of the search coil change, and an emf is induced in the coil. The coil is connected to a device, such as a ballistic galvanometer, which measures the net charge ΔQ passing through it. (See Prob. in Chap. 28.) This charge $\Delta Q = \int i_s\, dt$ corresponds to the induced current i_s in the search coil. Since i_s is proportional to the induced emf $\mathscr{E} = NS\, dB/dt$, the integral $\Delta Q = \int i_s\, dt$ is proportional to the integral $NS \int (dB/dt)dt = NS\, \Delta B$. In this way, the change ΔB in the magnitude of the magnetic field is determined by measuring the charge passing through the search coil.

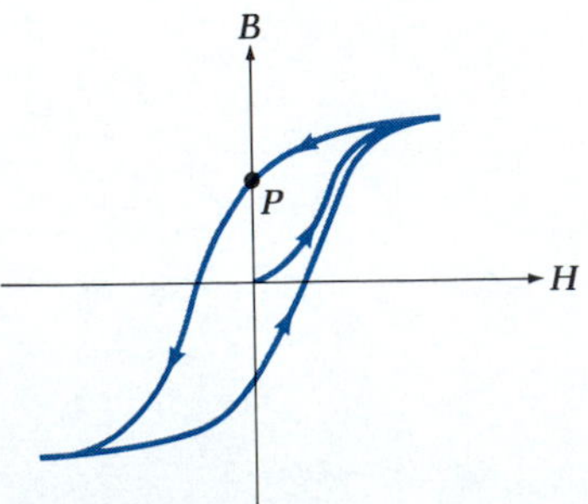

Figure 30-9. A ferromagnetic material exhibits hysteresis. The relation between **H** and **B** depends on the previous treatment of the sample.

By changing the toroid current in steps, one obtains pairs of values of components of **H** and **B**. Figure 30-9 shows a curve, typical of a hard ferromagnetic core, obtained by plotting such points on a graph. Conventionally, H is taken as the independent variable, since it is changed by adjusting the current in the toroid. The irreversibility of changes in the magnetic domain structure, called ***hysteresis,*** is evident in the figure. The magnetic intensity component H is changed such that the curve is traversed with the sense of the arrows. (Since H and B represent toroidal components of the fields in this discussion, negative values correspond to a reversal of direction. The sign of H is changed by reversing the sense of the current in the toroidal windings.) Note that $H = 0$, $B \neq 0$ at point P on the curve; the ferromagnetic core has a permanent magnetization in this state. You can also see that a value of B is not uniquely determined by a value of H. The state of the sample is determined by the history of its treatment.

30-6 THE MAGNETIC FIELD OF THE EARTH

A simple compass consists of a magnetized needle that can rotate freely in a plane. When used for navigation or direction-finding at the earth's surface, the compass is usually held with its plane horizontal, and the compass needle aligns with an approximately north-south orientation. The end of the needle that points north is identified as the north (north-seeking) pole of the needle. In the absence of other sources of magnetic field, the torque that orients a compass needle is provided by the magnetic field of the earth. The needle aligns with the magnetic field, and the direction of the field is from the south to the north pole of the needle.

The earth's magnetic field distribution outside its surface is shown schematically in Fig. 30-10. Note that the field has a component perpendicular to the earth's surface at most places. A compass can be used to map out the magnetic field of the earth. The direction of the needle when the plane of the compass is horizontal corresponds to the horizontal component of the field. If the plane of the compass is rotated about that direction until the plane is vertical, then the needle will give the direction of **B** at that point. (Try it using a small compass.) Notice from Fig. 30-10 that the lines representing **B** generally emerge from the earth's surface in the southern hemisphere and reenter in the northern hemisphere. These lines are suggestive of a magnetic dipole field. That is, outside the earth's surface, the magnetic field is essentially the same as that due to a magnetic dipole located at the center of the earth. Notice in the figure that lines entering the region near the north *geographic* pole correspond roughly to a *magnetic* south pole centered there. Likewise, the *magnetic* north pole is centered near the south *geographic* pole in Antarctica.

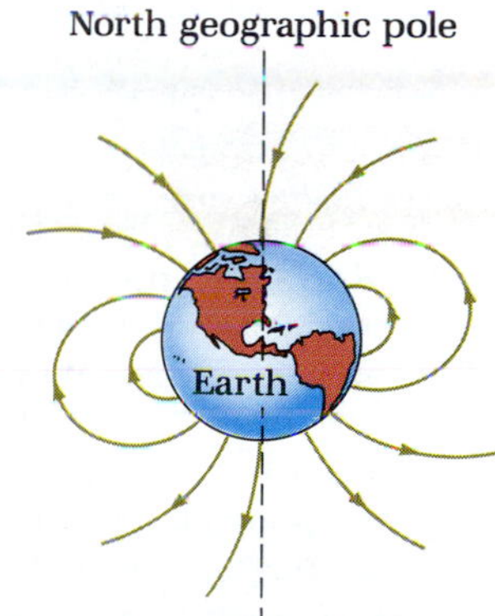

Figure 30-10. Outside the earth the magnetic field is approximately a dipole field. The region near the north geographic pole (toward the star Polaris) has lines similar to those near a south magnetic pole. The magnetic lines are continuous and close on themselves, although the behavior deep inside the earth is not known in detail.

At points several earth radii outside the surface, the earth's magnetic field is distorted by a contribution from the solar wind, a stream of charged particles from the sun. Some of the particles are trapped by the magnetic field around the earth. Light emitted by these particles in the upper atmosphere is responsible for the *auroras* that are sometimes visible in the higher latitudes.

The magnetic field distribution in the earth's interior is not known. This inaccessible region contains the source or cause of the earth's magnetic field. The mechanism by which the magnetic field is maintained is not understood. A successful theory of geomagnetism must account for features that are characterized by very different time scales. On a time scale measured in days or years, the earth's magnetic field seems static and, therefore, useful for navigation. But on a geologic time scale, the earth's magnetism is dynamically active. There are variations in the local magnetic field that occur over hundreds or thousands of years. And there is evidence from the direction of the magnetization in dated rocks that the direction of the earth's magnetic field undergoes abrupt reversals with intervals of up to a million years. The most recent reversal seems to have occurred somewhat over 10,000 years ago.

SUMMARY

Section 30-1. Atomic Currents, Magnetic Dipoles, and Magnetization

In a simple model, an orbiting electron has a magnetic moment proportional to its orbital angular momentum,

$$\mathbf{m} = -\frac{e}{2m_e}\mathbf{L} \qquad (30\text{-}1)$$

and a similar contribution due to the spin angular momentum,

$$\mathbf{m} = -\frac{e}{m_e}\mathbf{S} \qquad (30\text{-}2)$$

The magnetization in a material is the magnetic moment per unit volume:

$$\mathbf{M} = \frac{\langle \Sigma\, \mathbf{m}_i \rangle}{\Delta V} \qquad (30\text{-}3)$$

Section 30-2. Diamagnetism

In diamagnetic materials, magnetic dipole moments are induced in molecules by the magnetic field, and the vectors **M** and **B** have opposite directions.

Section 30-3. Paramagnetism

The permanent magnetic moment of an unpaired electron in a paramagnetic substance tends to become aligned with the magnetic field. The vectors **M** and **B** are parallel and are related by Curie's law

$$\mathbf{M} = \frac{C\mathbf{B}}{\mu_0 T} \qquad (30\text{-}4)$$

valid except at low temperatures and high fields.

Section 30-4. Ferromagnetism

Molecular magnetic dipoles in a magnetic domain tend to be aligned in a ferromagnetic material. If the domains are oriented preferentially by applying a magnetic field, the sample has a large magnetization. The magnetization can persist in hard magnetic materials to form a permanent magnet.

Section 30-5. Magnetic Intensity H

The magnetic intensity **H** is defined by the relation

$$\mathbf{B} = \mu_0(\mathbf{H} + \mathbf{M}) \qquad (30\text{-}6)$$

For a linear medium with permeability μ, the relation can be expressed as $\mathbf{B} = \mu\mathbf{H}$. For a Rowland ring, **H** is due to the macroscopic current in the windings. The relation between **B** and **H** for the ferromagnetic materials is nonlinear, and hysteresis effects are present.

Section 30-6. The Magnetic Field of the Earth

Outside its surface, the earth's magnetic field is approximately a dipole field. Large changes in the field occur over geological time intervals.

QUESTIONS

1. What is the direction of the orbital contribution to the magnetic moment **m** of an electron relative to the direction of its orbital angular momentum **L**?

2. A free electron at rest has a magnetic moment **m** due to its spin angular momentum **S**. What are the relative directions of these two vectors? A proton also has an intrinsic angular momentum and a magnetic moment. What are the relative directions of **m** and **S** for a free proton?

3. The electrons in an isolated He atom are paired so that the total angular momentum of the electrons is zero. Explain why you should expect that the net magnetic moment is also zero. Do you expect liquid helium to be diamagnetic or paramagnetic? Explain.

4. When isolated, a neutral Na atom and a neutral Cl atom are expected to have permanent magnetic moments. In table salt (NaCl), the *ions* Na^+ and Cl^- form an ionic crystal which is diamagnetic. Give a possible explanation of the diamagnetic behavior of NaCl.

5. An iron filing, released from rest near a stationary permanent magnet, accelerates toward the magnet. *(a)* What is the source of the increased kinetic energy of the filing as it moves toward the magnet? *(b)* What becomes of this kinetic energy as the filing strikes the magnet and sticks to it?

6. Why does a typical transformer have an iron core?

7. In a paramagnetic material, **M** and **B** are parallel and the permeability μ is positive. In a diamagnetic material, **M** and **B** have opposite directions. Why is the permeability μ not negative? From the value of μ, how can you distinguish a diamagnetic material from a paramagnetic one?

8. The core of a long solenoid can be filled with any of a variety of materials. A current in the windings causes a magnetic intensity **H** inside the solenoid parallel to the axis. Relative to the direction of **H**, what are the directions of **M** and **B** inside the solenoid if the core is *(a)* diamagnetic and *(b)* paramagnetic? *(c)* Explain why the question cannot be answered in general for a ferromagnetic core.

9. One property of a *superconductor* is that the magnetic field **B** is excluded from its interior. If the core of the solenoid in the previous question is a superconductor, then the magnetic field $\mathbf{B} = 0$ in the core. But **H** is not zero in the core if a current exists in the solenoid windings. *(a)* What is the direction of **M** in the core relative to the direction of **H**? *(b)* Explain why a superconductor is sometimes called "perfectly diamagnetic."

10. What is the magnetization **M** for vacuum? Explain.

11. A small nonferromagnetic sample is brought near the north pole of a strong magnet where the field is highly nonuniform. The sample is very weakly repelled by the north pole. *(a)* Is the sample diamagnetic or paramagnetic? *(b)* What happens to the sample if it is brought close to the south pole of the magnet?

12. What are the answers to the two parts of the previous question if the sample is weakly attracted to the north pole?

13. Consider the magnetic field **B** inside a long solenoid carrying current i *(a)* with and *(b)* without a soft iron core. In which case is B larger? Explain.

14. For the two cases in the previous question, which has the larger self-inductance? Explain.

15. You are given a small piece of wood, a glass of water, and a magnetized needle. Explain how a compass can be formed from these items.

16. Suppose that you have the compass from the previous question in an otherwise empty room. The magnetized needle

is unmarked, and the room has no windows or other openings. Is it possible under these circumstances to determine which way is north? Explain.

17. A compass needle points north when the plane of the compass is horizontal. If the plane is rotated through 90° about a north-south axis, the north-pole end of the needle points above the horizon. Are you more likely to be in Austria or in Australia? Explain.

18. Complete the following table:

Symbol	Represents	Type	SI Unit
M			
L	Orbital angular momentum		
S			$kg \cdot m^2 \cdot s^{-1}$
H			
μ		Scalar	

EXERCISES

Section 30-1. Atomic Currents, Magnetic Dipoles, and Magnetization

1. Suppose that the z component of the orbital angular momentum of an electron in an atom is $L_z = 1.06 \times 10^{-34}$ kg·m²·s⁻¹. Determine the z component of the orbital contribution to the magnetic dipole moment.

2. Verify, using Eq. 30-1, that the SI unit of magnetic moment is ampere-meters squared ($A \cdot m^2$) and, using Eq. 30-3, that the SI unit of magnetization is amperes per meter (A/m).

3. A free electron at rest has a magnetic moment component $m_z = 9.3 \times 10^{-24}\ A \cdot m^2$. Determine the corresponding component S_z of the electron's spin angular momentum.

4. Experiment shows that a component of angular momentum is *quantized*. That is, only certain discrete values occur, and for orbital angular momentum these are integer multiples of $h/2\pi$, where $h = 6.63 \times 10^{-34}\ kg \cdot m^2 \cdot s^{-1}$ is *Planck's constant*. (See Chap. 39.) *(a)* Show that the corresponding component of magnetic moment for an electron is also quantized in integer multiples of $m_B = eh/4\pi m_e$. This value, called the *Bohr magneton,* is a convenient unit of magnetic moment at the atomic level. *(b)* Determine the value of the Bohr magneton to three significant figures.

5. Some species of atomic nuclei have magnetic moments, and a convenient unit of nuclear magnetic moment is the *nuclear magneton* $m_N = eh/4\pi m_p$, where m_p is the proton mass. (See the previous exercise.) *(a)* Determine the value of the nuclear magneton to three significant figures. *(b)* Compare the values of the nuclear magneton and the Bohr magneton and explain why the nucleus of an atom usually contributes negligibly to the magnetic properties of materials.

6. Bismuth is one of the most diamagnetic of substances. If a 1-T magnetic field is applied parallel to the axis of a long Bi rod, the magnetization in the rod has magnitude $M = 1.7$ A/m. *(a)* What is the direction of **M** relative to the direction of **B**? *(b)* Determine the average magnetic moment per Bi atom. The density of Bi is $9.8 \times 10^3\ kg/m^3$, and the mass of a Bi atom is 3.5×10^{-25} kg. *(c)* What fraction of a Bohr magneton (see Exercise 4) is the average magnetic moment per atom?

7. In a strong permanent magnet, the average magnitude of the magnetic moment per atom is around $1 \times 10^{-23}\ A \cdot m^2$. Estimate the magnitude of the magnetization in such a magnet.

8. A permanently magnetized sphere of radius 25 mm has a uniform magnetization of magnitude 8400 A/m. *(a)* Determine the magnitude of the magnetic moment of the sphere. *(b)* Sketch the lines representing **B** inside and outside the sphere. Recall that a line representing **B** must close on itself and assume that **B** is uniform inside the sphere and parallel to **M**.

Section 30-2. Diamagnetism

9. No dissipation occurs for electronic currents in molecules, so they are resistanceless. To see how diamagnetism is related to Faraday's law, consider a *resistanceless* circular conducting filament. Initially there is no magnetic field and no current in the loop. A magnetic field is then applied perpendicular to the plane of the loop, and a current is induced in the loop. This current will persist even after the magnetic field stops changing. (Why?) *(a)* Construct a diagram showing the direction of the applied magnetic field and the sense of the induced current. *(b)* Determine the direction of the magnetic moment of the current loop and explain how this model is suggestive of diamagnetic behavior.

10. In an applied magnetic field of 1 T, the magnitude of the magnetization of water is about 8 A/m. What is the magnetization of water if *(a)* $B = 0$; *(b)* $B = 0.5$ T?

Section 30-3. Paramagnetism

11. Curie's constant for a paramagnetic salt is 1.8×10^{-3} K. *(a)* Determine the magnetization of this salt at room temperature ($T = 293$ K) in a 0.35-T magnetic field. *(b)* At what temperature would the magnetization have the same magnitude in a 0.25-T field?

12. Curie's constant for an ideal ($pV = nRT$) paramagnetic gas is proportional to the density of the gas. For O_2 at a density corresponding to 293 K and atmospheric pressure, the Curie constant is 5.5×10^{-4} K. *(a)* Determine the magnetization of O_2 with this density at 293 K in a 0.50-T magnetic field. *(b)* Determine the magnetization of the gas in the same field at 200 K at atmospheric pressure. (Notice that the density has changed.)

13. Curie's law is valid except for low temperatures or fields. One way to express this is in terms of the ratio of two energies: m_0B/kT. For a system at temperature T, the typical energy

change due to random thermal processes is about kT, where $k = 1.38 \times 10^{-23}$ J/K is the Boltzmann constant. (See Chap. 18.) *(a)* Consider a dipole, with a permanent dipole moment of magnitude m_0, that is aligned either parallel to or opposite the applied field **B**. Show that the difference in energy for these two orientations is $2m_0B$. *(b)* A typical molecular magnetic moment has $m_0 \approx 1 \times 10^{-23}$ A·m². Evaluate m_0B for such a molecule in a 1-T field. *(c)* Curie's law is valid if $m_0B/kT \ll 1$. Estimate a temperature range for the validity of Curie's law for the conditions in part *(b)*.

Section 30-4. Ferromagnetism

14. The magnetic moment per atom in Ni is about 6×10^{-24} A·m², and there are about 9×10^{28} atoms per cubic meter in the solid. *(a)* Determine the magnetization for a sample that has a single domain with virtually all of the magnetic moments aligned. *(b)* Determine the magnetization for a sample averaged over many domains oriented randomly. *(c)* What is the effective fraction of the many domains in a sample that are aligned with the average magnetization to give a magnitude $M = 2000$ A/m?

15. The magnetic moments per atom for Ni, Co, and Fe are 0.6×10^{-23} A·m², 1.6×10^{-23} A·m², and 2.1×10^{-23} A·m², respectively. Estimate the maximum (or saturation) magnetization that can exist in each of these solids.

Section 30-5. Magnetic Intensity H

16. A long solenoid has 2500 turns per meter and carries a 4.8-A current. Neglect end effects and determine B, M, and H inside the solenoid for *(a)* vacuum in the space, *(b)* a Pb core ($\mu = 0.999984\mu_0$), *(c)* air in the space ($\mu = 1.0000004\mu_0$).

17. For most practical purposes, the permeability of air is the same as for vacuum, $\mu \approx \mu_0$. Determine the magnitude H of the magnetic intensity at a point just above the earth's surface where $B = 42\ \mu$T.

18. If the relation $\mathbf{B} = \mu\mathbf{H}$ is used for a ferromagnetic material, then μ does not have a single value. A hysteresis curve for a ferromagnetic substance is shown in Fig. 30-11. For a point on the curve, the value of $\mu = B/H$ is the slope of the straight line from the origin to the point on the curve. Identify those portions (if any) for which μ is *(a)* positive, *(b)* negative, *(c)* zero, *(d)* infinite.

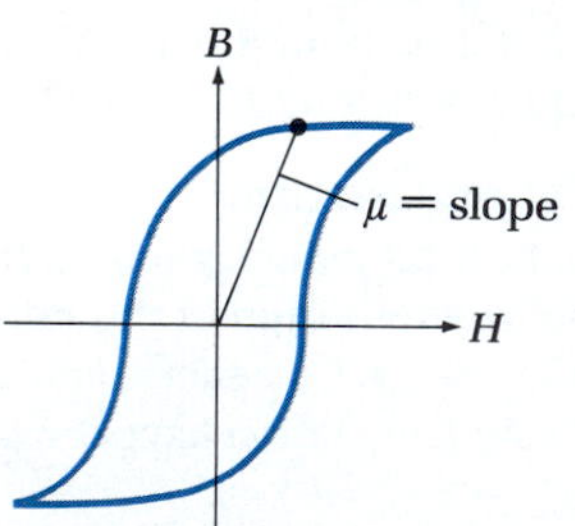

Figure 30-11. Exercise 18.

19. A soft iron rod forms the core of a long solenoid. Inside the rod the fields have magnitude $B = 1.10$ T, $H = 345$ A/m. *(a)* Determine a value of μ for this state. *(b)* Determine the magnetization in the iron rod, given that **H** and **B** are parallel to each other and to the solenoid axis. *(c)* Determine the percent error in evaluating B in this case from the approximate expression $B \approx \mu_0 M$.

20. Ferromagnetic iron forms the core of a Rowland ring. Measurements of H and B for the ring are shown in the following table:

H, A/m	*B*, mT	*H*, A/m	*B*, mT
0.0	0.0	32	16
0.8	0.2	48	30
4.0	1.5	64	150
8.0	2.6	80	540
16	6.3	160	1100
24	11	800	1600

Construct graphs of B versus H for *(a)* $0 \leq H \leq 24$ A/m and *(b)* $0 \leq H \leq 800$ A/m. *(c)* Estimate minimum and maximum values of $\mu = B/H$ from the two graphs. *(d)* Determine the magnitude of the magnetization when $H = 800$ A/m.

Section 30-6. The Magnetic Field of the Earth

21. Over much of the surface of the United States, the magnitude of the earth's magnetic field is around 6×10^{-5} T. The inclination (the angle between the direction of **B** and a horizontal plane tangent to the surface) averages around 70°. Estimate the vertical and horizontal components of **B** at a point in this region.

22. At points a large distance from a magnetic dipole in the perpendicular bisector plane of the dipole moment, the magnitude of the magnetic field is given by $B = \mu_0 m/4\pi r^3$, where m is the magnitude of the magnetic moment. The magnetic field outside the earth can be approximated by the field due to a magnetic dipole located at the earth's center. *(a)* Estimate the magnitude of this equivalent magnetic moment, given that $B = 3 \times 10^{-5}$ T at a point on the equator near the southern tip of India. *(b)* If the earth had a uniform magnetization, what magnitude M would correspond to the magnetic moment calculated above? *(c)* What direction should each of the vectors **m** and **M** have relative to the geographic poles?

PROBLEMS

1. ***Magnetic intensity inside an ideal solenoid.*** Ampere's law for steady current distributions, Eq. 27-8, is often written in terms of the magnetic intensity: $\oint \mathbf{H}\cdot d\boldsymbol{\ell} = \Sigma i$. Apply this form of Ampere's law and symmetry arguments to determine the magnetic intensity inside an ideal solenoid with n turns per unit length and current I. Note that the result is independent of the material in the core.

2. ***Finding H, B, and M inside a solenoid.*** A long solenoid has 2500 turns per meter and carries a steady 120-mA current. The core of the solenoid is ferromagnetic. *(a)* Determine **H** inside the solenoid core. *(b)* The core is in a state such that $\mu = 150\mu_0$. Determine **B** inside the solenoid core. *(c)* Determine **M** inside the solenoid core.

3. ***Magnetic force on a loop in a nonuniform field.*** A square current loop, shown in perspective in Fig. 30-12, lies in the xy plane perpendicular to a *nonuniform* magnetic field in the z direction. The z component of the field depends only on the y coordinate and ranges from B_{1z} to B_{2z} (with $B_{1z} < B_{2z}$) over a distance $\Delta y = a$. *(a)* Give the direction of the magnetic moment **m** of the loop relative to the direction of **B**. *(b)* Determine the direction of the net magnetic force on the current loop and explain why this result is suggestive of the force on a paramagnetic sample in a nonuniform magnetic field. *(c)* Reconsider parts *(a)* and *(b)* if the sense of the current is opposite that in the figure.

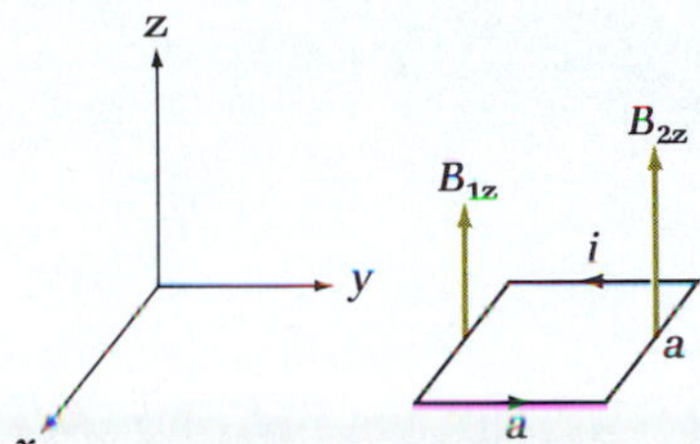

Figure 30-12. Problem 3.

4. ***The Larmor frequency.*** Consider an electron with speed v in a circular orbit of radius r about a nucleus. It is convenient to consider the angular speed, or angular frequency $\omega_0 = v/r$. The magnitude of the orbital angular momentum is $L = m_e vr = m_e r^2 \omega_0$. The centripetal force on the electron in this circular orbit, $m_e v^2/r$ or $m_e r \omega_0^2$, is provided by the electrostatic attraction F_E toward the nucleus, $m_e r \omega_0^2 = F_E$. Now suppose that an external magnetic field **B** is applied perpendicular to the plane of the orbit. For the orientation shown in Fig. 30-4*a*, the magnetic field exerts an additional force $-e\mathbf{v} \times \mathbf{B}$ on the electron. The force is directed radially inward and the electron increases its speed. Let $\omega_1 = \omega_0 + \omega_L$ represent the angular frequency for this case; the quantity ω_L is the small change in angular frequency due to the external magnetic field. *(a)* Assume that the radius of the orbit changes negligibly and show that $\omega_L = eB/2m_e$. This small change in frequency due to the magnetic field is called the *Larmor frequency*. *(b)* Show that the orbital angular momentum of the electron has changed by $\Delta \mathbf{L} = \frac{1}{2} er^2 \mathbf{B}$.

5. ***Magnetic moment of two paired electrons.*** Apply the ideas of the previous problem to a pair of electrons in orbits such as shown in Fig. 30-4. Show that this simplistic model leads to a net magnetic moment $\mathbf{m} = \mathbf{m}_1 + \mathbf{m}_2 = -e^2 r^2 \mathbf{B}/2m_e$. Why is this model suggestive of diamagnetic behavior?

6. ***Magnetic intensity due to a magnetized sphere.*** A permanently magnetized steel sphere of radius 8.5 mm has a uniform magnetization of magnitude 2500 A/m. At points inside the sphere, the magnetic field is given by $\mathbf{B} = 2\mu_0 \mathbf{M}/3$. Outside the sphere, **B** is a magnetic dipole field due to the dipole moment of the sphere. *(a)* Determine the magnitude and direction of **H** inside the sphere. *(b)* Determine the magnetic moment of the sphere. *(c)* What is **H** outside the sphere?

7. ***Magnetization in a neutron star.*** Very strong magnetic fields exist in a *neutron star*, the last stage in the evolution of stars somewhat more massive than our sun. In a model of a neutron star, the interior consists of a small sphere (radius $a \approx 20$ km) of an extremely dense fluid of neutrons (density $\rho \approx 10^{17}$ kg/m^3). The neutron, although neutral, has a magnetic moment of magnitude $m_N \approx 10^{-26}$ A·m^2. Suppose that the neutron star has a uniform magnetization similar to the sphere in the previous problem and that the magnetic field inside has magnitude $B \approx 10^8$ T. *(a)* Determine the magnitude of the magnetization inside the star. *(b)* Estimate the saturation magnetization which would correspond to the alignment of all the neutron magnetic moments. The neutron mass is $m_N = 1.67 \times 10^{-27}$ kg.

8. ***Hysteresis loss.*** Consider the material in the core of a long solenoid of cross-sectional area S, with N turns on a length ℓ. If the magnetic field **B** is changed by changing the current i in the windings, then an emf $\mathscr{E} = S\,dB/dt$ is induced in each turn. An external agent must supply energy at a rate $dW/dt = Ni\mathscr{E}$ to counter the induced emf in the N turns. *(a)* Show that the infinitesimal work can be expressed in terms of the fields and the volume of the region: $dW = \ell SH\,dB$. *(b)* The work per unit volume for a hysteresis loop can be expressed as $\oint H\,dB$. Explain how this integral can be interpreted as the area enclosed by a hysteresis loop such as that graphed in Fig. 30-11. This work corresponds to the dissipation of energy, called *hysteresis loss,* due to the irreversible changes in domain structure for such ferromagnetic materials.

31 ELECTROMAGNETIC OSCILLATIONS AND AC CIRCUITS

Satellite television antennas with rather ornate reflectors.

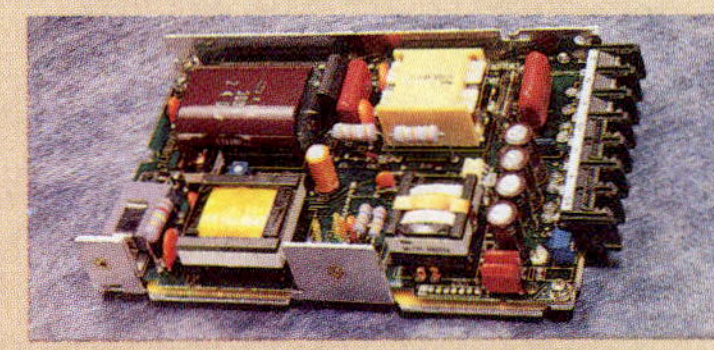

An electrical power supply reveals its complex circuitry.

Practically every day of our lives we use electrical devices that operate with alternating current (ac). Such devices include radios, television sets, computers, telephones, refrigerators, and on and on. The feature that makes ac electricity often more useful than dc is that ac can be more readily controlled. In this chapter we shall see how resistance, capacitance, and inductance in a circuit play a crucial role in the behavior of alternating currents and potential differences.

31-1 *LC* OSCILLATIONS

In Sec. 25-4 we discussed the behavior of a circuit containing a resistor and a capacitor, an *RC circuit.* When a capacitor is discharged through a resistor, the current in the circuit decreases exponentially to zero. The electric energy U_E stored in a capacitor is dissipated as heat in the resistor during the discharge. Similarly, in Sec. 29-2, we discussed the behavior of a circuit containing a resistor and an inductor, an *LR circuit,* and we found the magnetic energy U_B produced by the current. In an *LR* circuit the current decreases exponentially to zero while energy U_B is dissipated as heat in the resistor. Capacitors and inductors are energy-storage devices; a capacitor can store electric energy and an inductor can store magnetic energy. Resistance in a circuit causes energy to be dissipated as heat.

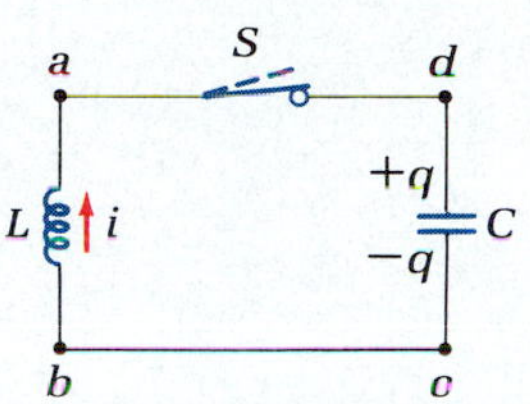

Figure 31-1. Applying the loop rule to an *LC* circuit.

We now examine the behavior of a circuit containing only a capacitance *C* and an inductance *L*, an ***LC* circuit** (Fig. 31-1). Ordinary circuits contain resistance, but for simplicity we begin by discussing an idealized circuit with negligible resistance. Suppose the capacitor in the circuit of Fig. 31-1 is charged by an external battery and then the battery is taken away. When the switch is closed, the capacitor will begin to discharge through the inductor so that at time *t* there will be a current *i* in the circuit and a charge *q* on the capacitor. The charge and the current are related because the current gives the rate at which charge is transferred from one plate to the other: $i = \pm\, dq/dt$. The choice of plus or minus depends on our sign convention for *i* and *q*. Suppose we let *i* be positive when the current is clockwise and let *q* be positive when the charge on the upper plate is positive. With this choice, *q* increases when *i* is positive: $i = dq/dt$.

From Kirchhoff's loop rule, the sum of the potential differences around the loop is zero:

$$(\mathcal{V}_b - \mathcal{V}_a) + (\mathcal{V}_c - \mathcal{V}_b) + (\mathcal{V}_d - \mathcal{V}_c) + (\mathcal{V}_a - \mathcal{V}_d) = 0$$

The potential difference across the inductor is $(\mathcal{V}_b - \mathcal{V}_a) = L(di/dt)$, where the algebraic sign is determined by Lenz's law. The induced emf opposes the change that causes it. Therefore, when $di/dt > 0$, $\mathcal{V}_b > \mathcal{V}_a$. With our sign convention for *q*, the potential difference across the capacitor is $\mathcal{V}_d - \mathcal{V}_c = q/C$. Thus

$$\mathrm{L}\frac{di}{dt} + \frac{q}{C} = 0$$

Since $i = dq/dt$, we have $di/dt = d^2q/dt^2$. Making this substitution and rearranging, we have

$$\frac{d^2q}{dt^2} = -\frac{1}{LC}q \qquad (31\text{-}1)$$

Differential equation for *LC* oscillations

This equation has the same mathematical form as the differential equation that describes a simple harmonic oscillator (Chap. 14). An example of a simple harmonic oscillator is the one-dimensional mass-spring system where frictional forces are negligible. For the mass-spring system, Newton's second law gives

$$\frac{d^2x}{dt^2} = -\frac{k}{m}x \qquad (14\text{-}11)$$

where *x* is the coordinate of an object of mass *m* connected to a spring of force constant *k*. The mass-spring harmonic oscillator provides a mechanical analog to the *LC* circuit. Table 31-1 lists some of the analogous quantities. From our experience with the motion of a simple harmonic oscillator, we expect that the charge on the capacitor in the *LC* circuit varies sinusoidally with time:

$$q = Q_m \cos(\omega_0 t + \phi) \qquad (31\text{-}2)$$

Oscillating charge

where Q_m is the maximum charge on the capacitor, ω_0 is the angular frequency of the oscillation, ϕ is the phase constant, and $(\omega_0 t + \phi)$ is the phase. The current is

$$i = \frac{dq}{dt} = \frac{d}{dt}[q_m \cos(\omega_0 t + \phi)] = -\omega_0 Q_m \sin(\omega_0 t + \phi)$$

or

$$i = -I_m \sin(\omega_0 t + \phi) \qquad (31\text{-}3)$$

Oscillating current

where $I_m = \omega_0 Q_m$ is the maximum current.

We now verify that Eq. 31-2 is a solution to Eq. 31-1. As a by-product of this

TABLE 31-1. ***Analogy between the Mechanical Block-Spring Harmonic Oscillator and the Electromagnetic LC Circuit***

Mechanical System	Electromagnetic System
Mass m	Inductance L
Force constant k	Reciprocal capacitance $1/C$
Coordinate x	Charge q
Velocity component $v_x = dx/dt$	Current $i = dq/dt$
Mechanical energy $E = \frac{1}{2}kx + \frac{1}{2}mv_x^2$	Electromagnetic energy $U = \frac{1}{2}q^2/C + \frac{1}{2}Li^2$

exercise, we shall find an expression for ω_0. To make this verification, we first determine d^2q/dt^2.

$$\frac{d^2q}{dt^2} = \frac{di}{dt} = \frac{d}{dt}[-\omega_0 Q_m \sin(\omega_0 t + \phi)] = -\omega_0^2 Q_m \cos(\omega_0 t + \phi)$$

Substituting this expression for d^2q/dt^2 and q from Eq. 31-2 into Eq. 31-1, we have

$$-\omega_0^2 Q_m \cos(\omega_0 t + \phi) = -\frac{1}{LC} Q_m \cos(\omega_0 t + \phi)$$

or

$$\omega_0^2 \cos(\omega_0 t + \phi) = \frac{1}{LC} \cos(\omega_0 t + \phi)$$

This equation is satisfied only if $\omega_0^2 = 1/LC$, or

Angular frequency of oscillation

$$\omega_0 = \frac{1}{\sqrt{LC}} \qquad (31\text{-}4)$$

Therefore, if $\omega_0 = 1/\sqrt{LC}$, then Eqs. 31-2 and 31-3 describe the oscillating charge and current. The frequency ν_0 of the oscillation is $\nu_0 = \omega_0/2\pi = 1/(2\pi\sqrt{LC})$.

The constants Q_m and ϕ in Eq. 31-2 are determined from the initial conditions. Suppose the capacitor was given a charge Q_0 while the switch was open, and the switch was closed at $t = 0$. Then the initial conditions are $q = Q_0$ and $i = 0$ at $t = 0$. Substituting $i = 0$ and $t = 0$ into Eq. 31-3, we have $0 = -I_m \sin\phi$. This condition can be satisfied by setting $\phi = 0$. Then we find Q_m by substituting $q = Q_0$ and $t = 0$ into Eq. 31-2: $Q_0 = Q_m \cos 0 = Q_m$. With these initial conditions, Eqs. 31-2 and 31-3 become

Definition of electromagnetic energy

$$q = Q_0 \cos(\omega_0 t) \qquad \text{and} \qquad i = -I_m \sin(\omega_0 t)$$

where $I_m = \omega_0 Q_0$. Figure 31-2 shows graphs of the oscillating charge and current versus time for the case with these initial conditions.

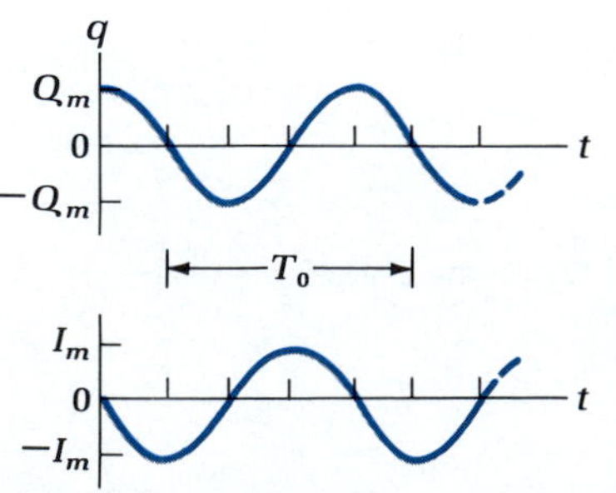

Figure 31-2. Oscillating charge and current versus time.

Now consider the energy of the LC circuit. From Sec. 23-3 the electric energy stored in a charged capacitor is $U_E = \frac{1}{2}q^2/C$, and from Sec. 29-3 the magnetic energy stored in a current-carrying inductor is $U_B = \frac{1}{2}Li^2$. The electromagnetic energy U of the LC circuit is

$$U = U_E + U_B$$

Using Eqs. 31-2 and 31-3, we find that the electric and magnetic energies at time t are

$$U_E = \frac{\frac{1}{2}q^2}{C} = \frac{\frac{1}{2}Q_m^2}{C} \cos^2(\omega_0 t + \phi)$$

and

$$U_B = \tfrac{1}{2}Li^2 = \tfrac{1}{2}LI_m^2 \sin^2(\omega_0 t + \phi)$$

We can use $I_m = \omega_0 Q_m$ and $\omega_0^2 = 1/LC$ to show that the factors that multiply

$\cos^2(\omega_0 t + \phi)$ and $\sin^2(\omega_0 t + \phi)$ in the two expressions above are equal:

$$\tfrac{1}{2}LI_m^2 = \tfrac{1}{2}L(Q_m\omega_0)^2 = \tfrac{1}{2}LQ_m^2\,\frac{1}{LC} = \frac{\tfrac{1}{2}Q_m^2}{C}$$

This means that the electromagnetic energy remains constant in an LC circuit:

$$\begin{aligned} U &= U_E + U_B \\ &= \frac{\tfrac{1}{2}Q_m^2}{C}\cos^2(\omega_0 t + \phi) + \tfrac{1}{2}LI_m^2\sin^2(\omega_0 t + \phi) \\ &= \frac{\tfrac{1}{2}Q_m^2}{C}[\cos^2(\omega_0 t + \phi) + \sin^2(\omega_0 t + \phi)] \end{aligned}$$

Since $\cos^2(\omega_0 t + \phi) + \sin^2(\omega_0 t + \phi) = 1$ for all t,

$$U = \frac{\tfrac{1}{2}Q_m^2}{C} = \tfrac{1}{2}LI_m^2$$

Thus the electromagnetic energy remains constant, continually changing back and forth between electric energy in the capacitor and magnetic energy in the inductor, as shown schematically in Fig. 31-3. Figure 31-4 shows graphs of the energies versus time for the same initial conditions as discussed above, $q = Q_0$ and $i = 0$ at $t = 0$ so that $Q_m = Q_0$, $I_m = \omega_0 Q_0$, and $\phi = 0$. The correlation with Fig. 31-3 is shown along the time axis in Fig. 31-4.

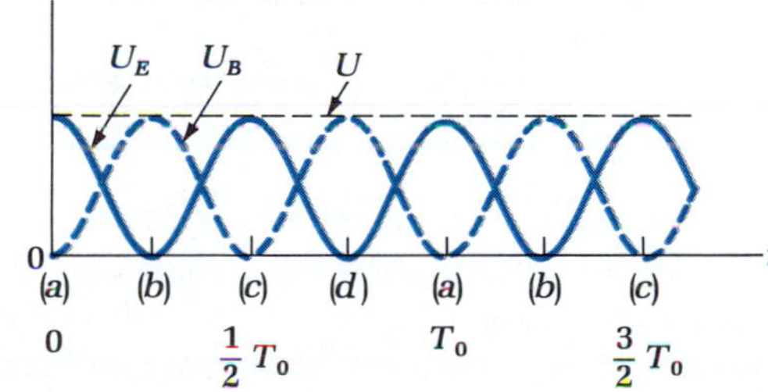

Figure 31-4. Energy in an LC circuit. The letters a, b, c, and d along the time axis correspond to the situations shown in Fig. 31-3.

EXAMPLE 31-1

***LC* oscillations.** With the switch open in Fig. 31-1, a 12-V battery is used to charge the capacitor and then the battery is removed. Given that $C = 3.7\ \mu\text{F}$ and $L = 96$ mH, determine *(a)* ω_0, *(b)* Q_m, *(c)* I_m, and *(d)* U for the oscillations that occur after the switch is closed.

Solution. *(a)* The angular frequency of the oscillations is

$$\omega_0 = \frac{1}{\sqrt{LC}} = \frac{1}{\sqrt{(96\text{ mH})(3.7\ \mu\text{F})}} = 1.7\text{ krad/s}$$

(b) The charge on the capacitor due to the battery is $Q_m = C\mathscr{E}$, where $\mathscr{E}$ is the emf of the battery (Sec. 25-4). Thus the amplitude of the oscillating charge is

$$Q_m = (3.7\ \mu\text{F})(12\text{ V}) = 44\ \mu\text{C}$$

(c) The amplitude of the oscillating current is

$$I_m = \omega_0 Q_m = (1.7\text{ krad/s})(44\ \mu\text{C}) = 74\text{ mA}$$

(d) The electromagnetic energy of the circuit is

$$U = \frac{\tfrac{1}{2}Q_m^2}{C} = \frac{\tfrac{1}{2}(44\ \mu\text{C})^2}{3.7\ \mu\text{F}} = 0.27\text{ mJ}$$

SELF-TEST 31-1. For the LC circuit in the above example, write expressions for the time dependence of *(a)* the charge q on the capacitor, *(b)* the current i in the circuit, *(c)* the electric energy U_E stored in the capacitor, and *(d)* the magnetic energy U_B stored in the inductor. Let $t = 0$ correspond to the instant the switch is closed. ***ANSWERS:*** *(a)* $q = (44\ \mu\text{C})\cos[(1.7\text{ krad/s})t]$; *(b)* $i = -(74\text{ mA})\sin[(1.7\text{ krad/s})t]$; *(c)* $U_E = (0.27\text{ mJ})\cos^2[(1.7\text{ krad/s})t]$; *(d)* $U_B = (0.27\text{ mJ})\sin^2[(1.7\text{ krad/s})t]$.

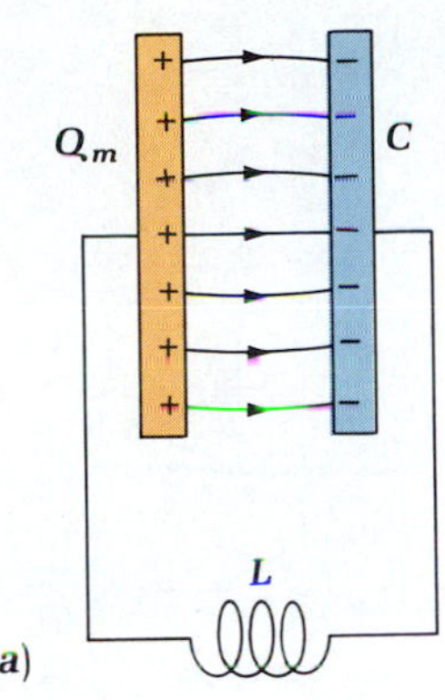

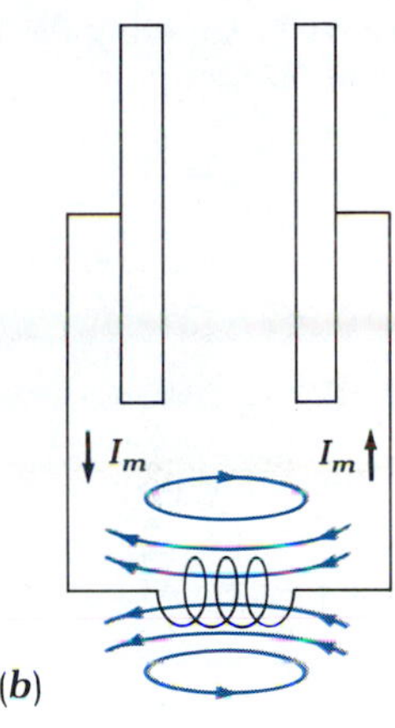

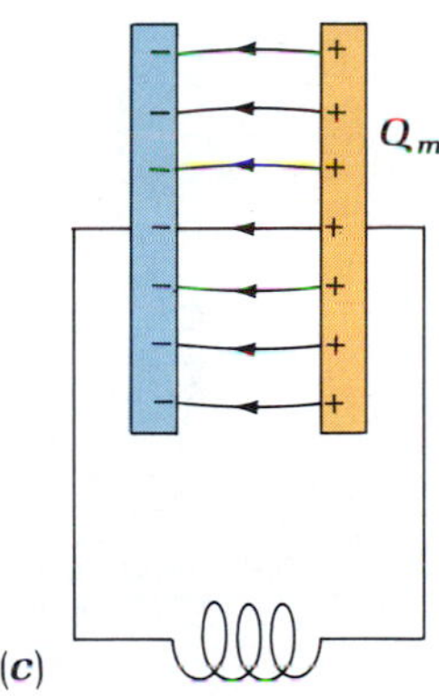

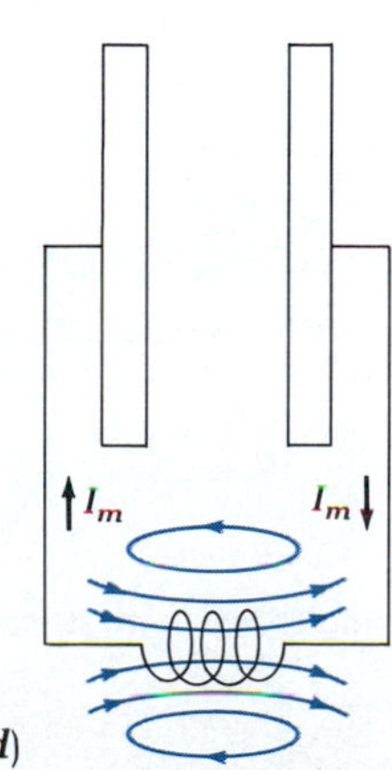

Figure 31-3. Electromagnetic energy is passed back and forth between the capacitor and the inductor in an LC circuit.

31-2 SERIES *RLC* CIRCUIT

In the previous section we made the simplifying assumption that the resistance in the circuit was negligible. With that assumption, we found that the electromagnetic energy of the circuit was constant. We now consider a circuit in which the resistance R is significant, an ***RLC circuit***. When a significant resistance is present in the circuit, the electromagnetic energy of the circuit decreases with time because energy is dissipated as heat from the resistor. Thus in an *RLC* circuit we expect that the charge on the capacitor and the current in the circuit will tend to approach zero as time goes by, but that oscillations in the charge and current may occur while they are dying out.

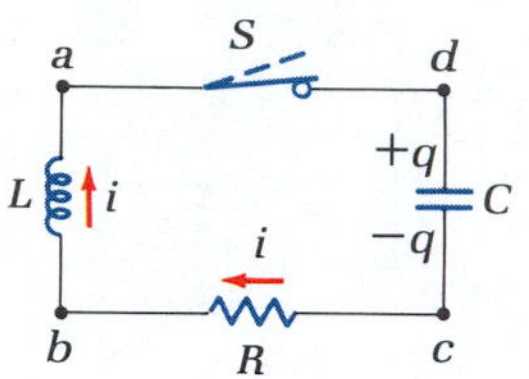

Figure 31-5. Applying the loop rule to an *RLC* circuit.

Figure 31-5 shows an *RLC* circuit. As before, the capacitor is charged by a battery, the battery is removed, and then the switch is closed. We let i be positive when the current is clockwise, and we let q be positive when the charge on the upper plate is positive so that $i = dq/dt$. Applying the loop rule, we have

$$(\mathcal{V}_b - \mathcal{V}_a) + (\mathcal{V}_c - \mathcal{V}_b) + (\mathcal{V}_d - \mathcal{V}_c) + (\mathcal{V}_a - \mathcal{V}_d) = 0$$

With our chosen sign convention for q and i, the loop rule gives

$$L\frac{di}{dt} + iR + \frac{q}{C} = 0$$

Rearranging and substituting dq/dt for i and d^2q/dt^2 for di/dt, we have

Differential equation for an *RLC* circuit

$$\frac{d^2q}{dt^2} + \frac{R}{L}\frac{dq}{dt} + \frac{1}{LC}q = 0 \tag{31-5}$$

This equation has the same mathematical form as the differential equation that describes a damped harmonic oscillator. The one-dimensional mass-spring system with a frictional force proportional to the object's speed is a damped harmonic oscillator (Sec. 14-6). Newton's second law applied to the mass-spring system with friction gives

$$\frac{d^2x}{dt^2} + \frac{\gamma}{m}\frac{dx}{dt} + \frac{k}{m}x = 0$$

where γ is the proportionality factor between the magnitude of the frictional force and the speed. If the damping factor γ is not too large, the object oscillates, and the amplitude of the oscillation decreases exponentially to zero. For example, the amplitude of a pendulum swinging in air gradually dies out. This is called the *underdamped* case. On the other hand, if the damping factor is larger than a certain critical amount, the object does not oscillate, and its displacement monotonically approaches zero. For example, suppose a pendulum is immersed in a viscous fluid, such as molasses, and the bob is displaced from its central position and released. The bob tends to return to its central position without overshooting. This is the *over*damped case. Because these two differential equations have the same form, we expect the *RLC* circuit to exhibit a behavior similar to the damped harmonic oscillator. From our discussion above and from a comparison of the two equations, you can see that the resistance R plays a role analogous to the damping factor γ.

From this mechanical analogy to the *RLC* circuit, we expect that a solution to Eq. 31-5 is

Oscillating charge with damping

$$q = Q_m e^{-t/\tau} \cos(\omega_d t + \phi) \tag{31-6}$$

The factor $e^{-t/\tau}$ provides the exponentially decreasing amplitude, and the factor $\cos(\omega_d t + \phi)$ provides the oscillations. (The subscript "*d*" on ω_d stands for *d*amped.) You can show that Eq. 31-6 is a solution by using it to substitute for q, dq/dt, and d^2q/dt^2 in Eq. 31-5. (See Prob. 3.) Further, this substitution will allow you to find values for τ and ω_d. The results are

$$\tau = \frac{2L}{R}$$

and

$$\omega_d = \sqrt{\frac{1}{LC} - \left(\frac{R}{2L}\right)^2} = \sqrt{\omega_0{}^2 - \frac{1}{\tau^2}}$$

Note that when $\omega_0 \gg 1/\tau$, $\omega_d \approx \omega_0$. In terms of R, L, and C this occurs when $(1/LC) \gg (R/2L)^2$ or $R \ll \sqrt{4L/C}$. This corresponds to the negligible-resistance case discussed in the previous section. Thus the criterion for the amount of damping caused by the resistance is the comparison between R and $\sqrt{4L/C}$. If $R > \sqrt{4L/C}$, then the circuit is overdamped. (Recall the pendulum in molasses.) Problem 2 provides further investigation of this case. For the particular case where $R = R_{crit} = \sqrt{4L/C}$, the circuit is said to be *critically damped*. Figure 31-6 shows a graph of q versus t for an underdamped circuit in which $R = 2.00 \times 10^2\ \Omega$, $L = 1.00$ mH, and $C = 1.00$ nF. The initial conditions are such that $Q_m = Q_0$ and $\phi = 0$ in Eq. 31-6.

Critical damping

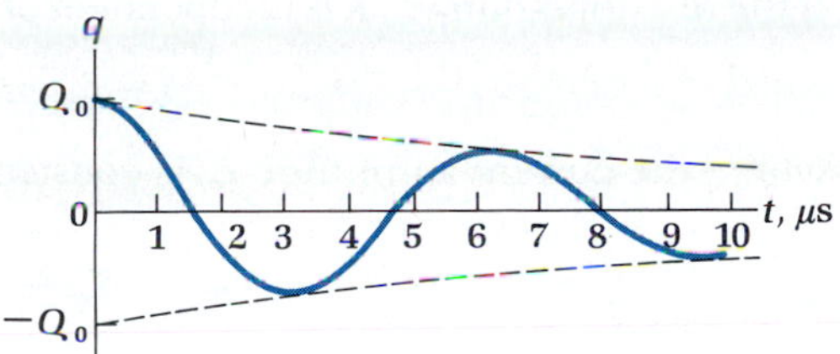

Figure 31-6. Charge on the capacitor in an underdamped *RLC* circuit.

EXAMPLE 31-2

***RLC* oscillations.** Using the values of R, L, and C given above for the graph in Fig. 31-6, find the expression for q.

Solution. To find the expression for q, we must determine τ and ω_d:

$$\tau = \frac{2L}{R} = \frac{2(1.00\text{ mH})}{2.00 \times 10^2\ \Omega} = 10.0\ \mu\text{s}$$

$$\omega_d = \sqrt{\frac{1}{LC} - \left(\frac{R}{2L}\right)^2}$$

$$= \sqrt{\frac{1}{(1.00\text{ mH})(1.00\text{ nF})} - \left[\frac{2.00 \times 10^2\ \Omega}{2(1.00\text{ mH})}\right]^2}$$

$$= 9.95 \times 10^5\text{ rad/s} = 995\text{ krad/s}$$

Since the initial conditions are such that $Q_m = Q_0$ and $\phi = 0$, we have

$$q = Q_0 e^{-t/10.0\mu s} \cos\left[(995\text{ krad/s})t\right]$$

SELF-TEST 31-2. From the above example, evaluate q in terms of Q_0 at *(a)* $t = 3.2\ \mu$s, *(b)* $t = 6.3\ \mu$s, and *(c)* $t = 9.5\ \mu$s. Use Fig. 31-6 to verify your answers. ***ANSWERS:*** *(a)* $-0.72\ Q_0$; *(b)* $0.53\ Q_0$; *(c)* $-0.39\ Q_0$.

31-3 AC SOURCE CONNECTED TO A RESISTOR

The circuits considered in earlier chapters had a source of emf, such as a battery, that was constant in time. We now investigate ac circuits, which have an alternating current. An ac current is sustained by an *ac source*. An ac source is a source of emf that produces an oscillating potential difference across its terminals; examples are the generator and the alternator discussed in Sec. 28-3. An electric outlet in your home is an ac source. The symbol for an ac source in a circuit diagram is ⊝. The

AC source

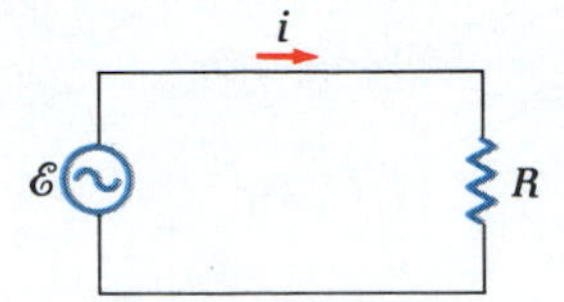

Figure 31-7. An ac source connected to a resistor.

potential difference $\mathcal{V}$ across the terminals of an ac source oscillates. For simplicity, we consider a sinusoidally varying potential difference across the source:

$$\mathcal{V} = \mathcal{V}_m \sin(\omega t) \tag{31-7}$$

where $\mathcal{V}_m$ is the amplitude of the oscillating potential difference and ω is its angular frequency. This oscillating potential difference is called an ***ac voltage***. The frequency ν of the ac voltage across the two connections to an electric outlet in your room is 60.0 Hz, so that $\omega = 2\pi\nu = 2\pi(60.0 \text{ Hz}) = 377$ rad/s.

In this section we consider a circuit that contains only a source and a resistor, a purely resistive ac circuit. Applying the loop rule to the circuit shown in Fig. 31-7, we find that the potential difference across the source is equal to the potential difference across the resistor:

$$\mathcal{V}_m \sin(\omega t) = iR$$

Solving for i gives

$$i = \frac{\mathcal{V}_m}{R}\sin(\omega t)$$

If the resistor is ohmic (R is independent of $\mathcal{V}$ or i), then the time dependence of i is

AC current in a resistive circuit

$$i = I_m \sin(\omega t) \tag{31-8}$$

where the current amplitude I_m is constant:

$$I_m = \frac{\mathcal{V}_m}{R} \tag{31-9}$$

Phasor diagram

Analysis of an ac circuit is facilitated by the use of a ***phasor diagram***. The phasor

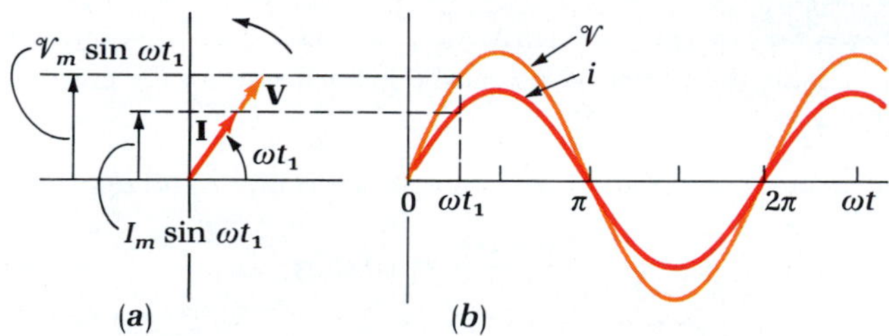

Figure 31-8. *(a)* A phasor diagram for the circuit in Fig. 31-7. *(b)* Graph of $\mathcal{V}$ and i versus ωt.

diagram for the circuit in Fig. 31-7 is shown in Fig. 31-8. A ***phasor*** is a vector that rotates about the origin with angular speed ω, as shown in the figure. The vertical components of phasors **V** and **I** represent the sinusoidally varying quantities $\mathcal{V}$ and i. The magnitudes of phasors **V** and **I** represent the amplitudes $\mathcal{V}_m$ and I_m of these oscillating quantities. The figure shows the relationship between a phasor diagram and a graph of the oscillating quantities versus ωt.

Voltage and current are in phase in a purely resistive circuit.

Figure 31-8 shows that for a purely resistive ac circuit, the voltage and current are in phase. Stated another way, the phase angle difference between the voltage and the current is zero. The phasor diagram shows this phase relationship with the phasors **V** and **I** being parallel.

31-4 AC SOURCE CONNECTED TO A CAPACITOR

Figure 31-9 shows an ac source connected to a capacitor only, a purely capacitive ac circuit. We let i be positive when the current is clockwise, and we let q be positive when the charge on the upper plate is positive so that $i = dq/dt$. We let Eq. 31-7 describe the voltage across the source: $\mathcal{V} = \mathcal{V}_m \sin(\omega t)$. From the loop rule, the voltages across the source and the capacitor are equal:

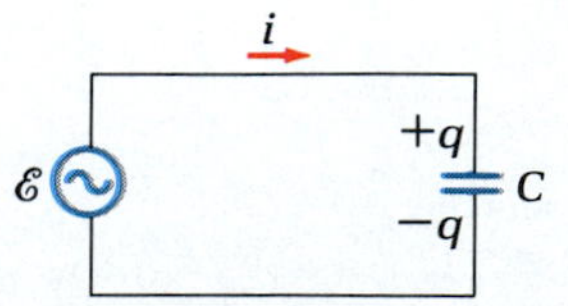

Fig. 31-9. An ac source connected to a capacitor.

$$\mathcal{V}_m \sin(\omega t) = \frac{q}{C}$$

To find the current, we solve for q and take its time derivative, $q = C\mathcal{V}_m \sin(\omega t)$, so that

$$i = \frac{dq}{dt} = \omega C\mathcal{V}_m \cos(\omega t)$$

Using $\cos(\omega t) = \sin(\omega t + \frac{1}{2}\pi)$ gives

$$i = I_m \sin(\omega t + \tfrac{1}{2}\pi) \qquad (31\text{-}10)$$

AC current in a capacitive circuit

where the amplitude of the oscillating current is $I_m = \omega C\mathcal{V}_m = \mathcal{V}_m/(1/\omega C)$. In a purely resistive circuit, the amplitude of the oscillating current was $I_m = \mathcal{V}_m/R$. By analogy to the resistance, we introduce the ***capacitive reactance*** X_C,

$$X_C = \frac{1}{\omega C} \qquad (31\text{-}11)$$

Capacitive reactance

so that the amplitude of the current is

$$I_m = \frac{\mathcal{V}_m}{X_C} \qquad (31\text{-}12)$$

Thus the current amplitude is inversely proportional to the capacitive reactance. Note that the dimension of capacitive reactance is the same as resistance, and its SI unit is the ohm (Ω).

The capacitive reactance limits the amplitude of the current in a purely capacitive circuit similar to the way resistance limits the amplitude of the current in a purely resistive circuit. But unlike resistance, capacitive reactance is frequency-dependent; it is inversely proportional to the frequency. The capacitive reactance of a capacitor is also inversely proportional to its capacitance. At a given frequency, a capacitor with a smaller capacitance impedes the current more than a capacitor with a larger capacitance.

A comparison between the expressions for $\mathcal{V}$ and i shows that these oscillating quantities are out of phase by $\frac{1}{2}\pi$ rad. From Eq. 31-7, $\mathcal{V} = \mathcal{V}_m \sin(\omega t)$, and from Eq. 31-10, $i = I_m \sin(\omega t + \frac{1}{2}\pi)$. This phase difference has been incorporated into Fig. 31-10, which shows the phasor diagram and graphs of $\mathcal{V}$ and i versus ωt for the purely capacitive circuit. In the phasor diagram, phasor **V** is $\frac{1}{2}\pi$ rad behind phasor **I** as they rotate counterclockwise. In the graphs of $\mathcal{V}$ and i versus ωt, the maxima in $\mathcal{V}$ are shifted $\frac{1}{2}\pi$ rad (or 90°) to the right of the maxima in i. This means that the voltage reaches its maximum value later than the current by one-fourth of a period $[T/4 = (\frac{1}{2}\pi)/\omega]$. This condition is described by saying, "The voltage lags the current by 90°," or "The current leads the voltage by 90°."

Voltage lags the current in a capacitive circuit.

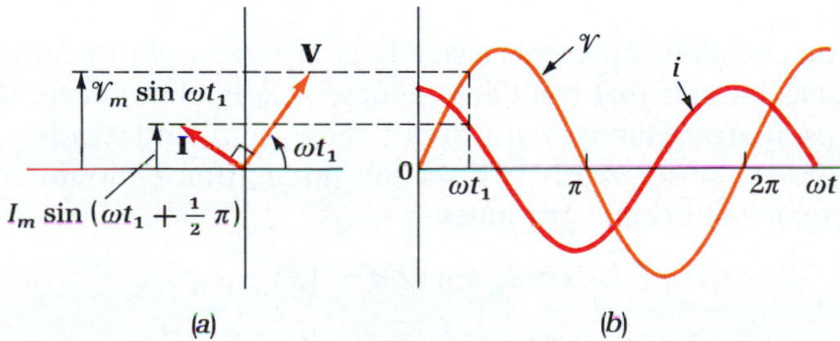

Figure 31-10. *(a)* A phasor diagram for the circuit in Fig. 31-9. *(b)* Graph of $\mathcal{V}$ and i versus ωt.

EXAMPLE 31-3

Capacitive reactance decreases with frequency. The terminals of a 650-nF capacitor are connected to an ac source with $\mathcal{V}_m = 158$ V. *(a)* If the frequency of the source is 20 kHz, determine the capacitive reactance of the capacitor and find the amplitude of the current. *(b)* If the frequency of the source is 20 Hz, determine the capacitive reactance of the capacitor and find the amplitude of the current.

Solution. *(a)* Using $\omega = 2\pi\nu$, we find that when $\nu = 20$ kHz, the capacitive reactance is

$$X_C = \frac{1}{\omega C} = \frac{1}{2\pi\nu C} = \frac{1}{(2\pi)(20\text{ kHz})(650\text{ nF})} = 12\ \Omega$$

The amplitude of the current in this case is

$$I_m = \frac{\mathcal{V}_m}{X_C} = \frac{158\text{ V}}{12\ \Omega} = 13\text{ A}$$

(b) When $\nu = 20$ Hz, the capacitive reactance is

$$X_C = \frac{1}{(2\pi)(20\text{ Hz})(650\text{ nF})} = 12\text{ k}\Omega$$

The amplitude of the current in this case is

$$I_m = \frac{158\text{ V}}{12\text{ k}\Omega} = 13\text{ mA}$$

The capacitive reactance is larger and the current amplitude is smaller for the case with the smaller frequency. The current-limiting ability of a capacitor increases with decreasing frequency. For a steady current, the frequency is zero, and a capacitor (ideally) stops all current.

SELF-TEST 31-3. An ac source with $\mathcal{V}_m = 20.0$ V and $\omega = 450$ rad/s is to be connected to a capacitor. If you want $I_m = 10.0$ mA in the circuit, what should the capacitance of the capacitor be? ***ANSWER:*** 1.1 μF.

31-5 AC SOURCE CONNECTED TO AN INDUCTOR

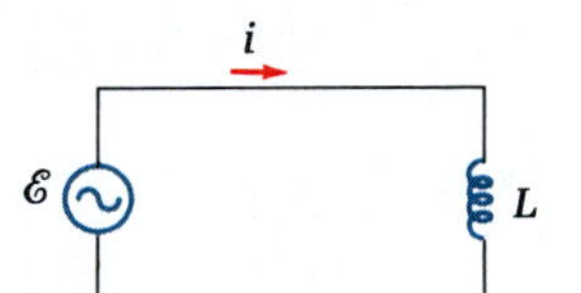

Figure 31-11. An ac source connected to an inductor.

Figure 31-11 shows an ac source connected to an inductor only, a purely inductive ac circuit. (Many inductors have appreciable resistance in their windings, but we make the simplifying assumption that this inductor has a negligibly small resistance.) Again let Eq. 31-7, $\mathcal{V} = \mathcal{V}_m \sin(\omega t)$, describe the voltage across the source. From the loop rule, the voltage across the source and the inductor are equal, $\mathcal{V}_m \sin(\omega t) = L\, di/dt$. To find the current, integrate di/dt with respect to time:

$$\int \frac{di}{dt}\, dt = \frac{\mathcal{V}_m}{L} \int \sin(\omega t)\, dt$$

Integration gives

$$i = -\frac{\mathcal{V}_m}{\omega L} \cos(\omega t) + \text{constant}$$

The integration constant represents a steady component of the current. Since the source produces an emf that oscillates symmetrically about zero, the current it sustains also oscillates symmetrically about zero, so that no steady component of the current exists. Consequently, we set the integration constant equal to zero. Using $-\cos(\omega t) = \sin(\omega t - \frac{1}{2}\pi)$ gives

AC current in an inductive circuit

$$i = I_m \sin(\omega t - \tfrac{1}{2}\pi) \tag{31-13}$$

where the amplitude of the current is $I_m = \mathcal{V}_m/(\omega L)$.

By analogy to the resistance and the capacitive reactance, we introduce the ***inductive reactance*** X_L:

Inductive reactance

$$X_L = \omega L \tag{31-14}$$

so that the amplitude of the current is

$$I_m = \frac{\mathcal{V}_m}{X_L} \tag{31-15}$$

The current amplitude is inversely proportional to the inductive reactance; the dimension of inductive reactance is the same as resistance and the same as capacitive reactance; and its SI unit is the ohm (Ω). The inductive reactance limits the current in a purely inductive circuit similar to the way the resistance limits the current in a purely resistive circuit or capacitive reactance limits the current in a purely capacitive circuit. The inductive reactance of an inductor is directly proportional to its inductance and to the frequency of the current. That is, contrary to capacitive reactance, inductive reactance increases with increasing frequency. An inductor weakly impedes a slowly varying current but it strongly impedes a rapidly varying current.

As with the capacitive circuit, a comparison between the expressions for $\mathscr{V}$ and i shows that these oscillating quantities are out of phase by $\frac{1}{2}\pi$ rad. But the phase difference has the opposite sign from the capacitive circuit. From Eq. 31-7, $\mathscr{V} = \mathscr{V}_m \sin(\omega t)$; and from Eq. 31-13, $i = I_m \sin(\omega t - \frac{1}{2}\pi)$. This phase difference has been incorporated into Fig. 31-12, which shows the phasor diagram and graphs of $\mathscr{V}$ and i versus ωt for the purely inductive circuit. In the phasor diagram, phasor **V** is $\frac{1}{2}\pi$ rad ahead of phasor **I** as they rotate counterclockwise. In the graphs of $\mathscr{V}$ and i versus ωt, the maxima in $\mathscr{V}$ are shifted $\frac{1}{2}\pi$ rad (or 90°) to the left of the maxima in i. This means that the voltage reaches its maximum value earlier than the current by one-fourth of a period $[T/4 = (\frac{1}{2}\pi)/\omega]$. Thus, the voltage leads the current by 90° or the current lags the voltage by 90°.

Voltage leads the current in an inductive circuit.

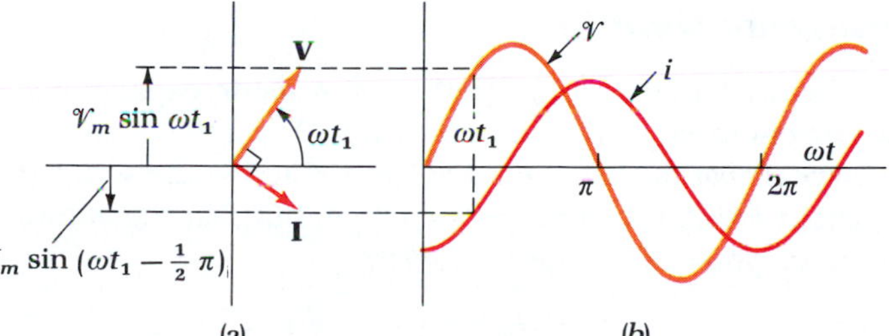

Figure 31-12. *(a)* A phasor diagram for the circuit in Fig. 31-11. *(b)* Graph of $\mathscr{V}$ and i versus ωt.

EXAMPLE 31-4

Inductive reactance increases with frequency. A 14-mH inductor is connected to an ac source with a voltage amplitude of 6.3 V and a variable frequency. *(a)* Determine the reactance of the inductor and the current amplitude in the circuit when $\omega = 340$ rad/s. *(b)* Determine the reactance of the inductor and the current amplitude in the circuit when $\omega = 340$ krad/s.

Solution. *(a)* When $\omega = 340$ rad/s,

$$X_L = \omega L = (340 \text{ rad/s})(14 \text{ mH}) = 4.8\ \Omega$$

and

$$I_m = \frac{\mathscr{V}_m}{X_L} = \frac{6.3 \text{ V}}{4.8\ \Omega} = 1.3 \text{ A}$$

(b) When $\omega = 340$ krad/s,

$$X_L = 340 \text{ krad/s})(14 \text{ mH}) = 4.8 \text{ k}\Omega$$

and

$$I_m = \frac{6.3 \text{ V}}{4.8 \text{ k}\Omega} = 1.3 \text{ mA}$$

SELF-TEST 31-4. An ac source with $\mathscr{V}_m = 55$ V and $\omega = 86$ krad/s is connected to an inductor. If you want $I_m = 380$ mA in the circuit, what should the inductance of the inductor be? Assume that the inductor's resistance is negligible. ***ANSWER:*** 1.7 mH.

31-6 SERIES *RLC* CIRCUIT DRIVEN BY AN AC SOURCE

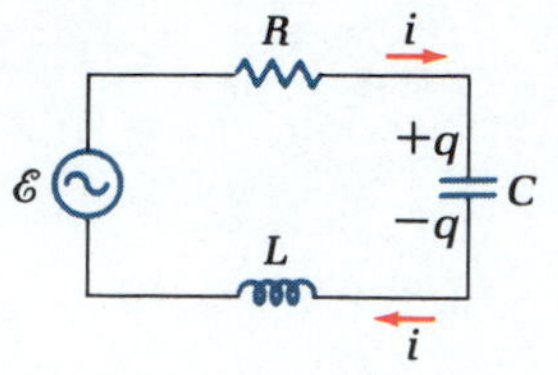

Figure 31-13. A series *RLC* circuit driven by an ac source.

A series *RLC* circuit driven by an ac source exhibits many properties that are common to ac circuits in general. The study of this circuit brings together the features of the previous three sections. Figure 31-13 shows the series combination of a resistor, inductor, capacitor, and source. From the loop rule, the sum of the voltages across the inductor, the resistor, and the capacitor is equal to the voltage across the source:

$$L\frac{di}{dt} + iR + \frac{q}{C} = \mathcal{V} \qquad (31\text{-}16)$$

where the voltage across the source is

$$\mathcal{V} = \mathcal{V}_m \sin(\omega t)$$

Five parameters characterize this circuit: L, R, C, $\mathcal{V}_m$, and ω. The values of these parameters determine the current in the circuit.

Equation 31-16 is analogous to the equation of motion for the forced, damped, harmonic oscillator discussed in Sec. 14-7. However, that analogy will not be used to help solve this circuit problem. Because of their extensive use in solving more complicated ac circuit problems, we shall use phasor diagrams.

Phasor-Diagram Solutions

Since the four elements in the circuit (inductor, resistor, capacitor, and source) are in series, the current in each element is the same. From the results of the past three sections, we expect that the oscillating voltage $\mathcal{V}$ across the source will sustain an oscillating current i with the same frequency ω, but that the voltage and the current may be out of phase. Therefore we write

$$i = I_m \sin(\omega t - \phi) \qquad (31\text{-}17)$$

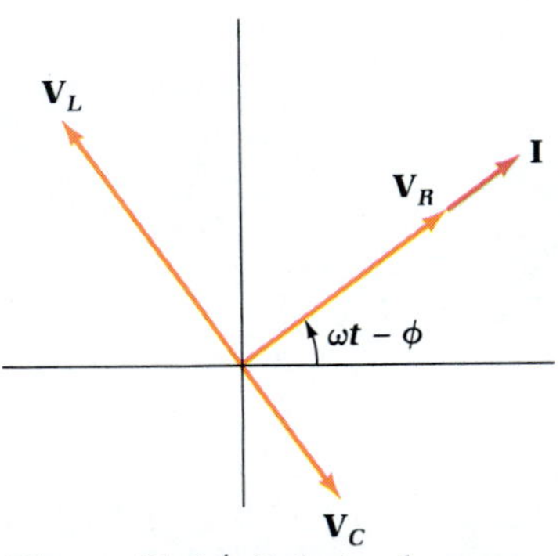

Figure 31-14. Relation between phasors $\mathbf{V}_L$, $\mathbf{V}_R$, $\mathbf{V}_C$, and **I** for the circuit in Fig. 31-13.

where ϕ is the phase difference between the voltage across the source and the current in the circuit. We need to construct a phasor diagram that will allow us to determine ϕ and the current amplitude I_m. In this diagram, a single phasor **I**, whose vertical component gives Eq. 31-17, represents the current in each element.

The voltage across each element is different in general, so we will need four phasors—$\mathbf{V}_L$, $\mathbf{V}_R$, $\mathbf{V}_C$, and **V**—to represent the voltage across the inductor, the resistor, the capacitor, and the source, respectively. Figure 31-14 shows a diagram with phasors $\mathbf{V}_L$, $\mathbf{V}_R$, $\mathbf{V}_C$, and **I**. Phasor $\mathbf{V}_R$ corresponds to the voltage phasor discussed in Sec. 31-3; $\mathbf{V}_C$ corresponds to the voltage phasor discussed in Sec. 31-4; $\mathbf{V}_L$ corresponds to the voltage phasor discussed in Sec. 31-5. This means that as these phasors rotate counterclockwise:

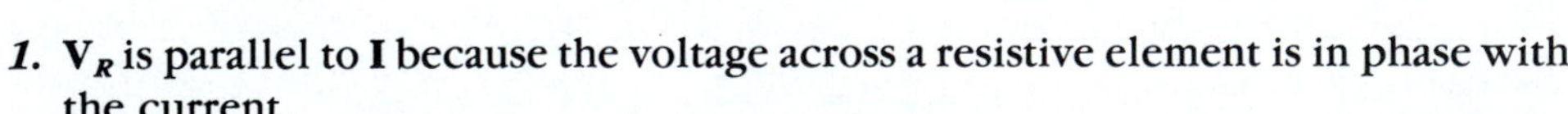

1. $\mathbf{V}_R$ is parallel to **I** because the voltage across a resistive element is in phase with the current.
2. $\mathbf{V}_C$ is $\frac{1}{2}\pi$ rad behind **I** because the voltage across a capacitive element lags the current by $\frac{1}{2}\pi$ rad.
3. $\mathbf{V}_L$ is $\frac{1}{2}\pi$ rad ahead of **I** because the voltage across an inductive element leads the current by $\frac{1}{2}\pi$ rad.

From Eqs. 31-9, 31-12, and 31-15, the lengths of these phasors are

$$\mathcal{V}_{Rm} = I_m R \qquad \mathcal{V}_{Cm} = I_m X_C \qquad \mathcal{V}_{Lm} = I_m X_L \qquad (31\text{-}18)$$

Now we use Eq. 31-16 to find the relation among phasors $\mathbf{V}_L$, $\mathbf{V}_R$, $\mathbf{V}_C$, and **V**. Rewriting Eq. 31-16, we have

$$\mathcal{V}_L + \mathcal{V}_R + \mathcal{V}_C = \mathcal{V} \qquad (31\text{-}19)$$

where $\mathcal{V}_L$, $\mathcal{V}_R$, $\mathcal{V}_C$, and $\mathcal{V}$ are the instantaneous voltages across the inductor, the resistor, the capacitor, and the source, respectively. The phasor relation whose vertical component gives Eq. 31-19 is

$$\mathbf{V}_L + \mathbf{V}_R + \mathbf{V}_C = \mathbf{V} \tag{31-20}$$

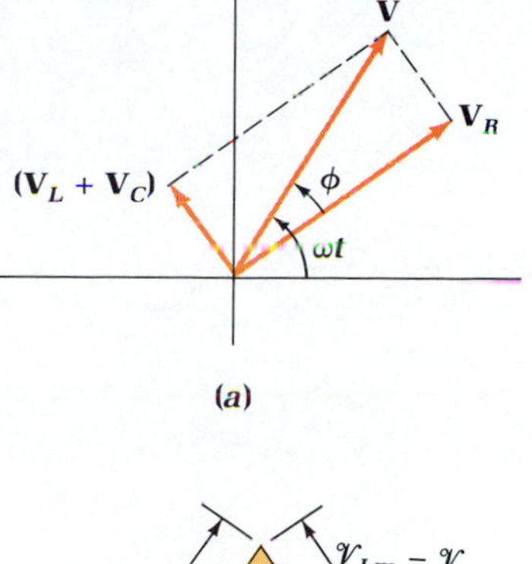

Figure 31-15 shows the phasor relation that represents Eq. 31-20. Since $\mathbf{V}_L$ and $\mathbf{V}_C$ are always along the same line and in opposite directions, we combine them into a single phasor $(\mathbf{V}_L + \mathbf{V}_C)$, which has magnitude $|\mathcal{V}_{Lm} - \mathcal{V}_{Cm}|$. Since $\mathbf{V}$ is represented as the hypotenuse of a right triangle whose sides are $\mathbf{V}_R$ and $(\mathbf{V}_L + \mathbf{V}_C)$, the phythagorean theorem gives

$$\mathcal{V}_m{}^2 = \mathcal{V}_{Rm}{}^2 + (\mathcal{V}_{Lm} - \mathcal{V}_{Cm})^2$$

Using Eqs. 31-18 to substitute for $\mathcal{V}_{Rm}$, $\mathcal{V}_{Cm}$, and $\mathcal{V}_{Lm}$, we have

$$\mathcal{V}_m{}^2 = (I_mR)^2 + (I_mX_L - I_mX_C)^2 = I_m{}^2[R^2 + (X_L - X_C)^2]$$

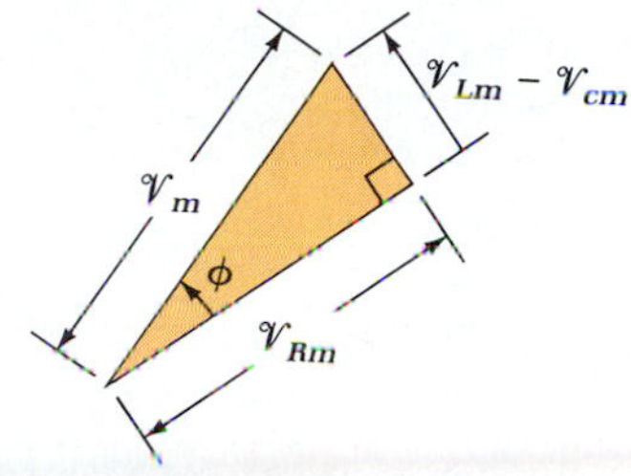

Figure 31-15. *(a)* Relation between phasors $\mathbf{V}$, $\mathbf{V}_R$, and $(\mathbf{V}_L + \mathbf{V}_C)$ for the circuit in Fig. 31-13. *(b)* Relation between the magnitudes of the phasors in *(a)*. The magnitude of phasor $\mathbf{V}$ is given by the pythagorean theorem because phasor $\mathbf{V}_R$ is perpendicular to phasor $(\mathbf{V}_L + \mathbf{V}_C)$.

Solving for I_m,

$$I_m = \frac{\mathcal{V}_m}{\sqrt{R^2 + (X_L - X_C)^2}} \tag{31-21}$$

By analogy to the resistance in a circuit, we introduce the ***impedance*** Z in an ac circuit:

$$\mathcal{V}_m = I_mZ \qquad \text{or} \qquad Z = \frac{\mathcal{V}_m}{I_m}$$

From Eq. 31-21, we see that the impedance of a series *RLC* circuit is

$$Z = \sqrt{R^2 + (X_L - X_C)^2} \tag{31-22}$$

Since phasor $\mathbf{I}$ is parallel to phasor $\mathbf{V}_R$, the phase difference ϕ between i and $\mathcal{V}$ can be determined from Fig. 31-15: $\tan\phi = (\mathcal{V}_{Lm} - \mathcal{V}_{Cm})/\mathcal{V}_{Rm}$. Using Eqs. 31-18, this can be written as

$$\tan\phi = \frac{X_L - X_C}{R} \tag{31-23}$$

Equations 31-22 and 31-23 are shown graphically in Fig. 31-16. This is an ***impedance diagram,*** a right triangle with Z as its hypotenuse. Note that if $X_L > X_C$, then ϕ is positive. In this case the circuit is predominantly inductive, and the voltage across the source leads the current. Figure 31-17 shows this case. On the other hand, if $X_L < X_C$ then ϕ is negative. In this case the circuit is predominantly capacitive, and the voltage across the source lags the current.

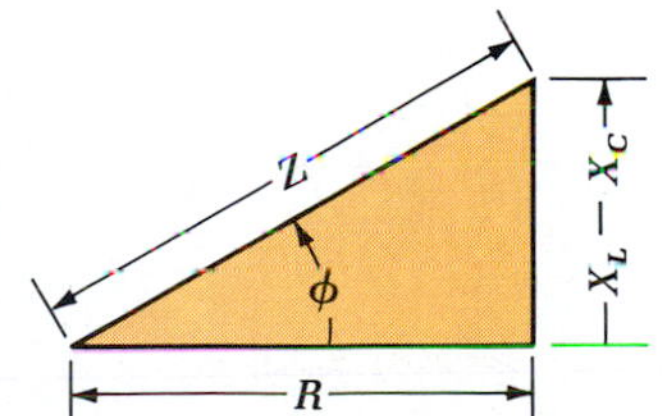

Figure 31-16. Impedance diagram.

Using phasor diagrams to solve a circuit problem is instructive, but causes us to bypass a feature that is sometimes significant. Notice that there is no mention of the initial conditions in our phasor-diagram discussion. The solution we obtained is called the ***steady-state solution.*** There is also a ***transient solution*** whose form is similar to that of the *RLC* circuit without a source (Sec. 31-2). The ***general solution*** is the sum of the transient and the steady-state solutions. After a sufficiently

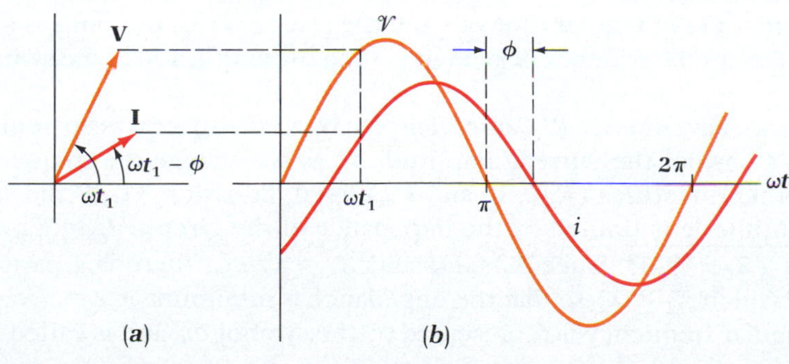

Figure 31-17. *(a)* Phasor diagram of $\mathbf{V}$ and $\mathbf{I}$ for a predominantly inductive series *RLC* circuit driven by an ac source. *(b)* Graphs of $\mathcal{V}$ and i versus ωt for this circuit. In this case, the voltage leads the current.

long time interval, the effects of the transient solution become insignificant, and the behavior of the circuit is described by the steady-state solution. (Recall Sec. 14-7).

EXAMPLE 31-5

Series RLC circuit with an ac source. A series *RLC* circuit with $R = 580\ \Omega$, $L = 31$ mH, and $C = 47$ nF is driven by an ac source. The amplitude and angular frequency of the source are 65 V and 33 krad/s. Determine *(a)* the reactance of the capacitor, *(b)* the reactance of the inductor, *(c)* the impedance of the circuit, *(d)* the phase difference between the voltage across the source and the current, and *(e)* the current amplitude. *(f)* Does the current lead or lag the voltage across the source?

Solution

(a) $$X_C = \frac{1}{\omega C} = \frac{1}{(33\text{ krad/s})(47\text{ nF})} = 640\ \Omega$$

(b) $$X_L = \omega L = (33\text{ krad/s})(31\text{ mH}) = 1000\ \Omega$$

(c) $$Z = \sqrt{R^2 + (X_L - X_C)^2}$$

$$= \sqrt{(580\ \Omega)^2 + (640\ \Omega - 1.0\text{ k}\Omega)^2} = 690\ \Omega$$

(d) $$\phi = \tan^{-1}\frac{X_L - X_C}{R}$$

$$= \tan^{-1}\frac{1000\ \Omega - 640\ \Omega}{580\ \Omega} = 0.58\text{ rad}$$

(e) $$I_m = \frac{V_m}{Z} = \frac{65\text{V}}{690\ \Omega} = 94\text{ mA}$$

(f) Since ϕ is positive, the voltage across the source leads the current.

SELF-TEST 31-5. Suppose, in the circuit in the above example, all circuit parameters remain the same except that the angular frequency of the source is changed to 24 krad/s. Determine each of the quantities asked for in that example. ***ANSWERS:*** *(a)* 890 Ω; *(b)* 740 Ω; *(c)* 600 Ω; *(d)* −0.24 rad; *(e)* 100 mA; *(f)* voltage across the source lags the current.

Resonance

An interesting and useful characteristic of the series *RLC* circuit driven by an ac source is the phenomenon of *resonance.* Resonance is a common feature of systems that have a tendency to oscillate at a particular frequency. This oscillation frequency is called the system's *natural frequency.* If such a system is driven by an energy source at a frequency that is near the natural frequency, then the amplitude of the oscillation is large. An example is a child on a playground swing. The child seated on the swing has a natural frequency for swinging back and forth. If the child pulls on the ropes at regular intervals and the frequency of the pulls is almost the same as the natural frequency of swinging, then the amplitude of the swinging will be large.

Suppose we have a series *RLC* circuit driven by an ac source whose frequency can be varied. Consider the current amplitude I_m as we change the frequency while keeping other quantities (R, L, C, and $\mathcal{V}_m$) fixed. Equation 31-20 shows that the current amplitude is limited by the impedance of the circuit, $I_m = \mathcal{V}_m/Z$, where $Z = \sqrt{R^2 + (X_L - X_C)^2}$. Since $X_L = \omega L$ and $X_C = 1/\omega C$, there is a particular frequency at which $X_L = X_C$ so that the impedance is minimum at $Z = \sqrt{R^2 + (0)^2} = R$. This angular frequency is represented by the symbol ω_0 and is called the ***reso-***

nant angular frequency. Using $X_L = X_C$ at $\omega = \omega_0$, we have $\omega_0 L = 1/\omega_0 C$ or

$$\omega_0 = \frac{1}{\sqrt{LC}} \tag{31-24}$$

Resonant angular frequency

When $\omega = \omega_0$, the current amplitude is maximum at $I_m = \mathcal{V}_m/R$. Notice that ω_0 is given by the same expression as the angular frequency of oscillation for the LC circuit with no resistor or source (Eq. 31-4 in Sec. 31-1). It is often called the circuit's ***natural angular frequency.***

Figure 31-18 shows graphs of I_m versus ω for two cases; in the upper curve, $R = 100\ \Omega$, and in the lower curve, $R = 200\ \Omega$. The other quantities are the same for each curve: $\mathcal{V}_m = 100$ V, $L = 1.00$ mH, $C = 1.00$ nF. Each curve exhibits a maximum current at a resonant frequency. Since the product LC is the same for each case, the resonant frequency is also the same: $\omega_0 = 1/\sqrt{LC} = 1/\sqrt{(1.00\text{ mH})(1.00\text{ nF})} = 1.00$ Mrad/s. At resonance, $I_m = \mathcal{V}_m/R$. Thus the current amplitude at resonance is twice as great for the circuit with $R = 100\ \Omega$ compared with the circuit with $R = 200\ \Omega$. At frequencies much less than ω_0, a circuit is predominantly capacitive, and the current is limited mainly by its capacitive reactance. At frequencies much greater than ω_0, a circuit is predominantly inductive, and the current is limited mainly by its inductive reactance.

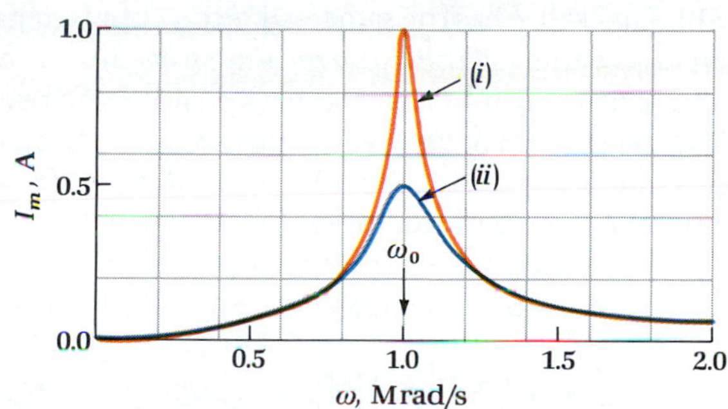

Figure 31-18. Current amplitude I_m versus ω for two cases: *(i)* $R = 100\ \Omega$ and *(ii)* $R = 200\ \Omega$. Other quantities are $\mathcal{V}_m = 100$ V, $L = 1.00$ mH, and $C = 1.00$ nF.

The tuning circuit of a radio or television set is an example of a circuit with a resonant frequency. The antenna of a radio accepts signals from many nearby stations. The antenna is the source in the tuning circuit, so the circuit is driven at many frequencies. However, the only large component of the current is the component that oscillates near the circuit's resonant frequency. The circuit discriminates against signals not near its resonant frequency. When you tune a radio, you vary the capacitance of a capacitor in the tuning circuit. This varies the resonant frequency of the circuit so that it matches the transmitting frequency of the station you wish to hear.

31-7 POWER FOR AN *RLC* CIRCUIT DRIVEN BY AN AC SOURCE

We now consider the rate at which energy is exchanged among the circuit elements of an *RLC* circuit driven by an ac source. That is, we examine the rate at which energy enters and leaves each of the elements. Ordinarily, the source frequency is too high for the time dependence of these energy exchanges to be of interest. Thus we are concerned chiefly with the average power $\overline{P}$, and this average is taken over an integral number of cycles. For simplicity, we continue to assume that the circuit's entire resistance is contained in the resistor, its entire capacitance is contained in the capacitor, and its entire inductance is contained in the inductor.

Consider the exchange of energy among the four elements of our circuit:

1. The source delivers electromagnetic energy to the circuit; it converts energy from some other form to electromagnetic energy.

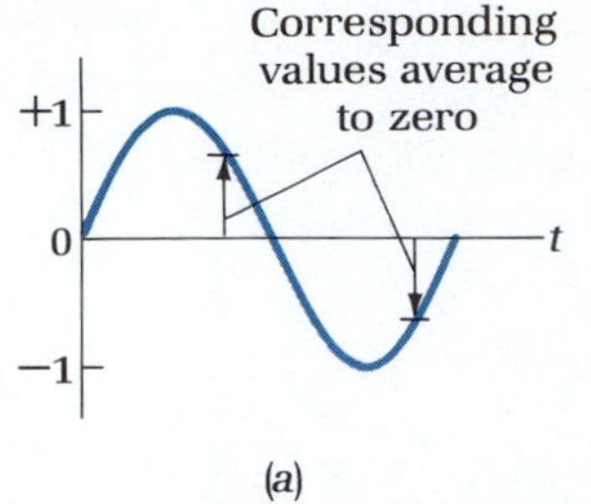

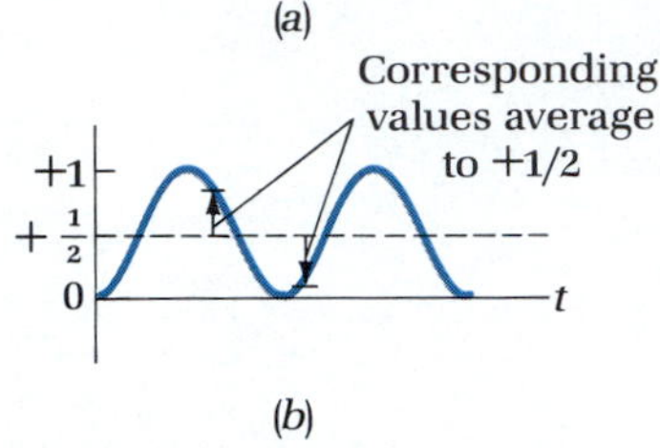

Figure 31-19. *(a)* The average value of sin (ωt) over an integral number of periods is zero. *(b)* The average value of $\sin^2(\omega t)$ over an integral number of periods is 1/2.

2. The resistor dissipates electromagnetic energy as heat; energy leaves the circuit through i^2R heating in the resistor (Sec. 25-2).

3. At any instant, electromagnetic energy may be entering or leaving the capacitor, depending on whether it is charging or discharging. Since the current oscillates sinusoidally, the energy that enters during the charging part of the cycle is equal to the energy that leaves during the discharging part of the cycle (Sec. 31-1). Consequently, the average power for the capacitor is zero.

4. As with the capacitor, the inductor is an energy-storage device. The average power for the inductor is zero.

For the circuit as a whole, energy enters at the source and leaves at the resistor. We shall refer to the average rate of this energy transfer as the ***average power*** $\overline{P}$ for the circuit.

The average power involves the average of the product of two sinusoidally varying quantities — for example, the square of the current. Therefore, it is convenient to introduce the ***root-mean-square (rms) values*** of the current and voltage. Because a sine function oscillates symmetrically about zero, the average value of a sinusoidally varying quantity, such as the current or voltage, is zero. For any instant that the function has a particular positive value, there is a corresponding instant in which its value has the same magnitude but is negative (Fig. 31-19*a*). However, the average value of the *square* of a sinusoidally varying quantity is *not* zero. The square of a sine function is always positive and oscillates symmetrically about +1/2 (Fig. 31-19*b*). For any instant that $\sin^2(\omega t)$ has a value greater than +1/2, there is a corresponding instant in which its value is the same amount smaller than +1/2. Thus the average value of $\sin^2(\omega t)$ is +1/2.

The root-mean-square value of a quantity is the square root of the average value of the square of the quantity. For example, the rms voltage is

$$\mathcal{V}_{\text{rms}} = \left(\overline{\mathcal{V}^2}\right)^{1/2} = \left\{\overline{[\mathcal{V}_m \sin(\omega t)]^2}\right\}^{1/2} = \mathcal{V}_m\left[\overline{\sin^2(\omega t)}\right]^{1/2} = \mathcal{V}_m(1/2)^{1/2}$$

or

RMS voltage

$$\mathcal{V}_{\text{rms}} = \frac{\mathcal{V}_m}{\sqrt{2}}$$

Similarly,

RMS current

$$I_{\text{rms}} = \frac{I_m}{\sqrt{2}}$$

When a value is given for an ac voltage or current, it is ordinarily the rms value. AC voltmeters and ammeters are calibrated to measure rms values. The voltage across the terminals of an outlet in your room is nominally 120 V. As shown in Fig. 31-20, this refers to the rms value of the voltage. The amplitude of this voltage is $\mathcal{V}_m = \sqrt{2}(120\text{ V}) = 170\text{ V}$.

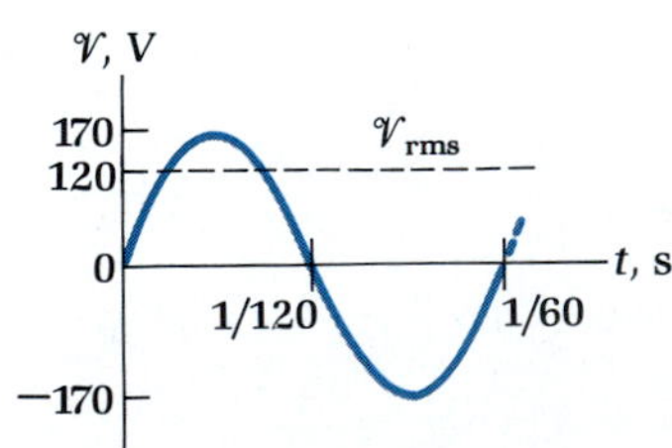

Figure 31-20. The voltage across the terminals of an electric outlet in your room.

Now we evaluate the average power $\overline{P}$ for an *RLC* circuit driven by an ac source. The instantaneous power P dissipated in the resistor is $P = i^2R$. Using Eq. 31-17 gives

$$P = [I_m \sin(\omega t + \phi)]^2 R$$

The average power $\overline{P}$ dissipated in the resistor is $\overline{P} = I_m^2 R\,\overline{\sin^2(\omega t + \phi)} = \frac{1}{2} I_m^2 R$, or

Average power

$$\overline{P} = (I_{\text{rms}})^2 R \tag{31-25}$$

This expression for the average power dissipated in the resistor resembles the expression for the power dissipated in a resistor in a dc circuit; the dc current is replaced by the rms current.

Another useful way to express $\overline{P}$ is in terms of the product of $\mathcal{V}_{\text{rms}}$ and I_{rms}, where

$\mathcal{V}_{\text{rms}}$ refers to the rms value of the voltage across the source. If we divide both sides of the equation $\mathcal{V}_m = I_m Z$ by $\sqrt{2}$, then we have

$$\mathcal{V}_{\text{rms}} = I_{\text{rms}} Z \quad \text{or} \quad I_{\text{rms}} = \frac{\mathcal{V}_{\text{rms}}}{Z}$$

Substitution into Eq. 31-25 gives

$$\overline{P} = \frac{\mathcal{V}_{\text{rms}}}{Z} I_{\text{rms}} R = \mathcal{V}_{\text{rms}} I_{\text{rms}} \frac{R}{Z}$$

From the impedance diagram, Fig. 31-16, we have

$$\cos\phi = \frac{R}{Z} = \frac{R}{\sqrt{R^2 + (X_L - X_C)^2}}$$

Power factor

where $\cos\phi$ is called the ***power factor.*** In terms of the power factor, the average power is

$$\overline{P} = \mathcal{V}_{\text{rms}} I_{\text{rms}} \cos\phi \tag{31-26}$$

Average power in terms of the power factor

If the circuit is driven by the source at the resonant frequency, then $X_C = X_L$, $Z = R$, and $\cos\phi = 1$. At resonance, the average power is $\mathcal{V}_{\text{rms}} I_{\text{rms}}$.

Figure 31-21 shows graphs of $\overline{P}$ versus ω for the same two cases that were used in Fig. 31-18: $R = 100\ \Omega$ and $R = 200\ \Omega$. For each case, $\mathcal{V}_m = 100$ V, $L = 1.00$ mH, and $C = 1.00$ nF.

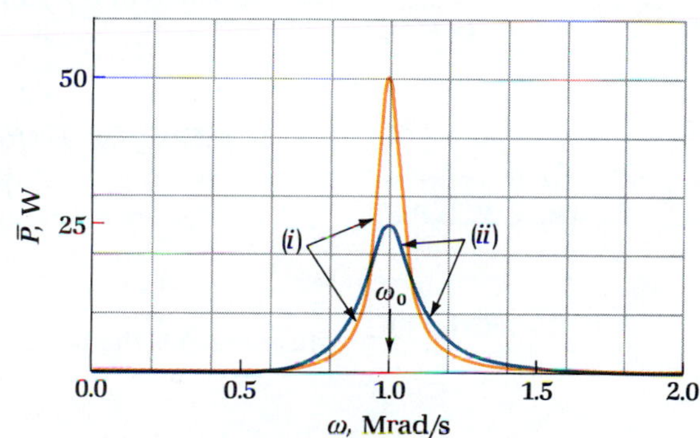

Figure 31-21. Average power $\overline{P}$ versus ω for two cases: *(i)* $R = 100\ \Omega$ and *(ii)* $R = 200\ \Omega$. Other quantities are $\mathcal{V}_m = 100$ V, $L = 1.00$ mH, and $C = 1.00$ nF.

EXAMPLE 31-6

Average power in terms of circuit parameters. The five parameters which describe the series *RLC* circuit driven by an ac source are R, L, C, $\mathcal{V}_m$, and ω. Develop an expression for $\overline{P}$ in terms of these parameters.

Solution. Using $I_{\text{rms}} = \mathcal{V}_{\text{rms}}/Z$ to substitute for $(I_{\text{rms}})^2$ in Eq. 31-25, we have

$$\overline{P} = \frac{(\mathcal{V}_{\text{rms}})^2 R}{Z^2} = \frac{(\mathcal{V}_{\text{rms}})^2 R}{R^2 + (X_L - X_C)^2}$$

Substituting for $\mathcal{V}_{\text{rms}}$, X_C, and X_L, we have

$$\overline{P} = \frac{\frac{1}{2}\mathcal{V}_m{}^2 R}{R^2 + [\omega L - (1/\omega C)]^2}$$

Note that if the circuit is driven by the source at the resonant frequency, then $\omega L = 1/\omega C$, and the average power is $\frac{1}{2}\mathcal{V}_m{}^2/R$.

SELF-TEST 31-6. Consider a series *RLC* circuit driven by an ac source with $R = 870\ \Omega$, $L = 24$ mH, $C = 390$ nF, $\mathcal{V}_m = 62$ V, and $\omega = 5.5$ krad/s. Use the expression in the above example to determine the average power dissipated by the resistor. ***ANSWER:*** 1.9 W.

SUMMARY

Section 31-1. *LC* Oscillations

In an LC circuit, the charge on the capacitor and the current in the circuit oscillate sinusoidally with the same angular frequency ω_0:

$$\omega_0 = \frac{1}{\sqrt{LC}} \tag{31-4}$$

The electromagnetic energy U, which is the sum of the electric energy U_E of the capacitor and the magnetic energy U_B of the inductor, remains constant. This energy is passed back and forth between the capacitor and the inductor.

Section 31-2. Series *RLC* Circuit

In an RLC circuit, if $R < \sqrt{4L/C}$, then the charge on the capacitor and the current in the circuit oscillate, and the circuit is said to be underdamped. These oscillations tend to die out as electromagnetic energy is dissipated in the resistor. If $R > \sqrt{4L/C}$, then no oscillations occur, and the circuit is said to be overdamped. A circuit is critically damped if $R = R_{crit} = \sqrt{4L/C}$.

Sections 31-3 through 31-6. AC Sources Connected to a Resistor, to a Capacitor, to an Inductor; Series *RLC* Circuit Driven by an AC Source

We considered circuits with an ac source in which the voltage across the source is given by

$$\mathcal{V} = \mathcal{V}_m \sin(\omega t) \tag{31-7}$$

The current sustained by this source is

$$i = I_m \sin(\omega t - \phi) \tag{31-17}$$

In a purely resistive circuit, $\phi = 0$ and $I_m = \mathcal{V}_m/R$. In a purely capacitive circuit, $\phi = -\frac{1}{2}\pi$ and $I_m = \mathcal{V}_m/X_C$, where X_C is the capacitive reactance:

$$X_C = \frac{1}{\omega C} \tag{31-11}$$

In a purely inductive circuit, $\phi = \frac{1}{2}\pi$ and $I_m = \mathcal{V}_m/X_L$, where X_L is the inductive reactance:

$$X_L = \omega L \tag{31-14}$$

In a series RLC circuit driven by an ac source,

$$\tan\phi = \frac{X_L - X_C}{R} \tag{31-23}$$

and

$$I_m = \frac{\mathcal{V}_m}{\sqrt{R^2 + (X_L - X_C)^2}} \tag{31-21}$$

The impedance of the circuit is

$$Z = \sqrt{R^2 + (X_L - X_C)^2} \tag{31-22}$$

If the angular frequency of the source is the same as the circuit's resonant angular frequency, $\omega = \omega_0 = 1/\sqrt{LC}$, then $X_C = X_L$, $\phi = 0$, the impedance has a minimum value $Z = R$, and the current amplitude has a maximum value $I_m = \mathcal{V}_m/R$.

Section 31-7. Power for an *RLC* Circuit Driven by an AC Source

The rms values of the current and resistance are

$$\mathcal{V}_{rms} = \frac{\mathcal{V}_m}{\sqrt{2}} \quad \text{and} \quad I_{rms} = \frac{I_m}{\sqrt{2}}$$

Two useful expressions for the average power are

$$\overline{P} = (I_{rms})^2 R \tag{31-25}$$

and

$$\overline{P} = \mathcal{V}_{rms} I_{rms} \cos\phi \tag{31-26}$$

where $\cos\phi$ is the power factor.

QUESTIONS

1. Once a current is started in an LC circuit (assumed to be resistanceless), why does it continue to oscillate despite the fact that no source is in the circuit?

2. By what factor will the frequency of the oscillations of an LC circuit change if the inductance is doubled while the capacitance is kept fixed? What is this factor if the capacitance is doubled while the inductance is kept fixed? What is this factor if both the capacitance and the inductance are doubled?

3. Using Lenz's law, explain why $\mathcal{V}_b - \mathcal{V}_a$ in Fig. 31-1 is given by $L(di/dt)$ rather than by $-L(di/dt)$.

4. In the analogy between the LC circuit and the harmonic oscillator, which energy—the electric or the magnetic—is analogous to the kinetic energy? Which is analogous to the potential energy?

5. What analogous quantities should be added to Table 31-1 to make a comparison between the RLC circuit and the damped harmonic oscillator?

6. In an LC circuit (resistance is negligible), the frequency depends on the product LC. Does the frequency of an underdamped RLC circuit depend only on the product LC? Suppose that for an underdamped RLC circuit you increase L and decrease C while keeping the product LC fixed. Will the frequency increase, remain the same, or decrease?

7. The current in a circuit is effectively stopped by an "open" and is not limited at all by a "short." At high frequencies, is a capacitor a short or an open? How about low frequencies? Consider whether an inductor is an open or a short at low frequencies and at high frequencies.

8. An inductor called a "ballast" is often used to limit the

current in fluorescent lights. Why is an inductor preferable to a resistor for this purpose?

9. A solenoid with a resistance of 1.2 Ω is connected across the terminals of an electric outlet in your room ($\mathcal{V}_{\rm rms} =$ 120 V). Despite the fact that a circuit breaker in the circuit will trip if $I_{\rm rms}$ in the circuit exceeds 15 A, the breaker does not trip. Explain why.

10. In some textbooks the oscillating current or voltage is taken to be the horizontal component of a phasor rather than the vertical component. How would such a convention affect the results we have given here? Find an equation in this chapter that would be different if we adopted this convention.

11. In some textbooks the expressions for the voltage and current are taken as $\mathcal{V} = \mathcal{V}_m \sin(\omega t)$ and $i = I_m \sin(\omega t + \phi)$. How would such a convention affect the results we have given here? Find an equation in this chapter that would be different if we adopted this convention.

12. In an *RLC* circuit driven by an ac source, the net energy delivered by the source during one cycle is 25 mJ. During one cycle, *(a)* what is the net energy that enters the inductor, *(b)* what is the net energy that enters the capacitor, *(c)* what is the energy dissipated as heat in the resistor?

13. The power rating of an element used in ac circuits refers to the element's average power rating. What is the maximum instantaneous power to a 60-W light bulb?

14. Consider the net power delivered by the source in an *RLC* circuit. Is the instantaneous power always positive? Does your answer depend on the frequency of the source relative to the resonant frequency? Explain.

15. The average current in the power line to your house is zero. Despite this fact, electric power is delivered to your house. Explain.

16. A resistor and an ac source are each inside unmarked boxes so that you cannot tell which is which by visual inspection. Two wires carry an alternating current between the two. Can you determine the direction of energy flow and so determine which box contains the resistor and which contains the source by measuring rms current and voltage values? By measuring instantaneous current and voltage values?

17. Capacitors have a maximum voltage rating. If this rating is exceeded, the dielectric between the plates may break down. In a series *RLC* circuit, it is possible to exceed the voltage rating of the capacitor (and damage the capacitor) even though this rating is higher than the amplitude of the voltage across the source. Explain. With a variable frequency source, is this more likely to happen at the resonant frequency, or when the circuit is predominantly capacitive?

18. Consider the "resonance curves" for the current amplitude (Fig. 31-18) and average power (Fig. 31-21). Discuss the appearance of these curves as *R* approaches zero while *L* and *C* are fixed.

19. Is the impedance diagram (Fig. 31-16) always a right triangle? Explain.

20. Is the circuit in Example 31-5 predominantly capacitive or predominantly inductive?

21. An *RLC* circuit with a resonant angular frequency in the range of 1 krad/s to 1 Mrad/s can be readily constructed. What practical problems make it difficult to construct a circuit with a resonant angular frequency of 1 rad/s?

22. What is a common characteristic between a radio playing the station you selected and a child swinging on a swing? What are some other systems that exhibit this characteristic?

23. Suppose an *RLC* series circuit is driven by an ac source at a particular frequency. If you know the power factor, can you determine whether the circuit is predominantly capacitive or predominantly inductive? Explain.

24. Electric power companies prefer to have the value of the power factor for their "load" be as nearly 1 as possible. Explain why this is so. (*Hint:* Take into account the i^2R losses on the transmission lines.)

25. In the latter part of the nineteenth century, prior to the advent of electric power transmission, George Westinghouse (1846–1914) and Thomas A. Edison (1847–1931) entered into a disagreement about whether ac or dc should be used to transmit electric energy. Westinghouse favored ac and Edison favored dc. Which system do you think is preferable? Explain. (*Hint:* Reexamine Sec. 29-5.)

26. Complete the following table:

Symbol	Represents	Type	SI Unit
ω_0			
I_m			
U	Electromagnetic energy		
$\mathbf{V}$			
X_C			
X_L			
Z			Ω
ϕ			
$I_{\rm rms}$		Scalar	
$\overline{P}$			
$\cos\phi$			

EXERCISES

Section 31-1. *LC* Oscillations

1. Show that $\sqrt{LC}$ has the dimension of time.

2. *(a)* What is the angular frequency of oscillation of the charge and current in an *LC* circuit with $L = 25$ mH and $C = 41$ nF? *(b)* What is the frequency of the oscillations? *(c)* What is the period of the oscillations?

3. In an *LC* circuit with $C = 58$ nF, the angular frequency of oscillation of the charge and current is 58 krad/s. What is the inductance of the inductor?

4. In an *LC* circuit with $L = 94$ mH, the frequency of oscillation of the charge and current is 130 kHz. What is the capacitance of the capacitor?

5. Consider an *LC* circuit in which $L = 5.3$ mH, $C = 17$ nF, the initial charge of the capacitor is 2.2 μC, and the initial current in the circuit is zero. Write expressions for q, i, U, U_E, and U_B as functions of t.

6. Consider an *LC* circuit in which $L = 71$ mH, $C = 130$ nF, the initial current in the circuit is 44 mA, and the initial charge on the capacitor is zero. Write expressions for q, i, U, U_E, and U_B as functions of t.

7. Consider an *LC* circuit in which $L = 31.4$ mH and $C = 159$ nF. At $t = 0$, the current in the circuit is 265 mA and the charge on the capacitor is 7.18 μC. Write expressions for q, i, U, U_E, and U_B as functions of t.

8. The current in an *LC* circuit is given by the expression $i = (27 \text{ mA}) \cos [(280 \text{ krad/s})t]$. *(a)* Write an expression for the charge on the capacitor. *(b)* Determine L if $C = 140$ nF. *(c)* Write expressions for U, U_E, and U_B.

9. The charge on the capacitor in an *LC* circuit is given by the expression $q = (71\ \mu\text{C}) \cos [(54 \text{ krad/s})t - \pi/4]$. *(a)* Write an expression for the current in the circuit. *(b)* Determine C if $L = 17$ mH. *(c)* Write expressions for U, U_E, and U_B.

10. The potential difference across the capacitor in an *LC* circuit is given by the expression $\mathscr{V}_C = (32 \text{ V}) \sin [(42 \text{ krad/s})t]$, and the inductance of the inductor is $L = 22$ mH. Write expressions for *(a)* the charge on the capacitor, *(b)* the current in the circuit, *(c)* the potential difference across the inductor, *(d)* U, *(e)* U_E, *(f)* U_B.

11. The potential difference across the inductor in an *LC* circuit is given by the expression $\mathscr{V}_L = (4.8 \text{ V}) \cos [(16 \text{ krad/s})t]$, and the capacitance of the capacitor is $C = 54$ nF. Write expressions for *(a)* the charge on the capacitor, *(b)* the current in the circuit, *(c)* the potential difference across the capacitor, *(d)* U, *(e)* U_E, *(f)* U_B.

12. Suppose switch S_2 in the circuit in Fig. 31-22 has been closed for a long enough time so that the potential difference across the capacitor is steady. At $t = 0$, switch S_1 is closed and switch S_2 is opened. Write expressions for the charge on the capacitor and the current in the inductor as functions of t.

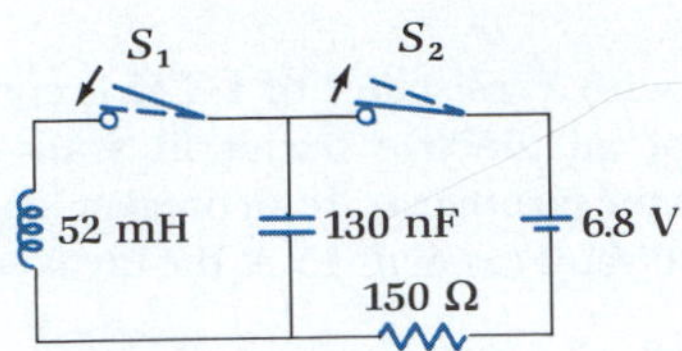

Figure 31-22. Exercise 12.

13. Suppose switch S_2 in the circuit of Fig. 31-23 has been closed for a long enough time so that the current in the inductor is steady. At $t = 0$, switch S_1 is closed and switch S_2 is opened. Write expressions for the charge on the capacitor and the current in the inductor as functions of t.

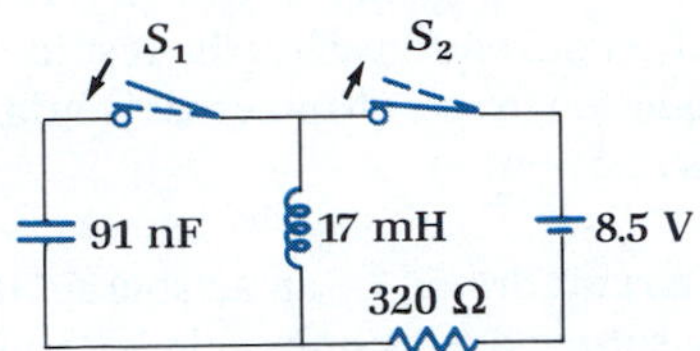

Figure 31-23. Exercise 13.

14. Consider an *LC* circuit at the instant the electric energy in the capacitor is equal to the magnetic energy in the inductor. *(a)* What is the charge on the capacitor in terms of its maximum charge? *(b)* What is the current in the circuit in terms of its maximum current?

15. Consider an *LC* circuit at an instant when 25 percent of the electromagnetic energy is stored in the capacitor and 75 percent is stored in the inductor. At this time, *(a)* what is the charge on the capacitor in terms of its maximum charge, and *(b)* what is the current in the circuit in terms of its maximum current?

16. Suppose you are given an inductor with $L = 38$ mH and two capacitors with $C_1 = 230$ nF and $C_2 = 510$ nF. What are the *LC* oscillation frequencies that you can produce with these elements?

17. Suppose you have a variable capacitor whose capacitance can be continuously varied in the range 0.14 to 3.2 nF. To produce a circuit whose oscillation frequency can be made to vary from 0.10 MHz to higher values, what value of inductance would you use in the circuit? What is the upper limit of the frequency for this circuit?

Section 31-2. Series *RLC* Circuit

18. Show that the SI unit of $\sqrt{4L/C}$ (or R_{crit}) is the ohm.

19. An *RLC* circuit has $R = 350\ \Omega$, $L = 16$ mH, and $C = 390$ nF. *(a)* Is this circuit underdamped or overdamped? If the circuit is underdamped, determine *(b)* ω_d and *(c)* τ.

20. Suppose you have an inductor with inductance 16 mH and a capacitor with capacitance 840 nF. Determine the value of the resistance you need in order to construct a critically damped *RLC* circuit.

21. The charge as a function of time for an *RLC* circuit is given by

$$q = (710\ \text{nC})e^{-t/(380\ \mu s)} \cos\left[(12.6\ \text{krad/s})t - 0.206\right]$$

The inductance of the inductor is 52 mH. *(a)* What is the resistance of the resistor? *(b)* What is the capacitance of the capacitor? *(c)* Determine the charge on the capacitor at $t = 230\ \mu s$.

22. *(a)* Show that if the charge on the capacitor in an *RLC* circuit is given by Eq. 31-6, then the current in the circuit is

$$i = -Q_m e^{-t/\tau}\left[\frac{1}{\tau}\cos(\omega_d t + \phi) + \omega_d \sin(\omega_d t + \phi)\right]$$

(b) Use the answer to part *(a)* to write an expression for the current in the circuit of the previous exercise. *(c)* Determine the value of the current at $t = 230\ \mu s$.

23. In an underdamped *RLC* circuit, the resistance is such that $\omega_d = \frac{1}{2}\omega_0$. Find this resistance in terms of L and C.

24. In an underdamped *RLC* circuit, the resistance is such that the time τ equals the period $2\pi/\omega_d$. Find this resistance in terms of L and C.

25. Consider an *RLC* circuit in which L and C are fixed, but the resistance can be varied in the range $0 < R < R_{crit}$, where R_{crit} is the resistance which produces critical damping: $R_{crit} = \sqrt{4L/C}$. *(a)* Show that the angular frequency of oscillation can be written $\omega_d = \omega_0\sqrt{1 - (R/R_{crit})^2}$. Make a graph of ω_d versus R for $R/R_{crit} = 0.01, 0.10, 0.20, 0.30, \ldots, 0.80, 0.90, 0.99$.

Section 31-3. AC Source Connected to a Resistor

26. An ac source with amplitude $\mathcal{V}_m = 170$ V and frequency 60 Hz is connected to a resistor with resistance 1.4 kΩ. *(a)* Determine the amplitude of the oscillating current i. Use Eqs. 31-7 and 31-8 to write expressions for *(b)* the voltage $\mathcal{V}$ across the resistor and *(c)* the current i in the circuit.

27. Consider the oscillating voltage and current calculated in the previous exercise. *(a)* What is the period of the oscillations? Determine $\mathcal{V}$ at *(b)* $t = \frac{1}{240}$ s and *(c)* $t = \frac{1}{120}$ s. Determine i at *(d)* $t = \frac{1}{240}$ s and *(e)* $t = \frac{1}{120}$ s.

Section 31-4. AC Source Connected to a Capacitor

28. Show that the SI unit for the capacitive reactance, $X_C = 1/\omega C$, is the ohm.

29. Determine the capacitive reactance of a 1.0-nF capacitor when the source frequency is *(a)* 100 Hz, *(b)* 100 kHz, *(c)* 100 MHz.

30. With a source frequency of 100 kHz, determine the capacitive reactance of a capacitor whose capacitance is *(a)* 1.0 pF, *(b)* 1.0 nF, *(c)* 1.0 μF.

31. What is the capacitance of a capacitor whose capacitive reactance is 2.5 kΩ when the source frequency is 3.8 krad/s?

32. Make a graph of X_C versus ω for a 1.0-nF capacitor. Plot points for $\omega = 1.0, 2.0, 3.0, 5.0, 7.0$, and 10.0 krad/s, and sketch a curve through the points. It is instructive to make this plot and that of Exercise 40 on the same graph.

33. A 2.6-nF capacitor is connected to a source in which $\mathcal{V}_m = 71$ V and $\omega = 360$ rad/s. What is the current amplitude?

34. The voltage across the terminals of a 230-pF capacitor is given by $\mathcal{V} = (27\ \text{V})\sin[(5.8\ \text{krad/s})t]$. *(a)* Write an expression for the current. *(b)* Determine the current at $t = 0.43$ ms.

35. A 2.1-μF capacitor is connected to a source whose voltage amplitude is 49 V and whose frequency can be varied. What value of the source angular frequency yields a current amplitude of 310 mA?

Section 31-5. AC Source Connected to an Inductor

36. Show that the SI unit for the inductive reactance, $X_L = \omega L$, is the ohm.

37. Determine the inductive reactance of a 1.0-mH inductor when the source frequency is *(a)* 100 Hz, *(b)* 100 kHz, *(c)* 100 MHz.

38. Determine the inductive reactance of an inductor with a source frequency of 100 kHz and with an inductance of *(a)* 1.0 μH, *(b)* 1.0 mH, *(c)* 1.0 H.

39. What is the inductance of an inductor whose inductive reactance is 420 Ω when the frequency of the source is 89 krad/s?

40. Make a graph of X_L versus ω for a 1.0-mH inductor. Plot points for $\omega = 0.0, 2.0, 4.0, 6.0, 8.0$, and 10.0 krad/s, and sketch a curve through the points. It is instructive to make this plot and that of Exercise 32 on the same graph.

41. A 65-mH inductor is connected to a source in which $\mathcal{V}_m = 130$ V and $\omega = 410$ rad/s. What is the current amplitude?

42. The voltage across the terminals of a 0.45-mH inductor is given by $\mathcal{V} = (8.1\ \text{V})\sin[(13\ \text{krad/s})t]$. *(a)* Write an expression for the current. *(b)* Determine the current at $t = 160\ \mu s$.

43. A 16-mH inductor is connected across a source whose voltage amplitude is 9.8 V and whose frequency can be varied. What value of the source angular frequency yields a current amplitude of 704 mA?

Section 31-6. Series *RLC* Circuit Driven by an AC Source

44. A series *RLC* circuit with $R = 510\ \Omega$, $L = 25$ mH, and $C = 240$ nF has an ac source with $\mathcal{V}_m = 17$ V and $\omega = 6.3$ krad/s. Determine the *(a)* capacitive reactance, *(b)* inductive reactance, *(c)* impedance, *(d)* current amplitude, *(e)* phase difference between $\mathcal{V}$ and i.

45. The voltage across an ac source is given by $\mathcal{V} = (5.4\ \text{V})\sin[(830\ \text{rad/s})t]$. The source is in a series *RLC* circuit with $R = 37\ \Omega$, $L = 85$ mH, and $C = 25\ \mu F$. *(a)* Write an expression for the current in the circuit. *(b)* On the same graph, plot $\mathcal{V}$ and i versus t from $t = 0$ to $t = 7.6$ ms.

46. Suppose the frequency of the source in the previous exercise can be varied. *(a)* What is the resonant frequency of the circuit? *(b)* What is the current amplitude at resonance?

47. Construct a graph similar to that of Fig. 31-17, except make it applicable to a predominantly capacitive circuit.

48. Verify some of the values for each of the resonance curves shown in Fig. 31-18. For the values of the parameters given in the caption, find I_m when $\omega = 0.5$, 0.9, 1.0, 1.1, and 1.5 Mrad/s. Use the figure to check your answers.

49. On the same graph, make plots of the phase difference ϕ versus ω for the two cases used in Fig. 31-18.

50. In a series *RLC* circuit driven by an ac source, $R = 140\ \Omega$, $L = 150$ mH, $C = 5.1\ \mu$F, and $\mathcal{V} = (14\text{ V})\sin[(530\text{ rad/s})t]$. Write expressions for the voltage across the *(a)* resistor, *(b)* capacitor, *(c)* inductor. *(d)* Draw a phasor diagram that includes $\mathbf{V}$, $\mathbf{V}_R$, $\mathbf{V}_C$, and $\mathbf{V}_L$.

51. Write Eq. 31-16 as a differential equation for the charge q on the capacitor. Compare your equation with Eq. 14-30 and make a table of analogous quantities.

52. By changing the capacitance of a capacitor, you can vary the resonant frequency of the tuning circuit in a radio from 500 to 1700 kHz. If you are to vary the resonant frequency of a series *RLC* circuit in this way, and the inductor has an inductance of 25 mH, what is the range of capacitance for the capacitor?

53. *(a)* Show that the current amplitude in a series *RLC* circuit driven by an ac source can be written

$$I_m = \frac{\mathcal{V}_m \omega}{\sqrt{\omega^2 R^2 + L^2(\omega^2 - \omega_0{}^2)^2}}$$

(b) Show that the phase angle difference can be written

$$\phi = \tan^{-1}\frac{L(\omega^2 - \omega_0{}^2)}{\omega R}$$

Section 31-7. Power for an *RLC* Circuit Driven by an AC Source

54. The current amplitude in an *RLC* series circuit driven by an ac source is $I_m = 260$ mA. *(a)* What is the rms current? *(b)* If the resistance in the circuit is 140 Ω, what is the average power delivered by the source?

55. In a series *RLC* circuit driven by an ac source, the voltage across the source is given by $\mathcal{V} = (17\text{ V})\sin[(230\text{ rad/s})t]$ and the current in the circuit is given by $i = (97\text{ mA})\sin[(230\text{ rad/s})t + 0.82\text{ rad}]$. What is the average power for the circuit?

56. The rms value of a time-dependent quantity $f(t)$ over the time interval T is

$$f_{\text{rms}} = \left\{\frac{1}{T}\int_0^T [f(t)]^2\,dt\right\}^{1/2}$$

Show that if $\mathcal{V} = \mathcal{V}_m \sin(\omega t)$, then $\mathcal{V}_{\text{rms}} = \mathcal{V}_m/\sqrt{2}$, where $T = 2\pi/\omega$. [*Hint:* Use the trigonometric identity $\sin^2\theta = \frac{1}{2} - \frac{1}{2}\cos(2\theta)$.]

57. A series *RLC* circuit is driven by an ac source at a frequency such that the circuit's impedance is 97 Ω. If the resistance in the circuit is 36 Ω and the rms voltage across the source is 6.2 V, what is average power for the circuit?

58. Suppose an ac source with a variable frequency drives a series *RLC* circuit in which $R = 1.5$ kΩ, $C = 10$ nF, and $L = 10$ mH. Determine the power factor when the source frequency is *(a)* 50 krad/s, *(b)* 100 krad/s, *(c)* 200 krad/s.

59. Verify some of the values for each of the resonance curves shown in Fig. 31-21. For the values of the parameters given in the caption, find $\bar{P}$ when $\omega = 0.5$, 0.9, 1.0, 1.1, and 1.5 Mrad/s. Use the figure to check your answers.

60. In a series *RLC* circuit driven by an ac source, the rms voltages across the source and the resistor are 9.4 and 2.7 V, respectively. Determine the power factor for the circuit.

61. Show that the average power for an *RLC* series circuit driven by an ac source can be written

$$\bar{P} = \frac{(\mathcal{V}_{\text{rms}})^2 R\omega^2}{R^2\omega^2 + L^2(\omega^2 - \omega_0{}^2)^2}$$

PROBLEMS

1. ***Solving the differential equation for the RLC circuit.*** By direct substitution, show that Eq. 31-6 is a solution to Eq. 31-5 if

$$\tau = \frac{2L}{R} \quad \text{and} \quad \omega_d = \sqrt{\frac{1}{LC} - \left(\frac{R}{2L}\right)^2}$$

[*Hint:* If the equation $A\sin(\omega t) + B\cos(\omega t) = 0$ is satisfied for all time t, then $A = 0$ and $B = 0$.]

2. ***The overdamped RLC circuit.*** When $R > \sqrt{4L/C}$, ω_d in Eq. 31-6 becomes imaginary:

$$\omega_d = j\sqrt{\left(\frac{R}{2L}\right)^2 - \frac{1}{LC}} = j\alpha$$

where $j = \sqrt{-1}$ and α is real. This corresponds to the overdamped circuit. *(a)* Show that

$$q = e^{-t/\tau}(C_1 e^{\alpha t} + C_2 e^{-\alpha t})$$

is a solution to Eq. 31-5, where C_1 and C_2 are constants that depend on the initial conditions. *(b)* Using typical values for the parameters, make a graph of the solution you found in part *(a)*.

3. ***LC circuit with two capacitors.*** In Fig. 31-24, capacitor C_2 is initially uncharged; C_1 is charged with a 12-V battery and then the battery is removed. Switches S_1 and S_2 are electronically controlled and can be opened and closed virtually instantaneously. *(a)* Describe a switching procedure that leaves C_1 with zero potential difference across it and C_2 with a potential difference of 36 V across it. *(b)* Determine the time interval between each change in switch settings. (*Hint:* Use energy considerations.)

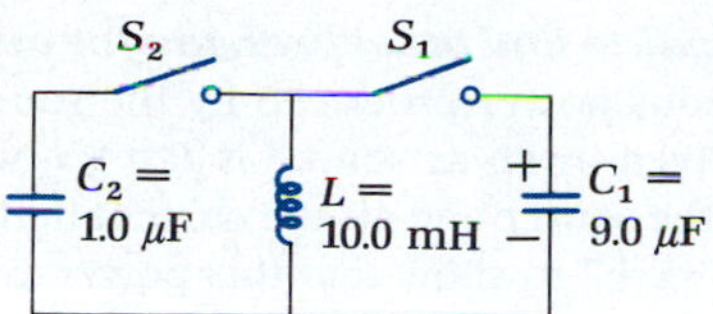

Figure 31-24. Problem 3.

4. ***Q value.*** Circuits that exhibit resonance are often characterized by their Q value. The Q value of a circuit is defined as $Q = L\omega_0/R$ and is related to the sharpness of the resonance peak in a graph of $\overline{P}$ versus ω (Fig. 31-21). *(a)* For a series *RLC* circuit driven by an ac source, show that $Q \approx \omega_0/\Delta\omega$, where $\Delta\omega$ is the width of the curve at half its maximum value. Assume that the resonance curve is sharply peaked ($\omega_0 \gg \Delta\omega$). *(b)* Using its definition, evaluate Q for each of the cases shown in Fig. 31-21. *(c)* Using a ruler to find $\Delta\omega$ for the curves in Fig. 31-21, determine Q for each case and compare your answers with those you found for part *(b)*.

5. ***Power in a series RLC circuit driven by an ac source.*** Equation 31-16 can be converted to an equation in which each term represents a power. *(a)* Multiply the equation by i and show that the resulting equation can be written

$$\frac{d}{dt}\left(\frac{1}{2}Li^2\right) + i^2R + \frac{d}{dt}\left(\frac{q^2}{2C}\right) = i\mathcal{V}$$

(b) Give a physical interpretation of each term in this expression.

6. ***Low-pass filter.*** Figure 31-25 shows an *RC* low-pass filter. Assume that the voltage across the input is a variable-frequency ac source in which $\mathcal{V} = \mathcal{V}_m \sin(\omega t)$. *(a)* Show that the amplitude of the output voltage is $\mathcal{V}_{Cm} = \mathcal{V}_m/\sqrt{(RC\omega)^2 + 1}$. *(b)* Make a graph of $\mathcal{V}_{Cm}$ versus ω. Plot points corresponding to $\omega = 0.0, 0.5, 1.0, \ldots, 3.5, 4.0$ in units of $1/RC$. *(c)* Explain the significance of the name of this circuit.

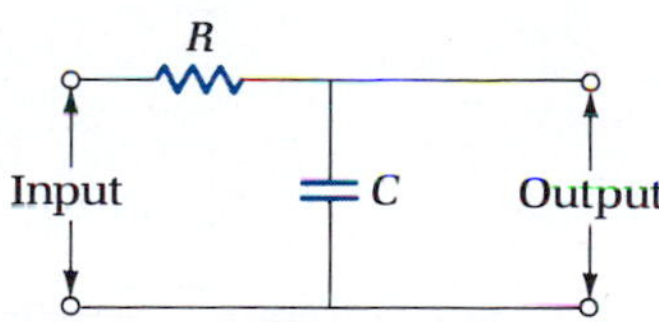

Figure 31-25. Problem 6: An *RC* low-pass filter.

7. ***High-pass filter.*** Figure 31-26 shows an *RC* high-pass filter. Assume that the voltage across the input is a variable-frequency ac source in which $\mathcal{V} = \mathcal{V}_m \sin(\omega t)$. *(a)* Show that the amplitude of the output voltage is $\mathcal{V}_{Rm} = \mathcal{V}_m/\sqrt{(1/RC\omega)^2 + 1}$. *(b)* Make a graph of $\mathcal{V}_{Rm}$ versus ω. Plot points corresponding to $\omega = 0.0, 0.5, 1.0, \ldots, 3.5, 4.0$ in units of $1/RC$. *(c)* Explain the significance of the name of this circuit.

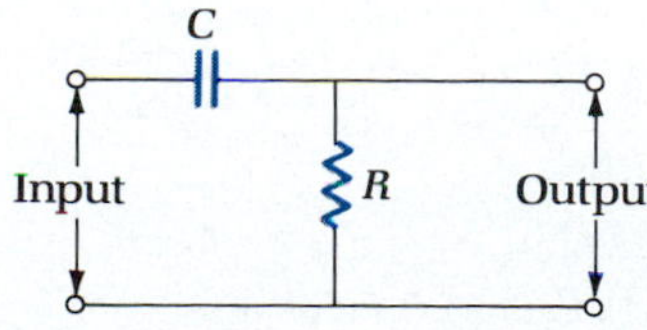

Figure 31-26. Problem 7: An *RC* high-pass filter.

8. ***Phase shifter.*** The circuit in Fig. 31-26 can be used as a phase shifter. Suppose $C = 14\ \mu F$ and the input voltage is $\mathcal{V}_i = (8.8\ V)\sin[(716\ rad/s)t]$. *(a)* Determine R such that the output voltage leads the input voltage by 0.56 rad. *(b)* Find the amplitude of the output voltage.

9. ***Another phase shifter.*** The circuit in Fig. 31-27 can be used as a phase shifter. Suppose $L = 86$ mH and the input voltage is $\mathcal{V}_i = (9.3\ V)\sin[(530\ rad/s)t]$. *(a)* Determine R such that the output voltage lags the input voltage by 0.70 rad. *(b)* Find the amplitude of the output voltage.

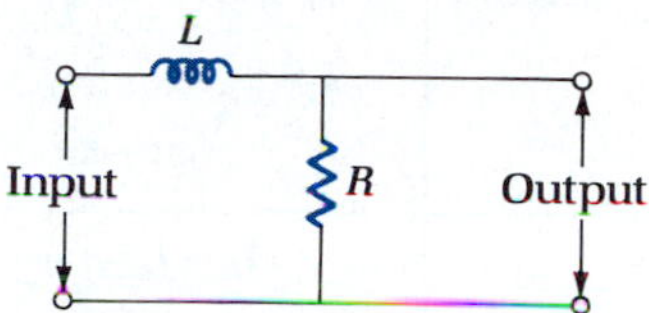

Figure 31-27. Problem 9.

10. ***Average power to a capacitor is zero.*** The instantaneous power for the capacitor in a series *RLC* circuit driven by an ac source is $P_C = i\mathcal{V}_C$, where $i = I_m \sin(\omega t - \phi)$ and $\mathcal{V}_C = X_C I_m \sin(\omega t - \phi - \frac{1}{2}\pi)$. Show that the average power for the capacitor is zero.

11. ***Average power to an inductor is zero.*** The instantaneous power for the inductor in a series *RLC* circuit driven by an ac source is $P_L = i\mathcal{V}_L$, where $i = I_m \sin(\omega t - \phi)$ and $\mathcal{V}_L = X_L I_m \sin(\omega t - \phi + \frac{1}{2}\pi)$. Show that the average power for the inductor is zero.

12. ***Triangular alternating current.*** Consider the nonsinusoidal alternating current shown in Fig. 31-28. Show that the rms value of the current is $I_{rms} = I_m/\sqrt{3}$. Refer to Exercise 56 for the definition of the rms value of a quantity.

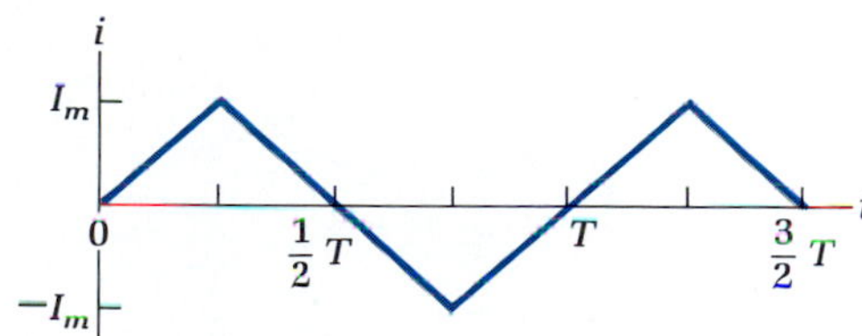

Figure 31-28. Problem 12.

13. ***Parallel RLC circuit driven by an ac source.*** Figure 31-29*a* shows a parallel *RLC* circuit driven by an ac source. Since the elements are in parallel, the voltage across each element is the same as the voltage across the source: $\mathcal{V} = \mathcal{V}_m \sin(\omega t)$. On the other hand, the current in each element is different. Figure 31-29*b* shows the phasor diagram for this circuit. *(a)* Explain the relative orientation of these phasors. *(b)* Show that the current in the source is $i = I_m \sin(\omega t - \phi)$, where

$$I_m = \mathcal{V}_m \sqrt{\left(\frac{1}{R}\right)^2 + \left(\frac{1}{X_L} - \frac{1}{X_C}\right)^2}$$

and

$$\phi = \tan^{-1}\left[R\left(\frac{1}{X_L} - \frac{1}{X_C}\right)\right]$$

(c) Discuss the frequency dependence of I_m.

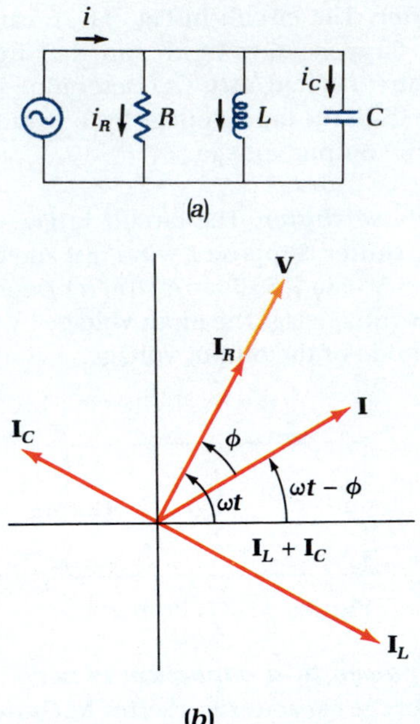

Figure 31-29. Problem 13; *(a)* Schematic diagram of a parallel *RLC* circuit driven by an ac source. *(b)* Phasor diagram for this circuit.

14. ***Another way to find the expression for average power.*** The instantaneous power produced by the source in a series *RLC* circuit driven by an ac source is $P = \mathcal{V} i$, where $\mathcal{V}$ is the voltage across the source and i is the current in the circuit. Use Eqs. 31-7 and 31-17 to show that this power averaged over a period is

$$\overline{P} = \mathcal{V}_{\text{rms}} I_{\text{rms}} \cos \phi$$

[*Hint:* Use the identity $\sin(\alpha + \beta) = \sin \alpha \cos \beta + \cos \alpha \sin \beta$. Also note that the product $\sin(\omega t) \cos(\omega t)$ averaged over a cycle is zero.]

WAVES 32

Waves on the surface of a lake.

As you read these words, the information comes to you in the form of light waves reflected from the page. When you go to class, the professor's lecture comes to you in the form of sound waves. Waves are important because a great deal of the contact we have with our environment comes to us as waves. But there is an additional reason to study waves. Matter in the size range of atoms and smaller exhibits an intrinsic wave behavior. For you to understand the nature of atoms, molecules, and nuclei, you must first learn about waves.

So that we may begin on familiar ground, this chapter is mostly about waves on a stretched rope, string, or spring. Waves on a rope are easy to visualize, and they exhibit many features that are common to all waves.

32-1 CHARACTERISTICS OF WAVES

We can separate waves into two categories. (i) *traveling waves* and (ii) *standing waves.* A wave propagating across the surface of water is an example of a traveling wave.

A traveling wave

A ***traveling wave*** can be defined as the propagation of energy without the propagation of matter.

A standing wave

By contrast, a ***standing wave*** is confined to a specific region of space by boundaries. For example, when you pluck a guitar string, you produce standing waves between the fixed ends of the string. For a standing wave, the energy associated with the wave remains between the boundaries. We begin by discussing traveling waves, and then we examine standing waves in Sec. 32-6.

Sound waves, waves on ropes and strings, and water waves are examples of mechanical waves. Mechanical waves exist in a *medium* and can be described with Newton's laws. Section 32-4 shows that there are two properties of a medium that govern the behavior of a mechanical wave: a restoring force and an inertial mass. In a water wave, for example, gravity provides a force that tends to restore the water to its equilibrium (flat) condition. That is, gravity pulls the wave crests down and fills in the wave troughs (Fig. 32-1). Because the water has inertial mass, it overshoots the equilibrium condition; the disturbance persists and the wave propagates.

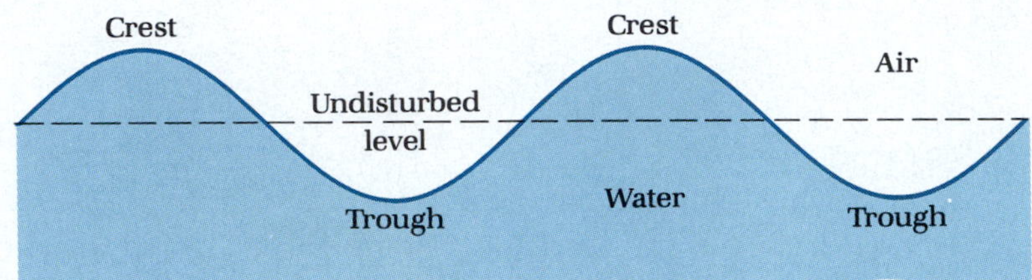

Figure 32-1. A water wave. Gravity (and surface tension) tends to restore the water surface to the flat equilibrium condition.

Waves involve two motions: that of the wave through the medium and that of the particles in the medium.

When a wave propagates through a medium, the particles of the medium do not move along with the wave. Suppose you tie one end of a rope to a post, stretch the rope out horizontally, and then wiggle the other end so that a wave moves along the rope. The wave moves along the length of the rope, but a particle of the rope oscillates about a central point. When we describe the motion associated with a wave, we must distinguish between two aspects of the motion: (i) the motion of the wave through the medium and (ii) the oscillatory motion of the particles of the medium.

A transverse wave

A longitudinal wave

One way to classify waves is according to the direction of the displacement of the particles relative to the propagation direction of the wave. A ***transverse wave*** is one in which the particles oscillate perpendicular to the propagation direction (Fig. 32-2*a*). A ***longitudinal wave*** is one in which the particles oscillate parallel to the propagation direction (Fig. 32-2*b*). A light wave is an example of a transverse wave. Rather than consisting of oscillating particles in a medium, a light wave

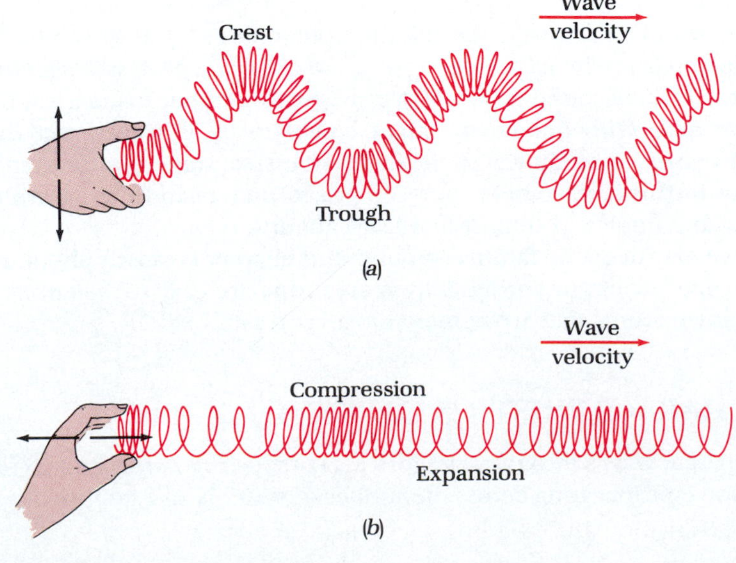

Figure 32-2. Waves on a spring. *(a)* A transverse wave. *(b)* A longitudinal wave.

Figure 32-3. A water wave is a combination transverse and longitudinal wave.

consists of oscillating electric and magnetic fields. In Chap. 34, we show that the directions of these wave fields are perpendicular to the wave velocity. In the next chapter, we find that sound in air (or any fluid) is a longitudinal wave. Some waves have both transverse and longitudinal components. For example, in a water wave the particles of the water follow elliptical paths so that the displacement of a particle can be resolved into components parallel and perpendicular to the wave velocity (Fig. 32-3).

32-2 WAVE PULSES

A ***wave pulse*** is a wave of relatively short extent. Because of this compactness, photographs of wave pulses can be used to demonstrate some important properties of waves. In this section, we use these photographs to discover the mathematical form for a traveling wave and to demonstrate the meanings of some of the terms used to describe waves.

Mathematical Expression for a Traveling Wave

In Fig. 32-4 we show a sequence of photographs of a wave pulse propagating toward the left along a stretched spring. The time interval between each photograph is the same. These photographs indicate that (i) the speed of a pulse is constant and (ii) the shape of the pulse remains nearly the same as the pulse moves along. Closer examination shows that the pulse gradually spreads out as it moves along; the pulse height decreases and the pulse width increases. This spreading out of the pulse is a result of *dispersion*. Dispersive effects are complex and depend on the properties of the medium. Also, dispersion is not of primary interest in the waves we wish to consider. Therefore we neglect dispersion.

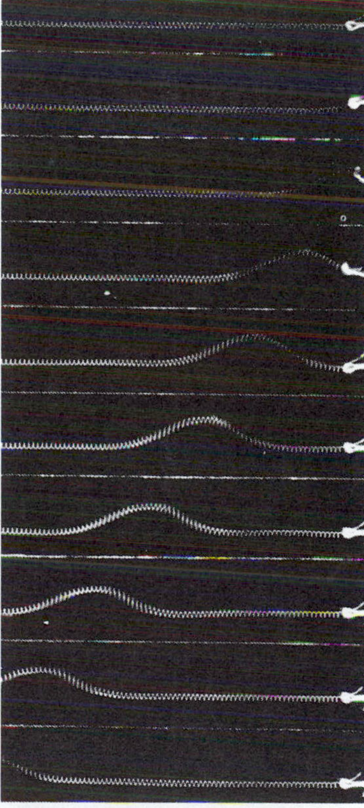

Figure 32-4. A wave pulse is generated by rapidly flipping the end of a long, stretched spring. The time interval between each subsequent photograph is the same. The pulse moves toward the left and, as far as we can tell from the photographs, travels at constant speed while maintaining its shape.

Figure 32-5 shows sketches of a pulse on a rope at two different times as the pulse moves to the right with speed v. A coordinate frame is shown as a backdrop, with the x axis along the undisturbed rope. Suppose the shape of the rope at $t = 0$ is given by the expression $y = f(x)$ (Fig. 32-5a). At a later time t, the pulse has moved to the right a distance vt (Fig. 32-5b). Recall that a function $f(x - a)$ has the same shape as the function $f(x)$, but $f(x - a)$ is displaced a distance a in the $+x$ direction. If we assume that the pulse maintains its shape as it moves, then the shape of the pulse at time t is given by

$$y(x, t) = f(x - vt) \tag{32-1}$$

A wave traveling in the $+x$ direction

A similar description with the pulse moving to the left at speed v gives

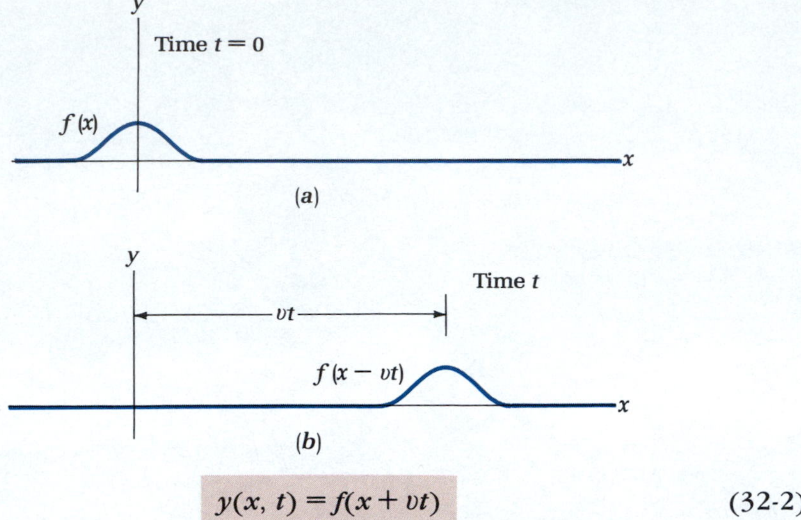

Figure 32-5. A pulse on a rope propagating in the $+x$ direction is shown at *(a)* time $t = 0$ and *(b)* time t. If $f(x)$ gives the pulse shape at $t = 0$ and the pulse maintains its shape as it moves with speed v, then the shape of the pulse at time t is given by $f(x - vt)$.

A wave traveling in the $-x$ direction

$$y(x, t) = f(x + vt) \tag{32-2}$$

To be specific, we usually discuss waves traveling in the $+x$ direction, Eq. 32-1.

Wave function

A function $y(x, t)$ that describes a wave is called a ***wave function.*** In the case of a wave on a rope, the wave function is the coordinate y of an element of the rope. Thus the wave function gives the displacement $y\mathbf{j}$ of the element from its equilibrium position at $y = 0$. A wave function depends on both x and t. This means that the displacement of an element of the rope depends on (i) the coordinate x of that element and (ii) the time t of the observation. Neglecting dispersion and leaving the shape of the wave unspecified, we have found that a wave function for a traveling wave has the form $f(x - vt)$ or $f(x + vt)$. That is, x and t must enter $y(x, t)$ in the combination $x - vt$ or $x + vt$. To specify the wave function, we must write it in terms of a particular function. For example, a specific wave function discussed in the next section is $y(x, t) = A \sin [k(x - vt)]$. Another specific wave function is given in the following example.

EXAMPLE 32-1

A wave pulse. Consider a wave pulse given by the wave function

$$y(x, t) = \frac{y_0}{[(x - vt)/x_0]^2 + 1}$$

where $y_0 = 10.0$ mm, $x_0 = 1.00$ m, and $v = 2.00$ m/s. *(a)* Make graphs of the pulse at times $t = 0.00$ s and $t = 2.50$ s. *(b)* The pulse is characterized by its height and width. The pulse height h is the magnitude of the maximum displacement due to the pulse, and the pulse width w is the distance between the two points on the pulse where the magnitude of the displacement is half the pulse height. Determine h and w for this pulse.

Solution. *(a)* Substituting the numerical values given for the parameters and $t = 0.00$ s into the expression for the pulse, we obtain

$$y(x, 0) = \frac{10.0 \text{ mm}}{[x/(1.00 \text{ m})]^2 + 1}$$

This function is shown as the solid curve in Fig. 32-6. To find y at $t = 2.50$ s, we set

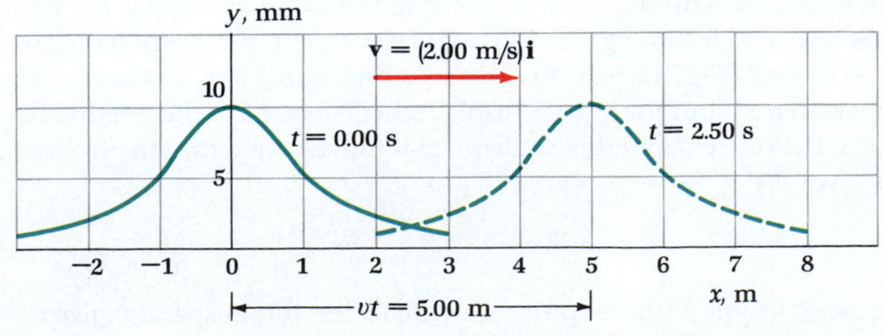

Figure 32-6. Example 32-1: A pulse traveling in the $+x$ direction at $t = 0.00$ s (solid line) and $t = 2.50$ s (dashed line). The pulse height is 10.0 mm and the pulse width is 2.00 m. Note the difference in scale between the x axis and the y axis.

$vt = (2.00\ \text{m/s})(2.50\ \text{s}) = 5.00\ \text{m}$:

$$y(x,\ 2.50\ \text{s}) = \frac{10.0\ \text{mm}}{[(x - 5.00\ \text{m})/(1.00\ \text{m})]^2 + 1}$$

This function is shown as the dashed curve in Fig. 32-6.

(b) By inspection of the graphs, you can see that the pulse height is $h = y_0 = 10.0$ mm and that the pulse width is $w = 2x_0 = 2.00$ m.

SELF-TEST 32-1. *(a)* For the wave pulse in the above example, write the expression for y as a function of x at the instant $t = 5.00$ s. *(b)* Where is the pulse centered at this instant?
ANSWERS: *(a)* $y(x,\ 5.00\ \text{s}) = 10.00\ \text{mm}/\{[(x - 10\ \text{m})/(1.00\ \text{m})]^2 + 1\}$; *(b)* $x = 10.00$ m.

Interference of Waves

When two or more waves encounter each other, we say that they *interfere.* Figure 32-7 shows the interference of two pulses with nearly the same size and shape. When the pulses come together, so that they occupy the same region of the spring, we say that they are *superposed.* From the figure you can see that when the pulses are superposed, the maximum displacement due to both pulses is the sum of the maximum displacements due to each pulse acting alone. To describe this result mathematically, let $f_1(x - vt)$ represent the pulse traveling to the right, $f_2(x + vt)$ represent the pulse traveling to the left, and $y(x, t)$ represent the shape of the rope due to both pulses. The photographs in Fig. 32-7 demonstrate that

$$y(x, t) = f_1(x - vt) + f_2(x + vt) \tag{32-3}$$

This result is an example of the *principle of superposition.*

Principle of superposition

The ***principle of superposition*** states that the resultant wave function due to two or more individual wave functions is the sum of the individual wave functions.

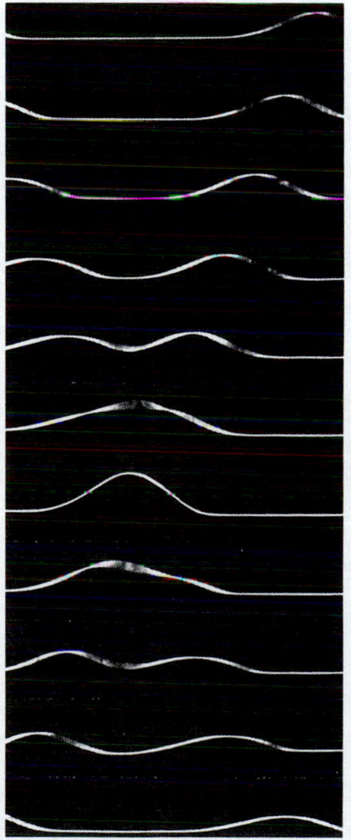

Figure 32-7. Two pulses traveling in opposite directions encounter each other. When the pulses are superposed (seventh photograph of the sequence), the maximum displacement of the rope is the sum of the maximum displacements due to each pulse acting alone.

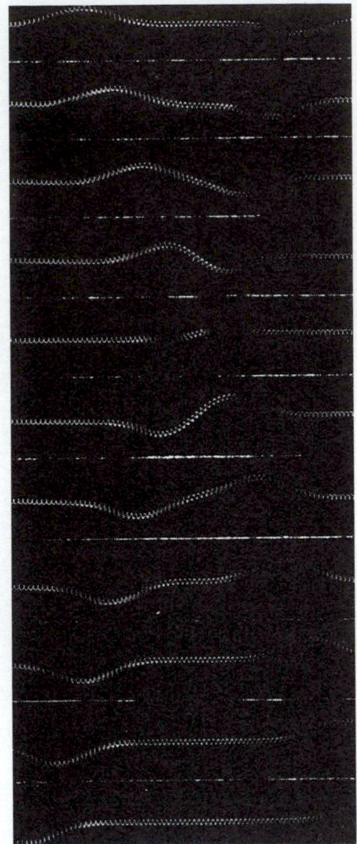

Figure 32-8. An "up" pulse moving to the right encounters a "down" pulse moving to the left. In accordance with the principle of superposition, the pulses nearly cancel when they are superposed (fifth photograph of the sequence).

According to the principle of superposition, the individual pulses act independently of each other. After the encounter, the size, shape, and speed of each pulse are the same as if there had been no encounter.

Figure 32-8 shows two pulses, one up and one down, encountering each other. As you can see from the photographs, when the pulses are superposed, they nearly cancel each other, giving a displacement of almost zero. This result is in accord with the principle of superposition. In this case y_1 is positive over the extent of pulse 1, and y_2 is negative over the extent of pulse 2. At the instant when both pulses occupy the same region of the spring, the sum of their displacements is nearly zero.

Reflection and Transmission

Waves can be reflected from boundaries and transmitted from one medium to another. Figure 32-9 shows a pulse on a spring traveling to the left and encountering a boundary where the end of the spring is fixed. The pulse, as it approaches the boundary, is referred to as an *incident pulse.* Note that the pulse is reflected from the boundary and that the *reflected pulse* is inverted relative to the incident pulse.

Figure 32-10 shows an incident pulse traveling to the left on a relatively light spring and encountering another spring that is relatively heavy. After the encounter, we see a reflected pulse on the light spring and a *transmitted pulse* on the heavy spring. The reflected pulse is inverted relative to the incident pulse but the transmitted pulse is not. Also, the speed of the pulse on the heavy spring is significantly less than the speed on the light spring. The speed of a wave depends on the medium in which it travels.

Figure 32-11 shows an incident pulse traveling to the right on a heavy spring and encountering another spring that is light. In this case, when the incident pulse is on the heavy spring, the reflected pulse is not inverted. Again, note that the speed of the pulse on the heavy spring is less than the speed of the pulse on the light spring.

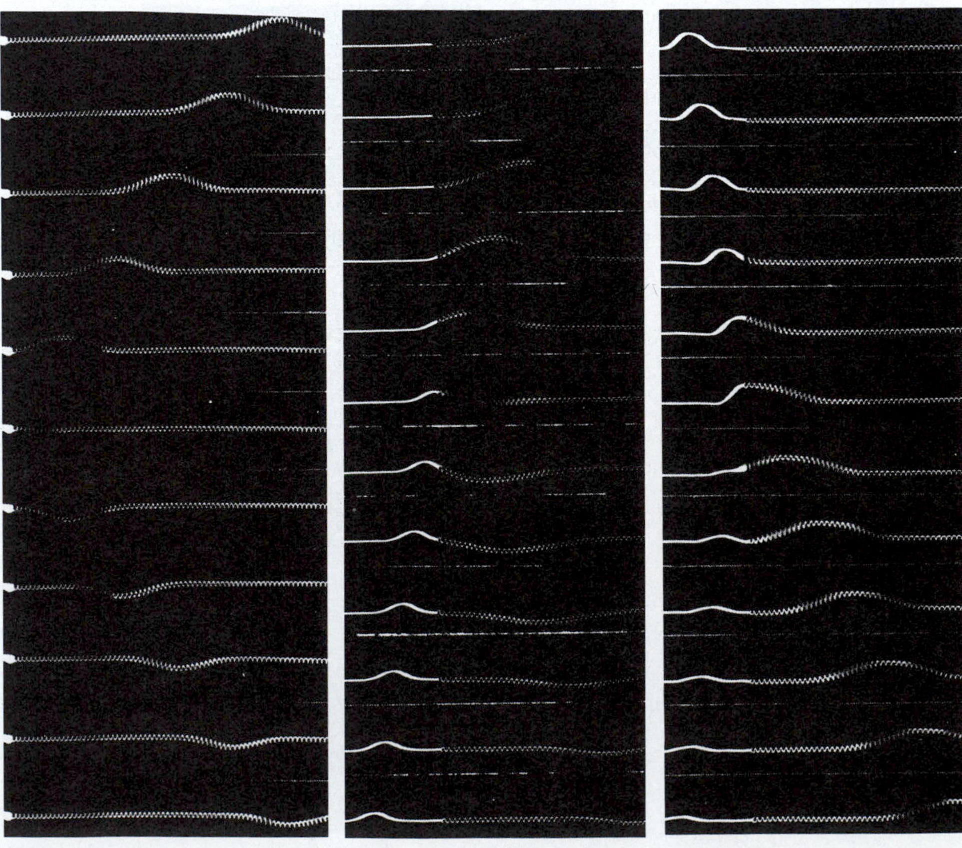

Figure 32-9. *(Left)* A pulse is reflected from a boundary where the end of the spring is fixed. The reflected pulse is inverted relative to the incident pulse.

Figure 32-10. *(Middle)* A pulse on a light spring is incident on a light-spring–heavy-spring boundary. The reflected pulse is inverted relative to the incident pulse, but the transmitted pulse is not inverted. Note that the pulse speed on the light spring is greater than it is on the heavy spring.

Figure 32-11. *(Right)* A pulse on a heavy spring incident on a light-spring–heavy-spring boundary. Neither the reflected pulse nor the transmitted pulse is inverted relative to the incident pulse. Note that the pulse speed on the light spring is greater than it is on the heavy spring.

32-3 HARMONIC WAVES

The mathematical description of waves is based primarily on the wave function for a ***harmonic wave.*** For a harmonic wave on a string, the string is shaped as a sine function at any particular instant. Figure 32-12 shows a "snapshot" of such a wave at two different instants—the solid curve shows the string at $t = 0$, and the dashed curve shows it a short time Δt later. At $t = 0$,

$$y = A \sin\left(\frac{2\pi}{\lambda}x\right)$$

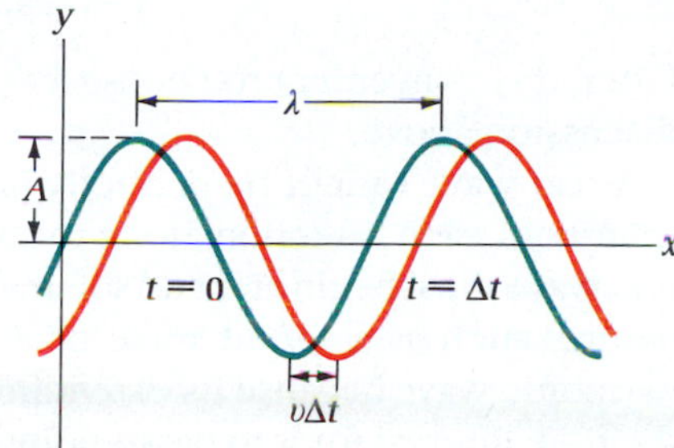

Figure 32-12. A graph of y versus x for a harmonic wave of amplitude A and wavelength λ. The solid curve shows the wave at $t = 0$, and the dashed curve shows the wave at a short time Δt later. Displacements along the y axis in this figure, and in others in this chapter, are exaggerated for purposes of illustration. Ordinarily, $\lambda \gg A$ for real waves.

Amplitude A and wavelength λ

The ***amplitude*** A is the maximum displacement of any element of the string from its equilibrium position at $y = 0$. The ***wavelength*** λ is the wave's repeat distance, for example, the distance between successive crests or successive troughs. In the previous section we found that for a wave moving to the right with speed v, the wave's time dependence can be developed from the expression for the wave's shape by simply replacing x with $x - vt$. Thus the wave function for a harmonic wave is

Wave function for a harmonic wave

$$y(x, t) = A \sin\left[\frac{2\pi}{\lambda}(x - vt)\right] \tag{32-4}$$

The argument of the sine function, $(2\pi/\lambda)(x - vt)$, is the ***phase*** of the wave.

Consider the motion of an element of the string as a harmonic wave moves along. For the element at $x = 0$, we have $y = A\sin[(-2\pi v/\lambda)t] = -A\sin[(2\pi v/\lambda)t]$. From Chap. 14 we know that simple harmonic motion (SHM) with period T is described by $y \propto \sin[(2\pi/T)t]$. Therefore the element executes SHM with period

$$T = \frac{\lambda}{v}$$

as shown in Fig. 32-13. Each element of the string executes SHM with this same period.

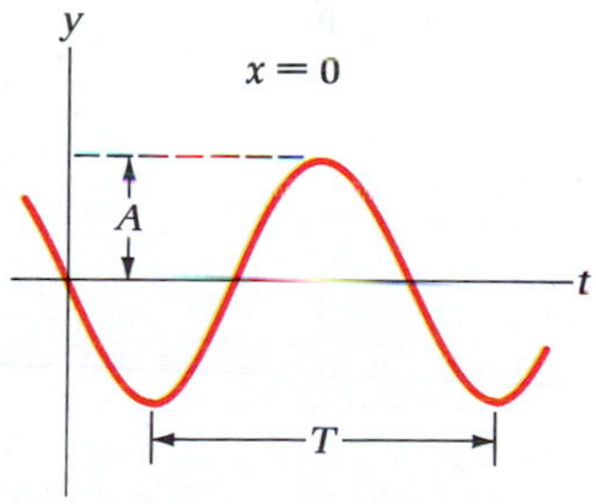

Figure 32-13. A graph of y versus t for a harmonic wave of amplitude A and period T. The motion of a particular element of the string is simple harmonic motion (SHM).

The above relation between T, λ, and v can be obtained another way. The ***period*** T is the time for a particular element to complete one oscillation, and it is also the time for a particular wave displacement (such as a crest) to move a distance of one wavelength λ. This means that the wave moves a distance λ in a time T, so its speed v is

Speed of a harmonic wave

$$v = \frac{\lambda}{T} \tag{32-5}$$

Frequency $\nu = 1/T$

Angular frequency $\omega = 2\pi/T$

Wave number $k = 2\pi/\lambda$

Traditionally, a number of parameters are used to describe a harmonic wave. The frequency ν of the wave is $\nu = 1/T$, and the angular frequency ω is $\omega = 2\pi/T = 2\pi\nu$. The ***wave number k*** is $k = 2\pi/\lambda$. A wave function can be written many different ways by using different combinations of these parameters, but a convenient way is in terms of k and ω:

$$y = A \sin (kx - \omega t)$$

The wave speed v can also be expressed a number of ways with these parameters. Two useful expressions are

$$v = \lambda \nu \qquad \text{and} \qquad v = \frac{\omega}{k}$$

The wave functions we have presented so far are not completely general because they require that $y = 0$ when $x = 0$ and $t = 0$. A more general expression includes a *phase constant* ϕ:

$$y = A \sin (kx - \omega t + \phi)$$

Often it is convenient to choose $x = 0$ and $t = 0$ such that $\phi = 0$, as we did in the discussion above.

A real wave cannot be perfectly harmonic because a harmonic wave extends to infinity in each direction along the x axis and has no beginning or ending time. A real wave must begin and end somewhere in space and in time. A wave that exists in nature, such as a sound wave or a light wave, often can be approximated as a harmonic wave because its extent in space is much larger than its wavelength, and the time interval for it to pass a point is much longer than its period. Such a wave is called a ***wave train.*** A harmonic wave is an idealized representation of a wave train.

Wave train

EXAMPLE 32-2

A harmonic wave. A harmonic wave on a string has an amplitude of 15 mm, a wavelength of 2.4 m, and a speed of 3.5 m/s. *(a)* Determine the period, the frequency, the angular frequency, and the wave number for the wave. *(b)* Write the wave function for this wave with the $+x$ direction as the direction of wave travel.

Solution. *(a)* Since the speed of a harmonic wave is given by $v = \lambda/T$, the period is

$$T = \frac{\lambda}{v} = \frac{2.4 \text{ m}}{3.5 \text{ m/s}} = 0.69 \text{ s}$$

The frequency is $\nu = 1/T = 1/0.69 \text{ s} = 1.5$ Hz, and the angular frequency is $\omega = 2\pi/T = 9.2$ rad/s. The wave number is $k = 2\pi/\lambda = 2\pi/2.4 \text{ m} = 2.6$ rad/m. *(b)* Using the concise form $y = A \sin (kx - \omega t)$, we have

$$y = (15 \text{ mm}) \sin [(2.6 \text{ rad/m})x - (9.2 \text{ rad/s})t]$$

Since the factor which involves x enters the expression with the opposite sign of the factor which involves t, the $+x$ direction is the direction of wave travel.

SELF-TEST 32-2. Write the wave function for a harmonic wave that has an amplitude of 12 mm and a wavelength of 3.1 m. The wave is traveling in the $-x$ direction at a speed of 44 m/s. ***ANSWER:*** $y = (12 \text{ mm}) \sin [(2.0 \text{ rad/m})x + (89 \text{ rad/s})t]$.

The Wave Equation

By investigating the derivatives of the wave function for a harmonic wave, we now introduce a differential equation that is called the ***wave equation.*** Later we shall find systems where the application of physical laws, such as Newton's second law, produces the wave equation. Such a finding constitutes a theoretical prediction that waves exist in the system.

First consider the derivative of y with respect to t, holding x constant—the partial derivative, $\partial y/\partial t$. This derivative gives the y component of the velocity of an element. Using $y(x, t) = A \sin (kx - \omega t)$, we have

$$\frac{\partial y}{\partial t} = \frac{\partial}{\partial t}[A \sin (kx - \omega t)] = -\omega A \cos (kx - \omega t)$$

A second time derivative of y, holding x constant, is the y component of the acceleration of the element:

$$\frac{\partial^2 y}{\partial t^2}=\frac{\partial}{\partial t}\frac{\partial y}{\partial t}=\frac{\partial}{\partial t}[-\omega A\cos(kx-\omega t)]=-\omega^2 A\sin(kx-\omega t)$$

Using $v=\omega/k$ in the form $\omega^2=v^2k^2$, we can write the acceleration component as

$$\frac{\partial^2 y}{\partial t^2}=-v^2k^2A\sin(kx-\omega t)$$

Since $y(x, t)=A\sin(kx-\omega t)$,

$$\frac{\partial^2 y}{\partial t^2}=-v^2k^2y(x, t) \qquad (32\text{-}6)$$

Next consider the derivative of y with respect to x, holding t constant—$\partial y/\partial x$:

$$\frac{\partial y}{\partial x}=\frac{\partial}{\partial x}[A\sin(kx-\omega t)]=kA\cos(kx-\omega t)$$

This derivative gives the slope of the string at a point x and time t. For example, it gives the slope at a point x for one of the graphs of y versus x shown in Fig. 32-12. The second derivative of y with respect to x, holding t constant is the change of the slope with x:

$$\frac{\partial^2 y}{\partial x^2}=\frac{\partial}{\partial x}\frac{\partial y}{\partial x}$$

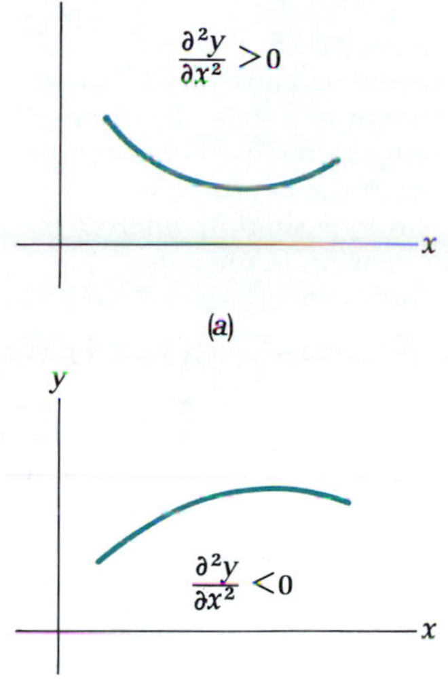

Figure 32-14. *(a)* A section of a curve with $\partial^2y/\partial x^2>0$. The curve bends upward with increasing x. *(b)* A section of a curve with $\partial^2y/\partial x^2<0$. The curve bends downward with increasing x. For a straight section, $\partial^2y/\partial x^2=0$.

This quantity gives a measure of the amount of bending in the string. The larger $|\partial^2y/\partial x^2|$, the tighter the bend in the string. If $\partial^2y/\partial x^2$ is positive, then the slope of the string increases with increasing x. We describe this case by saying that the string bends upward (Fig. 32-14*a*). If $\partial^2y/\partial x^2$ is negative, then the slope of the string decreases with increasing x and we say that the string bends downward (Fig. 32-14*b*). If $\partial^2y/\partial x^2=0$ at a point, then the string is straight at that point. For a harmonic wave,

$$\frac{\partial^2 y}{\partial x^2}=\frac{\partial}{\partial x}\frac{\partial y}{\partial x}=\frac{\partial}{\partial x}[kA\cos(kx-\omega t)]=-k^2A\sin(kx-\omega t)$$

Since $y(x, t)=A\sin(kx-\omega t)$,

$$\frac{\partial^2 y}{\partial x^2}=-k^2y(x, t) \qquad (32\text{-}7)$$

Combining Eqs. 32-6 and 32-7, we have

$$\frac{\partial^2 y}{\partial x^2}=\frac{1}{v^2}\frac{\partial^2 y}{\partial t^2} \qquad (32\text{-}8)$$

The wave equation

This differential equation is the wave equation. Because derivatives of a harmonic wave function produce this equation, we know that the wave function for a harmonic wave satisfies the wave equation, or is a solution to the wave equation. Indeed, the general traveling wave function $y(x, t)=f(x-vt)$ [or $y(x, t)=f(x+vt)$] is also a solution to the wave equation. (See Prob. 3.)

The wave equation expresses the result that the acceleration of an element of the string is related to the amount of bending of the string at that element. At points along the string where the bending is large ($|\partial^2y/\partial x^2|$ is large), the acceleration magnitude is large ($|\partial^2y/\partial t^2|$ is large). At points where the string is straight ($\partial^2y/\partial x^2=0$), the acceleration is zero ($\partial^2y/\partial t^2=0$). If $\partial^2y/\partial x^2$ is positive at a point (the string bends upward), the acceleration is directed upward; if $\partial^2y/\partial^2x^2$ is negative (the string bends downward), the acceleration is directed downward.

Why should the acceleration of an element depend on the bending of the string at that element? The answer comes from dynamics, the application of Newton's second law to each element of the string.

32-4 THE WAVE EQUATION FROM NEWTON'S SECOND LAW

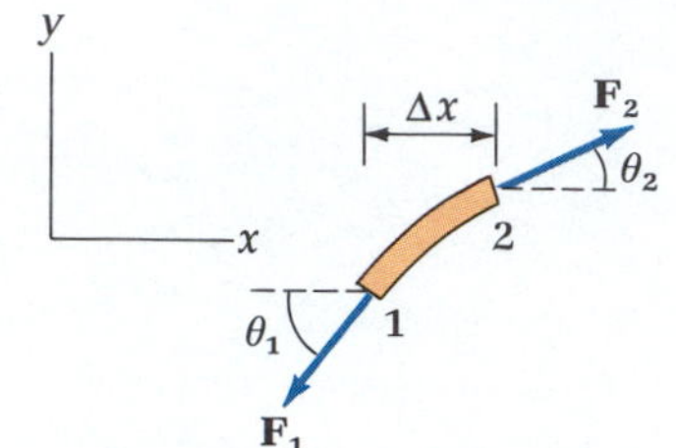

Figure 32-15. Forces $\mathbf{F}_1$ and $\mathbf{F}_2$ exerted on ends 1 and 2 of an element of a string by its neighboring elements. The y component of the net force is $F(\sin\theta_2 - \sin\theta_1)$, where F is the tension in the rope.

Newton's second law predicts that waves can occur in a medium with a linear elastic restoring force. As an example, consider an element of a uniform string. In equilibrium, the string is held taut along the x axis. Figure 32-15 shows an element of the string that is displaced from equilibrium by a wave. Forces $\mathbf{F}_1$ and $\mathbf{F}_2$ are exerted on ends 1 and 2 of the element by its neighboring elements. We assume that the effect of the wave is small enough that the tension F in the string is essentially uniform. This means that $|\mathbf{F}_1| = |\mathbf{F}_2| = F$. Also, we assume that the tension is great enough that the weight of the element can be neglected. With these approximations, the y component of the net force on the element is

$$\Sigma F_y = F_{y1} + F_{y2} = -F\sin\theta_1 + F\sin\theta_2 = F(\sin\theta_2 - \sin\theta_1)$$

Note that if the string is straight, then $\theta_1 = \theta_2$, and the net force on the element is zero. If the string is bent, $\theta_1 \neq \theta_2$, and there is a nonzero net force on the element.

Next we assume that the angles θ_1 and θ_2 are small so that $\sin\theta_1 \approx \tan\theta_1$ and $\sin\theta_2 \approx \tan\theta_2$. This approximation is useful because the slope of the string at a point equals the tangent of the angle between the string and the x axis at that point: $\tan\theta = \partial y/\partial x$. Accordingly, $\sin\theta \approx \partial y/\partial x$, and we can write the net force as

$$\Sigma F_y = F\left[\left(\frac{\partial y}{\partial x}\right)_2 - \left(\frac{\partial y}{\partial x}\right)_1\right]$$

The quantity $[(\partial y/\partial x)_2 - (\partial y/\partial x)_1]$ is the change in the slope between ends 1 and 2. If the element is small, then

$$\left(\frac{\partial y}{\partial x}\right)_2 - \left(\frac{\partial y}{\partial x}\right)_1 = \Delta\frac{\partial y}{\partial x} = \frac{\Delta(\partial y/\partial x)}{\Delta x}\Delta x \approx \frac{\partial}{\partial x}\left(\frac{\partial y}{\partial x}\right)\Delta x = \frac{\partial^2 y}{\partial x^2}\Delta x$$

where the approximation becomes exact as the size of the element approaches zero. The y component of the net force on a small element is

$$\Sigma F_y = F\frac{\partial^2 y}{\partial x^2}\Delta x$$

Let M and L represent the mass and length of the string, respectively. For a uniform string, the mass per unit length, or ***linear mass density*** μ, of the string is $\mu = M/L$. Using μ, we can write the mass m of the element in terms of Δx: $m = \mu\,\Delta x$. Applying the y component of Newton's second law, $\Sigma F_y = ma_y$, to the element gives

$$F\frac{\partial^2 y}{\partial x^2}\Delta x = \mu\,\Delta x\frac{\partial^2 y}{\partial t^2}$$

where we have used $a_y = \partial^2 y/\partial t^2$. Diving by Δx and rearranging, we obtain

Wave equation for waves on a string

$$\frac{\partial^2 y}{\partial x^2} = \frac{\mu}{F}\frac{\partial^2 y}{\partial t^2} \tag{32-9}$$

which is the wave equation. Thus Newton's second law predicts the existence of waves on a string.

Comparing Eq. 32-8 with Eq. 32-9, we see that $1/v^2 = \mu/F$, or

Speed of a wave on a string

$$v = \sqrt{\frac{F}{\mu}} \tag{32-10}$$

In addition to producing the wave equation, Newton's second law gives the speed of the waves in terms of the tension F in the string and the linear mass density μ of the string. For example, if a string of length 10 m and mass 1 kg ($\mu = M/L =$ 1 kg/10 m = 0.1 kg/m) is under a tension of 90 N, then the speed of a wave on this string is $v = \sqrt{F/\mu} = \sqrt{(90\text{ N})/(0.1\text{ kg/m})} = 30$ m/s. Experiment shows that Eq. 32-10 gives the correct value for the speed of waves on a string.

In the above derivation of the wave equation, we assumed that the angle between an element of the string and the x axis is small. For a harmonic wave, this corresponds to the wavelength being much larger than the amplitude, $\lambda \gg A$. The wave equation that is a result of this assumption, Eq. 32-9, is called a *linear* wave equation. A differential equation is linear if its terms involve y and derivatives of y to the first power. For example, a differential equation is linear if its terms involve y, $\partial y/\partial x$, or $\partial^2 y/\partial t^2$, but it is not linear if its terms involve $y(\partial y/\partial t)$, or $(\partial^2 y/\partial x^2)^2$, or y^2. An important feature of a linear wave equation is that the sum of individual wave functions is also a wave function. Suppose $y_1(x, t)$ and $y_2(x, t)$ are individual wave functions that satisfy a linear wave equation. Then

$$y(x, t) = y_1(x, t) + y_2(x, t) \tag{32-11}$$

The principle of superposition is obeyed by waves that satisfy a linear wave equation.

is also a wave function that satisfies the same linear wave equation. (See Exercise 24.) In other words, the principle of superposition is obeyed by waves that satisfy a linear wave equation.

In the next chapter we shall see another example in which Newton's second law yields a wave equation, a wave equation for sound in a fluid. For sound waves, the wave function represents the longitudinal displacement of the fluid caused by the wave. In Chap. 34 we shall find that the equations of electricity and magnetism predict the existence of electromagnetic waves, such as visible light. For electromagnetic waves, the wave function represents the oscillating electric and magnetic fields. In general, we can write the wave equation as

$$\frac{\partial^2 \psi}{\partial x^2} = \frac{1}{v^2}\frac{\partial^2 \psi}{\partial t^2} \tag{32-12}$$

Wave equation for a physical quantity ψ

where $\psi(x, t)$ represents the physical quantity that oscillates, or "waves," as the wave goes by. In the case of a wave on a string, ψ gives the transverse displacement of the string; in the case of a sound wave in a fluid, ψ gives the longitudinal displacement of the fluid; in the case of electromagnetic waves, ψ gives the electric or magnetic field.

When the wave equation is developed for a physical system, an expression for the wave speed v in terms of the properties of the medium is forthcoming. In the case of the string, we found that $v = \sqrt{F/\mu}$. For mechanical waves in a medium, the important characteristics of the medium are (i) a factor that characterizes the restoring force in the medium and (ii) a factor that characterizes the inertial mass of the medium. The expression for the wave speed has the form

$$v = \sqrt{\frac{\text{restoring force factor}}{\text{inertial mass factor}}}$$

The speed of a wave depends on the medium.

which shows how the speed of a wave depends on the properties of the medium through which the wave travels.

32-5 POWER OF A WAVE

As a wave moves along, it carries energy in the direction of wave travel. To determine the rate at which energy is propagated by a wave, or the power of a wave, we first find the wave's energy density. As we shall see, the power of a wave is given by the product of its energy density and its speed.

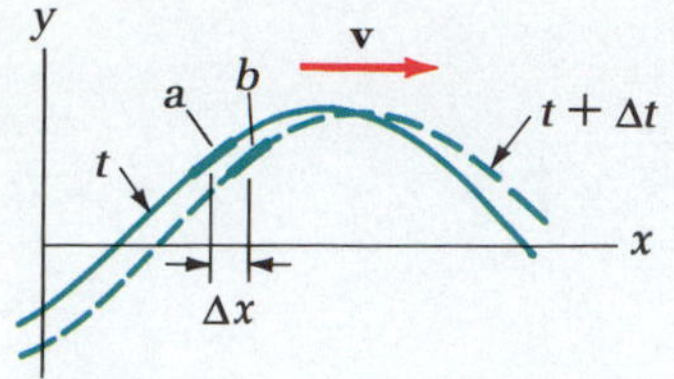

Figure 32-16. With a wave traveling in the $+x$ direction, the string is shown at time t and at time $t + \Delta t$. The energy of element a at t is equal to the energy of element b at $t + \Delta t$. Thus energy propagates along the string with speed $\Delta x/\Delta t = v$.

Consider element a of the string in Fig. 32-16 at some instant t as a wave moves along the string. Because of the wave, element a has both a kinetic energy due to its motion and a potential energy due to the amount it is stretched. The kinetic energy of the element is one-half its mass times its speed squared:

$$\Delta K = \frac{1}{2}(\mu\,\Delta x)\left(\frac{\partial y}{\partial t}\right)^2$$

Therefore the kinetic energy per unit length, or the kinetic energy density of the wave, is

$$\frac{\Delta K}{\Delta x} = \frac{1}{2}\mu\left(\frac{\partial y}{\partial t}\right)^2$$

By solving Prob. 7, you can show that the potential energy density is

$$\frac{\Delta U}{\Delta x} = \frac{1}{2}F\left(\frac{\partial y}{\partial x}\right)^2$$

The ***energy density*** of the wave is the sum of the kinetic and potential energy densities:

Energy density of a wave

$$\frac{\Delta E}{\Delta x} = \frac{1}{2}\mu\left(\frac{\partial y}{\partial t}\right)^2 + \frac{1}{2}F\left(\frac{\partial y}{\partial x}\right)^2$$

In Fig. 32-16, we show that the condition of element a at time t (its motion and the amount it is stretched) is the same as that of element b at time $t + \Delta t$. That is, the energy possessed by a is passed along to b in the time interval Δt, so that energy is propagating along the string with speed $\Delta x/\Delta t$, which is the same as the wave speed v. Thus the rate at which energy is propagating along the string, or the power of the wave, is $P = (\Delta E/\Delta x)(\Delta x/\Delta t)$, or

Power of a wave traveling in the $+x$ direction

$$P = \left[\frac{1}{2}\mu\left(\frac{\partial y}{\partial t}\right)^2 + \frac{1}{2}F\left(\frac{\partial y}{\partial x}\right)^2\right]v \qquad (32\text{-}13)$$

For a harmonic wave, $y = A \sin(kx - \omega t)$,

$$\frac{\partial y}{\partial t} = -\omega A \cos(kx - \omega t)$$

and

$$\frac{\partial y}{\partial x} = kA \cos(kx - \omega t)$$

Also, $v^2 = F/\mu = \omega^2/k^2$. With these expressions, we can write the power of a harmonic wave as

Power of a harmonic wave

$$P = \mu\omega^2 A^2 v \cos^2(kx - \omega t)$$

Figure 32-17 shows a graph of the power of a harmonic wave versus time at a fixed point on the string. Note that the power remains positive at all times, which indicates a continuous transfer of energy in the direction of wave travel.

Since the power of a harmonic wave oscillates between zero and a maximum value, a quantity of interest is the average power $\overline{P}$, and the average is taken over a whole number of cycles at a fixed point. The time dependence of P at a fixed point,

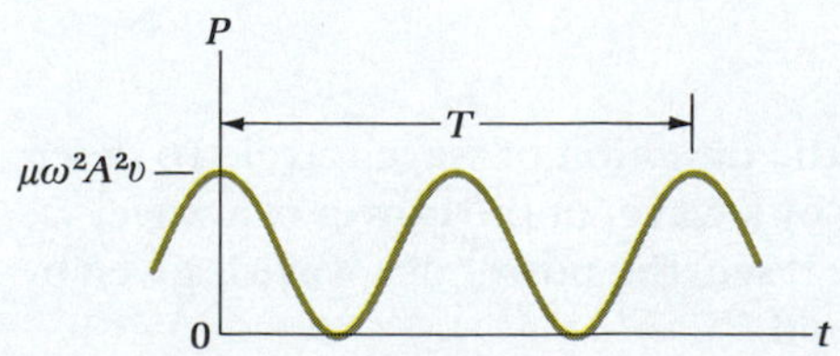

Figure 32-17. Power of a harmonic wave at a fixed point versus time. The time dependence of the power is $\cos^2(\omega t)$.

say $x = 0$, is $\cos^2(\omega t)$ [or equivalently, $\sin^2(\omega t)$]. As we have seen previously, (Sec. 31-7), the average of $\cos^2(\omega t)$ over a whole number of cycles is $\frac{1}{2}$. Thus

$$\bar{P} = \tfrac{1}{2}\mu\omega^2 A^2 v \qquad (32\text{-}14)$$

Average power of a harmonic wave

The dependence of the average power (and the power) on the amplitude squared, $P \propto A^2$, is a feature common to harmonic waves, whether they are sound waves, electromagnetic waves, or waves on a string.

According to Eq. 32-14, the average power of a wave is the same at each point along the string. That is, no energy is lost by the wave as it propagates along the string. When energy is lost by a wave as it propagates through a medium, we say that the wave is *attenuated.* Since our harmonic wave has a fixed amplitude, it describes an unattenuated wave. Real mechanical waves passing through a medium always exhibit some attenuation, but often the attenuation is small enough to be neglected. In our discussions, we neglect attenuation.

Attenuation

EXAMPLE 32-3

Power of a wave. A string with a linear mass density of 47 g/m is held taut so that the tension in the string is 75 N. A harmonic wave of amplitude $A = 13$ mm and frequency $\nu = 32$ Hz propagates along the string. What is the average power of the wave?

Solution. To use the expression for the average power given in Eq. 32-14, we must determine values for v and ω:

$$v = \sqrt{\frac{F}{\mu}} = \sqrt{\frac{75\ \text{N}}{47\ \text{g/m}}} = 40\ \text{m/s}$$

$$\omega = 2\pi\nu = 2\pi(32\ \text{Hz}) = 200\ \text{rad/s}$$

From Eq. 32-14,

$$\bar{P} = \tfrac{1}{2}\mu\omega^2 A^2 v = \tfrac{1}{2}(47\ \text{g/m})(200\ \text{rad/s})^2(13\ \text{mm})^2(40\ \text{m/s})$$
$$= 6.4\ \text{W}$$

SELF-TEST 32-3. Consider a particlar position on the string as the wave in the above example moves along. *(a)* What is the minimum value of the wave's power at any instant of time? *(b)* What is the maximum value of the wave's power at any instant of time? ***ANSWERS:*** *(a)* zero; *(b)* 12.7 W.

Waves in Three Dimensions: Wave Intensity

If you wiggle your finger up and down on the surface of water, waves propagate radially outward from the point of the disturbance. Each wave crest forms a circle whose radius continually increases as the wave propagates, and each of these expanding circles is called a ***wavefront.*** In Fig. 32-18 the wavefronts appear as a

Wavefront

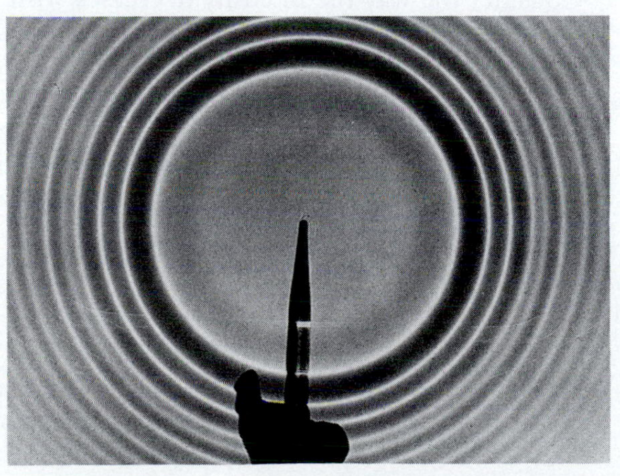

Figure 32-18. Circular wavefronts on water. The waves are caused by a periodic disturbance of the water surface; the disturbance behaves as a point source.

series of concentric circles with a separation of one wavelength. Waves on the surface of water are an example of waves propagating in two dimensions. You can use these two-dimensional waves to help visualize waves in three dimensions.

For a wave in three dimensions, such as a sound wave or a light wave, the wavefronts form surfaces. In Fig. 32-19*a* we show sections of concentric spherical surfaces. These spherical sections represent parts of the wavefronts of waves emanating from a point source located at the center of the spheres. In a uniform medium, each wavefront forms a complete spherical surface. Waves emanating from a point source in a uniform medium are often called ***spherical waves.*** The lines shown in the figure that are directed radially outward from the source are called ***rays.*** Rays are drawn perpendicular to wavefronts and are used to indicate the direction of propagation of a wave.

Spherical waves

Rays

At a large distance from a point source, the wavefronts become nearly planar.

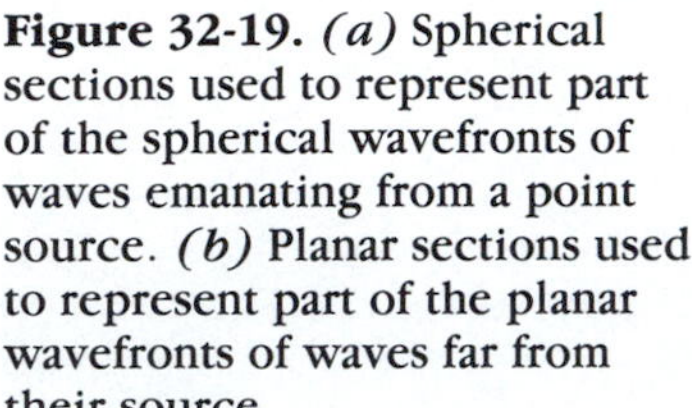

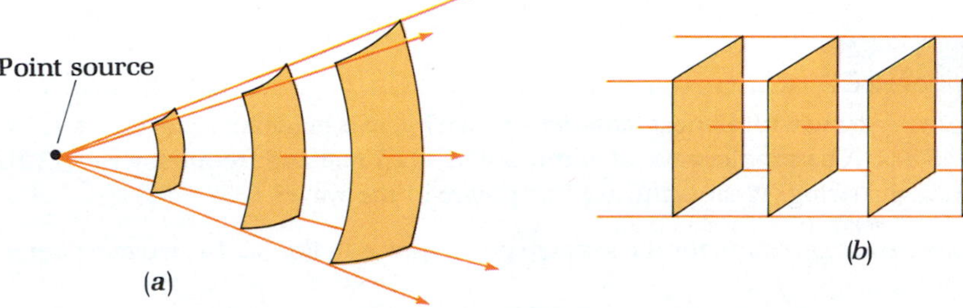

Figure 32-19. *(a)* Spherical sections used to represent part of the spherical wavefronts of waves emanating from a point source. *(b)* Planar sections used to represent part of the planar wavefronts of waves far from their source.

Then a section of a plane can be used to represent part of a wavefront, as shown in Fig. 32-19*b*. Such waves are referred to as ***plane waves.*** In the case of plane waves, the rays are straight lines that are parallel and equally spaced.

Plane waves

When sunlight comes through your window and into your room, the power to the room due to these electromagnetic waves depends on the area of the window —the larger the window area, the more power to the room. The effective area of the window is the projection of the window's area onto a plane perpendicular to the rays of the sunlight. The quantity that characterizes the energy flow due to the sunlight is the ***wave intensity.*** The intensity I of a wave is the power P propagated per unit area of a surface normal to the propagation direction. Since power is the rate of energy transfer, $P = \Delta E/\Delta t$, we have

Definition of wave intensity

$$I = \frac{P}{\Delta S} = \frac{\Delta E}{\Delta t\, \Delta S} \tag{32-15}$$

where ΔS is the area of a surface normal to the propagation direction and ΔE is the energy that passes through the surface in the time Δt. Since the dimension of power is energy divided by time, the dimension of intensity is (energy)/[(time)(area)]. The SI unit of intensity is the watt per square meter ($\mathrm{W/m^2}$).

Consider the way the intensity decreases with distance from a point source of waves. Let P_o represent the steady power output of the source, and assume that (i) the medium does not attenuate the waves and (ii) the waves emitted by the source travel uniformly in all directions. Since the intensity I is the power per unit area, the rate at which energy passes through a spherical surface of radius r (area $4\pi r^2$) centered at the source is $I(4\pi r^2)$. By conservation of energy, the power output of the source is equal to the rate at which energy passes through this spherical surface. That is, $P_o = 4\pi r^2 I$, or

Intensity of waves from a point source

$$I = \frac{P_o}{4\pi r^2} \tag{32-16}$$

Because the energy spreads out uniformly in three-dimensional space as it propagates from the point source, the intensity decreases as the inverse square of the distance r from the source.

EXAMPLE 32-4

Intensity of light from a light bulb. Determine the intensity of visible light waves at a distance of 1.5 m from a 60-W light bulb. Assume that 5 percent of the power to the bulb is emitted in the form of visible light, and treat the bulb as a point source that emits waves uniformly in all directions through a uniform medium.

Solution. The power output of the bulb in the form of visible light waves is $P_o = (0.05)(60\ \text{W}) = 3\ \text{W}$. Using Eq. 32-16, we find that the wave intensity at a distance of 1.5 m from the source is

$$I = \frac{P_o}{4\pi r^2} = \frac{3\ \text{W}}{4\pi(1.5\ \text{m})^2} = 0.1\ \text{W/m}^2$$

SELF-TEST 32-4. What is the intensity of visible light at a distance of 15 m from the light bulb in the above example? ***ANSWER:*** $0.001\ \text{W/m}^2$.

32-6 INTERFERENCE OF HARMONIC WAVES

If two or more waves exist in the same region, then the waves interfere. That is, when individual waves are superposed, they combine to produce a resultant wave. We examine two special cases of the interference of two harmonic waves. One of these cases involves phenomena called *constructive* and *destructive interference*. The other case reveals a phenomenon called *standing waves*.

Constructive and Destructive Interference

Consider the interference of two harmonic waves, waves 1 and 2:

$$y_1 = A \sin(kx - \omega t + \phi_1)$$

and

$$y_2 = A \sin(kx - \omega t + \phi_2)$$

Each wave is traveling in the same direction and has the same amplitude A, wave number k, and angular frequency ω, but their phase constants ϕ_1 and ϕ_2 may be different. The ***phase difference*** $\Delta\phi$ between these waves is

$$\Delta\phi = (kx - \omega t + \phi_2) - (kx - \omega t + \phi_1) = \phi_2 - \phi_1$$

Phase difference

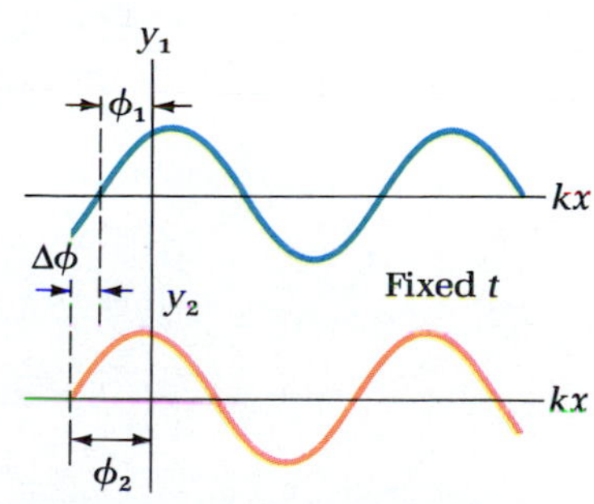

Figure 32-20. Graphs of y versus kx at fixed t for harmonic waves 1 and 2. The phase difference between the waves is $\Delta\phi$.

In Fig. 32-20, each wave is plotted on a separate graph, and the phase difference is shown. If $\phi_2 = \phi_1$ (or $\Delta\phi = 0$), we say that the waves are in phase, and if $\phi_2 \neq \phi_1$ (or $\Delta\phi \neq 0$), we say that the waves are out of phase by the phase difference $\Delta\phi$.

To find the resultant wave y due to the interference of y_1 and y_2, we use the principle of superposition:

$$y = y_1 + y_2 = A[\sin(kx - \omega t + \phi_1) + \sin(kx - \omega t + \phi_2)]$$

We can see the effect of the interference more clearly by using the trigonometric identity

$$\sin\alpha + \sin\beta = 2\sin[\tfrac{1}{2}(\alpha+\beta)]\cos[\tfrac{1}{2}(\alpha-\beta)] \tag{32-17}$$

Letting $\alpha = kx - \omega t + \phi_2$ and $\beta = kx - \omega t + \phi_1$, we find $\frac{1}{2}(\alpha+\beta) = kx - \omega t + \frac{1}{2}(\phi_1 + \phi_2)$ and $\frac{1}{2}(\alpha - \beta) = \frac{1}{2}\Delta\phi$, so that

$$y = [2A\cos(\tfrac{1}{2}\Delta\phi)]\sin[kx - \omega t + \tfrac{1}{2}(\phi_1 + \phi_2)] \tag{32-18}$$

Two features of the resultant wave are apparent from Eq. 32-18:

1. The resultant wave y is a harmonic wave with the same wave number k (or the same wavelength λ), the same angular frequency ω (or the same period T), and the

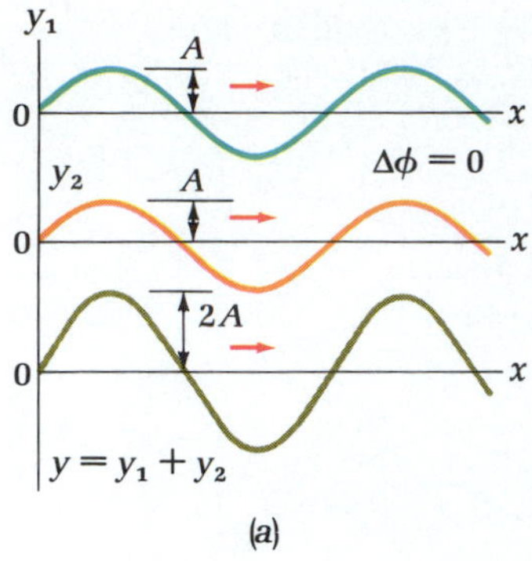

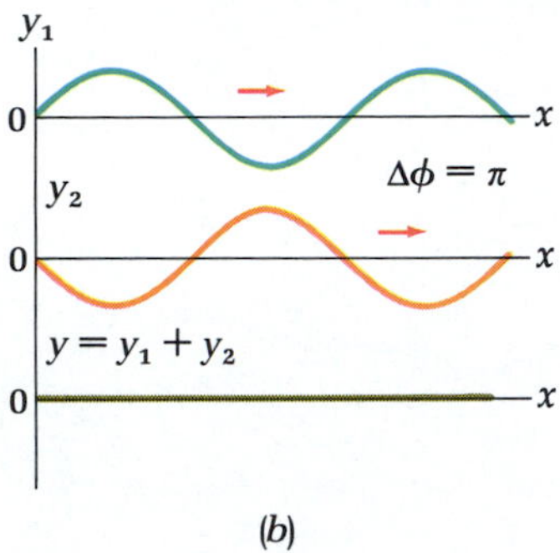

Figure 32-21. Interference of two harmonic waves. Each wave has the same A, k, ω, and propagation direction. *(a)* Constructive interference ($\Delta\phi = 0$). *(b)* Destructive interference ($\Delta\phi = \pi$).

same propagation direction ($+x$) that each individual wave (y_1 or y_2) would have if one of them were present alone.

2. The amplitude of y, which is $2A \cos(\frac{1}{2}\Delta\phi)$, depends on the phase difference $\Delta\phi$ between y_1 and y_2. Thus the phase difference plays an important role in the interference of these harmonic waves.

Suppose the phase difference between y_1 and y_2 is zero; the waves are in phase. Since $\Delta\phi = 0$, $\cos(\frac{1}{2}\Delta\phi) = \cos 0 = 1$, and the amplitude of y is $2A \cos 0 = 2A$. This type of interference is called ***constructive interference.*** When y_1 and y_2 constructively interfere, the resultant wave has twice the amplitude that either y_1 or y_2 would have if one of them were acting alone. In this case, y_1 and y_2 superpose crest on crest and trough on trough, as shown in Fig. 32-21*a*.

Suppose y_1 and y_2 are out of phase such that their phase difference is 180°, or π rad. Then $\cos(\frac{1}{2}\Delta\phi) = \cos(\frac{1}{2}\pi) = 0$, and the amplitude of y is $2A \cos(\frac{1}{2}\pi) = 0$. This type of interference is called ***destructive interference.*** The resultant wave is nonexistent because y_1 and y_2 superpose crest on trough and trough on crest, as shown in Fig. 32-21*b*. When harmonic waves are out of phase by 180°, or π rad, we say that the waves are completely out of phase.

For other values of the phase difference $\Delta\phi$, the resultant wave has an amplitude intermediate between $2A$ and zero. A graph of the case where $\Delta\phi = \pi/2 \text{ rad} = 90°$ is shown in Fig. 32-22. In this case the amplitude of y is $2A \cos(\pi/4) = 1.41A$.

These interference phenomena are peculiar to waves, and when we are investigating an effect, they can be used as evidence that waves are responsible for the effect. As we shall see in Chap. 36, the first convincing proof that light behaves as a wave was made with experiments in which light exhibited constructive and destructive interference.

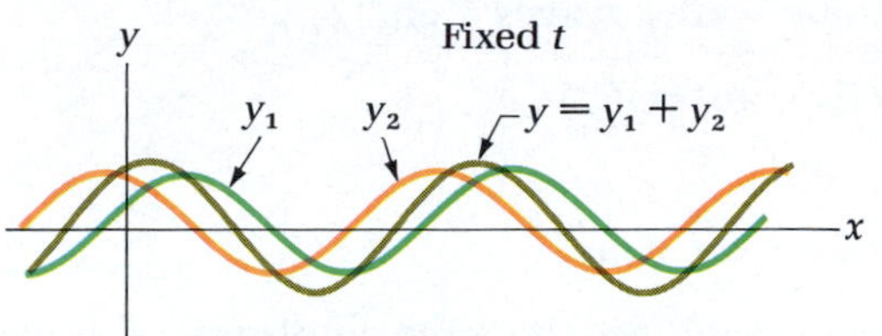

Figure 32-22. The interference of waves 1 and 2 on the same rope produce a resultant wave. In the case shown, waves 1 and 2 are out of phase by 90°.

Standing Waves

If a wave train encounters a boundary, the reflected part of the wave train interferes with the incident part of the wave train. This interference can create a stationary wave pattern called a *standing wave.* When waves are confined to a region of space by boundaries, waves reflected back and forth from the boundaries can establish standing waves. Standing waves are important in many aspects of science and engineering (see the Commentary). They are also important in the design of buildings, bridges, and musical instruments.

Figure 32-23 shows photographs of standing waves produced on a long rubber tube by wiggling one end of the tube. As we shall see, to establish a particular standing-wave pattern, like one of those shown in the figure, the end must be wiggled at a specific frequency.

To see how standing waves are formed, consider the interference of two harmonic waves that have the same amplitude, wave number, and angular frequency, but travel in opposite directions:

$$y_1 = A \sin(kx - \omega t) \qquad \text{and} \qquad y_2 = A \sin(kx + \omega t)$$

Wave 1 is traveling toward $+x$, wave 2 is traveling toward $-x$, and each has speed $v = \omega/k$. Let y_1 represent the incident wave and y_2 represent the reflected wave. Using the principle of superposition, we find that the resultant wave y is

$$y = y_1 + y_2 = A \sin(kx - \omega t) + A \sin(kx + \omega t)$$

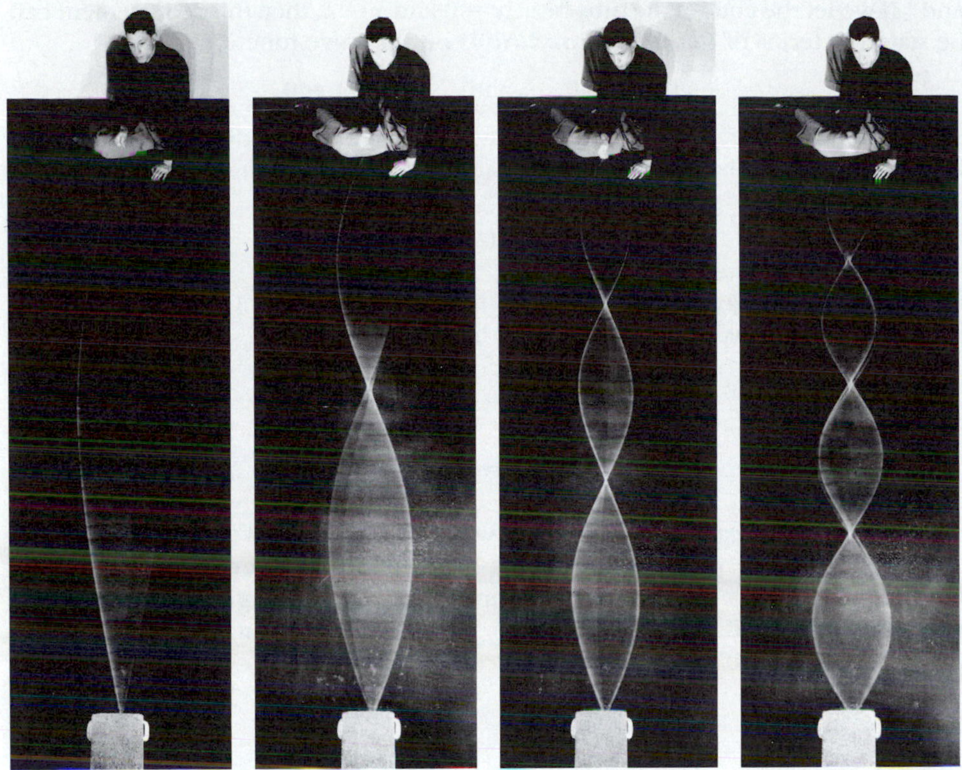

Figure 32-23. Standing waves on a long, rubber tube. In each case the end of the tube is wiggled gently at a frequency given by Eq. 32-23. The tube is moving so rapidly that it appears blurred over the extent of its motion.

The resultant wave can be expressed more simply by using the trigonometric identity given in Eq. 32-17. Letting $\alpha = kx + \omega t$ and $\beta = kx - \omega t$, we obtain

$$y(x, t) = 2A \cos(\omega t) \sin(kx) \qquad (32\text{-}19)$$

This is the wave function for a standing wave.

In a standing wave, the wave pattern does not move, but the elements of the rope do move. The motion is illustrated by the time sequence shown in Fig. 32-24. Writing the standing wave as $y = [2A \sin(kx)] \cos(\omega t)$, we can see that a particular element of the rope executes SHM with an amplitude equal to $2A|\sin(kx)|$. The amplitude of the SHM has its maximum value of $2A$ at positions where $|\sin(kx)| = 1$, or $kx = \pi/2, 3\pi/2, 5\pi/2$, and so on. These positions of maximum amplitude are called ***antinodes.*** Since $k = 2\pi/\lambda$, the antinodal positions are

$$x_n = \left(n + \frac{1}{2}\right)\frac{1}{2}\lambda \qquad (n = 0, 1, 2, \ldots) \qquad (32\text{-}20)$$

The antinodes are spaced one-half wavelength apart and are indicated by the letter A in Fig. 32-24.

Since $|\sin(kx)| = 0$ at values of x such that $kx = 0, \pi, 2\pi$, and so on, elements located at these positions do not move. The positions of these elements are called ***nodes.*** The nodal positions are

Nodes $$x_{n'} = n'\frac{1}{2}\lambda \qquad (n' = 0, 1, 2, \ldots) \qquad (32\text{-}21)$$

The nodes are spaced one-half wavelength apart and are indicated by the letter N in Fig. 32-24.

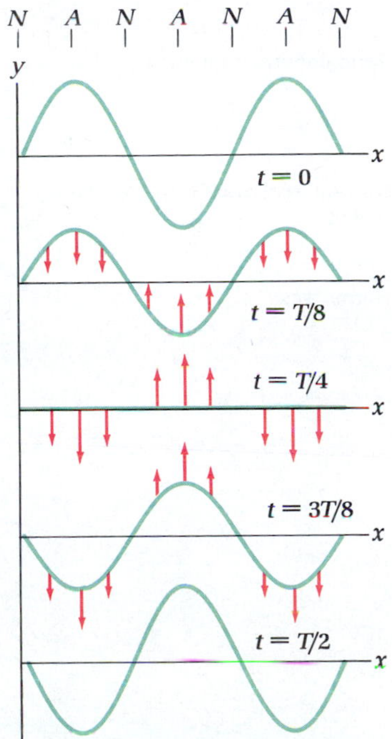

Figure 32-24. A standing wave at five instants of time. The arrows indicate the velocities of some representative elements. Note that at $t = 0$ and $t = T/2$, the entire rope is instantaneously at rest. The amplitude is exaggerated for purposes of illustration.

To produce standing waves of a single wavelength, like those shown in Fig. 32-23, the experimenter must wiggle the tube at a specific frequency. A node exists at each end of the tube, one at the experimenter's hand and the other at the fixed end.* If we let the ends of the tube be at $x = 0$ and $x = L$, then this requirement can be stated in terms of *boundary conditions* on the wave function:

$$y(0, t) = 0 \qquad \text{and} \qquad y(L, t) = 0$$

Since a node is at each end of the tube and since nodes are a distance $\frac{1}{2}\lambda$ apart, an integral number of half-wavelengths must fit along the length L of the tube, or

$$n\left(\frac{1}{2}\lambda_n\right) = L \qquad (n = 1, 2, 3, \ldots) \tag{32-22}$$

where n is an integer and λ_n represents the specific wavelengths that satisfy the boundary conditions. Solving for λ_n, we have

$$\lambda_n = \frac{2L}{n}$$

The allowed wavelengths for standing waves are $\lambda_1 = 2L$, $\lambda_2 = L$, $\lambda_3 = 2L/3$, and so on.

A standing wave cannot have just any wavelength whatsoever; it can only have one of the specific wavelengths λ_n that fit the boundary conditions. Since the frequency of a wave is related to its wavelength by the expression $v = \lambda\nu$, the frequency of a standing wave is similarly restricted to certain specific values ν_n. Using Eq. 32-22, we find that these values are

$$\nu_n = \frac{v}{\lambda_n} = n\left(\frac{v}{2L}\right)$$

These frequencies are called the ***natural frequencies.*** The lowest natural frequency, $\nu_1 = v/2L$, is called the ***fundamental frequency.*** Since $v = \sqrt{F/\mu}$, the fundamental frequency can be written

Fundamental frequency

$$\nu_1 = \frac{\sqrt{F/\mu}}{2L}$$

In terms of the fundamental frequency, the natural frequencies are

Natural frequencies for standing waves

$$\nu_n = n\nu_1 \qquad (n = 1, 2, 3, \ldots) \tag{32-23}$$

Harmonics

Thus the natural frequencies are whole multiples of the fundamental frequency. These natural frequencies are called ***harmonics;*** the fundamental frequency ν_1 is called the first harmonic, $\nu_2 = 2\nu_1$ is called the second harmonic, $\nu_3 = 3\nu_1$ is called the third harmonic, and so on. If the experimenter in Fig. 32-23 wiggles the end of the tube with SHM at one of the harmonic frequencies, then the standing wave corresponding to that frequency will exist on the tube.

Consider the fundamental frequency of a string in a musical instrument such as a guitar. Since $\nu_1 = \sqrt{F/\mu}/(2L)$, the frequency depends on (i) the tension F in the string, (ii) the linear mass density μ of the string, and (iii) the length L of the string between fixed ends. In the playing of a guitar, all three factors are taken into account. A guitar is tuned by changing F. Different strings have different values of μ. The length L between fixed ends of a string is changed by pushing the string against a fret.

EXAMPLE 32-5

Standing waves on a long rubber tube. For the long rubber tube in Fig. 32-23, suppose $F = 72$ N, $M = 0.84$ kg, and $L = 3.8$ m. *(a)* What is the fundamental frequency? *(b)* At what frequency should the experimenter wiggle the end of the tube in order to produce a standing wave into two antinodes?

* Actually, the experimenter must continuously wiggle the end of the tube slightly to compensate for attenuation. Thus the end held by the experimenter is only approximately a node.

Solution. *(a)* Since $\mu = M/L$, the fundamental frequency is

$$\nu_1 = \frac{\sqrt{F/\mu}}{2L} = \frac{\sqrt{FL/M}}{2L} = \frac{1}{2}\sqrt{\frac{F}{LM}}$$

$$= \frac{1}{2}\sqrt{\frac{72\text{ N}}{[(3.8\text{ m})(0.84\text{ kg})]}} = 2.4\text{ Hz}$$

(b) A standing wave with two antinodes is the second harmonic, $n = 2$. The frequency of the second harmonic is $\nu_2 = 2\nu_1 = 2(2.4\text{ Hz}) = 4.8\text{ Hz}$.

SELF-TEST 32-5. For the long rubber tube described in the above example, at what frequency should the experimenter wiggle the end of the tube to produce a standing wave with three antinodes? ***ANSWER:*** 7.2 Hz.

COMMENTARY: Atoms, Standing Waves, and Quantization

Suppose you have an account with a bank, and the bank has a rather strange rule. The rule is that your account is allowed to have only certain values, say \$0.00, \$17.40, \$34.02, \$52.87, and so on. If the account is at \$17.40, then you can deposit \$16.62 because \$16.62 + \$17.40 = \$34.02. This deposit puts your account at an allowed value. But you cannot deposit \$16.61 or \$16.63 or any other amount that would give the account a forbidden value. If your account is at \$34.02, then you can withdraw \$16.62 or \$34.02, but you cannot withdraw any other amount. We describe this strange rule by saying that your bank balance is quantized. (Of course, bank accounts really are quantized in units of one penny, but our strange rule provides a better analogy.)

Curiously, the energy of atoms is quantized. When an atom releases energy or accepts energy, it can release or accept only an amount that takes it from one allowed energy value to another. These allowed energy values are called *energy levels.*

What property of atoms is responsible for the quantization of atomic energies? The characteristics of standing waves provide a clue to the answer. In the previous section we found that when waves are confined to a region, standing waves can be established and the frequencies of these standing waves are quantized. On a string fixed at each end, the standing wave frequencies are $\nu_n = n\nu_1$, where ν_1 is the fundamental frequency and n is an integer. For example, if $\nu_1 = 2.1$ Hz, then the allowed frequencies are 2.1, 4.2, 6.3 Hz, and so on. A frequency of 4.1 Hz is forbidden; no standing wave with this frequency can exist on this string. These quantized frequencies are a result of boundary conditions placed on the wave function of the wave.

What do waves have to do with atoms? Atoms contain electrons, and electrons manifest a wave behavior. Indeed, an electron can be described by a wave function. Since the electrons in an atom are confined to that atom, the wave functions of atomic electrons must conform to boundary conditions. These boundary conditions cause the wave functions of atomic electrons to exhibit nodes and antinodes similar to those of standing waves on a string. Different atomic energy levels are associated with different standing wave patterns for the electron wave functions. The quantization of atomic energies is a result of the boundary conditions placed on the wave functions of atomic electrons.

SUMMARY

Section 32-1. Characteristics of Waves

A traveling wave is a disturbance that propagates from one position to another. In a transverse wave, the particles of the medium are displaced perpendicular to the propagation direction, and in a longitudinal wave, the particles of the medium are displaced parallel to the propagation direction.

Section 32-2. Wave Pulses

A wave is described by a wave function $y(x, t)$. In the case of a wave on a rope or string, the wave function gives the displacement $y\mathbf{j}$ of the string from equilibrium. For a wave traveling with speed v toward $+x$,

$$y(x, t) = f(x - vt) \tag{32-1}$$

and toward $-x$,

$$y(x, t) = f(x + vt) \tag{32-2}$$

When two or more waves interfere, the resultant wave function can be found with the principle of superposition.

Section 32-3. Harmonic Waves

A harmonic wave can be expressed as

$$y(x, t) = A \sin (kx - \omega t)$$

where $k = 2\pi/\lambda$ and $\omega = 2\pi\nu = 2\pi/T$. The speed of a harmonic wave is $v = \lambda/T = \omega/k$. Harmonic waves satisfy the wave equation:

$$\frac{\partial^2 y}{\partial x^2} = \frac{1}{v^2}\frac{\partial^2 y}{\partial t^2} \tag{32-8}$$

Section 32-4. The Wave Equation from Newton's Second Law

Newton's second law applied to an element of a rope or string yields the wave equation and gives the speed of the waves as $v = \sqrt{F/\mu}$. In general, the wave equation for a physical quantity ψ is

$$\frac{\partial^2 \psi}{\partial x^2} = \frac{1}{v^2}\frac{\partial^2 \psi}{\partial t^2} \tag{32-12}$$

This wave equation is a linear differential equation, which means that waves that satisfy this equation obey the principle of superposition.

Section 32-5. Power of a Wave

A wave's power is the product of its energy density and speed:

$$P = \left[\frac{1}{2}\mu\left(\frac{\partial y}{\partial t}\right)^2 + \frac{1}{2}F\left(\frac{\partial y}{\partial x}\right)^2\right] v \tag{32-13}$$

For a harmonic wave,

$$P = \mu\omega^2 A^2 v \cos^2 (kx - \omega t)$$

The average power of a harmonic wave at a particular point is

$$\bar{P} = \tfrac{1}{2}\mu\omega^2 A^2 v \tag{32-14}$$

In three dimensions, the intensity of a wave is the incident power per unit area for a surface normal to the propagation direction:

$$I = \frac{P}{\Delta S} \tag{32-15}$$

The intensity at a distance r from a point source is

$$I = \frac{P_o}{4\pi r^2} \tag{32-16}$$

Section 32-6. Interference of Harmonic Waves

When two harmonic waves with the same amplitude, wave number, and angular frequency interfere, the resultant wave is

$$y = [2A \cos (\tfrac{1}{2}\Delta\phi)] \sin [kx - \omega t + \tfrac{1}{2}(\phi_1 + \phi_2)] \tag{32-18}$$

If the phase difference $\Delta\phi$ is 0, the interference is constructive. If $\Delta\phi$ is π rad, the interference is destructive. When waves are confined to a region, a standing wave can be formed:

$$y = 2A \cos (\omega t) \sin (kx) \tag{32-19}$$

Because of the boundary conditions, only waves of certain wavelengths λ_n and frequencies ν_n can exist.

QUESTIONS

1. Discuss the similarities between the physical waves in this chapter and the waves in each of the following sentences:
(a) If your boss treats you unfairly, do not make waves by reporting it to the head of the company because you might be the one that gets hurt.
(b) Gene has wavy hair.
(c) Elda waved goodbye to Frank.
(d) The enemy attacked by sending forth waves of infantry.

2. List five different examples of mechanical waves. What is the medium in each case?

3. If you twirl one end of a horizontally stretched rope in a circle in a plane perpendicular to the rope, a wave will propagate along the rope. Is this wave transverse, longitudinal, some combination of the two, or none of these?

4. Suppose two wave pulses traveling in opposite directions interfere with each other. After they are no longer superposed, have their shapes been changed by the encounter? Have their speeds been changed by the encounter? If your answer to either question is yes, describe these changes.

5. When two wave pulses traveling in opposite directions encounter each other, do they bounce off one another like billiard balls in a head-on collision, or does each pulse pass through the other like a ghost in a cartoon? Which figure in this chapter best supports your answer?

6. When a wave pulse is reflected from a fixed barrier, as shown in Fig. 32-9, we say that the phase of the wave is reversed on reflection. When a pulse on a light spring is reflected from the light-spring–heavy-spring boundary, is the phase of the wave reversed on reflection? When a pulse on a heavy spring is reflected from a heavy-spring–light-spring boundary, is the phase of the pulse reversed on reflection?

7. Consider a harmonic wave of amplitude A on a horizontal string. What is y for an element with maximum upward acceleration? What is y for an element with maximum downward acceleration? What is y for an element with zero acceleration? What is y for an element that is instantaneously at rest? What is y for an element with maximum speed?

8. For a harmonic wave, what part of the wave has a positive $\partial^2 y/\partial x^2$? What part of the wave has a negative $\partial^2 y/\partial x^2$? What is y for an element that is straight?

9. Two harmonic waves are on different ropes and each rope has the same density and tension. The waves have the same

frequency, but wave 1 has twice the amplitude of wave 2. Which wave has the larger speed? Which wave causes the larger maximum speed for the elements of the rope on which it travels?

10. A graph of part of a wave pulse on a string at a particular instant is shown in Fig. 32-25, with the direction of propagation as indicated. Which, if any, of the elements marked with letters is instantaneously at rest? What is the sign of the velocity component of the elements that are moving? Do any of your answers depend on the direction of propagation?

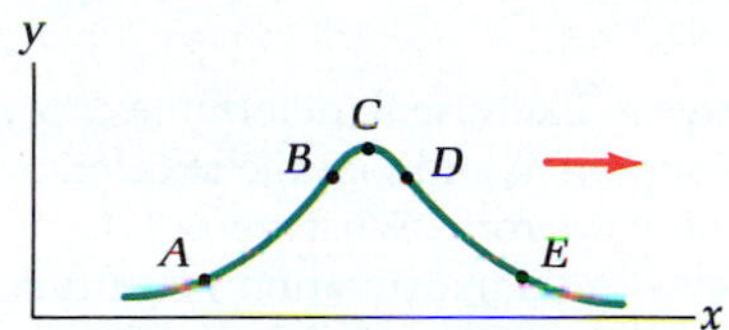

Figure 32-25. Question 10.

11. A graph of part of a wave pulse on a string at a particular instant is shown in Fig. 32-25, with the direction of propagation as indicated. The elements at B and D are inflection points —that is, $\partial^2 y/\partial x^2 = 0$ at these points. Which, if any, of the lettered elements has a zero acceleration? What is the sign of the acceleration component of those elements that have a nonzero acceleration? Do any of your answers depend on the propagation direction?

12. If the speed of a wave on a string is v_0 when the tension in the string is F_0, what is the speed of a wave when the tension is $2F_0$?

13. Two strings are under the same tension. The mass per unit length of string 1 is μ_0, and the mass per unit length of string 2 is $2\mu_0$. If the speed of a wave on string 1 is v_0, what is the speed of a wave on string 2?

14. If the amplitude of a harmonic wave is doubled, with other factors held fixed, how does the power of the wave change?

15. If the frequency of a harmonic wave on a rope is doubled and other factors are held fixed, how does the power of the wave change?

16. How does the amplitude of a harmonic wave depend on distance from a point source that emits waves uniformly in all directions in three-dimensional space?

17. The wavefronts of waves emitted uniformly from along the length of a long line source in a homogeneous medium are in the form of cylinders concentric with the line source, as shown in Fig. 32-26. If there is no attenuation, then the intensity of such waves fall off as $1/R$, where R is the perpendicular distance from the line source to the point where the intensity is measured. Explain this dependence on R. How does the amplitude of a harmonic wave depend on distance from a line source?

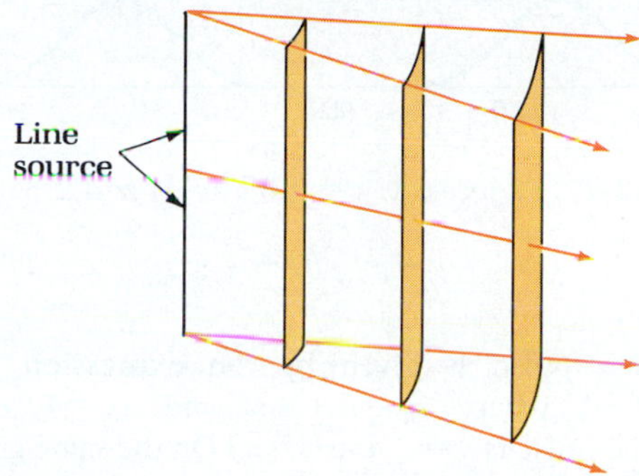

Figure 32-26. Question 17: Cylindrical sections used to represent cylindrical wavefronts emanating from a line source. Waves whose wavefronts have this shape are called *cylindrical waves.*

18. A wave propagates down a long rope that hangs freely from a support. As it propagates, what happens to the speed of the wave?

19. The distance from the sun to Mars is about $\frac{3}{2}$ that from the sun to the earth. Compare the intensity of sunlight at Mars with the intensity of sunlight at the earth.

20. Why do the strings on a guitar have different diameters? Which strings produce the lower notes?

21. Sometimes when an airplane flies near a house, the television signal received at the house fades periodically. Why?

22. Complete the following table:

Symbol	Represents	Type	SI Unit
$y(x, t)$		Component	
v			m/s
A			
λ	Wavelength		
k			
ν			
$\partial y/\partial t$			
$\Delta\phi$			
F			
μ			
P			
I			

EXERCISES

Section 32-2. Wave Pulses

1. Figure 32-27 shows a wave pulse on a string at two different times: (i) $t = 0.0$ s and (ii) $t = 2.3$ s. Use the figure to estimate *(a)* the speed of the pulse, *(b)* the height of the pulse, *(c)* the width of the pulse. Note that the vertical and horizontal scales in the figure are different.

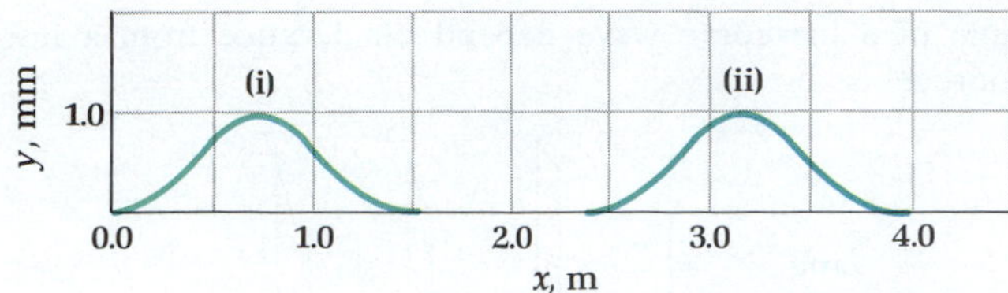

Figure 32-27. Exercise 1: $t = 0.0$ s at (i) and 2.3 s at (ii).

2. A wave pulse is given by the expression $y(x, t) = y_0 e^{-[(x-vt)/x_0]^2}$, where $y_0 = 4.1$ mm and $x_0 = 1.28$ m. The speed of the pulse is $v = 7.4$ m/s. *(a)* On the same graph, plot y versus x for the pulse at $t = 0.00$ s and $t = 0.50$ s. Exaggerate the vertical scale relative to the horizontal scale for purposes of illustration. *(b)* Determine the pulse height h and the pulse width w.

3. A wave pulse is given by the expression $y(x, t) = y_0 (2.00)^{-[(x-vt)/x_0]^4}$, where $y_0 = 10.0$ mm and $x_0 = 1.00$ m. The speed of the pulse is $v = 2.00$ m/s. *(a)* On the same graph, plot y versus x for the pulse at $t = 0.00$ s and $t = 1.00$ s. Exaggerate the vertical scale relative to the horizontal scale for purposes of illustration. *(b)* Determine the pulse height h and the pulse width w.

4. Plot y versus t at $x = 0$ for the pulse in Exercise 2.

5. Longitudinal and transverse waves produced by an earthquake travel at different speeds in the earth's crust, about 8 km/s for longitudinal and 5 km/s for transverse. If these two types of waves from an earthquake are initially received 250 s apart, what is the distance from the receiving site to the earthquake?

6. State which of the following functions have the form $y = f(x - vt)$ or $y = f(x + vt)$:
1 $y = y_0[(x + vt)/x_0]^{1/2}$
2 $y = y_0[(x^2 - 2vtx + v^2t^2)/x_0^2]$
3 $y = y_0[(x^2 - v^2t^2)/x_0^2]$
4 $y = y_0 \ln x/vt$
Could any of these functions be used to represent a traveling wave on a string? If not, why not?

Section 32-3. Harmonic Waves

7. A harmonic wave on a rope is given by the expression

$$y(x, t) = (4.3 \text{ mm}) \sin \left\{ \frac{2\pi}{0.82 \text{ m}} [x + (12 \text{ m/s})t] \right\}$$

What are the wave's *(a)* amplitude, *(b)* wavelength, *(c)* speed, *(d)* period, *(e)* wave number, *(f)* frequency, *(g)* angular frequency, *(h)* direction of propagation? *(i)* Determine y for the element located at $x = 0.58$ m at the instant $t = 0.41$ s.

8. A harmonic wave on a rope is given by the expression

$$y(x, t) = (6.8 \text{ mm}) \sin [(1.47 \text{ rad/m})x - (4.18 \text{ rad/s})t]$$

What are the wave's *(a)* amplitude, *(b)* wave number, *(c)* angular frequency, *(d)* speed, *(e)* wavelength, *(f)* frequency, *(g)* period, *(h)* direction of propagation? *(i)* Determine y for the element located at $x = 0.22$ m at the instant $t = 0.75$ s.

9. For the wave in Exercise 7, determine expressions for *(a)* the velocity component and *(b)* the acceleration component of each element of the rope. What are the *(c)* maximum speed and *(d)* maximum acceleration magnitude for each element of the rope? Determine *(e)* the velocity component and *(f)* the acceleration component of the element located at $x = 0.58$ m at the instant $t = 0.41$ s. *(g)* From your answer to part *(f)*, is the rope bending upward or downward at that element and at that instant?

10. For the wave in Exercise 8, determine expressions for *(a)* the velocity component and *(b)* the acceleration component of each element of the rope. What are *(c)* the maximum speed and *(d)* the maximum acceleration magnitude for each element of the rope? Determine *(e)* the velocity component and *(f)* the acceleration component of the element located at $x = 0.22$ m at the instant $t = 0.75$ s. *(g)* From your answer to part *(f)*, is the rope bending upward or downward at that element and at that instant?

11. For the wave in Exercise 7, determine expressions for *(a)* the slope of the rope and *(b)* $\partial^2y/\partial x^2$ as functions of x and t. Determine *(c)* the slope and *(d)* $\partial^2y/\partial x^2$ at the element located at $x = 0.58$ m at the instant $t = 0.41$ s. *(e)* From your answer to part *(d)*, is the rope accelerating upward or downward at that element and at that instant?

12. For the wave in Exercise 8, determine expressions for *(a)* the slope of the rope and *(b)* $\partial^2y/\partial x^2$ as functions of x and t. Determine *(c)* the slope and *(d)* $\partial^2y/\partial x^2$ at the element located at $x = 0.22$ m at the instant $t = 0.75$ s. *(e)* From your answer to part *(d)*, is the rope accelerating upward or downward at that element and at that instant?

13. Starting with the expression for the speed of a harmonic wave, $v = \lambda/T$, and using the relations between λ and k and between T, ν, and ω, show that *(a)* $v = \lambda\nu$ and *(b)* $v = \omega/k$.

14. The speed of electromagnetic waves in vacuum (or in air) is 3.0×10^8 m/s. The wavelengths of visible electromagnetic waves extend from about 750 nm (red light) to about 400 nm (violet light). Determine the frequency range of visible light.

15. The frequency range of the electromagnetic waves which correspond to commercial broadcasts (radio and television) extend from about 10^4 to 10^9 Hz. The speed of electromagnetic waves in vacuum (or air) is 3.0×10^8 m/s. What is the wavelength range of commercial broadcasts?

16. The frequency range of audible sound is from about 20 Hz to 20 kHz. The speed of sound in air is about 330 m/s. What is the wavelength range of audible sound in air?

17. *(a)* Write an expression for a wave on a string in which the amplitude is 25 mm, the wavelength is 0.72 m, and the frequency is 4.1 Hz. The propagation direction is toward $+x$. *(b)* What is the speed of the wave? What are the maximum *(c)* speed and *(d)* acceleration magnitude for an element of the string? What are the maximum *(e)* slope and *(f)* $\partial^2y/\partial x^2$ at each element of the string?

18. *(a)* Write an expression for a wave on a string in which the amplitude is 17 mm, the wave number is 5.3 rad/m, and the angular frequency is 19 rad/s. The propagation direction is toward $-x$. *(b)* What is the speed of the wave? What are the maximum *(c)* speed and *(d)* acceleration magnitude at each element of the string? What are the maximum *(e)* slope and *(f)* $\partial^2 y/\partial x^2$ at each element of the string?

19. Show that the dimensions are the same for each side of the wave equation

$$\frac{\partial^2 y}{\partial x^2} = \frac{1}{v^2}\frac{\partial^2 y}{\partial t^2}$$

20. The harmonic wave on a string shown in Fig. 32-28 has an amplitude of 25 mm, a speed of 46 m/s, an angular frequency of 160 rad/s and propagates toward $+x$. Determine *(a)* the velocity components and *(b)* the acceleration components of the lettered elements. Determine *(c)* the slope of the string and *(d)* $\partial^2 y/\partial x^2$ at the lettered elements.

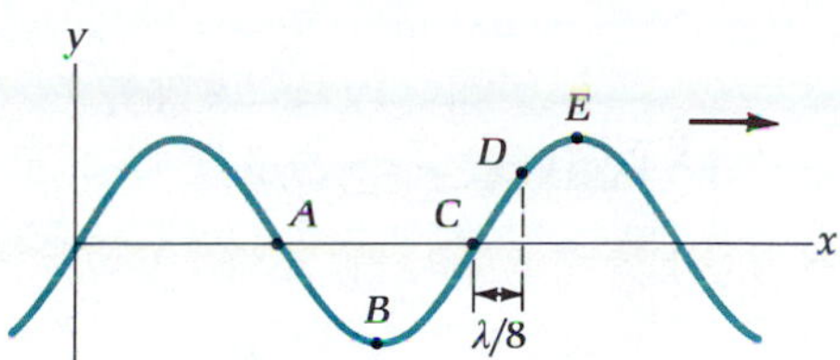

Figure 32-28. Exercise 20.

Section 32-4. The Wave Equation from Newton's Second Law

21. Show that the SI unit of $\sqrt{F/\mu}$ is meters per second.

22. A 5.5-m length of string has a mass of 0.34 kg. If the tension in the string is 77 N, what is the speed of a wave on the string?

23. The speed of a wave on a rope is 21 m/s when the tension in the rope is 92 N. What is the mass per unit length of the rope?

24. Suppose that $y_1(x, t)$ and $y_2(x, t)$ are solutions to the wave equation, $\partial^2 y/\partial x^2 = (1/v^2)(\partial^2 y/\partial t^2)$. Show that $y(x, t) = y_1(x, t) + y_2(x, t)$ is a solution to the wave equation. Thus solutions to the wave equation obey the principle of superposition.

25. The speed of a sound wave in a solid is given by $v = \sqrt{Y/\rho}$, where Y is Young's modulus (Chap. 15), and ρ is the mass density. Show that the SI unit of Y is newtons per square meter.

26. When the tension in a guitar string is 25 N, the speed of a wave on the string is 15 m/s. What tension would make the wave speed 30 m/s?

Section 32-5. Power of a Wave

27. A string with a mass per unit length of 0.15 kg/m is under a tension of 59 N. What is the average power of a wave of amplitude 42 mm and angular frequency of 130 rad/s that is propagated along the string?

28. The harmonic wave on a string shown in Fig. 32-28 has an amplitude of 25 mm, a speed of 46 m/s, and an angular frequency of 160 rad/s and propagates in the $+x$ direction. What is the power of the wave at each of the lettered elements? Let $\mu = 0.029$ kg/m.

29. *(a)* Explain why the power of a wave traveling in the $-x$ direction is

$$P = -\left[\frac{1}{2}\mu\left(\frac{\partial y}{\partial t}\right)^2 + \frac{1}{2}F\left(\frac{\partial y}{\partial x}\right)^2\right]v$$

(b) Show that the power of a harmonic wave traveling in the $-x$ direction is

$$P = -\mu\omega^2 A^2 v \cos^2(kx + \omega t)$$

(c) Explain the physical significance of the fact that this power is negative at all times and at any point.

30. A harmonic wave on a string has an amplitude of 32 mm, an angular frequency of 110 rad/s, and a speed of 28 m/s and propagates in the $+x$ direction. The mass per unit length of the string is 0.12 kg/m. At a particular element, plot the power of the wave versus time over an interval of $1.5T$.

31. *(a)* Show that for a harmonic wave, the kinetic energy density is equal to the potential energy density at any instant t and at any point x. *(b)* Use this result to show that the power of a harmonic wave $y = A \sin(kx - \omega t)$ is

$$P = \mu\omega^2 A^2 v \cos^2(kx - \omega t)$$

32. The intensity of the sunlight transmitted by a window of area 0.94 m² is 850 W/m². If the angle between the sun's rays and a normal to the window is 41°, what is the power to the room?

33. The intensity of sunlight just outside the earth's atmosphere is called the *solar constant* and has an average value of about 1.35 kW/m². What is the power incident on the earth due to sunlight?

34. A loudspeaker resting on the floor directs the sound upward such that the wave intensity is uniform over the surface of an imaginary hemisphere. If the power of the waves emitted from the loudspeaker is 12 W, what is the wave intensity at a distance of 1.4 m from this source?

Section 32-6. Interference of Harmonic Waves

35. Two waves with the same amplitude, wave number, angular frequency, and propagation direction are present on a rope. The phase difference between the waves is 0.65 rad, and the amplitude of each wave is 51 mm. What is the amplitude of the resultant wave?

36. Two waves 1 and 2 are present on a rope at the same time and are given by the expressions

$$y_1 = (14 \text{ mm}) \sin[(4.8 \text{ rad/m})x - (29 \text{ rad/s})t - 0.21 \text{ rad}]$$

$$y_2 = (14 \text{ mm}) \sin[(4.8 \text{ rad/m})x - (29 \text{ rad/s})t + 0.35 \text{ rad}]$$

(a) What is the phase difference between the waves? *(b)* What is the amplitude of the resultant wave $y = y_1 + y_2$. *(c)* Write an expression for the resultant wave.

37. A wave on a string is given by the expression

$$y_1 = (22 \text{ mm}) \sin[(3.4 \text{ rad/m})x + (36 \text{ rad/s})t + 0.16 \text{ rad}]$$

(a) Write an expression for a wave y_2 with the same amplitude

as y_1 that constructively interferes with y_1. *(b)* Write an expression for a wave y_2 that destructively interferes with y_1.

38. Two waves 1 and 2 have the same amplitude ($A_1 = A_2 = 46$ mm), angular frequency, and propagation direction, but they are out of phase with each other. Waves 1 and 2 are both present on a string, and the amplitude of the resultant wave is 31 mm. What is the phase difference between waves 1 and 2?

39. Two waves 1 and 2 are present on a string:

$$y_1 = (35 \text{ mm}) \sin [(8.4 \text{ rad/m})x - (15.7 \text{ rad/s})t]$$

$$y_2 = (35 \text{ mm}) \sin [(8.4 \text{ rad/m})x + (15.7 \text{ rad/s})t]$$

(a) Write the expression for the resultant wave $y = y_1 + y_2$ in the form of a wave function for a standing wave. *(b)* Give the x coordinates of the first two antinodes, starting at the origin and progressing toward $+x$. *(c)* What is the x coordinate of the node that is between the antinodes of part *(b)*? *(d)* What is the distance between the antinodes of part *(b)*?

40. The ends of a string are fixed such that the string is held taut with a tension of 122 N. The string is 2.4 m long and has a mass of 0.19 kg. What is the frequency of a standing wave with three antinodes?

41. On the same graph, plot the standing wave of Exercise 39 from $x = 0.00$ to 1.50 m at three different times: $t = 0.00$ s, $t = 0.10$ s, and $t = 0.20$ s.

42. A length L of string with mass per unit length μ is held taut with a tension F. Show that the frequency ν_n of a standing wave with n antinodes on the string is given by

$$\nu_n = \frac{n}{2L}\sqrt{\frac{F}{\mu}}$$

43. A standing wave with two antinodes exists on a 1.6-m-long string that is held fixed at each end. The string vibrates at a frequency of 7.2 Hz. *(a)* Write an expression for the wave function of this standing wave. *(b)* At what point on the string is the origin of the coordinates y and x? *(c)* Does $t = 0$ in your expression correspond to the string being straight or to each element being at its maximum displacement magnitude or to some other configuration?

44. A particular guitar string is in tune when the $n = 1$ standing wave has a frequency of 247 Hz. If the string has a mass per unit length of 1.3 g/m and the distance between its fixed ends is 0.58 m, what should the tension be?

45. The fundamental frequency for a standing wave on the rope in Fig. 32-29 is 16 Hz, and the linear mass density of the rope is 0.18 kg/m. What is the mass of the suspended block?

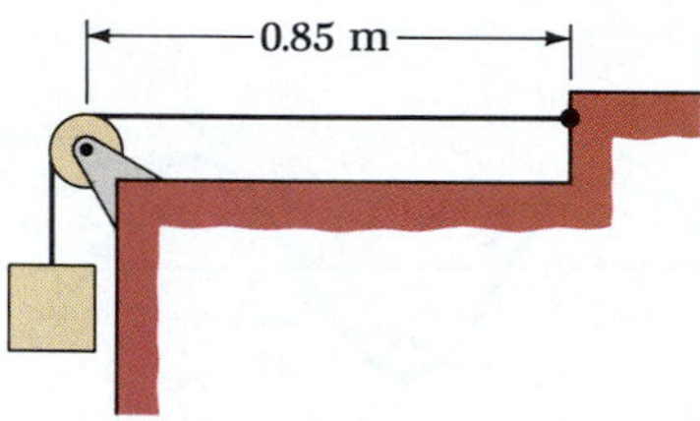

Figure 32-29. Exercise 45.

46. A beam of light with wavelength $\lambda = 630$ nm is normally incident on a mirror. The reflected light interferes with the incident light to form a standing wave that has a node at the mirror surface. *(a)* How far from the mirror is the nearest antinode. *(b)* How many nodes are within 1.0 mm of the mirror?

PROBLEMS

1. ***Comparison of space and time dependence.*** A pulse on a rope travels with a speed of 18 m/s in the $+x$ direction. The following data were taken from a photograph of the pulse at the time $t = 0$:

x, mm	−2.0	−1.0	0.0	1.0	2.0	3.0	4.0
y, mm	2.0	7.8	10.0	8.2	4.1	2.0	1.0

There was no measurable displacement of the string for $x < -3.0$ m or $x > 5.0$ m. *(a)* Make a graph of y versus x by plotting the above data and sketching a smooth curve between the points. For purposes of illustration, exaggerate the scale along the y axis relative to the x axis. *(b)* Make a similar graph of y versus t at the coordinate $x = 0$. Compare the shapes of these two graphs.

2. ***Distance to a thunderstorm.*** A procedure for finding the approximate distance to a thunderstorm is to measure the time interval between when a lightning flash is seen and when the subsequent thunderclap is heard. The time interval in seconds divided by 3 gives the distance in kilometers. *(a)* Justify this procedure and *(b)* estimate the percent error you might expect. Take the speed of light to be 3.0×10^8 m/s and the speed of sound to be 350 m/s. *(c)* Would you need to modify the procedure if the speed of sound were twice its given value? If the speed of light were twice its given value?

3. ***Traveling wave solution to the wave equation.*** *(a)* Show that the expression for a traveling wave, $y(x, t) = f(x - vt)$, is a solution to the wave equation,

$$\frac{\partial^2 y}{\partial x^2} = \frac{1}{v^2}\frac{\partial^2 y}{\partial t^2}$$

(*Hint:* Let $\xi = x - vt$ and note that

$$\frac{\partial f}{\partial x} = f'\frac{\partial \xi}{\partial x} \quad \text{and} \quad \frac{\partial f}{\partial t} = f'\frac{\partial \xi}{\partial t}$$

where $f' = df/d\xi$.) *(b)* Similarly show that the expression $y(x, t) = f(x + vt)$ is a solution to the wave equation.

4. ***Speed of waves on a stretched spring.*** *(a)* Show that the speed of transverse waves on a stretched spring is

$v = \sqrt{kL(L-\ell)/M}$, where k is the spring constant, ℓ is the unstretched length of the spring, L is the stretched length of the spring, and M is the spring's mass. *(b)* For the case where $L \gg \ell$, show that the time Δt required for a wave to travel from one end of the spring to the other is $\Delta t \approx \sqrt{M/k}$. The interesting feature of this result is that Δt is independent of the length L of the spring. The more the spring is stretched, the faster the wave travels, so the travel time remains nearly the same.

5. ***Waves on a rope hanging vertically.*** A rope of length L and mass M hangs freely from the ceiling. Show that the time Δt required for a transverse wave to travel the length of the rope is $\Delta t = 2\sqrt{L/g}$. (*Hint:* The wave speed varies with coordinate x measured from the rope's free end because the tension in the rope at a point is due to the weight of the rope below that point.)

6. ***A wave function for a wave pulse.*** Consider a wave pulse described by $y(x, t) = y_0 e^{-[(x-vt)/x_0]^2}$. Show explicitly that this wave function is a solution to the wave equation.

7. ***Potential energy density.*** Because a wave distorts the medium in which it travels, potential energy is associated with the wave. In Fig. 32-30, we show an element of a string that is stretched from its original length of Δx to a new length $\sqrt{(\Delta x)^2 + (\Delta y)^2}$ by a wave. The distance $\Delta\ell$ that the element is stretched by the wave is $\Delta\ell = \sqrt{(\Delta x)^2 + (\Delta y)^2} - \Delta x$. *(a)* Assuming that the element is small, show that

$$\Delta\ell = \left[\sqrt{1 + \left(\frac{\partial y}{\partial x}\right)^2} - 1\right]\Delta x$$

(b) Use the binomial expansion (App. M) and the assumption that $(\partial y/\partial x)^2 \ll 1$ to show that

$$\Delta\ell = \frac{1}{2}\left(\frac{\partial y}{\partial x}\right)^2 \Delta x$$

(c) The potential energy ΔU of the element due to the wave is the work done by the tension F in stretching the element: $\Delta U = F\,\Delta\ell$. Show that the potential energy per unit length $\Delta U/\Delta x$ of the wave is

$$\frac{\Delta U}{\Delta x} = \frac{1}{2} F \left(\frac{\partial y}{\partial x}\right)^2$$

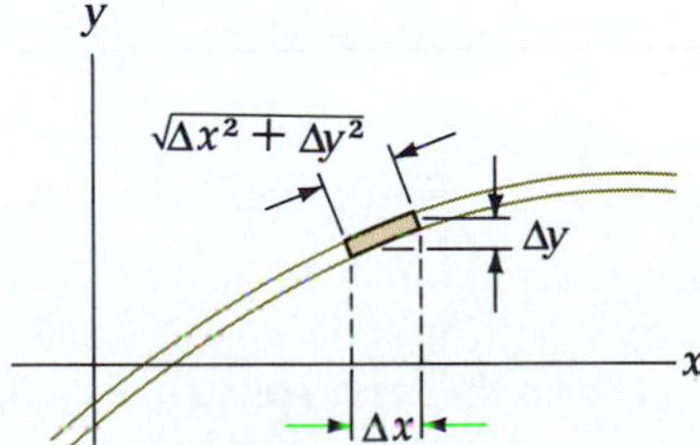

Figure 32-30. Problem 7.

8. ***Equality of kinetic and potential energy densities.*** The wave function for an arbitrarily shaped wave traveling in the $+x$ direction is $y = f(x - vt)$. Show that such a wave's kinetic energy density is equal to its potential energy density, so that

$$\frac{\Delta E}{\Delta x} = 2\frac{\Delta K}{\Delta x} = 2\frac{\Delta U}{\Delta x} = \mu\left(\frac{\partial y}{\partial t}\right)^2 = F\left(\frac{\partial y}{\partial x}\right)^2$$

(*Hint:* See Prob. 3.)

9. ***Energy of a standing wave.*** Consider a standing wave with n antinodes on a string of length L:

$$y = 2A \cos(\omega t) \sin(kx)$$

where $k = \pi n/L$. Show that the wave's average energy $\bar{E}$ (averaged over a whole number of cycles) is

$$\bar{E} = \frac{\pi^2 F A^2}{L} n^2$$

Thus, for standing waves of a given amplitude, the energy of the wave increases with the square of the number of antinodes. This result is analogous to an important problem in quantum mechanics called the "particle-in-a-box."

10. ***Amplitudes of reflected and transmitted waves.*** Strings a and b are tied together at one end of each string and then held taut such that they compose two media for waves. Waves can then encounter the boundary where the strings are joined. Three harmonic waves exist on the strings: an incident wave (wave 1), a reflected wave (wave 2), and a transmitted wave (wave 3). Let the boundary be at $x = 0$ and assume that these waves can be written

$$y_1 = A_1 \sin(k_a x - \omega t)$$

$$y_2 = A_2 \sin(k_a x + \omega t)$$

$$y_3 = A_3 \sin(k_b x - \omega t)$$

Note that since the frequency of a wave depends on the wave source, each wave has the same angular frequency ω. However, the wave number is different for waves on different strings (k_a for string a and k_b for string b) because the wave speed is different. We have chosen the $+x$ direction as the direction of travel for the incident wave, which means that the reflected wave travels in the $-x$ direction and the transmitted wave in the $+x$ direction. Show that the amplitudes of the reflected and transmitted waves are

$$A_2 = \frac{k_b - k_a}{k_b + k_a} A_1 \qquad A_3 = \frac{2k_a}{k_b + k_a} A_1$$

(*Hint:* Since the strings are smoothly joined at $x = 0$, we must have $y_1 + y_2 = y_3$ at $x = 0$ and $\partial y_1/\partial x + \partial y_2/\partial x = \partial y_3/\partial x$ at $x = 0$. That is, the strings and the slope of the string are continuous at the point where they are joined.)

11. ***Power of a wave pulse.*** *(a)* Show that the power of the wave pulse $y(x, t) = y_0 e^{-[(x-vt)/x_0]^2}$ is

$$P = 4\mu v^3 \left(\frac{y_0}{x_0}\right)^2 \left(\frac{x - vt}{x_0}\right)^2 e^{-2[(x-vt)/x_0]^2}$$

(b) What is the value of the power at the point of maximum displacement due to the pulse?

12. ***Standing waves.*** Two harmonic waves

$$y_1 = A \sin(kx - \omega t + \phi_1)$$

and

$$y_2 = A \sin(kx + \omega t + \phi_2)$$

combine to form a standing wave. Show that, if we adjust the coordinate origin and beginning time, the standing wave can be written

$$y = 2A \cos(\omega t') \sin(kx')$$

Determine the values of x' and t'.

HIGHLIGHTS OF MODERN PHYSICS

Solitons

In 1834 a new field of physics was born — the study of solitons. In the words of its creator, the Scottish engineer John Scott Russell:

> I was observing the motion of a boat which was rapidly drawn along a narrow channel by a pair of horses, when the boat suddenly stopped — not so the mass of water in the channel which it had put in motion; it accumulated round the prow of the vessel in a state of violent agitation, then suddenly leaving it behind, rolled forward with great velocity, assuming the form of a large solitary elevation, a rounded, smooth and well defined heap of water, which continued its course along the channel apparently without change of form or diminution of speed. I followed it on horseback and overtook it still rolling on at a rate of some eight or nine miles an hour, preserving its original figure some thirty feet long and a foot to a foot and a half in height. Its height gradually diminished and after a chase of one or two miles I lost it in the windings of the channel. Such in the month of August 1834 was my first chance interview with that singular and beautiful phenomenon which I have called the Wave of Translation

Russell reproduced this wave of translation in a series of careful experiments, which were ignored by the scientific world. No one saw the connection between this ***solitary wave,*** as Russell's wave is now known, and other phenomena, such as tidal bores (Fig. 1). Russell was able to show that the wave speed c depended on the wave amplitude A, the depth of the water h, and the acceleration of gravity g, according to

$$c = \sqrt{g(A+h)}$$

Figure 1. The tidal bore on the Petitcodiac River near the Bay of Fundy is an example of a soliton.

The fact that the speed depends on the wave's amplitude is significant because it says that the wave equation is nonlinear. Contrast this with the classical linear wave equation, Eq. 32-8:

$$\frac{\partial^2 u}{\partial t^2} = c^2 \frac{\partial^2 u}{\partial x^2} \tag{1}$$

Here, the wave speed c does not depend on the amplitude: if $u(x, t)$ is a solution, then so is $Au(x, t)$ for any value of A and with the same value of c. Apparently, the wave equation for Russell's waves has a nonlinear term, such as u^2 or $u\,\partial u/\partial x$. More than 60 years after Russell's discovery, the Dutch mathematicians D. J. Korteweg and G. de Vries finally obtained the wave equation describing zero-viscosity waves in shallow water, the "$K\,dV$ equation,"

$$\frac{1}{c_0}\frac{\partial u}{\partial t} + \left(1 + \frac{3}{2h}u\right)\frac{\partial u}{\partial x} + \frac{h^2}{6}\frac{\partial^3 u}{\partial x^3} = 0 \tag{2}$$

where $c_0 = \sqrt{gh}$. This wave equation is more complicated than Eq. 1, but we will dissect it piece by piece to understand its meaning.

Starting with the linear equation, Eq. 1, we note that if both terms are moved to the left side, the equation can be factored as

$$\left(\frac{\partial}{\partial t} + c\frac{\partial}{\partial x}\right)\left(\frac{\partial}{\partial t} - c\frac{\partial}{\partial x}\right)u = 0$$

or with the factors written in the opposite order. Therefore, Eq. 1 is satisfied if

$$\frac{\partial u}{\partial t} - c\frac{\partial u}{\partial x} = 0$$

for the factors written in the order shown above, or if

$$\frac{\partial u}{\partial t} + c\frac{\partial u}{\partial x} = 0 \tag{3}$$

for the opposite order. Those two possibilities correspond to the solutions $f(x \pm ct)$ described in Sec. 32-4.

The surprising thing about Russell's wave was that it did not quickly spread out, but rather maintained its shape. Most waves spread as they travel, a property called ***dispersion.*** Dispersion occurs when the wave speed depends on wavelength. Thus one way to produce a dispersive wave equation is to let c depend on λ — or equivalently on $k = 2\pi/\lambda$ — in Eq. 1 or 3, instead of its being a constant. But another simple way is to add a third term to Eq. 3:

$$\frac{\partial u}{\partial t} + c\frac{\partial u}{\partial x} + a\frac{\partial^3 u}{\partial x^3} = 0 \tag{4}$$

This equation is still linear but now the wave speed depends on k, as can be seen by substituting a trial solution, $u(x, t) = A\sin(kx - \omega t)$. We find that the wave speed $v = \omega/k$, is $(c - ak^2)$, so the wave disperses. Since the $K\,dV$ equation contains this dispersive term, we expect Russell's wave to disperse, but it does not.

To see what the nonlinear term in Eq. 2 does, we add it to Eq. 3:

$$\frac{\partial u}{\partial t} + c\frac{\partial u}{\partial x} + bu\frac{\partial u}{\partial x} = 0$$

or

$$\frac{\partial u}{\partial t} + (c + bu)\frac{\partial u}{\partial x} = 0 \tag{5}$$

A comparison with Eq. 3 suggests that a solution could be obtained by replacing c in $f(x - ct)$ with $c + bu$:

$$u(x, t) = f[x - (c + bu)t]$$

This is not a solution, of course, because f contains u itself, but it does suggest that the wave speed (the coefficient of t) depends on the displacement u, being greater for a larger (positive) displacement.

This dependence on displacement means that the crest of a shallow water wave will travel faster than the trough in front of it (Fig. 2), they will crowd together, the leading edge will steepen, and the crest will grow, eventually breaking into the trough, a phenomenon that is easy to see at the beach for ordinary waves, and more dramatically for tidal waves. This is just the opposite of dispersion, which causes the wave to *spread out* and become *less steep.* The question naturally arises as to whether these two effects might cancel each other since the $K\,dV$ equation contains both dispersive and nonlinear, antidispersive terms. The answer is that they can, and the result is a solitary wave like Russell's wave of translation. The solitary-wave solution to the $K\,dV$ equation is

$$u(x, t) = A \operatorname{sech}^2 \frac{x - ct}{L} \tag{6}$$

where $\operatorname{sech} x \equiv \dfrac{2}{e^x + e^{-x}}$, the wave speed is $c = c_o(1 + A/2h)$ and $L = (4h^3/3A)^{1/2}$. Therefore the speed *increases* linearly with A and the width of the wave L *decreases* inversely as the square root of A. Higher waves travel faster and are narrower.

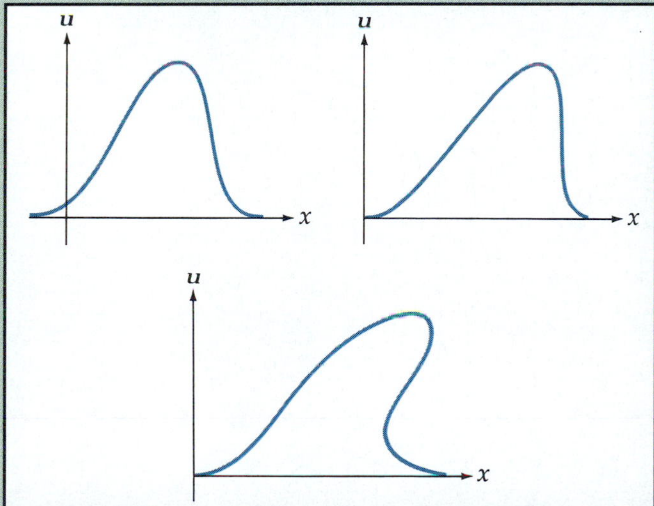

Figure 2. The effect of nonlinearities on wave shape. The initially symmetric wave becomes asymmetric as the wave crest, traveling faster, overtakes the trough in front and runs away from the one behind, steepening the leading edge slope and decreasing the slope behind.

What happens when a tall, fast solitary wave overtakes a short, slow one? Linear waves obey the superposition principle so two pulses simply add when they overlap; they pass right through each other unchanged (Sec. 32-2). When nonlinear pulses overlap, the event can more properly be described as a collision because the nonlinear term produces a strong interaction between the pulses, which distorts them both. The surprising thing about solitary solutions to the $K\,dV$ equation is

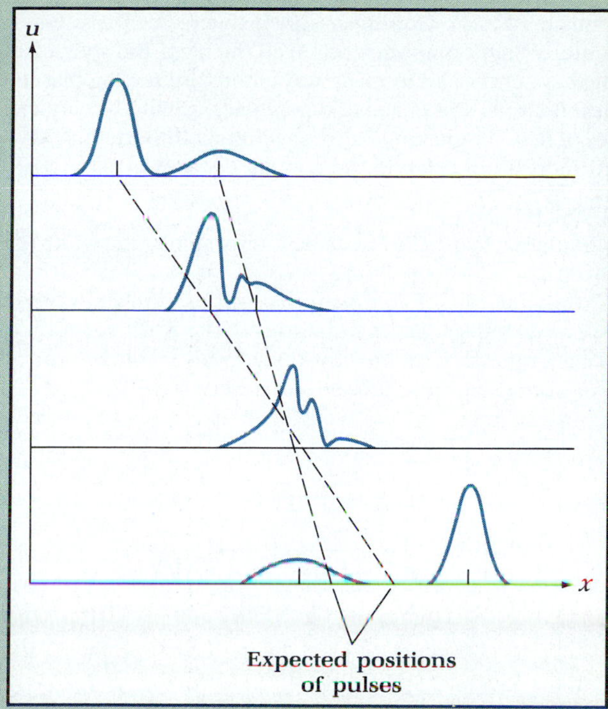

Figure 3. Solitons pass through each other without change in shape. The taller pulse travels faster than the shorter one and overtakes it. When they separate, it is as if they satisfied the superposition principle (they don't) except that they are not in their expected positions: the larger pulse reappears ahead of the position it would have had in the absence of a collision, while the smaller pulse is behind its expected position.

that they describe ***nonlinear*** waves that pass through each other without changing shape, just like wave solutions to linear equations (Fig. 3). It is this characteristic that led scientists to call these $K\,dV$ solutions ***solitons,*** a particlelike name that suggests a kind of permanence.

Today solitons are used in many places. In signal transmission of any kind it is important that the signal arrive without distortion and with enough strength to be reliably detected. Normal signals must be reconditioned and amplified with "repeaters" every so often because dispersion spreads and weakens the signal. Repeaters may be unnecessary in the near future because solitons maintain their strength much farther than ordinary waves. Solitons have been successfully transmitted through optical fibers for 10,000 km. Since they are unaffected when they collide, signals can move in both directions along an optical fiber. Nerve impulses have been shown to consist of solitons so that, once again, the signal reaches the destination with its original shape and with sufficient strength to be understood.

Any system will be linear for small amplitudes but will become nonlinear when the amplitude becomes large. Soliton formation requires an amplitude large enough for nonlinear terms to become important. If a spoon is placed in a cup of coffee, heat slowly diffuses up toward the hand, but if the spoon is placed in white hot coals, a soliton rapidly carries a heat pulse of energy to the other end. In optical-fiber communications, the light intensity must be large enough to drive the fiber into the nonlinear region.

There has been a role reversal in physics. Until about the middle of the twentieth century, physics was largely the study of linear systems. Nonlinear systems were seen as messy and uninteresting. Computers removed the mess and showed nonlinear systems to be in many ways more interesting that linear ones. Solitons and chaos (Commentary, Chap. 14) are examples of new, interesting, and exceedingly important fields that will increase in value in the coming decades.

QUESTIONS

1. Explain why L in Eq. 6 can be interpreted as the width of the soliton.
2. Apply the line of argument that was used to show how the dispersive and steepening qualities of the *K dV* equation cancel each other to explain why Russell was unable to produce a wave of translation in the form of a trough.

EXERCISES

1. Show that $Au(x, t)$ satisfies the classical wave equation, Eq. 1, for *any* A and the *same* wave speed.
2. Show that $f[x - (c + bu)t]$ satisfies Eq. 5.
3. Show that $f(x - ct)$ satsifies Eq. 3.
4. Show that the wave speed for harmonic solutions to Eq. 4 is $v = \omega/k = c - ak^2$.
5. *(a)* Verify that Eq. 6 is a solution to the *K dV* equation. *(b)* Show that the function obtained by reversing the sign of Eq. 6 is *not* a solution. Russell was easily able to produce waves of elevation but found that he could not produce the inverted wave of depression.

SOUND 33

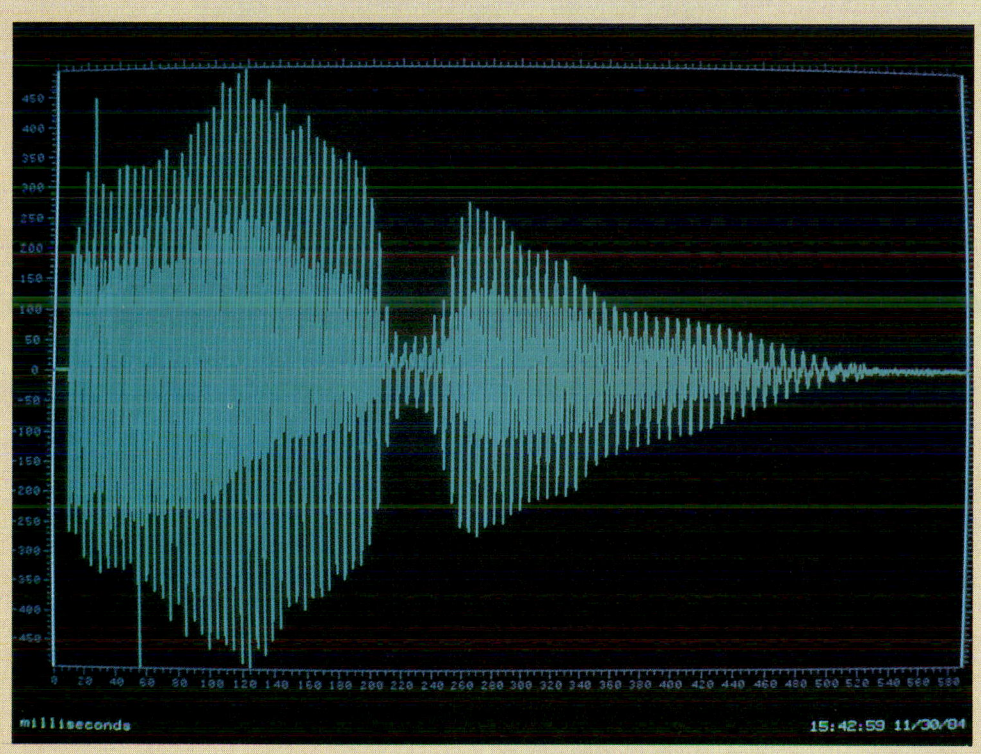

A sound wave displayed on an oscilloscope.

Mechanical waves with frequencies between about 20 Hz and 20 kHz are particularly important to us because these *sound waves* cause the sensation of hearing. Most of the sound that we hear is transmitted through air, but sound may also travel in liquids and solids. The wall of your room transmits the sound of your neighbor's radio, for example. Similar waves at higher frequencies, called *ultrasound,* have other uses — in medical diagnosis, for instance, and in detecting flaws in metal castings. For bats, dolphins, and submarines, these high-frequency waves are their means of finding their way about in the dark.

33-1 SOUND WAVES

A sound wave can be produced in a fluid, such as water or air, with the arrangement shown in Fig. 33-1. A long tube with a uniform cross section contains a fluid. If the piston remains at rest, as shown in Fig. 33-1*a*, the fluid is in equilibrium and has uniform density and pressure. The other end of the tube is assumed to be far away.

Suppose that the piston at one end of the tube moves back and forth, say in simple

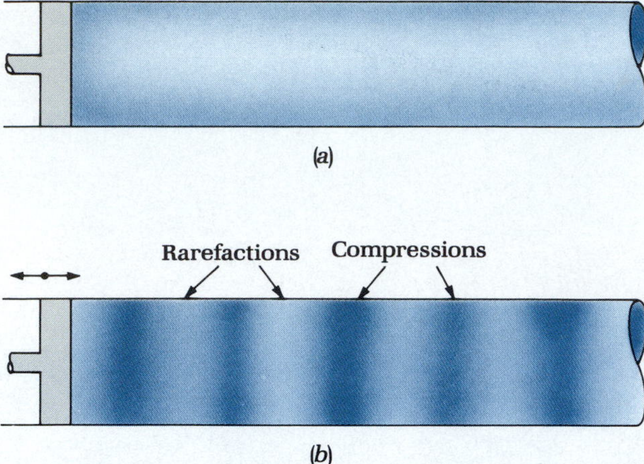

Figure 33-1. *(a)* No wave exists in the fluid. *(b)* The piston moves back and forth, causing regions of compression and of rarefaction to moving along the tube.

harmonic motion as sin (ωt). As the piston moves to the right, the layer of fluid next to it will also move in that direction. This element, or layer, of fluid exerts a force on the neighboring element of fluid, and a moving region of *compression* forms. In this region the density and pressure of the fluid are higher than the equilibrium values. The region of compression will continue to travel down the tube, even after the piston begins moving back to the left. During this return motion of the piston, a region of *rarefaction* of the fluid forms, where the density and pressure are lower than for equilibrium. This region of rarefaction will likewise move down the tube, sandwiched between one compression and the next compression, which is formed as the piston moves again to the right. Thus a pattern of compressions and rarefactions of the fluid moves along the tube, and this is a traveling sound wave.

A sound wave has compressions and rarefactions.

Although compressions and rarefactions move long distances along the tube, the fluid itself does not move very far. As the wave propagates along the tube, an element of the fluid moves back and forth along the tube. This motion is illustrated in Fig. 33-2, which shows velocities for elements of the fluid at an instant in a region containing two rarefactions and two compressions. Notice that the motion of the elements repeats for this wave, and the distance between consecutive compressions (or between consecutive rarefactions) is a wavelength λ. Figure 33-2 also indicates that the sound wave in a fluid is a longitudinal wave. The motion of the fluid is back and forth along the direction of propagation of the wave.

The sound wave in a fluid is longitudinal.

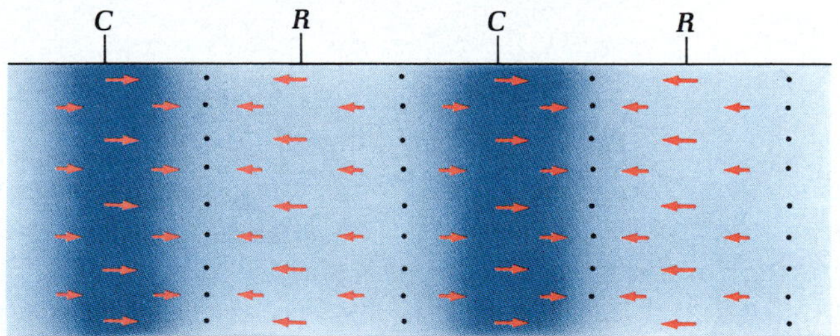

Figure 33-2. The arrows represent velocities of elements of the fluid. Regions of compression C and rarefaction R are indicated.

Consider an element of the fluid that is at x if no wave exists in the fluid. When a wave is propagating in the x direction, this element is displaced in the x direction by $\psi(x, t)$ at time t. (We use ψ instead of y to represent the longitudinal x component of the displacement of an element of the fluid.) A positive value of ψ corresponds to a displacement in the positive x direction.

A harmonic sound wave propagating in the positive x direction is described by

$$\psi(x, t) = A \cos (kx - \omega t) \tag{33-1}$$

where the amplitude A represents the maximum magnitude of the displacement, $k = 2\pi/\lambda$ is the wave number, and ω is the angular frequency. If v is the wave speed, then $\omega = kv$ (from Sec. 32-3).

If a sound wave exists in a fluid, the pressure of the fluid also varies. Let $p(x, t)$ represent the pressure of the fluid at position x at time t. If p_e is the equilibrium pressure when no wave exists, then $\Delta p(x, t) = p(x, t) - p_e$ is the pressure change due to the wave. For a harmonic wave, the pressure change Δp also varies sinusoidally:

Pressure change Δp for a wave in a fluid

$$\Delta p(x, t) = \Delta p_{max} \sin (kx - \omega t) \qquad (33\text{-}2)$$

where Δp_{max} is the maximum pressure change and occurs at a compression. At a rarefaction, $\Delta p = -\Delta p_{max}$; that is, $p(x, t)$ is less than the equilibrium value p_e. We now have two descriptions of a (harmonic) sound wave in a fluid, Eq. 33-1 for the longitudinal displacement and Eq. 33-2 for the pressure change. The connection between these descriptions is developed in Sec. 33-7.

Speed and Intensity of Sound Waves

We saw in Chap. 32 that the speed of a wave on a string (or a rope), $v = \sqrt{F/\mu}$, depends on the square root of the string tension divided by the mass per unit length. There is a similar expression for the speed of a sound wave in a fluid. The forcelike factor is related to the response of a fluid to a pressure change: If the pressure increases, the volume of the fluid decreases. For a given fluid, this response is expressed by the bulk modulus B. As defined in Chap. 15, the adiabatic bulk modulus is $B_s = -V(dp/dV)$, where dV is the small volume change that accompanies a small pressure change dp and V is the original volume. The adiabatic bulk modulus is used because, for a typical sound wave, the pressure changes (and temperature changes) in the fluid occur so rapidly that heat flow between neighboring elements of the fluid is negligible. See Sec. 17-5 for a discussion of an adiabatic process.

The adiabatic bulk modulus has the dimension of pressure, force per unit area. If we divide B_s by the equilibrium density ρ (the inertialike term) of the fluid, the ratio B_s/ρ has the dimension of speed squared. Thus the square root has the dimension of speed, and this dimensional analysis suggests that the wave speed v is proportional to $\sqrt{B_s/\rho}$. This result is confirmed in Sec. 33-7 where Newton's laws are applied to the fluid. The speed of sound in a fluid is given by

$$v = \sqrt{\frac{B_s}{\rho}} \qquad (33\text{-}3)$$

Speed of sound in a fluid

For sound waves in many gases, including air, the gas may be treated as an ideal gas, and we can determine the adiabatic bulk modulus. (See Prob. 3 for Newton's calculation of the speed of sound in air.) If an ideal gas undergoes an adiabatic process, the pressure and volume changes occur such that

$$pV^\gamma = \text{constant}$$

where $\gamma = c_p/c_V$ is the ratio of specific heats. Since the derivative of $pV^\gamma =$ constant is zero, we have

$$\frac{d}{dV} pV^\gamma = p\gamma V^{\gamma - 1} + V^\gamma \frac{dp}{dV} = 0$$

Rearranging and solving for $-V(dp/dV)$, we obtain

$$B_s = -V\frac{dp}{dV} = \gamma p$$

The density of the ideal gas can be expressed as $\rho = m/V = nM/V$, where V is the volume occupied by n mol and M is the molar mass (molecular weight). Then

$$\frac{B_s}{\rho} = \frac{\gamma p}{nM/V} = \frac{\gamma p V}{nM}$$

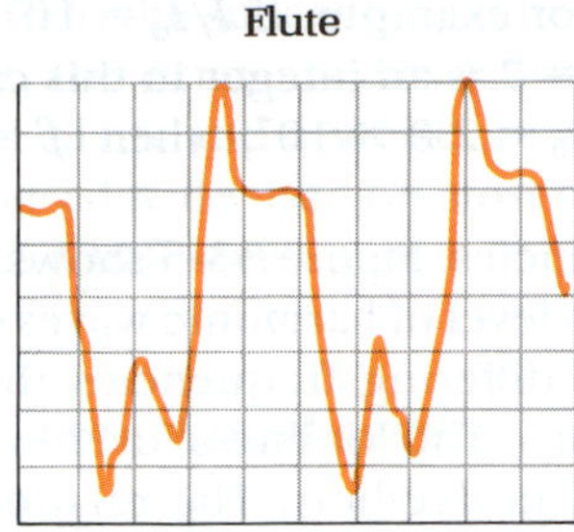

(a)

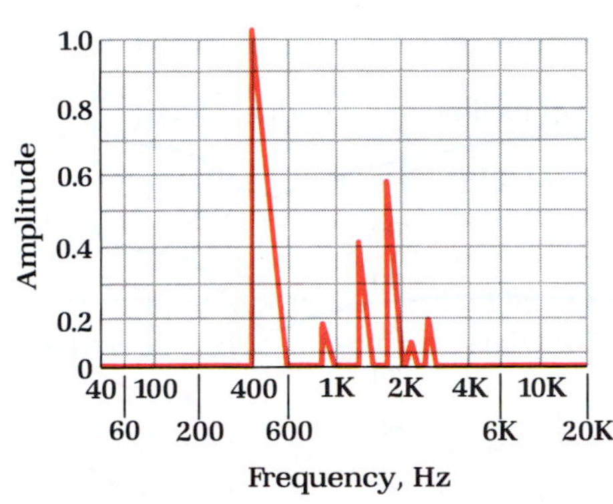

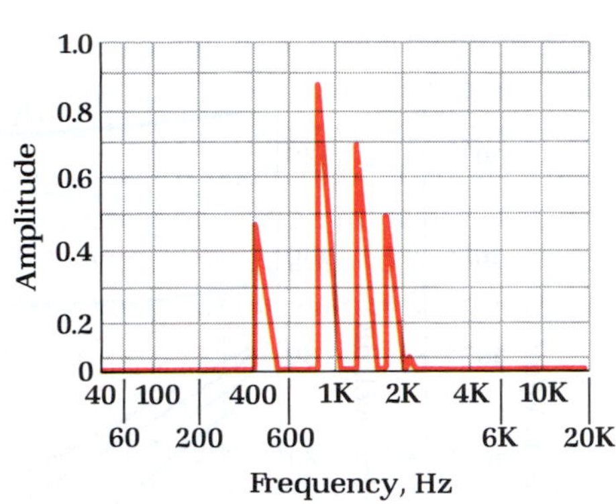

(b)

Figure 33-6. *(a)* Waveforms are shown for a flute and a bassoon playing the same note. Each waveform has a period that corresponds to a fundamental frequency of 440 Hz. *(b)* The amplitudes of the first few harmonics are shown for each waveform.

33-3 FOURIER ANALYSIS OF PERIODIC WAVES

Although harmonic (sinusoidal) waves are simple, most sounds of interest have more complicated periodic structures. Figure 33-6 shows the time dependence of the waveforms of two sounds that have the same period. These sounds have waveforms that are anharmonic (periodic but not sinusoidal). Both sounds are perceived subjectively to have the same pitch, but different quality. It appears that the pitch of these sounds has to do with the period of the waveform, which is the same for both sounds, and the quality has to do with the details of the waveform, which is different for each of these sounds.

Jean Baptiste Joseph Fourier (1768–1830) showed that complex periodic waveforms can be regarded as a sum of harmonic waves. Let $y(t)$ represent the periodic displacement of a wave at a certain position. If $y(t)$ and its derivative are continuous, then it can be shown that $y(t)$ can be represented by a sum of the form,

$$y(t) \approx \sum_{n=1}^{N} A_n \sin(n\omega t + \phi_n) \tag{33-9}$$

where $\omega = 2\pi/T$ and T is the period of the waveform. How large N must be to obtain a good representation depends on the waveform. The representations of some waveforms are shown in Fig. 33-7. As illustrated there, the use of just a few terms gives a reasonable facsimile of the desired waveform.

Fourier analysis of a waveform

The process of determining mathematically the coefficients A_n and phase constant ϕ_n for a given waveform is called *Fourier analysis* and is covered in more advanced texts. It is also possible to find the A_n's and ϕ_n's electronically. The waveforms of a flute and of a bassoon, both sounding the same pitch, are shown in Fig. 33-6*a*. The measured amplitudes A_n for the first few terms in the Fourier analysis of each waveform are indicated in Fig. 33-6*b*. In the analysis, the lowest frequency present is called the ***fundamental frequency*** and the multiples of this frequency are called ***higher harmonics,*** or ***overtones.*** The difference in the quality of the sounds produced by a bassoon and a flute playing the same note is due to the differences in the overtones.

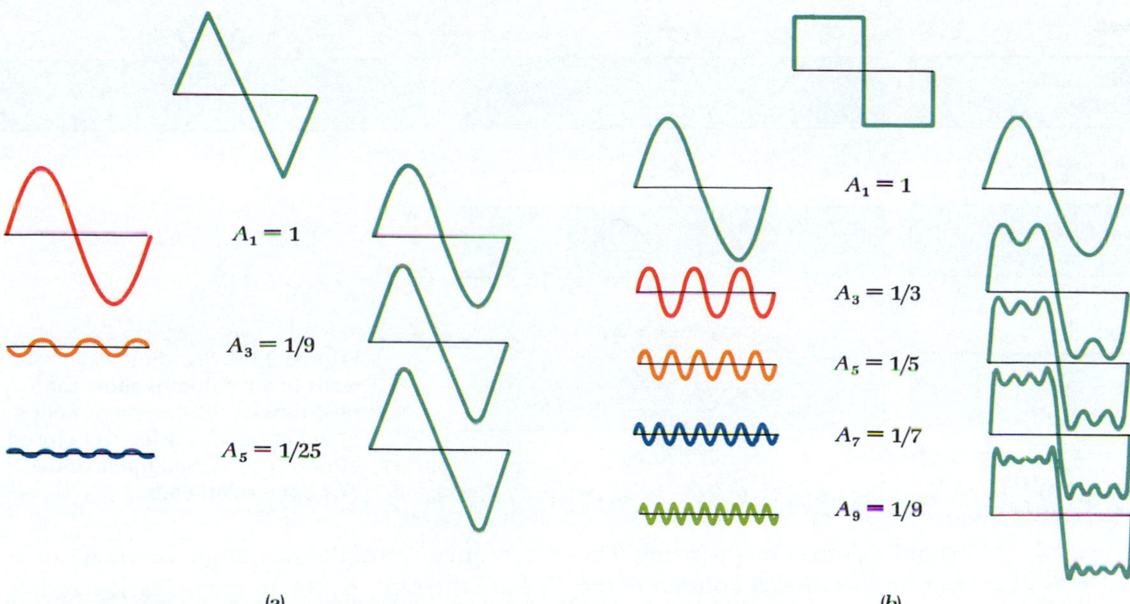

Figure 33-7. One cycle of a waveform is shown at the top of each diagram. Each sinusoidal wave that is added is shown on the left. On the right is the cumulative sum of the terms.

Just as a periodic waveform may be analyzed in terms of a Fourier series to give the relative amounts of the fundamental frequency and the higher harmonics present in the waveform, new periodic waveforms can be formed electronically by adding to a fundamental various amounts of its higher harmonics. This process is called *Fourier synthesis*. Some modern music is performed on a synthesizer rather than on a musical instrument that makes its sound mechanically.

33-4 SOURCES OF SOUND

Musical instruments are common sources of sound waves in air. A tuning fork is a particularly simple "musical instrument." When a tuning fork is struck, it will vibrate in very nearly simple harmonic motion and generate a harmonic sound wave in the surrounding air. The intensity of the sound wave in the air will depend on how much air the tuning fork is able to move — that is, on how well the tuning fork couples to the air.

A vibrating tuning fork is usually difficult to hear from a distance of a few meters or more. The area of the vibrating tines of the fork is small. Thus the tuning fork does not move much air, and the fork cannot deliver much power to the air. However, if the fork is held with its base in contact with a tabletop, it can be heard many meters away. The whole tabletop will vibrate with the tuning fork, so that much more air is moved and a sound wave with a larger intensity is generated.

If the tuning fork is put in a holder on top of a hollow wooden box of a suitable size, with one end of the box open, the coupling to the air that results is even better, and the sound can be heard at the back of a large lecture room. The hollow box is called a *sounding box*. The tuning fork is able to put much more power into the sound wave if its frequency of vibration is the same as the frequency of a standing wave of sound in the sounding box. This enhanced matching or coupling of a vibrating body with a sound wave is an example of resonance that was discussed in Sec. 14-7.

A sounding box enhances the coupling of a vibration and a sound wave.

Many musical instruments use such standing waves in a container of some sort to generate sound waves of sufficient intensity. The standing wave in the column of air in such a container is similar to the standing wave on a stretched string discussed in the previous chapter.

Consider a standing wave in a column of air in a container such as a tube or a pipe that is closed at both ends. Since the ends are closed, the displacement of the air must be zero at both ends. The standing wave has a node at each closed end. This condition is the same as for the transverse displacement at the ends of a string for

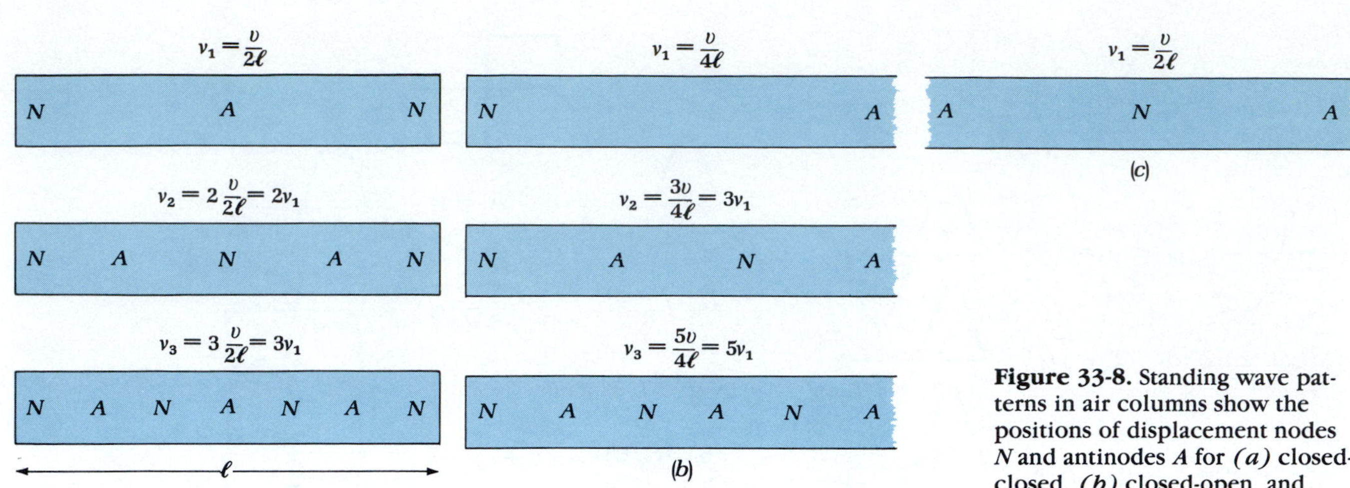

Figure 33-8. Standing wave patterns in air columns show the positions of displacement nodes N and antinodes A for *(a)* closed-closed, *(b)* closed-open, and *(c)* open-open ends.

standing waves on the string. Thus the frequencies of the standing waves that can be set up in a closed column of length ℓ are the same as the frequencies of standing waves on a string of length ℓ. These frequencies correspond to fitting a whole number n of half-wavelengths in the distance ℓ: $\frac{1}{2}\lambda n = \ell$, or $\lambda_n = 2\ell/n$. Using $\nu_n = v/\lambda_n$, we have

$$\nu_n = n\frac{v}{2\ell} \qquad (n = 1, 2, 3, \ldots) \qquad (33\text{-}10)$$

where v is the speed of sound waves in the air. Patterns of nodes N and antinodes A for several standing waves are shown in Fig. 33-8*a* for a closed-closed column.

Some musical instruments contain a column of air open to the atmosphere at one end and closed at the other, similar to the wooden box mentioned above that enabled the tuning fork to be heard at the back of a lecture room. It turns out that a standing sound wave in an open column has a node at the closed end and an antinode near the open end. Exactly where the antinode occurs depends on the details of the opening, such as whether the opening is in the side of the tube or in the end, and how the opening is shaped.

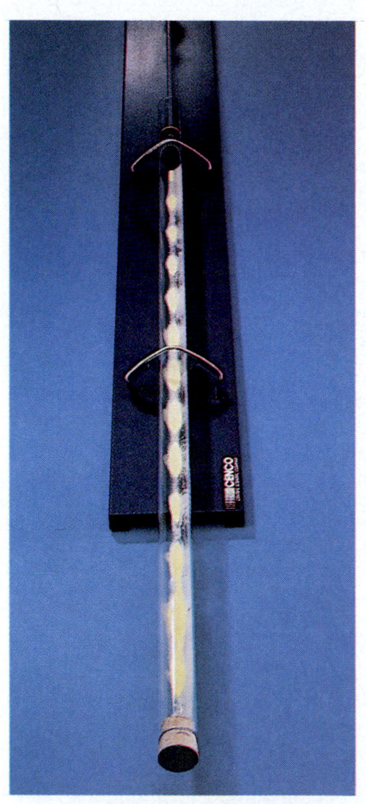

A Kundt's tube, named for the German physicist August Kundt, is used to measure the wavelength of standing sound waves. To make the measurement, a fine powder is spread uniformly inside the tube. A single-frequency sound wave is introduced into the tube such that standing waves are produced. The powder located at antinodes is agitated by the moving air, but the powder located at nodes is not. Thus, the powder accumulates at nodes and the distance between adjacent nodes is measured.

The patterns for several standing waves in a closed-open column are shown in Fig. 33-8*b*. In such a column, a standing wave can be regarded as being the sum of two traveling waves going in opposite directions. Each of these waves has the same wavelength, and their relative phase gives a node at the closed end and an antinode at the open end. For the lowest or fundamental frequency ν_1, the standing wave has a single node and a single antinode. This corresponds to fitting one-fourth of a wavelength in the length ℓ: $\frac{1}{4}\lambda_1 = \ell$, or $\lambda_1 = 4\ell$. Thus the frequency of this standing wave is $\nu_1 = v/\lambda_1 = v/4\ell$. The next-higher frequency occurs with an additional node and an additional antinode between the ends, as shown in the second case of Fig. 33-8*b*. This pattern corresponds to fitting three-fourths of a wavelength into the length ℓ. Thus the frequency is $\nu_2 = 3v/4\ell$. The general relation for the frequency of standing waves in a column open at one end and closed at the other is

$$\nu_{n'} = (2n' - 1)\frac{v}{4\ell} \qquad (n' = 1, 2, \ldots) \qquad (33\text{-}11)$$

Notice that only the odd harmonics ($\nu_1, 3\nu_1, 5\nu_1, \ldots$) occur for the closed-open column. Examples of musical instruments that use columns of air open at one end and closed at the other are the clarinet and the xylophone.

As you can see from Fig. 33-8*c*, the standing-wave pattern in a column open at both ends is similar to that in a column closed at both ends. The positions of the nodes and antinodes are just interchanged. Thus the standing-wave frequencies for open-open columns are the same as those for closed-closed columns and are given by Eq. 33-10. Examples of musical instruments that act as if they have open-open columns are the oboe, flute, and trombone.

The quality of the sound coming from an organ pipe closed at one end is different from the quality of the sound coming from an organ pipe open at both ends, even though the pitch of the two sounds is the same. This is due to the difference in the harmonics produced when the organ pipes are sounded. In a column open at both ends, both even and odd harmonics of the fundamental are present, but only the odd harmonics are present in the sound when the pipe is "stopped" to form a closed-open column. The absence of the even harmonics is readily apparent to the listener.

The quality of the sounds made by the human voice similarly depends on the various resonant cavities which are opened and closed to make the consonants and vowels of language (Fig. 33-9). The difference between the waveforms of the vowels "e" and "o" said at the same pitch by one of the authors is shown in Fig. 33-10. It has been proposed that among the differences between early and late forms of *Homo sapiens* is the presence of more of these cavities in late forms that enable more sounds to be made and hence more information to be transmitted in a given time by speech.

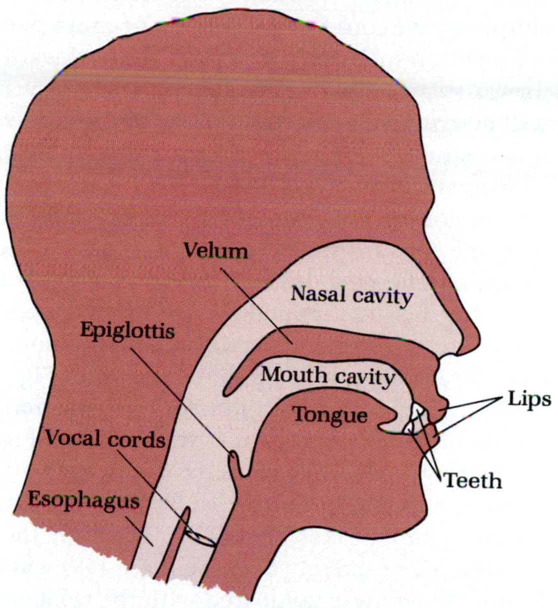

Figure 33-9. Some of the resonant chambers used in vocalization are shown schematically.

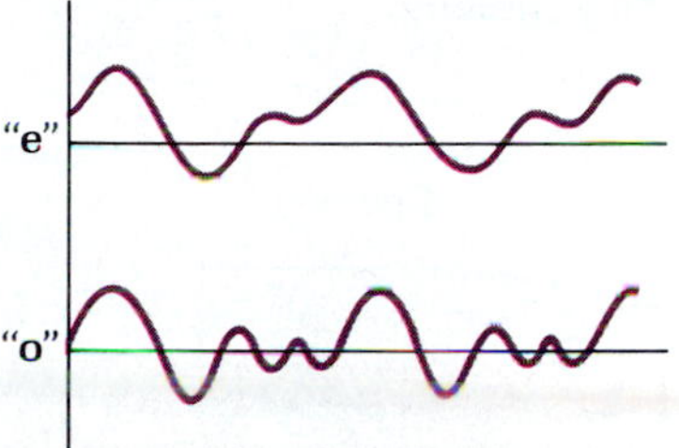

Figure 33-10. Waveforms are shown for the long vowel sounds "e" and "o" as voiced by one of the authors.

A common feature of modern life is the electrical reproduction of sound. Figure 33-11 is a diagram of a simple speaker in which sound is generated with essentially the same waveform as the electric current in the coil. The field of a permanent magnet surrounds the coil so that, when a current passes through the coil, a force is exerted that moves the coil back and forth. The coil is connected to a paper cone that moves in and out, pushing the air and generating a sound wave.

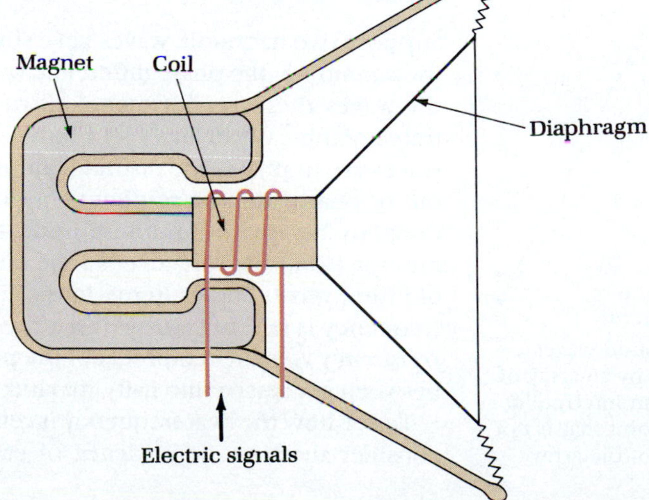

Figure 33-11. A simple speaker is shown schematically. Current in the voice coil causes a magnetic force that moves the paper diaphragm in and out.

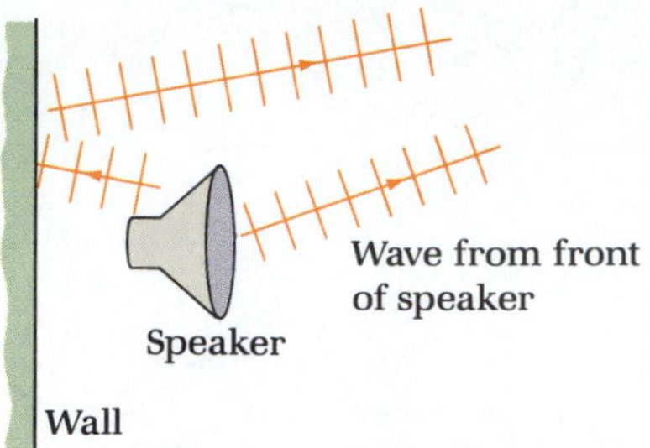

Figure 33-12. Sound waves coming from the front and back of a speaker interfere where they combine.

33-5 INTERFERENCE OF SOUND WAVES AND BEATS

When sound waves from two sources combine at a point, their displacements add, in accord with the principle of superposition. If the two waves are harmonic waves with the same frequency, then the resultant wave at a point where they combine will depend on the phase difference between them, as discussed in Sec. 32-6. Constructive interference occurs if the waves meet in phase, crest for crest, and the resultant amplitude is a maximum. Since the intensity is proportional to the square of the amplitude, the intensity is also a maximum if the waves are in phase. If the waves are out of phase by π rad, meeting crest for trough, then the resulting amplitude and intensity are at minimum.

It is important to consider interference effects when designing an auditorium, a recording studio, or audio components such as speakers and speaker cabinets. When the cone of a speaker moves, it not only creates sound waves in front of it, but also moves the air behind it and creates a backward-moving sound wave. That wave can reflect from a wall behind the speaker and lead to interference between the waves moving forward and backward from the speaker, as shown in Fig. 33-12. For simplicity we consider the speaker to emit a harmonic wave. Suppose the listener is at a position such that the path for the backward wave is longer than the path for the forward wave by half a wavelength. Then the waves will be out of phase by π rad and will destructively interfere. If one of these waves is not suppressed, then sound of this frequency will not be heard at such a position.

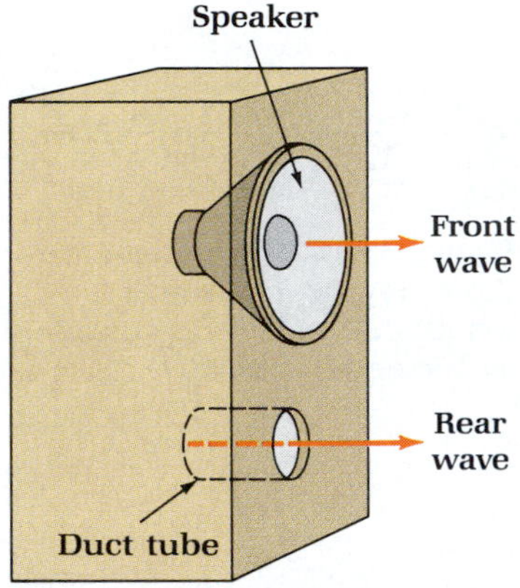

Figure 33-13. A duct-tube speaker enclosure is built so that the rear wave interferes constructively with the front wave for low frequencies.

To avoid such effects, speakers are put into enclosures of some kind. The enclosure is not just for decoration, but prevents the backward wave from interfering with the forward wave. Some enclosures, called *ducted-port enclosures,* have a means of reflecting the backward wave so that it will be in phase with the forward wave for the low frequencies (long wavelengths). This enables the speaker plus enclosure to reproduce low-frequency sounds more effectively, as if the speaker were larger. One such design is shown in Fig. 33-13.

Another effect of the interference of sound waves must be considered in the design of speaker arrays; it is illustrated in Fig. 33-14. The sound waves coming from an array of speakers interfere with each other. Consider the sound arriving at a point *P* that is not directly in front of the array. If the path difference between speaker 1 and speaker 5 is half a wavelength, the sound from these two speakers will interfere destructively because these two waves are out of phase by π rad. If the listener is far away compared with the separation of the speakers, then the waves from the pairs 2 – 6, 3 – 7, and 4 – 8 also interfere destructively. Thus no sound of this wavelength will be heard at position *P*. Effects of this type concentrate the sound in the forward direction, and make such arrays useful in projecting sound to a large audience.

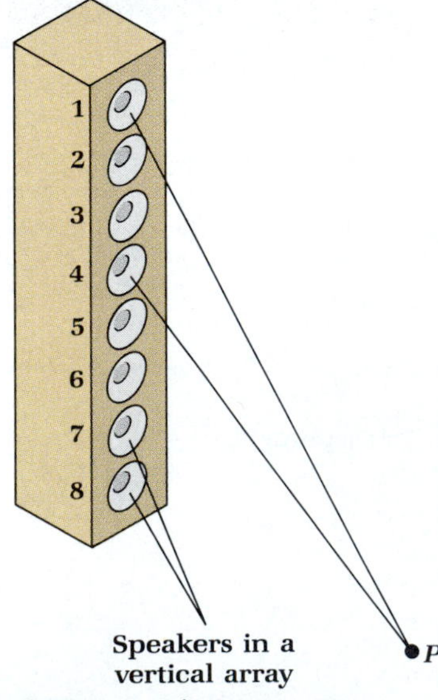

Figure 33-14. Sound waves emitted in phase by an array of small speakers can interfere destructively at a point that is not directly in front of the array.

Beats

Suppose two harmonic waves have slightly different frequencies. At a point where they combine, the phase difference will change with time, and the interference of the waves alternates between constructive and destructive. This behavior is illustrated graphically in Fig. 33-15 for waves of the same amplitude. At times when the waves are in phase, the resultant amplitude is large. At other times, the waves are out of phase, and the resultant amplitude is small. Since the intensity is proportional to the square of the amplitude, the intensity alternates between maxima and minima (loud and soft) over a time interval that is large compared with the period of either wave. These alternations of intensity are called ***beats,*** and a wave of one frequency is said to *beat* against a wave of a slightly different frequency. The *beat frequency* ν_b is the reciprocal of the period of the beats, which is the time interval between successive intensity maxima.

To see how the beat frequency is related to the frequencies of the two waves, we consider the time dependence of each wave at the point in space where they

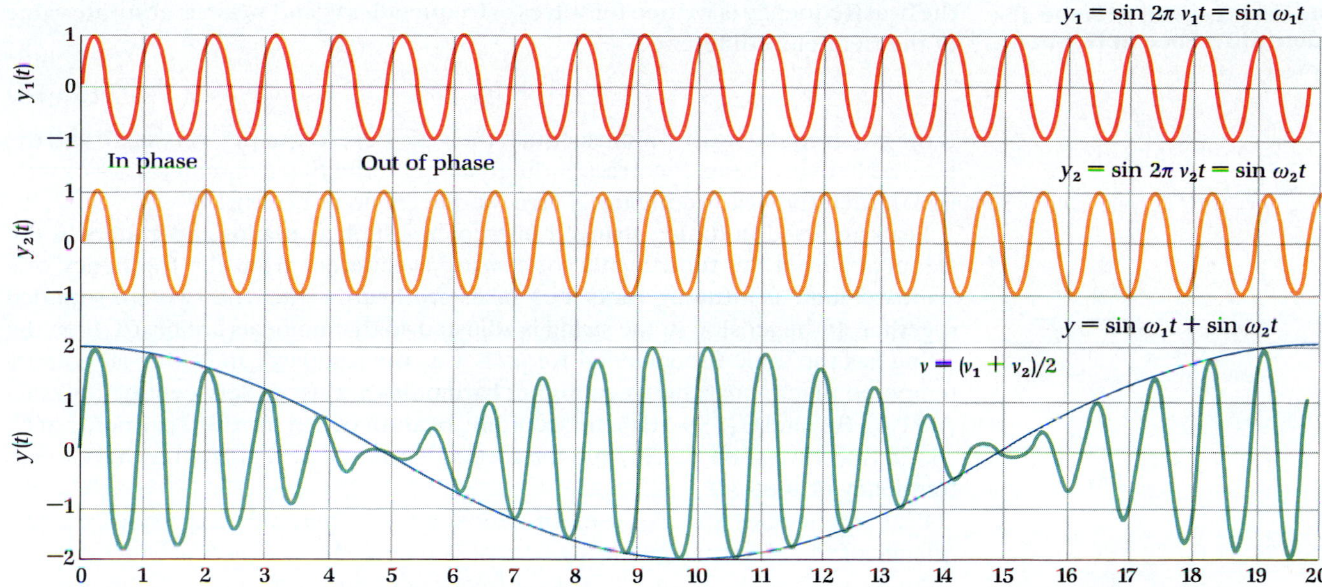

Figure 33-15. The time dependence is shown for two sinusoidal waves of slightly differing frequencies, ν_1 and $\nu_2 = 1.1\nu_1$. The resultant wave, $y = y_1 + y_2$, at a point oscillates with the average frequency and is modulated at the beat frequency, $\nu_b = |\nu_2 - \nu_1|$, which determines the envelope of the resultant wave.

combine. Let $y_1 = A \sin(\omega_1 t)$ and $y_2 = A \sin(\omega_2 t)$ represent these combining waves, where $\omega_1 = 2\pi\nu_1$ and $\omega_2 = 2\pi\nu_2$ are the angular frequencies. To obtain beats, we have $\omega_1 \approx \omega_2$ and, to be definite, we take $\omega_1 < \omega_2$. The resultant wave $y = y_1 + y_2$ at the point has a time dependence,

$$y = A[\sin(\omega_1 t) + \sin(\omega_2 t)] \tag{33-12}$$

It is convenient here to introduce the average angular frequency $\overline{\omega}$ and the angular frequency difference $\Delta\omega$ given by

$$\overline{\omega} = \tfrac{1}{2}(\omega_1 + \omega_2) \qquad \Delta\omega = \omega_2 - \omega_1$$

These equations can be solved for ω_1 and ω_2 in terms of $\overline{\omega}$ and $\Delta\omega$:

$$\omega_2 = \overline{\omega} + \tfrac{1}{2}\Delta\omega \qquad \omega_1 = \overline{\omega} - \tfrac{1}{2}\Delta\omega \tag{33-13}$$

Substituting Eqs. 33-13 into Eq. 33-12, we obtain

$$y = A\{\sin[(\overline{\omega} - \tfrac{1}{2}\Delta\omega)t] + \sin[(\overline{\omega} + \tfrac{1}{2}\Delta\omega)t]\}$$

This expression can be simplified by using the trigonometric identity, $\sin(\alpha = \beta) = \sin\alpha\cos\beta \pm \sin\beta\cos\alpha$. The result is

$$y = 2A\cos(\tfrac{1}{2}\Delta\omega\, t)\sin(\overline{\omega}t) \tag{33-14}$$

We interpret Eq. 33-14 as a sinusoidal term, $\sin(\overline{\omega}t)$ with frequency $\overline{\nu} = \overline{\omega}/2\pi$, the average frequency. This gives the pitch of the sound. It is multiplied or *modulated* by an overall amplitude factor, $2A\cos(\tfrac{1}{2}\Delta\omega\, t)$, which varies more slowly in time. This second factor is responsible for the envelope of the resultant wave in Fig. 33-15, which corresponds to beats. The beats that we hear are beats in intensity, and the intensity is proportional to the square of the amplitude factor, or to $[\cos(\tfrac{1}{2}\Delta\omega\, t)]^2$. The time interval Δt between successive intensity maxima corresponds to the cosine ranging from $+1$ to 0 to -1, so that the $(\text{cosine})^2$ factor ranges from $+1$ to 0 back to $+1$. Thus $\tfrac{1}{2}\Delta\omega \cdot \Delta t = \pi$, or $\Delta t = 2\pi/\Delta\omega$ is the period for beats. Since the beat frequency ν_b is the reciprocal of the beat period, we have

$$\nu_b = \frac{1}{\Delta t} = \frac{\Delta\omega}{2\pi} = \frac{\omega_2 - \omega_1}{2\pi} = \nu_2 - \nu_1$$

We took $\omega_1 < \omega_2$, or $\nu_1 < \nu_2$, in obtaining $\nu_b = \nu_2 - \nu_1$. It is preferable to deal with a positive beat-frequency value regardless of whether ν_1 or ν_2 is larger. Therefore

The beat frequency is the absolute difference in frequencies.

the beat frequency is written for waves of frequencies ν_1 and ν_2 as the absolute value of the frequency difference

$$\nu_b = |\nu_2 - \nu_1| \tag{33-15}$$

If the two frequencies are almost equal, then the beat frequency ν_b is small and the time Δt between beats is correspondingly large. If the two frequencies are exactly equal, then the beat frequency is zero, and beats do not occur.

Beats are used in tuning stringed instruments such as pianos and guitars. If the frequency from the fundamental of a string is different from the frequency of a standard such as a tuning fork, then beats are heard when the two are sounded together. If the tension in the string is adjusted so that no beats are heard, then the string has the same fundamental frequency as the standard. In tuning an equally tempered piano, beats between higher harmonics are used. Between the notes C_4 (261.63 Hz) and G_4 (392 Hz), for example, beats between the third harmonic of C_4 ($3 \cdot 261.63$ Hz $= 784.89$ Hz) and the second harmonic of G_4 (784 Hz) have a beat frequency of 0.89 Hz.

Beat frequencies are also used in radios, where "heterodyning" adds the frequency of the electromagnetic signal that is received to a signal produced in the radio by a "local oscillator." The beat frequency caused by this addition is much lower than the received frequency. Since lower-frequency signals are easier to amplify, the radio is made simpler and more effective by amplifying this beat frequency than by further amplifying the received radio frequency.

EXAMPLE 33-3

An out-of-tune cello. Two cellos bowed on their C_2 strings at the same time give rise to beats that have a minimum of intensity every $\frac{3}{4}$ s and thus a beat frequency of $\frac{4}{3}$ Hz. *(a)* If one cello is known to be properly tuned (65.406 Hz), by how much is the other out of tune? *(b)* By what fraction will the tension in the out-of-tune cello have to be changed to bring it into tune? *(c)* How can you tell if the tension in the out-of-tune cello needs to be increased or decreased?

Solution. *(a)* The difference in frequencies is the beat frequency:

$$|\nu_1 - \nu_2| = \nu_b = \tfrac{4}{3} \text{ Hz}$$

Thus the out-of-tune cello is off by $\frac{4}{3}$ Hz. *(b)* The fundamental frequency of standing waves on a string fixed at each end, as on a cello, is the same as that of standing waves in a closed-closed air column. We use Eq. 33-10, but with $v = \sqrt{F/\mu}$ from Chap. 32. Thus

$$\nu_1 = \frac{v}{2\ell} = \frac{\sqrt{F/\mu}}{2\ell} = C\sqrt{F}$$

where C is a constant for small changes in F. The change in the frequency can be obtained by differentiation. Thus

$$\Delta\nu = \frac{d\nu}{dF}\Delta F = \frac{\frac{1}{2}C}{\sqrt{F}}\Delta F$$

The *fractional* changes are obtained by dividing both sides by $\nu = C\sqrt{F}$:

$$\frac{\Delta\nu}{\nu} = \frac{\frac{1}{2}\Delta F}{F}$$

or

$$\frac{\Delta F}{F} = \frac{2\,\Delta\nu}{\nu} = \frac{2(\frac{4}{3}\text{ Hz})}{65.406\text{ Hz}} = 0.04 = 4\text{ percent}$$

(c) We cannot determine from the beats whether the out-of-tune cello string is above or below the frequency of the correctly tuned one. If the beat frequency increases with a small increase in tension, then the frequency was too high and the tension should be reduced to bring the string into tune. If the beat frequency decreases with a small increase in tension, then the frequency was too low and the tension should be increased.

SELF-TEST 33-3. In the above example, suppose you have determined that the sound from the out-of-tune cello is at a frequency that is $\frac{4}{3}$ Hz higher than the sound from the tuned cello. Thus the sound frequency from the tuned cello alone is 65.406 Hz, and the frequency from the out-of-tune cello alone is 65.406 Hz + 1.333 Hz = 66.739 Hz. What is the frequency of the sound when both cellos emit simultaneously? ***ANSWER:*** 66.072 Hz.

33-6 THE DOPPLER EFFECT

If you listen to the sound of an automobile horn as it passes by, you will hear a characteristic lowering in pitch of the sound. Such a shift in frequency, called the ***Doppler effect,*** or ***Doppler shift,*** is due to a difference in the number of oscillations per second reaching your ear because of the motion of the source. The Doppler shift was first discussed by Christian Doppler (1803–1853) in 1842 in connection with similar shifts in the frequency of light emitted by the stars revolving about each other in double-star systems. It is interesting to note that the speeds of travel in 1842 were such that observations of this shift in the frequency of sound were not common (but see Exercise 37).

The Doppler effect, or Doppler shift in frequency, is due to the motions of the source or the observer relative to the medium.

The Doppler shift occurs in other types of waves in addition to light and sound. Figure 33-16 shows the water waves produced by a vibrating tip that is moving to the right with respect to the water. The wavelength in front of the moving source is compressed, and the wavelength in back of the moving source is expanded. The speed v of the waves is the same in all directions relative to points fixed in the water. Since $\nu = v/\lambda$, the frequency of the waves reaching a point in front of the source will be greater than the frequency of waves reaching a point behind the source.

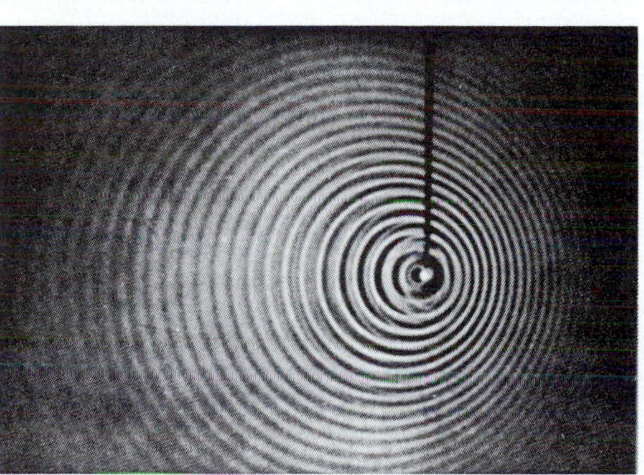

Figure 33-16. Water waves are generated by a vibrating tip moving to the right in the water.

When the source of a wave and an observer are in relative motion, there is a difference between the frequency ν_s emitted by a source and the frequency ν_o received by an observer. We now determine the relation between these frequencies. For simplicity, consider the case when the directions of the velocities of the observer and source lie along the line joining them, as seen in Fig. 33-17*a*. We work in an inertial reference frame in which the medium is at rest, and the speed of the wave in this medium is v. In Fig. 33-17 the positive x direction is from the source to the observer.

First we find the wavelength λ in the medium for the portion of the wave that will be received by the observer. Since the source emits a wave crest every period T_s of the source vibration, each wave crest moves a distance vT_s along the x axis before the source emits another wave crest. But the source is also moving through the medium along the positive x axis with speed v_s, so that the next wave crest is emitted a distance $vT_s - v_sT_s$ behind the one in front of it. Thus the wavelength λ of the wave in the medium is the distance moved by the wave in one period of the

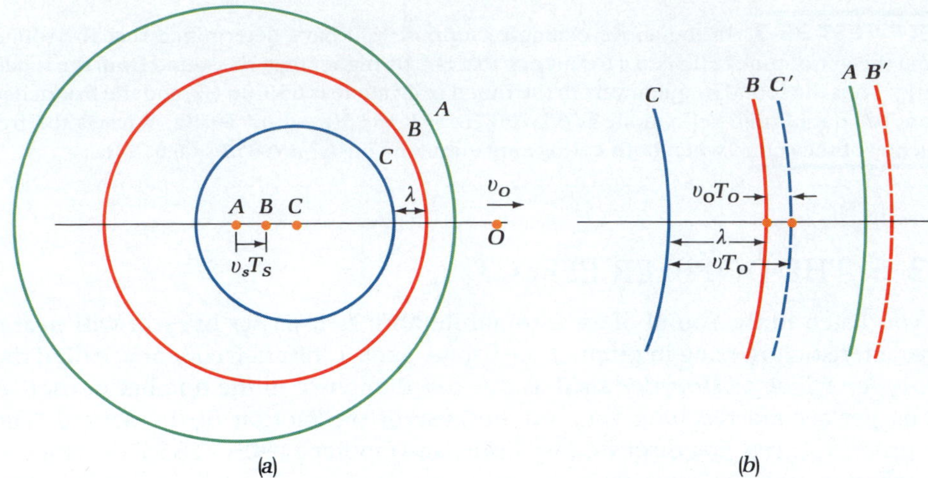

Figure 33-17. *(a)* Successive crests are emitted by the source as it moves a distance $v_s T_s$ toward an observer O. The wavelength in the forward direction is $\lambda = (v - v_s) T_s$. *(b)* A wave crest moves a distance vT_o from C to C' as the observer moves to the right a distance $v_o T_o$.

source vT_s minus the distance moved by the source in one period $v_s T_s$:

$$\lambda = vT_s - v_s T_s \tag{33-16}$$

Now we take the motion of the observer into account. Suppose the observer moves in the positive x direction (away from the source) with speed v_o, as seen in Fig. 33-17*b*. Then the time T_o for two successive wave crests to reach the observer is greater, because the observer is "running away" from the waves. In a time interval T_o, the observer moves a distance $v_o T_o$. In order for the wave crest labeled C in Fig. 33-17*b* to move to the position labeled C', the wave must travel a distance vT_o. This distance is a wavelength λ plus the distance moved by the observer, $vT_o = \lambda + v_o T_o$. Thus

$$\lambda = (v - v_o) T_o \tag{33-17}$$

Eliminating λ from Eqs. 33-16 and 33-17, we obtain

$$(v - v_s) T_s = (v - v_o) T_o$$

which relates the period of waves emitted by the source T_s to the period of waves received by the observer T_o. Since the frequency is the reciprocal of the period, we have

Doppler shift for sound

$$\nu_o = \nu_s \frac{v - v_o}{v - v_s} \tag{33-18}$$

which gives the Doppler shift for sound.

The Doppler shift expression in Eq. 33-18 was obtained for the case of all velocities along the positive x axis. It holds more generally, however, if v_o and v_s represent the x components of the velocities. Thus the components v_o and v_s could be negative. In Fig. 33-17*a*, for example, the velocity component v_o would be negative if the observer moved to the left toward the source, and the velocity component v_s would be negative if the source moved to the left away from the observer. If we set $v_s = v_o$ in Eq. 33-18, we find $\nu_s = \nu_o$. As expected, there is no frequency shift if source and observer move with the same velocity through the medium, or if neither source nor observer moves: $v_s = v_o = 0$.

EXAMPLE 33-4

Moving source, stationary observer. A foghorn on a ship vibrates at 69.3 Hz. The wind speed is zero, you are standing on the shore, and the ship is approaching you at 30.3 knots (15.6 m/s). The speed of sound is 345 m/s. *(a)* What is the wavelength of the sound in the air in front of the ship? *(b)* What frequency do you hear?

Solution. *(a)* Using Eq. 33-16 with $v_s = +15.6$ m/s, we have

$$\lambda = (v - v_s)T_s = \frac{v - v_s}{\nu_s} = \frac{345 \text{ m/s} - 15.6 \text{ m/s}}{69.3 \text{ Hz}}$$
$$= 4.75 \text{ m}$$

Note that if the ship were not moving, the wavelength would be

$$v/\nu_s = (345 \text{ m/s})/(69.3 \text{ Hz}) = 4.98 \text{ m}$$

(b) Using Eq. 33-18, we have

$$\nu_o = \frac{(69.3 \text{ Hz})(345 \text{ m/s} - 0)}{345 \text{ m/s} - 15.6 \text{ m/s}}$$
$$= 72.6 \text{ Hz}$$

This raises the pitch from C_2 sharp to almost D_2, a difference barely discernible by the human ear.

SELF-TEST 33-4. Rework the above example with the ship moving away from you at 15.6 m/s instead of toward you. ***ANSWERS:*** *(a)* 5.20 m; *(b)* 66.3 Hz.

EXAMPLE 33-5

Moving observer, stationary source. A warning foghorn on land vibrates at 69.3 Hz in still air. You approach the foghorn in a ship with a speed of 30.3 knots. The speed of sound is 345 m/s. *(a)* What is the wavelength of the sound in the air? *(b)* What frequency do you hear?

Solution. *(a)* The wavelength of the sound is just v/ν_s, because the source is stationary with respect to the air. Thus $\lambda = (345 \text{ m/s})/(69.3 \text{ Hz}) = 4.98$ m. *(b)* In this case the velocity of the source is zero, but the velocity component v_o of the observer is -15.6 m/s rather than zero as in the previous example. The sign of v_o is negative because the ship's velocity has a direction opposite to the direction from the source to the observer. Using Eq. 33-18, we have

$$\nu_o = 69.3 \text{ Hz} \frac{345 \text{ m/s} - (-15.6 \text{ m/s})}{345 \text{ m/s}}$$
$$= 72.4 \text{ Hz}$$

SELF-TEST 33-5. Rework the above example with you moving away from the foghorn rather than toward it. ***ANSWERS:*** *(a)* 4.98 m; *(b)* 66.2 Hz.

The relative speeds in the two examples above were the same, $|v_o| = |v_s|$, and the frequency shifts were almost the same. There is little difference in the frequency shift of sound if the source approaches the observer or the observer approaches the source with the same speed. It does make *some* difference, however. This difference makes it possible to determine whether the observer or the source is moving with respect to the medium in the foghorn examples above. As we will see in Chaps. 34 and 38, light does not have to have a medium in which to propagate, and only relative motions of source and observer affect measurements of the frequencies of light. There is a different expression for the Doppler shift for light, which we now quote. If c is the speed of light in vacuum and v_R is the relative velocity component, positive if the source and observer approach each other and negative if they recede from each other, then

$$\nu_o = \nu_s \left(\frac{c + v_R}{c - v_R}\right)^{1/2} \qquad (33\text{-}19)$$

Doppler shift for light

Exercise 43 asks you to show that the difference between using Eq. 33-18 and using Eq. 33-19 for light is immeasurably small for relative speeds that are small compared with the speed of light. Thus Doppler could not have known that his formula was correct when applied to sound, but incorrect when applied to the light waves in which he was interested.

There are further limitations on the Doppler shift for sound in Eq. 33-18. If the observer moves away from the source with a speed greater than the speed of sound, then the wave can never "catch up" with the observer, and the formula should not be applied. Another problem occurs if v_s exceeds v; the formula then predicts a negative frequency. Since this does not appear to make physical sense, we must look at the derivation and see where it goes wrong when $v_s \geq v$.

The problem comes in the calculation of λ. If $v_s \geq v$, then $\lambda \leq 0$. But a zero or negative wavelength does not make sense physically. Thus the Doppler formula for sound does not apply when the source moves toward the observer with a speed greater than the speed of sound.

A shock wave is formed if $\mathbf{v}_s > \mathbf{v}$.

What does happen when $v_s \geq v$? Among other things, a ***shock wave*** is formed. Figure 33-18 shows the conical envelope formed by the waves emitted by an object

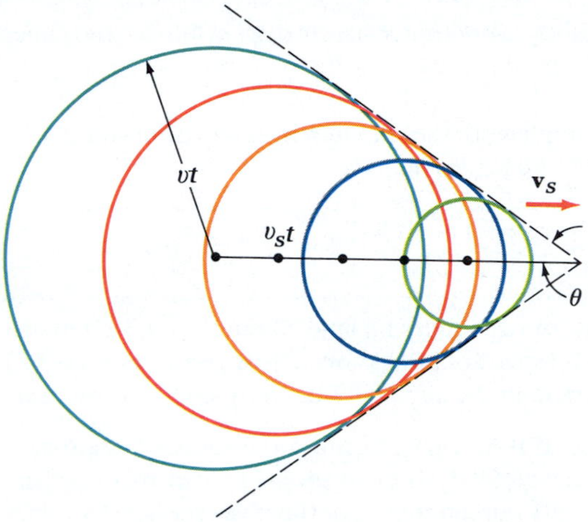

Figure 33-18. Wavefronts pile up on a cone and form a shock wave if the speed of the source exceeds the speed of sound in a medium.

moving faster than the speed of the wave. On this envelope the wave crests pile up, so that the wave amplitude becomes large. In air the resulting shock wave can increase the local pressure by enough to hurt an eardrum or break a window. A shock wave is responsible for the "sonic boom" that is heard when an aircraft passes by at a supersonic speed ($v_s > v$). Similar to the envelope for a shock wave in air is the bow wave produced by a boat or a ship that moves in the water with a speed greater than the speed of the water waves. Another similar effect is Cerenkov radiation, light that is produced when a charged particle moves in a medium such as water with a speed v_s greater than the speed of light v in that medium. In each case, the half-angle θ of the envelope in Fig. 33-18 is given by $\sin\theta = v/v_s$ (see Exercise 42).

33-7 THE WAVE EQUATION FOR SOUND

The wave equation for a wave on a string or a rope was obtained in Sec. 32-4 by considering the motion of an element due to the force exerted by the elements on either side. In a similar way, we develop the wave equation for a sound wave in a fluid by determining how an element of the fluid moves.

First we need the connection between the displacement $\psi(x, t)$ of an element of the fluid and the pressure change $\Delta p(x, t) = p(x, t) - p_e$. As discussed in Sec. 33-1, the adiabatic bulk modulus B_s relates a small pressure change Δp to a change in volume ΔV of an element of fluid, $B_s = -V(\Delta p/\Delta V)$, or

$$\Delta p = -B_s \frac{\Delta V}{V} \tag{33-20}$$

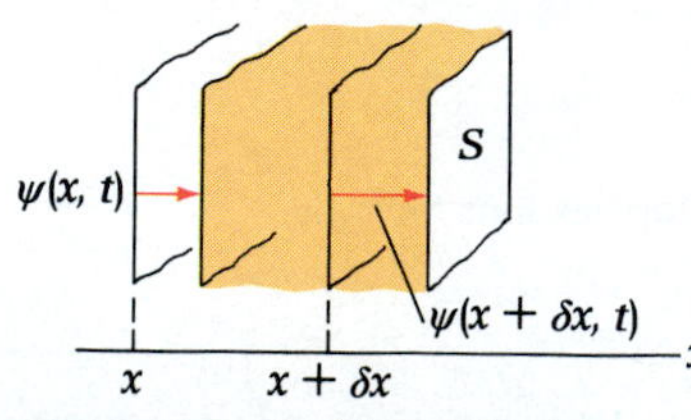

Figure 33-19. The volume $S\,\delta x$ between x and $x + \delta x$ changes to the volume $S[\delta x + \psi(x + \delta x, t) - \psi(x, t)]$ if a wave is present in the fluid.

Consider an element of fluid in the shape of a slab of area S that lies between faces at x and $x + \delta x$ when the fluid is in equilibrium. If a longitudinal wave is present, these faces are displaced by $\psi(x, t)$ and $\psi(x + \delta x, t)$, as shown in Fig. 33-19. Thus, because of the presence of the wave, the volume changes from $V = S\,\delta x$ to

$V + \Delta V = S[\delta x + \psi(x + \delta x, t) - \psi(x, t)]$. The change in volume is $\Delta V = S[\psi(x + \delta x, t) - \psi(x, t)]$, and the fractional change in volume, $\Delta V/V$ that appears in Eq. 33-20, is given by

$$\frac{\Delta V}{V} = \frac{S[\psi(x + \delta x, t) - \psi(x, t)]}{S\,\delta x} = \frac{\psi(x + \delta x, t) - \psi(x, t)}{\delta x}$$

In the limit as $\delta x \to 0$, this ratio approaches $\partial\psi/\partial x$. Thus Eq. 33-20 becomes

Pressure change and the derivative $\partial\psi/\partial x$

$$\Delta p(x, t) = -B_s \frac{\partial \psi}{\partial x} \tag{33-21}$$

We now consider the net force on the element shown in Fig. 33-20. The x component of the force exerted on the face at x due to the pressure change there is $\Delta p(x, t)S$. Similarly, the x component of the force exerted on the face at $x + \delta x$ is $-\Delta p(x + \delta x, t)S$. The x component of the net force on this element is

$$\delta F_x = S[\Delta p(x, t) - \Delta p(x + \delta x, t)]$$

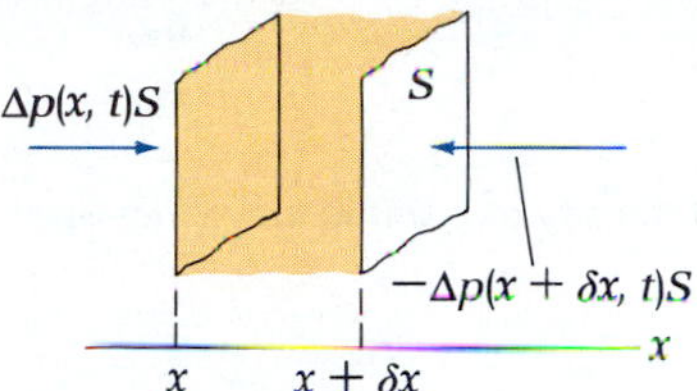

Figure 33-20. The x component of the net force on the element of fluid is $\delta F_x = S[\Delta p(x, t) - \Delta p(x + \delta x, t)]$.

Using Eq. 33-21, we express this force component in terms of the derivatives of ψ,

$$\delta F_x = B_s\, S \left(\left.\frac{\partial \psi}{\partial x}\right|_{x+\delta x} - \left.\frac{\partial \psi}{\partial x}\right|_{x} \right)$$

where the derivatives are evaluated at $x + \delta x$ and at x as indicated. The acceleration of this element is $a_x = \partial^2\psi/\partial t^2$, and the mass of the element is $\delta m = \rho S\,\delta x$, where ρ is the density of the fluid and $S\,\delta x$ is the volume of the element. Newton's second law gives $\delta m\, a_x = \delta F_x$, or

$$\rho S\,\delta x \frac{\partial^2 \psi}{\partial t^2} = B_s\, S \left(\left.\frac{\partial \psi}{\partial x}\right|_{x+\delta x} - \left.\frac{\partial \psi}{\partial x}\right|_{x} \right)$$

The area S divides out of the equation. Next we divide both sides of the equation by δx. The difference of first derivatives divided by δx becomes the second derivative $\partial^2\psi/\partial x^2$ as $\delta x \to 0$. Thus we obtain

$$\rho \frac{\partial^2 \psi}{\partial t^2} = B_s \frac{\partial^2 \psi}{\partial x^2} \tag{33-22}$$

Comparing this result with the form of the wave equation, Eq. 32-8, we see that sound waves in a fluid obey the wave equation and that the speed of sound in the fluid is given by

$$v = \sqrt{\frac{B_s}{\rho}}$$

which is Eq. 33-3.

Intensity of a Sound Wave in a Fluid

The intensity of a sound wave in a fluid can be obtained by considering the rate at which work is done on a part of the fluid by the surrounding fluid when a wave is present. Consider a wave traveling in one dimension, say in the positive x direction. Figure 33-21 shows in the fluid a plane surface of area S that is perpendicular to the direction of propagation of the wave. The pressure change due to the wave at this surface is $\Delta p(x, t)$ and the x component of the force exerted by the fluid on the surface is given by $F_x = \Delta p\, S$. The rate at which work is done is the power P, the rate

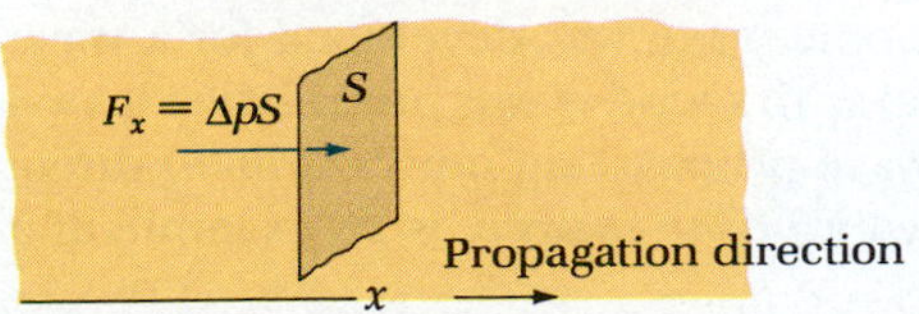

Figure 33-21. Work is done by the fluid to the left of the surface S as the surface moves when a wave is present.

at which energy passes in the x direction. Since $\partial\psi/\partial t$ is the x component of the velocity of the fluid at this surface, the power is $P = F_x\,\partial\psi/\partial t$. Substituting $\Delta p\, S$ for F_x, we obtain

$$P = \Delta p\, S\frac{\partial\psi}{\partial t}$$

The intensity I is the power per unit area and is obtained by dividing P by S:

$$I = \frac{P}{S} = \Delta p\frac{\partial\psi}{\partial t}$$

The intensity can be expressed in terms of the bulk modulus B_s and the derivatives of the displacement ψ by using Eq. 33-21, $\Delta p = -B_s(\partial\psi/\partial x)$. Thus

Intensity of a sound wave in a fluid

$$I = -B_s\frac{\partial\psi}{\partial x}\frac{\partial\psi}{\partial t} \qquad (32\text{-}23)$$

In Sec. 33-1, we displayed expressions for the average intensity of a harmonic wave, Eqs. 33-5 and 33-6. We now show how Eq. 33-23 leads to these expressions for a harmonic wave, say the wave given by Eq. 33-1, $\psi(x, t) = A\cos(kx - \omega t)$. The derivatives are

$$\frac{\partial\psi}{\partial x} = -kA\sin(kx - \omega t)$$

$$\frac{\partial\psi}{\partial t} = \omega A\sin(kx - \omega t)$$

and the intensity is

$$I = B_s\omega kA^2\sin^2(kx - \omega t)$$

Usually we are interested in the intensity averaged over a cycle of the wave. Since the average over a cycle of a sine squared is one-half, the average intensity is

$$\bar{I} = \tfrac{1}{2}B_s\omega kA^2$$

which is Eq. 33-5.

If the pressure change Δp is used to describe the harmonic wave, then the average intensity can be expressed in terms of Δp_{max}. Using $\psi = A\cos(kx - \omega t)$ and Eq. 33-21, we have

$$\Delta p = -B_s\frac{\partial\psi}{\partial x} = B_s kA\sin(kx - \omega t)$$

so that the maximum pressure change is

$$\Delta p_{max} = B_s kA \qquad (33\text{-}24)$$

Solving Eq. 33-24 for the displacement amplitude A and substituting into $\bar{I} = \frac{1}{2}B_s\omega kA^2$, gives

$$\bar{I} = \frac{(\Delta p_{max})^2\omega}{2B_s k}$$

Substituting $\omega/k = v$ and $B_s = \rho v^2$ from Eq. 33-3 gives

$$\bar{I} = \frac{(\Delta p_{max})^2}{2v\rho}$$

which is Eq. 33-6.

Our treatment of the intensity of a harmonic wave was developed for a wave propagating in one dimension. Often we deal with sound waves propagating in two and three dimensions. A spherical sound wave propagating outward in all directions from a small source, sometimes called a ***point source,*** is an example of a sound wave in three dimensions.

Consider a steady spherical wave passing through each of two spheres of radii r_1 and r_2 centered on the source, as shown in Fig. 33-22. By conservation of energy, the energy per second passing through an inner sphere of radius r_1 must equal the energy per second passing through an outer sphere of radius r_2. That is, the power at a distance r from the small source must be independent of r, $\bar{P} = \bar{I}4\pi r^2$, where $4\pi r^2$ is the surface area of the sphere. Thus the average intensity falls off as the inverse square of the distance from the point source:

Intensity of a spherical wave decreases as $1/r^2$.

$$\bar{I} = \frac{\bar{P}}{4\pi r^2} \tag{33-25}$$

We can still use Eqs. 33-5 and 33-6 for the average intensity of a harmonic spherical wave, but the amplitude A and the maximum pressure change Δp_{max} are then proportional to $1/r$. For example, if the amplitude is A_1 at a distance r_1 from a point source, then the amplitude at a distance r_2 from the source is $A_2 = A_1(r_1/r_2)$.

Amplitude of a spherical wave decreases as $1/r$.

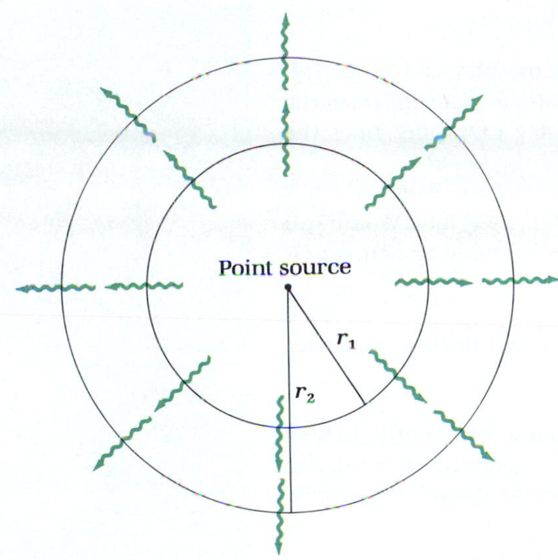

Figure 33-22. A point source produces a steady spherical wave. By conservation of energy, the energy per second, or power, passing through the inner sphere equals the energy per second passing through the outer sphere.

SUMMARY

Section 33-1. Sound Waves

A longitudinal sound wave in a fluid may be described by the displacement $\psi(x, t)$ of an element in the direction of propagation of the wave or by the pressure change from the equilibrium pressure, $\Delta p(x, t) = p(x, t) - p_e$. A harmonic wave is given by

$$\psi(x, t) = A \cos(kx - \omega t) \tag{33-1}$$

$$\Delta p(x, t) = \Delta p_{max} \sin(kx - \omega t) \tag{33-2}$$

The wave speed depends on the density and the adiabatic bulk modulus,

$$v = \sqrt{\frac{B_s}{\rho}} \tag{33-3}$$

The speed of sound in an ideal gas is given by

$$v = \sqrt{\frac{\gamma RT}{M}} \tag{33-4}$$

The average intensity of a harmonic wave is given by either

$$\bar{I} = \tfrac{1}{2}B_s\omega kA^2 \tag{33-5}$$

or

$$\bar{I} = \frac{(\Delta p_{max})^2}{2v\rho} \tag{33-6}$$

Equations 33-3, 33-5, and 33-6 are derived in Sec. 33-7.

Section 33-2. Hearing

Subjective judgments of equal steps in pitch correspond to equal multiples of frequency. Subjective judgments of equal steps in loudness level correspond to equal multiples of intensity. The intensity level is defined as

$$\beta = 10 \log_{10} \frac{I}{I_0} \tag{33-8}$$

where $I_0 = 10^{-12}$ W/m².

Section 33-3. Fourier Analysis of Periodic Waves

A periodic wave can be represented by a Fourier series, a linear combination of harmonic terms.

Section 33-4. Sources of Sound

Many musical instruments excite standing waves in an air column. The fundamental and higher harmonics depend on the length of the air column and the conditions at the ends of the column.

Section 33-5. Interference of Sound Waves and Beats

Sound waves exhibit interference, and the effect is important for acoustic design. Sounds with nearly equal frequencies produce beats with a beat frequency

$$\nu_b = |\nu_2 - \nu_1| \tag{33-15}$$

Section 33-6. The Doppler Effect
The frequency of sound received by a moving observer is Doppler-shifted from the frequency emitted by a moving source:

$$\nu_0 = \nu_s \frac{v - v_o}{v - v_s} \tag{33-18}$$

The Doppler shift for light depends only on the relative-velocity component of source and receiver:

$$\nu_o = \nu_s \left(\frac{c + v_R}{c - v_R} \right)^{1/2} \tag{33-19}$$

Section 33-7. The Wave Equation for Sound
The pressure change in a fluid is related to the fluid displacement by

$$\Delta p = -B_s \left(\frac{\partial \psi}{\partial x} \right) \tag{33-21}$$

Newton's second law applied to a fluid leads to the wave equation for $\psi(x, t)$. The wave speed is given by Eq. 33-3.

QUESTIONS

1. Water is denser than air, but the speed of sound in water is about 4 times faster than in air. Why?

2. When a sound wave goes from air to water, the frequency, which is determined by the source, is unchanged. Does the wave speed increase, decrease, or stay the same? Does the wavelength increase, decrease, or stay the same?

3. The average molecular mass of air is 29 g/mol. Would you expect the speed of sound to be higher in humid air or in dry air at the same temperature? Explain.

4. Would you expect the speed of sound in air to vary with altitude? Why? Where would it be greater?

5. If you see a lightning bolt and count seconds until you hear the thunder, you can then divide by 5 to determine the distance to the bolt in miles (dividing by 3 gives the distance in kilometers). Explain how this works.

6. By what fraction does the speed of sound in a dilute gas change when *(a)* the pressure p is increased by a factor of 2 at constant temperature T, *(b)* T is increased by a factor of 2 at constant p, *(c)* the gas changes from monatomic to diatomic while p and T are kept constant?

7. What is your answer to the old riddle, "If a tree falls in the forest with nobody around, is there any sound?" Explain.

8. Propose a mechanism whereby you can tell from which direction a sound wave comes.

9. Explain why you must be able to hear sounds of frequency higher than 1000 Hz if you are to distinguish between the vowel sounds "a" and "o" sung at 1000 Hz.

10. A tripling of the intensity of a sound increases the sound-intensity level from 70 to 75 dB. If the intensity is tripled yet again, what will the sound-intensity level be?

11. If doubling the intensity of a sound wave increases the sound-intensity level by about 3 dB, what does quadrupling the intensity do to the sound-intensity level?

12. The tubes below the bars of a xylophone or marimba (see Fig. 33-23) are sometimes called *resonant chambers.* The term "resonant" implies that an oscillator is in resonance with a driving frequency. What oscillator is in resonance with what driving frequency in these instruments? Why are the tubes of differing lengths?

Figure 33-23. Question 12: The vertical tubes of the xylophone are resonant air columns.

13. How does an increase in room temperature affect the pitch of an organ pipe?

14. The highest note played by a piano has a fundamental frequency of about 4 kHz. Why is it necessary that sounds with frequencies above 10 kHz be reproduced by the audio equipment in order to "capture" the sound of the piano on a recording?

15. Why does your voice sound different when you have a cold?

16. A loudspeaker giving off a constant-frequency sound is moved toward a wall at the front of a class. The class hears the sound coming directly from the speaker and the sound reflecting from the wall. The class reports that beats are heard. Explain how this result can be described either as a moving interference pattern or as beats between two Doppler-shifted waves.

17. Under what circumstances might you be able to hear beats from the sound waves given off by two tuning forks, both of which vibrate at 440 Hz?

18. Explain the V-shaped waves that a motorboat makes when moving fast on the water.

19. Two sound waves have the same amplitude, but one has twice the frequency of the other. Which has the greater intensity? How much greater?

20. A train blows its whistle as it approaches a tunnel cut into a sheer cliff, and the sound is reflected back toward the train. Compare the frequencies of the original sound and the reflected sound heard by *(a)* the engineer on the train, *(b)* a bystander near the tracks in front of the train, and *(c)* a bystander near the tracks in back of the train. Which listener hears the highest and which hears the lowest frequency? Does any listener hear the same frequency for the two sounds? Explain.

21. Complete the following table:

Symbol	Represents	Type	SI Unit
$\psi(x, t)$	Longitudinal displacement		
Δp_{max}			
B_s			Pa
M			
β			
v_n			
v_b		Scalar	
v_s			

EXERCISES

Section 33-1. Sound Waves*

1. A harmonic sound wave propagates in He in the positive x direction, with wave speed 950 m/s and wavelength 750 mm. Determine the *(a)* frequency, *(b)* angular frequency, *(c)* wave number for this wave. *(d)* Write an expression for the longitudinal displacement $\psi(x, t)$ if the wave amplitude is 5.0 μm.

2. A 1200-Hz sound wave travels in air at 348 m/s in the negative x direction. If the displacement amplitude is 3.0 μm, write expressions for *(a)* the longitudinal displacement, *(b)* the velocity component, *(c)* the acceleration component of an element of the air for this harmonic wave.

3. Determine the speed of sound in air at 0°C.

4. *(a)* Show that the bulk modulus has the dimension of pressure. *(b)* What is the percent change in pressure if the volume of a given mass of water is increased by 1 percent? (The bulk modulus of water is given in Table 15-1.) Repeat part *(b)* for air for which the volume change is *(c)* adiabatic and *(d)* isothermal.

5. Estimate the speed of sound in *(a)* ethyl alcohol and *(b)* glycerine.

6. What is the speed of sound at room temperature in *(a)* oxygen (O_2), *(b)* carbon monoxide (CO), *(c)* carbon dioxide (CO_2)? Assume γ is 7/5 for diatomic gases and 1.25 for CO_2.

7. A 100-m-long tube is filled with He gas at atmospheric pressure. If you shout at one end of the tube, how much sooner will your voice arrive at the other end through the tube than it will through the air outside the tube? Assume a uniform temperature of 300 K.

8. A ship emits a pulse of sound in water. The pulse is reflected off a submarine and returned to the ship in 5.2 s. How far away is the submarine?

9. Determine the speed of compressional sound waves in a rod of *(a)* Al, *(b)* Cu, *(c)* steel.

10. The sound of a train wheel hitting a pebble carries through the air and through the steel rails. How much sooner will the sound arrive through the rails at a position 3 km down the track than through the air?

11. Find the speed of compressional waves *(a)* in a rod of ice at 0°C and *(b)* in water at 0°C.

12. *(a)* Determine the average intensity of the wave of Exercise 2. Assume the wave travels in air of density 0.029 kg/m³. *(b)* What is the maximum pressure change for this wave?

13. A 1.0-kHz sound wave of intensity 8.8 nW/m² travels in water. Determine *(a)* the maximum pressure change and *(b)* the amplitude of the longitudinal displacement. *(c)* Repeat the calculations for a wave of the same frequency and intensity traveling in air at atmospheric pressure and 300 K.

14. Using Eqs. 33-5 and 33-6, show that $\Delta p_{max} = B_s kA$ for a harmonic wave.

Section 33-2. Hearing

15. The note A_4 ("middle A") is defined in modern music to be 440 Hz and corresponds to a white key on a piano. *(a)* What is the fundamental frequency of the next note, one half-step higher ($A_4^\#$, the next black key)? *(b)* What is the fundamental frequency of A_5, the note an octave higher than A_4?

16. What is the fundamental frequency of the sound five half-steps higher than A_4 (440 Hz)?

17. Sound B is louder than sound A. Sound C is perceived to be just as much louder than B as B is louder than A. If A has an intensity of 3.1 nW/m² and B has an intensity of 15 mW/m²,

* Useful data for some of the exercises in this set can be found in Tables 15-1 and 15-2.

what is the intensity of sound C? All sounds have a frequency of 1 kHz.

18. *(a)* A sound has an intensity of 9.2 $\mu W/m^2$. What is the sound-intensity level of this sound? *(b)* A sound has a sound-intensity level of 65.3 dB. What is the intensity of this sound?

19. Determine the change in sound-intensity level if the intensity changes by a factor of *(a)* 2, *(b)* 4, *(c)* 10, *(d)* 20, *(e)* 50, *(f)* 100, *(g)* 10^6.

Section 33-3. Fourier Analysis of Periodic Waves

20. Sum the first few terms in Eq. 33-9 graphically for $\omega = 1000$ rad/s. Take $A_n = (100 \text{ mm})/n^2$, $\phi_n = 0$, and graph $A_n \sin(n\omega t)$ for $0 \leq t \leq 2\pi/\omega$. Choose a scale for $y(t)$ on your graph so that you can add the graphs for $n = 1$, 2, and 3.

21. Repeat the previous exercise with $A_n = 100 \text{ mm}/n$, $\phi_n = 0$ for $n = 1, 3, 5, \ldots$, and $A_n = 0$ for $n = 2, 4, 6, \ldots$. Include terms through $n = 6$.

Section 33-4. Sources of Sound*

22. Determine the frequencies of the standing sound waves that can be set up in a 2.97-m organ pipe open at both ends.

23. Determine the frequencies of standing sound waves that can be set up in a 2.97-m organ pipe open at one end and closed at the other.

24. An organ pipe open at both ends and "tuned" to have a 440-Hz fundamental has its second harmonic ($n = 2$) with the same frequency as the third harmonic ($n' = 3$) of an organ pipe closed at one end and open at the other. How long is each pipe?

25. The finger holes on a flute effectively move the antinode from near the end to near the position of the first open hole. In the flutelike tube shown in Fig. 33-24, how far would the hole be from the open end for *(a)* "middle C" (262 Hz) and *(b)* the note that is a half-step lower than middle C?

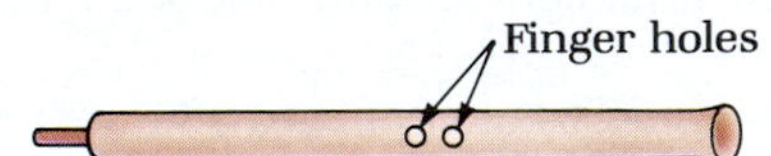

Figure 33-24. Exercise 25: Primitive flute.

26. How many of the harmonics of a 1.5-m-long organ pipe are within the hearing range of an average human if the pipe is *(a)* open at both ends and *(b)* open at one end and closed at the other?

27. A tuning fork is heard to resonate with a column of air in a soda bottle when the liquid level in the bottle is such that the air column is 50 mm long and again when it is 70 mm long. What is the frequency of vibration of the tuning fork? Assume the soda bottle is a straight tube like an organ pipe.

28. If you normally speak with a fundamental frequency of 280 Hz, what will be your fundamental frequency if you are breathing He gas?

* Take the speed of sound in air to be 348 m/s in the exercises for Sections 33-4 and 33-6.

29. By what fraction will the fundamental frequency of an open-open steel organ pipe change if the room temperature increases from 20 to 25°C? Take into account changes in both the steel (see Table 16-1) and the air. How would you answer change if the tube were open-closed?

Section 33-5. Interference of Sound Waves and Beats

30. Sound of frequency 1.16 kHz enters the arrangement of Fig. 33-25. For what values of x will the sound heard at the exit be *(a)* loudest and *(b)* faintest? Take $v = 348$ m/s.

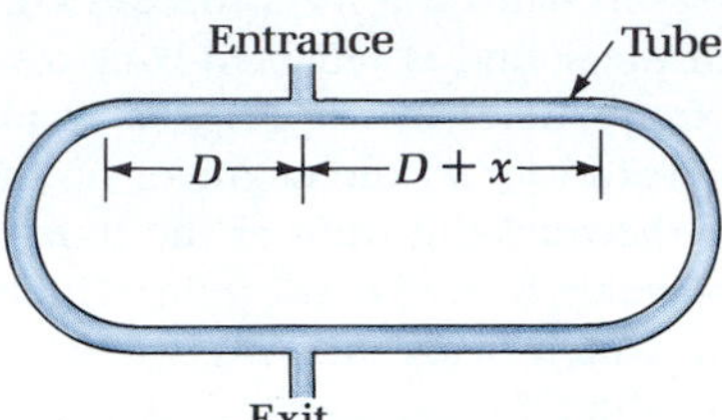

Figure 33-25. Exercise 30.

31. Three loudspeakers a, b, and c emit sound waves of the same frequency. When the sound waves arrive at point P, far from the loudspeakers, the sounds from the speakers have the same amplitude but different phases:

$$\psi_a = A \sin(\omega t - 2\pi/3)$$

$$\psi_b = A \sin(\omega t)$$

$$\psi_c = A \sin(\omega t + 2\pi/3)$$

Show that the sound intensity at point P is zero at all times.

32. Two loudspeakers emit sounds, one at 432.5 Hz and the other at 431.9 Hz. What is the frequency of beats heard where the two sounds combine?

33. Standing waves are set up in two open-ended tubes. One tube is 1.000 m long and the other is identical except that it is 1.002 m long. What is the frequency of beats heard when sound from both tubes is present? Take $v = 348$ m/s.

34. Standing waves are set up on two violin strings of the same length and density. The tensions are adjusted until the fundamental frequency of each string is 440 Hz. The tension in the string of one violin is then changed until the beat frequency heard between the violins is 5 Hz. By what fraction was the tension changed?

35. Two tuning forks are struck, and the sound coming from the two is found to have six beats per second. The first fork is labeled 880 Hz. The beat frequency decreases when the second tuning fork has putty stuck to one of its tines. What is the frequency of the second fork (but without the putty)?

Section 33-6. The Doppler Effect*

36. A sound source with a frequency of 8.46 kHz moves in the positive x direction with a speed of 34.8 m/s relative to an observer in still air. *(a)* What is the wavelength of the sound wave in front of the source along the x axis? *(b)* What frequency will be heard by the observer in front of the source along the x axis? *(c)* What frequency would be heard if the observer were in front of the source and moving in the negative x direction of 5.2 m/s relative to still air?

37. The first experimental confirmation of the effect Doppler predicted in 1842 was made by C. H. D. Buys Ballot in 1845. He compared the sound of stationary trumpeters and trumpeters approaching and receding on the Utrecht railway. How fast would the trains have to travel to have a difference of one half-step (a factor of $2^{1/12}$) for observers standing between the approaching and receding trumpeters?

38. An ambulance siren has a fundamental frequency of 261 Hz. If the ambulance travels at 100 km/h (27.8 m/s), what frequency is heard in still air by bystanders *(a)* in front of the ambulance? *(b)* In back of the ambulance?

39. Two police officers in separate cars head for the same wreck for which the ambulance of the previous exercise was called. One follows the ambulance at 90 km/h and the other approaches the wreck from the other direction at 90 km/h. What frequency does the ambulance siren have for each of the officers?

40. A bat chasing a moth emits a 55-kHz ultrasound. The bat is traveling at 13 m/s and the moth at 2.4 m/s in still air. *(a)* What frequency does the moth receive? *(b)* The ultrasound reflects from the moth and returns to the bat. What frequency does the bat hear from the reflected signal?

41. A train chases Wiley Coyote ***(Famisbus permanentus)*** down a track toward a tunnel cut into a shear cliff. The train's speed is 40 m/s and Wiley's is 30 m/s. The train blows its whistle, which has a fundamental frequency of 440 Hz. *(a)* What frequency does Wiley hear for the sound coming directly from the train? *(b)* What frequency does he hear for the sound reflected from the cliff?

42. Show that the half-angle of the cone for a shock wave is given by $\sin\theta = v/v_s$.

43. *(a)* Consider the Doppler shift for light (Eq. 33-19) when the speed of the observer relative to the source is much less than the speed of light, $|v_R| \ll c$. Use the binomial expansion, $(1+x)^n = 1 + nx + \cdots$, to show that

$$\nu_o \simeq \nu_s(1 + v_R/c)$$

where $v_R > 0$ when observer and source are approaching one another and $v_R < 0$ when they are receding. *(b)* Consider the Doppler shift for sound (Eq. 33-18) applied to the case where the speed of the observer relative to the source is $|v_R|$ and each is moving with speed $\frac{1}{2}|v_R|$ relative to the medium. Using the sign convention from part *(a)* for v_R, show that

$$\nu_o = \nu_s \frac{v + \frac{1}{2}v_R}{v - \frac{1}{2}v_R}$$

(c) Use the binomial expansion to show that when $v_R \ll v$, the result of part *(b)* gives

$$\nu_o \approx \nu_s(1 + v_R/v)$$

Notice the similarity between the expressions in parts *(a)* and *(c)*. To first order in the expansions, the Doppler shifts for light and for sound are the same.

44. An experimental police radar used to measure automobile speeds operates at 140 MHz. Suppose that the signal sent out from a police car traveling behind a suspect gives 5.5 beats per second when added to the signal reflected from the suspect car. How much faster is the suspect going than the police car? The speed of light is 3.00×10^8 m/s.

Section 33-7. The Wave Equation for Sound

45. A harmonic sound wave of amplitude A travels in the positive x direction in a fluid. *(a)* Show that the maximum pressure change is $\Delta p = B_s Ak$. *(b)* Write an expression for the pressure change $\Delta p(x, t)$ for such a wave of frequency 880 Hz and amplitude 25 nm in water.

46. A baby cries, putting 1 mW of power into a piercing wail at 1 kHz. The sound spreads out in all directions with equal intensity. The baby's mother can detect sound intensities of 10 pW/m² or greater. How far can the mother be from the baby and still hear its cry? Assume that there are no reflections and no absorption of the sound wave.

47. The intensity of the sound generated by a rock band at an open-air concert is 0.1 W/m² at a position 10 m from the band. Estimate the intensity of the sound 100 m from the band.

48. In 1 ms a volcanic eruption puts 10^{12} J of energy into a sound wave. Estimate the intensity of the wave when it reaches a point 1 km from the eruption.

PROBLEMS

1. ***Compressional waves in a solid.*** Two marks are made on a solid bar of cross-sectional area S at positions x and $x + \Delta x$. Suppose a compressional wave travels in the bar. *(a)* Show that the distance between marks is then $\Delta x + \psi(x + \Delta x) - \psi(x)$, where ψ is the displacement of the solid due to the wave. *(b)* Using Young's modulus $Y = (F/S)/(\Delta\ell/\ell)$, let $\Delta x \to 0$ and show that the stress F/S in the bar due to the wave is $-Y\partial\psi/\partial x$. *(c)* Show that the x component of the force on a small section of the bar δx long is $YS(\partial\psi/\partial x|_{x+\delta x} - \partial\psi/\partial x|_x)$. *(d)* Use Newton's second law for this small section of the bar to show that $Y\,\partial^2\psi/\partial x^2 = \rho\,\partial^2\psi/\partial t^2$, so that the speed of compressional waves in the solid is $\sqrt{Y/\rho}$.

2. ***Power into sound waves.*** Suppose the gas in the tube of Fig. 33-1 is He at a pressure of 0.11 MPa and a temperature of 297 K. If the piston has an area of 400 mm² and is moved sinusoidally with a frequency of 60 Hz, creating a wave with amplitude of 3.8 mm, what power goes into the sound waves formed?

3. ***Snitching on Newton.*** Newton first obtained the result $v = \sqrt{B/\rho}$ for the speed of sound in a fluid. He calculated the speed of sound in air by using Boyle's measurement of the ***isothermal*** bulk modulus B_T instead of the ***adiabatic*** bulk modulus B_s. Consequently, his value for v was too small. This error was not resolved until 1816 when Laplace noted that the changes in volume were adiabatic instead of isothermal. *(a)* Treat air as an ideal gas and show that $B_T = p$, the pressure. *(b)* Determine the percent error in the calculated speed of sound at 300 K if B_T is used instead of B_s.

4. ***Combining sounds in two different ways.*** *(a)* If one student yelling at a football game gives a sound-intensity level on the field of 35 dB, how many students must yell to give a sound-intensity level of 55 dB? Assume that the yells have random phases so that the net intensity is the sum of the intensities from each student. *(b)* If one loudspeaker gives a sound-intensity level of 35 dB on the field, how many loudspeakers producing identical sounds that arrive in phase at a point on the field would be necessary to give a sound-intensity level of 55 dB at that point?

5. ***Two ways to describe the same effect.*** Two loudspeakers are set up facing each other 20 m apart. They produce sounds with identical frequency, amplitude, and phase and of wavelength $\frac{1}{2}$ m. Quantitatively describe the minima and maxima of intensity you would hear as you walk with speed v_o from one speaker to the other in terms of *(a)* an interference phenomenon that leads to standing waves and *(b)* beats between Doppler-shifted frequencies received from the speakers.

6. ***More on the Doppler effect.*** Suppose that a source of sound with frequency ν_s moves with a velocity $\mathbf{v}_s$ with respect to a distant observer at rest in the medium. However, $\mathbf{v}_s$ does not lie along the line joining the source and observer. (See Fig. 33-26.) *(a)* Show that the wavelength according to the observer is $\lambda = (v - v_s \cos\alpha)/\nu_s$, where v is the speed of sound in the medium and α is the angle between $\mathbf{v}_s$ and the line directed from the source to the observer. *(b)* Show that the Doppler-shifted frequency received by the observer is

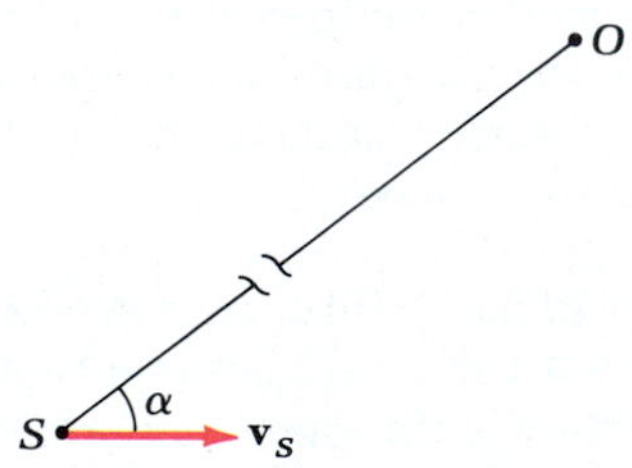

Figure 33-26. Problem 6.

7. ***Sound-wave interference.*** Two small loudspeakers 3.00 m apart emit sound waves of frequency 1.04 kHz with the same phase. Loudspeaker 1 produces 4.00 W of sound power, and loudspeaker 2 produces 2.56 W of sound power. Assume that there are no reflections from nearby surfaces and that the speed of sound is 348 m/s. *(a)* Show that the phase difference between the spherical waves arriving at a point is given by $2\pi d/\lambda$, where d is the difference in the distances from the speakers to the point. *(b)* Determine the average sound intensity at position P in Fig. 33-27. *(c)* Determine the average intensity at position Q in the figure. *(d)* Suppose speaker 2 is wired backwards so that its speaker cone moves out when that of speaker 1 moves in and vice versa. What are the intensities at point P and Q in this case?

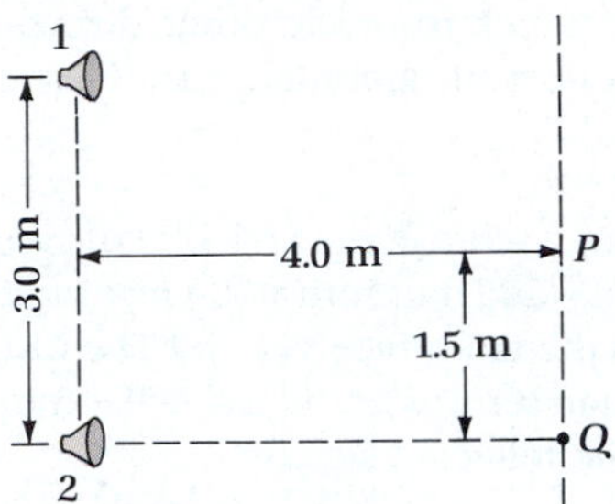

Figure 33-27. Problem 7.

8. ***A Fourier analysis.*** Write a spreadsheet or use a calculator to sum the first few terms of the right-hand side of Eq. 33-9, with $A_n = [\sin(n\pi/2)]/n$, $\phi_n = 0$, and $\omega = 400$ rad/s. Include enough terms, n from 1 to 10 perhaps, and enough values of t to sketch the form of $y(t)$ for a period.

9. ***Another Fourier analysis.*** Modify the procedure in the previous problem to sum the first few terms of the right-hand side of Eq. 33-9, with $A_n = e^{-2n^2}$, $\phi_n = 0$, and $\omega = 400$ rad/s. Include enough terms and enough values of t to sketch the form of $\psi(t)$ for a period. Was it necessary to include as many terms as in the previous problem? Explain.

MAXWELL'S EQUATIONS AND ELECTROMAGNETIC WAVES

34

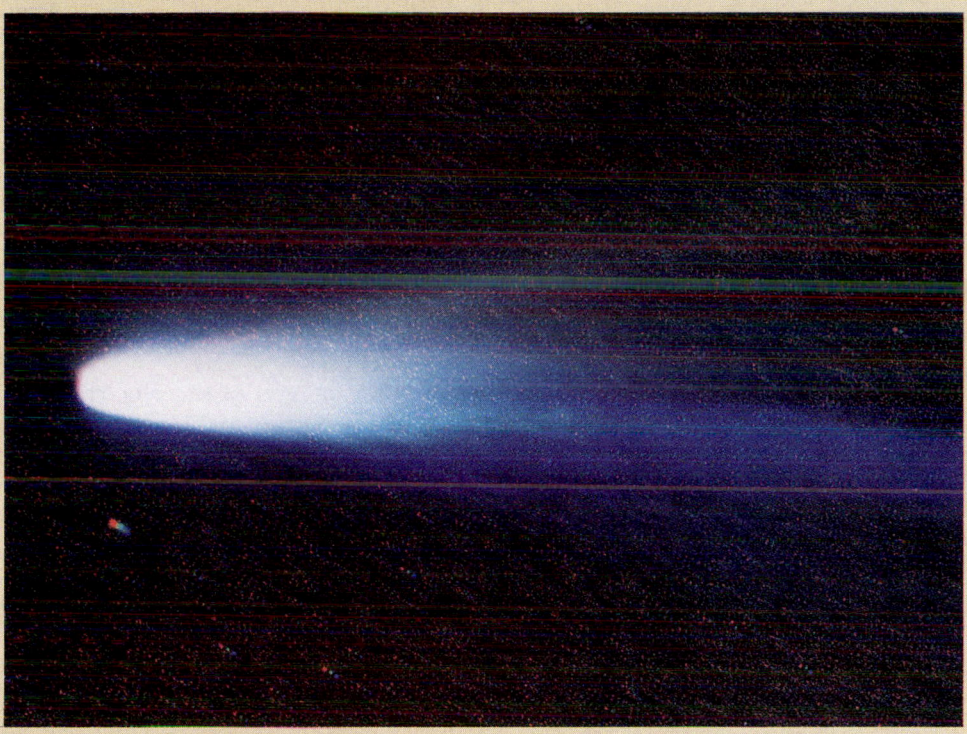

Halley's Comet. Comets have two tails. One tail is dust being blown from the comet by the radiation of sunlight, and the other is the ion tail which is caused by solar wind, a flow of charged particles (mostly protons and electrons) from the sun.

In 1864, the Scottish physicist James Clerk Maxwell published a paper entitled "Dynamical Theory of the Electromagnetic Field." In this paper, Maxwell presented equations that unified the electric and magnetic fields. In addition, he showed that these equations predict the existence of waves in the electric and magnetic fields — ***electromagnetic waves.*** Maxwell identified these electromagnetic waves as light. Therefore Maxwell's equations not only unified electric and magnetic phenomena, but optical phenomena as well. We now know that visible light is but one form of an electromagnetic wave; some other forms are radio waves, microwaves, and x-rays.

It is doubtful that Maxwell could have guessed the full impact of his findings on later human endeavors. However, the following quote from a letter to C. H. Hay on January 5, 1865, indicates that he believed his discoveries were quite significant: "I have also a paper afloat, with an electromagnetic theory of light, which, till I am convinced to the contrary, I hold to be great guns." In this chapter, we present a brief description of those "great guns."

34-1 MAXWELL'S EQUATIONS

You have already been introduced to the equations Maxwell used in developing his theory. The equations are

Gauss's law

$$\oint \mathbf{E}\cdot d\mathbf{S} = \frac{\Sigma q}{\epsilon_0} \tag{34-1}$$

Gauss's law for the magnetic field

$$\oint \mathbf{B}\cdot d\mathbf{S} = 0 \tag{34-2}$$

Faraday's law

$$\oint \mathbf{E}\cdot d\ell = -\frac{d}{dt}\int \mathbf{B}\cdot d\mathbf{S} \tag{34-3}$$

Ampere's law (modified form)

$$\oint \mathbf{B}\cdot d\ell = \mu_0 \Sigma I + \epsilon_0\mu_0 \frac{d}{dt}\int \mathbf{E}\cdot d\mathbf{S} \tag{34-4}$$

Let us briefly review each of these equations.

Equation 34-1 is Gauss's law. For static fields it is equivalent to Coulomb's law. It states that the flux of the electric field for a *closed* surface is proportional to the net charge contained in the volume enclosed by the surface. (See Sec. 21-3.)

Equation 34-2, Gauss's law for the magnetic field, states that the flux of the magnetic field for a *closed* surface is zero. Since this flux is zero, a magnetic counterpart to the electric charge does not exist. (See Sec. 27-5.)

Equation 34-3, Faraday's law, states that the line integral of the electric field around a *closed* path is proportional to the time rate of change of the magnetic flux for a surface bounded by that path. Thus a changing magnetic field is accompanied by an electric field. (See Sec. 28-4.)

Equation 34-4 is the modified form of Ampere's law. Maxwell modified the equation by adding the second term on the right-hand side, the displacement-current term which involves the flux of the electric field. The modified form of Ampere's law states that the line integral of the magnetic field around a *closed* path is proportional to the sum of two terms. The first term contains the net current which flows through a surface bounded by the closed path. The second term (Maxwell's modification) contains the time rate of change of the flux of the electric field for a surface bounded by the path. Because of Maxwell's modification, the equation states that a changing electric field is accompanied by a magnetic field. (See Sec. 27-6.)

34-2 THE WAVE EQUATION FOR E AND B

Maxwell's equations represent a complete and concise description of the electric and magnetic fields. Although these equations appear formidable, we shall use them in a simple way. For the surface integrals we shall choose flat surfaces with straight boundaries, and for the line integrals we shall choose straight-line paths. These simple applications will provide the result we seek—namely, to demonstrate that these equations predict the existence of an electromagnetic wave.

In Chap. 32 we showed that Newton's second law applied to an element of a rope yields the wave equation:

$$\frac{\partial^2 y}{\partial x^2} = \frac{\mu}{F}\frac{\partial^2 y}{\partial t^2}$$

The wave equation predicts the existence of waves in a system.

Thus Newton's second law predicts that a disturbance on a rope propagates as a wave. Even if we never had the opportunity to observe such waves, we would expect that they exist because of our confidence in Newton's second law and because we know that the wave equation is the theoretical harbinger of the existence of waves. That is, if we find some system that obeys the wave equation, then we expect waves to occur in that system. From Chap. 32, the wave equation is

$$\frac{\partial^2 \psi}{\partial x^2} = \frac{1}{v^2}\frac{\partial^2 \psi}{\partial t^2} \tag{34-5}$$

The wave equation

where ψ is the physical quantity that "waves" and v is the wave speed.

We now show that Maxwell's equations can be combined to produce two wave equations, one for the electric field and one for the magnetic field. Along the way, we discover some of the properties of these waves, and we determine the numerical value of their speed.

Plane-Wave Approximation

We simplify our discussion by anticipating the result. That is, we consider electric and magnetic fields that vary in a wavelike manner only. Any contribution to the fields that is uniform in space or constant in time is of no present interest. The space and time dependence of a wave field is oscillatory. For example, a harmonic wave in the electric field traveling in the $+x$ direction has the form $\mathbf{E} = \mathbf{E}_0 \sin(kx - \omega t)$.

Wave fields are oscillatory in space and time.

Also, we consider the fields in a region of free space, or vacuum, far from the source of the waves (point P in Fig. 34-1). The distance D from the source to P is much greater than the largest linear dimension d of the source. (We investigate the nature of the source in Sec. 34-6.) When $D \gg d$, the spatial variation of the wave fields depends only on a coordinate measured along the line from the source to the point P. It is independent of a coordinate measured perpendicular to this line. That is, a wave traveling along the x axis depends only on x, not on y or z. Therefore we orient our coordinate axes so that the x axis is along the direction of propagation (unit vector $\mathbf{i}$ points away from the source). With this orientation, the wave fields can be written

Far from the source, the wave fields depend only on x and t.

$$\mathbf{E} = \mathbf{E}(x, t) \qquad \text{and} \qquad \mathbf{B} = \mathbf{B}(x, t)$$

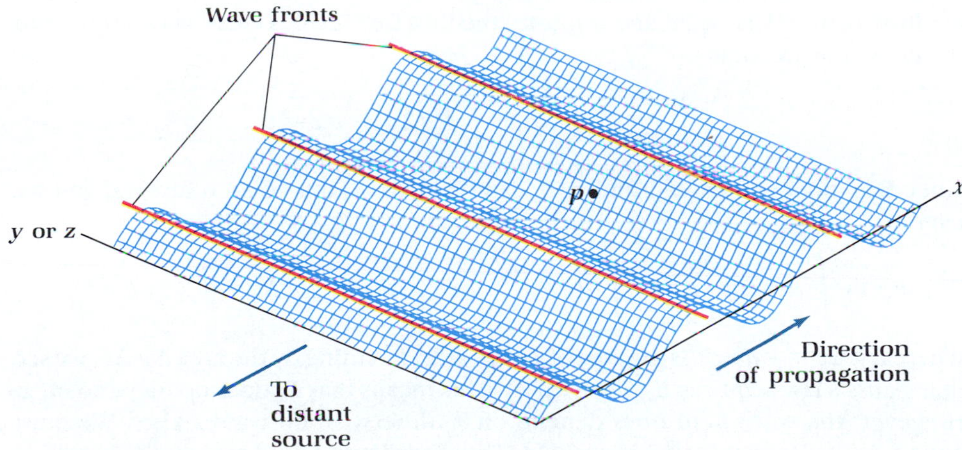

Figure 34-1. A grid-surface graph of a wavelike field plotted parallel to the x axis in a region far from the source. The field depends only on x; it is independent of y or z.

The important point here is that neither $\mathbf{E}$ nor $\mathbf{B}$ depends on y or z. Later in this section, we show that such fields have planar wavefronts, so that this assumption $(D \gg d)$ is called the *plane-wave approximation.*

Wave Fields E and B Are Transverse

We now find the direction of the wave's electric field by applying Gauss's law, $\oint \mathbf{E} \cdot d\mathbf{S} = (\Sigma q)/\epsilon_0$, to the cube shown in Fig. 34-2. The surface of the cube is our gaussian surface. Since free space has no charge, the net charge Σq enclosed by the

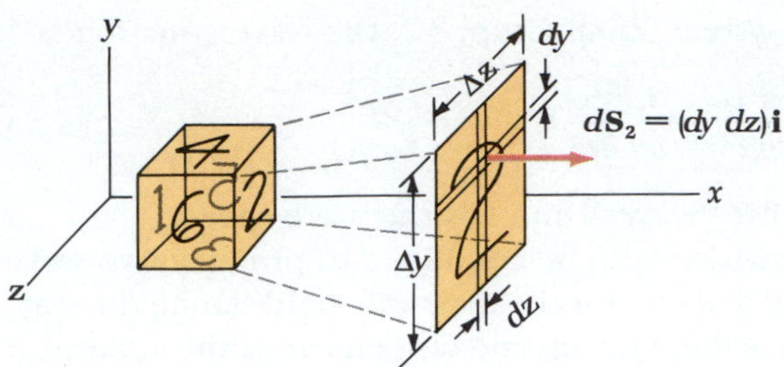

Figure 34-2. Gauss's law applied to a cube. The cube sides are numbered 1 through 6. An expanded view of side 2 shows that $d\mathbf{S}_2 = (dy\ dz)\mathbf{i}$. Can you write the expression for $d\mathbf{S}_4$ from examination of the figure?

gaussian surface is zero, so that $\oint \mathbf{E} \cdot d\mathbf{S} = 0$. Using the labeling of the cube sides shown in the figure, we have

$$\int \mathbf{E}(1) \cdot d\mathbf{S}_1 + \int \mathbf{E}(2) \cdot d\mathbf{S}_2 + \int \mathbf{E}(3) \cdot d\mathbf{S}_3 + \int \mathbf{E}(4) \cdot d\mathbf{S}_4 + \int \mathbf{E}(5) \cdot d\mathbf{S}_5 + \int \mathbf{E}(6) \cdot d\mathbf{S}_6 = 0$$

where $\mathbf{E}(n)$ is the electric field evaluated on the surface of side n and $d\mathbf{S}_n$ is the differential surface vector of side n. Recall that the surface vector for a closed surface points out of the enclosed volume, as shown by $d\mathbf{S}_2$ in the figure. The differential surface vectors are written: $d\mathbf{S}_1 = -(dy\ dz)\mathbf{i}$, $d\mathbf{S}_2 = +(dy\ dz)\mathbf{i}$, $d\mathbf{S}_3 = -(dx\ dz)\mathbf{j}$, $d\mathbf{S}_4 = +(dx\ dz)\mathbf{j}$, $d\mathbf{S}_5 = -(dx\ dy)\mathbf{k}$, and $d\mathbf{S}_6 = +(dx\ dy)\mathbf{k}$. Substituting into Gauss's law and performing the dot product, we obtain

$$-\int E_x(1)\ dy\ dz + \int E_x(2)\ dy\ dz - \int E_y(3)\ dx\ dz + \int E_y(4)\ dx\ dz - \int E_z(5)\ dx\ dy + \int E_z(6)\ dx\ dz = 0$$

Since $\mathbf{E}$ does not depend on y, $E_y(3) = E_y(4)$. Consequently, terms 3 and 4 cancel each other; the flux for side 3 is equal and opposite the flux for side 4. Similarly, $\mathbf{E}$ does not depend on z, so that $E_z(5) = E_z(6)$, and terms 5 and 6 cancel each other; the flux for side 5 is equal and opposite the flux for side 6. Gauss's law applied to the cube now becomes

$$-\int E_x(1)\ dy\ dz + \int E_x(2)\ dy\ dz = 0$$

Since $\mathbf{E}$ is independent of y and z, E_x can be factored out of each integral and we have

$$E_x(1) \int dy\ dz = E_x(2) \int dy\ dz \qquad \text{or} \qquad E_x(1)\ \Delta y\ \Delta z = E_x(2)\ \Delta y\ \Delta z$$

where $\Delta y\ \Delta z = \int dy\ dz$ is the area of a cube side. Dividing by the area $\Delta y\ \Delta z$, we see that Gauss's law requires $E_x(1) = E_x(2)$. This means that E_x does not depend on x. However, the wave field *does* depend on x; otherwise no wave exists. We must conclude that $E_x = 0$ for the wave field. Thus the electric field wave is transverse; it has no component along the direction of propagation. With a similar analysis using Eq. 34-2, you can show that the magnetic field wave is also transverse (Prob. 1).

Wave fields E and B are transverse to the direction of propagation.

Wave Fields E *and* B *Are Mutually Perpendicular*

We have oriented the x axis of our coordinate frame along the direction of propagation, but we are still free to choose a direction for the y (or z) axis. Since $\mathbf{E}$ is perpendicular to the x axis, it is customary to let the y axis be parallel to the oscillating $\mathbf{E}$ field so that $\mathbf{E}$ has neither an x nor a z component: $\mathbf{E} = E_y(x, t)\mathbf{j}$.

Now, with the y axis along $\mathbf{E}$, what is the direction of $\mathbf{B}$? We can find out by applying Faraday's law, $\oint \mathbf{E}\cdot d\boldsymbol{\ell} = -d/dt \int \mathbf{B}\cdot d\mathbf{S}$, to the small square path shown in Fig. 34-3. Since each differential displacement $d\boldsymbol{\ell}$ along this path is perpendicular to **E**, we have $\oint \mathbf{E}\cdot d\boldsymbol{\ell} = 0$, and Faraday's law gives $d/dt \int \mathbf{B}\cdot d\mathbf{S} = 0$, where the surface of integration is bounded by the square path. For this surface, $d\mathbf{S} = (dx\, dz)\mathbf{j}$ so that $\int \mathbf{B}\cdot d\mathbf{S} \approx B_y(\Delta x\, \Delta z)$, where B_y is evaluated at point P. Therefore

$$0 \approx \frac{d}{dt}[B_y(\Delta x\, \Delta z)] = (\Delta x\, \Delta z)\frac{\partial}{\partial t}B_y$$

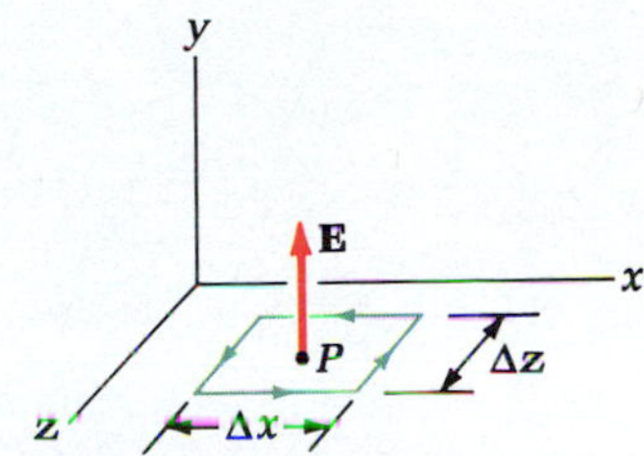

Figure 34-3. Faraday's law applied to a small square path contained in a plane parallel to the xz plane. Since **E** is along the y axis, each path element is perpendicular to **E**, so that $\oint \mathbf{E}\cdot d\boldsymbol{\ell} = 0$. Consequently, $d/dt \int \mathbf{B}\cdot d\mathbf{S} = 0$.

The equation becomes exact as the sides of the square converge on P. A partial derivative is indicated because **B** is a function of x as well as of t, and the point at which the derivative of B_y is evaluated is held fixed at P. Thus Faraday's law requires $\partial B_y/\partial t = 0$. That is, the y component of any time-varying magnetic field is zero. But a wave field *does* depend on time. Therefore $B_y = 0$ for the wave field. Since Gauss's law for the magnetic field requires that $B_x = 0$ for the wave field, the magnetic wave field can have a z component only: $\mathbf{B} = B_z(x, t)\mathbf{k}$. Since $\mathbf{E} = E_y(x, t)\mathbf{j}$, this means that the electric and magnetic fields are mutually perpendicular.

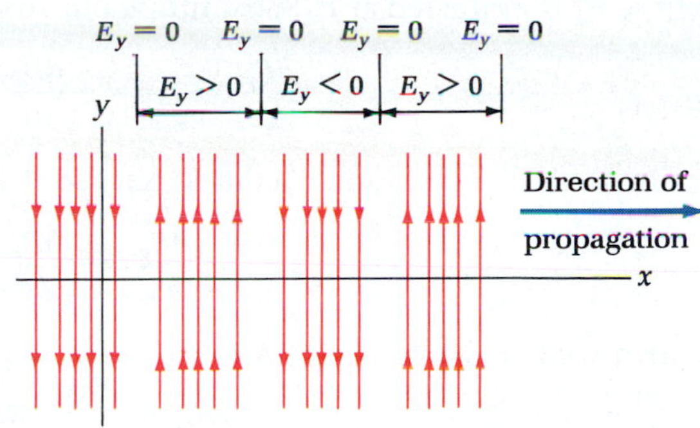

Figure 34-4. Lines of the **E** wave field in the xy plane at a particular instant. The pattern moves to the right as the wave propagates. Lines of the **B** wave field in the xz plane are similar in appearance.

We are now in a position to construct a schematic picture of the waves far from the source. Figure 34-4 shows some of the lines of **E** in the xy plane at a particular instant. The pattern moves to the right as the wave propagates. The lines of **B** in the xz plane have a similar appearance. From the figure you can see that the oscillating **E** field and the direction of propagation are contained in planes parallel to the xy plane. Similarly, the oscillating **B** field and the direction of propagation are contained in planes parallel to the xz plane. Such a wave is called a ***plane-polarized wave,*** and the ***plane of polarization*** is defined as the plane that contains **E** and the direction of propagation. For the case shown in Fig. 34-4, the xy plane is the plane of polarization.

The Wave Equation

Keeping in mind that **E** is along the y axis and **B** is along the z axis, we now apply Faraday's law, $\oint \mathbf{E}\cdot d\boldsymbol{\ell} = -d/dt \int \mathbf{B}\cdot d\mathbf{S}$, to the square path in Fig. 34-5. Again, point P is at the center of the square, and we consider the limit as the sides converge on P. With each side of the square path labeled as shown in the figure, the left-hand side of Faraday's law is

$$\oint \mathbf{E}\cdot d\boldsymbol{\ell} = \int \mathbf{E}(1)\cdot d\boldsymbol{\ell}_1 + \int \mathbf{E}(2)\cdot d\boldsymbol{\ell}_2 + \int \mathbf{E}(3)\cdot d\boldsymbol{\ell}_3 + \int \mathbf{E}(4)\cdot d\boldsymbol{\ell}_4$$

From the figure, $d\boldsymbol{\ell}_1 = dy\,\mathbf{j}$, $d\boldsymbol{\ell}_2 = -dx\,\mathbf{i}$, $d\boldsymbol{\ell}_3 = -dy\,\mathbf{j}$ and $d\boldsymbol{\ell}_4 = dx\,\mathbf{i}$. Since **E** has only a y component, **E** is perpendicular to $d\boldsymbol{\ell}_2$ and $d\boldsymbol{\ell}_4$ so that $\int \mathbf{E}(2)\cdot d\boldsymbol{\ell}_2 = 0$ and $\int \mathbf{E}(4)\cdot d\boldsymbol{\ell}_4 = 0$. This gives

$$\oint \mathbf{E}\cdot d\boldsymbol{\ell} = \int E_y(1)\, dy - \int E_y(3)\, dy = [E_y(1) - E_y(3)]\,\Delta y$$

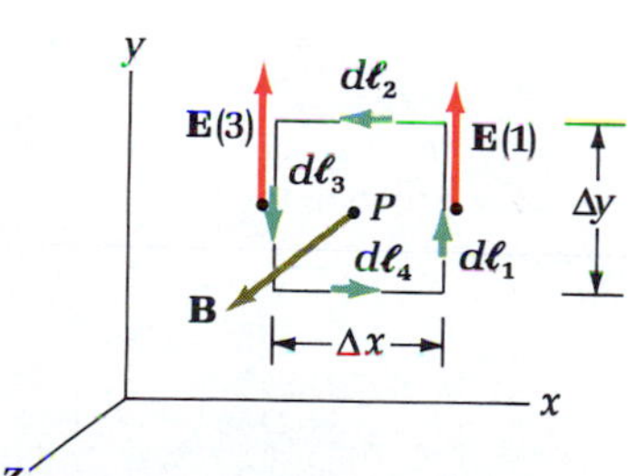

Figure 34-5. Faraday's law applied to a small square path contained in a plane parallel to the xy plane. The **E** wave field has a y component only, so that it is perpendicular to paths 2 and 4. The **B** wave field has a z component only, so that it is along a surface element $d\mathbf{S}$ on the flat surface enclosed by the square path. Apply the right-hand rule to the path shown, and determine whether $d\mathbf{S}$ is directed toward $+\mathbf{k}$ or $-\mathbf{k}$.

where we have factored E_y out of each integral because it is independent of y. Assuming Δx is small, we can write

$$E_y(1) - E_y(3) = \frac{E_y(1) - E_y(3)}{\Delta x}\,\Delta x \approx \frac{\partial E_y}{\partial x}\,\Delta x$$

which gives

$$\oint \mathbf{E}\cdot d\boldsymbol{\ell} \approx \frac{\partial E_y}{\partial x}\,\Delta x\,\Delta y$$

Now consider the right-hand side of Faraday's law. From the right-hand rule applied to the sense of the path of the line integral around the square, the differential surface vector for the plane surface bounded by this path is directed toward $+z$, so that $d\mathbf{S} = (dx\,dy)\mathbf{k}$. Therefore the approximate flux of the magnetic field for this surface is

$$\int \mathbf{B}\cdot d\mathbf{S} \approx B_z(\Delta x\,\Delta y)$$

where B_z is evaluated at P. Substituting our results into Faraday's law, we have

$$\frac{\partial E_y}{\partial x}\,\Delta x\,\Delta y \approx -\frac{\partial B_z}{\partial t}\,\Delta x\,\Delta y$$

In the limit, the equation becomes exact so that

$$\frac{\partial E_y}{\partial x} = -\frac{\partial B_z}{\partial t} \qquad (34\text{-}6)$$

With a similar analysis using Ampere's law (Eq. 34-4), you can show that

$$\frac{\partial B_z}{\partial x} = -\mu_0\epsilon_0\,\frac{\partial E_y}{\partial t} \qquad (34\text{-}7)$$

(See Prob. 2.) Combining Eqs. 34-6 and 34-7 gives the wave equations for E_y and B_z. Differentiating Eq. 34-6 with respect to x and Eq. 34-7 with respect to t yields

$$\frac{\partial^2 E_y}{\partial x^2} = -\frac{\partial}{\partial x}\frac{\partial B_z}{\partial t} \quad \text{and} \quad \frac{\partial}{\partial t}\frac{\partial B_z}{\partial x} = -\mu_0\epsilon_0\,\frac{\partial^2 E_y}{\partial t^2}$$

If we assume that the order of the x and t differentiation of B_z does not affect the result (see Exercises 6 and 7), then we can combine these two equations and obtain

The wave equation for E_y

$$\frac{\partial^2 E_y}{\partial x^2} = \mu_0\epsilon_0\,\frac{\partial^2 E_y}{\partial t^2} \qquad (34\text{-}8)$$

Similarly, differentiating Eq. 34-6 with respect to t and Eq. 34-7 with respect to x gives

The wave equation for B_z

$$\frac{\partial^2 B_z}{\partial x^2} = \mu_0\epsilon_0\,\frac{\partial^2 B_z}{\partial t^2} \qquad (34\text{-}9)$$

Equations 34-8 and 34-9 are the wave equations for E_y and B_z. Comparison with Eq. 34-5 shows that $1/v^2 = \mu_0\epsilon_0$, so that the wave speed is $v = 1/\sqrt{\mu_0\epsilon_0}$. Inserting the numerical values of μ_0 and ϵ_0, we have

$$\begin{aligned} v &= \frac{1}{\sqrt{(4\pi \times 10^{-7}\ \text{kg}\cdot\text{m/s}^2\cdot\text{A}^2)(8.85 \times 10^{-12}\ \text{s}^4\cdot\text{A}^2/\text{kg}\cdot\text{m}^3)}} \\ &= 3.00 \times 10^8\ \text{m/s} \end{aligned}$$

This speed has the same value as the speed c of light in vacuum or free space. (The traditional symbol for the speed of light in vacuum is c.)

As was Maxwell, we are led to the next logical step. We proclaim that we now know what light is. Light is a wave in the electric and magnetic fields, and its speed in vacuum depends on the electric and magnetic properties of vacuum:

$$c = \frac{1}{\sqrt{\mu_0 \epsilon_0}} \tag{34-10}$$

Speed of light in vacuum

In other words, light is a propagating wrinkle in the electric and magnetic fields.

EXAMPLE 34-1

Speed of light in a transparent medium. For electromagnetic waves traveling in a transparent dielectric, such as air or glass, the speed v of the wave is given by $v = 1/\sqrt{\mu_0 \kappa \epsilon_0}$, where κ is the dielectric constant of the material. That is, we replace ϵ_0 with $\kappa\epsilon_0$ in the formula which gives the speed. The value of κ depends on the wave frequency. The magnetic properties of transparent materials are usually such that a similar adjustment to μ_0 is too small to be significant. *(a)* For air at optical frequencies, $\kappa = 1.006$. Determine the speed of visible light in air. *(b)* Given that the speed of visible light in a particular type of glass is 2.0×10^8 m/s, determine κ at optical frequencies for this type of glass.

Solution. *(a)* Since $v = 1/\sqrt{\mu_0 \kappa \epsilon_0}$ and $c = 1/\sqrt{\mu_0 \epsilon_0}$, the speed of light in air is

$$v = \frac{c}{\sqrt{\kappa}} = \frac{3.00 \times 10^8 \text{ m/s}}{\sqrt{1.006}} = 2.99 \times 10^8 \text{ m/s}$$

The speed of visible light in air is nearly the same as in vacuum. *(b)* Solving for κ from $v = c/\sqrt{\kappa}$, we have

$$\kappa = \frac{c^2}{v^2}$$

The dielectric constant of this type of glass at optical frequencies is

$$\kappa = \frac{(3.0 \times 10^8 \text{ m/s})^2}{(2.0 \times 10^8 \text{ m/s})^2} = 2.2$$

SELF-TEST 34-1. The dielectric constant for water at optical frequencies and at room temperature is $\kappa = 1.78$. Determine the speed of light in water. ***ANSWER:*** 2.25×10^8 m/s.

34-3 ELECTROMAGNETIC WAVES

In Chap. 32 we studied solutions to the wave equation in one dimension, especially the harmonic or sinusoidal solution. The harmonic solutions to Eqs. 34-8 and 34-9 are

$$E_y = E_0 \sin (k_e x - \omega_e t) \tag{34-11}$$

and

$$B_z = B_0 \sin (k_b x - \omega_b t + \phi) \tag{34-12}$$

By placing subscripts on the wave numbers k_e and k_b, and on the angular frequencies ω_e and ω_b, we have allowed for the possibility that they may be different. Also, we allow for the possibility that the waves may be out of phase by inserting a phase constant ϕ in the expression for **B**. We know that both the electric field waves and the magnetic field waves have the same speed because the proportionality factor in both wave equations is the same: $c = 1/\sqrt{\mu_0 \epsilon_0}$. Thus $c = \omega_e/k_e = \omega_b/k_b$.

Using Eq. 34-6, we can determine ϕ and find relations between k_e and k_b, between ω_e and ω_b, and between E_0 and B_0. Differentiating Eqs. 34-11 and 34-12, we find

$$\frac{\partial E_y}{\partial x} = k_e E_0 \cos(k_e x - \omega_e t)$$

and

$$\frac{\partial B_z}{\partial t} = -\omega_b B_0 \cos(k_b x - \omega_b t + \phi) = -k_b c B_0 \cos(k_b x - \omega_b t + \phi)$$

Substitution into Eq. 34-6 gives

$$k_e E_0 \cos(k_e x - \omega_e t) = k_b c B_0 \cos(k_b x - \omega_b t + \phi) \tag{34-13}$$

For this equation to be valid for all x and t requires that $k_e = k_b$, $\omega_e = \omega_b$, and $\phi = 2\pi n$ ($n = 0$ or an integer). Since $k_e = k_b$, we let both be represented by k. Similarly, ω_e and ω_b can both be represented by ω. Also, we choose $\phi = 0$ for simplicity. The electric field waves and the magnetic field waves have the same wavelength λ ($\lambda = 2\pi/k$), the same frequency ν ($\nu = \omega/2\pi$), and they are in phase. Dividing Eq. 34-13 by k, we have

$$E_0 \cos(kx - \omega t) = cB_0 \cos(kx - \omega t)$$

For this equation to hold requires

$$E_0 = cB_0 \tag{34-14}$$

Consolidating these results, we now rewrite Eqs. 34-11 and 34-12:

Field components E_y and B_z for a harmonic electromagnetic wave

$$E_y = E_0 \sin(kx - \omega t) \tag{34-15}$$

and

$$B_z = B_0 \sin(kx - \omega t) \tag{34-16}$$

Also, as you can show (Exercise 17),

$$E_y = cB_z \tag{34-17}$$

There is but one wave, an electromagnetic wave.

We have found that the speed, wavelength, frequency, and phase of the electric field waves and magnetic field waves are the same, that their amplitudes are directly proportional (the proportionality constant is c), and that the fields are mutually perpendicular. Thus the electric field waves and magnetic field waves are not independent entities; the existence of one requires the existence of the other. There is but one wave, an ***electromagnetic wave.*** These features are shown in Fig. 34-6, which is a schematic representation of a plane-polarized electromagnetic wave at a particular instant.

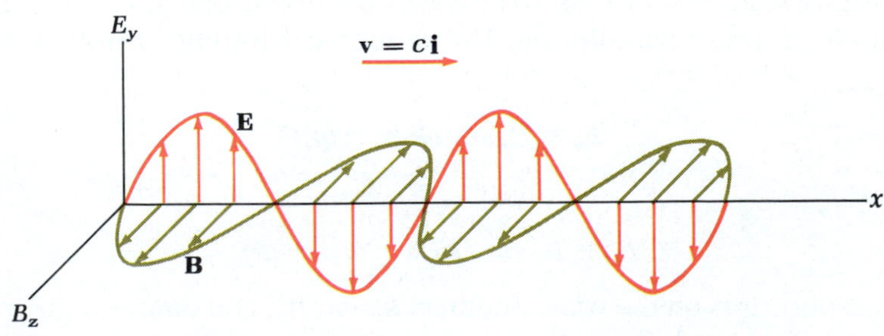

Figure 34-6. A schematic representation of a plane-polarized electromagnetic wave at a particular instant. The wave is propagating toward $+x$.

EXAMPLE 34-2

A particular electromagnetic wave. Suppose that the electric field amplitude of the wave shown in Fig. 34-6 is $E_0 = 120$ N/C and that its frequency is $\nu = 50.0$ MHz. *(a)* Determine B_0, ω, k, and λ. *(b)* Find expressions for **E** and **B**.

Solution. *(a)* From Eq. 34-14

$$B_0 = \frac{E_0}{c} = \frac{120\ \text{N/C}}{3.00 \times 10^8\ \text{m/s}} = 400\ \text{nT}$$

Using $\omega = 2\pi\nu$, we have

$$\omega = 2\pi(50.0\ \text{MHz}) = 3.14 \times 10^8\ \text{rad/s}$$

Since $k = \omega/c$,

$$k = \frac{3.14 \times 10^8\ \text{rad/s}}{3.00 \times 10^8\ \text{m/s}} = 1.05\ \text{rad/m}$$

Also, $\lambda = 2\pi/k$, so that

$$\lambda = \frac{2\pi}{1.05\ \text{rad/m}} = 6.00\ \text{m}$$

(b) Using the results from part *(a)*, we have

$$\mathbf{E} = \{(120\ \text{N/C}) \sin [(1.05\ \text{rad/m})x - (3.14 \times 10^8\ \text{rad/s})t]\}\mathbf{j}$$

$$\mathbf{B} = \{(400\ \text{nT}) \sin [(1.05\ \text{rad/m})x - (3.14 \times 10^8\ \text{rad/s})t]\}\mathbf{k}$$

Electromagnetic waves in this frequency range are used in television broadcasts.

SELF-TEST 34-2. A plane-polarized electromagnetic wave is propagating horizontally toward the south, and the plane of polarization is vertical. *(a)* Consider a point in space and an instant of time in which the wave's electric field **E** is directed vertically upward. What is the direction of the wave's magnetic field **B**? *(b)* Consider a point in space and an instant of time in which **E** is directed vertically downward. What is the direction of **B**? ***ANSWERS:*** *(a)* west; *(b)* east.

34-4 ELECTROMAGNETIC WAVE INTENSITY

Electromagnetic waves transport energy. For example, the sun emits electromagnetic radiation, and after traveling to the earth, a tiny fraction of this radiant energy is absorbed by green plants. By the process of photosynthesis, part of this absorbed energy is stored in the form of sugar molecules. This is how energy enters the life cycle of which we humans are a part. You used some of this energy when you picked up this book.

The energy transported by an electromagnetic wave consists of both electric energy and magnetic energy. In Sec. 23-3 we found that the energy density u_E associated with an electric field is $u_E = \frac{1}{2}\epsilon_0 E^2$, and in Sec. 29-3 we found that the energy density u_B associated with a magnetic field is $u_B = \frac{1}{2}B^2/\mu_0$. We now show that these energy densities are equal for plane electromagnetic waves. Using the relation $E_y = cB_z$ from the previous section and $c = 1/\sqrt{\mu_0\epsilon_0}$ or $\epsilon_0 = 1/\mu_0 c^2$, we have

$$u_E = \frac{1}{2}\epsilon_0 E^2 = \frac{1}{2}\epsilon_0 E_y{}^2 = \frac{1}{2}\left(\frac{1}{\mu_0 c^2}\right)(cB_z)^2 = \frac{\frac{1}{2}B^2}{\mu_0} = u_B$$

The sum of u_E and u_B is the electromagnetic energy density u:

$$u = u_E + u_B$$

Electromagnetic energy density

Since $u_E = u_B$, we have that $u = 2u_E = 2u_B$. By using $E_y = cB_z$ and $\epsilon_0\mu_0 = 1/c^2$, we can express u in several forms. A customary form is

$$u = \epsilon_0 E^2 \qquad (34\text{-}18)$$

In Fig. 34-7 a plane electromagnetic wave is shown passing through a slab-shaped region of space of thickness Δx and cross-sectional area $A = L^2$. We choose Δx to be much smaller than the wavelength so that the fields (and the energy density) in the

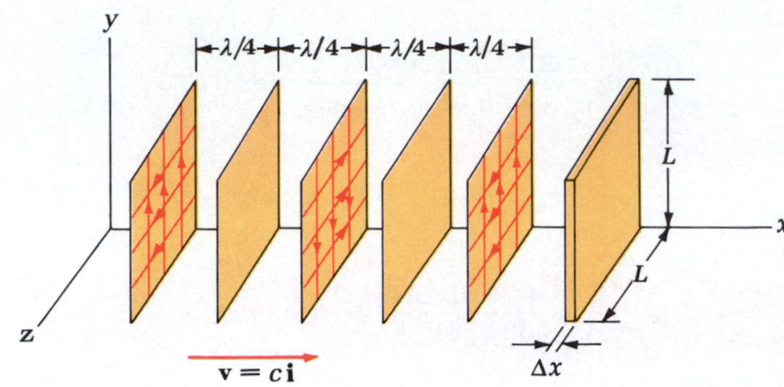

Figure 34-7. A plane electromagnetic wave passing through a slab-shaped volume element of area $A = L^2$ and thickness Δx. Because the wave speed is c, the time required for the energy $\Delta U = u(A\,\Delta x)$ contained in the slab to pass through the slab face is $\Delta t = \Delta x/c$.

volume are essentially uniform. Thus the electromagnetic energy ΔU within this volume is the product of the energy density and the volume:

$$\Delta U = u(A\,\Delta x)$$

Since the wave travels at speed c, the time Δt required for this energy to leave the slab-shaped volume and fill the adjoining volume on the right is $\Delta t = \Delta x/c$. Dividing ΔU by Δt gives the rate at which energy passes through a surface of area A perpendicular to the propagation direction:

$$\frac{\Delta U}{\Delta t} = uA\frac{\Delta x}{\Delta t} = uAc \qquad (34\text{-}19)$$

From Sec. 32-5, the wave ***intensity*** S is the rate at which energy passes through this area divided by the area:

Wave intensity

$$S = \frac{1}{A}\frac{\Delta U}{\Delta t} \qquad (34\text{-}20)$$

Inserting $\Delta U/\Delta t$ from Eq. 34-19 into Eq. 34-20, we have

$$S = uc \qquad (34\text{-}21)$$

The intensity is equal to the product of the energy density and the wave speed. We can express the intensity in terms of the magnitude of the electric field by substituting for u from Eq. 34-18:

$$S = \epsilon_0 E^2 c \qquad (34\text{-}22)$$

Suppose we construct an intensity vector $\mathbf{S}$ that points in the direction of propagation: $\mathbf{S} = S\mathbf{i}$. Note that the cross product $\mathbf{E} \times \mathbf{B}$ points in the propagation direction:

$$\mathbf{E} \times \mathbf{B} = (E_y\mathbf{j}) \times (B_z\mathbf{k}) = E_yB_z\mathbf{i}$$

You can show that S can be expressed in terms of the product E_yB_z (see Exercise 22): $S = E_yB_z/\mu_0$. Therefore $\mathbf{S} = (E_yB_z/\mu_0)\mathbf{i}$, or, more generally

Poynting vector

$$\mathbf{S} = \frac{1}{\mu_0}\mathbf{E} \times \mathbf{B} \qquad (34\text{-}23)$$

The vector $\mathbf{S}$ is called the ***Poynting vector,*** named for its originator, J. H. Poynting (1852–1914). The magnitude of $\mathbf{S}$ is the wave intensity, and its direction is the direction in which energy is propagated by the wave. (Be careful to avoid confusing the Poynting vector $\mathbf{S}$ with the differential surface vector $d\mathbf{S}$.)

For the case of a harmonic plane wave,

$$S = \frac{1}{\mu_0}E_0B_0\sin^2(kx - \omega t) \qquad (34\text{-}24)$$

Figure 34-8 shows graphs of E_y, B_z, and S versus the time at a particular point in space. The field components E_y and B_z depend on time as sin (ωt), whereas S depends on time as $\sin^2(\omega t)$. Therefore the direction of **S** does not oscillate, but its magnitude varies between zero and a maximum ($S_{max} = E_0B_0/\mu_0$) each quarter of a period.

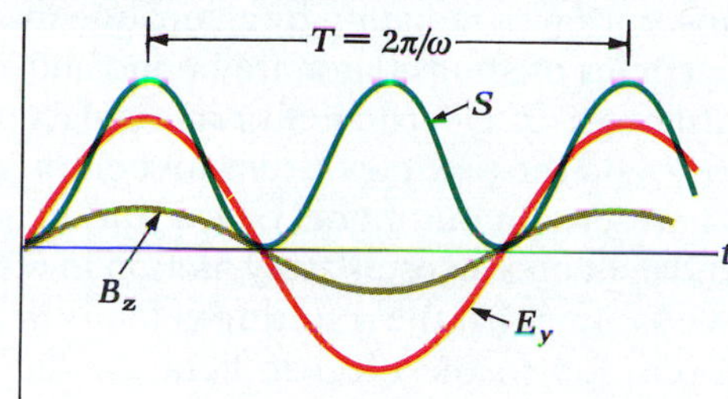

Figure 34-8. Graphs of E_y, B_z, and S versus time t. Each of these quantities has a different unit, so the scales shown here are arbitrary.

The time interval over which the intensity of an electromagnetic wave is measured or detected is usually much longer than the period of the wave. Therefore a quantity of more interest than the time-dependent value of S is its average value $\bar{S}$ over an integral number of half-periods. Since the average value of $\sin^2(\omega t)$ over an integral number of half-periods is 1/2, we have

$$\bar{S} = \frac{1}{2\mu_0} E_0B_0 \qquad (34\text{-}25)$$

Two alternative forms of $\bar{S}$ are

$$\bar{S} = \tfrac{1}{2}\epsilon_0 E_0{}^2 c \qquad \text{and} \qquad \bar{S} = \frac{cB_0{}^2}{2\mu_0}$$

The average intensity of a wave is proportional to the wave amplitude squared. Ordinarily when one speaks of the intensity of a wave, one is referring to the average intensity.

The average intensity of an electromagnetic wave is proportional to the square of the wave amplitude.

EXAMPLE 34-3

Poynting vector for an electromagnetic wave. *(a)* Determine the average intensity of the wave discussed in Example 34-2. *(b)* Write an expression for the Poynting vector of this wave.

Solution. *(a)* From Example 34-2, $E_0 = 120$ N/C. Using $\bar{S} = \frac{1}{2}\epsilon_0 E_0{}^2 c$, we have

$$\bar{S} = \tfrac{1}{2}(8.85 \times 10^{-12}\ \mathrm{C^2/N\cdot m^2})(120\ \mathrm{N/C})^2(3.00 \times 10^8\ \mathrm{m/s})$$
$$= 19.1\ \mathrm{W/m^2}$$

(b) From Eq. 34-25, we see that $E_0B_0/\mu_0 = 2\bar{S}$. Substitution into Eq. 34-24 gives

$$S = 2\bar{S}\sin^2(kx - \omega t)$$

Using the values of k and ω from Example 34-2 and $\bar{S}$ from part *(a)*, we have

$$\mathbf{S} = \{(38.2\ \mathrm{W/m^2})\sin^2[(1.05\ \mathrm{rad/m})x - (3.14 \times 10^8\ \mathrm{rad/s})t]\}\mathbf{i}$$

SELF-TEST 34-3. At a point in space and an instant of time, the electric field **E** of a plane-polarized electromagnetic wave is directed horizontally toward the north and the magnetic field **B** is directed vertically downward. What is the direction of the wave's Poynting vector? ***ANSWER:*** west.

34-5 RADIATION PRESSURE

We have discussed electromagnetic waves passing through free space. What happens when such a wave encounters a material object? For example, what happens when electromagnetic radiation is absorbed at the surface of an object? In Prob. 3, we present a model that is helpful in visualizing the absorption of radiation by

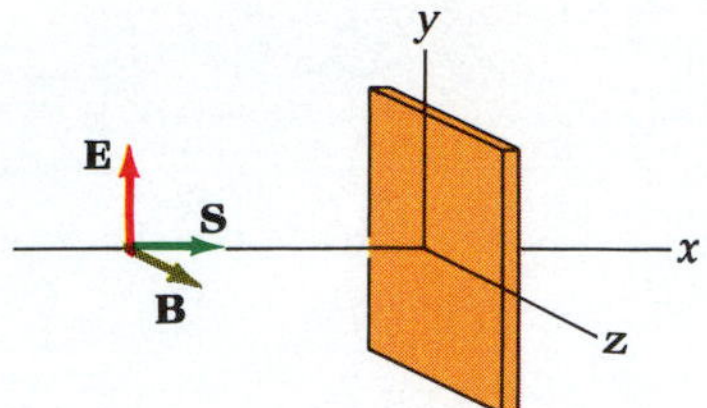

Figure 34-9. An electromagnetic wave is incident normally on a completely absorbing dielectric slab.

matter. This model introduces the concept of ***radiation pressure*** and provides the relation between absorbed intensity and radiation pressure. In this section, we give a qualitative discussion of the results you will get when you solve the problem.

When an electromagnetic wave is absorbed by an object, the wave's energy is transferred to some of the charged particles that compose the object. Consider the absorption of radiation by an opaque insulator (Fig. 34-9). Recall that the charged particles of an insulator are bound and are not free to move through the material. Although an electric field can displace the particles, the displacement is temporary, and the particles return to their original sites after the field is reduced to zero. Therefore, in our model of absorption, each charged particle is bound to a center and executes an oscillatory motion in response to the oscillating fields of the wave. We assume that the resulting velocity of a particle is parallel to the direction of the force due to the electric field, so the electric field does work on each of the particles. This work constitutes the energy transfer from the wave to the object. This electric force averaged over an integral number of cycles is zero because its direction changes every half-cycle. Thus the electric field does work on the charges but does not tend to displace the entire object.

In contrast, the magnetic field of the wave does no work on the particles because the magnetic force on a particle is directed perpendicular to its velocity ($\mathbf{F}_m = q\mathbf{v} \times \mathbf{B}$). Even though the oscillating magnetic field changes direction every half-cycle, the magnetic force on a particle averaged over an integral number of cycles is not zero. This average magnetic force is in the direction of propagation so that it tends to displace the entire object in that direction. In this model of absorption, the electric field part of the wave is responsible for energy being transferred to the object, and the magnetic field part of the wave is responsible for a force being exerted on the object in the direction of the propagation of the wave.

The solution to Prob. 3 shows that the relation between the pressure p exerted on a surface and the intensity S absorbed by the surface is

$$p = \frac{S}{c} \qquad \text{(total absorption)} \tag{34-26}$$

where $\mathbf{S}$ is normal to the surface. The force on the object due to this radiation pressure changes the momentum of the object. By conservation of momentum, the momentum transferred to the object must have come from the wave. Thus electromagnetic waves transport momentum as well as energy. When the waves encounter a material object, the waves transfer this momentum to the object.

Suppose the radiation incident on an object is reflected instead of absorbed. Recall the analogous case of a ball of mass m and speed v thrown against a wall such that the ball's initial velocity is normal to the wall's surface. The magnitude of the momentum Δp_w imparted to the wall is equal to the magnitude of the change in momentum of the ball by conservation of momentum. If the ball sticks to the wall, then $\Delta p_w = mv$ because the ball's initial momentum has magnitude mv and its final momentum is virtually zero. If the ball has an elastic collision with the wall, then $\Delta p_w = 2mv$ because the ball's final momentum is equal in magnitude but opposite in direction to its initial momentum.

Similarly, we can apply the principle of conservation of momentum to the interaction of radiation with a surface. The pressure exerted on a surface by a normally incident wave that is totally absorbed is $p = S/c$ from Eq. 34-26. The initial momentum of the wave was directed toward the surface and the wave was absorbed. Now suppose a normally incident wave is totally reflected by a surface. In this case the final momentum of the reflected wave is equal in magnitude but opposite in direction to the initial momentum of the incident wave. Thus the momentum imparted to the surface is double what it is for absorption, so the resulting pressure is doubled. In this case,

$$p = \frac{2S}{c} \qquad \text{(total reflection)} \tag{34-27}$$

The aurora australis, or southern lights. This view is from the British Antarctic Survey's Halley Station, Antarctica.

where S is the incident intensity of the wave.

The two expressions for radiation pressure, Eqs. 34-26 and 34-27, are valid for the two extreme cases of total absorption and total reflection. A real object partially reflects and partially absorbs radiation incident on its surface. The radiation pressure depends on the fraction of light that is reflected. Its value is in the range $S/c < p < 2S/c$.

Radiation pressure is difficult to detect. An ordinary beam of light that is readily observed because of its intensity will exert a radiation pressure that is minuscule. The first measurements of radiation pressure were made just after the turn of the century (1901 – 1903), about 30 years after the effect was predicted by Maxwell.

EXAMPLE 34-4

Radiation pressure from sunlight. When the sun is directly overhead on a clear day, the incident intensity on a horizontal surface at sea level is about 1 kW/m². *(a)* Assuming that 50 percent of this intensity is reflected and 50 percent is absorbed, determine the radiation pressure on this horizontal surface. *(b)* Find the ratio of this pressure to atmospheric pressure p_0 (about 1×10^5 Pa) at sea level.

Solution. *(a)* The half of the light that is absorbed exerts a radiation pressure of $\frac{1}{2}(S/c)$ and the half that is reflected exerts a pressure $\frac{1}{2}(2S/c)$. The total radiation pressure on the surface is

$$p_{\text{rad}} = \frac{\frac{3}{2}S}{c} = \frac{(1.5)(1\ \text{kW/m}^2)}{3 \times 10^8\ \text{m/s}} = 5 \times 10^{-6}\ \text{Pa}$$

(b)

$$\frac{p_{\text{rad}}}{p_0} = \frac{5 \times 10^{-6}\ \text{Pa}}{1 \times 10^5\ \text{Pa}} = 5 \times 10^{-11}$$

The radiation pressure due to sunlight at the earth's surface is negligibly small compared with the atmospheric pressure.

SELF-TEST 34-4. Rework the above example assuming that 75 percent of the light is absorbed and 25 percent is reflected. ***ANSWERS:*** *(a)* 4×10^{-6} Pa; *(b)* 4×10^{-11}.

34-6 EMISSION OF ELECTROMAGNETIC WAVES

An accelerating charge produces electromagnetic waves.

As electromagnetic waves propagate through space, the time-varying magnetic field induces an electric field and the time-varying electric field induces a magnetic field. According to Maxwell's equations, the existence of one of these time-varying fields requires the existence of the other. What is the source of these wave fields? We know that a stationary object with a static charge distribution produces a static electric field. Also, a wire carrying a steady current (the charge carriers have a constant average velocity) produces a static magnetic field. Therefore stationary charges or charges moving with constant velocity do not produce time-dependent wave fields. To produce a wave field, a charge must accelerate. An accelerating charge (or system of charges) is a source of electromagnetic waves.

To visualize how accelerated charges can lead to the generation of electromagnetic waves, consider the electric field of a dipole with point charges $+q$ at $(0, \frac{1}{2}\ell, 0)$ and $-q$ at $(0, -\frac{1}{2}\ell, 0)$ shown in Fig. 34-10. In Fig. 34-10*a* you are reminded that the field **E** is the vector sum of the individual fields due to each charge, so that at point O on the x axis the field is directed in the $-y$ direction. Now suppose the positions of the two charges are quickly interchanged around the time $t = 0$. Maxwell's equations show that the effect of this interchange propagates outward from the charges at speed c. Figure 34-10*b* shows the field at time t at points along the x axis in the vicinity of point O, far from the dipole. Since O is a distance ct from the dipole, the field at points farther from the dipole than point O is still directed toward $-y$, characteristic of the positions of the charges at times earlier than $t = 0$. At points closer to the dipole than O, the field is directed toward $+y$, which is characteristic of the positions of the charges at times later than $t = 0$. A wave pulse in the electric field propagates outward at speed c. If the two charges oscillate back and forth between the two positions, then an oscillating electric field propagates outward, as suggested in Fig. 34-10*c*.

This simple picture is useful to illustrate the emission of electromagnetic waves, but we cannot extend it further. We based this picture on the static field of a dipole, but at large distances r from the charges this field falls off with distance as $1/r^3$ (Sec. 20-5). From Sec. 32-5, the intensity of a wave decreases with distance from a

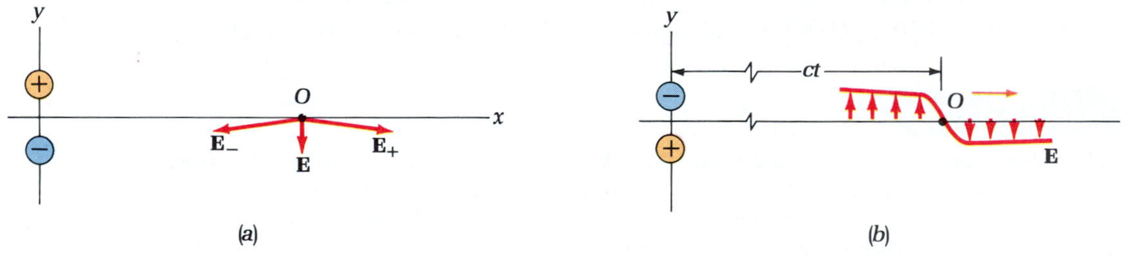

Figure 34-10. *(a)* A stationary electric dipole. *(b)* The charges exchange positions at time $t = 0$, which causes a wave pulse to propagate at speed c and pass point O at time t. *(c)* The charges continually exchange positions, producing an oscillating electric field at point O.

distant source as $1/r^2$. Also, the intensity of an electromagnetic wave depends on E^2. For the intensity to fall off as $1/r^2$, E^2 must fall off as $1/r^2$. Thus the wave fields that transport significant energy from the source must fall off as $1/r$.

The oscillating charges of Fig. 34-10 are similar to a common arrangement used to emit electromagnetic waves, an electric-dipole antenna. Figure 34-11 schematically shows an electric-dipole antenna used to emit electromagnetic waves, a transmitting antenna. The source of alternating current alternately places charges of first

Figure 34-11. A transmitting electric-dipole antenna.

one sign and then the other on each half of the antenna. An electromagnetic wave is emitted whose frequency ν is the same as the frequency of the ac source.

The distribution of the radiated intensity from a transmitting electric-dipole antenna located at the origin is shown in Fig. 34-12. On the scale of this figure, the antenna is too small to be seen. The (nearly) doughnut-shaped surface in Fig. 34-12*a* depicts the intensity pattern. The distance from the antenna to a point on the surface along a particular direction corresponds to the intensity emitted in that direction. The figure indicates that the intensity emitted in the perpendicular bisector plane of the antenna is maximum and that no energy is radiated along the axis of the antenna.

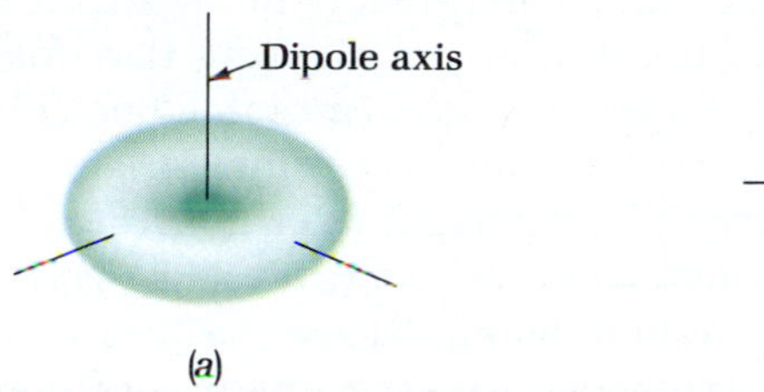

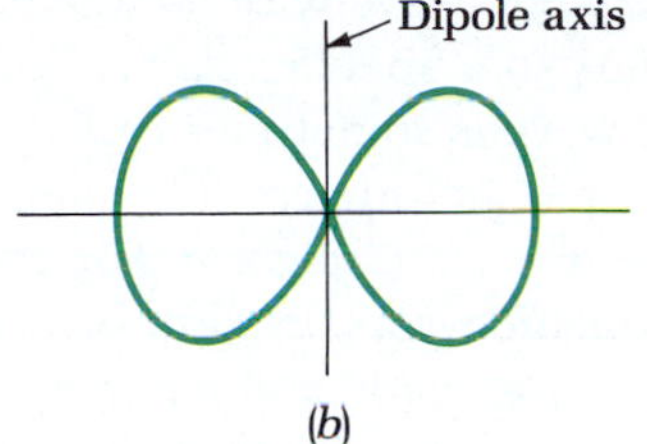

Figure 34-12. Intensity distribution for a transmitting electric-dipole antenna located at the origin. *(a)* Shown in three dimensions. *(b)* Shown in cross section.

During the period from 1887 to 1890, H. R. Hertz (1857–1894) conducted a series of experiments in which he generated and detected electromagnetic waves. He used an ac source to drive a transmitting antenna at a frequency of about 1 GHz. The receiving antenna was connected to a circuit turned to the same frequency. The distances over which he transmitted and detected waves were as great as 20 m.* Hertz showed that, similar to light, these waves could be polarized, reflected, and refracted, and he measured their speed to be the same as the speed of light. This direct verification of Maxwell's theory was performed about a decade after Maxwell's death.

Heinrich Hertz (1857–1894) was born in Hamburg, Germany, and graduated from the University of Berlin. In addition to his famous experiments on electromagnetic waves, he discovered the photoelectric effect (Chap. 39).

34-7 THE ELECTROMAGNETIC SPECTRUM

The wavelength λ and the frequency ν of electromagnetic waves in vacuum are related by the expression

$$c = \lambda\nu$$

where $c = 3.00 \times 10^8$ m/s. All frequency and wavelength values that satisfy $c = \lambda\nu$ are allowed. There are no intrinsic upper or lower limits to the wavelengths or frequencies. The ***electromagnetic spectrum,*** shown in Fig. 34-13, is the range of wavelengths or frequencies that are of most interest to us. To display conveniently the vast range of values of wavelength and frequency, scaling about 24 orders of magntitude, we use a logarithmic scale. Various intervals of the spectrum are given names that correspond to the origin or use of the waves — for instance, radio waves.

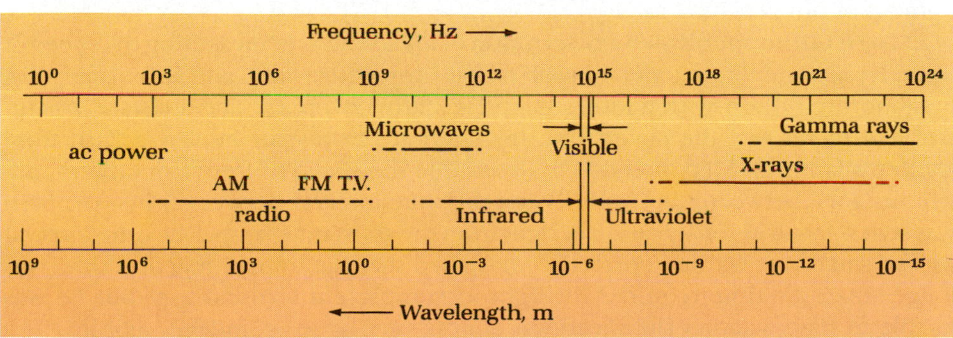

Figure 34-13. The electromagnetic spectrum.

* By 1901 Guglielmo Marconi (1874–1937) had detected electromagnetic waves that were transmitted from the other side of the Atlantic Ocean.

Figure 34-14. The wavelengths of the visible spectrum, in nanometers.

Figure 34-14 shows the range of wavelengths that can be readily detected by the human eye, the visible spectrum. Not only does the eye detect this radiation, but it also can distinguish different wavelengths by the sensation of color. The colors associated with various wavelength intervals are shown in the figure. A harmonic electromagnetic wave with its wavelength (or frequency) in the visible spectrum corresponds to a specific color. For this reason, visible light that consists of a harmonic wave is sometimes called *monochromatic* (meaning one color) *light*. The term "monochromatic" is often used to refer to any single-frequency or harmonic electromagnetic wave. For example, we might refer to harmonic x-rays as monochromatic x-rays. In addition to his many other contributions, Maxwell was an expert on color vision and made several important discoveries in that field.

Monochromatic light consists of a harmonic electromagnetic wave.

White light, such as sunlight, is a mixture of many wavelengths, or colors. Several phenomena can separate white light into its constituent colors; a rainbow is a beautiful example. The entire electromagnetic spectrum shown in Fig. 34-13 has been called "Maxwell's rainbow."

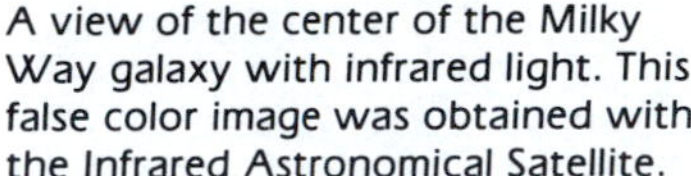
A view of the center of the Milky Way galaxy with infrared light. This false color image was obtained with the Infrared Astronomical Satellite.

COMMENTARY: *The Speed of Light*

The speed of light in vacuum is a fundamental property of nature. As such, an accurate and precise measurement of c has been of great importance to the development of physics.

In 1638 Galileo published a description of his attempts at measuring the speed of light. He reportedly stationed himself on a hilltop with a lamp and an assistant on a neighboring hilltop with another lamp (initially covered). The assistant was instructed to remove the cover from his lamp at the instant he saw a flash from Galileo's lamp. Galileo intended to determine the speed of light from the round-trip distance between the two hills and the time interval between when the flash was sent out and his observation of light from the assistant's lamp. The time interval was so small that Galileo correctly concluded that the human reaction time was longer than the time required for light to complete the round-trip. But he was unable to state whether the speed was simply very large or, indeed, infinite.

During the seventeenth century a great debate raged over the issue of whether light traveled at a finite speed. In 1676 Ole Roemer (1644 – 1710) reported that he had discovered a variation in the times at which the moon Io, in its orbit around

Jupiter, was eclipsed by Jupiter. Roemer correlated this variation to the relative positions of the earth and Jupiter in their orbits around the sun. He attributed the variation of the observed eclipse times to the different time intervals required for light to travel from Io to the earth. He found that light required 22 min to traverse the diameter of the earth's orbit. (From modern measurements, we now know that this time interval is about 17 min.)

Despite Roemer's evidence, the debate continued well into the next century. In 1729 James Bradley (1693–1762) effectively ended the controversy when he published his discovery of the aberration of light from stars due to the earth's orbital motion. The value of the speed of light determined by Bradley was in close agreement with values found using Roemer's method.

The definition of the meter is now based on the speed of light in a vacuum (Sec. 1-1). A meter is the distance traveled by light in a vacuum in a time interval of 1/299,792,458 s. Thus, from the definition of the meter, the speed of light in vacuum is exactly 299,792,458 m/s.

SUMMARY

Sections 34-1 through 34-3. Maxwell's Equations; the Wave Equation for E and B; Electromagnetic Waves

Maxwell's equations can be used to show that the **E** and **B** fields for a plane wave in vacuum (i) are perpendicular to the propagation direction, (ii) are perpendicular to each other, (iii) obey the wave equations with the speed $c = 1/\sqrt{\mu_0\epsilon_0}$, (iv) are in phase and have the same wavelength and frequency, and (v) have amplitudes related by $E_0 = cB_0$. Equations for a harmonic, plane-polarized electromagnetic wave are

$$E_y = E_0 \sin (kx - \omega t) \qquad (34\text{-}15)$$

$$B_z = B_0 \sin (kx - \omega t) \qquad (34\text{-}16)$$

where the wave propagates in the $+x$ direction with speed $c = \omega/k$ and the xy plane is the plane of polarization.

Section 34-4. Electromagnetic Wave Intensity

The Poynting vector **S** points in the direction of propagation and its magnitude is the wave intensity:

$$\mathbf{S} = \frac{1}{\mu_0}\mathbf{E}\times\mathbf{B} \qquad (34\text{-}23)$$

The average intensity $\bar{S}$ is proportional to the square of the wave amplitude:

$$\bar{S} = \tfrac{1}{2}\epsilon_0 E_0{}^2 c = \frac{B_0{}^2}{2\mu_0} c$$

Section 34-5. Radiation Pressure

Electromagnetic waves transport momentum as well as energy. If an electromagnetic wave with intensity S is incident normally on a surface and is totally absorbed by the surface, then the radiation pressure on the surface is $p = S/c$. If the wave is completely reflected by the surface, then $p = 2S/c$.

Section 34-6. Emission of Electromagnetic Waves

The source of electromagnetic waves is an accelerating charge or system of charges. If a charge oscillates with frequency ν, then it generates waves of the same frequency.

Section 34-7. The Electromagnetic Spectrum

The frequencies (or wavelengths) of electromagnetic waves span many orders of magnitude. Visible light corresponds to a small slice of the electromagnetic spectrum.

QUESTIONS

1. What is it that waves when an electromagnetic wave goes by?

2. Can an electric or magnetic field that is static in a region of free space have an effect on an electromagnetic wave passing through the region? Explain. Now reconsider this question for a transparent medium, assuming that κ depends on **E**.

3. When electric and magnetic fields were first introduced in this text (Chaps. 20 and 26), we could have regarded them as no more than computational tools for the calculation of electric and magnetic forces. In view of the electromagnetic waves described in this chapter, do you think that these fields are simply mathematical artifacts or do you think they really exist? Use experimental facts to support your contention that the fields do or do not exist.

4. When we discovered that the wavelike **E** and **B** fields far from their source were transverse (Sec. 34-2), we oriented the y axis of our coordinate system along **E** and later found that Faraday's law requires **B** to be along the z axis. Now suppose we let the y axis be along **B**, and proceed to find the direction of **E**. Which one of Maxwell's equations do we use? Describe the integration path that we should use.

5. Suppose an electromagnetic wave passes through a region that contains a volume charge density. Must the wavelike electric field be perpendicular to the direction of propagation? Explain.

6. If Ampere's law did not contain the term added by Maxwell, could Maxwell's equations still be combined to give wave equations for **E** and **B**? Suppose this added term is entered

into Ampere's law with a minus sign instead of a plus sign. Could Maxwell's equations still be combined to give wave equations?

7. In developing the wave equations for **E** and **B**, we assumed that the order of the x and t differentiation of B_z did not affect the result. Make up a function of x and t and take its derivative with respect to x and then its derivative with respect to t. Now take these derivatives of your function in reverse order—first t, then x. Are the derivatives equal? Attempt to find a function of x and t in which changing the order of differentiation changes the result.

8. Consider a comparison between sound waves and light waves. What are some similarities and what are some differences?

9. Can light waves travel through a perfect vacuum? Does vacuum have any physical properties? If so, name some of the properties of vacuum.

10. Recall that a wave in an elastic medium, such as a sound wave in air, has a speed given by the square root of the ratio of an elastic-force factor to an inertial-mass factor. Both of these factors are characteristic of the medium. If we regard vacuum as an elastic "medium" which supports light waves, then what can be said about this ratio for vacuum?

11. In a letter to William Thomson (Lord Kelvin) in 1861, Maxwell wrote: "I made out the equations in the country before I had any suspicion of the nearness of the two values of the velocity of propagation of magnetic effects and that of light, so that I think I have reason to believe that the magnetic and luminiferous media are identical." What did Maxwell mean by the magnetic medium? What did he mean by the luminiferous medium?

12. Consider a plane-polarized electromagnetic wave traveling horizontally toward the north. The plane of polarization is vertical. Determine which of the quantities, **E**, **B**, or **S**, is described by each of the following statements:
(a) It points continually northward with a magnitude that oscillates between zero and a maximum.
(b) Its direction oscillates along the vertical, pointing upward for half a cycle and then downward the other half.
(c) Its direction oscillates along the horizontal, pointing eastward for half a cycle and then westward for the other half.

13. In each part of Fig. 34-15, two of the three quantities **E**, **B**, and **S** that describe a plane electromagnetic wave are shown at a point in space. (The waves are different in each case.) In terms of the cartesian unit vectors, *(a)* what is the direction of **S** in Fig. 34-15*a*? *(b)* What is the direction of **B** in Fig. 34-15*b*? *(c)* What is the direction of **E** in Fig. 34-15*c*?

14. A number of parameters are associated with the description of a harmonic, plane-polarized electromagnetic wave: E_0, B_0, $\bar{S}$, k, λ, ω, and ν. To write an equation for a particular wave traveling in vacuum, what is the minimum number of these parameters that must be given? Of this minimum number, will any of the above suffice or do we need specific ones? For example, if you are given ν, do you also need k? If you are given $\bar{S}$, do you also need E_0?

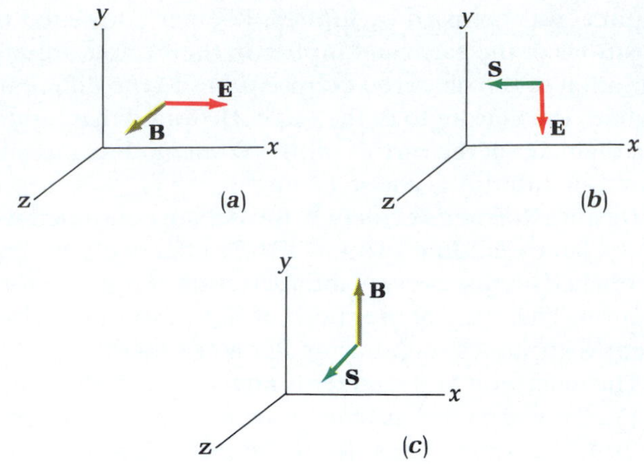

Figure 34-15. Question 13.

15. If the amplitude of the electric field part of a wave is tripled, by what factor is the average intensity changed? By what factor is the amplitude of the magnetic field changed?

16. Consider the flux of the Poynting vector for a surface: $\int \mathbf{S} \cdot d\mathbf{A}$, where $d\mathbf{A}$ is the differential surface vector for an element of area. What is the dimension of this flux? What is its SI unit? Explain the physical meaning of this flux.

17. When you turn on a flashlight, does the flashlight tend to recoil in your hand. If so, will you feel this recoil? Explain.

18. Would it be possible in principle to use a "sailing" spaceship to travel to Jupiter? To Venus? Should your sails be good reflectors or good absorbers?

19. In principle, is it possible for an electromagnetic wave to transfer momentum to an object but not transfer energy? Explain. Is it possible to transfer energy but not momentum? Explain.

20. A magnetic-dipole antenna is shown in Fig. 34-16. An ac source drives a current around the circular loop of wire first in one sense and then in the other. Sketch a representation of the magnetic field along the x axis in the vicinity of point P, which is far from the loop compared with the loop size or the wavelength of the waves. Explain why this antenna is called a magnetic-dipole antenna.

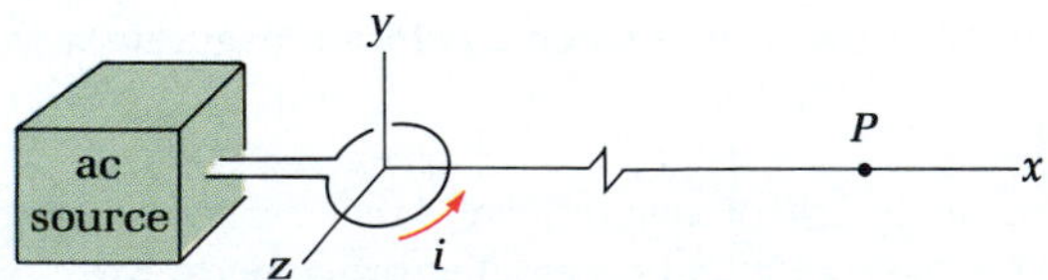

Figure 34-16. Question 20.

21. Consider the electromagnetic spectrum shown in Fig. 34-13. What type of waves have wavelengths about the length of a football field? The length of a finger? The width of a hair? The size of an atom? The size of an atomic nucleus?

22. Figure 34-12 indicates that no intensity is emitted from an electric-dipole transmitting antenna along the direction in which the antenna is aligned. Give a plausible argument for why this is so by using the transverse nature of electromagnetic waves.

23. In a comparison of infrared and ultraviolet light, which has the larger wavelength? Which has the larger frequency?

24. Describe how you could determine the speed of light by making electric and magnetic measurements. List the apparatus you would need. Is a clock necessary?

25. Complete the following table:

Symbol	Represents	Type	SI Unit
c	Speed of light in vacuum		
u			J/m^3
S		Vector	
p			

EXERCISES

Section 34-2. The Wave Equation for E and B

1. *(a)* Show that the dimension of $1/\sqrt{\epsilon_0\mu_0}$ is length/time. *(b)* Show by substitution that the SI unit for $1/\sqrt{\epsilon_0\mu_0}$ is meters per second.

2. *(a)* Show that $E_y = E_0 \sin[k(x - ct)]$ is a solution to Eq. 34-8 if $c = 1/\sqrt{\mu_0\epsilon_0}$. What is the direction of propagation of this wave? *(b)* Show that $E_y = E_0 \sin[k(x + ct)]$ is a solution to Eq. 34-8 if $c = 1/\sqrt{\mu_0\epsilon_0}$. What is the direction of propagation of this wave?

3. *(a)* Show that if the dependence of E_y on x and t is in the form $x - ct$, then E_y is a solution to the wave equation. That is, show that $E_y = f(x - ct)$ is a solution to Eq. 34-8, with $c = 1/\sqrt{\mu_0\epsilon_0}$. [*Hint:* Let $\xi = x - ct$. Note that $\partial E_y/\partial x = df/d\xi$ and $\partial E_y/\partial t = -c(df/d\xi)$.] What is the direction of propagation of this wave? *(b)* Show that $E_y = g(x + ct)$ is a solution to Eq. 34-8. What is the direction of propagation of this wave?

4. The speed of visible light in water is 2.25×10^8 m/s. What is the dielectric constant of water at optical frequencies?

5. The dielectric constant of ice at optical frequencies is 1.71. What is the speed of visible light in ice?

6. Suppose $B_z = B_0 \sin[k(x - ct)]$. Show that

$$\frac{\partial}{\partial x}\frac{\partial B_z}{\partial t} = \frac{\partial}{\partial t}\frac{\partial B_z}{\partial x}$$

7. Suppose the x and t dependence of B_z can be expressed $B_z = f(x - ct)$, where f and its first and second derivatives are continuous. Show that

$$\frac{\partial}{\partial x}\frac{\partial B_z}{\partial t} = \frac{\partial}{\partial t}\frac{\partial B_z}{\partial x}$$

(You may wish to use the hint from Exercise 3.)

8. To show that the wavelike **E** and **B** fields are mutually perpendicular, we oriented our y axis along the oscillating **E** field and used Faraday's law to show that the oscillating **B** field is along the z axis. Instead, let the y axis be along **B** and use the modified form of Ampere's law to find the direction of **E**. Note that the term involving ΣI is zero in vacuum.

9. To develop Eq. 34-6, we integrated around the square path in Fig. 34-5 in the counterclockwise sense. Develop Eq. 34-6 by integrating around this path in the clockwise sense.

10. Use Eqs. 34-6 and 34-7 to develop the wave equation for B_z, Eq. 34-9.

Section 34-3. Electromagnetic Waves

11. *(a)* What is the wavelength of the waves from an FM radio station with a frequency of 100 MHz? *(b)* What is the wavelength of the waves from an AM radio station with a frequency of 1000 kHz?

12. A harmonic electromagnetic wave passes by a football field. At an instant the spatial variation of the electric field part of the wave can be represented as shown in Fig. 34-17. *(a)* Estimate the frequency of this wave. *(b)* Use Fig. 34-13 to categorize this wave.

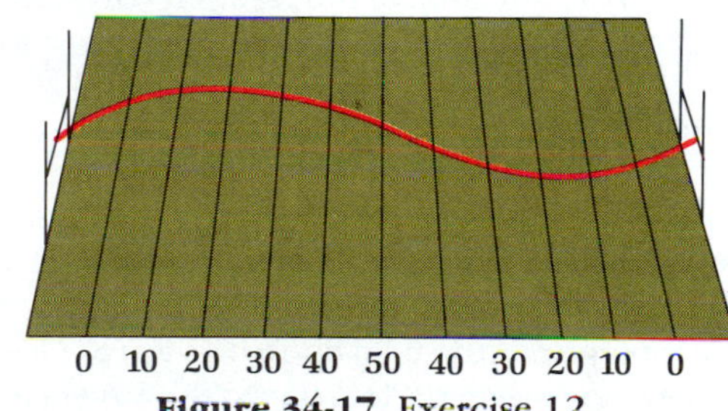

Figure 34-17. Exercise 12.

13. The amplitude of the magnetic field part of a harmonic electromagnetic wave in vacuum is $B_0 = 510$ nT. What is the amplitude of the electric field part of the wave? Would your answer be different if the wave were in air?

14. The angular frequency of a harmonic electromagnetic wave in vacuum is $\omega = 8.2 \times 10^{12}$ rad/s. Determine the wave number, wavelength, frequency, and period of this wave. Would any of your answers be different if the wave were in air? From Fig. 34-13, what type of wave is this?

15. Suppose the electric field part of an electromagnetic wave in vacuum is

$$\mathbf{E} = \{(31 \text{ N/C}) \cos[(1.8 \text{ rad/m})y + (5.4 \times 10^8 \text{ rad/s})t]\}\mathbf{i}$$

(a) What is the direction of propagation? *(b)* What is the wavelength λ? *(c)* What is the frequency ν? *(d)* What is the amplitude of the magnetic field part of the wave? *(e)* Write an expression for the magnetic field part of the wave.

16. Write expressions for the **E** and **B** fields of an electromagnetic wave that propagates in vacuum toward the $+z$ direction and has its plane of polarization parallel to the xz plane. The amplitude of the magnetic field part of the wave is $B_0 = 350$ nT

and the frequency of the wave is $\nu = 9.8$ GHz. Would your expressions be different if the wave were traveling in air? From Fig. 34-13, what type of wave is this?

17. Using Eqs. 34-14, 34-15, and 34-16, develop Eq. 34-17.

Section 34-4. Electromagnetic Wave Intensity

18. Equation 34-21 connects intensity with energy density, $S = uc$. Show that the dimension of intensity is the same as the dimension of the product of energy density and speed.

19. Suppose the electromagnetic wave intensity at a particular instant and at a point in vacuum is 1 kW/m². What is the energy density at that point and instant? Would your answer be different if the wave were in air?

20. The magnitude of the electric field due to an electromagnetic wave at a particular instant at a point in free space is 97 N/C. What is the energy density at that point and instant? Would your answer be different if the wave were in air?

21. *(a)* Show that the energy density due to an electromagnetic wave at a particular instant at a point in free space can be written $u = B^2/\mu_0$. *(b)* Show that the intensity can be written $S = B^2c/\mu_0$. *(c)* Determine u and S at a point and instant at which $B = 530$ nT.

22. Show that the intensity due to a plane-polarized electromagnetic wave in which the plane of polarization is parallel to the xy plane and the direction of propagation is toward $+x$ can be written $S = E_yB_z/\mu_0$.

23. Show that the average intensity of a harmonic wave can be written as *(a)* $\bar{S} = \frac{1}{2}\epsilon_0 E_0{}^2 c$ and *(b)* $\bar{S} = (B_0{}^2/2\mu_0)c$.

24. Suppose the average intensity $\bar{S}$ of a harmonic electromagnetic wave in vacuum or free space is 553 W/m². *(a)* What is the amplitude of the electric due to the wave? *(b)* What is the amplitude of the magnetic field due to the wave? *(c)* Determine these amplitudes when a wave of this average intensity travels in air.

25. The Poynting vector of an electromagnetic wave in vacuum is

$$\mathbf{S} = -\{(220\ \text{W/m}^2)\cos^2[(12\ \text{rad/m})z + (3.6 \times 10^9\ \text{rad/s})t]\}\mathbf{k}$$

(a) What is the direction of propagation? *(b)* What is the wavelength λ? *(c)* What is the frequency ν? *(d)* Write expressions for the **E** and **B** fields.

26. An electromagnetic wave in vacuum is traveling in the $+z$ direction, and its plane of polarization is parallel to the xz plane. The frequency of the wave is $\nu = 50$ MHz and its average intensity is $\bar{S} = 480$ W/m². Write expressions for **E**, **B**, and **S** as functions of z and t. Would these expressions be different if the wave were in air?

27. A light wave (in vacuum) with its plane of polarization parallel to the xy plane propagates in the $+\mathbf{i}$ direction. The wavelength of the wave is 580 nm, and its oscillating magnetic field has an amplitude of 86 nT. Write expressions for **E**, **B**, and **S**.

28. The average intensity of sunlight at the top of the earth's atmosphere is 1.35 kW/m². This radiation is unpolarized and consists of many frequencies, but, for purposes of this calculation, regard it as plane-polarized and harmonic or monochromatic. *(a)* What is the amplitude of the electric field part of the wave? *(b)* What is the amplitude of the magnetic field part of the wave? *(c)* What is the average electromagnetic energy density of the wave?

29. A helium-neon laser sends a beam of collimated, plane-polarized, monochromatic light into the air of a room. The beam has a circular cross section with a radius of 1.0 mm, and the intensity is essentially uniform within the beam. The average power of the beam is 3.5 mW and the wavelength of the light is 633 nm. *(a)* Determine $\bar{S}$ for the beam. *(b)* Determine the electromagnetic energy contained in a 1.0-m length of the beam. *(c)* Determine the amplitude of the electric field part of the wave. *(d)* Determine the amplitude of the magnetic field part of the wave. *(e)* If the beam direction is horizontal toward the north and the electric field oscillates along the horizontal east-west, what is the direction of the oscillating magnetic field? *(f)* Determine the frequency of the wave.

Section 34-5. Radiation Pressure

30. From Eq. 34-26, the radiation pressure is $p = S/c$. Show that the dimension of pressure is the same as the dimension of the ratio of intensity to speed.

31. The average power in a laser beam is 4.3 mW, and the beam has an essentially uniform intensity within the 1.2-mm radius of the beam. Suppose the beam is normally incident on a completely absorbing surface. *(a)* What is the pressure exerted by the beam on the part of the surface it strikes? *(b)* What is the force exerted on the surface by the beam?

32. The reflectivity r of a surface is the fraction of light intensity incident on the surface that is reflected. Show that the radiation pressure exerted on a surface of reflectivity r by a normally incident beam of intensity S is $p = (r + 1)S/c$.

33. The average intensity of sunlight at the top of the earth's atmosphere is 1.35 kW/m². Consider the force exerted on the earth by the absorption of the sun's radiation. *(a)* Explain why, for purposes of calculating this force, the earth may be regarded as a flat disk facing the sun. *(b)* Estimate this radiation force by assuming complete absorption. *(c)* Find the ratio of the radiation force to the gravitational force on the earth by the sun.

34. A 10,000-kg spaceship is drifting in interstellar space where all external forces on it are negligible. To propel the spaceship, a 30,000-W laser is turned on and directed into space. *(a)* What is the acceleration of the spaceship? *(b)* How many years will it take for the spaceship to change its speed by 1 m/s?

Section 34-6. Emission of Electromagnetic Waves

35. About 5 percent of the power of a 100-W light bulb is converted to visible radiation. *(a)* What is the average intensity of visible radiation at a distance of 1 m from the bulb? *(b)* At a distance of 10 m? Assume that the radiation is emitted isotropically and neglect reflections.

36. The average intensity of sunlight at the top of the earth's atmosphere is 1.35 kW/m². What is the radiative power of the sun?

37. Consider a long array of fluorescent tubes lined up end to end. Each 40-W tube is 1.22 m long and 20 percent of its power is emitted in the visible spectrum. *(a)* What is the average intensity in the visible spectrum at a perpendicular distance of 1 m from the tubes? *(b)* At a perpendicular distance of 10 m?

Section 34-7. The Electromagnetic Spectrum

38. What is the wavelength of the radiation from the 60-Hz alternating current in electric power lines?

39. Sodium arc lamps are often used in street lights and can be distinguished by their yellow light. The wavelength of this light is 590 nm. What is its frequency?

40. It is believed that the two hills used by Galileo in his speed-of-light experiment (see the Commentary) were about 1.5 km apart. What is the time required for light to traverse this round-trip distance?

41. The average earth-sun distance is 1.496×10^{11} m. What is the average time required for sunlight to reach the earth?

42. One light-year (ly) is the distance light travels in one year. *(a)* What is the conversion factor that changes meters to light years? *(b)* What is the conversion factor that changes light years to meters?

PROBLEMS

1. ***The magnetic-field wave is transverse.*** Use Gauss's law for the magnetic field to show that the magnetic field for a plane wave is transverse.

2. ***Developing the wave equation.*** Use the modified form of Ampere's law to develop Eq. 34-7. Note that the term that contains ΣI is zero in vacuum.

3. ***Radiation pressure.*** Consider the following model for the absorption of an electromagnetic wave. A plane-polarized electromagnetic wave propagating in the $+x$ direction is incident normally on the surface of an insulator and is completely absorbed by interacting with the charged particles of the material. These particles are bound to their lattice sites, but they oscillate parallel to the y axis because of the driving electric force $\mathbf{F}_e = (qE_y)\mathbf{j}$ resulting from the oscillating electric field of the wave (Fig. 34-18). Recall from Sec. 14-7 that the power absorbed by an oscillator from the driving force is maximum when the driving force is in phase with the particle's velocity $\mathbf{v}$ (at resonance). Therefore assume that $\mathbf{F}_e$ is in phase with $\mathbf{v}$. *(a)* Show that the magnetic force on the particle is $\mathbf{F}_m = (F_e v/c)\mathbf{i}$. (Assume that $v \ll c$ so that $F_m \ll F_e$ and the magnetic force has a negligible effect on the oscillations.) Explain why the magnetic force does *not* alternate in direction along the x axis, but varies between zero and a maximum while being directed toward $+x$ for charges of either sign. *(b)* Since $\mathbf{F}_e$ and $\mathbf{v}$ have the same direction, the power delivered to a particle by the wave is $F_e v$, and the power delivered to all such particles in the slab is $\Sigma(F_e v)$. Further, the pressure p on the slab of area A due to the magnetic forces on all such particles in the slab is $p = \Sigma(F_m)/A$. Use these results to show that $p = S/c$, where S is the absorbed intensity.

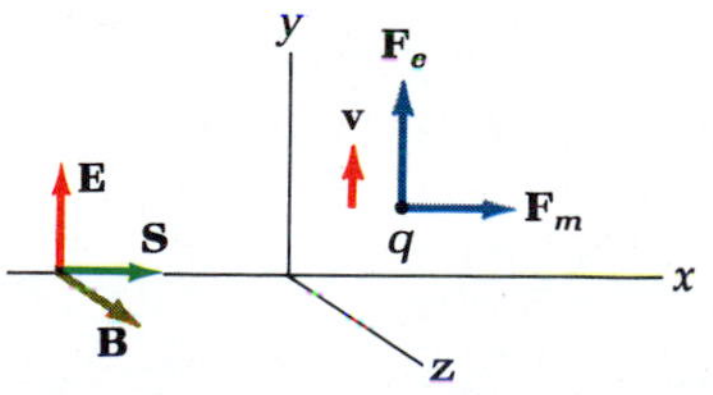

Figure 34-18. Problem 3: An electromagnetic wave interacting with a charged particle. The particle is bound to a central position by a linear restoring force, and we assume that its oscillation frequency is the same as the frequency of the wave. Thus the wave provides a driving force $\mathbf{F}_e = q\mathbf{E}$ in resonance with the particle so that $\mathbf{v}$ is parallel to $\mathbf{F}_e$. The magnetic force $\mathbf{F}_m$ exerted on the particle by the wave is in the direction of propagation.

4. ***A stationary (nonorbiting) space station.*** Consider placing a space station of mass m in the solar system without putting it in orbit around the sun. The station has a large reflecting surface which faces the sun so that it is in equilibrium; the force $\mathbf{F}_{rad}$ due to radiation pressure from the sun's radiated power P is equal and opposite the gravitational force $\mathbf{F}_G$ due to the sun's mass M_S. (Neglect forces due to the planets.) *(a)* Show that the area A of the station's reflecting surface is

$$A = \frac{2\pi G M_S m c}{P}$$

where G is the gravitational constant. Why is this equilibrium condition independent of the distance between the sun and the station? *(b)* Suppose the reflecting area is a square of side L and the mass of the station is 10^6 kg. Determine L. The power radiated by the sun is 3.77×10^{26} W.

5. ***An alternative view of i^2R heating.*** Consider a section of wire of length ℓ, resistance R, and radius a carrying a current i (Fig. 34-19). *(a)* Show that the magnitude of the Poynting vector at the surface of the wire is $S = (i^2R)/(2\pi a\ell)$. *(b)* Show that the direction of the Poynting vector at each point on the surface of the wire is normal to the surface and inward (toward the axis of the wire). *(c)* What is the flux of the Poynting vector for the surface of the wire? *(d)* Give a physical interpretation of these results.

6. ***Energy density and radiation pressure.*** A plane wave is normally incident on a surface. Show that the energy density just outside the surface is equal to the radiation pressure on the surface.

7. ***Comet tails.*** Consider a spherical object of radius a and uniform mass density ρ that is in equilibrium in the solar system under the action of the attractive gravitational force and the repulsive radiation force due to the sun. (Neglect forces due to the planets.) *(a)* Show that a completely absorbing object is in equilibrium if

$$a = \frac{3P}{16\pi c G M_S \rho}$$

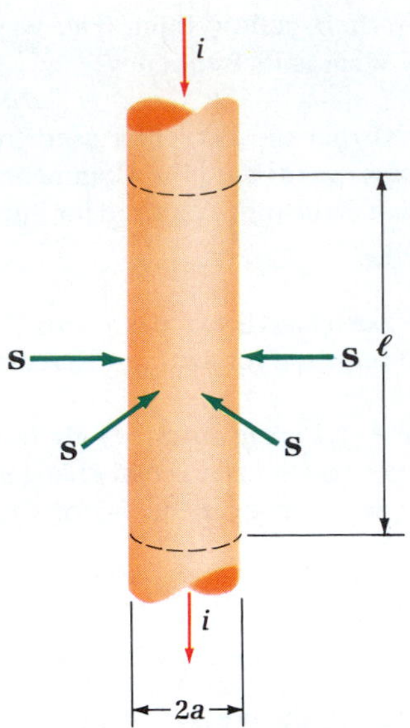

Figure 34-19. Problem 5.

where M_S is the sun's mass, P is the sun's radiated power, and G is the gravitational constant. *(b)* Why is this equilibrium condition independent of the distance between the object and the sun? *(c)* Determine a for a particle whose mass density is about the same as ice ($\rho \approx 10^3$ kg/m^3). The sun's radiated power is 3.77×10^{26} W. *(d)* What is the fate of particles of this density with radius smaller than a?

8. ***Solar intensity.*** The distance from the earth to the sun in early January is 1.446×10^{11} m, and in early July this distance is 1.543×10^{11} m. The power radiated by the sun is 3.77×10^{26} W. Determine the intensity at the top of the earth's atmosphere in *(a)* early January and *(b)* early July. *(c)* With this result, how do you account for the cold weather during January and the hot weather during July?

9. ***Measuring radiation pressure.*** Suppose we use the torsion balance shown in Fig. 34-20 to measure radiation pressure. Two coin-shaped mirrors, each of area A and centered a perpendicular distance ℓ from the axis are connected by a horizontal bar that is suspended by a fiber. The restoring torque τ when the fiber is twisted through an angle $\Delta\theta$ is $\tau = \kappa\,\Delta\theta$, where κ is the torque constant of the fiber. Light of known intensity $\bar{S}$ is normally incident on mirror 1 while mirror 2 is shaded, and the suspension comes to equilibrium after turning through an angle $\Delta\theta$. *(a)* Show that $\Delta\theta$ is given by

$$\Delta\theta = \frac{2\bar{S}A\ell}{c\kappa}$$

(b) Estimate the value required for κ in an experiment in which $\Delta\theta \approx 0.01$ rad, $A \approx 10^{-4}$ m^2, $\ell \approx 0.1$ m, and $\bar{S} \approx 10^5$ W/m^2.

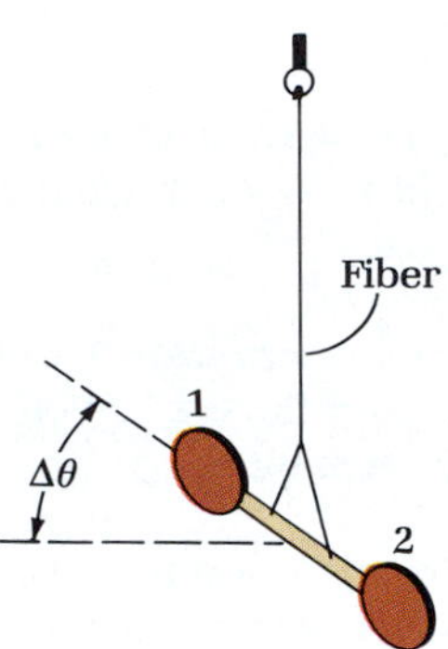

Figure 34-20. Problem 9: A torsion balance for measuring radiation pressure.

10. ***Energy from solar cells.*** Estimate the area required to produce electricity, using solar cells, for a typical family of four. Assume the cells have an efficiency of 10 percent for converting solar energy into electric energy. Will the area of a typical rooftop suffice?

GEOMETRICAL OPTICS 35

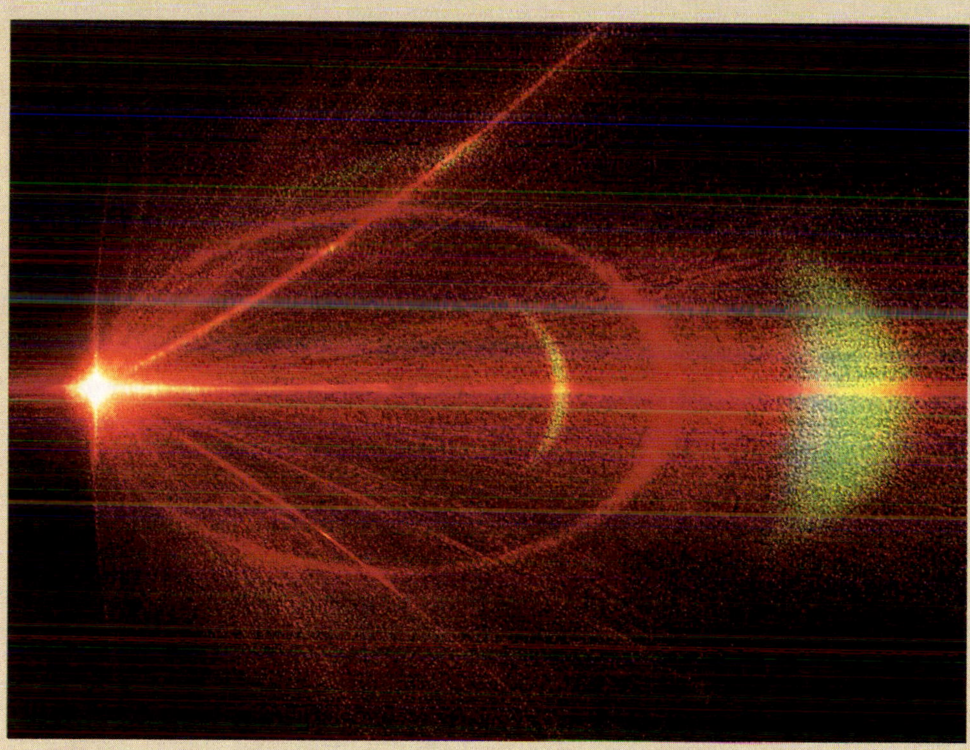

A laser beam strikes a crystal of barium titanate and scatters into a colorful fan of light due to the photorefractive effect. See *Scientific American,* October 1990.

The propagation of light waves is described by Maxwell's equations. The solution of these equations for the conditions of a given physical situation will determine **E** and **B** at every point—and thus the amplitude, polarization, and phase of the light wave at every point. Solving Maxwell's equations may be difficult, and often the detailed information they give is not needed. The information usually required is obtainable by a simpler method called ***geometrical optics,*** which was devised before light was known to be an electromagnetic wave. It has since been shown to approximate the results of Maxwell's equations when the wavelength of the light is much smaller than the objects that the light wave encounters.

35-1 GEOMETRICAL OPTICS

The crests of an electromagnetic wave progressing outward from a source are shown in Fig. 35-1. Two useful ways to represent the propagating wave are wavefronts and light rays. ***Wavefronts*** are surfaces of constant phase of the light wave, and can be likened to the crests of a water wave. The wavefronts of a spherical light

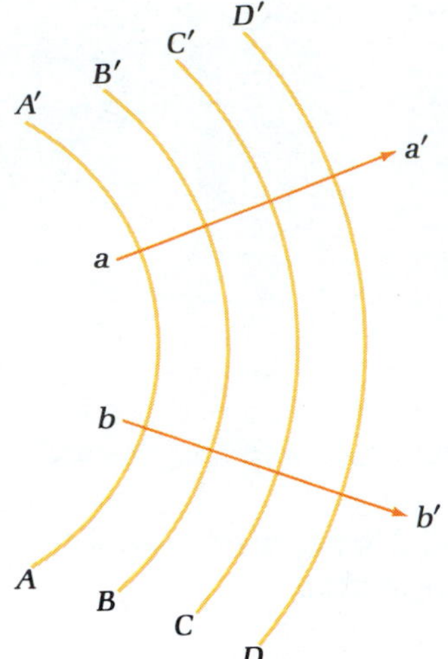

Figure 35-1. Wavefronts in an optical wave. AA' is a wavefront and represents positions that have the same phase at a given time, such as the positions of a certain crest of the wave. BB', CC', and DD' are other wavefronts. Lines aa' and bb' show the direction of propagation of the wave and are called *light rays*.

wave are shown as concentric circles in the two dimensions of Fig. 35-1. A ***light ray*** is a line pointing in the direction of wave propagation. If the speed of propagation is the same in all directions, the light rays are perpendicular to the wavefronts.

The path followed by a laser beam is a good example of the path of a ray of light. The beam can be observed by the light scattered off small particles in the path of the beam. When a laser beam hits a surface that bounds two different materials, several things can happen. The beam can be absorbed, as happens when a laser beam hits a piece of black paper. If the surface the laser beam hits is rough on a scale comparable to or greater than the wavelength of the light, then the light will be reflected in many directions, as happens when a laser beam hits a piece of white paper. If the laser beam hits a polished metal surface, it will be reflected in a single direction, a process called *specular reflection.* If a laser beam in air hits a glass surface it will be both reflected back into the air and transmitted, or ***refracted,*** into the glass. In this chapter we will be concerned with specular reflection, which we will just call reflection, and refraction.

When the wavelength of the light is much smaller than the dimensions of the physical system through which the light propagates, then the following three laws of geometrical optics apply:

1. The *law of rectilinear propagation.* Light rays in homogeneous media propagate in straight lines.

2. The *law of reflection.* At an interface between two media, an incident wave is partially reflected. The incident ray and the normal to the surface determine the plane of incidence, as seen in Fig. 35-2. If the incident ray makes an angle θ_1 with the normal, then the reflected ray lies in the plane of incidence on the other side of the normal and makes the same angle with it: $\theta_1 = \theta_{1r}$. In words, *the angle of incidence equals the angle of reflection.*

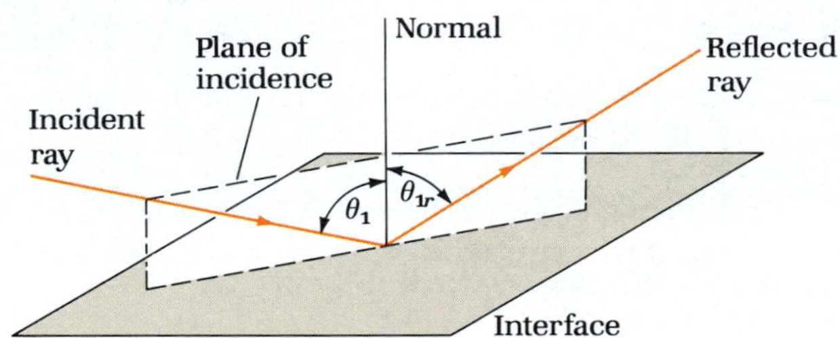

Figure 35-2. The geometry of reflection. The plane of incidence is the plane defined by the normal to the interface between the two media and the incident light ray. The reflected ray lies in this plane on the side of the normal opposite to the incident beam, such that $\theta_{1r} = \theta_1$.

Law of refraction

3. The *law of refraction.* The refracted ray is transmitted into the second medium, as shown in Fig. 35-3 (where the reflected ray has been omitted for clarity). The refracted ray also lies in the plane of incidence. It makes an angle θ_2 with the normal given by ***Snell's law:***

Snell's law of refraction

$$n_1 \sin \theta_1 = n_2 \sin \theta_2 \tag{35-1}$$

where n_1 and n_2 are properties of the two media. The ratio n_1/n_2 is called the *relative index of refraction* and is related to the speed of light rays in the media.

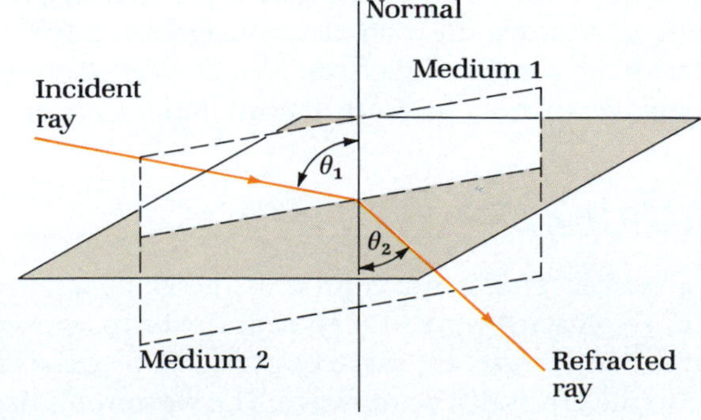

Figure 35-3. The geometry of refraction. The refracted ray lies in the incident plane in the second medium and makes an angle θ_2 with the normal to the interface.

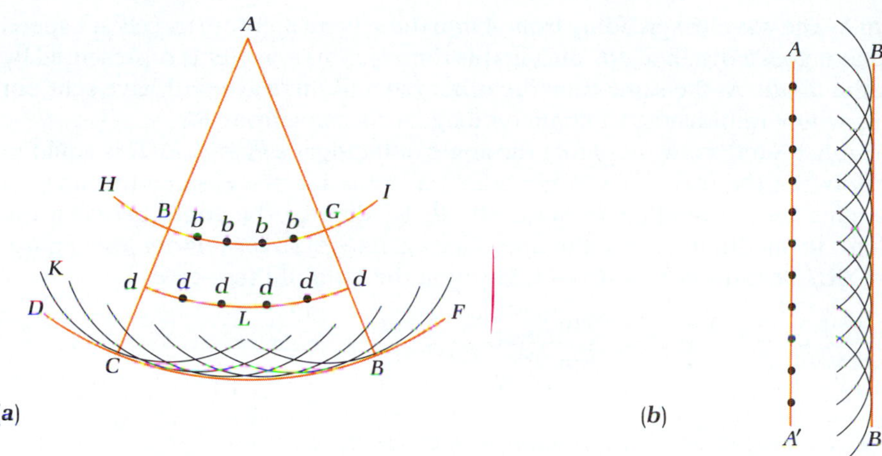

Figure 35-4. *(a)* Propagation of a spherical wave *(from Huygens' "Treatise on Light")*. On the wavefront *HI* each point *b* sends out a wavelet, such as the wavefront *KCL* sent out from *B*. The new wavefront *DF* is the tangent surface to the wavelets from all points (such as *b*) on the old wavefront *HI*. *(b)* A similar construction for a plane wave.

The first two of these laws were known to the ancient Greeks. The law of refraction was assiduously searched for in the early 1600s. Johannes Kepler, for example, found some 27 empirical rules to use in the design of lenses, but could not formulate a general rule. The law of refraction was probably first formulated by Willebrod Snell about 1620, and was first published by Descartes in 1637 (without mentioning Snell's name).

Pierre de Fermat (1601–1665) and Christian Huygens (1629–1695) formulated principles that led to the laws of ray optics. These principles have been shown to follow from Maxwell's equations in the approximation of ray optics—namely, that the wavelength of light is small compared with the dimensions of all of the parts of the optical system. We now discuss the theory of Huygens. Fermat's principle is illustrated in Probs. 1 and 2.

Huygens' Principle

Huygens, a contemporary of Newton, advocated a wave theory of light in contrast to Newton's corpuscular theory. Huygens' principle states that the propagation of a light wave can be determined by assuming that at every point on a wavefront there arises a spherical wavelet centered on that point. The next wavefront is the outward surface tangent to these wavelets. Figure 35-4*a*, from Huygens, shows that this predicts spherical waves from a point source. Similarly, a plane wavefront generates further plane waves, as shown in Fig. 35-4*b*. From this, the law of rectilinear propagation follows.

In Fig. 35-5 a plane wavefront is incident on a boundary. Following Huygens' principle, wavelets are constructed from the wavefront *AC*. The wavelet spreading from *C* will reach the boundary in time $t_0 = CB/v_1$, where v_1 is the speed of light in

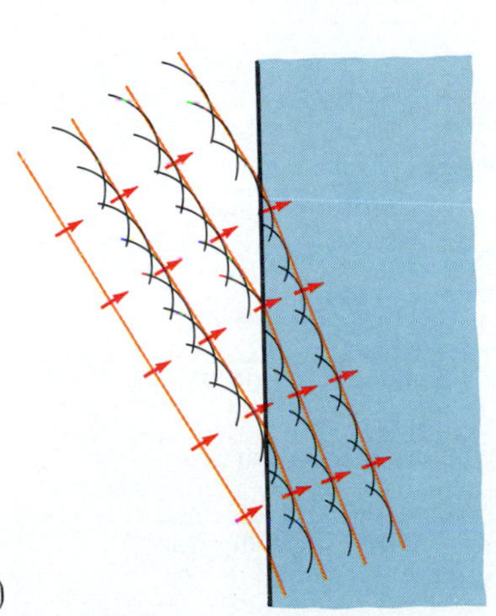

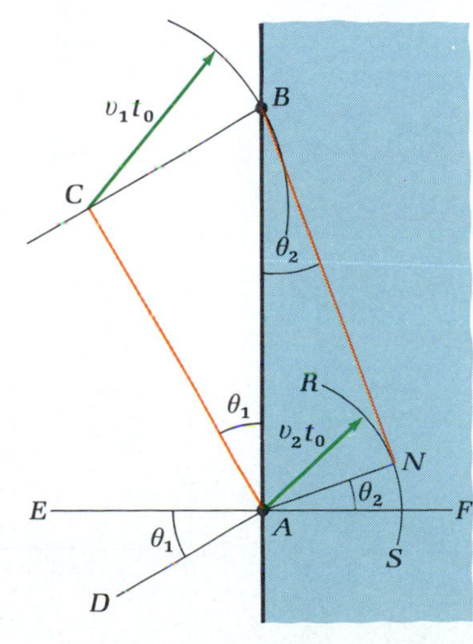

Figure 35-5. Refraction at a surface. *(a)* Huygens' construction for refraction at a surface between media with differing speeds for light. Successive positions of the wavefronts as they are refracted. *(b)* An enlargement of a portion of part *(a)*.

The refraction of water waves at a boundary between deep and shallow water obeys Snell's law.

medium 1. The wavelet spreading from A into the second medium travels at a speed v_2, so that it goes a distance $v_2 t_0$ during this time t_0. This wavelet is represented by SNR in the figure. At the same time the other parts of the wave will have sent out wavelets whose radii have as a common tangent the wavefront BN.

To establish Snell's law, note that the angle of incidence $\theta_1 = \angle EAD$ is equal to $\angle CAB$ between the wavefront in medium 1 and the interface, since the sides of these angles are perpendicular. Similarly, θ_2 is equal to the angle between the wavefront in medium 2 and the interface: $\angle FAN = \angle ABN$. From the figure, $\sin\theta_1 = BC/BA$ and $\sin\theta_2 = AN/BA$. Forming the ratio of these gives

$$\frac{\sin\theta_1}{\sin\theta_2} = \frac{BC}{AN} = \frac{v_1 t_0}{v_2 t_0}$$

or

$$v_1^{-1}\sin\theta_1 = v_2^{-1}\sin\theta_2$$

This is the same as Snell's law if $n_1/n_2 = v_2/v_1$. Thus not only does Snell's law follow from Huygens' principle, but Huygens' principle predicts that light travels slower in media of higher index. J. B. L. Foucault in 1850 showed that this was true by a direct measurement of the speed of light in water and in air.

Since only the ratio of speeds is involved in Snell's law, only the ratio n_1/n_2 is determined by measurements of refraction. The definition of the index of refraction of a single medium requires a convention. The convention used is that the index of refraction of a vacuum is exactly 1. Since the speed of light in vacuum is the constant c, the index of refraction n for a substance is given by $n/1 = c/v$, or

$$n = \frac{c}{v}$$

where v is the speed of light in the substance. The indices of several substances are given in Table 35-1.

The index of refraction of a substance depends somewhat on the wavelength of the light being used. If a ray composed of many wavelengths of light is refracted, it will be dispersed into rays whose directions depend on the index of refraction for the various wavelengths. The property whereby n varies with wavelength is called ***dispersion.*** For many substances at optical wavelengths, the index of refraction decreases as the wavelength increases. Figure 35-6 shows the index of refraction as a function of wavelength for a few substances.

At a boundary where the wave is reflected, Huygens' principle leads to the law of reflection, as shown in Fig. 35-7 and discussed in Exercise 3.

TABLE 35-1. ***The Approximate Index of Refraction for a Few Substances***

Substance	n^*
Gases (at 0°C)	
Air	1.000293
Ammonia	1.000376
Carbon dioxide	1.000451
Chlorine	1.000773
Hydrogen	1.000132
Methane	1.000444
Sulfur dioxide	1.000686
Liquids	
Benzene	1.501
Carbon disulfide	1.625
Ethyl alcohol	1.362
Methyl alcohol	1.329
Methylene iodide	1.726
Water	1.333
Solids	
Sapphire, ruby (Al_2O_3)	1.767
Diamond	2.417
Glasses: Fused quartz	1.458
Soda lime	1.512
Pyrex	1.474
Dense flint	1.655
Ice (0°C)	1.310
Lucite plastic	1.491
Rutile, E (470 nm)	3.095
Salt (NaCl)	1.544

* Values are at room temperature and atmospheric pressure for light of wavelength 589 nm unless otherwise indicated.

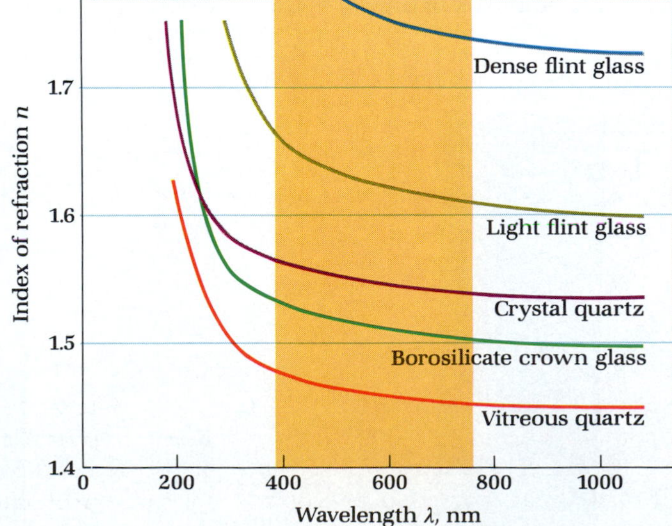

Figure 35-6. The index of refraction of some optical materials as a function of wavelength.

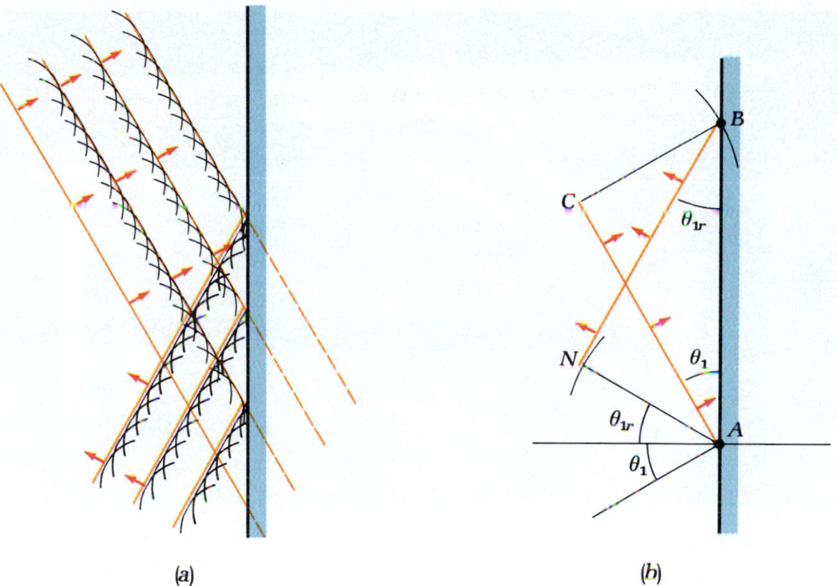

Figure 35-7. *(a)* Huygens' construction for reflection. *(b)* An enlargement of a portion of part *(a)*. The incident wavefronts are parallel to *AC*, and the reflected wavefronts are parallel to *NB*.

EXAMPLE 35-1

Refraction at an air-water boundary. *(a)* Measurements show that the speed of light in water is 225,000 km/s and in air 299,706 km/s. What are the indices of refraction of the two media? *(b)* A ray of light is incident from air onto a flat surface of water, as shown in Fig. 35-8. What is the angle of refraction, θ_2?

Solution. *(a)* The index of refraction for air is $n = (299{,}792 \text{ km/s})/(299{,}706 \text{ km/s}) = 1.00029$, and for water $n = (299{,}792 \text{ km/s})/(225{,}000 \text{ km/s}) = 1.33$.

(b) From the drawing, the angle of incidence is 28°. From Snell's law, $n_1 \sin \theta_1 = n_2 \sin \theta_2$; thus

$$\theta_2 = \sin^{-1}\left(\frac{n_1}{n_2}\sin \theta_1\right) = 21°$$

Notice that the path of the ray is bent toward the normal to the surface. When a ray goes from a medium of lower index to a medium of higher index, it is always bent toward the normal.

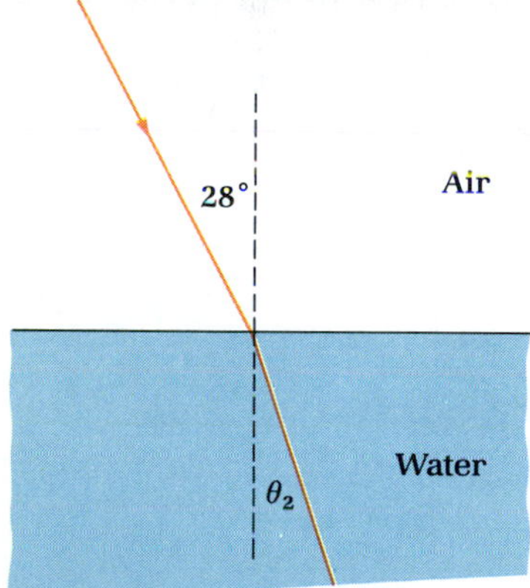

Figure 35-8. Example 35-1.

SELF-TEST 35-1. Consider the refraction of a light ray incident on a flat air-water surface when the light passes upward from water into air. If the angle of incidence is 21°, what is the angle of refraction? Compare your answer with that of the above example. This is an example of optical reversibility, which is discussed in the next section. ***ANSWER:*** 28°.

Total Internal Reflection

A ray of light proceeding from water to air cannot always get into air. Suppose a light ray comes up to the surface of water at an angle of incidence $\theta_1 = 50°$. Using Snell's law to solve for the angle θ_2 between the ray and the normal on the air side of the interface, we get

$$\theta_2 = \sin^{-1}\left(\frac{n_1}{n_2}\sin \theta_1\right) = \sin^{-1}\left(\frac{1.333}{1.000}\,0.766\right)$$
$$= \sin^{-1} 1.021$$

But there is no angle whose sine is greater than 1, so there is no solution to Snell's law for this situation! Why is there no solution? It is because no refracted ray exists for this angle of incidence. The only ray that leaves the interface is the reflected ray.

Figure 35-9 shows what happens to rays approaching such an interface at various angles. Since the rays are going from a medium of higher n to a medium of lower n,

(a)

(b)

Figure 35-9. *(a)* Demonstration of total internal reflection. A beam of light from above strikes three mirrors at the bottom of a fish tank. The light reflected from the mirrors forms three beams that strike the water-air interface at different angles. The two beams at angles less than the critical angle are both refracted and reflected at the water-air surface. By contrast, the beam at an angle greater than the critical angle is totally reflected back into the water. For total internal reflection to occur, the light must strike the surface from the medium with higher index n. *(b)* Light rays emerging from a point source S in a medium of index n_1 strike the surface with another medium of index n_2. The refracted angle θ_2 is greater than the incident angle θ_1 because $n_1 > n_2$. The critical angle θ_c is the value of θ_1 that corresponds to $\theta_2 = 90°$. Thus, for incident angles θ_1 equal to or greater than θ_c, the light undergoes total internal reflection.

each ray is bent away from the normal. The angle of incidence that produces an angle of refraction equal to 90° is called the ***critical angle*** θ_c. Suppose $n_2 < n_1$. Then from Snell's law

$$n_1 \sin \theta_c = n_2 \sin 90° = n_2$$

or

Critical angle

$$\sin \theta_c = \frac{n_2}{n_1} \tag{35-2}$$

Rays that approach an interface from the higher-index side with an angle of incidence less than the critical angle are partially reflected and partially refracted. A reflected ray and a refracted ray leave the interface, as seen in Fig. 35-9. Rays that approach the interface from the higher-index side at an angle of incidence greater than the critical angle are totally reflected. No refracted ray exists in this case. This phenomenon is called ***total internal reflection***. Note that there is no critical angle for light proceeding from a medium of lower index to a medium of higher index. In that case there is always a refracted beam.

Total internal reflection

Solving the same problem using Maxwell's equations shows that there is an *evanescent wave* in the medium of lower index. The evanescent wave is so called because it decays exponentially to a negligible amplitude a few wavelengths into the lower-index medium. Since ray optics is only valid for geometries in which all dimensions are much larger than a wavelength, ray optics cannot deal with this phenomenon. This evanescent wave can be detected by placing a third medium of index n_1 close to the interface, as shown in Fig. 35-10. If the third medium is very close (several wavelengths of light) to the interface, a ray will be found in the third medium, and the reflection will no longer be total. This phenomenon is called *frustrated total internal reflection.* (See Ques. 24.) There is a close relation between this phenomenon and the behavior of electrons at a similar barrier in tunnel diodes.

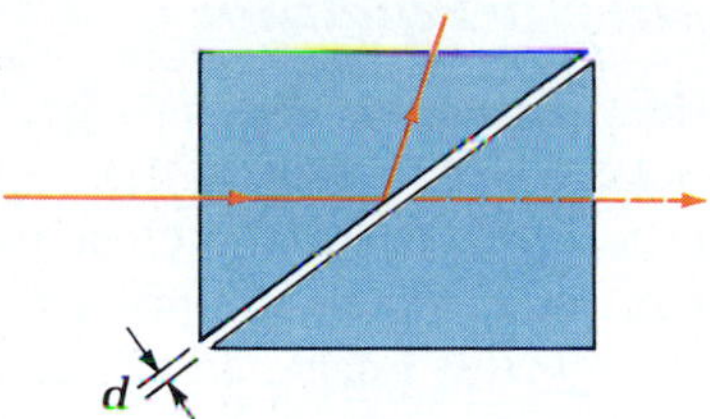

Figure 35-10. Schematic drawing of two prisms placed close together. A ray having an incident angle greater than the critical angle is shown approaching the boundary between the prisms. An evanescent wave is present several wavelengths into the medium of lower index between the prisms. If the prism to the right is pushed within several wavelengths of the prism to the left, a ray will propagate across the interface and into the second prism.

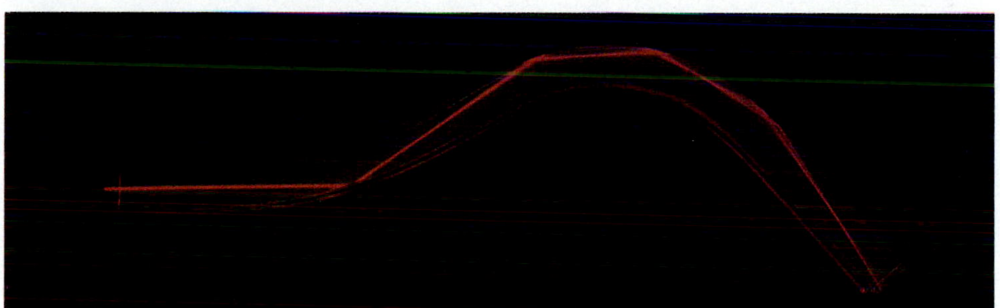

Demonstration of the path of a light beam in an optical fiber. Fiber optics is based on total internal reflection.

35-2 IMAGES FORMED BY REFLECTION

When you look into a plane mirror, as shown in Fig. 35-11, you see a world much like your own. By looking at the point P', your eye collects light that originated from point P on the candle flame; light from any other point on the candle is similarly reflected to your eye. The nonexistent candle flame that you see in the mirror is called an ***image,*** whereas the candle flame itself is called the ***object.*** The law of reflection shows that the line PP' is normal to the plane of the mirror and that P and P' are the same distance from the mirror. The image seen in a plane mirror is

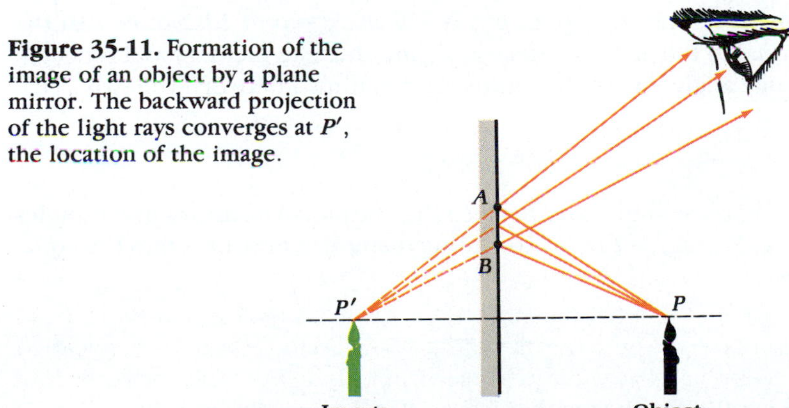

Figure 35-11. Formation of the image of an object by a plane mirror. The backward projection of the light rays converges at P', the location of the image.

always the same distance behind the mirror as the object is in front of the mirror (see Exercise 16).

Curved mirror surfaces also form images. For example, the curved surfaces of fun-house mirrors form images that appear elongated or shortened or otherwise distorted. The easiest curved surface to construct and analyze is a spherical surface. Further, spherical or nearly spherical mirrors are used in optical systems such as telescopes and solar collectors. We now discuss the images they form.

The Mirror Equation

Optic axis

Consider the rays that leave a point object P and travel toward the concave mirror of Fig. 35-12. The line joining the object and the center C of the spherical mirror is called the ***optic axis.*** Rays close to the optic axis, called ***paraxial rays,*** are all reflected close to the same point P', forming an image of P. The rays that are not paraxial blur this image, an effect called ***spherical aberration.*** Practical instruments using spherical mirrors minimize this blurring by allowing only rays that are nearly paraxial to be seen. In the following discussion, we will consider images formed by paraxial rays only.

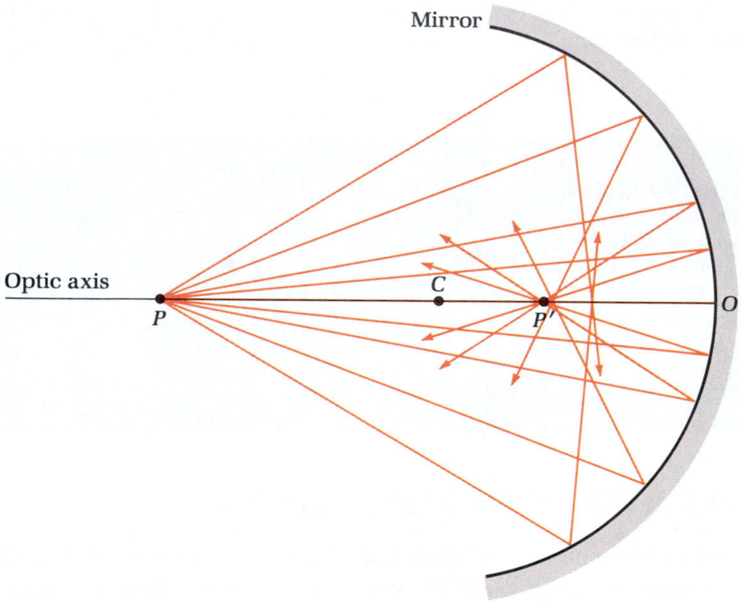

Figure 35-12. Reflection of rays from a concave spherical mirror. Rays from point P are reflected to P' if they are near the optic axis PCO.

Figure 35-13 shows a single ray leaving a point P on the optic axis and, after reflection from the mirror, crossing the optic axis at P'. The distance s is called the object distance, s' is called the image distance, R is the radius of the spherical mirror, and each distance is measured from the mirror ***vertex*** V. From plane geometry, the exterior angle of a triangle is the sum of the two opposite interior angles. For triangles PAC and PAP', this gives $\beta = \alpha + \theta$ and $\gamma = \alpha + 2\theta$. Notice that the radius CA is normal to the mirror surface at A, and thus the incident and reflected rays make the same angle θ with the radius. Eliminating θ between the two equations gives

$$\alpha + \gamma = 2\beta$$

For paraxial rays, all these angles are small, so that the radian measures of the angles are $\alpha \approx \ell/s$, $\beta = \ell/R$, and $\gamma \approx \ell/s'$. Substituting the radian measures into the equation above gives

$$\frac{1}{s} + \frac{1}{s'} = \frac{2}{R} \qquad (35\text{-}3)$$

This expression is valid for any ray that leaves P and passes through P'.

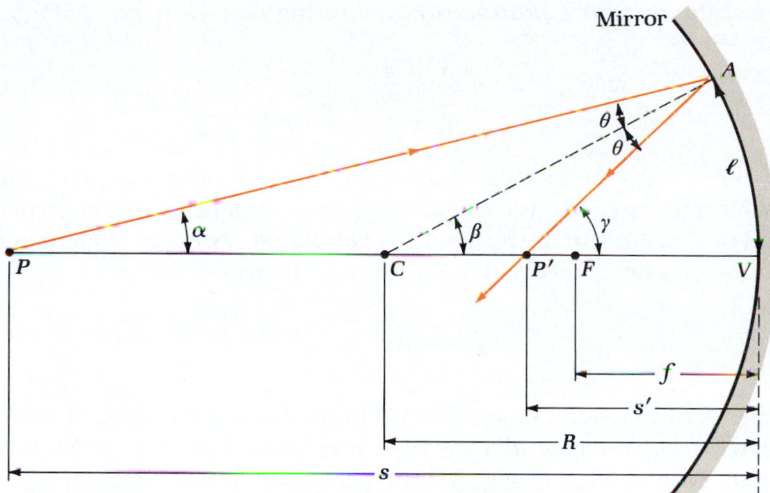

Figure 35-13. The geometry of reflection of paraxial rays by a spherical mirror. A ray leaving the object point P is reflected through the image point P'.

Suppose the point P is very far away from the mirror; that is, $s \gg R$. Substituting $s = \infty$ into Eq. 35-3 gives $1/s = 0$, so that $s' = \frac{1}{2}R$. This means that light from a distant point object crosses the optic axis at a point that is a distance $\frac{1}{2}R$ from the vertex V. This point is labeled F in Fig. 35-13 and is called the ***focal point*** of the mirror. *Any light ray that travels parallel to the optic axis and reflects from the mirror will then pass through the focal point.* Thus light from a distant object converges on the focal point and then diverges from the focal point, as shown in Fig. 35-14. This light comes to a *focus* at the focal point. With its blackened absorber located at the focal point, the solar heater in Fig. 35-15 collects light from the distant sun.

Definition of the focal point

The distance from the vertex to F is called the mirror's ***focal length*** f. For a concave mirror, we have found that

Focal length of a mirror

$$f = \tfrac{1}{2}R$$

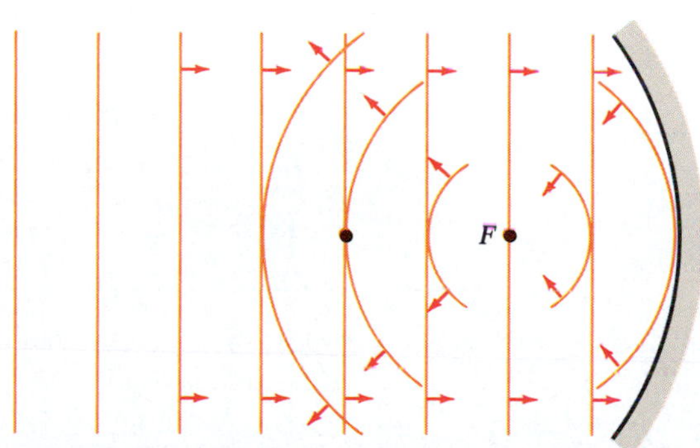

Figure 35-14. Plane wavefronts approaching a spherical mirror are reflected into spherical waves that converge to the focal point F of the mirror and then expand in spherical waves like those that would be given off by a point source at F.

Figure 35-15. An array of solar energy collectors. The parallel rays from the sun are reflected from reflectors and the energy is concentrated at the absorbers, which are located at the focal points.

We arrive at the ***mirror equation*** by substituting f for $\frac{1}{2}R$ in Eq. 35-3:

The mirror equation

$$\frac{1}{s}+\frac{1}{s'}=\frac{1}{f} \tag{35-4}$$

Optical reversibility

Notice that s and s' enter the mirror equation in the same way. The roles of the object and image can be interchanged simply by reversing the direction of the light rays. This is an example of the principle of *optical reversibility,* which states that if the direction of a ray is reversed, it will retrace its path.

Ray Diagrams

Principal rays

The image of an extended object can be located using a graphical construction called a ***ray diagram.*** Consider the light rays leaving point H at the head of the stick person (the object) in Fig. 35-16. The ***principal rays*** w, x, y, and z are shown. Ray w is parallel to the optic axis as it leaves H. After reflection from the concave mirror, it passes through the focal point F. Ray x strikes the vertex of the mirror, and its angle with the optic axis is the same after reflection as before. Ray y passes through the focal point F, and after reflection, it is parallel to the optic axis. Ray z passes through the center of the sphere and is reflected back on itself. All these rays, and all other rays from H that strike the mirror, converge on H' and then diverge from H'. Thus an image of H is formed at H'. Similarly, rays from any point on the object that strike the mirror are reflected and form a corresponding image point. An image of each point on the object from H to T is formed at H' to T'. Note that any two of the principal rays can be used to locate the image on a ray diagram. Also, the location of the image can be determined by solving the mirror equation for s':

$$s'=\frac{sf}{s-f}$$

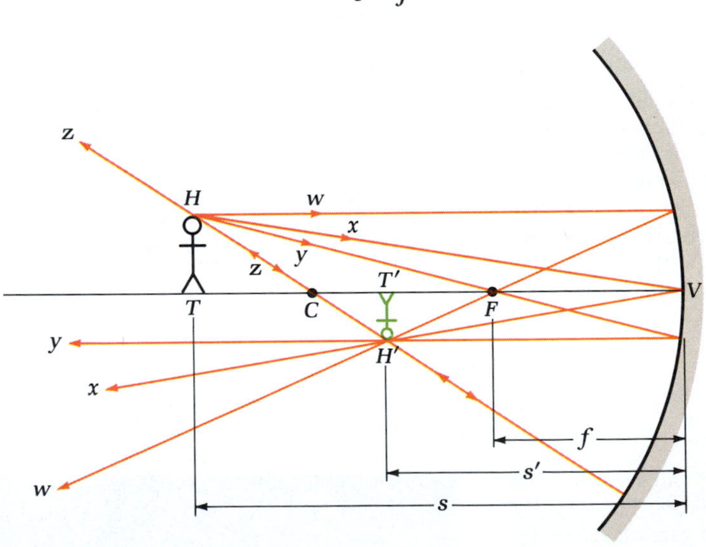

Figure 35-16. The principal rays for a mirror. Ray w is parallel to the optic axis CV, ray x hits the vertex V, ray y passes through the focal point F, and ray z passes through the center of curvature of the mirror C.

Erect and inverted images

An image produced by a spherical mirror may be right side up or upside down. An image that is right side up is called an *erect image,* and an image that is upside down is called an *inverted image.* A concave mirror can produce either an erect image or an inverted image, depending on the location of the object. When the object distance s is greater than the focal length f, as in Fig. 35-16, the image is inverted. The ray diagram in Fig. 35-17 shows that when s is less than f, then a concave mirror produces an erect image.

In Fig. 35-16, the image is formed by rays that converge after reflection from the mirror and then pass through the image. On the other hand, the image in Fig. 35-17

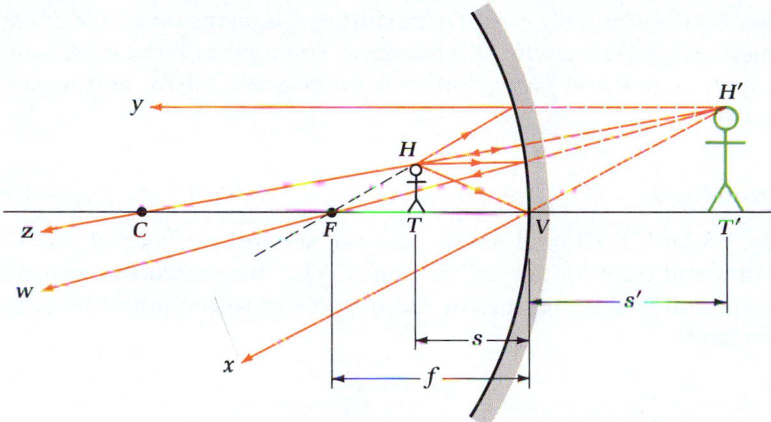

Figure 35-17. Reflection from a spherical mirror when the object is between the focal point and the mirror. Projecting the rays backward shows that they appear to be diverging from a virtual image located behind the mirror. By similar triangles, $FV/HT = (FV + VT')/H'T'$ and $VT/HT = VT'/H'T'$. After some algebra one gets $1/FV = (1/VT) - 1/VT')$ or $1/f = (1/s) + (1/s')$.

is formed by rays that diverge after reflection from the mirror. The location of the image in Fig. 35-17 is constructed by the extension of rays backward, behind the mirror, to a region where no rays exist. The image in Fig. 35-16 is called a ***real image,*** whereas the image in Fig. 35-17 is called a ***virtual image.*** A real image can be projected onto a viewing screen, whereas a virtual image cannot. If a sheet of white paper is placed at $H'T'$ in Fig. 35-16, a sharply focused image of the stick person is seen. In the case of the virtual image in Fig. 35-17, $H'T'$ is behind the mirror where no light exists, and a screen placed there shows no image.

Real and virtual images

Sign Convention

We developed the mirror equation by considering the case where $s > f$ (Fig. 35-13). This corresponds to the case in Fig. 35-16 where a real image is produced. Is this equation also valid for the case of the virtual image shown in Fig. 35-17? Yes, provided that you adhere to a sign convention (see Prob. 4). Also, this sign convention makes the mirror equation valid for a convex mirror (Fig. 35-18), as well as for a concave mirror. Indeed, as you shall see, the mirror equation also describes the behavior of a lens. (When applied to a lens, the equation is called the *lens equation.*) The sign convention applies to both mirrors and lenses; therefore some of its features will not become clear until you read the discussion of lenses. By the sign convention,

Sign convention

1. An object distance s is positive when the object is on the same side of the surface as the incoming light. Otherwise, the object distance is negative. When s is positive, the object is called a *real object,* and when s is negative, the object is called a *virtual object.*

2. An image distance s' is positive when the image is on the same side of the surface as the outgoing light. Otherwise, the image distance is negative. This means that s' is positive for a real image and negative for a virtual image.

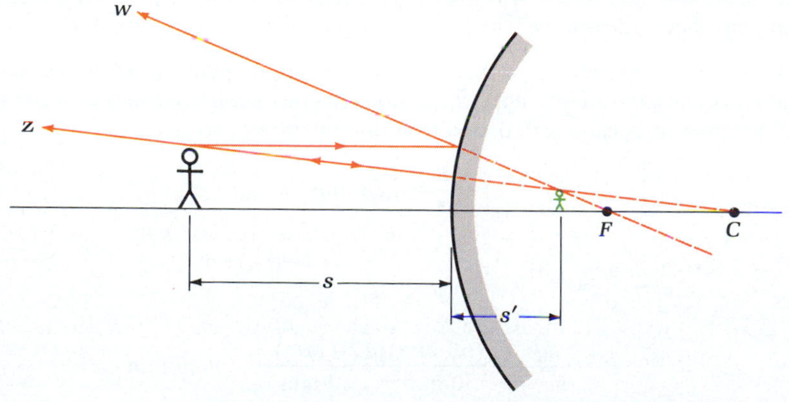

Figure 35-18. Reflection from a convex spherical mirror. Two principal rays define the image position. Ray w is parallel to the optic axis before reflection and diverges from F after reflection. Ray z is aimed at the center of the sphere and reflects back on itself. The image is found at the intersection of these rays. Problem 3 asks you to show that the mirror equation $1/f = 1/s + 1/s'$ works for this case if f is taken to be negative, in agreement with our convention.

3. A radius R is positive if the center of curvature C is on the same side of the surface as the outgoing light. Otherwise, R is negative. The sign of f is the same as the sign of R. This means that R and f are positive for a concave mirror and negative for a convex mirror.

Magnification

From Figs. 35-16, 35-17, and 35-18, you can see that, in general, the size of an image is different from the size of the object. The ***magnification*** m is defined so that its magnitude $|m|$ is the ratio of the image size to the object size. From Fig. 35-19, we have

$$|m| = \frac{H'T'}{HT}$$

If $|m| > 1$, the image is said to be magnified; if $|m| < 1$, the image is said to be diminished.

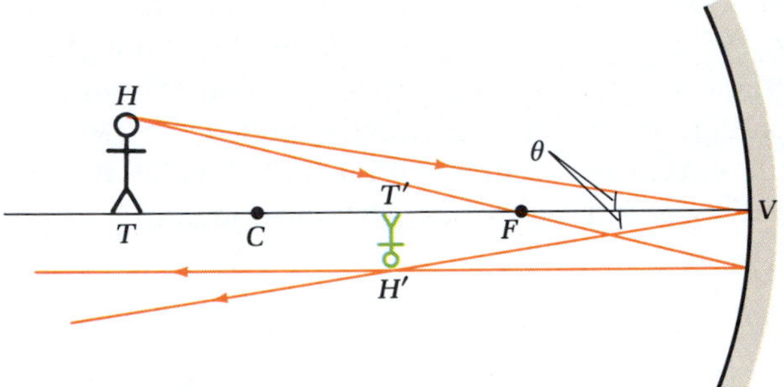

Figure 35-19. Geometry of image formation by a concave spherical mirror. Since triangles HTV and $H'T'V$ are similar, $m = -H'T'/HT = -VT'/VT = -s'/s$, where the negative sign indicates that the image is inverted.

Demonstration of the image produced by a concave mirror. (This mirror is cylindrical, not spherical.) For each object distance greater than the focal length, the image is real, inverted, and diminished. For each object distance less than the focal length, the image is virtual, erect, and magnified.

Since the triangles HTV and $H'T'V$ in Fig. 35-19 are similar, we have that $H'T'/HT = s'/s$. The values of s' and s are usually obtained more readily than $H'T'$ and HT, so we generally use these values to express the magnification. Also, the sign of m should tell us whether the image is erect or inverted. Thus the formula for magnification is written as

$$m = -\frac{s'}{s} \tag{35-5}$$

With the inclusion of the minus sign in this expression, m is positive when the image is erect and negative when it is inverted. Thus the magnitude of m characterizes the size of the image relative to the object, and its sign shows whether the image is erect or inverted.

EXAMPLE 35-2

A shaving mirror. A concave shaving mirror has a 240-mm radius. If you look into it from a distance of 60 mm, *(a)* where will your face appear to be and *(b)* how big will a 5-mm feature on your face appear to be?

Solution. Figure 35-20 shows a ray diagram of this example. (You are advised to sketch the ray diagram for each exercise you do.) Dimensions can be measured from a well-constructed sketch or, for more precision, calculated from the mirror equation:

(a)

$$f = \frac{R}{2} = 120 \text{ mm}$$

and

$$\frac{1}{s} + \frac{1}{s'} = \frac{1}{f}$$

$$s' = \frac{sf}{s-f} = \frac{(60 \text{ mm})(120 \text{ mm})}{60 \text{ mm} - 120 \text{ mm}} = -120 \text{ mm}$$

Demonstration of the image produced by a convex mirror. For any object distance, the image is virtual, erect, and diminished.

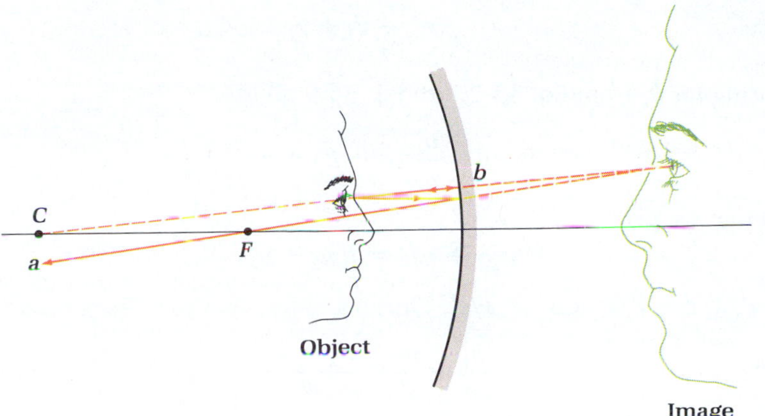

Figure 35-20. Example 35-2: Principal rays that pass through the focus and through the center locate the image.

The negative sign indicates that the image is 120 mm behind the mirror, and thus virtual.

(b) Using Eq. 35-5, we have

$$m = -\frac{s'}{s} = -\frac{(-120\text{ mm})}{(60\text{ mm})} = 2$$

Since m is positive, the image is erect. This means you see yourself right side up. The feature will appear to be twice as large as it is on your face, or 10 mm.

SELF-TEST 35-2. In Fig. 35-16, what are the signs of *(a)* f and *(b)* s'? In Fig. 35-17, what are the signs of *(c)* f and *(d)* s'? In Fig. 35-18, what are the signs of *(e)* f and *(f)* s'?
ANSWERS: *(a)* +; *(b)* +; *(c)* +; *(d)* −; *(e)* −; *(f)* −.

35-3 IMAGES FORMED BY REFRACTION

In preparation for the study of thin lenses, we first look at refraction at a single spherical interface. Consider the two rays shown leaving point P in Fig. 35-21. The ray propagating toward A will be refracted at the surface and meet the ray propagating along the optic axis at point P'. The ray propagating along the optic axis hits the interface normally and hence is not bent. An object at point P thus has its image at P'. If the rays are paraxial, then the angles α, β, γ, θ_1, and θ_2 are all small. From Snell's law, $n_1 \sin\theta_1 = n_2 \sin\theta_2$, or since the angles are small,

$$n_1\theta_1 = n_2\theta_2 \tag{35-6}$$

From plane geometry, the exterior angle of a triangle is equal to the sum of the two opposite angles. Thus in triangle PAC,

$$\theta_1 = \alpha + \beta \tag{35-7}$$

and in triangle $P'AC$,

$$\beta = \theta_2 + \gamma \tag{35-8}$$

The angles θ_1 and θ_2 can now be eliminated between these equations. Substituting for θ_2 from Eq. 35-6 into Eq. 35-8 gives

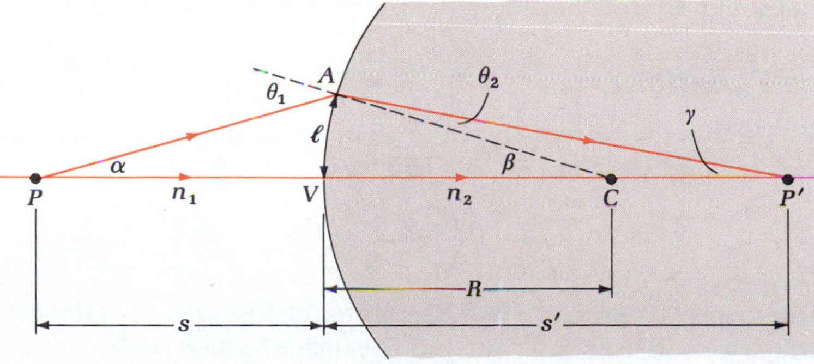

Figure 35-21. Refraction at a spherical surface.

$$\beta = \frac{n_1}{n_2}\theta_1 + \gamma \tag{35-9}$$

Substituting for θ_1 from Eq. 35-7 into Eq. 35-9 gives

$$\beta = \frac{n_1}{n_2}(\alpha + \beta) + \gamma$$

Simplifying, we get

$$n_1\alpha + n_2\gamma = \beta(n_2 - n_1)$$

But $\beta = \ell/R$, $\alpha \approx \ell/s$, and $\gamma \approx \ell/s'$. Thus

$$\frac{n_1}{s} + \frac{n_2}{s'} = \frac{n_2 - n_1}{R} \tag{35-10}$$

Sign convention

The sign convention stated in the previous section is applicable to Eq. 35-10: *A radius* R *is positive if the center of curvature* C *is on the same side of the surface as the outgoing light.* Thus, for a refracting surface, the radius R is positive if the surface is convex toward the object (as in Fig. 35-21), whereas R is negative if the surface is concave toward the object (see Prob. 4).

EXAMPLE 35-3

A fish in water. A fish viewed from directly above appears to be 1.5 m deep in the water. The index of refraction of air is 1.000 and of water is 1.333. What is the actual depth of the fish?

Solution. First draw Fig. 35-22. Notice that the image is on the incoming side of the surface. This means that the image distance is negative, so $s' = -1.5$ m. Since the surface is flat, the radius of curvature is infinite. Thus Eq. 35-10 becomes

$$\frac{1.333}{s} + \frac{1.000}{-1.5\text{ m}} = \frac{1.000 - 1.333}{\infty} = 0$$

$$s = (1.5\text{ m})\frac{1.333}{1.000} = 2.0\text{ m}$$

Thus the fish is deeper than it appears to be. Note that the image of the fish is formed by an extrapolation of the light rays; it is a virtual image.

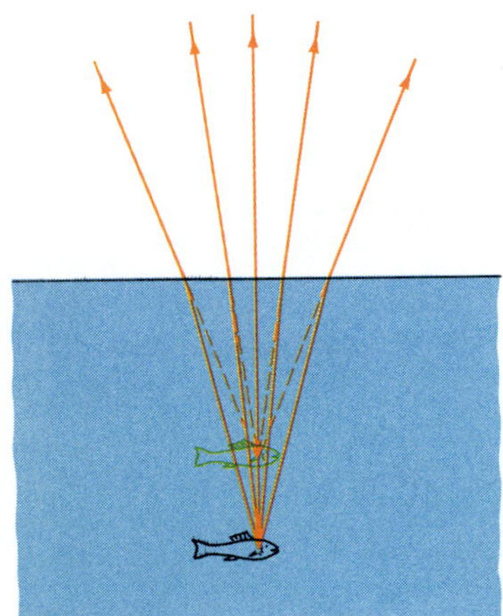

Figure 35-22. Example 35-3.

SELF-TEST 35-3. Suppose a fish is 1.00 m deep in water. How deep does it appear to be when viewed from above? ***ANSWER:*** 0.75 m.

35-4 LENSES

A piece of glass can be ground and polished to a smooth surface by rubbing it with a series of grits that have smaller and smaller particles. The nature of the polishing process makes spherical surfaces particularly easy to grind. When a slab of glass has a spherical surface ground on one or both sides, it forms a *lens.* "Burning glasses" were known to the Greeks and Romans, and lenses reappeared in Europe by the 1300s. The use of these lenses in spectacles became common after the invention of printing, although an understanding of how they worked came much later.

The Lens Equation

Figure 35-23 shows the path of rays from object P, through a lens, to image P'. At the first surface, application of Eq. 35-10 gives

$$\frac{n_1}{s_a} + \frac{n_2}{s'_a} = \frac{n_2 - n_1}{R_a} \tag{35-11}$$

Here the subscripts "*a*" on s_a, s'_a, and R_a refer to the first surface of the lens. For the second surface we will use subscript "*b*." The image formed by the first surface acts

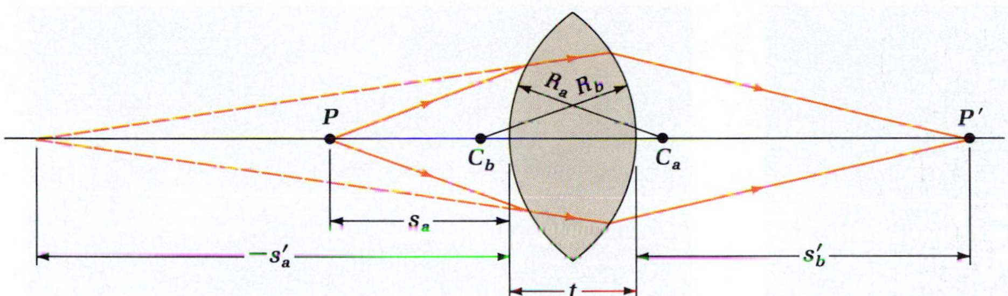

Figure 35-23. Path of light rays from an object at P through a lens to the image at P'.

as the object for the second surface of the lens: $s_b = -s'_a + t$, where t is the lens thickness. The negative sign comes from the convention we have adopted in which virtual objects on the outgoing-light side of a surface have negative object distances.

In the thin-lens approximation, the thickness t of the lens is small compared with the object and image distances. In this approximation, $s_b = -s'_a$. Applying Eq. 35-10 at this surface gives

$$\frac{n_2}{s_b} + \frac{n_1}{s'_b} = \frac{n_2}{-s'_a} + \frac{n_1}{s'_b} = \frac{n_1 - n_2}{R_b} = \frac{n_2 - n_1}{-R_b}$$

Using Eq. 35-11 and rearranging, we get

$$\frac{n_1}{s_a} + \frac{n_1}{s'_b} = (n_2 - n_1)\left(\frac{1}{R_a} - \frac{1}{R_b}\right) \qquad (35\text{-}12)$$

At this point, it is convenient to rename things. Considering the lens as a single entity, (i) let the object distance for the lens as a whole be $s = s_a$, (ii) let the image distance for the lens as a whole be $s' = s_b$, and (iii) let the focal length of the lens as a whole be f. The focal length f of the lens is defined to be equal to s' as $s \to \infty$. Substitution in Eq. 35-12 gives

$$\frac{n_1}{f} = (n_2 - n_1)\left(\frac{1}{R_a} - \frac{1}{R_b}\right) \qquad (35\text{-}13)$$

Lens-makers' equation

Equation (35-13) is known as the ***lens-makers' equation.*** It gives a prescription for making a lens with a given focal length. With these notation changes, Eq. 35-12

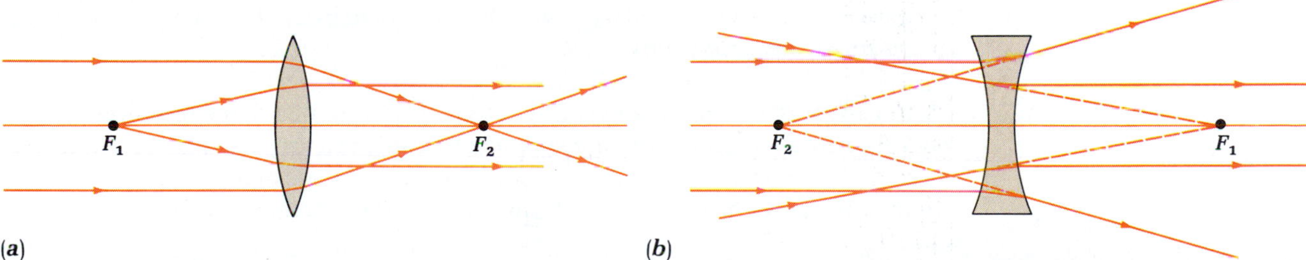

Figure 35-24. *(a)* Focal points of a converging lens. Rays diverging from the first focal point F_1 leave the lens as parallel rays. Parallel rays incident on the lens converge toward the second focal point F_2. *(b)* Focal points of a diverging lens. Rays converging toward the first focal point F_1 leave the lens as parallel rays. Parallel rays incident on the lens diverge from the second focal point F_2.

becomes the ***lens equation,***

$$\frac{1}{s} + \frac{1}{s'} = \frac{1}{f} \qquad (35\text{-}14)$$

Since the derivation of Eq. 35-14 used Eq. 35-10 for refraction at each surface, it is valid for objects and images on either side of the lens if the sign convention stated in Sec. 35-2 is followed.

A mirror has one focal point, whereas a lens has two, as shown in Fig. 35-24. The second focal point F_2 is the position where parallel light incident on the lens is focused. The first focal point F_1 is the position where an object produces an image at infinity. The focal points of a thin lens lie on opposite sides of the lens, each a

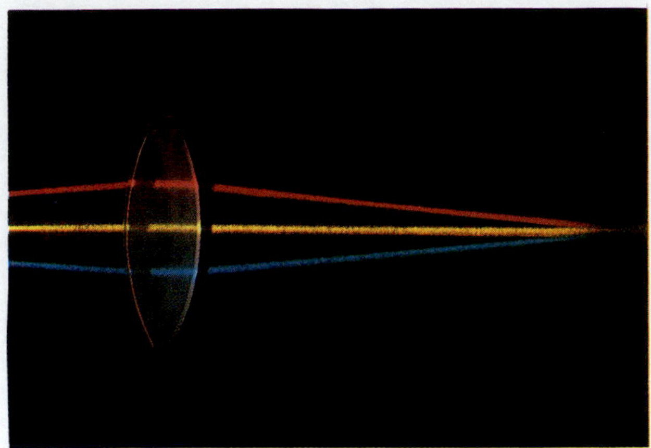

A converging, or positive, lens.

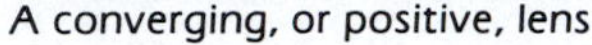

A diverging, or negative, lens.

distance f from the center of the lens. The focal points lie on the ***optic axis*** of the lens, which is the line that passes through the centers of curvature of the two surfaces of the lens.

Positive and Negative Lenses

The proper sign is given to f in Eq. 35-13 when we follow the sign convention for R. For a refractive surface, R is positive when the surface toward the object is convex, and it is negative when the surface toward the object is concave. We now apply the sign convention to the lenses shown in Fig. 35-24.

Positive lens

1. For an ordinary glass lens in air $n_2 > n_1$. A double convex lens (Fig. 35-24*a*) bends the ray toward the optic axis at each surface, so that the rays tend to converge. Equation 35-13 shows that the focal length f of a double convex glass lens in air is positive because R_a is positive, R_b is negative, and $n_2 > n_1$. A lens with a positive focal length is called a *converging,* or *positive, lens.*

Negative lens

2. Figure 35-24*b* shows the path of rays through a double concave lens. Both surfaces of the lens bend the rays away from the optic axis, so that the rays tend to diverge. The focal length of a double concave glass lens in air is negative because R_a is negative, R_b is positive, and $n_2 > n_1$. A lens with a negative focal length is called a *diverging,* or *negative, lens.*

You can measure f for a converging lens by standing in a darkened room near a wall opposite a window. Hold the lens near the wall and vary the distance until you can see a clear image projected on the wall. This is an image of the scene outside the window. Because the object is far from the lens, the distance from the lens to the wall is f.

Ray tracing and principal rays

Images formed by thin lenses can be found by tracing principal rays through the lens. The three principal rays of a lens are those that *(a)* pass through a focus of the lens, *(b)* pass through the center of the lens, or *(c)* are parallel to the optic axis of the lens. Figure 35-25 shows the simple paths these rays take for a positive lens.

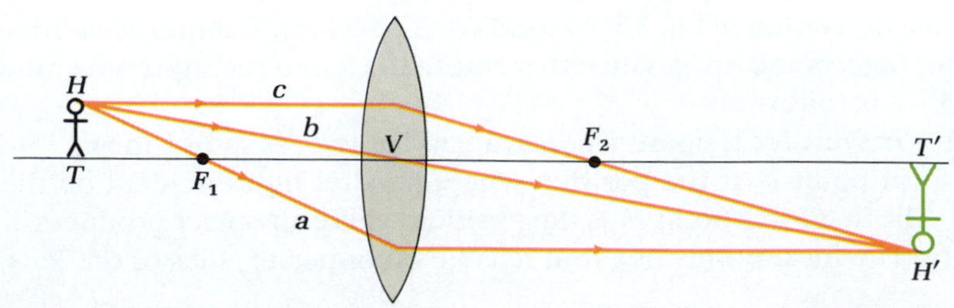

Figure 35-25. The principal rays of a lens. Ray *a* passes through the first focal point F_1 of the lens and is refracted by the lens parallel to the axis. Ray *b* passes through the center of the lens and is undeviated. Ray *c* enters the lens parallel to the optic axis and is refracted through the second focal point F_2.

Tracing the paths of these rays will locate the image position of an extended object and show whether the image is erect or inverted.

The magnification m can be determined using the principal ray that passes through the center of the lens, as shown in Fig. 35-25. The similar triangles HTV and $H'T'V$ show that $m = -H'T'/HT = T'V/TV$. But TV is the object distance s and $T'V$ is the image distance s'. Thus

$$m = -\frac{s'}{s} \tag{35-15}$$

Magnification of a lens

If $|m| > 1$, the image is magnified, but if $|m| < 1$, the image is diminished. If m is positive, the image is erect, but if m is negative, the image is inverted.

EXAMPLE 35-4

A positive lens. *(a)* Find the focal length of the planoconvex lens shown in Fig. 35-26. The spherical surface has a radius of curvature of 57.1 mm, and the index of refraction of the glass is 1.523. The lens is in air. An object is placed on the optic axis 50 mm in front of this lens. *(b)* Where does the lens form an image of this object? *(c)* Is the image real or virtual?

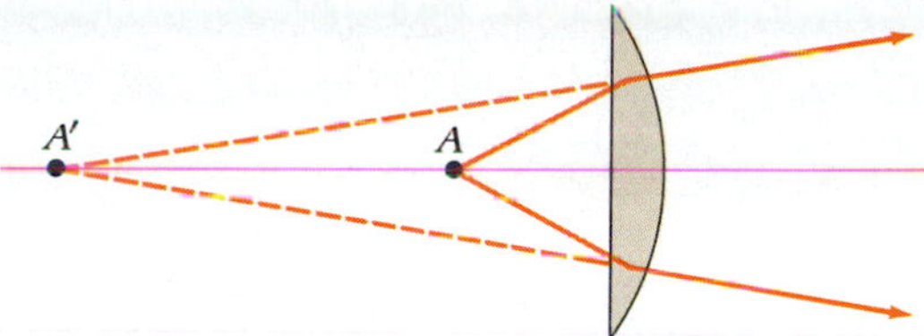

Figure 35-26. Example 35-4.

Solution. *(a)* The focal length is given by the lens-makers' equation:

$$\frac{1}{f} = (1.523 - 1)\left(0 - \frac{1}{-57.1 \text{ mm}}\right)$$

$$f = 109 \text{ mm}$$

Since the center of curvature of the second surface of the lens is not on the outgoing-light side of the interface, R_b is negative.

(b) The image distance is given by Eq. 35-14:

$$\frac{1}{50 \text{ mm}} + \frac{1}{s'} = \frac{1}{109 \text{ mm}}$$

$$s' = -150 \text{ mm}$$

(c) The negative sign indicates that the image is on the incoming-light side of the lens, and is thus a virtual image.

SELF-TEST 35-4. An object is placed on the optic axis of the lens in the above example at a distance of 185 mm from the lens. *(a)* Determine the image distance s'. *(b)* Is the image real or virtual? ***ANSWERS:*** *(a)* 265 mm; *(b)* real.

EXAMPLE 35-5

Ray diagrams. Use the ray-tracing method to find the image of a 7.5-mm-high object 35 mm in front of a lens of focal length *(a)* 20 mm and *(b)* −20 mm.

Solution. *(a)* The drawing is shown in Fig. 35-27*a*. Ray *a* passes through the first focal point of the lens and comes out on the other side parallel to the optic axis. Ray *b* passes through the center of the lens and is undeviated. Ray *c* is parallel to the optic axis and comes out on the other side along a line through the second focal point of the lens. Since light passes through the position of the image, the image is real. By a scaled measurement on the diagram, $s' = 46.5$ mm and the image is 10 mm high. For more precision, the algebraic solutions can be used.

(b) The drawing is shown in Fig. 35-27*b*. Ray *a* is aimed at F_1 and leaves the lens parallel to the optic axis. Ray *b* goes straight through the vertex of the lens. Ray *c* is parallel to the optic axis so that its path on the other side of the lens is directly away from F_2. The image is

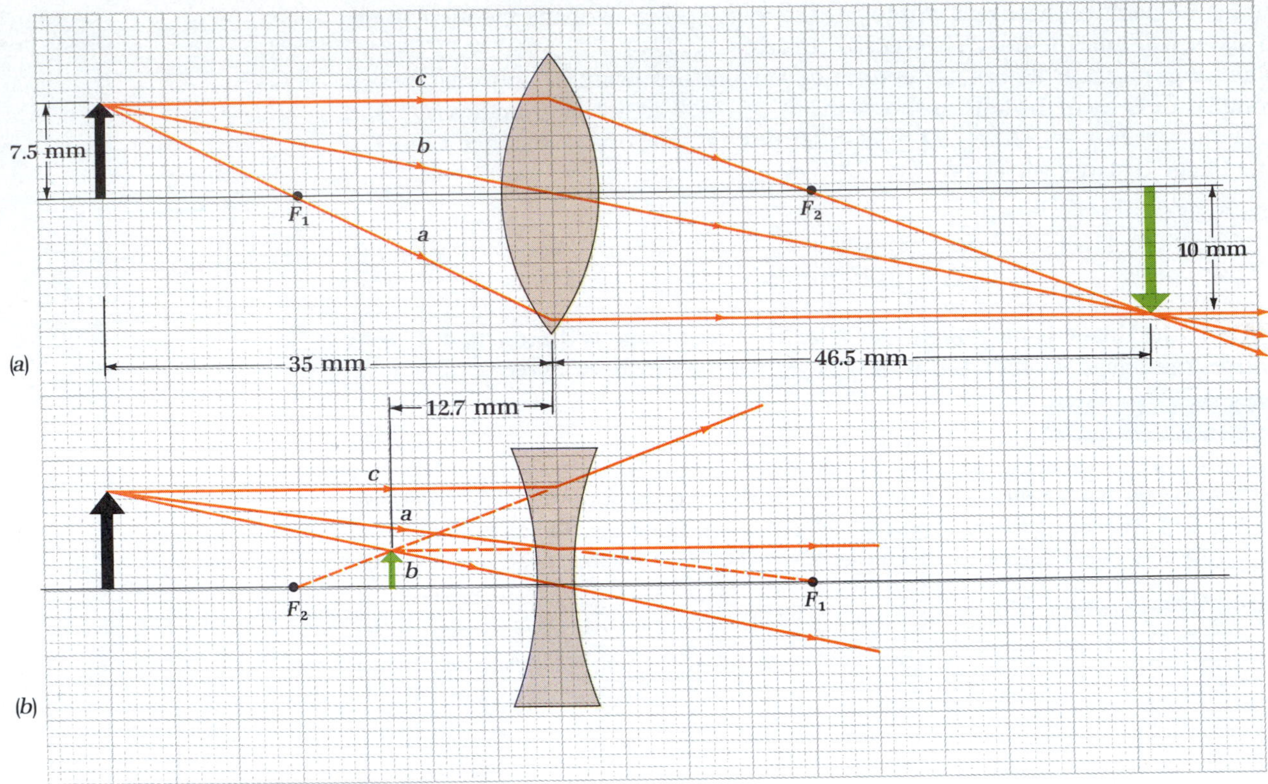

Figure 35-27. Example 35-5.

12.7 mm in front of the lens and 2.8 mm high. Since the image is on the same side of the lens with incoming light, the image is virtual.

SELF-TEST 35-5. For the positive lens in the above example, determine *(a)* the image distance s' and *(b)* the magnification m. For the negative lens in the above example, determine *(c)* the image distance and *(d)* the magnification. ***ANSWERS:*** *(a)* 47 mm; *(b)* -1.3; *(c)* -13 mm; *(d)* 0.36.

35-5 OPTICAL DEVICES

We have developed the theory of geometric optics to the point where it is possible to discuss the principles of some practical optical systems. These systems have extended the range of our senses to the very small and the very far away, making possible the sciences of microbiology and astronomy.

There are details in the design of optical instruments that we do not have space to discuss. In particular, we will not discuss the blurring of the image that occurs in finite-size lenses, the difficulties in forming an image of three-dimensional objects, or the techniques for minimizing the variation of the focal length with wavelength in lenses. The imperfections in the image caused by these effects are called ***aberrations.*** The design of lens combinations that minimize aberrations is a difficult procedure, one that does not need to be considered in the principles of these optical devices. One can build a serviceable telescope using the paraxial theory.

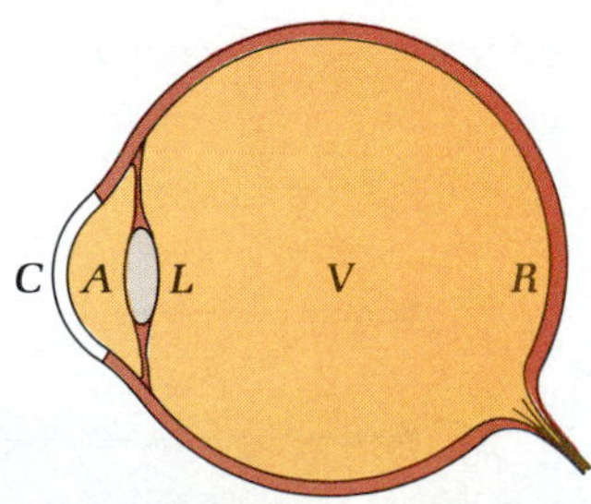

Figure 35-28. Schematic diagram of the eye. *C* is the cornea, *L* the lens, *A* the aqueous humor, *V* the vitreous humor, and *R* the retina.

The Eye

Many optical instruments are aids to the eye. Since the eye is part of the process, as well as an optical instrument itself, we discuss it first. The parts of the eye essential for our discussion are shown in Fig. 35-28. At the front of the eye is the sharply curved *cornea C*. Behind the cornea is a space filled with a liquid called the *aqueous humor A* ($n = 1.336$). Most of the refraction of incoming light occurs as light passes from air into this liquid. Next comes the lens *L*, which is made of material of average index $n = 1.396$. (The index is different at the center than at the edges.) The curvature of the lens can be adjusted by muscles attached to it in order to form images of objects at different distances. Behind the lens is a fluid similar to the aqueous humor called the *vitreous humor V*, and then the *retina R*. The retina contains photosensitive cells, which produce nerve signals when light hits them, and nerve cells that process these signals and send them to the brain for interpretation.

How can you best distinguish the details of a small object? The detail that you can distinguish in an object depends on the size of the object's image on your retina. In the unaided eye, the image size is determined by the angle subtended by the object at the eye. To enlarge the apparent size of the object, you bring it closer to your eye, making the angle that it subtends larger. Eventually you will reach a point where your eye can no longer comfortably focus the image on your retina. The closest distance at which you can comfortably bring an object into focus is called the *near point*. You can find your near point by bringing this text as close to your eyes as you can while still seeing a sharp image of the print.

Near point

When the muscles attached to the lens are relaxed, the normal eye is focused at infinity. If you tighten the muscles attached to the lens, in a process called *accommodation,* your eye can focus on objects at different distances. An infant can form clear images of objects as close as 70 mm. The lens of the average college student can accommodate objects as close as 100 mm. A 60-year-old professor may be able to accommodate only to 2 m, and be forced to wear glasses to focus on objects closer than 2 m. We will use an accommodation distance of 250 mm as an average near point.

The Simple Magnifier

If you place a converging lens in front of your eye, as shown in Fig. 35-29, it effectively increases the accommodation of your eye and decreases your near point. Your eye looks at a virtual object (not shown in figure) somewhere between the near point and infinity. The virtual object subtends a larger angle at your eye than the object would subtend at the near point. Thus the lens is called a *simple magnifier.*

To calculate the magnification, assume that you place the converging lens so that your eye is focused at infinity. You can then clearly see the object when it is placed just inside the focal point of the lens, a distance f from the lens. As shown in Fig. 35-29, the object subtends an angle θ_u when you see it with your unaided eye, but a larger angle θ_m when you see it through a simple magnifier. The *angular magnification M* is defined as the ratio of the angles subtended with and without the magnifier. Thus $M \equiv \theta_m/\theta_u = (h/f)/(h/250 \text{ mm})$, or

$$M = \frac{250 \text{ mm}}{f} \qquad (35\text{-}16)$$

Magnification of a simple magnifier

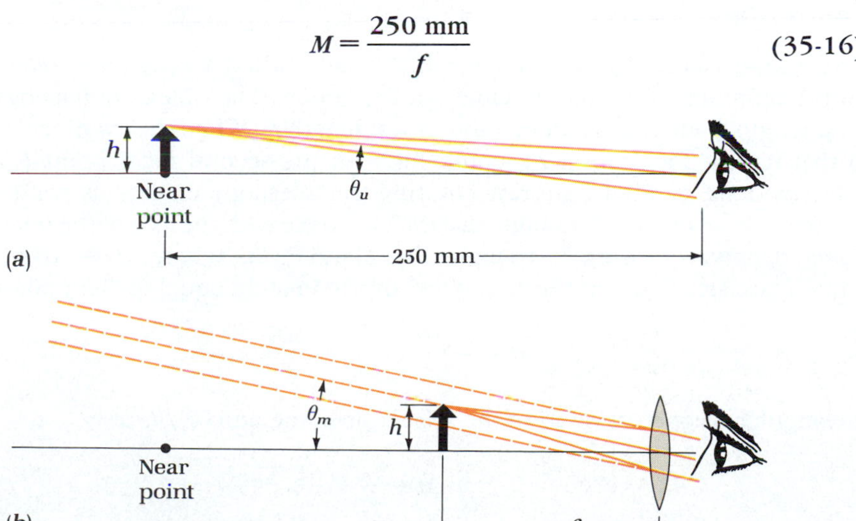

Figure 35-29. The simple magnifier. *(a)* An object of height h subtends an angle $\theta_u = h/250$ mm when seen by the eye at the near point of 250 mm. *(b)* The virtual image of this object when seen through a converging lens of focal length f subtends an angle $\theta_m = h/f$. Here the image is at infinity, so that $s = f$ and $s' = -\infty$.

Because of this relation, the ***power*** P of a lens is defined to be the inverse of its focal length f: $P = 1/f$. The power of a lens is measured in ***diopters*** (abbreviated D) with 1 D equal to 1 m^{-1}.

As the object is moved closer to the lens, the virtual image becomes closer to your eye. Exercise 42 asks you to show that if the lens is placed so that the virtual image is at 250 mm ($s' = -250$ mm, the average near point) then the angular magnification is

$$M = 1 + \frac{250 \text{ mm}}{f}$$

Simple magnifiers are generally used with the image at infinity because the normal eye is most comfortable when focused at infinity.

Notice that large magnifications M require small focal lengths f. Small focal lengths in turn require small radii of curvature of the lens. The small radius of curvature makes the lens thick and causes aberrations that blur the image. Simple single-lens magnifiers are limited by these aberrations to a magnification M of about 2.5, conventionally denoted as 2.5X. Simple magnifiers using more than one lens can decrease aberrations enough to reach a useful magnification of 15X.

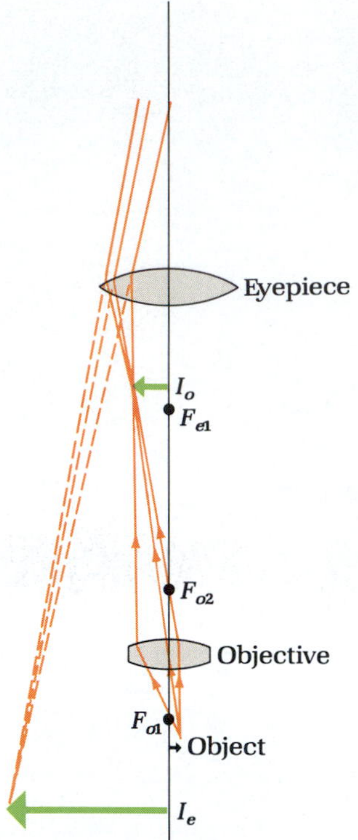

Figure 35-30. The compound microscope. An object just beyond the focal point F_{o1} of the objective lens is imaged by the objective lens at I_o. This real image forms the object of the eyepiece, which in turn forms an image at I_e.

The Compound Microscope

When higher angular magnifications are necessary, a compound microscope can be used. Invented around 1600 in Holland, it consists, in elemental form, of two lenses. The ***objective lens*** is placed so that the object to be examined is just beyond its first focal point, as shown in Fig. 35-30. An enlarged, real image I_o is formed beyond the second focal point of the objective lens. This image is viewed by a second lens, the ***eyepiece***. The eyepiece acts as a simple magnifier, producing a virtual image I_e of its object I_o.

The magnification m of the objective lens is given by Eq. 35-15: $m_o = -s'_o/s_o$. In this case the object is approximately at the focal point F_o of the objective lens so that $s_o \approx f_o$. The distance from the objective lens to its image is called the ***tube length T*** and is often set to 160 mm. In this case $m_o = -160 \text{ mm}/f_o$. The angular magnification M_e of the eyepiece using the simple magnifier relation is 250 mm$/f_e$, where f_e is the focal length of the eyepiece. The overall magnification M is thus

$$M = |M_e m_o| = \frac{(250 \text{ mm})(160 \text{ mm})}{f_e f_o} \tag{35-17}$$

Limitations due to the wave nature of light limit useful magnification to about 1000X when using visible light. These limitations will be discussed in Chap. 37.

Telescopes

A telescope whose primary element is a lens is called a ***refracting telescope***. A simple refracting telescope is shown in Fig. 35-31. The objective lens has a long focal length f_o and the eyepiece a short focal length f_e. When the eyepiece is placed so that its first focal point F_{e1} coincides with the second focal point F_{o2} of the objective lens, parallel light rays entering the telescope emerge as parallel light rays from the eyepiece. The angle that the rays make with the axis of the telescope is changed, however. Using the triangles $F_{o1}CD$ and $F_{e2}EG$, where $CD = AB = EG$ and θ and θ' are small so that the angles are approximately equal to their sines,

$$\theta = \frac{-AB}{f_o} \qquad \theta' = \frac{AB}{f_e}$$

The angular magnification M of the telescope is the ratio θ'/θ, or

Magnification of a simple refracting telescope

$$M = \frac{f_o}{f_e} \tag{35-18}$$

Figure 35-31. A simple keplerian refracting telescope. Parallel light rays from an object at infinity are focused by the objective lens to form an image at F_{o2}. When the eyepiece is adjusted so that its focal point F_{e2} coincides with F_{o2}, a virtual image is formed at infinity by the eyepiece.

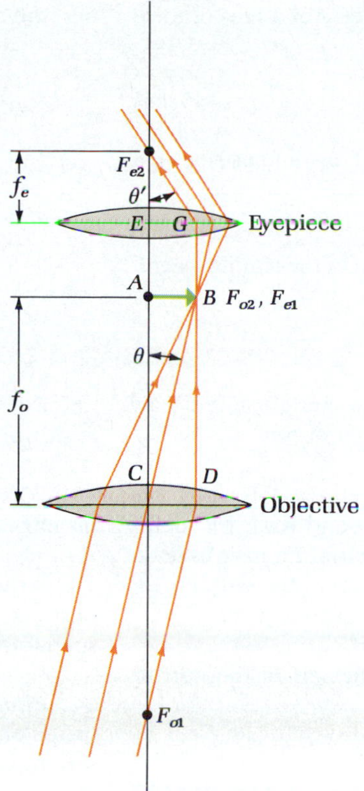

Galileo constructed a telescope with $M = 3$ as soon as he heard that someone had built a telescope using two lenses. His later telescopes had magnifications of around 30X. It is interesting to note that the Dutch government tried to keep the knowledge of the design of telescopes secret, but Galileo built a working model a very short time after he heard of the existence of the telescope. Historically it has often proved futile to keep the details of militarily and commercially important devices secret after the mere existence of the device is known.

The difficulties in making a large lens with acceptable aberrations limit the size of refracting telescopes. Large mirrors that are sufficiently free of aberrations are easier to make. A *reflecting telescope* has a mirror as its primary element. Figure 35-32 illustrates the path of the rays in a newtonian reflecting telescope.

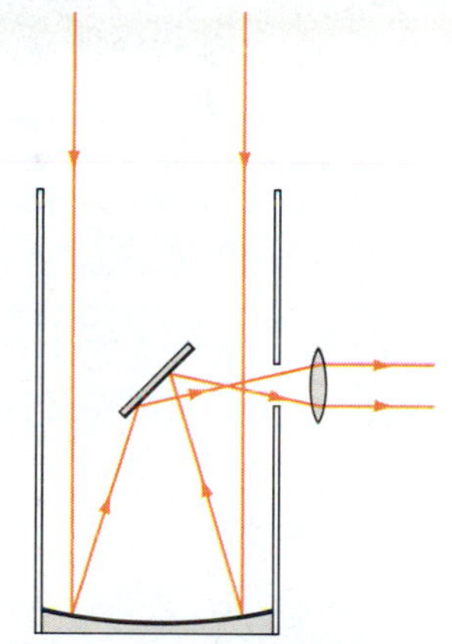

Figure 35-32. A newtonian reflecting telescope.

SUMMARY

Section 35-1. Geometrical Optics

Geometrical optics is an approximation to the results of Maxwell's equations that is valid when the dimensions of the system are much greater than the wavelength of the light used. The laws of geometrical optics are

1. Light travels in straight lines in a homogeneous medium.
2. The angle of reflection is equal to the angle of incidence.
3. Snell's law gives the angle of refraction:

$$n_1 \sin\theta_1 = n_2 \sin\theta_2 \tag{35-1}$$

When a light ray propagates from a medium of higher index n_1 to a medium of lower index n_2, there is a critical value of the angle of incidence $\theta_c = \sin^{-1}(n_2/n_1)$ above which a ray suffers total internal reflection.

Section 35-2. Images Formed by Reflection

The object distance s and the image distance s' of a spherical mirror are related by

$$\frac{1}{s} + \frac{1}{s'} = \frac{1}{f} \tag{35-4}$$

If the image has light rays that pass through it, it is called a real image. If the image is formed by the backward projection of light rays, it is called a virtual image. The magnification of a mirror (or lens) is given by $m = -s'/s$.

Section 35-3. Images Formed by Refraction

The image and object distances s and s' for a refracting surface of radius R between media of index n_1 and n_2 are related by

$$\frac{n_1}{s} + \frac{n_2}{s'} = \frac{n_2 - n_1}{R} \tag{35-10}$$

Section 35-4. Lenses

The focal length of a lens is given by the lens-makers' equation:

$$\frac{n_1}{f} = (n_2 - n_1)\left(\frac{1}{R_a} - \frac{1}{R_b}\right) \tag{35-13}$$

The image distance s' and object distance s of a lens of focal length f are related by

$$\frac{1}{s}+\frac{1}{s'}=\frac{1}{f} \tag{35-14}$$

All of the relations above assume paraxial rays and use the sign convention given in Sec. 35-2.

Section 35-5. Optical Devices

Expressions for the angular magnification of the simple magnifier, the compound microscope, and the telescope are

$$M=\frac{250\text{ mm}}{f}$$

$$M=\frac{(250\text{ mm})(160\text{ mm})}{f_o f_e}$$

$$M=\frac{f_o}{f_e}.$$

QUESTIONS

1. Suppose that light did not travel in straight lines. How would you tell whether a line was straight or not?

2. Are sound waves both reflected and refracted? Is it possible to make a lens or a mirror that focuses sound waves? If such a lens or mirror existed, about how large would it have to be so that the conditions of geometrical optics were satisfied for speed sounds, which have a wavelength around 1 m?

3. In Fig. 35-33, is Venus admiring herself in the mirror? Explain.

Figure 35-33. Question 3: ***The Rokeby Venus*** by Velásquez.

4. Huygens discussed the path of a light ray in a medium in which the index of refraction n varies with position. Will a light ray in such a medium be bent toward increasing or decreasing n? Explain.

5. At noon on a clear day, the air near the surface of a road is warmer than the air higher up. The index of refraction of air decreases with increasing temperature. Use these facts to explain the mirage one sees on such a road, in which the blue sky, when seen a long way ahead on the road, appears to be water.

6. Can a one-eyed fish see out of an aquarium in all directions? Does it have to look in all directions to see all the room around it? Explain.

7. Light has a wavelength, a frequency, and a speed. Which, if any, of these change when light goes from air to glass? Explain.

8. Does a mirror reverse left to right? Does a mirror reverse up to down? How about east to west?

9. What is the focal length of a plane mirror? What is the magnification of a plane mirror (including sign)?

10. Is the wide-angle rearview mirror used on many trucks and buses convex or concave? Estimate the radius of curvature of such mirrors.

11. Estimate the focal length of the mirror of Fig. 35-34. Is the image of the man's face virtual or real? Why is the image distorted at the edges of the sphere? What is the actual shape of the room reflected in the sphere?

Figure 35-34. Question 11: ***Hand with Reflecting Globe*** by Escher.

12. Suppose you have trouble seeing things far away when the muscles that control the shape of the lens in your eye are relaxed. Should you use eyeglasses which have diverging or converging lenses? What if you have trouble seeing close objects when your muscles have made the lens in your eye as converging as possible?

13. Would a double convex lens of index $n = 1.250$ be a converging or a diverging lens in air? In water ($n = 1.333$)? Explain why many people have difficulty seeing under water. Why do goggles help?

14. During a thunderstorm, the speed of sound in air is generally faster near the earth than above it. Consider the path of sound "rays" and concoct an explanation as to why thunder is usually not heard more than about 15 mi from a storm.

15. Is the image location changed when a planoconvex lens is turned around so that it is a "convexplano" lens? Explain.

16. Consider what happens to a light ray in a medium in which the index of refraction fluctuates with time. How can this explain "twinkle, twinkle, little star"?

17. Nonparaxial rays in a parallel beam of light reflected by a spherical concave mirror do not cross the optical axis at the focal point where paraxial rays do. Do the nonparaxial rays cross it between the mirror and its focus or beyond the focus? Use a diagram to check your answer.

18. As the object of a thin lens is moved from infinity to the first focal point of the lens, the image moves from the second focal point of the lens to infinity. What happens to the image as the object is moved closer to the lens?

19. Consider a two-lens system. Show that it is possible for the second lens to have a "virtual object"—that is, an object through which no rays pass and for which the object distance for the second lens is negative.

20. Can the magnification of a thin lens be infinite? Is this a practical possibility or a mathematical one?

21. A positive lens made of material of index $n = 2$ and a mirror have the same focal length in air. Which has the smaller focal length in water?

22*. The invisible man in H. G. Wells's novel of that name made himself invisible by changing the index of refraction of his body to that of air (Fig. 35-35). Could the invisible man see anything?

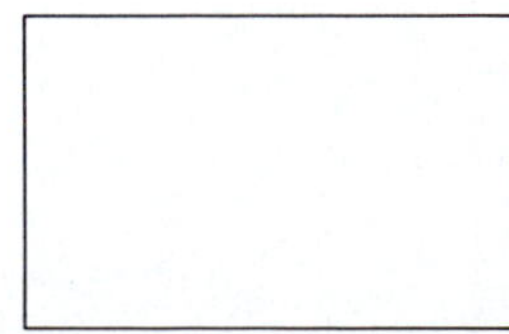

Figure 35-35. Question 22: The invisible man.

* Adapted from Jearl Walker, *The Flying Circus of Physics,* Wiley, New York, 1975.

23*. Can you explain why the shadow of a partially submerged pencil looks as it does in Fig. 35-36?

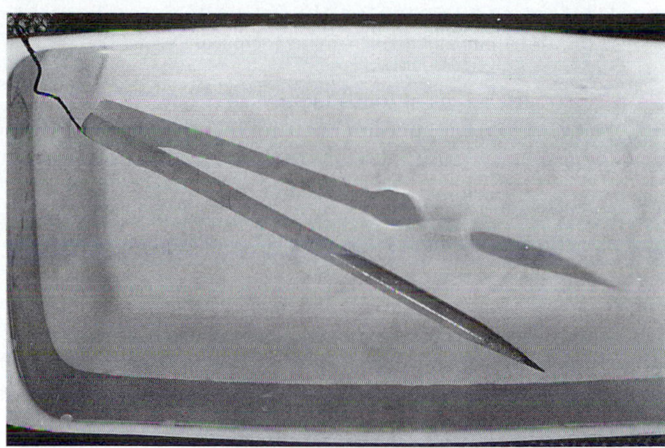

Figure 35-36. Question 23: Shadow of a partially submerged pencil.

24*. If you place a coin in a glass of water and look down at the proper angle, you will see the coin's image near the surface of the water (Fig. 35-37). Putting your hand on the opposite side of the glass has no effect on the image unless your hand is wet. Then the image of the coin disappears. Why?

Figure 35-37. Question 24: Image of a coin inside a glass of water.

25. How could you make a double convex lens that is diverging?

26. Is the image formed on the back of your retina right side up or upside down? (Experiments have been done with people wearing inverting prisms. After a relatively short time they were able to get around—and had trouble again when they took the inverting prisms off!)

27. The index of refraction of substances usually decreases with increasing wavelength. Will red or blue light bend more as a ray enters water from air? If the image of a star formed by a simple refracting telescope is a blue image on the retina of the eye, is the red image in front of the retina or behind it?

28. Can you use a lens to focus the rays of the sun to a point so that the point is at a higher temperature than the surface of the sun? (Consider the second law of thermodynamics.) Explain.

29. Where can a small length scale, etched in glass, be placed in a simple microscope so that the observer sees it coincident with the object of the microscope?

30. Some simple refracting telescopes have cross hairs. Where in the telescope are the hairs?

31. Complete the following table:

Symbol	Represents	Type	SI Units
n			
s			m
s'			
f		Scalar	
m			
M	Angular magnification		

EXERCISES

Section 35-1. Geometrical Optics

1. Two plane mirrors are joined so that their normals make an angle of 60°, as shown in Fig. 35-38. If an incoming ray makes an angle of 22° with one normal, by how much is the ray deviated by the mirrors; that is, what angle does the outgoing ray make with the incoming ray?

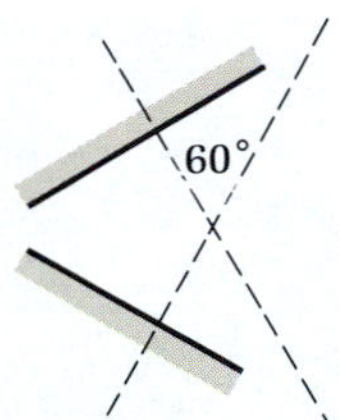

Figure 35-38. Exercise 1.

2. What angle must two mirrors make with each other if a ray in the plane formed by the normals to the mirrors is reflected by both mirrors so that it leaves on a path parallel to the path on which it entered?

3. Consider the reflection process shown in Fig. 35-7. Using Huygens' principle, show *(a)* that the length of *AN* is equal to the length of *BC*, *(b)* that the triangles *ABN* and *ABC* are congruent, *(c)* that $\theta_1 = \theta_{1r}$.

4. Parallel rays from the sun strike a window 3.0 m high by 2.0 m wide in a vertical wall of a building and are reflected onto the ground. If the normal to the window surface is north-south and the sun is due south 50° above the horizon, what is the size of the bright spot made by the parallel reflected rays on the flat ground?

5. Parallel light rays are incident on a glass prism as shown in Fig. 35-39. Show that the angle between the two reflected rays is twice the prism angle α.

6. A ray of light hits a mirror and is reflected. If the mirror is rotated by an angle α about an axis perpendicular to the plane of incidence, through what angle is the reflected ray rotated?

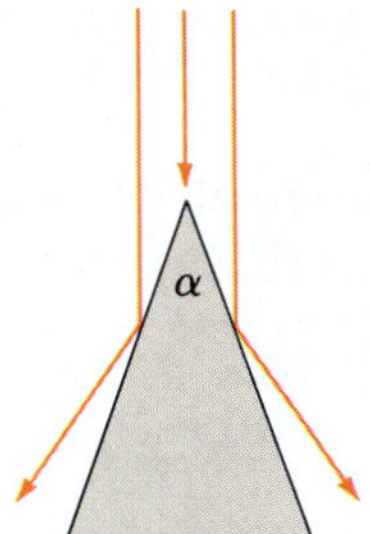

Figure 35-39. Exercise 5.

7. What is the speed of light in methylene iodide? See Table 35-1.

8. A ray of sunlight in air hits a water surface, making an angle of 40° with the normal to the water surface. What angle does the refracted ray make with the normal?

9. A ray of light in air makes an angle of 25° with the surface of a glass plate of index of refraction 1.525. *(a)* What angle does the incident ray make with the normal to the surface? *(b)* What angle does the refracted ray make with the surface of the glass?

10. A Pyrex glass prism has two faces which join at an angle $\alpha = 42°$. A ray of light hits one of these faces, making an angle $\theta = 18°$ with the normal to the surface, as shown in Fig. 35-40. A ray leaves the vertical surface. What angle does it make with the normal of that face? See Table 35-1 for indices of refraction.

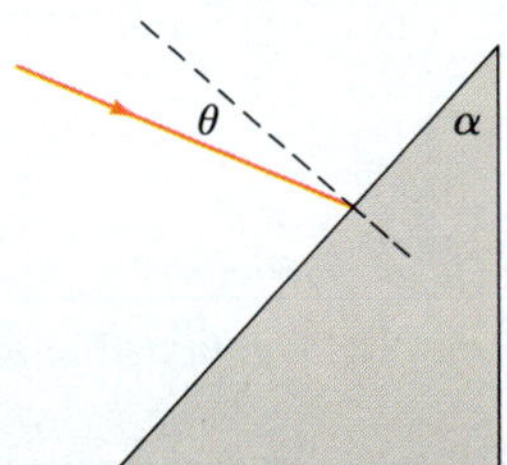

Figure 35-40. Exercise 10.

11. A light ray is refracted at an air-glass surface, making an angle of 75.00° with the normal on the air side and an angle of 39.71° with the normal on the glass side. Using Table 35-1, guess from which glass the prism is made.

12. When a ray of light passes through a plane-parallel sheet of material, it is displaced from its original path by an amount d, as shown in Fig. 35-41. A ray of light in air is incident on a plane-parallel sheet of index n' and thickness t oriented so that the normal to the surface makes an angle ϕ with the ray in the air and an angle ϕ' in the medium. Show that the displacement d of the ray is

$$d = t \sin\phi \left(1 - \frac{\cos\phi}{n' \cos\phi'}\right)$$

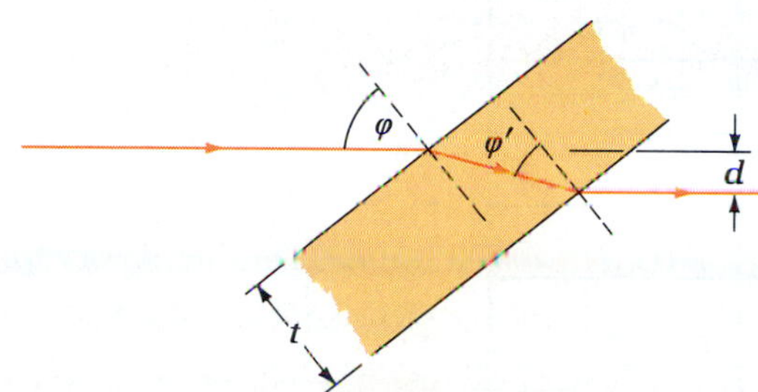

Figure 35-41. Exercise 12.

13. What is the critical angle for a ray leaving a diamond and entering air?

14. At what angle with respect to the vertical must a fish beneath water look to see the setting sun? Assume the fish is in a calm freshwater lake.

15. In Fig. 35-40, if $\alpha = 45°$ and the index of refraction of the prism is 1.655, what must θ be if the ray is to hit the right side of the prism at the critical angle?

Section 35-2. Images Formed by Reflection

16. *(a)* Use the law of reflection in Fig. 35-11 to show that triangle ABP is congruent to triangle ABP'. *(b)* Show that the distance from the plane of the mirror to the object is equal to the distance from the plane of the mirror to the image. *(c)* Show that this result is consistent with Eq. 35-4.

17. A student is 1.65 m tall, and her eyes are 120 mm below the top of her hair. She wishes to see her whole self in a vertical plane mirror. *(a)* How far above the floor can the bottom of the mirror be? *(b)* At least how far above the floor must the top of the mirror be? *(c)* Does it make any difference how close she stands to the mirror?

18. Two large plane mirrors are vertical, with an angle of 90° between them. *(a)* If you stand 1 m in front of their intersection on the bisector of the angle, how many images of yourself can you see? *(b)* How far from you is each one?

19. A 5-mm-high object is placed 250 mm in front of a convex mirror of radius of curvature $R = 400$ mm. *(a)* How far from the mirror will the image be? *(b)* Is the image erect or inverted? *(c)* What is the size of the image?

20. A concave shaving mirror has a radius of 335 mm. *(a)* What is the focal length of the mirror? *(b)* If your face is 105 mm from the mirror, how far from you is your image? *(c)* By what fraction is your image larger than you? *(d)* Is your image erect or inverted? *(e)* Is your image real or virtual?

21. In a convex spherical portion of a truck rearview mirror, the back of the truck, 20 m away, has a virtual image one-twentieth its real size. *(a)* What is the radius of curvature of the mirror? *(b)* Where is the image?

22. A *Cassegrain telescope* uses two mirrors, as shown in Fig. 35-42. Such a telescope is built with the mirrors 20 mm apart. If the radius of curvature of the large mirror is 220 mm and of the small mirror 140 mm, where will the final image of an object at infinity be?

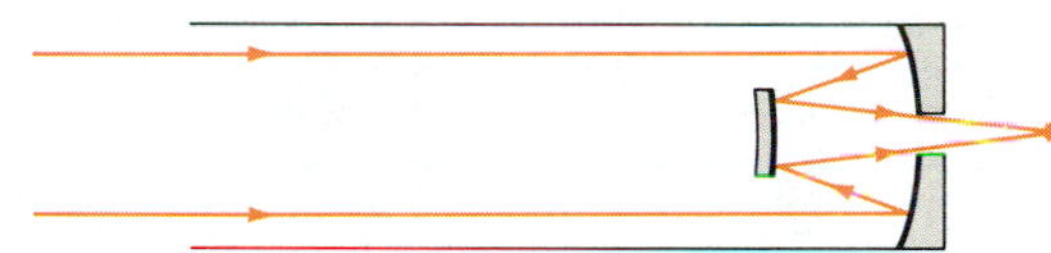

Figure 35-42. Exercise 23: A Cassegrain telescope.

23. Complete the table below for spherical mirrors with paraxial rays. All distances are in millimeters.

Type	Radius	Focal Length	Object Distance	Image Distance	Real Image?	Inverted Image?	m*
Convex	−90	−45	+15	−11	No	No	+0.75
Plane			+300				
Concave	120		+400				
	−480			−120			
		96		−36			
Convex			+160				0.5
	∞			−48			
Concave				+50	Yes		1.5

* Provide sign (+ or −) for all m values.

Section 35-3. Images Formed by Refraction

24. On its back surface, a 200-mm-diameter transparent sphere in air forms an image of paraxial rays from an object 1000 mm away. What is the index of refraction of the sphere?

25. A jar is filled with glycerine to a depth of 100.0 mm. The bottom appears to an observer to be raised 32.5 mm. Find the index of refraction of glycerine.

26. Complete the table below for spherical refracting surfaces. All distances are in millimeters. Let $n_1 = 1$ and $n_2 = 1.333$. Assume all rays are paraxial.

Type	Radius	Object Distance	Image Distance	Real Image?	Inverted Image?
Convex	+90	+15	−19	No	No
Plane		+300			
	+120		+480		
	−480	+120			
		+96	−96		

31. A planoconvex lens is to have a focal length of 240 mm and be made of glass of index 1.675. What radius of curvature should it have?

32. A polar explorer, out of matches, fashions a lens of ice to focus the sun's rays to start a fire. If he makes a planoconvex lens with the radius of curvature of $\frac{1}{4}$ m, how far from the tinder should he hold the lens?

33. A candle flame is 1.8 m from the side of a tent wall. A lens forms an image of the flame on the wall such that the image is inverted with a magnification of −6.0. What is the focal length of the lens?

Section 35-4. Lenses

27. Complete the table below for thin lenses. All distances are in millimeters. Assume paraxial rays.

Type of Lens	Focal Length	Object Distance	Image Distance	Real Image?	Inverted Image?	m*
Converging	+96	+144	+288	Yes	Yes	−2.0
Diverging		+300	−150			
	−120		−60			
Positive	+480	+120				
Negative		+96	3 (sign?)			
		−240		Yes		3.0
		+∞	+180		—	—
			−170			+2.5

* Provide sign (+ or −) for all m values.

28. The *newtonian* form of the lens equation is given in terms of the distance of the object and image from the first and second focal points. (The form we have used, $1/s + 1/s' = 1/f$, is called the *gaussian* form.) Let $x = s - f$ and $x' = s' - f$ and derive the newtonian form of the lens equation: $xx' = f^2$.

29. A double concave lens made of Lucite has both radii of curvature of magnitude 63 mm. What is its focal length? (See Table 35-1.)

30. An equiconcave lens is to have a focal length of −330 mm and be made of dense flint glass. (See Table 35-1.) What radii of curvature should it have?

34. An object on a 35-mm photographic slide is 20 mm high. The slide is 4.0 m from a screen. A lens of what focal length will be required to project an image of the slide that is 0.5 m high?

35. An object 15.7 mm high is 175 mm in front of a lens of focal length 85 mm. Fifty millimeters behind that lens is another lens of focal length −300 mm. *(a)* Where would the image of the first lens be if there were no second lens? *(b)* The image of the first lens acts as the (virtual) object for the second lens. Where does the second lens form its image? Is that image real or virtual?

36. Two thin lenses of focal length f_1 and f_2 are placed close together. Show that the combination of lenses acts like a single lens of focal length $f = f_1 f_2/(f_1 + f_2)$.

37. Prove that a diverging lens cannot form a real image of a real object. Note that real images have $s' > 0$.

38. *(a)* Prove that a positive lens forms a real image of a real object if and only if the object distance is greater than the focal length of the lens. *(b)* Show that if the object is virtual, the image formed by a positive lens is always real.

Section 35-5. Optical Devices

39. A certain glass has $n = 1.66650$ for light of wavelength 656.3 nm and $n = 1.68882$ for light of wavelength 434.0 nm. An equiconvex lens made of this glass has radii of magnitude 125.5 mm. For this lens, what is the difference between the focal lengths for light of these two wavelengths? Assume the lens is in air.

40. If the lens of the eye is treated as having an index $n_2 = 1.396$ in a medium of $n_1 = 1.336$ and as having radii of 10.0 mm and -6.0 mm, what is its focal length?

41. A quartz lens has a power of 4.5 D. What is its focal length?

42. Figure 35-43 shows a simple magnifier of focal length f, with the eye focused at a point corresponding to $s' = 250$ mm. *(a)* Show that the angle subtended at the eye by the object at the near point without the lens is $\theta_u \approx h/250$ mm and that the angle subtended by the image when the lens is in place is $\theta_m \approx h/s$, where $1/s - 1/s' = 1/f$. *(b)* Show that $M = \theta_m/\theta_u$ is given by

$$M = 1 + \frac{250 \text{ mm}}{f}$$

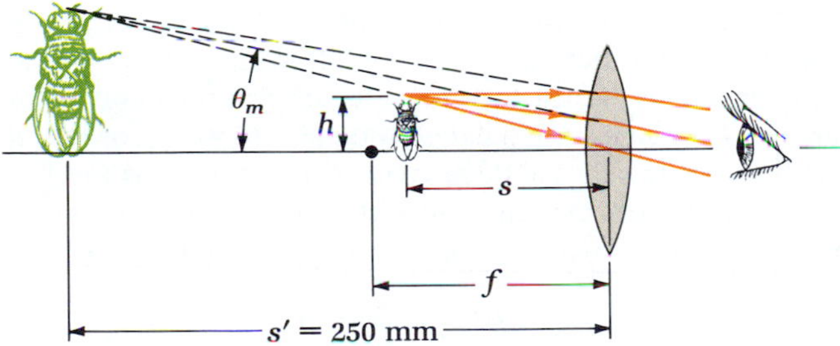

Figure 35-43. Exercise 42: A simple magnifier.

43. *(a)* What is the focal length of a 3X simple magnifier, where the 3X is measured with the eye focused at infinity? *(b)* What is the maximum magnification that a young child with an accommodation distance of 70 mm could obtain from this magnifier?

44. *(a)* If you use a 2X simple magnifier, with the magnification measured with your eye relaxed, how far from the magnifier is the object? *(b)* What is the magnification if you accommodate to a distance of 150 mm? *(c)* How far from the magnifier is the object in this case?

45. The objective and the eyepiece of a microscope have focal lengths of $+4.9$ and $+8.3$ mm, respectively. The tube length is 160 mm. Find *(a)* the distance from the objective to the object, *(b)* the linear magnification of the objective, *(c)* the overall magnification if the final (virtual) image is formed at infinity.

46. A microscope has an eyepiece marked 10X and an objective with a focal length of $+4.0$ mm. What is the overall magnification if the tube length is 160 mm?

47. A galilean telescope is shown in Fig. 35-44. Use a ray diagram to show that its image is virtual and erect.

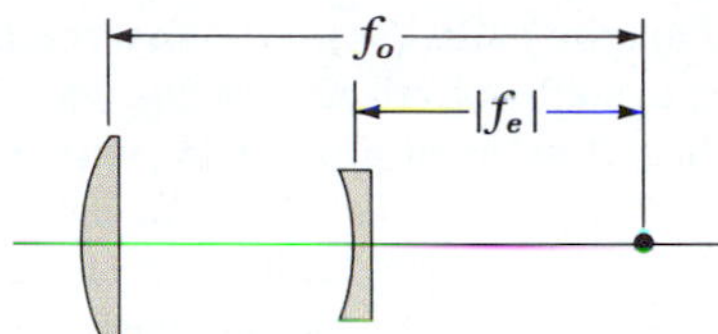

Figure 35-44. Exercise 47: A galilean telescope.

48. A simple refracting telescope of the keplerian design (Fig. 35-31) has an objective of focal length 850 mm and an eyepiece of focal length 25 mm. What is its angular magnification?

Additional Exercises

49. A circular raft 2.00 m in diameter floats on water of index $n = 1.333$. The sky is overcast so that light comes uniformly from all directions. What volume of water beneath the raft receives no direct illumination from the sky?

50. The diameter of the moon is 3.48 Mm and its distance from the earth is 384 Mm. What is the diameter of the image of the moon formed by a concave mirror of radius 0.415 m?

51. Draw a ray diagram of a stick person in front of a concave mirror of focal length 20 cm. Let $s = 30$ cm and let the stick person have a height of 5.0 cm. For simplicity, represent the mirror with a straight line and place the feet of the stick person on the optic axis. Show at least two principal rays; label the distances s and s'; and label the points C, F, and V.

52. Draw a ray diagram of a stick person in front of a concave mirror of focal length 30 cm. Let $s = 20$ cm and let the stick person have a height of 5.0 cm. For simplicity, represent the mirror with a straight line and place the feet of the stick person on the optic axis. Show at least two principal rays; label the distances s and s'; and label the points C, F, and V.

53. Draw a ray diagram of a stick person in front of a convex mirror of focal length 20 cm. Let $s = 15$ cm and let the stick person have a height of 5.0 cm. For simplicity, represent the mirror with a straight line and place the feet of the stick person on the optic axis. Show at least two principal rays; label the distances s and s'; label the points C, F, and V.

54. Draw a ray diagram of a stick person in front of a positive lens of focal length 20 cm. Let $s = 30$ cm and let the stick person have a height of 5.0 cm. For simplicity, represent the lens with a straight line and place the feet of the stick person on the optic axis. Show at least two principal rays; label the distances s and s'; and label the points F_1, F_2, and V.

55. Draw a ray diagram of a stick person in front of a positive lens of focal length 30 cm. Let $s = 20$ cm and let the stick person have a height of 5.0 cm. For simplicity, represent the lens with a straight line and place the feet of the stick person on the optic axis. Show at least two principal rays; label the distances s and s'; and label the points F_1, F_2, and V.

56. Draw a ray diagram of a stick person in front of a negative lens of focal length 30 cm. Let $s = 20$ cm and let the stick person have a height of 5.0 cm. For simplicity, represent the lens with a straight line and place the feet of the stick person on the optic axis. Show at least two principal rays; label the distances s and s'; and label the points F_1, F_2, and V.

PROBLEMS

1. ***Fermat's principle for reflection.*** Fermat's principle states that the path of a light ray between two points is such that the time for light to travel between the points is a minimum with respect to nearby paths (actually, that the time is stationary with respect to such variations). In Fig. 35-45 the path that a light ray might follow from point P to point P' is shown for rays reflected from a surface. Let the path length from P to P' be Δ and the position where the ray hits the mirror be x. According to Fermat's principle, the actual path will be the one for which the derivative $d\Delta/dx$ is zero, and the time taken for the light to go from P to P' is a minimum compared with that for neighboring paths. Show that the path length is $\Delta = (d^2 + x^2)^{1/2} + [d^2 + (\ell - x)^2]^{1/2}$ and that this path length (and therefore the time interval) is a minimum when $x = \ell/2$, so that $\theta = \theta_r$.

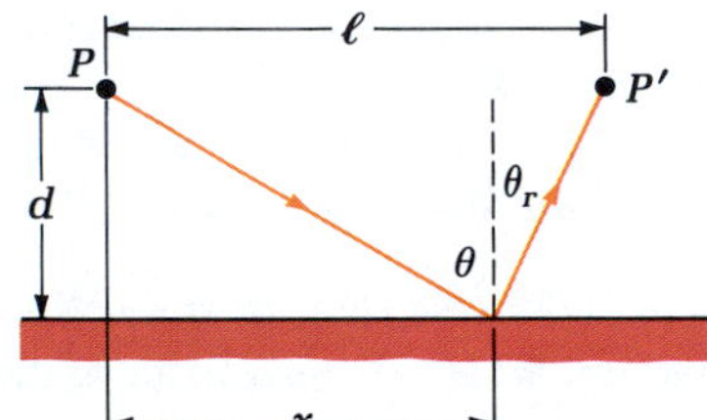

Figure 35-45. Problem 1.

2. ***Fermat's principle for refraction.*** In Fig. 35-46, the path a light ray might follow from P to P' is shown for rays refracted at a surface. Show that the time for the light to go from P to P' is a minimum if $n_1 \sin\theta_1 = n_2 \sin\theta_2$, and thus that Fermat's principle (see Prob. 1) leads to Snell's law. (*Hint:* Keep in mind that the speed of light in a medium of index n is $v = c/n$.)

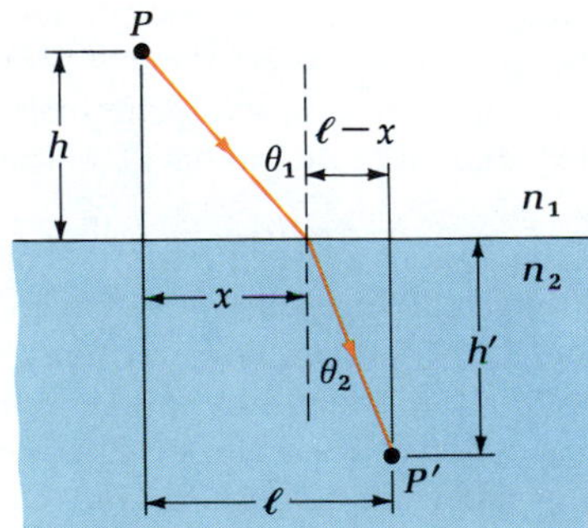

Figure 35-46. Problem 2.

3. ***Sign convention for mirrors.*** We developed the mirror equation, $1/s + 1/s' = 1/f$, by considering the concave mirror with $s > f$. *(a)* Show that the equation is valid for a concave mirror when $s < f$, provided that we let s' be negative. *(b)* Show that the mirror equation is valid for a convex mirror, provided that we let f and s' be negative. These results confirm the sign convention rules. (Although these rules are valid for virtual objects, for simplicity, assume that the object is real.

4. ***Sign convention for refraction at a surface.*** Using Fig. 35-47, show that Eq. 35-10 holds for paraxial rays if the radius R of the surface and the image distance s' are taken to be negative in accordance with the sign convention.

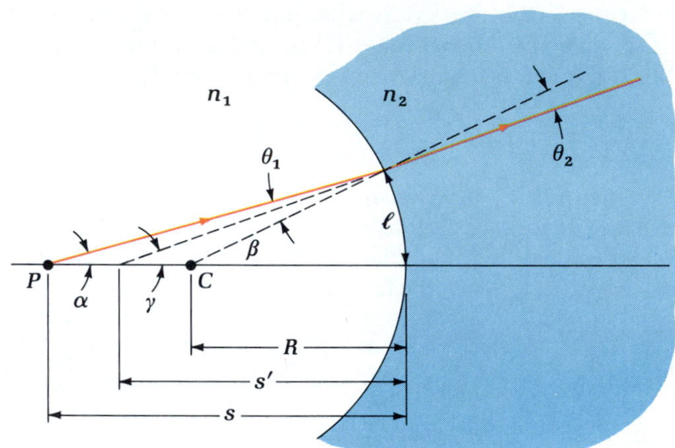

Figure 35-47. Problem 4.

5. ***Index of refraction of a liquid.*** The index of refraction of a liquid can be determined by the following method. A double convex lens of measured focal length f and radii of curvature $\pm R$ is brought near a flat glass plate, and a drop of liquid put between the lens and the plate. The liquid thus forms a planoconcave lens of radius $-R$. If the focal length of the combination of these two lenses is found to be f', show that

$$n = R\left(\frac{1}{f} - \frac{1}{f'}\right) + 1$$

6. ***Reflections from a parabolic mirror.*** A parabola can be defined as the locus of all points such that the distance from a point called the *focus* is equal to the distance $2p$ from a line called the *directrix*. *(a)* For the parabola $y^2 = 4p(x - p)$, show that the directrix is the y axis and the focus is at the point $(2p, 0)$. *(b)* Show that the tangent to the parabola is $dy/dx = 2p/y$, and that the tangent of the angle α in Fig. 35-48 is $y/(x - 2p)$ where x and y give the position where the ray hits the parabola. *(c)* Using the trigonometric formula for the tangent of the difference of angles, show that $\tan\gamma = \tan(\alpha - \beta) = \tan\beta$. Thus show that all rays perpendicular to the directrix are reflected to the focus of the parabola.

7. ***A combination of four lenses.*** Four thin lenses are placed along a common optic axis and spaced a distance d apart. Their

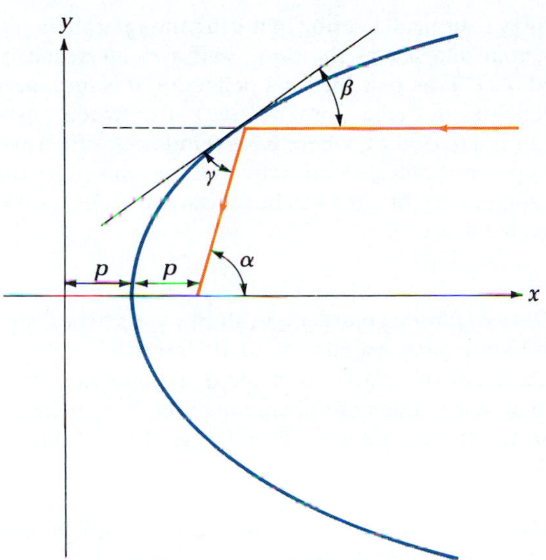

Figure 35-48. Problem 6.

focal lengths have identical magnitude f but alternating signs, with the first lens that the light encounters being positive. Show that if $f < d$, parallel light rays will all be brought to a focus on the axis by the combination.

8. ***Minimum real-object-to-real-image distance.*** Show that the minimum real-object-to-real-image distance for a converging lens is $4f$.

9. ***Geometric construction for refraction.*** The path of the refracted ray at a surface between media of indices n_1 and n_2 can be found with a geometric construction. Two arcs of circles centered at O, with radii of length proportional to n_1 and n_2, are drawn as shown in Fig. 35-49. Line OA is drawn parallel to the incident ray, and a line parallel to the normal to the surface is drawn through point B, where OA intersects the arc of radius n_1. This line intersects the arc of radius n_2 at C. Show that line OC is parallel to the refracted ray.

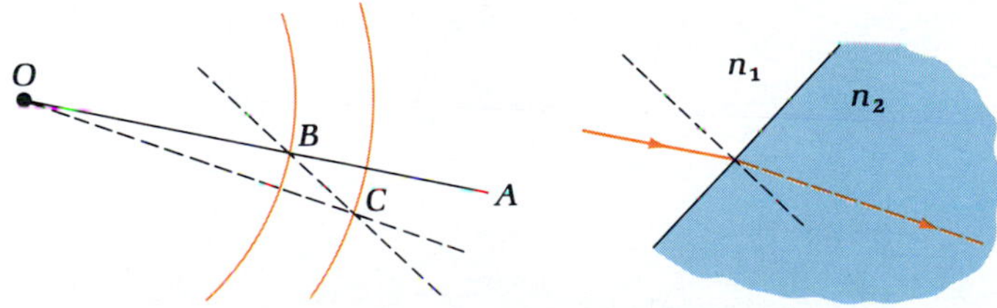

Figure 35-49. Problem 9.

10. ***Image superimposed on its object.*** A double convex lens of focal length f is sliced in half with a cut perpendicular to its optic axis. The plane surface of the cut is polished and silvered so that it forms a mirror. Show that an object $2f$ in front of this lens-mirror will be coincident with its own image. See Fig. 35-50.

11. ***GRIN.*** *A graded index lens (GRIN)* is made of glass of varying index of refraction. Use Fermat's principle (see Prob. 1) to show that the focal length of a 5.5-mm-high right cylinder of such glass whose index is cylindrically symmetric and varies

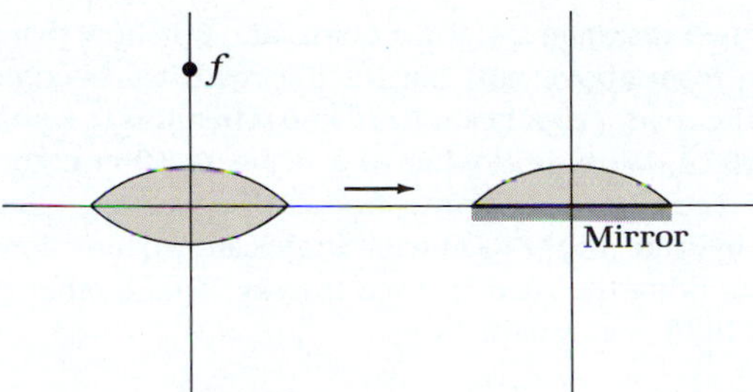

Figure 35-50. Problem 10.

linearly from 1.355 at the center to 1.300 at 15 mm from the center has a focal length of 409 mm.

12. ***Cone of light.*** A solid hemisphere of Pyrex glass of radius 150 mm lies with its flat face on a table. A beam of parallel light rays 5.15 mm in diameter is incident vertically on the center of the hemisphere. What is the diameter of the spot of light on the table?

13. ***Cube of light.*** A beam of light hits a cube of glass of side 20 mm, as shown in Fig. 35-51. The index of refraction of the glass is 1.555. Give the x, y, and z coordinates of the spot where the beam first exits the cube.

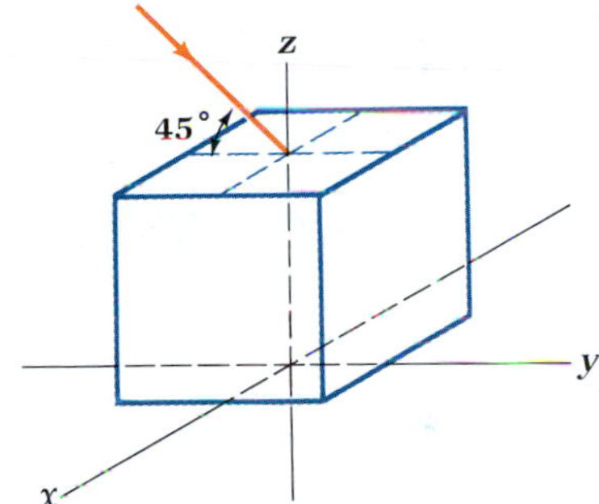

Figure 35-51. Problem 13.

14. ***Double depth.*** Show that a right cylinder of liquid of index n appears to be a fraction $1/n$ of its real depth when viewed from above, and that a right cylinder with a depth d_1 of liquid of index n_1 and a depth d_2 of liquid of index n_2 appears to have a depth $d_1/n_1 + d_2/n_2$.

15. ***Mirage 1.*** When air near the earth's surface is heated, its index of refraction varies vertically. Suppose the index of refraction is given by

$$n = 1.000293 - 0.000005(1 - e^{-y/10\,\text{mm}}),$$

where y is the height above the surface. Consider two rays of light propagating horizontally in this situation, one 5 mm above the surface and one 5.1 mm above the surface. *(a)* How much less than 1 mm does the upper beam travel in the time it takes the lower beam to travel exactly 1 m? *(b)* Through what angle (in radians) would a horizontal beam 0.1 mm high be bent upward after traveling 1 m? *(c)* Show that the change in the angle θ of a nearly horizontal beam as it propagates in the x direction through such a medium is $d\theta/dx = -(1/n)\, dn/dy$.

16. ***Mirage 2.*** Let θ be the angle that a beam of light has with the horizontal. *(a)* Write a spreadsheet that will find θ as a

function of distance x along a hot road. *(b)* Show that a beam coming from above will not hit the road if θ becomes zero above the road. *(c)* A beam has $\theta = 0$ when it is 0.1 mm above the surface. What is θ when the beam is 30 mm above the surface (and becoming free from those heating effects discussed in Prob. 15)? *(d)* At what angle can you look down from your car along the road and see the sky? (Remember that the path of light is reversible.)

17. ***Critical prism.*** Show that the condition $\alpha \leq 2\theta_c$ guarantees that a ray will emerge on the right side of the prism shown in Fig. 35-52 for some angle ϕ.

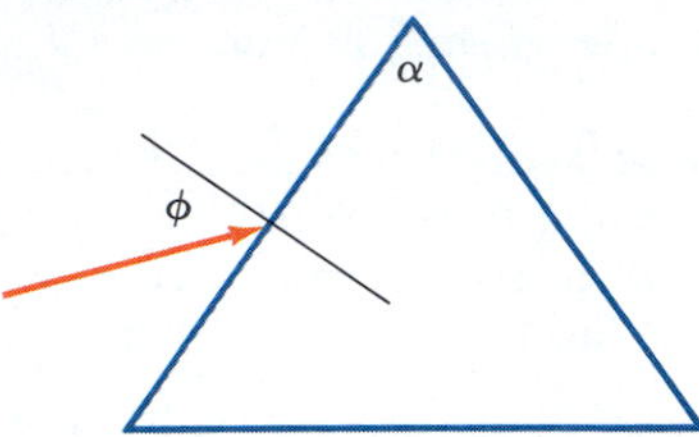

Figure 35-52. Problem 17.

18. ***Raindrops.*** A beam enters a raindrop and is reflected and refracted as shown in Fig. 35-53. *(a)* Show that $\alpha = \beta = \phi'$.

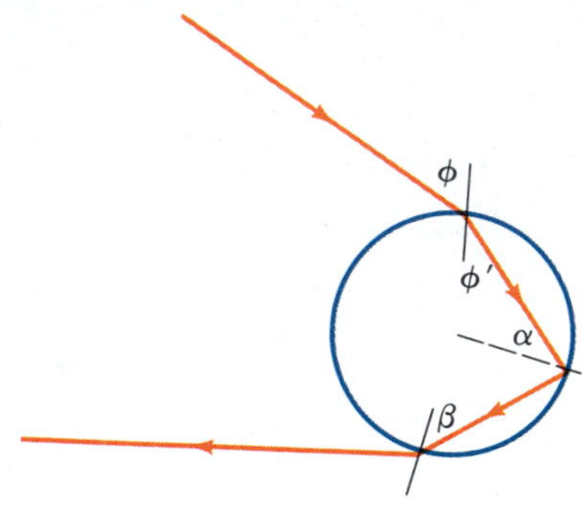

Figure 35-53. Problem 18.

(b) Show that the deflection of the incoming beam is $(\phi - \phi')$ as it enters and leaves the drop, and $\pi - 2\phi'$ when it is reflected. *(c)* Show that the total deflection θ is stationary with respect to ϕ (that is, $d\theta/d\phi = 0$) when $\phi = \phi_s = \cos^{-1}\sqrt{(1/3)(n^2 - 1)}$, where n is the index of refraction of the raindrop. Since more light is refracted into this angle than any other angle, a bright light will be seen at this angle. *(d)* What is this angle for water?

19. ***Optical fibers.*** Figure 35-54 shows an *optical fiber,* along which light is propagated as it totally internally reflects at the boundaries. If the cladding material has an index of 1.315, what would the index of the core material have to be to transmit all the light incident on its end, regardless of the incident angle?

20. ***Lens in water.*** A lens in air has a focal length of 89.4 mm. In water ($n = 1.333$) the lens has focal length of 277 mm. *(a)* What is the index of refraction of the glass that makes up the lens? *(b)* If the lens is planoconvex, what is the curvature of the convex side?

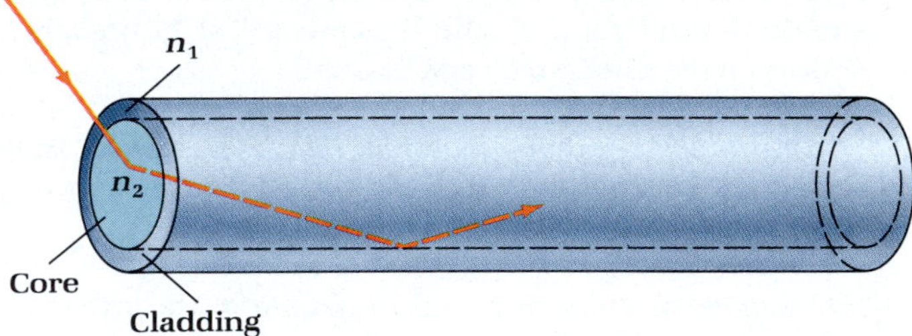

Figure 35-54. Problem 19.

HIGHLIGHTS OF MODERN PHYSICS

Precision, Accuracy, and the Hubble Space Telescope

All measurements involve error. ***Random errors*** arise because measurement instruments are not able to give results to an infinite number of significant figures. If you determine the value of pi by measuring a circle's circumference and diameter, each to four significant figures, the ratio will be known to only four figures, yet pi requires an infinite number of digits. Although they vary randomly, such measurements cluster around their average value; the more closely they cluster, the greater the *precision* of the measurement. ***Systematic errors,*** on the other hand, cause measured values to be consistently larger or smaller than the true value. If the ruler used to measure the circle's diameter is 10 percent too short, its measurements will be consistently 10 percent too large. A measurement with high precision can be very inaccurate if values are tightly clustered around an average that is far from the true value. The smaller the systematic error, the greater the *accuracy* of the measurement. Shortly after the Hubble Space Telescope (HST) was launched in 1990, it was discovered that its large mirror was the most precise but least accurate primary mirror in any major astronomical telescope. There had been a systematic error in measuring the shape of the mirror during manufacture.

The HST uses two hyperbolic mirrors to form images. To understand how it works, consider a simple reflecting telescope consisting of a single spherical mirror (Fig. 1). All rays from a point light source placed at the center of the sphere strike the mirror at right angles and reflect straight back to the same spot. They focus at the center of curvature, and a perfect image is formed (Fig. 1*a*). However, astronomical objects are not at the center of the sphere, but are so far away that their rays are essentially parallel. Rays striking close to the center of the mirror (paraxial rays) are reflected to a point halfway between the center of curvature and mirror's surface (Fig. 1*b*). Rays farther from the center of the mirror (marginal rays) focus closer to the mirror. The property of spherical mirrors in which different rays are focused at different points along the optic axis is called ***spherical aberration.*** For the marginal rays to focus at the same spot as the paraxial rays, the mirror must be less steep at the edges (Fig. 2). The desired surface is parabolic rather than spherical.

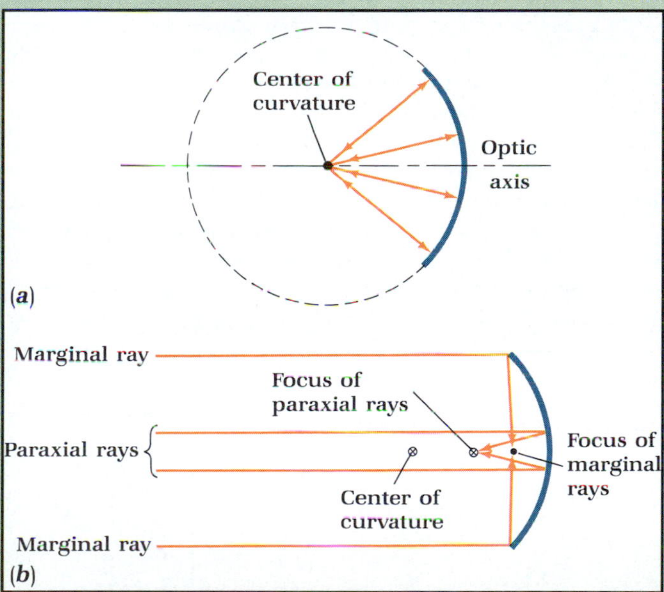

Figure 1. *(a)* Spherical mirror with the object at the center of curvature. All rays are focused back to the same point so the focus is at the center of curvature. The *optic axis* is the line through the center of curvature and the center of the mirror. *(b)* Spherical mirror with the object at infinity. The incoming rays are parallel, and those passing near the center of curvature (the paraxial rays) are focused at a point halfway between the mirror and the center of curvature. Marginal rays strike the increasingly steep slope of the mirror and are focused closer to the mirror. Because this lack of a single point of focus is associated with spherical mirrors, it is called ***spherical aberration,*** even though nonspherical mirrors can show it also.

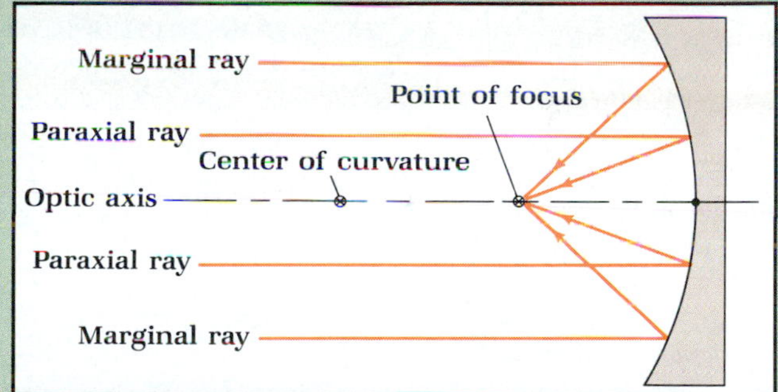

Figure 2. A parabolic mirror focuses all rays from an object at infinity and on the optic axis to a point, eliminating spherical aberration.

The focusing effect of a parabolic surface can be explained as follows. Light coming from one focus of an ellipsoid (obtained by rotating an ellipse about one of its axes) will converge at the other focus (Fig. 3). A sphere is an ellipsoid in which the two foci are at the center of curvature, which agrees with our earlier statement that rays coming from the center are focused back there. If we move one of the foci to infinity, the ellipse becomes a parabola, and rays from the focus at infinity will converge to a point at the other focus.

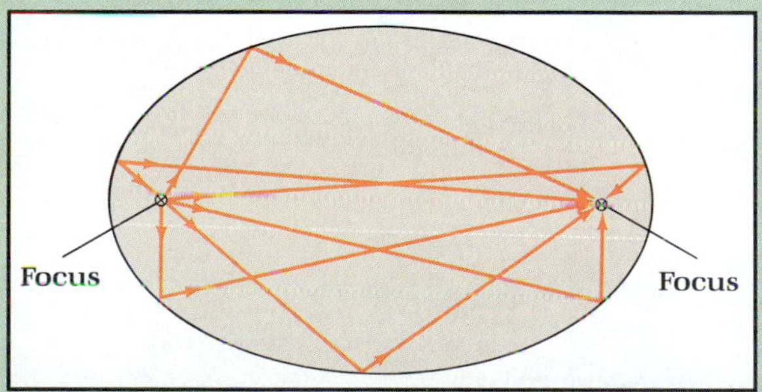

Figure 3. An ellipsoid focuses rays from one focus to the other. It is interesting to note that, since all points on an ellipse have the same sum of distances to the two foci, all light rays take the same time in going from one focus to the other. This is true for *all* optical systems, including systems with lenses: all rays take the same time to reach the focus.

Although stars in the center of the field of view will be imaged as sharp points, rays coming from an angle from stars near the edge of the field will not be parallel to the optic axis, and they produce tear-shaped images, an aberration called ***coma.*** To eliminate spherical aberration and reduce coma over a reasonably large field, two hyperbolic surfaces are commonly used for large telescopes such as the HST. This Ritchey-Chrêtien design, as it is called, is shown in Fig. 4. Although the design is straightforward, creating the mirror surfaces is not easy because the methods used to measure the shape of the surface are complicated.

For spherical surfaces there is a beautifully simple test called the Foucault knife-edge test (Fig. 5). If a knife edge is used to cut across the returning bundle of rays at point A, the mirror's surface will be seen to darken from bottom to top because the rays from the bottom will be cut off first, and conversely for point B. If cut at the center of curvature, all rays will be cut simultaneously and the mirror will darken all over uniformly. Any deviation from a perfect sphere will appear as a region of nonuniform darkening, and corrective polishing can be applied.

It is not so easy to measure a nonspherical surface because it does not darken uniformly. In both paraboloids and hyperboloids the radius of curvature is less at the edges than at the center so a knife edge cutting between the extreme focal points will produce top-to-bottom darkening at the center and bottom-to-top at the edges, assuming the same setup as in Fig. 5. It is hard to tell whether the surface is correct since all hyperboloids appear to darken alike except for subtle differences in contrast.

Telescope opticians use a null test in which uniform darkening certifies that the surface is correct. For a paraboloid, a null test is obtained by using a large flat mirror to produce a parallel beam of rays that simulates a star at infinity (Fig. 6). A knife-edge test at the focus will produce uniform darkening if the surface is a perfect paraboloid. Because accurate flat mirrors are as difficult to make as hyperboloidal ones, this approach was rejected for the large concave mirror of the HST. Such a flat mirror would have been very expensive to manufacture.

Instead, a small *reflective null corrector* was built to test the mirror during polishing. It let the hyperbolic mirror mimic a spherical surface by bending the rays from the light source so that when they exited the field lens they struck the hyperbolic mirror perpendicularly. Instead of using a point source of light and a knife edge, a laser was used in an interferometer arrangement so that a set of parallel bands was superimposed on the surface of the mirror, producing a sensitive and quantitative measure of the surface. For a perfect mirror the bands should be straight and evenly spaced. The function of the null corrector was to produce straight bands for the desired hyperbola rather than bands curved in some hard-to-interpret fashion. The finished primary mirror's precision was apparently accurate to at least 9 nm, even better than the specifications.

Unfortunately, after the HST was launched, it was discovered that the actual surface was accurate only to about 250 nm, one-thirtieth the accuracy expected. A group of optics experts was assigned to determine what had happened. The investiga-

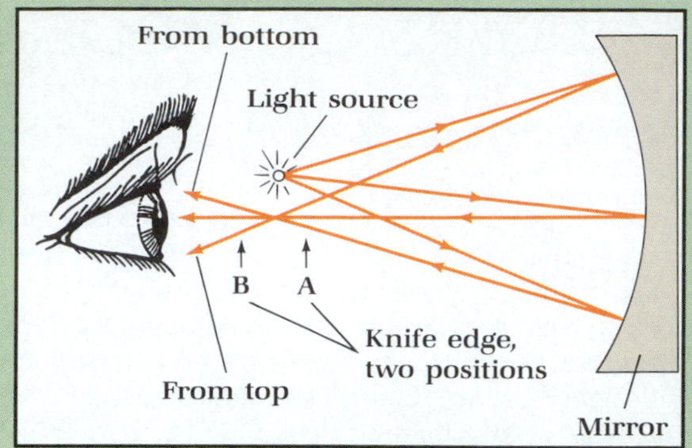

Figure 5. The Foucault knife-edge test. A pinhole light source is placed at the center of curvature of the mirror to be tested. If a knife edge cuts the return beam inside the center of curvature at point A as shown, the illuminated mirror surface will appear to darken from the same side, as those rays will be cut first. If the beam is cut at point B, the mirror will darken from the side opposite the knife edge. At the center of curvature a perfect sphere will darken all over simultaneously. Any deviation from sphericity will produce nonuniform darkening. This is called a *null test* because the partially darkened mirror of the desired shape — a sphere — shows no pattern.

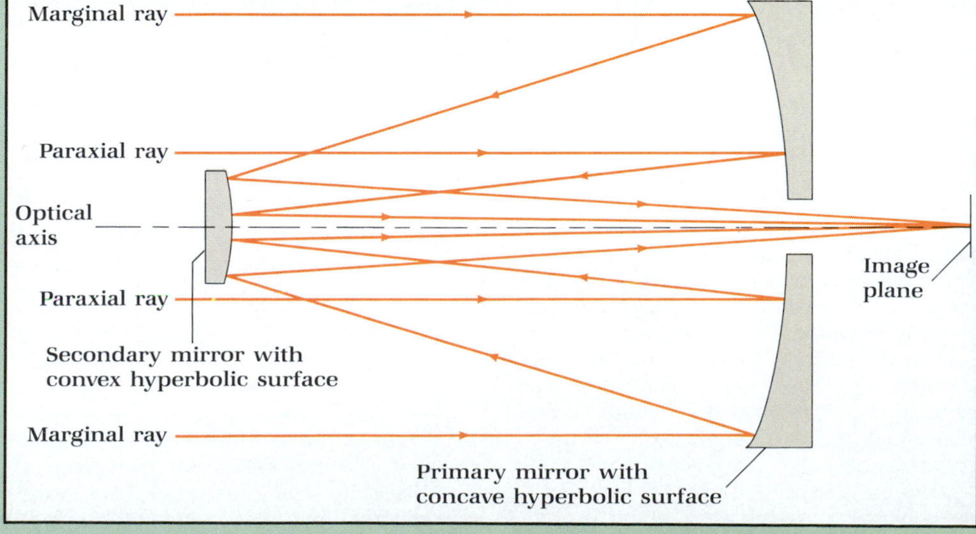

Figure 4. The Ritchey-Chrêtien telescope. It combines a short tube with a wide field of good focus. Both primary and secondary mirrors are hyperbolic, with the primary mirror concave and the secondary mirror convex.

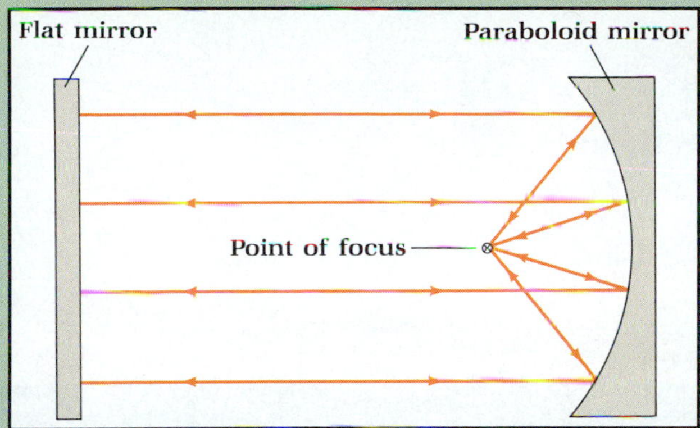

Figure 6. A null test for a paraboloid. The disadvantage of this method is that it requires an accurate flat mirror as large as the paraboloid. Accurate flats are themselves notoriously hard to make.

tion quickly focused on the reflective null corrector. The group found that the field lens was out of position by 1.3 mm and that this would explain the spherical aberration observed in the orbiting telescope. It was supposed to be positioned with an accuracy of 10 μm.

The field lens was positioned relative to the center of curvature of the lower mirror using a very accurate spacing rod (Fig. 7*a*). A laser beam was reflected off the top of the rod to fix its position. Since the end of the rod was rounded, there was concern that the beam might not reflect off the highest point, which would cause the field lens to be mispositioned. To prevent this problem, a cap was fitted over the end of the rod (Fig. 7*b*). It was painted flat black to keep light from being reflected off the cap itself, and a hole was drilled through its center to limit the laser beam to the exact top of the rod. Unfortunately, when the hole was drilled, some of the paint was chipped off the end cap. When the positioning procedure was run, the dominant reflection was from the chipped area of the cap, not the end of the rod. Apparently this reflection was used to position the spacing rod since this would misposition the field lens by 1.3 mm.

After it was determined that the primary mirror was misfigured, attention turned to how to correct it and to whether something could be done to improve the telescope's performance in the interim. About 20 options to correct the spherical aberration were considered, but it was finally decided to introduce corrective optics for four of the five instruments attached to the telescope during a repair mission in 1993. The fifth instrument, the high-speed photometer, will be sacrificed to provide room for a device to hold the corrective optics for two of the instruments. About 90 percent of the originally expected performance will be recovered for those instruments when this program is carried out.

Questions

1. Suppose the mirror in Fig. 5 is a paraboloid, and the knife edge cuts the beam from below at a point between the foci of the inner and outer parts of the mirror. Sketch the appearance of the mirror when the reflected beam is approximately half obstructed.

2. To correct spherical aberration (Fig. 1) material is removed from the outer part of the mirror to produce a paraboloid (Fig. 2). Consider the effect on the shape of the telescope mirror if the null corrector lens is positioned too far from the final focus of the null corrector mirrors (Fig. 6). That is, will the mirror be over-corrected (too much material removed from the outer part) or undercorrected (too little material removed)? Make a simple sketch to illustrate your answer. Hint: Recall that the mirror is polished so that its surface is perpendicular to all rays coming from the null lens.

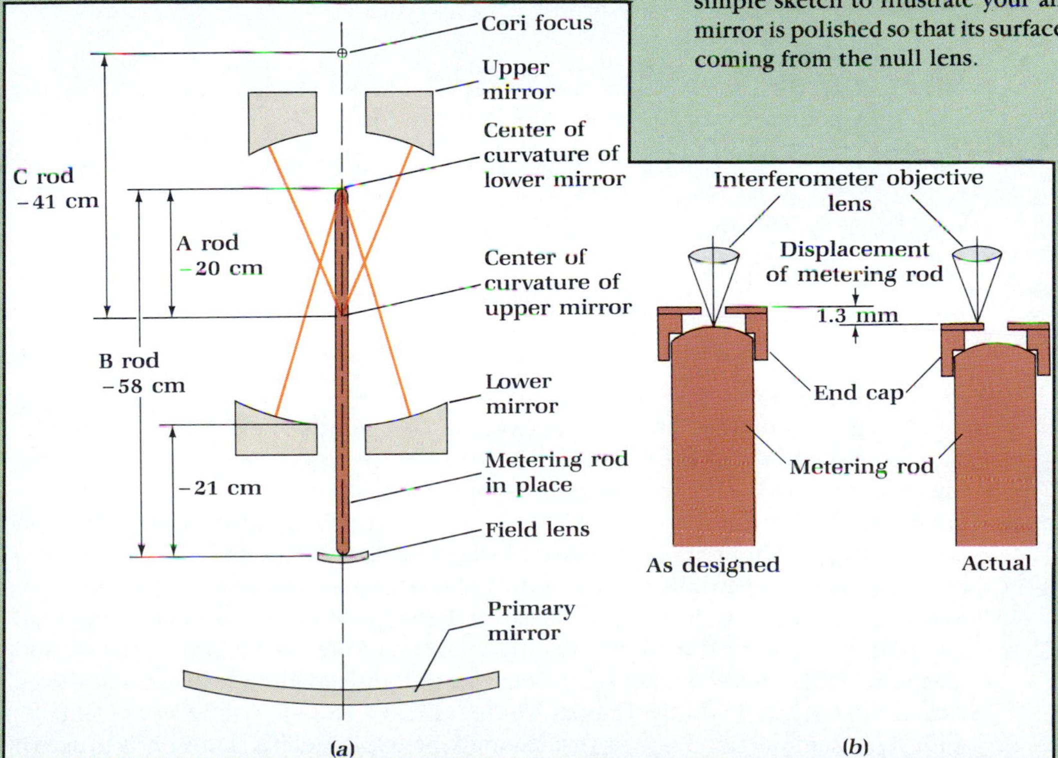

Figure 7. *(a)* Positioning the field lens. Light reflected off the top of the metering rod was supposed to allow accurate placement of that end. *(b)* An end cap was placed over the rod to limit reflected light to the center of the rod. Unfortunately, light reflected off the end cap (right-hand figure) instead of off the end of the positioning rod (left-hand figure).

36 INTERFERENCE AND DIFFRACTION

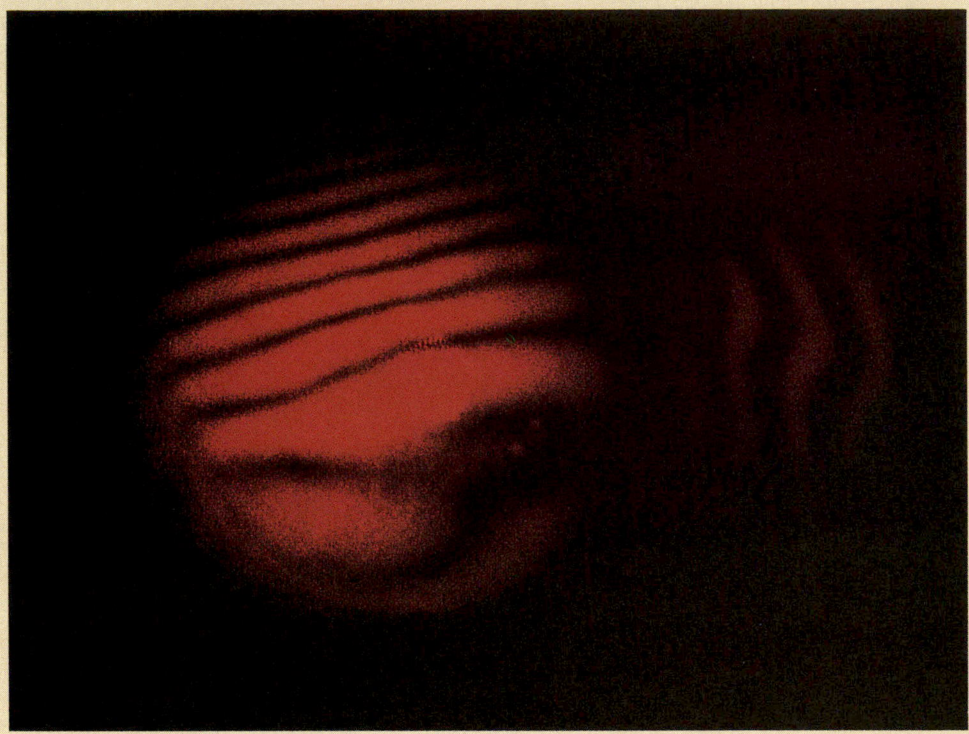

Interference fringes produced by passing a laser beam through a Michelson interferometer.

Circular waves on the surface of a pond. Notice how waves interfere when they occupy the same region.

What is light? This question has occupied the minds of many great scientists. Newton thought that light is a stream of particles, though he acknowledged some uncertainty. He was able to explain many optical effects using a particle, or corpuscular, theory of light. Christian Huygens (1629–1695), a contemporary of Newton's, believed that light is composed of waves. He initiated a wave theory of light, but his theory was partially unsuccessful because he assumed that light waves were longitudinal. We saw in Chap. 34 that in 1864 James Clerk Maxwell provided compelling theoretical evidence that light is a transverse wave in the electric and magnetic fields. However, the wave behavior of light had already been established experimentally in 1800 by Thomas Young (1773–1829). Young's experiments with light demonstrated effects that can only be explained in terms of destructive and constructive interference, effects that are exhibited only by waves.

In this chapter we discuss Young's famous double-slit experiment and several other important effects of the interference of light waves.

36-1 YOUNG'S DOUBLE-SLIT EXPERIMENT

Young's double-slit experiment provides a simple demonstration of the wave nature of light. Before discussing the double-slit experiment, we first consider an experiment where light passes through a single slit. The result may surprise you.

Suppose the light from an incandescent bulb passes through a narrow slit in a barrier and then impinges on a screen, as shown in Fig. 36-1. A color filter, such as a sheet of red cellophane, is placed over the bulb so that monochromatic light is incident on the slit. (Monochromatic light means light of one color or frequency or wavelength.) We are interested in the distribution of light intensity along the screen in the direction perpendicular to the slit, along the x direction in Fig. 36-1. If the slit is very narrow, then we observe that the light near the center of the screen (near $x = 0$) is almost uniform, gradually decreasing in intensity toward each side of the center. (In the next chapter we will see just how narrow we must make the slit.)

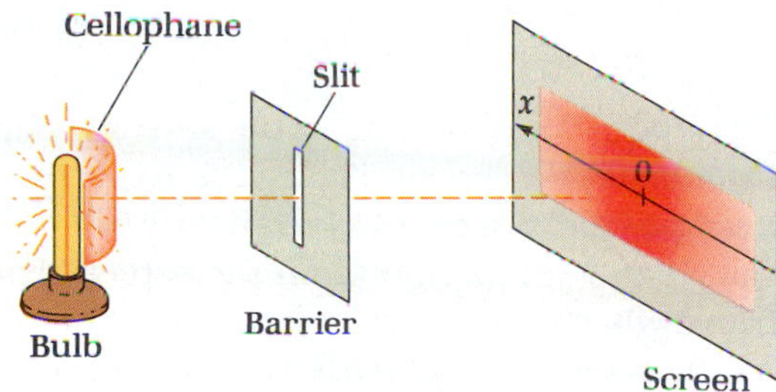

Figure 36-1. Monochromatic light from a bulb with a color filter passes through a slit and impinges on a screen.

The surprising feature about this experiment is that we do not see a narrow band of light on the screen. Instead, the light bends around the edges of the slit and illuminates part of the screen that is in the geometric shadow of the barrier. This phenomenon, whereby light spreads around the edge of a barrier, is called ***diffraction.*** We shall discuss diffraction and the single-slit experiment in more detail in the next chapter. For now we simply use the knowledge that when light passes through a very narrow slit, the light spreads out and is nearly uniformly distributed in the region close to $x = 0$ in Fig. 36-1. The narrow slit behaves as a line source of waves, sending out cylindrical wavefronts (Sec. 32-5).

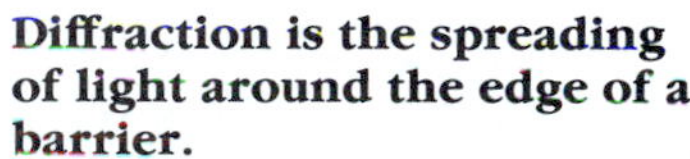
Diffraction is the spreading of light around the edge of a barrier.

Diffraction is a characteristic feature of wave behavior. As another example of diffraction, Fig. 36-2 shows a photograph of water waves spreading out as they pass through a small opening in a barrier.

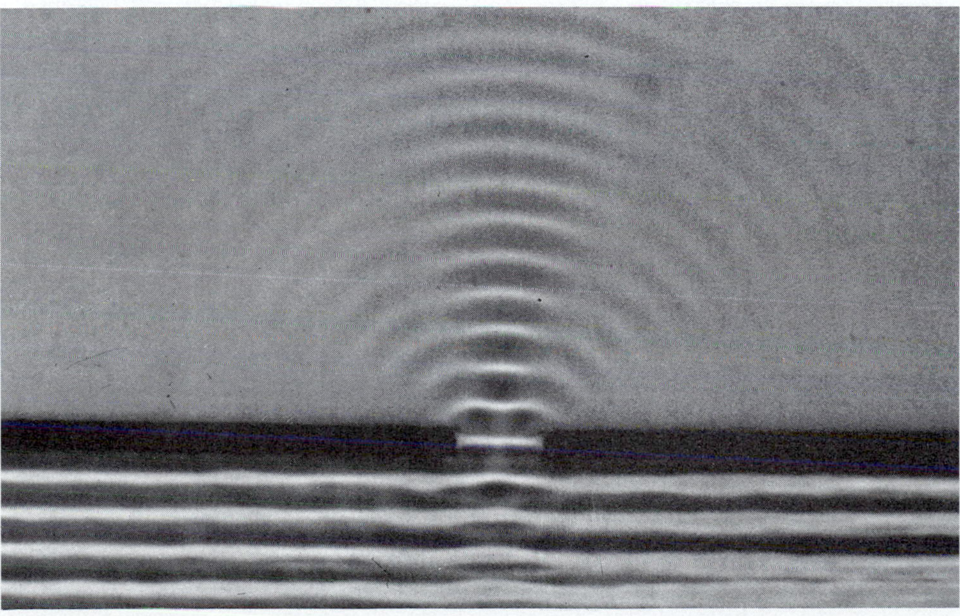

Figure 36-2. Water waves in a ripple tank. As the waves pass through the opening, they bend around the edge of the barrier. This is an example of diffraction.

The Double-Slit Experiment

In Fig. 36-3, monochromatic light passes through a collimating slit and then through two parallel slits S_1 and S_2 before impinging on a screen. The pattern of light seen on the screen consists of a series of brighter regions separated by darker regions, as shown in Fig. 36-4. These alternating bright and dark regions are called ***fringes,*** and this pattern of fringes is called the ***double-slit interference pattern.***

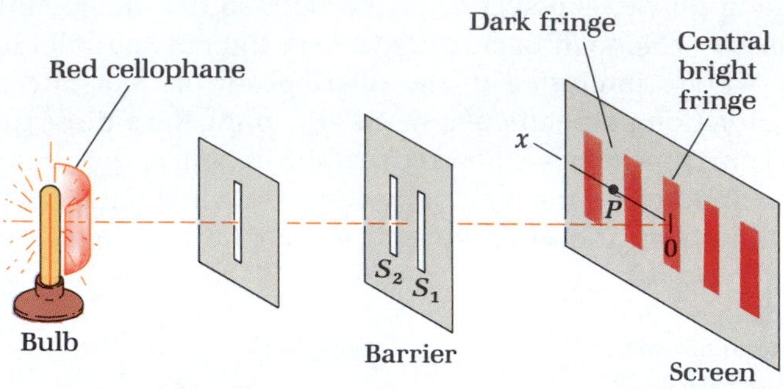

Figure 36-3. Young's double-slit experiment. Light from a bulb is made monochromatic by passing through a color filter. The light then passes through a collimating slit and parallel slits S_1 and S_2. The slit size and separation and the size of the double-slit interference pattern are exaggerated for clarity.

Figure 36-4. The double-slit interference pattern.

Suppose we cover one of the slits, say S_1, so that only light passing through S_2 illuminates the screen. Then we find that the fringes have disappeared and the light near the center of the screen is nearly uniform, as described previously (Fig. 36-1). Consider a specific point on the screen that corresponds to the center of a dark fringe when both slits are open, point P in Fig. 36-3. The light intensity at that point on the screen is essentially zero when both slits are open. However, if one of the slits is covered so that light from it cannot reach the screen, then the intensity at P is *not* zero.

How is it possible that light arrives at a point on the screen when only one slit is open, but when both slits are open (so that twice as much light passes through), no light arrives at that point? We can understand this result by assuming that light consists of waves. At the positions of the dark fringes, light waves from S_1 arrive out of phase with light waves from S_2. When two waves of equal intensity arrive out of phase at a point, they destructively interfere and the resultant wave is zero (Sec. 32-6). Thus the dark fringes are a consequence of the destructive interference between light waves from S_1 and light waves from S_2.

Dark fringes are a consequence of destructive wave interference.

Figure 36-5 shows a demonstration of interference effects from two sources of water waves. The interference of these waves can be used to visualize the way the double-slit interference pattern is formed.

Figure 36-5. Water waves in a ripple tank give a two-dimensional analog of the double-slit experiment. Two small spheres, vibrating up and down together, create waves of the same wavelength. Curved lines along which the waves destructively interfere are called *nodal lines.* The top edge of the photograph is analogous to the screen in the double-slit experiment. A point where a nodal line intersects the top edge is analogous to a dark fringe.

We can use the principle of superposition to determine the positions of the fringes in the double-slit pattern. In Fig. 36-6 we show slits S_1 and S_2 a perpendicular distance L from the screen. A point P on the screen can be located by the angle θ measured relative to the centerline. From the figure you can see that the waves from slits S_1 and S_2 travel different distances to reach point P; a wave from S_1 travels a distance r_1, and a wave from S_2 travels a distance r_2. For clarity in the figure, the separation d between the slits is greatly exaggerated compared with the distance L to the screen. In the experiment, d is much less than L, so the path difference $\Delta r = r_1 - r_2$ is approximately

$$r_1 - r_2 = d \sin \theta$$

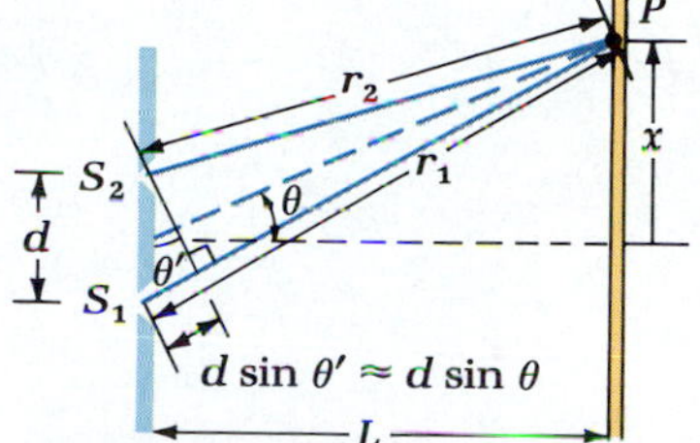

Figure 36-6. Light waves from S_1 and S_2 travel different distances to reach point P on the screen. In the actual experiment $L \gg d$, so the lines from each slit to P are essentially parallel and $\theta' \approx \theta$. Therefore $d \sin \theta' \approx d \sin \theta$ and $r_1 - r_2 \approx d \sin \theta$.

The interference of waves arriving at a point on the screen depends on this path difference. In Fig. 36-7*a* we show waves that produce the central bright fringe located at $\theta = 0$. The waves emerge from the slits in phase and arrive at the screen still in phase because they have the same wavelength and each wave travels the same distance to arrive at the screen: $r_1 - r_2 = 0$. Since the light waves arrive at the screen in phase, they constructively interfere and a bright fringe is produced.

In Fig. 36-7*b* we show light waves that produce the first dark fringe to one side of the central bright fringe. To reach the screen, the waves from S_1 travel half a wavelength farther than the waves from S_2: $r_1 - r_2 = \frac{1}{2}\lambda$. The waves emerge from the slits in phase, but, in traveling different distances, they arrive at the screen π rad (180°) out of phase. Thus the waves destructively interfere and a dark fringe is produced. Similarly, a dark fringe is centered at any point where the path difference is given by an odd integer times $\frac{1}{2}\lambda$. That is, path differences given by $r_1 - r_2 = \pm(m' + \frac{1}{2})\lambda$, where $m' = 0, 1, 2, \ldots$, correspond to the centers of the dark fringes. Positive and negative values of $\pm(m' + \frac{1}{2})\lambda$ refer to positions on either side of the central bright fringe. The angles $\theta_{m'}$ that locate the centers of the dark fringes are given by

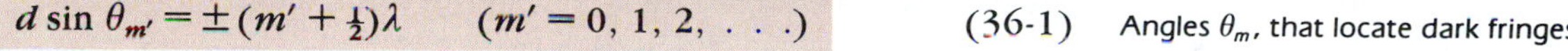

$$d \sin \theta_{m'} = \pm(m' + \tfrac{1}{2})\lambda \qquad (m' = 0, 1, 2, \ldots) \tag{36-1}$$

Angles $\theta_{m'}$ that locate dark fringes

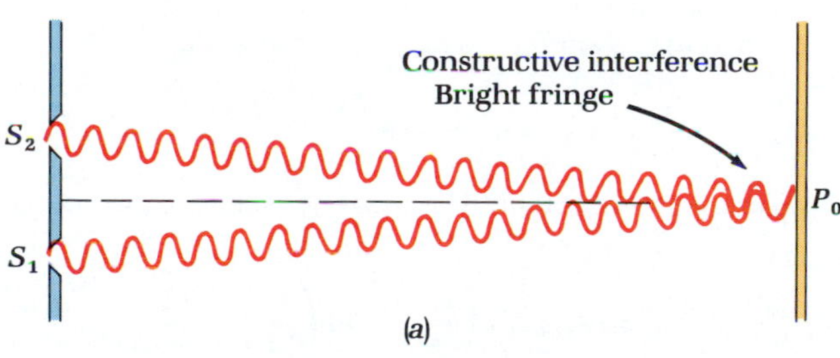

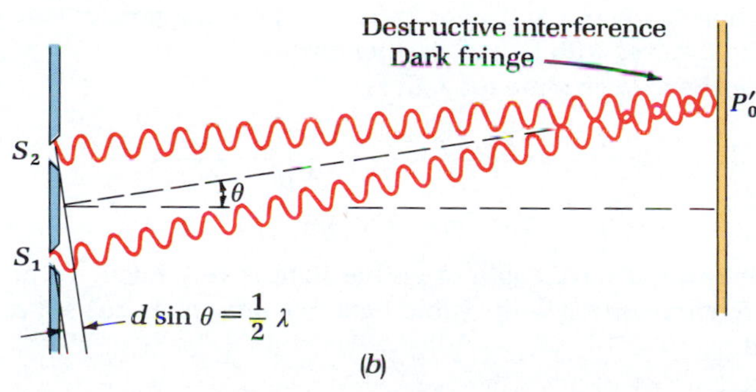

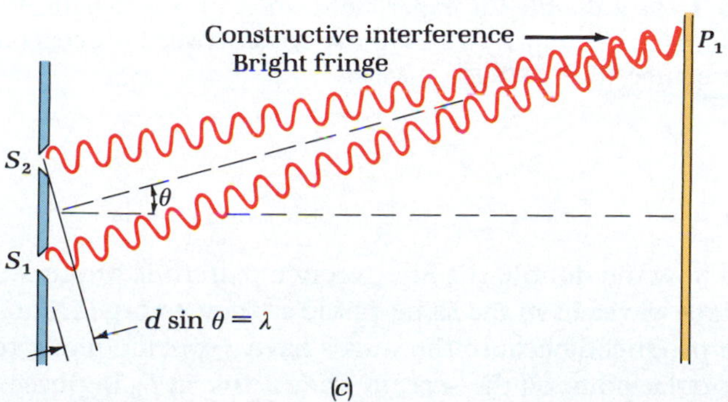

Figure 36-7. Light waves emerge from S_1 and S_2 in phase. *(a)* Waves from each slit arrive at P_0 in phase (crest on crest and trough on trough) so that they constructively interfere at the screen to produce the central bright fringe. *(b)* Waves arrive at P_0' out of phase (crest on trough and trough on crest) so that they destructively interfere to produce the $m' = 0$ dark fringe. *(c)* Waves arrive at P_1 in phase so that they constructively interfere to produce the $m = 1$ bright fringe.

Figure 36-7*c* shows how the first bright fringe to one side of the central bright fringe is produced. The light waves emerge from the slits in phase and arrive at the screen in phase because the waves from S_1 travel a full wavelength farther than the waves from S_2: $r_1 - r_2 = \lambda$. Thus the waves constructively interfere and a bright fringe is produced. Similarly, a bright fringe is centered at any point where the path difference is an integral number of wavelengths. That is, if $r_1 - r_2 = \pm m\lambda$, where $m = 0, 1, 2, \ldots$, then a bright fringe is produced. The angles θ_m that locate the centers of the bright fringes are given by

Angles θ_m that locate bright fringes

$$d \sin \theta_m = \pm m\lambda \qquad (m = 0, 1, 2, \ldots) \tag{36-2}$$

The bright fringes in the pattern are referred to according to their *order,* and the integer m in Eq. 36-2 labels the order of the fringe. The central bright fringe located at $\theta = 0$ corresponds to $m = 0$ and is called the ***zeroth-order maximum;*** the two bright fringes on either side of the central bright fringe correspond to $m = 1$ and are called the ***first-order maxima;*** and so on.

EXAMPLE 36-1

Spacing between adjacent fringes. ***(a)*** Find an expression for the spacing Δx between the centers of adjacent bright fringes for light of wavelength λ. Assume that $L \gg x_m$, where L is the distance from the slits to the screen and x_m is the coordinate of the center of a bright fringe relative to the center of the central bright fringe. (See Fig. 36-6.) ***(b)*** When red light is used in a double-slit experiment in which $L = 1.3$ m and $d = 0.12$ mm, we find that $\Delta x = 7.3$ mm. Determine the wavelength of this red light.

Solution. *(a)* Equation 36-2 gives the angles that locate the centers of the bright fringes. From Fig. 36-6, we see that $\sin \theta = x/\sqrt{x^2 + L^2}$. Since $L \gg x_m$, θ is small and $\sin \theta \approx \tan \theta = x/L$. Therefore $d \sin \theta_m \approx d(x_m/L) \approx \pm m\lambda$. Thus

$$x_m \approx \pm m \frac{L\lambda}{d}$$

The spacing between two adjacent bright fringes, say between the mth and the $(m + 1)$th for positive x_m, is

$$\Delta x \approx (m + 1)\left(\frac{L\lambda}{d}\right) - m\left(\frac{L\lambda}{d}\right) = \frac{L\lambda}{d}$$

Thus, in an experiment where L and d are fixed, the spacing is greater (the pattern is more spread out) for light waves with longer wavelengths.

(b) Solving the above expression for λ gives

$$\lambda \approx \frac{d\,\Delta x}{L} = \frac{(1.2 \times 10^{-4}\text{ m})(7.3 \times 10^{-3}\text{ m})}{1.3\text{ m}}$$
$$\approx 6.7 \times 10^{-7}\text{ m} = 670\text{ nm}$$

This result shows that the wavelength of visible light is very small. To produce an easily observable interference pattern with visible light, we must make the slit separation correspondingly small.

SELF-TEST 36-1. In a double-slit experiment, light of wavelength 585 nm is passed through slits spaced 0.23 mm apart. What is the spacing between adjacent bright fringes on a screen 0.88 m from the slits? ***ANSWER:*** 2.2 mm.

Coherence

To understand how the double-slit interference pattern is produced, we used the fact that the light waves have the same phase as they emerge from S_1 and S_2. The pattern is then produced because the waves have a specific phase relation as they arrive at a particular point on the screen. They arrive at P_0 in phase (bright), they arrive at P_0' out of phase (dark), they arrive at P_1 in phase (bright), and so on.

Suppose we arrange to have the waves emerge from S_1 out of phase by π rad with the waves emerging from S_2. (See Ques. 8.) Then, by a similar analysis, the waves arrive at P_0 out of phase (dark), they arrive at P_0' in phase (bright), they arrive at P_1 out of phase (dark), and so on. If the waves emerge from the slits π rad out of phase, then an interference pattern is still produced, but the positions of the bright and dark fringes are shifted from their positions when the waves emerge with a zero phase difference. Extending this argument to other phase differences, we can see that a pattern is produced for any particular phase difference between the waves emerging from the slits. Changing the phase difference of the emerging waves simply shifts the positions of the fringes. To produce the double-slit interference pattern, there is no need for the phase difference between the waves emerging from the two slits to be zero, but there is a need for this phase difference, whatever it is, to remain constant.

Now suppose that the phase difference of the emerging waves varies randomly. If the time of this variation in phase difference is short compared with the time of observation, then we would observe the average of many overlapping patterns. That is, *we would observe no double-slit interference pattern at all!* The fact that we do observe a stable pattern shows that the waves emerging from S_1 maintain a constant phase difference with the waves emerging from S_2.

If waves emerging from two sources maintain a constant phase difference, then the two sources are *coherent*. Only coherent sources can produce a stable interference pattern. The two vibrating spheres that produced the water waves in Fig. 36-5 are coherent sources because the spheres vibrate up and down together. The two slits in Young's double-slit experiment are coherent sources because the light from these sources originates from the same primary source, the collimating slit in Fig. 36-3.

Coherent sources maintain a constant phase difference.

In an ordinary light source, such as a light-bulb filament, the light is emitted from many individual atoms radiating independently. Consequently, the phase relation among these emissions is totally random. Thus two separate light-bulb filaments are incoherent sources, and they do not produce a stable interference pattern. In the double-slit experiment, the slits behave as two identical images of the same source. Any phase relation that exists for light emerging from one of the slits also exists for the other. Therefore the slits are coherent sources.

36-2 INTENSITY DISTRIBUTION IN THE DOUBLE-SLIT INTERFERENCE PATTERN

To find an expression for the distribution of light intensity in the double-slit interference pattern, we use the principle of superposition. That is, we add vectorially the electric fields due to the light waves from each slit. This addition is performed at each point P along the screen (Fig. 36-6). Then we find the intensity I at P by taking the square of the resultant electric field amplitude. (From Sec. 34-4, $I = \bar{S} = \frac{1}{2}\epsilon_0 E_0{}^2 c$.)

Let E_1 and E_2 be the electric field components of the light waves from S_1 and S_2 at point P on the screen. (For brevity, we drop the coordinate subscript on the symbols for the field components.) We assume that the slits are very narrow, so they may be treated as line sources of light. This means that the intensity distribution due to each slit acting alone is nearly uniform around the center of the screen, and the field components may be written as

$$E_1 = E_0 \sin(\omega t + \phi) \qquad \text{and} \qquad E_2 = E_0 \sin(\omega t)$$

Because the monochromatic light originates from the same source, each expression contains the same frequency ω. Also, since each slit has the same width and is nearly the same distance from P, we assume that the amplitude E_0 of each wave is the same. The crucial feature in the expressions for E_1 and E_2, as far as the interference pattern is concerned, is that the waves have a phase difference at P. The phase difference ϕ is due to the path difference $d \sin \theta$. In Fig. 36-7*b* the path difference is $d \sin \theta = \frac{1}{2}\lambda$, and the waves arrive at the screen with a phase difference $\phi = \pi$ rad. In Fig. 36-7*c* the path difference is $d \sin \theta = \lambda$, and the waves arrive at the screen with a phase difference $\phi = 2\pi$ rad. Thus the phase difference and the path difference are directly proportional:

$$\frac{\text{Phase difference}}{2\pi} = \frac{\text{path difference}}{\lambda} \quad \text{or} \quad \frac{\phi}{2\pi} = \frac{d \sin \theta}{\lambda}$$

Therefore

$$\phi = \frac{2\pi d \sin \theta}{\lambda} \tag{36-3}$$

To find the component E_{12} of the resultant electric field at P, we add E_1 and E_2:

$$E_{12} = E_1 + E_2 = E_0 \sin (\omega t + \phi) + E_0 \sin (\omega t)$$

Using the trigonometric identity

$$\sin \alpha + \sin \beta = 2 \cos [\tfrac{1}{2}(\alpha - \beta)] \sin [\tfrac{1}{2}(\alpha + \beta)]$$

with $\alpha = \omega t + \phi$ and $\beta = \omega t$, we obtain

$$E_{12} = [2E_0 \cos (\tfrac{1}{2}\phi)] \sin (\omega t + \tfrac{1}{2}\phi) \tag{36-4}$$

Thus E_{12} oscillates with an amplitude of $2E_0 \cos (\frac{1}{2}\phi)$.

Intensity is proportional to amplitude squared, so that if we let I_0 represent the intensity at P due to light from one of the slits acting alone (the other slit covered), then $I_0 \propto E_0{}^2$. From Eq. 36-4, the amplitude of the resultant wave is $2E_0 \cos (\frac{1}{2}\phi)$, so the intensity I_{12} due to light from both slits is

$$I_{12} \propto 4E_0{}^2 \cos^2 (\tfrac{1}{2}\phi)$$

Since the proportionality factor between intensity and amplitude squared is the same for both slits open as it is for each slit acting alone, we have

Double-slit intensity in terms of ϕ

$$I_{12} = 4I_0 \cos^2 (\tfrac{1}{2}\phi) \tag{36-5}$$

Using Eq. 36-3, $\phi = (2\pi d \sin \theta)/\lambda$, we can express this result in terms of the angle θ that locates the point P on the screen in Fig. 36-6:

Double-slit intensity in terms of θ

$$I_{12} = 4I_0 \cos^2 \left(\frac{\pi d \sin \theta}{\lambda}\right) \tag{36-6}$$

Let us check the agreement of this expression with Eqs. 36-1 and 36-2. From Eq. 36-1, dark fringes are located at angles θ_m, given by $d \sin \theta_{m'} = \pm(m' + \frac{1}{2})\lambda$. Substitution into Eq. 36-6 gives $I_{12} = 4I_0 \cos^2 [\pm(m' + \frac{1}{2})\pi]$. Since $\cos [\pm(m' + \frac{1}{2})\pi] = 0$ for $m' = 0$ or any integer, $I_{12} = 0$ at these angles. From Eq. 36-2, bright fringes are located at angles θ_m given by $d \sin \theta_m = \pm m\lambda$. Substitution into Eq. 36-6 gives $I_{12} = 4I_0 \cos^2 (\pm m\pi)$. Since $\cos^2 (\pm m\pi) = 1$ for $m = 0$ or any integer, $I_{12} = 4I_0$ at these angles. Thus Eq. 36-6 gives positions of dark fringes and bright fringes that are in agreement with Eqs. 36-1 and 36-2.

Often the experiment is performed such that the separation Δx of the fringes is much smaller than the distance L from the slits to the screen. In this case we are only interested in the intensity at points with $x \ll L$ in Fig. 36-6, and the approximation $\sin \theta \approx x/L$ is valid. Equation 36-6 then becomes

Double-slit intensity in terms of x

$$I_{12} \approx 4I_0 \cos^2 \left[\left(\frac{\pi d}{\lambda L}\right) x\right] \tag{36-7}$$

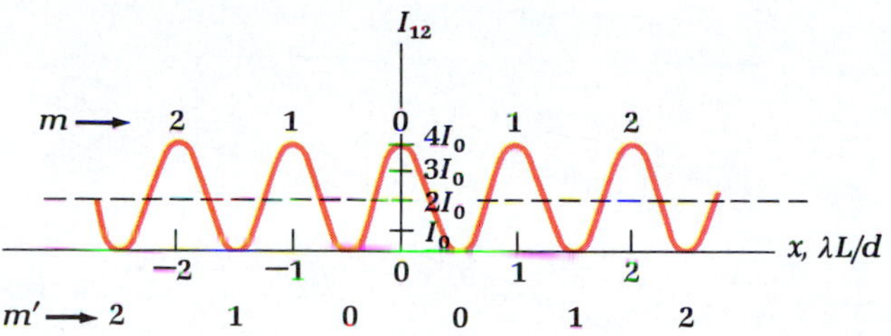

Figure 36-8. Intensity distribution of the double-slit interference pattern versus coordinate x in units of $\lambda L/d$. The slits are very narrow and $x \ll L$.

In Fig. 36-8, we show a graph of Eq. 36-7. With our assumption that the slits are very narrow, the intensity varies from zero to $4I_0$ as a cosine-squared function across the middle of the screen ($x \ll L$).

More about Coherence

To emphasize the distinction between coherent and incoherent sources, we recast Eq. 36-5 by using the trigonometric identity $2\cos^2(\frac{1}{2}\phi) = 1 + \cos\phi$:

$$I_{12} = 2I_0 + 2I_0 \cos\phi \qquad \text{(coherent sources)}$$

The phase difference ϕ corresponds to a particular point on the screen so that the average value of the second term over any integral number of fringes across the screen is zero. If the light from the slits originated from different sources, then the slits would be incoherent sources, and the phase difference ϕ would vary rapidly and randomly with time. In this case, the second term would have a time average of zero at each point on the screen, and the intensity would be

$$I_{12} = 2I_0 \qquad \text{(incoherent sources)}$$

That is, the intensity would be simply the sum of the intensities of two sources acting alone. The horizontal dashed line in Fig. 36-8 shows the intensity from two incoherent sources where I_0 is the intensity from each source acting alone.

To find the intensity due to combined waves from coherent sources, we add the electric field components and then square the resultant field. A stable interference pattern is a consequence of the combination of waves from coherent sources. To find the intensity due to the combined waves from incoherent sources, we first square the field components to determine the intensity of each source acting alone and then add the intensities. No stable interference pattern is produced by the combined effects of waves from incoherent sources.

36-3 DIFFRACTION GRATINGS

A ***diffraction grating*** is a device that can separate a beam of light into its constituent wavelengths or colors. A grating is an important tool for a scientist or engineer who performs research in optics. There are two types of gratings: *reflection gratings* and *transmission gratings.* A grating is constructed by forming parallel, evenly spaced grooves or scratches on a flat surface of a metal (reflection grating) or glass (transmission grating) plate. The action of a grating can be described in terms of a regular array of parallel slits. The grooves scatter the light and are effectively opaque, and the space between the grooves behaves as a slit. Typically the grating plate is rectangular, with dimensions of several centimeters on a side. The spacing d between the slits is very small, about 2 μm, and the number N of slits is usually large, about 10,000. A great deal of precision is required to assure that the closely spaced slits are parallel and of equal size and spacing.

Consider a beam of monochromatic light incident normally on a diffraction grating. (For nonnormal incidence see Prob. 6.) The light waves are in phase as they emerge from each of the slits, as shown in Fig. 36-9. To reach point P on the screen, the waves from adjacent slits travel a different distance by the amount $d \sin\theta$. For

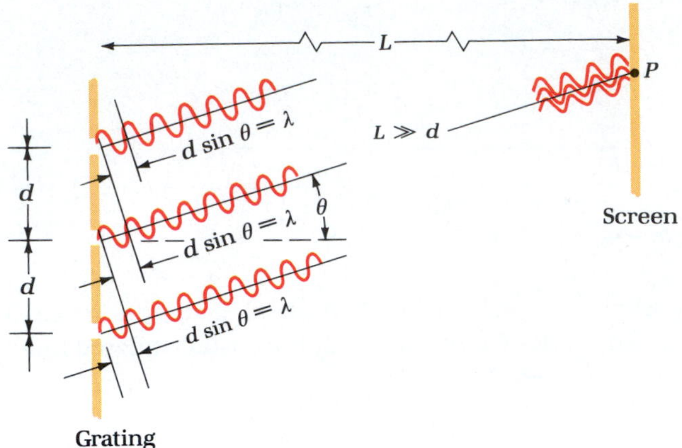

Figure 36-9. Waves from adjacent slits travel different distances by the amount $d \sin \theta$ to reach point P on the screen. In this particular case, $d \sin \theta = \lambda$, so an interference maximum is produced at P.

the particular case in the figure, the waves are shown directed such that $d \sin \theta = \lambda$. These waves arrive at the screen in phase and interfere constructively at P to form a first-order interference maximum. Waves will interfere constructively at the screen when the path difference $d \sin \theta$ is an integral number of wavelengths. That is, interference maxima are located at angles θ_m given by

Angles θ_m that locate interference maxima

$$d \sin \theta_m = \pm m\lambda \qquad (m = 0, 1, 2, \ldots) \tag{36-8}$$

where the integer m labels the order of the interference maximum. Note that this is the same equation as Eq. 36-2 for the bright fringes in the double-slit interference pattern.

The angle θ_m of an interference maximum for light of a given wavelength depends only on the spacing d. For example, consider light at each end of the visible spectrum passing through a grating with $d = 2.0\ \mu\text{m}$. The angles that locate the two first-order maxima ($m = 1$) for violet light ($\lambda \approx 400$ nm) are given by $\sin \theta_1 = \pm(1)(400\ \text{nm})/2.0\ \mu\text{m} = \pm 0.20$, so that $\theta_1 = \sin^{-1}(\pm 0.20) = \pm 12°$. Similarly, the first-order maxima for red light ($\lambda \approx 750$ nm) occurs at $\theta_1 = \pm 22°$.

Now let us examine the effect of adding slits to a grating, beginning with a double slit. From Eq. 36-4, the intensity distribution of the double-slit interference pattern is $I = 4I_0 \cos^2(\frac{1}{2}\phi)$. Using the trigonometric identity $2\cos\alpha = \sin(2\alpha)/\sin\alpha$, with $\alpha = \frac{1}{2}\phi$, we can write the double-slit ($N = 2$) intensity distribution as

$$I = I_0 \frac{\sin^2[2(\frac{1}{2}\phi)]}{\sin^2(\frac{1}{2}\phi)}$$

You can show (see Prob. 4) that for the case of three slits ($N = 3$),

$$I = I_0 \frac{\sin^2[3(\frac{1}{2}\phi)]}{\sin^2(\frac{1}{2}\phi)}$$

The general relation for the intensity distribution from a monochromatic source due to a grating with N slits (Prob. 13) is

$$I = I_0 \frac{\sin^2[N(\frac{1}{2}\phi)]}{\sin^2(\frac{1}{2}\phi)} \tag{36-9}$$

In each of these expressions, I_0 is the intensity at the screen due to one of the slits acting alone, and ϕ is the phase difference at the screen between waves from adjacent slits.

Figure 36-10 shows the progression of the intensity distribution from a double slit to a diffraction grating. The important feature of the interference pattern from a grating is that when N is large, the waves combine to give nearly complete cancellation at all angles except those that correspond to the interference maxima. The light is very intense at angles where $d \sin \theta_m = \pm m\lambda$, and much less intense at all other angles (Prob. 7).

Figure 36-10 shows two effects that result from adding more slits. First, the intensity of each interference maximum increases with N as N^2. That is, if we double the number of slits, keeping the slit size fixed, then the intensity of a maximum increases by a factor of 4. Second, the width of each interference maximum decreases with increasing N. The angular half-width $\Delta\theta_{1/2}$ of an interference maximum is defined as the angle between the center of the maximum and its adjacent minimum. The change $\Delta\phi$ in phase difference between the center of an interference maximum and its adjacent minimum is $\Delta\phi_{1/2} = 2\pi/N$. This result is suggested by the progression in Fig. 36-10. Also see Exercise 21. If we differentiate the expression $\phi = (2\pi d \sin \theta)/\lambda$ with ϕ and θ as variables, we find $d\phi = [(2\pi d \cos \theta)/\lambda]d\theta$. Thus a small change $\Delta\phi$ in phase difference corresponds to a small change $\Delta\theta$ in angle: $\Delta\theta = [\lambda/(2\pi d \cos \theta)]\Delta\phi$. Using $\Delta\theta = \Delta\theta_{1/2}$ and $\Delta\phi = \Delta\phi_{1/2} = 2\pi/N$, we find that the angular half-width of an interference maximum is

$$\Delta\theta_{1/2} = \frac{\lambda}{Nd \cos \theta_m} \qquad (36\text{-}10)$$

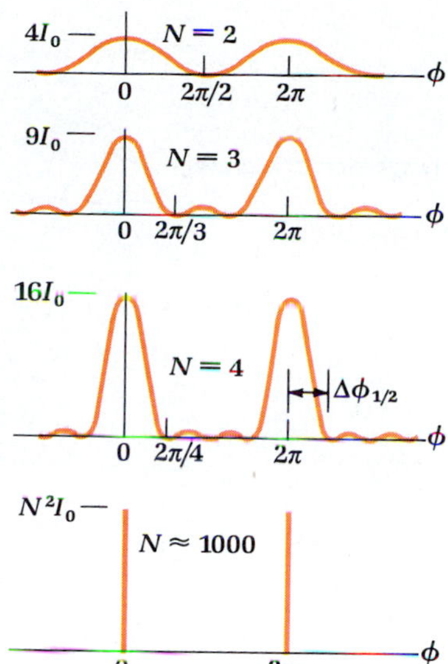

Figure 36-10. Progression of the intensity distribution with increasing N (d is fixed). The light is monochromatic and the slits are very narrow: $\phi = (2\pi d \sin \theta)/\lambda$.

The angular half-width of an interference maximum is proportional to $1/N$.

Suppose we have a source that emits light of two discrete wavelengths (or colors) $\lambda_v = 400$ nm (v for violet) and $\lambda_r = 750$ nm (r for red). For simplicity, we suppose each color is emitted with the same intensity. Figure 36-11 shows the $m = 0$ and $m = 1$ interference maxima in a graph of intensity versus θ for a grating with $d = 1.7$ μm. The light at the center ($\theta = 0$), which corresponds to the zeroth-order maximum, is a mixture of red and violet light, the same as the incident light. The first-order maxima on each side of $\theta = 0$ are spatially separated into violet and red.

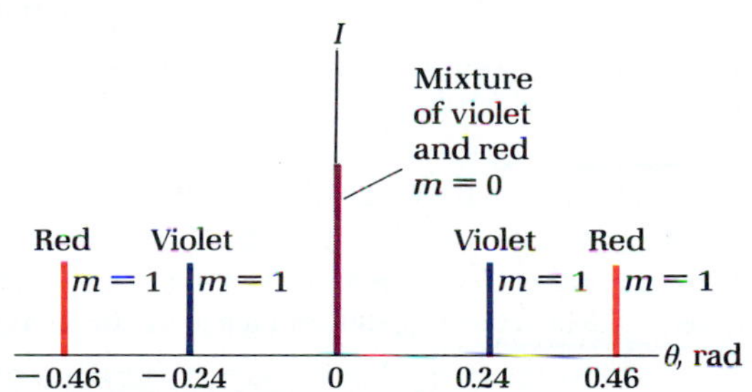

Figure 36-11. First-order interference maxima for light from a source that emits light with two discrete wavelengths of the same intensity.

Discrete wavelengths give a line spectrum.

When the emission from a source is in the form of discrete wavelengths, as in this case, the spectrum is called a ***line spectrum*** because the interference maxima appear as colored lines on the screen.

Continuous distribution of wavelengths gives a continuous spectrum.

Some light sources, such as the sun or an incandescent light bulb, emit light that contains a continuous distribution of wavelengths over a limited range. A grating will separate this light into a ***continuous spectrum*** for each order. When we view, say, a first-order spectrum from such a source, we see all the colors in the visible range of wavelengths spread out across the screen. For the grating in Fig. 36-11, we would see a "rainbow of colors" between $\theta = 0.24$ and 0.46 rad.

A quantity that characterizes the capability of a grating to spread a light beam spatially according to wavelength is called the grating's ***dispersion*** D. The dispersion of a grating is defined as

Definition of the dispersion of a grating

$$D = \frac{\Delta\theta_m}{\Delta\lambda}$$

where $\Delta\theta_m$ is the angular spacing between the interference maxima (of the same order) of waves with wavelengths that differ by an amount $\Delta\lambda$. If we differentiate the expression $d \sin \theta_m = m\lambda$, with θ_m and λ as variables, then we have $(d \cos \theta_m)\, d\theta_m = m\, d\lambda$. Thus two waves of nearly the same wavelength have their interference maxima separated by an amount $\Delta\theta_m = [m/(d \cos \theta_m)]\Delta\lambda$. Using $D = \Delta\theta_m/\Delta\lambda$, we have

Dispersion of a grating

$$D = \frac{m}{d \cos \theta_m} \tag{36-11}$$

Since d is in the denominator, Eq. 36-11 shows that the dispersion is larger for gratings with a smaller slit spacing d.

EXAMPLE 36-2

Angular half-width. *(a)* Determine an expression for $\Delta\theta_{1/2}$ for an interference maximum in terms of its order m. *(b)* What is the angular half-width of the first-order interference maximum when monochromatic light of wavelength 500 nm illuminates 1700 slits of a grating with $d = 1.8\ \mu m$?

Solution. *(a)* For the mth-order interference maximum, $\sin \theta_m = \pm m\lambda/d$. Using the identity $\sin^2 \theta + \cos^2 \theta = 1$, we have

$$\cos \theta_m = \sqrt{1 - \sin^2 \theta_m} = \sqrt{1 - \left(\frac{m\lambda}{d}\right)^2}$$

(We keep only the positive square root. Why?) Substitution into Eq. 36-10 gives

$$\Delta\theta_{1/2} = \frac{\lambda}{Nd\sqrt{1 - (m\lambda/d)^2}} = \frac{1}{N\sqrt{(d/\lambda)^2 - m^2}}$$

(b) Using the above expression, we have

$$\Delta\theta_{1/2} = \frac{1}{1700\sqrt{(1.8\ \mu m/500\ nm)^2 - 1^2}}$$
$$= 1.7 \times 10^{-4}\ \text{rad} = 0.0097°$$

The width of an interference maximum is very narrow for typical gratings because N is very large.

SELF-TEST 36-2. What is the dispersion D of the first-order spectrum for the grating in the above example for light with wavelength 500 nm? ***ANSWER:*** $D = 5.8 \times 10^5$ rad/m $= 5.8 \times 10^{-4}$ rad/nm $= 0.033°$/nm. This means that the angular separation between a maximum due to 500-nm light and a maximum due to 501-nm light is 0.033°.

EXAMPLE 36-3

The Rayleigh criterion. In a sodium vapor lamp, the sodium atoms are excited by an electric arc and emit radiation. A prominent part of the spectrum from this emission consists of two yellow lines with wavelengths 589.00 and 589.59 nm. Suppose a beam of light from a sodium lamp is incident normally on a grating with 12,000 slits and a spacing of 2.1 μm. The beam is narrow so that it illuminates a width of 1.7 mm of the grating. Determine whether the lines in the first-order spectra of these two wavelengths are resolved by the grating.

Solution. We use the *Rayleigh criterion* to decide whether the lines are resolved. According to this criterion, the lines are resolved if their angular separation $\Delta\theta_m$ is greater than the angular half-width $\Delta\theta_{1/2}$ of either line. (See Fig. 36-12.) From our discussion of dispersion, the angular separation $\Delta\theta_m$ of two lines with wavelengths that differ by $\Delta\lambda$ is

$$\Delta\theta_m = \frac{m}{d \cos \theta_m}\Delta\lambda$$

The Rayleigh criterion requires that $\Delta\theta_m > \Delta\theta_{1/2}$. Using $\Delta\theta_{1/2}$ from Eq. 36-11, we have that the lines are resolved when

$$\frac{m}{d \cos \theta_m}\Delta\lambda > \frac{\lambda}{Nd \cos \theta_m}$$

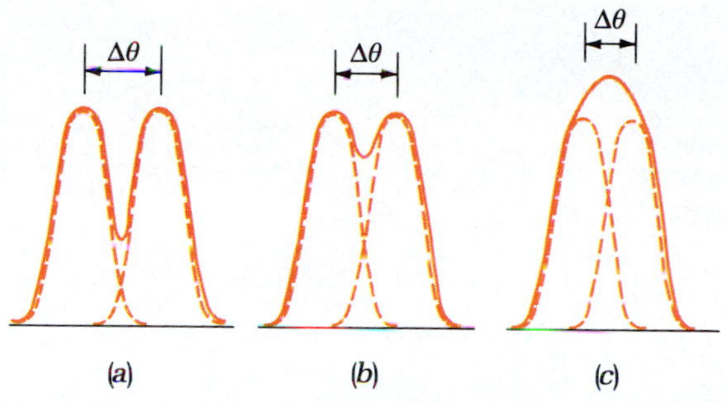

Figure 36-12. Example 36-3: Two spectral lines of equal intensity (shown dashed) separated by an angle $\Delta\theta$. *(a)* $\Delta\theta > \Delta\theta_{1/2}$. *(b)* $\Delta\theta = \Delta\theta_{1/2}$. *(c)* $\Delta\theta < \Delta\theta_{1/2}$. The Rayleigh criterion for resolution is $\Delta\theta > \Delta\theta_{1/2}$.

or

$$\Delta\lambda > \frac{\lambda}{Nm}$$

If the wavelengths of the two lines differ by more than λ/Nm, then the lines are separated enough so that they can be distinguished as two lines rather than one. In this expression, λ is the wavelength of either of the lines (they are nearly equal), N is the number of slits illuminated by the beam, and m is the order of the spectrum. For the beam of light in this example, the width w of the beam is too narrow to illuminate all the slits. The number of slits illuminated is $N = w/d = 1.7\ \text{mm}/2.1\ \mu\text{m} = 810$. For the first-order spectrum of the yellow lines from sodium, $\lambda/Nm = 589\ \text{nm}/[(810)(1)] = 0.73\ \text{nm}$. Since the wavelength difference for these lines is $\Delta\lambda = 589.59\ \text{nm} - 589.00\ \text{nm} = 0.59\ \text{nm}$, they are not resolved; they appear as one line. The lines could be resolved by either illuminating more slits, or by observing a higher-order spectrum.

SELF-TEST 36-3. For the grating in the above example, how many slits need to be illuminated to just resolve the two yellow lines from a sodium arc lamp? ***ANSWER:*** 1000 slits.

36-4 X-RAY DIFFRACTION BY CRYSTALS

X-rays were discovered in 1895 by Wilhelm Roentgen (1845–1923). They can be produced by accelerating electrons to high speeds through a large potential difference, 10 to 100 kV, and then causing these electrons to strike a metal target. X-rays are then emitted from the target (Fig. 36-13.) The identity of x-rays remained a mystery for some years after their discovery; that is why they were called "x." It was known that these rays were very penetrating, could darken photographic film, and could cause minerals to fluoresce. Since the deflection of a beam of x-rays by an electric or magnetic field could not be detected, it was assumed that the beam did not consist of charged particles. These properties led to the conclusion that x-rays consist of short-wavelength electromagnetic radiation—that is, light with wavelengths much smaller than visible light.

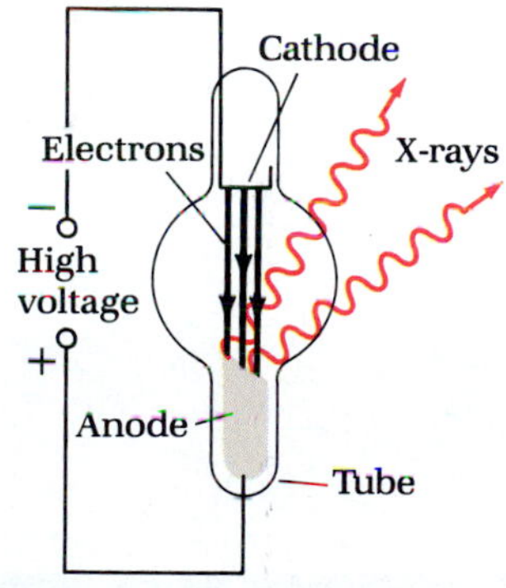

Figure 36-13. An x-ray tube. Electrons are emitted from the cathode and are accelerated to high speeds before striking the anode target. Rapid slowing (deceleration) of the electrons in the anode causes the emission of the x-rays.

In 1912 Max von Laue (1879–1960) suggested an experiment that verified the wave nature of x-rays. Von Laue pointed out that if x-rays have wavelengths λ that are about the same as the spacing d between planes of atoms in crystals, then x-ray waves impinging on crystals would exhibit interference effects. Recall that a transmission grating, because it consists of a regular array of slits, causes light waves to exhibit strong constructive interference at a few particular angles and almost complete cancellation at all other angles. To observe these interference effects, the slit spacing must be almost as small as the wavelength. Similarly, a crystalline solid consists of a regular array of atoms. When a beam of x-rays impinges on a crystal, strong constructive interference effects can be observed readily if the wavelength λ is somewhat smaller than the interplanar spacing d in the crystal.

Figure 36-14 shows a two-dimensional representation of a three-dimensional crystal; the rows of dots portray planes of atoms. X-rays of a single wavelength are in phase before being scattered from the atoms in plane A and the atoms in plane B. For constructive interference of the x-rays scattered from each plane of atoms, the angle of incidence turns out to be equal to the angle of reflection. To reach the detector,

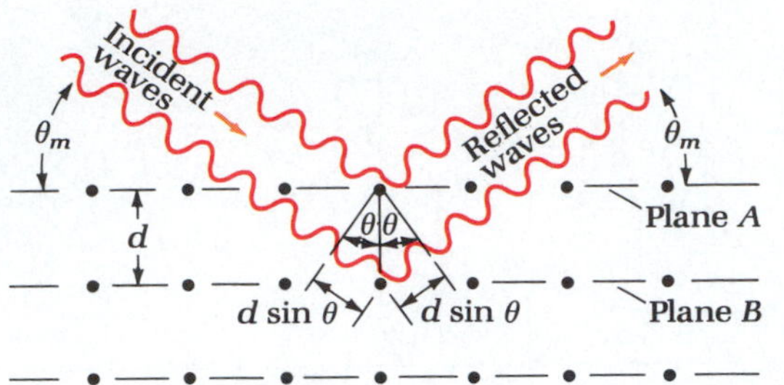

Figure 36-14. To reach the detector, x-ray waves reflected from plane *B* must travel a distance $2(d \sin \theta)$ farther than those reflected from plane *A*. The waves constructively interfere at the detector when $2d \sin \theta = m\lambda$.

the waves scattered from the atoms in plane *B* travel a greater distance than those scattered from the atoms in plane *A* by the amount $2(d \sin \theta)$. If the angle θ_m is given by the relation

Bragg's law

$$2d \sin \theta_m = m\lambda \qquad (m = 1, 2, \ldots) \tag{36-12}$$

then the waves scattered from the atoms in plane *A* will arrive at the detector in phase with the waves scattered from the atoms in plane *B*. Thus the waves will constructively interfere and produce an interference maximum. Similarly, constructive interference will occur for waves scattered from the atoms of each of the many planes that are parallel to planes *A* and *B*. This relation was first developed by W. L. Bragg (1890–1971) and is called ***Bragg's law.***

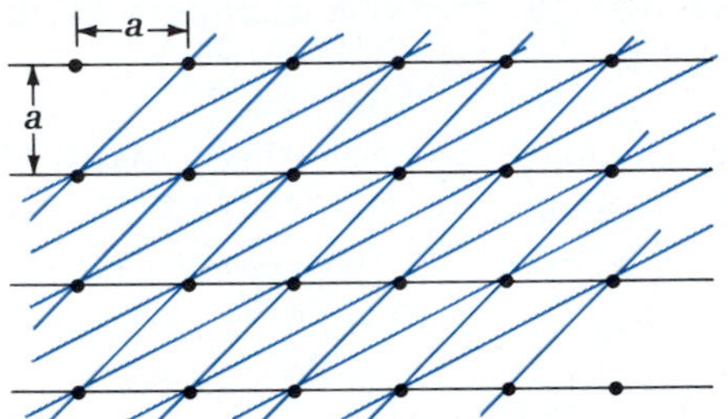

Figure 36-15. Two-dimensional analog of crystalline planes with closely packed atoms. Three such sets of planes are shown.

Bragg's law gives the angles that locate the maxima produced by the constructive interference of x-rays. The interference is caused by the scattering from the atoms in parallel sets of planes. Figure 36-15 shows the two-dimensional analog of three such sets of planes. Thus the distance d in Bragg's law refers to any of the many interplanar spacings that exist in a crystal. However, the intensity of the interference maxima for a set of planes depends on the density of atoms in those planes.

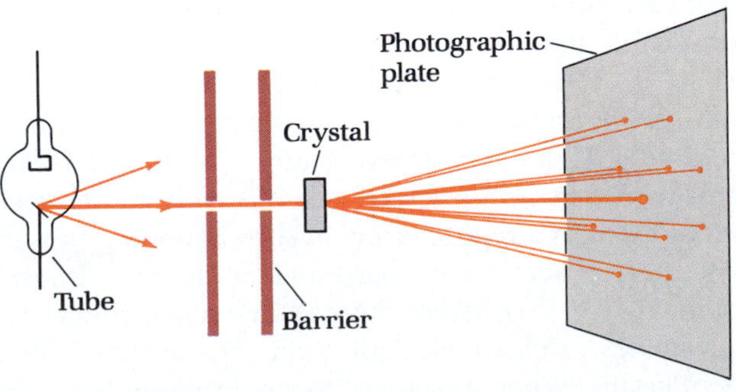

Figure 36-16. Experimental arrangement for x-ray diffraction.

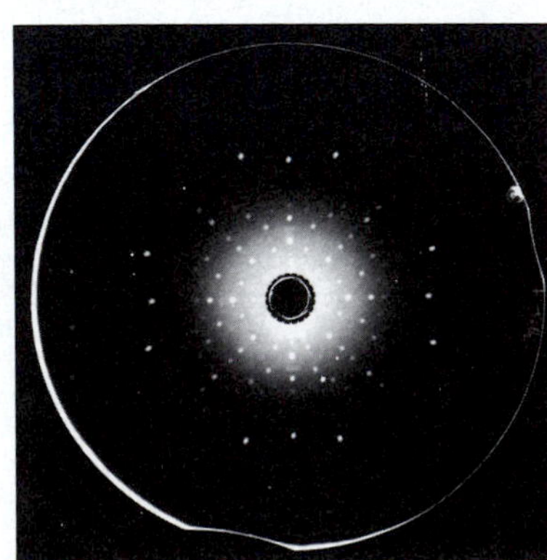

Figure 36-17. Photograph of the Laue pattern from a crystal. The symmetry of the crystal is indicated by the symmetry of the spots.

Figure 36-16 shows one experimental arrangement for x-ray diffraction. A collimated beam of x-rays that contains a continuous distribution of wavelengths strikes the crystal. Interference maxima are produced at angles that satisfy Bragg's law. These maxima form a pattern of spots on the film called a *Laue pattern* (Fig. 36-17). Properties of the crystal can be determined from the positions and intensities of the spots. Indeed, x-ray diffraction is one of the most powerful tools we have for studying the structure of solids.

Figure 36-18. Interference of reflected light from the front and back surfaces of a soap film. The film is very thin at the top, where it appears dark, and increases in thickness toward the bottom so that interference fringes are produced.

36-5 INTERFERENCE FROM THIN FILMS

An easily observed interference effect is that due to reflections from thin transparent films such as soap bubbles or oil films. The brilliant colors you often see reflected from such films are caused by interference.

Figure 36-18 shows a photograph of a soap film in a circular wire loop, the sort of arrangement used by children to blow bubbles. The loop is oriented vertically so that the film is slightly wedge-shaped because of its own weight; it is very thin at the top and becomes thicker toward the bottom. A source of monochromatic light was positioned behind the camera in taking the photograph in Fig. 36-18, and the incident light beam is directed nearly normal to the film surface. The light from the soap film arrived at the camera after reflection from the film's front and back surfaces. Two features of this reflected light are apparent from Fig. 36-18: (i) There is no reflected light from the top region of the film, which appears dark; and (ii) horizontal interference fringes occur below the dark region.

Consider the reflections that occur at the two surfaces of the soap film. Figure 36-19 shows a light beam at nearly normal incidence on a transparent film with air on each side. (The angle between the beam direction and the normal is exaggerated for clarity.) The incident beam (1) is split into two beams at the film's front surface: a reflected beam (2) and a transmitted beam (3). The transmitted beam (3) is then split into two beams at the film's back surface: a reflected beam (4) and a transmitted beam (5). The reflected beam (4) is further split into a transmitted beam (6) and a reflected beam (not shown) at the film's front surface, and so on. The superposition of beams (2) and (6) is observed at the camera. These beams are coherent because they originate from the same source. Also, their waves usually have about the same amplitude. (See Prob. 11.) It is interference between the waves in these two beams that causes the effects shown in Fig. 36-18.

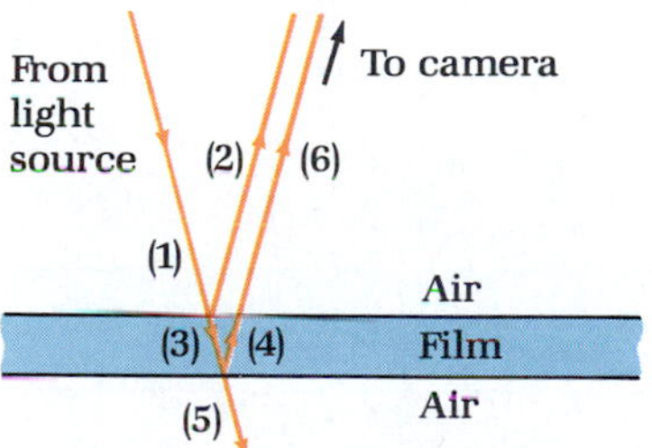

Figure 36-19. Reflections from the front and back surfaces of a film. Beams (2) and (6) have nearly the same intensity and are coherent.

There is an additional point to consider before we attempt to understand the effects shown in Fig. 36-18. Under certain conditions light waves undergo a phase change of π rad upon reflection; that is, they undergo a ***phase reversal***. Suppose light is incident on a boundary between transparent media. If the light is incident on the boundary from the medium with lower index of refraction, then the reflected waves undergo a phase reversal. If the light is incident on the boundary from the medium with higher index, then the reflected waves do not undergo a phase reversal. (A verse which may help you remember this is "low to high, phase change by π.") These phase relationships are predicted from the application of Maxwell's equations to the electric and magnetic fields at the boundary. Waves incident on the boundary between two stretched springs show similar phase relations (Sec. 32-2).

Reflection from a surface where the index changes from low to high causes a phase reversal.

Since the film in Fig. 36-19 has a larger index of refraction than air, the waves in beam (2) change phase by π rad when they are reflected, but the waves in beam (6) do not change phase when they are reflected.

Suppose the film in Fig. 36-19 has a thickness τ that is much less than the wavelength λ_n of the light. In this case, the waves in beams (2) and (6) will travel nearly the same distance to reach P; they will not become out of phase because of traveling different distances. However, because of the phase reversal of the waves in beam (2), the waves in beams (2) and (6) emerge from the film nearly π rad out of phase and interfere destructively. Therefore the top of the film appears dark. The dark appearance of the film in this region is verification of the phase reversal discussed above.

The film is thicker at positions farther down. Consider a position where the thickness is a quarter of a wavelength: $\tau = \lambda_n/4$. The waves in beam (6) travel farther than the waves in beam (2) by the amount $2\tau = 2(\lambda_n/4) = \frac{1}{2}\lambda_n$. This path difference for the two beams causes a phase difference of π rad. Since the phase reversal of beam (2) causes an additional phase difference of π rad, the waves emerge in phase. A bright fringe appears along the region of the film that has a thickness of $\lambda_n/4$. The relation between film thickness and wavelength that corresponds to constructive interference is

$$2\tau = (m + \tfrac{1}{2})\lambda_n \qquad (m = 0, 1, 2, \ldots) \tag{36-13}$$

Similarly, the relation that corresponds to destructive interference is

$$2\tau = m'\lambda_n \qquad (m' = 0, 1, 2, \ldots) \tag{36-14}$$

Thus, as the film in Fig. 36-18 becomes progressively thicker toward the bottom, interference fringes are produced in accordance with Eqs. 36-13 and 36-14.

Note that λ_n in Eqs. 36-13 and 36-14 refers to the wavelength in the film. If the film has an index of refraction n, the wavelength in the film is

$$\lambda_n = \frac{\lambda}{n}$$

where λ is the wavelength in vacuum (or, to a good approximation, in air).

Now we can understand the origin of the full spectrum of colors we often see reflected from a soap bubble or a film of oil. As you can see in Fig. 36-20, when light is reflected from a film at various angles of incidence, the extra distance $2\tau_{\text{eff}}$ that beam (6) must travel compared with beam (2) depends on the angle of incidence. (See Exercise 39.) This gives an angular dependence in the equations that describe interference maxima (and minima). When white light is incident on a film at a particular angle, the color corresponding to the wavelength that satisfies the relation describing constructive interference will be reflected with greater intensity than other colors. As we view different places on the film, we see light that is reflected at different angles. At each angle a given color is reflected with greater intensity than the other colors. This property, where the color of an object depends on the angle from which the object is viewed, is called ***iridescence.*** Thus the film appears as a rainbow as the incident white light is dispersed into its constituent colors.

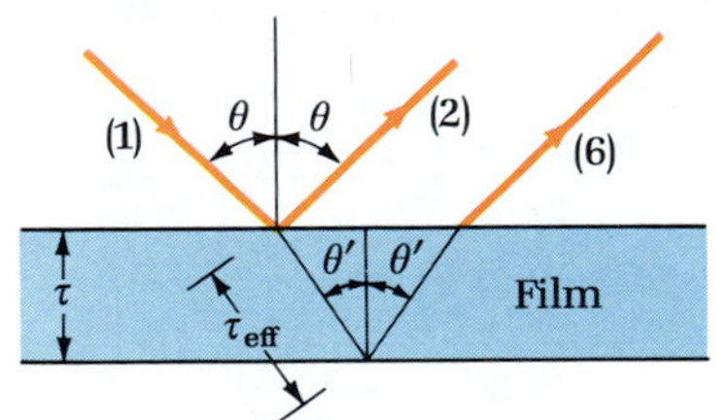

Figure 36-20. A beam of light incident at an angle θ to a normal to a film. Beam (6) must travel an extra distance $2\tau_{\text{eff}}$ compared with beam (2).

EXAMPLE 36-4

Nonreflecting coatings. To reduce reflection from an optical surface, such as the surface of a lens, the surface is often coated with a thin film. Suppose a glass lens ($n = 1.50$) is coated with a thin film of MgF_2 ($n = 1.38$). Determine the minimum thickness for a coating that will minimize reflection for normal incidence of light near the middle of the visible spectrum, say $\lambda = 550$ nm.

Solution. Equation 36-14 describes minimum reflection from a thin film (destructive interference between reflected beams), but it is not valid in this case. That equation was developed for the case where one of the beams had a phase reversal upon reflection and the other did not. For the case of a MgF_2 film on glass, the light reflected from the front surface of the film is incident from the medium of lower index (from air with $n = 1.00$ to MgF_2), so the reflected waves undergo a phase reversal. The light reflected from the back surface is also incident from the medium of lower index (from MgF_2 to glass), so these reflected waves also undergo a phase reversal. Since the effect of these two phase reversals cancels, the phase difference between waves reflected from the two surfaces is due to the path difference 2τ; the phase difference is π rad if 2τ equals an odd half-integral number of wavelengths. Thus Eq. 36-13 describes destructive interference (minimum reflection) in this case. Solving for τ, we have

$$\tau = \tfrac{1}{2}(m + \tfrac{1}{2})\lambda_n = \frac{\frac{1}{2}(m + \frac{1}{2})\lambda}{n}$$

Since we wish to find a minimum thickness for the film, we set $m = 0$. The extra distance that provides the phase difference is in the MgF_2 film, so $n = 1.38$. Thus

$$\tau = \frac{1}{2}\frac{1}{2}\frac{550 \text{ nm}}{1.38} = 100 \text{ nm}$$

This film thickness will minimize reflection of light with wavelengths near the middle of the visible spectrum. Lenses often have a purplish tint because the reflection contains only a small amount of light from the middle of the visible spectrum; most of the reflected light is from the red and violet ends of the spectrum. Thin films can also be used to maximize reflection. (See Exercise 38.)

SELF-TEST 36-4. For the coated glass surface in the above example, what wavelength of light nearest 550 nm will undergo a maximum reflection? ***ANSWER:*** 276 nm.

36-6 THE MICHELSON INTERFEROMETER

A device that uses wave-interference effects to make measurements is called an ***interferometer.*** Because the wavelength of visible light is very small, optical interferometers can be used to measure distance with great precision. The Michelson interferometer (named for A. A. Michelson, 1852–1931) is one of simple design and of historical importance.

Figure 36-21*a* shows the principal features of a Michelson interferometer. A beam of monochromatic light from the source impinges on mirror $M_{1/2}$, which is mounted at a 45° angle to the beam direction. The silver coating on this mirror is just the right thickness so that it reflects half of the incident beam and transmits the other half. Thus half of the incident beam is directed toward a movable mirror M_m [beam (1)], and the other half is directed toward a fixed mirror M_f [beam (2)]. Mirrors M_m and M_f then reflect the beams back to $M_{1/2}$, where half of each beam is again reflected and transmitted. We are interested in the parts of each beam that combine to form beam (3) and then propagate to the detector.

Beam (3) consists of waves that have traversed different distances. Beam (1) traversed a distance of $2\ell_1$ and beam (2) traversed a distance of $2\ell_2$. The difference in distance traversed by the beams is $|2\ell_2 - 2\ell_1|$. Since beams (1) and (2) are of equal intensity and are coherent (why?), they will constructively interfere if

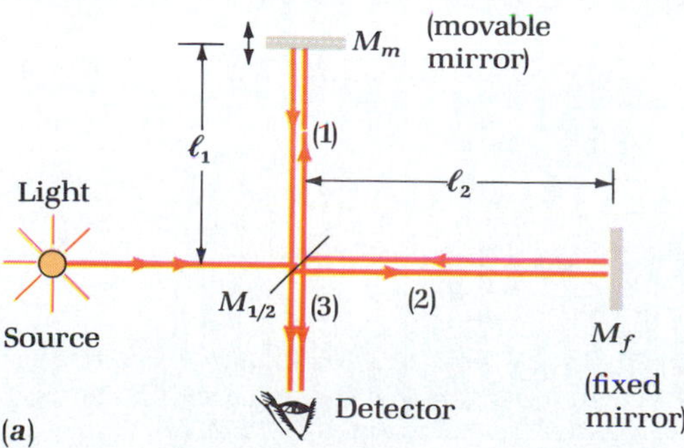

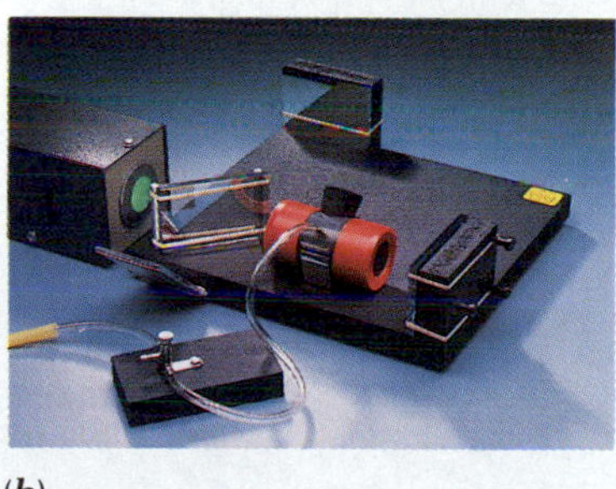

Figure 36-21. *(a)* A schematic diagram of the Michelson interferometer. *(b)* Michelson interferometer.

$2|\ell_2 - \ell_1|$ is an integral number of wavelengths, and they will destructively interfere if $2|\ell_2 - \ell_1|$ is an odd half-integral number of wavelengths. (For simplicity we ignore the thickness of $M_{1/2}$.) Therefore the intensity of the light measured at the detector depends on this distance difference.

Mirror M_m can be smoothly displaced small distances along the direction of beam (1) by turning a finely threaded screw. The light intensity measured by the detector can be caused to change from a maximum to a minimum by moving M_m a distance of $\lambda/4$. With a sensitive detector, distances as small as 1 percent of a wavelength (about 5 nm) can be readily detected.

At the time Michelson developed his interferometer, the standard of length, the meter, was defined as the distance between two fine scratches in a particular platinum-iridium bar kept at Sèvres, France. Using his interferometer, Michelson was able to measure the wavelength of an emission line from a cadmium light source with an accuracy of one part in 10^8. This capability ultimately led to a definition of the meter in terms of the wavelength of a certain spectral line from the element krypton.

COMMENTARY: Thomas Young

Thomas Young.

Thomas Young was born at Milverton, Somerset, England, in 1773. Blessed with a remarkable memory and extraordinary mechanical ability, Young applied his talents to physics, medicine, physiology, and languages. By the age of 14, he was familiar with Latin, Greek, Hebrew, Arabic, Persian, French, and Italian, and by the time he was 17, he had mastered Newton's *Principia* and *Opticks*.

Young's great-uncle, who was a prominent physician, persuaded him to study medicine so he could eventually acquire his practice. Young's studies of vision and hearing seem to have led to his more fundamental investigations of light and sound. He reported his definitive experiments on the interference of light in a paper entitled "Outlines of Experiments and Enquiries Respecting Sound and Light" (1800). There he compared the corpuscular theory of Newton with the wave theory of Huygens and showed that the experimental results could be explained with a wave theory, thus favoring Huygens. The suggestion that Newton could be wrong was not greeted with enthusiasm, especially in England, and it angered some scientists. Young showed how the wave theory correctly describes reflection and refraction, and he later explained the dispersive effects of gratings and the iridescence of thin films. Despite the fact that Young is generally credited with establishing the wave theory of light, he remained skeptical of his own theory.

Young published many important papers in physiology. He originated concepts that later developed into the three-color theory of vision. He investigated the functions of the heart and arteries and published a paper on capillarity and cohesion of fluids. He used his linguistic abilities to decipher Egyptian hieroglyphic inscriptions, and is especially remembered for his work on the famous Rosetta Stone.

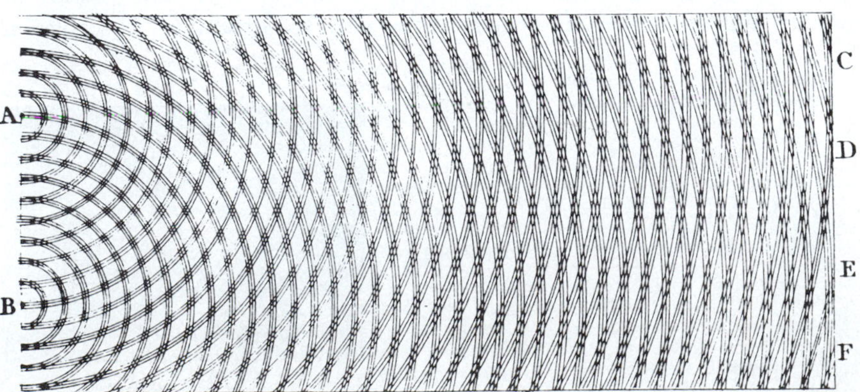

Young's original drawing illustrating the interference of light from two coherent sources. Compare this figure with Fig. 36-5.

SUMMARY

Section 36-1. Young's Double-Slit Experiment

The double-slit experiment demonstrates the wave nature of light because the pattern on the screen can be explained in terms of the interference of waves. Constructive interference gives bright fringes at angles θ_m:

$$d \sin \theta_m = \pm m\lambda \qquad (m = 0, 1, 2, \ldots) \qquad (36\text{-}2)$$

and destructive interference gives dark fringes at angles $\theta_{m'}$:

$$d \sin \theta_{m'} = \pm (m' + \tfrac{1}{2})\lambda \qquad (m' = 0, 1, 2, \ldots) \qquad (36\text{-}1)$$

Two coherent sources can produce a stable interference pattern. The two slits in the double-slit experiment are coherent sources because the light originates from a single source.

Section 36-2. Intensity Distribution in the Double-Slit Interference Pattern

The principle of superposition gives the intensity distribution for the double-slit interference pattern as

$$I_{12} = 4I_0 \cos^2\left(\frac{\pi d \sin \theta}{\lambda}\right) \qquad (36\text{-}6)$$

where I_0 is the intensity at the screen due to one of the slits acting alone.

Section 36-3. Diffraction Gratings

A diffraction grating is a device used to disperse light according to its wavelength. Angles which locate the interference maxima for light of a specific wavelength are given by

$$d \sin \theta_m = \pm m\lambda \qquad (m = 0, 1, 2, \ldots) \qquad (36\text{-}8)$$

where m is the order of the maximum.

Section 36-4. X-Ray Diffraction by Crystals

X-rays reflected from parallel sets of atomic planes exhibit interference effects. This demonstrates the wave nature of x-rays and provides a powerful tool for studying crystalline solids. Bragg's law gives the condition for constructive interference:

$$2d \sin \theta_m = m\lambda \qquad (m = 1, 2, \ldots) \qquad (36\text{-}12)$$

Section 36-5. Interference from Thin Films

Interference effects can be easily observed when light is reflected from thin transparent films. With these effects, we are able to use thin films on optical surfaces either to enhance or to inhibit reflection.

Section 36-6. The Michelson Interferometer

A Michelson interferometer splits a beam of light into two beams that travel different distances and then are rejoined. The observation of interference of the waves in these beams provides a technique for precisely measuring length.

QUESTIONS

1. What is meant by interference? What is the phase relationship between two waves that interfere constructively? What is the phase relationship between two waves that interfere destructively?

2. What is meant by diffraction? Do sound waves diffract? Would high frequencies or low frequencies be easier to hear around corners?

3. Why did we specify the use of monochromatic light in the double-slit experiment? Describe the appearance of the pattern if we used white light.

4. With the assumption that light consists of a beam of particles or corpuscles, develop a theory that describes the double-slit experiment. How do you account for the dark fringes in the pattern? Recall that these dark fringes occur at points on the screen that are not dark when one of the slits is covered. What property of the particles corresponds to different colors? From this property, how do you account for red light giving a pattern that is more spread out than violet light? Give up?

5. Explain how the double-slit experiment shows that different colors correspond to different wavelengths.

6. The intensity at the interference maxima in the double-slit pattern is $4I_0$, where I_0 is the intensity at that point due to one of the slits acting alone. Does this violate conservation of energy? If not, explain why not.

7. Suppose we perform the double-slit experiment under water. How would the pattern be affected?

8. Suppose the double-slit experiment is performed with a glass plate with index of refraction n covering the entrance to one of the slits (Fig. 36-22). Describe the effect of this plate on the resulting pattern on the screen.

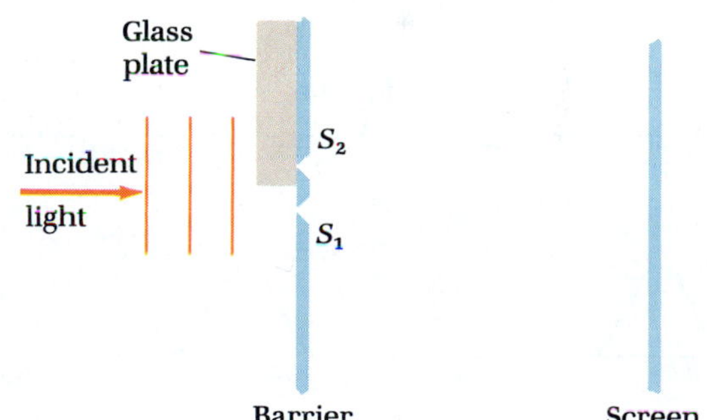

Figure 36-22. Question 8: A glass plate with index of refraction n covers the entrance to one of the slits.

9. Why are interference effects not more commonly observed? For example, why do we not observe interference from the light from two automobile headlamps?

10. In the arrangement shown in Fig. 36-3, suppose we let white light from the bulb fall on slits 1 and 2 and put a red color filter over slit 1 and a blue color filter over slit 2. Describe the resulting intensity pattern on the screen.

11. Consider water waves from two spheres vibrating in phase as seen in Fig. 36-5. If the separation between the spheres is 4λ, how many nodal lines will exist?

12. In his original experiment, Young used pinholes rather than slits and illuminated them with sunlight. Describe the pattern he saw on the screen. Why do you think we chose to discuss an experiment with slits and monochromatic light?

13. In developing the double-slit intensity distribution, we write the vector components of the electric field of light waves from slits 1 and 2 as $E_1 = E_0 \sin(\omega t + \phi)$ and $E_2 = E_0 \sin(\omega t)$. On what basis do we set the amplitude E_0 the same in these expressions? On what basis do we set the frequency ω the same in these expressions? Does ϕ represent the phase difference as the waves emerge from the slits or as they arrive at the screen? How does this phase difference arise?

14. Imagine observing the double-slit pattern for light of a given wavelength and gradually reducing the slit spacing d. What happens to the pattern? Is there a minimum spacing for observing a pattern? If so, what is this spacing?

15. What is the purpose of a diffraction grating? Why does a grating have a large number of slits? Why are the slits of a grating spaced very close together?

16. Figure 36-23 shows the interference maxima of the same two spectral lines as they appear in first-order diffraction by three different gratings: *A*, *B*, and *C*. (For simplicity we do not show the secondary maxima.) Which grating has the largest number of slits illuminated? Which grating has the smallest slit spacing? Which grating has the largest dispersion?

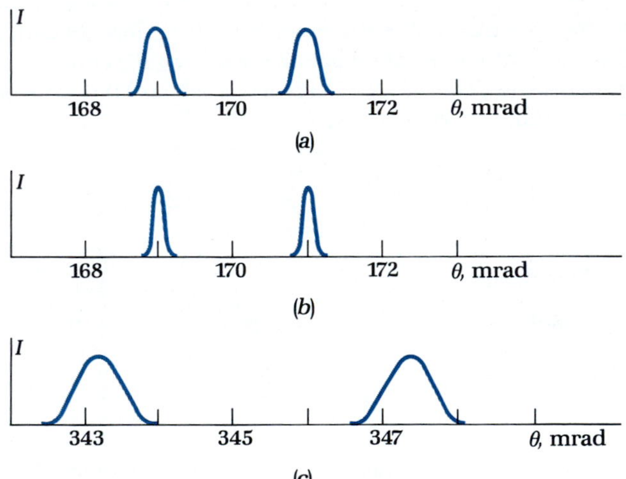

Figure 36-23. Question 16 and Exercise 26: First-order interference maxima of the same two spectral lines with *(a)* grating *A*, *(b)* grating *B*, and *(c)* grating *C*.

17. Bragg's law, $2d \sin \theta_m = m\lambda$, is closely related to the equation for the interference maxima due to a grating: $d \sin \theta_m = m\lambda$. Explain the origin of the factor 2 in Bragg's law.

18. If a beam of highly monochromatic (one wavelength) x-rays impinges on a single crystal with a random orientation, then, in general, no interference maxima will be observed. However, if the beam contains a continuous distribution of wavelengths (polychromatic, or "white," x-rays), then a Laue pattern is observed. Explain.

19. When light waves in air encounter the surface of water, is their phase reversed upon reflection?

20. Why must a film be thin to cause observable interference effects? Is it really important for the film to be thin, or is it only important that the film's opposite surfaces be very nearly parallel?

21. State which of the following objects are iridescent: a quartz crystal, a peacock's tail feather, the inside of an oyster shell, a TV screen, a soap bubble.

22. As a light wave passes from one transparent medium to another, say from air to glass, which, if any, of the wave's following characteristics change: speed, wavelength, frequency? Describe any change.

23. The index of refraction of air is very nearly 1, the value for vacuum. Explain how the Michelson interferometer could be used to accurately measure the index of refraction of air.

24. Complete the following table:

Symbol	Represents	Type	SI Unit
m, m'			
ϕ			rad
I_0			
N	Number of slits		
$\Delta\theta_{1/2}$			
D		Scalar	

EXERCISES

Section 36-1. Young's Double-Slit Experiment

1. Monochromatic light with a wavelength of 563 nm is used in a double-slit experiment in which the slit spacing is 0.18 mm. *(a)* What are the angles of the $m = 1$ and $m = 2$ maxima? *(b)* What are the angles of the $m' = 1$ and $m' = 2$ minima?

2. Monochromatic light with a wavelength of 713 nm is used in a double-slit experiment in which the slit spacing is 0.14 mm and the distance between the slits and the screen is 0.96 m. *(a)* What is the distance on the screen between the center of the zeroth-order maximum and one of the first-order maxima? *(b)* What is the distance between the center of the zeroth-order maximum and one of the $m' = 3$ minima?

3. Suppose you wish to project a double-slit pattern on a screen that is about 1 m from the slits. Your light source is a sodium lamp that has a bright yellow emission at about

590 nm. *(a)* If you make your slits with a spacing of about 2 mm, do you expect to have any trouble seeing a pattern? *(b)* Using $\lambda \approx 600$ nm and $L \approx 1$ m, develop a rough criterion for the maximum slit spacing that produces a pattern discernible with the naked eye.

4. In a double-slit experiment, $\lambda = 488$ nm, $L = 1.14$ m, and the spacing between dark fringes is $\Delta x = 6.1$ mm. What is the slit spacing.

5. Suppose you have a double slit with a spacing of 0.15 mm and a sodium lamp (which produces a bright yellow emission at 589 nm). How far from the slits should you place the screen to observe bright fringes that are 10 mm apart?

6. Monochromatic light of wavelength 730 nm is used in a double-slit experiment. Consider the distances the waves from each slit must travel to arrive at point P on the screen in Fig. 36-6. Waves from slit 1 travel a distance r_1, and waves from slit 2 travel a distance r_2. What is the distance difference, $r_1 - r_2$, when *(a)* point P is at the center of a second-order bright fringe, *(b)* point P is at the center of an $m' = 3$ dark fringe, *(c)* point P is halfway between a second-order bright fringe and an $m' = 2$ dark fringe?

7. Consider the acoustical equivalent of the double-slit experiment. Suppose you use two loudspeakers that emit sound of frequency 660 Hz. (The speed of sound in air is 330 m/s.) *(a)* To produce an easily measurable pattern, about how far apart should you place the loudspeakers? *(b)* At least how far away should you be with your microphone? *(c)* If the loudspeakers are 2.0 m apart, how many nodal lines will exist?

Section 36-2. Intensity Distribution in the Double-Slit Interference Pattern

8. In a double-slit experiment, the intensity at a particular point on the screen due to each slit acting alone is I_0. What is the intensity due to both slits when the waves have a phase difference of *(a)* zero, *(b)* $\pi/2$ rad, *(c)* π rad, *(d)* $3\pi/2$ rad, *(e)* 2π rad, *(f)* $5\pi/2$ rad?

9. Make a graph of Eq. 36-5. Plot I_{12} (in units of I_0) versus ϕ (in units of rad) from zero to 4π rad, with points every $\pi/4$ rad. Sketch a continuous curve through the points.

10. In a double-slit experiment, the intensity at the screen due to each of the slits acting alone is I_0, the slit spacing is 0.150 mm, and the wavelength of the light is 500 nm. What is the intensity due to both slits at points on the screen which correspond to angles θ of *(a)* zero, *(b)* 0.833 mrad, *(c)* 1.67 mrad, *(d)* 2.50 mrad, *(e)* 3.33 mrad, *(f)* 4.17 mrad? *(g)* What is the intensity at each of the above angles if the slits are replaced by two incoherent sources that each produce an intensity of I_0 when acting alone?

11. Make a graph of Eq. 36-6. Let $d = 0.150$ mm and $\lambda = 500$ nm. Plot I_{12} in units of I_0 versus θ in units of rad from zero to 5.00 mrad, with points every 0.833 mrad. Sketch a continuous curve through the points.

12. In a double-slit experiment, the intensity at the screen due to each of the slits acting alone is I_0, the slit spacing is 0.150 mm, the wavelength of the light is 500 nm, and the perpendicular distance from the slits to the screen is 1.00 m. What is the intensity due to both slits at positions x (see Fig. 36-6) of *(a)* zero, *(b)* 0.83 mm, *(c)* 1.7 mm, *(d)* 2.5 mm, *(e)* 3.3 mm, *(f)* 4.2 mm? *(g)* What is the intensity at each of the above points if the slits are replaced by incoherent sources, each of which produces intensity I_0 when acting alone?

13. Using the same data as Exercise 11, make a graph of Eq. 36-7. Plot I_{12} in units of I_0 versus x (see Fig. 36-6) in units of millimeters from zero to 4.2 mm, with points every 0.83 mm. Sketch a continuous curve through the points.

14. For a double-slit interference pattern, find the relationship similar to Eqs. 36-1 and 36-2 that gives the angles θ at which $I_{12} = 2I_0$.

15. Consider the first bright fringe on the positive x side of the zeroth bright fringe in a double-slit pattern. Let $d = 0.16$ mm, $\lambda = 550$ nm, and $L = 1.24$ m. Find the coordinate x of each of the two positions on each side of the fringe at which I_{12} is 75 percent its maximum.

Section 36-3. Diffraction Gratings

16. A hydrogen discharge tube emits light in the visible spectrum at four discrete wavelengths: the H_α line at $\lambda_\alpha = 656.3$ nm, the H_β line at $\lambda_\beta = 486.1$ nm, the H_γ line at $\lambda_\gamma = 434.1$ nm, and the H_δ line at $\lambda_\delta = 410.2$ nm. If a collimated beam of light from a hydrogen discharge tube is incident normally on a diffraction grating with a slit spacing of 1.9 μm, what are the angles that locate the first- and second-order maxima for these hydrogen lines?

17. A first-order maximum of the H_α line ($\lambda_\alpha = 656.3$ nm) from a hydrogen discharge tube is at $\theta = 18.3°$ for a particular grating. *(a)* What is the slit spacing of the grating? *(b)* A discrete line from another source has a maximum at $\theta = 15.7°$. What is the wavelength of this emission?

18. We have two gratings, A and B. The slit spacing of grating A is 1.86 μm, but we do not know the slit spacing of grating B. A monochromatic source gives a first-order line at $\theta = 19.4°$ when the light is incident normally on A, and the same source gives a first-order line at 22.1° when the same light is incident on B. What is the slit spacing of grating B?

19. Verify the angular positions of the lines in Fig. 36-11. The slit spacing is $d = 1.7$ μm, the wavelength of the red emission is $\lambda_r = 750$ nm, and the wavelength of the violet emission is $\lambda_v = 400$ nm.

20. Make a graph of Eq. 36-9, $I = I_0 \sin^2 [N(\frac{1}{2}\phi)]/\sin^2 (\frac{1}{2}\phi)$, between $\phi = 0.0001$ rad and $\phi = (2\pi - 0.0001)$ rad for the case where $N = 5$. (Note that I is indeterminate at $\phi = 0$ and 2π. See Prob. 5.) Plot points for $\phi = 0.0001$, $\pi/10$, $2\pi/10$, . . . , $19\pi/10$, and $(2\pi - 0.0001)$ rad. Sketch a continuous curve through the points. Compare your graph with Fig. 36-10.

21. Consider a minimum adjacent to a primary maximum in Eq. 36-9. To be specific, consider the minimum on the positive ϕ side of the central primary maximum ($m = 0$). *(a)* By direct substitution show that $I = 0$ when $\phi = 2\pi/N$. *(b)* Show that I is

nonzero in the range $0 < \phi < 2\pi/N$. (Note that I is indeterminate at $\phi = 0$. See Prob. 5.)

22. *(a)* Write Eq. 36-9 in terms of the angular position θ. *(b)* When N is odd, a secondary maximum occurs midway between primary maxima (see Fig. 36-10). Find the angular position of the secondary maximum that is midway between the zeroth-order and first-order primary maxima for the case where $N = 9$, $d = 40.0\ \mu m$, and $\lambda = 600$ nm. *(c)* Substitute your answer from part *(b)* into the expression you found in part *(a)* to find the intensity at this secondary maximum. *(d)* What is the ratio of this intensity to the intensity at a primary maximum?

23. A beam of monochromatic light with $\lambda = 532.8$ nm is incident normally on a grating with $d = 2.16\ \mu m$. *(a)* What are the angular positions of the first- and second-order interference maxima? *(b)* If 758 of the grating slits are illuminated by the beam, what is the angular half-width of each of these maxima?

24. Suppose the beam of light from the hydrogen discharge tube in Exercise 16 illuminates 844 slits of the grating. *(a)* What is the half-width of the first-order interference maxima of the H_α ($\lambda_\alpha = 656.3$ nm) and the H_δ ($\lambda_\delta = 410.2$ nm) lines? *(b)* What is the value of m such that the mth maximum of the H_α line has a larger angle than the $(m + 1)$th maximum of the H_δ line.

25. *(a)* By using Eq. 36-7, show that the angular half-width of a bright fringe or interference maximum in the double-slit pattern is $\Delta\theta_{1/2} \approx \lambda/2d$. [Note that Eq. 36-7 is valid only if $x \ll L$.] *(b)* Show that the result in part *(a)* is consistent with Eq. 36-10.

26. The two emission lines which gave the intensity distributions shown in Fig. 36-23 have wavelengths $\lambda_1 = 504.6$ nm and $\lambda_2 = 510.5$ nm. *(a)* From information given in the figure, determine the slit spacing for each of the gratings. *(b)* From information given in the figure, estimate the number of slits illuminated by the beam for each of the gratings.

27. Consider monochromatic light with $\lambda = 550$ nm incident normally on a grating with slit spacing $d = 2.11\ \mu m$. *(a)* What are the angular positions of the first-order interference maxima? *(b)* What is the dispersion of this grating at the angular positions of the maxima found in part *(a)*?

28. *(a)* Show that the expression for the dispersion of a grating can be written as

$$D = \frac{m}{\sqrt{d^2 - (m\lambda)^2}}$$

(b) Use the expression in part *(a)* to find the dispersion for the case where $m = 1$, $d = 2.11\ \mu m$, and $\lambda = 550$ nm. Compare your answer with that from the previous exercise.

29. Show that the expression for the dispersion of a grating can be written as $D = (\tan\theta_m)/\lambda$.

30. A beam of light from a source whose emission contains two lines of wavelengths 462.74 and 463.35 nm is incident normally on a grating, and the beam illuminates 1000 slits. According to the Rayleigh criterion, are the first-order interference maxima for these lines resolved?

31. For the grating in Example 36-3, determine the minimum number of slits that must be illuminated in order to satisfy the Rayleigh criterion.

32. In Fig. 36-12*b*, the spacing $\Delta\theta$ between two spectral lines of equal intensity is the same as their half-width $\Delta\theta_{1/2}$. Show that the ratio of the intensity at the point midway between the maxima and the intensity at the maxima is $2(2/\pi)^2 = 0.81$. [*Hint:* $\sin(\pi/2N) \approx \pi/2N$ for large N.]

33. ***Resolving power R.*** The resolving power of an optical device whose purpose is to disperse light according to wavelength is defined as

$$R = \frac{\lambda}{\Delta\lambda}$$

where $\Delta\lambda$ is the difference in wavelength of two spectral lines that can barely be resolved and λ is the wavelength of either line (they have nearly the same wavelength). Show that the resolving power of a grating for the mth-order spectrum is Nm.

Section 36-4. X-Ray Diffraction by Crystals

34. An interference maximum for the scattering of a beam of x-rays of wavelength 0.156 nm occurs when the angle between the beam and the surface of the crystal face is 12.8°. This maximum is due to scattering from the atoms in planes parallel to the crystal face. Assuming that this is a first-order maximum, determine the interplanar spacing for this set of planes of atoms.

35. A beam of x-rays gives a second-order interference maximum when it makes an angle of 24.1° with crystalline planes that have an interplanar spacing of 0.314 nm. What is the wavelength of these x-rays?

36. X-rays of wavelength 0.114 nm are scattered from the atoms in sets of crystalline planes with a spacing of 0.278 nm. At what angle will a first-order interference maximum occur?

Section 36-5. Interference from Thin Films

37. A thin film of a transparent material with $n = 1.29$ is to be placed on a glass ($n = 1.50$) surface. What is the minimum thickness for the film such that the reflection of normally incident light with $\lambda = 600$ nm is minimized?

38. A thin film of transparent material ($n = 1.27$) with air on each side is exposed to a normally incident beam of light that contains the full visible spectrum of colors. If the thickness of the film is 112 nm, what wavelength(s) of light will exhibit maximum reflection?

39. For nonnormal incidence of light on a thin film as shown in Fig. 36-20, show that

$$\tau_{\text{eff}} = \frac{\tau}{\cos\{\sin^{-1}[(\sin\theta)/n]\}}$$

where n is the index of refraction of the film.

40. A thin film of water ($n = 1.33$) on a flat glass ($n = 1.50$) surface is illuminated by a normally incident beam of light. The light in the beam is monochromatic but its wavelength can be varied. As the wavelength is varied continuously, the reflected intensity changes from a minimum at $\lambda = 530$ nm to a maximum at $\lambda = 790$ nm. What is the thickness of the film?

Section 36-6. The Michelson Interferometer

41. When the movable mirror of a Michelson interferometer is moved a distance Δl, 140.0 full fringes pass by the detector. (A full fringe consists of a maximum and a minimum in intensity.) The light used has a wavelength of 526.31 nm. Determine Δl.

42. Suppose we use a Michelson interferometer to measure the wavelength of a discrete emission from an arc lamp. When the movable mirror is moved a distance of 0.1724 mm, 628.00 fringes pass by the detector. What is the wavelength of the emission?

43. How many fringes move by the detector when the movable arm of a Michelson interferometer is moved a distance of 0.1152 mm? The light used has a wavelength of 754.1 nm.

44. Because it is nearly 1, the index of refraction of air is difficult to measure with a refraction experiment. Consider measuring n for air with a Michelson interferometer. A tube with glass end plates is aligned with its axis along one of the arms of an interferometer, say along ℓ_2 in Fig. 36-21. As the air in the tube is evacuated with a pump, 47.2 fringes of light with wavelength $\lambda = 589$ nm are observed to pass by the position of the detector. The length of the column of air that is evacuated is 47.9 mm. Determine the index of refraction of air.

Additional Exercises

45. Suppose a tiny glass plate of thickness 1.00 μm and index of refraction 1.458 is pushed behind one slit of a double slit with $\lambda = 488$ nm, $L = 1.14$ m, and $d = 0.150$ mm. By how much do the fringes shift as the plate is pushed behind the slit?

46. A beam of light emitted from a certain source is coherent with itself for about 2 ns. That is, the frequency and wavelength are only consistent for about 2 ns. By how much can ℓ_1 be greater than ℓ_2 if an interference pattern is still to be discerned in a Michelson interferometer?

47. The light emitted by heated cadmium vapor has two different visible wavelengths, 479.99 nm and 508.58 nm. Such light is incident on a double slit with $d = 0.55$ mm and $L = 1.25$ m. After how many fringes will the maximum of one set of fringes be coincident with the minimum of the other set?

48. Ten radio antennas are in a line spaced 0.30 km from each other. They form a radiotelescope used to detect radio stars, with the signals from the antennas fed to a single receiver. The length of the feeds is such that signals received simultaneously by the antennas arrive at the receiver simultaneously. Two 21-cm-wavelength point sources are observed near the highest point in the sky. What is the smallest angular separation these sources can have if they are to be seen as two sources by the radiotelescope?

49. A phased array radar has a number of antennas that send radar waves. The antennas send at the same frequency, but each has a different phase. This arrangement makes it possible to vary the angle at which the beam is sent out without physically moving the array. Suppose an array has 50 antennas spaced 75 mm apart along a straight line and sends out radar of 15-mm wavelength. What must the phase difference between the elements be to have the beam leave the array 35° from the perpendicular bisector of the line on which the antennas are placed?

50. There are three lines in the spectrum of the sun called the G lines and the G' line. One G line is due to Fe and has a wavelength of 430.7906 nm. The other is due to Ca and has a wavelength of 430.7741 nm. The G' line is due to H and has a wavelength of 434.0465 nm. *(a)* How large a grating would be needed to separate the G and G' lines in first order? *(b)* How large a grating would be needed to separate the two G lines in first order?

PROBLEMS

1. ***Lloyd's mirror.*** A pattern similar to the double-slit pattern can be produced using Lloyd's mirror (Fig. 36-24). At grazing angles almost 100 percent of the light that falls on the glass is reflected so that the reflected beam has nearly the same amplitude at P as the beam that passes directly from slit S to P. Slit S and its virtual image S' behave as coherent sources. Develop expressions similar to Eqs. 36-1 and 36-2 to give the positions of bright and dark fringes on the screen.

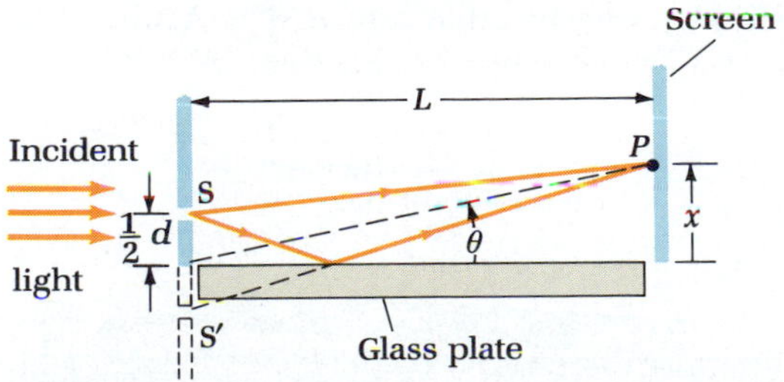

Figure 36-24. Problem 1: Lloyd's mirror. Slit S and its virtual image S' behave as coherent sources, and a pattern similar to the double-slit pattern is seen on the screen. Distance d is exaggerated for clarity.

2. ***Phasor diagram: Intensity from slits of different size.*** Suppose the slit openings in the double-slit experiment are different sizes so that the vector components of the electric field due to the waves must be written with different amplitudes. Thus

$$E_1 = E_{10} \sin(\omega t + \phi) \quad \text{and} \quad E_2 = E_{20} \sin(\omega t)$$

The intensities due to the slits acting alone are $I_1 \propto E_{10}{}^2$ and $I_2 \propto E_{20}{}^2$. *(a)* Use Fig. 36-25 to show that the component E_{12} of

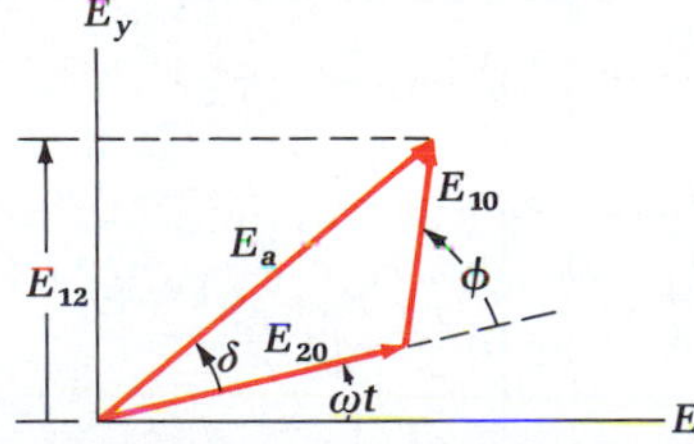

Figure 36-25. Problem 2: A phasor diagram, similar to those used in Chap. 31, can be used to find resultant wave amplitudes. The vector triangle rotates around the origin in the E_xE_y plane. The square of the resultant amplitude, E_a^2, is proportional to the resultant intensity I_{12}.

the resultant electric field is

$$E_{12} = E_a \sin (\omega t + \delta)$$

where E_a is the amplitude of the resultant wave and δ is a phase constant. *(b)* Use the law of cosiness to find E_a and show that

$$I_{12} = I_1 + I_2 + 2\sqrt{I_1 I_2} \cos \phi$$

(c) Show that your answer reduces to Eq. 36-5 when $E_{10} = E_{20} = E_0$.

3. ***Hyperbolic nodal lines.*** In Fig. 36-26, two coherent point sources of waves, S_1 and S_2, are separated by a distance d along the x axis. Show that the nodal lines are hyperbolas. That is, show that the coordinates of nodal lines satisfy the equation $(x/a)^2 - (y/b)^2 = 1$. (*Hint:* Recall that $|r_1 - r_2|$ is a constant on a nodal line.)

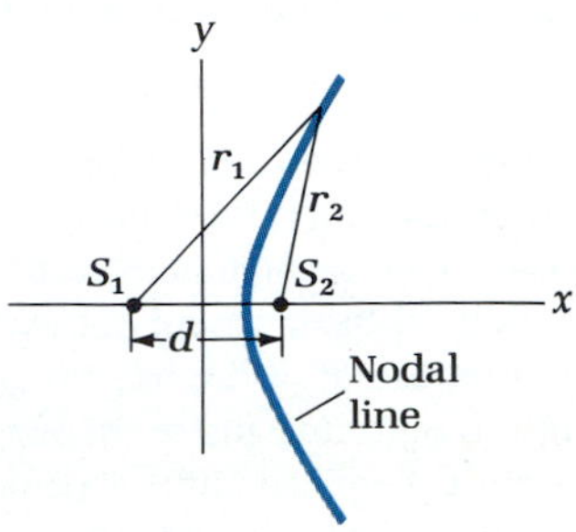

Figure 36-26. Problem 3: A nodal line due to two coherent sources of waves is a hyperbola.

4. ***Phasor diagram: Intensity due to three (and four) slits.*** Consider finding the intensity distribution due to light passing through three slits. The component E_c of the resultant electric field is $E_c = E_1 + E_2 + E_3$, where

$$E_1 = E_0 \sin (\omega t)$$

$$E_2 = E_0 \sin (\omega t + \phi)$$

$$E_3 = E_0 \sin (\omega t + 2\phi)$$

(a) Use Fig. 36-27 to show that

$$E_c = E_a \sin (\omega t + \phi)$$

where the amplitude E_a of the resultant wave is

$$E_a = E_0[2 \cos (\phi) + 1]$$

(b) Use trigonometric identities to show that

$$E_a = E_0 \frac{\sin [3(\frac{1}{2}\phi)]}{\sin (\frac{1}{2}\phi)}$$

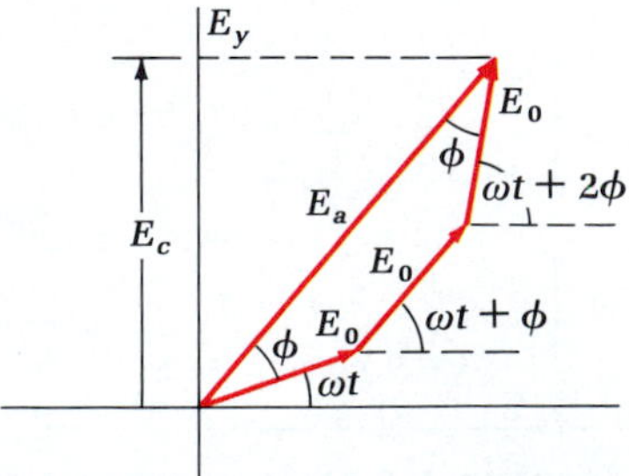

Figure 36-27. Problem 4: A phasor diagram, similar to those introduced in Chap. 31, can be used to find resultant wave amplitudes. The square of the resultant amplitude, E_a^2, is proportional to the resultant intensity.

(c) Show that your result in part *(b)* agrees with the general result given by Eq. 36-9 for the case where $N = 3$. *(d)* Draw the diagram similar to Fig. 36-27 that extends this argument to the case of four slits.

5. ***Intensity at the interference maxima from a grating.*** Show that the expression for the intensity distribution due to N slits, $I = I_0 \sin^2 [N(\frac{1}{2}\phi)]/\sin^2 (\frac{1}{2}\phi)$, is indeterminate for $\phi = \pm 2\pi m$, $m = 0$ 1, 2, ... Note that these values of ϕ correspond to the interference maxima. *(b)* Use l'Hospital's rule to show that $I = N^2 I_0$ when $\phi = \pm 2\pi m$, $m = 0, 1, 2, \ldots$.

6. ***A beam not at normal incidence.*** Consider a beam of light incident on a grating at an angle γ with respect to the normal, as shown in Fig. 36-28. Show that the interference maxima satisfy the relation

$$d(\sin \gamma + \sin \theta) = \pm m\lambda \qquad (m = 0, 1, 2, \ldots)$$

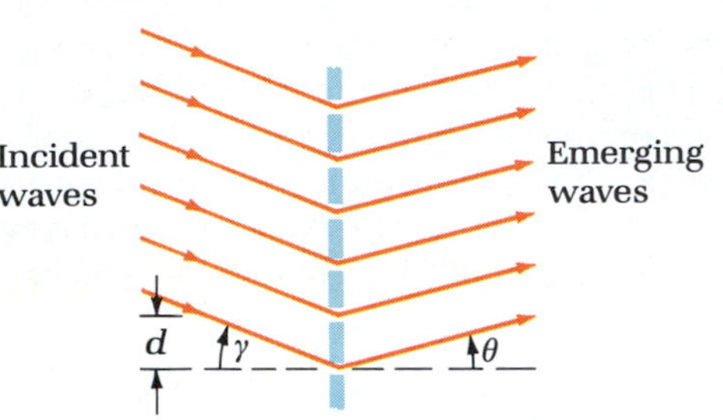

Figure 36-28. Problem 6: A light beam incident on a grating at an angle γ with respect to a normal to the grating face.

7. ***Intensities of secondary maxima from a grating.*** Consider the secondary maxima between two primary maxima in the intensity distribution due to a diffraction grating with a large number of slits. *(a)* Show that the intensity of the secondary maxima that are in the region midway between primary maxima is approximately I_0, the intensity due to one slit acting alone. *(b)* Show that the intensity of a secondary maximum adjacent to a primary maximum is approximately 4.5 percent of the intensity of the primary maximum. To be specific, consider the secondary maximum on the positive θ side of the $m = 0$ primary maximum. It is a valid approximation to assume that this secondary maximum occurs at $\phi = 3\pi/N$. Recall that the intensity of a primary maximum is $N^2 I_0$.

8. ***Fringes due to an air wedge.*** In Fig. 36-29, a flat glass plate is resting on a similar (horizontal) glass plate. The plates are in contact along one edge and are separated a distance d along the other edge by a thin wire: $d \ll L$. A thin wedge of air

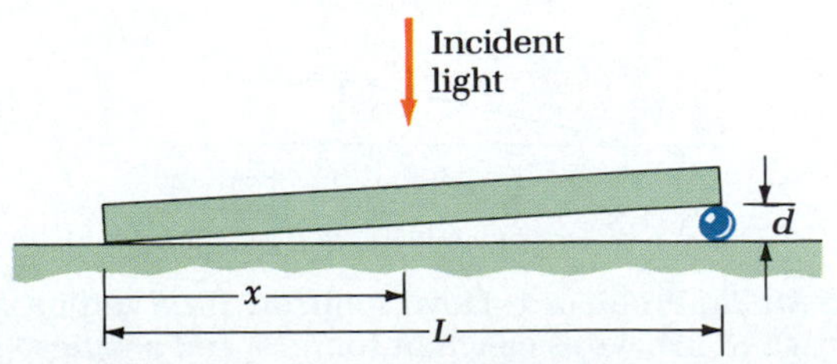

Figure 36-29. Problem 8: An air wedge is formed between two flat glass plates by placing a thin wire along one edge.

exists between the plates. Monochromatic light is incident from above and is nearly normal to the surface of both plates. Develop expressions for the positions x of *(a)* dark fringes and *(b)* bright fringes in the reflected light. *(c)* What is the spacing Δx between dark fringes for the case where $d = 31\ \mu\text{m}$, $L = 0.27$ m, $\lambda = 724$ nm?

9. ***Newton's rings.*** A planoconvex lens is shown with its spherical face down on a flat, horizontal glass plate in Fig. 36-30*a*. The planar side of the lens is horizontal and the spherical side has a radius of curvature R. A beam of monochromatic light of wavelength λ is directed vertically downward. When viewed from above, circular interference fringes of radius r, called *Newton's rings,* can be seen in the reflected light (Fig. 36-30*b*). Show that the radius of a bright fringe near the center of the pattern ($r \ll R$) is given by

$$r = \sqrt{(m + \tfrac{1}{2})\lambda R} \qquad (m = 0, 1, 2, \ldots)$$

10. ***Michelson-interferometer intensity.*** Show that the intensity at the detector in the Michelson interferometer (Fig. 36-21) varies with position x of the movable mirror as

$$I = 4I_0 \cos^2 \frac{2\pi x}{\lambda}$$

What are the interpretations of I_0 and of $x = 0$?

11. ***Comparison of intensities of beams reflected from opposite faces of a thin film.*** When light is reflected normally from the surface of a transparent material in air, the ratio of the intensities of the reflected and incident beams is $[(n-1)/(n+1)]^2$, where n is the index of refraction of the material. In Fig. 36-19, let I_1 represent the intensity of beam (1) acting alone and suppose that the film is water ($n = 1.33$). *(a)* Determine I_2 in terms of I_1. *(b)* Using your result from part *(a)*, estimate I_6 in terms of I_1. The comparison of the answers to parts *(a)* and *(b)* justifies our assumption that beams (2) and (6) acting alone would have about the same intensity.

12. ***The Fabry-Perot interferometer.*** A useful interferometer design is the Fabry-Perot interferometer. The essential elements of this design are shown in Fig. 36-31*a*. Show that the condition for constructive interference of the waves that reach the detector is

$$2d/\cos\theta = m\lambda$$

where m = an integer and d is the distance between the parallel mirrors.

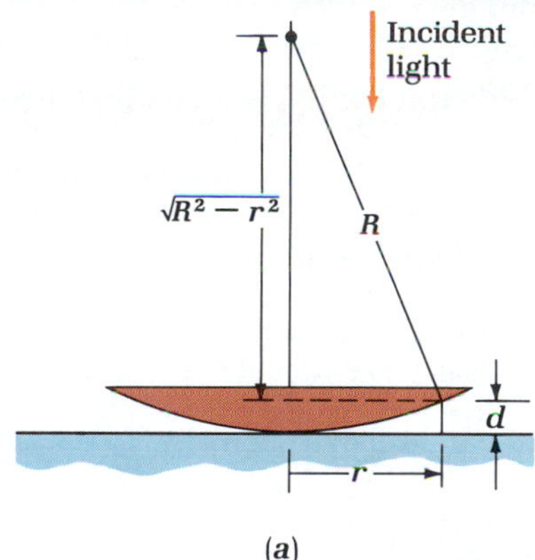

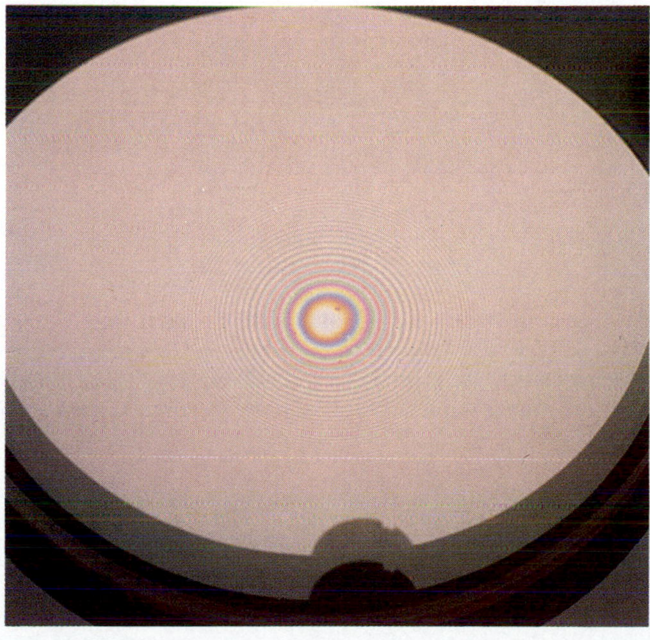

(b)

Figure 36-30. Problem 9: Newton's rings. *(a)* Light incident from above on a planoconvex lens. *(b)* Newton's rings.

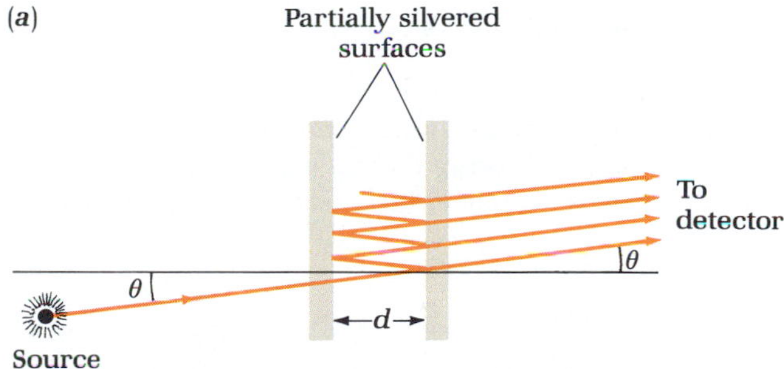

(b)

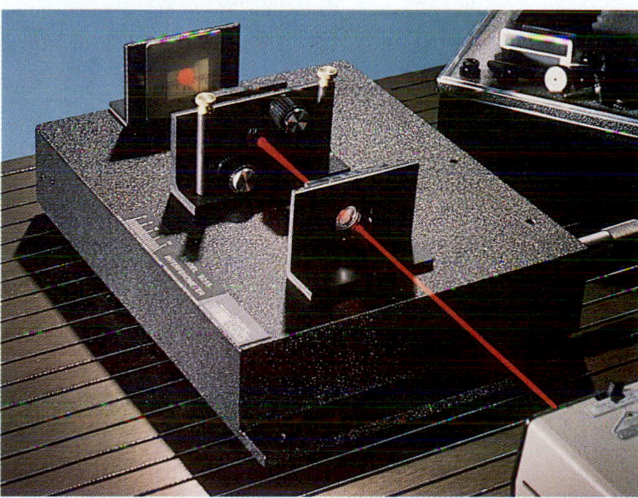

Figure 36-31. Problem 12: *(a)* Schematic diagram of the Fabry-Perot interferometer. Multiple reflections from partially silvered mirrors set up the conditions for interference fringes in the light that reaches the detector. *(b)* Fabry-Perot interferometer.

13. ***Complex variables: Intensity due to a grating.*** Consider using complex variables to develop the expression for the intensity due to a diffraction grating, Eq. 36-9. The resultant amplitude E_r may be written

$$E_r = E_0 \sum_{j=0}^{N-1} \cos(\omega t + j\phi)$$

where j is an index for summing the electric field components due to each of the N slits. Since $e^{i\alpha} = \cos\alpha + i\sin\alpha$, we can write this as

$$E_r = E_0 \, \mathbf{Re} \sum_{j=0}^{N-1} e^{i(\omega t + j\phi)}$$

where **Re** means "the real part of." *(a)* Use the sum of the geometric progression

$$a + ax + ax^2 + \cdots + ax^{N-1} = \frac{a(x^N - 1)}{x - 1}$$

to show that

$$E_r = E_0 \, \mathbf{Re} \, e^{i\omega t} \frac{e^{iN\phi} - 1}{e^{i\phi} - 1}$$

(b) Use the relation $\sin\alpha = (e^{i\alpha} - e^{-i\alpha})/2i$ to show that

$$E_r = E_0 \cos\left[\omega t + \tfrac{1}{2}(N-1)\phi\right] \frac{\sin\left[N(\tfrac{1}{2}\phi)\right]}{\sin(\tfrac{1}{2}\phi)}$$

[*Hint:* Note that $e^{i\alpha} - 1 = e^{i\alpha/2}(e^{i\alpha/2} - e^{-i\alpha/2})$.] *(c)* Use your answer from part *(b)* to develop Eq. 36-9.

14. ***Pohl's interferometer.*** In Pohl's interferometer, a thin sheet of mica is put behind a shielded light source, as in Fig. 36-32. Each side of the mica reflects light from the source, so that there are two coherent virtual sources along a line perpendicular to the screen. Show that the equation for the radius R of bright circular fringes on the screen is

$$\sqrt{R^2 + (\ell + 2d + 2t)^2} - \sqrt{R^2 + (\ell + 2d)^2} = (n - 1/2)\lambda.$$

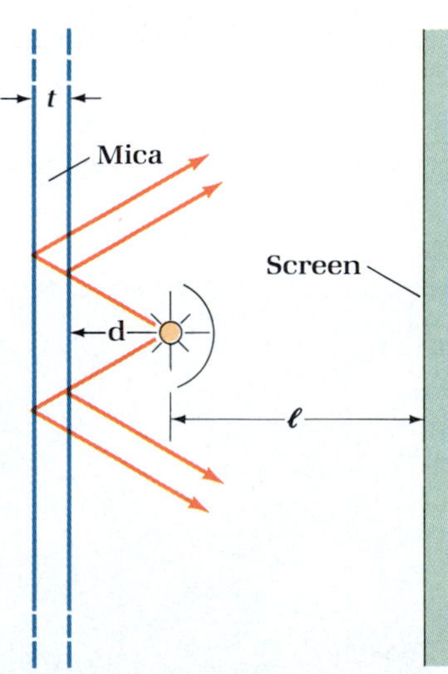

Figure 36-32. Problem 14.

15. ***Moving fringes or Doppler effect?*** *(a)* Suppose one mirror of a Michelson interferometer is moved at a constant speed of $v = 1.55\ \mu\text{m/s}$. If light of wavelength 546.1 nm is incident on the interferometer, how many fringes per second pass the detector? *(b)* Show that the Doppler shift of the light reflected from the moving mirror is given by $\nu' - \nu \approx \nu(1 - v/c)$, where c is the speed of light. *(c)* Show that the beat frequency between the beams reflected from the stationary and moving mirrors is $2v/\lambda = 5.68$ beats per second.

16. ***Four phasors.*** See Prob. 2 for the way phasor diagrams are used. *(a)* Show that the phasor diagrams for a four-slit interference pattern (Fig. 36-10) are a straight line of four phasors at the maxima, a square of phasors at the minima adjacent to the maxima, and two sets of antiparallel phasors at the minima centered between the maxima. *(b)* Draw the phasor diagram for the subsidiary maxima and show that the amplitude is one-sixteenth the amplitude at the center.

17. ***Uneven grating.*** A grating is made such that the even-numbered slits are twice as wide as the odd-numbered slits. How are the positions of the maxima affected? How are the minima midway between the maxima affected? (*Hint:* Look at the phasor diagram for a four-slit interference pattern.)

18. ***Double Newton.*** Two planoconvex lenses have radii of 2.00 and 3.00 m. They are brought in contact, and light of wavelength 650 nm is projected on them (Fig. 36-33). What is the radius of the tenth bright fringe?

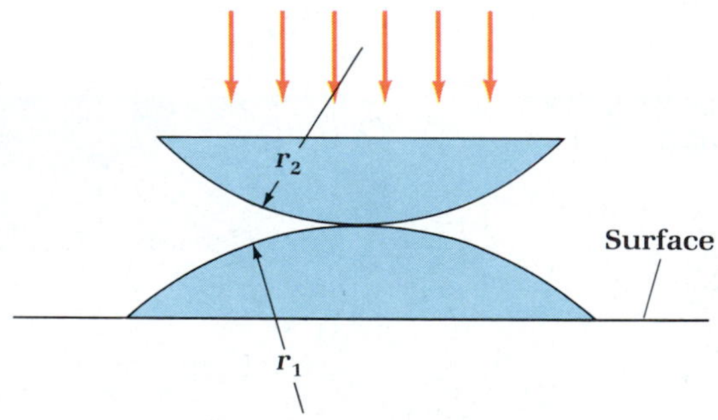

Figure 36-33. Problem 18.

19. ***Odd couple.*** One slit in a double slit is increased in size until it is 15 percent larger than the other. *(a)* By what fraction is the intensity at the maxima increased? *(b)* What is the intensity at the former minima in terms of the intensity at the maxima?

HIGHLIGHTS OF MODERN PHYSICS

The Aharonov-Bohm Effect

The discovery of a physics phenomenon is usually prompted by some very subtle effect, but sometimes the effect of the phenomenon is so obvious that its discovery merely waits for someone to ask the right question. In the theory of special relativity, for example, Einstein's 1905 paper showed that moving objects contract (Chap. 38), but it was not until 1959, when J. Terrell wondered how rapidly moving objects appear to the eye, that it was realized that in certain cases they look rotated, not contracted. This question could have been asked and answered decades earlier.

In 1959 two physicists, Yakir Aharonov and David Bohm, of the University of Bristol, England, asked a simple question: Which is more physically meaningful, the electric field **E** or the electric potential $\mathcal{V}$? The phenomenon they discovered as a result is called the ***Aharonov-Bohm effect.***

Most physicists thought the answer was obvious, that **E** was a real, physical quantity, while $\mathcal{V}$ was just a useful mathematical device. There were at least two reasons for this. First, Eq. 23-6 associates an energy density with the electric field so it is natural to attribute a physical existence to **E**. It is hard to see how a nonphysical quantity could produce energy. Second, the electric potential seems purely mathematical because, as we have seen (Sec. 22-3), it does not have a definite value. It is used to calculate **E** by taking derivatives,

$$E_x = -\frac{\partial \mathcal{V}}{\partial x} \qquad E_y = -\frac{\partial \mathcal{V}}{\partial y} \qquad E_z = -\frac{\partial \mathcal{V}}{\partial z}$$

so that adding a constant to $\mathcal{V}$ will leave **E** unchanged. Therefore both the energy density and forces on charges will be unaffected by changing $\mathcal{V}$ by a constant amount. There seemed little doubt that **E** was real and $\mathcal{V}$ was not until Aharonov and Bohm asked their question.

As we know from Chap. 36, monochromatic light produces an interference pattern of bright and dark fringes when it passes through Young's double slits. As shown in Fig. 1, the same interference pattern is produced if particles such as electrons are substituted for light because particles have wave properties similar to light (Chap. 40). Aharonov and Bohm suggested a variation on the two-slit experiment, again using electrons (Fig. 2). Since the only function of the slits in Fig. 1 is to provide two distinct paths for the particles, they suggested an arrangement in which the original beam is split into two (by an unspecified but irrelevant process). The two beams pass through long metal cylinders and are recombined at the screen. As for two-slit interference, if the path lengths are the same, the two beams will interfere constructively and many electrons will strike the screen there. At other points of the screen, the waves will be alternately in and out of phase, producing bright and dark fringes. Aharonov and Bohm asked what would happen if, *while electrons are inside the cylinders,* a potential difference is applied between the cylinders, and then is turned off before the particles emerge. They predicted that the fringes would be shifted to new positions.

Since **E** is zero inside a hollow, charged cylinder (Prob. 1), it was hard to see how the electrons could be affected if there was no electric field to exert forces on them. On the other hand, the electric potential inside the cylinders would change. In fact,

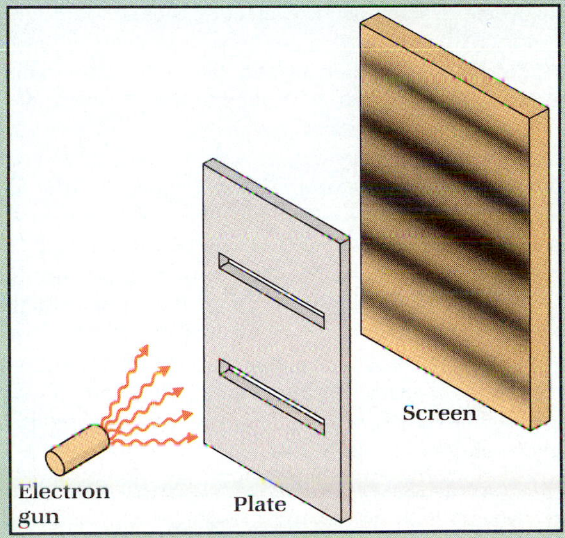

Figure 1. Electrons pass through a pair of closely spaced slits and produce interference fringes similar to the fringes produced by light. Bright fringes occur when the path lengths through the two slits differ by an integral number of wavelengths.

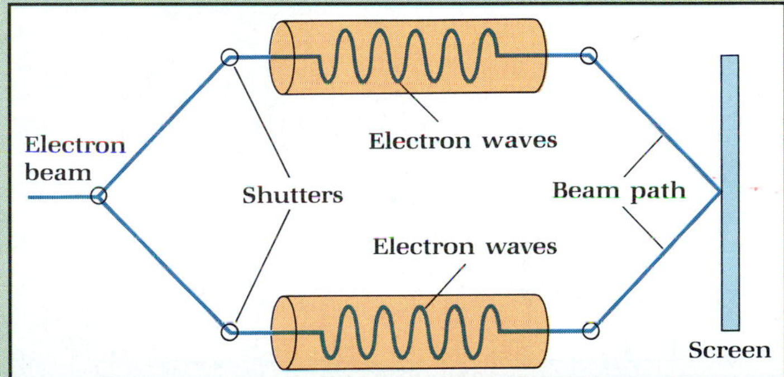

Figure 2. A variation of the two-slit experiment in which electrons follow two paths through metal cylinders from source to screen. If the potentials on and within the cylinders are changed, the interference fringes shift even though no **E** field penetrates into the cylinders where it could affect the electron motion.

Aharonov and Bohm calculated that the phase difference between the two beams depends on the potential difference between the two cylinders. They showed that the phases of the two beams changed by

$$\Delta\phi = \frac{2\pi e}{h} \int (\mathcal{V}_1 - \mathcal{V}_2)\, dt$$

relative to the phase difference with zero potential difference, where h is Planck's constant, 6.63×10^{-34} J·s. They concluded that the electrons were interacting with $\mathcal{V}$, not **E**, so that in some sense $\mathcal{V}$ is the more basic physical quantity. This is an effect of quantum mechanics (Chap. 40), not classical electromagnetism. (The presence of Planck's constant is the signature of a quantum-mechanical process.)

Their paper stimulated many reactions, some critical and some supportive. Some physicists felt that the calculation was incorrect and that, properly done, it would predict no effect at all; others tried to avoid the problem by reformulating electromagnetism to eliminate potentials from the theory. Experiments seemed to show that the effect does exist, but there was concern that stray fields had not been completely eliminated from the regions occupied by the electrons. To varying degrees, the controversy has continued to the present time. Not until 1986 did an experiment conclusively show that the effect exists as predicted. That experiment actually involved the magnetic field **B** rather than the electric field, but the idea is the same.

The experiment was a variation of the setup shown in Fig. 3. A solenoid (an electromagnet) is placed behind the double slits so the electron wave passing through the upper slit passes above the magnet, and the wave passing through the lower slit passes beneath it. A superconducting coating prevents stray magnetic fields from leaking out of the solenoid into the region of the electron waves. (Magnetic fields cannot penetrate certain superconductors; see Sec. 42-6.)

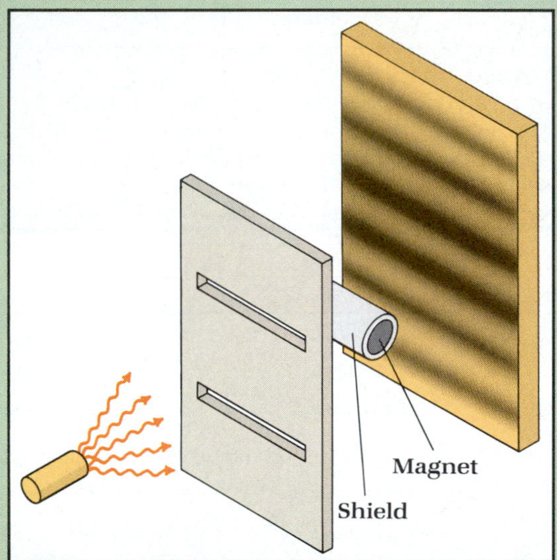

Figure 3. The interference fringes will shift when the magnet is inserted so electrons passing through the two slits pass on opposite sides of the magnet. The electrons are affected even though the **B** field is confined to the inside of the magnet and no electron penetrates there.

As in the **E**-field case, theory predicts a fringe shift purely in terms of a magnetic potential, called the ***vector potential* A**. The **B** field can be replaced in the equations of electromagnetism by derivatives of **A**, similar to the replacement of **E** by derivatives of $\mathcal{V}$. Aharonov and Bohm showed that the phase difference between waves passing through the two slits with the solenoid in place is

$$\Delta\phi = \frac{2\pi e}{h} \oint \mathbf{A} \cdot d\boldsymbol{\ell}$$

relative to the phase without the solenoid (Fig. 1), where the integral is around the loop formed by the two paths through the slits. Therefore, sending a current through the solenoid should produce a fringe shift. The **B** field outside a long thin solenoid is close to zero except for the weak return field, and even this field can be eliminated by bending the solenoid so its ends connect to form a torus. The experiment was performed in 1986 by Akiri Tonomura and his colleagues of the Hitachi Company of Japan. They substituted a permanent magnet for the solenoid.

The experiment confirmed the fringe shift predicted by Aharonov and Bohm. Since there is no **B** field in the region of the electron wave, it is difficult to see how **B** can cause the fringe shift. Yet, like $\mathcal{V}$, **A** is undetermined with respect to an additive term so it also seems to be a purely mathematical quantity. However, since $\mathbf{A} \neq 0$ while $\mathbf{B} = 0$ in the region of space accessible to the electrons, the experiment implies that **A** is the more physically significant quantity. We know **A** must be nonzero because

$$\oint \mathbf{A} \cdot d\boldsymbol{\ell} = \int\int \mathbf{B} \cdot d\mathbf{S}$$

for any path around the solenoid. That is, the line integral of **A** along any path around the solenoid equals the total magnetic flux through the solenoid.

However, not all physicists are ready to grant reality to the electric and magnetic potentials, since there are alternative interpretations. The concept of fields, such as electric and magnetic fields, was originally introduced to avoid the problem of ***action at a distance.*** The problem comes up in considering, for example, the force between two point charges (Fig. 4*a*). How is it possible for charges to exert forces on each other when they are not touching and nothing connects them? How is the force transmitted? The field concept solves the problem (Fig. 4*b*). In this interpretation each electric charge produces an electric field that extends outward in all directions. Each charge has a force exerted on it by the field, rather than directly by another charge. In this way there is no action at a distance because the force on a particle is exerted by the field at the particle's location. The force is said to be a *local interaction.*

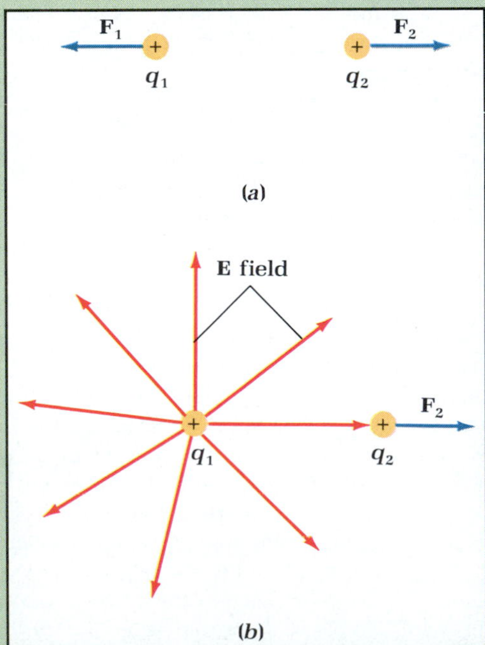

Figure 4. *(a)* The repulsion between like charges seems to require "action at a distance" since the charges repel each other without touching. *(b)* The field concept allows one to avoid action at a distance. The force $\mathbf{F}_2$ on q_2 can be thought of as the *local* interaction between q_1's **E** field and the charge q_2.

It is possible to interpret the Aharonov-Bohm effect without supposing that the potentials are real by letting the electromagnetic interactions be nonlocal—that is, by permitting action at a distance. Although physicists have traditionally resisted nonlocal theories, it turns out that nonlocal effects may be built into the quantum-mechanical description of nature. There are experiments for which the most natural explanation seems to require that an action at one location produce an instantaneous result at a distant location. This phenomenon is a subtle one in which the principle that signals cannot travel faster than light is not violated (Chap. 38), and it is surprising and poorly understood. It is a different kind of nonlocality from that suggested by the Aharonov-Bohm effect, but each situation hints that the quantum-mechanical universe, in some strange, unexpected way, may be a nonlocal one.

The Aharonov-Bohm effect is a rich phenomenon with numerous implications. As only one example, it suggests that in quantum mechanics the concept of force is no longer useful. The equations of quantum mechanics never involve forces, only potentials, so perhaps the Aharonov-Bohm effect should not have been so surprising. Nonetheless, the effect does seem to demonstrate that potentials are more fundamental than forces in the microscopic world.

QUESTIONS

1. In the **E**-field version of the Aharonov-Bohm effect, why is it important to apply the potential difference only while the electrons are inside the cylinder? Why should the electrons be far from the ends of the cylinders while the potential is applied?
2. Explain why the electric potential $\mathcal{V}$ must be constant inside the cylinders and equal to the potential of the respective cylinder, given that $\mathbf{E} = 0$ there.
3. Criticize the conclusion that the Aharonov-Bohm effect shows that the electric potential $\mathcal{V}$ is a physical quantity, assuming *local* interactions. (*Hint:* Consider what the effect actually depends on.)

EXERCISES

1. Electrons and other particles have wave properties, as mentioned in the passage above. For particles, the one-dimensional wave equation is

$$-\frac{\hbar^2}{2m}\frac{\partial^2\Psi(x, t)}{\partial x^2} + U(x)\,\Psi(x, t) = i\hbar\frac{\partial\Psi(x, t)}{\partial t}$$

and replaces the one we discussed in Chap. 32 ($\hbar \equiv h/2\pi$). Show by explicit substitution that

$$\Psi(x, t) = ce^{i(kx-\omega t)}$$

is a solution to the wave equation if $U = U_0 =$ constant. What relationship between k, ω, and U_0 must hold for the wave equation to be satisfied?
2. Use the results of Exercise 1 and the fact that the momentum of the electron is $p = \hbar k$ to give an interpretation to the quantity $\hbar\omega$.
3. Show that in the **B**-field Aharonov-Bohm effect there is no fringe shift for a total magnetic flux through the solenoid of $\Phi = nh/e$, where n is an integer.
4. Show that $\mathbf{E} = 0$ inside an infinitely long, hollow, uniformly charged metal cylinder. (*Hint:* Use Gauss's law and symmetry arguments.)

37 DIFFRACTION AND POLARIZATION

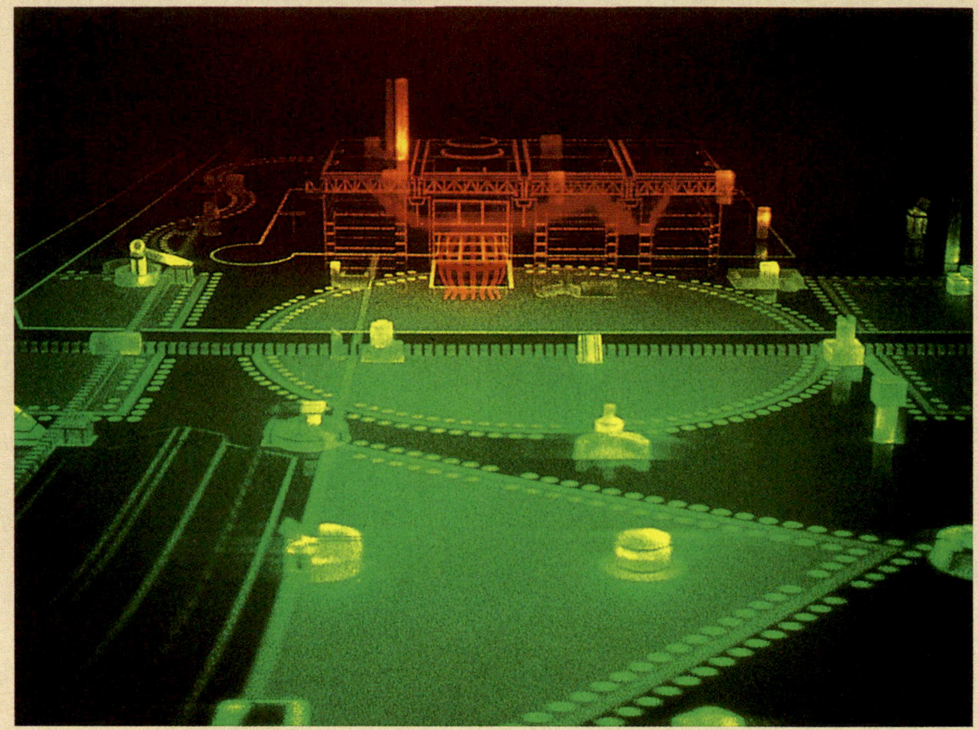

Transmission hologram of the Palace of Discovery at La Villette, Paris.

We are accustomed to the notion that light travels in a straight line. But light is a wave, and waves do not always propagate in a straight line. In particular, when waves pass near a barrier, they tend to bend around the barrier and spread into the region of the geometrical shadow. This phenomenon is called *diffraction*. In this chapter we consider a particularly simple example of diffraction, the diffraction of light as it passes through a slit.

From electromagnetic theory, light is a transverse wave in the electric and magnetic fields. In Chap. 34, we introduced the polarization of a light wave, which characterizes the directions of the oscillations of **E** and **B**. Here we describe how light can be polarized and how polarization can be measured.

37-1 DIFFRACTION

In the study of geometrical optics, light traveling in a homogeneous medium is assumed to follow a straight-line path. According to this assumption, the shadow cast by an object illuminated by a point source of light would have sharp edges.

Figure 37-1. The shadow of a razor blade. The blade was illuminated with monochromatic light from a small (nearly a point) source. Interference fringes are formed at the shadow's edge.

Figure 37-1 shows the shadow of a razor blade illuminated by a point source. As you can see, the edges of the shadow are not sharp; fringes occur around the shadow edges. This is evidence of ***diffraction***— the bending of light around the edge of a barrier. The amount of diffraction depends on wavelength; the longer the wavelength, the more noticeable the bending. The assumption that light follows a straight-line path is often valid because light has a very small wavelength. If we do not look too closely, diffraction may be neglected.

A curious phenomenon, called the *Arago spot* (sometimes called the *Poisson spot*), provides a startling example of diffraction. If a disk-shaped object, such as a penny, is illuminated with monochromatic light from a point source, a bright spot appears in the center of the shadow (Fig. 37-2). The Arago spot played an interesting role in the development of wave optics. In 1819, A. J. Fresnel (1788–1827) presented a wave theory of light that explained diffraction. Skeptical of this theory, S. D. Poisson (1781–1840) showed that it predicted a bright spot at the center of the shadow of a disk, a result so incredible that Poisson asserted that it discounted the wave theory. But the subsequent discovery of the spot by D. F. J. Arago (1786–1853) provided a striking confirmation of the wave nature of light.

Figure 37-2. The Arago spot.

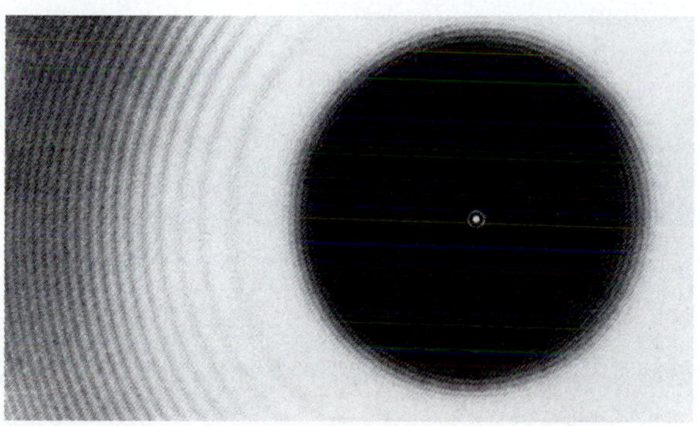

The study of diffraction is divided into two regimes: Fresnel diffraction and Fraunhofer diffraction. In Fresnel diffraction, the waves impinging on the barrier and the observing screen are not necessarily plane waves. In Fraunhofer diffraction, the source and the screen are effectively an infinite distance from the diffracting barrier so that the waves impinging on the barrier and on the screen are plane waves. Fraunhofer diffraction is a special case that is simpler to discuss than Fresnel diffraction. Fraunhofer diffraction can be achieved experimentally with the use of lenses or, to a good approximation, by placing the source and the viewing screen a large distance from the diffracting object.

37-2 DESCRIPTION OF THE SINGLE-SLIT DIFFRACTION PATTERN

In this section, we describe the diffraction pattern due to light passing through a narrow slit, and then we present a qualitative explanation of how the pattern is formed. For simplicity, we consider Fraunhofer diffraction; the source and the screen are far from the slit compared with the slit width.

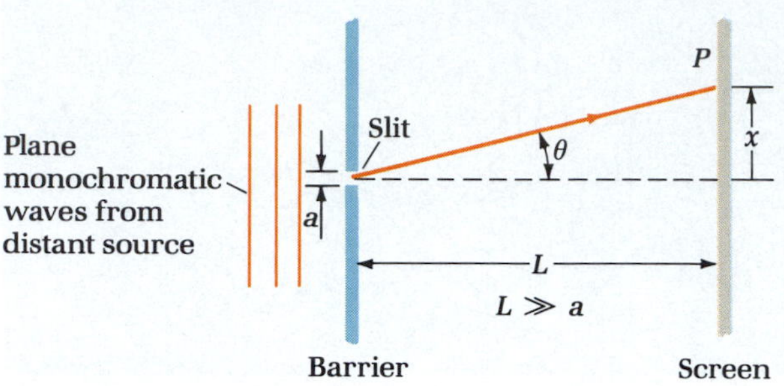

Figure 37-3. The arrangement used for the single-slit experiment (Fraunhofer diffraction).

In Fig. 37-3 we show a beam of monochromatic light of wavelength λ incident on a slit of width a. The light that passes through the slit and impinges on the screen produces the ***single-slit diffraction pattern*** shown in Fig. 37-4*a*. Figure 37-4*b* shows the intensity distribution as a function of $\sin\theta$, where θ is the angle that locates a point on the screen. The pattern consists of a bright central maximum flanked by secondary maxima. The intensity of each succeeding secondary maximum decreases with distance from the center. Between successive maxima, intensity minima occur at angles $\theta_{m'}$ given by

Intensity minima at angles $\theta_{m'}$

$$a \sin\theta_{m'} = \pm m'\lambda \qquad (m' = 1, 2, \ldots) \tag{37-1}$$

Note that $m' = 0$ is *not* included among the values of m' that give intensity minima. Indeed, $m' = 0$ corresponds to the center of the pattern, or the middle of the central maximum. This means that the width of the central maximum is twice that of the secondary maxima, as you can see in Fig. 37-4.

Figure 37-4. The single-slit diffraction pattern. *(a)* Photograph of the pattern. *(b)* The intensity distribution.

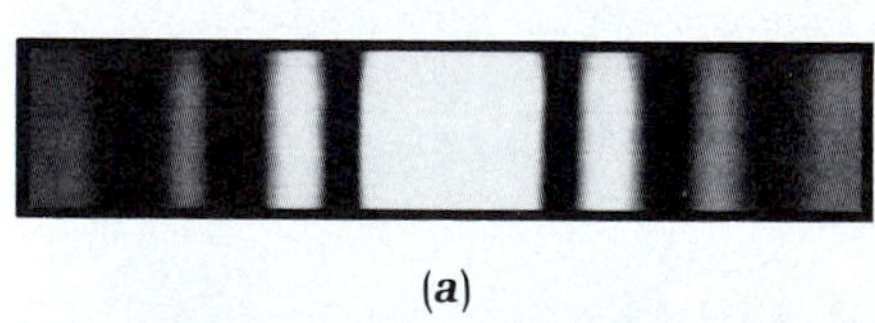

(a)

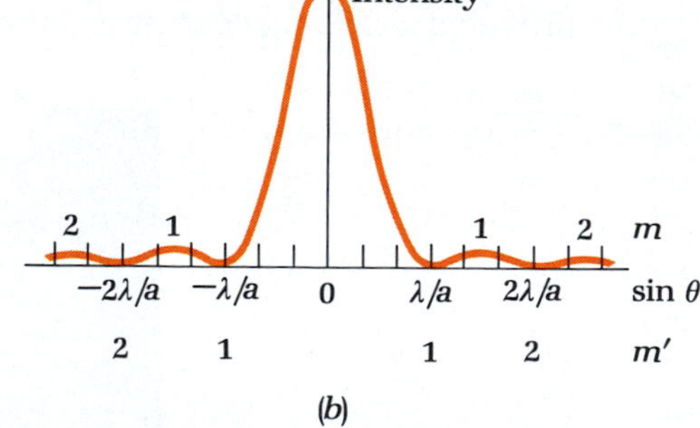

(b)

In Fig. 37-4, the secondary maxima are approximately midway between their adjacent minima. Thus the angles θ_m that locate the secondary maxima are given by

Secondary maxima at angles θ_m

$$a \sin\theta_m = \pm(m + \tfrac{1}{2})\lambda \qquad (m \approx 1, 2, \ldots) \tag{37-2}$$

Note that the value of m that designates a pair of secondary maxima is only approximately an integer because the secondary maxima are only approximately midway between their adjacent minima. (More accurate estimates of the two lowest values of m are $m = 0.93$ and $m = 1.96$. See Prob. 3.)

We can understand how the single-slit pattern is formed by replacing, in our mind's eye, the slit of width a with a number of parallel elementary slits, each of width Δy (Fig. 37-5). The more elementary slits we use, the smaller will be their width, and the more precise will be our conclusions. In the figure, we show 12 elementary slits so that $\Delta y = a/12$. Because the waves incident on the slit are plane

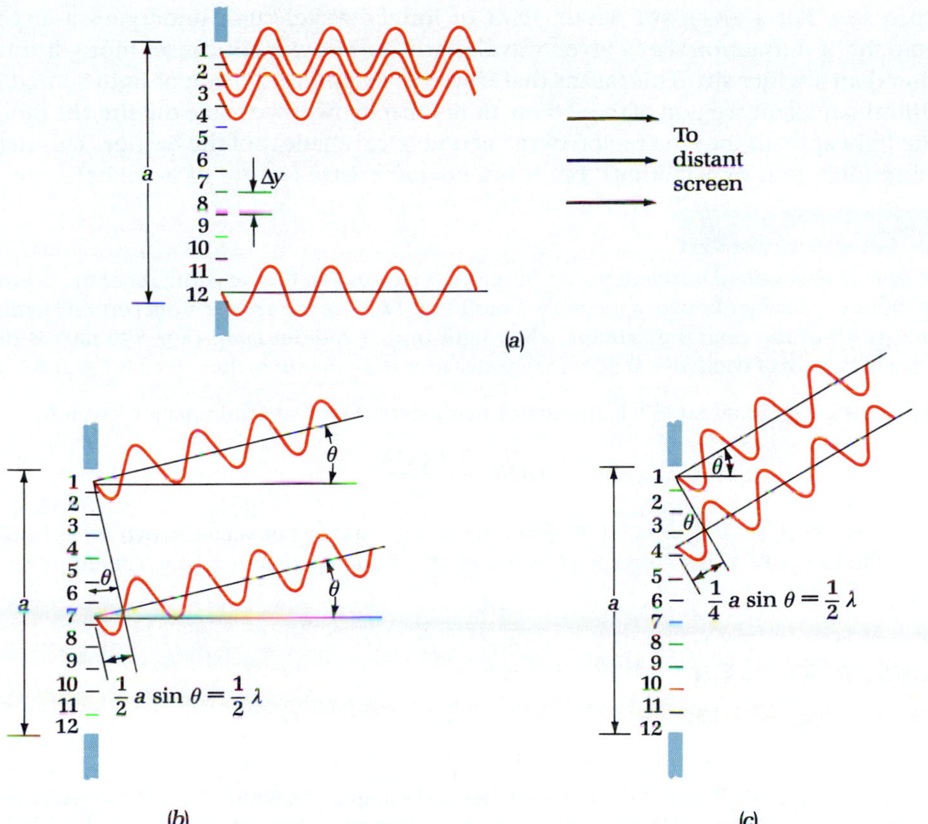

Figure 37-5. Waves emerging in phase from a single slit. The slit is shown divided into 12 elementary slits. *(a)* Waves that form the central maximum. *(b)* Waves that form one of the $m' = 1$ minima. *(c)* Waves that form one of the $m' = 2$ minima.

waves (Fraunhofer approximation), they emerge from each elementary slit in phase. Figure 37-5*a* shows schematically those waves that form the central maximum. Since $L \gg a$, the lines along the propagation direction of these waves are essentially parallel (Fraunhofer approximation), and each wave traverses approximately the same distance. Therefore they arrive at the center of the screen in phase, interfere constructively, and form an intensity maximum.

Figure 37-5*b* shows the waves that form one of the two minima adjacent to the central maximum. To arrive at the screen, the waves from elementary slit 7 travel half a wavelength farther than the waves from elementary slit 1; this pair of waves arrives at the screen with a phase difference $\phi = \pi$ rad and interferes destructively. Similarly, wave pairs from elementary slits 8 and 2, 9 and 3, 10 and 4, 11 and 5, and 12 and 6 arrive at the screen with $\phi = \pi$ rad and interfere destructively. This accounts for all 12 elementary slits. Thus the intensity is a minimum at this point on the screen. From Fig. 37-5*b* we see that the angle θ that locates this position on the screen is given by $\frac{1}{2}\lambda = \frac{1}{2}(a \sin\theta)$ or $a \sin\theta = +(1)\lambda$. This corresponds to $m' = 1$ in Eq. 37-1.

Figure 37-5*c* shows the waves that form one of the $m' = 2$ minima. To arrive at the screen, waves from elementary slit 4 travel half a wavelength farther than waves from elementary slit 1, so that this pair arrives with a phase difference $\phi = \pi$ rad and interferes destructively. Similarly, wave pairs from elementary slits 5 and 2, 6 and 3, 10 and 7, 11 and 8, and 12 and 9 arrive with $\phi = \pi$ rad and interfere destructively. From Fig. 37-5*c* we see that the angle θ that locates this point on the screen is given by $(a/4)\sin\theta = \frac{1}{2}\lambda$ or $a \sin\theta = +(2)\lambda$. This result corresponds to $m' = 2$ in Eq. 37-1.

If we consider an angle θ between the $m' = 1$ and $m' = 2$ minima, we find that this cancellation from pairs of elementary slits cannot account for the waves emerging from all the elementary slits. (See Ques. 10.) These remaining waves give rise to the light around an $m \approx 1$ secondary maximum.

We characterize the width of the single-slit pattern by the distance between the two $m' = 1$ minima. From Fig. 37-4, you can see that this width depends on the

A light beam cannot be confined by making the slit narrower and narrower.

ratio λ/a. For a given slit width, light of longer wavelength undergoes a larger amount of diffraction. For a given wavelength, a narrower slit causes more diffraction than a wider slit. This means that if we try to confine a beam of light so that it illuminates a tiny region of the screen, then the narrower we make the slit, the more the light spreads into the region of the geometrical shadow of the barrier. This may contradict your expectations, but it is a characteristic feature of wave behavior.

EXAMPLE 37-1

Width of the central maximum. *(a)* Develop an expression for the width Δx of the central maximum (distance between the $m' = 1$ minima). *(b)* Use the answer from part *(a)* to find the width of the central maximum when light from a sodium lamp ($\lambda = 590$ nm) is diffracted by a slit of width $a = 0.30$ mm. The distance from the slit to the screen is $L = 0.87$ m.

Solution. *(a)* From Eq. 37-1, the angles that locate the $m' = 1$ minima are given by

$$\sin \theta_1 = \frac{\pm(1)\lambda}{a}$$

From Fig. 37-3, $\sin \theta = x/\sqrt{x^2 + L^2}$. Since Δx is the distance between the two $m' = 1$ minima, the values of x that correspond to the $m' = 1$ minima are $x = \pm\frac{1}{2}\Delta x$. Therefore

$$\frac{\frac{1}{2}\Delta x}{\sqrt{(\frac{1}{2}\Delta x)^2 + L^2}} = \frac{\lambda}{a}$$

Solving for Δx, we find

$$\Delta x = \frac{2L\lambda}{\sqrt{a^2 - \lambda^2}}$$

Notice that if we make the slit width a smaller and smaller, the width Δx of the central maximum becomes larger and larger, approaching infinity as the slit width becomes as small as the wavelength. *(b)* For the value given in this example, $a \gg \lambda$, so we can use the approximation $\sqrt{a^2 - \lambda^2} \approx a$. Thus

$$\Delta x \approx \frac{2L\lambda}{a} = \frac{2(0.87 \text{ m})(590 \text{ nm})}{0.30 \text{ mm}} = 3.4 \text{ mm}$$

SELF-TEST 37-1. Repeat the above example except *(a)* let the slit size be doubled to 0.60 mm and *(b)* let the slit size be halved to 0.15 mm. ***ANSWERS:*** *(a)* 1.7 mm; *(b)* 6.4 mm. The width of the central maximum (and the entire pattern) becomes larger as the slit size becomes smaller.

37-3 THE SINGLE-SLIT INTENSITY DISTRIBUTION

In the previous section, we gave a qualitative explanation of the formation of the single-slit diffraction pattern by dividing the slit into 12 elementary slits. Now we develop an expression for the intensity distribution by dividing the slit into an infinite number of elementary slits, each with infinitesimal width. First we add vectorially the electric field contribution from each elementary slit to find the resultant field at a point on the screen. Then the intensity is determined by squaring the resultant field amplitude.

Figure 37-6 shows a wave emerging from an elementary slit of infinitesimal width dy. The coordinate y of the elementary slit is measured from the center of the actual slit. (By letting y be positive downward, we encounter fewer minus signs in the calculation.) At the screen, the component dE of the electric field* due to the wave emerging from this elementary slit may be written

$$dE = \left(E_c \frac{dy}{a}\right) \sin(\omega t + \phi) \tag{37-3}$$

The amplitude $E_c(dy/a)$ is proportional to the fraction dy/a of the slit occupied by

* As in the last chapter, for brevity we drop the coordinate subscript on the symbol for the field component.

Figure 37-6. Developing an expression for the single-slit intensity distribution. The slit is divided into an infinite number of infinitesimal elementary slits.

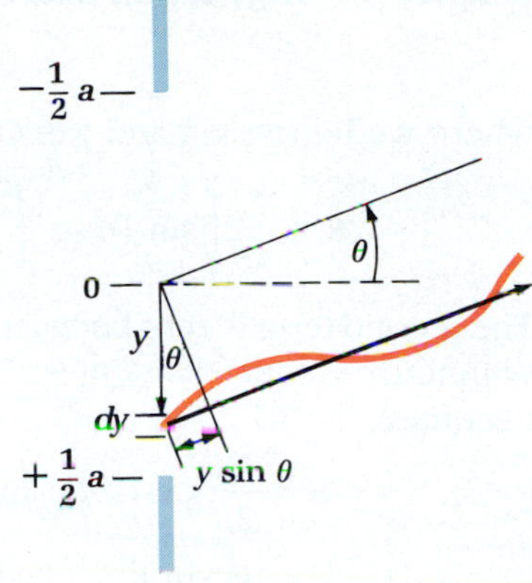

the elementary slit because, as the wave emerges from the slit, it is uniformly distributed along y. It turns out that E_c is the amplitude of the wave at the center of the screen (Prob. 2). Consistent with the Fraunhofer approximation, we neglect any θ dependence in the amplitude. The phase difference ϕ arises because waves emerging from different elementary slits arrive at the screen with different phases. The waves are in phase as they emerge from the slit, but become out of phase while traveling different distances to the screen. Thus ϕ is the phase difference at the screen between a wave that emerges at y and a wave that emerges at $y = 0$. To reach a point on the screen located by the angle θ (Fig. 37-6), a wave that emerges from the slit at y travels a distance $y \sin \theta$ farther than a wave that emerges at $y = 0$. Since a path difference of one wavelength causes a phase difference of 2π rad, we have

$$\frac{\text{Phase difference}}{2\pi} = \frac{\text{path difference}}{\lambda} \qquad \text{or} \qquad \frac{\phi}{2\pi} = \frac{y \sin \theta}{\lambda}$$

The relation between ϕ and y is

$$\phi = \left(\frac{2\pi}{\lambda} \sin \theta\right) y \tag{37-4}$$

To find the resultant electric field component E, we must add the contributions due to each elementary slit from $y = -\frac{1}{2}a$ to $y = +\frac{1}{2}a$. That is, we must integrate Eq. 37-3 with respect to y between these limits. The integration is performed at a specific instant (t fixed) and for a specific point on the screen (θ fixed). The value of E at a point on the screen is

$$E = \frac{E_c}{a} \int_{-\frac{1}{2}a}^{+\frac{1}{2}a} \sin (\omega t + \phi)\, dy \tag{37-5}$$

As we have seen, ϕ depends on y. To evaluate the integral we must either substitute for ϕ in terms of y or substitute for dy in terms of $d\phi$. We choose the latter. Solving Eq. 37-4 for y and differentiating (θ fixed), we obtain

$$dy = \frac{\lambda}{2\pi \sin \theta}\, d\phi$$

This substitution into Eq. 37-5 changes the variable of integration, so we must change the limits accordingly. Equation 37-4 shows that when $y = \pm\frac{1}{2}a$, $\phi = \pm(\pi a/\lambda) \sin \theta$. For brevity we let $\beta = (\pi a/\lambda) \sin \theta$, so that the limits on ϕ are from $-\beta$ to $+\beta$. [Note that β is the phase difference at the screen between a wave emerging from the center of the slit ($y = 0$) and a wave emerging from one edge of the slit ($y = \frac{1}{2}a$). Now

$$E = \frac{E_c}{2\beta} \int_{-\beta}^{+\beta} \sin (\omega t + \phi)\, d\phi$$

Time t is constant with respect to this integration, so we factor constants out of the integral with the trigonometric identity

$$\sin (\alpha + \gamma) = \sin \alpha \cos \gamma + \cos \alpha \sin \gamma$$

where we let $\alpha = \omega t$ and $\gamma = \phi$. This gives

$$E = \frac{E_c}{2\beta}\left[\sin(\omega t)\int_{-\beta}^{+\beta}\cos\phi\, d\phi + \cos(\omega t)\int_{-\beta}^{+\beta}\sin\phi\, d\phi\right]$$

The second term is zero because it involves the integral of an odd function between symmetric limits. (See Ques. 15 or Exercise 15.) After integration the first term becomes

$$E = \frac{E_c}{2\beta}\sin(\omega t)[\sin\phi]_{-\beta}^{+\beta} = \frac{E_c}{2\beta}\sin(\omega t)[\sin\beta - \sin(-\beta)]$$

$$= \frac{E_c \sin\beta}{\beta}\sin(\omega t)$$

The resultant amplitude at a point on the screen is $(E_c \sin\beta)/\beta$.

Light intensity is proportional to the square of the amplitude of the electric field component: $I \propto [(E_c \sin\beta)/\beta]^2$. Letting I_c be the intensity at the center of the screen (see Prob. 2), we obtain

Single-slit intensity distribution

$$I = I_c \frac{\sin^2\beta}{\beta^2} \qquad (37\text{-}6)$$

This is the *single-slit diffraction formula.* The dependence of the intensity on angular position θ is contained in $\beta : \beta = (\pi a/\lambda)\sin\theta$. A graph of I versus $\sin\theta$ is shown in Fig. 37-4*b* and was discussed in the previous section. Since $\sin\beta = 0$ when $\beta = \pm m'\pi$ ($m = 1, 2, \ldots$), Eq. 37-6 gives $I = 0$ when $(\pi a/\lambda)\sin\theta = \pm m'\pi$. This corresponds to the angles that locate minima given in Eq. 37-1, $a\sin\theta_{m'} = \pm m'\lambda$.

Figure 37-7 shows a comparison of I versus θ for two cases where the slit width is different by a factor of 2, but the wavelength and intensity of the incident beam are the same. When the slit width is reduced by a factor of 2, the amplitude of the wave at the center of the screen is reduced by a factor of 2, so the intensity at the center is reduced by a factor of 4.

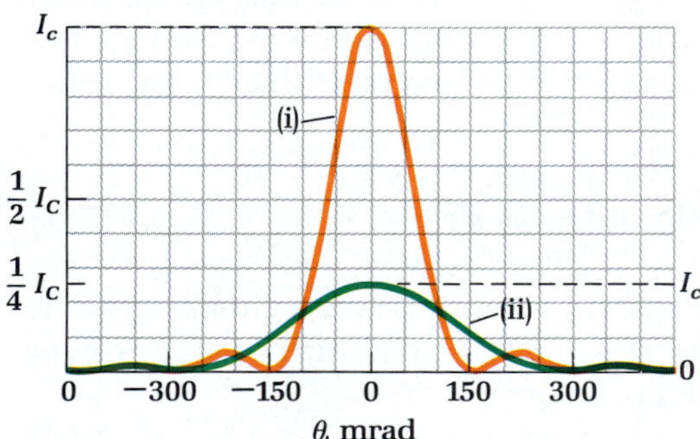

Figure 37-7. Comparison of intensity distributions for slits of widths different by a factor of 2. (i) Wide slit and (ii) narrow slit.

EXAMPLE 37-2

Finding a wavelength from an intensity minimum. Suppose the width of the wider slit for the intensity distribution shown in Fig. 37-7 is 4.0 μm. Determine the wavelength of the light.

Solution. From Fig. 37-7, the $m' = 1$ minima in the pattern for the wider slit are at $\theta = \pm 150$ mrad. Solving Eq. 37-1 for λ, with $m' = 1$, we have

$$\lambda = a\sin\theta = (4.0\ \mu\text{m})\sin(0.15\ \text{rad}) = 600\ \text{nm}$$

SELF-TEST 37-2. Suppose the intensity maximum at the center of the pattern produced by the wider slit in Fig. 37-7 is 8 W/m^2. What is the intensity maximum at the center of the pattern produced by the narrower slit? The wider slit has twice the width of the narrower slit. ***ANSWER:*** $(8\ \text{W/m}^2)/2^2 = 2\ \text{W/m}^2$.

EXAMPLE 37-3

Intensities at the secondary maxima. *(a)* Develop an expression for the intensity of the secondary maxima in terms of I_c. Employ the approximation that integers can be used for the values of m in Eq. 37-2. *(b)* Use your answer from part *(a)* to find the relative intensities of the $m \approx 1$ and $m \approx 2$ secondary maxima.

Solution. *(a)* From Eq. 37-2, angles θ_m, which locate secondary maxima, are given by $a \sin \theta_m = \pm(m + \frac{1}{2})\lambda$. Values of β that correspond to secondary maxima are

$$\beta_m = \frac{\pi a}{\lambda} \sin \theta_m = \pm(m + \tfrac{1}{2})\pi$$

Substitution into Eq. 37-6 gives the intensity I_m of a secondary maximum:

$$I_m = I_c \frac{\sin^2 [\pm(m + \frac{1}{2})\pi]}{[\pm(m + \frac{1}{2})\pi]^2}$$

If we use the approximation that the secondary maxima correspond to integral values of m, then we have $\sin^2 [\pm(m + \frac{1}{2})\pi] \approx 1$ for all m. Thus

$$I_m \approx \frac{I_c}{[(m + \frac{1}{2})\pi]^2}$$

where $m \approx 1, 2, \ldots$

(b) For the $m \approx 1$ maxima,

$$I_1 \approx \frac{I_c}{[(1 + \frac{1}{2})\pi]^2} = 0.045 I_c$$

For the $m \approx 2$ maxima,

$$I_2 \approx \frac{I_c}{[(2 + \frac{1}{2})\pi]^2} = 0.016 I_c$$

The $m \approx 1$ maxima have an intensity of about 4.5 percent that of the central maximum, and the $m \approx 2$ maxima have an intensity of about 1.6 percent that of the central maximum.

SELF-TEST 37-3. Write the expression, in terms of I_c, for the intensity I_3 at the $m \approx 3$ secondary maxima. ***ANSWER:*** $I_3 = I_c/[(3 + \frac{1}{2})\pi]^2 = 0.0083 I_c$

The Double-Slit Experiment Revisited

In discussing the double-slit pattern in the previous chapter, we made the simplifying assumption that the intensity distribution due to one slit acting alone is uniform across the screen. We now see, from Eq. 37-6, that this assumption is valid when the width of each slit is much less than the wavelength, $a \ll \lambda$. Then the central maximum would be spread across the screen. This situation is rarely achieved in an experiment because such narrow slits are difficult to make and would allow an exceedingly small amount of light to reach the screen. Using this assumption, we developed an expression for the intensity in the double-slit interference pattern (Eq. 36-5): $I_{12} = 4I_0 \cos^2 (\frac{1}{2}\phi)$, where $\phi = (2\pi d/\lambda) \sin \theta$.

We can now relax the condition that $a \ll \lambda$, and write a more general expression for the double-slit intensity distribution. Now the intensity due to one slit acting alone is not uniform; it is given by the single-slit diffraction formula. The expression for the double-slit pattern is found by replacing I_0 in Eq. 36-5 with $I_c \sin^2 \beta/\beta^2$:

$$I_{12} = 4I_c \left(\frac{\sin^2 \beta}{\beta^2}\right) \cos^2 (\tfrac{1}{2}\phi) \qquad (37\text{-}7)$$

Double-slit intensity distribution

where $\beta = (\pi a/\lambda) \sin \theta$ and $\phi = (2\pi d/\lambda) \sin \theta$. The factor $\sin^2 \beta/\beta^2$ depends on the slit width a and is usually called the *diffraction factor*. The factor $\cos^2 (\frac{1}{2}\phi)$ depends on the slit separation d and is usually called the *interference factor*. The intensity I_c is the intensity at the center of the single-slit pattern due to one of the slits acting alone. Figure 37-8 shows how the diffraction factor and the interference factor combine to give the double-slit pattern.

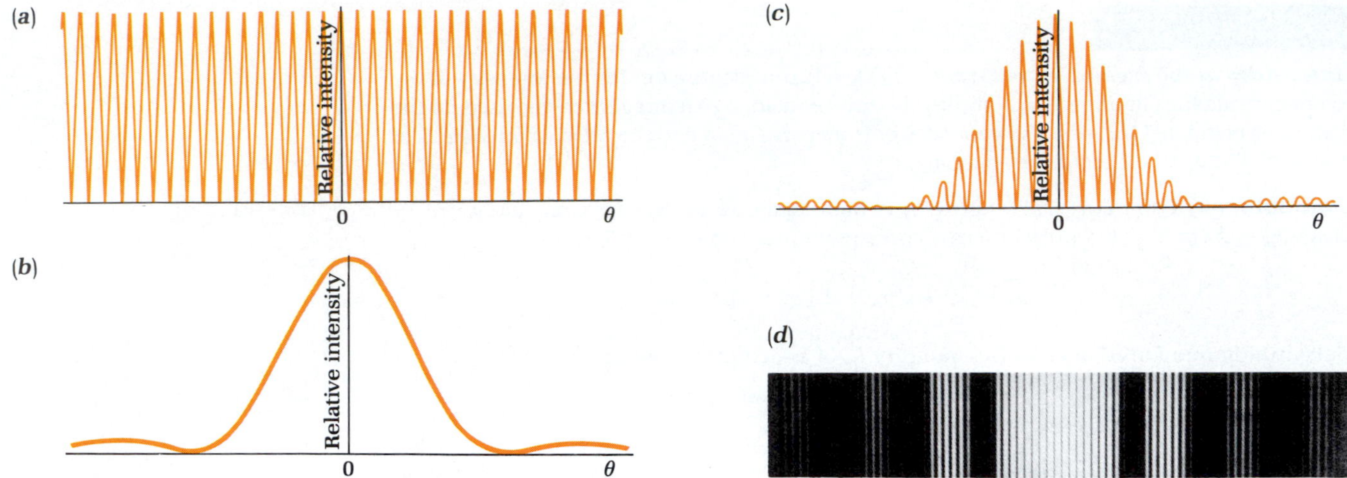

Figure 37-8. The double-slit intensity distribution. *(a)* A graph of the interference factor, $\cos^2(\frac{1}{2}\phi)$. *(b)* A graph of the diffraction factor, $\sin^2\beta/\beta^2$. *(c)* A graph of the double-slit intensity distribution, which is proportional to the product of parts *(a)* and *(b)*. *(d)* Photographs of a double-slit pattern.

37-4 THE LIMIT OF RESOLUTION

An important property of any optical instrument, such as a telescope, a camera, or an eye, is its resolving power. The ***resolving power*** of an optical instrument is a measure of the instrument's ability to produce separate images of two adjacent point objects. Often the measured resolving power of a particular instrument depends on imperfections in the lenses and/or mirrors, and sometimes on the properties of the surrounding medium. However, because of the wave nature of light there is an ***ultimate limit*** to the resolving power of all optical instruments. It is this ultimate limit we now discuss.

In the last section we described the diffraction of light by a slit. A slit is a rectangular aperture, very long and very narrow. Optical instruments, for example the pupil of an eye, usually have circular apertures. As you can see from Fig. 37-9, a circular aperture produces a diffraction pattern similar to that of a slit, except that it has circular symmetry. A mathematical analysis similar to the one we used in discussing a slit gives the expression for the angular position θ_1 of the first dark ring surrounding the bright central maximum in Fig. 37-9:

Figure 37-9. Fraunhofer diffraction by a circular aperture, the circular analogy to Fig. 37-4*a*. The central maximum is called the *Airy disk,* named for Sir George Airy (1801–1892).

$$\sin\theta_1 = 1.22\,\frac{\lambda}{d} \tag{37-8}$$

In this equation, d is the diameter of the circular aperture and θ_1 is the half-angle of a cone. As shown in Fig. 37-10, the cone has a circular base (of radius r_1) that is bounded by the first dark ring, and its apex is at the center of the aperture.

Ordinarily the angle θ_1 is small so that the approximations $\sin\theta_1 \approx \tan\theta_1 \approx \theta_1$ are valid, and we can use

$$\theta_1 \approx 1.22\,\frac{\lambda}{d} \tag{37-9}$$

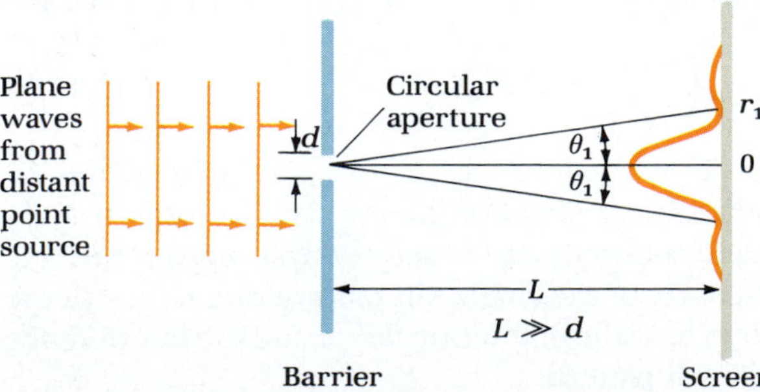

Figure 37-10. The angle θ_1 that locates the first dark ring in the Fraunhofer diffraction by a circular aperture is the half-angle of a cone. The base of the cone is bounded by the first dark ring, and the apex of the cone is at the center of the aperture.

Or, since $\tan \theta_1 = r_1/L$ (Fig. 37-10),

$$r_1 \approx 1.22 \frac{\lambda L}{d} \qquad (37\text{-}10)$$

Notice from Eqs. 37-9 and 37-10 that the size of the central maximum is directly proportional to λ/d. That is, the central maximum is more spread out for longer wavelengths and for smaller apertures.

Suppose we observe two distant stars with a telescope; the angular separation of the stars is very small and each star produces about the same intensity at the earth. In this case the circular aperture is the entrance mirror (or lens) of the telescope, and instead of a screen we have a solid-state detector. Each star is a point source, and its image on the film is the diffraction pattern due to the circular aperture. Since there are two stars, there will be two such diffraction patterns. No stable interference pattern will be produced by the interference of light from these sources because they are incoherent; the resultant intensity is the sum of the intensity from each diffraction pattern.

What must be the angular separation of the two stars for us to be able to tell that there are two and not one? To answer, we need a criterion for resolution. Figure 37-11 shows the images (diffraction patterns) of two point sources when they are unresolved, just resolved, and clearly resolved. The accepted criterion for resolution is the *Rayleigh criterion*. According to this criterion, two images are just resolved if the center of the central maximum of one pattern falls on the first dark ring of the other, the case shown in Fig. 37-11*b*. Therefore two point objects separated by an angle $\Delta\theta$ are resolved when $\Delta\theta > \Delta\theta_R$, where

$$\Delta\theta_R \approx 1.22 \frac{\lambda}{d} \qquad (37\text{-}11)$$

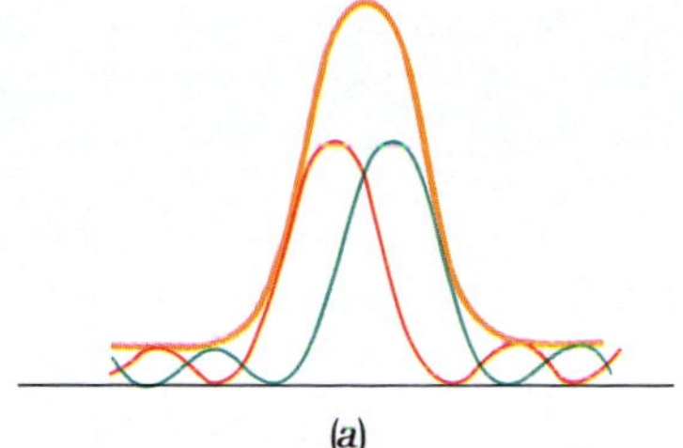
(a)

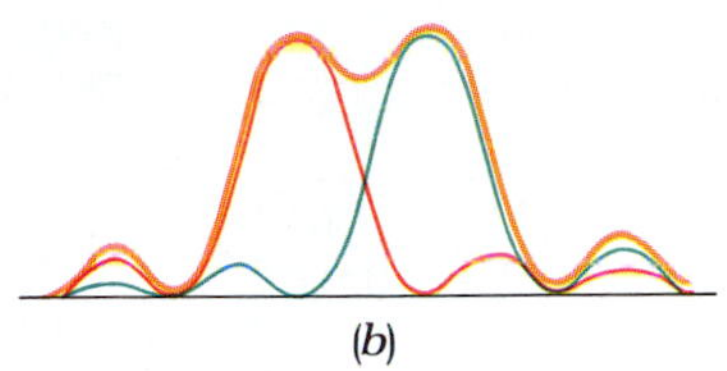
(b)

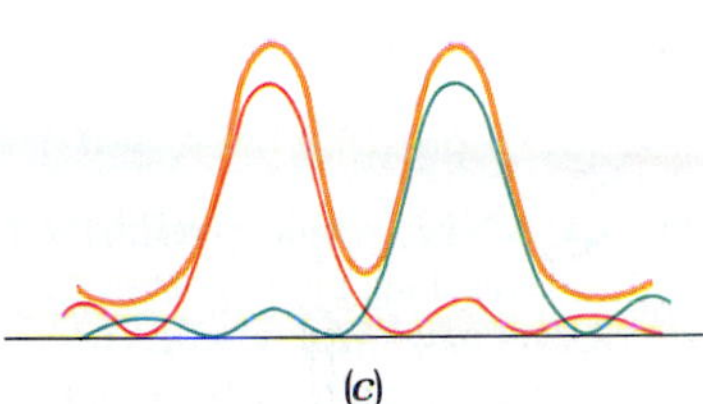
(c)

Figure 37-11. Showing the Rayleigh criterion. The diffraction patterns of two point sources by a circular aperture are *(a)* unresolved, *(b)* just resolved, with the maximum of one pattern at the minimum of the other, and *(c)* clearly resolved.

The angle $\Delta\theta_R$ is the limiting angle of resolution for an optical instrument with aperture diameter d. For the Rayleigh criterion to be useful, the intensity of the light that reaches the instrument from each object should be nearly the same. Regardless of the criterion we choose, the ultimate resolving power of an optical instrument depends on the wavelength of the light and the diameter of the instrument's aperture.

EXAMPLE 37-4

Limiting angle of resolution for a telescope. In an astronomical research telescope, the aperture diameter is usually large, and consequently, the telescope's limiting angle of resolution is small. The Mount Palomar telescope has a diameter of 5.1 m. Determine the limiting angle of resolution for this telescope when light of wavelength 550 nm is used.

Solution. From Eq. 37-11, the limiting angle is

$$\Delta\theta_R = 1.22 \frac{550 \text{ nm}}{5.1 \text{ m}} \approx 0.1\ \mu\text{rad}$$

This is about the same angle that is subtended by a dime at a distance of 14 km! This angle is smaller than the limiting angle due to "atmospheric blurring," which is from about 5 μrad to about 0.5 μrad, depending on conditions. Atmospheric problems can be avoided by placing a research telescope in earth orbit. However, even if we have a perfect telescope and ideal surroundings, the limit on resolution given by Eq. 37-11 cannot be avoided, since the limit is due to the wave nature of light.

SELF-TEST 37-4. What is the size of the smallest object on the moon that can be resolved when the angle of resolution is 0.1 μrad? The distance to the moon is 384 Mm. ***ANSWER:*** (384 Mm)(0.1 μrad) = 40 m

37-5 POLARIZATION

The polarization of a wave characterizes the direction of the wave oscillations. As we saw in Chap. 34, light is an electromagnetic wave in which the oscillating quantities are the electric and magnetic fields. In a plane wave, the directions of **E** and **B** are perpendicular to each other and perpendicular to the direction of propagation of the wave. Light is a transverse wave, and consequently, it can be polarized in different ways.

The simplest type of polarization to discuss is linear or plane polarization, shown schematically in Fig. 37-12. Recall from Chap. 34 that the propagation direction (the direction of the Poynting vector **S**) is given by the direction of **E** × **B**. In the figure, **E** is directed toward $+y$ at the same position and time that **B** is directed toward $+z$. Since **j** × **k** = **i**, the direction of propagation is toward $+x$. It is customary to let the polarization direction be defined as along **E** rather than **B**. In Fig. 37-12, the wave is polarized parallel to the y axis because **E** oscillates along the $\pm y$ directions. The plane that contains **E** and the propagation direction is called the ***plane of polarization*** (the xy plane for the case shown in the figure).

Plane-polarized light

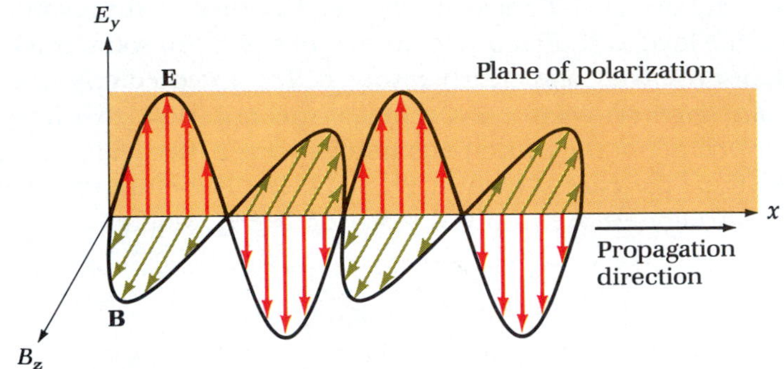

Figure 37-12. A plane-polarized light wave. The propagation direction is toward $+x$, and the plane of polarization is the xy plane.

At any point along a plane-polarized wave, **E** oscillates along a fixed line. Waves may also be *circularly* or *elliptically polarized.* At any point along a circularly polarized wave, **E** maintains a fixed magnitude, but its direction rotates in space with a constant angular frequency. Figure 37-13*a* shows the time dependence of **E** at a particular point due to a circularly polarized wave that is propagating along the $+x$ direction (out of the page). The variation of **E** may be represented by a vector of fixed magnitude that rotates about the x axis with a constant angular frequency ω. In this picture, the tip of the **E** vector traces out a circle, and the components of **E**—E_y and E_z—oscillate with the same amplitude and have a phase difference of $\frac{1}{2}\pi$ rad. If the tip of the **E** vector in Fig. 37-13*a* rotates in the clockwise sense, then the light is *right*-circularly polarized, whereas if the tip rotates in the counterclockwise sense, then the light is *left*-circularly polarized.

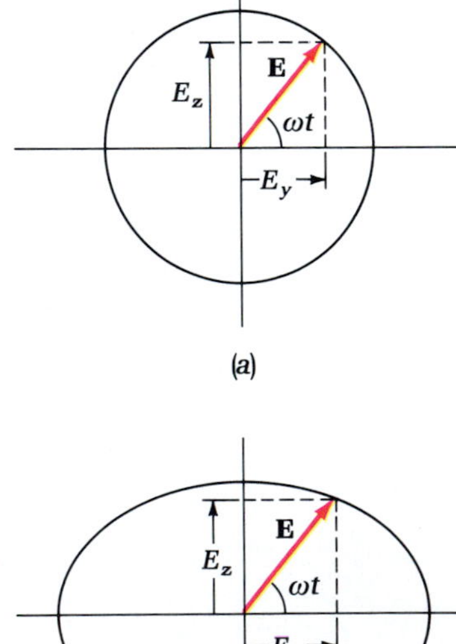

Figure 37-13. *(a)* The **E** vector of a circularly polarized wave at a particular point. The propagation direction is out of the page (toward $+x$). *(b)* An elliptically polarized wave.

An elliptically polarized wave (Fig. 37-13*b*) is similar to a circularly polarized wave except that, at a particular point, E_y and E_z have different amplitudes. In this case the tip of the **E** vector traces out an ellipse.

Ordinarily when one speaks of polarized light without qualifying it as plane, circular, or elliptical, one means *plane*-polarized light. Henceforth, we shall confine our attention to plane-polarized light.

Light emitted from an ordinary source, such as the filament of a light bulb, is *unpolarized.* Acting independently, the atoms and molecules emit wave trains of light, and the polarizations of these wave trains are unrelated. The resulting light consists of a random mixture of polarizations; it is unpolarized.

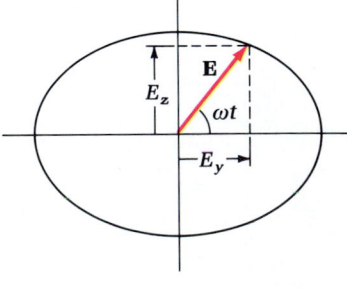

37-6 MEASUREMENT OF POLARIZATION

Light can be polarized by passing it through a *polarizer.* A familiar polarizer is the Polaroid film that is often used in sunglasses. A polarizer is an optical device that selectively transmits light having its plane of polarization parallel to the polarizer's

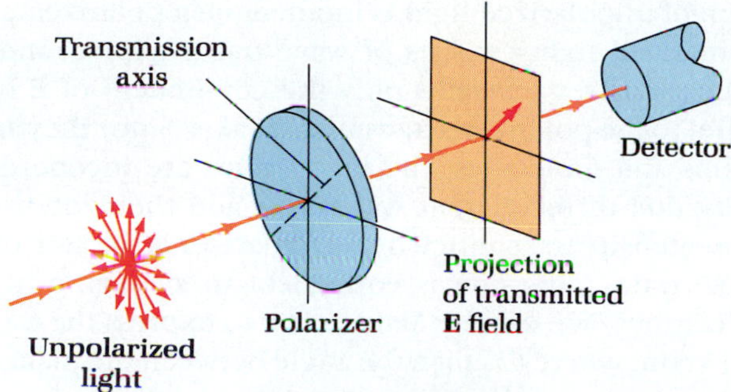

Figure 37-14. A polarizer transmits light with its plane of polarization parallel to the polarizer's transmission axis.

transmission axis. Light having its plane of polarization perpendicular to the transmission axis is blocked out by absorbtion or reflection. Figure 37-14 shows a beam of unpolarized light incident on a polarizer. The electric field of the transmitted wave at a particular point and time is parallel to the polarizer's transmission axis.

Malus's Law

Suppose we pass the polarized light from a polarizer through a second polarizer, as shown in Fig. 37-15. The second polarizer is often called the *analyzer* to distinguish it from the first polarizer. The azimuthal angle θ is a measure of the orientation of the polarizer's transmission axis relative to the analyzer's transmission axis. Let I_0 represent the intensity transmitted by the polarizer and incident on the analyzer. Notice that the detector can measure I_0 if the analyzer is removed temporarily. What light intensity I is transmitted by the analyzer and measured by the

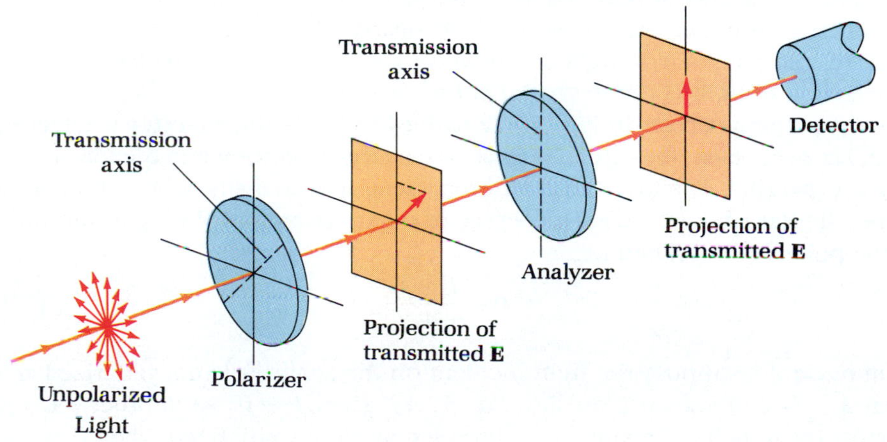

Figure 37-15. Plane-polarized light is transmitted by the polarizer. The analyzer transmits only the component of **E** parallel to its transmission axis.

detector? In Fig. 37-16 we show the projection of the field components of the light as it emerges from the polarizer with amplitude E_0 and from the analyzer with amplitude $E_0 \cos\theta$. That is, the analyzer transmits only the component of the wave that is parallel to its transmission axis. Since the intensity is proportional to the square of the wave amplitude, we have that $I_0 \propto (E_0)^2$ and $I \propto (E_0 \cos\theta)^2$. The proportionality constant is the same in each case so that

$$I = I_0 \cos^2\theta \tag{37-12}$$

This is *Malus's law* (named for E. L. Malus, 1775–1812). When $\theta = 0$ is substituted into Malus's law, we have $I = I_0 \cos^2 0 = I_0$. That is, the intensity transmitted by the analyzer is maximum when the transmission axes of the polarizer and analyzer are parallel. Further, when $\theta = \frac{1}{2}\pi$ rad, $I = I_0 \cos^2(\frac{1}{2}\pi) = 0$. That is, the intensity transmitted by the analyzer is minimum when the transmission axes are perpendicular. Because there is no positive sense associated with a transmission axis, all possible intensity values, $0 < I < I_0$, are realized in the interval from $\theta = 0$ to $\theta = \frac{1}{2}\pi$ rad.

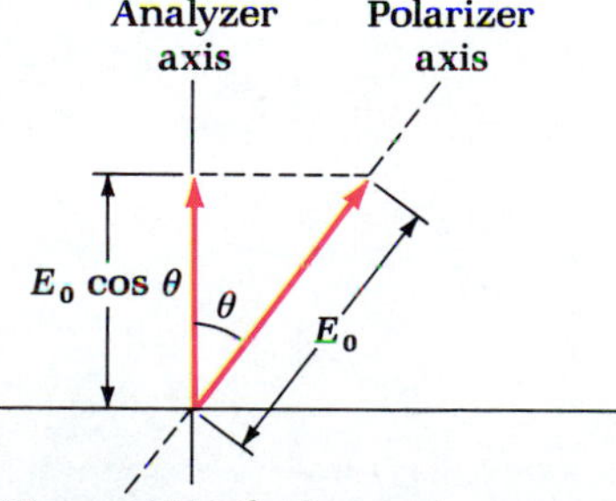

Figure 37-16. The projection of the field $\mathbf{E}_0$ transmitted by the polarizer onto the transmission axis of the analyzer is $E_0 \cos\theta$.

Suppose a beam of unpolarized light is incident on a polarizer, as shown in Fig. 37-14. The unpolarized light consists of wave trains with a random mixture of polarizations. The polarizer transmits only the component of **E** from each wave train that is parallel to the polarizer's transmission axis. Since they are emitted from independent atoms and molecules, the wave trains are incoherent. To find the resultant intensity due to incoherent waves, we add their intensities (not their amplitudes). The intensity transmitted by the polarizer is the sum of the intensities due to each wave train. This sum is equivalent to averaging over the random mixture of polarizations. We can use Malus's law to express the transmitted intensity due to a wave train, where θ is then the angle between the plane of polarization of the wave train and the polarizer's transmission axis. Since the average value of $\cos^2\theta$ over the interval from $\theta = 0$ to $\theta = \frac{1}{2}\pi$ rad is 1/2, the transmitted intensity is one-half the incident intensity. This intensity is independent of the orientation of the polarizer because of the random polarizations of the incident wave trains.

An unpolarized light beam consists of incoherent wave trains with randomly oriented planes of polarization. We can determine the contribution to the intensity due to a single wave train by resolving the field due to the wave train along two mutually perpendicular axes. This can be done for each wave train in the beam. Thus an unpolarized beam may be considered as two incoherent plane-polarized beams with perpendicular planes of polarization.

Degree of Polarization

In our development of Malus's law, we tacitly assumed that both the polarizer and the analyzer were ideal. That is, all light polarized parallel to the transmission axis was transmitted, and no light polarized perpendicular to the transmission axis was transmitted. Although polarizers that are nearly ideal can be constructed, real polarizers do not polarize light completely. The quantity used to characterize the polarization of light is the *degree of polarization P.*

Suppose the polarizer in Fig. 37-15 is not ideal, but the analyzer is effectively ideal. Let $I_\parallel$ represent the intensity measured by the detector when the transmission axes are parallel ($\theta = 0$), and let $I_\perp$ represent the intensity when the axes are perpendicular ($\theta = \frac{1}{2}\pi$ rad). The degree of polarization P of the light transmitted by the polarizer is defined as

Degree of polarization

$$P = \frac{I_\parallel - I_\perp}{I_\parallel + I_\perp} \tag{37-13}$$

As an example, suppose the light incident on the analyzer is not polarized at all. Then $I_\parallel = I_\perp$ and substitution into Eq. 37-13 gives $P = 0$. As another example, suppose the light incident on the analyzer is completely polarized. Then $I_\perp = 0$ and substitution into Eq. 37-13 gives $P = 1$. Thus the degree of polarization ranges from a minimum of zero for unpolarized light to a maximum of 1 for completely polarized light.

EXAMPLE 37-5

Degree of polarization. Suppose the degree of polarization of the light transmitted by a polarizer is checked by using the arrangement shown in Fig. 37-15. The intensity measured by the detector when the axes are parallel is I_m, and the intensity measured when the axes are perpendicular is $0.127I_m$. Assuming the analyzer is ideal, determine the degree of polarization due to the polarizer.

Solution. Using Eq. 37-13, we have

$$P = \frac{I_m - 0.127I_m}{I_m + 0.127I_m} = \frac{0.873}{1.127} = 0.775$$

Often this result is stated as a percentage, and we say that the light is 77.5 percent polarized.

SELF-TEST 37-5. What is the ratio $I_{\parallel}/I_{\perp}$ for a beam of light that has a degree of polarization of 50 percent? ***ANSWER:*** 3.

37-7 METHODS FOR POLARIZING LIGHT

When light undergoes a process such as reflection or scattering, it tends to become polarized. These processes can be utilized to produce polarized light. The particular processes that we shall discuss are (i) selective absorbtion, (ii) reflection and transmission, (iii) double refraction, and (iv) scattering.

Selective Absorbtion, or Dichroism

A common method of polarizing light is with a sheet of Polaroid, the material often used in sunglasses. Polaroid is an example of a *dichroic material.* A dichroic material transmits light that has its plane of polarization parallel to a particular alignment in the material and strongly absorbs light that has its plane of polarization perpendicular to this alignment. These alignments correspond to certain molecular or crystalline orientations. Figure 37-17 indicates the action of a dichroic material. A number of naturally occurring crystals are dichroic; an important example is tourmaline.

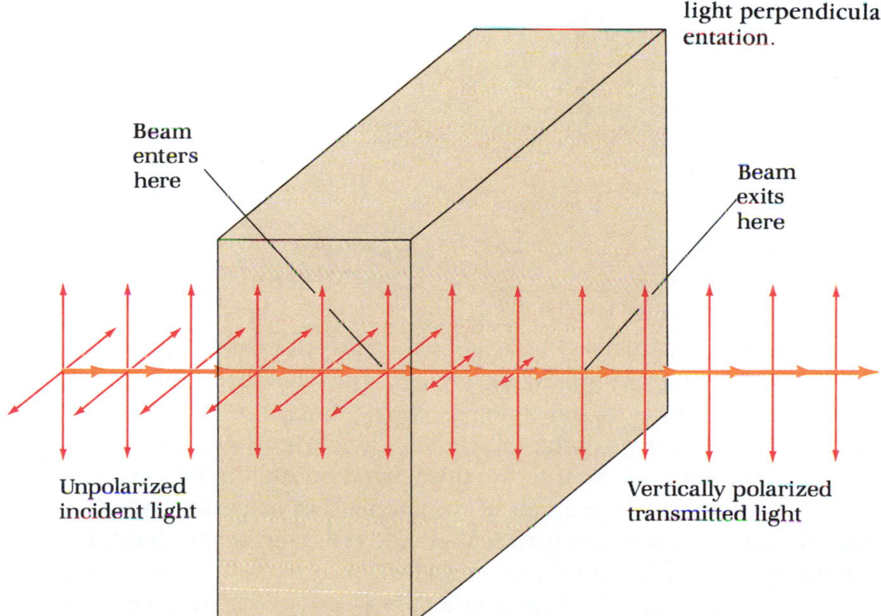

Figure 37-17. A dichroic material selectively transmits light with its plane of polarization parallel to a crystalline or molecular orientation and absorbs light perpendicular to this orientation.

Polaroid is produced by stretching plastic sheets that contain long-chain molecules and thereby aligning these molecules. When the plastic is stained with an ink that contains iodine, the material becomes dichroic. Polaroid is a convenient and inexpensive polarizer, but the transmitted light is colored because the absorbtion depends on wavelength as well as polarization. Also, the transmitted light is not completely polarized, and its degree of polarization depends somewhat on the wavelength. We have used Polaroid sunglasses to demonstrate Malus's law in Fig. 37-18.

(*a*)

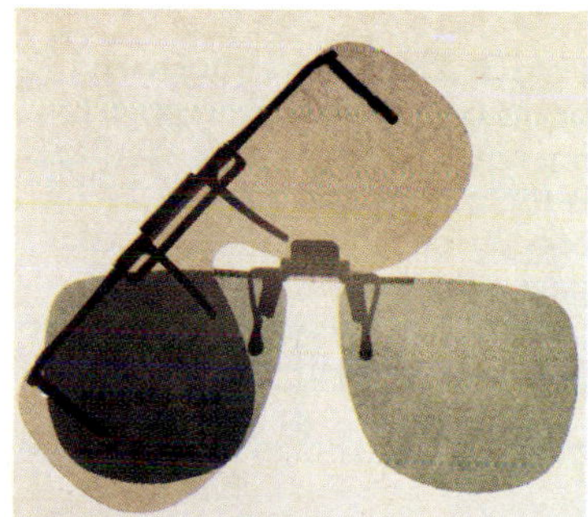

(*b*)

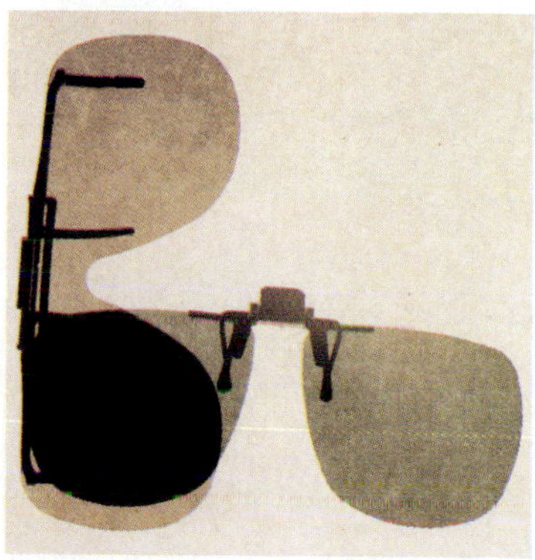

(*c*)

Figure 37-18. Demonstration of Malus's law with polariod sunglasses. *(a)* $\theta = 0$. *(b)* $\theta = \pi/4$. *(c)* $\theta = \pi/2$.

Reflection and Transmission

Using a pair of Polaroid sunglasses as an analyzer, you can observe the polarization of light by reflection. The next time you are wearing a pair of these glasses, tilt your head to one side and notice the greatly increased light intensity reflected from horizontal surfaces. Since the source of the light is the sun above, most of the light reaching your eyes is reflected from horizontal surfaces. When the light is reflected, it is also polarized, and the orientation of the plane of polarization depends on the orientation of the reflecting surface. The transmission axis of the Polaroid film in the glasses is aligned so that light that has been polarized by reflection from a horizontal surface is absorbed. With this alignment, these glasses reduce the bright glare from horizontal surfaces.

Consider an unpolarized light beam in air that is incident on a glass surface (Fig. 37-19). Except for a few particular angles, both the reflected and the refracted beams are partially polarized. If the beam is normally incident on the surface ($\theta = 0$), or if the beam grazes the surface ($\theta = \frac{1}{2}\pi$ rad), the reflected light is not polarized. In the discussion that follows, we consider angles of incidence between these two extremes, where the refracted beam is partially polarized at all angles. The reflected beam is also partially polarized at all angles except for one particular angle. At this angle, called the *polarizing angle* θ_p, the reflected beam is totally polarized.

The reflected beam is totally polarized at $\theta = \theta_p$.

When $\theta = \theta_p$, the reflected beam is completely polarized, with its plane of polarization perpendicular to the plane of incidence. Recall from Sec. 35-2 that the plane of incidence contains the incident beam, the reflected beam, and the normal to the surface. Thus the electric field of the reflected beam oscillates parallel to the reflecting surface (Fig. 37-19). For angles other than θ_p, the reflected beam is partially polarized perpendicular to the plane of incidence.

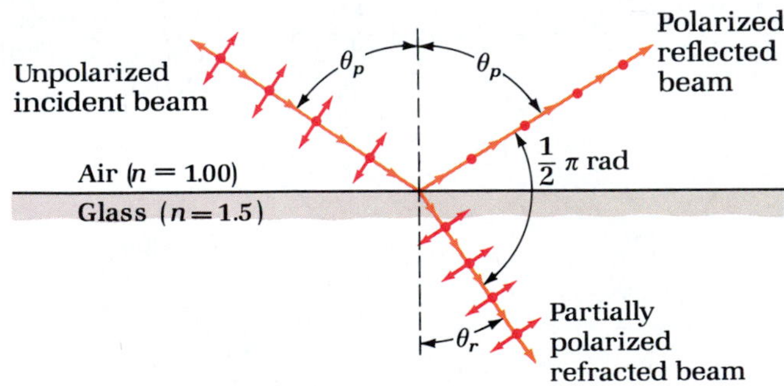

Figure 37-19. An unpolarized beam is incident at the polarizing angle. The plane of the page is the plane of incidence. Field oscillations in the plane of the page are marked ↕, and field oscillations perpendicular to the page are marked · .

In 1812 Sir David Brewster (1781 – 1868) experimentally discovered that when the angle of incidence θ_i is set at the polarizing angle θ_p, the reflected beam and the refracted beam are perpendicular, as shown in Fig. 37-19. Since $\theta_i = \theta_p$ in this case and since the angle of reflection equals the angle of incidence, Brewster's discovery may be written as $\theta_p + \theta_r = \frac{1}{2}\pi$ rad. We can use this result and Snell's law to find θ_p in terms of the relative index of refraction of the media. Let n_1 be the index of refraction of the medium that contains the incident and reflected beams; let n_2 be the index of refraction of the medium that contains the refracted beam. Using Snell's law, $n_1 \sin\theta_i = n_2 \sin\theta_r$, with $\theta_i = \theta_p$ and $\theta_r = \frac{1}{2}\pi \text{ rad} - \theta_p$, we have

$$n_1 \sin\theta_p = n_2 \sin(\tfrac{1}{2}\pi - \theta_p)$$

Since $\sin(\frac{1}{2}\pi - \theta p) = \cos\theta_p$,

$$\frac{n_2}{n_1} = \frac{\sin\theta_p}{\cos\theta_p} = \tan\theta_p$$

The ratio n_2/n_1 is the relative index of refraction n_{21}. Therefore

Brewster's law

$$\tan\theta_p = n_{21} \tag{37-14}$$

This equation is called *Brewster's law,* and sometimes the polarizing angle θ_p is called *Brewster's angle.* Brewster's law gives the value of the angle of incidence for which the reflected beam is completely polarized. Suppose medium 1 is air ($n_1 = 1.00$) and medium 2 is glass ($n_2 = 1.5$). Brewster's law gives $\tan\theta_p = 1.5$, or $\theta_p = \tan^{-1} 1.5 = 0.98\ \text{rad} = 56°$. Unpolarized light incident from air and reflected from this glass surface will be totally polarized if the angle of incidence is about 1 rad.

When unpolarized light is incident on the surface of a dielectric, the reflected beam is polarized. What about the refracted beam? The light in the reflected and refracted beams originated in the unpolarized incident beam. With the reflected beam polarized perpendicular to the plane of incidence, the refracted beam is left with light that is polarized parallel to the plane of incidence. This polarization is partial at all angles, even at $\theta = \theta_p$.

The refracted beam is partially polarized at all angles.

EXAMPLE 37-6

Polarization of reflected and refracted beams. For an unpolarized light beam incident from air onto glass ($n = 1.50$) at the polarizing angle, 7.4 percent of the incident intensity is reflected and 92.6 percent is refracted. Determine the degree of polarization of *(a)* the reflected beam and *(b)* the refracted beam.

Solution. *(a)* Since the light is incident at the polarizing angle, all the reflected light is polarized perpendicular to the plane of incidence: $P = 1$. *(b)* We may treat the unpolarized incident beam (of intensity I_i) as if half its intensity is polarized parallel to the plane of incidence and the other half perpendicular to the plane of incidence. All of the incident beam that is polarized parallel to the plane of incidence is refracted, because none of it is reflected. Therefore if we let $I_\parallel$ be the refracted beam intensity polarized parallel to the plane of incidence, then $I_\parallel = 0.500I_i$. Now let $I_\perp$ be the refracted beam intensity polarized perpendicular to the plane of incidence. Since all the reflected light (7.4 percent) is polarized perpendicular to the plane of incidence, $I_\perp = (0.500 - 0.074)I_i = 0.426I_i$. Using Eq. 37-13, we have

$$P = \frac{0.500I_i - 0.426I_i}{0.500I_i + 0.426I_i} = \frac{0.074}{0.926} = 0.080$$

The refracted beam is 8.0 percent polarized. Despite the 100 percent polarization of the reflected beam, the refracted beam has a rather small degree of polarization. The reason is that very little light is reflected; almost all of it is refracted.

SELF-TEST 37-6. An unpolarized beam is incident on a transparent material at the polarizing angle; 10.0 percent of the beam is reflected and 90.0 percent is refracted. What is the degree of polarization of the refracted beam? ***ANSWER:*** 11.1 percent.

Double Refraction, or Birefringence

Certain crystalline materials exhibit *double refraction,* or ***birefringence.*** Two important examples are calcite ($CaCO_3$) and crystalline quartz (SiO_2). In Fig. 37-20, we show a beam of unpolarized light incident on a calcite crystal. Within the crystal there are two refracted beams or rays, the *ordinary ray,* or *O* ray, and the *extraordinary ray,* or *E* ray. These two rays take divergent paths within the medium and are polarized with mutually perpendicular planes of polarization. If the angles of refraction are measured for a number of angles of incidence, the results show that Snell's law holds for the *O* ray but not for the *E* ray. (There is a special axis in birefringent crystals called the *optic axis.* If the plane of incidence is perpendicular to the optic axis, then Snell's law holds for the *E* ray as well as the *O* ray.) Birefringent materials are quite useful in optics, especially as polarizers.

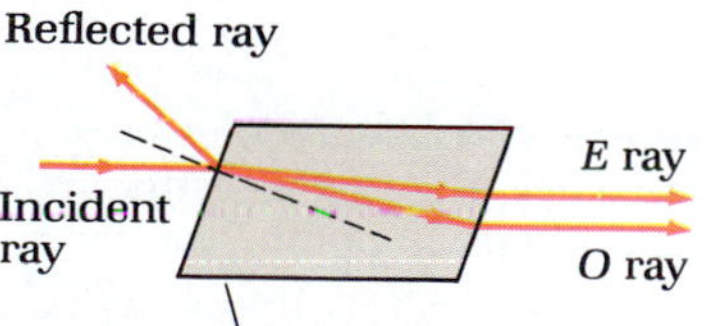

Figure 37-20. An unpolarized ray is incident on a birefringent crystal (say, calcite). The *O* ray and *E* ray are polarized, with mutually perpendicular planes of polarization.

Figure 37-21 shows the polarization of the *O* and *E* rays. For the normal incidence in the photograph, the light from the *O* ray is unrefracted, whereas the *E* ray is

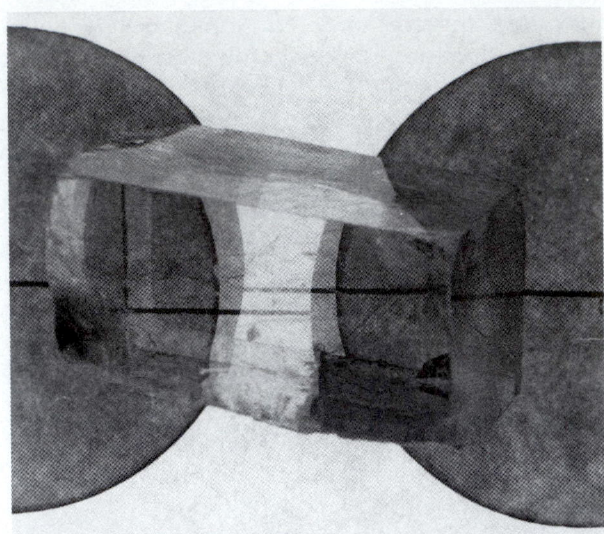

Figure 37-21. Demonstration of the polarization of the O and E rays. A straight horizontal line on a sheet of paper is viewed normally through circular Polaroid disks and through a calcite crystal. Since the light path is normal to the crystal faces, the O ray is unrefracted but the E ray is refracted. The transmission axis of the Polaroid disk on the left is horizontal and that of the disk on the right is vertical. You can see that the O ray and E ray are both polarized and that their planes of polarization are perpendicular.

refracted. The transmission axes of the polaroid sheets are given in the figure caption. Can you determine the plane of polarization of each ray?

Scattering

You can verify that scattered light is polarized by viewing the sky with a pair of Polaroid sunglasses on a clear day. Light comes to us from the sky because sunlight is scattered by the gas molecules in the atmosphere. If there were no atmosphere, the sky would be black and we could see stars during the day. The amount of scattering depends strongly on the wavelength; shorter wavelengths are scattered more than longer wavelengths. Thus the sky appears blue because the scattered light has a larger mixture of shorter wavelengths (violet and blue) than of longer wavelengths (yellow and red). Sunsets are red because this light comes to our eyes after passing through a large amount of atmosphere, and the short wavelengths have been selectively depleted from the light we see.

When you observe polarized skylight with your sunglasses, notice that the degree of polarization is maximum for light coming from the direction perpendicular to the sun's direction. Figure 37-22 shows the reason for this. A beam of unpolarized light propagating toward $+x$ is scattered by molecules in the vicinity of the origin. The direction of the scattered light that reaches an observer at point P on the z axis is at right angles to the incident beam direction. Since light is a transverse wave, the oscillating electric field of the waves incident on the scattering molecules has no component along the x axis. As a result, the oscillations in the beam scattered to the observer at P have no component along the x axis. The light detected at P is plane-polarized in the yz plane.

Some bees and ants can distinguish polarized sunlight, and they use it as a compass to navigate from their nest to a food source and back.

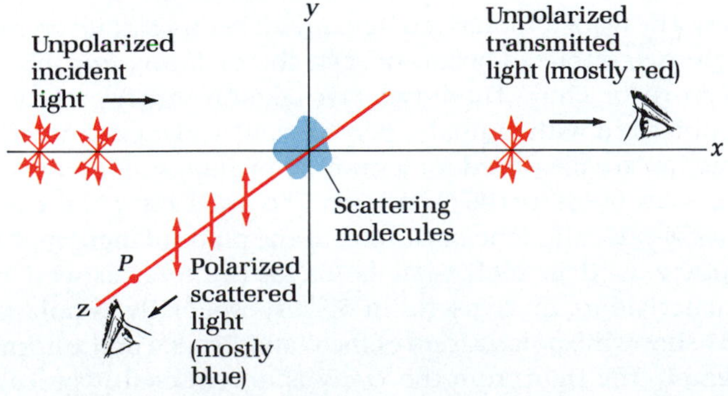

Figure 37-22. Molecules located near the origin scatter polarized light to point P on the z axis.

COMMENTARY: Holography

Holography is a technique for producing three-dimensional photographs by using the interference of light waves. A hologram of a scene gives a truly three-dimensional replica of the scene. Suppose we have a hologram of a saltshaker, a peppershaker, and a bottle. When viewed from a particular angle, the saltshaker is seen beside the bottle, but the peppershaker cannot be seen because it is behind the bottle. Then the viewer may change the angle of observation such that the saltshaker disappears behind the bottle and the peppershaker is seen on the other side.

One way in which the film is exposed in taking a hologram of an object is shown in Fig. 37-23*a*. A collimated beam of coherent, monochromatic light from a laser is incident on a mirror and on the object. The photographic film is exposed simultaneously to light reflected from the mirror and light scattered from the object. The light incident on the film from the mirror is called the *reference beam* and the light incident on the film from the object is called the *modified beam*. The incident light must be sufficiently coherent so that these two beams can produce a stable interference pattern on the surface of the film during exposure. The hologram is the photographic record of this interference pattern.

The viewing of a hologram is shown in Fig. 37-23*b*. The incident laser light is diffracted by the interference pattern of the hologram, which causes images of the object to be produced — a virtual image on the laser side of the hologram and a real image on the side opposite the laser.

We can get some understanding of how a hologram image is recorded and then viewed by simplifying the arrangement. In Fig. 37-24*a* we show a plane wave incident on a sheet of photographic film. (The wavelength is greatly exaggerated

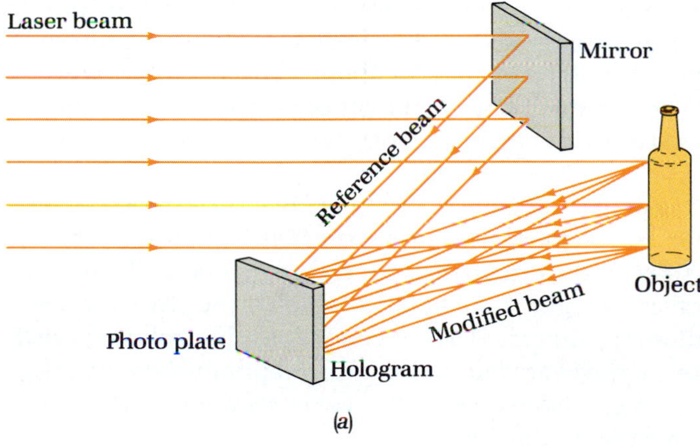

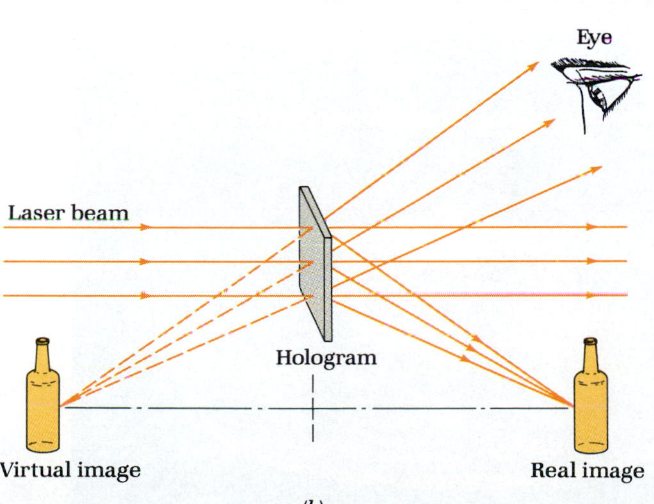

Figure 37-23. *(a)* A hologram is produced by exposing the film to a reference beam reflected from a mirror and a modified beam scattered from the object. (b) A hologram is viewed by observing the light that is diffracted by the hologram. The same type of laser light that was used in the exposure is also used in the viewing.

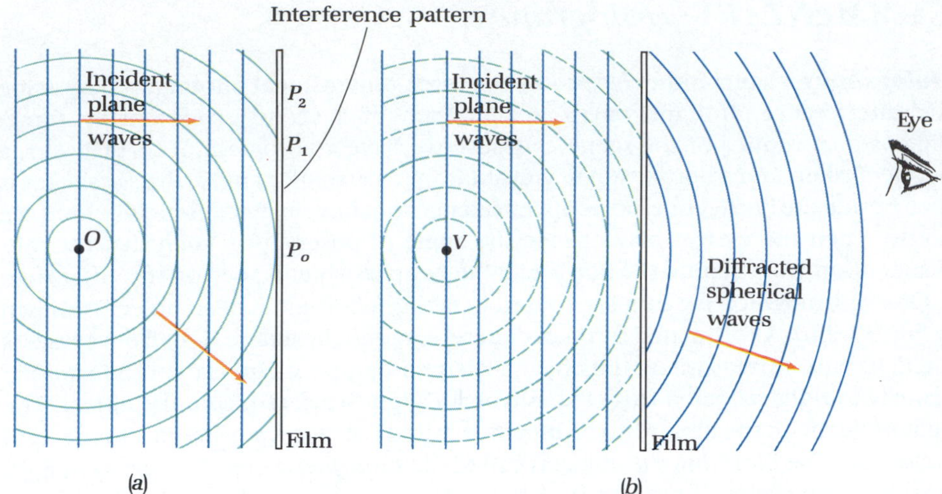

Figure 37-24. *(a)* Producing a hologram of a point object O. Plane waves incident on the film are scattered by the point object. Circular interference fringes are produced on the film during exposure. *(b)* Viewing the hologram of a point object. Plane waves incident on the film are diffracted by the interference pattern. Diffracted spherical waves diverge from virtual image V.

for purposes of illustration.) Waves scattered from the point object O are spherical waves centered at O's position. The incident plane waves and the scattered spherical waves produce a stable interference pattern on the film because they are coherent. At points where these waves arrive in phase they constructively interfere to produce interference maxima, and at points where they arrive out of phase they destructively interfere to produce minima. For the case shown in the figure, O is an integral number of wavelengths from the center of the screen at P_0. In this case interference maxima occur at P_0, P_1, and P_2. Thus interference fringes are formed and recorded on the film, and this exposed film is a hologram of a point object.

In the case of an extended object, there is interference between light scattered from each point on the object (within the modified beam in Fig. 37-23*a*) and the incident plane wave (the reference beam in Fig. 37-23*a*). The interference pattern produced on the film is due to the superposition of the interference of waves scattered from each point on the object and waves in the reference beam. Thus the hologram of an extended object consists of a very complex interference pattern, which is characteristic of the phases of waves scattered from the three-dimensional object.

The viewing of the hologram of the point object O is shown in Fig. 37-24*b*. Plane waves are incident on the hologram from the left. Although it is beyond the scope of this text to show it, the circular fringes recorded on the film diffract the light into spherical waves diverging from point V, which is the virtual image of point object O. (In addition, the diffracted waves consist of spherical waves that converge to a point, and plane waves are transmitted. For simplicity these waves are not shown.) In the case of an extended object, the viewer sees diffracted light that produces a complete three-dimensional image.

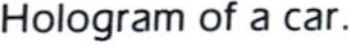
Hologram of a car.

SUMMARY

Section 37-1. Diffraction

Diffraction is the spreading of a wave around the edges of a barrier and into the region shaded by the barrier. In Fraunhofer diffraction the waves impinging on the barrier and the screen are plane waves, and in Fresnel diffraction the waves are not necessarily plane waves.

Section 37-2. Description of the Single-Slit Diffraction Pattern

The single-slit diffraction pattern is cast on a screen by monochromatic light passing through a narrow slit. The pattern is composed of fringes with a central maximum flanked by secondary maxima.

Section 37-3. The Single-Slit Intensity Distribution

The single-slit diffraction formula is

$$I = I_c \frac{\sin^2 \beta}{\beta^2} \qquad (37\text{-}6)$$

where $\beta = (\pi a/\lambda) \sin \theta$.

Section 37-4. The Limit of Resolution

The resolving power of optical instruments has an ultimate limit because of the wave nature of light. Using the Rayleigh criterion, we find that the limiting angle of resolution is

$$\Delta\theta_R \approx 1.22 \left(\frac{\lambda}{d}\right) \qquad (37\text{-}9)$$

where d is the aperture diameter of the instrument.

Section 37-5. Polarization

For a plane-polarized wave, the plane of polarization contains the oscillating electric field and the direction of propagation. Light may also be circularly or elliptically polarized.

Section 37-6. Measurement of Polarization

If plane-polarized light of intensity I_0 is incident on a polarizer, then the transmitted intensity I is given by Malus's law:

$$I = I_0 \cos^2 \theta \qquad (37\text{-}12)$$

where θ is the angle between the plane of polarization of the incident light and the polarizer's transmission axis.

Section 37-7. Methods for Polarizing Light

Light can be polarized by selective absorbtion, reflection and transmission, double refraction, and scattering. Light reflected from a surface is 100 percent polarized perpendicular to the plane of incidence if the angle of incidence is equal to the polarizing angle θ_p:

$$\tan \theta_p = n_{21} \qquad (37\text{-}14)$$

where n_{21} is the relative index of refraction.

QUESTIONS

1. Explain the distinction between Fresnel diffraction and Fraunhofer diffraction.

2. We have discussed three processes by which the propagation direction of waves may be altered: reflection, refraction, and diffraction. Which, if any, of these processes may be described by wave interference? Explain.

3. Use Fig. 37-25 to give a qualitative explanation for the formation of the Arago spot. Two rays are shown that pass the edge of the disk and impinge on the screen. What can you say about the path difference between these rays? What can you say about the phase difference between the light waves traveling along these rays? Extend your conclusions to include all light waves that similarly pass the edge of the disk.

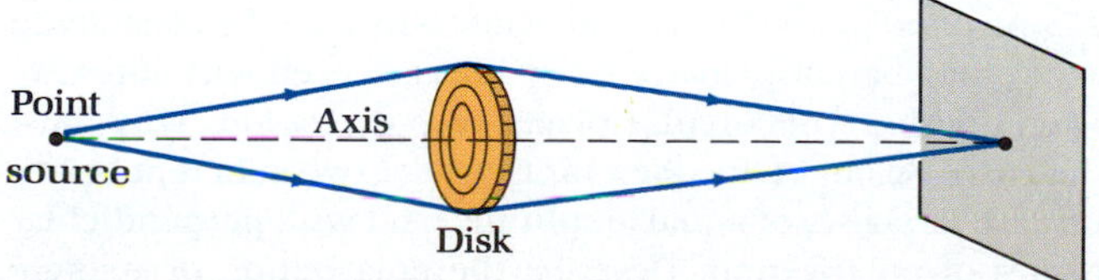

Figure 37-25. Question 3: A disk casts a shadow on a screen when illuminated by a point source of monochromatic light. The Arago spot is formed on the screen at the center of the shadow.

4. Suppose two astronauts are space-walking outside their spacecraft and each has a flashlight and walkie-talkie radio. When the spacecraft is between them, neither astronaut can see the other's flashlight, but they can communicate with their walkie-talkies. Explain this result. Assume that neither the radio waves nor the light waves can reach the other astronaut by reflection.

5. Consider the single-slit diffraction pattern. What is the effect on the width of the pattern, say the distance between the $m' = 1$ minima, when *(a)* the wavelength is doubled, *(b)* the slit width is doubled, *(c)* the incident intensity is doubled?

6. Consider the single-slit diffraction pattern. What is the effect on the intensity at the center of the central maximum when *(a)* the wavelength is doubled, *(b)* the slit width is doubled, *(c)* the incident intensity is doubled?

7. Consider the single-slit diffraction pattern. What is the effect on the ratio of the intensity at the first secondary maximum to the intensity at the central maximum when *(a)* the wavelength is doubled, *(b)* the slit width is doubled, *(c)* the incident intensity is doubled?

8. Suppose the single-slit diffraction experiment is performed in air and then the entire apparatus is immersed in water and the experiment is performed again. Describe any differences in the patterns for the two cases.

9. Suppose the single-slit diffraction experiment is performed with sunlight rather than monochromatic light. Describe the appearance of the pattern.

10. Use Fig. 37-26 to give a qualitative description of the formation of the $m \approx 1$ secondary maxima. (*Hint:* Recall the discussion of Fig. 37-5.) Extend your discussion to account for the $m \approx 2$ secondary maxima.

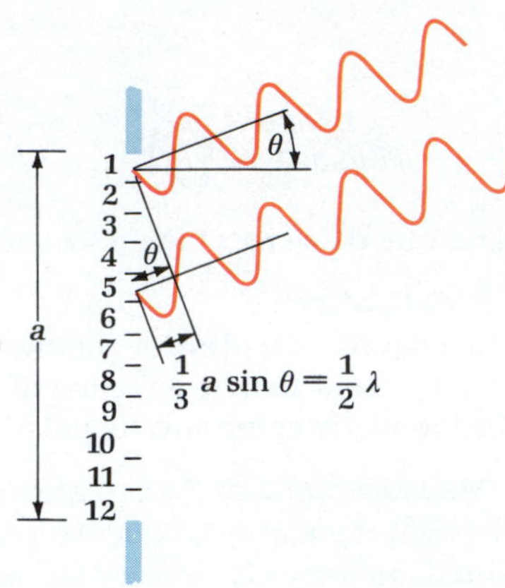

Figure 37-26. Question 10: Waves emerging in phase from a single slit. The slit is divided (in our mind's eye) into 12 elementary slits, and the waves shown arrive at the screen very near one of the $m \approx 1$ secondary maxima.

11. In our discussions of single-slit diffraction, we assumed that the length of the slit was much larger than its width, so we considered only the variation of light intensity in a direction perpendicular to the length of the slit. Suppose the slit's length is twice its width. Describe the way you think the pattern will appear.

12. In Figs. 37-5 and 37-6 we assumed that the waves emerged from each elementary slit in phase, and, to reach the same point on the screen, their directions of propagation were parallel. Explain the connection between these approximations and the assumption of Fraunhofer diffraction.

13. In Eq. 37-5, which of the factors inside the integral—ω, t, and/or ϕ—depends on the variable of integration y?

14. Note that the angle β entered the expression for the single-slit diffraction pattern as the limits to the variation of the phase difference ϕ. Give a physical interpretation of β.

15. A mathematical function $f(x)$ is an odd function of x if $f(-x) = -f(x)$. An example of an odd function is $f(x) = \sin x$. Use Fig. 37-27 to explain why the integral of an odd function between symmetric limits, say from $-a$ to $+a$, is always zero.

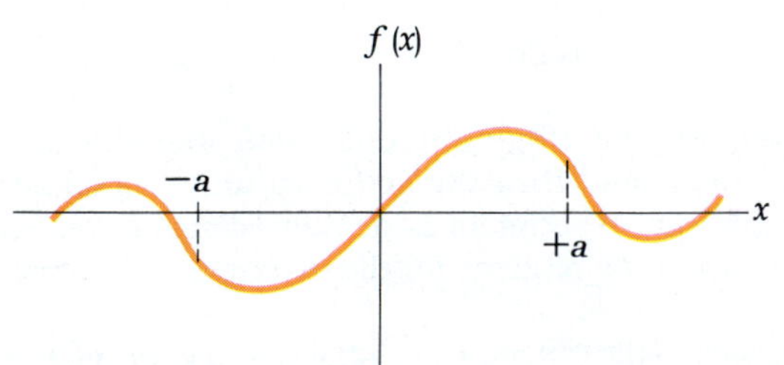

Figure 37-27. Question 15: Graph of an odd function. The area under the curve from $-a$ to $+a$ is zero.

16. If the width of a single slit is doubled, the power through the slit is doubled, but the intensity at the center of the pattern is quadrupled. Does this violate energy conservation? Explain.

17. Consider the double-slit intensity distribution, Eq. 37-7. Is it possible to have the $m' = 1$ minima in the interference term (see Eq. 36-1) to appear at the same point on the screen as the $m' = 1$ minima in the diffraction term (see Eq. 37-1)? Explain.

18. Suppose you are using a high-quality microscope whose resolution is determined solely by diffraction effects. Would higher resolution be attained with red light or blue light?

19. We have noted that a telescope with a larger aperture will have a better resolving power. Can you think of any other advantages to a large aperture? Are there disadvantages in having a large aperture? Explain.

20. In their dark and murky environment, sperm whales find their prey (squid) with echolocation. They emit high-frequency sound waves and then detect the reflected waves. Why do they use high- rather than low-frequency sound? Discuss the advantages of widely spaced ears for a whale.

21. How do Polaroid sunglasses "reduce glare"? Is the transmission axis in these glasses horizontal or vertical?

22. Fishers usually prefer Polaroid sunglasses because it helps them "see into water." Explain why this is so.

23. Some photographers use a polarizing filter over their camera lenses. Explain how such a filter could be useful in photography.

24. A light beam directed along the horizontal may be polarized in a variety of ways. For example, it might be plane-polarized in a horizontal plane or in a vertical plane. Can a similar thing be done to sound waves in air? Explain.

25. The polarization of light is considered as evidence that light is a transverse wave. But could it also have a longitudinal component? To answer, consider explaining Malus's law taking a longitudinal component into account.

26. Consider a double-slit experiment in which we polarize the light from the slits. Describe the pattern when the planes of polarization of the light from the two slits are *(a)* parallel and *(b)* perpendicular.

27. An unpolarized light beam of intensity I may be considered as two incoherent plane-polarized beams, each with intensity $\frac{1}{2}I$ and with perpendicular planes of polarization. Why must these two beams be incoherent? Consider two coherent beams that are in phase, of equal amplitude, and with perpendicular planes of polarization. Describe the polarization of a single beam that is equivalent to these two.

28. Suppose you are given a piece of plastic and asked to determine whether it is Polaroid. You do not have a polarizer at your disposal. Explain how you would make this determination. If it does turn out to be Polaroid, how would you determine the axis of transmission?

29. Viewed from the surface of the moon, the moon's sky is black. Explain.

30. Complete the following table:

Symbol	Represents	Type	SI Unit
β		Scalar	
$\Delta\theta_R$			rad
P	Degree of polarization		
$I_\perp$			
θ_p			

EXERCISES

Section 37-2. Description of the Single-Slit Diffraction Pattern

1. Monochromatic light is used in a single-slit diffraction experiment. The slit width is $a = 0.14$ mm, the distance to the screen is $L = 1.16$ m, and the width of the central maximum (distance between the $m' = 1$ minima) is 4.6 mm. What is the wavelength of the light?

2. A single-slit diffraction pattern is produced with the H_α line from a hydrogen source ($\lambda_\alpha = 656.3$ nm) onto a slit of width 0.041 mm. *(a)* What are the angles that locate the $m' = 1$ and $m' = 2$ minima? *(b)* What are the approximate angles that locate the $m \approx 1$ and $m \approx 2$ secondary maxima?

3. The width Δx of the central maximum (distance between the $m' = 1$ minima) in a single-slit pattern is 5.4 mm. The light has a wavelength $\lambda = 584$ nm, and the screen is 1.31 m from the slit. What is the width of the slit?

4. Determine the positions x_1 and x_2 (relative to the central maximum) of the $m' = 1$ and $m' = 2$ minima in the single-slit diffraction pattern of light of wavelength 450 nm. The slit width is 0.36 mm, and the screen is 1.03 m from the slit.

5. In the single-slit diffraction pattern of infrared light with wavelength 945 nm, the angular positions of the $m' = 1$ minima are $\theta_1 = \pm\pi/4$ rad. What is the width of the slit?

6. Consider the single-slit diffraction pattern due to a rather narrow slit: $a = 2.0\ \mu$m. Let $\lambda = 550$ nm. *(a)* Determine the angular width $2\theta_1$ of the central maximum (θ_1 represents the angular position of an $m' = 1$ minimum). *(b)* Determine the width Δx of the central maximum (distance between the $m' = 1$ minima) on a screen that is 1.00 m from the slit. Note that the approximation $\sin\theta \approx x/L$ is not valid in this case.

7. Consider the graph of the single-slit diffraction intensity, I versus $\sin\theta$, shown in Fig. 37-4*b*. What is the percent error in the values of $\sin\theta$ for the secondary maxima if you use $m = 1$ and $m = 2$ rather than the more accurate values—$m = 0.93$ and $m = 1.96$?

8. Suppose microwaves of wavelength 32 mm are incident normally on a slit in a metal barrier of width 25 mm. Is there a diffraction angle at which the intensity is zero? If so, what is this angle? If not, explain why not.

Section 37-3. The Single-Slit Intensity Distribution

9. Using Eq. 37-6, make a graph of I versus β. Plot points corresponding to $\beta = \pm\pi/4, \pm\pi/2, \pm 3\pi/4, \ldots, \pm 3\pi$, and use $I = I_c$ at $\beta = 0$. (See Prob. 2.) Sketch a smooth curve through the points.

10. Determine the intensity at a point midway between the center of the central maximum and an $m' = 1$ minimum on a graph of the single-slit diffraction formula, I versus β.

11. Divide the single-slit diffraction formula into two factors, $\sin^2\beta$ and $(1/\beta)^2$, and plot each factor on the same graph. Plot points corresponding to $\beta = 0, \pi/4, \pi/2, 3\pi/4, \ldots, 2\pi$, and sketch smooth curves for the two functions. Keep in mind that the single-slit pattern depends on the product of these two factors. From these curves, can you explain why the secondary maxima are not midway between their adjacent minima? Which do you expect to be closer to an integer, larger values of the index m or smaller values? The product of these factors is 1 at $\beta = 0$. (See Prob. 2.)

12. Using the approximation that the positions of secondary maxima correspond to integral values of m (see Example 37-3), determine the intensity of the $m \approx 3$ and $m \approx 4$ secondary maxima in terms of I_c.

13. Use the more accurate values of the two lowest indices of m, $m = 0.93$ and $m = 1.96$, in Eq. 37-6 to determine the intensity at these secondary maxima. From your answers, determine the percent error in the values of the intensity found in Example 37-3.

14. In developing the expression for the single-slit intensity distribution, Eq. 37-5 was integrated after replacing the variable y in terms of the variable ϕ. Accomplish the reverse procedure. Replace ϕ in terms of y in Eq. 37-5 and perform the integral using y as the variable of integration.

15. Many mathematical functions can be classified as either even functions or odd functions. For an odd function, $f(-x) = -f(x)$, and for an even function, $f(-x) = f(x)$. An example of an odd function is $f(x) = \sin x$, and an example of an even function is $f(x) = \cos x$. Show that the integral of an odd function between symmetric limits, say from $-a$ to $+a$, is zero.

16. In Eq. 37-7, we wrote the double-slit intensity distribution

for the more general case where the slit width is not necessarily much smaller than the wavelength. Make a similar modification to the intensity distribution due to a grating, Eq. 36-9.

17. Consider the double-slit intensity distribution (Eq. 37-7). What must be the ratio of the slit separation to the slit width, d/a, in order to have the $m = 2$ maxima in the interference term (see Eq. 36-2) occur at the same points on the screen as the $m' = 1$ minima in the diffraction term (Eq. 37-1)?

18. Consider the double-slit intensity distribution (Eq. 37-7). *(a)* What must be the ratio of the slit separation to the slit width, d/a, in order to have the $m' = 2$ minima in the interference term (see Eq. 36-1) occur at the same points on the screen as the $m' = 1$ minima in the diffraction term (Eq. 37-1)? *(b)* How many maxima in the interference term will occur "inside" the central maximum of the diffraction term (that is, between $\beta = -\pi$ and $\beta = +\pi$)? *(c)* How many interference maxima will occur "inside" one of the fringes corresponding to the $m \approx 1$ diffraction maxima? (For positive θ, this is between $\beta = +\pi$ and $\beta = +2\pi$.) *(d)* Make a sketch of I versus $\sin\theta$ for this case.

19. Determine the ratio of the slit separation to the slit width d/a for the double-slit intensity distribution shown in Fig. 37-8.

Section 37-4. The Limit of Resolution

20. Monochromatic light ($\lambda = 610$ nm) from a distant source impinges normally on a barrier that has a circular hole of diameter 0.50 mm. The light that passes through the hole forms a diffraction pattern on a screen. The screen is 1.0 m from the hole. What is the radius of the first dark ring in the pattern?

21. What is the limiting angle of resolution of a telescope that has an aperture diameter of 75 mm? The light from the objects being viewed has a wavelength of 500 nm.

22. Estimate the maximum distance a car can be from a person with perfect vision such that the person, viewing without the aid of a telescope, can tell whether both headlights are on. It is nighttime and the car is facing the person. (*Hint:* The diameter of the pupil of the human eye is several millimeters.

23. Two stars, each of which produces about the same intensity at the earth, have an angular separation of 7 μrad. If the light from the stars has an average wavelength of 600 nm, what is the minimum aperture diameter needed to resolve them?

Section 37-6. Measurement of Polarization

24. A polarizer and analyzer (both ideal) are arranged as shown in Fig. 37-15. Suppose the intensity at the detector is 0.65 W/m² when the azimuthal angle θ between the transmission axes is 0. *(a)* What is the intensity when $\theta = \pi/4$ rad? *(b)* What is the intensity when $\theta = 3\pi/4$ rad?

25. A polarizer and analyzer (both ideal) are arranged as shown in Fig. 37-15. The intensity at the detector is I_0 when the azimuthal angle θ between the transmission axes is zero. What is θ when $I = 0.25 I_0$?

26. Using two or more polarizers, one can rotate the plane of polarization of a light beam. Suppose three polarizers are placed in a beam of unpolarized light of intensity I_0 as shown in Fig. 37-28. Polarizer 1 polarizes the beam and polarizers 2 and 3 rotate the plane of polarization. Let θ be the azimuthal angle between the transmission axes of polarizers 1 and 2, and let α be the azimuthal angle between the axes of polarizers 2 and 3. *(a)* Assuming that all three polarizers are ideal, show that the intensity at the detector is $I = \frac{1}{2} I_0 (\cos\theta)^2 (\cos\alpha)^2$. Note that the emergent beam has had its plane of polarization rotated by the angle $\theta + \alpha$. *(b)* Determine I when $\theta = \alpha = \pi/4$ rad so that $\theta + \alpha = \pi/2$ rad.

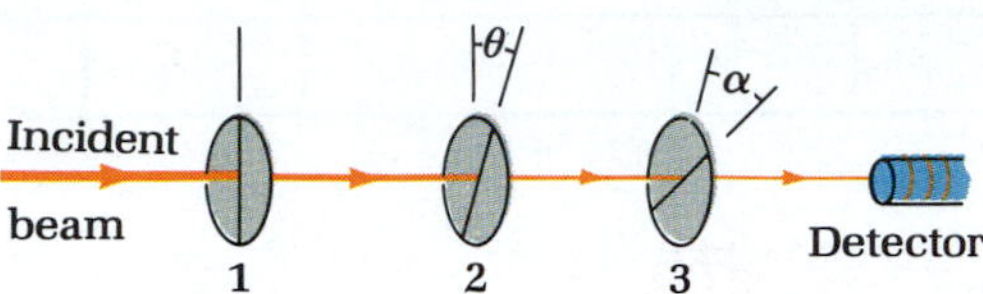

Figure 37-28. Exercise 26.

27. An effectively ideal polarizer is used to measure the polarization of a beam that is directed horizontally. As the polarizer is rotated, it transmits a maximum intensity I_m when the transmission axis is aligned with the vertical and a minimum intensity $0.10 i_m$ when aligned horizontally. *(a)* Is the plane of polarization horizontal or vertical? *(b)* What is the degree of polarization of the beam?

28. A partially polarized beam may be regarded as a combination of a completely unpolarized beam with intensity I_u and a completely polarized beam with intensity I_p. If this beam is incident on an ideal analyzer, then the maximum transmitted intensity is $I_{\parallel} = I_p + \frac{1}{2} I_u$ when the analyzer's axis is parallel to the plane of polarization of the polarized part of the beam. When the analyzer's axis is perpendicular to the plane of polarization of the polarized part of the beam, the transmitted intensity is $I_{\perp} = \frac{1}{2} I_u$. Show that the degree of polarization of the beam can be written $P = I_p/(I_p + I_u)$.

Section 37-7. Methods for Polarizing Light

29. What is the polarizing angle for a beam incident from air onto the surface of water ($n = 1.33$)?

30. A beam of light reflected from the surface of a transparent medium is completely polarized when the angle of reflection is 1.1 rad. What is the angle of refraction?

31. A flashlight beam is incident from air onto a sheet of plastic. The reflected light is found to be completely polarized when the angle of incidence is 0.89 rad. What is the index of refraction of the plastic?

32. A block of glass ($n = 1.5$) is immersed in water ($n = 1.3$). *(a)* What is the polarizing angle for a beam of light incident from the glass onto the glass-water surface? *(b)* What is the polarizing angle for a beam of light incident from the water onto the glass-water surface?

33. Brewster's law, $\tan\theta_p = n_2/n_1$, gives the angle of incidence for a beam in medium 1 onto the surface between media 1 and 2 such that the reflected beam is completely polarized. Let θ_{1p} represent this polarizing angle and θ_{2p} represent the angle of the refracted beam. Now suppose that a beam is incident from medium 2 onto the surface between media 1 and 2. Show that the polarizing angle for this beam is θ_{2p}.

34. *(a)* Show that if the incident beam is in the medium with the lower index of refraction, the polarizing angle must be between $\pi/4$ and $\pi/2$ rad. *(b)* Show that if the incident beam is in the medium with the higher index of refraction, the polarizing angle must be between 0 and $\pi/4$ rad.

35. Suppose the critical angle for the total internal reflection of light incident from medium 2 onto the surface between media 1 and 2 is 0.68 rad. *(a)* What is the polarizing angle for light incident from medium 2? *(b)* What is the polarizing angle for light incident from medium 1?

36. When light is reflected at the polarizing angle from a certain type of transparent plastic, 8.5 percent of the incident beam intensity is reflected. What is the degree of polarization of the refracted beam?

Additional Exercises

37. A long, narrow loudspeaker is set up with the long dimension vertical. If the narrow dimension of the speaker is 100 mm, what is the angle between the point at which sound comes straight out from the speaker and the point at which no sound can be heard (at a distance large compared to 100 mm) if the frequency of the sound is 20 kHz and the speed of sound in air is 1000 m/s?

38. Violet light ($\lambda = 400$ nm) incident on a slit produces a single-slit diffraction pattern that has its first minimum 24.5 mm from the center of the pattern on a screen 1.2 m from the slit. How far from the center of the pattern would the second minimum be if red light ($\lambda = 600$ nm) were used?

39. A single-slit pattern taken with light in air has the position of the $m' = 1$ minimum 5.4 mm from the central maxima and 1 m behind the slit. If the apparatus is moved into water, where will the position of the $m' = 1$ minimum be? The index of refraction of water is 1.3333.

40. Suppose your eye can just barely see as separate images two red lines that are 0.1 mm apart on a ruler 250 mm from your eye. Under the same conditions, how close can two violet lines be and still be resolved by your eye? (See Fig. 34-14.)

41. The sixth maximum from the center of a double-slit pattern is missing. What is the ratio of the slit separation to the slit width for the double slit?

42. What is the amplitude of the **E** vector of the light at the $m = -2$ maximum of a single-slit pattern in which the amplitude of the **E** vector at the center of the pattern is 10 V/m?

43. You are lost at sea in the northern Atlantic under a cloudy sky, but you have a polarizer and a watch. At noon the light from the sky is well polarized if viewed from the front or back of the ship, but nearly unpolarized if viewed from the port (left) side of the ship. In what direction is the ship headed?

44. Unpolarized light is incident on a plastic plate at the polarizing angle. The refracted light is 9.5 percent polarized. What percentage of the intensity of the incident light is reflected? Assume no absorption.

45. A beam of light propagating in the z direction has two coherent components. One has its **E** vector vibrating in the $\pm x$ direction with a maximum amplitude of 15 V/m, and the other has its **E** vector vibrating in the $\pm y$ direction with a maximum amplitude of 25 V/m. What is the degree of polarization of this beam?

46. Under what conditions of slit width and wavelength will the light passed by a single slit produce a diffraction pattern with no minima?

47. Light from a distant source of wavelength 550 nm is incident perpendicularly on a slit. The slit has a width of 2.5 μm. A screen is 1.5 m behind the slit. In the center of the diffraction pattern on the screen, the light intensity is 2.45 nW/m^2. What is the intensity 25 mm on either side of the center fringe?

48. Assuming your eye can distinguish fringes that are 200 times as dim as the center fringe of a single-slit pattern, what is the maximum number of fringes you could discern in a single-slit pattern?

49. A beam of parallel light having two wavelengths, 440 nm and 620 nm, is incident perpendicularly on a slit. The intensity of the two frequencies is equal. The slit is 3.0 μm wide. At the angle that gives the $m' = 1$ minimum of the 440-nm light, what is the intensity of the 620-nm light at the center of the pattern?

50. The photoreceptor cells in the back of an eye are 1 μm apart. Estimate the diameter of the pupil that will give a diffraction pattern for a distant point source that has its first minimum 1 μm from the center of the diffraction pattern.

PROBLEMS

1. ***Developing the single-slit formula from the grating formula.*** The single-slit diffraction formula can be developed by considering the slit to be a narrow grating of width a that has an infinite number of slits with infinitesimal separation. In Chap. 36 we gave the intensity distribution from a grating: $I = I_0 \sin^2 [N(\frac{1}{2}\phi)]/\sin^2 (\frac{1}{2}\phi)$, where $\phi = (2\pi d/\lambda) \sin \theta$, N is the number of slits, d is the slit separation, and I_0 is the intensity from one slit acting alone. Consider letting d approach zero as N approaches infinity such that the product $Nd = a$ remains fixed. Thus a is initially interpreted as the width of the grating, but in taking the limit this interpretation changes so that a becomes the width of a single slit. Also in this limit, I_0 approaches zero, but the product $N^2 I_0$ approaches I_c. Show that this limiting process leads to Eq. 37-6. Explain why $N^2 I_0 \rightarrow I_c$ rather than $N I_0 \rightarrow I_c$. [*Hint:* For large N, $\sin (\beta/N) \approx \beta/N$.]

2. ***Intensity at the center of the pattern.*** Use l'Hospital's rule to show that $[(\sin \beta)/\beta] \rightarrow 1$ as $\beta \rightarrow 0$. Note that this means E_c in Eq. 37-3 is the field amplitude and I_c in Eq. 37-6 is the intensity at the center of the single-slit diffraction pattern.

3. ***Accurate values of* m *for secondary maxima.*** Consider finding accurate values for the secondary maxima in the single-slit diffraction intensity distribution, Eq. 37-6. *(a)* Show that

the relation that gives the values of β for the maxima is $\tan\theta_m = \beta_m$. (*Hint:* Maxima in the function $\sin^2\beta/\beta^2$ occur at the same positions as extrema in the function $\sin\beta/\beta$.) *(b)* Verify that $\beta_m = \pm(m + \frac{1}{2})\pi$, with $m \approx 0.93$ and $m \approx 1.96$, give approximate solutions to the expression in part *(a)*. *(c)* By a trial-and-error procedure with your pocket calculator, determine the value of the index $m \approx 3$ that is accurate to three significant digits.

4. ***Graphical determination of values of β_m.*** Use a graphical procedure to find the solutions to the expression that gives maxima in the single-slit diffraction formula. From the previous problem, the maxima correspond to values of β given by $\tan\beta_m = \beta_m$. On the same graph plot $y_1 = \beta$ and $y_2 = \tan\beta$ from $\beta = 0$ to $\beta = 3\pi$. Points of intersection give values of β in which $\tan\beta = \beta$ so that they correspond to the values of β_m. Determine the two lowest values of β_m (other than 0).

5. ***Iterative procedure for finding m for secondary maxima.*** Using an iterative procedure with your hand calculator, you can find values of m in Eq. 37-2 to a high degree of accuracy. Since $\beta_m = \tan\beta_m$ (see Prob. 3), we also have $\beta_m = \tan^{-1}\beta_m$. To accurately find $m \approx 1(\beta_1 \approx 3\pi/2)$, make a guess at β_1 and call the guess β_{1a}. Try $\beta_{1a} = (3\pi/2 - 0.01)$ rad. Now calculate $\tan^{-1}\beta_{1a}$. The calculator finds the angle in the range $-\frac{1}{2}\pi$ to $+\frac{1}{2}\pi$, so you must add π to the value given by the calculator. Designate this angle β_{1b} and then find $\tan^{-1}\beta_{1b}$. Add π to the calculator's answer and call this β_{1c}. Continue this procedure until the calculator's answers no longer change with the next iteration. *(a)* Determine m for this value of β_m. Use this iterative procedure to find accurate values of m for *(b)* $m \approx 2$ and *(c)* $m \approx 3$.

6. ***Finding β at $I = \frac{1}{2}I_c$.*** Consider the values of β in Eq. 37-6 such that $I = \frac{1}{2}I_c$ and call these values $\beta_{1/2}$. Show that $\sin\beta_{1/2} = \beta_{1/2}/\sqrt{2}$. To find numerical values of $\beta_{1/2}$, use either a trial-and-error procedure similar to that used in Prob. 3 or a graphical procedure similar to that used in Prob. 4.

7. ***Degree of polarization of a refracted beam.*** A beam of unpolarized light with intensity I_i is incident on a boundary between transparent media. The reflected beam has an intensity of $0.060I_i$ and its degree of polarization is 0.90. What is the degree of polarization of the refracted beam?

8. ***Polarizing with a stack of glass plates.*** A light beam can be polarized by passing it through a "stack of glass plates," as shown in Fig. 37-29. The incident beam encounters the first plate at the polarizing angle. This causes the refracted beam inside the plate to encounter the opposite surface at the polarizing angle. (See Exercise 32.) Since the plates are parallel, each beam encounters each surface at the polarizing angle.

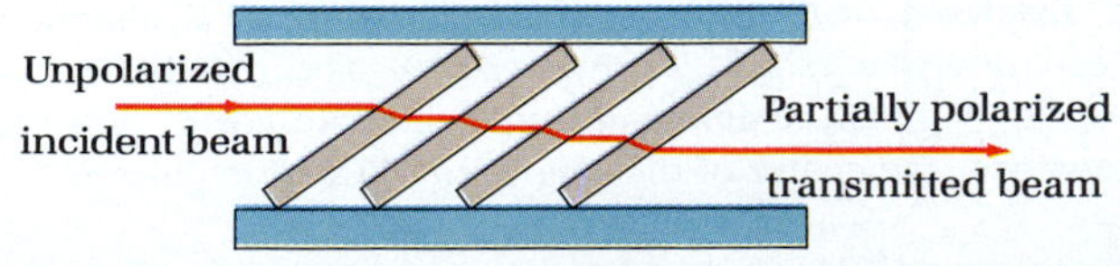

Figure 37-29. Problem 8: Polarization of a beam by passing it through a stack of glass plates. The reflected beams are not shown.

Determine the degree of polarization of the beam that passes through one of the plates. The fraction of the incident intensity reflected from each surface is 0.074. Neglect multiple reflections inside the plate.

9. ***Partially polarized beam incident on an ideal polarizer.*** A partially polarized beam may be regarded as two incoherent beams with different intensities and with perpendicular planes of polarization. Let I_1 represent the intensity of the more intense beam, and let I_2 represent the intensity of the less intense beam. *(a)* Show that when a partially polarized beam is incident on an ideal polarizer the transmitted intensity is

$$I = (I_1 - I_2)\cos^2\theta + I_2$$

where θ is the angle between the polarizer's transmission axis and the plane of polarization of the more intense beam. *(b)* Show that the intensity transmitted by the polarizer may be written

$$I = I_1 \frac{1 + P\cos(2\theta)}{1 + P}$$

where P is the degree of polarization of the beam.

10. ***Transmitted intensity through three polarizers.*** Suppose the angle between the transmission axes of polarizers 1 and 3 in Fig. 37-28 is held fixed at $\pi/2$ rad while the angle θ between the axes of 1 and 2 is allowed to vary. Show that the intensity at the detector is given by

$$I = I_0 \frac{1 - \cos(4\theta)}{16}$$

where I_0 is the intensity of the unpolarized beam incident on polarizer 1.

11. ***Optical activity.*** Suppose we have N (ideal) polarizers in the beam in Fig. 37-28 instead of three. The beam incident on polarizer 1 is plane-polarized, and polarizer 1 has its transmission axis rotated an angle θ/N to the plane of polarization of the incident beam. Each subsequent polarizer has its axis rotated in the same sense by an angle of θ/N to the one before it, so that the last polarizer has its axis rotated by an angle θ to the plane of polarization of the beam incident on polarizer 1. *(a)* Show that the intensity transmitted by the last polarizer is

$$I = I_0\left(\cos\frac{\theta}{N}\right)^{2N}$$

where I_0 is the intensity of the incident beam. Determine I/I_0 for the case where $\theta = \pi/2$ rad and *(b)* $N = 10$, *(c)* $N = 100$, *(d)* $N = 1000$. Some substances, for example, sugar dissolved in water, can rotate the plane of polarization of a polarized beam without noticeably reducing the beam intensity. This phenomenon is called *optical activity*.

12. ***Where did the energy go?*** Figure 37-7 shows that if the width of a slit is halved, the intensity at the center of the screen is quartered. From this it might seem that energy is not conserved, since half as much power is passed by the smaller slit, but a quarter as much intensity shows up at the center of the pattern. Show that the power delivered to the screen by the light in the central fringe (between the minima on either side of the central maximum) by the two slits shown in Fig. 37-7 is proportional to the width of the slit. You may do the integration necessary by counting squares or by a computer.

13. **Power to the fringes.** *(a)* Show that the total power delivered by the light in the region of a single-slit diffraction pattern between the first minima on either side of the central maximum is proportional to $\int_0^{\pi}[(\sin^2 \beta)/\beta^2]\, d\beta$ as long as $\sin\theta \approx \theta$. *(b)* Show that the ratio of the power delivered in the two first fringes on either side of central maximum is about 5 percent of the power delivered in the central maximum.

14. **Lighted from the side.** Light from a distant source is incident on a single slit at an angle γ (Fig. 37-30). Show that Eq. 37-6 still holds if β is replaced by $\pi a(\sin\gamma - \sin\theta)/\lambda$.

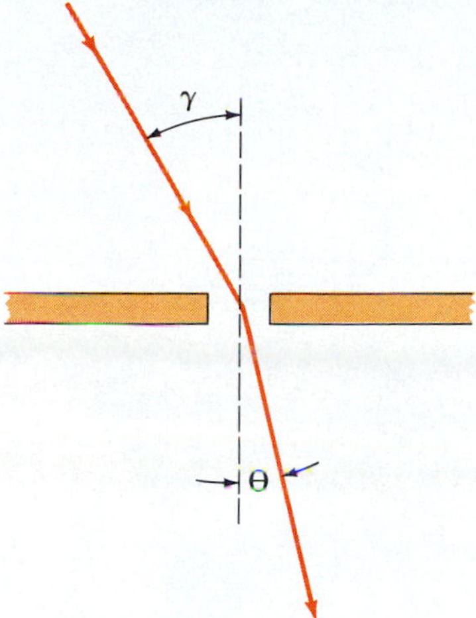

Figure 37-30. Problem 14.

15. **Unequal Rayleigh slits.** *(a)* Show that the Rayleigh criterion for the resolution of the diffraction patterns due to two sources of the same intensity passing through a single slit is $\Delta\theta_R \approx \lambda/a$, where a is the slit width (Fig. 37-31). *(b)* Show that this leads to an intensity at the center of the joint pattern of two fringes of 81 percent of the intensity at the peak on either side of the center.

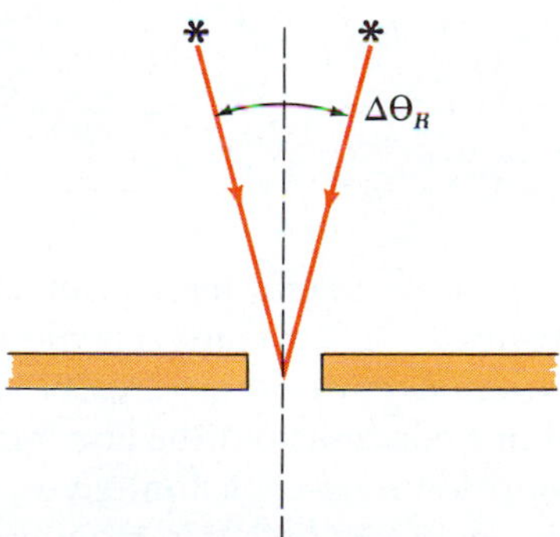

Figure 37-31. Problem 15.

16. **Adding to a circle.** Two coherent plane-polarized beams of light propagate along the z axis. They have the same frequency and amplitude. One is polarized along the x axis and the other along the y axis. The beam polarized along the x axis is advanced in phase by 90° relative to the beam polarized along the y axis. Show that the sum of these beams is a circularly polarized beam. Is the beam right- or left-circularly polarized?

17. **Elliptical intensities.** A beam of elliptically polarized light is passed through a perfect polarizer. As the polarizer is rotated, the maximum amplitude passed is 4 times the minimum amplitude passed. *(a)* Through what angle is the polarizer rotated as the amplitude goes from minimum to maximum? *(b)* Show that when the polarizer is set at an angle halfway between the minimum and maximum intensities the amplitude is two-fifths the maximum amplitude.

18. **Filtering a single slit.** Half of a single slit is covered by a filter (Fig. 37-32). The filter halves the amplitude of light passing through it, but does not appreciably affect the phase or polarization of the light it passes. Find the intensity at the following positions as a ratio of the intensity present at the center without the filter: *(a)* directly behind the slit; *(b)* at the position where the slit without the filter has the $m' = 1$ minimum; *(c)* at the position where the slit without the filter has the $m = 1$ maximum; *(d)* at the position where the slit without the filter has the $m' = 2$ minimum.

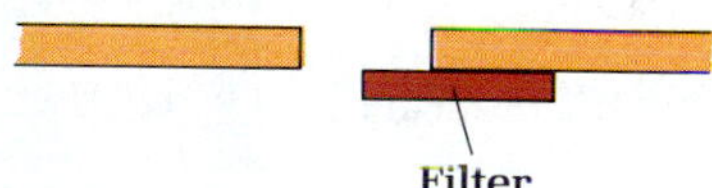

Figure 37-32. Problem 18.

19. **Doubling to no effect.** Show that at the positions where $\sin\theta = (2/3)\lambda/a$ the intensity in a single-slit pattern is unchanged when the slit width is halved.

20. **Imperfect polarizers.** Unpolarized light is passed through two polarizers, each of which absorbs 90 percent of the intensity of horizontally polarized light and 10 percent of the intensity of vertically polarized light. The intensity of the unpolarized light is 25 mW/m². *(a)* What are the intensity and the degree of polarization of the light after it has passed through the first polarizer? *(b)* What are the intensity and the degree of polarization of the light after it has passed through both polarizers? *(c)* The first polarizer is rotated by 30°. After the rotation, what are the intensity and the degree of polarization of the light after it has passed through both polarizers?

21. **Phasors in a circle.** *(a)* What is the phase difference between light from the two edges of a single slit when it arrives at the $m' = 1$ minimum? *(b)* Draw a phasor diagram similar to Fig. 36-27 in which you divide the slit into many sectors and add the contributions from each sector to find the resultant wave amplitude at the $m' = 1$ minimum, and thus show that the phasor diagram at this point is a circle.

38 RELATIVITY

VLA (Very-Large-Array) telescope at the National Radio Astronomy Laboratory, New Mexico.

Commenting on the close relationship between our description of nature and our direct experience, Albert Einstein wrote, ". . . this universe of ideas is just as little independent of the nature of our experiences as clothes are of the form of the human body. This is particularly true of our concepts of time and space. . . ."*

Our everyday experience, on which intuition is based, is limited to observations of slowly moving objects of ordinary size. An object is slowly moving in the sense that its speed is very small compared with the speed of light in vacuum. The orbital speed of an earth satellite, for example, is less than 10^4 m/s, much smaller than $c = 3 \times 10^8$ m/s. If an object moves with a speed that is comparable to the speed of light, the object is said to move *relativistically*. Most of us have had no experience with objects moving at such high speeds.

In this chapter we consider the physics of the very fast. As we extrapolate far beyond the realm of our ordinary experience, our intuition will be challenged and our most basic physical concepts must be modified. Be prepared for a profound change in the way you think of time and space.

* From Albert Einstein, *The Meaning of Relativity*, 3d ed., Princeton University Press, Princeton, N.J., 1950.

38-1 TRANSFORMATIONS

The basic process in the observation of phenomena is the cataloging of events. An event such as the collision of two particles is cataloged according to where and when it occurred. The location of the event can be specified by giving its coordinates (x, y, z) relative to a cartesian coordinate system. The time t of the event is assigned to be the simultaneous reading on a nearby clock. In discussing events, we often imagine an observer O noting the four space-time coordinates (x, y, z, t) for each event. The values of these space-time coordinates are relative—relative to that observer's set of coordinate axes and clock. A different observer O' may use a different clock and make a different choice of coordinate axes. Consequently, in this second observer's reference frame, the space-time coordinates for a given event (x', y', z', t') generally differ from the values (x, y, z, t) assigned by observer O to this same event.

Transformation Equations

By knowing the relationship between the reference frames of the two observers O and O', we can write the *transformation equations* that connect their observations. First we consider a simple transformation that illustrates the idea. We shall see how this transformation is related to the properties of space and time.

A transformation connects observations in different reference frames.

Suppose that two observers use coordinate systems with origins separated by a fixed distance a, as illustrated in Fig. 38-1. For convenience the observers use the same clock (so that $t' = t$). The transformation can be obtained by considering an arbitrary event such as the one shown in the figure. Since the origins of the coordinate systems lie on the xx' axes, we have that $x = x' + a$, $y = y'$, and $z = z'$. Therefore the transformation is

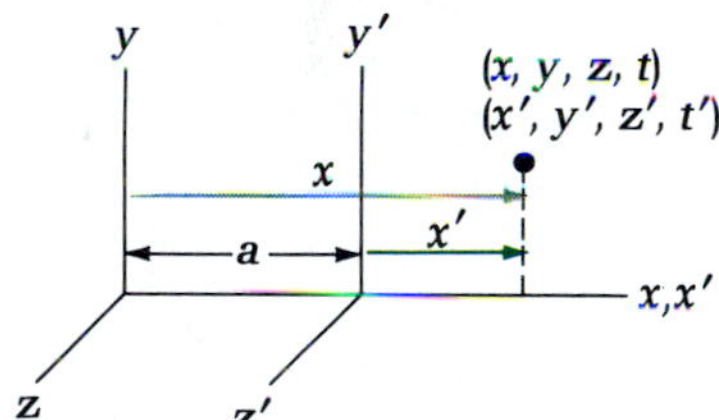

Figure 38-1. The origins of two coordinate systems are separated by a fixed distance *a*.

$$\begin{aligned} x' &= x - a \\ y' &= y \\ z' &= z \\ t' &= t \end{aligned} \qquad (38\text{-}1)$$

Although the transformation equations are simple in this case, they illustrate the general property: Given the space-time coordinates of an event for one observer, say O, the transformation gives the space-time coordinates of that event for the other observer, O'.

This transformation contains an implicit assumption about the nature of space. It can be illustrated by considering the spatial separation of two events. For simplicity we suppose the events occur on a line parallel to the xx' axes so that y and z coordinates can be ignored. Let one event, according to observer O, have space-time coordinates (x_1, t_1), while the second event occurs at (x_2, t_2). The spatial separation of the two events is $s_{12} = |x_2 - x_1|$ for observer O. The same two events have a spatial separation $s'_{12} = |x'_2 - x'_1|$ for the other observer, O'. How do these distances s_{12} and s'_{12} compare? The transformation in Eqs. 38-1 allows us to translate (or transform) one observer's results into the language of the other. We express the primed distance s'_{12} in terms of the unprimed quantities:

$$s'_{12} = |x'_2 - x'_1| = |(x_2 - a) - (x_1 - a)| = |x_2 - x_1| = s_{12}$$

Since the spatial separation of two events has the same value for both observers, it is *invariant* under this transformation. Put another way, the distance between two points in space does not depend on where we place the origin of the coordinate system—space is homogeneous. Of course, the homogeneity of space was tacitly assumed in the discussion that led to Eqs. 38-1. This is an example of how transformation equations contain our concepts of space and time.

Figure 38-2. Each observer describes the other as moving with constant speed v.

The Galilean Transformation

Of particular interest is the transformation connecting reference frames in relative uniform motion. Figure 38-2 shows observer O at rest relative to a train-station platform and observer O' at rest relative to a train on a straight track. If the train is moving at constant speed v past the station platform, then observer O' is moving to the right with speed v relative to observer O. But according to observer O', observer O is moving to the left with speed v. That is, each observer argues that the other is moving with relative speed v. Which one is *really* moving? Imagine that the argument is settled by an alien space traveler: "Silly humans. Both of them are moving as the earth rotates about its axis and revolves around the sun." The point is, of course, that motion is relative, relative to each observer. As we consider the transformation connecting such reference frames, we must remain flexible, identifying first with one observer, then with the other.

Suppose that the coordinate axes of these two observers are oriented as shown in Fig. 38-1 but with this essential difference: The distance a between origins is not fixed. Instead, there is relative motion along the xx' axes, with constant speed v. Because the observers are in relative motion, they do not use the same clock. Each observer has a clock for noting the time of events. For simplicity suppose that they synchronize their clocks so that $t = t' = 0$ at the instant that the origins of their coordinate systems coincide. This event is shown in Fig. 38-3a.

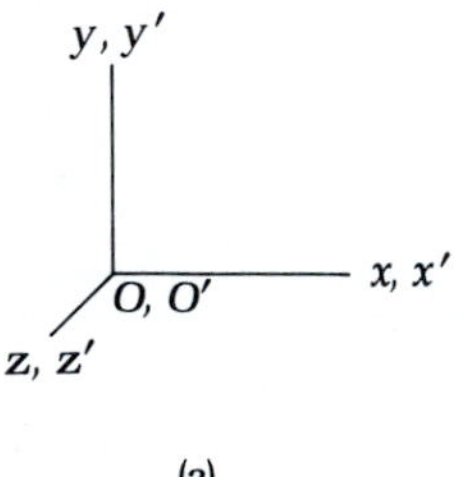

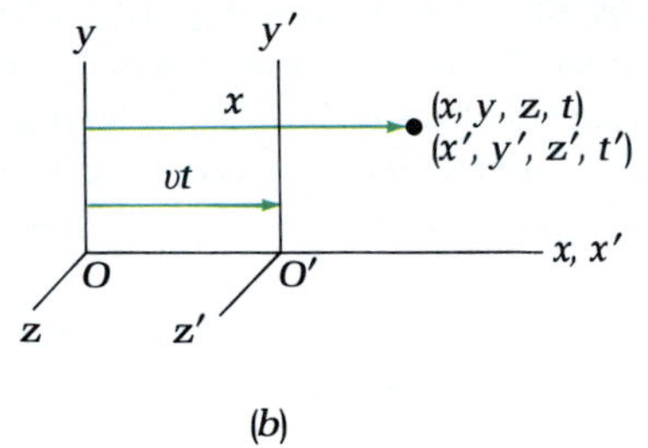

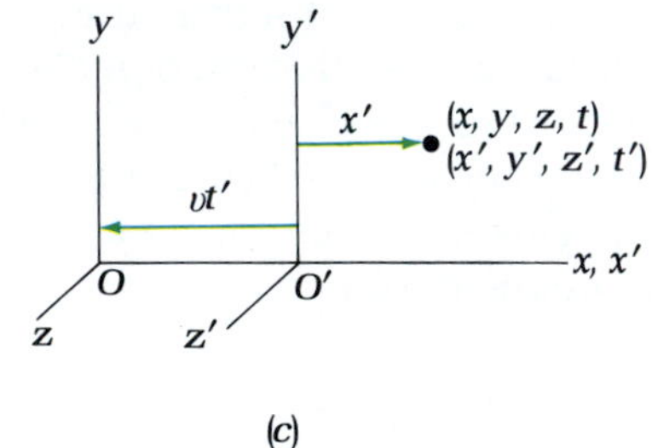

Figure 38-3. *(a)* The coordinate axes of two observers coincide at $t = t' = 0$. *(b)* An event occurs at (x, y, z, t) in the reference frame of observer O. *(c)* The same event occurs at (x', y', z', t') in the reference frame of observer O'.

Each observer now classifies an event that occurs later. The situation is shown as interpreted by observer O in Fig. 38-3b. At the time t that the event occurs, the origin of O' has moved a distance vt to the right along the positive x axis. A similar situation is shown in Fig. 38-3c as interpreted by observer O'. The event occurs at time t', and the origin of O has moved a distance vt' to the left along the negative x' axis.

Transformation contains some assumptions about space and time.

Writing down the transformation connecting these two reference frames involves some assumptions about the nature of space and time. Using Fig. 38-3b, you might think that the difference $x - vt$ should equal the coordinate x' that observer O' assigns to the event. Using Fig. 38-3c in the same way, you would interpret $x' + vt'$ as the coordinate x of the event for observer O. Notice that these two results, $x' = x - vt$ and $x = x' + vt'$, are not quite the same, since t appears in one of them and t' appears in the other. The expressions become the same if we make the further assumption that $t = t'$ for any event. This assumption embodies the newtonian concept of a universal, absolute time. Furthermore, a clock is assumed to run at a rate independent of its motion relative to another identically constructed clock. Similarly, the measurement of lengths (to determine the coordinates of a point) is assumed to be independent of relative motion. With these assumptions, the transformation (solved for the primed quantites) is

Galilean transformation

$$
\begin{aligned}
x' &= x - vt \\
y' &= y \\
z' &= z \\
t' &= t
\end{aligned}
\qquad (38\text{-}2)
$$

This ***galilean transformation,*** named after Galileo, was commonly (but incorrectly) thought, until near the beginning of this century, to reflect the essence of space and time. Indeed, the transformation provides an adequate interpretation for relative motion at speeds corresponding to trains, airplanes, and earth satellites ($v \ll c$ in these cases). Many of us make daily use of the transformation in walking, running, biking, and driving.

Velocity-Transformation Equation

Suppose that each observer measures the velocity of an object. Such a measurement involves two events. The position of the object is determined at each of two instants. For simplicity we consider only the x component of velocity. Let (x_1, t_1) and (x_2, t_2) represent the space-time coordinates of the events for observer O. The x component of the velocity u_x of the object is given by

$$u_x = \frac{x_2 - x_1}{t_2 - t_1}$$

(This gives the average velocity component; the limit as t_2 approaches t_1 can be taken at any stage.) Observer O' similarly evaluates the velocity component:

$$u'_x = \frac{x'_2 - x'_1}{t'_2 - t'_1}$$

We use the galilean transformation, Eqs. 38-2, to express u'_x in terms of the measurements of observer O. Using $t' = t$, we have $t'_2 - t'_1 = t_2 - t_1$ and

$$u'_x = \frac{x'_2 - x'_1}{t'_2 - t'_1} = \frac{(x_2 - vt_2) - (x_1 - vt_1)}{t_2 - t_1}$$

or, since $(x_2 - x_1)/(t_2 - t_1) = u_x$,

$$u'_x = u_x - v \qquad (38\text{-}3)$$

This result from kinematics was first encountered in Chap. 4. It is the galilean velocity-transformation formula.

EXAMPLE 38-1

A helicopter viewed from a train station and from a train. Observer O, while waiting for a train to pass the station, notices a helicopter traveling parallel to the tracks overtaking the train. She estimates the speed of the helicopter to be 100 km/h and that of the train to be 60 km/h. *(a)* What is the speed of the helicopter relative to an observer O' on the train? *(b)* If these speeds are maintained, how far ahead of the train is the helicopter after a half hour?

Solution. The galilean transformation applies since the speeds are small compared with the speed of light. *(a)* Selecting the xx' axes along the direction of motion and using Eq. 38-3, with $u_x = 100$ km/h and $v = 60$ km/h, we have

$$u'_x = 100 \text{ km/h} - 60 \text{ km/h} = 40 \text{ km/h}$$

(b) For the time interval $\Delta t' = 0.50$ h, the x' coordinate of the helicopter changes by $\Delta x' = u'_x \Delta t' = (40 \text{ km/h})(0.50 \text{ h}) = 20$ km according to observer O'. That is, the helicopter is 20 km ahead of the train at this time. What is the answer according to observer O?

SELF-TEST 38-1. As measured from a frame at rest relative to the helicopter in the above example, what is the speed of *(a)* the train and *(b)* the train station? ***ANSWERS:*** *(a)* 40 km/h; *(b)* 100 km/h.

38-2 THE PRINCIPLE OF RELATIVITY

One of the most fundamental tenets of science is the existence of an objective reality (nature) that can be described in logical terms. Furthermore, a description of nature, if correct, should be conceptually independent of a particular describer

Principle of relativity

or observer. These ideas are contained in a postulatelike statement called the ***principle of relativity:*** ***The laws of nature have the same mathematical form in all inertial reference frames.*** Put another way, the principle of relativity asserts that a valid law of nature cannot refer in any way to a special or particular reference frame. A law must have the same content in every reference frame. Consider as an example the law of conservation of momentum for an isolated system. The total momentum $\mathbf{P}$ evaluated in one reference frame may be different from the total momentum $\mathbf{P}'$ evaluated in another reference frame. But the law is stated (and obeyed) in the same way in each frame: $d\mathbf{P}/dt = 0$ in one frame and $d\mathbf{P}'/dt' = 0$ in the other frame. We shall restrict our attention to inertial reference frames except in the Commentary of this chapter. Remember that, in an inertial reference frame, a particle has zero acceleration unless a net force acts on it.

Galilean Relativity

Galilean relativity

The importance of the principle of relativity was recognized well before Einstein changed our views of space and time. Since the galilean transformation embodies earlier concepts of space-time, the pre-Einstein ideas of relative motion and relativity theory are often referred to as *galilean relativity.* We shall be discussing galilean relativity and problems associated with it in this section.

If we use the galilean transformation to connect inertial reference frames in relative motion, how do the laws of nature (as we have stated them so far) transform? Is Newton's second law, for example, in accord with the principle of relativity? Consider the acceleration component, $a_x = du_x/dt$, of a particle relative to an observer O. In the reference frame of observer O', moving with constant speed v relative to O, the acceleration component of the particle is $a'_x = du'_x/dt'$. Using the galilean transformation and Eq. 38-3, we have (note that $d/dt' = d/dt$ if $t' = t$)

$$a'_x = \frac{du'_x}{dt'} = \frac{d(u_x - v)}{dt} = \frac{du_x}{dt} = a_x$$

where $dv/dt = 0$ since v is the constant relative speed of the two observers. Thus the acceleration of a particle is invariant, $a'_x = a_x$, under a galilean transformation. If we assume that the mass of the particle is invariant, $m' = m$, then the product of mass and acceleration is also invariant: $m'a'_x = ma_x$. Suppose further that the net force component is the same in both frames, $\Sigma F'_x = \Sigma F_x$. (See Exercise 9.) Then, on combining the last pair of equalities, we find that if $\Sigma F'_x = m'a'_x$ is satisfied for observer O', then $\Sigma F_x = ma_x$ is satisfied for observer O. That is, Newton's second law has the same form in all inertial reference frames connected by galilean transformations and is in accord with the principle of relativity.

Maxwell's Equations and the Speed of Light

A problem arises, however, when we turn to Maxwell's equations, the laws of electromagnetism. As we saw in Chap. 34, Maxwell's equations in vacuum have wave solutions that we interpret as light. The waves travel with speed $c = (\mu_0\epsilon_0)^{-1/2}$, the speed of light in vacuum. These wave solutions and the value of their speed c come from the equations solved in *any* inertial reference frame. Therein lies a conflict in galilean relativity between electromagnetism and the galilean transformation. In the reference frame of observer O, the velocity component of a light wave in vacuum, propagating in the positive x direction, is $u_x = c$, the value obtained by solving Maxwell's equations in this frame. In the reference frame of observer O', related by the galilean transformation to the frame of O, the velocity component of the light is given by Eq. 38-3, $u'_x = c - v$. In contradiction, the wave solution of Maxwell's equations in this (and any) frame has velocity component $u'_x = c$.

One proposed resolution of the conflict was to regard light waves as propagating in a medium, and the name given to this supposed medium was *ether*. Light waves were assumed to propagate in this ether similar to the way mechanical waves propagate in a mechanical medium. In the wave equation that describes mechanical waves, the wave speed is the speed relative to the medium. So it was thought that Maxwell's equations were valid only in the *rest frame* of the ether, a reference frame in which the ether was at rest. The speed of light in vacuum, $c = (\mu_0\epsilon_0)^{-1/2}$, was the speed in the rest frame of the ether. (The ether supposedly permeated all of space.) In another reference frame, moving with speed v relative to the ether, the velocity of light was determined by the galilean transformation connecting this frame with the rest frame of the ether.

The Michelson-Morley Experiment

In spite of serious objections, the ether concept was widely accepted until an 1887 experiment by A. A. Michelson (1852–1931) and E. W. Morley (1838–1923) cast doubt on the whole idea. Michelson and Morley reasoned that in its orbital motion about the sun, the earth must be moving through the ether. Since the speed of light (measured on the earth) would depend on its direction of propagation relative to the direction of the earth's velocity through the ether, an experiment should determine how the earth moved through the ether.

The experiment utilized Michelson's interferometer (Sec. 36-6) in an attempt to detect phase shifts. These phase shifts were expected because of the dependence on direction of the speed of light in the earth frame. (See Exercise 11 and Prob. 1 for some details about the experiment.) Under a variety of conditions and during different times in the year, no phase shift was observed. Although this null result was puzzling at the time, we now recognize the implication of the Michelson-Morley experiment: ***The speed of light in vacuum is the same in all inertial reference frames. That is, the speed of light in vacuum is an invariant.***

The speed of light in vacuum is invariant.

Our modern interpretation of the Michelson-Morley experiment removes the need for the ether concept. The ether helped to explain how light could have different speeds in different reference frames. But light does not have different speeds in different frames; its speed in vacuum is the same in any inertial frame.

This "constancy" of the speed of light in different frames is devastating to galilean relativity. The speed of light is invariant, in contradiction to the galilean velocity-transformation formula (Eq. 38-3). We conclude that the galilean transformation and our ideas of space and time contained in the transformation are incorrect. It is important to note that the galilean transformation, although incorrect, does provide an adequate account of our everyday experience. It is only when high speeds are involved that the unusual or strange features of space and time are noticeable. But when they are noticeable, they can be jarringly so, as the following anecdotal example illustrates.

EXAMPLE 38-2

Speed of light according to two observers. A spaceship moves on a straight path past the earth at speed $v = 2 \times 10^8$ m/s. An earth-based observer directs a laser beam (light) along a parallel path and measures the speed of the light in the beam to be $u_x = c = 3 \times 10^8$ m/s. What is the speed of this beam of light as measured by an observer on the spaceship?

Solution. Since the speed of light in vacuum is invariant, the observer in the spaceship measures the speed of this same beam of light to be $u'_x = c = 3 \times 10^8$ m/s, the same as the earth-based observer. Notice that if Eq. 38-3 were applied to this case, the result (1×10^8 m/s) would be grossly incorrect.

SELF-TEST 38-2. State the definition of an inertial reference frame. ***ANSWER:*** If an object has zero net force acting on it, and if the object has zero acceleration relative to a reference frame, then that frame is an inertial frame.

38-3 THE LORENTZ TRANSFORMATION

In 1905 Albert Einstein proposed a new view of space-time and a new theory of relativity that we now call the ***special theory of relativity.*** It is special, or restricted, in the sense that it deals only with inertial reference frames and does not deal with gravitation. More general reference frames and the connection with gravity are considered in the ***general theory of relativity,*** which is described in the Commentary in this chapter.

Two Postulates

Einstein developed the special theory from the two postulates that are stated below. The first postulate is the principle of relativity, which presumably has a place in any meaningful physical theory. The second postulate formally acknowledges the invariance of the speed of light in vacuum. (It is not clear whether Einstein knew of the Michelson-Morley experiment. He likely was led to the second postulate on the basis of his belief in the correctness of Maxwell's equations.)

Postulates of the special theory of relativity

Postulate I. The laws of nature have the same mathematical form in all inertial reference frames.

Postulate II. The speed of light in vacuum is the same for all inertial reference frames.

The first task in developing the special theory is to determine the transformation connecting inertial reference frames in relative motion. This transformation replaces the galilean transformation, which is incorrect. We can be guided, however, by some features of the galilean transformation since it is adequate at low speeds. In particular, the correct transformation equations must reduce to the galilean transformation in the limit of low speeds.

The Lorentz Transformation Equations

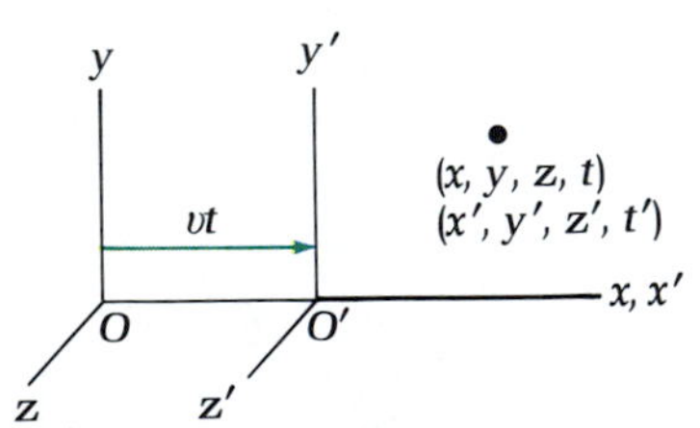

Figure 38-4. Two observers in uniform relative motion observe an event. This view is from the perspective of observer O.

We consider for simplicity only the transformation connecting two reference frames in relative motion along the xx' axes with constant speed v. Again we assume that observers O and O' associated with these reference frames set $t = t' = 0$ at the instant that the two coordinate systems coincide. The relationship later is shown in Fig. 38-4 (from the perspective of observer O). Each observer assigns space-time coordinates to events. Observer O assigns the set (x, y, z, t) to an event, and O' similarly assigns the set (x', y', z', t') to the same event. We seek the transformation connecting these sets for any event.

Since we assume that space-time is homogeneous and that there is a one-to-one correspondence of events for the observers, we consider only linear transformations. The origin of O' moves along the positive x axis with speed v relative to O, so that the equation for x' in terms of the unprimed quantities must be proportional to $x - vt$. In this way an event that occurs at $x' = 0$, the origin in the primed frame, occurs at $x = vt$ in the unprimed frame. Thus this transformation equation can be written in the form

$$x' = \gamma(x - vt) \tag{38-4}$$

where γ is independent of the space-time coordinates of the event.

An equation similar to Eq. 38-4 must hold for x expressed in terms of x' and t'. It is

$$x = \gamma(x' + vt') \tag{38-5}$$

The positive sign between x' and vt' corresponds to the origin of O moving along the negative x' axis with speed v relative to O'. The quantity γ must be the same in the two equations above. Consistent with the principle of relativity, we must not be able to single out one observer in favor of the other. All inertial observers have equally valid descriptions of nature. The value of γ does not characterize either observer but relates this pair of observers.

As with the galilean transformation, the primed and unprimed coordinates should be the same for axes perpendicular to the direction of the relative motion of the frames. This corresponds to the homogeneity and isotropy of space. For the case illustrated in Fig. 38-4, we have

$$y' = y$$
$$z' = z \qquad (38\text{-}6)$$

The remaining ingredient in the transformation involves time. What is the equation connecting t and t' for an event? It *cannot* be the galilean result $t' = t$. That would lead directly back to the galilean transformation. (See Exercise 15.) The form of the equation can be obtained by combining Eqs. 38-4 and 38-5. We substitute the expression in Eq. 38-4 for x' in Eq. 38-5 to obtain

$$x = \gamma\,[\gamma(x - vt) + vt']$$

and solve for t' in terms of t and x. The result is

$$t' = \gamma\left(t - \frac{\gamma^2 - 1}{\gamma^2 v}\,x\right) \qquad (38\text{-}7)$$

The set of four equations in Eqs. 38-4, 38-6, and 38-7 gives the form of the transformation connecting the reference frames of observers O and O'. The equations contain the quantity γ, which is yet to be determined. Its value must be consistent with postulate II: Both observers agree on the value of c, the speed of light in vacuum. Consider the following pair of events, which constitute a speed-of-light measurement:

1. At $t = t' = 0$ when the coordinate systems coincide, a pulse of light is emitted from a point source at the common origin. [If the space-time coordinates of this event are (0, 0, 0, 0) for observer O, what are they for observer O'?] The light propagates in vacuum.

2. The light pulse enters a detector located on the xx' axes. This event occurs at $(x, 0, 0, t)$ for observer O and at $(x', 0, 0, t')$ for observer O', as shown in Fig. 38-5. Since the speed of light in vacuum is the same for both observers, we have $x = ct$ and $x' = ct'$.

We substitute these expressions for x and x' in Eqs. 38-4 and 38-7, which results in

$$ct' = \gamma(ct - vt)$$

and

$$t' = \gamma\left(t - \frac{\gamma^2 - 1}{\gamma^2 v}\,ct\right)$$

Dividing one of these equations by the other eliminates both t and t', and we can solve for γ^2 (Exercise 16):

$$\gamma^2 = \frac{1}{1 - v^2/c^2}$$

On taking the positive square root (so that positive x and positive x' correspond to the same direction), we have

$$\gamma = \frac{1}{\sqrt{1 - v^2/c^2}} \qquad (38\text{-}8)$$

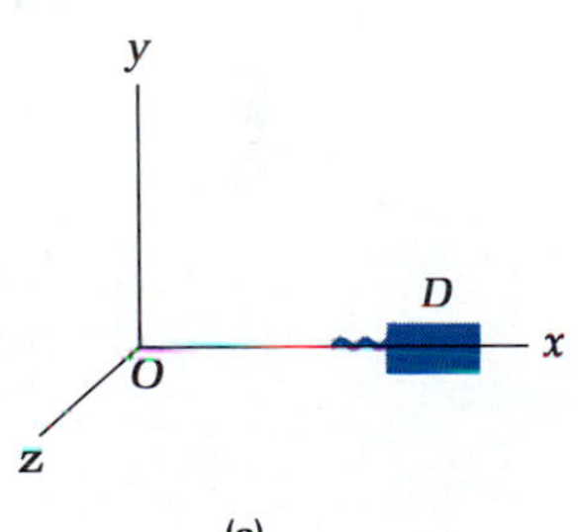

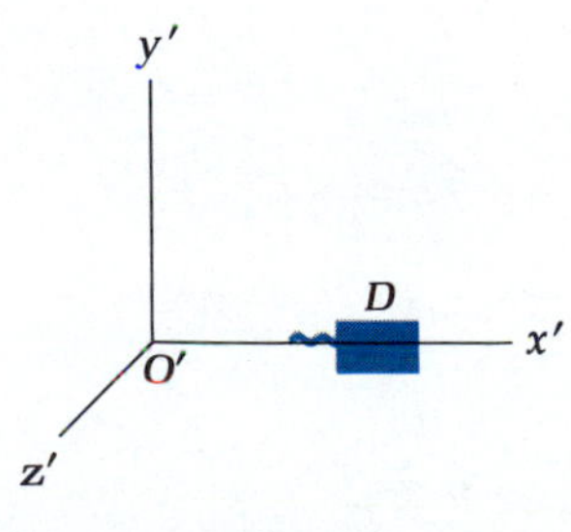

Figure 38-5. A light pulse enters a detector D, as seen in the reference frame of *(a)* observer O and *(b)* observer O'.

From the above expression for γ^2, we can express the factor that appears in Eq. 38-7 in terms of v and c: $(\gamma^2 - 1)/\gamma^2 v = v/c^2$.

We have now determined the transformation connecting the observers O and O'. The equations, solved for the primed space-time coordinates in terms of the unprimed ones, are

Lorentz transformation

$$\begin{aligned} x' &= \gamma(x - vt) \\ y' &= y \\ z' &= z \\ t' &= \gamma[t - (v/c^2)x] \end{aligned} \qquad (38\text{-}9)$$

This type of transformation is called a ***Lorentz transformation.*** The name honors H. A. Lorentz (see the Lorentz force in Sec. 26-4), who tried to explain the null result of the Michelson-Morley experiment in terms of a contraction of lengths by the factor $\sqrt{1 - v^2/c^2}$. But it was Einstein who first obtained this transformation from the postulates of special relativity.

The transformation equations can also be solved the other way, expressing the unprimed quantities in terms of the primed ones. You should do the algebra (Exercise 17) to obtain

$$\begin{aligned} x &= \gamma(x' + vt') \\ y &= y' \\ z &= z' \\ t &= \gamma[t' + (v/c^2)x'] \end{aligned} \qquad (38\text{-}9')$$

More simply, the result can be obtained from Eqs. 38-9 by interchanging the primed and unprimed quantities and changing the ($-$) sign in front of the vt and vx/c^2 terms to a ($+$). This change of sign corresponds to the observers describing each other as moving in opposite directions.

38-4 A NEW VIEW OF SPACE AND TIME

It is a remarkable fact that the speed of light in vacuum is independent of the relative motions of inertial reference frames in which the speed is measured. Two observers measuring the speed of light obtain the same value even though they may be moving relative to each other at a high speed. Are there other features of space-time that seem counter to our ordinary experience? What modifications in our everyday view of space and time are required in order to understand these features? To broaden our knowledge of space and time, we shall apply the Lorentz transformation to interpret some simple sets of events.

Time as a "Fourth Dimension"

An inspection of the Lorentz transformation equations shows that we must alter the concept of an absolute or universal time. From the equation for t',

$$t' = \gamma[t - (v/c^2)x]$$

we see that the value of the time t' assigned to an event by observer O' depends not only on the time t but also on the coordinate x assigned to the event by observer O. Thus we cannot draw a sharp distinction between space and time as separate concepts. Instead of three spatial coordinates (x, y, z) and a separate time (t) characterizing an event, there are four space-time coordinates (x, y, z, t) that are

Time as a "fourth dimension"

"mixed up" by a Lorentz transformation. In a mathematical way, time can be treated somewhat as if it were a fourth spatial coordinate. For this reason time is sometimes popularly referred to as the "fourth dimension."

It is important to understand how the factor $\gamma = 1/\sqrt{1 - v^2/c^2}$ in the Lorentz transformation equations depends on v, the relative speed of the two observers. This dependence is shown graphically in Fig. 38-6. If the relative speed v is small, $v/c \ll 1$, then $\gamma \approx 1$. For example, if $v = 0.001c$, then $\gamma = 1/\sqrt{1 - 0.001^2} = 1.0000005$. For increasing values of v, γ also increases. If $v = 0.70c$, then $\gamma = 1.4$; if $v = 0.99c$, then $\gamma = 7.1$. As v approaches c, the value of γ increases without bound.

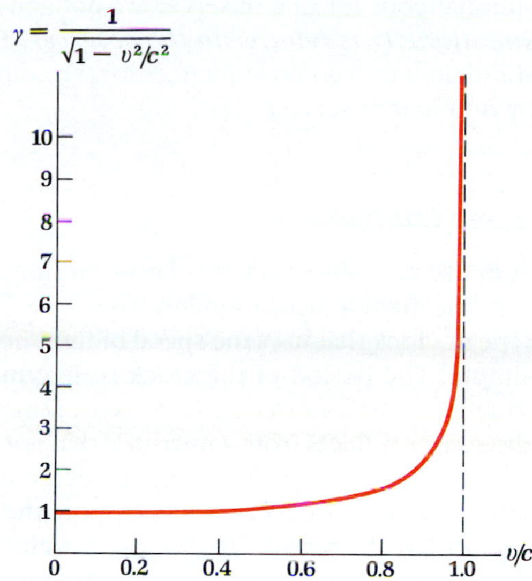

Figure 38-6. At low speeds, the factor $\gamma \approx 1$. As $v \to c$, the value of γ increases without bound.

The mixing of space and time is not apparent for transformations between reference frames with a low relative speed, $v \ll c$. Consider the time interval Δt between two events which are nearby for observer O so that the separation Δx is not large. From the Lorentz transformation, the time interval $\Delta t'$ between the events for observer O' is

$$\Delta t' = \gamma\left(\Delta t - \frac{v\,\Delta x}{c^2}\right)$$

If the relative speed is small compared with the speed of light, then $v/c \ll 1$ and $\gamma \approx 1$. Further, the term $v\,\Delta x/c^2 \ll \Delta t$. In this case the time-interval transformation equation becomes $\Delta t' \approx \Delta t$. That is, the observers agree on the time interval between the events. A similar analysis can be applied to the Lorentz transformation in Eqs. 38-9′. You should be able to show (Exercise 20) that the Lorentz transformation connecting reference frames can be approximated by the galilean transformation in the limit of a low relative speed, $v \ll c$.

Lorentz transformation approaches galilean transformation if $v \ll c$.

Simultaneity

In galilean relativity, time was considered to be absolute, the same for all observers ($t' = t$). If two events occurred at the same time, that is, simultaneously, in one reference frame, it was assumed that the two events would be simultaneous in all reference frames. The Lorentz transformation shows that this idea of absolute simultaneity is not valid. (See also Ques. 8.) Suppose that two events occur simultaneously in the reference frame of observer O. For example, a radioactive nucleus decays at (x_1, y_1, z_1, t_1) and a pair of molecules collide at (x_2, y_2, z_2, t_2), with $t_2 = t_1$. The time interval $\Delta t = t_2 - t_1$ being zero means that the events are simultaneous in this reference frame.

Are these events simultaneous in the reference frame of observer O', moving with speed v relative to O? We use the Lorentz transformation to determine the time interval $\Delta t' = t_2' - t_1'$:

$$\Delta t' = \gamma\left[(t_2 - t_1) - \frac{v(x_2 - x_1)}{c^2}\right] = \gamma\left(0 - \frac{v\,\Delta x}{c^2}\right)$$

Thus

$$\Delta t' = -(\gamma v/c^2)\,\Delta x \qquad \text{when } \Delta t = 0$$

Simultaneity is relative to the observer.

This means that two events that occur at the same time but at different locations ($\Delta x \neq 0$) for observer O occur at different times for observer O'. Events that are simultaneous for one observer are not generally simultaneous for the other. Thus *simultaneity is relative to the observer.* This lack of agreement on events being simultaneous has consequences for the comparison of lengths and of time intervals by different observers.

Time Dilation

Suppose two observers in relative motion compare the periods of the clocks that they use in their measurements. Figure 38-7 shows the essential features of a special type of clock that uses the speed of light in its operation, which makes the analysis simple. The period of the clock is determined by the time for light to travel (in vacuum) from a source S to a detector D next to the source. A light pulse emitted by the source reflects from a mirror M that is a fixed distance L from both S and D. The reception of the light pulse by the detector triggers the emission of another light pulse and the process repeats. Suppose the clock is at rest relative to O', as seen in Fig. 38-7*a*. The round-trip distance traveled by the light from S to D is $2L = c\,\Delta t'$, or the period of the clock according to O' is

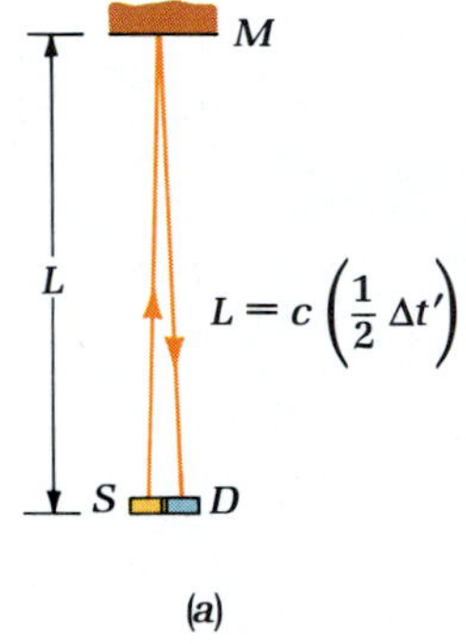

$$\Delta t' = \frac{2L}{c}$$

According to the other observer, the clock is moving with speed v as seen in Fig. 38-7*b*. The light travels from S to M along a path of length $c\,(\frac{1}{2}\Delta t) = \sqrt{L^2 + v^2(\frac{1}{2}\Delta t)^2}$, and an equal distance on the return to D. Squaring this expression and solving for Δt, the period of the moving clock according to observer O, we obtain

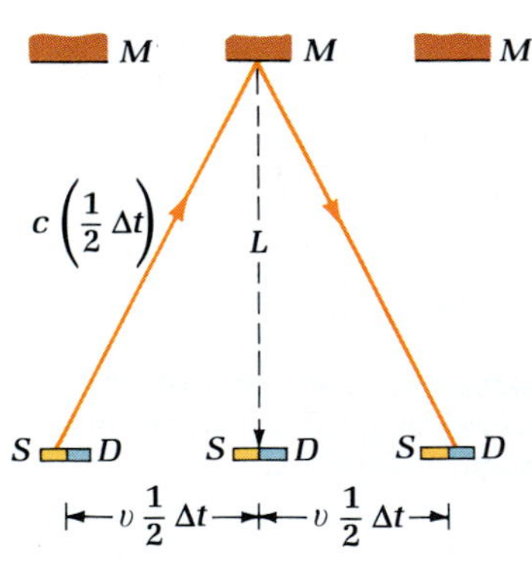

Figure 38-7. *(a)* A light pulse is emitted by source S, reflected by mirror M, and detected by detector D. As viewed in the rest frame of the clock by observer O', the time interval between emission and detection is such that $2L = c\,\Delta t'$. *(b)* For observer O, the clock is moving with speed v and $\sqrt{L^2 + v^2(\frac{1}{2}\,\Delta t)^2} = c(\frac{1}{2}\,\Delta t)$.

$$\Delta t = \frac{2L}{\sqrt{c^2 - v^2}}$$

We directly compare the periods $\Delta t'$ and Δt by forming the ratio $\Delta t/\Delta t'$ from the two expressions above. The result is $\Delta t/\Delta t' = c/\sqrt{c^2 - v^2}$, or

$$\Delta t = \frac{\Delta t'}{\sqrt{1 - v^2/c^2}}$$

The two observers disagree on the period of this clock!

The dependence of the period of a clock on its motion relative to an observer is a general feature of space-time; it is not a property of a specific type of clock, such as the one described above. To see this, suppose that the two observers use clocks that are identically constructed and that have the same period when at rest together. How do the periods compare if they are in relative motion?

While the clock mechanism is not important, we suppose that a device is attached to a clock that emits a flash of light at regular intervals, say at each "tick" of the clock. Then the period of the clock is just the time interval between flashes, which is easily measurable. We concentrate on flashes from the clock used by observer O'. This clock is at rest according to O', as illustrated in Fig. 38-8*a*, and this reference frame is called the *rest frame* of this clock. The period in this frame is denoted by $T_0' = t_2' - t_1'$, where the subscript on T_0' indicates the value measured in the rest

A clock in its rest frame has period T_0'.

frame. In the reference frame of observer O, this clock (belonging to O') is moving. The time interval between flashes as measured by O is $T = t_2 - t_1$. The flashes from the moving clock occur at different positions in this reference frame, as shown in Fig. 38-8*b*. Since the clock is moving in the positive x direction, with speed v, the spatial separation of the events is $\Delta x = x_2 - x_1 = v(t_2 - t_1) = vT$.

We substitute these results into the last equation in the Lorentz transformation of Eqs. 38-9:

$$T'_0 = t'_2 - t'_1 = \gamma\left[(t_2 - t_1) - \frac{v(x_2 - x_1)}{c^2}\right]$$
$$= \gamma\left[T - \frac{v(vT)}{c^2}\right] = T\gamma\left(1 - \frac{v^2}{c^2}\right)$$
$$= T\sqrt{1 - \frac{v^2}{c^2}}$$

since $\gamma = 1/\sqrt{1 - v^2/c^2}$. Solving for T, we obtain the time interval between flashes from the moving clock:

$$T = \frac{T'_0}{\sqrt{1 - v^2/c^2}} = \gamma T'_0 \tag{38-10}$$

Since the factor γ is greater than 1, the time interval T is greater than the time interval T'_0 in the rest frame of the clock. The interpretation of this result by observer O is that the moving clock is running slow. For example, suppose that T'_0 is 1 s. That is, the clock "ticks" once each second in its rest frame. But according to observer O, the time interval between the "ticks" of this *moving* clock is greater than 1 s. If $v = 0.600c$, then $\gamma = 1.25$ and $T = 1.25$ s. That is, the moving clock runs slow. This effect, the "stretching out" of time for a moving clock, is called ***time dilation.***

Our analysis of time dilation is independent of the mechanism of the clock. All physical processes in a moving system run at a slower rate. The biological clocks associated with the aging process should be no exception. See Prob. 3 for a discussion of the intriguing "twin paradox."

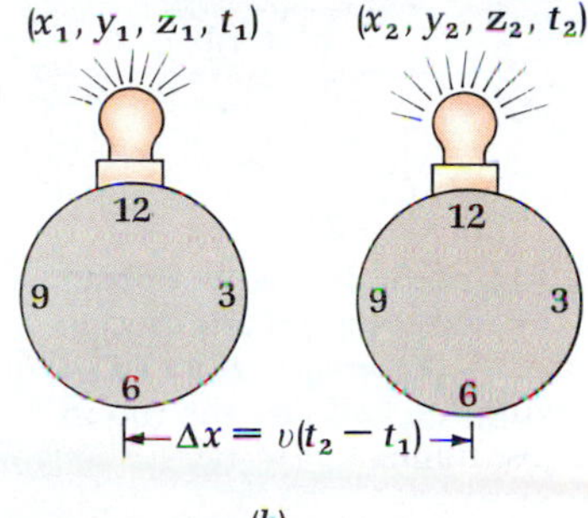

Figure 38-8. A light on a clock flashes at each "tick." *(a)* Observer O' is in the rest frame of the clock. Two ticks occur at the same place, and the period of the clock is $T'_0 = t'_2 - t'_1$. *(b)* The clock moves with speed v relative to observer O, and the period of the moving clock is $t_2 - t_1$.

EXAMPLE 38-3

Experimental verification of time dilation. The ***muon*** is an elementary particle similar to the electron. It is unstable and spontaneously disintegrates or decays, with a characteristic mean lifetime of 2.2 μs in its rest frame. High-speed muons are created by cosmic rays in the upper atmosphere. Determine the mean lifetime of muons with speed $v = 0.99c$.

Solution. Let a muon be at rest relative to observer O'; then $T'_0 = 2.2\ \mu\text{s}$. Since $\sqrt{1 - v^2/c^2} = \sqrt{1 - (0.99)^2} = 0.14$, we have $T = (2.2\ \mu\text{s})/0.14 = 16\ \mu\text{s}$ from Eq. 38-10. The observed increase in the lifetime of high-speed muons produced by cosmic rays was one of the first tests of time dilation.

SELF-TEST 38-3. Suppose an observer measures the mean lifetime of a type of unstable particle to be 35 ms when the particle moves at a speed of $0.85c$ relative to the observer. What is the mean lifetime of this type of particle in its rest frame? ***ANSWER:*** 18 ms.

Length Contraction

To measure a length, an observer compares the spatial interval between two endpoints with a standard such as a meter stick. It is straightforward to measure the length of an object in its rest frame, the reference frame in which the object is at rest. Suppose an object such as a rod is at rest in the reference frame of O' and that

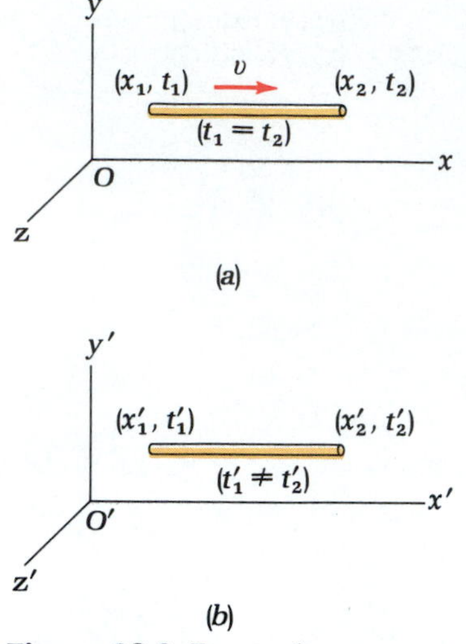

Figure 38-9. Events that occur at the ends of a rod are used to measure the length of *(a)* a moving rod and *(b)* the rod in its rest frame.

the length of the rod is parallel to the x' axis so that y' and z' are not needed. The position of one end is observed [event 1 at (x'_1, t'_1)], as is the position of the other end [event 2 at (x'_2, t'_2)]. Then the length $L'_0 = x'_2 - x'_1$ is obtained by laying off the distance with the meter stick. The subscript on L'_0 indicates the result of a measurement in the rest frame.

Measuring the length of an object that is moving is not so straightforward. Relative to observer O, the rod and the reference frame of O' are moving along the positive x axis with speed v. How does observer O determine the length of the rod? One method, outlined in Exercise 26, involves the measurement of a time interval and makes use of time dilation. Another procedure, illustrated in Fig. 38-9, is to determine *simultaneously* (in the reference frame of observer O) the endpoints x_1 and x_2 of the object. The length of the moving rod is the distance between its simultaneously observed endpoints, $L = x_2 - x_1$. How do L and L'_0 compare? Using the first equation in the Lorentz transformation, we have

$$L'_0 = x'_2 - x'_1 = \gamma[(x_2 - x_1) - v(t_2 - t_1)]$$

The ends were observed simultaneously by O so that $t_2 - t_1 = 0$. Thus

$$L'_0 = \gamma(x_2 - x_1) = \gamma L = \frac{L}{\sqrt{1 - v^2/c^2}}$$

Solving for L, the length of the moving rod in the direction of its velocity, we have

$$L = L'_0\sqrt{1 - \frac{v^2}{c^2}} = \frac{L'_0}{\gamma} \qquad (38\text{-}11)$$

Length contraction: The length of a moving object is shortened.

The length of a moving object, in the direction of its motion, is obtained by dividing its rest length L'_0 by γ. Since $\gamma > 1$, we see that $L < L'_0$. For example, if $v = 0.600c$, then $\gamma = 1.25$ and $L = L'_0/1.25$. There is a shortening, or *contraction,* of the length of an object in the direction of its motion. For a larger speed of the object, the factor γ is also larger, and the more contracted is the length. The analysis applies to the measurement by observer O of the length of any moving object, including the meter stick of observer O'.

Observer O' disagrees with the measurements performed by observer O. Suppose that the roles of the two observers are reversed so that O' measures the length of a moving object, an object at rest relative to observer O. The result (see Exercise 25) is that the length L' of the moving object is contracted from its rest length L_0 by the factor $\sqrt{1 - v^2/c^2}$—that is, $L' = L_0\sqrt{1 - v^2/c^2} = L_0/\gamma$. Thus each observer concludes that the length of a moving object is contracted.

EXAMPLE 38-4

A train just fits into a station. A superfast train, of rest length 1200 m, passes through an enclosed station. According to the stationmaster, the length of the station is 900 m and the train just fits into the station as it passes. That is, the stationmaster observes the train's back end just inside the entrance at the same time that its front end is just inside the exit (Fig. 38-10). What is the speed of the train?

Solution. To fit into the 900-m station, the train's moving length L must be 900 m. From Eq. 38-11, we have

$$900 \text{ m} = 1200 \text{ m}\sqrt{1 - \frac{v^2}{c^2}}$$

so that

$$1 - \frac{v^2}{c^2} = \left(\frac{900}{1200}\right)^2$$

Then

$$v = c\sqrt{1 - (\tfrac{9}{12})^2} = 2 \times 10^8 \text{ m/s}$$

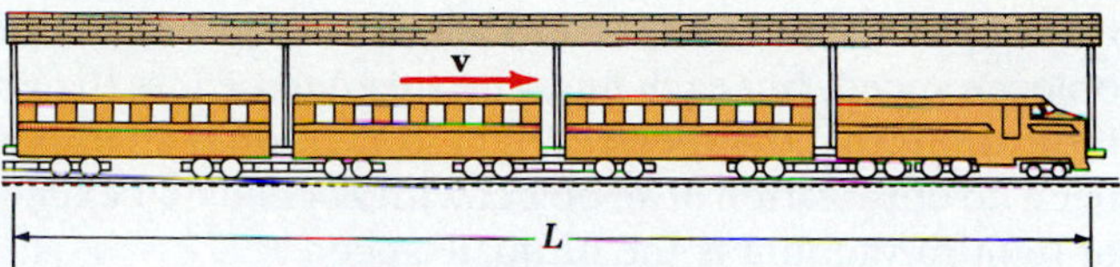

Figure 38-10. Example 38-4: Relative to the stationmaster, a moving train just fits into a station of length L.

What is the length of the station according to the conductor on the train?

Obviously, a train cannot travel at such a high speed relative to the stationmaster. Suppose that the train were almost supersonic: $v = 300$ m/s. What would be its contracted length?

SELF-TEST 38-4. In the above example, what is the length of the station according to the conductor on the train? ***ANSWER:*** 675 m.

38-5 THE ADDITION OF VELOCITIES

Although it is adequate for low speeds, the galilean formula for combining velocities, Eq. 38-3, is not correct for high speeds. Using the Lorentz transformation, we must reconsider the transformation of velocity components. Let events 1 and 2 correspond to observations of the position of a particle. The x component of the velocity of the particle according to observer O is

$$u_x = \Delta x/\Delta t = (x_2 - x_1)/(t_2 - t_1)$$

Similarly, observer O expresses the x' component of the velocity of the particle as $u'_x = (x'_2 - x'_1)/(t'_2 - t'_1)$. We transform the numerator and denominator using the first and last equations of the Lorentz transformation in Eqs. 38-9:

$$u'_x = \frac{\gamma[(x_2 - x_1) - v(t_2 - t_1)]}{\gamma[(t_2 - t_1) - v(x_2 - x_1)/c^2]}$$

Dividing numerator and denominator by $t_2 - t_1$ and canceling the common factor γ, we have

$$u'_x = \frac{(x_2 - x_1)/(t_2 - t_1) - v}{1 - (v/c^2)[(x_2 - x_1)/(t_2 - t_1)]}$$

Since $u_x = (x_2 - x_1)/(t_2 - t_1)$, we obtain

$$u'_x = \frac{u_x - v}{1 - vu_x/c^2} \qquad (38\text{-}12)$$

Lorentz velocity transformation

This *Lorentz velocity transformation* replaces the galilean result. In the limit of low speeds, $|vu_x/c^2| \ll 1$, the denominator in Eq. 38-12 approaches 1, and the galilean formula, $u'_x = u_x - v$, is recovered.

The transformation for the other components of the velocity of a particle can be obtained by a procedure similar to that above (see Exercise 29):

$$u'_y = \frac{u_y\sqrt{1 - v^2/c^2}}{1 - vu_x/c^2}$$

$$u'_z = \frac{u_z\sqrt{1 - v^2/c^2}}{1 - vu_x/c^2}$$

The transformation equations can also be solved for the unprimed velocity components. For example, Eq. 38-12 can be rearranged to give

$$u_x = \frac{u'_x + v}{1 + vu'_x/c^2} \qquad (38\text{-}13)$$

The mathematical structure of these velocity-transformation equations results in an important feature. If one observer, say O', determines the speed of an object to be less than the speed of light c, then the other observer O also determines the speed of

The speed of light in vacuum is the ultimate speed.

that object to be less than c. The two observers do not agree on the value of the object's speed, but each finds the speed to be less than c. This result is one of many indications that the speed of light in vacuum is an upper limit on speeds. There has been no observation of an object with a speed that exceeds c. In this sense the speed of light in vacuum is the ultimate speed.

EXAMPLE 38-5

Relative speeds. Observers in separate spaceships have relative speed $v = 2.4 \times 10^8$ m/s. Observer O' fires a space torpedo forward, as shown in Fig. 38-11, and measures its speed to be 1.8×10^8 m/s. *(a)* Determine the speed of the torpedo according to observer O. *(b)* Suppose O' fires a laser beam (light) forward instead of a torpedo. What is the speed of the light according to observer O?

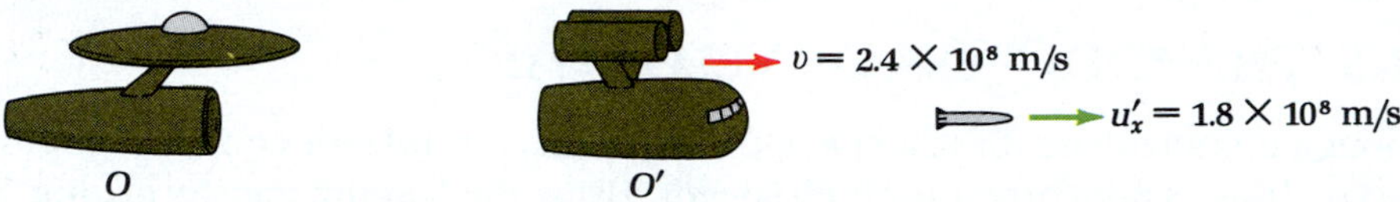

Figure 38-11. Example 38-5: Observer O', moving with speed v relative to observer O, fires a torpedo forward with speed $|u'_x|$.

Solution. *(a)* We use Eq. 38-13 to determine the velocity component u_x of the torpedo according to observer O (notice that $u'_x = 1.8 \times 10^8$ m/s):

$$u_x = \frac{(1.8 \times 10^8 \text{ m/s}) + (2.4 \times 10^8 \text{ m/s})}{1 + (2.4 \times 10^8 \text{ m/s})(1.8 \times 10^8 \text{ m/s})/c^2}$$

$$= \frac{4.2 \times 10^8 \text{ m/s}}{1 + (4.32 \times 10^{16})/(9.0 \times 10^{16})} = 2.8 \times 10^8 \text{ m/s}$$

The speed of the torpedo according to O is less than c. (The galilean formula would have incorrectly given the speed of the torpedo according to O to be greater than c.)

(b) Since the light has speed c according to O', we have $u'_x = c$. Substituting into Eq. 38-13 gives

$$u_x = \frac{c + v}{1 + vc/c^2} = \frac{c(1 + v/c)}{1 + v/c} = c$$

Observer O also determines the speed of light to be c. We should have expected this result because the Lorentz transformation was designed so that the speed of light in vacuum is invariant.

SELF-TEST 38-5. Suppose observer O' in the above example fires the space torpedo backward toward O at a speed of 1.8×10^8 m/s. *(a)* What is the torpedo's speed according to O? *(b)* According to O, is the torpedo traveling toward O or toward O'? ***ANSWERS:*** *(a)* 1.2×10^8 m/s; *(b)* toward O'.

38-6 MOMENTUM AND ENERGY

In everyday life we do not see relativistic effects such as length contraction or time dilation. The ordinary-sized objects around us do not move rapidly enough for the effects to be observable. Nor have we been in contact with another observer moving relative to us at a speed that is a significant fraction of the speed of light. Objects that do travel at speeds comparable to the speed of light are particles of atomic and subatomic sizes. For example, electrons in a television picture tube are accelerated to speeds of around $\frac{1}{2}c$. In dealing with such particles, our emphasis is usually placed on momentum and energy.

Along with revising our notions of space and time, we must modify the definitions of some dynamical quantities, if these are to be useful. To be useful concepts, momentum and energy should be defined such that they are conserved for an isolated system. In a collision of two particles, for example, the momentum gained by one particle should equal the momentum lost by the other particle. Further, the modified definition of momentum should reduce to our original definition, $\mathbf{p} = m\mathbf{v}$, for $v \ll c$.

Consider a particle of mass m with velocity $\mathbf{v}$ in some reference frame. The modified definition of the momentum $\mathbf{p}$ of the particle is

$$\mathbf{p} = \frac{m\mathbf{v}}{\sqrt{1 - v^2/c^2}} = \gamma m\mathbf{v} \qquad (38\text{-}14)$$

Definition of momentum

With this definition, momentum is conserved, in every inertial reference frame, in collisions between particles. Notice that the speed $v = |\mathbf{v}|$ that appears in the square root is the speed of the particle. If the speed is small compared with the speed of light, then $\gamma = 1/\sqrt{1 - v^2/c^2} \approx 1$, and the old definition of momentum, $\mathbf{p} = m\mathbf{v}$, is obtained. However, as the speed v approaches the speed of light, the square root approaches zero and the magnitude of the momentum increases without bound. This suggests that no particle can be accelerated to the speed of light. Its momentum change would be infinite in magnitude and that is not possible. The mass m of the particle is sometimes called its *rest mass.* It is the particle's mass in the rest frame of the particle—that is, in a frame for which the particle is at rest.

Rest mass of a particle

We must also reconsider energy and its conservation. In accounting for the conservation of energy, we must include every form of energy that is changing in a system. Einstein showed that there is a form of energy associated with mass and defined the ***total energy*** E of a particle of mass m and speed v to be

$$E = \frac{mc^2}{\sqrt{1 - v^2/c^2}} = \gamma mc^2 \qquad (38\text{-}15)$$

Total energy of a particle

This definition of the total energy of a particle results in the conservation of energy for an isolated system. (The term "total energy of a particle" does not include potential energy, which must be added in separately.)

Notice from Eq. 38-15 that the total energy of the particle is not zero if the particle is at rest. Setting $v = 0$ results in $E = mc^2$. This is the energy of a particle when it is at rest, and it is called the ***rest-mass energy*** of the particle. There are instances when the sum of the rest masses for an isolated system changes. The change Δm may be positive or negative as a result of energy transformations that are occurring in the system. The change in mass-energy, $\Delta E = \Delta mc^2$, is balanced by changes in other types of energy so that the total energy of the system is conserved. This inclusion of rest-mass energy into the total energy balance generalizes the law of conservation of energy. Often this generalized form is called the *law of conservation of mass-energy.* The conservation of mass-energy to other forms in nuclear fusion and fission processes has had political, economic, and military consequences that are controversial and well publicized.

Rest-mass energy, $E = mc^2$

Conservation of mass-energy

If a system is not isolated, then its energy changes by the amount ΔE added to it or taken from it. If we use the expression $\Delta E = \Delta mc^2$, then $\Delta m = \Delta E/c^2$ is interpreted as the change in the mass of the system. That is, *energy added to a system is equivalent to increasing the inertial mass of the system.* This result is often expressed as *the equivalence of mass and energy.*

Equivalence of mass and energy, $\Delta m = \Delta E/c^2$

The rest-mass energy mc^2 is the total energy of a particle at rest. When the particle is moving, its total energy is larger. The additional energy that a particle has because it is moving is defined to be the ***kinetic energy*** K of the particle. Subtracting mc^2 from the total energy of a particle in Eq. 38-15 gives the modified definition of the kinetic energy of a particle, $K = E - mc^2$, or

$$K = mc^2\left(\frac{1}{\sqrt{1 - v^2/c^2}} - 1\right) = (\gamma - 1)mc^2 \qquad (38\text{-}16)$$

Definition of kinetic energy

This same expression for kinetic energy is obtained by calculating the work done on the particle by the net force. (See Prob. 9.) At low speeds ($v/c \ll 1$), the term $\gamma = 1/\sqrt{1 - v^2/c^2} \approx 1 + \frac{1}{2}v^2/c^2$ (see App. M) and $K \approx \frac{1}{2}mv^2$. Only at low speeds is the kinetic-energy formula $\frac{1}{2}mv^2$ valid. As the speed of the particle approaches the speed of light, the square root in Eq. 38-16 approaches zero and the kinetic energy increases without bound. Again we see that no particle can have its speed increased to the speed of light. The particle would have gained an infinite kinetic energy, and this is not possible.

Often it is useful to have an expression connecting the total energy E of a particle with the magnitude p of its momentum. Exercise 36 and Prob. 6 suggest different ways to obtain the following result:

$$E = \sqrt{p^2c^2 + m^2c^4} \qquad (38\text{-}17)$$

Note that if the particle is at rest, $p = 0$ and $E = mc^2$. At the other extreme, for a particle with $p \gg mc$, then $E \approx pc$. The relation $E = pc$ holds exactly for a *photon*, the quantum of electromagnetic radiation discussed in the next chapter, whose rest mass $m = 0$.

EXAMPLE 38-6

An electron in an x-ray tube. An electron in a dentist's x-ray tube is accelerated from rest through a potential difference of 2.0×10^5 V. *(a)* Determine the resulting speed of the electron. *(b)* What is the magnitude of the momentum of an electron with this speed?

Solution. *(a)* In moving through a potential difference $\mathcal{V}$, the electron gains kinetic energy due to the change in electric potential energy. Thus, $K = e\mathcal{V}$ where e is the electronic charge. Solving Eq. 38-16 for the square root factor, we obtain

$$\sqrt{1 - \frac{v^2}{c^2}} = \frac{1}{1 + K/mc^2}$$

Squaring, rearranging, and solving for v gives

$$v = c\sqrt{1 - \frac{1}{(1 + K/mc^2)^2}}$$

The ratio

$$\frac{K}{mc^2} = \frac{e\mathcal{V}}{mc^2} = \frac{(1.6 \times 10^{-19}\ \text{C})(2.0 \times 10^5\ \text{V})}{(9.1 \times 10^{-31}\ \text{kg})(3.0 \times 10^8\ \text{m/s})^2} = 0.39$$

Then

$$v = c\sqrt{1 - \frac{1}{(1 + 0.39)^2}} = 0.69c = 2.1 \times 10^8\ \text{m/s}$$

(b) Taking the magnitude of Eq. 38-14, the magnitude of the momentum is $p = mv\sqrt{1 - v^2/c^2}$. Since $v = 0.69c$, the square root is $\sqrt{1 - v^2/c^2} = \sqrt{1 - 0.69^2} = 0.72$. Thus

$$p = \frac{(9.1 \times 10^{-31}\ \text{kg})(2.1 \times 10^8\ \text{m/s})}{0.72}$$
$$= 2.6 \times 10^{-22}\ \text{kg}\cdot\text{m/s}$$

SELF-TEST 38-6. When considering high-energy particles, we often give the energies in units of MeV. Give the following energies in MeV: *(a)* The kinetic energy of the electron in the above example. *(b)* The rest-mass energy of an electron. *(c)* The total energy of the electron in the above example. (*Hint:* The conversion factor for changing to joules from MeV is 1.602×10^{-13} J/MeV.) ***ANSWERS:*** *(a)* 0.2 MeV; *(b)* 0.511 MeV; *(c)* 0.7 MeV.

EXAMPLE 38-7

Nuclear-fusion reaction energy. In a nuclear-fusion reaction, a deuteron (^{2}H) of mass $m_d = 2.01355$ u and a triton (^{3}H) of mass $m_t = 3.01550$ u react to give a neutron (^{1}n) of mass $m_n = 1.00867$ u and an alpha particle (^{4}He) of mass 4.00150 u. Determine the mass energy released in this fusion reaction. [The mass values are expressed in unified atomic mass units (u) defined so that the mass of ^{12}C is exactly 12.00000 u. For conversion to the SI unit of mass, $1 \text{ u} = 1.66054 \times 10^{-27}$ kg.]

Solution. The rest mass of the system decreases from $m_i = 2.01355 \text{ u} + 3.01550 \text{ u} = 5.02905 \text{ u}$ to $m_f = 1.00867 \text{ u} + 4.00150 \text{ u} = 5.01017 \text{ u}$. The magnitude of the change (in kilograms) is

$$(5.02905 \text{ u} - 5.01017 \text{ u})(1.66054 \times 10^{-27} \text{ kg/u}) = 3.135 \times 10^{-29} \text{ kg}$$

The energy released in the fusion reaction is

$$|\Delta E| = |\Delta m|c^2 = (3.135 \times 10^{-29} \text{ kg})(2.998 \times 10^8 \text{ m/s})^2$$
$$= 2.818 \times 10^{-12} \text{ J} = 17.59 \text{ MeV}$$

SELF-TEST 38-7. In another nuclear-fusion reaction, two tritons react to form two neutrons and an alpha particle. Determine the mass-energy released in this reaction. ***ANSWER:*** $1.81 \times 10^{-12} \text{ J} = 11.3$ MeV.

COMMENTARY: General Relativity

Einstein's special theory applied the principle of relativity to *inertial* reference frames. In 1916, about 10 years after the special theory was introduced, Einstein extended the theory to include *noninertial* reference frames as well. This generalization is called the ***general theory of relativity.*** Since the mathematical structure of general relativity is too advanced for us to consider here, we shall instead discuss qualitatively some of the interesting aspects of space-time.

A noninertial reference frame is accelerated relative to an inertial reference frame. As an example of a noninertial reference frame, consider an observer, Aaron, in an enclosed box that has a constant acceleration **a** relative to an inertial frame. This situation is shown in Fig. 38-12*a* where, for simplicity, the box is imagined to be in intergalactic space where gravitational forces are negligible. Suppose Aaron drops a coin and watches it fall. He describes the coin as falling to the floor with a constant acceleration −**a**.

(*a*)

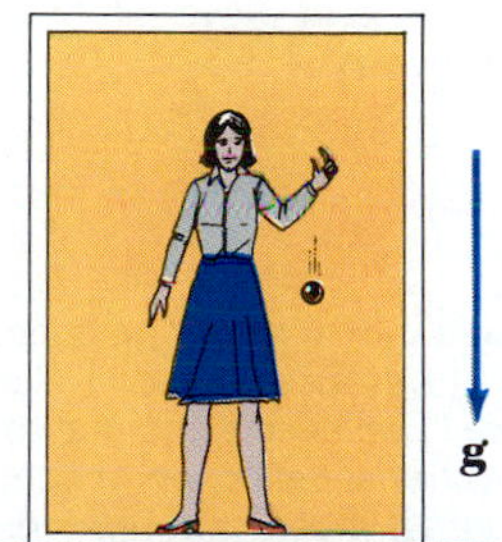

(*b*)

Figure 38-12. Measurements made inside the boxes cannot distinguish between *(a)* a uniformly accelerated reference frame and *(b)* a uniform gravitational field with **g** = −**a**.

The same kind of experiment is performed by Gloria in a similar box. As indicated in Fig. 38-12*b*, Gloria is in an inertial reference frame but in the presence of a uniform gravitational field **g**, with **g** = −**a**. She drops a coin and describes it as falling with a constant acceleration −**a**. Gloria observes her coin falling in exactly the same way that Aaron observes his coin falling.

Since the descriptions are identical, how can Gloria be sure that she is in an inertial reference frame with a gravitational field present? How can Aaron be sure that his reference frame is accelerated with no gravitational field present? A generalization of these ideas can be expressed as: *No experiment performed locally (inside such a box) can distinguish between a constantly accelerated frame and an inertial frame in a uniform gravitational field.* This is a statement of the ***principle of equivalence,*** which is a postulate of the general theory of relativity.

The principle of equivalence is related to the equivalence of inertial and gravitational mass (Sec. 7-3). The inertial mass of an object, such as the released coin, determines its response to a force in the inertial reference frame, and the gravitational mass of the object determines its response to the uniform gravitational field. Since the motion is indistinguishable from the corresponding motion in the accelerated reference frame, the two types of mass are equivalent.

The special theory of relativity caused us to merge separate concepts of space and time into a four-dimensional space-time. The general theory requires yet another

In his early life, Albert Einstein, born in Germany in 1879, showed little promise of his later accomplishments. After an undistinguished academic performance, he took his first job in the Swiss patent office, where he developed four profound papers on brownian motion, and the photoelectric effect, and the special theory of relativity. While Einstein did not seek fame or publicity, his name and features now have instant recognition. In this photograph, Einstein shows off his athletic prowess.

change in our view of space-time, one in which gravity is reduced to geometry! Instead of a "flat" space-time, it is curved or warped in the vicinity of a mass. The curvature is in the four-dimensional space-time, and it cannot be easily visualized by three-dimensional creatures like us.

The distinction between a flat space and a curved space can be illustrated in two dimensions. The surface of a plane is flat, whereas the surface of a sphere is curved. These two-dimensional surfaces have very different geometries. For example, the shortest path between two points in a plane is a straight line. But on the surface of a sphere, the shortest path between two points is along a great circle. (A *great circle* on a sphere has its center at the center of the sphere. Lines of longitude on a globe are great circles.)

Imagine being a two-dimensional creature constrained to move on the surface of a spherical earth. You know left from right and forward from backward, but you have no concept of up and down. Suppose you take the shortest path connecting two points on the surface (from San Francisco to Hong Kong by ship, for example). Since you are always going forward, bearing neither to the right nor to the left, you may think that you are traveling in a straight line. But viewed in three dimensions, the path is along a great circle; it is curved.

General relativity treats gravitation as a curvature of space-time in four dimensions. The curvature is determined by the presence of mass. Gravity is explained in terms of geometry. Consider the motion of the earth about the sun. Because of its large mass, the sun distorts space-time in its vicinity. The earth moves along the shortest path between two points in the curved space-time, "bearing neither to the right nor to the left," as it were. In this view, no force acts on the earth; the distortion of space-time *is* gravity. Of course, we can only see things from our three-dimensional, "flat-space" perspective; we see the path taken by the earth as an ellipse.

The effects of the curvature of space-time are important near a large distribution of mass such as a massive star or in cosmological theories. The consequences are dramatic near a black hole, which is thought to be the last stage in the evolution of a massive star. A black hole is very compact, as if the mass of our sun were in a sphere of radius 2 km, 3×10^{-5} of its actual radius. At the surface of a black hole, the space-time is so distorted that light does not emerge from within — hence the name "black hole."

In everyday situations, however, such extreme general relativistic effects are not noticeable. The effect of the earth's gravity on you, for example, is adequately described by Newton's law of universal gravitation, which is a limiting case of the general theory of relativity.

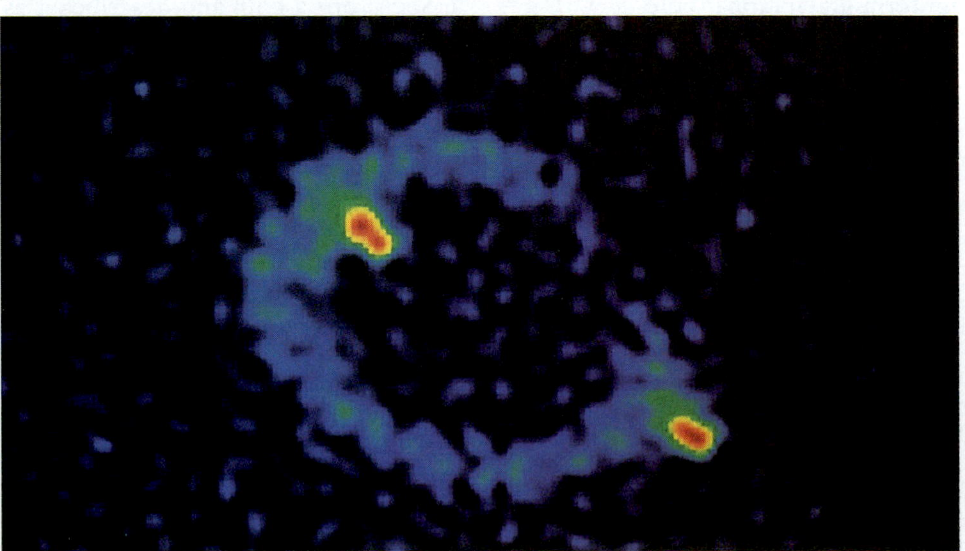

A gravitational lens? The appearance of this astronomical radio source is believed to be due to the gravitational deflection of light rays by a massive object in the foreground. The possibility of this general relativistic effect was suggested by Einstein in 1936.

SUMMARY

Section 38-1. Transformations

A transformation connects observations in different reference frames. The galilean transformation,

$$\begin{aligned} x' &= x - vt \\ y' &= y \\ z' &= z \\ t' &= t \end{aligned} \qquad (38\text{-}2)$$

which connects two inertial frames in relative motion with speed v, is adequate only for $v \ll c$. Velocity components in the direction of relative motion transform according to

$$u'_x = u_x - v \qquad (38\text{-}3)$$

where all speeds are small compared with c.

Section 38-2. The Principle of Relativity

The principle of relativity states that the laws of nature have the same form in all inertial reference frames. Maxwell's equations and the galilean transformation cannot both be in accord with the principle of relativity. The Michelson-Morley experiment showed that the speed of light in vacuum is independent of the motion of the earth.

Section 38-3. The Lorentz Transformation

Einstein's special theory of relativity is developed from two postulates:

I. The principle of relativity.
II. The speed of light in vacuum is the same for all inertial reference frames.

Observers in relative motion with speed v are connected by a Lorentz transformation,

$$\begin{aligned} x' &= \gamma(x - vt) \\ y' &= y \\ z' &= z \\ t' &= \gamma\left(t - \frac{vx}{c^2}\right) \end{aligned} \qquad (38\text{-}9)$$

with $\gamma = 1/\sqrt{1 - v^2/c^2}$.

Section 38-4. A New View of Space and Time

Simultaneity is relative to the observer. Moving clocks run slow by the factor $\gamma = 1/\sqrt{1 - v^2/c^2}$. The length of a moving object is contracted in the direction of motion by the factor $\sqrt{1 - v^2/c^2}$.

Section 38-5. The Addition of Velocities

Velocity components in the direction of relative motion transform under Lorentz transformation as

$$u'_x = \frac{u_x - v}{1 - vu_x/c^2} \qquad (38\text{-}12)$$

The speed of light in vacuum is the upper limit on speeds.

Section 38-6. Momentum and Energy

The momentum of a particle of rest mass m and velocity $\mathbf{v}$ is defined by

$$\mathbf{p} = \frac{m\mathbf{v}}{\sqrt{1 - v^2/c^2}} \qquad (38\text{-}14)$$

and the kinetic energy is defined by

$$K = mc^2(\gamma - 1) \qquad (38\text{-}16)$$

The rest-mass energy of the particle is mc^2, and the total energy is $E = K + mc^2 = \gamma mc^2$. The rest-mass energy of a system can change and must be included in the overall conservation-of-energy law.

QUESTIONS

1. What is the highest speed of an object that you have observed directly and visually?

2. Suppose that the speed of light were 3×10^3 m/s instead of 3×10^8 m/s. What are some everyday phenomena that would seem different from our present perceptions? See *Mr. Tompkins in Wonderland* by George Gamow (Macmillan, New York, 1940) for some interesting discussions of relativity.

3. Do you have any direct evidence to support the claim that the length of an object is independent of its speed? Explain.

4. What procedures could two widely separated observers (an earthling and Martian, for example) use to compare the rates at which their clocks run?

5. Imagine that a civilization in another galaxy is technologically comparable with our own. Is it reasonable to expect that their expressions for the laws of nature would be fundamentally different from ours? Explain.

6. An astronomer claims to have detected the light from a galaxy that is located 2×10^9 ly from us. When did the light received by the telescope originate? What answer should the astronomer give if asked, "Where is the galaxy now?"

7. How could you measure the length of a fire truck as it speeds by at 75 km/h? Describe in detail a procedure for doing so. Assume that you have ample opportunity to set up the necessary apparatus beforehand.

8. Einstein described the following situation to illustrate the relativity of simultaneity: A train moves with constant speed past a ground-based observer. Lightning bolts strike the train at each end, as shown in Fig. 38-13, leaving marks P', Q' on the train and marks P, Q on the ground. Observer O is positioned midway between the marks P and Q. Likewise, observer O' on the train is positioned midway between the marks P' and Q'. These observers note the light from the two events (the lightning bolts) reaching their eyes. For simplicity, assume that the light travels in vacuum. Observer O claims that the light from each bolt reaches her eyes at the same time. Are the events simultaneous for observer O? For observer O'? Explain.

9. Commercial airliners fly at an average speed relative to the ground of around 250 m/s. Should passengers adjust their watches after a flight to correct for time dilation? Explain.

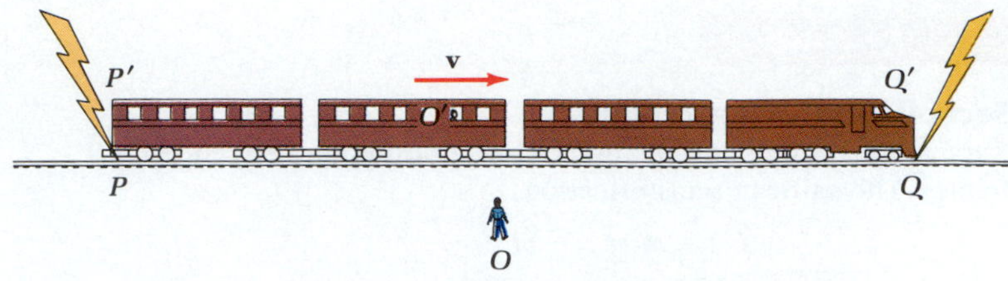

Figure 38-13. Question 8: Lightning strikes at each end of a train. The illustration shows the perspective of observer O, who stands on the ground beside the tracks.

10. The density of an object is defined as its rest mass divided by its volume. Will two observers in relative motion agree on the value of the rest mass of an object? Will they agree on the value of the density of the object? Explain.

11. Imagine that you speed by the earth toward the moon in a fast spaceship. If you measure the earth-moon separation, how does your measurement compare with the distance quoted in books such as this one? Explain.

12. Suppose that you are the earth-based observer of Example 38-2 who measures the speed of light. How can it be that the observer on the spaceship obtains the same value for the speed of the light as you do? Give a qualitative answer based on the length contraction of a measuring stick and time dilation for a moving clock.

13. Imagine that you catch the interstellar space shuttle to the nearby star Sirius. Will you find that your average pulse rate during the journey seems slower, faster, or the same as you found it to be on earth? Explain.

14. Suppose that you, at age 20, join the Space Cadet Corps. Your five-year training consists of a tour of nearby star systems aboard a fast starship. Explain why, at a celebration of your return to earth, your friends seem so much older than you.

15. The speed of a particle doubles, from 1×10^8 m/s to 2×10^8 m/s. Does the magnitude of the particle's momentum double? Does its kinetic energy quadruple? Explain.

16. There is an upper limit ($v < c$) on the speed of a particle such as an electron. Are there similar upper limits on its kinetic energy or on the magnitude of its momentum? Explain.

17. If you were responsible for programming the launch of a space shuttle into earth orbit, would you use $\mathbf{p} = m\mathbf{v}$ or $\mathbf{p} = \gamma m\mathbf{v}$ for the momentum of the shuttle? Explain.

18. Suppose that you add heat to a stationary object, such as a penny, and raise its temperature. Has the rest mass of the penny changed? Has its weight changed? Explain.

19. If a spring is compressed, its elastic potential energy increases. Can this increase be interpreted as an increase in rest-mass energy of the spring? Explain.

20. In the nuclear-fusion reaction of Example 38-7, why were the masses given with so many significant digits?

21. If an electron is moving fast so that Eqs. 38-14 and 38-16 must be used, then the electron is often described as a *relativistic electron*. What is meant by a *relativistic proton?*

22. Complete the following table:

Symbol	Represents	Type	SI Unit
γ		Scalar	
T'_0			s
L_0	Rest length of an object		
$\mathbf{p}$			
mc^2			
E			

EXERCISES

Section 38-1. Transformations

1. Consider two reference frames related by the transformation in Eqs. 38-1. Show that the spatial separation between any two points is invariant. Consider points that have different y and z coordinates as well as different x coordinates.

2. The transformation in Eqs. 38-1 connects reference frames that have origins at different points. Suppose that two observers have the same origin but use different clocks. Observer O uses standard time and observer O' uses daylight saving time. *(a)* Determine the transformation equations connecting these reference frames. *(b)* Show that the number of daylight hours (the time interval between sunrise and sunset) is invariant.

3. Show that the galilean transformation for space-time coordinates $x, x', y, \ldots, t'$ is also obeyed by space-time coordinate differences $\Delta x, \Delta x', \Delta y, \ldots, \Delta t'$.

4. Two reference frames are connected by the galilean transformation in Eqs. 38-2, with observer O' in a train traveling with speed $v = 60$ km/h relative to observer O at the station. At $t = 1.2$ min, observer O notices a bus crossing the track at $x = 3.0$ km. Where and when does the crossing occur according to observer O'?

5. Equation 38-3 can be generalized and written as a vector equation using the velocity vectors $\mathbf{u}$, $\mathbf{u}'$, and $\mathbf{v}$, where $\mathbf{v}$ is the relative velocity of observer O' with respect to observer O. *(a)* Write this vector galilean velocity-transformation formula. *(b)* What is the velocity of observer O relative to observer O'?

6. Two observers, O' on a train and O on the station platform, as illustrated in Fig. 38-2, notice a flying goose overtaking and passing the train parallel to the tracks. The goose has a speed of 2 m/s according to O' and a speed of 9 m/s according to O. *(a)*

What is the speed of the train relative to O? *(b)* What is the relative speed of the two observers if O describes the same goose seen before, but O' describes a goose flying *backwards* at 2 m/s?

7. Observers O and O' are connected by a galilean transformation with speed v, and observers O' and O'' are connected by the same type of transformation with relative speed v'. Show that observers O and O'' are also connected by a galilean transformation and determine their relative speed.

8. Relative to an observer standing at the roadside, you are traveling south in an automobile at 80 km/h and a truck is traveling north at 60 km/h. You observe a south-bound bus passing you at 10 km/h. Determine the velocity of the bus relative to *(a)* the bystander and *(b)* the truck driver.

Section 38-2. The Principle of Relativity

9. Suppose that the force on one particle due to another particle, such as the newtonian gravitational force, depends only on the instantaneous separation of the particles. *(a)* Show that this type of force is invariant with respect to galilean transformations. *(b)* List some forces discussed in this text that do *not* depend only on the instantaneous separation of the particles.

10. Observer O on a train-station platform tosses a ball straight up to a height of 2.6 m and catches it on its return. Observer O' is on a train passing the station at a constant speed of 12 m/s. Coordinate axes are oriented as shown in Fig. 38-3, with the y axis vertical. Determine for the motion of the ball according to observer O', *(a)* the path, *(b)* the time of flight, *(c)* the maximum height, *(d)* the velocity components u'_x and u'_y when the ball reaches the maximum height, *(e)* the acceleration components a'_x and a'_y. Neglect air-resistance effects.

11. In the ether theory, the earth should move at some time during the year relative to the ether with a speed at least equal to its orbital speed about the sun. *(a)* Estimate the fraction of the speed of light that the speed of the earth through the ether would be. *(b)* This fraction indicates the sensitivity required to detect such a speed in the Michelson-Morley experiment. Can you think of a speed larger than the orbital speed that was available or accessible at that time for this kind of experiment?

12. Suppose that the sun were at rest relative to the ether. *(a)* Using galilean relativity, estimate the expected maximum and minimum values of the speed of light in vacuum, depending on relative directions of the light and the earth's motion. *(b)* What are the results obtained from actual measurements?

Section 38-3. The Lorentz Transformation

13. Two events occur at the origin of the coordinate system of observer O at times $t_1 = 1.58$ s and $t_2 = 2.13$ s. Use the Lorentz transformation in Eqs. 38-9 to determine the spatial interval $\Delta x' = x'_2 - x'_1$ and the time interval $\Delta t' = t'_2 - t'_1$ between the events according to observer O' if the relative speed of the observers is *(a)* $0.0010c$, *(b)* $0.10c$, *(c)* $0.99c$.

14. The origins of the coordinate systems of two observers coincide at $t = t' = 0$. The observers are in relative motion along the yy' axes, with constant speed v. Write the Lorentz transformation connecting these reference frames.

15. Two of the equations used to obtain the Lorentz transformation were $x' = \gamma(x - vt)$ and $x = \gamma(x' + vt')$, where γ was to be determined. Show that if the assumption $t = t'$ is used with the above equations, then $\gamma = 1$ and the galilean transformation results. Since the galilean transformation is adequate only for low speeds, the assumption $t = t'$ is not valid.

16. *(a)* Review the discussion that led from Eq. 38-7 to Eq. 38-8 and show that $\gamma^2 = 1/(1 - v^2/c^2)$. *(b)* By considering the sign of spatial intervals Δx and $\Delta x'$, explain why the positive square root is taken to give $\gamma = 1/\sqrt{1 - v^2/c^2}$.

17. Starting with the Lorentz transformation in Eqs. 38-9, solve for the unprimed space-time coordinates in terms of the primed ones. This results in the *inverse* of the transformation.

18. Two intergalactic travelers O and O' are drifting with a relative velocity along their xx' axes of magnitude 2.5×10^8 m/s. Observer O catalogs an exploding star at $x_1 = -1.55 \times 10^{14}$ m, $t_1 = 1.68 \times 10^6$ s and an unrelated eclipse of a binary star system at $x_2 = 0.68 \times 10^{14}$ m, $t = 2.94 \times 10^6$ s. *(a)* Where and when did these events occur relative to O'? *(b)* Explain how $x'_2 - x'_1$ can be negative.

19. The symbol β is often used to represent the ratio v/c, where v is the relative speed of two reference frames. *(a)* Rewrite the Lorentz transformation in Eqs. 38-9, using β instead of v. *(b)* If $\beta \ll 1$, show that $\gamma = 1/\sqrt{1 - \beta^2} \approx 1 + \frac{1}{2}\beta^2$. Determine the percent error in using the approximation in part *(b)* for *(c)* $\beta = 0.010$, *(d)* $\beta = 0.20$, *(e)* $\beta = 0.50$.

Section 38-4. A New View of Space and Time

20. Show that the Lorentz transformation reduces to the galilean transformation in the limit of low relative speed, $v \ll c$. Justify your treatment of the term vx/c^2 in one of the Lorentz transformation equations.

21. An astronomer on earth determines that a volcanic eruption on Jupiter's moon Io, 8×10^{11} m from earth, occurred simultaneously with an eruption of a volcano in Mexico. These two events were also observed by a space traveler moving past the earth toward Jupiter at 2.5×10^8 m/s. According to the space traveler *(a)* which eruption occurred first and *(b)* what distance separated these two events? (See the next exercise also.)

22. *(a)* What is the distance between earth and Io according to the space traveler in the previous exercise? *(b)* Explain why this distance is different from the answer to part *(b)* of that exercise.

23. Two intergalactic observers are connected by the Lorentz transformation in Eqs. 38-9, with $v = 0.95c$. Observer O detects two supernova events at $x_1 = 30$ ly, $t_1 = 2.6$ y and $x_2 = 47$ ly, $t_2 = 5.4$ y. [One light-year (ly) is the distance light travels in one year. It equals cT, where $T = 1$ y.] *(a)* Where and when does each supernova occur according to observer O'? For which observer were the two events closer *(b)* in space and *(c)* in time?

24. Reconsider the previous exercise, changing only x_1, t_1 to the following values: $x_1 = 35$ ly, $t_1 = -6.0$ y. How do you account for the difference from the answers you found in the previous exercise?

25. Review the discussion around Eq. 38-11 and then consider a measurement by O' of the length L' of a moving object that is at rest relative to observer O. (Thus the rest length of the object is L_0.) Show that $L' = L_0\sqrt{1 - v^2/c^2}$.

26. Observer O measures the length of a stick moving with a known speed v by determining the time interval Δt required for the ends of the stick to pass a particular spatial point. The measured length is then $L = v\,\Delta t$. Use time dilation to show that the length of the moving stick is given by $L = L_0'\sqrt{1 - v^2/c^2}$, where L_0' is the rest length of the stick.

27. The mean lifetime of a free neutron is about 1×10^3 s in its rest frame. Estimate the mean lifetime in a reference frame in which the neutron speed *(a)* is $0.99c$, *(b)* is $0.80c$, *(c)* is $0.10c$, *(d)* corresponds to an average kinetic energy of $\frac{3}{2}kT$ in a nuclear reactor. (Assume $T = 600$ K; k is the Boltzmann constant.)

28. The rest length of a Klingon space cruiser is 1800 m and its chronometer (clock) utilizes a mechanism with a period of 4.77×10^{-7} s. The cruiser moves past the planet Xzalb at 2.95×10^8 m/s. Determine *(a)* the length of the cruiser and *(b)* the period of its chronometer relative to a Xzalbian observer.

Section 38-5. The Addition of Velocities

29. Consider the transformation of the y and z components of the velocity of an object for observers connected by the Lorentz transformation in Eqs. 38-9. Show that these velocity components transform as given in the equations following Eq. 38-12.

30. Two observers O and O' are in relative motion, with speed v. Observer O' moves in the positive x direction according to observer O. Both observers measure the speed of a proton moving along the xx' axes. According to observer O, the velocity component of the proton is $u_x = 2.76 \times 10^8$ m/s. What is the velocity (magnitude and direction) of the proton according to observer O' if the relative speed of the observers is *(a)* 300 m/s, *(b)* 1.00×10^8 m/s, *(c)* 2.76×10^8 m/s, *(d)* 2.95×10^8 m/s?

31. Observer O detects observer O' moving in the positive x direction with speed $v = 0.9990c$. Observer O' measures the velocity component of an electron to be $u_x' = 0.9999c$. What is the electron velocity component u_x according to observer O?

32. Use Eq. 38-13 to show that if $0 < u_x' < c$ and $0 < v < c$, then $0 < u_x < c$. That is, if the speed of an object is less than c for one observer, then the speed of the object is less than c for another observer.

Section 38-6. Momentum and Energy

33. Determine the kinetic energy and the magnitude of the momentum of an electron with speed *(a)* $0.99c$, *(b)* $0.50c$, *(c)* $0.10c$, *(d)* $0.001c$. *(e)* For which, if any, of these speeds can the nonrelativistic expressions $\frac{1}{2}mv^2$ and mv be used? Justify your answer.

34. For what speed v is the kinetic energy of a particle equal to *(a)* 10, *(b)* 1.0, *(c)* 0.10, *(d)* 0.01 times its rest-mass energy?

35. The nonrelativistic expression $\frac{1}{2}mv^2$ for the kinetic energy of a particle can be used if the speed v is small compared with c. For what speed does the use of this formula correspond to an error in the kinetic energy of *(a)* 1 percent and *(b)* 10 percent?

36. Eliminate the velocity $\mathbf{v}$ (and speed v) of the particle from Eqs. 38-14 and 38-16 to obtain the following relation between the total energy E of the particle and the magnitude p of its momentum:

$$E = \sqrt{p^2c^2 + m^2c^4}$$

(See Prob. 6 for another approach to this equation.)

37. In addition to his work on relativity, Einstein proposed a quantum picture of light as consisting of a stream of particles called *photons*. Each photon travels at the speed of light. To be consistent with special relativity, a particle traveling at speed c must have a zero rest mass. (Why?) Show that the energy and momentum of a photon are related by the simple expression $E = pc$.

38. An alpha particle (^{4}He nucleus with a rest mass of 6.6×10^{-27} kg) is emitted when ^{238}U spontaneously decays. The alpha particle has a kinetic energy of 4.2 MeV. Determine *(a)* the speed and *(b)* the magnitude of the momentum of the alpha particle. *(c)* What fraction of its rest-mass energy is the kinetic energy of the alpha particle?

39. *(a)* Determine the speed of an electron with kinetic energy equal to its rest-mass energy. *(b)* Repeat for a proton.

40. *(a)* Show that mc^2 has dimensions of energy. *(b)* The atomic mass unit ($1\ \text{u} = 1.6605 \times 10^{-27}$ kg) is a convenient mass unit for atomic and nuclear masses. Determine the equivalent rest-mass energy in J for 1 u. *(c)* What is the equivalent rest-mass energy of 1 u in MeV?

41. The ^{8}Be nucleus is unstable and spontaneously decays into two ^{4}He nuclei. *(a)* If the rest masses of the elements are $m_{\text{Be}} = 8.005308$ u and $m_{\text{He}} = 4.002603$ u, then what rest-mass energy is converted in the decay? *(b)* In a reference frame in which the Be nucleus is initially at rest, what kinetic energy does each He nucleus have after they are well separated? *(c)* What is the magnitude of the momentum of each He nucleus? *(d)* What is the speed of each He nucleus?

42. The rest mass M of a bound system such as an atom is less than the sum Σm_i of the rest masses of its constituent particles. The *binding energy* B of the system, the minimum energy required to separate the constituents, is given by the difference in rest-mass energies: $B = \Sigma m_i c^2 - Mc^2$. Determine the mass of *(a)* the deuteron (which consists of a proton and a neutron), given that $B = 2.23$ MeV, $m_p = 1.00728$ u, $m_n = 1.00867$ u and *(b)* the hydrogen atom (which consists of an electron and a proton), given that $B = 13.6$ eV, $m_p = 1.672648 \times 10^{-27}$ kg, $m_e = 9.1095 \times 10^{-31}$ kg.

Additional Exercises

43. Suppose a detective could travel at half the speed of light. While traveling at $c/2$, she looks out her window and sees a man even with her window shoot a phaser gun (which used light) at exactly noon. Further suppose she knows that a

woman 1 km down the tracks was hit by a phaser at noon plus 1 μs. Can she make a case against the man?

44. Passengers on a Trafalmadorian spaceship traveling by the solar system at $0.65c$ tune into a physics lecture from earth for laughs. The lecture lasts 50 minutes according to the students in the lecture hall on earth. How long must the Trafalmadorians watch to see the whole lecture?

45. The Trafalmadorians of the previous example also measure the distance across the United States. If the Coast and Geodetic survey says it is 4500 km, what distance will the Trafalmadorians measure?

46. The Trafalmadorians observe that the Super Bowl starts at exactly 4:00 Central Tralfamadorian Time on Los Angeles television. When do they see it start on New York television if they travel from west to east over Los Angeles and New York, and the distance between cities, as measured on earth, is 4500 km? Suppose that the network assures that the Super Bowl starts at the same earth time in both places.

47. Muons are observed traveling toward the earth at a height of 5 km with a velocity of $0.99c$. Half the muons will decay in 1.49×10^{-6} s (in the muon rest frame). *(a)* How close to the earth will the muons get before half of them decay? *(b)* How far will the muons travel in their reference frame?

48. An electron is accelerated in a television tube through a potential of 0.100 MV. *(a)* What is the rest-mass energy of an electron in MeV? (See App. C for conversions.) *(b)* What is the relativistic energy of the electron in the living room reference frame and in mega-electron volts? *(c)* What is the speed of the electron after it is accelerated?

49. A photon and a proton have the same total energy, 7 GeV. Which particle has the greater momentum? What is the ratio of their momenta?

50. If you wish to travel to a star that is 100 ly away, taking no more than the rest of your life expectancy (50 years?), how fast will your spaceship have to travel?

PROBLEMS

1. ***Michelson-Morley experiment.*** The design of the Michelson-Morley experiment involved the difference in path length for light traveling in different directions with respect to the velocity of the apparatus through the ether. For simplicity, consider the case where the interferometer (discussed in Chap. 36) moves relative to the ether (Fig. 38-14). *(a)* Consider the time $\Delta t_{\parallel}$ required for light to travel on the parallel path from $M_{1/2}$ to M_a and back to $M_{1/2}$. Show that $\Delta t_{\parallel} = 2Lc/(c^2 - v^2)$. *(b)* Similarly show, using Fig. 38-14*b*, that the time interval along the perpendicular path is $\Delta t_{\perp} = 2L/\sqrt{c^2 - v^2}$. The interference pattern is determined by these time differences and should change if the apparatus is rotated through 90° in the plane of the figure so as to interchange the parallel and perpendicular paths. No significant shift in the pattern occurred in the experiment, which implied that $v = 0$.

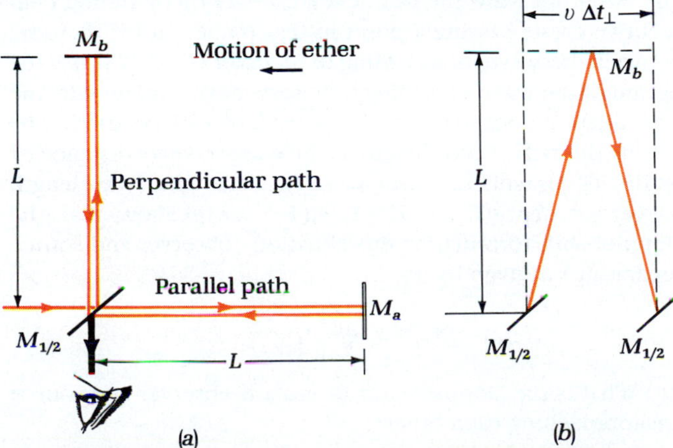

Figure 38-14. Problem 1: *(a)* The interferometer is shown in its rest frame. The light rays are assumed to travel from $M_{1/2}$ to M_a at speed $c - v$ and from M_a back to $M_{1/2}$ at speed $c + v$. *(b)* In the rest frame of the ether, the apparatus moves and the light traverses the perpendicular path with speed c.

2. ***Causality.*** Since simultaneity is a relative concept, what about causality? That is, if event 1 is the cause of event 2, then does the cause precede the effect for all observers? Suppose that these events occur at (x_1, t_1) and (x_2, t_2) in one reference frame, with $t_2 > t_1$. (We suppress the y and z coordinates for simplicity.) Suppose further that the events can be connected by a light signal: $|x_2 - x_1| < c(t_2 - t_1)$. Show that the events are also causal ($t_2' > t_1'$) in a second reference frame, connected to the first by the Lorentz transformation. This result suggests that the influences (forces) that one particle has on another cannot be instantaneous but are propagated at a speed no greater than c.

3. ***The twin paradox.*** Consider identical twins, Orestes and Opie. On their twenty-fifth birthday, Opie leaves earth on a spaceship, which quickly accelerates to a cruising speed of $v = 0.99c$ relative to Orestes. The cruise lasts for 12 years according to Opie, and he returns to celebrate his thirty-seventh birthday. *(a)* How old is Orestes at this celebration? *(b)* The paradox is that Opie could claim that Orestes (along with the solar system) accelerated and cruised at high speed; Orestes should return much younger than he. Suggest some ways to resolve the paradox.

4. ***A third observer.*** Suppose that three observers—O, O', and O''—are related in turn by the Lorentz transformations

$$x' = \gamma(x - vt) \qquad x'' = \gamma'(x' - v't')$$
$$y' = y \qquad y'' = y'$$
$$z' = z \qquad z'' = z'$$
$$t' = \gamma\left(t - \frac{vx}{c^2}\right) \qquad t'' = \gamma'\left(t' - \frac{v'x'}{c^2}\right)$$

where $\gamma = 1/\sqrt{1 - v^2/c^2}$, $\gamma' = 1/\sqrt{1 - v'^2/c^2}$, and v and v' are relative speeds of pairs of observers. *(a)* Show that the transformation directly connecting O and O'' is also a Lorentz transformation, with relative speed

$$u = \frac{v + v'}{1 + vv'/c^2}$$

(b) Can you explain why, if $v = v'$, then $u \neq 2v$? *(c)* What is the interpretation of these transformations for the case $v' = -v$? (Think of v and v' as velocity components rather than speeds in this instance.)

5. ***Proper time.*** Isotropy and homogeneity of space require that the distance $\Delta s = (\Delta x^2 + \Delta y^2 + \Delta z^2)^{1/2}$ between two points be invariant, independent of the orientation or origin of the three-dimensional coordinate system. There is a corresponding invariant quantity in four-dimensioanl space-time. It is convenient to express this invariant as a time interval called the *proper time interval* $\Delta\tau$ between two events. It is defined by the expression

$$c^2\,\Delta\tau^2 = c^2\,\Delta t^2 - (\Delta x^2 + \Delta y^2 + \Delta z^2)$$

Another observer (O') would evaluate the proper time interval for the same two events using

$$c^2\,\Delta\tau^2 = c^2\,\Delta t'^2 - (\Delta x'^2 + \Delta y'^2 + \Delta z'^2)$$

Use the Lorentz transformation in Eqs. 38-9 to show that the two expressions above are identically equal. A physical interpretation of the proper time comes from considering two events that occur at the same spatial point in the reference frame of one observer, say O. Then $\Delta\tau = \Delta t$, but notice that $\Delta\tau \neq \Delta t'$ in this case.

6. ***Energy-momentum relation.*** The three momentum components p_x, p_y, and p_z and the total energy E of a particle give four quantities (p_x, p_y, p_z, E/c^2) that transform under Lorentz transformation just as the space-time coordinates do. Thus

$$p_x' = \gamma\left(p_x - \frac{vE}{c^2}\right)$$

$$p_y' = p_y$$

$$p_z' = p_z$$

$$\frac{E'}{c^2} = \gamma\left(\frac{E}{c^2} - \frac{vp_x}{c^2}\right)$$

(a) Using the previous problem as a guide, show that the quantity $E^2 - p^2c^2$ is an invariant. Note that $p^2 = p_x{}^2 + p_y{}^2 + p_z{}^2$. *(b)* What is the value of the invariant?

7. ***Transformation of angles.*** Observers O and O' are connected by the Lorentz transformation in Eqs. 38-9. *(a)* Observer O detects a light ray traveling in the xy plane at an angle θ from the x axis, as shown in Fig. 38-15. Show that observer O' detects the ray traveling at angle θ' given by

$$\tan\theta' = \frac{\sin\theta}{\gamma(\cos\theta - v/c)}$$

(b) Suppose that observer O holds a rigid stick at angle θ from the x axis. Show that observer O' measures the angle θ' given by

$$\tan\theta' = \gamma\tan\theta$$

(c) Explain why the expressions in parts *(a)* and *(b)* are different.

8. ***Newton's second law.*** In the relativistic dynamics of a particle, Newton's second law is written in the form $\mathbf{F} = d\mathbf{p}/dt$,

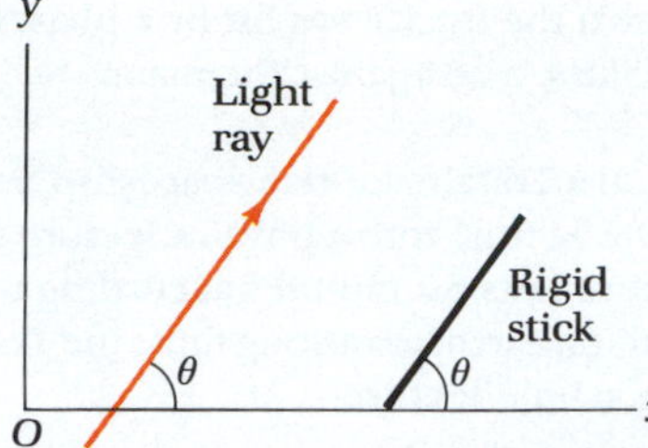

Figure 38-15. Problem 7: A propagating light ray and a stationary rigid stick are at the same angle θ from the x axis in the xy plane of observer O.

where $\mathbf{F}$ is the net force acting on the particle and $\mathbf{p} = \gamma m\mathbf{v}$, with $\gamma = 1/\sqrt{1 - v^2/c^2}$. *(a)* Show that the second law written in terms of the acceleration $\mathbf{a}$ is given in general by

$$\mathbf{F} = \frac{m\mathbf{a}}{\sqrt{1 - v^2/c^2}} + \frac{m\mathbf{v}(\mathbf{v}\cdot\mathbf{a})}{c^2(1 - v^2/c^2)^{3/2}}$$

(b) Consider the special case of uniform circular motion of a particle of charge q in a uniform magnetic field. At low speeds the particle circulates at the cyclotron frequency $\omega_0 = qB/m$, which is independent of the radius r of the path. (See Chap. 26.) Show that the angular frequency $\omega = v/r$ of the circular motion, correct for all speeds, is given by $\omega = \omega_0\sqrt{1 - v^2/c^2}$.

9. ***Work-energy.*** Consider the one-dimensional motion of a particle of rest mass m along the x axis, with a constant net force $F_x\mathbf{i}$ acting. The work dW for a small displacement $dx\,\mathbf{i}$ is given by $F_x\,dx = F_xv_x\,dt$, where v_x is the velocity component of the particle. *(a)* Use Newton's second law, $F_x = dp_x/dt$, and integrate to obtain the work done for a finite displacement. [*Hint:* Integrate $(dp_x/dt)v_x\,dt$ by parts, with $p_x = mv_x/\sqrt{1 - v_x^2/c^2}$.] *(b)* Apply the work-energy theorem to the case of a particle starting from rest and show that the kinetic energy of the particle is given by Eq. 38-16.

10. ***The Doppler shift.*** Although the speed of light in vacuum is the same in all inertial reference frames, the frequency of the light can be different for different observers. Suppose that observer O sends a light signal of frequency ν_s along the x axis. Let observer O', connected to O by the Lorentz transformation in Eqs. 38-9, measure the frequency $\nu_{O'} = 1/T_{O'}$ by timing consecutive crests passing a point in that frame. *(a)* If the time between these events according to observer O is T (T is *not* the period of the wave according to observer O), explain why the time-dilation formula $T = T_{O'}/\sqrt{1 - v^2/c^2}$ can be used. *(b)* During this time, according to O, the wave travels a distance cT while O' travels a distance vT. Thus the wavelength $\lambda_s = c/\nu_s = cT - vT$, or $T = 1/[\nu_s(1 - v/c)]$. Show that the Doppler-shift formula for this situation (observer and source separating) is given by

$$\nu_{O'} = \nu_s\sqrt{\frac{1 - v/c}{1 + v/c}}$$

(c) What is the Doppler-shift formula if observer and source are approaching each other?

11. ***Relativistic circles.*** Using the results of Prob. 8, show that the radius of curvature R of the path of a charged particle traveling perpendicular to a magnetic field B is $R = p/qB$, where p and q are the relativistic momentum and charge of the particle, respectively.

12. ***Identifying nuclei.*** A cosmic ray leaves a track in a detector. The density of the track indicates that the charge of the particle involved is +8 times that of an electron. The particle follows a circular path of radius 4.5 m in a magnetic field of 1.5 T. Assume the particle to be the bare nucleus of an atom. *(a)* What is its momentum (see Prob. 11)? *(b)* What is its energy? Appendix P has a periodic chart of the elements.

13. ***Nonradial Doppler.*** Show that the Doppler formula for the frequency ν of source radiating at a frequency ν' and traveling with speed v at an angle θ with respect to the line joining the source and the receiver is given by

$$\nu = \nu' \frac{1}{\gamma(1 + v \cos \theta / c)}$$

Remember to consider the time dilation and the radial component of the velocity of the source.

14. ***Transverse Doppler.*** Using the result of Prob. 13 and an expansion such as those shown in App. M, show that the radial Doppler shift (for a source moving away from the receiver) for $v \ll c$ is

$$(\nu - \nu')_R \approx v/c$$

while the transverse Doppler shift (for an object moving perpendicular to the line joining the source and the receiver) for $v \ll c$ is

$$(\nu - \nu')_T \approx (1/2)v^2/c^2$$

thus showing that the transverse Doppler effect is much smaller.

15. ***Equalizing protons.*** Protons from an accelerator are given a kinetic energy of 1 GeV. *(a)* Find the speed u of these protons as measured in the lab. They are incident on a target of solid hydrogen. *(b)* Show that in a reference frame moving toward the incoming protons at $v = (c^2/u)[(\gamma - 1)/\gamma]$, the momentum of the incoming protons is the same as that of those in the solid hydrogen. *(c)* What is this momentum?

16. ***Relativistic dissipation.*** How much energy do the protons in Prob. 15 have in the moving reference frame? If the protons were stationary in the moving reference frame after the collision, how much energy would be "dissipated" in the collision?

17. ***Higher-order terms in kinetic energy.*** Show that the kinetic energy of a particle can be expressed as

$$K = (mv^2/2)\ \Sigma[1 + (3/4)v^2/c^2 + (15/24)v^4/c^4 + \cdots]$$

18. ***Pions go to photons.*** A neutral pion of mass M moving with speed v decays into two photons whose paths make equal angles θ with the direction of motion of the pion. *(a)* Show that the photons have equal energy. *(b)* Show that $\cos \theta = v/c$.

19. ***Prohibited photon decay.*** Prove that the laws of conservation of relativistic energy and momentum prohibit an isolated photon from decaying into two particles with mass. Assume that the energy of the photon is greater than the rest-mass energy of the two particles.

39 QUANTIZATION OF ELECTROMAGNETIC RADIATION

Dueling prisms. White light is spatially separated into its constituent colors by each prism.

Imagine being a young scientist in 1899 and celebrating the turn of the century by reflecting on how well you understood the workings of the universe: Newtonian mechanics dealt with the motion of objects of all sizes — from planetary motions to a falling grain of sand. Thermodynamics had been developed from and had contributed to the improvements in the steam engine. Indeed, the steam locomotive provided fast, economical transportation across continents. Electricity and magnetism had been unified in Maxwell's equations. Further, since light consists of electromagnetic waves, optics was just a branch of electromagnetism. Of course, there were still some intriguing problems around. Many of these dealt with the structure of atoms and the light emitted and absorbed by them. You might be tempted to boast that these problems would soon be mopped up, that humanity was on the verge of mastery of the fundamental laws of nature. You would be greatly mistaken. As it happens, these problems led instead to a revolution of ideas in physics.

The basic outline of what is now called ***quantum mechanics,*** or ***quantum physics*** developed during a 30-year period beginning around 1900. Although the older, or classical, physics remains useful in its range of applicability, quantum physics deals with a whole new range of phenomena at the atomic and subatomic levels. We shall consider some of these phenomena in this chapter.*

39-1 INTERACTION OF LIGHT AND MATTER

Our perceptions of the physical world are dominated by the interaction of light and matter. One interaction occurs at the retina of the eye, leading physiologically to the sense of sight. The light that enters the eye and produces a rich variety of images is the result of the interaction of light and matter. As we mentioned in Chap. 34, our eyes are sensitive to only a small part of the electromagnetic spectrum. Henceforth, when we use the term "light," we mean electromagnetic radiation of any frequency—infrared, ultraviolet, microwave, whatever.

Light and matter seem to interact passively in some cases. For a nominally transparent substance such as glass, the main result of the interaction is the behavior at the surface—reflection and refraction (Chap. 35). The absorption of electromagnetic energy is usually negligible, at least in the visible spectrum.

In other cases the interaction of light and matter seems more active, as in the emission of light. Intense light is emitted from matter at high temperatures such as from the surface of the sun or from the flash of a photographer's light.

When light is incident on the surface of an opaque object, part of the light penetrates into the material and is absorbed. The remaining light is reflected from the surface. It is this reflected light that we observe; that is, we "see" the object by the light reflected from it. The fraction of light that is reflected from the surface depends on the wavelength. For example, the surface of an object may reflect most of the light in the blue portion of the spectrum and absorb strongly in the red portion. Thus if white light is incident on the surface, the reflected light contains a greater proportion of blue than red. As a result, the perceived color of the object is a shade of blue or perhaps green. The vast range of hues distinguishable by the human eye is a consequence of the sensitivity of the eye to the different wavelengths reflected.

Surfaces with darker colors are more strongly absorbing than surfaces with lighter colors. You have probably noticed this effect directly in the absorption of sunlight by your clothing. A dark shirt or blouse absorbs more of the incident sunlight than a light one, so that light clothing is cooler in summer. A white surface is one that reflects a large fraction of all frequencies of the incident light. Another type of surface that is a good reflector (and a poor absorber) is a polished, metallic surface of a substance such as aluminum or untarnished silver. At the other extreme are strongly absorbing (poorly reflecting) surfaces. Examples are tar, lampblack, and printer's ink. (How can you account for the high contrast between the paper and the printed symbols on this page?)

In addition to seeing objects by the light reflected from them, we can see some objects by the light that they emit. A glowing coal, for example, emits a significant fraction of its light in the visible spectrum; it can be seen in an otherwise darkened room. Other objects in the room, at much lower temperatures, do not emit with an appreciable intensity in the visible part of the spectrum. All objects emit electromagnetic radiation, and they do so at a rate that depends on the temperature of the surface. Only for temperatures well above room temperature does the surface emit enough in the visible spectrum to be easily observable.

The radiated power P (rate of emission of electromagnetic energy) from an area A of surface at temperature T is given by the Stefan-Boltzmann law from Chap. 16:

$$P = e\sigma AT^4 \qquad (16\text{-}12)$$

Stefan-Boltzmann law

where $\sigma = 5.67 \times 10^{-8}\ \mathrm{W \cdot m^{-2} \cdot K^{-4}}$ is the Stefan-Boltzmann constant. Josef Stefan (1835–1893) proposed the T^4 dependence in 1879 from an analysis of experi-

* In the "extended" edition of this text, the remaining chapters are largely devoted to discussion of quantum physics.

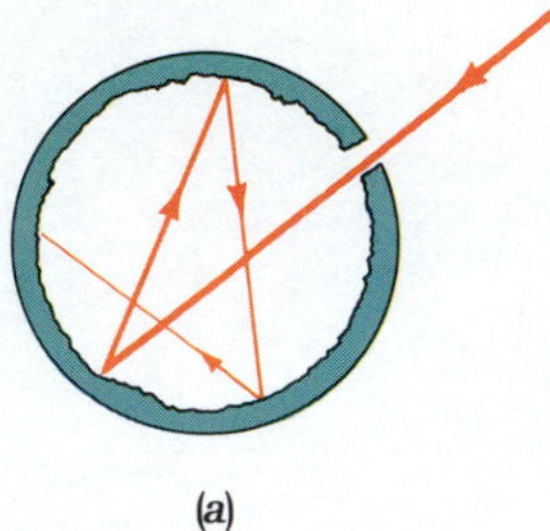

(a)

(b)

Figure 39-1. *(a)* Light entering a small opening to a cavity is absorbed after many reflections. *(b)* The light coming from the small opening to a cavity approximates blackbody radiation.

mental data. Five years later Ludwig Boltzmann derived the law theoretically. The ***emissivity*** e characterizes the emitting properties of the surface and is material-dependent. It is a dimensionless number with a range of $0 \leq e \leq 1$. The upper limit, $e = 1$, would correspond to a perfect, or ideal, emitter.

A blackbody is an ideal emitter and absorber.

A surface that is a good absorber of light is also a good emitter of light. Using the second law of thermodynamics, you can show (in Prob. 1) that the rate of emission of electromagnetic radiation of a given frequency must equal the rate of absorption at that frequency. Thus the ideal emitting surface with $e = 1$ would also be an ideal absorbing surface. That is, all of the radiation incident on the ideal absorbing surface would be absorbed and none would be reflected. Since no incident light would be reflected by the surface of an ideal emitter-absorber, such an object is sometimes called a ***blackbody.*** The electromagnetic radiation, or light, emitted by the ideal blackbody is often called ***blackbody radiation.***

A blackbody is an idealization in that no material has a surface with unit emissivity. Lampblack, with $e \approx 0.99$, closely approximates a blackbody. Only a small fraction of the light incident on such a surface is reflected.

Cavity radiation approximates blackbody radiation.

The light that would be emitted by a blackbody can be approximated as closely as desired by the radiation emerging from a small opening in a cavity (at temperatures below the melting point of the material that forms the walls of the cavity). Consider light entering the cavity through the opening shown in Fig. 39-1*a*. Part of the light is absorbed on each reflection from the interior cavity wall. After many reflections, virtually all of the energy that was incident on the opening has been absorbed. In this way the opening behaves as an ideal absorber, a blackbody. Thus the light emerging from the small opening of a cavity, as seen in Fig. 39-1*b*, is blackbody radiation. The emissivity $e = 1$ for a blackbody or for the opening of a cavity, and the radiated power is independent of material that forms the inside walls of the cavity.

The power in Eq. 16-12 includes all frequencies. In the next section we shall consider in detail how the radiation is distributed over a range of frequencies. This frequency dependence of blackbody radiation turns out to be universal, the same for all blackbodies. It is a feature of the interaction of light and matter and was one of the unsolved problems in classical physics that led to the development of quantum theory.

39-2 CAVITY RADIATION

The electromagnetic radiation that exists in a cavity is a mixture of standing waves. In analogy with one-dimensional standing waves that can fit on a string, three-dimensional standing wave modes of various frequencies ν can "fit" into the cavity.

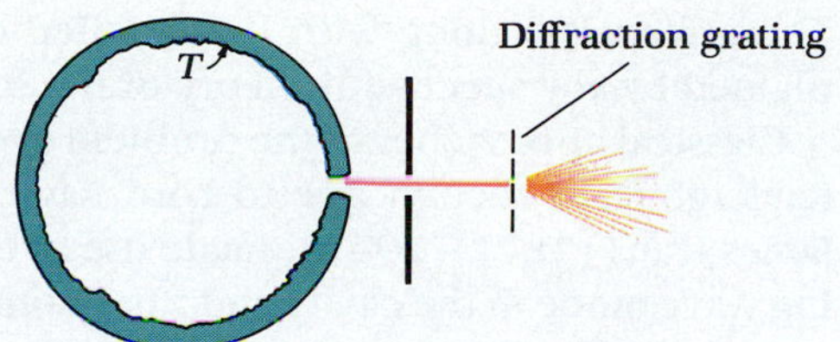

Figure 39-2. Light from a cavity passes through a collimating slit and is separated into its component wavelengths by a diffraction grating.

The electromagnetic energy is distributed among the standing wave modes such that at temperature T the radiation is in thermal equilibrium with the walls of the cavity. This energy distribution can be determined by sampling the radiation emerging from a small opening in the cavity wall. Since the small opening is equivalent to a blackbody, the power radiated from the opening is $P = \sigma A T^4$, where A is the area of the opening. We use the ***radiancy*** $R = P/A$ to denote the radiated power per unit area from the opening. Similarly, the ***spectral radiancy*** R_ν is the radiated power per unit area per unit frequency range. That is, $R_\nu\, d\nu$ is the radiated power per unit area with frequencies between ν and $\nu + d\nu$. The spectral radiancy R_ν summed or integrated over all frequencies gives the radiancy:

$$R = \int_0^\infty R_\nu\, d\nu$$

A related spectral radiancy R_λ is often used when wavelengths are measured. For a wavelength range $d\lambda$ between λ and $\lambda + d\lambda$, the radiated power per unit area is $R_\lambda\, d\lambda$. These two spectral radiancies are simply related (see Exercise 9) since $\nu\lambda = c$.

The spectral radiancy is measured by analyzing the radiation from the cavity with a spectrometer, as shown schematically in Fig. 39-2. The dependence on frequency or wavelength is shown for two different temperatures in Fig. 39-3. Notice the following general features:

1. At a given temperature, the spectral radiancy has a single peak or maximum.
2. If the temperature is increased, the spectral radiancy increases for every frequency or wavelength.
3. The peak wavelength, the wavelength at which the spectral radiancy is a maximum, shifts to smaller wavelengths (larger frequencies) at higher temperatures. (See Prob. 2.)

Figure 39-3. The spectral radiancy for a blackbody, or from an opening to a cavity, is shown for two temperatures. The frequency and wavelength ranges for visible light are indicated by the shaded regions. *(a)* The dependence of R_λ on λ is shown. *(b)* The dependence of R_ν on ν is shown.

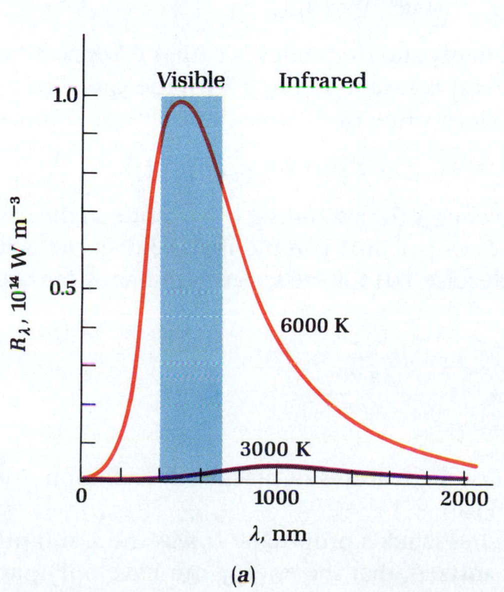

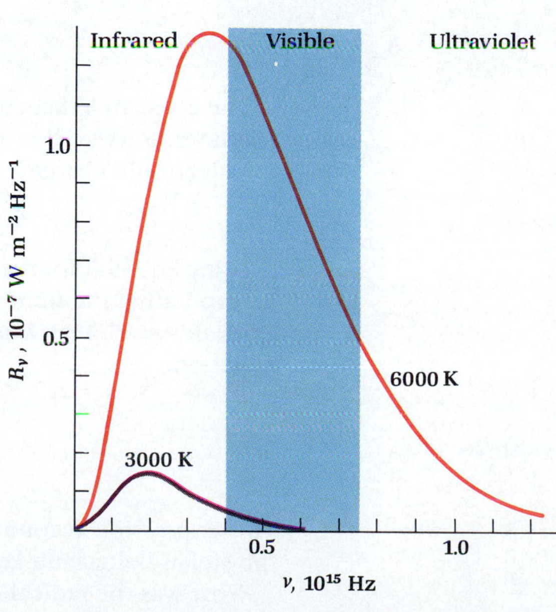

These features, along with the detailed dependence on frequency, must be explained by any successful theory of the interaction of light and matter.

Classical approaches to the problem of cavity radiation were not successful. The Rayleigh-Jeans theory, due to Lord Rayleigh (1842–1919) and modified by Sir James Jeans (1877–1946), made use of the mathematical equivalence of a standing wave mode in the cavity and a harmonic oscillator. The analogy may be seen by comparing the expressions $\frac{1}{2}\epsilon_0 E^2$ for the electric energy density of the standing wave and $\frac{1}{2}\alpha x^2$ for the potential energy of an oscillator with spring constant α. For a system of classical oscillators at temperature T, the average energy per oscillator is $\langle E \rangle = kT$ from the equipartition-of-energy theorem (Chap. 18), where k is the Boltzmann constant. Classically, a standing-wave mode should similarly have an average energy of $\langle E \rangle. = kT$.

The number of standing wave modes per unit volume with frequencies in a range $d\nu$ between ν and $\nu + d\nu$ can be calculated to be $8\pi\nu^2\, d\nu/c^3$. The radiancy, which is proportional to the energy per unit volume in the cavity, should be proportional to the product of the number of modes per unit volume and the average energy per mode. This reasoning gives the Rayleigh-Jeans formula, $R_\nu = 2\pi\nu^2 kT/c^2$. The Rayleigh-Jeans formula agrees with experiment only at low frequencies; at higher frequencies it is clearly wrong because it increases without bound (as ν^2). This mathematical behavior with increasing frequency of the Rayleigh-Jeans formula was called the "ultraviolet catastrophe."

In 1900 Max Planck (1858–1947) presented a formula that could describe the measured frequency distribution of blackbody radiation. To obtain a physical basis for some constants in the formula, Planck made some proposals that were thought to be unfounded and extremely radical at the time. These proposals can now be stated as follows:

Quantum of energy $h\nu$

The energy of an oscillator is quantized.

1. An oscillator (including a standing wave mode) of frequency ν can change its energy only by a whole multiple of a discrete amount, a quantum of energy $\Delta E = h\nu$, where h is a constant described below.
2. The energy of an oscillator is ***quantized;*** its energy is restricted to one of the values $E_n = n \cdot h\nu$, where the quantum number n is an integer.

Planck was able to show, as a consequence of energy quantization, that the average energy per oscillator for a collection of oscillators of frequency ν at temperature T is given by

$$\langle E \rangle = \frac{h\nu}{e^{h\nu/kT} - 1} \qquad (39\text{-}1)$$

The constant h that connects energy and frequency is called ***Planck's constant.*** It is considered to be a fundamental constant, ranking with the speed of light c or the electronic charge e. The modern value of Planck's constant in SI units is

Planck's constant

$$h = 6.626076 \times 10^{-34}\ \text{J}\cdot\text{s} \approx 6.63 \times 10^{-34}\ \text{J}\cdot\text{s}$$

Using Eq. 39-1 for the average energy for a standing wave mode in the cavity and incorporating the number of modes per unit volume from above, we can obtain ***Planck's radiation law*** (see Exercise 10) for the spectral radiancy for blackbody radiation:

Planck's radiation law

$$R_\nu = \frac{2\pi h\nu^3}{c^2(e^{h\nu/kT} - 1)} \qquad (39\text{-}2)$$

This expression accounts for all of the features of blackbody radiation, including the Stefan-Boltzmann law. (See Prob. 3.)

What was the radical element in Planck's proposals? It was the assumption that the energy of an oscillator is quantized, that the energy can have only particular, discrete values and not any value in between. For a classical mechanical or electric oscillator, the energy appears to be a continuously variable quantity. For example, the energy of a harmonic oscillator of mass m and angular frequency $\omega = 2\pi\nu$ can be expressed in terms of the amplitude A of the motion: $E = \frac{1}{2}m\omega^2A^2$. If the energy is quantized, then the amplitude likewise can have only particular, discrete values. However, no one had yet observed a discreteness in the amplitude of oscillators. The energy (and amplitude) of ordinary-sized oscillators does indeed seem to be continuously variable. The quantized values are closely spaced, as the following example shows, because of the small value of Planck's constant h.

EXAMPLE 39-1

Energy quantization is imperceptible for macroscopic oscillators. The 0.10-kg tip of a tuning fork vibrates harmonically at 440 Hz with amplitude $A = 1.2$ mm. *(a)* Determine the quantum number n for this state of the oscillating tip. *(b)* If the quantum number decreases from n to $n - 1$, determine the change ΔA in amplitude.

Solution. *(a)* The energy of the oscillator is

$$E = \tfrac{1}{2}m(2\pi\nu)^2A^2 = \tfrac{1}{2}(0.10\text{ kg})(2\pi \cdot 440\text{ Hz})^2(0.0012\text{ m})^2$$
$$= 0.55\text{ J}$$

Since $E = nh\nu$, we can solve for the quantum number n:

$$n = \frac{E}{h\nu} = \frac{0.55\text{ J}}{(6.63 \times 10^{-34}\text{ J}\cdot\text{s})(440\text{ Hz})}$$
$$= 1.9 \times 10^{30}$$

This large value of n for a typical state of a macroscopic oscillator is due to the small value of the quantum of energy, $h\nu = (6.63 \times 10^{-34}\text{ J}\cdot\text{s})(440\text{ Hz}) = 2.9 \times 10^{-31}$ J. *(b)* Since $E = nh\nu = \frac{1}{2}m(2\pi\nu)^2A^2$, we have $A^2 = nh/(2\pi^2m\nu)$. Taking the differential of this expression connects the change ΔA due to a change Δn: $2A\,\Delta A = h\,\Delta n/(2\pi^2m\nu)$. Dividing by $A^2 = nh/2\pi^2m\nu$, we obtain $2\,\Delta A/A = \Delta n/n$. Since $\Delta n = -1$, we have

$$\Delta A = \frac{\frac{1}{2}(1.2\text{ mm})(-1)}{1.9 \times 10^{30}} = -3.2 \times 10^{-34}\text{ m}$$

Such a small change in the amplitude is unobservable, and the amplitude of a macroscopic oscillator seems continuously variable.

SELF-TEST 39-1. In Chap. 34 we learned that a charged particle oscillating at some frequency emits electromagnetic radiation at that frequency. The middle of the visible spectrum is at a frequency of about 6×10^{14} Hz. What is the energy of an oscillator with this frequency that is in the $n = 1$ quantum state? Give your answer in joules and in electron volts. ***ANSWER:*** $4 \times 10^{-17}\text{ J} = 2$ eV.

39-3 THE PHOTOELECTRIC EFFECT

Planck's assumption that the energy of an oscillator is quantized did not at the time represent a new view of light. The oscillators discussed by Planck were thought to be atomic oscillators that formed the interior walls of the cavity, and not the standing wave modes. In 1906 Einstein extended the idea of quantization to light itself, as it propagated freely and interacted with matter. This new theory of light answered some puzzling questions about the photoelectric effect.

In the ***photoelectric effect,*** electrons are emitted from a material when light is incident on its surface. Ordinarily an electron is bound to the material and cannot escape from it unless energy is supplied. The light must supply each emitted electron with sufficient energy to escape from the surface. To be emitted from the surface, an electron must receive a minimum amount of energy ϕ, called the ***work function*** of the surface. The value of the work function depends on the material and on the condition of the surface. For example, the work function for aluminum (with a clean, unoxidized surface) is 4.2 eV.

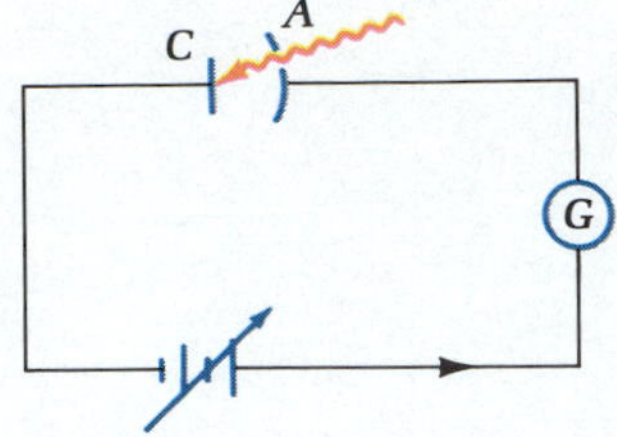

Figure 39-4. Light incident on the cathode C of a photocell causes electrons to be emitted. Electrons collected at the anode A contribute to the current in the circuit as measured by the galvanometer G. The potential difference across the photocell can be adjusted.

Quantitative studies of the photoelectric effect are made with an arrangement such as that shown in Fig. 39-4. Light is incident on the photosensitive surface, which is the cathode C of the photocell. Electrons emitted from the cathode are collected at the anode A at a rate that determines the current i in the galvanometer G. The potential difference across the photocell can be varied. If the anode is at a higher potential than the cathode, then the anode attracts the emitted electrons. If the polarity is reversed so that the anode is at a lower potential than the cathode, then the anode repels the electrons.

Suppose that monochromatic light of frequency ν is incident on the photocell. We list below several observed features of the photoelectric effect. These features have no reasonable explanation if we use the classical picture of electrons in the material interacting with the electromagnetic wave of the incident light.

1. There is a ***threshold frequency*** ν_0 below which no electrons are emitted from the surface. The current in the circuit in Fig. 39-4 is zero if the cathode is illuminated with light whose frequency ν is below the threshold value, $\nu < \nu_0$. The threshold frequency is characteristic of the material and the condition of its surface; it is independent of the intensity of the light. For example, $\nu_0 = 5.6 \times 10^{14}$ Hz for sodium with a clean surface. The classical wave theory provides no mechanism for a lowest, or threshold, frequency for the emission of electrons.

2. When the frequency of the incident light is above the threshold frequency, $\nu > \nu_0$, electrons are emitted with a distribution of kinetic energies. For electrons emitted from the surface, there is a maximum kinetic energy K_{max} that is independent of the intensity. The maximum value of the kinetic energy is measured by having the anode at a lower potential than the cathode. When electrons leave the cathode and move to a lower potential, they are slowed as their kinetic energy is transformed into electric potential energy. Electrons that are stopped before reaching the anode do not contribute to the current. An electron moving toward the anode with the maximum kinetic energy will be stopped before reaching the anode if the potential difference is adjusted to $\mathcal{V}_s$, the *stopping potential,* which stops all of the emitted electrons. Its value corresponds to the conversion of the maximum kinetic energy K_{max} into electric potential energy $e\mathcal{V}_s = K_{max}$. Contrary to experiment, the wave theory would have the distribution of kinetic energy dependent on intensity.

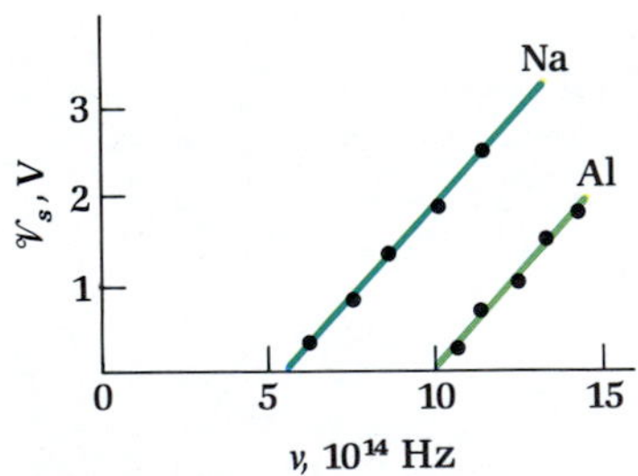

Figure 39-5. The maximum kinetic energy K_{max} of emitted electrons, and the stopping potential $\mathcal{V}_s$, $K_{max} = e\mathcal{V}_s$, depend linearly on the frequency of the incident light. The two lines have the same slope.

3. The maximum kinetic energy depends linearly on the frequency of the incident light. This behavior is shown graphically in Fig. 39-5 for two different materials. Notice that the intercept, which is the threshold frequency, is different for the two materials, but the two lines have the same slope. There is no basis for this kind of behavior in the wave theory.

4. Experiment shows that there is no appreciable time delay between the incidence of light on the surface and the emission of electrons. The wave theory requires that energy be continuously absorbed from the wave by an electron. If the intensity is low, there would be a time lag before the electrons could absorb enough energy to escape from the surface. (See Exercise 18.) The wave theory is completely at odds with the observation that electrons are emitted immediately.

The Photon

Einstein proposed a *corpuscular,* or particle, theory of light: Monochromatic light of frequency ν propagating in vacuum consists of a stream of particles or ***quanta,*** which we now call ***photons.*** Each photon travels at speed c and has a discrete amount or quantum of energy:

Energy of a photon

$$E = h\nu \tag{39-3}$$

where h is Planck's constant. Further, when a photon interacts "one on one" with an electron, the electron acquires all of the energy of the photon, which then exists no longer.

All of the features of the photoelectric effect can be simply explained by using the photon concept. Consider each of the properties listed above that could not be understood with the wave picture of light.

1. To escape from the surface of a material, an electron must gain an energy at least equal to the work function ϕ. If the energy acquired by absorbing a photon of energy $h\nu$ is less than the work function, then the electron will not be emitted. Thus there is a threshold frequency ν_0 such that $h\nu_0 = \phi$. Only for a frequency greater than the threshold frequency, $\nu > \nu_0$, can the electron obtain enough energy from a photon to escape from the surface.

Threshold frequency and work function, $h\nu_0 = \phi$

2. An electron receives a definite amount of energy $h\nu$ by absorbing a photon. It must give up at least an amount equal to the work function ϕ in escaping from the surface. Therefore there is a maximum kinetic energy for the emitted electrons.

3. The maximum kinetic energy K_{max} equals, by conservation of energy, the energy $h\nu$ absorbed from the photon, less the minimum energy ϕ required to escape from the surface. Thus

$$K_{max} = h\nu - \phi \qquad (39\text{-}4)$$

Einstein's photoelectric equation

which is known as ***Einstein's photoelectric equation.*** In terms of the stopping potential $\mathcal{V}_s$, where $K_{max} = e\mathcal{V}_s$, Eq. 39-4 can be expressed as

$$\mathcal{V}_s = \frac{h}{e}(\nu - \nu_0)$$

where $\nu_0 = \phi/h$. This result gives the linear dependence of the stopping potential on frequency. Each line in Fig. 39-5 has a threshold frequency ν_0 determined by the work function of the material. The slope of each line is independent of the material; its value is h/e.

4. Since an electron absorbs a photon and acquires the energy all at once, there need be no appreciable time delay between the incident of light and the emission of electrons. An electron can be emitted immediately.

The experimental features of the photoelectric effect discussed above were not clearly established when Einstein proposed the photon interpretation of light in 1905. The linear dependence of the stopping potential on frequency was confirmed by careful experiments in 1916 by R. A. Millikan (1868–1953). This work helped to gain the general acceptance of the photon picture of light and provided a value for the combination of fundamental constants h/e. Einstein was awarded the Nobel prize in 1921 for his work on the photoelectric effect.

The particlelike photon, the quantum of electromagnetic radiation, carries momentum as well as energy. Since the photon travels at the speed of light, its rest mass is zero. Using Eq. 38-17, $E = \sqrt{p^2c^2 + m^2c^4}$, with $m = 0$, we find that the magnitude of the momentum of the photon is given by $p = E/c$, where $E = h\nu$. Since $E/c = h\nu/c = h/\lambda$, we can write

$$p = \frac{h}{\lambda} \qquad (39\text{-}5)$$

Momentum of a photon

for the magnitude of the momentum of a photon.

EXAMPLE 39-2

Photoelectrons emitted from sodium. Monochromatic light of wavelength 450 nm is incident on a clean Na surface of work function $\phi = 3.7 \times 10^{-19}$ J $= 2.3$ eV. Determine *(a)* the energy of a photon of this light, *(b)* the maximum kinetic energy of emitted electrons, *(c)* the threshold frequency for Na, and *(d)* the magnitude of the momentum of a photon in the incident light.

Solution. *(a)* From Eq. 39-3, the energy of a photon is $E = h\nu$. Since $\nu = c/\lambda = 6.7 \times 10^{14}$ Hz,

$$E = (6.63 \times 10^{-34}\ \text{J}\cdot\text{s})(6.7 \times 10^{14}\ \text{Hz})$$
$$= 4.4 \times 10^{-19}\ \text{J} = 2.8\ \text{eV}$$

Notice that the photon energy is greater than the work function, so electrons are emitted from the surface.

(b) Einstein's photoelectric equation, Eq. 39-4, expresses conservation of energy for the interaction of an electron and a photon. An electron absorbs a photon and acquires its energy $h\nu$. If the electron gives up the minimum energy ϕ in escaping from the surface, it emerges with a maximum kinetic energy

$$K_{\text{max}} = h\nu - \phi = 2.8\ \text{eV} - 2.3\ \text{eV} = 0.5\ \text{eV}$$

(c) The threshold frequency is related to the work function by $h\nu_0 = \phi$, or

$$\nu_0 = \frac{\phi}{h} = \frac{3.7 \times 10^{-19}\ \text{J}}{6.63 \times 10^{-34}\ \text{J}\cdot\text{s}} = 5.6 \times 10^{14}\ \text{Hz}$$

(d) The magnitude of the momentum of a photon for the incident light is given by Eq. 39-5:

$$p = \frac{h}{\lambda} = \frac{6.63 \times 10^{-34}\ \text{J}\cdot\text{s}}{450\ \text{nm}}$$
$$= 1.5 \times 10^{-27}\ \text{kg}\cdot\text{m/s}$$

SELF-TEST 39-2. For the arrangement described in the above example, what stopping potential $\mathscr{V}_s$ is required to keep any emitted electrons from reaching the anode? ***ANSWER:*** 0.5 V.

39-4 PHOTONS AND ELECTRONS

In the photoelectric effect, the interaction of light and matter occurs as electron-photon encounters: A photon is absorbed by an electron, and the electron acquires the energy of the photon. If the photon energy $h\nu$ is greater than the work function of the material, then an electron, after absorbing a photon, has enough energy to escape from the surface. There are other processes that clearly involve the interaction of individual electrons and photons. We shall consider one of these to illustrate further the particle nature of light.

The Compton Effect

The scattering of an electromagnetic wave by a charged particle, such as an electron in an atom, had been described in classical terms in the following way: The incident plane wave, with electric and magnetic fields oscillating at frequency ν, exerts a sinusoidal driving force on the particle and causes it to oscillate at that same frequency. A charged particle oscillating with frequency ν emits electromagnetic radiation of that frequency, and this outgoing spherical wave is the scattered wave. Thus an incident wave of frequency ν is scattered by the charged particle, and the scattered wave has the same frequency as the incident wave.

In 1923 A. H. Compton (1892–1962) found that the frequency of some of the x-rays scattered by electrons was not the same as the frequency of the incident x-rays. This frequency change on scattering is called the ***Compton effect.*** Compton showed that the interaction can be interpreted as a collision of two particles, a photon and an electron. The final state is determined by applying the conservation laws for energy and momentum. The initial state is shown schematically in Fig. 39-6*a*. The initial momentum of the photon is along the x axis of a coordinate system, and the electron is initially at rest. As a result of the collision, the electron recoils with momentum **p** at angle ξ, as shown in Fig. 39-6*b*, and the photon scatters at angle θ from the incident direction. Since the electron has received kinetic energy in the collision, the energy of the scattered photon $h\nu'$ must be less than the energy of the incident photon $h\nu$. Relativistic expressions should be used for the electron's energy (Eq. 38-17) because its speed may be comparable to the speed of light.

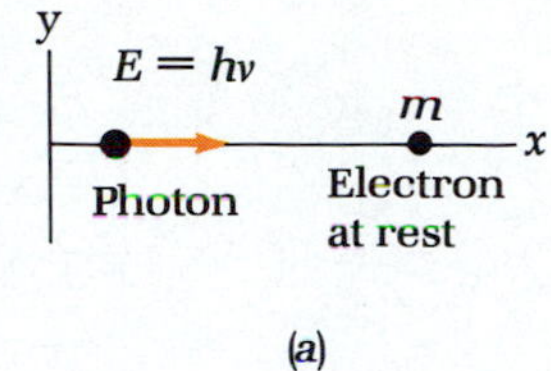

(a)

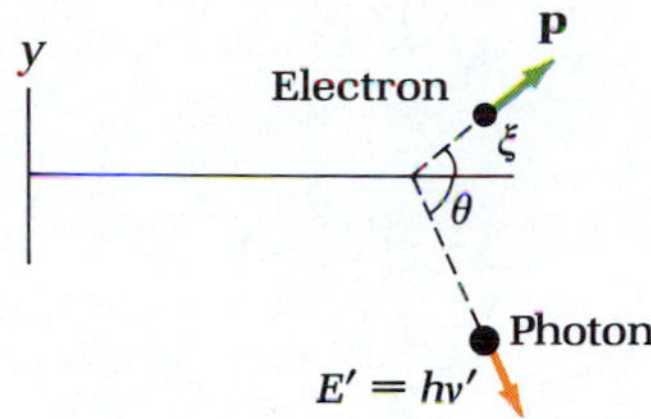

(b)

Figure 39-6. *(a)* Initial state: A photon of energy $h\nu$ and momentum $(h/\lambda)\mathbf{i}$ is incident on an electron at rest. The electron's energy is mc^2 and its momentum is zero. *(b)* Final state: The electron recoils with momentum **p** and energy $\sqrt{p^2c^2 + m^2c^4}$, and the scattered photon has energy $h\nu'$ and momentum of magnitude h/λ'.

The initial energy of the system is the photon's energy $h\nu$ plus the electron's rest-mass energy mc^2, and the final energy of the system is the photon's new energy $h\nu'$ plus the electron's total energy $\sqrt{p^2c^2 + m^2c^4}$. Thus conservation of energy gives

$$\frac{hc}{\lambda} + mc^2 = \frac{hc}{\lambda'} + \sqrt{p^2c^2 + m^2c^4} \qquad (39\text{-}6)$$

where we have used $\nu = c/\lambda$ and $\nu' = c/\lambda'$ because we are seeking an expression for the wavelengths. The system's initial x momentum is entirely due to the photon with magnitude h/λ. After the collision the photon is scattered at an angle θ so that its x momemtum is $(h/\lambda') \cos\theta$, and the electron moves off at angle ξ so that its x momentum is $p \cos\xi$. Conservation of momentum along the x axis gives

$$\frac{h}{\lambda} = \frac{h}{\lambda'} \cos\theta + p\cos\xi \qquad (39\text{-}7)$$

Similarly, conservation of momentum along the y axis gives

$$0 = p \sin\xi - \frac{h}{\lambda'} \sin\theta \qquad (39\text{-}8)$$

These three equations (Eqs. 39-6, 39-7, and 39-8) from conservation of energy and momentum can be used to find an expression for the photon's wavelength change $(\lambda' - \lambda)$ in terms of the scattering angle θ by eliminating p and ξ. We omit the details of solving the equations here (see Prob. 4) and just give the result:

Change in wavelength for Compton scattering

$$\lambda' - \lambda = \frac{h}{mc}(1 - \cos\theta) \qquad (39\text{-}9)$$

The wavelength λ' of the scattered radiation depends on the scattering angle θ. For the forward or incident direction ($\theta = 0$), the factor $(1 - \cos 0) = 0$, and there is no change in wavelength. At other angles there is a shift $\Delta\lambda = \lambda' - \lambda$ toward longer wavelengths. The maximum shift $\Delta\lambda_{max}$ occurs for backscattering ($\theta = 180°$) and from Eq. 39-9 $\Delta\lambda_{max} = (h/mc)[1 - (-1)] = 2h/mc$. Notice that the quantity h/mc, with dimensions of length, determines the shift in wavelength for scattering at a given angle. This length depends on the mass of the charged particle and is called the *Compton wavelength* for this type of particle. The Compton wavelength for an electron is $h/m_ec = 2.43 \times 10^{-12}$ m. Since this value is small compared with wavelengths in the visible spectrum (around 500 nm), the Compton effect is easily observable only for much shorter wavelengths, such as those in the x-ray region of the spectrum.

Compton wavelength

EXAMPLE 39-3

Compton scattering at 90°. X-rays of wavelength 1.14×10^{-11} m scatter from free electrons in a metal. *(a)* Determine the wavelength of x-rays scattered at $\theta = \pi/2$ from the incident direction. Assume that each electron is at rest initially. *(b)* What is the recoil kinetic energy of an electron?

Solution. *(a)* From Eq. 39-9, the change in wavelength is $\Delta\lambda = (h/mc)[1 - \cos(\pi/2)] = h/mc$. For an electron, the Compton wavelength is $h/mc = 2.43 \times 10^{-12}$ m. The wavelength of the scattered x-rays at this angle is

$$\lambda' = \lambda + \Delta\lambda = \lambda + \frac{h}{mc} = 1.14 \times 10^{-11}\text{ m} + 2.43 \times 10^{-12}\text{ m}$$
$$= 1.38 \times 10^{-11}\text{ m}$$

(b) The kinetic energy K received by the electron equals the difference in the energies of the incident and scattered photons. Thus

$$K = h\nu - h\nu' = hc\left(\frac{1}{\lambda} - \frac{1}{\lambda'}\right) = 3.07 \times 10^{-15}\text{ J}$$
$$= 19.2\text{ keV}$$

SELF-TEST 39-3. For the situation in the above example, determine the photon's momentum magnitude *(a)* before the encounter and *(b)* after the encounter. ***ANSWERS:*** *(a)* 5.8×10^{-23} kg·m/s; *(b)* 4.8×10^{-23} kg·m/s.

39-5 LINE SPECTRA

To see the structure or inner workings of a device such as a mechanical clock or an electric motor, we expose the object to light. With a good microscope we can resolve details of the structure that approach the limit of resolution, which is determined by the wavelength range of visible light, about 400 to 750 nm (Chap. 37). We cannot directly see the structure of an atom because its size, around 0.1 nm, is less than the wavelength of visible light. Nevertheless, the structure of an atom or molecule can be investigated through the interaction of light and matter. That is, atoms emit and absorb light and do so in a way that depends on their structure.

Continuous and line spectra

If the light from a cavity is analyzed with a spectrometer, as shown schematically in Fig. 39-2, then the intensity distribution contains all wavelengths in the range dispersed by the spectrometer. The spectrum is a ***continuous spectrum.*** In contrast, the light emitted from a "neon" sign contains intense light at a discrete set of wavelengths. Each wavelength from this source forms an image on a photographic plate of the narrow entrance slit of the spectrometer, so that the plate contains a sequence of lines (Fig. 39-7). Because of this appearance, a spectrum with a discrete set of wavelengths is called a ***line spectrum.***

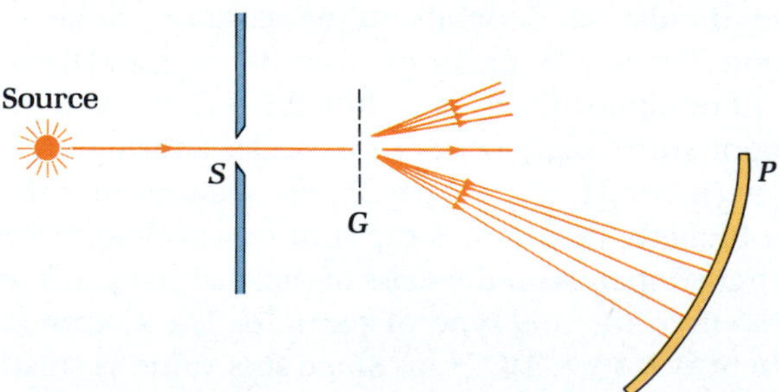

Figure 39-7. Light from a source passes through a slit S and is separated into its component wavelengths by a diffraction grating G. The position of each wavelength is recorded on the photographic plate P.

Emission and absorption spectra

A spectrum of the light emitted by a source, whether a continuous or a line spectrum, is called an ***emission spectrum.*** Figure 39-8 shows the visible portions of some emission spectra. An absorption spectrum can also show abrupt intensity changes for a discrete set of wavelengths, so it is a line spectrum. The ***absorption spectrum*** for a substance is obtained by passing light with a continuous spectrum through a substance. The transmitted light is analyzed with a spectrometer. Those wavelengths strongly absorbed by the substance are shown by their reduced intensity.

Whether an emission or an absorption spectrum, the line spectrum for a sub-

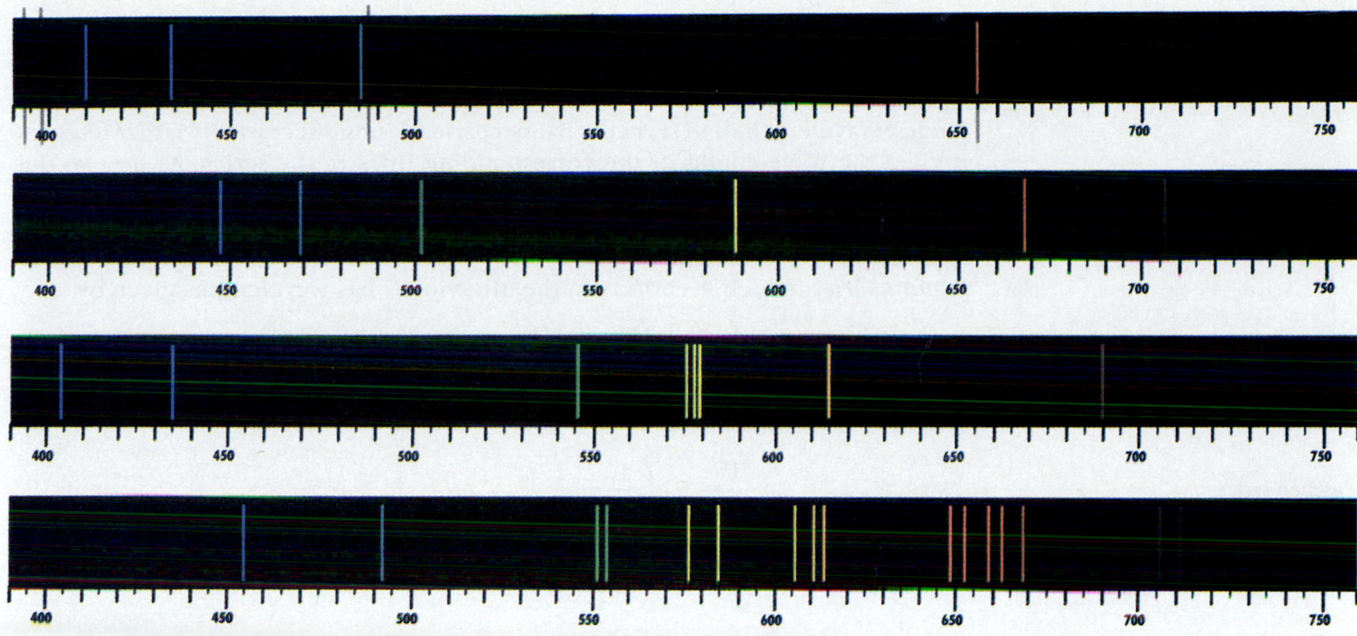

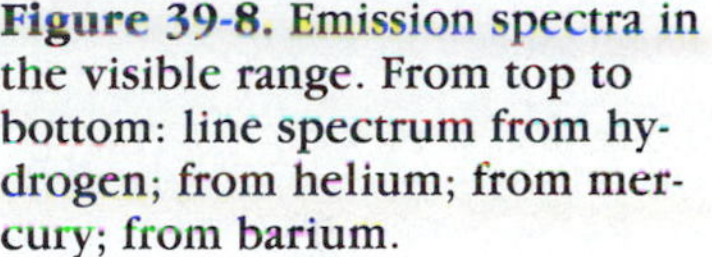
Figure 39-8. Emission spectra in the visible range. From top to bottom: line spectrum from hydrogen; from helium; from mercury; from barium.

stance is unique. It is analogous to a fingerprint, and can be used for identifying the presence of that substance in a sample. A spectrum for a substance is characteristic of its atomic or molecular structure and provides information about that structure. A successful model of an atom must account for its characteristic emission and absorption spectra.

The lightest atom, and presumably the simplest, is hydrogen. A portion of its emission spectrum is shown in Fig. 39-9. The four labeled lines—H_α, H_β, H_γ, and H_δ—are in the visible part of the spectrum. The remaining lines shown are in the ultraviolet. Notice the apparent regularity of the spacing of the lines and how the lines come to a wavelength limit at around 364.6 nm. They are members of a series of lines called the ***Balmer series.***

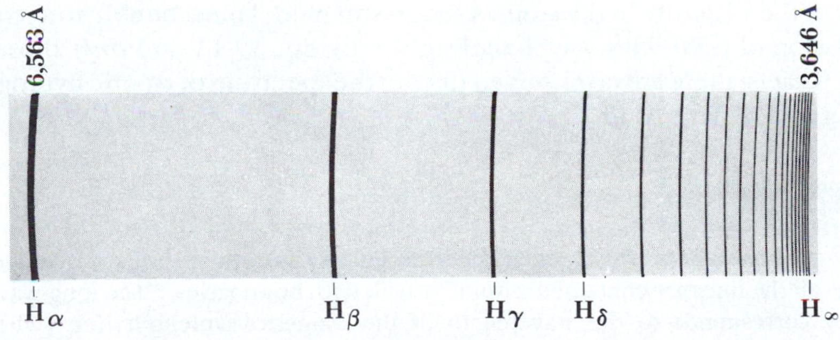

Figure 39-9. The Balmer series of spectral lines of hydrogen are shown. The four lines on the left are in the visible portion of the spectrum. The remaining lines are in the ultraviolet. The wavelengths are in angstrom (Å) units; 10 Å = 1 nm.

Johann Balmer (1825–1898), a Swiss schoolteacher, discovered in 1885 a simple mathematical formula for the wavelengths of the visible lines in the hydrogen spectrum. The Balmer formula had no theoretical basis but was instead an empirical relation that correctly described the regularity in the spectrum. In addition to describing the four visible lines known to Balmer, the formula also gave the wavelengths of lines in the ultraviolet part of the series that were observed thereafter. It is convenient to write the Balmer formula in a form that gives the reciprocal of the wavelength:

$$\frac{1}{\lambda} = R_H \left(\frac{1}{2^2} - \frac{1}{n^2} \right) \qquad (n = 3, 4, 5, \ldots) \qquad (39\text{-}10)$$

Balmer series formula

where $R_H = 1.097 \times 10^7\ \text{m}^{-1}$ is the ***Rydberg constant*** for hydrogen, named after J. R. Rydberg (1854–1919). If $n = 3$ is inserted into the Balmer formula, the calculated wavelength is $\lambda = 656.3$ nm. This is the wavelength of the first, or longest wavelength, line H_α in the Balmer series. Using successively larger integers n gives the wavelengths of the corresponding lines in the series. As $n \to \infty$, the Balmer formula gives the series limit, $\lambda = 364.6$ nm.

Series of spectral lines of hydrogen were later discovered in other parts of the spectrum. Each series is described by a formula similar to the Balmer formula. The *Lyman series,* which is entirely in the ultraviolet, has wavelengths given by

Lyman series

$$\frac{1}{\lambda} = R_H\left(\frac{1}{1^2} - \frac{1}{n^2}\right) \qquad (n = 2, 3, 4, \ldots)$$

There are several series that contain lines in the infrared with wavelengths given by

Paschen series

$$\frac{1}{\lambda} = R_H\left(\frac{1}{3^2} - \frac{1}{n^2}\right) \qquad (n = 4, 5, 6, \ldots)$$

Brackett series

$$\frac{1}{\lambda} = R_H\left(\frac{1}{4^2} - \frac{1}{n^2}\right) \qquad (n = 5, 6, 7, \ldots)$$

Pfund series

$$\frac{1}{\lambda} = R_H\left(\frac{1}{5^2} - \frac{1}{n^2}\right) \qquad (n = 6, 7, 8, \ldots)$$

Notice that the Rydberg constant R_H appears in all of these equations. From the common structure of the formulas for these series, it is evident that they can all be expressed by the single, more general expression:

Rydberg formula

$$\frac{1}{\lambda} = R_H\left(\frac{1}{n_f^2} - \frac{1}{n_i^2}\right) \qquad (39\text{-}11)$$

The integer n_f determines the series: $n_f = 1$ for the Lyman series, $n_f = 2$ for the Balmer series, ... For a given n_f, the values $n_i = n_f + 1$, $n_f + 2$, $n_f + 3$, . . . give the wavelengths of successive lines in the series.

It is remarkable that the simple expression in Eq. 39-11 correctly gives the wavelength of each line in the spectrum of hydrogen. It provides a stringent test for a model of the structure of the atom. A successful model must be able to account for the emission of light with wavelengths given by Eq. 39-11 and *only* those wavelengths. That is, there are no observed lines in the spectrum of atomic hydrogen that are not given by Eq. 39-11.

EXAMPLE 39-4

Wavelength boundaries for a spectral series. For any particular series of hydrogen spectral lines, all the lines are contained within "wavelength boundaries." The long-wavelength boundary corresponds to the wavelength of the longest-wavelength line (where $n_i = n_f + 1$) and the short-wavelength boundary corresponds to the series limit (where $n_i = \infty$). Determine these wavelength boundaries for the Balmer series and compare them with Fig. 39-9.

Solution. For the Balmer series, $n_f = 2$, so the longest-wavelength spectral line is given by

$$\frac{1}{\lambda} = (1.097 \times 10^7\ \text{m}^{-1})\left(\frac{1}{2^2} - \frac{1}{3^2}\right)$$

or

$$\lambda = 656.3\ \text{nm}$$

The wavelength of the series limit is given by

$$\frac{1}{\lambda} = (1.097 \times 10^7\ \text{m}^{-1})\left(\frac{1}{2^2} - \frac{1}{\infty^2}\right)$$

or

$$\lambda = 364.6 \text{ nm}$$

All the lines of the Balmer series are contained within these boundaries, as shown in Fig. 39-9.

SELF-TEST 39-4. Determine *(a)* the long-wavelength boundary and *(b)* the short-wavelength boundary for the Paschen series. All the spectral lines in this series are contained in the infrared. ***ANSWERS:*** *(a)* 1875 nm; *(b)* 820.4 nm.

39-6 THE BOHR MODEL OF HYDROGEN

In the period 1909–1913, two profound changes occurred in the accepted "picture" of the structure of an atom. It was (and is) widely believed that atoms contain electrons and that the size of an atom is around 0.1 nm. To be neutral, an atom must have, in addition to the negatively charged electrons, an equal amount of positive charge. In the "plum-pudding" model of J. J. Thomson (1856–1940), a continuous, positively charged medium extended throughout the atom with pointlike electrons arranged inside.

Rutherford's Planetary or Nuclear Model of an Atom

The first change in the picture of the atom concerned the distribution of positive charge and was developed by Ernest Rutherford (1871–1937) as a result of experiments performed under his direction. In these experiments, positively charged α particles (doubly ionized He ions from a naturally occurring radioactive source) were scattered by a thin gold foil. Rutherford was surprised to find that some of the particles were scattered backward from the foil. He later wrote: "It was almost as incredible as if you fired a 15-inch shell at a piece of tissue paper and it came back and hit you." Rutherford argued that the backscattering could occur only if the positive charge of the atom were concentrated in a region very much smaller than the size of that atom. It turns out that this region has a linear dimension of less than 10^{-14} m, much smaller than the atom's size of 10^{-10} m (see Prob. 6).

Rutherford proposed a ***nuclear*** model of the atom in which the positive charge and most of the mass of an atom is confined to a small ***nucleus.*** The electric force exerted by the positively charged nucleus on each electron is given by Coulomb's law. Thus the force on an electron is directed toward the nucleus, and its magnitude varies inversely with the square of the electron-nucleus separation. In analogy with the equivalent gravitational problem—planets attracted to a massive sun by the inverse-square gravitational force—the atom was pictured as a miniature solar system with electrons in orbit about the nucleus. The size of the atom was determined by the orbital motion of the electrons about the nucleus.

Rutherford's nuclear, or planetary, model of the atom

Consider the planetary model for the hydrogen atom. Hydrogen consists of a nucleus of charge $+e$ and a circulating electron of charge $-e$. The nucleus of hydrogen is called a proton, and its mass is nearly 2000 times the mass m of an electron. This means that, to a good approximation, we can neglect the motion of the proton. For simplicity, we assume that the electron is in a circular orbit of radius r about a fixed nucleus (Fig. 39-10). Also, we assume that the electron's speed v is much less than the speed of light so that we may use nonrelativistic mechanics. Thus the electron's acceleration magnitude is that of an object in uniform circular motion: $a = v^2/r$. The magnitude of the centripetal force that keeps the electron in its orbit is given by Coulomb's law as $e^2/4\pi\epsilon_0 r^2$. Newton's second law, $\Sigma\mathbf{F} = m\mathbf{a}$, applied to the electron is

$$\frac{e^2}{4\pi\epsilon_0 r^2} = m\frac{v^2}{r} \tag{39-12}$$

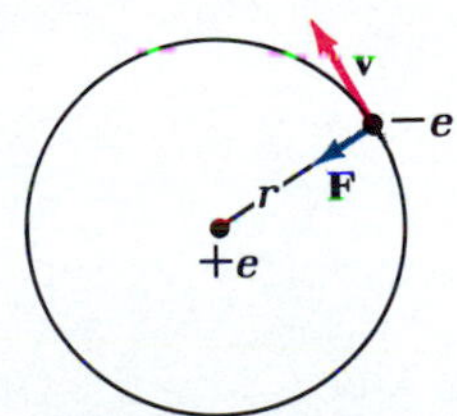

Figure 39-10. In a planetary model of the hydrogen atom, an electron is in a circular orbit about a proton.

If we multiply the right-hand side of this equation by $\frac{1}{2}r$, it becomes the kinetic energy: $\frac{1}{2}r(mv^2/r) = \frac{1}{2}mv^2 = K$. This gives the electron's kinetic energy K in terms of its orbital radius r:

$$K = \frac{e^2}{8\pi\epsilon_0 r} \tag{39-13}$$

The electric potential energy U of a pair of particles of charge $+e$ and $-e$ separated by a distance r is

$$U = -\frac{e^2}{4\pi\epsilon_0 r} \tag{39-14}$$

From Chap. 22, the reference level for this potential energy corresponds to the two particles being at an infinite separation; that is, $U = 0$ when $r = \infty$. This means that the potential energy is never positive. The farther the electron is from the nucleus, the greater is its potential energy, or the less negative is its potential energy.

The electron's mechanical energy, which we shall refer to as its energy, is $E = K + U$. Adding Eqs. 39-13 and 39-14, we find

$$E = -\frac{e^2}{8\pi\epsilon_0 r} \tag{39-15}$$

The energy turns out to be simply one-half of the potential energy. The energy is never positive; for larger separation distances r, the energy becomes larger, or less negative. Keep in mind that Eq. 39-15 is valid only for an electron in circular orbit because Eq. 39-13 was used in its development and Eq. 39-13 is valid only for a circular orbit. An electron can have a positive energy, but then it is not in circular orbit. An electron with positive energy corresponds to an ionized atom, and we are not presently concerned with an ionized atom. We are interested only in so-called *bound states,* where the electron remains associated with a single nucleus and the energy of the system is negative.

Using Eq. 39-15, we can discuss a serious difficulty with the planetary model of an atom. In classical electromagnetism, an accelerated charge emits electromagnetic radiation. Since an electron in orbit is accelerated ($a = v^2/r$), it should radiate electromagnetic energy. From conservation of energy, the atom would lose energy and, according to Eq. 39-15, the radius of the orbit would decrease as the energy decreases. Thus, in a classical planetary model, an electron orbit would be unstable, with the electron spiraling into the nucleus as it radiates electromagnetic energy. Contrary to the classical model, this catastrophic process does not occur: Stable atoms do exist.

The classical planetary model predicts an unstable atom.

EXAMPLE 39-5

Ionization energy of hydrogen. At the time Rutherford proposed the planetary model, it was known that the hydrogen atom has a diameter of about 0.1 nm, and it was known that the ionization energy of hydrogen is $I = 13.6$ eV. The ionization energy is the minimum energy required to remove the atom's electron ($r = \infty$). From our reference level for potential energy in Eq. 39-14, this corresponds to $U = 0$, and since this is the minimum energy, $K = 0$. Therefore the ionization energy is the energy required to raise the atom's energy from $E = -e^2/8\pi\epsilon_0 r$ to $E = 0$. Assume that the diameter of the orbit in the planetary model for hydrogen is 0.100 nm and calculate the ionization energy.

Solution. From the discussion above, the planetary model gives the ionization energy of hydrogen as

$$I = e^2/8\pi\epsilon_0 r$$

Since the atom's diameter is 0.100 nm, its radius is 0.050 nm. Substituting this value and the constants into this formula gives

$$I = e\,\frac{(9.0 \times 10^9\ \text{N}\cdot\text{m}^2/\text{C}^2)(1.6 \times 10^{-19}\ \text{C})}{2(0.050\ \text{nm})} = 14\ \text{eV}$$

As you can see, this is quite close to the measured value of I for hydrogen.

SELF-TEST 39-5. Use the planetary model and an orbital radius of 0.050 nm to determine (a) K, (b) U, and (c) E for hydrogen. State these energies in electron volts. *ANSWERS:* (a) 14 eV; (b) −29 eV; (c) −14 eV.

The Bohr Model

In 1913 Niels Bohr (1885–1962) modified the planetary model for hydrogen. This modified model contained features that conflicted with some of the principles of classical physics. It represented the beginning of a quantum theory of matter. To circumvent the classical problem of an electron radiating and spiraling into the nucleus, Bohr postulated the existence of certain *stationary* states of the atom. Such states are stationary in the sense that no radiation is emitted when an electron is in one of these states. That is, even though the electron in a stationary state is accelerating, it does not emit radiation.

An electron in a stationary state does not emit light.

Bohr recognized that discrete emission spectra demanded a new, nonclassical condition—that the mechanical properties of the atom must be discrete, or quantized. The quantization condition can be stated simply for the angular momentum of magnitude L:

$$L=\frac{h}{2\pi}n \qquad (n=1, 2, 3, \ldots) \tag{39-16}$$

Angular momentum is quantized.

where h is Planck's constant and n is a positive integer. Thus the Bohr model of hydrogen allows the electron to have only quantized values of angular momentum. The integer n is called the ***quantum number,*** and each integral value of the quantum number corresponds to a stationary state, or quantum state.

Quantum number n

Now we use the quantization condition for L in Eq. 39-16 to show how it leads to the quantization of other mechanical properties of the atom. From Sec. 13-2, an object of mass m and speed v in a circular orbit of radius r has an angular momentum of magnitude $L = mvr$. Applying Bohr's quantization condition to an electron in circular orbit, we have

$$mvr=\frac{nh}{2\pi} \tag{39-17}$$

Multiplying Eq. 39-12 by r^2, we find

$$mv^2r=\frac{e^2}{4\pi\epsilon_0} \tag{39-18}$$

If we divide Eq. 39-18 by Eq. 39-17, we arrive at the quantization condition for the electron's orbital speed:

$$v_n=\frac{e^2}{2h\epsilon_0}\left(\frac{1}{n}\right) \qquad (n=1, 2, 3, \ldots) \tag{39-19}$$

We use a subscript n on the symbol for the orbital speed to emphasize that the speed is quantized. The speed is quantized because it depends on n, and n can take on integer values only. It is common practice to label quantized variables with a quantum-number subscript.

Now we solve Eq. 39-17 for the orbital radius r,

$$r=nh/2\pi mv$$

and substitute v_n from Eq. 39-19 for the speed v. This gives the expression for the radius r_n of each quantized orbit:

$$r_n=\frac{\epsilon_0 h^2}{\pi m e^2}n^2 \qquad (n=1, 2, 3, \ldots) \tag{39-20}$$

The orbital radius increases as the quantum number increases. The smallest radius corresponds to $n=1$; it is called the ***Bohr radius*** and is denoted by the symbol a_0:

Bohr radius a_0

$$a_0 = \epsilon_0 h^2/\pi m e^2 = 0.053 \text{ nm}$$

The Bohr radius is an important quantity because it characterizes the size of atoms. In terms of the Bohr radius, the radii of the orbits are $r_1 = a_0$, $r_2 = 4a_0$, $r_3 = 9a_0$, and so on.

The quantized energy is found by substituting r_n for r in Eq. 39-15:

Energy of stationary states of hydrogen

$$E_n = -\frac{me^4}{8\epsilon_0^2 h^2}\left(\frac{1}{n^2}\right) \qquad (n = 1, 2, 3, \ldots) \tag{39-21}$$

The quantized energy for each stationary state is negative, and the energy increases as the quantum number increases. The lowest energy state, which is called the ***ground state,*** corresponds to $n = 1$:

$$E_1 = -me^4/8\epsilon_0^2 h^2 = -2.17 \times 10^{-18} \text{ J} = -13.6 \text{ eV}$$

States with higher energies are called ***excited states,*** and their energies are $E_2 = E_1/4 = -3.40$ eV, $E_3 = E_1/9 = -1.51$ eV, $E_4 = E_1/16 = -0.850$ eV, and so on.

The discrete nature of the energy for the stationary states is conveniently displayed on an *energy-level diagram.* A simple energy-level diagram for hydrogen is shown in Fig. 39-11. Each of the horizontal lines represents an allowed value of the energy of a stationary state. The vertical line shows the energy scale, with the zero of energy corresponding to an ionized atom. Notice that the quantum number n labels the energy of a stationary state.

Energy-level diagram

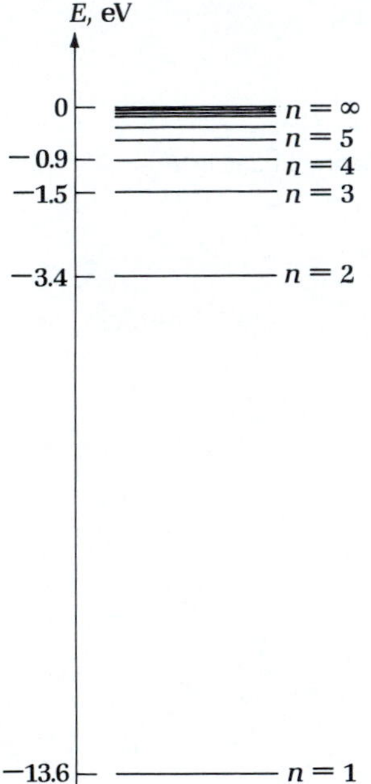

Figure 39-11. An energy-level diagram for hydrogen shows the energy of each of the stationary states.

Line Spectra from the Bohr Model

According to Bohr, an atom does not radiate when in a stationary state. Radiation is emitted, in the form of a photon, only when an atom undergoes a transition from one stationary state to another one of lower energy. To see how this idea results in the spectrum of hydrogen, consider an atom initially in a state with quantum number n_i. The energy of this state is given by inserting n_i for n in Eq. 39-21. Suppose the atom makes a transition to a state of lower energy with quantum number n_f, so that $n_f < n_i$. If a photon of energy $h\nu$ is emitted in the transition, then by conservation of energy, $E_i = E_f + h\nu$, or

$$h\nu = E_i - E_f \tag{39-22}$$

Since $\nu = c/\lambda$, Eq. 39-22 can be solved for $1/\lambda$ and compared with the formula containing the Rydberg constant in Eq. 39-11. As shown in Exercise 39, if the energies E_i and E_f are expressed in terms of n_i and n_f, then Eq. 39-22 can be rewritten as

$$\frac{1}{\lambda} = \frac{me^4}{8\epsilon_0^2 h^3 c}\left(\frac{1}{n_f^2} - \frac{1}{n_i^2}\right) \tag{39-23}$$

From Eq. 39-11, the Rydberg constant R_H is identified with the combination of fundamental constants that multiplies the term in parentheses in Eq. 39-23. Thus the Bohr model gives a theoretical basis to the Rydberg constant and

$$R_H = \frac{me^4}{8\epsilon_0^2 h^3 c} \tag{3-24}$$

Each of the lines in the spectrum of hydrogen can be associated with a transition between two stationary states. These transitions are conveniently displayed on the energy-level diagram in Fig. 39-12. The horizontal lines representing the energy levels have been extended so that a number of transitions can be shown. The transitions are arranged according to series. Notice that each line in the Lyman series is due to a transition to the ground state from a state of higher energy. Similarly, each line in the Balmer series corresponds to a transition to a state with

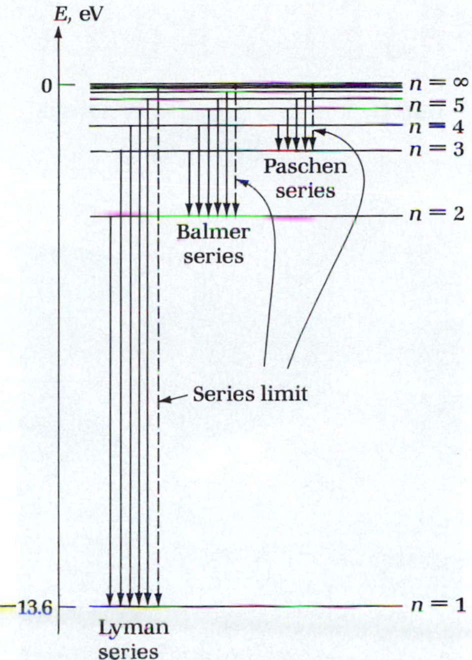

Figure 39-12. Transitions are shown for three of the series in the hydrogen emission spectrum.

$n = 2$ from a state of higher energy. You can also see from the diagram how a series limit occurs because of the concentration of levels near $E = 0$. If the atom is in the ground state ($n = 1$), there is no state of lower energy to which a downward transition can occur. Thus an atom in the ground state is stable.

The Bohr model of hydrogen represented a beginning step in the development of a quantum theory of atomic structure. It contained a mixture of the older classical physics and the newer quantum ideas that were being sorted out during the early part of this century. Some of the features of the Bohr model, such as a set of discrete circular orbits, are no longer taken seriously. However, many of Bohr's concepts are essential parts of the structure of quantum theory.

EXAMPLE 39-6

A hydrogen atom transition. An electron in a hydrogen atom makes a transition between the states with $n_i = 5$ and $n_f = 2$. *(a)* Determine the energy of the initial state and of the final state. *(b)* Determine the energy of the photon and the wavelength of the emitted radiation.

Solution. *(a)* Using Eq. 39-21 with $me^4/(8\epsilon_0{}^2h^2) = 13.6$ eV, we have

$$E_i = \frac{-13.6 \text{ eV}}{n_i^2} = -0.544 \text{ eV}$$

$$E_f = \frac{-13.6 \text{ eV}}{n_f^2} = -3.40 \text{ eV}$$

(b) Conservation of energy, as expressed in Eq. 39-22, gives the energy of the photon:

$$h\nu = E_i - E_f = (-0.54 \text{ eV}) - (-3.40 \text{ eV}) = 2.86 \text{ eV}$$

Solving for the frequency of the radiation, we obtain

$$\nu = \frac{(2.86 \text{ eV})(1.60 \times 10^{-19} \text{ J/eV})}{6.63 \times 10^{-34} \text{ J}\cdot\text{s}} = 6.89 \times 10^{14} \text{ Hz}$$

The wavelength, $\lambda = c/\nu$, is

$$\lambda = \frac{3.00 \times 10^8 \text{ m/s}}{6.89 \times 10^{14} \text{ Hz}} = 435 \text{ nm}$$

SELF-TEST 39-6. Repeat the above example except let $n_i = 4$ and $n_f = 1$. ***ANSWERS:*** *(a)* -0.850 eV and -13.6 eV; *(b)* 12.7 eV and 97 nm.

American Institute of Physics poster celebrating the centennial of Niels Bohr's birth.

COMMENTARY: Neils Bohr and His Model of the Hydrogen Atom

Niels Bohr, born in Copenhagen, Denmark, on October 7, 1885, began his study of physics in 1903 at the University of Copenhagen. Four years later, he was awarded the gold medal of the Royal Danish Academy for his careful work on the surface tension of liquids. For his graduate work, Bohr considered electric, magnetic, and thermal properties of metals and noted the inadequacy of the existing theory of electrons for understanding these properties.

After receiving the doctoral degree in 1911, Bohr obtained an appointment at Cambridge University and worked in the Cavendish Laboratory under the direction of J. J. Thomson. There was a notable and unresolved language problem. He wrote to his mother, "You have no idea of the confusion reigning in the Cavendish laboratory, and a poor foreigner, who does not even know what the different things he cannot find are called, is in a very awkward position. . . ."

Niels Bohr in 1917.

Early in 1912, Bohr left Cambridge and joined the Manchester Laboratory under the leadership of Ernest Rutherford. Only the year before, Rutherford had proposed his nuclear, or planetary, model of the atom. Bohr was interested in the model because of its success in describing alpha-particle scattering, and he wanted to work on the apparent lack of stability in the planetary model.

Although the eventual picture that emerged of the atom is familiar to us and seems obvious, there was no general agreement in 1912 about some of the most elementary features of the atom. For example, it was not at all clear how many electrons were contained in an atom. J. J. Thomson had earlier supposed that each atom contained thousands of electrons arranged in coplanar, revolving rings inside a uniform sphere of positive charge. One of Thomson's major goals was to determine the number of electrons in an atom. His experiments led him to revise the number downward. It was Rutherford's alpha-particle experiments that provided the evidence that the number of electrons was approximately half the mass number of an atom.

Returning to Copenhagen, Bohr attempted to find the final, stable configuration of electrons revolving about a central nucleus. He was not successful, but his calculations did lead to the suggestion that hydrogen should have one electron, helium should have two electrons, lithium should have three electrons, ...

Since he had been concentrating on determining a final, stable electronic configuration, Bohr had not yet attempted to deal with the characteristic spectra emitted by atoms. In February of 1913, a colleague of Bohr's called his attention to Balmer's formula for the spectral lines of hydrogen. Bohr later related, "As soon as I saw Balmer's formula, everything became clear to me." In less than a month, he was able to put together the first of his famous papers on the structure of the atom.

Bohr was still a very young man when he developed his model of atomic hydrogen. He made many important contributions to atomic and nuclear physics. Perhaps even more important was the leadership that he provided as the quantum theory developed over the next 20 years.

For further reading, see *Niels Bohr—His Life and Work as Seen by His Friends and Colleagues,* edited by Stefan Rozental (North-Holland, Amsterdam, 1967) and "Bohr's First Theories of the Atom" by John L. Heilbron in *Physics Today* (vol. 38, no. 10, 1985, pp. 28–36).

SUMMARY

Section 39-1. Interaction of Light and Matter

A surface that is a good absorber (poor reflector) is a good emitter of light, and a surface that is a poor absorber (good reflector) is a poor emitter of light. The power radiated by a surface is given by the Stefan-Boltzmann law:

$$P = e\sigma AT^4$$

The ideal absorber, a blackbody with emissivity $e = 1$, is an ideal emitter. Blackbody radiation can be approximated by radiation emerging from a cavity.

Section 39-2. Cavity Radiation

The energy of an oscillator of frequency ν is quantized;

$$E = nh\nu$$

Spectral radiancy for blackbody radiation is given by the Planck expression,

$$R_\nu = \frac{2\pi h\nu^3}{c^2(e^{h\nu/kT} - 1)} \tag{39-2}$$

Section 39-3. The Photoelectric Effect

Einstein proposed that light consists of a stream of photons; each photon has energy

$$E = h\nu \tag{39-3}$$

In the photoelectric effect, an ejected electron has a maximum kinetic energy:

$$K_{\text{max}} = h\nu - \phi \tag{39-4}$$

The work function ϕ for a material is related to the threshold frequency ν_0 by $\phi = h\nu_0$. A photon for light of wavelength λ has momentum of magnitude

$$p = \frac{h}{\lambda} \tag{39-5}$$

Section 39-4. Photons and Electrons

In the Compton effect, electrons scatter radiation with a change in the wavelength of the radiation that depends on the scattering angle,

$$\lambda' - \lambda = \frac{h}{mc}(1 - \cos\theta) \tag{39-9}$$

Section 39-5. Line Spectra

The spectrum of hydrogen consists of series of lines described by

$$\frac{1}{\lambda} = R_{\text{H}}\left(\frac{1}{n_f^2} - \frac{1}{n_i^2}\right) \tag{39-11}$$

Each pair of integers, n_i and n_f with $n_i > n_f$, corresponds to a line in the spectrum.

Section 39-6. The Bohr Model of Hydrogen

In the Rutherford-Bohr model of the hydrogen atom, the angular momentum and the energy of stationary states are quantized:

$$L = \frac{nh}{2\pi} \tag{39-16}$$

$$E_n = -\frac{me^4}{8\epsilon_0{}^2h^2n^2} \tag{39-21}$$

In a transition between stationary states, a photon is emitted with energy

$$h\nu = E_i - E_f \tag{39-22}$$

QUESTIONS

1. If all objects emit radiation, then why can we not see in the dark? What *is* "the dark"?

2. A surface absorbs more strongly in the red than in the blue part of the spectrum. What would you expect the color of the surface to be if illuminated by *(a)* the sun, *(b)* a red incandescent bulb, *(c)* a blue incandescent bulb?

3. The fraction of light reflected by a surface can depend on the wavelength of the radiation. When you look at your image in a mirror, are all colors faithfully imaged? How can you be sure?

4. Develop an analogy between the quantized energy of an oscillator and the currency of the United States. What is the discrete amount or quantum of money? Is it possible to make change for an item costing $1.50? 1.5¢?

5. The light coming from the surface of the sun approximates blackbody radiation for $T = 6000$ K. Would the temperature of the surface of a red giant star be higher or lower than 6000 K? Explain.

6. As a result of the "big bang," the universe seems to contain background electromagnetic radiation as if it were a cavity. How can a measurement of this radiation be related to the background temperature (about 3 K) of the universe?

7. Use a mirror to look at the pupil (the innermost circle) of your eye, or look at the pupil of a friend's eye. What color is it? Explain.

8. Consider electrons ejected from a surface due to monochromatic incident light. How does the maximum kinetic energy of ejected electrons change if *(a)* the frequency is changed, *(b)* the intensity is doubled, *(c)* the exposure time is doubled?

9. Explain how a circuit similar to that in Fig. 39-4 could be used in a burglar alarm. What are some other applications of the photoelectric effect?

10. A chemical bond can be broken if a sufficient amount of energy is suddenly supplied. It is observed that most plastic storage bags are stable when exposed to ordinary house lights but deteriorate when exposed to sunlight. Explain.

11. Photosynthesis proceeds by chemical reactions that are initiated by the absorption of light. Explain why a plant can thrive if exposed to light containing visible and ultraviolet frequencies but cannot if exposed only to infrared radiation.

12. A beam of monochromatic x-ray photons is scattered by free electrons in a metal foil. Is the frequency of the scattered radiation greater or less than that of the incident radiation? How do the wavelengths compare? Explain.

13. A shift in x-ray wavelength due to Compton scattering by free electrons in a foil can be detected. Would you expect a detectable shift due to Compton scattering by the positively charged ions in the foil? Explain.

14. Which series in the emission spectrum of hydrogen has the highest frequencies? In what part of the spectrum are these lines?

15. The lines in the Lyman series do not overlap those of the Balmer series. Is there a series that overlaps the Paschen series? If so, what series?

16. Which series in the spectrum of hydrogen has lines in the visible part of the spectrum? How many lines are there in the visible spectrum? In what part of the spectrum are the remainder of the lines of this series?

17. The masses of the electron, the alpha particle, and the gold atom are $m_e = 9.1 \times 10^{-31}$ kg, $m_\alpha = 6.4 \times 10^{-27}$ kg, $m_{Au} = 3.3 \times 10^{-25}$ kg. Why could Rutherford neglect the effect on the alpha particle of the electrons in a gold atom while claiming that the positive charge of the atom is confined to a very small region?

18. In the Bohr model, is an electron in a stationary state of hydrogen at rest? Explain.

19. List and explain some similarities and differences between the Bohr model of hydrogen and the orbital motion of a planet about the sun.

20. When an electron undergoes a transition from a state of higher energy to one of lower energy, how is the energy difference accounted for? What is the interpretation if the transition is from lower to higher energy?

21. Consider the transitions shown in Fig. 39-12 for the lines of the Balmer series. Is one electron in one atom responsible for all of these transitions, or are these lines formed by electrons making transitions in many different atoms? Explain.

22. Complete the following table:

Symbol	Represents	Type	SI Unit
R			
R_ν			
R_λ			
h			
$\mathcal{V}_s$	Stopping potential		
K_{max}		Scalar	
R_H			m^{-1}

EXERCISES

Section 39-1. Interaction of Light and Matter

1. A sphere of radius 250 mm has a layer of lampblack on its surface. Determine the radiated power if the surface temperature is *(a)* $T = 300$ K, *(b)* $T = 600$ K, *(c)* $T = 1500$ K.

2. Treating the sun as a blackbody at 6000 K, estimate the intensity of solar radiation incident on the earth. The radius of the sun is 7×10^8 m, and the mean radius of the earth's orbit is 1.5×10^{11} m.

3. A thin-walled Al sphere of radius $R = 280$ mm has a circular opening in its surface of radius $a = 2.8$ mm. The sphere is kept at 1100 K, and the emissivity of its surface under these conditions is $e = 0.32$. *(a)* Determine the energy radiated from the opening in an hour. *(b)* What surface area of the sphere would radiate the same energy in that time interval? *(c)* What differences do you expect in the visual appearances of the opening and the surface?

Section 39-2. Cavity Radiation

4. Show that Planck's constant has dimensions of *(a)* (energy)(time), *(b)* (momentum)(length), *(c)* (angular momentum)(angle).

5. A quantum oscillator has frequency $\nu = 10^{13}$ Hz. *(a)* What is the minimum change in energy of the oscillator? *(b)* For what frequency would the minimum energy change be 1 eV? 1 J?

6. Consider a cavity in the form of a cube of edge 200 mm. *(a)* Using standing waves on a string (Chap. 32) as a guide, estimate the lowest frequency for standing-wave modes in the cavity. *(b)* What is the minimum change in the energy of this mode? *(c)* Compare the answer for part *(b)* with the average energy of a *classical* standing-wave mode at $T = 300$ K.

7. Determine the average energy of a quantum oscillator of frequency 2×10^{12} Hz in equilibrium at *(a)* $T = 0.42$ K, *(b)* $T = 4.2$ K (liquid helium temperature), *(c)* $T = 300$ K, *(d)* $T = 1000$ K.

8. *(a)* For each case in the preceding exercise, compare the average energy of the quantum oscillator with the average energy of a classical oscillator at that temperature. *(b)* Using Eq. 39-1, show that $\langle E \rangle \approx kT$ if $h\nu \ll kT$. Note that $e^x \approx 1 + x$ if $x \ll 1$.

9. The spectral radiancy is often expressed for a differential range $d\lambda$ of wavelength such that $R_\lambda\, d\lambda = -R_\nu\, d\nu$, or $R_\lambda = -R_\nu\, d\nu/d\lambda$, with R_ν given by Eq. 39-2. Determine the expression for R_λ. This is the form used by Planck.

10. *(a)* Obtain Eq. 39-2 by taking 1/4 of the product of Eq. 39-1, the average energy of an oscillator, and $8\pi\nu^2\, d\nu/c^3$, the number of standing-wave modes per unit volume with frequencies in the range $d\nu$. *(b)* Show from Eq. 39-2 that $R_\nu\, d\nu$ has dimensions of power/area.

Section 39-3. The Photoelectric Effect

11. The visible part of the spectrum has wavelengths in the range of 400 to 750 nm. What is the range of energies of photons in this part of the spectrum? Express your answers both in electron-volts and joules.

12. One of the intense spectral lines emitted by a mercury vapor light has a wavelength $\lambda = 546.1$ nm. For light of this wavelength, determine *(a)* the frequency, *(b)* the energy of a photon, *(c)* the magnitude of the momentum of a photon. *(d)* Suppose that a beam of this monochromatic light has intensity 1 W/m². Determine the number of photons per second that pass through a 1-m² area oriented perpendicular to the direction of the beam.

13. The work function for a clean surface of Na is 2.5 eV. *(a)* Determine the photoelectric threshold frequency. *(b)* The surface is illuminated with monochromatic light of wavelength 550 nm. Will electrons be emitted from the surface? Explain.

14. Monochromatic light of frequency 6.77×10^{14} Hz is incident on a Na surface with work function 2.46 eV. *(a)* Determine the maximum kinetic energy of the photoelectrons. *(b)* What potential must be applied between cathode and anode to reduce the photocurrent to zero?

15. In a photoelectric experiment, the stopping potential is determined for each set of frequencies. The data for one run are summarized in the table below. *(a)* Construct a graph similar to that in Fig. 39-5. *(b)* Use the graph to determine a value for the work function of the cathode surface and a value for Planck's constant:

ν, 10^{14} Hz	$\mathcal{V}_s$, V
7.8	0.11
7.9	0.16
8.1	0.25
8.6	0.46
8.7	0.49

16. For the surface described in the previous exercise, determine the stopping potential for electrons emitted when the surface is illuminated with light of wavelength *(a)* 200 nm, *(b)* 400 nm, *(c)* 600 nm.

17. When light of wavelength $\lambda_1 = 620$ nm is incident on a photocell surface, electrons are ejected with a maximum kinetic energy of 0.14 eV. Determine *(a)* the work function and *(b)* the threshold frequency for this surface. What is the maximum kinetic energy of ejected electrons if the surface is illuminated with light of wavelength *(c)* $\lambda_2 = \frac{1}{2}\lambda_1$ and *(d)* $\lambda_3 = 2\lambda_1$?

18. Monochromatic light of intensity 1 W/m² illuminates at normal incidence at 1-cm² surface with work function 3 eV. Suppose that an electron bound to an atom near the surface can continuously absorb all of the energy from the *classical wave*

incident on an area of 1 nm by 1 nm. *(a)* What is the power absorbed by the electron? *(b)* Estimate the time necessary for an electron to absorb enough energy from the classical wave to escape from the surface. *(c)* Using the photon picture, explain how an electron can be ejected from the surface with no noticeable time delay.

19. The surface of a metal alloy is illuminated with 280-nm light in the presence of oxygen. As the surface gradually becomes corroded, the stopping potential changes from 1.3 to 0.7 V. Determine the corresponding changes, if any, in *(a)* the maximum kinetic energy of electrons ejected from the surface, *(b)* the work function, *(c)* the threshold frequency, *(d)* Planck's constant.

20. For monochromatic light of wavelength λ and frequency ν, show that *(a)* the energy of a photon is $E = hc/\lambda$ and *(b)* the magnitude of the momentum of a photon is $p = h\nu/c$. *(c)* What is the value of E/p for a photon?

Section 39-4. Photons and Electrons

21. Show that the meter is the SI unit of the Compton wavelength h/mc for a particle.

22. Determine the Compton wavelength for *(a)* an electron, *(b)* a proton, *(c)* an O_2 molecule, *(d)* a 75-kg student.

23. X-rays of wavelength 71 pm are scattered by free electrons in a metal foil. Determine the wavelength of x-rays Compton-scattered through an angle of *(a)* π rad, *(b)* $\frac{1}{2}\pi$ rad, *(c)* 0.1 rad.

24. When a beam of x-rays is Compton-scattered through 60° by free electrons, the wavelength is measured to be 12.6 pm. Determine the wavelength of *(a)* the incident radiation and *(b)* the radiation scattered through 120°.

25. X-ray wavelengths can be measured in a certain spectrometer with a resolution of 1 pm. An incident x-ray beam of wavelength 71 pm is scattered by free electrons. At what minimum angle of scatter can the Compton-scattered wavelength be resolved from the wavelength of the incident beam?

26. X-rays of wavelength 0.1542 nm are scattered by free electrons. *(a)* Determine the energy of a photon in the incident beam. *(b)* If an electron, initially at rest, scatters a photon through π rad, determine the recoil energy of the electron.

27. An 8.3-MeV gamma ray (electromagnetic radiation from a nuclear process) is Compton-scattered through $\frac{1}{2}\pi$ rad by a free electron initially at rest. Determine *(a)* the energy of the scattered photon and *(b)* the magnitude and direction of the recoil momentum of the electron.

Section 39-5. Line Spectra

28. Calculate *(a)* the wavelength and *(b)* the frequency of each of the four lines of hydrogen that are in the visible part of the spectrum.

29. *(a)* Show that the Balmer series formula can be written in the form $\lambda_m = bm^2/(m^2 - n^2)$, where b is a constant, $n = 2$, and $m = 3, 4, 5, \ldots$. This is the form originally proposed by Balmer. *(b)* Express b in terms of the Rydberg constant R_H. *(c)* What is the numerical value of b?

30. Determine the values of the longest wavelength and the series limit for *(a)* the Lyman series, *(b)* the Balmer series, *(c)* the Paschen series, *(d)* the Brackett series. *(e)* Which, if any, of these series have lines in overlapping regions of the spectrum?

31. Show that the frequency ν_{31} of the hydrogen line for $n_i = 3$, $n_f = 1$ equals the sum $\nu_{32} + \nu_{21}$, where ν_{32} corresponds to $n_i = 3$, $n_f = 2$, and ν_{21} corresponds to $n_i = 2$, $n_f = 1$.

32. One of the lines in the hydrogen spectrum has a frequency $\nu = 1.6 \times 10^{14}$ Hz. Determine the series to which this line belongs and identify the line by giving the values of n_i and n_f.

Section 39-6. The Bohr Model of Hydrogen

33. *(a)* Show that the Bohr radius, $a_0 = (\epsilon_0 h^2/\pi m e^2)$, has dimensions of length. *(b)* Substitute in the values of the constants and determine the value of a_0 to three significant figures.

34. A hydrogen atom is in the state $n = 2$. Assume that the electron is in a circular orbit and determine *(a)* the radius of the orbit, *(b)* the electric potential energy, *(c)* the kinetic energy, *(d)* the total energy of the electron in this orbit.

35. An electron in a hydrogen atom makes a transition from an initial state with $n_i = 5$ to a final state with $n_f = 3$. *(a)* Identify the series and the line for the radiation emitted for this transition. *(b)* Determine the initial and the final energies of the electron. *(c)* Determine the frequency and wavelength of the radiation emitted for this transition.

36. An electron undergoes successive transitions in a hydrogen atom. Initially the electron is in the state $n_a = 6$. In the first transition, to an intermediate state n_b, a photon of energy 1.13 eV is emitted. After the second transition, the electron is in the ground state $n_c = 1$. *(a)* Determine the quantum number n_b for the intermediate state. *(b)* Determine the energy of the photon emitted in the second transition. *(c)* Show these two transitions on an energy-level diagram for hydrogen.

37. *(a)* Evaluate the orbital speed v_1 of the electron in the $n = 1$ quantum state of hydrogen. *(b)* Determine $\gamma = 1/\sqrt{1 - v^2/c^2}$ for this speed. Since this speed is much less than the speed of light, and the orbital speeds for all other quantum states are smaller than v_1, this justifies our use of nonrelativistic formulas for the Bohr model of hydrogen.

38. The mechanical energy, $E = K + U$, of an electron in a circular orbit in the Bohr model of hydrogen is given by Eq. 39-15. Use the radius of the nth Bohr orbit in Eq. 39-20 to obtain the energy of a stationary state of hydrogen, Eq. 39-21.

39. *(a)* Show that Eqs. 39-21 and 39-22 lead to the Balmer formula and express the Rydberg constant in terms of the fundamental constants. *(b)* Substitute in the values of the constants and determine the value of R_H to five significant figures.

40. *(a)* Suppose that the orbital angular momentum of the earth about the sun is quantized as in Eq. 39-17. Assume a

circular orbit of radius 1.5×10^{11} m and estimate the value of n for this orbit. *(b)* Develop the counterpart of Eq. 39-20 by replacing the Coulomb force of the proton on the electron by the gravitational force of the sun on the earth. *(c)* Determine the radius of an orbit with $n = 1$. Is this a physically meaningful value?

41. Consider an electron in the ground state of a hydrogen atom. It can undergo a transition by absorbing a photon. *(a)* Develop an expression, similar to the Rydberg formula for emission, for the inverse wavelengths $(1/\lambda)$ ***absorbed*** by atomic hydrogen. *(b)* Show these transitions on an energy-level diagram. *(c)* To which series in the ***emission*** spectrum of hydrogen does this absorption series correspond? *(d)* Explain why lines corresponding to the other emission series are not readily observed in the absorption spectrum.

42. A large number of hydrogen atoms are initially in the state with $n = 6$. Consider all possible combinations of transitions originating from atoms in this state as the electrons undergo radiative transitions. Be sure to include transitions to and from intermediate states. *(a)* Show each of these transitions on an energy-level diagram. *(b)* Identify the series and line in the emission spectrum of hydrogen that corresponds to each transition. *(c)* What is the total number of distinct transitions in this sequence?

Additional Exercises

43. Estimate the power radiated by your body. Assume that it acts like a blackbody and that it stays at 37°C. Remember the Stefan-Boltzmann law.

44. If the earth had no atmosphere and had $e = 1$, it would absorb energy from the sun and reradiate it into space. In equilibrium, the energy absorbed and the energy radiated would be equal. What would the temperature of the earth be if this were true? Note that the earth absorbs on one side and radiates in all directions. (See Exercise 2.)

45. A tungsten filament in a bulb is 3 mm in diameter and 10 mm long and has an emissivity of essentially 1. It is kept at 2500 K. What is the bulb's power in watts?

46. A sensitive, dark-adapted eye has a quantum efficiency of about 1/6; that is, it takes about six photons to cause a photoreceptor to detect the presence of light. An eye collects light from an area of about 1 cm². Estimate the distance at which you could barely see a 10-mm-radius black sphere kept at a temperature of 6000 K. Assume the eye collects photons in the range 500 to 600 nm and that R_λ is constant at the value for 500 nm (see Fig. 39-3*a*).

47. *(a)* How hot would the surface of a star have to be so that the average thermal energy $\langle E \rangle \approx kT$ was enough to promote electrons in hydrogen to the first excited state? *(b)* Why would a star have to be approximately this hot before the Balmer absorption lines would be seen in the spectrum of the star?

48. The work function of tungsten is about 4.5 eV. *(a)* What is the stopping potential for tungsten? *(b)* How hot must a tungsten filament be to have the average energy of an electron in the tungsten be 0.1 times the work function (and thus have a small chance of being emitted)?

49. A photon of yellow light ($\lambda = 550$ nm) is Compton-scattered through π rad by a free electron. Could you notice the color change of the photon? If such π-rad scattering caused a 55-nm change in wavelength, what would the mass of the scattering object have to be?

50. An 83-MeV gamma ray photon is Compton-scattered through π rad by a free proton initially at rest. Determine *(a)* the energy of the scattered photon and *(b)* the magnitude of the recoil momentum of the proton.

PROBLEMS

1. ***Rate of emission equals rate of absorption.*** Consider an object in thermal equilibrium with its surroundings at temperature T, as shown in Fig. 39-13. Prove that the rate at which the object emits radiation equals the rate at which it absorbs radiation. The proof consists of assuming that the rates are unequal and showing that this assumption would lead to a violation of the second law of thermodynamics. The space between the object and the surrounding walls is evacuated so that there is no convection or conduction of heat. How can the proof be extended to show that the rates of emission and absorption are the same at each frequency?

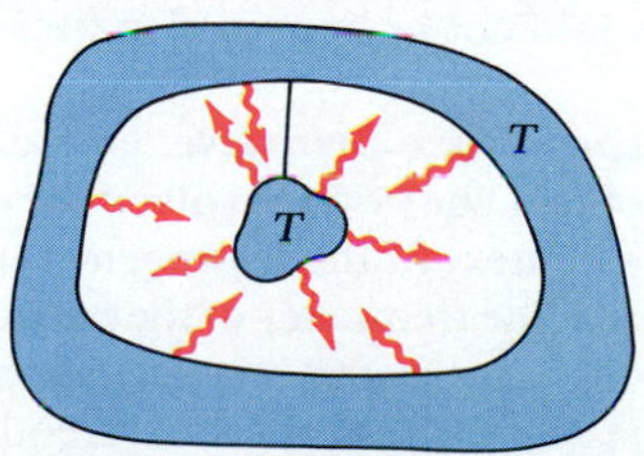

Figure 39-13. Problem 1: A body at temperature T is suspended by an insulating thread and is surrounded by walls at the same temperature.

2. ***Wien's displacement law.*** *(a)* Obtain, as in Exercise 9, the blackbody spectral radiancy R_λ for a differential wavelength range $d\lambda$. *(b)* Show that the wavelength λ_m at which the spectral radiancy is maximum depends on the temperature such that $\lambda_m T =$ constant and show that the value of this constant is 2.90×10^{-3} K·m. This result is known as *Wien's displacement law.* (*Hint:* $e^{hc/\lambda kT} \gg 1$ for $\lambda = \lambda_m$.) *(c)* Determine λ_m for a blackbody at $T = 6000$ K, the surface temperature of the sun.

3. ***Connection between Planck's law and the Stefan-Boltzmann law.*** Show that Planck's radiation law leads to *(a)* the Rayleigh-Jeans result, $R_\nu = 2\pi\nu^2 kT/c^2$, at low frequencies ($h\nu \ll kT$), where $e^{h\nu/kT} - 1 \approx h\nu/kT$, and to *(b)* the Stefan-Boltzmann law, $R = \sigma T^4$. Determine the constant σ in terms of fundamental constants. *Hint:* You should encounter an integral that can be put in the form

$$\int_0^\infty \frac{x^3}{e^x - 1}\,dx = \frac{\pi^4}{15}$$

4. ***The Compton effect.*** Apply the conservation laws for energy and momentum to the Compton effect and obtain Eq. 39-9.

5. ***Nonrelativistic Compton effect.*** If the incident photon energy is not too high, nonrelativistic expressions for the energy ($\frac{1}{2}mv^2$) and magnitude of momentum (mv) of the electron may be used. *(a)* Set up the equations similarly to those in Sec. 39-4 and show that

$$\lambda' - \lambda = \frac{h}{2mc}\left(\frac{\lambda'}{\lambda} + \frac{\lambda}{\lambda'} - 2\cos\theta\right)$$

(b) For a low photon energy, $h\nu = hc/\lambda \ll mc^2$ and $\lambda' \approx \lambda$; show that the right-hand side of the above expression then reproduces Eq. 39-9, which is valid for any photon energy.

6. ***Maximum size of a nucleus.*** Consider a head-on collision between an alpha particle, treated as a point particle of charge $+2e$, and a spherical, positively charged part of an atom of charge $+Ze$, where Z is the atomic number and R is the radius of the sphere. Assume that the massive atom remains essentially at rest and that the alpha particle has initial kinetic energy K, with $K \ll m_\alpha c^2$. If the alpha particle is scattered backward as a result of the Coulomb force, then it must come momentarily to rest at a minimum distance r_0 from the center of the atom. *(a)* Show that this minimum distance is $r_0 = Ze^2(2\pi\epsilon_0 K)$ if the alpha particle does not penetrate the sphere of charge—that is, if $r_0 > R$. *(b)* Use this result to estimate an upper limit on the size of the sphere of charge for gold ($Z = 79$) if $K = 5$ MeV. *(c)* If the positively charged sphere were larger, so that an alpha particle penetrated the sphere of charge, explain why the alpha particle would not be expected to backscatter.

7. ***The correspondence principle.*** One of the arguments used by Bohr to obtain Eq. 39-21 involved the *correspondence principle.* In the context of the hydrogen-atom problem, this principle states that the frequency of the radiation emitted in a transition between state n and state $n - 1$, where n is very large, is the same as the classical orbital frequency of an electron in an orbit of radius r_n. *(a)* Show that the Rydberg formula, coupled with the expression $h\nu = E_{n_i} - E_{n_f}$ implies that $E_n = -hcR_H/n^2 = -e^2/(8\pi\epsilon_0 r_n)$ from Eq. 39-15. Use the last equality to obtain $r_n = n^2e^2/(8\pi\epsilon_0 hcR_H)$. *(b)* Use the Rydberg formula to show that the frequency emitted in the transition from state n to state $n - 1$, with $n \gg 1$, is $\nu = 2cR_H/n^3$. *(c)* Classically, the frequency emitted by a charged particle in a circular orbit is equal to the particle's orbital frequency, $\nu = v_n/2\pi r_n$. Express this frequency in terms of the radius r_n of the orbit and show that its square is $\nu^2 = e^2/(16\pi^3\epsilon_0 m r_n^3)$. *(d)* Equate expressions for ν^2 from parts *(b)* and *(c)* and solve for r_n^3. *(e)* Eliminate r_n from parts *(a)* and *(d)* to show that $R_H = me^4/(8\epsilon_0^2h^3c)$, and that $E_n = -me^4/(8\epsilon_0^2h^2n^2)$.

8. ***Reduced mass.*** The mass m that appears in Eqs. 39-20, 39-21, and 39-24 is the electron mass m_e, based on the assumption that the nucleus is at rest. To correct for the motion of the nucleus of mass M, the *reduced mass* $m = m_eM/(m_e + M)$ of the system should be used in those expressions. Determine the values, to five significant figures, of the Rydberg constant and the wavelength of the H_α line for *(a)* normal hydrogen with a single proton as the nucleus and *(b)* deuterium, or heavy hydrogen, with the deuteron as the nucleus ($m_d = 3.3441 \times 10^{-27}$ kg). *(c)* Determine the difference in wavelengths of the H_α line for normal and heavy hydrogen. This difference can be observed in the spectrum and corresponds to the 0.015 percent natural abundance of the heavy-hydrogen isotope.

9. ***Singly ionized helium.*** Singly ionized He has one remaining electron and a nuclear charge of $2e$. *(a)* Show that the Rydberg constant for singly ionized He is 4 times the Rydberg constant for H. *(b)* Identify the quantum numbers for the transitions in singly ionized He that have virtually the same wavelengths as the wavelengths of the Balmer series lines in H. How is this part of the singly ionized He spectrum different from the H spectrum? Bohr's explanation of these spectral lines of singly ionized He was an early success (in 1913) of his model of the atom.

10. ***Maximum wavelength.*** Show from the results of Exercise 9 that the wavelength at which the radiated power per unit area is a maximum is inversely proportional to the absolute temperature.

11. ***Series guessing.*** Each of the following spectral lines of hydrogen belongs to a different series. Identify the series in each case: *(a)* 469 nm; *(b)* 122 nm; *(c)* 188 nm; *(d)* 65.6 nm.

12. ***Impotent "passion."*** How many wavelengths in the Paschen series cannot cause the ejection of electrons from Cd, which has a work function of 4.07 V?

13. ***Equality among -ons.*** *(a)* An electron has a speed $0.75c$, where c is the speed of light. What is its energy? *(b)* What is the wavelength of a photon with the same energy? *(c)* This electron and this photon are viewed from a frame moving at $0.5c$ in the direction of the electron's motion. What are the energies of the electron and photon in this reference frame?

14. ***Hubble red shift.*** A quasi-stellar source of radiation moves away from the earth so fast that the H_β line in the spectrum of Fig. 39-9 is Doppler-shifted to the position of the H_α line. *(a)* What is the recessional speed of the source? *(b)* The Hubble constant H relates the distance of an astronomic object from the earth to its recessional speed according to the expression $v = Hd$, where v is the speed and d is the distance to the source. If $H = 1.6 \times 10^{-18}\ s^{-1}$, how far from the earth is the quasi-stellar object?

15. ***Relativistic corrections to Bohr.*** The speed of an electron in the ground state of a hydrogen atom according to the Bohr theory is $v = e^2/(2\epsilon_0 h)$. *(a)* Estimate the size of the relativistic corrections to the kinetic energy of the electron in the ground state of the hydrogen atom according to the Bohr theory. The results of Prob. 18 in Chap. 38 may help. Compare the size of this correction to that due to the motion of the nucleus, found in Prob. 8 above. *(b)* The results of Prob. 9 show that this speed increases proportional to the charge of the nucleus. How large would the relativistic correction be for Sn? (See App. P.) Note that modern theory does not have the electron moving around the nucleus in the ground state.

16. ***Photon recoil.*** When a stationary atom emits a photon, the atom must recoil to conserve momentum. Consider a H atom in the first excited state. It has an energy of 10.2 eV above the ground state. *(a)* Write the equations that express conservation of momentum (zero initially) and conservation of energy (10.2 eV initially). Show that the energy of the photon is $hv = -Mc^2 \pm (Mc^2)\sqrt{[1 + (2E/Mc^2)]}$ or, to a good approximation,

$$hv \approx E[1 - (E/Mc^2)]$$

where M is the mass of the hydrogen atom. *(b)* Appendix F shows that R_H is known to one part in 10^{10}. Is the recoil correction observable?

17. ***Countering recoil with Doppler.*** Because of the recoil correction discussed in Prob. 16, the photon emitted by one stationary H atom as the electron falls from the first excited state to the ground state cannot be used to promote the electron of another stationary H atom from the ground state to the first excited state. But if one H atom approaches the other, the Doppler shifting of the photon will allow the process. What must the relative velocity between two H atoms be in order for the photon emitted by one to promote the electron of the other? Remember that both atoms must recoil.

18. ***Positronium.*** "Positronium" is a quasi-atom formed for a short time by an electron and a positron. Using the reduced mass suggested by Prob. 8, find the Rydberg constant for positronium.

19. ***Semiconducting Bohr.*** Electrons in semiconductors are attracted by negatively charged defects. The problem is similar to that of a hydrogen atom except that the dielectric constant is not that of free space, but that of the semiconducting material. The dielectric constant of diamond is 5.5. What is the wavelength of the line equivalent to the first Balmer line for an electron in diamond near a defect with a charge of $+e$?

APPENDIX A
ASTRONOMICAL DATA

Body	Body orbited	Mean radius of orbit, m	Radius of body, m	Period of orbit, s	Mass of body, kg
Sun	Galaxy	5.6×10^{20}	6.96×10^{8}	8×10^{15}	1.99×10^{30}
Mercury	Sun	5.79×10^{10}	2.42×10^{6}	7.60×10^{6}	3.35×10^{23}
Venus	Sun	1.08×10^{11}	6.10×10^{6}	1.94×10^{7}	4.89×10^{24}
Earth	Sun	1.50×10^{11}	6.37×10^{6}	3.16×10^{7}	5.97×10^{24}
Mars	Sun	2.28×10^{11}	3.38×10^{6}	5.94×10^{7}	6.46×10^{23}
Jupiter	Sun	7.78×10^{11}	7.13×10^{7}	3.74×10^{8}	1.90×10^{27}
Saturn	Sun	1.43×10^{12}	6.04×10^{7}	9.35×10^{8}	5.69×10^{26}
Uranus	Sun	2.87×10^{12}	2.38×10^{7}	2.64×10^{9}	8.73×10^{25}
Neptune	Sun	4.50×10^{12}	2.22×10^{7}	5.22×10^{9}	1.03×10^{26}
Pluto	Sun	5.91×10^{12}	3×10^{6}	7.82×10^{9}	5.4×10^{24}
Moon	Earth	3.84×10^{8}	1.74×10^{6}	2.36×10^{6}	7.35×10^{22}
Phobos	Mars	9×10^{6}	6×10^{3}	3×10^{4}	1×10^{16}
Deimos	Mars	2.3×10^{7}	3×10^{3}	1.09×10^{5}	2×10^{15}
Io	Jupiter	4.22×10^{8}	1.67×10^{6}	1.53×10^{5}	7.3×10^{22}
Europa	Jupiter	6.71×10^{8}	1.46×10^{6}	3.07×10^{5}	4.75×10^{22}
Ganymede	Jupiter	1.07×10^{9}	2.55×10^{6}	6.18×10^{5}	1.54×10^{23}
Callisto	Jupiter	1.88×10^{9}	2.36×10^{6}	1.44×10^{6}	9.5×10^{22}
Mimas	Saturn	1.86×10^{8}	3×10^{5}	8.12×10^{4}	4×10^{19}
Enceladus	Saturn	2.38×10^{8}	3×10^{5}	1.18×10^{5}	7×10^{19}
Tethys	Saturn	2.95×10^{8}	5×10^{5}	1.63×10^{5}	6.5×10^{20}
Dione	Saturn	3.77×10^{8}	5×10^{5}	2.37×10^{5}	1.0×10^{21}
Rhea	Saturn	5.27×10^{8}	7×10^{5}	3.91×10^{5}	2.3×10^{21}
Titan	Saturn	1.22×10^{9}	2.44×10^{7}	1.38×10^{6}	1.37×10^{23}
Iapetus	Saturn	1.48×10^{9}	5×10^{5}	6.85×10^{6}	1×10^{21}
Ariel	Uranus	1.92×10^{8}	3×10^{5}	2.18×10^{5}	1.2×10^{21}
Umbriel	Uranus	2.67×10^{8}	2×10^{5}	3.58×10^{5}	5×10^{20}
Titania	Uranus	4.38×10^{8}	5×10^{5}	7.53×10^{5}	4×10^{21}
Oberon	Uranus	5.86×10^{9}	4×10^{5}	1.16×10^{6}	2.6×10^{21}
Triton	Neptune	3.53×10^{8}	2×10^{6}	4.82×10^{5}	1.40×10^{23}
Nereid	Neptune	5.6×10^{9}	1×10^{5}	3.11×10^{7}	3×10^{19}

APPENDIX C
CONVERSION FACTORS

Length

	m	inch	ft	mi
1 meter	1	39.37	3.281	6.214×10^{-4}
1 inch	2.540×10^{-2}	1	8.333×10^{-2}	1.578×10^{-5}
1 foot	0.3048	12	1	1.894×10^{-4}

1 fermi = 10^{-15} m
1 Bohr radius = 5.292×10^{-11} m
1 angstrom = 10^{-10} m
1 light-year = 9.460×10^{15} m
1 parsec = 3.084×10^{16} m
1 mil = 10^{-3} inch
1 yard = 3 ft
1 fathom = 6 ft
1 nautical mile = 1852 m

Time

	s	min	h	d	yr
1 second	1	1.667×10^{-2}	2.778×10^{-4}	1.157×10^{-5}	3.169×10^{-8}
1 minute	60	1	1.667×10^{-2}	6.944×10^{-4}	1.901×10^{-6}
1 hour	3600	60	1	4.167×10^{-2}	1.141×10^{-4}
1 day	8.640×10^{4}	1440	24	1	2.738×10^{-3}
1 year	3.156×10^{7}	5.260×10^{5}	8.766×10^{3}	365.2	1

Mass

Quantities in the colored areas are weights, not masses, but are commonly equated to masses. For example, 1 kg has a weight of 2.205 lb in a region in which $g = 9.80665$ m/s².

	kg	u	slug	oz	lb	ton
1 kilogram	1	6.022×10^{26}	6.852×10^{-2}	35.27	2.205	1.102×10^{-3}
1 atomic mass unit	1.661×10^{-27}	1	1.138×10^{-28}	5.857×10^{-27}	3.661×10^{-27}	1.830×10^{-30}
1 slug	14.59	8.788×10^{27}	1	514.8	32.17	1.609×10^{-2}

1 metric tonne = 1000 kg

Area

	m²	in²	ft²
1 square meter	1	1550	10.76
1 square inch	6.452×10^{-4}	1	6.944×10^{-3}
1 square foot	9.290×10^{-2}	144	1

1 barn = 10^{-28} m²
1 hectare = 10^4 m² = 2.471 acres
1 acre = 4.356×10^4 ft²
1 square mile = 640 acres = 2.788×10^7 m²

Volume

	m³	cm³	L	inch³	ft³
1 cubic meter	1	10^6	10^3	6.102×10^{4}	35.31
1 cubic centimeter	10^{-6}	1	10^{-3}	6.102×10^{-2}	3.531×10^{-5}
1 liter	10^{-3}	10^3	1	61.02	3.531×10^{-2}
1 cubic inch	1.639×10^{-5}	16.39	1.639×10^{-2}	1	5.787×10^{-4}
1 cubic foot	2.832×10^{-2}	2.831×10^{4}	28.32	1728	1

1 U.S. fluid gallon = 4 U.S. fluid qt = 8 U.S. pt = 128 U.S. fluid oz = 231 inches³
1 British Imperial gallon = 277.42 inches³
1 U.S. barrel = $31\frac{1}{2}$ gal (Other definitions of the barrel exist.)

Speed

	m/s	km/h	ft/s	mi/h
1 meter per second	1	3.600	3.281	2.237
1 kilometer per hour	0.2778	1	0.9113	0.6214
1 foot per second	0.3048	1.097	1	0.6818
1 mile per hour	0.4470	1.609	1.467	1

Force

	N	dyn	lb
1 newton	1	10^5	0.2248
1 dyne	10^{-5}	1	2.248×10^{-6}
1 pound	4.448	4.448×10^5	1

Power

	W	cal/s	hp	ft · lb/s	Btu/h
1 watt	1	0.2390	1.341×10^{-3}	0.7376	3.414
1 calorie* per second	4.184	1	5.611×10^{-3}	3.086	14.29
1 horsepower	745.7	178.2	1	550	2546
1 foot pound per second	1.356	0.3240	1.818×10^{-3}	1	4.629
1 British thermal unit per hour	0.2929	7.000×10^{-2}	3.928×10^{-4}	0.2160	1

* The thermochemical calorie is defined to be 4.184 J. The Calorie used in human diets is 10^3 cal.

Density The pound per cubic foot is weight density, the others mass density. See Mass table.

	kg/m³	g/cm³	lb/ft³
1 kilogram per cubic meter	1	10^{-3}	6.243×10^{-2}
1 gram per cubic centimeter	10^3	1	62.43
1 pound per cubic foot	16.02	1.602×10^{-2}	1

Pressure

	Pa	dyn/cm²	atm	mmHg (torr)	lb/inch²	inch of water
1 pascal (1 N/m²)	1	10	9.869×10^{-6}	7.501×10^{-3}	1.450×10^{-4}	4.015×10^{-3}
1 dyne per square centimeter	0.1	1	9.869×10^{-7}	7.501×10^{-4}	1.450×10^{-5}	4.015×10^{-4}
1 atmosphere	1.013×10^5	1.013×10^6	1	760	14.70	406.8
1 millimeter of mercury	133.3	1.333×10^3	1.316×10^{-3}	1	1.934×10^{-2}	0.5352
1 pound per square inch	6895	6.895×10^4	0.6805	51.71	1	27.68
1 inch of water	249.1	2491	2.458×10^{-3}	1.868	3.613×10^{-2}	1

Energy

	J	erg	eV	cal	kW · h	ft · lb	hp · h	Btu
1 joule	1	10^7	6.242×10^{18}	0.2390	2.778×10^{-7}	0.7376	3.725×10^{-7}	9.484×10^{-4}
1 erg	10^{-7}	1	6.242×10^{11}	2.390×10^{-8}	2.778×10^{-14}	7.376×10^{-8}	3.725×10^{-14}	9.484×10^{-11}
1 electron volt	1.602×10^{-19}	1.602×10^{-12}	1	3.829×10^{-20}	4.450×10^{-26}	1.182×10^{-19}	5.968×10^{-26}	1.520×10^{-22}
1 calorie*	4.184	4.184×10^7	2.611×10^{19}	1	1.162×10^{-6}	3.086	1.559×10^{-6}	3.968×10^{-3}
1 kilowatt-hour	3.6×10^6	3.6×10^{13}	2.247×10^{25}	8.604×10^5	1	2.655×10^6	1.341	3414
1 foot pound	1.356	1.356×10^7	8.462×10^{18}	0.3240	3.766×10^{-7}	1	5.051×10^{-7}	1.286×10^{-3}
1 horsepower-hour	2.685×10^6	2.685×10^{13}	1.676×10^{25}	6.416×10^5	0.7457	1.980×10^6	1	2546
1 British thermal unit	1054	1.054×10^{10}	6.581×10^{21}	252	2.929×10^{-4}	7.777×10^2	3.928×10^{-4}	1

* The thermochemical calorie is defined to be 4.186 J. The Calorie used in human diets is 10^3 cal.

APPENDIX D DIFFERENTIAL CALCULUS

The derivative of $y = f(x)$ is defined to be the limit of the slope $\Delta y/\Delta x$ of the y-versus-x curve:

$$\frac{dy}{dx} = \lim_{\Delta x \to 0} \frac{\Delta y}{\Delta x} = \lim_{\Delta x \to 0} \frac{f(x+\Delta x) - f(x)}{\Delta x}$$

Some general relations about derivatives:

Sums of functions: $$\frac{d}{dx}[f(x) \pm g(x)] = \frac{df}{dx} \pm \frac{dg}{dx}$$

Products of functions: $$\frac{d}{dx}[f(x)g(x)] = f\frac{dg}{dx} + g\frac{df}{dx}$$

Quotient of two functions: $$\frac{d(f/g)}{dx} = \frac{g\,df/dx - f\,dg/dx}{g^2}$$

Chain rule: If $y = f(x)$ and $x = g(z)$, then

$$\frac{df(x)}{dz} = \frac{df(x)}{dx}\frac{dx}{dz}$$

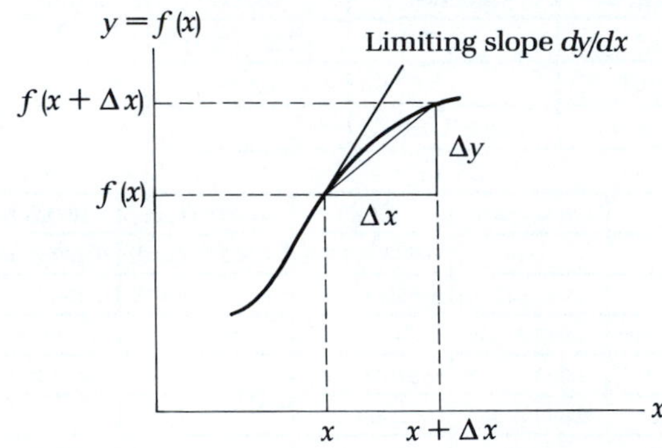

Derivatives of some particular functions (a and n are constants):

$$\frac{d(ax^n)}{dx} = nax^{n-1}$$

$$\frac{d}{dx}\sin ax = a\cos ax \qquad \frac{d}{dx}a^{nx} = na^x \ln a \qquad \frac{d}{dx}\sin^{-1} ax = \frac{a}{\sqrt{1-a^2x^2}}$$

$$\frac{d}{dx}\cos ax = -a\sin ax \qquad \frac{d}{dx}e^{ax} = ae^{ax} \qquad \frac{d}{dx}\cos^{-1} ax = \frac{-a}{\sqrt{1-a^2x^2}}$$

$$\frac{d}{dx}\tan ax = a\sec^2 ax \qquad \frac{d}{dx}\ln ax = \frac{1}{x} \qquad \frac{d}{dx}\tan^{-1} ax = \frac{a}{1+a^2x^2}$$

APPENDIX F
FUNDAMENTAL CONSTANTS

Summary of the 1986 recommended values of the fundamental physical constants.

Quantity	Symbol	Value	Units	Relative uncertainty, ppm
Speed of light in vacuum	c	299 792 458	$m \cdot s^{-1}$	(Exact)
Triple-point temperature	T_t	273.16	K	(Exact)
Permeability of vacuum	μ_0	$4\pi \times 10^{-7}$	$N \cdot A^{-2}$	
		= 12.566 370 614. . .	$10^{-7}\ N \cdot A^{-2}$	(Exact)
Permittivity of vacuum, $1/\mu_0 c^2$	ϵ_0	8.854 187 817. . .	$10^{-12}\ F \cdot m^{-1}$	(Exact)
Newtonian constant of gravitation	G	6.672 59(85)	$10^{-11}\ m^3 \cdot kg^{-1} \cdot s^{-2}$	128
Planck constant	h	6.626 075 5(40)	$10^{-34}\ J \cdot s$	0.60
$h/2\pi$	$\hbar$	1.054 572 66(63)	$10^{-34}\ J \cdot s$	0.60
Elementary charge	e	1.602 177 33(49)	10^{-19} C	0.30
Magnetic flux quantum, $h/2e$	Φ_0	2.067 834 61(61)	10^{-15} Wb	0.30
Electron mass	m_e	9.109 389 7(54)	10^{-31} kg	0.59
Proton mass	m_p	1.672 623 1(10)	10^{-27} kg	0.59
Proton-electron mass ratio	m_p/m_e	1836.152 701(37)		0.020
Neutron mass	m_n	1.674 928 6(10)	10^{-27} kg	0.59
Compton wavelength, $h/m_e c$	λ_c	2.426 310 58(22)	10^{-12} m	0.089
Fine-structure constant, $\mu_0 ce^2/2h$	α	7.297 353 08(33)	10^{-3}	0.045
Inverse fine-structure constant	α^{-1}	137.035 989 5(61)		0.045
Rydberg constant, $m_e c\alpha^2/2h$	R_∞	10 973 731.534(13)	m^{-1}	0.0012
Avogadro constant	N_A, L	6.022 136 7(36)	$10^{23}\ mol^{-1}$	0.59
Faraday constant, $N_A e$	F	96 485.309(29)	$C \cdot mol^{-1}$	0.30
Molar gas constant	R	8.314 510(70)	$J \cdot mol^{-1} \cdot K^{-1}$	8.4
Boltzmann constant, R/N_A	k	1.380 658(12)	$10^{-23}\ J \cdot K^{-1}$	8.5
Stefan-Boltzmann constant, $(\pi^2/60)k^4/\hbar^3 c^2$	σ	5.670 51(19)	$10^{-8}\ Wm^{-2} \cdot K^{-4}$	34
		Non-SI units used with SI		
Electron volt, $(e/C)J = \{e\}J$	eV	1.602 177 33(49)	10^{-19} J	0.30
Atomic mass unit (unified), $1\ u = m_u = \frac{1}{12}m(^{12}C)$	u	1.660 540 2(10)	10^{-27} kg	0.59

From E. Richard Cohen and B. N. Taylor, *Reviews of Modern Physics*, vol. 59, No. 4, October 1987, p 1139.

APPENDIX I
INTEGRAL CALCULUS

The *integral I* of the function $f(x)$ between the limits a and b is written

$$I = \int_a^b f(x)\, dx$$

and is equal to the area under the curve $f(x)$ between the lines $x = a$ and $x = b$, as shown in Fig. I-1. The fundamental theorem of calculus shows that if the upper limit is a variable w, then

$$I(w) = \int_a^w f(x)\, dx$$

$$\frac{d}{dw} I(w) = \frac{d}{dw} \int_a^w f(x)\, dx = f(w)$$

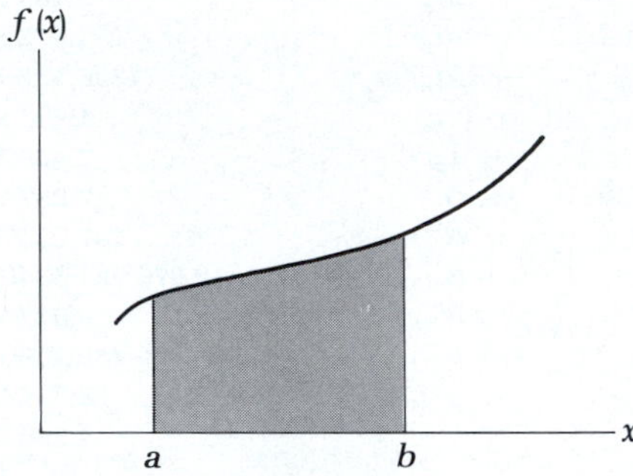

Thus we think of integration as the inverse of differentiation. The *indefinite integral* $I(x)$ of $f(x)$ is the function whose differential is $f(x)$; for example, the indefinite integral of $ax^2 + bx + c$ is $\frac{1}{3}ax^3 + \frac{1}{2}bx^2 + cx + d$.

Some general rules about indefinite integrals (f, g, u, and v are functions; a, b and c are constants) can be expressed as

$$\int dx = x + c$$

$$\int \frac{d[f(x)]}{dx}\, dx = f(x) + c$$

$$\int af(x)\, dx = a \int f(x)\, dx$$

$$\int [af(x) + bg(x)]\, dx = a \int f(x)\, dx + b \int g(x)\, dx$$

$$\int u\, dv = uv - \int v\, du$$

Below are tables of some indefinite and definite integrals.

Indefinite integrals An arbitrary constant should be added to each integral. a, b, and n represent constants.

$$\int x^n\, dx = \frac{x^{n+1}}{n+1}$$

$$\int (a+bx)^n\,dx = \frac{(a+bx)^{n+1}}{b(n+1)} \qquad \text{(provided } n \neq -1\text{)}$$

$$\int \frac{dx}{x} = \ln x$$

$$\int \frac{dx}{a+bx} = \frac{1}{b}\ln(a+bx)$$

$$\int \frac{dx}{a+bx^2} = \frac{1}{\sqrt{ab}}\tan^{-1}\left(\frac{\sqrt{b}}{\sqrt{a}}\,x\right) \qquad \text{(provided } ab > 0\text{)}$$

$$\int \frac{dx}{a+bx^2} = \frac{1}{2\sqrt{|ab|}}\ln\left(\frac{a-x\sqrt{|ab|}}{a+x\sqrt{|ab|}}\right) \qquad \text{(provided } ab < 0\text{)}$$

$$\int \frac{x\,dx}{(a+bx^2)^n} = -\frac{1}{2b(n-1)(a+bx^2)^{n-1}} \qquad \text{(provided } n \neq 1\text{)}$$

$$\int \frac{x\,dx}{a+bx^2} = \frac{1}{2b}\ln(a+bx)$$

Let $u = \sqrt{a+cx^2}$, $I = \dfrac{1}{\sqrt{c}}\ln(x\sqrt{c}+u)$ if $c > 0$

$$= \frac{1}{\sqrt{-c}}\sin^{-1}\left(x\sqrt{\frac{-c}{a}}\right) \qquad \text{if } c < 0 \text{ and } a > 0$$

then

$$\int u\,dx = \tfrac{1}{2}(xu + aI)$$

$$\int \frac{dx}{u} = I$$

$$\int xu\,dx = \frac{u^3}{3c}$$

$$\int \frac{x\,dx}{u} = \frac{u}{c}$$

$$\int e^{ax}\,dx = \frac{e^{ax}}{a}$$

$$\int xe^{ax}\,dx = \frac{e^{ax}}{a^2}(ax-1)$$

$$\int x^2e^{ax}\,dx = \frac{e^{ax}}{a^3}(a^2x^2 - 2ax + 2)$$

$$\int \frac{dx}{a+be^{nx}} = \frac{x}{a} - \frac{\ln(a+be^{ax})}{an}$$

$$\int \ln ax\,dx = (x\ln ax) - x$$

$$\int \sin ax\,dx = -\frac{\cos ax}{a}$$

$$\int \cos ax\,dx = \frac{\sin ax}{a}$$

$$\int \tan ax\,dx = -\frac{\ln(\cos ax)}{a}$$

$$\int \sin^2 ax\, dx = \frac{x}{2} - \frac{\sin 2ax}{4a}$$

$$\int \cos^2 ax\, dx = \frac{x}{2} + \frac{\sin 2ax}{4a}$$

$$\int \tan^2 ax\, dx = \frac{\tan ax}{a} - x$$

$$\int \sin^{-1}\left(\frac{x}{a}\right) dx = x \sin^{-1}\left(\frac{x}{a}\right) + \sqrt{a^2 + x^2}$$

$$\int \cos^{-1}\left(\frac{x}{a}\right) dx = x \cos^{-1}\left(\frac{x}{a}\right) - \sqrt{a^2 - x^2}$$

$$\int \tan^{-1}\left(\frac{x}{a}\right) dx = x \tan^{-1}\left(\frac{x}{a}\right) - \left(\frac{a}{2}\right) \ln (a^2 + x^2)$$

Definite integrals
($a > 0$)

$$\int_0^\infty e^{-ax}\, dx = \frac{1}{a}$$

$$\int_0^\infty x^n e^{-ax}\, dx = n!a^{-n-1}$$

$$\int_0^\infty \frac{dx}{1 + e^{ax}} = \frac{\ln 2}{a}$$

$$\int_0^\infty e^{-a^2x^2}\, dx = \frac{\sqrt{\pi}}{2a}$$

$$\int_0^\infty x e^{-ax^2}\, dx = \frac{1}{2a}$$

$$\int_0^\infty x^2 e^{-ax^2}\, dx = \frac{1}{4}\sqrt{\frac{\pi}{a^3}}$$

$$\int_0^\infty x^3 e^{-ax^2}\, dx = \frac{1}{2a^2}$$

$$\int_0^\infty x^4 e^{-ax^2}\, dx = \frac{3}{8}\sqrt{\frac{\pi}{a^5}}$$

$$\int_0^\infty \frac{\sin ax}{x} = \frac{\pi}{2}$$

APPENDIX M MATHEMATICAL APPROXIMATIONS, FORMULAS, AND SYMBOLS

Expansions

$(1+x)^n = 1 + nx + \frac{n(n-1)}{2!}x^2 + \cdots$	$\lvert x\rvert < 1$
$\sin\theta = \theta - \frac{\theta^3}{3!} + \frac{\theta^5}{5!} - \cdots$	θ in rad
$\cos\theta = 1 - \frac{\theta^2}{2!} + \frac{\theta^4}{4!} - \cdots$	θ in rad
$\tan\theta = \theta + \frac{\theta^3}{3} + \frac{2}{15}\theta^5 + \cdots$	θ in rad
$\sin^{-1} x = x + \frac{1}{6}x^3 + \frac{3}{40}x^5 + \cdots$	$\lvert x\rvert < 1$, angle in rad
$\cos^{-1} x = \frac{\pi}{2} - \sin^{-1} x$	
$\tan^{-1} x = x - \frac{x^3}{3} + \frac{x^5}{5} - \cdots$	$x^2 < 1$
$\quad = \frac{\pi}{2} - \frac{1}{x} + \frac{1}{3x^3} - \frac{1}{5x^5} + \cdots$	$x^2 > 1$
$e^x = 1 + x + \frac{x^2}{2!} + \frac{x^3}{3!} + \cdots$	
$\ln(1+x) = x - \frac{1}{2}x^2 + \frac{1}{3}x^3 - \cdots$	$x < 1$

Areas

Square of side a	a^2
Rectangle with sides a and b	ab
Triangle of base b and height h	$\frac{1}{2}bh$
Parallelogram of base b and height h	bh
Circle of radius r	πr^2
Ellipse with semimajor axis a and semiminor axis b	πab
Sphere of radius r	$4\pi r^2$
Circular cylinder of radius r and height h	$2\pi r^2 + 2\pi rh$ (top and bottom) + (side)

Volumes

Cube of side a	a^3

Parallelepiped with base area A and height h	Ah
Sphere of radius r	$\frac{4}{3}\pi r^3$
Cylinder of base area A and height h	Ah
Cone of base area A and height h	$\frac{1}{3}Ah$

Equations of curves

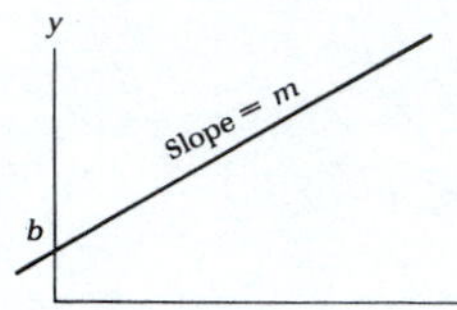

Straight line of slope m and intercept b $\quad y = mx + b$

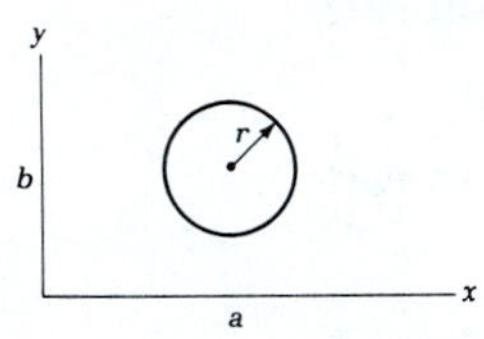

Circle of radius r centered at (a, b) $\quad (x - a)^2 + (y - b)^2 = r^2$

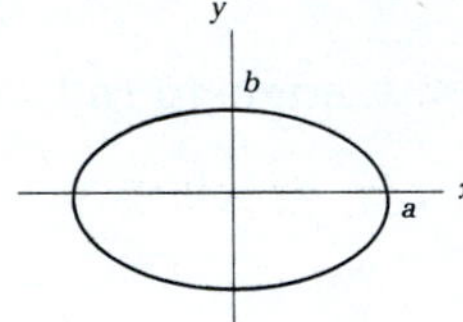

Ellipse with semiaxes a and b and center at (0, 0) $\quad \dfrac{x^2}{a^2} + \dfrac{y^2}{b^2} = 1$

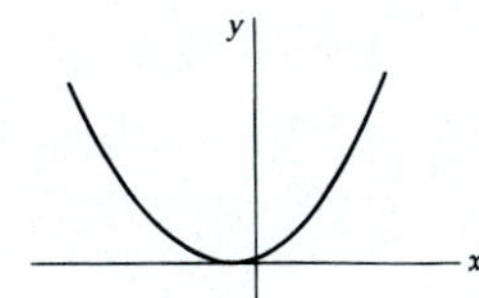

Parabola with vertex at origin branching upward $\quad y = Ax^2$

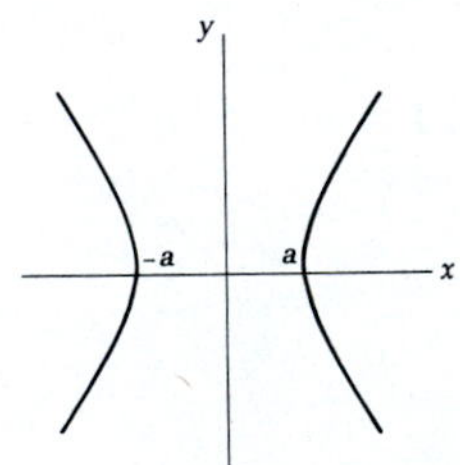

Hyperbola with vertices at $(\pm a, 0)$ and eccentricity e given by $e^2 = (b^2/a^2) + 1$ $\quad \dfrac{x^2}{a^2} - \dfrac{y^2}{b^2} = 1$

Quadratic equations

The solutions to the equation $ax^2 + bx + c = 0$ are given by

$$x = \frac{-b \pm \sqrt{b^2 - 4ac}}{2a}$$

Logarithms

If $x = a^y$, then $y = \log_a x$.

The base of the natural logarithms is $e = 2.718281828 \ldots$, so that $y = \log_e x = \ln x$.

$\log 1 = 0$

$\log_a a = 1$

$\log (uv) = \log u + \log v$

$\log (u/v) = \log u - \log v$

$\log u^n = n \log u$

$\ln e = 1$

$\ln e^n = n$

$\ln 10 = 2.303$

$\ln 2 = 0.693$

Mathematical symbols

Symbol	Definition
$=$	is equal to
$\neq$	is not equal to
$\approx$	is approximately equal to
$\propto$	is proportional to
$>$	is greater than
$\gg$	is much greater than
$<$	is less than
$\ll$	is much less than
Δx	change in x
$n!$	$n(n-1)(n-2)\cdots 1$
Σ	sum
lim	limit
$\Delta t \rightarrow 0$	Δt approaches zero
$\int$	integral
$\dfrac{df}{dx}$	derivative of f with respect to x
$\dfrac{\partial f}{\partial x}$	partial derivative of f with respect to x

The Greek alphabet

Character	Upper case	Lower case	Character	Upper case	Lower case
Alpha	A	α	Nu	N	ν
Beta	B	β	Xi	Ξ	ξ
Gamma	Γ	γ	Omicron	O	o
Delta	Δ	δ	Pi	Π	π
Epsilon	E	ϵ	Rho	P	ρ
Zeta	Z	ζ	Sigma	Σ	σ
Eta	H	η	Tau	T	τ
Theta	Θ	θ	Upsilon	Υ	υ
Iota	I	ι	Phi	Φ	ϕ, φ
Kappa	K	κ	Chi	X	χ
Lambda	Λ	λ	Psi	Ψ	ψ
Mu	M	μ	Omega	Ω	ω

APPENDIX P
PERIODIC TABLE OF THE ELEMENTS

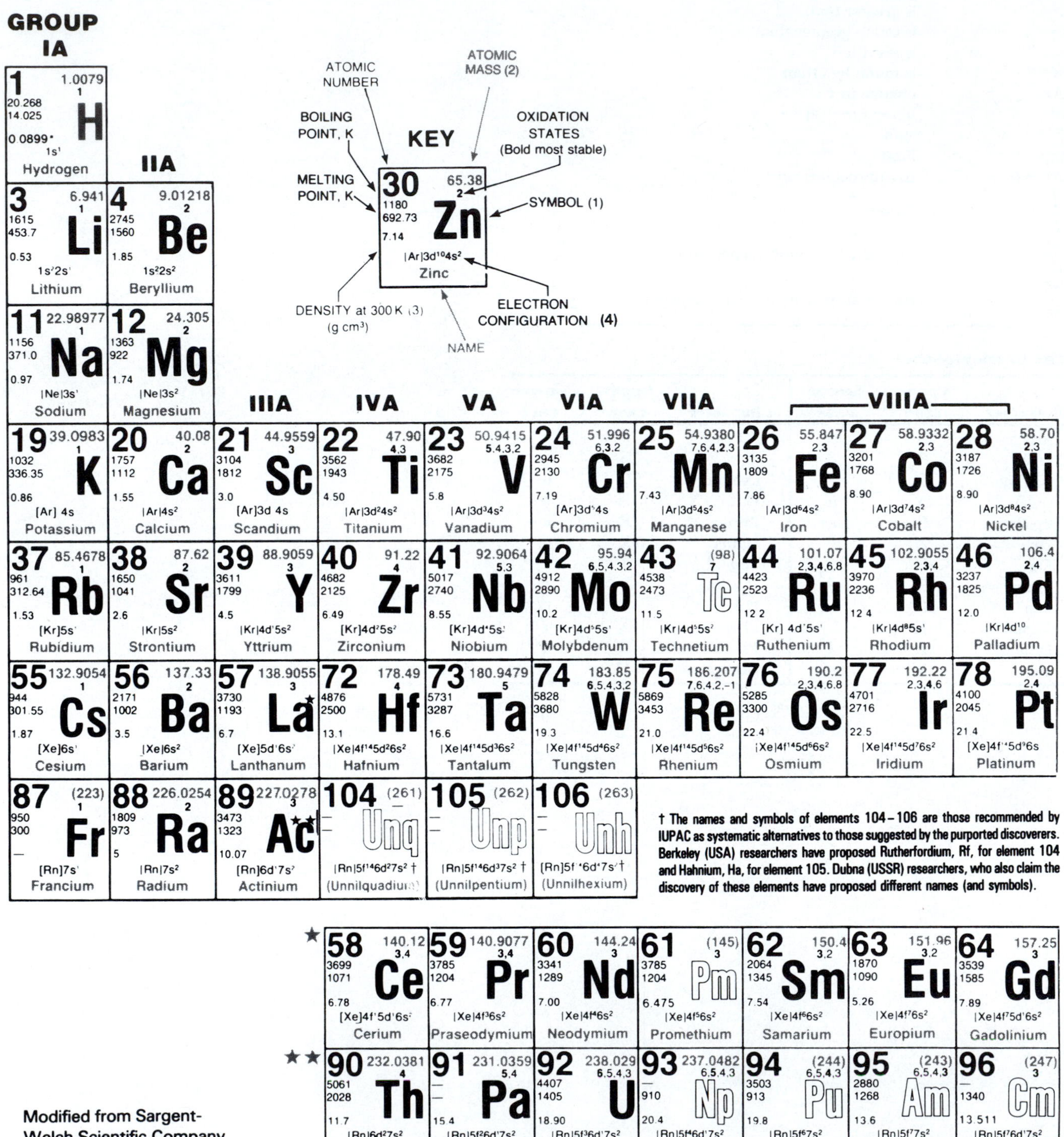

IB	IIB	IIIB	IVB	VB	VIB	VIIB	VIII
							2 4.00260 4.215 0.95 0.1787* **He** $1s^2$ Helium
		5 10.81 3 4275 2300 2.34 **B** $1s^22s^22p^1$ Boron	**6** 12.011 ±4,2 4470* 4100* 2.62 **C** $1s^22s^22p^2$ Carbon	**7** 14.0067 ±3,5,4,2 77.35 63.14 1.251* **N** $1s^22s^22p^3$ Nitrogen	**8** 15.9994 −2 90.18 50.35 1.429* **O** $1s^22s^22p^4$ Oxygen	**9** 18.998403 −1 84.95 53.48 1.696* **F** $1s^22s^22p^5$ Fluorine	**10** 20.179 27.096 24.553 0.901* **Ne** $1s^22s^22p^6$ Neon
		13 26.98154 3 2793 933.25 2.70 **Al** $[Ne]3s^23p^1$ Aluminum	**14** 28.0855 4 3540 1685 2.33 **Si** $[Ne]3s^23p^2$ Silicon	**15** 30.97376 ±3,5,4 550 317.30 1.82 **P** $[Ne]3s^23p^3$ Phosphorus	**16** 32.06 ±2,4,6 717.75 388.36 2.07 **S** $[Ne]3s^23p^4$ Sulfur	**17** 35.453 ±1,3,5,7 239.1 172.16 3.17* **Cl** $[Ne]3s^23p^5$ Chlorine	**18** 39.948 87.30 83.81 1.784* **Ar** $[Ne]3s^23p^6$ Argon
29 63.546 2,1 2836 1357.6 8.96 **Cu** $[Ar]3d^{10}4s^1$ Copper	**30** 65.38 2 1180 692.73 7.14 **Zn** $[Ar]3d^{10}4s^2$ Zinc	**31** 69.72 3 2478 302.90 5.91 **Ga** $[Ar]3d^{10}4s^24p^1$ Gallium	**32** 72.59 4 3107 1210.4 5.32 **Ge** $[Ar]3d^{10}4s^24p^2$ Germanium	**33** 74.9216 ±3,5 876 (subl.) 1081 (28 atm) 5.72 **As** $[Ar]3d^{10}4s^24p^3$ Arsenic	**34** 78.96 −2,4,6 958 494 4.80 **Se** $[Ar]3d^{10}4s^24p^4$ Selenium	**35** 79.904 ±1,5 332.25 265.90 3.12 **Br** $[Ar]3d^{10}4s^24p^5$ Bromine	**36** 83.80 119.80 115.78 3.74* **Kr** $[Ar]3d^{10}4s^24p^6$ Krypton
47 107.868 1 2436 1234 10.5 **Ag** $[Kr]4d^{10}5s^1$ Silver	**48** 112.41 2 1040 594.18 8.65 **Cd** $[Kr]4d^{10}5s^2$ Cadmium	**49** 114.82 3 2346 429.76 7.31 **In** $[Kr]4d^{10}5s^25p^1$ Indium	**50** 118.69 4,2 2876 505.06 7.30 **Sn** $[Kr]4d^{10}5s^25p^2$ Tin	**51** 121.75 ±3,5 1860 904 6.68 **Sb** $[Kr]4d^{10}5s^25p^3$ Antimony	**52** 127.60 −2,4,6 1261 722.65 6.24 **Te** $[Kr]4d^{10}5s^25p^4$ Tellurium	**53** 126.9045 ±1,5,7 458.4 386.7 4.92 **I** $[Kr]4d^{10}5s^25p^5$ Iodine	**54** 131.30 165.03 161.36 5.89* **Xe** $[Kr]4d^{10}5s^25p^6$ Xenon
79 196.9665 3,1 3130 1337.58 19.3 **Au** $[Xe]4f^{14}5d^{10}6s^1$ Gold	**80** 200.59 2,1 630 234.28 13.53 **Hg** $[Xe]4f^{14}5d^{10}6s^2$ Mercury	**81** 204.37 3,1 1746 577 11.85 **Tl** $[Xe]4f^{14}5d^{10}6s^26p^1$ Thallium	**82** 207.2 4,2 2023 600.6 11.4 **Pb** $[Xe]4f^{14}5d^{10}6s^26p^2$ Lead	**83** 208.9804 3,5 1837 544.52 9.8 **Bi** $[Xe]4f^{14}5d^{10}6s^26p^3$ Bismuth	**84** (209) 4,2 1235 527 9.4 **Po** $[Xe]4f^{14}5d^{10}6s^26p^4$ Polonium	**85** (210) ±1,3,5,7 610 575 — **At** $[Xe]4f^{14}5d^{10}6s^26p^5$ Astatine	**86** (222) 211 202 9.91* **Rn** $[Xe]4f^{14}5d^{10}6s^26p^6$ Radon

The A & B subgroup designations, applicable to elements in rows 4, 5, 6, and 7, are those recommended by the International Union of Pure and Applied Chemistry. It should be noted that some authors and organizations use the opposite convention in distinguishing these subgroups.

*** Estimated Values**

65 158.9254 3,4 3496 1630 8.27 **Tb** $[Xe]4f^96s^2$ Terbium	**66** 162.50 3 2835 1682 8.54 **Dy** $[Xe]4f^{10}6s^2$ Dysprosium	**67** 164.9304 3 2968 1743 8.80 **Ho** $[Xe]4f^{11}6s^2$ Holmium	**68** 167.26 3 3136 1795 9.05 **Er** $[Xe]4f^{12}6s^2$ Erbium	**69** 168.9342 3,2 2220 1818 9.33 **Tm** $[Xe]4f^{13}6s^2$ Thulium	**70** 173.04 3,2 1467 1097 6.98 **Yb** $[Xe]4f^{14}6s^2$ Ytterbium	**71** 174.967 3 3668 1936 9.84 **Lu** $[Xe]4f^{14}5d^16s^2$ Lutetium
97 (247) 4,3 — — — **Bk** $[Rn]5f^97s^2$ Berkelium	**98** (251) 3 — 900 — **Cf** $[Rn]5f^{10}7s^2$ Californium	**99** (252) — — — — **Es** $[Rn]5f^{11}7s^2$ Einsteinium	**100** (257) — — — — **Fm** $[Rn]5f^{12}7s^2$ Fermium	**101** (258) — — — — **Md** $[Rn]5f^{13}7s^2$ Mendelevium	**102** (259) — — — — **No** $[Rn]5f^{14}7s^2$ Nobelium	**103** (260) — — — — **Lr** $[Rn]5f^{14}6d^17s^2$ Lawrencium

APPENDIX T TRIGONOMETRY

The sine, cosine, and tangent of θ (see Fig. T-1) are given by

$$\sin\theta = \frac{y}{r} \qquad \cos\theta = \frac{x}{r} \qquad \tan\theta = \frac{y}{x} = \frac{\sin\theta}{\cos\theta}$$

The cosecant, secant, and cotangent of θ are given by

$$\csc\theta = \frac{r}{y} = \frac{1}{\sin\theta} \qquad \sec\theta = \frac{r}{x} = \frac{1}{\cos\theta} \qquad \cot\theta = \frac{x}{y} = \frac{1}{\tan\theta}$$

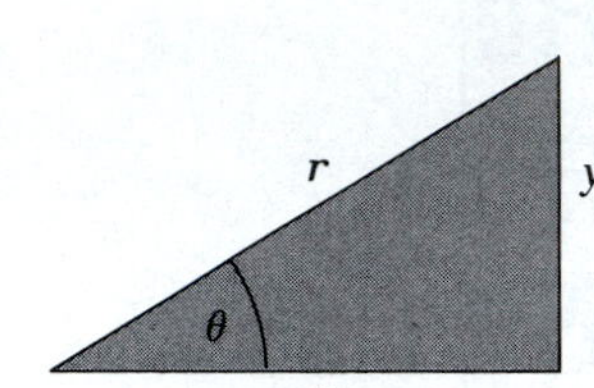

Figure T-1.

From Fig. T-1,

$$\sin\left(\theta \pm \frac{\pi}{2}\right) = \pm\cos\theta \qquad \cos\left(\theta \pm \frac{\pi}{2}\right) = \mp\sin\theta \qquad \tan\left(\theta - \frac{\pi}{2}\right) = \mp\cot\theta$$

From the pythagorean theorem, $x^2 + y^2 = r^2$, and

$$\sin^2\theta + \cos^2\theta = 1 \qquad \sec^2\theta - \tan^2\theta = 1 \qquad \csc^2\theta - \cot^2\theta = 1$$

The following related identities are derived in trigonometry:

$$\sin(\alpha \pm \beta) = \sin\alpha\cos\beta \pm \sin\beta\cos\alpha$$

$$\cos(\alpha \pm \beta) = \cos\alpha\cos\beta \mp \sin\alpha\sin\beta$$

$$\tan(\alpha \pm \beta) = \frac{\tan\alpha \pm \tan\beta}{1 \mp \tan\alpha\tan\beta}$$

$$\sin\alpha \pm \sin\beta = 2\sin\tfrac{1}{2}(\alpha \pm \beta)\cos\tfrac{1}{2}(\alpha \mp \beta)$$

$$\cos\alpha + \cos\beta = 2\cos\tfrac{1}{2}(\alpha + \beta)\cos\tfrac{1}{2}(\alpha - \beta)$$

$$\cos\alpha - \cos\beta = 2\sin\tfrac{1}{2}(\alpha + \beta)\sin\tfrac{1}{2}(\beta - \alpha)$$

$$\tan\alpha \pm \tan\beta = \frac{\sin(\alpha \pm \beta)}{\cos\alpha\cos\beta}$$

$$\sin\alpha\sin\beta = \tfrac{1}{2}[\cos(\alpha - \beta) - \cos(\alpha + \beta)]$$

$$\cos\alpha\cos\beta = \tfrac{1}{2}[\cos(\alpha + \beta) + \cos(\alpha - \beta)]$$

$$\sin\alpha\cos\beta = \tfrac{1}{2}[\sin(\alpha + \beta) + \sin(\alpha - \beta)]$$

$$\sin^2\alpha - \sin^2\beta = \sin(\alpha + \beta)\sin(\alpha - \beta) = \cos^2\beta - \cos^2\alpha$$

$$\cos^2\alpha - \sin^2\beta = \cos(\alpha + \beta)\cos(\alpha - \beta) = \cos^2\beta - \sin^2\alpha$$

$$\sin 2\theta = 2\sin\theta\cos\theta$$

$$\cos 2\theta = \cos^2\theta - \sin^2\theta = 2\cos^2\theta - 1 = 1 - 2\cos\theta$$

$$\tan 2\theta = \frac{2\tan\theta}{1 - \tan^2\theta}$$

$$\sin^2\frac{\theta}{2} = \frac{1}{2}(1 - \cos\theta) \qquad \cos^2\frac{\theta}{2} = \frac{1}{2}(1 + \cos\theta)$$

$$\sin^3\theta = \tfrac{1}{4}(-\sin 3\theta + 3\sin\theta) \qquad \cos^3\theta = \tfrac{1}{4}(\cos^3\theta + 3\cos\theta)$$

$$\sin 3\theta = 3\sin\theta - 4\sin^3\theta \qquad \cos 3\theta = 4\cos^3\theta - 3\cos\theta$$

$$\sin^4\theta = \tfrac{1}{8}(\cos 4\theta - 4\cos 2\theta + 3) \qquad \cos^4\theta = \tfrac{1}{8}(\cos 4\theta + 4\cos 2\theta + 3)$$

$$\sin 4\theta = \cos\theta\,(4\sin\theta - 8\sin^3\theta) \qquad \cos 4\theta = 8\cos^4\theta - 8\cos^2\theta + 1$$

$$\sin(-\theta) = -\sin\theta \qquad \cos(-\theta) = \cos\theta \qquad \tan(-\theta) = -\tan\theta$$

For any triangle (see Fig. T-2),

$$\alpha + \beta + \gamma = \pi \text{ rad} = 180^\circ$$

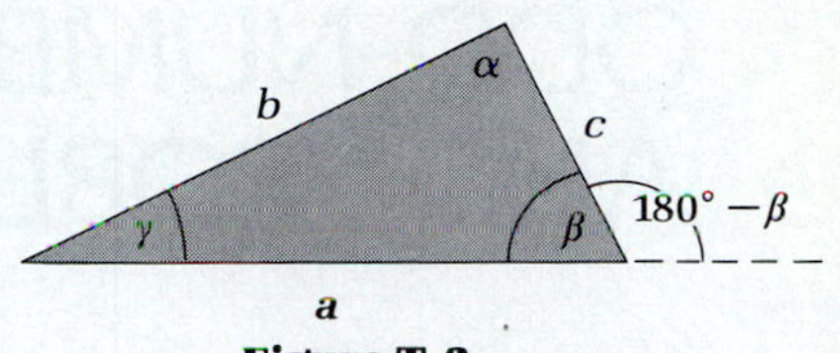

Figure T-2.

Law of cosines:

$$\begin{aligned} a^2 &= b^2 + c^2 - 2bc \cos \alpha \\ &= a^2 + b^2 + 2bc \cos (180^\circ - \alpha) \end{aligned}$$

Law of sines:

$$\frac{a}{\sin \alpha} = \frac{b}{\sin \beta} = \frac{c}{\sin \gamma}$$

ANSWERS TO SELECTED ODD-NUMBERED EXERCISES AND PROBLEMS

CHAPTER 1

Exercises

1. [length]/[time]²; acceleration
3. [mass][length]²/[time]².
5. $b = 1$, $c = -1$, $d = 1$.
7. 0.2778 (m/s)/(km/h).
9. *(a)* 0.447 (m/s²)/(mi/h·s); *(b)* 5.4 m/s²
11. 62.4 pound-mass/ft³.
13. 1.86×10^5 mi/s.
15. 7.69×10^{17}.
17. 3.3×10^{24} kg.
19. 1.89×10^5.
21. *(a)* 10^{15} kg; *(b)* 10^4 kg.
23. 10^{21} kg.
25. 10^6 kg.

CHAPTER 2

Exercises

1. *(a)* 2.2 m; *(b)* 2.2 m; *(c)* 3.7 m.
7. *(a)* **d** and **e** opposite; *(b)* **d** and **e** parallel; *(c)* **d** and **e** opposite; *(d)* **d** and **e** perpendicular.
9. *(b)* 58 mm, 22°; *(c)* 54 mm, 22 mm.
11. *(a)* −260 m, 220 m; *(b)* 260 m, −220 m; *(c)* −260 m, 220 m, 35 m.
13. *(a)* $\sqrt{x^2 + y^2}$; *(b)* $(x/r)\mathbf{i} + (y/r)\mathbf{j}$.
15. *(a)* $\mathbf{i} - 2\mathbf{j} + 5\mathbf{k}$; *(b)* $5\mathbf{i} + 10\mathbf{j} - 5\mathbf{k}$; *(c)* $-5\mathbf{i} - 10\mathbf{j} + 5\mathbf{k}$; *(d)* $\mathbf{i} - 2\mathbf{j} + 5\mathbf{k}$.
19. *(a)* 1.9 km, 80° north of east.
21. *(a)* (25 mm, 0), (18 mm, 18 mm), (0, 25 mm), (−18 mm, 18 mm), (−25 mm, 0), (−18 mm, −18 mm), (0, −25 mm), (18 mm, −18 mm); *(b)* (**a**: 61 mm, 45°), (**b**: 66 mm, 112°), (**c**: 47 mm, 160°).
23. 170°.
25. *(b)* 180°; *(c)* 0.
27. $(-0.6)\mathbf{i} + (0.8)\mathbf{j}$.
29. $(4\text{ m})\mathbf{i} + (4\text{ m})\mathbf{j} + (4\text{ m})\mathbf{k}$.

Problems

1. *(c)* $F_x = F\cos\alpha$, $F_y = F\cos\beta$, $F_z = F\cos\gamma$.
3. *(a)* 90°; *(b)* 60°; *(c)* 180°; *(d)* 0°.
5. *(b)* x/r, y/r, z/r; $r = \sqrt{x^2 + y^2 + z^2}$.
7. *(a)* $(3\text{ m})\mathbf{i} - (4\text{ m})\mathbf{j}$; *(b)* $(0.6)\mathbf{i} - (0.8)\mathbf{j}$; *(c)* $(0.8)\mathbf{i} + (0.6)\mathbf{j}$; *(d)* $\mathbf{k}$.
9. *(a)* 3.7 m; *(b)* 74°, 58°, 37°.

CHAPTER 3

Exercises

1. *(a)* −16 m; *(b)* 37 m; *(c)* $(-16\text{ m})\mathbf{i}$; *(d)* $(37\text{ m})\mathbf{i}$; *(e)* $(53\text{ m})\mathbf{i}$.
3. *(a)* For example, $x(2.0\text{ s}) = 40$ mm; *(b)* 153 mm; *(c)* $(-49\text{ mm})\mathbf{i}$.
5. *(a)* 0.3048 (m/s)/(ft/s); *(b)* 82 ft/s.
7. *(a)* 500 s; *(b)* 9 light-minutes.
9. 200 m.
11. *(a)* 29 m/s; *(b)* 26 m/s.
13. *(a)* 0.87 m/s; *(b)* 1.3 m/s; *(c)* 1.7 m/s; *(d)* 2.0 m/s; *(e)* 2.5 m/s.
15. 4×10^4 m/s.
17. *(a)* $x(t) = -0.02\text{ m} + (1.3\text{ m/s})t$; *(b)* $x(t) = 0.76\text{ m} - (1.3\text{ m/s})t$.
19. 0.50 s.
21. *(a)* 0.447 (m/s²); *(b)* (mi/h·s); *(c)* 5.4 m/s².
23. *(a)* −0.9 m/s²; *(b)* +0.9 m/s².
25. *(a)* 0.43 m/s²; *(b)* 0.63 m/s²; *(c)* 0.85 m/s²; *(d)* 1.1 m/s²; *(e)* 1.4 m/s².
27. *(a)* $a(t) = (6.4\text{ m/s}^3)t$; *(b)* 17 m/s²; *(c)* zero.
29. *(a)* $v(t) = (3.6\text{ m/s}^2)t$; *(b)* 86 m/s; *(c)* $x(t) = (1.8\text{ m/s}^2)t^2$; *(d)* 1.0 km.
31. *(a)* 12 m/s; *(b)* 16 m/s.
33. 1.8 m/s².
35. *(a)* 1.6 s; *(b)* 2.3 s.
37. 4.2 s.
39. 2×10^5 m/s².
41. *(a)* 3.0 m; *(b)* 3.2 m/s²; *(c)* 2.9 m/s²; *(d)* $x(t) = 3.0\text{ m} + (2.9\text{ m/s})t + (1.6\text{ m/s}^2)t^2$.
43. 0.64 g.
45. 10 m/s².
47. *(a)* $h_m = 9.1$ m, $t_m = 1.2$ s; *(b)* 0.71 s and 1.7 s; *(c)* 7.8 m and 7.8 m.
49. 10 m/s.
51. 10 m/s².
53. 6.5 m/s².
55. −7 m/s; negative; no.
57. $t = 2$ s.

Problems

1. *(a)* 11.5 m/s; *(b)* 2.6 s; *(c)* 7.4 s; *(d)* 4.4 m/s².
3. *(a)* 7.8 s; *(b)* 140 m; *(c)* 36 m/s.
5. 1.6 km.
11. Tails up.
15. *(a)* $b = 6.0\text{ m/s}^2$, $c = 1.0\text{ m/s}^3$; *(b)* $t = 2.0$ s, $x = 16$ m.
17. *(b)* 9.815 m/s².
19. *(a)* $v = (\lambda/\tau)(1 - e^{-t/\tau})$; *(b)* $a = (\lambda/\tau^2)e^{-t/\tau}$.
21. *(a)* 6.3 m/s; *(b)* 4.4 m/s; *(c)* 9.8×10^3 m/s² upward; *(d)* 4.9×10^3 m/s² upward.

CHAPTER 4

Exercises

1. *(a)* $\mathbf{r} = (31.8\text{ m})\mathbf{i} + (31.8\text{ m})\mathbf{j}$; *(b)* $\Delta\mathbf{r} = -(45.0\text{ m})\mathbf{i} + (45.0\text{ m})\mathbf{j}$; *(c)* 70.7 m.
3. *(a)* $\overline{\mathbf{v}} = -(1.34\text{ m/s})\mathbf{i} + (1.34\text{ m/s})\mathbf{j}$; *(b)* $\overline{\mathbf{v}} = -(1.45\text{ m/s})\mathbf{i} + (1.45\text{ m/s})\mathbf{j}$.
5. *(a)* $\mathbf{v} = (16\text{ m/s})\mathbf{j}$; *(b)* $\mathbf{a} = (0.88\text{ m/s})\mathbf{j}$; *(c)* $\mathbf{v} = (16\text{ m/s})\mathbf{i} + (16\text{ m/s})\mathbf{j}$; *(d)* $\overline{\mathbf{a}} = (1.7\text{ m/s}^2)\mathbf{i} - (1.7\text{ m/s}^2)\mathbf{j}$.

7. *(a)* $\mathbf{v} = (3.5 \text{ m/s})\mathbf{i} + (5.1 \text{ m/s})\mathbf{j}$.
9. *(a)* $v_h = 14$ m/s; *(b)* $v_v = 7.8$ m/s.
11. *(a)* $a_x = 1.7 \text{ m/s}^2$, $a_y = -0.47 \text{ m/s}^2$, $v_x = (1.7 \text{ m/s}^2)t$, $v_y = -(0.47 \text{ m/s}^2)t$, $x = (0.87 \text{ m/s}^2)t^2$, $y = -(0.23 \text{ m/s}^2)t^2$; *(b)* $a_x = 1.81 \text{ m/s}^2$, $a_y = 0$, $v_x = (1.81 \text{ m/s}^2)t$, $v_y = 0$, $x = (0.905 \text{ m/s}^2)t^2$, $y = 0$.
13. $v_x = 17$ m/s, $v_y = 32 \text{ m/s} - (9.8 \text{ m/s}^2)t$, $x = (17 \text{ m/s})t$, $y = (32 \text{ m/s})t - (4.9 \text{ m/s}^2)t^2$.
15. 16 m/s.
17. *(a)* 13°; *(b)* 77°.
19. *(a)* 5.1 s; *(b)* 130 m; *(c)* 140 m.
21. $v_0 = 28$ m/s, $\theta_0 = 45°$.
23. *(a)* 1.6 m/s^2; *(b)* 2.9 m/s^2.
25. *(a)* 23 m/s northeast; *(b)* 23 m/s southeast; *(c)* $\mathbf{a}_D = (1.9 \text{ m/s}^2)\mathbf{i} - (1.9 \text{ m/s}^2)\mathbf{j}$, $\mathbf{a}_H = -(1.2 \text{ m/s}^2)\mathbf{i} - (1.2 \text{ m/s}^2)\mathbf{j}$.
27. *(a)* 5.5 m/s; *(b)* 31 m/s^2.
29. *(a)* $3.37 \times 10^{-2} \text{ m/s}^2 = 3.44 \times 10^{-3}\, g$; *(b)* $5.9 \times 10^{-3} \text{ m/s}^2 = 6.1 \times 10^{-4}\, g$; *(c)* $2.2 \times 10^{-10} \text{ m/s}^2 = 2.2 \times 10^{-11}\, g$.
31. *(b)* 14 m/s^2.
33. *(a)* 6 m/s; *(b)* 6 m/s south; *(c)* 6 m/s; *(d)* 6 m/s north.
35. 4.3 m/s 11° north of east.
37. *(a)* $\mathbf{v}_{BW} = -(2.3 \text{ m/s})\mathbf{i} + (7.5 \text{ m/s})\mathbf{j}$; *(b)* 4.0 min; *(c)* $\mathbf{v}_{BW} = -(4.6 \text{ m/s})\mathbf{i} + (6.3 \text{ m/s})\mathbf{j}$, 4.8 min.
41. 1.1×10^{-11} m.
43. 16.1°, 0.26 s.
45. *(a)* $\mathbf{v}_{GM} = 0$, $\mathbf{v}_{GB} = (1.3 \text{ m/s})\mathbf{j}$, $\mathbf{a}_{GM} = \mathbf{a}_{GB} = -(9.8 \text{ m/s}^2)\mathbf{j}$; *(b)* $\mathbf{v}_{GM} = -(1.3 \text{ m/s})\mathbf{j}$, $\mathbf{v}_{GB} = 0$, $\mathbf{a}_{GM} = \mathbf{a}_{GB} = -(9.8 \text{ m/s}^2)\mathbf{j}$.
47. 7.4 m/s.
49. 16 m.
51. 8×10^5 m/s.

Problems

1. *(e)* $\mathbf{v} = -(1.49 \text{ m/s})\mathbf{i} + (1.49 \text{ m/s})\mathbf{j}$, $\mathbf{a} = -(6.95 \times 10^{-2} \text{ m/s}^2)\mathbf{i} - (6.95 \times 10^{-2} \text{ m/s}^2)\mathbf{j}$; *(f)* uniform circular motion.
3. *(b)* 38°; *(c)* 76°; *(d)* $h_m = R_m/4$.
5. 130 m.
9. *(b)* $(0.900)v^2/R$; *(c)* $(0.974)v^2/R$; *(d)* $(0.996)v^2/R$; *(e)* $(1.000)v^2/R$; *(f)* 1/2.
11. 3.8 m/s^2.
13. *(a)* 108 s; *(b)* 87 s; *(c)* parallel to the current took longer by 21 s.
15. 70 m/s.
17. *(a)* 8.4 m/s; *(b)* 13 m/s.
19. *(b)* $v_x = v - v\cos(vt/R)$, $v_y = v\sin(vt/R)$; *(c)* $a_x = (v^2/R)\sin(vt/R)$, $a_y = (v^2/R)\cos(vt/R)$.
23. *(b)* 20 m/s.
25. *(a)* $3.21 \times 10^{-3} \text{ m/s}^2$; *(b)* $8.65 \times 10^{-3} \text{ m/s}^2$.

CHAPTER 5

Exercises

1. 4.9 sl.
3. *(a)* 14 kN; *(b)* 630 tons.
5. *(a)* $F_1 = 6.6$ N, F_2 13.0 N; *(b)* $\theta_1 = 111°$, $\theta_2 = -49°$; *(d)* $|\Sigma\mathbf{F}| = 7.1$ N, $\theta = 31°$.
7. 300 N, 19° east of north.
9. *(b)* $\mathbf{F}_{air} = 720$ N up, $\mathbf{F}_e = 720$ N down.
11. 1.8 kN.
13. *(a)* 10 N; *(b)* 10^{-21} s.
15. 50 N.
17. 2 kN.
19. *(a)* 7×10^{28} N; *(b)* 3.5×10^{22} N.
21. 1.24 kg.
23. *(a)* $\mathbf{F}_{12} = (4 \text{ N})\mathbf{i}$, $\mathbf{F}_{21} = -(4 \text{ N})\mathbf{i}$; *(b)* $\mathbf{F}_{12} = (8 \text{ N})\mathbf{i}$, $\mathbf{F}_{21} = -(8 \text{ N})\mathbf{i}$.
25. 13 m/s^2.
27. *(a)* 8.9×10^{-30} N.
29. *(a)* 970 N; *(b)* 520 N; *(c)* 750 N.
31. 1.7 s.
33. *(b)* 270 N up; *(c)* 4.2 m/s^2; *(d)* 3.5 m.
35. *(a)* 21 N; *(b)* 6.5 N.
37. *(a)* 2.2 N; *(b)* 1.0 N, west.
39. *(b)* 1.5 N; *(c)* 15 N.
41. *(a)* 1.8 kN; *(b)* 9.2 kN; *(c)* 9.4 kN, 79° above horizontal.
43. 24 N.
45. $F_2 = 220$ N, $F_3 = 270$ N.
47. *(a)* 190 N; *(b)* 540 N; *(c)* 570 N.
49. *(a)* $F_1 = 4.1$ N, $F_2 = 12.3$ N; *(b)* $F_1 = 8.2$ N, $F_2 = 12.3$ N.
51. 0.45 m.

Problems

1. $6.5 \times 10^{-3} \text{ m/s}^2$; no; an inertial frame.
3. *(b)* 16 kg.
5. 7.6 m/s^2.
7. $F_{T1} = 10$ N, $F_{T2} = 6$ N, $F_{T3} = 4$ N.
11. *(a)* $a_A = 0.10 \text{ m/s}^2$, $a_B = 0.20 \text{ m/s}^2$; *(b)* 0.50; *(c)* $d_A = 2.0$ m, $d_B = 4.0$ m; *(d)* parts *(b)* and *(c)*.
15. *(b)* Upward; *(c)* 1 kN; *(d)* larger, factors m and d.
17. *(b)* 20 kg.
21. *(a)* $m_C g \sin\theta$; *(b)* $(m_C + m_B)g\sin\theta$; *(c)* $m_B\cos\theta$.

CHAPTER 6

Exercises

1. *(a)* 310 N; *(b)* 190 N; *(c)* zero.
3. *(a)* 180 N; *(b)* 160 N.
5. *(a)* 1.1; *(b)* 0.75.
7. Yes, $v = 30$ m/s.
9. 11 m.
11. *(c)* $F_N = 310$ N, $F_c = 370$ N.
13. 4.2 m/s^2.
15. *(a)* 77 N; *(b)* 71 N.
17. 83 N.
19. 0.25 m/s^2.
21. 0.061 m/s^2.
23. 12°.
25. 1.1 km.
27. *(a)* 3.7 m/s^2; *(b)* 3.3 kN; *(c)* 8.6 kN; *(d)* 9.2 kN; *(e)* 21°.
29. 0.11.
31. *(a)* $v_m = \sqrt{gR}$; *(b)* 3.1 m/s.
33. *(a)* 1.0 kN; *(b)* 6.7 m/s; *(c)* 6.6 s.
35. 12.79 N.
37. *(a)* 12.8 m/s^2; *(b)* 5.2 Mm.
39. 880 N.
41. With a force in this direction, the block cannot slide up the wall.
43. 20 m.
45. 57°.
47. 19 m/s.
49. *(a)* 3.7 N; *(b)* 24 N.

Problems

3. $F_{a,\min} = \dfrac{mg}{\mu_s}\left(1 + \dfrac{m}{M}\right)$.
5. (c) 290 N; (d) 35°; (e) 350 N.
7. 33 m.
9. (a) 0.64 m/s²; (b) 46 N.
13. (a) 2.9 m/s, south; (b) 2.2 m/s², west; (c) 1.5 N, west.
17. (a) 1.5 kN, inward; (b) 1.2 kN, outward.
19. 6.9 m/s².

CHAPTER 7

Exercises

3. (a) 3.86×10^7 s; (b) 5.69×10^7 s.
5. 1.3×10^{-10} N.
7. 0.64 m.
9. 20 Mm.
11. (a) 4.18×10^{15} N; (b) 1.80×10^{15} N; (c) 8.69×10^{15} N.
13. See Table 7-1.
15. (a) 1.984×10^{-29} C/kg; (b) 39.5 C.
17. 259 Mm, 0.175 percent.
19. (a) 281 N; (b) 69.4 N; (c) 27.8 N.
21. 3.18×10^{-5}.
23. $F = \sqrt{3}Gm^2/a^2$.
25. 9.75 m/s².
27. 0.003.
29. (a) 3.7 N/kg; (b) 260 N.
31. $\mathbf{g} = (2.2 \times 10^{-11}\ \text{N/kg})\mathbf{i} - (5.5 \times 10^{-11}\ \text{N/kg})\mathbf{j}$.
33. (a) 7.44×10^3 m/s; (b) 6.07×10^3 s.
35. (a) 1.37×10^3 m/s.
37. (b) $x = 0$, $y = 14$ Gm; (c) $x = 0$, $y = 100$ Gm; (d) $v_x = -130$ km/s, $v_y = 0$; (e) $v_x = 18$ km/s, $v_y = 0$.
39. $r^3/T^2 = 3.21 \times 10^{15}$ m³/s².
41. See Table 7-1.
43. (a) 6.02×10^{24} kg; (b) 1 percent.
45. (a) $(3.3 \times 10^{-11}\ \text{N})\mathbf{j}$;
 (b) $(1.6 \times 10^{-11}\ \text{N})\mathbf{i} + (2.1 \times 10^{-11}\ \text{N})\mathbf{j}$;
 (c) $(1.6 \times 10^{-11}\ \text{N})\mathbf{i} + (5.5 \times 10^{-11}\ \text{N})\mathbf{j}$;
 (d) $(0.80 \times 10^{-11}\ \text{N})\mathbf{i} + (2.7 \times 10^{-11}\ \text{N})\mathbf{j}$.
47. (a) 5.93×10^{-3} m/s²; (b) 33.7×10^{-3} m/s²; (c) 27.8×10^{-3} m/s²; (d) 39.6×10^{-3} m/s²; (e) midnight.
49. 1×10^6 N/kg.

Problems

1. (a) $G \approx 14 \times 10^{-11}$ N·m²/kg² (in error by more than a factor of 2); (b) 5.51×10^3 kg/m³.
3. (a) 2.908×10^{25} kg; (b) 8.00×10^3 kg/m³; (c) 21.3 N/kg; (d) 17.8 N.
5. (a) $x_n = x_c/(1 + \sqrt{m_c/m_b})$; (b) $x_n = 2$ m.
13. (a) 17.3 km/s; (b) 13.1 km/s; (c) opposite; (d) 4.2 km/s.
15. (f) 2×10^{-6}.
19. (b) 7×10^3 s; (c) 7×10^{-4} s.

CHAPTER 8

Exercises

1. (a) 40 N, up; (b) 80 J.
5. (a) −15 J; (b) 0.030.
9. (a) 3 J; (b) 87°.
13. (a) −21 J; (b) minimum = 0, maximum = 290 N.
17. $C(1/z_f - 1/z_i)$.
21. (a) 22 J; (b) 0.
23. (a) 86 kJ; (b) 170 kJ.
25. 11 m/s.
27. (a) −2.9 J; (b) 0.050; (c) 4.6 N; (d) 0.
29. (a) −5.8 J; (b) 12 m/s.
31. 4.8 m/s.
33. 210 J.
35. 410 N.
37. (a) 2100 hp; (b) 690 kN; (c) 2.1 MN.
39. (a) 100 J; (b) 10 J; (c) 1 mJ.
41. (a) −20 J; (b) 110°.
43. (a) zero; (b) −60 J; (c) +60 J.
45. (a) $-(A/4)(x_f^4 - x_i^4)$; (b) -4×10^{-18} J.
47. (a) 12 mW; (b) −12 mW.
49. (a) 3.1×10^{-16} J; (b) 2.6×10^7 m/s.

Problems

1. (a) −160 J; (b) −100 J; (c) 0; (d) −60 J; (e) 1.4 m.
3. 0.
5. (a) 14 kJ; (b) −14 kJ; (c) 1.0 kW; (d) 1.2 kW; (e) 0.5 kW.
7. (a) 4.7 m/s; (b) 42 N.
9. (a) 1×10^7 J; (b) 3×10^7 J; (c) 2 kW.
11. (a) 3 m/s; (b) 300 N; (c) 1.2 kJ in each case.
13. (a) −300 kJ; (b) 7 kW.
15. $W = \frac{1}{2}k(x_i^2 - x_f^2) + \frac{1}{2}k(y_i^2 - y_f^2)$.

CHAPTER 9

Exercises

1. (a) 54 J; (b) 54 J; (c) 10 m.
3. (a) 0.21 m; (b) 11 m/s; (c) 0.18 m.
5. 4.2 MN/m.
7. (a) 3.5 kJ; (b) no; (c) 9.6 m/s.
9. (a) 60 J; (b) 4.9 m/s; (c) 4.2 m/s.
11. (a) 2.0 m; (b) 200 J; (c) 1.0 m; (d) $-(100\ \text{N})\mathbf{j}$.
13. (a) 0.18×10^{-20} J; (b) −0.067 nm.
15. (a) −2 kJ; (b) 2 kJ; (c) 0.
17. (a) $-(3\ \text{N})(x_f - x_i) - (4\ \text{N})(y_f - y_i)$; (b) $-(3\ \text{N})x - (4\ \text{N})y$; (c) −48 J; (d) 0.
19. (a) 87 J; (b) 26 m/s; (c) 44 J.
21. (a) 1.1 m/s; (b) $mg(3 - 2\cos 30°)$.
23. $\sqrt{3gr}$.
25. (a) 2.7 J if y is measured from the initial position on the countertop; (b) 8.6 kN/m; (c) 6.2 m.
27. (a) −4200 J.
29. (a) 6 kJ.
31. (a) $0.05\ mg \approx 500$ N; (b) 10 kW.
33. (a) $\Delta K = -59$ J, $\Delta U = 45$ J, $\Delta E_{int} = 14$ J.
35. (a) -1.7×10^{11} J; (b) 8.5×10^{10} J; (c) -8.5×10^{10} J; (d) kinetic energy.
37. (a) -3.8×10^{32} J; (b) 2.3×10^{32} J.
39. (a) 7.0 km/s; (b) 7.5 km/s.
41. 40 mJ.
43. (a) 50 kN/m; (b) 60 J, 90 J.
45. (a) −80 kJ; (b) +70 kJ; (c) 5 m/s.
47. (a) $+3 \times 10^{-11}$ m, -3×10^{-11} m; (b) $+4 \times 10^{-11}$ m, -3×10^{-20} J; (c) -4×10^{-11} m, -3×10^{-20} J; (d) -4.5×10^{-11} m, 4×10^{-20} J.
49. −6 kJ.
51. (a) 7.6×10^3 m/s; (b) 5.8×10^3 s; (c) 1.1×10^4 m/s.

Problems

1. *(a)* 1.6 m; *(b)* 7.7 m/s.
3. *(a)* $U = mgy + \frac{1}{2}ky^2$; *(b)* $F_y = -mg - ky$.
7. *(b)* $\sqrt{2ghm_1/(m_1 + m_2)}$.
13. *(a)* $-A/(x^2 - a^2)$.
15. *(a)* -10^{39} J; *(b)* 2×10^5 m/s.
17. 1.8 m.
19. 1×10^5 kg/s, 1×10^{-4} m/s.
23. 2×10^{-14} J.

CHAPTER 10

Exercises

1. $x_{cm} = -1.1$ m, $y_{cm} = 0.79$ m.
3. $x_{cm} = 25$ mm, $y_{cm} = 25$ mm.
5. $x_{cm} = \ell/6$, $y_{cm} = 2\ell/3$; $\mathbf{r}_{cm} = (\ell/6)\mathbf{i} + (2\ell/3)\mathbf{j}$.
7. 0.50 m/s.
9. *(a)* 1.7 m/s², south; *(b)* zero.
11. *(a)* $g[(m_2 - m_1)^2/(m_1 + m_2)^2]$; *(b)* $g[(m_2 - m_1)^2/(m_1 + m_2)]$; *(c)* the earth and the pulley.
13. 1.8×10^{29} kg·m/s.
15. *(a)* 1.1×10^{-23} kg·m/s, north; *(b)* 1.1×10^{-23} kg·m/s, south; *(c)* 2.1×10^{-23} kg·m/s, south.
17. *(a)* 2200 kg·m/s; *(b)* 50 kN.
19. 0.53 N·s, upward; 0.53 N·s, downward.
21. *(a)* 4.02 m/s; *(b)* 29 J.
23. 0.23 m/s.
25. *(a)* 0.20 m/s; *(b)* zero; *(c)* yes.
27. 0.68 m/s.
29. 51 kg.
31. 0.91 m/s.
33. 5.0 m/s.
35. 1.2×10^5 m/s.
37. 420 m/s.
39. $(-1.2$ m/s$)\mathbf{i}$. The initially stationary cart has the larger mass.
41. 18 km/h, 11° east of south.
43. $(-6.0$ m/s$)\mathbf{i} + (3.0$ m/s$)\mathbf{j}$.
45. *(a)* 0.85 m/s; *(b)* 15°; *(c)* 75°.
49. 736 kg.
51. *(a)* 20 N; *(b)* 1000 s; *(c)* 10 kg.
53. 1300 bullets/min.
55. 1.1 m/s.
57. *(a)* $v = Mv_m/(m + M)$; *(b)* $V = mv_m/(m + M)$; *(c)* 99 m/s; *(d)* 0.99 m/s.
59. 200 N.

Problems

11. *(a)* $v_{1f} = -1.0$ m/s, $v_{2f} = 2.0$ m/s; *(b)* $v_{1f} = 1.3$ m $v_{2f} = 3.0$ m/s; *(c)* $v_{1f} = -0.3$ m/s, $v_{2f} = 1.0$ m/s.
13. 10^{-9} m.
19. $(58\text{ N})\mathbf{i} - (58\text{ N})\mathbf{j}$.

CHAPTER 11

Exercises

1. *(a)* 7.8 N·m, clockwise; *(b)* 0; *(c)* 7.8 N·m, counterclockwise; *(d)* 0.
3. *(a)* 20.0 N; *(b)* 20.0 N·m, assuming a 1-m moment arm.
5. *(a)* at 20.9 cm; *(b)* 14 N.
7. 400 N.
9. *(a)* 0 vertical, 24 kN horizontal, 31 kN tension; *(b)* along the boom to the right.
11. *(a)* 10 kN front, 40 kN rear; *(b)* 36 kN.
13. *(a)* 470 N; *(b)* 400 N; *(c)* 120 N.
15. *(a)* 0.81 F_e; *(b)* 0.33 F_e horizontal to the left, 0.19 F_e vertically up; *(c)* toward the center of the disk.
17. *(a)* $F_a/\sqrt{2}$ tension, $\frac{1}{2}F_a$ friction, $\frac{1}{2}F_a$ normal.
19. $x = 0.56$ m, $y = 3.44$ m relative to the pin.
21. 45°.
23. $(369\text{ N·m})\mathbf{i} + (224\text{ N·m})\mathbf{j} + (660\text{ N·m})\mathbf{k}$.
25. 1 m³.
27. *(b)* $\mathbf{i} + 7\mathbf{j} - 5\mathbf{k}$.
31. *(a)* $\tau_{z1} = 5.0$ N·m, $\tau_{z2} = 5.0$ N·m, $\tau_z = 10.0$ N·m; *(b)* $\tau_{z1} = 0.0$ N·m, $\tau_{z2} = 10.0$ N·m, $\tau_z = 10.0$ N·m; *(c)* $\tau_{z1} = 10.0$ N·m, $\tau_{z2} = 0.0$ N·m, $\tau_z = 10.0$ N·m.
33. *(a)* 1.5 kN·m, out of page; *(b)* 1.2 kN, up; *(c)* 740 N, down.
35. 300 N.
37. 0.4 m.
39. *(a)* $2F$; *(b)* $\mathbf{j}$; *(c)* $a/2\sqrt{2}$.
41. $x = 1.8$ m, $y = 2.0$ m; no.

Problems

3. $\frac{1}{2}F_e$, $\mu_s \geq \frac{1}{2}$.
5. *(a)* 150 N on the left, 130 N on the right; *(b)* 120 N; *(c)* 120 N at 15° with the horizontal.
7. *(a)* 28-kN tension, $F_{px} = 28$ kN, $F_{py} = 18$ kN; *(b)* 12-kN tension, $F_{px} = 11$ kN, $F_{py} = 21$ kN.
11. 10^{-9} N·m.
13. *(a)* $F_{MV} = 200$ N, $F_{LV} = 200$ N, $F_{MH} = 250$ N, $F_{Lh} = 250$ N; *(b)* $F_{UV} = 200$ N, $F_{LV} = 200$ N, $F_{UH} = 130$ N, $F_{LH} = 130$ N.
15. 6.0 kN.
17. *(a)* $F_{Ax} = 0$, $F_{Ay} = 150$ N; *(b)* $F_{Bx} = 77$ N, $F_{By} = 0$; *(c)* $F_{cx} = -150$ N, $F_{cy} = 0$; *(d)* $F_{Dx} = -77$ N, $F_{Dy} = 0$; *(e)* $F_{Ex} = 0$, $F_{Ey} = 150$ N.
19. $F_{Ax} = Mg/\cos(30°)$, $F_{Bx} = -Mg/\cos(30°)$, $F_{Ay} + F_{By} = Mg$.
21. 280 N horizontal, 310 N tension, 350 N vertical.

CHAPTER 12

Exercises

1. *(a)* 2.1 rad; *(b)* 120°; *(c)* 0.33 rev.
3. *(a)* 0.51 m; *(b)* 0.51 m.
5. 2.6×10^9 m.
7. 8.2 rad/s.
9. *(a)* $-(4.2\text{ rad/s}^3)t^2$; *(b)* -19 rad/s; *(d)* 19 rad/s; *(c)* -6.2 rad.
11. $\theta(t) = -(3.5\text{ rad/s})t$.
13. *(a)* $\alpha_z = -2.2$ rad/s²; *(b)* $\theta(t) = (5.8\text{ rad/s})t - (1.1\text{ rad/s}^2)t^2$; *(c)* $t_q = 2.6$ s; *(d)* north before t_q, south after t_q; *(e)* $\omega_z^2 = (5.8\text{ rad/s})^2 - (4.4\text{ rad/s}^2)\theta$.
15. *(a)* $\omega_z(t) = -(1.4\text{ rad/s}^2)t$; *(b)* $\theta(t) = -(0.7\text{ rad/s}^2)t^2$; *(c)* $\omega_z^2 = -(2.7\text{ rad/s}^2)\theta$.
17. 9.7 mm/s.
19. *(a)* $|a_t| = 0.25$ m/s², $|a_R| = 0.16$ m/s²; *(b)* $v = 0.48$ m/s, $a = 0.30$ m/s².
21. *(a)* $(ML^2)(T^{-2}) = ML^2T^{-2}$; *(b)* $(\text{kg·m}^2)(\text{s}^{-2}) = \text{kg(m/s)}^2 = \text{J}$.

23. $0.34\ \text{kg}\cdot\text{m}^2$.
25. *(a)* $20\ \text{kg}\cdot\text{m}^2$; *(b)* $9\ \text{kg}\cdot\text{m}^2$; *(c)* $29\ \text{kg}\cdot\text{m}^2$.
27. $34\ \text{kg}\cdot\text{m}^2$.
29. $Ma^2/6$.
31. $13Mr_0^2/20$.
33. *(a)* 4.1 rad/s; *(b)* 2.8 m/s.
35. *(a)* 3.9 m/s; *(b)* 41 rad/s.
37. 5 J, 1/2.
39. *(a)* 62 s; *(b)* 1200 rad.
41. $\omega_z(t) = (0.93\ \text{rad/s}^3)t^2 - 2.9\ \text{rad/s}$, $\alpha_z(t) = (1.86\ \text{rad/s}^3)t$.
43. *(a)* 2.00×10^{-7} rad/s; *(b)* 2.99×10^4 m/s; *(c)* $5.96 \times 10^{-3}\ \text{m/s}^2$.
45. 160 kJ.
47. *(a)* $(4.0\ \text{m/s})\mathbf{i}$; *(b)* $(2.0\ \text{m/s})\mathbf{i} - (2.0\ \text{m/s})\mathbf{j}$.
49. *(a)* $\sqrt{16gh/13}$; *(b)* 5/13.

Problems

1. *(a)* $\alpha_z = 2.6\ \text{rad/s}^2$, $\omega_{z0} = -5.1$ rad/s; *(b)* $\omega_z = -5.1\ \text{rad/s} + (2.6\ \text{rad/s}^2)t$, $\theta = -(5.1\ \text{rad/s})t + (1.3\ \text{rad/s}^2)t^2$.
3. $K_{\text{orbit}} = 2.7 \times 10^{33}$ J, $K_{\text{spin}} = 2.6 \times 10^{29}$ J.
5. *(b)* $v = \sqrt{2gh/(1 + K^2/R^2)}$.
7. $I = MR^2/4$.
9. $6.0 \times 10^{-5}\ \text{kg}\cdot\text{m}^2$.
11. *(a)* $v = \sqrt{2gh/[1 + (M/2m)]}$; *(b)* $\omega = (1/R_0)\sqrt{2gh/[1 + (M/2m)]}$.
13. *(a)* $(2.5\ \text{rad/s})\mathbf{k}$; *(b)* $-(1.8\ \text{rad/s}^2)\mathbf{k}$; *(c)* $-(32\ \text{m/s}^2)\mathbf{i} - (9.8\ \text{m/s}^2)\mathbf{j}$.
17. $4m[(a^4 + 2a^2b^2)/(2a^2 + b^2)]$.
19. 0.31 rad/s.

CHAPTER 13

Exercises

1. *(a)* $6.10 \times 10^6\ \text{kg}\cdot\text{m}^2/\text{s}$, down; *(b)* $6.10 \times 10^6\ \text{kg}\cdot\text{m}^2/\text{s}$, down.
3. $(100\ \text{kg}\cdot\text{m}^2/\text{s})\mathbf{k}$.
5. $-(0.25\ \text{kg}\cdot\text{m}^2/\text{s})\mathbf{k}$.
7. $-(2.4\ \text{kg}\cdot\text{m}^2/\text{s})\mathbf{i} + (1.6\ \text{kg}\cdot\text{m}^2/\text{s})\mathbf{j} - (6.1\ \text{kg}\cdot\text{m}^2/\text{s})\mathbf{k}$.
11. $7.8 \times 10^{-3}\ \text{kg}\cdot\text{m}^2/\text{s}$, down.
13. $\alpha_z = 120\ \text{rad/s}^2$, $\omega_z = (120\ \text{rad/s}^2)t$, $\theta = \theta_0 + (60\ \text{rad/s}^2)t^2$.
15. *(a)* $\alpha = 64\ \text{rad/s}^2$; *(b)* $30\ \text{rad/s}^2$.
17. $0.016\ \text{kg}\cdot\text{m}^2$.
19. *(a)* $m_cg/[m_c + m_b + (I_0/R_0^2)]$; *(b)* $m_bm_cg/[m_c + m_b + (I_0/R_0^2)]$; *(c)* $m_cg[m_b + (I_0/R_0^2)]/[m_c + m_b + (I_0/R_0^2)]$.
21. 720 N.
25. 0.19 m/s.
27. *(a)* 15 rad/s; *(b)* $K_f - K_i = 44$ J.
29. *(c)* $\omega = MD(v_i + v_f)/(Mw^2/3)$; *(d)* 4.7 rad/s.
31. *(b)* 27 N.
33. *(a)* $2v/D$; *(b)* $\Delta K = 0$.
35. *(a)* $(2.3\ \text{N}\cdot\text{m})$, out of page; $(12\ \text{rad/s}^2)$, out of page.
37. 2.1 rad/s.
39. *(a)* 10^{-4} rad/s; *(b)* 2.9×10^{-6} rad/s.
41. *(b)* $1.1\ \text{m/s}^2$.
43. $\boldsymbol{\ell} = (-9.8\ \text{kg}\cdot\text{m/s})\mathbf{i} + (-21\ \text{kg}\cdot\text{m/s})\mathbf{j} + (-4.2\ \text{kg}\cdot\text{m/s})\mathbf{k}$.
45. 3.1 rad/s.
47. $\omega_f = \omega_i/[1 + (m/2M)]$; *(b)* 4.0 rad/s.
49. 3.8 W.

Problems

3. $h = 2r_0/5$.
7. $h = 27R_0/10$.
11. *(b)* $W = -\frac{1}{2}mv_i^2(1 - R_i^2/R_f^2)$.
15. *(b)* $Mg\sqrt{1 + [M^4g^2D^6/(I_s\omega_s)^4]}$; *(c)* $\tan^{-1}[M^2gD^3/(I_s\omega_s)^2]$.
21. $90° - \tan^{-1}(R_1/R_0)$.

CHAPTER 14

Exercises

1. 1.1 s, 0.92 Hz.
3. *(a)* $x(t) = (0.063\ \text{m})\cos[(4.1\ \text{rad/s})t]$, $v_x(t) = -(0.26\ \text{m/s})\sin[(4.1\ \text{rad/s})t]$, $a_x(t) = -(1.1\ \text{m/s}^2)\cos[(4.1\ \text{rad/s})t]$; *(b)* $x(1.7\ \text{s}) = 0.049$ m, $v_x(1.7\ \text{s}) = -0.16$ m/s, $a_x(1.7\ \text{s}) = -0.82\ \text{m/s}^2$.
5. *(a)* $\omega = 7.1$ rad/s, $A = 0.25$ m, $\nu = 1.1$ Hz, $T = 0.88$ s, $\phi = \pi$ rad; *(b)* $x(t) = -(0.25\ \text{m})\cos[(7.1\ \text{rad/s})t]$, $a_x(t) = (13\ \text{m/s}^2)\cos[(7.1\ \text{rad/s})t]$; *(c)* $x(0.25\ \text{s}) = 0.051$ m, $v_x(0.25\ \text{s}) = 1.8$ m/s, $a_x(0.25\ \text{s}) = -2.6\ \text{m/s}^2$.
7. *(a)* $x(t) = (0.29\ \text{m})\cos[(6.7\ \text{rad/s})t + \pi/2]$, $v_x(t) = -(1.9\ \text{m/s})\sin[(6.7\ \text{rad/s})t + \pi/2]$, $a_x(t) = -(13\ \text{m/s}^2)\cos[(6.7\ \text{rad/s})t + \pi/2]$; *(b)* $x(0.54\ \text{s}) = 0.13$ m, $v_x(0.54\ \text{s}) = 1.7$ m/s, $a_x(0.54\ \text{s}) = -5.9\ \text{m/s}^2$.
9. $A = 0.49$ m, $v_{\text{max}} = 1.3$ m/s, $a_{\text{max}} = 3.4\ \text{m/s}^2$.
11. *(a)* 7.2 rad/s; *(b)* 1.1 Hz; *(c)* 0.88 s.
13. 32 N/m.
15. 0.34 kg.
17. 0.083 J.
19. *(a)* $U = (24\ \text{mJ})\cos^2[(6.5\ \text{rad/s})t]$; *(b)* $K = (24\ \text{mJ})\sin^2[(6.5\ \text{rad/s})t]$.
21. *(a)* 0.047 m; *(b)* 0.33 m/s; *(c)* 0.32 m/s; *(d)* 0.030 m.
23. *(a)* 1.67 kg; *(b)* 19.3 N/m.
25. *(a)* 1.00 J; *(b)* 1.38 J; *(c)* at $y = 0$, 0.39 m/s; *(d)* 1.2 J.
27. *(a)* 3 s; *(b)* 0.3 Hz, 0.4 m/s.
29. *(a)* 1.6 s; *(b)* 1.5 s.
31. *(a)* $2.65 \times 10^{-5}\ \text{N}\cdot\text{m}$; *(b)* $284\ \text{rad/s}^2$.
33. *(a)* 2.1 s; (b) 0.13 mJ.
35. $x = (150\ \text{mm})\cos[(3.5\ \text{rad/s})t]$, $v_x = -(0.52\ \text{m/s})\sin[(3.5\ \text{rad/s})t]$, $a_x = -(1.8\ \text{m/s}^2)\cos[(3.5\ \text{rad/s})t]$.
37. *(b)* $L\sin\theta$, $\sqrt{g/(L\sin\theta)}$.
39. *(a)* After an additional 2.4 min; *(b)* $4.8 \times 10^{-3}\ \text{s}^{-1}$.
41. *(a)* $1.2 \times 10^{-3}\ \text{s}^{-1}$; *(b)* 9:31.
47. *(a)* 9.3 mm; *(b)* $\tan^{-1}(-4.0) = -1.3$ rad $[+\pi] =$ 1.8 rad; *(c)* 9.6 mm for $\omega_E = \omega$.
49. *(a)* $v_x = \omega A\cos(\omega t + \delta)$; *(b)* $a_x = -\omega^2 A\sin(\omega t + \delta)$; *(c)* $\delta = \tan^{-1}(x_0/\omega v_{x0})$; *(d)* $A = \sqrt{x_0^2 + (v_{x0}/\omega)^2}$.
51. *(a)* 0.27 m; *(b)* 9.3 rad/s; *(c)* 1.5 Hz; *(d)* 0.68 s; *(e)* $(0.21\ \text{m})\mathbf{i}$.
53. *(a)* 23 mJ; *(b)* 7.2 m/s; *(c)* 13 mJ; *(d)* 5.4 m/s.
57. $190\ \text{N}\cdot\text{m/rad}$.
59. 100 N/m.
61. 2 s.
63. $2\pi\sqrt{5D/3g}$.

65. *(a)* $x = (1.5 \times 10^{11}\ \text{m}) \cos[(200\ \text{nrad/s})t]$,
$y = (1.5 \times 10^{11}\ \text{m}) \sin[(200\ \text{nrad/s})t]$;
(b) $v_x = -(30\ \text{km/s}) \sin[(200\ \text{nrad/s})t]$,
$v_y = (30\ \text{km/s}) \cos[(200\ \text{nrad/s})t]$;
(c) $a_x = -(6.0 \times 10^{-3}\ \text{m/s}^2) \cos[(200\ \text{nrad/s})t]$,
$a_y = -(6.0 \times 10^{-3}\ \text{m/s}^2) \sin[(200\ \text{nrad/s})t]$.

Problems

1. $T/6$.
3. *(a)* 0.09 s; *(b)* 0.11 s; *(c)* 0.16 s.
7. 0.75 s.
9. *(a)*
$$P(t) = -\frac{F_0{}^2\omega_E}{m\sqrt{(\omega_E{}^2 - \omega^2)^2 + 4\gamma^2\omega_E{}^2}} \cos \omega_E t \sin(\omega_E t - \phi_E)$$
13. *(a)* $2\pi\sqrt{3R/2g}$; *(b)* $2\pi\sqrt{2R/g}$; *(c)* $2\pi\sqrt{7R/4g}$.

CHAPTER 15

Exercises

1. 1.33 MPa.
3. *(a)* 34 kPa; *(b)* 30 kPa.
5. *(a)* 1.21×10^5; *(b)* 154 nm.
7. *(a)* 8.4×10^{-7}; *(b)* 8.4×10^{-7} rad; *(c)* 0.43 μm.
9. $1.98 \times 10^4\ \text{kg/m}^3$.
11. *(a)* $1.01 \times 10^3\ \text{kg/m}^3$; *(b)* $11.3 \times 10^3\ \text{kg/m}^3$.
13. 100 MPa.
15. 37.1 kPa.
17. 27 N.
19. 840 m^3.
23. 1.00129.
25. *(a)* $3.3 \times 10^{-3}\ \text{m}^3/\text{s}$, $1.7 \times 10^{-3}\ \text{m}^3/\text{s}$; *(b)* 3.4 m/s; *(c)* 71 kPa.
29. -2.5 kPa.
31. *(a)* 7.7 m/s; *(b)* 0.19 m^3/s; *(c)* 2.3×10^4 Pa, 1.8×10^4 Pa.
33. 0.32 N.
35. 0.544 m^3.
37. 0.96.
39. 920 kg/m^3.
41. 15 kPa.
43. 1.22 kg/m^3.
45. 97 N.
47. 4.2 $\text{N}\cdot\text{s/m}^2$.

Problems

1. *(a)* 4.8 kg; *(b)* $6.8 \times 10^3\ \text{kg/m}^3$; *(c)* 0.47 mm.
3. $(\rho_c - \rho_0)/(\rho - \rho_0)$.
5. 1.06 km, 2.6 km.
7. $80h_1$.
9. 90 min.
11. *(b)* 12.3°.
17. 16 km.
19. 1600 kg/m^3.

CHAPTER 16

Exercises

1. 10^6 years.
3. *(a)* 0.282 m; *(b)* 8.03 kN.
5. *(a)* 3650 Pa; *(b)* 77.1 K.
7. *(a)* 17.9 mm_{Hg}; *(b)* 17.9 mm_{Hg}; *(c)* 20.4 mm_{Hg}.
9. 35°C.
11. *(a)* $-40°$; *(b)* 575 K; *(c)* 0; *(d)* none.
13. *(a)* 9.997 mm; *(b)* 0.03%; *(c)* 30 mm.
15. *(a)* -22°C; *(b)* no.
17. *(a)* -2.8×10^{-5} s; *(b)* 2.4 s gain.
19. 280 mm.
23. 600 W.
25. *(a)* 1700°C/m; *(b)* 0.070 $\text{W}\cdot\text{K}^{-1}\cdot\text{m}^{-1}$; *(c)* insulator.
27. *(a)* 3800 Btu/h; *(b)* 0.77; *(c)* 0.14 $\text{W}\cdot\text{K}^{-1}\cdot\text{m}^{-1}$.
31. *(a)* 45 W; *(b)* 160 kJ.
33. *(a)* 70 MW/m^2; *(b)* 500 W/m^2; *(c)* 100 W/m^2; *(d)* 5 $\mu\text{W/m}^2$.
35. *(a)* 0.021 kg; *(b)* 1.2×10^{24}.
37. 26 mm.
39. *(a)* 23°C; *(b)* 1.0044 m.
41. *(a)* 0.163 m^2; *(b)* 365 mm $\times$ 450 mm; *(c)* 0.164 m^2.
43. 5×10^{26} W.
45. 460 K/m in the glass and 1400 K/m in the air.

Problems

5. *(a)* 86 W; *(b)* $T = 140°\text{C} - (124°\text{C}) \ln(r/12\ \text{mm})$; *(d)* -6.2×10^3 C°/m.
9. $T = T_2 - \dfrac{T_2 - T_1}{L}x$.
13. 280 K.
17. 3600 K.
19. *(a)* 310 K, 400 K, 1300 K; *(b)* 10 km.

CHAPTER 17

Exercises

1. *(a)* 2/3.
3. *(a)* 3.3×10^{-3} mol; *(b)* 0.49 kPa.
5. 0.0823 $\text{atm}\cdot\text{L}\cdot\text{mol}^{-1}\cdot\text{K}^{-1}$.
7. 2.7×10^{25} molecules/m^3.
9. 30 J.
11. *(a)* 24°C; *(b)* 11 kJ.
13. 1.1 kg.
15. *(a)* 58.1°C.
17. *(a)* 8.08 kJ; *(b)* -8.08 kJ.
19. *(a)* 60 kJ; *(b)* -55 kJ; *(c)* 5 kJ.
21. *(a)* $\frac{1}{2}(p_i + p_f)(V_f - V_i)$.
23. *(a)* 3.1 kJ; *(b)* 121 kPa.
25. *(a)* 13.0 kJ; *(b)* 24.0 kJ; *(c)* 25.6 kJ.
27. *(a)* 6.0 kJ; *(b)* 9.0 kJ; *(c)* 15.0 kJ.
29. 7.53 kJ.
33. 0.3 K.
35. 1.6 kg.
37. *(b)* -160 J.
39. \$8.
41. 340 K.
43. 20°C.
45. *(a)* 0; *(b)* 12 kJ; 12 kJ for the entire process.
47. 3.13 kJ.
49. *(a)* 450 J; *(b)* 650 J; *(c)* -200 J; *(d)* -850 J; *(e)* heat is rejected.

Problems

1. *(a)* $20.7\ \text{J}\cdot\text{mol}^{-1}\cdot\text{K}^{-1} + (0.0123\ \text{J}\cdot\text{mol}^{-1}\cdot\text{K}^{-2})T$; *(c)* 13 kJ.
3. *(a)* $1/p$; *(b)* $\dfrac{1 - b/V}{p - a/V^2 + 2ab/V^3}$.

11. *(a)* 250 K, 1000 K; *(b)* 1.08 kJ; *(c)* −1.38 kJ; *(d)* 255 J.

13.

Process	W (J)	Q (J)	ΔU (J)
a	0	416	416
b	−270	−502	−232
c	184	0	−184
Cycle	−86	−86	0

15. *(b)* 3.1 kJ.

CHAPTER 18

Exercises

1. *(a)* 298 m/s; *(b)* 1.01×10^5 m²/s²; *(c)* $\langle v\rangle^2 = 0.89 \times 10^5$ m²/s².
5. 500.
9. *(a)* 520 m/s; *(b)* 480 m/s; *(c)* 410 m/s; *(d)* 6.2×10^{-21} J.
11. *(a)* 1.2; *(b)* 0.8; *(c)* 2.3.
13. *(a)* 0.01 eV; *(b)* 0.04 eV; *(c)* 0.8 eV.
15. *(a)* $\langle v_x\rangle = 80$ km/h, $\langle v_y\rangle = \langle v_z\rangle = 0$; *(b)* $\langle v_x\rangle = \langle v_y\rangle = \langle v_z\rangle = 0$.
17. *(a)* 6.0 kJ.
19. *(a)* 6.0 kJ, 0; *(b)* 29.4 J·mol⁻¹·K⁻¹; *(c)* 8.4 kJ, 2.4 kJ.
21. *(a)* 23.1 J·mol⁻¹·K⁻¹.
23. 1.7, 1.4, 1.3.
25. *(a)* 0, approximately; *(b)* −1.1 kJ.
27. *(a)* 160 kPa; *(b)* 330 K, 250 K.
29. $T_i/V_i = T_f/V_f$ for an isobaric process.
31. *(a)* 1.3 kJ; *(b)* 1.3 kJ; *(c)* 1.3 kJ.
37. *(a)* 260 m/s; *(b)* 280 m/s; *(c)* 250 m/s.
39. 0.80, 074.

Problems

3. *(a)* 940 m/s; *(b)* 1×10^{-89} ($v_{\text{escape}} = 1.1 \times 10^4$ m/s); *(c)* yes.
5. *(a)* 5×10^{-46} kg·m²; *(b)* 5×10^{12} rad/s.

CHAPTER 19

Exercises

1. *(a)* 120 J; *(b)* 320 J; *(c)* 0.
3. *(a)* 0.28; *(b)* 1600 J; *(c)* 67 kW, 49 kW, 18 kW.
5. *(a)* 272 K, 544 K, 1088 K, 544 K; *(b)* 14.7 kJ; *(c)* 2.3 kJ; *(d)* 12.4 kJ; *(e)* 15 percent.
7. *(a)* 170 J; *(b)* 2.1.
9. *(a)* 7.7 kW; *(b)* 4.2 kW; *(c)* \$0.35.
13. *(a)* $|Q_C|/|Q_H| = 0.732$; *(b)* 9.15 mJ; *(c)* no.
17. *(a)* $\Delta\eta = +0.05$; *(b)* 0.04; *(c)* 0.08, 0.07.
19. *(a)* 5 percent; *(b)* 40 MW; *(c)* 10 m³/s.
21. 1 kW.
23. *(a)* Q/T; *(b)* 0.1 J/K.
25. 6.1 kJ/K.
27. *(a)* 51°C; *(b)* $\Delta S_{250} = 490$ J/K, $\Delta S_{950} = -400$ K; *(c)* 90 J/K; *(d)* irreversible.
31. *(a)* −0.1 J/K; *(b)* −0.1 J/K; *(c)* greater than 0.1 J/K.
33. 1070 MW.
35. *(a)* 2.11 kW; *(b)* 2.2.
37. *(a)* −6.3 J/K; *(b)* +6.2 J/K; *(c)* +0.2 J/K.
39. *(a)* 301 K; *(b)* 1252 J; *(c)* 626 J; *(d)* 626 J.
41. *(a)* 60 J; *(b)* 40 J; *(c)* 75 J; *(d)* 15 J; *(e)* 0; *(f)* 0.25 J/K.

Problems

1. Q_H/T_H, 0, $-|Q_C|/T_C$, 0.
7. *(a)* 0.40 J/K; *(b)* 140 J additional.
9. 23 J/K.

CHAPTER 20

Exercises

3.

r, m	0.10	0.20	0.30	0.40	0.50
F, μN	4.0	1.0	0.44	0.25	0.16

5. *(a)* 6.25×10^9; *(b)* 1.875×10^{10}.
7. *(a)* 9.63×10^4 C/faraday; *(b)* 4×10^3 C.
9. *(a)* $F_{ab} = 0.23$ mN, $F_{cb} = 0.19$ mN, $F_b = 0.04$ mN toward a; *(b)* $F_{ac} = 0.034$ mN, $F_{bc} = 0.19$ mN, $F_c = 0.16$ mN toward b.
11. $\sqrt{3}q^2/4\pi\epsilon_0 d^2$ directed away from the center of the triangle.
13. 420 nC.
15. *(a)* 220 N/C up; *(b)* 2.9 μN down.
17. *(a)* −0.065 C/kg; *(b)* 1.4×10^{-9}.
19. *(a)* $E_x = 2300$ N/C, $E_y = 0$, $E_z = 0$; $E_x = 820$ N/C, $E_y = 820$ N/C, $E_z = 0$; $E_x = 450$ N/C, $E_y = 450$ N/C, $E_z = 450$ N/C; $E_x = 470$ N/C, $E_y = 930$ N/C, $E_z = 0$; *(b)* $E = 2300$ N/C, $E = 1160$ N/C, $E = 770$ N/C, $E = 1040$ N/C.
21. *(a)* 1400 N/C to the right; *(b)* 7500 N/C to the left.
23. *(a)* $q/2\pi\epsilon_0 d^2$ toward vacant corner (positive q); *(b)* $(\sqrt{2} + \frac{1}{2})q/4\pi\epsilon_0 d^2$ away from center (positive q).
25. $\mathbf{E} = 2\mathbf{p}/4\pi\epsilon_0 z^3$.
29. 71 N/C, −140 N/C.
31. 0.19×10^{-10} m.
35. $Q/2\pi^2\epsilon_0 a^2$.
39. *(a)* -1.76×10^{11} C/kg, 9.58×10^7 C/kg; *(b)* 1.76×10^{11} m/s², 9.58×10^7 m/s²; *(c)* 5.46×10^{-4}.
41. *(a)* 9.7×10^7 C/kg; *(b)* proton.
43. 20°.
45. *(a)* (11 N/C)**i** − (19 N/C)**j**; *(b)* (22 nN)**i** − (38 nN)**j**.
47. *(a)* 2.6 kN/C; *(b)* toward rod.
49. *(a)* 2.3 kN/C; *(b)* toward disk.
51. 18°.

Problems

1. *(a)* $q = -Q/2\sqrt{2}$; *(b)* $7q^2/8\pi\epsilon_0 a^2$.
7. *(a)* −(3100 N/C)**i**; *(b)* 0.
9. $\mathbf{E} = \dfrac{2\lambda\ell x}{\pi\epsilon_0(x^2 + \ell^2)\sqrt{x^2 + 2\ell^2}}\,\mathbf{i}$
11. *(a)* Oscillatory; *(b)* $T = 2\pi\sqrt{4\pi\epsilon_0 a^3 m/Q|q|}$.
15. $E_x = \dfrac{Qx}{4\pi\epsilon_0(x^2 + a^2)^{3/2}}$, $E_\perp = \dfrac{Qa}{2\pi^2\epsilon_0(x^2 + a^2)^{3/2}}$

CHAPTER 21

Exercises

1. *(a)* 2100 N·m²/C; *(b)* 0; *(c)* 1800 N·m²/C.
3. *(a)* 440 N·m²/C; *(b)* 200 N·m²/C.
5. $\Delta S_x = 9.0$ m², $\Delta S_y = 12.0$ m².

7. 67°.
9. *(a)* 13 m^2; *(b)* 9.0 N/C; *(c)* 110 N·m^2/C; *(d)* 50 m^2, 2.2 N/C, 110 N·m^2/C.
11. −2900 N·m^2/C.
15. 4.4×10^{-12} C/m^3.
17. $\Phi_E = -480$ N·m^2/C for the entire cube; $\Phi_E = 0$ for each of the three sides in the *xy*, *xz*, and *yz* planes; $\Phi_E = -160$ N·m^2/C for each of the other three sides.
21. *(a)* −37 nC/m; *(b)* 26×10^3 N/C toward rod.
23. *(a)* 66 nC/m^2; *(b)* 3.7×10^3 N/C in each case.
25. *(a)* −2.3 μC/m^2; *(b)* 0, 0, 140 kN/C, 78 kN/C.
27. *(a)* 130 μC/m^3; *(b)* 120 kN/C, 240 kN/C, 60 kN/C.
29. *(a)* None of the quantities can be determined; *(b)* $Q = 37$ nC, $r_0 = 26$ mm, $\rho = 50$ μC/m^3.
31. 8.3 nC/m^2.
33. 5.5 nC/m^2.
37. 3.1 N·m^2/C.
39. $(0.22\ m^2)\mathbf{i} + (0.14\ m^2)\mathbf{j}$ or $(-0.22\ m^2)\mathbf{i} + (-0.14\ m^2)\mathbf{j}$.
41. *(a)* zero; *(b)* zero; *(c)* 444 N/C; *(d)* 250 N/C.

Problems

3. *(a)* $\sigma_{inner} = 160$ nC/m^2, $\sigma_{outer} = -110$ nC/m^2.
5. *(a)* $E = 0$, $r < a$; $E = Q/4\pi\epsilon_0 r^2$, $a < r < b$; $E = 0$, $r > b$.
7. *(a)* $\sigma_{inner} = -2.5$ μC/m^2, $\sigma_{outer} = 0.89$ μC/m^2; *(b)* $E = q/4\pi\epsilon_0 r^2$, $r < 30$ mm; $E = 0$, 30 mm $< r <$ 50 mm; $E = q/4\pi\epsilon_0 r^2$, $r > 50$ mm.

CHAPTER 22

Exercises

1. *(a)* $U_a = 87$ μJ, $U_b = 40$ μJ; *(b)* $U_b - U_a = -47$ μJ; *(c)* $W = 47$ μJ.
3. *(a)* 54 μJ; *(b)* 0.022 m/s.
5. *(a)* 65 μJ; *(b)* 130 μJ.
7. *(a)* 33 kV; *(b)* 33 kV; *(c)* 33 kV.
9. 12 μJ.
15. *(a)* 330 V; *(b)* 330 V; *(c)* −330 V; *(d)* 660 V; *(e)* 330 V.
17. *(a)* 1.3×10^6 V/m; *(b)* 430 V.
21. *(a)* 18 kV; *(b)* 10 kV.
23. 4×10^6 V.
25. $E = 0$.
27. *(a)* 800 V; *(b)* $[4Ax/(x^2 + a^2)^3]\mathbf{i}$; *(c)* (3.2 kV/m)**i**.
29. $E_x = \dfrac{\sigma}{2\epsilon_0}\left(\dfrac{x}{\sqrt{x^2}} - \dfrac{x}{\sqrt{x^2 + R^2}}\right)$.
31. −290 V.
33. 17 V.
39. *(a)* 4.5 nC; *(b)* 62 nC/m^2; *(c)* 7.0 kV/m.

Problems

3. *(c)* 0.053 nm; *(d)* 2.2×10^6 m/s.
5. 8×10^{-15} m.
13. *(a)* $\mathcal{V} = \dfrac{Q_a + Q_b}{4\pi\epsilon_0 r_{bo}}$

(b) $\mathcal{V}_a - \mathcal{V}_b = \dfrac{Q_a}{4\pi\epsilon_0}\left(\dfrac{1}{r_a} - \dfrac{1}{r_{bi}}\right)$

(c) $\mathcal{V}_a = \dfrac{Q_a}{4\pi\epsilon_0}\left(\dfrac{1}{r_a} - \dfrac{1}{r_{bi}}\right) + \dfrac{Q_a + Q_b}{4\pi\epsilon_0 r_{bo}}$

15. $\mathcal{V} = \dfrac{Q}{4\pi\epsilon_0 r}$, $r > r_0$

$\mathcal{V} = \dfrac{Q}{4\pi\epsilon_0}\left[\dfrac{1}{r} + \dfrac{\frac{1}{2}(r_0^2 - r^2) + r_0^3(r_0^{-1} - r^{-1})}{r_0^3 - r_i^3}\right]$, $r_0 > r > r_i$;

$\mathcal{V} = \dfrac{Q}{4\pi\epsilon_0}\left[\dfrac{1}{r_i} + \dfrac{\frac{1}{2}(r_0^2 - r_i^2) + r_0^3(r_0^{-1} - r_i^{-1})}{r_0^3 - r_i^3}\right]$, $r < r_i$.

23. 2,500 MeV.
25. 3 mC.
27. 600 MeV.
29. 6.8×10^5 C.
31. *(a)* 10^9 V; *(b)* 8.85×10^6 N/m^2.
33. 4.0×10^{-15} kg.
35. 1.8×10^6 C.

CHAPTER 23

Exercises

3. 0.58 μF.
5. 34 km.
9. *(b)* 46 mF.
11. *(a)* 5.5 μF; *(b)* 6.1 V; *(c)* 15 μC, 19 μC.
13. 5.8 μF, in series.
15. *(a)* 740 nC and 460 nC; *(b)* 12 V; *(c)* 179 nC and 109 nC; *(d)* 2.9 V.
17. *(b)* $C_{eq} = \dfrac{C_1C_2C_3C_4}{C_2C_3C_4 + C_3C_4C_1 + C_4C_1C_2 + C_1C_2C_3}$.
19. *(a)* 7.4 μF; *(b)* $\mathcal{V}_3 = 65$ V, $\mathcal{V}_1 = 39$ V, $\mathcal{V}_2 = 26$ V; *(c)* $Q_3 = 320$ μC, $Q_1 = 160$ μC, $Q_2 = 160$ μC.
21. *(a)* 50 μJ; *(b)* 200 μJ.
23. *(a)* 40 nJ/m^3; *(b)* 2×10^{12} J.
25. 16 mJ.
31. 1 MeV.
33. *(b)* $W = -Q^2d/\epsilon_0 A$; *(c)* $\Delta E = -W$.
35. 5.6 nF.
37. 5.6.
39. 3000 V.
41. 0.3 m^2.
43. *(a)* 0.6 μC/m^2; *(b)* 0.5 μC/m^2; *(c)* 10,000 V/m; *(d)* 60,000 V/m; *(e)* 50,000 V/m.
45. 2×10^{-15} m.

Problems

3. C_1.
5. *(b)* 1.1×10^{-12} F.
7. *(a)* 2.5 kV; *(c)* 3.0×10^4 V/m.
9. 2 mm.
11. *(b)* $F_x = \dfrac{Q^2d(\kappa - 1)}{2\epsilon_0 w_2[w_1 + (\kappa - 1)x]^2}$.
13. $\epsilon_0(48\pi^2 KZ^x/4\pi)^{1/3}$
15. 2×10^{-11} C.
17. 707 μF.
19. $[V_0\sqrt{2\epsilon_0(\kappa - 1)/\rho}]^{1/2}/d$.
21. Either way; $5CV_{max}^2$.
23. No.

CHAPTER 24

Exercises

1. *(a)* 750 C; *(b)* 4.7×10^{21}.
3. *(a)* $I(t) = (13.0\ C/s^2)t$; *(b)* $I(3.4\ s) = 44$ A.
5. *(a)* $I_+ = 0.645$ A, $I_- = 0.215$ A; *(b)* $J_+ = 1400\ A/m^2$, $J_- = 480\ A/m^2$; *(c)* $v_{d+} = 0.16$ mm/s, $v_{d-} = 0.052$ mm/s.
9. *(a)* 0.29 A; *(b)* 4.1 V/m.
11. *(a)* 3.2 mΩ; *(b)* 5.0 mΩ.
13. *(a)* 7.7×10^{-9} m; *(b)* 19 Ω. Resistance is independent of the ratio of length to width.
15. 57 mΩ.
17. 360 mΩ.
19. $1.1 \times 10^{-7}\ \Omega\cdot$m.
21. −19°C, or 254 K.
23. *(a)* 6.1 mV/m; *(b)* $2.8 \times 10^4\ A/m^2$; *(c)* $4.6 \times 10^6\ \Omega^{-1}\cdot m^{-1}$; *(d)* $2.2 \times 10^{-7}\ \Omega\cdot$m.
25. *(a)* 7.4 Ω; *(b)* $I_{16} = 0.87$ A, $\mathcal{V}_{16} = 14$ V; $I_{25} = 0.56$ A, $\mathcal{V}_{25} = 14$ V; $I_{31} = 0.45$ A, $\mathcal{V}_{31} = 14$ V.
29. *(a)* 0; *(b)* 4.0 Ω; *(c)* $\mathcal{V}_{ca} = \mathcal{V}_{da} = 5.0$ V, $\mathcal{V}_{bc} = \mathcal{V}_{bd} = 3.0$ V; $I = 1.0$ A in all resistors except resistor 3; $I = 0$ in resistor 3.
33. *(c)* 27 Ω.
35. 2×10^{-12} s.
37. *(a)* 2.7 mA; *(b)* −1.0 μA; *(c)* $R_f = 74$ Ω, $R_r = 200$ kΩ.
39. *(a)* 208 Ω; *(b)* 140 mA.
41. *(a)* 6.57×10^{28} carriers/m³; *(b)* 3.6×10^{-4} m/s.

Problems

3. *(b)* $\alpha = -E_g/2kT_0^2 = -7.4 \times 10^{-2}\ K^{-1}$; *(c)* $\rho_0 = \rho_1 e^{E_g/2kT_0}$.
9. *(b)* 900 MΩ.
11. *(b)* 160 kΩ.
13. $R/2$; no.
15. 200 Ω
17. 150 Ω.
19. 14 Ω.
21. $(\rho/2\pi\ell)\ln(R/r)$.

CHAPTER 25

Exercises

1. 1.4 V.
3. $\mathcal{E} = 6.3$ V, $r = 3$ Ω.
5. $\mathcal{E} = 6.1$ V, $r = 0.2$ Ω.
7. *(a)* 0.46 W; *(b)* 4.6×10^{-4} kW·h $= 1.7 \times 10^3$ J.
9. *(a)* 240 Ω; *(b)* 0.50 A; *(c)* 50 W.
11. *(a)* 16 mA; *(b)* 16 V.
13. 75 mA, 10 V.
15. *(a)* 3.2×10^5 C; *(b)* 3.9×10^6 J.
17. *(a)* 88 W; *(b)* 88 W; *(c)* 0.59 W.
21. *(a)* 4 V; *(b)* 36 V.
23. $i_2 = -0.5$ A, $i_3 = 1$ A.
25. *(a)* 0.0 A; *(b)* 0.5 A; *(c)* 0.3 A; *(d)* 0.2 A.
27. 2.2 Ω.
29. $i_5 = 0.1$ A, $i_2 = i_4 = 0.6$ A, $i_3 = 0.9$ A, $i_7 = 0.5$ A; $\mathcal{V}_5 = 1$ V, $\mathcal{V}_2 = 1$ V, $\mathcal{V}_4 = 2$ V, $\mathcal{V}_3 = 3$ V, $\mathcal{V}_7 = 3$ V.
31. *(a)* 0.063 s; *(b)* 6.4 μC; *(c)* 0.19 mA; *(d)* 84 μA; *(e)* 12 μC.
33. *(a)* 28 μC; *(b)* 0.14 mJ; *(c)* 0.58 mJ; *(d)* 0.44 mJ.
35. *(a)* 0.32 mA; *(b)* 11 mW; *(c)* 6.4 mW; *(d)* 5 mW.
39. 0.72 μF.
41. 8.65 s.

Problems

3. 92 percent.
9. $i_1 = 1.1$ A, $i_2 = 0.87$ A, $i_3 = 0.73$ A, $i_4 = 0.36$ A, $i_5 = 0.15$ A, $i_6 = 0.22$ A.
11. 2 ms.
13. *(a)* 3 A; *(b)* 2×10^4 C.
15. $[RC\ ln(2)]/2$; $\mathcal{E}^2C/4$.
17. 1/4.
19. 0.19 cents.

CHAPTER 26

Exercises

1. *(b)* 3.30×10^{-14} N; *(c)* 3.30×10^{-14} N.
3. *(a)* $F_x = 6.1 \times 10^{-15}$ N, $F_y = 8.5 \times 10^{-15}$ N, $F_z = 8.5 \times 10^{-15}$ N.
5. $F_g/F_B = 5.6 \times 10^{-12}$.
7. *(a)* $F_g = 0.02$ N, $F_E = 8 \times 10^{-10}$ N, $F_B = 2 \times 10^{-14}$ N.
9. *(a)* 26 mA.
11. *(a)* Positive z direction if $A > 0$; *(b)* $AI \ln 3$.
13. *(b)* 0.
15. *(a)* 6.1×10^{-3} N·m; *(b)* 7.1×10^{-3} N·m; *(c)* 6.1×10^{-3} N·m.
17. 1.0×10^{-4} A.
19. *(a)* 0.026 A·m²; *(b)* 0.031 N·m; *(c)* ±30°, ±150°.
21. *(a)* 5 μJ; *(b)* 5 μN·m; *(c)* 0, 5 μJ.
23. *(a)* 0.004 A·m²; *(b)* 7×10^{-5} m²; *(c)* −1 mJ; *(d)* 3 mJ.
25. *(a)* 4.8×10^{-20} kg·m/s; *(b)* 7.1×10^6 m/s; *(c)* 1.1 MeV.
27. *(a)* 8.2×10^7 rad/s; *(b)* 13 mm; *(c)* 21 MeV; *(d)* 2.1×10^4.
29. *(a)* Outer part; *(b)* positive; *(c)* 6×10^{-22} to 45×10^{-22} kg·m/s.
31. *(a)* From upper to lower plate; *(b)* 80 μT into the plane of the figure; *(c)* 5×10^7 m/s, 10 MeV.
33. *(a)* 1×10^{13} rad/s; *(b)* 1×10^7 m/s; *(c)* 3×10^8 m/s, or $c(1 - 5 \times 10^{-10})$.
35. 48×10^6 C/kg.
37. 430.
39. 74 μJ.
41. *(b)* 1.60×10^{-19} A·m²·T/eV.

Problems

3. *(b)* 0.41 T.
5. *(b)* 11.01 u.
7. *(b)* $mg/(2NLB)$; *(c)* 90°; *(d)* 30°; *(e)* stable.
13. 37.4 N; the forces oppose one another.
15. 2 kPa; no.
17. Tellurium.
19. 36 mT.

CHAPTER 27

Exercises

3. *(a)* 8.6×10^{-5} T.
5. *(a)* $2\mu_0 I/(\pi a)$ to the right; *(b)* $\mu_0 I/(2\pi a)$ to the right; *(c)* 3.8×10^{-5} T at P, 9.6×10^{-6} T at Q.
7. *(a)* $D/3$; *(b)* none.
11. *(a)* $\dfrac{\mu_0 I}{8ab}(b - a)$; *(b)* 42 μT.
13. *(a)* $\dfrac{\mu_0 IL}{4\pi R\sqrt{R^2 + L^2}}$; *(b)* $\mu_0 I/(4\pi R)$.

15. *(a)* 1.5×10^{-7} A·m²; *(b)* 2.9×10^{-14} T; *(c)* 1.9×10^{-6} T.
17. $-\mu_0\mathbf{m}/4\pi r^3$.
19. *(a)* 10.0 A; *(b)* $4\pi \times 10^{-6}$ T·m.
21. *(a)* 1.25 mm.
25. *(a)* $B = 0$, $R \le b$; *(b)* $B = \dfrac{\mu_0 I_0(R^2 - b^2)}{2\pi R(c^2 - b^2)}$, $b \le R \le c$; *(c)* $B = \mu_0 I_0/(2\pi R)$, $c \le R$.
27. *(a)* 2.6 mT; *(b)* 170; *(c)* 1.2 mm.
29. *(a)* 290; *(b)* 16 m; *(c)* 1.3 Ω; *(d)* 0.95 V.
31. *(a)* $\dfrac{\mu_0 Ix}{2\pi a^2} + \dfrac{\mu_0 I}{2\pi(D - x)}$; *(b)* $\dfrac{\mu_0 I}{2\pi}\left(\dfrac{1}{x} + \dfrac{1}{D - x}\right)$; *(c)* $\dfrac{\mu_0 I}{2\pi}\left(\dfrac{1}{x} + \dfrac{D - x}{a^2}\right)$, all directed toward the top of the page.
33. *(a)* 17 A; *(b)* 1/4.
35. 0.055 Wb.
37. *(a)* $(\mu_0 Ic/2\pi)\ln(b/a)$; *(b)* $(\mu_0 Ic/2\pi)\ln(b/a)$.
39. *(a)* ρI; *(b)* for $d(\rho I)/dt \neq 0$.
41. *(a)* 460 μC; *(b)* 2.2×10^{11} N·C⁻¹·s⁻¹.

Problems

3. $\mu_0 ni$.
5. $\dfrac{\mu_0 nIL}{\sqrt{L^2 + 4a^2}}$.
7. $B = 0$ above the top sheet, $B = \mu_0 K$ between the sheets, $B = 0$ below the bottom sheet.
13. $(\mu_0 cI^2/2\pi)[(1/a) - (1/b)]$.
15. $4\pi\mu_0\omega\sigma R$.
17. Replace ϵ_0 with $\kappa\epsilon_0$.
19. $B_{\text{circle}}/B_{\text{square}} = \pi^2\sqrt{2}/16$.

CHAPTER 28

Exercises

1. *(a)* 1.5 mWb, 2.3 mWb; *(b)* 6.7 mV.
3. *(a)* 23 μWb; *(b)* 7.8 mV; *(c)* 7.9 μWb, 2.7 mV.
5. *(a)* 0.030 Wb/s; *(b)* 0.47 T/s.
7. *(a)* 45 mV; *(b)* 1.5 s; *(c)* 3.0 μA.
9. *(a)* 0.17 V; *(b)* 0.22 mA; *(c)* 14 μN, opposite the velocity.
11. *(a)* 0; *(b)* 0; *(c)* $B\ell v\cos\theta$.
13. *(a)* $2B\ell v$; *(b)* clockwise.
15. *(a)* 71 ms; *(b)* 170 mV; *(c)* counterclockwise; *(d)* 4.2 μN, opposite the velocity; *(e)* no.
17. *(a)* $B\ell^2\omega/2R$; *(b)* 900 μA; *(c)* counterclockwise; *(d)* 3 μN·m.
21. *(a)* 20 mWb; *(b)* 5 mA; *(c)* 4 ms, 12 ms, ...
23. *(a)* $\omega C_1 \sin\omega t + 3\omega C_3 \sin 3\omega t$; *(b)* 2.2 V; *(c)* 1.7 V.
25. *(a)* $300\,\omega_1 C \sin\omega_2 t$; *(b)* 20 V; *(c)* 100 V.
27. *(a)* 0; *(b)* 0.7 mN/C; *(c)* 0.8 mN/C; *(d)* the same.
31. 55.
33. *(a)* CCW; *(b)* CW.
35. *(a)* 1 mV; *(b)* 1 mV.
37. 1.5 V.
39. 1.4 mW.

Problems

3. *(a)* $B\ell vt(1 + t/t_1)$; *(b)* $B\ell v(1 + 2t/t_1)$; *(c)* 0.62 V.
5. *(b)* Clockwise; *(c)* to the left; *(d)* counterclockwise, to the right.
7. *(a)* $\dfrac{\mu_0 i\ell}{2\pi}\ln\dfrac{w + R_0 + vt}{R_0 + vt}$; *(b)* $\dfrac{\mu_0 i\ell wv}{2\pi(R_0 + vt)(R_0 + w + vt)}$; *(c)* clockwise.
11. *(a)* Opposite the sense of the current; into the page on the left, out of the page on the right.
13. $\mathscr{E} - \mathscr{E}_0 e^{-t/\tau}$, $\tau - mR_L/B^2R^2$, $\mathscr{E}_0 = BR^2\omega_0/2$.
15. 2.2 V across wings.

CHAPTER 29

Exercises

3. *(a)* 850 μV; *(b)* 850 μV; *(c)* 850 μV, with opposite sense from *(a)* and *(b)*; *(d)* 0.
5. *(a)* 15 A/s; *(b)* either.
9. *(b)* $\dfrac{\mu_0 i\ell}{\pi}\ln\dfrac{D - a}{a}$.
11. *(a)* 19 ms; *(b)* 0.35 A; *(c)* 0.48 A; *(d)* 0.48 A.
13. Many possibilities consistent with $\mathscr{E}_0/R = 0.80$ A, $R \gg 2.3$ Ω; for example, $\mathscr{E}_0 = 24$ V, $R = 30$ Ω.
17. *(a)* 0, 74 mA, 110 mA, 130 mA; *(b)* 0, 5.3 V, 8.1 V, 9.2 V; *(c)* 9.2 V, 3.9 V, 1.1 V, 0.
19. *(a)* 94 ms; *(b)* 47 mH.
21. *(a)* $\frac{1}{2}Li_0^2 e^{-2t/\tau_L}$.
23. *(a)* $i_0^2 Re^{-2t/\tau_L}$; *(b)* energy dissipated $= U_0$.
25. $u_B \approx 0.6$ J/m³, $u_E \approx 0.04$ J/m³.
27. *(b)* 60 μH.
29. *(a)* To the left; *(b)* to the right; *(c)* to the left.
31. *(a)* 0.011; *(b)* 1.1 A; *(c)* 24 kW.
33. *(a)* 14 A; *(b)* 4 kW; *(c)* 0.2 percent; *(d)* 10 mm.
35. *(a)* $N_s/N_p = 50$: if $N_p = 10$, then $N_s = 500$; *(b)* 0.09 A; *(c)* 70 kΩ.
37. *(a)* 8.5 Ω; *(b)* 0.14 ms; *(c)* 0.098 ms.
39. 2×10^{12} J.
41. 9.5×10^{-12} J.

Problems

1. *(a)* $\dfrac{\mu_0 Nib}{2\pi}\ln\dfrac{c + a}{c}$; *(b)* $\dfrac{\mu_0 N^2 b}{2\pi}\ln\dfrac{c + a}{c}$.
3. *(a)* $\dfrac{\mu_0 i\ell}{2\pi}\ln b/a$; *(b)* $\dfrac{\mu_0}{2\pi}\ln b/a$; *(c)* the same.
7. *(a)* 200 mm; *(b)* 16 m; *(c)* 0.34 Ω; *(d)* 0.12 mH.
9. *(b)* $i_1 = \dfrac{\mathscr{E}_0}{R_1}(1 - e^{-R_1 t/L_1})$, $i_2 = \dfrac{M\mathscr{E}_0}{L_1 R_2}(e^{-R_2 t/L_2} - e^{-R_1 t/L_1})$; *(d)* 48 V across R_1 asymptotically, 30 V across R_2 at $t = 0.01$ s.

CHAPTER 30

Exercises

1. -9.31×10^{-24} A·m².
3. -5.3×10^{-35} kg·m²/s.
5. *(a)* 5.05×10^{-27} A·m²; *(b)* $m_N/m_B = 5.46 \times 10^{-4}$.
7. 1×10^6 A·m², assuming 1×10^{29} atoms per cubic meter.
11. *(a)* 1.7 A/m; *(b)* 210 K.
13. *(b)* 1×10^{-23} J; *(c)* $m_0 B/kT \le 0.1$ for $T > 7$ K.
15. 0.6×10^6 A/m, 1.6×10^6 A/m, 2.1×10^6 A/m, assuming 1×10^{29} atoms per cubic meter.
17. 33 A/m.
19. *(a)* 3.19×10^{-3} T·m/A; *(b)* 8.75×10^5 A/m; *(c)* 0.04 percent.
21. 20 μT, horizontal; 56 μT vertical (down).

Problems

1. $H = nI$.
3. *(a)* Parallel; *(b)* positive y direction, toward the stronger field region; *(c)* negative y direction.
7. *(a)* 1×10^{14} A/m; *(b)* 6×10^{17} A/m.

CHAPTER 31

Exercises

3. 5.1 mH.
5. $q(t) = (2.2\ \mu C) \cos[(0.11\ \text{Mrad/s})t]$, $i(t) = (230\ \text{mA}) \sin[(0.11\ \text{Mrad/s})t]$, $U = 140\ \mu J$, $U_E(t) = (140\ \mu J) \cos^2[(0.11\ \text{Mrad/s})t]$, $U_B(t) = (140\ \mu J) \sin^2[(0.11\ \text{Mrad/s})t]$.
7. $q(t) = (20\ \mu C) \cos[(14\ \text{krad/s})t - 1.2\ \text{rad}]$, $i(t) = -(280\ \text{mA}) \sin[(14\ \text{krad/s})t - 1.2\ \text{rad}]$, $U = 1.3$ mJ, $U_E(t) = (1.3\ \text{mJ}) \cos^2[(14\ \text{krad/s})t - 1.2\ \text{rad}]$, $U_B(t) = (1.3\ \text{mJ}) \sin^2[(14\ \text{krad/s})t - 1.2\ \text{rad}]$.
9. *(a)* $i(t) = -(3.8\ \text{A}) \sin[(54\ \text{krad/s})t - \pi/4]$; *(b)* 20 nF; *(c)* $U = 130$ mJ, $U_E(t) = (130\ \text{mJ}) \cos^2[(54\ \text{krad/s})t - \pi/4]$, $U_B(t) = (130\ \text{mJ}) \sin^2[(54\ \text{krad/s})t - \pi/4]$.
11. *(a)* $q(t) = -(260\ \text{nC}) \cos[(16\ \text{krad/s})t]$; *(b)* $i(t) = (4.2\ \text{mA}) \sin[(16\ \text{krad/s})t]$; *(c)* $\mathcal{V}_C = (4.8\ \text{V}) \cos[(16\ \text{krad/s})t]$; *(d)* $U = 630$ nJ; *(e)* $U_E = (630\ \text{nJ}) \cos^2[(16\ \text{krad/s})t]$; *(f)* $U_B = (630\ \text{nJ}) \sin^2[(16\ \text{krad/s})t]$.
13. $q(t) = (1.1\ \mu C) \cos[(25\ \text{krad/s})t - \pi/2]$; $i(t) = -(27\ \text{mA}) \sin[(25\ \text{krad/s})t - \pi/2]$.
15. *(a)* $q = Q_m/2$; *(b)* $i = -\sqrt{3}I_m/2$.
17. 790 μH, 3.0 Mrad/s or 0.48 MHz.
19. *(a)* Underdamped; *(b)* 6.3 krad/s; *(c)* 91 μs.
21. *(a)* 270 Ω; *(b)* 120 nF.
23. $R = \sqrt{3L/C}$.
27. *(a)* 17 ms; *(b)* 170 V; *(c)* 0; *(d)* 0.12 A; *(e)* 0.
29. *(a)* 1.6 MΩ; *(b)* 1.6 kΩ; *(c)* 1.6 Ω.
31. 0.11 μF.
33. 66 μA.
35. 3.0 krad/s.
37. *(a)* 0.63 Ω; *(b)* 630 Ω; *(c)* 630 kΩ.
39. 4.7 mH.
41. 4.9 A.
43. 820 rad/s.
45. *(a)* $i(t) = (120\ \text{mA}) \sin[(830\ \text{rad/s})t + 0.54\ \text{rad}]$.
55. 560 mW.
57. 9.5 mW.

Problems

9. *(a)* 54 Ω; *(b)* 7.1 V.

CHAPTER 32

Exercises

1. *(a)* 1.1 m/s; *(b)* 1 mm; *(c)* 0.7 m.
3. *(b)* $h = 10.0$ mm, $w = 2.00$ m.
5. 3.3 Mm.
7. *(a)* 4.3 mm; *(b)* 0.82 m; *(c)* 12 m/s; *(d)* 68 ms; *(e)* 7.7 rad/m; *(f)* 15 Hz; *(g)* 92 rad/s; *(h)* $-\mathbf{i}$; *(i)* -4.1 mm.
9. *(a)* $\partial y/\partial t = (0.40\ \text{m/s}) \cos[(7.7\ \text{rad/m})x + (92\ \text{rad/s})t]$; *(b)* $\partial^2 y/\partial t^2 = -(36\ \text{m/s}^2) \sin[(7.7\ \text{rad/m})x + (92\ \text{rad/s})t]$; *(c)* 0.40 m/s; *(d)* 36 m/s²; *(e)* -0.11 m/s; *(f)* 35 m/s²; *(g)* upward.
11. *(a)* $\partial y/\partial x = (0.033) \cos[(7.7\ \text{rad/m})x + (92\ \text{rad/s})t]$; *(b)* $\partial^2 y/\partial x^2 = -(0.25\ \text{m}^{-1}) \sin[(7.7\ \text{rad/m})x + (92\ \text{rad/s})t]$; *(c)* -0.0087; *(d)* 0.24 m⁻¹; *(e)* upward.
15. 0.3 m to 30 km.
17. *(a)* $y = (25\ \text{mm}) \sin\{[2\pi/(0.72\ \text{m})]x + [2\pi(4.1\ \text{Hz})]t\}$; *(b)* 3.0 m/s; *(c)* 0.64 m/s; *(d)* 17 m/s²; *(e)* 0.22; *(f)* 1.9 m⁻¹.
23. 0.21 kg/m.
27. 44 W.
33. 1.7×10^{17} W.
35. 97 mm.
37. *(a)* $y_2 = (22\ \text{mm}) \sin[(3.4\ \text{rad/m})x + (36\ \text{rad/s})t + 0.16\ \text{rad}]$; *(b)* $y_2 = (22\ \text{mm}) \sin[(3.4\ \text{rad/m})x + (36\ \text{rad/s})t + 3.30\ \text{rad}]$.
39. *(a)* $y = (70\ \text{mm}) \cos[(15.7\ \text{rad/s})t] \sin[(8.4\ \text{rad/m})x]$; *(b)* 0.37 m and 0.75 m; *(c)* 0.56 m, 0.37 m.
43. *(a)* $y = 2A \cos[(45\ \text{rad/s})t] \sin[(3.9\ \text{rad/m})x]$. *(b)* The origin is at one of the fixed ends. *(c)* Each element has its maximum displacement at $t = 0$.
45. 14 kg.

Problems

11. *(b)* $4\mu v^3 y_0{}^2/x_0{}^2$.

CHAPTER 33

Exercises

1. *(a)* 1.3 kHz; *(b)* 8.0×10^3 rad/s; *(c)* 8.4 m⁻¹; *(d)* $(5.0\ \mu\text{m}) \cos[(8.4\ \text{rad/m})x - (8.0 \times 10^3\ \text{rad/s})t]$.
3. 330 m/s.
5. *(a)* 1.1×10^3 m/s; *(b)* 2.0×10^3 m/s.
7. 0.2 s.
9. *(a)* 5.1×10^3 m/s; *(b)* 3.5×10^3 m/s; *(c)* 5.1×10^3 m/s.
11. *(a)* 3.9×10^3 m/s; *(b)* 1.5×10^3 m/s.
13. *(a)* 0.16 Pa; *(b)* 1.7×10^{-11} m; *(c)* 2.7×10^{-3} Pa, 1.0×10^{-9} m.
15. *(a)* 466 Hz; *(b)* 880 Hz.
17. 73 nW/m².
19. *(a)* 3; *(b)* 6; *(c)* 10; *(d)* 13; *(e)* 17; *(f)* 20; *(g)* 60.
23. 29 Hz, 88 Hz, 146 Hz, ...
25. *(a)* 0.66 m; *(b)* 0.70 m.
27. 8.7 kHz.
29. 8×10^{-3}.
33. 0.3 Hz.
35. 886 Hz.
37. 10 m/s.
39. 260 Hz following, 300 Hz approaching.
41. *(a)* 450 Hz; *(b)* 540 Hz.
45. *(b)* $(210\ \text{Pa}) \cos[(3.73\ \text{rad/m})x - (5530\ \text{rad/s})t]$.
47. 1 mW/m².

Problems

3. *(b)* 15 percent.
5. *(b)* $v_b = v_s(2v_0/v) = v_0/(0.25\ \text{m})$.
7. *(b)* 0.057 W/m²; *(c)* 0.051 W/m²; *(d)* 7.0×10^{-4} W/m², 0.

CHAPTER 34

Exercises

3. *(a)* $+\mathbf{i}$; *(b)* $-\mathbf{i}$.
5. 2.29×10^8 m/s.
11. *(a)* 3 m; *(b)* 300 m.
13. 150 N/C.

15. *(a)* $-\mathbf{j}$; *(b)* 3.5 m; *(c)* 86 MHz; *(d)* 100 nT; *(e)* (100 nT) cos [(1.8 rad/m)y + (5.4 × 10^8 rad/s)t]**k**.
19. 3 μJ/m³.
25. *(a)* $-\mathbf{k}$; *(b)* 0.52 m; *(c)* 570 MHz; *(d)* **E** = (290 N/C) cos [(12 rad/m)z + (3.6 Grad/s)t]**i**, **B** = −(960 nT) cos [(12 rad/m)z + (3.6 Grad/s)t]**j**. This answer is not unique.
27. **E** = (26 N/C) cos [(1.1 × 10^7 rad/m)x − (3.2 × 10^{15} rad/s)t]**j**, **B** = (85 nT) cos [(1.1 × 10^7 rad/m)x − (3.2 × 10^{15} rad/s)t]**k**, **S** = (1.7 W/m²) cos² [(1.1 × 10^7 rad/m) x − (3.2 × 10^{15} rad/s)t]**i**.
29. *(a)* 1.1 kW/m²; *(b)* 1.2 × 10^{-11} J; *(c)* 920 N/C; *(d)* 3.1 μT; *(e)* up and down; *(f)* 4.7 × 10^{14} Hz.
31. *(a)* 3.2 mN/m²; *(b)* 1.4 × 10^{-11} N.
33. *(b)* 5.8 × 10^5 N; *(c)* 1.6 × 10^{-17}.
35. *(a)* 0.4 W/m²; *(b)* 0.004 W/m².
37. *(a)* 1 W/m²; *(b)* 0.1 W/m².
39. 5.1 × 10^{14} Hz.
41. 499.0 s = 8.317 min.

Problems

7. *(c)* 0.56 μm.
9. *(b)* 6.7 × 10^{-7} N·m/rad.

CHAPTER 35

Exercises

1. 120°.
3. *(a)* AN and BC are traveled by the light in the same time; *(b)* side AB is common, $AN = AC$ as in *(a)* and AC and BN are bounded by the same parallel rays; *(c)* they are equivalent angles in congruent triangles.
7. 1.734 × 10^8 m/s.
9. *(a)* 65°; *(b)* 36.5°.
11. Soda lime.
13. 24.44°.
15. 7.827°.
17. *(a)* 0.765 m; *(b)* 1.59 m; *(c)* any distance.
19. *(a)* 111 mm; *(b)* erect; *(c)* 2 mm.
21. *(a)* 2 m; *(b)* 1 m behind mirror.
23.

Type	f	s	s'	Real?	Inverted?	m
Convex	−45	+15	−11	No	No	+0.75
Plane	∞	+300	−300	No	No	+1.0
Concave	+60	+400	+71	Yes	Yes	−0.18
Convex	−240	+240	−120	No	No	+0.5
Concave	+96	+26	−36	No	No	+1.4
Convex	−160	+160	−80	No	No	+0.5
Plane	∞	+48	−48	No	No	+1.0
Concave	+20	+33	+50	Yes	Yes	−1.5

25. 1.48.
27.

Type	f	s	s'	Real?	Inverted?	m
Converging	96	144	288	Yes	Yes	−2.0
Diverging	−300	300	−150	No	No	1/2
Diverging	−120	120	−60	No	No	1/2
Positive	480	120	−160	No	No	4/3
Negative	−58	96	−36	No	No	1.5/4
Diverging	−360	−240	720	Yes	No	+3
Positive	180	+∞	180	Yes	—	0
Positive	113	68	−170	No	No	2.5

29. 64 mm.
31. −162 mm.
33. 220 mm.
35. *(a)* 170 mm; *(b)* s' for second lens is 186 mm, real.
39. 3.05 mm.
41. 220 mm.
43. *(a)* 80 mm; *(b)* 7.
45. *(a)* 4.9 mm; *(b)* 33; *(c)* 980.
49. 7.4 m³.

Problems

13. x = 10 mm, y = 19.1 mm, z = 0.
15. *(a)* 3.1 × 10^{-8} m; *(b)* 6.2 × 10^{-8} m.
19. 1.650.

CHAPTER 36

Exercises

1. *(a)* $\theta_1 = \pm 0.18°$, $\theta_2 = \pm 0.36°$; *(b)* $\theta_1 = \pm 0.27°$, $\theta_2 = \pm 0.45°$.
5. 2.5 m.
15. 5.0 mm and 3.5 mm.
17. *(a)* 2.09 μm; *(b)* 566 nm.
23. *(a)* ±14.3° and ±29.6°; *(b)* 336 and 374 μrad.
27. *(a)* ±15.1°; *(b)* 4.91 × 10^5 m⁻¹.
31. 998 slits.
35. 0.13 nm.
37. 116 nm.
41. 36.84 μm.
43. 305.5 fringes.
45. 3.5 mm.
47. 9.
49. 18 rad.

Problems

1. $d \sin \theta_m = (m + \frac{1}{2})\lambda$ and $d \sin \theta_{m'} = m'\lambda$.
11. *(a)* $I_2 = (0.0200)I_1$; *(b)* $I_6 = (0.0200)(1 - 0.0200)^2 I_1 = (0.0193)I_1$.
17. *(a)* Unaffected; *(b)* They no longer have zero intensity, but rather 1/9th the amplitude at the center.
19. *(a)* 16 percent; *(a)* 0.49 percent.

CHAPTER 37

Exercises

1. 280 nm (ultraviolet).
3. 0.280 mm.
5. 1.3 μm.
7. 4.9 and 1.6 percent.
13. 4.6 and 1.6 percent.
17. 2.
19. 10.
21. 8.1 μrad.
23. 0.10 m.
25. 60°.
27. *(a)* Vertical; *(b)* 82 percent.
29. 53.1°.
31. 1.23.
35. *(a)* 0.56 rad; *(b)* 1.0 rad.
37. 30°.
39. 4.1 mm.
41. d/a = 6.
43. West.
45. 47 percent.

47. 2.40 nW/m^2.
49. 0.126.

Problems

5. *(c)* 2.97.
7. −0.057.
11. *(b)* 0.781; *(c)* 0.976: *(d)* 0.998.
17. *(a)* 90°.
21. *(a)* 2π (or zero).

CHAPTER 38

Exercises

5. *(a)* $\mathbf{u}' = \mathbf{u} - \mathbf{v}$; *(b)* $-\mathbf{v}$.
7. $v'' = v + v'$.
11. *(a)* 1×10^{-4}.
13. *(a)* -1.65×10^5 m, 0.55 s; *(b)* -1.66×10^7 m, 0.55 s; *(c)* -1.16×10^9 m, 3.9 s.
17. $x = \gamma(x' + vt')$, $y = y'$, $z = z'$, $t = \gamma(t + vx'/c^2)$ with $\gamma = 1/\sqrt{1 - v^2/c^2}$.
19. *(a)* $x' = \gamma(x - \beta ct)$, $y' = y$, $z' = z$, $t' = \gamma(t - \beta x/c)$ with $\gamma = 1/\sqrt{1 - \beta^2}$; *(c)* negligible; *(d)* 0.06 percent; *(e)* 2.6 percent.
21. *(a)* Eruption in Io; *(b)* 14×10^{11} m.
23. *(a)* $x_1' = 88$ ly, $t_1' = -83$ years; $x_2' = 134$ ly, $t_2' = -126$ years; *(b)* observer *O*; *(c)* observer *O*.
27. *(a)* 7×10^3 s; *(b)* 2×10^3 s; *(c)* 1×10^3 s; *(d)* 1×10^3 s.
31. $0.99999995c$.
33. *(a)* 3.1 MeV, 1.9×10^{-21} kg·m/s; *(b)* 79 keV, 1.6×10^{-22} kg·m/s; *(c)* 2.6 keV, 2.7×10^{-23} kg·m/s; *(d)* 0.26 eV, 2.7×10^{-25} kg·m/s.
35. *(a)* $0.12c$; *(b)* $0.36c$.
39. *(a)* $0.866c$; *(b)* $0.866c$.
41. *(a)* 1.0×10^{-4} u; *(b)* 7.6×10^{-15} J; *(c)* 1.0×10^{-20} kg·m/s; *(d)* 1.5×10^6 m/s.
43. No.
45. 3400 km.
47. *(a)* 4.6 km; *(b)* 0.65 km.
49. *(a)* For given *E*, photon always has more momentum; *(b)* 0.991.

Problems

3. *(a)* 110 years old.
15. *(a)* $0.77c$; *(c)* 3.0×10^{-19} kg·m/s.

CHAPTER 39

Exercises

1. *(a)* 360 W; *(b)* 5.8 kW; *(c)* 230 kW.
3. *(a)* 7.4 kJ; *(b)* 7.7×10^{-5} m^2.
5. *(a)* 7×10^{-21} J; *(b)* 2×10^{14} Hz, 2×10^{33} Hz.
7. *(a)* 7×10^{-121} J; *(b)* 2×10^{-31} J; *(c)* 4×10^{-21} J; *(d)* 1×10^{-20} J.
11. 1.7 to 3.1 eV, 2.7×10^{-19} to 5.0×10^{-19} J.
13. *(a)* 6.0×10^{14} Hz; *(b)* no.
17. *(a)* 1.87 eV; *(b)* 4.5×10^{14} Hz; *(c)* 2.14 eV; *(d)* none.
19. *(a)* 1.3 to 0.7 eV; *(b)* 3.1 to 3.7 eV; *(c)* 7.6×10^{14} to 8.9×10^{14} Hz; *(d)* no change.
23. *(a)* 76 pm; *(b)* 73 pm; *(c)* 71 pm.
25. 53°.
27. *(a)* 0.48 MeV; *(b)* $p = 4.4 \times 10^{-21}$ kg·m/s at 3.3° from the initial direction, 93.3° from the scattered photon.
29. *(b)* $b = n^2/R_H = 4/R_H$; *(c)* $b = 3.646 \times 10^{-7}$ m.
33. *(b)* 5.29×10^{-11} m.
35. *(a)* Paschen series, second longest wavelength; *(b)* −0.54 eV, −1.51 eV; *(c)* 2.3×10^{14} Hz, 1.3 μm.
37. *(a)* 2.2×10^6 m/s; *(b)* 1.00003.
39. *(b)* $R_H = 1.0974 \times 10^7$ m^{-1}.
41. *(a)* $1/\lambda = R_H(1 - 1/n_f^2)$; *(c)* Lyman.
43. 100 W.
45. 200 W.
47. *(a)* 118,000 K
49. *(a)* 0.005 nm, no; *(b)* 8×10 kg.

Problems

11. *(a)* 2nd Balmer; *(b)* 1st Lyman; *(c)* 1st Paschen; *(d)* 2nd Pfund.
13. *(a)* 0.773 meV; *(b)* 1.69×10^{-12} M. *(c)* 0.558 MeV, 0.446 MeV.
15. *(a)* 4×10^{-5}; *(b)* Approximately $\frac{1}{2}$, must use relativistic expression.
17. 4.35 m/s.
19. 22 nm.

PHOTO CREDITS

Page numbers are shown below in boldface italic.

Chapter 1 ***1*** Nano Scope AMF/Digital Instruments, Inc. ***2*** Museo Archeologico, Italy. ***3*** *Left,* National Institute of Standards & Technology, Boulder Laboratories, US Dept. of Commerce; *right,* National Institute of Standards & Technology, Boulder Laboratories, US Dept. of Commerce; *bottom,* National Institute of Standards & Technology, Boulder Laboratories, US Dept. of Commerce. ***7*** Dave Schaefer/The Picture Cube.

Chapter 2 ***13*** *Top,* Porterfield/Chickering/Photo Researchers; *bottom,* Dan Overcash. ***22*** AIP Niels Bohr Library.

Chapter 3 ***27*** Loren M. Winters. ***28*** *Top,* NASA; *bottom,* Loren M. Winters. ***35*** Loren M. Winters. ***41*** Loren M. Winters. ***42*** Fig. 3-16, Tom Richard. ***43*** Physics Today Collection/AIP Niels Bohr Library. ***44*** Fig. 3-17, James Sugar/Black Star. ***50*** Fig. 3-27, Dr. Harold Edgerton/M.I.T., Cambridge, Mass. ***51*** Fig. 3-28, National Center for Atmospheric Research; Fig. 3-29, Tom Richard. ***56*** Fig. 3-38, Loren M. Winters.

Chapter 4 ***57*** Loren M. Winters. ***58*** Caltech Jet Propulsion Laboratory/NASA. ***64*** Fig. 4-10, Dr. Harold Edgerton/M.I.T., Cambridge, Mass. ***67*** Fig. 4-13, Loren M. Winters. ***71*** Loren M. Winters. ***73*** Denis Milon. ***74*** New Mexico State University. ***84*** Rob Crandall/Stock, Boston.

Chapter 5 ***87*** *Top,* Brad McDonald; *bottom,* AIP Niels Bohr Library. ***88*** Fig. 5-1, CENCO. ***89*** Fig. 5-2, *right,* Joe Strunk and Loren M. Winters. Fig. 5-3, Loren M. Winters. ***91*** Fig. 5-4, PASCO. ***92*** Fig. 5-6, Loren M. Winters. ***94*** Fig. 5-7, Dan Overcash; Fig. 5-8, Bill Belknap/Science Source/Photo Researchers. ***105*** Ward's Scientific/Photo Researchers. ***109*** Mike J. Howell/Stock, Boston. ***113*** Margaret Durrance/Photo Researchers.

Chapter 6 ***116*** *Top,* Focus On Sports; *bottom,* Safra Nimrod. ***117*** Fig. 6-2, Dan Overcash. ***118*** *Left,* The Bettmann Archive; Fig. 6-4, Joan Hudson. ***126*** Fig. 6-16, The Smithsonian Institution. ***128*** Giraudon/Art Resource. ***130*** Fig. 6-18, Richard Pasley/Stock, Boston.

Chapter 7 ***138*** NASA. ***139*** Fig. 7-1, AMNH/Haydn Planetarium. ***140*** Fig. 7-2, NGC. ***144*** Fig. 7-6a, CENCO. ***152*** Fig. 7-13. AIP Niels Bohr Library. ***153*** NASA. ***155*** AIP Niels Bohr Library. ***156*** AIP Niels Bohr Library. ***158*** Fig. 7-21, Courtesy of Stanford Linear Accelerator, Stanford Univ., Calif. ***159*** Fig. 7-22, Argonne National Laboratory. ***160*** *Top and bottom,* Courtesy of Fermilab. ***161*** NASA. ***162*** 7-24, U.S. Naval Observatory; *bottom,* NASA. ***164*** Tass/Sovfoto.

Chapter 8 ***170*** *Top,* John G. Ross/Photo Researchers; *bottom,* NASA. ***181*** Fig. 8-12*b*[a], Stanley Rowin/The Picture Cube. ***184*** Barbara Alper/Stock, Boston. ***185*** Fig. 8-14*a*, Tom Richard.

Chapter 9 ***194*** *Top,* Focus on Sports; *bottom,* NASA. ***212*** AIP Niels Bohr Library. ***213*** Photo by Alan Richards/AIP Niels Bohr Library. ***219*** NASA.

Chapter 10 ***224*** *Top,* NASA; *bottom,* Tom Branch/Science Source/Photo Researchers. ***225*** Fig. 10-1*a* and *b,* Manfred Bucher and Randy Dotta-Dovidio. ***231*** Fig. 10-8, Dr. Harold Edgerton/M.I.T., Cambridge, Mass. ***234*** Fig. 10-12, J. Williamson/Photo Researchers. ***237*** Fig. 10-16, Dan Overcash. ***240*** Fig. 10-24, Science Photo Library/Photo Researchers; Fig. 10-25, Dan Overcash. ***250*** Fig. 10-41. Dan Overcash.

Chapter 11 ***257*** Margot Granitsas/The Image Works. ***259*** Dan Overcash. ***266*** Dan Overcash. ***271*** Fig. 11-20, Spencer Grant/Stock, Boston. ***279*** Fig. 11-50, Dan Overcash.

Chapter 12 ***280*** Richard Hutchings/Photo Researchers. ***281*** Fig. 12-1*b*, Loren M. Winters. ***294*** Fig. 12-19, Richard Megna/Fundamental Photographs. ***309*** NASA.

Chapter 13 ***318*** *Top,* CENCO; Fig. 13-17, Dan Overcash. ***321*** Dan Overcash.

Chapter 14 ***331*** Loren M. Winters. ***332*** Figs. 14-1 and 14-2, Loren M. Winters. ***340*** NASA. ***341*** Fig. 14-10*a*, Loren M. Winters. ***349*** *Both,* Nancy Dudley/Stock, Boston. ***351*** Fig. 14-22*a* and *b, top,* Ed Gettys; Fig. 14-23, Ed Gettys. ***352*** Fig. 14-24 S. Neil Rasband, Professor of Physics & Astronomy, Brigham Young University.

Chapter 15 ***361*** Bob Daemmrich/The Image Works. ***369*** Loren M. Winters. ***373*** Fig. 15-12, Tom Richard. ***374*** Fig. 15-13*a*, Illustrated Experiments in Fluid Mechanics (The NCFMF Book of Film Notes) National Committee for Fluid Mechanics Films, Educational Development Center, Inc., MIT Press, Cambridge, MA, 1972. Fig. 15-13*b*, Tecquipment Ltd., Nottingham, England. ***379*** Fig. 15-20*e*, and *middle,* CENCO. Fig. 15-21, Wolf D. Seufert. ***381*** Culver Pictures. ***383*** Fig. 15-27, CENCO. Fig. 15-28*a* and *b*, Dan Overcash. ***384*** Fig. 15-30, Haags Gementemuseum.

Chapter 16 ***391*** Daedalus Enterprises, Inc. ***396*** Dan Overcash. ***397*** The Bettmann Archive. ***400*** AP/Wide World. ***401*** CENCO. ***401*** CENCO.

Chapter 17 ***416*** *Top,* Kent & Donna Dannen/Photo Researchers; *bottom,* CENCO. ***426*** The Bettmann Archive. ***433*** Culver Pictures.

Chapter 18 ***442*** Safra Nimrod. ***443*** CENCO. ***447*** AIP Niels Bohr Library. ***458*** University of Virginia/AIP Niels Bohr Library.

Chapter 19 ***469*** Gianni Tortoli/Photo Researchers. ***472*** Collier/Condit/Stock, Boston. ***473*** *Above,* Max Planck Institute. *Below,* CENCO, CENCO. ***477*** AIP Niels Bohr Library. ***480*** The Bettmann Archive.

Chapter 20 ***501*** Arjen Verkaik/The Stock Market. ***503*** Dan Overcash. ***517*** *All,* Dan Overcash.

Chapter 21 ***527*** Andre Baranowski. ***528*** From Albert Betlex, The Discovery of Nature, Simon & Schuster, New York, 1965/AIP Niels Bohr Library. ***541*** Dan Overcash. ***543*** New York Public Library. ***545*** Fig. 21-27, Tom Richard.

Chapter 22 ***551*** Mula & Haramati/Phototake. ***560*** CENCO. ***566*** *Top,* CENCO; Fig. 22-21, Dan Overcash ***567*** Hank Morgan/Rainbow.

Chapter 23 ***576*** Dan McCoy/Rainbow. ***577*** *Top,* Dan Overcash; *bottom,* The Image Works Archives. ***579*** Fig. 23-4*c*, The Image Works Archives.

Chapter 24 ***600*** The Image Works. ***613*** Gerd Binnig & Heinrich Rohrer/IBM.

Chapter 25 ***624*** C. Falco/Photo Researchers. ***625*** Dan Overcash. ***647*** Srinivias Manne and Scot A.C. Gould, Department of Physics, UCSB. ***648*** IBM Corporation, Research Division, Almaden Research Center.

Chapter 26 ***650*** Michael Giannechini/Photo Researchers. ***651*** Fig. 26-1*a* and *b*, Dan Overcash. ***652*** Fig. 23-3*b*, CENCO. ***654*** CENCO. ***660*** CENCO. ***661*** Fig. 26-12, Brookhaven National Laboratory; bottom, Brenneis/Photo Researchers. ***666*** Fig. 26-16, CENCO.

Chapter 27 ***673*** Courtesy of the Princeton Plasma Physics Laboratory. ***676*** Fig. 27-3*b*, Richard Megna/Fundamental Photographs. ***684*** *Both,* Richard Megna/Fundamental Photographs. ***692*** New York Public Library Collection.

Chapter 28 ***701*** Porterfield/Chickering/Photo Researchers. ***702*** Richard Megna/Fundamental Photographs. ***703*** CENCO. ***720*** Deena Cloud. ***721*** Will & Deni McIntyre/Photo Researchers.

Chapter 29 ***723*** Ray Ellis/Photo Researchers. ***724*** The Image Works Archives. ***733*** CENCO. ***735*** Coco McCoy/Rainbow. ***736*** Culver Pictures.

Chapter 30 ***743*** CNRI/Science Photo Library/Photo Researchers. ***744*** NASA/Mark Marten/Science Source/Photo Researchers. ***752*** Fig. 30-8, Tom Richard.

Chapter 31 ***758*** *Top,* John G. Ross/Photo Researchers; *bottom,* Courtesy of Kepco, Inc.

Chapter 32 ***781*** Judith Canty/Stock, Boston. ***783*** Fig. 32-3, M. Serraillier/Photo Researchers. Fig. 32-4, Physics, 2nd ed., Physical Science Study Committee, 1965. Reprinted by permission of D. C. Heath & Co. and Educational Development Center. ***785*** Fig. 32-7, ***Physics,*** 2d ed., Physical Science Study Committee, 1965. Reprinted by permission of D. C. Heath & Co. and Educational Development Center. ***786*** Figs. 32-8 and 32-9, ***Physics,*** 2d ed., Physical Science Study Committee, 1965. Reprinted by permission of D. C. Heath & Co. and Educational Development Center. Figs. 32-10 and 32-11, ***Physics,*** 2d ed., Physical Science Study Committee, 1965. Reprinted by permission of D. C. Heath & Co. and Educational Development Center. ***793*** Fig. 32-18, Berenice Abbott/Photo Researchers. ***797*** Fig. 32-23, ***Physics,*** 2d ed., Physical Science Study Committee, 1965. Reprinted by permission of D. C. Heath & Co. and Educational Development Center. ***806*** J. H. Robinson/Photo Researchers.

Chapter 33 ***809*** Hank Morgan/Science Source/Photo Researchers. ***818*** CENCO. ***823*** Fig. 33-16, ***Physics,*** 2d ed., Physical Science Study Committee, 1965. Reprinted by permission of D. C. Heath & Co., and Educational Development Center. ***830*** Fig. 33-23, Courtesy of Ludwig and Musser Industries.

Chapter 34 ***835*** Royal Observatory, Edinburgh. ***847*** Science Photo Library/Photo Researchers. ***849*** AIP Niels Bohr Library. ***850*** *Top,* Fig. 34-14, R. D. Black/Wabash Instrument Corporation; *bottom,* Courtesy, COBE Science Working Group.

Chapter 35 ***857*** Jack Feinberg. ***860*** From PSSC Physics, 3e, by Uri Haber-Schaim et al., D. C. Heath & Co., 1971, p. 117. ***862*** Fig. 35-9*a*, Ken Kay/Fundamental Photographs. ***863*** CENCO. ***865*** Fig. 35-15, Mark Antman/The Image Works. ***866*** Richard Megna/Fundamental Photographs. ***867*** Richard Megna/Fundamental Photographs. ***872*** *Both,* Richard Megna/Fundamental Photographs. ***878*** Fig. 35-33, National Gallery, London/Bridgeman/Art Resource; Fig. 35-34, Cordon Art-Baarn-Holland. ***879*** Figs. 35-36 and 35-37, Dan Overcash.

Chapter 36 ***890*** CENCO. ***891*** Fig. 36-2, Reprinted by permission of D. C. Heath & Company and Educational Development Center. ***892*** Fig. 36-4, Photo Researchers; Fig.

36-5. Berenice Abbott/Photo Researchers. ***902*** Fig. 36-17, Photo Researchers. ***903*** Fig. 36-18, Richard Megna/Fundamental Photographs. ***905*** Fig. 36-21*b*, CENCO. ***906*** *Above,* New York Public Library. *Below,* Thomas Young, A Course of Lectures on Natural Philosophy on the Mechanical Arts, Taylor & Walton, London, 1845/AIP Niels Bohr Library. ***913*** Fig. 36-30*b*, CENCO; Fig. 36-31*b*, CENCO.

Chapter 37 ***918*** Philippe Phailly/Science Photo Library/Photo Researchers. ***919*** Fig. 37-1, Ken Kay/Fundamental Photographs; Fig. 37-2, Philip M. Rinaro, Los Alamos National Laboratory. ***920*** Fig. 37-4*a*, Cagnet, Francon, Phrierr, Atlas of Optical Phenomena, Springer Verlag, Berlin, 1962. ***926*** Fig. 37-8*d*, Cagnet, Francon, Phrierr, Atlas of Optical Phenomena, Springer Verlag, Berlin, 1962; Fig. 37-9, Photo Researchers. ***931*** Fig. 37-18*a,b* and *c*, Dan Overcash. ***936*** CENCO.

Chapter 38 ***944*** Stan Osolinski/The Stock Market. ***962*** *Bottom,* Courtesy NRAO/AUI; *top,* California Institute Millikan Library.

Chapter 39 ***970*** David Parker/Science Photo Library/Photo Researchers. ***972*** Fig. 39-1*b*, Ed Gettys ***980*** Martin Dohrn/Science Photo Library/Photo Researcher. ***981*** Fig. 39-8, R. D. Black/Wabash Instrument Corporation; Fig. 39-9, From Gerhard Herzberg, Atomic Spectra and Atomic Structures, Prentice-Hall Englewood Cliffs, N.J., 1937. ***988*** *Both,* W. F. Meggers Collection/AIP Niels Bohr Library.

INDEX

SOME SYMBOLS USED IN THIS TEXT

a	acceleration
A	area; amplitude; mass number of nucleus
B	magnetic field
B	bulk modulus
c	speed of light; specific heat (per unit mass)
C	molar heat capacity; capacitance
e	base of natural logs, 2.71828. . . ; emissivity; magnitude of charge of electron
E	energy
E	electric field
$\mathcal{E}$	emf
f	focal length
F	force
g	acceleration of gravity
G	gravitational constant
h	Planck's constant
$\hbar$	$h/2\pi$
H	heat current
H	magnetic intensity
i	current (signed)
i	unit vector in x direction
I	moment of inertia; current; intensity
j	unit vector in y direction
J	impulse, current density
k	Boltzmann's constant; spring constant; thermal conductivity
k	unit vector in z direction
K	kinetic energy; coefficient of performance
ℓ	angular momentum of particle; orbitial angular momentum quantum number
L	latent heat; inductance
L	angular momentum of system
m	mass of particle; order of intensity maximum
m	magnetic dipole moment
M	mass of system; mutual inductance
M	magnetic moment per unit volume
n	number of moles; index of refraction; number density of charge carriers; number of turns per unit length
N	neutron number of nucleus; number of turns in a coil
p	pressure
p	momentum of particle; electric dipole moment
P	power
$\mathcal{P}$	momentum of a system of particles
q	charge of particle
Q	volume flow rate; heat; charge of system
r	position vector
R	universal gas constant; resistance, radiancy
s	object distance; s' image distance
S	surface area; shear modulus; entropy
S	surface vector; spin angular momentum; Poynting vector
t	time
T	temperature; period
u	energy density
U	potential energy; internal energy
v	speed
v	velocity
$\mathcal{V}$	electric potential; potential difference
W	work
X	reactance
Y	Young's modulus
Z	impedence; atomic number of nucleus
α	magnitude of angular acceleration; coefficient of linear expansion; alpha particle
β	coefficient of volume expansion; sound intensity level; electron; v/c
γ	ratio of specific heats, c_p/c_v; $1/\sqrt{1-v^2/c^2}$; photon
ϵ	strain; permittivity
η	viscocity; efficiency of an engine
κ	dielectric constant